MEISTERPRO reference guides

professional tools to grow with!

PUT UP TO DATE TECHNOLOGY IN THE PALM OF YOUR HAND!

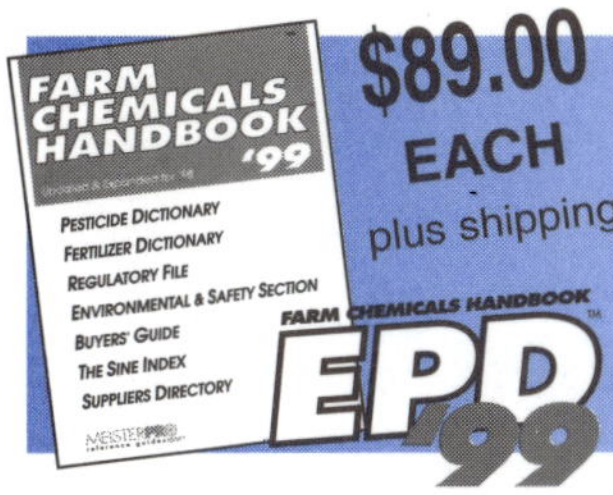

$89.00 EACH plus shipping

1999 Farm Chemicals Handbook and Electronic Pesticide Dictionary

Almost 1000 pages completely updated with the latest chemical information from around the world. More than 8000 products are referenced in the Pesticide Dictionary. Also has the Fertilizer Dictionary, Regulatory File, Environmental and Safety section, and a 200-page Buyers' Guide. The Electronic Pesticide Dictionary software is fully indexed and searchable, perfect for your PC or Macintosh. CD-Rom only!

New for 1999!

1999 Insect and Disease Control Guide and IDCG-E

600-plus pages of pest and disease control information for all major crops grown in the U.S. More than 300 insects and 250 diseases are covered. Also includes a photo identification guide and Worker Protection Standard information. Have a PC or a Macintosh? Then you need our IDCG-E software. Fully indexed and searchable by crop, insect, disease, and product. CD-Rom only!

$54.00 EACH plus shipping

IDCG-E

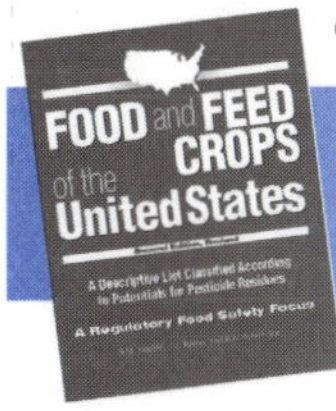

Brand New!

$34.95 plus shipping

Food and Feed Crops of the Unite

More than 500 pages of crop monograph … s, crop classifications, plus a complete index section w … ames.

AF854573

Weed Control Manual and WCM-E

If you have weeds, you need the Weed Control Manual! With 600-plus pages of weed control information, you can look up detailed product information for almost every crop grown in the U.S.. The Weed Identification Guide alone makes this worth the price. Our PC-compatible WCM-E software is a perfect companion to the book.

$59.00 EACH plus shipping

WCM-E

$54.00 Hard cover*
$40.00 Soft cover*
*plus shipping

Vegetable Insect Management

A practical guide to Integrated Pest Management. Has 190-plus pages packed with photos, charts, and detailed description of insects and IPM.

The More You Buy The More You $ave!

TITLES		QTY	PRICE	SHIPPING*	TOTAL PRICE
Farm Chemicals Handbook	$89.00				
Electronic Pesticide Dictionary (CD-Rom)	$89.00				
Insect & Disease Control Guide	$54.00				
IDCG-E Software (CD-Rom)	$54.00				
Food and Feed Crop of the U.S.	$34.95				
Weed Control Manual	$59.00				
WCM-E Software	$59.00				
☐ CD-Rom ☐ Diskette					
Vegetable Insect Management					
Hard Cover	$54.00				
Soft Cover	$40.00				

The More You Buy The More You $ave!

- Buy 1 product. . . . . .Save $5.00
- Buy 2 products. . . . . .Save $10.00
- Buy 3 products. . . . . .Save $15.00
- Buy 4 or more. . . . . Save $20.00

Subtotal ______
Quantity Discount ______
New Subtotal After Discount ______
Residents of CA, FL, IL, NY, OH, TN, WA add Applicable Sales Tax ______
Canada add 7% GST ______
Total ______

Please Indicate Method Of Payment

☐ Check Enclosed* ☐ Visa ☐ MasterCard ☐ AMEX ☐ Discover

Acct # ______ Exp. ___ ___/ ______

Signature Required ✘ ______

***Shipping per copy:**
US $8.00 All other countries $35.00 Airmail
For courier delivery rates:
Call: 440/942-2000 Fax: 440/942-0662
Email fchb_circ@meisterpubl.com
Canada must be in U.S. funds
International must be pre-paid and in U.S. funds drawn on a U.S. bank or by credit card

Name ______
Co. Name ______
Occupation and Title ______
Address ______
City ______
State ______ Zip ______
Country ______
Phone (____) ______
Fax (____) ______
E-Mail ______

Meister Publishing Company • 37733 Euclid Avenue. • Willoughby, Ohio 44094-5992 • Fax 440-942-0662 • Phone 440-942-2000 • 1-800-572-7740 • E-mail: fchb_circ@meisterpubl.com • Website: www.meisterpro.com

DIN99

Insect and Disease Control Guide

EXECUTIVE OFFICE
37733 Euclid Ave., Willoughby, OH 44094
Phone: 440-942-2000 • Fax: 440-942-0662
e-mail: icg_ads@meisterpubl.com
e-mail: icg_circ@meisterpubl.com
e-mail: icg_edit@meisterpubl.com
e-mail: icg_prod@meisterpubl.com

REGIONAL OFFICES

Eastern: Jack Driscoll
1811 Rt. 52, Suite 213, Hopewell, NY 12533
Phone: 914-221-0330 • Fax: 914-221-0370

Central: Alan C. Strohmaier, Ric Abernethy
37333 Euclid Ave., Willoughby, OH 44094
Phone: 440-942-2000 • Fax: 440-942-0662

Southern: John Mayfield
65 Germantown Court, Suite 220, Cordova, TN 38018
Phone: 901-756-8822 • Fax: 901-756-8879

MEISTER PUBLISHING COMPANY
Richard T. Meister, Chairman
Gary T. Fitzgerald, President
Charlotte Sine, Executive Vice President
William J. Miller II, Vice President

ISBN 1-892829-01-0
Library of Congress Card Number: 98-88129

Foreword

Volume 1, 1999

Welcome to the first edition of the new ***Insect and Disease Control Guide***. It's a critical time for crop protection. New regulations spawned by the Food Quality Protection Act are already changing the nature of pest management, with more to come in the next couple of years. That's why it's important to have this new book on your desk.

Combining insecticides and fungicides in one volume makes this a one-stop source for your pest and disease problems. With this book, you can immediately determine which insecticides and fungicides are cleared for any crop and what restrictions apply. Our exclusive listing format makes it easy to compare products and make the right decision quickly. Every product is indexed and cross-referenced by trade name and common name.

We've added and revamped sections you are familiar with. Besides our exclusive crop listings, we've added charts to each section describing Bt. products available as well as Enhanced Seed products. We've expanded our photo identification section (pages 9-28) to include more than 45 insects and 60 diseases. Finally, we include our exclusive Worker Protection Standard chart (pages 39-51). It allows you to quickly determine what personal protective equipment should be worn when applying a product, as well as EPA numbers, common names, signal words, and restricted entry intervals.

Add to this insect and disease photos, Worker Protection information, complete indexing, and our Suppliers Directory, and there is no other book you need to turn to.

Don't forget about our software. If you haven't ordered it yet, it's the perfect companion to this book. Simply put the book in your truck for quick reference in the field and have the ***Insect and Disease Control Guide-Electronic*** on your computer in your office. It will make it quick and easy to figure out which product you should use in the field.

Editorial Advisory Board

Dr. Edgar R. Butts

Dr. Edgar R. Butts is founder and president of Taensa, Inc., a firm commercializing biopesticides and other low-risk pesticides that present minimal safety and environmental risks. He received his B.S. in agronomy and Ph.D. in weed science from Louisiana State University, his M.S. in plant physiology from Virginia Polytechnic Institute and State University, and M.B.A. from St. Mary's College of California.

Dr. Bill Curran

Dr. Curran is Associate Professor of Weed Science at Penn State University, joining the faculty in 1990, after receiving his Ph.D. from the University of Illinois. He received his bachelor's degree at Colorado State University and his master's degree from Washington State University. At Penn State, he has statewide responsibilities for weed science extension and research in agronomic crops. He is also an active member of the American Society of Agronomy, the Weed Science Society of America (WSSA), the North Central Society of Weed Science, and the Northeastern Weed Science Society. He is the Northeastern representative to the WSSA board of directors.

Dr. Rick Foster

Dr. Foster is an associate professor with responsibility for extension and research of insect pests of vegetables and fruit in the Department of Entomology at Purdue University. He received his B.A. from Southern Illinois University, his M.S. from the University of Illinois, and his Ph.D. from Iowa State University. He served on the faculty at the University of Florida for 4 years before accepting the position at Purdue in 1988. Dr. Foster's research deals with all aspects of pest management. He is actively involved in the development and implementation of IPM programs on vegetables and fruit.

Robert H. McCarty

Robert McCarty is director of the bureau of plant industry, Mississippi Department of Agriculture and Commerce. His duties include supervision of personnel responsible for administering state laws regulating pesticides and pesticide applications. He is responsible for all pesticide regulatory programs and pesticide applicator certification and licensing programs in the state. A former president of the Association of American Pesticide Control Officials, he has served on the States/FIFRA issues Research and Evaluation Group working group on enforcement and certification since 1975. McCarty received his B.S. and M.S. from Mississippi State University.

Dr. John J. Mortvedt

Dr. Mortvedt is a retired soil fertility specialist. Recently he was interim extension soil specialist with Colorado State University. He was formerly with the National Fertilizer and Environmental Research Center (NFERC) of the Tennessee Valley Authority (TVA). With NFERC from 1962-1993, he was involved mainly in research and development of secondary and micronutrient fertilizers, and heavy metal contaminants in phosphate fertilizers and industrial by-products. Dr. Mortvedt is a co-editor of Micronutrients in Agriculture, and is a former president and editor-in-chief of the Soil Science Society of America. He received his B.S. and M.S. in soil science from South Dakota State University and his Ph.D. in soil chemistry from the University of Wisconsin.

Doug Soper

Doug Soper is Senior Technical Information Specialist for the Agriculture Division of Bayer Corp. He coordinates the dissemination of technical product information pertaining to Bayer's Crop Protection and Garden and Professional Care business unit both inside Bayer and externally. Soper is active in industry committees developing standards to deliver product information to distributors, dealers, and growers and electronically submit research studies to regulatory authorities. He received his bachelor's and master's degrees from the University of Missouri.

Table of Contents

Insect and Disease Control Guide

UPFRONT

Insect and Disease Identification Photos
- Insects 10
- Diseases 18

- FQPA 29
- PPE Chart 32
- Formulation Guide 34
- Nozzle Guide 36
- WPS Chart 39

1. BIOCONTROLS

- Product Listings 54
  - Biocontrol Agents 54
  - Beneficial Organisms 62

2. FIELD CORN, POPCORN

- Enhanced Seed Products 72
- Bt. products 76
- Product Listings 77
  - Insecticides 77
  - Fungicides 85

3. SOYBEANS

- Enhanced Seed Products 90
- Bt. Products 94
- Product Listings 95
  - Insecticides 95
  - Fungicides 101

4. COTTON

- Enhanced Seed Products 108
- Bt. Products 108
- Product Listings 109
  - Insecticides 109
  - Fungicides 122

5. SMALL GRAINS *(Barley, Oats, Rye, Triticale, Wheat)*

- Bt. Products 128
- Product Listings 129
  - Insecticides 129
  - Fungicides 134

6. OTHER FIELD CROPS

- Bt. Products 144
- Product Listings 146
  - Insecticides 146
    - Hops 146
    - Mint 148
    - Oil Seed Crops 150
    - Peanuts 150
    - Rice 154
    - Safflower 156
    - Sorghum 157
    - Sugar Beets 161
    - Sugarcane 166
    - Sunflower 166
    - Tobacco 169
  - Fungicides 172
    - Castor Beans 172
    - Flax 172
    - Hops 173
    - Mint 174
    - Mung Beans 175
    - Peanuts 175
    - Rice 181
    - Safflower 184
    - Sorghum 185
    - Sugar Beets 187
    - Sugarcane 191
    - Sunflower 191
    - Tobacco 192

7. VEGETABLES

- Enhanced Seed Products 196
- Bt. Products 196
- Product Listings 198
  - Insecticides 198
    - Artichokes 198
    - Asparagus 199
    - Beans 202
    - Carrots, etc. 209
    - Celery 213
    - Cole Crops 217
    - Cucurbits 225
    - Eggplant 231
    - Garlic 236
    - Greens 238
    - Herbs & Spices 243
    - Horseradish 245
    - Lentils 246
    - Lettuce 248
    - Mushrooms 254
    - Okra 255
    - Onions 256
    - Peas 260
    - Peppers 264
    - Potatoes 271
    - Radishes 278
    - Rhubarb 281

Southern Peas 282
Spinach....284
Sweet Corn 288
Sweet Potatoes, Yams 294
Table Beets 297
Tomatoes....300
Watercress....307
Fungicides....308
Artichokes....308
Asparagus....308
Beans....309
Carrots, etc. 316
Celery 319
Cole Crops....322
Cucurbits 328
Eggplant 334
Garbanzos....336
Garlic 336
Greens 338
Herbs & Spices....341
Horseradish 342
Lentils....342
Lettuce....343
Mushrooms 345
Okra....345
Onions 346
Peas....351
Peppers 353
Potatoes 357
Radishes....362
Rhubarb....363
Southern Peas 364
Spinach....365
Sweet Corn 367
Sweet Potatoes, Yams 370
Table Beets 370
Tomatoes....372
Watercress....378

8. FRUITS & NUTS

Bt. Products....380
Product Listings 381
Insecticides 381
Small Fruits 381
Tree Fruits, Citrus 392
Tree Fruits, Deciduous....401
Tree Fruits, Subtropical....418
Nuts....423
Fungicides....433
Small Fruits 433
Tree Fruits, Citrus 446
Tree Fruits, Decidous 452
Tree Fruits, Subtropical....470
Nuts....477

9. TURF & ORNAMENTALS

Bt. Products....488
Product Listings 489
Insecticides 489
Lawns & Turf....489
Ornamentals 494
Fungicides....509
Lawns & Turf....509
Ornamentals 519

10. OTHER USES

Enhanced Seed Products 536
Bt. Products....536
Product Listings 537
Insecticides 537
Forage (Alfalfa, etc.) 537
Forage (Grasses, Hay, etc.)....543
Pastures & Rangeland 544
Noncropland....546
Fungicides....547
Forage (Alfalfa, etc.) 547
Forage (Grasses, Hay, etc.)....549

INDEXES

Crop Index 597
Disease Index 590
Pest Index....571
Product Index....552
Suppliers Directory....599

Insect & Disease Photo Identification

Insects

Alfalfa weevil 10
Aphids 12
Apple maggot fly 12
Beet armyworm 15
Black cutworm 15
Boll weevil 10
Boxelder bug 12
Citrus mealybug 13
Citrus red mite 10
Codling moth 16
Corn earworm 16
Cottonycushion scale 13
Euonymus scale 13
European corn borer 16
European red mite 10
Fall armyworm 16
Green peach aphid 13
Greenhouse whitefly 13
Japanese beetle 11
Medfly 12
Northern corn rootworm . . 11
Pink bollworm 15
Plum curculio 11
Potato leafhopper 13
Red headed sawfly 15
Soybean looper 16
Spotted cucumber beetle . . 11
Squash vine borer 17
Striped cucumber beetle . . 11
Sweet potato whitefly 14
Tarnished plant bug 12
Tea scale 14
Thrips 17
Tobacco hornworm 17
Tomato hornworm 17
Western corn rootworm . . 11
White peach scale 14
Whiteflies 14

Diseases

Apple
Black spot canker 18
Fire blight 18
Powdery mildew 18

Beans
Blight 18
Gray mold 18
Leaf blotch 18

Berries
Blossom blight 19
Cane spot 19
Downy mildew 19
Gray mold 19
Powdery mildew 19

Citrus
Anthracnose 20
Botrytis blight 20
Fusarium rot 20
Greasy spot 20
Leaf spot 20

Cucumber
Anthracnose 20
Cottony leak 21

Grape
Bird's eye rot 21
Botrytis 21
Downy mildew 22
Powdery mildew 21

Lettuce
Downy mildew 22
Gray mold 22
Powdery mildew 22
Rhizoctonia 22
Shot hole 23

Melon
Bacterial wilt 23
Downy mildew 23
Fusarium rot 23

Onion
Basal plate rot 23
Soil line rot 23

Ornamentals
Colletotrichum leaf spot . . 24
Leaf spot 24
Powdery mildew 24

Peanut
Botrytis blight 24
Pod rot 24

Pear
Blossom end-rot 24

Potato
Late blight 25

Rose
Black spot 25
Botrytis blight 25
Downy mildew 25
Powdery mildew 25

Stone fruit
Anthracnose 25
Bacterial spot 26
Leaf spot 26

Small grains
Head blight 26
Phytophthora 26
Summer black stem 26

Tobacco
Anthracnose 27
Verticillium wilt 27

Tomato
Common rot 27
Early blight 27
Leaf mold 27
Pythium 27

Turf
Anthracnose 28
Blight 28
Crown & root rot 28
Snow mold 28

Watermelon
Alternaria 28

IDENTIFICATION IS START FOR PEST CONTROL PROGRAM

Find out what pest or disease is in your field with this photo identification section.

Identifying an insect or disease is the first step in a pest control program. This section contains more than 100 common insect and disease photos and short descriptions. These photos are only a starting point for identifying your pest problem.

Descriptions are condensed from three sources, The American Phytopathological Society; *Westcott's Plant Disease Handbook*, by R. Kenneth Horst; and *Destructive and Useful Insects*, by Robert L. Metcalf and Robert A. Metcalf.

Since determining your pest or disease problem can be difficult, be sure to discuss it with your county extension agent, consultant, or other experts before embarking on a pest control program.

How Insects Are Classified

Phyla...

Are divided into Classes.

Insects make up the Class Hexapoda. Mites are not insects and belong to the Class Arachnida, which also includes spiders and ticks.

Classes...

Are divided into Orders.

For insects, there are 24 or more orders, of which 8 are major:

Lepidoptera

Caterpillars, butterflies, and moths.

Includes corn earworm, cotton bollworm, armyworm, and cabbage looper.

Homoptera

Aphid, scale, leafhopper and whitefly.

Coleoptera

Beetles: Colorado potato beetle, boll weevil, flea beetle.

Diptera

Flies: Leafminer, maggots.

Thysanoptera

Thrips: Western flower thrips.

Orthoptera

Grasshoppers, crickets.

Hemiptera

True bugs: Plant bug, lace bug.

Hymenoptera

Wasps, ants, bees.

Orders...

Are divided into Families.

The family name is always found by adding the suffix "idae" to the stem of the name of the typical genus. For instance, the order Lepidoptera has many families of which the family Noctuidae includes many important crop pests like armyworm and bollworm.

Families...

Are divided into Genera.

Most families have a number of genera based on common characteristics.

Genera...

Are divided into Species.

Members of a species generally interbreed. Members of different species seldom interbreed.

All of the individuals of a species look alike, act alike, eat the same kind of food in the same manner, and are controllable in the same way.

Every species is given a scientific name consisting of the genus name and the species name.

Scientific names are written in Latin form and are italicized.

The corn earworm/cotton bollworm species is the order Lepidoptera, and family Noctuidae. The genus and species are *Helicoverpa zea*.

Disease Identification

Determining which disease is lurking in your field is not simple. Many variables are involved including weather, field conditions, fertilization, and herbicide injury. For example, some diseases become a problem only after temperatures rise above a certain level or humidity remains high for a period of time.

There are soil-borne diseases, water-borne diseases that spread to the plant when splashed by rain or irrigation, air-borne diseases, and insect-borne diseases. Other diseases carry over from year to year in infected tree limbs, canes, or stems. Finally, some diseases are only a problem if trash is left in the field after harvest and the disease is allowed to overwinter. Some plants can exhibit disease symptoms when there is no disease present and it's due to drought stress or overwatering.

Because of all of the factors involved, the disease photos in this book should not be used to make a final determination. Instead, you can try to narrow it down to a few possibilities and then work with your consultant or county extension agent to make a final determination.

Insect Identification

Arachnida, Coleoptera

Phil Phillips, University of California.

Citrus red mite

Citrus red mite. *Panonychus citri*
The citrus red mite is most prevalent in Florida and California. The female lays 20-50 eggs, at the rate of 2-3 per day on leaves, twigs, or fruit. Hatching occurs in about 1 week. Larvae migrate over the plant and begin feeding.

European red mite. *Panonychus ulmi*
An important pest of deciduous trees. Leaf damage ranges from specking to bronzing. Mites are usually rust in color, newly emerged females, bright velvety red. Adult females live about 19 days, depositing about 20 summer eggs.

University of California.

European red mite

APHIS.

Alfalfa weevil

Alfalfa weevil. *Hypera poaica*
Generally overwinters in the adult stage. Beetles are grayish brown or nearly black, with short grayish hairs giving a spotted appearance, 1/8"-1/4" long. Females make a cavity in the alfalfa and lay 1-40 eggs, 600-800 eggs in its 10-14 month life span. Larvae are white, becoming green with a prominent, white, mid-dorsal stripe and a black head. They feed on the plant stems, working to the new growth below the tip. Larvae become full-grown about time of first cutting, when the pest goes into the soil, spins a nearly spherical cocoon, pupates, and emerges as an adult about 10 days later.

APHIS.

APHIS.

Boll weevil larva

Boll weevil. *A*
The boll wee
cotton pests.
adults. Adult
feed and lay e
and bolls, d
Adults are br
1/4" long. Fem
ing in 3-4 day
days. Matur
curved body.
the adult eati
is 25 days, w

Errata

The **Boll weevil adult** photo shown on this page is incorrect. The correct photo is shown here.

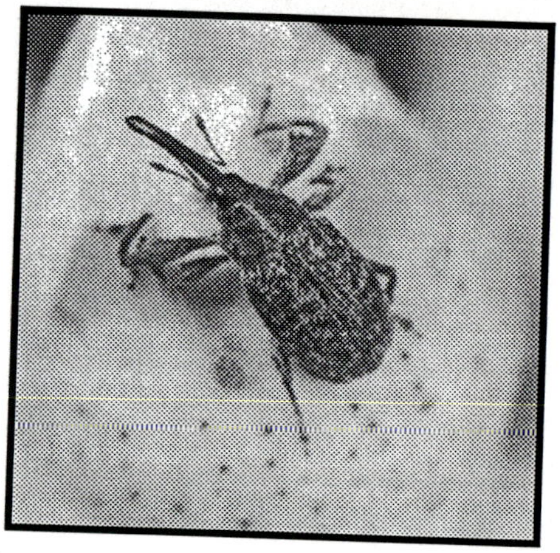

Coleoptera

Rhone-Poulenc.

Northern corn rootworm

Northern corn rootworm. *Diabrotica barberi*
Larvae are distinguished by pattern on terminal segment on body. After hatching, larvae feed on corn roots. Adults emerge after pupating and feed on silks and pollen of corn. Adult is green to yellowish-green and $^1/_5$" long.

Japanese beetles. *Popillia japonica*
Japanese beetles are metallic green or greenish bronze $^1/_3$"-$^5/_8$" long. It has reddish wing covers and two prominent and several smaller, white spots near the tip of the abdomen and along the sides. Adults feed on foliage, flowers, cornsilks, and fruits on warm, sunny days.

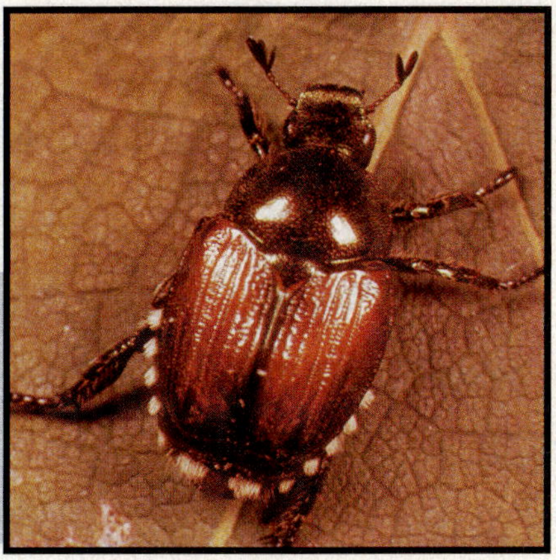

Clemson University.

Japanese beetle

New York State Agricultural Experiment Station, Geneva.

Plum curculio

Plum curculio. *Conotrachelus nenuphar*
Adult is $^1/_4$" long with grayish or whitish patches on its back and four humps on the wing covers. Females deposit eggs in holes in the fruit. Grayish-white larva feed on the fruit. Spends pupal stage in the ground, emerging after 1 month as adult.

Western corn rootworm. *Diabrotica virgifera*
Larvae feed on roots, adults feed on newly formed corn silk and leaves. Eggs are deposited in late summer and hatch in the spring. Larvae go through 3 instars and a prepupal stage. Pests typically are full-grown by early July.

USDA/ARS

Western corn rootworm

Gerald M. Ghidiu, Rutgers University.

Striped cucumber beetle

Striped cucumber beetle. *Acalymma vittatum*
The most serious pest of cucurbits in the U.S. east of the Rockies. Adults are $^1/_5$" long with black-yellow stripes. Orange-yellow eggs are laid at the base of plants and the whitish larvae feed on the roots and underground stems.

Spotted cucumber beetle. *Diabrotica undecimpunctata*
Also known as Southern corn rootworm. Larvae feed on plant roots and underground parts of stem. Adult is $^1/_4$" long with 12 conspicuous black spots on wing covers.

Gerald M. Ghidiu, Rutgers University.

Striped & Spotted cucumber beetle

Photo Identification

Diptera, Heteroptera, Homoptera

Apple maggot fly

New York State Agricultural Experiment Station, Geneva.

Apple maggot fly. *Rhagoletis pomonella*
Pupae overwinter in the soil, emerging as adult flies in the summer. Females lay their eggs singly in punctures in the apple skin. Apple maggots develop slowly in the green fruit, not fully developed until fruit drops to the ground.

Mediterranean fruit fly. *Ceratitis capitata*
Subtropical insect pest. Adult is about the size of a housefly with a mosaic of yellow and black markings. Wings are transparent. Female lays up to 40 eggs per day under fruit skin. Larvae bore into the fruit flesh, destroying the fruit.

Medfly adult

Phil Phillips, University of California.

Tarnished plant bug (nymph)

Cornell University.

Tarnished plant bug. *Lygus lineolaris*
Tarnished plant bug adults are $\frac{1}{4}$" long, flattened, oval in outline, with the small head projecting in front. They have a general brown color mottled with small, irregular splotches of white, yellow, reddish-brown, and black. Nymphs are pale green. Pests become active in the early spring, attacking terminal shoots and fruits. Females lay eggs on herbaceous weeds, vegetables, and flowers. After about 10 days, a small, yellowish-green nymph $\frac{1}{25}$" long emerges. As pest feeds, it introduces a toxic saliva into the plant. Fruit can become pitted, leaves deformed, or stems scarred and discolorerd.

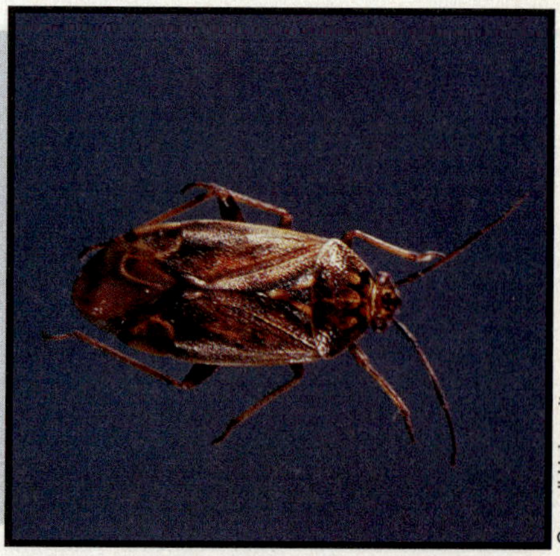

Tarnished plant bug (adult)

Cornell University.

Boxelder bug

Entomology Dept., Clemson University.

Boxelder bug. *Leptocoris trivittatus*
The boxelder bug sucks sap from the leaves, tender twigs, and developing seeds of the boxelder tree. Adults are $\frac{1}{2}$" long, brownish-black with three longitudinal red stripes on the thorax and red veins on the wings. Nymphs are bright red.

Aphids. Aphidae
Also known as plant lice, aphids will attack every vegetable crop. They feed by puncturing leaves and sucking out sap. They overwinter as eggs, hatching as small nymphs. Females can reproduce without mating.

Aphids

Entomology Dept., Clemson University.

Ohio Florists Association.

Citrus mealybug (nymph)

Citrus mealybug. *Plannococcus citri*
Pest can be found in all stages of development throughout the year. Young insect feeds by sucking sap or juices from leaves or fruit. Infested citrus trees have masses of white, cotton-appearing insects on leaves, twigs, and angles where fruit touches.

Cottony cushion scale. *Icerya purchasi*
Scale is large, red-brown, and covered with white or pale yellow waxy filaments. Egg masses contain up to 1000 bright red eggs. Red nymphs begin to feed, soon becoming covered with waxy, cottony filaments and powder.

Entomology Dept., Clemson Unviesity.

Cottonycushion scale

Entomology Dept., Clemson University.

Euonymus scale

Euonymus scale. *Unaspis euonymi*
This ornamental pest usually attaches itself to the plant for most of its life cycle. Scales reduce plant vigor by removing plant juices. Females pass the winter fully grown and pale-yellow crawlers appear in May and June.

Green peach aphid. *Myzus persicae*
Passes the winter as black shining eggs on the bark of peach, plum, apricot, and cherry trees. Young aphids are pale yellowish-green with three dark lines on the back of the abdomen. It is a vector of more than 50 plant virus diseases.

Gerald M. Ghidiu, Rutgers University.

Green peach aphid

Lance Osborne, University of Florida.

Greenhouse whitefly

Greenhouse whitefly. *Trialeurodes vaporariorum*
The pest is tiny, $^{1}/_{16}$" long. Plants are covered, especially on the underside, with small, snow-white, four-winged flies and very small, oval, flat, pale-green nymphs. Infested plants lack vigor and leaves are coated with a sticky material.

Potato leafhopper. *Empoasca fabae*
Adult is pale-green wedge-shaped insect about $^{1}/_{8}$" long. Eggs are thrust into main veins or petioles of underside of leaves. Insects feed on underside of leaves, sucking out the sap from the veins.

Rick Foster, Purdue University.

Potato leafhopper

Photo Identification

Homoptera

Ohio Florists Association.

Sweet potato whitefly (pupa)

Sweet potato whitefly. *Bemisia tabaci*
The sweet potato whitefly has become a serious pest of cotton, lettuce, and numerous other crops. The pest is a vector of numerous plant virus diseases, including lettuce infectious yellows and squash silver leaf. Adults are four-winged about $^1/_{16}$" long. Pests suck sap from plants and excrete sticky honeydew on the leaves.

USDA/ARS

Sweet potato whitefly (adult)

Whitefly nymphs and eggs

Whiteflies. Aleyrodidae
Nymphs are light-green, flattened, oval-shaped, about the size of a pinhead with fringed white threads radiating from the body. Nymphal period lasts about 1 month, passing through 4 instars. After emerging as adults, the insects begin to mate, attaching their eggs to the underside of the leaves. Both adults and nymphs feed on the underside of leaves, leaving a honeydew secretion which covers the leaves. Adults are about $^1/_8$", and winged. Is a major pest of citrus, greenhouse plants, cotton, and vegetables.

Whiteflies

Entomology Dept., Clemson University.

Tea scale

Tea scale. *Fiorinia theae*
Tea scale attach to underside of tea and camellia leaves. Females are dark brown to black. Males are snow-white with a ridge down the middle. When abundant, they give the underside of leaves a frosty appearance.

White peach scale. *Pseudaulacaspis pentagona*
Mainly found in the South. It attacks stone fruit and ornamental plants. On peaches, sometimes as injurious as San Jose scale. Females are circular and white tinged with brown. Males are elongated and white.

Entomology Dept., Clemson University

White peach scale

Hymenoptera, Lepidoptera

Entomology Dept., Clemson University.

Redheaded sawfly

Redheaded sawfly. *Neodiprion rugifrons*
Some sawflies feed on plant leaves while others bore in stems, fruits, or leaves. Some species are slug like and covered with slime. The adults range in size from $^{1}/_{4}$" to 1" long. The larvae consume the foliage of trees and shrubs.

Black cutworm. *Agrotis ipsilon*
Black cutworm adults lay eggs singly or a few together in early spring on winter annual weeds. Larvae move to crop plants when weed hosts are destroyed by tillage. Larvae is greasy gray to brown with faint lighter stripes.

Rick Foster, Purdue University.

Black cutworm

Lee Townsend.

Beet armyworm (larva)

Beet armyworm. *Spodoptera exigua*
The beet armyworm is a general feeder that attacks foliage, stems, and roots of field and vegetable crops. Wingspan on the adult is $1^{1}/_{4}$" with grayish-brown forewings and a pale spot in the mid-front. Hind wings are white with a dark anterior margin. There are 4 generations a year. Larvae hatch in 2-5 days, and feed for 3 weeks. Mature larvae are green with prominent, dark lateral stripes about $1^{1}/_{4}$" long. Life cycle is 36 days at 80°F with 4 generations a year in warmer areas.

Ric Bessin.

Beet armyworm (adult)

USDA/ARS

Pink bollworm (larva)

Pink bollworm. *Pectinophora gossypiella*
The bollworm is a major cotton pest. Larvae bore into the squares, causing them to fall or blossoms fail to open. Adult is a small ($^{3}/_{4}$"-wingspan), dark-brown moth. Female begins laying eggs all over the cotton plant 1 or 2 days after emergence. Eggs hatch in 3-5 days, with larvae promptly boring into the squares. Mature larvae are $^{1}/_{2}$" long with a prominent pinkish color. There can be 4-6 generations per season.

DuPont

Pink bollworm (adult)

Photo Identification

Lepidoptera

Entomology Dept., Clemson University.

Codling moth (larva)

Codling moth. *Cydia pomonella*
The codling moth is a major apple pest. Worms tunnel into fruit, with frass or castings protruding, especially at blossom end. Codling moth spins cocoons under loose scales of bark on apple trees. After overwintering, grayish moths (wingspan 1/2"-3/4") emerge in early-/mid-May. After mating, females lay eggs, usually a short distance from an apple cluster. Larvae hatch in 6-20 days, working their way to the fruit, burrowing to the core. Moths are most active when temperatures are between 50°-86°F.

USDA/ARS

Codling moth (adult)

Entomology Dept., Clemson University.

Corn earworm

Corn earworm. *Helicoverpa zea*
Considered the worst corn pest. Worms range in color from light green or pink to brown or nearly black, 2" long when full-grown. Adults have 1 1/2" wingspan. Worms bore directly into the tip of the ear. Usually one worm per ear is found.

European corn borer. *Ostrinia nubilalis*
Insect passes the winter as full-grown caterpillar in the stems of plants where they were feeding. Size is 3/4" to 1" long, flesh-colored with small, round, brown spots.

European corn borer

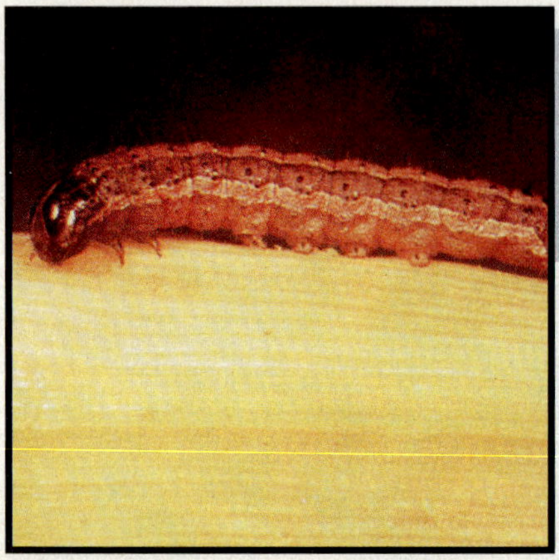

USDA

Fall armyworm

Fall armyworm. *Spodoptera frugiperda*
Female moths lay about 1000 eggs in masses of 150 eggs. Larvae are voracious feeders, consuming a wide variety of crops. Full-grown larvae are 1"-1 1/2" long. Color ranges from light tan or green to nearly black.

Soybean looper. *Pseudoplusia includens*
Major defoliator of soybeans. Overwinters in the pupal stage, emerging as adult (1 1/2" wingspan) in the spring. Forewings are mottled brown with two silver spots, hindwings dusky brown. Larvae are green with light stripes.

Gerald M. Ghidiu, Rutgers University.

Soybean looper

Lepidoptera, Thysanoptera

Gerald M. Ghidiu, Rutgers University.

Squash vine borer (larva)

Squash vine borer. *Melittia cucurbitae*
Squash vine borer larvae tunnel into stems of squash and gourds, often killing the plant. They burrow 1-2" into the ground 4-6 weeks after hatching, emerging as adults in the spring. Adults have clear wings 1"-1$^1/_2$" wide. Females lay eggs on stems and leaf stalks. The borers enter the stem 1-2 weeks later, tunneling through the plant, eating out the inner tissue for a month.

Gerald M. Ghidiu, Rutgers University.

Squash vine borer (adult)

Gerald M. Ghidiu, Rutgers University.

Tobacco hornworm (instars)

Tobacco hornworm. *Manduca sexta*
The tobacco hornworm is 3-4" in length with seven diagonal stripes on each side of the body with a slightly curved, red horn projecting from near the rear end. Moths are 4-5" wide and grayish or brownish in color, with white and dark mottlings. The moths do no damage. The larvae feed on the leaves. Female adults lay eggs on the underside of leaves. The larvae feed on leaves for 3-4 weeks before burrowing into the soil and pupate, emerging in the spring.

Gerald M. Ghidiu, Rutgers University.

Tobacco hornworm (adult)

Gerald M. Ghidiu, Rutgers University.

Tomato hornworm (adult)

Tomato hornworm. *Manduca quinquemaculata*
This pest feeds on tomato foliage, also serious tobacco pest. The worms are large, green, white-barred, 3-4" long. A slender horn projects from the rear end. The adult has zig-zag, dark stripes across the hind wings and orange-yellow spots on the abdomen.

Thrips. Thysanoptera
Eggs are laid on plant tissues, sometimes inserted into slits on plants. Pest is very active when disturbed. Pest is $^1/_8$" long and feeds on plant sap. Is major pest in onion, pears, and citrus.

Ohio Florists Association.

Thrips

Disease Identification

Apple, Beans

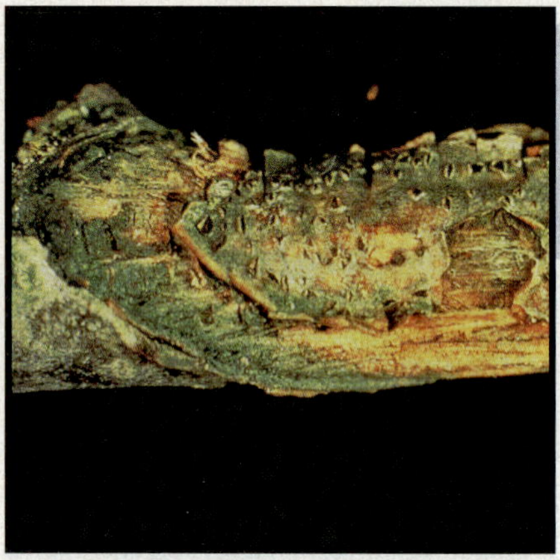

Apple,
Black spot canker

Black spot canker. *Neofabraea malicorticis*
In late fall, cankers form on younger branches through pruning wounds. Appears as reddish-brown, circular spots on bark. Fruit infected from pustules on bark, but may not show until after harvest.

Fire blight. *Erwinia amylovora*
On infected twigs, leaves and blossoms suddenly wilt, turning dark brown to black, shrivel, and die. Bark is shrunken, dark brown to purplish and sometimes blistered with gum oozing out. Treat with fungicide spray during bloom.

Apple,
Fire blight

Apple,
Powdery mildew

Powdery mildew. *Podosphaera leucotricha*
Fruit affected during bloom become russetted, stunted. Leaves become crinkled and curl with gray to white "felty" patches, become brittle and die. Overwinters on twigs and buds. Control with fungicide treatments.

Blight. *Pythium* spp.
Starts at lower node then spreads up and down stem. Has water-soaked then brownish look. Infected tissues are soft, slimy, eventually wilting and dying. Complete defoliation can take place. Infection results in thin stands. Crop rotation can help in disease control.

Beans,
Blight

Beans,
Gray mold

Gray mold. *Botryotinia fuckeliana*
Starts in old blossoms then invades pod or any part of plant the flower falls on. Gray lesions form, often with concentric zonation and yellowish margins. Pods first appear water-soaked, then grayish-brown. Timely fungicide sprays necessary in varieties with history of disease.

Cercospora leaf blotch, leaf spot.
Cercospora canescens
Leaves show brown or rust-colored lesions, circular to angular, 2-10 mm in diameter. Lesions can appear on branches, stems, and leaves. Rotation with nonhost crop can help control disease.

Beans,
Leaf blotch

Blackberry, Powdery mildew

Powdery mildew. *Sphaerotheca macularis*
Infected leaves develop light green blotches on upper surface. Lower surface becomes covered with white fungus growth. If severely infected, shoots become long and spindly with curled leaves. Control with fungicide sprays.

Cane spot. *Elsinoe veneta*
Reddish-brown spots with purple margins and light gray center found mostly on canes. Canes can become girdled, drying developing fruit. Control with improved air circulation in plants.

Black raspberry, Cane spot

Blueberry, Blossom blight

Blossom blight, fruit rot. *Botrytis cinerea*
Leaves become infected after contact with infected flowers. First symptom is chlorotic area, becoming necrotic, turning light brown. Attacks all parts of plant. Rarely attacks older plant parts. Rainy weather during bloom causes disease persistence. Annual pruning will help air movement.

Downy mildew. *Peronospora sparsa*
Affects leaves and berries. On berries, fruit appears dull. Early infection causes fruit to shrivel and harden. On leaves, it appears as yellow discoloration, turning red-purple. Control with sucker removal, destroying canes.

Boysenberry, Downy mildew

Red raspberry, Gray mold

Gray mold, blossom blight. *Botrytis cinerea*
As flowers open, fungus attacks, and blighting occurs. During ripening, drupelets develop a watery rot and appear tan or golden tan. It soon becomes covered by a grayish-brown dust mass. Control with fungicides, cultural practices.

Gray mold, ash mold. *Botrytis cinerea*
Found at all fruit stages. Most frequently found on calyx end or sides of fruit touching rotten fruits. Tissue turns brown, remaining firm. Whole rotted berries retain general shape, becoming tough and dry, with little or no leak. Apply protective fungicides beginning at bloom.

Strawberry, Gray mold

Disease Identification

Citrus, Cucumber

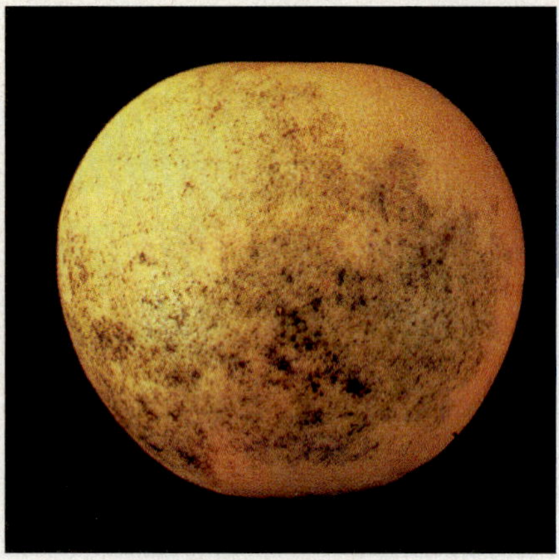

Grapefruit, Greasy spot

Greasy spot. *Mycosphaerella citra*
On fruit, necrotic specks appear 3-6 months after infection. Most are too small to see, but coloring on surrounding tissue may be delayed, giving it a green tint. Lesions can expand on fruit. Fungicides control this disease.

Botrytis blight. *Botrytis cinerea*
Most frequently found on flowers and young fruit, appearing first as yellowish-buff spot. Infected petals are main source of infection of fruit and twigs. Cool, wet weather can cause epidemic.

Lemon, Botrytis blight

Lime, Anthracnose

Lime anthracnose, weathertip.
Gloeosporium limetticola
Young shoots, leaves, and fruits infected. Twigs wither and shrivel from 1-inch to several inches back from tip. Leaves are distorted or have dead areas. Young fruit usually shed.

Fusarium rot. *Fusarium moniliforme*
Postharvest disease. Develops on fruit in long storage. Skin is leathery, beige to light or dark brown, and sunken. If stored under humid conditions, white surface mycelium appears.

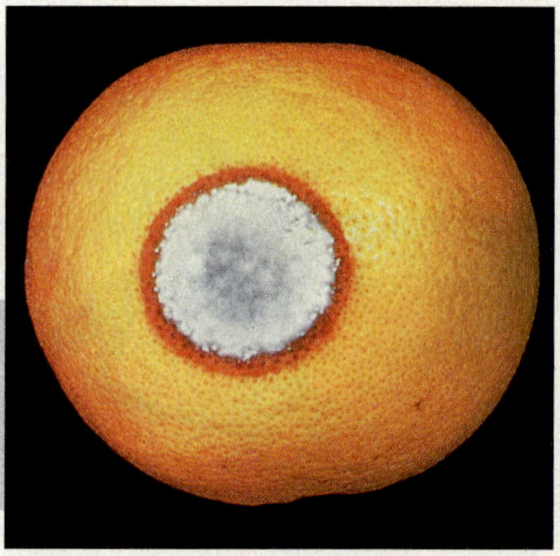

Orange, Fusarium rot

Orange, Leaf spot

Leaf spot, fruit spot. *Cercospora angolensis*
Affects leaves and fruit. Produces grayish-brown circular spots, up to 4 mm in diameter. Spots have a shriveled center and brown border, surrounded by yellow halo. Control with fungicide treatments.

Anthracnose. *Colletotrichum orbiculare*
Lesions appear near leaf veins first, color ranges from light brown to reddish, up to 1 cm in diameter. Fruit lesions are circular, sunken, water-soaked. Control by planting resistant varieties, using protective fungicides, crop rotation.

Cucumber, Anthracnose

Cucumber, Grapes

Cucumber, Cottony leak

Pythium cottony leak, fruit rot. *Pythium* spp. Common soil disease. On fruit, starts as brownish discoloration, spreading throughout, with soft, watery, yellow-brown, oozing areas. Cottony growth then appears on fruit. Enters fruit in contact with soil.

Grape powdery mildew. *Uncinula necator* Leaves, canes, and young fruits covered with white patches. Young leaves become stunted, distorted with whitish-gray, dusty, or powdery appearance. In fruit, berries split and rot. Control with fungicide treatments.

Grape, Powdery mildew

Grape, Bird's eye rot

Bird's eye rot. *Elsinoe ampelina*
Lesions form small sunken spots with dark margins and light centers on fruits, shoots, and leaves. Dead tissue drops out of lesion, leaving leaves with a ragged appearance. Spores develop on overwintering infected berries. Disease develops during periods of heavy rains.
For control, dormant season fungicide sprays recommended. Also apply foliar fungicides at 2-week intervals during growing season.

Grape, Bird's eye rot

Grape, Botrytis

Botrytis. *Botrytis cinerea*
Infection occurs throughout season. From fruit ripening onward, grapes are infected through wounds or the epidermis. Rot then invades the rest of the cluster. Infected white grapes turn brown. Table grapes in cold storage develop circular brown lesions that gradually cover entire fruit. For control, limit vegetation, and improve air flow through vines and increase exposure to sun.

Grape, Botrytis

Disease Identification

Grape, Lettuce

Grape, Downy mildew

Downy mildew. *Plasmopara viticola*
Leaves have yellow-reddish-brown, oily lesions. Infected leaves are source for subsequent infection. Young berries are very susceptible. These berries stop growing, turn dark, and dry with much grayish growth. Older fruits have brown rot without the mildew effect.
Control: Reduce inoculum source by cleaning vineyard floor. Also use fungicide sprays.

Grape, Downy mildew

Lettuce, Downy mildew

Downy mildew. *Bremia lactucae*
A major lettuce disease. Light green or yellowish areas develop on upper leaf surfaces with downy patches on the underside. Leaf turns brown and dies. Disease is worse in cool, damp conditions. For control, fungicides important. Check for resistant varieties.

Gray mold. *Botrytis cinerea*
Has water-soaked, brownish-gray, soft-rot on damaged or older leaves. Older leaves in contact with soil especially susceptible. Progresses to younger part of plant, decaying main stem and detached leaves.

Lettuce, Gray mold

Lettuce, Powdery mildew

Powdery mildew. *Erysiphe cichoracearum*
Affects older, outer leaves, covering both upper and lower surfaces with white, powdery spores. Leaves become slightly yellow, then brown. Heads are small and poor quality. Control with sulfur applications at first sign of infection.

Bottom rot. *Rhizoctonia solani*
Appears on lower leaves in direct contact with soil as small, brown spots. Advanced symptoms are seen at heading. Amber-colored droplets can ooze from lesions on infected midribs. Plant on high beds to improve air circulation.

Lettuce, Rhizoctonia (bottom rot)

Lettuce, Melon, Onion

Lettuce, Shot hole

Shot hole, rust. *Microdochium panattonianum* Small, rounded, water-soaked spots on leaves, later turning brown. Centers often fall out. Can cause heavy losses during prolonged periods of wet weather. Control by crop rotation and discing field debris.

Bacterial wilt. *Erwinia tracheiphila* Serious threat to melons and cucumbers. Distributed by the cucumber beetle. Dull-green patches form on leaves, followed by sudden wilting and shriveling. Diagnose by cutting apart a runner, rejoining, then pulling apart. Slime strands extend from surface. Pest control is key.

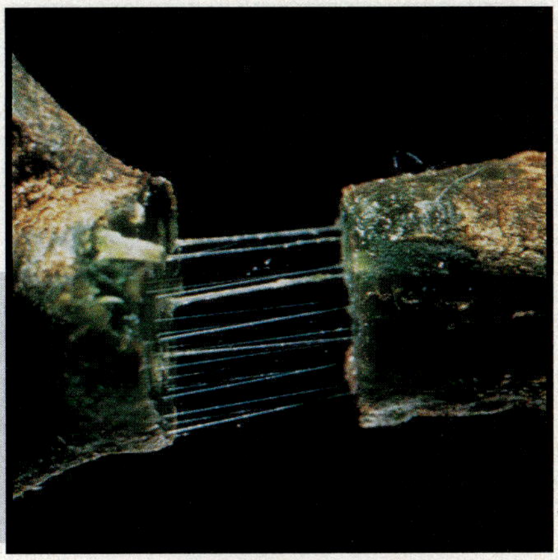

Melon, Bacterial wilt

Melon, Downy mildew

Downy mildew. *Pseudoperonospora cubensis* Affects cucurbit leaves, rarely affects fruit. Irregular yellow spots appear on upper leaves, brown on underside. Under favorable conditions, leaf undersides become downy covered, light gray to deep purple. Leaves die in a few days, exposing fruit to sunscald. Control with fungicides.

Fusarium rot. *Fusarium acuminatum* Most infections occur in the field. Produces reddish-purple pigmentation in infected area. It is easy to detect on fruit. Control with hot water immersion for >1 minute with postharvest fungicide treatment.

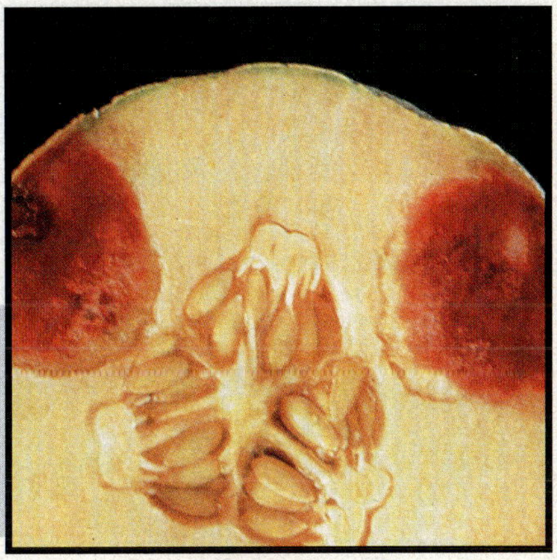

Melon, Fusarium rot

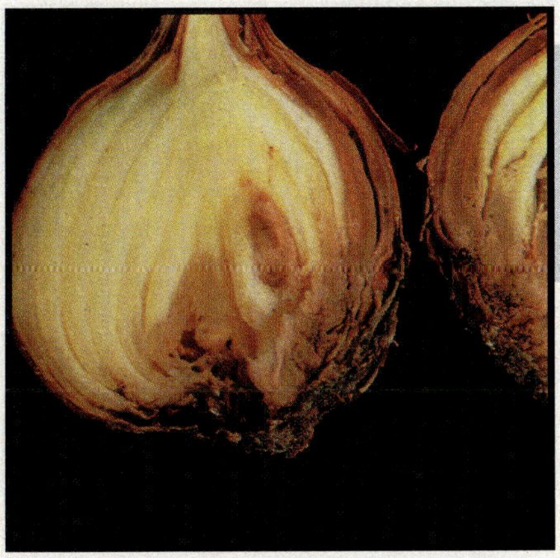

Onion, Basal plate rot

Basal plate rot. *Fusarium oxysporum cepae* Bulb affected any time in field. There is progressive yellowing and dying back from tips. Roots turn pink and gradually decay. Infected bulbs may appear discolored with tissue brown and watery. Rot may not appear until storage.

Soil-line rot. *Botrytis allii* More severe in transplanted onions during cool, moist weather. Fungus may penetrate a number of outer scales and secondary rotting by bacteria and other organisms is common.

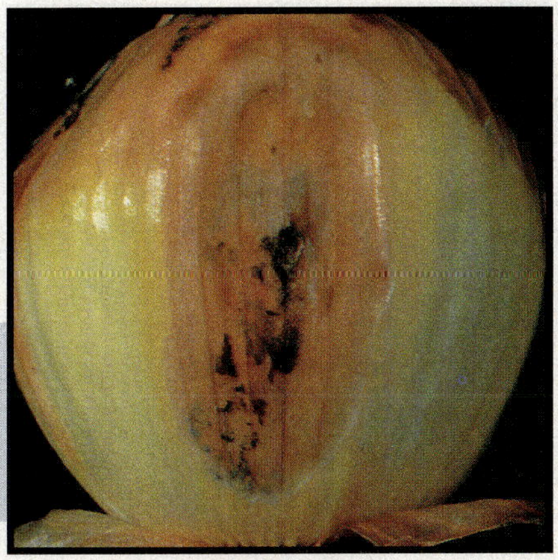

Onion, Soil-line rot

Disease Identification

Ornamentals, Peanut, Pear

Ornamentals, Colletotrichum leaf spot

Colletotrichum leaf spot.
Colletotrichum gloeosporiodes
Usually tan to dark brown, circular lesions. Often accompanied by dieback. Wounding leads to infection. Spread by splashing infected water onto plants.

Cercospora leaf spot. *Cercospora* spp.
Typically starts as tiny, raised area on leaf, becoming scabby, and changing from clear to dark brown, about 1 mm in diameter. Fungicide applications help disease control.

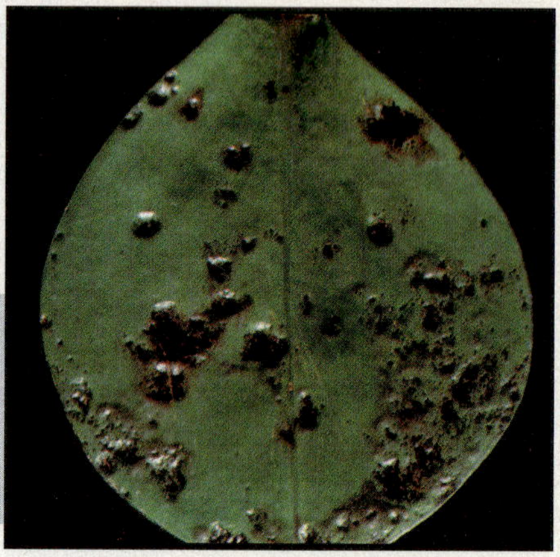

Ornamentals, Leaf spot

Grape Ivy, Powdery mildew

Powdery mildew. *Oidium* spp.
Affects African violet and grape ivy (shown here). Lesions appear on flowers, petioles, and leaves, 1 cm in diameter. A powdery, white mass can cover entire flower or leaf. Fungicide applications can help with disease control.

Botrytis blight. *Botrytis cinerea*
Soilborne fungus attacks all part of plant. Leaflets sometimes infected without evidence of disease anywhere else. For infection, temperatures < 68°F with heavy dew or rain needed. Frost-injured or mechanical-injured plants prone to infection. Fungicides offer some protection.

Peanut, Botrytis blight

Peanut, Pod rot

Pod rot. *Pythium myriotylum*
Widespread throughout world. Symptoms include a black root rot, with plant collapse. Leaves are light green, quickly dying. Roots are water-soaked and clumped. Plant disease-resistant varieties for control.

Blossom end-rot, gray mold. *Botrytis cinerea*
A postharvest disease. First appears as pale tan areas. As it enlarges, older portions of decay may turn darker brown, with edges often remaining a pale color. Control with postharvest fungicides.

Pear, Blossom end-rot

Potato, Rose, Stone Fruit

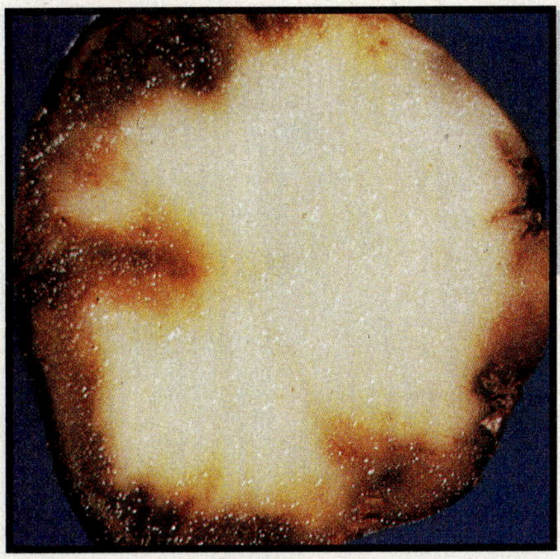

Photo courtesy American Vegetable Grower.

Potato, Late blight

Late blight. *Phytophthora infestans*
Large, dark-green, water-soaked spots appear on lower leaves in wet weather. Spot enlarges and is shriveled, dark brown to black with downy, whitish growth on underside of leaf. Tubers have small, brown-to-purple discoloration on skin. When cut, a reddish-brown dry rot is seen.

Black spot, leaf blotch. *Diplocarpon rosae*
Most important disease of rose. One to a dozen black spots appear, up to 1/2" diameter on upper leaf surface. Leaves turn yellow then fall off. Is spread by splashing inoculum by rain or watering. Do not allow leaves to remain wet for more than 7-12 hours.

Rose, Black spot

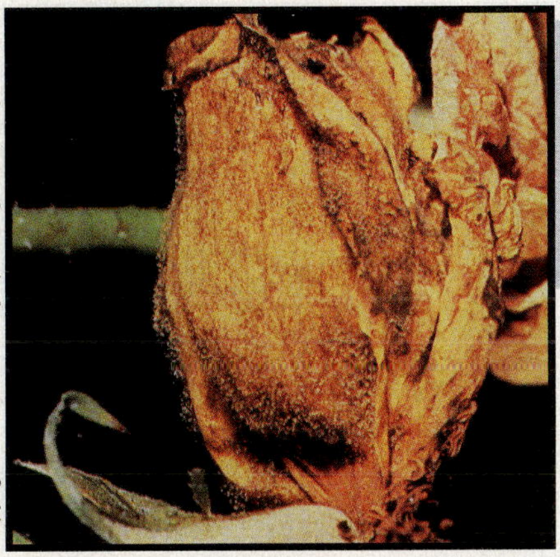

Rose, Botrytis blight

Botrytis blight, gray mold blight, blossom blight.
Botrytis cinerea
During continued wet weather and cool temperatures, infected buds won't open and develop grayish-brown fungus growth. Buds droop and grayish-black lesions found down the stem. For control, remove and destroy all infected parts of plant.

Downy mildew. *Peronospora sparsa*
Young leaves are spotted and drop. Purplish-red to dark brown irregular spots develop on leaf. Leaflets may turn yellow. Control by lowering humidity in greenhouse and raising temperature to 81°F during the day.

Rose, Downy mildew

Rose, Powdery mildew

Powdery mildew, rose mildew.
Sphaerotheca pannosa var. *rosae*
First symptom is slight curling of leaves followed by a white coating, which can expand to flower buds. Usually affects younger leaves. On canes, mildew is heavier and more "felty," especially near thorns. Control with sulfur dusting.

Anthracnose.
Colletotrichum acutatum, C. gloeosporioides
Occurs on ripe or nearly ripe fruit. Has circular, tan to brown necrotic lesions with sunken concentric circles, 4-5 cm in diameter, and firm to touch. Early fungicide treatment is important.

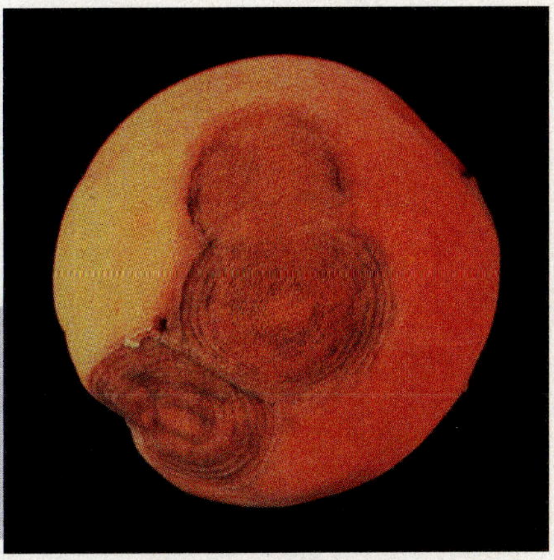

Stone fruit, Peach Anthracnose

Disease Identification

Stone Fruit, Small Grains

Stone fruit, Bacterial spot

Bacterial spot, canker. *Xanthomonas pruni*
Affects all parts of tree. One of most destructive diseases on stone fruit. On fruit, spots turn into brown to black, saucer-shaped depressions with small masses of gummy, yellow exudation. Trees can become defoliated when leaves infected.

Leaf spot, shot hole. *Cercospora circumscissa*
Lesions are round, reddish-brown necrotic spots, enlarging to 4-5 mm and changing to light brown with brownish-red edges. Can defoliate tree by early summer. Fungicide treatments recommended for control.

Stone fruit, Leaf spot

Alfalfa, Phytophthora

Root rot. *Phytophthora megasperma*
Plants wilt and foliage turns yellow to reddish-brown. Regrowth is slow after cutting. Disease occurs in poorly drained soils. Deep plowing, land leveling help in disease control.

Summer black stem, leaf spot. *Cercospora medicaginis*
Leaf spots appear first, then stem lesions. Spots are brown, 2-6 mm in diameter, becoming silvery-brown. 2-3 spots cause leaf to fall off in a few days.

Alfalfa, Summer black stem

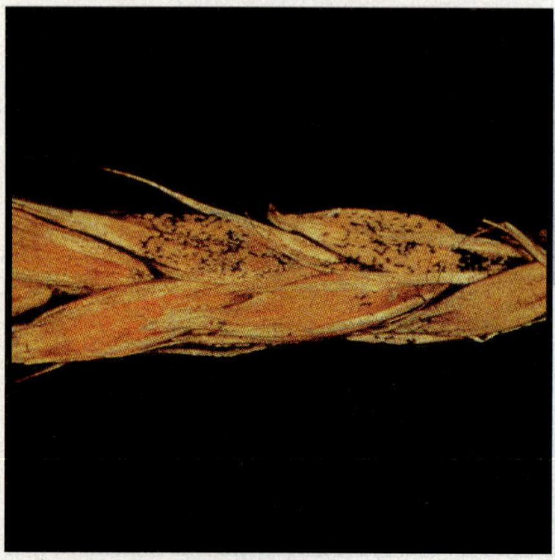

Barley, Head blight

Head blight. *Fusarium graminearum*
Infection starts with small, water-soaked brownish spot at base or in the middle of the glumes or on rachis. Spreads from point of infection. Kernels are shrunken and grayish-brown with interior floury and discolored. Control with fungicide seed treatment, crop rotation.

Head blight. *Fusarium moniliforme*
Kills large areas of the panicle. Causes premature death when panicle or rachis branches infected. Initial infection is in axillaries of panicle branches, spreading throughout panicle causing collapse and drooping of rachis branches.

Sorghum, Head blight

Tobacco, Anthracnose

Anthracnose. *Colletotrichum gloeosporioides* Lesions start as small, light green, water-soaked pinpoint spots, soon enlarging to form oily circular areas up to 3 mm in diameter. Spots dry out, becoming thin and gray-white.

Verticillium wilt. *Verticillium dahliae* Infected leaves become orange-colored and wilt. As it progresses, infected tissue becomes necrotic, leaving orange border between dead and living tissue. Control with field fumigation.

Tobacco, Verticillium wilt

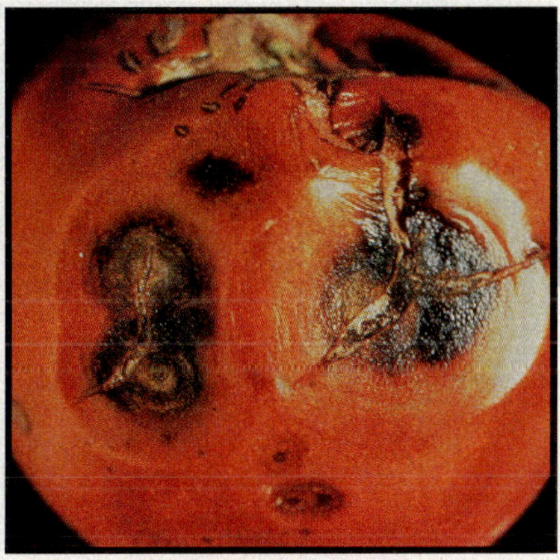

Tomato, Common rot

Common rot. *Colletotrichum phomoides* Small, circular, sunken spots to 1" diameter appear on ripe fruit. Lesion starts with water-soaked look, spots turning dark, with pinkish, cream, or brown spore masses in depressed centers. Fungicide protection as crop ripens is important.

Early blight. *Alternaria solani* Leaves have dark brown, circular to oval spots. Spots grow together and eventually kill off leaves, exposing fruit to sun. Diseased fruit become leathery and covered with a velvety mass of black spores.

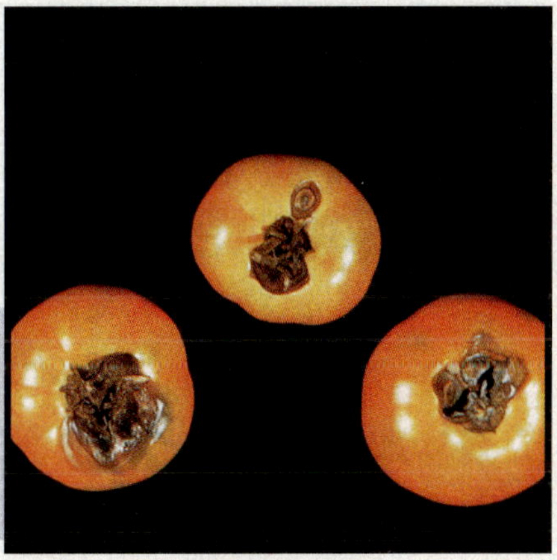

Tomato, Early blight

Tomato, Leaf mold

Cercospora leaf mold. *Pseudocercospora fuligena* Lesions develop on young leaves as indistinct discolorations with no definite margins. A faint halo appears as lesion size increases. Appears on upper and lower leaf surfaces. Destroy crop residue after harvest to limit disease carryover.

Pythium. *Pythium* spp. In emerging plants, begins as dark-colored, water-soaked lesion extending 2-4 cm above soil line. Plants wilt and die. On fruit, starts as small, water-soaked lesion, spreads over entire fruit, skin ruptures and fruit's contents spill out.

Tomato, Pythium

Disease Identification

Turf, Watermelon

Turf,
Anthracnose

Anthracnose. *Colletotrichum graminicola*
Causes root decay and stem rot. In warm weather, leaf spots are small, circular to elliptical, reddish purple which enlarge and fade with age. Control with fungicides.

Blight, grease spot. *Pythium* spp.
Small spots appear, spreading rapidly in hot, humid weather. Infected leaves become light tan to brown, shriveled, and matted. Patches can coalesce, forming large disease areas. Equipment can spread disease.

Turf,
Blight

Turf,
Crown and root rot

Crown and root rot of *Poa pratensis*.
Fusarium spp.
Diseased plants become light green, fading to tan and straw-colored, then dying in hot, dry weather. Roots are black to dark brown. Infection can easily spread in rewetted thatch. Control by removing thatch. Water thoroughly but infrequently.

Snow mold. *Coprinus psychromorbidus*
Initially seen as conspicuous white knots, turning light brown. Matures into brownish-black, irregular-shaped areas. Infected plants have rotted, water-soaked leaves. When dry, leaves are pale-brown with dark red-brown margins.

Turf,
Snow mold

Watermelon,
Alternaria

Alternaria leaf blight. *Alternaria cucumerina*
Lesions appear on older leaves first. Initially yellow-brown and small, expanding to brown, necrotic areas up to 20 mm in diameter. Leaves become cupped, then die, exposing fruit to sun. Spreads rapidly in warm, humid weather.

FOOD QUALITY PROTECTION ACT PUTS FARMING AT RISK

This act could make it difficult to produce crops in the United States if it's not implemented properly.

By Laurie Moses
Editor

Folks, farming as we know it is in danger. It's not because of drought. It's not because of floods. No, it's because of a regulation EPA is implementing, the Food Quality Protection Act (FQPA).

This regulation, passed in 1996, will do what no anti-pesticide person has been able to do, take away pest control products you've been able to legally use for years in integrated pest management programs for all crops grown in the United States.

Farmers and growers are not opposed to food safety laws. In fact, if abundant crops could be grown without the use of pest control products, farmers would be the first to do it. However, since this isn't possible, the next best thing is to have the safest products available.

What makes FQPA such a concern? Mainly that decisions will be made arbitrarily and uses banned without the benefit of input from agricultural interests. Many products do not have an alternate product growers and farmers can turn to.

OPs, Carbamates First To Be Evaluated

Congress passed FQPA. It requires all existing pesticide tolerances to be re-evaluated for risk, using a much stricter set of scientific standards. It forces EPA to re-assess all existing chemical tolerances within 10 years. Organophosphates (OPs) (such as diazinon, malathion, chlorpyrifos, and dimethoate), and carbamates are the two classes that must be re-evaluated by August 1999. There are 39 total, registered for agricultural and non-agricultural uses.

Just one problem: These two chemical classes are very important to farming in the U.S. While growers and farmers have been using them for years to produce the safest, most abundant food supply in the world, EPA is now determined to eliminate these uses – without the benefit of finding alternatives first.

Farming isn't going to give in without a fight. A coalition of farm groups has banded together to create a "roadmap" for EPA to follow. This roadmap gives some concrete, reasonable steps EPA could follow to implement FQPA fairly (see sidebar next page).

Why has this issue become such a hot topic? To get to this point, we need to go back in time, to the 1950s. Back then, Congress passed the Delaney Clause which required no detectable residues in processed food, and if detected, the product be banned from use. This law was fine back then. But technology raced ahead of the law. Instead of being able to detect residues down to parts per thousands, we're able to detect down to parts per million and as low as parts per billion or parts per trillion. One part per million is equivalent to one drop of water in 21.7 gallons. One part per billion is equivalent to one drop of water in 21,700 gallons.

Now, zoom ahead to 1996. Congress finally acted and passed FQPA. EPA must look at all uses for a product: agricultural and non-agricultural and decide whether a use should continue to be registered. Non-agricultural includes mosquito control, home lawn and garden use, etc. This means there are a lot of people interested in the outcome of this debate as it continues. It's not an easy process for EPA to undertake (see sidebar).

Proper Procedure Is Key

Why should farmers and growers care about this? Because it isn't just OPs and carbamates potentially on the chopping block. They are only the first of many chemical classes EPA must evaluate in the next few years. Whatever procedure is set up now will be used to evaluate chemical classes in the future. If this process is set up fairly and properly and actual use evaluated accurately, it will be easy for EPA to evaluate other products. If it isn't, farmers and growers are going to lose other crop protection products down the road.

What can be done? Contact your congressional representatives and let them know you're very concerned about the process EPA is undertaking. Proper implementation and sound scientific principles will keep pest control products on the tractor and in the field for years to come. □

Risk Cup Can't Run Over

As EPA looks at a product, it determines the "risk" it has based on the amount of active ingredient a person can be exposed to each day without that ingredient becoming a health threat. What makes this so difficult for farming is EPA is evaluating exposure to the active ingredient in food, water, and in the home. All of this exposure is put into an imaginary "risk cup." If this cup overflows, uses have to be dropped to stop it from overflowing.

What fills this cup? Use data. For some active ingredients, there is no actual use data (information from growers that states specifically the actual amount of a pesticide being used). Without this real data, EPA assumes farmers and growers are using the maximum amount of a product on a crop every season. This is rarely the case. Add to the cup consumer uses (where use data is more difficult to determine), and the cup will overflow.

If EPA assumes the maximum amount is being used, then the risk cup fills quite quickly, and something has to give. What will give could be an active ingredient for all registered uses, or specific uses dropped, such as an organophosphate apple growers use in their orchards.

Want one more complication? There is a possibility that EPA will add into the cup "phantom residues." This means even if there are no detectable residues on a crop, EPA is going to assume there has to be some sort of residue present, therefore it's added into the risk cup.

With all of these factors on the table, EPA will make their decisions. The first chemical classes on the docket are organophosphates and carbamates (see chart, page 31). These classes have been used by farmers and growers for years now and are an important part of integrated pest management programs (see page 30). Whatever process EPA sets up now will be used for other chemical classes.

IPM AND RESISTANCE MANAGEMENT ON THE ROPES

Organophosphates and carbamates are essential elements of effective crop management.

By Jennifer Speth
Assistant Editor

Over the years, farmers and researchers have built IPM and resistance management programs around the cultural and chemical tools available. With the likelihood of losing organophosphates (OPs) and carbamates looming, growers are gravely concerned about their ability to manage weeds, insects, and diseases in the future without significant impact on the quality of the food and fiber they grow.

Resistance to insecticides is an ever increasing problem. It happens as individually resistant insects survive, breed, and pass on their resistant genes to their offspring. Trying to determine how EPA will finally interpret FQPA has made this problem even more frustrating.

Resistance management programs use tactics to slow resistance to a chemical, while also reducing the impact of pesticides on beneficial predators and parasites. The only proven method to control resistance problems is to rotate different chemical classes or families of insecticides.

Says *Insect and Disease Control Guide* advisor, Dr. Rick Foster, associate professor, Purdue University, "One of the good things about OPs in apple production is that some of our predator mites are resistant to them. If we replace the OPs with something else that kills the predators, we wind up applying more pesticides to control the mites."

If OPs and carbamates are banned or highly restricted, more pressure is placed on remaining products to perform, and at this point there are not many viable options. Russell Mizell, IPM coordinator in Florida, points out, "There are a lot of crops and pests that do not have any alternatives to OPs and carbamates."

According to Leonard Gianessi from the National Center for Food and Ag Policy, "When you talk about reducing the number of classes of chemistries in an arsenal, you have the potential for pests developing resistance to the remaining classes, that's a very scary scenario."

While many new products either presently available or in the pipeline are useful, not only because they work, but because they don't kill beneficial insects, Michigan State University entomologist Blake Layton, believes that "these products are so specific they provide opportunities for new pests to increase in status. OPs and carbamates provide us alternative methods of control."

According to Bob Nichols of Cotton Incorporated, "It's very clear. Reduce flexibility in the system in any way, and you increase the probability of failure." Mizell agrees, "It boils down to we grow the crops and use the pesticides, or we don't grow the crops." And that can have many in agriculture with a noose around the neck. □

Implementing FQPA Requires Grower Guidance

The Implementation Working Group (IWG), a coalition of organizations and individuals from all walks of agriculture, directly addressed the issue of Food Quality Protection Act (FQPA) implementation through the creation of its historic "Road Map." This visionary document provides guiding principles for the fair and efficient implementation of FQPA, and state what the industry desires from EPA:

1. Sound Science. Implementation must be based on sound science and reliable information. All major science policies and science assessment methodologies must be peer-reviewed before they are adopted. In the absence of reliable scientific information, the use of unrealistic default assumptions or unclear decision criteria will jeopardize the credibility of the decision-making process.

2. Transparency. Implementation must be transparent: The public must be informed of the criteria used to assess risk and the process by which decisions are reached. It is essential that the very complex and evolving science and regulatory policies and guidelines that govern decisions be formally communicated to the public in a clear, timely, and open manner. Opportunities for public review and comment must be provided prior to the adoption of all major science and regulatory policies and processes.

3. Balance. Implementation must be balanced: Responsible use of pesticides contributes to public health by ensuring an abundant, wholesome, and affordable supply of food. Further, many pesticide products are essential in the fight against direct public health threats such as malaria and Lyme disease.

As EPA considers canceling some older pesticide uses as a result of the tolerance reassessment and reregistration processes, the Agency must give priority to the review and approval of new products. Balance is critical to sustaining the global competitiveness of American agriculture and for developing reasonable policies that will avoid trade inequities or barriers.

4. Workability. Implementation must be workable: The law must be administered in a practical and reasonable way. For example, if it is decided that certain existing products do not meet the FQPA standard, it is important to provide users and producers with opportunities to provide input to the final decisions, as well as sufficient transition time for users to adapt to new or alternative products and practices. Failure to implement the law reasonably and consistently will result in excessive litigation and potential major disruption of agricultural production and public health protection.

CROPS AFFECTED BY OP AND CARBAMATE USE

Organophosphates and carbamates are used on many crops in the United States. This list gives a quick overview of those uses. Find out more by turning to the crop listings in this book.

	Organophosphates																						Carbamates						
	Acephate	Azinphos-Methyl	Chlorpyrifos	Diazinon	Dicrotophos	Dimethoate	Disulfoton	Ethion	Ethoprop	Fenamiphos	Fonofos	Malathion	Methamidophos	Methidathion	Methyl Parathion	Naled	Oxydemeton-Methyl	Phorate	Phosmet	Profenofos	Sulprofos	Terbufos	Aldicarb	Carbaryl	Carbofuran	Formetanate HCL	Methomyl	Oxamyl	Thiodicarb
Almonds		✔	✔	✔										✔					✔					✔					
Apples		✔	✔	✔		✔		✔				✔		✔	✔				✔					✔		✔	✔	✔	
Blackberries	✔	✔		✔								✔												✔					
Blueberries		✔		✔								✔							✔					✔			✔		
Cherries		✔	✔	✔		✔						✔		✔	✔				✔					✔					
Citrus		✔	✔	✔		✔		✔		✔		✔		✔		✔							✔	✔		✔	✔		
Cranberries	✔	✔	✔	✔								✔												✔	✔				
Grapes		✔	✔	✔		✔				✔		✔			✔	✔			✔					✔	✔		✔		
Nectarines		✔	✔	✔						✔				✔	✔				✔					✔		✔	✔		
Peaches		✔	✔	✔						✔		✔		✔	✔				✔					✔		✔	✔	✔	
Pears		✔	✔	✔		✔								✔	✔				✔					✔		✔	✔	✔	
Pecans		✔	✔	✔		✔	✔	✔				✔		✔	✔								✔	✔			✔		
Raspberries		✔		✔						✔		✔					✔							✔			✔		
Strawberries		✔	✔	✔							✔	✔			✔	✔	✔							✔	✔		✔		
Walnuts		✔	✔	✔								✔		✔		✔	✔							✔					
Asparagus			✔				✔				✔	✔												✔					
Broccoli			✔	✔		✔	✔			✔	✔	✔	✔		✔	✔	✔							✔			✔		
Cabbage		✔	✔	✔		✔	✔		✔	✔	✔	✔	✔		✔	✔	✔							✔			✔		
Carrots				✔								✔			✔									✔			✔	✔	
Celery	✔	✔		✔									✔		✔	✔								✔			✔	✔	
Cucumbers		✔		✔		✔			✔			✔	✔		✔		✔							✔	✔		✔	✔	
Green Beans	✔		✔	✔		✔	✔		✔		✔	✔			✔	✔	✔	✔						✔			✔		
Lettuce	✔			✔		✔	✔					✔	✔		✔		✔							✔			✔		
Melons		✔		✔		✔	✔	✔				✔	✔		✔	✔	✔							✔	✔		✔	✔	
Onions		✔	✔	✔		✔		✔			✔	✔			✔		✔							✔			✔	✔	
Peppers	✔	✔		✔		✔	✔				✔	✔	✔		✔		✔							✔	✔		✔	✔	
Potatoes		✔		✔		✔	✔		✔		✔	✔	✔		✔		✔	✔	✔					✔	✔		✔	✔	
Sweet Corn			✔	✔		✔			✔		✔	✔			✔		✔	✔				✔		✔	✔		✔		✔
Tomatoes		✔	✔	✔		✔	✔				✔	✔	✔		✔									✔			✔	✔	
Alfalfa		✔	✔	✔		✔						✔	✔	✔	✔	✔			✔					✔	✔	✔	✔		
Field Corn			✔	✔		✔	✔				✔	✔			✔		✔	✔				✔		✔	✔		✔		
Cotton	✔	✔	✔	✔	✔	✔	✔			✔		✔	✔	✔	✔	✔	✔	✔		✔	✔		✔	✔			✔	✔	✔
Soybeans	✔		✔			✔	✔								✔									✔	✔		✔		✔
Wheat			✔	✔		✔	✔					✔			✔			✔						✔	✔		✔		
Ornamentals	✔	✔	✔	✔								✔				✔								✔	✔				

Personal Protective Equipment

QUICK GUIDE TO WORKER PROTECTION

This guide is intended for APPLICATION clothing & equipment, it is ***not***

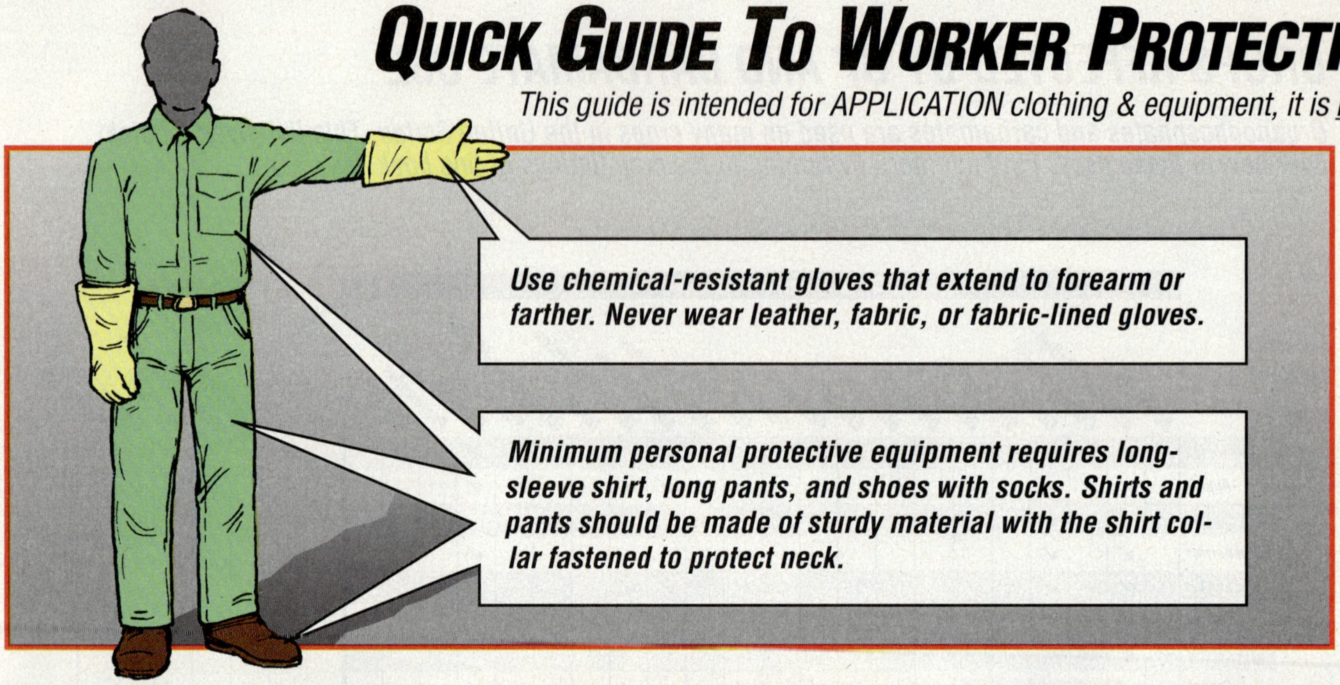

NOTE: ***Always refer to label for exact protective clothing information.***

CLOTHING & EQUIPMENT

intended for EARLY ENTRY or MIXING & LOADING.

IMPORTANT

Always refer to product label for exact protective clothing and equipment requirements. This chart is intended only as a guide.

GREEN

These are the minimum requirements – long-sleeve shirt, long pants, shoes with socks, and waterproof or chemical-resistant gloves.

YELLOW

In addition to shirt, pants, and gloves, requires varying combinations of chemical resistant footwear, apron, hat, and eye protection as specified by label. Respirator protection is not required.

BLUE

Shirt, pants, gloves, and varying combinations of chemical-resistant footwear, apron, hat, eye protection, and respirator as specified by label.

PURPLE

Requires coveralls over shirt and pants, chemical-resistant gloves, and varying combinations of chemical-resistant footwear, apron, hat, and eye protection as label specifies. Respirator protection not required.

ORANGE

Respirator protection required, plus coveralls over shirt and pants, chemical-resistant gloves and varying combinations of chemical-resistant footwear, apron, hat, and eye protection as label specifies.

RED

Requires chemical-resistant protective suit, chemical-resistant gloves, chemical-resistant footwear, apron, hat, eye protection, and respirator as label requires.

PURPLE

Respirator protection not required.

Coveralls over shirt and pants, and a combination of chemical-resistant gloves, footwear, apron, hat, and eye protection.

ORANGE

Same as purple, except requires respirator protection.

RED

Chemical resistant protective suit, chemical-resistant gloves, and footwear, apron, hat, eye protection, respirator.

NOTE: ***Always refer to label for exact protective clothing information.***

KNOW YOUR FORMULATION

This guide will help you match formulation to type of application being made.

The active ingredient (a.i.) is the agent in a formulation which has a specific effect on a pest, weed, or plant. A single active ingredient often is sold in several different kinds of formulations. You must choose the formulation that will be best for each use. In making your choice, consider:

- The plant, animal, or surface to be protected (phytotoxicity, animal absorption, pitting or marring surface);
- Application machinery available and best suited for the job;
- Hazard of drift and runoff (nearness to sensitive areas, likelihood of wind or rain);
- Safety to applicator, helpers, other humans, and pets likely to be exposed;
- Habits or growth patterns of the pest (bait vs. broadcast spray, granular vs. foliar spray);
- Cost; and
- Type of environment in which the application must be made (agricultural, aquatic, forest, urban, etc.).

The amount of active ingredient and the kind of formulation are listed on the label. For instance, a 50W contains 50% by weight of a.i. and is a wettable powder. If it is a 10 lb. bag, it contains 5 lb. of a.i. and 5 lb. of inert ingredients. Liquid formulations indicate the amount of a.i. in pounds per gallon. For example, a 4E means 4 lb. per gallon of the a.i. in an emulsifiable concentrate formulation.

Aerosol

1. Ready-to-use type such as household sprays. Commercial type holds 5 to 10 pounds of formulation and can be refillable.

2. For smoke or fog generators that break the liquid formulation into a fine mist or fog using a rapidly whirling disc or a heated surface.

DF or WDG - Dry Flowables

Dry flowables, also known as water-dispersible granules, are like wettable powders except the active ingredient is formulated on a granule instead of a powder. Requires agitation. Easier to pour and mix than wettable powders because there is less dust.

D - Dusts

Low percentage of a.i. on a very fine dry inert carrier like talc, chalk or clay. Most are ready to use. Danger of drift.

EC or E - Emulsifiable Concentrate

A liquid formulation containing the a.i., one or more solvents, and an emulsifier which allows mixing with water.

Advantages:

- High concentration means price per pound of a.i. is relatively low and product is easy to handle, transport, and store.

Suffixes of Chemical Brand Names

Suffix	Meaning
Formulations:	
AF	Aqueous Flowable
AS	Aqueous Suspension
D	Dust
DF	Dry Flowable, also Water-Dispersible Granule
EC or E	Emulsifiable Concentrate
ES	Emulsifiable Solution
F, FL or LF	Flowable, Liquid Flowable
G	Granular
OL	Oil-Soluble liquid
P or PS	Pelleted
S	Solution
SL	Slurry
SP	Soluble Powder
SG	Sand Granules
ULV	Ultra-Low Volume Concentrate
W or WP	Wettable Powder
WDG	Water-Dispersible Granules, also Dry Flowable
WSB	Water-soluble Bag
WSP	Water-soluble Pak or Packet
How A Pesticide Is Used:	
AG	Agricultural
GS	For Treatment of Grass Seed
LSR	For Leaf Spot and Rust
PM	For Powdery Mildew
RP	For Range and Pasture
RTU	Ready to Use
SD	For Use as a Side Dressing
TC	Termiticide Concentrate
TGF	Turf Grass Fungicide
WK	To Be Used with Weed Killers
Describe Characteristics Of The Formulation:	
BE	The Butyl Ester of 2,4-D
D	An Ester of 2,4-D
K	A Potassium Salt of the Active Ingredient
LO	Low Odor
LV	Low Volatility
MF	Modified Formulation
T	A Triazole
2X	Double Strength

- Little agitation required; not abrasive; will not settle out or separate when equipment is running.
- Little visible residue on fresh fruits and vegetables and on finished surfaces.

Disadvantages:
- High concentration makes it easy to overdose or underdose through mixing or calibration errors.
- Phytotoxicity hazard usually greater.
- Easily absorbed through skin of humans or animals.
- Solvents may cause rubber or plastic hoses, gaskets, and pump parts and surfaces to deteriorate.

F or FL - Flowable

A liquid formulation consisting of a finely ground active ingredient suspended in a liquid. Mixed with water for application.

Advantages:
- Seldom clogs nozzles.
- Easy to handle and apply.

Disadvantages:
- Require moderate agitation.
- May leave a visible residue.

G - Granules

Most often used for soil applications. The a.i. is coated or absorbed onto coarse particles like clay, ground walnut shells, or ground corn cobs.

Advantages:
- Ready to use; no mixing.
- Drift hazard is low because particles settle quickly.
- No spray, little dust mean low hazard to applicator.
- Weight carries the formulation through foliage to soil target.
- Simple application equipment, often seeders or fertilizer spreaders.

Disadvantages:
- Won't stick to foliage.
- More expensive than WPs or ECs.
- May need to be incorporated into soil.
- May need moisture to activate pesticidal action.

Invert Emulsion

A water soluble pesticide dispersed in an oil carrier. Forms large droplets which do not drift easily.

Low Concentrate Solution

Small amounts of a.i., 1% or less, used without dilution for structural pests, space sprays in barns, mosquito control, etc.

Microencapsulation

Particles of a pesticide, either liquid or dry, surrounded by a plastic coating. Mixed with water and applied as a spray. Encapsulation makes timed release possible.

SP - Soluble Powder

A dry formulation which, when mixed with water, dissolves readily and forms a true solution. When thoroughly mixed, no agitation necessary. Not many formulations of this type are available because few active ingredients are soluble in water.

S - Solution

For those active ingredients which dissolve readily in water, of which there are relatively few in number. The formulation is a liquid and usually consists of the a.i. and additives. When mixed with water will not settle out or separate.

Water-Soluble Packet

Water-soluble packets are used to reduce the mixing and handling hazards of some highly toxic pesticides. Preweighed amounts of wettable powder or soluble powder formulations are packaged in water-soluble plastic bags. When the bags are dropped into a filled spray tank, they dissolve and release their contents to mix with the water. There are no risks of inhaling or coming into contact with the undiluted pesticide during mixing as long as the packets are not opened.

WP or W - Wettable Powders

Dry, finely ground formulations which look like dusts. The a.i. is combined with a finely ground dry carrier, usually mineral clay, along with other ingredients that enhance the ability of the powder to suspend in water. The powder is mixed with water for application as a spray. Wettable powders are one of the most widely used pesticide formulations.

Advantages:
- Low cost.
- Easy to store, transport, and handle.
- Lower phytotoxicity hazard than ECs and other liquid formulations.
- Easily measured and mixed.
- Less skin and eye absorption than ECs and other liquid formulations.

Disadvantages:
- Inhalation hazard to applicator while pouring and mixing.
- Require good and constant agitation (usually mechanical) in the spray tank.
- Abrasive to many pumps and nozzles.
- Residues may be visible. □

Source: Applying Pesticides Correctly, Ohio State University; Safe & Effective Use of Pesticides, University of California.

KNOW YOUR NOZZLES

The right nozzle selection will help minimize drift, improve droplet size, and make application more efficient.

By Alvin R. Womac, Ph.D.

It's well known that drift can be minimized — and spray efficiency maximized — through selection of the proper sprayer nozzle. Droplet size also comes into play when selecting a nozzle. Typically, the larger the droplet size, the less chance there will be for drift.

But current pesticide labels usually do not recommend specific nozzles for specific pesticides for two reasons:

1. They wouldn't allow applicators any flexibility to best match the nozzle with the spray condition.

2. They wouldn't allow for use of new designs in nozzles without having to change labels as well.

Nonetheless, we know that nozzle type, flow rate, and operating pressure all influence the suitability of a nozzle application relating to droplet size — and that droplet size, in turn, may affect product performance, application efficiency, and drift. We also know that some products require varying levels of application droplet sizes, depending on the situation.

Label Language

Because of these facts, there is a trend toward specifying information for nozzle selection on labels. But, in practical terms, how specific can recommendations be?

The American Society of Agricultural Engineers — working with nozzle manufacturers, regulators, and the Spray Drift Task Force (with EPA and USDA) — has determined that the most practical approach would be to create general droplet size categories (such as very fine, which is the smallest; fine; medium; coarse; very coarse; and ultra coarse, which is the largest), then specify which nozzles would fall into each. This classification system would allow applicators more freedom to select nozzle type, size, and pressure — so long as the com-

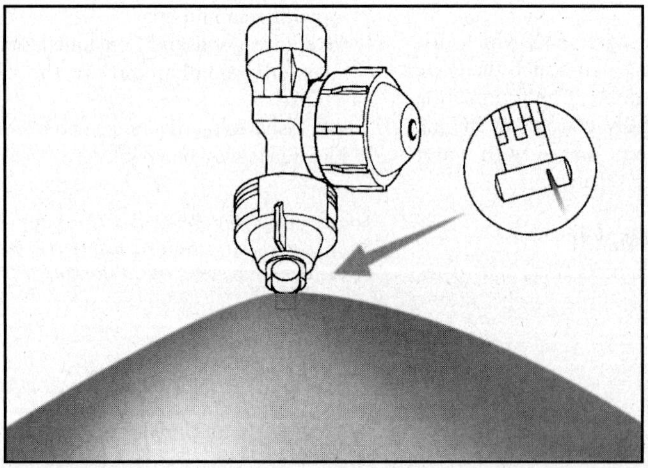

Turbo flat fan front

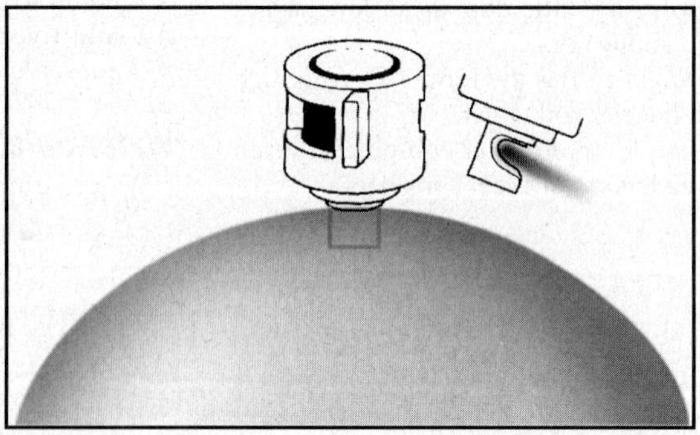

Flooding flat fan

bination they selected produces the droplet size that falls within the specified category.

While a system of nozzle classification has not yet been fully developed and is not yet in effect, it likely will be in the next several years. The wise applicator will be well prepared. Here's a sneak preview of how it might work.

More Informed Choices

Currently, selecting a general nozzle type (such as a hollow-cone, flood, flat-fan, etc.) is the main consideration given to droplet size. For instance, cone types and flat-fans are selected for applications needing uniform foliage coverage from small droplets, while flood nozzles are selected for applications that can use large droplets, such as soil incorporation.

These basic selections will remain unchanged — except that the nozzle user will have more information to see how much the droplets vary in size. Nozzle classification will add a degree of refinement to nozzle selection. The main reason: Nozzle tips at different pressures can be compared for droplet size quality. Therefore, a nozzle can fall into two or more categories because operating pressure affects droplet size.

Under a classification system, a chemical label would specify the droplet size range, then allow the applicator to choose a spray nozzle within that parameter. □

TYPES OF NOZZLES AVAILABLE

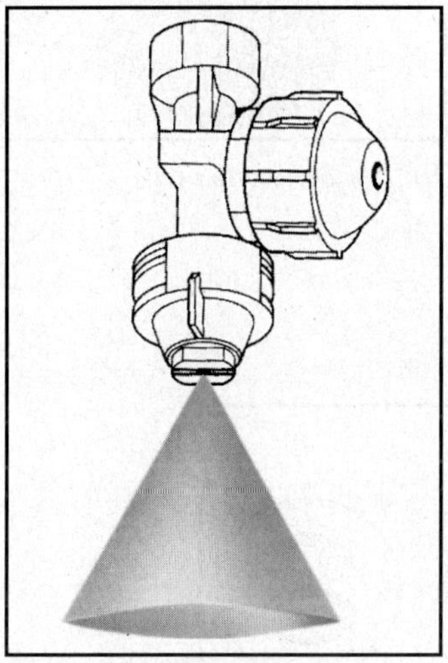

Flat fan extended range

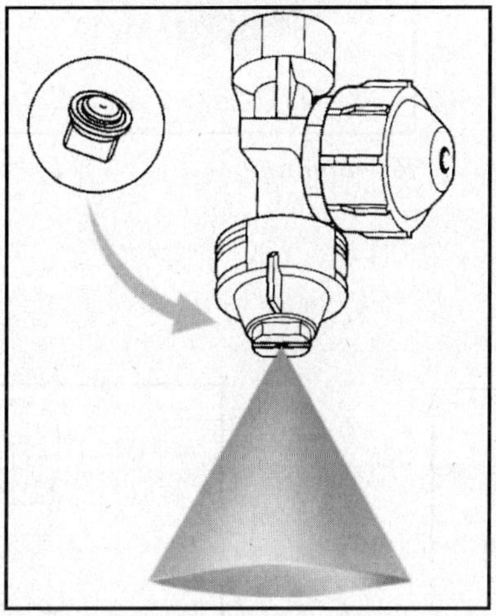

Drift reduction

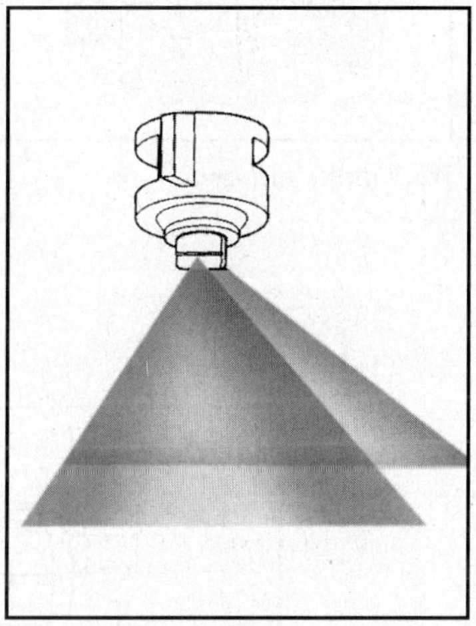

Twin orifice flat fan

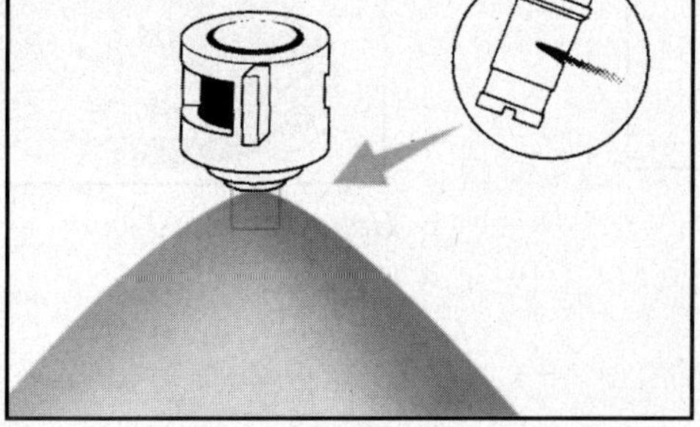

Turbo flood flat fan

TYPES OF NOZZLES AVAILABLE

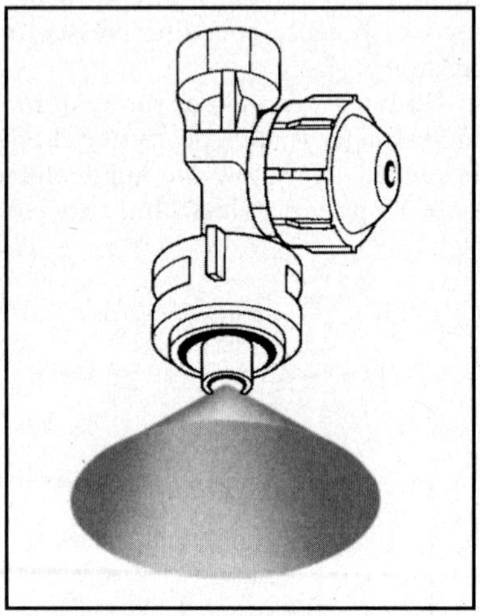

Wide angle full cone

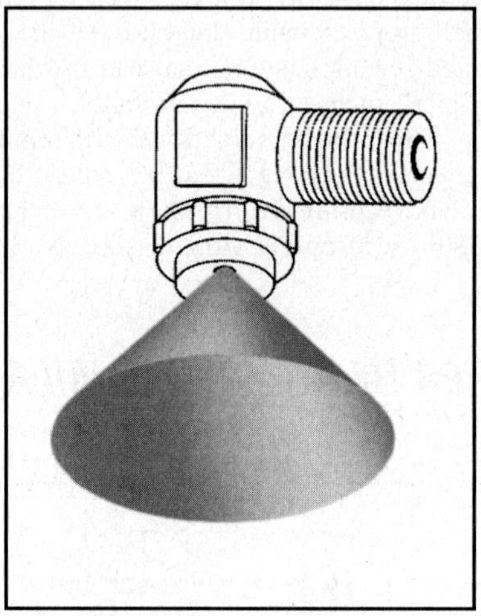

RA-raindrop

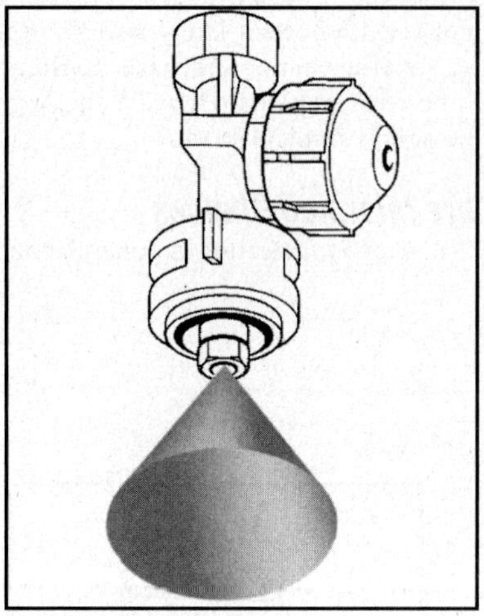

Hollow cone one piece

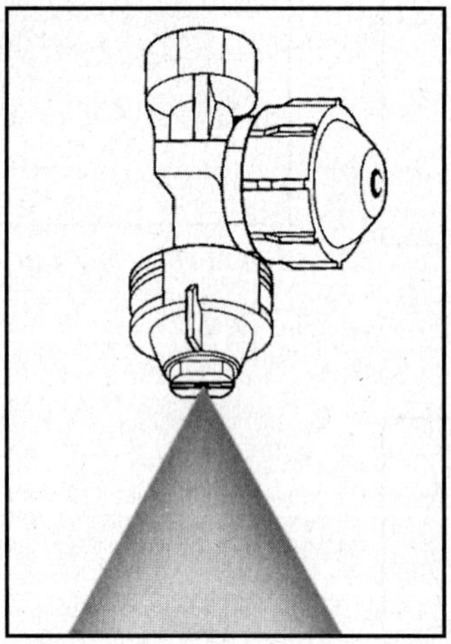

Even flat fan

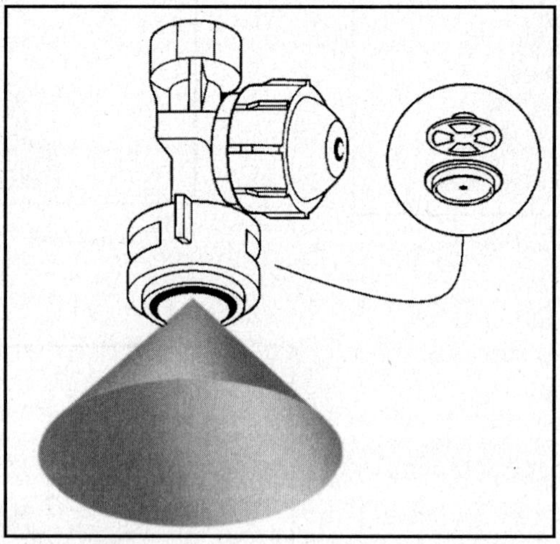

Hollow core disk-core

Illustration source: 1997 Illinois Agricultural Pest Management Handbook.

Understanding Worker Protection Standard

This information will help you keep up-to-date on federal WPS requirements you must follow

The following summary is intended to help you comply with the U.S. EPA's Worker Protection Standard (WPS). Pesticide products that are subject to the WPS must carry updated labeling. Look for the "Agricultural Use Requirement" box on the label. All products being distributed or sold must bear labeling that complies with the WPS.

For major manufacturers, this list contains pesticide products that are used in agricultural production and are subject to the WPS. For other companies' products, check the label. Unless noted, the products in this chart have a federal registration. Individual states may restrict the use of certain products, however, so be sure to read the product label carefully.

This list has been compiled from information supplied to the *Farm Chemicals Handbook*, the *Weed Control Manual*, and the *Insect and Disease Control Guide*, as well as from individual product labels. This list is a helpful reference, but should not be regarded as a substitute for the product label. Always refer to the label to ensure that you are using a pesticide properly and are in compliance with the WPS.

How To Use This Table

For each product, you'll find the following information:

Product name — Information in the chart applies to specific formulated products. Check your product label before using this chart.

Manufacturer — Products in this chart are grouped by manufacturer. If you don't know a product's manufacturer, look for the product name in the Product Index.

EPA registration number — Issued by the EPA.

Active ingredient(s) — If the product is a premix, two or more active ingredients will be given.

Restricted Use Pesticide (RUP) status — In some cases, only certain uses of a product may be restricted. Always check the product label for complete information. This information is not required for WPS information posting, but it indicates which products are subject to the federal recordkeeping rule for restricted use pesticides. Individual states may restrict other products, so review local regulations before using any pesticide.

Signal Word — Pesticides are labeled with a signal word based on their levels of toxicity. Class I pesticides carry the word "Danger" or "Danger-Poison." Class II pesticides are labeled "Warning," while Class III or IV products bear the signal word "Caution."

Action(s) — The product's major pesticidal activity.

Worker notification — Notification of a pesticide application must be given to all workers who will be within 1/4-mile of a field during the application or before the restricted-entry interval expires. This notification may be either oral or written, e.g., a posted sign at the entrances to treated sites. For some products, both oral and written notification is necessary.

Restricted-entry Interval (REI) — The time, in hours, before it is safe to reenter a field that has been treated with the product without wearing appropriate personal protective equipment and clothing for early entry. For some products, the REI may be different in arid areas, or it may vary depending on the crop or method of application. If the REI is footnoted in the chart with a "†", refer to the product label for clarification.

Personal protective equipment (PPE) — The minimum protective clothing and equipment that is needed to safely handle or load the product.

Early entry PPE — The minimum protective clothing and equipment that is needed to safely enter an area that has been sprayed with this pesticide before the REI has elapsed. □

Worker Protection Standard

Product Name	EPA Number	Active Ingredient(s)	RUP	Signal Word	Action	Notification	REI	PPE	Early Entry PPE
Abbott Laboratories									
Biobit* HP	275-120	*Bt.* spp. *kurstaki*	No	Caution	IN	Oral	4	a e g	b e g
Biobit* XL	275-116	*Bt.* spp. *kurstaki*	No	Caution	IN	Oral	4	a e g	b e g
DiPel* DF	275-103	*Bt.* spp. *kurstaki*	No	Caution	IN	Oral	4	a e g m	b e g
DiPel* ES	275-65	*Bt.* spp. *kurstaki*	No	Caution	IN	Oral	4	a f g	b f g
DiPel* ES-CPI	275-65	*Bt.* spp. *kurstaki*	No	Caution	IN	Oral	4	a f g	b f g
DiPel* 10G	275-55	*Bt.* spp. *kurstaki*	No	Caution	IN	Oral	4	a g	b e g
DiPel* SG Plus	275-96	*Bt.* spp. *kurstaki*	No	Caution	IN	Oral	4	a g	b e g
DiTera* ES	275-110	*Myrothecium verrucaria*	No	Caution	NE	Oral	4	a f g	b f g
DiTera* G	275-133	*Myrothecium verrucaria*	No	Caution	NE	Oral	4	a e g	b e g
DiTera* WDG	275-133	*Myrothecium verrucaria*	No	Caution	NE	Oral	4	a e g	b e g
DiTera* WP	275-100	*Myrothecium verrucaria*	No	Caution	NE	Oral	4	a e g	b e g
Foray* 48B	275-116	*Bt.* spp. *kurstaki*	No	Caution	IN	Oral	4	a e g	b e g
Foray* 76B	275-119	*Bt.* spp. *kurstaki*	No	Caution	IN	Oral	4	a e g	b e g
Foray* 48F	275-98	*Bt.* spp. *kurstaki*	No	Caution	IN	Oral	12†	a e g	b e g
Gnatrol*	275-52	*Bt.* spp. *israelensis*	No	Caution	IN	Oral	4	a e g	b e g
Novodor*	275-118	*Bt.* spp. *tenebrionis*	No	Caution	IN	Oral	12	a f g	b f g
XenTari*	275-85	*Bt.* spp. *aizawai*	No	Caution	IN	Oral	4	a e g m	b e g m
AgrEvo Environmental Health, Inc.									
Pyrenone*	4816-490	pyrethrins	No	Caution	IN	Oral	12	a f g	b f g
AgrEvo USA Co.									
Apollo*	45639-135	clofentezine	No	Caution	OV/MI	Oral	12	a f g	b f g
Banol*	45639-88	propamocarb hydrochloride	No	Caution	FN	Oral	12	a e g	b e g
Carzol* SP	45639-163	formetanate hydrochloride	No	Danger-Poison	IN/MI	Oral	48	a e g j	b e g j
Decis*	34147-12-45639	deltamethrin	Yes	Danger-Poison	IN	Oral	12	c f h j k l	c f h j l
Mitac* W	45639-164	amitraz	No	Danger	IN	Oral	24	d e h j k l	d e h j l
Moncut* 50WP - Rice	45639-203	flutolanil	Yes	Caution	FN	Oral	12	a e g	b e g
Moncut* 50WP - Peanuts	45639-195	flutolanil	No	Caution	FN	Oral	12	a e g	b e g
Ovasyn*	45639-146	amitraz	No	Danger	IN/MI	Oral	24	d f h j k l	d f h j l
Phaser* 3EC	45639-169	endosulfan	No	Danger-Poison	IN	Oral	24	d f h j k l m	d f h j l
Phaser* 50WSB	45639-194	endosulfan	No	Danger-Poison	IN	Oral	24	c e h j k l m	c e h j l
ProStar* 50WP	45639-153	flutolanil	No	Caution	FN	Oral	12	a e g	b e g
Proxol* 80SP	45639-123	trichlorfon	No	Warning	IN	Oral	24	a e g j k l m	b e g j
Scout X-TRA*	34147-3-45639	tralomethrin	Yes	Danger	IN	Oral	24	a f g j	b f g j
Scout X-TRA* Gel	34147-11-45639	tralomethrin	Yes	Danger	IN	Oral	24	a f g j k	b f g j
Turcam* 2 1/2 G	45639-100	bendiocarb	Yes	Caution	IN	Oral	12	a g	b e g
Turcam*	45639-59	bendiocarb	Yes	Warning	IN	Oral	12	c e h k l	c e h l
Agriculture Sciences Inc. (AGSCI)									
Sincocin*	59714-2	plant extract	No	Caution	NE	Oral	12	a e g	a e g
AGSCO, Inc.									
DB-Green*	554-142	maneb + lindane	No	Warning	ST/IN	Oral	12†	d e g j k m	d e g j
DB-Green L*	554-140	maneb + lindane	No	Warning	ST/IN	Oral	12†	d f g j k m	d f g j
DB-Green* + Vitavax*	7501-37-554	carboxin + maneb + lindane	No	Warning	ST/IN	Oral	12†	d e h j k l m	d e g j l
Double R*	43813-3-554	imazalil	No	Danger	ST	Oral	48†	a f g j	b f g j
Double R* II	43813-14-554	imazalil	No	Danger	ST/FN	Oral	48	a f g j	b f g j
Dustret* A	554-141	maneb + streptomycin sulfate	No	Caution	ST/FN	Oral	12†	d e g j k m	d e g j
Dustret* LD	554-141	maneb + streptomycin sulfate	No	Caution	ST/FN	Oral	12†	d e g j k m	d e g j
Dustret* Maxim LD	100-821-554	fludioxonil	No	Caution	ST/FN	Oral	12	a e g	b e g
Dustret* T	7501-32-554	thiophanate-methyl	No	Caution	ST/FN	Oral	12	a e g m	b e g
Lindane ST 40	554-144	lindane	No	Warning	IN	Oral	12	c e h m	b e g
Agtrol International									
Champ* FL	55146-41	copper hydroxide	No	Caution	FN/IN	Oral	24	a e g	b e g
Champ* Formula 2 FL	55146-64	copper hydroxide	No	Caution	FN/IN	Oral	24	a e g	b e g
Champion*	55146-1	copper hydroxide	No	Danger	FN/IN	Oral	24	a e g j	b e g j
American Cyanamid Co.									
Amdro* Granular	241-322	hydramethylnon	No	Caution	IN	Oral	12	a e g	b e g
Counter* CR	241-314	terbufos	Yes	Danger-Poison	IN/NE	Oral, Posted	48	d e h j k l m	d e h j l
Counter* 15G	241-238	terbufos	Yes	Danger-Poison	IN/NE	Oral, Posted	48	d e h j k l m	d e h j l
Thimet* 20-G	241-257	phorate	Yes	Danger-Poison	IN	Oral, Posted	48	d e h j k l m	d e h j l
Amvac Chemical Corp.									
Bidrin* 8	5481-488	dicrotophos	Yes	Danger-Poison	IN	Oral, Posted	72	d f h j l m	d f h j l
PCNB 10% Granules	5481-211	PCNB	No	Caution	FN	Oral	12	a e g	b e g
PCNB 75DG	5481-441	PCNB	No	Caution	FN	Oral	12	a e g m	b e g
PCNB 2-E	5481-212	PCNB	No	Danger	FN	Oral	12	d f h j k l m	d f h j

Product Name	EPA Number	Active Ingredient(s)	RUP	Signal Word	Action	Notification	REI	PPE	Early Entry PPE
PCNB 2FL Turf & Ornamental	5481-443	PCNB	No	Caution	FN	Oral	12	a f g m	b f g
PCNB 10G Turf & Ornamental	5481-444	PCNB	No	Caution	FN	Oral	12	a e g m	b e g
PCNB 2LF	5481-215	PCNB	No	Caution	FN/ST	Oral	†	a g	b g
PCNB 75W Turf & Ornamental	5481-419	PCNB	No	Caution	FN	Oral	12	a e g m	b e g
PCNB 75WP	5481-279	PCNB	No	Caution	FN	Oral	12	a e g m	b e g
TurfPro* WSP	5481-457	PCNB	No	Caution	FN	Oral	12	a e g m	b e g
Vapam*	5481-466	metam-sodium	No	Danger	FM	Oral, Posted	48	d e h j k l m	d e h j k l m
Vapam* HL	5481-468	metam-sodium	No	Danger	FM	Oral, Posted	48	d e h j k l m	d e h j k l m
Win-Flo* 4F	5481-472	PCNB	No	Caution	FN	Oral	12	a f g m	a f g
Win-Flo* 6F	5481-471	PCNB	No	Caution	FN	Oral	12	a f g m	a f g
BASF Corp.									
Basamid* Granular	7969-99	dazomet	No	Warning	FM	Oral, Posted	24	c e h j m	c e h
Curalan* DF	7969-85	vinclozolin	No	Caution	FN	Oral	12	d e h j k l m	d e h j l
Curalan* EG	7969-85	vinclozolin	No	Caution	FN	Oral	12	d e h j k l m	d e h j l
Curalan* FL	7969-62	vinclozolin	No	Caution	FN	Oral	12	d f h j k l m	d f h j l
Kumulus* DF	7969-61	sulfur	No	Caution	FN/IN	Oral	24	a e g j	b e g j
Nexter*	7969-106	pyridaben	No	Caution	IN/MI	Oral	12	a e g j m	b e g m
Polyram* 80DF	7969-105	metiram	No	Caution	FN	Oral	12	d e g j k	d f h j l
Pyramite*	7969-125	pyridaben	No	Caution	IN/MI	Oral	12	a e g j m	b e g m
Ronilan* DF	7969-85	vinclozolin	No	Caution	FN	Oral	12	d e h j k l m	d e h j l
Ronilan* EG	7969-85	vinclozolin	No	Caution	FN	Oral	12	d e h j k l m	d e h j l
Ronilan* FL	7969-62	vinclozolin	No	Caution	FN	Oral	12	d f h j k l m	d f h j l
Sanmite*	7969-106	pyridaben	No	Danger	IN/MI	Oral	12	a e g j m	b e g m
Bayer Corp.									
Admire* 2F	3125-422	imidacloprid	No	Caution	IN	Oral	12†	a e g	b e g
Aztec* 2.1%	3125-412	cyfluthrin + tebupirimphos	Yes	Warning	IN	Oral	†	a e g	b e g
Bayleton* 50%	3125-320	triadimefon	No	Caution	FN	Oral	12	a e g l m	b e g l
Baythroid* 2	3125-351	cyfluthrin	Yes	Danger	IN	Oral	12	a f g j	b f g j
Bolstar* 6	3125-321	sulprofos	Yes	Warning	IN	Oral, Posted	24	c f h k l m	c f h l
Di-Syston* 8	3125-307	disulfoton	Yes	Danger	IN	Oral, Posted	48	d f h j k l m	d f h j l
Di-Syston* 15%	3125-172	disulfoton	Yes	Danger	IN	Oral, Posted	48	c e h k l m	c e h l
Elite* 45 DF	3125-388	tebuconazole	No	Warning	FN	Oral	12	a e g j	a e g j
Folicur* 3.6F	3125-394	tebuconazole	No	Caution	FN	Oral	12	a f g	b f g
Guthion* 2L	3125-102	azinphos-methyl	Yes	Danger	IN	Oral	48	f h j l m n	f h j l n
Guthion* Solupak	3125-301	azinphos-methyl	No	Danger	IN	Oral	48	e h j l m n	e h l n
Monitor* 4	3125-280	methamidophos	Yes	Danger	IN	Oral, Posted	48	c f h j k l m	c f h j l
Nemacur* 3	3125-283	fenamiphos	Yes	Danger	IN/NE	Oral	48	f h j l m n	f h j l n
Nemacur* 15%	3125-236	fenamiphos	Yes	Danger	IN/NE	Oral	48	e h j l m n	e h j l n
Provado* 1.6	3125-457	imidacloprid	No	Caution	IN	Oral	12	a e g	b e g
Provado* Solupak	3125-428	imidacloprid	No	Caution	IN	Oral	12	a e g	b e g
Becker Microbial Products, Inc.									
BMP 123* 64ES	62637-10	*Bt.* spp. *kurstaki*	No	Caution	LA	Oral	4	a e g	b e g
BMP 123* 10G	62637-8	*Bt.* spp. *kurstaki*	No	Caution	LA	Oral	4	a e g	b e g
BMP 123* 2X WDG	62637-5	*Bt.* spp. *kurstaki*	No	Caution	LA	Oral	4	a e g	b e g
BMP 123* 2X WP	62637-5	*Bt.* spp. *kurstaki*	No	Caution	LA	Oral	4	a e g	b e g
Best Sulfur Products									
Lime-Sulfur Solution	66196-2	calcium polysulfide	No	Danger	FN/IN	Oral	48	d e h j k l m	d e h j l
Sulforix*	64348-2	calcium polysulfide	No	Danger	FN/IN	Oral	48	d e h j k l m	d e h j l
Brandt Consolidated, Inc.									
SAF-T-SIDE*	48813-1	petroleum oil	No	Caution	IN	Oral	4	a g	b e g
Cheminova, Inc.									
Cheminova Dimethoate 4E	51036-110-67760	dimethoate	No	Warning	IN	Oral	12	a f g j l m	b f g j l
Cheminova Methyl 4EC	67760-29	methyl parathion	Yes	Danger	IN	Oral	48	d f h j k l m	d f h j l
Cyren 2E	67760-6	chlorpryifos	No	Warning	IN	Oral	12	c f h j	c f h j

Action Word Abbreviations

BA......Bactericide
BI.......Biostimulant
FM.....Fumigant
FN......Fungicide
IGR....Insect Growth Regulator
IN.......Insecticide
LA......Larvicide
MI......Miticide
MO....Molluscicide
NE......Nematicide
OV.....Ovicide
RE......Repellent
SP......Seed Protectant
ST......Seed Treatment

Personal Protective Equipment (PPE) Abbreviations

a... Long-sleeved shirt and long pants
b .. Coveralls
c... Coveralls over short-sleeved shirt and short pants
d .. Coveralls over long-sleeved shirt and long pants
e... Waterproof gloves
f ... Chemical-resistant gloves
g .. Shoes plus socks
h .. Chemical-resistant footwear plus socks
j ... Protective eyewear
k... Chemical-resistant apron when cleaning equipment, mixing, or loading
l ... Chemical-resistant headgear for overhead exposure
m.. Approved respirator
n .. Chemical-resistant protective suit

* = Trade Name/R/TM † May vary; please refer to product label for complete information.
Information presented herein is for preliminary planning only. Exclusive reliance must be placed on information/directions supplied by the manufacturer.

Product Name	EPA Number	Active Ingredient(s)	RUP	Signal Word	Action	Notification	REI	PPE	Early Entry PPE
Cyren 4E	67760-7	chlorpyrifos	No	Warning	IN	Oral	12	d f g j	b f g j
Nufos* 4E-SG	67760-27	chlorpyrifos	Yes	Warning	IN	Oral	24	c f h j k l	c f h j l
Nufos* 4E	67760-28	chlorpyrifos	No	Warning	IN	Oral	24	c f h j k l	c f h j l
Cleary Chemical Corp.									
3336* F	1001-69	thiophanate-methyl	No	Caution	FN	Oral	12	a f g m	b e g
3336* G	1001-70	thiophanate-methyl	No	Caution	FN	Oral	12	a g	b e g
3336* WP	1001-63	thiophanate-methyl	No	Caution	FN	Oral	12	a e g	b e g
Defend* 10G	5481-444-1001	PCNB	No	Caution	FN	Oral	12	a e g m	b e g
Defend* 75WP	5481-419-1001	PCNB	No	Caution	FN	Oral	12	a e g m	b e g
KnoxOut GH	4581-379-1001	diazinon	Yes	Caution	IN	Oral	12	a g	b f g
Protect* T/O	1001-65	mancozeb	No	Caution	FN	Oral	24	a e g j k	d e g
Diatect International Inc.									
Diatect* Multipurpose Insecticide II	42850-4	pyrethrins + piperonyl butoxide	No	Caution	IN	Oral	12	a e g m	b e g
Dow AgroSciences									
Conserve* SC	62719-291	spinosad	No	Caution	IN	Oral	4	a g	b e g
Dursban* 50W	62719-72	chlorpyrifos	No	Warning	IN	Oral	12	a e g l	b e g l
Lock-On*	62719-79	chlorpyrifos	No	Warning	IN	Oral	12	a f g j	b f g j
Lorsban*-4E	62719-220	chlorpyrifos	No	Warning	IN	Oral	24	c f h j k l	c f h j l
Lorsban* 4E-SG	62719-245	chlorpyrifos	Yes	Warning	IN	Oral	24	c f h j k l	c f h j l
Lorsban* 15G	62719-34	chlorpyrifos	No	Caution	IN	Oral	12	a g	b e g
Rubigan* AS	62719-142	fenarimol	No	Caution	FN	Oral	12	a f g j	b f g j
Rubigan* EC	62719-134	fenarimol	No	Warning	FN	Oral	12	c f h j k l	c f h j l
Success* Naturalyte*	62719-292	spinosad	No	Caution	IN	Oral	4	a e g	b e g
Telone* C-17	62719-12	dichloropropene + chloropicrin	Yes	Danger	FM	Oral	72	c f h j l k m	b f h j m
Telone* II	62719-32	dichloropropene	Yes	Warning	FM	Oral	72	c h j k l m	b f h j m
Tracor* Naturalyte*	62719-267	spinosad	No	Caution	IN	Oral	4	a e g	b e g
Drexel Chemical Co.									
435 Oil 98.8	19713-394	petroleum oil	No	Caution	IN/MI	Oral	4	a f g	b f g
455 Oil 98.8	19713-396	petroleum oil	No	Caution	IN/MI	Oral	4	a f g	b f g
Basic Copper Sulfate	19713-72	copper sulfate, basic	No	Warning	FN	Oral	24	a g j	b e g j
Captan 4L	19713-156	captan	No	Danger	FN	Oral	96†	a e g j k	b e g j
Captan Plus Molybdenum	19713-126	captan	No	Danger	FN	Oral	96†	a e g j k	b e g j
Captan 50W	19713-235	captan	No	Danger	FN	Oral	96†	a e g j k m	b e g j
Captan 80W	19713-362	captan	No	Danger	FN	Oral	96†	a e g j k m	b e g j
Captan-50/10 Seed Protectant	19713-261	captan	No	Danger	FN/SP	Oral	48	a e g j m	b e g j
Carbaryl 4L	19713-49	carbaryl	No	Caution	IN	Oral	12	a e g m	b e g
Carbaryl 50W	19713-52	carbaryl	No	Caution	IN	Oral	12	a e g m	b e g
Carbaryl 80S	19713-50	carbaryl	No	Warning	IN	Oral	12	a e g m	b e g
D-264 14G	19713-95	diazinon	Yes	Caution	IN	Oral	†	a e g j	b e g
Damoil*	19713-123	refined petroleum distillate	No	Caution	IN/MI	Oral	4	a g	b e g
Diazinon Insecticide	19713-91	diazinon	Yes	Caution	IN	Oral	†	a f g j	b f g
Diazinon 50W	19713-492	diazinon	Yes	Caution	IN	Oral	†	a e g	b e g
Dimethoate 4EC	19713-231	dimethoate	No	Warning	IN/MI	Oral	48	a f h j l m	b f h j l
Endosulfan 2EC	19713-99	endosulfan	No	Warning	IN	Oral	24	d f h j k l m	d f h j l
Endosulfan 3EC	51036-92-19713	endosulfan	No	Danger-Poison	IN	Oral	24	c f h j k l m	c f h j l
Endosulfan 50WP	51036-91-19713	endosulfan	No	Danger-Poison	IN	Oral	24	a e g j k l m	b e g l
KOP* 300	19713-81	copper sulfate, basic	No	Warning	FN	Oral	12	a e g j	b e g j
KOP* Oxy-85	59716-1-19713	copper oxychloride	No	Danger	FN	Oral	24	a e g j	b e g j
KOP*-Hydroxide	19713-301	copper hydroxide	No	Danger	FN	Oral	48	a e g j m	b e g j
KOP*-Hydroxide 50	19713-321	copper hydroxide	No	Danger	FN	Oral	48	a e g j m	b e g j
Malathion 5EC	19713-217	malathion	No	Warning	IN	Oral	12	a f g	b f g
Soy Gro*	19713-197	captan	No	Danger	FN	Oral	96†	a e g j k m	b e g j
Suffa*	19713-39	sulfur	No	Caution	FN/MI	Oral	24	a e g j	b e g j
Suffa* 8	19713-87	sulfur	No	Caution	FN/MI	Oral	24	a e g j	b e g j
Sulfa* 80	19713-295	sulfur	No	Caution	FN	Oral	24	a e g	b e g j
Sulfur 90W	19713-238	sulfur	No	Caution	FN/MI	Oral	24	a e g j	b e g j
Sumulsion*	19713-489	petroleum oil	No	Caution	IN/MI	Oral	4	a f g	b f g
Ziram 4L	19713-93	ziram	No	Danger	FN	Oral	48	a e g j l	b e g j l
Ziram 76	19713-68	ziram	No	Danger	FN	Oral	48	a e g j l	b e g j l
DuPont Agricultural Products									
Asana* XL	352-515	esfenvalerate	Yes	Warning	IN	Oral	12	a f g j	b f g j
Benlate* SP	352-564	benomyl	No	Caution	FN	Oral	24	a e h k m	b e h
Benlate* WP	352-354	benomyl	No	Caution	FN	Oral	24	a e h k m	b e h
Fortress* 2.5G	352-579	chlorethoxyfos	No	Danger-Poison	IN	Oral	48†	a e g j m	b e g
Fortress* 5G	352-552	chlorethoxyfos	No	Danger-Poison	IN	Oral	48†	d e g j m	b e g

Product Name	EPA Number	Active Ingredient(s)	RUP	Signal Word	Action	Notification	REI	PPE	Early Entry PPE
Lannate*	352-342	methomyl	Yes	Danger-Poison	IN	Oral	48†	a e g j	b e g j
Vydate* C-LV	352-532	oxamyl	Yes	Danger-Poison	IN	Oral	48	c f h j k l	c f h j l
Vydate* L	352-372	oxamyl	Yes	Danger-Poison	IN/NE	Oral	48	c f h j k l	c f h j l

Ecogen Inc.

Product Name	EPA Number	Active Ingredient(s)	RUP	Signal Word	Action	Notification	REI	PPE	Early Entry PPE
AQ10*	55638-16	*Ampelomyces quisqualis*	No	Caution	FN	Oral	4	a g m	b e g
AQ10* t/o	55638-16	*Ampelomyces quisqualis*	No	Caution	FN	Oral	4	a g m	b e g
Condor*	55638-7	*Bt.* spp. *kurstaki*	No	Caution	IN	Oral	4	a g	b e g
Condor* t/o	55638-7	*Bt.* spp. *kurstaki*	No	Caution	IN	Oral	4	a g	a e g
Condor* WP	55638-12	*Bt.* spp. *kurstaki*	No	Caution	IN	Oral	4	a g	b e g
Condor* XL	55638-33	*Bt.* spp. *kurstaki*	No	Caution	IN	Oral	4	a g	a e g
CRYMAX* WDG	55638-34	*Bt.* spp. *kurstaki*	No	Caution	IN	Oral	4	a g	b e g
Cutlass* WP	55638-8	*Bt.* spp. *kurstaki*	No	Caution	IN	Oral	4	a g	b e g
Lepinox* G	55638-36	*Bt.* spp. *kurstaki*	No	Caution	IN	Oral	4	a g	b e g
Lepinox* WDG	55638-36	*Bt.* spp. *kurstaki*	No	Warning	IN	Oral	8	a g j	b e g j
Raven*	55638-27	*Bt.* spp. *kurstaki*	No	Caution	IN	Oral	4	a g	b g

Elf Atochem North America, Inc.

Product Name	EPA Number	Active Ingredient(s)	RUP	Signal Word	Action	Notification	REI	PPE	Early Entry PPE
Kryocide*	4581-116	cryolite	No	Caution	IN	Oral	12	a e g	b e g
Maneb 75DF	4581-371	maneb	No	Caution	FN	Oral	24	d e g j k	d e g
Maneb 80WP	4581-255	maneb	No	Caution	FN	Oral	24	d e g j k	d e g
Microthiol* Special	4581-373	sulfur	No	Caution	FN/MI	Oral	24	a e g	b e g
Penncap-M*	4581-292	methyl parathion	Yes	Warning	IN	Oral	48	d e h j l	d e h j l
Penncozeb* 75DF	4581-370	mancozeb	No	Caution	FN	Oral	24	d e g j k	d e g
Penncozeb* 80WP	4581-358	mancozeb	No	Caution	FN	Oral	24	d e g j k	d e g
Topsin* M 70W	4581-322	thiophanate-methyl	No	Caution	FN	Oral	12	a g	b e g
Topsin* M WSB	4581-377	thiophanate-methyl	No	Caution	FN	Oral	12	a g	b e g
Ziram 76DF	4581-140	ziram	No	Danger	FN	Oral	48	a e g j m	b e g j

Entek Corp.

Product Name	EPA Number	Active Ingredient(s)	RUP	Signal Word	Action	Notification	REI	PPE	Early Entry PPE
Enzone*	68891-2	sodium tetrathiocarbonate	No	Danger	FM	Oral, Posted	96	d e h j k l	d e h j l

FMC Corp.

Product Name	EPA Number	Active Ingredient(s)	RUP	Signal Word	Action	Notification	REI	PPE	Early Entry PPE
Ammo* 2.5EC	279-3027	cypermethrin	Yes	Caution	IN	Oral	12	c f g j k	c f g j
Ammo* WSB	279-3084	cypermethrin	Yes	Warning	IN	Oral	12	a e g	b e g
Astro*	279-3141	permethrin	No	Caution	IN	Oral	24	a f g	b f g
Astro* 25WP	279-3051	permethrin	Yes	Warning	IN	Oral	24	a e g	b e g
Brigade* WSB	279-3108	bifenthrin	No	Warning	IN/MI	Oral	12	a e g j	b e g
Capture* 2 EC (Calif.)	279-3114	bifenthrin	Yes	Warning	IN/MI	Oral	†	a f g j	b f g
Capture* 2 EC	279-3069	bifenthrin	Yes	Warning	IN/MI	Oral	12	a f g j	b f g
Carbamate WDG	45728-7-279	ferbam	No	Caution	FN	Oral	24	a e g j l	b e g j l
Ethion 4 Miscible	279-1254	ethion	No	Warning	IN/MI	Oral	48	c f h j k m	c f h
Furadan* 3G	279-2792	carbofuran	Yes	Warning	IN/NE	Oral, Posted	48	a e g	b e g
Furadan* 5G	279-2874	carbofuran	Yes	Warning	IN/NE	Oral, Posted	48†	a e g m	b e g
Furadan* 4F	279-2876	carbofuran	Yes	Danger-Poison	IN/NE	Oral, Posted	48†	a f g j m	b f g
Furadan* 15G	279-3023	carbofuran	Yes	Warning	IN/NE	Oral, Posted	48†	a e g m	b e g
Fury* 1.5EC	279-3125	*zeta*-cypermethrin	Yes	Warning	IN	Oral	12	a f g	b f g
Golden Leaf*	279-2924	endosulfan	No	Danger-Poison	IN	Oral	24	d f h j k l m	d f h j l
Mustang* 1.5 EW	279-3126	*zeta*-cypermethrin	Yes	Warning	IN	Oral	12	a f g j	b f g
Pounce* 3.2EC	279-3014	permethrin	Yes	Caution	IN	Oral	12	a f g	b f g
Pounce* 1.5G	279-3059	permethrin	Yes	Caution	IN	Oral	12	a e g	b e g
Pounce* 25WP	279-3051	permethrin	Yes	Warning	IN	Oral	12	a e g	b e g
Pounce* WSB	279-3083	permethrin	Yes	Warning	IN	Oral	12	a e g	b e g
Talstar* Nursery Flowable	279-3155	bifenthrin	Yes	Caution	IN/MI	Oral	12	a f g j	b f g
Talstar* Nursery Granular	279-3130	bifenthrin	No	Caution	IN/MI	Oral	12†	a f g	b f g
Thiodan* Cottonseed Oil	279-2659	endosulfan	No	Danger-Poison	IN	Oral	24	c f h j k l m	c f h l
Thiodan* 3EC	279-2924	endosulfan	No	Danger-Poison	IN	Oral	24	d f h j k l m	d f h j l
Thiodan* 50WP	279-1380	endosulfan	No	Danger-Poison	IN	Oral	24	a e g j k l m	b e g l
Thiodan* WSB	279-3120	endosulfan	No	Danger-Poison	IN	Oral	24	a e g j k l m	b e g l

Action Word Abbreviations

BA......Bactericide
BI.......Biostimulant
FM .....Fumigant
FN......Fungicide
IGR ....Insect Growth Regulator
IN.......Insecticide
LA......Larvicide
MI......Miticide
MO ....Molluscicide
NE......Nematicide
OV .....Ovicide
RE......Repellent
SP......Seed Protectant
ST......Seed Treatment

Personal Protective Equipment (PPE) Abbreviations

a... Long-sleeved shirt and long pants
b .. Coveralls
c... Coveralls over short-sleeved shirt and short pants
d .. Coveralls over long-sleeved shirt and long pants
e... Waterproof gloves
f ... Chemical-resistant gloves
g .. Shoes plus socks
h .. Chemical-resistant footwear plus socks
j ... Protective eyewear
k... Chemical-resistant apron when cleaning equipment, mixing, or loading
l ... Chemical-resistant headgear for overhead exposure
m.. Approved respirator
n .. Chemical-resistant protective suit

* = Trade Name/R/TM † May vary; please refer to product label for complete information.
Information presented herein is for preliminary planning only. Exclusive reliance must be placed on information/directions supplied by the manufacturer.

Product Name	EPA Number	Active Ingredient(s)	RUP	Signal Word	Action	Notification	REI	PPE	Early Entry PPE
Gowan Company									
Azinphos-M 50-W	10163-78	azinphos-methyl	Yes	Danger-Poison	IN	Oral	48†	d e h j k l m	d e h l
Botran* 75-W	10163-189	DCNA	No	Caution	FN	Oral	12†	a g	b e g
Diazinon 14G	10163-104	diazinon	Yes	Caution	IN	Oral	12†	a e g	b e g
Diazinon 50WSB	10163-163	diazinon	Yes	Caution	IN	Oral	12†	a e g j	b e g j l
Diazinon 4E	10163-116	diazinon	Yes	Caution	IN	Oral	12†	a f g	b f g
Dimethoate E267	10163-56	dimethoate	No	Warning	IN	Oral	48	c f h j k l	c f h j l
Endosulfan 3EC	10163-100	endosulfan	No	Danger-Poison	IN	Oral	24	d f h j k l m	d f h j l
Endosulfan 50WSB	10163-130	endosulfan	No	Danger-Poison	IN	Oral	24	d e h j k l m	d e h j l
Hexygon* 50-WP	10163-208	hexythiazox	No	Caution	OV/MI	Oral	12	a e g	b e g
Imidan* 70-W	10163-169	phosmet	No	Warning	IN	Oral	24	a e g l m	a e g l
Lorsban* 50-W	62719-221-10163	chlorpyrifos	No	Warning	IN	Oral	24†	a e g l	b e g l
Malathion 8	10163-21	malathion	No	Caution	IN	Oral	12†	a f g	b f g
Mesurol* 75W	3125-288-10163	methiocarb	Yes	Warning	IN	Oral	24	c e h j l k m	c e h j l
Metasystox-R* SC	10163-220	oxydemeton-methyl	Yes	Warning	IN	Oral	48†	d f h j k l m	d f h l
Prokil Cryolite 96	10163-41	cryolite	No	Caution	IN	Oral	12	a e g	b e g
Savey*	10163-208	hexythiazox	No	Caution	OV/MI	Oral	12	a e g	b e g
Sevin Bait	10163-32	carbaryl	No	Caution	IN	Oral	12	a e g	b e g
Supracide* 2E	100-501	methidathion	Yes	Danger	IN/MI	Oral	48†	a f g j m	b f g j
Supracide* 25WP	100-754	methidathion	Yes	Warning	IN/MI	Oral	48†	a e g m	b e g
Great Lakes Chemical									
Brom-O-Gas*	5785-4	methyl bromide + chloropicrin	Yes	Danger-Poison	FM	Oral, Posted	48†	a g j m	a g j
Chlor-O-Pic* 100	5785-17	chloropicrin	Yes	Danger-Poison	FM	Oral, Posted	48†	a g j m	a g j
Terr-O-Gas*	5785-24	methyl bromide + chloropicrin	Yes	Danger-Poison	FM	Oral, Posted	48†	a g j m	a g j
Griffin L.L.C.									
Basicop*	1812-300	copper	No	Warning	FN	Oral	24	a e g j	b e g j
Blue Viking* CSP Instant	1812-313	copper sulfate pentahydrate	No	Danger	FN	Oral	24	a e g j	b e g j
Chlorfos 15G	66222-18-1812	chlorpyrifos	No	Caution	IN	Oral	12	a e g	b e g
Declare* 4EC	1812-405	methyl parathion	Yes	Danger-Poison	IN	Oral, Posted	48	c f h j k l m	c f h j l
Dicofol 4E	10163-97-1812	dicofol	No	Caution	IN	Oral	12	a f g	b f g
Kocide* 101	1812-288	copper hydroxide	No	Danger	FN	Oral	24	a e g j	b e g j
Kocide* 20/20	1812-305	copper hydroxide	No	Danger	FN	Oral	24	a e g j	b e g j
Kocide* 2000	1812-358	copper hydroxide	No	Warning	FN	Oral	24	a e g j	b e g j
Kocide* DF	1812-334	copper hydroxide	No	Danger	FN	Oral	24	a e g j	b e g j
Kocide* LF	1812-338	copper hydroxide	No	Caution	FN/BA	Oral	24	a e h j l	b e h j l
Kocide* 4.5LF	1812-303	copper hydroxide	No	Caution	FN/BA	Oral	24	a e g j	b e g j
Lannate* LV	352-384	methomyl	Yes	Danger-Poison	IN	Oral	48†	a f g j	b f g j
Manex*	1812-251	maneb	No	Caution	FN	Oral	24	d e g j k	d e g
ManKocide*	1812-360	mancozeb + copper hydroxide	No	Danger	FN	Oral	48	a e g j k	b e g j
Manzate* 200 DF	352-449	mancozeb	No	Caution	FN	Oral	24	d e g j k	d e g
Orbit*/SuperTin*	1812-395	propiconazole + triphenyltin hydroxide	Yes	Orbit: Caution SuperTin: Danger	FN	Oral, Posted	24	d e h j k l m	d e h j l
Pro-tex*	1812-351	maneb + triphenyltin hydroxide	Yes	Danger-Poison	FN	Oral, Posted	48	d e h j k l	d e h j l
Super Tin* 80WP	1812-350	triphenyltin hydroxide	Yes	Danger-Poison	FN	Oral, Posted	48	d e h j k l	d e h j l
Tennessee Brand* CSP Instant	1812-313	copper sulfate pentahydrate	No	Danger	FN	Oral	24	a e g j	b e g j
Vendex* 50WP	352-480	fenbutatin-oxide	Yes	Danger-Poison	MI	Oral	48	a e g j	b e g j
Gustafson, Inc.									
Agricultural Streptomycin	618-100-7501	streptomycin sulfate	No	Caution	BI/ST	Oral	12	a e g j m	b e g j
Allegiance*-FL	7501-42	metalaxyl	No	Warning	ST/FN	Oral	24†	c f h	b f j
Apron* FL	7501-42	metalaxyl	No	Warning	FN/ST	Oral	12	a f g	b f g
Baytan 30	7501-91	triadimenol	No	Caution	FN	Oral	12	a f g	b f g
Flo-Pro IMZ*	7501-127	imazalil	No	Warning	FN	Oral	48†	a f g j	b f g j
Kodiak*	7501-144	*Bacillus subtillis*	No	Caution	ST	Oral	12†	a e g m	b e g
Lindane 30C	7501-34	lindane	No	Warning	IN	Oral	12	a f g m	b f g
Raxil*-Thiram	7501-151	tebuconazole + thiram	No	Caution	ST/FN	Oral	24†	a f g j m	b f g j
RTU* FL	7501-135	thiram + thiabendazole	No	Caution	FN	Oral	24†	a f g j	b f g j
RTU*-Vitavax-Extra	7501-166	carboxin + imazalil + thiabendazole	No	Caution	FN	Oral	48†	a f g	b f g
RTU-Vitavax*-Thiram	7501-114	carboxin + thiram	No	Caution	FN	Oral	24	a e g j m	b e g j
42-S* Thiram	7501-14	thiram	No	Caution	FN	Oral	24	a f g j	b f g j
Thiram 50WP Dyed	7501-105	thiram	No	Caution	FN	Oral	24	a e g j m	b e g j
TOPS* 5	7501-149	thiophanate-methyl	No	Caution	ST/FN	Oral	12†	a e g m	b e g
TOPS 2.5D*	7501-32	thiophanate-methyl	No	Caution	FN	Oral	12	a e g m	b e g
TOPS* MZ	7501-157	thiophanate-methyl	No	Caution	ST/FN	Oral	24†	d e g j k m	d e g j
Vitavax*-Thiram-Lindane	7501-141	carboxin + thiram + lindane	No	Caution	FN	Oral	24†	a f g m	d f g

Product Name	EPA Number	Active Ingredient(s)	RUP	Signal Word	Action	Notification	REI	PPE	Early Entry PPE
Helena Chemical Co.									
Bravo* S	5905-472	chlorothalonil + sulfur	No	Caution	FN/MI	Oral	48	a e g j	b e g j
Diazinon AG500	5905-248	diazinon	No	Caution	IN	Oral	24†	a f g	b f g
Dimethoate 4EC	5905-493	dimethoate	No	Warning	IN	Oral	48	a f h j l m	b f h j l
Fyfanon* 5EC	5905-196	malathion	No	Caution	IN	Oral	12	a f g j l	b f g j l
Fyfanon* 8E	5905-250	malathion	No	Caution	IN	Oral	12	a f g	b f g
Liquid Sulfur Six	5905-437-AA	sulfur	No	Caution	FN	Oral	24	a e g	b e g
4 lb. Methyl Parathion	5905-528	methyl parathion	Yes	Danger-Poison	IN	Oral, Posted	48	d f h j k l m	d f h j l
Omni Supreme Spray	5705-368	petroleum oil	No	Caution	IN/MI	Oral	4	a f g	b f g
System3*	7501-148-38167	PCNB + metalaxyl + *Bacillus subtillis*	No	Caution	FN/ST	Oral	12	a e g m	b e g
Hot Pepper Wax, Inc.									
Hot Pepper Wax*	67238-1	capsaicin	No	Caution	IN/RE	Oral	0	a e g	b e g
International Sulphur, Inc.									
Red Ball* Commercial Dusting	49668-1	sulfur	No	Caution	FN/IN	Oral	24	a e g	b e g
Red Ball* EM-53	49668-4	sulfur	No	Caution	FN/IN	Oral	24	a e g j	b e g j
Red Ball* Superfine	49668-2	sulfur	No	Caution	FN/IN	Oral	24	a e g	b e g
Red Ball* Lightning	49668-7	sulfur	No	Caution	IN	Oral	24	a e g	b e g
Kincaid Enterprises, Inc.									
Demosan* FL	41014-8	chloroneb	No	Caution	FN/ST	Oral	12	b f g	b f g
Demosan* 65W	41014-4	chloroneb	No	Caution	FN/ST	Oral	12	b f g	b f g
Marlate* 50	41014-2	methoxychlor	No	Caution	IN	Oral	12	a e g	a e g
Terraneb* SP	41014-6	chloroneb	No	Caution	FN	Oral	12	b f g	b f g
Terraneb* SP FL	41014-8	chloroneb	No	Caution	FN	Oral	12	b f g	b f g
Micro Flo Co.									
Acephate 75SP	51036-236	acephate	No	Caution	IN	Oral	24	a e g l	b e g l
Acephate 90SP	51036-238	acephate	No	Caution	IN	Oral	24	a e g l	b e g l
Adios*	51036-185	carbaryl	No	Caution	IN	Oral	12	a e g	b e g
Azinphosmethyl 2EC	51036-76	azinphos-methyl	Yes	Danger-Poison	IN	Oral	48†	c f h j k l m	c f h j l
Azinphosmethyl 50W	51036-164	azinphos-methyl	No	Danger-Poison	IN	Oral	48†	c e h j k l m	c e h j l
Basic Copper 53	51036-24	basic copper sulfate	No	Warnin	FN/BA	Oral	24	a e g j l m	b e g j l
Captan 50WP	10182-145-51036	captan	No	Danger	FN	Oral, Posted	96†	a e g j	b e g j
Captan 80WP	10182-164-51036	captan	No	Danger	FN	Oral, Posted	96†	a e g j	b e g j
Captec 4L	51036-181	captan	No	Danger	FN	Oral	96†	a e g j k m	b e g j
Chlorpyrifos 15G AG	51036-300	chlorpyrifos	No	Caution	IN	Oral	12	a e g	b e g
Chlorpyrifos 4E AG	51036-291	chlorpyrifos	No	Warning	IN	Oral	24	c f h j k l	c f h j k l
Chlorpyrifos 4E AG-Wheat	51036-294	chlorpyrifos	Yes	Warning	IN	Oral	24	a f g j k l	b f g j
Chlorpyrifos Pro 2	51036-152	chlorpyrifos	Yes	Warning	IN	Oral	12	c f h j	c f h j
Chlorpyrifos Pro 4	51036-154	chlorpyrifos	Yes	Warning	IN	Oral	12	a f g j	b f g j
Diazinon 14G	51036-70	diazinon	Yes	Caution	IN	Oral	24	a e g	b e g
Diazinon 50W	51036-108	diazinon	Yes	Caution	IN	Oral	24	a e g	b e g l
Diazinon AG500	51036-71	diazinon	Yes	Caution	IN	Oral	24	a f g	b f g
Dimethoate 267	51036-198	dimethoate	No	Warning	IN	Oral	48	a f g j l m	b f g j l
Dimethoate 4E	51036-110	dimethoate	No	Warning	IN	Oral	48	a f g j l m	b f g j l
Endosulfan 3EC	51036-92	endosulfan	No	Danger-Poison	IN	Oral	24	c f h j k l m	c f h j l
Endosulfan 50WP	51036-91	endosulfan	No	Danger-Poison	IN	Oral	24	a e g j k l m	b e g l
Ethion 4EC	51036-89	ethion	No	Warning	IN	Oral	48	c f h j k l m	c f h j l
Malathion 5EC	51036-104	malathion	No	Warning	IN	Oral	12	a f g l	b f g l
Malathion ULV	51036-103	malathion	No	Caution	IN	Oral	12	a e g l	b e g l
Microsperse* COC 53WP	51036-125	copper oxychloride	No	Danger	FN/BA	Oral	24	c e h j l k	c e h j l
Microsperse* Wettable Sulfur	51036-14	sulfur	No	Caution	FN/MI	Oral	24	a e g l	b e g l
Nu-Cop* 3L	51036-268	copper hydroxide	No	Warning	FN/BA	Oral	24	a e g j	b e g j
Nu-Cop* 50DF	51036-269	copper hydroxide	No	Danger	FN/BA	Oral	48	a e g j	b e g j
Nu-Cop* 50WP	51036-270	copper hydroxide	No	Danger	FN/BA	Oral	48	a e g j	b e g j
Slam*	51036-185	carbaryl	No	Caution	IN	Oral	12	a e g	b e g
Sulfur 6L	51036-16	sulfur	No	Caution	FN/MI	Oral	24	a e g j l	b e g j l

Action Word Abbreviations

BA......Bactericide
BI.......Biostimulant
FM......Fumigant
FN......Fungicide
IGR....Insect Growth Regulator
IN.......Insecticide
LA......Larvicide
MI......Miticide
MO.....Molluscicide
NE......Nematicide
OV.....Ovicide
RE......Repellent
SP......Seed Protectant
ST......Seed Treatment

Personal Protective Equipment (PPE) Abbreviations

a... Long-sleeved shirt and long pants
b .. Coveralls
c... Coveralls over short-sleeved shirt and short pants
d .. Coveralls over long-sleeved shirt and long pants
e... Waterproof gloves
f ... Chemical-resistant gloves
g .. Shoes plus socks
h .. Chemical-resistant footwear plus socks
j ... Protective eyewear
k... Chemical-resistant apron when cleaning equipment, mixing, or loading
l ... Chemical-resistant headgear for overhead exposure
m.. Approved respirator
n .. Chemical-resistant protective suit

* = Trade Name/R/TM † May vary; please refer to product label for complete information.
Information presented herein is for preliminary planning only. Exclusive reliance must be placed on information/directions supplied by the manufacturer.

Product Name	EPA Number	Active Ingredient(s)	RUP	Signal Word	Action	Notification	REI	PPE	Early Entry PPE

Mineral Research & Development, Div. Chemical Specialties Inc.

Copper-Count*-N	10465-3	copper ammonium	No	Caution	FN	Oral	12	a f g	b f g

Mycogen Crop Protection

M-Pede*	53219-6	potassium salts of fatty acids	No	Warning	FN/IN	Oral	12	c f h j k l	b e g
M-Peril*	53219-9	*Bt.* spp. *kurstaki*	No	Caution	IN	Oral	4	a e g	b e g
Mattch*	53219-10	*Bt.* spp. *kurstaki* + *Bt.* spp. *aizawai*	No	Caution	IN	Oral	4	a e g	b e g
MVP II*	53219-12	*Bt.* spp. *kurstaki*	No	Caution	IN	Oral	4	a e g	b e g

Mycotech Corp.

BotainGard* ES	65626-8	*Beauveria bassiana*	No	Warning	IN	Oral	12	a g	d e g j
BotaniGard* 22WP	65626-10	*Beauveria bassiana*	No	Caution	IN	Oral	12	a e g m	b e g m
Mycotrol* ES	65626-8	*Beauveria bassiana*	No	Warning	IN	Oral	12	a g	d e g j
Mycotrol* 22WP	65626-11	*Beauveria bassiana*	No	Caution	IN	Oral	12	a e g m	b e g m

Novartis

Agri-Mek* 0.15EC	618-98	abamectin	Yes	Warning	IN/MI	Oral	12	c f h j k l	c f h j l
Agri-mycin* 17	100-899	streptomycin sulfate	No	Caution	FN	Oral	12	a e g m	b e g
Apron* 50W	100-738	metalaxyl	No	Warning	FN	Oral	12	a e g j	b e g j
Apron XL* LS	100-799	mefenoxam	No	Warning	FN	Oral	48	a e g j	b e g j
Avid* 0.15EC	618-96	abamectin	No	Warning	IN	Oral	12	c f g j k l	c f j l
Award*	100-722	fenoxycarb	No	Caution	IGR	Oral	12	a g	b e g
Banner* GL	100-736	propiconazole	No	Warning	FN	Oral	24	a f g j	b f g j
Banner* MAXX*	100-741	propiconazole	No	Warning	FN	Oral	24	a f g j	b f g j
Break* EC	100-702	propiconazole	No	Warning	FN	Oral	24	a f g j	b f g j
Citation*	100-667	cyromazine	No	Caution	IN	Oral	12	d e h k	d e h
Curacron* 8E	100-669	profenofos	Yes	Warning	IN/MI	Oral, Posted	48†	c f h j k	c f h j
D-z-n* diazinon AG500	100-461	diazinon	Yes	Caution	IN	Oral	12†	a f g	b f g
D-z-n* diazinon AG600 WBC	100-784	diazinon	Yes	Caution	IN	Oral	24	a e g m	b e g
D-z-n* diazinon 14G	100-469	diazinon	Yes	Caution	IN	Oral	12	a e g	b e g
D-z-n* diazinon 50W	100-460	diazinon	Yes	Caution	IN	Oral	12†	a e g	b e g
Dividend* XL RTA	100-826	difenoconazole	No	Caution	FN/ST	Oral	48	a g	b e g
Logic*	100-722	fenoxycarb	No	Caution	IGR	Oral	12	a g	b e g
Medallion*	100-769	fludioxonil	No	Caution	FN	Oral	12	a e g	b e g
Mycoshield*	100-900	oxytetracycline	No	Caution	FN	Oral	12	a e g j m	b e g j
Orbit*	100-702	propiconazole	No	Warning	FN	Oral	24	a f g j	b f g j
Precision*	100-750	fenoxycarb	No	Caution	IGR	Oral	12	a e g	b e g
Ridomil Gold* EC	100-801	mefenoxam	No	Warning	FN	Oral	48	a e g j	b e g j
Ridomil Gold* GR	100-798	mefenoxam	No	Caution	FN	Oral	48	a e g	b e g
Ridomil Gold* MZ	100-803	mefenoxam + mancozeb	No	Caution	FN	Oral	48	d e g j k	d e g
Ridomil Gold* PC	100-792	mefenoxam + PCNB	No	Warning	FN	Oral	48	a e g j	b e g j
Ridomil Gold* PC GR	100-823	mefenoxam + PCNB	No	Warning	FN	Oral	48	a e g j	b e g j
Ridomil Gold* WSP	100-802	mefenoxam	No	Caution	FN	Oral	48	a e g	b e g
Ridomil Gold*/Bravo*	100-800	mefenoxam + chlorothalonil	No	Danger	FN	Oral	48	a e g j m	b e g j
Ridomil Gold*/Bravo* Liquid	100801 + 50534-188	Ridomil Gold: mefenoxam + Bravo 720: chlorothalonil	No	Warning	FN	Oral	48	a e g j	b e g j
Ridomil Gold*/Copper	100-804	mefenoxam + copper hydroxide	No	Danger	FN	Oral	48	a e g j m	b e g j
Ridomil MZ* 72	100-767	metalaxyl + mancozeb	No	Caution	FN	Oral	24	d e g j k	d e g
Ridomil PC* 11G	100-664	PCNB + metalaxyl	No	Caution	FN	Oral	12	a e g	b e g
Ridomil PC* Liquid	100607 + 5481-212	Ridomil 2E: metalaxyl + PCNB 2-E: PCNB	No	Danger	FN	Oral	12	d f h j k l m	d f h j
Sentinel* 40WG	100-863	cyproconazole	No	Caution	FN	Oral	12	a e g	b e g
Subdue* 2E	100-619	metalaxyl	No	Warning	FN	Oral	12	a f g j	b f g j
Subdue* 2X WSP	100-795	mefenoxam	No	Caution	FN	Oral		a e g	b e g
Tilt*	100-617	propiconazole	No	Warning	FN	Oral	24	a f g j	b f g j
Trigard*	100-654	cyromazine	No	Caution	IN	Oral	12	a e h	b e h
Triumph*	100-643	isazofos	Yes	Warning	IN	Oral	24	c f h k m	c f h
Vangard* WG	100-828	cyprodinil	No	Caution	FN	Oral	12	a e g	b e g
Zephyr* 0.15EC	618-97	abamectin	Yes	Warning	IN/MI	Oral	12	c f h j k l	c f h j l

Olympic Horticultural Products

Azatin* EC	62552-1-59107	azadirachtin	No	Warning	IN	Oral	12	a f g j	a f g j
Azatin* XL	62552-10-59107	azadirachtin	No	Caution	IN	Oral	4	a f g j	a f g j
Decathlon*	3125-430-59807	cyfluthrin	No	Caution	IN	Oral	12	a e g	b e g
Joust*	3125-437-59807	oxythioquinox	No	Caution	MI	Oral	24	a f g	b f g
Marathon* 1%G	3125-452-59807	imidacloprid	No	Caution	IN	Oral	12†	a e g	b e g
Marathon* 60WP	3125-492-59807	imidacloprid	No	Caution	IN	Oral	12	a e g	b e g
Strike* 25 WDG	3125-436-59807	triadimefon	No	Caution	FN	Oral	12	a g	b e g

Product Name	EPA Number	Active Ingredient(s)	RUP	Signal Word	Action	Notification	REI	PPE	Early Entry PPE
Pace International LP									
Deadline*	64864-2	metaldehyde	No	Warning	MO	Oral	12	b e g	b g
Platte Chemical Co.									
Clean Crop* C-O-C-S* WDG	34704-326	copper oxychloride + basic copper sulfate	No	Warning	FN	Oral	24	a e g j	b e g j
Clean Crop* Captan 5%	34704-30	captan	No	Danger	ST	Oral	96†	a e g j k m	b e g j
Clean Crop* Captan 7.5 Dust	34704-22	captan	No	Danger	FN	Oral	48†	a e g j k m	b e g j
Clean Crop* Dimethoate 400	34704-207	dimethoate	No	Warning	IN/MI	Oral	48	a f h j l m	b f h j l
Clean Crop* Malathion 8EC	34704-119	malathion	No	Caution	IN	Oral	12	a e g	b e g
Clean Crop* Malathion 57EC	34704-108	malathion	No	Caution	IN	Oral	12	a f g j	b f g
Clean Crop* Nemasol* 42%	34704-769	metam-sodium	No	Danger	FM	Oral, Posted	48	d e h j k l m	d e h j m
Clean Crop* Phorate 20G	34704-259	phorate	Yes	Danger-Poison	IN	Oral, Posted	48†	d e h j k l m	d e h j l
Clean Crop* Sevin Bait	34704-23	carbaryl	No	Caution	IN	Oral	12	a e g	b e g
Dusting Sulfur	34704-735	sulfur	No	Caution	FN/IN	Oral	24	a e g	b e g
Polyram* 80DF	7969-105-34704	metiram	No	Caution	FN	Oral	24	d e g j k	d e g
Sniper* 2-E	34704-691	azinphos-methyl	Yes	Danger-Poison	IN	Oral, Posted	48†	d f h j k l m	d f h l
Sniper* 50 PVA	10163-180-34704	azinphos-methyl	Yes	Danger-Poison	IN	Oral	48†	d e h j k l m	d e h l
Sniper* 50W	66222-11-34704	azinphos-methyl	Yes	Danger-Poison	IN	Oral	48†	d e h j k l m	d e h l
Syllit* 65W	264-508-34704	dodine	No	Danger	FN	Oral, Posted	48	d e h j k l m	d e h j l
Thiolux*	55947-48-34704	sulfur	No	Caution	FN/IN	Oral	24	a e g j	b e g j
TOPS* 5	7501-149-34704	thiophanate-methyl	No	Caution	FN	Oral	12†	a e g m	b e g
TOPS* 2.5D	7501-32-34704	thiophanate-methyl	No	Caution	FN	Oral	12	a e g m	b e g
Wettable Sulfur	34704-734	sulfur	No	Caution	FN/IN	Oral	24	a e g	b e g
Prentiss Incorporated									
Prentox* 2 lb. Methoxychlor Spray	655-742	methoxychlor	No	Caution	IN	Oral	12	a f g	a f g
Prentox* Diazinon AG500	655-459	diazinon	Yes	Caution	IN	Oral	12†	b f g	b f g
Prentox* Diazinon 50W	655-456	diazinon	Yes	Caution	IN	Oral	12†	b e g	b e g
Prentox* Fogging Insecticide	655-643	pyrethrins	No	Caution	IN	Oral	12†	b f g	b f g
Prentox* 5 lb. Malathion Spray	655-777	malathion	No	Warning	IN	Oral	12	a f g j	b f g j
Rhone-Poulenc Ag Co.									
Aliette*/Maneb 2+2	264-516 & 4581-371	fosetyl-aluminum + maneb	No	Caution	FN	Oral	24	a e g j k	d e g j
Aliette* WDG	264-516	fosetyl-aluminum	No	Caution	FN	Oral	12	a e g j	b e g j
Aliette* WSP	264-516	fosetyl-aluminum	No	Caution	FN	Oral	12	a e g j	b e g j
Chipco* Aliette* Signature* WDG	264-515	fosetyl-aluminum	No	Caution	FN	Oral	12	a e g j	b e g j
Chipco* Aliette* WDG	264-515	fosetyl-aluminum	No	Caution	FN	Oral	12	a e g j	b e g j
Chipco* 26019 FLO	264-480	iprodione	No	Caution	FN	Oral	12	a f g	b f g
Chipco* Mocap* 10G	264-546	ethoprop	No	Warning	IN/NE	Oral	†	c e g l m	c e g l
Chipco* Sevin* SL	264-335	carbaryl	No	Warning	IN	Oral	12	a e g l	b e g l
Chipco* Sevin* 80WSP	264-526	carbaryl	No	Warning	IN	Oral	12	a e g l	b e g l
Chipco* 26019 WP	264-481	iprodione	No	Caution	FN	Oral	12	a e g	b e g
Larvin* 3.2	264-379	thiodicarb	No	Warning	IN/OV	Oral	12	a e g l	b e g l
Mocap* EC	264-458	ethoprop	Yes	Danger-Poison	IN/NE	Oral, Posted	48†	d f h j k l m	d f h j l
Mocap* 10%G	264-465	ethoprop	No	Warning	IN/NE	Oral, Posted	48†	c e h k l m	c e h l
Mocap* 20%G	264-469	ethoprop	Yes	Danger-Poison	IN/NE	Oral, Posted	48†	c e h k l m	c e h l
Mocap* Plus 4-2 EC	264-464	ethoprop + disulfoton	Yes	Danger-Poison	IN/NE	Oral, Posted	48†	f h j m n	f h j l n
Rovral*	264-453	iprodione	No	Caution	FN	Oral	12	a e g j	b e g j
Rovral* 4 Flowable	264-482	iprodione	No	Caution	FN	Oral	12	a e g j	b e g j
Rovral* WG	264-524	iprodione	No	Warning	FN	Oral	12	a e g j	b e g j
Sevin* 4F	264-349	carbaryl	No	Caution	IN	Oral	12	a f g l	b f g l
Sevin* 80S	264-316	carbaryl	No	Warning	IN	Oral	12	a e g l	b e g l
Sevin* 80WSP	264-526	carbaryl	No	Warning	IN	Oral	12	a e g l	b e g l
Sevin* XLR Plus	264-333	carbaryl	No	Caution	IN	Oral	12	a f g l	b f g l
Temik* 15G	264-330	aldicarb	Yes	Danger-Poison	IN/NE	Oral, Posted	48†	c e h j k l m	c e h j l
Temik* 15G (Calif.)	264-426	aldicarb	Yes	Danger-Poison	IN/NE	Oral, Posted	48†	c e h j k l m	c e h j l
Temik* 15G Citrus	264-417	aldicarb	Yes	Danger-Poison	IN/NE	Oral, Posted	48†	c e h j k l m	c e h j l

Action Word Abbreviations

- BA......Bactericide
- BI.......Biostimulant
- FM.....Fumigant
- FN......Fungicide
- IGR....Insect Growth Regulator
- IN.......Insecticide
- LA......Larvicide
- MI......Miticide
- MO....Molluscicide
- NE......Nematicide
- OV.....Ovicide
- RE......Repellent
- SP......Seed Protectant
- ST......Seed Treatment

Personal Protective Equipment (PPE) Abbreviations

- a... Long-sleeved shirt and long pants
- b.. Coveralls
- c... Coveralls over short-sleeved shirt and short pants
- d.. Coveralls over long-sleeved shirt and long pants
- e... Waterproof gloves
- f... Chemical-resistant gloves
- g.. Shoes plus socks
- h.. Chemical-resistant footwear plus socks
- j... Protective eyewear
- k... Chemical-resistant apron when cleaning equipment, mixing, or loading
- l... Chemical-resistant headgear for overhead exposure
- m.. Approved respirator
- n.. Chemical-resistant protective suit

* = Trade Name/R/TM † May vary; please refer to product label for complete information.
Information presented herein is for preliminary planning only. Exclusive reliance must be placed on information/directions supplied by the manufacturer.

Product Name	EPA Number	Active Ingredient(s)	RUP	Signal Word	Action	Notification	REI	PPE	Early Entry PPE
RohMid L.L.C.									
Mach 2*	69075-4	halofenozide	No	Caution	IN	Oral	12	a e g	b e g
Rohm and Haas Co.									
Confirm* 2F	707-238	tebufenozide	No	Caution	IN	Oral	4	a e g	b e g
Dithane* DF	707-180	mancozeb	No	Caution	FN	Oral	24	d e g j k	d e g
Dithane* F-45	707-156	mancozeb	No	Caution	FN	Oral	24	d e g j k	d e g
Dithane* M-45	707-78	mancozeb	No	Caution	FN	Oral	24	d e g j k	d e g
Dithane* ST	707-156	mancozeb	No	Caution	FN	Oral	24	d e g j k	d e g
Dithane* T&O	707-180	mancozeb	No	Caution	FN	Oral	24	d e g j k	d e g
Dithane* WF	707-156	mancozeb	No	Caution	FN	Oral	24	d e g j k	d e g
Dithane* WSP	707-241	mancozeb	No	Caution	FN	Oral	24	d e g j k	d e g
Eagle* WSP	707-232	myclobutanil	No	Warning	FN	Oral	24	a e g j l	b e g j l
Enable* 2F	707-231	fenbuconazole	No	Caution	FN	Oral	12	a e g	b e g
Fore* FL	707-156	mancozeb	No	Caution	FN	Oral	24	d e g j k	d e g
Fore* FloXL	707-156	mancozeb	No	Caution	FN	Oral	24	d e g j k	d e g
Fore* T&O	707-87	mancozeb	No	Caution	FN	Oral	24	d e g j k	d e g
Fore* WSP	707-240	mancozeb	No	Caution	FN	Oral	24	d e g j k	d e g
Indar* 75WSP	707-239	fenbuconazole	No	Caution	FN	Oral	12	a e g	b e g
Kelthane* 35	707-205	dicofol	No	Warning	MI	Oral	12	a e g j	b e g j
Kelthane* 50	707-229	dicofol	No	Danger	MI	Oral	12	d e g j k l m	b e g
Kelthane* MF	707-202	dicofol	No	Caution	MI	Oral	12	d f h k l m	b f g
Kelthane* T/O	707-229	dicofol	No	Caution	MI	Oral	12	d e g j k l m	b e g
Nova* 40W	707-221	myclobutanil	No	Warning	FN	Oral	24	a e g j l	b e g j l
Rally* 40W	707-215	myclobutanil	No	Warning	FN	Oral	24	a e g j l	b e g j l
Rally* 40W	707-221	myclobutanil	No	Warning	FN	Oral	24	a e g j l	b e g j l
Systhane* WSP	707-253	myclobutanil	No	Warning	FN	Oral	24	a e g j l	b e g j l
The Scotts Company									
Fungo* 50WSB	58185-30	thiophanate-methyl	No	Warning	FN	Oral	12	a e g j m	b e g j
Koban* 30	58185-5	etridiazole	No	Warning	FN	Oral	12	c e h j k l m	c e h j l
Penstar* Flo	400-454-538	PCNB	No	Caution	FN	Oral	12	a f g	b f g
Penstar* 10G	5481-211-58185	PCNB	No	Caution	FN	Oral	12	a e g	b e g
ProTurf* Fluid Fungicide	538-183	iprodione + thiophanate-methyl	No	Caution	FN	Oral	12	a f g	b f g
ProTurf* Insecticide III	538-111	chlorpyrifos	No	Caution	IN	Oral	12	a f g l	b f g l
ProTurf* Pythium Control	538-185	metalaxyl	No	Caution	FN	Oral	24	a e g j m	b f g j
Vorlan* DF	7969-85-5815	vinclozalin	No	Caution	FN	Oral	12	d f h j k l m	d f h j l
SePRO Corp.									
Pipron*	67690-1	piperalin	No	Warning	FN	Oral	12	b f g	b f g
Rubigan* T&O	62719-134-67690	fenarimol	No	Warning	FN	Oral	12	c f h l k	c f h j l
Soil Technologies Corp.									
Garlic Gard AG*	66352-2-67638	garlic juice	No	Caution	IN	Oral	4	a g	b e g
Nemastop*	59714-2-67638	plant extract	No	Caution	NE	Oral	12	a e g	a e g
Sostram Corp.									
Echo* 500	60063-4	chlorothalonil	No	Warning	FN	Oral	48	a e g j m	b e g j
Echo* 720	60063-7	chlorothalonil	No	Warning	FN	Oral	48	a e g m	b e g
Echo* 90DF	60063-10	chlorothalonil	No	Danger	FN	Oral	48	a e g j m	b e g j
Stoller Enterprises, Inc.									
THAT*	57538-5	sulfur	No	Caution	AC/FN	Oral	24	a e g j	b e g j
THAT* Big 8	57528-8	sulfur	No	Caution	AC/IN	Oral	24	a e g j	b e g j
TOP COP* Tri-Basic*	57538-7	basic copper sulfate	No	Caution	FN/BA	Oral	24	a e g	b e g
TOP COP* w/Sulfur	57538-6	sulfur + tribasic copper sulfate	No	Caution	FN/BA	Oral	24	a e h l	b e h
Sun Company, Inc.									
SunSpray* 6C	862-21	refined petroleum distillate	No	Caution	IN	Oral	12	a g	b e g
SunSpray* 9C	862-24	refined petroleum distillate	No	Caution	IN/FN	Oral	4	a g	b e g
SunSpray* 11C	862-22	refined petroleum distillate	No	Caution	IN	Oral	12	a g	b e g
SunSpray* 6E	862-11	refined petroleum distillate	No	Caution	IN	Oral	4	a g	b e g
SunSpray* 7E	862-8	refined petroleum distillate	No	Caution	IN	Oral	4	a g	b e g
SunSpray* 9E	862-19	refined petroleum distillate	No	Caution	IN	Oral	4	a g	b e g
SunSpray* 11E	862-9	refined petroleum distillate	No	Caution	IN	Oral	4	a g	b e g
SunSpray* Ultra-Fine Spray Oil	862-23	paraffinic oil	No	Caution	IN/FN	Oral	4	a g	b e g
Terra Industries Inc.									
Bean Treater* LW	42056-10-9779	thiram	No	Warning	ST	Oral	24	a e g j	b e g j
Captain Moly*	42036-3-9779	captan	No	Danger	ST	Oral	48	a e g j k	b e g j
Dimate* 4E	9779-273	dimethoate	No	Warning	IN	Oral	48	a f h j l m	b f h j l
Kernel Guard*	42056-3-9779	diazinon + lindane	No	Danger	ST	Oral	12	c e h j k m	b e g j

Product Name	EPA Number	Active Ingredient(s)	RUP	Signal Word	Action	Notification	REI	PPE	Early Entry PPE
Riverside* Diazinon 4AG	9779-210	diazinon	Yes	Caution	IN	Oral	12†	a f g	b f g
Riverside* Diazinon 5WB	100-784-9779	diazinon	Yes	Caution	IN	Oral	24	a e g	b e g
Riverside* Malathion 5	9779-5	malathion	No	Warning	IN	Oral	12	a f g j	b f g j
Riverside* Malathion ULV	4787-8-9779	malathion	No	Caution	IN	Oral	12	a f g	b f g
Riverside* Methomyl 5G	9779-331	methomyl	Yes	Warning	IN	Oral	48	c e h j k l m	c e h j l
Riverside* Phorate 20G	9779-293	phorate	Yes	Danger-Poison	IN	Oral, Posted	48†	e h j l m n	e h j l n
Riverside* Sulfur 6L	51036-16-9779	sulfur	No	Caution	FN	Oral	24	a e g j l	b e g j l
Terranil* Cu	9779-339	chlorothalonil + copper oxychloride	No	Danger	FN	Oral	48	a e g j m	b e g j
Terranil* 90DF	9779-280	chlorothalonil	No	Danger	FN	Oral	48	a e g j m	b e g j
Terranil* 90DF WSP	9779-328	chlorothalonil	No	Danger	FN	Oral	48	a e g j m	b e g j
Terranil* 6L	9779-320	chlorothalonil	No	Warning	FN	Oral	48	a e g j m	b e g j
Terranil* S	9779-337	chlorothalonil + sulfur	No	Warning	FN	Oral	48	a e g j m	b e g j
Terranil* Zn	9779-333	chlorothalonil	No	Warning	FN	Oral	48	a e g j m	b e g j
Thalonil* 4L	9779-270	chlorothalonil	No	Warning	FN	Oral	48	a e g j m	b e g j
Thalonil* 90DF	9779-280	chlorothalonil	No	Danger	FN	Oral	48	a e g j m	b e g j
Thalonil* 90DF WSP	9779-328	chlorothalonil	No	Danger	FN	Oral	48	a e g j m	b e g j
Thirethrin*	9779-330	endosulfan + piperonyl butoxide	No	Warning	IN	Oral	24	c f h j k l m	c f h j l

Thermo Trilogy

Agree* WG	70051-47	*Bt.* spp. *aizawai*	No	Caution	IN	Oral	4	a e g j m	b e g j
Azatin* XL Plus	70051-27	azadirachtin	No	Caution	IN	Oral	4	a f g j	a f g j
Javelin* WG	70051-66	*Bt.* spp. *kurstaki*	No	Caution	IN	Oral	4	a e g	b e g
Neemazad* 0.25EC	70051-5	azadirachtin	No	Caution	IN	Oral	12	a f g	b f g
Neemazad* 4.5EC	70051-9	azadirachtin	No	Warning	IN	Oral	12	a f g	b f g
Neemix*	70051-5	azadirachtin	No	Caution	IN	Oral	12	a f g	b f g
Neemix* 4.5	70051-9	azadirachtin	No	Caution	IN	Oral	12	a f g	b f g
Soil Gard* 12G	70051-3	*Gliocladium virens*	No	Caution	FN	Oral	4	a f g j	b f g j
Triact* 90EC	70051-8	neem oil	No	Caution	FN/IN	Oral	4	a f g	b f g
Trilogy*	70051-2	neem oil	No	Caution	FN/IN	Oral	4	a f g	b f g
Trilogy* 90EC	70051-12	neem oil	No	Caution	FN/IN	Oral	4	a f g	b f g

Trace Chemicals, Inc.

Apron* Dry	7501-132-42056	metalaxyl	No	Caution	FN	Oral	12	a e g	b e g
Bean Guard*	7501-43-42056	captan + carboxin	No	Danger	FN	Oral	48†	a e g j m	b e g j
Captan-Diazinon	42056-18	captan + diazinon	No	Danger	FN/IN	Oral	48†	a e g j	b e g j
Enhance* Plus	7501-37-42056	carboxin + maneb + lindane	No	Warning	FN/IN	Oral	12†	d e h j k l m	d e f j l
Enhance* Vitavax*-Captan 20-20	7501-36-42056	captan + carboxin	No	Danger	FN/IN	Oral	48	a e g j k m	b e g j
Germate Plus*	7501-112-42056	carboxin + diazinon + lindane	No	Caution	FN/IN	Oral	12	a e g m	b e g
Grain Guard*	42056-6	mancozeb	No	Caution	FN/IN	Oral	12†	d e g j k	d e g j
Grain Guard* Plus	42056-15	mancozeb + lindane	No	Warning	FN/ST	Oral	12†	d e g j k	d e g j
Hi Moly/Captan*	42056-1	captan	No	Danger	FN/ST	Oral	48†	a f g j k	b f g j
Hi Moly/Captan*-D	42056-1	captan	No	Danger	FN/ST	Oral	48†	a f g j k	b f g j
Kernel Guard*	42056-11	captan + diazinon + lindane	No	Danger	FN/IN	Oral	12	a e g j k m	b e g j
Kernel Guard* Supreme	42056-21	carboxin + permethrin	No	Caution	FN	Oral	12†	a e g m	b e g
Prevail*	7501-145-42056	PCNB + carboxin + metalaxyl	No	Caution	FN	Oral	12	a e g m	b e g
Protector-L*	42056-10	thiram	No	Warning	FN	Oral	24†	a e g j	b e g j
Sorghum Guard*	42056-14	captan + lindane	No	Danger	FN/IN	Oral	12	a e g j k m	b e g j
Vitavax* T-L	7501-114-42056	carboxin + thiram	No	Caution	FN/SP	Oral	24	a e g j	b e g j

Troy Biosciences, Inc.

Larvo-Bt*	53871-6	*Bt.* spp. *kurstaki*	No	Caution	IN	Oral	12	a f g	b f g
Naturalis-O*	53871-9	*Beauveria bassiana*	No	Caution	IN	Oral	4	a f g m	b f g
Naturalis-T*	53871-9	*Beauveria bassiana*	No	Caution	IN	Oral	4	a f g m	b f g
Stirrup M*	53871-2	trimethyl dodectriene	No	Caution	IN	Oral	12	a f g	b f g
Troy-Bt*	53871-6	*Bt.* spp. *kurstaki*	No	Caution	IN	Oral	12	a f g	b f g

UCB Agrochemicals

Ferbam Granuflo*	45728-7	ferbam	No	Caution	FN	Oral	24	a e g j l m	b e g j l
Metam CLR*	45728-16	metam-sodium	No	Danger	FM	Oral	48	d e h k j l m	d e h j m

Action Word Abbreviations

BA......Bactericide
BI.......Biostimulant
FM.....Fumigant
FN......Fungicide
IGR....Insect Growth Regulator
IN.......Insecticide
LA......Larvicide
MI......Miticide
MO....Molluscicide
NE......Nematicide
OV.....Ovicide
RE......Repellent
SP......Seed Protectant
ST......Seed Treatment

Personal Protective Equipment (PPE) Abbreviations

a... Long-sleeved shirt and long pants
b .. Coveralls
c... Coveralls over short-sleeved shirt and short pants
d .. Coveralls over long-sleeved shirt and long pants
e... Waterproof gloves
f ... Chemical-resistant gloves
g .. Shoes plus socks
h .. Chemical-resistant footwear plus socks
j ... Protective eyewear
k... Chemical-resistant apron when cleaning equipment, mixing, or loading
l ... Chemical-resistant headgear for overhead exposure
m.. Approved respirator
n .. Chemical-resistant protective suit

* = Trade Name/R/TM † May vary; please refer to product label for complete information.
Information presented herein is for preliminary planning only. Exclusive reliance must be placed on information/directions supplied by the manufacturer.

Product Name	EPA Number	Active Ingredient(s)	RUP	Signal Word	Action	Notification	REI	PPE	Early Entry PPE
Thiram 65*	45728-24	thiram	No	Caution	FN	Oral	24	a e g l	a e g l
Thiram Granuflo*	45728-21	thiram	No	Caution	FN/SP	Oral	24	a e g l	a e g l
Ziram Granuflo*	45728-12	ziram	No	Danger	FN	Oral	48	a e g j l m	b e g j l
Uniroyal Chemical Co., Inc.									
Adept*	400-469	diflubenzuron	No	Caution	IGR	Oral	12	ag	b e g
Comite*	400-104	propargite	No	Danger	MI	Oral, Posted	48†	a f g j k l m	b f g j l
Comite* II	400-154	propargite	No	Danger	MI	Oral, Posted	48†	a f g j k l m	b f g j l
Dimilin* 2F	400-473	diflubenzuron	Yes	Caution	IGR	Oral	12	a g	b e g
Dimilin* 2L	400-474	diflubenzuron	Yes	Caution	IGR	Oral	12	a f g	b f g
Dimilin* 4L	400-461	diflubenzuron	Yes	Caution	IGR	Oral	12	a e g	b e g
Dimilin* 25W	400-465	diflubenzuron	Yes	Caution	IGR	Oral	12	a g	b e g
Dimilin* 25W (Mushrooms)	400-469	diflubenzuron	No	Caution	IGR	Oral	12	a g	b e g
Fireban*	10182-365-400	tefluthrin	No	Caution	IN	Oral, Posted	48	a e g	b e g
Micromite* 25WS	400-471	diflubenzuron	Yes	Caution	IGR	Oral	12	a g	b e g
Omite*-CR (Calif.)	400-425	propargite	No	Danger	MI	Oral, Posted	48†	a e g j k l	b e g j l
Omite*-6E	400-89	propargite	No	Warning	MI	Oral, Posted	48†	c f h j k l m	c f h j l
Omite*-30W	400-82	propargite	No	Danger	MI	Oral, Posted	48†	a e g j k l	b e g j l
Omite*-30WS	400-427	propargite	No	Danger	MI	Oral, Posted	48†	a e g j k l	b e g j l
Omite*-CR	400-426	propargite	No	Danger	MI	Oral, Posted	48†	a e g j k l	b e g j l
Ornamite*	400-426	propargite	No	Danger	MI	Oral, Posted	48†	a e g j k l	b e g j l
Plantvax* 75W	400-144	oxycarboxin	No	Caution	FN	Oral	12	a g	b e g
Terraclor* 400	400-454	PCNB	No	Caution	FN	Oral	12†	a f g m	b f g
Terraclor* 75% WP	400-399	PCNB	No	Caution	FN	Oral	12†	a e g	b e g
Terraguard* 50W	400-433	triflumizole	No	Caution	FN	Oral	12†	a e g	b e g
Terrazole* 35% WP	400-416	etridiazole	No	Warning	FN	Oral	12†	a e g j m	b e g j
Topcide*	10182-379-400	*lambda*-cyhalothrin	No	Warning	IN	Oral	24	a e g j	b e g j
Topcide* O/S	10182-417-400	*lambda*-cyhalothrin	Yes	Warning	IN	Oral	24	a e g j	b e g j
Turfcide* 400 Flowable	400-454	PCNB	No	Caution	FN	Oral	12†	a f g m	b f g
Turfcide* 10% Granular	400-407	PCNB	No	Caution	FN	Oral	12†	a e g	b e g
Valent U.S.A. Corp.									
Danitol* 2.4 EC	59639-35	fenpropathrin	Yes	Danger	IN/MI	Oral	24	a f g j	b f g j
Distance*	59639-96	pyriproxyfen	No	Caution	IGR	Oral	12	c f h k l	c f h l
Knack*	59639-95	pyriproxyfen	No	Caution	IGR	Oral	12	c f h k l	c f h l
Monitor* 4 Spray	59639-56	methamidophos	Yes	Danger-Poison	IN	Oral, Posted	48†	c f h j k l m	c f h j l
Orthene* 75 S	59639-26	acephate	No	Caution	IN	Oral	24	a e g l	b e g l
Orthene* 90 S	59639-33	acephate	No	Caution	IN	Oral	24	a e g l	b e g l
Orthene* 97 S	59639-91	acephate	No	Caution	IN	Oral	24	a e g l	b e g l
Orthene* Turf, T & O Spray	59639-26	acephate	No	Caution	IN	Oral	24	a g l	b e g l
Orthene* Turf, T & O WSP	59639-88	acephate	No	Caution	IN	Oral	24	a g l	b e g l
Orthene* 75 WSP	59639-89	acephate	No	Caution	IN	Oral	24	a e g l	b e g l
Payload* 15 G	59639-75	acephate	No	Caution	IN	Oral	12	a e g	b e g
Pinpoint* 15 G	59639-87	acephate	No	Caution	IN	Oral	12	a e g	b e g
Tame* 2.4 EC	59639-77	fenpropathrin	Yes	Danger	IN/MI	Oral	24	a f g j	b f g j
Velocity*	59639-87	acephate	No	Caution	IN	Oral	12	a e g	b e g
Volck* Supreme Spray	59639-20	petroleum oil	No	Caution	IN/MI	Oral	4	a g	b e g
Webb Wright Corp.									
Pyrellin* E.C.	30573-2	pyrethrins + rotenone	No	Caution	IN	Oral	12	a f g	b f g
Wellmark International									
Enstar II*	2724-476	*s*-kinoprene	No	Warning	IGR	Oral	4	a f g j	b f g j
Mavrik Aquaflow*	2724-478	*tau*-fluvalinate	No	Caution	IN/MI	Oral	12	a f g m	b f g
Wilbur-Ellis Co.									
Agrox Premiere*	2935-525	captan + diazinon + lindane + metalaxyl	No	Danger	IN	Oral	96†	d e h j l m	d e h j l
Bio-Trek* 22G	68539-3-2935	*Trichoderma harzianum*	No	Caution	FN	Oral	4	a g m	a e g
Bio-Trek* HB	68539-4-2935	*Trichoderma harzianum*	No	Caution	FN	Oral	4	a g m	a e g
Bio-Trek* Nursery Drench	68539-4-2935	*Trichoderma harzianum*	No	Caution	FN	Oral	4	a g m	a e g
Botran* 6 Dust	10951-13-2935	DCNA	No	Caution	FN	Oral	12	a e g	b e g
Botran* 6 Sulfur 25 Dust	10951-14-2935	DCNA	No	Warning	FN	Oral	12	a e g j	b e g j
Bt 320 Sulfur 25 Dust	2935-398	Bt var. *berliner* + sulfur	No	Caution	IN	Oral	24	a e g	b e g
Bt 320 Sulfur 50 Dust	2935-485	Bt var. *berliner* + sulfur	No	Caution	IN	Oral	24	a e g	b e g
Citri-Oil*	2935-461	petroleum oil	No	Caution	IN	Oral	12	a f g	b f g
Copper 3 Dust	WA-880002	copper	No	Caution	FN	Oral	12	a e g	b e g
Digon* 400	2935-520	dimethoate	No	Warning	IN/MI	Oral	48	a f h j l m	b f h j l
Nu-Gro* Isotox*	10182-311-2935	captan + lindane	No	Danger	ST	Oral	96†	d e h j k l m	d e h j l
PCNB 2 Spray	2935-208	PCNB	No	Caution	FN	Oral	12	a f g	b f g
PCNB 10 Granular	2935-357	PCNB	No	Caution	FN	Oral	12	a e g j	b e g

Product Name	EPA Number	Active Ingredient(s)	RUP	Signal Word	Action	Notification	REI	PPE	Early Entry PPE
Py-Rin* 60-6EC	1021-1091-2935	pyrethrins	No	Caution	IN	Oral	12	a f g j	b f g j
Sevin* 5 Dust	2935-193	carbaryl	Yes	Caution	IN	Oral	12	a e g	b e g
Sevin* 10 Dust	2935-320-ZA	carbaryl	Yes	Caution	IN	Oral	12	a e g	b e g
Wilbur-Ellis Diazinon 4 Spray	2935-388-AA	diazinon	Yes	Caution	IN	Oral	24	a e g	b e g
Wilbur-Ellis Diazinon 14 Spray	2935-408-ZA	diazinon	Yes	Caution	IN	Oral	12	a e g	b e g
Wilbur-Ellis Dimethoate 267	2749-41-AA-2935	dimethoate	No	Warning	IN/MI	Oral	48	a f h j l m	b f h j l
Wilbur-Ellis Malathion 8	2935-83-ZA	malathion	No	Caution	IN	Oral	12	a f g	b f g
Wilbur-Ellis Methyl Parathion 4	2935-528	methyl parathion	Yes	Danger-Poison	IN	Oral, Posted	48	d f h j k l m	d f h j l
Wilbur-Ellis Methyl Parathion 5	2935-527	methyl parathion	Yes	Danger-Poison	IN	Oral, Posted	48	d f h j k l m	d f h j l
Wilbur-Ellis Methyl Parathion 7.5	2935-421	methyl parathion	Yes	Danger-Poison	IN	Oral, Posted	48	d f h j k l m	d f h j l

ZENECA Ag Products

Product Name	EPA Number	Active Ingredient(s)	RUP	Signal Word	Action	Notification	REI	PPE	Early Entry PPE
Abound*	10182-415	azoxystrobin	No	Caution	FN	Oral	12	a e g	b e g
Ambush*	10182-18	permethrin	Yes	Warning	IN	Oral	12	a f g j m	b f g j
Ambush* 25WP	10182-35	permethrin	Yes	Warning	IN	Oral	12	a e g j	b e g j
Ambush* 25WSP	10182-110	permethrin	Yes	Warning	IN	Oral	12	a e g j	b e g j
Bravo* 500	50534-8	chlorothalonil	No	Warning	FN	Oral	48	a e g j	b e g j
Bravo* 720	50534-188	chlorothalonil	No	Warning	FN	Oral	48	a e g j m	b e g j
Bravo Ultrex*	50534-201	chlorothalonil	No	Danger	FN	Oral	48	a e g j m	b e g j
Bravo Weather Stik*	50534-188	chlorothalonil	No	Warning	FN	Oral	48	a e g j m	b e g j
Bravo Zn*	50534-204	chlorothalonil	No	Warning	FN	Oral	48	a e g j	b e g j
Daconil 2787*	50534-9	chlorothalonil	No	Warning	FN	Oral	48	a e g j m	b e g j
Daconil Ultrex*	50534-202	chlorothalonil	No	Danger	FN	Oral	48	a e g j m	b e g j
Daconil Weather Stik*	50534-209	chlorothalonil	No	Warning	FN	Oral	48	a e g j m	b e g j
Daconil Zn*	50534-211	chlorothalonil	No	Warning	FN	Oral	48	a e g j m	b e g j
Force* 3G	10182-373	tefluthrin	Yes	Caution	IN	Oral	0†	a e g	b e g
Heritage*	10182-408	azoxystrobin	No	Caution	FN	Oral	12	a e g	b e g
Karate*	10182-96	*lambda*-cyhalothrin	Yes	Danger	IN	Oral	24	d f h j k l m	d f h j l
Pirimor* 50-DF	10182-370	pirimicarb	No	Warning	IN	Oral	12	a e g	b e g
Quadris*	10182-415	azoxystrobin	No	Caution	FN	Oral	12	a e g	b e g
Warrior*	10182-96	*lambda*-cyhalothrin	Yes	Danger	IN	Oral	24	d f h j k l m	d f h j l

Action Word Abbreviations

BA......Bactericide
BI.......Biostimulant
FM .....Fumigant
FN......Fungicide
IGR ....Insect Growth Regulator
IN.......Insecticide
LA......Larvicide
MI......Miticide
MO ....Molluscicide
NE......Nematicide
OV .....Ovicide
RE......Repellent
SP......Seed Protectant
ST......Seed Treatment

Personal Protective Equipment (PPE) Abbreviations

a... Long-sleeved shirt and long pants
b .. Coveralls
c... Coveralls over short-sleeved shirt and short pants
d .. Coveralls over long-sleeved shirt and long pants
e... Waterproof gloves
f ... Chemical-resistant gloves
g .. Shoes plus socks
h .. Chemical-resistant footwear plus socks
j ... Protective eyewear
k... Chemical-resistant apron when cleaning equipment, mixing, or loading
l ... Chemical-resistant headgear for overhead exposure
m.. Approved respirator
n .. Chemical-resistant protective suit

* = Trade Name/R/TM † May vary; please refer to product label for complete information.
Information presented herein is for preliminary planning only. Exclusive reliance must be placed on information/directions supplied by the manufacturer.

Notes • Notes • Notes

Insect and Disease Control Guide

Section 1

Biocontrols

The information in this 1999 Insect and Disease Control Guide is updated to August 15, 1998.
Because manufacturers' product recommendations are constantly being updated and revised as additional product registrations and label changes are approved by EPA, there may be changes after this date. However, every effort has been made to give manufacturers the opportunity to include in the Guide their latest information. This Guide is for preliminary planning – not a use guide. Be sure to follow label directions, notwithstanding information contained herein.

Biocontrols

Biocontrol Agents

AGRI-MYCIN 17

Fungicide/bactericide for control of bacterial plant pathogens.

Agri-mycin 17 (Streptomycin sulfate) *(Novartis)*

Crop: Tobacco.
Rate: Preventive:
0.5 lb. in 100 gal./A
Curative:
1 lb. in 100 gal./A
Disease:
blue mold
wildfire
Time: Apply when plants are in 2-leaf stage or when disease first appears in area; repeat at 5- to 7-day intervals until plants are set in field; additional protection may be obtained by maintaining a weekly spray schedule.

Crop: Peppers, tomatoes.
Rate: 1 lb. in 100 gal./A
Disease:
bacterial spot
Time: Apply when seedlings are in 2-leaf stage (when first true leaves appear); repeat at 4- to 5-day intervals when transplanted in field.

Crop: Celery—Florida.
Rate: 1 lb. in 100 gal./A
Disease:
bacterial blight
Time: Apply when first true leaves appear; repeat when transplanted in field.

Crop: Potatoes.
Rate: 0.5 lb. in 100 gal./A
Disease:
blackleg
soft rot
Time: Seedpiece treatment. Soak cut seed pieces for several minutes; plant as usual.

Crop: Apples, pears.
Rate: 24-48 oz. in 100 gal./A
West Coast:
28.8 oz. in 100 gal./A
Disease:
fire blight
Time: Apply at 20-30% bloom; repeat at 3- to 4-day intervals during blossom time, and 10- to 14-day intervals after petal fall. West Coast: For apples, apply at petal fall, and late secondary bloom; continue at 5- to 7-day intervals. West Coast: For pears, apply at 10% bloom; repeat at 5-day intervals until late bloom is over; continue at 5- to 7-day intervals to control shoot and fruit infections.

Rate: 1 lb. in 100 gal./A
Disease:
bacterial leaf rot (philodendron)
bacterial stem rot (dieffenbachia)
crown gall (roses) (New Jersey)
Time: For dieffenbachia, soak cuttings in solution for 20 minutes; plant in sterilized rooting soil. For crown gall, remove infected plant, cut out gall tissue, then soak infected plant root system and cut surfaces in solution for 15 minutes; replant in soil free of crown gall organisms. For bacterial leaf rot preventive, apply at first signs of water-soaked areas on leaf; repeat at 4- to 5-day intervals, curative, remove all rotted leaves and spray every 4 days.

Crop: Ornamentals, tobacco, tree fruits deciduous (apples, pears), vegetables (celery, peppers, potatoes, tomatoes).
Rate: 4 oz. in 100 gal./A
Disease:
bacterial wilt (chrysanthemum)
crown gall (roses) (New Jersey)
Time: For bacterial wilt, soak plant cuttings in solution for 4 hr.; plant as usual. For crown gall, apply weekly in watering solution or in foliar sprays starting week after planting.
Remarks: Suitable fungicide (dithiocarbamate) should be used as an adjunct to seedpiece treatment for potatoes. Do not apply through any type of irrigation system.
Days To Harvest: Pears-30; apples-50.

- - - - - - - - - - - - - - - - -

Signal Word/Toxicity Class: Caution.

AQ10

Fungus that parasitizes powdery mildew. AQ-10 spores germinate into powdery mildew mycelia and parasitize it.

AQ10 (WDG) or AQ10 t/o (*Ampelomyces quisqualis*) *(Ecogen)*

Crop: Carrots, cucurbits, grapes, pome fruit, stone fruit, strawberries, sugar beets, tomatoes.
Rate: 0.5-1 oz. WDG/A

Crop: Bedding and flowering plants, nursery and landscape plants, potted flowering plants, shade and flowering trees, woody ornamentals.
Rate: 1 oz. t/o/100 gal. spray volume
Disease:
powdery mildew
Remarks: Begin application as soon as susceptible host tissue emerges. Apply at least 2 sequential sprays 7-14 days apart. AQ10 germination and parasitism occur under high humidity conditions.
Days To Harvest: 0.

- - - - - - - - - - - - - - - - -

Signal Word/Toxicity Class: Caution.
REI: 4 hr.

BEAUVERIA BASSIANA

Insect specific fungus. Exposure to nontarget insects should be avoided. Works primarily through contact action, so good spray coverage is essential.

◆ **BotaniGard ES or 22WP (strain GHA)** *(Mycotech)*

Crop: Herbs and spices, fruits, ornamentals, turf and forestry, vegetables.
Rate: 1-4 pt. ES, 22WP/1000 sq. ft.
Low volume:
0.5-2 lb. 22 WP/100 gal. spray volume
Cuttings:
0.25-0.5 oz. 22WP/gal. water
Pest:

Altenius*	green peach aphid	potato psylla*
banded wing whitefly	greenbug	rose aphid
bean aphid	greenhouse thrips	Russian wheat aphid
black vine weevil*	greenhouse whitefly	silverleaf whitefly
buffalo grass mealybug	hop aphid	spotted alfalfa aphid
cabbage aphid	leafhoppers*	sweetpotato whitefly
citrus blackfly	melon aphid	*Thrips palmi*
citrus mealybug	onion thrips	tomato psylla*
citrus root weevil*	pea aphid	variegated leafhopper*
citrus whitefly	pear thrips	Virginia creeper leafhopper*
cowpea aphid	pepper weevil*	western flower thrips
giant whitefly*	plant hoppers*	white grubs*
grape leafhopper*	potato aphid	
grape mealybug	potato leafhopper*	
green June beetle*		

* *BotaniGard 22WP*

Remarks: Apply when insects first appear; repeat at 7- to 10-day intervals as needed. Intense insect pressure may require more frequent applications. *Beauveria bassiana* not compatible with fungicides in tank mixtures. Fungicides will kill spores.
Days To Harvest: 0.

- - - - - - - - - - - - - - - - -

Signal Word/Toxicity Class: Caution (22WP); Warning/II (ES).
REI: 12 hr.

◆ - **new product** • ai-active ingredient • DF-dry flowable • E/EC-emulsifiable concentrate • EW-emulsion in water
F/FL-flowable • G-granule • L/LC-liquid • SP/WSP-soluble packet • W/WP-wettable powder • WSB-water soluble bag

◆ **Mycotrol ES or 22WP (strain GHA)** *(Mycotech)*

Crop: Corn, cotton, fruits, forage, herbs and spices, hops, nuts, oil seed crops, peanuts, safflower, sorghum, soybeans, small grains, sugar beets, sugarcane, sunflower, vegetables.

Rate: 1-6 pt. ES/A
0.25-1 lb. 22WP/A
Orchards:
1-4 oz. 22WP/1000 sq. ft.

Pest:

alfalfa weevil	flea beetles	pecan weevil
apple curculio	fleahopper	plant bug
banded wing whitefly	Fuller rose weevil	plum curculio
bean aphid	giant whitefly	potato aphid
bean leaf beetle	grape leafhopper	potato leafhopper
billbugs	grape mealybug	potato psylla*
black vine weevil	grasshoppers	root weevil*
cabbage aphid	green peach aphid	rose aphid
cereal leaf beetle	greenbug	rose curculio
chinch bugs	greenhouse thrips	Russian wheat aphid
citrus blackfly	greenhouse whitefly	seed bug
citrus mealybug	hop aphid	silverleaf whitefly
citrus root weevil*	locusts	spotted alfalfa aphid
citrus whitefly	longtailed mealybug*	strawberry root weevil
Colorado potato beetle	Lygus bug	sweetpotato weevil
corn rootworm	melon aphid	sweetpotato whitefly
cotton boll weevil	mole crickets	tarnished plant bug
cowpea aphid	Mormon crickets	*Thrips palmi*
Cuban laurel thrips*	onion thrips	tomato psylla*
cucumber beetle	pea aphid	variegated leafhopper
diamondback moth*	pear psylla*	vegetable weevil
elm leaf beetle	pear thrips	western flower thrips

* *Mycotrol 22WP*

Remarks: Apply when insects first appear; repeat at 7- to 10-day intervals as needed. Intense insect pressure may require more frequent applications and a combination of Mycotrol with a compatible insecticide. *Beauveria bassiana* not compatible with fungicides in tank mixtures. Fungicides will kill spores. Maximum rate 3 lb. 22WP/A. Thorough spray coverage essential.

Days To Harvest: 0.

Signal Word/Toxicity Class: Caution (22WP); Warning/II (ES).

REI: 12 hr.

Naturalis-O *(Troy Biosciences)*

Crop: Ornamentals (foliage, flowering plants, shrubs, shade and ornamental trees).

Rate: 30-100 fl. oz./100 gal. water

Pest:

aphids	flea beetles	psyllids
black vine root weevil	leafhoppers	strawberry root weevil
caterpillars	Lygus bugs	thrips
cucumber beetle	mites	whiteflies
elm leaf beetle	obscure root weevil	

Remarks: Apply at 3- to 5-day intervals when insects first appear.

Signal Word/Toxicity Class: Caution.

REI: 12 hr.

Naturalis-T *(Troy Biosciences)*

Crop: Lawns and turf.

Rate: 1-3 oz./1000 sq. ft.

Pest:

ants	chinch bugs	millipedes
armyworms	European chafer	mole cricket
bermudagrass mite	fire ants	northern masked chafer
billbugs	grasshoppers	sod webworms
black turfgrass ataenius	green June beetle	sowbugs
chiggers	Japanese beetle	white grubs

Remarks: Apply at 3- to 5-day intervals when insect first appear.

Signal Word/Toxicity Class: Caution.

REI: 12 hr.

BIO-TREK

Bio-Trek 22G (*Trichoderma harzianum* strain KRL-AG2) *(Wilbur-Ellis)*

Crop: Turf.

Rate: 1.5 lb. 22G/1000 sq. ft.

Disease:

Pythium — *Sclerotinia homeocarpa*

Time: Apply to new spring growth.

Remarks: Can be used in conjunction with certain chemical fungicides, consult detailed information on compatibility.

Signal Word/Toxicity Class: Caution.

REI: 12 hr.

Bio-Trek HB (*Trichoderma harzianum* strain KRL-AG2) *(Wilbur-Ellis)*

Crop: Ornamental crops, turf.

Rate: 6-8 oz. HB/100 lb. seed

Disease:

Fusarium — Pythium — Rhizoctonia

Time: At planting.

Remarks: For maximum protection, especially in cold soils, apply Bio-Trek HB to commercially treated seed such as seed treated with Apron and/or Captan for stand establishment.

Signal Word/Toxicity Class: Caution.

REI: 12 hr.

Bio-Trek Nursery Drench (*Trichoderma harzianum* strain KRL-AG2) *(Wilbur-Ellis)*

Crop: Cabbage, cucumbers, ornamentals, tomatoes.

Rate: 3-6 oz. Drench/cu. yd.

Disease:

Fusarium — Pythium — Rhizoctonia

Time: At planting.

Remarks: Suspend in sufficient water to achieve uniform application.

Signal Word/Toxicity Class: Caution.

REI: 12 hr.

BotaniGard — see *Beauveria bassiana*

CheckMate — see pheromones

CLANDOSAN

Dried, granular material isolated from tough polymer matrix of crustacean exoskeletons. Product acts in soils to stimulate growth of normal soil microorganisms, which produce chitinase and other enzymes to destroy plant-pathogenic nematodes and their eggs.

ClandoSan 618 (chitin-protein) *(IGENE Biotechnology)*

Crop: Cotton, field corn, forage, greenhouse, nuts, ornamentals, other field crops (flax, mint, peanuts, sorghum, sunflower), small fruits, small grains, soybeans, tree fruits (citrus, deciduous, subtropical), turf, vegetables.

Rate: 1-3 tons/A broadcast

Pest:

plant-pathogenic nematode

Remarks: Single annual application usually sufficient. Soil analysis for nematode typing and counts of adult forms and larvae should be undertaken both before and after each crop season. For preplant soil application, till on broadcast or band basis, thoroughly mixing into soil 2-4 weeks before planting. For broadcast and band application, till to 6-8" depth below surface where plant roots commonly extend and where nematodes tend to proliferate. For bulk soil treatment, soil should be sieved and blended with ClandoSan. For lawns and turfgrasses, apply during growing season using conventional application and immediately irrigate. Do not use on crops which may be adversely affected by high rates of urea nitrogen.

Days To Harvest: 0.

Signal Word/Toxicity Class: Caution/III.

REI: 12 hr.

Decoy — see pheromones

DIATOMACEOUS EARTH

Fossilized shells of minuscule water-dwelling organisms known as diatoms. Microscopically fine, sharp edges desiccate insects on contact.

Diatomaceous Earth *(Nature's Control)*

Crop: Ornamentals, vegetables.

Rate: Determined on individual basis.

Pest:

crawling insects

Remarks: Apply as dust early in the morning, prior to or at daybreak, or preferably at sunset to minimize harm to beneficials. Most effective when dry. May be sprayed, dusted, or used as a barrier for crawling insects. Wear a mask when applying large amounts.

NOTICE The information on these pages is for preliminary planning — not a guide for use. Be sure to follow the manufacturer's directions, notwithstanding information contained here.

For personal protective equipment and EPA registration numbers, please turn to page 39.

Disrupt — see pheromones

DITERA

Biological nematicide that suppresses plant nematodes parasitizing food, fiber, and ornamental crops.

DiTera G, ES, WDG (*Myrothecium verrucaria*) *(Abbott Laboratories)*

Crop: Greenhouse crops, ornamental flowers and bedding plants, small fruits, tree fruits (citrus, subtropical), turf, vegetables.

Rate: 10-40 gal. ES/A
13-100 lb. G, WDG/A (broadcast)

Pest:

burrowing nematode	lesion nematode	root-knot nematode
citrus nematode	needle nematode	spiral nematode
cyst nematode	pin nematode	sting nematode
dagger nematode	ring nematode	stunt nemtode
lance nematode	reniform nematode	stunt root nematode

Remarks: For preplant or preseeding, apply as row or band treatment. Use 2:1 ratio water to product. For postplanting on established crops, apply as band treatment on soil surface. Apply with enough water for penetration. Pests and rates vary by crop; refer to label.

Signal Word/Toxicity Class: Caution.
REI: 4 hr.

ECOSAFE

Essential oil.

◆ **Ecosafe (EC) (eugenol)** *(Arbico)*

Crop: Fruits and vegetables, indoor and outdoor ornamental plants.

Rate: 1 qt. EC/A

Pest:

aphids	cutworms	mites
armyworms	grasshoppers	weevils
beetles	loopers	

Remarks: Apply when eggs, larvae, or adults appear; repeat as needed to maintain control. For best results, thoroughly wet all plant surfaces.

Signal Word/Toxicity Class: Caution.
REI: 0 hr.

ENTICE

Insect feeding stimulant.

Entice *(Custom Chemicides)*

Crop: Cotton, field corn, ornamentals, small fruits, small grains, soybeans, tree fruits, vegetables.

Rate: 1-10 lb./A

Pest: Lepidoptera.

Remarks: Apply in recommended amount of water needed to apply insecticide. Time application according to label. When applied with a stomach poison, increases effectiveness of insecticides. Also improves efficiency of *Bacillus thuringiensis* and nuclear polyhedrosis viral insecticides.

Frustrate — see pheromones

GALLTROL

Bacterial inoculant which aids in the control of crown gall.

Galltrol (*Agrobacterium radiobacter*) *(AgBioChem)*

Crop: Almonds, apples, apricots, blueberries, bushberries (blackberries, boysenberries, raspberries, youngberries), cherries, chrysanthemums, clematis, euonymus, filberts, grapes, gypsophila, kiwi, nectarines, peaches, pears, pecans, plums, poplar, prunes, roses, walnuts, willows, etc.

Rate: Preplant dip or spray: 1 agar plate (6" diameter)/gal. unchlorinated water.

Disease:

crown gall

Time: Apply during fall, winter, and spring planting season.

Remarks: Nonbearing. For use in field or greenhouse. Do not mix or combine with Captan, Vydate, Nemacur, and other pesticides or fertilizers. Generally one preventive preplant application will give control. Treat as soon as possible.

Signal Word/Toxicity Class: Caution.
REI: 12 hr.

GARLIC GARD

Insect Repellent.

◆ **Garlic Gard AG (garlic juice)** *(Soil Technologies)*

Crop: Cotton.

Pest:

aphids	cutworms	saltmarsh caterpillar
armyworms	fleahoppers	thrips
boll weevils	loopers	webworms
bollworms	Lygus bugs	whiteflies
cotton leafperforator	mites	

Crop: Field corn, popcorn, and small grains (barley, oats, rye, wheat).

Pest:

aphids	cutworms	mites
armyworms	earworms	rootworms
beetles	grasshoppers	white grubs
borers	maggots	wireworms

Crop: Peanuts.

Pest:

aphids	earworms	rootworms
armyworms	leafhoppers	thrips
cutworms	mites	

Crop: Vegetables.

Pest:

aphids	hornworms	rust flies
armyworms	leafhoppers	slugs
beetles	leafminers	sowbugs
borers	leafrollers	symphylans
caterpillars	loopers	thrips
chinch bugs	Lygus bugs	tuberworms
curculios	maggots	webworms
cutworms	mealybugs	weevils
earworms	mites	white grubs
fruitworms	plant bugs	whiteflies
grasshoppers	psyllids	wireworms

Crop: Fruits and Nuts.

Pest:

ants	fruit moths	sawflies
aphids	fruitworms	scales
appleworms	gall wasps	shuckworms
beetles	grubs	slugs
borers	Japanese beetle	spittlebugs
bud moths	leafhoppers	symphylans
cambium curculio	leafminers	thrips
carpenterworms	leafrollers	tortrix
casebearers	maggots	tree crickets
catocales	mealybugs	twig girdlers
chafers	mite scale	walnut caterpillar
codling moths	mites	webworms
curculios	Phylloxera	weevils
cutworms	plant bugs	white grubs
earwigs	psylla	whiteflies
fruit flies	rootworms	wireworms

Crop: Ornamentals.

Pest:

aphids	leafhoppers	plant bugs
bagworms	leafminers	sawflies
beetles	leafrollers	scales
borers	maggots	spittlebugs
chafers	mealybugs	webworms
curculios	mites	whiteflies
cutworms		

Crop: Forage crops (alfalfa, clovers, grasses, timothy, vetch).

Pest:

aphids	grasshoppers	spittlebugs
armyworms	Japanese beetles	thrips
caterpillars	leafhoppers	webworms
chinch bugs	Lygus bugs	weevils
cutworms	mites	

Rate: 5 gal. diluted spray/A

Remarks: Mix one part Garlic Gard to 100 parts water for diluted spray. Make no more than 14 applications per season. Refer to label.

Signal Word/Toxicity Class: Caution.
REI: 4 hr.

Gemstar — see nuclear polyhedrosis virus

◆ - **new product** • ai-active ingredient • DF-dry flowable • E/EC-emulsifiable concentrate • EW-emulsion in water
F/FL-flowable • G-granule • L/LC-liquid • SP/WSP-soluble packet • W/WP-wettable powder • WSB-water soluble bag

GOLDEN NATUR'L SPRAY OIL

Soybean oil.

Golden Natur'l Spray Oil *(Stoller)*

Crop: Ornamentals, vegetables (lima beans, peppers, radishes).
Rate: 2.5-5 Tbl. 93%/gal. water
Pest:

aphids	leafminers	spider mites
fungus gnats	mealybugs	whiteflies
lace bugs	scales	

Remarks: Greenhouse application.

- - - - - - - - - - - - - - - - - -

Crop: Cotton.
Rate: 2.5 oz. 93%/gal. water
Pest:

aphids	spider mites	whiteflies

Crop: Corn (field, sweet).
Rate: 2 gal. 93%/100 gal. water
Pest:

corn rootworm	earworms	fall armyworm

Crop: Sugar beets.
Rate: 2 gal. 93%/A
Pest:

leafminers	mites

Crop: Soybeans, vegetables (asparagus, beans, beets, cabbage, cauliflower, celery, corn, cucurbits, eggplant, lettuce, melon, peas, peppers, potatoes, squash, tomatoes).
Rate: 1.2 gal. 93%/100 gal. water
Pest:

adelgids	leafhoppers	psyllids
aphids	leafminers	soft scales
armored scale	leafrollers	thrips
beetles (larvae)	mealybugs	webworms
cankerworms	mites	whiteflies
caterpillars	plant bugs	

Remarks: For cotton, apply when insects appear; repeat weekly as needed. For sugar beets, apply as needed. For field corn, apply early spring. For soybeans and vegetables (including sweet corn), apply in summer.

- - - - - - - - - - - - - - - - - -

Crop: Deciduous (apples, apricots, cherries, nectarines, peaches, pears, plums, prunes), nuts (pecans), small fruits (blueberries).
Rate: 2-3 gal. 93%/100 gal. water
Pest:

apple red bug	European red mite	pear psylla
cottony peach scale	fruittree leafroller	pearleaf blister mite
European fruit lecanium scale	mites	San Jose scale
	obscure scale	scurfy scale

Crop: Nuts (almonds).
Rate: 1-1.5 gal. 93%/100 gal. water
Pest:

aphid eggs	scales	spider mites

Crop: Subtropical.
Rate: Figs:
2-3 gal. 93%/A
Olives:
1.5 gal. 93%/100 gal. water
Pest:
scales

Crop: Small fruits (grapes).
Rate: 1 gal. 93%/A
Pest:
mealybugs

Crop: Citrus.
Rate: 2.75-5.5 gal. 93%/A
Pest:

citrus red mite	citrus rust mite	Texas citrus mite

Remarks: For citrus, apply postbloom, summer or fall spray application.

- - - - - - - - - - - - - - - - - -

Crop: Ornamentals (flowers, foliage plants, shade trees, shrubs).
Rate: 2-3 gal. 93%/100 gal. water
Pest:

adelgids	mealybugs	scales (immature)
aphids	plant bugs (immature)	spider mites
caterpillars	psyllids (immature)	whiteflies (immature)
gall mites	sawflies (larvae)	
lace bugs		
leaf beetle (larvae)		

Remarks: Summer application.

- - - - - - - - - - - - - - - - - -

Crop: Forage (alfalfa).
Rate: 1 qt. 93%/A
Pest:
spider mites
Remarks: Seed crop. Apply as needed.

- - - - - - - - - - - - - - - - - -

Signal Word/Toxicity Class: Caution.
REI: 12 hr.

GOOD BUG POWER MEAL

Beneficial insect attractant.

Good Bug Power Meal *(Arbico)*

Crop: Cotton, field corn, fruits, nuts, ornamentals, small grains, soybeans, vegetables, noncropland.
Rate: Varies.
Pest: Any insect pest that can be controlled by a beneficial insect.
Remarks: Increases beneficial insect activity by 50%; provides food source for beneficials as development continues; helps beneficials stay in place. Use indoors or outdoors, mixing with water. May be sprayed directly on crop or mixed into paste or slurry.

HOT PEPPER WAX

Insect Repellent.

◆ **Hot Pepper Wax (C) (capsaicin)** *(Hot Pepper Wax)*

Crop: Including but not limited to: vegetables (bulb, cole, cucurbit, fruiting, leafy, legumes, root and tuber), tree fruits (citrus), indoor and outdoor ornamental plants.
Rate: 6 gal. C/A in 100 gal. water
Pest:

aphids	scales	thrips
lace bugs	spider mites	whiteflies
leafhoppers		

Remarks: Apply until foliage is wet. Repeat if necessary in 1-2 weeks to establish control. Apply 2 gal. C/A in 100 gal. water every 2-4 weeks for maintenance.

- - - - - - - - - - - - - - - - - -

Signal Word/Toxicity Class: Caution.
REI: 0 hr.

Isomate — see pheromones

KONSUME

Insect feeding stimulant.

Konsume *(Troy Biosciences)*

Crop: Cotton, field corn, fruits, small grains, soybeans, vegetables.
Rate: 1-8 pt./A
California:
1 pt./5 gal. spray solution
Pest:

beet armyworm	cutworms	sod webworm
Colorado potato beetle	grasshoppers	tobacco budworm
corn earworm	loopers	tomato fruitworm
	mole crickets	

Remarks: Apply by air or ground equipment in tank mix with registered insecticide to stimulate insects to feed on treated foliage. May also be combined with biological controls such as *Bacillus thuringiensis.* For California, with agitation system running, add insecticide after all Konsume has been suspended. Provide sufficient agitation during mixing and application to ensure sprayable suspension.

- - - - - - - - - - - - - - - - - -

Signal Word/Toxicity Class: Caution.

NOTICE The information on these pages is for preliminary planning — not a guide for use. Be sure to follow the manufacturer's directions, notwithstanding information contained here.	For personal protective equipment and EPA registration numbers, please turn to page 39.

MYCOSHIELD

Agricultural terramycin, controls fire blight on pears and bacterial spot on nectarines and peaches.

Mycoshield (oxytetracycline-agricultural terramycin) *(Novartis)*

Crop: Pears.

Rate: 1 lb. product/100 gal. water

Disease:
fire blight

Time: Apply at 10% bloom; repeat at 4- to 6-day intervals for 8-10 applications; apply to point of runoff.

Crop: Nectarines, peaches.

Rate: 11.33 oz./100 gal.

Disease:
bacterial spot

Time: Apply on weekly schedule starting with shuck split; apply to point of runoff.

Remarks: Nectarines, peaches, pears. Spray application schedules are based on shuck split of peaches which varies by region and varieties. Do not apply through any type of irrigation system.

Days To Harvest: Pears-60; peaches, nectarines-21.

- - - - - - - - - - - - - - - - -

Signal Word/Toxicity Class: Caution.

REI: 12 hr.

Mycotrol — see *Beauveria bassiana*

Naturalis-O — see *Beauveria bassiana*

Naturalis-T — see *Beauveria bassiana*

NEEM OIL

Multi-purpose organic insecticide/fungicide/miticide. Kills eggs, larval, and adult stages of insects.

Trilogy 90EC *(Thermo Trilogy)*

Crop: Aspargus, cotton, field corn, forage (alfalfa, etc.), hops, mushrooms, okra, popcorn, rice, safflower, small grains, sorghum, soybeans, sugarcane, sunflower, sweet corn, tobacco.

Pest:
greenhouse whitefly, silverleaf whitefly, sweetpotato whitefly, thrips, whiteflies

Disease:
rust

Crop: Cucurbits.

Pest:
aphids, calico scale, frosted scales, greenhouse whitefly, leafcurl aphids, lecanium scales, mites, San Jose scale, scales, silverleaf whitefly, spider mites, squash bugs, sweet potato whitefly, whiteflies, woolly aphid

Disease:
downy mildew, powdery mildew

Crop: Bulb and cole vegetables.

Pest:
aphids, greenhouse whitefly, rust mites, silverleaf whitefly, spider mites, whiteflies, woolly aphid

Disease:
Alternaria, Botrytis, downy mildew, powdery mildew, rust, white rust

Crop: Legume and fruiting vegetables.

Pest:
aphids, greenhouse whitefly, mites, psyllids, rust mites, San Jose scale, silverleaf whitefly, spider mites, whiteflies, woolly aphid

Disease:
blight, gray mold, leaf blight, leaf spot, mold, powdery mildew, rust

Crop: Root and tuber vegetables.

Pest:
aphids, calico scale, frosted scales, greenhouse whitefly, lecanium scales, mites, rust mite, scales, silverleaf whitefly, spider mites, sweetpotato whitefly

Disease:
blight, powdery mildew, rust

Crop: Herbs and spices.

Pest:
aphids, calico scale, chestnut weevil, frosted scale, fruittree leafroller, leafrollers, lecanium scale, mites, rust mite, San Jose scale, scales, spider mites, weevils, woolly aphid

Disease:
leaf spot, powdery mildew

Rate: Pest:
0.5-2 gal. 90EC/A in 100 gal. water
Disease:
0.5-1 gal. 90EC/A in 100 gal. water

Crop: Nuts, small fruits.

Pest:
aphids, calico scale, chestnut weevil, frosted scale, fruittree leafroller, leafrollers, lecanium scale, mites, rust mite, San Jose scale, scales, spider mites, weevils, woolly aphid

Disease:
downy mildew (small fruits), leaf spot (nuts), powdery mildew (small fruits)

Crop: Tree fruits (citrus, deciduous, subtropical).

Pest:
aphids, apple grain aphid, black cherry aphid, calico scale, citrus whitefly, fruit flies, frosted scale, green apple aphid, greenhouse whitefly, leafcurl aphids, leafminers, lecanium scale, mealy plum aphid, mealybugs, mites, psyllids, rosy apple aphid, rust mite, San Jose scale, scales, silverleaf whitefly, spider mites, sweetpotato whitefly, whiteflies, woolly aphid

Disease:
Alternaria, Anthracnose, downy mildew, greasy spot, fruit drop (postbloom), powdery mildew, scab

Rate: Pest:
1.25-5 gal. 90EC/A in 250 gal. water
Disease:
1.25-2.5 gal. 90EC/A in 250 gal. water

Remarks: Apply at first signs of crop damage. Repeat at 7- to 10-day intervals as needed. For walnuts, do not apply after husk split.

Days To Harvest: Lemons-60.

- - - - - - - - - - - - - - - - -

Signal Word/Toxicity Class: Caution.

REI: 4 hr.

◆ **Triact 90EC** *(Thermo Trilogy)*

Crop: Ornamental flowering plants, trees, shrubs, foliage.

Rate: Pest:
1-2 gal. 90EC/A in 100 gal. water
Disease:
1 gal. 90EC/A in 100-200 gal. water

Pest:
aphids, leafhoppers, mealybugs, mites, scales, whiteflies

Disease:
Anthracnose, black spot, Botrytis, downy mildew, flower blight, leaf spot, needle rust, powdery mildew, rust, scab, tip blight, twig blight

Remarks: Do not use on food crops. Do not apply to stressed plants, or known sensitive species. Repeat at 7- to 14-day intervals as needed.

Signal Word/Toxicity Class: Caution.
REI: 4 hr.

Nemastop — see plant extracts

Nolo Bait — see *Nosema locustae*

NoMate — see pheromones

NOSEMA LOCUSTAE

Grasshopper pathogen; single-celled protozoan.

Nolo Bait *(M & R Durango)*
Crop: Cotton, field corn, fruits, noncropland, nuts, ornamentals, pastures and rangeland, small grains, soybeans, vegetables.
Rate: 1 lb./A
Pest:
grasshoppers, locusts, Mormon crickets
Remarks: For best knockdown results, apply when grasshoppers are 2nd to 3rd instar. Older insects may be treated as an inoculative measure. Temperature should be between 60-85°F. Do not use during rain or excessively damp conditions.

Nosema locustae *(Arbico; Harmony Farm Supply)*
Crop: Cotton, field corn, fruits, noncropland, nuts, ornamentals, pastures and rangeland, small grains, soybeans, vegetables.
Rate: 1-2 lb./A
Pest:
crickets, grasshoppers, locusts
Remarks: Apply early in season as overwintering hoppers emerge. Broadcast around entire infected area as well as outlying areas such as drying grasses and fields; second application may be necessary to treat heavily infected areas 4-6 weeks later.

NUCLEAR POLYHEDROSIS VIRUS

Naturally occuring insect-specific pathogenic viruses which infect lepidopterous larvae.

Gemstar LC (*Helicoverpa zea*) *(Thermo Trilogy)*
Crop: Row crops, vegetables.
Rate: 10 oz. LC/A
Pest:
corn earworm, tobacco budworm, tomato fruitworm, cotton bollworm
Remarks: Apply when larvae are in 1st and 2nd instar. Larvae must ingest sprayed plant material for product to be effective.
Days To Harvest: 0.

Signal Word/Toxicity Class: Caution.
REI: 4 hr.

◆ **SPOD-X LC (*Spodoptera exigua*)** *(Thermo Trilogy)*
Crop: Ornamentals, row crops, vegetables.
Rate: 1.7-3.4 oz. EC/A
Pest:
beet armyworm
Remarks: Apply when larvae are in 1st and 2nd instar. Larvae must ingest sprayed plant material for product to be effective.
Days To Harvest: 0.

Signal Word/Toxicity Class: Caution.
REI: 4 hr.

PB-ROPE L

Sex pheromone which prevents mating of pink bollworm moths.

PB-Rope L *(Pacific Biocontrols)*
Crop: Cotton.
Rate: 100-150 dispensers/A
Pest:
pink bollworm
Remarks: Attach dispensers at pinsquare. Twist dispensers loosely around mainstem of cotton near bottom of plant. Apply immediately prior to moth emergence in field or adjoining area. Dispensers must be applied uniformly throughout treated acreage to obtain a reduction in mating. If a major source of mated female moths is present adjacent to field, migration of these moths may significantly reduce level of control achieved. Sources are likely to be unsprayed cotton or other host species within 300 yd. of treated field. Solutions include treatment of entire blocks rather than selected sections, or treatment of infestation sources with PB-Rope L or effective insecticide. Other pests must be monitored so timely intervention with insecticides is possible.

Signal Word/Toxicity Class: Caution.

PHEROMONES

Pheromone-based insect mating disruptant products.

CheckMate CM *(Consep)*
Crop: Apples, pears, walnuts.
Rate: 120-200 dispensers/A
Pest:
codling moth
Remarks: Apply by hand. Single application controls for up to 70 days.

CheckMate OFM or OFM-F *(Consep)*
Crop: Nectarines, peaches.
Rate: 108 dispensers OFM/A
1.32-2.93 fl. oz. OFM-F/A
Pest:
Oriental fruit moth
Remarks: Apply dispensers by hand. Single application controls for up to 90 days. For OFM-F, apply in sufficient water for uniform coverage. Do not exceed 22 fl. oz. OFM-F/A per yr. Reapply as needed.

Signal Word/Toxicity Class: OFM-F: Caution.

◆ **CheckMate OLR or OLR-F** *(Consep)*
Crop: Grapes, stone fruits, and other tree crops.
Rate: 100 dispensers OLR/A
0.37-1.46 fl. oz. OLR-F/A
Pest:
omnivorous leafroller
Remarks: Apply dispensers by hand. Single application controls for up to 100 days. For OLR-F apply in sufficient water for uniform coverage. Do not exceed 22 fl. oz. OLR-F/A per yr. Reapply as needed.

Signal Word/Toxicity Class: Caution.

CheckMate PBW or PBW-F *(Consep)*
Crop: Cotton.
Rate: 6-8 g. PBW/A
0.31-1.55 fl. oz. PBW-F/A
Pest:
pink bollworm
Remarks: For PBW, application will control for 20-30 days; compatible with most spray equipment. Spreader and sticker is recommended for maximum field life. For PBW-F, apply in sufficient water for uniform coverage. Do not exceed 23 fl. oz. PBW-F/A per yr.

Signal Word/Toxicity Class: Caution.

CheckMate PTB-XL *(Consep)*
Crop: Almonds, apricots, nectarines, peaches.
Rate: 150 dispensers/A
Pest:
peach twig borer
Remarks: Apply by hand for 90 day control.

NOTICE	The information on these pages is for preliminary planning — not a guide for use. Be sure to follow the manufacturer's directions, notwithstanding information contained here.	For personal protective equipment and EPA registration numbers, please turn to page 39.

CheckMate TPW or TPW-F *(Consep)*
Crop: Tomatoes.
Rate: 200 dispensers TPW/A
1.2-2.4 fl. oz. TPW-F/A
Pest:
tomato pinworm
Remarks: Apply dispensers by hand. Single application controls for up to 75 days. For TPW-F, apply in sufficient water for uniform coverage. Do not exceed 22 fl. oz. TPW-F/A per yr. Reapply as needed.

Signal Word/Toxicity Class: Caution.

Decoy P.B.W. Sprayable *(AgriSense)*
Frustrate PBW Sprayable *(Thermo Trilogy)*
Crop: Cotton.
Rate: 2.6-14.3 g P.B.W./A
17-70 g PBW/A
Pest:
pink bollworm
Remarks: Apply by air or ground equipment upon emergence of adults and before larvae infestation occurs. Rate depends on infestation level and time of application. Can be incorporated into IPM program, applied alone or tank mixed with variety of insecticides.

◆ **Disrupt CM, CM-Xtra** *(Hercon Environmental)*
Crop: Fruit and nuts crops including: apples, nectarines, peaches, pears, pecans, plums, prunes, quinces.
Rate: 200 dispensers/A
Pest:
codling moth
Remarks: Attach dispensers to lateral branches in top 2-3 ft. of tree early in season prior to moth emergence. Distribute dispensers evenly throughout area at given rate. If adjacent areas are heavily infested, additional control measures may be needed. Disrupt CM prevents or suppresses mating for 75-90 days; CM-Xtra releases pheromones for 120-140 days. Reapply as needed based on local conditions, insect monitoring data, and field scouting observations.

Signal Word/Toxicity Class: Caution.

◆ **Disrupt OFM** *(Hercon Environmental)*
Crop: Fruit and nuts crops including: almonds, apples, macadamias, nectarines, peaches, plums, quinces, walnuts.
Rate: 108 dispensers/A
Pest:
Oriental fruit moth
Remarks: Attach dispensers to lateral branches in top 1/3 of tree early in season prior to moth emergence. Distribute dispensers evenly throughout area at given rate. If adjacent areas are heavily infested, additional control measures may be needed. Prevents or suppresses mating for 90-105 days. Reapply as needed based on local conditions, insect monitoring data, and field scouting observations.

Signal Word/Toxicity Class: Caution.

◆ **Disrupt PTB** *(Hercon Environmental)*
Crop: Fruit and nuts crops including: almonds, apples, cherries, nectarines, peaches, plums, prunes.
Rate: 200 dispensers/A
Pest:
peach twig borer
Remarks: Attach dispensers to lateral branches in top 1/3 of tree early in season prior to moth emergence. Distribute dispensers evenly throughout area at given rate. If adjacent areas are heavily infested, additional control measures may be needed. Prevents or suppresses mating for 90-105 days. Reapply as needed based on local conditions, insect monitoring data, and field scouting observations.

Signal Word/Toxicity Class: Caution.

◆ **Disrupt II GM** *(Hercon Environmental)*
Crop: Hardwood and evergreen trees in forests, parks, and parkways, municipal and residential shade tree areas, recreational areas, tree nurseries, rights of way.
Rate: 168-224 g/A
Pest:
gypsy moth
Remarks: Apply single application by air or ground equipment at least 2 weeks prior to moth emergence. Inert sticker should be used with product to hold flakes to leaves.

Signal Word/Toxicity Class: Caution.

◆ **Disrupt PBW** *(Hercon Environmental)*
Crop: Cotton.
Rate: 100 dispensers/A
Pest:
pink bollworm
Remarks: Registration pending. Make first application no later than 2 weeks after transplant or when monitoring traps capture 2-4 moths per day. Attach to plant vine or stalk of suitable size near top of plant. Distribute dispensers evenly at indicated rate. Reapply as needed based on local conditions, insect monitoring data, and field scouting observations.

Signal Word/Toxicity Class: Caution.

◆ **Disrupt TPW** *(Hercon Environmental)*
Crop: Tomatoes.
Rate: 400 dispensers/A
Pest:
tomato pinworm
Remarks: Registration pending. Make first application no later than 2 weeks after transplant or when monitoring traps capture 2-4 moths per day. Attach to plant vine or stalk of suitable size near top of plant. Distribute dispensers evenly at indicated rate. Reapply as needed based on local conditions, insect monitoring data, and field scouting observations.

Signal Word/Toxicity Class: Caution.

Frustrate PBW Bands *(Thermo Trilogy)*
Crop: Cotton.
Rate: 100 bands/A
Pest:
pink bollworm
Remarks: Hand-applied product that controls pink bollworm for 120 days. Apply evenly throughout cotton field by placing them over terminal portion of plant before flowering and emergence of adults. The bands release pheromone throughout season, disrupting adult moth mating.

Signal Word/Toxicity Class: Caution.

Isomate-C Plus *(Pacific Biocontrol)*
Crop: Deciduous (apples, nectarines, peaches, pears, plums, prunes, quinces), nuts (pecans, walnuts).
Rate: 400 dispensers/A
Pest:
codling moth hickory shuckworm
Remarks: Apply dispensers securely to lateral branches 2-3 ft. from top of tree prior to moth emergence in spring. Dispensers release pheromone for 120-140 days. Reapply dispensers to crops with long field seasons or in orchards with high populations. If a major source of mated female moths is present adjacent to field, migration of these moths may significantly reduce level of control achieved. These sources are unsprayed apple, pear, crabapple, plum, prune, quince, or walnut trees or other host species within 300 yd. of treated field. Treat entire orchards rather than selected sections or treat infestation sources with either Isomate-C Plus or effective insecticide. Other pests must be monitored and controlled by timely intervention with conventional insecticides.

Signal Word/Toxicity Class: Warning/II.

Isomate-DBM (EUP; registration pending) *(Pacific Biocontrol)*
Crop: Vegetables (broccoli, Brussels sprouts, cabbage, carrots, cauliflower, lettuce, radish).
Rate: 400-425 dispensers/A
Pest:
diamondback moth
Remarks: Attach after transplant establishment of each field. Reapply dispensers to crops with long season (more than 90 days). Use with releases of biocontrol agents such as predators and parasites and in conjunction with Bt products.

Signal Word/Toxicity Class: Caution.

Isomate-GBM *(Pacific Biocontrol)*
Crop: Grapes.
Rate: 250-400 dispensers/A
Pest:
grape berry moth
Remarks: Attach to upper training wire of vines before moths emerge in spring. One application sufficient for season-long control. If a major source of mated female moths is present adjacent to field, migration of these moths may significantly reduce level of control. Sources are likely to be unsprayed vineyards or wild grapevine species within 300 yd. of treated field.

◆ - **new product** • ai-active ingredient • DF-dry flowable • E/EC-emulsifiable concentrate • EW-emulsion in water
F/FL-flowable • G-granule • L/LC-liquid • SP/WSP-soluble packet • W/WP-wettable powder • WSB-water soluble bag

Treat entire blocks of vineyards (not just large sections) or treat infestation sources with GBM-Rope or effective insecticide. Other pests must be monitored so that timely intervention with conventional insecticides is possible.

Signal Word/Toxicity Class: Caution.

Isomate-M *(Pacific Biocontrol)*
Isomate-M-100 *(Pacific Biocontrol)*
Crop: Deciduous (apples, apricots, nectarines, peaches, plums, quinces), nuts (almonds, macadamias).
Rate: 300-400 dispensers M/A
100-150 dispensers M-100/A
Pest:
koa seedworm macademia nut borer Oriental fruit moth
Remarks: Attach to top 1/3 of tree before moths emerge in spring. Protects orchards for 3 months, after which second application may be made if necessary. Where Isomate-M is used in orchards adjacent to untreated trees or sprayed orchards that can serve as sources of mated female moths, fruit damage from migrating females is possible. Treat whole orchard units as all conventionally sprayed orchards serve as sources of some mated females. Tests in single rows or small corners of conventionally sprayed orchard must be avoided. Other pests need to be carefully monitored so timely control measures can be taken as necessary.

Signal Word/Toxicity Class: Caution.

Isomate-P *(Pacific Biocontrol)*
Crop: Deciduous (apricots, cherries, nectarines, peaches, plums, prunes), nuts (almonds).
Rate: 100-250 dispensers/A; 1 dispenser/tree
Pest:
peachtree borer
Remarks: Attach onto branches above mid-point of tree before emergence of borers in spring. Apply once per season. If a major source of mated female borers is present adjacent to field, migration of these moths may significantly reduce level of control achieved. Sources are likely to be untreated host trees within 300 yd. of treated field. Treat entire orchards rather than selected sections, or treat infestation sources with either Isomate-P or an effective insecticide. Other pests must be monitored and controlled by timely intervention with conventional insecticides.

Signal Word/Toxicity Class: Caution.

◆ **NoMate LRX MEC** *(Ecogen)*
Crop: Small fruit, tree fruit, and vine crops.
Rate: 2-2.5 oz. MEC/A
Pest:
leafroller
Remarks: Apply when populations are relatively low, repeat no more than 4 weeks after last application. Can also be used in conjunction with Cutlass for control of other lepidopterous pests.
Days To Harvest: 0.

Signal Word: Caution.

◆ **NoMate OLR Spiral (tetradecen acetate)** *(Ecogen)*
Crop: Grapes, kiwi, tree fruit.
Rate: 130 spirals/A
Pest:
omnivorous leafroller
Remarks: Wrap spirals around branches at least 1-2 ft. within canopy and in upper portion. Apply 2-3 days after first male moth is captured in a "wing-style" trap.
Days To Harvest: 0.

Signal Word: Caution.

NoMate PBW MEC *(Ecogen)*
Crop: Cotton.
Rate: 0.33-1.66 fl. oz. MEC/A
Pest:
pink bollworm
Remarks: Apply at pinsquare and continue throughout growing season. Can also be used in conjunction with *Bacillus thuringiensis* or beneficials in control programs or in combination with insecticides as bioirritant. Mixing with suitable adhesive such as Bio-Tac is recommended to hold individual fibers on foliage. Can also be used in conjunction with Cutlass or Condor for control of other lepidopterous pests. Do not apply through any type of irrigation system.

Signal Word/Toxicity Class: Caution.

NoMate TPW Fiber, MEC, Spiral *(Ecogen)*
Crop: Tomatoes.
Rate: 1.34-2.68 oz. MEC/A
Pest:
tomato pinworm
Remarks: Apply when populations are relatively low, economic threshold is 2-5 moths/day per trap catch; repeat no more than 4 weeks after last application. Can also be used in conjunction with Cutlass for control of other lepidopterous pests. Do not apply through any type of irrigation system.

Signal Word: Caution.

Stirrup M *(Troy Biosciences)*
Crop: Cotton, field corn, fruits, nuts, ornamentals, small grains, soybeans, vegetables.
Rate: 2-6 oz./A
Pest: Suppression:
tetranychid mites
Remarks: Use higher rate as pest population increases, when applying over 10 gal./A, or for ornamentals and nurseries. Make at least 2 applications at intervals recommended by label. Reapply as needed to maintain control and/or after rainstorms.

Signal Word/Toxicity Class: Caution.

Stirrup PBW *(Troy Biosciences)*
Crop: Cotton.
Rate: 3-6 oz./A
Pest: Suppression:
pink bollworm
Remarks: Use higher rates as pest population increases or when applying spray volumes over 3 gal./A. Repeat as necessary to maintain suppression and after rainstorms.

Signal Word/Toxicity Class: Caution.

PLANT EXTRACTS

Biologically derived agent for control of nematodes and associated pathogens.

◆ **Nemastop (extracts of *Quercus falcata, Opuntia lindheimeri, Rhus aromatica,* and *Rhizophoria mangle*)** *(Soil Technologies)*
◆ **Sincocin (extracts of *Quercus falcata, Opuntia lindheimeri, Rhus aromatica,* and *Rhizophoria mangle*)** *(Agricultural Sciences [AGSCI])*
Crop: Food crops, orchards.
Rate: 26 fl. oz./A
Pest:
nematodes
Remarks: Apply in minimum 90 gal. water/A. Make first application during initial leaf flush; repeat at 60-day intervals. No more than 6 applications/yr. For orchards, if a systemic fungicide or nematicide is used, wait 60 days before applying Sincocin.

Crop: Ornamentals, turf.
Rate: 87 fl. oz./A
Pest:
nematodes
Remarks: Apply in minimum 90 gal. water/A. Make first application during initial leaf flush; repeat at 30- to 60-day intervals. Make no more than 10 applications/yr. If a systemic fungicide or nematicide is used, wait 14-28 days before applying Sincocin.

Signal Word/Toxicity Class: Caution.
REI: 12 hr.

PRED-FEED

Sprayable powder containing food supplements to attract and increase number of predators including green lacewings, ladybugs, and Syrphid flies. Will encourage predators to remain in crops and reproduce as much as tenfold.

PRED-FEED I.P.M. *(Custom Chemicides)*
Crop: Cotton, field corn, fruits, nuts, ornamentals, small grains, soybeans, vegetables.
Rate: 4-10 lb. in 5-10 gal. water/A
Pest:
aphids bollworms mites

NOTICE The information on these pages is for preliminary planning — not a guide for use. Be sure to follow the manufacturer's directions, notwithstanding information contained here.

For personal protective equipment and EPA registration numbers, please turn to page 39.

Remarks: Width of swath treated area should be 40-50 ft. and only 1/2 of gross acreage should be treated by skipping every other swath width. Make 1-3 applications prior to influx of pests. Higher rates are used to support higher numbers of predators. Do not use with or within 1 week of toxic pesticide applications. Spray equipment that has been contaminated with pesticides should be thoroughly cleaned and neutralized. In situations where pesticide use is needed, consult entomologists or other qualified personnel.

PROSPER NEMA

Pathogenic fungi.

◆ ***Prosper nema*** *(Arbico)*
Crop: Fruits and vegetables, indoor and outdoor ornamental plants.
Rate: 3.2 oz./A
Pest:
nematodes
Remarks: Apply as needed to maintain control.

Sincocin — see plant extracts

SOILGARD

Naturally occurring soil fungus which is antagonistic to plant pathogenic fungi.

SoilGard 12G (*Gliocladium virens* GL-21) *(Thermo Trilogy)*
Crop: Ornamentals, vegetables.
Rate: 1-1.5 lb. 12G/yd³.
Disease:
root rot
Remarks: Thoroughly incorporate into soil or soilless growing media prior to planting.
Days To Harvest: 0.

- - - - - - - - - - - - - - - - - -

Signal Word/Toxicity Class: Caution.
REI: 4 hr.

SPOD-X — see nuclear polyhedrosis virus

Stirrup — see pheromones

Triact — see neem oil

Trilogy — see neem oil

Beneficial Organisms

APHIDIUS SPP.

Parasitic wasp.

Aphidius colmani *(Florikan)*
Crop: Greenhouse vegetables, ornamentals.
Rate: Preventive:
0.5/sq. yd.
Established populations:
2/sq. yd.
Pest:
aphids
Environment: Humidity should be 70-85%; temperature 65°-77°F.
Remarks: The wasp pierces back of aphid, laying egg inside. Aphid will live for some time as larvae is feeding. Larvae glues aphid to leaf spinning a cocoon from which it pupates. Parasitized aphids are swollen, papery, and silvery brown, often called mummies. Open bottle in aphid-infested area, allow adult aphids to fly out. Sprinkle mummies onto aphid-infested plants. Release for 3 consecutive weeks. Check release area for ants which can protect aphid colonies from natural predators.

◆ ***Aphidius matricariae*** *(Arbico; Hydro-Gardens)*
Crop: Greenhouse vegetables, ornamentals.
Rate: 500-3000/A
Pest:
aphids
Remarks: Parasitizes aphids in same manner as *A. colmani.* Open bottle in aphid-infested area, allow adult aphids to fly out. This parasite can be used in greenhouses throughout the year as it is not affected by short-day lengths.

APHIDOLETES APHIDIMYZA

Predator midge.

Aphidoletes aphidimyza *(International Technology Services; IPM Laboratories; Nature's Control)*
Crop: Greenhouse.
Rate: Determined on individual basis.
Pest:
aphids
Environment: Humidity should be 50-90%; temperature 60°-80°F (International Technology Services) or 68°-81°F (IPM Laboratories).
Remarks: For preventive control, release in low numbers. When aphids are first observed, higher number of predators must be introduced. Apply every 2 weeks. Lays its eggs near aphid colonies; larvae emerging from eggs feed on aphids. To reach full development, larvae need about 10 aphids, however, where aphid densities are high they will destroy many more. Killed aphids remain attached to plant leaves and will dry up eventually. Adult predators are very sensitive to chemical pesticides, although larvae have greater resistance. It is important that other pests be controlled by biological methods.

Aphidoletes aphidimyza *(Arbico; Hydro-Gardens)*
Crop: Greenhouse.
Rate: Arbico:
1-3 midges/10 sq. ft.
Hydro-Gardens:
1 midge/10 sq. ft.
Pest:
aphids
Remarks: Apply at first sign of aphids. Active at night and during dusk.

APHYTIS MELINUS

Scale parasite.

Aphytis melinus *(Arbico; Sespe Creek Insectary)*
Crop: Citrus.
Rate: 10,000-100,000/A
Pest:
citrus red scale, citrus yellow scale, ivy scale, oleander scale, San Jose scale
Environment: Humidity should be 20-80%; temperature 65°-95°F.
Remarks: Primary target is red scale. Target 2-6 releases with maximum mating using pheromone traps. Apply by hand, air blowers, or aerially. Release where infestation is heaviest at first to guarantee parasites will establish greatest population to control pests. Releases usually begin within 2 weeks of full bloom, continuing through Sept.-Oct., utilizing pheromone traps to visually detect if wasps are attacking female scales awaiting fertilization.

Aphytis melinus *(Caltec)*
Crop: Citrus.
Rate: 10,000/A
Pest:
California red scale, Dictyosperium scale, oleander scale, yellow scale
Remarks: Release 4 times per season.

Beneficial Nematodes — see *Steinernema* spp.

◆ - **new product** • ai-active ingredient • DF-dry flowable • E/EC-emulsifiable concentrate • EW-emulsion in water
F/FL-flowable • G-granule • L/LC-liquid • SP/WSP-soluble packet • W/WP-wettable powder • WSB-water soluble bag

BENE-NEM

Beneficial nematodes.

◆ **Bene-Nem (*Steinernema* spp. & *Heterhabditis* spp.)** (Florikan)
Crop: Greenhouse.
Rate: 26 million/A
Pest:
banana moth, fungus gnats, shore fly, cutworms
Remarks: Apply in evening, directly to soil or growing media. Do not apply during daytime, as ultraviolet rays kill these parasites. Watering in will help them migrate to protected area within root zone.

BioVector — see *Steinernema* spp.

Caterpillar parasites — see *Trichogramma* spp.

COLEOMEGILLA IMACULATA

Pink ladybird beetle.

◆ ***Coleo imaculata*** (Arbico)
Crop: Vegetable and flower crops.
Rate: 1/sq. ft.
Pest:
aphids, mites, thrips, caterpillars, scales, whiteflies
Remarks: Shipped as larvae and eggs.

COTESIA SPP.

Larval parasite.

Cotesia marginiventris (Arbico; Caltec)
Crop: Cotton, soybeans, vegetables.
Rate: 150 pair/A
Pest:
cabbage looper, fall armyworm, *Heliothis* spp.
Remarks: Release weekly.

Cotesia plutella (Arbico; Caltec)
Crop: Vegetables.
Rate: 150 pair/A
Pest:
diamondback moth
Remarks: Release weekly.

Cruiser — see *Heterohabditis* spp.

CRYPTOLAEMUS MONTROUZIERI

Predator.

Cryptolaemus montrouzieri (Arbico; Caltec; Florikan; International Technology Services; IPM Laboratories; Natural Pest Controls; Nature's Control)
Crop: Citrus, ornamentals (greenhouse), soybeans, vegetables, vine crops.
Rate: Arbico: 500-5000/A
Caltec: 500-2000/A
International Technology Services: Determined on individual basis
Florikan; IPM; Natural Pest Controls; Nature's Control: 2-5/infested plant
Pest: Soft bodied insects including:
aphids, mealybugs, scales
Environment: Humidity should be 70-80%; temperature 70°-80°F.
Remarks: Adults and young larvae prefer host eggs and nymphs; larger larvae will consume mealybugs of any size. Larvae are cannibalistic and should not be confined together without food. Repeat as necessary. Do not wear white while distributing.

DACNUSA SIBIRICA/ DIGLYPHUS ISAEA

Parasites.

Dacnusa sibirica* and *Diglyphus isaea (Arbico; International Technology Services; Natural Pest Controls)
Crop: Ornamentals, vegetables (beans, eggplant, peppers, tomatoes).
Rate: Arbico: 500-1000/A
International Technology Services; Natural Pest Controls: Determined on individual basis.
Pest:
leafminers
Environment: Humidity should be 50-90%; temperature 60°-85°F. *Dacnusa sibirica* suited to cooler climates; *Diglyphus isaea* to warmer climates.
Remarks: Outdoors and greenhouses. Introduce when leafminers are first observed. *Dacnusa sibirica* deposits eggs inside leafminer larvae, killing larvae. *Diglyphus isaea* kills leafminers in tunnel and deposits eggs next to them, using dead larvae as food. Also kills small larvae, which are used solely as food. To determine parasite activity, examine leaf samples for type of leafminer, any parasites which are present and level of infestation. If sufficient parasites are noted, no action is advised. If parasites are of insufficient quantity, one of the two parasites should be introduced.

DELPHASTUS PUSILLUS

Predatory beetle.

Delphastus pusillus (Arbico; IPM Laboratories; Nature's Control)
Crop: Ornamentals, vegetables.
Rate: 2000/3000 sq. ft.
Nature's Control:
1/15-50 sq. ft.
Pest:
greenhouse whitefly, sweetpotato whitefly
Environment: Temperature 60-85°F.
Remarks: For use in greenhouses. Will eat at least 150 whitefly eggs a day. Will feed on spider mites if no whiteflies are available. Should be used with *Encarsia formosa* in greenhouses, along with traps.

Devour — see *Steinernema* spp.

DIADIGMA INSULARE

Diamondback moth parasite; pre-fed fertile adults.

Diadigma insulare (Arbico; Caltec)
Crop: Vegetables.
Rate: 150 pair/A
Pest:
diamondback moth
Remarks: Release weekly.

DIARETIELLA RAPAE

Aphid parasite.

Diaretiella rapae (Arbico)
Crop: Cotton, field corn, fruits, nuts, ornamentals, small grains, soybeans, vegetables.
Rate: Determined on individual basis.
Pest:
aphids
Remarks: Kills aphids by mummifying them. Release rates vary by type and stage of infestation.

Ecomask — see *Steinernema* spp.

En-Strip — see whitefly parasites

NOTICE The information on these pages is for preliminary planning — not a guide for use. Be sure to follow the manufacturer's directions, notwithstanding information contained here.

For personal protective equipment and EPA registration numbers, please turn to page 39.

FIRE MITE

Ant control; natural enemy of fire ants and other ants.

Fire mite for ant control *(Caltec)*
Crop: Vine crops.
Rate: 2-3 oz./mound
Pest:
ants fire ant
Remarks: Mixed with wheat bran. Must be introduced to mound so it can reach queens. Probe mound, introduce and cover. Shipped with sitratroge eggs for food.

FLY PARASITES/PREDATORS

Destroy pest flies in their immature larval and pupal stages.

Fly Parasites *(Beneficial Insectary; Hydro-Gardens)*
Fly Predators (*Muscidifurax Raptorellus, Spanglia Endius, Spanglia Cameroni*) *(Spalding Laboratories)*
Crop: Mushrooms.
Rate: 5000 parasites or predators/1000 sq. ft. growing area
Pest:
flies
Remarks: Release every 4 weeks of fly season. Recommended starting point is to receive first shipment in early spring when flies first appear and continue monthly shipments throughout fly season or until first frost in the fall.

Fungus Gnat Predators— see *Hypoaspis miles*

GALENDROMUS SPP.

Predator mite.

Galendromus occidentalis *(Arbico; Harmony Farm Supply)*
Crop: Cotton, field corn, fruits, nuts, ornamentals, small grains, soybeans, vegetables.
Rate: Determined on individual basis.
Pest:
spider mites
Environment: Can tolerate wide range of weather conditions (humidity 50% minimum; temperature 70°-95°F), well-adapted to outdoor use.
Remarks: Introduce at first sign of spider mites. Tolerant to carbaryl and Guthion. Will actively search out and destroy spider mites throughout growing season.

◆ ***Galendromus helvelous*** *(Arbico)*
Crop: Avocados, citrus,
Rate: 5000 predators/A
Pest:
avocado mite sixspotted spider mite
Remarks: These predators are specific for the avocado mites in Southern California and the sixspotted spider mite in Florida. It is best to release predators immediately, although they may be stored for 2-3 days at 45°-50°F.

GONIOZUS LEGNERI

Navel orangeworm parasite.

Goniozus legneri *(Caltec; M & R Durango)*
Crop: Nuts (almonds, walnuts).
Rate: Caltec: 500/A
M & R Durango: 1000-2000/A/release
Pest:
navel orangeworm (larvae)
Remarks: Make up to 2-3 releases/yr.

GREEN LACEWING

Predator.

Chrysopa carnea *(Beneficial Insectary; Caltec; Natural Pest Controls)*
Chrysoperla rufilabris *(Arbico; Beneficial Insectary; IPM Laboratories)*
Crop: Cotton, field corn, fruits, nuts, ornamentals, small grains, soybeans, vegetables.
Rate: 1 lacewing/5-30 aphids
Pest: Small, soft-bodied insects and moth eggs including:
aphids scales thrips
caterpillars (small) spider mites whiteflies
mealybugs
Remarks: Lacewing eggs produce tiny larvae that attack and consume aphids for 2-3 weeks. Called aphid lions, these larvae also feed on other insects that cross their path. Each aphid lion will consume 200-300 aphids during its lifetime. Just prior to pupation, larvae will spin a silk cocoon. Depending on remaining aphid population, may be useful to make another release of lacewings to keep active predation going. Adult lacewing emerges after about 5 days to feed on nectar and pollen and produce new aphid lions. For best results, release lacewing eggs upon arrival or hatching, whichever occurs later. Larger pre-fed larvae are also available. These will consume aphids at a faster rate than tiny hatchlings during first week after release.

Chrysopa carnea *(Hydro-Gardens)*
Crop: Cotton, field corn, fruits, nuts, ornamentals, soybeans, vegetables.
Rate: 1 lacewing/10 aphids
Pest:
aphids soft-bodied insects
Remarks: Lacewing eggs should be sprinkled around aphid-infested plants. Make 2-3 releases, 2 weeks apart.

***Chrysoperla* spp.** *(Florikan)*
Crop: Orchard trees, ornamentals, row crops.
Rate: 1 or more/plant depending on level of infestation
Pest:
aphids mealybugs spider mites
caterpillars (eggs and small larvae) moth (eggs) thrips
leafhoppers scales whiteflies
Environment: Egg development and larval hatching: humidity should be 75%, temperature 75°-79°F. Larvae: humidity should be 60%, temperature should be 68°-88°F.
Remarks: Apply any time during plant growth cycle or at first appearance of pests. Release during cooler time of day when there is little or no wind. Check release area for ants which can protect aphid colonies natural predators. Do not refrigerate eggs. May be use in combination with *Bacillus thuringiensis*.

***Chrysoperla* spp.** *(M & R Durango)*
Crop: Cotton, field corn, forage (hay), fruits, nuts, ornamentals, small grains, soybeans, vegetables.
Rate: Determined on individual basis.
Pest:
aphids moth (eggs) thrips
caterpillars (eggs) spider mites whiteflies
mealybugs
Remarks: Predator insect in larval stage. Pierces insect with mouth parts and sucks body fluid from host.

Chrysoperla carnea *(Arbico)*
Crop: Cotton, field corn, fruits, nuts, ornamentals, small grains, soybeans, vegetables.
Rate: 1000 eggs/200 sq. ft.
10,000-50,000/A (weekly-monthly distribution)
Pest: Eggs and soft bodies of:
aphids mealybugs thrips
caterpillars (eggs) scales whiteflies
leafhoppers spider mites
Remarks: Larvae will feed for 1-3 weeks before becoming adults (eating only nectar and honeydew). If infestation is not controlled after 5-7 days, additional releases may be necessary. Shipped in bran and freshly hatched, eggs are packed with moth eggs for food. Pre-fed adult lacewings also available.

Chrysoperla rufilabris *(Caltec)*
Crop: Small grains.
Rate: 1000 eggs/200 sq. ft.
5000-50,000/A per year (weekly-monthly distribution)
Pest:
Russian wheat aphid
Remarks: Larvae will feed for 1-3 weeks before becoming adults (eating only nectar and honeydew). If infestation is not controlled after 5-7 days, additional releases may be necessary. Shipped in bran while freshly hatched, eggs are packed with moth eggs for food. Pre-fed adult lacewings also available.

◆ - **new product** • ai-active ingredient • DF-dry flowable • E/EC-emulsifiable concentrate • EW-emulsion in water
F/FL-flowable • G-granule • L/LC-liquid • SP/WSP-soluble packet • W/WP-wettable powder • WSB-water soluble bag

Chrysopa carnea *(A-1 Unique Insect Control)*
Chrysopa rufilabris *(A-1 Unique Insect Control; Nature's Control)*

Crop: Cotton, field corn, fruits, nuts, ornamentals, soybeans, vegetables.

Rate: A-1 Unique Insect Control: 5000 eggs/A
Nature's Control: 4000-50,000 eggs/A

Pest: Many insects and eggs including:

aphids	red spider mites

Remarks: Lacewings come as eggs ready to hatch (also available as hatched larvae), and will immediately begin eating pests. They grow to almost 1/2" long over 3-4 weeks, consuming 400 or more aphids or other pests in the process. Adults feed only on nectar and pollen. Repeat every 2 weeks throughout growing season.

Guardian — see *Steinernema* spp.

HETEROHABDITIS SPP.

Beneficial nematodes.

Cruiser (*H. bacteriophora*) *(Ecogen)*

Crop: Cranberries

Rate: 1-1.5 billion nematodes/A

Pest:

black vine weevil	cranberry root grub	Oriental beetles
cranberry girdler	*Hoplia* spp.	white grubs

Remarks: Apply when grubs are active, before damage occurs, usually in August. Bogs should be not flooded for at least 2 weeks after nematodes are applied. Only one application required per season.

– – – – – – – – – – – – – – – – – –

Crop: Turf.

Rate: 1-1.5 billion nematodes/A

Pest:

annual bluegrass weevil	cutworms	rose chafer (grubs)
armyworms	fleas (larvae, pupae)	sod webworms
billbugs	Japanese beetle (grubs)	tropical sod webworm
chinch bugs	masked chafer (grubs)	

Remarks: Rate varies by pest. For grubs, apply following egg-laying of adult beetles, usually mid- to late-Aug. or at latest Sept. If grub pressure is very heavy, retreat after 4 weeks. For fleas, larvae, and pupae, apply to all areas frequented by dogs and cats. For billbugs, annual bluegrass weevils, and chinch bugs, apply as soon as damage is detected. For webworms and cutworms, apply when early instar caterpillars surface feeding is first noticed. Reapply after 14-21 days if needed.

– – – – – – – – – – – – – – – – – –

Crop: Ornamentals.

Rate: 1 billion nematodes/A

Pest:

black vine weevil

Heterohabditis bacteriophora *(Arbico; Harmony Farm Supply; M & R Durango)*

Crop: Field crops, greenhouse, orchards, turf.

Rate: Determined by size of area to be treated, soil type, climatic factors, type of pest and life cycle, infestation levels, pest tolerance threshold, stage of plant development, and available research information.

Pest:

beetles (larvae)	grubs	rootworms
billbug	Japanese beetles	wireworms
crown borers	mint flea beetle	

Environment: Needs moist environment to survive and move through soil.

Remarks: Apply in the evening. Soil should be damp prior to application. Keeping soil moist for several day after application recommended.

Heteromask (*H. bacteriophora*) *(BioLogic)*

Crop: Ornamentals.

Rate: 35,000 active units/sq. ft.

Pest:

black vine weevil

Lawn Patrol (*H. bacteriophora*) *(Hydro-Gardens)*

Crop: Greenhouse.

Rate: 1 million/3000 sq. ft.

Pest: Over 250 insect pests with resting periods in the soil including:

corn rootworm	fungus gnats	thrips
cutworms	sod webworms	white grubs
European corn borer	strawberry root weevil	

Remarks: Apply in evening, directly to soil or growing media. Do not apply during daytime, as ultraviolet rays kill these parasites. Watering-in will help them migrate to protected area within root zone. Does not control other nematodes.

Heteromask — see *Heterohabditis* spp.

HYPOASPIS MILES

Controls soil organisms.

◆ **Fungus Gnat Predators** *(Nature's Control)*

Crop: Cotton, field corn, fruits, nuts, ornamentals, small grains, soybeans, vegetables.

Rate: 5000/200 sq. ft.

Pest:

fungus gnats	mites	springtails

Remarks: Live and breed in the top 0.5" of soil.

◆ ***Hypoaspis miles*** *(Florikan; Harmony Farm Supply)*

Crop: Greenhouse crops.

Rate: Preventive:
25/pot
Curative:
55/pot

Pest:

sciarid flies

Rate: 100-300/sq. meter

Pest:

thrips

Remarks: Will also attack other moving organisms in the soil.

LADYBUGS

Predators.

Ladybugs (*Hippodamia convergens*) *(A-1 Unique Insect Control; Arbico; Caltec; Harmony Farm Supply; Hydro-Gardens; IPM Laboratories; Natural Pest Controls; Nature's Control)*

Crop: Cotton, field corn, fruits, nuts, ornamentals, small grains, soybeans, vegetables.

Rate: Determined on individual basis.

Pest: Soft-bodied insects and eggs:

aphids	mites	whiteflies

Remarks: During both larval and adult stages, ladybugs feed on aphids and other pests. Adult females lay clusters of orange, bullet-shaped eggs which hatch into tiny alligator-shaped larvae. Larvae feed on aphids and grow to adult size over 3-4 weeks before entering into pupal stage for about 1 week. To keep ladybug adults near release area, release at dusk, near aphids that will provide immediate food source. Spraying plants with water prior to release will provide them with needed moisture. Once ladybugs have been released, spraying area with pesticides may kill their young.

Lawn Patrol — see *Heterohabditis* spp.

LEPTOMASTIX DACTYLOPII

Parasitic wasp.

Leptomastix dactylopii *(Arbico; IPM Labortories)*

Crop: Citrus.

Rate: 5/plant.

Pest:

citrus mealybug

Environment: Sunny, warm (75°-80°F), humid environment.

Remarks: Tiny yellowish-brown wasp with long antennae. The female lays her eggs in third-stage mealybug nymphs and adults. Monitor for new mealybugs weekly on new growth on susceptible plants. Good control should be achieved in 2-3 months. Repeat releases as necessary, usually once or twice a year.

LINDORUS LOPHANTHAE

Black ladybeetle.

◆ ***Lindorus lophanthae*** *(Arbico)*

Crop: Greenhouse.

Rate: 3-6/sq. yd.
1000-2000/A

Pest:

black scale	brown scale	red scale

Environment: 59°-77°F, 20-90% relative humidity.

Remarks: Primary prey are soft scales, such as black, brown, and red. Excellent search and detection capabilities.

NOTICE The information on these pages is for preliminary planning — not a guide for use. Be sure to follow the manufacturer's directions, notwithstanding information contained here. For personal protective equipment and EPA registration numbers, please turn to page 39.

MESOSEIULUS LONGIPES

Predator. Also known as Phytoseiulus longipes.

Mesoseiulus longipes *(Arbico; Harmony Farm Supply; Nature's Control)*
Crop: Greenhouse strawberries, vegetables (vine crops).
Rate: Determined on individual basis.
Pest:
spider mites
Environment: Can tolerate hot, dry conditions (minimum humidity 40%; optimum temperature 70°-90°F). Requires increasing humidity as temperature rises; tolerates extremes of temperature and humidity better than other mite predators.
Remarks: Introduce at first sign of spider mites. Will control spider mites throughout growing season, actively searching mites out and destroying up to 20 immature or 5 mature spider mites per day.

METAPHYCUS HELVOLUS

Black scale parasite.

Metaphycus helvolus *(Arbico; Harmony Farm Supply; Natural Pest Controls; Sespe Creek Insectary)*
Crop: Citrus.
Rate: 1000/A
Pest:
citrus black scale
Remarks: Release on infested trees in early spring.

Millenium — see *Steinernema spp.*

MOSQUITO FISH

Mosquito control.

◆ **Mosquito fish (*Gambusis affinis*)** *(Natural Pest Controls)*
Crop: Rice.
Rate: 200-400 fish/A
Pest:
mosquitoes
Remarks: Apply after frost.

NEOSEIULUS SPP.

Predator. Also known as Amblyseius.

◆ ***Ambylesius cucumeris*** *(Florikan)*
Crop: Greenhouse crops.
Rate: 5/sq. ft.
Pest:
broad mite, tetramychid mite, thrips, cyclamen mite
Environment: 70% relative humidity, temperature 68°-80°F.
Remarks: Can survive on pollen when prey is not available.

Neoseiulus barkeri *(IPM Laboratories)*
Crop: Greenhouse crops.
Rate: 10-30/plant per week.
Pest:
broad mites, thrips
Remarks: Predators can consume 15-26 broad mites per day.

Neoseiulus californicus *(Arbico; Beneficial Insectary; Caltec; IPM Laboratories; Nature's Control)*
Crop: Corn, fruits, nuts, ornamentals, vegetables.
Rate: Determined on individual basis.
Pest:
spider mites
Environment: Can tolerate hot conditions (humidity 60% minimum; temperature 70°-90°F).
Remarks: Introduce at first sign of spider mites. Will actively search out and destroy spider mites throughout growing season. Although slow-acting, can survive absence of prey longer than other predator mites.

Neoseiulus cucumeris *(Arbico; Natural Pest Controls)*
Crop: Sweet peppers.
Rate: Determined on individual basis.
Pest:
thrips
Remarks: Greenhouse.

N. cucumeris* and *N. barkeri *(Hydro-Gardens)*
Crop: Greenhouse crops.
Rate: 1 predator/sq. ft.
Pest:
thrips
Remarks: Predator population must be built up and established early in crop. Introduce 2 weeks after transplanting. Repeat every month during periods of warm, dry weather.

Neoseiulus cucumeris *(IPM Laboratories)*
Crop: Greenhouse crops.
Rate: 10-200/plant
Pest:
aphids, mites, thrips
Environment: Minimum humidity moderate: temperature 70°F.
Remarks: Rate varies with level of infestation and plant size.

Thrips Predator Mites (*N. cucumeris*) *(Harmony Farm Supply; International Technology Services; Nature's Control)*
Crop: Greenhouse ornamentals, vegetables.
Rate: Determined on individual basis.
Pest:
thrips
Environment: Humidity should be 70-90%; temperature 50°-85°F.
Remarks: Introduce when thrips are first observed. Especially recommended for cucumbers. Will actively search for thrips; adults will destroy 1-2 thrips larvae per day. Because *Neoseiulus* predators do not reproduce as rapidly as thrips, it is important to tackle infestation as soon as possible. Timely and regular introductions of large numbers of predator mites are necessary.

ORIUS INSIDIOSUS

Minute Pirate Bug.

Orius insidiosus *(Arbico; Florikan; Harmony Farm Supply)*
Crop: Greenhouse vegetable and ornamentals.
Rate: Variable depending on crop, pest, and level of infestation.
Pest:
aphids, insect eggs, thrips, caterpillars (small), mites, whiteflies
Environment: Humidity 70-85%; temperature 68°-78°F.
Remarks: Foliage should be dry at time of application. Typical rates are 1-5/sq. yd. repeated every 2-3 weeks.

Orius insidiosus *(IPM Laboratories)*
Crop: Ornamentals.
Rate: Preventive:
1/10 sq. ft.
Thrips present:
1/2 sq. ft.
Pest:
aphids, thrips (preferred), whiteflies (pupae), mites
Environment: Temperature 70°-90°F, long days.
Remarks: Small, predaceous bugs that kill their prey by attacking with their piercing mouthparts and sucking out body contents. Both adults and nymphs are predaceous. Adults lay eggs beneath plant tissue surface, often on stems or near leaf veins. Eggs hatch into small, yellow nymphs that develop into adults in 9-25 days, depending on temperature. Adults can eat 5-20 thrips per day. They can survive on pollen in the absence of prey. Dormant Sept.-April.

PEDIOBIUS FOVEOLATUS

Bean beetle parasite.

Pediobius foveolatus *(Arbico)*
Crop: Beans.
Rate: 1 unit/400 sq. ft.
Pest:
Mexican bean beetle
Remarks: Parasite deposits its eggs in larvae of beetle. Parasites must be released during larval stage of beetle.

◆ - **new product** • ai-active ingredient • DF-dry flowable • E/EC-emulsifiable concentrate • EW-emulsion in water
F/FL-flowable • G-granule • L/LC-liquid • SP/WSP-soluble packet • W/WP-wettable powder • WSB-water soluble bag

PENTILITOMASTIZ SP.

Navel orangeworm parasite.

***Pentilitomastiz* sp.** *(Arbico; Caltec)*
Crop: Nuts.
Rate: 500/A
Pest:
navel orangeworm
Remarks: Egg is polyembronic, producing 500-1300 parasites from 1 navel orangeworm.

PHYTOSEIULUS SPP.

Predator mite.

Phytoseiulus persimilis *(Arbico)*
Crop: Greenhouse ornamentals, strawberries, vegetables (cucumber, tomatoes).
Rate: Cucumbers, tomatoes:
1 mite/plant + 1-2 mites/infested leaf
Strawberries:
1000-10,000 mites/A
Greenhouse:
2000 mites/3000 sq. ft.
Pest:
spider mites
Remarks: Apply at first sign of spider mites. Upon release, spider mite infestations can be reduced to very low numbers within 2-3 weeks. Soon afterward they will die from lack of food. If pests reinfest plants, re-introduce predators every 3-5 weeks or in combination with either *Galendromus occidentalis* or *Neoseiulus californicus*. Also available in *Phytoseiulus persimilis* Ht, for higher temperature, which can be used in desert regions.

Phytoseiulus persimilis *(Beneficial Insectary; Harmony Farm Supply; Hydro-Gardens; International Technology Services; Natural Pest Controls; Nature's Control)*
Crop: Greenhouse ornamentals, strawberries, vegetables.
Rate: Determined on individual basis.
Pest:
spider mites
Environment: Humidity should be 60-90%; temperature 60°-85°F (strawberries 50°-85°F). Nature's Control *P. persimilis* is a cool- and high-temperature-adapted strain (tolerates temperature to 105°).
Remarks: Introduce at first sign of spider mites. Will control spider mites throughout growing season, actively searching out and destroying up to 20 immature or 5 mature spider mites in 1 day. Adults are bright orange and slightly larger than spider mites.

Phytoseiulus persimilis *(Caltec)*
Crop: Fruits, greenhouse strawberries, nuts, vegetables (vine crops).
Rate: 5000-20,000/A
Pest:
spider mites
Remarks: Apply at first sighting of spider mites. Upon release, infestations can be reduced to very low numbers within 2-3 weeks. They will soon die from lack of food. If pests reinfest plants, re-introduce predators every 3-5 weeks or in combination with either *Galendromus occidentalis* or *Neoseiulus californicus*. Also available in *Phytoseiulus persimilis* Ht, for higher temperature, which can be used in desert regions.

Phytoseiulus persimilis *(Florikan)*
Crop: Greenhouse ornamentals, strawberries, vegetables.
Rate: 1-2/sq. ft.
Pest:
spider mites
Environment: Temperature should be 40°-95°F; humidity greater than 60%.
Remarks: Rate depends on density of infestation. Apply anytime during plant growing cycle or at first appearance of spider mites. May be used in combination with *Bacillus thuringiensis* or Enstar.

Phytoseiulus persimilis *(Hydro-Gardens)*
Crop: Greenhouse.
Rate: Preventive:
1/5 sq. ft.
Existing:
1/25-50 mites
Pest:
spider mites
Remarks: For preventive, introduce soon after transplanting; repeat in 2 weeks.

Phytoseiulus persimilis *(IPM Laboratories)*
Crop: Greenhouse.
Rate: Determined on individual basis.
Pest:
twospotted spider mite
Environment: Humidity should be 60% minimum; temperature 65°-80°F.
Remarks: Adults are bright orange and slightly larger than spider mites. Will actively seek out and consume 20 young or 5 adult spider mites per day.

PODISUS MACULIVENTRIS

"Spined soldier bug." Predator.

Podisus Maculiventris *(Arbico)*
Crop: Field corn, fruits, nuts, ornamentals, small grains, soybeans, vegetables.
Rate: 1 unit/10 ft. row
Pest:
cabbage looper
imported cabbageworm
Mexican bean beetle
Remarks: Larvae begin feeding 4-5 days after hatching and continue to attack prey throughout next 3 weeks of development. Depending on conditions, will provide 30-85 days of control from time of release. Additional releases may be necessary if adult population has diminished or if subsequent plantings are desired. Timing is important; shipments begin around April 1.

RHIZOBIUS LOPHANTHAE

Black ladybug.

◆ ***Rhizobius lophanthae*** *(Harmony Farm Supply)*
Crop: Citrus, interiorscapes.
Rate: Citrus:
150-500/a
Interiorscapes:
1-2/sq. ft.
Pest:
armored scales
Remarks: Release as soon as scale infestation is detected. After hatching *Rhizobius lophanthae* consumes scale eggs, as a larvae will attack scale crawlers. Adults consume all life stages of scales as well as mealybugs.

RUMINA DECOLLATA

Predatory snail.

Rumina decollata *(Arbico; Sespe Creek Insectary)*
Crop: Ornamentals.
Rate: 1000-1500 snails/A
Pest:
brown garden snail
Remarks: Release in infested areas. Among the best places for release are under irrigated perennial shrubs or ground covers where there is a rich supply of organic matter. Do not release in areas where chemical snail bait has been used within past 6-8 weeks, vegetable gardens, or annual flower beds.
State Restrictions: Arbico not available in California north of the Tehachapa Mountains.

Scanmask — see *Steinernema* spp.

SPIDER MITE DESTROYER

Ladybug.

◆ **Spider Mite Destroyer (*Stethorus punctillum*)** *(Nature's Control)*
Crop: Cotton, field corn, fruits, nuts, ornamentals, small grains, soybeans, vegetables.
Rate: 100 destroyers begin a colony
Pest:
twospotted spider mite
Remarks: Each spider mite destroyer eats more than 40 mites per day, living 4-5 weeks. Life cycle takes 18 days at 70°.

NOTICE	The information on these pages is for preliminary planning — not a guide for use. Be sure to follow the manufacturer's directions, notwithstanding information contained here.	For personal protective equipment and EPA registration numbers, please turn to page 39.

STEINERNEMA SPP.

Beneficial nematodes.

Beneficial Nematodes *(Nature's Control)*
Guardian *(Hydro-Gardens)*
Crop: Greenhouse.
Rate: 1 million/3000 sq. ft.
Pest: Over 250 insect pests with resting periods in the soil including:

corn rootworm	fungus gnats	thrips
cutworms	sod webworms	white grubs
European corn borer	strawberry root weevil	

Remarks: Apply in evening, directly to soil or growing media. Do not apply during daytime, as ultraviolet rays kill these parasites. Watering in will help them migrate to protected area within root zone. Does not control other nematodes.

BioVector 355 for Citrus (*S. riobravis*) *(Thermo Trilogy)*
Crop: Citrus.
Rate: 1 container/20 grove A
1 container/1.33 A nursery plantings
Pest: Larvae:

apopka weevil	blue green weevil	citrus root weevil

Environment: Apply when temperature exceeds 55°F.
Remarks: For best results apply to moist soil. Compatible with most registered citrus products; see compatibility chart on label.

Signal Word/Toxicity Class: Caution.

Devour (*S. riobravis*) *(Thermo Trilogy)*
Crop: Turf.
Rate: 1 billion nematodes/A
Pest:
mole crickets

Rate: 2 billion nematodes/A
Pest:

ring nematode	root-knot nematode	sting nematode

Environment: Apply when temperature is between 55°-100°F.
Remarks: Apply early spring for overwintering adult mole crickets and late summer/fall to control adult and last instar nymphs. For nematodes, apply when soil sampling shows unacceptable levels. Compatible with most registered turf products; see compatibility chart on label.

Signal Word/Toxicity Class: Caution.

Ecomask (*S. carpocapsae*) *(BioLogic)*
Scanmask (*S. feltiae*) *(BioLogic)*
Crop: Fruits, nuts, ornamentals, turf, vegetables.
Rate: 11,000-35,000 active units/sq. ft. (depends on insect)
Pest:

artichoke moth (larvae)	cutworms	raspberry crown borer
banana moth	flea beetles	root maggots
black currant borer	fungus gnats (larvae)	sod webworm
black vine weevil	gypsy moth (larvae)	strawberry root weevil
cabbage root maggot	Japanese beetle	tobacco budworm
codling moth (larvae)	mole crickets	white grubs
corn earworm	onion maggot	wireworms
corn rootworm	pine weevil	
cucumber beetle	poplar clearwing borer	

Environment: Works in moist soil at temperatures from 50°-85°F. Ecomask is inactive below 60°F.
Remarks: For best results, apply at dusk. Ecomask is less active against fly larvae and Scarab larvae.

◆ **Millenium (*S. carpocapse*)** *(Thermo Trilogy)*
Crop: Turf.
Rate: 1 billion nematodes/A
Pest:

annual bluegrass weevil	billbugs	sod webworm
armyworm	chinch bugs	tropical sod webworm
	cutworms	

Remarks: For billbugs and worms, apply when small larvae or damage is first detected. For annual bluegrass weevil, apply against early instars and again 2-3 weeks later. For chinch bugs, apply with silicone-type surfactant. Compatible with most registered turf products; see compatibility chart on label.

Crop: Ornamentals.
Rate: 1 billion nematodes/14,500 sq. ft.
Pest:

black vine weevil	strawberry root weevil

Rate: 1.2 oz./1 gal. water
Pest:

clearwing borer	dogwood borer	peachtree borer

Rate: 1 billion nematodes/500 plants
Pest:

banana moth	giant palmetto weevil	rotten sugarcane borer

Rate: 1 billion nematodes/A
Pest:

fungus gnats	shore flies

Environment: Apply when temperature is between 55°-100°F.
Remarks: For black vine and strawberry root weevils, apply when larvae are active, but before damage occurs. For clearwing, dogwood, and peachtree borers, apply between May-Sept. when larvae are present, but before insect entry hole has been plugged with debris. For banana moth, giant palmetto weevil, and rotten sugarcane borer, apply 3 applications at 6- to 8-week intervals as a crown drench to point of runoff. For fungus gnat and shore flies, make 3 applications at 7-day intervals. Compatible with most registered ornamental products; see compatibility chart on label.

Signal Word/Toxicity Class: Caution.

Steinernema carpocapsae *(Arbico)*
Crop: Fruits, nuts, ornamentals, vegetables.
Rate: Conventional rows:
7 million nematodes (1 package)/600 sq. ft.
Raised beds or wide rows:
14 million nematodes/200-300 sq. ft.
Pest: White grubs and larval or grub stage of:

armyworms	fungus gnats	northern masked chafer
billbug	girdlers	Oriental beetle
black vine weevil	iris borer	root maggot
cabbage root maggot	Japanese beetle	sciarid larvae
carrot weevil	June beetle	sod webworms
citrus weevil	maggots	strawberry root weevil
cutworms	mole crickets	
European chafer		

Remarks: Parasites enter host through body apertures or by penetration of body wall. Once inside, they release a bacterium that kills host within 48 hr. Mix with water and spray or sprinkle on soil along crop rows or around larger plants. Apply before setting out transplants. For Japanese beetle grubs, apply in late summer or mid-spring or when insect damage is detected. Water first if soil is dry. During hot weather, release at night when temperatures are cooler. Parasites are shipped in juvenile stage of their life cycle and can be stored in the refrigerator for up to 2 months.

Steinernema carpocapsae *(Harmony Farm Supply; M & R Durango)*
Steinernema feltiae *(M & R Durango)*
Crop: Field crops, greenhouse, orchards, turf, mushrooms (feltiae only)
Rate: Determined by size of area to be treated, soil type, climatic factors, type of pest and life cycle, infestation levels, pest tolerance threshold, stage of plant development, and available research information.
Pest:

caterpillars	fungus gnats	sciarid fly (feltiae)
cutworms		

Environment: Needs moist environment to survive and move through soil.
Remarks: Apply in the evening. Soil should be damp prior to application. Keeping soil moist for several day after application recommended.

TENODERA ARIDIFOLIA SINENSIS

Praying mantis. Predator.

Tenodera aridifolia sinensis *(A-1 Unique Insect Control; Arbico; Natural Pest Controls; Nature's Control)*
Crop: Fruits, nuts, ornamentals, vegetables.
Rate: A-1 Unique Insect Control: 3 cases/5000 ft. area
Arbico: 3 cases/5000 sq. ft. or 10-100 cases/A
Natural Pest Controls: 2 cases/ft.
Nature's Control: 10-20 cases/A
Pest: Most pest insects, mites, and eggs including:

grasshoppers	Japanese beetles	mosquitoes
flies	locusts	

Remarks: Each egg case hatches about 200 (300 for Natural Pest Controls). Allow 2-8 weeks of warm weather for hatching. Will not bother lady bugs. Attach to fences, twigs, leaves, other vegetation, or in the crotch of a bush or tree. Do not place on ground, as they become easy prey for ants. Release should begin during early spring. May be used in conjunction with all other beneficial insect releases.

◆ - **new product** • ai-active ingredient • DF-dry flowable • E/EC-emulsifiable concentrate • EW-emulsion in water
F/FL-flowable • G-granule • L/LC-liquid • SP/WSP-soluble packet • W/WP-wettable powder • WSB-water soluble bag

THRIPOBIUS SEMILUTEUS

Thrips parasite.

Thripobius semiluteus *(Arbico; Nature's Control)*
Crop: Greenhouse crops including citrus, avocados.
Rate: 2000/A
Pest: Greenhouse thrips.
Remarks: Attach vial to tree horizontally.

Thrips Predator Mites — see *Neoseiulus* spp.

TRICHOGRAMMA SPP.

Parasitic wasp.

Caterpillar Parasites *(Nature's Control)*
Crop: Corn, orchards, ornamentals, vegetables.
Rate: 5000-50,000/A
Pest: Over 200 kinds of caterpillars.
Remarks: Release when pest moths are first seen. Continue releases weekly or bi-weekly until moths are no longer present.

***Trichogramma* spp.** *(A-1 Unique Insect Control; Arbico; Beneficial Insectary; Caltec; Harmony Farm Supply; M & R Durango; Natural Pest Controls)*
Crop: Avocados, corn, orchards, ornamentals, vegetables.
Rate: A-1 Unique Insect Control: 40,000/A
Arbico: 12,000/500 sq. ft./40,000-200,000/A
Beneficial Insectary; Caltec; Harmony Farm Supply; M & R Durango: Determined on individual basis
Natural Pest Controls: 35,000-40,000.
Pest: Many insect species including:

armyworms	codling moth	leafworms
bollworms	corn borers	moth eggs
cabbage looper	corn earworm	tomato hornworm
cabbageworm	cutworms	webworms
cane borers	fruitworms	

Remarks: Available in 4 species: *T. pretiosum* recommended for field crops and vegetables, *T. planterni* for avocados, *T. minutum* for ornamentals and orchard crops, *T. brassicae* for European corn borer control.

WHITEFLY PARASITES

Parasitic wasp.

Encarsia formosa *(Arbico)*
Crop: Ornamentals, vegetables.
Rate: 2-4 parasites/sq. ft. greenhouse
Cucumbers, tomatoes:
8 parasites/plant
Pest:
whiteflies
Remarks: For greenhouses. When low densities of whiteflies are present, release 1 parasite/sq. ft. weekly for 3 weeks. Should be used in conjunction with traps for trapping and monitoring.

Encarsia formosa *(Florikan; Harmony Farm Supply; Hydro-Gardens; Nature's Control)*
Crop: Ornamentals, vegetables.
Rate: Determined on individual basis.
Pest:
greenhouse whitefly sweetpotato whitefly
Environment: Temperature should be at least 64°F (average 68°F).
Remarks: For greenhouses. Introduce at intervals when pests are first observed. Deposits an egg in whitefly larva which turns black after approximately 10 days (with sweetpotato whitefly, eggs turn a darker shade but not jet-black). Parasitized larvae die; instead of a whitefly, a tiny wasp will emerge and continue beneficial process, which includes feeding on young whitefly larvae. Do not throw away pruned leaves without checking for black parasite scales. Leave these under plant for approximately 1 week until wasps have emerged. May be used in combination with other beneficials. *Encarsia formosa* is very susceptible to chemicals; use extreme caution when treating other diseases or pests.

Encarsia formosa *(IPM Laboratories)*
Crop: Cucumbers, tomatoes, poinsettias.
Rate: **Cucumbers:**
20,000/A weekly (8 weeks)
Tomatoes:
10,000/A biweekly (4-6 applications)
Poinsettias:
2-4 parasites/plant weekly (8 weeks)
Pest:
whiteflies
Environment: Temperature should be at least 64°F.
Remarks: For greenhouses. Tiny wasp that attacks only whiteflies. Very susceptible to chemicals.

Encarsia formosa *(International Technology Services)*
En-Strip *(Natural Pest Controls)*
Crop: Ornamentals, vegetables.
Rate: 1 parasite/1 sq. ft.
Pest:
whiteflies
Remarks: Greenhouse. Release at first sign of infestations, repeat at 2-week intervals. Introduce before whitefly populations become too large.

Eretmocerus californicus *(Arbico; Beneficial Insectary; Florikan; Hydro-Gardens)*
Crop: Cotton, field corn, fruits, greenhouse crops, nuts, other field crops, ornamentals, small grains, soybeans, vegetables.
Rate: Determined on individual basis.
Pest:
greenhouse whitefly silverleaf whitefly sweetpotato whitefly
Environment: Capable of handling hot, dry temperatures.
Remarks: Introduce at intervals when whiteflies are first observed. Deposits an egg in whitefly nymphs. Instead of a whitefly, a tiny wasp will emerge and continue the beneficial process, which includes killing the young nymphs by feeding on them. May be used in combination with other beneficials such as green lacewing. *Eretmocerus* is more tolerant to pesticides than *Encarsia formosa*.

For personal protective equipment and EPA registration numbers, please turn to page 39.

Notes • Notes • Notes

Insect and Disease Control Guide

Section 2

Field Corn

Enhanced Seed Chart 72
Bt Chart 76
Field Corn crop listings 77
Insecticides 77
Fungicides 85

The information in this 1999 Insect and Disease Control Guide is updated to August 15, 1998.
Because manufacturers' product recommendations are constantly being updated and revised as additional product registrations and label changes are approved by EPA, there may be changes after this date. However, every effort has been made to give manufacturers the opportunity to include in the Guide their latest information. This Guide is for preliminary planning – not a use guide. Be sure to follow label directions, notwithstanding information contained herein.

Enhanced Seed Products Chart

Field Corn

Find enhanced seed products quickly and easily with this enhanced seed chart. Simply find the enhanced seed qualities you want, and follow across to the marked spot to find the company name you can contact. Turn to the Suppliers Directory on page 599 for addresses and phone numbers.

	AgriPro Seeds Inc.	Asgrow Seed Co.	Beck's Superior Hybrids, Inc.	Campbell Seed, Inc.	Cargill Hybrid Seeds	Chemgro Seeds	Countrymark Co-op/Growmark, Inc.	DEKALB Genetics Corp.	Garst Seed Co.	Golden Harvest Seeds, Inc.	Hoegemeyer Hybrids	Hoffman Seeds Inc.	Interstate Payco Seed	Merschman Seeds	Midwest Seed Genetics	Mycogen Seeds	NC+ Hybrids	Novartis Seeds	Patriot Seed Co.	Pioneer Hi-Bred	PSA Genetics	Rupp Seed, Inc.	Sands of Iowa	Sieben Hybrids, Inc.	Stine Seed Co.	Terra Industries Inc.	Trisler Seed Farms, Inc.
Bt Corn	■	■	■	■		■	■	■	■	■		■	■	■	■	■	■	■	■	■	■	■	■	■	■	■	■
Bt Corn + eyespot resistance/tolerance																											
Bt Corn + gray leaf spot resistance/ tolerance			■	■	■	■		■	■			■			■			■						■			
Bt Corn + Liberty Link									■				■														
Bt Corn + Liberty Link + gray leaf spot resistance/tolerance																		■									
Bt Corn + Northern leaf blight resistance/ tolerance + eyespot resistance/tolerance					■																						
Bt Corn + Northern leaf blight resistance/ tolerance + eyespot resistance/tolerance + Helminthosporium resistance/tolerance					■																						
Bt Corn + Southern leaf blight resistance/tolerance																	■										
Bt Corn + Southern leaf blight resistance/ tolerance + Helminthosporium leaf spot resistance/tolerance					■																						
Bt Corn + Southern leaf blight resistance/ tolerance + Stewart's wilt resistance/ tolerance																	■										
Bt Corn + Southern rust resistance/ tolerance		■																									
IMI Corn + Bt							■		■	■			■			■		■									
IMI Corn + Bt + gray leaf spot resistance/ tolerance																	■										
IMI Corn + Bt + Liberty Link									■				■														
IMI Corn + Bt + Liberty Link + gray leaf spot resistance/tolerance									■																		
IMI Corn + Bt + Liberty Link + gray leaf spot resistance/tolerance + eyespot resistance/tolerance									■																		
IMI Corn + Bt + Liberty Link + gray leaf spot resistance/tolerance + Northern leaf blight resistance/tolerance									■																		
IMI Corn + Bt + Liberty Link + Northern leaf blight resistance/tolerance + eyespot resistance/tolerance									■																		
IMI Corn + Bt + Northern leaf blight resistance/tolerance									■																		
IMI Corn + eyespot resistance/tolerance																	■							■			

	AgriPro Seeds Inc.	Asgrow Seed Co.	Beck's Superior Hybrids, Inc.	Campbell Seed, Inc.	Cargill Hybrid Seeds	Chemgro Seeds	Countrymark Co-op/Growmark, Inc.	DEKALB Genetics Corp.	Garst Seed Co.	Golden Harvest Seeds, Inc.	Hoegemeyer Hybrids	Hoffman Seeds Inc.	Interstate Payco Seed	Merschman Seeds	Midwest Seed Genetics	Mycogen Seeds	NC+ Hybrids	Novartis Seeds	Patriot Seed Co.	Pioneer Hi-Bred	PSA Genetics	Rupp Seed, Inc.	Sands of Iowa	Sieben Hybrids, Inc.	Stine Seed Co.	Terra Industries Inc.	Trisler Seed Farms, Inc.
IMI Corn + eyespot resistance/tolerance + gray leaf spot resistance/tolerance									■																		
IMI Corn + Goss' wilt resistance/tolerance																	■										
IMI Corn + Goss' wilt resistance/tolerance + Southern leaf blight resistance/tolerance																	■										
IMI Corn + gray leaf spot resistance/ tolerance			■	■				■			■				■	■	■							■			■
IMI Corn + Helminthosporium leaf spot resistance/tolerance + Stewart's wilt resistance/tolerance																								■			
IMI Corn + Northern leaf blight resistance/ tolerance									■																		
IMI Corn + Northern leaf blight resistance/ tolerance + Anthracnose resistance/ tolerance + Helminthosporium leaf spot resistance/tolerance + Stewart's wilt resistance/tolerance																								■			
IMI Corn + Northern leaf blight resistance/ tolerance + eyespot resistance/tolerance					■				■																		
IMI Corn + Northern leaf blight resistance/ tolerance + eyespot resistance/tolerance + Goss' wilt resistance/tolerance																		■						■			
IMI Corn + Northern leaf blight resistance/ tolerance + gray leaf spot resistance/ tolerance									■																		
IMI Corn + Northern leaf blight resistance/ tolerance + gray leaf spot resistance/ tolerance + eyespot resistance/tolerance									■																		
IMI Corn + Northern leaf blight resistance/ tolerance + gray leaf spot resistance/ tolerance + Helminthosporium leaf spot resistance/tolerance + eyespot resistance/tolerance					■																						
IMI Corn + Northern leaf blight resistance/tolerance + gray leaf spot resistance/tolerance + Stewart's wilt resistance/tolerance																	■							■			
IMI Corn + Northern leaf blight resistance/tolerance + Helminthosporium leaf spot resistance/tolerance					■																						
IMI Corn + Northern & Southern leaf blight resistance/tolerance									■																		
IMI Corn + Northern & Southern leaf blight resistance/tolerance + Anthracnose resistance/tolerance																								■			

Enhanced Seed Products Chart, continued

Field Corn

	AgriPro Seeds Inc.	Asgrow Seed Co.	Beck's Superior Hybrids, Inc.	Campbell Seed, Inc.	Cargill Hybrid Seeds	Chemgro Seeds	Countrymark Co-op/Growmark, Inc.	DEKALB Genetics Corp.	Garst Seed Co.	Golden Harvest Seeds, Inc.	Hoegemeyer Hybrids	Hoffman Seeds Inc.	Interstate Payco Seed	Merschman Seeds	Midwest Seed Genetics	Mycogen Seeds	NC+ Hybrids	Novartis Seeds	Patriot Seed Co.	Pioneer Hi-Bred	PSA Genetics	Rupp Seed, Inc.	Sands of Iowa	Sieben Hybrids, Inc.	Stine Seed Co.	Terra Industries Inc.	Trisler Seed Farms, Inc.
IMI Corn + Northern & Southern leaf blight resistance/tolerance + Anthracnose resistance/tolerance + Goss' wilt resistance/tolerance + Stewart's wilt resistance/tolerance																								■			
IMI Corn + Southern leaf blight resistance/tolerance		■			■			■									■							■			
IMI Corn + Southern rust resistance/tolerance		■																									
IMI Corn + Southern rust resistance/tolerance + gray leaf spot resistance/tolerance									■															■			
IMI Corn + Stewart's wilt resistance/tolerance																	■										
Liberty Link + corn borer resistance/tolerance													■														■
Liberty Link + eyespot resistance/tolerance		■																						■			
Liberty Link + Goss' wilt resistance/tolerance																	■										
Liberty Link + gray leaf spot resistance/tolerance		■															■							■			
Liberty Link + gray leaf spot resistance/tolerance + Anthracnose resistance/tolerance																								■			
Liberty Link + gray leaf spot resistance/tolerance + Southern rust resistance/tolerance																								■			
Liberty Link + Northern leaf blight resistance/tolerance + eyespot resistance/tolerance					■																						
Liberty Link + Northern leaf blight resistance/tolerance + eyespot resistance/tolerance + Helminthosporium leaf spot resistance/tolerance					■																						
Liberty Link + Northern leaf blight resistance/tolerance + Helminthosporium leaf spot resistance/tolerance					■																						
Liberty Link + Northern & Southern leaf bligth resistance/tolerance + gray leaf spot resistance/tolerance + Anthracnose resistance/tolerance																								■			
Liberty Link + Southern leaf blight resistance/tolerance		■																						■			

	AgriPro Seeds Inc.	Asgrow Seed Co.	Beck's Superior Hybrids, Inc.	Campbell Seed, Inc.	Cargill Hybrid Seeds	Chemgro Seeds	Countrymark Co-op/Growmark, Inc.	DEKALB Genetics Corp.	Garst Seed Co.	Golden Harvest Seeds, Inc.	Hoegemeyer Hybrids	Hoffman Seeds Inc.	Interstate Payco Seed	Merschman Seeds	Midwest Seed Genetics	Mycogen Seeds	NC+ Hybrids	Novartis Seeds	Patriot Seed Co.	Pioneer Hi-Bred	PSA Genetics	Rupp Seed, Inc.	Sands of Iowa	Sieben Hybrids, Inc.	Stine Seed Co.	Terra Industries Inc.	Trisler Seed Farms, Inc.
Liberty Link + Southern leaf blight resistance/tolerance + Helminthosporium leaf spot resistance/tolerance					■																						
Liberty Link + Southern rust resistance/tolerance		■																									
Liberty Link + Stewart's wilt resistance/tolerance + Nothern leaf blight resistance/tolerance + Northern leaf spot resistance/tolerance																	■										
GR corn + eyespot resistance/tolerance								■																			
GR corn + gray leaf spot resistance/tolerance								■																			
GR corn + gray leaf spot resistance/tolerance + Southern rust resistance/tolerance								■																			
GR corn + Southern leaf blight resistance/tolerance								■																			
Poast resistance/tolerance + eyespot resistance/tolerance								■																			
Poast resistance/tolerance + gray leaf spot resistance/tolerance + Southern rust resistance/tolerance								■																			
Poast resistance/tolerance + Northern leaf blight resistance/tolerance					■																						
Poast resistance/tolerance + Northern leaf blight resistance/tolerance + eyespot resistance/tolerance + Helminthosporium leaf spot resistance/tolerance					■																						
Roundup Ready + Bt			■																								
Roundup Ready + corn borer resistance/tolerance								■																			
Gray leaf spot resistance/tolerance								■						■			■							■			■
Gray leaf spot resistance/tolerance + rust resistance/tolerance + eyespot resistance/tolerance								■									■							■			
Maize dwarf mosaic virus resistance/tolerance + gray leaf spot resistance/tolerance								■																			
Southern leaf blight resistance/tolerance + Stewart's wilt resistance/tolerance								■																■			

Bt Products Chart

Field Corn

Find Bt products quickly and easily with this Bt products chart. Simply find the insect you want to control at the top of the chart, and follow down to the marked spot to find the Bt product you can use to control that pest. Company address information can be found on page 599.

	Armyworms	Beet armyworm	Corn earworm	European corn borer	Fall armyworm	Fusarium	Headworms	Pythium	Southwestern corn borer	Variegated cutworm	Webworms	Western bean cutworm
Agree				■								
Bacillus subtilis						■		■				
Biobit	■			■					■			
BMP 123				■	■				■			
Bt. ssp. aizawai	■			■			■					
Bt.* ssp. *kurstaki	■	■	■	■	■		■		■	■	■	■
Condor		■		■					■			
Crymax	■			■					■			
Cutlass	■			■					■			
DiPel	■		■	■	■		■		■	■	■	■
Javelin	■								■			
Kodiak						■		■				
Lepinox	■		■	■	■				■			
M-Peril				■	■				■			
XenTari	■			■			■					

Manufacturers List

These Bt products are manufactured by the following companies. Look for their addresses, phone numbers, and fax numbers in the Suppliers Directory beginning on page 599.

Agree, Javelin – Thermo Trilogy
Biobit, DiPel, XenTari – Abbott Laboratories
BMP 123 – Becker Microbial
Condor, Crymax, Cutlass, Lepinox – Ecogen
Kodiak – Gustafson, Trace
M-Peril – Mycogen

Field Corn, Popcorn

Insecticides

435 or 455 Oil — see petroleum oil

Adios — see carbaryl

Ambush — see permethrin

ASANA

Broad-spectrum synthetic pyrethroid insecticide. May be tank mixed with herbicides and other insecticides. Restricted Use Pesticide due to toxicity to fish and aquatic organisms.

Asana XL (esfenvalerate) *(DuPont)*

Rate: 0.015-0.03 lb. ai/A
2.9-5.8 fl. oz. XL/A

Pest:
western bean cutworm

Rate: 0.03-0.05 lb. ai/A
5.8-9.6 fl. oz. XL/A

Pest:
chinch bug, corn earworm, corn leaf aphid, corn rootworm (adult), cutworms, flea beetles, grasshoppers, oat birdcherry aphid, southwestern corn borer, stalk borers, true armyworm

Remarks: Field corn (including seed), popcorn. Repeat as necessary to maintain control. For western bean cutworm, apply before larvae enter ear. For aphids, direct spray at aphid population to achieve maximum coverage of exposed insects. Aphids not contacted by spray, such as in whorls and leaf axis, may not be adequately controlled. For chinch bug, over-the-top sprays are not adequate. Direct spray at plant base by drop nozzles or some other mechanism. For corn earworm, first application should be at or before silking; repeat at 3- to 5-day intervals until silking is completed. For corn rootworm, spray at first sign of silk feeding. For flea beetles and stalk borers, apply early in migration from grassy areas to corn, before borers enter plant. For southwestern corn borer, higher rates recommended for moderate to heavy populations.

- - - - - - - - - - - - - - - -

Rate: 0.04-0.05 lb. ai/A
7.8-9.6 fl. oz. XL/A

Pest:
European corn borer

Remarks: Field corn (including seed), popcorn. First brood: Spray while eggs are in blackhead stage or before larvae enter whorl. Application by ground equipment suggested; good coverage essential. Second brood: Apply when sufficient egg masses are found. Spray when eggs are in blackhead stage or starting to hatch. When egg laying is prolonged or 3rd generation is present, additional sprays may be required. Do not exceed 0.25 lb. ai/A per season.

- - - - - - - - - - - - - - - -

Signal Word/Toxicity Class: Warning/II.
REI: 12 hr.

AZTEC (Premix)

Organophosphate chemistry of tebupirimphos combined with cyfluthrin provides two modes of action to control soil insect pests. Restricted Use Pesticide due to toxicity to aquatic invertebrates.

Aztec (2.1%) (2% tebupirimphos & 0.1% cyfluthrin) *(Bayer)*

Rate: 0.12-0.15 lb. ai/A
5.5-7.3 lb. 2.1%/A

Pest:
cutworms, northern corn rootworm (larvae), seedcorn beetles, seedcorn maggots, southern corn rootworm (larvae), western corn rootworm (larvae), white grubs, wireworms

Remarks: Field corn (including seed, silage), popcorn. Apply band, T-band, or in-furrow. Do not apply more than 7.3 lb./A per crop season. Do not use on other crops grown for food or forage. Refer to label for recommended applications.

State Restrictions: For use in Arizona, Arkansas, Colorado, Idaho, Illinois, Indiana, Iowa, Kansas, Kentucky, Michigan, Minnesota, Missouri, Montana, Nebraska, Nevada, New Mexico, North Dakota, Ohio, Oklahoma, Pennsylvania, South Dakota, Tennessee, Utah, Vermont, West Virginia, Wisconsin, Wyoming.

- - - - - - - - - - - - - - - -

Signal Word/Toxicity Class: Warning/II.
REI: 0.

CAPTURE

Synthetic pyrethroid insecticide with broad-spectrum activity. Formulated as an emulsifiable concentrate. Toxic to fish and aquatic organisms. Restricted Use Pesticide.

Capture 2EC (bifenthrin) *(FMC)*

Rate: 0.08-0.1 lb. ai/A
5.1-6.4 fl. oz. 2EC/A

Pest:
Banks grass mite, carmine mite, corn earworm, corn rootworm (adult), European corn borer, southwestern corn borer, twospotted spider mite, western bean cutworm

Remarks: Field (grain, seed, silage), popcorn. Apply in minimum 10 gal./A by ground; 2 gal./A by air. May be tank mixed with other products approved for use on corn (except clay-based fungicides). Do not graze treated areas or cut treated crops for feed within 30 days of last application. Straw may not be used for food or feed. Refer to label for spray drift precautions.

Days To Harvest: 30.

Crop Rotation: Anytime for crops for which bifenthrin tolerances exist; leafy vegetables and root crops-30 days; all other crops following final bifenthrin application-7 months.

State Restrictions: For use in Colorado, Kansas, Nebraska, New Mexico, Oklahoma, and Texas.

- - - - - - - - - - - - - - - -

Signal Word/Toxicity Class: Warning/II.
REI: 12 hr.

CARBARYL

Broad-spectrum carbamate pesticidal bait. Moderately toxic to humans. Extremely toxic to estuarine and aquatic invertebrates. Highly toxic to bees exposed to direct treatment or residues on blooming crops or weeds.

Adios or Slam (13%) *(Micro Flo)*

Rate: 0.03-0.1 lb. ai/A
0.25-0.75 pt. 13%/A

Pest:
corn rootworm

Remarks: Field (seed), popcorn. Do not apply through any type of irrigation system. Do not apply more than 3.75 lb./A per season.

Days To Harvest: 0.

- - - - - - - - - - - - - - - -

Signal Word/Toxicity Class: Caution/III.
REI: 12 hr.

NOTICE The information on these pages is for preliminary planning — not a guide for use. Be sure to follow the manufacturer's directions, notwithstanding information contained here.	For personal protective equipment and EPA registration numbers, please turn to page 39.

Clean Crop Sevin 5 Bait *(Platte)*

Rate: 1-2 lb. ai/A
20-40 lb. Bait/A

Pest:
ants, armyworms, cutworms, darkling beetle, field crickets, grasshoppers, sowbugs

Remarks: Field corn, popcorn. Apply with ground or air equipment.
Days To Harvest: 0.

Signal Word/Toxicity Class: Caution/III.
REI: 12 hr.

Drexel Carbaryl 4L, 50W, 80S *(Drexel)*
Sevin 4F, XLR Plus, 80S, 80WSP *(Rhone-Poulenc)*

Rate: 0.5-1.5 lb. ai/A
1-3 pt. 4L, 4F, XLR/A
1-3 lb. 50W/A
0.66-1.88 lb. 80S, 80WSP/A

Pest:
grasshoppers

Rate: 1-2 lb. ai/A
2-4 pt. 4L, 4F, XLR/A
1.25-2.5 lb. 80S, 80WSP/A

Pest:
armyworms, chinch bug, corn earworm, corn rootworm, European corn borer, fall armyworm, flea beetles, Japanese beetle, leafhoppers, sap beetle, southwestern corn borer, western bean cutworm (high rate)

Rate: 2-6.5 lb. ai/A
4-13 pt. 4L/A
2-3 lb. ai/A
4-6 pt. 4F, XLR/A
2 lb. ai/A
4 lb. 50W/A
2-6.4 lb. ai/A
2.5-8 lb. 80S, 80WSP/A

Pest:
cutworms

Remarks: Apply when insects or their damage appear. For insects attacking silks and ears, apply at 1- to 6-day intervals when first silks appear and continue until silks begin to dry. For larvae in whorl and foliage feeders, apply as necessary. For western bean cutworm, treat when infestation averages 15% and at 90-100% tassel emergence; treatment after 100% silk emergence will reduce effectiveness. For cutworms, apply in 12" band over row, using sufficient water to obtain thorough coverage; broadcast, use at least 20 gal. water/A by ground or 5 gal. water/A by air. Do not apply when foliage is wet or when rain or excessive humidity is expected during next 2 days.
Days To Harvest: 0.

Signal Word/Toxicity Class: Caution (4F, 4L, XLR); Warning/II (50W, 80S, 80WSP).
REI: 12 hr.

◆ **Gowan Sevin Bait** *(Gowan)*

Rate: 2 lb. ai/A
40 lb. Bait/A

Pest:
armyworms, crickets, cutworms, darkling beetle, grasshoppers, sowbugs

Remarks: Corn forage. Apply when pests appear in damaging numbers; repeat at 7- to 14-day intervals if necessary.
Days To Harvest: 1.
State Restrictions: For use in Arizona and California.

Signal Word/Toxicity Class: Caution/III.
REI: 12 hr.

◆ **Wilbur-Ellis Sevin 5 or 10 Dust** *(Wilbur-Ellis)*

Rate: 2 lb. ai/A
40 lb. 5/A
20 lb. 10/A

Pest:
armyworms, corn earworm, Japanese beetle, sap beetle

Remarks: Make 3-4 applications at 2- to 3-day intervals; repeat immediately after rain.
Days To Harvest: 7.
State Restrictions: Sevin 5 not registered in California.

Signal Word/Toxicity Class: Caution.
REI: 12 hr.

Chlorfos — see chlorpyrifos

CHLORPYRIFOS

Broad-spectrum organophosphate insecticide available in granular and liquid formulations for soil and foliar treatments. Very toxic to mammals, birds, fish, and aquatic organisms.

◆ **Chlorfos 15G** *(Griffin)*
Lorsban 15G *(Dow AgroSciences)*
◆ **Micro Flo Chlorpyrifos 15G AG** *(Micro Flo)*

Rate: 1-2 lb. ai/A
6.7-13.5 lb. 15G/A

Pest:
cutworms

Rate: 2 lb. ai/A
13.5 lb. 15G/A

Pest:
billbugs, flea beetles (larvae), grubs, wireworms

Remarks: Preplant broadcast. Apply and incorporate 4-6". Do not exceed 13.5 lb. 15G/A per crop season.

Rate: 8 oz. 15G/1000 ft. row

Pest:
cutworms, northern corn rootworm, seedcorn maggots, southern corn rootworm, western corn rootworm

Rate: 8-16 oz. 15G/1000 ft. row

Pest:
billbugs, grubs, lesser cornstalk borer, seedcorn beetles, symphylans, wireworms

Remarks: At planting. Apply at planting in 6-7" T-band over open seed furrow or behind planter shoe and ahead of press wheel with additional incorporation. Do not apply more than 16 oz./1000 ft. row per season.
Days To Harvest: Forage, silage-14; fodder, grain-35.

Rate: Aerial application:
0.75-1 lb. ai/A
5-6.5 lb. 15G/A
Band treatment:
6-8 oz. 15G/1000 ft. row

Pest:
European corn borer, southwestern corn borer

Remarks: Apply postplant in row over the whorl or broadcast by ground or air. Do not exceed 13 lb./A per crop season.
Days To Harvest: Forage or silage-14; fodder or grain-35.

Signal Word/Toxicity Class: Caution.
REI: 12 hr.

Lorsban-4E *(Dow AgroSciences)*
◆ **Micro Flo Chlorpyrifos 4E AG** *(Micro Flo)*
Nufos 4E *(Cheminova)*

Rate: 1-3 lb. ai/A
2-6 pt. 4E/A

Pest:
billbugs, corn rootworm (larvae), cutworms, flea beetles (larvae), grubs, lesser cornstalk borer, seedcorn beetles, seedcorn maggots, symphylans, wireworms

Remarks: Field corn (including seed), popcorn. Preplant incorporated. Apply to soil surface in minimum 10 gal. water/A; incorporate same day into soil 2-4". May be tank mixed; refer to label.
Days To Harvest: 35.

Rate: 0.5-1 lb. ai/A
1-2 pt. 4E/A

Pest:
armyworm, cutworms

Remarks: Field corn (including seed), popcorn. Preplant, at planting, preemergence treatment in conservation tillage. Use higher rates for residual control. May be tank mixed; refer to label.
Days To Harvest: 35.

◆ - **new product** • ai-active ingredient • DF-dry flowable • E/EC-emulsifiable concentrate • EW-emulsion in water
F/FL-flowable • G-granule • L/LC-liquid • SP/WSP-soluble packet • W/WP-wettable powder • WSB-water soluble bag

Rate: 1 lb. ai/A
2 pt. 4E/A

Pest:
corn rootworm (larvae)

Remarks: Field corn (including seed), popcorn. Cultivation treatment. Apply as water emulsion on both sides of row at base of plants just ahead of cultivator shovels; cover with soil around brace roots. Another application may be made in addition to at-planting application of chlorpyrifos 15G. Refer to both labels for complete use directions.

Days To Harvest: 35.

Rate: 0.25-1.5 lb. ai/A
0.5-3 pt. 4E/A

Pest:

aphids	corn rootworm (adult)	lesser cornstalk borer
armyworms	cutworms	southwestern corn borer
billbugs	European corn borer	webworms
chinch bug	flea beetles (adult)	western bean cutworm
common stalk borer	grasshoppers	
corn earworm		

Remarks: Field corn (including seed), popcorn. Postemergence. Rate varies by pest species. For common stalk borer, apply approximately 11 days after glyphosate application or 3-5 days after complete burndown by paraquat. Do not use in combination with a burndown herbicide.

Days To Harvest: 35.

Signal Word/Toxicity Class: Warning/II.
REI: 24 hr.

Comite — see propargite

COUNTER

Organophosphate insecticide and nematicide. Counter CR Lock'n Load formulation is a closed handling system. May be applied at planting time to control soil pests or over the whorl to control corn borer. Translocated from soil into plants providing systemic activity. Extremely toxic to fish and wildlife. Restricted Use Pesticide due to acute oral and dermal toxicity.

Counter 15G or CR (terbufos) *(American Cyanamid)*

Rate: 8 oz. 15G/1000 ft. row
6 oz. CR/1000 ft. row

Pest: At planting (banded or in-furrow):

corn flea beetle	maize billbug	stubby-root nematode
corn rootworm	root-knot nematode	stunt nematode
cutworms*	seedcorn beetles	symphylans
dagger nematode	seedcorn maggots	thrips
lance nematode	southern corn billbug	white grubs
lesion nematode	spiral nematode	wireworms
lesser cornstalk borer*	sting nematode	

Postemergence incorporated (banded):

maize billbug	southern corn billbug

At cultivation:

corn rootworm	European corn borer

***Suppression**

Remarks: Field corn, field corn grown for seed, popcorn. Make only 1 application per year. Refer to label for use directions and precautions.

Days To Harvest: Grazing or cutting for forage-30.

Signal Word/Toxicity Class: Danger-Poison/I.
REI: 48 hr.

lambda-CYHALOTHRIN

Broad-spectrum synthetic pyrethroid insecticide formulated as an emulsifiable concentrate. Restricted Use Pesticide due to toxicity to fish and aquatic organisms.

Karate or Warrior (1E) *(ZENECA)*

Rate: 0.015-0.03 lb. ai/A
1.92-3.84 fl. oz. 1E/A

Pest:

armyworms**	cutworm	meadow spittlebug
bean leaf beetle	English grain aphid*	oat birdcherry aphid*
beet armyworm**	European corn borer	southwestern corn borer
cereal leaf beetle	flea beetles	
chinch bug	grasshoppers	stalk borer
corn earworm	green cloverworm	tobacco budworm
corn leaf aphid*	greenbug*	webworm
corn rootworm beetle	Japanese beetle	western bean cutworm

Suppression **1st & 2nd instar

Remarks: Field corn (including seed), popcorn. For cutworm, apply before, during, or after planting. For chinch bug, apply as directed spray to base of plants when bugs migrate from small grains or grass weeds to small corn; repeat at 3- to 5-day intervals if needed. For corn borer, earworm, budworm, stalk borer, and western bean cutworm, apply before larvae bores into plant stalk or ear. Do not apply more than 0.96 pt./A per season. Refer to label for spray drift precautions and state and county restrictions.

Days To Harvest: 21.

Signal Word/Toxicity Class: Danger/I.
REI: 24 hr.

DB-Green — see lindane premixes

DEADLINE

Packaged as Bullets, M-Ps (mini-pellets), and granules.

Deadline Bullets, M-Ps, Granules (4%) (metaldehyde) *(Pace)*

Rate: 0.4-1 lb. ai/A
10-40 lb. 4%/A

Pest:
slugs   snails

Remarks: Apply every 3-4 weeks during growing season, or as needed. Do not apply directly to edible portion of plant.

Signal Word/Toxicity Class: Warning/II (granule); Caution (bullets, M-P).
REI: 12 hr.

Declare — see methyl parathion

Diatect — see pyrethrins

DIAZINON (Premix)

Seed treatment. Contains graphite.

◆ **Captan-Diazinon (25% diazinon & 36.67% captan)** *(Trace)*
◆ **Germate Plus (15% diazinon & 25% lindane & 14% carboxin)** *(Trace)*
◆ **Kernel Guard (15% diazinon & 25% lindane & 14.67% captan)** *(Trace)*

Rate: 2 oz./bu. seed

Pest:
seedcorn beetles   seedcorn maggots   wireworms

Remarks: Apply in planter box before planting. Do not use Germate Plus or Kernel Guard on Foundation seed stock. Treat only those seeds needed for immediate use. Do not use treated seed for food, feed, or oil purposes. Do not store excess treated seeds beyond planting time. Refer to label for use directions.

Signal Word/Toxicity Class: Danger/I.
REI: 12 hr.; Captan-Diazinon 4 days with exceptions.

Digon — see dimethoate

Dimate — see dimethoate

DIMETHOATE

Broad-spectrum organophosphate insecticide/acaricide with systemic action. Environmental hazards include toxicity to birds, fish, and bees.

Cheminova Dimethoate 4E *(Cheminova)*
Clean Crop Dimethoate 400 *(Platte)*
Dimate 4E (4EC) *(Terra)*
Drexel Dimethoate 4EC *(Drexel)*
Gowan Dimethoate E267 *(Gowan)*
Helena Dimethoate 4E *(Helena)*
Micro Flo Dimethoate 267 or 4E *(Micro Flo)*
Wilbur-Ellis Dimethoate 267 or Digon 400 *(Wilbur-Ellis)*

Rate: 0.33-0.5 lb. ai/A
0.66-1 pt. 4E, 4EC, 400/A
1-1.5 pt. E267/A
0.22-0.5 lb. ai/A
0.66-1 pt. 267/A

NOTICE The information on these pages is for preliminary planning — not a guide for use. Be sure to follow the manufacturer's directions, notwithstanding information contained here.	For personal protective equipment and EPA registration numbers, please turn to page 39.

Pest:

aphids, Banks grass mite (except Trans-Pecos area of TX), bean beetle, corn rootworm (adult) (except CA), grasshoppers, twospotted spider mite

Remarks: Do not apply during pollen-shed period. Use high rate for grasshoppers. Do not apply more than 3 applications per year.
Days To Harvest: Feeding, grazing-14.

- - - - - - - - - - - - - - - -

Signal Word/Toxicity Class: Warning/II.
REI: 48 hr.

DI-SYSTON

Systemic organophosphate insecticide that is very effective against sucking insects. Very toxic to mammals, birds, fish, aquatic organisms, and other wildlife. Restricted Use Pesticide.

Di-Syston 8 (disulfoton) *(Bayer)*

Rate: 1 lb. ai/A
1 pt. 8/A (40" row)

Pest:

aphids, leafhoppers, mites, thrips

Remarks: Field corn, popcorn. At planting: As water emulsion, spray in 4-6" band directly behind planter shoe in front of press wheel or apply with liquid fertilizer in split bands on each side of row. Do not apply directly to seed. Postplant sidedress: Alternative treatment for early-season control of aphids, leafhoppers, mites, and thrips may be made in water emulsion or with liquid fertilizer as a soil injection on each side of furrow after emergence.
Days To Harvest: Forage or cut treated corn for food, feed, or ensilage-28.

- - - - - - - - - - - - - - - -

Rate: 0.5-1 lb. ai/A
0.5-1 pt. 8/A

Pest:

aphids, mites

Remarks: Field corn, popcorn. Foliar: Spray in sufficient water for thorough coverage but not less than 1 gal. total volume/A. Apply only 1 soil and 1 foliar treatment per crop season.
Days To Harvest: Forage or cut treated corn for food, feed, or ensilage-28.

- - - - - - - - - - - - - - - -

Signal Word/Toxicity Class: Danger/I.
REI: 48 hr.

Enhance Plus — see lindane premixes

FORCE

Synthetic pyrethroid insecticide with activity on a broad spectrum of soil insects. Due to low water solubility and low soil mobility, risk of leaching or groundwater contamination is low. Shows low toxicity to mammals and birds; no toxic effects on earthworms have been observed. Toxic to fish and aquatic organisms. Restricted Use Pesticide.

Force 3G (tefluthrin) *(ZENECA)*

Rate: 4-5 oz. 3G/1000 ft. row

Pest:

black cutworm, lesser cornstalk borer, northern corn rootworm, seedcorn maggots, seedcorn beetles, southern corn rootworm, western corn rootworm, white grubs (suppression), wireworms (suppression)

Rate: 3 oz. 3G/1000 ft. row

Pest:

cutworms (1st yr. corn)

Remarks: Field corn (including seed), popcorn. At planting band or in-furrow application. For band, incorporate into top 1" of soil.
Crop Rotation: No restrictions after 30 days.

- - - - - - - - - - - - - - - -

Signal Word/Toxicity Class: Caution/III.
REI: 48 hr.

FORTRESS

Fortress 5G is applied only with SMARTBOX system, a closed handling and application system. Toxic to wild mammals, birds, fish, and aquatic invertebrates. Restricted Use Pesticide.

Fortress 2.5G or 5G (chlorethoxyfos) *(DuPont)*

Rate: 6 oz. 2.5G/1000 ft. row
3 oz. 5G/1000 ft. row

Pest:

cutworms, northern corn rootworm, seedcorn maggots, southern corn rootworm, symphylans, western corn rootworm, white grubs, wireworms

Remarks: Field corn (including seed). Apply T-band or in-furrow. Do not make more than 1 application per year. Do not apply through any type of irrigation system.
Days To Harvest: 21.
Crop Rotation: Corn–anytime; other crops–30 days.

- - - - - - - - - - - - - - - -

Signal Word/Toxicity Class: Danger-Poison.
REI: 48 hr.; 72 hr. if average rainfall less than 25" per year.

FURADAN

Broad-spectrum carbamate used as an insecticide and nematicide. Relatively nonphytotoxic enabling application in-furrow at planting. Provides systemic activity. Breakdown of product by soil microorganisms has been a problem in some soils. Toxic to bees, birds, fish, and other wildlife. Restricted Use Pesticide due to acute oral and inhalation toxicity.

Furadan 4F (carbofuran) *(FMC)*

Rate: Foliar:
0.125-1 lb. ai/A
0.25-2 pt. 4F/A

Pest:

Banks grass mite (suppression), European corn borer, grasshoppers, southwestern corn borer

Remarks: Field corn. Apply in minimum 10 gal. finished spray by ground; 1 gal. by air. For grasshoppers, apply when insects appear or feeding is noticed. For corn borers, apply broadcast foliar spray. For ground equipment, direct spray into corn whorl for 1st brood and into ear zone for 2nd brood and repeat if necessary. For Banks grass mite, apply in sufficient water for thorough coverage. Rate varies by pest species and infestation level; refer to label for rates and time of application. Do not apply on seed corn less than 14 days prior to detasseling or rogueing. Do not make more than 2 applications per season at 1.5-2 pt. rate; 4 applications per season at 1 pt. rate. Do not apply through any type of irrigation system.
Days To Harvest: Harvest/forage/cutting: 30.
Crop Rotation: Do not rotate with any crop on soil treated with more than 3 lb. ai/A for at least 10 months or plant with any crop for which carbofuran is not registered for at least 10 months.
State Restrictions: Not for use in Long Island, New York.

- - - - - - - - - - - - - - - -

Rate: At planting:
1 lb. ai/A
2 pt. 4F/A (40" row)

Pest:

armyworms*, corn rootworm, European corn borer (1st generation)*, flea beetles, seedcorn maggots, wireworms

***Aids in control for 4-6 weeks after planting**

Rate: Postplant:
1 lb. ai/A
2 pt. 4F/A (40" row)

Pest:

northern corn rootworm, southern corn rootworm, western corn rootworm

Remarks: Field corn, popcorn. At-planting: Apply as 7" band over row; inject on each side of row by mixing with water or liquid fertilizers, or apply directly into seed furrow. Postplant: Apply as postemergent spray by banding over the row, by side-dressing, or basal spraying both sides of row after corn emerges. Control will generally be improved if treatment is cultivated into soil.
Days To Harvest: Do not feed forage within 30 days of last application.
Crop Rotation: Do not rotate with any crop on soil treated with more than 3 lb. ai/A for at least 10 months. Do not plant with any crop for which carbofuran is not registered for at least 10 months.
State Restrictions: Not for use in Long Island, New York.

- - - - - - - - - - - - - - - -

Signal Word/Toxicity Class: Danger-Poison/I.
REI: 48 hr.

Fyfanon — see malathion

Germate Plus — see diazinon premixes

◆ - **new product** • ai-active ingredient • DF-dry flowable • E/EC-emulsifiable concentrate • EW-emulsion in water
F/FL-flowable • G-granule • L/LC-liquid • SP/WSP-soluble packet • W/WP-wettable powder • WSB-water soluble bag

Grain Guard Plus — see lindane premixes

Karate — see *lambda*-cyhalothrin

Kernel Guard — see diazinon premixes

Kernel Guard Supreme — see permethrin premix

Kumulus — see sulfur

LANNATE

Broad-spectrum carbamate insecticide primarily used as treatment of foliar and ear pests in corn. Highly toxic to mammals; also toxic to fish and bees. Restricted Use Pesticide.

Lannate SP or LV (methomyl) *(DuPont)*

Rate: 0.225-0.45 lb. ai/A
0.25-0.5 lb. SP/A
0.75-1.5 pt. LV/A

Pest:

aphids, armyworms, beet armyworm, corn rootworm (adult), earworm, fall armyworm, flea beetles, picnic beetle, variegated cutworm

Rate: 0.45 lb. ai/A
0.5 lb. SP/A
1.5 pt. LV/A

Pest:

beet armyworm, variegated cutworm

Remarks: Apply when insects first appear; repeat at 5- to 7-day intervals. Do not apply more than 2.25 lb. ai/A per crop or make more than 10 applications per crop.

Days To Harvest: Ears/fodder 21; forage-3.

Signal Word/Toxicity Class: Danger-Poison/I.
REI: 48 hr.

LINDANE (Premixes)

Dual purpose insecticide and fungicide seed treatment. Grain Guard Plus and Sorghum Guard contain graphite.

◆ **DB-Green (18.75% lindane & 50% maneb)** *(AGSCO)*

Rate: 2.75 oz./bu. seed

Pest:

seedcorn maggots, wireworms

Remarks: Drill or planter box application. Treat only seeds needed for immediate use, minimizing interval between treatment and planting. Do not use treated seed for food, feed, or oil purposes. Do not store excess treated seeds beyond planting time. Do not graze or feed livestock on treated areas for 6 weeks after planting.

Signal Word/Toxicity Class: Warning/II.
REI: 12 hr.

◆ **DB-Green L (8.6% lindane & 25.6% maneb)** *(AGSCO)*

Rate: 4 fl. oz./bu. seed

Pest:

wireworms

Remarks: Drill or planter box application. Treat only seeds needed for immediate use, minimizing interval between treatment and planting. Do not use treated seed for food, feed, or oil purposes. Do not store excess treated seeds beyond planting time.

Signal Word/Toxicity Class: Warning/II.
REI: 12 hr.

◆ **Enhance Plus (18.75% lindane & 35% maneb & 20% carboxin)** *(Trace)*
◆ **Grain Guard Plus (18.75% lindane & 50% mancozeb)** *(Trace)*

Rate: 3 oz./bu. seed

Pest:

wireworms

Remarks: Drill box treatment. Treat only those seeds needed for immediate use. Do not use treated seed for food, feed, or oil purposes. Do not store excess treated seeds beyond planting time. Do not graze or feed livestock on treated areas for 6 weeks after planting.

Signal Word/Toxicity Class: Danger/I.
REI: 12 hr.

◆ **Sorghum Guard (16.6% lindane & 32.75% captan)** *(Trace)*

Rate: 4 oz./50 lb. seed

Pest:

wireworms

Remarks: Apply in planter or drill box before planting. Treat only those seeds needed for immediate use. Do not use treated seed for food, feed, or oil purposes. Do not store excess treated seeds beyond planting time. Do not graze or feed livestock on treated areas for 6 weeks after planting.

Signal Word/Toxicity Class: Danger/I.
REI: 12 hr.

Lorsban — see chlorpyrifos

MALATHION

Broad-spectrum organophosphate insecticide which was one of the first organophosphate insecticides developed. Has relatively low toxicity to mammals, highly toxic to fish and bees.

Drexel Malathion 5EC *(Drexel)*
Fyfanon (5EC) *(Helena)*
◆ **Micro Flo Malathion 5EC** *(Micro Flo)*
Prentox 5 lb. Malathion Spray (5EC) *(Prentiss)*

Rate: 1 lb. ai/A
1.5 pt. 5EC/A

Pest:

aphids, corn earworm, corn rootworm (adult), grasshoppers, sap beetle, thrips

Rate: 1-1.25 lb. ai/A
1.5-2 pt. 5EC/A

Pest:

armyworm (except Fyfanon)

Remarks: **Field corn (forage, grain).** For earworm and sap beetle, apply when 10% of ears show silk, repeat 4-5 times at 3- to 5-day intervals. Injury may occur in whorl and silk stages.

Days To Harvest: Harvest/forage-5.

Signal Word/Toxicity Class: Warning/II (Caution/III for Fyfanon).
REI: 12 hr.

◆ **Drexel Malathion ULV** *(Drexel)*
Fyfanon ULV *(Cheminova)*
◆ **Micro Flo Malathion ULV** *(Micro Flo)*

Rate: 4 fl. oz. ULV/A

Pest:

corn rootworm (adult)

Remarks: Apply low rate for corn rootworm; high rate when grasshoppers are present at an economic threshold.

Days To Harvest: 5.

Signal Word/Toxicity Class: Caution.
REI: 12 hr.

Fyfanon (8E) *(Helena)*
◆ **Gowan Malathion 8 (8E)** *(Gowan*

Rate: 1.25 lb. ai/A
1.25 pt. 8E/A

Pest:

aphids, corn rootworm (adult), grasshoppers, sap beetle, thrips

Remarks: Injury may occur in whorl and silk stages.

Days To Harvest: 5.

Signal Word/Toxicity Class: Caution/III.
REI: 12 hr.

NOTICE	The information on these pages is for preliminary planning — not a guide for use. Be sure to follow the manufacturer's directions, notwithstanding information contained here.	For personal protective equipment and EPA registration numbers, please turn to page 39.

METHOXYCHLOR

Organochlorine insecticide that acts as a stomach and contact poison on a wide range of pests. Primarily used for flea beetle control and sometimes other foliar pests such as armyworm and Japanese beetle. Low toxicity to mammals and birds. Toxic to bees, fish, and various aquatic organisms.

Prentox Methoxychlor 2EC *(Prentiss)*

Rate: 1.3-1.7 lb. ai/A
75-100 fl. oz. 2EC/A

Pest:
corn earworm, fall armyworm, flea beetles, imported cabbageworm, Japanese beetle

Remarks: Apply at first sign of infestations; repeat at 7- to 14-day intervals as needed.
Days To Harvest: 7.

Signal Word/Toxicity Class: Caution.
REI: 12 hr.

METHYL PARATHION

Highly toxic organophosphate insecticide/acaricide with both contact and stomach poison activity. Penncap-M is a microencapsulated formulation. Restricted Use Pesticide due to residual effects to avian species and hazard to bees.

Cheminova Methyl 4EC *(Cheminova)*
◆ Declare (4EC) *(Griffin)*
Helena 4 lb. Methyl Parathion (4EC) *(Helena)*

Rate: 0.25 lb. ai/A
0.5 pt. 4EC/A

Pest:
aphids, armyworms, corn leaf aphid, corn rootworm, fall armyworm, flea beetles, mites, stink bugs

Rate: 0.5 lb. ai/A
1 pt. 4EC/A

Pest:
grasshoppers

Remarks: Apply when insects appear; repeat as needed. Do not apply during pollen shed.
Days To Harvest: 12.

Signal Word/Toxicity Class: Danger-Poison/I.
REI: 48 hr.

Penncap-M (2FL) *(Elf Atochem)*

Rate: 0.25-0.5 lb. ai/A
1-2 pt. 2FL/A

Pest:
corn rootworm (adult)

Rate: 0.25-0.75 lb. ai/A
1-3 pt. 2FL/A

Pest:
stink bugs*

Rate: 0.5-0.75 lb. ai/A
2-3 pt. 2FL/A

Pest:
aphids, flea beetles, grasshoppers, true armyworm*

Rate: 0.5-1 lb. ai/A
2-4 pt. 2FL/A

Pest:
corn earworm, European corn borer, Japanese beetle*, silk fly*, southwestern corn borer, western bean cutworm

Rate: 1 lb. ai/A
4 pt./ 2FLA

Pest:
black cutworm

**Except California*

Remarks: Do not apply during pollen shed if bees are foraging in areas to be treated. For black cutworm, apply under moist soil conditions. For grasshoppers, use high rate when weather is cool and/or hoppers are large. Do not apply chemigate in California. Refer to label for additional use information.
Days To Harvest: 12.

Signal Word/Toxicity Class: Warning/II.
REI: 48 hr.

Microthiol Special — see sulfur

MOCAP

Organophosphate insecticide/nematicide with broad-spectrum soil pest activity. Extremely toxic to mammals, birds, and aquatic organisms. Mocap EC is a Restricted Use Pesticide.

Mocap EC or 10%G (ethoprop) *(Rhone-Poulenc)*

Rate: Broadcast:
3 lb. ai/A
2 qt. EC/A
30 lb. 10%G/A

Pest:
cutworms

Rate: Broadcast:
6 lb. ai/A
4 qt. EC/A
60 lb. 10%G/A
Row:
1.5-2 lb. ai/A
1-1.33 qt. EC/A
15-20 lb. 10%G/A

Pest:
nematodes

Rate: Row treatment:
1 lb. ai/A
0.66 qt. EC/A (1.66 oz./1000 ft. row)
10 lb. 10%G/A (12 oz./1000 ft. row)

Pest:
corn rootworm, garden symphylans, white grub (suppression)

Remarks: Broadcast: Apply from 3 days before planting to at-planting; mix with 2-4" soil (2" for cutworms) right after application. Row treatment: Apply at planting in 6-7" band on row (40" spacing) over closed seed furrow; mix with top 1/2" soil.
State Restrictions: Do not apply in Long Island, New York.

Signal Word/Toxicity Class: Danger-Poison/I (EC); Warning/II (10%G).
REI: 48 hr.; 72 hr. if average rainfall less than 25" per year.

Nufos — see chlorpyrifos

Omite — see propargite

Penncap-M — see methyl parathion

PERMETHRIN

Broad-spectrum synthetic pyrethroid insecticide. Readily breaks down in most soils except organic types. Has low toxicity to mammals and birds, but is toxic to fish and very toxic to bees. Restricted Use Pesticide due to possible adverse effects on aquatic organisms.

Ambush 2E, 25WP or 25WSP *(ZENECA)*

Rate: 0.1-0.2 lb. ai/A
6.4-12.8 fl. oz. 2EC/A
6.4-12.8 oz. 25WP, 25WSP/A

Pest:
armyworms, corn earworm*, corn rootworm*, cutworms, European corn borer, flea beetles, southwestern corn borer, stalk borer, tobacco budworm*

**Except California*

Remarks: Field corn (including grown for seed), popcorn. For preemergent application on armyworm and cutworm, apply from 5 days prior to planting to crop emergence. For foliar use, apply prior to brown silk stage. For stalk borer, apply when, or shortly before stalk borer larvae are moving into corn from surrounding weeds and grasses.

Signal Word/Toxicity Class: Warning/II.
REI: 24 hr.

◆ - **new product** • ai-active ingredient • DF-dry flowable • E/EC-emulsifiable concentrate • EW-emulsion in water
F/FL-flowable • G-granule • L/LC-liquid • SP/WSP-soluble packet • W/WP-wettable powder • WSB-water soluble bag

Pounce 3.2EC, 1.5G, 25WP or WSB *(FMC)*
Rate: 0.1-0.2 lb. ai/A
4-8 oz. 3.2EC/A (0.3-0.6 oz./1000 lin. ft. row)
6.7-13.3 lb. 1.5G/A (8-16 oz./1000 lin. ft. row)
6.4-12.8 oz. 25WP, WSB/A (0.5-1 oz./1000 lin. ft. row)
Pest: At planting:
armyworms, cutworms
Preemergence:
armyworms, cutworms, stalk borers
Foliar:
armyworms, corn earworm, corn rootworm beetle, cutworms, European corn borer, flea beetles, hop vine borer, southwestern corn borer, stalk borer

Rate: 0.05-0.1 lb. ai/A
2-4 oz. 3.2EC/A
3.2-6.4 oz. 25WP, WSB/A
Pest:
western bean cutworm
Remarks: Field corn (including seed), popcorn. Do not apply less than 6 days apart. At planting, apply in-furrow, band, or T-band using minimum 4" band. For preemergence application, apply 14 days prior to planting up to crop emergence. For foliar, apply when eggs begin to hatch. For stalk borer, apply when or shortly before larvae are moving into corn from surrounding weeds and grasses; mowing or burndown herbicide suggested to initiate movement.
Days To Harvest: Grain, fodder-30; forage-0.

Signal Word/Toxicity Class: Caution (3.2EC, 1.5G); Warning/II (25WP, WSB).
REI: 24 hr.

PERMETHRIN (Premix)

Seed treatment.

Kernel Guard Supreme (10.42% permethrin & 14% carboxin) *(Trace)*
Rate: 2 oz./bu. seed
Pest:
seedcorn beetles, seedcorn maggots, wireworms
Remarks: Field corn, popcorn. Refer to label for use directions.

Signal Word/Toxicity Class: Caution.
REI: 12 hr. with exceptions.

PETROLEUM OIL

Petroleum based spray oil. Mode of action is through suffocation of eggs, larvae, and nymphs of insects and adults of soft bodied insects. This mode of action necessitates total spray coverage. Toxic to fish.

◆ **435 or 455 Oil 98.8** *(Drexel)*
◆ **SunSpray 6E or 7E** *(Sun)*
Rate: 2 gal./100 gal. water
Pest:
corn rootworm, earworms, fall armyworm
Remarks: Field corn. Do not apply in combination with or immediately preceding or following applications of captan, carbaryl, Dyrene, folpet, Morestan, propargite, and sulfur compounds.

Signal Word/Toxicity Class: Caution.
REI: 4 hr.

SAF-T-SIDE *(Brandt Consolidated)*
◆ **Sumulsion** *(Drexel)*
Rate: 1-2 gal./100 gal. water
Pest:
aphids, leafminers, mites, whiteflies
Remarks: Field corn (including seed), popcorn. Apply as needed. Compatible with most insecticides and fungicides. Do not use with carbaryl, dimethoate, or any product whose label recommends the use of no oils. Do not use in combination with or immediately before or after spraying with fungicides such as captan, folpet, Morestan or any sulfur products.

Signal Word/Toxicity Class: Caution.
REI: 4 hr.

◆ **SunSpray Ultra-Fine** *(Sun)*
Rate: 2 gal./100 gal. water
Pest:
aphids, armyworms, corn earworm, leafminers, mites, rootworms, whiteflies
Remarks: Field corn. Apply at 2-week intervals. Do not exceed 4 applications per season.

Signal Word/Toxicity Class: Caution.
REI: 4 hr.

PHORATE

Organophosphate insecticide that controls pests by contact and systemic activity. May be applied at planting time to control soil pests. Restricted Use Pesticide due to acute oral and dermal toxicity and aquatic and avian hazards.

Clean Crop Phorate 20G *(Platte)*
Riverside Phorate 20G *(Terra)*
Rate: 6 oz. 20G/1000 ft. row (min. 30" spacing)
Pest: At planting:
corn rootworm, mites, seedcorn beetles (IA, IL), wireworms (North Central, Southeast)
At cultivation:
corn rootworm

Rate: 1 lb. ai/A
5 lb. 20G/A
Pest: Application over plant:
corn rootworm, European corn borer (1st brood)
Remarks: At-planting application: Place granules in 7" band over row; 1 additional application can be applied at cultivation. Do not place in direct contact with seed. Cultivation application: Apply granules to base of plants just ahead of cultivator shovels. Do not apply after cultivation. For application over plant, apply granules into whorl of plant by ground or air. Do not make more than 1 application over plant.
Days To Harvest: Harvest/grazing/cutting for forage-30.

Signal Word/Toxicity Class: Danger-Poison/I.
REI: 48 hr.

Thimet 20-G *(American Cyanamid)*
Rate: 6 oz. 20-G/1000 ft. row (min. 30" spacing)
Pest: At planting:
corn rootworm, flea beetles, mites, seedcorn beetles, seedcorn maggots, white grubs, wireworm
At cultivation:
chinch bug (nymphs) (AR, CO, KS, LA, MS, NE, OK, TN, TX), corn rootworm
Remarks: Not for use on popcorn. At-planting: Place granules in 7" band over row; 1 additional application can be applied at cultivation. Do not use in-furrow. Cultivation application: Apply granules to base of plants just ahead of cultivator shovels. Do not apply after cultivation. For chinch bug, apply as soon as nymphs are observed at base of plants. Do not make more than 1 cultivation treatment per season.
Days To Harvest: Harvest/grazing/cutting for forage-30.

Signal Word/Toxicity Class: Danger-Poison/I.
REI: 48 hr.

Pounce — see permethrin

PROPARGITE

Organic cyclo compound used as an acaricide that controls mites by contact activity. Nonsystemic, but has moderately long residual activity. Used for control of twospotted spider mite and Banks grass mite. Nontoxic to bees and allows survival of beneficial predatory mites. Toxic to fish and shrimp.

Comite or Comite II *(Uniroyal)*
Rate: 1.64-2.46 lb. ai/A
2-3 pt./A
1.845 lb. ai/A
2.25 pt. II/A
Pest:
Banks grass mite, twospotted spider mite

NOTICE The information on these pages is for preliminary planning — not a guide for use. Be sure to follow the manufacturer's directions, notwithstanding information contained here. For personal protective equipment and EPA registration numbers, please turn to page 39.

Remarks: Apply early to building populations, usually prior to tasseling. Do not make more than 1 application per year. Do not apply through any type of irrigation system.
Days To Harvest: 30.

Signal Word/Toxicity Class: Danger-Eye/I.
REI: 48 hr.

Omite-6E *(Uniroyal)*
Rate: 1.5 lb. ai/A
2 pt. 6E/A
Pest:
twospotted spider mite
Remarks: Apply in 20-50 gal. spray solution by ground; minimum 10 gal. spray solution/A by air. Do not make more than 1 application per year. Do not apply through any type of irrigation system.
Days To Harvest: 56.
State Restrictions: For use in California.

Signal Word/Toxicity Class: Warning/II.
REI: 48 hr.

PYRELLIN (Premix)

Premix containing pyrethrins and rotenone. May be combined with other insecticides to enhance knockdown or as an insect "exciter" to help facilitate contact with other spray residues.

Pyrellin E.C. (0.042 lb./gal. pyrethrins & 0.035 lb./gal. rotenone) *(Webb Wright)*
Rate: 1-2 pt. E.C./A
Pest:
aphids, armyworms, chinch bug, mites, thrips
Remarks: Apply at 7-day intervals; repeat as needed.
Days To Harvest: 0.

Signal Word/Toxicity Class: Caution/III.
REI: 12 hr.

Pyrenone — see pyrethrins

PYRETHRINS

Contains piperonyl butoxide as a synergist. Diatect also contains silicon dioxide (diatomaceous earth). May be combined in tank mix with other insecticides for quicker and more complete kill where insect resistance may be a problem, and as an "exciter" to flush insects out of hiding and into contact with other spray residues.

Diatect Multipurpose Insecticide II *(Diatect)*
Pyrenone (0.5EC) *(AgrEvo Environmental Health)*
Rate: 1-6 lb. II/A
12 oz. 0.5EC/A
Pest: Including but not limited to:
aphids, armyworms, caterpillars, chinch bug, corn earworm, flea beetles, mites, thrips
Remarks: Field corn, popcorn. Apply in sufficient water for thorough coverage; repeat as needed. Apply Diatect as dust or wettable powder.
Days To Harvest: 0.

Signal Word/Toxicity Class: Caution.
REI: 12 hr.

Redball Lightning — see sulfur

REGENT

Broad-spectrum phenylpyrazol insecticide formulated as a wettable granule. Regent interferes with the GABA receptor in the central nervous system to kill insects. The contact and systemic activity controls corn rootworm while the systemic activity helps controls 1st generation European corn borer. Must be used with the 1-PASS System. Not for use on popcorn. Restricted Use Pesticide.

◆ **Regent (80WG) (fipronil)** *(Rhone-Poulenc)*
Rate: 0.13 lb. ai/A
2.6 oz. 80WG/A (30" row spacing)
Pest:
chinch bug, common stalk borer, corn rootworm, European corn borer (1st generation), grubs, seedcorn beetles, seedcorn maggots, wireworms
Remarks: Field corn (including seed). Apply single application as a liquid in-furrow at-planting.
Days To Harvest: 90.
Crop Rotation: Leafy vegetables-1 month; root crops-5 months; small grains and other crops-12 months following application.

Signal Word/Toxicity Class: Warning/II.
REI: 24 hr.

SAF-T-SIDE — see petroleum oil

Sevin — see carbaryl

Slam — see carbaryl

Sorghum Guard — see lindane premixes

SULFUR

Stirrup Mylox combines flowable sulfur with pheromones. Microthiol and Thiolux are micronized sulfurs. Redball Lightning is a dustless granule formulated to breakdown into smaller particles when wetted by water to provide effectiveness of a powder with the handling ease of a granule.

◆ **Kumulus DF** *(BASF)*
Rate: 4.8-12 lb. ai/A
6-15 lb. DF/A
Pest:
Banks grass mite, Pacific mite, red spider mite, twospotted spider mite
Remarks: Do not apply if temperatures during or within 3 days after application are expected to exceed 90°F. Do not apply through any type of irrigation system.

Signal Word/Toxicity Class: Caution.
REI: 24 hr.

Microthiol Special (80W) *(Elf Atochem)*
Rate: 6.4-8 lb. ai/A
8-10 lb. 80W/A
Pest:
Atlantic mite, Pacific mite, red spider mite, twospotted spider mite (except CA)
Remarks: Apply when mites first appear; repeat as needed. Do not chemigate in California.

Signal Word/Toxicity Class: Caution.
REI: 24 hr.

◆ **Redball Lightning Degradable (G)** *(International Sulphur)*
Rate: 5-15 lb. G/A
Pest:
fleahoppers, mites, thrips
Remarks: Do not apply within 3 weeks of oil spray.
Days To Harvest: 0.

Signal Word/Toxicity Class: Caution.
REI: 24 hr.

THAT (6FL) or THAT Big 8 (8FL) *(Stoller)*
Rate: 9 lb. ai/A
12 pt. 6FL/A
8 lb. ai/A
8 pt. 8FL/A
Pest:
Banks grass mite
Remarks: Apply when mites appear on bottom 2 leaves. Can be tank mixed with other miticides.
State Restrictions: Not for use in California.

Signal Word/Toxicity Class: Caution.
REI: 12 hr.

◆ **- new product** • ai-active ingredient • DF-dry flowable • E/EC-emulsifiable concentrate • EW-emulsion in water
F/FL-flowable • G-granule • L/LC-liquid • SP/WSP-soluble packet • W/WP-wettable powder • WSB-water soluble bag

Thiolux (DF) *(Platte)*
Rate: 4.8 lb. ai/A
6 lb. DF/A

Pest: Suppression:
Banks grass mite
Pacific spider mite
red spider mite
twospotted spider mite

Remarks: Apply when mites first appear; repeat as necessary.

Signal Word/Toxicity Class: Caution.
REI: 24 hr.

Sumulsion — see petroleum oil

SunSpray — see petroleum oil

THAT or THAT Big 8 — see sulfur

Thimet — see phorate

Thiolux — see sulfur

Warrior — see *lambda*-cyhalothrin

Fungicides

Allegiance — see metalaxyl

Apron — see metalaxyl

Apron XL — see mefenoxam

BAYTAN 30

Triazole systemic fungicide for use only through commercial seed treating equipment which are closed application systems.

Baytan 30 (FL) (triadimenol) *(Gustafson)*
Rate: 3 fl. oz. FL/100 lb. seed
Disease:
head smut
Time: Seed treatment.
Remarks: Use with EPA-approved dye that imparts unnatural color to seed. Do not use for feed, food, or oil purposes. Green forage may be grazed 45 days after seeding. Refer to label for use directions.

Signal Word/Toxicity Class: Caution.

Bravo — see chlorothalonil

CAPTAN

Gustafson Captan 30-DD contains the toxic inert Rhodamine B.

Drexel Captan 4FL *(Drexel)*
Gustafson Captan 30-DD or 400 *(Gustafson)*
Rate: 1.5 fl. oz. 30-DD/100 lb. seed
1.25-2.375 fl. oz. 400/100 lb. seed
1.75 fl. oz. 4FL/100 lb. seed
Disease:
damping-off
mold
seed decay
seedling blight
seedling disease
Time: Liquid or slurry seed treatment.
Remarks: Use with EPA-approved dye that imparts unnatural color to seed. Do not use treated seed for food, feed, or oil purposes. Refer to label for use directions.

Signal Word/Toxicity Class: Danger/I.

Drexel Captan 50/10 or Captan-75 *(Drexel)*
Rate: 2.4 oz. 50/10/100 lb. seed
1.66 oz. 75/100 lb. seed
Planter box:
1.33 oz. 50/10/bu. seed
1 oz. 75/bu. seed
Disease:
damping-off
seed decay
seedling blight
seedling disease
soilborne disease
Time: Seed treatment.
Remarks: Do not use treated seed for food, feed, or oil purposes. Refer to label for use directions.

Signal Word/Toxicity Class: Danger/I.
REI: 4 days (Captan-75); 48 hr. (Captan 50/10).

CAPTAN (Premixes)

Premix combining a fungicide with an insecticide. Contains graphite. Toxic to fish, birds, and wildlife.

Captan-Diazinon (36.67% captan & 25% diazinon) *(Trace)*
Nu-Gro Isotox (12.2% captan & 25% lindane) *(Wilbur-Ellis)*
Seed Mate Isotox (12.2% captan & 25% lindane) *(Loveland)*
Sorghum Guard (32.75% captan & 16.6% lindane) *(Trace)*
Rate: 2 oz./bu. seed (Captan-Diazinon)
4-8 oz./100 lb. seed (Isotox)
4 oz./50 lb. seed (Sorghum Guard)
Disease:
damping-off
seed decay
seedling blight
seedling disease
Time: Planter box seed treatment.
Remarks: Do not mix with bare hands. Do not use treated seed for food, feed, or oil purposes. Do not store excess treated seeds beyond planting time. Refer to label for use directions.

Signal Word/Toxicity Class: Danger/I.

CARBOXIN

Ready-to-use seed treatment applied as slurry treatment by machine. Contains toxic inert Rhodamine B. Toxic to fish.

Vitavax-34 *(Gustafson)*
Rate: 2-4 fl. oz./100 lb. seed
Disease:
Rhizoctonia solani
seedborne disease
Sphacelotheca reiliana
Time: Slurry seed treatment.
Remarks: Incorporation of dye into treating slurry is recommended to distinguish seed as treated. Do not graze or feed livestock on treated areas for six weeks after planting. Do not apply this product through any type of irrigation system. Refer to label for use directions.

Signal Word/Toxicity Class: Caution.

NOTICE	The information on these pages is for preliminary planning — not a guide for use. Be sure to follow the manufacturer's directions, notwithstanding information contained here.	For personal protective equipment and EPA registration numbers, please turn to page 39.

CARBOXIN (Premixes)

Germate Plus contains graphite. Prevail is a premix of the active ingredients of Apron, Terraclor, and Vitavax. Kernel Guard Supreme contains the synthetic pyrethroid insecticide permethrin. Vitaflo-280 is a combination of a systemic fungicide (Vitavax) and a contact fungicide (thiram). Toxic to fish. Germate Plus is also toxic to birds and other wildlife.

Germate Plus (14% carboxin & 15% diazinon & 25% lindane) *(Trace)*
Prevail (15% carboxin & 15% PCNB & 3.12% metalaxyl) *(Trace)*

Rate: 2 oz./bu. seed (Germate Plus)
3 oz./bu. seed (Prevail)

Disease:
seedling diseases

Time: Planter box seed treatment.

Remarks: Treat only those seeds needed for immediate use. Do not use treated seed for food, feed, or oil purposes. For Prevail, do not graze or feed livestock on treated areas for 6 weeks after planting. Refer to label for use directions.

Signal Word/Toxicity Class: Caution.
REI: 12 hr.

Kernel Guard Supreme (14% carboxin & 10.42% permethrin) *(Trace)*

Rate: 2 oz./bu. seed

Disease:
damping-off, seedling blight, seedling disease, seed decay

Time: Seed treatment.

Remarks: Field corn, popcorn. Treat only those seeds needed for immediate use. Do not use treated seed for food, feed, or oil purposes. Do not store excess treated seeds beyond planting time. Refer to label for use directions.

Signal Word/Toxicity Class: Caution.
REI: 12 hr.

Vitaflo-280 (14.9% carboxin & 13.2% thiram) *(Gustafson)*

Rate: 4.5 fl. oz./100 lb. seed
Corn head smut:
8.5-11.5 fl. oz./100 lb. seed

Disease:
corn head smut, damping-off, seed decay

Time: Seed treatment.

Remarks: Field corn, popcorn. Thorough seed coverage is required. Use higher rate under conditions of heavy, expected disease pressure. Avoid planting corn under cold, wet conditions. Stress conditions during first few weeks after planting may increase incidence of head smut. Will not control soilborne head smut. Do not use treated seed for food, feed or oil purposes. Do not graze or feed livestock on treated areas for 6 weeks after planting. Refer to label for use directions.

Signal Word/Toxicity Class: Caution.

CHLOROTHALONIL

Broad-spectrum contact fungicide. Toxic to fish, aquatic invertebrates and marine/estuarine organisms.

Bravo 500, 720/Weather Stik, Ultrex (WDG) *(ZENECA)*

Rate: 0.56-1.4 lb. ai/A
1.125-2.75 pt. 500/A
0.56-1.5 lb. ai/A
0.75-2 pt. 720/A
0.7-1.8 lb. WDG/A

Disease:
Helminthosporium leaf blight, rust

Time: Apply when conditions favor disease development; repeat at 4- to 7-day intervals or as needed.

Remarks: Corn grown for seed, popcorn. Apply in sufficient water to obtain adequate coverage. Do not allow livestock to graze treated fields. Do not ensile treated corn or use as livestock forage. Refer to label for use directions.

Days To Harvest: 14.

Signal Word/Toxicity Class: Danger/I (WDG); Warning/II (500, 720).
REI: 48 hr.

DB-Green — see maneb premixes

DB-Green + Vitavax — see maneb premixes

Dithane — see mancozeb

Enhance Plus — see maneb premixes

FLUDIOXONIL

Contact fungicide for control of seedborne and soilborne fungi. Toxic to fish and aquatic invertebrates.

Maxim 4FS *(Novartis)*

Rate: 0.08-0.16 fl. oz. 4FS/100 lb. seed

Disease:
damping-off, seedborne disease, seedling disease, seed decay, seedling blight, soilborne disease

Time: Seed treatment.

Remarks: Under severe disease conditions, use high rate. May be tank mixed. Do not use treated seed for food, feed, or oil purposes. Do not graze on forage until 30 days after planting. Use EPA-approved dye or colorant that imparts unnatural color to seed.

Signal Word/Toxicity Class: Caution.

Germate Plus — see carboxin premixes

GRAIN GUARD PLUS (Premix)

Contains graphite. Toxic to fish, birds, and other wildlife.

Grain Guard Plus (50% mancozeb & 18.75% lindane) *(Trace)*

Rate: 3 oz./bu. seed

Disease:
damping-off, seed decay

Time: Drill box or planter box seed treatment.

Remarks: Treat only those seeds needed for immediate use. Do not use treated seed for food, feed, or oil purposes. Do not store excess treated seeds beyond planting time. Refer to label for use directions.

Signal Word/Toxicity Class: Warning/II.

Kernel Guard Supreme — see carboxin premixes

Kumulus — see sulfur

MANCOZEB

Dithiocarbamate fungicide with broad-spectrum activity with no resistance problems. Coordination product of zinc ion and manganese ethylene bisdithiocarbamate. Toxic to fish.

Dithane (DF, F-45, M-45, ST, WSP) *(Rohm and Haas)*
Manzate 200DF *(Griffin)*
Penncozeb 75DF or 80WP *(Elf Atochem)*

Rate: 1.12 lb. ai/A
1.5 lb. DF, 200DF/A
1.2 lb. ai/A
1.5 lb. M-45, WSP/A
2.4 pt. F-45/A
0.75-1.12 lb. ai/A
1-1.5 lb. 75DF/A
0.8-1.2 lb. ai/A
1-1.5 lb. 80WP/A

Disease:
common rust, gray leaf spot (Manzate, Penncozeb), Helminthosporium leaf blight

Time: Apply when disease threatens; repeat at 4- to 14-day intervals.

Remarks: Field corn, seed corn, popcorn. Addition of surfactant will improve performance. Do not graze or feed forage or fodder to livestock. Refer to label for use directions.

Days To Harvest: Popcorn-7; field corn, seed corn-40.

State Restrictions: Manzate and Penncozeb: For use east of the Mississippi (including Arkansas and Louisiana); west of the Mississippi (except Arkansas and Louisiana). Refer to label for California precautions.

◆ - **new product** • ai-active ingredient • DF-dry flowable • E/EC-emulsifiable concentrate • EW-emulsion in water
F/FL-flowable • G-granule • L/LC-liquid • SP/WSP-soluble packet • W/WP-wettable powder • WSB-water soluble bag

Rate: 2.9-5.8 oz. DF, 75DF/100 lb. seed
2.7-5.4 oz. M-45, WSP, 200DF, 80WP/100 lb. seed
4.3-8.6 fl. oz. F-45, ST/100 lb. seed
Disease:
damping-off　seed rot　seedling disease
Time: Seed treatment.
Remarks: Incorporation of dye into treating slurry is recommended to distinguish seed as treated. Do not use treated seed for food, feed, or oil purposes. Refer to label for use directions.
State Restrictions: Refer to label for California precautions.

Signal Word/Toxicity Class: Caution.
REI: 24 hr.

MANEB

Dithiocarbamate contact fungicide. Toxic to fish.

Maneb 75DF or 80WP *(Elf Atochem)*
Manex (FL) *(Griffin)*
Rate: 1.12 lb. ai/A
1.5 lb. 75DF/A
1.2 lb. ai/A
1.2 qt. FL/A
1.5 lb. 80WP/A
Disease:
common rust　Helminthosporium leaf blight
Time: Begin when disease threatens; repeat at 3- to 10-day intervals.
Remarks: Popcorn. Use sufficient water for thorough coverage. Add surfactant or spreader/sticker if needed for better coverage. Do not feed treated forage to livestock. Refer to label for use directions.
Days To Harvest: 7.
State Restrictions: For use east of the Mississippi (including Arkansas, Louisiana), and west of the Mississippi (except Arkansas, Louisiana).

Signal Word/Toxicity Class: Caution.
REI: 24 hr.

MANEB (Premixes)

Dual purpose ready-to-use fungicide and insecticide. DB-Green is a dust-based formulation seed treatment. DB-Green L is a liquid formulation seed treatment. Seed Mate contains graphite. Toxic to fish, birds, and other wildlife.

DB-Green (dust) (50% maneb & 18.75% lindane) *(AGSCO)*
DB-Green L (liquid) (25.6% maneb & 8.6% lindane) *(AGSCO)*
Rate: 2.75 oz. dust/bu. seed
4 fl. oz. L/bu. seed
Disease:
seedling diseases　soilborne diseases
Time: Seed treatment.
Remarks: Drill or planter box application. Treat only seeds needed for immediate use, minimizing interval between treatment and planting. Do not store excess treated seeds beyond planting time. Treated seed must not be used for or mixed with food of animal feed, or processed for oil. Care must be exercised in handling treated seed. Do not graze or feed livestock on treated areas for 6 weeks after planting.

Signal Word/Toxicity Class: Warning/II.
REI: 12 hr.

DB-Green + Vitavax (35% maneb & 20% carboxin & 18.75% lindane) *(AGSCO)*
Enhance Plus (35% maneb & 20% carboxin & 18.75% lindane) *(Trace)*
Rate: 3 oz./bu. seed
Disease:
seed decay　seedling diseases　soilborne diseases
Time: Seed treatment.
Remarks: Drill box application. Treat only seeds needed for immediate use, minimizing interval between treatment and planting. Do not store excess treated seeds beyond planting time. Treated seed must not be used for or mixed with food or animal feed, or processed for oil. Care must be exercised in handling treated seed. Do not graze or feed livestock on treated areas for 6 weeks after planting.

Signal Word/Toxicity Class: Warning/II.
REI: 12 hr.

Seed Mate Maneb-Lindane (50% maneb & 18.75% lindane) *(Loveland)*
Rate: 3 oz./bu. seed
Disease:
damping-off　seed decay
Time: Apply as drill box or planter box treatment.
Remarks: Treat only seeds needed for immediate use, minimizing interval between treatment and planting. Do not store excess treated seeds beyond planting time. Do not use treated seed for food, feed, or oil purposes. Refer to label for use directions.

Signal Word/Toxicity Class: Caution.
REI: 24 hr.

Manex — see maneb

Manzate — see mancozeb

Maxim — see fludioxonil

MEFENOXAM

Systemic fungicide having a specific mode of action.

Apron XL LS (mefenoxam) *(Novartis)*
Rate: 0.32-0.64 fl. oz. LS/100 lb. seed
Disease:
damping-off
Time: Seed treatment.
Remarks: Do not use treated seed for food, feed, or oil purposes.

Signal Word/Toxicity Class: Warning/II.

METALAXYL

Systemic fungicide having a specific mode of action.

Allegiance or Apron (FL) *(Gustafson)*
Rate: 0.375-0.75 fl. oz. Allegiance/100 lb. seed
0.75-1.5 fl. oz. Apron/100 lb. seed
Disease:
damping-off　Phytophthora (early-season)
Time: Seed treatment.
Remarks: Field corn, popcorn. Use with EPA-approved dye that imparts unnatural color to seed. Do not use for feed, food, or oil purposes. Refer to label for use directions.

Signal Word/Toxicity Class: Warning/II.

Apron Dry *(Trace)*
Rate: 3-4 oz./100 lb. seed
Disease:
damping-off
Time: Seed treatment.
Remarks: Treat only those seeds needed for immediate use. Do not use treated seed for food, feed, or oil purposes. Do not store excess treated seeds beyond planting time. Refer to label for use directions.

Signal Word/Toxicity Class: Caution.
REI: 12 hr.

Micro Flo — see sulfur

Nu-Gro Isotox — see captan premixes

PCNB

RTU-PCNB contains toxic inert Rhodamine B. Toxic to fish and aquatic organisms.

PCNB 2-E or 2LF *(Amvac)*
RTU-PCNB *(Gustafson)*
Rate: 3.35 fl. oz. 2-E/100 lb. seed
4 fl. oz. 2LF/100 lb. seed
3 fl. oz. RTU/100 lb. seed
Disease:
covered smut　Fusarium　Rhizoctonia
Time: Seed treatment.

NOTICE	The information on these pages is for preliminary planning — not a guide for use. Be sure to follow the manufacturer's directions, notwithstanding information contained here.	For personal protective equipment and EPA registration numbers, please turn to page 39.

Remarks: Do not use treated seed for food, feed, or oil purposes. Refer to label for use directions.
State Restriction: PCNB 2-E and RTU-PCNB not for use in California.

Signal Word/Toxicity Class: Danger/I (2-E); Caution.
REI: 12 hr. (PCNB 2-E, 2LF).

Penncozeb — see mancozeb

Prevail — see carboxin premixes

PROPICONAZOLE

Triazole sterol inhibitor. Toxic to fish.

Tilt (EC) *(Novartis)*
Rate: 2-4 fl. oz. EC/A
Disease:
eyespot; gray leaf spot; Helminthosporium leaf blight; rust
Time: Apply when disease first appears. For leaf blight and rust, apply at 7- to 14-day intervals. For eyespot and gray leaf spot, apply at 14-day intervals.
Remarks: Field corn, field corn for seed, popcorn. Do not apply to field corn and field corn for seed after silking. Do not apply more than 16 fl. oz./crop per season. Do not graze or feed livestock treated forage or cut green crop for hay or silage.
Days To Harvest: 30.

Signal Word/Toxicity Class: Warning/II.
REI: 24 hr.

Redball — see sulfur

Seed Mate Isotox — see captan premixes

Seed Mate Maneb-Lindane — see maneb premixes

Sorghum Guard — see captan premixes

SULFUR

Nonsystemic fungicide with protective action. Microthiol Special, Sulfa 80, and Thiolux are micronized sulfurs.

Drexel Suffa (6FL), Sulfa 80 (80DF) or Sulfur (90W) *(Drexel)*
Micro Flo Sulfur (6L) *(Micro Flo)*
Terra Sulfur (6L) *(Terra)*
Rate: 4-6 pt. 6FL/A
7.5-12.5 lb. 80DF/A
3-5 lb. 90W/A
5.3-20 pt. 6L/A
Disease:
powdery mildew
Time: Apply when disease appears; repeat as necessary.
Remarks: Do not use within 2 weeks of oil application (4 weeks 6FL). Some plants are susceptible to injury from sulfur under certain climatic conditions. Do not make sulfur applications at excessively high temperatures. Do not apply 6FL, 80DF, or 90W through any type of irrigation system. Refer to label for use directions.

Signal Word/Toxicity Class: Caution.
REI: 24 hr.

Kumulus DF (sulfur) *(BASF)*
Rate: 4.8-12 lb. ai/A
6-15 lb. DF/A
Disease:
powdery mildew
Time: Apply when disease first appears and repeat as necessary.
Remarks: Do not apply if temperatures during or within 3 days after application are expected to exceed 90°F. Do not apply through any type of irrigation system. Refer to label for use directions.

Signal Word/Toxicity Class: Caution.
REI: 24 hr.

Redball EM-53 (6L), Superfine Wettable Sulphur (WS) or Commercial Dusting Sulphur (DS) *(International Sulphur)*
Rate: 4-20 pt. 6L/A
5-10 lb. WS/100 gal. water
5-15 lb. DS/A
Disease:
powdery mildew
Time: Apply when disease first appears; repeat as necessary.
Remarks: Apply in sufficient water for thorough coverage. Do not apply when temperatures exceed 90°F. Refer to label for use directions.

Signal Word/Toxicity Class: Caution.
REI: 12 hr.

SULFUR (Premix)

Fungicide/bactericide in flowable formulation. Toxic to fish and aquatic organisms.

Top Cop with Sulfur (6.25 lb./gal. sulfur & 1 lb./gal. tribasic copper sulfate) *(Stoller)*
Rate: 2 qt./A
Disease:
Helminthosporium leaf blight
Time: Apply when disease threatens; repeat at 4- to 7-day intervals.
Remarks: Refer to label for use directions.

Signal Word/Toxicity Class: Caution.
REI: 24 hr.

SYSTEM[3] (Premix)

Biological and chemical fungicide containing the active ingredients of Apron, Terraclor, and Kodiak. Contains graphite. Toxic to fish and aquatic organisms.

System[3] (0.1% *Bacillus subtilis* & 14.25% metalaxyl & 16.67% PCNB) *(Helena)*
Rate: 2-3 oz./bu.
Disease:
Pythium; *Rhizoctonia solani*
Time: Apply at planting.
Remarks: A non-oil lubricant has been added for seeds to flow more easily and evenly from planter boxes. Incorporation of dye is recommended to distinguish seed as treated. Do not use treated seed for food, feed, or oil purposes. Refer to label for use directions.

Signal Word/Toxicity Class: Caution.
REI: 12 hr.

THIRAM

Broad-spectrum fungicide. Toxic to fish. Thiram 50WP contains the toxic inert Rhodamine B and is also toxic to birds and wildlife.

42-S Thiram or Thiram 50WP Dyed *(Gustafson)*
Thiram Granuflo (WDG) *(UCB Agrochemicals)*
Rate: 1.5 fl. oz. 42-S/bu. seed
3 oz. 50WP/100 lb. seed
2 oz. WDG/100 lb. seed
Disease:
damping-off; seed decay; seedling blight; seedling disease
Time: Seed treatment.
Remarks: Use with EPA-approved dye that imparts unnatural color to seed. Do not use for feed, food, or oil purposes. Do not apply Thiram Granuflo through any type of irrigation system. Refer to label for use directions.
State Restrictions: Thiram Granuflo not for use in California.

Signal Word/Toxicity Class: Caution.
REI: 24 hr. (WDG).

Tilt — see propiconazole

Top Cop with Sulfur — see sulfur premix

Vitaflo — see carboxin premixes

Vitavax — see carboxin

◆ - **new product** • ai-active ingredient • DF-dry flowable • E/EC-emulsifiable concentrate • EW-emulsion in water
F/FL-flowable • G-granule • L/LC-liquid • SP/WSP-soluble packet • W/WP-wettable powder • WSB-water soluble bag

Section 3

Soybeans

Enhanced Seed Chart .90
Bt Chart 94
Soybeans crop listings 95
Insecticides 95
Fungicides 101

The information in this 1999 Insect and Disease Control Guide is updated to August 15, 1998.
Because manufacturers' product recommendations are constantly being updated and revised as additional product registrations and label changes are approved by EPA, there may be changes after this date. However, every effort has been made to give manufacturers the opportunity to include in the Guide their latest information. This Guide is for preliminary planning – not a use guide. Be sure to follow label directions, notwithstanding information contained herein.

Enhanced Seed Products Chart

Soybeans

Find enhanced seed products quickly and easily with this enhanced seed chart. Simply find the insect you want to control on the top of the chart, and follow down to the marked spot to find the Bt product you can use to control that pest.

	Beck's Superior Hybrids, Inc.	Bergmann-Taylor, Inc.	Campbell Seed, Inc.	CHEMGRO Seeds	Dairyland Seed Co., Inc.	DEKALB Genetics Corp.	Deltapine Seed	Farm Science Genetics	Garst Seed Co.	Gutwein Seeds	Interstate Payco Seed	Latham Seeds	Merschman Seeds	Midwest Seed Genetics	Mycogen Seeds	NC+ Hybrids	Novartis Seeds	Pioneer Hi-Bred	Rupp Seed, Inc.	Sands of Iowa	Sieben Hybrids, Inc.	Stine Seed Co.	Terra Industries Inc.	Trisler Seed Farms, Inc.
Roundup Ready + brown stem rot resistance/tolerance	■		■	■	■	■						■	■	■	■	■				■	■	■	■	
Roundup Ready + brown stem rot resistance/tolerance + white mold resistance/tolerance											■			■	■	■			■		■			■
Roundup Ready + iron chlorosis resistance/tolerance					■																			
Roundup Ready + Phytophthora root rot resistance/tolerance	■		■	■	■	■		■			■	■	■	■	■	■			■	■	■	■	■	■
Roundup Ready + Phytophthora root rot resistance/tolerance + brown stem rot resistance/tolerance	■		■			■		■		■	■		■		■	■			■	■	■		■	■
Roundup Ready + Phytophthora root rot resistance/tolerance + brown stem rot resistance/tolerance + charcoal rot resistance/tolerance										■														
Roundup Ready + Phytophthora root rot resistance/tolerance + brown stem rot resistance/tolerance + white mold resistance/tolerance										■										■				
Roundup Ready + Phytophthora root rot resistance/tolerance + charcoal rot resistance/tolerance																					■			
Roundup Ready + Phytophthora root rot resistance/tolerance + charcoal rot resistance/tolerance + white mold resistance/tolerance																					■			
Roundup Ready + Phytophthora root rot resistance/tolerance + iron chlorosis resistance/tolerance					■										■									
Roundup Ready + Phytophthora root rot resistance/tolerance + soybean cyst nematode resistance/tolerance	■		■	■		■		■					■	■	■	■					■		■	
Roundup Ready + Phytophthora root rot resistance/tolerance + stem canker resistance/tolerance + frogeye leaf spot resistance/tolerance																							■	
Roundup Ready + Phytophthora root rot resistance/tolerance + sudden death syndrome resistance/tolerance						■							■				■				■			■

	Beck's Superior Hybrids, Inc.	Bergmann-Taylor, Inc.	Campbell Seed, Inc.	CHEMGRO Seeds	Dairyland Seed Co., Inc.	DEKALB Genetics Corp.	Deltapine Seed	Farm Science Genetics	Garst Seed Co.	Gutwein Seeds	Interstate Payco Seed	Latham Seeds	Merschman Seeds	Midwest Seed Genetics	Mycogen Seeds	NC+ Hybrids	Novartis Seeds	Pioneer Hi-Bred	Rupp Seed, Inc.	Sands of Iowa	Sieben Hybrids, Inc.	Stine Seed Co.	Terra Industries Inc.	Trisler Seed Farms, Inc.
Roundup Ready + Phytophthora root rot resistance/tolerance + white mold resistance/tolerance										■					■	■			■		■			■
Roundup Ready + soybean cyst nematode resistance/tolerance		■	■	■		■	■	■	■				■	■	■	■					■	■	■	■
Roundup Ready + soybean cyst nematode resistance/tolerance + brown stem rot resistance/tolerance	■		■													■					■		■	
Roundup Ready + soybean cyst nematode resistance/tolerance + charcoal rot resistance/tolerance													■											
Roundup Ready + soybean cyst nematode resistance/tolerance + Phytophthora root rot resistance/tolerance + brown stem rot resistance/tolerance			■										■											
Roundup Ready + soybean cyst nematode resistance/tolerance + stem canker resistance/tolerance																							■	
Roundup Ready + soybean cyst nematode resistance/tolerance + sudden death syndrome resistance/tolerance + Phythophthora root rot resistance/tolerance + charcoal rot resistance/tolerance																					■			■
Roundup Ready + stem canker resistance/tolerance							■																	
Roundup Ready + stem canker resistance/tolerance + aerial blight resistance/tolerance + frogeye leaf spot resistance/tolerance + soybean cyst nematode resistance/tolerance							■																	
Roundup Ready + stem canker resistance/tolerance + frogeye leaf spot resistance/tolerance							■																	
Roundup Ready + stem canker resistance/tolerance + javanese root knot nematode resistance/tolerance							■																	
Roundup Ready + stem canker resistance/tolerance + soybean mosaic virus resistance/tolerance							■																	

Enhanced Seed Products Chart, continued

Soybeans

	Beck's Superior Hybrids, Inc.	Bergmann-Taylor, Inc.	Campbell Seed, Inc.	CHEMGRO Seeds	Dairyland Seed Co., Inc.	DEKALB Genetics Corp.	Deltapine Seed	Farm Science Genetics	Garst Seed Co.	Gutwein Seeds	Interstate Payco Seed	Latham Seeds	Merschman Seeds	Midwest Seed Genetics	Mycogen Seeds	NC+ Hybrids	Novartis Seeds	Pioneer Hi-Bred	Rupp Seed, Inc.	Sands of Iowa	Sieben Hybrids, Inc.	Stine Seed Co.	Terra Industries Inc.	Trisler Seed Farms, Inc.
Roundup Ready + stem canker resistance/tolerance + frogeye leaf spot resistance/tolerance + soybean cyst nematode resistance/tolerance							■																	
Roundup Ready + stem canker resistance/tolerance + frogeye leaf spot resistance/tolerance + soybean mosaic virus resistance/tolerance + soybean cyst nematode resistance/tolerance							■																	
Roundup Ready + STS + Phytophthora root rot resistance/tolerance																								■
Roundup Ready + sudden death syndrome resistance/tolerance						■							■											■
Roundup Ready + white mold resistance/tolerance												■		■	■	■	■		■		■		■	
STS + brown stem rot resistance/tolerance	■		■		■							■		■		■					■		■	
STS + Phytophthora root rot resistance/tolerance	■		■	■	■			■				■		■	■	■			■	■	■		■	
STS + Phytophthora root rot resistance/tolerance + brown stem rot resistance/tolerance				■	■									■		■				■	■			■
STS + Phytophthora root rot resistance/tolerance + brown stem rot resistance/tolerance + white mold resistance/tolerance				■				■												■	■		■	
STS + Phytophthora root rot resistance/tolerance + frogeye leaf spot resistance/tolerance																					■			
STS + soybean cyst nematode resistance/tolerance		■						■						■										
STS + soybean cyst nematode resistance/tolerance + stem canker resistance/tolerance								■															■	
STS + stem canker resistance/tolerance + frogeye leaf spot resistance/tolerance + sudden death syndrome + soybean cyst nematode resistance/tolerance + aerial blight resistance/tolerance							■																	

	Beck's Superior Hybrids, Inc.	Bergmann-Taylor, Inc.	Campbell Seed, Inc.	CHEMGRO Seeds	Dairyland Seed Co., Inc.	DEKALB Genetics Corp.	Deltapine Seed	Farm Science Genetics	Garst Seed Co.	Gutwein Seeds	Interstate Payco Seed	Latham Seeds	Merschman Seeds	Midwest Seed Genetics	Mycogen Seeds	NC+ Hybrids	Novartis Seeds	Pioneer Hi-Bred	Rupp Seed, Inc.	Sands of Iowa	Sieben Hybrids, Inc.	Stine Seed Co.	Terra Industries Inc.	Trisler Seed Farms, Inc.
STS + white mold resistance/tolerance				■																	■		■	
Brown stem rot resistance/tolerance													■					■						
Brown stem rot resistance/tolerance + iron chlorosis resistance/tolerance + white mold resistance/tolerance													■											
Iron chlorosis resistance/tolerance + yellow hilum resistance/tolerance													■											
Phytophthora root rot resistance/tolerance										■			■					■						
Phytophthora root rot resistance/tolerance + brown stem rot resistance/tolerance										■														
Phytophthora root rot resistance/tolerance + brown stem rot resistance/tolerance + white mold resistance/tolerance										■														
Phytophthora root rot resistance/tolerance + brown stem rot resistance/tolerance + charcoal rot resistance/tolerance										■														
Sclerotinia white mold resistance/tolerance									■															
Soybean cyst nematode resistance/tolerance	■		■	■		■			■		■			■		■	■	■			■	■	■	■
Soybean cyst nematode resistance/tolerance + brown stem rot resistance/tolerance	■					■						■				■	■				■		■	■
Soybean cyst nematode resistance/tolerance + Phytophthora resistance/tolerance	■		■			■					■	■				■	■				■		■	■
Soybean cyst nematode resistance/tolerance + sudden death syndrome resistance/tolerance													■											
Sudden death syndrome resistance/tolerance													■											

Bt Products Chart

Soybeans

Find Bt products quickly and easily with this Bt products chart. Simply find the insect you want to control at the top of the chart, and follow down to the marked spot to find the Bt product you can use to control that pest. Company address information can be found on page 599.

	Armyworms	Bollworms	Budworms	Caterpillars	Cutworms	Diamondback moth	Earworms	European corn borer	Fusarium	Green cloverworm	Hornworms	Imported cabbageworm	Loopers	Podworms	Rhizoctonia	Saltmarsh caterpillar	Velvetbean caterpillar
Agree	■												■	■			■
Bacillus subtilis									■						■		
Biobit	■				■	■		■		■	■	■	■	■		■	■
BMP 123	■				■	■		■		■	■	■	■	■		■	■
Bt. ssp. *aizawai*	■									■			■	■			■
Bt. ssp. *aizawai* + *Bt.* ssp. *kurstaki*	■		■	■			■	■					■				
Bt. ssp. *kurstaki*	■	■	■	■	■	■	■	■		■	■	■	■	■		■	■
Condor	■									■			■	■			■
Crymax										■			■	■			■
Cutlass										■			■				■
DiPel	■				■	■		■		■	■	■	■	■		■	■
Javelin										■			■				■
Kodiak									■						■		
Lepinox										■			■				■
Mattch	■		■	■			■	■					■				
MVP II	■		■	■			■	■					■				
Troy-Bt	■	■	■							■	■		■			■	■
XenTari	■									■			■	■			■

Manufacturers List

These Bt products are manufactured by the following companies. Look for their addresses, phone numbers, and fax numbers in the Suppliers Directory beginning on page 599.

Agree, Javelin – Thermo Trilogy
Biobit, DiPel, XenTari – Abbott Laboratories
BMP 123 – Becker Microbial
Condor, Crymax, Cutlass, Lepinox – Ecogen
Kodiak – Gustafson
Mattch, MVP II – Mycogen
Troy-Bt – Troy Biosciences

Soybeans

Insecticides

ACEPHATE

Contact and systemic organophosphate insecticide. Compatible with other pesticides. Practically nontoxic to fish, moderately toxic to birds, and toxic to bees.

Orthene 75 S *(Valent)*

Rate: 0.2 lb. ai/A
4 oz. 75 S/A

Pest:
thrips

Remarks: Apply when eggs or insects appear; repeat as needed to maintain control. Do not apply through any type of irrigation system. Do not graze or cut vines for hay or forage.

Days To Harvest: 14.

State Restrictions: SLN-Arkansas, Mississippi.

- - - - - - - - - - - - - - - - -

Rate: Seed treatment:
0.5 lb. ai/A
0.66 lb. 75 S/A

Pest:
black cutworm thrips

Remarks: Apply in planter box as dry powder. Even coverage required for acceptable insect control. Loss of Orthene 75 S will occur if air planters are used. Plants grown from improperly treated seed will not exhibit insect control. Reduction of black cutworm for about 14 days after planting and thrips for about 3 weeks after planting. When planting seed into fields where large cutworms are present (5th instar and larger), economic damage may occur. Do not use treated seed for food or feed purposes. Do not graze or cut vines for feed or forage.

State Restrictions: SLN-Mississippi.

- - - - - - - - - - - - - - - - -

Signal Word/Toxicity Class: Caution.

REI: 24 hr.

Orthene 90 S *(Valent)*

Rate: 0.25-1 lb. ai/A
0.28-1.1 lb. 90 S/A

Pest:
armyworms, cabbage looper, corn earworm, grasshoppers, green cloverworm, soybean looper, stink bugs, threecornered alfalfa hopper, velvetbean caterpillar

Remarks: Rate varies by pest species; refer to label. Apply in 5-10 gal. water by air; 10-50 gal. water by ground. Repeat as necessary to maintain control. Do not apply through any type of irrigation system. Do not graze or cut vines for hay or forage.

Days To Harvest: 14.

State Restrictions: SLN-Louisiana.

- - - - - - - - - - - - - - - - -

Signal Word/Toxicity Class: Caution.

REI: 24 hr.

Adios — see carbaryl

Ambush — see permethrin

ASANA

Broad-spectrum pyrethroid insecticide with contact and stomach poison action. May be tank mixed with herbicides and other insecticides. Restricted Use Pesticide due to toxicity to fish and aquatic organisms.

Asana XL (esfenvalerate) *(DuPont)*

Rate: 0.03-0.05 lb. ai/A
5.8-9.6 fl. oz. XL/A

Pest:
bean leaf beetle, beet armyworm (aids in control), cabbage looper, corn earworm, cutworms, grasshoppers, Japanese beetle (adult), southern green stink bug, threecornered alfalfa hopper

Rate: 0.015-0.03 lb. ai/A
2.9-5.8 fl. oz. XL/A

Pest:
green cloverworm, Mexican bean beetle, potato leafhopper, saltmarsh caterpillar, velvetbean caterpillar, woollybear caterpillar

Remarks: Repeat as necessary to maintain control. When applying in nonvolatile vegetable oils, use total spray volume of 2 pt./A or more. Rate varies by pest species; refer to label. Do not feed or graze livestock. Do not exceed 0.2 lb. ai/A per season.

Days To Harvest: 21.

- - - - - - - - - - - - - - - - -

Signal Word/Toxicity Class: Warning/II.

REI: 12 hr.

AZADIRACHTIN

Extracted from seed of neem tree. Insect growth regulator that controls targeted insect larvae when they ingest or come in contact with it, by interfering with the insect's ability to molt and shutting off feeding behavior. Affected larvae slowly starve and die at next molt.

◆ **NEEMAZAD 4.5EC or NEEMIX 4.5** *(Thermo Trilogy)*

Rate: 0.125-1 pt. 4.5/A
High Volume:
2.25-4.5 oz. 4.5EC/100 gal. water
Low Volume:
0.5 pt. 4.5EC/A

Pest:
armyworms, bean leaf beetle, cabbage looper, caterpillars, Colorado potato beetle, cutworms, garden webworm, greenhouse whitefly, hornworms, leafminers, loopers, Mexican bean beetle, psyllids, soybean looper, weevils

Remarks: Apply in sufficient water for adequate coverage. Low rate should be used as preventative, or if used with adulticide products. For low insect pressure, apply on preventative 7-day schedule; high insect pressure, every 3-4 days. Do not apply through any type of irrigation system.

- - - - - - - - - - - - - - - - -

Signal Word/Toxicity Class: Warning.

REI: 12 hr.

NOTICE	The information on these pages is for preliminary planning — not a guide for use. Be sure to follow the manufacturer's directions, notwithstanding information contained here.	For personal protective equipment and EPA registration numbers, please turn to page 39.

AZINPHOS-METHYL

Toxic to fish, wildlife, and to bees exposed to direct treatment or residues on crops or blooming weeds. Restricted Use Pesticide.

Gowan Azinphos-M 50-W *(Gowan)*
◆ **Micro Flo Azinphosmethyl 2EC or 50W** *(Micro Flo)*
Sniper 2-E, 50PVA, 50W *(Platte)*

Rate: 0.375-0.75 lb. ai/A
1.5-3 pt. 2EC, 2-E/A
0.75-1.5 lb. 50-W, 50PVA, 50W/A

Pest:

aphids	leafhoppers	Mexican bean beetle
bean leaf beetle	leafminers	stink bugs
green cloverworm	leafrollers	velvetbean caterpillar

Remarks: Apply in sufficient water to give complete coverage; repeat as needed. Rate varies by pest species, refer to label. In some areas, aphids may have developed resistance to organophosphate insecticide. Do not graze or feed treated vines to livestock.
Days To Harvest: 45.
State Restrictions: Sniper 50PVA not registered in California.

Signal Word/Toxicity Class: Danger-Poison/I.
REI: 48 hr.; 72 hr. if average rainfall less than 25" per year.

CARBARYL

Carbamate insecticide with activity against a broad spectrum of insects. Mode of attack is as contact and stomach poison with slight systemic properties. Compatible with most insecticides and acaricides, and toxic to fish and bees.

Adios or Slam (13%) *(Micro Flo)*

Rate: 0.065-0.1 lb. ai/A
0.5-0.75 lb. 13%/A

Pest:
cucumber beetle

Remarks: Do not apply through any type of irrigation system. Do not apply more than 3.75 lb./A per season.
Days To Harvest: 0.

Signal Word/Toxicity Class: Caution/III.
REI: 12 hr.

Clean Crop Sevin 5 Bait *(Platte)*

Rate: 1-2 lb. ai/A
20-40 lb. Bait/A

Pest:

ants	darkling beetle	grasshoppers
armyworms	field crickets	sowbugs
cutworms		

Remarks: Apply with ground or air equipment.
Days To Harvest: 0.

Signal Word/Toxicity Class: Caution/III.
REI: 12 hr.

Drexel Carbaryl 4L, 80S, 50W *(Drexel)*
Sevin 4F, XLR Plus, 80S, 80WSP *(Rhone-Poulenc)*

Rate: 0.5-1.5 lb. ai/A
1-3 pt. 4L, 4F, XLR/A
1-3 lb. 50W/A
0.66-1.875 lb. 80S, 80WSP/A

Pest:
grasshoppers

Rate: 0.5-2 lb. ai/A
1-4 pt. 4F, XLR/A
1-4 lb. 50W/A
0.5-2.5 lb. ai/A
1-5 pt. 4L/A
0.5-2.4 lb. ai/A
0.66-3 lb. 80S, 80WSP/A

Pest:

alfalfa caterpillar	grape colaspis	threecornered alfalfa hopper
armyworms	green cloverworm	thrips
bean leaf beetle	Japanese beetle	velvetbean caterpillar
blister beetle	leafhoppers	webworms
corn earworm	Mexican bean beetle	woollybear caterpillar
cucumber beetles	painted lady	yellowstriped armyworm
cutworms	saltmarsh caterpillar	
fall armyworm	thistle caterpillar	

Remarks: Rate varies by pest species and infestation level; refer to label. Apply when insects or their damage appear; repeat at 7- to 14-day intervals or as necessary unless shorter interval is specified. Do not apply when foliage is wet or when rain or excessive humidity is expected during the next 2 days. Do not apply in combination with 2,4-DB herbicides.
Days To Harvest: 0.

Signal Word/Toxicity Class: Caution (4F, 4L, XLR); Warning/II (50W, 80S, 80WSP).
REI: 12 hr.

Chlorfos — see chlorpyrifos

CHLORPYRIFOS

Nonsystemic, organophosphate insecticide with broad spectrum of activity. Mode of action is as contact and stomach poison. Toxic to birds, fish, and bees.

Lorsban-4E *(Dow AgroSciences)*
◆ **Micro Flo Chlorpyrifos 4E AG** *(Micro Flo)*

Rate: 0.25-1 lb. ai/A
0.5-2 pt. 4E/A

Pest:

armyworms	grasshoppers	southern green stink bug
bean leaf beetle	green cloverworm	spider mites
corn earworm	lesser cornstalk borer	velvetbean caterpillar
cutworms	Mexican bean beetle	woollybears
European corn borer	saltmarsh caterpillar	

Remarks: For soil treatment of cutworm or lesser cornstalk borer, apply at planting in 4-6" band or postemergence in 9-12" band. For foliar, apply when insects become a problem and repeat when necessary. For plants under 6" high, apply over the top as 6-12" band. Use minimum 10 gal. spray/A. May also be applied through center pivot irrigation systems. Do not apply in-furrow. Do not apply last 2 applications closer than 14 days. On determinate soybeans, do not apply more than 1 application after pod set. Do not apply more than 6 pt./A per season. Rate varies by pest species; refer to label. Do not graze treated areas or feed treated forage, hay, and straw to livestock.
Days To Harvest: 28.

Signal Word/Toxicity Class: Warning/II.
REI: 24 hr.

◆ **Chlorfos 15G** *(Griffin)*
Lorsban 15G *(Dow AgroSciences)*
◆ **Micro Flo Chlorpyrifos 15G** *(Micro Flo)*

Rate: 8 oz. 15G/1000 ft. row
Southeast—at planting:
4-8 oz. 15G/1000 ft. row

Pest:

cutworms	lesser cornstalk borer (larvae)

Remarks: Apply at planting or postemergence as band treatment and incorporate 1". Do not apply in-furrow. Do not make more than 1 application per season.

Signal Word/Toxicity Class: Caution.
REI: 12 hr.

lambda-CYHALOTHRIN

Synthetic pyrethroid insecticide with broad-spectrum activity, formulated as an emulsifiable concentrate. Restricted Use Pesticide due to toxicity to fish and aquatic organisms.

Karate or Warrior (1E) *(ZENECA)*

Rate: 0.015-0.03 lb. ai/A
1.92-3.84 fl. oz. 1E/A

Pest:

bean leaf beetle	green cloverworm	stink bugs
beet armyworm**	Japanese beetle (adult)	threecornered alfalfa hopper
blister beetles	lesser cornstalk borer*	thrips
cabbage looper	Mexican bean beetle	tobacco budworm
corn earworm	plant bugs	velvetbean caterpillar
cucumber beetle	potato leafhopper	webworms
cutworms	saltmarsh caterpillar	woollybear caterpillar
European corn borer	silverspotted skipper	yellowstriped armyworm**
fall armyworm**	soybean looper*	
grasshoppers	spider mites*	

Suppression **1st & 2nd instar

◆ **- new product** • ai-active ingredient • DF-dry flowable • E/EC-emulsifiable concentrate • EW-emulsion in water
F/FL-flowable • G-granule • L/LC-liquid • SP/WSP-soluble packet • W/WP-wettable powder • WSB-water soluble bag

Remarks: Rate varies by pest species. Apply as required by scouting, usually at intervals of 5 or more days. Do not apply more than 0.48 pt./A per season. Do not graze or harvest treated soybean forage, straw, or hay for livestock feed. Refer to label for spray drift precautions and state and county restrictions.
Days To Harvest: 45.

Signal Word/Toxicity Class: Danger/I.
REI: 24 hr.

DB-Green — see lindane premixes

Declare — see methyl parathion

Diatect — see pyrethrins

Digon — see dimethoate

Dimate — see dimethoate

DIMETHOATE

Broad-spectrum organophosphate insecticide and acaricide with systemic action. Mode of action is as contact and stomach poison. Compatible with most pesticides but is incompatible with alkaline materials. Toxic to birds, fish, and bees.

Cheminova Dimethoate 4E *(Cheminova)*
Clean Crop Dimethoate 400 *(Platte)*
Digon 400 *(Wilbur-Ellis)*
Dimate 4E *(Terra)*
Drexel Dimethoate 4EC *(Drexel)*
Gowan Dimethoate E267 *(Gowan)*
Micro Flo Dimethoate 267 or 4E *(Micro Flo)*

Rate: 0.5 lb. ai/A
1.5 pt. E267/A
1 pt. 4E, 4EC, 400/A
1.38 pt. 267/A

Pest:
bean leaf beetle, grasshoppers, leafhoppers, Mexican bean beetle, spider mites

Remarks: Do not feed or graze within 5 days of last application.
Days To Harvest: 21.

Signal Word/Toxicity Class: Warning/II.
REI: 48 hr.

DIMILIN

Member of benzoylphenylurea chemical family. An insect growth regulator that interferes with chitin formation and molting process of larvae; therefore does not provide immediate kill. Nontoxic to bees and most parasitic and predatory insects; relatively environmentally safe and has low toxicity to fish and birds. Restricted Use Pesticide due to toxicity to aquatic invertebrates.

Dimilin 2F, 2L, 25W (diflubenzuron) *(Uniroyal)*

Rate: 0.03-0.06 lb. ai/A
2-4 oz. 2F, 2L, 25W/A

Pest:
beet armyworm, fall armyworm, green cloverworm, Mexican bean beetle, soybean looper, velvetbean caterpillar

Remarks: Apply 9-35 gal./A by ground or 1-3 gal./A by air. Apply when larvae are small (less than 1/2"); repeat if damaging numbers reappear. Do not make more than 2 applications per year. Do not cut for hay or allow milk or meat animals to graze treated fields.
Days To Harvest: 21.
Crop Rotation: Do not plant food or feed crops in treated soils within 6 months following last application unless Dimilin is authorized for use on these crops.
State Restrictions: Not for use in California.

Signal Word/Toxicity Class: Caution/III.
REI: 12 hr.

DI-SYSTON

Broad-spectrum organophosphate insecticide and acaricide. Has both contact and systemic action and can be absorbed by the roots and translocated throughout the plant. Compatible with most pesticides. Toxic to birds, fish, and bees. Restricted Use Pesticide.

Di-Syston 15% (disulfoton) *(Bayer)*

Rate: 1 lb. ai/A
6.7 lb. 15%/A (less than 39" row spacing)
8 oz. 15%/1000 ft. (at least 39" row spacing)

Pest:
aphids, leafhoppers, Mexican bean beetle, mites, thrips (except flower thrips)

Remarks: Place granules in 4" band over seed furrow behind planter shoe and in front of press wheel or in a band on each side of seed furrow at planting time. Do not apply directly on seed.

Signal Word/Toxicity Class: Danger/I.
REI: 48 hr.; 72 hr. if average rainfall less than 25" per year.

FURADAN

Broad-spectrum carbamate insecticide, acaricide, and nematicide. Has systemic activities; mode of action is as contact and stomach poison. Compatible with most pesticides except those that are alkaline. Toxic to bees, fish, birds, and other wildlife. Restricted Use Pesticide due to acute oral and inhalation toxicity.

Furadan 4F (carbofuran) *(FMC)*

Rate: 1.5-2 lb. ai/A
3-4 pt. 4F/A (40" row spacing)
3.75-5 fl. oz. 4F/1000 lin. ft. row

Pest:
cyst nematode, dagger nematode, lance nematode, lesion nematode, ring nematode, root-knot nematode, spiral nematode, sting nematode, stubby-root nematode, stunt nematode

Remarks: Apply at planting as 12" surface band. Soybeans may be grazed or cut for forage 30 days or later following at-planting application.

Rate: 0.125-0.5 lb. ai/A
0.25-1 pt. 4F/A

Pest:
grasshoppers, Mexican bean beetle

Remarks: Foliar application. Use low rate for grasshoppers and high rate for Mexican bean beetles. Do not make more than 2 applications per season. Do not apply through any type of irrigation system. Do not graze or feed treated forage to livestock or cut for silage or hay.
Days To Harvest: 21.
Crop Rotation: Do not rotate with soil treated with more than 3 lb. ai/A or plant with any crop for which carbofuran is not registered for at least 10 months.
State Restrictions: Not for use in Long Island, New York.

Signal Word/Toxicity Class: Danger-Poison/I.
REI: 48 hr.

Fyfanon — see malathion

Karate — see *lambda*-cyhalothrin

Kernel Guard Supreme — see permethrin premix

Kumulus — see sulfur

LANNATE

Carbamate insecticide and acaricide that has systemic properties and kills both by contact and stomach action. Effective against a broad spectrum of pests, and compatible with other pesticides. Toxic to fish and bees. Restricted Use Pesticide.

Lannate SP or LV (methomyl) *(DuPont)*

Rate: 0.1125-0.45 lb. ai/A
0.125-0.5 lb. SP/A
0.375-1.5 pt. LV/A

For personal protective equipment and EPA registration numbers, please turn to page 39.

Pest:

bean leaf beetle, beet armyworm, corn earworm, fall armyworm, green cloverworm, Mexican bean beetle, saltmarsh caterpillar, silverspotted skipper, thrips, velvetbean caterpillar

Remarks: Rate varies by pest species; refer to label. Apply when economic thresholds are reached; repeat as needed. Do not apply more than 1.35 lb. ai/A per crop or make more than 3 applications per crop.

Days To Harvest: 14. Feeding/grazing under 0.45 lb. ai: forage-3, hay-7; more than 0.45 lb. ai: forage-10, hay-12.

- - - - - - - - - - - - - - - - -

Signal Word/Toxicity Class: Danger-Poison/I.

REI: 48 hr.

LARVIN

Broad-spectrum carbamate insecticide/ovicide that kills predominantly as a stomach poison after ingestion, and has some limited contact action. Known to have systemic action as a seed treatment. Moderately toxic to bees during direct application, but nontoxic after spray has dried.

Larvin 3.2 (thiodicarb) *(Rhone-Poulenc)*

Rate: Standard spray volume:
0.45-0.75 lb. ai/A
18-30 fl. oz. 3.2/A

Pest:

bean leaf beetle, cabbage looper, Mexican bean beetle, soybean looper, stink bugs (suppression), threecornered alfalfa hopper (suppression)

Rate: Standard & low spray volume:
0.25-0.75 lb. ai/A
10-30 fl. oz. 3.2/A

Pest:

beet armyworm, bollworm, corn earworm, fall armyworm, green cloverworm, podworms, southern armyworm, tobacco budworm, velvetbean caterpillar, woollybear caterpillar, yellowstriped armyworm

Remarks: Rate varies by pest species; refer to label. Use lower rates for low to moderate populations and maximum protection of beneficials. When applying low volume applications at 1 gal. spray volume/A by air, use 1 qt. emulsified crop or vegetable oil; water alone may be used as diluent for spray volumes 2 gal./A or higher. Only use low volume aerial application where spray volume will provide adequate coverage of crop. Do not feed forage, hay, or straw to livestock.

Days To Harvest: 28.

State Restrictions: Low volume application for use in Alabama, Arkansas, Delaware, Georgia, Louisiana, Maryland, Mississippi, Missouri, North Carolina, South Carolina, Virginia.

- - - - - - - - - - - - - - - - -

Rate: Postemergence rescue treatment:
0.5-0.75 lb. ai/A
20-30 fl. oz. 3.2/A

Pest:

cutworms

Remarks: Apply as broadcast spray when cutworms are actively feeding. If cutworms are feeding below ground, control will be more difficult. Light cultivation with rotary hoe before spraying may enhance control. Do not feed forage, hay, or straw to livestock.

Days To Harvest: 28.

- - - - - - - - - - - - - - - - -

Signal Word/Toxicity Class: Warning/II.

REI: 12 hr.

LINDANE (Premixes)

Dual purpose insecticide and fungicide seed treatment. Sorghum Guard contains graphite.

◆ **DB-Green (18.75% lindane & 50% maneb)** *(AGSCO)*

Rate: 2.75 oz./bu. seed

Pest:

seedcorn maggots, wireworms

Remarks: Drill or planter box application. Treat only seeds needed for immediate use, minimizing interval between treatment and planting. Do not use treated seed for food, feed, or oil purposes. Do not store excess treated seeds beyond planting time. Do not graze or feed livestock on treated areas for 6 weeks after planting.

- - - - - - - - - - - - - - - - -

Signal Word/Toxicity Class: Warning/II.

REI: 12 hr.

◆ **DB-Green L (8.6% lindane & 25.6% maneb)** *(AGSCO)*

Rate: 4 fl. oz./bu. seed

Pest:

wireworms

Remarks: Drill or planter box application. Treat only seeds needed for immediate use, minimizing interval between treatment and planting. Do not use treated seed for food, feed, or oil purposes. Do not store excess treated seeds beyond planting time.

- - - - - - - - - - - - - - - - -

Signal Word/Toxicity Class: Warning/II.

REI: 12 hr.

◆ **Sorghum Guard (16.6% lindane & 32.75% captan)** *(Trace)*

Rate: 1 oz./50 lb. seed

Pest:

wireworms

Remarks: Apply in planter or drill box before planting. Treat only those seeds needed for immediate use. Do not use treated seed for food, feed, or oil purposes. Do not store excess treated seeds beyond planting time.

Signal Word/Toxicity Class: Danger/I.

REI: 12 hr.

Lorsban — see chlorpyrifos

MALATHION

Organophosphate insecticide/acaricide with contact, stomach, and respiratory killing action. Effective against a wide variety of insects and compatible with many pesticides, but incompatible with alkaline materials. Often used where safety to mammals is a concern, but is toxic to fish and bees.

Clean Crop Malathion 57EC *(Platte)*

Rate: 1.6-1.9 lb. ai/A
2.5-3 pt. 57EC/A

Pest:

green cloverworm, Mexican bean beetle

Remarks: Apply when insects first appear and repeat as necessary. Use high rate for cloverworm.

Days To Harvest: 3.

- - - - - - - - - - - - - - - - -

Signal Word/Toxicity Class: Caution/III.

REI: 12 hr.

Drexel Malathion 5EC *(Drexel)*
◆ **Micro Flo Malathion 5EC** *(Micro Flo)*

Rate: 1.88 lb. ai/A
3 pt. 5EC/A

Pest:

green cloverworm, Mexican bean beetle

Remarks: Do not apply through any type of irrigation system.

Days To Harvest: Harvest or forage-0.

- - - - - - - - - - - - - - - - -

Signal Word/Toxicity Class: Warning/II.

REI: 12 hr.

◆ **Drexel Malathion ULV** *(Drexel)*
Fyfanon ULV *(Cheminova)*
◆ **Micro Flo Malathion ULV** *(Micro Flo)*

Rate: 8 fl. oz. ULV/A

Pest:

grasshoppers, green cloverworm, Japanese beetle, Mexican bean beetle

Remarks: Apply when pests are present at economically damaging levels.

Days To Harvest: Harvest or forage-7.

- - - - - - - - - - - - - - - - -

Signal Word/Toxicity Class: Caution.

REI: 12 hr.

◆ **Gowan Malathion 8 (8E)** *(Gowan)*

Rate: 2 lb. ai/A
2 pt. 8E/A

Pest:

grasshoppers, Japanese beetle, Mexican bean beetle

Remarks: Do not apply through any type of irrigation system.

Days To Harvest: 0.

- - - - - - - - - - - - - - - - -

Signal Word/Toxicity Class: Caution.

REI: 12 hr.

◆ - **new product** • ai-active ingredient • DF-dry flowable • E/EC-emulsifiable concentrate • EW-emulsion in water
F/FL-flowable • G-granule • L/LC-liquid • SP/WSP-soluble packet • W/WP-wettable powder • WSB-water soluble bag

METHYL PARATHION

Highly toxic organophosphate insecticide with stomach, contact, and some respiratory killing capacity. Compatible with other insecticides but not alkaline materials. Penncap-M is a microencapsulated formulation. Restricted Use Pesticide due to residual effects to avian species and hazard to bees.

Cheminova Methyl 4EC *(Cheminova)*
◆ **Declare (4EC)** *(Griffin)*

Rate: 0.375-1 lb. ai/A
0.75-2 pt. 4EC/A

Pest:

aphids	corn earworm*	plant bug*
armyworms	fall armyworm	stink bugs
bean leaf beetle*	garden webworm	threecornered alfalfa
blister beetles	grasshoppers	hopper
bollworms	green cloverworm	twospotted spider mite
climbing cutworm	Mexican bean beetle	velvetbean caterpillar

***Declare only.**

Remarks: Apply when insects first appear and repeat as needed. Rate varies by pest species; refer to label.
Days To Harvest: Up to 1 pt.-15; up to 2 pt.-20.

Signal Word/Toxicity Class: Danger-Poison/I.
REI: 48 hr.; 72 hr. if average rainfall less than 25" per year.

Penncap-M (2FL) *(Elf Atochem)*

Rate: 0.25-0.75 lb. ai/A
1-3 pt. 2FL/A

Pest:
stink bugs*

Rate: 0.5-0.75 lb. ai/A
2-3 pt. 2FL/A

Pest:

bean leaf beetle	potato leafhopper	threecornered alfalfa
grasshoppers	silverspotted skipper	hopper
green cloverworm	soybean looper*	thrips*
Mexican bean beetle		velvetbean caterpillar

Rate: 0.75-1 lb. ai/A
3-4 pt. 2FL/A

Pest:
Japanese beetle*

***Except CA**

Remarks: Apply when insect damage threatens. Do not apply more than twice per season. For grasshoppers, use higher rate if majority of insects are large or weather is cool. Do not chemigate in California. Refer to label for additional use information.
Days To Harvest: Harvest or grazing-20.

Signal Word/Toxicity Class: Warning/II.
REI: 48 hr.

M-PEDE

Contact insecticide and fungicide for control of soft-bodied insects and mites.

M-Pede (potassium salts of fatty acids) *(Mycogen)*

Rate: 4 pt./A in 3-25 gal. water
2% v/v spray solution/A in 25 gal. water or more

Pest:

aphids	fleahoppers	spider mites
caterpillars (tank mix)	greenbugs	thrips
	plant bugs	whiteflies

Remarks: Apply at first sign of infestation; repeat at 7- to 14-day intervals. May be tank mixed with other insecticides; refer to both labels for use directions and precautions.
Days To Harvest: 0.

Signal Word/Toxicity Class: Warning/II.
REI: 12 hr.

NEEMAZAD — see azadirachtin

NEEMIX — see azadirachtin

Orthene — see acephate

Penncap-M — see methyl parathion

PERMETHRIN

Nonsystemic, pyrethroid insecticide with a broad spectrum of activity. Kills by contact and stomach action and is compatible with many insecticides and herbicides. Very low in toxicity to mammals and birds but is highly toxic to bees. Restricted Use Pesticide due to toxicity to fish and aquatic organisms.

Ambush 2EC or 25WP *(ZENECA)*
Pounce 3.2EC, 25WP or WSB *(FMC)*

Rate: 0.05-0.2 lb. ai/A
2-8 oz. 3.2EC/A
3.2-12.8 fl. oz. 2EC/A
3.2-12.8 oz. 25WP/A
1-2 bags WSB/A

Pest:

bean leaf beetle	green cloverworm	saltmarsh caterpillar
beet armyworm	Japanese beetles (except CA**)	soybean looper
cabbage looper		thistle caterpillar*
corn earworm	Mexican bean beetle	velvetbean caterpillar
cutworms*	potato leafhopper	webworms*

***Pounce **Ambush**

Remarks: Rate varies by pest species; refer to label. Apply by air or ground. Do not apply more than 0.4 lb. ai/A per season. Do not graze or feed soybean forage or hay.
Days To Harvest: 60.

Signal Word/Toxicity Class: Caution (3.2EC); Warning (2EC, 25WP, WSB).
REI: 24 hr. (Ambush); 12 hr. (Pounce).

PERMETHRIN (Premix)

Seed treatment.

Kernel Guard Supreme (10.42% permethrin & 14% carboxin) *(Trace)*

Rate: 1.8 oz./bu. seed

Pest:

seedcorn beetles	seedcorn maggots	wireworms

Remarks: Refer to label for use directions.

Signal Word/Toxicity Class: Caution.
REI: 12 hr. with exceptions.

PHORATE

Organophosphate insecticide and nematicide. Highly toxic to birds. Restricted Use Pesticide.

Clean Crop Phorate 20G *(Platte)*
Riverside Phorate 20G *(Terra)*
Thimet 20-G *(American Cyanamid)*

Rate: Early-season control:
9 oz. 20G, 20-G/1000 ft. row (min. 30" spacing)

Pest:

aphids	Mexican bean beetle	seedcorn maggots
leafhoppers	mites	thrips
Lygus bugs		

Remarks: At planting. Drill granules to side of seed. For seedcorn maggots, place granules in 7" band over row directly behind planter shoe and in front of press wheel. Do not place granules in direct contact with seed. Crop injury may result when preplant incorporated or preemergence applications of metribuzin herbicides are used in conjunction with this product or other soil-applied organophosphate pesticides. Do not feed foliage of treated soybeans.

Signal Word/Toxicity Class: Danger-Poison/I.
REI: 48 hr.; 72 hr. if average rainfall less than 25" per year.

Pounce — see permethrin

3

Soybeans

NOTICE The information on these pages is for preliminary planning — not a guide for use. Be sure to follow the manufacturer's directions, notwithstanding information contained here.	For personal protective equipment and EPA registration numbers, please turn to page 39.

PYRELLIN (Premix)

Premix containing pyrethrins and rotenone. Pyrethrins are among the oldest, most widely used insecticides, derived from plants belonging to the genus Chrysanthemum. Rotenone is also derived from plants. Both are nonsystemic with contact and stomach poison action, and are highly toxic to fish.

Pyrellin E.C. (0.042 lb./gal. pyrethrins & 0.035 lb./gal. rotenone) *(Webb Wright)*
Rate: 1-2 pt. E.C./A
Pest:
aphids, bean beetle, beet webworm, caterpillars, loopers, Mexican bean beetle, mites, thrips
Remarks: Apply as necessary to maintain control.
Days To Harvest: 0.

Signal Word/Toxicity Class: Caution/III.
REI: 12 hr.

Pyrenone — see pyrethrins

PYRETHRINS

Contains piperonyl butoxide as a synergist. Diatect also contains silicon dioxide (diatomaceous earth). May be combined in tank mix with other insecticides for quicker and more complete kill where insect resistance may be a problem, and as an "exciter" to flush insects out of hiding and into contact with other spray residues.

Diatect Multipurpose Insecticide II *(Diatect)*
Pyrenone (0.5EC) *(AgrEvo Environmental Health)*
◆ **Py-Rin 60-6 EC** *(Wilbur-Ellis)*
Rate: 1-6 lb. II/A
1-12 fl. oz. 0.5EC/A
2-16 oz. 6-60 EC/A
Pest: Including but not limited to:
aphids, armyworms, caterpillars, corn earworm, flea beetles, loopers, Mexican bean beetle, mites, thrips
Remarks: Apply in sufficient water for thorough coverage; repeat as needed. Apply Diatect as dust or wettable powder.
Days To Harvest: 0.

Signal Word/Toxicity Class: Caution.
REI: 12 hr.

Py-Rin — see pyrethrins

Redball Lightning — see sulfur

SCOUT X-TRA

Nonsystemic pyrethroid insecticide that kills by contact and stomach action and is effective against a wide variety of insects. Scout X-TRA Gel is packaged in water-soluble bags. Low toxicity to mammals and birds, highly toxic to bees. Restricted Use Pesticide due to toxicity to fish and aquatic organisms.

Scout X-TRA (0.9EC) or Gel (WSB) (tralomethrin) *(AgrEvo USA)*
Rate: 0.012-0.024 lb. ai/A
1.71-3.41 fl. oz. 0.9EC/A
1 WSB/16.7-8.3 A
Pest:
beet armyworm, cabbage looper, corn earworm, cutworms, fall armyworm, grasshoppers, green cloverworm, leafhoppers, Mexican bean beetle, soybean looper (suppression), stink bugs, threecornered alfalfa hopper, velvetbean caterpillar
Remarks: Dilute in water or once-refined crop oil and use as foliar spray. Make first application before insect populations increase to economic significance. Additional applications may be made if insect populations rebuild. Do not apply more than 0.12 lb. ai/A per growing season. Do not apply through any type of irrigation system. Do not graze or harvest treated soybean forage, straw, or hay for livestock.
Days To Harvest: 21.
State Restrictions: Not for use in California.

Signal Word/Toxicity Class: Danger/I.
REI: 24 hr.

Sevin — see carbaryl

Slam — see carbaryl

Sniper — see azinphos-methyl

Sorghum Guard — see lindane premixes

SULFUR

Thiolux is micronized sulfur. Redball Lightning is a dustless granule formulated to break down into smaller particles when wetted by water to provide effectiveness of a powder with the handling ease of a granule.

◆ **Kumulus DF** *(BASF)*
Rate: 4.8-12 lb. ai/A
6-15 lb. DF/A
Pest:
Banks grass mite, Pacific mite, red spider mite, twospotted spider mite
Remarks: Do not apply if temperatures during or within 3 days after application are expected to exceed 90°F. Do not apply through any type of irrigation system.

Signal Word/Toxicity Class: Caution.
REI: 24 hr.

◆ **Redball Lightning Degradable (G)** *(International Sulphur)*
Rate: 5-15 lb. G/A
Pest:
fleahoppers, mites, thrips
Remarks: Do not apply within 3 weeks of oil spray.
Days To Harvest: 0.

Signal Word/Toxicity Class: Caution.
REI: 24 hr.

THAT (6FL) or THAT Big 8 (8FL) *(Stoller)*
Rate: 9 lb. ai/A
12 pt. 6FL/A
8 lb. ai/A
8 pt. 8FL/A
Pest:
Banks grass mite
Remarks: Apply when mites appear on bottom 2 leaves. Can be tank mixed with other miticides.
State Restrictions: Not for use in California.

Signal Word/Toxicity Class: Caution.
REI: 12 hr.

Thiolux (DF) *(Platte)*
Rate: 6-10 lb. ai/A
7.5-12.5 lb. DF/A
Pest: Suppression:
Banks grass mite, Pacific spider mite, red spider mite, twospotted spider mite
Remarks: Apply when mites first appear; repeat as necessary.

Signal Word/Toxicity Class: Caution.
REI: 24 hr.

THAT or THAT Big 8 — see sulfur

Thimet — see phorate

◆ - **new product** • ai-active ingredient • DF-dry flowable • E/EC-emulsifiable concentrate • EW-emulsion in water
F/FL-flowable • G-granule • L/LC-liquid • SP/WSP-soluble packet • W/WP-wettable powder • WSB-water soluble bag

Thiolux — see sulfur

VYDATE

Systemic carbamate insecticide/nematicide that is highly toxic to humans and other mammals, and toxic to fish and bees. Restricted Use Pesticide.

Vydate L (oxamyl) *(DuPont)*

Rate: In-furrow or incorporated band:
0.5-1 lb. ai/A
2-4 pt. L/A in 10-20 gal. water (36" row)
Incorporated broadcast:
2-4 lb. ai/A
8-16 pt. L/A in 10-20 gal. water

Pest:
cyst nematode, lesion nematode, root-knot nematode (except javanese), lance nematode, ring nematode, stunt nematode

Remarks: In-furrow band treatment: Spray over open drill row in seed zone to ensure incorporation with soil. Incorporated band treatment: Apply in 7-10" band in drill area at planting or preplant up to 1 week before planting. Incorporate 2-4" into soil in seed zone. Will not control severe nematode infestations such as where injury to plants is manifested by severe stunting and chlorosis. Do not apply more than 16 pt./A per season. Do not cut for hay or feed treated forage to livestock.
State Restrictions: Not registered for use in California. Do not use in Suffolk and Nassau counties, Long Island, New York.

Signal Word/Toxicity Class: Danger-Poison/I.
REI: 48 hr.

Warrior — see *lambda*-cyhalothrin

Fungicides

Allegiance — see metalaxyl

Apron — see metalaxyl

Apron XL — see mefenoxam

Bean Guard — see captan premixes

BENLATE

Benzimidazole systemic fungicide. Toxic to fish.

Benlate (SP, WP) (benomyl) *(DuPont)*

Rate: 0.25-0.5 lb. ai/A
0.5-1 lb. SP, WP/A

Disease:
aerial blight, Anthracnose, Diaporthe pod blight, Diaporthe stem blight, frogeye leaf spot, purple seed stain, Septoria brown spot

Time: Apply at early pod set (pods 1/8-1/4" long at 1 of 4 main stem upper nodes); repeat 14- to 21-day intervals as needed.
Remarks: Do not tank mix or alternate with benzimidazole or thiophanate products. Do not graze livestock in treated groves. Refer to label for use directions.
Days To Harvest: 35 days.
State Restrictions: Benlate SP not registered for use in California.

Signal Word/Toxicity Class: Caution.
REI: 24 hr.

Bravo — see chlorothalonil

CAPTAN

Some products contain molybdenum for plant nutrition and nitrogen fixation, graphite as a lubricant, and the toxic inert Rhodamine B.

Drexel Captan 4FL *(Drexel)*
Gustafson Captan 30-DD or 400 *(Gustafson)*

Rate: 2 fl. oz. 30-DD/100 lb. seed
1.5-2.5 fl. oz. 400/100 lb. seed
2.5 fl. oz. 4FL/100 lb. seed

Disease:
damping-off, mold, seed decay, seedling blight, seedling disease

Time: Liquid or slurry seed treatment.
Remarks: Do not use treated seed for food, feed, or oil purposes. Refer to label for use directions.

Signal Word/Toxicity Class: Danger/I.

Drexel Captan 50/10 or Captan-75 *(Drexel)*
Drexel Captan Plus Molybdenum (FL) *(Drexel)*
Soy Gro *(Drexel)*

Rate: 3.5 oz. 50/10/100 lb. seed
2.33 oz. 75/100 lb. seed
Planter box:
1 fl. oz. FL/bu. seed
2 oz. 50/10/bu. seed
1.33 oz. 75/bu. seed
3 oz. Soy Gro/bu. seed

Disease:
damping-off, seed decay, seedling blight, seedling disease, soilborne diseases

Time: Seed treatment.
Remarks: Do not use treated seed for food, feed, or oil purposes. Refer to label for use directions.

Signal Word/Toxicity Class: Danger/I.
REI: 4 days; 48 hr. (Captan 50/10).

Hi Moly/Captan (18.86%) or Hi Moly/Captan-D (50%) *(Trace)*

Rate: 5 fl. oz. 18.86%/bu. seed
2 oz. 50%/bu. seed

Disease:
damping-off, seed decay, seedling blight, seedling disease

Time: Planter box seed treatment.
Remarks: Treat only those seeds needed for immediate use. Do not use treated seed for food, feed, or oil purposes. Do not store excess treated seeds beyond planting time. Refer to label for use directions.

Signal Word/Toxicity Class: Danger/I.

Seed Mate Soybean Mo (2X) *(Loveland)*

Rate: 3.33 oz. 2X/100 lb. seed

Disease:
seed rot, seedling blight, seedling disease, soilborne diseases

Time: Seed protectant.
Remarks: Do not use treated seed for food, feed, oil, or feed forage grown from treated seed. Refer to label for use directions.

NOTICE	The information on these pages is for preliminary planning — not a guide for use. Be sure to follow the manufacturer's directions, notwithstanding information contained here.	For personal protective equipment and EPA registration numbers, please turn to page 39.

CAPTAN (Premixes)

Bean Guard contains molybdenum for plant nutrition and nitrogen fixation. Toxic to fish, birds, and other wildlife.

Bean Guard (23.9% captan & 12.5% carboxin) *(Trace)*
Nu-Gro Isotox (12.2% captan & 25% lindane) *(Wilbur-Ellis)*
Seed Mate Isotox (12.2% captan & 25% lindane) *(Loveland)*
Sorghum Guard (32.75% captan & 16.6% lindane) *(Trace)*

Rate: 2 oz./bu. seed (Bean Guard)
2 oz./100 lb. seed (Isotox)
1 oz./50 lb. seed (Sorghum Guard)

Disease:
damping-off, seedling blight, seedling disease, seed decay

Time: Planter box treatment.

Remarks: Treated seed must not be used for or mixed with food or animal feed, or processed for oil. Treat only those seeds needed for immediate use, minimizing the interval between treatment and planting. Do not store excess treated seeds beyond planting time. Refer to label for use directions.

Signal Word/Toxicity Class: Danger/I.

Rival (19.8% captan & 8.4% PCNB & 1% thiabendazole) *(Gustafson)*

Rate: 4 fl. oz./100 lb. seed

Disease:
damping-off, seed decay, seedling disease, Diaporthe pod blight, seedling blight, stem blight

Time: Seed treatment

Remarks: Use with EPA-approved dye that imparts unnatural color to seed. Do not use for feed, food, or oil purposes. Refer to label for use directions.

Signal Word/Toxicity Class: Danger/I.

CARBOXIN

Ready-to-use seed treatment applied as slurry treatment by machine. Contains toxic inert Rhodamine B.

Vitavax-34 *(Gustafson)*

Rate: 3-4 fl. oz./100 lb. seed

Disease:
seedling diseases

Time: Slurry seed treatment.

Remarks: Incorporation of dye into treating slurry is recommended to distinguish seed as treated. Do not use treated seed for food, feed, or oil purposes. Do not graze or feed livestock on forage or hay on treated areas. Do not apply this product through any type of irrigation system. Refer to label for use directions.

Signal Word/Toxicity Class: Caution.

CARBOXIN (Premixes)

Carboxin is a systemic fungicide and seed protectant. Kernel Guard Supreme contains the synthetic pyrethroid insecticide permethrin. Prevail is a premix of the active ingredients of Apron, Terraclor, and Vitavax. Toxic to fish.

Enhance Vitavax-Captan 20-20 (20% carboxin & 19.1% captan) *(Trace)*
Seed Mate Vitavax-Captan (20% carboxin & 19.1% captan) *(Loveland)*

Rate: 3 oz. Enhance/bu. seed

Disease:
damping-off, seedling blight, soilborne disease, seed decay, seedling disease

Rate: 4 oz. Seed Mate/100 lb. seed

Disease:
seedling diseases

Time: Apply as planter box treatment before planting.

Remarks: Best results are obtained by premixing seed with product in a suitable container, then pouring treated seed into planter box. Contains graphite. Do not use treated seed for food, feed, or oil purposes. Refer to label for use directions.

Signal Word/Toxicity Class: Danger/I (Enhance).

Kernel Guard Supreme (14% carboxin & 10.42% permethrin) *(Trace)*

Rate: 1.8 oz./bu. seed

Disease:
damping-off, seedling blight, seedling disease, seed decay

Time: Seed treatment.

Remarks: Treat only those seeds needed for immediate use. Do not use treated seed for food, feed, or oil purposes. Do not store excess treated seeds beyond planting time. Refer to label for use directions.

Signal Word/Toxicity Class: Caution.
REI: 12 hr.

Prevail (15% carboxin & 15% PCNB & 3.12% metalaxyl) *(Trace)*

Rate: 4 oz./bu. seed

Disease:
damping-off, seed rot, seedling disease

Time: Planter box seed treatment.

Remarks: Treat only those seeds needed for immediate use. Do not use treated seed for food, feed, or oil purposes. Do not store excess treated seeds beyond planting time. All seed treated with this product (except planter box) must be colored with EPA approved dye which imparts an unnatural color to seed. Do not graze or feed livestock hay grown from treated seed. Refer to label for use directions.

Signal Word/Toxicity Class: Caution.
REI: 12 hr.

RTU-Vitavax-Thiram (10% carboxin & 10% thiram) *(Gustafson)*
Vitavax T-L (10% carboxin & 10% thiram) *(Trace)*

Rate: 6.8 fl. oz./100 lb. seed

Disease: RTU-Vitavax-Thiram:
Fusarium, Pythium, seedling diseases, Helminthosporium, *Rhizoctonia solani*

Time: Seed treatment.

Disease: Vitavax T-L:
Rhizoctonia solani, seedling disease

Time: Hopper box seed treatment.

Remarks: Use with EPA-approved dye that imparts unnatural color to seed. Do not use for feed, food, or oil purposes. Do not graze or feed livestock on forage and hay in treated areas. Refer to label for use directions.

Signal Word/Toxicity Class: Danger/I (T-L); Caution (RTU).

Vitaflo-280 (14.9% carboxin & 13.2% thiram) *(Gustafson)*
Vitavax-PCNB (17% carboxin & 17% PCNB) *(Gustafson)*

Rate: 4 fl. oz./100 lb. seed

Disease:
damping-off, seed decay, seedling diseases

Time: Seed treatment.

Remarks: Use higher rate under conditions of heavy expected disease pressure. Use with EPA-approved dye that imparts unnatural color to seed. Do not use treated seed for food, feed, or oil purposes. Do not graze or feed livestock in treated areas for 6 weeks after planting. Do not graze or feed livestock on vines grown from treated seed. Refer to label for use directions.

Signal Word/Toxicity Class: Caution.

CHLORONEB

Supplemental seed treatment; use in conjunction with a standard fungicide seed treatment.

Demosan Flowable or 65W *(Kincaid)*

Rate: 7 oz. FL/100 lb. seed
4 oz. 65W/100 lb. seed

Disease:
damping-off, Rhizoctonia, seedling diseases, Pythium blight, Sclerotium blight, sore shin

Time: Seed treatment or hopper box treatment.

Remarks: Refer to label for use directions. Do not use excess treated seed for food, feed, or oil purposes. Do not graze within 45 days of planting.

Signal Word/Toxicity Class: Caution.
REI: 12 hr.

◆ - **new product** • ai-active ingredient • DF-dry flowable • E/EC-emulsifiable concentrate • EW-emulsion in water
F/FL-flowable • G-granule • L/LC-liquid • SP/WSP-soluble packet • W/WP-wettable powder • WSB-water soluble bag

CHLOROTHALONIL

Broad-spectrum contact fungicide. Toxic to fish, aquatic invertebrates and marine/estuarine organisms.

Bravo 500, 720/Weather Stik, Ultrex (WDG) *(ZENECA)*
Echo 500, 720 or 90DF *(Sostram)*
Terranil (6L, 90DF or 90DF WSP) *(Terra)*

Rate: 2 application program:
1-1.8 lb. ai/A
2-3.5 pt. 500/A
1.5-2.5 pt. 720, 6L/A
1.4-2.2 lb. WDG/A
1.2-2 lb. 90DF/A
1.125-2 lb. 90DF, 90DF WSP/A
3 application program (Northern varieties):
0.75-1.5 lb. ai/A
1.5-2.75 pt. 500/A
1-2 pt. 720, 6L/A
0.9-1.8 lb. WDG/A
0.8-1.6 lb. 90DF/A
0.875-1.5 lb. 90DF, 90DF WSP/A

Disease:
Anthracnose; leaf blight; Septoria brown spot; Diaporthe pod blight; purple seed stain; stem blight; frogeye leaf spot (*Cercospora sojina*)

Time: Determinate (Southern) varieties: 2 application program-early pod set stage when majority of pods are 1/8-3/8" in length; beginning of seed formation which occurs about 14 days later. 3 application program-beginning of flowering; early pod set; beginning of seed formation. Indeterminate (Northern) varieties: 2 application program-when largest pods are 1-1 1/2" in length; 14 days later; 3 application program-1 week after first flowering and repeat at 14-day intervals.

Remarks: Apply in sufficient water to obtain complete coverage, using at least 5 gal. water/A for aerial application. Use 3 application program in areas having a history of moderate to severe disease intensity. Do not exceed 3 applications per season. Do not graze treated fields or feed treated parts to livestock. May be applied through sprinkler irrigation equipment. Refer to label for use directions and restrictions.

Days To Harvest: 6 weeks.

- - - - - - - - - - - - - - - - - -

Rate: 0.75 lb. ai/A
1.5 pt. 500/A
1 pt. 720, 6L/A
0.9 lb. WDG/A
0.875 lb. 90DF, 90DF WSP/A

Disease:
stem canker

Time: Apply at emergence of second trifoliate leaves. If conditions favor disease, make a second and third application at 10- to 14-day intervals.

Remarks: Determinate (Southern) varieties. Apply as a band treatment in 10-20 gal. water directing spray to cover entire plant. Refer to label for use directions and restrictions.

- - - - - - - - - - - - - - - - - -

Signal Word/Toxicity Class: Danger/I (WDG, WSP, 90DF); Warning/II (6L, 500, 720).
REI: 48 hr.

DB-Green — see maneb premixes

Demosan — see chloroneb

Echo — see chlorothalonil

Enhance Vitavax-Carboxin — see carboxin premixes

Kernel Guard Supreme — see carboxin premises

Kumulus — see sulfur

MANEB (Premixes)

Dual purpose ready-to-use fungicide/insecticide. DB-Green is a dust-based seed treatment. DB-Green L is a liquid seed treatment. Seed Mate contains graphite. Toxic to fish, birds, and other wildlife.

DB-Green (dust) (50% maneb & 18.75% lindane) *(AGSCO)*
DB-Green L (liquid) (25.6% maneb & 8.6% lindane) *(AGSCO)*

Rate: 2.75 oz. dust/bu. seed
4 fl. oz. L/bu seed.

Disease:
seedling diseases; soilborne diseases

Time: Seed treatment.

Remarks: Drill or planter box application. Treat only seeds needed for immediate use, minimizing interval between treatment and planting. Do not store excess treated seeds beyond planting time. Treated seed must not be used for or mixed with food or animal feed, or processed for oil. Do not graze or feed livestock on treated areas for 6 weeks after planting.

- - - - - - - - - - - - - - - - - -

Signal Word/Toxicity Class: Warning/II.
REI: 12 hr.

Seed Mate Maneb-Lindane (50% maneb & 18.75% lindane) *(Loveland)*

Rate: 3 oz./bu. seed

Disease:
damping-off; seed decay

Time: Drill box or planter box treatment.

Remarks: Seed treatment. Treat only seeds needed for immediate use, minimizing interval between treatment and planting. Do not store excess treated seeds beyond planting time. Do not use treated seed for food, feed, or oil purposes. Refer to label for use directions.

MEFENOXAM

Systemic fungicide having a specific mode of action.

Apron XL LS *(Novartis)*

Rate: 0.16-0.64 fl. oz. LS/100 lb. seed

Disease:
damping-off; Phytophthora (early-season)

Time: Seed treatment at planting.

Remarks: For hopper box treatment, refer to label. For soilborne diseases such as Rhizoctonia, apply in combination with other registered seed dressing treatments. Do not use treated seed for food, feed, or oil purposes.

- - - - - - - - - - - - - - - - - -

Signal Word/Toxicity Class: Warning/II.

Ridomil Gold EC, GR, or WSP *(Novartis)*

Rate: Surface Application:
0.63 lb. ai/A
1.25 pt. EC/A
0.56 lb. ai/A
1.25 lb. WSP/A
In-Furrow Application:
0.25 fl. oz. EC/1000 lin. ft. row
6 oz. GR/1000 lin. ft. row
0.25 oz. WSP/1000 lin. ft. row

Disease:
damping-off; Phytophthora root rot; Phytophthora stem rot

Time: At planting.

Remarks: Use higher rates in areas with a history of heavy Phytophthora damage. May be applied broadcast, banded, or in-furrow. For banded application, 7" band recommended. Do not plant any crop which is not registered for use with mefenoxam in treated soil for 12 months.

- - - - - - - - - - - - - - - - - -

Signal Word/Toxicity Class: Warning/II (EC); Caution (GR, WSP).
REI: 48 hr.

METALAXYL

Systemic fungicide having a specific mode of action.

Allegiance or Apron (FL) *(Gustafson)*

Rate: 0.375-0.75 fl. oz. Allegiance/100 lb. seed
0.75-1.5 fl. oz. FL/100 lb. seed

Disease:
damping-off; Phytophthora (early-season)

Time: Seed treatment.

Remarks: Use with EPA-approved dye that imparts unnatural color to seed. Do not use for feed, food, or oil purposes. Refer to label for use directions.

- - - - - - - - - - - - - - - - - -

Signal Word/Toxicity Class: Warning/II.

NOTICE The information on these pages is for preliminary planning — not a guide for use. Be sure to follow the manufacturer's directions, notwithstanding information contained here.	For personal protective equipment and EPA registration numbers, please turn to page 39.

Apron Dry *(Trace)*
Rate: 3-4 oz./100 lb. seed
Disease:
damping-off Phytophthora
Time: Seed treatment.
Remarks: Treat only those seeds needed for immediate use. Do not use treated seed for food, feed, or oil purposes. Do not store excess treated seeds beyond planting time. Refer to label for use directions.

Signal Word/Toxicity Class: Caution.
REI: 12 hr.

Micro Flo — see sulfur

Microthiol — see sulfur

Nu-Gro Isotox — see captan premixes

PCNB

RTU-PCNB contains toxic inert Rhodamine B. Toxic to fish and aquatic organisms.

PCNB 2-E or 2LF *(Amvac)*
Rate: 3.35-6.70 fl. oz. 2-E/100 lb. seed
4-8 fl. oz. 2LF/100 lb. seed
Disease:
covered smut (2-E) Fusarium Rhizoctonia
Time: Seed treatment.
Remarks: Do not apply through any type of irrigation system. Do not use for food, feed or oil purposes. Do not plant root crops within 12 months of application unless PCNB is registered for use on those crops.

Signal Word/Toxicity Class: Danger/I (2-E); Caution (2LF).
REI: 12 hr.

RTU-PCNB *(Gustafson)*
Rate: 3-6 fl. oz. RTU/100 lb. seed
Disease:
Fusarium Rhizoctonia
Time: Seed treatment.
Remarks: Do not use treated seed for food, feed, or oil purposes. Do not feed treated seed to animals. Refer to label for use directions.
State Restrictions: Not for use in California.

Signal Word/Toxicity Class: Caution.

Prevail — see carboxin premixes

Protector-L — see thiram

Redball — see sulfur

Ridomil Gold — see mefenoxam

Rival — see captan premixes

RTU (Premix)

Ready-to-use flowable fungicide containing the active ingredients of thiram and thiabendazole (Mertect).

RTU (FL) (12.60% thiram & 0.34% thiabendazole) *(Gustafson)*
Rate: 4 fl. oz. FL/bu. seed
Disease:
damping-off seed decay seedling disease
Diaporthe pod blight seedling blight stem blight
Time: Slurry seed suspension
Remarks: May be applied undiluted or with water when additional volumes are needed to insure uniform seed coverage. Do not tank mix unless testing or prior experience insures compatability. Use with EPA-approved dye that imparts unnatural color to seed. Do not use for feed, food, or oil purposes. Do not graze or feed livestock on forage or hay from treated areas. Refer to label for use directions.

Signal Word/Toxicity Class: Caution.

RTU-Vitavax-Thiram — see carboxin premixes

Seed Mate Isotox — see captan premixes

Seed Mate Soybean Mo — see captan

Seed Mate Vitavax-Captan — see carboxin premixes

Sorghum Guard — see captan premixes

Soy Gro — see captan

SULFUR

Nonsystemic fungicide with protective action. Microthiol Special, Sulfa 80, and Thiolux are micronized sulfurs.

Drexel Suffa (6FL), Sulfa 80 (80DF) or Sulfur (90W) *(Drexel)*
Rate: 3-4.5 lb. ai/A
4-6 pt. 6FL/A
6-10 lb. ai/A
7.5-12.5 lb. 80DF/A
2.7-4.5 lb. ai/A
3-5 lb. 90W/A
Disease:
powdery mildew
Time: Apply when disease appears; repeat as necessary.
Remarks: Do not use within 2 weeks of oil application (4 weeks 6FL). Some plants are susceptible to injury from sulfur under certain climatic conditions. Do not make sulfur applications at excessively high temperatures. Do not apply this product through any type of irrigation system. Refer to label for use directions.

Signal Word/Toxicity Class: Caution.
REI: 24 hr.

Kumulus DF (sulfur) *(BASF)*
Rate: 4.8-12 lb. ai/A
6-15 lb. DF/A
Disease:
powdery mildew
Time: Apply when disease first appears and repeat as necessary.
Remarks: Do not apply if temperatures during or within 3 days after application are expected to exceed 90°F. Do not apply through any type of irrigation system. Refer to label for use directions.

Signal Word/Toxicity Class: Caution.
REI: 24 hr.

Micro Flo Sulfur (6L), Dual Spraying & Dusting Sulfur (DS) or Microsperse Wettable Sulfur (WS) *(Micro Flo)*
Microthiol Special (80WP) *(Elf Atochem)*
Rate: 3-12 lb. ai/A
4-16 pt. 6L/A
11.3 lb. ai/A
12.5 lb. DS, WS/A
8-12 lb. ai/A
10-15 lb. 80WP/A
Disease:
leaf spot powdery mildew
Time: Apply at early leaf stage; repeat at 14-day intervals as needed.
Remarks: Do not apply DS and WS through any type of irrigation system. Refer to label for use directions.

Signal Word/Toxicity Class: Caution.
REI: 24 hr.

◆ - **new product** • ai-active ingredient • DF-dry flowable • E/EC-emulsifiable concentrate • EW-emulsion in water
F/FL-flowable • G-granule • L/LC-liquid • SP/WSP-soluble packet • W/WP-wettable powder • WSB-water soluble bag

Redball EM-53 (6L), Superfine Wettable Sulphur (WS) or Commercial Dusting Sulphur (DS) *(International Sulphur)*
Rate: 4-20 pt. 6L/A
5-10 lb. WS/100 gal. water
5-15 lb. DS/A
Disease:
powdery mildew
Time: Apply when disease first appears; repeat as necessary.
Remarks: Apply in sufficient water for thorough coverage. Do not apply when temperatures exceed 90°F. Refer to label for use directions.

Signal Word/Toxicity Class: Caution.
REI: 12 hr.

Terra Sulfur (6L) *(Terra)*
Rate: 12-16 pt. 6L/A
Disease:
leaf spot, powdery mildew
Time: Apply at early leaf stage; repeat at 14-day intervals.
Remarks: Do not use within 2 weeks of oil application. Do not make sulfur applications at excessively high temperatures. Refer to label for directions.
Days To Harvest: 30.

Signal Word/Toxicity Class: Caution.
REI: 24 hr.

Thiolux (DF) *(Platte)*
Rate: 7.5-12.5 lb. DF/A
Disease:
powdery mildew
Time: Apply when disease first appears; repeat as needed.
Remarks: Do not make applications within 2 weeks of oil spray. Refer to label for use directions.

Signal Word/Toxicity Class: Caution.
REI: 24 hr.

SULFUR (Premix)

Fungicide/bactericide in flowable formulation. Toxic to fish and aquatic organisms.

Top Cop with Sulfur (6.25 lb./gal. sulfur & 1 lb./gal. tribasic copper sulfate) *(Stoller)*
Rate: 4 pt./A
Disease:
Anthracnose, Cercospora leaf spot, stem blight, brown spot, Diaporthe pod blight
Time: Apply when pods are 1/8-1/2" long (early pod set); repeat at 10- to 14-day intervals throughout growing season.

Rate: Seed treatment:
8-12 fl. oz./100 lb. seed
Disease:
damping-off, seedling disease
Time: Seed treatment.
Remarks: Do not use treated seed for food, feed or oil purposes. Refer to label for use directions.

Signal Word/Toxicity Class: Caution.
REI: 24 hr.

SYSTEM³ (Premix)

Biological and chemical fungicide containing the active ingredients of Apron, Terraclor, and Kodiak. Contains graphite. Toxic to fish and aquatic organisms.

System³ (0.1% *Bacillus subtilis* & 14.25% metalaxyl & 16.67% PCNB) *(Helena)*
Rate: 4-6 oz./bu. seed
Disease:
Phytophthora, Pythium, *Rhizoctonia solani*
Time: Apply at planting.
Remarks: A non-oil lubricant has been added for seeds to flow more easily and evenly from planter boxes. Incorporation of dye is recommended to distinguish seed as treated. Do not use treated seed for food, feed or oil purposes. Refer to label for use directions.

Signal Word/Toxicity Class: Caution.
REI: 12 hr.

Terranil — see chlorothalonil

Thiolux — see sulfur

THIOPHANATE-METHYL

Benzimidazole systemic fungicide. Toxic to fish.

Topsin M (WSB, 70W) *(Elf Atochem)*
Rate: 0.35-0.7 lb. ai/A
0.5-1 lb. WSB, 70W/A
Disease:
aerial blight (suppression), brown spot, purple seed stain, frogeye leaf spot, Sclerotinia, Anthracnose, Diaporthe pod blight, stem blight
Time: Apply from full bloom to when pods are 1/8-1/4" in length; repeat 14- to 21-days later. Do not make second application later than 14 days after pods average 1/4" in length or when beans form in pod. Use high rate under severe disease pressure. For Sclerotinia, make first application at early bloom; repeat 7- to 14-days later if conditions are favorable for continued disease pressure.
Remarks: For seed beans, make single application at high rate when beans form in pod. Do not make more than 2 applications per year. Do not graze or feed treated vines or hay to livestock. Refer to label for directions.

Signal Word/Toxicity Class: Caution.
REI: 12 hr.

THIRAM

Broad-spectrum fungicide. Protector-L contains molybdenum for plant nutrition and nitrogen fixation. Protector-L and Thiram 50WP contain the toxic inert Rhodamine B. Toxic to fish; Thiram 50WP also toxic to birds and wildlife.

Protector-L *(Trace)*
Rate: 4 fl. oz./bu. seed
Disease:
damping-off, seedling blight, seedling disease, seed decay
Time: Planter box seed treatment.
Remarks: Treat only those seeds needed for immediate use. Do not use treated seed for food, feed, or oil purposes. Do not store excess treated seeds beyond planting time. Refer to label for use directions.

Signal Word/Toxicity Class: Warning/II.

42-S Thiram or Thiram 50WP Dyed *(Gustafson)*
Thiram Granuflo (WDG) *(UCB Agrochemicals)*
Rate: 2 fl. oz. 42-S/100 lb. seed
3.3 oz. 50WP/100 lb. seed
2.2 oz. WDG/100 lb. seed
Disease:
damping-off, seedling blight, seedling disease, seed decay
Time: Seed treatment.
Remarks: Use with EPA-approved dye that imparts unnatural color to seed. Do not use for feed, food, or oil purposes. Do not apply Thiram Granuflo through any type of irrigation system. Refer to label for use directions.
State Restrictions: Thiram Granuflo not for use in California.

Signal Word/Toxicity Class: Caution.
REI: 24 hr. (WDG).

Top Cop with Sulfur — see sulfur premix

Topsin M — see thiophanate-methyl

Vitaflo — see carboxin premixes

Vitavax — see carboxin

Vitavax-PCNB — see carboxin premixes

Vitavax T-L — see carboxin premixes

NOTICE The information on these pages is for preliminary planning — not a guide for use. Be sure to follow the manufacturer's directions, notwithstanding information contained here. For personal protective equipment and EPA registration numbers, please turn to page 39.

Notes • Notes • Notes

Section 4

Cotton

Enhanced Seed Chart 108
Bt Chart 108
Cotton crop listings 109
Insecticides 109
Fungicides 122

The information in this 1999 Insect and Disease Control Guide is updated to August 15, 1998. Because manufacturers' product recommendations are constantly being updated and revised as additional product registrations and label changes are approved by EPA, there may be changes after this date. However, every effort has been made to give manufacturers the opportunity to include in the Guide their latest information. This Guide is for preliminary planning – not a use guide. Be sure to follow label directions, notwithstanding information contained herein.

Enhanced Seed & Bt Products Chart

Cotton

Find enhanced seed and Bt products quickly and easily with these charts. For enhanced seed, find the product description along the left-hand side of the chart and follow over to the marked spot. For Bt products, find the insect you want to control at the top of the chart, and follow down to the marked spot to find the Bt product you can use to control that pest. Company address information can be found on page 599.

Enhanced Seed Products

Enhanced Seed Products	Stoneville Pedigreed Seeds	Deltapine Seed	Paymaster Cottonseed
BXN cotton + Bollgard	■		
Bollgard + Fusarium wilt resistance/tolerance		■	
Bollgard		■	■
Bollgard + Fusarium wilt resistance/tolerance + Verticillium wilt resistance/tolerance		■	
Roundup Ready + Fusarium wilt resistance/tolerance		■	
Roundup Ready + Fusarium wilt resistance/ tolerance + Verticillium wilt resistance/tolerance		■	
Roundup Ready + Fusarium wilt resistance/ tolerance + Verticillium wilt resistance/tolerance + Bollgard		■	
Roundup Ready + Fusarium wilt resistance/ tolerance + root-knot nematode resistance/tolerance		■	
Roundup Ready + Bollgard			■
Roundup Ready + Verticillium wilt resistance/tolerance			■

Manufacturers List

These Cotton Bt products are manufactured by the following companies. Look for their addresses, phone numbers, and fax numbers in the Suppliers Directory beginning on page 599.

Javelin – Thermo Trilogy
DiPel, XenTari – Abbott Laboratories
BMP 123 – Becker Microbial
Condor, Crymax, Cutlass, Lepinox – Ecogen
Kodiak – Gustafson
Mattch, MVP II – Mycogen
Troy-Bt – Troy Biosciences

Bt Products

Bt Products	Armyworms	Cotton bollworm	Cotton leafperforator	Cotton leafworm	Fusarium	Green cloverworm	Helicoverpa/Heliothis	Hornworms	Loopers	Rhizoctonia	Saltmarsh caterpillar	Tobacco budworm	Velvetbean caterpillar	Yellowstriped caterpillar
Bacillus subtilis					■					■				
BMP 123	■	■	■	■					■		■	■		
Bt. ssp. *aizawai*	■	■							■		■	■		
Bt. ssp. *aizawai* + *Bt.* ssp. *kurstaki*	■	■							■			■		
Bt. ssp. *kurstaki*	■	■	■	■		■	■	■	■		■	■	■	■
Condor	■	■	■						■		■	■		
Crymax	■	■	■						■		■	■		
Cutlass	■	■	■						■		■	■		
DiPel	■	■					■		■		■	■		
Javelin	■	■	■	■					■		■	■		
Kodiak					■					■				
Lepinox	■		■			■			■		■		■	■
Mattch	■	■							■			■		
MVP II	■	■							■			■		
Troy-Bt	■	■				■		■	■		■	■	■	
XenTari	■	■							■		■	■		

Cotton

Insecticides

ABAMECTIN

Emulsifiable concentrate for control of spider mites in cotton. Highly toxic to humans, other mammals, fish, and wildlife. Restricted Use Pesticide.

Zephyr 0.15EC *(Novartis)*

Rate: Cotton under 10" tall (early-season):
4-16 fl. oz. 0.15EC/A
Cotton over 10" tall:
8-16 fl. oz. 0.15EC/A

Pest:

carmine spider mite, Pacific spider mite, strawberry spider mite, twospotted spider mite

Remarks: Rate depends on plant size and foliage density. Apply when mites are first noticed; repeat if necessary. Apply in minimum 5 gal. water/A with ground sprayers or aerially by fixed wing aircraft or helicopter. For early-season cotton (under 10"), apply by ground. Do not graze or feed cotton foliage.

Days To Harvest: 20.

State Restrictions: Not for use east of the Mississippi.

Signal Word/Toxicity Class: Warning/II.

REI: 12 hr.

ACEPHATE

Systemic organophosphate with several uses in cotton. Often used as seed treatment or applied in-furrow for thrips control. As a foliar spray, effective against several other pests. Frequently tank mixed with other insecticides. Toxic to birds.

Micro Flo Acephate 75SP or 90SP *(Micro Flo)*
Orthene 75 S or 90 S *(Valent)*

Rate: 0.1-1 lb. ai/A
2 oz.-1.33 lb. 75SP, 75 S/A
0.18-0.9 lb. ai/A
3.2 oz.-1 lb. 90SP, 90 S/A

Pest:

armyworms (except beet armyworm), bollworms, cabbage looper, cotton aphid (except AZ, CA), fleahoppers, Lygus bugs, pink bollworm (AZ, CA), plant bugs, stink bugs (AZ, CA), thrips, tobacco budworm, whiteflies

Remarks: Apply when eggs or insects appear and repeat as needed to maintain control. For Lygus bugs, use higher rate for adults that have migrated into cotton. May be tank mixed with Lorsban; refer to label for rates and directions. Do not feed gin trash to livestock or allow animals to graze treated areas.

Days To Harvest: 21.

Rate: Seed hopper box (except AZ, CA, Blacklands of TX):
0.15-0.2 lb. ai/A
3-4 oz. 75SP, 75 S/A
2.5-3.25 oz. 90 S/A
In-furrow:
0.5-1 lb. ai/A
0.66-1.33 lb. 75 S/A
0.56-1.1 lb. 90SP, 90 S/A

Pest:

black cutworm, cotton aphid, thrips

Remarks: For hopper box treatment, apply in planter box as dry powder. Loss of acephate will occur if air planters are used. Plants grown from improperly treated seed will not exhibit insect control. Do not use treated seed for food or feed purposes or process for oil.

State Restrictions: Not for use in Arizona, California, and Blacklands of Texas for hopper box treatment. For Blacklands, use 1 lb. ai/A in-furrow.

Signal Word/Toxicity Class: Caution.

REI: 24 hr.

Orthene 80 or 80WSP *(Valent)*

Rate: 0.4 lb. ai/100 lb. seed
8 oz. 80, 80WSP/100 lb. seed

Pest:

black cutworm, cotton aphid, seedcorn maggots, thrips

Remarks: Seed protectant. Reduction of aphids and thrips up to 3 weeks after planting; reduction of black cutworm from planting through 3- to 4-leaf stage. Do not use treated seed for food, feed, or oil.

Signal Word/Toxicity Class: Caution.

Payload (15 G) *(Valent)*

Rate: 0.75-1 lb. ai/A
5-6.66 lb. 15 G/A

Pest:

aphids, cutworms, thrips

Remarks: Apply in-furrow at planting.

Signal Word/Toxicity Class: Caution.

REI: 12 hr.

Admire — see imidacloprid

AMITRAZ

Insecticide/miticide with ovicidal and synergistic activity. Not phytotoxic to young cotton. Toxic to fish.

Ovasyn (EC) *(AgrEvo USA)*

Rate: 0.125-0.25 lb. ai/A
0.66-1.33 pt. EC/A

Pest:

aphids, beet armyworm, bollworms, pink bollworm, tobacco budworm, whiteflies

Rate: 0.25-1 lb. ai/A
1.33-5.33 pt. EC/A

Pest:

aphids, mites

Remarks: Do not apply after cotton bolls open. Do not apply more than 5.33 pt./A per growing season. Do not feed foliage from treated plants to livestock.

Signal Word/Toxicity Class: Danger/I.

REI: 24 hr.

Ammo — see cypermethrin

NOTICE	The information on these pages is for preliminary planning — not a guide for use. Be sure to follow the manufacturer's directions, notwithstanding information contained here.	For personal protective equipment and EPA registration numbers, please turn to page 39.

ASANA

Synthetic pyrethroid insecticide formulated in an emulsifiable concentrate. Usually recommended to control cotton bollworm, pink bollworm, and tobacco budworm, and will give some control of various other pests. Has low toxicity to humans and other mammals. Restricted Use Pesticide due to toxicity to fish and aquatic organisms.

Asana XL (esfenvalerate) *(DuPont)*

Rate: 0.02-0.05 lb. ai/A
3.9-9.6 fl. oz. XL/A

Pest:

beet armyworm (aids in control)	cotton leafworm	plant bugs
boll weevil	cutworms	saltmarsh caterpillar
cabbage looper	fleahoppers	thrips (seedling cotton)
cotton aphid	grasshoppers	tobacco budworm
cotton bollworm	leafhoppers	whiteflies (suppression)
cotton leafperforator	Lygus bugs	
	pink bollworm	

Remarks: May be applied in water or nonvolatile vegetable oils. Rate varies by pest species; refer to label. For boll weevil, 3- to 5-day interval between sprays may be required. Apply no more than 0.5 lb./A per growing season. Do not graze livestock on treated fields or feed treated trash. Refer to label for spray drift precautions.

Days To Harvest: 21.

Signal Word/Toxicity Class: Warning/II.
REI: 12 hr.

AZADIRACHTIN

Extracted from seed of neem tree. Insect growth regulator that controls targeted insect larvae when they ingest or come in contact with it, by interfering with the insect's ability to molt and shutting off feeding behavior. Affected larvae slowly starve and die at next molt.

◆ **NEEMAZAD 4.5EC or NEEMIX 4.5** *(Thermo Trilogy)*

Rate: 0.125-1 pt. 4.5/A
High Volume:
2.25-4.5 oz. 4.5EC/100 gal. water
Low Volume:
0.5 pt. 4.5EC/A

Pest:

armyworms	cutworms	loopers
boll weevil	leafhoppers	pink bollworm
cotton bollworm	leafminers	thrips
cotton leafperforator	leafrollers	tobacco budworm

Remarks: For low volume, apply in sufficient water for adequate coverage. For low insect pressure, apply preventative at 7-day intervals or at first sign of insects; high insect pressure, apply at 3- to 4-day intervals.

Signal Word/Toxicity Class: Warning.
REI: 12 hr.

AZINPHOS-METHYL

Organophosphate insecticide usually targeted toward boll weevil but also effective against various other pests. Highly toxic to humans and other mammals. Restricted Use Pesticide.

Gowan Azinphos-M 50-W *(Gowan)*
◆ **Micro Flo Azinphosmethyl 50W** *(Micro Flo)*

Rate: 0.125-1 lb. ai/A
0.25-2 lb. 50-W, 50W/A

Pest:

boll weevil	cotton leafhopper	rapid plant bug
brown cotton leafworm	cotton leafworm	tarnished plant bug
	pink bollworm	thrips

Remarks: Apply by ground or air in sufficient water for complete coverage but not less than 5 gal. water/A; repeat as needed. Rate varies by pest species; refer to label. Cotton receiving late-season applications should not be pastured.

Days To Harvest: 1.

Signal Word/Toxicity Class: Danger-Poison/I.
REI: 48 hr.; 72 hr. if average rainfall less than 25" per year.

Guthion 2L *(Bayer)*
◆ **Micro Flo Azinphosmethyl 2EC** *(Micro Flo)*

Rate: 0.125-1.5 lb. ai/A
0.5-3 pt. 2L/A
0.125-2 lb. ai/A
0.25-4 pt. 2EC/A

Pest:

boll weevil	cotton leafworm	stink bugs
brown cotton leafworm	Lygus bugs	tarnished plant bug
cotton leafhopper	pink bollworm	thrips
	rapid plant bug	

Remarks: Conventional or low volume spray: Apply by air or ground in sufficient water for complete coverage, but not less than 1 gal./A. Rate varies by pest species; refer to label. May be tank mixed with DEF defoliant in 1 application; refer to labels. Ultra low volume spray: May be used undiluted with ground or aerial spray equipment that has been adapted and calibrated for ultra low volume spraying. For early- and mid-season control of boll weevil, apply in accordance with local recommendations. For diapause weevils, use 1 pt./A. Repeat as necessary, but not within 2 days of hand-picking. Do not apply more than 12 applications per season. Do not graze livestock in treated areas.

Days To Harvest: Conventional or LV spray (hand-picking)-1; ULV spray (hand-picking)-2; machine harvested-0.

State Restrictions: Pink bollworm in California only for 2L.

Signal Word/Toxicity Class: Danger/I.
REI: 48 hr.; 72 hr. if average rainfall less than 25" per year.

BAYTHROID

Synthetic pyrethroid formulated in an emulsifiable concentrate. Frequently recommended to control cotton bollworm, pink bollworm, and tobacco budworm, and gives good control of boll weevil. Toxic to fish and aquatic organisms. Restricted Use Pesticide.

Baythroid (2EC) (cyfluthrin) *(Bayer)*

Rate: 0.0125-0.05 lb. ai/A
0.8-3.2 fl. oz. 2EC/A

Pest:

aphids	flea beetles	soybean looper
beet armyworm (1st & 2nd instar)	fleahoppers	stink bugs
boll weevil	garden webworm	threecornered alfalfa hopper
cabbage looper	grasshoppers	thrips
cotton aphid	leafminers	tobacco budworm
cotton bollworm	leafworm	whiteflies (suppression)
cotton leafperforator	Lygus bugs	yellowstriped armyworm
cutworms	pink bollworm	
European corn borer	saltmarsh caterpillar	
fall armyworm	southern garden leafhopper	

Remarks: Apply in minimum 1 gal./A as needed for thorough coverage. Applications may also be made with ground or aerial equipment adapted and calibrated for ULV application. May be mixed with once-refined vegetable oil and applied in total volume at least 1 qt./A. For sprinkler irrigation systems, follow chemigation directions on label. Timing of applications should be based on careful scouting and local economic thresholds. A 3-5 day application schedule may be required to maintain control of heavy boll weevil infestations. Lower rates adequate for low to moderate insect pressure, but may require more frequent applications. For armyworms, scout carefully for early detection. Subsequent applications may be necessary to maintain control. For cotton aphid, thorough coverage under light infestation pressure necessary for control; for heavy infestation, systemic insecticide may be necessary. For maximum contact ovicidal effect on bollworm/budworm, time application to correspond with peak egg deposition. Up to 32 fl. oz./A may be applied per crop season. Do not make more than 10 applications per crop season. Do not graze treated fields. Refer to label for spray drift precautions.

Days To Harvest: 0.

Signal Word/Toxicity Class: Danger/I.
REI: 12 hr.

BIDRIN

Organophosphate insecticide formulated as an emulsifiable concentrate. Extremely toxic to humans, other mammals, fish, birds, bees, and other wildlife. Restricted Use Pesticide.

Bidrin 8 (dicrotophos) *(Amvac)*

Rate: 0.1-0.5 pt. ai/A
1.6-8.0 fl. oz. 8/A

Pest:

aphids	grasshoppers	saltmarsh caterpillar
black fleahopper	leafperforators	stink bugs
boll weevil	Lygus bugs	thrips
cotton fleahopper	mites	

◆ - **new product** • ai-active ingredient • DF-dry flowable • E/EC-emulsifiable concentrate • EW-emulsion in water
F/FL-flowable • G-granule • L/LC-liquid • SP/WSP-soluble packet • W/WP-wettable powder • WSB-water soluble bag

Remarks: Apply to give uniform coverage. Repeat as necessary but do not apply more than 3 times per season. Rate varies by pest species and time of season; refer to label. Do not graze livestock on treated fields or feed treated gin trash.
Days To Harvest: 30.

Signal Word/Toxicity Class: Danger-Poison/I.
REI: 72 hr.

BOLSTAR

Organophosphate insecticide formulated as an emulsifiable concentrate. Restricted Use Pesticide due to acute avian toxicity.

Bolstar 6 (sulprofos) *(Bayer)*
Rate: 0.75-1.5 lb. ai/A
1-2 pt. 6/A
Pest:

beet armyworm*	fleahoppers	twospotted spider mite**
carmine spider mite**	Lygus bugs (adult**, nymph)	whiteflies** (except sweetpotato)
cotton bollworm*	pink bollworm	
fall armyworm*	tobacco budworm*	

** 1st & 2nd instar larvae ** Suppression*

Remarks: Apply by air, ground, or sprinkler irrigation equipment, alone, in sufficient water, or refined vegetable oil. For best results during mid- to late-season or under hot, dry conditions, use 3-5 gal. water by air. For bollworm and budworm, apply at 5-7 day intervals when treatment thresholds are reached. For light infestations, apply 0.75 pt./A (1.33 pt./A in AZ, CA); moderate infestations 1.66 pt./A; heavy infestations 2 pt./A. In areas of frequent rainfall, higher rates and closer spray intervals (5-day minimum interval required) may be needed for adequate control. For armyworms, make 2 applications 5 days apart when larvae first appear. For light infestations, apply 1.33 pt./A; moderate infestations 1.66 pt./A; heavy infestations 2 pt./A. Space applications at least 5 days apart and do not apply more than 8 pt. 6E/A. May be tank mixed with Guthion 2L or 2S. Rate varies by pest species; refer to label and follow all directions, limitations, and use precautions. Do not graze treated fields.
Days To Harvest: 14; Arizona, California-21.
Crop Rotation: Crops to be grown in rotation with cotton should not be planted within 60 days after last application.

Rate: 0.25 lb. ai/A
0.33 pt. 6/A
Pest:

fleahoppers	plant bugs	thrips (except western flower)

Remarks: Apply early- to mid-season as needed when insects reach economic threshold. Space applications at least 5 days apart and do not apply more than 8 pt. 6E/A. Refer to label and follow all directions, limitations, and use precautions. Do not graze treated fields.
State Restrictions: For use in Southern areas only.

Signal Word/Toxicity Class: Warning/II.
REI: 48 hr.; 72 hr. if average rainfall less than 25" per year.

Bt 320 Sulfur 25 or 50 Dust (Premix)

Bt 320 Sulfur 25 Dust (0.064% Bt & 25% sulfur) or Bt 320 Sulfur 50 Dust (0.064% Bt & 50% sulfur) *(Wilbur-Ellis)*
Rate: 25-50 lb. Dust/A
Pest:
cabbage looper
Remarks: Do not mix or dilute with other products. Thorough coverage essential.

Signal Word/Toxicity Class: Caution.
REI: 24 hr.

CAPTURE

Synthetic pyrethroid insecticide with broad spectrum activity against most cotton pests. Formulated as an emulsifiable concentrate and frequently recommended for pest combinations. Restricted Use Pesticide.

Capture 2EC (bifenthrin) *(FMC)*
Rate: 0.02-0.1 lb. ai/A
1.3-6.4 fl. oz. 2EC/A
Pest:

banded soybean thrips	cotton leafperforator	southern garden leafhopper
beet armyworm	cutworms	stink bugs
boll weevil	European corn borer	tobacco budworm
bollworms	fall armyworm	tobacco thrips
cabbage looper	Lygus bugs	twospotted spider mite
carmine spider mite	pink bollworm	whiteflies
cotton aphid	plant bugs	yellowstriped armyworm
cotton fleahopper	saltmarsh caterpillar	

Remarks: Apply in minimum 5 gal./A by ground; 1 gal./A by air. For ULV applications, apply in refined vegetable oil in minimum 1 qt. finished spray/A with aircraft calibrated to give adequate coverage. Rate varies by pest species. For boll weevils, apply at 3- to 4-day intervals until pest numbers are reduced to acceptable levels. For aphids and mites, apply when pests first appear; repeat as necessary to maintain control. Higher rates required once a damaging threshold is established. Do not apply more than 0.5 lb. ai/A per season. May be tank mixed with other products approved for use on cotton; observe all label restrictions. Do not graze treated areas. Refer to label for spray drift precautions.
Days To Harvest: 14.
Crop Rotation: Not within 30 days after last application.

Signal Word/Toxicity Class: Warning/II.
REI: 12 hr.

CARBARYL

Carbamate insecticide with some activity on a wide variety of pests. Has many formulations and is relatively safe to humans and other higher vertebrates, but is highly toxic to bees.

Clean Crop Sevin 5 Bait *(Platte)*
◆ **Gowan Sevin Bait** *(Gowan)*
Rate: 1.5 lb. ai/A
30 lb. Bait/A
Pest:

armyworm	cutworms	grasshoppers
crickets	darkling beetle	sowbugs

Remarks: Apply with ground or air equipment; repeat at 7- to 14-day intervals as needed. Do not use when bees are present.
Days To Harvest: 7.

Signal Word/Toxicity Class: Caution/III.
REI: 12 hr.

Drexel Carbaryl 4L or 80S *(Drexel)*
Sevin 4F, XLR Plus, 80S, 80WSP *(Rhone Poulenc)*
Rate: 0.5-2.5 lb. ai/A
1-5 pt. 4L, 4F, XLR/A
0.66-3 lb. 80S, 80WSP/A
Pest:

boll weevil	fall armyworm	saltmarsh caterpillar
bollworms	flea beetles*	stink bugs
cotton fleahopper*	grasshoppers	striped blister beetle*
cotton leafperforator (4L, 80S only)	leafhoppers	tarnished plant bug
	leafrollers	thrips*
cotton leafworm*	Lygus bugs	yellowstriped armyworm
cutworms	pink bollworm	

** Early-season control*

Remarks: Aphid population will be suppressed by repeated application. Rate varies by pest species, infestation level, and time of season; refer to label. To avoid possible injury on tender foliage, do not apply when foliage is wet or excessive humidity is expected during the next 2 days.
Days To Harvest: 7.

Signal Word/Toxicity Class: Caution (4F, 4L, XLR); Warning/II (80S, 80WSP).
REI: 12 hr.

◆ **Wilbur-Ellis Sevin 5 or 10 Dust** *(Wilbur-Ellis)*
Rate: 2-2.5 lb. ai/A
40-50 lb. 5/A
2.5 lb. ai/A
25 lb. 10/A
Pest:

bollworms	leafperforators	Lygus bugs
fall armyworm	leafhoppers	saltmarsh caterpillar
flea beetles	leafrollers	stink bugs
fleahoppers	leafworm	thrips
grasshoppers		

Remarks: Do not graze treated fields or feed cotton trash to livestock.
Days To Harvest: 7.
State Restrictions: Sevin 5 not registered in California.

Signal Word/Toxicity Class: Caution.
REI: 12 hr.

For personal protective equipment and EPA registration numbers, please turn to page 39.

CHLORPYRIFOS

Organophosphate insecticide/miticide with broad spectrum activity on cotton pests. Can be tank mixed with other cotton insecticides to control Helicoverpa spp. Lorsban 30 and 50-SL for use only by commercial seed treaters.

Lock-On *(Dow AgroSciences)*

Rate: 0.5 lb. ai/A
2 pt./A

Pest:

armyworms	cotton leafperforator (suppression)	saltmarsh caterpillar
boll weevil		spider mites (suppression)
cotton aphid	cutworms	
cotton bollworm	Lygus bugs	thrips
cotton fleahopper	pink bollworm	tobacco budworm

Remarks: Apply when field counts indicate damaging insect populations are developing or present; repeat when necessary. Do not graze treated areas or feed treated forage to livestock. Do not apply when bees are foraging or clustered outside of hives.

Days To Harvest: 40.

State Restrictions: For use in Arizona, California.

- - - - - - - - - - - - - - - - -

Signal Word/Toxicity Class: Warning/II.

REI: 12 hr.

Lorsban-4E *(Dow AgroSciences)*
◆ **Micro Flo Chlorpyrifos 4E AG** *(Micro Flo)*
Nufos 4E *(Cheminova)*

Rate: Except Arizona, California:
0.19-1 lb. ai/A
0.375-2 pt. 4E/A

Pest:

beet armyworm	fall armyworm	saltmarsh caterpillar
boll weevil	grasshoppers	spider mites
bollworms	Lygus bugs	thrips
cotton aphid	mirids	yellowstriped armyworm
cotton fleahopper	pink bollworm	
cutworms	plant bugs	

Rate: Arizona, California:
0.5-1 lb. ai/A
1-2 pt. 4E/A

Pest:

armyworms	cotton leafperforator (suppression)	saltmarsh caterpillar
boll weevil		spider mites (suppression)
bollworms	cutworms	
cotton aphid	Lygus bugs	thrips
cotton fleahopper	pink bollworm	

Remarks: Rate varies by pest species; refer to label. Do not make more than 6 applications per season. Do not graze treated areas or feed treated gin trash or forage to livestock.

Days To Harvest: 14.

- - - - - - - - - - - - - - - - -

Signal Word/Toxicity Class: Warning/II.

REI: 24 hr.

Lorsban 30, 50-SL *(Gustafson)*

Rate: Delinted seed:
5.5 fl. oz. 30/100 lb. seed
4 oz. 50-SL/100 lb. seed

Pest:
seedcorn maggots

Remarks: Apply prior to planting as a slurry treatment. Treated seed must not be used for or mixed with food or animal feed or processed for oil. Commercially treated seed must be labeled in accordance with all applicable requirements of Federal Seed Act. Do not allow livestock to graze or feed bean hay or pea hay grown from treated seed.

- - - - - - - - - - - - - - - - -

Signal Word/Toxicity Class: Caution (30); Warning/II (50-SL).

Comite — see propargite

CURACRON

Organophosphate insecticide/miticide formulated as emulsifiable concentrate. Frequently recommended to control cotton bollworm, tobacco budworm, and certain other insects and mites on cotton. Use tank mix to broaden spectrum of insect control and make a more useful spray application if multiple pests occur on cotton. Restricted Use Pesticide.

Curacron 8E (profenofos) *(Novartis)*

Rate: 0.125-1 lb. ai/A
0.125-1 pt. 8E/A

Pest:

aphids	fall armyworm	plant bugs (AZ, CA suppression only)
beet armyworm	Lygus bugs	
cotton bollworm	mites	tobacco budworm
cotton leafperforator		whiteflies

Remarks: Rate varies by pest species and infestation level. May be tank mixed with Ammo, Guthion, methomyl, and methyl parathion in all cotton growing states in accordance with label limitations and precautions. For suppression of plant bugs in Arizona and California. Do not apply more than 3 qt./A per season. Do not apply through any type of irrigation system. Do not graze treated cotton plants or feed gin trash or foliage from treated plants to livestock. Refer to label for further directions.

Days To Harvest: 14.

- - - - - - - - - - - - - - - - -

Signal Word/Toxicity Class: Warning/II.

REI: 48 hr; 72 hr. if average rainfall less than 25" per year.

lambda-CYHALOTHRIN

Synthetic pyrethroid insecticide with broad-spectrum activity, formulated as an emulsifiable concentrate. Restricted Use Pesticide due to toxicity to fish and aquatic organisms.

Karate (1E) *(ZENECA)*

Rate: 0.01-0.04 lb. ai/A
1.28-5.12 fl. oz. 1E/A

Pest:

bandedwing whitefly*	cotton fleahopper	southern green stink bug
	cotton leafperforator	
beet armyworm (1st & 2nd instar)	cutworms	soybean thrips
	European corn borer	sweetpotato whitefly*
boll weevil	fall armyworm	thrips
brown stink bug	green stink bug	tobacco budworm
cabbage looper	Lygus bugs	tobacco thrips
cotton aphid*	pink bollworm	twospotted spider mite*
cotton bollworm	saltmarsh caterpillar	

*** Suppression**

Remarks: Apply with ground or air equipment using sufficient water for full coverage of foliage. Rate varies according to pest controlled and insect pressure. For boll weevils, spray on a 3-5 day schedule. Do not apply more than 1.6 pt. per season or make more than 10 synthetic pyrethroid applications per crop season. Timing and frequency of applications should be based on scouting reports. Under light bollworm/budworm pressure, 0.02 lb. ai/A may be used with intense field monitoring. Ovicidal control of unhatched *Heliothis* spp. eggs obtained at larvicidal rates. Do not graze livestock in treated areas. Refer to label for spray drift precautions and further use restrictions.

Days To Harvest: 21.

State Restrictions: For use in all cotton producing states except California.

- - - - - - - - - - - - - - - - -

Signal Word/Toxicity Class: Danger/I.

REI: 24 hr.

CYPERMETHRIN

Synthetic pyrethroid with broad-spectrum activity on cotton pests. Formulated as emulsifiable concentrate, and generally recommended for control of pink bollworms but will also control boll weevils, cutworms, fleahoppers, plant bugs and other insect pests. Low toxicity to humans and other mammals, but is highly toxic to fish. Restricted Use Pesticide.

Ammo 2.5 EC or WSB *(FMC)*

Rate: 0.025-0.1 lb. ai/A
1.3-5 fl. oz. 2.5 EC/A
0.05-0.1 lb. ai/A
1-2 bags WSB/A

Pest:

beet armyworm	fall armyworm	tobacco budworm
boll weevil	Lygus bugs	tobacco thrips
cabbage looper	pink bollworm	whiteflies (aids in control)
cotton bollworm	plant bugs	
cotton fleahopper	saltmarsh caterpillar	yellowstriped armyworm
cotton leafperforator	soybean thrips	
cutworms	tarnished plant bug	

Remarks: Apply by ground or air in water or refined vegetable oil. When water is used, apply a minimum of 5 gal. finished spray/A by ground; 1 gal. finished spray/A by air. For boll weevils, apply at 3-4 day intervals until pest numbers are reduced to acceptable levels. Apply no more than 0.6 lb. ai/A per season. Do not graze or feed cotton for forage. Refer to label for spray drift precautions.

Days To Harvest: 14.

Crop Rotation: Not within 30 days after last application.

- - - - - - - - - - - - - - - - -

Signal Word/Toxicity Class: Caution/III (2.5 EC); Warning/II (WSB).

REI: 12 hr.

◆ - **new product** • ai-active ingredient • DF-dry flowable • E/EC-emulsifiable concentrate • EW-emulsion in water
F/FL-flowable • G-granule • L/LC-liquid • SP/WSP-soluble packet • W/WP-wettable powder • WSB-water soluble bag

zeta-CYPERMETHRIN

Synthetic pyrethroid with broad-spectrum activity. Extremely toxic to fish and aquatic invertebrates; highly toxic to bees. Restricted Use Pesticide.

Fury 1.5 EC or Mustang 1.5 EW *(FMC)*

Rate: 0.016-0.024 lb. ai/A
1.37-2.05 fl. oz. EC, EW/A

Pest:
banded soybean thrips, cutworms, tobacco thrips

Rate: 0.033-0.045 lb. ai/A
2.82-3.83 fl. oz. EC, EW/A

Pest:
boll weevil, cabbage looper, cotton bollworm, cotton fleahopper, cotton leafperforator, European corn borer, fall armyworm, pink bollworm, saltmarsh caterpillar, tarnished plant bug, tobacco budworm, whiteflies, yellowstriped armyworm

Rate: 0.035-0.050 lb. ai/A
2.99-4.26 fl. oz. EC, EW/A

Pest:
beet armyworm, cotton aphid, Lygus bugs, whiteflies

Remarks: Apply in water or refined vegetable oil. In water, apply a minimum of 5 gal. finished spray/A by ground; 1 gal. finished spray/A by air (1 qt. emulsifiable oil may be substituted for 1 qt. water in finished spray). Rate varies by pest species, insect pressure, timing of spray, and field scouting. For boll weevils, apply at 3- to 4-day intervals until pest numbers are reduced to acceptable levels. Up to 0.3 lb. ai/A may be applied per season. May be tank mixed with other products approved for use on cotton; observe all label restrictions. Do not graze or feed treated cotton for forage. Refer to label for spray drift precautions.

Days To Harvest: 14.

Crop Rotation: Not within 30 days after last application.

State Restrictions: For control of beet armyworm only in the high plains of Arizona, California, and Texas.

Signal Word/Toxicity Class: Warning/II.

REI: 12 hr.

DANITOL

Restricted Use Pesticide due to toxicity to fish and aquatic organisms. Also toxic to wildlife.

Danitol 2.4EC (fenpropathrin) *(Valent)*

Rate: 0.2-0.3 lb. ai/A
10.66-16 fl. oz. 2.4EC/A

Pest:
beet armyworm, bollworms, cotton leafperforator, leafhoppers, pink bollworm, plant bugs (Lygus), saltmarsh caterpillar, sweetpotato whitefly, tobacco budworm

Remarks: For bollworm and whitefly, tank mix with 9 oz. Orthene 90S. Do not exceed 2.66 pt./A per season. Do not feed gin trash or treated forage to livestock or allow animals to graze treated fields. Do not make more than 10 synthetic pyrethroid applications to cotton crop in one growing season.

Days To Harvest: 21.

Crop Rotation: No restrictions.

Signal Word/Toxicity Class: Danger/I.

REI: 24 hr.

DECIS

Restricted Use Pesticide.

Decis 1.5 EC (deltamethrin) *(AgrEvo USA)*

Rate: 0.013-0.019 lb. ai/A
1.11-1.62 fl. oz. EC/A

Pest:
cotton fleahopper, cotton leafperforator, cotton leafworm, cutworm, leafhoppers, Lygus bugs, plant bugs, thrips

Rate: 0.019-0.03 lb. ai/A
1.62-2.56 fl. oz. EC/A

Pest:
aphids*, beet armyworm*, boll weevil, cabbage looper, cotton bollworm, European corn borer, fall armyworm*, mites*, pink bollworm, saltmarsh caterpillar, soybean looper*, stink bugs, tobacco budworm, whiteflies*

**Suppression*

Remarks: Apply before insect populations increase to economic significance. Do not make more than 10 applications or apply more than 0.3 lb. ai/A per season. Allow at least 5 days between applications. Do not apply through any type of irrigation system. Refer to label for spray drift precautions.

Days To Harvest: 21.

Signal Word/Toxicity Class: Danger-Poison/I.

REI: 12 hr.

Declare — see methyl parathion

Diatect — see pyrethrins

DIBROM

Organophosphate insecticide formulated as an emulsifiable concentrate. Short residual contact insecticide for kill of insects and mites (adult, larvae) present at time of application. Toxic to fish, aquatic invertebrates, and wildlife. Highly toxic to bees exposed to direct treatment.

Dibrom 8 (8E) (naled) *(Amvac)*

Rate: 1 lb. ai/A
1 pt. 8E/A
Lygus bugs:
0.5-1 lb. ai/A
0.5-1 pt. 8E/A

Pest:
aphids, armyworms, bollworms, fleahoppers, leafperforators, loopers, Lygus bugs, mites (suppression)

Remarks: Apply in 3-20 gal. water by ground; 3-10 gal. water by air at 7- to 10-day intervals while pests persist. For heavy populations, use 0.5-0.75 pt./A in tank mix with other insecticides registered for use on cotton. Do not apply after first bolls open. Do not use with highly alkaline materials such as lime or Bordeaux. Do not apply through any type of irrigation system. Do not graze livestock in treated fields. Do not apply more than 5 pt./A per season. Apply at 7-day intervals.

Signal Word/Toxicity Class: Danger/I.

REI: 48 hr.

DICOFOL

Chlorinated hydrocarbon miticide available in several formulations, but emulsified concentrate is most frequently used in cotton. Toxic to fish, has low toxicity to humans and other mammals, and is nontoxic to bees.

◆ **Griffin Dicofol 4E** *(Griffin)*

Kelthane MF (4E) *(Rohm and Haas)*

Rate: 1-1.5 lb. ai/A
2-3 pt. 4E/A

Pest:
desert mite, Pacific mite, twospotted spider mite

Remarks: Apply at first sign of mite buildup. Rate depends on size of plant and density of foliage. Do not feed cotton trash to livestock. Do not apply through any type of irrigation system.

Days To Harvest: 14.

Signal Word/Toxicity Class: Caution.

REI: 12 hr.

Digon — see dimethoate

Dimate — see dimethoate

NOTICE The information on these pages is for preliminary planning — not a guide for use. Be sure to follow the manufacturer's directions, notwithstanding information contained here.

For personal protective equipment and EPA registration numbers, please turn to page 39.

DIMETHOATE

Organophosphate insecticide with foliar systemic activity, frequently recommended for control of aphids, fleahoppers, plant bugs, and thrips.

Cheminova Dimethoate 4E *(Cheminova)*
Clean Crop Dimethoate 400 *(Platte)*
Dimate 4E *(Terra)*
Drexel Dimethoate 2.67 or 4EC *(Drexel)*
Gowan Dimethoate E267 *(Gowan)*
Helena Dimethoate 4E *(Helena)*
Micro Flo Dimethoate 267 or 4E *(Micro Flo)*
Wilbur-Ellis Dimethoate 267 or Digon 400 *(Wilbur-Ellis)*

Rate: 0.125-0.25 lb. ai/A
0.33-0.66 pt. 2.67, E267/A
0.25-0.5 pt. 4E, 4EC, 400/A
0.125-0.5 lb. ai/A
0.33-1.5 pt. 267/A

Pest:
aphids; Lygus bugs (high rate); mites; fleahoppers; thrips

Rate: Arizona, California:
0.25-0.5 lb. ai/A
0.75-1.5 pt. 2.67, E267/A
0.5-1 pt. 4E, 4EC, 400/A

Pest:
black fleahopper; leafhoppers (Wilbur-Ellis); Lygus bugs; fleahoppers

Remarks: Repeat at 14-day intervals. In Arizona and California, do not apply more than twice per season. Do not graze treated fields or feed treated forage to livestock.

Days To Harvest: 14.

- - - - - - - - - - - - - - - - - -

Signal Word/Toxicity Class: Warning/II.
REI: 48 hr.

DIMILIN

Substituted benzoylphenylurea which acts as a growth regulator by inhibiting chitin formation in insect cuticles. Can be effective in control of boll weevils and is used as a preventive treatment on armyworms. Relatively nontoxic to humans and other mammals. Restricted Use Pesticide due to toxicity to aquatic invertebrates.

Dimilin 2F, 2L, 25W (diflubenzuron) *(Uniroyal)*

Rate: 0.03-0.12 lb. ai/A
2-8 fl. oz. 2F, 2L/A
0.06-0.12 lb. ai/A
4-8 oz. 25W/A

Pest:
beet armyworm; boll weevil; fall armyworm

2F, 2L only:
cabbage looper; southern armyworm; yellowstriped armyworm; saltmarsh caterpillar

Remarks: For armyworms, apply during early stage of larval development; 2-6 pt. emulsifiable vegetable or paraffinic crop oil may be added using sufficient water for adequate coverage. For ground application, apply 10-20 gal. water/A; for ULV application, use a total volume of 20-64 fl. oz./A. For aerial application, apply 3-5 gal. total volume/A; for ULV application use a total volume of 20-48 fl. oz./A. For boll weevil, apply in 4-8 pt. emulsifiable vegetable or paraffinic crop oil with or without water. In California, do not apply in oil without water. For ground application, apply 10-20 gal. water/A. For aerial application, apply 3-5 gal. total volume/A. For maximum control of weevil reproduction, make first application at pinhead square stage of cotton growth when overwintering weevils are entering fields. Repeat at 5- to 7 -day intervals. Do not exceed 6 applications or apply more than 24 fl. oz. 2F, 2L or 1.5 lb. 25W/A per growing season. Do not exceed 3 applications or apply more than 12 fl. oz. 2F, 2L or 12 oz. 25W after bolls begin to open.

Crop Rotation: Do not plant food or feed crops in treated soils within 6 months following last application unless Dimilin authorized for use on these crops.

Days To Harvest: 14.

- - - - - - - - - - - - - - - - - -

Signal Word/Toxicity Class: Caution/III.
REI: 12 hr.

DI-SYSTON

Organophosphate insecticide, often applied at planting for control of thrips and other early-season insects. Soil systemic, and highly toxic to humans, birds, and wildlife. Restricted Use Pesticide.

Di-Syston 8 (disulfoton) *(Bayer)*

Rate: 3-9 oz. ai/A
3-9 fl. oz. 8/A

Pest:
aphids; thrips

Remarks: Foliar application prior to bloom. Spray using aircraft or ground equipment in sufficient water for thorough coverage but not less than 1 gal. total volume/A. Repeat as necessary but do not apply more than 3 times per season with last application no later than bloom. If foliar applications are made, do not apply soil treatments within same crop year. Do not graze treated fields.

- - - - - - - - - - - - - - - - - -

Rate: 9-16 oz. ai/A
9-16 fl. oz. 8/A (40" row spacing)

Pest:
aphids; mites; thrips

Remarks: Preplant or planting application may be followed by postplant application or by 1 or 2 sidedress or broadcast-over-foliage soil applications of Di-Syston 15G. Do not make more than 3 soil applications per crop season regardless of method of application or formulation used. For preplant, apply in 16" band as water emulsion. At planting, apply as water emulsion or with liquid fertilizer; spray in-furrow or soil inject on each side of furrow. For postplant, apply in water emulsion or with liquid fertilizer as soil injection on each side of row. Do not apply more than twice per season; allow minimum of 21 days between applications. If preplant, planting, or postplant applications are made, no foliar application may be made within same crop season.

Days To Harvest: Single postplant application at rates up to 1 pt./A-28; single postplant application at rates up to 2 pt./A-90.

- - - - - - - - - - - - - - - - - -

Rate: 1.6-3.2 fl. oz. 8/A

Pest:
aphids

Remarks: None.

State Restrictions: SLN-Texas.

- - - - - - - - - - - - - - - - - -

Signal Word/Toxicity Class: Danger/I.
REI: 48 hr.

Di-Syston 15% (disulfoton) *(Bayer)*

Rate: At planting:
0.6-1 lb. ai/A
4-6.7 lb. 15%/A
Hill-dropped:
0.3-0.4 lb. ai/A
2-2.7 lb. 15%A

Pest:
aphids; mites; thrips

Remarks: At planting: Place granules in seed furrow or in a band on each side of furrow. Postplant (irrigated cotton only): If additional treatment is necessary, make one soil incorporated sidedress application. Do not apply more than twice per season. Hill-dropped cotton: Apply with seed at planting. Use lower rate where hill spacing is greater than 14" apart or where application is made to light, sandy soil.

Days To Harvest: 28.

- - - - - - - - - - - - - - - - - -

Signal Word/Toxicity Class: Danger/I.
REI: 48 hr.

Drexel Suffa — see sulfur

ENDOSULFAN

Chlorinated bicyclic sulfite insecticide/miticide with broad-spectrum activity. Highly toxic to fish, and moderately to highly toxic to mammals.

Drexel Endosulfan 3EC or 50WP *(Drexel)*
Gowan Endosulfan 3EC or 50WSB *(Gowan)*
◆ **Micro Flo Endosulfan 3EC or 50WP** *(Micro Flo)*
Phaser 3EC or 50WSB *(AgrEvo USA)*
Thiodan 3EC or 50WP *(FMC)*

Rate: 0.375-1.5 lb ai/A
0.5-2 qt. 3EC/A
0.5-1.5 lb. ai/A
1-3 lb. 50WP, 50WSB/A

◆ - **new product** • ai-active ingredient • DF-dry flowable • E/EC-emulsifiable concentrate • EW-emulsion in water
F/FL-flowable • G-granule • L/LC-liquid • SP/WSP-soluble packet • W/WP-wettable powder • WSB-water soluble bag

Pest:

boll weevil, cotton leafperforator, stink bugs
bollworms, fleahoppers, thrips
cabbage looper, Lygus bugs, tobacco budworm

Remarks: Rate varies by pest species; refer to label. Do not apply after bolls open or graze livestock in treated fields. Do not exceed 3 lb. ai/A per year.

Signal Word/Toxicity Class: Danger-Poison/I.
REI: 24 hr.

FURADAN

Carbamate insecticide for control of thrips. Toxic to fish, birds, and other wildlife. Restricted Use Pesticide due to acute oral and inhalation toxicity.

Furadan 4F (carbofuran) *(FMC)*

Rate: 1 lb. ai/A
2 pt. 4F/A in 40" row
2.5 fl. oz. 4F/1000 lin. ft. row

Pest:
thrips

Remarks: Apply at planting in the seed furrow. May be mixed with water or liquid fertilizer. Do not feed cotton forage. Do not apply through any type of irrigation system.

Crop Rotation: Do not rotate with any crop on soil treated at greater than 3 lb. ai/A for at least 10 months. Do not plant with any crop for which carbofuran treatment is not registered for at least 10 months.

Signal Word/Toxicity Class: Danger-Poison/I.
REI: 48 hr.

Fury — see *zeta*-cypermethrin

Fyfanon — see malathion

Gaucho — see imidacloprid

Guthion — see azinphos-methyl

IMIDACLOPRID

Admire 2F *(Bayer)*

Rate: 1.3 fl. oz. 2F/1000 row ft.

Pest:

aphids, thrips, whiteflies
plant bugs

Remarks: Early-season control. Apply as narrow band directly below seed row 7 or fewer days before planting or as in-furrow spray during planting. Do not apply more then 0.33 lb. ai/A as soil application.

Signal Word/Toxicity Class: Caution.
REI: 12 hr.

Gaucho 75 ST or 480 *(Gustafson)*

Rate: 5.3 oz. 75 ST/100 lb. seed
Slurry treatment:
8 fl. oz. 480/100 lb. seed

Pest:

aphids, thrips

Remarks: Delinted seed only. Early-season control. Apply prior to planting; ensure thorough coverage. Treated seed must not be used for or mixed with food or animal feed or processed for oil.

Crop Rotation: May be replanted with any labeled crop, or any crop for which ai tolerance exists; 12-month plantback interval for other crops.

Signal Word/Toxicity Class: Caution.

Provado 1.6 *(Bayer)*

Rate: 0.05 lb. ai/A
3.75 fl. oz. 1.6/A

Pest:

aphids, plant bugs, sweetpotato whitefly*
fleahoppers, silverleaf whitefly*, whiteflies*

****Suppression***

Remarks: Foliar. For aphids and plant bugs, apply as pests begin to build; 2 applications at 7- to 10-day intervals may be required. For whiteflies, apply when adults appear, prior to development of nymphs; repeat at 7-day intervals as long as pest pressure continues.

Days To Harvest: 14.

Signal Word/Toxicity Class: Caution.
REI: 12 hr.

IMIDAN

Organophosphate insecticide that is most effective against boll weevils in cotton. Toxic to fish and bees, and moderately toxic to humans and other mammals.

Imidan 70-W (phosmet) *(Gowan)*

Rate: 0.25-0.5 lb. ai/A
0.33-0.75 lb. 70-W/A

Pest:
overwintering boll weevil

Rate: 0.75-1 lb. ai/A
1-1.33 lb. 70-W/A

Pest: 1st, 2nd, or 3rd generation:
boll weevil

Remarks: Apply in 5 gal. water by ground equipment or 1 lb./A in 3 gal. water by air. For overwintering boll weevil, make 2 applications: first application, one- third grown square stage; second, 5-7 days later. For 1st, 2nd, or 3rd generation boll weevils, apply at 3- to 7-day intervals depending on weevil population and migration in fields. Use higher rate for heavy infestations; check infestations regularly. Do not exceed 14 lb./A per season. Do not graze or feed forage to livestock. Do not apply within 1 mile of any coastal or estuarine waters; 100 ft. of aquatic habitats.

Days To Harvest: 21.

State Restrictions: Do not use in Lauderdale and Madison counties, Alabama; Copiah and Clairborne counties, Mississippi; Lawrence County, Tennessee; and San Joaquin Valley, CA.

Signal Word/Toxicity Class: Warning/II.
REI: 24 hr.

Karate — see *lambda*-cyhalothrin

Kelthane — see dicofol

KNACK

Insect growth regulator which suppresses embryogenesis within insect egg and inhibits metamorphosis and adult emergence. No activity on adult insects, but hatching of eggs laid by treated adults will be suppressed.

◆ **Knack (pyriproxyfen)** *(Valent)*

Rate: 8-10 fl. oz./100 gal. water

Pest:

silverleaf whitefly, sweetpotato whitefly

Remarks: Apply by air or ground when whiteflies reach threshold levels (3-5 adults per leaf). Under severe pressure, use higher rate. Make only 1 application per season.

Days To Harvest: 28.

Signal Word/Toxicity Class: Caution.
REI: 12 hr.

Kumulus — see sulfur

LANNATE

Broad-spectrum carbamate type insecticide/ovicide. Formulated as a liquid in various amounts of active ingredient per gallon. Toxic to fish and bees. Highly toxic to humans and other mammals. Restricted Use Pesticide.

Lannate SP or LV (methomyl) *(DuPont)*

Rate: 0.1125-0.225 lb. ai/A
0.125-0.25 lb. SP/A
0.4-0.75 pt. LV/A

Pest:

bollworms, Lygus bugs, tobacco budworm
cotton fleahopper

NOTICE The information on these pages is for preliminary planning — not a guide for use. Be sure to follow the manufacturer's directions, notwithstanding information contained here. For personal protective equipment and EPA registration numbers, please turn to page 39.

Rate: 0.225-0.45 lb. ai/A
0.25-0.5 lb. SP/A
0.75-1.5 pt. LV/A

Pest:

aphids	cotton leafworm	thrips

Remarks: Ovicide/larvicide. For bollworm and tobacco budworm, apply when significant numbers of eggs are present; repeat at 5- to 7-day intervals while eggs are present and larval control adequate. If significant larvae survive, use higher rate. For Lygus bugs, start treatment on low level population for suppression. Use may redden cotton. If excessive, stop or alternate with other insecticides. Do not apply more than 1.8 lb. ai/A per crop or make more than 8 applications per crop. Do not graze or feed to livestock.

Days To Harvest: 15.

- - - - - - - - - - - - - - - - -

Rate: East of the Rockies:
0.45 lb. ai/A
0.5 lb. SP/A
1.5 pt. LV/A
West of the Rockies:
0.3-0.675 lb. ai/A
0.33-0.75 lb. SP/A
0.875-2.25 pt. LV/A

Pest:

beet armyworm	cotton leafperforator	Lygus bugs
bollworms	fall armyworm	tobacco budworm

Remarks: West of the Rockies rate varies by pest species. Use may redden cotton, if excessive, stop or alternate with other insecticides. Do not apply more than 1.8 lb. ai/A per crop or make more than 8 applications per crop. Do not graze or feed to livestock.

Days To Harvest: 15.

- - - - - - - - - - - - - - - - -

Signal Word/Toxicity Class: Danger-Poison/I.
REI: 72 hr.

LARVIN

Carbamate type insecticide/ovicide most active against armyworms and Helicoverpa spp. Toxic to fish, moderately toxic to mammals and bees.

Larvin 3.2 (thiodicarb) *(Rhone-Poulenc)*

Rate: 0.125-0.25 lb. ai/A
5-10 fl. oz. 3.2/A

Pest:

cotton bollworm (eggs)	tobacco budworm (eggs)

Remarks: Apply at 3- to 5-day intervals when moth flights begin or sufficient eggs are present. Thorough spray coverage necessary; eggs laid on untreated foliage or not contacted by Larvin will not be controlled. If significant number of larvae survive, use higher recommended rates or apply in tank mix combinations with pyrethroid or organophosphate insecticides at their standard rate. In Arizona, do not exceed total 1.8 lb. ai/A per season or apply after bolls have opened. Do not allow livestock to graze treated fields.

Days To Harvest: 28.

- - - - - - - - - - - - - - - - -

Rate: 0.6-0.9 lb. ai/A
24-36 fl. oz. 3.2/A

Pest:

beet armyworm	cotton leafperforator	plant bugs*
boll weevil*	cotton leafworm	soybean looper
cabbage looper	fall armyworm	tobacco budworm (larvae)
cotton bollworm (larvae)	fleahoppers*	
	pink bollworm*	

*** Suppression**

Remarks: As pest pressure increases, reduce intervals between applications and/or increase rate. Initiate sprays when sufficient newly hatched larvae are present. Rate of 24 fl. oz. should be used when maximum survival of beneficial insects and spiders is desired; 30-36 fl. oz. should be used when insect populations are heavy or infestation is predominately large larvae. Larvae in bolls will not be controlled. For pink bollworm, application should be timed for presence of adult moths. For light boll weevil populations when spraying on 5-7 day worm control schedule, boll weevil levels will usually be maintained below established economic threshold. For moderate to heavy boll weevil populations, intersprays of a registered boll weevil product and/or tank mix of Larvin 3.2 plus registered product should be used; 3-5 day spray intervals may be required under heavy infestations. Do not allow livestock to graze treated fields.

Days To Harvest: 28.

- - - - - - - - - - - - - - - - -

Rate: 0.5-0.75 lb. ai/A
20-30 fl. oz. 3.2/A

Pest:
cutworms

Remarks: Apply in a minimum 15 gal./A finished spray by ground; 3 gal./A finished spray by air. For postemergence rescue treatment, apply as broadcast spray when cutworms are actively feeding. If cutworms are feeding below ground, control will be more difficult. Light cultivation with rotary hoe before spraying may enhance control. Do not feed forage, hay, or straw to livestock.

Days To Harvest: 28.

- - - - - - - - - - - - - - - - -

Signal Word/Toxicity Class: Warning/II.
REI: 12 hr.

LINDANE

Lindane 30C contains toxic inert Rhodamine B. Toxic to fish, birds, and wildlife.

Lindane 30C *(Gustafson)*
◆ Lindane ST 40 *(AGSCO)*

Rate: Acid delinted or reginned:
5.4 fl. oz. 30C/100 lb. seed
4 fl. oz. 40/100 lb. seed
Fuzzy (not delinted):
6.1 fl. oz. 30C/100 lb. seed
4.5 fl. oz. 40/100 lb. seed

Pest:

seedcorn maggots	wireworms

Remarks: Do not use treated seed for food, feed, or oil purposes. Exposed treated seed may be hazardous to birds and other wildlife. Dispose of excess seed by burial away from bodies of water.

- - - - - - - - - - - - - - - - -

Signal Word/Toxicity Class: Warning/II.
REI: 12 hr.

Lock-On — see chlorpyrifos

Lorsban — see chlorpyrifos

MALATHION

Organophosphate insecticide with broad spectrum of activity but mostly targeted toward aphids and boll weevils in cotton. Available in many formulations. Toxic to fish and bees, low toxicity to mammals.

Clean Crop Malathion 57EC *(Platte)*

Rate: 1-1.25 lb. ai/A
1.5-2 pt. 57EC/A

Pest:

aphids	cotton leafperforator	thrips
brown cotton leafworm	leafhoppers	whiteflies
	Lygus bugs	

Rate: 1.25-4 lb. ai/A
2-6 pt. 57EC/A

Pest:
boll weevil

Remarks: Rate varies by pest species; refer to label.

Days To Harvest: 0.

- - - - - - - - - - - - - - - - -

Signal Word/Toxicity Class: Caution/III.
REI: 12 hr.

Clean Crop Malathion 8EC *(Platte)*

Rate: 1 lb. ai/A
1 pt. 8EC/A

Pest:

cotton aphid	leafhoppers	thrips
cotton leafperforator	Lygus bugs	whiteflies
cotton leafworm		

Remarks: Apply when insects first appear; repeat as necessary to maintain control.

Days To Harvest: 0.

- - - - - - - - - - - - - - - - -

Signal Word/Toxicity Class: Caution/III.
REI: 12 hr.

◆ **- new product** • ai-active ingredient • DF-dry flowable • E/EC-emulsifiable concentrate • EW-emulsion in water
F/FL-flowable • G-granule • L/LC-liquid • SP/WSP-soluble packet • W/WP-wettable powder • WSB-water soluble bag

Fyfanon (8E) *(Helena)*
Rate: 0.75-2 lb. ai/A
0.75-2 pt. 8E/A
Pest:
boll weevil, brown cotton leafworm, cotton aphid, cotton fleahoppers, cotton leafperforator, cotton leafworm, fall armyworm, garden webworm, grasshoppers, leafhoppers, Lygus bugs, spider mites, thrips, whiteflies
Remarks: Rate varies by pest species; refer to label.
Days To Harvest: 0.

Signal Word/Toxicity Class: Caution/III.
REI: 12 hr.

Drexel Malathion 5EC *(Drexel)*
◆ **Fyfanon (5EC)** *(Helena)*
◆ **Micro Flo Malathion 5EC** *(Micro Flo)*
Prentox 5 lb. Malathion Spray (5EC) *(Prentiss)*
Rate: 0.3-2.5 lb. ai/A
0.5-4 pt. 5EC/A
Pest:
boll weevil, brown cotton leafworm, cotton aphid, cotton fleahoppers, cotton leafperforator, cotton leafworm, desert spider mite, fall armyworm, garden webworm, grasshoppers, leafhoppers, Lygus bugs, thrips, whiteflies
Remarks: Rate varies by pest species; refer to label.
Days To Harvest: 0.

Signal Word/Toxicity Class: Warning/II.
REI: 12 hr.

◆ **Drexel Malathion ULV** *(Drexel)*
Fyfanon ULV *(Cheminova)*
◆ **Micro Flo Malathion ULV** *(Micro Flo)*
Rate: 4-8 fl. oz. ULV/A
Pest:
boll weevil, cotton aphid, fleahoppers, grasshoppers, green peach aphid, leafhoppers, Lygus bugs, tarnished plant bugs, thrips
Remarks: Rate varies by pest species and time of season; refer to label. For heavy migrating populations of Lygus bugs, use high rate.
Days To Harvest: 0.

Signal Word/Toxicity Class: Caution/III.
REI: 12 hr.

◆ **Gowan Malathion 8 (8E)** *(Gowan)*
Rate: 1-4 lb. ai/A
1-4 pt. 8E/A
Pest:
aphids, boll weevil, brown cotton leafworm, cotton leafperforator, fall armyworm, fleahoppers, garden webworm, grasshoppers, leafhoppers, Lygus bugs, spider mites, stink bugs, whiteflies
Remarks: Do not apply through any type of irrigation system.
Days To Harvest: 0.

Signal Word/Toxicity Class: Caution/III.
REI: 12 hr.

◆ **Wilbur-Ellis Malathion 8 Spray** *(Wilbur-Ellis)*
Rate: 1-4 lb. ai/A
1-4 pt. 8/A
Pest:
boll weevils, cotton aphid, cotton leafperforator, cotton leafworm, leafhoppers, Lygus bugs, thrips, whiteflies
Remarks: Do not feed or graze treated crop. In California, do not apply more than 1 pt./A.

Signal Word/Toxicity Class: Caution.
REI: 12 hr.

METASYSTOX-R

Systemic organophosphate insecticide/miticide normally formulated as emulsifiable concentrate. Recommended for use on cotton against aphids and mites. Toxic to fish, bees, and mammals. Restricted Use Pesticide.

Metasystox-R (SC) (oxydemeton-methyl) *(Gowan)*
Rate: Except California:
0.125-0.25 lb. ai/A
0.5-1 pt. SC/A
Pest:
aphids, fleahoppers, thrips

Rate: Except California:
0.375-0.75 lb. ai/A
1.5-3 pt. SC/A
Pest:
mites

Rate: California:
0.375-0.5 lb. ai/A
1.5-2 pt. SC/A
Pest:
mites, aphids, Lygus bugs
Remarks: Apply in sufficient water for complete coverage but not less than 1 gal./A. Do not apply more than 3 times per season. May be applied through sprinkler irrigation systems. Do not graze or feed gin trash to dairy or meat animals.
Days To Harvest: 14.

Signal Word/Toxicity Class: Warning/II.
REI: 48 hr.; 72 hr. if average rainfall less than 25" per year.

METHAMIDOPHOS

Systemic organophosphate insecticide/miticide normally formulated as an emulsifiable concentrate. Toxic to fish, bees, and mammals. Restricted Use Pesticide.

Monitor 4 *(Bayer)*
Rate: 0.1-1 lb. ai/A
0.2-2 pt. 4/A
Pest:
aphids, beet armyworm, cabbage looper, fleahoppers, Lygus bugs, mites, stink bugs, thrips, western flower thrips, whiteflies (except sweetpotato)
Remarks: Apply as needed before bolls open. Rate varies by pest species, infestation level and time of season; refer to label. Do not feed gin trash to livestock or allow animals to graze on treated areas.
Days To Harvest: 50.

Rate: 0.1-1.25 lb. ai/A
0.2-2.5 pt. 4/A
Pest:
aphids, beet armyworm, cabbage looper, fleahoppers, Lygus bugs, mites, thrips, western flower thrips, whiteflies
Remarks: Rate varies by pest species, infestation level and time of season; refer to label.
State Restrictions: SLN-Arkansas, California, Mississippi.

Signal Word/Toxicity Class: Danger/I.
REI: 48 hr.; 72 hr. if average rainfall less than 25" per year.

Monitor 4 Spray *(Valent)*
Rate: 0.25-0.5 lb. ai/A
0.5-1 pt. 4S/A
Pest:
whiteflies

Rate: 0.5-1 lb. ai/A
1-2 pt. 4S/A
Pest:
aphids, beet armyworm, cabbage looper, Lygus bugs (AZ, CA), mites, thrips
Remarks: Apply before bolls open. For whiteflies, a second application may be needed for heavy infestations. Do not apply through any type of irrigation system. Do not feed gin trash to livestock or allow animals to graze treated areas.
Days To Harvest: 50.

4 Cotton

NOTICE The information on these pages is for preliminary planning — not a guide for use. Be sure to follow the manufacturer's directions, notwithstanding information contained here. For personal protective equipment and EPA registration numbers, please turn to page 39.

Rate: 0.1-1 lb. ai /A
0.2-2 pt. 4S/A

Pest:

aphids	Lygus bugs (CA)	western flower thrips (AR, LA)
beet armyworm	mites	
cabbage looper	thrips	

Remarks: Do not apply after 65% of the bolls are open. Do not apply through any type of irrigation system. Do not feed gin trash to livestock or allow animals to graze treated areas.
State Restrictions: SLN-Arkansas, California, Louisiana, Mississippi, Tennessee.

- - - - - - - - - - - - - - - - - -

Signal Word/Toxicity Class: Danger-Poison/I.
REI: 48 hr.

METHYL PARATHION

Highly toxic organophosphate insecticide. Used mostly against boll weevils, but will give some control of other pests. Penncap-M is a microencapsulated formulation. Restricted Use Pesticide due to residual effects to avian species and hazard to bees.

Cheminova Methyl 4EC *(Cheminova)*

Rate: 0.125-3 lb. ai/A
0.25-6 pt. 4EC/A

Pest:

aphids	cotton leafworm	grasshoppers
armyworms	cutworms	leafhoppers
boll weevil	fall armyworm	saltmarsh caterpillar
bollworm	false chinch bug	stink bugs
cotton leafperforator	garden webworm	thrips

Remarks: Rate varies by pest species; refer to label. Applications should be made at 4-5 day intervals until control is obtained.
Days To Harvest: Mechanical pickers-1; hand pickers-5.

- - - - - - - - - - - - - - - - - -

Signal Word/Toxicity Class: Danger-Poison/I.

◆ **Declare (4EC)** *(Griffin)*
Helena 4 lb. Methyl Parathion (4EC) *(Helena)*

Rate: 0.125-3 lb. ai/A
0.25-6 pt. 4EC/A

Pest:

aphids	cutworms	mites
armyworm	desert spider mite	saltmarsh caterpillar
boll weevil	fall armyworm	southern green stink bug
bollworm	false chinch bug	
cabbage looper	fleahoppers	stink bugs
cotton aphid	garden webworm	thrips
cotton fleahopper	grasshoppers	tobacco budworm
cotton leafperforator	leafrollers	twospotted spider mite
cotton leafworm	Lygus bugs	

Remarks: Rate varies by pest species; refer to label. For aphids, apply when pests appear in buds and on leaves of seedlings or when honeydew appears on older plants. For thrips, apply before or during 4-leaf stage. For early-season control of boll weevils, apply as early as 2-4 leaf stage. Repeat 1-3 times at 7- to 10-day intervals. For mid-season control of boll weevils, apply at start of squaring or after 10%-25% of first squares are punctured and repeat at 3- to 8-day intervals as needed or until top bolls are full grown. Repeat at 3-day intervals during major fruiting season and periods of heavy weevil migration.

- - - - - - - - - - - - - - - - - -

Signal Word/Toxicity Class: Danger-Poison/I.
REI: 48 hr.

Penncap-M (2FL) *(Elf Atochem)*

Rate: 0.125-0.25 lb. ai/A
0.5-1 pt. 2FL/A

Pest:

fleahoppers	plant bugs	thrips*

Rate: 0.125-1 lb. ai/A
0.5-4 pt. 2FL/A

Pest:
Lygus bugs

Rate: 0.25-0.75 lb. ai/A
1-3 pt. 2FL/A

Pest:

boll weevil	stink bugs*

Rate: 0.5-0.75 lb. ai/A
2-3 pt. 2FL/A

Pest:
soybean looper

Rate: 0.25-1 lb. ai/A
1-4 pt. 2FL/A

Pest:
aphids (suppression)

Rate: 0.5-1 lb. ai/A
2-4 pt. 2FL/A

Pest:

pink bollworm	whiteflies

**Except CA*

Remarks: Do not apply if bees are foraging in areas to be treated. For whiteflies, may be tank mixed with amitraz or non-organophosphate insecticides. Do not chemigate in California. Refer to label for additional use information.
Days To Harvest: Harvest or handpicking-7.

- - - - - - - - - - - - - - - - - -

Rate: 0.25-0.5 lb. ai/1-2 pt. water
1-2 pt. 2FL/1-2 pt. water

Pest:
boll weevil

Remarks: ULV aerial application. Do not apply if bees are foraging in areas to be treated. Apply when damaging levels of boll weevil appear; repeat as necessary. Refer to label for additional use information.
Days To Harvest: Machine harvested-anytime.
State Restrictions: Not for use in California.

- - - - - - - - - - - - - - - - - -

Signal Word/Toxicity Class: Warning/II.
REI: 48 hr.

Wilbur-Ellis Methyl Parathion 4 or 5 Spray *(Wilbur-Ellis)*

Rate: 1.5-3 lb. ai/A
3-6 pt. 4/A
1.9-3 lb. ai/A
3-4.75 pt.5/A

Pest:

aphids	bollworms	pink bollworm
armyworms (3rd instar)	cotton leafperforator	spider mites
	cotton leafworm	thrips
boll weevil	leafhoppers	tobacco budworm

Remarks: Rate varies by pest species; refer to label.
Days To Harvest: Handpicking-15.

- - - - - - - - - - - - - - - - - -

Signal Word/Toxicity Class: Danger-Poison/I.
REI: 48 hr.

METHYL PARATHION (Premix)

Methyl Parathion 2 Thiodan 3EC (2 lb./gal. methyl parathion & 3 lb./gal. endosulfan) *(FMC)*

Rate: 2.5 lb. ai/A
2 qt. product/A

Pest:

aphids	cabbage looper	Lygus bugs
armyworms	cotton leafperforator	stink bugs
boll weevil	cotton leafworm	thrips
bollworms	fleahoppers	

Remarks: Apply when insects first appear and repeat as necessary. Do not apply after bolls open. Do not graze in treated fields.

- - - - - - - - - - - - - - - - - -

Signal Word/Toxicity Class: Danger-Poison/I.
REI: 48 hr.

Microthiol Special — see sulfur

Monitor — see methamidophos

M-PEDE

M-Pede (potassium salts of fatty acids) *(Mycogen)*

Rate: 4 pt./A in 3-25 gal. water
2% v/v spray solution/A in 25 gal. water or more

Pest:

aphids	fleahoppers	spider mites
caterpillars (tank mix)	green bugs	thrips
	plant bugs	whiteflies

Remarks: Apply at first sign of infestation; repeat at 7- to 14-day intervals. May be tank mixed with other insecticides; refer to both labels for use directions and precautions.

◆ - **new product** • ai-active ingredient • DF-dry flowable • E/EC-emulsifiable concentrate • EW-emulsion in water
F/FL-flowable • G-granule • L/LC-liquid • SP/WSP-soluble packet • W/WP-wettable powder • WSB-water soluble bag

Days To Harvest: 0.

- - - - - - - - - - - - - - - - -

Signal Word/Toxicity Class: Warning/II.
REI: 12 hr.

Mustang — see *zeta*-cypermethrin

NEMACUR

Organophosphate systemic nematicide. Frequently used at or before planting for nematode control. Toxic to humans and other mammals. Restricted Use Pesticide.

Nemacur 3 or 15% (fenamiphos) *(Bayer)*

Rate: 0.75-2.2 lb. ai/A
1-2.9 qt. 3/A
5-9.8 lb. 15%/A (40" rows)

Pest:
nematodes thrips

Remarks: In furrow/covering soil: Apply in furrow or band in covering soil behind the seed drop and in front of covering devices. Band: Apply in a 6-12" band over the row. Incorporate thoroughly to ensure uniform distribution.

- - - - - - - - - - - - - - - - -

Rate: 3 lb. ai/A
4 qt. 3/A (40" rows)

Pest:
nematodes thrips

Remarks: Apply by soil injection in water emulsion or liquid fertilizer in an 18" band.
State Restrictions: For use in California only.

- - - - - - - - - - - - - - - - -

Rate: 0.825-1.65 lb. ai/A
1.1-2.2 qt. 3/A (40" rows)

Pest:
nematodes thrips (early season)

Remarks: Apply as water emulsion spray or as tank mix with Treflan 4EC and/or liquid fertilizer in 12-18" band over the row.
State Restrictions: SLN-Texas.

- - - - - - - - - - - - - - - - -

Signal Word/Toxicity Class: Danger/I.
REI: 48 hr.

NEEMAZAD — see azadirachtin

NEEMIX — see azadirachtin

Nufos — see chlorpyrifos

Orthene — see acephate

Ovasyn — see amitraz

Payload — see acephate

Penncap-M — see methyl parathion

PETROLEUM OIL

Petroleum based spray oil. Mode of action is through suffocation of eggs, larvae, and nymphs of insects and adults of soft bodied insects. This mode of action necessitates total spray coverage. Toxic to fish.

SAF-T-SIDE *(Brandt Consolidated)*
◆ **Sumulsion** *(Drexel)*

Rate: 1-2 gal./100 gal. water

Pest:
aphids caterpillars mites
bollworms leafhoppers thrips
(eggs and larvae) loopers whiteflies

Remarks: Apply as needed. May be tank mixed with most commonly used insecticides and fungicides. Do not use in combination with or immediately before or after spraying with fungicides such as captan, Dyrene, folpet, Morestan, or any sulfur products. Do not use with carbaryl or dimethoate.

- - - - - - - - - - - - - - - - -

Signal Word/Toxicity Class: Caution.
REI: 4 hr.

Phaser — see endosulfan

PHORATE

Systemic organophosphate frequently applied in granular formulation at planting to control thrips, mites, aphids, and other early-season insects. Restricted Use Pesticide.

Clean Crop Phorate 20G *(Platte)*
Riverside Phorate 20G *(Terra)*
Thimet 20-G *(American Cyanamid)*

Rate: 6-9 oz. 20G, 20-G/1000 ft. row (min. 36" spacing)

Pest:
aphids leafminers thrips
black cutworm mites whiteflies
leafhoppers

Remarks: Apply evenly in furrow at planting. Under adverse conditions plant stand may be affected. Recommended for areas requiring extended period of mite control as well as early-season protection against other cotton pests. Higher rate for heavier soils. Do not apply more than 1.6 lb. ai/A. Do not graze or feed treated hay or forage to livestock.

- - - - - - - - - - - - - - - - -

Rate: 3 oz. 20G, 20-G/1000 ft. row (min. 30" spacing)

Pest: Early-season control:
aphids mites thrips
leafminers

Remarks: Apply evenly in furrow at planting. Suggested for areas where limited period of protection against early-season pests is necessary for good cotton growth. Do not apply more than 0.66 lb. ai/A. Do not graze or feed treated hay or forage to livestock.

- - - - - - - - - - - - - - - - -

Rate: Irrigated-cotton:
9-12 oz. 20G, 20-G/1000 ft. row (min. 36" spacing)

Pest: Spring-early summer:
mites

Remarks: Incorporate granules into soil as sidedress application deep enough to avoid disturbance by future cultivation. Irrigate as soon as possible after treatment. Do not apply more than 2.1 lb. ai/A. Do not graze or feed treated hay or forage to livestock.
Days To Harvest: 60.

- - - - - - - - - - - - - - - - -

Signal Word/Toxicity Class: Danger-Poison/I.
REI: 48 hr.

PROPARGITE

Sulfite ester miticide formulated as an emulsifiable concentrate, frequently used to control spider mites in cotton. Nontoxic to bees and allows survival of beneficial predatory mites. Toxic to fish and shrimp.

Comite or Comite II *(Uniroyal)*

Rate: 0.82-1.64 lb. ai/A
1-2 pt./A
1.025-1.845 lb. ai/A
1.25-2.25 pt. II/A

Pest:
Pacific spider mite strawberry spider mite twospotted spider mite

Remarks: Apply in 15-50 gal./A by ground or 5-15 gal./A by air after plants have hardened, usually after plants are 6-12" high. Do not apply more than 3 times per year or after bolls open. Do not apply through any type of irrigation system. Do not feed treated foliage or cotton trash to livestock.
State Restrictions: Comite II for use east of the Rockies.

- - - - - - - - - - - - - - - - -

Signal Word/Toxicity Class: Danger-Eye/I.
REI: 48 hr.

Provado — see imidacloprid

NOTICE	The information on these pages is for preliminary planning — not a guide for use. Be sure to follow the manufacturer's directions, notwithstanding information contained here.	For personal protective equipment and EPA registration numbers, please turn to page 39.

PYRELLIN (Premix)

Premix containing pyrethrins and rotenone. May be combined with other insecticides to enhance knockdown or as an insect "exciter" to help facilitate contact with other spray residues.

Pyrellin E.C. (0.042 lb./gal. pyrethrins & 0.035 lb./gal. rotenone) *(Webb Wright)*
Rate: 1-2 pt. E.C./A
Pest:
aphids, fleahoppers, loopers, mites, pink bollworm, thrips
Remarks: Apply in intervals of 7 days or less; repeat as needed.
Days To Harvest: 0.

Signal Word/Toxicity Class: Caution/III.
REI: 12 hr.

Pyrenone — see pyrethrins

PYRETHRINS

Contains piperonyl butoxide as a synergist. Diatect also contains silicon dioxide (diatomaceous earth). May be combined in tank mix with other insecticides for quicker and more complete kill where insect resistance may be a problem, and as an "exciter" to flush insects out of hiding and into contact with other spray residues.

Diatect Multipurpose Insecticide II *(Diatect)*
Pyrenone (0.5EC) *(AgrEvo Environmental Health)*
Rate: 1-6 lb. II/A
12 oz. 0.5EC/A
Pest: Including but not limited to:
aphids, bollworms, boll weevil, mites, pink bollworm, thrips, whiteflies
Remarks: Apply in sufficient water for thorough coverage; repeat as needed. Apply Diatect as dust or wettable powder.
Days To Harvest: 0.

Signal Word/Toxicity Class: Caution.
REI: 12 hr.

Redball Lightning — see sulfur

SAF-T-SIDE — see petroleum oil

SCOUT X-TRA

Synthetic pyrethroid insecticide formulated as an emulsifiable concentrate. Scout X-TRA Gel is packaged in water-soluble bags. Low toxicity to humans and other mammals, highly toxic to bees. Restricted Use Pesticide due to toxicity to fish and aquatic organisms.

Scout X-TRA (0.9EC) or Gel (WSB) (tralomethrin) *(AgrEvo USA)*
Rate: 0.016-0.024 lb. ai/A
2.28-3.41 fl. oz. 0.9EC/A
1 WSB/12.5-8.3 A
Pest:
aphids*, beet armyworm*, boll weevil, cabbage looper, cotton bollworm, cotton fleahopper, cotton leafperforator, cotton leafworm, cutworms (except CA), European corn borer, fall armyworm*, leafhoppers, Lygus bugs, mites* (except CA), pink bollworm, plant bugs, saltmarsh caterpillar, soybean looper*, stink bugs, thrips (except CA), tobacco budworm, whiteflies* (except CA)
****Suppression***
Remarks: Uniform coverage essential. Do not apply more than 0.19 lb. ai/A per growing season. Refer to label for spray drift precautions.
Days To Harvest: 28.
State Restrictions: Gel not for use in California.

Signal Word/Toxicity Class: Danger/I.
REI: 24 hr.

Sevin — see carbaryl

SPINOSAD

Fermentation-derived insect control agent. Does not significantly impact predaceous and parasitic insect populations. Highly toxic to bees and aquatic invertebrates.

◆ **Tracer Naturalyte (SC) (spinosyn A & spinosyn D)** *(Dow AgroSciences)*
Rate: 0.045-0.089 lb. ai/A
1.4-2.9 oz. SC/A
Pest:
cotton bollworm (prebloom), cotton leafperforator, European corn borer, tobacco budworm

Rate: 0.067-0.089 lb. ai/A
2.14-2.9 oz. SC/A
Pest:
armyworms, beet armyworm, cabbage looper, cotton bollworm (postbloom), fall armyworm, loopers, soybean looper, saltmarsh caterpillar
Remarks: For tobacco budworm and cotton bollworm, apply when majority of insect population is within blackhead egg stage to 1/8" larval length. For beet armyworm, apply when scouting reveals 3 or more occurrences of egg hatch or larval feeding. For loopers, apply when there is more than 4 larvae/1 ft. row or 25% defoliation. Do not apply more than 14.4 fl. oz./A.
Days To Harvest: 28.

Signal Word/Toxicity Class: Caution.
REI: 4 hr.

SULFUR

Microthiol and Thiolux are micronized sulfurs. Stirrup Mylox combines flowable sulfur with pheromones. Redball Lightning is a dustless granule formulated to break down into smaller particles when wetted by water to provide effectiveness of a powder with the handling ease of a granule. Relatively nontoxic.

Drexel Suffa 6 or 8, Sulfur 90W *(Drexel)*
Rate: 1.5-6 lb. ai/A
2-8 pt. 6/A
1.5-6 pt. 8/A
1.6-6.6 lb. 90W/A
Pest:
Atlantic mite, Lygus bugs
Remarks: Use low rate for early-season control; high rate for mid- to late-season.

Signal Word/Toxicity Class: Caution.
REI: 12 hr. (Suffa 8); 24 hr. (Suffa 6, Sulfur 90W).

◆ **Kumulus DF** *(BASF)*
Rate: 2.4-4 lb. ai/A
3-5 lb. DF/A
Pest:
red spider mite
Remarks: Apply at early leaf stage; repeat at 10- to 14-day intervals with or without other pesticides immediately after a rain. Do not apply if temperatures during or within 3 days after application are expected to exceed 90°F. Do not apply through any type of irrigation system.

Signal Word/Toxicity Class: Caution.
REI: 24 hr.

Microthiol Special (80W) *(Elf Atochem)*
Rate: 4-8 lb. ai/A
5-10 lb. 80W/A
California:
4 lb. ai/A
5 lb. 80W/A
Pest:
Atlantic mite (except CA), red spider mite, twospotted spider mite (except CA)
Remarks: Apply as needed throughout season. Do not chemigate in California.

Signal Word/Toxicity Class: Caution.
REI: 24 hr.

◆ - **new product** • ai-active ingredient • DF-dry flowable • E/EC-emulsifiable concentrate • EW-emulsion in water
F/FL-flowable • G-granule • L/LC-liquid • SP/WSP-soluble packet • W/WP-wettable powder • WSB-water soluble bag

◆ **Redball Lightning Degradable (G)** *(International Sulphur)*
Rate: 5-15 lb. G/A
Pest:
fleahoppers mites thrips
Remarks: Do not apply within 3 weeks of oil spray.
Days To Harvest: 0.

- - - - - - - - - - - - - - - - -

Signal Word/Toxicity Class: Caution.
REI: 24 hr.

THAT (6FL) or THAT Big 8 (8FL) *(Stoller)*
Rate: 1.5-6 lb. ai/A
2-8 pt. 6FL/A
1.33-5 lb. ai/A
1.33-5 pt. 8FL/A
Pest:
Atlantic mite
Remarks: Apply as needed throughout season. Use low rate for early-season control; high rate for mid- to late-season.

- - - - - - - - - - - - - - - - -

Signal Word/Toxicity Class: Caution.
REI: 24 hr.

Thiolux (DF) *(Platte)*
Rate: 2.4-4 lb. ai/A
3-5 lb. DF/A
Pest:
Atlantic mite red spider mite
Remarks: Apply as needed.

- - - - - - - - - - - - - - - - -

Signal Word/Toxicity Class: Caution.
REI: 24 hr.

Sumulsion — see petroleum oil

SUPRACIDE

Organophosphate insecticide/miticide. Gives broad spectrum control of insects and mites. Toxic to humans, other mammals, fish, and bees. Restricted Use Pesticide.

Supracide 25WP (methidathion) *(Gowan)*
Rate: 1 lb. ai/A
4 lb. 25WP/A
Pest:
boll woovil bollworms spider mites
Remarks: Apply when infestations warrant and repeat at 5-7 day intervals as needed. Minimum 10 gal. water by ground equipment; 3 gal. water by air. Do not apply more than 4 lb. ai/A during any one growing season. Do not apply more than 0.5 lb. ai/A to open bolls. Do not apply through any type of irrigation system. Do not graze treated plants or feed gin trash to livestock.
Days To Harvest: 14.

- - - - - - - - - - - - - - - - -

Rate: 0.5-1 lb. ai/A
2-4 lb. 25WP/A
Pest:
Lygus bugs pink bollworm
Remarks: Apply in minimum 10 gal. water by ground equipment or 3 gal. water by air. Do not apply more than 4 lb. ai/A during any one growing season. Do not apply more than 0.5 lb. ai/A to open bolls. For pink bollworm, use low rate and repeat at 5- to 7-day intervals if needed. For Lygus bugs, use low rate for light to moderate infestations; high rate for heavier infestations; repeat as necessary. Do not graze treated plants or feed gin trash to livestock.
Days To Harvest: 14.
State Restrictions: For use in Arizona, California.

- - - - - - - - - - - - - - - - -

Rate: 0.25-0.5 lb. ai/A
1-2 lb. 25WP/A
Pest:
bandedwing whitefly
Remarks: Apply in minimum 10 gal. water by ground equipment or 3 gal. water by air. Apply low rate for light to moderate infestations; high rate for moderate to heavy infestations. Repeat at 5- to 7-day intervals if necessary. Do not apply more than 4 lb. ai/A during any one growing season. Do not apply more than 0.5 lb. ai/A to open bolls. Do not graze treated plants or feed gin trash to livestock.
Days To Harvest: 14.
State Restrictions: For use in the mid-South, Southeast.

- - - - - - - - - - - - - - - - -

Signal Word/Toxicity Class: Warning/II.
REI: 48 hr.

TEMIK

Carbamate type insecticide/nematicide. Usually applied at planting as a granular formulation for control of early-season insects and nematodes. Highly toxic to humans, other mammals, birds, and other wildlife. Restricted Use Pesticide.

Temik 15G (aldicarb) *(Rhone-Poulenc)*
Rate: Except Arizona:
0.5-1.5 lb. ai/A
3.5-10 lb. 15G/A
Arizona only:
2-4 lb. ai/A
14-27 lb. 15G/A
Pest:
nematodes
Remarks: Drill granules just below seed line or place in seed furrow and cover with soil. If rate exceeds 7 lb., apply in 4-6" band and work into soil or cover with soil; plant seed in or above treated zone. For Arizona, apply in 4-6" band 3-6" below seed line or cover with 3-6" of soil; plant seed above treated zone. Do not exceed a total of 33 lb./A. Refer to label for environmental precautions and soil restrictions table.
Days To Harvest: 90.

- - - - - - - - - - - - - - - - -

Rate: Texas, Oklahoma, New Mexico:
0.3-0.75 lb. ai/A
2-5 lb. 15G/A
Except Texas, Oklahoma, New Mexico:
0.5-0.75 lb. ai/A
3.5-5 lb. 15G/A
Pest:
aphids thrips

- - - - - - - - - - - - - - - - -

Rate: 0.6-1 lb. ai/A
4-7 lb. 15G/A
Pest:
boll weevil leafminers mites
fleahoppers Lygus bugs plant bugs
Remarks: Drill granules just below seed line or place in seed furrow and cover with soil. Do not make more than 1 application at planting and 1 postemergence application per crop. If Temik and seeds are hill-dropped, rate may be reduced by half. Do not exceed total 33 lb./A. Refer to label for environmental precautions and soil restrictions table.
Days To Harvest: 90.

- - - - - - - - - - - - - - - - -

Rate: 1-2 lb. ai/A
7-14 lb. 15G/A
Pest:
fleahoppers leafhoppers

Rate: 2 lb. ai/A
14 lb. 15G/A
Pest:
leafperforator

Rate: 1-3 lb. ai/A
7-20 lb. 15G/A
Pest:
boll weevil Lygus bugs whiteflies
mites plant bugs
Remarks: Side-dress granules 8-16" to one side of plant row and 2-6" deep. For whiteflies, apply before populations reach an economic threshold. Apply for control of moderate populations. Do not exceed total 33 lb./A. Refer to label for environmental precautions and soil restrictions table.
Days To Harvest: 90.

- - - - - - - - - - - - - - - - -

Signal Word/Toxicity Class: Danger-Poison/I.
REI: 48 hr.

THAT or THAT Big 8 — see sulfur

Thimet — see phorate

NOTICE The information on these pages is for preliminary planning — not a guide for use. Be sure to follow the manufacturer's directions, notwithstanding information contained here.

For personal protective equipment and EPA registration numbers, please turn to page 39.

Thiodan — see endosulfan

Thiolux — see sulfur

Tracer Naturalyte — see spinosad

VYDATE

Systemic carbamate insecticide/nematicide that is highly toxic to humans and other mammals, and toxic to fish and bees. Restricted Use Pesticide.

Vydate C-LV (oxamyl) *(DuPont)*
Rate: 0.125-0.5 lb. ai/A
4.25-17 fl. oz. C-LV/A
Pest:
boll weevil cotton fleahopper tarnished plant bug

Rate: 0.25-1 lb. ai/A
8.5-34 fl. oz. C-LV/A
Pest:
cotton aphids *Lygus hesperus* pink bollworm
cotton leafperforator
Remarks: Apply in water or refined vegetable oil (min. 3 pt./A). Do not apply more than 2.5 lb. ai/A for boll weevil, fleahopper, plant bug; 4 lb. ai/A for aphids, bollworm, leafperforator, and Lygus per season. Do not graze or feed treated cotton to livestock.
Days To Harvest: 14.
Crop Rotation: Do not plant crops other than bell peppers, cotton, carrots, celery, cucumbers, eggplant, melons (cantaloupe, honeydew, watermelon), peanuts, potatoes, pumpkins, squash, sweet potatoes, tobacco, or tomatoes within 4 months after last application.

- - - - - - - - - - - - - - - - -

Signal Word/Toxicity Class: Danger-Poison/I.
REI: 48 hr.

Vydate L (oxamyl) *(DuPont)*
Rate: 0.125-0.25 lb. ai/A
0.5-1 pt. L/A
Pest:
boll weevil cotton fleahopper tarnished plant bug

Rate: 0.25-1 lb. ai/A
1-4 pt. L/A
Pest:
cotton leafperforator
Remarks: Apply when damaging populations appear; repeat as needed. For leafperforator, repeat at 6- to 8-day intervals for up to 4 applications. Do not apply more than 10 pt./A (16 pt./A for leafperforator) per season. Do not graze or feed treated cotton to livestock.
Days To Harvest: 21.
Crop Rotation: Do not plant crops other than registered on Vydate L label within 4 months after last application.

- - - - - - - - - - - - - - - - -

Signal Word/Toxicity Class: Danger-Poison/I.
REI: 48 hr.

Zephyr — see abamectin

Fungicides

Allegiance — see metalaxyl

Apron — see metalaxyl

Apron XL — see mefenoxam

BAYTAN 30

Triazole systemic fungicide for use only through commercial seed treating equipment which are closed application systems.

Baytan 30 (FL) (triadimenol) *(Gustafson)*
Rate: 0.25-0.75 fl. oz. FL/100 lb. seed
Disease:
Rhizoctonia

Rate: 1-3 fl. oz. FL/100 lb. seed
Disease:
black root rot
Time: Seed treatment.
Remarks: Acid delinted cotton. Use high rates only for areas with severe history of disease caused by black root rot. Use with EPA-approved dye that imparts unnatural color to seed. Do not use for feed, food, or oil purposes. Refer to label for use directions.

- - - - - - - - - - - - - - - - -

Signal Word/Toxicity Class: Caution.

CAPTAN

Gustafson Captan 30-DD contains the toxic inert Rhodamine B.

Drexel Captan 4FL *(Drexel)*
Rate: Acid delinted:
3 fl. oz. 4FL/100 lb. seed
Reginned, fuzzy:
4.5 fl. oz. 4FL/100 lb. seed
Disease:
seed decay seedling disease soilborne disease
seedling blight
Time: Liquid or slurry seed treatment.
Remarks: Do not use treated seed for food, feed, or oil purposes. Refer to label for use directions.

- - - - - - - - - - - - - - - - -

Signal Word/Toxicity Class: Danger/I.

Gustafson Captan 30-DD or 400 *(Gustafson)*
Rate: Acid delinted:
2.33 fl. oz. 30-DD/100 lb. seed
2.5-5 fl. oz. 400/100 lb. seed
Reginned, fuzzy:
3.5 fl. oz. 30-DD/100 lb. seed
3.5-7 fl. oz. 400/100 lb. seed
Disease:
damping-off seed decay seedling disease
mold seedling blight
Time: Seed treatment.
Remarks: Do not combine colloidal suspensions, wettable powders, or emulsifiable concentrates unless previous experience assures compatibility. Use with EPA-approved dye that imparts unnatural color to seed. Do not use for feed, food, or oil purposes. Refer to label for use directions.

- - - - - - - - - - - - - - - - -

Signal Word/Toxicity Class: Danger/I.

CAPTAN (Premixes)

Premix combining a fungicide with an insecticide. Toxic to fish, birds, and wildlife.

Nu-Gro Isotox (12.2% captan & 25% lindane) *(Wilbur-Ellis)*
Seed Mate Isotox (12.2% captan & 25% lindane) *(Loveland)*
Rate: 8 oz./100 lb. seed
Disease:
damping-off seedling blight seedling disease
seed decay
Time: Planter box treatment.

◆ - **new product** • ai-active ingredient • DF-dry flowable • E/EC-emulsifiable concentrate • EW-emulsion in water
F/FL-flowable • G-granule • L/LC-liquid • SP/WSP-soluble packet • W/WP-wettable powder • WSB-water soluble bag

Remarks: Treated seed must not be used for or mixed with food or animal feed, or processed for oil. Treat only those seeds needed for immediate use, minimizing the interval between treatment and planting. Do not store excess treated seeds beyond planting time. Refer to label for use directions.

Signal Word/Toxicity Class: Danger/I.
REI: 4 days.

CARBOXIN

Vitavax-30C is a concentrate formulation. Vitavax-34 is a ready-to-use seed treatment applied as slurry treatment by machine containing toxic inert Rhodamine B. Toxic to fish.

Vitavax-30C or Vitavax-34 *(Gustafson)*

Rate: 3.3-6.6 fl. oz. 30C/100 lb. seed
4-8 fl. oz. 34/100 lb. seed

Disease:
Rhizoctonia solani

Time: Slurry seed treatment.

Remarks: Incorporation of dye into treating slurry is recommended to distinguish seed as treated. Mix with enough water to apply at 20 fl. oz./100 lb. seed. For early-season planting in cold soils, use high rate. Under severe planting conditions, use approved in-furrow granule. Do not use treated seed for food, feed, or oil purposes. Do not graze or feed livestock hay grown from treated seed. Do not apply this product through any type of irrigation system. Refer to label for use directions.

Signal Word/Toxicity Class: Warning/II (30C); Caution (34).

CARBOXIN (Premixes)

Prevail is a premix of the active ingredients of Apron, Terraclor, and Vitavax. Vitavax-PCNB is a ready-to-use formulation combining the systemic action of carboxin with the surface action of PCNB. Toxic to fish.

Prevail (15% carboxin & 15% PCNB & 3.12% metalaxyl) *(Trace)*

Rate: 8-16 oz./100 lb. seed

Disease:
damping-off, seed rot, seedling disease

Time: Planter box seed treatment.

Remarks: Treat only those seeds needed for immediate use. Do not use treated seed for food, feed, or oil purposes. Do not store excess treated seeds beyond planting time. All seed treated with this product (except planter box) must be colored with EPA approved dye which imparts unnatural color to seed. Do not graze or feed livestock hay grown from treated seed. Refer to label for use directions.

Signal Word/Toxicity Class: Caution.
REI: 12 hr.

RTU-Vitavax-Thiram (10% carboxin & 10% thiram) *(Gustafson)*

Rate: 11-13.5 fl. oz./100 lb. seed

Disease:
damping-off, seed rot, sore shin, *Rhizoctonia solani*, seedling diseases

Time: Seed treatment.

Remarks: Use high rates under heavy disease pressure. Use with EPA-approved dye that imparts unnatural color to seed. Do not use for feed, food, or oil purposes. Do not graze or feed livestock on treated areas for 6 weeks after planting. Refer to label for use directions.

Signal Word/Toxicity Class: Caution.

Vitavax-PCNB (17% carboxin & 17% PCNB) *(Gustafson)*

Rate: 5-7 fl. oz./100 lb. seed

Disease:
seed decay, seedling diseases

Time: Slurry seed treatment.

Remarks: Use high rates under heavy disease pressure. Use with EPA-approved dye that imparts unnatural color to seed. Do not use treated seed for food, feed, or oil purposes. Do not graze or feed livestock on hay grown from treated seed. Refer to label for use directions.

Signal Word/Toxicity Class: Caution.

CHLORONEB

Supplemental seed treatment; use in conjunction with a standard fungicide seed treatment.

Demosan Flowable or 65W *(Kincaid)*

Rate: East of Rockies:
10.5 oz. FL/100 lb. seed
6 oz. 65W/100 lb. seed
West of Rockies:
17 oz. FL/100 lb. seed
10 oz. 65W/100 lb. seed
Texas high plains:
7 oz. FL/100 lb. seed
4 oz. 65W/100 lb. seed

Disease:
damping-off, Rhizoctonia, seedling diseases, Pythium blight, Sclerotium blight, sore shin

Time: Seed treatment or hopper box treatment.

Remarks: Refer to label for use directions. Do not use excess treated seed for food, feed, or oil purposes. Do not graze within 45 days of planting.

Signal Word/Toxicity Class: Caution.
REI: 12 hr.

Demosan — see chloroneb

Dithane — see mancozeb

MANCOZEB

Dithiocarbamate fungicide with broad-spectrum activity with no resistance problems. Coordination product of zinc ion and manganese ethylene bisdithiocarbamate. Toxic to fish.

Dithane (DF, F-45, M-45, ST, WSP) *(Rohm and Haas)*
Manzate 200DF *(Griffin)*
Penncozeb 75DF or 80WP *(Elf Atochem)*

Rate: 0.9-1.5 lb. ai/A
1.25-2 lb. DF, 200DF/A
1-1.6 lb. ai/A
1.25-2 lb. M-45, WSP/A
2-3.2 pt. F-45/A
1.12-1.5 lb. ai/A
1.5-2 lb. 75DF/A
1.2-1.6 lb. ai/A
1.5-2 lb. 80WP/A

Disease:
rust

Time: Apply when disease appears; repeat at 10- to 14-day intervals until disease threat is past.

Remarks: Do not apply after bolls open. Do not graze livestock in treated areas or feed gin trash to livestock. Refer to label for use directions.

Days To Harvest: 45; Penncozeb-60.

State Restrictions: For use in the Southwest only. Refer to label for California precautions.

Rate: 3.2-6.4 oz. DF, 75DF/100 lb. seed
3-6 oz. M-45, WSP, 200DF, 80WP/100 lb. seed
4.8-9.6 fl. oz. F-45, ST/100 lb. seed

Disease:
damping-off, seedling blight, seedling disease

Time: Seed treatment.

Remarks: Use low rate for acid delinted; high rate for reginned. Incorporation of dye into treating slurry recommended to distinguish seed as treated. Do not use for food, feed, or oil purposes. Refer to label for use directions.

State Restrictions: Refer to label for California precautions.

Signal Word/Toxicity Class: Caution.
REI: 24 hr.

Manzate — see mancozeb

NOTICE The information on these pages is for preliminary planning — not a guide for use. Be sure to follow the manufacturer's directions, notwithstanding information contained here.

For personal protective equipment and EPA registration numbers, please turn to page 39.

MEFENOXAM

Systemic fungicide having a specific mode of action.

Apron XL LS *(Novartis)*
Rate: 0.32-0.64 fl. oz. LS/100 lb. seed
Disease:
damping-off   seed rot
Time: Seed treatment at planting.
Remarks: Incorporation of dye into treating slurry is recommended to distinguish seed as treated. Do not use treated seed for food, feed, or oil purposes.

Signal Word/Toxicity Class: Warning/II.

Ridomil Gold EC, GR, or WSP *(Novartis)*
Rate: 0.075-0.15 fl. oz. EC/1000 lin. ft. row
1.5-3 oz. GR/1000 lin. ft. row
0.075-0.15 oz. WSP/1000 lin. ft. row
Disease:
seed rot   seedling diseases
Time: In-furrow at planting.
Remarks: May be tank mixed. Do not plant any crop which is not registered for use with mefenoxam in treated soil for 12 months.

Signal Word/Toxicity Class: Warning/II (EC); Caution (GR, WSP).
REI: 48 hr.

METALAXYL

Systemic fungicide having a specific mode of action.

Allegiance or Apron (FL) *(Gustafson)*
Apron Dry *(Trace)*
Rate: 0.75 fl. oz. Allegiance/100 lb. seed
0.75-1.5 fl. oz. Apron FL/100 lb. seed
3-4 oz. Dry/100 lb. seed
Disease:
damping-off   seed rot
Time: Seed treatment.
Remarks: Apply to machine delinted or acid delinted cotton. Use with EPA-approved dye that imparts unnatural color to seed. Do not use for feed, food, or oil purposes. Refer to label for use directions.

Signal Word/Toxicity Class: Warning/II (FL); Caution (Dry).
REI: 12 hr. (Dry).

Nu-Gro Isotox — see captan premixes

PCNB

Toxic to fish and aquatic organisms.

PCNB 2-E, 2LF, 10%, 75DG or 75WP *(Amvac)*
Rate: Row treatment:
1-2 lb. ai/13,000 lin. ft. row (40" spacing)
0.5-1 gal. 2-E/13,000 lin. ft. row (40" spacing)
Furrow treatment (1300 lin. ft. row):
1-2 lb. ai/A
10-20 lb. 10%/A
1.33-2.75 lb. 75DG, 75WP/A (40" spacing)
Disease:
Rhizoctonia solani
Time: At planting.
Remarks: Apply in 15 gal. water. Use lower rate for sandy soil; higher rate for organic soil. Do not apply through any type of irrigation system. Do not allow livestock to graze or feed on foliage in treated fields. Do not plant root crops within 12 months of application unless PCNB is registered for use on those crops.

Rate: 12-16.20 fl. oz. 2-E/100 lb. seed
14.25-19 fl. oz. 2LF/100 lb. seed
Disease:
Fusarium   *Rhizoctonia solani*
Time: Seed treatment.
Remarks: Use lower rate for acid delinted; higher rate for mechanically delinted. Do not apply through any type of irrigation system. Do not use for food, feed or oil purposes. Do not plant root crops within 12 months of application unless PCNB is registered for use on those crops.

Signal Word/Toxicity Class: Danger/I (2-E); Caution (2LF, 10%, 75DG, 75WP).
REI: 12 hr.

PCNB 2 Spray or 10 Granular *(Wilbur-Ellis)*
Rate: 4-8 pt. 2S/A (13,000 lin. ft. 40" row)
10 lb. 10G/A (12,400 lin. ft. furrow)
Disease:
damping-off
Time: Spray over seed and surrounding soil at time of planting.
Remarks: Do not allow feeding or grazing of cotton foliage by livestock. Do not plant root crops within 12 months of application unless PCNB is registered for use on those crops. Do not apply PCNB 2S through any type of irrigation system. Refer to label for use directions.

Signal Word/Toxicity Class: Caution.
REI: 12 hr.

RTU-PCNB or Terra-Coat LT-2N *(Gustafson)*
Rate: 10.75-14.5 fl. oz. RTU/100 lb. seed
12-16 fl. oz. LT-2N/100 lb. seed
Disease: RTU:
Fusarium   Rhizoctonia
LT-2N:
damping-off   seed rot
Time: Seed treatment.
Remarks: Use low rate for acid delinted; high rate for mechanically delinted. Do not use treated seed for food, feed, or oil purposes. Do not feed treated seed to animals. Refer to label for use directions.

Signal Word/Toxicity Class: Caution.

Win-Flo 4F or 6F *(Amvac)*
Rate: 0.75-1.5 lb. ai/A
1.5-3 pt. 4F/A
1-2 pt. 6F/A
Disease:
damping-off   *Rhizoctonia solani*
Time: Furrow treatment.
Remarks: Apply in 5-15 gal. water with 40" row spacing. Use higher rates on heavy soils, when weather conditions are expected to be unfavorable to rapid germination, on fields having a history of disease problems, or when minimum or low till programs are utilized. Do not apply through any type of irrigation system. Do not allow livestock to graze or feed on foliage in treated fields. Do not plant root crops within 12 months of application unless PCNB is registered for use on those crops.

Signal Word/Toxicity Class: Caution.
REI: 12 hr.

Penncozeb — see mancozeb

Prevail — see carboxin premixes

Ridomil Gold — see mefenoxam

RIDOMIL GOLD PC (Premix)

Systemic fungicide having a specific mode of action. Toxic to fish and aquatic organisms.

Ridomil Gold PC or PC GR (0.5% mefenoxam & 10% PCNB) *(Novartis)*
Rate: 0.73-1.05 lb. ai/13,000 lin. ft. row
7-10 lb./13,000 lin. ft. row
Disease:
damping-off   seed rot
Time: At planting.
Remarks: Do not allow livestock to feed or graze foliage. Do not plant any crop not registered for use with mefenoxam in treated soil for 12 months.

Signal Word/Toxicity Class: Warning/II.
REI: 48 hr.

◆ - **new product** • ai-active ingredient • DF-dry flowable • E/EC-emulsifiable concentrate • EW-emulsion in water
F/FL-flowable • G-granule • L/LC-liquid • SP/WSP-soluble packet • W/WP-wettable powder • WSB-water soluble bag

RTU-BAYTAN-THIRAM (Premix)

Ready-to-use seed treatment combining the systemic action of Baytan with the surface action of thiram. Systemic activity of Baytan provides early-season control of infections by powdery mildew and rusts. Toxic to fish.

RTU-Baytan-Thiram (5% triadimenol & 15.3% thiram) *(Gustafson)*

Rate: 3-6 fl. oz. FL/100 lb. seed
Disease:
Rhizoctonia

Rate: 6-12 fl. oz. FL/100 lb. seed
Disease:
black root rot

Time: Seed treatment.
Remarks: Acid delinted cotton. Use high rates only for areas with severe history of disease caused by black root rot. Use with EPA-approved dye that imparts unnatural color to seed. Do not use for feed, food, or oil purposes. Refer to label for use directions.

- - - - - - - - - - - - - - - - -

Signal Word/Toxicity Class: Caution.

RTU-Vitavax-Thiram — see carboxin premixes

Seed Mate Isotox — see captan premixes

SULFUR (Premix)

Fungicide/bactericide in flowable formulation. Toxic to fish and aquatic organisms.

Top Cop with Sulfur (6.25 lb./gal. sulfur & 1 lb./gal. tribasic copper sulfate) *(Stoller)*

Rate: 1-3 qt./A
Disease:
boll rot   rust
Time: Apply early bloom, or when disease threatens; repeat at 7- to 14-day intervals until defoliation.

Rate: 8-12 fl. oz./100 lb. seed
Disease:
damping-off   seedling disease
Time: Seed treatment.
Remarks: Do not use treated seed for food, feed or oil purposes. Refer to label for use directions.

- - - - - - - - - - - - - - - - -

Signal Word/Toxicity Class: Caution.
REI: 24 hr.

SYSTEM³ (Premix)

Biological and chemical fungicide containing the active ingredients of Apron, Terraclor, and Kodiak. Contains graphite. Toxic to fish and aquatic organisms.

System³ (0.1% *Bacillus subtilis* & 14.25% metalaxyl & 16.67% PCNB) *(Helena)*

Rate: 12 oz./100 lb.
Disease:
Pythium   *Rhizoctonia solani*
Time: Apply at planting.
Remarks: A non-oil lubricant has been added for seeds to flow more easily and evenly from planter boxes. Incorporation of dye is recommended to distinguish seed as treated. Do not use treated seed for food, feed or oil purposes. Refer to label for use directions.

- - - - - - - - - - - - - - - - -

Signal Word/Toxicity Class: Caution.
REI: 12 hr.

TELONE

Liquid fumigant for preplant soil treatment. Restricted Use Pesticide.

Telone II (dichloropropene) *(Dow AgroSciences)*

Rate: 12 gal. II/A
Disease:
Fusarium wilt
Time: Row treatment.
Remarks: Use lower rate for mineral soils; higher rate for heavier soils. Do not apply through any type of irrigation system. Refer to label for use directions.

- - - - - - - - - - - - - - - - -

Signal Word/Toxicity Class: Warning/II.
REI: 72 hr.

Terra-Coat — see PCNB

THIRAM

Broad-spectrum fungicide. Toxic to fish. Thiram 50WP contains the toxic inert Rhodamine B and is also toxic to birds and wildlife.

42-S Thiram or Thiram 50WP Dyed *(Gustafson)*
Thiram Granuflo (WDG) *(UCB Agrochemicals)*

Rate: 4.5 fl. oz. 42-S/100 lb. seed
4.5 oz. 50WP/100 lb. seed
3 oz. WDG/100 lb. seed
Disease:
damping-off   seedling blight   seedling disease
seed decay
Time: Seed treatment.
Remarks: Use with EPA-approved dye that imparts unnatural color to seed. Do not use for feed, food, or oil purposes. Do not apply Thiram Granuflo through any type of irrigation system. Refer to label for use directions.
State Restrictions: Thiram Granuflo not for use in California.

- - - - - - - - - - - - - - - - -

Signal Word/Toxicity Class: Caution.
REI: 24 hr. (WDG).

Top Cop with Sulfur — see sulfur premix

Vitavax — see carboxin

Vitavax-PCNB — see carboxin premixes

Win-Flo — see PCNB

4

Cotton

NOTICE	The information on these pages is for preliminary planning — not a guide for use. Be sure to follow the manufacturer's directions, notwithstanding information contained here.	For personal protective equipment and EPA registration numbers, please turn to page 39.

Notes • Notes • Notes

Section 5

Small Grains

Bt Chart *128*
Small Grains crop listings *129*
Insecticides *129*
Fungicides *134*

The information in this 1999 Insect and Disease Control Guide is updated to August 15, 1998. *Because manufacturers' product recommendations are constantly being updated and revised as additional product registrations and label changes are approved by EPA, there may be changes after this date. However, every effort has been made to give manufacturers the opportunity to include in the Guide their latest information. This Guide is for preliminary planning – not a use guide. Be sure to follow label directions, notwithstanding information contained herein.*

Bt Products Chart

Small Grains

Find Bt products quickly and easily with this Bt products chart. Simply find the insect you want to control at the top of the chart, and follow down to the marked spot to find the Bt product you can use to control that pest. Company address information can be found on page 599.

	Armyworms	Bollworms	Budworms	Corn earworm	Diamondback moth	Fusarium	Green cloverworm	Heliothis spp.	Hornworms	Loopers	Rangeland caterpillar	Rhizoctonia	Saltmarsh caterpillar	Skippers	Variegated cutworm	Velvetbean caterpillar	Webworms
Bacillus subtilis						■						■					
Biobit	■									■							
BMP 123	■				■		■	■	■	■			■			■	
Bt. ssp. *aizawai*	■									■							
Bt. ssp. *kurstaki*	■	■	■	■	■		■	■	■	■	■		■	■	■	■	■
Condor	■									■							
Crymax	■									■							
Cutlass	■									■							
DiPel	■			■						■					■		■
Javelin	■																
Kodiak						■						■					
Lepinox	■									■							
Troy-Bt	■	■	■				■		■	■	■			■			
XenTari	■									■							

Manufacturers List

These Bt products are manufactured by the following companies. Look for their addresses, phone numbers, and fax numbers in the Suppliers Directory beginning on page 599.

Javelin – Thermo Trilogy
Biobit, DiPel, XenTari – Abbott Laboratories
BMP 123 – Becker Microbial
Condor, Crymax, Cutlass, Lepinox – Ecogen
Kodiak – Gustafson
Troy-Bt – Troy Biosciences

Small Grains
(Barley, Oats, Rye, Triticale, Wheat)

Insecticides

AZINPHOS-METHYL

Restricted Use Pesticide.

Gowan Azinphos-M 50-W *(Gowan)*
Guthion 2L *(Bayer)*
◆ Micro Flo Azinphosmethyl 2EC or 50W *(Micro Flo)*
Sniper 2-E, 50PVA, 50W *(Platte)*

Rate: 0.375-0.5 lb. ai/A
1.5-2 pt. 2EC, 2-E, 2L/A
0.75-1 lb. 50-W, 50PVA, 50W/A

Pest:
cereal leaf beetle

Remarks: Barley, oats, rye, wheat. Apply in sufficient water to give complete coverage. Do not apply more than once per season.
Days To Harvest: Harvest/grazing/feeding-30.
State Restrictions: Sniper 50PVA not registered in California.

- - - - - - - - - - - - - - - - - -

Signal Word/Toxicity Class: Danger/I.
REI: 48 hr.; 72 hr. if average rainfall less than 25" per year.

CARBARYL

Nonrestricted, broad-spectrum carbamate insecticide registered for use on many crops. Available in a number of formulations. Sevin XLR Plus is residual in action and affords excellent control of chewing insect pests.

Clean Crop Sevin 5 Bait *(Platte)*

Rate: **Wheat:**
1.5 lb. ai/A
30 lb. Bait/A
Barley, oats, rye:
1-2 lb. ai/A
20-40 lb. Bait/A

Pest:

armyworms	darkling beetles	grasshoppers
cutworms	field crickets	sowbugs

Remarks: Barley, oats, rye, wheat. Do not make more than 2 applications to wheat after grain heads emerge from boot. Do not feed treated fodder to dairy animals or animals being finished for slaughter or allow them to graze treated crops.
Days To Harvest: Wheat: forage, pasture-0; fall or spring (grain)-21.

- - - - - - - - - - - - - - - - - -

Signal Word/Toxicity Class: Caution/III.
REI: 12 hr.

Drexel Carbaryl 4L *(Drexel)*
Sevin 4F, XLR Plus, 80S, 80WSP *(Rhone-Poulenc)*

Rate: 1-1.5 lb. ai/A
2-3 pt. 4L, 4F, XLR/A
1.25-1.875 lb. 80S, 80WSP/A

Pest:

armyworms	flea beetles	grasshoppers
cereal leaf beetle		

Remarks: Wheat. Apply when insects or their damage appears. Rate varies by pest species; refer to label. Do not make more than 2 applications of 80S after grain heads emerge from boot. For cereal leaf beetle, effective against eggs, larvae, and adults. To avoid possible injury to tender foliage, do not apply when foliage is wet or when rain or excessive humidity is expected during next 2 days.
Days To Harvest: Forage-0; grain-21.
State Restrictions: Not for use in California.

- - - - - - - - - - - - - - - - - -

Signal Word/Toxicity Class: Caution (4F, 4L, XLR); Warning/II (80S, 80WSP).
REI: 12 hr.

CHLORPYRIFOS

Restricted Use Pesticide due to acute toxicity to avian and aquatic organisms.

Lorsban 4E-SG *(Dow AgroSciences)*
Micro Flo Chlorpyrifos 4E-Wheat *(Micro Flo)*
Nufos 4E-SG *(Cheminova)*

Rate: 0.25-0.5 lb. ai/A
0.5-1 pt. 4E/A

Pest:

aphids	cutworms	Russian wheat aphid
brown wheat mite	grasshoppers	wheat midge

Remarks: Wheat. Do not apply more than twice per season. Do not apply directly to bodies of water.
State Restrictions: For use in Arizona, California, Colorado, Idaho, Kansas, Montana, Nebraska, Nevada, New Mexico, Oklahoma, Oregon, South Dakota, Texas, Utah, Washington, and Wyoming.
Days To Harvest: 28; Graze or forage-14.

- - - - - - - - - - - - - - - - - -

Signal Word/Toxicity Class: Warning/II.
REI: 24 hr.

lambda-CYHALOTHRIN

Synthetic pyrethroid insecticide with broad-spectrum activity, formulated as an emulsifiable concentrate. Restricted Use Pesticide due to toxicity to fish and aquatic organisms.

Karate or Warrior (1E) *(ZENECA)*

Rate: 0.015-0.03 lb. ai/A
1.92-3.84 fl. oz. 1E/A

Pest:

army cutworm	English grain aphid	oat birdcherry aphid
armyworms	fall armyworm	Russian wheat aphid
cereal leaf beetle	flea beetles	stink bugs
chinch bug	grasshoppers	yellowstriped armyworm
cutworms	greenbug	

Remarks: Wheat, triticale. Apply as required by scouting, usually at intervals of 5 or more days. For chinch bugs, repeat at 3- to 5-day intervals if needed. For aphids and greenbug, apply before insects begin to roll leaves; once wheat has started to boot, may provide suppression only. Do not apply more than 0.48 pt./A per season. Refer to label for spray drift precautions and state and county restrictions.
Days To Harvest: 30.

- - - - - - - - - - - - - - - - - -

Signal Word/Toxicity Class: Danger/I.
REI: 24 hr.

DB-Green — see lindane premixes

Declare — see methyl parathion

Diatect — see pyrethrins

Digon — see dimethoate

Dimate — see dimethoate

For personal protective equipment and EPA registration numbers, please turn to page 39.

5 Small Grains

DIMETHOATE

Broad-spectrum, organophosphate insecticide systemic in action for insect and mite control. Excellent for most aphid species and for grasshopper control on wheat.

Cheminova Dimethoate 4E *(Cheminova)*
Clean Crop Dimethoate 400 *(Platte)*
Dimate 4E *(Terra)*
Drexel Dimethoate 2.67 or 4EC *(Drexel)*
Gowan Dimethoate E267 *(Gowan)*
Micro Flo Dimethoate 267 or 4E *(Micro Flo)*
Wilbur-Ellis Dimethoate 267 or Digon 400 *(Wilbur-Ellis)*

Rate: 0.25-0.33 lb. ai/A
0.75-1 pt. 2.67, E267/A
0.16-0.375 lb. ai/A
0.33-1.125 pt. 267/A
0.33-0.75 pt. 4E, 4EC, 400/A

Pest:
aphids; brown wheat mite; grasshoppers (except 2.67); greenbug

Remarks: Wheat. Do not apply more than twice per season or within 14 days of grazing immature plant. Rate varies by pest species; refer to label.
Days To Harvest: 35; 267, 4E (California)-60.

Signal Word/Toxicity Class: Warning/II.
REI: 48 hr.

DI-SYSTON

Organophosphate insecticide and acaricide, systemic in action, that shows good aphid control and has been used extensively on wheat against the Russian aphid. Insecticidal efficacy on aphid, grasshopper, and Hessian fly is excellent. Restricted Use Pesticide.

Di-Syston 8 (disulfoton) *(Bayer)*

Rate: Barley:
Foliar:
0.5-1 lb. ai/A
0.5-1 pt. 8/A
Soil:
1 lb. ai/A
1 pt. 8/A (0.25 fl. oz./1000 ft. row)

Pest:
aphids; greenbug; Hessian fly; mites (except wheat curl); thrips

Rate: Wheat:
Foliar—spring or fall:
0.25-0.75 lb. ai/A
4-12 fl. oz. 8/A

Pest:
aphids; greenbug; mites

Rate: Fall wheat:
Soil:
1 lb. ai/A
1 pt. 8/A (0.25 fl. oz./1000 ft. row)

Pest:
aphids; greenbug; Hessian fly; oat birdcherry aphid (except CA)

Remarks: Barley, wheat. Foliar: Apply in sufficient water to obtain thorough coverage but not less than 1 gal. total volume/A. Use lower rate on plants up to tillering (stooling); higher rate after plants begin to tiller. Applications may be made following an at-planting application. Do not apply more than 2 pt./A per crop season. Soil: Apply as a soil injection in water emulsion or with liquid fertilizer at planting time. Due to greenbug resistance in certain areas, Di-Syston alone may not provide satisfactory control. Consult agricultural advisor or extension service for recommendations.
Days To Harvest: Foliar-30; do not graze treated fields. Soil-60; do not graze or cut for forage within 30 days.

Rate: 0.5 lb. ai/A
0.5 pt. 8/A
Pest:
aphids
Remarks: Oats.
State Restrictions: SLN-Maine.

Rate: 0.25-0.75 lb. ai/A
4-12 oz. 8/A
Pest:
aphids; mites
Remarks: Triticale, wheat.
State Restrictions: SLN-Wyoming: triticale; SLN-New Mexico, Oklahoma, Texas: wheat.

Signal Word/Toxicity Class: Danger/I.
REI: 48 hr.

Di-Syston 15% (disulfoton) *(Bayer)*

Rate: 1 lb. ai/A
6.7 lb. 15%/A

Pest:
aphids; greenbug (wheat); grasshoppers (wheat); Hessian fly (wheat); mites (barley)

Remarks: Barley, fall wheat. For barley, apply by drilling or broadcast at planting time, or as broadcast after emergence. Do not apply more than twice per season. Allow minimum of 21 days between applications. For fall wheat, apply by drilling or broadcast at planting time in the fall. Make only 1 application. Due to greenbug resistance in certain areas, Di-Syston alone may not provide satisfactory control. Consult agricultural advisor or extension service for recommendations.
Days To Harvest: Grazing or cut forage barley-30; wheat-75.

Rate: 1 lb. ai/A
6.7 lb. 15%/A
Pest:
aphids; grasshoppers (except CA); Hessian fly (except CA)
Remarks: Barley (except California), wheat.
State Restrictions: SLN-California, Idaho, Montana, Oregon, Utah, Washington.

Signal Word/Toxicity Class: Danger/I.
REI: 48 hr.

ENDOSULFAN

For army cutworms at higher rates. On cereal leaf beetle at reduced rates when beetle larvae are small.

Drexel Endosulfan 3EC or 50WP *(Drexel)*
Gowan Endosulfan 3EC or 50WSB *(Gowan)*
◆ Micro Flo Endosulfan 3EC or 50WP *(Micro Flo)*
Phaser 3EC or 50WSB *(AgrEvo USA)*
Thiodan 3EC or 50WP *(FMC)*

Rate: 0.25-0.75 lb. ai/A
0.33-1 qt. 3EC/A
0.5-1.5 lb. 50WP, 50WSB/A

Pest:
aphids; army cutworm; cereal leaf beetle

Remarks: Barley, oats, rye, wheat. For aphids, apply when insects appear or feeding is noticed. For cereal leaf beetle, apply when small larvae are readily found in field. Do not apply after heads begin to form. Do not exceed 1 lb. ai/A per year or 2 applications per year. Do not feed treated forage to livestock.
State Restrictions: For use on cereal leaf beetle in Illinois, Indiana, Michigan, Ohio.

Signal Word/Toxicity Class: Danger-Poison/I.
REI: 24 hr.

Enhance Plus — see lindane premixes

FURADAN

Broad-spectrum carbamate insecticide, nematicide, and miticide. Effective against a number of crop pests applied both as soil and foliar insecticide. Shows good residual efficacy as an insecticide against chewing and sucking insects. Toxic to bees, fish, birds, and other wildlife. Restricted Use Pesticide due to acute oral and inhalation toxicity.

Furadan 4F (carbofuran) *(FMC)*

Rate: 0.125-0.25 lb. ai/A
0.25-0.5 pt. 4F/A
Pest:
cereal leaf beetle; grasshoppers

◆ - **new product** • ai-active ingredient • DF-dry flowable • E/EC-emulsifiable concentrate • EW-emulsion in water
F/FL-flowable • G-granule • L/LC-liquid • SP/WSP-soluble packet • W/WP-wettable powder • WSB-water soluble bag

Remarks: Barley, oats, wheat. Apply in minimum 10 gal. finished spray/A by ground; 2 gal./A by air when insects appear and before heads emerge from boot. Do not apply more than twice per season. Apply high rate for cereal leaf beetle. For waterfowl protection, do not apply on fields in proximity to waterfowl nesting areas or on fields where waterfowl are known to repeatedly feed. Do not feed treated forage to livestock. Do not apply through any type of irrigation system.

Crop Rotation: Do not rotate with any crop on soil treated at greater than 3 lb. ai/A for at least 10 months or plant with any crop for which carbofuran treatment is not registered for at least 10 months.

State Restrictions: Not for use in Long Island, New York.

Signal Word/Toxicity Class: Danger-Poison/I.
REI: 48 hr.

Fyfanon — see malathion

Gaucho — see imidacloprid

Grain Guard Plus — see lindane premixes

Guthion — see azinphos-methyl

IMIDACLOPRID

For use by commercial seed treaters only.

Gaucho 75 ST, 480 *(Gustafson)*

Rate: 0.67-2 oz. 75 ST/100 lb. seed
Slurry treatment:
1-3 fl. oz. 480/100 lb. seed

Pest:

aphids	Hessian fly	Russian wheat aphid
English grain aphid	oat birdcherry aphid	wireworm
greenbug		

Remarks: Barley, wheat. Apply prior to planting; ensure thorough coverage. Do not graze or feed livestock on treated areas for 45 days after planting. Treated seed must not be used for or mixed with food or animal feed or processed for oil.

Crop Rotation: May be replanted with any labeled crop or any crop for which ai tolerance exists; 12-month plant-back interval for other crops.

Signal Word/Toxicity Class: Caution.
REI: 12 hr.

Karate — see *lambda*-cyhalothrin

Kumulus — see sulfur

LANNATE

Broad-spectrum carbamate insecticide effective against a number of insect species on field crops. Very efficacious on armyworm, aphids, and cereal leaf beetle on cereal grains. Restricted Use Pesticide.

Lannate SP or LV (methomyl) *(DuPont)*

Rate: 0.225-0.45 lb. ai/A
0.25-0.5 lb. SP/A
0.75-1.5 pt. LV/A

Pest:

aphids	armyworm	cereal leaf beetle

Remarks: Barley, oats, rye, wheat. Apply by air or ground when insects first appear; repeat at 5- to 7-day intervals or as needed. Do not apply more than 1.8 lb. ai/A per crop or make more than 4 applications per crop.

Days To Harvest: 7; grazing-10.

Signal Word/Toxicity Class: Danger-Poison/I.
REI: 48 hr.

LINDANE

Lindane 30 C contains toxic inert Rhodamine B. Toxic to fish, birds, and wildlife.

Lindane 30C *(Gustafson)*
◆ **Lindane ST 40** *(AGSCO)*

Rate: Barley:
1.35 fl. oz. 30C/100 lb. seed
1 fl. oz. 40/100 lb. seed
Oats, rye, wheat:
1.4 fl. oz. 30C/100 lb. seed
1 fl. oz. 40/100 lb. seed

Pest:

seedcorn maggots	wireworms

Remarks: Barley, oats, rye, wheat. Do not use treated seed for food, feed, or oil purposes. Exposed treated seed may be hazardous to birds and other wildlife. Dispose of excess seed by burial away from bodies of water.

Signal Word/Toxicity Class: Warning/II.
REI: 12 hr.

LINDANE (Premixes)

Dual purpose insecticide and fungicide seed treatment. Grain Guard Plus contains graphite.

◆ **DB-Green (18.75% lindane & 50% maneb)** *(AGSCO)*

Rate: 2 oz./bu. seed

Pest:

seedcorn maggots	wireworms

Remarks: Barley, oats, rye, wheat. Drill or planter box application. Treat only seeds needed for immediate use, minimizing interval between treatment and planting. Do not use treated seed for food, feed, or oil purposes. Do not store excess treated seeds beyond planting time. Do not graze or feed livestock on treated areas for 6 weeks after planting.

Signal Word/Toxicity Class: Warning/II.
REI: 12 hr.

◆ **DB-Green L (8.6% lindane & 25.6% maneb)** *(AGSCO)*

Rate: 3 fl. oz./bu. seed

Pest:
wireworms

Remarks: Barley, oats, rye, wheat. Drill or planter box application. Treat only seeds needed for immediate use, minimizing interval between treatment and planting. Do not use treated seed for food, feed, or oil purposes. Do not store excess treated seeds beyond planting time.

Signal Word/Toxicity Class: Warning/II.
REI: 12 hr.

◆ **Enhance Plus (18.75% lindane & 35% maneb & 20% carboxin)** *(Trace)*

Rate: 2 oz./bu. seed

Pest:
wireworms

Remarks: Barley, oats, wheat. Apply as drill box treatment at planting. Treat only those seeds needed for immediate use. Do not use treated seed for food, feed, or oil purposes. Do not store excess treated seeds beyond planting time. Do not graze or feed livestock on treated areas for 6 weeks after planting.

Signal Word/Toxicity Class: Danger/I.
REI: 12 hr.

◆ **Grain Guard Plus (18.75% lindane & 50% mancozeb)** *(Trace)*

Rate: 2 oz./bu. seed

Pest:
wireworms

Remarks: Barley, oats, rye, wheat. Apply in planter or drill box as seed treatment before planting. Treat only those seeds needed for immediate use. Do not use treated seed for food, feed, or oil purposes. Do not store excess treated seeds beyond planting time.

Signal Word/Toxicity Class: Warning/II.
REI: 12 hr.

5

Small Grains

NOTICE The information on these pages is for preliminary planning — not a guide for use. Be sure to follow the manufacturer's directions, notwithstanding information contained here.	For personal protective equipment and EPA registration numbers, please turn to page 39.

Lorsban — see chlorpyrifos

MALATHION

Organic phosphate insecticide with numerous uses on many crops. Shows good efficacy against both sucking and chewing insects. Its advantage on cereal grains and alfalfa is the short harvest interval.

Clean Crop Malathion 57EC or 8EC *(Platte)*
Fyfanon (8E) *(Helena)*
Gowan Malathion 8 (8E) *(Gowan)*

Rate: 1-1.25 lb. ai/A
1-1.25 pt. 8E, 8EC/A
1.5-2 pt. 57EC/A

Pest:
armyworms, grasshoppers, greenbug, English grain aphid

Remarks: Barley, oats (except Fyfanon), rye, wheat. For 57EC, use high rate for armyworm.
Days To Harvest: 7.

Signal Word/Toxicity Class: Caution/III.
REI: 12 hr.

Drexel Malathion 5EC *(Drexel)*
◆ **Micro Flo Malathion 5EC** *(Micro Flo)*
Prentox 5 lb. Malathion Spray (5EC) *(Prentiss)*

Rate: 0.63-1.25 lb. ai/A
1-2 pt. 5EC/A

Pest:
cereal leaf beetle, grasshoppers, winter grain mite (except Cythion), English grain aphid, greenbug

Remarks: Barley, oats, rye, wheat. Apply when insects first appear and repeat as necessary. Rate varies by pest species; refer to label.
Days To Harvest: Harvest or forage-7.

Signal Word/Toxicity Class: Warning/II.
REI: 12 hr.

◆ **Drexel Malathion ULV** *(Drexel)*
Fyfanon ULV *(Cheminova)*
◆ **Micro Flo Malathion ULV** *(Micro Flo)*

Rate: 4-8 fl. oz. ULV/A

Pest:
cereal leaf beetle (except rye), grasshoppers (high rate)

Remarks: Barley, oats, rye, wheat. Apply high rate when grasshoppers are present at an economic threshold.
Days To Harvest: 7.

Signal Word/Toxicity Class: Caution.
REI: 12 hr.

◆ **Wilbur-Ellis Malathion 8 Spray** *(Wilbur-Ellis)*

Rate: 1.25 lb. ai/A
1.25 pt. 8/A

Pest:
aphids, armyworms

Remarks: Barley, oats, rye, wheat. Apply in minimum 5 gal. by air; 10 gal. by ground.
Days To Harvest: 7.

Signal Word/Toxicity Class: Caution.
REI: 12 hr.

METHYL PARATHION

Broad-spectrum organophosphate insecticide. Highly effective for control of foliar insect pests on cereal grains and alfalfa. Penncap-M is a microencapsulated formulation. Restricted Use Pesticide due to residual effects to avian species and hazard to bees.

Cheminova Methyl 4EC *(Cheminova)*
◆ **Declare (4EC)** *(Griffin)*
Helena 4 lb. Methyl Parathion (4EC) *(Helena)*

Rate: 0.25-0.375 lb. ai/A
0.5-1.5 pt. 4EC/A
0.125-0.75 lb. ai/A
0.25-1.5 pt. 4EC/A (Helena)

Pest:
aphids, armyworms, barley thrips, black grass bug, black mite, brown wheat mite, chinch bug, climbing cutworm, fall armyworm, false chinch bug, grasshoppers, greenbugs, leafhoppers, Say stink bug, spider mites, stink bugs, thrips, western wheat mite, winter grain mite

Remarks: Barley, oats, rye, wheat. Apply when insects or their damage appears. Rate varies by pest species; refer to label.
Days To Harvest: 15.

Signal Word/Toxicity Class: Danger-Poison/I.
REI: 48 hr.; 72 hr. if average rainfall less than 25" per year.

Penncap-M (2FL) *(Elf Atochem)*

Rate: 0.25-0.75 lb. ai/A
1-3 pt. 2FL/A

Pest:
stink bugs*

Rate: 0.5-0.75 lb. ai/A
2-3 pt. 2FL/A

Pest:
aphids, grasshoppers, greenbug, true armyworm*

****Except CA***

Remarks: Barley, oats, wheat. For grasshoppers, use higher rate if majority of insects are large or weather is cool. Do not chemigate in California.
Days To Harvest: Harvest or grazing-15.

Signal Word/Toxicity Class: Warning/II.
REI: 48 hr.

Wilbur-Ellis Methyl Parathion 4 or 5 Spray *(Wilbur-Ellis)*

Rate: 0.5 lb. ai/A
1 pt. 4/A
0.63 lb. ai/A
1 pt. 5/A

Pest:
aphids, armyworms, greenbug, leafhoppers

Remarks: Barley, oats, rye, wheat.
Days To Harvest: 15.

Signal Word/Toxicity Class: Danger-Poison/I.
REI: 48 hr.; 72 hr. if annual rainfall is less than 25" per year.

Wilbur-Ellis Methyl Parathion 7.5 Spray *(Wilbur-Ellis)*

Rate: Barley:
0.23-0.7 lb. ai/A
0.25-0.75 pt. 7.5/A
Wheat:
0.117-0.7 lb. ai/A
0.125-0.75 pt. 7.5/A

Pest:
aphids, armyworms, barley thrips, black grass bug, brown wheat mite, chinch bug, climbing cutworm, cutworms, false chinch bug, greenbug, leafhoppers, mites, Say stink bug, stink bugs, thrips, winter grain mite

Remarks: Rates vary by pest.
Days To Harvest: 15.

Signal Word/Toxicity Class: Danger-Poison/I.
REI: 48 hr.; 72 hr. if annual rainfall is less than 25" per year.

Microthiol Special — see sulfur

M-PEDE

M-Pede (potassium salts of fatty acids) *(Mycogen)*

Rate: 4 pt./A in 3-25 gal. water
2% v/v spray solution in at least 25 gal. water/A

Pest:
aphids, caterpillars (tank mix only), greenbug, fleahoppers, Lygus bugs, plant bugs, spider mites, stink bugs, thrips, whiteflies

◆ **- new product** • ai-active ingredient • DF-dry flowable • E/EC-emulsifiable concentrate • EW-emulsion in water
F/FL-flowable • G-granule • L/LC-liquid • SP/WSP-soluble packet • W/WP-wettable powder • WSB-water soluble bag

Remarks: Wheat. Apply at first sign of infestation; repeat at 7- to 14-day intervals. May be tank-mixed with other insecticides; refer to both labels for use directions and precautions.
Days To Harvest: 0.
State Restrictions: Not for use in California.

Signal Word/Toxicity Class: Warning/II.
REI: 12 hr.

Nufos — see chlorpyrifos

Penncap-M — see methyl parathion

Phaser — see endosulfan

PHORATE

Organophosphate insecticide, systemic in action, used on wheat for Hessian fly, aphids, and grasshopper control. Restricted Use Pesticide.

Clean Crop Phorate 20G *(Platte)*
Riverside Phorate 20G *(Terra)*
Thimet 20-G *(American Cyanamid)*
Rate: 1.2 oz. 20G, 20-G/1000 ft. row (min. 8" spacing)
Pest:

aphids (North Central, Western states)	grasshoppers (winter wheat)	Hessian fly

Remarks: Wheat. At planting time, apply granules in seed furrow. Do not apply after planting time.
Days To Harvest: 70.

Signal Word/Toxicity Class: Danger-Poison/I.
REI: 48 hr.

Prentox — see malathion

PYRELLIN (Premix)

Pyrethrins and rotenone are botanically derived, general use, broad-spectrum insecticides used on a wide variety of crops to control a wide range of insect pests. Pyrethrins are effective against a number of sucking insects while both are effective against many chewing insects. These insecticides have little residual activity. Rotenone is highly toxic to fish.

Pyrellin E.C. (0.042 lb./gal. pyrethrins & 0.035 lb./gal. rotenone) *(Webb Wright)*
Rate: 1-2 pt. E.C./A
Pest:

aphids	caterpillars	mites
armyworms	chinch bug	thrips

Remarks: Barley, wheat. Apply in intervals of 7 days or less; repeat as needed.
Days To Harvest: 0.

Signal Word/Toxicity Class: Caution/III.
REI: 12 hr.

Pyrenone — see pyrethrins

PYRETHRINS

Contains piperonyl butoxide as a synergist and silicon dioxide (diatomaceous earth).

Diatect Multipurpose Insecticide II *(Diatect)*
Pyrenone (0.5EC) *(AgrEvo Environmental Health)*
Rate: 1-6 lb. II/A
12 oz. 0.5EC/A
Pest: Including but not limited to:

aphids	caterpillars	mites
armyworms	chinch bug	thrips

Remarks: Barley, oats, rye, wheat. Apply in sufficient water for thorough coverage; repeat as needed. Apply Diatect as dust or wettable powder.
Days To Harvest: 0.

Signal Word/Toxicity Class: Caution.
REI: 12 hr.

Redball Lightning — see sulfur

Sevin — see carbaryl

Sniper — see azinphos-methyl

SULFUR

Microthiol and Thiolux are micronized sulfurs. Redball Lightning is a dustless granule formulated to break down into smaller particles when wetted by water to provide effectiveness of a powder with the handling ease of a granule.

◆ **Kumulus DF** *(BASF)*
Rate: 4.8-12 lb. ai/A
6-15 lb. DF/A
Pest:

Atlantic mite	Pacific mite	twospotted spider mite
Banks grass mite	red spider mite	

Remarks: Do not apply if temperatures during or within 3 days after application are expected to exceed 90°F. Do not apply through any type of irrigation system.

Signal Word/Toxicity Class: Caution.
REI: 24 hr.

Microthiol Special (80W) *(Elf Atochem)*
Rate: 6.4-8 lb. ai/A
8-10 lb. 80W/A
Pest:

Atlantic mite	red spider mite	twospotted spider mite (except CA)
Pacific mite		

Remarks: Barley, oats, rye, wheat. Apply when mites first appear; repeat as necessary. Do not chemigate in California.

Signal Word/Toxicity Class: Caution.
REI: 24 hr.

◆ **Redball Lightning Degradable (G)** *(International Sulphur)*
Rate: 5-15 lb. G/A
Pest:

fleahoppers	mites	thrips

Remarks: Oats, wheat. Do not apply within 3 weeks of oil spray.
Days To Harvest: 0.

Signal Word/Toxicity Class: Caution.
REI: 24 hr.

Thiolux (DF) *(Platte)*
Rate: 4.8 lb. ai/A
6 lb. DF/A
Pest:

Atlantic spider mite (suppression)	Banks grass mite	red spider mite
	Pacific spider mite	twospotted spider mite

Remarks: Barley, oats, rye, wheat. Apply when mites first appear; repeat as necessary.

Signal Word/Toxicity Class: Caution.
REI: 24 hr.

Thimet — see phorate

Thiodan — see endosulfan

Thiolux — see sulfur

Warrior — see *lambda*-cyhalothrin

5 Small Grains

NOTICE The information on these pages is for preliminary planning — not a guide for use. Be sure to follow the manufacturer's directions, notwithstanding information contained here.	For personal protective equipment and EPA registration numbers, please turn to page 39.

Fungicides

Allegiance — see metalaxyl

Apron — see metalaxyl

Apron XL — see mefenoxam

Bayleton — see triadimefon

BAYTAN 30

Triazole systemic fungicide for use only through commercial seed treating equipment which are closed application systems.

Baytan 30 (FL) (triadimenol) *(Gustafson)*

Rate: 0.75-1.5 fl. oz. FL/100 lb. seed

Disease: Barley:

barley rust, loose smut, stripe rust, covered smut, net blotch*, Typhula rot*, foot rot*, powdery mildew

Oats:

crown rust, powdery mildew, stripe rust, loose smut

Rye:

powdery mildew, stripe rust, stripe smut, rye rust

Wheat:

flag smut, leaf rust, stinking smut, foot rot*, loose smut, stripe rust, glume blotch, powdery mildew, take-all*

* *Suppression.*

Time: Seed treatment.

Remarks: Barley, oats, rye, wheat. Use with EPA-approved dye that imparts unnatural color to seed. Do not use for feed, food, or oil purposes. Green forage may be grazed 40 days after seeding. Refer to label for use directions.

Signal Word/Toxicity Class: Caution.

BENLATE

Benzimidazole systemic fungicide. Toxic to fish.

Benlate (SP, WP) (benomyl) *(DuPont)*

Rate: 0.13 lb. ai/A
0.25 lb. SP, WP/A

Disease:

glume blotch, powdery mildew, Septoria leaf blotch, leaf rust

Time: Apply at boot stage; repeat 14 days later.

Rate: 0.13-0.25 lb. ai/A
0.25-0.5 lb. SP, WP/A

Disease:

Helminthosporium leaf blight, leaf rust, powdery mildew

Time: Apply at boot stage to early heading; repeat 14 days later.

Rate: 0.375-1 lb. ai/A
0.75-2 lb. SP, WP/A

Disease:

strawbreaker

Time: Apply once at full tillering but before elongation.

Remarks: Wheat. Do not tank mix or alternate with benzimidazole or thiophanate products. Do not allow livestock to graze in treated fields. Refer to label for use directions.

Days To Harvest: 21.

State Restrictions: Benlate SP not registered for use in California.

Rate: 1-2 oz. SP, WP/bu. seed

Disease:

bunt, covered smut, loose smut, common bunt, flag smut

Time: Dry or slurry treatment.

Remarks: Barley, oats, rye, wheat. Use with EPA-approved dye that imparts unnatural color to treated seed. Do not tank mix or alternate with benzimidazole or thiophanate products. Do not allow livestock to graze on plants grown from treated seed. Do not use treated seed for food, feed, or oil purposes. Refer to label for use directions.

State Restrictions: Benlate SP not registered for use in California.

Signal Word/Toxicity Class: Caution.

REI: 24 hr.

CAPTAN

Gustafson Captan 30-DD contains the toxic inert Rhodamine B.

Drexel Captan 4FL *(Drexel)*

Rate: Barley:
2.25 fl. oz. 4FL/100 lb. seed
Oats:
3.25 fl. oz. 4FL/100 lb. seed
Wheat:
1.75 fl. oz. 4Fl/100 lb. seed

Disease:

seed decay, seedling blight, soilborne disease, seed rot

Time: Liquid or slurry seed treatment.

Remarks: Will not control bunt or smut. Do not use treated seed for food, feed, or oil purposes. Refer to label for use directions.

Signal Word/Toxicity Class: Danger/I.

Gustafson Captan 30-DD or 400 *(Gustafson)*

Rate: Barley, Rye:
2 fl. oz. 30-DD/100 lb. seed
2-3 fl. oz. 400/100 lb. seed
Oats:
2.33 fl. oz. 30-DD/100 lb. seed
2-4 fl. oz. 400/100 lb. seed
Wheat:
1.5 fl. oz. 30-DD/100 lb. seed
1.5-4 fl. oz. 400/100 lb. seed

Disease:

damping-off, seed decay, seedling blight, mold

Time: Seed treatment.

Remarks: Barley, oats, rye, wheat. Will not control bunt or smut. Do not combine colloidal suspensions, wettable powders, or emulsifiable concentrates unless previous experience assures compatibility. Use with EPA-approved dye that imparts unnatural color to seed. Do not use for feed, food, or oil purposes. Refer to label for use directions.

Signal Word/Toxicity Class: Danger/I.

CAPTAN (Premixes)

Toxic to fish, birds and other wildlife.

Nu-Gro Isotox (12.2% captan & 25% lindane) *(Wilbur-Ellis)*
Seed Mate Isotox (12.2% captan & 25% lindane) *(Loveland)*

Rate: Oats, rye, wheat:
4 oz./100 lb. seed
Barley:
4 oz./100 lb. seed

Disease:

damping-off, seed decay, seedling blight

◆ - **new product** • ai-active ingredient • DF-dry flowable • E/EC-emulsifiable concentrate • EW-emulsion in water
F/FL-flowable • G-granule • L/LC-liquid • SP/WSP-soluble packet • W/WP-wettable powder • WSB-water soluble bag

Time: Planter box treatment.
Remarks: Treated seed must not be used for or mixed with food or animal feed, or processed for oil. Treat only those seeds needed for immediate use, minimizing the interval between treatment and planting. Do not store excess treated seeds beyond planting time. Refer to label for use directions.

Signal Word/Toxicity Class: Danger/I.
REI: 4 days.

Rival (19.8% captan & 8.4% PCNB & 1% thiabendazole) *(Gustafson)*
Rate: 4 fl. oz./100 lb. seed
Disease:
damping-off; seedling blight; stem blight; seed decay
Time: Seed treatment
Remarks: Wheat. Use with EPA-approved dye that imparts unnatural color to seed. Do not use for feed, food, or oil purposes. Refer to label for use directions.

Signal Word/Toxicity Class: Danger/I.

CARBOXIN

Ready-to-use seed treatment applied as slurry treatment by machine. Contains toxic inert Rhodamine B.

Vitavax-34 *(Gustafson)*
Rate: 2-3 fl. oz./100 lb. seed
Disease:
common smut (wheat); covered smut (barley, oats); loose smut
Time: Slurry seed treatment.
Remarks: Barley, oats, wheat. Incorporation of dye into treating slurry is recommended to distinguish seed as treated. Do not use treated seed for food, feed, or oil purposes. Do not graze or feed livestock on forage or hay on treated areas. Do not apply this product through any type of irrigation system. Refer to label for use directions.

Signal Word/Toxicity Class: Caution.

CARBOXIN (Premixes)

Vitavax-Captan contains graphite. Toxic to fish.

Enhance Vitavax-Captan 20-20 (20% carboxin & 19.1% captan) *(Trace)*
Seed Mate Vitavax-Captan (20% carboxin & 19.1% captan) *(Loveland)*
Rate: 4 oz./100 lb. seed
Disease:
covered smut; seed rot; seedling blight; loose smut (barley, wheat)
Time: Planter box seed treatment.
Remarks: Barley, oats, wheat. Do not use treated seed for food, feed, or oil purposes. Refer to label for use directions.

Signal Word/Toxicity Class: Danger/I.

Prevail (15% carboxin & 15% PCNB & 3.12% metalaxyl) *(Trace)*
Rate: Barley, oats:
2-4 oz./bu. seed
Wheat:
3 oz./bu. seed
Disease:
common bunt (wheat); loose smut (barley, wheat); seed rot; covered smut (barley); seedling disease; damping-off; oat smut; stinking smut (wheat)
Time: Planter box seed treatment.
Remarks: Barley, oats, wheat. Treat only those seeds needed for immediate use. Do not use treated seed for food, feed, or oil purposes. Do not store excess treated seeds beyond planting time. All seed treated with this product (except planter box) must be colored with EPA approved dye which imparts unnatural color to seed. Do not graze or feed livestock on treated areas for 6 weeks after planting. Refer to label for use directions.

Signal Word/Toxicity Class: Caution.
REI: 12 hr.

RTU-Vitavax-Extra (1.5 lb. carboxin & 0.1 lb. imazalil & 0.13 lb. thiabendazole) *(Gustafson)*
Rate: 5 fl. oz./100 lb. seed
Disease:
common bunt; loose smut; seedling blight; glume blotch; seed rot; *Septoria nodorum*
Time: Seed treatment.
Remarks: Wheat. Use high rate where smut and bunt are severe. Use with EPA-approved dye that imparts unnatural color to seed. Do not use for feed, food, or oil purposes. Do not graze or feed livestock on treated areas for 6 weeks after planting. Refer to label for use directions.

Signal Word/Toxicity Class: Caution.

RTU-Vitavax-Thiram (10% carboxin & 10% thiram) *(Gustafson)*
Rate: 5-6.8 fl. oz./100 lb. seed
Disease:
common bunt; flag smut; loose smut; covered smut
Time: Seed treatment.
Remarks: Barley, oats, wheat. Use with EPA-approved dye that imparts unnatural color to seed. Do not use for feed, food, or oil purposes. Do not graze or feed livestock on treated areas for 6 weeks after planting. Refer to label for use directions.

Signal Word/Toxicity Class: Caution.

Vitaflo-280 (14.9% carboxin & 13.2% thiram) *(Gustafson)*
Rate: 3.5-5 fl. oz./100 lb. seed
Disease:
common bunt (wheat); covered smut (barley, oats); loose smut; net blotch (barley)
Time: Seed treatment.
Remarks: Barley, oats, wheat. Do not use treated seed for food, feed, or oil purposes. Do not graze or feed livestock on treated areas for 6 weeks after planting. Refer to label for use directions.

Signal Word/Toxicity Class: Caution.

Vitavax-PCNB (17% carboxin & 17% PCNB) *(Gustafson)*
Rate: 3-4 fl. oz./100 lb. seed
Disease:
common bunt; loose smut; seedling diseases; covered smut; seed decay
Remarks: Barley, oats, wheat. Use with EPA-approved dye that imparts unnatural color to seed. Do not use treated seed for food, feed, or oil purposes. Do not graze or feed livestock hay grown from treated seed for 6 weeks after planting. Refer to label for use directions.

Signal Word/Toxicity Class: Caution.

Vitavax-Thiram-Lindane (14% carboxin & 12% thiram & 8% lindane) *(Gustafson)*
Rate: Barley:
6 fl. oz./100 lb. seed
Oats, wheat:
5 fl. oz./100 lb. seed
Disease:
common bunt (wheat); covered smut (barley); loose smut
Time: Slurry seed treatment.
Remarks: Barley, oats, wheat. Use with EPA-approved dye that imparts unnatural color to seed. Do not use treated seed for food, feed, or oil purposes. Do not graze or feed livestock hay grown from treated seed for 6 weeks after planting. Refer to label for use directions.

Signal Word/Toxicity Class: Caution.

Vitavax T-L (10% carboxin & 10% thiram) *(Trace)*
Rate: 5-6.8 fl. oz./100 lb. seed
Disease:
common bunt; *Rhizoctonia solani*; seedling disease; loose smut
Time: Hopper box seed treatment.
Remarks: Wheat. Do not use treated seed for food, feed, or oil purposes. Do not graze or feed livestock on treated areas for 6 weeks after planting. Refer to label for use directions.

Signal Word/Toxicity Class: Danger/I.

Champ — see copper compounds

Champion — see copper compounds

NOTICE	The information on these pages is for preliminary planning — not a guide for use. Be sure to follow the manufacturer's directions, notwithstanding information contained here.	For personal protective equipment and EPA registration numbers, please turn to page 39.

Clean Crop COCS — see copper compounds

COPPER COMPOUNDS

Fungicide/bactericide.

Champ (FL or Formula 2) or Champion (WP) (copper hydroxide) *(Agtrol)*

Rate: 2-2.66 pt. FL/A
1-1.33 pt. Formula 2/A
1.5-2 lb. WP/A

Disease:
Helminthosporium spot blotch; Septoria leaf blotch

Time: Apply at early heading; repeat 10 days later.

Rate: 4 fl. oz. FL/100 lb. seed
2 fl. oz. Formula 2/100 lb. seed

Disease:
bacterial leaf blight; bacterial leaf streak; common bunt

Time: Apply as seed dressing.
Remarks: Barley, wheat. Refer to label for use directions.

Signal Word/Toxicity Class: Danger/I (Champion); Caution (Champ).
REI: 24 hr.

Clean Crop C-O-C-S (WDG) (74.8% copper oxychloride & 14.2% basic copper sulfate) *(Platte)*
TOP COP Tri-Basic (FL) (basic copper sulfate) *(Stoller)*

Rate: 1-2 lb. WDG/A
3-4 pt. FL/A

Disease:
Helminthosporium spot blotch; Septoria leaf blotch

Time: Apply at early heading; repeat 10 days later.
Remarks: Barley, wheat. Do not apply WDG through any type of irrigation system. Refer to label for use directions.
State Restrictions: WDG not for use on barley in California.

Signal Word/Toxicity Class: Warning/II; Caution (TOP COP).
REI: 24 hr.

Kocide (101, DF, 2000, 4.5LF, LF) (copper hydroxide) *(Griffin)*

Rate: 1.5-2 lb. 101, DF/A
1.25-1.5 lb. 2000/A
1-1.33 pt. 4.5LF/A
1.33-2.66 pt. LF/A

Disease:
Helminthosporium spot blotch; Septoria leaf blotch

Time: Apply at early heading; repeat 10 days later.
Remarks: Barley, oats, wheat. Use higher rate when conditions favor disease. Do not apply in a spray solution having a pH of less than 6.5 as phytotoxicity may occur. Refer to label for use directions.

Rate: 2 fl. oz. 4.5LF/100 lb. of seed

Disease:
Pseudomonas syringae; *Tilletia caries*; *Xanthomonas translucens*

Time: Seed treatment.
Remarks: Barley, wheat. Do not apply in a spray solution having a pH less than 6.5 as phytotoxicity could occur. Refer to label for use directions.
State Restrictions: Not for use in California.

Signal Word/Toxicity Class: Danger/I (101, DF); Warning/II (2000); Caution (4.5LF, LF).
REI: 24 hr.

KOP-Hydroxide (FL, 50WP) (copper hydroxide) *(Drexel)*

Rate: 2-2.66 pt. FL/A
1.5-2 lb. 50WP/A

Disease:
Helminthosporium spot blotch; Septoria leaf blotch

Time: Apply at early boot stage; repeat at early head stage.
Remarks: Barley, wheat. KOP-Hydroxide 50WP should not be applied in spray solution having pH less than 6.5 or greater than 9.0 as phytotoxicity may occur. Refer to label for use directions.

Signal Word/Toxicity Class: Danger/I.
REI: 48 hr.

Microsperse COC 53WP (copper oxychloride) *(Micro Flo)*

Rate: 2-3 lb. 53WP/A

Disease:
glume blotch; Helminthosporium leaf blotch; Helminthosporium spot blotch; Septoria leaf blotch

Time: Apply at early heading; again 10-days later.
Remarks: Barley, oats, wheat. Refer to label for use directions.

Signal Word/Toxicity Class: Danger/I.
REI: 24 hr.

Nu-Cop (3L, 50DF, 50WP) (copper hydroxide) *(Micro Flo)*

Rate: Barley, oats, wheat:
1-1.33 pt. 3L/A
1.5-2 lb. 50DF, 50WP/A

Disease:
Helminthosporium spot blotch; Septoria leaf blotch

Time: Apply at early heading; repeat 10 days later.

Rate: Barley, wheat:
2 fl. oz. 3L/100 lb. seed

Disease:
bacterial leaf blight; bacterial leaf streak; common bunt

Time: Seed treatment.
Remarks: Barley, oats, wheat. Refer to label for use directions.

Signal Word/Toxicity Class: Danger/I (50DF, 50WP); Warning/II (3L).
REI: 48 hr. (50DF, 50WP); 24 hr. (3L).

DB-Green — see maneb premixes

Dithane — see mancozeb

DIVIDEND XL (Premixes)

Systemic seed treatment containing active ingredients of Apron XL and Dividend. Apron XL provides damping-off activity, Dividend provides activity on the remaining diseases. EPA approved coloring agent has been added to formulation. Toxic to fish and aquatic invertebrates.

Dividend XL (1.54 lb. difenoconazole & 0.13 lb. mefenoxam) *(Novartis)*

Rate: Spring, winter wheat:
1-2 fl. oz./100 lb. seed

Disease:
common bunt; common root rot*; damping-off; dwarf bunt (winter wheat); flag smut; Fusarium crown rot* (winter wheat); Fusarium root rot* (winter wheat); Fusarium seed scab; loose smut; Rhizoctonia* (spring wheat); seed rot; Septoria; take-all*

*** Partial control**

Rate: Winter wheat (fall season foliar):
2 fl. oz./100 lb. seed

Disease:
leaf rust; powdery mildew; Septoria leaf blotch

Time: Apply as seed dressing.
Remarks: Winter wheat, spring wheat. Do not use treated seed for feed, food, or oil. Green forage may not be grazed until 55 days after planting.
Crop Rotation: Do not plant any crop other than wheat within 30 days to fields in which treated seeds were planted.

Signal Word/Toxicity Class: Caution.

Dividend XL RTA (0.31 lb. difenoconazole & 0.025 lb. mefenoxam) *(Novartis)*

Rate: 5-10 fl. oz. XL RTA/100 lb. seed

Disease:
common bunt; common root rot*; damping off; dwarf bunt (winter wheat); flag smut; Fusarium crown rot* (winter wheat); Fusarium root rot* (winter wheat); Fusarium seed scab; loose smut; Rhizoctonia* (spring wheat); seed rot; Septoria; take-all*

Fall season foliar diseases:
leaf rust; powdery mildew; Septoria leaf blotch

*** Partial control**

◆ - **new product** • ai-active ingredient • DF-dry flowable • E/EC-emulsifiable concentrate • EW-emulsion in water
F/FL-flowable • G-granule • L/LC-liquid • SP/WSP-soluble packet • W/WP-wettable powder • WSB-water soluble bag

Time: Apply as seed dressing.
Remarks: Winter wheat, spring wheat. Rates vary by disease. Do not use treated seed for food, feed, or oil purposes. Green forage may not be grazed until 55 days after planting.

Signal Word/Toxicity Class: Caution.

Double R — see imazalil

Enhance — see carboxin premixes

Enhance Plus — see maneb premixes

Flo-Pro — see imazalil

Grain Guard — see mancozeb

GRAIN GUARD PLUS (Premix)

Contains graphite. Toxic to fish, birds, and other wildlife.

Grain Guard Plus (50% mancozeb & 18.75% lindane) *(Trace)*
Rate: 2 oz./bu. seed
Disease:

covered smut (barley)	loose smut (barley)	seed decay
damping-off	oat smut	stinking smut (wheat)
false loose smut (barley)		

Time: Drill box or planter box seed treatment.
Remarks: Barley, oats, rye, wheat. Treat only those seeds needed for immediate use. Do not use treated seed for food, feed, or oil purposes. Do not store excess treated seeds beyond planting time. Refer to label for use directions.

Signal Word/Toxicity Class: Warning/II.

GUSTAFSON LSP

Systemic fungicide formulated as a liquid suspension seed protectant. Toxic to fish.

Gustafson LSP (FL) (thiabendazole) *(Gustafson)*
Rate: Spring, winter wheat:
2-4 fl. oz. FL/bu. seed
Disease:
common bunt — stinking smut
Time: Spray mist or slurry seed treatment.

Rate: Winter wheat:
4-6 fl. oz. FL/bu. seed
Disease:
dwarf bunt
Time: For best results, plant treated wheat as late in growing season as practical. Use only on normal or late winter wheat plantings.
Remarks: Winter wheat. Use low rate for late plantings. Apply in equal amount of water. Use with EPA-approved dye that imparts unnatural color to seed. Do not use for feed, food, or oil purposes. Refer to label for use directions.
State Restrictions: For use in Colorado, Idaho, Montana, Oregon, Utah, Washington, Wyoming only on winter wheat for dwarf bunt.

Signal Word/Toxicity Class: Caution.

IMAZALIL

Flowable systemic fungicide.

Double R or Double R II *(AGSCO)*
Rate: 0.8-1.5 fl. oz./100 lb. seed
Disease:

barley leaf stripe	Helminthosporium	seedling blight
common root rot	net blotch	*Septoria nodorum*
Fusarium		

Time: Seed treatment.
Remarks: Barley, wheat. Do not use treated seed for food, feed, or oil purposes. Do not graze or feed livestock on treated areas for 6 weeks after planting. Refer to label for use directions.

Signal Word/Toxicity Class: Danger/I.
REI: 48 hr.

Flo-Pro IMZ *(Gustafson)*
Rate: 0.25-0.5 fl. oz./100 lb. seed
Disease:
barley leaf stripe — common root rot
Time: Slurry seed treatment.
Remarks: Barley, wheat. Use with EPA-approved dye that imparts unnatural color to seed. Do not use for feed, food, or oil purposes. Do not graze or feed livestock on treated areas for 6 weeks after planting. Refer to label for use directions.

Signal Word/Toxicity Class: Warning/II.

Kocide — see copper compounds

KOP-Hydroxide — see copper compounds

Kumulus — see sulfur

MANCOZEB

Dithiocarbamate fungicide with broad-spectrum activity. Coordination product of zinc ion and manganese ethylene bisdithiocarbamate. Grain Guard contains graphite. Toxic to fish.

Dithane (DF, F-45, M-45, ST, WSP) *(Rohm and Haas)*
Manzate 200DF *(Griffin)*
Penncozeb 75DF or 80WP *(Elf Atochem)*
Rate: 1.6 lb. ai/A
2 lb. M-45, WSP/A
2.1 lb. DF/A
3.2 pt. F-45/A
1.5 lb. ai/A
2 lb. 200DF/A
0.75-1.5 lb. ai/A
1-2 lb. 75DF/A
0.8-1.6 lb. ai/A
1-2 lb. 80WP/A
Disease:

Helminthosporium leaf spot	leaf rust	Septoria leaf spot
	Septoria glume blotch	tan spot

Time: Apply at onset of disease or when plants are in tillering to jointing stage; repeat at 7- to 10-day intervals.
Remarks: Barley, oats, rye, wheat. Addition of surfactant will improve performance. Do not make more than 3 applications per season. Do not graze livestock in treated areas prior to harvest. Refer to label for use directions.
Days To Harvest: 26.
State Restrictions: Refer to label for California precautions.

Rate: Barley:
4.3-6.7 fl. oz. ST/100 lb. seed
2.7-4.2 oz. M-45, WSP, 200DF, 80WP/100 lb. seed
2.9-4.5 oz. 75DF/100 lb.
Oats:
6.4-10 fl. oz. ST/100 lb. seed
4-6.3 oz. M-45, WSP, 200DF, 80WP/100 lb. seed
4.3-6.7 oz. 75DF/100 lb.
Rye:
3.6-5.7 fl. oz. ST/100 lb. seed
2.3-3.6 oz. M-45, WSP, 200DF, 80WP/100 lb. seed
2.5-3.8 oz. 75DF/100 lb.
Wheat:
3.5-5.2 fl. oz. ST/100 lb. seed
2.2-3.3 oz. M-45, WSP, 200DF, 80WP/100 lb. seed
2.3-3.5 oz. 75DF/100 lb.
Disease: Barley:

covered smut	false loose smut	seed rot
damping-off	(barley)	seedling blight

Time: Seed treatment.

5 Small Grains

NOTICE	The information on these pages is for preliminary planning — not a guide for use. Be sure to follow the manufacturer's directions, notwithstanding information contained here.	For personal protective equipment and EPA registration numbers, please turn to page 39.

Remarks: Barley, oats, rye, wheat. Dye must be added to treating slurry so that unnatural color will distinguish seed as treated. Do not use for food, feed, or oil purposes. Refer to label for use directions.
State Restrictions: Refer to label for California precautions.

Signal Word/Toxicity Class: Caution.
REI: 24 hr.

Grain Guard *(Trace)*
Rate: 2 oz./bu. seed
Disease:
bunt (rye, wheat), damping-off, seedling blight, covered smut (barley, oats), seed rot
Time: Drill box seed treatment.
Remarks: Barley, oats, rye, wheat. Treat only those seeds needed for immediate use. Do not use treated seed for food, feed, or oil purposes. Do not store excess treated seeds beyond planting time. Refer to label for use directions.

Signal Word/Toxicity Class: Caution.
REI: 12 hr.

MANEB (Premixes)

Dual purpose ready-to-use fungicide/insecticide; ManKocide is a fungicide/bactericide. DB-Green is a dust-based formulation seed treatment; DB-Green L is a liquid formulation. Seed Mate contains graphite. Toxic to fish, birds, and other wildlife.

DB-Green (dust) (50% maneb & 18.75% lindane) *(AGSCO)*
DB-Green L (liquid) (25.6% maneb & 8.6% lindane) *(AGSCO)*
Rate: 2 oz. dust/bu. seed
3 fl. oz. L/bu. seed
Disease: Barley:
covered smut, false loose smut, seedling blight
Oats:
oat smut, seedling blight
Rye:
bunt, seedling blight
Wheat:
seedling blight, stinking smut
Time: Seed treatment.
Remarks: Barley, oats, rye, wheat. Drill or planter box application. Treat only seeds needed for immediate use, minimizing interval between treatment and planting. Do not store excess treated seeds beyond planting time. Treated seed must not be used for or mixed with food or animal feed, or processed for oil. Do not graze or feed livestock on treated areas for 6 weeks after planting.

Signal Word/Toxicity Class: Warning/II.
REI: 12 hr.

DB-Green + Vitavax (35% maneb & 20% carboxin & 18.75% lindane) *(AGSCO)*
Enhance Plus (35% maneb & 20% carboxin & 18.75% lindane) *(Trace)*
Rate: 2 oz./bu. seed
Disease: Barley:
covered smut, false loose smut, seedling blight
Oats:
oat smut, seedling blight
Wheat:
bunt, seed decay, stinking smut
Time: For DB-Green, use drill box treatment. For Enhance, use planter box seed treatment.
Remarks: Barley, oats, wheat. Treat only those seeds needed for immediate use. Do not use treated seed for food, feed, or oil purposes. Do not store excess treated seeds beyond planting time. Do not graze or feed livestock on treated areas for 6 weeks after planting. Refer to label for use directions.

Signal Word/Toxicity Class: Danger/I (Enhance); Warning/II (DB-Green).
REI: 12 hr. (DB-Green).

ManKocide (DF) (15% maneb & 46.1% copper hydroxide) *(Griffin)*
Rate: 2-2.5 lb. DF/A
Disease:
glume blotch, Helminthosporium leaf spot, Septoria leaf spot
Time: Apply at early heading; repeat 10 days later.
Remarks: Barley, oats, wheat. Do not graze livestock in treated areas prior to harvest. Refer to label for use directions.
Days To Harvest: 26.

Signal Word/Toxicity Class: Danger/I.
REI: 48 hr.

Seed Mate Maneb-Lindane (50% maneb & 18.75% lindane) *(Loveland)*
Rate: 2 oz./bu. seed
Disease:
bunt (wheat), false loose smut (barley), seed decay, covered smut, stinking smut (wheat), damping-off, oat smut (oats)
Time: Apply as drill box or planter box treatment.
Remarks: Barley, oats, rye, wheat. Treat only seeds needed for immediate use, minimizing interval between treatment and planting. Do not store excess treated seeds beyond planting time. Do not use treated seed for food, feed, or oil purposes. Refer to label for use directions.

ManKocide — see maneb premixes

Manzate — see mancozeb

MEFENOXAM

Systemic fungicide having a specific mode of action.

Apron XL LS (mefenoxam) *(Novartis)*
Rate: 0.32-0.64 fl. oz. LS/100 lb. seed
Disease:
damping-off
Time: Seed treatment at planting.
Remarks: Barley, oats, rye, wheat. Do not use treated seed for food, feed, or oil purposes. Refer to label for use directions.

Signal Word/Toxicity Class: Warning/II.

METALAXYL

Systemic fungicide having a specific mode of action.

Allegiance or Apron (FL) *(Gustafson)*
Rate: 0.375-0.75 fl. oz. Allegiance/100 lb. seed
0.75-1.5 fl. oz. Apron/100 lb. seed
Disease:
damping-off, Phytophthora (early-season)
Time: Seed treatment.
Remarks: Barley, oats, rye, wheat. Use with EPA-approved dye that imparts unnatural color to seed. Do not use for feed, food, or oil purposes. Refer to label for use directions.

Signal Word/Toxicity Class: Warning/II.

Apron Dry *(Trace)*
Rate: 3-4 oz./100 lb. seed
Disease:
damping-off
Time: Seed treatment.
Remarks: Barley, oats, rye, wheat. Do not use treated seed for food, feed, or oil purposes. Do not store excess treated seeds beyond planting time. Refer to label for use directions.

Signal Word/Toxicity Class: Caution.
REI: 12 hr.

Metam CLR — see metam-sodium

METAM-SODIUM

Fumigant for soil preparation only. Toxic to fish.

Metam CLR (42%) *(UCB Agrochemicals)*
Rate: 1.85-7.5 gal. 42%/A
Disease:
soil fungi

◆ - **new product** • ai-active ingredient • DF-dry flowable • E/EC-emulsifiable concentrate • EW-emulsion in water
F/FL-flowable • G-granule • L/LC-liquid • SP/WSP-soluble packet • W/WP-wettable powder • WSB-water soluble bag

Time: Preplant. Apply 14-21 days before planting.
Remarks: Barley, wheat. Use only in areas that receive more than 15" rain. Can be diluted with water or nonacidic liquid fertilizer immediately before applying. Do not mix with acidic fertilizer or other acidic solution. Soil temperature should be in 40°-90°F range. Refer to label for use directions.
State Restrictions: In California, application must be in compliance with Technical Information Bulletin "Guidelines for All Application Methods for Metam-sodium in California." This bulletin may be obtained from your local pesticide dealer or a metam-sodium registrant.

Signal Word/Toxicity Class: Danger/I.
REI: 48 hr.

Micro Flo — see sulfur

Microsperse — see copper compounds

Nu-Cop — see copper compounds

Nu-Gro Isotox — see captan premixes

PCNB

RTU-PCNB contains toxic inert Rhodamine B. Toxic to fish and aquatic organisms.

PCNB 2-E or 2LF *(Amvac)*
Rate: Wheat:
3.35 fl. oz. 2-E/100 lb. seed
4 fl. oz. 2LF/100 lb. seed
Barley:
4.2-8.4 fl. oz. 2-E/100 lb. seed
5-10 fl. oz. 2LF/100 lb. seed
Oats:
6.13-12.26 fl. oz. 2-E/100 lb. seed
7.25-14.5 fl. oz. 2LF/100 lb. seed
Disease:
bunt (wheat), common smut, covered smut (barley), oat smut (oats)
Time: Seed treatment.
Remarks: Barley, oats, wheat. Do not apply through any type of irrigation system. Do not use for food, feed, or oil purposes. Do not plant root crops within 12 months of application unless PCNB is registered for use on those crops.

Signal Word/Toxicity Class: Danger/I (2-E); Caution (2LF).
REI: 12 hr.

RTU-PCNB or Terra-Coat LT-2N *(Gustafson)*
Rate: Wheat:
3 fl. oz. RTU/100 lb. seed
2 fl. oz. LT-2N/bu. seed
Disease:
bunt, common smut, damping-off

Rate: Barley:
3.75-7.5 fl. oz. RTU/100 lb. seed
2-4 fl. oz. LT-2N/bu. seed
Oats:
5.5-11 fl. oz. RTU/100 lb. seed
2-4 fl. oz. LT-2N/bu. seed
Disease:
covered smut, oat smut
Time: Seed treatment.
Remarks: Barley, oats, wheat. Do not use treated seed for food, feed, or oil purposes. Refer to label for use directions.

Signal Word/Toxicity Class: Caution.

Penncozeb — see mancozeb

Prevail — see carboxin premixes

PROPICONAZOLE

Triazole sterol inhibitor fungicide. Toxic to fish.

Tilt (EC) *(Novartis)*
Rate: 4 fl. oz. EC/A
Disease:
barley scald, foot rot (wheat), glume blotch, leaf blight, net blotch, powdery mildew, rust, tan spot
Time: Apply to emerging flag leaf. Do not apply after ligule of flag leaf emerges. For foot rot, apply at tillering before elongation.
Remarks: Barley, oats, rye, wheat. Make only 1 application per season. Do not graze or feed livestock treated forage or cut green crop for hay or silage. Do not apply this product through any type of irrigation system.

Signal Word/Toxicity Class: Warning/II.
REI: 24 hr.

RAXIL

Toxic to estuarine invertebrates.

Raxil 2.6F (tebuconazole) *(Gustafson)*
Rate: 0.1 fl. oz. 2.6F/100 lb. seed
Disease:
barley leaf stripe, barley rust* **, common root rot**, covered smut, crown rust* **, damping-off, flag smut, foot rot**, leaf rust* **, loose smut, powdery mildew* **, Rhizoctonia**, Septoria**, stinking smut
***Suppression **Early-season.**
Time: Seed treatment.
Remarks: Barley, oats, wheat. Use with EPA-approved dye that imparts unnatural color to seed. Do not use for feed, food, or oil purposes. Do not graze barley for 31 days, wheat for 38 days, and oats for 51 days after seeding. Refer to label for use directions.

Signal Word/Toxicity Class: Caution.

RAXIL-THIRAM (Premix)

Combines the systemic action of Raxil with the surface action of thiram. Toxic to Fish.

Raxil-Thiram (FL) (0.055 lb. tebuconazole & 1.72 lb. thiram) *(Gustafson)*
Rate: 3.5-4.6 fl. oz. FL/100 lb. seed
Disease:
barley leaf stripe*, barley rust* **, common root rot**, covered smut, crown rust* **, damping-off, flag smut, foot rot**, Fusarium scab, leaf rust* **, loose smut, powdery mildew* **, Rhizoctonia**, seed rot, Septoria**, stinking smut
***Suppression **Early-season.**
Time: Seed treatment.
Remarks: Barley, oats, wheat. Use with EPA-approved dye that imparts unnatural color to seed. Do not use for feed, food, or oil purposes. Do not graze barley or wheat for 31 days, oats for 30 days after seeding. Refer to label for use directions.

Signal Word/Toxicity Class: Caution.

Redball — see sulfur

Rival — see captan premixes

RTU (Premix)

Ready-to-use flowable fungicide containing the active ingredients of thiram and thiabendazole (Mertect).

RTU (FL) (12.60% thiram & 0.34% thiabendazole) *(Gustafson)*
Rate: 6.6 fl. oz. FL/100 lb. seed
Disease:
damping-off, seed decay, seedling blight, stinking smut
Time: Slurry seed suspension

NOTICE	The information on these pages is for preliminary planning — not a guide for use. Be sure to follow the manufacturer's directions, notwithstanding information contained here.	For personal protective equipment and EPA registration numbers, please turn to page 39.

Remarks: May be applied undiluted or with water when additional volumes are needed to insure uniform seed coverage. Do not tank mix unless testing or prior experience insures compatability. Use with EPA-approved dye that imparts unnatural color to seed. Do not use for feed, food, or oil purposes. Do not graze or feed livestock on forage or hay from treated areas. Refer to label for use directions.

Signal Word/Toxicity Class: Caution.

RTU-BAYTAN-THIRAM (Premix)

Ready-to-use seed treatment combining the systemic action of Baytan with the surface action of thiram. Systemic activity of Baytan provides early-season control of infections by powdery mildew and rusts. Toxic to fish.

RTU-Baytan-Thiram (5% triadimenol & 15.3% thiram) *(Gustafson)*

Rate: 4.5-9 fl. oz. FL/100 lb. seed

Disease: Barley:

barley rust	net blotch*	scab
covered smut	powdery mildew	stripe rust
foot rot*	Pythium rot*	Typhula rot*
loose smut		

Oats:

crown rust	powdery mildew	scab
loose smut	Pythium rot*	stripe rust

Rye:

powdery mildew	rye rust	stripe rust
Pythium rot*	scab	stripe smut

Wheat:

flag smut	loose smut	stinking smut
foot rot*	powdery mildew	stripe rust
glume blotch	Pythium rot*	take-all*
leaf rust	scab	

*** Suppression.**

Time: Seed treatment.

Remarks: Barley, oats, rye, wheat. Use low rate for smuts. Use with EPA-approved dye that imparts unnatural color to seed. Do not use for feed, food, or oil purposes. Green forage may be grazed 40 days after seeding. Refer to label for use directions.

Signal Word/Toxicity Class: Caution.

RTU-PCNB — see PCNB

RTU-Vitavax-Extra — see carboxin premixes

RTU-Vitavax-Thiram — see carboxin premixes

Redball — see sulfur

Seed Mate Isotox — see captan premixes

Seed Mate Vitavax-Captan — see carboxin premixes

SULFUR

Nonsystemic fungicide with protective action. Microthiol Special, Sulfa 80, and Thiolux are micronized sulfurs.

Drexel Suffa (6FL) or Sulfur (90W) *(Drexel)*

Rate: 3-4.5 lb. ai/100 gal. water
4-6 pt. 6FL/100 gal. water
5.4 lb. ai/100 gal. water
6 lb. 90W/100 gal. water

Disease:
powdery mildew

Time: Apply when disease appears; repeat as necessary.

Remarks: Barley, oats, rye, wheat. Apply as dust or spray. Do not use within 2 weeks of oil application (4 weeks 6FL). Do not make sulfur applications at excessively high temperatures. Do not apply this product through any type of irrigation system. Refer to label for use directions.

Signal Word/Toxicity Class: Caution.
REI: 24 hr.

Kumulus DF (sulfur) *(BASF)*

Rate: 4.8-12 lb. ai/A
6-15 lb. DF/A

Disease:
powdery mildew

Time: Apply when disease first appears and repeat as necessary.

Remarks: Do not apply if temperatures during or within 3 days after application are expected to exceed 90°F. Do not apply through any type of irrigation system. Refer to label for use directions.

Signal Word/Toxicity Class: Caution.
REI: 24 hr.

Micro Flo Sulfur (6L) *(Micro Flo)*
Terra Sulfur (6L) *(Terra)*

Rate: 4-15 lb. ai/A
5.3-20 pt. 6L/A

Disease:
powdery mildew

Time: Apply at first sign of disease; repeat as necessary.

Remarks: Barley, oats, rye, wheat. Refer to label for use directions.

Signal Word/Toxicity Class: Caution.
REI: 24 hr.

Redball EM-53 (6L), Superfine Wettable Sulphur (WS) or Commercial Dusting Sulphur (DS) *(International Sulphur)*

Rate: 4-20 pt. 6L/A
5-10 lb. WS/100 gal. water
5-15 lb. DS/A

Disease:
powdery mildew

Time: Apply when disease first appears; repeat as necessary.

Remarks: Barley, oats, rye, wheat. Do not apply when temperatures exceed 90°F. Refer to label for use directions.

Signal Word/Toxicity Class: Caution.
REI: 12 hr.

THAT (6FL) or THAT Big 8 (8FL) *(Stoller)*

Rate: 2.25-3.75 lb. ai/A
3-5 pt. 6FL/A
2.5-4 lb. ai/A
2.5-4 pt. 8FL/A

Disease:
powdery mildew

Time: Apply when disease appears; repeat at 2-week intervals.

Remarks: Barley, oats, rye, wheat. Refer to label for use directions.

State Restrictions: Not for use in California.

Signal Word/Toxicity Class: Caution.
REI: 24 hr.

SULFUR (Premix)

Fungicide/bactericide in flowable formulation. Toxic to fish and aquatic organisms.

Top Cop with Sulfur (6.25 lb./gal. sulfur & 1 lb./gal. tribasic copper sulfate) *(Stoller)*

Rate: Barley:
2 qt./A

Disease:

Helminthosporium spot blotch	Septoria leaf blotch

Time: Apply in foliar spray at early heading; repeat 10 days later.

Rate: Wheat:
2 qt./A

Disease:

leaf rust	powdery mildew

Time: Apply when disease first appears; repeat at 2 week intervals if needed.

Remarks: Barley, wheat. Some wheat varieties may be sensitive to copper. Do not apply at excessively high temperatures. Refer to label for use directions.

Signal Word/Toxicity Class: Caution.
REI: 24 hr.

◆ - **new product** • ai-active ingredient • DF-dry flowable • E/EC-emulsifiable concentrate • EW-emulsion in water
F/FL-flowable • G-granule • L/LC-liquid • SP/WSP-soluble packet • W/WP-wettable powder • WSB-water soluble bag

SYSTEM3 (Premix)

Biological and chemical fungicide containing the active ingredients of Apron, Terraclor, and Kodiak. Contains graphite. Toxic to fish and aquatic organisms.

System3 (0.1% *Bacillus subtilis* & 14.25% metalaxyl & 16.67% PCNB) *(Helena)*

Rate: **Barley:**
2-4 oz./bu.
Wheat:
2-3 oz./bu.

Disease:
common bunt (wheat), covered smut (barley), *Rhizoctonia solani*, common smut (wheat), Pythium

Time: Apply at planting.

Remarks: Barley, wheat. Incorporation of dye is recommended to distinguish seed as treated. Do not use treated seed for food, feed or oil purposes. Refer to label for use directions.

- - - - - - - - - - - - - - - - -

Signal Word/Toxicity Class: Caution.
REI: 12 hr.

Terra-Coat — see PCNB

THAT or THAT Big 8 — see sulfur

THIOPHANATE-METHYL

Benzimidazole systemic fungicide with broad-spectrum control. Toxic to fish.

Topsin M (WSB, 70W) *(Elf Atochem)*

Rate: 0.7 lb. ai/A
1 lb. WSB, 70W/A

Disease:
eyespot, foot rot, strawbreaker

Time: Apply after wheat is tillered but before stem elongation has begun.

Remarks: Fall-seeded wheat. Do not cut for hay within 90 days of application. Do not allow livestock to graze treated areas before harvest. Refer to label for use directions.

State Restrictions: For use in Idaho, Oregon, Washington.

- - - - - - - - - - - - - - - - -

Signal Word/Toxicity Class: Caution.
REI: 12 hr.

THIRAM

Broad-spectrum fungicide. Toxic to fish. Thiram 50WP contains the toxic inert Rhodamine B and is also toxic to birds and wildlife.

42-S Thiram or Thiram 50WP Dyed *(Gustafson)*
Thiram Granuflo (WDG) *(UCB Agrochemicals)*

Rate: **Barley, oats, rye, wheat:**
2 fl. oz. 42-S/bu. seed
Barley:
4.1 oz. 50WP/100 lb. seed
2.7 oz. WDG/100 lb. seed
Rye:
3.7 oz. 50WP/100 lb. seed
2.5 oz. WDG/100 lb. seed
Wheat:
3.3 oz. 50WP/100 lb. seed
Oats, wheat:
2.2 oz. WDG/100 lb. seed

Disease:
bunt, damping-off, seedling blight, covered smut, seed decay

Time: Seed treatment.

Remarks: Barley, oats, rye, wheat. Use with EPA-approved dye that imparts unnatural color to seed. Do not use for feed, food, or oil purposes. Do not apply Thiram Granuflo through any type of irrigation system. Refer to label for use directions.

State Restrictions: 42-S Thiram or Thiram 50WP not for use in California on oats. Thiram Granuflo not for use in California.

- - - - - - - - - - - - - - - - -

Signal Word/Toxicity Class: Caution.
REI: 24 hr. (WDG).

Tilt — see propiconazole

Top Cop with Sulfur — see sulfur premix

Topsin M — see thiophanate-methyl

TOP COP Tri-Basic — see copper compounds

TRIADIMEFON

Triazole sterol inhibitor systemic fungicide compatible with many insecticides, miticides, and fungicides.

Bayleton 50% DF *(Bayer)*

Rate: 2-6 oz. 50% DF/A

Disease:
glume blotch, powdery mildew, tan spot, leaf blotch, rust

Time: Apply when disease symptoms first appear on leaves or stems; additional applications should be made if new symptoms appear.

Remarks: Wheat. Do not apply this product through any type of irrigation system. Do not use on other crops grown for food or forage. Refer to label for use directions.

Days To Harvest: 35.

Crop Rotation: Small grains, corn, sorghum, soybeans, beans, peas and cucurbits may be planted 35 days after last application; forage or vines from these crops may not be used for food or feed. Root crops may be planted 120 days after last application but tops must not be used for food or feed. All crops may be planted 12 months or later after last application without restrictions.

- - - - - - - - - - - - - - - - -

Signal Word/Toxicity Class: Caution.
REI: 12 hr.

Vitaflo — see carboxin premixes

Vitavax — see carboxin

Vitavax-PCNB — see carboxin premixes

Vitavax T-L — see carboxin premixes

Vitavax-Thiram-Lindane — see carboxin premixes

5 Small Grains

NOTICE The information on these pages is for preliminary planning — not a guide for use. Be sure to follow the manufacturer's directions, notwithstanding information contained here.

For personal protective equipment and EPA registration numbers, please turn to page 39.

Notes • Notes • Notes

Insect and Disease Control Guide

Section 6

Other Field Crops

Bt Chart 144

Insecticide Listings

Hops. 146
Mint 148
Oil Seed Crops 150
Peanuts. 150
Rice 154
Safflower. 156
Sorghum. 157
Sugar beets. 161
Sugarcane. 166
Sunflower 166
Tobacco 169

Fungicide Listings

Castor Beans. 172
Flax. 172
Hops. 173
Mint 174
Mung Beans 175
Peanuts 175
Rice 181
Safflower 184
Sorghum. 185
Sugar beets. 187
Sugarcane. 191
Sunflower 191
Tobacco 192

The information in this 1999 Insect and Disease Control Guide is updated to August 15, 1998.
Because manufacturers' product recommendations are constantly being updated and revised as additional product registrations and label changes are approved by EPA, there may be changes after this date. However, every effort has been made to give manufacturers the opportunity to include in the Guide their latest information. This Guide is for preliminary planning – not a use guide. Be sure to follow label directions, notwithstanding information contained herein.

Bt Products Chart

Other Field Crops

Find Bt products quickly and easily with this Bt products chart. Simply find the insect you want to control on the top of the chart, and follow down to the marked spot to find the Bt product you can use to control that pest. Company address information can be found on page 599.

	Alfalfa caterpillar	Armyworms	Artichoke plume moth	Aspergillis	Banded sunflower moth	Bollworms	Borers	Budworms	Caterpillars	Cross-striped cabbageworm	Cutworms	Diamondback moth	Earworms	Fusarium	Green cloverworm	Headmoth	Headworm	Helicoverpa	Heliothis
Agree		■								■	■	■							
Bacillus subtilis				■										■					
Bactimos																			
Biobit	■	■					■			■	■	■			■	■	■		■
BMP 123		■					■			■	■	■			■	■			■
Bt. ssp. *aizawai*		■	■				■			■	■	■			■	■	■		■
Bt. ssp. *aizawai* + *Bt.* ssp. *kurstaki*		■					■	■	■			■	■						
Bt. ssp. *israelensis*																			
Bt. ssp. *kurstaki*	■	■			■	■	■	■	■	■	■	■	■		■	■	■	■	■
Condor		■			■		■				■	■			■	■	■		
Crymax		■			■		■				■	■			■	■	■		
Cutlass		■			■		■				■	■			■	■	■		
DiPel		■			■	■	■			■	■	■	■		■	■	■	■	■
Javelin		■			■		■				■	■			■	■	■		■
Kodiak				■										■					
Lepinox		■					■								■		■		
M-Peril																			
Mattch		■					■	■	■			■	■						
MVP II		■					■	■	■			■	■						
Troy-Bt		■				■		■			■	■			■				
VectoBac																			
XenTari		■	■				■			■	■	■			■	■	■		■

Manufacturers List

These Bt products are manufactured by the following companies. Look for their addresses, phone numbers, and fax numbers in the Suppliers Directory beginning on page 599.

Agree, Javelin – Thermo Trilogy
Bactimos, Biobit, DiPel, VectoBac, XenTari – Abbott Laboratories
BMP 123 – Becker Microbial
Condor, Crymax, Cutlass, Lepinox – Ecogen
Kodiak – Gustafson, Trace
Mattch, M-Peril, MVP II – Mycogen
Troy-Bt – Troy Biosciences

	Hornworms	Imported cabbageworm	Leafrollers	Loopers	Melonworm	Mosquitoes	Omnivorous leaftier	Pickleworm	Podworms	Rhizoctonia	Rindworm	Saltmarsh caterpillar	Sunflower moth	Tobacco budworm	Tobacco hornworm	Tomato fruitworm	Velvetbean caterpillar	Webworms
Agree	■	■		■					■					■			■	
Bacillus subtilis										■								
Bactimos						■												
Biobit	■	■	■	■			■		■			■		■	■		■	■
BMP 123	■	■	■	■			■		■			■		■	■		■	■
Bt.* ssp. *aizawai	■	■	■	■					■			■		■			■	■
Bt.* ssp. *aizawai* + *Bt* ssp. *kurstaki		■		■										■		■		
Bt.* ssp. *israelensis						■												
Bt.* ssp. *kurstaki	■	■	■	■	■		■	■	■		■	■	■	■	■	■	■	■
Condor		■	■	■			■		■			■	■	■	■		■	
Crymax		■	■	■			■		■			■	■	■	■		■	
Cutlass		■	■	■			■		■			■	■	■	■		■	
DiPel	■	■	■	■					■			■	■	■	■		■	■
Javelin		■	■	■			■		■			■			■		■	
Kodiak										■								
Lepinox				■					■			■		■	■		■	
M-Peril														■				
Mattch		■		■										■		■		
MVP II		■		■										■		■		
Troy-Bt	■	■		■	■			■			■	■		■		■	■	■
VectoBac						■												
XenTari	■	■	■	■					■			■		■			■	■

Other Field Crops

Insecticides

Hops

AZADIRACHTIN

Extracted from seed of neem tree. Insect growth regulator that controls targeted insect larvae when they ingest or come in contact with it, by interfering with the insect's ability to molt and shutting off feeding behavior. Affected larvae slowly starve and die at next molt.

◆ **Azatin XL Plus** *(Thermo Trilogy)*
Rate: 5-21 oz. XL Plus/A
Pest:
aphids, armyworms, cutworms, leafhoppers, leafminers, loopers, mealybugs, thrips, whiteflies
Remarks: Rate varies by species. Apply when pests first appear; repeat at 7-day intervals. When pest populations are high, use higher rates. Do not apply more that 21 fl. oz./A.
Days To Harvest: 0.

Signal Word/Toxicity Class: Caution.
REI: 4 hr.

Azatin — see azadirachtin

BRIGADE

Synthetic pyrethroid insecticide with broad-spectrum activity. Formulated as a wettable powder and packaged in a water soluble bag. Restricted Use Pesticide.

Brigade WSB (bifenthrin) *(FMC)*
Rate: 0.06-0.1 lb. ai/A
9.6-16 oz. WSB/A
Pest:
aphids, armyworms, cutworms, loopers, root weevil, twospotted spider mite (high rate)
Remarks: Apply at 21-day intervals. May be tank mixed with other products approved for use on hops; observe all label restrictions. Do not exceed 0.1 lb. ai (2 bags)/A per application or 3 lb. ai (6 bags)/A per season. Refer to label for spray drift precautions.
Days To Harvest: 14.

Signal Word/Toxicity Class: Warning/II.
REI: 12 hr.

Diatect — see pyrethrins

DIAZINON

Restricted Use Pesticide due to avian and aquatic toxicity.

D•z•n diazinon AG600 WBC, 50W *(Novartis)*
Drexel Diazinon Insecticide (4EC), 50W *(Drexel)*
Gowan Diazinon 4E, 50WSB *(Gowan)*
Helena Diazinon AG500 *(Helena)*
Micro Flo Diazinon AG500, 50W *(Micro Flo)*
Prentox Diazinon AG500, 50W *(Prentiss)*
◆ **Wilbur-Ellis Diazinon 4 Spray (4E)** *(Wilbur-Ellis)*
Rate: 1 lb. ai/A
2 pt. 4E, 4EC, AG500/A
27 fl. oz. AG600/A
2 lb. 50W, 50WSB/A
Pest:
aphids, mites
Remarks: Apply as pests occur; repeat every 14 days as necessary.
Days To Harvest: 14.

Signal Word/Toxicity Class: Caution/III; Warning/II (Drexel 50W).
REI: 24 hr.

DIBROM

Dibrom 8 (8E) (naled) *(Amvac)*
Rate: 1 lb. ai/A
1 pt. 8E/A
Pest:
armyworms, hop aphid
Remarks: Apply up to 5 applications at 14-day intervals. Do not use with highly alkaline materials such as lime or Bordeaux. Do not apply through any type of irrigation system. Refer to label for directions and precautions.
Days To Harvest: 7.

Signal Word/Toxicity Class: Danger/I.
REI: 48 hr.

DICOFOL

◆ **Griffin Dicofol 4E** *(Griffin)*
Kelthane MF (4E) *(Rohm and Haas)*
Rate: 1-1.17 lb. ai/A
2-2.33 pt. 4E/A
Pest:
Pacific mite, tropical mite, twospotted spider mite
Remarks: Apply at first sign of mite buildup. Do not feed treated vines or crop residues to livestock. Do not apply through any type of irrigation system.
Days To Harvest: 7.

Signal Word/Toxicity Class: Caution.
REI: 12 hr.

D•z•n — see diazinon

IMIDACLOPRID

Systemic insecticide for control of aphids. Toxic to bees and aquatic invertebrates.

Provado 1.6 *(Bayer)*
Rate: 0.1 lb. ai/A
8 fl. oz. 1.6/A
Pest:
aphids
Remarks: Foliar. Allow at least 21 days between applications. Do not apply more than 24 fl. oz./A per crop season.
Days To Harvest: 28.

Signal Word/Toxicity Class: Caution.
REI: 12 hr.

◆ - **new product** • ai-active ingredient • DF-dry flowable • E/EC-emulsifiable concentrate • EW-emulsion in water
F/FL-flowable • G-granule • L/LC-liquid • SP/WSP-soluble packet • W/WP-wettable powder • WSB-water soluble bag

Kelthane — see dicofol

Kumulus — see sulfur

MALATHION

Clean Crop Malathion 57EC *(Platte)*
Drexel Malathion 5EC *(Drexel)*
◆ **Micro Flo Malathion 5EC** *(Micro Flo)*
Prentox 5 lb. Malathion Spray (5EC) *(Prentiss)*

Rate: 0.625 lb. ai/A
1 pt. 5EC, 57EC/A

Pest:
aphids mites

Remarks: Apply when insects first appear and repeat as necessary.
Days To Harvest: 57EC-7; 5EC-10.

Signal Word/Toxicity Class: Caution/III (57EC); Warning/II (5EC).
REI: 12 hr.

◆ **Gowan Malathion 8 (8E)** *(Gowan)*

Rate: 0.5-1.25 lb. ai/A
0.5-1.25 pt. 8E/A

Pest:
aphids

Remarks: Apply at first sign of infestation; repeat at 7- to 10-day intervals.
Days To Harvest: 7.
State Restrictions: Not for use in California.

Signal Word/Toxicity Class: Caution/III.
REI: 12 hr.

Microthiol Special — see sulfur

MORESTAN

Morestan Solupak 25 (oxythioquinox) *(Bayer)*

Rate: 0.5 lb ai/A
2 lb. 25/A

Pest:
mite eggs spider mites

Remarks: Refer to label for use directions and precautions.
State Restrictions: SLN-Oregon, Washington.

Signal Word/Toxicity Class: Warning/II.
REI: 24 hr.

M-PEDE

M-Pede (potassium salts of fatty acids) *(Mycogen)*

Rate: 1-2% v/v solution

Pest:
aphids, leafhoppers, mealybugs, plant bugs, scales, thrips, whiteflies

Remarks: Apply at first sign of infestation; repeat at 7- to 14-day intervals. May be tank mixed with other insecticides. Refer to both labels for use directions and precautions.
Days To Harvest: 0.
State Restrictions: Not for use in California.

Signal Word/Toxicity Class: Warning/II.
REI: 12 hr.

Omite — see propargite

PROPARGITE

Omite 6E, 30W, 30WS or CR *(Uniroyal)*

Rate: 1.5 lb. ai/A
2 pt. 6E/A
5 lb. 30W, 30WS/A
1.35 lb. ai/A
3 bags CR/A

Pest:
twospotted spider mite

Remarks: Apply by ground in 100-200 gal. spray solution (200 gal. water for CR). Do not make more than 2 applications per year. Do not apply through any type of irrigation system.
Days To Harvest: 14.

Signal Word/Toxicity Class: Warning/II (6E); Danger/I (30W, 30WS, CR).
REI: 7 days. After 48 hr., workers can enter treated areas to perform hand labor tasks if they wear early-entry personal protective equipment.

Provado — see imidacloprid

PYRELLIN (Premix)

Pyrellin E.C. (0.042 lb./gal. pyrethrins & 0.035 lb./gal. rotenone) *(Webb Wright)*

Rate: 1-2 pt. E.C./A

Pest:
aphids, leafhoppers, leafminers, loopers, Lygus bugs, mites, plant bugs, thrips, whiteflies

Remarks: Apply as necessary to maintain control.
Days To Harvest: 0.

Signal Word/Toxicity Class: Caution/III.
REI: 12 hr.

PYRETHRINS

Contains piperonyl butoxide as a synergist and silicon dioxide (diatomaceous earth).

Diatect Multipurpose Insecticide II *(Diatect)*

Rate: 1-6 lb. II/A

Pest:
aphids, loopers, mites, plant bugs, thrips, whiteflies

Remarks: Apply as dust or wettable powder. Repeat as needed.

Signal Word/Toxicity Class: Caution.
REI: 12 hr.

SULFUR

Microthiol is micronized sulfur.

◆ **Kumulus DF** *(BASF)*

Rate: 1.6-3.2 lb. ai/A
2-4 lb. DF/A

Pest:
red spider mite (suppression)

Remarks: Do not apply in less than 10 gal. water/A. Do not apply if temperatures during or within 3 days after application are expected to exceed 90°F. Do not apply through any type of irrigation system.

Signal Word/Toxicity Class: Caution.
REI: 24 hr.

Microthiol Special (80W) *(Elf Atochem)*

Rate: 1.6-4.8 lb. ai/A
2-6 lb. 80W/A

Pest:
red spider mites

Remarks: Apply in early leaf stage; repeat as needed.
State Restrictions: Not for use in California.

Signal Word/Toxicity Class: Caution.
REI: 24 hr.

NOTICE The information on these pages is for preliminary planning — not a guide for use. Be sure to follow the manufacturer's directions, notwithstanding information contained here.

For personal protective equipment and EPA registration numbers, please turn to page 39.

Mint

(Peppermint, Spearmint)

ACEPHATE

Organophosphate systemic insecticide with contact and stomach poison activity. Effective against many aphids, thrips, caterpillars, and plant bugs. Has moderate residual effects and is compatible with most pesticides.

Acephate 75SP AG *(Micro Flo)*
Orthene 75 S *(Valent)*
Rate: 1 lb. ai/A
1.33 lb. 75SP, 75 S/A
Pest:
alfalfa looper aphids cutworms
Remarks: Peppermint, spearmint. Apply when eggs or insects first appear. Repeat once if necessary to maintain control. Do not apply more than 2.66 lb./A per season. Do not use spent hay for feed for dairy animals.
Days To Harvest: 14.

Rate: 1 lb. ai/A
1.33 lb. 75SP, 75 S/A
Pest:
strawberry root weevil (adult)
Remarks: Peppermint, spearmint. Early spring application timed at 90% adult emergence. Make applications at dusk or in the evening when weevils are active on foliage. Do not apply more than 2.66 lb./A per season. Do not apply through any type of irrigation system.
Days To Harvest: 14.
State Restrictions: SLN-Oregon.

- - - - - - - - - - - - - - - - -

Signal Word/Toxicity Class: Caution.
REI: 24 hr.

AZADIRACHTIN

Extracted from seed of neem tree. Insect growth regulator that controls targeted insect larvae when they ingest or come in contact with it, by interfering with the insect's ability to molt and shutting off feeding behavior. Affected larvae slowly starve and die at next molt.

◆ **Azatin XL Plus** *(Thermo Trilogy)*
Rate: 5-21 oz. XL Plus/A
Pest:
aphids, armyworm, cutworms, flea beetles, leafhoppers, loopers, thrips, whiteflies
Remarks: Rate varies by species. Apply when pests first appear; repeat at 7-day intervals. When pest populations are high, use higher rates. Do not apply more that 21 fl. oz./A.
Days To Harvest: 0.

- - - - - - - - - - - - - - - - -

Signal Word/Toxicity Class: Caution.
REI: 4 hr.

◆ **NEEMAZAD 4.5EC or NEEMIX 4.5** *(Thermo Trilogy)*
Rate: 0.125-1 pt. 4.5/A
High Volume:
2.25-4.5 oz. 4.5EC/100 gal. water
Low Volume:
0.5 pt. 4.5EC/A
Pest:
aphids flea beetles leafhoppers
Remarks: Peppermint, spearmint, wintergreen. For low volume, apply in sufficient water for adequate coverage. For low insect pressure, apply preventative at 7-day intervals or at first sign of insects; high insect pressure, at 3- to 4-day intervals.

- - - - - - - - - - - - - - - - -

Signal Word/Toxicity Class: Warning.
REI: 12 hr.

Azatin — see azadirachtin

CHLORPYRIFOS

Lorsban-4E *(Dow AgroSciences)*
◆ **Micro Flo Chlorpyrifos 4E AG** *(Micro Flo)*
Rate: 1-2 lb. ai/A
2-4 pt. 4E/A
Pest:
cutworms mint root borer
Remarks: For cutworms, apply as rescue treatment in minimum 10 gal. water. Apply high rate for cutworms over 3/4" in length. Make 1 application per year. For mint root borer, apply high rate postharvest; follow with approximately 1 acre inch of sprinkler irrigation to incorporate insecticide. Make only 1 postharvest application per season.
Days To Harvest: 90.

- - - - - - - - - - - - - - - - -

Signal Word/Toxicity Class: Warning/II.
REI: 24 hr.

DEADLINE

Packaged as Bullets, M-Ps (mini-pellets), and granules.

Deadline Bullets, M-Ps, Granules (4%) (metaldehyde) *(Pace)*
Rate: 0.4-1 lb. ai/A
10-40 lb. 4%/A
Pest:
slugs snails
Remarks: Apply every 3-4 weeks during growing season, or as needed. Do not apply directly to edible portion of plant.

- - - - - - - - - - - - - - - - -

Signal Word/Toxicity Class: Warning/II (granule); Caution (bullets, M-P).
REI: 12 hr.

Diatect — see pyrethrins

DICOFOL

◆ **Griffin Dicofol 4E** *(Griffin)*
Kelthane MF (4E) *(Rohm and Haas)*
Rate: 0.88-1.25 lb. ai/A
1.75-2.5 pt. 4E/A
Pest:
twospotted spider mite
Remarks: Peppermint, spearmint. Commercial oil production. Apply at first sign of mite buildup. Do not make more than 1 application per crop season. Do not feed treated fresh or spent hay to livestock. Do not apply through any type of irrigation system.
Days To Harvest: 30.

- - - - - - - - - - - - - - - - -

Signal Word/Toxicity Class: Caution.
REI: 12 hr.

Fyfanon — see malathion

Kelthane — see dicofol

LANNATE

Restricted Use Pesticide.

Lannate SP or LV (methomyl) *(DuPont)*
Rate: 0.675-0.9 lb. ai/A
0.75-1 lb. SP/A
2.25-3 pt. LV/A
Pest:
alfalfa looper flea beetles variegated cutworm
Remarks: Apply in sufficient water to obtain thorough, uniform coverage when insects first appear; repeat at 5- to 7-day intervals or as needed. Rate varies by pest species; refer to label. Do not apply more than 1.8 lb. ai/A per crop or make more than 4 applications per crop.
Days To Harvest: 14.

- - - - - - - - - - - - - - - - -

Signal Word/Toxicity Class: Danger-Poison/I.
REI: 48 hr.

◆ - **new product** • ai-active ingredient • DF-dry flowable • E/EC-emulsifiable concentrate • EW-emulsion in water
F/FL-flowable • G-granule • L/LC-liquid • SP/WSP-soluble packet • W/WP-wettable powder • WSB-water soluble bag

Lorsban — see chlorpyrifos

MALATHION

Clean Crop Malathion 57EC *(Platte)*
Rate: 0.7-1 lb. ai/A
1-1.5 pt. 57EC/A
Pest:
aphids, caterpillars, flea beetles (adult), leafhoppers, spider mites
Remarks: Apply when insects first appear and repeat as necessary.
Days To Harvest: 7.

Signal Word/Toxicity Class: Caution/III.
REI: 12 hr.

Drexel Malathion 5EC *(Drexel)*
◆ Micro Flo Malathion 5EC *(Micro Flo)*
Prentox 5 lb. Malathion Spray (5EC) *(Prentiss)*
Rate: 0.94 lb. ai/A
1.5 pt. 5EC/A
Pest:
aphids, flea beetles (adult), leafhoppers, spider mites
Remarks: Peppermint, spearmint. Apply when insects first appear and repeat as necessary.
Days To Harvest: 7.

Signal Word/Toxicity Class: Warning/II.
REI: 12 hr.

Fyfanon (8E) *(Helena)*
Gowan Malathion 8 (8E) *(Gowan)*
Rate: 1 lb. ai/A
1 pt. 8E/A
Pest:
aphids (Fyfanon), flea beetles (adult), leafhoppers (Gowan), spider mites (Fyfanon)
Remarks: Peppermint, spearmint. Apply at first sign of infestation; repeat at 7- to 10-day intervals.
Days To Harvest: 7.

Signal Word/Toxicity Class: Caution/III.
REI: 12 hr.

◆ Wilbur-Ellis Malathion 8 Spray *(Wilbur-Ellis)*
Rate: 1 lb. ai/A
1 pt. 8/A
Pest:
aphids, flea beetles (adults), leafhoppers, spider mites
Remarks: Peppermint, spearmint. Apply in minimum 5 gal. by air; 10 gal. by ground.
Days To Harvest: 7.

Signal Word/Toxicity Class: Caution
REI: 12 hr.

METAM-SODIUM

Metam CLR (42%) *(UCB Agrochemicals)*
Rate: 15 gal. 42%/A
Pest:
symphylids
Remarks: Peppermint. Maintain adequate soil moisture in the spring. Use injector blade or thin shank injector rig with injectors spaced 5" apart. Refer to label for use directions.
State Restrictions: In California, application must be in compliance with Technical Information Bulletin "Guidelines for All Application Methods for Metam-sodium in California." This bulletin may be obtained from your local pesticide dealer or a metam-sodium registrant.

Signal Word/Toxicity Class: Danger/I.
REI: 48 hr.

Metam CLR — see metam-sodium

METASYSTOX-R

Restricted Use Pesticide.

Metasystox-R (SC) (oxydemeton-methyl) *(Gowan)*
Rate: 0.75 lb. ai/A
3 pt. SC/A
Pest:
aphids, mites
Remarks: Apply with ground equipment in at least 20 gal. water/A. For established infestations, make 2 applications 10-14 days apart.
Days To Harvest: 14.

Signal Word/Toxicity Class: Warning/II.
REI: 48 hr.; 72 hr. if average rainfall less than 25" per year.

NEEMAZAD — see azadirachtin

NEEMIX — see azadirachtin

Omite — see propargite

Orthene — see acephate

PROPARGITE

Omite 6E *(Uniroyal)*
Rate: 1.5-2.25 lb. ai/A
2-3 pt. 6E/A
Pest:
twospotted spider mite
Remarks: Apply in 20-50 gal. spray solution by ground; minimum 10 gal. spray solution/A by air. Do not apply more than twice per year. Do not apply through any type of irrigation system.
Days To Harvest: 14.

Signal Word/Toxicity Class: Warning/II.
REI: 7 days. After 48 hr., workers can enter treated areas to perform hand labor tasks if they wear early-entry personal protective equipment.

PYRELLIN (Premix)

Pyrellin E.C. (0.042 lb./gal. pyrethrins & 0.035 lb./gal. rotenone) *(Webb Wright)*
Rate: 1-2 pt. E.C./A
Pest:
aphids, loopers, mites, plant bugs, thrips, whiteflies
Remarks: Apply in intervals of 7 days or less; repeat as needed.
Days To Harvest: 0.

Signal Word/Toxicity Class: Caution/III.
REI: 12 hr.

PYRETHRINS

Contains piperonyl butoxide as a synergist and silicon dioxide (diatomaceous earth).

Diatect Multipurpose Insecticide II *(Diatect)*
Rate: 1-6 lb. II/A
Pest:
aphids, loopers, mites, plant bugs, thrips, whiteflies
Remarks: Apply as dust or wettable powder; repeat as needed.

Signal Word/Toxicity Class: Caution.
REI: 12 hr.

6 Other Field Crops

NOTICE The information on these pages is for preliminary planning — not a guide for use. Be sure to follow the manufacturer's directions, notwithstanding information contained here.

For personal protective equipment and EPA registration numbers, please turn to page 39.

Oil Seed Crops

Declare — see methyl parathion

METHYL PARATHION

Restricted Use Pesticide.

Cheminova Methyl 4EC *(Cheminova)*
◆ **Declare (4EC)** *(Griffin)*
Rate: 0.5 lb. ai/A
1 pt. 4EC/A
Pest:
cabbage aphid, cabbage seedpod weevil, diamondback moth (larvae), flea beetles
Remarks: Canola, rapeseed. Apply by ground or air in sufficient water. Do not apply more than 2 applications per season. Do not graze treated fields or feed treated forage or threshing to livestock. In Oregon, apply only in the evening (3 hr. before sunset until midnight of same day). Refer to label for use directions and precautions.
Days To Harvest: 28.
State Restrictions: Do not apply in Idaho and Washington.

- - - - - - - - - - - - - - - - -

Signal Word/Toxicity Class: Danger-Poison/I.
REI: 48 hr.; 72 hr. if average rainfall less than 25" per year.

M-PEDE

M-Pede (potassium salts of fatty acids) *(Mycogen)*
Rate: 4 pt./A in 3-25 gal. water
2% v/v spray solution more than 25 gal. water/A
Pest:
aphids, caterpillars (tank mix only), fleahoppers, greenbug, Lygus bugs, plant bugs, spider mites, stink bugs, thrips, whiteflies
Remarks: Canola. Apply at first sign of infestation; repeat at 7- to 14-day intervals. May be tank mixed with other insecticides; refer to both labels for use directions and precautions.
Days To Harvest: 0.
State Restrictions: Not for use in California.

- - - - - - - - - - - - - - - - -

Signal Word/Toxicity Class: Warning/II.
REI: 12 hr.

PYRELLIN (Premix)

Pyrellin E.C. (0.042 lb./gal. pyrethrins & 0.035 lb./gal. rotenone) *(Webb Wright)*
Rate: 1-2 pt. E.C./A
Pest:
aphids, armyworms, chinch bug, mites, thrips
Remarks: Jojoba. Apply at 7-day intervals or less; repeat as needed.
Days To Harvest: 0.

- - - - - - - - - - - - - - - - -

Signal Word/Toxicity Class: Caution/III.
REI: 12 hr.

Peanuts

ACEPHATE

Organophosphate insecticide with systemic activity. Used on peanuts for thrips and leafhopper control. Also effective for major foliage feeders of peanuts. Has low toxicity to humans.

Micro Flo Acephate 75SP AG *(Micro Flo)*
Orthene 75 S *(Valent)*
Rate: 0.25-1 lb. ai/A
0.33-1.33 lb. 75SP, 75S/A
Pest:
corn earworm, fall armyworm, grasshoppers, leafhoppers, loopers, thrips, velvetbean caterpillar
Remarks: Rate varies by pest species; refer to label. Apply when insects first appear and repeat as needed to maintain control. Do not apply through any type of irrigation system. Do not feed treated forage or hay to livestock or allow animals to graze treated areas.
Days To Harvest: 14 days of digging.

- - - - - - - - - - - - - - - - -

Rate: 0.75-1 lb. ai/A
1-1.33 lb. 75 S/A (36" row spacing)
Pest:
thrips
Remarks: Apply in-furrow spray at planting. Apply only 1 application per year. Do not feed treated forage or hay to livestock or allow animals to graze treated areas.
State Restrictions: SLN-North Carolina, Virginia.

- - - - - - - - - - - - - - - - -

Rate: Hopper box treatment:
0.2 lb. ai/A
4 oz. 75 S/100 lb. seed
Pest:
thrips
Remarks: Apply to seed in planter box as dry powder. Lay in powder as seeds are filled into planter box. Treated seeds which become wet or moist due to exposure to rain or heavy dew may exhibit a decrease in germination. Do not mix with peanut seed by stirring or other agitation. Do not use treated seed for food, feed purposes or process for oil.
Days To Harvest: 14.
State Restrictions: SLN-Alabama, Georgia, Oklahoma, Virginia.

- - - - - - - - - - - - - - - - -

Signal Word/Toxicity Class: Caution.
REI: 24 hr.

Payload (15 G) *(Valent)*
Rate: SLN-New Mexico, Texas.
0.75-1 lb. ai/A
5-6.66 lb. 15 G/A
SLN-Oklahoma:
0.75 lb. ai/A
0.5 lb. 15 G/A
Pest:
thrips
Remarks: Apply in-furrow at planting. In New Mexico and Texas, some delay in emergence, stunting of seedlings, or stand reduction may occur under adverse conditions when used at 6.66 lb/A rate. Do not feed treated forage to livestock or allow animals to graze treated areas.
State Restrictions: SLN-New Mexico, Oklahoma, Texas.

- - - - - - - - - - - - - - - - -

Signal Word/Toxicity Class: Caution.
REI: 12 hr.

Adios — see carbaryl

ASANA

Synthetic pyrethroid insecticide with low mammalian toxicity. Restricted Use Pesticide due to toxicity to fish and aquatic organisms.

Asana XL (esfenvalerate) *(DuPont)*
Rate: 0.015-0.05 lb. ai/A
2.9-9.6 fl. oz. XL/A
Pest:
beet armyworm*, corn earworm, cutworms, granulate cutworm, grasshoppers, potato leafhopper, rednecked peanutworm, velvetbean caterpillar

Rate: 0.05 lb. ai/A
9.6 fl. oz. XL/A
Pest:
fall armyworm*, lesser cornstalk borer*
*** Aids in control**
Remarks: Rate varies by pest species. Repeat as necessary to maintain control. Do not feed or graze livestock on treated vines. Do not exceed 0.15 lb. ai/A per season. Refer to label for directions and precautions.
Days To Harvest: 21.

- - - - - - - - - - - - - - - - -

Signal Word/Toxicity Class: Warning/II.
REI: 12 hr.

◆ - **new product** • ai-active ingredient • DF-dry flowable • E/EC-emulsifiable concentrate • EW-emulsion in water
F/FL-flowable • G-granule • L/LC-liquid • SP/WSP-soluble packet • W/WP-wettable powder • WSB-water soluble bag

AZADIRACHTIN

Extracted from seed of neem tree. Insect growth regulator that controls targeted insect larvae when they ingest or come in contact with it, by interfering with the insect's ability to molt and shutting off feeding behavior. Affected larvae slowly starve and die at next molt.

◆ **NEEMAZAD 4.5EC or NEEMIX 4.5** *(Thermo Trilogy)*

Rate: 0.125-1 pt. 4.5/A
High Volume:
2.25-4.5 oz. 4.5EC/100 gal. water
Low Volume:
0.5 pt. 4.5EC/A

Pest:

armyworms	cutworms	loopers
bean leaf beetle	garden webworm	Mexican bean beetles
cabbage loopers	hornworms	soybean loopers
caterpillars	leafminers	weevils

Remarks: For low volume, apply in sufficient water for adequate coverage. For low insect pressure, apply preventative at 7-day intervals or at first sign of insects; high insect pressure, apply at 3- to 4-day intervals.

Signal Word/Toxicity Class: Warning/II.
REI: 12 hr.

CARBARYL

Carbamate insecticide with broad activity on many insects. Low mammalian toxicity, but highly toxic to bees.

Adios (13%) *(Micro Flo)*

Rate: 0.06-0.1 lb. ai/A
0.5-0.75 lb. 13%/A

Pest:
cucumber beetle

Remarks: Do not apply through any type of irrigation system. Do not apply more than 3.75 lb./A per season.
Days To Harvest: 0.

Signal Word/Toxicity Class: Caution/III.
REI: 12 hr.

Clean Crop Sevin 5 Bait *(Platte)*

Rate: 1 lb. ai/A
20 lb. Bait/A

Pest:

ants	darkling beetle	grasshoppers
armyworms	field crickets	sowbugs
cutworms		

Remarks: Apply with ground or air equipment.
Days To Harvest: 0.

Signal Word/Toxicity Class: Caution/III.
REI: 12 hr.

Drexel Carbaryl 4L or 50W *(Drexel)*
Sevin 4F, XLR Plus, 80S, 80WSP *(Rhone-Poulenc)*

Rate: 0.5-2 lb. ai/A
1-4 pt. 4L, 4F, XLR/A
1-4 lb. 50W/A
0.66-2.5 lb. 80S, 80WSP/A

Pest:

alfalfa caterpillar	grasshoppers	threecornered alfalfa hopper
armyworms	green cloverworm	thrips
bean leaf beetle	Japanese beetle	velvetbean caterpillar
blister beetle	leafhoppers	webworms
corn earworm	Mexican bean beetle	whitefringed beetle (adult)
cucumber beetle	rednecked peanutworm	
cutworms	stink bugs	
fall armyworm		

Remarks: Rate varies by pest species; refer to label. Do not apply when foliage is wet or when rain or excessive humidity is expected during next 2 days.
Days To Harvest: 0.

Signal Word/Toxicity Class: Caution (4F, 4L, XLR); Warning/II (50W, 80S, 80WSP).
REI: 12 hr.

Chlorfos — see chlorpyrifos

CHLORPYRIFOS

Organophosphate insecticide formulated as a granule and an emulsifiable concentrate. Used primarily for control of soil insects on peanuts.

Lorsban-4E *(Dow AgroSciences)*
◆ **Micro Flo Chlorpyrifos 4E AG** *(Micro Flo)*

Rate: 2 lb. ai/A
4 pt. 4E/A

Pest:
wireworms (suppression)

Remarks: Apply preplant broadcast to soil surface in minimum 10 gal. total spray/A Immediately incorporate to depth of 3-4". Apply only once per season. Do not feed treated peanut forage or hay to livestock.
Days To Harvest: 21.

Signal Word/Toxicity Class: Warning/II.
REI: 24 hr.

◆ **Chlorfos 15G** *(Griffin)*
Lorsban 15G *(Dow AgroSciences)*
◆ **Micro Flo Chlorpyrifos 15G** *(Micro Flo)*

Rate: Broadcast:
1.5-1.95 lb. ai/A
10-13 lb. 15G/A
Band:
7.5-15 oz. 15G/1000 ft. row

Pest:
lesser cornstalk borer

Remarks: Broadcast rescue treatment: When lesser cornstalk borer first appears, uniformly broadcast by air. Care should be taken to ensure proper equipment calibration and delivery. Band rescue treatment: Apply just at or prior to pegging in 10-18" band over fruiting zone.

Rate: 7.5-15 oz. 15G/1000 ft. row

Pest:

cutworms	southern corn	wireworms
lesser cornstalk borer	rootworm (larvae)	

Remarks: At planting: Apply in 6-12" band over row behind planter shoe and in front of press wheel. If 7.5 oz. rate is used at planting time, repeat at same rate postplant to maintain control. Postplant: Apply at early flowering in a 6-12" band over row or during early pegging in 10-18" band over fruiting zone. Irrigation or rain following application will enhance activity. Do not apply more than 15 oz./1000 ft. row or 13.6 lb./crop season. Do not feed treated peanut forage or hay to livestock.
Days To Harvest: 21.

Signal Word/Toxicity Class: Caution.
REI: 12 hr.

Comite — see propargite

lambda-CYHALOTHRIN

Synthetic pyrethroid insecticide with broad-spectrum activity, formulated as an emulsifiable concentrate. Restricted Use Pesticide due to toxicity to fish and aquatic organisms.

Karate or Warrior (1EC) *(ZENECA)*

Rate: 0.015-0.03 lb. ai/A
1.92-3.84 fl. oz. 1EC/A

Pest:

aphids*	grasshoppers	soybean looper*
bean leaf beetle	green cloverworm	spider mites*
beet armyworm (1st & 2nd instar)	lesser cornstalk borer*	stink bugs
	potato leafhopper	tobacco thrips
corn earworm	rednecked peanutworm	vegetable weevil
cutworms		velvetbean caterpillar
fall armyworm (1st & 2nd instar)	southern corn rootworm (adult)	whitefringed beetle (adult)

* *Suppression*

Remarks: Apply as required by scouting, usually at least 7-day intervals. Do not apply more than 0.96 pt./A per season. Do not graze livestock in treated areas or use treated vines or hay for animal feed. Refer to label for spray drift precautions and state and county restrictions.
Days To Harvest: 14.

Signal Word/Toxicity Class: Danger/I.
REI: 24 hr.

NOTICE The information on these pages is for preliminary planning — not a guide for use. Be sure to follow the manufacturer's directions, notwithstanding information contained here.

For personal protective equipment and EPA registration numbers, please turn to page 39.

DANITOL

Restricted Use Pesticide due to toxicity to fish and aquatic organisms. Also toxic to wildlife.

Danitol 2.4EC (fenpropathrin) *(Valent)*

Rate: 0.2-0.3 lb. ai/A
10.66-16 fl. oz. 2.4EC/A

Pest:
corn earworm, fall armyworm, potato leafhopper, silverleaf whitefly, sweetpotato whitefly, thrips, twospotted spider mite

Remarks: Apply in 5-10 gal. water/A by air; 10-50 gal. water/A by ground in seasonal program when pest activity is first noticed. Repeat at 7-day intervals as needed. Under severe pest pressure use high rate. For whiteflies, tank mix with Orthene 75 S. Do not graze or feed treated peanut vine forage or dried hay within 14 days of last application. Do not exceed 2.66 pt. 2.4EC/A per season.
Days To Harvest: Do not apply within 14 days of digging.

Signal Word/Toxicity Class: Danger/I.
REI: 24 hr.

DI-SYSTON

Organophosphate insecticide used at planting on peanuts for thrips control. Highly toxic to mammals and fish. Restricted Use Pesticide.

Di-Syston 15% (disulfoton) *(Bayer)*

Rate: 1-2 lb. ai/A
6.7-13.3 lb. 15%/A (36" row spacing)

Pest:
aphids, leafhoppers, mites, thrips

Remarks: Place granules in soil incorporated band on each side of seed furrow at planting, or as sidedressing after plants are established. Do not use treated vines for feed.

Rate: 7.4-14.7 oz. 15%1000 ft.

Pest:
aphids, leafhoppers, mites, thrips

Remarks: None.
State Restrictions: SLN-North Carolina, Virginia.

Signal Word/Toxicity Class: Danger/I.
REI: 48 hr.

Karate — see *lambda*-cyhalothrin

Kumulus — see sulfur

LANNATE

Carbamate insecticide available in water soluble formulations. Usually recommended for foliage feeding caterpillars on peanuts. Short residual, quick acting and highly toxic to humans, fish, and bees. Restricted Use Pesticide.

Lannate SP or LV (methomyl) *(DuPont)*

Rate: 0.225-0.9 lb. ai/A
0.25-1 lb. SP/A
0.75-3 pt. LV/A

Pest:
beet armyworm, corn earworm, fall armyworm, granulate cutworm, green cloverworm, loopers, potato leafhopper, thrips, velvetbean caterpillar

Remarks: Apply in sufficient water to obtain thorough, uniform coverage when insects first appear; repeat at 5- to 7-day intervals or as needed. Rate varies by pest species; refer to label. Do not feed treated vines. Do not apply more than 3.6 lb. ai/A per crop or make more than 8 applications per crop.
Days To Harvest: 21.

Signal Word/Toxicity Class: Danger-Poison/I.
REI: 48 hr.

Lorsban — see chlorpyrifos

MALATHION

Organophosphate insecticide with low mammalian toxicity. Toxic to fish and bees. Liquid sprays used in peanuts to control leafhoppers and thrips.

Clean Crop Malathion 57EC *(Platte)*
Drexel Malathion 5EC *(Drexel)*
◆ **Micro Flo Malathion 5EC** *(Micro Flo)*

Rate: 1 lb. ai/A
1.5 pt. 5EC, 57EC/A

Pest:
leafhoppers, thrips

Remarks: Apply when insects first appear and repeat as necessary.
Days To Harvest: 0.

Signal Word/Toxicity Class: Caution/III (57EC); Warning/II (5EC).
REI: 12 hr.

Gowan Malathion 8 (8E) *(Gowan)*

Rate: 1 lb. ai/A
1 pt. 8E/A

Pest:
leafhoppers

Remarks: Apply at first sign of infestation; repeat at 7- to 10-day intervals.
Days To Harvest: 0.
State Restrictions: Not for use in California.

Signal Word/Toxicity Class: Caution/III.
REI: 12 hr.

Microthiol Special — see sulfur

MOCAP

Organophosphate insecticide used on peanuts to control soil insects and nematodes. Hazardous to birds and fish. Mocap EC is a Restricted Use Pesticide.

Mocap EC or 10%G (ethoprop) *(Rhone-Poulenc)*

Rate: Broadcast:
4-6 lb. ai/A
3-4 qt. EC/A
40-60 lb. 10%G/A
Row treatment:
2-4 lb. ai/A
1.33-2.66 qt. EC/A (2.9-5.9 fl. oz./1000 lin. ft. row)
20-40 lb. 10%G/A

Pest:
nematodes

Remarks: Broadcast: Apply 1 week before planting to at-planting time. Row treatment: Apply at planting in band 12" wide (36" spacing). Mix with top 2-4" soil. Do not use as seed furrow treatment or allow spray to contact seed. Use high rate for control of peanut root-knot nematode and in fields where nematode infestations have been especially severe.

Rate: Row treatment:
2.5-3 lb. ai/A
25-30 lb. 10%G/A (1.7-2.1 lb./1000 lin. ft. row)

Pest:
corn rootworm, lesser cornstalk borer (suppression), nematodes

Remarks: Apply at pegging in band 15-18" wide on row (36" spacing). Incorporate lightly into soil.

Signal Word/Toxicity Class: Danger-Poison/I (EC); Warning/II (10%G).
REI: 48 hr.; 72 hr. if average rainfall less than 25" per year.

M-PEDE

M-Pede (potassium salts of fatty acids) *(Mycogen)*

Rate: 4 pt./A in 3-25 gal. water
2% v/v spray solution more than 25 gal. water/A

Pest:
aphids, caterpillars (tank mix only), fleahoppers, greenbug, Lygus bugs, plant bugs, spider mites, stink bugs, thrips, whiteflies

◆ - **new product** • ai-active ingredient • DF-dry flowable • E/EC-emulsifiable concentrate • EW-emulsion in water
F/FL-flowable • G-granule • L/LC-liquid • SP/WSP-soluble packet • W/WP-wettable powder • WSB-water soluble bag

Remarks: Apply at first sign of infestation; repeat at 7- to 14-day intervals. May be tank mixed with other insecticides; refer to both labels for use directions and precautions.
Days To Harvest: 0.
State Restrictions: Not for use in California.

Signal Word/Toxicity Class: Warning/II.
REI: 12 hr.

NEMACUR

Restricted Use Pesticide.

Nemacur 3 or 15% (fenamiphos) *(Bayer)*

Rate: 1.5-2.5 lb. ai/A
2-3.3 qt. 3/A (36" row)
10-17 lb. 15%/A (36" row)
4.5-7.3 fl. oz. 3/1000 ft. row (any row spacing)
11-18.7 oz. 15%/1000 ft. row (any row spacing)

Pest:
nematodes
thrips (early-season)

Remarks: Apply in 12" band and incorporate. Refer to label for further use information.

Signal Word/Toxicity Class: Danger/I.
REI: 48 hr.

NEEMAZAD — see azadirachtin

NEEMIX — see azadirachtin

Omite — see propargite

Orthene — see acephate

Payload — see acephate

PETROLEUM OIL

Petroleum-based spray oil. Mode of action is through suffocation of eggs, larvae, and nymphs of insects and adults of soft bodied insects. This mode of action necessitates total spray coverage. Toxic to fish.

SAF-T-SIDE *(Brandt Consolidated)*
◆ **Sumulsion** *(Drexel)*

Rate: 1-2 gal./100 gal. water

Pest:
aphids
beetles (larvae)
leafhoppers
leafminers
mites
spiders
thrips
whiteflies

Remarks: Apply as needed. May be tank mixed with most commonly used insecticides and fungicides. Do not use in combination with or immediately before or after spraying with fungicides such as captan, Dyrene, folpet, Morestan, or any sulfur products. Do not use with carbaryl or dimethoate.

Signal Word/Toxicity Class: Caution.
REI: 4 hr.

PHORATE

Systemic organophosphate insecticide applied in-furrow at planting for thrips control or banded over the row at pegging for soil insect control. Restricted Use Pesticide.

Clean Crop Phorate 20G *(Platte)*
Riverside Phorate 20G *(Terra)*
Thimet 20-G *(American Cyanamid)*

Rate: 5.13 oz. 20 G/1000 ft. row (min. 24" spacing)
5.5 oz. 20-G/1000 ft. row (min. 24" spacing)

Pest:
leafhoppers
thrips

Remarks: Apply at planting. Distribute granules evenly in furrow at planting time. Do not graze or feed treated hay or foliage.
Days To Harvest: 90 (Thimet).

Rate: 11 oz. 20G, 20-G/1000 ft. row (min. 24" spacing)

Pest:
leafhoppers
southern corn rootworm

Remarks: Apply at pegging. Distribute granules as a band over fruiting zone at pegging time. Work into top few inches of soil immediately. Do not graze or feed treated hay or foliage.
Days To Harvest: 90 (Thimet).

Signal Word/Toxicity Class: Danger-Poison/I.
REI: 48 hr.

PROPARGITE

Miticide formulated as emulsifiable concentrate or wettable powder. Used to control spider mites on peanuts. Toxic to fish.

Comite or Comite II *(Uniroyal)*

Rate: 1.64 lb. ai/A
2 pt./A
1.845 lb. ai/A
2.25 pt. II/A

Pest:
twospotted spider mite

Remarks: Apply in a minimum 20 gal./A by ground or 5 gal./A by air when mites first appear. Do not apply more than once per year. Do not apply through any type of irrigation system. Do not graze or feed livestock on treated areas or cut treated forage for hay.
Days To Harvest: 14.

Signal Word/Toxicity Class: Danger/I.
REI: 7 days. After 48 hr., workers can enter treated areas to perform hand labor tasks if they wear early-entry personal protective equipment.

Omite 30W or 30WSB *(Uniroyal)*

Rate: 0.9-1.5 lb. ai/A
3-5 lb. 30W, 30WSB/A in a minimum 20 gal. water

Pest:
twospotted spider mite

Remarks: Apply in sufficient water to provide thorough coverage. Do not apply more than 2 applications per year. Do not apply through any type of irrigation system.
Days To Harvest: 14.

Signal Word/Toxicity Class: Danger/I.
REI: 7 days. After 48 hr., workers can enter treated areas to perform hand labor tasks if they wear early-entry personal protective equipment.

PYRELLIN (Premix)

Premix containing pyrethrins and rotenone. May be combined with other insecticides to enhance knockdown or as an insect "exciter" to help facilitate contact with other spray residues.

Pyrellin E.C. (0.042 lb./gal. pyrethrins & 0.035 lb./gal. rotenone) *(Webb Wright)*

Rate: 1-2 pt. E.C./A

Pest:
aphids
bean beetles
caterpillars
loopers
mites
thrips

Remarks: Apply as needed to maintain control.
Days To Harvest: 0.

Signal Word/Toxicity Class: Caution/III.
REI: 12 hr.

SAF-T-SIDE — see petroleum oil

Sevin — see carbaryl

NOTICE The information on these pages is for preliminary planning — not a guide for use. Be sure to follow the manufacturer's directions, notwithstanding information contained here.	For personal protective equipment and EPA registration numbers, please turn to page 39.

SULFUR

Microthiol and Thiolux are micronized sulfur.

◆ Kumulus DF *(BASF)*
Thiolux (DF) *(Platte)*
Rate: 2.4-4 lb. ai/A
3-5 lb. DF/A
Pest:
red spider mite
Remarks: Apply at early leaf stage; repeat at 10- to 14-day intervals with or without other pesticides immediately after a rain. Do not apply if temperatures during or within 3 days after application are expected to exceed 90°F. Do not apply through any type of irrigation system.

- - - - - - - - - - - - - - - - -

Signal Word/Toxicity Class: Caution.
REI: 24 hr.

Microthiol Special (80W) *(Elf Atochem)*
Rate: 4-6.4 lb. ai/A
5-8 lb. 80W/A
Pest:
red spider mite twospotted spider mite
Remarks: Apply at early leaf stage; repeat as needed. Do not chemigate in California.

- - - - - - - - - - - - - - - - -

Signal Word/Toxicity Class: Caution.
REI: 24 hr.

Sumulsion — see petroleum oil

TEMIK

Carbamate insecticide/nematicide. Applied on peanuts at planting, in-furrow, or in a band for thrips and nematode control. Some states allow split applications for spider mite and nematode control. Highly toxic. Restricted Use Pesticide.

Temik 15G (aldicarb) *(Rhone-Poulenc)*
Rate: 1-2 lb. ai/A
7-14 lb. 15G/A
Pest:
thrips

Rate: 2-3 lb. ai/A
14-20 lb. 15G/A
Pest:
nematodes
Remarks: At planting. For thrips, apply in 6-12" band and work into soil or cover with soil to a depth of 2-4"; plant seed in treated zone. In Southwest, use high rate only. For nematodes, apply in seed furrow and cover with soil. Do not exceed a total 20 lb./A. Do not make more than 1 application per crop, hog off treated fields, or feed hay or vines to livestock. Refer to label for environmental precautions and soil restrictions table.
Days To Harvest: 90.

- - - - - - - - - - - - - - - - -

Signal Word/Toxicity Class: Danger-Poison/I.
REI: 48 hr.

Thimet — see phorate

Thiolux — see sulfur

VYDATE

Systemic carbamate insecticide/nematicide. Highly toxic to humans, wildlife, bees, and fish. Restricted Use Pesticide.

Vydate L (oxamyl) *(DuPont)*
Rate: 1-3 lb. ai/A
4-12 pt. L/A in 10 gal. water
Pest:
lesion nematode, ring nematode, root-knot nematode (except javanese), sting nematode, thrips
Remarks: Soil treatment. Preplant: Apply in 7-12" band directly to soil and incorporate 3-5" into soil immediately after application. For severe infestations, use high rate and wider band. Plant peanuts within 24 hr. after application. At-planting: Apply in 7" band immediately after planting. Do not apply more than 20 pt./A per season.
State Restrictions: Not registered for use in California.

- - - - - - - - - - - - - - - - -

Rate: 0.5-1 lb. ai/A
2-4 pt. L/A in 20-40 gal. water
Pest:
lesion nematode, ring nematode, root-knot nematode (except javanese), sting nematode, thrips
Remarks: Foliar supplement. Use only after soil fumigation, preplant, or at-planting application of Vydate L or other contact nematicides. Apply 3 weeks after emergence; second application 3 weeks later. For best results, concentrate spray on row using 3 cone-type nozzles positioned over and to each side of row. Thorough coverage important.
State Restrictions: Not registered for use in California.

- - - - - - - - - - - - - - - - -

Signal Word/Toxicity Class: Danger-Poison/I.
REI: 48 hr.

Warrior — see *lambda*-cyhalothrin

Rice

CARBARYL

Carbamate insecticide. Extremely toxic to aquatic and estuarine invertebrates.

Clean Crop Sevin 5 Bait *(Platte)*
Rate: 1-2 lb. ai/A
20-40 lb. Bait/A
Pest:
ants, armyworms, cutworms, darkling beetle, field crickets, grasshoppers, sowbugs
Remarks: Apply with ground or air equipment. Do not use on rice fields in which crayfish or catfish farming is included in the cultural practice.
Days To Harvest: 14.

- - - - - - - - - - - - - - - - -

Signal Word/Toxicity Class: Caution/III.
REI: 12 hr.

Drexel Carbaryl 4L or 50W *(Drexel)*
Sevin 4F, XLR Plus, 80S, 80 WSP *(Rhone-Poulenc)*
Rate: 0.5-1.5 lb. ai/A
1-3 pt. 4L, 4F, XLR/A
0.66-1.875 lb. 80S, 80WSP/A
Pest:
grasshoppers
Remarks: Do not apply when foliage is wet or excessive humidity is expected in next 2 days.
Days To Harvest: 14.

- - - - - - - - - - - - - - - - -

Rate: 1-1.5 lb. ai/A
2-3 pt. 4L, 4F, XLR/A
2-3 lb. 50W/A
1.25-1.875 lb. 80S, 80WSP/A
Pest:
armyworms, chinch bug, stink bugs, tall armyworm
Remarks: Do not use on rice fields in which crayfish or catfish farming is included in cultural practice. Do not apply when foliage is wet or excessive humidity is expected in next 2 days. Do not apply propanil within 15 days before or after carbaryl application.
Days To Harvest: 14.
State Restrictions: For use in Mississippi Delta, Texas.

- - - - - - - - - - - - - - - - -

Rate: 2 lb. ai/A
4 pt. 4L, 4F, XLR/A
4 lb. 50W/A
2.5 lb. 80S, 80WSP/A
Pest:
armyworms, leafhoppers, tadpole shrimp

◆ - **new product** • ai-active ingredient • DF-dry flowable • E/EC-emulsifiable concentrate • EW-emulsion in water
F/FL-flowable • G-granule • L/LC-liquid • SP/WSP-soluble packet • W/WP-wettable powder • WSB-water soluble bag

Remarks: For optimum tadpole shrimp control, apply to water when pests first appear. Do not apply when foliage is wet or excessive humidity is expected in next 2 days.
Days To Harvest: 14.
State Restrictions: For use in California.

Signal Word/Toxicity Class: Caution (4F, 4L, XLR); Warning/II (50W, 80S, 80WSP).
REI: 12 hr.

Declare — see methyl parathion

Diatect — see pyrethrins

Fyfanon — see malathion

MALATHION

Clean Crop Malathion 57EC *(Platte)*
Drexel Malathion 5EC *(Drexel)*
Fyfanon (5EC) *(Helena)*
◆ **Micro Flo Malathion 5EC** *(Micro Flo)*
Prentox 5 lb. Malathion Spray (5EC) *(Prentiss)*

Rate: 0.63-1 lb. ai/A
1-1.5 pt. 5EC, 57EC/A
Pest:
rice stink bug

Rate: 1.6 lb. ai/A
2.5 pt. 5EC, 57EC/A
Pest:
rice leafminer

Remarks: For stink bugs, apply early milk and dough stage of rice. For leafminers, apply when first rice blades appear on water surface; repeat if necessary. Do not apply propanil within 15 days of malathion application.
Days To Harvest: 7.

Signal Word/Toxicity Class: Caution (Clean Crop, Fyfanon); Warning/II (Drexel, Micro Flo, Prentox).
REI: 12 hr.

◆ **Clean Crop malathion 8EC** *(Platte)*
Rate: 0.63 lb. ai/A
10 oz. 8EC/A
Pest:
grasshoppers rice stink bug

Rate: 1.25 lb. ai/A
1.25 pt. 8EC/A
Pest:
rice leafminers

Rate: 2 lb. ai/A
2 pt. 8EC/A
Pest:
tadpole shrimp

Remarks: Restrict water spill from treated fields for 48 hr. after application.
Days To Harvest: 7.
State Restrictions: For use on tadpole shrimp in California.

Signal Word/Toxicity Class: Caution.
REI: 12 hr.

◆ **Drexel Malathion ULV** *(Drexel)*
Fyfanon ULV *(Cheminova)*
◆ **Micro Flo Malathion ULV** *(Micro Flo)*
Rate: 8 fl. oz. ULV/A
Pest:
grasshoppers rice stink bug

Remarks: For rice stink bug, apply by aircraft during early milk and dough stage of growing rice.
Days To Harvest: 7.
State Restrictions: For use on rice stink bug in Louisiana and Texas.

Signal Word/Toxicity Class: Caution.
REI: 12 hr.

Fyfanon (8E) *(Helena)*
Rate: 1 lb. ai/A
1 pt. 8E/A
Pest:
stink bugs

Rate: 1.5 lb. ai/A
1.5 pt. 8E/A
Pest:
leafminers

Remarks: For stink bugs, apply by aircraft during early milk and dough stage of growing rice. For leafminers, apply when first rice blades appear on water surface; repeat if necessary.
Days To Harvest: 7.

Signal Word/Toxicity Class: Caution.
REI: 12 hr.

Gowan Malathion 8 *(Gowan)*
Rate: 1.5 lb. ai/A
1.5 pt. 8E/A
Pest:
rice leafminer rice stink bug

Remarks: Do not apply propanil within 15 days of malathion application.
Days To Harvest: 7.

Signal Word/Toxicity Class: Caution.
REI: 12 hr.

◆ **Wilbur-Ellis Malathion 8 Spray** *(Wilbur-Ellis)*
Rate: 1.5 lb. ai/A
1.5 pt. 8/A
Pest:
rice leafminer

Remarks: Apply when first rice plants appear on surface of water.
Days To Harvest: 7.

Signal Word/Toxicity Class: Caution
REI: 12 hr.

METHYL PARATHION

Highly toxic organophosphate insecticide. Penncap-M is a microencapsulated formulation. Restricted Use Pesticide due to residual effects to avian species and hazard to bees.

Cheminova Methyl 4EC *(Cheminova)*
Rate: 0.5-0.75 lb. ai/A
1-1.5 pt 4EC/A
Pest:
grasshoppers rice stink bug

Remarks: Apply when insects appear; repeat as needed. Apply high rate for grasshoppers. Do not apply within 14 days after propanil application.
Days To Harvest: 15.

Signal Word/Toxicity Class: Danger-Poison/I.
REI: 48 hr.; 72 hr. if average rainfall less than 25" per year.

◆ **Declare (4EC)** *(Griffin)*
Helena 4 lb. Methyl Parathion (4EC) *(Helena)*
Rate: 0.25-0.5 lb. ai/A
0.5-1 pt. 4EC/A
Pest:
rice stink bug (Helena)

Rate: 0.5-0.75 lb. ai/A
1-1.5 pt. 4EC/A
Pest:
armyworm rice stink bug (Declare)

Rate: 0.5 lb. ai/A
1 pt. 4EC/A
Pest:
grasshoppers

Remarks: Do not apply where crabs, crayfish, and shrimp are important resources; they may be killed.
Days To Harvest: 15.

Rate: 0.75 lb. ai/A
1.5 pt. 4EC/A
Pest:
rice leafminer tadpole shrimp

6 Other Field Crops

NOTICE The information on these pages is for preliminary planning — not a guide for use. Be sure to follow the manufacturer's directions, notwithstanding information contained here. For personal protective equipment and EPA registration numbers, please turn to page 39.

Remarks: Apply after planting by air in 5-10 gal. water at first sign of infestation. Restrict spill from rice fields for 3 days. Do not use within 14 days of propanil application.
Days To Harvest: 15.
State Restrictions: For use in California only.

Signal Word/Toxicity Class: Danger-Poison/I.
REI: 48 hr.

Penncap-M *(Elf Atochem)*
Rate: 0.25-0.5 lb. ai/A
1-2 pt./A
Pest:
stink bugs

Rate: 0.25-0.75 lb. ai/A
1-3 pt./A
Pest:
grasshoppers
Remarks: For grasshoppers, use higher rate if majority of insects are large or weather is cool. Do not apply within 14 days of propanil application. Do not chemigate in California.
Days To Harvest: 15.

Signal Word/Toxicity Class: Warning/II.
REI: 48 hr.

Wilbur-Ellis Methyl Parathion 5 Spray *(Wilbur-Ellis)*
Rate: 0.2-0.63 lb. ai/A
0.33-1 pt. 5/A
Pest:
rice leafminer tadpole shrimp
Remarks: Hold water in check for 3 days after application. Do not apply more than 1 pt./A.
Days To Harvest: 15.

Signal Word/Toxicity Class: Danger-Poison/I.
REI: 48 hr.

Penncap-M — see methyl parathion

PYRELLIN (Premix)

Pyrellin E.C. (0.042 lb./gal. pyrethrins & 0.035 lb./gal. rotenone) *(Webb Wright)*
Rate: 1-2 pt. E.C./A
Pest: Including but not limited to:
armyworm chinch bug thrips
Remarks: Apply at 7-day intervals or less; repeat as needed.
Days To Harvest: 0.

Signal Word/Toxicity Class: Caution/III.
REI: 12 hr.

Pyrenone — see pyrethrins

PYRETHRINS

Contains piperonyl butoxide as a synergist. May be combined in tank mix with other insecticides for quicker and more complete kill where insect resistance may be a problem, and as an "exciter" to flush insects out of hiding and into contact with other spray residues.

Diatect Multipurpose Insecticide II *(Diatect)*
Pyrenone (0.5EC) *(AgrEvo Environmental Health)*
◆ **Py-Rin 60-6 EC** *(Wilbur-Ellis)*
Rate: 1-6 lb. II/A
1-12 fl. oz. 0.5EC/A
2-16 oz. 60-6 EC/A
Pest: Including but not limited to:
armyworm chinch bug stink bugs
Remarks: Apply in sufficient water for thorough coverage; repeat as needed. Apply Diatect as dust or wettable powder.
Days To Harvest: 0.

Signal Word/Toxicity Class: Caution.
REI: 12 hr.

Py-Rin — see pyrethrins

Sevin — see carbaryl

Safflower

DIBROM

Short residual contact insecticide for kill of insects and mites present at the time of application.

Dibrom 8 (8E) (naled) *(Amvac)*
Rate: 0.75 lb. ai/A
0.75 pt. 8E/A
Pest:
aphids Lygus bugs thrips
Remarks: Apply at first sign of insects; repeat at 7-day intervals as needed. Apply in 30-250 gal. water/A by ground; 3-10 gal. spray/A by air. Do not use with highly alkaline materials such as lime or Bordeaux. Do not apply through any type of irrigation system. Do not apply more than 2.25 pt./A per season.
Days To Harvest: 30.
State Restrictions: For use in Arizona, California.

Signal Word/Toxicity Class: Danger/I.
REI: 48 hr.

Digon — see dimethoate

Dimate — see dimethoate

DIMETHOATE

Organophosphate insecticide toxic to wildlife and aquatic invertebrates.

Cheminova Dimethoate 4E *(Cheminova)*
Clean Crop Dimethoate 400 *(Platte)*
Dimate 4E *(Terra)*
Drexel Dimethoate 2.67 or 4EC *(Drexel)*
Gowan Dimethoate E267 *(Gowan)*
Helena Dimethoate 4E *(Helena)*
Micro Flo Dimethoate 267 or 4E *(Micro Flo)*
Wilbur-Ellis Dimethoate 267 or Digon 400 *(Wilbur-Ellis)*
Rate: 0.25-0.5 lb. ai/A
0.75-1.5 pt. 2.67, 267/A
0.5-1 pt. 4E, 4EC/A
0.67 lb. ai/A
2 pt. E267/A
Pest:
aphids Lygus bugs thrips
leafhoppers plant bugs
Remarks: Apply when insects or their damage appears; repeat at 14-day intervals. Make no more than 2 applications per season at higher rate.
Days To Harvest: 14.
State Restrictions: For use in Arizona, California.

Signal Word/Toxicity Class: Warning/II.
REI: 48 hr.

ENDOSULFAN

Drexel Endosulfan 3EC or 50WP *(Drexel)*
Gowan Endosulfan 3EC or 50WSB *(Gowan)*
◆ **Micro Flo Endosulfan 3EC or 50WP** *(Micro Flo)*
Phaser 3EC or 50WSB *(AgrEvo USA)*
Thiodan 3EC or 50WP *(FMC)*
Rate: 1 lb. ai/A
1.33 qt. 3EC/A
2 lb. 50WP, 50WSB/A
Pest:
green peach aphid whiteflies

◆ - **new product** • ai-active ingredient • DF-dry flowable • E/EC-emulsifiable concentrate • EW-emulsion in water
F/FL-flowable • G-granule • L/LC-liquid • SP/WSP-soluble packet • W/WP-wettable powder • WSB-water soluble bag

Remarks: Apply when insects appear or feeding is noticed. Do not apply after flower heads open. Do not exceed 2 lb. ai/A per year or 2 applications per year. Do not apply through any type of irrigation system.

Signal Word/Toxicity Class: Danger-Poison/I.
REI: 24 hr.

Fyfanon — see malathion

LINDANE

Lindane 30C contains toxic inert Rhodamine B. Toxic to fish, birds, and wildlife.

Lindane 30C *(Gustafson)*
◆ **Lindane ST 40** *(AGSCO)*
Rate: 1.4 fl. oz. 30C/100 lb. seed
1 fl. oz. 40/100 lb. seed
Pest:
seedcorn maggots wireworms
Remarks: Do not use treated seed for food, feed, or oil purposes. Exposed treated seed may be hazardous to birds and other wildlife. Dispose of excess seed by burial away from bodies of water.

Signal Word/Toxicity Class: Warning/II.
REI: 12 hr.

MALATHION

Organophosphate insecticide. Toxic to fish and bees.

Clean Crop Malathion 57EC *(Platte)*
Rate: 1-1.25 lb. ai/A
1.5-2 pt. 57EC/A
Pest:
aphids grasshoppers Lygus bugs
Remarks: Apply when insects first appear and repeat as necessary.
Days To Harvest: 3.

Signal Word/Toxicity Class: Caution/III.
REI: 12 hr.

Fyfanon ULV *(Cheminova)*
◆ **Micro Flo Malathion ULV** *(Micro Flo)*
Rate: 0.5 pt. ULV/A
Pest:
grasshoppers Lygus bugs
Remarks: None.
Days To Harvest: Harvesting seeds-3.

Signal Word/Toxicity Class: Caution.
REI: 12 hr.

Phaser — see endosulfan

SUPRACIDE

Restricted Use Pesticide.

Supracide 25WP (methidathion) *(Gowan)*
Rate: 0.5 lb. ai/A
2 lb. 25WP/A
Pest:
Lygus bugs thrips
Remarks: Apply in minimum 20 gal. water by ground; 5 gal. water by air. Apply at bud stage, 12-14 days before onset of bloom, or as infestations develop; repeat at 7- to 14-day intervals or as needed. Do not make more than 3 applications per year. Do not apply through any type of irrigation system. Do not graze treated areas.
Days To Harvest: 28.

Signal Word/Toxicity Class: Warning/II.
REI: 48 hr.

Thiodan — see endosulfan

Sorghum

Adios — see carbaryl

BAYTHROID

Synthetic pyrethroid formulated as emulsifiable concentrate. Toxic to fish and aquatic organisms. Restricted Use Pesticide.

Baythroid (2EC) (cyfluthrin) *(Bayer)*
Rate: 0.015-0.044 lb. ai/A
1-2.8 fl. oz. 2EC/A
Pest:
chinch bug, cutworms, fall armyworm, false chinch bug, grasshoppers, headworm, sorghum midge, sorghum webworm, stink bugs
Remarks: Apply by air or ground in sufficient water for complete coverage. Allow 10 days between applications. Up to 3 applications may be made per crop season. If 1-2 applications are made, green forage may be fed or grazed on day of treatment; 3 applications, allow at least 15 days between last application and grazing. Refer to label for spray drift precautions.
Days To Harvest: 14.

Signal Word/Toxicity Class: Danger/I.
REI: 12 hr.

CARBARYL

Adios (13%) *(Micro Flo)*
Rate: 0.03-0.1 lb. ai/A
0.25-0.75 lb. 13%/A
Pest:
cucumber beetle
Remarks: Field, grain sorghum. Not for use on sweet sorghum. Do not apply through any type of irrigation system. Do not apply more than 3.75 lb./A per season.
Days To Harvest: Forage-0; grain-21.
State Restrictions: Not for use in California.

Signal Word/Toxicity Class: Caution/III.
REI: 12 hr.

Clean Crop Sevin 5 Bait *(Platte)*
Rate: 1-2 lb. ai/A
20-40 lb. Bait/A
Pest:
ants, armyworms, cutworms, darkling beetle, field crickets, grasshoppers, sowbugs
Remarks: Forage, milo, grain, sweet sorghum, and hybrids. Apply with ground or air equipment.
Days To Harvest: 0; grain-21.

Signal Word/Toxicity Class: Caution/III.
REI: 12 hr.

Sevin 4F, XLR Plus, 80S, 80WSP *(Rhone-Poulenc)*
Rate: 0.5-2 lb. ai/A
1-4 qt. 4F or XLR/A
0.66-2.5 lb. 80S, 80WSP/A
Pest:
armyworms, chinch bug, corn earworm, cutworms, fall armyworm, grasshoppers, sorghum midge, southwestern corn borer, stink bugs
Remarks: Milo, grain, sweet sorghum. Rate varies by pest species; refer to label. For sorghum midge, treat when 25-30% of heads have emerged from boot and are in bloom; repeat 3-5 days later if adults are still active (third application may be necessary in late planted sorghum or if midges are abundant). For chinch bug, use high gallonage ground application directed at base of plants. Do not apply when foliage is wet or excessive humidity is expected in next 2 days.
Days To Harvest: Forage-0; grain-21.

Signal Word/Toxicity Class: Caution (4F, XLR); Warning/II (80S, 80WSP).
REI: 12 hr.

6 Other Field Crops

NOTICE The information on these pages is for preliminary planning — not a guide for use. Be sure to follow the manufacturer's directions, notwithstanding information contained here.

For personal protective equipment and EPA registration numbers, please turn to page 39.

Chlorfos — see chlorpyrifos

CHLORPYRIFOS

◆ **Chlorfos 15G** *(Griffin)*
Lorsban 15G *(Dow AgroSciences)*
◆ **Micro Flo Chlorpyrifos 15G** *(Micro Flo)*

Rate: 4-12 oz. 15G/1000 ft. row

Pest:

chinch bug*	cutworms	lesser cornstalk borer
corn rootworm	fire ants*	

* *Suppression*

Remarks: Apply in 6-8" band over row and lightly incorporate by using press wheel or drag chain. Rate varies by pest species; refer to label. Necessary to incorporate, especially at lower rates.

Signal Word/Toxicity Class: Caution.
REI: 12 hr.

Lorsban-4E *(Dow AgroSciences)*
◆ **Micro Flo Chlorpyrifos 4E** *(Micro Flo)*
Nufos 4E *(Cheminova)*

Rate: 0.25-1 lb. ai/A
0.5-2 pt. 4E/A

Pest:

aphids	European corn borer	southwestern corn borer
armyworms	grasshoppers	
chinch bug	greenbug	webworms
corn earworm	lesser cornstalk borer	yellow sugarcane aphid
cutworms	sorghum midge	

Remarks: Not for use on sweet sorghum. Rate varies by pest species; refer to label. For midge control, apply when 30-50% of seedheads are in bloom; repeat at 3 day intervals if necessary. Do not apply to drought-stressed sorghum within 3 days of irrigation or rain except if applied in irrigation water. Do not apply more than 1 pt./A to seed sorghum if risk of crop injury is unacceptable.

Days To Harvest: Grain, forage, fodder, hay, silage: 1 pt.-30; over 1 pt.-60.

Signal Word/Toxicity Class: Warning/II.
REI: 24 hr.

Comite — see propargite

COUNTER

Counter CR Lock'n Load formulation is a closed handling system. Extremely toxic to fish and wildlife. Restricted Use Pesticide due to acute oral and dermal toxicity.

Counter 15G or CR (terbufos) *(American Cyanamid)*

Rate: Banded (at planting):
8 oz. 15G/1000 ft. row (min. 20" row)
6 oz. CR/1000 ft. row (min. 20" row)

Pest:

chinch bug	greenbug	white grubs
corn leaf aphid	nematodes	wireworms
corn rootworm		

Rate: Knifed-in (at bedding, at planting):
8-16 oz. 15G/1000 ft. row (min. 20" row)
6-12 oz. CR/1000 ft. row (min. 20" row)

Pest:

corn leaf aphid	greenbug

Remarks: Grain sorghum. Apply at planting. Knifed-in application: Drill granules 1-4" directly below seed or 1-4" below seed and up to 5" to side of seed. Banded application: Apply in 5-7" band directly behind planter shoe in front of press wheel and lightly incorporate with drag chains or tines. Do not place granules in direct contact with seed. Only 1 application per year may be used.

Days To Harvest: Grazing/forage-50; fodder/grain-100.

State Restrictions: Do not use banded application for aphid control in New Mexico, west Texas, and panhandle of Oklahoma. Refer to labels for chinch bug restrictions.

Signal Word/Toxicity Class: Danger-Poison/I.
REI: 48 hr.

lambda-CYHALOTHRIN

Synthetic pyrethroid insecticide with broad-spectrum activity, formulated as an emulsifiable concentrate. Restricted Use Pesticide due to toxicity to fish and aquatic organisms.

Karate or Warrior (1EC) *(ZENECA)*

Rate: 0.015-0.03 lb. ai/A
1.92-3.84 fl. oz. 1EC/A

Pest:

armyworms	European corn borer	sorghum midge
beet armyworm (1st & 2nd instar)	fall armyworm (1st & 2nd instar)	southwestern corn borer
chinch bug	flea beetles	stink bugs
corn earworm	grasshoppers	webworms
cutworms	lesser cornstalk borer	

Remarks: Grain sorghum. Apply as required by scouting, usually at 5-day intervals or more. For corn borers, apply before larvae bores into plant stalk. For sorghum midge, apply when 25% of sorghum heads have emerged and are in tip bloom. For chinch bug, direct spray to base of plant when bugs migrate from small grains or grass weeds to small sorghum; repeat at 3-5 day intervals if needed. Do not apply more than 0.64 pt./A per season or more than 0.48 pt./A per season after crop emergence. Do not graze treated areas or harvest for fodder, silage, or hay. Refer to label for spray drift precautions and state and county restrictions.

Days To Harvest: 30.

Signal Word/Toxicity Class: Danger/I.
REI: 24 hr.

Diatect — see pyrethrins

Digon — see dimethoate

Dimate — see dimethoate

DIMETHOATE

Cheminova Dimethoate 4E *(Cheminova)*
Clean Crop Dimethoate 400 *(Platte)*
Dimate 4E *(Terra)*
Drexel Dimethoate 4EC *(Drexel)*
Gowan Dimethoate E267 *(Gowan)*
Helena Dimethoate 4EC *(Helena)*
Micro Flo Dimethoate 267 or 4E *(Micro Flo)*
Wilbur-Ellis Dimethoate 267 or Digon 400 *(Wilbur-Ellis)*

Rate: 0.1-0.5 lb. ai/A
0.33-1.5 pt. 267, E267/A
0.25-1 pt. 4E, 4EC, 400/A

Pest:

aphids	grasshoppers	spider mites
Banks grass mite (except Trans-Pecos area of TX)	sorghum midge	

Remarks: Milo sorghum. Apply when insects are present. Do not apply more than 3 applications per season. Do not apply after heading.

Days To Harvest: Feeding or grazing-28.

Signal Word/Toxicity Class: Warning/II.
REI: 48 hr.

Drexel Dimethoate 2.67 EC *(Drexel)*

Rate: 0.25-0.5 lb. ai/A
0.75-1.5 pt. 2.67/A

Pest:

aphids	greenbug

Remarks: Apply when insects or their damage appear. Do not apply more than 3 applications per season. Do not apply after heading.

Days To Harvest: Feeding or grazing-28.

Signal Word/Toxicity Class: Warning/II.
REI: 48 hr.

◆ - **new product** • ai-active ingredient • DF-dry flowable • E/EC-emulsifiable concentrate • EW-emulsion in water
F/FL-flowable • G-granule • L/LC-liquid • SP/WSP-soluble packet • W/WP-wettable powder • WSB-water soluble bag

DI-SYSTON

Restricted Use Pesticide.

Di-Syston 8 (disulfoton) *(Bayer)*

Rate: At planting (furrow):
0.75-1 lb. ai/A
12-16 fl. oz. 8/A (40" row)
At planting (band), postplant (sidedress):
1 lb. ai/A
1 pt. 8/A (40" row)

Pest:
aphids greenbug

Rate: Postplant (foliar):
0.25-0.5 lb. ai/A
4-8 fl. oz. 8/A

Pest:
aphids
Banks grass mite (OK, TX except Trans-Pecos area)
greenbug
sorghum midge (AR, LA, OK, TX)

Remarks: Use high rate for Banks grass mite. Furrow: Apply as water emulsion or in liquid fertilizer in seed furrow approximately 1/2" above seed. Do not apply directly on seed. Band: Apply as water emulsion or in liquid fertilizer in 4-6" band directly behind planter shoe in front of the press wheel. Do not apply directly on seed. Sidedress: Apply as water emulsion or in liquid fertilizer injected into the row on both sides of the furrow at any time up to boot stage. Foliar: Apply in sufficient water for thorough coverage. For aerial application, use not less than 1 gal. water/A; 5 gal. water/A with ground equipment. For sorghum midge, apply when 50% of seed heads have emerged from boot; repeat 3-5 days later. A maximum of 3 foliar applications may be made at rates of 8 fl. oz./A or less. Due to greenbug resistance in certain areas, Di-Syston used alone may not provide satisfactory control. Consult agricultural advisor or extension service for recommendations.

Days To Harvest: Varies from 7-60 days depending on number and type of applications made; refer to labels.

- - - - - - - - - - - - - - - - - -

Signal Word/Toxicity Class: Danger/I.
REI: 48 hr.

Di-Syston 15% (disulfoton) *(Bayer)*

Rate: At planting (in-furrow):
0.75-1 lb. ai/A
5-6.7 lb. 15%/A (40" row)
At planting (band), postplant (whorl):
1 lb. ai/A
6.7 lb. 15%/A (40" row)

Pest:
aphids greenbugs

Remarks: In-furrow: Place granules in seed furrow approximately 1/2" above seed. Do not apply directly on seed. Band: Place granules in 4-6" band directly behind planter shoe in front of the press wheel. Do not apply directly on seed. Whorl: Apply by broadcasting into whorl. Do not apply more than once per season. Due to greenbug resistance in certain areas, Di-Syston alone may not provide satisfactory control. Consult agricultural advisor or Extension Service for recommendations.

Days To Harvest: Varies from 7-60 days depending on number and type of applications made; refer to labels.

- - - - - - - - - - - - - - - - - -

Signal Word/Toxicity Class: Danger/I.
REI: 48 hr.

Enhance Plus — see lindane premixes

Fyfanon — see malathion

Gaucho — see imidacloprid

Grain Guard Plus — see lindane premixes

IMIDACLOPRID

Gaucho 75 ST, 480 *(Gustafson)*

Rate: 5.3 oz. 75 ST/100 lb. seed
Slurry treatment:
8 fl. oz. 480/100 lb. seed

Pest:
aphids
chinch bug
corn leaf aphid
English grain aphid
fire ants
greenbug
wireworm
yellow sugarcane aphid

Remarks: Early-season protection. Apply prior to planting; ensure thorough coverage. Do not graze or feed livestock on treated areas for 45 days after planting. Treated seed must not be used for or mixed with food or animal feed or processed for oil.

Crop Rotation: Any labeled crop or any crop for which an ai tolerance exists; 12-month plant-back interval for other crops.

- - - - - - - - - - - - - - - - - -

Signal Word/Toxicity Class: Caution.
REI: 12 hr.

Karate — see *lambda*-cyhalothrin

Kumulus — see sulfur

LANNATE

Restricted Use Pesticide.

Lannate SP or LV (methomyl) *(DuPont)*

Rate: 0.225-0.45 lb. ai/A
0.25-0.5 lb. SP/A
0.75-1.5 pt. LV/A

Pest:
armyworms
beet armyworm
corn earworm
fall armyworm
sorghum midge
sorghum webworm

Remarks: Apply when insects first appear in significant numbers; repeat at 5- to 7-day intervals or as needed. For sorghum midge, apply at 50% bloom; repeat 3-5 days later if necessary. Apply high rate for sorghum webworm. Do not apply more than 0.9 lb. ai/A per crop or make more than 2 applications per crop.

Days To Harvest: 14; livestock grazing/feeding-14.

- - - - - - - - - - - - - - - - - -

Signal Word/Toxicity Class: Danger-Poison/I.
REI: 48 hr.

LINDANE (Premixes)

Seed treatment. Grain Guard Plus and Sorghum Guard contain graphite.

Enhance Plus (18.75% lindane & 35% maneb & 20% carboxin) *(Trace)*
Grain Guard Plus (18.75% lindane & 50% mancozeb) *(Trace)*
Sorghum Guard (16.6% lindane & 32.75% captan) *(Trace)*

Rate: 3 oz./bu. seed

Pest:
wireworms

Remarks: Apply in drill box for Enhance Plus; planter or drill box for Grain Guard Plus and Sorghum Guard. Treat only those seeds needed for immediate use. Do not use treated seed for food, feed, or oil purposes. Do not store excess treated seeds beyond planting time. Do not graze or feed livestock on treated areas for 6 weeks after planting.

- - - - - - - - - - - - - - - - - -

Signal Word/Toxicity Class: Danger/I.
REI: 12 hr.

Lorsban — see chlorpyrifos

MALATHION

Clean Crop Malathion 57EC *(Platte)*

Rate: 1 lb. ai/A
1.5 pt. 57EC/A

Pest:
greenbug

6 Other Field Crops

NOTICE	The information on these pages is for preliminary planning — not a guide for use. Be sure to follow the manufacturer's directions, notwithstanding information contained here.	For personal protective equipment and EPA registration numbers, please turn to page 39.

Remarks: Grain sorghum. Apply full coverage application; repeat as needed.
Days To Harvest: Harvest or forage-7.

Signal Word/Toxicity Class: Caution/III.
REI: 12 hr.

Fyfanon ULV *(Cheminova)*
◆ **Micro Flo Malathion ULV** *(Micro Flo)*
Rate: 0.5-0.75 pt. ULV/A
Pest:
grasshoppers sorghum midge
Remarks: Grain sorghum. Apply low rate for grasshoppers. For sorghum midge, apply during bloom stage.
Days To Harvest: 7.

Signal Word/Toxicity Class: Caution.
REI: 12 hr.

METHYL PARATHION

Wilbur-Ellis Methyl Parathion 5 Spray *(Wilbur-Ellis)*
Rate: 0.94 lb. ai/A
1.5 pt. 5/A
Pest:
greenbug
Remarks: Milo. Leaf injury may occur on some sorghum varieties. Do not apply through any type of irrigation system.
Days To Harvest: 21.
State Restrictions: For use in California.

Signal Word/Toxicity Class: Danger-Poison/I.
REI: 48 hr.

Wilbur-Ellis Methyl Parathion 7.5 Spray *(Wilbur-Ellis)*
Rate: 0.47-0.94 lb. ai/A
0.5-1 pt. 7.5/A
Pest:
corn leaf aphid sorghum midge spider mites
greenbug
Remarks: Apply when 90% of heads have completely emerged from boot. Leaf injury may occur on some hybrid varieties. Do not use wetting agents as severe injury may occur.
Days To Harvest: 21.

Signal Word/Toxicity Class: Danger-Poison/I.
REI: 48 hr.; 72 hr. if annual rainfall is less than 25" per year.

Microthiol Special — see sulfur

Nufos — see chlorpyrifos

PHORATE

Restricted Use Pesticide.

Clean Crop Phorate 20G *(Platte)*
Riverside Phorate 20G *(Terra)*
Rate: 6 oz. 20G/1000 ft. row (min. 30" row)
Pest: At planting:
aphids (greenbug)
At cultivation (Riverside):
chinch bug (nymphs)
(CO, KS, NE)
Remarks: For aphids, place granules in 7" band over row or drill 1" below and 2" to side of seed. Do not place granules in direct contact with seed. For chinch bug, apply granules at time of cultivation in a band at base of plants or over row just ahead of cultivator so as to cover granules with soil. Apply as soon as nymphs are observed at base of plants. Granules must be placed at infested region of plants and adequately covered with soil. Later infestations of chinch bugs may not be controlled. Do not make any applications after cultivation treatment. Do not feed foliage before grain harvest.

Rate: Postemergence-over the plant:
1 lb. ai/A
4.9 lb. 20G/A
Pest:
aphids Banks grass mite (except Trans-Pecos area of Texas) greenbug
Remarks: Broadcast granules into the whorls of plants by air or ground. Make only one application after plant emergence.
Days To Harvest: 28.

Signal Word/Toxicity Class: Danger-Poison/I.
REI: 48 hr.

Thimet 20-G *(American Cyanamid)*
Rate: 6 oz. 20-G/1000 ft. row (min. 30" row)
Pest: At planting:
aphids (greenbug)
At cultivation:
chinch bug (nymphs)
(CO, KS, NE)
Remarks: For aphids, place granules in 7" band over row or drill 1" below and 2" to side of seed. Do not place granules in direct contact with seed. Second application of granules may be made at cultivation. For chinch bug, apply granules at time of cultivation in a band at base of plants or over row just ahead of cultivator so as to cover granules with soil. Apply as soon as nymphs are observed at base of plants. Granules must be placed at infested region of plants and adequately covered with soil. Later infestations of chinch bugs may not be controlled. Do not make any applications after cultivation treatment. Do not feed foliage before grain harvest.
Days To Harvest: 30.

Signal Word/Toxicity Class: Danger-Poison/I.

PROPARGITE

Comite or Comite II *(Uniroyal)*
Rate: 1.2-1.64 lb. ai/A
1.5-2 pt./A
1.23-1.845 lb. ai/A
1.5-2.25 pt. II/A
Pest:
Banks grass mites twospotted spider mite
Remarks: Grain, silage sorghum. Apply when mites first appear in minimum 20 gal./A by ground; 5 gal./A by air. Apply no more than once per year. Do not apply through any type of irrigation system.
Days To Harvest: Comite: silage-30; grain-60; Comite II: grain-30.
State Restrictions: Comite II for use east of the Rockies.

Signal word/Toxicity Class: Danger/I.
REI: 7 days. After 48 hr., workers can enter treated areas to perform hand labor tasks if they wear early-entry personal protective equipment.

PYRELLIN (Premix)

Premix containing pyrethrins and rotenone. May be combined with other insecticides to enhance knockdown or as an insect "exciter" to help facilitate contact with other spray residues.

Pyrellin E.C. (0.042 lb./gal. pyrethrins & 0.035 lb./gal. rotenone) *(Webb Wright)*
Rate: 1-2 pt. E.C./A
Pest:
aphids chinch bug thrips
armyworms mites
Remarks: Apply in intervals of 7 days or less; repeat as needed.
Days To Harvest: 0.

Signal Word/Toxicity Class: Caution/III.
REI: 12 hr.

PYRETHRINS

Contains piperonyl butoxide as a synergist and silicon dioxide (diatomaceous earth).

Diatect Multipurpose Insecticide II *(Diatect)*
Rate: 1-6 lb. II/A
Pest: Including but not limited to:
aphids chinch bug mites
armyworms corn earworm thrips
caterpillars flea beetles
Remarks: Milo. Apply as dust or wettable powder. Apply in sufficient water for thorough coverage; repeat as needed.
Days To Harvest: 0.

Signal Word/Toxicity Class: Caution.
REI: 12 hr.

◆ - **new product** • ai-active ingredient • DF-dry flowable • E/EC-emulsifiable concentrate • EW-emulsion in water
F/FL-flowable • G-granule • L/LC-liquid • SP/WSP-soluble packet • W/WP-wettable powder • WSB-water soluble bag

Redball Lightning — see sulfur

Sevin — see carbaryl

Sorghum Guard — see lindane premixes

SULFUR

Microthiol and Thiolux are micronized sulfurs. Relatively non-toxic. Redball Lightning is a dustless granule formulated to breakdown into smaller particles when wetted by water to provide effectiveness of a powder with the handling ease of a granule.

◆ **Kumulus DF** *(BASF)*
Rate: 4.8-12 lb. ai/A
6-15 lb. DF/A
Pest:
Banks grass mite, Pacific mite, red spider mite, twospotted spider mite
Remarks: Do not apply if temperatures during or within 3 days after application are expected to exceed 90°F. Do not apply through any type of irrigation system.

Signal Word/Toxicity Class: Caution.
REI: 24 hr.

Microthiol Special (80W) *(Elf Atochem)*
Rate: 6.4-8 lb. ai/A
8-10 lb. 80W/A
Pest:
Atlantic mite, Pacific mite, red spider mite, twospotted spider mite (except CA)
Remarks: Apply when mites first appear; repeat as needed. Do not chemigatechemigate in California.

Signal Word/Toxicity Class: Caution.
REI: 24 hr.

◆ **Redball Lightning Degradable (G)** *(International Sulphur)*
Rate: 5-15 lb. G/A
Pest:
fleahoppers, mites, thrips
Remarks: Do not apply within 3 weeks of oil spray.
Days To Harvest: 0.

Signal Word/Toxicity Class: Caution.
REI: 24 hr.

THAT (6FL) or THAT Big 8 (8FL) *(Stoller)*
Rate: 9 lb. ai/A
12 pt. 6FL/A
8 lb. ai/A
8 pt. 8FL/A
Pest:
Banks grass mite
Remarks: Apply when mites appear on bottom 2 leaves. Can be tank mixed with other miticides.
State Restrictions: Not for use in California.

Signal Word/Toxicity Class: Caution.
REI: 24 hr.

Thiolux (DF) *(Platte)*
Rate: 4.8 lb. ai/A
6 lb. DF/A
Pest:
Banks grass mite, Pacific spider mite, red spider mite, twospotted spider mite
Remarks: Apply when mites first appear; repeat as necessary.

Signal Word/Toxicity Class: Caution.
REI: 24 hr.

TEMIK

Restricted Use Pesticide.

Temik 15G (aldicarb) *(Rhone-Poulenc)*
Rate: 1 lb. ai/A
7 lb. 15G/A
Pest:
chinch bug, greenbug, nematodes
Rate: Oklahoma, Texas:
0.5-1 lb. ai/A
3.5-7 lb. 15G/A
Pest:
nematodes
Remarks: Grain sorghum. Apply granules at planting in seed furrow and cover with soil. Do not apply more than 1 application per year or exceed a total of 7 lb./A. Do not feed green forage to livestock. Refer to label for environmental precautions and soil restrictions table.
Days To Harvest: 90.

Signal Word/Toxicity Class: Danger-Poison/I.
REI: 48 hr.

THAT or THAT Big 8 — see sulfur

Thimet — see phorate

Thiolux — see sulfur

Warrior — see lambda-cyhalothrin

Sugar Beets

435 or 455 Oil — see petroleum oil

AZADIRACHTIN

Extracted from seed of neem tree. Insect growth regulator that controls targeted insect larvae when they ingest or come in contact with it, by interfering with the insect's ability to molt and shutting off feeding behavior. Affected larvae slowly starve and die at next molt.

◆ **Azatin XL Plus** *(Thermo Trilogy)*
Rate: 5-21 oz. XL Plus/A
Pest:
aphids, armyworms, beet armyworm, beetles, caterpillars, cutworms, fall armyworm, flea beetles, green peach aphid, leafhoppers, leafminers, loopers, thrips, webworms, whiteflies
Remarks: Rate varies by species. Apply when pests first appear; repeat at 7-day intervals. When pest populations are high, use higher rates. Do not apply more that 21 fl. oz./A.
Days To Harvest: 0.

Signal Word/Toxicity Class: Caution.
REI: 4 hr.

Azatin — see azadirachtin

Bt 320 SULFUR 50 DUST (Premix)

Bt 320 Sulfur 50 Dust (0.064% Bt & 50% sulfur) *(Wilbur-Ellis)*
Rate: 35-50 lb. Dust/A
Pest:
Pacific mite, strawberry mite, twospotted spider mite

NOTICE The information on these pages is for preliminary planning — not a guide for use. Be sure to follow the manufacturer's directions, notwithstanding information contained here.

For personal protective equipment and EPA registration numbers, please turn to page 39.

Remarks: Do not mix or dilute with other products. Thorough coverage essential.

Signal Word/Toxicity Class: Caution.
REI: 24 hr.

CARBARYL

Clean Crop Sevin 5 Bait *(Platte)*
Rate: 1-2 lb. ai/A
20-40 lb. Bait/A
Pest:
ants, armyworms, cutworms, darkling beetle, field crickets, grasshoppers, sowbugs
Remarks: Apply with ground or air equipment.
Days To Harvest: 14.

Signal Word/Toxicity Class: Caution/III.
REI: 12 hr.

Drexel Carbaryl 4L or 50W *(Drexel)*
Sevin 4F, XLR Plus, 80S, 80WSP *(Rhone-Poulenc)*
Rate: 0.5-1.5 lb. ai/A
1-3 pt. 4L, 4F, XLR/A
2-4 lb. 50W/A
0.66-1.875 lb. 80S, 80WSP/A
Pest:
armyworms, beet leaf beetle, cutworms, fall armyworm, flea beetles, grasshoppers, leafhoppers, webworms
Remarks: Rate varies by pest species; refer to label. To avoid possible injury on tender foliage, do not apply when foliage is wet or excessive humidity is expected in the next 2 days.
Days To Harvest: 14.

Signal Word/Toxicity Class: Caution (4F, 4L, XLR); Warning/II (50W, 80S, 80WSP).
REI: 12 hr.

Gowan Sevin Bait *(Gowan)*
Rate: 2 lb. ai/A
40 lb. Bait/A
Pest:
armyworms, crickets, cutworms, darkling beetle, grasshoppers, sowbugs
Remarks: Apply when pests appear in damaging numbers; repeat at 7- to 14-day intervals if necessary. Do not use when bees are present.
Days To Harvest: 14.
State Restrictions: For use in Arizona and California.

Signal Word/Toxicity Class: Caution/III.
REI: 12 hr.

Chlorfos — see chlorpyrifos

CHLORPYRIFOS

◆ **Chlorfos 15G** *(Griffin)*
Lorsban 15G *(Dow AgroSciences)*
◆ **Micro Flo Chlorpyrifos 15G** *(Micro Flo)*
Rate: Band:
4.5-9 oz. 15G/1000 ft. row
Postplant:
6.5-9 oz. 15G/1000 ft. row
Pest:
sugarbeet root maggot

Rate: Band:
6.5-9 oz. 15G/1000 ft. row
Pest:
cutworms, wireworms
Remarks: Apply at planting in front of press wheel in 3-5" band and incorporate into top 1-2" soil. Can also be applied postemergence if incorporated 1/2-1"; refer to label for use directions. Make no more than 1 application per year.

Signal Word/Toxicity Class: Caution.
REI: 12 hr.

Lorsban-4E *(Dow AgroSciences)*
◆ **Micro Flo Chlorpyrifos 4E** *(Micro Flo)*
Rate: 0.75-1 lb. ai/A
1.5-2 pt. 4E/A
Pest:
beet armyworm, cutworms, fall armyworm, flea beetles, grasshoppers, spider mites, sugarbeet root maggot, webworms, yellowstriped armyworm
Remarks: Do not apply more than 8 pt./A or make more than 4 applications per season. Refer to label for use directions.
Days To Harvest: Roots, tops, grazing-30.

Signal Word/Toxicity Class: Warning/II.
REI: 24 hr.

COUNTER

Restricted Use Pesticide due to acute oral and dermal toxicity.

Counter 15G or CR (terbufos) *(American Cyanamid)*
Rate: Modified in-furrow or banded:
4-8 oz. 15G/1000 ft. row (min. 20" row)
3-6 oz. CR/1000 ft. row (min. 20" row)
Pest: At planting:
aphids*, sugarbeet leafhopper*, sugarbeet root maggot, white grubs, wireworms
Postemergence:
sugarbeet root maggot
Modified in-furrow only at 8 oz. rate

Rate: Banded:
8 oz. 15G/1000 ft. row (min. 20" row)
6 oz. CR/1000 ft. row (min. 20" row)
Pest: At planting:
cutworms (suppression)
Postemergence:
aphids, sugarbeet leafhopper, sugarbeet root aphid

Rate: Knifed-in application:
18 oz. 15G/1000 ft. row (min. 20" row)
12 oz. CR/1000 ft. row (min. 20" row)
Pest: At planting:
aphids, sugarbeet cyst nematode (suppression), sugarbeet leafhopper
Remarks: Banded: Apply in 5-7" band; lightly incorporate into soil. Modified in-furrow: Apply in furrow 2-3" behind seed drop zone after some soil has covered seed. Knifed-in: Drill granules 2" to side of seed and 2-4" below seed. At planting: Do not place granules in direct contact with seed or incorporate deeper than 2"; power incorporation may be used.

Signal Word/Toxicity Class: Danger-Poison/I.
REI: 48 hr.

DEADLINE

Packaged as Bullets, M-Ps (mini-pellets), and granules.

Deadline Bullets, M-Ps, Granules (4%) (metaldehyde) *(Pace)*
Rate: 0.4-1 lb. ai/A
10-40 lb. 4%/A
Pest:
slugs, snails
Remarks: Apply every 3-4 weeks during growing season, or as needed. Do not apply directly to edible portion of plant.

Signal Word/Toxicity Class: Warning/II (granule); Caution (bullets, M-P).
REI: 12 hr.

Declare — see methyl parathion

Diatect — see pyrethrins

◆ - **new product** • ai-active ingredient • DF-dry flowable • E/EC-emulsifiable concentrate • EW-emulsion in water
F/FL-flowable • G-granule • L/LC-liquid • SP/WSP-soluble packet • W/WP-wettable powder • WSB-water soluble bag

DIAZINON

Restricted Use Pesticide due to avian and aquatic toxicity.

D•z•n diazinon AG600 WBC, 50W *(Novartis)*
Drexel Diazinon Insecticide (4EC) *(Drexel)*
Gowan Diazinon 4E, 50WSB *(Gowan)*
Helena Diazinon AG500 *(Helena)*
Micro Flo Diazinon AG500, 50W *(Micro Flo)*
Prentox Diazinon AG500, 50W *(Prentiss)*
◆ **Wilbur-Ellis Diazinon 4 Spray (4E)** *(Wilbur-Ellis)*

Rate: 0.375-0.5 lb. ai/A
0.75-1 pt. 4E, 4EC, AG500/A
9.5-12.75 fl. oz. AG600/A
0.75-1 lb. 50W, 50WSB/A

Pest:
aphids; beet leafhopper; dipterous leafminer; grasshoppers (high rate)

Rate: 3-4 lb. ai/A
6-8 pt. 4E, 4EC, AG500/A
76.5-102 fl.oz. AG600/A
6-8 lb. 50WSB/A

Pest:
wireworms

Remarks: Apply when insects appear; repeat at 7-day intervals as needed. For wireworms, broadcast just before planting; incorporate into soil 4-6". Do not apply more than 5 times per season. Tops may be fed to livestock.

Signal Word/Toxicity Class: Caution.
REI: 24 hr.

Drexel D-264 14G *(Drexel)*
◆ **Micro Flo Diazinon 14G** *(Micro Flo)*
◆ **Wilbur-Ellis Diazinon 14G** *(Wilbur-Ellis)*

Rate: 1-2 lb. ai/A
7-14 lb. 14G/A
4.7-9.4 oz. 14G/1000 lin. ft. row

Pest:
sugarbeet root maggot

Remarks: Apply in 5-7" band (22" row spacing) incorporated 2-3" at planting. Refer to label for use directions and precautions.

Rate: 3-4 lb. ai/A
21-28 lb. 14G/A

Pest:
wireworms

Remarks: Broadcast and incorporate 4-8" into soil before planting. Refer to label for use directions and precautions.

Signal Word/Toxicity Class: Caution.
REI: 12 hr. (24 hr. for Micro Flo).

DIBROM

Short residual contact insecticide for kill of insects and mite adults and larvae present at time of application.

Dibrom 8 (8E) (naled) *(Amvac)*

Rate: 1 lb. ai/A
1 pt. 8E/A

Pest:
leafhoppers; spider mites

Remarks: Apply in 30-250 gal. water/A by ground; 1-5 gal. water/A by air. Do not use with highly alkaline materials such as lime or Bordeaux. Do not apply through any type of irrigation system. Do not apply more than 5 pt./A per season. Apply at 7-day intervals.
Days To Harvest: 2.

Signal Word/Toxicity Class: Danger/I.
REI: 48 hr.

D•z•n — see diazinon

ENDOSULFAN

Drexel Endosulfan 3EC or 50WP *(Drexel)*
Gowan Endosulfan 3EC or 50WSB *(Gowan)*
◆ **Micro Flo Endosulfan 3EC or 50WP** *(Micro Flo)*
Phaser 3EC or 50WSB *(AgrEvo USA)*
Thiodan 3EC or 50WP *(FMC)*

Rate: 0.5-1 lb. ai/A
0.66-1.33 qt. 3EC/A
1-2 lb. 50WP, 50WSB/A

Pest:
green peach aphid; meadow spittlebug; sugarbeet webworm

Remarks: Apply 1-2 applications per year as needed when insects appear or feeding is noticed. Do not exceed 2 lb. ai/A per year or 2 applications per year. Do not feed treated tops to livestock.
Days To Harvest: 30.

Signal Word/Toxicity Class: Danger-Poison/I.
REI: 24 hr.

Fyfanon — see malathion

Kumulus — see sulfur

LANNATE

Restricted Use Pesticide.

Lannate SP or LV (methomyl) *(DuPont)*

Rate: 0.225-0.9 lb. ai/A
0.25-1 lb. SP/A
0.75-3 pt. LV/A

Pest:
aphids; beet armyworm; beet webworm; carrion beetles; flea beetles; variegated cutworm; western yellowstriped armyworm

Remarks: Rate varies by pest species; refer to label. Apply in sufficient water to obtain thorough, uniform coverage when insects first appear; repeat at 5- to 7-day intervals or as needed. Do not apply more than 4.5 lb. ai/A per crop or make more than 10 applications per crop.
Days To Harvest: 7; livestock grazing or feeding: tops-30.

Signal Word/Toxicity Class: Danger-Poison/I.
REI: 48 hr.

Lorsban — see chlorpyrifos

MALATHION

Clean Crop Malathion 57EC *(Platte)*
Drexel Malathion 5EC *(Drexel)*
◆ **Micro Flo Malathion 5EC** *(Micro Flo)*

Rate: 1-1.3 lb. ai/A
1.5-2 pt. 5EC, 57EC/A

Pest:
aphids; spider mites

Rate: 2 lb. ai/A
3 pt. 5EC, 57EC/A

Pest:
grasshoppers

Remarks: None.
Days To Harvest: Clean Crop-3; Drexel, Prentox: tops used for feed or forage-7.

Signal Word/Toxicity Class: Caution/III (57EC); Warning/II (5EC).
REI: 12 hr.

◆ **Drexel Malathion ULV** *(Drexel)*
Fyfanon ULV *(Cheminova)*
◆ **Micro Flo Malathion ULV** *(Micro Flo)*

Rate: 8 fl. oz. ULV/A

6 Other Field Crops

NOTICE	The information on these pages is for preliminary planning — not a guide for use. Be sure to follow the manufacturer's directions, notwithstanding information contained here.	For personal protective equipment and EPA registration numbers, please turn to page 39.

Pest:
grasshoppers sugarbeet root maggot (adult)
Remarks: None.
Days To Harvest: 0; tops used for food or feed-7.

Signal Word/Toxicity Class: Caution/III.
REI: 12 hr.

◆ Gowan Malathion 8 (8E) *(Gowan)*
Rate: 1-2 lb. ai/A
1-2 pt. 8E/A
Pest:
aphids

Rate: 2 lb. ai/A
2 pt. 8E/A
Pest:
grasshoppers
Remarks: Apply at first sign of infestation; repeat at 7- to 10-day intervals.
Days To Harvest: 3; tops used for food or feed-7.

Signal Word/Toxicity Class: Caution/III.
REI: 12 hr.

METASYSTOX-R

Restricted Use Pesticide.

Metasystox-R (SC) (oxydemeton-methyl) *(Gowan)*
Rate: 0.375-0.75 lb. ai/A
1.5-3 pt. SC/A
Pest:
aphids leafhoppers mites
Remarks: Apply in sufficient water for complete coverage but not less than 1 gal./A. Do not apply more than 2 times per season.
Days To Harvest: Tops for feed or forage-30.

Signal Word/Toxicity Class: Warning/II.
REI: 48 hr.; 72 hr. if average rainfall less than 25" per year.

METHYL PARATHION

Restricted Use Pesticide.

Cheminova Methyl 4EC *(Cheminova)*
◆ Declare (4EC) *(Griffin)*
Helena 4 lb. Methyl Parathion (4EC) *(Helena)*
Wilbur-Ellis Methyl Parathion 7.5 Spray *(Wilbur-Ellis)*
Rate: 0.25-0.375 lb. ai/A
0.5-0.75 pt. 4EC/A
0.21-0.35 lb. ai/A
0.25-0.375 pt. 7.5/A
Pest:
aphids (Declare), armyworms, cutworms, fall armyworm, flea beetles, grasshoppers, leafhoppers, Lygus bugs, mites, stink bugs, sugarbeet webworm, webworms
Remarks: Apply when insects appear; repeat as needed.
Days To Harvest: Roots-20; tops fed to livestock-60.

Signal Word/Toxicity Class: Danger-Poison/I.
REI: 48 hr.

Wilbur-Ellis Methyl Parathion 5 Spray *(Wilbur-Ellis)*
Rate: 0.32 lb. ai/A
0.5 pt. 5/A
Pest:
aphids, armyworms, flea beetles, webworms
Remarks: Do not apply through any type of irrigation system.
Days To Harvest: 20; 60 days if tops are to be fed to livestock.

Signal Word/Toxicity Class: Danger-Poison/I.
REI: 48 hr.

Microthiol Special — see sulfur

M-PEDE

M-Pede (potassium salts of fatty acids) *(Mycogen)*
Rate: 1-2% v/v solution
Pest:
aphids*, broad mites, green peach aphid*, leafhoppers, leafminers*, plant bugs, russet mites, spider mites, thrips, whiteflies
** Tank mix only*
Remarks: Apply at first sign of infestation; repeat at 7- to 14-day intervals. May be tank mixed with other insecticides. Refer to both labels for use directions and precautions.
Days To Harvest: 0.

Signal Word/Toxicity Class: Warning/II.
REI: 12 hr.

PETROLEUM OIL

Petroleum-based spray oil. Mode of action is through suffocation of eggs, larvae, and nymphs of insects and adults of soft bodied insects. This mode of action necessitates total spray coverage. Toxic to fish.

SAF-T-SIDE *(Brandt Consolidated)*
◆ Sumulsion *(Drexel)*
Rate: 1-2 gal./100 gal. water
Pest:
aphids, caterpillars, corn rootworm, earworm, fall armyworm, leafminers, mites, whiteflies
Remarks: Apply as needed. May be tank mixed with most commonly used insecticides and fungicides. Do not use in combination with or immediately before or after spraying with dinitro compounds, fungicides such as Captan, Folpet, Dyrene, Morestan or any other product containing sulfur. Do not use with carbaryl or dimethoate.

Signal Word/Toxicity Class: Caution.
REI: 4 hr.

◆ 435 or 455 Oil 98.8 *(Drexel)*
◆ SunSpray 6E or 7E *(Drexel)*
Rate: 2 gal./A
Pest:
leafminers mites
Remarks: Apply as needed. May be tank mixed with most commonly used insecticides and fungicides. Do not use in combination with or immediately before or after spraying with fungicides such as captan, Dyrene, folpet, Morestan or any sulfur products. Do not use with carbaryl or dimethoate.

Signal Word/Toxicity Class: Caution.
REI: 4 hr.

◆ SunSpray Ultra-Fine *(Sun)*
Rate: 2 gal./100 gal. water
Pest:
aphids, armyworms, corn earworm, leafminers, mites, rootworms, whiteflies
Remarks: Apply at 2-week intervals. Do not exceed 4 applications per season.

Signal Word/Toxicity Class: Caution.
REI: 4 hr.

Phaser — see endosulfan

PHORATE

Restricted Use Pesticide.

Clean Crop Phorate 20G *(Platte)*
Riverside Phorate 20G *(Terra)*
Thimet 20-G *(American Cyanamid)*
Rate: At planting:
4.5 oz. 20G, 20-G/1000 ft. row (min. 20" spacing)
Pest:
aphids, beet root maggot, leafhoppers, leafminers, mites

◆ - **new product** • ai-active ingredient • DF-dry flowable • E/EC-emulsifiable concentrate • EW-emulsion in water
F/FL-flowable • G-granule • L/LC-liquid • SP/WSP-soluble packet • W/WP-wettable powder • WSB-water soluble bag

Rate: Postemergence:
0.9-1.5 lb. ai/A
4.9-7.5 lb. 20G, 20-G/A

Pest:
aphids · beet root maggot · mites

Remarks: Apply to foliage when plants are dry. For lower rate, application should be made by drilling to side of seed or banding over row at seeding time; higher rate to dense foliage. Do not place granules in direct contact with seed. Do not feed beet tops or silage to dairy cattle. Consult state experiment station or state extension service for proper timing of application in your area.

Days To Harvest: 30.

Signal Word/Toxicity Class: Danger-Poison/I.
REI: 48 hr.

PYRELLIN (Premix)

Pyrellin E.C. (0.042 lb./gal. pyrethrins & 0.035 lb./gal. rotenone) *(Webb Wright)*

Rate: 1-2 pt. E.C./A

Pest:
aphids · bean beetle · beet webworm · caterpillars · loopers · Mexican bean beetle · mites · thrips

Remarks: Apply as necessary to maintain control.

Days To Harvest: 0.

Signal Word/Toxicity Class: Caution/III.
REI: 12 hr.

Pyrenone — see pyrethrins

PYRETHRINS

Contains piperonyl butoxide as a synergist. Diatect also contains silicon dioxide (diatomaceous earth). May be combined in tank mix with other insecticides for quicker and more complete kill where insect resistance may be a problem, and as an "exciter" to flush insects out of hiding and into contact with other spray residues.

Diatect Multipurpose Insecticide II *(Diatect)*
Pyrenone (0.5EC) *(AgrEvo Environmental Health)*
◆ **Py-Rin 60-6 EC** *(Wilbur-Ellis)*

Rate: 1-6 lb. II/A
1-12 fl. oz. 0.5EC/A
2-16 oz. 60-6 EC/A

Pest: Including but not limited to:
aphids · beet webworm · caterpillars · loopers · Mexican bean beetle · mites · thrips

Remarks: Apply in sufficient water for thorough coverage; repeat as needed. Apply Diatect as dust or wettable powder.

Days To Harvest: 0.

Signal Word/Toxicity Class: Caution.
REI: 12 hr.

Py-Rin — see pyrethrins

SAF-T-SIDE — see petroleum oil

Sevin — see carbaryl

SULFUR

Microthiol and Thiolux are micronized sulfur.

◆ **Kumulus DF** *(BASF)*

Rate: 8-16 lb. ai/A
10-20 lb. DF/A

Pest:
red spider mite

Remarks: Apply at first indication of problem with or without other pesticides; repeat as needed. Do not apply if temperatures during or within 3 days after application are expected to exceed 90°F. Do not apply through any type of irrigation system.

Signal Word/Toxicity Class: Caution.
REI: 24 hr.

Microthiol Special (80W) *(Elf Atochem)*

Rate: 4-8 lb. ai/A
5-10 lb. 80W/A

Pest:
red spider mite

Remarks: Apply at first appearance; repeat every 2-3 weeks, or as needed. Do not chemigate in California.

Signal Word/Toxicity Class: Caution.
REI: 24 hr.

Thiolux DF *(Platte)*

Rate: 4-8 lb. ai/A
5-10 lb. DF/A

Pest:
Banks grass mite · Pacific spider mite · red spider mite · twospotted spider mite

Remarks: Apply at 10-30 day intervals.

Signal Word/Toxicity Class: Caution.
REI: 24 hr.

Sumulsion — see petroleum oil

SunSpray — see petroleum oil

TEMIK

Restricted Use Pesticide.

Temik 15G (aldicarb) *(Rhone-Poulenc)*

Rate: At planting or 1 week before planting:
4-5 lb. ai/A
27-33 lb. 15G/A
Postemergence (within 60 days after planting):
4 lb. ai/A
27 lb. 15G/A
Split application (at planting plus postemergence):
2-3 lb. ai/A
14-20 lb. 15G/A

Pest:
nematodes

Rate: At planting, postemergence:
1-3 lb. ai/A
7-20 lb. 15G/A

Pest:
aphids · leafhoppers · leafminers · sugarbeet root maggot

Remarks: Do not make any postemergence application if 27-33 lb./A were applied at planting or 1 week before planting. Do not make more than 1 at-planting and 2 postemergence applications per crop. Total use not to exceed 33 lb./A. Do not use tops for human food. Refer to label for environmental precautions and soil restrictions table.

Days To Harvest: 90; if tops are to be fed to livestock-120.

Signal Word/Toxicity Class: Danger-Poison/I.
REI: 48 hr.

Thimet — see phorate

Thiodan — see endosulfan

Thiolux — see sulfur

6 Other Field Crops

NOTICE	The information on these pages is for preliminary planning — not a guide for use. Be sure to follow the manufacturer's directions, notwithstanding information contained here.	For personal protective equipment and EPA registration numbers, please turn to page 39.

Sugarcane

ASANA

Restricted Use Pesticide.

Asana XL (esfenvalerate) *(DuPont)*
Rate: 0.03-0.05 lb. ai/A
5.8-9.6 fl. oz. XL/A
Pest:
sugarcane borer
Remarks: Repeat as necessary to maintain control. Do not exceed 0.2 lb. ai/A per season.
Days To Harvest: 21.

- - - - - - - - - - - - - - - - - -

Signal Word/Toxicity Class: Warning/II.
REI: 12 hr.

AZADIRACHTIN

Extracted from seed of neem tree. Insect growth regulator that controls targeted insect larvae when they ingest or come in contact with it, by interfering with the insect's ability to molt and shutting off feeding behavior. Affected larvae slowly starve and die at next molt.

◆ **Azatin XL Plus** *(Thermo Trilogy)*
Rate: 5-21 oz. XL Plus/A
Pest:
beetles
Remarks: Rate varies by species. Apply when pests first appear; repeat at 7-day intervals. When pest populations are high, use higher rates. Do not apply more that 21 fl. oz./A.
Days To Harvest: 0.

- - - - - - - - - - - - - - - - - -

Signal Word/Toxicity Class: Caution.
REI: 4 hr.

Azatin — see azadirachtin

BAYTHROID

Synthetic pyrethroid formulated in an emulsifiable concentrate. Toxic to fish and aquatic organisms. Restricted Use Pesticide.

Baythroid (2EC) (cyfluthrin) *(Bayer)*
Rate: 0.033-0.044 lb. ai/A
2.1-2.8 fl. oz. 2ECA
Pest:
rice borer sugarcane borer
Remarks: Apply in sufficient water for complete coverage of foliage. Up to 6 applications may be made per crop season. Allow 7 days between applications. Refer to label for application timing and spray drift precautions.
Days To Harvest: 15.

- - - - - - - - - - - - - - - - - -

Signal Word/Toxicity Class: Danger/I.
REI: 12 hr.

FURADAN

Restricted Use Pesticide.

Furadan 4F (carbofuran) *(FMC)*
Rate: 0.5-0.75 lb. ai/A
1-1.5 pt. 4F/A
Pest:
sugarcane borer
Remarks: Apply by ground or air. Check fields weekly, early June-Aug. Make first application only after visible joints form and 5% or more of plants are infested with young larvae feeding in or under leaf sheath and which have not bored into stalks. Repeat whenever field checks indicate infestation exceeds 5%. Do not apply through any type of irrigation system.
Days To Harvest: 17.
Crop Rotation: Do not rotate with any crop on soil treated at greater than 3 lb. ai/A for at least 10 months or plant with any crop for which carbofuran treatment is not registered for at least 10 months.
State Restrictions: Not for use in Hawaii.

- - - - - - - - - - - - - - - - - -

Signal Word/Toxicity Class: Danger-Poison/I.
REI: 48 hr.

MOCAP

Mocap EC and 20%G are Restricted Use Pesticides.

Mocap EC, 10%G or 20%G (ethoprop) *(Rhone-Poulenc)*
Rate: 2-4 lb. ai/A
1.33-2.66 qt. EC/A (5-8 fl. oz./1000 lin. ft. row)
20-40 lb. 10%G/A (2.4-3.2 lb./1000 lin. ft. row)
10-20 lb. 20%G/A (1.4-2.8 lb. 20%/1000 lin. ft. row)
Pest:
nematodes wireworms
Remarks: Furrow treatment. Apply evenly at planting in band 12-15" wide, centered over seed pieces in opened furrow immediately before closing. Use higher rates where nematodes and wireworms have been especially severe.

- - - - - - - - - - - - - - - - - -

Signal Word/Toxicity Class: Danger-Poison/I (EC, 20%G); Warning/II (10%G).
REI: 48 hr.; 72 hr. if average rainfall less than 25" per year.

PHORATE

Restricted Use Pesticide.

Clean Crop Phorate 2G *(Platte)*
Riverside Phorate 2G *(Terra)*
Thimet 20-G *(American Cyanamid)*
Rate: 3.9 lb. ai/A
19.5 lb. 20G, 20-G/A
Pest:
wiroworms
Remarks: Apply in 10-12" band directly on seed piece and surrounding soil in open furrow immediately before covering. Do not graze or feed treated hay or forage to livestock.
State Restrictions: For use in Florida.

- - - - - - - - - - - - - - - - - -

Signal Word/Toxicity Class: Danger-Poison/I.
REI: 48 hr.

TEMIK

Restricted Use Pesticide.

Temik 15G (aldicarb) *(Rhone-Poulenc)*
Rate: 2-3 lb. ai/A
14-20 lb. 15G/A
Pest:
nematodes
Remarks: Apply at planting. Apply granules in open row on top of newly planted cane and cover immediately with at least 6" of soil. Do not make more than 1 application per crop or exceed a total of 20 lb./A. Refer to label for environmental precautions and soil restrictions table.
Days To Harvest: 120.
State Restrictions: For use in Louisiana.

- - - - - - - - - - - - - - - - - -

Signal Word/Toxicity Class: Danger-Poison/I.
REI: 48 hr.

Thimet — see phorate

Sunflower

ASANA

Restricted Use Pesticide.

Asana XL (esfenvalerate) *(DuPont)*
Rate: 0.015-0.03 lb. ai/A
2.9-5.8 fl. oz. XL/A

◆ - **new product** • ai-active ingredient • DF-dry flowable • E/EC-emulsifiable concentrate • EW-emulsion in water
F/FL-flowable • G-granule • L/LC-liquid • SP/WSP-soluble packet • W/WP-wettable powder • WSB-water soluble bag

Pest:
sunflower beetle

Rate: 0.03-0.05 lb. ai/A
5.8-9.6 fl. oz. XL/A

Pest:

banded sunflower moth	cutworms	sunflower maggot
beet armyworm (aids in control)	grasshoppers	sunflower moth
	Heliothis	sunflower seed weevil
	leafhoppers	sunflower stem weevil

Remarks: Repeat as necessary to maintain control. Do not exceed 0.2 lb. ai/A per season.
Days To Harvest: 28.

Signal Word/Toxicity Class: Warning/II.
REI: 12 hr.

BAYTHROID

Synthetic pyrethroid formulated in an emulsifiable concentrate. Toxic to fish and aquatic organisms. Restricted Use Pesticide.

Baythroid (2EC) (cyfluthrin) *(Bayer)*
Rate: 0.0125-0.044 lb. ai/A
0.5-2.8 fl. oz. 2EC/A

Pest:

banded sunflower moth	sunflower beetle	sunflower midge
cutworm	sunflower bud moth	sunflower moth
grasshoppers	sunflower headclipper weevil	sunflower seed weevil
		sunflower stem weevil

Remarks: Apply in sufficient water for complete coverage of foliage and heads. Use lower rate for low to moderate insect pressure; higher rate for moderate to heavy pest pressure and longer residual control. Allow 7 days between applications. Up to 8.4 fl. oz./A may be applied per crop season. Do not graze or feed forage to livestock within 30 days after application. Refer to label for timing of applications and spray drift precautions.
Days To Harvest: 30.

Signal Word/Toxicity Class: Danger/I.
REI: 12 hr.

CARBARYL

Clean Crop Sevin 5 Bait *(Platte)*
Rate: 1-2 lb. ai/A
20-40 lb. Bait/A

Pest:

ants	darkling beetle	grasshoppers
armyworms	field crickets	sowbugs
cutworms		

Remarks: Apply with ground or air equipment. Do not allow animals to graze treated sunflower crops.
Days To Harvest: Seeding purposes-0; crop used for oil, food, or feed-60.

Signal Word/Toxicity Class: Caution/III.
REI: 12 hr.

Drexel Carbaryl 4L *(Drexel)*
Sevin 4F, XLR Plus, 80S, 80WSP *(Rhone-Poulenc)*
Rate: 0.5-2 lb. ai/A
1-4 pt. 4L, 4F, XLR/A
0.66-2.5 lb. 80S, 80WSP/A

Pest:

armyworms	grasshoppers	sunflower beetle
cutworms	stem weevil	sunflower moth
fall armyworm		

Remarks: Do not apply when foliage is wet or excessive humidity is expected during next 2 days. Do not allow animals to graze treated crops.
Days To Harvest: 60.
State Restrictions: Not for use in California.

Signal Word/Toxicity Class: Caution (4F, 4L, XLR); Warning/II (80S, 80WSP).
REI: 12 hr.

Chlorfos — see chlorpyrifos

CHLORPYRIFOS

Lorsban-4E *(Dow AgroSciences)*
◆ **Micro Flo Chlorpyrifos 4E** *(Micro Flo)*
Rate: 0.5-0.75 lb. ai/A
1-1.5 pt. 4E/A

Pest:

banded sunflower moth	seed weevil	sunflower moth
grasshoppers	stem weevil	woollybears
	sunflower beetle	

Rate: Preplant:
1-2 lb. ai/A
2-4 pt. 4E/A
Foliar:
1-1.5 lb. ai/A
2-3 pt. 4E/A

Pest:
cutworms

Remarks: Do not apply more than 9 pt./A per season. For grasshoppers, apply low rate. For cutworms, soil treatment, apply preplant incorporated into top 2-4" soil using not less than 10 gal. water/A; foliar, apply by air or ground as broadcast spray when insects become a problem. Do not graze treated areas.
Days To Harvest: 42.

Signal Word/Toxicity Class: Warning/II.
REI: 24 hr.

◆ **Chlorfos 15G** *(Griffin)*
Lorsban 15G *(Dow AgroSciences)*
◆ **Micro Flo Chlorpyrifos 15G** *(Micro Flo)*
Rate: 8 oz. 15G/1000 ft. row
Pest:
cutworms

Remarks: Apply at planting in 7" band behind planter shoe and in front of press wheel. Do not make more than 1 application per season.

Signal Word/Toxicity Class: Caution.
REI: 12 hr.

lambda-CYHALOTHRIN

Synthetic pyrethroid insecticide with broad-spectrum activity, formulated as an emulsifiable concentrate. Restricted Use Pesticide due to toxicity to fish and aquatic organisms.

Karate or Warrior (1EC) *(ZENECA)*
Rate: 0.01-0.03 lb. ai/A
1.28-3.84 fl. oz. 1EC/A

Pest:

banded sunflower moth (adult)	headworms	spotted cabbage looper
beet armyworm (1st & 2nd instar)	Japanese beetle	stem weevil
cutworms	leafhoppers	stink bugs
fall armyworm (1st & 2nd instar)	meadow spittlebug	sunflower beetle
grasshoppers	painted lady	sunflower maggot
headclipper weevil	seed weevil	sunflower moth
	spider mites (suppression)	thistle caterpillar
		woollybear caterpillar

Remarks: Apply as required by scouting, usually at intervals of 5 or more days. Use sufficient water to obtain full coverage of sunflower heads and/or foliage. When applying by air, apply in a minimum of 2 gal./A per season. Do not apply more than 0.72 pt./A after bloom initiation. Do not apply more than 0.96 pt./A per season. Refer to label for spray drift precautions and state and county restrictions.
Days To Harvest: 45.

Signal Word/Toxicity Class: Danger/I.
REI: 24 hr.

DB-Green — see lindane premixes

Declare — see methyl parathion

NOTICE The information on these pages is for preliminary planning — not a guide for use. Be sure to follow the manufacturer's directions, notwithstanding information contained here.

For personal protective equipment and EPA registration numbers, please turn to page 39.

Diatect — see pyrethrins

ENDOSULFAN

Drexel Endosulfan 3EC or 50WP *(Drexel)*
Gowan Endosulfan 3EC or 50WSB *(Gowan)*
◆ **Micro Flo Endosulfan 3EC or 50WP** *(Micro Flo)*
Phaser 3EC or 50WSB *(AgrEvo USA)*
Thiodan 3EC or 50WP *(FMC)*

Rate: 1 lb. ai/A
1.33 qt. 3EC/A
2 lb. 50WP, 50WSB/A

Pest:
sunflower moth

Remarks: Apply when insects appear or feeding is noticed. Make first application at onset of bloom; repeat at 7-day intervals. Do not exceed 3 lb. ai/A per year or 3 applications per year. Do not feed treated forage to livestock.

Days To Harvest: 1.

- - - - - - - - - - - - - - - - -

Signal Word/Toxicity Class: Danger-Poison/I.
REI: 24 hr.

FURADAN

Restricted Use Pesticide.

Furadan 4F (carbofuran) *(FMC)*

Rate: At planting:
1.4-2.8 lb. ai/A
1.4-2.8 qt. 4F/A (30" row)
2.5-5 fl. oz. 4F/1000 lin. ft.

Pest:
grasshoppers, sunflower beetle, sunflower stem weevil

Rate: Foliar:
0.125-0.5 lb. ai/A
0.25-1 pt. 4F/A

Pest:
banded sunflower moth, seed weevil, sunflower beetle, grasshoppers, stem weevil, sunflower moth

Remarks: Confectionery and oil. Use higher rate for heavy infestations. At planting: Apply directly into seed furrow or in 7" band. May be mixed with water or liquid fertilizer. Maintain agitation after mixing and during application. Do not mix until ready to use. Foliar: Apply when insects appear. Do not make more than 2 foliar applications per season. Do not apply through any type of irrigation system.

Days To Harvest: Foliar-28.

Crop Rotation: Do not rotate with any crop on soil treated at greater than 3 lb. ai/A for at least 10 months or plant with any crop for which carbofuran treatment is not registered for at least 10 months.

State Restrictions: Not for use in Long Island, New York.

- - - - - - - - - - - - - - - - -

Signal Word/Toxicity Class: Danger-Poison/I.
REI: 48 hr.

Karate — see *lambda*-cyhalothrin

LINDANE

Lindane 30C contains toxic inert Rhodamine B. Toxic to fish, birds, and wildlife.

Lindane 30C *(Gustafson)*
◆ **Lindane ST 40** *(AGSCO)*

Rate: 2.7 fl. oz. 30C/100 lb. seed
2 fl. oz. 40/100 lb. seed

Pest:
seedcorn maggots, wireworms

Remarks: Do not use treated seed for food, feed, or oil purposes. Exposed treated seed may be hazardous to birds and other wildlife. Dispose of excess seed by burial away from bodies of water.

- - - - - - - - - - - - - - - - -

Signal Word/Toxicity Class: Warning/II.

LINDANE (Premixes)

Dual purpose insecticide and fungicide seed treatment. Sorghum Guard contains graphite.

◆ **DB-Green L (8.6% lindane & 25.6% maneb)** *(AGSCO)*

Rate: 4 fl. oz./bu. seed

Pest:
wireworms

Remarks: Drill or planter box application. Treat only seeds needed for immediate use, minimizing interval between treatment and planting. Do not use treated seed for food, feed, or oil purposes. Do not store excess treated seeds beyond planting time.

- - - - - - - - - - - - - - - - -

Signal Word/Toxicity Class: Warning/II.
REI: 12 hr.

◆ **Sorghum Guard (16.6% lindane & 32.75% captan)** *(Trace)*

Rate: 2 oz./50 lb. seed

Pest:
wireworms

Remarks: Apply in planter or drill box as seed treatment before planting. Treat only those seeds needed for immediate use. Do not use treated seed for food, feed, or oil purposes. Do not store excess treated seeds beyond planting time.

- - - - - - - - - - - - - - - - -

Signal Word/Toxicity Class: Danger/I.
REI: 12 hr.

Lorsban — see chlorpyrifos

METHYL PARATHION

Restricted Use Pesticide.

Cheminova Methyl 4EC *(Cheminova)*
◆ **Declare (4EC)** *(Griffin)*
Helena 4 lb. Methyl Parathion (4EC) *(Helena)*

Rate: 1 lb. ai/A
2 pt. 4EC/A

Pest:
sunflower maggot, sunflower moth, sunflower seed weevil

Remarks: Make no more than 3 applications at 5-day intervals. Do not feed seed to birds.

Days To Harvest: 30.

- - - - - - - - - - - - - - - - -

Signal Word/Toxicity Class: Danger-Poison/I.
REI: 48 hr.

Wilbur-Ellis Methyl Parathion 5 Spray *(Wilbur-Ellis)*

Rate: 0.63-1.3 lb. ai/A
1-2 pt. 5/A

Pest:
seed weevil, stem weevil, sunflower beetle

Remarks: Apply at set of flowering before 10% of plants begin to flower. Repeat 2-3 times at 5-day intervals.

- - - - - - - - - - - - - - - - -

Signal Word/Toxicity Class: Danger-Poison/I.
REI: 48 hr.

Phaser — see endosulfan

PYRELLIN (Premix)

Pyrellin E.C. (0.042 lb./gal. pyrethrins & 0.035 lb./gal. rotenone) *(Webb Wright)*

Rate: 1-2 pt. E.C./A

Pest:
aphids, caterpillars, mites, bean beetle, loopers, thrips

Remarks: Apply as necessary to maintain control.

Days To Harvest: 0.

- - - - - - - - - - - - - - - - -

Signal Word/Toxicity Class: Caution/III.
REI: 12 hr.

◆ - **new product** • ai-active ingredient • DF-dry flowable • E/EC-emulsifiable concentrate • EW-emulsion in water
F/FL-flowable • G-granule • L/LC-liquid • SP/WSP-soluble packet • W/WP-wettable powder • WSB-water soluble bag

PYRETHRINS

Contains piperonyl butoxide as a synergist and silicon dioxide (diatomaceous earth).

Diatect Multipurpose Insecticide II *(Diatect)*
Rate: 1-6 lb. II/A
Pest: Including but not limited to:
aphids, mites, thrips, loopers
Remarks: Apply as dust or wettable powder. Repeat as needed.

- - - - - - - - - - - - - - - -

Signal Word/Toxicity Class: Caution.
REI: 12 hr.

SCOUT X-TRA

Synthetic pyrethroid insecticide formulated as an emulsifiable concentrate. Scout X-TRA Gel is packaged in water-soluble bags. Low toxicity to humans and other mammals, highly toxic to bees. Restricted Use Pesticide due to toxicity to fish and aquatic organisms.

Scout X-TRA (0.9EC) or Gel (WSB) (tralomethrin) *(AgrEvo USA)*
Rate: 0.005-0.019 lb. ai/A
0.71-2.67 fl. oz. 0.9EC/A
1 WSB/40-20 A
Pest:
banded sunflower moth, cabbage looper, corn earworm, cutworms, grasshoppers (except lubber grasshopper), leafhoppers, potato flea beetle, sunflower beetle, sunflower maggot, sunflower moth, sunflower seed weevil, sunflower stem weevil (moth)
Remarks: Uniform coverage essential. Do not apply as ULV spray. Do not apply more than 0.06 lb. ai/A per growing season. Do not graze or harvest treated sunflower foliage. Refer to label for spray drift precautions.
Days To Harvest: 21.
State Restrictions: Not for use in California.

- - - - - - - - - - - - - - - -

Signal Word/Toxicity Class: Danger/I.
REI: 24 hr.

Sevin — see carbaryl

Sorghum Guard — see lindane premixes

Thiodan — see endosulfan

Warrior — see *lambda*-cyhalothrin

Tobacco

ACEPHATE

Broad spectrum, general use organophosphate insecticide. May be applied to transplant water or as a foliar spray.

Acephate 75SP AG *(Micro Flo)*
Orthene 75 S or 97 S *(Valent)*
Rate: 0.25-0.75 lb. ai/A
0.33-1 lb. 75SP, 75 S/A
0.24-0.73 lb. ai/A
0.25-0.75 lb. 97 S/A
Pest:
budworms, cabbage looper, cutworms, flea beetles, grasshoppers, green peach aphid, hornworms, stink bugs, tobacco aphid, tobacco thrips, vegetable weevils
Remarks: Air-cured, flue-cured, dark fire-cured tobacco. Foliar application. Apply in 10-50 gal. water/A by ground; minimum 3 gal. water/A by air. Apply at 7-day intervals as needed. Rate varies by pest species; refer to label. Do not apply through any type of irrigation system.
Days To Harvest: 3.
State Restrictions: Not for use in California.

- - - - - - - - - - - - - - - -

Rate: 0.75 lb. ai/A
1 lb. 75SP, 75 S/A
0.73 lb. ai/A
0.75 lb. 97 S/A
Pest:
cutworms, flea beetles, green peach aphid, tobacco aphid, tobacco thrips (transplant water)
Remarks: Greenhouse, plant bed, transplant water application. For greenhouse or plant bed treatment, apply 1 Tbs./1 gal. water/1000 sq. ft. bed evenly to foliage. For transplant water application, apply in minimum 100 gal. transplant water/A. Provides control of early-season insects for approximately 3-4 weeks after transplanting. Do not apply more than 1 lb./A in transplant water as some phytotoxicity may occur. Do not apply through any type of irrigation system.
State Restrictions: Not for use in California.

- - - - - - - - - - - - - - - -

Rate: Mound treatment—Drench:
1 oz. 75SP, 75 S/A in 5 gal. water
0.75 oz. 97 S/A in 5 gal. water
Pest:
ants, harvester ants, imported fire ants
Remarks: Soil application. Mount treatment-drench: Apply 1 gal. mix to each mound area by sprinkling mound until wet and treat 4 ft. diameter circle around mound. For best results, apply early morning or late afternoon when ants are most active. Applications made after prolonged hot and dry conditions may be ineffective due to location of ants deep in the nest. Grass in treated area may be injured. Treat maximum 13 mounds/A. Do not treat mound more than once per season.
Days To Harvest: 3.
State Restrictions: Not for use in California.

- - - - - - - - - - - - - - - -

Signal Word/Toxicity Class: Caution.
REI: 12 hr.; 97S: 24 hr.

Admire — see imidacloprid

AZADIRACHTIN

Extracted from seed of neem tree. Insect growth regulator that controls targeted insect larvae when they ingest or come in contact with it, by interfering with the insect's ability to molt and shutting off feeding behavior. Affected larvae slowly starve and die at next molt.

◆ **Azatin XL Plus** *(Thermo Trilogy)*
Rate: 5-21 oz. XL Plus/A
Pest:
aphids, armyworms, black cutworm, budworms, caterpillar, cutworms, fall armyworm, flea beetles, green peach aphid, Japanese beetle, June beetle, leafhoppers, leafminers, loopers, thrips, tobacco budworm, tobacco hornworm
Remarks: Rate varies by species. Apply when pests first appear; repeat at 7-day intervals. When pest populations are high, use higher rates. Do not apply more that 21 fl. oz./A.
Days To Harvest: 0.

- - - - - - - - - - - - - - - -

Signal Word/Toxicity Class: Caution.
REI: 4 hr.

Azatin — see azadirachtin

AZINPHOS-METHYL

Restricted Use Pesticide.

Gowan Azinphos-M 50-W *(Gowan)*
◆ **Micro Flo Azinphosmethyl 2EC or 50W** *(Micro Flo)*
Sniper 2-E, 50PVA, 50W *(Platte)*
Rate: 0.5-0.75 lb. ai/A
2-3 pt. 2EC, 2-E/A
1-1.5 lb. 50-W, 50 PVA, 50W/A
Pest:
aphids, grasshoppers, tobacco budworm, tobacco flea beetle, tobacco hornworm

6 Other Field Crops

For personal protective equipment and EPA registration numbers, please turn to page 39.

Remarks: Apply in sufficient water to give complete coverage; repeat as needed. Rate varies by pest species, refer to label. Prime before treating.
Days To Harvest: 6.
State Restrictions: Sniper 50PVA not registered in California.

- - - - - - - - - - - - - - - -

Signal Word/Toxicity Class: Danger-Poison/I.
REI: 48 hr.; 72 hr. if average rainfall less than 25" per year.

CARBARYL

General use, broad spectrum carbamate insecticide that provides good control of insects with chewing mouthparts. May be phytotoxic to small plants in the tobacco bed. Highly toxic to bees.

Clean Crop Sevin 5 Bait *(Platte)*

Rate: 1-2 lb. ai/A
20-40 lb. Bait/A

Pest:
ants, armyworms, cutworms, darkling beetle, field crickets, grasshoppers, sowbugs

Remarks: Apply with ground or air equipment.
Days To Harvest: Tobacco beds-0; field tobacco-3.

- - - - - - - - - - - - - - - -

Signal Word/Toxicity Class: Caution/III.
REI: 12 hr.

Drexel Carbaryl 4L or 50W *(Drexel)*
Sevin 4F, XLR Plus, 80S, 80WSP *(Rhone-Poulenc)*

Rate: 0.5-1.5 lb. ai/A
1-3 pt. 4L, 4F, XLR/A
1-3 lb. 50W/A
0.66-1.875 lb. 80S, 80WSP/A

Pest:
grasshoppers

Rate: Plant beds:
1 lb. ai/A
2 pt. 4L, 4F, XLR/A
1 lb. ai/A
2 lb. 50W/A in 50 gal. water
1.25 lb. 80S, 80WSP/A in 50 gal. water

Pest:
tobacco flea beetles

Rate: Plant beds:
8 lb. ai/A
8 qt. 4L, 4F, XLR/A
0.5 lb. ai/A
1 lb. 50W/A in 100 gal. water
0.66 lb. 80S, 80WSP/A in 200 gal.water

Pest:
green June beetle grubs

Remarks: For grubs, apply only to areas that larvae have uprooted by sprinkling mixture as drench treatment. Do not apply when foliage is wet or excessive humidity is expected in next 2 days.
Days To Harvest: 0.

- - - - - - - - - - - - - - - -

Rate: Field:
1-2 lb. ai/A
2-4 pt. 4L, 4F or XLR/A
2-4 lb. 50W/A
1.25-2.5 lb. 80S, 80WSP/A

Pest:
budworms, fall armyworm, flea beetles, hornworms, Japanese beetle, June beetle, suckfly

Remarks: Use lower rate on young plants (up to knee height). Use at least 10 gal. prepared spray/A. Apply when worms are small. If late-season application necessary, allow 3 days before priming or cutting. Do not apply when foliage is wet or excessive humidity is expected in next 2 days.
Days To Harvest: 0.

- - - - - - - - - - - - - - - -

Signal Word/Toxicity Class: Caution (4F, 4L, XLR); Warning/II (50W, 80S, 80WSP).
REI: 12 hr.

Chlorfos — see chlorpyrifos

CHLORPYRIFOS

General use organophosphate insecticide-nematicide. Extremely toxic to fish and other aquatic organisms.

◆ **Chlorfos 15G** *(Griffin)*
Lorsban-4E or 15G *(Dow AgroSciences)*
◆ **Micro Flo Chlorpyrifos 4E or 15G** *(Micro Flo)*

Rate: 2-3 lb. ai/A
4-6 pt. 4E/A
13.5-20 lb. 15G/A
North Carolina, South Carolina, Virginia:
5 lb. ai/A
10 pt. 4E/A

Pest:
cutworms, flea beetles, mole crickets (4E), root-knot nematode (low to moderate populations), root maggot, wireworms

Remarks: Apply preplant broadcast (overall spray) 1 week before transplanting. Do not make more than 1 application per season. Will also suppress movement of red imported fire ants into treated fields. For nematodes, 4E may be tank mixed with Nemacur 3. Refer to both labels for rates and use directions.

Signal Word/Toxicity Class: Warning/II (4E); Caution (15G).
REI: 24 hr (4E); 12 hr. (15G).

DI-SYSTON

Organophosphate insecticide with systemic activity. Highly toxic. Granules may be applied to tobacco bed before seeding. Liquid or granular formulations may be applied to soil prior to transplant. Restricted Use Pesticide.

Di-Syston 8 (disulfoton) *(Bayer)*

Rate: Field:
4 lb. ai/A
4 pt. 8/A

Pest:
aphids, flea beetles

Remarks: Apply as water emulsion broadcast spray in sufficient water for thorough coverage. Disk or till soil 2-3" before setting plants. Do not make more than 1 application per crop season.

- - - - - - - - - - - - - - - -

Signal Word/Toxicity Class: Danger/I.
REI: 48 hr.

Di-Syston 15% (disulfoton) *(Bayer)*

Rate: Field:
2-4 lb. ai/A
13.3-26.7 lb. 15%/A (48" row or broadcast)
Transplant beds:
1 oz. 15%100 sq. ft. of bed

Pest:
aphids, flea beetles

Remarks: Field: Apply in 6-12" band over row or broadcast immediately before transplanting. Incorporate granules by working into top 2-3" soil. Transplant beds: Apply evenly over bed area. Work granules into top 2-3" soil and seedbeds in normal manner. Alternative treatment may be made by broadcasting granules evenly over plants after emergence and watering thoroughly. Do not make more than 1 application per season or to transplant beds.

- - - - - - - - - - - - - - - -

Signal Word/Toxicity Class: Danger/I.
REI: 48 hr.

ENDOSULFAN

Broad spectrum, general use chlorinated bicyclic sulfite insecticide. Should not be applied within 300 ft. of lakes, ponds, streams, or estuaries due to risk of runoff or drift.

Drexel Endosulfan 3EC or 50WP *(Drexel)*
Gowan Endosulfan 3EC or 50WSB *(Gowan)*
◆ **Micro Flo Endosulfan 3EC or 50WP** *(Micro Flo)*
Phaser 3EC or 50WSB *(AgrEvo USA)*
Thiodan 3EC, 50WP or Golden Leaf (3EC) *(FMC)*

Rate: Plant bed:
0.25 lb. ai/100 gal. water
0.33 qt. 3EC/100 gal. water
0.5 lb. 50WP, 50WSB/100 gal. water

Pest:
green June bug (larvae)

◆ **- new product** • ai-active ingredient • DF-dry flowable • E/EC-emulsifiable concentrate • EW-emulsion in water
F/FL-flowable • G-granule • L/LC-liquid • SP/WSP-soluble packet • W/WP-wettable powder • WSB-water soluble bag

Rate: Field:
0.5 lb. ai/A
0.66 qt. 3EC/A
1-1.5 lb. ai/100 gal. water
2-3 lb. 50WP, 50WSB/100 gal. water
Pest:
stink bugs

Rate: Seedbed:
0.5 lb. ai/100 gal. water
0.66 qt. 3EC/100 gal. water
1 lb. 50WP, 50WSB/100 gal. water
Field:
0.5-1 lb. ai/A
0.66-1.33 qt. 3EC/A
1-2 lb. 50WP, 50WSB/A
Pest:
budworms, flea beetles, hornworms, cabbage looper, green peach aphid, tobacco aphid
Remarks: Apply when insects appear or feeding is noticed. For seedbed, apply 6 gal. finished spray/100 sq. yd. For plant bed, drench at rate of 1 gal./100 sq. yd.
Days To Harvest: 5.

Signal Word/Toxicity Class: Danger-Poison/I.
REI: 24 hr.

FURADAN

Restricted Use carbamate insecticide-nematicide due to acute oral and inhalation toxicity.

Furadan 4F (carbofuran) *(FMC)*
Rate: 4 lb. ai/A
1 gal. 4F/A
Pest:
flea beetles
Remarks: **Burley tobacco.** Apply as broadcast spray over soil surface prior to transplanting; incorporate into top 3" of soil with suitable device.

Rate: 4-6 lb. ai/A
1-1.5 gal. 4F/A
Pest:
budworms, root-knot nematode, wireworms, flea beetles
Remarks: **Flue-cured tobacco.** Apply as broadcast spray over soil surface prior to forming beds; incorporate into top 3" soil. Form beds and plant as usual. May induce flecking of bottom or lower leaves. Do not apply through any type of irrigation system.
Crop Rotation: Do not rotate with any crop on soil treated at greater than 3 lb. ai/A for at least 10 months. Do not plant with any crop for which carbofuran treatment is not registered for at least 10 months.
State Restrictions: Not for use in Long Island, New York.

Signal Word/Toxicity Class: Danger-Poison/I.
REI: 48 hr.

Golden Leaf — see endosulfan

IMIDACLOPRID

Admire (2F) *(Bayer)*
Rate: 0.016 lb. ai/1000 plants
1 fl. oz. 2F/1000 plants
Pest:
aphids, flea beetles

Rate: 0.02-0.04 lb. ai/1000 plants
1.4-2.8 fl. oz. 2F/1000 plants
Pest:
mole crickets, wireworms
Remarks: Soil application. Apply in transplant water, broadcast, or as a drench to trays or flats prior to transplants. Water by overhead irrigation to wash Admire from foliage into potting media. Failure to wash from foliage may result in control reduction. Adverse growing conditions may cause a delay in uptake of Admire into plant and control. Do not apply more than 0.5 lb. ai/A per season.
Days To Harvest: 14.

Signal Word/Toxicity Class: Caution.
REI: 12 hr.

LANNATE

Carbamate insecticide/ovicide. Highly toxic. Lannate LV contains methanol. Restricted Use Pesticide.

Lannate SP or LV (methomyl) *(DuPont)*
Rate: 0.225-0.45 lb. ai/A
0.25-0.5 lb. SP/A
0.75-1.5 pt. LV/A
Pest:
aphids, flea beetles, loopers, fall armyworm, hornworms, tobacco budworm
Remarks: Apply in sufficient water for thorough, uniform coverage. Low rate effective only for flea beetles and hornworms. Do not apply more than 2.25 lb. ai/A per crop or make more than 5 applications per crop.
Days To Harvest: Flue-cured-5; air- or fire-cured-14.

Signal Word/Toxicity Class: Danger-Poison/I.
REI: 48 hr.

Lorsban — see chlorpyrifos

MALATHION

General use organophosphate insecticide labeled for aphid control. Highly toxic to bees.

Clean Crop Malathion 57EC *(Platte)*
Gowan Malathion 8 (8E) *(Gowan)*
Rate: 1-1.5 lb. ai/A
1.5-2.5 pt. 5EC, 57EC/A
1-1.5 pt. 8E/A
Pest:
aphids
Remarks: None.
Days To Harvest: 7.

Signal Word/Toxicity Class: Caution/III.
REI: 12 hr.

MOCAP

Organophosphate insecticide-nematicide labeled for wireworm and nematode control. Highly toxic. Mocap EC is a Restricted Use Pesticide.

Mocap EC or 10%G (ethoprop) *(Rhone-Poulenc)*
Rate: Broadcast
2 lb. ai/A
1.33 qt. EC/A
20 lb. 10%G/A
Pest:
wireworms
Remarks: Apply from 2 weeks before planting to at-planting time. Mix with top 2-4" soil right after application.
State Restrictions: Not for use in Long Island, New York or Florida.

Rate: 6-12 lb. ai/A
1-2 gal. EC/A
60-120 lb. 10%G/A
Pest:
nematodes, wireworms
Remarks: Apply from 1 week before planting to at-planting time. Mix with top 2-4" soil right after application. For row treatment, apply in 18-24" wide band on 42" row. Use higher rate in fields in which nematode injury to tobacco has been especially severe or where soil is sandy.
State Restrictions: Not for use in Long Island, New York or Florida.

Signal Word/Toxicity Class: Danger-Poison/I (EC); Warning/II (10%G).
REI: 48 hr.; 72 hr. if rainfall is less than 25" per yr.

M-PEDE

M-Pede (potassium salts of fatty acids) *(Mycogen)*
Rate: 4 pt./A in 3-25 gal. water
2% v/v spray solution more than 25 gal. water/A
Pest:
aphids, greenbug, stink bugs, caterpillars (tank mix), Lygus bugs, thrips, plant bugs, whiteflies, fleahoppers, spider mites

NOTICE The information on these pages is for preliminary planning — not a guide for use. Be sure to follow the manufacturer's directions, notwithstanding information contained here.

For personal protective equipment and EPA registration numbers, please turn to page 39.

Remarks: Apply at first sign of infestation; repeat at 7- to 14-day intervals. May be tank mixed with other insecticides; refer to both labels for use directions and precautions.
Days To Harvest: 0.
State Restrictions: Not for use in California.

- - - - - - - - - - - - - - - - -

Signal Word/Toxicity Class: Warning/II.
REI: 12 hr.

NEMACUR

Organophosphate systemic nematicide applied to soil and incorporated prior to transplant. Not for use on shade-grown tobacco. Restricted Use Pesticide.

Nemacur 3 (fenamiphos) *(Bayer)*
Rate: 4-6 lb. ai/A
1.33-2 gal. 3/A
Pest:
aphids (suppression) nematodes
Remarks: Broadcast and incorporate 2-4" deep. May be tank mixed with products registered and recommended for tobacco, including Lorsban 4E and Mocap EC. Refer to labels for rates and further use directions.

- - - - - - - - - - - - - - - - -

Signal Word/Toxicity Class: Danger/I.
REI: 48 hr.

Orthene — see acephate

PETROLEUM OIL

Petroleum based spray oil. Mode of action is through suffocation of eggs, larvae, and nymphs of insects and adults of soft bodied insects. This mode of action necessitates total spray coverage. Toxic to fish.

SAF-T-SIDE *(Brandt Consolidated)*
◆ **Sumulsion** *(Drexel)*
◆ **SunSpray Ultra-Fine** *(Sun)*
Rate: 1-2 gal./100 gal. water
Pest:
leafminers mites whiteflies
Remarks: May be tank mixed with most commonly used insecticides and fungicides. Do not use in combination with or immediately before or after spraying with fungicides such as captan, Dyrene, folpet, Morestan or any sulfur products. Do not use with carbaryl or dimethoate.

- - - - - - - - - - - - - - - - -

Signal Word/Toxicity Class: Caution.
REI: 4 hr.

Phaser — see endosulfan

SAF-T-SIDE — see petroleum oil

Sevin — see carbaryl

Sniper — see azinphos-methyl

Sumulsion — see petroleum oil

SunSpray — see petroleum oil

Thiodan — see endosulfan

VYDATE

Carbamate insecticide/nematicide. May be applied as preplant incorporated treatment. Restricted Use Pesticide due to acute toxicity to birds and mammals.

Vydate L (oxamyl) *(DuPont)*
Rate: 2 lb. ai/A
1 gal. L/A
Pest:
flea beetles lesion nematodes root-knot nematodes (except Javanese)
Remarks: Soil treatment: Incorporate 4-6" into soil. For broadcast and bed treatment, use only treated soil for beds. Do not transplant tobacco for 48 hr. after soil treatment. Row treatment: Apply in 18-24" band (12,000 ft. row) in at least 20 gal. water. Do not apply more than 1 gal./A per growing season.
State Restrictions: Do not use in Suffolk and Nassau counties, Long Island, New York.

- - - - - - - - - - - - - - - - -

Signal Word/Toxicity Class: Danger-Poison/I.
REI: 48 hr.

Fungicides

Castor Beans

THIRAM

42-S Thiram or Thiram 50WP Dyed *(Gustafson)*
Thiram Granuflo (WDG) *(UCB Agrochemicals)*
Rate: 4.5 fl. oz. 42-S/100 lb. seed
4.5 oz. 50WP/100 lb. seed
3 oz. WDG/100 lb. seed
Disease:
damping-off seed decay seedling blight
Time: Seed treatment.
Remarks: Use with EPA-approved dye that imparts unnatural color to seed. Do not use for feed, food, or oil purposes. Do not apply Thiram Granuflo through any type of irrigation system. Refer to label for use directions.
State Restrictions: Thiram Granuflo not for use in California.

- - - - - - - - - - - - - - - - -

Signal Word/Toxicity Class: Caution.
REI: 24 hr. (Thiram Granuflo).

Flax

CAPTAN

Gustafson Captan 30-DD or 400 *(Gustafson)*
Rate: 3 fl. oz. 30-DD/100 lb. seed
2-3.75 fl. oz. 400/100 lb. seed
Disease:
damping-off seed decay seedling blight mold
Time: Seed treatment.
Remarks: Do not combine colloidal suspensions, wettable powders, or emulsifiable concentrates unless previous experience assures compatibility. Use with EPA-approved dye that imparts unnatural color to seed. Do not use for feed, food, or oil purposes. Refer to label for use directions.

- - - - - - - - - - - - - - - - -

Signal Word/Toxicity Class: Danger/I.

◆ - **new product** • ai-active ingredient • DF-dry flowable • E/EC-emulsifiable concentrate • EW-emulsion in water
F/FL-flowable • G-granule • L/LC-liquid • SP/WSP-soluble packet • W/WP-wettable powder • WSB-water soluble bag

CAPTAN (Premix)

Nu-Gro Isotox (12.2% captan & 25% lindane) *(Wilbur-Ellis)*
Seed Mate Isotox (12.2% captan & 25% lindane) *(Loveland)*
Rate: 2 oz./100 lb. seed
Disease:
damping-off seed decay seedling blight
Time: Planter box treatment.
Remarks: Treated seed must not be used for or mixed with food or animal feed, or processed for oil. Treat only those seeds needed for immediate use, minimizing the interval between treatment and planting. Do not store excess treated seeds beyond planting time. Refer to label for use directions.

- - - - - - - - - - - - - - - - -

Signal Word/Toxicity Class: Danger/I.
REI: 4 days.

Dithane — see mancozeb

MANCOZEB

Dithane (DF, F-45, M-45, ST, WSP) *(Rohm and Haas)*
Manzate 200DF *(Griffin)*
Penncozeb 75DF or 80WP *(Elf Atochem)*
Rate: 3.8-7.6 oz. DF, 75DF/100 lb. seed
3.6-7.1 oz. M-45, WSP, 200DF, 80WP/100 lb. seed
5.7-11.3 fl. oz. F-45, ST/100 lb. seed
Disease:
damping-off seed rot seedling blight
Time: Seed treatment.
Remarks: Incorporation of dye into the treating slurry is recommended to distinguish seed as treated. Do not use for food, feed, or oil purposes. Refer to label for use directions.
State Restrictions: Refer to label for California precautions.

- - - - - - - - - - - - - - - - -

Signal Word/Toxicity Class: Caution.
REI: 24 hr.

Manzate — see mancozeb

Nu-Gro Isotox — see captan premix

Penncozeb — see mancozeb

Seed Mate Isotox — see captan premix

SULFUR

Micro Flo Sulfur (6L) *(Micro Flo)*
Terra Sulfur (6L) *(Terra)*
Rate: 20-30 lb. ai/A
3.33-5 gal. 6L/A
Disease:
powdery mildew
Time: Apply when disease first appears; repeat at 7- to 10-day intervals.
Remarks: Do not use within 2 weeks of oil application. Some plants are susceptible to injury from sulfur under certain climatic conditions. Do not make sulfur applications at excessively high temperatures. Refer to label for use directions.

- - - - - - - - - - - - - - - - -

Signal Word/Toxicity Class: Caution.
REI: 24 hr.

THIRAM

42-S Thiram or Thiram 50WP Dyed *(Gustafson)*
Thiram Granuflo (WDG) *(UCB Agrochemicals)*
Rate: 3 fl. oz. 42-S/bu. seed
5.3 oz. 50WP/100 lb. seed
3.5 oz. WDG/100 lb. seed
Disease:
damping-off seed decay seedling blight
Time: Seed treatment.
Remarks: Use with EPA-approved dye that imparts unnatural color to seed. Do not use for feed, food, or oil purposes. Do not apply Thiram Granuflo through any type of irrigation system. Refer to label for use directions.
State Restrictions: Thiram Granuflo not for use in California.

- - - - - - - - - - - - - - - - -

Signal Word/Toxicity Class: Caution.
REI: 24 hr. (Thiram Granuflo).

Hops

Aliette — see fosetyl-aluminum

BASAMID

Granular soil fumigant. Toxic to fish.

Basamid Granular (dazomet) *(BASF)*
Rate: 6-8 lb./1000 sq ft.
Time: Preplant.
Disease:
soilborne diseases
Remarks: Till to desired depth and follow suggested cultural practices; refer to label. Do not use if soil temperature is below 43°F.
Days To Harvest: 1 year.

- - - - - - - - - - - - - - - - -

Signal Word/Toxicity Class: Warning/II.
REI: 24 hr.

Champ — see copper compounds

Champion — see copper compounds

Clean Crop C-O-C-S — see copper compounds

COPPER COMPOUNDS

Champ (FL or Formula 2) or Champion (WP) copper hydroxide) *(Agtrol)*
Kocide (101, DF, 2000, 4.5LF, LF) (copper hydroxide) *(Griffin)*
Rate: 2.66 pt. FL, LF/A
1.33 pt. Formula 2, 4.5LF/A
1.5 lb. 2000/A
2 lb. WP, 101, DF/A
Disease:
downy mildew
Time: Apply crown treatment after pruning, but before training. After training, repeat at 10-day intervals.
Remarks: Kocide should not be applied in spray solution having pH less than 6.5 as phytotoxicity may occur. Refer to label for use directions.
Days To Harvest: 2 weeks.

- - - - - - - - - - - - - - - - -

Signal Word/Toxicity Class: Varies by formulation.
REI: Varies by formulation.

Clean Crop C-O-C-S (WDG) (74.8% copper oxychloride & 14.2% copper sulfate, basic) *(Platte)*
Rate: 2-3 lb. ai/A
4-6 lb. WDG/A
Disease:
downy mildew
Time: Apply when growth is approximately 18" long; repeat at 7- to 10-day intervals throughout growing season.
Remarks: Do not apply this product through any type of irrigation system. Refer to label for use directions.
State Restrictions: Not for use in California.

- - - - - - - - - - - - - - - - -

Signal Word/Toxicity Class: Warning/II.
REI: 24 hr.

6 Other Field Crops

NOTICE	The information on these pages is for preliminary planning — not a guide for use. Be sure to follow the manufacturer's directions, notwithstanding information contained here.	For personal protective equipment and EPA registration numbers, please turn to page 39.

KOP-Hydroxide (FL, 50WP) (copper hydroxide) *(Drexel)*
Nu-Cop (3L, 50DF, 50WP) (copper hydroxide) *(Micro Flo)*
TOP COP Tri-Basic (FL) (basic copper sulfate) *(Stoller)*

Rate: 2.66 pt. FL/A (Drexel)
1.33-2.66 pt. 3L/A
2 lb. 50DF, 50WP/A
2 pt. FL/A (Stoller)

Disease:
downy mildew

Time: Apply crown treatment after pruning, but before training; repeat at 10-day intervals.

Remarks: KOP-Hydroxide 50WP should not be applied in spray solution having pH less than 6.5 or greater than 9.0 as phytotoxicity may occur. Refer to label for use directions.

Days To Harvest: 2 weeks.

Signal Word/Toxicity Class: Varies by formulation.
REI: Varies by formulation.

FOSETYL-ALUMINUM

Systemic fungicide/bactericide. Toxic to estuarine invertebrates.

Aliette (WDG, WSP) *(Rhone-Poulenc)*

Rate: 2 lb. ai/A
2.5 lb. WDG, WSP/A

Disease:
downy mildew

Time: Apply when warm and humid conditions favor disease development. Apply when shoots are 6-12" high, after training when vines are 5-6 ft. tall, 3-weeks after second application, and during bloom.

Remarks: Apply using ground equipment. Do not apply more than 10 lb./A per season. With moderate disease pressure, high rate and shorter interval should be used. Do not tank mix with copper compounds. Do not feed refuse to animals. Refer to label for use directions.

Days To Harvest: 24.

State Restrictions: Not registered for use in California.

Signal Word/Toxicity Class: Caution.
REI: 12 hr.

Kocide — see copper compounds

KOP-Hydroxide — see copper compounds

MEFENOXAM

Ridomil Gold EC or WSP *(Novartis)*

Rate: Soil Drench:
0.25 lb. ai/A
0.5 pt. EC/A
0.23 lb. ai/A
0.5 lb. WSP/A

Disease:
seed rots — seedling diseases

Time: Apply before shoots are 6" long.

Remarks: Apply in minimum 20 gal. water. If primary infection persists or secondary infection occurs, Ridomil Gold PC can be tank mixed as foliar spray with Kocide 101 or another contact copper fungicide; refer to label for rates. Do not make more than 3 applications per season. Do not plant any crop which is not registered for use with mefenoxam in treated soil for 12 months.

Days To Harvest: 45.

Signal Word/Toxicity Class: Warning/II (EC); Caution (WSP).
REI: 48 hr.

Nu-Cop — see copper compounds

Ridomil Gold — see mefenoxam

SULFUR (Premix)

Top Cop with Sulfur (6.25 lb./gal. sulfur & 1 lb./gal. tribasic copper sulfate) *(Stoller)*

Rate: 4 qt./100 gal. water

Disease:
downy mildew — powdery mildew

Time: Apply after pruning but before training; after training, repeat at 10-day intervals.

Remarks: Refer to label for use directions.

Days To Harvest: 2 weeks.

Signal Word/Toxicity Class: Caution.
REI: 24 hr.

TOP COP — see copper compounds

Mint
(Peppermint, Spearmint)

Bravo — see chlorothalonil

CHLOROTHALONIL

Bravo 500, 720/Weather Stik, Ultrex (WDG) *(ZENECA)*
Echo 500, 720 or 90DF *(Sostram)*
Terranil (6L, 90DF or 90DF WSP) *(Terra)*

Rate: 1 lb. ai/A
2 pt. 500/A
1.375 pt. 720, 6L/A
1.25 lb. WDG/A
1.2 lb. 90DF/A (Echo)
1.125 lb. 90DF, 90DF WSP/A (Terranil)

Disease:
rust — Septoria leaf spot

Time: Apply when emerging plants are 4-8" high; repeat at 7- to 10-day intervals as necessary.

Remarks: Mint. Use in sufficient water to obtain adequate coverage. Do not apply more than 3 times per season. Do not feed fresh or extracted mint hay from treated fields to livestock. Refer to label for directions and restrictions.

Days To Harvest: 80.

State Restrictions: For use in Indiana, Michigan, Wisconsin. SLN-Oregon (Bravo 500).

Signal Word/Toxicity Class: Danger/I (WDG, 90DF); Warning/II (6L, 500, 720).
REI: 48 hr.

Clean Crop Nemasol — see metam-sodium

Echo — see chlorothalonil

Kumulus — see sulfur

METAM-SODIUM

Water-soluble liquid that when applied to properly prepared soil is converted into a gaseous fumigant.

Clean Crop Nemasol (42%) *(Platte)*
Metam CLR (42%) *(UCB Agrochemicals)*
Vapam (WS) or HL *(Amvac)*

Rate: 75 gal. 42%, HL/A
100 gal. WS/A
74.5 gal. 42%/A (UCB)

Disease:
Verticillium

Time: Preplant.

Remarks: Peppermint. Use injector blade or thin shank injector rig. Refer to label for use directions.

◆ - **new product** • ai-active ingredient • DF-dry flowable • E/EC-emulsifiable concentrate • EW-emulsion in water
F/FL-flowable • G-granule • L/LC-liquid • SP/WSP-soluble packet • W/WP-wettable powder • WSB-water soluble bag

State Restrictions: In California, application must be in compliance with Technical Information Bulletin "Guidelines for All Application Methods for Metam-sodium in California." This bulletin may be obtained from your local pesticide dealer or a metam-sodium registrant.

Signal Word/Toxicity Class: Danger/I.
REI: 48 hr.

Microthiol Special — see sulfur

Sulfa — see sulfur

SULFUR

Kumulus DF (sulfur) *(BASF)*
Rate: 3.2 lb. ai/A
4 lb. DF/A
Disease:
powdery mildew
Time: Apply when disease appears; repeat at 3 week intervals as often as necessary to cover new growth or until growth ceases in the fall.
Remarks: Mint. Do not apply if temperatures during or within 3 days after application are expected to exceed 90°F. Do not apply through any type of irrigation system. Refer to label for use directions.

Signal Word/Toxicity Class: Caution.
REI: 24 hr.

Micro Flo Sulfur (6L) *(Micro Flo)*
Microthiol Special (80WP) *(Elf Atochem)*
Sulfa 80 (80DF) *(Drexel)*
Terra Sulfur (6L) *(Terra)*
Thiolux (DF) *(Platte)*
Rate: 3-4.5 lb. ai/A
4-6 pt. 6L/A
3.2-4.8 lb. ai/A
4-6 lb. 80WP, 80DF, DF/A
Disease:
powdery mildew
Time: Apply when mint is 5-6" tall or when disease appears; repeat twice, at 30-day intervals.
Remarks: Peppermint, spearmint. Refer to label for use directions.
Days To Harvest: 30.
State Restrictions: Not for use in California for Microthiol Special.

Signal Word/Toxicity Class: Caution.
REI: 24 hr.

THAT (6FL) or THAT Big 8 (8FL) *(Stoller)*
Rate: 2.4-4.7 lb. ai/A
3.25-6.25 pt. 6FL/A
3-5 lb. ai/A
3-5 pt. 8FL/A
Disease:
powdery mildew
Time: Apply when first sign of infection appears or when plants have 4 to 5 leaves followed by additional spray at 2-week intervals, until pink bud stage, or 4 weeks before harvest.
Remarks: Peppermint, spearmint. Obtain authorization from buyer of oil before applying. Refer to label for use directions.

Signal Word/Toxicity Class: Caution.
REI: 24 hr.

TELONE

Liquid fumigant for preplant soil treatment. Restricted Use Pesticide.

Telone II (dichloropropene) *(Dow AgroSciences)*
Rate: 25-30 gal. II/A
Disease:
Verticillium
Time: Apply in spring, or preferably in fall.
Remarks: Mint. Use lower rate for mineral soils; higher rate for heavier soils. Do not apply through any type of irrigation system. Refer to label for use directions.

Signal Word/Toxicity Class: Warning/II.
REI: 72 hr.

TELONE (Premix)

Liquid fumigant for preplant soil treatment. Restricted Use Pesticide.

Telone C-17 (8.2 lb./gal. dichloropropene & 1.7 lb./gal. chloropicrin) *(Dow AgroSciences)*
Rate: 27.5 gal. C-17/A
Disease:
Verticillium
Time: Preplant soil treatment. Apply whenever soil conditions permit.
Remarks: Mint. Rate varies by soil type. Do not use in extremely heavy clay or muck soils. Do not apply this product through any type of irrigation system. Refer to label for use directions, precautions, and restrictions.

Signal Word/Toxicity Class: Danger/I.
REI: 72 hr.

Terranil — see chlorothalonil

THAT or THAT Big 8 — see sulfur

Thiolux — see sulfur

Vapam — see metam-sodium

Mung Beans

MEFENOXAM

Ridomil Gold EC or WSP *(Novartis)*
Rate: 0.25-0.5 lb. ai/A
0.5-1 pt. EC/A
0.23-0.45 lb. ai/A
0.5-1 lb. WSP/A
Disease:
damping-off root rot
Time: Apply preplant incorporated or as surface application just after planting.
Remarks: For banded applications, 7" band recommended. Do not use in greenhouse or on field grown vegetable bedding plants. May be tank mixed. Do not plant any crop which is not registered for use with mefenoxam in treated soil for 12 months.
Days To Harvest: 1.

Signal Word/Toxicity Class: Warning/II.
REI: 48 hr.

Ridomil Gold — see mefenoxam

Peanuts

Abound — see azoxystrobin

Allegiance — see metalaxyl

Apron XL — see mefenoxam

6 Other Field Crops

For personal protective equipment and EPA registration numbers, please turn to page 39.

AZOXYSTROBIN

Flowable broad-spectrum fungicide with preventive, systemic, and curative properties. Toxic to freshwater and estuarine/ marine fish and aquatic invertebrates.

Abound (FL) *(ZENECA)*
Rate: 0.3-0.4 lb. ai/A
18.5-24.6 fl. oz. FL/A
Disease:
early leaf spot, late leaf spot, pod rot, Pythium stem rot, web blotch, white mold
Time: Apply 60-90 days after planting.
Remarks: Apply foliar application by ground or air. Do not apply more than 1.54 qt./A per yr. or more than 2 applications per yr. Do not apply through any type of irrigation system or ultra-low volume spray system. Do not feed peanut hay. Refer to label for use directions.
Days To Harvest: 50.

Signal Word/Toxicity Class: Caution.
REI: 12 hr.

Basicop — see copper compounds

BENLATE

Benlate (SP, WP) (benomyl) *(DuPont)*
Rate: 0.13 lb. ai/A
0.25 lb. SP, WP/A
Disease:
Ascochyta web blotch, blackhull, rust
Time: Apply 35-40 days after planting or when disease first appears; repeat at 7- to 10-days for rust, 7- to 14-days for web blotch. For blackhull, apply in-furrow at planting.
Remarks: Do not tank mix or alternate with benzimidazole or thiophanate products. Do not graze or feed treated vines, hay, or hulls to livestock. Refer to label for use directions.
Days To Harvest: 14.
State Restrictions: Benlate SP not registered for use in California.

Signal Word/Toxicity Class: Caution.
REI: 24 hr.

Bravo — see chlorothalonil

Bravo S — see sulfur premix

CAPTAN

Drexel Captan 4FL *(Drexel)*
Rate: 6 fl. oz. 4FL/100 lb. seed
Disease:
seed decay, seedling blight, soilborne disease
Time: Liquid or slurry seed treatment.
Remarks: Do not use treated seed for food, feed, or oil purposes. Refer to label for use directions.

Signal Word/Toxicity Class: Danger/I.

Gustafson Captan 30-DD or 400 *(Gustafson)*
Rate: 6 fl. oz. 30-DD/100 lb. seed
3-6 fl. oz. 400/100 lb. seed
Disease:
damping-off mold, seed decay, seedling blight
Time: Seed treatment.
Remarks: Do not combine colloidal suspensions, wettable powders, or emulsifiable concentrates unless previous experience assures compatibility. Use with EPA-approved dye that imparts unnatural color to seed. Do not use for feed, food, or oil purposes. Refer to label for use directions.

Signal Word/Toxicity Class: Danger/I.

CAPTAN (Premix)

Special formulation for dry application to peanut seed. Toxic to fish and aquatic organisms.

Vitavax PC (45% captan & 15% PCNB & 10% carboxin) *(Gustafson)*
Rate: 4-5 oz./100 lb. seed
Disease:
Aspergillus damping-off, *Rhizoctonia solani*, Rhizopus seed rot
Time: Seed treatment.
Remarks: Use higher rate under conditions of heavy pressure. Do not use treated seed for food, feed, or oil purposes. Refer to label for use directions.

Signal Word/Toxicity Class: Danger/I.

CARBOXIN

Vitavax-30C or Vitavax-34 *(Gustafson)*
Rate: 3 fl. oz. 30C/100 lb. seed
3-6 fl. oz./100 lb. seed
Disease:
seed rot
Time: Slurry seed treatment.
Remarks: Incorporation of dye into treated slurry is recommended to distinguish seed as treated. Do not use treated seed for food, feed, or oil purposes. Do not graze or feed livestock on treated areas or hay grown from treated seed. Do not apply through any type of irrigation system. Refer to label for use directions.

Signal Word/Toxicity Class: Warning/II (30C); Caution (34).

CARBOXIN (Premix)

Prevail (15% carboxin & 15% PCNB & 3.12% metalaxyl) *(Trace)*
Rate: 4-8 oz./100 lb. seed
Disease:
damping-off, seed rot, seedling disease
Time: Planter box seed treatment.
Remarks: Treat only those seeds needed for immediate use. Do not use treated seed for food, feed, or oil purposes. Do not store excess treated seeds beyond planting time. All seed treated with this product (except planter box) must be colored with EPA approved dye which imparts unnatural color to seed. Do not graze or feed livestock hay grown from treated seed. Refer to label for use directions.

Signal Word/Toxicity Class: Caution.
REI: 12 hr.

Champ — see copper compounds

Champion — see copper compounds

CHLOROTHALONIL

Bravo 500, 720/Weather Stik, Ultrex (WDG) *(ZENECA)*
Echo 500, 720 or 90DF *(Sostram)*
Terranil (6L, 90DF or 90DF WSP) *(Terra)*
Rate: 0.75-1.1 lb. ai/A
1.5-2.125 pt. 500/A
1-1.5 pt. 720, 6L/A
0.9-1.4 lb. WDG/A
0.8-1.2 lb. 90DF/A (Echo)
0.9-1.1 lb. ai/A
1-1.25 lb. 90DF, 90DF WSP/A (Terranil)
Disease:
early leaf spot (Cercospora), late leaf spot (Cercosporidium), rust, web blotch
Time: Apply when leaf wetness first occurs and before disease symptoms appear (30-40 days after planting); repeat at 10- to 14-day intervals. When late leaf spot prevails or when rust or web blotch occurs, apply high rate at 10-day intervals for remainder of season.

◆ - **new product** • ai-active ingredient • DF-dry flowable • E/EC-emulsifiable concentrate • EW-emulsion in water
F/FL-flowable • G-granule • L/LC-liquid • SP/WSP-soluble packet • W/WP-wettable powder • WSB-water soluble bag

Remarks: Use in sufficient water to obtain adequate coverage. Do not graze treated areas or feed hay or threshings from treated fields to livestock. Can be applied through sprinkler irrigation equipment; refer to label for use directions and type of equipment.
Days To Harvest: 14.

Signal Word/Toxicity Class: Danger/I (WDG, 90DF); Warning/II (6L, 500, 720).
REI: 48 hr.

Clean Crop Nemasol — see metam-sodium

COPPER COMPOUNDS

Basicop (WP) (basic copper sulfate) *(Griffin)*
Rate: 1.5 lb. ai/A
3 lb. WP/A
Disease:
Cercospora leaf spot
Time: Apply when disease occurs, repeat at 14-day intervals.
Remarks: Do not apply through any type of irrigation system. Refer to label for use directions.

Signal Word/Toxicity Class: Warning/II.
REI: 24 hr.

Champ (FL or Formula 2) or Champion (WP) (copper hydroxide) *(Agtrol)*
Kocide (101, DF, 2000, 4.5LF, LF) (copper hydroxide) *(Griffin)*
KOP-Hydroxide (FL, 50WP) (copper hydroxide) *(Drexel)*
TOP COP Tri-Basic (FL) (basic copper sulfate) *(Stoller)*
Rate: 2-4 pt. FL, LF/A
1-2 pt. Formula 2, 4.5LF/A
1.5-3 lb. WP, 101, DF, 50WP/A
1.25-2.25 lb. 2000/A
Disease:
Cercospora leaf spot
Time: Apply after planting or when disease symptoms first appear; repeat at 10- to 14-day intervals; 7-day intervals during humid weather.
Remarks: Use sufficient water for thorough coverage. Refer to label for use directions.
State Restrictions: Kocide LF is not for use in California.

Signal Word/Toxicity Class: Varies by formulation.
REI: Varies by formulation.

Copper-Count-N (L) (copper ammonium) *(Mineral Research)*
Rate: 1.5-3 qt. L/A
Disease:
Cercospora leaf spot
Time: Apply on appearance of disease at 10- to 14-day intervals.

Rate: North Carolina:
12 qt. L/A
Disease:
Pythium *Rhizoctonia solani* *Sclerotium rolfsii*
Time: Apply at pegging in a 12-15" band over the row.
Remarks: Applications may be made by ground or air; refer to label for use directions. Do not apply through any type of irrigation system unless supplemental labeling on chemigation is followed.

Signal Word/Toxicity Class: Caution.
REI: 12 hr.

Drexel Basic Copper Sulfate (WP) or KOP 300 (3FL) (basic copper sulfate) *(Drexel)*
Rate: 2-3 lb. WP/A
2-4 pt. 3FL/A
Disease:
Cercospora leaf spot
Time: Repeat at 10- to 14-day intervals.
Remarks: Do not apply through any type of irrigation system. Refer to label for use directions.

Signal Word/Toxicity Class: Warning/II.
REI: 24 hr. (WP); 12 hr. (3FL).

Micro Flo Basic Copper "53" (basic copper sulfate) *(Micro Flo)*
Rate: 2-3 lb. 53/A
Disease:
early leaf spot late leaf spot
Time: Repeat at 10- to 14-day intervals as needed.
Remarks: Refer to label for use directions.

Signal Word/Toxicity Class: Warning/II.
REI: 24 hr.

Microsperse COC 53WP (copper oxychloride) or Nu-Cop (3L, 50DF, 50WP) (copper hydroxide) *(Micro Flo)*
Rate: 1-4 pt. 3L/A
2-3 lb. 53WP/A
1.5-3 lb. 50DF, 50WP/A
Disease:
Cercospora leaf spot
Time: Apply 25-40 days (35-40 days for 3L) after planting or when disease symptoms appear; repeat at 10- to 14-day intervals as needed.
Remarks: Refer to label for use directions.

Signal Word/Toxicity Class: Danger/I (50DF, 50WP, 53WP); Warning/II (3L).
REI: 48 hr. (50DF, 50WP); 24 hr. (53WP, 3L).

Copper-Count-N — see copper compounds

Dithane — see mancozeb

Echo — see chlorothalonil

FLUTOLANIL

Systemic fungicide in water-soluble package. Toxic to shrimp.

Moncut 50WP *(Gowan)*
Rate: 1-2 lb. ai/A
2-4 lb. 50WP/A
Disease:
pod rot *Sclerotium rolfsii* white mold
Rhizoctonia solani stem rot
Time: Broadcast foliage application.
Remarks: May be applied by ground, air, or chemigation. A split application may be used in fields with a history of high disease, or under prolonged conditions conducive to disease infection and development. Refer to label for use directions. Do not apply more than 4 lb. 50WP/A per single growing season. Do not feed hay or threshing from rotational crops of soybeans or cotton planted following harvest of treated peanuts (more than 2 lb. 50WP/A). Do not allow cattle to graze soybean or cotton forage in previously treated fields.
Crop Rotation: Do not plant rotation crops other than peanuts for 12 months following last application except: corn, soybeans, or cotton may be planted 240 or more days following the last application.

Signal Word/Toxicity Class: Caution.
REI: 12 hr.

Folicur — see tebuconazole

IPRODIONE

Dicarboximide fungicide. Toxic to aquatic invertebrates.

Rovral (4FL, WG, WP) *(Rhone-Poulenc)*
Rate: 1 lb. ai/A
2 pt. 4FL/A
2 lb. WG, WP/A
Disease:
Sclerotinia blight
Time: Apply when conditions favor disease development. Up to 2 subsequent applications should be made at 14- to 21-day intervals.
Remarks: Up to 3 applications may be applied per season. Do not apply by air. Do not feed peanut hay to livestock. Refer to label for use directions.
Days To Harvest: 10.
Crop Rotation: Beans, broccoli, carrots, garlic, lettuce, onions (dry bulb), peanuts, potatoes, and rice can be rotated after harvest.
State Restrictions: Not registered for use in California.

Signal Word/Toxicity Class: Warning/II (WG); Caution (4FL, WP).
REI: 12 hr.

NOTICE	The information on these pages is for preliminary planning — not a guide for use. Be sure to follow the manufacturer's directions, notwithstanding information contained here.	For personal protective equipment and EPA registration numbers, please turn to page 39.

Kocide — see copper compounds

KOP-Hydroxide — see copper compounds

Kumulus — see sulfur

MANCOZEB

Dithane (DF, F-45, M-45, ST, WSP) *(Rohm and Haas)*
Manzate 200DF *(Griffin)*
Penncozeb 75DF or 80WP *(Elf Atochem)*

Rate: 0.75-1.5 lb. ai/A
1-2 lb. DF, 200DF, 75DF/A
0.8-1.6 lb. ai/A
1-2 lb. M-45, WSP, 80WP/A
1.6-3.2 pt. F-45/A

Disease:
Cercospora leaf spot rust

Time: Apply when disease first appears; repeat at 7- to 14-day intervals.
Remarks: Do not feed treated vines to livestock. Refer to label for use directions.
Days To Harvest: 14.
State Restrictions: Refer to label for California precautions.

Rate: 8.5-17 oz. DF, 75DF/100 lb. seed
8-16 oz. M-45, WSP, 200DF, 80WP/100 lb. seed
12.8-25.6 fl. oz. F-45, ST/100 lb. seed

Disease:
damping-off seed rot seedling blight

Time: Seed treatment.
Remarks: **Peanuts (shelled).** Dye must be added to the treating slurry so that unnatural color will distinguish the seed as treated. Do not use for food, feed or oil purposes. Refer to label for use directions.
State Restrictions: Refer to label for California precautions.

Signal Word/Toxicity Class: Caution.
REI: 24 hr.

MANEB (Premix)

Fungicide/bactericide. Toxic to fish, birds, and other wildlife.

ManKocide (DF) (15% maneb & 46.1% copper hydroxide) *(Griffin)*

Rate: 2-4 lb. DF/A

Disease:
Cercospora leaf spot

Time: Apply when disease first appears; repeat at 7- to 14-day intervals. Reduce sprays to 7-day intervals during humid weather.
Remarks: High rate when conditions favor disease. Do not feed treated vines to livestock. Refer to label for use directions.
Days To Harvest: 14.

Signal Word/Toxicity Class: Danger/I.
REI: 48 hr.

ManKocide — see maneb premix

Manzate — see mancozeb

MEFENOXAM

Apron XL LS *(Novartis)*

Rate: 0.32-0.64 fl. oz. LS/100 lb. seed

Disease:
damping-off seed rot

Time: Seed treatment at planting.
Remarks: Do not use treated seed for food, feed, or oil purposes. Refer to label for use directions.

Signal Word/Toxicity Class: Warning/II.

Ridomil Gold EC, GR, or WSP *(Novartis)*

Rate: 0.25-1 lb. ai/A
0.5-1 pt. EC/A
0.5-1 lb. WSP/A
13 oz. GR/1000 lin. ft. row

Disease:
pod rot

Time: Apply at early pod set or pegging.

Rate: 0.125 lb. ai/A
0.25 pt. EC/A
0.25 lb. WSP/A
6.5 oz. GR/1000 lin. ft. row

Disease:
seedling diseases

Time: Apply in-furrow or at planting.
Remarks: Where pathogens other than Pythium are present, use fungicides that contol those pathogens in combination with Ridomil Gold. Do not plant any crop which is not registered for use with mefenoxam in treated soil for 12 months. Refer to label for use directions.

Signal Word/Toxicity Class: Warning/II (EC); Caution (GR, WSP).
REI: 48 hr.

METALAXYL

Allegiance (FL) *(Gustafson)*

Rate: 0.75 fl. oz. FL/100 lb. seed

Disease:
damping-off seed rot

Time: Seed treatment.
Remarks: Use with EPA-approved dye that imparts unnatural color to seed. Do not use for feed, food, or oil purposes. Refer to label for use directions.
State Restrictions: Allegiance not for use in California.

Signal Word/Toxicity Class: Warning/II.

METAM-SODIUM

Fumigant for soil preparation only. Toxic to fish.

Clean Crop Nemasol (42%) *(Platte)*

Rate: 7.5 gal. 42%/A

Disease:
Cylindrocladium black rot

Time: Preplant.
Remarks: Plant peanuts in center of treated beds no earlier than 14 days following application. Soil temperature must be in 60°-90°F range. Refer to label for use directions.
State Restrictions: In California, application must be in compliance with Technical Information Bulletin "Guidelines for All Application Methods for Metam-sodium in California." This bulletin may be obtained from your local pesticide dealer or a metam-sodium registrant.

Signal Word/Toxicity Class: Danger/I.
REI: 48 hr.

Metam CLR (42%) *(UCB Agrochemicals)*

Rate: Resistant cultivar:
7.5 gal. 42%/A
Susceptible cultivar:
15 gal. 42%/A

Disease:
Cylindrocladium black rot

Time: Preplant.
Remarks: Do not use on highly susceptible cultivars. Plant peanuts in center of treated beds no earlier than 14 days following application. Soil temperature must be in 60°-90°F range. Refer to label for use directions.
State Restrictions: In California, application must be in compliance with Technical Information Bulletin "Guidelines for All Application Methods for Metam-sodium in California." This bulletin may be obtained from your local pesticide dealer or a metam-sodium registrant.

Signal Word/Toxicity Class: Danger/I.
REI: 48 hr.

Micro Flo — see copper compounds

◆ - **new product** • ai-active ingredient • DF-dry flowable • E/EC-emulsifiable concentrate • EW-emulsion in water
F/FL-flowable • G-granule • L/LC-liquid • SP/WSP-soluble packet • W/WP-wettable powder • WSB-water soluble bag

Microsperse — see copper compounds

Microthiol Special — see sulfur

Moncut — see flutolanil

Nu-Cop — see copper compounds

PCNB

PCNB 2LF, 10%, 75DG or 75WP *(Amvac)*

Rate: 10 lb. ai/A
100 lb. 10%/12,400 lin. ft. row
13.3 lb. 75DG, 75WP/A in 8-10 gal. water
Time: Preplant soil, surface band at-pegging.

Rate: 3.3 lb. ai/A
33 lb. 10%/12,400 lin. ft. row
4.33 lb. 75DG, 75WP/A
Time: At cultivation.
Disease:
root rot
Remarks: Preplant soil mix: Apply in 8-12" band centered on row; incorporate to a depth of 2-4". Application may be made simultaneously with planting, using suitable equipment. At-pegging band application: Apply in 12" band centered on row so as to cover crown and pegging zones of plant. At cultivation: Attach 2 spray nozzles per row of inside cultivator sweeps and adjust so as to treat a total band of 12" centered on row. Do not apply more than 3 times per crop season. Do not pile untreated soil on crown of plant. Do not exceed 10 lb./A in any one season. Do not apply through any type of irrigation system. Do not feed treated vines to livestock. Do not allow hogging down of peanuts. Do not plant root crops within 12 months of application unless PCNB is registered for use on those crops.
Days To Harvest: 45.

- - - - - - - - - - - - - - - - -

Rate: 2.33-4.66 fl. oz. 2LF/100 lb. seed
Disease:
Fusarium Rhizoctonia
Time: Seed treatment.
Remarks: Do not dilute with water. Do not apply through any type of irrigation system. Do not feed treated vines to livestock. Do not plant root crops within 12 months of application unless PCNB is registered for use on those crops.

- - - - - - - - - - - - - - - - -

Signal Word/Toxicity Class: Caution.
REI: 12 hr.

RTU-PCNB *(Gustafson)*

Rate: 1.75-3.5 fl. oz. RTU/100 lb. seed
Disease:
Fusarium Rhizoctonia
Time: Seed treatment.
Remarks: Do not dilute with water. Do not use treated seed for food, feed, or oil purposes. Do not feed treated seed to animals. Refer to label for use directions.

- - - - - - - - - - - - - - - - -

Signal Word/Toxicity Class: Caution.

Penncozeb — see mancozeb

PROPICONAZOLE

Tilt (EC) *(Novartis)*

Rate: 4 fl. oz. EC/A
Disease:
late leaf spot
Time: Apply 35-40 days after planting or when disease is first noticed; repeat at 10- to 14-day intervals.

Rate: 2.5-4 fl. oz. EC/A
Disease:
early leaf spot
Time: Apply 35-40 days after planting or when disease is first noticed; repeat at 14-day intervals.
Remarks: Do not apply more than 16 fl. oz./crop per season. May be tank mixed. Do not graze or feed livestock treated forage or cut green crop for hay or silage.

- - - - - - - - - - - - - - - - -

Signal Word/Toxicity Class: Warning/II.
REI: 24 hr.

Prevail — see carboxin premix

Redball — see sulfur

Ridomil Gold — see mefenoxam

RIDOMIL GOLD PC (Premix)

Ridomil Gold PC or PC GR (0.5% mefenoxam & 10% PCNB) *(Novartis)*

Rate: 12.5-25 lb./14,520 lin. ft. row
Disease:
seed rot
Time: Apply at planting.

Rate: 50 lb./14,520 lin. ft. row
Disease:
pod rot
Time: Apply at early pegging.
Remarks: Do not apply more than 75 lb./14,520 lin. ft. row per season. Do not allow livestock to feed or graze foliage or harvest residues. Do not plant any crop which is not registered for use with mefenoxam in treated soil for 12 months. Refer to label for use directions.
Days To Harvest: 75.

- - - - - - - - - - - - - - - - -

Signal Word/Toxicity Class: Warning/II.
REI: 48 hr.

Rovral — see iprodione

SULFUR

Drexel Suffa (6FL, 8FL), Sulfa 80 (80DF) or Sulfur (90W) *(Drexel)*

Rate: 1.5-3 lb. ai/A
2-4 pt. 6FL/A
1.5-3 pt. 8FL/A
2-4 lb. aiA
3-5 lb. 80DF/A
2.7-4.5 lb. ai/A
3-5 lb. 90W/A
Disease:
leaf spot (except 8FL) powdery mildew (except 8FL) rust
Time: Apply early leaf stage; repeat at 10- to 24-day intervals as necessary. For Suffa-8, apply 6 weeks after planting; repeat at 7- to 10-day intervals as necessary.
Remarks: Do not use within 2 weeks of oil application (4 weeks FL). Some plants are susceptible to injury from sulfur under certain climatic conditions. Do not make sulfur applications at excessively high temperatures. Do not apply this product through any type of irrigation system. Refer to label for use directions.

- - - - - - - - - - - - - - - - -

Signal Word/Toxicity Class: Caution.
REI: 24 hr.

Kumulus DF (sulfur) *(BASF)*

Rate: 2.4-4 lb. ai/A
3-5 lb. DF/A
Disease:
powdery mildew
Time: Apply in early leaf stage; repeat applications as necessary, usually 10- to 24-days.
Remarks: Do not apply if temperatures are expected to exceed 90°F during or within 3 days after application. Do not apply through any type of irrigation system. Refer to label for use directions.

- - - - - - - - - - - - - - - - -

Signal Word/Toxicity Class: Caution.
REI: 24 hr.

6 Other Field Crops

NOTICE	The information on these pages is for preliminary planning — not a guide for use. Be sure to follow the manufacturer's directions, notwithstanding information contained here.	For personal protective equipment and EPA registration numbers, please turn to page 39.

Liquid Sulfur Six (6FL) *(Helena)*
THAT (6FL) or THAT Big 8 (8FL) *(Stoller)*

Rate: 1.5-3 lb. ai/A
2-4 pt. 6FL/A
1.3-2.5 lb. ai/A
1.3-2.5 pt. 8FL/A

Disease:
rust

Time: Apply when disease threatens; repeat at 7- to 10-day intervals.
Remarks: Do not apply Liquid Sulfur Six through any type of irrigation system. Refer to label for use directions.

Signal Word/Toxicity Class: Caution.
REI: 24 hr.

Micro Flo Sulfur (6L), Dual Spraying & Dusting Sulfur (DS) or Microsperse Wettable Sulfur (WS) *(Micro Flo)*

Rate: 4.5-6.4 lb. ai/A
5.33-8.5 pt. 6L/A
4.5 lb. ai/A
5 lb. DS, WS/A

Disease:
leaf spot — powdery mildew — rust (6L)

Time: Apply at early leaf stage; repeat at 18-day intervals.
Remarks: Do not apply DS and WS through any type of irrigation system. Refer to label for use directions.

Signal Word/Toxicity Class: Caution.
REI: 24 hr.

Microthiol Special (80WP) *(Elf Atochem)*

Rate: 4-6.4 lb. ai/A
5-8 lb. 80WP/A

Disease:
powdery mildew — rust

Time: Apply at early leaf stage; repeat as necessary.
Remarks: Refer to label for use directions.
State Restrictions: Not for use in California for rust.

Signal Word/Toxicity Class: Caution.
REI: 24 hr.

Redball EM-53 (6L), Superfine Wettable Sulphur (WS) or Commercial Dusting Sulphur (DS) *(International Sulphur)*

Rate: 2-4 pt. 6L/A
6-16 lb. WS/A
20-35 lb. DS/A

Disease:
leaf spot (WS, DS) — powdery mildew (WS, DS) — rust (6L)

Time: 6L: Apply 40-60 days after planting; repeat at 7- to 14-day intervals as needed; WS, DS: Apply early leaf stage; repeat at 10- to 24-day intervals as needed.
Remarks: Apply in sufficient water for thorough coverage. Do not apply when temperatures exceed 90°F. Refer to label for use directions.

Signal Word/Toxicity Class: Caution.
REI: 12 hr.

Terra Sulfur (6L) *(Terra)*

Rate: 5.33-8.5 pt. 6L/A

Disease:
leaf spot — powdery mildew — rust

Time: Apply at early leaf stage; repeat at 18-day intervals.
Remarks: Do not use within 2 weeks of oil application. Some plants are susceptible to injury from sulfur under certain climatic conditions. Do not make sulfur applications at excessively high temperatures. Refer to label for use directions.

Signal Word/Toxicity Class: Caution.
REI: 24 hr.

Thiolux (DF) *(Platte)*

Rate: 3-5 lb. DF/A

Disease:
leaf spot — rust

Time: Apply early leaf stage; repeat at 10- to 24-day intervals.
Remarks: Do not make applications within 2 weeks of oil spray. Refer to label for use directions.

Signal Word/Toxicity Class: Caution.
REI: 24 hr.

SULFUR (Premix)

Bravo S (27.25% sulfur & 19.15% chlorothalonil) *(Helena)*

Rate: 3-4.5 pt./A

Disease:
early leaf spot (Cercospora) — late leaf spot (Cercosporidium) — rust — web blotch

Time: Apply when disease first appears; repeat at 10- to 14-day intervals as needed.
Remarks: Under severe conditions, use high rate. Do not allow livestock to graze treated areas, or feed hay or threshings from treated fields to livestock. Refer to label for use directions.
Days To Harvest: 14.

Signal Word/Toxicity Class: Caution.
REI: 48 hr.

Terranil S (19.15% chlorothalonil & 27.25% sulfur) *(Terra)*

Rate: 3-4.5 pt. S/A

Disease:
Cercospora leaf spot — Cercosporidium leaf spot — rust — web blotch

Time: Apply when leaf wetness first occurs or 30-40 days after planting; repeat at 10- to 14-day intervals.
Remarks: When conditions favor late leaf spot or when rust or web blotch occur, apply high rate and shorter interval for remainder of season. Use in sufficient water to obtain adequate coverage. Do not allow livestock to graze in treated areas or feed hay or threshing from treated fields to livestock. Refer to label for use directions.
Days To Harvest: 14.

Signal Word/Toxicity Class: Warning/II.
REI: 48 hr.

Top Cop with Sulfur (6.25 lb./gal. sulfur & 1 lb./gal. tribasic copper sulfate) *(Stoller)*

Rate: 6 pt./A

Disease:
early leaf spot — late leaf spot — rust

Time: Apply 40-60 days after planting; repeat at 10- to 14-day intervals.
Remarks: Refer to label for use directions.

Signal Word/Toxicity Class: Caution.
REI: 24 hr.

SYSTEM³

System³ (0.1% *Bacillus subtilis* & 14.25% metalaxyl & 16.67% PCNB) *(Helena)*

Rate: 4-6 oz./100 lb.

Disease:
Pythium — *Rhizoctonia solani*

Time: Apply at planting.
Remarks: A non-oil lubricant has been added for seeds to flow more easily and evenly from planter boxes. Incorporation of dye is recommended to distinguish seed as treated. Do not use treated seed for food, feed or oil purposes. Refer to label for use directions.

Signal Word/Toxicity Class: Caution.
REI: 12 hr.

TEBUCONAZOLE

Triazole sterol inhibitor. Toxic to estuarine/marine invertebrates.

Folicur 3.6F *(Bayer)*

Rate: 7.2 fl. oz. 3.6F/A

Disease:
early leaf spot — late leaf spot — Rhizoctonia limb rot — rust — Sclerotium stem rot — white mold

Time: Apply 4 application spray program for early and late leaf spots and rust; 7 application for Rhizoctonia limb rot and white mold.

◆ - **new product** • ai-active ingredient • DF-dry flowable • E/EC-emulsifiable concentrate • EW-emulsion in water
F/FL-flowable • G-granule • L/LC-liquid • SP/WSP-soluble packet • W/WP-wettable powder • WSB-water soluble bag

Remarks: Chlorothalonil should be applied prior to and following Folicur application to discourage development of resistant strains of fungi. For control of early and late leaf spots, spray surfactant should be used. Do not feed hay or threshing or allow livestock to graze treated areas. Do not apply through any type of irrigation system. Refer to label for use directions.
Days To Harvest: 14.
Crop Rotation: Treated areas may be replanted with peanuts as soon as practical after last application; any other crop may be planted 120 days after last application.

Signal Word/Toxicity Class: Caution.
REI: 12 hr.

TELONE (Premix)

Restricted Use Pesticide.

Telone C-17 (8.2 lb./gal. dichloropropene & 1.7 lb./gal. chloropicrin) *(Dow AgroSciences)*
Rate: 10.8-21.6 gal. C-17/A
Disease:
pod rot
Time: Preplant soil treatment. Apply whenever soil conditions permit.
Remarks: Rate varies by soil type. Do not use in extremely heavy clay or muck soils. Do not apply this product through any type of irrigation system. Refer to label for use directions, precautions, and restrictions.

Signal Word/Toxicity Class: Danger/I.
REI: 72 hr.

Terranil — see chlorothalonil

TERRANIL Cu (Premix)

Toxic to fish, aquatic invertebrates, and marine/estuarine organisms.

Terranil Cu (24% chlorothalonil & 24% copper oxychloride) *(Terra)*
Rate: 2.25-3.4 pt. Cu/A
Disease:
Cercospora leaf spot, Cercosporidium leaf spot, rust, web blotch
Time: Apply when leaf wetness first occurs or 30-40 days after planting; repeat at 10- to 14-day intervals.
Remarks: When conditions favor late leaf spot or when rust or web blotch occur, apply high rate and shorter interval for remainder of season. Use in sufficient water to obtain adequate coverage. Do not allow livestock to graze in treated areas or feed hay or threshing from treated fields to livestock. Refer to label for use directions.
Days To Harvest: 14.

Signal Word/Toxicity Class: Danger/I.
REI: 48 hr.

THAT or THAT Big 8 — see sulfur

Thiolux — see sulfur

THIOPHANATE-METHYL

Topsin M (WSB, 70W) *(Elf Atochem)*
Rate: 0.35 lb. ai/A
0.5 lb. WSB, 70W/A
Disease:
Ascochyta web blotch, leaf spot, limb rot, rust
Time: Begin applications 35 days after planting or when disease first appears; repeat at 7- to 14-day intervals as needed (14- to 21-day for 70W). Use 7-day interval (14-day for 70W) under severe disease pressure.
Remarks: Refer to label for use directions.
Days To Harvest: 14.

Signal Word/Toxicity Class: Caution.
REI: 12 hr.

THIRAM

42-S Thiram or Thiram 50WP Dyed *(Gustafson)*
Thiram Granuflo (WDG) *(UCB Agrochemicals)*
Rate: 3 fl. oz. 42-S/100 lb. seed
4.5 oz. 50WP/100 lb. seed
3 oz. WDG/100 lb. seed
Disease:
damping-off, seed decay, seedling blight
Time: Seed treatment.
Remarks: Peanuts (shelled, unshelled). Do not dilute with water. Use with EPA-approved dye that imparts unnatural color to seed. Do not use for feed, food, or oil purposes. Do not apply Thiram Granuflo through any type of irrigation system. Refer to label for use directions.
State Restrictions: Thiram Granuflo not for use in California.

Signal Word/Toxicity Class: Caution.
REI: 24 hr. (Thiram Granuflo).

Tilt — see propiconazole

TOP COP — see copper compounds

Top Cop with Sulfur — see sulfur premix

Topsin M — see thiophanate-methyl

Vitavax — see carboxin

Vitavax PC — see captan premix

Rice

Allegiance — see metalaxyl

Apron — see metalaxyl

Apron XL — see mefenoxam

BENLATE

Benlate (SP, WP) (benomyl) *(DuPont)*
Rate: 0.5-1 lb. ai/A
1-2 lb. SP, WP/A
Disease:
blast, leaf smut, narrow brown leaf spot, sheath blight, stem rot
Time: Apply at booting; repeat at heading.
Remarks: Do not apply to stubble rice. Do not apply to fields where crayfish or catfish farming is practiced. Drain water from treated areas into areas where such farming is practiced. Water from treated areas must not be used to irrigate other crops. Do not tank mix or alternate with benzimidazole or thiophanate products. Refer to label for use directions.
Days To Harvest: 21 days.
State Restrictions: Not for use in California.

Signal Word/Toxicity Class: Caution.
REI: 24 hr.

For personal protective equipment and EPA registration numbers, please turn to page 39.

CAPTAN

Drexel Captan 4FL *(Drexel)*
Rate: 3.4 fl. oz. 4FL/100 lb. seed
Disease:
seed decay   seedling blight   soilborne disease
Time: Liquid or slurry seed treatment.
Remarks: Do not use treated seed for food, feed, or oil purposes. Refer to label for use directions.
- - - - - - - - - - - - - - - - -
Signal Word/Toxicity Class: Danger/I.

CARBOXIN (Premix)

RTU-Vitavax-Thiram (10% carboxin & 10% thiram) *(Gustafson)*
Rate: 5-6.8 fl. oz./100 lb. seed
Disease:
Fusarium   Pythium   seedling diseases
Helminthosporium oryzae   *Rhizoctonia solani*
Time: Seed treatment.
Remarks: Use high rates under heavy disease pressure. Use with EPA-approved dye that imparts unnatural color to seed. Do not use for feed, food, or oil purposes. Refer to label for use directions.
- - - - - - - - - - - - - - - - -
Signal Word/Toxicity Class: Caution.

Champ — see copper compounds

COPPER COMPOUNDS

Champ (FL or Formula 2) (copper hydroxide) *(Agtrol)*
Rate: 4-8 fl. oz. FL/100 lb. seed
2-4 fl. oz. Formula 2/100 lb. seed
Disease:
seed rot   water mold
Time: Apply as seed dressing.
Remarks: Consult State Agricultural Experiment Station for specific recommendations.
- - - - - - - - - - - - - - - - -
Signal Word/Toxicity Class: Caution.
REI: 24 hr.

Kocide (4.5LF) (copper hydroxide) *(Griffin)*
Rate: 2-4 fl. oz. 4.5LF/100 lb. seed
Disease:
Achlya   Pythium
Time: Seed treatment.
Remarks: Do not apply in a spray solution having pH less than 6.5 as phytotoxicity could occur. Refer to label for use directions.
State Restrictions: Not for use in California.
- - - - - - - - - - - - - - - - -
Signal Word/Toxicity Class: Caution.
REI: 24 hr.

Nu-Cop (3L) (copper hydroxide) *(Micro Flo)*
Rate: Barley, wheat:
2 fl. oz. 3L/100 lb. seed
Disease:
bacterial blight   common bunt

Rate: Rice:
2-4 fl. oz. 3L/100 lb. seed
Disease:
seed rot   water mold
Time: Seed treatment.
Remarks: Refer to label for use directions.
- - - - - - - - - - - - - - - - -
Signal Word/Toxicity Class: Warning/II.
REI: 24 hr.

Dithane — see mancozeb

FLUTOLANIL

Moncut 50WP *(Gowan)*
Rate: 0.35-0.5 lb. ai/A
0.7-1 lb. 50WP/A in 5-10 gal. water
Disease:
Rhizoctonia solani
Time: Apply at first internode elongation; repeat 10- to 14-days later. Use higher rate where disease pressure is expected to be heavy.

Rate: 0.35-0.7 lb. ai/A
0.7-1.4 lb. 50WP/A
Disease:
sheath blight
Time: In fields where close scouting is practiced, apply when 5-10% of tillers of a susceptible variety or 10-15% of moderately susceptible variety have sheath blight lesions. Continue scouting and reapply if disease begins to move up the stem again.
Remarks: Do not apply more than 2 lb. 50WP/A per year. Do not apply beyond 75% heading development stage. Do not use oil as a carrier or add other additives to finished spray.
Days To Harvest: 30.
Crop Rotation: Do not plant rotational crops other than rice or peanuts for 12 months following last application except: soybeans or grain sorghum may be planted 240 days or more following last application.
State Restrictions: Not for use in California, and in certain counties in Arkansas (with restrictions); refer to label.
- - - - - - - - - - - - - - - - -
Signal Word/Toxicity Class: Caution.
REI: 12 hr.

Grain Guard — see mancozeb

IPRODIONE

Rovral (4FL) *(Rhone-Poulenc)*
Rate: 0.5 lb. ai/A
1 pt. 4FL/A
Disease:
brown spot   narrow brown leaf spot   sheath blight
Time: Make first application between joint movement and booting; second 14-days after first application, but no later than 75% heading.
Remarks: Apply as broadcast spray by air. Up to 2 applications may be applied per season. Do not apply in areas where catfish and crayfish are commercially cultivated. Refer to label for use directions.
Crop Rotation: Beans, broccoli, carrots, garlic, lettuce, onions (dry bulb), peanuts, potatoes, and rice can be rotated after harvest.
State Restrictions: Not registered for use in California. Restrictions apply to various counties in Arkansas; refer to label.
- - - - - - - - - - - - - - - - -
Signal Word/Toxicity Class: Caution.
REI: 12 hr.

Kocide — see copper compounds

MANCOZEB

Grain Guard contains graphite.

Dithane (DF, F-45, M-45, ST, WSP) *(Rohm and Haas)*
Manzate 200DF *(Griffin)*
Penncozeb 75DF or 80WP *(Elf Atochem)*
Rate: 2.1-4.3 oz. DF, 75DF/100 lb. seed
2-4 oz. M-45, WSP, 200DF, 80WP/100 lb. seed
3.2-6.4 fl. oz. F-45, ST/100 lb. seed
Disease:
damping-off   seed rot   seedling blight
Time: Seed treatment. Apply before, during, or after soaking in water.
Remarks: Incorporation of dye into the treating slurry is recommended to distinguish seed as treated. Do not use treated seed for food, feed, or oil purposes. Refer to label for use directions.
State Restrictions: For Manzate, refer to label for California use restrictions.
- - - - - - - - - - - - - - - - -
Signal Word/Toxicity Class: Caution.
REI: 24 hr.

◆ - **new product** • ai-active ingredient • DF-dry flowable • E/EC-emulsifiable concentrate • EW-emulsion in water
F/FL-flowable • G-granule • L/LC-liquid • SP/WSP-soluble packet • W/WP-wettable powder • WSB-water soluble bag

Grain Guard *(Trace)*
Rate: 4 oz./100 lb. seed
Disease:
damping-off seed rot seedling blight
Time: Drill box seed treatment.
Remarks: Treat only those seeds needed for immediate use. Do not use treated seed for food, feed, or oil purposes. Do not store excess treated seeds beyond planting time. Refer to label for use directions.

Signal Word/Toxicity Class: Caution.
REI: 12 hr.

Manzate — see mancozeb

MEFENOXAM

Apron XL LS *(Novartis)*
Rate: 0.32-0.64 fl. oz. LS/100 lb. seed
Disease:
damping-off
Time: Seed treatment at planting.
Remarks: Do not use treated seed for food, feed, or oil purposes. Incorporation of dye into treating slurry is recommended to distinguish seed as treated.

Signal Word/Toxicity Class: Warning/II.

METALAXYL

Allegiance or Apron (FL) *(Gustafson)*
Rate: 0.375-0.75 fl. oz. Allegiance/100 lb. seed
0.75-1.5 fl. oz. Apron/100 lb. seed
Disease:
damping-off Phytophthora (early-season)
Time: Seed treatment.
Remarks: Do not use Ridomil for control of Pythium in water-seeded rice if using Allegiance or Apron (FL). Use with EPA-approved dye that imparts unnatural color to seed. Do not use treated seed for food, feed, or oil purposes. Refer to label for use directions.
State Restrictions: Not for use in California.

Signal Word/Toxicity Class: Warning/II.

Apron Dry *(Trace)*
Rate: 3-4 oz./100 lb. seed
Disease:
damping-off
Time: Seed treatment.
Remarks: Incorporation of dye into treating slurry recommended to distinguish seed as treated. Do not use treated seed for food, feed, or oil purposes. Refer to label for use directions.

Signal Word/Toxicity Class: Caution.

Moncut — see flutolanil

Nu-Cop — see copper compounds

PCNB

PCNB 2-E or 2LF *(Amvac)*
Rate: 4.45-8.90 fl. oz. 2-E/100 lb. seed
5.25-10.5 fl. oz. 2LF/100 lb. seed
Disease:
covered smut (2-E) Fusarium Rhizoctonia
Time: Seed treatment.
Remarks: Do not apply through any type of irrigation system. Do not use treated seed for food, feed, or oil purposes. Do not plant root crops within 12 months of application unless PCNB is registered for use on those crops.
State Restriction: PCNB 2-E not for use in California.

Signal Word/Toxicity Class: Danger/I (2-E); Caution (2LF).
REI: 12 hr.

Penncozeb — see mancozeb

PROPICONAZOLE

Tilt (EC) *(Novartis)*
Rate: 6-10 fl. oz. EC/A (in 5-10 gal. water)
Disease:
brown blotch leaf smut sheath spot
brown leaf spot narrow brown leaf spot stem rot (suppression)
Time: Apply 6 fl. oz. at first internode elongation; repeat at swollen boot. Make second application 10-14 days after first application, but before boot splits and head emerges or apply 10 fl. oz. (if greater than 10% of tillers are infected with sheath blight) at first internode elongation. If disease reappears, use another registered fungicide for second application.

Rate: Wild rice:
6-8 fl. oz. EC/A (in 5-10 gal. water)
Disease:
brown blotch Helminthsporium leaf blight
Time: Apply low rate at both booting and heading, or high rate at booting.
Remarks: Aerial application. Do not apply to stubble or ratoon crop rice. Do not use in rice fields where commercial farming of crayfish will be practiced. Do not drain water from treated fields to irrigate other crops.
State Restrictions: Not for use in California. Refer to label for endangered species restrictions in certain counties of Arkansas. For use on wild rice in Minnesota only.

Signal Word/Toxicity Class: Warning/II.
REI: 24 hr.

Rovral — see iprodione

RTU-Vitavax-Thiram — see carboxin premix

Sulfa — see sulfur

SULFUR

Sulfa 80 (80DF) *(Drexel)*
Rate: 6-10 lb. ai/A
7.5-12.5 lb. 80DF/A
Disease:
powdery mildew
Time: Apply when disease appears; repeat as necessary.
Remarks: Do not use within 2 weeks of oil application. Some plants are susceptible to injury from sulfur under certain climatic conditions. Do not make sulfur applications at excessively high temperatures. Do not apply this product through any type of irrigation system. Refer to label for directions.

Signal Word/Toxicity Class: Caution.
REI: 24 hr.

SULFUR (Premix)

Top Cop with Sulfur (6.25 lb./gal. sulfur & 1 lb./gal. tribasic copper sulfate) *(Stoller)*
Rate: 2 qt./A
Disease:
brown leaf spot leaf spot sheath spot
leaf smut sheath blight stem rot
Time: Apply at panicle initiation; repeat at 14-day intervals.
Remarks: Rice. Refer to label for use directions.

Signal Word/Toxicity Class: Caution.
REI: 24 hr.

SYSTEM³

System³ (0.1% *Bacillus subtilis* & 14.25% metalaxyl & 16.67% PCNB) *(Helena)*
Rate: 2-4 oz./bu.
Disease:
Pythium *Rhizoctonia solani*

6 Other Field Crops

NOTICE	The information on these pages is for preliminary planning — not a guide for use. Be sure to follow the manufacturer's directions, notwithstanding information contained here.	For personal protective equipment and EPA registration numbers, please turn to page 39.

Time: Apply at planting.
Remarks: A non-oil lubricant has been added for seeds to flow more easily and evenly from planter boxes. Incorporation of dye is recommended to distinguish seed as treated. Do not use treated seed for food, feed, or oil purposes. Refer to label for use directions.

Signal Word/Toxicity Class: Caution.
REI: 12 hr.

THIRAM

42-S Thiram or Thiram 50WP Dyed *(Gustafson)*
Thiram Granuflo (WDG) *(UCB Agrochemicals)*
Rate: 1.5 fl. oz. 42-S/bu. seed
4 fl. oz. 42-S/100 lb. seed (California)
4.3 oz. 50WP/100 lb. seed
2.9 oz. WDG/100 lb. seed
Disease:
damping-off seed decay seedling blight
Time: Seed treatment; slurry treatment (California).
Remarks: Use with EPA-approved dye that imparts unnatural color to seed. Do not use treated seed for food, feed, or oil purposes. Do not apply Thiram Granuflo through any type of irrigation system. Refer to label for use directions.
State Restrictions: Thiram Granuflo not for use in California.

Signal Word/Toxicity Class: Caution.
REI: 24 hr. (Thiram Granuflo).

Tilt — see propiconazole

Top Cop with Sulfur — see sulfur premix

Safflower

CAPTAN (Premix)

Nu-Gro Isotox (12.2% captan & 25% lindane) *(Wilbur-Ellis)*
Seed Mate Isotox (12.2% captan & 25% lindane) *(Loveland)*
Rate: 2 oz./100 lb. seed
Disease:
damping-off seed decay seedling blight
Time: Planter box treatment.
Remarks: Treated seed must not be used for or mixed with food or animal feed, or processed for oil. Treat only those seeds needed for immediate use, minimizing the interval between treatment and planting. Do not store excess treated seeds beyond planting time. Refer to label for use directions.

Signal Word/Toxicity Class: Danger/I.
REI: 4 days.

CARBOXIN

Vitavax-34 *(Gustafson)*
Rate: 2 fl. oz./100 lb. seed
Disease:
seedling diseases
Time: Slurry seed treatment.
Remarks: Incorporation of dye into treating slurry is recommended to distinguish seed as treated. Do not use treated seed for food, feed, or oil purposes. Do not graze or feed livestock on forage or hay in treated areas. Do not apply this product through any type of irrigation system. Refer to label for use directions.
State Restrictions: Not for use in California.

Signal Word/Toxicity Class: Caution.

Dithane — see mancozeb

Grain Guard — see mancozeb

MANCOZEB

Grain Guard contains graphite.

Dithane (DF, F-45, M-45, ST, WSP) *(Rohm and Haas)*
Manzate 200DF *(Griffin)*
Penncozeb 75DF or 80WP *(Elf Atochem)*
Rate: 2.1 oz. DF, 75DF/100 lb. seed
2 oz. M-45, WSP, 200DF, 80WP/100 lb. seed
3.2 fl. oz. F-45, ST/100 lb. seed
Disease:
seedborne disease
Time: Seed treatment.
Remarks: Incorporation of dye into treating slurry recommended to distinguish seed as treated. Do not use treated seed for food, feed, or oil purposes. Refer to label for use directions.
State Restrictions: For Manzate, refer to label for California use restrictions.

Signal Word/Toxicity Class: Caution.
REI: 24 hr.

Grain Guard *(Trace)*
Rate: 3 oz./100 lb. seed
Disease:
damping-off seedborne disease seedling blight
seed rot
Time: Drill box seed treatment.
Remarks: Treat only those seeds needed for immediate use. Do not use treated seed for food, feed, or oil purposes. Do not store excess treated seeds beyond planting time. Refer to label for use directions.

Signal Word/Toxicity Class: Caution.
REI: 12 hr.

Manzate — see mancozeb

Nu-Gro Isotox — see captan premix

PCNB

PCNB 2-E or 2LF *(Amvac)*
Rate: 1.95-3.90 fl. oz. 2-E/100 lb. seed
2.33-4.66 fl. oz. 2LF/100 lb. seed
Disease:
Fusarium Rhizoctonia seedborne disease
Time: Seed treatment.
Remarks: Do not apply through any type of irrigation system. Do not use treated seed for food, feed, or oil purposes. Do not plant root crops within 12 months of application unless PCNB is registered for use on those crops.
State Restriction: PCNB 2-E not for use in California.

Signal Word/Toxicity Class: Danger/I (2-E); Caution (2LF).
REI: 12 hr.

Penncozeb — see mancozeb

Seed Mate Isotox — see captan premix

THIRAM

42-S Thiram or Thiram 50WP Dyed *(Gustafson)*
Thiram Granuflo (WDG) *(UCB Agrochemicals)*
Rate: 2 fl. oz. 42-S/100 lb. seed
4 oz. 50WP/100 lb. seed
2.7 oz. WDG/100 lb. seed
Disease:
damping-off seed decay seedling blight

◆ - **new product** • ai-active ingredient • DF-dry flowable • E/EC-emulsifiable concentrate • EW-emulsion in water
F/FL-flowable • G-granule • L/LC-liquid • SP/WSP-soluble packet • W/WP-wettable powder • WSB-water soluble bag

Time: Seed treatment.
Remarks: Use with EPA-approved dye that imparts unnatural color to seed. Do not use treated seed for food, feed, or oil purposes. Do not apply Thiram Granuflo through any type of irrigation system. Refer to label for use directions.
State Restrictions: Thiram Granuflo not for use in California.

- - - - - - - - - - - - - - - - - -

Signal Word/Toxicity Class: Caution.
REI: 24 hr. (Thiram Granuflo).

Vitavax — see carboxin

Sorghum

(Forage, Grain [Milo], Sweet)

Allegiance — see metalaxyl

Apron — see metalaxyl

Apron XL — see mefenoxam

CAPTAN

Drexel Captan 4FL *(Drexel)*
Rate: 4 fl. oz. 4FL/100 lb. seed
Disease:
kernel smut
Time: Liquid or slurry seed treatment.
Remarks: Do not use treated seed for food, feed, or oil purposes. Refer to label for use directions.

- - - - - - - - - - - - - - - - - -

Signal Word/Toxicity Class: Danger/I.

Gustafson Captan 30-DD or 400 *(Gustafson)*
Rate: 3.5 fl. oz. 30-DD/100 lb. seed
1.75-4.5 fl. oz. 400/100 lb. seed
Disease:
damping-off, mold, seedling blight, kernel smut, seed decay
Time: Seed treatment.
Remarks: Hulled sorghum. Do not combine colloidal suspensions, wettable powders, or emulsifiable concentrates unless previous experience assures compatibility. Use with EPA-approved dye that imparts unnatural color to seed. Do not use for feed, food, or oil purposes. Refer to label for use directions.

- - - - - - - - - - - - - - - - - -

Signal Word/Toxicity Class: Danger/I.

CAPTAN (Premix)

Sorghum Guard (32.75% captan & 16.6% lindane) *(Trace)*
Rate: 3 oz./bu. seed
Disease:
damping-off, seed decay, seedling blight
Time: Planter box seed treatment.
Remarks: Treat only those seeds needed for immediate use. Do not use treated seed for food, feed, or oil purposes. Do not store excess treated seeds beyond planting time. Refer to label for use directions.

- - - - - - - - - - - - - - - - - -

Signal Word/Toxicity Class: Danger/I.
REI: 12 hr.

Dithane — see mancozeb

FLUDIOXONIL

Maxim 4FS *(Novartis)*
Rate: 0.08-0.16 fl. oz. 4FS/100 lb. seed
Disease:
damping-off, seedborne disease, soilborne disease, seed decay, seedling blight
Time: Seed treatment.
Remarks: Under severe disease conditions, use high rate. May be tank mixed. Do not use treated seed for food, feed, or oil purposes. Do not graze on forage until 30 days after planting. Use EPA-approved dye or colorant that imparts unnatural color to seed.

- - - - - - - - - - - - - - - - - -

Signal Word/Toxicity Class: Caution.

Grain Guard — see mancozeb

Grain Guard Plus — see mancozeb premix

Kumulus — see sulfur

MANCOZEB

Grain Guard contains graphite.

Dithane (DF, F-45, M-45, ST, WSP) *(Rohm and Haas)*
Manzate 200DF *(Griffin)*
Penncozeb 75DF or 80WP *(Elf Atochem)*
Rate: 2.9-4.8 oz. DF, 75DF/100 lb. seed
2.7-4.5 oz. M-45, WSP, 200DF, 80WP/100 lb. seed
4.3-7.2 fl. oz. F-45, ST/100 lb. seed
Disease:
covered kernel smut, seed rot, seedling blight, damping-off
Time: Seed treatment.
Remarks: Incorporation of dye into treating slurry recommended to distinguish seed as treated. Do not use treated seed for food, feed, or oil purposes. Refer to label for use directions.
State Restrictions: For Manzate, refer to label for California use restrictions.

- - - - - - - - - - - - - - - - - -

Signal Word/Toxicity Class: Caution.
REI: 24 hr.

Grain Guard *(Trace)*
Rate: 3 oz./100 lb. seed
Disease:
covered kernel smut, seed rot, seedling blight, damping-off, seedborne disease
Time: Drill box seed treatment.
Remarks: Treat only those seeds needed for immediate use. Do not use treated seed for food, feed, or oil purposes. Do not store excess treated seeds beyond planting time. Refer to label for use directions.

- - - - - - - - - - - - - - - - - -

Signal Word/Toxicity Class: Caution.
REI: 12 hr.

MANCOZEB (Premix)

Grain Guard Plus contains graphite.

Grain Guard Plus (50% mancozeb & 18.75% lindane) *(Trace)*
Rate: 3 oz./bu. seed
Disease:
damping-off, seed decay
Time: Drill box or planter box seed treatment.
Remarks: Treat only those seeds needed for immediate use. Do not use treated seed for food, feed, or oil purposes. Do not store excess treated seeds beyond planting time. Refer to label for use directions.

- - - - - - - - - - - - - - - - - -

Signal Word/Toxicity Class: Warning/II.

6 Other Field Crops

NOTICE The information on these pages is for preliminary planning — not a guide for use. Be sure to follow the manufacturer's directions, notwithstanding information contained here.

For personal protective equipment and EPA registration numbers, please turn to page 39.

Manzate — see mancozeb

Maxim — see fludioxonil

MEFENOXAM

Apron XL LS *(Novartis)*
Rate: 0.32-0.64 fl. oz. LS/100 lb. seed
Disease:
damping-off

Rate: 1.28 fl. oz. LS/100 lb. seed
Disease:
downy mildew
Time: Seed treatment at planting.
Remarks: Treat only high grade sorghum (90% germ or above). Consult label for rates for downy mildew in heavily infected areas with recent severe history or for cultivars which do not possess resistance to a particular race. Do not use treated seed for food, feed, or oil purposes. Refer to label for use directions.
Signal Word/Toxicity Class: Warning/II.

METALAXYL

Allegiance or Apron (FL) *(Gustafson)*
Rate: 0.375-1.5 fl. oz. Allegiance/100 lb. seed
0.75-3 fl. oz. Apron/100 lb. seed
Disease:
damping-off downy mildew
Time: Seed treatment.
Remarks: Consult label for rates. For downy mildew in heavily infected areas with recent severe history or for cultivars which do not possess resistance to a particular race. Treat only high grade sorghum (90% germ or above). Use with EPA-approved dye that imparts unnatural color to seed. Do not use treated seed for food, feed, or oil purposes. Refer to label for use directions.

Signal Word/Toxicity Class: Warning/II.

Apron Dry *(Trace)*
Rate: 4-8 oz./100 lb. seed
Disease:
damping-off downy mildew
Time: Seed treatment.
Remarks: Treat only high grade sorghum (90% germ or above). Consult label for rates for downy mildew in heavily infected areas with recent severe history or for cultivars which do not possess resistance to a particular race. Incorporation of dye into the treating slurry recommended to distinguish seed as treated. Do not use treated seed for food, feed, or oil purposes. Refer to label for use directions.

Signal Word/Toxicity Class: Caution.

PCNB

PCNB 2-E or 2LF *(Amvac)*
RTU-PCNB *(Gustafson)*
Rate: 1.95 fl. oz. 2-E/100 lb. seed
2.5 fl. oz. 2LF/100 lb. seed
1.75 fl. oz. RTU/100 lb. seed
Disease:
covered smut (2-E) Fusarium Rhizoctonia
Time: Seed treatment.
Remarks: Do not apply 2-E or 2LF through any type of irrigation system. Do not use for food, feed or oil purposes. Do not plant root crops within 12 months of application unless PCNB is registered for use on those crops. Do not feed treated seed to animals. Refer to label for use directions.
State Restrictions: RTU-PCNB not for use in California.

Signal Word/Toxicity Class: Danger/I (2-E); Caution (2LF, RTU).
REI: 12 hr. (2-E, 2LF).

Penncozeb — see mancozeb

Redball — see sulfur

Sorghum Guard — see captan premix

SULFUR

Drexel Suffa (6FL), Sulfa 80 (80DF) or Sulfur (90W) *(Drexel)*
Rate: 3-4.5 lb. ai/A
4-6 pt. 6FL/A
6-10 lb. ai/A
7.5-12.5 lb. 80DF/A
2.7-4.5 lb. ai/A
3-5 lb. 90W/A
Disease:
powdery mildew
Time: Apply when disease appears; repeat as necessary.
Remarks: Do not use within 2 weeks of oil application (4 weeks 6FL). Some plants are susceptible to injury from sulfur under certain climatic conditions. Do not make sulfur applications at excessively high temperatures. Do not apply this product through any type of irrigation system. Refer to label for use directions.

Signal Word/Toxicity Class: Caution.
REI: 24 hr.

Kumulus DF (sulfur) *(BASF)*
Micro Flo Sulfur (6L) *(Micro Flo)*
Terra Sulfur (6L) *(Terra)*
Rate: 6-15 lb. DF/A
5.3-20 pt. 6L/A
Disease:
powdery mildew
Time: Apply at first sign of disease; repeat as necessary.
Remarks: Do not use within 2 weeks of oil application. Do not apply if temperatures during or within 3 days after application are expected to exceed 90°F. Do not apply Kumulus through any type of irrigation system. Refer to label for use directions.

Signal Word/Toxicity Class: Caution.
REI: 24 hr.

Redball EM-53 (6L), Superfine Wettable Sulphur (WS) or Commercial Dusting Sulphur (DS) *(International Sulphur)*
Rate: 4-20 pt. 6L/A
5-10 lb. WS/100 gal. water
5-15 lb. DS/A
Disease:
powdery mildew
Time: Apply when disease first appears; repeat as necessary.
Remarks: Apply in sufficient water for thorough coverage. Do not apply when temperatures exceed 90°F. Refer to label for use directions.

Signal Word/Toxicity Class: Caution.
REI: 12 hr.

THIRAM

42-S Thiram or Thiram 50WP Dyed *(Gustafson)*
Thiram Granuflo (WDG) *(UCB Agrochemicals)*
Rate: 2 fl. oz. 42-S/bu. seed
2.7 oz. 50WP/100 lb. seed
1.8 oz. WDG/100 lb. seed
Disease:
covered kernel smut seed decay seedling blight
damping-off
Time: Seed treatment.
Remarks: Use with EPA-approved dye that imparts unnatural color to seed. Do not use treated seed for food, feed, or oil purposes. Do not apply Thiram Granuflo through any type of irrigation system. Refer to label for use directions.
State Restrictions: Thiram Granuflo not for use in California.

Signal Word/Toxicity Class: Caution.
REI: 24 hr. (Thiram Granuflo).

◆ - **new product** • ai-active ingredient • DF-dry flowable • E/EC-emulsifiable concentrate • EW-emulsion in water
F/FL-flowable • G-granule • L/LC-liquid • SP/WSP-soluble packet • W/WP-wettable powder • WSB-water soluble bag

Sugar Beets

Allegiance — see metalaxyl

Apron — see metalaxyl

Apron XL — see mefenoxam

Basicop — see copper compounds

Bayleton — see triadimefon

Bravo — see chlorothalonil

BT 320 SULFUR 50 DUST (Premix)

Microbial insecticide/fungicide containing Bacillus thuringiensis and sulfur.

Bt 320 Sulfur 50 Dust (0.064% Bt & 50% sulfur) *(Wilbur-Ellis)*
Rate: 35-50 lb. Dust/A
Disease:
powdery mildew
Time: Apply at 7- to 10-day intervals.
Remarks: Refer to label for use directions.

Signal Word/Toxicity Class: Caution.
REI: 24 hr.

CAPTAN

Drexel Captan 4FL *(Drexel)*
Rate: Western States:
6 fl. oz. 4FL/100 lb. seed
Eastern States:
12 fl. oz. 4FL/100 lb. seed
Disease:
seed decay — seedling blight — soilborne disease
Time: Liquid or slurry seed treatment.
Remarks: Do not use treated seed for food, feed, or oil purposes. Refer to label for use directions.

Signal Word/Toxicity Class: Danger/I.

Gustafson Captan 30-DD or 400 *(Gustafson)*
Rate: Western States:
4.75 fl. oz. 30-DD/100 lb. seed
6 fl. oz. 400/100 lb. seed
Eastern States:
9.5 fl. oz. 30-DD/100 lb. seed
12 fl. oz. 400/100 lb. seed
Disease:
damping-off — seed decay — seedling blight
mold
Time: Seed treatment.
Remarks: Do not combine colloidal suspensions, wettable powders, or emulsifiable concentrates unless previous experience assures compatibility. Use with EPA-approved dye that imparts unnatural color to seed. Do not use treated seed for food, feed, or oil purposes. Refer to label for directions.

Signal Word/Toxicity Class: Danger/I.

CAPTAN (Premix)

Nu-Gro Isotox (12.2% captan & 25% lindane) *(Wilbur-Ellis)*
Seed Mate Isotox (12.2% captan & 25% lindane) *(Loveland)*
Rate: 16 oz./100 lb. seed
Disease:
damping-off — seed decay — seedling blight
Time: Planter box treatment.
Remarks: Treated seed must not be used for or mixed with food or animal feed, or processed for oil. Treat only those seeds needed for immediate use, minimizing the interval between treatment and planting. Do not store excess treated seeds beyond planting time. Refer to label for use directions.

Signal Word/Toxicity Class: Danger/I.
REI: 4 days.

Champ — see copper compounds

Champion — see copper compounds

CHLORONEB

Supplemental seed treatment; use in conjunction with a standard fungicide seed treatment.

Demosan Flowable or 65W *(Kincaid)*
Rate: 7.8 oz. FL/100 lb. seed
4.5 oz. 65W/100 lb. seed
Disease:
damping-off — Rhizoctonia — seedling diseases
Pythium — Sclerotium blight — sore shin
Time: Seed treatment or hopper box treatment.
Remarks: Refer to label for use directions. Do not use excess treated seed for food, feed, or oil purposes. Do not graze within 45 days of planting.

Signal Word/Toxicity Class: Caution.
REI: 12 hr.

CHLOROTHALONIL

Bravo 500 *(ZENECA)*
Rate: 1.25 lb. ai/A
2.5 pt. 500/A
Disease:
downy mildew
Time: Make first application in mid-Oct. when disease conditions become favorable; repeat as conditions warrant through April of following year. Do not apply after bolting and seedhead emergence has occurred.
Remarks: Sugar beets grown for seed. Use in sufficient water to obtain adequate coverage. Do not graze treated areas or feed treated crop residues to livestock. Refer to label for use directions and restrictions.
Days To Harvest: 14.
State Restrictions: SLN-Oregon.

Signal Word/Toxicity Class: Warning/II.
REI: 48 hr.

Clean Crop C-O-C-S — see copper compounds

COPPER COMPOUNDS

Basicop (WP) (basic copper sulfate) *(Griffin)*
Drexel Basic Copper Sulfate (WP) or KOP 300 (3FL) (basic copper sulfate) *(Drexel)*
Rate: 2 lb. ai/A
4 lb. WP/A
5.3 pt. 3FL/A
Disease:
Cercospora leaf spot
Time: Apply when disease appears; repeat at 7- to 10-day intervals.
Remarks: Do not apply through any type of irrigation system. Refer to label for use directions.

Signal Word/Toxicity Class: Warning/II.
REI: 24 hr. (WP); 12 hr. (3FL).

Champ (FL or Formula 2) or Champion (WP) (copper hydroxide) *(Agtrol)*
KOP-Hydroxide (FL, 50WP) (copper hydroxide) *(Drexel)*
Nu-Cop (3L, 50DF, 50WP) (copper hydroxide) *(Micro Flo)*
Rate: 2.66-6.66 pt. FL/A
1.33-3.33 pt. Formula 2/A
1.33-6.66 pt. 3L/A
2-5 lb. WP, 50DF, 50WP/A

NOTICE The information on these pages is for preliminary planning — not a guide for use. Be sure to follow the manufacturer's directions, notwithstanding information contained here.

For personal protective equipment and EPA registration numbers, please turn to page 39.

Disease:
Cercospora leaf spot
Time: Apply when disease threatens; repeat at 10- to 14-day intervals.
Remarks: Refer to label for use directions.

Signal Word/Toxicity Class: Varies by formulation.
REI: Varies by formulation.

Clean Crop C-O-C-S (WDG) (74.8% copper oxychloride & 14.2% copper sulfate, basic) *(Platte)*
Rate: 1-2 lb. ai/A
2-4 lb. WDG/A
Disease:
Cercospora leaf spot
Time: Apply when disease first threatens; repeat at 7- to 10-day intervals to protect new growth throughout growing season.
Remarks: May be tank mixed with sulfur to enhance control. Do not apply this product through any type of irrigation system. Refer to label for use directions.

Signal Word/Toxicity Class: Warning/II.
REI: 24 hr.

Copper-Count-N (L) (copper ammonium) *(Mineral Research)*
Rate: 1.5-3 qt. L/A
Disease:
Cercospora leaf spot
Time: Apply on disease appearance, 3-6 sprays at 10- to 14-day intervals, more frequently under severe conditions.
Remarks: Applications may be made by ground or air; refer to label for use directions. Do not apply through any type of irrigation system unless supplemental labeling on chemigation is followed.

Signal Word/Toxicity Class: Caution.
REI: 12 hr.

Kocide (101, DF, 2000, 4.5LF, LF) (copper hydroxide) *(Griffin)*
Rate: 2-5 lb. 101, DF/A
1.5-3.75 lb. 2000/A
1.33-2.66 4.5LF/A
2.66-6.66 pt. LF/A
Disease:
Cercospora leaf spot
Time: Apply when disease threatens; repeat every 10- to 14-days.
Remarks: Use higher rate when disease is severe. Addition of 2 qt./A agricultural spray oil is recommended. Do not apply in a spray solution having a pH less than 6.5 as phytotoxicity may occur. Refer to label for use directions.

Signal Word/Toxicity Class: Danger/I (101, DF); Warning/II (2000); Caution (4.5LF, LF).
REI: 24 hr.

KOP Oxy-85 (WP) (copper oxychloride) *(Drexel)*
TOP COP Tri-Basic (FL) (basic copper sulfate) *(Stoller)*
Rate: 6 lb. WP/A
5 pt. FL/A
Disease:
Cercospora leaf spot
Time: Apply when disease first appears; repeat at 7- to 10-day intervals.
Remarks: Use in sufficient water and thoroughly wet plants. Refer to label for use directions.

Signal Word/Toxicity Class: Danger/I (KOP); Caution (TOP COP).
REI: 24 hr.

Micro Flo Basic Copper "53" (basic copper sulfate) or Microsperse COC (53WP) (copper oxychloride) *(Micro Flo)*
Rate: 2-4 lb. 53/A
3-5 lb. 53WP/A
Disease: 53:
downy mildew Cercospora leaf spot
53WP:
leaf spot
Time: Apply when disease first appears; repeat at 7- to 10-day intervals as needed.
Remarks: Refer to label for use directions.

Signal Word/Toxicity Class: Danger/I (53WP); Warning/II (53).
REI: 24 hr.

Copper-Count-N — see copper compounds

Demosan — see chloroneb

Dithane — see mancozeb

Kocide — see copper compounds

KOP-Hydroxide — see copper compounds

KOP Oxy — see copper compounds

Kumulus — see sulfur

MANCOZEB

Dithane (DF, F-45, M-45, WSP) *(Rohm and Haas)*
Manzate 200DF *(Griffin)*
Penncozeb 75DF or 80WP *(Elf Atochem)*
Rate: 1.12-1.5 lb. ai/A
1.5-2 lb. DF, 200DF, 75DF/A
1.2-1.6 lb. ai/A
1.5-2 lb. M-45, WSP, 80WP/A
2.4-3.2 pt. F-45/A
Disease:
Cercospora leaf spot rust (80WP)
Time: Apply when disease first threatens; repeat at 7- to 10-day intervals as needed.
Remarks: Addition of surfactant will improve performance. Do not feed treated tops to livestock. Refer to label for use directions.
Days To Harvest: 14.
State Restrictions: For Manzate, refer to label for California use restrictions.

Signal Word/Toxicity Class: Caution.
REI: 24 hr.

MANEB

Maneb 75DF or 80WP *(Elf Atochem)*
Manex (FL) *(Griffin)*
Rate: 1.12-1.5 lb. ai/A
1.5-2 lb. 75DF/A
1.2-1.6 lb. ai/A
1.5-2 lb. 80WP/A
1.2-1.6 qt. FL/A
Disease:
Cercospora leaf spot
Time: Apply when disease appears; repeat at 7- to 10-day intervals.
Remarks: For Manex, do not use any other product containing a different EBDC ai during the same growing season. Do not feed treated tops to livestock. Refer to label for use directions.
Days To Harvest: 14.

Signal Word/Toxicity Class: Caution.
REI: 24 hr.

MANEB (Premix)

Pro-Tex is a Restricted Use Pesticide.

Pro-Tex (3.5 lb./gal. maneb & 0.5 lb./gal. triphenyltin hydroxide) *(Griffin)*
Rate: 1-1.8 lb. ai/A
1-1.8 qt./A
Disease:
Cercospora leaf spot
Time: Apply with appearance of blight weather conditions; repeat at 7-day intervals.

◆ - **new product** • ai-active ingredient • DF-dry flowable • E/EC-emulsifiable concentrate • EW-emulsion in water
F/FL-flowable • G-granule • L/LC-liquid • SP/WSP-soluble packet • W/WP-wettable powder • WSB-water soluble bag

Remarks: For ground application (closed cabs only), apply in at least 15 gal. water; aerial application in 3-10 gal. water. Use lower rate early in season and high rate mid to late season or when blight infection is in area. Do not exceed 6 qt./A per season. Do not graze or feed treated beet tops to livestock. Refer to label for use directions.
Days To Harvest: 21.

Signal Word/Toxicity Class: Danger-Poison/I.
REI: 48 hr.

Manex — see maneb

Manzate — see mancozeb

MEFENOXAM

Apron XL LS *(Novartis)*
Rate: 0.32-0.64 fl. oz. LS/100 lb. seed
Disease:
damping-off
Time: Seed treatment at planting.
Remarks: Do not use treated seed for food, feed, or oil purposes.

Signal Word/Toxicity Class: Warning/II.

Ridomil Gold EC, GR, or WSP *(Novartis)*
Rate: 0.5-1 lb. ai/A
1-2 pt. EC/A
20-40 lb. GR/A
0.45-0.9 lb. ai/A
1-2 lb. WSP/A
Disease:
Pythium
Time: Apply as preplant incorporated or as a surface spray at planting.
Remarks: For banded applications, 7" band recommended. Do not plant any crop which is not registered for use with mefenoxam in treated soil for 12 months.

Signal Word/Toxicity Class: Warning/II (EC); Caution (GR, WSP).
REI: 48 hr.

METALAXYL

Allegiance or Apron (FL) *(Gustafson)*
Rate: 0.75 fl. oz. Allegiance/100 lb. seed
1.5 fl. oz. Apron/100 lb. seed
Disease:
damping-off
Time: Seed treatment.
Remarks: Use with EPA-approved dye that imparts unnatural color to seed. Do not use treated seed for food, feed, or oil purposes. Refer to label for use directions.

Signal Word/Toxicity Class: Warning/II.

Apron Dry *(Trace)*
Rate: 4 oz./100 lb. seed
Disease:
damping-off
Time: Seed treatment.
Remarks: Do not use treated seed for food, feed, or oil purposes. Incorporation of dye into the treating slurry recommended to distinguish seed as treated. Refer to label for use directions.

Signal Word/Toxicity Class: Caution.

Micro Flo — see copper compounds

Microsperse — see copper compounds

Microthiol Special — see sulfur

Nu-Cop — see copper compounds

Nu-Gro Isotox — see captan premix

PCNB

PCNB 2-E or 2LF *(Amvac)*
RTU-PCNB *(Gustafson)*
Rate: 6.13-12.0 fl. oz. 2-E/100 lb. seed
7.25-14.5 fl. oz. 2LF/100 lb. seed
5.5-10.75 fl. oz. RTU/100 lb. seed
Disease:
Fusarium Rhizoctonia
Time: Seed treatment.
Remarks: Do not apply 2-E or 2LF through any type of irrigation system. Do not use treated seed for food, feed, or oil purposes. Do not feed treated seed to animals. Do not plant root crops within 12 months of application unless PCNB is registered for use on those crops. Refer to label for use directions.
State Restrictions: RTU-PCNB not for use in California.

Signal Word/Toxicity Class: Danger/I (2-E); Caution (2LF, RTU).
REI: 12 hr. (2-E, 2LF).

Penncozeb — see mancozeb

Pro-Tex — see maneb premix

Redball — see sulfur

Ridomil Gold — see mefenoxam

RIDOMIL GOLD MZ (Premix)

Premix containing the active ingredients of Dithane and Manzate. Packaged in Accu-Pak, a wettable powder in a water-soluble bag. Toxic to fish.

Ridomil Gold MZ (WP) (4% mefenoxam & 64% mancozeb) *(Novartis)*
Rate: 1.7 lb. ai/A
2.5 lb. WP/A
Disease:
downy mildew
Time: Apply when conditions are favorable for disease development, but before infection; repeat at 14-day intervals until disease threat is over.
Remarks: Up to 4 applications may be made. If conditions for disease development are still favorable, use other fungicides for control. If other registered fungicides containing EBDCs are used during growing season, do not apply more than 11.2 lb. EBDC /A. If tank mixed with other EBDC products, apply no more than 1.6 lb. EBDC. Do not plant any crop which is not registered for use with mefenoxam in treated soil for 12 months. Do not feed treated tops to livestock.
Days To Harvest: 7.

Signal Word/Toxicity Class: Caution.
REI: 48 hr.

Seed Mate Isotox — see captan premix

Sulfa — see sulfur

SULFUR

Drexel Suffa (6FL, 8FL), Sulfa 80 (80DF) or Sulfur (90W) *(Drexel)*
Rate: 0.75-2.25 lb. ai/A
1-3 pt. 6FL/A
0.75-2.25 pt. 8FL/A (aerial)
4-8 lb. ai/A
5-10 lb. 80DF/A
2.7-4.5 lb. ai/A
3-5 lb. 90W/A

6 Other Field Crops

NOTICE The information on these pages is for preliminary planning — not a guide for use. Be sure to follow the manufacturer's directions, notwithstanding information contained here.	For personal protective equipment and EPA registration numbers, please turn to page 39.

Disease:
powdery mildew
Time: Apply when disease first appears; repeat at 3- to 4-week intervals; 10- to 30-day intervals for Sulfa 80.
Remarks: Do not use within 2 weeks of oil application (4 weeks FL). Some plants are susceptible to injury from sulfur under certain climatic conditions. Do not make sulfur applications at excessively high temperatures. Do not apply this product through any type of irrigation system. Refer to label for use directions.

Signal Word/Toxicity Class: Caution.
REI: 24 hr.

Kumulus DF (sulfur) *(BASF)*
Thiolux (DF) *(Platte)*
Rate: 8-16 lb. ai/A
10-20 lb. DF/A
Disease:
powdery mildew
Time: Apply when disease first appears; repeat as necessary with or without pesticides immediately after rain.
Remarks: Do not apply if temperatures during or within 3 days after application are expected to exceed 90°F. Do not make applications within 2 weeks of oil spray. Do not apply Kumulus through any type of irrigation system. Refer to label for use directions.

Signal Word/Toxicity Class: Caution.
REI: 24 hr.

Liquid Sulfur Six *(Helena)*
Rate: 4-8 pt./100 gal. water
Disease:
powdery mildew
Time: Apply at first sign of infection; repeat at 2-week intervals for 2-4 applications.
Remarks: Do not apply this product through any type of irrigation system. Refer to label for use directions.

Signal Word/Toxicity Class: Caution.
REI: 24 hr.

Micro Flo Sulfur (6L), Dual Spraying & Dusting Sulfur (DS) or Microsperse Wettable Sulfur (WS) *(Micro Flo)*
Terra Sulfur (6L) *(Terra)*
Rate: 4 lb. ai/A
5.3 pt. 6L/A
3.6 lb. ai/A
4 lb. DS, WS/A
Disease:
powdery mildew
Time: Apply at 18-day intervals.
Remarks: Do not use within 2 weeks of oil application. Some plants are susceptible to injury from sulfur under certain climatic conditions. Do not make sulfur applications at excessively high temperatures. Do not apply DS and WS through any type of irrigation system. Refer to label for use directions.
Days To Harvest: 30.

Signal Word/Toxicity Class: Caution.
REI: 24 hr.

Microthiol Special (80WP) *(Elf Atochem)*
Rate: 4-8 lb. ai/A
5-10 lb. 80WP/A
Disease:
leaf spot powdery mildew
Time: Apply as soon as symptoms appear; repeat every 2-3 weeks as needed.
Remarks: Refer to label for use directions.

Signal Word/Toxicity Class: Caution.
REI: 24 hr.

Redball EM-53 (6L), Superfine Wettable Sulphur (WS) or Commercial Dusting Sulphur (DS) *(International Sulphur)*
THAT (6FL) or THAT Big 8 (8FL) *(Stoller)*
Rate: 6-12 lb. ai/A
8-16 pt. 6L/A
5-10 lb. WS/100 gal. water
5-15 lb. DS/A
6-12 pt. 8FL/A
Disease:
powdery mildew
Time: Apply at first sign of disease; repeat at 10- to 30-day intervals as needed throughout season.
Remarks: Apply in sufficient water for thorough coverage. Do not apply when temperatures exceed 90°F. Refer to label for use directions.
Days To Harvest: 0.

Signal Word/Toxicity Class: Caution.
REI: 24 hr. (THAT); 12 hr. (Redball).

SULFUR (Premix)

Top Cop with Sulfur (6.25 lb./gal. sulfur & 1 lb./gal. tribasic copper sulfate) *(Stoller)*
Rate: 4-8 pt./A
Disease:
Cercospora leaf spot downy mildew powdery mildew
Time: Apply before or at first appearance of disease; repeat at 10- to 14-day intervals for 3-6 treatments.
Remarks: Refer to label for use directions.

Signal Word/Toxicity Class: Caution.
REI: 24 hr.

SUPER TIN

Packaged in AgPak, a water-soluble pack. Toxic to fish and wildlife. Restricted Use Pesticide.

Super Tin (80WP) (triphenyltin hydroxide) *(Griffin)*
Rate: 2.5-5 oz. 80WP/A
Disease:
Cercospora leaf spot
Time: Apply when disease appears; repeat at 10- to 14-day intervals.
Remarks: For ground application (closed cabs only), apply in at least 15 gal. water; for aerial, apply in 5-10 gal. water when plants are small and increase volume with plant size. Use lower rate for protective sprays and higher rates later in season or during high infection periods. Do not exceed 15 oz. 80WP/A per season. Do not mix with pesticides, micronutrients, or adjuvants unless testing or prior experience has shown the mixture to be nonphytotoxic to crop. Do not graze or feed beet tops to livestock. Refer to label for use directions.
Days To Harvest: 21.
State Restrictions: Not for use in California.

Signal Word/Toxicity Class: Danger-Poison/I.
REI: 48 hr.

TELONE

Restricted Use Pesticide.

Telone II (dicloropropene) *(Dow AgroSciences)*
Rate: 10-18 gal. II/A
Disease:
Rhizomania
Time: Preplant broadcast.
Remarks: Use higher rate for heavier soils and/or higher levels of disease infestation. Do not apply through any type of irrigation system. Refer to label for use directions.

Signal Word/Toxicity Class: Warning/II.
REI: 72 hr.

THAT or THAT Big 8 — see sulfur

Thiolux — see sulfur

THIOPHANATE-METHYL

Topsin M (WSB, 70W) *(Elf Atochem)*
Rate: 0.26-0.35 lb. ai/A
0.375-0.5 lb. WSB, 70W/A
Disease:
leaf spot
Time: Begin applications when disease first appears; repeat at 14-day intervals (14- to 21-day intervals 70W) as needed.

◆ - **new product** • ai-active ingredient • DF-dry flowable • E/EC-emulsifiable concentrate • EW-emulsion in water
F/FL-flowable • G-granule • L/LC-liquid • SP/WSP-soluble packet • W/WP-wettable powder • WSB-water soluble bag

Remarks: Refer to label for use directions.
Days To Harvest: 21.

Signal Word/Toxicity Class: Caution.
REI: 12 hr.

THIRAM

42-S Thiram or Thiram 50WP Dyed *(Gustafson)*
Thiram Granuflo (WDG) *(UCB Agrochemicals)*

Rate: 8 fl. oz. 42-S/100 lb. seed
8 oz. 50WP/100 lb. seed
5.3 oz. WDG/100 lb. seed

Disease:
damping-off seed decay seedling blight

Time: Seed treatment.
Remarks: Use with EPA-approved dye that imparts unnatural color to seed. Do not use treated seed for food, feed, or oil purposes. Do not apply Thiram Granuflo through any type of irrigation system. Refer to label for use directions.
State Restrictions: Thiram Granuflo not for use in California.

Signal Word/Toxicity Class: Caution.
REI: 24 hr. (Thiram Granuflo).

TOP COP — see copper compounds

Top Cop with Sulfur — see sulfur premix

Topsin M — see thiophanate-methyl

TRIADIMEFON

Bayleton 50% DF *(Bayer)*

Rate: 4-8 oz. 50% DF/A
Disease:
powdery mildew

Time: Broadcast: Apply as foliar spray when disease first appears; repeat at 10- to 14-day intervals. Band: Apply spray directed into leaf whorl in 2" band.
Remarks: Use higher rate for severe disease conditions. Do not apply through any type of irrigation system. Refer to label for use directions.
Days To Harvest: Broadcast-15; band-30.
Crop Rotation: Small grains, corn, sorghum, soybeans, beans, peas, and cucurbits may be planted 35 days after the last application; forage or vines from these crops may not be used for food or feed. Root crops may be planted 120 days after the last application but tops must not be used for food or feed. All crops may be planted 12 months or later after last application without restrictions.

Signal Word/Toxicity Class: Caution.
REI: 12 hr.

Sugarcane

PROPICONAZOLE

Tilt (EC) *(Novartis)*

Rate: 0.75 fl. oz. EC/100 gal. water
Disease:
pineapple disease

Time: Apply to cut seed pieces as cold or hot dip. For cold dip, immerse seed pieces to provide thorough coverage; remove and allow to drain. For hot dip, maintain temperature of dip at 52°C and soak seed pieces for 20 minutes; remove and allow to drain.
Remarks: Do not use treated seed pieces for food, feed, or fodder. Do not graze or feed treated forage to livestock.
State Restrictions: For use in Hawaii only.

Signal Word/Toxicity Class: Warning/II.
REI: 24 hr.

SULFUR

Micro Flo Sulfur (6L) or Microsperse Wettable Sulfur (WS) *(Micro Flo)*
Terra Sulfur (6L) *(Terra)*

Rate: 6 lb. ai/A
8 pt. 6L/A
5.4 lb. ai/A
6 lb. WS/A

Disease:
rust

Time: Apply as required by disease pressure.
Remarks: Apply in sufficient water for thorough coverage. Do not use within 2 weeks of oil application. Some plants are susceptible to injury from sulfur under certain climatic conditions. Do not make sulfur applications at excessively high temperatures. Do not apply WS through any type of irrigation system. Refer to label for use directions.

Signal Word/Toxicity Class: Caution.
REI: 24 hr.

Tilt — see propiconazole

Sunflower

Allegiance — see metalaxyl

Apron — see metalaxyl

Apron XL — see mefenoxam

CAPTAN

Gustafson Captan 30-DD or 400 *(Gustafson)*

Rate: 4 fl. oz. 30-DD/100 lb. seed
2-4 fl. oz. 400/100 lb. seed

Disease:
damping-off seed decay seedling blight
mold

Time: Seed treatment.
Remarks: Do not combine colloidal suspensions, wettable powders, or emulsifiable concentrates unless previous experience assures compatibility. Use with EPA-approved dye that imparts unnatural color to seed. Do not use treated seed for food, feed, or oil purposes. Refer to label for use directions.

Signal Word/Toxicity Class: Danger/I.

CAPTAN (Premixes)

Nu-Gro Isotox (12.2% captan & 25% lindane) *(Wilbur-Ellis)*
Seed Mate Isotox (12.2% captan & 25% lindane) *(Loveland)*
Sorghum Guard (32.75% captan & 16.6% lindane) *(Trace)*

Rate: 4 oz./100 lb. seed
2 oz./50 lb. seed

Disease:
damping-off seed decay seedling blight

Time: Planter box treatment.
Remarks: Treated seed must not be used for or mixed with food or animal feed, or processed for oil. Treat only those seeds needed for immediate use, minimizing the interval between treatment and planting. Do not store excess treated seeds beyond planting time. Refer to label for use directions.

Signal Word/Toxicity Class: Danger/I.
REI: 4 days; (12 hr. Trace).

DB-Green — see maneb premix

6 Other Field Crops

NOTICE The information on these pages is for preliminary planning — not a guide for use. Be sure to follow the manufacturer's directions, notwithstanding information contained here. For personal protective equipment and EPA registration numbers, please turn to page 39.

MANEB (Premix)

Dual purpose ready-to use fungicide and insecticide. DB-Green is a dust-based formulation seed treatment. DB-Green L is a liquid formulation seed treatment.

DB-Green (dust) (50% maneb & 18.75% lindane) *(AGSCO)*
DB-Green L (liquid) (25.6% maneb & 8.6% lindane) *(AGSCO)*
Rate: 2.75 oz. dust/bu.
4 fl. oz. L/bu.
Disease:
seedling diseases soilborne diseases
Time: Seed treatment.
Remarks: Drill or planter box application. Treat only seeds needed for immediate use, minimizing interval between treatment and planting. Do not store excess treated seeds beyond planting time. Treated seed must not be used for or mixed with food of animal feed, or processed for oil. Care must be exercised in handling treated seed. Do not graze or feed livestock on treated areas for 6 weeks after planting.

Signal Word/Toxicity Class: Warning/II.
REI: 12 hr.

MEFENOXAM

Apron XL LS *(Novartis)*
Rate: 1.28 fl. oz. LS/100 lb. seed
Disease:
downy mildew
Time: Seed treatment at planting.
Remarks: Do not use treated seed for food, feed, or oil purposes.

Signal Word/Toxicity Class: Warning/II.

METALAXYL

Allegiance or Apron (FL) *(Gustafson)*
Rate: 1.5 fl. oz. Allegiance/100 lb. seed
3 fl. oz. Apron/100 lb. seed
Disease:
downy mildew
Time: Seed treatment.
Remarks: Use with EPA-approved dye that imparts unnatural color to seed. Do not use for feed, food, or oil purposes. Refer to label for use directions.

Signal Word/Toxicity Class: Warning/II.

Apron Dry *(Trace)*
Rate: 8 oz./100 lb. seed
Disease:
downy mildew
Time: Seed treatment.
Remarks: Incorporation of dye recommended to distinguish seed as treated. Treat only those seeds needed for immediate use. Do not use treated seed for food, feed, or oil purposes. Refer to label for use directions.

Signal Word/Toxicity Class: Caution.

Nu-Gro Isotox — see captan premix

Seed Mate Isotox — see captan premix

Sorghum Guard — see captan premix

THIRAM

42-S Thiram *(Gustafson)*
Thiram Granuflo (WDG) *(UCB Agrochemicals)*
Rate: 2 fl. oz. 42-S/100 lb. seed
5.4 oz. WDG/100 lb. seed
Disease:
damping-off seed decay seedling blight
Time: Seed treatment.
Remarks: Use with EPA-approved dye that imparts unnatural color to seed. Do not use treated seed for food, feed, or oil purposes. Do not apply Thiram Granuflo through any type of irrigation system. Refer to label for use directions.
State Restrictions: Thiram Granuflo not for use in California.

Signal Word/Toxicity Class: Caution.
REI: 24 hr. (Thiram Granuflo).

Tobacco

Brom-O-Gas — see methyl bromide

CHLOR-O-PIC

Restricted Use Pesticide.

Chlor-O-Pic (chloropicrin) *(Great Lakes Chemical)*
Rate: Nontarp:
150-500 lb./A
Disease:
Cylindrocladium, Fusarium, Phytophthora, *Pseudomonas solanacearum*, Pythium, Rhizoctonia, Sclerotinia, Sclerotium, Verticillium
Time: Apply up to 2 weeks prior to planting when soil moisture is good and temperature is 60°-85°F.
Remarks: Work upwind of product and never fumigate without 2 qualified applicators. Do not fumigate when there is no or little air movement. Refer to label for use directions.

Signal Word/Toxicity Class: Danger/I.

Clean Crop Nemasol — see metam-sodium

COPPER COMPOUNDS

Microsperse COC 53WP (copper oxychloride) *(Micro Flo)*
Rate: 0.25-0.66 lb. 53WP/10 gal. water
Disease:
angular leaf spot damping-off root rot
Time: Begin at plant emergence using lower rate and increase as seedlings grow by applying as a spray every 15 yd. of bed; repeat at 10- to 14-day intervals.
Remarks: Cold frames, greenhouses, etc. Refer to label for use directions.

Signal Word/Toxicity Class: Danger/I.
REI: 24 hr.

FERBAM

Broad-spectrum fungicide. Toxic to fish.

Ferbam Granuflo (WDG) *(UBC Agrochemicals)*
Rate: 1.5-2.3 lb. ai/100 gal. water
2-3 lb. WDG/100 gal. water
Disease:
blue mold
Time: Begin applications when plants are size of a dime, or when blue mold is reported in area; repeat twice weekly until plants are transplanted.
Remarks: Seedbeds. Apply 3 gal./100 sq. yd. when plants are small, increasing to 6 gal. when plants are ready for transplanting. Refer to label for use directions.

Signal Word/Toxicity Class: Caution.
REI: 24 hr.

◆ - **new product** • ai-active ingredient • DF-dry flowable • E/EC-emulsifiable concentrate • EW-emulsion in water
F/FL-flowable • G-granule • L/LC-liquid • SP/WSP-soluble packet • W/WP-wettable powder • WSB-water soluble bag

MEFENOXAM

Ridomil Gold EC or WSP *(Novartis)*

Rate: Flue-Cured:
0.5-4 lb. ai/A
1-8 pt. EC/A
0.45-0.9 ai/A
1-2 lb. WSP/A
Burley and Other:
4-4.5 lb. ai/A
8-9 pt. EC/A
0.9-1.35 lb. ai/A
2-3 lb. WSP/A
No-till:
0.25-0.5 lb. ai/A
0.5-1 pt. EC/A
0.225-0.45 lb. ai/A
0.5-1 lb. WSP/A

Disease:
black shank blue mold

Time: Apply before transplanting. Supplemental applications may be made at lay-by or last cultivation (no-till 30-35 days after planting).
Remarks: Not for use for greenhouse or tobacco plant beds. Do not dip plants in solutions containing Ridomil Gold or crop injury may occur. Do not use in transplant water on field tobacco. Do not use as a foliar spray. Do not plant any crop which is not registered for use with mefenoxam in treated soil for 12 months. Refer to label for use directions.
State Restrictions: Do not use Ridomil Gold EC for black shank control in Pennsylvania.

- - - - - - - - - - - - - - - - -

Signal Word/Toxicity Class: Warning/II (EC); Caution (WSP).
REI: 48 hr.

Mertect — see thiabendazole

METAM-SODIUM

Water-soluble liquid when applied to properly prepared soil is converted into a gaseous fumigant.

Clean Crop Nemasol (42%) *(Platte)*
Metam CLR (42%) *(UCB Agrochemicals)*

Rate: Tarp method:
30-75 gal. Nemasol/A
0.75-1.1 gal. CLR/100 sq. yd.
Drench method:
1.875 gal. Nemasol/100 sq. yd.
1.85 gal. CLR/100 sq. yd.

Disease:
damping-off root rot wilt

Time: Preplant. Fall application. In the South, apply before Nov. 30.
Remarks: Plant beds. Tarp method: Apply uniformly over entire bed; cover immediately with a plastic cover for 1-2 days; 7 days after application loosen treated soil to 2" depth. Drench method: Application may be made with sprinklers, sprayers with nozzles, or any suitable equipment. Do not seed tobacco earlier than 21 days after application. Refer to label for use directions.
State Restrictions: In California, application must be in compliance with Technical Information Bulletin "Guidelines for All Application Methods for Metam-sodium in California." This bulletin may be obtained from your local pesticide dealer or a metam-sodium registrant.

- - - - - - - - - - - - - - - - -

Signal Word/Toxicity Class: Danger/I.
REI: 48 hr.

Vapam (WS) or HL *(Amvac)*

Rate: Tarp method:
1-1.5 gal. WS/100 sq. yd.
0.5-1 gal. HL/100 sq. yd.
Drench method:
2.5 gal. WS/100 sq. yd.
2 gal. HL/100 sq. yd

Disease:
soilborne diseases

Time: Fall application. In the South, apply before Nov. 30.
Remarks: Prepare bed 5- to 7-days before application to ensure best conditions for weed seed germination and fumigant action. Loosen treated soil to a depth of 2" within 7 days after application. Do not seed tobacco earlier than 21 days after application. Refer to label for use directions.

- - - - - - - - - - - - - - - - -

Signal Word/Toxicity Class: Danger/I.
REI: 48 hr.

METHYL BROMIDE

2% Chloropicrin added as warning agent. Classified as an ozone depleter under the Clean Air Act. Use to be eliminated starting Jan. 1, 2001. Restricted Use Pesticide.

Brom-O-Gas *(Great Lakes Chemical)*

Rate: Injected with chisels + covered with tarp:
872 lb./A
Raised tarp:
2 lb./100 ft.

Disease:
Fusarium, Pythium, Sclerotinia
oak root fungus, Rhizoctonia, Sclerotium
Phytophthora

Time: Preplant.
Remarks: Refer to label for use directions.
State Restrictions: Brom-O-Gas not for use in Arizona and not for use in California for oak root fungus (Armillaria root rot) control.

- - - - - - - - - - - - - - - - -

Signal Word/Toxicity Class: Danger/I.

Microsperse — see copper compounds

Ridomil Gold — see mefenoxam

TELONE (Premix)

Restricted Use Pesticide.

Telone C-17 (8.2 lb./gal. dichloropropene & 1.7 lb./gal. chloropicrin) *(Dow AgroSciences)*

Rate: 10.8-21.6 gal. C-17/A

Disease:
black root rot black shank Granville bacterial wilt

Time: Preplant soil treatment. Apply whenever soil conditions permit.
Remarks: Rate varies by soil type. Do not use in extremely heavy clay or muck soils. Do not apply this product through any type of irrigation system. Refer to label for use directions, precautions, and restrictions.

- - - - - - - - - - - - - - - - -

Signal Word/Toxicity Class: Danger/I.
REI: 72 hr.

TERR-O-GAS (Premix)

Classified as an ozone depleter under the Clean Air Act. Use to be eliminated starting Jan. 1, 2001. Restricted Use Pesticide.

Terr-O-Gas (67% methyl bromide & 33% chloropicrin) *(Great Lakes Chemical)*

Rate: Field-nontarp:
60-130 lb./A
Seedbed:
1.5 lb./100 ft.

Disease:
Fusarium, Pythium, Sclerotinia
Phytophthora, Rhizoctonia

Time: Apply up to 2 weeks before planting when soil moisture is good and temperature is 60°-85°F.
Remarks: Work upwind of product and never fumigate without 2 qualified applicators. Do not fumigate when there is no or little air movement. Tarping is not used after field application. Refer to label for use directions.

- - - - - - - - - - - - - - - - -

Signal Word/Toxicity Class: Danger/I.

THIABENDAZOLE

Antimycotic to retard or control mold growth in reconstituted tobacco leaf. Toxic to fish.

Mertect 40 *(Novartis)*

Rate: 2.4 oz. 40/100 lb. tobacco leaf

Disease:
mold

Remarks: Apply to bone-dry tobacco by tobacco water stream.

- - - - - - - - - - - - - - - - -

Signal Word/Toxicity Class: Caution.

Vapam — see metam-sodium

NOTICE The information on these pages is for preliminary planning — not a guide for use. Be sure to follow the manufacturer's directions, notwithstanding information contained here.

For personal protective equipment and EPA registration numbers, please turn to page 39.

Section 7

Vegetables

Enhanced Seed products 196
Bt Chart 196

Insecticide Listings
Artichokes 198
Asparagus 199
Beans 202
Carrots, etc. 209
Celery 213
Cole Crops 217
Cucurbits 225
Eggplant 231
Garlic 236
Greens 238
Herbs & Spices 243
Horseradish 245
Lentils 246
Lettuce 248
Mushrooms 254
Okra 255
Onions 256
Peas 260
Peppers 264
Potatoes 271
Radishes 278
Rhubarb 281
Southern Peas 282
Spinach 284
Sweet Corn 288
Sweet Potatoes, Yams . .294
Table Beets 297
Tomatoes 300
Watercress 307

Fungicide Listings
Artichokes 308
Asparagus 308
Beans 309
Carrots, etc. 316
Celery 319
Cole Crops 322
Cucurbits 328
Eggplant 334
Garbanzos 336
Garlic 336
Greens 338
Herbs & Spices 341
Horseradish 342
Lentils 342
Lettuce 343
Mushrooms 345
Okra 345
Onions 346
Peas 351
Peppers 353
Potatoes 357
Radishes 362
Rhubarb 363
Southern Peas 364
Spinach 365
Sweet Corn 367
Sweet Potatoes, Yams . .370
Table Beets 370
Tomatoes 372
Watercress 378

The information in this 1999 Insect and Disease Control Guide is updated to August 15, 1998.
Because manufacturers' product recommendations are constantly being updated and revised as additional product registrations and label changes are approved by EPA, there may be changes after this date. However, every effort has been made to give manufacturers the opportunity to include in the Guide their latest information. This Guide is for preliminary planning – not a use guide. Be sure to follow label directions, notwithstanding information contained herein.

Enhanced Seed/Bt Products Chart

Vegetables

Vegetable Enhanced Seed Varieties

NatureMark Potatoes has three enhanced seed varieties:
Bt potatoes for Colorado potato beetle resistance.
Bt potatoes + potato virus Y
Bt potatoes + potato leafroll virus.
Find company address information on page 599.

Find Bt products quickly and easily with this Bt products chart. Simply find the insect you want to control on the top of the chart, and follow down to the marked spot to find the Bt product you can use to control that pest. Company address information can be found on page 599.

	Alfalfa caterpillar	Armyworms	Artichoke plume moth	Cabbage budworm	Celery leaftier	Colorado potato beetle	Corn borers	Corn earworm	Cross-striped cabbageworm	Cutworms	Diamondback moth	Fungus gnats	Fusarium	Green cloverworm	Headworms	Heliothis
Agree		■	■						■	■	■			■		
Bacillus subtilis													■			
Biobit	■	■	■				■			■	■			■		■
BMP 123	■	■	■				■			■	■			■		■
Bt.* ssp. *aizawai		■	■						■	■	■			■	■	■
Bt.* ssp. *aizawai* + *Bt.* ssp. *kurstaki		■					■	■			■					
Bt.* ssp. *israelensis												■				
Bt.* ssp. *kurstaki	■	■	■	■	■	■	■	■	■	■	■			■	■	■
Bt.* ssp. *tenebriosis						■										
Condor		■	■	■	■		■		■	■	■			■		
Crymax		■	■	■	■		■		■	■	■			■		
Cutlass		■	■	■	■		■		■	■	■			■		
DiPel	■	■					■	■	■	■	■			■	■	■
Gnatrol												■				
Javelin		■	■	■	■		■		■	■	■			■		
Kodiak													■			
Lepinox		■					■	■								
M-Peril		■					■									
Mattch		■					■	■			■					
MVP II		■					■	■			■					
Novodor						■										
Raven						■										
Troy-Bt		■						■		■	■			■		
XenTari		■	■						■	■	■			■	■	■

	Hornworms	Imported cabbageworm	Loopers	Melonworm	Omnivorous leafroller	Pickleworm	Pinworm	Podworms	Rhizobium	Rhizoctonia	Saltmarsh caterpillar	Tobacco budworm	Tomato fruitworm	Velvetbean caterpillar	Webworms
Agree	■	■	■	■							■		■	■	
Bacillus subtilis									■	■					
Biobit	■	■	■	■	■	■	■	■			■		■	■	■
BMP 123	■	■	■		■	■		■			■		■	■	■
Bt.* ssp. *aizawai	■	■	■	■	■			■			■		■	■	■
Bt.* ssp. *aizawai* + *Bt.* ssp. *kurstaki		■	■									■	■		
Bt.* ssp. *israelensis															
Bt.* ssp. *kurstaki	■	■	■	■	■	■	■	■			■	■	■	■	■
Bt.* ssp. *tenebriosis															
Condor	■	■	■	■	■	■	■	■			■		■	■	■
Crymax	■	■	■	■	■	■	■	■			■		■	■	■
Cutlass	■	■	■	■	■	■	■	■			■		■	■	■
DiPel	■	■	■	■	■	■	■	■			■		■	■	■
Gnatrol															
Javelin	■	■	■	■	■	■	■	■			■		■	■	■
Kodiak									■	■					
Lepinox															
M-Peril											■				
Mattch		■	■								■		■	■	
MVP II		■	■										■	■	
Novodor															
Raven															
Troy-Bt	■	■	■	■		■					■	■	■		■
XenTari	■	■	■	■	■			■			■		■	■	■

Manufacturers List

These Bt products are manufactured by the following companies. Look for their addresses, phone numbers, and fax numbers in the Suppliers Directory beginning on page 599.

Agree, Javelin – Thermo Trilogy
Biobit, DiPel, Gnatrol, Novodor, XenTari – Abbott Laboratories
BMP 123 – Becker Microbial
Condor, Crymax, Cutlass, Lepinox, Raven – Ecogen
Kodiak – Gustafson, Trace
Mattch, M-Peril, MVP II – Mycogen
Troy-Bt – Troy Biosciences

Vegetables

Insecticides

Artichokes

Ambush — see permethrin

ASANA

Pyrethroid insecticide that has both contact and stomach poison activity. Restricted Use Pesticide due to toxicity to fish and aquatic organisms.

Asana XL (esfenvalerate) *(DuPont)*

Rate: 0.03-0.05 lb. ai/A
5.8-9.6 fl. oz. XL/A

Pest:
artichoke plume moth

Remarks: Apply in minimum 10 gal. water by air; 50-200 gal. water by ground; repeat at 14-day intervals as needed. Use sufficient water to obtain coverage without excessive runoff. Apply no more than 0.075 lb. ai/A between bud formation and harvest of individual artichokes.

Days To Harvest: 7.

- - - - - - - - - - - - - - - - -

Signal Word/Toxicity Class: Warning/II.
REI: 12 hr.

AZADIRACHTIN

Extracted from seed of neem tree. Insect growth regulator that controls targeted insect larvae when they ingest or come in contact with it, by interfering with the insect's ability to molt and shutting off feeding behavior.

◆ **Azatin XL Plus** *(Thermo Trilogy)*

Rate: 5-21 oz. XL Plus/A

Pest:

aphids	leafhoppers	thrips
artichoke plume moth	leafminers	whiteflies
green peach aphid	loopers	

Remarks: Rate varies by species. Apply when pests first appear; repeat at 7-day intervals. When pest populations are high, use higher rates. Do not apply more that 21 fl. oz./A.

Days To Harvest: 0.

- - - - - - - - - - - - - - - - -

Signal Word/Toxicity Class: Caution.
REI: 4 hr.

◆ **NEEMAZAD 4.5EC or NEEMIX 4.5** *(Thermo Trilogy)*

Rate: 0.125-1 pt. 4.5/A
High Volume:
2.25-4.5 oz. 4.5EC/100 gal. water
Low Volume:
0.5 pt. 4.5EC/A

Pest:

aphids	caterpillars	leafhoppers
armyworms	Colorado potato beetle	potato flea beetle
artichoke plume moth	diamondback moth	psyllids
aster leafhopper	greenhouse whitefly	sweetpotato whitefly
cabbage loopers	hornworms	webworms

Remarks: For low volume, apply in sufficient water for adequate coverage. For low insect pressure, apply preventive at 7-day intervals or at first sign of insects; high insect pressure, at 3- to 4-day intervals.

- - - - - - - - - - - - - - - - -

Signal Word/Toxicity Class: Warning/II.
REI: 12 hr.

Azatin — see azadirachtin

AZINPHOS-METHYL

Restricted Use Pesticide.

◆ **Micro Flo Azinphosmethyl 2EC or 50W** *(Micro Flo)*
Sniper 2-E, 50PVA, 50W *(Platte)*

Rate: 1.5 lb. ai/A
6 pt. 2EC, 2-E/A
3 lb. 50PVA, 50W/A

Pest:
plume moth

Remarks: Apply in sufficient water for complete coverage; repeat as needed (3 times for 50PVA). Do not feed or ensile treated forage.

Days To Harvest: 30.

- - - - - - - - - - - - - - - - -

Signal Word/Toxicity Class: Danger-Poison/I.
REI: 48 hr.; 72 hr. if average rainfall less than 25" per year.

DEADLINE

Packaged as Bullets, M-Ps (mini-pellets), and granules. Resistant to rain and watering.

Deadline Bullets, M-Ps, Granules (4%) (metaldehyde) *(Pace)*

Rate: 1-3 lb. ai/A
25-75 lb. 4%/A

Pest:
slugs snails

Remarks: Apply by air or ground before tight bud stage. After tight bud stage, apply by ground only. Do not apply directly to edible portion of plant.

- - - - - - - - - - - - - - - - -

Signal Word/Toxicity Class: Warning/II (granule); Caution (bullets, M-P).
REI: 12 hr.

Diatect — see pyrethrins

ENDOSULFAN

Toxic to birds, fish, and wildlife; toxic to bees exposed to direct application.

Drexel Endosulfan 3EC or 50WP *(Drexel)*
Gowan Endosulfan 3EC or 50WSB *(Gowan)*
◆ **Micro Flo Endosulfan 3EC or 50WP** *(Micro Flo)*
Phaser 3EC or 50WSB *(AgrEvo USA)*
Thiodan 3EC or 50WP *(FMC)*

Rate: 0.75-1 lb. ai/A
1-1.33 qt. 3EC/A
1.5-2 lb. 50WP, 50WSB/A

Pest:

artichoke aphid	green peach aphid	whiteflies
artichoke plume moth		

Remarks: Apply when insects appear or feeding is noticed. Use high rate for artichoke plume moth and whiteflies. Do not exceed 2 applications per year or 2 lb. ai/A per year.

Days To Harvest: 7.

- - - - - - - - - - - - - - - - -

Signal Word/Toxicity Class: Danger-Poison/I.
REI: 24 hr.

◆ - **new product** • ai-active ingredient • DF-dry flowable • E/EC-emulsifiable concentrate • EW-emulsion in water
F/FL-flowable • G-granule • L/LC-liquid • SP/WSP-soluble packet • W/WP-wettable powder • WSB-water soluble bag

METHYL PARATHION

Organophosphate insecticide toxic to bees exposed to direct treatment or residues on blooming crops or weeds. Restricted Use Pesticide due to very high acute toxicity to humans and birds.

Wilbur-Ellis Methyl Parathion 4 Spray *(Wilbur-Ellis)*
Rate: 0.5 lb. ai/A
1 pt. 4/A
Pest:
aphids, flea beetles, spider mites, armyworms, leafhoppers
Remarks: Do not apply through any type of irrigation system.
Days To Harvest: 15.

Signal Word/Toxicity Class: Danger-Poison/I.
REI: 24 hr.

M-PEDE

Effective contact insecticide/miticide for control of soft-bodied insects and mites.

M-Pede (potassium salts of fatty acids) *(Mycogen)*
Rate: 1-2% v/v solution
Pest:
aphids*, leafminers*, spider mites, broad mite, plant bugs, thrips, green peach aphid*, russet mite, whiteflies, leafhoppers
** Tank mix only*
Remarks: Apply at first sign of infestation; repeat at 7- to 14-day intervals. May be tank mixed with other insecticides. Refer to both labels for use directions and precautions.
Days To Harvest: 0.
State Restrictions: Not for use in California.

Signal Word/Toxicity Class: Warning/II.
REI: 12 hr.

NEEMAZAD — see azadirachtin

NEEMIX — see azadirachtin

PERMETHRIN

Restricted Use Pesticide.

Ambush 2E or 25W *(ZENECA)*
Rate: 0.1-0.3 lb. ai/A
6.4-19.2 fl. oz. 2E/A
6.4-19.2 oz. 25W/A
Pest:
artichoke plume moth
Remarks: Apply by ground as needed using sufficient water to obtain uniform coverage. Do not apply more than 1.5 lb. ai/A per season.
Days To Harvest: 0.

Signal Word/Toxicity Class: Warning/II.
REI: 24 hr.

Phaser — see endosulfan

PYRELLIN (Premix)

Premix containing pyrethrins and rotenone. May be combined with other insecticides to enhance knockdown or as an insect "exciter" to help facilitate contact with other spray residues.

Pyrellin E.C. (0.042 lb./gal. pyrethrins & 0.035 lb./gal. rotenone) *(Webb Wright)*
Rate: 1-2 pt. E.C./A
Pest:
aphids, loopers, plant bugs, leafhoppers, Lygus bugs, thrips, leafminers, mites, whiteflies
Remarks: Apply as necessary to maintain control.
Days To Harvest: 0.

Signal Word/Toxicity Class: Caution/III.
REI: 12 hr.

PYRETHRINS

Contains piperonyl butoxide as a synergist and silicon dioxide (diatomaceous earth).

Diatect Multipurpose Insecticide II *(Diatect)*
Rate: 1-6 lb. II/A
Pest: Including but not limited to:
aphids, mites, thrips, loopers, plant bugs, whiteflies
Remarks: Apply as dust or wettable powder. Repeat as needed.
Days To Harvest: 0.

Signal Word/Toxicity Class: Caution.
REI: 12 hr.

Sniper — see azinphos-methyl

SUPRACIDE

Restricted Use Pesticide.

Supracide 25WP (methidathion) *(Gowan)*
Rate: 1 lb. ai/A
4 lb. 25WP/A
Pest:
artichoke plume moth, chrysanthemum leafminer
Remarks: Apply in minimum 50 gal. water with ground equipment; 10 gal. water by air. Apply when insects appear in newly planted fields or when shoots emerge following complete cutback of plants; repeat at 14-day intervals until buds begin to appear. Do not apply after buds begin to form or under conditions which favor drift to adjacent artichoke fields where buds are present. Do not exceed 8 applications per crop cycle. Do not plant crops other than those listed on label within 1 year after last application. Do not apply through any type of irrigation system.

Signal Word/Toxicity Class: Warning/II.
REI: 48 hr.

Thiodan — see endosulfan

Asparagus

Ambush — see permethrin

AZADIRACHTIN

Extracted from seed of neem tree. Insect growth regulator that controls targeted insect larvae when they ingest or come in contact with it, by interfering with the insect's ability to molt and shutting off feeding behavior. Affected larvae slowly starve and die at next molt.

◆ **Azatin XL Plus** *(Thermo Trilogy)*
Rate: 5-21 oz. XL Plus/A
Pest:
aphids, flea beetles, leafrollers, armyworms, green peach aphid, loopers, beet armyworm, Japanese beetle, thrips, beetles, leafhoppers, webworms, caterpillars, leafminers, whiteflies, cutworms
Remarks: Rate varies by species. Apply when pests first appear; repeat at 7-day intervals. When pest populations are high, use higher rates. Do not apply more that 21 fl. oz./A.
Days To Harvest: 0.

Signal Word/Toxicity Class: Caution.
REI: 4 hr.

For personal protective equipment and EPA registration numbers, please turn to page 39.

Azatin — see azadirachtin

CARBARYL

Carbamate insecticide with broad activity on many insects. Low mammalian toxicity, but highly toxic to bees.

Clean Crop Sevin 5 Bait *(Platte)*

Rate: 2 lb. ai/A
40 lb. Bait/A

Pest:
ants, armyworms, cutworms, darkling beetle, field crickets, grasshoppers, sowbugs

Remarks: Apply with ground or air equipment.
Days To Harvest: 1.

- - - - - - - - - - - - - - - - -

Signal Word/Toxicity Class: Caution/III.
REI: 12 hr.

Drexel Carbaryl 4L or 50W *(Drexel)*
Sevin 4F, XLR Plus, 80S, 80WSP *(Rhone-Poulenc)*

Rate: 0.5-1.5 lb. ai/A
1-3 pt. 4L, 4F, XLR/A
1-3 lb. 50W/A
0.66-1.875 lb. 80S, 80WSP/A

Pest:
grasshoppers

Remarks: Use high rate for adult grasshoppers or applications to dense vegetation. Observe plant response precautions.
Days To Harvest: 1.

- - - - - - - - - - - - - - - - -

Rate: 1-2 lb. ai/A
2-4 pt. 4L, 4F, XLR/A
2-4 lb. 50W/A
1.25-2.5 lb. 80S, 80WSP/A

Pest:
asparagus beetle

Rate: Postharvest application
2-4 lb. ai/A
4-8 pt. 4L, 4F, XLR/A
4-8 lb. 50W/A
2.5-5 lb. 80S, 80WSP/A

Pest:
Apache cicada, asparagus beetle

Remarks: Ferns or brush growth. Do not treat more than once every 3 days. Observe plant response precautions.
Days To Harvest: Asparagus beetle (up to 2 lb. ai/A)-1.

- - - - - - - - - - - - - - - - -

Signal Word/Toxicity Class: Caution (4F, 4L, XLR); Warning/II (50W, 80S, 80WSP).
REI: 12 hr.

Gowan Sevin Bait *(Gowan)*

Rate: 2 lb. ai/A
40 lb. Bait/A

Pest:
armyworm, crickets, cutworms, darkling beetle, grasshoppers, sowbugs

Remarks: Apply when pests appear in damaging numbers; repeat at 7- to 14-day intervals if necessary. Do not use when bees are present.
Days To Harvest: 1.
State Restrictions: For use in Arizona and California.

- - - - - - - - - - - - - - - - -

Signal Word/Toxicity Class: Caution/III.
REI: 12 hr.

CHLORPYRIFOS

Organophosphate insecticide toxic to birds and wildlife, extremely toxic to fish and aquatic organisms.

Lorsban 4E *(Dow AgroSciences)*
◆ **Micro Flo Chlorpyrifos 4E AG** *(Micro Flo)*

Rate: 1 lb. ai/A
2 pt. 4E/A

Pest:
asparagus aphid, asparagus beetle, cutworms

Remarks: For cutworms, apply when soil is moist and worms are active near surface. For beetles and aphids, apply during fern stage. Do not make more than 1 preharvest application or 2 postharvest applications during fern stage.
Days To Harvest: 1.
State Restrictions: For use in the Midwest, Pacific Northwest.

- - - - - - - - - - - - - - - - -

Signal Word/Toxicity Class: Warning/II.
REI: 24 hr.

DEADLINE

Packaged as Bullets, M-Ps (mini-pellets), and granules.

Deadline Bullets, M-Ps, Granules (4%) (metaldehyde) *(Pace)*

Rate: 0.8-1 lb. ai/A
20-40 lb. 4%/A

Pest:
slugs, snails

Remarks: Apply every 3-4 weeks during growing season or as needed. Do not apply directly to edible portion of plant.

- - - - - - - - - - - - - - - - -

Signal Word/Toxicity Class: Warning/II (granule); Caution (bullets, M-P).
REI: 12 hr.

Diatect — see pyrethrins

DI-SYSTON

Restricted Use Pesticide.

Di-Syston 8 (disulfoton) *(Bayer)*

Rate: 1 lb. ai/A
1 pt. 8/A

Pest:
asparagus aphid, thrips (except NC)

Remarks: None.
State Restrictions: SLN-Arizona, California, North Carolina, Oregon, and Washington.

- - - - - - - - - - - - - - - - -

Signal Word/Toxicity Class: Danger/I.
REI: 48 hr.

LANNATE

Restricted Use Pesticide.

Lannate SP or LV (methomyl) *(DuPont)*

Rate: 0.45-0.9 lb. ai/A
0.5-1 lb. SP/A
1.5-3 pt. LV/A

Pest:
asparagus beetle, beet armyworm, redbacked cutworm, spotted asparagus beetle, variegated cutworm, western yellowstriped armyworm, white cutworm

Remarks: Apply in sufficient water to obtain thorough, uniform coverage when insects first appear; repeat at 5- to 7-day intervals or as needed. Do not apply more than 4.5 lb. ai/A per crop or make more than 8 applications per crop.
Days To Harvest: 1.

- - - - - - - - - - - - - - - - -

Signal Word/Toxicity Class: Danger-Poison/I.
REI: 48 hr.

Lorsban — see chlorpyrifos

MALATHION

Organophosphate insecticide toxic to fish, aquatic invertebrates, aquatic life stages of amphibians, and bees exposed to direct treatment.

Clean Crop Malathion 57EC *(Platte)*

Rate: 1-1.25 lb. ai/A
1.5-2 pt. 57EC/A

Pest:
thrips

◆ **- new product** • ai-active ingredient • DF-dry flowable • E/EC-emulsifiable concentrate • EW-emulsion in water
F/FL-flowable • G-granule • L/LC-liquid • SP/WSP-soluble packet • W/WP-wettable powder • WSB-water soluble bag

Rate: 1.25 lb. ai/A
2 pt. 57EC/A
Pest:
asparagus aphid asparagus beetle
Remarks: Apply when insects first appear and repeat as necessary.
Days To Harvest: 1.

Signal Word/Toxicity Class: Caution/III.
REI: 12 hr.

◆ **Gowan Malathion 8 (8E)** *(Gowan)*
Rate: 1.25 lb. ai/A
1.25 pt. 8E/A
Pest:
asparagus beetle thrips
Remarks: Apply at first sign of infestation; repeat at 7- to 10-day intervals.
Days To Harvest: 1.

Signal Word/Toxicity Class: Caution/III.
REI: 12 hr.

METHOXYCHLOR

Organochlorine pesticide toxic to fish and aquatic invertebrates.

Prentox Methoxychlor 2EC *(Prentiss)*
Rate: 1.3-17 lb. ai/A
75-100 fl. oz. 2EC/A
Pest:
asparagus beetle
Remarks: Apply as thorough cover spray at first sign of infestation.
Days To Harvest: 3.

Signal Word/Toxicity Class: Caution.
REI: 12 hr.

M-PEDE

M-Pede (potassium salts of fatty acids) *(Mycogen)*
Rate: 1-2% v/v solution
Pest:
aphids* leafminers* spider mites
broad mite plant bugs thrips
green peach aphid* russet mite whiteflies
leafhoppers
*** Tank mix only**
Remarks: Apply at first sign of infestation, repeat at 7- to 14-day intervals. May be tank mixed with other insecticides. Refer to both labels for use directions and precautions.
Days To Harvest: 0.

Signal Word/Toxicity Class: Warning/II.
REI: 12 hr.

NEMACUR

Restricted Use Pesticide.

Nemacur 3 (fenamiphos) *(Bayer)*
Rate: 2 lb. ai/A
5.33 pt. 3/A
Pest:
nematodes
Remarks: Apply in incorporated band centered on the row using band width of 50% row spacing. Newly planted crowns: Apply before planting crowns but after making furrows. Apply following spring before spear emergence and incorporate by reshaping beds. Postharvest: Apply after all spears have been removed and incorporate by reshaping beds.
Days To Harvest: 9 months.
State Restrictions: For use in Connecticut, Delaware, Maine, Maryland, Massachusetts, New Hampshire, New Jersey, New York, Pennsylvania, and Rhode Island only.

Signal Word/Toxicity Class: Danger/I.
REI: 48 hr.

Omni Supreme Spray — see petroleum oil

PERMETHRIN

Restricted Use Pesticide.

Ambush 2EC or 25W *(ZENECA)*
Rate: 0.05-0.1 lb. ai/A
3.2-6.4 fl. oz. 2EC/A
3.2-6.4 oz. 25W/A
Pest:
asparagus beetle cutworms
Remarks: Apply as needed by ground using sufficient water to obtain uniform coverage. Do not apply more than 0.4 lb. ai/A per season.
Days To Harvest: 1.

Rate: 0.1 lb. ai/A
6.4 fl. oz. 2EC/A
6.4 oz. 25W/A
Pest:
asparagus beetle Japanese beetle Lygus bugs
Remarks: Apply to fern stage of asparagus after spear harvest when larvae and adult stage are present.

Signal Word/Toxicity Class: Warning/II.
REI: 24 hr.

Pounce 3.2EC, 25WP or WSB *(FMC)*
Rate: 0.05-0.1 lb./A
2-4 oz. 3.2EC/A
3.2-6.4 oz. 25WP/A
1 bag WSB/A
Pest:
asparagus beetle cutworms

Rate: Postharvest:
0.1 lb. ai/A
4 oz. 3.2EC/A
6.4 oz. 25WP/A
1 bag WSB/A
Pest:
asparagus beetle Lygus bugs tarnished plant bug
Japanese beetle (adult)*
*** Except California**
Remarks: Apply by ground as needed. Use sufficient water to obtain uniform coverage. Do not apply more than 0.4 lb. ai/A per season. For postharvest application, apply to fern stage of asparagus plant after spear harvest when larval and adult stage are present.

Signal Word/Toxicity Class: Caution (3.2 EC), Warning/II (25WP, WSB).
REI: 24 hr.

PETROLEUM OIL

Petroleum-based spray oil. Mode of action is through suffocation of eggs, larvae, and nymphs of insects and adults of soft bodied insects. This mode of action necessitates total spray coverage. Toxic to fish.

Omni Supreme Spray *(Helena)*
Rate: 1-2 gal./100 gal. water
Pest:
aphids caterpillars mites
beetles (larvae) leafminers
Remarks: None.

Signal Word/Toxicity Class: Caution.
REI: 4 hr.

SAF-T-SIDE *(Brandt Consolidated)*
◆ **Sumulsion** *(Drexel)*
◆ **SunSpray Ultra-Fine** *(Sun)*
Rate: 1-2 gal./100 gal. water
Pest:
aphids leafminers thrips
beetles (larvae) mites whiteflies
leafhoppers spiders
Remarks: Apply as needed. May be tank mixed with most commonly used insecticides and fungicides. Do not use in combination with or immediately before or after spraying with fungicides such as captan, Dyrene, folpet, Morestan, or any sulfur products. Do not use with carbaryl or dimethoate.

Signal Word/Toxicity Class: Caution.
REI: 4 hr.

7 Vegetables

NOTICE	The information on these pages is for preliminary planning — not a guide for use. Be sure to follow the manufacturer's directions, notwithstanding information contained here.	For personal protective equipment and EPA registration numbers, please turn to page 39.

Pounce — see permethrin

PYRELLIN (Premix)

Premix containing pyrethrins and rotenone. May be combined with other insecticides to enhance knockdown or as an insect "exciter" to help facilitate contact with other spray residues.

Pyrellin E.C. (0.042 lb./gal. pyrethrins & 0.035 lb./gal. rotenone) *(Webb Wright)*

Rate: 1-2 pt. E.C./A

Pest: Including but not limited to:

aphids	leafminers	plant bugs
asparagus beetle	loopers	thrips
leafhoppers	mites	whiteflies

Remarks: Apply at 7-day intervals or less; repeat as needed.

Days To Harvest: 0.

Pyrenone — see pyrethrins

PYRETHRINS

Contains piperonyl butoxide as a synergist. Diatect also contains silicon dioxide (diatomaceous earth). May be combined in tank mix with other insecticides for quicker and more complete kill where insect resistance may be a problem, and as an "exciter" to flush insects out of hiding and into contact with other spray residues.

Diatect Multipurpose Insecticide II *(Diatect)*
Pyrenone (0.5EC) *(AgrEvo Environmental Health)*
◆ **Py-Rin 60-6 EC** *(Wilbur-Ellis)*

Rate: 1-6 lb. II/A
1-12 fl. oz. 0.5EC/A
2-16 oz. 6-60 EC/A

Pest: Including but not limited to:

aphids	loopers	thrips
asparagus beetle	mites	whiteflies
leafminers	plant bugs	

Remarks: Repeat as necessary. Apply Diatect as dust or wettable powder.

Days To Harvest: 0.

- - - - - - - - - - - - - - - -

Signal Word/Toxicity Class: Caution.

REI: 12 hr.

Py-Rin — see pyrethrins

SAF-T-SIDE — see petroleum oil

Sevin — see carbaryl

Sumulsion — see petroleum oil

SunSpray — see petroleum oil

Beans

(Dry, Green, Kidney, Lima, Navy, Pole, Snap, Wax)

ACEPHATE

Organophosphate systemic insecticide with contact and stomach poison activity. Effective against many aphids, thrips, caterpillars, and plant bugs. Has moderate residual effects and is compatible with most pesticides.

Orthene 75 S *(Valent)*

Rate: 0.25-1 lb. ai/A
0.33-1.33 lb. 75 S/A

Pest:

aphids	cutworms	Lygus bugs
armyworms	European corn borer	Mexican bean beetle
bean leaf beetle	fleahoppers	plant bugs
bean leafroller	grasshoppers	soybean looper
cabbage looper	green cloverworm	thrips
corn earworm	leafhoppers	whiteflies

Remarks: Dry, succulent beans, lima beans. Apply at 7- to 10-day intervals as needed when eggs or insects first appear. Rate varies by pest species and infestation level; refer to label. Do not apply through any type of irrigation system. Do not feed treated vines.

Days To Harvest: Lima beans (succulent)-0; dry, snap beans-14.

- - - - - - - - - - - - - - - -

Signal Word/Toxicity Class: Caution.

REI: 24 hr.

Adios — see carbaryl

ASANA

Pyrethroid insecticide that has both contact and stomach poison activity. Restricted Use Pesticide due to toxicity to fish and aquatic organisms.

Asana XL (esfenvalerate) *(DuPont)*

Rate: 0.015-0.03 lb. ai/A
2.9-5.8 fl. oz. XL/A

Pest:

Mexican bean beetle	saltmarsh caterpillar

Rate: 0.03-0.05 lb. ai/A
5.8-9.6 fl. oz. XL/A

Pest:

beet armyworm (aids in control)	cutworms	painted lady butterfly (larvae) (dry beans)
cabbage looper	European corn borer (snap beans)	pea aphid
corn earworm	flea beetles (snap beans)	potato leafhopper
corn rootworm (adult)	grasshoppers	velvetbean caterpillar
cowpea curculio (dry beans)	green cloverworm	western bean cutworm
cucumber beetle	leafhoppers	

Remarks: Dry, snap beans. Repeat as needed to maintain control. Do not exceed 0.2 lb. ai/A per season. Do not allow livestock to feed or graze treated fields. For snap beans, do not harvest treated vines for livestock forage, fodder, or hay.

Days To Harvest: Snap bean-3; dry bean-21.

- - - - - - - - - - - - - - - -

Signal Word/Toxicity Class: Warning/II.

REI: 12 hr.

AZADIRACHTIN

Extracted from seed of neem tree. Insect growth regulator that controls targeted insect larvae when they ingest or come in contact with it, by interfering with the insect's ability to molt and shutting off feeding behavior. Affected larvae slowly starve and die at next molt.

◆ **NEEMAZAD 4.5EC or NEEMIX 4.5** *(Thermo Trilogy)*

Rate: 0.125-1 pt. 4.5/A
High Volume:
2.25-4.5 oz. 4.5EC/100 gal. water
Low Volume:
0.5 pt. 4.5EC/A

Pest:

armyworms	garden webworm	psyllids
bean leaf beetle	greenhouse whitefly	soybean loopers
cabbage loopers	hornworms	sweetpotato whitefly
caterpillars	leafminers	tomato fruitworms
Colorado potato beetle	loopers	tomato pinworms
cutworms	Mexican bean beetles	weevils
	pepper weevils	

Remarks: For low volume, apply in sufficient water for adequate coverage. For low insect pressure, apply preventative at 7-day intervals or at first sign of insects; high insect pressure, at 3- to 4-day intervals.

- - - - - - - - - - - - - - - -

Signal Word/Toxicity Class: Warning/II.

REI: 12 hr.

◆ - **new product** • ai-active ingredient • DF-dry flowable • E/EC-emulsifiable concentrate • EW-emulsion in water
F/FL-flowable • G-granule • L/LC-liquid • SP/WSP-soluble packet • W/WP-wettable powder • WSB-water soluble bag

AZINPHOS-METHYL

Restricted Use Pesticide.

Gowan Azinphos-M 50-W *(Gowan)*
◆ Micro Flo Azinphosmethyl 2EC or 50W *(Micro Flo)*
Sniper 2-E, 50PVA, 50W *(Platte)*

Rate: 0.25-0.5 lb. ai/A
1-2 pt. 2EC, 2-E/A
0.5-1 lb. 50-W, 50PVA, 50W/A

Pest:

aphids	Mexican bean beetle	striped cucumber beetle
bean leaf beetle	mites	tarnished plant bug
green cloverworm	spotted cucumber beetle	velvetbean caterpillar
leafhoppers	stink bugs	western striped cucumber beetle
leafminers		
leafrollers		

Remarks: Dry, snap beans. Apply in sufficient water to give complete coverage; repeat as needed. Rate varies by pest species, refer to label. Do not exceed 4 applications. Do not feed or ensile treated forage.
Days To Harvest: Dry beans-30; snap beans-7.
State Restrictions: Sniper 50PVA not registered in California.

Signal Word/Toxicity Class: Danger-Poison/I.
REI: 48 hr.; 72 hr. if average rainfall less than 25" per year.

Bravo S — see sulfur premix

CARBARYL

Adios (13%) *(Micro Flo)*

Rate: 0.065-0.1 lb. ai/A
0.5-0.75 lb. 13%/A

Pest:
cucumber beetle

Remarks: Dry, green, lima, navy, snap beans. Do not apply through any type of irrigation system. Do not apply more than 3.75 lb./A per season.
Days To Harvest: 0.

Signal Word/Toxicity Class: Caution/III.
REI: 12 hr.

Clean Crop Sevin 5 Bait *(Platte)*

Rate: 1-2 lb. ai/A
20-40 lb. Bait/A

Pest:

ants	darkling beetle	grasshoppers
armyworms	field crickets	sowbugs
cutworms		

Remarks: Apply with ground or air equipment.
Days To Harvest: 0.

Signal Word/Toxicity Class: Caution/III.
REI: 12 hr.

Drexel Carbaryl 4L or 50W *(Drexel)*
Sevin 4F, XLR Plus, 80S, 80WSP *(Rhone-Poulenc)*

Rate: 0.5-1.5 lb. ai/A
1-3 pt. 4L, 4F, XLR/A
1-3 lb. 50W/A
0.66-1.875 lb. 80S, 80WSP/A

Pest:

alfalfa caterpillar	European corn borer	stink bugs
alfalfa looper	fall armyworm	tarnished plant bug
armyworms	flea beetles	threecornered alfalfa hopper
bean leaf beetle	grasshoppers	thrips
blister beetle	green cloverworm	velvetbean caterpillar
corn earworm	Japanese beetle	webworms
cucumber beetle	leafhoppers	western bean cutworm
cutworms	Mexican bean beetle	

Rate: 2 lb. ai/A
4 pt. 4L, 4F, XLR/A
4 lb. 50W/A
2.5 lb. 80S, 80WSP/A

Pest:

corn earworm*	limabean pod borer*	southern green stink bug*
cowpea curculio	Lygus bugs*	stink bugs*

*** California**

Remarks: Dry, green, lima, navy, snap beans. Apply when insects or their damage appear; repeat at 7- to 14-day intervals or as needed. Rate varies by pest species; refer to label. Observe plant response precautions.
Days To Harvest: 0.

Signal Word/Toxicity Class: Caution (4F, 4L, XLR); Warning/II (50W, 80S, 80WSP).
REI: 12 hr.

Gowan Sevin Bait *(Gowan)*

Rate: 2 lb. ai/A
40 lb. Bait/A

Pest:

armyworms	cutworms	grasshoppers
crickets	darkling beetle	sowbugs

Remarks: Apply when pests appear in damaging numbers; repeat at 7- to 14-day intervals if necessary. Do not use when bees are present.
Days To Harvest: 1.
State Restrictions: For use in Arizona and California.

Signal Word/Toxicity Class: Caution/III.
REI: 12 hr.

Wilbur-Ellis Sevin 5 or 10 Dust *(Wilbur-Ellis)*

Rate: 2 lb. ai/A
40 lb. 5/A
20 lb. 10/A

Pest:

armyworm	cutworms	Mexican bean beetle
bean leaf beetle	flea beetles	stink bugs
corn earworm	leafhoppers	velvetbean caterpillar
cucumber beetle	Lygus bugs	western bean cutworm

Remarks: Do not feed treated vines to livestock.
Days To Harvest: 0.
State Restrictions: Sevin 5 not registered in California.

Signal Word/Toxicity Class: Caution.
REI: 12 hr.

CHLORPYRIFOS

For use only by commercial seed treaters.

Lorsban 30, 50-SL *(Gustafson)*

Rate: 2.7 fl. oz. 30/100 lb. seed
2 oz. 50-SL/100 lb. seed

Pest:
seedcorn beetles — seedcorn maggots

Remarks: Field, green, kidney, lima, navy, snap, string, wax beans. Apply as slurry treatment before planting. Do not use treated seed for food, feed, or oil purposes. Commercially treated seed must be labeled in accordance with all applicable requirements of Federal Seed Act. Do not allow livestock to graze or feed bean hay or pea hay grown from treated seed.

Signal Word/Toxicity Class: Warning (50-SL); Caution (30).
REI: 24 hr.

Comite — see propargite

DEADLINE

Packaged as Bullets, M-Ps (mini-pellets), and granules.

Deadline Bullets, M-Ps, Granules (4%) (metaldehyde) *(Pace)*

Rate: 0.8-1 lb. ai/A
20-40 lb. 4%/A

Pest:
slugs — snails

Remarks: Apply every 3-4 weeks during growing season or as needed. Do not apply directly to edible portion of plant.

Signal Word/Toxicity Class: Warning/II (granule); Caution (bullets, M-P).
REI: 12 hr.

Declare — see methyl parathion

Diatect — see pyrethrins

NOTICE The information on these pages is for preliminary planning — not a guide for use. Be sure to follow the manufacturer's directions, notwithstanding information contained here.

For personal protective equipment and EPA registration numbers, please turn to page 39.

DIAZINON

Restricted Use Pesticide due to avian and aquatic toxicity.

D•z•n diazinon AG600 WBC, 50W *(Novartis)*
Drexel Diazinon Insecticide (4EC) *(Drexel)*
Gowan Diazinon 4E, 50WSB *(Gowan)*
Helena Diazinon AG500 *(Helena)*
Micro Flo Diazinon AG500, 50W *(Micro Flo)*
Prentox Diazinon AG500, 50W *(Prentiss)*
◆ **Wilbur-Ellis Diazinon 4 (4E)** *(Wilbur-Ellis)*

Rate: 0.375 lb. ai/A
0.75 pt. 4E, 4EC, AG500/A in 100 gal. water
9.5 fl. oz. AG600/A in 100 gal. water
0.25-0.375 lb. ai/A
0.5-0.75 lb. 50W, 50WSB/A in 100 gal. water

Pest:
cucumber beetle

Rate: 0.5-0.75 lb. ai/A
1-1.5 pt. 4E, 4EC, AG500/A
12.75-19 fl. oz. AG600/A
1-1.5 lb. 50W, 50WSB/A

Pest:
black bean aphid · dipterous leafminer · Mexican bean beetle

Rate: 2-4 lb. ai/A
4-8 pt. 4E, 4EC, AG500/A
51-102 fl. oz. AG600/A
4-8 lb. 50W, 50WSB/A

Pest:
cutworms

Rate: 3-4 lb. ai/A
6-8 pt. 4E, 4EC, AG500/A
76.5-102 fl. oz. AG600/A
6-8 lb. 50W, 50WSB/A

Pest:
wireworm

Rate: **Lima beans only:**
0.5 lb. ai/A
1 pt. 4E, 4EC, AG500/A
12.75 fl. oz. AG600/A

Pest:
leafhoppers · spider mites

Remarks: Lima, pole, snap beans. Succulent beans only. For cutworms and wireworms, broadcast just before planting and incorporate into soil immediately. Will not control organophosphate-resistant leafminers. Bean forage or bean hay may be fed to dairy and beef cattle and sheep. Wait 4 days following treatment if forage is to be cut for hay.
Days To Harvest: 7.

Signal Word/Toxicity Class: Caution/III.
REI: 24 hr.

Drexel D-264 14G *(Drexel)*
Gowan Diazinon 14G *(Gowan)*
◆ **Micro Flo Diazinon 14G** *(Micro Flo)*
◆ **Wilbur-Ellis Diazinon 14G** *(Wilbur-Ellis)*

Rate: 1-2 lb. ai/A
7-14 lb. 14G/A (6" band over row)

Pest:
lesser cornstalk borer

Rate: 2-4 lb. ai/A
14-28 lb. 14G/A

Pest:
cutworms

Rate: 3-4 lb. ai/A
21-28 lb. 14G/A

Pest:
wireworms

Remarks: Lima, pole, snap beans. Succulent beans only. Broadcast and incorporate just before planting.

Signal Word/Toxicity Class: Caution.
REI: 12 hr. (24 hr. for Micro Flo).

DIBROM

Short residual contact insecticide for control of insect and mite adults and larvae present at time of application.

Dibrom 8 (8E) (naled) *(Amvac)*

Rate: 1 lb. ai/A
1 pt. 8E/A

Pest:
alfalfa looper · aphids · leafhoppers · leafminers · loopers · Lygus bugs · nitidulid beetles · spider mites

Rate: 1.5 lb. ai/A
1.5 pt. 8E/A

Pest:
large loopers · stink bugs

Remarks: Beans, lima beans. Apply in sufficient water to cover (100 gal. water/A preferable) at first sign of insects. For aerial application (California only), apply in minimum 10 gal. water/A by air. Make only 1 application per 7-day period. Do not use with highly alkaline materials such as lime or Bordeaux. Do not apply through any type of irrigation system. Do not apply less than 1 day before cutting or pulling plants. Do not apply more than 4.5 pt./A per season.
Days To Harvest: 1.

Signal Word/Toxicity Class: Danger/I.
REI: 48 hr. for 1 pt./A or less; 72 hr. for more than 1 pt./A.

DICOFOL

Chlorinated hydrocarbon acaricide with contact activity, but no systemic or insecticidal activity. Undersides of leaves must be covered for maximum effectiveness. Long residual effects, but incompatible with lime compounds. Toxic to fish.

◆ **Griffin Dicofol 4E** *(Griffin)*
Kelthane MF (4E) *(Rohm and Haas)*

Rate: 0.38-0.5 lb. ai/A
0.75-1 pt. 4E/A

Pest:
Pacific mite · tropical mite · twospotted spider mite

Rate: California:
1.5 lb. ai/A
3 pt. 4E/A

Pest:
Pacific mite · tomato russet mite · tropical mite · twospotted spider mite

Remarks: Dry, green, lima beans. Apply at first sign of mite buildup. Do not feed treated vines or crop residues to livestock. Do not apply through any type of irrigation system.
Days To Harvest: 0.75-1 pt. rate-7; 3 pt. rate-21.

Signal Word/Toxicity Class: Caution.
REI: 12 hr.

Digon — see dimethoate

Dimate — see dimethoate

DIMETHOATE

Highly toxic to bees.

Cheminova Dimethoate 4E *(Cheminova)*
Clean Crop Dimethoate 400 *(Platte)*
Dimate 4E *(Terra)*
Drexel Dimethoate 2.67 or 4EC *(Drexel)*
Gowan Dimethoate E267 *(Gowan)*
Helena Dimethoate 4EC *(Helena)*
Micro Flo Dimethoate 267 or 4E *(Micro Flo)*
Wilbur-Ellis Dimethoate 267 or Digon 400 *(Wilbur-Ellis)*

Rate: 0.25-0.5 lb. ai/A
0.75-1.5 pt. 2.67, 267, E267/A
0.5-1 pt. 4E, 4EC, 400/A

◆ - **new product** • ai-active ingredient • DF-dry flowable • E/EC-emulsifiable concentrate • EW-emulsion in water
F/FL-flowable • G-granule • L/LC-liquid • SP/WSP-soluble packet • W/WP-wettable powder • WSB-water soluble bag

Pest:

aphids	leafhoppers	Mexican bean beetle
bean leaf beetle	leafminers	mites
grasshoppers	Lygus bugs	

Remarks: Dry, green, lima, snap beans. Do not apply if bees are visiting areas to be treated when crop or weeds are in bloom. Do not feed treated vines.

Days To Harvest: 0; Helena-7.

Signal Word/Toxicity Class: Warning/II.
REI: 48 hr.

DI-SYSTON

Restricted Use Pesticide.

Di-Syston 8 or 15% (disulfoton) *(Bayer)*

Rate: 1-2 lb. ai/A
1-2 pt. 8/A
6.7-13.3 lb. 15%/A (30" row spacing)

Pest:

aphids	Mexican bean beetle (larvae)	mites
leafhoppers		thrips

Remarks: Dry, green, lima, snap beans. Lima, snap beans: For Di-Syston 8, apply in water emulsion or with liquid fertilizer as soil injection on each side of seed furrow at planting time. For Di-Syston 15%, place granules in soil incorporated band on each side of seed furrow at planting time. Dry beans: Apply as above at planting time or inject as side-dressing (soil incorporated for 15%) on each side of row. Do not apply directly on seed. Do not apply more than once per season. Do not feed vines or hay. Refer to label for additional use information.

Days To Harvest: Dry beans-60.

Signal Word/Toxicity Class: Danger/I.
REI: 48 hr.

Drexel D-264 — see diazinon

ENDOSULFAN

Toxic to birds, fish, and wildlife; toxic to bees exposed to direct application.

Drexel Endosulfan 3EC or 50WP *(Drexel)*
Gowan Endosulfan 3EC or 50WSB *(Gowan)*
◆ **Micro Flo Endosulfan 3EC or 50WP** *(Micro Flo)*
Phaser 3EC or 50WSB *(AgrEvo USA)*
Thiodan 3EC or 50WP *(FMC)*

Rate: 0.5-1 lb. ai/A
0.66-1.33 qt. 3EC/A
1-2 lb. 50WP, 50WSB/A

Pest:

aphids	cowpea curculio	leafhoppers
armyworms	cucumber beetle	Mexican bean beetle
bean leafskeletonizer	flea beetle	western bean cutworm
black bean aphid	green stink bug	whiteflies

Remarks: Dry beans (except lima), succulent beans. Do not use on lima beans that are to be removed from field for processing. Apply when insects appear or feeding is noticed. For cowpea curculio, apply at 5-day intervals starting when pods are 1/2" long. Do not exceed 3 applications per year or 3 lb. ai/A per year. Do not feed treated threshings to livestock or allow livestock to graze treated fields.

Days To Harvest: 3.

Signal Word/Toxicity Class: Danger-Poison/I.
REI: 24 hr.

Fyfanon — see malathion

Kelthane — see dicofol

Kumulus — see sulfur

LANNATE

Restricted Use Pesticide.

Lannate SP or LV (methomyl) *(DuPont)*

Rate: 0.225-0.9 lb. ai/A
0.25-1 lb. SP/A
0.75-3 pt. LV/A

Pest:

aphids	loopers (except AL, GA)	thrips
beet armyworm	Lygus bugs	variegated cutworm
corn earworm	Mexican bean beetle	western yellowstriped armyworm
European corn borer	saltmarsh caterpillar	
fall armyworm	spotted cucumber beetle	yellowstriped armyworm
leafhoppers		

Remarks: Dry, succulent beans. Apply in sufficient water to obtain thorough, uniform coverage when insects first appear; repeat at 5- to 7-day intervals or as needed. Do not apply more than 4.5 lb. ai/A per crop or make more than 10 applications per crop.

Days To Harvest: Succulent beans: up to 0.45 lb. ai-1; over 0.45 lb. ai-3. Dry beans-14 (from cutting). Grazing/feeding: vines, hay-14 (from cutting).

Signal Word/Toxicity Class: Danger-Poison/I.
REI: 48 hr.

LINDANE

Lindane 30C contains toxic inert Rhodamine B. Toxic to fish, birds, and wildlife.

Lindane 30C *(Gustafson)*
◆ **Lindane ST 40** *(AGSCO)*

Rate: 0.7 fl. oz. 30C/100 lb. seed
0.5 fl. oz. 40/100 lb. seed

Pest:

seedcorn maggots wireworms

Remarks: Do not use treated seed for food, feed, or oil purposes. Exposed treated seed may be hazardous to birds and other wildlife. Dispose of excess seed by burial away from bodies of water.

Signal Word/Toxicity Class: Warning/II.
REI: 12 hr.

Lorsban — see chlorpyrifos

MALATHION

Organophosphate insecticide toxic to fish, aquatic invertebrates, aquatic life stages of amphibians, and bees exposed to direct treatment on blooming crops or weeds.

Clean Crop Malathion 57EC *(Platte)*
Drexel Malathion 5EC *(Drexel)*
◆ **Micro Flo Malathion 5EC** *(Micro Flo)*
Prentox 5 lb. Malathion Spray (5EC) *(Prentiss)*

Rate: Green, lima, kidney, navy, snap, wax beans:
1-1.25 lb. ai/A
1.5-2 pt. 5EC, 57EC/A

Pest:

aphids	cucumber beetle	Mexican bean beetle
bean leaf beetle	Japanese beetle	potato leafhopper
blister beetle	Lygus bugs	red spider mites

Rate: Dry beans—California, Northwest:
1-1.25 lb./A
1.5-2 pt. 5EC/A

Pest:

Lygus bugs

Remarks: Rate varies by pest species; refer to label.

Days To Harvest: 1.

Signal Word/Toxicity Class: Caution/III (57EC); Warning/II (5EC).
REI: 12 hr.

7

Vegetables

NOTICE The information on these pages is for preliminary planning — not a guide for use. Be sure to follow the manufacturer's directions, notwithstanding information contained here.

For personal protective equipment and EPA registration numbers, please turn to page 39.

◆ **Clean Crop Malathion 8EC** *(Platte)*
◆ **Wilbur-Ellis Malathion 8 Spray** *(Wilbur-Ellis)*

Rate: 1-1.5 lb. ai/A
1-1.5 pt. 8ECA

Pest:
aphids, leafhoppers, Mexican bean beetle, pea weevil, spider mites, thrips

Remarks: Dry, succulent beans. Apply by air or ground.
Days To Harvest: 1.

Signal Word/Toxicity Class: Caution/III.
REI: 12 hr.

Fyfanon (8E) *(Helena)*
◆ **Gowan Malathion 8 (8E)** *(Gowan)*

Rate: 1.5 lb. ai/A
1.5 pt. 8E/A

Pest:
aphids, cucumber beetle, Japanese beetle, Mexican bean beetle, nitidulid beetle, pea leaf weevil, potato leafhopper, spider mites

Rate: Dry beans—California:
1-15 lb./A
1-1.5 8E/A

Pest:
Lygus bugs

Remarks: Do not graze or feed treated foliage to livestock.
Days To Harvest: 1.

Signal Word/Toxicity Class: Caution/III.

Fyfanon ULV *(Cheminova)*
◆ **Micro Flo Malathion ULV** *(Micro Flo)*

Rate: 0.5 pt. ULV/A

Pest:
green cloverworm, Japanese beetle, leafhoppers, Lygus bugs, Mexican bean beetle

Remarks: Dry, green, kidney (red), lima, navy, snap, wax beans. Apply when insects are present.
Days To Harvest: 1.

Signal Word/Toxicity Class: Caution.
REI: 12 hr.

Marlate — see methoxychlor

METASYSTOX-R

Restricted Use Pesticide.

Metasystox-R (SC) (oxydemeton-methyl) *(Gowan)*

Rate: 0.5 lb. ai/A
2 pt. SC/A in 4 gal. water

Pest:
leafhoppers, Lygus bugs (CA only), mites

Remarks: Lima beans. Up to 3 applications may be made per picking season.
Days To Harvest: Grazing or cutting treated vines for feed or forage-21.

Signal Word/Toxicity Class: Warning/II.
REI: 48 hr.; 72 hr. if average rainfall less than 25" per year.

METHOXYCHLOR

Organochlorine pesticide toxic to fish and aquatic invertebrates.

Marlate 50 (WP) *(Kincaid)*
Prentox Methoxychlor 2EC *(Prentiss)*

Rate: 1.3-1.7 lb. ai/A
75-100 fl. oz. 2EC/A
2 Tbs. 50WP/gal. water

Pest:
bean leaf beetle, corn earworm, cucumber beetle, fall armyworm, flea beetles, garden webworm, Japanese beetle, Mexican bean beetle, potato leafhopper

Remarks: Apply at first sign of infestation; repeat at 7- to 14-day intervals as needed.
Days To Harvest: 3; if vines are used for feed or forage-7.

Signal Word/Toxicity Class: Caution.
REI: 12 hr.

METHYL PARATHION

Organophosphate insecticide toxic to bees exposed to direct treatment or residues on blooming crops or weeds. Penncap-M is a microencapsulated formulation. Restricted Use Pesticide due to very high acute toxicity to humans and birds.

Cheminova Methyl 4EC *(Cheminova)*
◆ **Declare (4EC)** *(Griffin)*

Rate: 0.5-1.5 lb. ai/A
1-3 pt. 4EC/A

Pest:
aphids, armyworms, climbing cutworm, cowpea curculio, cutworms, flea beetles, leafhoppers, Lygus bugs, mites, stink bugs, surface cutworm

Remarks: Dry, green beans. Apply when insects first appear; repeat as needed. Rate varies by pest species; refer to label for use information.
Days To Harvest: Up to 1 pt.-15; above 1 pt.-21.

Signal Word/Toxicity Class: Danger-Poison/I.
REI: 48 hr.; 72 hr. if average rainfall less than 25" per year.

Helena 4 lb. Methyl Parathion (4EC) *(Helena)*

Rate: 0.5-1.25 lb. ai/A
1-2.5 pt. 4EC/A

Pest:
aphids, armyworms (to 3rd instar), cabbage looper, climbing cutworm, cutworms, flea beetles, leafhoppers, plant bugs, stink bugs, surface cutworm

Remarks: Rate varies by pest species; refer to label.
Days To Harvest: 15; green and lima beans over 0.5 lb. ai-21.

Signal Word/Toxicity Class: Danger-Poison/I.
REI: 48 hr.

Penncap-M (2FL) *(Elf Atochem)*

Rate: 0.5 lb. ai/A
2 pt. 2FL/A

Pest:
aphids (except CA), cucumber beetle, Diabrotica, green cloverworm, leafhoppers, Lygus bugs, Mexican bean beetle

Rate: 0.5-1 lb. ai/A
2-4 pt. 2FL/A

Pest:
cowpea curculio, European corn borer, stink bugs (except CA)

Remarks: Dry, succulent beans. Apply at 7-day intervals or more. Do not apply more than 12 pt. per season. Do not chemigate in California.
Days To Harvest: Dry beans-15. Succulent beans: Average rainfall exceeds 25" -3; average rainfall less than 25"-7; California-15.

Signal Word/Toxicity Class: Warning/II.
REI: 48 hr.

Wilbur-Ellis Methyl Parathion 4 or 5 Spray *(Wilbur-Ellis)*

Rate: 0.5 lb. ai/A
1 pt. 4/A
0.95-1.3 lb. ai/A
1.5-2 pt. 5/A

Pest:
aphids, armyworms, flea beetles, spider mites

Remarks: Do not apply if crop or weed in treatment area is in bloom. Do not apply through any type of irrigation system.
Days To Harvest: 21; Dry beans-15.

Signal Word/Toxicity Class: Danger-Poison/I.
REI: 24 hr.

◆ - **new product** • ai-active ingredient • DF-dry flowable • E/EC-emulsifiable concentrate • EW-emulsion in water
F/FL-flowable • G-granule • L/LC-liquid • SP/WSP-soluble packet • W/WP-wettable powder • WSB-water soluble bag

MOCAP

Mocap EC is a Restricted Use Pesticide.

Mocap EC or 10%G (ethoprop) *(Rhone-Poulenc)*

Rate: Broadcast:
6-8 lb. ai/A
1-1.33 gal. EC/A
60-80 lb. 10%/A
Row treatment:
2-3 lb. ai/A
1.33-2 qt. EC/A (2-4.4 fl. oz./1000 lin. ft. row)
20-30 lb. 10%G/A (1.4-2 lb./1000 lin. ft. row)

Pest:
garden symphylans, nematodes

Remarks: Lima, snap beans. Broadcast: Apply from 3 days before planting to at-planting time. Row treatment: Apply in band 12-15" wide on the row; do not use as seed furrow treatment or allow spray to contact seed. Mix with top 2-4" of soil. Beans grown in treated soil may remain green longer than other beans and mature 1-2 weeks later.

State Restrictions: Do not apply in Long Island, New York.

Signal Word/Toxicity Class: Danger-Poison/I (EC); Warning/II (10%G).
REI: 48 hr.; 72 hr. if average rainfall less than 25" per year.

M-PEDE

M-Pede (potassium salts of fatty acids) *(Mycogen)*

Rate: 1-2% v/v solution

Pest:
aphids*, broad mite, green peach aphid*, leafhoppers, leafminers*, plant bugs, russet mite, spider mites, thrips, whiteflies

** Tank mix only*

Remarks: Apply at first sign of infestation; repeat at 7- to 14-day intervals. May be tank mixed with other insecticides. Refer to both labels for use directions and precautions.

Days To Harvest: 0.

Signal Word/Toxicity Class: Warning/II.
REI: 12 hr.

NEEMAZAD — see azadirachtin

NEEMIX — see azadirachtin

Omni Supreme Spray — see petroleum oil

Orthene — see acephate

Penncap-M — see methyl parathion

PETROLEUM OIL

Petroleum based spray oil. Mode of action is through suffocation of eggs, larvae, and nymphs of insects and adults of soft bodied insects. This mode of action necessitates total spray coverage. Toxic to fish.

Omni Supreme Spray *(Helena)*

Rate: 1-2 gal./100 gal. water

Pest:
aphids, beetles (larvae), caterpillars, leafminers, mites

Remarks: None.

Signal Word/Toxicity Class: Caution.
REI: 4 hr.

SAF-T-SIDE *(Brandt Consolidated)*
◆ **Sumulsion** *(Drexel)*
◆ **SunSpray Ultra-Fine** *(Sun)*

Rate: 1-2 gal./100 gal. water

Pest:
aphids, beetles (larvae), leafhoppers, leafminers, mites, spiders, thrips, whiteflies

Remarks: Apply as needed. May be tank mixed with most commonly used insecticides and fungicides. Do not use in combination with or immediately before or after spraying with fungicides such as captan, Dyrene, folpet, Morestan or any sulfur products. Do not use with carbaryl or dimethoate.

Signal Word/Toxicity Class: Caution.
REI: 4 hr.

Phaser — see endosulfan

PHORATE

Restricted Use Pesticide.

Clean Crop Phorate 20G *(Platte)*
Thimet 20-G *(American Cyanamid)*

Rate: 4.5-7 oz. 20G, 20-G/1000 ft. row (min. 30" spacing)

Pest:
aphids, leafhoppers, Lygus bugs, Mexican bean beetle (up to 9.4 oz.), mites, seedcorn maggots, thrips

Remarks: Apply at planting. Drill granules to side of seed. For seedcorn maggots, apply granules in band over row. Do not place granules in direct contact with seed. Rate varies by pest species; refer to label.

Days To Harvest: 60.

Signal Word/Toxicity Class: Danger-Poison/I.
REI: 48 hr.

PROPARGITE

For use on dry beans only.

Comite *(Uniroyal)*

Rate: 1.64-2.46 lb. ai/A
2-3 pt./A

Pest:
twospotted spider mite

Remarks: Dry beans. Apply when mites first appear in at least 20 gal./A by ground; 5 gal./A by air. Apply no more than twice per year. Do not apply through any type of irrigation system. Do not feed or forage treated bean vines or bean trash after harvesting.

Days To Harvest: 14.

State Restrictions: For use west of the Rockies.

Signal Word/Toxicity Class: Danger/I.
REI: 7 days. After 48 hr., workers can enter treated areas to perform hand labor tasks if they wear early-entry personal protective equipment.

PYRELLIN (Premix)

Premix containing pyrethrins and rotenone. May be combined with other insecticides to enhance knockdown or as an insect "exciter" to help facilitate contact with other spray residues.

Pyrellin E.C. (0.042 lb./gal. pyrethrins & 0.035 lb./gal. rotenone) *(Webb Wright)*

Rate: 1-2 pt. E.C./A

Pest: Including but not limited to:
alfalfa webworm, aphids, bean beetle, cabbageworms, leafhoppers, loopers, Lygus bugs, mites, plant bugs, stink bugs, thrips, whiteflies

Remarks: Dry, green, lima, snap beans. Apply at 7-day intervals or less; repeat as needed.

Days To Harvest: 0.

Signal Word/Toxicity Class: Caution/III.
REI: 12 hr.

7 Vegetables

NOTICE The information on these pages is for preliminary planning — not a guide for use. Be sure to follow the manufacturer's directions, notwithstanding information contained here.

For personal protective equipment and EPA registration numbers, please turn to page 39.

Pyrenone — see pyrethrins

PYRETHRINS

Contains piperonyl butoxide as a synergist. Diatect also contains silicon dioxide (diatomaceous earth). May be combined in tank mix with other insecticides for quicker and more complete kill where insect resistance may be a problem, and as an "exciter" to flush insects out of hiding and into contact with other spray residues.

Diatect Multipurpose Insecticide II *(Diatect)*
Pyrenone (0.5EC) *(AgrEvo Environmental Health)*
◆ **Py-Rin 60-6 EC** *(Wilbur-Ellis)*

Rate: 1-6 lb. II/A
1-12 fl. oz. 0.5EC/A
2-16 oz. 6-60 EC/A

Pest: Including but not limited to:

aphids	caterpillar	mites
armyworms	flea beetles	plant bugs
bean beetle	leafhoppers	stink bugs
cabbage looper	loopers	thrips
cabbageworms	Lygus bugs	whiteflies

Remarks: Kidney, lima, navy, pinto, snap, wax beans, etc. Apply in sufficient water for thorough coverage; repeat as needed. Apply Diatect as dust or wettable powder.
Days To Harvest: 0.

Signal Word/Toxicity Class: Caution.
REI: 12 hr.

Py-Rin — see pyrethrins

SAF-T-SIDE — see petroleum oil

Sevin — see carbaryl

Sniper — see azinphos-methyl

SULFUR

Thiolux is micronized sulfur.

◆ **Kumulus DF** *(BASF)*
Rate: 2.4-4 lb. ai/A
3-5 lb. DF/A
Pest:
red spider mite
Remarks: Apply early leaf stage; repeat at 10- to 14-day intervals with or without other pesticides immediately after a rain. Do not apply if temperatures during or within 3 days after application are expected to exceed 90°F. Do not apply through any type of irrigation system.

Signal Word/Toxicity Class: Caution.
REI: 24 hr.

Liquid Sulfur Six *(Helena)*
Rate: 3-4 pt./8 gal. water
Pest:
red spider mite
Remarks: Apply by air. Do not apply through any type of irrigation system. Refer to label for use directions.

Signal Word/Toxicity Class: Caution.
REI: 24 hr.

Thiolux (DF) *(Platte)*
Rate: 2.4-8 lb. ai/A
3-10 lb. DF/A
Pest:
red spider mite
Remarks: Apply early leaf stage.

Signal Word/Toxicity Class: Caution.
REI: 24 hr.

SULFUR (Premix)

Miticide/fungicide. Toxic to fish, aquatic invertebrates, and marine/estuarine organisms.

Bravo S (27.25% sulfur & 19.15% chlorothalonil) *(Helena)*
Rate: 8 pt./A
Pest:
red spider mite
Remarks: Do not graze treated areas or feed treated plant parts to livestock.
Days To Harvest: 7.

Signal Word/Toxicity Class: Caution.
REI: 48 hr.

Sumulsion — see petroleum oil

SunSpray — see petroleum oil

TEMIK

Organophosphate soil insecticide with systemic activity, used for many soil- and foliage-feeding insect pests and mites; ineffective against caterpillars. Has long residual effects. Incompatible with alkaline compounds, highly toxic to animals, and rapidly absorbed through the skin. For use on dry beans only. Restricted Use Pesticide.

Temik 15G (aldicarb) *(Rhone-Poulenc)*
Rate: 0.75-2 lb. ai/A
5-14 lb. 15G/A
Pest:

aphids	Mexican bean beetle	nematodes
leafhoppers	mites	

Remarks: Dry beans. Apply at planting. Rate varies by pest species; refer to label. For nematodes, apply in 6" band and work into or cover with soil. Plant seed in treated zone. For all other pests, drill granules either 2-3" below seed or 2-3" to side of seed and 2-3" deep. Granules may also be placed directly in seed furrow if rate does not exceed 5 lb./A. Do not make more than 1 application per crop or exceed a total of 14 lb./A. Do not feed green forage, hay, or straw to livestock or use green pods as food for humans. Refer to label for environmental precautions and soil restrictions.
Days To Harvest: 90.

Rate: 0.5-0.75 lb. ai/A
3.5-5 lb. 15G/A
Pest:
seedcorn maggot
Remarks: Dry beans. Apply at planting in seed furrow and cover with soil. Do not make more than 1 application per crop or exceed a total of 14 lb./A. Do not feed green forage, hay, or straw to livestock or use green pods as food for humans. Refer to label for environmental precautions and soil restrictions table.
Days To Harvest: 90.
State Restrictions: For use in Michigan only.

Signal Word/Toxicity Class: Danger-Poison/I.
REI: 48 hr.

Thimet — see phorate

Thiodan — see endosulfan

Thiolux — see sulfur

THIRETHRIN (Premix)

Broad-spectrum insecticide/aphicide. Contains piperonyl butoxide. Toxic to fish, shrimp, crab, birds, and other wildlife.

Thirethrin (2 lb./gal. endosulfan & 0.045 lb./gal. pyrethrins) *(Terra)*
Rate: 1-2 qt./A

◆ - **new product** • ai-active ingredient • DF-dry flowable • E/EC-emulsifiable concentrate • EW-emulsion in water
F/FL-flowable • G-granule • L/LC-liquid • SP/WSP-soluble packet • W/WP-wettable powder • WSB-water soluble bag

Pest:

aphids, bean leafskeletonizer, black bean aphid, cowpea curculio, cucumber beetle, green stink bugs, Mexican bean beetles

Remarks: Dry (except lima), succulent beans. Do not make more than 3 applications. Do not feed treated threshing to livestock or allow livestock to graze treated fields.

Days To Harvest: 3.

Signal Word/Toxicity Class: Warning/II.
REI: 24 hr.

Carrots, Parsley, Parsnips, Salsify

ASANA

Pyrethroid insecticide that has both contact and stomach poison activity. Restricted Use Pesticide due to toxicity to fish and aquatic organisms.

Asana XL (esfenvalerate) *(DuPont)*

Rate: 0.03-0.05 lb. ai/A
5.8-9.6 fl. oz. XL/A

Pest:

aster leafhopper, carrot weevil, cutworms, leafhoppers

Remarks: Carrots. Apply as needed, but do not exceed 0.5 lb. ai/A per season. For aerial application, apply in minimum 5 gal. water/A. For carrot weevils, apply high rate (ground application recommended) when weevils become active; thorough spray coverage of crown area essential.

Days To Harvest: 7.

Signal Word/Toxicity Class: Warning/II.
REI: 12 hr.

AZADIRACHTIN

Extracted from seed of neem tree. Insect growth regulator that controls targeted insect larvae when they ingest or come in contact with it, by interfering with the insect's ability to molt and shutting off feeding behavior. Affected larvae slowly starve and die at next molt.

◆ **Azatin XL Plus** *(Thermo Trilogy)*

Rate: 5-21 oz. XL Plus/A

Pest:

aphids, beet armyworm, beetles, black cutworm, cabbage looper, caterpillars, Colorado potato beetle, cutworms, diamondback moth, fall armyworm, flea beetles, green peach aphid, imported cabbageworm, Japanese beetle, leafhoppers, leafminers, leafrollers, loopers, Mexican bean beetle, southern armyworm, thrips, tomato fruitworm, webworms, whiteflies

Remarks: Carrots, parsley, parsnips, salsify. Rate varies by species. Apply when pests first appear; repeat at 7-day intervals. When pest populations are high, use higher rates. Do not apply more that 21 fl. oz./A.

Signal Word/Toxicity Class: Caution.
REI: 4 hr.

◆ **NEEMAZAD 4.5EC or NEEMIX 4.5** *(Thermo Trilogy)*

Rate: 0.125-1 pt. 4.5/A
High Volume:
2.25-4.5 oz. 4.5EC/100 gal. water
Low Volume:
0.5 pt. 4.5EC/A

Pest:

aphids, armyworms, artichoke plume moth, aster leafhopper, cabbage loopers, caterpillars, Colorado potato beetle, diamondback moth, greenhouse whitefly, hornworms, leafhoppers, potato flea beetle, psyllids, sweetpotato whitefly, webworms

Remarks: Carrots, parsnips. For low volume, apply in sufficient water for adequate coverage. For low insect pressure, apply preventative at 7-day intervals or at first sign of insects; high insect pressure, at 3- to 4-day intervals.

Signal Word/Toxicity Class: Warning/II.
REI: 12 hr.

Azatin — see azadirachtin

BAYTHROID

Synthetic pyrethroid formulated in an emulsifiable concentrate. Toxic to fish and aquatic organisms. Restricted Use Pesticide.

Baythroid (2EC) (cyfluthrin) *(Bayer)*

Rate: 0.025-0.044 lb. ai/A
1.6-2.8 fl. oz. 2EC/A

Pest:

aster leafhopper, black cutworm, carrot weevil

Remarks: Carrots. Do not apply to crops grown for seed. Apply in sufficient water for good coverage by ground only. Up to 5 applications may be made per crop season; allow at least 7 days between applications. Refer to label for spray drift precautions.

Days To Harvest: 0.

Signal Word/Toxicity Class: Danger/I.
REI: 12 hr.

CARBARYL

Clean Crop Sevin 5 Bait *(Platte)*

Rate: 1-2 lb. ai/A
20-40 lb. Bait/A

Pest:

ants, armyworms, cutworms, darkling beetle, field crickets, grasshoppers, sowbugs

Remarks: Carrots, parsley, parsnips, salsify. Apply by ground or air.

Days To Harvest: Carrots-0; parsnips, salsify (roots)-3; parsley, salsify (tops)-14.

Signal Word/Toxicity Class: Caution/III.
REI: 12 hr.

Drexel Carbaryl 4L or 50W *(Drexel)*
Sevin 4F, XLR Plus, 80S, 80WSP *(Rhone-Poulenc)*

Rate: 0.5-2 lb. ai/A
1-4 pt. 4L, 4F, XLR/A
1-4 lb. 50W/A
0.66-2.5 lb. 80S, 80WSP/A

Pest:

armyworms, aster leafhopper, corn earworm, fall armyworm, flea beetles, grasshoppers, harlequin bug (salsify), imported cabbageworm (salsify), leafhoppers, Lygus bugs, spittlebugs, stink bugs, tarnished plant bug

Remarks: Carrots, parsley, parsnips, salsify. Treat at 5- to 7-day intervals. For salsify, treat after heads begin to form. To avoid possible injury to tender foliage, do not apply to wet foliage or when rain or high humidity is expected during next 2 days. Rate varies by pest species; refer to label.

Days To Harvest: Carrots-0; parsnips, salsify (roots)-3; parsley, salsify (tops)-14.

Signal Word/Toxicity Class: Caution (4F, 4L, XLR); Warning/II (50W, 80S, 80WSP).
REI: 12 hr.

Gowan Sevin Bait *(Gowan)*

Rate: 2 lb. ai/A
10 lb. Bait/A

Pest:

armyworms, crickets, cutworms, darkling beetle, grasshoppers, sowbugs

Remarks: Carrots, parsley, parsnips. Apply when pests appear in damaging numbers; repeat at 7- to 14-day intervals if necessary. Do not use when bees are present.

Days To Harvest: Carrots-1; parsnips-3; parsley-14.

State Restrictions: For use in Arizona and California.

Signal Word/Toxicity Class: Caution/III.
REI: 12 hr.

For personal protective equipment and EPA registration numbers, please turn to page 39.

DEADLINE

Packaged as Bullets, M-Ps (mini-pellets), and granules.

Deadline Bullets, M-Ps, Granules (4%) (metaldehyde) *(Pace)*

Rate: 0.8-1 lb. ai/A
20-40 lb. 4%/A

Pest:
slugs snails

Remarks: Carrots, parsnips, salsify. Apply every 3-4 weeks during growing season or as needed. Do not apply directly to edible portion of plant.

Signal Word/Toxicity Class: Warning/II (granule); Caution (bullets, M-P).
REI: 12 hr.

Declare — see methyl parathion

Diatect — see pyrethrins

DIAZINON

Restricted Use Pesticide due to avian and aquatic toxicity.

D•z•n diazinon AG600 WBC, 50W *(Novartis)*
Drexel Diazinon Insecticide (4EC), D-264 14G *(Drexel)*
Gowan Diazinon 4E, 14G, 50WSB *(Gowan)*
Helena Diazinon AG500 *(Helena)*
Micro Flo Diazinon AG500, 14G, 50W *(Micro Flo)*
Prentox Diazinon AG500 or 50W *(Prentiss)*
◆ **Wilbur-Ellis Diazinon 4 (4E), 14G** *(Wilbur-Ellis)*

Rate: Carrots:
0.5 lb. ai/A
1 pt. 4E, 4EC, AG500/A
12.75 fl. oz. AG600/A
1 lb. 50W 50WSB /A

Pest:
aphids

Rate: Parsnips:
0.25-0.5 lb. ai/A
0.5-1 pt. 4E, 4EC, AG500/A
6.5-12.75 fl. oz. AG600/A
0.5-1 lb. 50W, 50WSB/A

Pest:
aphids dipterous leafminer flea beetles

Remarks: Carrots, parsnips. Apply as insects occur; repeat at 7-day intervals as needed. Do not apply more than 5 times per season. Will not control organophosphate-resistant leafminers.
Days To Harvest: 14.

Rate: 1 lb. ai/A
2 pt. 4E, 4EC, AG500/A
25.5 fl. oz. AG600/A
2 lb. 50W, 50WSB/A
7 lb. 14G/A

Pest:
mole crickets

Remarks: Carrots. Apply to plant beds or fields 1-2 days before sowing of seed or transplanting and immediately incorporate into top 1-2" of soil.

Rate: 2-4 lb. ai/A
4-8 pt. 4E, 4EC, AG500/A
51-102 fl. oz. AG600/A
4-8 lb. 50W, 50WSB/A
14-28 lb. 14G/A

Pest:
cutworms

Rate: 3-4 lb. ai/A
6-8 pt. 4E, 4EC, AG500/A
76.5-102 fl. oz. AG600/A
6-8 lb. 50W, 50WSB/A
21-28 lb. 14G/A

Pest:
wireworm

Remarks: Carrots, parsley. Broadcast just before planting and immediately incorporate into soil.

Signal Word/Toxicity Class: Caution/III.
REI: 24 hr.; 12 hr. (Drexel/Gowan/Wilbur-Ellis 14G).

ENDOSULFAN

Toxic to birds, fish, and wildlife; toxic to bees exposed to direct application.

Drexel Endosulfan 3EC or 50WP *(Drexel)*
Gowan Endosulfan 3EC or 50WSB *(Gowan)*
◆ **Micro Flo Endosulfan 3EC or 50WP** *(Micro Flo)*
Phaser 3EC or 50WSB *(AgrEvo USA)*
Thiodan 3EC or 50WP *(FMC)*

Rate: 0.5-1 lb. ai/A
0.66-1.33 qt. 3EC/A
1-2 lb. 50WP, 50WSB/A

Pest:
armyworms, flea beetle, green peach aphid, leafhoppers, whiteflies

Remarks: Carrots. Apply when insects appear or feeding is noticed. Do not exceed 1 application per year or 1 lb. ai/A per year. Do not use tops for food or feed.
Days To Harvest: 7.

Signal Word/Toxicity Class: Danger-Poison/I.
REI: 24 hr.

Fyfanon — see malathion

Kumulus — see sulfur

LANNATE

Restricted Use Pesticide.

Lannate (90) or LV (methomyl) *(DuPont)*

Rate: 0.225-0.9 lb. ai/A
0.25-1 lb. 90/A
0.75-3 pt. LV/A

Pest:
armyworms, aster leafhopper, beet armyworm, variegated cutworm

Remarks: Carrots. Apply in sufficient water to obtain thorough, uniform coverage when insects first appear; repeat at 5- to 7-day intervals or as needed. Do not apply more than 6.3 lb. ai/A per crop or make more than 10 applications per crop.
Days To Harvest: 1.

Signal Word/Toxicity Class: Danger-Poison/I.
REI: 48 hr.

LARVIN

Broad spectrum carbamate insecticide that has stomach poison activity and some limited contact activity. Known to have systemic action as a seed treatment. Toxic to fish and birds, and moderately toxic to bees as direct treatment.

Larvin 3.2 (thiodicarb) *(Rhone-Poulenc)*

Rate: 0.4-0.75 lb. ai/A
16-30 fl. oz. 3.2/A

Pest:
alfalfa looper, beet armyworm, bollworm, corn earworm, fall armyworm, southern armyworm, tomato fruitworm

Rate: 0.6-0.75 lb. ai/A
24-30 fl. oz. 3.2/A

Pest:
cabbage looper

Remarks: Parsley. Spray as needed. Use higher rates for heavier infestations of larger larvae. Do not exceed 1.5 lb. ai/A per season.
Days To Harvest: 14.

Signal Word/Toxicity Class: Warning/II.
REI: 12 hr.

◆ - **new product** • ai-active ingredient • DF-dry flowable • E/EC-emulsifiable concentrate • EW-emulsion in water
F/FL-flowable • G-granule • L/LC-liquid • SP/WSP-soluble packet • W/WP-wettable powder • WSB-water soluble bag

LINDANE

Lindane 30C contains toxic inert Rhodamine B. Toxic to fish, birds, and wildlife.

Lindane 30C *(Gustafson)*
◆ **Lindane ST 40** *(AGSCO)*
Rate: 2.7 fl. oz. 30C/100 lb. seed
2 fl. oz. 40/100 lb. seed
Pest:
seedcorn maggots wireworms
Remarks: Do not use treated seed for food, feed, or oil purposes. Exposed treated seed may be hazardous to birds and other wildlife. Dispose of excess seed by burial away from bodies of water.

Signal Word/Toxicity Class: Warning/II.
REI: 12 hr.

MALATHION

Organophosphate insecticide toxic to fish, aquatic invertebrates, aquatic life stages of amphibians, and bees exposed to direct treatment on blooming crops or weeds.

Clean Crop Malathion 57EC or 8EC *(Platte)*
Rate: Carrots:
1-1.6 lb. ai/A
1.5-2.5 pt. 57EC/A
Pest:
aphids leafhoppers

Rate: Parsley, parsnips, salsify:
0.63-1.25 lb. ai/A
1 -2 pt. 57EC/A
0.5-1 lb. ai/A
0.5-1 pt. 8EC/A
Pest:
aphids cabbage looper imported cabbageworm
Remarks: Carrots, parsley, parsnips, salsify.
Days To Harvest: Carrots, parsnip, salsify-7; parsley-21.

Signal Word/Toxicity Class: Caution/III.
REI: 12 hr.

Drexel Malathion 5EC *(Drexel)*
Rate: Carrots, parsley, parsnips:
1-1.25 lb. ai/A
1.5-2 pt. 5EC/A
Pest:
aphids

Rate: Carrots:
1.6 lb. ai/A
2.5 pt. 5EC/A
Pest:
leafhoppers
Remarks: Carrots, parsley, parsnips.
Days To Harvest: Carrots, parsnips-7; parsley-21.

Signal Word/Toxicity Class: Warning/II.
REI: 12 hr.

Fyfanon (8E) *(Helena)*
◆ **Gowan Malathion 8 (8E)** *(Gowan)*
Rate: 2 lb. ai/A
2 pt. 8E/A
Pest:
aphids flea beetles leafhoppers
diamondback moth
Remarks: Carrots, parsley, parsnips, salsify.
Days To Harvest: 7.

Signal Word/Toxicity Class: Caution/III.
REI: 12 hr.

◆ **Micro Flo Malathion 5EC** *(Micro Flo)*
Rate: 1-1.3 lb. ai/A
1.5-2 pt. 5EC/A
Pest:
aphids
Remarks: Parsley, parsnips.
Days To Harvest: Parsnip-7; parsley-21.

Signal Word/Toxicity Class: Warning/II.
REI: 12 hr.

Marlate — see methoxychlor

METHOXYCHLOR

Organochlorine pesticide toxic to fish and aquatic invertebrates.

Marlate 50 (WP) *(Kincaid)*
Prentox Methoxychlor 2EC *(Prentiss)*
Rate: 1.3-1.7 lb. ai/A
75-100 fl. oz. 2EC/A
2 Tbs. 50WP/gal. water
Pest:
alfalfa looper, flea beetles, Japanese beetle, blister beetle, imported cabbageworm, leafhoppers, fall armyworm
Remarks: Apply at first sign of infestation; repeat at 7- to 14-day intervals as needed.
Days To Harvest: Marlate: 14. Prentox: 7; 14 if tops are use for food.

Signal Word/Toxicity Class: Caution.
REI: 12 hr.

METHYL PARATHION

Organophosphate insecticide toxic to bees exposed to direct treatment or residues on blooming crops or weeds. Restricted Use Pesticide due to very high acute toxicity to humans and birds.

Cheminova Methyl 4EC *(Cheminova)*
Rate: 1 lb. ai/A
2 pt. 4EC/A
Pest:
aphids, cutworms, leafhoppers, armyworms, flea beetles, mites
Remarks: Carrots. Apply when insects appear; repeat as needed. Do not feed tops.
Days To Harvest: 15.

Signal Word/Toxicity Class: Danger-Poison/I.
REI: 48 hr.; 72 hr. if average rainfall less than 25" per year.

◆ **Declare (4EC)** *(Griffin)*
Helena 4 lb. Methyl Parathion (4EC) *(Helena)*
Rate: 0.47-0.94 lb. ai/A
1-2 pt. 4EC/A
Pest:
aphids, cutworms, mites, armyworms, flea beetles, plant bugs, cabbage looper, leafhoppers
Remarks: Rate varies by pest species; refer to label. Do not use tops for food or feed.
Days To Harvest: 15.

Signal Word/Toxicity Class: Danger-Poison/I.
REI: 48 hr.

Wilbur-Ellis Methyl Parathion 4 or 5 Spray *(Wilbur-Ellis)*
Rate: 0.5 lb. ai/A
1 pt. 4/A
0.63 lb. ai/A
1 pt. 5/A
Pest:
aphids, flea beetles, spider mites, armyworms, leafhoppers
Remarks: Do not use treated tops for food or feed. Do not apply through any type of irrigation system.
Days To Harvest: 15; tops to be used as food-21 (5 Spray).

Signal Word/Toxicity Class: Danger-Poison/I.
REI: 24 (4 Spray); 48 (5 Spray).

Microthiol Special — see sulfur

7 Vegetables

NOTICE The information on these pages is for preliminary planning — not a guide for use. Be sure to follow the manufacturer's directions, notwithstanding information contained here. For personal protective equipment and EPA registration numbers, please turn to page 39.

M-PEDE

M-Pede (potassium salts of fatty acids) *(Mycogen)*
Rate: 1-2% v/v solution
Pest:

aphids*	leafminers*	spider mites
broad mite	plant bugs	thrips
green peach aphid*	russet mite	whiteflies
leafhoppers		

** Tank mix only*

Remarks: Apply at first sign of infestation; repeat at 7- to 14-day intervals. May be tank mixed with other insecticide products. Refer to both labels for use directions and precautions.
Days To Harvest: 0.

Signal Word/Toxicity Class: Warning/II.
REI: 12 hr.

NEEMAZAD — see azadirachtin

NEEMIX — see azadirachtin

PERMETHRIN

Restricted Use Pesticide.

Pounce 3.2 EC, 25WP or WSB *(FMC)*
Rate: 0.05-0.2 lb. ai/A
2-8 oz. 3.2 EC/A
3.2-12.8 oz. 25WP/A
1-2 bags WSB/A
Pest.

alfalfa looper	cutworms	green cloverworm
beet armyworm	European corn borer	leafhoppers
cabbage looper	fall armyworm	vegetable leafminer
corn earworm		

Remarks: Parsley. Apply every 3-5 days or as needed. Use sufficient water to obtain full coverage of foliage. Rate varies by pest species. Do not graze treated areas or feed crop refuse to livestock.
Days To Harvest: 1.

Signal Word/Toxicity Class: Caution (3.2 EC); Warning/II (25WP, WSB).
REI: 24 hr.

Phaser — see endosulfan

Pounce — see permethrin

PYRELLIN (Premix)

Premix containing pyrethrins and rotenone. May be combined with other insecticides to enhance knockdown or as an insect "exciter" to help facilitate contact with other spray residues.

Pyrellin E.C. (0.042 lb./gal. pyrethrins & 0.035 lb./gal. rotenone) *(Webb Wright)*
Rate: 1-2 pt. E.C./A
Pest:

alfalfa webworm	harlequin bug	mites
aphids	Japanese beetle	plant bugs
bean beetle	leafhoppers	stink bugs
cabbageworms	leafminers	thrips
celery leaftier	loopers	whiteflies
garden fleahopper	Lygus bugs	

Remarks: Carrots, parsnips, salsify. For carrots, apply at 7-day intervals or less; repeat as necessary. For parsnips and salsify, apply as necessary to maintain control.
Days To Harvest: 0.

Rate: 1.5-2 pt. E.C./A
Pest:

Colorado potato beetle	cucumber beetle	flea beetles
	European corn borer	vegetable weevil

Remarks: Carrots. Make 3-4 applications or more as eggs start to hatch; repeat at 5- to 7-day intervals.
Days To Harvest: 0.

Signal Word/Toxicity Class: Caution/III.
REI: 12 hr.

Pyrenone — see pyrethrins

PYRETHRINS

Contains piperonyl butoxide as a synergist. Diatect also contains silicon dioxide (diatomaceous earth). May be combined in tank mix with other insecticides for quicker and more complete kill where insect resistance may be a problem, and as an "exciter" to flush insects out of hiding and into contact with other spray residues.

Diatect Multipurpose Insecticide II *(Diatect)*
Pyrenone (0.5EC) *(AgrEvo Environmental Health)*
◆ **Py-Rin 60-6 EC** *(Wilbur-Ellis)*
Rate: 1-6 lb. II/A
1-12 fl. oz. 0.5EC/A
2-16 oz. 6-60 EC/A
Pest: Including but not limited to:

aphids	chinch bug	flea beetles
carrot rust fly	cutworms	leafhoppers
carrot weevil	diamondback moth	

Remarks: Carrots, parsley, parsnips, salsify. Apply in sufficient water for thorough coverage; repeat as needed. Apply Diatect as dust or wettable powder.
Days To Harvest: 0.

Signal Word/Toxicity Class: Caution.
REI: 12 hr.

Py-Rin — see pyrethrins

Sevin — see carbaryl

SULFUR

Microthiol and Thiolux are micronized sulfur.

◆ **Kumulus DF** *(BASF)*
Microthiol Special (80W) *(Elf Atochem)*
Thiolux (DF) *(Platte)*
Rate: 2.4-8 lb. ai/A
3-10 lb. 80W, DF/A
Pest:
petrobia mite
Remarks: Apply in early leaf stage; repeat at 10- to 14-day intervals or as needed. Do not apply if temperatures during or within 3 days after application are expected to exceed 90°F. Refer to label for use directions.

Signal Word/Toxicity Class: Caution.
REI: 24 hr.

Thiodan — see endosulfan

Thiolux — see sulfur

THIRETHRIN (Premix)

Broad-spectrum insecticide/aphicide. Contains 0.45 lb. piperonyl butoxide. Toxic to bees, fish, shrimp, crabs, birds, and other wildlife.

Thirethrin (2 lb./gal. endosulfan & 0.045 lb./gal. pyrethrins) *(Terra)*
Rate: 1-2 qt./A
Pest:

cabbage looper	green peach aphid

Remarks: Do not make more than 3 applications per season.
Days To Harvest: 1 qt.-7; 1-2 qt.-4.

Signal Word/Toxicity Class: Warning/II.
REI: 24 hr.

◆ - **new product** • ai-active ingredient • DF-dry flowable • E/EC-emulsifiable concentrate • EW-emulsion in water
F/FL-flowable • G-granule • L/LC-liquid • SP/WSP-soluble packet • W/WP-wettable powder • WSB-water soluble bag

VYDATE

Restricted Use Pesticide.

Vydate L (oxamyl) *(DuPont)*
Rate: Preplant:
4-8 lb. ai/A
2-4 gal. L/A
In-furrow:
2-4 lb. ai/A
1-2 gal. L/A
Pest:

lesion nematode	spiral nematode	stunt nematode
root-knot nematode (except javanese)	sting nematode	

Rate: Foliar:
0.5-1 lb. ai/A
2-4 pt. L/A
Pest:
carrot weevil

Remarks: Carrots. Apply in at least 20 gal. water. Preplant: Apply within 1 week of planting and incorporate 4-6" into soil. In-furrow: Apply in seed furrow during planting. Foliar: Apply 3 treatments at 2- to 3- week intervals. Do not apply more than 4 gal./A per season.
State Restrictions: Not registered for use in California. Do not use in Suffolk and Nassau counties, Long Island, New York.

Signal Word/Toxicity Class: Danger-Poison/I.
REI: 48 hr.

Celery

ABAMECTIN

Natural product produced by soil microorganism, Streptomyces avermitilis. Unrelated to other miticides. Toxic to fish, mammals, and aquatic organisms; highly toxic to honeybees. Restricted Use Pesticide.

Agri-Mek 0.15EC *(Novartis)*
Rate: 8-16 fl. oz. 0.15EC/A
Pest:
leafminers mites
Remarks: Apply when mites first appear; repeat as necessary. For leafminers, apply when adults are first observed; repeat at 7-day intervals as necessary. Thorough coverage essential. Do not apply in less than 20 gal. water/A. Do not make more than 2 sequential applications after transplanting. Do not apply through any type of irrigation system or by aircraft.
Days To Harvest: 7.

Signal Word/Toxicity Class: Warning/II.
REI: 12 hr.

ACEPHATE

Organophosphate systemic insecticide with contact and stomach poison activity. Effective against many aphids, thrips, caterpillars, and plant bugs. Has moderate residual effects and is compatible with most pesticides.

Orthene 75 S *(Valent)*
Rate: 0.5-1 lb. ai/A
0.66-1.33 lb. 75S/A
Pest:
green peach aphid

Rate: 1 lb. ai/A
1.33 lb. 75S/A
Pest:
beet armyworm cabbage looper fall armyworm
Remarks: Apply when eggs or insects appear; repeat at 3- to 10-day intervals as needed to maintain control. Plants should be trimmed (tops removed) before shipping or use. Do not apply through any type of irrigation system. Tops should not be used for food or feed.
Days To Harvest: 21.

Signal Word/Toxicity Class: Caution.
REI: 24 hr.

Agri-Mek — see abamectin

Ambush — see permethrin

AZADIRACHTIN

Extracted from seed of neem tree. Insect growth regulator that controls targeted insect larvae when they ingest or come in contact with it, by interfering with the insect's ability to molt and shutting off feeding behavior. Affected larvae slowly starve and die at next molt.

◆ **Azatin EC or XL** *(Olympic)*
◆ **Azatin XL Plus** *(Thermo Trilogy)*
Rate: 8-21 oz. EC/A
5-21 oz. XL, XL Plus/A
Pest:

aphids	fall armyworm	loopers
beet armyworm	flea beetles	Mexican bean beetle
beetles	green peach aphid	serpentine leafminer
cabbage looper	imported cabbageworm	southern armyworm
caterpillars	Japanese beetle	thrips
Colorado potato beetle	leafhoppers	tomato fruitworm
cutworms	leafminers	webworms
diamondback moth	leafrollers	whiteflies

Remarks: Rate varies by species. Apply when pests first appear; repeat at 7-day intervals. When pest populations are high, use higher label rates. Do not apply more that 21 fl. oz./A.
Days To Harvest: 0.

Signal Word/Toxicity Class: Caution (XL); Warning/II (EC).
REI: 4 hr. (XL), 12 hr. (EC).

◆ **NEEMAZAD 4.5EC or NEEMIX 4.5** *(Thermo Trilogy)*
Rate: 0.125-1 pt. 4.5/A
High Volume:
2.25-4.5 oz. 4.5EC/100 gal. water
Low Volume:
0.5 pt. 4.5EC/A
Pest:

aphids	diamondback moth	onion maggot
armyworms	greenhouse whitefly	sweetpotato whitefly
cabbage loopers	hornworms	thrips
caterpillars	imported cabbageworm	webworms
cutworms	leafminers	weevils

Remarks: For low volume, apply in sufficient water for adequate coverage. For low insect pressure, apply preventative at 7-day intervals or at first sign of insects; high insect pressure, at 3- to 4-day intervals.

Signal Word/Toxicity Class: Warning/II.
REI: 12 hr.

Agri-Mek — see abamectin

Azatin — see azadirachtin

AZINPHOS-METHYL

Restricted Use Pesticide.

Gowan Azinphos-M 50-W *(Gowan)*
◆ **Micro Flo Azinphosmethyl 2EC or 50W** *(Micro Flo)*
Sniper 2-E, 50PVA, 50W *(Platte)*
Rate: 0.5 lb. ai/100 gal. water
2 pt. 2EC, 2-E/100 gal. water
1 lb. 50-W, 50PVA, 50W/100 gal. water
Pest:

aphids	leafminers	tarnished plant bug
leafhoppers	spittlebug	

Remarks: Apply in sufficient water to give complete coverage.
Days To Harvest: 14.
State Restrictions: Sniper 50PVA not registered in California.

Signal Word/Toxicity Class: Danger-Poison/I.
REI: 48 hr.; 72 hr. if average rainfall less than 25" per year.

NOTICE The information on these pages is for preliminary planning — not a guide for use. Be sure to follow the manufacturer's directions, notwithstanding information contained here.

For personal protective equipment and EPA registration numbers, please turn to page 39.

BT 320 SULFUR 25 or 50 DUST (Premix)

Bt 320 Sulfur 25 Dust (0.064% Bt & 25% sulfur) or Bt 320 Sulfur 50 Dust (0.064% Bt & 50% sulfur) *(Wilbur-Ellis)*

Rate: 20-50 lb. Dust/A

Pest:
cabbage looper

Remarks: Do not mix or dilute with other products. Thorough coverage essential when using this product.

Signal Word/Toxicity Class: Caution.
REI: 24 hr.

CARBARYL

Carbamate insecticide with broad activity on many insects. Low mammalian toxicity, but highly toxic to bees.

Clean Crop Sevin 5 Bait *(Platte)*

Rate: 1-2 lb. ai/A
20-40 lb. Bait/A

Pest:
ants, armyworms, cutworms, darkling beetle, field crickets, grasshoppers, sowbugs

Remarks: Apply with ground or air equipment.
Days To Harvest: 14.

Signal Word/Toxicity Class: Caution/III.
REI: 12 hr.

Drexel Carbaryl 4L *(Drexel)*
Sevin 4F, XLR Plus, 80S, 80WSP *(Rhone-Poulenc)*

Rate: 0.5-2 lb. ai/A
1-4 pt. 4L, 4F, XLR/A
0.66-2.5 lb. 80S, 80WSP

Pest:
armyworms, aster leafhopper, corn earworm, fall armyworm, flea beetles, grasshoppers, leafhoppers, Lygus bugs, spittlebugs, stink bugs, tarnished plant bug

Remarks: Treat at 5- to 7-day intervals. Rate varies by pest species; refer to label.
Days To Harvest: 14.
State Restrictions: Not for use in California.

Signal Word/Toxicity Class: Caution (4F, 4L, XLR); Warning/II (80S, 80WSP).
REI: 12 hr.

DEADLINE

Packaged as Bullets, M-Ps (mini-pellets), and granules.

Deadline Bullets, M-Ps, Granules (4%) (metaldehyde) *(Pace)*

Rate: 0.8-1 lb. ai/A
20-40 lb. 4%/A

Pest:
slugs, snails

Remarks: Apply every 3-4 weeks during growing season, or as needed. Do not apply directly to edible portion of plant.

Signal Word/Toxicity Class: Warning/II (granule); Caution (bullets, M-P).
REI: 12 hr.

Declare — see methyl parathion

Diatect — see pyrethrins

DIAZINON

Restricted Use Pesticide due to avian and aquatic toxicity.

Drexel Diazinon D-264 14G *(Drexel)*
Gowan Diazinon 14G *(Gowan)*
◆ Micro Flo Diazinon 14G *(Micro Flo)*
◆ Wilbur-Ellis Diazinon 14G *(Wilbur-Ellis)*

Rate: 2-4 lb. ai/A
14-28 lb. 14G/A

Pest:
cutworms

Rate: 3-4 lb. ai/A
21-28 lb. 14G/A

Pest:
wireworms

Remarks: Broadcast and incorporate just before planting.

Signal Word/Toxicity Class: Caution.
REI: 12 hr. (24 hr. for Micro Flo).

Digon — see dimethoate

Dimate — see dimethoate

DIMETHOATE

Clean Crop Dimethoate 400 *(Platte)*
Digon 400 *(Wilbur-Ellis)*
Dimate 4E *(Terra)*
Drexel Dimethoate 4EC *(Drexel)*
Micro Flo Dimethoate 4E *(Micro Flo)*

Rate: 0.5 lb. ai/A
1 pt. 4E, 4EC, 400/A

Pest:
Carmine mite, leafminers, twospotted spider mite

Remarks: Mites for Clean Crop only.
Days To Harvest: 7.
State Restrictions: For use in Florida.

Signal Word/Toxicity Class: Warning/II.
REI: 48 hr.

ENDOSULFAN

Drexel Endosulfan 3EC or 50WP *(Drexel)*
Gowan Endosulfan 3EC or 50WSB *(Gowan)*
◆ Micro Flo Endosulfan 3EC or 50WP *(Micro Flo)*
Phaser 3EC or 50WSB *(AgrEvo USA)*
Thiodan 3EC or 50WP *(FMC)*

Rate: 0.5-1 lb. ai/A
0.66-1.33 qt. 3EC/A
1-2 lb. 50WP, 50WSB/A

Pest:
armyworm, cabbage looper, flea beetles, green peach aphid, leafhoppers, whiteflies

Remarks: Apply when insects appear or feeding is noticed. Do not exceed more than 1 application per season (2 for green peach aphid at 0.5 lb. ai rate) or 1 lb. ai/A per year. Do not use tops for food or feed.
Days To Harvest: 4; green peach aphid at 0.5 lb. ai rate-7.

Signal Word/Toxicity Class: Danger-Poison/I.
REI: 24 hr.

LANNATE

Restricted Use Pesticide.

Lannate SP or LV (methomyl) *(DuPont)*

Rate: 0.225-0.9 lb. ai/A
0.25-1 lb. SP/A
0.75-3 pt. LV/A

Pest:
armyworms, aster leafhopper, beet armyworm, loopers, variegated cutworm

Remarks: Apply in sufficient water to obtain thorough, uniform coverage when insects first appear; repeat at 5- to 7-day intervals or as needed. Do not apply more than 9 lb. ai/A per crop or make more than 10 applications per crop.
Days To Harvest: 7.

Signal Word/Toxicity Class: Danger-Poison/I.
REI: 48 hr.

◆ - **new product** • ai-active ingredient • DF-dry flowable • E/EC-emulsifiable concentrate • EW-emulsion in water
F/FL-flowable • G-granule • L/LC-liquid • SP/WSP-soluble packet • W/WP-wettable powder • WSB-water soluble bag

LARVIN

Broad spectrum carbamate insecticide that has stomach poison activity and some limited contact activity. Known to have systemic action as a seed treatment. Toxic to fish and birds, and moderately toxic to bees as direct treatment.

Larvin 3.2 (thiodicarb) *(Rhone-Poulenc)*

Rate: 0.4-0.75 lb. ai/A
16-30 fl. oz. 3.2/A

Pest:
alfalfa looper, corn earworm, southern armyworm, beet armyworm, fall armyworm, tomato fruitworm, bollworm

Rate: 0.6-0.75 lb. ai/A
24-30 fl. oz. 3.2/A

Pest:
cabbage looper

Remarks: Celery. Spray as needed. Use higher rates for heavier infestations of larger larvae. Do not exceed 1.5 lb. ai/A per season.

Days To Harvest: 14.

Signal Word/Toxicity Class: Warning/II.
REI: 12 hr.

MALATHION

Organophosphate insecticide toxic to fish, aquatic invertebrates, aquatic life stages of amphibians, and bees exposed to direct treatment on blooming crops or weeds.

Clean Crop Malathion 57EC or 8EC *(Platte)*
Drexel Malathion 5EC *(Drexel)*
◆ **Gowan Malathion 8 (8E)** *(Gowan)*
◆ **Micro Flo Malathion 5EC** *(Micro Flo)*
Prentox 5 lb. Malathion Spray (5EC) *(Prentiss)*
◆ **Wilbur-Ellis Malathion 8 Spray** *(Wilbur-Ellis)*

Rate: 1 lb. ai/A
1.5 pt. 5EC, 57EC/A
1.5 lb. ai/A
1.5 pt. 8EC/A
1-1.5 lb. ai/A
1-1.5 pt. 8E/A

Pest:
aphids, spider mites

Remarks: None.
Days To Harvest: 7.

Signal Word/Toxicity Class: Caution/III; Warning/II (Drexel, Prentox).
REI: 12 hr.

METHYL PARATHION

Organophosphate insecticide toxic to bees exposed to direct treatment or residues on blooming crops or weeds. Restricted Use Pesticide due to very high acute toxicity to humans and birds.

Cheminova Methyl 4EC *(Cheminova)*

Rate: 1 lb. ai/A
2 pt. 4EC/A

Pest:
aphids, leafhoppers, mites, armyworms, Lygus bugs, stink bugs, flea beetles

Remarks: Apply when insects first appear and repeat as needed.
Days To Harvest: 15.

Signal Word/Toxicity Class: Danger-Poison/I.
REI: 48 hr.; 72 hr. if average rainfall less than 25" per year.

◆ **Declare (4EC)** *(Griffin)*
Helena 4 lb. Methyl Parathion (4EC) *(Helena)*

Rate: 0.5-1 lb. ai/A
1-2 pt. 4EC/A

Pest:
aphids, flea beetles, mites, armyworms, leafhoppers, plant bugs, cabbage looper, Lygus bugs, stink bugs

Remarks: Rate varies by pest species; refer to label.
Days To Harvest: 15.

Signal Word/Toxicity Class: Danger-Poison/I.
REI: 48 hr.

Wilbur-Ellis Methyl Parathion 4 or 5 Spray *(Wilbur-Ellis)*

Rate: 0.5 lb. ai/A
1 pt. 4/A
0.63 lb. ai/A
1 pt. 5/A

Pest:
aphids, flea beetles, spider mites, armyworms, leafhoppers

Remarks: Do not apply through any type of irrigation system.
Days To Harvest: 15.

Signal Word/Toxicity Class: Danger-Poison/I.
REI: 24 (4 Spray); 48 (5 Spray).

M-PEDE

M-Pede (potassium salts of fatty acids) *(Mycogen)*

Rate: 1-2% v/v solution

Pest:
aphids*, leafminers*, spider mites, broad mite, plant bugs, thrips, green peach aphid*, russet mite, whiteflies, leafhoppers

* *Tank mix only*

Remarks: Apply at first sign of infestation; repeat at 7- to 14-day intervals. May be tank mixed with other insecticides. Refer to labels for use directions and precautions.
Days To Harvest: 0.

Signal Word/Toxicity Class: Warning/II.
REI: 12 hr.

NEEMAZAD — see azadirachtin

NEEMIX— see azadirachtin

Orthene — see acephate

PERMETHRIN

Restricted Use Pesticide.

Ambush 2EC or 25W *(ZENECA)*

Rate: 0.1-0.2 lb. ai/A
6.4-12.8 fl. oz. 2EC/A
6.4-12.8 oz. 25W/A

Pest:
beet armyworm, cutworms, green cloverworm, cabbage looper, European corn borer, leafhoppers, corn earworm, fall armyworm, vegetable leafminer

Remarks: Apply in sufficient water by air or ground at 3- to 5-day intervals or as needed. Do not apply more than 2 lb. ai/A per season.
Days To Harvest: 1.

Signal Word/Toxicity Class: Warning/II.

Pounce 3.2 EC, 25WP or WSB *(FMC)*

Rate: 0.05-0.2 lb. ai/A
2-8 oz. 3.2 EC/A
3.2-12.8 oz. 25WP/A
1-2 bags WSB/A

Pest:
alfalfa looper, cutworms, leafhoppers, beet armyworm, European corn borer, leafminers, cabbage looper, fall armyworm, southern armyworm, corn earworm, green cloverworm

Remarks: Apply every 3-5 days or as needed. Use sufficient water to obtain full coverage. Rate varies by pest species. Do not apply more than 2 lb. ai/A per season in California. Do not graze celery forage or feed trimmings from treated fields to livestock.
Days To Harvest: 1.

Signal Word/Toxicity Class: Caution (3.2 EC); Warning/II (25WP, WSB).
REI: 24 hr.

7 Vegetables

NOTICE The information on these pages is for preliminary planning — not a guide for use. Be sure to follow the manufacturer's directions, notwithstanding information contained here.	For personal protective equipment and EPA registration numbers, please turn to page 39.

PETROLEUM OIL

Petroleum based spray oil. Mode of action is through suffocation of eggs, larvae, and nymphs of insects and adults of soft bodied insects. This mode of action necessitates total spray coverage. Toxic to fish.

SAF-T-SIDE *(Brandt Consolidated)*
◆ **Sumulsion** *(Drexel)*
◆ **SunSpray Ultra-Fine** *(Sun)*

Rate: 1-2 gal./100 gal. water
Pest:

aphids	leafminers	plant bugs
caterpillars	loopers	thrips
leafhoppers	mites	whiteflies

Remarks: Apply as needed. May be tank mixed with most commonly used insecticides and fungicides. Do not use in combination with or immediately before or after spraying with fungicides such as captan, Dyrene, folpet, Morestan or any sulfur products. Do not use with carbaryl or dimethoate.

- - - - - - - - - - - - - - - - -

Signal Word/Toxicity Class: Caution.
REI: 4 hr.

Phaser — see endosulfan

Pounce — see permethrin

PYRELLIN (Premix)

Premix containing pyrethrins and rotenone. May be combined with other insecticides to enhance knockdown or as an insect "exciter" to help facilitate contact with other spray residues.

Pyrellin E.C. (0.042 lb./gal. pyrethrins & 0.035 lb./gal. rotenone) *(Webb Wright)*

Rate: 1-2 pt. E.C./A
Pest:

aphids	Japanese beetle	mites
bean beetle	leafhoppers	plant bugs
cabbageworms	leafminers	stink bugs
celery leaftier	loopers	thrips
garden fleahopper	Lygus bugs	whiteflies
harlequin bug		

Remarks: Apply at 7-day intervals or less; repeat as needed.
Days To Harvest: 0.

- - - - - - - - - - - - - - - - -

Rate: 1.5-2 pt. E.C./A
Pest:

Colorado potato beetle	cucumber beetle	flea beetles
	European corn borer	vegetable weevil

Remarks: Make 3-4 applications or more as eggs start to hatch; repeat at 5- to 7-day intervals.

- - - - - - - - - - - - - - - - -

Signal Word/Toxicity Class: Caution/III.
REI: 12 hr.

Pyrenone — see pyrethrins

PYRETHRINS

Contains piperonyl butoxide as a synergist. Diatect also contains silicon dioxide (diatomaceous earth). May be combined in tank mix with other insecticides for quicker and more complete kill where insect resistance may be a problem, and as an "exciter" to flush insects out of hiding and into contact with other spray residues.

Diatect Multipurpose Insecticide II *(Diatect)*
Pyrenone (0.5EC) *(AgrEvo Environmental Health)*
◆ **Py-Rin 60-6 EC** *(Wilbur-Ellis)*

Rate: 1-6 lb. II/A
1-12 fl. oz. 0.5EC/A
2-16 oz. 6-60 EC/A
Pest: Including but not limited to:

aphids	cutworms	loopers
carrot weevil	leafhoppers	mites
celery leaftier	leafminers	tarnished plant bug

Remarks: Apply in sufficient water for thorough coverage; repeat as needed. Apply Diatect as dust or wettable powder.
Days To Harvest: 0.

- - - - - - - - - - - - - - - - -

Signal Word/Toxicity Class: Caution.
REI: 12 hr.

Py-Rin — see pyrethrins

SAF-T-SIDE — see petroleum oil

Sevin — see carbaryl

Sniper — see azinphos-methyl

Sumulsion — see petroleum oil

SunSpray — see petroleum oil

Thiodan — see endosulfan

THIRETHRIN (Premix)

Broad-spectrum insecticide/aphicide. Contains 0.45 lb. piperonyl butoxide. Toxic to bees, fish, shrimp, crab, birds, and other wildlife.

Thirethrin (2 lb./gal. endosulfan & 0.045 lb./gal. pyrethrins) *(Terra)*

Rate: 1 qt./A
Pest:
cabbage looper — green peach aphid
Remarks: Repeat as needed.
Days To Harvest: 7.

- - - - - - - - - - - - - - - - -

Rate: 1-2 qt./A
Pest:
cabbage looper — green peach aphid
Remarks: Do not make more than 3 applications.
Days To Harvest: 4.

- - - - - - - - - - - - - - - - -

Signal Word/Toxicity Class: Warning/II.
REI: 24 hr.

TRIGARD

Triazine compound used as insect growth regulator for control of leafminers. Ineffective against other insect species. Toxic to bees.

Trigard (WSP) (cyromazine) *(Novartis)*

Rate: 0.125 lb. ai/A
0.167 lb. WSP/A (1 packet)
Pest:
leafminers
Remarks: Apply as foliar spray when leafminers first appear; repeat at 7-day intervals or as necessary. Do not make more than 6 applications per crop. Do not apply through any type of irrigation system.
Days To Harvest: 7.
Crop Rotation: All crops on label may be planted following harvest of treated crop. Do not rotate to any other crop except sweet corn or radishes. Do not plant sweet corn or radishes within 3 months after last application.

- - - - - - - - - - - - - - - - -

Signal Word/Toxicity Class: Caution.
REI: 12 hr.

◆ - **new product** • ai-active ingredient • DF-dry flowable • E/EC-emulsifiable concentrate • EW-emulsion in water
F/FL-flowable • G-granule • L/LC-liquid • SP/WSP-soluble packet • W/WP-wettable powder • WSB-water soluble bag

VYDATE

Restricted Use Pesticide.

Vydate L (oxamyl) *(DuPont)*

Rate: Foliar:
0.5-1 lb. ai/A
2-4 pt. L/A

Pest:
serpentine leafminers (except *Liriomyza trifolii*)

Remarks: Apply when insects first appear; repeat at 5- to 7-day intervals or as needed. Use low rates for light infestations, high rates for severe infestations. Do not apply more than 24 pt./A per season.

Days To Harvest: 14.

State Restrictions: For use in Arizona, California, Florida.

- - - - - - - - - - - - - - - - -

Rate: Transplant:
1-2 lb. ai/A
4-8 pt. L/A in 100 gal. water
Foliar:
2 lb. ai/A
8 pt. L/A in 100 gal. water
Preplant (8-16" row):
4 lb. ai/A
16 pt. L/A in 20 gal. water
Foliar preplant extension:
1 lb. ai/A
4 pt. L/A in 200 gal. water

Pest:
pin nematode, root knot nematode

Rate: Foliar alone or preplant extension:
1 lb. ai/A
4 pt. L/A in 200 gal. water

Pest:
carrot weevil, pin nematode, root knot nematode

Remarks: Transplant: Apply immediately after transplanting seedlings. Foliar: Apply first spray 3 weeks after transplanting; second spray 3 weeks later. Preplant row: Incorporate 4" into soil. Foliar preplant extension: Apply 2 treatments 2-3 weeks apart beginning 2-3 weeks after transplanting.

Days To Harvest: 14.

State Restrictions: For use in Florida, Michigan, Ohio, Pennsylvania, and Texas.

- - - - - - - - - - - - - - - - -

Rate: Band treatment/shank injection:
1 lb. ai/A
4 pt. L/A in 100 gal. water

Pest:
root knot nematode, stubby root nematode

Remarks: Apply as 1-2" band over plant line or near base of transplants after seedling or tranplanting. After initial treatment, apply as band spray or by shank injection at 21- to 30-day intervals.

Days To Harvest: 14.

State Restrictions: For use in California.

- - - - - - - - - - - - - - - - -

Signal Word/Toxicity Class: Danger-Poison/I.
REI: 48 hr.

Cole Crops

(Bok Choy, Broccoli, Brussels Sprouts, Cabbage, Cauliflower, Kohlrabi, Rutabagas, Turnips)

ACEPHATE

Organophosphate systemic insecticide with contact and stomach poison activity. Effective against many aphids, thrips, caterpillars, and plant bugs. Has moderate residual effects and is compatible with most pesticides.

Orthene 75S *(Valent)*

Rate: 0.5-1 lb. ai/A
0.66-1.33 lb. 75S/A

Pest:
green peach aphid

Rate: 1 lb. ai/A
1.33 lb. 75S/A

Pest:
cabbage looper, diamondback moth, imported cabbageworm

Remarks: Brussels sprouts, cauliflower. Apply when insects or eggs first appear; repeat as needed. Use higher rate for heavy aphid infestations. Do not apply more than 8 lb./A per season to Brussels sprouts. Do not apply through any type of irrigation system. Do not feed trimmings to livestock or allow animals to graze in treated areas.

Days To Harvest: 14.

- - - - - - - - - - - - - - - - -

Signal Word/Toxicity Class: Caution.
REI: 24 hr.

Admire — see imidacloprid

Ambush — see permethrin

Ammo — see cypermethrin

ASANA

Pyrethroid insecticide that has both contact and stomach poison activity. Restricted Use Pesticide due to toxicity to fish and aquatic organisms.

Asana XL (esfenvalerate) *(DuPont)*

Rate: 0.015-0.03 lb. ai/A
2.9-5.8 fl. oz. XL/A

Pest:
imported cabbageworm

Rate: 0.03-0.05 lb. ai/A
5.8-9.6 fl. oz. XL/A

Pest:
alfalfa looper, beet armyworm (aids in control), cabbage looper, cutworms, flea beetles, grasshoppers

Remarks: Broccoli, cabbage, cauliflower, Chinese cabbage (tight-headed varieties). Repeat as necessary to maintain control. Do not exceed 0.4 lb. ai/A per season.

Days To Harvest: 3.

- - - - - - - - - - - - - - - - -

Signal Word/Toxicity Class: Warning/II.
REI: 12 hr.

AZADIRACHTIN

Extracted from seed of neem tree. Insect growth regulator that controls targeted insect larvae when they ingest or come in contact with it, by interfering with the insect's ability to molt and shutting off feeding behavior. Affected larvae slowly starve and die at next molt.

◆ **Azatin EC or XL** *(Olympic)*
◆ **Azatin XL Plus** *(Thermo Trilogy)*

Rate: 8-21 oz. EC/A
5-21 oz. XL, XL Plus/A

Pest:
aphids, beet armyworm, beetles, cabbage looper, caterpillars, Colorado potato beetle, cutworms, diamondback moth, fall armyworm, flea beetles, green peach aphid, imported cabbageworm, Japanese beetle, leafhoppers, leafminers, leafrollers, loopers, Mexican bean beetle, silverleaf whitefly, soybean looper, sweetpotato whitefly, thrips, tobacco budworm, webworms, whiteflies

Remarks: Broccoli, Brussels sprouts, bok choy, Chinese cabbage, cauliflower, kohlrabi, rutabaga, turnip. Rate varies by species. Apply when pests first appear; repeat at 7 day intervals. When pest populations are high, use higher rates. Do not apply more that 21 fl. oz./A.

Days To Harvest: 0.

- - - - - - - - - - - - - - - - -

Signal Word/Toxicity Class: Caution (XL); Warning/II (EC).
REI: 4 hr. (XL), 12 hr. (EC).

7 Vegetables

NOTICE The information on these pages is for preliminary planning — not a guide for use. Be sure to follow the manufacturer's directions, notwithstanding information contained here.

For personal protective equipment and EPA registration numbers, please turn to page 39.

◆ **NEEMAZAD 4.5EC or NEEMIX 4.5** *(Thermo Trilogy)*
Rate: 0.125-1 pt. 4.5/A
High Volume:
2.25-4.5 oz. 4.5EC/100 gal. water
Low Volume:
0.5 pt. 4.5EC/A
Pest:

aphids	diamondback moth	onion maggot
armyworm	greenhouse whitefly	sweetpotato whitefly
cabbage loopers	hornworms	thrips
caterpillars	imported cabbageworm	webworms
cutworms	leafminers	weevils

Remarks: Broccoli, bok choy, Brussels sprouts, cabbage, cauliflower, kohlrabi. For low volume, apply in sufficient water for adequate coverage. For low insect pressure, apply preventative at 7-day intervals or at first sign of insects; high insect pressure, at 3- to 4-day intervals.

Signal Word/Toxicity Class: Warning/II.
REI: 12 hr.

Azatin — see azadirachtin

AZINPHOS-METHYL

Restricted Use Pesticide.

Gowan Azinphos-M 50-W *(Gowan)*
◆ **Micro Flo Azinphosmethyl 2EC or 50W** *(Micro Flo)*
Sniper 2-E, 50PVA or 50W *(Platte)*
Rate: 0.25-0.75 lb. ai/A
0.5-3 pt. 2EC, 2-E/A
1-1.5 lb. 50-W, 50PVA, 50W/A
Pest:

aphids	cabbage maggot	imported
cabbage looper	diamondback moth	cabbageworm

Remarks: Broccoli, Brussels sprouts, cabbage, cauliflower, Chinese cabbage (includes tight-headed). Apply in sufficient water for complete coverage; repeat as needed. For cabbage maggot, see label for rate and California use directions. Do not apply more than 3 times per season.
Days To Harvest: Brussels sprouts-7; broccoli, cauliflower-15; cabbage-21.
State Restrictions: California labels may vary. Sniper 50PVA not registered in California.

Signal Word/Toxicity Class: Danger-Poison/I.
REI: 48 hr.; 72 hr. if average rainfall less than 25" per year.

Bravo S — see sulfur premix

BT 320 SULFUR 25 or 50 DUST (Premix)

Bt 320 Sulfur 25 Dust (0.064% Bt & 25% sulfur) or Bt 320 Sulfur 50 Dust (0.064% Bt & 50% sulfur) *(Wilbur-Ellis)*
Rate: 20-50 lb. Dust/A
Pest:

cabbage looper	imported cabbageworm (cauliflower)

Remarks: Broccoli, cabbage, cauliflower. Do not mix or dilute with other products. Thorough coverage essential.

Signal Word/Toxicity Class: Caution.
REI: 24 hr.

CARBARYL

Clean Crop Sevin 5 Bait *(Platte)*
Rate: 1-2 lb. ai/A
20-40 lb. Bait/A
Pest:

ants	darkling beetle	grasshoppers
armyworms	field crickets	sowbugs
cutworms		

Remarks: Broccoli, Brussels sprouts, cabbage, cauliflower, Chinese cabbage, kohlrabi, rutabagas, turnips. Apply by ground or air.
Days To Harvest: 3; Chinese cabbage-14.

Signal Word/Toxicity Class: Caution/III.
REI: 12 hr.

Drexel Carbaryl 4L or 50W *(Drexel)*
Sevin 4F, XLR Plus, 80S, 80WSP *(Rhone-Poulenc)*
Rate: 0.5-2 lb. ai/A
1-4 pt. 4L, 4F, XLR/A
1-4 lb. 50W/A
0.66-2.5 lb. 80S, 80WSP/A
Pest:

armyworms	fall armyworm	leafhoppers
aster leafhopper	flea beetles	stink bugs
corn earworm	harlequin bug	tarnished plant bug
diamondback moth	imported cabbageworm	

Remarks: Broccoli, brussels sprouts, cabbage, cauliflower, Chinese cabbage, kohlrabi, rutabagas, turnips. Apply when insects or their damage appears; repeat at 7- to 14-day intervals or as necessary unless shorter interval is specified.
Days To Harvest: Broccoli, brussels sprouts, cabbage, cauliflower, kohlrabi, rutabagas, turnips-3; Chinese cabbage-14.

Signal Word/Toxicity Class: Caution (4F, 4L, XLR); Warning/II (50W, 80S, 80WSP).
REI: 12 hr.

Gowan Sevin Bait *(Gowan)*
Rate: 2 lb. ai/A
40 lb. Bait/A
Pest:

armyworms	cutworms	grasshoppers
crickets	darkling beetle	sowbugs

Remarks: Broccoli, Brussels sprouts, cabbage, cauliflower, rutabagas, turnips. Apply when pests appear in damaging numbers; repeat at 7- to 14-day intervals if necessary. Do not use when bees are present.
Days To Harvest: 3.
State Restrictions: For use in Arizona and California.

Signal Word/Toxicity Class: Caution/III.
REI: 12 hr.

Chlorfos — see chlorpyrifos

CHLORPYRIFOS

Organophosphate insecticide toxic to birds and wildlife, extremely toxic to fish and aquatic organisms.

Lorsban-4E *(Dow AgroSciences)*
◆ **Micro Flo Chlorpyrifos 4E** *(Micro Flo)*
Rate: **Broccoli, Brussels sprouts, cabbage, kohlrabi, turnips:**
1.6-2.75 fl. oz. 4E/1000 lin. ft. row
Cauliflower:
1.6-2.4 fl. oz. 4E/1000 lin. ft. row
Rutabagas:
1.6-3.3 fl. oz. 4E/1000 lin. ft. row
Pest:
root maggot
Remarks: Broccoli, Brussels sprouts, cabbage (including Chinese), kohlrabi, turnips. Apply to direct-seeded crops as 4" band over row at planting time and to transplants directed to base of plants immediately after setting. For transplants, do not apply as foliar application. Do not apply more than once per season. Do not use rutabaga tops for food or feed. To avoid phytotoxicity in vegetable crops, do not mix with other pesticide products or treat plants that are under extreme heat or drought stress. Refer to label for use directions.
Days To Harvest: 30.

Rate: 1 lb. ai/A
2 pt. 4E/A
Pest:

aphids	imported	striped flea beetle
cutworms	cabbageworm	

Remarks: Brussels sprouts. Apply when insects appear as full coverage foliar spray in 20-150 gal. water/A; repeat at 7- to 14-day intervals. Consult state agricultural experiment station, extension service specialist, or integrated pest control advisor for proper time to treat depending on area. To avoid phytotoxicity, do not mix with other pesticides or treat plants that are under extreme heat and drought stress. Refer to label for use directions.
Days To Harvest: 21.

Signal Word/Toxicity Class: Warning/II.
REI: 24 hr.

◆ - **new product** • ai-active ingredient • DF-dry flowable • E/EC-emulsifiable concentrate • EW-emulsion in water
F/FL-flowable • G-granule • L/LC-liquid • SP/WSP-soluble packet • W/WP-wettable powder • WSB-water soluble bag

◆ **Chlorfos 15G** *(Griffin)*
Lorsban 15G *(Dow AgroSciences)*
◆ **Micro Flo Chlorpyrifos15G** *(Micro Flo)*

Rate: 4.6-9.2 oz. 15G/1000 ft. row

Pest:
root maggot

Remarks: Broccoli, Brussels sprouts, cabbage (including Chinese), cauliflower, kohlrabi, rutabagas, turnips. Apply at planting across seed row in 4" band behind planter shoe and in front of press wheel. Do not apply more than once per season. Do not use rutabaga tops for food or feed. Refer to label for use directions.

Signal Word/Toxicity Class: Caution/III.
REI: 12 hr.

Lorsban 50-W *(Gowan)*

Rate: 2 lb. 50-W/A (two 1 lb. bags)

Pest:

armyworm	cutworms	striped flea beetle (adult)
beet armyworm	diamondback moth	
cabbage aphid	imported cabbageworm	

Remarks: Broccoli, Brussels sprouts, cabbage, cauliflower, kohlrabi. Will not control organophosphate-resistant diamondback moths. Do not apply more than 6 applications per season. Do not treat plants under stress from extreme heat or lack of moisture.
Days To Harvest: 21.

Signal Word/Toxicity Class: Warning/II.
REI: 24 hr.

CRYOLITE

Kryocide (96%) *(Elf Atochem)*

Rate: 7.7-15.4 lb. ai/A
8-16 lb. 96%/A

Pest:

cabbage looper	Diabrotica beetles	imported cabbageworm
corn earworm	diamondback moth	yellowstriped armyworm
cutworms	flea beetles	

Remarks: Broccoli, Brussels sprouts, cabbage, cauliflower, kohlrabi. Apply at 14-day intervals as needed. Do not exceed 96 lb./A per season. Do not use in combination with lime or compounds containing free lime. Do not use chemigation in California.
Days To Harvest: 14.
State Restrictions: Not registered for use on Brussels sprouts and kohlrabi in California.

Signal Word/Toxicity Class: Caution.
REI: 12 hr.

◆ **Prokil Cryolite 96** *(Gowan)*

Rate: 9.6-15.4 lb. ai/A
10-16 lb. 96/A

Pest:

cabbage looper	Diabrotica beetles	imported cabbageworm
corn earworm	diamondback moth	yellowstriped armyworm
cutworms	flea beetles	

Remarks: Broccoli, Brussels sprouts, cauliflower. Apply at 7-day intervals as needed. Do not exceed 96 lb./A per season. Do not use in combination with lime or compounds containing free lime. Do not apply through any type of irrigation system.
Days To Harvest: 7.

Signal Word/Toxicity Class: Caution.
REI: 12 hr.

lambda-CYHALOTHRIN

Synthetic pyrethroid insecticide with broad-spectrum activity, formulated as an emulsifiable concentrate. Restricted Use Pesticide due to toxicity to fish and aquatic organisms.

Karate or Warrior (1E) *(ZENECA)*

Rate: 0.015-0.03 lb. ai/A
1.92-3.84 fl. oz. 1E/A

Pest:

alfalfa looper	diamondback moth	plant bugs
aphids*	fall armyworm (1st & 2nd instar)	southern cabbageworm
armyworms	flea beetles	spider mites*
beet armyworm (1st & 2nd instar)	grasshoppers	stink bugs
cabbage looper	imported cabbageworm	thrips
cabbage webworm	Japanese beetle (adult)	vegetable weevil (adult)
corn earworm	leafhoppers	whiteflies*
cutworms	meadow spittlebug	yellowstriped armyworm

* *Suppression*

Remarks: Broccoli, cabbage. Apply as required by scouting, usually at 5-day intervals or more. Do not apply more than 1.92 pt./A per season. Refer to label for spray drift precautions and state and county restrictions.
Days To Harvest: 1.

Signal Word/Toxicity Class: Danger/I.
REI: 24 hr.

CYPERMETHRIN

Synthetic pyrethroid insecticide with both contact and stomach poison activity. No systemic activity and is ineffective against soil pests. Controls many caterpillars, beetles, flies, plant bugs, and other insect pests. Fast knockdown effects. Toxic to fish and bees, and may cause eye and skin irritation. Restricted Use Pesticide.

Ammo 2.5 EC or WSB *(FMC)*

Rate: 0.05-0.1 lb. ai/A
2.5-5 fl. oz. 2.5 EC/A
1-2 bags WSB/A

Pest:

aphids (aids in control)	diamondback moth	onion thrips
armyworms	flea beetles	saltmarsh caterpillar
corn earworm	ground beetles	stink bugs
crickets	imported cabbageworm	tobacco budworm
cucumber beetle	leafhoppers	whiteflies (aids in control)
cutworms	loopers	wireworms (adult)
	Lygus bugs	

Remarks: Cabbage. Apply in water using minimum 15 gal. finished spray/A by ground or 5 gal./A by air. Lower rates should be used for light to moderate insect pressure, higher rates for heavy to extremely heavy insect populations. Rate varies by pest species; refer to label. Apply no more than 0.6 lb. ai/A per season.
Days To Harvest: 1.
Crop Rotation: Not within 30 days after last application.

Signal Word/Toxicity Class: Caution (2.5 EC); Warning/II (WSB).
REI: 12 hr.

zeta-CYPERMETHRIN

Synthetic pyrethroid with broad spectrum activity. Extremely toxic to fish and aquatic invertebrates; highly toxic to bees. Restricted Use Pesticide.

Fury 1.5 EC or Mustang 1.5 EW *(FMC)*

Rate: 0.028-0.05 lb. ai/A
2.39-4.26 fl. oz. EC, EW/A

Pest:

aphids*	diamondback moth	Lygus bugs
corn earworm	flea beetles	saltmarsh caterpillar
cucumber beetle	imported cabbageworm	tobacco budworm
cutworms	leafhoppers	whiteflies*

* *Aids in control*

Rate: 0.04-0.05 lb. ai/A
3.41-4.26 fl. oz. EC, EW/A

Pest:

armyworms	loopers	stink bugs
crickets	onion thrips	wireworm (adult)
ground beetles		

Remarks: Cabbage. Apply in water with minimum 15 gal. finished spray/A by ground; 5 gal. finished spray/A by air. Up to 0.3 lb. ai/A may be applied per season. In arid climatic areas, such as Arizona and California, higher minimum rates may be required.
Days To Harvest: 1.

Signal Word/Toxicity Class: Warning/II.
REI: 12 hr.

For personal protective equipment and EPA registration numbers, please turn to page 39.

DEADLINE

Packaged as Bullets, M-Ps (mini-pellets), and granules.

Deadline Bullets, M-Ps, Granules (4%) (metaldehyde) *(Pace)*

Rate: 0.8-1 lb. ai/A
20-40 lb. 4%A

Pest:
slugs snails

Remarks: Broccoli, Brussels sprouts, cabbage, cauliflower, kohlrabi, rutabagas, turnips. Apply every 3-4 weeks during growing season, or as needed. Do not apply directly to edible portion of plant.

Signal Word/Toxicity Class: Warning/II (granule); Caution (bullets, M-P).
REI: 12 hr.

Declare — see methyl parathion

Diatect — see pyrethrins

DIAZINON

Restricted Use Pesticide due to avian and aquatic toxicity.

D•z•n diazinon AG600 WBC, 50W *(Novartis)*
Drexel Diazinon Insecticide (4EC), D-264 14G *(Drexel)*
Gowan Diazinon 4E, 14G, 50WSB *(Gowan)*
Helena Diazinon AG500 *(Helena)*
Micro Flo Diazinon AG500, 14G, 50W *(Micro Flo)*
Prentox Diazinon AG500 or 50W *(Prentiss)*
◆ **Wilbur-Ellis Diazinon 4 (4E), 14G** *(Wilbur-Ellis)*

Rate: **Broccoli, cabbage, cauliflower, turnips:**
0.25-0.5 lb. ai/A
0.5-1 pt. 4E, 4EC, AG500/A
6.5-12.75 fl. oz. AG600/A
0.5-1 lb. 50W, 50WSB/A
Brussels sprouts:
0.25-0.5 lb. ai/A
0.5-1 pt. 4E, 4EC, AG500/A
6.5-12.75 fl. oz. AG600/A
1 lb. 50W, 50WSB/A

Pest:
aphids
diamondback moth (except turnips)
dipterous leafminer
flea beetles (turnips)
imported cabbageworm (except turnips)

Rate: 1 lb. ai/A
2 pt. 4E, 4EC, AG500/A
25.5 fl. oz. AG600/A
2 lb. 50W, 50WSB/A
7 lb. 14G/A

Pest:
mole crickets

Rate: 2-4 lb. ai/A
4-8 pt. 4E, 4EC, AG500/A
51-102 fl. oz. AG600/A
4-8 lb. 50W, 50WSB/A
14-28 lb. 14G/A

Pest:
cutworms

Rate: 3-4 lb. ai/A
6-8 pt. AG500/A
76.5-102 fl. oz. AG600/A
6-8 lb. 50W, 50WSB/A
21-28 lb. 14G/A

Pest:
wireworms

Remarks: Broccoli, Brussels sprouts, cabbage, cauliflower, turnips. Apply every 7 days as needed. Will not control organophosphate-resistant leafminers.
Days To Harvest: Broccoli, Brussels sprouts, cauliflower-7; turnips-14; cabbage-21.

Rate: Broadcast:
2-3 lb. ai/A
4-6 pt. 4E, 4EC, AG500/A
51-76.5 fl. oz. AG600/A
4-6 lb. 50W, 50WSB/A
14-21 lb. 14G/A
Transplant water (drench application):
0.125-0.25 lb. ai/A
3.25-6.5 fl. oz. AG600/in 50 gal. water
4-8 oz. 4E, 4EC, AG500, 50W, 50WSB/in 50 gal. water

Pest:
root maggots

Remarks: Broccoli, Brussels sprouts, cabbage, cauliflower. Broadcast just before sowing of seeds or transplanting and immediately incorporate into top 3-4" soil. For transplant water application, apply per plant by hand or by tractor-mounted sprayer equipped with drop nozzles to direct spray to base of plant. Transplant water treatments may result in stand reduction due to plant stress at time of transplants.

Signal Word/Toxicity Class: Caution/III.
REI: 24 hr.; 12 hr. (Drexel/Gowan/Wilbur-Ellis 14G).

DIBROM

Organophosphate insecticide with both contact and stomach poison activity. Provides broad spectrum control of insect pests and mites with fast knockdown and short residual effects. Can be phytotoxic to some bean and melon varieties. Causes eye injury and is absorbed through the skin.

Dibrom 8 (8E) (naled) *(Amvac)*

Rate: 1 lb. ai/A
1 pt. 8E/A

Pest:
aphids diamondback caterpillar imported cabbageworm

Rate: 2 lb. ai/A
2 pt. 8E/A

Pest:
loopers

Remarks: Broccoli, Brussels sprouts, cabbage, Chinese cabbage (tightheaded), cauliflower. Apply at first sign of insects; repeat as necessary. Apply in 30-250 gal. water by ground; 3-10 gal. water by air. Do not use with highly alkaline materials such as lime or Bordeaux. Do not apply through any type of irrigation system. Apply at 7-day intervals. Do not apply more than 1 pt./A in Florida. Do not apply more than 10 pt./A per season. Do not apply when temperature exceeds 90°F.
Days To Harvest: 1.

Signal Word/Toxicity Class: Danger/I.
REI: 48 hr. for 1 pt./A or less-; 72 hr. over 1 pt./A.

Digon — see dimethoate

Dimate — see dimethoate

DIMETHOATE

Organophosphate insecticide/acaricide with both contact and stomach poison activity. Systemic with residual effects. Controls a broad spectrum of insect pests on many vegetable crops. Can be phytotoxic to some bean and tomato varieties. Causes eye irritation and is absorbed through the skin.

Cheminova Dimethoate 4E *(Cheminova)*
Clean Crop Dimethoate 400 *(Platte)*
Dimate 4E *(Terra)*
Drexel Dimethoate 2.67 or 4EC *(Drexel)*
Gowan Dimethoate E267 *(Gowan)*
Helena Dimethoate 4E *(Helena)*
Micro Flo Dimethoate 267 or 4E *(Micro Flo)*
Wilbur-Ellis Dimethoate 267 or Digon 400 *(Wilbur-Ellis)*

Rate: 0.25-0.5 lb. ai/A
0.75-1.5 pt. 2.67, 267, E267/A
0.5-1 pt. 4E, 4EC, 400/A

Pest:
aphids

Remarks: Broccoli, Brussels sprouts, cabbage, cauliflower, turnips.
Days To Harvest: Broccoli, cabbage, cauliflower-7; Brussels sprouts-10; turnips-14.
State Restrictions: Brussels sprouts for California aerial application only.

Signal Word/Toxicity Class: Warning/II.
REI: 48 hr.

◆ **- new product** • ai-active ingredient • DF-dry flowable • E/EC-emulsifiable concentrate • EW-emulsion in water
F/FL-flowable • G-granule • L/LC-liquid • SP/WSP-soluble packet • W/WP-wettable powder • WSB-water soluble bag

DI-SYSTON

Organophosphate insecticide/acaricide with systemic and contact activity. Controls broad spectrum of insect pests and mites on a wide range of vegetable crops with long residual effects. Especially effective against sapsucking insect pests. Can be mixed with fertilizers. Can cause plant injury if in contact with the seed or if used in conjunction with some preemergence herbicides. Restricted Use Pesticide.

Di-Syston 8 or 15% (disulfoton) *(Bayer)*

Rate: Broccoli, Brussels sprouts, cabbage, cauliflower:
1 lb. ai/A
1 pt. 8/A (36" row spacing)
6.7 lb. 15%/A (36" row spacing)

Pest:
aphids, flea beetles, leafhoppers, mites (Brussels sprouts), thrips (Brussels sprouts)

Rate: Cabbage:
1.7 fl. oz. 8/1000 ft. row (any row spacing)
11.3 oz. 15%/1000 ft. row (any row spacing)

Pest:
root aphid

Remarks: Transplanting seedbeds: For Di-Syston 8, apply evenly over bed area prior to seeding, work into top 2-3" soil. For Di-Syston 15%, thoroughly incorporate into top 2-3" soil. Seed beds in normal manner. Field applications: Apply Di-Syston 8 as injection with or without liquid fertilizer or Di-Syston 15% as soil incorporated band on each side of seed furrow or transplanted row at planting, or sidedress (soil incorporated for 15%) after plants are established. Do not apply directly on seed. Do not apply more than 1 field application to broccoli and cabbage; 2 field applications to Brussels sprouts and cauliflower. Refer to label for additional information.

Days To Harvest: Broccoli-14; Brussels sprouts-30; cauliflower-40; cabbage-42.

Signal Word/Toxicity Class: Danger/I.
REI: 48 hr.

ENDOSULFAN

Drexel Endosulfan 3EC or 50WP *(Drexel)*
Gowan Endosulfan 3EC or 50WSB *(Gowan)*
◆ Micro Flo Endosulfan 3EC or 50WP *(Micro Flo)*
Phaser 3EC or 50WSB *(AgrEvo USA)*
Thiodan 3EC or 50WP *(FMC)*

Rate: 0.75-1 lb. ai/A
1-1.33 qt. 3EC/A
1.5-2 lb. 50WP, 50WSB/A

Pest:
armyworms, cabbage aphid, cabbage looper, cross-striped cabbageworm, cutworms, diamondback moth (larvae), flea beetles, harlequin bug, imported cabbageworm, leafhoppers, stink bugs, whiteflies

Remarks: Broccoli, Brussels sprouts, cabbage, cauliflower. Apply when insects appear or feeding is noticed. Rate varies by pest species; refer to label. Do not exceed 4 applications per year or 3 lb. ai/A per year.

Days To Harvest: Broccoli, cabbage-7; Brussels sprouts, cauliflower-14.

Signal Word/Toxicity Class: Danger-Poison/I.
REI: 24 hr.

Fury — see *zeta*-cypermethrin

IMIDACLOPRID

Systemic insecticide for control of whiteflies. Toxic to bees and aquatic invertebrates. Restricted Use Pesticide as required for emergency exemptions.

Admire 2F *(Bayer)*
Provado 1.6F *(Bayer)*

Rate: 0.16-0.375 lb. ai/A
10-24 fl. oz. 2F/A
0.05 lb. ai/A
3.75 fl. oz. 1.6F/A

Pest:
aphids, silverleaf whitefly, sweetpotato whitefly, whiteflies

Remarks: Broccoli, Brussels sprouts, cabbage, cauliflower, Chinese cabbage, kohlrabi. Direct seed or transplant. Soil application. Do not apply more than 24 fl. oz. 2F/A or 18.75 fl. oz. 1.6F/A per year.

Days To Harvest: 2F-21; 1.6F-7.

Signal Word/Toxicity Class: Caution.
REI: 12 hr.

Karate — see *lambda*-cyhalothrin

Kryocide — see cryolite

Kumulus — see sulfur

LANNATE

Restricted Use Pesticide.

Lannate SP or LV (methomyl) *(DuPont)*

Rate: 0.225-0.9 lb. ai/A
0.25-1 lb. SP/A
0.75-3 pt. LV/A

Pest:
beet armyworm, diamondback moth, fall armyworm (cabbage), imported cabbageworm, loopers, variegated cutworm

Remarks: Broccoli, Brussels sprouts, cabbage, cauliflower, Chinese cabbage. Apply in sufficient water for thorough, uniform coverage when insects first appear; repeat at 5- to 7-day intervals or as needed. Add wetting agent (except Chinese cabbage). For Chinese cabbage, apply by ground only. Refer to label for restrictions and precautions.

Days To Harvest: Cabbage-1; broccoli, Brussels sprouts, cauliflower-3; Chinese cabbage-10.

Signal Word/Toxicity Class: Danger-Poison/I.
REI: 48 hr.

LARVIN

Broad spectrum carbamate insecticide that has stomach poison activity and some limited contact activity. Known to have systemic action as a seed treatment. Toxic to fish and birds, and moderately toxic to bees as direct treatment.

Larvin 3.2 (thiodicarb) *(Rhone-Poulenc)*

Rate: 0.4-0.8 lb. ai/A
16-32 fl. oz. 3.2/A

Pest:
alfalfa looper, beet armyworm, cabbage flea beetle, cutworms (DF only), imported cabbageworm

Rate: 0.6-1 lb. ai/A
24-40 fl. oz. 3.2/A

Pest:
cabbage looper, diamondback moth

Remarks: Broccoli, cabbage, cauliflower. Spray as needed. Use higher rates for heavier infestations or large larvae. Do not exceed 6 lb. ai/A per season. Regional differences have been noted in susceptibility of certain strains of diamondback moth to thiodicarb. Postemergence rescue treatment for cutworms: Apply as broadcast spray when cutworms are actively feeding. If cutworms are feeding below ground, control will be more difficult. Light cultivation with rotary hoe before spraying may enhance control.

Days To Harvest: 7.

Signal Word/Toxicity Class: Warning/II.
REI: 12 hr.

LINDANE

Contains toxic inert Rhodamine B. Toxic to fish, birds, and wildlife.

Lindane 30C *(Gustafson)*
◆ Lindane ST 40 *(AGSCO)*

Rate: 4.9 fl. oz. 30C/100 lb. seed
3.6 fl. oz. 40/100 lb. seed

7 Vegetables

NOTICE The information on these pages is for preliminary planning — not a guide for use. Be sure to follow the manufacturer's directions, notwithstanding information contained here.	For personal protective equipment and EPA registration numbers, please turn to page 39.

Pest:

seedcorn maggots wireworms

Remarks: Broccoli, Brussels sprouts, cabbage, cauliflower. Do not use treated seed for food, feed, or oil purposes. Exposed treated seed may be hazardous to birds and other wildlife. Dispose of excess seed by burial away from bodies of water.

Signal Word/Toxicity Class: Warning/II.
REI: 12 hr.

Lorsban — see chlorpyrifos

MALATHION

Organophosphate insecticide toxic to fish, aquatic invertebrates, aquatic life stages of amphibians, and bees exposed to direct treatment on blooming crops or weeds.

Clean Crop Malathion 57EC *(Platte)*
Drexel Malathion 5EC *(Drexel)*
◆ **Micro Flo Malathion 5EC** *(Micro Flo)*
Prentox 5 lb. Malathion Spray (5EC) *(Prentiss)*

Rate: 0.625-1.25 lb. ai/A
1-2 pt. 5EC, 57EC/A

Pest:

aphids, caterpillar, imported cabbageworm, cabbage looper, diamondback moth

Remarks: Broccoli, Brussels sprouts, cabbage, cauliflower, kohlrabi, rutabagas, turnips. Apply when insects appear. For caterpillars on cabbage and turnips (summer and fall plantings in the South), apply when true leaves appear.
Days To Harvest: Broccoli, rutabagas-3; Brussels sprouts, cabbage, cauliflower, kohlrabi, turnips (tops used for food or feed)-7.

Signal Word/Toxicity Class: Caution (57EC); Warning/II (5EC).
REI: 12 hr.

◆ **Clean Crop Malathion 8EC** *(Platte)*
◆ **Wilbur-Ellis Malathion 8 Spray** *(Wilbur-Ellis)*

Rate: 1-2 lb. ai/A
1-2 pt. 8EC/A

Pest:

aphids, cabbage looper, imported cabbageworm

Remarks: Broccoli, Brussels sprouts, cabbage, turnips.
Days To Harvest: Broccoli, turnips-3; Brussels sprouts, cabbage-7.

Signal Word/Toxicity Class: Caution/III.
REI: 12 hr.

◆ **Gowan Malathion 8 (8E)** *(Gowan)*

Rate: 1.5-2.5 lb. ai/A
1.5-2.5 pt. 8E/A

Pest:

aphids, flea beetles, imported cabbageworm, cabbage looper

Remarks: Broccoli, Brussels sprouts, cabbage, cauliflower, kohlrabi, rutabagas. Rutabagas for aphids only.
Days To Harvest: Broccoli, rutabagas-3; Brussels sprouts, cabbage, cauliflower, kohlrabi-7.

Signal Word/Toxicity Class: Caution/III.
REI: 12 hr.

Marlate — see methoxychlor

METASYSTOX-R

Organophosphate systemic insecticide with contact and stomach poison activity. Controls variety of insect pests and mites, especially sapsucking pests. Incompatible with alkaline compounds, and toxic to wildlife. Restricted Use Pesticide.

Metasystox-R (SC) (oxydemeton-methyl) *(Gowan)*

Rate: Broccoli, Brussels sprouts, cauliflower:
0.375-0.5 lb. ai/A
1.5-2 pt. SC/A

Pest:

aphids

Rate: Cabbage, Chinese cabbage:
0.375-0.75 lb. ai/A
1.5-3 pt. SC/A

Pest:

aphids, thrips

Remarks: Broccoli, Brussels sprouts, cabbage, cauliflower, Chinese cabbage (includes tight-headed varieties). Apply in sufficient water for complete coverage but not less than 1 gal./A. Do not apply more than 3 times per season for cabbage.
Days To Harvest: Broccoli, cabbage, cauliflower-7; Brussels sprouts-10.

Signal Word/Toxicity Class: Warning/II.
REI: 48 hr.; 72 hr. if average rainfall less than 25" per year.

METHOXYCHLOR

Organochlorine pesticide toxic to fish and aquatic invertebrates.

Marlate 50 (WP) *(Kincaid)*

Rate: 2 Tbs. 50WP/gal. water

Pest:

alfalfa looper, flea beetles, Japanese beetle, blister beetle, imported cabbageworm, leafhoppers, fall armyworm

Remarks: Broccoli, Brussels sprouts, cabbage, cauliflower, kohlrabi, rutabagas, turnips. Apply at first sign of infestation; repeat at 7- to 14-day intervals as needed.
Days To Harvest: Cabbage, cauliflower, rutabaga-3; kohlrabi-7; broccoli, Brussels sprouts, turnips-14.

Signal Word/Toxicity Class: Caution.
REI: 12 hr.

Prentox Methoxychlor 2EC *(Prentiss)*

Rate: 1.3-1.7 lb. ai/A
75-100 fl. oz. 2EC/A

Pest:

bean leaf beetle, corn earworm, Japanese beetle, blister beetle, cross-striped cabbageworm, leafhoppers, Colorado potato beetle, tomato hornworm

Remarks: Broccoli, Brussels sprouts, cabbage, cauliflower, kohlrabi, rutabagas, turnips. Apply at first sign of infestation; repeat at 7- to 14-day intervals as needed.
Days To Harvest: Kohlrabi-1; cabbage-3; cauliflower, rutabagas, turnips-7; broccoli, Brussels sprouts-14.

Signal Word/Toxicity Class: Caution.
REI: 12 hr.

METHYL PARATHION

Toxic to bees exposed to direct treatment or residues on blooming crops or weeds. Restricted Use Pesticide due to very high acute toxicity to humans and birds.

Cheminova Methyl 4EC *(Cheminova)*

Rate: 0.5-1.5 lb. ai/A
1-3 pt. 4EC/A

Pest:

aphids, flea beetles, leafminers (turnips), armyworms, leafhoppers, mites

Rate: 1-1.5 lb. ai/A
2-3 pt. 4EC/A

Pest:

stink bugs

Remarks: Broccoli, Brussels sprouts, cabbage, cauliflower, turnips. Apply when insects or their damage first appear and repeat as needed. Rate varies by pest species; refer to label.
Days To Harvest: Broccoli, Brussels sprouts, cauliflower-7; cabbage, turnips-10; rates over 1 pt. and turnips tops used for food or feed-21.

Signal Word/Toxicity Class: Danger-Poison/I.
REI: 48 hr.; 72 hr. if average rainfall less than 25" per year.

◆ **Declare (4EC)** *(Griffin)*

Rate: Broccoli:
0.25-1.5 lb. ai/A
0.5-3 pt. 4EC/A
Brussels sprouts, cabbage, cauliflower, turnips:
0.5-1.5 lb. ai/A
1-3 pt. 4EC/A

◆ **- new product** • ai-active ingredient • DF-dry flowable • E/EC-emulsifiable concentrate • EW-emulsion in water
F/FL-flowable • G-granule • L/LC-liquid • SP/WSP-soluble packet • W/WP-wettable powder • WSB-water soluble bag

Pest:

aphids, armyworms, cabbage looper, flea beetles, imported cabbageworm, leafhoppers, mites, plant bugs, stink bugs, thrips (broccoli, Brussels sprouts)

Remarks: Broccoli, Brussels sprouts, cabbage, cauliflower, turnips. Rate varies by pest species.

Days To Harvest: Up to 1 pt.-7 days (cabbage 10 days); over 1 pt. or turnips tops used as food or feed-21 days.

Signal Word/Toxicity Class: Danger-Poison/I.
REI: 48 hr.

Helena 4 lb. Methyl Parathion (4EC) *(Helena)*
Rate: Broccoli, Brussels sprouts:
0.25-1.25 lb. ai/A
0.5-2.5 pt. 4EC/A
Cabbage:
0.5-1.25 lb. ai/A
1-2.5 pt. 4EC/A
Cauliflower:
0.5-1.5 lb. ai/A
1-3 pt. 4EC/A
Kohlrabi:
0.5-1.5 lb. ai/A
1-3 pt. 4EC/A
Turnips:
0.25-0.75 lb. ai/A
0.5-1.5 pt. 4/A

Pest:

aphids, armyworms, cabbage looper, flea beetles, imported cabbageworm, leafhoppers, mites, plant bugs, stink bugs, thrips (broccoli, Brussels sprouts)

Remarks: Broccoli, Brussels sprouts, cabbage, cauliflower, kohlrabi, turnips. Rate varies by pest species.

Days To Harvest: Up to 1 pt.-7 days (cabbage 10 days); over 1 pt.-21 days.

Signal Word/Toxicity Class: Danger-Poison/I.
REI: 48 hr.

Wilbur-Ellis Methyl Parathion 4 or 5 Spray *(Wilbur-Ellis)*
Rate: 0.5 lb. ai/A
1 pt. 4/A
0.95-1.3 lb. ai/A
1.5-2 pt. 5/A

Pest:

aphids, armyworms, flea beetles, spider mites

Remarks: Broccoli, Brussels sprouts, cabbage, cauliflower. Do not apply through any type of irrigation system.

Days To Harvest: 21.

Signal Word/Toxicity Class: Danger-Poison/I.
REI: 24 (4 Spray); 48 (5 Spray).

Microthiol Special — see sulfur

MOCAP

Mocap EC is a Restricted Use Pesticide.

Mocap EC or 10%G (ethoprop) *(Rhone-Poulenc)*
Rate: Broadcast:
5 lb. ai/A
3.33 qt. EC/A
50 lb. 10%G/A
Row treatment:
2 lb. ai/A
1.33 qt. EC/A (2.9 fl. oz./1000 lin. ft. row)
20 lb. 10%G/A (1.38 lb./1000 lin. ft. row)

Pest:

nematodes, symphylans

Remarks: Cabbage. Broadcast: Apply from 1 week before planting to at-planting time; mix with top 2-4" of soil. Row treatment: Apply at planting in a band 15" wide on the row; do not use as a seed furrow treatment or allow spray to contact the seed.

State Restrictions: Do not apply in Long Island, New York.

Signal Word/Toxicity Class: Danger-Poison/I (EC); Warning/II (10%G).
REI: 48 hr.; 72 hr. if average rainfall less than 25" per year.

M-PEDE

M-Pede (potassium salts of fatty acids) *(Mycogen)*
Rate: 1-2% v/v solution

Pest:

aphids*, broad mite, green peach aphid*, leafhoppers, leafminers*, plant bugs, russet mite, spider mites, thrips, whiteflies

* *Tank mix only*

Remarks: Broccoli, Brussels sprouts, cabbage, Chinese cabbage (Bok choy). Apply at first sign of infestation; repeat at 7- to 14-day intervals. May be tank mixed with other insecticides. Refer to labels for use directions and precautions.

Days To Harvest: 0.

Signal Word/Toxicity Class: Warning/II.
REI: 12 hr.

Mustang — see *zeta*-cypermethrin

NEEMAZAD — see azadirachtin

NEEMIX — see azadirachtin

NEMACUR

Organophosphate systemic nematicide with contact activity. Applied broadcast, in-furrow, in band, or by drench for control of many species of nematodes. Long residual effects and systemic activity against certain foliage-feeding insect pests. Toxic to fish. Restricted Use Pesticide.

Nemacur 3 (fenamiphos) *(Bayer)*
Rate: 1.69 lb. ai/A
2.25 qt. 3/A (36" row spacing)
5 fl. oz. 3/1000 ft. row

Pest:

nematodes

Remarks: Cabbage (transplanted). Apply in 10" band as postplant drench within 24 hr. after transplanting. Refer to label for use information.

State Restrictions: SLN-Florida.

Signal Word/Toxicity Class: Danger/I.
REI: 48 hr.

Nemacur 15% (fenamiphos) *(Bayer)*
Rate: 1.8-3 lb. ai/A
12-20 lb. 15G/A (38-40" row spacing)
7.3-18.4 oz. 15%/1000 ft. row (any row spacing)

Pest:

nematodes

Remarks: Brussels sprouts (transplanted), cabbage (direct-seeded, transplanted). Apply in 6-15" band and incorporate into soil. Refer to label for further use information.

Signal Word/Toxicity Class: Danger/I.
REI: 48 hr.

Orthene — see acephate

PERMETHRIN

Restricted Use Pesticide.

Ambush 2EC or 25W *(ZENECA)*
Rate: Broccoli, brussels sprouts, cauliflower, Chinese broccoli, turnips:
0.05-0.1 lb. ai/A
3.2-6.4 fl. oz. 2EC/A
3.2-6.4 oz. 25W/A
Cabbage, Chinese cabbage (tight-headed):
0.05-0.2 lb. ai/A
3.2-12.8 fl. oz. 2EC/A
3.2-12.8 oz. 25W/A

NOTICE The information on these pages is for preliminary planning — not a guide for use. Be sure to follow the manufacturer's directions, notwithstanding information contained here. For personal protective equipment and EPA registration numbers, please turn to page 39.

Pest:

cabbage aphid (suppression)	cabbage looper	imported
	diamondback moth	cabbageworm

Remarks: Broccoli, brussels sprouts, cabbage, cauliflower, Chinese broccoli, Chinese cabbage, turnips. Use sufficient water to obtain full coverage of foliage. For turnips, apply by ground only; do not exceed 4 applications in Florida, Georgia, South Carolina, Washington; 8 applications in Illinois, Indiana, Oklahoma, Texas. Do not apply more than 0.8 lb. ai (1 lb. ai on cabbage, Chinese cabbage) per season. Turnips: Do not graze treated areas or feed crop refuse to livestock.

Days To Harvest: 1.

State Restrictions: For use on turnips in Florida, Georgia, Illinois, Indiana, Oklahoma, South Carolina, Texas, and Washington.

- - - - - - - - - - - - - - - - -

Signal Word/Toxicity Class: Warning/II.
REI: 24 hr.

Pounce 3.2 EC, 25WP or WSB *(FMC)*

Rate: 0.05-0.1 lb. ai/A
2-4 oz. 3.2 EC/A
3.2-6.4 oz. 25WP/A
1 bag WSB/A

Pest:

beet armyworm	diamondback	imported
cabbage looper	moth	cabbageworm

Remarks: Broccoli, Brussels sprouts, cabbage, cauliflower, Chinese cabbage. Apply when insects appear or feeding is noticed; repeat as needed. Higher rate should be used as pest populations increase. Apply in minimum of 20 gal. finished spray/A by ground; 3 gal. finished spray/A (1 gal. on cabbage, Chinese cabbage) by air. Do not apply more than 0.8 lb. ai (1 lb. ai on cabbage, Chinese cabbage) per season.

Days To Harvest: 1.

- - - - - - - - - - - - - - - - -

Signal Word/Toxicity Class: Caution (3.2 EC), Warning/II (25WP, WSB).
REI: 24 hr.

PETROLEUM OIL

Petroleum based spray oil. Mode of action is through suffocation of eggs, larvae, and nymphs of insects and adults of soft bodied insects. This mode of action necessitates total spray coverage. Toxic to fish.

SAF-T-SIDE *(Brandt Consolidated)*
◆ **Sumulsion** *(Drexel)*
◆ **SunSpray Ultra-Fine** *(Sun)*

Rate: 1-2 gal./100 gal. water

Pest:

aphids	leafminers	plant bugs
caterpillars	loopers	thrips
leafhoppers	mites	whiteflies

Remarks: Cabbage, cauliflower. Apply as needed. May be tank mixed with most commonly used insecticides and fungicides. Do not use in combination with or immediately before or after spraying with fungicides such as captan, Dyrene, folpet, Morestan or any sulfur products. Do not use with carbaryl or dimethoate.

- - - - - - - - - - - - - - - - -

Signal Word/Toxicity Class: Caution.
REI: 4 hr.

Phaser — see endosulfan

Pounce — see permethrin

Provado — see imidacloprid

PYRELLIN (Premix)

Premix containing pyrethrins and rotenone. May be combined with other insecticides to enhance knockdown or as an insect "exciter" to help facilitate contact with other spray residues.

Pyrellin E.C. (0.042 lb./gal. pyrethrins & 0.035 lb./gal. rotenone) *(Webb Wright)*

Rate: 1-2 pt. E.C./A

Pest:

aphids	harlequin bug	mites
bean beetle	Japanese beetle	plant bugs
beet webworm	leafhoppers	stink bugs
cabbageworms	leafminers	thrips
celery leaftier	loopers	whiteflies
garden fleahopper	Lygus bugs	

Remarks: Broccoli, Brussels sprouts, cabbage, cauliflower, Chinese cabbage, kohlrabi, rutabagas, turnips. Apply at 7-day intervals or less; repeat as needed.

Days To Harvest: 0.

- - - - - - - - - - - - - - - - -

Rate: 1.5-2 pt. E.C./A

Pest:

Colorado potato beetle	cucumber beetle	flea beetles
	European corn borer	vegetable weevil

Remarks: Cabbage, Chinese cabbage. Make 3-4 applications or more as eggs start to hatch; repeat at 5- to 7-day intervals.

Days To Harvest: 0.

- - - - - - - - - - - - - - - - -

Signal Word/Toxicity Class: Caution/III.
REI: 12 hr.

Pyrenone — see pyrethrins

PYRETHRINS

Contains piperonyl butoxide as a synergist. Diatect also contains silicon dioxide (diatomaceous earth). May be combined in tank mix with other insecticides for quicker and more complete kill where insect resistance may be a problem, and as an "exciter" to flush insects out of hiding and into contact with other spray residues.

Diatect Multipurpose Insecticide II *(Diatect)*
Pyrenone (0.5EC) *(AgrEvo Environmental Health)*
◆ **Py-Rin 60-6 EC** *(Wilbur-Ellis)*

Rate: 1-6 lb.II/A
1-12 fl. oz. 0.5EC/A
2-16 oz. 6-60 EC/A

Pest: Including but not limited to:

aphids	diamondback moth	leafminers
armyworms	European corn borer	loopers
cabbage looper	flea beetles	Lygus bugs
cabbageworms	garden fleahoppers	Mexican bean beetle
caterpillars	harlequin bug	mites
celery leaftier	imported cabbageworm	plant bugs
Colorado potato beetles	Japanese beetle	thrips
		vegetable weevil
cucumber beetle	leafhoppers	whiteflies

Remarks: Broccoli, Brussels sprouts, cabbage, cauliflower, kohlrabi, rutabagas, turnips. Apply in sufficient water for thorough coverage; repeat as needed. Apply Diatect as dust or wettable powder.

Days To Harvest: 0.

- - - - - - - - - - - - - - - - -

Signal Word/Toxicity Class: Caution.
REI: 12 hr.

Py-Rin — see pyrethrins

SAF-T-SIDE — see petroleum oil

SCOUT X-TRA

Synthetic pyrethroid insecticide formulated as an emulsifiable concentrate. Scout X-TRA Gel is packaged in water-soluble bags. Low toxicity to humans and other mammals, highly toxic to bees. Restricted Use Pesticide due to toxicity to fish and aquatic organisms.

Scout X-TRA (0.9EC) or Gel (WSB) (tralomethrin) *(AgrEvo USA)*

Rate: 0.016-0.024 lb. ai/A
2.28-3.41 fl. oz. 0.9EC/A
1 WSB/12.5-8.3 A

◆ - **new product** • ai-active ingredient • DF-dry flowable • E/EC-emulsifiable concentrate • EW-emulsion in water
F/FL-flowable • G-granule • L/LC-liquid • SP/WSP-soluble packet • W/WP-wettable powder • WSB-water soluble bag

Pest:

alfalfa looper
aphids (suppression)
beet armyworm
cabbage looper
corn earworm
cutworms
diamondback moth
European corn borer
fall armyworm
flea beetle
grasshoppers
green cloverworm
imported cabbageworm
leafhoppers
saltmarsh caterpillar
stink bug

Remarks: Broccoli. Uniform coverage essential. Refer to label for spray drift precautions and state restrictions on minimum spray carrier volumes. Do not apply more than 0.19 lb. ai/per growing season.

Days To Harvest: 5.

- - - - - - - - - - - - - - - - -

Signal Word/Toxicity Class: Danger/I.
REI: 24 hr.

Sevin — see carbaryl

Sniper — see azinphos-methyl

SULFUR

Microthiol Special and Thiolux are micronized sulfurs.

◆ **Kumulus DF** *(BASF)*

Rate: 2.4-4 lb. ai/A
3-5 lb. DF/A

Pest:
red spider mite

Remarks: Broccoli, Brussels sprouts, cabbage, cauliflower, rutabagas, turnips. Apply at early leaf stage; repeat at 10- to 14-day intervals with or without other pesticides immediately after a rain. Do not apply if temperatures during or within 3 days after application are expected to exceed 90°F. Do not apply through any type of irrigation system.

- - - - - - - - - - - - - - - - -

Signal Word/Toxicity Class: Caution.
REI: 24 hr.

Liquid Sulfur Six *(Helena)*

Rate: 3-4 pt./8 gal. water

Pest:
red spider mite

Remarks: Apply by air. Do not apply this product through any type of irrigation system. Refer to label for use directions.

- - - - - - - - - - - - - - - - -

Signal Word/Toxicity Class: Caution.
REI: 24 hr.

Microthiol Special (80W) *(Elf Atochem)*

Rate: 2.4-8 lb. ai/A
3-10 lb. 80W/A
California—rutabagas, turnips:
4 lb. ai/A
5 lb. 80W/A

Pest:
red spider mite

Remarks: Broccoli, Brussels sprouts, cabbage, cauliflower, rutabagas, turnips. Apply early leaf stage; repeat at 10- to 14-day intervals or as needed. Do not use chemigation in California.

- - - - - - - - - - - - - - - - -

Signal Word/Toxicity Class: Caution.
REI: 24 hr.

Thiolux (DF) *(Platte)*

Rate: Broccoli, Brussels sprouts, cabbage:
2.4-4 lb. ai/A
3-5 lb. DF/A
Rutabagas, turnips:
2.4-8 lb. ai/A
3-10 lb. DF/A

Pest:
red spider mite

Remarks: Broccoli, Brussels sprouts, cabbage, rutabagas, turnips. Apply early leaf stage; repeat at 10- to 14-day intervals as needed.

- - - - - - - - - - - - - - - - -

Signal Word/Toxicity Class: Caution.
REI: 24 hr.

SULFUR (Premix)

Miticide/fungicide. Toxic to fish, aquatic invertebrates, and marine/estuarine organisms.

Bravo S (27.25% sulfur & 19.15% chlorothalonil) *(Helena)*

Rate: 5.5 pt./A

Pest:
red spider mite

Remarks: Broccoli, Brussels sprouts, cabbage, cauliflower. Apply in 8 gal. by air.

- - - - - - - - - - - - - - - - -

Signal Word/Toxicity Class: Caution.
REI: 48 hr.

Sumulsion — see petroleum oil

SunSpray — see petroleum oil

Thiodan — see endosulfan

Thiolux — see sulfur

THIRETHRIN (Premix)

Broad-spectrum insecticide/aphicide. Contains 0.45 lb. piperonyl butoxide. Toxic to bees, fish, shrimp, crab, birds, and other wildlife.

Thirethrin (2 lb./gal. endosulfan & 0.045 lb./gal. pyrethrins) *(Terra)*

Rate: 1.5-2 qt./A

Pest:

armyworms
cabbage aphid
cabbage looper
cross-striped cabbageworm
cutworms
diamondback moth (larvae)
flea beetles
harlequin bugs
imported cabbageworm
stink bugs

Remarks: Broccoli, Brussels sprouts, cabbage, cauliflower.

Days To Harvest: Broccoli, cabbage-7; Brussels sprouts, cauliflower-14.

- - - - - - - - - - - - - - - - -

Signal Word/Toxicity Class: Warning/II.
REI: 24 hr.

Warrior — see *lambda*-cyhalothrin

Cucurbits

(Cucumbers, Melons, Pumpkins, Squash)

Adios — see carbaryl

Ambush — see permethrin

ASANA

Pyrethroid insecticide that has both contact and stomach poison activity. Restricted Use Pesticide due to toxicity to fish and aquatic organisms.

Asana XL (esfenvalerate) *(DuPont)*

Rate: 0.03-0.05 lb. ai/A
5.8-9.6 fl. oz. XL/A

Pest:

cabbage looper
corn earworm
cucumber beetle (adult)
cutworms (seedling spray)
grasshoppers
leafhoppers
pickleworm
plant bugs
rindworm
squash bug
squash vine borer

For personal protective equipment and EPA registration numbers, please turn to page 39.

Remarks: Cucumbers, melons (cantaloupe, honeydew, muskmelon, watermelon), pumpkins, squash (summer, winter). Repeat as necessary to maintain control. Do not exceed 0.25 lb. ai/A per season.
Days To Harvest: 3.

Signal Word/Toxicity Class: Warning/II.
REI: 12 hr.

AZADIRACHTIN

Extracted from seed of neem tree. Insect growth regulator that controls targeted insect larvae when they ingest or come in contact with it, by interfering with the insect's ability to molt and shutting off feeding behavior. Affected larvae slowly starve and die at next molt.

◆ **Azatin XL Plus** *(Thermo Trilogy)*
Rate: 5-21 oz. XL Plus/A
Pest:

aphids	flea beetles	Mexican bean beetle
beet armyworm	green peach aphid	potato aphid
beetles	imported cabbageworm	psyllids
cabbage looper	Japanese beetle	serpentine leafminer
caterpillars	leafhoppers	thrips
Colorado potato beetle	leafminers	tobacco budworm
cutworms	leafrollers	webworms
diamondback moth	loopers	whiteflies
fall armyworm	melon aphid	yellowstriped armyworm

Remarks: Cucumbers, gherkins, gourds, melons (cantaloupe, casaba, crenshaw, honeydew, honeyball, watermelon), pumpkins, squash. Rate varies by species. Apply when pests first appear; repeat at 7-day intervals. When pest populations are high, use higher rates. Do not apply more that 21 fl. oz./A.
Days To Harvest: 0.

Signal Word/Toxicity Class: Caution.
REI. 4 hr.

◆ **NEEMAZAD 4.5EC or NEEMIX 4.5** *(Thermo Trilogy)*
Rate: 0.125-1 pt. 4.5/A
High Volume:
2.25-4.5 oz. 4.5EC/100 gal. water
Low Volume:
0.5 pt. 4.5EC/A
Pest:

budworm	melonworm	squash bug
fall armyworm	pickleworm	squash vine borer
greenhouse whitefly	rindworms	sweetpotato whitefly
leafminers		

Remarks: Cucumbers, gherkins, gourds, melons (cantaloupe, honeydew, watermelon), pumpkins, squash. For low volume, apply in sufficient water for adequate coverage. For low insect pressure, apply preventative at 7-day intervals or at first sign of insects; high insect pressure at 3- to 4-day intervals.

Signal Word/Toxicity Class: Warning/II.
REI: 12 hr.

Azatin — see azadirachtin

AZINPHOS-METHYL

Restricted Use Pesticide (except Guthion Solupak).

Gowan Azinphos-M 50-W *(Gowan)*
◆ **Micro Flo Azinphosmethyl 2EC or 50W** *(Micro Flo)*
Guthion 2L, Solupak (50WSP) *(Bayer)*
Sniper 2-E, 50PVA, 50W *(Platte)*
Rate: Cucumbers:
0.5 lb. ai/A
2 pt. 2EC, 2-E, 2L/A
1 lb. 50-W, 50PVA, 50W, 50WSP/A
Pest:

spotted cucumber beetle	striped cucumber beetle	western striped cucumber beetle

Rate: Melons:
0.375-0.5 lb. ai/A
1.5-2 pt. 2EC, 2-E, 2L/A
0.75-1 lb.50-W, 50PVA, 50W, 50WSP/A
Pest:

leafhoppers	spotted cucumber beetle	western striped cucumber beetle
leafminers	striped cucumber beetle	
rindworms		

Remarks: Cucumbers, melons (cantaloupe, honeydew, muskmelon, watermelon). Apply by air or ground in sufficient water for complete coverage. Do not apply more than 3 times per season. Allow 7 days between applications for cucumbers; 5 days for melons.
Days To Harvest: Cucumbers-1; melons-7.
State Restrictions: California labels may vary. Sniper 50PVA not registered in California.

Signal Word/Toxicity Class: Danger/I.
REI: 48 hr.; 72 hr. if average rainfall less than 25" per year.

CARBARYL

Carbamate insecticide with broad activity on many insects. Highly toxic to bees.

Adios (13%) *(Micro Flo)*
Rate: 0.065-0.1 lb. ai/A
0.5-0.75 lb. 13%/A
Pest:
cucumber beetle
Remarks: Cucumbers, melons (cantaloupe, crenshaw, honeydew, muskmelon, watermelon), pumpkins, squash. Do not apply through any type of irrigation system. Do not apply more than 3.75 lb./A per season.
Days To Harvest: 0.

Signal Word/Toxicity Class: Caution/III.
REI: 12 hr.

Clean Crop Sevin 5 Bait *(Platte)*
Rate: 1 lb. ai /A
20 lb. Bait/A
Pest:

ants	darkling beetle	grasshoppers
armyworms	field crickets	sowbugs
cutworms		

Remarks: Cucumbers, melons, pumpkins, squash. Apply by ground or air.
Days To Harvest: 0.

Signal Word/Toxicity Class: Caution/III.
REI: 12 hr.

Drexel Carbaryl 4L or 50W *(Drexel)*
Sevin 4F, XLR Plus 80S, 80WSP *(Rhone-Poulenc)*
Rate: 0.5-1.5 lb. ai/A
1-3 pt. 4L, 4F, XLR/A
1-3 lb. 50W/A
0.66-1.875 lb. 80S, 80WSP/A
Pest:

cucumber beetle	leafhoppers	pickleworm
flea beetles	melonworm	squash bug
grasshoppers		

Remarks: Cucumbers, melons, pumpkins, squash. Rate varies by pest species. Do not apply to wet foliage or when rain or high humidity is expected during next 2 days.
Days To Harvest: 0.

Signal Word/Toxicity Class: Caution (4F, 4L, XLR); Warning/II (50W, 80S, 80WSP).
REI: 12 hr.

Gowan Sevin Bait *(Gowan)*
Rate: 1 lb. ai/A
20 lb. Bait/A
Pest:

armyworm	cutworms	grasshoppers
crickets	darkling beetle	sowbugs

Remarks: Cucumbers, melons, squash. Apply when pests appear in damaging numbers; repeat at 7- to 14-day intervals if necessary.
Days To Harvest: 1.
State Restrictions: For use in Arizona and California.

Signal Word/Toxicity Class: Caution/III.
REI: 12 hr.

◆ - **new product** • ai-active ingredient • DF-dry flowable • E/EC-emulsifiable concentrate • EW-emulsion in water
F/FL-flowable • G-granule • L/LC-liquid • SP/WSP-soluble packet • W/WP-wettable powder • WSB-water soluble bag

◆ **Wilbur-Ellis Sevin 5 Dust** *(Wilbur-Ellis)*
Rate: 2 lb. ai/A
40 lb. 5/A
Pest:
cucumber beetle | leafhoppers | pickleworm
flea beetles | melonworm | squash bug
Remarks: Cucumbers, summer squash. Do not feed treated vines to livestock.
Days To Harvest: 0.
State Restrictions: Not registered in California.

Signal Word/Toxicity Class: Caution.
REI: 12 hr.

CHLORPYRIFOS

For use only by commercial seed treaters.

Lorsban 30, 50-SL *(Gustafson)*
Rate: 2.7 fl. oz. 30/100 lb. seed
2 oz. 50-SL/100 lb. seed
Pest:
seedcorn beetles | seedcorn maggots
Remarks: Cucumbers, pumpkins. Apply prior to planting as slurry treatment. Treated seed must not be used for or mixed with food or animal feed or processed for oil. Commercially treated seed must be labeled in accordance with all applicable requirements of Federal Seed Act. Do not allow livestock to graze or feed bean hay or pea hay grown from treated seed.

Signal Word/Toxicity Class: Warning/II (50-SL); Caution (30).
REI: 24 hr.

CRYOLITE

Kryocide (96%) *(Elf Atochem)*
Rate: Cucumbers:
7.7-11.5 lb. ai/A
8-12 lb. 96%/A
Melons, squash:
7.7-15.4 lb. ai/A
8-16 lb. 96%/A
Pest:
cabbage looper | Diabrotica beetles | melonworm
cucumber beetle | flea beetles | pickleworm
Remarks: Cucumbers, melons (cantaloupe, watermelon), squash. For cucumbers, apply at 10-day intervals as needed. Do not exceed 48 lb./A per season. For melons and squash, apply at 7-day intervals as needed. Do not exceed 64 lb./A per season. Remove excess residues on edible portions by washing, brushing, or other effective means. Do not use in combination with lime or compounds containing free lime. Do not use chemigation in California.
Days To Harvest: 14.

Signal Word/Toxicity Class: Caution.
REI: 12 hr.

◆ **Prokil Cryolite 96** *(Gowan)*
Rate: 9.6-15.4 lb. ai/A
108-16 lb. 96/A
Pest:
cabbage looper | Diabrotica beetles | melonworm
Colorado potato beetle | flea beetles | pickleworm
cucumber beetle
Remarks: Melons (cantaloupe, watermelon), squash (summer, winter). Apply at 7-day intervals as needed. Do not apply more than 80 lb./A per season (64 lb. for summer squash). Do not use in combination with lime or compounds containing free lime. Do not apply through any type of irrigation system.
Days To Harvest: 14; summer squash-7.

Signal Word/Toxicity Class: Caution.
REI: 12 hr.

DEADLINE

Packaged as Bullets, M-Ps (mini-pellets), and granules.

Deadline Bullets, M-Ps, Granules (4%) (metaldehyde) *(Pace)*
Rate: 0.8-1 lb. ai/A
20-40 lb. 4%/A
Pest:
slugs | snails
Remarks: Cucumbers, melons (cantaloupe, watermelon), pumpkins, squash. Apply every 3-4 weeks during growing season, or as needed. Do not apply directly to edible portion of plant.

Signal Word/Toxicity Class: Warning/II (granule); Caution (bullets, M-P).
REI: 12 hr.

Diatect — see pyrethrins

DIAZINON

Restricted Use Pesticide due to avian and aquatic toxicity.

D•z•n diazinon AG600 WBC, 50W *(Novartis)*
Drexel Diazinon Insecticide (4EC), D-264 14G *(Drexel)*
Gowan Diazinon 4E, 14G, 50WSB *(Gowan)*
Helena Diazinon AG500, 50W *(Helena)*
Micro Flo Diazinon AG500, 14G, 50W *(Micro Flo)*
Prentox Diazinon AG500, 50W *(Prentiss)*
◆ **Wilbur-Ellis Diazinon 4 (4E), 14G** *(Wilbur-Ellis)*
Rate: Cucumbers:
0.5 lb. ai/A
1 pt. 4E, 4EC, AG500/A
12.75 fl. oz. AG600/A
1 lb. 50W, 50WSB/A
Melons, squash:
0.25-0.75 lb. ai/A
0.5-1.5 pt. 4E, 4EC, AG500/A
1-1.5 lb. 50W, 50WSB/A
Pest:
aphids | melonworm | spider mites (melons, squash)
dipterous leafminer (melons) | serpentine leafminers (cucumber, squash) | striped cucumber beetle
leafhoppers | | thrips

Rate: Cucumbers, melons, squash:
2-4 lb. ai/A
4-8 pt. 4E, 4EC, AG500/A
51-102 fl. oz. AG600/A
4-8 lb. 50W, 50WSB/A
14-28 lb. 14G/A
Pest:
cutworms

Rate: Cucumbers, melons, squash:
3-4 lb. ai/A
6-8 pt. 4E, 4EC, AG500/A
76.5-102 fl. oz. AG600/A
6-8 lb. 50W, 50WSB/A
21-28 lb. 14G/A
Pest:
wireworms
Remarks: Cucumbers, melons (cantaloupe, casaba, crenshaw, honeydew, muskmelon, Persian melons, watermelon), squash (summer, winter). Apply as insects occur; repeat every 7 days as needed. Will not control organophosphate-resistant leafminers. For cutworms and wireworms, broadcast and incorporate into soil just before planting.
Days To Harvest: Melons, winter squash-3; cucumbers, summer squash-7.

Signal Word/Toxicity Class: Caution/III.
REI: 24 hr.; 12 hr. (Drexel/Gowan/Wilbur-Ellis 14G).

DIBROM

Dibrom 8 (8E) (naled) *(Amvac)*
Rate: Melons:
1 lb. ai/A
1 pt. 8E/A
Pest:
aphids | leafhoppers | loopers
armyworms | leafminers

Rate: Squash:
1-2 lb. ai/A
1-2 pt. 8E/A
Pest:
aphids | leafminers | spider mites
armyworms | loopers | thrips
cucumber beetle | pickleworms

NOTICE The information on these pages is for preliminary planning — not a guide for use. Be sure to follow the manufacturer's directions, notwithstanding information contained here. | For personal protective equipment and EPA registration numbers, please turn to page 39.

7 Vegetables

Remarks: Melons (cantaloupe, honeydew, muskmelon, watermelon), summer squash. Apply by air or ground at 7-day spray intervals. For melons, do not apply more than 2 pt./A per season; squash, 6 pt./A per season. Do not apply when temperature exceeds 90°F.
Days To Harvest: 1.

Signal Word/Toxicity Class: Danger/I.
REI: 48 hr. for 1 pt./A or less; 72 hr. for more than 1 pt./A.

DICOFOL

◆ **Griffin Dicofol 4E** *(Griffin)*
Rate: 0.38 lb. ai/A
0.75 pt. 4E/A
Pest:
Pacific mite, tomato russet mite, tropical mite, twospotted spider mite
Remarks: Cucumbers, squash. Apply at first sign of mite buildup. Do not make more than 2 applications. Do not feed crop residues to livestock. Do not apply through any type of irrigation system.
Days To Harvest: 2.

Signal Word/Toxicity Class: Caution.
REI: 12 hr.

Kelthane 35 or 50 *(Rohm and Haas)*
Rate: 0.35-0.6 lb. ai/A
1-1.66 lb. 35/A
0.75 lb. ai/A
1.25 lb. 50/A
Pest:
Pacific mite, tomato russet mite, tropical mite, twospotted spider mite
Remarks: Cucumbers, melons (cantaloupe, watermelon), pumpkins, squash. Add 2 qt. adjuvant/100 gal. spray. Do not make more than 2 applications. Do not feed treated vines, forage, or crop residues to livestock. Do not apply through any type of irrigation system.
Days To Harvest: 2.

Signal Word/Toxicity Class: Warning/II (35); Caution (50).
REI: 12 hr.

Digon — see dimethoate

Dimate — see dimethoate

DIMETHOATE

Cheminova Dimethoate 4E *(Cheminova)*
Clean Crop Dimethoate 400 *(Platte)*
Dimate 4E *(Terra)*
Drexel Dimethoate 2.67 or 4EC *(Drexel)*
Gowan Dimethoate E267 *(Gowan)*
Helena Dimethoate 4E *(Helena)*
Micro Flo Dimethoate 267 or 4E *(Micro Flo)*
Wilbur-Ellis Dimethoate 267 or Digon 400 *(Wilbur-Ellis)*
Rate: **Melons except watermelon:**
0.5 lb. ai/A
1.5 pt. 2.67, 267, E267/A
1 pt. 4E, 4EC, 400/A
Watermelon:
0.25-0.5 lb. ai/A
0.75-1.5 pt. 2.67, 267/A
0.5-1 pt. 4E, 4EC, 400/A
Pest:
aphids, leafhoppers, leafminers, thrips (4E, 4EC, 400) (except watermelon)
Remarks: Melons.
Days To Harvest: 3.

Signal Word/Toxicity Class: Warning/II.
REI: 48 hr.

ENDOSULFAN

Toxic to birds, fish, and wildlife; toxic to bees exposed to direct application.

Drexel Endosulfan 3EC or 50WP *(Drexel)*
Gowan Endosulfan 3EC or 50WSB *(Gowan)*
◆ **Micro Flo Endosulfan 3EC or 50WP** *(Micro Flo)*
Phaser 3EC or 50WSB *(AgrEvo USA)*
Thiodan 3EC or 50WP *(FMC)*
Rate: 0.5-1 lb. ai/A
0.66-1.33 qt. 3EC/A
1-2 lb. 50WP, 50WSB/A
Pest:
aphids, cabbage looper, cucumber beetle, melonworm, omnivorous leafroller, pickleworm, rindworm, squash beetle, squash bug, squash vine borer, striped flea beetle, whiteflies
Remarks: Cucumbers, melons, pumpkins, squash (summer, winter). Apply when insects appear or feeding is noticed. For squash vine borer, apply weekly to flower buds, stems, and vines beginning when moths first appear. Do not exceed 3 lb. ai/A per year or 6 applications per year.
Days To Harvest: 2.

Signal Word/Toxicity Class: Danger-Poison/I.
REI: 24 hr.

Fyfanon — see malathion

Guthion — see azinphos-methyl

Kelthane — see dicofol

Kryocide — see cryolite

LANNATE

Restricted Use Pesticide.

Lannate SP or LV (methomyl) *(DuPont)*
Rate: 0.45-0.09 lb. ai/A
0.5-1 lb. SP/A
1.5-3 pt. LV/A
Pest:
beet armyworm, cucumber beetle, fall armyworm, flea beetles, granulate cutworm, loopers, melon aphids, melonworm, pickleworm, tobacco budworm, variegated cutworm, yellowstriped armyworm
Remarks: Cucumbers, melons, summer squash. Apply in sufficient water to obtain thorough, uniform coverage when insects first appear; repeat at 5- to 7-day intervals or as needed. Do not apply more than 5.4 lb. ai/A per crop or make more than 12 applications per crop.
Days To Harvest: 90: 0.5 lb.-1; over 1 lb.-3; LV: 1.5 pt.-1; over 1.5 pt.-3.

Signal Word/Toxicity Class: Danger-Poison/I.
REI: 48 hr.

LINDANE

Lindane 30C contains toxic inert Rhodamine B. Toxic to fish, birds, and wildlife.

Lindane 30C *(Gustafson)*
◆ **Lindane ST 40** *(AGSCO)*
Rate: 4.9 fl. oz. 30C/100 lb. seed
3.6 fl. oz. 40/100 lb. seed
Pest:
seedcorn maggots, wireworms
Remarks: Cucumbers, melons (cantaloupe, watermelon), squash. Do not use treated seed for food, feed, or oil purposes. Exposed treated seed may be hazardous to birds and other wildlife. Dispose of excess seed by burial away from bodies of water.

Signal Word/Toxicity Class: Warning/II.
REI: 12 hr.

◆ - **new product** • ai-active ingredient • DF-dry flowable • E/EC-emulsifiable concentrate • EW-emulsion in water
F/FL-flowable • G-granule • L/LC-liquid • SP/WSP-soluble packet • W/WP-wettable powder • WSB-water soluble bag

Lorsban — see chlorpyrifos

MALATHION

Organophosphate insecticide toxic to fish, aquatic invertebrates, aquatic life stages of amphibians, and bees exposed to direct treatment on blooming crops or weeds.

◆ **Clean Crop Malathion 8EC** *(Platte)*
◆ **Wilbur-Ellis Malathion 8 Spray** *(Wilbur-Ellis)*

Rate: **Cucumbers:**
1.5 lb. ai/A
1.5 pt. 8, 8EC/A

Pest:
aphids — cucumber beetle — spider mites

Rate: **Squash:**
1.5-1.75 lb. ai/A
1.5-1.75 pt. 8, 8EC/A

Pest:
aphids — leafhoppers — spider mites — cucumber beetle — pickleworm

Remarks: Apply in minimum 5 gal. by air; 10 gal. by ground.
Days To Harvest: 1.

Signal Word/Toxicity Class: Caution/III.
REI: 12 hr.

Drexel Malathion 5EC *(Drexel)*
Fyfanon (5EC) *(Helena)*
◆ **Micro Flo Malathion 5EC** *(Micro Flo)*
Prentox 5 lb. Malathion Spray (5EC) *(Prentiss)*

Rate: 1-2 lb. ai/A
1.5-3 pt. 5EC/A

Pest:
aphids — leafminers — spider mites — cucumber beetle — pickleworm — squash vine borer

Remarks: Cucumbers, melons (cantaloupe, casaba, crenshaw, honeyball, honeydew, muskmelon, Persian, watermelon, etc.), pumpkins, squash. Rate varies by pest species; refer to label. Do not apply unless plants are dry.
Days To Harvest: Cucumbers, melons, squash-1; pumpkins-3.

Signal Word/Toxicity Class: Warning/II.
REI: 12 hr.

Fyfanon (8E) *(Helena)*

Rate: 1.75 lb. ai/A
1.75 pt. 8E/A

Pest:
spider mites — squash vine borer — thrips

Remarks: Cucumbers, squash. Do not apply unless plants are dry.
Days To Harvest: 1

Signal Word/Toxicity Class: Caution/III.
REI: 12 hr.

◆ **Gowan Malathion 8** *(Gowan)*

Rate: 1.75 lb. ai/A
1.75 pt. 8E/A

Pest:
aphids — darkling beetle — spider mites — cucumber beetle — leafhoppers — squash vine borer — cutworms — pickleworm — thrips

Remarks: Cucumbers, melons, pumpkins, squash. For vine borer, apply weekly to vines and stems at base of plant. For pumpkins, apply up to 1.75 pt. Do not apply unless plants are dry.
Days To Harvest: Cucumbers, melons, squash-1; pumpkins-3.

Signal Word/Toxicity Class: Caution/III.
REI: 12 hr.

Marlate — see methoxychlor

METASYSTOX-R

Restricted Use Pesticide.

Metasystox-R (SC) (oxydemeton-methyl) *(Gowan)*

Rate: 0.375-0.5 lb. ai/A
1.5-2 pt. SC/A

Pest:
aphids — mites

Remarks: Cucumbers, melons (cantaloupe, muskmelon, watermelon), pumpkins, squash (summer, winter). Apply in sufficient water for complete coverage but not less than 1 gal./A. Do not apply more than once per season for pumpkins and squash, 2 times for cucumbers and watermelons, and 3 times for other melons.
Days To Harvest: Cucumbers, summer squash-3; watermelons-7; other melons, pumpkins, winter squash-14.

Signal Word/Toxicity Class: Warning/II.
REI: 48 hr.; 72 hr. if average rainfall less than 25" per year.

METHOXYCHLOR

Organochlorine pesticide toxic to fish and aquatic invertebrates.

Marlate 50 (WP) *(Kincaid)*

Rate: 2 Tbs. 50WP/gal. water

Pest:
cucumber beetle — flea beetles — squash vine borer — fall armyworm

Remarks: Cucumbers, melons, pumpkins, squash (including summer). Apply at first sign of infestation; repeat at 7- to 14-day intervals as needed.
Days To Harvest: 7.

Signal Word/Toxicity Class: Caution.
REI: 12 hr.

Prentox Methoxychlor 2EC *(Prentiss)*

Rate: 1.3-1.7 lb. ai/A
75-100 fl. oz. 2EC/A

Pest:
cucumber beetle — flea beetles — squash vine borer — fall armyworm

Remarks: Cucumbers, melons, pumpkins, squash. Apply at first sign of infestation; repeat at 7- to 14-day intervals as needed.
Days To Harvest: Cucumbers, pumpkins, squash-1; melons-7.

Signal Word/Toxicity Class: Caution.
REI: 12 hr.

MOCAP

Mocap EC is a Restricted Use Pesticide.

Mocap EC or 10%G (ethoprop) *(Rhone-Poulenc)*

Rate: Row treatment:
2 lb. ai/A
1.33 qt. EC/A (6.8 fl. oz./1000 lin. ft. row)
20 lb. 10%G/A (3.2 lb./1000 lin. ft. row)

Pest:
nematodes

Remarks: Cucumbers. Apply in a 12-15" band (7 ft. spacing) at or just before planting. Prepare bed and knock off top to provide level area for treatment; mix with top 2-4" of soil right after application. Do not use as seed furrow treatment or allow spray to contact seed.
State Restrictions: Do not apply in Long Island, New York.

Signal Word/Toxicity Class: Danger-Poison/I (EC); Warning/II (10%G).
REI: 48 hr.; 72 hr. if average rainfall less than 25" per year.

MORESTAN

Morestan Solupak 25 (oxythioquinox) *(Bayer)*

Rate: 0.375 lb. ai/A
1.5 lb. 25/A

Pest:
mite eggs — mites

Remarks: Seed crop. Refer to label for use directions and precautions.
State Restrictions: SLN-California.

Signal Word/Toxicity Class: Warning/II.
REI: 24 hr.

7 Vegetables

For personal protective equipment and EPA registration numbers, please turn to page 39.

M-PEDE

M-Pede (potassium salts of fatty acids) *(Mycogen)*

Rate: 1-2% v/v solution

Pest:

aphids*	leafminers*	spider mites
broad mite	plant bugs	thrips
green peach aphid*	russet mite	whiteflies
leafhoppers		

** Tank mix only*

Remarks: Cucumbers, melons, pumpkins, summer squash. Apply at first sign of infestation; repeat at 7- to 14-day intervals. May be tank mixed with other insecticides. Refer to labels for use directions and precautions.

Days To Harvest: 0.

Signal Word/Toxicity Class: Warning/II.

REI: 12 hr.

NEEMAZAD — see azadirachtin

NEEMIX — see azadirachtin

Omni Supreme Spray — see petroleum oil

PERMETHRIN

Restricted Use Pesticide.

Ambush 2EC or 25W *(ZENECA)*

Rate: 0.1-0.2 lb. ai/A
6.4-12.8 fl. oz. 2EC/A
6.4-12.8 oz. 25W/A

Pest:

cabbage looper	Lygus bugs	rindworms
cucumber beetle (adult)	melonworm	squash bug
cutworms	pickleworm	squash vine borer
leafminers	plant bugs	stink bugs

Remarks: Cucumbers, melons (including watermelon), pumpkins, squash. Do not apply more than 1.6 lb. ai/A per season.

Days To Harvest: 0.

Signal Word/Toxicity Class: Warning/II.

Pounce 3.2 EC, 25WP or WSB *(FMC)*

Rate: 0.1-0.2 lb. ai/A
4-8 oz. 3.2 EC/A
6.4-12.8 oz. 25WP/A
1-2 bags WSB/A

Pest:

cabbage looper	Lygus bugs	rindworms
cucumber beetle	melonworm	squash bug
cutworms	pickleworm	squash vine borer
leafminers	plant bugs	stink bugs

Remarks: Melons (including watermelon), pumpkins, squash. Apply in a minimum 20 gal. finished spray by ground; 4 gal. by air. Rate varies by pest species. Do not apply more than 1.6 lb. ai/A per season.

Days To Harvest: 0.

Signal Word/Toxicity Class: Caution (3.2 EC); Warning/II (25WP, WSB).

REI: 24 hr.

PETROLEUM OIL

Petroleum based spray oil. Mode of action is through suffocation of eggs, larvae, and nymphs of insects and adults of soft bodied insects. This mode of action necessitates total spray coverage. Toxic to fish.

Omni Supreme Spray *(Helena)*

Rate: 1-2 gal./100 gal. water

Pest:

aphids	caterpillars	mites
beetles (larvae)	leafminers	

Remarks: None.

Signal Word/Toxicity Class: Caution.

REI: 4 hr.

SAF-T-SIDE *(Brandt Consolidated)*
◆ **Sumulsion** *(Drexel)*
◆ **SunSpray Ultra-Fine** *(Sun)*

Rate: 1-2 gal./100 gal. water

Pest:

aphids	leafminers	thrips
beetles (larvae)	mites	whiteflies
leafhoppers	spiders	

Remarks: Cucumbers, melons, pumpkins, squash. Apply as needed. May be tank mixed with most commonly used insecticides and fungicides. Do not use in combination with or immediately before or after spraying with fungicides such as captan, Dyrene, folpet, Morestan, or any sulfur products. Do not use with carbaryl or dimethoate.

Signal Word/Toxicity Class: Caution.

REI: 4 hr.

Phaser — see endosulfan

Pounce — see permethrin

PYRELLIN (Premix)

Premix containing pyrethrins and rotenone. May be combined with other insecticides to enhance knockdown or as an insect "exciter" to help facilitate contact with other spray residues.

Pyrellin E.C. (0.042 lb./gal. pyrethrins & 0.035 lb./gal. rotenone) *(Webb Wright)*

Rate: 1-2 pt. E.C./A

Pest:

aphids	harlequin bug	mites
bean beetle	Japanese beetle	plant bugs
beet webworm	leafhoppers	stink bugs
cabbageworms	leafminers	thrips
celery leaftier	loopers	whiteflies
garden fleahopper	Lygus bugs	

Remarks: Cucumbers, melons (cantaloupe, casaba, crenshaw, honeydew, muskmelon, Persian, watermelon), pumpkins, squash (summer, winter, acorn, butternut, zucchini). For cucumbers and squash, apply at 7-day intervals or less; repeat as needed. For melons and pumpkins, apply as necessary to maintain control

Days To Harvest: 0.

Rate: 1.5-2 pt. E.C./A

Pest:

Colorado potato beetle	cucumber beetle	flea beetles
	European corn borer	vegetable weevil

Remarks: Cucumbers, squash (summer, winter, acorn, butternut, zucchini). Make 3-4 applications or more as eggs start to hatch; repeat at 5- to 7-day intervals.

Days To Harvest: 0.

Signal Word/Toxicity Class: Caution/III.

REI: 12 hr.

Pyrenone — see pyrethrins

PYRETHRINS

Contains piperonyl butoxide as a synergist. Diatect also contains silicon dioxide (diatomaceous earth). May be combined in tank mix with other insecticides for quicker and more complete kill where insect resistance may be a problem, and as an "exciter" to flush insects out of hiding and into contact with other spray residues.

Diatect Multipurpose Insecticide II *(Diatect)*
Pyrenone (0.5EC) *(AgrEvo Environmental Health)*
◆ **Py-Rin 60-6 EC** *(Wilbur-Ellis)*

Rate: 1-6 lb. II/A
1-12 fl. oz. 0.5EC/A
2-16 oz. 6-60 EC/A

◆ - **new product** • ai-active ingredient • DF-dry flowable • E/EC-emulsifiable concentrate • EW-emulsion in water
F/FL-flowable • G-granule • L/LC-liquid • SP/WSP-soluble packet • W/WP-wettable powder • WSB-water soluble bag

Pest: Including but not limited to:

aphids	harlequin bug	plant bugs
cabbage looper	Japanese beetle	spotted cucumber beetle
cabbageworms	leafhoppers	
caterpillars	leafminers	squash vine borer
cucumber beetle	loopers	stink bugs
diamondback moth	Lygus bugs	thrips
flea beetles	Mexican bean beetle	vegetable weevil
garden fleahoppers	mites	whiteflies

Remarks: Cucumbers, melons (including watermelon), pumpkins, squash. Apply in sufficient water for thorough coverage; repeat as needed. Apply Diatect as dust or wettable powder.

Days To Harvest: 0.

- - - - - - - - - - - - - - - -

Signal Word/Toxicity Class: Caution.
REI: 12 hr.

Py-Rin — see pyrethrins

SAF-T-SIDE — see petroleum oil

Sevin — see carbaryl

Sniper — see azinphos-methyl

Sumulsion — see petroleum oil

SunSpray — see petroleum oil

Thiodan — see endosulfan

THIRETHRIN (Premix)

Broad-spectrum insecticide/aphicide. Contains 0.45 lb. piperonyl butoxide. Toxic to bees, fish, shrimp, crab, birds, and other wildlife.

Thirethrin (2 lb./gal. endosulfan & 0.045 lb./gal. pyrethrins) *(Terra)*

Rate: 1-2 qt./A

Pest:

aphids	omnivorous leafroller	squash beetle
cabbage looper	pickleworm	squash bug
cucumber beetle	rindworm (watermelons)	squash vine borer
melonworm		striped flea beetle

Remarks: Cucumber, melons (cantaloupe, muskmelon, watermelon), pumpkins, squash (summer, winter). For squash vine borers, apply weekly to flower buds, stems, and vines when moths first appear.

Days To Harvest: 0.

- - - - - - - - - - - - - - - -

Signal Word/Toxicity Class: Warning/II.
REI: 24 hr.

TRIGARD

Triazine compound used as insect growth regulator for control of leafminers. Ineffective against other insect species. Toxic to bees.

Trigard (WSP) (cyromazine) *(Novartis)*

Rate: 0.125 lb. ai/A
0.167 lb. WSP/A (1 packet)

Pest:
leafminers

Remarks: Cucumbers, melons (including cantaloupe, casaba, crenshaw, honeydew, muskmelon, Persian, watermelon), pumpkins, squash (summer, winter). Apply when leafminers first appear as foliar spray in minimum 5 gal. water by air; 50 gal. water by ground. Repeat at 7-day intervals or as necessary. Do not make more than 6 applications per growing season. Do not apply through any type of irrigation system.

Days To Harvest: 0.

Crop Rotation: All crops on label may be planted following harvest of treated crop. Do not rotate to any other crop except sweet corn or radishes. Do not plant sweet corn or radishes within 3 months of last application.

- - - - - - - - - - - - - - - -

Signal Word/Toxicity Class: Caution.
REI: 12 hr.

VYDATE

Restricted Use Pesticide.

Vydate L (oxamyl) *(DuPont)*

Rate: Preplant and planting:
2-4 lb. ai/A
8-16 pt. L/A
Foliar:
0.5-1 lb. ai/A
2-4 pt. L/A

Pest:

aphids	ring nematode	sting nematode
leafminers	root-knot nematode (except javanese)	stunt nematode
lesion nematode		

Remarks: Cucumbers, melons (cantaloupe, honeydew, watermelon), pumpkins, squash. Preplant, planting treatment: Apply broadcast (band treatment use proportionately less). Following application and before planting, incorporate 2-4" into soil. Foliar treatment: Apply in sufficient water to obtain uniform coverage of foliage. Make first application 2-4 weeks after planting and repeat 2-3 weeks later. Use low rate for light infestations. Where leafminer infestations occur annually, initiate treatment schedule 2-4 weeks after planting and repeat weekly or as needed. Do not apply more than 24 pt./A per season.

Days To Harvest: 1.

State Restrictions: Do not use in Suffolk and Nassau counties, Long Island, New York.

- - - - - - - - - - - - - - - -

Signal Word/Toxicity Class: Danger-Poison/I.
REI: 48 hr.

Eggplant

Adios — see carbaryl

Admire — see imidacloprid

Ambush — see permethrin

ASANA

Pyrethroid insecticide that has both contact and stomach poison activity. Restricted Use Pesticide due to toxicity to fish and aquatic organisms.

Asana XL (esfenvalerate) *(DuPont)*

Rate: 0.03-0.05 lb. ai/A
5.8-9.6 fl. oz. XL/A

Pest:

Colorado potato beetle	corn earworm	flea beetles
	European corn borer	loopers

Remarks: Apply when insects or their damage are observed and repeat at 7- to 10-day intervals. Do not exceed 0.35 lb. ai/A per season.

Days To Harvest: 7.

- - - - - - - - - - - - - - - -

Signal Word/Toxicity Class: Warning/II.
REI: 12 hr.

AZADIRACHTIN

Extracted from seed of neem tree. Insect growth regulator that controls targeted insect larvae when they ingest or come in contact with it, by interfering with the insect's ability to molt and shutting off feeding behavior. Affected larvae slowly starve and die at next molt.

◆ **Azatin EC or XL** *(Olympic)*
◆ **Azatin XL Plus** *(Thermo Trilogy)*

Rate: 8-21 oz. EC/A
5-21 oz. XL, XL Plus/A

7 Vegetables

NOTICE The information on these pages is for preliminary planning — not a guide for use. Be sure to follow the manufacturer's directions, notwithstanding information contained here.

For personal protective equipment and EPA registration numbers, please turn to page 39.

Pest:

aphids	flea beetles	Mexican bean beetle
beet armyworm	green peach aphid	psyllids
beetles	imported cabbageworm	silverleaf whitefly
cabbage looper	Japanese beetle	sweetpotato whitefly
caterpillars	leafhoppers	thrips
Colorado potato beetle	leafminers	tomato fruitworm
cutworms	leafrollers	tomato pinworm
diamondback moth	loopers	webworms
fall armyworm		whiteflies

Remarks: Rate varies by species. Apply when pests first appear; repeat at 7-day intervals. When pest populations are high, use higher rates. Do not apply more that 21 fl. oz./A.
Days To Harvest: 0.

Signal Word/Toxicity Class: Caution (XL); Warning/II (EC).
REI: 4 hr. (XL), 12 hr. (EC).

◆ **NEEMAZAD 4.5EC or NEEMIX 4.5** *(Thermo Trilogy)*
Rate: 0.125-1 pt. 4.5/A
High Volume:
2.25-4.5 oz. 4.5EC/100 gal. water
Low Volume:
0.5 pt. 4.5EC/A
Pest:

armyworms	garden webworm	psyllids
bean leaf beetle	greenhouse whitefly	soybean loopers
cabbage looper	hornworms	sweetpotato whitefly
caterpillars	leafminers	tomato fruitworm
Colorado potato beetle	loopers	tomato pinworm
cutworms	Mexican bean beetle	weevils
	pepper weevil	

Remarks: For low volume, apply in sufficient water for adequate coverage. For low insect pressure, apply preventive at 7-day intervals or at first sign of insects; high insect pressure, at 3- to 4-day intervals.

Signal Word/Toxicity Class: Warning/II.
REI: 12 hr.

Azatin — see azadirachtin

AZINPHOS-METHYL

Restricted Use Pesticide.

Gowan Azinphos-M 50-W *(Gowan)*
◆ **Micro Flo Azinphosmethyl 2EC or 50W** *(Micro Flo)*
Sniper 2-E, 50PVA, 50W *(Platte)*
Rate: 0.375-0.5 lb. ai/A
1.5-2 pt. 2EC, 2-E/A
0.75-1 lb. 50-W, 50PVA, 50W/A
Pest:
European corn borer, flea beetles, leafminers
Remarks: Apply at 7-day intervals in sufficient water to give complete coverage; repeat as needed. Do not apply after fruit set. Except 2-E: Do not apply more than 3 times per crop season.
State Restrictions: Sniper 50PVA not registered in California.

Signal Word/Toxicity Class: Danger-Poison/I.
REI: 48 hr.; 72 hr. if average rainfall less than 25" per year.

CARBARYL

Carbamate insecticide with broad activity on many insects. Highly toxic to bees.

Adios (13%) *(Micro Flo)*
Rate: 0.065-0.1 lb. ai/A
0.5-0.75 lb. 13%/A
Pest:
cucumber beetle
Remarks: Do not apply through any type of irrigation system. Do not apply more than 3.75 lb./A per season.
Days To Harvest: 0.
State Restrictions: Not for use in California.

Signal Word/Toxicity Class: Caution/III.
REI: 12 hr.

Clean Crop Sevin 5 Bait *(Platte)*
Rate: 1-2 lb. ai/A
20-40 lb. Bait/A
Pest:

ants	darkling beetle	grasshoppers
armyworms	field crickets	sowbugs
cutworms	flea beetles	

Remarks: Apply with ground or air equipment.
Days To Harvest: 0.

Signal Word/Toxicity Class: Caution/III.
REI: 12 hr.

Drexel Carbaryl 4L or 50W *(Drexel)*
Sevin 4F, XLR Plus, 80S, 80WSP *(Rhone-Poulenc)*
Rate: 0.5-2 lb. ai/A
1-4 pt. 4L, 4F, XLR/A
1-4 lb. 50W/A
0.66-2.5 lb. 80S, 80WSP/A
Pest:

Colorado potato beetle	flea beetles	tarnished plant bug
cutworms	grasshoppers	tomato fruitworm
European corn borer	lace bug	tomato hornworm
fall armyworm	leafhoppers	tomato pinworm
	stink bugs	

Remarks: Use higher rate for mature grasshoppers or dense vegetation.
Days To Harvest: 0.

Signal Word/Toxicity Class: Caution (4F, 4L, XLR); Warning/II (50W, 80S, 80WSP).
REI: 12 hr.

Gowan Sevin Bait *(Gowan)*
Rate: 2 lb. ai/A
40 lb. Bait/A
Pest:

armyworms	cutworms	grasshoppers
crickets	darkling beetle	sowbugs

Remarks: Apply when pests appear in damaging numbers; repeat at 7- to 14-day intervals if necessary. Do not use when bees are present.
Days To Harvest: 1.
State Restrictions: For use in Arizona and California.

Signal Word/Toxicity Class: Caution/III.
REI: 12 hr.

◆ **Wilbur-Ellis Sevin 5 or 10 Dust** *(Wilbur-Ellis)*
Rate: 2 lb. ai/A
40 lb. 5/A
20 lb. 10/A
Pest:

Colorado potato beetle	flea beetles	stink bugs
European corn borer	leafhoppers	tomato fruitworm
fall armyworm	Lygus bugs	tomato hornworm

Remarks: Do not feed treated vines to livestock.
Days To Harvest: 0.
State Restrictions: Sevin 5 not registered in California.

Signal Word/Toxicity Class: Caution.
REI: 12 hr.

DEADLINE

Packaged as Bullets, M-Ps (mini-pellets), and granules.

Deadline Bullets, M-Ps, Granules (4%) (metaldehyde) *(Pace)*
Rate: 0.8-1 lb. ai/A
20-40 lb. 4%/A
Pest:
slugs, snails
Remarks: Apply every 3-4 weeks during growing season or as needed. Do not apply directly to edible portion of plant.

Signal Word/Toxicity Class: Warning/II (granule); Caution (bullets, M-P).
REI: 12 hr.

Diatect — see pyrethrins

◆ - **new product** • ai-active ingredient • DF-dry flowable • E/EC-emulsifiable concentrate • EW-emulsion in water
F/FL-flowable • G-granule • L/LC-liquid • SP/WSP-soluble packet • W/WP-wettable powder • WSB-water soluble bag

DIBROM

Dibrom 8 (8E) (naled) *(Amvac)*

Rate: 1-2 lb. ai/A
1-2 pt. 8E/A

Pest:
aphids, blister beetle, flea beetles, leafminers, spider mites

Remarks: Apply by air or ground at 7-day spray intervals in minimum 20 gal. water by ground/A; 10 gal. water by air/A. Do not apply more than 6 pt./A per season. Do not apply more than 1 pt./A in Florida. Do not apply when temperatures exceed 90°F.

Days To Harvest: 1.

Signal Word/Toxicity Class: Danger/I.
REI: 48 hr. for 1 pt./A or less; 72 hr. for more than 1 pt./A.

ENDOSULFAN

Toxic to birds, fish, and wildlife; toxic to bees exposed to direct application.

Drexel Endosulfan 3EC or 50WP *(Drexel)*
Gowan Endosulfan 3EC or 50WSB *(Gowan)*
◆ **Micro Flo Endosulfan 3EC or 50WP** *(Micro Flo)*
Phaser 3EC or 50WSB *(AgrEvo USA)*
Thiodan 3EC or 50WP *(FMC)*

Rate: 0.5-1 lb. ai/A
0.66-1.33 qt. 3EC/A
1-2 lb. 50WP, 50WSB/A

Pest:
blister beetle, Colorado potato beetle, flea beetle, green peach aphid, green stink bug, whiteflies

Remarks: Apply when insects appear or feeding is noticed. Do not exceed 2 applications per year or 0.5 lb. ai/A per year.

Days To Harvest: 1.

Signal Word/Toxicity Class: Danger-Poison/I.
REI: 24 hr.

Fyfanon — see malathion

IMIDACLOPRID

Admire 2F *(Bayer)*

Rate: 0.25-0.375 lb. ai/A
16-24 fl. oz. 2F/A

Pest:
aphids, Colorado potato beetle, flea beetles, silverleaf whitefly, sweetpotato whitefly, thrips, whiteflies

Remarks: Direct seed or transplant. Soil application. Use high rate where infestations occur later in crop development, or where pest pressure is continuous. Do not apply more than 24 fl. oz./A per year.

Days To Harvest: 21.

Signal Word/Toxicity Class: Caution.
REI: 12 hr.

Provado 1.6 *(Bayer)*

Rate: 0.05 lb. ai/A
3.75 fl. oz. 1.6/A

Pest:
aphids, Colorado potato beetle, silverleaf whitefly, sweetpotato whitefly, whiteflies

Remarks: For aphids and Colorado potato beetle, good foliage coverage necessary. Apply before damaging population becomes established. For whiteflies, apply at first appearance; repeat at 5- to 7-day intervals until no longer in crop. Do not apply more than 18.75 fl. oz./A per year.

Days To Harvest: 0.

Signal Word/Toxicity Class: Caution.
REI: 12 hr.

LANNATE

Restricted Use Pesticide.

Lannate SP or LV (methomyl) *(DuPont)*

Rate: 0.225-0.9 lb. ai/A
0.25-1 lb. SP/A
0.75-3 pt. LV/A

Pest:
beet armyworm, green peach aphid, tomato pinworm

Remarks: Apply in sufficient water to obtain thorough, uniform coverage when insects first appear; repeat at 5- to 7-day intervals or as needed. For tomato pinworm, use ground application only and apply minimum 0.45 lb. ai/A. Do not apply more than 4.5 lb. ai/A per crop or make more than 10 applications per crop.

Days To Harvest: 5.

Signal Word/Toxicity Class: Danger-Poison/I.
REI: 48 hr.

MALATHION

Organophosphate insecticide toxic to fish, aquatic invertebrates, aquatic life stages of amphibians, and bees exposed to direct treatment on blooming crops or weeds.

Clean Crop Malathion 57EC *(Platte)*

Rate: 0.63-0.94 lb. ai/A
1-1.5 pt. 57EC/A

Pest:
aphids

Rate: 1.9 lb. ai/A
3 pt. 57EC/A

Pest:
lace bugs

Remarks: Apply when insects first appear and repeat as necessary.

Days To Harvest: 3.

Signal Word/Toxicity Class: Caution/III.
REI: 12 hr.

Clean Crop Malathion 8EC *(Platte)*
◆ **Wilbur-Ellis Malathion 8 Spray** *(Wilbur-Ellis)*

Rate: 0.5-1 lb. ai/A
0.5-1 pt. 8EC/A

Pest:
aphids, leafhoppers, Mexican bean beetle, spider mites, thrips

Remarks: Apply when insects first appear and repeat as necessary.

Days To Harvest: 3.

Signal Word/Toxicity Class: Caution/III.
REI: 12 hr.

Drexel Malathion 5EC *(Drexel)*
◆ **Gowan Malathion 8 (8E)** *(Gowan)*
◆ **Micro Flo Malathion 5EC** *(Micro Flo)*
Prentox 5 lb. Malathion Spray (5EC) *(Prentiss)*

Rate: 0.63 lb. ai/A
1 pt. 5EC/A
0.75-3.5 lb. ai/A
0.75-3.5 pt. 8E/A

Pest:
aphids, spider mites

Rate: 2 lb. ai/A
3 pt. 5EC/A
2-3.5 lb. ai/A
2-3.5 pt. 8E/A

Pest:
lace bug

Remarks: Apply when insects first appear and repeat as necessary.

Days To Harvest: 3.

Signal Word/Toxicity Class: Caution/III (8E); Warning/II (5EC).
REI: 12 hr.

For personal protective equipment and EPA registration numbers, please turn to page 39.

Fyfanon (8E) *(Helena)*
Rate: 1.25-2 lb. ai/A
1.25-2 pt. 8E/A
Pest:
aphids | lace bug | spider mites
Remarks: Use higher rate for lace bugs.
Days To Harvest: 3.

- - - - - - - - - - - - - - - - -

Signal Word/Toxicity Class: Caution/III.
REI: 12 hr.

Marlate — see methoxychlor

METASYSTOX-R

Restricted Use Pesticide.

Metasystox-R (SC) (oxydemeton-methyl) *(Gowan)*
Rate: 0.5 lb. ai/A
2 pt. SC/A
Pest:
aphids | mites
Remarks: Apply in sufficient water for complete coverage but not less than 1 gal./A. Do not apply more than 3 times per season.
Days To Harvest: 7.

- - - - - - - - - - - - - - - - -

Signal Word/Toxicity Class: Warning/II.
REI: 48 hr.; 72 hr. if average rainfall less than 25" per year.

METHOXYCHLOR

Organochlorine pesticide toxic to fish and aquatic invertebrates.

Marlate 50 (WP) *(Kincaid)*
Prentox Methoxychlor 2EC *(Prentiss)*
Rate: 2 Tbs. 50WP/gal. water
Pest:
alfalfa looper | flea beetles | Japanese beetle
blister beetle | imported cabbageworm | leafhoppers
fall armyworm
Remarks: Apply at first sign of infestation; repeat at 7- to 14-day intervals as needed.
Days To Harvest: Marlate-7; Prentox-1.

- - - - - - - - - - - - - - - - -

Signal Word/Toxicity Class: Caution.
REI: 12 hr.

M-PEDE

M-Pede (potassium salts of fatty acids) *(Mycogen)*
Rate: 1-2% v/v solution
Pest:
aphids* | leafminers* | spider mites
broad mite | plant bugs | thrips
green peach aphid* | russet mite | whiteflies
leafhoppers
*** Tank mix only***
Remarks: Apply at first sign of infestation; repeat at 7- to 14-day intervals. May be tank mixed with other insecticides. Refer to both labels for use directions and precautions.
Days To Harvest: 0.

- - - - - - - - - - - - - - - - -

Signal Word/Toxicity Class: Warning/II.
REI: 12 hr.

NEEMAZAD — see azadirachtin

NEEMIX — see azadirachtin

NEMACUR

Restricted Use Pesticide.

Nemacur 3 or 15% (fenamiphos) *(Bayer)*
Rate: 2 lb. ai/A
2.66 qt. 3/A (36" row)
13.4 lb. 15%/A (36" row)
5.9 fl. oz. 3/1000 ft. row in 12" band
14.7 oz. 15%/1000 ft. row in 12" band
Pest:
nematodes
Remarks: Apply at transplanting and immediately incorporate. On narrow rows, do not allow bands to overlap.

- - - - - - - - - - - - - - - - -

Signal Word/Toxicity Class: Danger/I.
REI: 48 hr.

PERMETHRIN

Phytotoxicity has occurred on eggplant. Restricted Use Pesticide.

Ambush 2E or 25W *(ZENECA)*
Rate: 0.1-0.2 lb. ai/A
6.4-12.8 fl. oz. 2E/A
6.4-12.8 oz. 25W/A
Pest:
cabbage looper | flea beetles | vegetable leafminers
Colorado potato beetle
Remarks: Apply as needed by air or ground equipment using sufficient water to obtain uniform coverage. Apply high rate for Colorado potato beetle. Do not apply more than 3 lb. ai/A per season.
Days To Harvest: 3.

- - - - - - - - - - - - - - - - -

Signal Word/Toxicity Class: Warning/II.
REI: 24 hr.

Pounce 3.2 EC, 25WP or WSB *(FMC)*
Rate: 0.1-0.2 lb. ai/A
4-8 oz. 3.2 EC/A
6.4-12.8 oz. 25WP/A
1-2 bags WSB/A
Pest:
cabbage looper | Lygus bugs | rindworms
cucumber beetle | melonworm | squash bug
cutworms | pickleworm | squash vine borer
leafminers | plant bugs | stink bugs
Remarks: Rate varies by pest species; refer to label. Apply in minimum 20 gal. finished spray by ground; 4 gal. by air. Do not apply more than 1.6 lb. ai/A per season.
Days To Harvest: 0.

- - - - - - - - - - - - - - - - -

Signal Word/Toxicity Class: Caution (3.2 EC); Warning/II (25WP, WSB).
REI: 24 hr.

PETROLEUM OIL

Petroleum based spray oil. Mode of action is through suffocation of eggs, larvae, and nymphs of insects and adults of soft bodied insects. This mode of action necessitates total spray coverage. Toxic to fish.

SAF-T-SIDE *(Brandt Consolidated)*
◆ **Sumulsion** *(Drexel)*
◆ **SunSpray Ultra-Fine** *(Sun)*
Rate: 1-2 gal./100 gal. water
Pest:
aphids | leafminers | thrips
beetles (larvae) | mites | whiteflies
leafhoppers | spiders
Remarks: Apply as needed. May be tank mixed with most commonly used insecticides and fungicides. Do not use in combination with or immediately before or after spraying with fungicides such as captan, Dyrene, folpet, Morestan, or any sulfur products. Do not use with carbaryl or dimethoate.

- - - - - - - - - - - - - - - - -

Signal Word/Toxicity Class: Caution.
REI: 4 hr.

◆ - **new product** • ai-active ingredient • DF-dry flowable • E/EC-emulsifiable concentrate • EW-emulsion in water
F/FL-flowable • G-granule • L/LC-liquid • SP/WSP-soluble packet • W/WP-wettable powder • WSB-water soluble bag

Phaser — see endosulfan

Pounce — see permethrin

Provado — see imidacloprid

PYRELLIN (Premix)

Premix containing pyrethrins and rotenone. May be combined with other insecticides to enhance knockdown or as an insect "exciter" to help facilitate contact with other spray residues.

Pyrellin E.C. (0.042 lb./gal. pyrethrins & 0.035 lb./gal. rotenone) *(Webb Wright)*

Rate: 1-2 pt. E.C./A

Pest:

aphids	harlequin bug	mites
bean beetle	Japanese beetle	plant bugs
beet webworm	leafhoppers	stink bugs
cabbageworm	leafminers	thrips
celery leaftier	loopers	whiteflies
garden fleahopper	Lygus bugs	

Remarks: Apply at 7 day intervals or less; repeat as needed.

Days To Harvest: 0.

Rate: 1.5-2 pt. E.C./A

Pest:

Colorado potato beetle	cucumber beetle	flea beetles
	European corn borer	vegetable weevil

Remarks: Make 3-4 applications or more as eggs start to hatch; repeat at 5- to 7-day intervals.

Days To Harvest: 0.

Signal Word/Toxicity Class: Caution/III.

REI: 12 hr.

Pyrenone — see pyrethrins

PYRETHRINS

Contains piperonyl butoxide as a synergist. Diatect also contains silicon dioxide (diatomaceous earth). May be combined in tank mix with other insecticides for quicker and more complete kill where insect resistance may be a problem, and as an "exciter" to flush insects out of hiding and into contact with other spray residues.

Diatect Multipurpose Insecticide II *(Diatect)*
Pyrenone (0.5EC) *(AgrEvo Environmental Health)*
Py-Rin 60-6 EC *(Wilbur-Ellis)*

Rate: 1-6 lb. II/A
1-12 fl. oz. 0.5EC/A
2-16 oz. 6-60 EC/A

Pest: Including but not limited to:

aphids	leafminers	stink bugs
caterpillars	loopers	tarnished plant bug
Colorado potato beetle	Lygus bugs	thrips
cutworms	mites	tomato fruitworm
European corn borer	plant bugs	vegetable weevil
flea beetles	psyllids	whiteflies
green peach aphid		

Remarks: Apply in sufficient water for thorough coverage; repeat as needed. Apply Diatect as dust or wettable powder.

Days To Harvest: 0.

Signal Word/Toxicity Class: Caution.

REI: 12 hr.

Py-Rin — see pyrethrins

SAF-T-SIDE — see petroleum oil

Sevin — see carbaryl

Sniper — see azinphos-methyl

Sumulsion — see petroleum oil

SunSpray — see petroleum oil

Thiodan — see endosulfan

THIRETHRIN (Premix)

Contains 0.45 lb. piperonyl butoxide.

Thirethrin (2 lb./gal. endosulfan & 0.045 lb./gal. pyrethrins) *(Terra)*

Rate: 1 qt./A

Pest:

blister beetles	flea beetles	green stink bug
Colorado potato beetle	green peach aphid	whiteflies

Remarks: Repeat as needed.

Days To Harvest: 1.

Signal Word/Toxicity Class: Warning/II.

REI: 24 hr.

VENDEX

Vendex 50WP (fenbutatin-oxide) *(Griffin)*

Rate: 1.5 lb. ai/A
3 lb. 50WP/A

Pest:

clover mite	twospotted spider mite

Remarks: Apply when mites appear; repeat as necessary. Do not make more than 3 applications per season or apply more than 4.5 lb. ai/A per year.

Days To Harvest: 3.

Signal Word/Toxicity Class: Danger/I.

REI: 48 hr.

VYDATE

Restricted Use Pesticide.

Vydate L (oxamyl) *(DuPont)*

Rate: Foliar:
0.5-1 lb. ai/A
2-4 pt. L/A
Soil—nematodes only:
2 lb. ai/A
8 pt. L/A

Pest:

aphids	leafminers	nematodes
Colorado potato beetle	mites	

Remarks: Foliar: Apply when insects first appear; repeat at 1- to 3-week intervals as needed. For nematodes, apply twice by ground at 1- to 2-week intervals 2-4 weeks after second soil treatment. Soil: Apply 2- to 3-weeks after transplanting; repeat 4 weeks after first application. Do not apply more than 24 pt./A per season.

Days To Harvest: 1; nematodes-7.

State Restrictions: Do not use in Suffolk and Nassau Counties, Long Island, New York. Not for use on nematodes in California.

Signal Word/Toxicity Class: Danger-Poison/I.

REI: 48 hr.

NOTICE The information on these pages is for preliminary planning — not a guide for use. Be sure to follow the manufacturer's directions, notwithstanding information contained here.

For personal protective equipment and EPA registration numbers, please turn to page 39.

Garlic

Ambush — see permethrin

Ammo — see cypermethrin

AZADIRACHTIN

Extracted from seed of neem tree. Insect growth regulator that controls targeted insect larvae when they ingest or come in contact with it, by interfering with the insect's ability to molt and shutting off feeding behavior. Affected larvae slowly starve and die at next molt.

◆ **Azatin EC or XL** *(Olympic)*
◆ **Azatin XL Plus** *(Thermo Trilogy)*

Rate: 8-21 oz. EC/A
5-21 oz. XL, XL Plus/A

Pest:

aphids	cutworms	leafhoppers
armyworms	diamondback moth	leafminers
beet armyworm	flea beetles	loopers
beetles	green peach aphid	thrips
caterpillars	Japanese beetle	whiteflies
Colorado potato beetle		

Remarks: Rate varies by species. Apply when pests first appear; repeat at 7-day intervals. When pest populations are high, use higher rates. Do not apply more that 21 fl. oz./A.
Days To Harvest: 0.

- - - - - - - - - - - - - - - - -

Signal Word/Toxicity Class: Caution (XL); Warning/II (EC).
REI: 4 hr. (XL), 12 hr. (EC).

◆ **NEEMAZAD 4.5EC or NEEMIX 4.5** *(Thermo Trilogy)*

Rate: 0.125-1 pt. 4.5/A
High Volume:
2.25-4.5 oz. 4.5EC/100 gal. water
Low Volume:
0.5 pt. 4.5EC/A

Pest:

aphids	diamondback moth	onion maggot
armyworms	greenhouse whitefly	sweetpotato whitefly
cabbage loopers	hornworms	thrips
caterpillars	imported cabbageworm	webworms
cutworms	leafminers	weevils

Remarks: For low volume, apply in sufficient water for adequate coverage. For low insect pressure, apply preventive at 7-day intervals or at first sign of insects; high insect pressure, at 3- to 4-day intervals.

- - - - - - - - - - - - - - - - -

Signal Word/Toxicity Class: Warning/II.
REI: 12 hr.

Azatin — see azadirachtin

lambda-CYHALOTHRIN

Synthetic pyrethroid insecticide with broad-spectrum activity, formulated as an emulsifiable concentrate. Restricted Use Pesticide due to toxicity to fish and aquatic organisms.

Karate or Warrior (1E) *(ZENECA)*

Rate: 0.015-0.03 lb. ai/A
1.92-3.84 fl. oz. 1E/A

Pest:

aphids*	leafminers (adult)	seedcorn maggots
armyworms	onion maggot (adult)	(adult)
(1st & 2nd instar)	onion thrips	stink bugs
cutworms	plant bugs	tobacco thrips
flower thrips*		western flower thrips*

* ***Suppression***

Remarks: Apply as required by scouting, usually at intervals of 5 or more days. Do not apply more than 1.92 pt./A per season. Use higher rates as thrips population increase. Refer to label for spray drift precautions and state and county restrictions.
Days To Harvest: 14.

- - - - - - - - - - - - - - - - -

Signal Word/Toxicity Class: Danger/I.
REI: 24 hr.

CYPERMETHRIN

Synthetic pyrethroid insecticide with both contact and stomach poison activity. No systemic activity and is ineffective against soil pests. Controls many caterpillars, beetles, flies, plant bugs, and others and has fast knockdown effects. Toxic to fish and bees, and may cause eye and skin irritation. Restricted Use Pesticide.

Ammo 2.5 EC or WSB *(FMC)*

Rate: 0.04-0.1 lb. ai/A
2-5 fl. oz. 2.5 EC/A
1-2 bags WSB/A

Pest:

aphids	cutworms	onion maggot (adult)
armyworms	leafminers	stink bugs

Rate: 0.08-0.1 lb. ai/A
4-5 fl. oz. 2.5 EC/A
1-2 bags WSB/A

Pest:
onion thrips

Remarks: Apply in water as necessary using minimum 20 gal. finished spray/A by ground; 3 gal./A by air. Rate varies by pest species. Apply no more than 0.5 lb. ai/A per season. Do not graze treated areas or cut treated crops for feed. Refer to label for spray drift precautions.
Days To Harvest: 7.
Crop Rotation: Not within 30 days after last application.

- - - - - - - - - - - - - - - - -

Signal Word/Toxicity Class: Caution (2.5 EC); Warning/II (WSB).
REI: 12 hr.

DEADLINE

Packaged as Bullets, M-Ps (mini-pellets), and granules.

Deadline Bullets, M-Ps, Granules (4%) (metaldehyde) *(Pace)*

Rate: 0.8-1 lb. ai/A
20-40 lb. 4%/A

Pest:
slugs snails

Remarks: Apply every 3-4 weeks during growing season, or as needed. Do not apply directly to edible portion of plant.

- - - - - - - - - - - - - - - - -

Signal Word/Toxicity Class: Warning/II (granule); Caution (bullets, M-P).
REI: 12 hr.

Diatect — see pyrethrins

Karate — see *lambda*-cyhalothrin

Kumulus — see sulfur

LANNATE

Restricted Use Pesticide.

Lannate SP or LV (methomyl) *(DuPont)*

Rate: 0.45 lb. ai/A
0.5 lb. SP/A
1.5 pt. LV/A

Pest:
beet armyworm

Remarks: Apply in sufficient water to obtain thorough, uniform coverage when insects first appear; repeat at 5- to 7-day intervals or as needed. Add wetting agent. Do not apply more than 3.6 lb. ai/A per crop or make more than 6 applications per crop.
Days To Harvest: 7.

- - - - - - - - - - - - - - - - -

Signal Word/Toxicity Class: Danger-Poison/I.
REI: 48 hr.

◆ - **new product** • ai-active ingredient • DF-dry flowable • E/EC-emulsifiable concentrate • EW-emulsion in water
F/FL-flowable • G-granule • L/LC-liquid • SP/WSP-soluble packet • W/WP-wettable powder • WSB-water soluble bag

MALATHION

Organophosphate insecticide toxic to fish, aquatic invertebrates, aquatic life stages of amphibians, and bees exposed to direct treatment on blooming crops or weeds.

Clean Crop Malathion 57EC or 8EC *(Platte)*
Drexel Malathion 5EC *(Drexel)*
◆ **Gowan Malathion 8 (8E)** *(Gowan)*
◆ **Micro Flo Malathion 5EC** *(Micro Flo)*
Prentox 5 lb. Malathion Spray (5EC) *(Prentiss)*
◆ **Wilbur-Ellis Malathion 8 Spray** *(Wilbur-Ellis)*

Rate: 1-1.25 lb. ai/A
1.5-2 pt. 5EC, 57EC/A
0.5-1 lb. ai/A
0.5-1 pt. 8EC/A
1-2 lb. ai/A
1-2 pt. 8E/A

Pest:
aphids
thrips

Remarks: None.
Days To Harvest: 3.

Signal Word/Toxicity Class: Caution/III; Warning/II (5EC).
REI: 12 hr.

Microthiol Special — see sulfur

M-PEDE

M-Pede (potassium salts of fatty acids) *(Mycogen)*
Rate: 1-2% v/v solution
Pest:
aphids*
broad mite
green peach aphid*
leafhoppers
leafminers*
plant bugs
russet mite
spider mites
thrips
whiteflies

* *Tank mix only*

Remarks: Apply at first sign of infestation; repeat at 7- to 14-day intervals. May be tank mixed with other insecticides; refer to labels for use directions and precautions.
Days To Harvest: 0.

Signal Word/Toxicity Class: Warning/II.
REI: 12 hr.

NEEMAZAD — see azadirachtin

NEEMIX — see azadirachtin

NEMACUR

Restricted Use Pesticide.

Nemacur 15% (fenamiphos) *(Bayer)*
Rate: 2.25-4.5 lb. ai/A
15-30 lb. 15%/A (40" row)
9.2-18.4 oz. 15%/1000 ft. row (any row spacing)

Pest:
bulb nematode
stem nematode

Remarks: Apply in-furrow with garlic seed (cloves) at planting. Refer to label for further use information.

Signal Word/Toxicity Class: Danger/I.
REI: 48 hr.

PERMETHRIN

Restricted Use Pesticide.

Ambush 2E or 25W *(ZENECA)*
Rate: 0.1-0.3 lb. ai/A
6.4-19.2 fl. oz. 2E/A
6.4-19.2 oz. 25W/A

Pest:
armyworms
cutworms
leafminers
onion maggot (adult)
onion thrips
stink bugs

Remarks: Apply when insects appear and repeat as necessary. For foliar application, apply in minimum 20 gal. water/A with ground equipment and minimum 6 gal. water/A by air. Use higher rates on onion thrips and avoid rescue situations. Apply no more than 2.4 lb. ai/A per season. Do not graze livestock in treated areas or feed forage to livestock.
Days To Harvest: 1.

Signal Word/Toxicity Class: Warning/II.
REI: 24 hr.

PYRELLIN (Premix)

Premix containing pyrethrins and rotenone. May be combined with other insecticides to enhance knockdown or as an insect "exciter" to help facilitate contact with other spray residues.

Pyrellin E.C. (0.042 lb./gal. pyrethrins & 0.035 lb./gal. rotenone) *(Webb Wright)*
Rate: 1-2 pt. E.C./A
Pest:
aphids
leafhoppers
leafminers
loopers
Lygus bugs
mites
plant bugs
thrips
whiteflies

Remarks: Apply as necessary to maintain control.
Days To Harvest: 0.

Signal Word/Toxicity Class: Caution/III.
REI: 12 hr.

Pyrenone — see pyrethrins

PYRETHRINS

Contains piperonyl butoxide as a synergist. Diatect also contains silicon dioxide (diatomaceous earth). May be combined in tank mix with other insecticides for quicker and more complete kill where insect resistance may be a problem, and as an "exciter" to flush insects out of hiding and into contact with other spray residues.

Diatect Multipurpose Insecticide II *(Diatect)*
Pyrenone (0.5EC) *(AgrEvo Environmental Health)*
Rate: 1-6 lb. II/A
1-12 fl. oz. 0.5EC/A

Pest: Including but not limited to:
aphids
onion maggot
thrips

Remarks: Apply in sufficient water for thorough coverage; repeat as necessary. Apply Diatect as dust or wettable powder.
Days To Harvest: 0.

Signal Word/Toxicity Class: Caution.
REI: 12 hr.

SULFUR

Microthiol and Thiolux are micronized sulfurs.

◆ **Kumulus DF** *(BASF)*
Microthiol Special (80W) *(Elf Atochem)*
Thiolux (DF) *(Platte)*
Rate: 2.4-8 lb. ai/A
3-10 lb. 80W, DF/A
California:
4.8 lb. ai/A
6 lb. 80W/A

Pest:
petrobia mite

Remarks: Apply when mites first appear; repeat as needed. Do not apply if temperatures during or within 3 days after application are expected to exceed 90°F. Refer to label for use directions.
State Restrictions: Kumulus not registered in California.

Signal Word/Toxicity Class: Caution.
REI: 24 hr.

Thiolux — see sulfur

Warrior — see *lambda*-cyhalothrin

7 Vegetables

NOTICE The information on these pages is for preliminary planning — not a guide for use. Be sure to follow the manufacturer's directions, notwithstanding information contained here.

For personal protective equipment and EPA registration numbers, please turn to page 39.

Greens

(Collards, Dandelion, Hanover Salad, Kale, Mustard Greens, Swiss Chard, Turnip Greens)

Admire — see imidacloprid

Ambush — see permethrin

ASANA

Pyrethroid insecticide that has both contact and stomach poison activity. Restricted Use Pesticide due to toxicity to fish and aquatic organisms.

Asana XL (esfenvalerate) *(DuPont)*

Rate: 0.03-0.05 lb. ai/A
5.8-9.6 fl. oz. XL/A

Pest:

alfalfa looper	cabbage looper	grasshoppers
beet armyworm (aids in control)	cutworms	imported cabbageworm
	flea beetle	

Remarks: Collards. Apply by ground application as needed to maintain control. Do not exceed 0.2 lb. ai/A per season Do not feed livestock on treated plant parts.

Days To Harvest: 7.

Signal Word/Toxicity Class: Warning/II.
REI: 12 hr.

AZADIRACHTIN

Extracted from seed of neem tree. Insect growth regulator that controls targeted insect larvae when they ingest or come in contact with it, by interfering with the insect's ability to molt and shutting off feeding behavior. Affected larvae slowly starve and die at next molt.

◆ **Azatin XL Plus** *(Thermo Trilogy)*

Rate: 5-21 oz. XL Plus/A

Pest:

aphids	flea beetles	moths
armyworms	green peach aphid	psyllids
beet armyworm	imported cabbageworm	silverleaf whitefly
beetles	Japanese beetle	southern armyworm
cabbage looper	leafhoppers	sweetpotato whitefly
caterpillars	leafminers	thrips
Colorado potato beetle	leafrollers	tomato fruitworm
cutworms	loopers	webworms
diamondback moth	Mexican bean beetle	yellowstriped armyworm
fall armyworm		

Remarks: Collard greens, kale, mustard greens, Swiss chard, turnip greens. Rate varies by species. Apply when pests first appear; repeat at 7-day intervals. When pest populations are high, use higher rates. Do not apply more that 21 fl. oz./A.

Days To Harvest: 0.

Signal Word/Toxicity Class: Caution.
REI: 4 hr.

◆ **NEEMAZAD 4.5EC or NEEMIX 4.5** *(Thermo Trilogy)*

Rate: 0.125-1 pt. 4.5/A
High Volume:
2.25-4.5 oz. 4.5EC/100 gal. water
Low Volume:
0.5 pt. 4.5EC/A

Pest:

aphids	diamondback moth	onion maggot
armyworms	greenhouse whitefly	sweetpotato whitefly
cabbage loopers	hornworms	thrips
caterpillars	imported cabbageworm	webworms
cutworms	leafminers	weevils

Remarks: Collards, kale, mustard greens, Swiss chard, turnip greens. For low volume, apply in sufficient water for adequate coverage. For low insect pressure, apply preventive at 7-day intervals or at first sign of insects; high insect pressure, at 3- to 4-day intervals.

Signal Word/Toxicity Class: Warning/II.
REI: 12 hr.

Azatin — see azadirachtin

BT 320 SULFUR 25 or 50 DUST (Premix)

Bt 320 Sulfur 25 Dust (0.064% Bt & 25% sulfur) or Bt 320 Sulfur 50 Dust (0.064% Bt & 50% sulfur) *(Wilbur-Ellis)*

Rate: 20-35 lb. Dust/A

Pest:
imported cabbageworm

Remarks: Collards, kale, mustard greens, turnip greens. Do not mix or dilute with other products. Thorough coverage essential.

Signal Word/Toxicity Class: Caution.
REI: 24 hr.

CARBARYL

Carbamate insecticide with broad activity on many insects. Highly toxic to bees.

Clean Crop Sevin 5 Bait *(Platte)*

Rate: 1-2 lb. ai/A
20-40 lb. Bait/A

Pest:

ants	darkling beetle	grasshoppers
armyworms	field crickets	sowbugs
cutworms		

Remarks: Collards, dandelion, Hanover salad, kale, mustard greens, Swiss chard, turnip greens. Apply with ground or air equipment.

Days To Harvest: 14.

Signal Word/Toxicity Class: Caution/III.
REI: 12 hr.

Drexel Carbaryl 4L or 50W *(Drexel)*
Sevin 4F, XLR Plus, 80S, 80WSP *(Rhone-Poulenc)*

Rate: 0.5-2 lb. ai
1-4 pt. 4L, 4F, XLR/A
1-4 lb. 50W/A
0.66-2.5 lb. 80S, 80WSP/A

Pest:

armyworms	grasshoppers	Lygus bugs (dandelion)
aster leafhopper	harlequin bug	
corn earworm	imported cabbageworm (except Swiss chard)	spittlebugs (dandelion)
fall armyworm		stink bugs
flea beetles	leafhoppers	tarnished plant bug

Remarks: Collards, dandelion, Hanover salad, kale, mustard greens, Swiss chard, turnip greens. For Swiss chard, treat at 5- to 7-day intervals. Do not apply to wet foliage or when rain or high humidity is expected during next 2 days. Rate varies by pest species; refer to label.

Days To Harvest: 14.

Signal Word/Toxicity Class: Caution (4F, 4L, XLR); Warning/II (50W, 80S, 80WSP).
REI: 12 hr.

Gowan Sevin Bait *(Gowan)*

Rate: 2 lb. ai/A
40 lb. Bait/A

Pest:

armyworms	cutworms	grasshoppers
crickets	darkling beetle	sowbugs

Remarks: Collards, kale, Swiss chard, turnip greens. Apply when pests appear in damaging numbers; repeat at 7- to 14-day intervals if necessary.

Days To Harvest: 14.
State Restrictions: For use in Arizona and California.

Signal Word/Toxicity Class: Caution/III.
REI: 12 hr.

◆ - **new product** • ai-active ingredient • DF-dry flowable • E/EC-emulsifiable concentrate • EW-emulsion in water
F/FL-flowable • G-granule • L/LC-liquid • SP/WSP-soluble packet • W/WP-wettable powder • WSB-water soluble bag

Chlorfos — see chlorpyrifos

CHLORPYRIFOS

Organophosphate insecticide toxic to birds and wildlife, extremely toxic to fish and aquatic organisms.

◆ **Chlorfos 15G** *(Griffin)*
Lorsban-4E or 15G *(Dow AgroSciences)*
◆ **Micro Flo Chlorpyrifos 4E or 15G** *(Micro Flo)*

Rate: 1.6-2.75 fl. oz. 4E/1000 lin. ft. row
3.6-9.2 oz. 15G/1000 ft. row (4.6-9.2 oz. for Dow)

Pest:
root maggot

Remarks: Collards, kale. Do not apply more than 2.6 pt. 4E/A or 7.5 lb. 15G/A on 40" rows; more than 4.5 pt. 4E/A or 15 lb. 15G/A on 20" rows. Use proportional amounts on other row spacing not to exceed 4.5 pt. 4E/A or 15 lb. 15G/A. Do not apply more than once per season. Do not mix with other pesticides or treat plants under extreme heat or drought stress. Refer to label for use directions.

Days To Harvest: 30.

Signal Word/Toxicity Class: Warning/II (4E); Caution/III (15G).
REI: 24 hr. (4E); 12 hr. (15G).

Lorsban 50-W *(Gowan)*

Rate: 2 lb. 50-W/A (two 1 lb. bags)

Pest:

armyworms	cutworms	striped flea beetle (adult)
beet armyworm	diamondback moth	
cabbage aphid	imported cabbageworm	

Remarks: Collards, kale. Will not control organophosphate-resistant diamondback moths. Do not apply more than 6 applications per season. Do not treat plants under stress from extreme heat or lack of moisture.

Days To Harvest: 21.

Signal Word/Toxicity Class: Warning/II.
REI: 24 hr.

CRYOLITE

Kryocide (96%) *(Elf Atochem)*

Rate: 7.7-15.4 lb. ai/A
8-16 lb. 96%/A

Pest:

cabbage looper	Diabrotica beetles	imported cabbageworm
corn earworm	diamondback moth	yellowstriped armyworm
cutworms	flea beetles	

Remarks: Collards. Apply at 10-day intervals as needed. Remove excess residues on edible portions by washing, brushing, or other effective means. Do not exceed 96 lb./A per season. Do not use in combination with lime or compounds containing free lime. Do not use chemigation in California.

Days To Harvest: 14.

Signal Word/Toxicity Class: Caution.
REI: 12 hr.

DEADLINE

Packaged as Bullets, M-Ps (mini-pellets), and granules.

Deadline Bullets, M-Ps, Granules (4%) (metaldehyde) *(Pace)*

Rate: 0.8-1 lb. ai/A
20-40 lb. 4%/A

Pest:
slugs snails

Remarks: Collards, kale, mustard greens, Swiss chard. Apply every 3-4 weeks during growing season, or as needed. Do not apply directly to edible portion of plant.

Signal Word/Toxicity Class: Warning/II (granule); Caution (bullets, M-P).
REI: 12 hr.

Declare — see methyl parathion

Diatect — see pyrethrins

DIAZINON

Restricted Use Pesticide due to avian and aquatic toxicity.

D•z•n diazinon AG600 WBC, 50W *(Novartis)*
Drexel Diazinon Insecticide (4EC), D-264 14G *(Drexel)*
Gowan Diazinon 4E, 14G, 50WSB *(Gowan)*
Micro Flo Diazinon AG500, 14G, 50W *(Micro Flo)*
Prentox Diazinon AG500, 50W *(Prentiss)*
◆ **Wilbur-Ellis Diazinon 4 (4E) or 14G** *(Wilbur-Ellis)*

Rate: 0.25-0.5 lb. ai/A
0.5-1 pt. 4E, 4EC/A
6.5-12.75 fl. oz. AG600/A
0.5-1 lb. 50W, 50WSB/A

Pest:

aphids	diamondback moth	harlequin bug
corn earworm	flea beetles	imported cabbageworm

Rate: 1 lb. ai/A
2 pt. 4E, 4EC, AG500/A
25.5 fl. oz. AG600/A
2 lb. 50W, 50WSB/A
7 lb. 14G/A

Pest:
mole crickets

Rate: 2-4 lb. ai/A
4-8 pt. 4E, 4EC, AG500/A
51-102 fl. oz. AG600/A
4-8 lb. 50W, 50WSB/A
14-28 lb. 14G/A

Pest:
cutworms

Rate: 3-4 lb. ai/A
6-8 pt. 4E, 4EC, AG500/A
76.5-102 fl. oz. AG600/A
6-8 lb. 50W, 50WSB/A
21-28 lb. 14G/A

Pest:
wireworms

Remarks: Collards, kale, mustard, Swiss chard, turnip tops. Apply as insects occur; repeat at 7-day intervals. Do not apply more than 5 times per season. For crickets, cutworms, and wireworms, apply just before planting and immediately incorporate into soil.

Days To Harvest: Collards, kale, mustard-10; Swiss chard, turnip tops-14.

Signal Word/Toxicity Class: Caution/III.
REI: 24 hr.; 12 hr. (Drexel/Gowan/Wilbur-Ellis 14G).

DIBROM

Short residual contact insecticide for kill of insect and mite adults and larvae present at time of application.

Dibrom 8 (8E) (naled) *(Amvac)*

Rate: 1 lb. ai/A
1 pt. 8E/A

Pest:
aphids diamondback caterpillar imported cabbageworm

Rate: 2 lb. ai/A
2 pt. 8E/A

Pest:
loopers

Remarks: Collards, kale. Apply at first sign of insects; repeat as necessary. Apply in 30-250 gal. water by ground; 3-10 gal. water by air. Do not use with highly alkaline materials such as lime or Bordeaux. Do not apply through any type of irrigation system. Do not apply more than 1 pt./A in Florida. Do not apply more than 10 pt./A per season. Do not apply when temperatures exceed 90°F.

Days To Harvest: 1.

Signal Word/Toxicity Class: Danger/I.
REI: 48 hr. for 1 pt./A or less; 72 hr. for more than 1 pt./A.

7 Vegetables

NOTICE The information on these pages is for preliminary planning — not a guide for use. Be sure to follow the manufacturer's directions, notwithstanding information contained here.	For personal protective equipment and EPA registration numbers, please turn to page 39.

Digon — see dimethoate

Dimate — see dimethoate

DIMETHOATE

Cheminova Dimethoate 4E *(Cheminova)*
Clean Crop Dimethoate 400 *(Platte)*
Dimate 4E *(Terra)*
Drexel Dimethoate 2.67 or 4EC *(Drexel)*
Gowan Dimethoate E267 *(Gowan*
Helena Dimethoate 4E *(Helena)*
Micro Flo Dimethoate 267 or 4E *(Micro Flo)*
Wilbur-Ellis Dimethoate 267 or Digon 400 *(Wilbur-Ellis)*

Rate: 0.25 lb. ai/A
0.75 pt. 2.67, 267, E267/A
0.5 pt. 4E, 4EC, 400/A

Pest:
aphids leafhoppers leafminers

Remarks: Collards, kale, mustard greens, Swiss chard, turnip greens.
Days To Harvest: 14.

Signal Word/Toxicity Class: Warning/II.
REI: 48 hr.

D•z•n — see diazinon

ENDOSULFAN

Toxic to birds, fish, and wildlife; toxic to bees exposed to direct application.

Drexel Endosulfan 3EC or 50WP *(Drexel)*
Gowan Endosulfan 3EC or 50WSB *(Gowan)*
◆ **Micro Flo Endosulfan 3EC or 50WP** *(Micro Flo)*
Phaser 3EC or 50WSB *(AgrEvo USA)*
Thiodan 3EC or 50WP *(FMC)*

Rate: 0.75-1 lb. ai/A
1-1.33 qt. 3EC/A
1.5-2 lb. 50WP, 50WSB/A

Pest: Collards:
cabbage looper harlequin bug imported cabbageworm
diamondback moth

Kale:
cabbage flea beetles harlequin bug imported cabbageworm

Mustard greens:
aphids cabbage looper imported cabbageworm

Remarks: Apply once per season when insects appear and feeding is noticed.
Days To Harvest: 21.

Signal Word/Toxicity Class: Danger-Poison/I.
REI: 24 hr.

IMIDACLOPRID

Admire 2F *(Bayer)*

Rate: 0.16-0.375 lb. ai/A
10-24 fl. oz. 2F/A

Pest:
aphids silverleaf whitefly sweetpotato whitefly

Remarks: Collards, kale, mustard greens, rape greens. Direct seed or transplant. Soil application. Use higher rates where infestations occur later in crops development, or where pest pressure is continuous. Do not apply more than 24 fl. oz./A per year.
Days To Harvest: 21.

Signal Word/Toxicity Class: Caution.
REI: 12 hr.

Provado 1.6 *(Bayer)*

Rate: 0.05 lb. ai/A
3.75 fl. oz. 1.6/A

Pest:
aphids sweetpotato whitefly whiteflies
silverleaf whitefly

Remarks: Collards, kale, mustard greens, rape greens. Apply before damaging population becomes established. For whiteflies, apply at first appearance; repeat at 5- to 7-day intervals until no longer in crop. Do not apply more than 18.75 fl. oz./A per year.
Days To Harvest: 7.

Signal Word/Toxicity Class: Caution.
REI: 12 hr.

Kryocide — see cryolite

Kumulus — see sulfur

LANNATE

Restricted Use Pesticide.

Lannate SP or LV (methomyl) *(DuPont)*

Rate: 0.45-0.9 lb. ai/A
0.5-1 lb. SP/A
1.5-3 pt. LV/A

Pest:
beet armyworm diamondback moth (except collards) imported cabbageworm
cabbage looper (except collards)

Rate: **Collards only:**
0.45 lb. ai/A
0.5 lb. SP/A
1.5 pt. LV/A

Pest:
diamondback moth loopers variegated cutworm

Remarks: Beet tops, collards (fresh market only), dandelions, kale, mustard greens, Swiss chard, turnip greens. Apply in sufficient water to obtain thorough, uniform coverage when insects first appear; repeat at 5- to 7-day intervals or as needed. Do not apply more than 3.6 lb. ai/A (5.4 lb. ai/A for collards) per crop or make more than 8 applications per crop. For collards, do not apply when temperature is less than 50°F or crop is less than 10" tall.
Days To Harvest: 10.
State Restrictions: Not for use on loopers and cabbage looper in Alabama and Georgia.

Signal Word/Toxicity Class: Danger-Poison/I.
REI: 48 hr.

LARVIN

Broad spectrum carbamate insecticide that has stomach poison activity and some limited contact activity. Known to have systemic action as a seed treatment. Toxic to fish and birds, and moderately toxic to bees as direct treatment.

Larvin 3.2 (thiodicarb) *(Rhone-Poulenc)*

Rate: 0.4-0.75 lb. ai/A
16-30 fl. oz. 3.2/A

Pest:
alfalfa looper corn earworm southern armyworm
beet armyworm fall armyworm tomato fruitworm
bollworm

Rate: 0.6-0.75 lb. ai/A
24-30 fl. oz. 3.2/A

Pest:
cabbage looper

Remarks: Dandelion, Swiss chard. Use higher rate for heavier infestations of larger larvae. Spray as needed. Do not exceed 1.5 lb. ai/A per season.
Days To Harvest: 14.

Signal Word/Toxicity Class: Warning/II.
REI: 12 hr.

Lorsban — see chlorpyrifos

◆ - **new product** • ai-active ingredient • DF-dry flowable • E/EC-emulsifiable concentrate • EW-emulsion in water
F/FL-flowable • G-granule • L/LC-liquid • SP/WSP-soluble packet • W/WP-wettable powder • WSB-water soluble bag

MALATHION

Organophosphate insecticide toxic to fish, aquatic invertebrates, aquatic life stages of amphibians, and bees exposed to direct treatment on blooming crops or weeds.

Clean Crop Malathion 57EC *(Platte)*

Rate: Collards:
0.32-1.75 lb. ai/A
0.5-2.8 pt. 57EC/A

Pest:
cabbage looper, caterpillars, diamondback moth, harlequin bug, imported cabbageworm, leafhoppers, leafminers

Rate: Kale, mustard greens, Swiss chard:
0.7-1.4 lb. ai/A
1-2 pt. 57EC/A

Pest:
aphids, cabbage looper, imported cabbageworm

Rate: Dandelion:
1-1.4 lb. ai/A
1.5-2 pt. 57EC/A

Pest:
aphids

Remarks: Collards, dandelion, kale, mustard greens, Swiss chard.
Days To Harvest: 7.

Signal Word/Toxicity Class: Caution/III.
REI: 12 hr.

Clean Crop Malathion 8EC *(Platte)*
◆ **Wilbur-Ellis Malathion 8 Spray** *(Wilbur-Ellis)*

Rate: Dandelions, Swiss chard:
0.5-1 lb. ai/A
0.5-1 pt. 8, 8EC/A

Pest:
aphids, leafhoppers, Mexican bean beetle, spider mites, thrips

Rate: Kale, mustard greens, turnip greens:
1-2 lb. ai/A
1-2 pt. 8, 8EC/A

Pest:
aphids, cabbage looper, imported cabbageworm

Remarks: Apply in minimum 5 gal. by air; 10 gal. by ground.
Days To Harvest: Dandelions, kale, mustard greens, Swiss chard-7; turnip greens-3.

Signal Word/Toxicity Class: Caution/III.
REI: 12 hr.

Drexel Malathion 5EC *(Drexel)*
◆ **Micro Flo Malathion 5EC** *(Micro Flo)*
Prentox 5 lb. Malathion Spray (5EC) *(Prentiss)*

Rate: Collards:
0.63 lb. ai/A
1 pt. 5EC/A

Pest:
harlequin bug

Rate: Collards, dandelion, Swiss chard:
1-1.3 lb. ai/A
1.5-2 pt. 5EC/A

Pest:
aphids

Rate: Kale:
0.63-1.3 lb. ai/A
1-2 pt. 5EC/A

Pest:
aphids, cabbage looper, diamondback moth, imported cabbageworm, webworm

Rate: Mustard:
1-1.3 lb. ai/A
1.5-2 pt. 5EC/A

Pest:
aphids, cabbage looper, flea beetles, imported cabbageworm

Remarks: Collards, dandelion, kale, mustard greens, Swiss chard. Apply when insects first appear and repeat as necessary.
Days To Harvest: 7.

Signal Word/Toxicity Class: Warning/II.
REI: 12 hr.

◆ **Gowan Malathion 8 (8E)** *(Gowan)*

Rate: Dandelion, Swiss chard:
1-2 lb. ai/A
1-2 pt. 8E/A

Pest:
aphids

Rate: Collards, kale, mustard greens:
1.5-2.5 lb. ai/A
1.5-2.5 pt. 8E/A

Pest:
aphids, cabbage looper, flea beetles, imported cabbageworm

Remarks: Collards, dandelion, kale, mustard greens, Swiss chard.
Days To Harvest: 7.

Signal Word/Toxicity Class: Caution/III.
REI: 12 hr.

Marlate — see methoxychlor

METHOXYCHLOR

Organochlorine pesticide toxic to fish and aquatic invertebrates.

Marlate 50 (WP) *(Kincaid)*
Prentox Methoxychlor 2EC *(Prentiss)*

Rate: 1.3-1.7 lb. ai/A
75-100 fl. oz. 2EC/A
2 Tbs. 50WP/gal. water

Pest:
alfalfa looper, blister beetle, fall armyworm, flea beetles, imported cabbageworm, Japanese beetle, leafhoppers

Remarks: Collards, kale. Apply at first sign of infestation; repeat at 7- to 14-day intervals as needed.
Days To Harvest: 14.

Signal Word/Toxicity Class: Caution.
REI: 12 hr.

METHYL PARATHION

Organophosphate insecticide toxic to bees exposed to direct treatment or residues on blooming crops or weeds. Restricted Use Pesticide due to very high acute toxicity to humans and birds.

Cheminova Methyl 4EC *(Cheminova)*

Rate: 0.5-1.5 lb. ai/A
1-3 pt. 4EC/A

Pest:
aphids, armyworms, flea beetles, leafhoppers, mites

Rate: 1-1.5 lb. ai/A
2-3 pt. 4EC/A

Pest:
stink bugs

Remarks: Collards, kale, mustard greens. Apply when insects appear; repeat as needed.
Days To Harvest: 10; rates over 1 pt.-21.

Signal Word/Toxicity Class: Danger-Poison/I.
REI: 48 hr.; 72 hr. if average rainfall less than 25" per year.

◆ **Declare (4EC)** *(Griffin)*
Helena 4 lb. Methyl Parathion (4EC) *(Helena)*

Rate: 0.23-1.4 lb. ai/A
0.5-3 pt. 4EC/A

Pest:
aphids, armyworms, cabbage looper, flea beetles, imported cabbageworm, leafhoppers, mites, plant bugs, stink bugs, thrips (collards)

NOTICE The information on these pages is for preliminary planning — not a guide for use. Be sure to follow the manufacturer's directions, notwithstanding information contained here.	For personal protective equipment and EPA registration numbers, please turn to page 39.

7 Vegetables

Remarks: Collards, kale, mustard greens. Rate varies by pest species; refer to label.
Days To Harvest: Up to 1 pt.–10; over 1 pt.–21.

- - - - - - - - - - - - - - - - -

Signal Word/Toxicity Class: Danger-Poison/I.
REI: 48 hr.

Microthiol Special — see sulfur

M-PEDE

M-Pede (potassium salts of fatty acids) *(Mycogen)*
Rate: 1-2% v/v solution
Pest:

aphids*	leafminers*	spider mites
broad mite	plant bugs	thrips
green peach aphid*	russet mite	whiteflies
leafhoppers		

** Tank mix only*

Remarks: Collards, kale, mustard greens, Swiss chard. Apply at first sign of infestation; repeat at 7- to 14-day intervals. May be tank mixed with other insecticides. Refer to label for use directions and precautions.
Days To Harvest: 0.

- - - - - - - - - - - - - - - - -

Signal Word/Toxicity Class: Warning/II.
REI: 12 hr.

NEEMAZAD — see azadirachtin

NEEMIX — see azadirachtin

PERMETHRIN

Restricted Use Pesticide.

Ambush 2E or 25W *(ZENECA)*
Rate: 0.05-0.1 lb. ai/A
3.2-6.4 fl. oz. 2E/A
3.2-6.4 oz. 25W/A
Pest:

cabbage aphid (suppression)	cabbage looper	imported cabbageworm
	diamondback moth	

Remarks: Collards. Apply by ground only in sufficient water to obtain uniform coverage. Do not make more than 8 applications per season. Do not graze livestock in treated areas or feed crop refuse to livestock.
State Restrictions: For use on collards in Arizona, Arkansas, Georgia, Illinois, North Carolina, Oklahoma, South Carolina, and Texas.

- - - - - - - - - - - - - - - - -

Rate: 0.1-0.2 lb. ai/A
6.4-12.8 fl. oz. 2E/A
6.4-12.8 oz. 25W/A
Pest:

beet armyworm	cutworms	green cloverworm
cabbage looper	European corn borer	leafhoppers
corn earworm	fall armyworm	vegetable leafminer

Remarks: Dandelion, Swiss chard. Refer to label for complete list of minor crop leafy vegetables. Apply every 3-5 days or as needed by ground or air. Use sufficient water to obtain full coverage of foliage. Do not apply more than 2 lb. ai per season.
Days To Harvest: 1.

- - - - - - - - - - - - - - - - -

Signal Word/Toxicity Class: Warning/II.
REI: 24 hr.

Pounce 3.2 EC, 25WP or WSB *(FMC)*
Rate: 0.05-0.2 lb. ai/A
2-8 oz. 3.2 EC/A
3.2-12.8 oz. 25WP/A
1-2 bags WSB/A
Pest:

alfalfa looper	cutworms	green cloverworm
beet armyworm	European corn borer	leafhoppers
cabbage looper	fall armyworm	vegetable leafminer
corn earworm		

Remarks: Dandelion, Swiss chard. Rate varies by pest species; refer to label. Apply every 3-5 days or as needed in sufficient water to obtain full foliage coverage. Do not graze treated areas or feed crop refuse to livestock.
Days To Harvest: 1.

- - - - - - - - - - - - - - - - -

Signal Word/Toxicity Class: Caution (3.2 EC); Warning/II (25WP, WSB).
REI: 12 hr.

Phaser — see endosulfan

Pounce — see permethrin

Provado — see imidacloprid

PYRELLIN (Premix)

Premix containing pyrethrins and rotenone. May be combined with other insecticides to enhance knockdown or as an insect "exciter" to help facilitate contact with other spray residues.

Pyrellin E.C. (0.042 lb./gal. pyrethrins & 0.035 lb./gal. rotenone) *(Webb Wright)*
Rate: 1-2 pt. E.C./A
Pest:

aphids	loopers	plant bugs
leafhoppers	Lygus bugs	thrips
leafminers	mites	whiteflies

Remarks: Collards, dandelion, kale, mustard greens, Swiss chard, turnip greens. Apply as necessary to maintain control.

- - - - - - - - - - - - - - - - -

Signal Word/Toxicity Class. Caution/III.
REI: 12 hr.

Pyrenone — see pyrethrins

PYRETHRINS

Contains piperonyl butoxide as a synergist. Diatect also contains silicon dioxide (diatomaceous earth). May be combined in tank mix with other insecticides for quicker and more complete kill where insect resistance may be a problem, and as an "exciter" to flush insects out of hiding and into contact with other spray residues.

Diatect Multipurpose Insecticide II *(Diatect)*
Pyrenone (0.5EC) *(AgrEvo Environmental Health)*
Py-Rin 60-6 EC *(Wilbur-Ellis)*
Rate: 1-6 lb. II/A
1-12 fl. oz. 0.5EC/A
2-16 oz. 6-60 EC/A
Pest:

aphids	diamondback moth	loopers
armyworms	flea beetles	Lygus bugs
blister beetle	garden fleahoppers	Mexican bean beetle
cabbage looper	harlequin bugs	mites
cabbageworms	imported cabbageworm	plant bugs
caterpillars		stink bugs
Colorado potato beetle	Japanese beetles	thrips
	leafhoppers	vegetable weevils
corn earworm	leafminers	webworms
cucumber beetle	leaftiers	whiteflies

Remarks: Collards, dandelion, kale, mustard greens, Swiss chard, turnip greens, etc. Apply in sufficient water for thorough coverage; repeat as necessary. Apply Diatect as dust or wettable powder.
Days To Harvest: 0.

- - - - - - - - - - - - - - - - -

Signal Word/Toxicity Class: Caution.
REI: 12 hr.

Py-Rin — see pyrethrins

Sevin — see carbaryl

◆ - **new product** • ai-active ingredient • DF-dry flowable • E/EC-emulsifiable concentrate • EW-emulsion in water
F/FL-flowable • G-granule • L/LC-liquid • SP/WSP-soluble packet • W/WP-wettable powder • WSB-water soluble bag

SULFUR

Microthiol and Thiolux are micronized sulfurs.

◆ **Kumulus DF** *(BASF)*
Thiolux (DF) *(Platte)*
Rate: 2.4-4 lb. ai/A
3-5 lb. DF/A
Pest:
red spider mite
Remarks: Collards, kale. Apply at early leaf stage. Repeat at 10- to 14-day intervals with or without other pesticides immediately after a rain. Do not apply if temperatures during or within 3 days after application are expected to exceed 90°F. Refer to label for use directions.

Signal Word/Toxicity Class: Caution.
REI: 24 hr.

Microthiol Special (80W) *(Elf Atochem)*
Rate: 2.4-8 lb. ai/A
3-10 lb. 80W/A
Pest:
red spider mite
Remarks: Collards, kale, mustard greens. Apply at early leaf stage; repeat every 14 days or as needed.

Signal Word/Toxicity Class: Caution.
REI: 24 hr.

Thiodan — see endosulfan

Thiolux — see sulfur

THIRETHRIN (Premix)

Contains 0.45 lb. piperonyl butoxide.

Thirethrin (2 lb./gal. endosulfan & 0.045 lb./gal. pyrethrins) *(Terra)*
Rate: 1-1.5 qt./A
Pest:
aphids, cabbage looper, diamondback moth, harlequin bugs, imported cabbageworm
Remarks: Collards, mustard greens, turnip greens. Do not apply more than once per season.
Days To Harvest: 21.

Signal Word/Toxicity Class: Warning/II.
REI: 24 hr.

Herbs & Spices

(Anise, Basil, Chive, Dill, Fennel, Marjoram, Sage)

Ambush — see permethrin

AZADIRACHTIN

Extracted from seed of neem tree. Insect growth regulator that controls targeted insect larvae when they ingest or come in contact with it, by interfering with the insect's ability to molt and shutting off feeding behavior. Affected larvae slowly starve and die at next molt.

◆ **Azatin EC or XL** *(Olympic)*
◆ **Azatin XL Plus** *(Thermo Trilogy)*
Rate: 8-21 oz. EC/A
5-21 oz. XL, XL Plus/A
Pest:
aphids, beet armyworm, cutworms, fall armyworm, green peach aphid, imported cabbageworm, leafhoppers, leafminers, loopers, moths, southern armyworm, thrips, tomato fruitworm, whiteflies
Remarks: Rate varies by species. Apply when pests first appear; repeat at 7-day intervals as needed. When pest populations are high, use higher label rates. Do not apply more that 21 fl. oz./A.
Days To Harvest: 0.

Signal Word/Toxicity Class: Caution (XL); Warning/II (EC).
REI: 4 hr. (XL), 12 hr. (EC).

◆ **NEEMAZAD 4.5EC or NEEMIX 4.5** *(Thermo Trilogy)*
Rate: 0.125-1 pt. 4.5/A
High Volume:
2.25-4.5 oz. 4.5EC/100 gal. water
Low Volume:
0.5 pt. 4.5EC/A
Pest:
aphids, flea beetles, leafhoppers
Remarks: Anise, basil, dill, fennel, rosemary, sage, thyme, etc. For low volume, apply in sufficient water for adequate coverage. For low insect pressure, apply preventive at 7-day intervals or at first sign of insects; high insect pressure, at 3- to 4-day intervals.

Signal Word/Toxicity Class: Warning/II.
REI: 12 hr.

Azatin — see azadirachtin

Diatect — see pyrethrins

LANNATE

For use on anise only. Restricted Use Pesticide.

Lannate SP or LV (methomyl) *(DuPont)*
Rate: 0.45-0.9 lb. ai/A
0.5-1 lb. SP/A
1.5-3 pt. LV/A
Pest:
beet armyworm, cabbage looper
Remarks: Anise. Apply in sufficient water for thorough coverage at 5- to 7-day intervals or as needed when insects first appear. Apply high rate for cabbage looper. Do not apply more than 4.5 lb. ai/A per crop or make more than 10 applications per crop.
Days To Harvest: 7.

Signal Word/Toxicity Class: Danger-Poison/I.
REI: 48 hr.

LARVIN

Broad spectrum carbamate insecticide that has stomach poison activity and some limited contact activity. Known to have systemic action as a seed treatment. Toxic to fish and birds, and moderately toxic to bees as direct treatment.

Larvin 3.2 (thiodicarb) *(Rhone-Poulenc)*
Rate: 0.4-0.75 lb. ai/A
16-30 fl. oz. 3.2/A
Pest:
alfalfa looper, beet armyworm, bollworm, corn earworm, fall armyworm, southern armyworm, tomato fruitworm

Rate: 0.6-0.75 lb. ai/A
24-30 fl. oz. 3.2/A
Pest:
cabbage looper
Remarks: Fennel. Spray as needed. Use higher rates for heavier infestations of larger larvae. Do not exceed 1.5 lb. ai/A per season.
Days To Harvest: 14.

Signal Word/Toxicity Class: Warning/II.
REI: 12 hr.

For personal protective equipment and EPA registration numbers, please turn to page 39.

MALATHION

Organophosphate insecticide toxic to fish, aquatic invertebrates, aquatic life stages of amphibians, and bees exposed to direct treatment on blooming crops or weeds.

Clean Crop Malathion 57EC *(Platte)*
Drexel Malathion 5EC *(Drexel)*
◆ **Micro Flo Malathion 5EC** *(Micro Flo)*
Prentox 5 lb. Malathion Spray (5EC) *(Prentiss)*

Rate: 0.94 lb. ai/A
1.5 pt. 5EC, 57EC/A

Pest:
aphids, spider mites

Remarks: Anise. Fresh leaves and stalks. Do not use on crops grown for seed and oil. Apply when insects first appear and repeat as necessary.
Days To Harvest: 7.

Signal Word/Toxicity Class: Caution/III (57EC); Warning/II (5EC).
REI: 12 hr.

M-PEDE

M-Pede (potassium salts of fatty acids) *(Mycogen)*

Rate: 1-2% v/v solution

Pest:
aphids*, broad mite, green peach aphid*, leafhoppers, leafminers*, plant bugs, russet mite, spider mites, thrips, whiteflies

**** Tank mix only***

Remarks: Basil, chive, dill, marjoram, sage. Apply at first sign of infestation; repeat at 7- to 14-day intervals. May be tank mixed with other insecticides. Refer to both labels for use directions and precautions.
Days To Harvest: 0.

Signal Word/Toxicity Class: Warning/II.
REI: 12 hr.

NEEMAZAD — see azadirachtin

NEEMIX — see azadirachtin

PERMETHRIN

Restricted Use Pesticide.

Ambush 2E or 25W *(ZENECA)*

Rate: 1-2 lb. ai/A
6.4-12.8 fl. oz. 2E/A
6.4-12.8 oz. 25W/A

Pest:
beet armyworm, cabbage looper, corn earworm, cutworms, European corn borer, fall armyworm, green cloverworm, leafhoppers, vegetable leafminer

Remarks: Refer to label for complete list of herbs and spices. Apply at 3- to 5-day intervals or as needed by air or ground. Use sufficient water to obtain full coverage of foliage. Do not apply more than 2 lb. ai/A per season. Do not graze or feed crop refuse to livestock.
Days To Harvest: 1.

Signal Word/Toxicity Class: Warning/II.
REI: 24 hr.

Pounce 3.2 EC, 25WP or WSB *(FMC)*

Rate: 0.05-0.2 lb. ai/A
2-8 oz. 3.2 EC/A
3.2-12.8 oz. 25WP/A
1-2 bags WSB/A

Pest:
alfalfa looper, beet armyworm, cabbage looper, corn earworm, cutworms, European corn borer, fall armyworm, green cloverworm, leafhoppers, vegetable leafminer

Remarks: Anise, fennel. Apply every 3-5 days or as needed. Use sufficient water to obtain full coverage of foliage. Rate varies by pest species; refer to label. Do not graze treated areas or feed crop refuse to livestock.
Days To Harvest: 1.

Signal Word/Toxicity Class: Caution (3.2 EC); Warning/II (25WP, WSB).
REI: 24 hr.

PETROLEUM OIL

Petroleum based spray oil. Mode of action is through suffocation of eggs, larvae, and nymphs of insects and adults of soft bodied insects. This mode of action necessitates total spray coverage. Toxic to fish.

◆ **SunSpray Ultra-Fine** *(Sun)*

Rate: 1-2 gal./100 gal. water

Pest:
aphids, caterpillars, leafhoppers, leafminers, loopers, mites, plant bugs, thrips, whiteflies

Remarks: Apply as needed. May be tank mixed with most commonly used insecticides and fungicides. Do not use in combination with or immediately before or after spraying with fungicides such as captan, Dyrene, folpet, Morestan, or any sulfur products. Do not use with carbaryl or dimethoate.

Signal Word/Toxicity Class: Caution.
REI: 4 hr.

Pounce — see permethrin

PYRELLIN (Premix)

Premix containing pyrethrins and rotenone. Broad spectrum, botanically derived product with a blend of essential oils and soya oil.

Pyrellin E.C. (0.042 lb./gal. pyrethrins & 0.035 lb./gal. rotenone) *(Webb Wright)*

Rate: 1-2 pt. E.C./A

Pest:
aphids, loopers, mites, plant bugs, thrips, whiteflies

Remarks: Anise, basil, dill, fennel, ginger, etc. Apply at 7-day intervals or less; repeat as needed. Refer to label for minor herb and spice crops.
Days To Harvest: 0.

Signal Word/Toxicity Class: Caution/III.
REI: 12 hr.

Pyrenone — see pyrethrins

PYRETHRINS

Contains piperonyl butoxide as a synergist. Diatect also contains silicon dioxide (diatomaceous earth). May be combined in tank mix with other insecticides for quicker and more complete kill where insect resistance may be a problem, and as an "exciter" to flush insects out of hiding and into contact with other spray residues.

Diatect Multipurpose Insecticide II *(Diatect)*
Pyrenone (0.5EC) *(AgrEvo Environmental Health)*
Py-Rin 60-6 EC *(Wilbur-Ellis)*

Rate: 1-6 lb. II/A
1-12 fl. oz. 0.5EC/A
2-16 oz. 6-60 EC/A

Pest:
aphids, loopers, mites, plant bugs, thrips, whiteflies

Remarks: Anise, basil, dill, fennel, etc. Apply in sufficient water for thorough coverage; repeat as needed. Apply Diatect as dust or wettable powder.
Days To Harvest: 0.

Signal Word/Toxicity Class: Caution.
REI: 12 hr.

Py-Rin — see pyrethrins

SunSpray — see petroleum oil

◆ **- new product** • ai-active ingredient • DF-dry flowable • E/EC-emulsifiable concentrate • EW-emulsion in water
F/FL-flowable • G-granule • L/LC-liquid • SP/WSP-soluble packet • W/WP-wettable powder • WSB-water soluble bag

Horseradish

Ambush — see permethrin

AZADIRACHTIN

Extracted from seed of neem tree. Insect growth regulator that controls targeted insect larvae when they ingest or come in contact with it, by interfering with the insect's ability to molt and shutting off feeding behavior. Affected larvae slowly starve and die at next molt.

◆ **Azatin XL Plus** *(Thermo Trilogy)*
Rate: 5-21 oz. XL Plus/A
Pest:

aphids	cutworms	Japanese beetle
armyworms	diamondback moth	leafhoppers
beetles	fall armyworm	leafminers
black cutworm	flea beetles	loopers
caterpillars	green peach aphid	thrips
corn earworm	imported cabbageworm	webworms

Remarks: Rate varies by species. Apply when pests first appear; repeat at 7-day intervals. When pest populations are high, use higher rates. Do not apply more that 21 fl. oz./A.
Days To Harvest: 0.

Signal Word/Toxicity Class: Caution.
REI: 4 hr.

◆ **NEEMAZAD 4.5EC or NEEMIX 4.5** *(Thermo Trilogy)*
Rate: 0.125-1 pt. 4.5/A
High Volume:
2.25-4.5 oz. 4.5EC/100 gal. water
Low Volume:
0.5 pt. 4.5EC/A
Pest:

aphids	caterpillars	leafhoppers
armyworms	Colorado potato beetle	potato flea beetle
artichoke plume moth	diamondback moth	psyllids
aster leafhopper	greenhouse whitefly	sweetpotato whitefly
cabbage loopers	hornworms	webworms

Remarks: For low volume, apply in sufficient water for adequate coverage. For low insect pressure, apply preventive at 7-day intervals or at first sign of insects; high insect pressure, apply at 3- to 4-day intervals.

Signal Word/Toxicity Class: Warning/II.
REI: 12 hr.

Azatin — see azadirachtin

CARBARYL

Carbamate insecticide with broad activity on many insects. Highly toxic to bees.

Clean Crop Sevin 5 Bait *(Platte)*
Rate: 1-2 lb. ai/A
20-40 lb. Bait/A
Pest:

ants	darkling beetle	grasshoppers
armyworms	field crickets	sowbugs
cutworms		

Remarks: Apply with ground or air equipment.
Days To Harvest: 3.

Signal Word/Toxicity Class: Caution/III.
REI: 12 hr.

Drexel Carbaryl 4L *(Drexel)*
Sevin 4F, XLR Plus, 80S, 80WSP *(Rhone-Poulenc)*
Rate: 0.5-2 lb. ai/A
1-4 pt. 4L, 4F, XLR/A
0.66-2.5 lb. 80S, 80WSP/A
Pest:

armyworms	flea beetles	leafhoppers
aster leafhopper	grasshoppers	stink bugs
corn earworm	harlequin bug	tarnished plant bug
fall armyworm	imported cabbageworm	

Remarks: Use high rate for adult grasshoppers or dense vegetation.
Days To Harvest: 3.

Signal Word/Toxicity Class: Caution (4F, 4L, XLR); Warning/II (80S, 80WSP).
REI: 12 hr.

Gowan Sevin Bait *(Gowan)*
Rate: 2 lb. ai/A
40 lb. Bait/A
Pest:

armyworms	cutworms	grasshoppers
crickets	darkling beetle	sowbugs

Remarks: Apply when pests appear in damaging numbers; repeat at 7- to 14-day intervals if necessary.
Days To Harvest: 3.
State Restrictions: For use in Arizona and California.

Signal Word/Toxicity Class: Caution/III.
REI: 12 hr.

DEADLINE

Packaged as Bullets, M-Ps (mini-pellets), and granules.

Deadline Bullets, M-Ps, Granules (4%) (metaldehyde) *(Pace)*
Rate: 0.8-1 lb. ai/A
20-40 lb. 4%/A
Pest:

slugs	snails

Remarks: Apply every 3-4 weeks during growing season or as needed. Do not apply directly to edible portion of plant.

Signal Word/Toxicity Class: Warning/II (granule); Caution (bullets, M-P).
REI: 12 hr.

Diatect — see pyrethrins

Fyfanon — see malathion

LANNATE

Restricted Use Pesticide.

Lannate SP or LV (methomyl) *(DuPont)*
Rate: 0.45 lb. ai/A
0.5 lb. SP/A
1.5 pt. LV/A
Pest:

aphids	thrips

Remarks: Ground application only. Apply in sufficient water to obtain thorough, uniform coverage when insects first appear; repeat at 5- to 7-day intervals or as needed. Do not apply more than 1.8 lb. ai/A per crop or make more than 4 applications per crop.
Days To Harvest: 65.

Signal Word/Toxicity Class: Danger-Poison/I.
REI: 48 hr.

MALATHION

Organophosphate insecticide toxic to fish, aquatic invertebrates, aquatic life stages of amphibians, and bees exposed to direct treatment on blooming crops or weeds.

Clean Crop Malathion 57EC or 8EC *(Platte)*
Rate: 0.7-1.4 lb. ai/A
1-2 pt. 57EC/A
0.5-1 lb. ai/A
0.5-1 pt. 8EC/A
Pest:

aphids	cabbage looper	imported cabbageworm

Remarks: Apply when insects first appear and repeat as necessary.
Days To Harvest: 7.

Signal Word/Toxicity Class: Caution/III.
REI: 12 hr.

For personal protective equipment and EPA registration numbers, please turn to page 39.

Fyfanon (8E) *(Helena)*
◆ **Gowan Malathion 8 (8E)** *(Gowan)*
◆ **Wilbur-Ellis Malathion 8 Spray** *(Wilbur-Ellis)*

Rate: 2 lb. ai/A
2 pt. 8E/A (Gowan, Wilbur-Ellis)
2.5 lb. ai/A
2.5 pt. 8E/A (Fyfanon)

Pest:
aphids, flea beetles, leafhoppers, diamondback moth

Remarks: None.
Days To Harvest: 7.

Signal Word/Toxicity Class: Caution/III.
REI: 12 hr.

Drexel Malathion 5EC *(Drexel)*
◆ **Micro Flo Malathion 5EC** *(Micro Flo)*
Prentox 5 lb. Malathion Spray (5EC) *(Prentiss)*

Rate: 1-1.25 lb. ai/A
1.5-2 pt. 5EC, 57EC/A

Pest:
aphids

Remarks: Apply when insects first appear and repeat as necessary.
Days To Harvest: 7.

Signal Word/Toxicity Class: Warning/II.
REI: 12 hr.

M-PEDE

M-Pede (potassium salts of fatty acids) *(Mycogen)*

Rate: 1-2% v/v solution

Pest:
aphids*, leafminers*, spider mites, broad mite, plant bugs, thrips, green peach aphid*, russet mite, whiteflies, leafhoppers

**** Tank mix only***

Remarks: Apply at first sign of infestation; repeat at 7- to 14-day intervals. May be tank mixed with other insecticides; refer to labels for use directions and precautions.
Days To Harvest: 0.

Signal Word/Toxicity Class: Warning/II.
REI: 12 hr.

NEEMAZAD — see azadirachtin

NEEMIX— see azadirachtin

PERMETHRIN

Restricted Use Pesticide.

Ambush 2E *(ZENECA)*
Pounce 3.2 EC, 25WP or WSB *(FMC)*

Rate: Foliar:
0.2 lb. ai/A
12.8 fl. oz. 2E/A (0.1% ai solution as spring preplant dip)
16 oz. 3.2 EC/A
12.8 oz. 25WP/A
2 bags WSB/A

Pest:
imported crucifer weevil

Remarks: Apply by ground equipment in minimum spray volume of 20 gal./A. Make up to 3 foliar applications as needed to control weevil adults during oviposition. For 0.1% ai solution, soak sets for 30 minutes and air dry before planting.
Days To Harvest: 22.

Signal Word/Toxicity Class: Caution (3.2 EC); Warning/II (2E, 25WP, WSB).
REI: 24 hr.

Pounce — see permethrin

PYRELLIN (Premix)

Premix containing pyrethrins and rotenone. May be combined with other insecticides to enhance knockdown or as an insect "exciter" to help facilitate contact with other spray residues.

Pyrellin E.C. (0.042 lb./gal. pyrethrins & 0.035 lb./gal. rotenone) *(Webb Wright)*

Rate: 1-2 pt. E.C./A

Pest:
aphids, loopers, plant bugs, leafhoppers, Lygus bugs, thrips, leafminers, mites, whiteflies

Remarks: Apply as necessary to maintain control.
Days To Harvest: 0.

Signal Word/Toxicity Class: Caution/III.
REI: 12 hr.

Pyrenone — see pyrethrins

PYRETHRINS

Contains piperonyl butoxide as a synergist. Diatect also contains silicon dioxide (diatomaceous earth). May be combined in tank mix with other insecticides for quicker and more complete kill where insect resistance may be a problem, and as an "exciter" to flush insects out of hiding and into contact with other spray residues.

Diatect Multipurpose Insecticide II *(Diatect)*
Pyrenone (0.5EC) *(AgrEvo Environmental Health)*
Py-Rin 60-6 EC *(Wilbur-Ellis)*

Rate: 1-6 lb. II/A
1-12 fl. oz. 0.5EC/A
2-16 oz. 6-60 EC/A

Pest: Including but not limited to:
aphids, harlequin bugs, mites, diamondback moth, leafhoppers, thrips, cucumber beetle, leafminers, vegetable weevil, flea beetles, loopers, whiteflies

Remarks: Apply in sufficient water for thorough coverage; repeat as needed. Apply Diatect as dust or wettable powder.
Days To Harvest: 0.

Signal Word/Toxicity Class: Caution.
REI: 12 hr.

Py-Rin — see pyrethrins

Sevin — see carbaryl

Lentils

AZADIRACHTIN

Extracted from seed of neem tree. Insect growth regulator that controls targeted insect larvae when they ingest or come in contact with it, by interfering with the insect's ability to molt and shutting off feeding behavior. Affected larvae slowly starve and die at next molt.

◆ **Azatin EC or XL** *(Olympic)*
◆ **Azatin XL Plus** *(Thermo Trilogy)*

Rate: 8-21 oz. EC/A
5-21 oz. XL, XL Plus/A

Pest:
aphids, cutworms, Japanese beetle, armyworms, diamondback moth, leafhoppers, beetles, fall armyworm, leafminers, black cutworm, flea beetles, loopers, caterpillars, green peach aphid, thrips, corn earworm, imported cabbageworm, webworms

◆ - **new product** • ai-active ingredient • DF-dry flowable • E/EC-emulsifiable concentrate • EW-emulsion in water
F/FL-flowable • G-granule • L/LC-liquid • SP/WSP-soluble packet • W/WP-wettable powder • WSB-water soluble bag

Remarks: Rate varies by species. Apply when pests first appear; repeat at 7-day intervals. When pest populations are high, use higher rates. Do not apply more that 21 fl. oz./A.
Days To Harvest: 0.

Signal Word/Toxicity Class: Caution (XL); Warning (EC).
REI: 4 hr. (XL), 12 hr. (EC).

◆ NEEMAZAD 4.5EC or NEEMIX 4.5 *(Thermo Trilogy)*
Rate: 0.125-1 pt. 4.5/A
High Volume:
2.25-4.5 oz. 4.5EC/100 gal. water
Low Volume:
0.5 pt. 4.5EC/A
Pest:
aphids, armyworms, artichoke plume moth, aster leafhopper, cabbage loopers, caterpillars, Colorado potato beetle, diamondback moth, greenhouse whitefly, hornworms, leafhoppers, potato flea beetle, psyllids, sweetpotato whitefly, webworms
Remarks: For low volume, apply in sufficient water for adequate coverage. For low insect pressure, apply preventive at 7-day intervals or at first sign of insects; high insect pressure, apply at 3- to 4-day intervals.

Signal Word/Toxicity Class: Warning/II.
REI: 12 hr.

Azatin — see azadirachtin

CARBARYL

Carbamate insecticide with broad activity on many insects. Highly toxic to bees.

Sevin 4F, XLR Plus, 80S, 80WSP *(Rhone-Poulenc)*
Rate: 0.5-1.5 lb. ai/A
1-3 pt. 4F, XLR/A
0.66-1.875 lb. 80S, 80WSP/A
Pest:
alfalfa caterpillar, alfalfa looper, armyworms, bean leaf beetle, blister beetle, corn earworm, cucumber beetle, cutworms, European corn borer, fall armyworm, flea beetles, grasshoppers, green cloverworm, Japanese beetle, leafhoppers, Mexican bean beetle, stink bugs, tarnished plant bug, threecornered alfalfa hopper, thrips, velvetbean caterpillar, webworms, western bean cutworm

Rate: 2 lb. ai/A
2.5 lb. 80S, 80WSP/A
Pest:
cowpea curculio
Remarks: Do not apply to wet foliage or when rain or high humidity is expected during next 2 days.
Days To Harvest: 7.

Signal Word/Toxicity Class: Caution (4F, XLR); Warning/II (80S, 80WSP).
REI: 12 hr.

Digon — see dimethoate

Dimate — see dimethoate

DIMETHOATE

Cheminova Dimethoate 4E *(Cheminova)*
Digon 400 *(Wilbur-Ellis)*
Rate: 0.17-0.5 lb. ai/A
0.33-1 pt. 4E, 4EC, 400/A
Pest:
aphids, lygus bugs
Remarks: Do not apply if crop or weeds in treatment area are in bloom. Do not make more than 1 application per season. Do not feed or graze hay within 21 days after last application.
Days To Harvest: 0.
State Restrictions: Not for use in California.

Signal Word/Toxicity Class: Warning/II.
REI: 48 hr.

Clean Crop Dimethoate 400 (4EC) *(Platte)*
Dimate 4E *(Terra)*
Gowan Dimethoate E267 *(Gowan)*
Rate: 0.25-0.5 lb. ai/A
0.75-1.5 pt. E267/A
0.5-1 pt. 4EC/A
Pest:
aphids, Lygus bugs
Remarks: Do not apply if crop or weeds in treatment area are in bloom. Do not feed or graze hay.
Days To Harvest: 14.
State Restrictions: Dimate 4E: SLN-Idaho, Washington.

Signal Word/Toxicity Class: Warning/II.
REI: 48 hr.

DI-SYSTON

Restricted Use Pesticide.

Di-Syston 8 (disulfoton) *(Bayer)*
Rate: 1-2.5 lb. ai/A
1-2.5 pt. 8/A
Pest:
aphids, leafhoppers
Remarks: Apply in water emulsion or with liquid fertilizer as in-furrow spray, injection on each side of seed furrow at planting, or injected as a sidedressing after plants are established. Do not apply more than once per season. Do not feed vines or hay.
Days To Harvest: 50.

Signal Word/Toxicity Class: Danger/I.
REI: 48 hr.

LANNATE

Restricted Use Pesticide.

Lannate SP or LV (methomyl) *(DuPont)*
Rate: 0.45-0.9 lb. ai/A
0.5-1 lb. SP/A
1.5-3 pt. LV/A
Pest:
western yellowstriped armyworm
Remarks: Apply in sufficient water to obtain thorough, uniform coverage when insects first appear; repeat at 5- to 7-day intervals or as needed. Do not apply more than 0.9 lb. ai/A per crop or make more than 2 applications per crop.
Days To Harvest: 21; hay-7; forage-3.

Signal Word/Toxicity Class: Danger-Poison/I.
REI: 48 hr.

LINDANE

Lindane 30C contains toxic inert Rhodamine B. Toxic to fish, birds, and wildlife.

Lindane 30C *(Gustafson)*
◆ Lindane ST 40 *(AGSCO)*
Rate: 2.7 fl. oz. 30C/100 lb. seed
2 fl. oz. 40/100 lb. seed
Pest:
seedcorn maggots, wireworms
Remarks: Do not use treated seed for food, feed, or oil purposes. Exposed treated seed may be hazardous to birds and other wildlife. Dispose of excess seed by burial away from bodies of water.

Signal Word/Toxicity Class: Warning/II.
REI: 12 hr.

7 Vegetables

NOTICE The information on these pages is for preliminary planning — not a guide for use. Be sure to follow the manufacturer's directions, notwithstanding information contained here.

For personal protective equipment and EPA registration numbers, please turn to page 39.

MALATHION

Organophosphate insecticide toxic to fish, aquatic invertebrates, aquatic life stages of amphibians, and bees exposed to direct treatment on blooming crops or weeds.

Clean Crop Malathion 57EC *(Drexel)*
Drexel Malathion 5EC *(Drexel)*
◆ **Gowan Malathion 8** *(Gowan)*
◆ **Micro Flo Malathion 5EC** *(Micro Flo)*
Prentox 5 lb. Malathion Spray (5EC) *(Prentiss)*
◆ **Wilbur-Ellis Malathion 8** *(Wilbur-Ellis)*

Rate: 0.94-1.25 lb. ai/A
1-2 pt. 57EC/A
0.94 lb. ai/A
1.5 pt. 5EC/A
1 lb. ai/A
1 pt. 8/A

Pest:
aphids

Remarks: None.
Days To Harvest: 3.

Signal Word/Toxicity Class: Caution/III (57EC, 8); Warning/II (5EC).
REI: 12 hr.

NEEMAZAD — see azadirachtin

NEEMIX — see azadirachtin

PYRELLIN (Premix)

Premix containing pyrethrins and rotenone. May be combined with other insecticides to enhance knockdown or as an insect "exciter" to help facilitate contact with other spray residues.

Pyrellin E.C. (0.042 lb./gal. pyrethrins & 0.035 lb./gal. rotenone) *(Webb Wright)*

Rate: 1-2 pt. E.C./A

Pest:
aphids, leafhoppers, leafminers, loopers, Lygus bugs, mites, plant bugs, thrips, whiteflies

Remarks: Apply as necessary to maintain control.
Days To Harvest: 0.

Signal Word/Toxicity Class: Caution/III.
REI: 12 hr.

Pyrenone — see pyrethrins

PYRETHRINS

Contains piperonyl butoxide as a synergist. May be combined in tank mix with other insecticides for quicker and more complete kill where insect resistance may be a problem, and as an "exciter" to flush insects out of hiding and into contact with other spray residues.

Diatect Multipurpose Insecticide II *(Diatect)*
Pyrenone (0.5EC) *(AgrEvo Environmental Health)*
Py-Rin 60-6 EC *(Wilbur-Ellis)*

Rate: 1-6 lb. II/A
1-12 fl. oz. 0.5EC/A
2-16 oz. 6-60 EC/A

Pest: Including but not limited to:
aphids, cabbage looper, cabbageworm, diamondback moth, flea beetles, harlequin bug, imported cabbageworm, leafhoppers, thrips

Remarks: Apply in sufficient water for thorough coverage; repeat as needed. Apply Diatect as dust or wettable powder.
Days To Harvest: 0.

Signal Word/Toxicity Class: Caution.
REI: 12 hr.

Py-Rin — see pyrethrins

Sevin — see carbaryl

Lettuce
(Chicory, Endive, Escarole)

ABAMECTIN

Natural product produced by soil microorganism, Streptomyces avermitilis. Unrelated to other miticides. Toxic to fish, mammals, aquatic organisms, and highly toxic to honeybees. For use on head lettuce only. Restricted Use Pesticide.

Agri-Mek 0.15EC *(Novartis)*

Rate: 8-16 fl. oz. 0.15EC/A

Pest:
leafminers

Remarks: Head lettuce. Apply when adults are first observed; repeat as necessary. Thorough coverage essential. Do not apply in less than 20 gal. water/A. Do not exceed 48 fl. oz./A per growing season. Do not make more than 2 sequential applications after transplanting. Do not apply through any type of irrigation system or by aircraft.
Days To Harvest: 7.

Signal Word/Toxicity Class: Warning/II.
REI: 12 hr.

Admire — see imidacloprid

Agri-Mek — see abamectin

Ambush — see permethrin

Ammo — see cypermethrin

AZADIRACHTIN

Extracted from seed of neem tree. Insect growth regulator that controls targeted insect larvae when they ingest or come in contact with it, by interfering with the insect's ability to molt and shutting off feeding behavior. Affected larvae slowly starve and die at next molt.

◆ **Azatin XL Plus** *(Thermo Trilogy)*

Rate: 5-21 oz. XL Plus/A

Pest:
aphids, armyworms, beet armyworm, beetles, black cutworm, cabbage looper, caterpillars, Colorado potato beetle, cutworms, diamondback moth, fall armyworm, flea beetles, green peach aphid, imported cabbageworm, Japanese beetle, leafhoppers, leafminers, leafrollers, loopers, Mexican bean beetle, pea aphid, psyllids, silverleaf whitefly, southern armyworm, sweetpotato whitefly, tent caterpillar, thrips, tobacco budworm, tomato fruitworm, webworms, whiteflies

Remarks: Lettuce, endive. Rate varies by species. Apply when pests first appear; repeat at 7-day intervals. When pest populations are high, use higher rates. Do not apply more that 21 fl. oz./A.
Days To Harvest: 0.

Signal Word/Toxicity Class: Caution.
REI: 4 hr.

◆ - **new product** • ai-active ingredient • DF-dry flowable • E/EC-emulsifiable concentrate • EW-emulsion in water
F/FL-flowable • G-granule • L/LC-liquid • SP/WSP-soluble packet • W/WP-wettable powder • WSB-water soluble bag

◆ **NEEMAZAD 4.5EC or NEEMIX 4.5** *(Thermo Trilogy)*

Rate: 0.125-1 pt. 4.5/A
High Volume:
2.25-4.5 oz. 4.5EC/100 gal. water
Low Volume:
0.5 pt. 4.5EC/A

Pest:

aphids	diamondback moth	onion maggot
armyworms	greenhouse whitefly	sweetpotato whitefly
cabbage loopers	hornworms	thrips
caterpillars	imported cabbageworm	webworms
cutworms	leafminers	weevils

Remarks: Endive, lettuce. For low volume, apply in sufficient water for adequate coverage. For low insect pressure, apply preventive at 7-day intervals or at first sign of insects; high insect pressure, at 3- to 4-day intervals.

Signal Word/Toxicity Class: Warning/II.
REI: 12 hr.

Azatin — see azadirachtin

BT 320 SULFUR 25 or 50 DUST (Premix)

Bt 320 Sulfur 25 Dust (0.064% Bt & 25% sulfur) or 50 Dust (0.064% Bt & 50% sulfur) *(Wilbur-Ellis)*

Rate: 20-50 lb. Dust/A
Pest:
cabbage looper

Remarks: Do not mix or dilute with other products. Thorough coverage essential.

Signal Word/Toxicity Class: Caution.
REI: 24 hr.

CARBARYL

Carbamate insecticide with broad activity on many insects. Highly toxic to bees.

Clean Crop Sevin 5 Bait *(Platte)*

Rate: 1-2 lb. ai/A
20-40 lb. Bait/A

Pest:

ants	darkling beetle	grasshoppers
armyworms	field crickets	sowbugs
cutworms		

Remarks: Endive (escarole), lettuce (head, leaf). Apply with ground or air equipment.
Days To Harvest: Head lettuce-3; leaf lettuce-14.

Signal Word/Toxicity Class: Caution/III.
REI: 12 hr.

Drexel Carbaryl 4L *(Drexel)*
Sevin 4F, XLR Plus, 80S or 80WSP *(Rhone-Poulenc)*

Rate: 0.5-2 lb. ai/A
1-4 pt. 4L, 4F, XLR/A
0.66-2.5 lb. 80S, 80WSP/A

Pest:

armyworms	grasshoppers	Lygus bugs
aster leafhopper	harlequin bug	spittlebugs
corn earworm	imported cabbageworm	stink bugs
fall armyworm		tarnished plant bug
flea beetles	leafhoppers	

Endive (escarole):

Lygus bugs	spittlebugs

Remarks: Endive (escarole), lettuce (head, leaf). Rate varies by pest species; refer to label. Apply at 5- to 7-day intervals after heads begin to form. Do not apply to wet foliage or when rain or high humidity is expected during next 2 days.
Days To Harvest: Head lettuce-3; endive, leaf lettuce-14.

Signal Word/Toxicity Class: Caution (4F, 4L, XLR); Warning/II (80S, 80WSP).
REI: 12 hr.

Gowan Sevin Bait *(Gowan)*

Rate: 2 lb. ai/A
40 lb. Bait/A

Pest:

armyworms	cutworms	grasshoppers
crickets	darkling beetle	sowbugs

Remarks: Endive, lettuce (head, leaf). Apply when pests appear in damaging numbers; repeat at 7- to 14-day intervals if necessary.
Days To Harvest: Head lettuce-3; endive, leaf lettuce-14.
State Restrictions: For use in Arizona and California.

Signal Word/Toxicity Class: Caution/III.
REI: 12 hr.

CRYOLITE

Kryocide (96%) *(Elf Atochem)*

Rate: 7.7-19.2 lb. ai/A
8-20 lb. 96%/A

Pest:

armyworms	corn earworm	tobacco budworm
cabbage looper		

Remarks: Lettuce (head, leaf). Apply at 7-day intervals as needed. Remove excess residues on edible portions by washing, brushing, or other effective means. Do not exceed 160 lb./A per season. Do not use in combination with lime or compounds containing free lime. Do not use chemigation in California.
Days To Harvest: 14.

Signal Word/Toxicity Class: Caution.
REI: 12 hr.

lambda-CYHALOTHRIN

Synthetic pyrethroid insecticide with broad-spectrum activity, formulated as an emulsifiable concentrate. Restricted Use Pesticide due to toxicity to fish and aquatic organisms.

Karate or Warrior (1E) *(ZENECA)*

Rate: 0.015-0.03 lb. ai/A
1.92-3.84 fl. oz. 1E/A

Pest:

alfalfa looper	fall armyworm (1st & 2nd instar)	meadow spittlebug
aphids*		plant bugs
armyworms	flea beetles	saltmarsh caterpillar
beet armyworm (1st & 2nd instar)	grasshoppers	southern armyworm
	green cloverworm	spider mites*
cabbage looper	imported cabbageworm	stink bugs
corn earworm		tobacco budworm
cutworms	Japanese beetle (adult)	vegetable weevil (adult)
diamondback moth		
European corn borer	leafhoppers	whiteflies*

* *Suppression*

Remarks: Head lettuce. Apply as required by scouting, usually at intervals of 5 or more days. Do not apply more than 2.4 pt./A per season. Refer to label for spray drift precautions and state and county restrictions.
Days To Harvest: 1.

Signal Word/Toxicity Class: Danger/I.
REI: 24 hr.

CYPERMETHRIN

Synthetic pyrethroid insecticide with both contact and stomach poison activity. No systemic activity and is ineffective against soil pests. Controls many caterpillars, beetles, flies, and plant bugs, and has fast knockdown effects. Toxic to fish and bees, and may cause eye and skin irritation. Restricted Use Pesticide.

Ammo 2.5 EC or WSB *(FMC)*

Rate: 0.05-0.1 lb. ai/A
2.5-5 fl. oz. 2.5 EC/A
1-2 bags WSB/A

Pest:

armyworms	diamondback moth	Lygus bugs
corn earworm	flea beetles	onion thrips
crickets	imported cabbageworm	saltmarsh caterpillar
cucumber beetle	leafhoppers	stink bugs
cutworms	loopers	tobacco budworm

7 Vegetables

NOTICE The information on these pages is for preliminary planning — not a guide for use. Be sure to follow the manufacturer's directions, notwithstanding information contained here.

For personal protective equipment and EPA registration numbers, please turn to page 39.

Remarks: Head lettuce. Rate varies by pest species. Apply in water using minimum of 15 gal. finished spray/A by ground; 5 gal./A by air. Lower rates should be used for light to moderate insect pressure, higher rates for heavy to extremely heavy insect populations. Apply no more than 0.6 lb. ai/A per season.

Days To Harvest: 5.

Crop Rotation: Not within 30 days after last application.

Signal Word/Toxicity Class: Caution (2.5 EC); Warning/II (WSB).

REI: 12 hr.

zeta-CYPERMETHRIN

Synthetic pyrethroid with broad spectrum activity. Extremely toxic to fish and aquatic invertebrates; highly toxic to bees. Restricted Use Pesticide.

Fury 1.5 EC or Mustang 1.5 EW *(FMC)*

Rate: 0.028-0.05 lb. ai/A
2.39-4.26 fl. oz./A

Pest:

corn earworm, cucumber beetles, cutworms, diamondback moth, flea beetles, imported cabbageworm, leafhoppers, Lygus bugs, saltmarsh caterpillar, tobacco budworm

Rate: 0.040-0.05 lb. ai/A
3.41-4.26 fl. oz./A

Pest:

armyworms, crickets, loopers, onion thrips, stink bugs

Remarks: Head lettuce. Apply in water as necessary using minimum of 15 gal. finished spray/A by ground; 5 gal. finished spray/A by air. Rate varies by pest species, insect pressure, timing of spray, and field scouting. Up to 0.3 lb. ai/A may be applied per season. In arid climatic areas, such as Arizona and California, higher than minimum recommended rates may be required.

Days To Harvest: 5.

Crop Rotation: Not within 30 days after last application.

Signal Word/Toxicity Class: Warning/II.

REI: 12 hr.

DEADLINE

Packaged as Bullets, M-Ps (mini-pellets), and granules.

Deadline Bullets, M-Ps, Granules (4%) (metaldehyde) *(Pace)*

Rate: 0.8-1 lb. ai/A
20-40 lb. 4%A

Pest:

slugs, snails

Remarks: Endive, lettuce. Apply every 3-4 weeks during growing season or as needed. Do not apply directly to edible portion of plant.

Signal Word/Toxicity Class: Warning/II (granule); Caution (bullets, M-P).

REI: 12 hr.

Declare — see methyl parathion

Diatect — see pyrethrins

DIAZINON

Restricted Use Pesticide due to avian and aquatic toxicity.

D•z•n diazinon AG600 WBC or 50W *(Novartis)*
Drexel Diazinon Insecticide (4EC), D-264 14G *(Drexel)*
Gowan Diazinon 4E, 14G, 50WSB *(Gowan)*
Micro Flo Diazinon AG500, 14G, 50W *(Micro Flo)*
Prentox Diazinon AG500, 50W *(Prentiss)*
◆ **Wilbur-Ellis Diazinon 4 (4E), 14G** *(Wilbur-Ellis)*

Rate: 0.25-0.5 lb. ai/A
0.5-1 pt. 4E, 4EC, AG500/A
6.5-12.75 fl. oz. AG600/A
0.5-1 lb. 50W, 50WSB/A

Pest:

aphids, dipterous leafminer

Rate: 1 lb. ai/A
2 pt. 4E, 4EC, AG500/A
25.5 fl. oz. AG600/A
2 lb. 50W, 50WSB/A
7 lb. 14G/A

Pest:

mole crickets

Rate: 2-4 lb. ai/A
4-8 pt. 4E, 4EC, AG500/A
51-102 fl. oz. AG600/A
4-8 lb. 50W, 50WSB/A
14-28 lb. 14G/A

Pest:

cutworms

Rate: 3-4 lb. ai/A
6-8 pt. 4E, 4EC/A
76.5-102 fl. oz. AG600/A
6-8 lb. 50W, 50WSB/A
21-28 lb. 14G/A

Pest:

wireworms

Remarks: Endive (escarole), lettuce. For aphids and leafminers, apply as insects appear; repeat at 7-day intervals as needed. Do not apply more than 5 times per season. Will not control organophosphate-resistant leafminers. For mole crickets, apply to plant beds 1-2 days before sowing or transplanting, immediately work into soil. For cutworms and wireworms, apply just before planting, immediately incorporate into soil.

Days To Harvest: 14.

Signal Word/Toxicity Class: Caution/III.

REI: 24 hr.; 12 hr. (Drexel/Gowan/Wilbur-Ellis 14G).

Digon — see dimethoate

Dimate— see dimethoate

DIMETHOATE

Cheminova Dimethoate 4E *(Cheminova)*
Clean Crop Dimethoate 400 *(Platte)*
Dimate 4E *(Terra)*
Drexel Dimethoate 2.67 or 4EC *(Drexel)*
Gowan Dimethoate E267 *(Gowan)*
Helena Dimethoate 4E *(Helena)*
Micro Flo Dimethoate 267 or 4E *(Micro Flo)*
Wilbur-Ellis Dimethoate 267 or Digon 400 *(Wilbur-Ellis)*

Rate: 0.25 lb. ai/A
0.75 pt. 2.67, 267, E267/A
0.5 pt. 4E, 4EC, 400/A

Pest:

aphids, leafhoppers, leafminers

Remarks: Endive (escarole), lettuce (head, leaf).

Days To Harvest: 14; head lettuce-7.

Signal Word/Toxicity Class: Warning/II.

REI: 48 hr.

DI-SYSTON

Restricted Use Pesticide.

Di-Syston 8 or 15% (disulfoton) *(Bayer)*

Rate: 1-2 lb. ai/A
1-2 pt. 8/A
6.7-13.3 15%/A (20" row spacing)

Pest:

aphids, leafhoppers, mites, root aphids (8), thrips

Remarks: Spring- or summer-seeded lettuce: For Di-Syston 8, apply in water emulsion or with liquid fertilizer as an injection on each side of seed furrow at seeding time only. For Di-Syston 15%, place granules as soil incorporated band on each side of row at seeding time only. Use higher rate on heavy organic soils. Do not apply directly on seed or to transplanted lettuce.Fall- or winter-seeded lettuce: For Di-Syston 8, apply in water emulsion or as soil injected sidedressing with liquid fertilizer at thinning time. Refer to label for additional information.

Days To Harvest: 60.

◆ - **new product** • ai-active ingredient • DF-dry flowable • E/EC-emulsifiable concentrate • EW-emulsion in water
F/FL-flowable • G-granule • L/LC-liquid • SP/WSP-soluble packet • W/WP-wettable powder • WSB-water soluble bag

Rate: 1.2 oz. 8/1000 ft. row
Pest:
lettuce root aphid
Remarks: None.
State Restrictions: SLN-California.

Signal Word/Toxicity Class: Danger/I.
REI: 48 hr.

D•z•n — see diazinon

ENDOSULFAN

Toxic to birds, fish, and wildlife; toxic to bees exposed to direct application.

Drexel Endosulfan 3EC or 50WP *(Drexel)*
Gowan Endosulfan 3EC or 50WSB *(Gowan)*
◆ **Micro Flo Endosulfan 3EC or 50WP** *(Micro Flo)*
Phaser 3EC or 50WSB *(AgrEvo USA)*
Thiodan 3EC or 50WP *(FMC)*

Rate: 0.75-1 lb. ai/A
1-1.33 qt. 3EC/A
1.5-2 lb. 50WP, 50WSB/A
Pest:
cabbage looper, green peach aphid, imported cabbageworm, diamondback moth
Remarks: Lettuce (head, leaf). Apply in sufficient water for thorough coverage when insects first appear; repeat as necessary. Coverage of upper and lower leaf surfaces essential for good control. Do not exceed 2 applications on leaf lettuce; 3 applications after thinning head lettuce. Remove outside leaves at harvest. Do not feed crop refuse to livestock. Do not exceed 3 lb. ai/A per year.
Days To Harvest: 14.

Signal Word/Toxicity Class: Danger-Poison/I.
REI: 24 hr.

Fury — see *zeta*-cypermethrin

Fyfanon — see malathion

IMIDACLOPRID

Admire 2F *(Bayer)*
Rate: 0.16-0.375 lb. ai/A
10-24 fl. oz. 2F/A
Pest:
aphids, silverleaf whitefly, sweetpotato whitefly
Remarks: Lettuce (head, leaf). Direct seed or transplant. Soil application. Use higher rates where infestations occur later in crop development, or where pest pressure is continuous. Do not apply more than 24 fl. oz./A per year.
Days To Harvest: 21.

Signal Word/Toxicity Class: Caution.
REI: 12 hr.

Provado 1.6 *(Bayer)*
Rate: 0.05 lb. ai/A
3.75 fl. oz. 1.6/A
Pest:
aphids, silverleaf whitefly, sweetpotato whitefly
Remarks: Lettuce (head, leaf). Apply before damaging population becomes established. For whiteflies, apply at first appearance; repeat at 5- to 7-day intervals until no longer in crop. For aphids, good coverage of foliage necessary. Do not apply more than 18.75 fl. oz./A per year.
Days To Harvest: 7.

Signal Word/Toxicity Class: Caution.
REI: 12 hr.

Karate — see *lambda*-cyhalothrin

Kryocide — see cryolite

LANNATE

Restricted Use Pesticide.

Lannate SP or LV (methomyl) *(DuPont)*
Rate: 0.225-0.9 lb. ai/A
0.25-1 lb. SP/A
0.75-3 pt. LV/A
Pest:
aphids, aster leafhopper, beet armyworm, corn earworm, loopers, thrips, variegated cutworm
Remarks: Lettuce (head, leaf). Apply in sufficient water to obtain thorough, uniform coverage when insects first appear; repeat at 5- to 7-day intervals or as needed. For head lettuce, do not apply more than 9 lb. ai/A per crop or make more than 15 applications per crop. For leaf lettuce, do not apply more than 3.6 lb. ai/A per crop or make more than 8 applications per crop.
Days To Harvest: 0.225-0.45 lb. ai-7, over 0.45 lb. ai-10.
State Restrictions: Leaf lettuce for Arizona, California only.

Rate: 0.45-0.9 lb. ai/A
0.5-1 lb. SP/A
1.5-3 pt. LV/A
Pest:
beet armyworm, leafhoppers, variegated cutworm
Remarks: Chicory. Apply in sufficient water to obtain thorough, uniform coverage when insects first appear at 5- to 7-day intervals or as needed. Do not apply more than 1.8 lb. ai/A per crop or make more than 2 applications per crop.
Days To Harvest: 80.

Signal Word/Toxicity Class: Danger-Poison/I.
REI: 48 hr.

LARVIN

Broad-spectrum carbamate insecticide that has stomach poison activity and some limited contact activity. Known to have systemic action as a seed treatment. Toxic to fish and birds, and moderately toxic to bees as direct treatment.

Larvin 3.2 (thiodicarb) *(Rhone-Poulenc)*
Rate: 0.4-0.75 lb. ai/A
16-30 fl. oz. 3.2/A
Pest:
alfalfa looper, beet armyworm, bollworm, corn earworm, fall armyworm, southern armyworm, tomato fruitworm

Rate: 0.6-0.75 lb. ai/A
24-30 fl. oz. 3.2/A
Pest:
cabbage looper
Remarks: Endive, lettuce (head, leaf). Spray as needed. Use higher rates for heavier infestations of larger larvae. Do not exceed 1.5 lb. ai/A per season.
Days To Harvest: 14.

Signal Word/Toxicity Class: Warning/II.
REI: 12 hr.

MALATHION

Organophosphate insecticide toxic to fish, aquatic invertebrates, aquatic life stages of amphibians, and bees exposed to direct treatment on blooming crops or weeds.

Clean Crop Malathion (57EC) *(Platte)*
◆ **Wilbur-Ellis Malathion 8** *(Wilbur-Ellis)*
Rate: Lettuce:
0.63-1.25 lb. ai/A
1-2 pt. 57EC/A
1-2 lb. ai/A
1-2 pt. 8/A
Pest:
aphids, leafhoppers

NOTICE	The information on these pages is for preliminary planning — not a guide for use. Be sure to follow the manufacturer's directions, notwithstanding information contained here.	For personal protective equipment and EPA registration numbers, please turn to page 39.

7 Vegetables

Rate: Endive:
0.32-0.63 lb. ai/100 gal.
0.5-2 pt. 57EC/100 gal.
0.5-1 lb. ai/A
0.5-1 pt. 8/A

Pest:
aphids mites

Remarks: Endive, lettuce (head, leaf). Apply in minimum 5 gal. by air; 10 gal. by ground.
Days To Harvest: Endive, head lettuce-7; leaf lettuce-14.

Signal Word/Toxicity Class: Caution/III.
REI: 12 hr.

Drexel Malathion 5EC *(Drexel)*
◆ **Micro Flo Malathion 5EC** *(Micro Flo)*
Prentox 5 lb. Malathion Spray (5EC) *(Prentiss)*

Rate: Lettuce:
1.25-1.9 lb. ai/A
2-3 pt. 5EC/A

Pest:
aphids leafhoppers spider mites
cabbage looper

Rate: Endive:
0.94-0.63 lb. ai/A
1.5-2 pt. 57EC/A

Pest:
aphids spider mites

Remarks: Endive, lettuce (head, leaf). Apply in minimum 5 gal. by air; 10 gal. by ground.
Days To Harvest: Endive, head lettuce-7; leaf lettuce-14.

Signal Word/Toxicity Class: Warning/II.
REI: 12 hr.

Fyfanon (8E) *(Helena)*
◆ **Gowan Malathion 8 (8E)** *(Gowan)*

Rate: 2 lb. ai/A
2 pt. 8E/A

Pest:
aphids cabbage looper (Fyfanon) leafhoppers
alfalfa looper mites

Remarks: Endive, lettuce (head, leaf).
Days To Harvest: Head lettuce, endive-7; leaf lettuce-14.

Signal Word/Toxicity Class: Caution/III.
REI: 12 hr.

Marlate — see methoxychlor

METASYSTOX-R

Restricted Use Pesticide.

Metasystox-R (SC) (oxydemeton-methyl) *(Gowan)*

Rate: 0.375-0.5 lb. ai/A
1.5-2 pt. SC/A

Pest:
aphids mites

Remarks: Head lettuce. Apply in sufficient water for complete coverage but not less than 1 gal./A. Do not apply more than 3 times per season.
Days To Harvest: 21; Arizona-28; California-refer to label.

Signal Word/Toxicity Class: Warning/II.
REI: 48 hr.; 72 hr. if average rainfall less than 25" per year.

METHOXYCHLOR

Organochlorine pesticide toxic to fish and aquatic invertebrates.

Marlate 50 (WP) *(Kincaid)*
Prentox Methoxychlor 2EC *(Prentiss)*

Rate: 1.3-1.7 lb. ai/A
75-100 fl. oz. 2EC/A
2 Tbs. 50WP/ gal. water

Pest:
alfalfa looper flea beetles Japanese beetle
blister beetle imported cabbageworm leafhoppers
fall armyworm

Remarks: Apply at first sign of infestation; repeat at 7- to 14-day intervals as needed.
Days To Harvest: 14.

Signal Word/Toxicity Class: Caution.
REI: 12 hr.

METHYL PARATHION

Organophosphate insecticide toxic to bees exposed to direct treatment or residues on blooming crops or weeds. Restricted Use Pesticide due to very high acute toxicity to humans and birds.

Cheminova Methyl 4EC *(Cheminova)*

Rate: 1 lb. ai/A
2 pt. 4EC/A

Pest:
aphids flea beetles mites
armyworms leafhoppers

Remarks: Apply when insects first appear; repeat as needed.
Days To Harvest: 21.

Signal Word/Toxicity Class: Danger-Poison/I.
REI: 48 hr.; 72 hr. if average rainfall less than 25" per year.

◆ **Declare (4EC)** *(Griffin)*
Helena 4 lb. Methyl Parathion (4EC) *(Helena)*
Wilbur-Ellis Methyl Parathion 7.5 Spray *(Wilbur-Ellis)*

Rate: 1 lb. ai/A
2 pt. 4E/A
0.47-0.94 lb. ai/A
0.5-1 pt. 7.5/A

Pest:
aphids flea beetles leafhoppers
armyworms imported cabbageworm mites
cabbage looper plant bugs

Remarks: None.
Days To Harvest: 21.

Signal Word/Toxicity Class: Danger-Poison/I.
REI: 48 hr.; 72 hr. if annual rainfall is less than 25" per year (Wilbur-Ellis).

Wilbur-Ellis Methyl Parathion 4 or 5 Spray *(Wilbur-Ellis)*

Rate: 0.5 lb. ai/A
1 pt. 4/A
0.63 lb. ai/A
1 pt. 5/A
Cabbage looper (5 Spray):
1 lb. ai/A
1.6 pt. 5/A

Pest:
aphids flea beetles leafhoppers
armyworms imported cabbageworm spider mites
cabbage looper

Remarks: Do not apply through any type of irrigation system.
Days To Harvest: 21.

Signal Word/Toxicity Class: Danger-Poison/I.
REI: 24 (4 Spray); 48 (5 Spray).

Microthiol Special — see sulfur

M-PEDE

M-Pede (potassium salts of fatty acids) *(Mycogen)*

Rate: 1-2% v/v solution

Pest:
aphids* leafminers* spider mites
broad mite plant bugs thrips
green peach aphid* russet mite whiteflies
leafhoppers

*** Tank mix only**

Remarks: Endive (escarole), lettuce. Apply at first sign of infestation; repeat at 7- to 14-day intervals. May be tank mixed with other insecticides; refer to labels for use directions and precautions.
Days To Harvest: 0.

Signal Word/Toxicity Class: Warning/II.
REI: 12 hr.

◆ - **new product** • ai-active ingredient • DF-dry flowable • E/EC-emulsifiable concentrate • EW-emulsion in water
F/FL-flowable • G-granule • L/LC-liquid • SP/WSP-soluble packet • W/WP-wettable powder • WSB-water soluble bag

Mustang — see *zeta*-cypermethrin

NEEMAZAD — see azadirachtin

NEEMIX — see azadirachtin

PERMETHRIN

Ambush 25 W is wettable powder or water-soluble pack formulation. Restricted Use Pesticide due to toxicity to fish and aquatic organisms.

Ambush (2E, 25W) (WP or WSP) *(ZENECA)*

Rate: 0.1-0.2 lb. ai/A
6.4-12.8 fl. oz. 2E/A
6.4-12.8 oz. 25W/A

Pest:

beet armyworm	cutworms	green cloverworm
cabbage looper	European corn borer	leafhoppers
corn earworm	fall armyworm	vegetable leafminer

Remarks: Endive (escarole), lettuce (head, leaf). Apply by air or ground in sufficient water to obtain full coverage at 3- to 5-day intervals or as needed by air or ground. Do not apply more than 2 lb. ai/A per season.

Days To Harvest: 1.

Signal Word/Toxicity Class: Warning/II.
REI: 24 hr.

PETROLEUM OIL

Petroleum based spray oil. Mode of action is through suffocation of eggs, larvae, and nymphs of insects and adults of soft bodied insects. This mode of action necessitates total spray coverage. Toxic to fish.

SAF-T-SIDE *(Brandt Consolidated)*
◆ **Sumulsion** *(Drexel)*
◆ **SunSpray Ultra-Fine** *(Sun)*

Rate: 1-2 gal./100 gal. water

Pest:

aphids	leafminers	plant bugs
caterpillars	loopers	thrips
leafhoppers	mites	whiteflies

Remarks: Lettuce. Apply as needed. May be tank mixed with most commonly used insecticides and fungicides. Do not use in combination with or immediately before or after spraying with fungicides such as captan, Dyrene, folpet, Morestan or any sulfur products. Do not use with carbaryl or dimethoate.

Signal Word/Toxicity Class: Caution.
REI: 4 hr.

Phaser — see endosulfan

Provado — see imidacloprid

PYRELLIN (Premix)

Premix containing pyrethrins and rotenone. May be combined with other insecticides to enhance knockdown or as an insect "exciter" to help facilitate contact with other spray residues.

Pyrellin E.C. (0.042 lb./gal. pyrethrins & 0.035 lb./gal. rotenone) *(Webb Wright)*

Rate: 1-2 pt. E.C./A

Pest:

aphids	harlequin bug	mites
bean beetle	leafhoppers	plant bugs
beet webworm	leafminers	stink bugs
cabbageworm	loopers	thrips
celery leaftier	Lygus bugs	whiteflies

Remarks: Lettuce (head, leaf, Romaine). Apply in intervals of 7 days or less; repeat as needed to maintain control.

Days To Harvest: 0.

Rate: 1.5-2 pt. E.C./A

Pest:

Colorado potato beetle	cucumber beetle	flea beetles
	European corn borer	vegetable weevil

Remarks: Lettuce (head, leaf, Romaine). Make 3-4 applications or more as eggs start to hatch; repeat at 5-7 day intervals.

Days To Harvest: 0.

Signal Word/Toxicity Class: Caution/III.
REI: 12 hr.

Pyrenone — see pyrethrins

PYRETHRINS

Contains piperonyl butoxide as a synergist. Diatect also contains silicon dioxide (diatomaceous earth). May be combined in tank mix with other insecticides for quicker and more complete kill where insect resistance may be a problem, and as an "exciter" to flush insects out of hiding and into contact with other spray residues.

Diatect Multipurpose Insecticide II *(Diatect)*
Pyrenone (0.5EC) *(AgrEvo Environmental Health)*
Py-Rin 60-6 EC *(Wilbur-Ellis)*

Rate: 1-6 lb. II/A
1-12 fl. oz. 0.5EC/A
2-16 oz. 6-60 EC/A

Pest: Including but not limited to:

aphids	cucumber beetle	mites
cabbage looper	diamondback moth	plant bugs
loopers	flea beetles	thrips
cabbageworm	leafhoppers	vegetable weevil
caterpillars	leafminers	webworms
corn earworm	loopers	whiteflies

Remarks: Lettuce, endive (escarole). Apply in sufficient water for thorough coverage; repeat as needed. Apply Diatect as dust or wettable powder.

Days To Harvest: 0.

Signal Word/Toxicity Class: Caution.
REI: 12 hr.

Py-Rin — see pyrethrins

SAF-T-SIDE — see petroleum oil

SCOUT X-TRA

Synthetic pyrethroid insecticide formulated as an emulsifiable concentrate. Scout X-TRA Gel is packaged in water-soluble bags. Low toxicity to humans and other mammals, highly toxic to bees. Restricted Use Pesticide due to toxicity to fish and aquatic organisms.

Scout X-TRA (0.9EC) or Gel (WSB) (tralomethrin) *(AgrEvo USA)*

Rate: 0.016-0.024 lb. ai/A
2.28-3.41 fl. oz. 0.9EC/A
1 WSB/12.5-8.3 A

Pest:

alfalfa looper	diamondback moth	imported cabbageworm
aphids	European corn borer	
beet armyworm	fall armyworm	leafhopper
cabbage looper	flea beetles	saltmarsh caterpillar
corn earworm	grasshoppers	stink bug
cutworms	green cloverworm	

Remarks: Lettuce (head, leaf). Uniform coverage essential. Refer to label for spray drift precautions and state restrictions on minimum spray carrier volumes. Do not apply more than 0.19 lb. ai/A per growing season.

Days To Harvest: 3.

Signal Word/Toxicity Class: Danger/I.
REI: 24 hr.

Sevin — see carbaryl

NOTICE The information on these pages is for preliminary planning — not a guide for use. Be sure to follow the manufacturer's directions, notwithstanding information contained here.	For personal protective equipment and EPA registration numbers, please turn to page 39.

SULFUR

Micronized sulfur.

Microthiol Special (80W) *(Elf Atochem)*
Rate: 4-4.8 lb. ai/A
5-6 lb. 80W/A
Pest:
red spider mite
Remarks: Lettuce (head, leaf). Apply in early leaf stage; repeat at 14-day intervals or as needed. Thorough coverage required. Do not use chemigation in California.

Signal Word/Toxicity Class: Caution.
REI: 24 hr.

Sumulsion — see petroleum oil

SunSpray — see petroleum oil

Thiodan — see endosulfan

THIRETHRIN (Premix)

Contains 0.45 lb. piperonyl butoxide.

Thirethrin (2 lb./gal. endosulfan & 0.045 lb./gal. pyrethrins) *(Terra)*
Rate: 1-2 qt. 2.9L/A
Pest:
cabbage loopers, diamondback moth, green peach aphid, imported cabbageworm
Remarks: Head, leaf lettuce. For head lettuce, remove wrapper leaves at harvest; do not apply more than 3 applications. For leaf lettuce, do not apply more than 2 applications.
Days To Harvest: 14.

Signal Word/Toxicity Class: Warning/II.
REI: 24 hr.

TRIGARD

Triazine compound used as insect growth regulator for control of leafminers. Ineffective against other insect species. Very low toxicity to mammals and safe to beneficial insects.

Trigard (WSP) (cyromazine) *(Novartis)*
Rate: 0.125 lb. ai/A
0.167 lb. WSP/A (1 packet)
Pest:
leafminers
Remarks: Lettuce (head, leaf). Apply as foliar spray in minimum 5 gal. water by air; 50 gal. water by ground when leafminers first appear; repeat at 7-day intervals. Do not make more than 6 applications per crop to head lettuce; 5 applications per crop to leaf lettuce. Do not apply through any type of irrigation system.
Days To Harvest: 7.
Crop Rotation: All crops on label may be planted following harvest of treated crop. Do not rotate to any other crop except sweet corn or radishes. Do not plant sweet corn or radishes within 3 months after last application.

Signal Word/Toxicity Class: Caution.
REI: 12 hr.

Warrior — see *lambda*-cyhalothrin

Mushrooms

Ambush — see permethrin

Diatect — see pyrethrins

DIAZINON

Restricted Use Pesticide due to avian and aquatic toxicity.

Prentox Diazinon AG500 *(Prentiss)*
Micro Flo Diazinon AG500 *(Micro Flo)*
Rate: 2-4 lb. ai/50 gal. water
4-8 pt. AG500/50 gal. water
Pest:
phorid mushroom fly, sciarid mushroom fly
Remarks: Spray on outside and inside walls and ventilators of mushroom houses after compost has been pasteurized by heating. Do not spray compost in beds or after spawning, or illegal residues may result. To treat wooden sides of beds, post, doors and door frames after pasteurizing, mix 0.5 lb. in 6 gal. water and apply with a brush. Do not apply this treatment after spawning. Do not treat compost.

Signal Word/Toxicity Class: Caution.
REI: 24 hr.

DIMILIN

Dimilin 25W (diflubenzuron) *(Uniroyal)*
Rate: Compost:
0.6-1 lb. ai/1000 sq. ft.
2.4-4 lb. 25W/1000 sq. ft.
Casing:
3.4 oz. ai/1000 sq. ft.
13.5 oz. 25W/1000 sq. ft.
Pest:
sciarid mushroom fly
Remarks: Compost: Thoroughly incorporate at spawning time with spawning machine. Casing: Incorporated or drench.

Signal Word/Toxicity Class: Caution/III.
REI: 12 hr.

D•z•n — see diazinon

MALATHION

Organophosphate insecticide toxic to fish, aquatic invertebrates, aquatic life stages of amphibians, and bees exposed to direct treatment on blooming crops or weeds.

Clean Crop Malathion 57EC or 8EC *(Platte)*
Drexel Malathion 5EC *(Drexel)*
◆ **Gowan Malathion 8** *(Gowan)*
◆ **Micro Flo Malathion 5EC** *(Micro Flo)*
Prentox 5 lb. Malathion Spray (5EC) *(Prentiss)*
◆ **Wilbur-Ellis Malathion 8** *(Wilbur-Ellis)*
Rate: 1.56 lb. ai/A
2.5 pt. 5EC, 57EC/A in 130 gal. water
1.5 pt. 8, 8EC/A in 130 gal. water
Pest:
mites (except Gowan), phorid flies, sciarid flies
Remarks: Apply as soon as possible after picking and repeat as necessary, usually twice per week.
Days To Harvest: 1.
State Restrictions: Gowan not for use in California.

Signal Word/Toxicity Class: Caution/III (57EC, 8, 8EC); Warning/II (5EC).
REI: 12 hr.

PERMETHRIN

Restricted Use Pesticide.

Ambush 2E or 25W *(ZENECA)*
Rate: 0.05-0.0625 lb. ai/A
3.2-4 oz. 2E/30 oz. water or suitable diluent
1 qt. solution/standard double (35,000 cu. ft.: 8000 sq. ft.)
6 oz. 25W/11 gal. water
8 level Tbs. 25W/3 gal. water
Pest:
phorid mushroom fly (adult), sciarid mushroom fly (adult)

◆ **- new product** • ai-active ingredient • DF-dry flowable • E/EC-emulsifiable concentrate • EW-emulsion in water
F/FL-flowable • G-granule • L/LC-liquid • SP/WSP-soluble packet • W/WP-wettable powder • WSB-water soluble bag

Remarks: For 2E, apply as fogging or aerosol treatment. For 25W, apply using 1 gal. spray/750 sq. ft. as a guide; spray directly to walls and ceilings (to point of runoff) as residual surface treatment only. Do not use when mushrooms are present. Treat once daily or as needed when flies appear. Do not make more than 20 applications prior to pinning of first break; no more than 2 applications between each break. Do not apply more than 30 applications per crop of 5 breaks. Refer to label for use information.
Days To Harvest: 3.

Signal Word/Toxicity Class: Warning/II.
REI: 12 hr.

Prentox Fogging Insecticide PF — see pyrethrins

Pyrenone — see pyrethrins

PYRETHRINS

Contains piperonyl butoxide as a synergist. Diatect also contains silicon dioxide (diatomaceous earth). May be combined in tank mix with other insecticides for quicker and more complete kill where insect resistance may be a problem, and as an "exciter" to flush insects out of hiding and into contact with other spray residues.

Diatect Multipurpose Insecticide II *(Diatect)*
Pyrenone (0.5EC) *(AgrEvo Environmental Health)*
Py-Rin 60-6 EC *(Wilbur-Ellis)*
Rate: 5 oz. II/1000 cu. ft.
1-12 fl. oz. 0.5EC/A
2-16 oz. 6-60 EC/A
Pest:
fungus gnats, mushroom flies
Remarks: Apply as dust or wettable powder. Repeat as necessary. Rate varies with insect population.

Signal Word/Toxicity Class: Caution.
REI: 12 hr.

Prentox Fogging Insecticide PF *(Prentiss)*
Rate: 1 oz. product/1000 cu. ft.
Pest:
fungus gnats, mushroom flies
Remarks: Fogger should be operated 8-10 ft. away from beds and directed slightly upward over them. Circulating fans should be left in operation, but for best results, refrigeration should be turned off.

Signal Word/Toxicity Class: Caution.
REI: 12 hr.

Py-Rin — see pyrethrins

Okra

AZADIRACHTIN

Extracted from seed of neem tree. Insect growth regulator that controls targeted insect larvae when they ingest or come in contact with it, by interfering with the insect's ability to molt and shutting off feeding behavior. Affected larvae slowly starve and die at next molt.

◆ **Azatin EC or XL** *(Olympic)*
◆ **Azatin XL Plus** *(Thermo Trilogy)*
Rate: 8-21 oz. EC/A
5-21 oz. XL, XL Plus/A
Pest:
aphids, armyworms, beetles, caterpillars, cutworms, diamondback moth, flea beetles, green peach aphid, imported cabbageworm, Japanese beetle, leafhoppers, leafminers, loopers, thrips, webworms, whiteflies
Remarks: Rate varies by species. Apply when pests first appear; repeat at 7-day intervals. When pest populations are high, use higher rates. Do not apply more that 21 fl. oz./A.
Days To Harvest: 0.

Signal Word/Toxicity Class: Caution (XL); Warning (EC).
REI: 4 hr. (XL), 12 hr. (EC).

Azatin — see azadirachtin

CARBARYL

Carbamate insecticide with broad activity on many insects. Highly toxic to bees.

Clean Crop Sevin 5 Bait *(Platte)*
Rate: 1-2 lb. ai/A
20-40 lb. Bait/A
Pest:
ants, armyworms, cutworms, darkling beetle, field crickets, grasshoppers, sowbugs
Remarks: Apply with ground or air equipment.
Days To Harvest: 0.

Signal Word/Toxicity Class: Caution/III.
REI: 12 hr.

Drexel Carbaryl 4L or 50W *(Drexel)*
Sevin 4F, XLR Plus, 80S or 80WSP *(Rhone-Poulenc)*
Rate: 0.5-1.5 lb. ai/A
1-3 pt. 4L, 4F, XLR/A
1-3 lb. 50W/A
0.66-1.875 lb. 80S, 80WSP/A
Pest:
grasshoppers
Remarks: Use high rate for mature grasshoppers or dense vegetation.
Days To Harvest: 0.

Rate: 1-2 lb. ai/A
2-4 pt. 4L, 4F, XLR/A
2-4 lb. 50W/A
1.25-2.5 lb. 80S, 80WSP/A
Pest:
corn earworms, stink bugs
Remarks: Treat on a 5- to 7-day schedule.
Days To Harvest: 0.

Signal Word/Toxicity Class: Caution (4F, 4L, XLR); Warning/II (50W, 80S, 80WSP).
REI: 12 hr.

Gowan Sevin Bait *(Gowan)*
Rate: 2 lb. ai/A
40 lb. Bait/A
Pest:
armyworms, crickets, cutworms, darkling beetle, grasshoppers, sowbugs
Remarks: Apply when pests appear in damaging numbers; repeat at 7- to 14-day intervals if necessary.
Days To Harvest: 1.
State Restrictions: For use in Arizona and California.

Signal Word/Toxicity Class: Caution/III.
REI: 12 hr.

DEADLINE

Packaged as Bullets, M-Ps (mini-pellets), and granules.

Deadline Bullets, M-Ps, Granules (4%) (metaldehyde) *(Pace)*
Rate: 0.8-1 lb. ai/A
20-40 lb. 4%/A
Pest:
slugs, snails
Remarks: Apply every 3-4 weeks during growing season or as needed. Do not apply directly to edible portion of plant.

Signal Word/Toxicity Class: Warning/II (granule); Caution (bullets, M-P).
REI: 12 hr.

Diatect — see pyrethrins

For personal protective equipment and EPA registration numbers, please turn to page 39.

MALATHION

Organophosphate insecticide toxic to fish, aquatic invertebrates, aquatic life stages of amphibians, and bees exposed to direct treatment on blooming crops or weeds.

Drexel Malathion 5EC *(Drexel)*
◆ **Micro Flo Malathion 5EC** *(Micro Flo)*
Prentox 5 lb. Malathion Spray (5EC) *(Prentiss)*

Rate: 0.94-1.25 lb. ai/A
1.5-2 pt. 5EC/A

Pest:

aphids	Japanese beetle

Remarks: Apply when insects first appear and repeat as necessary. Use high rate for Japanese beetles. Do not apply after pods begin to form.
Days To Harvest: 1.

Signal Word/Toxicity Class: Warning/II.
REI: 12 hr.

◆ **Gowan Malathion 8 (8E)** *(Gowan)*

Rate: 1.5 lb. ai/A
1.5 pt. 8E/A

Pest:

aphids	Japanese beetle

Remarks: Apply at first sign of infestation; repeat at 7- to 10-day intervals.
Days To Harvest: 1.
State Restrictions: Not for use in California.

Signal Word/Toxicity Class: Caution/III.
REI: 12 hr.

M-PEDE

M-Pede (potassium salts of fatty acids) *(Mycogen)*

Rate: 1-2% v/v solution

Pest:

aphids*	leafminers*	spider mites
broad mite	plant bugs	thrips
green peach aphid*	russet mite	whiteflies
leafhoppers		

* *Tank mix only*

Remarks: Apply at first sign of infestation; repeat at 7- to 14-day intervals. May be tank mixed with other insecticides. Refer to label for use directions and precautions.
Days To Harvest: 0.

Signal Word/Toxicity Class: Warning/II.
REI: 12 hr.

NEMACUR

Restricted Use Pesticide.

Nemacur 15% (fenamiphos) *(Bayer)*

Rate: Band:
2-2.5 lb. ai/ A
13.4-16.7 lb. 15%/A (36" row)
14.7-18.4 oz. 15%/1000 ft. row (any row spacing)

Pest:
nematodes

Remarks: Apply in 12-15" band and incorporate into soil. Refer to label for further use information.
State Restrictions: Not for use in California.

Signal Word/Toxicity Class: Danger/I.
REI: 48 hr.

PYRELLIN (Premix)

Premix containing pyrethrins and rotenone. May be combined with other insecticides to enhance knockdown or as an insect "exciter" to help facilitate contact with other spray residues.

Pyrellin E.C. (0.042 lb./gal. pyrethrins & 0.035 lb./gal. rotenone) *(Webb Wright)*

Rate: 1-2 pt. E.C./A

Pest:

aphids	loopers	plant bugs
leafhoppers	Lygus bugs	thrips
leafminers	mites	whiteflies

Remarks: Apply as necessary to maintain control.
Days To Harvest: 0.

Signal Word/Toxicity Class: Caution/III.
REI: 12 hr.

PYRETHRINS

Contains piperonyl butoxide as a synergist and silicon dioxide (diatomaceous earth). May be combined in tank mix with other insecticides for quicker and more complete kill where insect resistance may be a problem, and as an "exciter" to flush insects out of hiding and into contact with other spray residues.

Diatect Multipurpose Insecticide II *(Diatect)*

Rate: 1-6 lb. II/A

Pest: Including but not limited to:

aphids	imported cabbageworm	mites
cabbageworms	leafhoppers	plant bugs
caterpillars	leafminers	stink bugs
cucumber beetles	loopers	thrips
diamondback moth	Lygus bugs	vegetable weevils
flea beetles		whiteflies

Remarks: Apply as dust or wettable powder. Repeat as necessary.

Signal Word/Toxicity Class: Caution.
REI: 12 hr.

Sevin — see carbaryl

Onions

Ambush — see permethrin

Ammo — see cypermethrin

AZADIRACHTIN

Extracted from seed of neem tree. Insect growth regulator that controls targeted insect larvae when they ingest or come in contact with it, by interfering with the insect's ability to molt and shutting off feeding behavior. Affected larvae slowly starve and die at next molt.

◆ **Azatin EC or XL** *(Olympic)*
◆ **Azatin XL Plus** *(Thermo Trilogy)*

Rate: 8-21 oz. EC/A
5-21 oz. XL, XL Plus/A

Pest:

aphids	cutworms	leafminers
armyworms	diamondback moth	leafrollers
beet armyworm	flea beetles	loopers
beetles	green peach aphid	thrips
caterpillars	Japanese beetle	webworms
Colorado potato beetle	June beetle	whiteflies
corn earworm	leafhoppers	

Remarks: Leeks, onions, shallots. Rate varies by species. Apply when pests first appear; repeat at 7-day intervals. When pest populations are high, use higher rates. Do not apply more that 21 fl. oz./A.
Days To Harvest: 0.

Signal Word/Toxicity Class: Caution (XL); Warning/II (EC).
REI: 4 hr. (XL), 12 hr. (EC).

◆ **NEEMAZAD 4.5EC or NEEMIX 4.5** *(Thermo Trilogy)*

Rate: 0.125-1 pt. 4.5/A
High Volume:
2.25-4.5 oz. 4.5EC/100 gal. water
Low Volume:
0.5 pt. 4.5EC/A

◆ - **new product** • ai-active ingredient • DF-dry flowable • E/EC-emulsifiable concentrate • EW-emulsion in water
F/FL-flowable • G-granule • L/LC-liquid • SP/WSP-soluble packet • W/WP-wettable powder • WSB-water soluble bag

Pest:

aphids	diamondback moth	onion maggot
armyworms	greenhouse whitefly	sweetpotato whitefly
cabbage loopers	hornworms	thrips
caterpillars	imported cabbageworm	webworms
cutworms	leafminers	weevils

Remarks: Leeks, onions, shallots. For low volume, apply in sufficient water for adequate coverage. For low insect pressure, apply preventive at 7-day intervals or at first sign of insects; high insect pressure, at 3- to 4-day intervals.

Signal Word/Toxicity Class: Warning/II.
REI: 12 hr.

Azatin — see azadirachtin

AZINPHOS-METHYL

Restricted Use Pesticide (except Guthion Solupak).

Gowan Azinphos-M 50-W *(Gowan)*
Guthion 2L, Solupak (50WSP) *(Bayer)*
◆ **Micro Flo Azinphosmethyl 2EC or 50W** *(Micro Flo)*
Sniper 2-E, 50PVA, 50W *(Platte)*

Rate: 0.5-0.75 lb. ai/A
2-3 pt. 2EC, 2-E, 2L/A
1-1.5 lb. 50-W, 50PVA, 50W, 50WSP/A

Pest:
thrips

Remarks: Dry, green onions. Apply by air or ground in sufficient water for complete coverage. Do not apply more than 3 times per season. Allow 7 days between applications for dry bulb onions; 10 days for green onions.
Days To Harvest: Dry onions-28; green onions-14 (Gowan-7).
State Restrictions: California labels may vary. Sniper 50PVA not registered in California.

Signal Word/Toxicity Class: Danger/I.
REI: 48 hr.; 72 hr. if average rainfall less than 25" per year.

Chlorfos — see chlorpyrifos

CHLORPYRIFOS

◆ **Chlorfos 15G** *(Griffin)*
Lorsban-4E or 15G *(Dow AgroSciences)*
◆ **Micro Flo Chlorpyrifos 4E or 15G** *(Micro Flo)*

Rate: 1.1 fl. oz. 4E/1000 lin. ft. row
3.7 oz. 15G/1000 lin. ft. row

Pest:
onion maggot

Remarks: Dry bulb onions. In-furrow treatment. For 4E, apply as drench on 18" row spacing (minimum 40 gal. drench/A, incorporate 1-2"). Apply no more than once per year. Refer to label for use directions.

Signal Word/Toxicity Class: Warning/II (4E); Caution (15G).
REI: 4E: 24 hr.; 15G: 12 hr.

lambda-CYHALOTRHIN

Synthetic pyrethroid insecticide with broad-spectrum activity, formulated as an emulsifiable concentrate. Restricted Use Pesticide due to toxicity to fish and aquatic organisms.

Karate or Warrior (1E) *(ZENECA)*

Rate: 0.015-0.03 lb. ai/A
1.92-3.84 fl. oz. 1E/A

Pest:

aphids*	leafminers (adult)	seedcorn maggot (adult)
armyworms (1st & 2nd instar)	onion maggot (adult)	stink bugs
cutworms	onion thrips	tobacco thrips
flower thrips*	plant bugs	western flower thrips*

* *Suppression*

Remarks: Apply as required by scouting, usually at intervals of 5 or more days. Do not apply more than 1.92 pt./A per season. Use higher rates as thrips population increase. Refer to label for spray drift precautions and state and county restrictions.
Days To Harvest: 14.

Signal Word/Toxicity Class: Danger/I.
REI: 24 hr.

CYPERMETHRIN

Synthetic pyrethroid insecticide with both contact and stomach poison activity. No systemic activity and is ineffective against soil pests. Controls many caterpillars, beetles, flies, plant bugs, and other insect pests and has fast knockdown effects. Toxic to fish and bees. May cause eye and skin irritation. Restricted Use Pesticide.

Ammo 2.5 EC or WSB *(FMC)*

Rate: 0.04-0.1 lb. ai/A
2-5 fl. oz. 2.5 EC/A
1-2 bags WSB/A

Pest:

aphids	cutworms	onion maggot (adult)
armyworms	leafminers	stink bugs

Rate: 0.08-0.1 lb. ai/A
4-5 fl. oz. 2.5 EC/A
1-2 bags WSB/A

Pest:
onion thrips

Remarks: Onions, shallots. Apply as needed in water using minimum 20 gal. finished spray/A by ground; 3 gal./A by air. Rate varies by pest species. Apply no more than 0.5 lb. ai/A per season. Do not graze treated areas or cut treated crops for feed. Refer to label for spray drift precautions.
Days To Harvest: 7.
Crop Rotation: Not within 30 days after last application.

Signal Word/Toxicity Class: Caution (2.5 EC); Warning/II (WSB).
REI: 12 hr.

zeta-CYPERMETHRIN

Synthetic pyrethroid with broad spectrum activity. Extremely toxic to fish and aquatic invertebrates; highly toxic to bees. Restricted Use Pesticide.

Fury 1.5 EC or Mustang 1.5 EW *(FMC)*

Rate: 0.0375-0.05 lb. ai/A
3.20-4.26 fl. oz./A

Pest:
onion thrips

Rate: 0.028-0.05 lb. ai/A
2.39-4.26 fl. oz./A

Pest:

aphids	cutworms	onion maggot (adult)
armyworms	leafminers	stink bugs

Remarks: Apply in minimum of 20 gal./A with ground equipment or 3 gal./A by air when pests appear; repeat as necessary. Do not apply more than 0.25 lb. ai/A per season. Do not graze treated areas or cut treated crops for feed.
Days To Harvest: 7.

Signal Word/Toxicity Class: Warning/II.
REI: 12 hr.

DEADLINE

Packaged as Bullets, M-Ps (mini-pellets), and granules.

Deadline Bullets, M-Ps, Granules (4%) (metaldehyde) *(Pace)*

Rate: 0.8-1 lb. ai/A
20-40 lb. 4%/A

Pest:
slugs snails

Remarks: Leeks, onions, shallots. Apply every 3-4 weeks during growing season or as needed. Do not apply directly to edible portion of plant.

Signal Word/Toxicity Class: Warning/II (granule); Caution (bullets, M-P).
REI: 12 hr.

Declare — see methyl parathion

Diatect — see pyrethrins

NOTICE	The information on these pages is for preliminary planning — not a guide for use. Be sure to follow the manufacturer's directions, notwithstanding information contained here.	For personal protective equipment and EPA registration numbers, please turn to page 39.

DIAZINON

Organophosphate insecticide. Restricted Use Pesticide due to avian and aquatic toxicity.

D•z•n diazinon AG600 WBC, 50W *(Novartis)*
Drexel Diazinon Insecticide (4EC), D-264 14G *(Drexel)*
Gowan Diazinon 4E, 14G, 50WSB *(Gowan)*
Helena Diazinon AG500 *(Helena)*
Micro Flo Diazinon AG500, 14G, 50W *(Micro Flo)*
Prentox Diazinon AG500 *(Prentiss)*
◆ Wilbur-Ellis Diazinon 4 (4E), 14G *(Wilbur-Ellis)*

Rate: 0.5 lb. ai/A
1 pt. 4E, 4EC, AG500/A
12.75 fl. oz. AG600/A
1 lb. 50W, 50WSB/A

Pest:
onion thrips

Rate: 2-4 lb. ai/A
2-4 qt. 4E, 4EC, AG500/A
51-102 fl. oz. AG600/A
4-8 lb. 50W, 50WSB/A
14-28 lb. 14G/A

Pest:
onion maggot

Rate: 3-4 lb. ai/A
3-4 qt. 4E, 4EC, AG500/A
76.5-102 fl. oz. AG600/A
6-8 lb. 50W, 50WSB/A
21-28 lb. 14G/A

Pest:
wireworm

Remarks: Bulb, green onions. Apply as thrips appear and repeat after 7 days if necessary. Do not apply more than 3 times per season. For maggots and wireworms, apply just before planting and immediately incorporate into soil. Will not control organophosphate-resistant onion maggot.

Days To Harvest: 14.

Signal Word/Toxicity Class: Caution.
REI: 24 hr.; 12 hr. (Drexel/Gowan/Wilbur-Ellis 14G).

D•z•n — see diazinon

Fury — see *zeta*-cypermethrin

Fyfanon — see malathion

Guthion — see azinphos-methyl

Kumulus — see sulfur

LANNATE

Restricted Use Pesticide.

Lannate SP or LV (methomyl) *(DuPont)*

Rate: 0.45-0.9 lb. ai/A
0.5-1 lb. SP/A
1.5-3 pt. LV/A

Pest:
beet armyworm   thrips   variegated cutworm

Remarks: Dry, bulb, green onions. Apply in sufficient water for uniform coverage when insects first appear; repeat at 5- to 7-day intervals or as needed. Add wetting agent. Rate varies by pest species; refer to label. For green onions, do not apply more than 5.4 lb. ai/A per crop or make more than 8 applications per crop. For dry bulb onions, do not apply more than 3.6 lb. ai/A per crop or make more than 8 applications per crop.

Days To Harvest: 7.

Signal Word/Toxicity Class: Danger-Poison/I.
REI: 48 hr.

LINDANE

Lindane 30C contains toxic inert Rhodamine B. Toxic to fish, birds, and wildlife.

Lindane 30C *(Gustafson)*
◆ Lindane ST 40 *(AGSCO)*

Rate: 2.7 fl. oz. 30C/100 lb. seed
2 fl. oz. 40/100 lb. seed

Pest:
seedcorn maggots   wireworms

Remarks: Do not use treated seed for food, feed, or oil purposes. Exposed treated seed may be hazardous to birds and other wildlife. Dispose of excess seed by burial away from bodies of water.

Signal Word/Toxicity Class: Warning/II.
REI: 12 hr.

Lorsban — see chlorpyrifos

MALATHION

Organophosphate insecticide toxic to fish, aquatic invertebrates, aquatic life stages of amphibians, and bees exposed to direct treatment on blooming crops or weeds.

Clean Crop Malathion 57EC *(Platte)*
Fyfanon (5EC) *(Helena)*

Rate: Leeks, shallots:
0.94-1.25 lb. ai/A
1.5-2 pt. 57EC, 5EC/A

Pest:
aphids   thrips

Rate: Onions:
0.94 lb. ai/A
1.5 pt. 57EC, 5EC/A

Pest:
onion thrips

Rate: Onions:
1.6-1.9 lb. ai/A
2.5-3 pt. 57EC
1.6 lb. ai/A
2.5 pt. 5EC/A

Pest:
onion maggot

Remarks: Leeks, onions, shallots.

Days To Harvest: 3; shallots-7 (Gowan).

Signal Word/Toxicity Class: Caution/III.
REI: 12 hr.

Clean Crop Malathion 8EC *(Platte)*

Rate: Leeks, shallots:
0.5-1 lb. ai/A
0.5-1 pt. 8EC/A

Pest:
aphids   thrips

Rate: Onions:
1.5-2 lb. ai/A
1.5-2 pt. 8EC/A

Pest:
onion maggot   thrips

Remarks: Leeks, onions, shallots.

Days To Harvest: 3.

Signal Word/Toxicity Class: Caution/III.
REI: 12 hr.

Drexel Malathion 5EC *(Drexel)*
◆ Gowan Malathion 8 (8E) *(Gowan)*
◆ Micro Flo Malathion 5EC *(Micro Flo)*
Prentox 5 lb. Malathion Spray (5EC) *(Prentiss)*

Rate: 1-1.6 lb. ai/A
1.5-3 pt. 5EC/A
1-2 lb. ai/A
1-2 pt. 8E/A

◆ - new product • ai-active ingredient • DF-dry flowable • E/EC-emulsifiable concentrate • EW-emulsion in water
F/FL-flowable • G-granule • L/LC-liquid • SP/WSP-soluble packet • W/WP-wettable powder • WSB-water soluble bag

Pest:

aphids onion maggot thrips

Remarks: Leeks, onions (including green), shallots.

Days To Harvest: 3.

Signal Word/Toxicity Class: Caution/III (8E); Warning/II (5EC).

REI: 12 hr.

Fyfanon (8E) *(Helena)*

Rate: 2 lb. ai/A
2 pt. 8E/A

Pest:

onion maggot thrips

Remarks: Onions. For onion maggot, apply at 4-day intervals when flies first appear.

Days To Harvest: 3.

Signal Word/Toxicity Class: Caution/III.

REI: 12 hr.

METHYL PARATHION

Organophosphate insecticide toxic to bees exposed to direct treatment or residues on blooming crops or weeds. Penncap-M is a microencapsulated formulation. Restricted Use Pesticide due to very high acute toxicity to humans and birds.

Cheminova Methyl 4EC *(Cheminova)*

◆ **Declare (4EC)** *(Griffin*

Rate: 0.5 lb. ai/A
1 pt. 4EC/A

Pest:

thrips

Remarks: Apply when insects appear; repeat as needed.

Days To Harvest: 15.

Signal Word/Toxicity Class: Danger-Poison/I.

REI: 48 hr.; 72 hr. if average rainfall less than 25" per year.

Penncap-M (2FL) *(Elf Atochem)*

Rate: 0.5 lb. ai/A
2 pt. 2FL/A

Pest:

thrips

Remarks: Do not apply when onions are blooming or when bees are foraging areas to be treated. Do not chemigate in California.

Days To Harvest: 15.

Signal Word/Toxicity Class: Warning/II.

REI: 48 hr.

Microthiol Special — see sulfur

M-PEDE

M-Pede (potassium salts of fatty acids) *(Mycogen)*

Rate: 1-2% v/v solution

Pest:

aphids* leafminers* spider mites
broad mite plant bugs thrips
green peach aphid* russet mite whiteflies
leafhoppers

**** Tank mix only***

Remarks: Apply at first sign of infestation; repeat at 7- to 14-day intervals. May be tank mixed with other insecticides; refer to labels for use directions and precautions.

Days To Harvest: 0.

Signal Word/Toxicity Class: Warning/II.

REI: 12 hr.

Mustang — see *zeta*-cypermethrin

NEEMAZAD — see azadirachtin

NEEMIX — see azadirachtin

Penncap-M — see methyl parathion

PERMETHRIN

Restricted Use Pesticide due to toxicity to fish and aquatic organisms.

Ambush 2E or 25W *(ZENECA)*

Rate: 0.1-0.3 lb. ai/A
6.4-19.2 fl. oz. 2E/A
6.4-19.2 oz. 25W/A

Pest:

armyworm leafminers onion thrips
cutworms onion maggot (adult) stink bugs

Remarks: Dry, bulb onions. Apply when insects appear and repeat as necessary. For foliar application, apply in minimum 20 gal. water/A by ground; 6 gal. water/A by air. Use higher rates on onion thrips and avoid rescue situations. Apply no more than 2.4 lb. ai/A per season. Do not graze livestock in treated areas or feed forage to livestock.

Days To Harvest: 1.

Signal Word/Toxicity Class: Warning/II.

REI: 24 hr.

PETROLEUM OIL

Petroleum based spray oil. Mode of action is through suffocation of eggs, larvae, and nymphs of insects and adults of soft bodied insects. This mode of action necessitates total spray coverage. Toxic to fish.

SAF-T-SIDE *(Brandt Consolidated)*

◆ **Sumulsion** *(Drexel)*

Rate: 1-2 gal./100 gal. water

Pest:

aphids leafminers plant bugs
caterpillars loopers thrips
leafhoppers mites whiteflies

Remarks: Apply as needed. May be tank mixed with most commonly used insecticides and fungicides. Do not use in combination with or immediately before or after spraying with fungicides such as captan, Dyrene, folpet, Morestan, or any sulfur products. Do not use with carbaryl or dimethoate.

Signal Word/Toxicity Class: Caution.

REI: 4 hr.

PYRELLIN (Premix)

Premix containing pyrethrins and rotenone. May be combined with other insecticides to enhance knockdown or as an insect "exciter" to help facilitate contact with other spray residues.

Pyrellin E.C. (0.042 lb./gal. pyrethrins & 0.035 lb./gal. rotenone) *(Webb Wright)*

Rate: 1-2 pt. E.C./A

Pest:

aphids loopers stink bugs
bean beetle Lygus bugs thrips
leafhoppers mites whiteflies
leafminers plant bugs

Remarks: Green, dry, bulb onions. Apply at 7-day intervals or less; repeat as needed to maintain control.

Days To Harvest: 0.

Rate: 1.5-2 pt. E.C./A

Pest:

Colorado potato beetle European corn borer flea beetles

Remarks: Green, dry, bulb onions. Make 3-4 applications or more as eggs start to hatch; repeat at 5- to 7 day-intervals.

Days To Harvest: 0.

Signal Word/Toxicity Class: Caution/III.

REI: 12 hr.

Pyrenone — see pyrethrins

7

Vegetables

NOTICE The information on these pages is for preliminary planning — not a guide for use. Be sure to follow the manufacturer's directions, notwithstanding information contained here.

For personal protective equipment and EPA registration numbers, please turn to page 39.

PYRETHRINS

Contains piperonyl butoxide as a synergist and silicon dioxide (diatomaceous earth). May be combined in tank mix with other insecticides for quicker and more complete kill where insect resistance may be a problem, and as an "exciter" to flush insects out of hiding and into contact with other spray residues.

Diatect Multipurpose Insecticide II *(Diatect)*
Pyrenone (0.5EC) *(AgrEvo Environmental Health)*
Py-Rin 60-6 EC *(Wilbur-Ellis)*

Rate: 1-6 lb. II/A
1-12 fl. oz. 0.5EC/A
2-16 oz. 6-60 EC/A

Pest: Including but not limited to:
aphids, leafhoppers, leafminers, nematodes, onion maggot, thrips

Remarks: Apply in sufficient water for thorough coverage; repeat as needed. Apply Diatect as dust or wettable powder.
Days To Harvest: 0.

Signal Word/Toxicity Class: Caution.
REI: 12 hr.

Py-Rin — see pyrethrins

SAF-T-SIDE — see petroleum oil

Sniper — see azinphos-methyl

SULFUR

Microthiol and Thiolux are micronized sulfurs.

◆ **Kumulus DF** *(BASF)*
Microthiol Special (80W) *(Elf Atochem)*
Thiolux (DF) *(Platte)*

Rate: 2.4-8 lb. ai/A
3-10 lb. 80W, DF/A
California:
4.8 lb. ai/A
6 lb. 80W/A

Pest:
petrobia mite

Remarks: Onions (including dry), dry shallots. Apply when insects first appear; repeat as needed. Do not apply if temperatures during or within 3 days after application are expected to exceed 90°F. Refer to label for use directions.
State Restrictions: Kumulus not registered in California.

Signal Word/Toxicity Class: Caution.
REI: 24 hr.

Sumulsion — see petroleum oil

Thiolux — see sulfur

Warrior — see *lambda*-cyhalothrin

Peas

Adios — see carbaryl

ASANA

Pyrethroid insecticide that has both contact and stomach poison activity. Restricted Use Pesticide due to toxicity to fish and aquatic organisms.

Asana XL (esfenvalerate) *(DuPont)*

Rate: 0.015-0.05 lb. ai/A
2.9-9.6 fl. oz. XL/A

Pest:
beet armyworm (aids in control), cabbage looper, corn earworm, corn rootworm (adult), cowpea curculio, cucumber beetle, cutworms, grasshoppers, green cloverworm, leafhoppers, Mexican bean beetle, painted lady butterfly (larvae), pea aphid, potato leafhopper, saltmarsh caterpillar, velvetbean caterpillar, western bean cutworm

Remarks: Dry peas. Repeat as necessary to maintain control. Rate varies by pest species; refer to label. Do not exceed 0.2 lb. ai/A per season. Do not feed or graze livestock on treated vines.
Days To Harvest: 21.

Rate: 0.015-0.05 lb. ai/A
2.9-9.6 fl. oz. XL/A

Pest:
alfalfa caterpillar, alfalfa looper, armyworms, cabbage looper, celery looper, corn earworm, cutworms, green cloverworm, imported cabbageworm, pea aphid

Remarks: Green peas. Repeat as necessary to maintain control. Rate varies by pest species; refer to label. Do not exceed 0.1 lb. ai/A per season. Do not feed treated vines to livestock.
Days To Harvest: 3.

Signal Word/Toxicity Class: Warning/II.
REI: 12 hr.

AZADIRACHTIN

Extracted from seed of neem tree. Insect growth regulator that controls targeted insect larvae when they ingest or come in contact with it, by interfering with the insect's ability to molt and shutting off feeding behavior. Affected larvae slowly starve and die at next molt.

◆ **NEEMAZAD 4.5EC or NEEMIX 4.5** *(Thermo Trilogy)*

Rate: 0.125-1 pt. 4.5/A
High Volume:
2.25-4.5 oz. 4.5EC/100 gal. water
Low Volume:
0.5 pt. 4.5EC/A

Pest:
armyworms, bean leaf beetle, cabbage loopers, caterpillars, Colorado potato beetle, cutworms, garden webworm, greenhouse whitefly, hornworms, leafminers, loopers, Mexican bean beetles, pepper weevils, psyllids, soybean loopers, sweetpotato whitefly, tomato fruitworms, tomato pinworms, weevils

Remarks: For low volume, apply in sufficient water for adequate coverage. For low insect pressure, apply preventive at 7-day intervals or at first sign of insects; high insect pressure, at 3- to 4-day intervals.

Signal Word/Toxicity Class: Warning/II.
REI: 12 hr.

CARBARYL

Carbamate insecticide with broad activity on many insects. Highly toxic to bees.

Adios (13%) *(Micro Flo)*

Rate: 0.065-0.1 lb. ai/A
0.5-0.75 lb. 13%/A

Pest:
cucumber beetle

Remarks: For use on peas with pods. Do not apply through any type of irrigation system. Do not apply more than 3.75 lb./A per season.
Days To Harvest: 3.
State Restrictions: Not for use in California.

Signal Word/Toxicity Class: Caution/III.

Clean Crop Sevin 5 Bait *(Platte)*

Rate: 1.5 lb. ai/A
30 lb. Bait/A

Pest:
ants, armyworms, cutworms, darkling beetle, field crickets, grasshoppers, sowbugs

◆ - **new product** • ai-active ingredient • DF-dry flowable • E/EC-emulsifiable concentrate • EW-emulsion in water
F/FL-flowable • G-granule • L/LC-liquid • SP/WSP-soluble packet • W/WP-wettable powder • WSB-water soluble bag

Remarks: Apply with ground or air equipment.
Days To Harvest: 3.

Signal Word/Toxicity Class: Caution/III.
REI: 12 hr.

Drexel Carbaryl 4L or 50W *(Drexel)*
Sevin 4F, XLR Plus, 80S, 80WSP *(Rhone-Poulenc)*
Rate: 0.5-1.5 lb. ai/A
1-3 pt. 4L, 4F, XLR/A
1-3 lb. 50W/A
0.66-1.875 lb. 80S, 80WSP/A
Pest:
alfalfa caterpillar, armyworms, Colorado potato beetle, cutworms, grasshoppers, leafhoppers, pea leaf weevil, pea weevil, yellowstriped armyworm

Rate: 2.5 lb. ai/A
5 pt. 4L, 4F, XLR/A
3 lb. 80S, 80WSP/A
Pest:
alfalfa looper (Washington)
Remarks: Rate varies by pest species; refer to label.
Days To Harvest: 3.

Signal Word/Toxicity Class: Caution (4F, 4L, XLR); Warning/II (50W, 80S, 80WSP).
REI: 12 hr.

Gowan Sevin Bait *(Gowan)*
Rate: 1.5 lb. ai/A
30 lb. Bait/A
Pest:
armyworms, crickets, cutworms, darkling beetle, grasshoppers, sowbugs
Remarks: Apply when pests appear in damaging numbers; repeat at 7- to 14-day intervals if necessary.
Days To Harvest: 3.
State Restrictions: For use in Arizona and California.

Signal Word/Toxicity Class: Caution/III.
REI: 12 hr.

CHLORPYRIFOS

For use only by commercial seed treaters.

Lorsban 30, 50-SL *(Gustafson)*
Rate: 2.7 fl. oz. FL/100 lb. seed
2 oz. WP/100 lb. seed
Pest:
seedcorn beetles, seedcorn maggots
Remarks: Field, garden peas. Apply prior to planting as slurry treatment. Do not use treated seed for food, feed, or oil purposes. Commercially treated seed must be labeled in accordance with all applicable requirements of Federal Seed Act. Do not allow livestock to graze or feed bean hay or pea hay grown from treated seed.

Signal Word/Toxicity Class: Warning/II (50-SL); Caution (30).
REI: 12 hr.

DEADLINE

Packaged as Bullets, M-Ps (mini-pellets), and granules.

Deadline Bullets, M-Ps, Granules (4%) (metaldehyde) *(Pace)*
Rate: 0.8-1 lb. ai/A
20-40 lb. 4G/A
Pest:
slugs, snails
Remarks: Apply every 3-4 weeks during growing season or as needed. Do not apply directly to edible portion of plant.

Signal Word/Toxicity Class: Warning/II (granule); Caution (bullets, M-P).
REI: 12 hr.

Declare — see methyl parathion

Diatect — see pyrethrins

DIAZINON

Organophosphate insecticide. Restricted Use Pesticide due to avian and aquatic toxicity.

D•z•n diazinon AG600 WBC, 50W *(Novartis)*
Drexel Diazinon Insecticide (4EC), D-264 14G *(Drexel)*
Gowan Diazinon 4E, 14G, 50WSB *(Gowan)*
Helena Diazinon AG50 *(Helena)*
Micro Flo Diazinon AG500, 14G, 50W *(Micro Flo)*
Prentox Diazinon AG500, 50W *(Prentiss)*
◆ **Wilbur-Ellis Diazinon 4 (4E), 14G** *(Wilbur-Ellis)*
Rate: 0.38-0.5 lb. ai/A
0.75-1 pt. 4E, 4EC, AG500/A
9.5-12.75 fl. oz. AG600/A
0.75-1 lb. 50W, 50WSB/A
0.3-0.4 oz. 50W/1000 sq. ft.
Pest:
aphids, dipterous leafminer

Rate: 2-4 lb. ai/A
4-8 pt. 4E, 4EC, AG500/A
51-102 fl. oz. AG600/A
4-8 lb. 50W, 50WSB/A
14-28 lb. 14G/A
Pest:
cutworms

Rate: 3-4 lb. ai/A
6-8 pt. 4E, 4EC, AG500/A
76.5-102 fl. oz. AG600/A
6-8 lb. 50W, 50WSB/A
21-28 lb. 14G/A
Pest:
wireworms
Remarks: Succulent peas. Apply as insects occur; repeat at 5-day intervals as needed. Do not apply more than 3 times per season. Will not control organophosphate-resistant leafminers. For cutworms and wireworms, broadcast just before planting and immediately incorporate into soil. Pea vines and hay may be fed to dairy and beef cattle and sheep; wait 7 days following treatment if forage is to be cut for hay.
Days To Harvest: 7.

Signal Word/Toxicity Class: Caution.
REI: 24 hr.; 12 hr. (Drexel/Gowan/Wilbur-Ellis 14G).

DIAZINON (Premix)

Seed treatment. Contains graphite.

Captan-Diazinon (25% diazinon & 36.67% captan) *(Trace)*
Kernel Guard (15% diazinon & 25% lindane & 14.67% captan) *(Trace)*
Rate: 2 oz./bu. seed
Pest:
seedcorn beetles, seedcorn maggots, wireworms
Remarks: Apply in planter box as seed treatment before planting. Treat only those seeds needed for immediate use. Do not use treated seed for food, feed, or oil purposes. Do not store excess treated seeds beyond planting time. Refer to label for use directions.

Signal Word/Toxicity Class: Danger/I.
REI: 12 hr.; Captan-Diazinon 4 days with exceptions.

DIBROM

Short residual contact insecticide for kill of insects and mite adults and larvae present at time of application.

Dibrom 8 (8E) (naled) *(Amvac)*
Rate: 1 lb. ai/A
1 pt. 8E/A
California:
1 lb. ai/A
1 pt. 8E/A (min. 10 gal. water by air)
Pest:
alfalfa looper, aphids, leafhoppers, leafminers, loopers, Lygus bugs, nitidulid beetles, spider mites

7 Vegetables

NOTICE The information on these pages is for preliminary planning — not a guide for use. Be sure to follow the manufacturer's directions, notwithstanding information contained here.	For personal protective equipment and EPA registration numbers, please turn to page 39.

Remarks: Dry, succulent peas (except field peas grown for livestock feeding). Apply at first sign of insects. Make only 1 application per 7-day period. Do not use highly alkaline materials such as lime or Bordeaux. Do not apply through any type of irrigation system. Do not apply more than 4.5 pt./A per season.
Days To Harvest: 1.

- - - - - - - - - - - - - - - -

Signal Word/Toxicity Class: Danger/I.
REI: 48 hr.

Digon — see dimethoate

Dimate — see dimethoate

DIMETHOATE

Highly toxic to bees.

Cheminova Dimethoate 4E *(Cheminova)*
Clean Crop Dimethoate 400 *(Platte)*
Dimate 4E *(Terra)*
Drexel Dimethoate 2.67 or 4EC *(Drexel)*
Gowan Dimethoate E267 *(Gowan)*
Helena Dimethoate 4E *(Helena)*
Micro Flo Dimethoate 267 or 4E *(Micro Flo)*
Wilbur-Ellis Dimethoate 267 or Digon 400 *(Wilbur-Ellis)*

Rate: 0.16 lb. ai/A
0.5 pt. 2.67, 267, E267/A
0.33 pt. 4E, 4EC, 400/A

Pest:
aphids

Remarks: Do not make more than 1 application per season. Do not apply if crop or weeds in treatment area are in bloom. Do not feed or graze hay within 21 days after last application when stationary viner is used. Do not feed or graze when mobile viner is used.
Days To Harvest: 0; Helena-7.

- - - - - - - - - - - - - - - -

Signal Word/Toxicity Class: Warning/II.
REI: 48 hr.

DI-SYSTON

Restricted Use Pesticide.

Di-Syston 8 or 15% (disulfoton) *(Bayer)*

Rate: 1-2.5 lb. ai/A
1-2.5 pt. 8/A
6.7-16.7 lb. 15G/A

Pest:
aphids — leafhoppers

Remarks: For Di-Syston 8, apply in water emulsion or with liquid fertilizer as in-furrow spray, injection on each side of seed furrow at planting, or injected as sidedressing after plants are established. For Di-Syston 15%, apply by drilling or broadcasting at planting time or as soil incorporated sidedressing on each side of row after emergence. Do not apply more than once per season. Do not feed vines or hay.
Days To Harvest: 50.

- - - - - - - - - - - - - - - -

Signal Word/Toxicity Class: Danger/I.
REI: 48 hr.

D•z•n — see diazinon

ENDOSULFAN

Toxic to birds, fish, and wildlife; toxic to bees exposed to direct application.

Drexel Endosulfan 3EC or 50WP *(Drexel)*
Gowan Endosulfan 3EC or 50WSB *(Gowan)*
◆ **Micro Flo Endosulfan 3EC or 50WP** *(Micro Flo)*
Phaser 3EC or 50WSB *(AgrEvo USA)*
Thiodan 3EC or 50WP *(FMC)*

Rate: 0.5-1 lb. ai/A
0.66-1.33 qt. 3EC/A
1-2 lb. 50WP, 50WSB/A

Pest:
pea aphid — pea weevil

Remarks: Seed crop. Use only on peas to be harvested by combine. Apply when insects first appear or feeding is noticed. Do not exceed 2 applications per year or 3 lb. ai/A per year. Do not feed treated vines to livestock or allow livestock to graze in treated fields.
Days To Harvest: 1.

- - - - - - - - - - - - - - - -

Signal Word/Toxicity Class: Danger-Poison/I.
REI: 24 hr.

Fyfanon — see malathion

IMIDAN

Imidan 70-W (phosmet) *(Gowan)*

Rate: 0.75-1 lb. ai/A
1-1.33 lb. 70-W/A

Pest:
pea leaf weevil — pea weevil

Remarks: Dry, fresh peas. Apply between emergence and early pod formation when adult populations are present but before eggs are laid. Consult county agent or extension service representative regarding proper timing of application. Do not cut treated fresh pea forage for hay within 10 days of application.
Days To Harvest: Harvest, feeding, or grazing-7.
State Restrictions: For use in the Pacific Northwest only.

- - - - - - - - - - - - - - - -

Signal Word/Toxicity Class: Warning/II.
REI: 24 hr.

Kernel Guard — see diazinon premixes

Kumulus — see sulfur

LANNATE

Restricted Use Pesticide.

Lannate (90) or LV (methomyl) *(DuPont)*

Rate: 0.225-0.9 lb. ai/A
0.25-1 lb. 90/A
0.75-3 pt. LV/A

Pest:

alfalfa caterpillar	green cloverworm	pea aphid
armyworms	loopers	saltmarsh caterpillar
beet armyworm	(except AL, GA)	variegated cutworm

Remarks: Succulent peas. Apply in sufficient water to obtain thorough, uniform coverage when insects first appear at 5- to 7-day intervals or as needed. Rate varies by pest species; refer to label. Do not apply more than 2.7 lb. ai/A per crop or make more than 6 applications per crop.
Days To Harvest: 1; livestock grazing or feeding: forage-5, hay-14.

- - - - - - - - - - - - - - - -

Signal Word/Toxicity Class: Danger-Poison/I.
REI: 48 hr.

LINDANE

Lindane 30C contains toxic inert Rhodamine B. Toxic to fish, birds, and wildlife.

Lindane 30C *(Gustafson)*
◆ **Lindane ST 40** *(AGSCO)*

Rate: 2.7 fl. oz. 30C/100 lb. seed
2 fl. oz. 40/100 lb. seed

Pest:
seedcorn maggots — wireworms

Remarks: Do not use treated seed for food, feed, or oil purposes. Exposed treated seed may be hazardous to birds and other wildlife. Dispose of excess seed by burial away from bodies of water.

- - - - - - - - - - - - - - - -

Signal Word/Toxicity Class: Warning/II.
REI: 12 hr.

Lorsban — see chlorpyrifos

◆ - **new product** • ai-active ingredient • DF-dry flowable • E/EC-emulsifiable concentrate • EW-emulsion in water
F/FL-flowable • G-granule • L/LC-liquid • SP/WSP-soluble packet • W/WP-wettable powder • WSB-water soluble bag

MALATHION

Organophosphate insecticide toxic to fish, aquatic invertebrates, aquatic life stages of amphibians, and bees exposed to direct treatment on blooming crops or weeds.

Clean Crop Malathion 57EC *(Platte)*
Rate: 1-1.25 lb. ai/A
1.5-2 pt. 57EC/A
Pest:
aphids
Remarks: Apply when insects first appear and repeat as necessary. Forage may be fed to livestock.
Days To Harvest: 3.

Signal Word/Toxicity Class: Caution/III.
REI: 12 hr.

Drexel Malathion 5EC *(Drexel)*
◆ **Micro Flo Malathion 5EC** *(Micro Flo)*
Prentox 5 lb. Malathion Spray (5EC) *(Prentiss)*
Rate: 0.8-2.5 lb. ai/A
1.25-4 pt. 5EC/A
Pest:
aphids, grasshoppers, leafhoppers
Remarks: Rate varies by pest species; refer to label.
Days To Harvest: Vines not to be fed-3; vines to be fed-7.

Signal Word/Toxicity Class: Warning/II.
REI: 12 hr.

Fyfanon ULV *(Cheminova)*
◆ **Micro Flo Malathion ULV** *(Micro Flo)*
Rate: 0.5 pt. ULVA
Pest:
pea weevil
Remarks: None.
Days To Harvest: 14.
State Restrictions: For use in the Northwest.

Signal Word/Toxicity Class: Caution.
REI: 12 hr.

◆ **Gowan Malathion 8 (8E)** *(Gowan)*
Rate: 1-2.5 lb. ai/A
1-2.5 pt. 8E/A
Pest:
aphids, pea weevil
Remarks: If forage is fed to livestock, use 1-1.25 pt. 8E/A.
Days To Harvest: 3; forage-7.

Signal Word/Toxicity Class: Caution/III.
REI: 12 hr.

◆ **Wilbur-Ellis Malathion 8 Spray** *(Wilbur-Ellis)*
Rate: 0.5-1 lb. ai/A
0.5-1 pt. 8/A
Pest:
aphids, Mexican bean beetle, spider mites, leafhoppers, pea weevil, thrips
Remarks: Apply in minimum 5 gal. by air; 10 gal. by ground.
Days To Harvest: 3.

Signal Word/Toxicity Class: Caution/III.
REI: 12 hr.

Marlate — see methoxychlor

METHOXYCHLOR

Marlate 50 (WP) *(Kincaid)*
Prentox Methoxychlor 2EC *(Prentiss)*
Rate: 1.3-1.7 lb. ai/A
75-100 fl. oz. 2EC/A
2 Tbs. 50WP/gal. water
Pest:
alfalfa looper, bean leaf beetle, pea weevil
Remarks: Apply at first sign of infestation; repeat at 7- to 14-day intervals as needed. For pea weevil, apply to foliage early bloom before eggs are laid; repeat 5 days later.
Days To Harvest: 7.

Signal Word/Toxicity Class: Caution.
REI: 12 hr.

Wilbur-Ellis Methoxychlor 2 *(Wilbur-Ellis)*
Rate: 1.5-2 lb. ai/A
3-4 qt. 2/A
Pest:
armyworms, pea weevil
Remarks: Do not apply within 14 days of sulfur or sulfur containing sprays.
Days To Harvest: 7.

Signal Word/Toxicity Class: Danger-Poison/I.
REI: 48 hr.

METHYL PARATHION

Organophosphate insecticide toxic to bees exposed to direct treatment or residues on blooming crops or weeds. Penncap-M is a microencapsulated formulation. Restricted Use Pesticide due to very high acute toxicity to humans and birds.

Cheminova Methyl 4EC *(Cheminova)*
Rate: 0.5-1 lb. ai/A
1-2 pt. 4E, 4EC/A
Pest:
aphids, flea beetles, mites, armyworms, leafhoppers, stink bugs, climbing cutworm, Lygus bugs, surface cutworm, cowpea curculio
Remarks: Do not apply if crops or weeds in treatment area are in bloom. Rate varies by pest species; refer to label.
Days To Harvest: Up to 1 pt.: Clean Crop-7; Cheminova-10. Above 1 pt.-15.

Signal Word/Toxicity Class: Danger-Poison/I.
REI: 48 hr.; 72 hr. if average rainfall less than 25" per year.

◆ **Declare (4EC)** *(Griffin)*
Helena 4 lb. Methyl Parathion *(Helena)*
Rate: 0.5-1 lb. ai/A
1-2 pt. 4EC/A
Pest:
aphids, climbing cutworm, mites, armyworms (to 3rd instar), cowpea curculio, plant bugs, flea beetles, stink bugs, cabbage looper, leafhoppers, surface cutworm
Remarks: Rate varies by pest species; refer to label.
Days To Harvest: Up to 1 pt.-10 days; 1-2 pt.-15 days.

Signal Word/Toxicity Class: Danger-Poison/I.
REI: 48 hr.

Penncap-M (2LF) *(Elf Atochem)*
Rate: 0.5 lb. ai/A
2 pt. 2FL/A
Pest:
pea aphid, pea leaf weevil, pea weevil
Remarks: **Austrian winter, white blooming peas.** Do not apply more than 4 pt./A per year. Do not apply if peas are blooming and bees are foraging area to be treated. Do not chemigate in California.
Days To Harvest: Harvest of pods-10; forage or grazing-15.

Signal Word/Toxicity Class: Warning/II.
REI: 48 hr.

Wilbur-Ellis Methyl Parathion 5 Spray *(Wilbur-Ellis)*
Rate: 0.63 lb. ai/A
1 pt. 5/A
Pest:
aphids, flea beetles, spider mites, armyworms, leafhoppers
Remarks: Do not apply through any type of irrigation system.
Days To Harvest: 15.

Signal Word/Toxicity Class: Danger-Poison/I.
REI: 48 hr.

NOTICE The information on these pages is for preliminary planning — not a guide for use. Be sure to follow the manufacturer's directions, notwithstanding information contained here.

For personal protective equipment and EPA registration numbers, please turn to page 39.

M-PEDE

M-Pede (potassium salts of fatty acids) *(Mycogen)*
Rate: 1-2% v/v solution
Pest:

aphids*	leafminers*	spider mites
broad mite	plant bugs	thrips
green peach aphid*	russet mite	whiteflies
leafhoppers		

*** Tank mix only**

Remarks: Apply at first sign of infestation; repeat at 7- to 14-day intervals. May be tank mixed with other insecticides; refer to labels for use directions and precautions.
Days To Harvest: 0.

Signal Word/Toxicity Class: Warning/II.
REI: 12 hr.

NEEMAZAD — see azadirachtin

NEEMIX — see azadirachtin

Penncap-M — see methyl parathion

Phaser — see endosulfan

PYRELLIN (Premix)

Premix containing pyrethrins and rotenone. May be combined with other insecticides to enhance knockdown or as an insect "exciter" to help facilitate contact with other spray residues.

Pyrellin E.C. (0.042 lb./gal. pyrethrins & 0.035 lb./gal. rotenone) *(Webb Wright)*
Rate: 1-2 pt. E.C./A
Pest:

aphids	loopers	plant bugs
leafhoppers	Lygus bugs	thrips
leafminers	mites	whiteflies

Remarks: Apply as necessary to maintain control.
Days To Harvest: 0.

Signal Word/Toxicity Class: Caution/III.
REI: 12 hr.

Pyrenone — see pyrethrins

PYRETHRINS

Contains piperonyl butoxide as a synergist. Diatect also contains silicon dioxide (diatomaceous earth). May be combined in tank mix with other insecticides for quicker and more complete kill where insect resistance may be a problem, and as an "exciter" to flush insects out of hiding and into contact with other spray residues.

Diatect Multipurpose Insecticide *(Diatect)*
Rate: 1-6 lb. product/A
Pest:

aphids	garden fleahoppers	Lygus bugs
armyworms	imported cabbageworm	mites
cabbage looper	Japanese beetles	plant bugs
caterpillars	leafhoppers	stink bugs
cucumber beetle	leafminers	thrips
diamondback moth	loopers	vegetable weevils
flea beetles		whiteflies

Remarks: Apply as dust or wettable powder. Repeat as necessary. Rate varies with insect population.

Signal Word/Toxicity Class: Caution.
REI: 12 hr.

Pyrenone (0.5EC) *(AgrEvo Environmental Health)*
◆ **Py-Rin 60-6 EC** *(Wilbur-Ellis)*
Rate: 1-12 fl. oz. 0.5EC/A
2-16 oz. 6-60 EC/A
Pest: Including but not limited to:

alfalfa caterpillar	cabbage looper	green cloverworm
aphids	cutworm	ground beetles
armyworms	flea beetle	leafhoppers
bean leaf beetle	grasshoppers	stink bugs

Remarks: Apply in sufficient water for thorough coverage of upper and lower leaf surfaces and repeat as necessary to maintain control.
Days To Harvest: 0.

Signal Word/Toxicity Class: Caution.
REI: 12 hr.

Py-Rin — see pyrethrins

Sevin — see carbaryl

SULFUR

Thiolux is micronized sulfur.

◆ **Kumulus DF** *(BASF)*
Rate: 2.4-4 lb. ai/A
3-5 lb. DF/A
Pest:
red spider mite
Remarks: Apply at early leaf stage. Repeat at 10- to 14-day intervals with or without other pesticides immediately after a rain. Do not apply if temperatures during or within 3 days after application are expected to exceed 90°F. Do not apply through any type of irrigation system. Refer to label for use directions.

Signal Word/Toxicity Class: Caution.
REI: 24 hr.

Liquid Sulfur Six *(Helena)*
Rate: 3-4 pt./8 gal. water
Pest:
red spider mite
Remarks: Apply by air. Do not apply this product through any type of irrigation system. Refer to label for use directions.

Signal Word/Toxicity Class: Caution.
REI: 24 hr.

Thiolux (DF) *(Platte)*
Rate: 2.4-8 lb. ai/A
3-10 lb. DF/A
Pest:
red spider mite
Remarks: Apply in early leaf stage.

Signal Word/Toxicity Class: Caution.
REI: 24 hr.

Thiodan — see endosulfan

Thiolux — see sulfur

Peppers

ABAMECTIN

Natural product produced by soil microorganism, Streptomyces avermitilis. Unrelated to other miticides. Toxic to fish, mammals, aquatic organisms and highly toxic to honey bees. Restricted Use Pesticide.

Agri-Mek 0.15EC *(Novartis)*
Rate: 8-16 fl. oz. 0.15EC/A

◆ **- new product** • ai-active ingredient • DF-dry flowable • E/EC-emulsifiable concentrate • EW-emulsion in water
F/FL-flowable • G-granule • L/LC-liquid • SP/WSP-soluble packet • W/WP-wettable powder • WSB-water soluble bag

Pest:

leafminers mites

Remarks: For mites, apply at first appearance; leafminers when adults are first observed. Repeat as necessary. Do not apply through any type of irrigation system or by aircraft.

Days To Harvest: 7.

State Restrictions: For use in Texas only.

Signal Word/Toxicity Class: Warning/II.

REI: 12 hr.

ACEPHATE

Organophosphate systemic insecticide with contact and stomach poison activity. Effective against many aphids, thrips, caterpillars, and plant bugs. Has moderate residual effects and is compatible with most pesticides.

Orthene 75S *(Valent)*

Rate: 0.25-1 lb. ai/A
0.33-0.66 lb. 75S/A

Pest:

cabbage looper, grasshoppers, tobacco hornworm, European corn borer, green peach aphid

Remarks: Sweet bell peppers. Apply when eggs or insects appear and apply as needed until insects have been reduced below economic numbers. Rate varies by pest species; refer to label. Do not apply through any type of irrigation system.

Days To Harvest: 7.

Rate: 0.5 lb. ai/A
0.66 lb. 75S/A

Pest:

aphids

Remarks: Non-bell peppers. Apply at 7- to 10-day intervals throughout growing season. Do not exceed 7 applications per season. Do not apply through any type of irrigation system.

Days To Harvest: 7.

State Restrictions: For use in Midwestern and Eastern states, Puerto Rico.

Signal Word/Toxicity Class: Caution.

REI: 24 hr.

Payload (15G) *(Valent)*

Rate: 1 lb. ai/A
6.66 lb. 15G/A

Pest:

aphids, cutworms, thrips

Remarks: In-furrow or sidedress treatment. Sidedress application should be used to extend residual control in areas which experience prolonged insect pressure. Sidedress applications should be followed by irrigation.

State Restrictions: SLN-New Mexico.

Signal Word/Toxicity Class: Caution.

REI: 12 hr.

Adios — see carbaryl

Admire — see imidacloprid

Agri-Mek — see abamectin

Ambush — see permethrin

ASANA

Pyrethroid insecticide that has both contact and stomach poison activity. Restricted Use Pesticide due to toxicity to fish and aquatic organisms.

Asana XL (esfenvalerate) *(DuPont)*

Rate: 0.03-0.05 lb. ai/A
5.8-9.6 fl. oz. XL/A

Pest:

beet armyworm (aids in control), cucumber beetle (adult), loopers, pepper weevil (aids in control), Colorado potato beetle, European corn borer, corn earworm, flea beetles, southern armyworm

Remarks: Apply when insects or their damage are observed and repeat at 7- to 10-day intervals. Do not exceed 0.35 lb. ai/A per season.

Days To Harvest: 7.

Signal Word/Toxicity Class: Warning/II.

REI: 12 hr.

AZADIRACHTIN

Extracted from seed of neem tree. Insect growth regulator that controls targeted insect larvae when they ingest or come in contact with it, by interfering with the insect's ability to molt and shutting off feeding behavior. Affected larvae slowly starve and die at next molt.

◆ **Azatin EC or XL** *(Olympic)*
◆ **Azatin XL Plus** *(Thermo Trilogy)*

Rate: 8-21 oz. EC/A
5-21 oz. XL, XL Plus/A

Pest:

aphids, armyworms, beet armyworm, beetles, cabbage looper, caterpillars, cutworms, diamondback moth, fall armyworm, flea beetles, green peach aphid, imported cabbageworm, Japanese beetle, leafhoppers, leafminers, leafrollers, loopers, psyllids, silverleaf whitefly, southern armyworm, sweetpotato whitefly, thrips, tobacco hornworm, tomato fruitworm, tomato pinworm, webworms, whiteflies

Remarks: Rate varies by species. Apply when pests first appear; repeat at 7-day intervals. When pest populations are high, use higher rates. Do not apply more that 21 fl. oz./A.

Days To Harvest: 0.

Signal Word/Toxicity Class: Caution (XL); Warning/II (EC).

REI: 4 hr. (XL), 12 hr. (EC).

◆ **NEEMAZAD 4.5EC or NEEMIX 4.5** *(Thermo Trilogy)*

Rate: 0.125-1 pt. 4.5/A
High Volume:
2.25-4.5 oz. 4.5EC/100 gal. water
Low Volume:
0.5 pt. 4.5EC/A

Pest:

armyworms, bean leaf beetle, cabbage looper, caterpillars, Colorado potato beetle, cutworms, garden webworm, greenhouse whitefly, hornworms, leafminers, loopers, Mexican bean beetle, pepper weevil, psyllids, soybean loopers, sweetpotato whitefly, tomato fruitworm, tomato pinworm, weevils

Remarks: For low volume, apply in sufficient water for adequate coverage. For low insect pressure, apply preventive at 7-day intervals or at first sign of insects; high insect pressure, at 3- to 4-day intervals.

Signal Word/Toxicity Class: Warning/II.

REI: 12 hr.

Azatin — see azadirachtin

AZINPHOS-METHYL

Restricted Use Pesticide.

Gowan Azinphos-M 50-W *(Gowan)*
◆ **Micro Flo Azinphosmethyl 2EC or 50W** *(Micro Flo)*
Sniper 2-E, 50PVA, 50W *(Platte)*

Rate: 0.375-0.5 lb. ai/A
1.5-2 pt. 2EC, 2-E/A
0.75-1 lb. 50-W, 50PVA, 50W/A

Pest:

European corn borer, flea beetles, leafminers

Remarks: Apply in sufficient water to give complete coverage. Do not apply more than 3 times per season. Allow at least 7 days between applications.

NOTICE The information on these pages is for preliminary planning — not a guide for use. Be sure to follow the manufacturer's directions, notwithstanding information contained here. For personal protective equipment and EPA registration numbers, please turn to page 39.

Days To Harvest: 2-E, 50W: Up to 4 applications-3; over 4 applications-14; 50-W (except CA), 50PVA-7.
State Restrictions: Sniper 50PVA not registered in California.

Signal Word/Toxicity Class: Danger-Poison/I.
REI: 48 hr.; 72 hr. if average rainfall less than 25" per year.

BAYTHROID

Synthetic pyrethroid formulated in emulsifiable concentrate. Toxic to fish and aquatic organisms. Restricted Use Pesticide.

Baythroid (2EC) (cyfluthrin) *(Bayer)*
Rate: 0.025-0.044 lb. ai/A
1.6-2.8 fl. oz. 2EC/A
Pest:
beet armyworm, cabbage looper, celery leaftier, corn earworm, European corn borer, garden webworm, leafminers (suppression), pepper weevil, potato leafhopper, thrips (except *Thrips palmi*), western yellowstriped armyworm
Remarks: Do not apply to crops grown for seed. Apply by ground only in sufficient water for complete coverage. Allow at least 7 days between applications. Up to 16.8 fl. oz./A may be applied per crop season. Refer to label for spray drift precautions.
Days To Harvest: 7.

Signal Word/Toxicity Class: Danger/I.
REI: 12 hr.

CARBARYL

Carbamate insecticide with broad activity on many insects. Highly toxic to bees.

Adios (13%) *(Micro Flo)*
Rate: 0.065-0.1 lb. ai/A
0.5-0.75 lb. 13%/A
Pest:
cucumber beetle
Remarks: Do not apply through any type of irrigation system. Do not apply more than 3.75 lb./A per season.
Days To Harvest: 0.
State Restrictions: Not for use in California.

Signal Word/Toxicity Class: Caution/III.
REI: 12 hr.

Clean Crop Sevin 5 Bait *(Platte)*
Rate: 1-2 lb. ai/A
20-40 lb. Bait/A
Pest:
ants, armyworms, cutworms, darkling beetle, field crickets, grasshoppers, sowbugs
Remarks: Apply with ground or air equipment.
Days To Harvest: 0.

Signal Word/Toxicity Class: Caution/III.
REI: 12 hr.

Drexel Carbaryl 4L or 50W *(Drexel)*
Sevin 4F, XLR Plus, 80S, 80WSP *(Rhone-Poulenc)*
Rate: 0.5-2 lb. ai/A
1-4 pt. 4L, 4F, XLR/A
1-4 lb. 50W/A
0.66-1.25 lb. 80S, 80WSP/A
Pest:
Colorado potato beetle, cutworms, European corn borer, fall armyworm, flea beetles, grasshoppers, lace bugs, leafhoppers, stink bugs, tarnished plant bug, tomato fruitworm, tomato hornworm, tomato pinworm
Remarks: Rate varies by pest species; refer to label. Thorough coverage essential to effectively suppress stink bugs. When disease transmission is suspected, monitor field following application and repeat if infestation occurs.
Days To Harvest: 0.

Signal Word/Toxicity Class: Caution (4F, 4L, XLR); Warning/II (50W, 80S, 80WSP).
REI: 12 hr.

Gowan Sevin Bait *(Gowan)*
Rate: 2 lb. ai/A
40 lb. Bait/A
Pest:
armyworms, crickets, cutworms, darkling beetle, grasshoppers, sowbugs
Remarks: Apply when pests appear in damaging numbers; repeat at 7- to 14-day intervals if necessary.
Days To Harvest: 1.
State Restrictions: For use in Arizona and California.

Signal Word/Toxicity Class: Caution/III.
REI: 12 hr.

◆ **Wilbur-Ellis Sevin 5 or 10 Dust** *(Wilbur-Ellis)*
Rate: 2 lb. ai/A
40 lb. 5/A
20 lb. 10/A
Pest:
Colorado potato beetle, European corn borer, fall armyworm, flea beetles, leafhoppers, Lygus bugs, stink bugs, tomato fruitworm, tomato hornworm
Remarks: Do not feed treated vines to livestock.
Days To Harvest: 0.
State Restrictions: Sevin 5 not registered in California.

Signal Word/Toxicity Class: Caution.
REI: 12 hr.

CRYOLITE

Kryocide (96%) *(Elf Atochem)*
Rate: 7.7-11.5 lb. ai/A
8-12 lb. 96%/A
Pest:
hornworms, omnivorous leafroller, pepper weevil
Remarks: Apply at 7-day intervals as needed. Do not exceed 24 lb./A per season. Do not use in combination with lime or compounds containing free lime. Do not chemigate in California.
Days To Harvest: 14.

Signal Word/Toxicity Class: Caution.
REI: 12 hr.

Prokil Cryolite 96 *(Gowan)*
Rate: 9.6-11.5 lb. ai/A
10-12 lb. 96%/A
Pest:
armyworms, blister beetle, cabbage looper, Colorado potato beetle, flea beetles, hornworms
Remarks: Apply at 7-day intervals as needed. Do not exceed 24 lb./A per season. Do not use in combination with lime or compounds containing free lime. Do not apply through any type of irrigation system.
Days To Harvest: 14.

Signal Word/Toxicity Class: Caution.
REI: 12 hr.

DEADLINE

Packaged as Bullets, M-Ps (mini-pellets), and granules.

Deadline Bullets, M-Ps, Granules (4%) (metaldehyde) *(Pace)*
Rate: 0.8-1 lb. ai/A
20-40 lb. 4%/A
Pest:
slugs, snails
Remarks: Apply every 3-4 weeks during growing season or as needed. Do not apply directly to edible portion of plant.

Signal Word/Toxicity Class: Warning/II (granule); Caution (bullets, M-P).
REI: 12 hr.

Diatect — see pyrethrins

◆ **- new product** • ai-active ingredient • DF-dry flowable • E/EC-emulsifiable concentrate • EW-emulsion in water
F/FL-flowable • G-granule • L/LC-liquid • SP/WSP-soluble packet • W/WP-wettable powder • WSB-water soluble bag

DIAZINON

Restricted Use Pesticide due to avian and aquatic toxicity.

D•z•n diazinon AG600 WBC, 50W *(Novartis)*
Drexel Diazinon Insecticide (4EC), D-264 14G *(Drexel)*
Gowan Diazinon 4E, 14G, 50WSB *(Gowan)*
Helena Diazinon AG500 *(Helena)*
Micro Flo Diazinon AG500, 14G, 50W *(Micro Flo)*
Prentox Diazinon AG500, 50W *(Prentiss)*
◆ **Wilbur-Ellis Diazinon 4 (4E), 14G** *(Wilbur-Ellis)*

Rate: 0.25-0.5 lb. ai/A
0.5-1 pt. 4E, 4EC, AG500/A
6.5-12.75 fl. oz. AG600/A
0.5 lb. ai/A
1 lb. 50W, 50WSB/A

Pest:
aphids — serpentine leafminer

Rate: 1 lb. ai/A
2 pt. 4E, 4EC, AG500/A
25.5 fl. oz. AG600/A
2 lb. 50W, 50WSB/A
7 lb. 14G/A

Pest:
mole crickets

Rate: 2-4 lb. ai/A
4-8 pt. 4E, 4EC, AG500/A
51-102 fl. oz. AG600/A
4-8 lb. 50W, 50WSB/A
14-28 lb. 14G/A

Pest:
cutworms

Rate: 3-4 lb. ai/A
6-8 pt. 4E, 4EC, AG500/A
76.5-102 fl. oz. AG600/A
6-8 lb. 50W, 50WSB/A
21-28 lb. 14G/A

Pest:
wireworms

Remarks: Apply as insects appear; repeat after 7 days if needed. Do not make more than 5 applications per season. For mole crickets, broadcast to plant beds 1-2 days before sowing seed or transplanting and immediately incorporate into top 4-8" soil. For cutworms and wireworms, broadcast just before planting and immediately incorporate into soil.
Days To Harvest: 5.

Signal Word/Toxicity Class: Caution/III.
REI: 24 hr.; 12 hr. (Drexel/Gowan/Wilbur-Ellis 14G).

DIBROM

Dibrom 8 (8E) (naled) *(Amvac)*

Rate: 1-2 lb. ai/A
1-2 pt. 8E/A

Pest:
aphids — blister beetles — flea beetles — leafminers — spider mites

Remarks: Apply at 7-day intervals in minimum 10 gal. water by air/A; 20 gal. water by ground/A. Do not apply more than 6 pt. product/A per season. Do not apply more than 1 pt./A in Florida. Do not apply when temperatures exceed 90°F.
Days To Harvest: 1.

Signal Word/Toxicity Class: Danger/I.
REI: 48 hr. for 1 pt./A or less; 72 hr. for more than 1 pt./A.

DICOFOL

◆ **Griffin Dicofol 4E** *(Griffin)*
Kelthane MF (4E) *(Rohm and Haas)*

Rate: 0.38-0.75 lb. ai/A
0.75-1.5 pt. 4E/A

Pest:
Pacific mite — tomato russet mite — tropical mite — twospotted spider mites

Remarks: Apply at first sign of mites. Do not make more than 2 applications or exceed 1.6 pt. per year. Do not feed treated plants, forage, or crop residues to livestock. Do not apply through any type of irrigation system.
Days To Harvest: 2.

Signal Word/Toxicity Class: Caution.
REI: 12 hr.

Digon — see dimethoate

Dimate — see dimethoate

DIMETHOATE

Cheminova Dimethoate 4E *(Cheminova)*
Clean Crop Dimethoate 400 *(Platte)*
Dimate 4E *(Terra)*
Drexel Dimethoate 2.67 EC or 4EC *(Drexel)*
Gowan Dimethoate E267 *(Gowan)*
Helena Dimethoate 267 or 4E *(Helena)*
Micro Flo Dimethoate 267 or 4E *(Micro Flo)*
Wilbur-Ellis Dimethoate 267 or Digon 400 *(Wilbur-Ellis)*

Rate: 0.25-0.33 lb. ai/A
0.75-1 pt. 2.67, 267, E267/A
0.5-0.66 pt. 4E, 4EC, 400/A

Pest:
aphids — leafminers — maggots

Remarks: Repeat as necessary.
Days To Harvest: 0.

Signal Word/Toxicity Class: Warning/II.
REI: 48 hr.

DI-SYSTON

Restricted Use Pesticide.

Di-Syston 8 (disulfoton) *(Bayer)*

Rate: 2 lb. 8/A

Pest:
aphids

Remarks: None.
State Restrictions: SLN-California.

Signal Word/Toxicity Class: Danger/I.
REI: 12 hr.

Di-Syston 15% (disulfoton) *(Bayer)*

Rate: 6.7-13.3 lb. 15%/A (32" row spacing)

Pest:
aphids

Remarks: Place granules in soil incorporated band on each side of seed furrow at planting or at transplanting. Do not apply more than once per crop season. Do not apply directly on seed.
Days To Harvest: 90.

Signal Word/Toxicity Class: Danger/I.
REI: 48 hr.

D•z•n — see diazinon

ENDOSULFAN

Toxic to birds, fish, and wildlife; toxic to bees exposed to direct application.

Drexel Endosulfan 3EC or 50WP *(Drexel)*
Gowan Endosulfan 3EC or 50WSB *(Gowan)*
Phaser 3EC or 50WSB *(AgrEvo USA)*
◆ **Micro Flo Endosulfan 3EC or 50WP** *(Micro Flo)*
Thiodan 3EC or 50WP *(FMC)*

Rate: 0.5-1 lb. ai/A
0.66-1.33 qt. 3EC/A
1-2 lb. 50WP, 50WSB/A

NOTICE The information on these pages is for preliminary planning — not a guide for use. Be sure to follow the manufacturer's directions, notwithstanding information contained here.

For personal protective equipment and EPA registration numbers, please turn to page 39.

Pest:

armyworms* | hornworms | pepper maggot
flea beetles | leafhoppers* | whiteflies (high rate)
green peach aphid

****Except California***

Remarks: Apply when insects first appear or feeding is noticed. Do not exceed 2 applications per year or 2 lb. ai/A per year.

Days To Harvest: 0.5 lb.ai; over 0.5 lb. ai-4.

Signal Word/Toxicity Class: Danger-Poison/I.
REI: 24 hr.

Fyfanon — see malathion

IMIDACLOPRID

Systemic insecticide for control of Thrips palmi. Toxic to bees and aquatic invertebrates.

Admire 2F *(Bayer)*

Rate: 0.25-0.5 lb. ai/A
16-32 fl. oz. 2F/A

Pest:

aphids, Colorado potato beetle, flea beetles, silverleaf whitefly, sweetpotato whitefly, thrips, whiteflies

Remarks: Bell, chili, cooking, pimentos, sweet peppers. Direct seed or transplant. Soil application. Use high rate where infestations occur later in crop development, or where pest pressure is continuous. Do not apply more than 32 fl. oz./A per year.

Days To Harvest: 21.

Signal Word/Toxicity Class: Caution.
REI: 12 hr.

Provado 1.6 *(Bayer)*

Rate: 0.05 lb. ai/A
3.75 fl. oz. 1.6/A

Pest:

aphids, Colorado potato beetle, silverleaf whitefly, sweetpotato whitefly, whiteflies

Remarks: Bell, chili, cooking, pimentos, sweet peppers. For aphids and Colorado potato beetle, good coverage of foliage necessary. Apply before damaging population becomes established. For whiteflies, apply at first appearance; repeat at 5- to 7-day intervals until no longer in crop. Do not apply more than 18.75 fl. oz./A per year.

Days To Harvest: 0.

Signal Word/Toxicity Class: Caution.
REI: 12 hr.

Kelthane — see dicofol

Kryocide — see cryolite

Kumulus — see sulfur

LANNATE

Restricted Use Pesticide.

Lannate SP or LV (methomyl) *(DuPont)*

Rate: 0.225-0.45 lb. ai/A
0.25-0.5 lb. SP/A
0.75-1.5 pt. LV/A

Pest:

armyworms, beet armyworm, fall armyworm, green peach aphid, loopers, variegated cutworm

Remarks: Apply in sufficient water to obtain thorough, uniform coverage when insects first appear; repeat at 5- to 7-day intervals or as needed. Rate varies by pest species; refer to label. Do not apply more than 4.05 lb. ai/A per crop or make more than 10 applications per crop.

Days To Harvest: 3.

Signal Word/Toxicity Class: Danger-Poison/I.
REI: 48 hr.

MALATHION

Organophosphate insecticide toxic to fish, aquatic invertebrates, aquatic life stages of amphibians, and bees exposed to direct treatment on blooming crops or weeds.

Clean Crop Malathion 57EC or 8EC *(Platte)*
Drexel Malathion 5EC *(Drexel)*
Fyfanon (8E) *(Helena)*
Gowan Malathion 8 (8E) *(Gowan)*
◆ **Micro Flo Malathion 5EC** *(Micro Flo)*
Prentox 5 lb. Malathion Spray (5EC) *(Prentiss)*

Rate: 0.63-1.5 lb. ai/A
1-2.5 pt. 5EC, 57EC/A
1.5 lb. ai/A
1.5 pt. 8E, 8EC/A

Pest:

aphids, pepper maggots

Remarks: Including bell peppers.

Days To Harvest: 3.

Signal Word/Toxicity Class: Caution/III (57EC, 8E, 8EC); Warning/II (5EC).
REI: 12 hr.

◆ **Wilbur-Ellis Malathion 8 Spray** *(Wilbur-Ellis)*

Rate: 0.5-1 lb. ai/A
0.5-1 pt. 8/A

Pest:

aphids, leafhoppers, Mexican bean beetle, spider mites, thrips

Remarks: Apply in minimum 5 gal. by air; 10 gal. by ground.

Days To Harvest: 3.

Signal Word/Toxicity Class: Caution/III.
REI: 12 hr.

Marlate — see methoxychlor

METASYSTOX-R

Restricted Use Pesticide.

Metasystox-R (SC) (oxydemeton-methyl) *(Gowan)*

Rate: 0.5 lb. ai/A
2 pt. SC/A

Pest:

aphids

Remarks: Apply in sufficient water for complete coverage but not less than 1 gal./A. Do not apply more than twice per year.

Days To Harvest: 3.

Signal Word/Toxicity Class: Warning/II.
REI: 48 hr.; 72 hr. if average rainfall less than 25" per year.

METHOXYCHLOR

Marlate 50 (WP) *(Kincaid)*
Prentox Methoxychlor 2EC *(Prentiss)*

Rate: 1.3-1.7 lb. ai/A
75-100 fl. oz. 2EC/A
2 Tbs. 50WP/gal. water

Pest:

alfalfa looper, blister beetle, fall armyworm, flea beetles, imported cabbageworm, Japanese beetle, leafhoppers

Remarks: Apply at first sign of infestation; repeat at 7- to 14-day intervals as needed.

Days To Harvest: Marlate: 7; Prentox: 1.

Signal Word/Toxicity Class: Caution.
REI: 12 hr.

Microthiol Special — see sulfur

◆ - **new product** • ai-active ingredient • DF-dry flowable • E/EC-emulsifiable concentrate • EW-emulsion in water
F/FL-flowable • G-granule • L/LC-liquid • SP/WSP-soluble packet • W/WP-wettable powder • WSB-water soluble bag

M-PEDE

M-Pede (potassium salts of fatty acids) *(Mycogen)*

Rate: 1-2% v/v solution

Pest:

aphids*	leafminers*	spider mites
broad mite	plant bugs	thrips
green peach aphid*	russet mite	whiteflies
leafhoppers		

*** Tank mix only**

Remarks: Apply at first sign of infestation; repeat at 7- to 14-day intervals. May be tank mixed with other insecticides; refer to labels for use directions and precautions.

Days To Harvest: 0.

Signal Word/Toxicity Class: Warning/II.

REI: 12 hr.

NEEMAZAD — see azadirachtin

NEEMIX — see azadirachtin

NEMACUR

Restricted Use Pesticide.

Nemacur 15% (fenamiphos) *(Bayer)*

Rate: California, Georgia, Puerto Rico:
1.35-2 lb. ai/A
9-13.3 lb. 15%/A (36" row)
10-14.7 oz. 15%/1000 ft. row (any row spacing)

Rate: SLN-New Mexico:
1.5-2.2 oz. ai
10-14.7 oz. 15%/1000 ft. row (any row spacing)
1.35-2 lb. ai/A
9-13.3 lb. 15%/A (36" row)

Pest:

nematodes

Remarks: Non-bell peppers. Apply in a 12" band and incorporate.

Signal Word/Toxicity Class: Danger/I.

REI: 48 hr.

Omni Supreme Spray — see petroleum oil

Orthene — see acephate

Payload — see acephate

PERMETHRIN

Restricted Use Pesticide.

Ambush 2EC or 25W *(ZENECA)*

Pounce 3.2 EC, 25WP or WSB *(FMC)*

Rate: 0.1-0.2 lb. ai/A
6.4-12.8 fl. oz. 2EC/A
6.4-12.8 oz. 25W, 25WP/A
4-8 oz. 3.2 EC/A
1-2 bags WSB/A

Pest:

cabbage looper	European corn borer	pepper weevil
corn earworm	flea beetles	vegetable leafminer

Remarks: Bell peppers. Apply as needed by ground or air using sufficient water to obtain uniform coverage. Apply high rate for European corn borer. Do not apply more than 1.6 lb. ai/A per season.

Days To Harvest: 3.

Signal Word/Toxicity Class: Caution (3.2 EC); Warning/II (2EC, 25W, 25WP, WSB).

PETROLEUM OIL

Petroleum based spray oil. Mode of action is through suffocation of eggs, larvae, and nymphs of insects and adults of soft bodied insects. This mode of action necessitates total spray coverage. Toxic to fish.

Omni Supreme Spray *(Helena)*

Rate: 1-2 gal./100 gal. water

Pest:

aphids	caterpillars	mites
beetles (larvae)	leafminers	

Remarks: None.

Signal Word/Toxicity Class: Caution.

REI: 4 hr.

SAF-T-SIDE *(Brandt Consolidated)*

◆ **Sumulsion** *(Drexel)*

◆ **SunSpray Ultra-Fine** *(Sun)*

Rate: 1-2 gal./100 gal. water

Pest:

aphids	leafminers	thrips
beetles (larvae)	mites	whiteflies
leafhoppers	spiders	

Remarks: Apply as needed. May be tank mixed with most commonly used insecticides and fungicides. Do not use in combination with or immediately before or after spraying with fungicides such as captan, Dyrene, folpet, Morestan or any sulfur products. Do not use with carbaryl or dimethoate.

Signal Word/Toxicity Class: Caution.

REI: 4 hr.

Phaser — see endosulfan

Pounce — see permethrin

Provado — see imidacloprid

PYRELLIN (Premix)

Premix containing pyrethrins and rotenone.

Pyrellin E.C. (0.042 lb./gal. pyrethrins & 0.035 lb./gal. rotenone) *(Webb Wright)*

Rate: 1-2 pt. E.C./A

Pest:

aphids	harlequin bug	mites
bean beetle	leafhoppers	plant bugs
cabbageworm	leafminers	stink bugs
celery leaftier	loopers	thrips
garden fleahopper	Lygus bugs	whiteflies

Remarks: Apply at 7-day intervals or less; repeat as needed.

Days To Harvest: 0.

Rate: 1.5-2 pt. E.C./A

Pest:

Colorado potato beetle	cucumber beetle	flea beetles
	European corn borer	vegetable weevil

Remarks: Make 3-4 applications or more as eggs start to hatch; repeat at 5- to 7-day intervals.

Days To Harvest: 0.

Signal Word/Toxicity Class: Caution/III.

REI: 12 hr.

Pyrenone — see pyrethrins

7 Vegetables

NOTICE The information on these pages is for preliminary planning — not a guide for use. Be sure to follow the manufacturer's directions, notwithstanding information contained here.

For personal protective equipment and EPA registration numbers, please turn to page 39.

PYRETHRINS

Contains piperonyl butoxide as a synergist and silicon dioxide (diatomaceous earth). May be combined in tank mix with other insecticides for quicker and more complete kill where insect resistance may be a problem, and as an "exciter" to flush insects out of hiding and into contact with other spray residues.

Diatect Multipurpose Insecticide II *(Diatect)*
Pyrenone (0.5EC) *(AgrEvo Environmental Health)*
Py-Rin 60-6 EC *(Wilbur-Ellis)*

Rate: 1-6 lb. II/A
1-12 fl. oz. 0.5EC/A
2-16 oz. 6-60 EC/A

Pest: Including but not limited to:

aphids	leafminers	stink bugs
caterpillars	loopers	tarnished plant bug
Colorado potato beetle	Lygus bugs	thrips
cutworms	mites	tomato fruitworm
European corn borer	plant bugs	vegetable weevil
flea beetles	psyllids	whiteflies
green peach aphid		

Remarks: Apply in sufficient water for thorough coverage; repeat as needed. Apply Diatect as dust or wettable powder.
Days To Harvest: 0.

Signal Word/Toxicity Class: Caution.
REI: 12 hr.

Py-Rin — see pyrethrins

SAF-T-SIDE — see petroleum oil

Sevin — see carbaryl

Sniper — see azinphos-methyl

SULFUR

Microthiol Special and Thiolux are micronized sulfurs.

◆ **Kumulus DF** *(BASF)*
Microthiol Special (80W) *(Elf Atochem)*
Thiolux (DF) *(Platte)*

Rate: 2.4-4 lb. ai/A
3-5 lb. DF/A

Pest:
broad mite (80W only) (except CA)
red spider mite

Remarks: Apply at early leaf stage. Repeat at 10- to 14-day intervals with or without other pesticides immediately after a rain. Do not apply if temperatures during or within 3 days after application are expected to exceed 90°F. Refer to label for use directions.

Signal Word/Toxicity Class: Caution.
REI: 24 hr.

Liquid Sulfur Six *(Helena)*
Rate: 3-4 pt./8 gal. water
Pest:
red spider mite
Remarks: Apply by air. Do not apply through any type of irrigation system. Refer to label for use directions.

Signal Word/Toxicity Class: Caution.
REI: 24 hr.

Sumulsion — see petroleum oil

SunSpray — see petroleum oil

Thiodan — see endosulfan

Thiolux — see sulfur

THIRETHRIN (Premix)

Thirethrin (2.9L) (2 lb./gal. endosulfan & 0.45 lb./gal. piperonyl butoxide & 0.045 lb./gal. pyrethrins) *(Terra)*
Rate: 1-2 qt. 2.9L/A
Pest:

flea beetles	hornworms	pepper maggots
green peach aphids		

Remarks: Apply when insects first appear and repeat as required.
Days To Harvest: 4.

Signal Word/Toxicity Class: Warning/II.
REI: 24 hr.

TRIGARD

Triazine compound used as insect growth regulator for control of leafminers. Ineffective against other insect species. Very low toxicity to mammals and safe to beneficial insects.

Trigard (WSP) (cyromazine) *(Novartis)*
Rate: 0.125 lb. ai/A
0.167 lb. WSP/A (1 packet)
Pest:
leafminers
Remarks: Apply as foliar spray in minimum 5 gal. water by air or 50 gal. water by ground when leafminers first appear; repeat at 7-day intervals. Do not make more than 6 applications per crop. Do not apply through any type of irrigation system.
Days To Harvest: 7.
Crop Rotation: All crops on label may be planted following harvest of treated crop. Do not rotate to any other crop except sweet corn or radishes. Do not plant sweet corn or radishes within 3 months after last application.

Signal Word/Toxicity Class: Caution.
REI: 12 hr.

VYDATE

Restricted Use Pesticide.

Vydate L (oxamyl) *(DuPont)*
Rate: Transplant water:
0.5 lb. ai/A
2 pt. L/A
Foliar:
0.5-1 lb. ai/A
2-4 pt. L/A
Pest:

green peach aphid	root-knot nematode (except javanese)	stubby root nematode
pepper weevil		stunt nematode
ring nematode	sting nematode	vegetable leafminer

Remarks: Bell, nonbell peppers. Transplant water treatment: Apply during transplanting. For high nematode populations, begin with transplant water treatment and supplement with drip irrigation or foliar sprays. Do not apply transplant water treatment during periods of slow plant growth, such as when temperatures fall below 45°F. Foliar: Apply when insects first appear; repeat at 1- to 2-week intervals or as needed. Apply low rates for light infestations; high rates at shorter intervals for severe infestations. Do not apply more than 24 pt./A per season.
Days To Harvest: 7.
State Restrictions: Do not use in Suffolk and Nassau counties, Long Island, New York. Not for use in California on nematodes.

Signal Word/Toxicity Class: Danger-Poison/I.
REI: 48 hr.

◆ - **new product** • ai-active ingredient • DF-dry flowable • E/EC-emulsifiable concentrate • EW-emulsion in water
F/FL-flowable • G-granule • L/LC-liquid • SP/WSP-soluble packet • W/WP-wettable powder • WSB-water soluble bag

Potatoes

Adios — see carbaryl

Admire — see imidacloprid

Ambush — see permethrin

ASANA

Pyrethroid insecticide that has both contact and stomach poison activity. Restricted Use Pesticide due to toxicity to fish and aquatic organisms.

Asana XL (esfenvalerate) *(DuPont)*
Rate: 0.015-0.05 lb. ai/A
2.9-9.6 fl. oz. XL/A
Pest:

beet armyworm (aids in control)	cutworms	potato psyllid
buckthorn aphid	European corn borer	potato tuberworm
cabbage looper	flea beetles	tarnished plant bug
cucumber beetle (adult)	grasshoppers	western yellowstriped armyworm
	potato aphid	
	potato leafhopper	

Rate: 0.03-0.05 lb. ai/A
5.8-9.6 fl. oz. XL/A
Long Island, NY:
0.05 lb. ai/A
9.6 fl. oz. XL/A
Pest:
Colorado potato beetle
Remarks: Repeat as necessary to maintain control. Rate varies by pest species; refer to label. Do not exceed 0.35 lb. ai/A per season.
Days To Harvest: 7.

- - - - - - - - - - - - - - - -

Signal Word/Toxicity Class: Warning/II.
REI: 12 hr.

AZADIRACHTIN

Extracted from seed of neem tree. Insect growth regulator that controls targeted insect larvae when they ingest or come in contact with it, by interfering with the insect's ability to molt and shutting off feeding behavior. Affected larvae slowly starve and die at next molt.

◆ **Azatin XL Plus** *(Thermo Trilogy)*
Rate: 5-21 oz. XL Plus/A
Pest:

aphids	cutworms	leafhoppers
armyworms	diamondback moth	leafminers
beet armyworm	fall armyworm	leafrollers
beetles	flea beetles	potato aphid
cabbage looper	green peach aphid	psyllids
caterpillars	imported cabbageworm	southern armyworm
Colorado potato beetle	Japanese beetle	thrips
		webworms

Remarks: Rate varies by species. Apply when pests first appear; repeat at 7-day intervals. When pest populations are high, use higher rates. Do not apply more that 21 fl. oz./A.
Days To Harvest: 0

- - - - - - - - - - - - - - - -

Signal Word/Toxicity Class: Caution.
REI: 4 hr.

◆ **NEEMAZAD 4.5EC or NEEMIX 4.5** *(Thermo Trilogy)*
Rate: 0.125-1 pt. 4.5/A
High Volume:
2.25-4.5 oz. 4.5EC/100 gal. water
Low Volume:
0.5 pt. 4.5EC/A
Pest:

aphids	caterpillars	leafhoppers
armyworms	Colorado potato beetle	potato flea beetle
artichoke plume moth	diamondback moth	psyllids
aster leafhopper	greenhouse whitefly	sweetpotato whitefly
cabbage loopers	hornworms	webworms

Remarks: For low volume, apply in sufficient water for adequate coverage. For low insect pressure, apply preventive at 7-day intervals or at first sign of insects; high insect pressure, at 3- to 4-day intervals.

- - - - - - - - - - - - - - - -

Signal Word/Toxicity Class: Warning/II.
REI: 12 hr.

Azatin — see azadirachtin

AZINPHOS-METHYL

Restricted Use Pesticide (except Guthion Solupak).

Gowan Azinphos-M 50-W *(Gowan)*
Guthion 2L, Solupak (50WSP) *(Bayer)*
◆ **Micro Flo Azinphosmethyl 2EC or 50W** *(Micro Flo)*
Sniper 2-E, 50PVA, 50W *(Platte)*
Rate: 0.375-0.75 lb. ai/A
1.5-3 pt. 2EC, 2-E, 2L/A
0.75-1.5 lb. 50-W, 50PVA, 50W, 50WSP/A
Pest:

banded cucumber beetle	flea beetles	spittlebugs
Colorado potato beetle	leafhoppers	tarnished plant bug
European corn borer	leafminers	tuberworm

Remarks: Apply in sufficient water for compete coverage. Up to 3 applications may be made per crop season. Allow at least 7 days between applications.
Days To Harvest: 7.
State Restrictions: California labels may vary. Sniper 50PVA not registered in California.

- - - - - - - - - - - - - - - -

Signal Word/Toxicity Class: Danger/I.
REI: 48 hr.; 72 hr. if average rainfall less than 25" per year.

BT 320 SULFUR 25 or 50 DUST (Premix)

Bt 320 Sulfur 25 Dust (0.064% Bt & 25% sulfur) or Bt 320 Sulfur 50 Dust (0.064% Bt & 50% sulfur) *(Wilbur-Ellis)*
Rate: 20-40 lb. Dust/A
Pest:
cabbage looper
Remarks: Do not mix or dilute with other products. Thorough coverage essential.

- - - - - - - - - - - - - - - -

Signal Word/Toxicity Class: Caution.
REI: 24 hr.

CARBARYL

Carbamate insecticide with broad activity on many insects. Highly toxic to bees.

Adios (13%) *(Micro Flo)*
Rate: 0.065-0.1 lb. ai/A
0.5-0.75 lb. 13%/A
Pest:
cucumber beetle
Remarks: Irish, white potatoes. Do not apply through any type of irrigation system. Do not apply more than 3.75 lb./A per season.
Days To Harvest: 0.

- - - - - - - - - - - - - - - -

Signal Word/Toxicity Class: Caution/III.

Clean Crop Sevin 5 Bait *(Platte)*
Rate: 1-2 lb. ai/A
20-40 lb. Bait/A
Pest:

ants	darkling beetle	grasshoppers
armyworms	field crickets	sowbugs
cutworms		

Remarks: Apply with ground or air equipment.
Days To Harvest: 0.

- - - - - - - - - - - - - - - -

Signal Word/Toxicity Class: Caution/III.
REI: 12 hr.

NOTICE The information on these pages is for preliminary planning — not a guide for use. Be sure to follow the manufacturer's directions, notwithstanding information contained here.	For personal protective equipment and EPA registration numbers, please turn to page 39.

Drexel Carbaryl 4L or 50W *(Drexel)*
Sevin 4F, XLR Plus, 80S or 80WSP *(Rhone-Poulenc)*
Rate: 0.5-2 lb. ai/A
1-4 pt. 4L, 4F, XLR/A
1-4 lb. 50W/A
0.66-1.25 lb. 80S, 80WSP/A
Pest:
Colorado potato beetle, cutworms, European corn borer, fall armyworm, flea beetles, grasshoppers, lace bug, leafhoppers, stink bugs, tarnished plant bug, tomato fruitworm, tomato hornworm, tomato pinworm
Remarks: Rate varies by pest species; refer to label. Thorough coverage essential to effectively suppress stink bugs.
Days To Harvest: 0.
- - - - - - - - - - - - - - - - -
Signal Word/Toxicity Class: Caution (4F, 4L, XLR); Warning/II (50W, 80S, 80WSP).
REI: 12 hr.

Gowan Sevin Bait *(Gowan)*
Rate: 2 lb. ai/A
40 lb. Bait/A
Pest:
armyworms, crickets, cutworms, darkling beetle, grasshoppers, sowbugs
Remarks: Apply when pests appear in damaging numbers; repeat at 7- to 14-day intervals if necessary.
Days To Harvest: 1.
State Restrictions: For use in Arizona and California.
- - - - - - - - - - - - - - - - -
Signal Word/Toxicity Class: Caution/III.
REI: 12 hr.

◆ **Wilbur-Ellis Sevin 5 or 10 Dust** *(Wilbur-Ellis)*
Rate: 1.5-2 lb. ai/A
30-40 lb. 5/A
15-20 lb. 10/A
Pest:
Colorado potato beetle, flea beetles, leafhoppers
Remarks: Apply to foliage only.
Days To Harvest: 0.
State Restrictions: Sevin 5 not registered in California.
- - - - - - - - - - - - - - - - -
Signal Word/Toxicity Class: Caution.
REI: 12 hr.

Champ — see copper compounds

Champion — see copper compounds

Comite — see propargite

COPPER COMPOUNDS

◆ **Champ (4.6FL), Champ Formula 2 (3FL), or Champion (WP) (copper hydroxide)** *(Agtrol)*
Rate: 1.33-5.33 pt. 4.6FL/A
0.66-2.66 pt. 3FL/A
1-4 lb. WP/A
Pest:
Colorado potato beetle (suppression)
Time: Apply when plants are 6" high; repeat at 7- to 10-day intervals.
Remarks: Refer to label for use directions.
Days To Harvest: 2 weeks.
- - - - - - - - - - - - - - - - -
Signal Word/Toxicity Class: Danger/I (Champion); Caution (Champ).
REI: 24 hr.

CRYOLITE

Kryocide (96%) *(Elf Atochem)*
◆ **Prokil Cryolite 96** *(Gowan)*
Rate: 9.6-11.5 lb. ai/A
10-12 lb. 96%, 96/A
Pest:
Colorado potato beetle

Remarks: Apply at 7-day intervals as needed. Application to exposed tubers may result in excess residues. Do not exceed 96 lb./A per season. Do not use in combination with lime or compounds containing free lime.
Days To Harvest: 0.
State Restrictions: Not registered in California.
- - - - - - - - - - - - - - - - -
Signal Word/Toxicity Class: Caution.
REI: 12 hr.

DEADLINE

Packaged as Bullets, M-Ps (mini-pellets), and granules.

Deadline Bullets, M-Ps, Granules (4%) (metaldehyde) *(Pace)*
Rate: 0.8-1 lb. ai/A
20-40 lb. 4%/A
Pest:
slugs, snails
Remarks: Apply every 3-4 weeks during growing season or as needed. Do not apply directly to edible portion of plant.
- - - - - - - - - - - - - - - - -
Signal Word/Toxicity Class: Warning/II (granule); Caution (bullets, M-P).
REI: 12 hr.

Declare — see methyl parathion

Diatect — see pyrethrins

DIAZINON

Organophosphate insecticide. Restricted Use Pesticide due to avian and aquatic toxicity.

D•z•n diazinon AG600 WBC, 50W *(Novartis)*
Drexel Diazinon Insecticide (4EC) *(Drexel)*
Gowan Diazinon 4E, 50WSB *(Gowan)*
Helena Diazinon AG500 *(Helena)*
Micro Flo Diazinon AG500, 50W *(Micro Flo)*
Prentox Diazinon AG500, 50W *(Prentiss)*
◆ **Wilbur-Ellis Diazinon 4 (4E)** *(Wilbur-Ellis)*
Rate: 0.25-0.375 lb. ai/A
0.5-0.75 pt. 4E, 4EC, AG500/A
0.5-0.75 lb. 50W, 50WSB/A
6.5-9.5 fl. oz. AG600/A
Pest:
aphids, Colorado potato beetle, dipterous leafminer, flea beetles

Rate: 0.375-0.5 lb. ai/A
0.75-1 pt. 4E, 4EC, AG500/A
0.75-1 lb. 50W, 50WSB/A
9.5-12.75 fl. oz. AG600/A
Pest:
banded cucumber beetle, leafhoppers, southern armyworm

Rate: 2-4 lb. ai/A
4-8 pt. 4E, 4EC, AG500/A
4-8 lb. 50W, 50WSB/A
51-102 fl. oz. AG600/A
Pest:
cutworms, southern potato wireworm (Southeast)

Rate: 3-4 lb. ai/A
6-8 pt. 4E, 4EC, AG500/A
6-8 lb. 50W, 50WSB/A
76.5-102 fl. oz. AG600/A
Pest:
wireworms
Remarks: Do not use on commercially grown potatoes that will be hand-harvested. Apply when insects appear; repeat at 7-day intervals as needed. Do not apply more than 5 times per season. Will not control organophosphate-resistant leafminers. For cutworms and wireworms, broadcast just before planting and immediately incorporate into soil. Refer to label for use directions and precautions.
Days To Harvest: 35.
- - - - - - - - - - - - - - - - -
Signal Word/Toxicity Class: Caution.
REI: 24 hr.

◆ - **new product** • ai-active ingredient • DF-dry flowable • E/EC-emulsifiable concentrate • EW-emulsion in water
F/FL-flowable • G-granule • L/LC-liquid • SP/WSP-soluble packet • W/WP-wettable powder • WSB-water soluble bag

Drexel Diazinon D-264 14G *(Drexel)*
Gowan Diazinon 14G *(Gowan)*
◆ **Micro Flo Diazinon 14G** *(Micro Flo)*
◆ **Wilbur-Ellis Diazinon 14G** *(Wilbur-Ellis)*

Rate: 2-4 lb. ai/A
14-28 lb. 14G/A

Pest:
cutworms

Rate: 3-4 lb. ai/A
21-28 lb. 14G/A

Pest:
wireworms

Remarks: Do not use on commercially grown potatoes that will be hand-harvested. Broadcast and incorporate into soil just before planting. Refer to label for use directions and precautions.

Signal Word/Toxicity Class: Caution.
REI: 12 hr. (24 hr. for Micro Flo).

Digon — see Dimethoate

Dimate — see Dimethoate

DIMETHOATE

Cheminova Dimethoate 4E *(Cheminova)*
Clean Crop Dimethoate 400 *(Platte)*
Dimate 4E *(Terra)*
Drexel Dimethoate 2.67 or 4EC *(Drexel)*
Gowan Dimethoate E267 *(Gowan)*
Helena Dimethoate 4E *(Helena)*
◆ **Micro Flo Dimethoate 267 or 4E** *(Micro Flo)*
Wilbur-Ellis Dimethoate 267 or Digon 400 *(Wilbur-Ellis)*

Rate: 0.25-0.5 lb. ai/A
0.75-1.5 pt. 2.67, 267, E267/A
0.5-1 pt. 4E, 4EC, 400/A

Pest:
aphids
grasshoppers (4E, 4EC, 400)
leafhoppers
leafminers

Remarks: Do not apply when weather conditions favor drift.
Days To Harvest: 0.

Signal Word/Toxicity Class: Warning/II.
REI: 48 hr.

DI-SYSTON

Restricted Use Pesticide.

Di-Syston 8 (disulfoton) *(Bayer)*

Rate: 3-4 lb. ai/A
3-4 pt. 8/A

Pest:
aphids

Remarks: Soil application-preplant broadcast. Incorporate to depth of 2-3". Do not make more than 2 soil applications per crop season. Do not tank mix with phosalone.

Rate: 2-3 lb. ai/A
2-3 pt. 8/A (38" row spacing)

Pest:
aphids
Colorado potato beetle (early-season reduction)
flea beetles (except tuber flea beetle)
leafhoppers
potato psyllid
southern potato wireworm

Remarks: In-furrow soil application. Apply undiluted in specially designed equipment for low volume soil injection, or apply in a water emulsion or with liquid fertilizer as in-furrow spray at planting. Additional sidedress application may be made as water emulsion injection after plants become established. Do not make more than 2 soil applications per crop season. Higher rate to be used in organic soils. Do not tank mix with phosalone.
Days To Harvest: 75.

Rate: 0.375-1 lb. ai/A
0.375-1 pt. 8/A

Pest:
green peach aphid

Remarks: Foliar application. Apply by air or ground in sufficient water (not less than 1 gal. total volume/A) when aphids first appear. Use higher rates for heavy infestations. Repeat as necessary but not more than 3 times per season. Do not tank mix with phosalone.
Days To Harvest: 30.
State Restrictions: For use east of the Rockies.

Rate: 3 lb. ai/A
3 pt. 8/A

Pest:
Colorado potato beetle
green peach aphid

Remarks: Sprinkler irrigated only. Apply as sprinkler injection. Do not apply more than once per crop season.
State Restrictions: For use in Idaho, Oregon, Utah, and Washington.

Signal Word/Toxicity Class: Danger/I.
REI: 48 hr.

Di-Syston 15% (disulfoton) *(Bayer)*

Rate: Row treatment:
2-3 lb. ai/A
13.5-20.5 lb. 15%/A (38" spacing)
Preplant broadcast:
3-4 lb. ai/A
20-26.7 lb. 15%/A

Pest:
aphids
Colorado potato beetle (early-season reduction)
flea beetles (except tuber flea beetle)
leafhoppers
potato psyllid
southern potato wireworm

Remarks: Row treatment: Place granules in seed furrow or in soil incorporated band on each side of seed furrow at planting. Additional application may also be made as soil incorporated sidedressing after plants become established. Do not make more than 2 applications per season. Use higher rate in organic soils. Broadcast application: Apply preplant. Work granules into soil to depth of 2-3" using disc or other suitable equipment. Postemergence sidedress may also be made at 15-23 oz./1000 ft. row (any row spacing) after plants become established. Do not make more than 2 applications per season.
Days To Harvest: 75.

Rate: 3 lb. ai/A
20 lb. 15%/A

Pest:
aphids
green peach aphid (except Idaho)

Remarks: None.
State Restrictions: SLN-Idaho, Oregon, and Washington.

Signal Word/Toxicity Class: Danger/I.
REI: 48 hr.

D•z•n — see diazinon

ENDOSULFAN

Toxic to fish, birds, and other wildlife.

Drexel Endosulfan 3EC or 50WP *(Drexel)*
Gowan Endosulfan 3EC or 50WSB *(Gowan)*
◆ **Micro Flo Endosulfan 3EC or 50WP** *(Micro Flo)*
Phaser 3EC or 50WSB *(AgrEvo USA)*
Thiodan 3EC or 50WP *(FMC)*

Rate: 0.5-1 lb. ai/A
0.66-1.33 qt. 3EC/A
1-2 lb. 50WP, 50WSB/A

Pest:
aphids
armyworms
Colorado potato beetle
European corn borer
false chinch bug
green stink bug
leaffooted bug
plant bugs
potato flea beetle
potato leafhopper
potato psyllid
potato tuberworm
threelined potato beetle
whiteflies

Remarks: Rate varies by pest species. Apply when insects appear or feeding is noticed. Do not exceed 3 lb. ai/A per year or 6 applications per year.
Days To Harvest: 1.

Signal Word/Toxicity Class: Danger-Poison/I.
REI: 24 hr.

NOTICE	The information on these pages is for preliminary planning — not a guide for use. Be sure to follow the manufacturer's directions, notwithstanding information contained here.	For personal protective equipment and EPA registration numbers, please turn to page 39.

FURADAN

Restricted Use Pesticide due to acute oral and inhalation toxicity. Toxic to bees, fish, birds, and other wildlife.

Furadan 4F (carbofuran) *(FMC)*
Rate: 0.5-1 lb. ai/A
1-2 pt. 4F/A
Pest:
Colorado potato beetle, potato flea beetle, potato tuberworm (Virginia only), European corn borer, potato leafhopper
Remarks: Apply in 10 gal. spray volume by ground; 3 gal. by air when insects first appear; repeat as necessary. Do not apply more than 2 pt./application or make more than 8 foliar applications per season. For waterfowl protection, do not apply immediately before or during furrow irrigation, on fields in proximity of waterfowl nesting area, or on fields where waterfowl are known to repeatedly feed.
Days To Harvest: 14.
Crop Rotation: Do not rotate with any crop on soil treated at greater than 3 lb. ai/A for at least 10 months or plant with any crop for which carbofuran is not registered for at least 10 months.
State Restrictions: Not for use in Long Island, New York.

Signal Word/Toxicity Class: Danger-Poison/I.
REI: 48 hr.

Fyfanon — see malathion

Guthion — see azinphos-methyl

IMIDACLOPRID

Admire 2F *(Bayer)*
Rate: 0.9-1.3 fl. oz. 2F/1000 row ft.
Pest:
aphids, flea beetles, psyllids, Colorado potato beetle, potato leafhopper
Remarks: Apply as narrow band, in-furrow, or sidedress. Narrow band directly below seed row up to 7 days before planting or as in-furrow spray during planting. Higher rate recommended when planting into muck soils, or when extended length of control is needed. Do not apply more then 0.31 lb. ai/A per season.
Days To Harvest: 21.

Signal Word/Toxicity Class: Caution.
REI: 12 hr.

Provado 1.6 *(Bayer)*
Rate: 3.75 fl. oz. 1.6/A
Pest:
aphids, flea beetles, psyllids, Colorado potato beetle, potato leafhopper
Remarks: Apply before damaging population becomes established. Apply 2 applications at 7-day intervals. For aphids, leafhoppers, or potato beetles west of the Rockies, apply by ground only. For potato beetles east of the Rockies, apply by ground or air. Good foliage coverage necessary for optimal control. Do not apply more than 15 fl. oz. per year.
Days To Harvest: 7.

Signal Word/Toxicity Class: Caution.
REI: 12 hr.

IMIDAN

Imidan 70-W (phosmet) *(Gowan)*
Rate: 1 lb. ai/A
1.33 lb. 70-W/A
Pest:
Colorado potato beetle, potato flea beetle, potato leafhopper
Remarks: Repeat as necessary throughout growing season. Use only on potatoes to be harvested by machine.
Days To Harvest: 7.
State Restrictions: Not for use in California.

Signal Word/Toxicity Class: Warning/II.
REI: 24 hr.

Kryocide — see cryolite

Kumulus — see sulfur

LANNATE

Restricted Use Pesticide.

Lannate SP or LV (methomyl) *(DuPont)*
Rate: 0.45-0.9 lb. ai/A
0.5-1 lb. 90/A
1.5-3 pt. LV/A
Pest:
aphids, flea beetles, tuberworm, beet armyworm, leafhoppers, variegated cutworm, fall armyworm, loopers
Remarks: Rate varies by pest species; refer to label. Apply in sufficient water to obtain thorough, uniform coverage when insects first appear; repeat at 5-7 day intervals or as needed. Do not apply more than 4.5 lb. ai/A per crop or 10 applications per crop.
Days To Harvest: 6.

Signal Word/Toxicity Class: Danger-Poison/I.
REI: 48 hr.

MALATHION

Organophosphate insecticide toxic to fish, aquatic invertebrates, aquatic life stages of amphibians, and bees exposed to direct treatment on blooming crops or weeds.

Clean Crop Malathion 57EC *(Platte)*
Rate: 0.63-0.94 lb. ai/A
1-1.5 pt. 57EC/A
Pest:
aphids, leafhoppers
Remarks: Apply when insects first appear and repeat as necessary.
Days To Harvest: 0.

Signal Word/Toxicity Class: Caution/III.
REI: 12 hr.

Clean Crop Malathion 8EC *(Platte)*
◆ **Gowan Malathion 8 (8E)** *(Gowan)*
Rate: 1-3 lb. ai/A
1-3 pt. 8E, 8EC/A
Pest:
aphids, false chinch bug, mealybugs, blister beetle, leafhoppers
Remarks: Apply when insects first appear.
Days To Harvest: 0.

Signal Word/Toxicity Class: Caution/III.
REI: 12 hr.

Drexel Malathion 5EC *(Drexel)*
◆ **Micro Flo Malathion 5EC** *(Micro Flo)*
Prentox 5 lb. Malathion Spray (5EC) *(Prentiss)*
Rate: 0.94-1.6 lb. ai/A
1.5-2.5 pt. 5EC/A
Pest:
aphids, grasshoppers, mealybugs, false chinch bug, leafhoppers
Remarks: Rate varies by pest species; refer to label.
Days To Harvest: 0.

Signal Word/Toxicity Class: Warning/II.
REI: 12 hr.

Fyfanon (5EC or 8E) *(Helena)*
Rate: 0.63-0.94 lb. ai/A
1-1.5 pt. 5EC/A
1-2.5 lb. ai/A
1-2.5 pt. 8E/A
Pest:
aphids, false chinch bug, mealybugs, blister beetle (8E), leafhoppers
Remarks: Rate varies by pest species; refer to label.
Days To Harvest: 0.

Signal Word/Toxicity Class: Caution/III.
REI: 12 hr.

◆ - **new product** • ai-active ingredient • DF-dry flowable • E/EC-emulsifiable concentrate • EW-emulsion in water
F/FL-flowable • G-granule • L/LC-liquid • SP/WSP-soluble packet • W/WP-wettable powder • WSB-water soluble bag

Wilbur-Ellis Malathion 8 *(Wilbur-Ellis)*
Rate: 1-1.5 lb. ai/A
1-1.5 pt. 8/A
Pest:
aphids, leafhoppers, spider mites, false chinch bug
Remarks: Apply in minimum 5 gal. by air; 10 gal. by ground.
Days To Harvest: 0.

Signal Word/Toxicity Class: Caution/III.
REI: 12 hr.

Marlate — see methoxychlor

METHAMIDOPHOS

Restricted Use Pesticide due to acute dermal toxicity and residue effects on avian species.

Monitor 4 *(Bayer)*
Monitor 4 Spray *(Valent)*
Rate: 0.75-1 lb. ai/A
1.5-2 pt. 4, 4S/A
Pest:
aphids (except melon aphid), armyworms* (Bayer), cabbage looper, Colorado potato beetle (except NY), cutworms* (foliage feeding), European corn borer*, flea beetles, Lygus bugs* (Bayer), potato leafhopper, potato psyllid (Bayer), potato tuberworm
**Except CA for Bayer*
Remarks: Apply at 7- to 10-day intervals as necessary. Resistance to Colorado potato beetle has occurred in some areas of New York; consult local extension service.
Days To Harvest: 14.

Signal Word/Toxicity Class: Danger/I.
REI: 48 hr.

METHOXYCHLOR

Organochlorine pesticide toxic to fish and aquatic invertebrates.

Marlate 50 (WP) *(Kincaid)*
Prentox Methoxychlor 2EC *(Prentiss)*
Rate: 1.3-1.7 lb. ai/A
75-100 fl. oz. 2EC/A
2 Tbs. 50WP/gal. water
Pest:
blister beetle, Colorado potato beetle, fall armyworm, flea beetles, leafhoppers
Remarks: **Irish potatoes (Marlate).** Apply at first sign of infestation; repeat at 7- to 14-day intervals as needed.
Days To Harvest: 0.

Signal Word/Toxicity Class: Caution.
REI: 12 hr.

METAM-SODIUM

Metam CLR (42%) *(UCB Agrochemicals)*
Rate: 37.25-74.5 gal. 42%/A
Oregon:
29.75 gal. 42%/A
Pest:
root-knot nematode
Remarks: Fall application. Inject into sprinkler system and apply in at least 1" water. Can be applied where crop stubble or vegetation exists without prior tillage, if there is adequate soil penetration. In Oregon, use thin shank injector rig with shanks spaced 5" apart. Soil temperature should be in 40°-90°F range. Refer to label for use directions.
State Restrictions: In California, application must be in compliance with Technical Information Bulletin "Guidelines for All Application Methods for Metam-sodium in California."

Signal Word/Toxicity Class: Danger/I.
REI: 48 hr.

Metam CLR — see metam-sodium

METHYL PARATHION

Organophosphate insecticide toxic to bees exposed to direct treatment or residues on blooming crops or weeds. Penncap-M is a microencapsulated formulation. Restricted Use Pesticide due to very high acute toxicity to humans and birds.

Cheminova Methyl 4EC *(Cheminova)*
Rate: 1.5 lb. ai/A
3 pt. 4EC/A
Pest:
aphids, armyworms, false chinch bug, flea beetles, leafhoppers, mites, shield bug
Remarks: Apply when insects first appear and repeat as needed.
Days To Harvest: 5.

Signal Word/Toxicity Class: Danger-Poison/I.
REI: 48 hr.; 72 hr. if average rainfall less than 25" per year.

◆ **Declare (4EC)** *(Griffin)*
Helena 4 lb. Methyl Parathion (4EC) *(Helena)*
Rate: 0.25-1.5 lb. ai/A
0.5-3 pt. 4EC/A
Pest:
aphids, armyworms, cabbage looper, false chinch bug, flea beetles, leafhoppers, Lygus bugs, mites, plant bugs, shield bug, stink bug, thrips
Remarks: Rate varies by pest species; refer to label.
Days To Harvest: 6.

Signal Word/Toxicity Class: Danger-Poison/I.
REI: 48 hr.

Penncap-M (2FL) *(Elf Atochem)*
Rate: 0.5-0.75 lb. ai/A
2-3 pt. 2FL/A
Pest:
grasshoppers

Rate: 0.5-1 lb. ai/A
2-4 pt. 2FL/A
Pest:
cutworms, European corn borer, flea beetles, potato leafhopper, tarnished plant bug

Rate: 0.5-1.5 lb. ai/A
2-6 pt. 2FL/A
Pest:
Colorado potato beetle
Remarks: For grasshoppers, use higher rate if majority of insects are large or weather is cool. Do not chemigate in California.
Days To Harvest: 5.

Signal Word/Toxicity Class: Warning/II.
REI: 48 hr.

Wilbur-Ellis Methyl Parathion 4 or 5 Spray *(Wilbur-Ellis)*
Rate: 0.75 lb. ai/A
1.5 pt. 4/A
0.95 lb. ai/A
1.5 pt. 5/A
Pest:
aphids (5 Spray), armyworms, flea beetles, leafhoppers (5 Spray), spider mites
Remarks: Do not apply through any type of irrigation system.
Days To Harvest: 5.

Signal Word/Toxicity Class: Danger-Poison/I.
REI: 24 (4 Spray); 48 (5 Spray).

Microthiol Special — see sulfur

NOTICE The information on these pages is for preliminary planning — not a guide for use. Be sure to follow the manufacturer's directions, notwithstanding information contained here.

For personal protective equipment and EPA registration numbers, please turn to page 39.

7 Vegetables

MOCAP

Mocap EC is a Restricted Use Pesticide.

Mocap EC or 10%G (ethoprop) *(Rhone-Poulenc)*

Rate: Broadcast:
6-12 lb. ai/A
4-8 qt. EC/A
60-120 lb. 10%G/A
Wireworms, symphylans:
4-6 lb. ai/A
2.66-4 qt. EC/A
40-60 lb. 10%G/A
Row treatment:
3 lb. ai/A
2 qt. EC/A (4.4 fl. oz./1000 lin. ft. row)
30 lb. 10%G/A (2.1 lb./1000 lin. ft. row)

Pest:

garden symphylans	root-knot nematode	stubby-root nematode
lesion nematode	spiral nematode	stunt nematode
ring nematode	sting nematode	wireworms

Remarks: Broadcast: Apply from 2 weeks before planting to at-planting time. Row treatment: Apply in 12" wide band on 36" row spacing at planting. Mix with top 2-4" soil. For suppression of Columbia root-knot nematodes, apply following fall or spring application of registered fumigant nematicide. Do not exceed 12 lb. ai/A per season.

State Restrictions: Not for use in Long Island, New York.

\- - - - - - - - - - - - - - - - -

Signal Word/Toxicity Class: Danger-Poison/I (EC); Warning/II (10%G).
REI: 48 hr.; 72 hr. if average rainfall less than 25" per year.

Monitor — see methamidophos

M-PEDE

M-Pede (potassium salts of fatty acids) *(Mycogen)*

Rate: 1-2% v/v solution

Pest:

aphids*	leafminers	spider mites
broad mite	plant bugs	thrips
green peach aphid*	russet mite	whiteflies
leafhoppers		

*** Tank mix only**

Remarks: Apply at first sign of infestation; repeat at 7- to 14-day intervals. May be tank mixed with other insecticides. Refer to both labels for use directions and precautions.

Days To Harvest: 0.

\- - - - - - - - - - - - - - - - -

Signal Word/Toxicity Class: Warning/II.
REI: 12 hr.

NEEMAZAD — see azadirachtin

NEEMIX — see azadirachtin

Omite — see propargite

Penncap-M — see methyl parathion

PERMETHRIN

Restricted Use Pesticide.

Ambush 2EC or 25W *(ZENECA)*

Rate: 0.05-0.2 lb. ai/A
3.2-12.8 fl. oz. 2EC/A
3.2-12.8 oz. 25W/A

Pest:

aster leafhopper	potato aphid	potato psyllid
cabbage looper	potato flea beetle	potato tuberworm
Colorado potato beetle	potato leafhopper	

Remarks: Apply as needed by air or ground. Use sufficient water to obtain full foliage coverage. Do not apply more than 2.4 lb. ai/A per season.

Days To Harvest: 7.

\- - - - - - - - - - - - - - - - -

Signal Word/Toxicity Class: Warning/II.
REI: 24 hr.

Pounce 3.2 EC, 25WP or WSB *(FMC)*

Rate: 0.1-0.2 lb. ai/A
4-8 oz. 3.2 EC/A
6.4-12.8 oz. 25WP/A
1-2 bags WSB/A

Pest:

aster leafhopper	cutworms	potato psyllid
beet armyworm	European corn borer	potato tuberworm
cabbage looper	potato flea beetle	tarnished plant bug
Colorado potato beetle	potato leafhopper	

Remarks: Apply in minimum 20 gal. finished spray by ground; 3 gal. by air. Do not make more than 12 applications per season. Do not graze or feed potato forage.

Days To Harvest: 7.

\- - - - - - - - - - - - - - - - -

Signal Word/Toxicity Class: Caution (3.2 EC); Warning/II (25WP, WSB).
REI: 24 hr.

PETROLEUM OIL

Petroleum based spray oil. Mode of action is through suffocation of eggs, larvae, and nymphs of insects and adults of soft bodied insects. This mode of action necessitates total spray coverage. Toxic to fish.

SAF-T-SIDE *(Brandt Consolidated)*
◆ **Sumulsion** *(Drexel)*
◆ **SunSpray Ultra-Fine** *(Sun)*

Rate: 1-2 gal./100 gal. water

Pest:

aphids	mites	whiteflies
leafminers		

Remarks: Apply as needed. May be tank mixed with most commonly used insecticides and fungicides. Do not use in combination with or immediately before or after spraying with fungicides such as captan, Dyrene, folpet, Morestan, or any sulfur products. Do not use with carbaryl or dimethoate.

\- - - - - - - - - - - - - - - - -

Signal Word/Toxicity Class: Caution.
REI: 4 hr.

Phaser — see endosulfan

PHORATE

Restricted Use Pesticide.

Clean Crop Phorate 20G *(Platte)*
Riverside Phorate 20G *(Terra)*
Thimet 20-G *(American Cyanamid)*

Rate: Light or sandy soils:
11.3 oz. 20G, 20-G/1000 ft. row (min. 32" spacing)
Heavy or clay soils:
17.3 oz. 20G, 20-G/1000 ft. row (min. 32" spacing)

Pest:

aphids	flea beetles	psyllids
Colorado potato beetle (early-season control)	leafhoppers	wireworms
	leafminers	

Remarks: Apply at planting. Distribute granules evenly in furrow or granules may be banded on each side of the row at planting time.

Days To Harvest: 90.

State Restrictions: Thimet not for Colorado potato beetle control on Del Marva Peninsula.

\- - - - - - - - - - - - - - - - -

Signal Word/Toxicity Class: Danger-Poison/I.
REI: 48 hr.

Pounce — see permethrin

◆ - **new product** • ai-active ingredient • DF-dry flowable • E/EC-emulsifiable concentrate • EW-emulsion in water
F/FL-flowable • G-granule • L/LC-liquid • SP/WSP-soluble packet • W/WP-wettable powder • WSB-water soluble bag

PROPARGITE

Comite *(Uniroyal)*
Omite 6E *(Uniroyal)*
Rate: 1.2-2 lb. ai/A
1.5-2.5 pt. Comite/A
1.5-2.25 lb. ai/A
2-3 pt. Omite/A
Pest:
twospotted spider mite
Remarks: Apply when mites first appear. Apply 20-50 gal./A by ground or minimum 10 gal./A by air. Apply no more than twice per year. Do not apply through any type of irrigation system.
Days To Harvest: 14.
State Restrictions: For use west of the Rockies.

Signal Word/Toxicity Class: Danger/I (Comite); Warning/II (Omite 6E).
REI: 7 days. After 48 hr., workers can enter treated areas to perform hand labor tasks if they wear early-entry personal protective equipment.

Provado — see imidacloprid

PYRELLIN (Premix)

Premix containing pyrethrins and rotenone. May be combined with other insecticides to enhance knockdown or as an insect "exciter" to help facilitate contact with other spray residues.

Pyrellin E.C. (0.042 lb./gal. pyrethrins & 0.035 lb./gal. rotenone) *(Webb Wright)*
Rate: 1-2 pt. E.C./A
Pest:
alfalfa webworm, aphids, bean beetle, beet webworm, cabbageworm, garden fleahopper, leafhoppers, leafminers, loopers, Lygus bugs, mites, plant bugs, stink bugs, thrips, whiteflies
Remarks: Apply in intervals of 7 days or less; repeat as needed.
Days To Harvest: 0.

Rate: 1.5-2 pt. E.C./A
Pest:
Colorado potato beetle, European corn borer, flea beetles
Remarks: Make 3-4 applications or more as eggs start to hatch; repeat at 5- to 7-day intervals.
Days To Harvest: 0.

Signal Word/Toxicity Class: Caution/III.
REI: 12 hr.

Pyrenone — see pyrethrins

PYRETHRINS

Contains piperonyl butoxide as a synergist. Diatect also contains silicon dioxide (diatomaceous earth). May be combined in tank mix with other insecticides for quicker and more complete kill where insect resistance may be a problem, and as an "exciter" to flush insects out of hiding and into contact with other spray residues.

Diatect Multipurpose Insecticide *(Diatect)*
Pyrenone (0.5EC) *(AgrEvo Environmental Health)*
Py-Rin 60-6 EC *(Wilbur-Ellis)*
Rate: 1-6 lb. II/A
1-12 fl. oz. 0.5EC/A
2-16 oz. 6-60 EC/A
Pest: Including but not limited to:
aphids, Colorado potato beetle, cutworms, European corn borer, flea beetles, green peach aphid, leafhoppers, plant bugs
Remarks: Apply in sufficient water for thorough coverage; repeat as needed. Apply Diatect as dust or wettable powder.
Days To Harvest: 0.

Signal Word/Toxicity Class: Caution.
REI: 12 hr.

Py-Rin — see pyrethrins

SAF-T-SIDE — see petroleum oil

Sevin — see carbaryl

Sniper — see azinphos-methyl

SULFUR

Microthiol and Thiolux are micronized sulfurs.

◆ **Kumulus DF** *(BASF)*
Rate: 2.4-4 lb. ai/A
3-5 lb. DF/A
Pest:
red spider mite
Remarks: Apply at early leaf stage. Repeat at 10- to 14-day intervals with or without other pesticides immediately after a rain. Do not apply if temperatures during or within 3 days after application are expected to exceed 90°F. Do not apply through any type of irrigation system. Refer to label for use directions.

Signal Word/Toxicity Class: Caution.
REI: 24 hr.

Liquid Sulfur Six *(Helena)*
Rate: 3-4 pt./8 gal. water
Pest:
red spider mite
Remarks: Apply by air. Do not apply this product through any type of irrigation system. Refer to label for use directions.

Signal Word/Toxicity Class: Caution.
REI: 24 hr.

Microthiol Special (80W) *(Elf Atochem)*
Thiolux (DF) *(Platte)*
Rate: 4 lb. ai/A
5 lb. 80W/A
2.4-8 lb. ai/A
3-10 lb. DF/A
Pest:
red spider mite
Remarks: Apply early leaf stage. Do not chemigate for Microthiol Special in California.

Signal Word/Toxicity Class: Caution.
REI: 24 hr.

Sumulsion — see petroleum oil

SunSpray — see petroleum oil

Thimet — see phorate

Thiodan — see endosulfan

Thiolux — see sulfur

THIRETHRIN (Premix)

Thirethrin 2.9L (2 lb./gal. endosulfan & 0.45 lb. /gal. piperonyl butoxide & 0.045 lb./gal. pyrethrins) *(Terra)*
Rate: 1-2 qt. 2.9L/A
Pest:
aphids, armyworms, Colorado potato beetle, European corn borer, false chinch bug, green stink bug, leaffooted plant bugs, potato flea beetle, potato leafhopper, potato psyllid, potato tuberworm, threelined potato beetle, whiteflies

7 Vegetables

NOTICE The information on these pages is for preliminary planning — not a guide for use. Be sure to follow the manufacturer's directions, notwithstanding information contained here. For personal protective equipment and EPA registration numbers, please turn to page 39.

Remarks: Repeat as needed.
Days To Harvest: 0.

Signal Word/Toxicity Class: Warning/II.
REI: 24 hr.

VYDATE

Restricted Use Pesticide.

Vydate L (oxamyl) *(DuPont)*

Rate: In-furrow:
2-4 lb. ai/A
1-2 gal. L/A
Broadcast:
4-8 lb. ai/A
2-4 gal. L/A
Foliar:
0.5-1 lb. ai/A
2-4 pt. L/A

Pest:

aphids	lesion nematode	root-knot nematode (except javanese)
Colorado potato beetle	potato leafhopper	sting nematode
flea beetles	ring nematode	tarnished plant bug

Remarks: Preplant in-furrow: Apply to seed furrow during planting. Preplant broadcast: Apply within 1 week of planting and incorporate 4-6" into soil. Foliar: Apply after scouting fields to determine that early-season control has diminished; repeat at 5- to 7-day intervals or as needed. Apply low rate for light infestations; high rate for severe infestations. In the Midwest and Northeast, 1-4 pt./A may be used for Colorado potato beetle. Do not apply more than 4.5 gal./A per season or make more than 6 foliar applications/A per crop.
Days To Harvest: 7.
State Restrictions: Do not use in Suffolk and Nassau counties, Long Island, New York. Not for use in California. Not for use on nematodes west of the Rockies.

Signal Word/Toxicity Class: Danger-Poison/I.
REI: 48 hr.

Radishes

ACEPHATE

Organophosphate systemic insecticide with contact and stomach poison activity. Effective against many aphids, thrips, caterpillars, and plant bugs. Has moderate residual effects and is compatible with most pesticides.

Orthene 75S *(Valent)*

Rate: 1 lb. ai/A
1.33 lb. 75S/A

Pest:

cabbage aphid	diamondback moth	imported cabbageworm
cabbage looper	green peach aphid	

Remarks: Seed crop. Do not use on Daikon radish variety. Apply during early morning or late evening only as necessary. Do not apply before or during peak bloom period. Do not apply through any type of irrigation system. Do not feed crop residues to livestock or graze treated areas.
State Restrictions: SLN-Washington.

Signal Word/Toxicity Class: Caution.
REI: 24 hr.

ASANA

Pyrethroid insecticide that has both contact and stomach poison activity. Restricted Use Pesticide due to toxicity to fish and aquatic organisms.

Asana XL (esfenvalerate) *(DuPont)*

Rate: 0.03-0.05 lb. ai/A
5.8-9.6 fl. oz. XL/A

Pest:

armyworms	beetles

Remarks: Repeat as necessary. Do not exceed 0.1 lb. ai/A per season.
Days To Harvest: 7.

Signal Word/Toxicity Class: Warning/II.
REI: 12 hr.

AZADIRACHTIN

Extracted from seed of neem tree. Insect growth regulator that controls targeted insect larvae when they ingest or come in contact with it, by interfering with the insect's ability to molt and shutting off feeding behavior. Affected larvae slowly starve and die at next molt.

◆ **Azatin XL Plus** *(Thermo Trilogy)*

Rate: 5-21 oz. XL Plus/A

Pest:

aphids	fall armyworm	leafrollers
armyworms	flea beetles	loopers
beetles	green peach aphid	moths
cabbage looper	imported cabbageworm	psyllids
caterpillar	Japanese beetle	thrips
corn earworm	leafhoppers	webworms
cutworms	leafminers	whiteflies
diamondback moth		

Remarks: Controls larval and pupal stages of insects. Rate varies by species. Apply when pests first appear; repeat at 7-day intervals. When pest populations are high, use higher rates. Do not apply more that 21 fl. oz./A.
Days To Harvest: 0.

Signal Word/Toxicity Class: Caution.
REI: 4 hr.

◆ **NEEMAZAD 4.5EC or NEEMIX 4.5** *(Thermo Trilogy)*

Rate: 0.125-1 pt. 4.5/A
High Volume:
2.25-4.5 oz. 4.5EC/100 gal. water
Low Volume:
0.5 pt. 4.5EC/A

Pest:

aphids	caterpillars	leafhoppers
armyworms	Colorado potato beetle	potato flea beetle
artichoke plume moth	diamondback moth	psyllids
aster leafhopper	greenhouse whitefly	sweetpotato whitefly
cabbage loopers	hornworms	webworms

Remarks: For low volume, apply in sufficient water for adequate coverage. For low insect pressure, apply preventive at 7-day intervals or at first sign of insects; high insect pressure, at 3- to 4-day intervals.

Signal Word/Toxicity Class: Warning/II.
REI: 12 hr.

Azatin — see azadirachtin

Baythroid — see Cyfluthrin

BT 320 SULFUR 25 or 50 DUST (Premix)

Bt 320 Sulfur 25 Dust (0.064% Bt & 25% sulfur) or Bt 320 Sulfur 50 Dust (0.064% Bt & 50% sulfur) *(Wilbur-Ellis)*

Rate: 20-40 lb. Dust/A

Pest:
cabbage looper

Remarks: Do not mix or dilute with other products. Thorough coverage essential when using this product.

Signal Word/Toxicity Class: Caution.
REI: 24 hr.

CARBARYL

Clean Crop Sevin 5 Bait *(Platte)*

Rate: 1-2 lb. ai/A
20-40 lb. Bait/A

Pest:

ants	darkling beetle	grasshoppers
armyworms	field crickets	sowbugs
cutworms		

◆ - **new product** • ai-active ingredient • DF-dry flowable • E/EC-emulsifiable concentrate • EW-emulsion in water
F/FL-flowable • G-granule • L/LC-liquid • SP/WSP-soluble packet • W/WP-wettable powder • WSB-water soluble bag

Remarks: Apply with ground or air equipment.
Days To Harvest: 3.

Signal Word/Toxicity Class: Caution/III.
REI: 12 hr.

Drexel Carbaryl 4L or 50W *(Drexel)*
Sevin 4F, XLR Plus, 80S, 80WSP *(Rhone-Poulenc)*

Rate: 0.5-2 lb. ai/A
1-4 pt. 4L, 4F, XLR/A
0.66-2.5 lb. 80S, 80WSP/A

Pest:

armyworms	flea beetles	leafhoppers
aster leafhopper	grasshoppers	stink bugs
corn earworm	harlequin bug	tarnished plant bug
fall armyworm	imported cabbageworm	

Remarks: Rate varies by pest species; refer to label.
Days To Harvest: 3.

Signal Word/Toxicity Class: Caution (4F, 4L, XLR); Warning/II (50W, 80S, 80WSP).
REI: 12 hr.

Gowan Sevin Bait *(Gowan)*

Rate: 2 lb. ai/A
40 lb. Bait/A

Pest:

armyworms	cutworms	grasshoppers
crickets	darkling beetle	sowbugs

Remarks: Do not use when bees are present.
Days To Harvest: 3.
State Restrictions: For use in Arizona and California.

Signal Word/Toxicity Class: Caution/III.
REI: 12 hr.

Chlorfos — see chlorpyrifos

CHLORPYRIFOS

◆ **Chlorfos 15G** *(Griffin)*
Lorsban-4E or 15G *(Dow AgroSciences)*
◆ **Micro Flo Chlorpyrifos 4E or 15G** *(Micro Flo)*

Rate: 1 fl. oz. 4E/1000 lin. ft. row
3.3 oz. 15G/1000 lin. ft. row

Pest:
root maggot

Remarks: Apply at planting in-furrow with seed at planting time. For 4E, apply as water-based drench (minimum 40 gal. total drench/A). Do not apply more than 5.5 pt. 4E or 18.3 lb. 15G/A or make more than 1 application per season. Refer to label for complete use directions.

Signal Word/Toxicity Class: Warning/II (4E); Caution/III (15G).
REI: 4E: 24 hr.; 15G: 12 hr.

CYFLUTHRIN

Synthetic pyrethroid formulated in an emulsifiable concentrate. Toxic to fish and aquatic organisms. Restricted Use Pesticide.

Baythroid (2EC) *(Bayer)*

Rate: 0.025-0.044 lb. ai/A
1.6-2.8 fl. oz. 2EC/A

Pest:

cutworms	flea beetles

Remarks: Do not apply to crops grown for seed. Apply by ground only in sufficient water for complete foliage coverage. Up to 5 applications may be made per crop season; allow at least 7 days between applications. Refer to label for spray drift precautions.

Signal Word/Toxicity Class: Danger/I.
REI: 12 hr.

DEADLINE

Packaged as Bullets, M-Ps (mini-pellets), and granules.

Deadline Bullets, M-Ps, Granules (4%) (metaldehyde) *(Pace)*

Rate: 0.8-1 lb. ai/A
20-40 lb. 4%/A

Pest:

slugs	snails

Remarks: Apply every 3-4 weeks during growing season, or as needed. Do not apply directly to edible portion of plant.

Signal Word/Toxicity Class: Warning/II (granule); Caution (bullets, M-P).
REI: 12 hr.

Diatect — see pyrethrins

DIAZINON

Organophosphate insecticide. Restricted Use Pesticide due to avian and aquatic toxicity.

D•z•n diazinon AG600 WBC, 50W *(Novartis)*
Drexel Diazinon Insecticide (4EC) *(Drexel)*
Gowan Diazinon 4E, 14G, 50WSB *(Gowan)*
Helena Diazinon AG500 *(Helena)*
Micro Flo Diazinon AG500, 50W *(Micro Flo)*
Prentox Diazinon AG500, 50W *(Prentiss)*
◆ **Wilbur-Ellis Diazinon 4 (4E), 14G** *(Wilbur-Ellis)*

Rate: 0.25-0.5 lb. ai/A
0.5-1 pt. 4E, 4EC, AG500/A
0.5-1 lb. 50W, 50WSB/A
6.5-9.5 fl. oz. AG600/A

Pest:

aphids	dipterous leafminer	flea beetles

Rate: 1 lb. ai/A
2 pt. 4E, 4EC, AG500/A
2 lb. 50W, 50WSB/A
7 lb. 14G/A
25.5 fl. oz. AG600/A

Pest:
mole crickets

Rate: 2-4 lb. ai/A
4-8 pt. 4E, 4EC, AG500/A
4-8 lb. 50W, 50WSB/A
51-102 fl. oz. AG600/A

Pest:
cutworms

Rate: 3-4 lb. ai/A
6-8 pt. 4E, 4EC, AG500/A
6-8 lb. 50W, 50WSB/A
76.5-102 fl. oz. AG600/A

Pest:
wireworms

Remarks: Apply as insects occur; repeat at 7-day intervals if needed. Do not apply more than 3 times per season. Will not control organophosphate-resistant leafminers. For mole crickets, apply to plant beds or fields 1-2 days before sowing of seeds or transplanting; immediately incorporate into top 1-2" soil. For cutworms and wireworms, broadcast just before planting; immediately incorporate into top 1-2" soil. Refer to label for directions and precautions.
Days To Harvest: 14.

Signal Word/Toxicity Class: Caution.
REI: 24 hr.

Digon — see dimethoate

Dimate — see dimethoate

7 Vegetables

NOTICE	The information on these pages is for preliminary planning — not a guide for use. Be sure to follow the manufacturer's directions, notwithstanding information contained here.	For personal protective equipment and EPA registration numbers, please turn to page 39.

DIMETHOATE

Cheminova Dimethoate 4E *(Cheminova)*
Clean Crop Dimethoate 400 *(Platte)*
Digon 400 *(Wilbur-Ellis)*
Dimate 4E *(Terra)*
Drexel Dimethoate 2.67 or 4EC *(Drexel)*
Gowan Dimethoate E267 *(Gowan)*
Helena Dimethoate 4E *(Helena)*
Micro Flo Dimethoate 267 or 4E *(Micro Flo)*

Rate: 0.25 lb. ai/A
0.75 pt. 2.67, 267, E267/A
0.5 pt. 4E, 4EC, 400/A

Pest:
aphids; leafhoppers; leafminers

Remarks: None.
Days To Harvest: 14.

Signal Word/Toxicity Class: Warning/II.
REI: 48 hr.

D•z•n — see diazinon

Fyfanon — see malathion

LINDANE

Lindane 30C contains toxic inert Rhodamine B. Toxic to fish, birds, and wildlife.

Lindane 30C *(Gustafson)*
◆ **Lindane ST 40** *(AGSCO)*

Rate: 1.4 fl. oz. 30C/100 lb. seed
1 fl. oz. 40/100 lb. seed

Pest:
seedcorn maggots; wireworms

Remarks: Do not use treated seed for food, feed, or oil purposes. Exposed treated seed may be hazardous to birds and other wildlife. Dispose of excess seed by burial away from bodies of water.

Signal Word/Toxicity Class: Warning/II.
REI: 12 hr.

Lorsban — see chlorpyrifos

MALATHION

Organophosphate insecticide toxic to fish, aquatic invertebrates, aquatic life stages of amphibians, and bees exposed to direct treatment on blooming crops or weeds.

Clean Crop Malathion 57EC or 8EC *(Platte)*

Rate: 0.7-1.4 lb. ai/A
1-2 pt. 57EC/A
0.5-1 lb. ai/A
0.5-1 pt. 8EC/A

Pest:
aphids; cabbage looper; imported cabbageworm

Remarks: Apply when insects first appear and repeat as necessary.
Days To Harvest: 7.

Signal Word/Toxicity Class: Caution/III.
REI: 12 hr.

Fyfanon (5EC) *(Helena)*

Rate: 0.94-1.25 lb. ai/A
1.5-2 pt. 5EC/A

Pest:
aphids

Remarks: None.
Days To Harvest: 7.

Signal Word/Toxicity Class: Caution/III.
REI: 12 hr.

Fyfanon (8E) *(Helena)*
◆ **Gowan Malathion 8 (8E)** *(Gowan)*
◆ **Wilbur-Ellis Malathion 8** *(Wilbur-Ellis)*

Rate: 2 lb. ai/A
2 pt. 8E, 8/A

Pest:
aphids; flea beetles; leafhoppers; diamondback moth

Remarks: None.
Days To Harvest: 7.

Signal Word/Toxicity Class: Caution/III.
REI: 12 hr.

Marlate — see methoxychlor

METHOXYCHLOR

Organochlorine pesticide toxic to fish and aquatic invertebrates.

Marlate 50 (WP) *(Kincaid)*
Prentox Methoxychlor 2EC *(Prentiss)*

Rate: 1.3-1.7 lb. ai/A
75-100 fl. oz. 2EC/A
2 Tbs. 50WP/gal. water

Pest:
alfalfa looper; flea beetles; Japanese beetle; blister beetle; imported cabbageworm; leafhoppers; fall armyworm

Remarks: Apply at first sign of infestation; repeat at 7- to 14-day intervals as needed.
Days To Harvest: 14.

Signal Word/Toxicity Class: Caution.
REI: 12 hr.

M-PEDE

M-Pede (potassium salts of fatty acids) *(Mycogen)*

Rate: 1-2% v/v solution

Pest:
aphids*; leafminers; spider mites; broad mite; plant bugs; thrips; green peach aphid*; russet mite; whiteflies; leafhoppers

** Tank mix only*

Remarks: Apply at first sign of infestation; repeat at 7- to 14-day intervals. May be tank mixed with other insecticides. Refer to label for use directions and precautions.
Days To Harvest: 0.

Signal Word/Toxicity Class: Warning/II.
REI: 12 hr.

NEEMAZAD — see azadirachtin

NEEMIX — see azadirachtin

Orthene — see acephate

PETROLEUM OIL

Petroleum based spray oil. Mode of action is through suffocation of eggs, larvae, and nymphs of insects and adults of soft bodied insects. This mode of action necessitates total spray coverage. Toxic to fish.

SAF-T-SIDE *(Brandt Consolidated)*
◆ **Sumulsion** *(Drexel)*
◆ **SunSpray Ultra-Fine** *(Sun)*

Rate: 1-2 gal./100 gal. water

Pest:
aphids; leafminers; thrips; beetle (larvae); mites; whiteflies; leafhoppers; spiders

◆ - **new product** • ai-active ingredient • DF-dry flowable • E/EC-emulsifiable concentrate • EW-emulsion in water
F/FL-flowable • G-granule • L/LC-liquid • SP/WSP-soluble packet • W/WP-wettable powder • WSB-water soluble bag

Remarks: Apply as needed. May be tank mixed with most commonly used insecticides and fungicides. Do not use in combination with or immediately before or after spraying with fungicides such as captan, Dyrene, folpet, Morestan, or any sulfur products. Do not use with carbaryl or dimethoate.

Signal Word/Toxicity Class: Caution.
REI: 4 hr.

PYRELLIN (Premix)

Premix containing pyrethrins and rotenone. May be combined with other insecticides to enhance knockdown or as an insect "exciter" to help facilitate contact with other spray residues.

Pyrellin E.C. (0.042 lb./gal. pyrethrins & 0.035 lb./gal. rotenone) *(Webb Wright)*

Rate: 1-2 pt. E.C./A
Pest:

aphids	loopers	plant bugs
leafhoppers	Lygus bugs	thrips
leafminers	mites	whiteflies

Remarks: Apply as necessary to maintain control.
Days To Harvest: 0.

Signal Word/Toxicity Class: Caution/III.
REI: 12 hr.

Pyrenone — see pyrethrins

PYRETHRINS

Contains piperonyl butoxide as a synergist. Diatect also contains silicon dioxide (diatomaceous earth). May be combined in tank mix with other insecticides for quicker and more complete kill where insect resistance may be a problem, and as an "exciter" to flush insects out of hiding and into contact with other spray residues.

Diatect Multipurpose Insecticide II *(Diatect)*
Pyrenone (0.5EC) *(AgrEvo Environmental Health)*
Py-Rin 60-6 EC *(Wilbur-Ellis)*

Rate: 1-6 lb. II/A
1-12 fl. oz. 0.5EC/A
2-16 oz. 6-60 EC/A

Pest: Including but not limited to:

aphids	diamondback moth	leafminers
cabbage looper	flea beetles	mites
cross-striped cabbageworm	harlequin bug	plant bugs
cucumber beetles	imported cabbageworm	stink bugs
cutworms	leafhoppers	thrips
		whiteflies

Remarks: Apply in sufficient water for thorough coverage; repeat as needed. Apply Diatect as dust or wettable powder.
Days To Harvest: 0.

Signal Word/Toxicity Class: Caution.
REI: 12 hr.

Py-Rin — see pyrethrins

SAF-T-SIDE — see petroleum oil

Sevin — see carbaryl

Sumulsion — see petroleum oil

SunSpray — see petroleum oil

Rhubarb

AZADIRACHTIN

Extracted from seed of neem tree. Insect growth regulator that controls targeted insect larvae when they ingest or come in contact with it, by interfering with the insect's ability to molt and shutting off feeding behavior. Affected larvae slowly starve and die at next molt.

◆ **Azatin XL Plus** *(Thermo Trilogy)*

Rate: 5-21 oz. XL Plus/A
Pest:

aphids	corn earworm	leafhoppers
armyworms	cutworms	leafminers
beet armyworm	diamondback moth	loopers
beetles	fall armyworm	southern armyworm
black cutworm	flea beetles	thrips
cabbage looper	imported cabbageworm	tomato fruitworm
caterpillars		webworms
Colorado potato beetle	Japanese beetle	whiteflies

Remarks: Rate varies by species. Apply when pests first appear; repeat at 7-day intervals. When pest populations are high, use higher rates. Do not apply more that 21 fl. oz./A.
Days To Harvest: 0.

Signal Word/Toxicity Class: Caution.
REI: 4 hr.

◆ **NEEMAZAD 4.5EC or NEEMIX 4.5** *(Thermo Trilogy)*

Rate: 0.125-1 pt. 4.5/A
High Volume:
2.25-4.5 oz. 4.5EC/100 gal. water
Low Volume:
0.5 pt. 4.5EC/A

Pest:

aphids	diamondback moth	leafminers
armyworms	greenhouse whitefly	thrips
cabbage looper	hornworms	webworms
caterpillars	imported cabbageworm	weevils
cutworms		

Remarks: For low volume, apply in sufficient water for adequate coverage. For low insect pressure, apply preventive at 7-day intervals or at first sign of insects; high insect pressure, at 3- to 4-day intervals.

Signal Word/Toxicity Class: Warning/II.
REI: 12 hr.

Azatin — see azadirachtin

DEADLINE

Packaged as Bullets, M-Ps (mini-pellets), and granules.

Deadline Bullets, M-Ps, Granules (4%) (metaldehyde) *(Pace)*

Rate: 0.8-1 lb. ai/A
20-40 lb. 4%/A

Pest:
slugs snails

Remarks: Apply every 3-4 weeks during growing season, or as needed. Do not apply directly to edible portion of plant.

Signal Word/Toxicity Class: Warning/II (granule); Caution (bullets, M-P).
REI: 12 hr.

Diatect — see pyrethrins

LARVIN

Broad spectrum carbamate insecticide that has stomach poison activity and some limited contact activity. Known to have systemic action as a seed treatment. Toxic to fish and birds, and moderately toxic to bees as direct treatment.

Larvin 3.2 (thiodicarb) *(Rhone-Poulenc)*

Rate: 0.4-0.75 lb. ai/A
16-30 fl. oz. 3.2/A

7 Vegetables

NOTICE	
The information on these pages is for preliminary planning — not a guide for use. Be sure to follow the manufacturer's directions, notwithstanding information contained here.	For personal protective equipment and EPA registration numbers, please turn to page 39.

Pest:

alfalfa looper, beet armyworm, bollworm, corn earworm, fall armyworm, southern armyworm, tomato fruitworm

Rate: 0.6-0.75 lb. ai/A
24-30 fl. oz. 3.2/A

Pest:

cabbage looper

Remarks: Spray as needed. Use higher rates for heavier infestations of larger larvae. Do not exceed 1.5 lb. ai/A per season.

Days To Harvest: 14.

Signal Word/Toxicity Class: Warning/II.
REI: 12 hr.

NEEMAZAD — see azadirachtin

NEEMIX — see azadirachtin

PERMETHRIN

Restricted Use Pesticide.

Pounce 3.2 EC, 25WP or WSB *(FMC)*

Rate: 0.05-0.2 lb. ai/A
2-8 oz. 3.2 EC/A
3.2-12.8 oz. 25WP/A
1-2 bags WSB/A

Pest:

alfalfa looper, beet armyworm, cabbage looper, corn earworm, cutworms, European corn borer, fall armyworm, green cloverworm, leafhoppers, leafminers, southern armyworm

Remarks: Apply every 3-5 days or as needed. Use sufficient water to obtain full coverage of foliage. Rate varies by pest species. Do not graze treated areas or feed crop refuse to livestock.

Days To Harvest: 1.

Signal Word/Toxicity Class: Caution (3.2 EC); Warning/II (25WP, WSB).
REI: 24 hr.

Pounce — see permethrin

PYRELLIN (Premix)

Premix containing pyrethrins and rotenone. May be combined with other insecticides to enhance knockdown or as an insect "exciter" to help facilitate contact with other spray residues.

Pyrellin E.C. (0.042 lb./gal. pyrethrins & 0.035 lb./gal. rotenone) *(Webb Wright)*

Rate: 1-2 pt. E.C./A

Pest:

aphids, leafhoppers, leafminers, loopers, Lygus bugs, mites, plant bugs, thrips, whiteflies

Remarks: Apply as necessary to maintain control.

Days To Harvest: 0.

Signal Word/Toxicity Class: Caution/III.
REI: 12 hr.

Pyrenone — see pyrethrins

PYRETHRINS

Contains piperonyl butoxide as a synergist. Diatect also contains silicon dioxide (diatomaceous earth). May be combined in tank mix with other insecticides for quicker and more complete kill where insect resistance may be a problem, and as an "exciter" to flush insects out of hiding and into contact with other spray residues.

Diatect Multipurpose Insecticide II *(Diatect)*
Pyrenone (0.5EC) *(AgrEvo Environmental Health)*
Py-Rin 60-6 EC *(Wilbur-Ellis)*

Rate: 1-6 lb. II/A
1-12 fl. oz. 0.5EC/A
2-16 oz. 6-60 EC/A

Pest: Including but not limited to:

aphids, cutworms, flea beetle, leafminers, loopers

Remarks: Apply in sufficient water for thorough coverage; repeat as need. Apply Diatect as dust or wettable powder.

Days To Harvest: 0.

Signal Word/Toxicity Class: Caution.
REI: 12 hr.

Py-Rin — see pyrethrins

Southern Peas

(Blackeyed Peas, Cowpeas, Crowder Peas)

Adios — see carbaryl

AZINPHOS-METHYL

Organophosphate insecticide. Restricted Use Pesticide.

Gowan Azinphos-M 50-W *(Gowan)*
◆ **Micro Flo Azinphosmethyl 2EC or 50W** *(Micro Flo)*
Sniper 2-E, 50PVA or 50W *(Platte)*

Rate: 0.375-1 lb. ai/A
1.5-4 pt. 2EC, 2-E/A
0.75-2 lb. 50-W, 50PVA, 50W/A

Pest:

corn earworm, cowpea curculio, leafminers, stink bugs

Remarks: Blackeyed peas, crowder peas. Rate varies by pest species, refer to label. Apply in sufficient water to give complete coverage. Do not apply more than 4 times per season. Do not use vines for feed or forage nor pasture treated areas.

Days To Harvest: 7.

State Restrictions: Sniper 50PVA not registered in California.

Signal Word/Toxicity Class: Danger-Poison/I.
REI: 48 hr.; 72 hr. if average rainfall less than 25" per year.

CARBARYL

Carbamate insecticide with broad activity on many insects. Highly toxic to bees.

Adios (13%) *(Micro Flo)*

Rate: 0.065-0.1 lb. ai/A
0.5-0.75 lb. 13%/A

Pest:

cucumber beetle

Remarks: Cowpeas. Do not apply through any type of irrigation system. Do not apply more than 3.75 lb./A per season.

Days To Harvest: 0.

State Restrictions: Not for use in California.

Signal Word/Toxicity Class: Caution/III.

Drexel Carbaryl 4L or 50W *(Drexel)*
Sevin 4F, XLR Plus, 80S, 80WSP *(Rhone-Poulenc)*

Rate: 0.5-1.5 lb. ai/A
1-3 pt. 4L, 4F, XLR/A
1-3 lb. 50W/A
0.66-1.875 lb. 80S, 80WSP/A

Pest:

alfalfa caterpillar, alfalfa looper, armyworms, bean leaf beetle, blister beetle, corn earworm, cucumber beetle, cutworms, European corn borer, fall armyworm, flea beetles, grasshoppers, green cloverworm, Japanese beetle, leafhoppers, Mexican bean beetle, stink bugs, tarnished plant bug, threecornered alfalfa hopper, thrips, velvetbean caterpillar, webworms, western bean cutworm

◆ - **new product** • ai-active ingredient • DF-dry flowable • E/EC-emulsifiable concentrate • EW-emulsion in water
F/FL-flowable • G-granule • L/LC-liquid • SP/WSP-soluble packet • W/WP-wettable powder • WSB-water soluble bag

Rate: 2 lb. ai/A
4 pt. 4L, 4F, XLR/A
4 lb. 50W/A
2.5 lb. 80S, 80WSP/A

Pest:
cowpea curculio
California only:
corn earworm, Lygus bugs, stink bugs, limabean pod borer

Remarks: Blackeyed peas, cowpeas, crowder peas, southern peas. Rate varies by pest species; refer to label. Do not apply to wet foliage or when rain or high humidity is expected during next 2 days. Observe plant response precautions.
Days To Harvest: 0; except cowpeas-3.

Signal Word/Toxicity Class: Caution (4F, 4L, XLR); Warning/II (50W, 80S, 80WSP).
REI: 12 hr.

CHLORPYRIFOS

Organophosphate insecticide. For use only by commercial seed treaters.

Lorsban 30, 50-SL *(Gustafson)*
Rate: 2.7 fl. oz. 30/100 lb. seed
2 oz. 50-SL/100 lb. seed

Pest:
seedcorn beetles, seedcorn maggost

Remarks: Blackeyed peas. Apply prior to planting as slurry treatment. Do not use treated seed for food, feed, or oil purposes. Commercially treated seed must be labeled in accordance with all applicable requirements of Federal Seed Act. Do not allow livestock to graze or feed bean hay or pea hay grown from treated seed.

Signal Word/Toxicity Class: Warning/II (50-SL); Caution (30).
REI: 12 hr.

DEADLINE

Packaged as Bullets, M-Ps (mini-pellets), and granules.

Deadline Bullets, M-Ps, Granules (4%) (metaldehyde) *(Pace)*
Rate: 0.8-1 lb. ai/A
20-40 lb. 4%/A

Pest:
slugs, snails

Remarks: Blackeyed peas, cowpeas. Apply every 3-4 weeks during growing season or as needed. Do not apply directly to edible portion of plant.

Signal Word/Toxicity Class: Warning/II (granule); Caution (bullets, M-P).
REI: 12 hr.

Diatect — see pyrethrins

ENDOSULFAN

Toxic to fish, birds, and other wildlife.

Drexel Endosulfan 3EC or 50WP *(Drexel)*
Gowan Endosulfan 3EC or 50WSB *(Gowan)*
◆ Micro Flo Endosulfan 3EC or 50WP *(Micro Flo)*
Phaser 3EC or 50WSB *(AgrEvo USA)*
Thiodan 3EC or 50WP *(FMC)*
Rate: 0.5-1 lb. ai/A
0.66-1.33 qt. 3EC/A
1-2 lb. 50WP, 50WSB/A

Pest:
aphids, bean leafskeletonizer, black bean aphid, cucumber beetle, flea beetles, green stink bug, leafhoppers, Mexican bean beetle, western bean cutworm, whiteflies

Remarks: Apply when insects appear or feeding is noticed. Do not apply more than 3 lb. ai/A per year or make more than 3 applications per year. Do not feed treated threshings or allow livestock to graze treated fields.
Days To Harvest: 3.

Signal Word/Toxicity Class: Danger-Poison/I.
REI: 24 hr.

Lorsban — see chlorpyrifos

MALATHION

Organophosphate insecticide toxic to fish, aquatic invertebrates, aquatic life stages of amphibians, and bees exposed to direct treatment on blooming crops or weeds.

Drexel Malathion 5EC *(Drexel)*
◆ Micro Flo Malathion 5EC *(Micro Flo)*
Prentox 5 lb. Malathion Spray (5EC) *(Prentiss)*
Rate: 1-1.25 lb. ai/A
1.5-2 pt. 5EC/A

Pest:
aphids, bean leaf beetle, blister beetle, cucumber beetle, Japanese beetle, Lygus bugs, Mexican bean beetle, potato leafhopper, red spider mite

Remarks: Blackeyed peas, cowpeas. Rate varies by pest species; refer to label.
Days To Harvest: 1.

Signal Word/Toxicity Class: Warning/II.
REI: 12 hr.

Marlate — see methoxychlor

METHYL PARATHION

Organophosphate insecticide toxic to bees exposed to direct treatment or residues on blooming crops or weeds. Penncap-M is a microencapsulated formulation. Restricted Use Pesticide due to very high acute toxicity to humans and birds.

Penncap-M *(Elf Atochem)*
Rate: 0.5 lb. ai/A
2 pt. 2FL/A

Pest:
aphids (except CA), cucumber beetle, Diabrotica, green cloverworm, leafhoppers, Lygus bugs, Mexican bean beetle

Rate: 0.5-1 lb. ai/A
2-4 pt. 2FL/A

Pest:
cowpea curculio, European corn borer, stink bugs (except CA)

Remarks: Apply when blooms are first observed and repeat at 3- to 5-day intervals. Do not chemigate in California.
Days To Harvest: 15.

Signal Word/Toxicity Class: Warning/II.
REI: 48 hr.

METHOXYCHLOR

Organochlorine pesticide toxic to fish and aquatic invertebrates.

Marlate 50 (WP) *(Kincaid)*
Rate: 2 Tbs. 50WP/gal. water

Pest:
alfalfa webworm, bean leaf beetle, corn earworm, cucumber beetles, fall armyworm, flea beetles, garden webworm, Japanese beetle, Mexican bean beetle, potato leafhopper

Remarks: Blackeyed peas. Apply at first sign of infestation; repeat at 7- to 14-day intervals as needed.
Days To Harvest: 3; if vines used for feed or forage-7.

Signal Word/Toxicity Class: Caution.
REI: 12 hr.

Prentox Methoxychlor 2EC *(Prentiss)*
Rate: 0.66-2 lb. ai/A
1-4 qt. 2EC/A

Pest:
alfalfa webworm, bean leaf beetle, corn earworm, cucumber beetles, fall armyworm, flea beetles, garden webworm, Japanese beetle, Mexican bean beetle, potato leafhopper

For personal protective equipment and EPA registration numbers, please turn to page 39.

Remarks: Cowpeas. Apply at first sign of infestation; repeat at 7- to 14-day intervals as needed. Rate varies by pest species; refer to label.
Days To Harvest: Harvest or grazing-7.

Signal Word/Toxicity Class: Caution.
REI: 12 hr.

Penncap-M — see methyl parathion

Phaser — see endosulfan

PYRELLIN (Premix)

Premix containing pyrethrins and rotenone. May be combined with other insecticides to enhance knockdown or as an insect "exciter" to help facilitate contact with other spray residues.

Pyrellin E.C. (0.042 lb./gal. pyrethrins & 0.035 lb./gal. rotenone) *(Webb Wright)*

Rate: 1-2 pt. E.C./A

Pest:

aphids	loopers	plant bugs
leafhoppers	Lygus bugs	thrips
leafminers	mites	whiteflies

Remarks: Blackeyed peas. Apply as necessary to maintain control.
Days To Harvest: 0.

Signal Word/Toxicity Class: Caution/III.
REI: 12 hr.

Pyrenone — see pyrethrins

PYRETHRINS

Contains piperonyl butoxide as a synergist. Diatect also contains silicon dioxide (diatomaceous earth). May be combined in tank mix with other insecticides for quicker and more complete kill where insect resistance may be a problem, and as an "exciter" to flush insects out of hiding and into contact with other spray residues.

Diatect Multipurpose Insecticide II *(Diatect)*
Pyrenone (0.5EC) *(AgrEvo Environmental Health)*
Py-Rin 60-6 EC *(Wilbur-Ellis)*

Rate: 1-6 lb. II/A
1-12 fl. oz. 0.5EC/A
2-16 oz. 6-60 EC/A

Pest: Including but not limited to:

alfalfa caterpillar	flea beetles	Lygus bugs
aphids	grasshoppers	mites
armyworms	green cloverworm	plant bugs
bean leaf beetle	leafhoppers	stink bugs
caterpillars	loopers	thrips

Remarks: Apply in sufficient water for thorough coverage; repeat as needed. Apply Diatect as dust or wettable powder.
Days To Harvest: 0.

Signal Word/Toxicity Class: Caution.
REI: 12 hr.

Py-Rin — see pyrethrins

Sevin — see carbaryl

Sniper — see azinphos-methyl

Thiodan — see endosulfan

Spinach

Ambush — see permethrin

AZADIRACHTIN

Extracted from seed of neem tree. Insect growth regulator that controls targeted insect larvae when they ingest or come in contact with it, by interfering with the insect's ability to molt and shutting off feeding behavior. Affected larvae slowly starve and die at next molt.

◆ **Azatin XL Plus** *(Thermo Trilogy)*

Rate: 5-21 oz. XL Plus/A

Pest:

aphids	fall armyworm	Mexican bean beetle
armyworms	flea beetles	psyllids
beet armyworm	green peach aphid	silverleaf whitefly
beetles	imported cabbageworm	southern armyworm
black cutworm		sweetpotato whitefly
cabbage looper	Japanese beetle	thrips
caterpillars	leafhoppers	tobacco budworm
corn earworm	leafminers	tomato fruitworm
cutworms	leafrollers	webworms
diamondback moth	loopers	whiteflies

Remarks: Rate varies by species. Apply when pests first appear; repeat at 7-day intervals. When pest populations are high, use higher rates. Do not apply more that 21 fl. oz./A.
Days To Harvest: 0.

Signal Word/Toxicity Class: Caution.
REI: 4 hr.

◆ **NEEMAZAD 4.5EC or NEEMIX 4.5** *(Thermo Trilogy)*

Rate: 0.125-1 pt. 4.5/A
High Volume:
2.25-4.5 oz. 4.5EC/100 gal. water
Low Volume:
0.5 pt. 4.5EC/A

Pest:

aphids	diamondback moth	onion maggot
armyworms	greenhouse whitefly	sweetpotato whitefly
cabbage loopers	hornworms	thrips
caterpillars	imported cabbageworm	webworms
cutworms	leafminers	weevils

Remarks: For low volume, apply in sufficient water for adequate coverage. For low insect pressure, apply preventive at 7-day intervals or at first sign of insects; high insect pressure, at 3- to 4-day intervals.

Signal Word/Toxicity Class: Warning/II.
REI: 12 hr.

Azatin — see azadirachtin

AZINPHOS-METHYL

Organophosphate insecticide. Restricted Use Pesticide.

Gowan Azinphos-M 50-W *(Gowan)*
◆ **Micro Flo Azinphosmethyl 2EC or 50W** *(Micro Flo)*
Sniper 2-E, 50PVA or 50W *(Platte)*

Rate: 0.375-05 lb. ai/A
1.5-2 pt.2EC, 2-E/A
0.75-1 lb. 50-W, 50PVA, 50W/A

Pest:

aphids	leafminers	mites

Remarks: Apply in sufficient water to give complete coverage; repeat as needed. Do not apply more than 3 times per season.
Days To Harvest: 14.
State Restrictions: Sniper 50PVA not registered in California.

Signal Word/Toxicity Class: Danger-Poison/I.
REI: 48 hr.; 72 hr. if average rainfall less than 25" per year.

◆ - **new product** • ai-active ingredient • DF-dry flowable • E/EC-emulsifiable concentrate • EW-emulsion in water
F/FL-flowable • G-granule • L/LC-liquid • SP/WSP-soluble packet • W/WP-wettable powder • WSB-water soluble bag

BT 320 SULFUR 25 or 50 DUST (Premix)

Bt 320 Sulfur 25 Dust (0.064% Bt & 25% sulfur) or Bt 320 Sulfur 50 Dust (0.064% Bt & 50% sulfur) *(Wilbur-Ellis)*
Rate: 20-40 lb. Dust/A
Pest:
cabbage looper
Remarks: Do not mix or dilute with other products. Thorough coverage essential.

Signal Word/Toxicity Class: Caution.
REI: 24 hr.

CARBARYL

Clean Crop Sevin 5 Bait *(Platte)*
Rate: 1-2 lb. ai/A
20-40 lb. Bait/A
Pest:
ants, darkling beetle, grasshoppers, armyworms, field crickets, sowbugs, cutworms
Remarks: Apply with ground or air equipment.
Days To Harvest: 14.

Signal Word/Toxicity Class: Caution/III.
REI: 12 hr.

Drexel Carbaryl 4L or 50W *(Drexel)*
Sevin 4F, XLR Plus, 80S, 80WSP *(Rhone-Poulenc)*
Rate: 0.5-2 lb. ai/A
1-4 pt. 4L, 4F, XLR/A
1-4 lb. 50W/A
0.66-2.5 lb. 80S, 80WSP/A
Pest:
armyworms, flea beetles, leafhoppers, aster leafhopper, grasshoppers, stink bugs, corn earworm, harlequin bug, tarnished plant bug, fall armyworm
Remarks: Rate varies by pest species; refer to label. Apply at 5- to 7-day intervals. Do not apply to wet foliage or when rain or high humidity is expected during next 2 days.
Days To Harvest: 14.

Signal Word/Toxicity Class: Caution (4F, 4L, XLR); Warning/II (50W, 80S, 80WSP).
REI: 12 hr.

Gowan Sevin Bait *(Gowan)*
Rate: 2 lb. ai/A
40 lb. Bait/A
Pest:
armyworms, cutworms, grasshoppers, crickets, darkling beetle, sowbugs
Remarks: Do not use when bees are present.
Days To Harvest: 14.
State Restrictions: For use in Arizona and California.

Signal Word/Toxicity Class: Caution/III.
REI: 12 hr.

DEADLINE

Packaged as Bullets, M-Ps (mini-pellets), and granules.

Deadline Bullets, M-Ps, Granules (4%) (metaldehyde) *(Pace)*
Rate: 0.8-1 lb. ai/A
20-40 lb. 4%/A
Pest:
slugs, snails
Remarks: Apply every 3-4 weeks during growing season or as needed. Do not apply directly to edible portion of plant.

Signal Word/Toxicity Class: Warning/II (granule); Caution (bullets, M-P).
REI: 12 hr.

Declare — see methyl parathion

Diatect — see pyrethrins

DIAZINON

Restricted Use Pesticide due to avian and aquatic toxicity.

D•z•n diazinon AG600 WBC, 50W *(Novartis)*
Drexel Diazinon Insecticide (4EC), D-264 14G *(Drexel)*
Gowan Diazinon 4E, 14G, 50WSB *(Gowan)*
Micro Flo Diazinon AG500, 14G, 50W *(Micro Flo)*
Prentox Diazinon AG500, 50W *(Prentiss)*
◆ Wilbur-Ellis Diazinon 4 (4E), 14G *(Wilbur-Ellis)*
Rate: 0.25-0.5 lb. ai/A
0.5-1 pt. 4E, 4EC, AG500/A
6.5-12.75 fl. oz. AG600/A
0.5-1 lb. 50W, 50WSB/A
Pest:
aphids, dipterous leafminer
Rate: 2-4 lb. ai/A
4-8 pt. 4E, 4EC, AG500/A
51-102 fl. oz. AG600/A
4-8 lb. 50W, 50WSB/A
14-28 lb. 14G/A
Pest:
cutworms
Rate: 3-4 lb. ai/A
6-8 pt. 4E, 4EC, AG500/A
76.5-102 fl. oz. AG600/A
6-8 lb. 50W, 50WSB/A
21-28 lb. 14G/A
Pest:
wireworms
Remarks: Broadcast just before planting and immediately incorporate into soil. Refer to label for use directions and precautions.
Days To Harvest: 14.

Signal Word/Toxicity Class: Caution/III.
REI: 24 hr.; 12 hr. (Drexel/Gowan/Wilbur-Ellis 14G).

Digon — see dimethoate

Dimate — see dimethoate

DIMETHOATE

Cheminova Dimethoate 4E *(Cheminova)*
Clean Crop Dimethoate 400 *(Platte)*
Digon 400 *(Wilbur-Ellis)*
Dimate 4E *(Terra)*
Drexel Dimethoate 2.67 or 4EC *(Drexel)*
Gowan Dimethoate E267 *(Gowan)*
Helena Dimethoate 4E *(Helena)*
Micro Flo Dimethoate 267 or 4E *(Micro Flo)*
Rate: 0.25 lb. ai/A
0.75 pt. 2.67, 267, E267/A
0.5 pt. 4E, 4EC, 400/A
Pest:
aphids, leafhoppers, leafminers
Remarks: None.
Days To Harvest: 14.

Signal Word/Toxicity Class: Warning/II.
REI: 48 hr.

D•z•n — see diazinon

ENDOSULFAN

Toxic to fish, birds, and other wildlife.

Drexel Endosulfan 3EC or 50WP *(Drexel)*
Gowan Endosulfan 3EC or 50WSB *(Gowan)*
◆ Micro Flo Endosulfan 3EC or 50WP *(Micro Flo)*
Phaser 3EC or 50WSB *(AgrEvo USA)*
Thiodan 3EC or 50WP *(FMC)*
Rate: 0.75-1 lb. ai/A
1-1.33 qt. 3EC/A
1.5-2 lb. 50WP, 50WSB/A

7 Vegetables

NOTICE The information on these pages is for preliminary planning — not a guide for use. Be sure to follow the manufacturer's directions, notwithstanding information contained here. For personal protective equipment and EPA registration numbers, please turn to page 39.

Pest:

armyworms, crown mite, flea beetles, green peach aphid, leafhoppers, whiteflies (high rate)

Remarks: Apply when insects appear or feeding is noticed. Do not apply more than 1 lb. ai/A per year or 1 application per year.

Days To Harvest: 21.

Signal Word/Toxicity Class: Danger-Poison/I.

REI: 24 hr.

LANNATE

Restricted Use Pesticide.

Lannate (90) or LV (methomyl) *(DuPont)*

Rate: 0.45-0.9 lb. ai/A
0.5-1 lb. 90/A
1.5-3.0 pt. LV/A

Pest:

beet armyworm, fall armyworm, loopers, variegated cutworm

Remarks: Apply in sufficient water to obtain thorough, uniform coverage when insects first appear; repeat at 5- to 7-day intervals or as needed. Do not apply when minimum daily temperature is 32°F or lower. Do not apply to seedlings less than 3" diameter. Do not apply more than 3.6 lb. ai/A per crop or 8 applications per crop.

Days To Harvest: 7.

Signal Word/Toxicity Class: Danger-Poison/I.

REI: 48 hr.

LARVIN

Broad spectrum carbamate insecticide that has stomach poison activity and some limited contact activity. Known to have systemic action as a seed treatment. Toxic to fish and birds, and moderately toxic to bees as direct treatment.

Larvin 3.2 (thiodicarb) *(Rhone-Poulenc)*

Rate: 0.4-0.75 lb. ai/A
16-30 fl. oz. 3.2/A

Pest:

alfalfa looper, beet armyworm, bollworm, corn earworm, fall armyworm, southern armyworm, tomato fruitworm

Rate: 0.6-0.75 lb. ai/A
24-30 fl. oz. 3.2/A

Pest:

cabbage looper

Remarks: Spray as needed. Use higher rates for heavier infestations of larger larvae. Do not exceed 1.5 lb. ai/A per season.

Days To Harvest: 14.

Signal Word/Toxicity Class: Warning/II.

REI: 12 hr.

MALATHION

Organophosphate insecticide toxic to fish, aquatic invertebrates, aquatic life stages of amphibians, and bees exposed to direct treatment on blooming crops or weeds.

Clean Crop Malathion 57EC or 8EC *(Platte)*
Drexel Malathion 5EC *(Drexel)*
◆ **Gowan Malathion 8 (8E)** *(Gowan)*
◆ **Micro Flo Malathion 5EC** *(Micro Flo)*
Prentox 5 lb. Malathion Spray (5EC) *(Prentiss)*
◆ **Wilbur-Ellis Malathion 8** *(Wilbur-Ellis)*

Rate: 1.25 lb. ai/A
2 pt. 5EC, 57EC/A
1.5 lb. ai/A
1.5 pt. 8E/A
0.5-1 lb. ai/A
0.5-1 pt. 8EC, 8/A

Pest:

aphids

Remarks: Apply when insects first appear and repeat as necessary.

Days To Harvest: 7.

Signal Word/Toxicity Class: Caution/III; Warning/II (5EC).

REI: 12 hr.

Marlate — see methoxychlor

METHOXYCHLOR

Organochlorine pesticide toxic to fish and aquatic invertebrates.

Marlate 50 (WP) *(Kincaid)*
Prentox Methoxychlor 2EC *(Prentiss)*

Rate: 1.3-1.7 lb. ai/A
75-100 fl. oz. 2EC/A
2 Tbs. 50WP/gal. water

Pest:

alfalfa looper, blister beetle, fall armyworm, flea beetles, imported cabbageworm, Japanese beetle, leafhoppers

Remarks: Apply at first sign of infestation; repeat at 7- to 14-day intervals as needed.

Days To Harvest: 14.

Signal Word/Toxicity Class: Caution.

REI: 12 hr.

METHYL PARATHION

Organophosphate insecticide toxic to bees exposed to direct treatment or residues on blooming crops or weeds. Restricted Use Pesticide due to very high acute toxicity to humans and birds.

Cheminova Methyl 4EC *(Cheminova)*
◆ **Declare (4EC)** *(Griffin)*

Rate: 0.5-1 lb. ai/A
1-2 pt. 4EC/A

Pest:

aphids, armyworms, crown mite, flea beetles, leafhoppers, mites, seedcorn maggots (in crown)

Remarks: Apply when insects appear; repeat as needed. For stink bugs, apply a minimum of 1 lb. ai/A.

Days To Harvest: 15; rates above 1 pt.-21.

Signal Word/Toxicity Class: Danger-Poison/I.

REI: 48 hr.; 72 hr. if average rainfall less than 25" per year.

Helena 4 lb. Methyl Parathion (4EC) *(Helena)*

Rate: 0.5-1 lb. ai/A
1-2 pt. 4EC/A

Pest:

aphids, armyworms, cabbage looper, crown mite, flea beetles, mites, plant bugs, seedcorn maggots (in crown)

Remarks: Rate varies by pest species; refer to label.

Days To Harvest: Up to 1 pt-15; over 1 pt.-21.

Signal Word/Toxicity Class: Danger-Poison/I.

REI: 48 hr.; 72 hr. if annual rainfall is less than 25" per year (Wilbur-Ellis).

Wilbur-Ellis Methyl Parathion 4 or 5 Spray *(Wilbur-Ellis)*

Rate: 0.5 lb. ai/A
1 pt. 4/A
0.63 lb. ai/A
1 pt. 5/A
Cabbage looper (5 Spray):
1 lb. ai/A
1.6 pt. 5/A

Pest:

aphids, armyworms, cabbage looper, flea beetles, imported cabbageworm, leafhoppers, spider mites

Remarks: Do not apply through any type of irrigation system.

Days To Harvest: 21.

Signal Word/Toxicity Class: Danger-Poison/I.

REI: 24 (4 Spray); 48 (5 Spray).

◆ - **new product** • ai-active ingredient • DF-dry flowable • E/EC-emulsifiable concentrate • EW-emulsion in water
F/FL-flowable • G-granule • L/LC-liquid • SP/WSP-soluble packet • W/WP-wettable powder • WSB-water soluble bag

M-PEDE

M-Pede (potassium salts of fatty acids) *(Mycogen)*

Rate: 1-2% v/v solution

Pest:

aphids*	leafminers*	spider mites
broad mite	plant bugs	thrips
green peach aphid*	russet mite	whiteflies
leafhoppers		

* *Tank mix only*

Remarks: Apply at first sign of infestation; repeat at 7- to 14-day intervals. May be tank mixed with other insecticides. Refer to both labels for use directions and precautions.

Days To Harvest: 0.

- - - - - - - - - - - - - - - - - -

Signal Word/Toxicity Class: Warning/II.

REI: 12 hr.

NEEMAZAD — see azadirachtin

NEEMIX — see azadirachtin

PERMETHRIN

Restricted Use Pesticide.

Ambush 2E or 25W *(ZENECA)*

Rate: 0.1-0.2 lb. ai/A
6.4-12.8 fl. oz. 2E/A
6.4-12.8 oz. 25W/A

Pest:

beet armyworm	cutworms	green cloverworm
cabbage looper	European corn borer	leafhoppers
corn earworm	fall armyworm	vegetable leafminer

Remarks: Fine spinach, New Zealand spinach. Apply by ground or air at 3- to 5-day intervals or as needed. Use sufficient water to obtain full coverage of foliage. Do not apply more than 2 lb. ai/A per season. Refer to label for complete list of minor crop leafy vegetables.

Days To Harvest: 1.

- - - - - - - - - - - - - - - - - -

Signal Word/Toxicity Class: Warning/II.

REI: 24 hr.

Pounce 3.2 EC, 25WP or WSB *(FMC)*

Rate: 0.05-0.2 lb. ai/A
2-8 oz. 3.2 EC/A
3.2-12.8 oz. 25WP/A
1-2 bags WSB/A

Pest:

alfalfa looper	cutworms	green cloverworm
beet armyworm	European corn borer	leafhoppers
cabbage looper	fall armyworm	vegetable leafminer
corn earworm		

Remarks: Apply in sufficient water at 3- to 5-day intervals or as needed to obtain full foliage coverage. Rate varies by pest species; refer to label. Do not graze treated areas or feed crop refuse to livestock.

Days To Harvest: 1.

- - - - - - - - - - - - - - - - - -

Signal Word/Toxicity Class: Caution (3.2 EC); Warning/II (25WP, WSB).

REI: 24 hr.

Phaser — see endosulfan

Pounce — see permethrin

PYRELLIN (Premix)

Premix containing pyrethrins and rotenone. May be combined with other insecticides to enhance knockdown or as an insect "exciter" to help facilitate contact with other spray residues.

Pyrellin E.C. (0.042 lb./gal. pyrethrins & 0.035 lb./gal. rotenone) *(Webb Wright)*

Rate: 1-2 pt. E.C./A

Pest:

aphids	harlequin bug	mites
bean beetle	leafhoppers	plant bugs
cabbageworms	leafminers	stink bugs
celery leaftier	loopers	thrips
garden fleahopper	Lygus bugs	whiteflies

Remarks: Apply at 7-day intervals or less; repeat as needed.

Days To Harvest: 0.

- - - - - - - - - - - - - - - - - -

Rate: 1.5-2 pt. E.C./A

Pest:

Colorado potato beetle	cucumber beetle	flea beetles
	European corn borer	vegetable weevil

Remarks: Make 3-4 applications or more as eggs start to hatch; repeat at 5- to 7-day intervals.

Days To Harvest: 0.

- - - - - - - - - - - - - - - - - -

Signal Word/Toxicity Class: Caution/III.

REI: 12 hr.

Pyrenone — see pyrethrins

PYRETHRINS

Contains piperonyl butoxide as a synergist. Diatect also contains silicon dioxide (diatomaceous earth). May be combined in tank mix with other insecticides for quicker and more complete kill where insect resistance may be a problem, and as an "exciter" to flush insects out of hiding and into contact with other spray residues.

Diatect Multipurpose Insecticide II *(Diatect)*
Pyrenone (0.5EC) *(AgrEvo Environmental Health)*
Py-Rin 60-6 EC *(Wilbur-Ellis)*

Rate: 1-6 lb. II/A
1-12 fl. oz. 0.5EC/A
2-16 oz. 6-60 EC/A

Pest: Including but not limited to:

aphids	flea beetles	Lygus bugs
cabbage looper	imported cabbageworm	mites
caterpillars		plant bugs
corn earworm	leafhoppers	stink bugs
cucumber beetles	leafminers	thrips
cutworms	loopers	whiteflies

Remarks: Apply in sufficient water for thorough coverage; repeat as needed. Apply Diatect as dust or wettable powder.

Days To Harvest: 0.

- - - - - - - - - - - - - - - - - -

Signal Word/Toxicity Class: Caution.

REI: 12 hr.

Py-Rin — see pyrethrins

Sevin — see carbaryl

Sniper — see azinphos-methyl

Thiodan — see endosulfan

THIRETHRIN (Premix)

Contains 0.45 lb. piperonyl butoxide.

Thirethrin (2 lb./gal. endosulfan & 0.045 lb./gal. pyrethrins) *(Terra)*

Rate: 1.5 qt./A

Pest:

crown mite	green peach aphid

Remarks: Do not apply more than once per season.

Days To Harvest: 21.

- - - - - - - - - - - - - - - - - -

Signal Word/Toxicity Class: Warning/II.

REI: 24 hr.

For personal protective equipment and EPA registration numbers, please turn to page 39.

TRIGARD

Triazine compound used as insect growth regulator for control of leafminers. Ineffective against other insect species. Toxic to bees.

Trigard (WSP) (cyromazine) *(Novartis)*

Rate: 0.125 lb. ai/A
0.167 lb. WSP/A (1 packet)

Pest:
leafminers

Remarks: Apply as foliar spray in minimum 5 gal. water by air or 50 gal. water by ground when leafminers first appear. Repeat at 7-day intervals or as necessary. Do not make more than 5 applications per crop. Do not apply through any type of irrigation system.

Days To Harvest: 7.

Crop Rotation: All crops on label may be planted following harvest of treated crop. Do not rotate to any other crop except sweet corn or radishes. Do not plant sweet corn or radishes within 3 months after last application.

Signal Word/Toxicity Class: Caution.
REI: 12 hr.

Sweet Corn

435 or 455 Oil — see petroleum oil

Adios — see carbaryl

Ambush — see permethrin

ASANA

Pyrethroid insecticide that has both contact and stomach poison activity. Restricted Use Pesticide due to toxicity to fish and aquatic organisms.

Asana XL (esfenvalerate) *(DuPont)*

Rate: 0.015-0.03 lb. ai/A
2.9-5.8 fl. oz. XL/A

Pest:
western bean cutworm

Rate: 0.03-0.05 lb. ai/A
5.8-9.6 fl. oz. XL/A

Pest:

armyworm	corn leaf aphid	oat birdcherry aphid
banded cucumber beetle	corn rootworm (adult)	sap beetles (adults)
beet armyworm (aids in control)	cutworm	southwestern corn borer
chinch bug	European corn borer	stalk borers
corn earworm	flea beetles	tarnished plant bug
	grasshoppers	

Remarks: Repeat as necessary to maintain control, but do not exceed 0.5 lb. ai/A per season. For western bean cutworm, apply before larvae enter ear. For aphids, direct spray at aphid population to achieve maximum coverage of exposed insects. Aphids not contacted by spray, such as in whorls and leaf axis, may not be adequately controlled. For chinch bug, over-the-top sprays not adequate for control. Direct spray at base of plant through use of drop nozzles or some other mechanism. For corn earworm, first application should be at or before silking. For corn rootworm, spray at first sign of silk feeding. For flea beetles and stalk borers, apply early in migration from grassy areas to corn, before borers enter plant. For southwestern corn borer, higher rates recommended for moderate to heavy populations.

Days To Harvest: 1.

Signal Word/Toxicity Class: Warning/II.
REI: 12 hr.

AZTEC (Premix)

The organophosphate chemistry of tebupirimphos combined with the effectiveness of a proven pyrethroid (cyfluthrin) provides two modes of action to control soil insect pests. Restricted Use Pesticide due to toxicity to aquatic invertebrates.

Aztec 2.1% (2% tebupirimphos & 0.1% cyfluthrin) *(Bayer)*

Rate: 0.12-.015 lb. ai/A
5.5-7.3 lb. 2.1%/A

Pest:

cutworms	seedcorn maggots	western corn rootworm (larvae)
northern corn rootworm (larvae)	southern corn rootworm (larvae)	white grubs
seedcorn beetles		wireworms

Remarks: Apply band, T-band, or in-furrow. Up to 7.3 lb./A may be applied per crop season. Do not use on other crops grown for food or forage. Refer to label for recommended applications.

State Restrictions: For use in Arizona, Arkansas, Colorado, Idaho, Illinois, Indiana, Iowa, Kansas, Kentucky, Michigan, Minnesota, Missouri, Montana, Nebraska, Nevada, New Mexico, North Dakota, Ohio, Oklahoma, Pennsylvania, South Dakota, Tennessee, Utah, Vermont, West Virginia, Wisconsin, and Wyoming.

Signal Word/Toxicity Class: Warning/II.
REI: 0 hr.

Baythroid — see cyfluthrin

CARBARYL

Carbamate insecticide with both contact and stomach poison activity against broad spectrum of insect pests. Compatible with other pesticides, but incompatible with lime or alkaline compounds. Toxic to bees.

Adios or Slam (13%) *(Micro Flo)*

Rate: 0.03-0.1 lb. ai/A
0.25-0.75 pt. 13%/A

Pest:
corn rootworm

Remarks: Do not apply through any type of irrigation system. Do not apply more than 3.75 lb./A per season.

Days To Harvest: 0.

Signal Word/Toxicity Class: Caution/III.
REI: 12 hr.

Clean Crop Sevin 5 Bait *(Platte)*

Rate: 1-2 lb. ai/A
20-40 lb. Bait/A

Pest:

ants	darkling beetle	grasshoppers
armyworms	field crickets	sowbugs
cutworms		

Remarks: Apply with ground or air equipment.

Days To Harvest: 0.

Signal Word/Toxicity Class: Caution/III.
REI: 12 hr.

Drexel Carbaryl 4L, 80S or 50W *(Drexel)*
Sevin 4F, XLR Plus, 80S, 80WSP *(Rhone-Poulenc)*

Rate: 0.5-1.5 lb. ai/A
1-3 pt. 4L, 4F, XLR/A
1-3 lb. 50W/A
0.66-1.875 lb. 80S, 80WSP/A

Pest:
grasshoppers

Rate: 1-2 lb. ai/A
2-4 pt. 4L, 4F, XLR/A
2-4 lb. 50W/A
1.25-2.5 lb. 80S, 80WSP/A

Pest:

armyworms	fall armyworm	southwestern corn borer
chinch bug	flea beetles	
corn earworm	Japanese beetle	western bean cutworm (high rate)
corn rootworm (adult)	leafhoppers	
European corn borer	sap beetle	

◆ - **new product** • ai-active ingredient • DF-dry flowable • E/EC-emulsifiable concentrate • EW-emulsion in water
F/FL-flowable • G-granule • L/LC-liquid • SP/WSP-soluble packet • W/WP-wettable powder • WSB-water soluble bag

Rate: 2-6.5 lb. ai/A
4-13 pt. 4L/A
4-13 lb. 50W/A
2-3 lb. ai/A
4-6 pt. 4F, XLR/A
2.5-8 lb. 80S, 80WSP/A

Pest:
cutworms

Remarks: Apply when insects or their damage appear. For insects attacking silks and ears, apply at 1- to 6-day intervals starting when first silks appear and continuing until silks begin to dry. For larvae in whorl and foliage feeders, apply as necessary. For chinch bug, use high gallonage ground application directed at base of plants (optimum timing and good coverage essential for effective control). For western bean cutworm, use high rate and treat when infestation averages 15% and at 90-100% tassel emergence; treatment after 100% silk emergence will reduce effectiveness. For cutworms, apply in 12" band over row.

Days To Harvest: 0.

Signal Word/Toxicity Class: Caution (4F, 4L, XLR); Warning/II (50W, 80S, 80WSP).

REI: 12 hr.

Gowan Sevin Bait *(Gowan)*

Rate: 2 lb. ai/A
40 lb. Bait/A

Pest:

armyworms	cutworms	grasshoppers
crickets	darkling beetle	sowbugs

Remarks: Apply when pests appear in damaging numbers; repeat at 7- to 14-day intervals if necessary. Do not use when bees are present.

Days To Harvest: 1.

State Restrictions: For use in Arizona and California.

Signal Word/Toxicity Class: Caution/III.

REI: 12 hr.

Chlorfos — see chlorpyrifos

CHLORPYRIFOS

Organophosphate insecticide with both contact and stomach poison activity. Controls broad spectrum of soil and foliar insect pests. Has no systemic activity and short residual on foliage, but strongly resists leaching when in the soil. Lorsban 30 and 50-SL for use only by commercial seed treaters.

Lorsban 4E *(Dow AgroSciences)*
◆ **Micro Flo Chlorpyrifos 4E** *(Micro Flo)*
Nufos 4E *(Cheminova)*

Rate: 1-3 lb. ai/A
2-6 pt. 4E/A

Pest:

billbugs	grubs	seedcorn maggots
corn rootworm (larvae)	lesser cornstalk borer	symphylans
cutworms	seedcorn beetles	wireworms
flea beetles (larvae)		

Remarks: Preplant incorporated. Apply in minimum 10 gal. water/A to soil surface and incorporate 2-4". Rate varies by pest species. Can be tank mixed; refer to label.

Days To Harvest: 35.

Rate: 0.5-1 lb. ai/A
1-2 pt. 4E/A

Pest:
armyworms cutworms

Remarks: Preplant, at-plant, or preemergence treatment in conservation tillage. Use higher rates for residual control. Refer to label for use directions.

Days To Harvest: 35.

Rate: 1 lb. ai/A
2 pt. 4E/A

Pest:
corn rootworm (larvae)

Remarks: Cultivation time. Apply as water emulsion on both sides of row at base of plants just ahead of cultivator shovels. Cover insecticide with soil around brace roots. Application may be made in addition to at-planting application of 15G. Refer to label for use directions.

Days To Harvest: 35.

Rate: 0.25-1.5 lb. ai/A
0.5-3 pt. 4E/A

Pest:

armyworms	corn rootworm (adult)	lesser cornstalk borer
billbugs	cutworms	southwestern corn borer
chinch bug	European corn borer	
common stalk borer	flea beetles (adult)	webworms
corn earworm	grasshoppers	western bean cutworm

Remarks: Postemergence. Apply when insects become a problem. Refer to label for use directions.

Days To Harvest: 35.

Rate: 0.5-1 lb. ai/A
1-2 pt. 4E/A

Pest:

beet armyworm	corn earworm	fall armyworm

Remarks: Apply as broadcast foliar spray in sufficient water to ensure thorough coverage when damaging populations are developing or present. Repeat as necessary but apply no more than 22 pt. per season. Refer to label for use directions.

Days To Harvest: 21.

State Restrictions: For use in Florida and Georgia only.

Signal Word/Toxicity Class: Warning/II.

REI: 24 hr.

◆ **Chlorfos 15G** *(Griffin)*
Lorsban 15G *(Dow AgroSciences)*
◆ **Micro Flo Chlorpyrifos 15G** *(Micro Flo)*

Rate: Preplant broadcast:
1-2 lb. ai/A
6.7-13.5 lb. 15G/A

Pest:

billbugs	flea beetles	wireworms
cutworms	grubs	

Rate: At planting-banded:
8-16 oz. 15G/1000 ft. row

Pest:

billbugs	seedcorn beetles	symphylans
grubs	seedcorn maggots	western corn rootworm
lesser cornstalk borer	southern corn rootworm	wireworms
northern corn rootworm		

Rate: Postplant-banded:
6-8 oz. 15G/1000 ft. row
Postplant-aerial:
5-6 lb. 15G/A

Pest:
European corn borer southwestern corn borer

Remarks: Do not graze treated areas or harvest treated corn silage as feed. Refer to label for use directions.

Days To Harvest: Forage, silage-14; fodder, grain-35.

Signal Word/Toxicity Class: Caution/III.

REI: 12 hr.

Lorsban 30, 50-SL *(Gustafson)*

Rate: 2.7 fl. oz. 30/100 lb. seed
2 oz. 50-SL/100 lb. seed

Pest:
seedcorn beetles seedcorn maggots

Remarks: Apply prior to planting as slurry treatment. Do not use treated seed for food, feed, or oil purposes. Commercially treated seed must be labeled in accordance with all applicable requirements of Federal Seed Act. Do not allow livestock to graze or feed bean hay or pea hay grown from treated seed.

Signal Word/Toxicity Class: Warning/II (50-SL); Caution (30).

REI: 24 hr.

Comite — see propargite

7 Vegetables

NOTICE The information on these pages is for preliminary planning — not a guide for use. Be sure to follow the manufacturer's directions, notwithstanding information contained here.	For personal protective equipment and EPA registration numbers, please turn to page 39.

COUNTER

Organophosphate insecticide/nematicide with systemic activity. Effective against broad spectrum of soil- and foliage-feeding insect pests. Extremely toxic to fish and wildlife. Restricted Use Pesticide due to acute oral and dermal toxicity.

Counter 15G or CR (terbufos) *(American Cyanamid)*

Rate: 8 oz. 15G/1000 ft. row
6 oz. CR/1000 ft. row

Pest: At planting (banded or in-furrow):

corn flea beetle, corn rootworm, cutworms*, dagger nematode, lance nematode, lesion nematode, lesser cornstalk borer*, maize billbug, root-knot nematode, seedcorn beetles, seedcorn maggots, southern corn billbug, spiral nematode, sting nematode, stubby-root nematode, stunt nematode, symphylans, thrips, white grubs, wireworms

Postemergence incorporated (banded):

maize billbug, southern corn billbug

At cultivation:

corn rootworm, European corn borer

***Suppression**

Remarks: Refer to label for use directions and precautions.

Days To Harvest: CR: 60.

- - - - - - - - - - - - - - - - -

Signal Word/Toxicity Class: Danger-Poison/I.

REI: 48 hr.

CYFLUTHRIN

Synthetic pyrethroid formulated in an emulsifiable concentrate. Toxic to fish and aquatic organisms. Restricted Use Pesticide.

Baythroid (2EC) *(Bayer)*

Rate: 0.8-2.8 fl. oz. 2EC/A

Pest:

black cutworm, chinch bug, common stalk borer, corn earworm, corn rootworm, European corn borer, fall armyworm, granulate cutworm, sandhill cutworm, southwestern corn borer, true armyworm, western bean cutworm

Remarks: Apply in sufficient water for complete coverage of foliage. For cutworms, apply as needed; lower rate for black cutworms under low or no trash conditions; higher rate for other cutworms and high trash conditions, such as those existing under low or no till culture. Up to 10 applications may be applied per crop and a total of 28 fl. oz./A may be applied per crop season. Refer to label for spray drift precautions.

Days To Harvest: 0.

- - - - - - - - - - - - - - - - -

Signal Word/Toxicity Class: Danger/I.

REI: 12 hr.

lambda-CYHALOTHRIN

Synthetic pyrethroid insecticide with broad-spectrum activity, formulated as an emulsifiable concentrate. Restricted Use Pesticide due to toxicity to fish and aquatic organisms.

Karate or Warrior (1E) *(ZENECA)*

Rate: 0.02-0.03 lb. ai/A
2.54-3.84 fl. oz. 1E/A

Pest:

aphids*, aster leafhopper, beet armyworm**, chinch bug, common cornstalk borer, corn earworm, corn rootworm (adult), cucumber beetle (adult), cutworms, European corn borer, fall armyworm**, flea beetles, grasshoppers, Japanese beetle (adult), sap beetle (adult), southern armyworm**, southwestern corn borer, spider mites*, tarnished plant bug, webworms, western bean cutworm, yellowstriped armyworm**

***Suppression **1st & 2nd instar**

Remarks: Do not apply more than 3.84 pt./A per season. Refer to label for spray drift precautions and state and county restrictions.

Days To Harvest: 1.

- - - - - - - - - - - - - - - - -

Signal Word/Toxicity Class: Danger/I.

REI: 24 hr.

DEADLINE

Packaged as Bullets, M-Ps (mini-pellets), and granules.

Deadline Bullets, M-Ps, Granules (4%) (metaldehyde) *(Pace)*

Rate: 0.8-1 lb. ai/A
20-40 lb. 4%/A

Pest:

slugs, snails

Remarks: Apply every 3-4 weeks during growing season or as needed. Do not apply directly to edible portion of plant.

- - - - - - - - - - - - - - - - -

Signal Word/Toxicity Class: Warning/II (granule); Caution (bullets, M-P).

REI: 12 hr.

DIAZINON

Organophosphate insecticide/nematicide with both contact and stomach poison activity. Controls broad spectrum of soil insects, nematodes, and foliage insect pests on wide range of vegetables with long residual effects. Incompatible with copper compounds. Restricted Use Pesticide due to avian and aquatic toxicity.

D•z•n diazinon AG600 WBC, 50W *(Novartis)*
Drexel Diazinon Insecticide (4EC) *(Drexel)*
Gowan Diazinon 4E, 14G, 50WSB *(Gowan)*
Helena Diazinon AG500 *(Helena)*
Micro Flo Diazinon AG500, 50W *(Micro Flo)*
Prentox Diazinon AG500, 50W *(Prentiss)*
◆ Wilbur-Ellis Diazinon 4 Spray (4E), 14G *(Wilbur-Ellis)*

Rate: 0.25-1.25 lb. ai/A
0.5-2.5 pt. 4E, 4EC, AG500/A
0.5-2.5 lb. 50W, 50WSB/A
6.5-32 fl. oz. AG600/A

Pest:

corn earworm, corn leaf aphid, corn rootworm (adult), flea beetles, grasshoppers, spider mites, sap beetle

Rate: 2-4 lb. ai/A
4-8 pt. 4E, 4EC, AG500/A
4-8 lb. 50W, 50WSB/A
14-28 lb. 14G/A
51-102 fl. oz. AG600/A

Pest:

cutworms, seedcorn maggots

Rate: 3-4 lb. ai/A
6-8 pt. 4E, 4EC, AG500/A
6-8 lb. 50W, 50WSB/A
21-28 lb. 14G/A
76.5-102 fl. oz. AG600/A

Pest:

wireworms

Remarks: Rate varies by pest species. Repeat at 7 day-intervals as needed. Do not apply more than 5 times per season. For corn earworm, apply when silks first appear. For sap beetle, consult state agricultural experiment stations or extension service specialists for proper timing of sprays. For cutworms, maggots, and wireworms, broadcast just before planting and immediately incorporate into soil. Corn forage may be fed to dairy and beef cattle and sheep.

Days To Harvest: 7.

- - - - - - - - - - - - - - - - -

Signal Word/Toxicity Class: Caution/III.

REI: 24 hr.

DIAZINON (Premix)

Seed treatment. Contains graphite.

Captan-Diazinon (25% diazinon & 36.67% captan) *(Trace)*
Germate Plus (15% diazinon & 25% lindane & 14% carboxin) *(Trace)*
Kernel Guard (15% diazinon & 25% lindane & 14.67% captan) *(Trace)*

Rate: 2 oz./bu. seed

Pest:

seedcorn beetles, seedcorn maggots, wireworms

◆ **- new product** • ai-active ingredient • DF-dry flowable • E/EC-emulsifiable concentrate • EW-emulsion in water
F/FL-flowable • G-granule • L/LC-liquid • SP/WSP-soluble packet • W/WP-wettable powder • WSB-water soluble bag

Remarks: Apply in planter box before planting. Do not use Germate Plus on Foundation seed stock. Treat only those seeds needed for immediate use. Do not use treated seed for food, feed, or oil purposes. Do not store excess treated seeds beyond planting time. Refer to label for use directions.

Signal Word/Toxicity Class: Danger/I.
REI: 12 hr.; Captan-Diazinon 4 days with exceptions.

D•z•n — see diazinon

ENDOSULFAN

Nonsystemic, chlorinated hydrocarbon insecticide/acaricide with contact and stomach poison activity. Controls broad spectrum of foliage-feeding insect pests. Toxic to fish, birds, and other wildlife.

Drexel Endosulfan 3EC or 50WP *(Drexel)*
Gowan Endosulfan 3EC or 50WSB *(Gowan)*
◆ Micro Flo Endosulfan 3EC or 50WP *(Micro Flo)*
Phaser 3EC or 50WSB *(AgrEvo USA)*
Thiodan 3EC or 50WP *(FMC)*

Rate: 1 lb. ai/A
1.33 qt. 3EC/A
2 lb. 50WP, 50WSB/A
Pest:
corn leaf aphid

Rate: 1.5 lb. ai/A
2 qt. 3EC/A
3 lb. 50WP, 50WSB/A
Pest:
corn earworm

Remarks: Fresh vegetable use only. Do not apply to sweet corn to be processed. Apply when insects appear or feeding is noticed. For corn earworm, apply when silks first appear and continue until they begin to dry; allow 5 days between applications. Do not exceed 3 lb. ai/A per year or 3 applications per year. Do not feed treated forage or ensilage to livestock or allow to graze in treated fields.
Days To Harvest: 1.

Signal Word/Toxicity Class: Danger-Poison/I.
REI: 24 hr.

FORTRESS

Fortress 5C is applied only with SMARTBOX system, a closed handling and application system. Toxic to mammals, birds, fish, and aquatic invertebrates. Restricted Use Pesticide.

Fortress 2.5G or 5G (chlorethoxyfos) *(DuPont)*

Rate: 0.15 oz. ai/1000 ft. row
6 oz. 2.5G/1000 ft. row
3 oz. 5G/1000 ft. row
Pest:

cutworms	southern corn rootworm	western corn rootworm
northern corn rootworm	symphylans	white grubs
seedcorn maggots		wireworms

Remarks: Apply by T-band or in-furrow. Do not make more than one application per year. Do not apply through any type of irrigation system.
Days To Harvest: 21.
Crop Rotation: Corn-anytime; other crops-30 days.

Signal Word/Toxicity Class: Danger-Poison.
REI: 48 hr.; 72 hr. if average rainfall less than 25" per year.

FURADAN

Carbamate insecticide/nematicide used as soil systemic or foliar applied material. Has both contact and stomach poison activity, and controls a wide range of insect pests and nematodes. Half-life in soil is 40-60 days. Toxic to fish and other wildlife. To prevent bird kill, leave no granules exposed. Restricted Use Pesticide.

Furadan 4F (carbofuran) *(FMC)*

Rate: Foliar:
0.5 lb. ai/A
1 pt. 4F/A
Pest:
European corn borer (2nd generation)

Remarks: Machine harvested only. Apply in minimum 10 gal. finished spray/A by ground; 2 gal./A by air. Make first application just prior to first silking and repeat at weekly intervals. Do not make more than 4 applications per season.
Days To Harvest: 7; grazing or harvesting stalks-21.
Crop Rotation: Do not rotate with any crop on soil treated at greater than 3 lb. ai/A for at least 10 months or plant with any crop for which carbofuran is not registered for at least 10 months.
State Restrictions: Not for use in Long Island, New York.

Rate: At planting:
1 lb. ai/A
2 pt. 4F/A (40" row)
Pest:

armyworms*	European corn borer (1st generation)*	seedcorn maggots
corn rootworm	flea beetles	wireworms

***Aids in control for 4-6 weeks after planting**

Remarks: Apply as 7" band over row; inject on each side of row by mixing with water or liquid fertilizers, or apply directly into seed furrow.
Days To Harvest: Do not feed forage within 30 days of last application.
Crop Rotation: Do not rotate with any crop on soil treated at greater than 3 lb. ai/A for at least 10 months or plant with any crop for which carbofuran is not registered for at least 10 months.
State Restrictions: Not for use in Long Island, New York.

Signal Word/Toxicity Class: Danger-Poison/I.
REI: 48 hr.

Fyfanon — see malathion

Germate Plus — see diazinon premixes

Karate — see *lambda*-cyhalothrin

Kernel Guard — see diazinon premixes

Kernel Guard Supreme — see permethrin premix

Kumulus — see sulfur

LANNATE

Carbamate insecticide with contact and stomach poison activity. Controls broad spectrum of insect pests. Phytotoxic to certain hybrid sweet corn varieties. Restricted Use Pesticide.

Lannate SP or LV (methomyl) *(DuPont)*

Rate: 0.3-0.45 lb. ai/A
0.33-0.5 lb. SP/A
1-1.5 pt. LV/A
Pest:
earworms (whorl as needed)

Rate: 0.225-0.45 lb. ai/A
0.25-0.5 lb. SP/A
0.75-1.5 pt. LV/A
Pest:

aphids	earworms	flea beetles
armyworms	European corn borer	picnic beetles
corn rootworm (adult)	fall armyworm	

Rate: 0.45 lb. ai/A
0.5 lb. SP/A
1.5 pt. LV/A
Pest:
beet armyworm variegated cutworm

Remarks: Apply when insects first appear; repeat at 1-day intervals or as needed. Certain hybrid varieties of sweet corn are susceptible to methomyl injury; treat small area to determine crop safety before full scale spraying. Do not apply more than 7.2 lb. ai/A per crop or make more than 28 applications per crop.
Days To Harvest: Ears-0; forage-3.

Signal Word/Toxicity Class: Danger-Poison/I.
REI: 48 hr.

7 Vegetables

NOTICE The information on these pages is for preliminary planning — not a guide for use. Be sure to follow the manufacturer's directions, notwithstanding information contained here.	For personal protective equipment and EPA registration numbers, please turn to page 39.

LARVIN

Broad-spectrum carbamate insecticide that has stomach poison activity and some limited contact activity. Known to have systemic action as a seed treatment. Toxic to fish and birds, and moderately toxic to bees as direct treatment.

Larvin 3.2 (thiodicarb) *(Rhone-Poulenc)*

Rate: 0.5-0.75 lb. ai/A
20-30 fl. oz. 3.2/A

Pest:

beet armyworm	cutworms	fall armyworm
corn earworm	European corn borer	western bean cutworm

Remarks: Fresh market only. For insects attacking silks and ears, apply at 1- to 7-day intervals starting when silks first appear and continuing until silks begin to dry or infestation potential ceases. As insect populations increase and/or silk growth is rapid, spray interval should be shorter. Follow local recommendations for proper timing. For whorl or foliage feeders, apply when insect populations reach recognized economic threshold levels. After silk initiation, do not exceed 7.5 lb. ai/A per season. For postemergence rescue treatments for cutworms, apply as broadcast spray when cutworms are actively feeding. If cutworms are feeding below ground, control will be more difficult. Light cultivation with rotary hoe before spraying may enhance control. Do not allow livestock to graze treated field or feed treated corn silage (green plant) or fodder to livestock.

Days To Harvest: 0.

State Restrictions: For use in Florida only.

Signal Word/Toxicity Class: Warning/II.

REI: 12 hr.

Lorsban — see chlorpyrifos

MALATHION

Organophosphate insecticide used to control a broad spectrum of insect pests and mites on many vegetable varieties. Has less residual effects when mixed with alkaline compounds, although toxicity is not decreased. Does not persist in the soil.

Drexel Malathion 5EC *(Drexel)*
◆ Micro Flo Malathion 5EC *(Micro Flo)*
Prentox 5 lb. Malathion Spray (5EC) *(Prentiss)*

Rate: 1.25 lb. ai/A
2 pt. 5EC/A

Pest:
Japanese beetle

Remarks: Injury may occur in whorl and silk stages.

Days To Harvest: 5.

Signal Word/Toxicity Class: Warning/II.

REI: 12 hr.

Fyfanon (5EC or 8E) *(Helena)*
◆ Gowan Malathion 8 (8E) *(Gowan)*

Rate: 0.94 lb. ai/A
1.5 lb. 5EC/A
1.25 lb. ai/A
1.25 pt. 8E/A

Pest:

aphids	grasshoppers	sap beetle
corn rootworm (adult)		thrips

Remarks: Injury may occur in whorl and silk stages.

Days To Harvest: 5.

Signal Word/Toxicity Class: Caution/III.

REI: 12 hr.

Marlate — see methoxychlor

METASYSTOX-R

Restricted Use Pesticide.

Metasystox-R (SC) (oxydemeton-methyl) *(Gowan)*

Rate: 0.375-0.5 lb. ai/A
1.5-2 pt. SC/A

Pest:

aphids	leafhoppers	thrips
corn rootworm (beetle)	mites	

Remarks: Apply in sufficient water for complete coverage but not less than 1 gal./A. Do not apply more than 3 times per season.

Days To Harvest: 1 application-7; 2-3 applications-21.

Signal Word/Toxicity Class: Warning/II.

REI: 48 hr.; 72 hr. if average rainfall less than 25" per year.

METHOXYCHLOR

Organochlorine pesticide toxic to fish and aquatic invertebrates.

Marlate 50 (WP) *(Kincaid)*
Prentox Methoxychlor 2EC *(Prentiss)*

Rate: 1.3-1.7 lb. ai/A
75-100 fl. oz. 2EC/A
2 Tbs. 50WP/gal. water

Pest:

armyworms	flea beetles	Japanese beetle
fall armyworm		

Remarks: Apply at first sign of infestation; repeat at 7- to 14-day intervals as needed.

Days To Harvest: 7.

Signal Word/Toxicity Class: Caution.

REI: 12 hr.

MOCAP

Organophosphate soil insecticide/nematicide with contact activity. Must be incorporated or carried into soil by water to be effective. Long residual effects with no fumigant activity. Toxic to fish and other wildlife. Mocap EC is a Restricted Use Pesticide.

Mocap EC or 10%G (ethoprop) *(Rhone-Poulenc)*

Rate: Broadcast:
3 lb. ai/A
2 qt. EC/A
30 lb. 10%G/A

Pest:
cutworms

Rate: Broadcast treatment:
6 lb. ai/A
4 qt. 6EC/A
60 lb. 10%G/A
Row treatment:
1.5-2 lb. ai/A
1-1.33 qt. EC/A
15-20 lb. 10%G/A

Pest:
nematodes

Rate: Row treatment:
1 lb. ai/A
0.66 qt. EC/A (1.66 oz./1000 ft. row)
10 lb. 10%G/A (12 oz./1000 ft. row)

Pest:

corn rootworm	garden symphylan	white grubs (suppression)

Remarks: Broadcast: Apply from 3 days before planting to at-planting time; mix with 2-4" of soil (2" for cutworms) right after application. Row treatment: Apply at planting in 6-7" band on 40" row spacing over a closed seed furrow; mix with top 1/2" of soil.

State Restrictions: Do not apply in Long Island, New York.

Signal Word/Toxicity Class: Danger-Poison/I (EC); Warning/II (10%G).

REI: 48 hr.; 72 hr. if average rainfall less than 25" per year.

Nufos — see chlorpyrifos

Omni Supreme Spray — see petroleum oil

◆ **- new product** • ai-active ingredient • DF-dry flowable • E/EC-emulsifiable concentrate • EW-emulsion in water
F/FL-flowable • G-granule • L/LC-liquid • SP/WSP-soluble packet • W/WP-wettable powder • WSB-water soluble bag

PERMETHRIN

Restricted Use Pesticide.

Ambush 2E or 25W *(ZENECA)*

Rate: 0.1-0.2 lb. ai/A
6.4-12.8 fl. oz. 2E/A
6.4-12.8 oz. 25W/A

Pest:
aster leafhopper, cutworms, fall armyworm, corn earworm, European corn borer, flea beetle (except CA), corn rootworm (adult)

Remarks: Apply every 5 days or as needed by air or ground. Do not apply more than 1.2 lb. ai/A per season.

Days To Harvest: 1.

- - - - - - - - - - - - - - - - -

Rate: 0.1-0.25 lb. ai/A
6.4-16 fl. oz. 2E/A
6.4-16 oz. 25W/A

Pest:
aster leafhopper, cutworms, fall armyworm, corn earworm, European corn borer, southern armyworm, corn rootworm (adult)

Remarks: Fresh market. Apply every 5 days or as needed by air or ground. Do not apply more than 3.25 lb. ai/A per crop per season. Do not graze livestock in treated areas or harvest for forage.

State Restrictions: For use in Florida.

- - - - - - - - - - - - - - - - -

Signal Word/Toxicity Class: Warning/II.
REI: 24 hr.

Pounce 1.5G *(FMC)*

Rate: 0.1-0.2 lb. ai/A
6.7-13.3 lb. 1.5G/A
8-16 oz. 1.5G/1000 linear ft. row (band)

Pest:
armyworms, southwestern corn borer, stalk borer, cutworms, western bean cutworm, European corn borer

Remarks: Preemergence: Broadcast by ground or air. Foliar application: Apply when eggs begin to hatch. For stalk borer, apply when or shortly before larvae are moving into corn from surrounding weeds and grass. Mowing or burndown herbicide suggested to initiate movement of stalk borer larvae. Apply by aircraft by broadcasting over corn plants or with ground equipment by directing granules into whorl. Do not apply more than 0.4 lb. ai/A after brown silk stage or exceed 1 lb. ai/A per season.

Signal Word/Toxicity Class: Caution.
REI: 24 hr.

Pounce 3.2 EC, 25WP or WSB *(FMC)*

Rate: 0.1-0.2 lb. ai/A
4-8 oz. 3.2 EC/A
6.4-12.8 oz. 25WP/A
1-2 bags WSB/A

Pest:
corn earworm, European corn borer, leafhoppers, corn rootworm beetle (except CA), fall armyworm, southern armyworm, flea beetle, stalk borers, cutworms

Remarks: All markets. Apply every 5 days or as needed. Do not apply more than 1.2 lb. ai/A per season.

Days To Harvest: Ears or livestock grazing-1.

- - - - - - - - - - - - - - - - -

Rate: 0.1-0.2 lb. ai/A
4-10 oz. 3.2 EC/A
6.4-12.8 oz. 25WP/A
1-2 bags WSB/A

Pest:
corn earworm, fall armyworm, southern armyworm, European corn borer

Remarks: Fresh market use. Apply every 5 days or as needed. Do not apply more than 3.25 lb. ai/A per crop season. Do not graze livestock in treated area or harvest for forage.

State Restrictions: For use in Florida.

- - - - - - - - - - - - - - - - -

Signal Word/Toxicity Class: Caution (3.2 EC); Warning/II (25WP, WSB).
REI: 24 hr.

PERMETHRIN (Premix)

Seed treatment.

Kernel Guard Supreme (10.42% permethrin & 14% carboxin) *(Trace)*

Rate: 2 oz./bu. seed

Pest:
seedcorn beetles, seedcorn maggots, wireworms

Remarks: Refer to label for use directions.

- - - - - - - - - - - - - - - - -

Signal Word/Toxicity Class: Caution.
REI: 12 hr. with exceptions.

PETROLEUM OIL

Petroleum based spray oil. Mode of action is through suffocation of eggs, larvae, and nymphs of insects and adults of soft bodied insects. This mode of action necessitates total spray coverage. Toxic to fish.

◆ **435 or 455 Oil 98.8** *(Drexel)*
◆ **SunSpray 6E or 7E** *(Sun)*

Rate: 2 gal./100 gal. water

Pest:
corn rootworm (435), earworms, fall armyworm (435)

Remarks: Do not apply in combination with or immediately preceding or following applications of captan, carbaryl, Dyrene, folpet, Morestan, propargite, and sulfur compounds.

Signal Word/Toxicity Class: Caution.
REI: 4 hr.

Omni Supreme Spray *(Helena)*

Rate: 1-2 gal./100 gal. water

Pest:
aphids, caterpillars, mites, beetles (larvae), leafminers

Remarks: None.

- - - - - - - - - - - - - - - - -

Signal Word/Toxicity Class: Caution.
REI: 4 hr.

SAF-T-SIDE *(Brandt Consolidated)*
◆ **Sumulsion** *(Drexel)*

Rate: 1-2 gal./100 gal. water

Pest:
aphids, mites, whiteflies, leafminers

Remarks: Apply as needed. May be tank mixed with most commonly used insecticides and fungicides. Do not use in combination with or immediately before or after spraying with fungicides such as captan, Dyrene, folpet, Morestan, or any sulfur products. Do not use with carbaryl or dimethoate.

- - - - - - - - - - - - - - - - -

Signal Word/Toxicity Class: Caution.
REI: 4 hr.

◆ **SunSpray Ultra-Fine** *(Sun)*

Rate: 2 gal./100 gal. water

Pest:
aphids, leafminer, rootworm, armyworms, mites, whiteflies, corn earworm

Remarks: Apply at 2-week intervals. Do not exceed 4 applications per season.

- - - - - - - - - - - - - - - - -

Signal Word/Toxicity Class: Caution.
REI: 4 hr.

Phaser — see endosulfan

NOTICE	The information on these pages is for preliminary planning — not a guide for use. Be sure to follow the manufacturer's directions, notwithstanding information contained here.	For personal protective equipment and EPA registration numbers, please turn to page 39.

PHORATE

Restricted Use Pesticide.

Clean Crop Phorate 20G *(Platte)*
Riverside Phorate 20G *(Terra)*
Thimet 20-G *(American Cyanamid)*

Rate: 6 oz. 20G, 20-G/1000 ft. row (min. 30" row spacing)
Pest: At planting:

corn rootworm	seedcorn beetles (IA, IL*)	wireworms (reduction North Central, Southeast *)
flea beetles	seedcorn maggots	
mites	white grubs	

At cultivation:
corn rootworm

*** *Clean Crop, Riverside***

Remarks: At planting application: Place granules in 7" band over row; 1 additional application can be applied at cultivation. Do not use in-furrow. Cultivation application: Apply granules to base of plants just ahead of cultivator shovels. Do not apply after cultivation. Do not make more than 1 cultivation treatment per season.
Days To Harvest: 30.

- - - - - - - - - - - - - - - - -

Signal Word/Toxicity Class: Danger-Poison/I.
REI: 48 hr.

Pounce — see permethrin

PROPARGITE

Nonsystemic organic material with contact activity used as acaricide with residual killing action. Controls mites when applied before populations reach destructive numbers, but has no insecticide activity. Incompatible with oils, petroleum solvents, alkaline materials, or in a spray solution with a pH above 10. Can cause eye damage.

Comite *(Uniroyal)*

Rate: 1.64-2.46 lb. ai/A
2-3 pt./A
Pest:

Banks grass mite	twospotted spider mite

Remarks: Apply when mite populations are low and just beginning to build, and corn is still small enough (usually 3-4 ft.) to allow good coverage. Apply only when corn leaves are dry, especially for white corn varieties which are most susceptible to plant injury. Apply no more than once per year. Do not apply through any type of irrigation system.
Days To Harvest: 30.
State Restrictions: For use in California.

- - - - - - - - - - - - - - - - -

Signal Word/Toxicity Class: Danger/1.
REI: 7 days. After 48 hr., workers can enter treated areas to perform hand labor tasks if they wear early-entry personal protective equipment.

PYRELLIN (Premix)

Premix containing pyrethrins and rotenone. May be combined with other insecticides to enhance knockdown or as an insect "exciter" to help facilitate contact with other spray residues.

Pyrellin E.C. (0.042 lb./gal. pyrethrins & 0.035 lb./gal. rotenone) *(Webb Wright)*

Rate: 1-2 pt. E.C./A
Pest:

alfalfa webworms	garden fleahopper	Lygus bugs
aphids	harlequin bug	mites
bean beetle	Japanese beetle	plant bugs
beet webworm	leafhoppers	stink bugs
cabbageworm	leafminers	thrips
celery leaftier	loopers	whiteflies

Remarks: Apply in intervals of 7 days or less; repeat as needed.
Days To Harvest: 0.

- - - - - - - - - - - - - - - - -

Rate: 1.5-2 pt. E.C./A
Pest:

Colorado potato beetle	cucumber beetle	flea beetles
	European corn borer	vegetable weevil

Remarks: Make 3-4 applications or more as eggs start to hatch; repeat at 5-7 day intervals.
Days To Harvest: 0.

- - - - - - - - - - - - - - - - -

Signal Word/Toxicity Class: Caution/III.
REI: 12 hr.

SAF-T-SIDE — see petroleum oil

Sevin — see carbaryl

Slam — see carbaryl

SULFUR

◆ **Kumulus DF** *(BASF)*

Rate: 4.8-12 lb. ai/A
6-15 lb. DF/A
Pest:

Banks grass mite	red spider mite	twospotted spider mite
Pacific mite		

Remarks: Do not apply if temperatures during or within 3 days after application are expected to exceed 90°F. Do not apply through any type of irrigation system. Refer to label for use directions.

- - - - - - - - - - - - - - - - -

Signal Word/Toxicity Class: Caution.
REI: 24 hr.

Sumulsion — see petroleum oil

SunSpray — see petroleum oil

Thimet — see phorate

Thiodan — see endosulfan

Warrior — see lambda-cyhalothrin

Sweet Potatoes, Yams

Adios — see carbaryl

AZADIRACHTIN

Extracted from seed of neem tree. Insect growth regulator that controls targeted insect larvae when they ingest or come in contact with it, by interfering with the insect's ability to molt and shutting off feeding behavior. Affected larvae slowly starve and die at next molt.

◆ **Azatin XL Plus** *(Thermo Trilogy)*

Rate: 5-21 oz. XL Plus/A
Pest:

aphids	cutworms	leafminers
armyworms	diamondback moth	leafrollers
beetles	flea beetles	loopers
cabbage looper	green peach aphid	psyllids
caterpillars	imported cabbageworm	thrips
Colorado potato beetle	Japanese beetle	webworms
corn earworm	leafhoppers	whiteflies

Remarks: Sweet potatoes, yams. Rate varies by species. Apply when pests first appear; repeat at 7-day intervals. When pest populations are high, use higher rates. Do not apply more that 21 fl. oz./A.
Days To Harvest: 0.

- - - - - - - - - - - - - - - - -

Signal Word/Toxicity Class: Caution.
REI: 4 hr.

◆ - **new product** • ai-active ingredient • DF-dry flowable • E/EC-emulsifiable concentrate • EW-emulsion in water
F/FL-flowable • G-granule • L/LC-liquid • SP/WSP-soluble packet • W/WP-wettable powder • WSB-water soluble bag

◆ **NEEMAZAD 4.5EC or NEEMIX 4.5** *(Thermo Trilogy)*
Rate: 0.125-1 pt. 4.5/A
High Volume:
2.25-4.5 oz. 4.5EC/100 gal. water
Low Volume:
0.5 pt. 4.5EC/A
Pest:

aphids	caterpillars	leafhoppers
armyworms	Colorado potato beetle	potato flea beetle
artichoke plume moth	diamondback moth	psyllids
aster leafhopper	greenhouse whitefly	sweetpotato whitefly
cabbage loopers	hornworms	webworms

Remarks: Sweet potatoes, yams. For low volume, apply in sufficient water for adequate coverage. For low insect pressure, apply preventive at 7-day intervals or at first sign of insects; high insect pressure, at 3- to 4-day intervals.

Signal Word/Toxicity Class: Warning/II.
REI: 12 hr.

Azatin — see azadirachtin

CARBARYL

Carbamate insecticide with broad activity on many insects. Highly toxic to bees.

Adios (13%) *(Micro Flo)*
Rate: 0.065-0.1 lb. ai/A
0.5-0.75 lb. 13%/A
Pest:
cucumber beetle
Remarks: Sweet potatoes. Do not apply through any type of irrigation system. Do not apply more than 3.75 lb./A per season.
Days To Harvest: 0.

Signal Word/Toxicity Class: Caution/III.

Clean Crop Sevin 5 Bait *(Platte)*
Rate: 1-2 lb. ai/A
20-40 lb. Bait/A
Pest:

ants	darkling beetle	grasshoppers
armyworms	field crickets	sowbugs
cutworms		

Remarks: Apply with ground or air equipment.
Days To Harvest: 0.

Signal Word/Toxicity Class: Caution/III.
REI: 12 hr.

Drexel Carbaryl 4L *(Drexel)*
Sevin 4F, XLR Plus, 80S or 80WSP *(Rhone-Poulenc)*
Rate: 0.5-2 lb. ai/A
1-4 pt. 4L, 4F, XLR/A
0.66-2.5 lb. 80S, 80WSP/A
Pest:

corn earworm	grasshoppers	tortoise beetle
cucumber beetle	sweetpotatohornworm	yellowstriped
flea beetles	sweetpotato weevil	armyworm

Remarks: Rate varies by pest species; refer to label. Apply as foliar spray as needed for all pests except grasshoppers. For sweetpotato weevil, full coverage of plants essential.
Days To Harvest: 0.
State Restrictions: Sevin XLR Plus or 4F not for use in California.

Rate: 9 Tbs. 80S, 80WSP/1 gal. water
Pest:
sweetpotato weevil
Remarks: Preplant dip. Just prior to planting, dip sweet potato cuttings in a suspension of this mixture.
State Restrictions: Not for use in California.

Signal Word/Toxicity Class: Caution (4F, 4L, XLR); Warning/II (80S, 80WSP).
REI: 12 hr.

Chlorfos — see chlorpyrifos

CHLORPYRIFOS

◆ **Chlorfos 15G** *(Griffin)*
Lorsban-4E or 15G *(Dow AgroSciences)*
◆ **Micro Flo Chlorpyrifos 4E or 15G** *(Micro Flo)*
Rate: 2 lb. ai/A
4 pt. 4E/A
13.5 lb. 15G/A
Pest:
flea beetles (larvae) sweetpotato flea beetle wireworms
Remarks: Preplant broadcast. Incorporate 4-6". Do not apply more than once per season or plant crop later than 14 days after treatment. Will not control false wireworms, whitefringed beetles or other grubs that attack sweet potatoes. Refer to label for use directions.
Days To Harvest: 125.

Signal Word/Toxicity Class: Warning/II (4E); Caution/III (15G).
REI: 4E: 24 hr.; 15G: 12 hr.

DEADLINE

Packaged as Bullets, M-Ps (mini-pellets), and granules.

Deadline Bullets, M-Ps, Granules (4%) (metaldehyde) *(Pace)*
Rate: 0.8-1 lb. ai/A
20-40 lb. 4%A
Pest:
slugs snails
Remarks: Apply every 3-4 weeks during growing season, or as needed. Do not apply directly to edible portion of plant.

Signal Word/Toxicity Class: Warning/II (granule); Caution (bullets, M-P).
REI: 12 hr.

Diatect — see pyrethrins

DIAZINON

Restricted Use Pesticide due to avian and aquatic toxicity.

D•z•n diazinon AG600 WB, 50W *(Novartis)*
Drexel Diazinon Insecticide (4EC), D-264 14G *(Drexel)*
Gowan Diazinon 4E, 14G, 50WSB *(Gowan)*
Micro Flo Diazinon AG500, 14, 50W *(Micro Flo)*
Prentox Diazinon AG500 50W *(Prentiss)*
◆ **Wilbur-Ellis Diazinon 4 Spray (4E), 14G** *(Wilbur-Ellis)*
Rate: 3-4 lb. ai/A
6-8 pt. 4E, 4EC, AG500/A
6-8 lb. 50W, 50WSB /A
76.5-102 fl. oz. AG600/A
3 lb. ai/A
21 lb. 14G/A
Pest:
flea beetles wireworms
Remarks: Broadcast before planting and immediately incorporate into soil 4-8". Refer to label for use directions and precautions.

Signal Word/Toxicity Class: Caution.
REI: 24 hr.; 12 hr. (Drexel/Gowan/Wilbur-Ellis 14G).

D•z•n — see diazinon

ENDOSULFAN

Toxic to fish, birds, and other wildlife.

Drexel Endosulfan 3EC or 50WP *(Drexel)*
Gowan Endosulfan 3EC or 50WSB *(Gowan)*
◆ **Micro Flo Endosulfan 3EC or 50WP** *(Micro Flo)*
Phaser 3EC or 50WSB *(AgrEvo USA)*
Thiodan 3EC or 50WP *(FMC)*
Rate: 0.5 lb. ai/A
0.66 qt. 3EC/A
1 lb. 50WP, 50WSB/A
Pest:
sweetpotato flea beetle sweetpotato weevil whiteflies

7 Vegetables

NOTICE The information on these pages is for preliminary planning — not a guide for use. Be sure to follow the manufacturer's directions, notwithstanding information contained here.

For personal protective equipment and EPA registration numbers, please turn to page 39.

Remarks: Sweet potatoes. Apply shortly after transplanting or as soon as flea beetles appear. Do not exceed 3 applications per year or 3 lb. ai/A per year.
Days To Harvest: 1.

Rate: 1-2 lb. ai/A
1.33-2.66 qt. 3EC/A
2-4 lb. 50WP, 50WSB/A
Pest:
banded cucumber beetle (larvae)
Remarks: Sweet potatoes. Broadcast on 16" band over 48" row; work into soil to a depth of approximately 3". Apply just prior to plant set. Do not feed cull potatoes to livestock or allow livestock to graze treated fields.
Days To Harvest: 1.
State Restrictions: For use in South Central states and Puerto Rico only.

Signal Word/Toxicity Class: Danger-Poison/I.
REI: 24 hr.

Fyfanon — see malathion

Lorsban — see chlorpyrifos

MALATHION

Organophosphate insecticide toxic to fish, aquatic invertebrates, aquatic life stages of amphibians, and bees exposed to direct treatment on blooming crops or weeds.

Clean Crop Malathion 57EC *(Platte)*
Drexel Malathion 5EC *(Drexel)*
Fyfanon (5EC) *(Helena)*
◆ **Gowan Malathion 8 (8E)** *(Gowan)*
◆ **Micro Flo Malathion 5EC** *(Micro Flo)*
Prentox 5 lb. Malathion Spray (5EC) *(Prentiss)*
Rate: 1-1.9 lb. ai/A
1.5-3 pt. 5EC, 57EC/A
1-1.75 lb. ai/A
1-1.75 pt. 8E/A
Pest:
leafhoppers, morningglory leafminer
Remarks: Sweet potatoes. Rate varies by pest species.
Days To Harvest: 3.

Signal Word/Toxicity Class: Varies by formulation.
REI: 12 hr.

Marlate — see methoxychlor

METHOXYCHLOR

Organochlorine pesticide toxic to fish and aquatic invertebrates.

Marlate 50 (WP) *(Kincaid)*
Rate: 2 Tbs. 50WP/gal. water
Pest:
fall armyworm, flea beetles
Remarks: Sweet potatoes, yams. Apply at first sign of infestation; repeat at 7- to 14-day intervals as needed.
Days To Harvest: 0.

Signal Word/Toxicity Class: Caution.
REI: 12 hr.

MOCAP

Mocap EC is a Restricted Use Pesticide.

Mocap EC or 10%G (ethoprop) *(Rhone-Poulenc)*
Rate: Broadcast:
6-8 lb. ai/A
1-1.33 gal. EC/A
60-80 lb. 10%G/A
Row treatment:
3-4 lb. ai/A
2-2.66 qt. EC/A (5.1-6.9 fl. oz./1000 lin. ft. row)
30-40 10%G/A (2.4-3.2 lb./1000 lin. ft. row)
Pest:
banded cucumber beetle, flea beetle larvae, white grubs, nematodes, wireworms
Remarks: Broadcast: Apply evenly on top of soil 2-3 weeks before planting. Row treatment: Apply in a band 12-15" wide on row (min. 42") 2-3 weeks before planting. Mix with top 2-4" soil immediately after application.
State Restrictions: Do not apply in Long Island, New York.

Signal Word/Toxicity Class: Danger-Poison/I (EC); Warning/II (10%G).
REI: 48 hr.; 72 hr. if average rainfall less than 25" per year.

M-PEDE

M-Pede (potassium salts of fatty acids) *(Mycogen)*
Rate: 1-2% v/v solution
Pest:
aphids*, leafminers*, spider mites, broad mite, plant bugs, thrips, green peach aphid*, russet mite, whiteflies, leafhoppers
** Tank mix only*
Remarks: Yams. Apply at first sign of infestation; repeat at 7- to 14-day intervals. May be tank mixed with other insecticides. Refer to label for use directions and precautions.
Days To Harvest: 0.

Signal Word/Toxicity Class: Warning/II.
REI: 12 hr.

NEEMAZAD — see azadirachtin

NEEMIX — see azadirachtin

PETROLEUM OIL

Petroleum based spray oil. Mode of action is through suffocation of eggs, larvae, and nymphs of insects and adults of soft bodied insects. This mode of action necessitates total spray coverage. Toxic to fish.

SAF-T-SIDE *(Brandt Consolidated)*
◆ **Sumulsion** *(Drexel)*
Rate: 1-2 gal./100 gal. water
Pest:
aphids, mites, whiteflies, leafminers
Remarks: Apply as needed. May be tank mixed with most commonly used insecticides and fungicides. Do not use in combination with or immediately before or after spraying with fungicides such as captan, Dyrene, folpet, Morestan, or any sulfur products. Do not use with carbaryl or dimethoate.

Signal Word/Toxicity Class: Caution.
REI: 4 hr.

Phaser — see endosulfan

PYRELLIN (Premix)

Premix containing pyrethrins and rotenone. May be combined with other insecticides to enhance knockdown or as an insect "exciter" to help facilitate contact with other spray residues.

Pyrellin E.C. (0.042 lb./gal. pyrethrins & 0.035 lb./gal. rotenone) *(Webb Wright)*
Rate: 1-2 pt. E.C./A
Pest:
aphids, loopers, plant bugs, leafhoppers, Lygus bugs, thrips, leafminers, mites
Remarks: Yams. Apply as necessary to maintain control.
Days To Harvest: 0.

Signal Word/Toxicity Class: Caution/III.
REI: 12 hr.

◆ **- new product** • ai-active ingredient • DF-dry flowable • E/EC-emulsifiable concentrate • EW-emulsion in water
F/FL-flowable • G-granule • L/LC-liquid • SP/WSP-soluble packet • W/WP-wettable powder • WSB-water soluble bag

Pyrenone — see pyrethrins

PYRETHRINS

Contains piperonyl butoxide as a synergist. Diatect also contains silicon dioxide (diatomaceous earth). May be combined in tank mix with other insecticides for quicker and more complete kill where insect resistance may be a problem, and as an "exciter" to flush insects out of hiding and into contact with other spray residues.

Diatect Multipurpose Insecticide II *(Diatect)*
Pyrenone (0.5EC) *(AgrEvo Environmental Health)*
Py-Rin 60-6 EC *(Wilbur-Ellis)*

Rate: 1-6 lb. II/A
1-12 fl. oz. 0.5EC/A
2-16 oz. 6-60 EC/A

Pest: Including but not limited to:

aphids	leafhoppers	plant bugs
cucumber beetle	leafminers	stink bugs
cutworms	loopers	weevils
flea beetles	mites	whiteflies

Remarks: Apply in sufficient water for thorough coverage; repeat as needed. Apply Diatect as dust or wettable powder.
Days To Harvest: 0.

Signal Word/Toxicity Class: Caution.
REI: 12 hr.

Py-Rin — see pyrethrins

Sevin — see carbaryl

SAF-T-SIDE — see petroleum oil

Sumulsion — see petroleum oil

TEMIK

Restricted Use Pesticide.

Temik 15G (aldicarb) *(Rhone-Poulenc)*

Rate: 1.5-3 lb. ai/A
10-20 lb. 15G/A

Pest:
nematodes

Remarks: Apply at planting in 12" band in open row. Cover immediately with soil by hilling up 8-10". Plant in center of treated zone. Apply with positive displacement applicators; refer to label. Do not make more than 1 application per crop or exceed 20 lb./A. Do not feed vines to livestock. Refer to label for environmental precautions and soil restrictions table.
Days To Harvest: 120.

Signal Word/Toxicity Class: Danger-Poison/I.
REI: 48 hr.

Thiodan — see endosulfan

VYDATE

Restricted Use Pesticide.

Vydate L (oxamyl) *(DuPont)*

Rate: Preplant:
4-6 lb. ai/A
2-3 gal. L/A
In-furrow:
2-4 lb. ai/A
1-2 gal. L/A

Pest:

root-knot nematode (except javanese)	spiral nematode

Remarks: Sweet potatoes. Preplant: Apply within 1 week of planting; thoroughly incorporate to a depth of 4-6". In-furrow: Apply during planting of slips. Do not apply as transplant water treatment during periods of slow plant growth such as when temperatures fall below 45°F, or crop injury may result. Do not apply more than 3 gal./A per season.
State Restrictions: Do not use in Suffolk and Nassau counties, Long Island, New York. Not for use in California.

Rate: Seed piece dip:
1200-2400 ppm ai/100 gal. water
4-8 pt. L/100 gal. water
Foliar:
0.5 lb. ai/A
2 pt. L/A

Pest:
nematodes

Remarks: Yams. Commercial farming. Seed piece dip: Dip seed pieces into solution for 15 minutes; allow to dry for 24 hr., then plant. Foliar: Apply as supplement to seed piece dip at 2-week intervals. Apply approximately 2 months after planting and continue for up to 12 applications. Do not allow spray to drift away from target crop.
Days To Harvest: 60.
State Restrictions: For use in Puerto Rico only.

Signal Word/Toxicity Class: Danger-Poison/I.
REI: 48 hr.

Table Beets

AZADIRACHTIN

Extracted from seed of neem tree. Insect growth regulator that controls targeted insect larvae when they ingest or come in contact with it, by interfering with the insect's ability to molt and shutting off feeding behavior. Affected larvae slowly starve and die at next molt.

◆ **Azatin XL Plus** *(Thermo Trilogy)*

Rate: 5-21 oz. XL Plus/A

Pest:

aphids	cutworms	leafminers
armyworms	diamondback moth	leafrollers
beet armyworm	fall armyworm	loopers
beetles	flea beetles	Mexican bean beetle
cabbage looper	green peach aphid	psyllids
caterpillars	imported cabbageworm	thrips
Colorado potato beetle	Japanese beetle	webworms
corn earworm	leafhoppers	whiteflies

Remarks: Controls larval and pupal stages of insects. Rate varies by species. Apply when pests first appear; repeat at 7-day intervals. When pest populations are high, use higher rates. Do not apply more that 21 fl. oz./A.
Days To Harvest: 0.

Signal Word/Toxicity Class: Caution.
REI: 4 hr.

◆ **NEEMAZAD 4.5EC or NEEMIX 4.5** *(Thermo Trilogy)*

Rate: 0.125-1 pt. 4.5/A
High Volume:
2.25-4.5 oz. 4.5EC/100 gal. water
Low Volume:
0.5 pt. 4.5EC/A

Pest:

aphids	caterpillars	leafhoppers
armyworms	Colorado potato beetle	potato flea beetle
artichoke plume moth	diamondback moth	psyllids
aster leafhopper	greenhouse whitefly	sweetpotato whitefly
cabbage looper	hornworms	webworms

Remarks: For low volume, apply in sufficient water for adequate coverage. For low insect pressure, apply preventive at 7-day intervals or at first sign of insects; high insect pressure, apply at 3- to 4-day intervals.

Signal Word/Toxicity Class: Warning/II.
REI: 12 hr.

Azatin — see azadirachtin

NOTICE The information on these pages is for preliminary planning — not a guide for use. Be sure to follow the manufacturer's directions, notwithstanding information contained here.

For personal protective equipment and EPA registration numbers, please turn to page 39.

CARBARYL

Clean Crop Sevin 5 Bait *(Platte)*
Rate: 1-2 lb. ai/A
20-40 lb. Bait/A
Pest:
ants, darkling beetle, grasshoppers, armyworm, field crickets, sowbugs, cutworms
Remarks: Apply with ground or air equipment.
Days To Harvest: Beet roots-3; tops-14.

Signal Word/Toxicity Class: Caution/III.
REI: 12 hr.

Drexel Carbaryl 4L *(Drexel)*
Sevin 4F, XLR Plus, 80S, 80WSP *(Rhone-Poulenc)*
Rate: 0.5-2 lb. ai/A
1-4 pt. 4L, 4F, XLR/A
1-4 lb. 50W/A
0.66-2.5 lb. 80S, 80WSP/A
Pest:
armyworms, flea beetles, leafhoppers, aster leafhopper, grasshoppers, stink bugs, corn earworm, harlequin bug, tarnished plant bug, fall armyworm
Remarks: Treat on a 5- to 7-day schedule.
Days To Harvest: Beet roots-3; tops-14.

Signal Word/Toxicity Class: Caution (4F, 4L, XLR); Warning/II (80S, 80WSP).
REI: 12 hr.

Gowan Sevin Bait *(Gowan)*
Rate: 2 lb. ai/A
40 lb. Bait/A
Pest:
armyworms, cutworms, grasshoppers, crickets, darkling beetle, sowbugs
Remarks: Do not use when bees are present.
Days To Harvest: Beet roots-3; tops-14.
State Restrictions: For use in Arizona and California.

Signal Word/Toxicity Class: Caution/III.
REI: 12 hr.

DEADLINE

Packaged as Bullets, M-Ps (mini-pellets), and granules.

Deadline Bullets, M-Ps, Granules (4%) (metaldehyde) *(Pace)*
Rate: 0.8-1 lb. ai/A
20-40 lb. 4%/A
Pest:
slugs, snails
Remarks: Apply every 3-4 weeks during growing season or as needed. Do not apply directly to edible portion of plant.

Signal Word/Toxicity Class: Warning/II (granule); Caution (bullets, M-P).
REI: 12 hr.

Diatect — see pyrethrins

DIAZINON

Organophosphate insecticide. Restricted Use Pesticide due to avian and aquatic toxicity.

D•z•n diazinon AG600 WBC, 50W *(Novartis)*
Drexel Diazinon Insecticide (4EC), D-264 14G *(Drexel)*
Gowan Diazinon 4E, 14G, 50WSB *(Gowan)*
Micro Flo Diazinon AG500, 14G, 50W *(Micro Flo)*
Prentox Diazinon AG500, 50W *(Prentiss)*
◆ **Wilbur-Ellis Diazinon 4 Spray (4E), 14G** *(Wilbur-Ellis)*
Rate: 0.25-0.5 lb. ai/A
0.5-1 pt. 4E, 4EC, AG500/A
6.5-12.75 fl. oz. AG600/A
0.5-1 lb. 50W, 50WSB/A
Pest:
dipterous leafminer

Rate: 1 lb. ai/A
1 qt. 4E, 4EC, AG500/A
25.5 fl. oz. AG600/A
2 lb. 50W, 50WSB/A
7 lb. 14G/A
Pest:
mole crickets

Rate: 2-4 lb. ai/A
4-8 pt. 4E, 4EC, AG500/A
51-102 fl. oz. AG600/A
4-8 lb. 50W, 50WSB/A
14-28 lb. 14G/A
Pest:
cutworms

Rate: 3-4 lb. ai/A
6-8 pt. 4E, 4EC, AG500/A
76.5-102 fl. oz. AG600/A
4-8 lb. 50W, 50WSB/A
21-28 lb. 14G/A
Pest:
wireworms
Remarks: For leafminers, repeat at 7-da intervals as needed. Will not control organophosphate-resistant leafminers. For mole crickets, apply to plant beds or fields 1-2 days before planting and immediately incorporate into top 1-2" soil. For cutworms and wireworms, broadcast just before planting and immediately incorporate into soil. Do not apply more than 5 times per season. Refer to label for use directions and precautions.
Days To Harvest: 14.

Signal Word/Toxicity Class: Caution.
REI: 24 hr.; 12 hr. (Drexel/Gowan/Wilbur-Ellis 14G).

Declare — see methyl parathion

Digon — see dimethoate

Dimate — see dimethoate

DIMETHOATE

Cheminova Dimethoate 4E *(Cheminova)*
Clean Crop Dimethoate 400 *(Platte)*
Dimate 4E *(Terra)*
Drexel Dimethoate 2.67 or 4EC *(Drexel)*
Gowan Dimethoate E267 *(Gowan)*
Helena Dimethoate 4E *(Helena)*
Micro Flo Dimethoate 267 or 4E *(Micro Flo)*
Wilbur-Ellis Dimethoate 267 or Digon 400 *(Wilbur-Ellis)*
Rate: 0.25-0.5 lb. ai/A
0.75-1.5 pt. 2.67, 267, E267/A
0.5-1 pt. 4E, 4EC, 400/A
Pest:
aphids, leafhoppers, leafminers
Remarks: None.
Days To Harvest: 7.

Signal Word/Toxicity Class: Warning/II.
REI: 48 hr.

D•z•n — see diazinon

Kumulus — see sulfur

◆ - **new product** • ai-active ingredient • DF-dry flowable • E/EC-emulsifiable concentrate • EW-emulsion in water
F/FL-flowable • G-granule • L/LC-liquid • SP/WSP-soluble packet • W/WP-wettable powder • WSB-water soluble bag

LANNATE

Restricted Use Pesticide.

Lannate SP or LV (methomyl) *(DuPont)*

Rate: 0.225-0.9 lb. ai/A
0.25-1 lb. 90/A
0.75-3 pt. LV/A

Pest:

beet armyworm, cabbage looper, cucumber beetles, diamondback moth, imported cabbageworm, variegated cutworm

Remarks: Apply in sufficient water to obtain thorough, uniform coverage when insects first appear; repeat at 5- to 7-day intervals or as needed. Rate varies by pest species; refer to label. Do not apply more than 3.6 lb. ai/A per crop or make more than 8 applications per crop.

Days To Harvest: Roots-0; tops-14.

Signal Word/Toxicity Class: Danger-Poison/I.
REI: 48 hr.

MALATHION

Organophosphate insecticide toxic to fish, aquatic invertebrates, aquatic life stages of amphibians, and bees exposed to direct treatment on blooming crops or weeds.

Clean Crop Malathion 57EC *(Platte)*
Drexel Malathion 5EC *(Drexel)*
◆ **Micro Flo Malathion 5EC** *(Micro Flo)*
Prentox 5 lb. Malathion Spray (5EC) *(Prentiss)*

Rate: 1-1.25 lb. ai/A
1.5-2 pt. 5EC, 57EC/A

Pest:

aphids

Remarks: Garden, table. Apply when insects first appear; repeat as needed.

Days To Harvest: If tops are to be used as feed-7.

Signal Word/Toxicity Class: Caution/III (57EC); Warning/II (5EC).
REI: 12 hr.

Clean Crop Malathion 8EC *(Platte)*
◆ **Wilbur-Ellis Malathion 8** *(Wilbur-Ellis)*

Rate: 0.5-1 lb. ai/A
0.5-1 pt. 8EC, 8/A

Pest:

aphids, leafhoppers, Mexican bean beetle, spider mites, thrips

Remarks: Apply when insects first appear and repeat as necessary.

Days To Harvest: 7.

Signal Word/Toxicity Class: Caution/III.
REI: 12 hr.

◆ **Gowan Malathion 8 (8E)** *(Gowan)*

Rate: Garden—seed crop:
1.25 lb. ai/A
1.25 pt. 8E/A

Pest:

Lygus bugs

Rate: Table:
2.5 lb. ai/A
2.5 pt. 8E/A

Pest:

aphids, beet armyworm, blister beetle, flea beetle

Remarks: For Lygus bugs, apply in seedball stage to hard seed stage; repeat as needed.

Days To Harvest: 7.

Signal Word/Toxicity Class: Caution/III.
REI: 12 hr.

Marlate — see methoxychlor

METHOXYCHLOR

Organochlorine pesticide toxic to fish and aquatic invertebrates.

Marlate 50 (WP) *(Kincaid)*
Prentox Methoxychlor 2EC *(Prentiss)*

Rate: 1.3-1.7 lb. ai/A
75-100 fl. oz. 2EC/A
2 Tbs. 50WP/gal. water

Pest:

alfalfa webworm, bean leaf beetle, corn earworm, cucumber beetles, fall armyworm, flea beetles, garden webworm, Japanese beetle, Mexican bean beetle, potato leafhopper

Remarks: Apply at first sign of infestation; repeat at 7- to 14-day intervals as needed.

Days To Harvest: 14.

Signal Word/Toxicity Class: Caution.
REI: 12 hr.

METHYL PARATHION

Organophosphate insecticide toxic to bees exposed to direct treatment or residues on blooming crops or weeds. Restricted Use Pesticide due to very high acute toxicity to humans and birds.

◆ **Declare (4EC)** *(Griffin*
Helena 4 lb. Methyl Parathion (4EC) *(Helena)*

Rate: 0.47-1.18 lb. ai/A
1-2 pt. 4EC/A

Pest:

aphids, armyworms, cabbage looper, flea beetles, leafhoppers, Lygus bugs, mites, plant bugs, stink bugs

Remarks: Rate varies by pest species; refer to label.

Days To Harvest: 20; if tops used for food-60.

Signal Word/Toxicity Class: Danger-Poison/I.
REI: 48 hr.

Wilbur-Ellis Methyl Parathion 4 or 5 Spray *(Wilbur-Ellis)*

Rate: 0.5 lb. ai/A
1 pt. 4/A
0.63 lb. ai/A
1 pt. 5/A

Pest:

aphids, armyworms, flea beetles, leafhoppers, spider mites

Remarks: Do not apply through any type of irrigation system.

Days To Harvest: 15; tops used for food-21.

Signal Word/Toxicity Class: Danger-Poison/I.
REI: 24 (4 Spray); 48 (5 Spray).

M-PEDE

M-Pede (potassium salts of fatty acids) *(Mycogen)*

Rate: 1-2% v/v solution

Pest:

aphids*, broad mite, green peach aphid*, leafhoppers, leafminers, plant bugs, russet mite, spider mites, thrips, whiteflies

*** Tank mix only**

Remarks: Apply at first sign of infestation; repeat at 7- to 14-day intervals. May be tank mixed with other insecticides. Refer to label for use directions and precautions.

Days To Harvest: 0.

Signal Word/Toxicity Class: Warning/II.
REI: 12 hr.

NEEMAZAD — see azadirachtin

NEEMIX — see azadirachtin

7 Vegetables

NOTICE The information on these pages is for preliminary planning — not a guide for use. Be sure to follow the manufacturer's directions, notwithstanding information contained here.

For personal protective equipment and EPA registration numbers, please turn to page 39.

NEMACUR

Restricted Use Pesticide.

Nemacur 3 (fenamiphos) *(Bayer)*
Rate: 4-6 fl. oz. 3/1000 ft. row (any row spacing)
Pest:
cyst nematodes
Remarks: Apply as band treatment before or at seeding. Center-treated band on row using band width of 33%-50% row spacing. Do not make more than 1 application per season. Apply by ground equipment only. Refer to label for use directions and precautions.
Days To Harvest: 90.
State Restrictions: For use in Illinois, Indiana, Michigan, New York, Ohio, and Pennsylvania.

Signal Word/Toxicity Class: Danger/I.
REI: 48 hr.

Omni Supreme Spray — see petroleum oil

PETROLEUM OIL

Petroleum based spray oil. Mode of action is through suffocation of eggs, larvae, and nymphs of insects and adults of soft bodied insects. This mode of action necessitates total spray coverage. Toxic to fish.

Omni Supreme Spray *(Helena)*
Rate: 1-2 gal./100 gal. water
Pest:

aphids	caterpillars	mites
beetles (larvae)	leafminers	

Remarks: None.

Signal Word/Toxicity Class: Caution.
REI: 4 hr.

◆ **SunSpray Ultra-Fine** *(Sun)*
Rate: 1-2 gal./100 gal. water
Pest:

aphids	leafminers	thrips
beetle (larvae)	mites	whiteflies
leafhoppers	spiders	

Remarks: Do not exceed 4 applications per growing season.

Signal Word/Toxicity Class: Caution.
REI: 4 hr.

PYRELLIN (Premix)

Premix containing pyrethrins and rotenone. May be combined with other insecticides to enhance knockdown or as an insect "exciter" to help facilitate contact with other spray residues.

Pyrellin E.C. (0.042 lb./gal. pyrethrins & 0.035 lb./gal. rotenone) *(Webb Wright)*
Rate: 1-2 pt. E.C./A
Pest:

aphids	garden fleahopper	mites
bean beetle	leafhoppers	plant bugs
beet webworm	leafminers	stink bugs
cabbageworm	loopers	thrips
celery leaftier	Lygus bugs	whiteflies

Remarks: Apply at 7-day intervals s or less; repeat as needed.
Days To Harvest: 0.

Rate: 1.5-2 pt. E.C./A
Pest:

Colorado potato beetle	cucumber beetle	flea beetles
	European corn borer	vegetable weevil

Remarks: Make 3-4 applications or more as eggs start to hatch; repeat at 5- to 7-day intervals.
Days To Harvest: 0.

Signal Word/Toxicity Class: Caution/III.
REI: 12 hr.

Pyrenone — see pyrethrins

PYRETHRINS

Contains piperonyl butoxide as a synergist. Diatect also contains silicon dioxide (diatomaceous earth). May be combined in tank mix with other insecticides for quicker and more complete kill where insect resistance may be a problem, and as an "exciter" to flush insects out of hiding and into contact with other spray residues.

Diatect Multipurpose Insecticide II *(Diatect)*
Pyrenone (0.5EC) *(AgrEvo Environmental Health)*
Py-Rin 60-6 EC *(Wilbur-Ellis)*
Rate: 1-6 lb. II/A
1-12 fl. oz. 0.5EC/A
2-16 oz. 6-60 EC/A
Pest: Including but not limited to:

aphids	beet webworm	flea beetles
armyworms	cutworm	leafminers

Remarks: Apply in sufficient water for thorough coverage; repeat as needed. Apply Diatect as dust or wettable powder.
Days To Harvest: 0.

Signal Word/Toxicity Class: Caution.
REI: 12 hr.

Py-Rin — see pyrethrins

Sevin — see carbaryl

SULFUR

◆ **Kumulus DF** *(BASF)*
Rate: 2.4-4 lb. ai/A
3-5 lb. DF/A
Pest:
red spider mite
Remarks: Apply at first indication of problem with or without other pesticides; repeat as necessary. Repeat at 10- to 14-day intervals with or without other pesticides immediately after a rain. Do not apply if temperatures during or within 3 days after application are expected to exceed 90°F. Do not apply through any type of irrigation system. Refer to label for use directions.

Signal Word/Toxicity Class: Caution.
REI: 24 hr.

SunSpray — see petroleum oil

Tomatoes

ABAMECTIN

Natural product produced by soil microorganism, Streptomyces avermitilis. Unrelated to other miticides. Toxic to fish, mammals, aquatic organisms and highly toxic to honey bees. Restricted Use Pesticide.

Agri-Mek 0.15EC *(Novartis)*
Rate: 8-16 fl. oz. 0.15EC/A
Pest:

Colorado potato beetle	leafminers	tomato pinworm
	spider mites	tomato russet mite

Remarks: Apply at first appearance; repeat as necessary. Thorough coverage essential. For tomato pinworm, use high rate. Do not apply in less than 20 gal. water/A. Do not exceed 48 fl. oz./A per growing season. Do not make more than 2 sequential applications. Do not apply through any type of irrigation system or by aircraft.
Days To Harvest: 7.

Signal Word/Toxicity Class: Warning/II.
REI: 12 hr.

◆ - **new product** • ai-active ingredient • DF-dry flowable • E/EC-emulsifiable concentrate • EW-emulsion in water
F/FL-flowable • G-granule • L/LC-liquid • SP/WSP-soluble packet • W/WP-wettable powder • WSB-water soluble bag

Adios — see carbaryl

Admire — see imidacloprid

Agri-Mek — see abamectin

Ambush — see permethrin

ASANA

Pyrethroid insecticide that has both contact and stomach poison activity. Restricted Use Pesticide due to toxicity to fish and aquatic organisms.

Asana XL (esfenvalerate) *(DuPont)*

Rate: 0.015-0.05 lb. ai/A
2.9-9.6 fl. oz. XL/A

Pest:

beet armyworm (aids in control)	grasshoppers	tomato pinworm
cabbage looper	potato aphid	vegetable leafminer
Colorado potato beetle	southern armyworm	western yellowstriped armyworm
cutworms	tobacco hornworm	whiteflies
flea beetles	tomato fruitworm	
	tomato hornworm	

Remarks: Repeat as necessary to maintain control. Rate varies by pest species; refer to label. Do not exceed 0.5 lb. ai/A per season.
Days To Harvest: 1.
State Restrictions: Not recommended on vegetable leafminer in Florida.

- - - - - - - - - - - - - - - -

Signal Word/Toxicity Class: Warning/II.
REI: 12 hr.

AZADIRACHTIN

Extracted from seed of neem tree. Insect growth regulator that controls targeted insect larvae when they ingest or come in contact with it, by interfering with the insect's ability to molt and shutting off feeding behavior. Affected larvae slowly starve and die at next molt.

◆ **Azatin EC or XL** *(Olympic)*
◆ **Azatin XL Plus** *(Thermo Trilogy)*

Rate: 8-21 oz. EC/A
5-21 oz. XL, XL Plus/A

Pest:

aphids	flea beetles	serpentine leafminer
armyworms	green peach aphid	silverleaf whitefly
beet armyworm	imported cabbageworm	southern armyworm
beetles	Japanese beetle	sweetpotato whitefly
cabbage looper	leafhoppers	thrips
caterpillar	leafminers	tobacco hornworm
Colorado potato beetle	leafrollers	tomato fruitworm
cutworms	loopers	tomato pinworm
diamondback moth	potato aphid	webworms
fall armyworm	psyllids	whiteflies

Remarks: Ground cherry, tomato, tomatillo. Controls larval and pupal stages of insects. Rate varies by species. Apply when pests first appear; repeat at 7-day intervals. When pest populations are high, use higher rates. Do not apply more that 21 fl. oz./A.
Days To Harvest: 0.

- - - - - - - - - - - - - - - -

Signal Word/Toxicity Class: Caution (XL); Warning/II (EC).
REI: 4 hr. (XL), 12 hr. (EC).

◆ **NEEMAZAD 4.5EC or NEEMIX 4.5** *(Thermo Trilogy)*

Rate: 0.125-1 pt. 4.5/A
High Volume:
2.25-4.5 oz. 4.5EC/100 gal. water
Low Volume:
0.5 pt. 4.5EC/A

Pest:

armyworms	garden webworm	psyllids
bean leaf beetle	greenhouse whitefly	soybean looper
cabbage looper	hornworms	sweetpotato whitefly
caterpillars	leafminers	tomato fruitworm
Colorado potato beetle	loopers	tomato pinworm
	Mexican bean beetle	weevils
cutworms	pepper weevil	

Remarks: Ground cherries, tomatoes. For low volume, apply in sufficient water for adequate coverage. For low insect pressure, apply preventative at 7-day intervals or at first sign of insects; high insect pressure, at 3- to 4-day intervals.

- - - - - - - - - - - - - - - -

Signal Word/Toxicity Class: Warning/II.
REI: 12 hr.

Azatin — see azadirachtin

AZINPHOS-METHYL

Restricted Use Pesticide (except Guthion Solupak).

Gowan Azinphos-M 50-W *(Gowan)*
Guthion 2L, Solupak (50WSP) *(Bayer)*
◆ **Micro Flo Azinphosmethyl 2EC or 50W** *(Micro Flo)*
Sniper 2-E, 50PVA, 50W *(Platte)*

Rate: 0.375-1.5 lb. ai/A
1.5-6 pt. 2EC, 2-E, 2L/A
0.75-3 lb. 50-W, 50PVA, 50W, 50WSP/A

Pest:

aphids	European corn borer	leafminers
banded cucumber beetle	flea beetles	pinworm
	fruitworms	thrips
Colorado potato beetle	grasshoppers	tuberworm
	green stink bug	whiteflies
corn earworm	hornworms	yellowstriped armyworm
Drosophila	leafhoppers	

Remarks: Apply at 7-day intervals by air or ground in sufficient water for complete coverage. Rate varies by pest species; refer to label. Higher rates should be used where heavy infestations of large worms and pinworms are present. Do not apply more than 4 times per season. Colorado potato beetle has developed resistance in some areas; consult extension service or sales representative for details.
Days To Harvest: Up to 1.5 lb. ai-0; above 1.5 lb. ai-14.
State Restrictions: California labels may vary. Sniper 50PVA not registered in California.

- - - - - - - - - - - - - - - -

Signal Word/Toxicity Class: Danger/I.
REI: 48 hr.; 72 hr. if average rainfall less than 25" per year.

Baythroid — see cyfluthrin

Bravo S — see sulfur premix

BT 320 SULFUR 25 or 50 DUST (Premix)

Bt 320 Sulfur 25 Dust (0.064% Bt & 25% sulfur) or Bt 320 Sulfur 50 Dust (0.064% Bt & 50% sulfur) *(Wilbur-Ellis)*

Rate: 20-40 lb. Dust/A

Pest:
cabbage looper — tomato hornworm

Remarks: Do not mix or dilute with other products. Thorough coverage essential.

- - - - - - - - - - - - - - - -

Signal Word/Toxicity Class: Caution.
REI: 24 hr.

CARBARYL

Carbamate insecticide with broad activity on many insects. Highly toxic to bees.

Adios (13%) *(Micro Flo)*

Rate: 0.065-0.1 lb. ai/A
0.5-0.75 lb. 13%/A

Pest:
cucumber beetle

Remarks: Irish, white potatoes. Do not apply through any type of irrigation system. Do not apply more than 3.75 lb./A per season.
Days To Harvest: 0.
State Restrictions: Not for use in California.

- - - - - - - - - - - - - - - -

Signal Word/Toxicity Class: Caution/III.

NOTICE The information on these pages is for preliminary planning — not a guide for use. Be sure to follow the manufacturer's directions, notwithstanding information contained here.	For personal protective equipment and EPA registration numbers, please turn to page 39.

Clean Crop Sevin 5 Bait *(Platte)*
Rate: 1-2 lb. ai/A
20-40 lb. Bait/A
Pest:

ants	cutworms	grasshoppers
armyworms	darkling beetle	sowbugs
crickets	field crickets	

Remarks: Apply with ground or air equipment.
Days To Harvest: 0.

Signal Word/Toxicity Class: Caution/III.
REI: 12 hr.

Drexel Carbaryl 4L or 50W *(Drexel)*
Sevin 4F, XLR Plus, 80S or 80WSP *(Rhone-Poulenc)*
Rate: 0.5-2 lb. ai/A
1-4 pt. 4L, 4F, XLR/A
1-4 lb. 50W/A
0.66-2.5 lb. 80S, 80WSP/A
Pest:

Colorado potato beetle	flea beetles	tarnished plant bug
cutworms	grasshoppers	tomato fruitworm
European corn borer	lace bugs	tomato hornworm
fall armyworm	leafhoppers	tomato pinworm
	stink bugs	

Remarks: Thorough coverage essential to effectively suppress stink bugs. When disease transmission is suspected, monitor fields following application and re-treat if reinfestation occurs.
Days To Harvest: 0.

Signal Word/Toxicity Class: Caution (4F, 4L, XLR); Warning/II (50W, 80S, 80WSP).
REI: 12 hr.

Gowan Sevin Bait *(Gowan)*
Rate: 2 lb. ai/A
40 lb. Bait/A
Pest:

armyworms	cutworms	grasshoppers
crickets	darkling beetle	sowbugs

Remarks: Apply when pests appear in damaging numbers; repeat at 7- to 14-day intervals if necessary.
Days To Harvest: 1.
State Restrictions: For use in Arizona and California.

Signal Word/Toxicity Class: Caution/III.
REI: 12 hr.

◆ **Wilbur-Ellis Sevin 5 or 10 Dust** *(Wilbur-Ellis)*
Rate: 2 lb. ai/A
40 lb. 5/A
20 lb. 10/A
Pest:

Colorado potato beetle	flea beetles	stink bugs
European corn borer	leafhoppers	tomato fruitworm
fall armyworm	Lygus bugs	tomato hornworm

Remarks: Do not feed treated vines to livestock.
Days To Harvest: 0.
State Restrictions: Sevin 5 not registered in California.

Signal Word/Toxicity Class: Caution.
REI: 12 hr.

CRYOLITE

Kryocide (96%) *(Elf Atochem)*
Rate: 7.7-15.4 lb. ai/A
8-16 lb. 96%/A
Pest:

blister beetle	flea beetles	hornworms
cabbage looper	fruitworms	tomato pinworm
Colorado potato beetle (except CA)		

Remarks: Apply at 7-day intervals as needed. Remove excess residues on edible portions by washing, brushing, or other effective means. Do not exceed 64 lb./A per season. Do not use in combination with lime or compounds containing free lime. Do not chemigate in California.
Days To Harvest: 14.

Signal Word/Toxicity Class: Caution.
REI: 12 hr.

CYFLUTHRIN

Synthetic pyrethroid formulated in an emulsifiable concentrate. Toxic to fish and aquatic organisms. Restricted Use Pesticide.

Baythroid (2EC) *(Bayer)*
Rate: 1.6-2.8 fl. oz. 2EC/A
Pest:

beet armyworm	potato aphid	tomato pinworm
cabbage looper	southern armyworm	variegated cutworm
Colorado potato beetle	stink bugs	western flower thrips
dipterous leafminer	tomato fruitworm	western yellowstriped
European corn borer	tomato hornworm	armyworm

Remarks: Apply in sufficient water for complete foliage coverage by ground only; repeat as needed to maintain control. Up to 16.8 fl. oz./A may be applied per crop season. Allow at least 7 days between applications. Do not apply to crops grown for seed. Colorado potato beetle has developed resistance in some areas; consult agricultural advisor or cooperative extension service for recommendations. Refer to label for spray drift precautions.

Signal Word/Toxicity Class: Danger/I.
REI: 12 hr.

lambda-CYHALOTHRIN

Synthetic pyrethroid insecticide with broad-spectrum activity, formulated as an emulsifiable concentrate. Restricted Use Pesticide due to toxicity to fish and aquatic organisms.

Karate or Warrior (1E) *(ZENECA)*
Rate: 0.015-0.03 lb. ai/A
1.92-3.84 fl. oz. 1E/A
Pest:

aphids*	flea beetles	stink bugs
beet armyworm (1st & 2nd instar)	grasshoppers	tobacco budworm
cabbage looper	hornworm	tomato fruitworm
Colorado potato beetle	leafhoppers	tomato pinworm
cutworms	leafminers*	whiteflies*
European corn borer	meadow spittlebug	yellowstriped armyworm (1st & 2nd instar)
fall armyworm (1st & 2nd instar)	plant bugs	
	southern armyworm (1st & 2nd instar)	

* *Suppression*
Remarks: Not for use on varieties in which mature tomatoes will be less than 1" in diameter (such as cherry tomatoes). Apply as required by scouting, usually at least 5-day intervals. Do not apply more than 2.88 pt./A per season. Refer to label for spray drift precautions and state and county restrictions.
Days To Harvest: 5.

Signal Word/Toxicity Class: Danger/I.
REI: 24 hr.

DANITOL

Danitol 2.4EC (fenpropathrin) *(Valent)*
Rate: 0.2 lb. ai/A
10.66 fl. oz. 2.4EC/A
Pest:

fruitworms	sweetpotato whitefly	yellowstriped armyworm
potato aphid	tomato pinworm	
silverleaf whitefly		

Remarks: Repeat at 7-day intervals as needed. Do not apply more than 2.66 pt./A per season. For aphids, pinworms, and whiteflies, tank mix with Monitor.
Days To Harvest: 3; tank mix-7.

Signal Word/Toxicity Class: Danger/I.
REI: 24 hr.

DEADLINE

Packaged as Bullets, M-Ps (mini-pellets), and granules.

Deadline Bullets, M-Ps, Granules (4%) (metaldehyde) *(Pace)*
Rate: 0.8-1 lb. ai/A
20-40 lb. 4G/A
Pest:

slugs	snails

◆ - **new product** • ai-active ingredient • DF-dry flowable • E/EC-emulsifiable concentrate • EW-emulsion in water
F/FL-flowable • G-granule • L/LC-liquid • SP/WSP-soluble packet • W/WP-wettable powder • WSB-water soluble bag

Remarks: Apply every 3-4 weeks during growing season or as needed. Do not apply directly to edible portion of plant.

Signal Word/Toxicity Class: Warning/II (granule); Caution (bullets, M-P).
REI: 12 hr.

Diatect — see pyrethrins

DIAZINON

Organophosphate insecticide. Restricted Use Pesticide due to avian and aquatic toxicity.

D•z•n diazinon AG600 WBC, 50W *(Novartis)*
Drexel Diazinon Insecticide (4EC), D-264 14G *(Drexel)*
Gowan Diazinon 4E, 14G, 50WSB *(Gowan)*
Helena Diazinon AG500 *(Helena)*
Micro Flo Diazinon AG500, 14G, 50W *(Micro Flo)*
Prentox Diazinon AG500, 50W *(Prentiss)*
◆ Wilbur-Ellis Diazinon 4 Spray (4E), 14G *(Wilbur-Ellis)*

Rate: 0.25-0.75 lb. ai/A
0.5-1.5 pt. 4E, 4EC, AG500/A
0.5-1.5 lb. 50W, 50WSB/A
6.5-19 fl. oz. AG600/A

Pest:
aphids, banded cucumber beetle, beet armyworm, dipterous leafminer, fall armyworm, southern armyworm, vinegar flies

Rate: 1 lb. ai/A
2 pt. 4E, 4EC, AG500/A
2 lb. 50W, 50WSB/A
25.5 fl. oz. AG600/A
7 lb. 14G/A

Pest:
mole crickets

Rate: 2-4 lb. ai/A
4-8 pt. 4E, 4EC, AG500/A
4-8 lb. 50W, 50WSB/A
51-102 fl. oz. AG600/A
14-28 lb. 14G/A

Pest:
cutworms

Rate: 3-4 lb. ai/A
6-8 pt. 4E, 4EC, AG500/A
6-8 lb. 50W, 50WSB/A
76.5-102 fl. oz. AG600/A
21-28 lb. 14G/A

Pest:
wireworms

Remarks: Rate varies by pest species. Apply when aphids and leafminers appear; repeat at 7-day intervals as needed. Will not control organophosphate-resistant leafminers. For mole crickets, apply to plant beds 1-2 days before sowing of seed or transplanting and immediately incorporate into top 1-2" soil. For cutworms and wireworms, broadcast just before planting and immediately incorporate into soil. Do not apply more than 5 times per season. Refer to label for directions and precautions.
Days To Harvest: 1.

Signal Word/Toxicity Class: Caution/III.
REI: 24 hr.; 12 hr. (Drexel/Gowan/Wilbur-Ellis 14G).

DICOFOL

◆ Griffin Dicofol 4E *(Griffin)*
Kelthane MF (4E) *(Rohm and Haas)*

Rate: 0.38-0.75 lb. ai/A
0.75-1.5 pt. 4E/A

Pest:
Pacific mite, tomato russet mite, tropical mite, twospotted spider mite

Remarks: Apply at first sign of mite buildup. Do not make more than 2 applications or exceed 1.6 pt. per year. Do not feed treated plants, forage, or crop residues to livestock. Do not apply through any type of irrigation system.
Days To Harvest: 2.

Signal Word/Toxicity Class: Caution.
REI: 12 hr.

Digon — see dimethoate

Dimate — see dimethoate

DIMETHOATE

Cheminova Dimethoate 4E *(Cheminova)*
Clean Crop Dimethoate 400 (4EC) *(Platte)*
Digon 400 (4EC) *(Wilbur-Ellis)*
Dimate 4E *(Terra)*
Drexel Dimethoate 2.67 or 4EC *(Drexel)*
Gowan Dimethoate E267 *(Gowan)*
Helena Dimethoate 4E *(Helena)*
◆ Micro Flo Dimethoate 267 or 4E *(Micro Flo)*

Rate: 0.25-0.5 lb. ai/A
0.75-1.5 pt. 2.67, 267, E267/A
0.5-1 pt. 4E, 4EC/A

Pest:
aphids, leafhoppers, leafminers

Remarks: None.
Days To Harvest: 7.

Signal Word/Toxicity Class: Warning/II.
REI: 48 hr.

DI-SYSTON

Restricted Use Pesticide.

Di-Syston 8 (disulfoton) *(Bayer)*

Rate: 1-3 lb. ai/A
1-3 pt. 8/A (38" row)

Pest:
aphids, Colorado potato beetle (early season reduction), flea beetles, leafhoppers, leafminers, mites

Remarks: Transplant seed beds: Apply evenly over bed area in water emulsion as broadcast spray prior to seeding. Work into top 2-3" of soil and plant seed in normal manner. Field applications: Apply in water emulsion or with liquid fertilizer as an injection on each side of seed furrow or transplanted row at planting time. Do not apply directly to seed. Allow minimum of 21 days between applications. Refer to label for additional information and special precautions.
Days To Harvest: 30.

Signal Word/Toxicity Class: Danger/I.
REI: 48 hr.

Drexel Suffa — see sulfur

D•z•n — see diazinon

ENDOSULFAN

Toxic to fish, birds, and other wildlife.

Drexel Endosulfan 3EC or 50WP *(Drexel)*
Gowan Endosulfan 3EC or 50WSB *(Gowan)*
◆ Micro Flo Endosulfan 3EC or 50WP *(Micro Flo)*
Phaser 3EC or 50WSB *(AgrEvo USA)*
Thiodan 3EC or 50WP *(FMC)*

Rate: 0.5-1 lb. ai/A
0.66-1.33 qt. 3EC/A
1-2 lb. 50WP, 50WSB/A

Pest:
blister beetle, cabbage looper, Colorado potato beetle, flea beetle, green peach aphid, hornworms, stink bug, tomato fruitworm, tomato russet mite, whiteflies, yellowstriped armyworm

Remarks: Field, greenhouse. Apply when insects appear or feeding is noticed. Do not exceed 3 lb. ai/A per year or 6 applications per year. Rate varies by pest species; refer to label.
Days To Harvest: 2.

Signal Word/Toxicity Class: Danger-Poison/I.
REI: 24 hr.

NOTICE	The information on these pages is for preliminary planning — not a guide for use. Be sure to follow the manufacturer's directions, notwithstanding information contained here.	For personal protective equipment and EPA registration numbers, please turn to page 39.

7 Vegetables

Fyfanon — see malathion

Guthion — see azinphos-methyl

IMIDACLOPRID

Admire 2F *(Bayer)*
Rate: 16-24 fl. oz. 2F/A
Pest:
aphids | flea beetles | sweetpotato whitefly
Colorado potato beetle | silverleaf whitefly | thrips
Remarks: Tomatoes, tomatillos. Apply direct seed or transplant. Do not apply to vegetables grown for seed. Use high rate where infestations occur later in crop development, or where pest pressure is continuous. Do not apply more than 24 fl. oz./A per year as soil application.
Days To Harvest: 21.

Signal Word/Toxicity Class: Caution.
REI: 12 hr.

Provado 1.6 *(Bayer)*
Rate: 3.75 fl. oz. 1.6/A
Pest:
aphids | silverleaf whitefly | sweetpotato whitefly
Colorado potato beetle
Remarks: Tomatoes, tomatillos. For aphids and Colorado potato beetle, good coverage of foliage necessary for optimal control. Apply before damaging population becomes established. For whiteflies, apply at first appearance; repeat at 5- to 7-day intervals until no longer in crop. Do not apply more than 18.75 fl. oz. per year.
Days To Harvest: 0.

Signal Word/Toxicity Class: Caution.
REI: 12 hr.

Karate — see lambda-cyhalothrin

Kelthane — see dicofol

Kryocide — see cryolite

Kumulus — see sulfur

LANNATE

Restricted Use Pesticide.

Lannate (90) or LV (methomyl) *(DuPont)*
Rate: 0.225-0.9 lb. ai/A
0.25-1 lb. 90/A
0.75-3 pt. LV/A
Pest:
aphids | hornworms | southern armyworm
armyworms | loopers | tomato fruitworm
beet armyworm | pinworms | variegated cutworm
fall armyworm
Remarks: Apply in sufficient water to obtain thorough, uniform coverage when insects first appear; repeat at 5- to 7-day intervals or as needed. Rate varies by pest species; refer to label. Do not apply more than 7.2 lb. ai/A per crop or make more than 16 applications per crop.
Days To Harvest: 1.

Signal Word/Toxicity Class: Danger-Poison/I.
REI: 48 hr.

MALATHION

Organophosphate insecticide toxic to fish, aquatic invertebrates, aquatic life stages of amphibians, and bees exposed to direct treatment on blooming crops or weeds.

Clean Crop Malathion 57EC *(Platte)*
Rate: 0.63-1.9 lb. ai/A
1-3 pt. 57EC/A
Pest:
aphids | tomato russet mite

Rate: 1.6-1.9 lb. ai/A
2.5-3 pt. 57EC/A
Pest:
Drosophila flies

Rate: California:
3.5 lb. ai/A
5.5 pt. 57EC/A
Pest:
armyworms | tomato fruitworm
Remarks: Apply when insects first appear and repeat as necessary.
Days To Harvest: 1; California-3.

Signal Word/Toxicity Class: Caution/III.
REI: 12 hr.

◆ **Clean Crop Malathion 8EC** *(Platte)*
Fyfanon (8E) *(Helena)*
◆ **Gowan Malathion 8 (8E)** *(Gowan)*
Rate: 1.5 lb. ai/A
1.5 pt. 8E, 8EC/A
Pest:
aphids | spider mites

Rate: 2 lb. ai/A
2 pt. 8E, 8EC/A
Pest:
Drosophila flies
Remarks: Apply full coverage to fruit and foliage.
Days To Harvest: 1; rates over 2 pt.-5; California higher rates-3.

Signal Word/Toxicity Class: Caution/III.
REI: 12 hr.

◆ **Wilbur-Ellis Malathion 8 Spray** *(Wilbur-Ellis)*
Rate: 0.5-1 lb. ai/A
0.5-1 pt. 8/A
Pest:
aphids | leafhoppers | spider mites
false chinch bugs

Rate: California:
3.5 lb. ai/A
3.5 pt. 8/A
Pest:
tomato fruitworm
Remarks: Apply in minimum 5 gal. by air; 10 gal. by ground.
Days To Harvest: 1; tomato fruitworm-5.

Signal Word/Toxicity Class: Caution/III.
REI: 12 hr.

Marlate— see methoxychlor

METHAMIDOPHOS

Restricted Use Pesticide.

Monitor 4 *(Bayer)*
Monitor 4 Spray *(Valent)*
Rate: 0.25-0.75 lb. ai/A
0.5-1.5 pt. 4/A
Pest: SLN-Georgia:
aphids | spider mites | thrips

Rate: 0.75-1 lb. ai/A
1.5-2 pt. 4/A
Pest: SLN-Arkansas California, Delaware, Indiana, Maryland, New Jersey, Virginia:
aphids | fruitworms | stink bugs
beet armyworm | hornworms | thrips (except CA)
black cutworm | leafminers | variegated cutworm
cabbage looper

SLN-Tennessee:
aphids | fruitworms | stink bugs
beet armyworm | hornworms | tomato pinworm
cabbage looper | leafminers | whiteflies
cutworms

◆ - **new product** • ai-active ingredient • DF-dry flowable • E/EC-emulsifiable concentrate • EW-emulsion in water
F/FL-flowable • G-granule • L/LC-liquid • SP/WSP-soluble packet • W/WP-wettable powder • WSB-water soluble bag

SLN-Indiana, North Carolina, Ohio—Fresh fruit only:

aphids	Colorado potato beetle (IN, OH)	stink bugs (NC)
beet armyworm (NC, OH)	fruitworms	tomato pinworm (NC)
black cutworm (NC, OH)	hornworms	variegated cutworm (IN, OH)
cabbage looper	leafminers	whiteflies (NC)

SLN-Florida—Fresh fruit only:

aphids	leafminers	tomato pinworms
fruitworms	sweetpotato whitefly	

SLN-South Carolina—Fresh fruit only:

leafminers

SLN-Texas—Fresh fruit only:

aphids	leafminers	thrips
beet armyworm	stink bugs	variegated cutworm
cabbage looper	sweetpotato whitefly (suppression)	
fruit hornworm		

Rate: 0.5-0.75 lb. ai/A
1-1.5 pt. 4/A

Pest: SLN-Puerto Rico—Fresh market only:

fruitworms	leafminers	pinworms
granulate cutworm		

Remarks: Refer to label for use directions and precautions.

Signal Word/Toxicity Class: Danger/I.
REI: 48 hr.

METHOXYCHLOR

Organochlorine pesticide toxic to fish and aquatic invertebrates.

Marlate 50 (WP) *(Kincaid)*
Prentox Methoxychlor 2EC *(Prentiss)*

Rate: 1.3-1.7 lb. ai/A
75-100 fl. oz. 2EC/A
2 Tbs. 50WP/gal. water

Pest:

blister beetle	fall armyworm	leafhoppers
Colorado potato beetle	flea beetles	

Remarks: Apply at first sign of infestation; repeat at 7- to 14-day intervals as needed.
Days To Harvest: 7.

Signal Word/Toxicity Class: Caution.
REI: 12 hr.

METHYL PARATHION

Organophosphate insecticide toxic to bees exposed to direct treatment or residues on blooming crops or weeds. Penncap-M is a microencapsulated formulation. Restricted Use Pesticide due to very high acute toxicity to humans and birds.

Penncap-M (2FL) *(Elf Atochem)*

Rate: 0.25-0.5 lb. ai/A
1-2 pt. 2FL/A
Pest:
thrips*

Rate: 0.5-0.75 lb. ai/A
2-3 pt. 2FL/A
Pest:
stink bugs

Rate: 0.5-1 lb. ai/A
2-4 pt. 2FL/A
Pest:

aphids* (suppression)	flea beetles	leafminers
European corn borer*	grasshoppers*	Lygus bugs*
	leafhoppers*	true armyworm*

Rate: 0.75-1 lb. ai/A
3-4 pt. 2FL/A
Pest:
Japanese beetle*

Rate: 1 lb. ai/A
4 pt. 2FL/A
Pest:

Colorado potato beetle	cutworms*	tomato fruitworm
	hornworms	tomato pinworm

Except CA

Remarks: Do not apply to cherry tomatoes. Do not use chemigation in California.
Days To Harvest: 15.

Signal Word/Toxicity Class: Warning/II.
REI: 48 hr.

Microthiol Special — see sulfur

Monitor — see methamidophos

M-PEDE

M-Pede (potassium salts of fatty acids) *(Mycogen)*

Rate: 1-2% v/v solution
Pest:

aphids*	leafminers	spider mites
broad mite	plant bugs	thrips
green peach aphid*	russet mite	whiteflies
leafhoppers		

** Tank mix only*

Remarks: Apply at first sign of infestation; repeat at 7- to 14-day intervals. May be tank mixed with other insecticides. Refer to both labels for use directions and precautions.
Days To Harvest: 0.

Signal Word/Toxicity Class: Warning/II.
REI: 12 hr.

NEEMAZAD — see azadirachtin

NEEMIX — see azadirachtin

Omni Supreme Spray — see petroleum oil

Penncap-M — see methyl parathion

PERMETHRIN

Restricted Use Pesticide

Ambush 2E or 25W *(ZENECA)*
Pounce 3.2 EC, 25WP or WSB *(FMC)*

Rate: 0.05-0.2 lb. ai/A
3.2-12.8 fl. oz. 2E/A
3.2-12.8 oz. 25W, 25WP/A
2-8 oz. 3.2 EC/A
1-2 bags WSB/A

Pest:

beet armyworm	granulated cutworm	tomato fruitworm
cabbage looper	hornworm	tomato pinworm
Colorado potato beetle	southern armyworm	vegetable leafminer

Remarks: Do not apply to cherry tomatoes or other varieties which produce fruit less than 1" in diameter. Apply by air or ground in sufficient water to obtain uniform coverage. Do not apply more than 1.2 lb. ai/A per season.
Days To Harvest: 0.

Signal Word/Toxicity Class: Caution (3.2 EC); Warning/II (2E, 25W, 25WP, WSB).
REI: 24 hr.

PETROLEUM OIL

Petroleum based spray oil. Mode of action is through suffocation of eggs, larvae, and nymphs of insects and adults of soft bodied insects. This mode of action necessitates total spray coverage. Toxic to fish.

Omni Supreme Spray *(Helena)*

Rate: 1-2 gal./100 gal. water
Pest:

aphids	caterpillars	mites
beetles (larvae)	leafminers	

NOTICE The information on these pages is for preliminary planning — not a guide for use. Be sure to follow the manufacturer's directions, notwithstanding information contained here.

For personal protective equipment and EPA registration numbers, please turn to page 39.

Remarks: None.

Signal Word/Toxicity Class: Caution.
REI: 4 hr.

SAF-T-SIDE *(Brandt Consolidated)*
◆ **Sumulsion** *(Drexel)*
◆ **SunSpray Ultra-Fine** *(Sun)*
Rate: 1-2 gal./100 gal. water
Pest:

aphids	leafminers	thrips
beetle (larvae)	mites	whiteflies
leafhoppers	spiders	

Remarks: Apply as needed. May be tank mixed with most commonly used insecticides and fungicides. Do not use in combination with or immediately before or after spraying with fungicides such as captan, Dyrene, folpet, Morestan, or any sulfur products. Do not use with carbaryl or dimethoate.

Signal Word/Toxicity Class: Caution.
REI: 4 hr.

Phaser — see endosulfan

Pounce — see permethrin

Provado — see imidacloprid

PYRELLIN (Premix)

Premix containing pyrethrins and rotenone. May be combined with other insecticides to enhance knockdown or as an insect "exciter" to help facilitate contact with other spray residues.

Pyrellin E.C. (0.042 lb./gal. pyrethrins & 0.035 lb./gal. rotenone) *(Webb Wright)*
Rate: 1-2 pt. E.C./A
Pest:

aphids	Japanese beetle	mites
bean beetle	leafhoppers	plant bugs
cabbageworms	leafminers	stink bugs
garden fleahopper	loopers	thrips
harlequin bug	Lygus bugs	whiteflies

Remarks: Apply at 7-day intervals or less; repeat as needed.
Days To Harvest: 0.

Rate: 1.5-2 pt. E.C./A
Pest:

Colorado potato beetle	cucumber beetle	flea beetles
	European corn borer	vegetable weevil

Remarks: Make 3-4 applications or more as eggs start to hatch; repeat at 5- to 7-day intervals.
Days To Harvest: 0.

Signal Word/Toxicity Class: Caution/III.
REI: 12 hr.

Pyrenone — see pyrethrins

PYRETHRINS

Contains piperonyl butoxide as a synergist. Diatect also contains silicon dioxide (diatomaceous earth). May be combined in tank mix with other insecticides for quicker and more complete kill where insect resistance may be a problem, and as an "exciter" to flush insects out of hiding and into contact with other spray residues.

Diatect Multipurpose Insecticide II *(Diatect)*
Pyrenone (0.5EC) *(AgrEvo Environmental Health)*
Py-Rin 60-6 EC *(Wilbur-Ellis)*
Rate: 1-6 lb. II/A
1-12 fl. oz. 0.5EC/A
2-16 oz. 6-60 EC/A
Pest: Including but not limited to:

aphids	leafminers	stink bugs
caterpillars	loopers	tarnished plant bug
Colorado potato beetle	Lygus bugs	thrips
cutworms	mites	tomato fruitworm
European corn borer	plant bugs	vegetable weevil
flea beetles	psyllids	whiteflies
green peach aphid		

Remarks: Apply in sufficient water for thorough coverage; repeat as needed. Apply Diatect as dust or wettable powder.
Days To Harvest: 0.

Signal Word/Toxicity Class: Caution.
REI: 12 hr.

Py-Rin — see pyrethrins

SAF-T-SIDE — see petroleum oil

Sevin — see carbaryl

Sniper — see azinphos-methyl

SULFUR

Microthiol and Thiolux are micronized sulfurs.

Drexel Suffa 6 or 8, Sulfur 90W *(Drexel)*
Rate: 3-6 lb. ai/A
4-8 pt. 6/A
3-6 pt. 8/A
3.3-6.6 lb. 90W/A
Pest:
tomato russet mite
Remarks: Use as full coverage spray when insects first appear.

Signal Word/Toxicity Class: Caution.
REI: 12 hr. (Suffa 8); 24 hr. (Suffa 6, Sulfur 90W).

◆ **Kumulus DF** *(BASF)*
Thiolux (DF) *(Platte)*
Rate: 8-16 lb. ai/A
10-20 lb. DF/A
Pest:
russet mite
Remarks: Thorough coverage required. Do not apply if temperatures during or within 3 days after application are expected to exceed 90°F. Refer to label for use directions.
State Restrictions: Not registered in California.

Signal Word/Toxicity Class: Caution.
REI: 24 hr.

Microthiol Special (80W) *(Elf Atochem)*
Rate: 4-8 lb. ai/A
5-10 lb. 80W/A
Pest:

tomato russet mite	twospotted spider mite (except CA)

Remarks: Apply when mites first appear; repeat as needed. Thorough coverage required. Do not chemigate in California.

Signal Word/Toxicity Class: Caution.
REI: 24 hr.

THAT (6FL) or THAT Big 8 (8FL) *(Stoller)*
Rate: 3-6 lb. ai/A
4-8 pt. 6FL/A
2.5-5 lb. ai/A
2.5-5 pt. 8FL/A
Pest:
tomato russet mite
Remarks: Rate depends on size of plants.

Signal Word/Toxicity Class: Caution.
REI: 24 hr.

◆ - **new product** • ai-active ingredient • DF-dry flowable • E/EC-emulsifiable concentrate • EW-emulsion in water
F/FL-flowable • G-granule • L/LC-liquid • SP/WSP-soluble packet • W/WP-wettable powder • WSB-water soluble bag

SULFUR (Premixes)

Bravo S (27.25% sulfur & 19.15% chlorothalonil) *(Helena)*

Rate: 8.5 pt./A

Pest:

tomato russet mite

Remarks: Under severe conditions, use high rate and short interval. Refer to label for use directions.

Signal Word/Toxicity Class: Caution.
REI: 48 hr.

Sumulsion — see petroleum oil

SunSpray — see petroleum oil

THAT or THAT Big 8 — see sulfur

Thiodan — see endosulfan

Thiolux — see sulfur

TRIGARD

Triazine compound used as insect growth regulator for control of leafminers. Ineffective against other insect species. Toxic to bees. Supplemental label: must be in possession of user at time of pesticide application.

Trigard (WSP) (cyromazine) *(Novartis)*

Rate: 0.125 lb. ai/A
0.167 lb. WSP/A (1 packet)

Pest:

leafminers

Remarks: Apply foliar spray in a minimum 50 gal. water by ground when leafminers first appear; repeat at 7-day intervals or as necessary. Do not make more than 6 applications per crop. Do not apply through any type of irrigation system.

Days To Harvest: 7.

Crop Rotation: All crops on label may be planted following harvest of treated crop. Do not rotate to any other crop except sweet corn or radishes. Do not plant sweet corn or radishes within 3 months after last application.

Signal Word/Toxicity Class: Caution.
REI: 12 hr.

VYDATE

Restricted Use Pesticide.

Vydate L (oxamyl) *(DuPont)*

Rate: 0.5-1 lb. ai/A
2-4 pt. L/A

Pest:

aphids	root-knot nematode (except javanese)	sting nematode
Colorado potato beetle	serpentine leafminer (except *Liriomyza trifolii*)	stubby root nematode
nematodes		stunt nematode
reniform nematode		
ring nematode		

Remarks: Apply in sufficient water to obtain uniform coverage when insects first appear; repeat at 5- to 7-day intervals or as needed. Rates vary by infestation level. For nematodes, apply as soon as plants become established; repeat at 1- to 2-week intervals throughout season. Under high nematode populations, other effective soil treatments at or before planting may be necessary, followed by foliar sprays of Vydate L to extend and maintain control. Supplemental sprays should begin when nematode populations start to recover. Do not apply more than 6 gal./A per season.

Days to Harvest: 1.

Signal Word/Toxicity Class: Danger-Poison/I.
REI: 48 hr.

Warrior — see *lambda*-cyhalothrin

Watercress

AZADIRACHTIN

Extracted from seed of neem tree. Insect growth regulator that controls targeted insect larvae when they ingest or come in contact with it, by interfering with the insect's ability to molt and shutting off feeding behavior. Affected larvae slowly starve and die at next molt.

◆ **Azatin XL Plus** *(Thermo Trilogy)*

Rate: 5-21 oz. XL Plus/A

Pest:

aphids	leafhoppers	thrips
green peach aphid	leafminers	whiteflies

Remarks: Rate varies by species. Apply when pests first appear; repeat at 7-day intervals. When pest populations are high, use higher rates. Do not apply more that 21 fl. oz./A.

Days To Harvest: 0.

Signal Word/Toxicity Class: Caution.
REI: 4 hr.

Azatin — see azadirachtin

MALATHION

Organophosphate insecticide toxic to fish, aquatic invertebrates, aquatic life stages of amphibians, and bees exposed to direct treatment on blooming crops or weeds.

Clean Crop Malathion 8EC *(Platte)*
◆ **Wilbur-Ellis Malathion 8** *(Wilbur-Ellis)*

Rate: 0.5-1 lb. ai/A
0.5-1 pt. 8EC, 8/A

Pest:

aphids	spider mites	thrips
leafhoppers		

Remarks: Apply when insects first appear and repeat as necessary.

Days To Harvest: 7.

Signal Word/Toxicity Class: Caution/III.
REI: 12 hr.

M-PEDE

Contact insecticide and fungicide for control of soft-bodied insects and mites.

M-Pede (potassium salts of fatty acids) *(Mycogen)*

Rate: 1-2% v/v solution

Pest:

aphids*	leafminers	spider mites
broad mite	plant bugs	thrips
green peach aphid*	russet mite	whiteflies
leafhoppers		

* ***Tank mix only***

Remarks: Apply at first sign of infestation; repeat at 7- to 14-day intervals. May be tank mixed with other insecticides. Refer to both labels for use directions and precautions.

Days To Harvest: 0.

State Restrictions: Not for use in California.

Signal Word/Toxicity Class: Warning/II.
REI: 12 hr.

7 Vegetables

NOTICE The information on these pages is for preliminary planning — not a guide for use. Be sure to follow the manufacturer's directions, notwithstanding information contained here.	For personal protective equipment and EPA registration numbers, please turn to page 39.

Fungicides

Artichokes

SULFUR

Micro Flo Sulfur (6L) *(Micro Flo)*
Terra Sulfur (6L) *(Terra)*
Rate: 4-5.66 gal. 6L/A
Disease:
leaf spot
Time: Apply when disease appears; repeat at 7- to 10-day intervals.
Remarks: Globe artichokes. Do not use Terra Sulfur within 2 weeks of oil application. Some plants are susceptible to injury from sulfur under certain climatic conditions. Do not make sulfur applications at excessively high temperatures. Refer to label for use directions.

Signal Word/Toxicity Class: Caution.
REI: 24 hr.

Asparagus

Aliette — see fosetyl-aluminum

Brom-O-Gas — see methyl bromide

Dithane — see mancozeb

FOSETYL-ALUMINUM

Aliette (WDG, WSP) *(Rhone-Poulenc)*
Rate: 4 lb. ai/A
5 lb. WDG, WSP/A
Disease:
asparagus spear slime crown rot
Time: Apply once per season to fully expanded ferns. Do not apply to ferns that are beginning to senesce.
Remarks: Thorough coverage is required. Refer to label for use directions.
Days To Harvest: 110.
State Restrictions: For use in California.

Signal Word/Toxicity Class: Caution.
REI: 12 hr.

Kumulus — see sulfur

MANCOZEB

Dithane (DF, F-45, M-45, WSP) *(Rohm and Haas)*
Manzate 200DF *(Griffin)*
Penncozeb 75DF or 80WP *(Elf Atochem)*
Rate: 0.75 lb. ai/100 gal.
1 lb. DF, 200DF/100 gal.
0.8 lb. ai/100 gal.
1 lb. M-45, WSP/100 gal.
1.6 pt. F-45, ST/100 gal.
Disease:
crown rot
Remarks: Place loosely packed crowns into burlap bag; soak using gentle agitation in solution for 5 minutes. Remove bag, drain well, plant crowns as soon as possible. A tank large enough to hold a single burlap bag will treat 2 bags of crowns. Clean dipping suspension should then be prepared in a clean tank. Dirty crowns should be prewashed to remove excess soil. Refer to label for use directions.

Rate: 1.5 lb. ai/A
2 lb. DF, 200DF, 75DF/A
1.6 lb. ai/A
2 lb. M-45, WSP, 80WP/A
3.2 pt. F-45/A
Disease:
Cercospora leaf spot rust
Time: Apply when disease first appears; repeat at 10-day intervals (4 applications usually sufficient).
Remarks: Apply only on ferns after spears have been harvested. Do not use any other product containing a different EBDC ai during the same growing season when using Manzate. Refer to label for use directions.
Days To Harvest: 180; Arizona, California-120.
State Restrictions: Refer to label for California precautions.

Signal Word/Toxicity Class: Caution.
REI: 24 hr.

Manzate — see mancozeb

MEFENOXAM

Ridomil Gold EC or WSP *(Novartis)*
Rate: 0.5 lb. ai/A
1 pt. EC/A
0.45 lb. ai/A
1 lb. WSP/A
Disease:
crown rot spear rot
Time: For cutting beds, apply 30-60 days before first cutting, make another application just before harvesting. For new plantings, apply after planting seedlings or after covering one-year-old crowns.
Remarks: Do not plant any crop which is not registered for use with mefenoxam in treated soil for 12 months.
Days To Harvest: 1.

Signal Word/Toxicity Class: Warning/II (EC); Caution (WSP).
REI: 48 hr.

METHYL BROMIDE

2% Chloropicrin added as warning agent. Restricted Use Pesticide. Classified as an ozone depleter under the Clean Air Act. Use is to be completely eliminated starting Jan. 1, 2001.

Brom-O-Gas *(Great Lakes Chemical)*
Rate: Injected with chisels + covered with tarp:
180-240 lb./A
Raised tarp with topical application:
0.4-0.55 lb./100 ft.
Disease:
Fusarium, oak root fungus, Phytophthora, Plasmodiophora, Pythium, Rhizoctonia, Sclerotinia, Sclerotium
Time: Preplant.
Remarks: Do not fumigate when soil temperature is below 50°F. For best results, fumigate when temperature is 60°-80°F at depth of 6". Refer to label for use directions.
State Restrictions: Brom-O-Gas not for use in Arizona and not for use in California for oak root fungus (Armillaria root rot) control.

Signal Word/Toxicity Class: Danger/I.

◆ - **new product** • ai-active ingredient • DF-dry flowable • E/EC-emulsifiable concentrate • EW-emulsion in water
F/FL-flowable • G-granule • L/LC-liquid • SP/WSP-soluble packet • W/WP-wettable powder • WSB-water soluble bag

Penncozeb — see mancozeb

Redball — see sulfur

Ridomil Gold — see mefenoxam

SULFUR

Drexel Suffa (6FL), Sulfa 80 (80DF) or Sulfur (90W) *(Drexel)*
Rate: 9-18 lb. ai/A
12-14 pt. 6FL/A
10-20 lb. 90W/A
8-24 lb. ai/A
10-30 lb. 80DF/A
Disease:
rust
Time: Apply after cutting stops; repeat at 7- to 10-day intervals throughout season. During fern period, apply at 2- to 3-week intervals or with a regular insect program.
Remarks: Do not use within 2 weeks of oil application (4 weeks 6FL). Some plants are susceptible to injury from sulfur under certain climatic conditions. Do not make sulfur applications at excessively high temperatures. Do not apply this product through any type of irrigation system. Refer to label for use directions.

Signal Word/Toxicity Class: Caution.
REI: 24 hr.

Kumulus DF *(BASF)*
Micro Flo Sulfur (6L) *(Micro Flo)*
Terra Sulfur (6L) *(Terra)*
Thiolux (DF) *(Platte)*
Rate: 8-24 lb. ai/A
1.33-4 gal. 6L/A
10-30 lb. DF/A
Disease:
rust
Time: Apply after cutting stops; repeat at 7- to 10-day intervals through season.
Remarks: Do not use within 2 weeks of oil application. Do not apply Kumulus through any type of irrigation system. Refer to label for use directions.
State Restrictions: Kumulus not registered in California.

Signal Word/Toxicity Class: Caution.
REI: 24 hr.

Redball Superfine Wettable Sulphur (WS) or Commercial Dusting Sulphur (DS) *(International Sulphur)*
Rate: 6-16 lb. WS/A
10-30 lb. DS/A
Disease:
rust
Time: Apply after cutting stops; repeat at 7- to 14-day intervals (WS), 7- to 10-day intervals (DS).
Remarks: Apply in sufficient water for thorough coverage. Do not apply when temperatures exceed 90°F. Refer to label for use directions.

Signal Word/Toxicity Class: Caution.
REI: 24 hr.

SULFUR (Premix)

Top Cop with Sulfur (6.25 lb./gal. sulfur & 1 lb./gal. tribasic copper sulfate) *(Stoller)*
Rate: 2 qt./A
Disease:
rust
Time: Apply when disease first appears; repeat at 10-day intervals for 4 applications.
Remarks: Refer to label for use directions.

Signal Word/Toxicity Class: Caution.
REI: 24 hr.

TERR-O-GAS (Premix)

Classified as an ozone depleter under the Clean Air Act. Use to be eliminated starting Jan. 1, 2001. Restricted Use Pesticide.

Terr-O-Gas (67% methyl bromide & 33% chloropicrin)
(Great Lakes Chemical)
Rate: Nontarp:
270-355 lb./A
Disease:
Fusarium, Pythium, Sclerotinia, Phytophthora, Rhizoctonia, Sclerotium, Plasmodiophora
Time: Apply up to 2 weeks prior to planting when soil moisture is good and temperature is 60°-85°F.
Remarks: Work upwind of product and never fumigate without 2 qualified applicators. Do not fumigate when there is little or no air movement. Tarping is not used after field application. Refer to label for use directions.

Signal Word/Toxicity Class: Danger/I.

Thiolux — see sulfur

Beans

(Dry, Great Northern, Kidney, Lima, Navy, Pink, Pinto, Red, Snap, White)

Allegiance — see metalaxyl

Apron — see metalaxyl

Apron XL — see mefenoxam

Basicop — see copper compounds

BENLATE

Benzimidazole systemic fungicide. Toxic to fish.

Benlate (SP, WP) (benomyl) *(DuPont)*
Rate: 0.375-0.5 lb. ai/A
0.75-1 lb. SP, WP/A
Disease:
Fusarium root rot
Time: Apply in-furrow at planting.

Rate: 0.25-0.75 lb. ai/A
0.5-1.5 lb. SP, WP/A
Disease:
Anthracnose

Rate: 0.75-1 lb. ai/A
1.5-2 lb. SP, WP/A
Disease:
Botrytis gray mold, Sclerotinia white mold
Time: Apply at 10-50% bloom; repeat 7- to -10 days later. Pacific NW: Apply prior to first petal fall/row fill; repeat 7- to 10-days later. Colorado, Montana, Nebraska, Wyoming for narrow-row irrigated dry beans: Apply at initial bloom; repeat 7- to 10-days later.
Remarks: Dry, snap, succulent beans. Do not tank mix or alternate with benzimidazole or thiophanate products. Refer to label for use directions.
Days To Harvest: Dry, snap (except Fusarium)-14; succulent-28.
State Restrictions: Not for use on Anthracnose in California. Benlate SP not registered for use in California.

Rate: 6 oz. WP/50 lb. seed
Disease:
Fusarium root rot
Time: Seed treatment.

NOTICE The information on these pages is for preliminary planning — not a guide for use. Be sure to follow the manufacturer's directions, notwithstanding information contained here.	For personal protective equipment and EPA registration numbers, please turn to page 39.

Remarks: Snap beans. Apply dry in planter box. Do not tank mix or alternate with benzimidazole or thiophanate products. Do not use treated seed for food, feed, or oil purposes. Refer to label for use directions.
State Restrictions: Benlate SP not registered for use in California.

Signal Word/Toxicity Class: Caution.
REI: 24 hr.

Botran — see dicloran

Bravo — see chlorothalonil

Bravo S — see sulfur premix

CAPTAN

Captan 30-DD contains the toxic inert Rhodamine B.

Gustafson Captan 30-DD or 400 *(Gustafson)*
Rate: 2.25 fl. oz. 30-DD/100 lb. seed
2-3 fl. oz. 400/100 lb. seed
Disease:
damping-off, seed decay, seedling blight, mold
Time: Seed treatment.
Remarks: Dry, snap beans. Do not combine colloidal suspensions, wettable powders, or emulsifiable concentrates unless previous experience assures compatibility. Use with EPA-approved dye that imparts unnatural color to seed. Do not use treated seed for food, feed, or oil purposes. Refer to label for use directions.

Signal Word/Toxicity Class: Danger/I.

Drexel Captan 4FL *(Drexel)*
Rate: 2.5 fl. oz. 4FL/100 lb. seed
Disease:
seed decay, seedling blight, soilborne disease
Time: Liquid or slurry seed treatment.
Remarks: Dry, lima, snap beans. Do not use treated seed for food, feed, or oil purposes. Refer to label for use directions.

Signal Word/Toxicity Class: Danger/I.

CAPTAN (Premixes)

Captan-Diazinon contains graphite.

Captan-Diazinon (36.67% captan & 25% diazinon) *(Trace)*
Kernel Guard (14.67% captan & 15% diazinon & 25% lindane) *(Trace)*
Rate: 2 oz./bu. seed
Disease:
damping-off, seed decay, seedling blight
Time: Planter box seed treatment.
Remarks: Do not use treated seed for food, feed, or oil purposes. Refer to label for use directions.

Signal Word/Toxicity Class: Danger/I.
REI: 12 hr. (Kernel Guard).

Nu-Gro Isotox (12.2% captan & 25% lindane) *(Wilbur-Ellis)*
Seed Mate Isotox (12.2% captan & 25% lindane) *(Loveland)*
Rate: Limas (Fordhook, standard, large):
1 oz./100 lb. seed
Baby bush limas, bush, snap:
2 oz./100 lb. seed
Disease:
damping-off, seed decay, seedling blight
Time: Planter box treatment.
Remarks: Bush, limas (baby bush, large, standard, Fordhook), snap beans. Use with protective fungicide. Treated seed must not be used for or mixed with food or animal feed, or processed for oil. Refer to label for use directions.

Signal Word/Toxicity Class: Danger/I.
REI: 4 days.

Rival (19.8% captan & 8.4% PCNB & 1% thiabendazole) *(Gustafson)*
Rate: 4 fl. oz./100 lb. seed
Disease:
damping-off, seed decay, stem blight, pod blight, seedling blight
Time: Seed treatment
Remarks: Dry beans. Use with EPA-approved dye that imparts unnatural color to seed. Do not use treated seed for food, feed, or oil purposes. Refer to label for use directions.

CARBOXIN

Vitavax-30C is a concentrate formulation. Vitavax-34 is a ready-to-use seed treatment applied as slurry treatment by machine containing toxic inert Rhodamine B. Toxic to fish.

Vitavax-30C or Vitavax-34 *(Gustafson)*
Rate: 3.3-6.6 fl. oz. 30C/100 lb. seed
4-8 fl. oz. 34/100 lb. seed
Disease:
Rhizoctonia solani
Time: Slurry seed treatment.
Remarks: Incorporation of dye into treating slurry is recommended to distinguish seed as treated. Mix with enough water to apply at 20 fl. oz./100 lb. seed. For early-season planting in cold soils, use high rate. Under severe planting conditions, use approved in-furrow granule to aid in disease control. Do not use treated seed for food, feed, or oil purposes. Do not graze or feed livestock hay grown from treated seed. Do not apply this product through any type of irrigation system. Refer to label for use directions.

Signal Word/Toxicity Class: Warning/II (30C); Caution (34).

CARBOXIN (Premixes)

Germate Plus contains graphite. Prevail is a premix of the active ingredients of Apron, Terraclor, and Vitavax. Vitaflo-280 is a combination of a systemic fungicide (Vitavax) and a contact fungicide (thiram). Toxic to fish. Germate Plus is also toxic to birds and other wildlife.

Germate Plus (14% carboxin & 15% diazinon & 25% lindane) *(Trace)*
Rate: 2 oz./bu. seed
Disease:
seedling diseases
Time: Planter box seed treatment.
Remarks: Do not use treated seed for food, feed, or oil purposes. Refer to label for use directions.

Signal Word/Toxicity Class: Caution.
REI: 12 hr.

Prevail (15% carboxin & 15% PCNB & 3.12% metalaxyl) *(Trace)*
Rate: 6-8 oz./100 lb. seed
Disease:
damping-off, seed rot, seedling disease
Time: Planter box seed treatment.
Remarks: Do not use treated seed for food, feed, or oil purposes. Use with EPA-approved dye that imparts unnatural color to seed (except planter box). Do not graze or feed livestock hay grown from treated seed. Refer to label for use directions.

Signal Word/Toxicity Class: Caution.
REI: 12 hr.

RTU-Vitavax-Thiram (10% carboxin & 10% thiram) *(Gustafson)*
Rate: 6.8 fl. oz./100 lb. seed
Disease:
Fusarium, Pythium, seedling diseases, Helminthosporium, *Rhizoctonia solani*
Time: Seed treatment.
Remarks: Dry beans. Use high rates under heavy disease pressure. Use with EPA-approved dye that imparts unnatural color to seed. Do not use treated seed for food, feed, or oil purposes. Do not graze or feed livestock on bean forage until 60 days after treated seed is planted. Refer to label for use directions.

Signal Word/Toxicity Class: Caution.

◆ - **new product** • ai-active ingredient • DF-dry flowable • E/EC-emulsifiable concentrate • EW-emulsion in water
F/FL-flowable • G-granule • L/LC-liquid • SP/WSP-soluble packet • W/WP-wettable powder • WSB-water soluble bag

Vitaflo-280 (14.9% carboxin & 13.2% thiram) *(Gustafson)*
Rate: 4 fl. oz./100 lb. seed
Disease:
damping-off   seed decay   seedling diseases
Time: Seed treatment.
Remarks: Thorough seed coverage is required. Use higher rate under conditions of heavy expected disease pressure. Do not use treated seed for food, feed, or oil purposes. Do not graze or feed livestock on treated areas for 6 weeks after planting. Do not graze or feed livestock on vines grown from treated seed. Refer to label for use directions.

Signal Word/Toxicity Class: Caution.

Champ — see copper compounds

Champion — see copper compounds

CHLORONEB

Supplemental seed treatment; use in conjunction with a standard fungicide seed treatment.

Demosan (65W or FL) *(Kincaid)*
Rate: 2.6 oz. ai/cwt seed
4 oz. 65W/cwt seed
7 oz. FL/cwt seed
Disease:
damping-off   Rhizoctonia   seedling diseases
Pythium blight   Sclerotium blight   sore shin
Time: Seed treatment or hopper box treatment.
Remarks: Field, green, kidney, lima, navy, pole, snap, string, wax beans. Do not use treated seed for food, feed, or oil purposes. Do not graze within 45 days of planting. Refer to label for use directions.

Signal Word/Toxicity Class: Caution.
REI: 12 hr.

CHLOROTHALONIL

Broad-spectrum contact fungicide. Toxic to fish, aquatic invertebrates, and marine/estuarine organisms.

Bravo 500, 720/Weather Stik, Ultrex (WDG) *(ZENECA)*
Echo 500, 720 or 90DF *(Sostram)*
Rate: 1-1.5 lb. ai/A
2-3 pt. 500/A
1.375-2 pt. 720/A
1.25-1.8 lb. WDG/A
1.2-1.7 lb. 90DF/A
Disease:
Anthracnose   downy mildew   rust
Time: Apply during early bloom stage; repeat at 7- to 10-day intervals.
Remarks: Dry, kidney, lima, navy, pinto beans. For use only on beans harvested dry with pods removed. Use in sufficient water to obtain adequate coverage. Do not graze treated areas or feed treated plant parts to livestock. Refer to label for use directions and restrictions.
Days To Harvest: 6 weeks.
State Restrictions: SLN-Colorado, Kansas, Nebraska (Bravo 720).

Rate: 1-2.25 lb. ai/A
2-4.25 pt. 500/A
1.375-3 pt. 720/A
1.25-2.7 lb. WDG/A
1.2-2.4 lb. 90DF/A
Disease:
Botrytis blight   rust
Time: Apply during early bloom stage or when disease first threatens; repeat at weekly intervals as necessary.
Remarks: Snap beans. Use in sufficient water to obtain adequate coverage. Use high rate for Botrytis blight. Do not graze treated areas or feed treated plant parts to livestock. Refer to label for use directions and restrictions.
Days To Harvest: 7.

Signal Word/Toxicity Class: Danger/I (WDG, 90DF); Warning/II (500, 720).
REI: 48 hr.

Bravo Zn *(ZENECA)*
Rate: 1-1.5 lb. ai/A
2-3 pt. Zn/A
Disease:
Anthracnose   Cercospora leaf spot   rust
Ascochyta blight   downy mildew
Time: Apply during early bloom stage; repeat at 7- to 10-day intervals.
Remarks: Dry beans. For use only on beans to be harvested dry with pods removed. Use in sufficient water to obtain adequate coverage. Do not apply more than 4 times per growing season. Do not allow livestock to graze treated areas or feed treated plant parts to livestock. Refer to label for use directions and restrictions.
Days To Harvest: 14.

Signal Word/Toxicity Class: Warning/II.
REI: 48 hr.

Terranil (6L, 90DF or 90DF WSP) or Terranil Zn *(Terra)*
Rate: Dry:
1-1.6 lb. ai/A
2-3 pt. Zn/A
1.375-2 pt. 6L/A
1.125-1.75 lb. 90DF, 90DF WSP/A
Disease:
Anthracnose   downy mildew   rust
Time: Apply during early bloom stage; repeat at 7- to 10-day intervals. For use only on beans harvested dry with pods removed.

Rate: Snap:
1-2.25 ai/A
1.375-3 pt. 6L/A
1.125-2.5 lb. 90DF, 90DF WSP/A
Disease:
Botrytis blight   rust
Time: Apply during early bloom stage or when disease first threatens; repeat at weekly intervals or as necessary.
Remarks: Dry, kidney, lima, navy, pinto, snap beans. Use in sufficient water to obtain adequate coverage. Do not graze treated areas or feed treated plant parts to livesotck. Refer to label for use directions.
Days To Harvest: Dry-6 weeks; snap-7.

Signal Word/Toxicity Class: Danger/I (90DF, 90DF WSP); Warning/II (6L, Zn).
REI: 48 hr.

Clean Crop C-O-C-S — see copper compounds

COPPER COMPOUNDS

Fungicide/bactericide.

Basicop (WP) (basic copper sulfate) *(Griffin)*
Rate: 1 lb. ai/A
2 lb. WP/A
Disease:
Anthracnose   leaf spot   powdery mildew
downy mildew
Time: Apply when plants are 5" tall; repeat every 7- to 10-days.

Rate: 2 lb. ai/A
4 lb. WP/A
Disease:
bacterial blight
Time: Apply when disease appears, repeat at 7-day intervals.
Remarks: Do not apply through any type of irrigation system. Refer to label for use directions.

Signal Word/Toxicity Class: Warning/II.
REI: 24 hr.

Champ (FL or Formula 2) or Champion (WP) (copper hydroxide) *(Agtrol)*
Rate: 1.33-4 pt. FL/A
0.66-2 pt. Formula 2/A
1-3 lb. WP/A
Disease:
bacterial blight   common blight   halo blight
Time: Apply when plants are 6" high; repeat at 7- to 14-day intervals.
Remarks: Refer to label for use directions.

Signal Word/Toxicity Class: Danger/I (Champion); Caution (Champ).
REI: 24 hr.

7 Vegetables

NOTICE	The information on these pages is for preliminary planning — not a guide for use. Be sure to follow the manufacturer's directions, notwithstanding information contained here.	For personal protective equipment and EPA registration numbers, please turn to page 39.

Clean Crop C-O-C-S (WDG) (74.8% copper oxychloride & 14.2% copper sulfate, basic) *(Platte)*
Rate: 1-2 lb. ai/A
2-4 lb. WDG/A
Disease:
Anthracnose Cercospora leaf spot downy mildew
bacterial blight
Time: Apply when plants are 5-6" tall; repeat as needed to protect new growth throughout growing season.
Remarks: Dry and succulent beans. Do not apply this product through any type of irrigation system. Refer to label for use directions.

Signal Word/Toxicity Class: Warning/II.
REI: 24 hr.

Copper-Count-N (L) (copper ammonium) *(Mineral Research)*
Rate: 1-3 qt. L/A
Disease:
bacterial blight common blight halo blight
Time: Apply when plants are 3-5" high and before diseases appear; repeat at 7- to 10-day intervals (5- to 7-days under severe conditions).
Remarks: Dry, snap beans. Apply by ground or air; refer to label for use directions. Do not apply through any type of irrigation system unless supplemental labeling on chemigation is followed.

Signal Word/Toxicity Class: Caution.
REI: 12 hr.

Drexel Basic Copper Sulfate (WP) or KOP 300 (3FL) (basic copper sulfate) *(Drexel)*
Rate: 2-4 lb. WP/A
0.16-0.5 gal. 3FL/A
Disease:
angular leaf spot bacterial blight downy mildew
Anthracnose
Time: Apply when plants are 5" tall; repeat at 5- to 7-day intervals.
Remarks: Dry, green beans. Do not apply through any type of irrigation system. Refer to label for use directions.

Signal Word/Toxicity Class: Warning/II.
REI: 24 hr. (WP); 12 hr. (3FL).

Kocide (101, DF, 2000, 4.5LF, LF) (copper hydroxide) *(Griffin)*
Rate: 1-3 lb. 101, DF/A
0.75-2.25 lb. 2000/A
0.66-2 pt. 4.5LF/A
1.33-4 pt. LF/A
Disease:
brown spot common blight halo blight
Time: Apply when plants are 6" high on a 7- to 14-day schedule.
Remarks: Dry, green beans. Refer to label for use directions.

Signal Word/Toxicity Class: Danger/I (101, DF); Warning/II (2000); Caution (4.5LF, LF).
REI: 24 hr.

KOP-Hydroxide (FL, 50WP) (copper hydroxide) *(Drexel)*
Rate: 1.33-4 pt. FL/A
1-3 lb. 50WP/A
Disease:
bacterial blight common blight halo blight
Time: Apply when plants are 5" high; repeat at 5- to 10-day intervals as needed.
Remarks: KOP-Hydroxide 50WP should not be applied in spray solution having pH less than 6.5 or greater than 9.0 as phytotoxicity may occur. Refer to label for use directions.

Signal Word/Toxicity Class: Danger/I.
REI: 48 hr.

Micro Flo Basic Copper "53" (basic copper sulfate) or Microsperse COC (53WP) (copper oxychloride) *(Micro Flo)*
Rate: 2-4 lb. 53, 53WP/A
Disease:
angular leaf spot bacterial blight powdery mildew (53)
Anthracnose downy mildew
Time: Apply when plants are 5" high; repeat at 5- to 10-day intervals as needed.
Remarks: Dry, green beans. Refer to label for use directions.

Signal Word/Toxicity Class: Danger/I (53WP); Warning/II (53).
REI: 24 hr.

Nu-Cop (3L, 50DF, 50WP) (copper hydroxide) *(Micro Flo)*
Rate: 0.66-4 pt. 3L/A
1-3 lb. 50DF, 50WP/A
Disease:
bacterial blight brown spot (DF, WP)
Time: Apply when plants have second trifoliate leaves or are about 6" high; repeat at 7- to 14-day intervals as needed.
Remarks: Refer to label for use directions.

Signal Word/Toxicity Class: Danger/I (50DF, 50WP); Warning/II (3L).
REI: 48 hr. (50DF, 50WP); 24 hr. (3L).

TOP COP Tri-Basic (FL) (basic copper sulfate) *(Stoller)*
Rate: 1-1.5 pt. FL/A
Disease:
angular leaf spot bacterial blight downy mildew
Anthracnose
Time: Apply when plants are about 5" tall; repeat at 5- to 7-day intervals.
Remarks: Refer to label for use directions.

Signal Word/Toxicity Class: Caution.
REI: 24 hr.

Copper-Count-N — see copper compounds

Demosan — see chloroneb

DICLORAN

Protective fungicide. Toxic to fish.

Botran 75-W *(Gowan)*
Rate: Bush:
1.7 lb. ai/A
2.25 lb. 75-W/A
Pole:
3 lb. ai/A
4 lb. 75-W/A
Disease:
white mold
Time: Apply when disease is anticipated; repeat at 7-day intervals during periods favorable to disease development.
Remarks: Snap beans. Do not feed treated foliage to livestock. Refer to label for use directions.
Days To Harvest: 2.

Signal Word/Toxicity Class: Caution.
REI: 12 hr.

Echo — see chlorothalonil

Germate Plus — see carboxin premixes

IPRODIONE

Dicarboximide fungicide. Toxic to aquatic invertebrates.

Rovral (4FL, WG, WP) *(Rhone-Poulenc)*
Rate: 0.75-1 lb. ai/A
1.5-2 pt. 4FL/A
1.5-2 lb. WG, WP/A
Disease:
gray mold white mold
Time: Apply at first bloom to when 10% of plants have one open bloom; repeat at 5- to 7-day intervals up to peak bloom if conditions are favorable for disease development.
Remarks: Dry, lima, snap beans. Under severe disease conditions, use high rate and shorter spray intervals. Maximum of 2 applications per season, with last application made no later than peak bloom. Do not forage for 14 days after last application. Do not feed snap or succulent bean hay to livestock. Do not feed dry bean hay to livestock until 45 days after last application. Refer to label for use directions.

◆ - **new product** • ai-active ingredient • DF-dry flowable • E/EC-emulsifiable concentrate • EW-emulsion in water
F/FL-flowable • G-granule • L/LC-liquid • SP/WSP-soluble packet • W/WP-wettable powder • WSB-water soluble bag

Crop Rotation: Beans, broccoli, carrots, garlic, lettuce, onions (dry bulb), peanuts, potatoes, and rice can be rotated after harvest.
State Restrictions: Aerial application not registered for use in California.

- - - - - - - - - - - - - - - -

Signal Word/Toxicity Class: Warning/II (WG); Caution (4FL, WP).
REI: 12 hr.

Kernel Guard — see captan premixes

Kocide — see copper compounds

KOP-Hydroxide — see copper compounds

Kumulus — see sulfur

MANEB

Dithiocarbamate contact fungicide. Toxic to fish.

Maneb 75DF or 80WP *(Elf Atochem)*
Manex (FL) *(Griffin)*

Rate: 1.12-1.5 lb. ai/A
1.5-2 lb. 75DF/A
1.2-1.6 lb. ai/A
1.5-2 lb. 80WP/A
1.2-1.6 qt. FL/A

Disease:
Anthracnose downy mildew rust

Time: Apply when plants are small; repeat at 5- to 7-day intervals.
Remarks: Dry beans. Do not use any other product containing a different EBDC ai during the same growing season if using Manex. Refer to label for use directions.
Days To Harvest: 30.

- - - - - - - - - - - - - - - - - - - -

Signal Word/Toxicity Class: Caution.
REI: 24 hr.

Manex — see maneb

MEFENOXAM

Systemic fungicide having a specific mode of action.

Apron XL LS *(Novartis)*
Rate: 0.32-0.64 fl. oz. LS/100 lb. seed
Disease:
damping-off Phytophthora (early season)
Time: Seed treatment at planting.
Remarks: Field, green, kidney, lima, navy, pinto, pole, snap, string, wax beans. Do not use treated seed for food, feed, or oil purposes.

- - - - - - - - - - - - - - - -

Signal Word/Toxicity Class: Warning/II.

Ridomil Gold EC or WSP *(Novartis)*
Rate: 0.25-0.5 lb. ai/A
0.5-1 pt. EC/A
0.225-0.45 lb. ai/A
0.5-1 lb. WSP/A
Disease:
damping-off root and stem rot
Time: Preplant incorporated or surface application just after planting.
Remarks: Field, French, kidney, lima, navy, pinto, runner, snap, wax beans. For banded applications, 7" band recommended. Do not use in greenhouse or on field grown vegetable bedding plants. May be tank mixed. Do not plant any crop which is not registered for use with mefenoxam in treated soil for 12 months.
Days To Harvest: 1.

- - - - - - - - - - - - - - - -

Signal Word/Toxicity Class: Warning/II (EC); Caution (WSP).
REI: 48 hr.

METALAXYL

Systemic fungicide having a specific mode of action.

Allegiance or Apron (FL) *(Gustafson)*
Rate: 0.75 fl. oz. Allegiance/100 lb. seed
0.75-1.5 fl. oz. Apron/100 lb. seed
Disease:
damping-off Phytophthora (early season)
Time: Seed treatment.
Remarks: Field, green, kidney, lima, navy, pole, snap, string, wax beans. Use with EPA-approved dye that imparts unnatural color to seed. Do not use for feed, food, or oil purposes. Refer to label for use directions.

- - - - - - - - - - - - - - - -

Signal Word/Toxicity Class: Warning/II.

Apron Dry *(Trace)*
Rate: 4 oz./100 lb. seed
Disease:
damping-off Phytophthora
Time: Seed treatment.
Remarks: Do not use treated seed for food, feed, or oil purposes. Refer to label for use directions.

- - - - - - - - - - - - - - - -

Signal Word/Toxicity Class: Caution.
REI: 12 hr.

Micro Flo — see copper compounds

Microsperse COC — see copper compounds

Microthiol Special — see sulfur

Nu-Cop — see copper compounds

Nu-Gro Isotox — see captan premixes

PCNB

RTU-PCNB contains toxic inert Rhodamine B. Toxic to fish and aquatic organisms.

PCNB 2-E, 2LF, 10%, 75DG or 75WP *(Amvac)*
PCNB 2 Spray *(Wilbur-Ellis)*
Rate: 1-1.5 lb. ai/8-10 gal. water
4-6 pt. 2-E, 2S/8-10 gal. water
10-15 lb. 10%
1.33-2 lb. 75DG, 75WP/A in 8-10 gal. water
Arizona, California:
0.75-1 lb. ai/8-10 gal. water
3-4 pt. 2-E, 2S/8-10 gal. water
Disease:
root and stem rot
Time: Apply over seed at planting time in the furrow.
Remarks: Bush, pole beans. For bush beans apply in 14,500 lin. ft. row; pole beans in 8340 lin. ft. row. Do not apply through any type of irrigation system. Do not feed treated vines to livestock. Do not plant root crops within 12 months of application unless PCNB is registered for use on those crops.

- - - - - - - - - - - - - - - -

Rate: Band application (8"):
2 lb. ai/8-10 gal. water
1 gal. 2-E, 2S/8-10 gal. water
2.75 lb. 75DG, 75WP/A in 8-10 gal. water
Disease:
white mold
Time: Apply immediately after or at planting. If disease is severe, repeat at 2- to 3-week intervals. Do not apply after first bloom appears.
Remarks: Dry, snap beans. Do not apply through any type of irrigation system. Do not feed treated vines to livestock. Do not plant root crops within 12 months of application unless PCNB is registered for use on those crops.
State Restrictions: Postplant application not for use in California.

- - - - - - - -

NOTICE	The information on these pages is for preliminary planning — not a guide for use. Be sure to follow the manufacturer's directions, notwithstanding information contained here.	For personal protective equipment and EPA registration numbers, please turn to page 39.

Rate: 3.35 fl. oz. 2-E/100 lb. seed
4 fl. oz. 2LF/100 lb. seed

Disease:
Fusarium Rhizoctonia

Time: Seed treatment.

Remarks: Do not apply through any type of irrigation system. Do not use treated seed for food, feed, or oil purposes. Do not plant root crops within 12 months of application unless PCNB is registered for use on those crops.

Signal Word/Toxicity Class: Danger/I (2-E); Caution.
REI: 12 hr.

RTU-PCNB *(Gustafson)*

Rate: 3 fl. oz. RTU/100 lb. seed

Disease:
Fusarium Rhizoctonia

Time: Seed treatment.

Remarks: Do not use treated seed for food, feed, or oil purposes. Do not feed treated seed to animals. Refer to label for use directions.

Signal Word/Toxicity Class: Caution.

Terraclor (75WP) *(Uniroyal)*

Rate: 4-8 oz. 75WP/100 gal. water

Disease:
damping-off root and stem rot

Time: Soil drench. One repeat application may be made 4-6 weeks later if necessary.

Remarks: Bedding plants in greenhouse or nursery. Refer to label for use directions.

Crop Rotation: Do not plant root crops in PCNB treated fields within 12 months of broadcast and banding applications unless PCNB registered for use on those crops.

Signal Word/Toxicity Class: Caution.

Prevail — see carboxin premixes

Redball — see sulfur

Ridomil Gold — see mefenoxam

RIDOMIL GOLD PC (Premix)

Systemic fungicide having a specific mode of action. Toxic to fish and aquatic organisms.

Ridomil Gold PC or PC GR (0.5% mefenoxam & 10% PCNB) *(Novartis)*

Rate: 0.08 lb. ai/A
0.75 lb./A

Disease:
damping-off seed rot

Time: At planting.

Remarks: Beans (bush, pole). Do not allow livestock to feed treated vines or graze treated foliage. Do not plant any crop which is not registered for use with mefenoxam in treated soil for 12 months.

Signal Word/Toxicity Class: Warning/II.
REI: 48 hr.

Rival — see captan premixes

Rovral — see iprodione

RTU-Vitavax-Thiram — see carboxin premixes

Seed Mate — see captan premixes

SULFUR

Nonsystemic fungicide with protective action. Microthiol Special, Sulfa 80, and Thiolux are micronized sulfurs.

Drexel Suffa (6FL), Sulfa 80 (80DF) or Sulfur (90W) *(Drexel)*
Thiolux (DF) *(Platte)*

Rate: 2.6-5.3 lb. ai/A
3.5-7 pt. 6FL/A
2.4-8 lb. ai/A
3-10 lb. 80DF, DF/A
2.7-5.4 lb. ai/A
3-6 lb. 90W/A

Disease:
leaf spot (6FL, 90W) powdery mildew rust

Time: Apply at first signs of infection; repeat at 7- to 14-day intervals as needed. For 80DF and DF, apply early leaf stage; repeat at 10- to 14-day intervals with or without pesticides immediately after rain.

Remarks: Do not use within 2 weeks of oil application (4 weeks 6FL). Some plants are susceptible to injury from sulfur under certain climatic conditions. Do not make sulfur applications at excessively high temperatures. Do not apply this product through any type of irrigation system. Refer to label for use directions.

Signal Word/Toxicity Class: Caution.
REI: 24 hr.

Dusting Sulfur (DS) *(Platte)*

Rate: 15-30 lb. DS/A

Disease:
powdery mildew rust

Time: Apply as soon as first true leaves form; repeat at weekly intervals until blossom time.

Remarks: May cause injury to some varieties. Do not make applications within 2 weeks of oil spray. Refer to label for use directions.

Signal Word/Toxicity Class: Caution.
REI: 24 hr.

Kumulus DF *(BASF)*

Rate: 2.4-4 lb. ai/A
3-5 lb. DF/A

Disease:
powdery mildew Septoria leaf spot

Time: Apply at early leaf stage; repeat at 10- to 14-day intervals with or without other pesticides immediately after a rain.

Remarks: Do not apply if temperatures during or within 3 days after application are expected to exceed 90°F. Do not apply through any type of irrigation system. Refer to label for use directions.

Signal Word/Toxicity Class: Caution.
REI: 24 hr.

Liquid Sulfur Six *(Helena)*

Rate: 3-4 pt./8 gal. water

Disease:
powdery mildew

Time: Apply at first sign of infection; repeat at 3-week intervals.

Remarks: Apply by air. Do not apply this product through any type of irrigation system. Refer to label for use directions.

Signal Word/Toxicity Class: Caution.
REI: 24 hr.

Micro Flo Sulfur (6L), Dual Spraying & Dusting Sulfur (DS) or Microsperse Wettable Sulfur (WS) *(Micro Flo)*
Terra Sulfur (6L) *(Terra)*

Rate: 2.25-6.75 lb. ai/A
3-9 pt. 6L/A
6.3 lb. ai/A
7 lb. DS/A
2.7-4.5 lb. ai/A
3-5 lb. WS/A

Disease:
leaf spot (DS, 6L) powdery mildew rust (6L)

Time: Apply at early leaf stage; repeat at 14-day intervals (10- to 14-day intervals (WS)) as necessary.

Remarks: Do not use within 2 weeks of oil application. Do not apply DS and WS through any type of irrigation system. Refer to label for use directions.

Signal Word/Toxicity Class: Caution.
REI: 24 hr.

◆ - **new product** • ai-active ingredient • DF-dry flowable • E/EC-emulsifiable concentrate • EW-emulsion in water
F/FL-flowable • G-granule • L/LC-liquid • SP/WSP-soluble packet • W/WP-wettable powder • WSB-water soluble bag

Microthiol Special (80WP) *(Elf Atochem)*
Rate: 5.6 lb. ai/A
7 lb. 80WP/A
Disease:
leaf spot powdery mildew
Time: Apply at early leaf stage; repeat at 14-day intervals or as needed.
Remarks: Refer to label for use directions.

Signal Word/Toxicity Class: Caution.
REI: 24 hr.

Redball EM-53 (6L), Superfine Wettable Sulphur (WS) or Commercial Dusting Sulphur (DS) *(International Sulphur)*
Rate: 8 pt. 6L/A
6-16 lb. WS/A
10-25 lb. DS/A
Disease:
powdery mildew rust
Time: Apply at first sign of disease; repeat at 7- to 14-day intervals (6L, WS), 7- to 10-day intervals (DS).
Remarks: Dry, green, lima, snap beans. Sulfur may injure certain varieties of beans, check tolerance prior to application. Do not apply when temperatures exceed 90°F. Refer to label for use directions.

Signal Word/Toxicity Class: Caution.
REI: 24 hr.

THAT (6FL) *(Stoller)*
Rate: 3-6 lb. ai/A
4-8 pt. 6FL/A
Disease:
leaf spot powdery mildew rust
Time: Apply at early leaf stage; repeat at 14-day intervals as necessary.
Remarks: Dry, colored, green, snap, lima beans. Refer to label for use directions.

Signal Word/Toxicity Class: Caution.
REI: 24 hr.

SULFUR (Premixes)

Fungicide/bactericide in flowable formulation. Toxic to fish and aquatic organisms.

Bravo S (27.25% sulfur & 19.15% chlorothalonil) *(Helena)*
Terranil S (27.25% sulfur & 19.15% chlorothalonil) *(Terra)*
Rate: 4-8 pt./A
Disease:
Botrytis blight powdery mildew rust
Time: Apply during early bloom or when disease first threatens; repeat at weekly intervals as needed.
Remarks: Snap beans. Do not graze treated areas or feed treated plant parts to livestock. Refer to label for use directions.
Days To Harvest: 7.

Signal Word/Toxicity Class: Warning/II (Terranil S); Caution (Bravo S).
REI: 48 hr.

Top Cop with Sulfur (6.25 lb./gal. sulfur & 1 lb./gal. tribasic copper sulfate) *(Stoller)*
Rate: 2 qt./A
Disease:
downy mildew powdery mildew rust
halo blight
Time: Apply early bloom or when disease first threatens.
Remarks: Refer to label for use directions.

Signal Word/Toxicity Class: Caution.
REI: 24 hr.

SYSTEM³

Biological and chemical fungicide containing the active ingredients of Apron, Terraclor, and Kodiak. Contains graphite. Toxic to fish and aquatic organisms.

System³ (0.1% *Bacillus subtilis* & 14.25% metalaxyl & 16.67% PCNB) *(Helena)*
Rate: 2-3 oz./bu.
Disease:
Pythium *Rhizoctonia solani*
Time: Apply at planting.
Remarks: A non-oil lubricant has been added for seeds to flow more easily and evenly from planter boxes. Incorporation of dye is recommended to distinguish seed as treated. Do not use treated seed for food, feed or oil purposes. Refer to label for use directions.

Signal Word/Toxicity Class: Caution.
REI: 12 hr.

Terraclor — see PCNB

Terranil — see chlorothalonil

TERRANIL Cu (Premix)

Toxic to fish, aquatic invertebrates, and marine/estuarine organisms.

Terranil Cu (24% chlorothalonil & 24% copper oxychloride) *(Terra)*
Rate: Dry:
3-4.5 pt. Cu/A
Disease:
Anthracnose downy mildew rust
Time: Apply during early bloom stage; repeat at 7- to 10-day intervals.

Rate: Snap:
3-6.75 pt. Cu/A
Disease:
Botrytis blight rust
Time: Apply during early bloom stage or when disease first threatens; repeat at weekly intervals or as necessary.
Remarks: Dry, kidney, lima, navy, pinto, snap beans. For use only on beans harvested dry with pods removed. Do not graze treated areas or feed treated plant parts to livesotck. Refer to label for use directions.
Days To Harvest: Dry-6 weeks; snap-7.

Signal Word/Toxicity Class: Danger/I.
REI: 48 hr.

Terranil S — see sulfur premixes

THAT — see sulfur

Thiolux — see sulfur

THIOPHANATE-METHYL

Benzimidazole systemic fungicide. Toxic to fish.

Topsin M (WSB, 70W) *(Elf Atochem)*
Rate: Single:
1-1.4 lb. ai/A
1.5-2 lb. WSB, 70W/A
Split:
0.7-1 lb. ai/A
1-1.5 lb. WSB, 70W/A
Disease:
gray mold white mold
Time: Single application: Apply at 50%-70% of full bloom. Split application: Apply first application at 10%-30% of full bloom; second 4-7 days later or at peak bloom.
Remarks: Dry, lima, snap bean. Refer to label for use directions.
Days To Harvest: Dry, snap beans-14 days; lima beans-28.

Signal Word/Toxicity Class: Caution.
REI: 12 hr.

NOTICE The information on these pages is for preliminary planning — not a guide for use. Be sure to follow the manufacturer's directions, notwithstanding information contained here.

For personal protective equipment and EPA registration numbers, please turn to page 39.

THIRAM

Broad-spectrum fungicide. Toxic to fish. Thiram 50WP contains the toxic inert Rhodamine B and is also toxic to birds and wildlife.

42-S Thiram or Thiram 50WP Dyed *(Gustafson)*
Thiram Granuflo (WDG) *(UCB Agrochemicals)*
Rate: 2-3 fl. oz. 42-S/100 lb. seed
2-3 oz. 50WP/100 lb. seed
1.3-2.2 oz. WDG/100 lb. seed
Disease:
damping-off seed decay seedling blight
Time: Seed treatment.
Remarks: Dry, lima, snap beans. Use high rate for lima beans, low rate for dry and snap beans. Use with EPA-approved dye that imparts unnatural color to seed. Do not use for feed, food, or oil purposes. Do not apply Thiram Granuflo through any type of irrigation system. Refer to label for use directions.
State Restrictions: Thiram Granuflo not for use in California.

Signal Word/Toxicity Class: Caution.
REI: 24 hr. (Thiram Granuflo).

TOP COP — see copper compounds

Top Cop with Sulfur — see sulfur premix

Topsin M — see thiophanate-methyl

Vitaflo — see carboxin premixes

Vitavax — see carboxin

ZIRAM

Liquid formulation with excellent adhesive properties and safe on a wide range of plants. Toxic to fish and wildlife.

Ziram 4L *(Drexel)*
Rate: 1 lb. ai/100 gal.
2 pt. 4L/100 gal.
Disease:
Anthracnose rust
Time: Apply when disease first appears; repeat at 4- to 7-day intervals as needed.
Remarks: Do not apply this product through any type of irrigation system. Refer to label for use directions.
Days To Harvest: Snap beans-4.
State Restrictions: Not for use in California.

Signal Word/Toxicity Class: Danger/I.
REI: 48 hr.

Carrots, Parsley, Parsnips, Salsify

Aliette — see fosetyl-aluminum

Allegiance — see metalaxyl

Apron — see metalaxyl

Apron XL — see mefenoxam

Basicop — see copper compounds

BENLATE

Benlate (SP, WP) (benomyl) *(DuPont)*
Rate: 0.13-0.5 lb. ai/A
0.25-1 lb. SP, WP/A
Disease:
Sclerotinia white mold
Time: Apply when disease first appears; repeat at 7- to 10-day intervals.
Remarks: Do not tank mix or alternate with benzimidazole or thiophanate products. Refer to label for use directions.
Days To Harvest: 4.
State Restrictions: Benlate SP not registered for use in California.

Signal Word/Toxicity Class: Caution.
REI: 24 hr.

Bravo — see chlorothalonil

CAPTAN (Premix)

Nu-Gro Isotox (12.2% captan & 25% lindane) *(Wilbur-Ellis)*
Seed Mate Isotox (12.2% captan & 25% lindane) *(Loveland)*
Rate: 4 oz./100 lb. seed
Disease:
damping-off seed decay seedling blight
Time: Planter box treatment.
Remarks: Treated seed must not be used for or mixed with food or animal feed, or processed for oil. Refer to label for use directions.

Signal Word/Toxicity Class: Danger/I.
REI: 4 days.

Champ — see copper compounds

Champion — see copper compounds

CHLOROTHALONIL

Bravo 500, 720/Weather Stik, Ultrex (WDG) *(ZENECA)*
Echo 500, 720 or 90DF *(Sostram)*
Terranil (6L, 90DF or 90DF WSP) *(Terra)*
Rate: Carrots:
1.1-1.4 lb. ai/A
2.25-2.75 pt. 500/A
1.1-1.5 lb. ai/A
1.5-2 pt. 720, 6L/A
1.4-1.8 lb. WDG/A
1.3-1.5 lb. 90DF, 90DF WSP/A
Disease:
Alternaria blight Cercospora blight
Time: Apply when disease threatens; repeat at 7- to 10-day intervals as necessary.

Rate: Parsnips:
1-1.5 lb. ai/A
2-3 pt. 500/A
1.5-2 pt. 720, 6L/A
1.4-1.8 lb. WDG/A
1.2-1.7 lb. 90DF, 90DF WSP/A
Disease:
Alternaria leaf spot Botrytis blight downy mildew
Anthracnose bottom rot
Time: Apply at first sign of disease or when conditions are favorable for infection; repeat at 7- to 10-day intervals.
Remarks: Carrots, parsnips. Use in sufficient water to obtain adequate coverage. Do not graze treated areas or feed treated plant parts to livestock. Refer to label for use directions and restrictions.
Days To Harvest: Carrots-6 weeks; parsnips-10 days.

Signal Word/Toxicity Class: Danger/I (WDG, 90DF); Warning/II (6L, 500, 720).
REI: 48 hr.

◆ - **new product** • ai-active ingredient • DF-dry flowable • E/EC-emulsifiable concentrate • EW-emulsion in water
F/FL-flowable • G-granule • L/LC-liquid • SP/WSP-soluble packet • W/WP-wettable powder • WSB-water soluble bag

Clean Crop C-O-C-S — see copper compounds

COPPER COMPOUNDS

Basicop (WP) (basic copper sulfate) *(Griffin)*
Rate: 1.5-2 lb. ai/A
3-4 lb. WP/A
Disease:
Alternaria leaf blight Cercospora leaf blight downy mildew
Time: Apply when disease appears; repeat at 7- to 10-day intervals.
Remarks: Use higher rate when disease is heavy. Do not apply through any type of irrigation system. Refer to label for use directions.

Signal Word/Toxicity Class: Warning/II.
REI: 24 hr.

Champ (FL or Formula 2) or Champion (WP) (copper hydroxide) *(Agtrol)*
Rate: Carrots:
2.66 pt. FL/A
1.33 pt. Formula 2/A
2 lb. WP/A
Disease:
carrot blight
Time: Apply when disease threatens; repeat at 7- to 14-day intervals.

Rate: Parsley:
2 pt. Formula 2/A
Disease:
bacterial blight
Time: Apply when plants are first established in field; repeat at 5- to 7-day intervals.
Remarks: Refer to label for use directions.

Signal Word/Toxicity Class: Danger/I (Champion); Caution (Champ).
REI: 24 hr.

Clean Crop C-O-C-S (WDG) (74.8% copper oxychloride & 14.2% copper sulfate, basic) *(Platte)*
Rate: 1-2 lb. ai/A
2-4 lb. WDG/A
Disease:
leaf blight leaf spot
Time: Repeat applications are necessary to protect new growth throughout growing season.
Remarks: Do not apply this product through any type of irrigation system. Refer to label for use directions.

Signal Word/Toxicity Class: Warning/II.
REI: 24 hr.

Copper-Count-N (L) (copper ammonium) *(Mineral Research)*
Rate: 2-3 qt. L/A
Disease:
early blight late blight
Time: Apply when plants are 6" high; 3-5 applications at 7- to 10-day intervals.
Remarks: Applications may be made by ground or air; refer to label for use directions. Do not apply through any type of irrigation system unless supplemental labeling on chemigation is followed.

Signal Word/Toxicity Class: Caution.
REI: 12 hr.

Drexel Basic Copper Sulfate (WP) or KOP 300 (3FL) (basic copper sulfate) *(Drexel)*
Rate: 3-4 lb. WP/A
0.33 gal. 3FL/A
Disease:
Alternaria leaf blight Cercospora leaf blight downy mildew
Time: Apply when disease first appears; repeat at 7- to 10-day intervals.
Remarks: Do not apply through any type of irrigation system. Refer to label for use directions.

Signal Word/Toxicity Class: Warning/II.
REI: 24 hr. (WP); 12 hr. (3FL).

Kocide (101, DF, 2000, 4.5LF, LF) (copper hydroxide) *(Griffin)*
Rate: Carrots:
2 lb. 101, DF/A
1.5 lb. 2000/A
1.33 pt. 4.5LF/A
2.66 pt. LF/A
Disease:
Alternaria leaf spot (2000, 4.5LF) Cercospora leaf spot
Time: Apply when disease threatens; repeat every 7- to 14-days depending on disease severity.

Rate: Parsley:
3 lb. 101, DF/A
2.25 lb. 2000/A
2 pt. 4.5LF/A
6 pt. LF/A
Disease:
bacterial blight
Time: Apply when plants are first established in field; repeat at 5- to 7-day intervals depending upon disease severity.
Remarks: Do not apply in a spray solution having a pH less than 6.5 as phytotoxicity could occur. Refer to label for use directions.
State Restrictions: Kocide 101, LF is not for use on parsley in California.

Signal Word/Toxicity Class: Danger/I (101, DF); Warning/II (2000); Caution (4.5LF, LF).
REI: 24 hr.

KOP Oxy-85 (WP) (copper oxychloride) *(Drexel)*
Rate: 4 lb. WP/A
Disease:
leaf spot (Cercospora) leaf spot (Macrosporium)
Time: Apply when disease is reported to be present in area; repeat at 7- to 10-day intervals as needed.
Remarks: Use in sufficient water and thoroughly wet plants. Refer to label for use directions.

Signal Word/Toxicity Class: Danger/I.
REI: 24 hr.

Micro Flo Basic Copper "53" (basic copper sulfate) or Microsperse COC (53WP) (copper oxychloride) *(Micro Flo)*
Rate: 2-4 lb. 53/A
3-6 lb. 53WP/A
Disease: 53:
late light (Alternaria) early blight (Cercospora) downy mildew
53WP:
leaf blight leaf spot
Time: Apply when disease first appears; repeat at 7- to 10-day intervals as needed.
Remarks: Carrots. Refer to label for use directions.

Signal Word/Toxicity Class: Danger/I (53WP); Warning/II (53).
REI: 24 hr.

Nu-Cop (3L, 50DF, 50WP) (copper hydroxide) *(Micro Flo)*
Rate: Carrots:
1.33-2.66 pt. 3L/A
2 lb. 50DF, 50WP/A
Disease:
Cercospora blight
Time: Apply when diseases threatens; repeat at 7- to 14-day intervals as needed.

Rate: Parsley:
4 pt. 3L/A
Disease:
bacterial blight
Time: Apply when plants are first established in field; repeat at 5- to 7-day intervals as needed.
Remarks: Carrots, parsley. Refer to label for use directions.

Signal Word/Toxicity Class: Danger/I (50DF, 50WP); Warning/II (3L).
REI: 48 hr. (50DF, 50WP); 24 hr. (3L).

NOTICE The information on these pages is for preliminary planning — not a guide for use. Be sure to follow the manufacturer's directions, notwithstanding information contained here.

For personal protective equipment and EPA registration numbers, please turn to page 39.

TOP COP Tri-Basic (FL) (basic copper sulfate) *(Stoller)*
Rate: 2-4 pt. FL/A
Disease:
Alternaria leaf blight Cercospora leaf spot downy mildew
Time: Apply when disease first appears; repeat at 7- to 10-day intervals.
Remarks: Refer to label for use directions.

Signal Word/Toxicity Class: Caution.
REI: 24 hr.

Copper-Count-N — see copper compounds

Echo — see chlorothalonil

FOSETYL-ALUMINUM

Aliette (WDG, WSP) *(Rhone-Poulenc)*
Rate: 1.6-4 lb. ai/A
2-5 lb. WDG, WSP/A
Disease:
downy mildew
Time: Apply when conditions favor disease development (high moisture and moderate temperatures); repeat at 7- to 21-day intervals.
Remarks: Parsley. Apply in sufficient water for good coverage. Do not apply less than 10 gal./A or make more than 7 applications per season. Do not tank mix with copper compounds. With moderate disease pressure, high rate and shorter interval should be used. Refer to label for use directions.
Days To Harvest: 3.

Signal Word/Toxicity Class: Caution.
REI: 12 hr.

IPRODIONE

Rovral (4FL, WG, WP) *(Rhone-Poulenc)*
Rate: 0.5-1 lb. ai/A
1-2 pt. 4FL/A
1-2 lb. WG, WP/A
Disease:
Alternaria blight black crown rot
Time: Apply when conditions favor disease development; repeat at 7- to 14-day intervals (7- to 10-day if tank mixed).
Remarks: Under severe disease conditions, use high rate and shorter spray intervals. May be tank mixed with another fungicide. Do not make more than 4 applications per season (10 applications if tank mixed). Do not drench. Refer to label for use directions.
Days To Harvest: 0.
Crop Rotation: Beans, broccoli, carrots, garlic, lettuce, onions (dry bulb), peanuts, potatoes, and rice can be rotated after harvest.

Signal Word/Toxicity Class: Warning/II (WG); Caution (4FL, WP).
REI: 12 hr.

Kocide — see copper compounds

KOP Oxy — see copper compounds

Kumulus — see sulfur

MEFENOXAM

Apron XL LS *(Novartis)*
Rate: 0.32-0.64 fl. oz. LS/100 lb. seed
Disease:
damping-off
Time: Seed treatment at planting.
Remarks: Do not use treated seed for food, feed, or oil purposes.

Signal Word/Toxicity Class: Warning/II.

Ridomil Gold EC or WSP *(Novartis)*
Rate: 0.5-1 lb. ai/A
1-2 pt. EC/A
0.45-0.9 lb. ai/A
1-2 lb. WSP/A
Disease:
Phytophthora Pythium
Time: Preplant incorporated or surface application just after planting.
Remarks: Carrots, parsley, parsnips. For banded applications, 7" band recommended. Do not use in greenhouse or on field grown vegetable bedding plants. Do not plant any crop which is not registered for use with mefenoxam in treated soil for 12 months.

MEFENOXAM (Premixes)

Systemic fungicide having a specific mode of action. Both are packaged in Accu-Pak, a wettable powder in a water-soluble bag. Toxic to fish; Ridomil Gold/Copper is also toxic to aquatic organisms.

Ridomil Gold/Bravo (WP) (4.5% mefenoxam & 72% chlorothalonil) *(Novartis)*
Ridomil Gold/Copper (WSB) (5% mefenoxam & 60% copper hydroxide) *(Novartis)*
Rate: 1.1-1.5 lb. ai/A
1.5-2 lb. WP/A
1.3 lb. ai/A
2 lb. WSB/A
Disease:
cavity spot Cercospora blight (WP only)
Time: May be applied 40-50 days after Ridomil Gold EC or WSP at planting application. Apply 2-4 applications at 14-day intervals.
Remarks: Carrots. For season-long control of cavity spot, a preplant or at-planting application of Ridomil Gold EC or WSP is recommended. Do not plant any crop which is not registered for use with mefenoxam in treated soil for 12 months.
Days To Harvest: 7.

Signal Word/Toxicity Class: Danger/I.
REI: 48 hr.

Mertect — see thiabendazole

METALAXYL

Allegiance or Apron (FL) *(Gustafson)*
Rate: 0.75 fl. oz. Allegiance/100 lb. seed
0.75-1.5 fl. oz. Apron/100 lb. seed
Disease:
damping-off
Time: Seed treatment.
Remarks: Use with EPA-approved dye that imparts unnatural color to seed. Do not use treated seed for food, feed, or oil purposes. Refer to label for use directions.
State Restrictions: Not for use in California.

Signal Word/Toxicity Class: Warning/II.

Micro Flo — see sulfur

Microthiol Special — see sulfur

Nu-Cop — see copper compounds

Nu-Gro Isotox — see captan premixes

Redball — see sulfur

Ridomil Gold — see mefenoxam

◆ - **new product** • ai-active ingredient • DF-dry flowable • E/EC-emulsifiable concentrate • EW-emulsion in water
F/FL-flowable • G-granule • L/LC-liquid • SP/WSP-soluble packet • W/WP-wettable powder • WSB-water soluble bag

Ridomil Gold/Bravo — see mefenoxam premixes

Ridomil Gold/Copper — see mefenoxam premixes

Rovral — see iprodione

Seed Mate Isotox — see captan premixes

SULFUR

Drexel Suffa (6FL), Sulfa 80 (80DF) or Sulfur (90W) *(Drexel)*
Kumulus DF *(BASF)*
Microthiol Special (80WP) *(Elf Atochem)*
Thiolux (DF) *(Platte)*

Rate: 4.5-9 lb. ai/A
6-12 pt. 6FL/A
5-10 lb. 90W/A
2.4-8 lb. ai/A
3-10 lb. 80DF, DF, 80WP/A

Disease:
powdery mildew

Time: Apply before disease appears; repeat as necessary.
Remarks: Do not use within 2 weeks of oil application (4 weeks 6FL). Some plants are susceptible to injury from sulfur under certain climatic conditions. Do not make sulfur applications at excessively high temperatures. Do not apply Drexel or Kumulus through any type of irrigation system. Refer to label for use directions.

Signal Word/Toxicity Class: Caution.
REI: 24 hr. (except Kumulus).

Micro Flo Sulfur (6L) or Dual Spraying & Dusting Sulfur (DS) *(Micro Flo)*
Terra Sulfur (6L) *(Terra)*

Rate: 10 lb. ai/A
13 pt. 6L/A
9 lb. ai/A
10 lb. DS/A

Disease:
powdery mildew

Time: Apply at early leaf stage; repeat at 14-day intervals as necessary.
Remarks: Do not use within 2 weeks of oil application. Do not apply DS through any type of irrigation system. Refer to label for use directions.

Signal Word/Toxicity Class: Caution.
REI: 24 hr.

Redball EM-53 (6L), Superfine Wettable Sulphur (WS) or Commercial Dusting Sulphur (DS) *(International Sulphur)*

Rate: 2-12 pt. 6L/A
5-10 lb. WS/100 gal. water
10-30 lb. DS/A

Disease:
powdery mildew

Time: Apply at first sign of disease; repeat as necessary.
Remarks: Apply in sufficient water for thorough coverage. Do not apply when temperatures exceed 90°F. Refer to label for use directions.

Signal Word/Toxicity Class: Caution.
REI: 12 hr.

SULFUR (Premix)

Top Cop with Sulfur (6.25 lb./gal. sulfur & 1 lb./gal. tribasic copper sulfate) *(Stoller)*

Rate: 2 qt./A

Disease:
Alternaria leaf blight   Cercospora blight   downy mildew

Time: Apply when disease first appears; repeat at 7- to 10-day intervals.
Remarks: Refer to label for use directions.

Signal Word/Toxicity Class: Caution.
REI: 24 hr.

Terranil — see chlorothalonil

TERRANIL Cu (Premix)

Terranil Cu (24% chlorothalonil & 24% copper oxychloride) *(Terra)*

Rate: 3.4-4.5 pt./A

Disease:
Alternaria blight   Cercospora blight

Time: Apply when disease threatens; repeat at 7- to 10-day intervals.
Remarks: Refer to label for use directions.

Signal Word/Toxicity Class: Danger/I.
REI: 48 hr.

THIABENDAZOLE

Toxic to fish.

Mertect 340-F *(Novartis)*

Rate: 41 fl. oz. 340-F/100 gal. water

Disease:
gray mold   Sclerotium

Time: Dip before storage, 5-10 seconds.
Remarks: Do not apply through any type of irrigation system.

Signal Word/Toxicity Class: Caution.

Thiolux — see sulfur

THIRAM

42-S Thiram or Thiram 50WP Dyed *(Gustafson)*

Rate: 8 fl. oz. 42-S/100 lb. seed
8 oz. 50WP/100 lb. seed

Disease:
damping-off   seed decay   seedling blight

Time: Seed treatment.
Remarks: Use with EPA-approved dye that imparts unnatural color to seed. Do not use treated seed for food, feed, or oil purposes. Refer to label for use directions.

Signal Word/Toxicity Class: Caution.

TOP COP — see copper compounds

Top Cop with Sulfur — see sulfur premix

Celery

Aliette — see fosetyl-aluminum

Basicop — see copper compounds

BENLATE

Benlate (SP, WP) (benomyl) *(DuPont)*

Rate: 0.13-0.25 lb. ai/A
0.25-0.5 lb. SP, WP/A

Disease:
early blight (Cercospora)   late blight (Septoria)

Time: Apply when disease first appears; repeat at 7- to 10-day intervals.
Remarks: Do not tank mix or alternate Benlate with benzimidazole or thiophanate products. Refer to label for use directions.
Days To Harvest: 7.
State Restrictions: Benlate SP not registered for use in California.

Signal Word/Toxicity Class: Caution.
REI: 24 hr.

 For personal protective equipment and EPA registration numbers, please turn to page 39.

Botran — see dicloran

Bravo — see chlorothalonil

Champ — see copper compounds

Champion — see copper compounds

CHLOROTHALONIL

Bravo 500, 720/Weather Stik, Ultrex (WDG) *(ZENECA)*
Echo 500, 720 or 90DF *(Sostram)*

Rate: Transplants:
0.75-2.25 lb. ai/A
1.5-4.25 pt. 500/A
1-3 pt. 720/A
0.9-2.7 lb. WDG/A
0.8-2.4 lb. 90DF/A

Disease:
basal rot; late blight; Septoria late blight; early blight (Cercospora); pink rot (suppression)

Time: Apply low rate at 3- to 5-day intervals; high rate at weekly intervals. Start applications shortly after crop emergence. For suppression of pink rot, use high rate on a weekly schedule.

Remarks: Can be applied through sprinkler irrigation equipment; refer to label for use directions and type of equipment.

Days To Harvest: 7.

- - - - - - - - - - - - - - - - -

Rate: Seedbeds:
1.1-1.4 lb. ai/100 gal.
2.25-2.75 pt. 500/100 gal.
1.5-2 pt. 720/100 gal.
1.4-1.8 lb. WDG/100 gal.
1.3-1.6 lb. 90DF/100 gal.

Disease:
early blight; late blight

Time: Start applications shortly after crop emergence. Apply 125 gal./A twice weekly or as needed to maintain control.

Remarks: Use higher rate under severe disease conditions. Refer to label for use directions and restrictions.

- - - - - - - - - - - - - - - - -

Signal Word/Toxicity Class: Danger/I (WDG, 90DF); Warning/II (500, 720).
REI: 48 hr.

Terranil (6L, 90DF or 90DF WSP) *(Terra)*

Rate: 3-5 day schedule:
0.78-1.12 lb. ai/A
1-1.5 pt. 6L/A
0.875-1.25 lb. 90DF, 90DF WSP/A
7 day schedule:
1.6-2.25 lb. ai/A
2-3 pt. 6L/A
1.75-2.5 lb. 90DF, 90DF WSP/A

Disease:
basal rot; Cercospora blight; Septoria blight

Time: Apply when transplants are set in field.

Rate: 1.12-1.4 lb. ai/A
1.5-2 pt. 6L/A
1.25-1.5 lb. 90DF, 90DF WSP/100 gal.

Disease:
early blight; late blight

Time: For seedbeds, apply 125 gal./A twice weekly or as needed starting shortly after crop emergence.

Remarks: Use in sufficient water to obtain adequate coverage. Use high rate under severe disease conditions. Refer to label for use directions.

Days To Harvest: 7.

- - - - - - - - - - - - - - - - -

Signal Word/Toxicity Class: Danger/I (90DF, 90DF WSP); Warning/II (6L).
REI: 48 hr.

Clean Crop C-O-C-S — see copper compounds

COPPER COMPOUNDS

Basicop (WP) (basic copper sulfate) *(Griffin)*

Rate: 1.5-2 lb. ai/A
3-4 lb. WP/A

Disease:
early blight; late blight; leaf spot

Time: Apply when plants are 6" tall, or when disease appears; repeat every 5- to 14-days as necessary.

Remarks: Use higher rate when disease is heavy. Do not apply through any type of irrigation system. Refer to label for use directions.

- - - - - - - - - - - - - - - - -

Signal Word/Toxicity Class: Warning/II.
REI: 24 hr.

Champ (FL or Formula 2) or Champion (WP) (copper hydroxide) *(Agtrol)*
KOP-Hydroxide (FL, 50WP) (copper hydroxide) *(Drexel)*

Rate: 2.66 pt. FL/A
1.33 pt. Formula 2/A
2 lb. WP, 50WP/A

Disease:
bacterial blight; early blight; late blight

Time: Apply when plants are established; repeat at 5- to 7-day intervals.

Remarks: KOP-Hydroxide 50WP should not be applied in spray solution having pH less than 6.5 or greater than 9.0 as phytotoxicity may occur. Refer to label for use directions.

- - - - - - - - - - - - - - - - -

Signal Word/Toxicity Class: Danger/I (Champion, KOP); Caution (Champ).
REI: 48 hr. (KOP); 24 hr. (Champ, Champion).

Clean Crop C-O-C-S (WDG) (74.8% copper oxychloride & 14.2% copper sulfate, basic) *(Platte)*

Rate: 1-2 lb. ai/A
2-4 lb. WDG/A

Disease:
blights; downy mildew

Time: Apply when disease first threatens; repeat at 7- to 10-day intervals to protect new growth throughout growing season.

Remarks: Do not apply this product through any type of irrigation system. Refer to label for use directions.

- - - - - - - - - - - - - - - - -

Signal Word/Toxicity Class: Warning/II.
REI: 24 hr.

Copper-Count-N (L) (copper ammonium) *(Mineral Research)*

Rate: 2-3 qt. L/A

Disease:
bacterial blight; early blight; late blight

Time: Apply as soon as plants are established; repeat at 5- to 7-day intervals.

Remarks: Apply by ground or air; refer to label for use directions. Do not apply through any type of irrigation system unless supplemental labeling on chemigation is followed.

- - - - - - - - - - - - - - - - -

Signal Word/Toxicity Class: Caution.
REI: 12 hr.

Drexel Basic Copper Sulfate (WP) or KOP 300 (3FL) (basic copper sulfate) *(Drexel)*

Rate: 3-4 lb. WP/A
2.6 pt. 3FL/A

Disease:
bacterial blight; early blight; late blight

Time: Treat every 7-days after emergence.

Remarks: Do not apply through any type of irrigation system. Refer to label for use directions.

- - - - - - - - - - - - - - - - -

Signal Word/Toxicity Class: Warning/II.
REI: 24 hr. (WP); 12 hr. (3FL).

Kocide (101, DF, 20/20, 2000, 4.5LF, LF) (copper hydroxide) *(Griffin)*

Rate: 2 lb. 101, DF/A
5 lb. 20/20/A
1.5 lb. 2000/A
1.33 pt. 4.5LF/A
2.66 pt. LF/A

Disease:
bacterial blight; early blight (Cercospora); late blight (Septoria)

◆ - **new product** • ai-active ingredient • DF-dry flowable • E/EC-emulsifiable concentrate • EW-emulsion in water
F/FL-flowable • G-granule • L/LC-liquid • SP/WSP-soluble packet • W/WP-wettable powder • WSB-water soluble bag

Time: Apply when plants are firmly established; repeat every 5- to 7-days as needed.
Remarks: Do not apply in a spray solution having a pH less than 6.5 as phytotoxicity could occur. Refer to label for use directions.

Signal Word/Toxicity Class: Danger/I (101, DF, 20/20); Warning/II (2000); Caution (4.5LF, LF).
REI: 24 hr.

KOP Oxy-85 (WP) (copper oxychloride) *(Drexel)*
Rate: 4 lb. WP/A
Disease:
early blight late blight
Time: Apply when disease is reported to be present in area; repeat at 7- to 10-day intervals as needed.
Remarks: Refer to label for use directions.

Signal Word/Toxicity Class: Danger/I.
REI: 24 hr.

Micro Flo Basic Copper "53" (basic copper sulfate) or Microsperse COC (53WP) (copper oxychloride) *(Micro Flo)*
Rate: 3-4 lb. 53/A
3-6 lb. 53WP/A
Disease:
bacterial blight early blight late blight
Time: Apply in plant bed or field when plants are 6" high or when disease first appears; repeat at 5- to 14-day intervals (5- to 10-day intervals 53WP).
Remarks: Refer to label for use directions.

Signal Word/Toxicity Class: Danger/I (53WP); Warning/II (53).
REI: 24 hr.

Nu-Cop (3L, 50DF, 50WP) (copper hydroxide) *(Micro Flo)*
Rate: 1.3-2.66 pt. 3L/A
1-2 lb. 50DF, 50WP/A
Disease:
bacterial blight early blight late blight
Time: Apply when plants are first established in field; repeat at 3- to 7-day intervals as needed.
Remarks: Refer to label for use directions.

Signal Word/Toxicity Class: Danger/I (50DF, 50WP); Warning/II (3L).
REI: 48 hr. (50DF, 50WP); 24 hr. (3L).

TOP COP Tri-Basic (FL) (basic copper sulfate) *(Stoller)*
Rate: 2-4 pt. FL/A
Disease:
bacterial blight early blight late blight
Time: Apply every 7 days after emergence.
Remarks: Refer to label for use directions.

Signal Word/Toxicity Class: Caution.
REI: 24 hr.

Copper-Count-N — see copper compounds

DICLORAN

Botran 75-W *(Gowan)*
Rate: 1.5 lb. ai/A
2 lb. 75-W/A
Disease:
pink rot
Time: Apply when disease is anticipated or approximately 10 weeks before harvest; repeat at 7-day intervals in summer or 14-day intervals in fall and winter. If early disease is unlikely, apply single application of 5.33 lb./A 4-8 weeks before harvest.
Remarks: Refer to label for use directions.
Days To Harvest: 7.

Signal Word/Toxicity Class: Caution.
REI: 12 hr.

Echo — see chlorothalonil

FOSETYL-ALUMINUM

Aliette (WDG, WSP) *(Rhone-Poulenc)*
Rate: 1.6-4 lb. ai/A
2-5 lb. WDG, WSP/A
Disease:
downy mildew
Time: Apply when conditions favor disease development (high moisture and moderate temperatures); repeat at 7- to 21-day intervals.
Remarks: Apply in sufficient water for good coverage. Do not apply less than 10 gal./A or make more than 7 applications per season. Do not tank mix with copper compounds. With moderate disease pressure, high rate and shorter interval should be used. Refer to label for use directions.
Days To Harvest: 3.

Signal Word/Toxicity Class: Caution.
REI: 12 hr.

Kocide — see copper compounds

KOP-Hydroxide — see copper compounds

KOP Oxy — see copper compounds

MEFENOXAM

Ridomil Gold EC, GR, or WSP *(Novartis)*
Rate: 0.5-1 lb. ai/A
1-2 pt. EC/A
20-40 lb. GR/A
0.45-0.9 lb. ai/A
1-2 lb. WSP/A
Disease:
damping-off
Time: Apply at planting.
Remarks: For banded applications, 7" band recommended. Do not use in greenhouse or on field grown vegetable bedding plants. Do not plant any crop which is not registered for use with mefenoxam in treated soil for 12 months.

Signal Word/Toxicity Class: Warning/II (EC); Caution (GR, WSP).
REI: 48 hr.

Micro Flo — see copper compounds

Microsperse COC — see copper compounds

Nu-Cop — see copper compounds

PROPICONAZOLE

Tilt (EC) *(Novartis)*
Rate: 4 fl. oz. EC/A
Disease:
early blight late blight
Time: Apply at 7-day intervals.
Remarks: Apply no more than 16 fl. oz./crop per season. May be tank mixed with appropriate spreader/sticker.
Days To Harvest: 14.

Signal Word/Toxicity Class: Warning/II.
REI: 24 hr.

Ridomil Gold — see mefenoxam

Sulfa — see sulfur

7 Vegetables

NOTICE The information on these pages is for preliminary planning — not a guide for use. Be sure to follow the manufacturer's directions, notwithstanding information contained here.

For personal protective equipment and EPA registration numbers, please turn to page 39.

SULFUR

Sulfa 80 (80DF) *(Drexel)*
Rate: 2-8 lb. ai/A
3-10 lb. 80DF/A
Disease:
powdery mildew rust
Time: Apply early leaf stage; repeat at 10- to 14-day intervals as needed, or immediately after rain.
Remarks: Do not use within 2 weeks of oil application. Some plants are susceptible to injury from sulfur under certain climatic conditions. Do not make sulfur applications at excessively high temperatures. Do not apply this product through any type of irrigation system. Refer to label for use directions.
- - - - - - - - - - - - - - - - - -
Signal Word/Toxicity Class: Caution.
REI: 24 hr.

SULFUR (Premix)

Top Cop with Sulfur (6.25 lb./gal. sulfur & 1 lb./gal. tribasic copper sulfate) *(Stoller)*
Rate: 2 qt./A
Disease:
bacterial blight late blight leaf spot
early blight
Time: Apply as foliar spray at weekly intervals beginning in plant bed; repeat at 7- to 10-day intervals in field when plants are established.
Remarks: Refer to label for use directions.
- - - - - - - - - - - - - - - - - -
Signal Word/Toxicity Class: Caution.
REI: 24 hr.

Terranil — see chlorothalonil

TERRANIL Cu (Premix)

Terranil Cu (24% chlorothalonil & 24% copper oxychloride) *(Terra)*
Rate: 3-5 day schedule:
2.25-3.4 pt./A
7 day schedule:
4.5-6.75 pt./A
Disease:
basal rot Cercospora blight Septoria blight
Time: Apply when transplants are set in field. For suppression of pink rot, use 7 day schedule rate.

Rate: 3.4-4.5 pt./100 gal.
Disease:
early blight late blight
Time: For seedbeds, apply 125 gal./A twice weekly or as needed starting shortly after crop emergence.
Remarks: Use in sufficient water to obtain adequate coverage. Use high rate under severe disease conditions. Refer to label for use directions.
Days To Harvest: 7.
- - - - - - - - - - - - - - - - - -
Signal Word/Toxicity Class: Danger/I.
REI: 48 hr.

Tilt — see propiconazole

TOP COP — see copper compounds

Top Cop with Sulfur — see sulfur premix

ZIRAM

Ziram 4L or 76 *(Drexel)*
Rate: 1-1.5 lb. ai/100 gal.
2-3 pt. 4L/100 gal.
0.76-1.5 lb. ai/100 gal.
1-2 lb. 76/100 gal.
Disease:
early blight late blight
Time: Apply before disease first appears; repeat at 5- to 10-day intervals as needed.
Remarks: Do not apply this product through any type of irrigation system. Refer to label for use directions.
State Restrictions: Not for use in California.
- - - - - - - - - - - - - - - - - -
Signal Word/Toxicity Class: Danger/I.
REI: 48 hr.

Cole Crops

(Bok Choy, Broccoli, Brussels Sprouts, Cabbage, Cauliflower, Kohlrabi, Rutabagas, Turnips)

ALIETTE/MANEB 2+2 (Co-Pack)

Twin pack product containing a systemic and a contact fungicide in attached water soluble packets. Each twin pack contains 2 lbs. of Aliette and 2 lbs. of Maneb.

Aliette/Maneb 2+2: Aliette (80% fosetyl-aluminum) *(Rhone-Poulenc)* **& Maneb (75% maneb)** *(Elf Atochem)*
Rate: 4 lb./A
Disease:
Alternaria leaf spot downy mildew
Time: Apply when conditions favor disease development; repeat at 7- to 10-day intervals.
Remarks: Broccoli, Brussels sprouts, cabbage, cauliflower, Chinese cabbage (tightheaded), kohlrabi. Apply in sufficient water for good coverage. Do not exceed 6 applications per season. Do not apply by air in less than 10 gal./A. Do not tank mix with copper fungicides. Do not mix with surfactants or foliar fertilizers. Refer to label for use directions and restrictions.
Days To Harvest: 7.
State Restrictions: Chemigation application not allowed in California.
- - - - - - - - - - - - - - - - - -
Signal Word/Toxicity Class: Caution.
REI: 24 hr.

Aliette — see fosetyl-aluminum

Basicop — see copper compounds

BENLATE

Benzimidazole systemic fungicide. Toxic to fish.

Benlate (SP, WP) (benomyl) *(DuPont)*
Rate: 0.25-1 lb. ai/A
0.5-2 lb. SP, WP/A
Disease:
Anthracnose ring spot Sclerotinia white mold
Botrytis gray mold
Time: Apply at first petal fall; repeat at 14-day intervals.
Remarks: Broccoli, Brussels sprouts, cabbage, cauliflower, Chinese cabbage, kohlrabi, rutabagas, turnips. Do not tank mix or alternate with benzimidazole or thiophanate products. Do not graze treated areas. Do not use seed or plant parts for food or feed purposes. Refer to label for use directions.
Days To Harvest: Brussels sprouts-7.
State Restrictions: Benlate SP not registered for use in California.
- - - - - - - - - - - - - - - - - -
Rate: 8 oz. SP, WP/100 lb. seed in 0.5-3 pt. water
Disease:
Phoma lingam
Time: Slurry seed treatment.
Remarks: Broccoli, Brussels sprouts, cabbage, Chinese cabbage, cauliflower, kohlrabi, rutabagas, turnips. Use EPA-approved dye that imparts unnatural color to treated seed. Allow seeds to dry before planting; seeds may be planted any time after being well cured. Do not tank mix or alternate with benzimidazole or thiophanate products. Do not use treated seed for food, feed, or oil purposes. Refer to label for use directions.
State Restrictions: Benlate SP not registered for use in California.
- - - - - - - - - - - - - - - - - -
Signal Word/Toxicity Class: Caution.
REI: 24 hr.

◆ - **new product** • ai-active ingredient • DF-dry flowable • E/EC-emulsifiable concentrate • EW-emulsion in water
F/FL-flowable • G-granule • L/LC-liquid • SP/WSP-soluble packet • W/WP-wettable powder • WSB-water soluble bag

Bravo — see chlorothalonil

Bravo S — see sulfur premix

Brom-O-Gas — see methyl bromide

CAPTAN

Captan 30-DD contains the toxic inert Rhodamine B.

Gustafson Captan 30-DD or 400 *(Gustafson)*
Rate: 1.25 fl. oz. 30-DD/100 lb. seed
1-2 fl. oz. 400/100 lb. seed
Disease:
damping-off, seed decay, seedling blight, mold
Time: Seed treatment.
Remarks: Broccoli, Brussels sprouts, cabbage, cauliflower, turnips. Do not combine with colloidal suspensions, wettable powders, or emulsifiable concentrates unless previous experience assures compatibility. Use with EPA-approved dye that imparts unnatural color to seed. Do not use treated seed for food, feed, or oil purposes. Refer to label for use directions.

Signal Word/Toxicity Class: Danger/I.

Drexel Captan 4FL *(Drexel)*
Rate: 1.5 fl. oz. 4FL/100 lb. seed
Disease:
seed decay, seedling blight, soilborne disease
Time: Liquid or slurry seed treatment.
Remarks: Broccoli, Brussels sprouts, cabbage, cauliflower. Do not use treated seed for food, feed, or oil purposes. Refer to label for use directions.

Signal Word/Toxicity Class: Danger/I.

CAPTAN (Premix)

Premix combining a fungicide with an insecticide. Toxic to fish, birds, and wildlife.

Nu-Gro Isotox (12.2% captan & 25% lindane) *(Wilbur-Ellis)*
Seed Mate Isotox (12.2% captan & 25% lindane) *(Loveland)*
Rate: 2 oz./100 lb. seed
Disease:
damping-off, seed decay, seedling blight
Time: Planter box treatment.
Remarks: Treated seed must not be used for or mixed with food or animal feed, or processed for oil. Refer to label for use directions.

Signal Word/Toxicity Class: Danger/I.
REI: 4 days.

Champ — see copper compounds

Champion — see copper compounds

CHLOROTHALONIL

Broad-spectrum contact fungicide. Toxic to fish, aquatic invertebrates, and marine/estuarine organisms.

Bravo 500, 720/Weather Stik, Ultrex (WDG) *(ZENECA)*
Echo 500, 720 or 90DF *(Sostram)*
Rate: 1.12 lb. ai/A
2.25 pt. 500/A
1.5 pt. 720/A
1.4 lb. WDG/A
1.3 lb. 90DF/A
Disease:
Alternaria leaf spot, downy mildew
Time: Apply after transplants are set in the field, shortly after emergence of field-seeded crop, or when conditions favor disease development; repeat at 7- to 10-day intervals as necessary.

Rate: Field-seeded Brussels sprouts:
1.35 lb. ai/A
2.75 pt. 500/A
1.5 lb. ai/A
1.5 lb. 90DF/A
2 pt. 720/A
1.12 lb. ai/A
1.4 lb. WDG/A
Disease:
ring spot (California)
Time: Apply at time of early sprout development or when conditions favor disease development; repeat at 7- to 10-day intervals as necessary.
Remarks: Broccoli, Brussels sprouts (field-seeded), cabbage, cauliflower, Chinese broccoli, Chinese cabbage (tightheaded varieties). Use in sufficient water to obtain adequate coverage. Refer to label for use directions and restrictions.
Days To Harvest: Chinese broccoli, Chinese cabbage-7.

Signal Word/Toxicity Class: Danger/I (WDG, 90DF); Warning/II (500, 720).
REI: 48 hr.

Terranil (6L, 90DF or 90DF WSP) *(Terra)*
Rate: 1.12-1.4 lb. ai/A
1.5-2 pt. 6L/A
1.25-1.5 lb. 90DF, 90DF WSP/A
Disease:
Alternaria leaf spot, downy mildew, ring spot (CA only)
Time: Apply after transplants are set in field, or shortly after emergence of field-seeded crops, or when conditions favor disease development; repeat at 7- to 10-day intervals as needed. For powdery mildew, apply at first sign of infection; repeat at 3-week intervals. For field-seeded Brussels sprouts, apply at time of early sprout development or when conditions favor disease development; repeat at 7- to 10-day intervals.
Remarks: Broccoli, Brussels sprouts, cabbage, cauliflower, Chinese broccoli, Chinese cabbage (tightheaded). Use in sufficient water to obtain adequate coverage. Do not graze treated areas or feed treated plant parts to livestock. Refer to label for use directions.
Days To Harvest: Chinese broccoli, Chinese cabbage-7.

Signal Word/Toxicity Class: Danger/I (90DF, 90DF WSP); Warning/II (6L).
REI: 48 hr.

Clean Crop C-O-C-S — see copper compounds

COPPER COMPOUNDS

Fungicide/bactericide.

Basicop (WP) (basic copper sulfate) *(Griffin)*
Micro Flo Basic Copper "53" (basic copper sulfate) *(Micro Flo)*
Rate: 0.5-1.5 lb. ai/A
1-3 lb. WP, 53/A
Disease:
downy mildew, leaf spot
Time: Apply when disease appears, repeat at 7- to 10-day intervals.
Remarks: Broccoli, Brussels sprouts, cabbage, cauliflower. Use higher rate when disease is heavy. Do not apply through any type of irrigation system. Refer to label for use directions.

Signal Word/Toxicity Class: Warning/II.
REI: 24 hr.

Champ (FL or Formula 2) or Champion (WP) (copper hydroxide) *(Agtrol)*
Rate: Broccoli, Brussels sprouts, cabbage, cauliflower:
0.66-1.33 pt. FL/A
0.33-0.66 pt. Formula 2/A
0.5-1 lb. WP/A
Disease:
downy mildew
Rate: Cabbage:
2.66 pt. FL/A
1.33 pt. Formula 2/A
2 lb. WP/A
Disease:
black leaf spot, black rot
Time: Apply after transplanting or shortly after emergence of field-seeded crops; repeat at 7- to 10-day intervals.

7 Vegetables

NOTICE	The information on these pages is for preliminary planning — not a guide for use. Be sure to follow the manufacturer's directions, notwithstanding information contained here.	For personal protective equipment and EPA registration numbers, please turn to page 39.

Remarks: Broccoli, Brussels sprouts, cabbage, cauliflower. Use high rate when conditions favor disease. Reddening of older leaves may occur on broccoli and a flecking of wrapper leaves may occur on cabbage. Refer to label for use directions.

Signal Word/Toxicity Class: Danger/I (Champion); Caution (Champ).
REI: 24 hr.

Clean Crop C-O-C-S (WDG) (74.8% copper oxychloride & 14.2% copper sulfate, basic) *(Platte)*
Rate: 1.5-2 lb. ai/A
3-4 lb. WDG/A
Disease:
black rot downy mildew
Time: Apply after transplants are set in field or after emergence of seedlings; repeat at 7- to 10-day intervals to protect new growth throughout growing season.
Remarks: Broccoli, Brussels sprouts, cabbage, cauliflower. Reddening of older leaves may occur on broccoli and flecking may occur on wrapper leaves of cabbage. Do not apply this product through any type of irrigation system. Refer to label for use directions.

Signal Word/Toxicity Class: Warning/II.
REI: 24 hr.

Copper-Count-N (L) (copper ammonium) *(Mineral Research)*
Rate: Broccoli, Brussels sprouts, cauliflower:
2 qt. L/A
Disease:
Alternaria black rot Xanthomonas
black leaf spot

Rate: Cabbage:
0.5-1 qt. L/A
Disease:
downy mildew
Time: Apply in minimum of 25 gal./A; repeat at 7-day intervals for cabbage, 7- to 10-day intervals for broccoli, Brussels sprouts, cauliflower.
Remarks: Broccoli, Brussels sprouts, cabbage, cauliflower. Applications may be made by ground or air; refer to label for use directions. Do not apply through any type of irrigation system unless supplemental labeling on chemigation is followed.
State Restrictions: Not for use on downy mildew in California.

Signal Word/Toxicity Class: Caution.
REI: 12 hr.

Drexel Basic Copper Sulfate (WP) or KOP 300 (3FL) (basic copper sulfate) *(Drexel)*
TOP COP Tri-Basic (FL) (basic copper sulfate) *(Stoller)*
Rate: 1-3 lb. WP/A
0.33 gal. 3FL/A
2-4 pt. FL/A
Disease:
downy mildew leaf spot
Time: Apply when plants are above ground in plant bed or before disease appears; repeat at 7- to 10-day intervals.
Remarks: Broccoli, Brussels sprouts, cabbage, cauliflower; TOP COP only: kohlrabi, turnips. Do not apply through any type of irrigation system. Refer to label for use directions.

Signal Word/Toxicity Class: Warning/II; Caution (FL).
REI: 24 hr. (WP, FL); 12 hr. (3FL).

Kocide (101, DF, 2000, 4.5LF, LF) (copper hydroxide) *(Griffin)*
Rate: 1-2 lb. 101, DF/A
0.75-1.5 lb. 2000/A
0.66-1.33 pt. 4.5LF/A
0.66-2.66 pt. LF/A
Disease:
black leaf spot black rot downy mildew
Time: Apply after transplanting or shortly after emergence of field-seeded crops or when conditions favor disease development; repeat at 7- to 10-day intervals.
Remarks: Broccoli, Brussels sprouts, cabbage, cauliflower. Reddening of older leaves could occur on broccoli and flecking of wrapper leaves could occur on cabbage. Refer to label for use directions.

Signal Word/Toxicity Class: Danger/I (101, DF); Warning/II (2000); Caution (4.5LF, LF).
REI: 24 hr.

KOP-Hydroxide (FL, 50WP) (copper hydroxide) *(Drexel)*
Rate: Broccoli, Brussels sprouts, cauliflower:
0.66-1.33 pt. FL/A
0.5-1 lb. 50WP/A
Disease:
downy mildew
Time: Apply at 7-day intervals.

Rate: Cabbage:
2.66 pt. FL/A
2 lb. 50WP/A
Disease:
black leaf spot black rot
Time: Apply at 7- to 10-day intervals.
Remarks: Broccoli, Brussels sprouts, cabbage, cauliflower. A slight reddening of older leaves may occur on broccoli, and slight flecking of wrapper leaves may occur on cabbage. KOP-Hydroxide 50WP should not be applied in spray solution having pH less than 6.5 or greater than 9.0 as phytotoxicity may occur. Refer to label for use directions.

Signal Word/Toxicity Class: Danger/I.
REI: 48 hr.

Nu-Cop (3L, 50DF, 50WP) (copper hydroxide) *(Micro Flo)*
Rate: Broccoli, Brussels sprouts, cauliflower:
0.33-1.33 pt. 3L/A
0.5-1 lb. 50DF, 50WP/A
Disease:
downy mildew

Rate: Cabbage:
1.33-2.66 pt. 3L/A
2 lb. 50DF, 50WP/A
Disease:
black leaf spot black rot
Time: Apply after transplants are set in field, or shortly after emergence of field seeded crops; repeat at 7-day intervals (7- to 10-day intervals for cabbage).
Remarks: Broccoli, Brussels sprouts, cabbage, cauliflower. Refer to label for use directions.

Signal Word/Toxicity Class: Danger/I (50DF, 50WP); Warning/II (3L).
REI: 48 hr. (50DF, 50WP); 24 hr. (3L).

Copper-Count-N — see copper compounds

Echo — see chlorothalonil

FOSETYL-ALUMINUM

Systemic fungicide/bactericide. toxic to estuarine invertebrates.

Aliette (WDG, WSP) *(Rhone-Poulenc)*
Rate: 1.6-4 lb. ai/A
2-5 lb. WDG, WSP/A
Disease:
downy mildew
Time: Apply when conditions favor disease development; repeat at 7- to 21-day intervals.
Remarks: Bok choy, broccoli, Brussels sprouts, cabbage, cauliflower, Chinese broccoli, kohlrabi. Do not apply less than 10 gal./A or make more than 7 applications per season. Do not tank mix with copper compounds. With moderate disease pressure, high rate and shorter interval should be used. Refer to label for use directions.
Days To Harvest: 3.

Signal Word/Toxicity Class: Caution.
REI: 12 hr.

IPRODIONE

Dicarboximide fungicide. Toxic to aquatic invertebrates.

Rovral (4FL, WG, WP) *(Rhone-Poulenc)*
Rate: 1 lb. ai/A
2 pt. 4FL/A
2 lb. WG, WP/A
Disease:
blackleg

◆ - **new product** • ai-active ingredient • DF-dry flowable • E/EC-emulsifiable concentrate • EW-emulsion in water
F/FL-flowable • G-granule • L/LC-liquid • SP/WSP-soluble packet • W/WP-wettable powder • WSB-water soluble bag

Time: Apply immediately after thinning (2-4 leaf stage). If disease conditions persist or recur, second application may be made up to day of harvest.
Remarks: Broccoli. Do not make more than 2 applications per crop. Do not drench. Refer to label for use directions.
Days To Harvest: 0.
Crop Rotation: Beans, broccoli, carrots, garlic, lettuce, onions (dry bulb), peanuts, potatoes, and rice.

Signal Word/Toxicity Class: Warning/II (WG); Caution (4FL, WP).
REI: 12 hr.

Kocide — see copper compounds

KOP-Hydroxide — see copper compounds

Kumulus — see sulfur

MANEB

Dithiocarbamate contact fungicide. Toxic to fish.

Maneb 75DF or 80WP *(Elf Atochem)*
Manex (FL) *(Griffin)*

Rate: 1.12-1.5 lb. ai/A
1.5-2 lb. 75DF/A
1.2-1.6 lb. ai/A
1.5-2 lb. 80WP/A
1.2-1.6 qt. FL/A

Disease:
Alternaria leaf spot   downy mildew

Time: Plant beds and direct seeded fields, spray on 7- to 10-day intervals.
Days To Harvest: 7.

Rate: California, Hawaii-looseheaded Chinese cabbage:
0.75-1.12 lb. ai/A
1-1.5 lb. 75DF/A
0.8-1.2 lb. ai/A
1-1.5 lb. 80WP/A
0.8-1.2 qt. FL/A

Disease:
Alternaria leaf spot   downy mildew

Time: Apply at 7- to 10-day intervals.
Remarks: Broccoli, Brussels sprouts, cabbage, cauliflower, kohlrabi, Chinese cabbage (tightheaded, looseheaded). Do not use any other product containing a different EBDC ai during the same growing season if using Manex. Refer to label for use directions.
Days To Harvest: 7; loose-head Chinese cabbage: Hawaii-10.
State Restrictions: 80WP for use in California and Hawaii only.

Signal Word/Toxicity Class: Caution.
REI: 24 hr.

Manex — see maneb

MEFENOXAM

Systemic fungicide having a specific mode of action.

Ridomil Gold EC or WSP *(Novartis)*

Rate: Broccoli, cabbage, cauliflower:
0.25-0.5 lb. ai/A
0.5-1 pt. EC/A
0.45-0.9 lb. ai/A
1-2 lb. WSP/A

Disease:
basal rot   damping-off

Rate: Rutabagas, turnips:
0.5-1 lb. ai/A
1-2 pt. EC/A
0.45-0.9 lb. ai/A
1-2 lb. WSP/A

Disease:
Phytophthora   Pythium

Time: Preplant incorporated or surface application just after planting.
Remarks: Broccoli, cabbage, cauliflower, rutabagas, turnips. For banded applications, 7" band recommended. Do not use in greenhouse or on field grown vegetable bedding plants. May be tank mixed. Do not plant any crop which is not registered for use with mefenoxam in treated soil for 12 months.
Days To Harvest: Broccoli, cabbage, cauliflower-1.

Signal Word/Toxicity Class: Warning/II (EC); Caution (WSP).
REI: 48 hr.

MEFENOXAM (Premix)

Systemic fungicide having a specific mode of action. Packaged in Accu-Pak, a wettable powder in a water-soluble bag. Toxic to fish.

Ridomil Gold/Bravo (WP) (4.5% mefenoxam & 72% chlorothalonil) *(Novartis)*

Rate: 1.1 lb. ai/A
1.5 lb. WP/A

Disease:
Alternaria leaf spot   downy mildew

Time: Apply when conditions are favorable for disease development, but before infection, repeat at 14-day intervals until disease threat is over.
Remarks: Broccoli, Brussels sprouts, cabbage, cauliflower. Up to 4 applications per crop may be applied. Under severe Alternaria leaf spot pressure, use other EPA-registered products between applications. Do not plant any crop which is not registered for use with mefenoxam in treated soil for 12 months.
Days To Harvest: 7.

Signal Word/Toxicity Class: Danger/I.
REI: 48 hr.

METHYL BROMIDE

2% Chloropicrin added as warning agent. Restricted Use Pesticide. Classified as an ozone depleter under the Clean Air Act. Use is to be completely eliminated starting Jan. 1, 2001.

Brom-O-Gas *(Great Lakes Chemical)*

Rate: Injected with chisels + covered with tarp:
180-240 lb./A
Raised tarp with topical application:
0.4-0.55 lb./100 ft.

Disease:
Fusarium   Plasmodiophora   Sclerotinia
oak root fungus   Pythium   Sclerotium
Phytophthora   Rhizoctonia

Time: Preplant.
Remarks: Broccoli, cauliflower. Do not fumigate when soil temperature is below 50°F. For best results, fumigate when temperature is 60°-80°F at depth of 6". Refer to label for use directions.
State Restrictions: Brom-O-Gas not for use in Arizona and not for use in California for oak root fungus (Armillaria root rot) control.

Signal Word/Toxicity Class: Danger/I.

Micro Flo — see copper compounds

Microthiol Special — see sulfur

Nu-Cop — see copper compounds

Nu-Gro Isotox — see captan premixes

PCNB

Toxic to fish and aquatic organisms.

PCNB 10%, 75DG or 75WP *(Amvac)*

Rate: Broadcast:
30 lb. ai/A
40 lb. 75DG, 75WP/A in 30 gal. water
300 lb. 10%/A
Band:
22.5 lb. ai/A
30 lb. 75DG, 75WP/A in 25 gal. water

NOTICE The information on these pages is for preliminary planning — not a guide for use. Be sure to follow the manufacturer's directions, notwithstanding information contained here.

For personal protective equipment and EPA registration numbers, please turn to page 39.

Transplants:
1.5-4.5 lb. ai/A
2-6 lb. 75DG, 75WP/A in 100 gal. water
Pretransplant:
20 lb. ai
200 lb. 10%/13,100 ft. (cauliflower–10,900 lin. ft.)

Disease:
club root

Time: For band (12-15") and broadcast application, apply immediately prior to planting.

Rate: Broadcast drench:
1.5-4.5 lb. ai/A
15-20 lb. 75DG, 75WP/A in 50 gal. water
1.5-4.5 lb. ai/A
10-15 lb. 75DG, 75WP/A in 35 gal. water
Band-**Cauliflower:**
110-150 lb. 10%/10,900 lin. ft. row

Disease:
black rot wire stem

Time: For soil drench or 8" band centered on row drench, apply at time of or immediately after seeding.

Remarks: Broccoli, Brussels sprouts, cabbage, cauliflower. Use 40" row spacing; (48" row spacing for cauliflower). For club root, thoroughly mix PCNB into soil. For band or broadcast, incorporate to a depth of 4-6". Do not apply through any type of irrigation system. Do not plant root crops within 12 months of application unless PCNB is registered for use on those crops.

Signal Word/Toxicity Class: Caution.
REI: 12 hr.

Terraclor (75WP) or Terraclor 400 (FL) *(Uniroyal)*
Rate: 4-8 oz. 75WP/100 gal. water
6-12 fl. oz. FL/100 gal. water

Disease:
damping-off root and stem rot

Time: Soil drench. One repeat application may be made 4-6 weeks later if necessary.

Remarks: Broccoli, Brussel sprouts, cabbage, cauliflower. Bedding plants in greenhouse or nursery. Refer to label for use directions.

Crop Rotation: Do not plant root crops in PCNB treated fields within 12 months of broadcast and banding applications unless PCNB registered for use on those crops.

Signal Word/Toxicity Class: Caution.

Turfcide (10G) *(Uniroyal)*
Rate: 1.5 oz. 10G/100 ft. row

Disease:
root and stem rot

Rate: 11 oz. 10G/100 sq. ft.

Disease:
club root

Time: Apply at planting time.

Remarks: Broccoli, Brussel sprouts, cabbage, cauliflower bedding plants container grown. Direct material into planting row and cover soil. Refer to label for use directions.

Crop Rotation: Do not plant root crops in PCNB treated fields within 12 months of broadcast and banding applications unless PCNB registered for use on those crops.

Signal Word/Toxicity Class: Caution.

Redball — see sulfur

Ridomil Gold — see mefenoxam

Ridomil Gold/Bravo — see mefenoxam premix

Rovral — see iprodione

Seed Mate — see captan premixes

SULFUR

Nonsystemic fungicide with protective action. Microthiol Special, Sulfa 80, and Thiolux are micronized sulfurs.

Drexel Suffa (6FL), Sulfa 80 (80DF) or Sulfur (90W) *(Drexel)*
Rate: 2.3-4.5 lb. ai/A
3-6 pt. 6FL/A
2-8 lb. ai/A
3-10 lb. 80DF/A
2.7-4.5 lb. ai/A
3-5 lb. 90W/A

Disease:
powdery mildew rust (80DF)

Time: Apply early leaf stage; repeat at 10- to 14-day intervals as needed, or immediately after rain.

Remarks: Broccoli, Brussels sprouts, cabbage, cauliflower; also rutabagas, turnips for 80DF. Do not use within 2 weeks of oil application (4 weeks 6FL). Some plants are susceptible to injury from sulfur under certain climatic conditions. Do not make sulfur applications at excessively high temperatures. Do not apply this product through any type of irrigation system. Refer to label for use directions.

Signal Word/Toxicity Class: Caution.
REI: 24 hr.

Dusting Sulfur (DS) *(Platte)*
Rate: 15-20 lb. DS/A

Disease:
powdery mildew

Time: Apply at 7- to 14-day intervals.

Remarks: Rutabagas, turnips. Do not make applications within 2 weeks of oil spray. Refer to label for use directions.

Signal Word/Toxicity Class: Caution.
REI: 24 hr.

Kumulus DF *(BASF)*
Rate: 2.4-4 lb. ai/A
3-5 lb. DF/A

Disease:
powdery mildew Septoria leaf spot

Time: Apply at early leaf stage; repeat at 10- to 14-day intervals with or without other pesticides immediately after a rain.

Remarks: Do not apply if temperatures during or within 3 days after application are expected to exceed 90°F. Do not apply through any type of irrigation system. Refer to label for use directions.

Signal Word/Toxicity Class: Caution.
REI: 24 hr.

Liquid Sulfur Six *(Helena)*
Rate: 3-4 pt./8 gal. water

Disease:
powdery mildew

Time: Apply at first sign of infection; repeat at 3-week intervals.

Remarks: Broccoli, Brussels sprouts, cabbage, cauliflower. Apply by air. Do not apply this product through any type of irrigation system. Refer to label for use directions.

Signal Word/Toxicity Class: Caution.
REI: 24 hr.

Micro Flo Sulfur (6L), Dual Spraying & Dusting Sulfur (DS) or Microsperse Wettable Sulfur (WS) *(Micro Flo)*
Terra Sulfur (6L) *(Terra)*
Rate: 2.25-8 lb. ai/A
3-10.66 pt. 6L/A
4.5 lb. ai/A
5 lb. DS/A
2.7-4.5 lb. ai/A
3-5 lb. WS/A

Disease:
powdery mildew rust (6L)

Time: Apply early leaf stage; repeat at 10- to 14-day intervals (14-day intervals-DS) as necessary.

Remarks: Broccoli, Brussels sprouts, cabbage, cauliflower, rutabagas, turnips. Do not apply DS and WS through any type of irrigation system. Refer to label for use directions.

Signal Word/Toxicity Class: Caution.
REI: 24 hr.

◆ - **new product** • ai-active ingredient • DF-dry flowable • E/EC-emulsifiable concentrate • EW-emulsion in water
F/FL-flowable • G-granule • L/LC-liquid • SP/WSP-soluble packet • W/WP-wettable powder • WSB-water soluble bag

Microthiol Special (80WP) *(Elf Atochem)*
Rate: 2.4-8 lb. ai/A
3-10 lb. 80WP/A
Rutabagas, turnips-CA:
4 lb. ai/A
5 lb. 80WP/A
Disease:
powdery mildew
Time: Apply early leaf stage; repeat at 10- to 14-day intervals as needed. In California, repeat at 14-day intervals as necessary for rutabagas, turnips.
Remarks: Broccoli, Brussels sprouts, cabbage, cauliflower, rutabagas, turnips. Refer to label for use directions.

Signal Word/Toxicity Class: Caution.
REI: 24 hr.

Redball EM-53 (6L), Superfine Wettable Sulphur (WS) or Commercial Dusting Sulphur (DS) *(International Sulphur)*
Rate: 2-7 pt. 6L/A
6-16 lb. WS/A
10-25 lb. DS/A
Disease:
powdery mildew
Time: 6L: Apply early leaf stage; repeat as necessary. WS, DS: Apply when disease first appears; repeat at 7- to 14-day intervals (WS), 7- to 10-day intervals (DS).
Remarks: Broccoli, Brussels sprouts, cabbage, cauliflower; WS, DS: rutabagas, turnips. Apply in sufficient water for thorough coverage. Do not apply when temperatures exceed 90°F. Refer to label for use directions.

Signal Word/Toxicity Class: Caution.
REI: 12 hr.

Thiolux (DF) *(Platte)*
Rate: 3-5 lb. DF/A
Rutabagas, turnips:
3-10 lb. DF/A
Disease:
powdery mildew; rust
Time: Apply early leaf stage; repeat at 10- to 14-day intervals with or without pesticides immediately after rain.
Remarks: Broccoli, Brussels sprouts, cabbage, cauliflower, rutabagas. Do not make applications within 2 weeks of oil spray. Refer to label for use directions.

Signal Word/Toxicity Class: Caution.
REI: 24 hr.

SULFUR (Premixes)

Fungicide/bactericide in flowable formulation. Toxic to fish and aquatic organisms. Terranil S is also toxic to marine/estuarine organisms.

Bravo S (27.25% sulfur & 19.15% chlorothalonil) *(Helena)*
Rate: 4.5-5.5 pt./A
Disease:
Alternaria leaf spot; powdery mildew; ring spot (California only); downy mildew
Time: Alternaria leaf spot and downy mildew: Apply after transplants are set in field, or shortly after emergence of field seeded crop, or when conditions favor disease development; repeat at 7- to 10-day intervals as needed. Powdery mildew: Apply at first sign of infection; repeat at 3-week intervals. Ring spot (California only, field-seeded Brussels sprouts): Apply at time of early sprout development, or when conditions favor disease development; repeat at 7- to 10-day intervals as needed.
Remarks: Broccoli, Brussels sprouts, cabbage, cauliflower. For powdery mildew, apply in 8 gal. water by air. Refer to label for use directions.
Days To Harvest: 7

Signal Word/Toxicity Class: Caution.
REI: 48 hr.

Terranil S (27.25% sulfur & 19.15% chlorothalonil) *(Terra)*
Rate: 4.5-5.5 pt./A
Disease:
Alternaria leaf spot; powdery mildew; ring spot (CA only); downy mildew
Time: Apply after transplants are set in field, shortly after emergence of field-seeded crops, or when conditions favor disease development; repeat at 7- to 10-day intervals as needed. For powdery mildew, apply at first sign of infection; repeat at 3-week intervals. For field-seeded Brussels sprouts, apply at time of early sprout development or when conditions favor disease development; repeat at 7- to 10-day intervals.
Remarks: Broccoli, Brussels sprouts, cabbage, cauliflower, Chinese broccoli, Chinese cabbage. Do not graze treated areas or feed treated plant parts to livestock. Refer to label for use directions.
Days To Harvest: Chinese broccoli, Chinese cabbage-7.

Signal Word/Toxicity Class: Warning/II.
REI: 48 hr.

Top Cop with Sulfur (6.25 lb./gal. sulfur & 1 lb./gal. tribasic copper sulfate) *(Stoller)*
Rate: 2 qt./A
Disease:
black rot; downy mildew; leaf spot
Time: Apply when disease threatens. Repeat at 7- to 10-day intervals; 3-day intervals in plant beds.
Remarks: Broccoli, Brussels sprouts, cabbage, cauliflower, Chinese cabbage, kohlrabi, turnips. Refer to label for use directions.

Signal Word/Toxicity Class: Caution.
REI: 24 hr.

Terraclor — see PCNB

Terranil — see chlorothalonil

Terranil S — see sulfur premixes

TERR-O-GAS (Premix)

Restricted Use Pesticide. Classified as an ozone depleter under the Clean Air Act. Use is to be completely eliminated starting Jan. 1, 2001.

Terr-O-Gas (67% methyl bromide & 33% chloropicrin) *(Great Lakes Chemical)*
Rate: Nontarp:
270-355 lb./A
Disease:
Fusarium; Pythium; Sclerotinia; Phytophthora; Rhizoctonia; Sclerotium; Plasmodiophora
Time: Apply up to 2 weeks prior to planting when soil moisture is good and temperature is 60°-85°F.
Remarks: Broccoli, cauliflower. Work upwind of product and never fumigate without 2 qualified applicators. Do not fumigate when there is no or little air movement. Tarping is not used after field application. Refer to label for use directions.

Signal Word/Toxicity Class: Danger/I.

Thiolux — see sulfur

THIRAM

Broad-spectrum fungicide. Toxic to fish. Thiram 50WP contains the toxic inert Rhodamine B and is also toxic to birds and wildlife.

42-S Thiram or Thiram 50WP Dyed *(Gustafson)*
Thiram Granuflo (WDG) *(UCB Agrochemicals)*
Rate: 8 fl. oz. 42-S/100 lb. seed
8 oz. 50WP/100 lb. seed
5.3 oz. WDG/100 lb. seed
Disease:
damping-off; seed decay; seedling blight
Time: Seed treatment.

7
Vegetables

NOTICE The information on these pages is for preliminary planning — not a guide for use. Be sure to follow the manufacturer's directions, notwithstanding information contained here.	For personal protective equipment and EPA registration numbers, please turn to page 39.

Remarks: Broccoli, Brussels sprouts, cabbage, cauliflower, kohlrabi, turnips. Use with EPA-approved dye that imparts unnatural color to seed. Do not use treated seed for food, feed, or oil purposes. Do not apply Thiram Granuflo through any type of irrigation system. Refer to label for use directions.
State Restrictions: Thiram Granuflo not for use in California.

Signal Word/Toxicity Class: Caution.
REI: 24 hr. (Thiram Granuflo).

TOP COP — see copper compounds

Top Cop with Sulfur — see sulfur premix

Turfcide — see PCNB

Cucurbits

(Cucumbers, Melons, Pumpkins, Squash)

ALIETTE/MANEB 2+2 (Co-Pack)

Twin pack product containing a systemic and a contact fungicide in attached water soluble packets. Each twin pack contains 2 lbs. of Aliette and 2 lbs. of Maneb.

Aliette/Maneb 2+2: Aliette (80% fosetyl-aluminum) *(Rhone-Poulenc)* **& Maneb (75% maneb)** *(Elf Atochem)*
Rate: 4 lb./A
Disease: Cucumbers:
Alternaria leaf spot, Anthracnose, Pythium fruit rot, angular leaf spot, downy mildew
Melons:
Alternaria leaf spot, Cercospora leaf spot, gummy stem blight, Anthracnose, downy milew
Pumpkins:
angular leaf spot, downy mildew
Squash:
Anthracnose, downy mildew
Time: Apply when conditions favor disease development; repeat at 7- to 10-day intervals.
Remarks: Cucumbers, melons (cantaloupe, casaba, crenshaw, honeydew, watermelon), pumpkins, squash (summer, winter). Do not exceed 7 applications per season. Do not tank mix with copper fungicides. Do not mix with surfactants or foliar fertilizers. Refer to label for use directions and restrictions.
Days To Harvest: 5.
State Restrictions: Chemigation application not allowed in California.

Signal Word/Toxicity Class: Caution.
REI: 24 hr.

Aliette — see fosetyl-aluminum

Allegiance — see metalaxyl

Basicop — see copper compounds

Bayleton — see triadimefon

BENLATE

Benlate (SP, WP) (benomyl) *(DuPont)*
Rate: 0.13-0.25 lb. ai/A
0.25-0.5 lb. SP, WP/A
Disease:
Anthracnose, powdery mildew, target spot, Cephalosporium wilt
Time: Apply when plants begin to run or when disease first appears; repeat at 7- to 14-day intervals as needed.
Remarks: Cucumbers, melons, pumpkins, squash. Do not tank mix or alternate with benzimidazole or thiophanate products. Do not graze livestock in treated fields. Refer to label for use directions.
Days To Harvest: 1 (except Cephalosporium wilt).
State Restrictions: Benlate SP not registered for use in California.

Signal Word/Toxicity Class: Caution.
REI: 24 hr.

Botran — see dicloran

Bravo — see chlorothalonil

Brom-O-Gas — see methyl bromide

CAPTAN

Gustafson Captan 30-DD or 400 *(Gustafson)*
Rate: Muskmelon, pumpkin, squash, watermelon:
1.66 fl. oz. 30-DD/100 lb. seed
1.5-2 fl. oz. 400/100 lb. seed
Cucumber, cantaloupe:
2.33 fl. oz. 30-DD/100 lb. seed
2.5-3 fl. oz. 400/100 lb. seed
Disease:
damping-off, seed decay, seedling blight, mold
Time: Seed treatment.
Remarks: Cucumbers, melons (cantaloupe, muskmelon, watermelon), pumpkin, squash. Do not combine colloidal suspensions, wettable powders, or emulsifiable concentrates unless previous experience assures compatibility. Use with EPA-approved dye that imparts unnatural color to seed. Do not use treated seed for food, feed, or oil purposes. Refer to label for use directions.

Signal Word/Toxicity Class: Danger/I.

Drexel Captan 4FL *(Drexel)*
Rate: Muskmelon, pumpkin, squash, watermelon:
2 fl. oz. 4FL/100 lb. seed
Cucumbers, cantaloupe:
3 fl. oz. 4FL/100 lb. seed
Disease:
seed decay, seedling blight, soilborne disease
Time: Liquid or slurry seed treatment.
Remarks: Cucumber, melons (cantaloupe, muskmelon, watermelon), pumpkins, squash. Do not use treated seed for food, feed, or oil purposes. Refer to label for use directions.

Signal Word/Toxicity Class: Danger/I.

CAPTAN (Premix)

Nu-Gro Isotox (12.2% captan & 25% lindane) *(Wilbur-Ellis)*
Seed Mate Isotox (12.2% captan & 25% lindane) *(Loveland)*
Rate: 4 oz./100 lb. seed
Disease:
damping-off, seed decay, seedling blight
Time: Planter box treatment.
Remarks: Cucumbers, melons (cantaloupe, watermelon), pumpkins, squash. Treated seed must not be used for or mixed with food or animal feed, or processed for oil. Refer to label for use directions.

Signal Word/Toxicity Class: Danger/I.
REI: 4 days.

Champ — see copper compounds

Champion — see copper compounds

◆ - **new product** • ai-active ingredient • DF-dry flowable • E/EC-emulsifiable concentrate • EW-emulsion in water
F/FL-flowable • G-granule • L/LC-liquid • SP/WSP-soluble packet • W/WP-wettable powder • WSB-water soluble bag

CHLOR-O-PIC

Restricted Use Pesticide.

Chlor-O-Pic (chloropicrin) *(Great Lakes Chemical)*

Rate: Nontarp:
300-500 lb./A

Disease:

Cylindrocladium, Fusarium, Phytophthora, Plasmodiophora, Pseudomonas, Pythium, Rhizoctonia, Sclerotinia, Sclerotium, Verticillium

Time: Apply up to 2 weeks prior to planting when soil moisture is good and temperature is 60°-85°F.

Remarks: Cucumbers, melons. Work upwind of product and never fumigate without 2 qualified applicators. Do not fumigate when there is no or little air movement. Refer to label for use directions.

Signal Word/Toxicity Class: Danger/I.

CHLOROTHALONIL

Bravo 500, 720/Weather Stik, Ultrex (WDG) *(ZENECA)*
Echo 500, 720 or 90DF *(Sostram)*
Terranil (6L, 90DF or 90DF WSP) *(Terra)*

Rate: 1.12-2.25 lb. ai/A
2.25-4.25 pt. 500/A
1.5-3 pt. 720, 6L/A
1.4-2.7 lb. WDG/A
1.3-2.4 lb. 90DF/A
1.25-2.5 90DF, 90DF WSP/A

Disease:

Alternaria leaf blight, Anthracnose, Cercospora leaf spot, downy mildew, gummy stem blight, powdery mildew, scab, target spot

Rate: 6 lb. ai/A
12 pt. 500/A
8.25 pt. 720/A
7.6 lb. WDG/A
6.9 lb. 90DF/A
7 lb. 90DF, 90DF WSP/A

Disease:

belly rot

Time: Apply when plants are in first true leaf stage or when conditions are favorable for disease development; repeat at 7-day intervals (for severe disease conditions shorten spray interval). For cucumber belly rot, make single application when vines begin to form.

Remarks: Cucumbers, melons (cantaloupe, honeydew, muskmelon, watermelon), pumpkins, squash. For cucumber belly rot, use sufficient water to obtain runoff to soil surface. May be applied through sprinkler irrigation equipment. Refer to label for use directions and restrictions.

Days To Harvest: 7.

Signal Word/Toxicity Class: Danger/I (WDG, 90DF); Warning/II (500, 720, 6L).
REI: 48 hr.

Clean Crop C-O-C-S — see copper compounds

COPPER COMPOUNDS

Basicop (WP) (basic copper sulfate) *(Griffin)*

Rate: 1 lb. ai/A
2 lb. WP/A

Disease:

angular leaf spot, Anthracnose, downy mildew

Time: Apply when disease appears, repeat at 7- to 10-day intervals.

Remarks: Cucumbers, melons (cantaloupe), squash. Do not apply through any type of irrigation system. Refer to label for use directions.

Signal Word/Toxicity Class: Warning/II
REI: 24 hr.

Champ (FL or Formula 2) or Champion (WP) (copper hydroxide) *(Agtrol)*
Kocide (101, DF, 2000, 4.5LF, LF) (copper hydroxide) *(Griffin)*
KOP-Hydroxide (FL, 50WP) (copper hydroxide) *(Drexel)*

Rate: Cucumbers:
2-2.66 pt. FL, LF/A
1-1.33 pt. Formula 2/A
1.5-2 lb. WP, 101, DF/A
2 lb. 50WP/A

Disease:

angular leaf spot, downy mildew

Time: Apply weekly when plants begin to vine.

Rate: Cantaloupe, honeydew, muskmelon:
2.66 pt. FL, LF/A
1.33 pt. Formula 2/A
2 lb. WP, 101, DF, 50WP/A

Disease:

downy mildew

Time: Apply when conditions are favorable for disease development; repeat at 5- to 7-day intervals as needed.

Rate: Pumpkins, squash:
2-4 pt. FL, LF/A
1-2 pt. Formula 2/A
1.5-3 lb. WP, 101, DF/A
2 lb. 50WP/A

Disease:

powdery mildew

Time: Apply when plants are 3 weeks old or when disease symptoms first appear; repeat at weekly intervals.

Rate: Watermelon:
2-4 pt. FL, LF/A
1-2 pt. Formula 2/A (0.66-1.5 pt./A bacterial fruit blotch)
1.5-3 lb. WP, 101, 50WP/A
2 lb. DF/A

Disease:

Anthracnose, bacterial fruit blotch (3FL, DF), downy mildew

Time: Apply as soon as plants become established in the field; repeat at weekly intervals depending on disease severity.

Remarks: Cucumbers, melons (cantaloupe, honeydew, muskmelon, watermelon), pumpkins, squash. Refer to label for use directions.

State Restrictions: Kocide DF is not for use in California on watermelons for bacterial fruit blotch.

Signal Word/Toxicity Class: Varies by formulation.
REI: Varies by formulation.

Clean Crop C-O-C-S (WDG) (74.8% copper oxychloride & 14.2% copper sulfate, basic) *(Platte)*

Rate: 1-2 lb. ai/A
2-4 lb. WDG/A

Disease:

Anthracnose, bacterial wilt (aids in control), downy mildew, leaf spot

Time: Apply when disease first threatens; repeat at 7- to 10-day intervals to protect new growth throughout growing season.

Remarks: Cucumbers, melons (watermelons), pumpkins, squash (summer, winter). Do not apply this product through any type of irrigation system. Refer to label for use directions.

Signal Word/Toxicity Class: Warning/II.
REI: 24 hr.

Copper-Count-N (L) (copper ammonium) *(Mineral Research)*

Rate: 2-3 qt. L/A

Disease:

Alternaria, angular leaf spot, downy mildew, powdery mildew, scab

Time: Apply after vining and continue until mid-harvest.

Remarks: Cucumbers, melons (including cantaloupe), squash. Applications may be made by ground or air; refer to label for use directions. Do not apply through any type of irrigation system unless supplemental labeling on chemigation is followed.

Signal Word/Toxicity Class: Caution.
REI: 12 hr.

Drexel Basic Copper Sulfate (WP) or KOP 300 (3FL) (basic copper sulfate) *(Drexel)*
Micro Flo Basic Copper "53" (basic copper sulfate) *(Micro Flo)*

Rate: 2 lb. WP, 53/A
0.33 gal. 3FL/A

Disease:

Alternaria leaf spot, angular leaf spot, Anthracnose, bacterial wilt, downy mildew, gummy stem blight, leaf spot, powdery mildew, scab

Time: Apply when plants begin to vine; repeat at 7- to 10-day intervals (5- to 10-day for 53).

NOTICE	The information on these pages is for preliminary planning — not a guide for use. Be sure to follow the manufacturer's directions, notwithstanding information contained here.	For personal protective equipment and EPA registration numbers, please turn to page 39.

Remarks: Cucumbers, melons (including cantaloupe), pumpkins, squash. Do not apply WP or 3FL through any type of irrigation system. Refer to label for use directions.

Signal Word/Toxicity Class: Warning/II.
REI: 24 hr. (WP, 53); 12 hr. (3FL).

Kocide (2000, 4.5LF) (copper hydroxide) *(Griffin)*
Rate: **Cucumber, melons, pumpkins, squash, watermelon:**
1.5 lb. 2000/A
1.33 pt. 4.5LF/A
Disease:

Alternaria leaf spot	bacterial fruit blotch	gummy stem blight
angular leaf spot	downy mildew	powdery mildew
Anthracnose		

Time: Apply when conditions are favorable for disease development; repeat at 5- to 7-day intervals as needed.
Remarks: Cucumbers, melons (cantaloupe, honeydew, muskmelon, watermelon), pumpkins, squash. Refer to label for use directions.

Signal Word/Toxicity Class: Warning/II (2000); Caution (4.5LF).
REI: 24 hr.

KOP Oxy-85 (WP) (copper oxychloride) *(Drexel)*
Rate: 4 lb. WP/A
Disease:

angular leaf spot	downy mildew	scab
Anthracnose		

Time: Apply when disease is reported to be present in area; repeat at 7- to 10-day intervals as needed.
Remarks: Cucumbers, melons. Refer to label for use directions.

Signal Word/Toxicity Class: Danger/I.
REI: 24 hr.

Microsperse COC (53WP) (copper oxychloride) *(Micro Flo)*
Rate: 3-4 lb. 53WP/A
Disease:

Alternaria leaf spot	Anthracnose	downy mildew
angular leaf spot	Cercospora leaf spot	powdery mildew

Time: Cucumbers: Apply weekly once plants start to vine. Melons: Apply before disease first appears. Pumpkins, squash: Apply when plants are 3 weeks old or when symptoms first appear; repeat at weekly intervals as needed.
Remarks: Cucumbers, melons (cantaloupe, honeydew, muskmelon), pumpkins, squash. Refer to label for use directions.

Signal Word/Toxicity Class: Danger/I.
REI: 24 hr.

Nu-Cop (3L) (copper hydroxide) *(Micro Flo)*
Rate: 1-4 pt. 3L/A
Disease:

Alternaria leaf spot	bacterial fruit blotch (suppression)	gummy stem blight
angular leaf spot	downy mildew	powdery mildew
Anthracnose		

Time: Apply when conditions are favorable for disease development; repeat at 5- to 10-day intervals.
Remarks: Cucumbers, melons (cantaloupe, honeydew, muskmelon, watermelon), pumpkins, squash. Refer to label for use directions.

Signal Word/Toxicity Class: Warning/II.
REI: 24 hr.

Nu-Cop (50DF, 50WP) (copper hydroxide) *(Micro Flo)*
Rate: 1.5-3 lb. 50DF, 50WP/A
Disease:

angular leaf spot (cucumbers)	Anthracnose (watermelons)	downy mildew
		powdery mildew (pumpkins, squash)

Time: Apply weekly once plants begin to vine or disease first appears.
Remarks: Cucumbers, melons (including watermelon), pumpkins, squash. Refer to label for use directions.

Signal Word/Toxicity Class: Danger/I.
REI: 48 hr.

TOP COP Tri-Basic (FL) (basic copper sulfate) *(Stoller)*
Rate: 2-4 pt. FL/A
Disease:

angular leaf spot	downy mildew	powdery mildew
Anthracnose	gummy stem blight	scab
bacterial wilt	leaf spot	

Time: Apply when plants begin to vine; repeat at 7- to 10-day intervals. In dense plants, applications every 3-4 days may be required.
Remarks: Cucumbers, melons (cantaloupe, honeydew, muskmelon, watermelon), pumpkins, squash. Refer to label for use directions.

Signal Word/Toxicity Class: Caution.
REI: 24 hr.

Copper-Count-N — see copper compounds

DICLORAN

Botran 75-W *(Gowan)*
Rate: 1 lb. ai/A
1.33 lb. 75-W/A
Disease:
white mold
Time: Apply when disease first appears; repeat at 14-day intervals.
Remarks: Cucumbers (greenhouse/hothouse). Refer to label for use directions.
Days To Harvest: 1.

Signal Word/Toxicity Class: Caution.
REI: 12 hr.

Dithane — see mancozeb

Echo — see chlorothalonil

FOSETYL-ALUMINUM

Aliette (WDG, WSP) *(Rhone-Poulenc)*
Rate: 1.6-4 lb. ai/A
2-5 lb. WDG, WSP/A
Disease:

downy mildew	Phytophthora fruit rot	Phytophthora root rot

Time: Apply when conditions favor disease development (high moisture and moderate temperatures); repeat at 7- to 14-day intervals.
Remarks: Cucumber, melons (muskmelon, watermelon), pumpkins, squash (summer, winter). Do not apply less than 10 gal./A or make more than 7 applications per season. Do not tank mix with copper compounds. With moderate disease pressure, high rate and shorter interval should be used. Refer to label for use directions.
Days To Harvest: 0.
State Restrictions: Use rate limited to maximum 3.75 lb. product/A in certain counties of Alabama, Arkansas, Illinois, Indiana, Iowa, Kentucky, Louisiana, Maryland, Minnesota, Mississippi, Missouri, North Carolina, Ohio, Tennessee, Virginia, Vermont; refer to label for specific counties.

Signal Word/Toxicity Class: Caution.
REI: 12 hr.

Kocide — see copper compounds

KOP-Hydroxide — see copper compounds

KOP Oxy — see copper compounds

Kumulus — see sulfur

◆ - **new product** • ai-active ingredient • DF-dry flowable • E/EC-emulsifiable concentrate • EW-emulsion in water
F/FL-flowable • G-granule • L/LC-liquid • SP/WSP-soluble packet • W/WP-wettable powder • WSB-water soluble bag

MANCOZEB

Dithane (DF, F-45, M-45, WSP) *(Rohm and Haas)*
Manzate 200DF *(Griffin)*

Rate: 1.5-2.25 lb. ai/A
2-3 lb. DF, 200DF/A
1.6-2.4 lb. ai/A
2-3 lb. M-45, WSP/A
3.2-4.8 pt. F-45/A

Disease:
Alternaria leaf spot, Cercospora leaf spot, gummy stem blight, Anthracnose, downy mildew, scab

Time: Apply when plants are 2-leaf stage; repeat at 7- to 10-day intervals.
Remarks: Cucumbers, melons (cantaloupe, casaba, crenshaw, honeydew, muskmelon, watermelon), summer squash. Use sufficient water to provide thorough coverage of both upper and lower leaf surfaces. Refer to label for use directions.
Days To Harvest: 5.
State Restrictions: Refer to label for California precautions.

Signal Word/Toxicity Class: Caution.
REI: 24 hr.

Penncozeb 75DF or 80WP *(Elf Atochem)*

Rate: Cucumbers:
1.12-2.25 lb. ai/A
1.5-3 lb. 75DF/A
1.2-2.4 lb. ai/A
1.5-3 lb. 80WP/A

Disease:
Alternaria leaf spot, Anthracnose, Pythium fruit rot, angular leaf spot, downy mildew

Rate: Melons, squash:
1.5-2.25 lb. ai/A
2-3 lb. 75DF/A
1.6-2.4 lb. ai/A
2-3 lb. 80WP/A

Disease:
Alternaria leaf spot, Cercospora leaf spot, gummy stem blight, Anthracnose, downy mildew

Time: Apply when disease threatens or plants begin to run; spray at 7- to 10-day intervals using higher rate and/or shorter intervals under severe disease pressure.
Remarks: Cucumbers, melons (cantaloupe, casaba, crenshaw, honeydew, watermelon), squash (summer). Use sufficient water to provide thorough coverage of both upper and lower leaf surfaces. Refer to label for use directions.
Days To Harvest: 5.

Signal Word/Toxicity Class: Caution.
REI: 24 hr.

MANEB

Maneb 75DF or 80WP *(Elf Atochem)*
Manex (FL) *(Griffin)*

Rate: 1.12-1.5 lb. ai/A
1.5-2 lb. 75DF/A
1.2-1.6 lb. ai/A
1.5-2 lb. 80WP/A
1.2-1.6 qt. FL/A

Disease: Cucumbers:
Alternaria leaf spot, Anthracnose, Pythium fruit rot, angular leaf spot, downy mildew

Pumpkins:
angular leaf spot, downy mildew

Squash:
Anthracnose, downy mildew

Rate: 0.75-1.5 lb. ai/A
1-2 lb. 75DF/A
0.8-1.6 lb. ai/A
1-2 lb. 80WP/A
1.2-1.6 lb. ai/A
1.2-1.6 qt. FL/A

Disease: Melons:
Alternaria leaf spot, Cercospora leaf spot, gummy stem blight, Anthracnose, downy mildew

Time: Apply when disease threatens or plants begin to run; repeat at 7- to 10-day intervals.
Remarks: Cucumbers, melons (cantaloupe, casaba, crenshaw, honeydew, watermelon), pumpkins, squash (summer, winter). Do not use any other product containing a different EBDC ai during same growing season if using Manex. Refer to label for use directions.
Days To Harvest: 5.

Signal Word/Toxicity Class: Caution.
REI: 24 hr.

MANEB (Premix)

Fungicide/bactericide. Toxic to fish, birds, and other wildlife.

ManKocide (DF) (15% maneb & 46.1% copper hydroxide) *(Griffin)*

Rate: Cucumbers:
2-2.5 lb. DF/A

Disease:
angular leaf spot, downy mildew, gummy stem blight, Anthracnose

Time: Apply when plants begin to vine; repeat weekly.

Rate: Melons:
2.5 lb. DF/A

Disease:
Alternaria leaf spot, bacterial fruit blotch, downy mildew, Anthracnose, Cercospora leaf spot, gummy stem blight

Time: Apply when plants are in 2-leaf stage; repeat at 7-day intervals.
Remarks: Cucumbers, melons (cantaloupe, casaba, crenshaw, honeydew, muskmelon, watermelon). Use sufficient water and direct spray to provide thorough coverage of both upper and lower leaf surfaces. Refer to label for use directions.
Days To Harvest: 5.

Signal Word/Toxicity Class: Danger/I.
REI: 48 hr.

Manex — see maneb

ManKocide — see maneb premix

Manzate — see mancozeb

MEFENOXAM

Systemic fungicide having a specific mode of action.

Ridomil Gold EC or WSP *(Novartis)*

Rate: 0.5-1 lb. ai/A
1-2 pt. EC/A
0.45-0.9 lb. ai/A
1-2 lb. WSP/A

Disease:
cottony leak, damping-off

Time: Preplant incorporated or surface spray at planting.
Remarks: Cucumbers, gherkins, melons (cantaloupe, casaba, crenshaw, honedew, honeyball, muskmelon, Persian), pumpkins, squash (summer, winter). For banded application, 7" band recommended. Do not use in greenhouse or on field grown vegetable bedding plants. Do not dip plants in solutions containing Ridomil Gold or crop injury may occur. Do not plant any crop which is not registered for use with mefenoxam in treated soil for 12 months.

Signal Word/Toxicity Class: Warning/II (EC); Caution (WSP).
REI: 48 hr.

MEFENOXAM (Premixes)

Ridomil Gold/Bravo (WP) (4.5% mefenoxam & 72% chlorothalonil) *(Novartis)*
Ridomil Gold/Copper (WSB) (5% mefenoxam & 60% copper hydroxide) *(Novartis)*

Rate: 1.5 lb. ai/A
2 lb. WP/A
1.3 lb. ai/A
2 lb. WSB/A

Disease:
downy mildew

NOTICE The information on these pages is for preliminary planning — not a guide for use. Be sure to follow the manufacturer's directions, notwithstanding information contained here. For personal protective equipment and EPA registration numbers, please turn to page 39.

Time: Apply when conditions are favorable for disease development, but before infection; repeat at 14-day intervals until disease threat is over.

Rate: 1.5 lb. ai/A
2 lb. WP/A

Disease:
Anthracnose
Cercospora leaf spot
gummy stem blight (black rot)
leaf blight
scab

Time: Apply when conditions are favorable for disease development, but before infection, repeat at 10- to 14-day intervals until disease threat is over.
Remarks: Cucumber, melons (cantaloupe, casaba, crenshaw, honeyball, honeydew, muskmelon, Persian), pumpkins, squash (summer, winter). Up to 4 applications of Ridomil Gold/Bravo per crop may be applied. Avoid late season applications when plants reach full maturity or begin senescence. See label for additional information. Do not plant any crop which is not registered for use with mefenoxam in treated soil for 12 months.

- - - - - - - - - - - - - - - - -

Signal Word/Toxicity Class: Danger/I.
REI: 48 hr.

Ridomil Gold MZ (WP) (4% mefenoxam & 64% mancozeb) *(Novartis)*
Rate: 1.7 lb. ai/A
2.5 lb. WP/A
Disease:
downy mildew
Time: Apply when conditions are favorable for disease development before infection; repeat at 14-day intervals until disease threat is over.
Remarks: Cucumbers, melons (cantaloupe, casaba, crenshaw, honeydew, watermelon), summer squash. Make up to 4 applications. If conditions for disease development are still favorable after 4 applications, use other fungicides for control. Avoid late season applications when plants reach full maturity or begin senescence. If other registered fungicides containing EBDCs are used during growing season, do not apply more than 12.8 lb. EBDC/A. If tank mixed with other products containing mancozeb, apply no more than 2.4 lb. mancozeb. If tank mixed with maneb, apply no more than 1.6 lb. EBDC. Do not plant any crop which is not registered for use with mefenoxam in treated soil for 12 months.
Days To Harvest: 5.

- - - - - - - - - - - - - - - - -

Signal Word/Toxicity Class: Caution.
REI: 48 hr.

METALAXYL

Allegiance (FL) *(Gustafson)*
Rate: 0.75 fl. oz. FL/100 lb. seed
Disease:
damping-off
Time: Seed treatment.
Remarks: Cucumbers. Use with EPA-approved dye that imparts unnatural color to seed. Do not use treated seed for food, feed, or oil purposes. Refer to label for use directions.
State Restrictions: Not for use in California.

- - - - - - - - - - - - - - - - -

Signal Word/Toxicity Class: Warning/II.

METHYL BROMIDE

2% Chloropicrin added as warning agent. Restricted Use Pesticide. Classified as an ozone depleter under the Clean Air Act. Use is to be completely eliminated starting Jan. 1, 2001.

Brom-O-Gas *(Great Lakes Chemical)*
Rate: Injected with chisels + covered with tarp:
180-240 lb./A
Raised tarp with topical application:
0.4-0.55 lb./100 ft.
Disease:
Fusarium
oak root fungus
Phytophthora
Plasmodiophora
Pythium
Rhizoctonia
Sclerotinia
Sclerotium
Time: Preplant.
Remarks: Muskmelon. Do not fumigate when soil temperature is below 50°F. For best results, fumigate when temperature is 60°-80°F at depth of 6". Refer to label for use directions.
State Restrictions: Brom-O-Gas not for use in Arizona and not for use in California for oak root fungus (Armillaria root rot) control.

- - - - - - - - - - - - - - - - -

Signal Word/Toxicity Class: Danger/I.

Microsperse — see copper compounds

Microthiol Special — see sulfur

M-PEDE

Effective contact insecticide, miticide, and fungicide . Toxic to aquatic invertebrates.

M-Pede (potassium salts of fatty acids) *(Mycogen)*
Rate: 2% v/v solution
Disease:
powdery mildew
Time: Apply when disease appears; repeat at 7- to 10-day intervals.
Remarks: Cucumbers (greenhouse). May be tank mixed at 1-2% v/v solution. Do not apply through any type of irrigation system. Refer to label for use directions.

- - - - - - - - - - - - - - - - -

Signal Word/Toxicity Class: Warning/II.
REI: 12 hr.

Nu-Cop — see copper compounds

Nu-Gro Isotox — see captan premixes

Penncozeb — see mancozeb

Ridomil Gold — see mefenoxam

Ridomil Gold MZ — see mefenoxam premix

Ridomil Gold/Bravo — see mefenoxam premix

Ridomil Gold/Copper — see mefenoxam premix

Seed Mate — see captan premixes

SULFUR

Drexel Suffa (6FL), Sulfa 80 (80DF) or Sulfur (90W) *(Drexel)*
Rate: **Cucumbers:**
1.6-3 lb. ai/A
2-4 lb. 80DF/A
Melons, squash:
8-12 lb. ai/A
10-15 lb. 80DF/A
Melons:
13.5-18 lb. ai/A
18-24 pt. 6FL/A
15-20 lb. 90W/A
Disease:
powdery mildew
Time: Apply when first true leaves appear; repeat at weekly intervals.
Remarks: Cucumbers, melons (cantaloupe, watermelon), squash. Do not use within 2 weeks of oil application (4 weeks 6FL). Some plants are susceptible to sulfur injury under certain climatic conditions. Do not make sulfur applications at excessively high temperatures. Do not apply this product through any type of irrigation system. Refer to label for use directions.

- - - - - - - - - - - - - - - - -

Signal Word/Toxicity Class: Caution.
REI: 24 hr.

◆ - **new product** • ai-active ingredient • DF-dry flowable • E/EC-emulsifiable concentrate • EW-emulsion in water
F/FL-flowable • G-granule • L/LC-liquid • SP/WSP-soluble packet • W/WP-wettable powder • WSB-water soluble bag

Kumulus DF *(BASF)*
Rate: 16-20 lb. ai/A
20-25 lb. DF/A
Disease:
powdery mildew
Time: Apply when disease first appears; repeat as necessary.
Remarks: Melons. When the temperature reaches 95°F, do not use on sulfur-sensitive varieties due to possible injury. Do not apply through any type of irrigation system. Refer to label for use directions.

Signal Word/Toxicity Class: Caution.
REI: 24 hr.

Micro Flo Sulfur (6L), Dual Spraying & Dusting Sulfur (DS) or Microsperse Wettable Sulfur (WS) *(Micro Flo)*
Terra Sulfur (6L) *(Terra)*
Rate: 24 lb. ai/A
4 gal. 6L/A
22.5 lb. ai/A
25 lb. DS, WS/A
Disease:
powdery mildew
Time: Apply when disease first appears; repeat as necessary.
Remarks: Cucumbers, melons, squash. Do not use within 2 weeks of oil application. Do not apply DS and WS through any type of irrigation system. Refer to label for use directions.

Signal Word/Toxicity Class: Caution.
REI: 24 hr.

Microthiol Special (80WP) *(Elf Atochem)*
Rate: **Cucumbers:**
1.6-3.2 lb. ai/A
2-4 lb. 80WP/A
Melons, squash:
4-8 lb. ai/A
5-10 lb. 80WP/A
Disease:
powdery mildew
Time: Cucumbers, melons, squash. Apply when disease first appears; repeat as needed.
Remarks: Do not use on sulfur sensitive varieties. Refer to label for directions.

Signal Word/Toxicity Class: Caution.
REI: 24 hr.

Thiolux (DF) *(Platte)*
Rate: **Cucumbers:**
4-6 lb. DF/A
Melons:
6-15 lb. DF/A
Squash:
4-10 lb. DF/A
Disease:
powdery mildew
Time: Apply when disease first appears; repeat at 10- to 14-day intervals as necessary throughout growing season.
Remarks: Cucumbers, melons, squash. Sulfur can injure plants, especially when temperatures reach 95°F. Do not use on sulfur-sensitive varieties. Do not make applications within 2 weeks of oil spray. Refer to label for use directions.

Signal Word/Toxicity Class: Caution.
REI: 24 hr.

SULFUR (Premix)

Top Cop with Sulfur (6.25 lb./gal. sulfur & 1 lb./gal. tribasic copper sulfate) *(Stoller)*
Rate: 0.5-1 qt./A
Disease:
downy mildew powdery mildew
Time: Apply when disease first appears; repeat at 7- to 10-day intervals through season.
Remarks: Cucumbers, melons (cantaloupe, muskmelon, watermelon), pumpkins, squash. Some plants are susceptible to injury from sulfur under certain climatic conditions. Use lower rates over 75°F. Do not apply when temperatures exceed 90°F. Refer to label for use directions.

Signal Word/Toxicity Class: Caution.
REI: 24 hr.

Terranil — see chlorothalonil

TERRANIL Cu (Premix)

Terranil Cu (24% chlorothalonil & 24% copper oxychloride) *(Terra)*
Rate: 3.4-4.5 pt./A
Disease:
Anthracnose downy mildew target spot

Rate: 4.5-6.75 pt./A
Disease:
Alternaria leaf blight gummy stem blight scab
Cercospora leaf spot powdery mildew

Rate: 8.25 pt./A
Disease:
belly rot
Time: Apply when plants are in first true leaf stage or when conditions are favorable for disease development; repeat at 7-day intervals. For cucumber belly rot, apply a single application when vines begin to form.
Remarks: Cucumbers, melons (cantaloupe, honeydew, muskmelon), pumpkin, squash. Under severe disease conditions, shorten spray interval. Refer to label for use directions.
Days To Harvest: 7.

Signal Word/Toxicity Class: Danger/I.
REI: 48 hr.

TERR-O-GAS (Premix)

Classified as an ozone depleter under the Clean Air Act. Use to be eliminated starting Jan. 1, 2001. Restricted Use Pesticide.

Terr-O-Gas (67% methyl bromide & 33% chloropicrin) *(Great Lakes Chemical)*
Rate: Nontarp:
270-355 lb./A
Disease:
Fusarium Pythium Sclerotinia
Phytophthora Rhizoctonia Sclerotium
Plasmodiophora
Time: Apply up to 2 weeks prior to planting when soil moisture is good and temperature is 60°-85°F.
Remarks: Muskmelon. Work upwind of product and never fumigate without 2 qualified applicators. Do not fumigate when there is little or no air movement. Tarping is not used after field application. Refer to label for use directions.

Signal Word/Toxicity Class: Danger/I.

Thiolux — see sulfur

THIOPHANATE-METHYL

Topsin M (WSB, 70W) *(Elf Atochem)*
Rate: 0.18-0.35 lb. ai/A
0.25-0.5 lb. WSB, 70W/A
Disease:
Anthracnose gummy stem blight target spot
belly rots powdery mildew
Time: Apply when plants begin to run or when disease first appears; repeat at 7-14 days intervals as needed (7 days for target spot).
Remarks: Cucumbers, melons, pumpkins, squash (summer, winter). Apply by ground or air (high rate). Refer to label for use directions.
State Restrictions: For use only on powdery mildew in California.

Signal Word/Toxicity Class: Caution.
REI: 12 hr.

THIRAM

42-S Thiram or Thiram 50WP Dyed *(Gustafson)*
Thiram Granuflo (WDG) *(UCB Agrochemicals)*
Rate: 4.5 fl. oz. 42-S/100 lb. seed
4.5 oz. 50WP/100 lb. seed
3 oz. WDG/100 lb. seed
Disease:
damping-off seed decay seedling blight

7 Vegetables

NOTICE The information on these pages is for preliminary planning — not a guide for use. Be sure to follow the manufacturer's directions, notwithstanding information contained here. For personal protective equipment and EPA registration numbers, please turn to page 39.

Time: Seed treatment.
Remarks: Cucumbers, melons (cantaloupe, watermelon), pumpkins, squash. Use with EPA-approved dye that imparts unnatural color to seed. Do not use treated seed for food, feed, or oil purposes. Do not apply Thiram Granuflo through any type of irrigation system. Refer to label for use directions.
State Restrictions: Thiram Granuflo not for use in California.

Signal Word/Toxicity Class: Caution.
REI: 24 hr. (Thiram Granuflo).

TOP COP — see copper compounds

Top Cop with Sulfur — see sulfur premix

Topsin M — see thiophanate-methyl

TRIADIMEFON

Bayleton 50% DF *(Bayer)*
Rate: 2-4 oz. 50% DF/A
Disease:
powdery mildew
Time: Apply spray when disease first appears; repeat at 14-day intervals.
Remarks: Cucumbers, melons (cantaloupe, casaba, crenshaw, honeydew, honey ball, mango melon, muskmelon, Persian, watermelon), pumpkins, squash (summer, winter). Do not apply through any type of irrigation system. Refer to label for use directions.
Days To Harvest: 0.
Crop Rotation: Small grains, corn, sorghum, soybeans, beans, peas, and cucurbits may be planted 35 days after last application; forage or vines from these crops may not be used for food or feed. Root crops may be planted 120 days after last application but tops must not be used for food or feed. All crops may be planted 12 months or later after last application without restrictions.

Signal Word/Toxicity Class: Caution.
REI: 12 hr.

ZIRAM

Ziram 4L or 76 *(Drexel)*
Rate: 1-1.5 lb. ai/100 gal.
2-3 pt. 4L/100 gal.
0.76-1.5 lb. ai/100 gal.
1-2 lb. 76/100 gal.
Disease:
Alternaria leaf blight, Anthracnose, downy mildew, angular leaf spot, bacterial wilt, scab
Time: Apply before disease first appears; repeat at 7- to 10-day intervals as needed.
Remarks: Cucumbers, melons (cantaloupe), pumpkins, squash. Do not apply this product through any type of irrigation system. Refer to label for use directions.
State Restrictions: Not for use in California.

Signal Word/Toxicity Class: Danger/I.
REI: 48 hr.

Eggplant

Basicop — see copper compounds

Brom-O-Gas — see methyl bromide

Champ — see copper compounds

Champion — see copper compounds

CHLOR-O-PIC

Restricted Use Pesticide.

Chlor-O-Pic (chloropicrin) *(Great Lakes Chemical)*
Rate: Nontarp:
300-500 lb./A
Disease:
Cylindrocladium, Pseudomonas, Sclerotinia, Fusarium, Pythium, Sclerotium, Phytophthora, Rhizoctonia, Verticillium, Plasmodiophora
Time: Apply up to 2 weeks prior to planting when soil moisture is good and temperature is 60°-85°F.
Remarks: Work upwind of product and never fumigate without 2 qualified applicators. Do not fumigate when there is no or little air movement. Refer to label for use directions.

Signal Word/Toxicity Class: Danger/I.

Clean Crop C-O-C-S — see copper compounds

COPPER COMPOUNDS

Basicop (WP) (basic copper sulfate) *(Griffin)*
Champ (FL or Formula 2) or Champion (WP) (copper hydroxide) *(Agtrol)*
Drexel Basic Copper Sulfate (WP) or KOP 300 (3FL) (basic copper sulfate) *(Drexel)*
Kocide (101, DF, 2000, 4.5LF, LF) (copper hydroxide) *(Griffin)*
KOP-Hydroxide (FL, 50WP) (copper hydroxide) *(Drexel)*
Micro Flo Basic Copper "53" (basic copper sulfate) or Microsperse COC (53WP) (copper oxychloride) or Nu-Cop (3L, 50DF, 50WP) (copper hydroxide) *(Micro Flo)*
Rate: 3-4 lb. WP, 53, 53WP/A
2.66 pt. 3FL, FL, LF/A
1.33 pt. Formula 2, 3L, 4.5LF/A
2 lb. WP, 50WP, 50DF, 101, DF/A
1.5 lb. 2000/A
Disease:
Alternaria leaf blight, Anthracnose, Phomopsis
Time: Apply when disease appears; repeat at 7- to 10-day intervals.
Remarks: Use high rate when disease is heavy. Do not apply Basicop, Drexel Basic Copper Sulfate, KOP 300, Micro Flo Basic Copper "53", or Microsperse COC through any type of irrigation system. Refer to label for use directions, restrictions, and precautions.
State Restrictions: Champ, Champion, KOP-Hydroxide 50WP, and Microsperse COC not for use in California.

Signal Word/Toxicity Class: Varies by formulation.
REI: Varies by formulation.

Clean Crop C-O-C-S (WDG) (74.8% copper oxychloride & 14.2% copper sulfate, basic) *(Platte)*
Rate: 1-2 lb. ai/A
2-4 lb. WDG/A
Disease:
Alternaria, downy mildew, Phytophthora blight, Anthracnose
Time: Apply when disease first threatens; repeat at 7- to 10-day intervals to protect new growth throughout growing season.
Remarks: Do not apply this product through any type of irrigation system. Refer to label for use directions.

Signal Word/Toxicity Class: Warning/II.
REI: 24 hr.

Copper-Count-N (L) (copper ammonium) *(Mineral Research)*
TOP COP Tri-Basic (FL) (basic copper sulfate) *(Stoller)*
Rate: 4 pt. L/A
3-4 pt. FL/A
Disease:
Alternaria blight, Anthracnose, Phomopsis
Time: Apply before disease appears; repeat at 7- to 10-day intervals.
Remarks: Applications may be made by ground or air; refer to label for use directions. Do not apply through any type of irrigation system unless supplemental labeling on chemigation is followed.
State Restrictions: Copper-Count-N not for use in California.

Signal Word/Toxicity Class: Caution.
REI: 24 hr. (FL); 12 hr. (L).

◆ - **new product** • ai-active ingredient • DF-dry flowable • E/EC-emulsifiable concentrate • EW-emulsion in water
F/FL-flowable • G-granule • L/LC-liquid • SP/WSP-soluble packet • W/WP-wettable powder • WSB-water soluble bag

Copper-Count-N — see copper compounds

Kocide — see copper compounds

KOP-Hydroxide — see copper compounds

MANEB

Maneb 75DF or 80WP *(Elf Atochem)*
Manex (FL) *(Griffin)*
Rate: 1.12-1.5 lb. ai/A
1.5-2 lb. 75DF/A
1.2-1.6 lb. ai/A
1.5-2 lb. 80WP/A
1.2-1.6 qt. FL/A
Disease:
Anthracnose early blight late blight
Cladosporium leaf mold gray leaf spot Septoria leaf spot
Time: Begin at first fruit cluster; repeat at 7- to 10-day intervals.
Remarks: Do not use any other product containing a different EBDC ai during the same growing season if using Manex. Refer to label for use directions.
Days To Harvest: 5.

Signal Word/Toxicity Class: Caution.
REI: 24 hr.

Manex — see maneb

MEFENOXAM

Ridomil Gold EC or WSP *(Novartis)*
Rate: 0.5 lb. ai/A
1 pt. EC/A
0.45 lb. ai/A
1 lb. WSP/A
Disease:
damping-off root and stem rot
Time: Preventative. Apply at planting; up to 2 applications may be made at 30-day intervals.
Remarks: Do not apply more than 3 pt. EC/A or 3 lb. WSP/A per season. Do not use in greenhouse or on field grown vegetable bedding plants. May be tank mixed. Do not plant any crop which is not registered for use with mefenoxam in treated soil for 12 months.
Days To Harvest: 7.

Signal Word/Toxicity Class: Warning/II (EC); Caution (WSP).
REI: 48 hr.

METHYL BROMIDE

2% Chloropicrin added as warning agent. Classified as an ozone depleter under the Clean Air Act. Use to be eliminated starting Jan. 1, 2001. Restricted Use Pesticide.

Brom-O-Gas *(Great Lakes Chemical)*
Rate: Injected with chisels + covered with tarp:
180-240 lb./A
Raised tarp with topical application:
0.4-0.55 lb./100 ft.
Disease:
Fusarium Plasmodiophora Sclerotinia
oak root fungus Pythium Sclerotium
Phytophthora Rhizoctonia
Time: Preplant.
Remarks: Do not fumigate when soil temperature is below 50°F. For best results, fumigate when temperature is 60°-80°F at depth of 6". Refer to label for use directions.
State Restrictions: Brom-O-Gas not for use in Arizona and not for use in California for oak root fungus (Armillaria root rot) control.

Signal Word/Toxicity Class: Danger/I.

Micro Flo — see copper compounds

Microsperse COC — see copper compounds

Nu-Cop — see copper compounds

Ridomil Gold — see mefenoxam

SULFUR

Micro Flo Sulfur (6L) *(Micro Flo)*
Terra Sulfur (6L) *(Terra)*
Rate: 20-34 lb. ai/A
3.33-5.66 gal. 6L/A
Disease:
powdery mildew
Time: Apply when first true leaves appear; repeat at weekly intervals.
Remarks: Do not use within 2 weeks of oil application. Some plants are susceptible to sulfur injury under certain climatic conditions. Do not make sulfur applications at excessively high temperatures. Refer to label for use directions.

Signal Word/Toxicity Class: Caution.
REI: 24 hr.

SULFUR (Premix)

Top Cop with Sulfur (6.25 lb./gal. sulfur & 1 lb./gal. tribasic copper sulfate) *(Stoller)*
Rate: 2 qt./A
Disease:
Alternaria blight Anthracnose Phomopsis
Time: Apply when disease appears; repeat at 7- to 10-day intervals until harvest begins.
Remarks: Refer to label for use directions.

Signal Word/Toxicity Class: Caution.
REI: 24 hr.

TERR-O-GAS (Premix)

Classified as an ozone depleter under the Clean Air Act. Use to be eliminated starting Jan. 1, 2001. Restricted Use Pesticide.

Terr-O-Gas (67% methyl bromide & 33% chloropicrin) *(Great Lakes Chemical)*
Rate: Nontarp:
270-355 lb./A
Disease:
Fusarium Pythium Sclerotinia
Phytophthora Rhizoctonia Sclerotium
Plasmodiophora
Time: Apply up to 2 weeks prior to planting when soil moisture is good and temperature is 60°-85°F.
Remarks: Work upwind of product and never fumigate without 2 qualified applicators. Do not fumigate when there is no or little air movement. Tarping is not used after field application. Refer to label for use directions.

Signal Word/Toxicity Class: Danger/I.

THIRAM

42-S Thiram or Thiram 50WP Dyed *(Gustafson)*
Thiram Granuflo (WDG) *(UCB Agrochemicals)*
Rate: 6.5 fl. oz. 42-S/100 lb. seed
6 oz. 50WP/100 lb. seed
4 oz. WDG/100 lb. seed
Disease:
damping-off seed decay seedling blight
Time: Seed treatment.
Remarks: Use with EPA-approved dye that imparts unnatural color to seed. Do not use treated seed for food, feed, or oil purposes. Do not apply Thiram Granuflo through any type of irrigation system. Refer to label for directions.
State Restrictions: Thiram Granuflo not for use in California.

Signal Word/Toxicity Class: Caution.
REI: 24 hr. (Thiram Granuflo).

NOTICE The information on these pages is for preliminary planning — not a guide for use. Be sure to follow the manufacturer's directions, notwithstanding information contained here.

For personal protective equipment and EPA registration numbers, please turn to page 39.

TOP COP — see copper compounds

Top Cop with Sulfur — see sulfur premix

Garbanzos
(Chickpeas)

Allegiance — see metalaxyl

Apron — see metalaxyl

Apron XL — see mefenoxam

BENLATE

Benlate (SP, WP) (benomyl) *(DuPont)*
Rate: 10 oz. SP, WP/100 lb. seed
Disease:
Ascochyta rabiei
Time: Slurry treatment.
Remarks: Chickpeas. Use with EPA-approved dye that imparts unnatural color to treated seed. Allow seeds to dry before planting; seeds may be planted any time after being well cured. Do not tank mix or alternate with benzimidazole or thiophanate products. Do not use treated seed for food, feed, or oil purposes. Refer to label for use directions.
State Restrictions: For use in Idaho and Washington only. Benlate SP not registered for use in California.

Signal Word/Toxicity Class: Caution.
REI: 24 hr.

MEFENOXAM

Apron XL LS *(Novartis)*
Rate: 0.32-0.64 fl. oz. LS/100 lb. seed
Disease:
damping-off   Phytophthora (early season)
Time: Seed treatment at planting.
Remarks: Do not use treated seed for food, feed, or oil purposes.

Signal Word/Toxicity Class: Warning/II.

Ridomil Gold EC or WSP *(Novartis)*
Rate: 0.25-0.5 lb. ai/A
0.5-1 pt. EC/A
0.225-0.45 lb. ai/A
0.5-1 lb. WSP/A
Disease:
damping-off   root and stem rot
Time: Preplant incorporated or surface application just after planting.
Remarks: For banded application, 7" band recommended. Do not use in greenhouse or on field grown vegetable bedding plants. May be tank mixed. Do not plant any crop which is not registered for use with mefenoxam in treated soil for 12 months.
Days To Harvest: 1.

Signal Word/Toxicity Class: Warning/II (EC); Caution (WSP).
REI: 48 hr.

METALAXYL

Allegiance or Apron (FL) *(Gustafson)*
Rate: 0.75 fl. oz. Allegiance/100 lb. seed
0.75-1.5 fl. oz. Apron/100 lb. seed
Disease:
damping-off   Phytophthora (early season)
Time: Seed treatment.
Remarks: Chickpeas. Use with EPA-approved dye that imparts unnatural color to seed. Do not use treated seed for food, feed, or oil purposes. Refer to label for use directions.

Signal Word/Toxicity Class: Warning/II.

Apron Dry *(Trace)*
Rate: 4 oz./100 lb. seed
Disease:
damping-off   Phytophthora
Time: Seed treatment.
Remarks: Do not use treated seed for food, feed, or oil purposes. Refer to label for use directions.

Signal Word/Toxicity Class: Caution.
REI: 12 hr.

Ridomil Gold — see mefenoxam

Garlic

Botran — see dicloran

Bravo — see chlorothalonil

CHLOROTHALONIL

Bravo 500, 720/Weather Stik, Ultrex (WDG) *(ZENECA)*
Echo 500, 720 or 90DF *(Sostram)*
Rate: 1-2 lb. ai/A
2-4.25 pt. 500/A
1.5-3 pt. 720/A
1.4-2.7 lb. WDG/A
1.2-2.4 lb. 90DF/A
Disease:
Botrytis leaf blight   downy mildew (suppression)   purple blotch
Time: Apply prior to favorable infection periods; repeat at 7- to 10-day intervals as necessary. Use high rate and 7-day interval when heavy dew or rain persists.
Remarks: Do not apply more than 3 times per season. If additional disease control is needed before harvest, use another registered fungicide. Refer to label for use directions and restrictions.
Days To Harvest: 7.

Signal Word/Toxicity Class: Danger/I (WDG, 90DF); Warning/II (500, 720).
REI: 48 hr.

Terranil (6L, 90DF or 90DF WSP) *(Terra)*
Rate: 1-2 lb. ai/A
1.5-3 pt. 6L/A
1.125-2.25 lb. 90DF, 90DF WSP/A
Disease:
Botrytis leaf blight   downy mildew (suppression)   purple blotch
Time: Apply at first sign of disease or when dew or rain occurs; repeat at 7- to 10-day intervals as needed.
Remarks: Use in sufficient water to obtain adequate coverage. Use high rate and 7-day schedule when heavy dew or rain persists. Do not apply more than 3 times per season. Refer to label for use directions.
Days To Harvest: 7.

Signal Word/Toxicity Class: Danger/I (90DF, 90DF WSP); Warning/II (6L).
REI: 48 hr.

COPPER COMPOUNDS

Microsperse COC 53WP (copper oxychloride) *(Micro Flo)*
Rate: 3-4 lb. 53WP/A
Disease:
downy mildew   purple blotch
Time: Apply when plants are 4-6" high; repeat at 7- to 10-day intervals as needed.
Remarks: Spreader/sticker may be added. Refer to label for use directions.

Signal Word/Toxicity Class: Danger/I.
REI: 24 hr.

◆ - **new product** • ai-active ingredient • DF-dry flowable • E/EC-emulsifiable concentrate • EW-emulsion in water
F/FL-flowable • G-granule • L/LC-liquid • SP/WSP-soluble packet • W/WP-wettable powder • WSB-water soluble bag

DICLORAN

Botran 6 Dust *(Wilbur-Ellis)*

Rate: 15-24 lb. ai/A
250-400 lb. Dust/A

Disease:
white rot

Time: For direct seeding, apply 1-2 weeks before; for planting sets or buds, apply broadcast.

Remarks: Refer to label for use directions.

Crop Rotation: Do not plant spinach as a follow-up crop in treated soil. If lettuce is planted as follow-up crop, plow soil 8" and cross disk area prior to seeding.

State Restrictions: Not for use in North Central states.

Signal Word/Toxicity Class: Caution.
REI: 12 hr.

Botran 75-W *(Gowan)*

Rate: 4 lb. ai/A
5.33 lb. 75-W/A

Disease:
white rot

Time: For direct-seeded, apply 5" band over seed row/bed and incorporate in top 1 1/2" of soil 1-2 weeks before garlic is seeded. For planting sets or buds, apply in furrow application.

Rate: 1.1-2 lb. ai/A
1.5-2.66 lb. 75-W/A

Disease:
Botrytis rot leaf blight

Time: Apply when disease is anticipated; repeat at 14-day intervals if necessary.

Remarks: Refer to label for use directions.

Days To Harvest: 14.

Crop Rotation: Do not plant spinach as a follow-up crop in treated soil.

State Restrictions: Direct-seeded application for use in eastern Oregon, Washington, and western Idaho only. Not for use in California.

Signal Word/Toxicity Class: Caution.
REI: 12 hr.

Echo — see chlorothalonil

IPRODIONE

Rovral (4FL, WG, WP) *(Rhone-Poulenc)*

Rate: 2 lb. ai/A
4 pt. 4FL/A
4 lb. WG, WP/A

Disease:
white rot

Time: Apply in furrow at planting.

Remarks: Rate on 38-40" row spacing. Do not make more than 1 application per year. Refer to label for use directions.

Crop Rotation: Beans, broccoli, carrots, garlic, lettuce, onions (dry bulb), peanuts, potatoes, and rice can be rotated after harvest.

Signal Word/Toxicity Class: Warning/II (WG); Caution (4FL, WP).
REI: 12 hr.

Kumulus — see sulfur

MEFENOXAM

Ridomil Gold EC or WSP *(Novartis)*

Rate: 0.25-0.5 lb. ai/A
0.5-1 pt. EC/A
0.225-0.45 lb. ai/A
0.5-1 lb. WSP/A

Disease:
damping-off

Time: Preplant incorporated or surface application just after planting.

Remarks: For banded application, 7" band recommended. Not for use in greenhouse or on field grown vegetable bedding plants. May be tank mixed. Do not plant any crop not registered for use with mefenoxam in treated soil for 12 months.

Signal Word/Toxicity Class: Warning/II (EC); Caution (WSP).
REI: 48 hr.

MEFENOXAM (Premixes)

Ridomil Gold/Bravo (WP) (4.5% mefenoxam & 72% chlorothalonil) *(Novartis)*

Rate: 1.5 lb. ai/A
2 lb. WP/A

Disease:
Botrytis leaf blight downy mildew purple blotch

Time: Apply when conditions are favorable for disease development, but before infection. For downy mildew, apply at 14-day intervals until disease threat is over. For Botrytis leaf blight and purple blotch, apply at 7-day intervals or at 14-day intervals when other effective EPA-registered fungicides are applied between Ridomil Gold/Bravo applications.

Remarks: Up to 4 applications per crop may be applied. Do not plant any crop which is not registered for use with mefenoxam in treated soil for 12 months.

Days To Harvest: 7.

Signal Word/Toxicity Class: Danger/I.
REI: 48 hr.

Ridomil Gold/Copper (WSB) (5% mefenoxam & 60% copper hydroxide) *(Novartis)*
Ridomil Gold MZ (WP) (4% mefenoxam & 64% mancozeb) *(Novartis)*

Rate: 1.3 lb. ai/A
2 lb. WSB/A
1.7 lb. ai/A
2.5 lb. WP/A

Disease:
downy mildew

Time: Apply when conditions are favorable for disease development before infection; repeat at 14-day intervals until disease threat is over.

Remarks: Make up to 4 applications. If conditions for disease development are still favorable after 4 applications, use other fungicides for control. Do not apply more than 2.4 lb. ai/A. Do not plant any crop which is not registered for use with mefenoxam in treated soil for 12 months. Refer to label for restrictions and precautions.

Days To Harvest: 7.

Signal Word/Toxicity Class: Danger/I (WSB); Caution (WP).
REI: 48 hr.

Microsperse — see copper compounds, sulfur

Microthiol Special — see sulfur

PCNB

PCNB 2 Spray *(Wilbur-Ellis)*

Rate: 20 lb. ai/100 gal. water
10 gal./100 gal. water (21,800 lin. ft.)

Disease:
white rot

Time: Apply at planting time.

Remarks: Do not apply this product through any type of irrigation system. Do not plant root crops within 12 months of application unless PCNB registered for use on those crops. Refer to label for use directions.

Signal Word/Toxicity Class: Caution.
REI: 12 hr.

PCNB 2-E, 75DG or 75WP *(Amvac)*

Rate: In-furrow:
20 lb. ai/100 gal.w ater
10 gal. 2-E/100 gal. water (21,800 lin. ft. row)
27.5 lb. 75DG, 75WP/A
Clove treatment:
10 lb. ai/1000 lb. seed
13.5 lb. 75DG, 75WP/1000 lb. seed

Disease:
white rot

7 Vegetables

NOTICE	The information on these pages is for preliminary planning — not a guide for use. Be sure to follow the manufacturer's directions, notwithstanding information contained here.	For personal protective equipment and EPA registration numbers, please turn to page 39.

Time: Furrow treatment: at planting.
Remarks: For clove treatment, mix in sufficient water to apply as uniform mist-spray for complete coverage of treated cloves. Add commercial sticker to spray solution. Do not apply through any type of irrigation system. Do not allow livestock to graze or feed on foliage in treated fields. Do not plant root crops within 12 months of application unless PCNB is registered for use on those crops.
State Restrictions: Not for use in California.

Signal Word/Toxicity Class: Danger/I (2-E); Caution (75DG, 75WP).
REI: 12 hr.

Redball — see sulfur

Ridomil Gold — see mefenoxam

Ridomil Gold/Bravo — see mefenoxam premixes

Ridomil Gold/Copper — see mefenoxam premixes

Ridomil Gold MZ — see mefenoxam premixes

Rovral — see iprodione

SULFUR

Drexel Sulfa (6FL), Sulfa 80 (80DF) or Sulfur (90W) *(Drexel)*
Kumulus DF *(BASF)*
Microthiol Special (80WP) *(Elf Atochem)*
Thiolux (DF) *(Platte)*

Rate: 4.5-9 lb. ai/A
6-12 pt. 6FL/A
5-10 lb. 90W/A
2.4-8 lb. ai/A
3-10 lb. 80DF, 80WP, DF/A
California:
4.8 lb. ai/A
6 lb. 80WP/A

Disease:
powdery mildew

Time: Apply when disease appears; repeat as necessary.
Remarks: Do not use within 2 weeks of oil application (4 weeks 6FL). Some plants are susceptible to injury from sulfur under certain climatic conditions. Do not make sulfur applications at excessively high temperatures. Refer to label for use directions.

Signal Word/Toxicity Class: Caution.
REI: 24 hr.

Micro Flo Sulfur (6L), Dual Spraying & Dusting Sulfur (DS) or Microsperse Wettable Sulfur (WS) *(Micro Flo)*
Terra Sulfur (6L) *(Terra)*

Rate: 3-8 lb. ai/A
4-10.6 pt. 6L/A
5.4 lb. ai/A
6 lb. DS, WS/A

Disease:
powdery mildew

Time: Apply when disease first appears; repeat as necessary.
Remarks: Do not use within 2 weeks of oil application. Do not apply DS and WS through any type of irrigation system. Refer to label for use directions.

Signal Word/Toxicity Class: Caution.
REI: 24 hr.

Redball Superfine Wettable Sulphur (WS) or Commercial Dusting Sulphur (DS) *(International Sulphur)*

Rate: 5-10 lb. WS/100 gal. water
10-30 lb. DS/A

Disease:
downy mildew powdery mildew

Time: Apply at first sign of disease; repeat as necessary.
Remarks: Apply in sufficient water for thorough coverage. Do not apply when temperatures exceed 90°F. Refer to label for use directions.

Signal Word/Toxicity Class: Caution.
REI: 12 hr.

Terranil — see chlorothalonil

Thiolux — see sulfur

Greens

(Collards, Kale, Mustard, Swiss Chard, Turnip Greens)

ALIETTE/MANEB 2+2 (Co-Pack)

Twin pack product containing a systemic and a contact fungicide in attached water soluble packets. Each twin pack contains 2 lbs. of Aliette and 2 lbs. of Maneb.

Aliette/Maneb 2+2: Aliette (80% fosetyl-aluminum) *(Rhone-Poulenc)* **& Maneb (75% maneb)** *(Elf Atochem)*

Rate: 4 lb./A

Disease:
Alternaria leaf spot downy mildew

Time: Apply when conditions favor disease development; repeat at 7- to 10-day intervals.
Remarks: Kale. Do not exceed 2 applications per cutting; 7 applications per season. Do not apply by air in less than 10 gal./A. Do not tank mix with copper fungicides. Do not mix with surfactants or foliar fertilizers. Refer to label for use directions and restrictions.
Days To Harvest: 10.
State Restrictions: Chemigation application not allowed in California.

Signal Word/Toxicity Class: Caution.
REI: 24 hr.

Aliette — see fosetyl-aluminum

BENLATE

Benzimidazole systemic fungicide. Toxic to fish.

Benlate (SP, WP) (benomyl) *(DuPont)*

Rate: 0.25-1 lb. ai/A
0.5-2 lb. SP, WP/A

Disease: Collards, kale, mustard:
Anthracnose ring spot Sclerotinia white mold
Botrytis gray mold

Turnip greens:
Anthracnose leaf spots powdery mildew

Time: Collards, kale, mustard: Apply at first petal fall (when disease appears for turnip greens); repeat at 14-day intervals.
Remarks: Collards, kale, mustard, turnip greens. Do not tank mix or alternate with benzimidazole or thiophanate products. Do not graze treated areas. Do not use seed or plant parts for food or feed purposes. Refer to label for use directions.
Days To Harvest: Turnip greens-14.
State Restrictions: Benlate SP not registered for use in California. Benlate WP for use only on turnip greens in Alabama, Arkansas, Florida, Georgia, Kentucky, Louisiana, Mississippi, North Carolina, Oklahoma, South Carolina, Tennessee, Texas, and Virginia.

Rate: 8 oz. SP, WP/100 lb. seed in 0.5-3 pt. water

Disease:
Phoma lingam

Time: Slurry seed treatment.
Remarks: Collards, kale, mustard. Use with EPA-approved dye that imparts unnatural color to treated seed. Allow seeds to dry before planting; seeds may be planted any time after being well cured. Do not tank mix or alternate with benzimidazole or thiophanate products. Do not use treated seed for food, feed, or oil purposes. Refer to label for use directions.
State Restrictions: Benlate SP not registered for use in California.

Signal Word/Toxicity Class: Caution.
REI: 24 hr.

◆ - **new product** • ai-active ingredient • DF-dry flowable • E/EC-emulsifiable concentrate • EW-emulsion in water
F/FL-flowable • G-granule • L/LC-liquid • SP/WSP-soluble packet • W/WP-wettable powder • WSB-water soluble bag

CAPTAN

Captan 30-DD contains the toxic inert Rhodamine B.

Gustafson Captan 30-DD or 400 *(Gustafson)*
Rate: **Collards, kale, mustard, rape greens:**
1.25 fl. oz. 30-DD/100 lb. seed
Mustard, rape greens:
1-2 fl. oz. 400/100 lb. seed
Swiss chard:
9.5 fl. oz. 30-DD/100 lb. seed
10-12 fl. oz. 400/100 lb. seed
Disease:
damping-off seed decay seedling blight mold
Time: Seed treatment.
Remarks: Collards, kale, mustard, rape greens, Swiss chard. Do not combine colloidal suspensions, wettable powders, or emulsifiable concentrates unless previous experience assures compatibility. Use with EPA-approved dye that imparts unnatural color to seed. Do not use treated seed for food, feed, or oil purposes. Refer to label for use directions.

Signal Word/Toxicity Class: Danger/I.

Drexel Captan 4FL *(Drexel)*
Rate: 12 fl. oz. 4FL/100 lb. seed
Disease:
seed decay seedling blight soilborne disease
Time: Liquid or slurry seed treatment.
Remarks: Swiss chard. Do not use treated seed for food, feed, or oil purposes. Refer to label for use directions.

Signal Word/Toxicity Class: Danger/I.

COPPER COMPOUNDS

Fungicide/bactericide.

Kocide (101, DF, 2000, 4.5LF, LF) (copper hydroxide) *(Griffin)*
Rate: 1-2 lb. 101, DF/A
0.75-1.5 lb. 2000/A
0.66-1.33 pt. 4.5LF/A
2.66 pt. LF/A
Disease:
black leaf spot black rot downy mildew (DF)
Time: Apply at 7- to 10-day intervals after transplants are set in the field, shortly after emergence of field-seeded crops, or when conditions favor disease development.
Remarks: Collards, mustard, and turnip greens (except 101, LF). Reddening of older leaves may occur on broccoli and flecking of wrapper leaves may occur on cabbage. Do not apply in a spray solution having a pH of less than 6.5 as phytotoxicity could occur. Refer to label for use directions.

Signal Word/Toxicity Class: Danger/I (101, DF); Warning/II (2000); Caution (4.5LF, LF).
REI: 24 hr.

Nu-Cop (3L, 50DF, 50WP) (copper hydroxide) *(Micro Flo)*
Rate: 0.33-1.33 pt. 3L/A
0.5-1 lb. 50DF, 50WP/A
Disease:
downy mildew
Time: Apply after transplants are set in field, or shortly after emergence of field seeded crops; repeat at 7-day intervals.
Remarks: Collards. Refer to label for use directions.

Signal Word/Toxicity Class: Danger/I (50DF, 50WP); Warning/II (3L).
REI: 48 hr. (50DF, 50WP); 24 hr. (3L).

TOP COP Tri-Basic (FL) (basic copper sulfate) *(Stoller)*
Rate: 2-4 pt. FL/A
Disease:
downy mildew leaf spot
Time: Apply when plants are above-ground in plant bed or before disease appears; repeat at 7- to 10-day intervals in plant bed and fields.
Remarks: Collards, mustard. Refer to label for use directions.

Signal Word/Toxicity Class: Caution.
REI: 24 hr.

FOSETYL-ALUMINUM

Systemic fungicide/bactericide. Toxic to estuarine invertebrates.

Aliette (WDG, WSP) *(Rhone-Poulenc)*
Rate: 1.6-4 lb. ai/A
2-5 lb. WDG, WSP/A
Disease:
downy mildew
Time: Apply when conditions favor disease development (high moisture and moderate temperatures); repeat at 7- to 21-day intervals.
Remarks: Collards, kale, mustard greens, rape greens, Swiss chard. Do not apply less than 10 gal./A or make more than 7 applications per season. May be tank mixed with another fungicide labeled for downy mildew using lower rate; when used alone, use higher rate. With moderate disease pressure (curative treatment), high rate and shorter interval should be used. Do not tank mix with copper compounds. Refer to label for use directions.
Days To Harvest: 3.

Signal Word/Toxicity Class: Caution.
REI: 12 hr.

Kocide — see copper compounds

Kumulus — see sulfur

MANEB

Dithiocarbamate contact fungicide. Toxic to fish.

Maneb 75DF or 80WP *(Elf Atochem)*
Rate: **Collards, mustard, turnip tops:**
1.2 lb. ai/A
1.6 lb. 75DF/A
1.5 lb. 80WP/A
Disease:
Alternaria leaf spot Cercospora leaf spot downy mildew
Time: Begin when disease threatens; repeat on 14-day intervals.

Rate: **Kale:**
1.12-1.5 lb. ai/A
1.5-2 lb. 75DF/A
1.2-1.6 lb. ai/A
1.5-2 lb. 80WP/A
Disease:
Alternaria leaf spot downy mildew
Time: Begin when disease threatens; repeat at 7- to 10-day intervals.
Remarks: Collards, kale, mustard, turnip tops. Refer to label for use directions.
Days To Harvest: Collards, mustard, turnip tops-14; kale-10.
State Restrictions: Not for use on kale in California. For use on collards, mustard, and turnip tops in Georgia and Tennessee only.

Signal Word/Toxicity Class: Caution.
REI: 24 hr.

Manex (FL) *(Griffin)*
Rate: **Kale:**
1.2-1.6 lb. ai/A
2.4-3.2 pt. FL/A
Disease:
Alternaria leaf spot downy mildew
Time: Begin when disease threatens; repeat at 7- to 10-day intervals.
Remarks: Kale. Do not use any other product containing a different EBDC ai during the same growing season. Refer to label for use directions.
Days To Harvest: Kale-10.
State Restrictions: Not for use on kale in California.

Signal Word/Toxicity Class: Caution.
REI: 24 hr.

Manex — see maneb

7 Vegetables

NOTICE The information on these pages is for preliminary planning — not a guide for use. Be sure to follow the manufacturer's directions, notwithstanding information contained here.	For personal protective equipment and EPA registration numbers, please turn to page 39.

MEFENOXAM

Systemic fungicide having a specific mode of action.

Ridomil Gold EC, GR, or WSP *(Novartis)*
Rate: 0.5-1 lb. ai/A
1-2 pt. EC/A
20-40 lb. GR/A
0.45-0.9 lb. ai/A
1-2 lb. WSP/A
Disease:
damping-off
Time: Apply at planting.
Remarks: Swiss chard. For banded application, 7" band recommended. Not for use in greenhouse or on field grown vegetable bedding plants. Do not plant any crop which is not registered for use with mefenoxam in treated soil for 12 months.

Signal Word/Toxicity Class: Warning/II (EC); Caution (GR, WSP).
REI: 48 hr.

Microthiol Special — see sulfur

Nu-Cop — see copper compounds

Ridomil Gold — see mefenoxam

SULFUR

Nonsystemic fungicide with protective action. Microthiol Special, Sulfa 80, and Thiolux are micronized sulfurs.

Drexel Suffa (6FL), Sulfa 80 (80DF) or Sulfur (90W) *(Drexel)*
Rate: 2.3-4.5 lb. ai/A
3-6 pt. 6FL/A
2-8 lb. ai/A
3-10 lb. 80DF/A
2.7-4.5 lb. ai/A
3-5 lb. 90W/A
Disease:
powdery mildew rust (80DF)
Time: Apply early leaf stage; repeat at 10- to 14-day intervals as needed, or immediately after rain.
Remarks: Collards, kale; mustard (80DF). Do not use within 2 weeks of oil application (4 weeks 6FL). Some plants are susceptible to injury from sulfur under certain climatic conditions. Do not make sulfur applications at excessively high temperatures. Do not apply this product through any type of irrigation system. Refer to label for use directions.

Signal Word/Toxicity Class: Caution.
REI: 24 hr.

Kumulus DF *(BASF)*
Rate: 2.4-4 lb. ai/A
3-5 lb. DF/A
Disease:
powdery mildew Septoria leaf spot
Time: Apply at early leaf stage; repeat at 10- to 14-day intervals with or without other pesticides immediately after a rain.
Remarks: Do not apply if temperatures during or within 3 days after application are expected to exceed 90°F. Do not apply through any type of irrigation system. Refer to label for use directions.

Signal Word/Toxicity Class: Caution.
REI: 24 hr.

Micro Flo Sulfur (6L), Dual Spraying & Dusting Sulfur (DS) or Microsperse Wettable Sulfur (WS) *(Micro Flo)*
Rate: 2.25-8 lb. ai/A
3-10.66 pt. 6L/A
4.5 lb. ai/A
5 lb. DS/A
2.7-4.5 lb. ai/A
3-5 lb. WS/A
Disease:
powdery mildew rust (6L)
Time: Apply at early leaf stage; repeat at 10- to 14-day intervals (14-day intervals-DS) as necessary.
Remarks: Collards, kale, mustard. Do not apply DS and WS through any type of irrigation system. Refer to label for use directions.

Signal Word/Toxicity Class: Caution.
REI: 24 hr.

Microthiol Special (80WP) *(Elf Atochem)*
Rate: 2.4-8 lb. ai/A
3-10 lb. 80WP/A
Disease:
powdery mildew
Time: Apply at early leaf stage; repeat at 14-day intervals or as needed.
Remarks: Collards, kale, mustard. Refer to label for use directions.

Signal Word/Toxicity Class: Caution.
REI: 24 hr.

Terra Sulfur (6L) *(Terra)*
Thiolux (DF) *(Platte)*
Rate: 3-10.66 pt. 6L/A
3-5 lb. DF/A
Disease:
powdery mildew rust
Time: Apply at early leaf stage; repeat at 10- to 14-day intervals with or without pesticides immediately after rain.
Remarks: Collards, kale, mustard (6L). Do not use within 2 weeks of oil application. Some plants are susceptible to injury from sulfur under certain climatic conditions. Do not make sulfur applications at excessively high temperatures. Refer to label for use directions.

Signal Word/Toxicity Class: Caution.
REI: 24 hr.

SULFUR (Premix)

Fungicide/bactericide in flowable formulation. Toxic to fish and aquatic organisms.

Top Cop with Sulfur (6.25 lb./gal. sulfur & 1 lb./gal. tribasic copper sulfate) *(Stoller)*
Rate: 2 qt./A
Disease:
black rot downy mildew leaf spot
Time: Apply when disease threatens. Repeat at 7- to 10-day intervals; 3-day intervals in plant beds.
Remarks: Collards, kale, mustard. Refer to label for use directions.

Signal Word/Toxicity Class: Caution.
REI: 24 hr.

Thiolux — see sulfur

THIRAM

Broad-spectrum fungicide. Toxic to fish. Thiram 50WP contains the toxic inert Rhodamine B and is also toxic to birds and wildlife.

42-S Thiram or Thiram 50WP Dyed *(Gustafson)*
Thiram Granuflo (WDG) *(UCB Agrochemicals)*
Rate: 8 fl. oz. 42-S/100 lb. seed
8 oz. 50WP/100 lb. seed
5.3 oz. WDG/100 lb. seed
Disease:
damping-off seed decay seedling blight
Time: Seed treatment.
Remarks: Collards, kale, mustard, Swiss chard. Use with EPA-approved dye that imparts unnatural color to seed. Do not use treated seed for food, feed, or oil purposes. Do not apply Thiram Granuflo through any type of irrigation system. Refer to label for use directions.
State Restrictions: Thiram Granuflo not for use in California.

Signal Word/Toxicity Class: Caution.
REI: 24 hr. (Thiram Granuflo).

TOP COP — see copper compounds

Top Cop with Sulfur — see sulfur premix

◆ - **new product** • ai-active ingredient • DF-dry flowable • E/EC-emulsifiable concentrate • EW-emulsion in water
F/FL-flowable • G-granule • L/LC-liquid • SP/WSP-soluble packet • W/WP-wettable powder • WSB-water soluble bag

Herbs & Spices

Aliette — see fosetyl-aluminum

Allegiance — see metalaxyl

Apron — see metalaxyl

Botran — see dicloran

COPPER COMPOUNDS

Kocide (101, DF, 2000, 4.5LF, LF) (copper hydroxide) *(Griffin)*

Rate: **Chive:**
2 lb. 101, DF/A
1.5 lb. 2000/A
1.33 pt. 4.5LF/A
4 pt. LF/A

Disease:
downy mildew

Rate: **Dill:**
2 lb. 2000/A
1.33 pt. 4.5LF/A

Disease:
foliage blight Phoma leaf spot Rhizoctonia

Time: Apply when plants are established in field; repeat every 7- to 10-days as dictated by disease conditions.

Remarks: Chive, dill. If disease pressure is high, use shorter spray interval. Do not apply in a spray solution having a pH less than 6.5 as phytotoxicity may occur. Refer to label for use directions.

State Restrictions: Kocide 101, LF is not for use in California.

Signal Word/Toxicity Class: Danger/I (101, DF); Warning/II (2000); Caution (4.5LF, LF).

REI: 24 hr.

Nu-Cop (3L) (copper hydroxide) *(Micro Flo)*

Rate: **Chive:**
2.66 pt. 3L/A

Disease:
downy mildew

Rate: **Dill:**
2.66-4 pt. 3L/A

Disease:
Phoma leaf spot Rhizoctonia foliage blight

Time: Apply when plants are established in field; repeat at 7- to 10-day intervals.

Remarks: Chive, dill. Refer to label for use directions.

Signal Word/Toxicity Class: Warning/II.

REI: 24 hr.

DICLORAN

Botran 75-W *(Gowan)*

Rate: 1.5 lb. ai/A
2 lb. 75-W/A

Disease:
pink rot

Time: Apply when disease is anticipated or approximately 10 weeks before harvest; repeat at 7-day intervals in summer or 14 day intervals in fall and winter. If early disease is unlikely, apply single application of 5.33 lb./A 4-8 weeks before harvest.

Remarks: Fennel. Refer to label for use directions.

Days To Harvest: 7.

Signal Word/Toxicity Class: Caution.

REI: 12 hr.

Dithane — see mancozeb

FOSETYL-ALUMINUM

Aliette (WDG, WSP) *(Rhone-Poulenc)*

Rate: 1.6-4 lb. ai/A
2-5 lb. WDG, WSP/A

Disease:
downy mildew

Time: Apply when conditions favor disease development (high moisture and moderate temperatures); repeat at 7- to 21-day intervals.

Remarks: Fennel. Do not apply less than 10 gal./A or make more than 7 applications per season. Do not tank mix with copper compounds. With moderate disease pressure, high rate and shorter interval should be used. Refer to label for use directions.

Days To Harvest: 3.

Signal Word/Toxicity Class: Caution.

REI: 12 hr.

Kocide — see copper compounds

MANCOZEB

Dithane (DF, F-45, M-45, WSP) *(Rohm and Haas)*
Manzate 200DF *(Griffin)*
Penncozeb 75DF or 80WP *(Elf Atochem)*

Rate: 1.5 lb. ai/A
2 lb. DF, 200DF/A
1.6 lb. ai/A
2 lb. M-45, WSP/A
3.2 pt. F-45/A
1.12-1.5 lb. ai/A
1.5-2 lb. 75DF/A
1.2-1.6 lb. ai/A
1.5-2 lb. 80WP/A

Disease:
early blight late blight

Time: Apply when disease first appears; repeat at 7- to 10-day intervals.

Remarks: Fennel. Refer to label for use directions.

Days To Harvest: 14.

State Restrictions: Refer to label for California precautions.

Signal Word/Toxicity Class: Caution.

REI: 24 hr.

Manzate — see mancozeb

MEFENOXAM

Ridomil Gold EC, GR, or WSP *(Novartis)*

Rate: 0.5-1 lb. ai/A
1-2 pt. EC/A
20-40 lb. GR/A
0.45-0.9 lb. ai/A
1-2 lb. WSP/A

Disease:
damping-off

Time: Apply at planting.

Remarks: Fennel. For banded application, 7" band recommended. Not for use in greenhouse or on field grown vegetable bedding plants. Do not plant any crop which is not registered for use with mefenoxam in treated soil for 12 months.

Signal Word/Toxicity Class: Warning/II (EC); Caution (GR, WSP).

REI: 48 hr.

METALAXYL

Allegiance or Apron (FL) *(Gustafson)*

Rate: 0.75 fl. oz. Allegiance/100 lb. seed
0.75-1.5 fl. oz. Apron/100 lb. seed

Disease:
damping-off Phytophthora (early season)

Time: Seed treatment.

Remarks: Dill. Use with EPA-approved dye that imparts unnatural color to seed. Do not use treated seed for food, feed, or oil purposes. Refer to label for use directions.

Signal Word/Toxicity Class: Warning/II.

For personal protective equipment and EPA registration numbers, please turn to page 39.

Apron Dry *(Trace)*
Rate: 4 oz./100 lb. seed
Disease:
damping-off Phytophthora
Time: Seed treatment.
Remarks: Dill. Do not use treated seed for food, feed, or oil purposes. Refer to label for use directions.

Signal Word/Toxicity Class: Caution.
REI: 12 hr.

Nu-Cop — see copper compounds

Penncozeb — see mancozeb

Ridomil Gold — see mefenoxam

Horseradish

MEFENOXAM

Ridomil Gold EC or WSP *(Novartis)*
Rate: 0.5-1 lb. ai/A
1-2 pt. EC/A
0.45-0.9 lb. ai/A
1-2 lb. WSP/A
Disease:
Phytophthora Pythium
Time: Preplant incorporated or surface application just after planting.
Remarks: For banded application, 7" band recommended. Not for use in greenhouse or on field grown vegetable bedding plants. Do not plant any crop which is not registered for use with mefenoxam in treated soil for 12 months.
Days To Harvest: 1.

Signal Word/Toxicity Class: Warning/II (EC); Caution (WSP).
REI: 48 hr.

Ridomil Gold — see mefenoxam

Lentils

Allegiance — see metalaxyl

Apron — see metalaxyl

Apron XL — see mefenoxam

CAPTAN

Gustafson Captan 30-DD or 400 *(Gustafson)*
Rate: 2 fl. oz. 30-DD/100 lb. seed
2.5 fl. oz. 400/100 lb. seed
Disease:
damping-off seed decay seedling blight
mold
Time: Seed treatment.
Remarks: Do not combine colloidal suspensions, wettable powders, or emulsifiable concentrates unless previous experience assures compatibility. Use with EPA-approved dye that imparts unnatural color to seed. Do not use treated seed for food, feed, or oil purposes. Refer to label for use directions.

Signal Word/Toxicity Class: Danger/I.

MEFENOXAM

Apron XL LS *(Novartis)*
Rate: 0.32-0.64 fl. oz. LS/100 lb. seed
Disease:
damping-off Phytophthora (early season)
Time: Seed treatment at planting.
Remarks: Do not use treated seed for food, feed, or oil purposes.

Signal Word/Toxicity Class: Warning/II.

Ridomil Gold EC or WSP *(Novartis)*
Rate: 0.25-0.5 lb. ai/A
0.5-1 pt. EC/A
0.225-0.45 lb. ai/A
0.5-1 lb. WSP/A
Disease:
damping-off root and stem rot
Time: Preplant incorporated or surface application just after planting.
Remarks: For banded application, 7" band recommended. Not for use in greenhouse or on field grown vegetable bedding plants. May be tank mixed. Do not plant any crop which is not registered for use with mefenoxam in treated soil for 12 months.
Days To Harvest: 1.

Signal Word/Toxicity Class: Warning/II (EC); Caution (WSP).
REI: 48 hr.

METALAXYL

Allegiance or Apron (FL) *(Gustafson)*
Rate: 0.75 fl. oz. Allegiance/100 lb. seed
0.75-1.5 fl. oz. Apron/100 lb. seed
Disease:
damping-off Phytophthora (early season)
Time: Seed treatment.
Remarks: Use with EPA-approved dye that imparts unnatural color to seed. Do not use for feed, food, or oil purposes. Refer to label for use directions.

Signal Word/Toxicity Class: Warning/II.

Apron Dry *(Trace)*
Rate: 4 oz./100 lb. seed
Disease:
damping-off Phytophthora
Time: Seed treatment.
Remarks: Do not use treated seed for food, feed, or oil purposes. Refer to label for use directions.

Signal Word/Toxicity Class: Caution.
REI: 12 hr.

Ridomil Gold — see mefenoxam

THIRAM

Thiram Granuflo (WDG) *(UCB Agrochemicals)*
Rate: 2 oz. WDG/100 lb. seed
Disease:
damping-off seed decay seedling blight
Time: Seed treatment.
Remarks: Do not use treated seed for food, feed, or oil purposes. Treat only seeds needed for immediate use. Do not store excess treated seeds beyond planting time. Do not apply this product through any type of irrigation system. Refer to label for use directions.
State Restrictions: Not for use in California.

Signal Word/Toxicity Class: Caution.
REI: 24 hr.

◆ - **new product** • ai-active ingredient • DF-dry flowable • E/EC-emulsifiable concentrate • EW-emulsion in water
F/FL-flowable • G-granule • L/LC-liquid • SP/WSP-soluble packet • W/WP-wettable powder • WSB-water soluble bag

Lettuce

(Chicory, Endive, Escarole)

ALIETTE/MANEB 2+2 (Co-Pack)

Twin pack product containing a systemic and a contact fungicide in attached water soluble packets. Each twin pack contains 2 lbs. of Aliette and 2 lbs. of Maneb.

Aliette/Maneb 2+2: Aliette (80% fosetyl-aluminum) *(Rhone-Poulenc)* **& Maneb (75% maneb)** *(Elf Atochem)*

Rate: 4 lb./A
Disease:
downy mildew
Time: Apply when conditions favor disease development; repeat at 7- to 10-day intervals.
Remarks: Endive, lettuce (head, leaf). Do not exceed 6 applications per season (4 applications in California). Do not apply by air in less than 10 gal./A. Do not mix with surfactants or foliar fertilizers. Refer to label for use directions and restrictions.
Days To Harvest: 10, California-14.
State Restrictions: Chemigation application not allowed in California.

Signal Word/Toxicity Class: Caution.
REI: 24 hr.

Aliette — see fosetyl-aluminum

Botran — see dicloran

Brom-O-Gas — see methyl bromide

Champ — see copper compounds

Champion — see copper compounds

Clean Crop C-O-C-S — see copper compounds

COPPER COMPOUNDS

Champ (FL or Formula 2) or Champion (WP) (copper hydroxide) *(Agtrol)*
KOP-Hydroxide (FL) (copper hydroxide) *(Drexel)*

Rate: 1.33-2.66 pt. FL/A
0.66-1.33 pt. Formula 2/A
1.5-2 lb. WP/A
Disease:
downy mildew
Time: Apply when disease first appears; repeat at 7- to 10-day intervals as needed.
Remarks: Endive, lettuce. Refer to label for use directions.

Signal Word/Toxicity Class: Danger/I (Champion, KOP); Caution (Champ).
REI: 48 hr. (KOP); 24 hr.

Clean Crop C-O-C-S (WDG) (74.8% copper oxychloride & 14.2% copper sulfate, basic) *(Platte)*

Rate: 0.5-1.5 lb. ai/A
1-3 lb. WDG/A
Disease:
downy mildew
Time: Apply when disease first threatens; repeat at 7- to 10-day intervals to protect new growth throughout growing season.
Remarks: Lettuce (head, leaf). Under adverse weather conditions flecking of wrapper leaves may occur. Do not apply this product through any type of irrigation system. Refer to label for use directions.
State Restrictions: Not for use in California.

Signal Word/Toxicity Class: Warning/II.
REI: 24 hr.

Micro Flo Basic Copper "53" (basic copper sulfate) or Nu-Cop (3L, 50DF, 50WP) (copper hydroxide) *(Micro Flo)*

Rate: 1-3 lb. 53/A
0.66-2.66 pt. 3L/A
1-2 lb. 50DF, 50WP/A
Disease:
downy mildew
Time: Apply when disease first appears; repeat at 7- to 10-day intervals; (3- to 7-day intervals for 53) as needed.
Remarks: Endive, lettuce. Refer to label for use directions.

Signal Word/Toxicity Class: Danger/I (50DF, 50WP); Warning/II (53, 3L).
REI: 48 hr. (50DF, 50WP); 24 hr. (53, 3L).

TOP COP Tri-Basic (FL) (basic copper sulfate) *(Stoller)*

Rate: 2-3 pt. FL/A
Disease:
downy mildew
Time: Apply when disease appears; repeat at 7- to 10-day intervals.
Remarks: Refer to label for use directions.

Signal Word/Toxicity Class: Caution.
REI: 24 hr.

DICLORAN

Botran 6 Dust *(Wilbur-Ellis)*

Rate: Leaf (greenhouse):
1.8 lb. ai/A
30 lb. Dust/A
Head (field):
2 lb. ai/A
33.3 lb. Dust/A
Disease:
Botrytis rot — Sclerotinia wilt (head)
Time: For leaf, apply 7 days after transplanting; repeat when plants are half mature. For head, apply at thinning; repeat at 7- to 10-day intervals.
Remarks: Lettuce (head, leaf). For leaf lettuce, do not apply to wilted plants or seedlings. Refer to label for use directions.
Days To Harvest: 14.

Signal Word/Toxicity Class: Caution.
REI: 12 hr.

Botran 75-W *(Gowan)*

Rate: 1.5-4 lb. ai/A
2-5.33 lb. 75-W/A
Disease:
drop rot — Sclerotinia minor — watery soft rot
gray mold
Time: For drop rot and watery soft rot, apply preemergence in 4-6" band over transplant or seed row; make second application immediately after thinning. For Sclerotinia minor, apply single application immediately after thinning. For gray mold, apply prior to thinning (2-6 leaf stage) and repeat immediately after thinning; greenhouse and hothouse lettuce, apply 7-days following transplanting and repeat when plants are half mature.
Remarks: Endive, greenhouse/hothouse lettuce (leaf), lettuce (head, leaf). Some leaf bronzing may be noted soon after application, but is temporary. Do not apply to wilted plants or seedlings. Refer to label for use directions.
Days To Harvest: 14.

Signal Word/Toxicity Class: Caution.
REI: 12 hr.

FOSETYL-ALUMINUM

Aliette (WDG, WSP) *(Rhone-Poulenc)*

Rate: 1.6-4 lb. ai/A
2-5 lb. WDG, WSP/A
Disease:
downy mildew
Time: Apply when conditions favor disease development (high moisture and moderate temperatures); repeat at 7- to 21-day intervals.
Remarks: Endive, lettuce (head, leaf). Do not apply less than 10 gal./A or make more than 7 applications per season. Do not tank mix with copper compounds. With moderate disease pressure, high rate and shorter interval should be used. Refer to label for use directions.
Days To Harvest: 3.

Signal Word/Toxicity Class: Caution.
REI: 12 hr.

NOTICE	
The information on these pages is for preliminary planning — not a guide for use. Be sure to follow the manufacturer's directions, notwithstanding information contained here.	For personal protective equipment and EPA registration numbers, please turn to page 39.

IPRODIONE

Rovral (4FL, WG, WP) *(Rhone-Poulenc)*
Rate: 0.75-1 lb. ai/A
1.5-2 pt. 4FL/A
1.5-2 lb. WG, WP/A
Disease:
bottom rot lettuce drop
Time: Apply from 3-leaf stage to just after thinning; repeat 10 days later. If conditions still favor disease development, make third application 10 days after second application.
Remarks: Lettuce (head, leaf). Do not make more than 3 applications per crop. Do not cultivate after application; if necessary, make application during or immediately after cultivation. Do not drench. Refer to label for use directions.
Days To Harvest: 14.
Crop Rotation: Beans, broccoli, carrots, garlic, lettuce, onions (dry bulb), peanuts, potatoes, and rice can be rotated after harvest.
State Restrictions: Chemigation application not registered in California.

Signal Word/Toxicity Class: Warning/II (WG); Caution (4FL, WP).
REI: 12 hr.

KOP-Hydroxide — see copper compounds

MANEB

Maneb 75DF or 80WP *(Elf Atochem)*
Manex (FL) *(Griffin)*
Rate: 1.12-1.5 lb. ai/A
1.5-2 lb. 75DF/A
1.2-1.6 lb. ai/A
1.5-2 lb. 80WP/A
1.2-1.6 qt. FL/A
Disease:
downy mildew
Time: Apply when disease appears; repeat at 7- to 10-day intervals.
Remarks: Endive, lettuce (head, leaf). Do not use any other product containing a different EBDC ai during the same growing season if using Manex. Refer to label for use directions.
Days To Harvest: 10; California-14.

Signal Word/Toxicity Class: Caution.
REI: 24 hr.

Manex — see maneb

MEFENOXAM

Ridomil Gold EC, GR, or WSP *(Novartis)*
Rate: 0.5-1 lb. ai/A
1-2 pt. EC/A
20-40 lb. GR/A
0.45-0.9 lb. ai/A
1-2 lb. WSP/A
Disease:
damping-off
Time: Apply at planting.
Remarks: Endive, lettuce (head, leaf). For banded application, 7" band recommended. Not for use in greenhouse or on field grown vegetable bedding plants. Do not plant any crop which is not registered for use with mefenoxam in treated soil for 12 months.

Signal Word/Toxicity Class: Warning/II (EC); Caution (GR, WSP).
REI: 48 hr.

METHYL BROMIDE

2% Chloropicrin added as warning agent. Restricted Use Pesticide. Classified as an ozone depleter under the Clean Air Act. Use is to be completely eliminated starting Jan. 1, 2001.

Brom-O-Gas *(Great Lakes Chemical)*
Rate: Injected with chisels + covered with tarp:
180-240 lb./A
Raised tarp with topical application:
0.4-0.55 lb./100 ft.
Disease:
Fusarium, oak root fungus, Phytophthora, Plasmodiophora, Pythium, Rhizoctonia, Sclerotinia, Sclerotium
Time: Preplant.
Remarks: Do not fumigate when soil temperature is below 50°F. For best results, fumigate when temperature is 60°-80°F at depth of 6". Refer to label for use directions.
State Restrictions: Brom-O-Gas not for use in Arizona and not for use in California for oak root fungus (Armillaria root rot) control.

Signal Word/Toxicity Class: Danger/I.

Micro Flo — see copper compounds

Microthiol Special — see sulfur

Nu-Cop — see copper compounds

Ridomil Gold — see mefenoxam

Ronilan — see vinclozolin

Rovral — see iprodione

Sulfa — see sulfur

SULFUR

Micro Flo Sulfur (6L) *(Micro Flo)*
Microthiol Special (80WP) *(Elf Atochem)*
Terra Sulfur (6L) *(Terra)*
Rate: 4-4.5 lb. ai/A
5.3-6 pt. 6L/A
4-4.8 lb. ai/A
5-6 lb. 80WP/A
Disease:
powdery mildew rust
Time: Apply at early leaf stage; repeat at 14-day intervals as needed.
Remarks: Lettuce (head, leaf). Thorough coverage is required. Refer to label for use directions and precautions.

Signal Word/Toxicity Class: Caution.
REI: 24 hr.

Sulfa 80 (80DF) *(Drexel)*
Rate: 2.4-8 lb. ai/A
3-10 lb. 80DF/A
Disease:
powdery mildew rust
Time: Apply early leaf stage; repeat at 10- to 14-day intervals as needed, or immediately after rain.
Remarks: Do not use within 2 weeks of oil application. Refer to label for use directions and precautions.

Signal Word/Toxicity Class: Caution.
REI: 24 hr.

SULFUR (Premix)

Top Cop with Sulfur (6.25 lb./gal. sulfur & 1 lb./gal. tribasic copper sulfate) *(Stoller)*
Rate: 2 qt./A
Disease:
Anthracnose, leaf spot, powdery mildew, downy mildew
Time: Apply when disease threatens; repeat at 7- to 10-day intervals.
Remarks: Refer to label for use directions.

Signal Word/Toxicity Class: Caution.
REI: 24 hr.

◆ - **new product** • ai-active ingredient • DF-dry flowable • E/EC-emulsifiable concentrate • EW-emulsion in water
F/FL-flowable • G-granule • L/LC-liquid • SP/WSP-soluble packet • W/WP-wettable powder • WSB-water soluble bag

TERR-O-GAS (Premix)

Classified as an ozone depleter under the Clean Air Act. Use to be eliminated starting Jan. 1, 2001. Restricted Use Pesticide.

Terr-O-Gas (67% methyl bromide & 33% chloropicrin) *(Great Lakes Chemical)*

Rate: Nontarp:
270-355 lb./A

Disease:
Fusarium, Phytophthora, Plasmodiophora, Pythium, Rhizoctonia, Sclerotinia, Sclerotium

Time: Apply up to 2 weeks prior to planting when soil moisture is good and temperature is 60°-85°F.

Remarks: Work upwind of product and never fumigate without 2 qualified applicators. Do not fumigate when there is little or no air movement. Tarping is not used after field application. Refer to label for use directions.

Signal Word/Toxicity Class: Danger/I.

THIRAM

42-S Thiram or Thiram 50WP Dyed *(Gustafson)*
Thiram Granuflo (WDG) *(UCB Agrochemicals)*

Rate: Endive, lettuce:
8 fl. oz. 42-S/100 lb. seed
8 oz. 50WP/100 lb. seed
5.3 oz. WDG/100 lb. seed
Chickory:
5.2 fl. oz. 42-S/100 lb. seed
3.5 oz. WDG/100 lb. seed

Disease:
damping-off, seed decay, seedling blight

Time: Seed treatment.

Remarks: Chickory, endive, lettuce. Use with EPA-approved dye that imparts unnatural color to seed. Do not use treated seed for food, feed, or oil purposes. Do not apply Thiram Granuflo through any type of irrigation system. Refer to label for use directions.

State Restrictions: 42-S not for chickory in California. Thiram Granuflo not for use in California.

Signal Word/Toxicity Class: Caution.
REI: 24 hr. (Thiram Granuflo).

TOP COP — see copper compounds

Top Cop with Sulfur — see sulfur premix

VINCLOZOLIN

Dicarboximide fungicide. Toxic to fish.

Ronilan FL, DF, EG *(BASF)*

Rate: 0.5-1 lb. ai/A
1-2 pt. FL/A
1-2 lb. DF, EG/A

Disease:
Rhizoctonia bottom rot, Sclerotinia drop

Time: Apply within 2 days of thinning for direct seeded; 7- to 10-days after transplanting. Second application should be made 14 days after first if cool, wet conditions occur for periods lasting 48 hrs. or more. A third application should be made 14 days after second if conditions favoring disease occurs for direct seeded only.

Remarks: Lettuce (direct seeded, transplanted). Do not apply more than 3 lb. ai/A in one season. Do not use as plant dip or injury may occur. Do not apply this product through any type of irrigation system. Refer to label for use directions.

Days To Harvest: 28.

Signal Word/Toxicity Class: Caution.
REI: 12 hr.

Mushrooms

BENLATE

Benlate (SP, WP) (benomyl) *(DuPont)*

Rate: 0.5 lb. ai/100 gal.
1 lb. SP, WP/100 gal.

Disease:
Verticillium spot

Time: New bed: Apply after casing; repeat 3- to 4-days before harvest. Production bed: Apply after picking; repeat 10-days later.

Remarks: Do not tank mix or alternate with benzimidazole or thiophanate products. Refer to label for use directions.

Days To Harvest: 2.

State Restrictions: Benlate SP not registered for use in California.

Signal Word/Toxicity Class: Caution.
REI: 24 hr.

Bravo — see chlorothalonil

CHLOROTHALONIL

Bravo 500, 720/Weather Stik, Ultrex (WDG) *(ZENECA)*

Rate: 4-8 fl. oz. 500/1000 sq. ft.
2.75-5.5 fl. oz. 720/1000 sq. ft.
2.5-5 oz. WDG/1000 sq. ft.

Disease:
Verticillium spot

Time: Make 2 drench applications to bed surface. First application should be made within 2 days of top-dressing spawn-colonized compost with casing layer using high rate; second, use low rate at pinning.

Remarks: Use in sufficient water to obtain adequate coverage. Do not graze treated areas or feed treated crop residues to livestock. Refer to label for use directions and restrictions.

Days To Harvest: 5.

Signal Word/Toxicity Class: Danger/I (WDG); Warning/II (500, 720).
REI: 48 hr.

Mertect — see thiabendazole

THIABENDAZOLE

Mertect 340-F *(Novartis)*

Rate: 4-8 fl. oz. 340-F/1000 sq. ft.

Disease:
Verticillium spot

Time: Apply at casting, fuzzing, pinning.

Remarks: First application may be up to 8 fl. oz./1000 sq. ft.; subsequent not more than 4 fl. oz./1000 sq. ft.; maximum 20 fl. oz./1000 sq. ft. per crop. Do not apply through any type of irrigation system.

Days To Harvest: 12 hr.

Signal Word/Toxicity Class: Caution.

Okra

Allegiance — see metalaxyl

Apron — see metalaxyl

Apron XL — see mefenoxam

NOTICE The information on these pages is for preliminary planning — not a guide for use. Be sure to follow the manufacturer's directions, notwithstanding information contained here.

For personal protective equipment and EPA registration numbers, please turn to page 39.

CAPTAN (Premix)

Nu-Gro Isotox (12.2% captan & 25% lindane) *(Wilbur-Ellis)*
Seed Mate Isotox (12.2% captan & 25% lindane) *(Loveland)*

Rate: 4 oz./100 lb. seed
Disease:
damping-off seed decay seedling blight
Time: Planter box treatment.
Remarks: Do not use treated seed for food, feed, or oil purposes. Refer to label for use directions.

Signal Word/Toxicity Class: Danger/I.
REI: 4 days.

MEFENOXAM

Apron XL LS *(Novartis)*

Rate: 0.32-0.64 fl. oz. LS/100 lb. seed
Disease:
damping-off Phytophthora (early season)
Time: Seed treatment at planting.
Remarks: Do not use treated seed for food, feed, or oil purposes.

Signal Word/Toxicity Class: Warning/II.

METALAXYL

Allegiance or Apron (FL) *(Gustafson)*

Rate: 0.75 fl. oz. Allegiance/100 lb. seed
0.75-1.5 fl. oz. Apron/100 lb. seed
Disease:
damping-off Phytophthora (early season)
Time: Seed treatment.
Remarks: Use with EPA-approved dye that imparts unnatural color to seed. Do not use treated seed for food, feed, or oil purposes. Refer to label for use directions.

Signal Word/Toxicity Class: Warning/II.

Apron Dry *(Trace)*

Rate: 4 oz./100 lb. seed
Disease:
damping-off Phytophthora
Time: Seed treatment.
Remarks: Do not use treated seed for food, feed, or oil purposes. Do not store excess treated seed beyond planting time. Refer to label for use directions.

Signal Word/Toxicity Class: Caution.
REI: 12 hr.

Nu-Gro Isotox — see captan premix

Seed Mate — see captan premix

SULFUR

Micro Flo Sulfur (6L) *(Micro Flo)*
Terra Sulfur (6L) *(Terra)*

Rate: 1.5-3 lb. ai/A
2-4 pt. 6L/A
Disease:
powdery mildew
Time: Apply when disease first appears; repeat at 7-day intervals as needed.
Remarks: Head, leaf lettuce. Do not use within 2 weeks of oil application. Some plants are susceptible to injury from sulfur under certain climatic conditions. Do not make sulfur applications at excessively high temperatures. Refer to label for use directions.

Signal Word/Toxicity Class: Caution.
REI: 24 hr.

THIRAM

42-S Thiram or Thiram 50WP Dyed *(Gustafson)*
Thiram Granuflo (WDG) *(UCB Agrochemicals)*

Rate: 6 fl. oz. 42-S/100 lb. seed
6 oz. 50WP/100 lb. seed
4 oz. WDG/100 lb. seed
Disease:
damping-off seed decay seedling blight
Time: Seed treatment.
Remarks: Use with EPA-approved dye that imparts unnatural color to seed. Do not use treated seed for food, feed, or oil purposes. Do not apply Thiram Granuflo through any type of irrigation system. Refer to label for use directions.
State Restrictions: 42-S Thiram and Thiram Granuflo not for use in California.

Signal Word/Toxicity Class: Caution.
REI: 24 hr. (Thiram Granuflo).

Onions

ALIETTE/MANEB 2+2 (Co-Pack)

Twin pack product containing a systemic and a contact fungicide in attached water soluble packets. Each twin pack contains 2 lbs. of Aliette and 2 lbs. of Maneb.

Aliette/Maneb 2+2: Aliette (80% fosetyl-aluminum) *(Rhone-Poulenc)* **& Maneb (75% maneb)** *(Elf Atochem)*

Rate: 4 lb./A
Disease:
Botrytis leaf blight neck rot purple blotch
downy mildew
Time: Apply when conditions favor disease development; repeat at 7-day intervals.
Remarks: Dry bulb onions. Do not apply to exposed bulbs. Do not exceed 7 applications per season. Do not tank mix with copper fungicides. Do not mix with surfactants or foliar fertilizers. Refer to label for use directions and restrictions.
Days To Harvest: 7.
State Restrictions: Chemigation application not allowed in California.

Signal Word/Toxicity Class: Caution.
REI: 24 hr.

Aliette — see fosetyl-aluminum

Botran — see dicloran

Bravo — see chlorothalonil

Brom-O-Gas — see methyl bromide

CAPTAN

Drexel Captan 4FL *(Drexel)*

Rate: 1.5 pt. 4FL/100 lb. seed
Disease:
seed decay seedling blight soilborne disease
Time: Liquid or slurry seed treatment.
Remarks: Pelleting onions. Do not use treated seed for food, feed, or oil purposes. Refer to label for use directions.

Signal Word/Toxicity Class: Danger/I.

◆ - **new product** • ai-active ingredient • DF-dry flowable • E/EC-emulsifiable concentrate • EW-emulsion in water
F/FL-flowable • G-granule • L/LC-liquid • SP/WSP-soluble packet • W/WP-wettable powder • WSB-water soluble bag

CAPTAN (Premix)

Nu-Gro Isotox (12.2% captan & 25% lindane) *(Wilbur-Ellis)*
Seed Mate Isotox (12.2% captan & 25% lindane) *(Loveland)*
Rate: 4 oz./100 lb. seed
Disease:
damping-off seed decay seedling blight
Time: Planter box treatment.
Remarks: Treated seed must not be used for or mixed with food or animal feed, or processed for oil. Refer to label for use directions.

Signal Word/Toxicity Class: Danger/I.
REI: 4 days.

Champ — see copper compounds

Champion — see copper compounds

CHLOR-O-PIC

Restricted Use Pesticide.

Chlor-O-Pic (chloropicrin) *(Great Lakes Chemical)*
Rate: Nontarp:
200-400 lb./A
Disease:
Cylindrocladium, Fusarium, Phytophthora, Plasmodiophora, Pseudomonas, Pythium, Rhizoctonia, Sclerotinia, Sclerotium, Verticillium
Time: Apply up to 2 weeks prior to planting when soil moisture is good and temperature is 60°-85°F.
Remarks: Dry bulb onions. Work upwind of product and never fumigate without 2 qualified applicators. Do not fumigate when there is no or little air movement. Refer to label for use directions.

Signal Word/Toxicity Class: Danger/I.

CHLOROTHALONIL

Bravo 500, 720/Weather Stik, Ultrex (WDG) *(ZENECA)*
Echo 500, 720 or 90DF *(Sostram)*
Terranil (6L, 90DF or 90DF WSP) *(Terra)*
Rate: 0.75-1.5 lb. ai/A
1.5-3 pt. 500/A
1-2 pt. 720, 6L/A
0.9-1.8 lb. WDG/A
1.2-1.7 lb. 90DF/A (Echo)
0.875-1.75 lb. 90DF, 90DF WSP/A (Terranil)
Disease:
Botrytis leaf blight Botrytis neck rot (suppression) purple blotch
Time: Apply at 7- to 10-day intervals according to disease hazard. For suppression of neck rot during storage, at least 3 weekly applications prior to lifting recommended.
Remarks: Onions (dry bulb). Use in sufficient water to obtain thorough coverage of tops. Refer to label for use directions and restrictions.
Days To Harvest: 7.

Rate: 1-2 lb. ai/A
2-4.25 pt. 500/A
1.5-3 pt. 720, 6L/A
1.4-2.7 lb. WDG/A
1.2-2.4 lb. 90DF/A (Echo)
1.125-2.25 lb. 90DF, 90DF WSP/A (Terranil)
Disease:
Botrytis leaf blight downy mildew (suppression) purple blotch
Time: Apply prior to favorable infection periods; repeat at 7- to 10-day intervals as necessary. Use high rate and 7-day interval when heavy dew or rains persist.
Remarks: Onions (green bunching, grown for seed), leeks, shallots. Use in sufficient water to obtain thorough coverage of tops. Do not apply more than 3 times per season. If additional disease control is needed before harvest, use another registered fungicide. Refer to label for use directions and restrictions.
Days To Harvest: Dry bulb-7; green bunching, leeks, shallots-14.

Signal Word/Toxicity Class: Danger/I (WDG, 90DF); Warning/II (500, 720, 6L).
REI: 48 hr.

Clean Crop C-O-C-S — see copper compounds

COPPER COMPOUNDS

Champ (FL or Formula 2) or Champion (WP) (copper hydroxide) *(Agtrol)*
Copper-Count-N (L) (copper ammonium) *(Mineral Research)*
Kocide (101, DF, 2000, 4.5LF, LF) (copper hydroxide) *(Griffin)*
KOP-Hydroxide (FL, 50WP) (copper hydroxide) *(Drexel)*
Rate: 2.66 pt. FL, LF/A
1.33 pt. Formula 2, 4.5LF/A
2 lb. WP, 50WP, 101, DF/A
4 pt. L/A
0.75-1.5 lb. 2000/A
Disease:
bacterial blight (2000, 4.5LF) downy mildew purple blotch
Time: Apply when plants are 4-6" high; repeat at 7- to 10-day intervals.
Remarks: Do not apply Copper-Count-N through any type of irrigation system unless supplemental labeling on chemigation is followed. Refer to label for use directions and precautions.

Signal Word/Toxicity Class: Varies by formulation.
REI: Varies by formulation.

Clean Crop C-O-C-S (WDG) (74.8% copper oxychloride & 14.2% copper sulfate, basic) *(Platte)*
Rate: 1.5-2 lb. ai/A
3-4 lb. WDG/A
Disease:
downy mildew purple blotch
Time: Apply when plants are 4-8" high; repeat at 7- to 10-day intervals to protect new growth throughout growing season.
Remarks: If disease is severe, increase dosage up to 7 lb./A on larger plants. Do not apply this product through any type of irrigation system. Refer to label for use directions.
State Restrictions: Not for use in California.

Signal Word/Toxicity Class: Warning/II.
REI: 24 hr.

Drexel Basic Copper Sulfate (WP) or KOP 300 (3FL) (basic copper sulfate) *(Drexel)*
Rate: 3-4 lb. WP/A
2.6 pt. 3FL
Disease:
downy mildew purple blotch (California)
Time: Apply when plants are 4-6"; repeat at 7- to 10-day intervals.
Remarks: Do not apply through any type of irrigation system. Refer to label for use directions.

Signal Word/Toxicity Class: Warning/II.
REI: 24 hr. (WP); 12 hr. (3FL).

Micro Flo Basic Copper "53" (basic copper sulfate) or Microsperse COC (53WP) (copper oxychloride) or Nu-Cop (3L, 50DF, 50WP) (copper hydroxide) *(Micro Flo)*
Rate: 3-4 lb. 53, 53WP/A
1.33-2.66 pt. 3L/A
2 lb. 50DF, 50WP/A
Disease:
downy mildew purple blotch
Time: Apply when plants are 4-6" high; repeat at 7- to 10-day intervals as needed.
Remarks: Leeks (53WP), onions. Addition of spreader/sticker may be desirable. Refer to label for use directions.

Signal Word/Toxicity Class: Danger/I (50DF, 50WP, 53WP); Warning/II (53, 3L).
REI: 48 hr. (50DF, 50WP); 24 hr. (53, 53WP, 3L).

TOP COP Tri-Basic (FL) (basic copper sulfate) *(Stoller)*
Rate: 4 pt. FL/A
Disease:
downy mildew purple blotch
Time: Apply when plants are 4-6" high; repeat at 7- to 10-day intervals.
Remarks: Refer to label for use directions.

Signal Word/Toxicity Class: Caution.
REI: 24 hr.

NOTICE The information on these pages is for preliminary planning — not a guide for use. Be sure to follow the manufacturer's directions, notwithstanding information contained here.

For personal protective equipment and EPA registration numbers, please turn to page 39.

Copper-Count-N — see copper compounds

DICLORAN

Botran 6 Dust *(Wilbur-Ellis)*

Rate: 15-24 lb. ai/A
250-400 lb. Dust/A

Disease:
white rot

Time: For direct seeding, apply 1-2 weeks before; for planting sets or buds apply broadcast.

Remarks: Refer to label for use directions.

Crop Rotation: Do not plant spinach as follow-up crop on treated soil. If lettuce is planted as follow-up crop, plow soil 8" and cross disc area prior to seeding.

State Restrictions: Not for use in North Central states.

Signal Word/Toxicity Class: Caution.
REI: 12 hr.

Botran 75-W *(Gowan)*

Rate: 4 lb. ai/A
5.33 lb. 75-W/A

Disease:
white rot

Time: For direct-seeded, apply 5" band over seed row/bed and incorporate in top 1 1/2" of soil 1-2 weeks before onions are seeded. For planting sets or buds, apply in-furrow application.

Rate: 1.1-2 lb. ai/A
1.5-2.66 lb. 75-W/A

Disease:
Botrytis rot leaf blight

Time: Apply when disease is anticipated; repeat at 14-day intervals if necessary.

Remarks: Onions, shallots. Refer to label for use directions.

Days To Harvest: 14.

Crop Rotation: Do not plant spinach as a follow-up crop in treated soil.

State Restrictions: Direct-seeded application for use in eastern Oregon, Washington, and western Idaho only. Not for use in California.

Signal Word/Toxicity Class: Caution.
REI: 12 hr.

Dithane — see mancozeb

Echo — see chlorothalonil

FOSETYL-ALUMINUM

Aliette (WDG, WSP) *(Rhone-Poulenc)*

Rate: 1.6-2.4 lb. ai/A
2-3 lb. WDG, WSP/A

Disease:
downy mildew purple blotch (suppression)

Time: Preventive: Apply when conditions (high humidity and cool evening temperatures) favor disease development; repeat at 7- to 14-day intervals. If disease is already present, use high rate at 7-day intervals.

Remarks: Dry bulb onions. Apply in sufficient water for good coverage. Do not apply less than 10 gal./A by air (20 gal./A by ground). May be tank mixed with other foliar-applied fungicides and insecticides registered for use on onions except copper materials or acid type compatibility adjuvants. Do not apply more than 7 applications per season. Refer to label for use directions.

Days To Harvest: 7.

Signal Word/Toxicity Class: Caution.
REI: 12 hr.

IPRODIONE

Rovral (4FL, WG, WP) *(Rhone-Poulenc)*

Rate: 0.5-0.75 lb. ai/A
1-1.5 pt. 4FL/A
1-1.5 lb. WG, WP/A

Disease:
Botrytis leaf blight Botrytis neck rot purple blotch

Time: Apply when conditions favor disease development; repeat at 7- to 14-day intervals (7- to 10-day if tank mixed).

Remarks: Dry bulb onions. May be tank mixed with another fungicide. Do not make more than 5 applications per season (10 applications if tank mixed). Do not drench. Refer to label for use directions.

Days To Harvest: 7.

Crop Rotation: Beans, broccoli, carrots, garlic, lettuce, onions (dry bulb), peanuts, potatoes, and rice can be rotated after harvest.

Signal Word/Toxicity Class: Warning/II (WG); Caution (4FL, WP).
REI: 12 hr.

Kocide — see copper compounds

KOP-Hydroxide — see copper compounds

Kumulus — see sulfur

MANCOZEB

Dithane (DF, F-45, M-45, WSP) *(Rohm and Haas)*
Manzate 200DF *(Griffin)*

Rate: 2.25 lb. ai/A
3 lb. DF, 200DF/A
2.4 lb. ai/A
3 lb. M-45, WSP/A
4.8 pt. F-45/A

Disease: Dry bulb:
Botrytis leaf blight neck rot purple blotch
downy mildew

Furrow drench (onion seeds):
damping-off seedling blight smut
seed rot

Time: Apply when disease first reported in area; repeat at 7-day intervals throughout season. For furrow drench, apply in 75-125 gal. water/A at time of planting seeds.

Remarks: Onions (dry bulb, seed). Addition of surfactant will improve performance. Do not apply to exposed bulbs. Do not allow spray or drift to contact bulbs after lifting from soil. For furrow drench, use 18" row spacing. Refer to label for use directions.

Days To Harvest: 7.

State Restrictions: Refer to label for California precautions.

Signal Word/Toxicity Class: Caution.
REI: 24 hr.

Penncozeb 75DF or 80WP *(Elf Atochem)*

Rate: Dry bulb:
1.5-2.25 lb. ai/A
2-3 lb. 75DF/A
1.6-2.4 lb. ai/A
2-3 lb. 80WP/A

Disease:
Botrytis leaf blight neck rot purple blotch
downy mildew

Time: Begin applications when diseases are first reported in area at 7-day intervals throughout season.

Rate: Onion seeds:
2.25 lb. ai/A
3 lb. 75DF/A
2.4 lb. ai/A
3 lb. 80WP/A

Disease:
smut

Time: Apply as furrow drench at time of planting seeds.

Remarks: Do not apply to exposed bulbs. Refer to label for use directions.

Days To Harvest: Dry bulb onions-7.

State Restrictions: Refer to label for California precautions.

Signal Word/Toxicity Class: Caution.
REI: 24 hr.

◆ - **new product** • ai-active ingredient • DF-dry flowable • E/EC-emulsifiable concentrate • EW-emulsion in water
F/FL-flowable • G-granule • L/LC-liquid • SP/WSP-soluble packet • W/WP-wettable powder • WSB-water soluble bag

MANEB

Maneb 75DF or 80WP *(Elf Atochem)*
Manex (FL) *(Griffin)*

Rate: **Dry bulb, green:**
1.5-2.25 lb. ai/A
2-3 lb. 75DF/A
1.6-2.4 lb. ai/A
2-3 lb. 80WP/A
1.6-2.4 qt. FL/A

Disease:
Botrytis leaf blight neck rot purple blotch
downy mildew

Time: Begin applications when diseases are first reported in area at 7-day intervals throughout the season.

Rate: **Onion seeds:**
2.25 lb. ai/A
3 lb. 75DF/A
2.4 lb. ai/A
3 lb. 80WP/A
1.6-2.4 lb. ai/A
1.6-2.4 qt. FL/A

Disease:
smut

Time: Apply as furrow drench at time of planting seeds.
Remarks: Dry bulb, green and seed onions. Do not apply to exposed bulbs. Refer to label for use directions.
Days To Harvest: 7.
State Restrictions: Furrow drench not for use in California.

- - - - - - - - - - - - - - - - -

Signal Word/Toxicity Class: Caution.
REI: 24 hr.

MANEB (Premix)

ManKocide (DF) (15% maneb & 46.1% copper hydroxide) *(Griffin)*

Rate: 1.5-2.25 lb. DF/A
Disease:
bacterial blight

Rate: 2.5 lb. DF/A
Disease:
Botrytis leaf blight downy mildew purple blotch

Time: Apply protective spray starting when diseases are first reported in area; repeat at 7-day intervals throughout season.
Remarks: Dry bulb onions. Do not apply to exposed bulbs. Refer to label for use directions.
Days To Harvest: 7.

- - - - - - - - - - - - - - - - -

Signal Word/Toxicity Class: Danger/I.
REI: 48 hr.

Manex — see maneb

ManKocide — see maneb premix

Manzate — see mancozeb

MEFENOXAM

Ridomil Gold EC or WSP *(Novartis)*

Rate: 0.25-0.5 lb. ai/A
0.5-1 pt. EC/A
0.225-0.45 lb. ai/A
0.5-1 lb. WSP/A

Disease:
damping-off

Time: Preplant incorporated or surface application just after planting.
Remarks: Onions (bulb, dry, green, spring onions, Japanese bunching), leeks, scallions, green shallots, green eschalots. For banded applications, 7" band recommended. Do not use in greenhouse or on field grown vegetable bedding plants. May be tank mixed. Do not plant any crop which is not registered for use with mefenoxam in treated soil for 12 months.

- - - - - - - - - - - - - - - - -

Signal Word/Toxicity Class: Warning/II (EC); Caution (WSP).
REI: 48 hr.

MEFENOXAM (Premixes)

Ridomil Gold/Bravo (WP) (4.5% mefenoxam & 72% chlorothalonil) *(Novartis)*

Rate: 1.5 lb. ai/A
2 lb. WP/A

Disease:
Botrytis leaf blight downy mildew purple blotch

Time: Apply when conditions are favorable for disease development, but before infection. For downy mildew, repeat at 14-day intervals until disease threat is over. For Botrytis leaf blight and purple blotch, apply at 7-day intervals or at 14-day intervals when other effective EPA-registered fungicides are applied between applications.
Remarks: Onions (bulb, dry, green, spring, Japanese bunching), leeks, scallions, green shallots, green eschalots. Up to 4 applications per crop may be applied to bulb, dry or seed onions; 3 applications to green onions. If conditions for disease development are still favorable after 3 applications to green onions, use other registered fungicides. Do not plant any crop which is not registered for use with mefenoxam in treated soil for 12 months. Refer to label for use directions.
Days To Harvest: Bulb, dry: 7; green onions: 21.

- - - - - - - - - - - - - - - - -

Signal Word/Toxicity Class: Danger/I.
REI: 48 hr.

Ridomil Gold/Copper (WSB) (5% mefenoxam & 60% copper hydroxide) *(Novartis)*

Rate: 1.3 lb ai/A
2 lb. WSB/A

Disease:
downy mildew

Time: Apply when conditions are favorable for disease development, but before infection; repeat at 14-day intervals until disease threat is over.
Remarks: Onions (bulb, dry, green, spring, Japanese bunching), leeks, scallions, green shallots, green eschalots. Up to 4 applications per crop may be applied to bulb, dry, or seed onions; 3 applications to green onions. If conditions for disease development are still favorable after 3 applications, use other fungicides registered for disease control. Do not apply more than 0.4 lb. mefenoxam/A. Do not plant any crop which is not registered for use with mefenoxam in treated soil for 12 months.
Days To Harvest: Bulb, dry: 7; green onions: 21.

- - - - - - - - - - - - - - - - -

Signal Word/Toxicity Class: Danger/I.
REI: 48 hr.

Ridomil Gold MZ (WP) (4% mefenoxam & 64% mancozeb) *(Novartis)*

Rate: 1.7 lb. ai/A
2.5 lb. WP/A

Disease:
downy mildew

Time: Apply when conditions are favorable for disease development, but before infection; repeat at 14-day intervals until disease threat is over.
Remarks: Onion (bulb, dry), shallots (bulb, dry). Make up to 4 applications. If conditions for disease development are still favorable after 4 applications, use other fungicides for control. If other registered fungicides containing EBDCs are used during growing season, do not apply more than 24 lb. EDDC/A. If tank mixed with other EDDC products, apply no more than 2.4 lb. EBDC. Do not plant any crop which is not registered for use with mefenoxam in treated soil for 12 months.
Days To Harvest: 7.

- - - - - - - - - - - - - - - - -

Signal Word/Toxicity Class: Caution.
REI: 48 hr.

METHYL BROMIDE

2% Chloropicrin added as warning agent. Restricted Use Pesticide. Classified as an ozone depleter under the Clean Air Act. Use is to be completely eliminated starting Jan. 1, 2001.

Brom-O-Gas *(Great Lakes Chemical)*

Rate: Injected with chisels + covered with tarp:
180-240 lb./A
Raised tarp with topical application:
0.4-0.55 lb./100 ft.

Disease:
Fusarium Plasmodiophora Sclerotinia
oak root fungus Pythium Sclerotium
Phytophthora Rhizoctonia

Time: Preplant.

NOTICE The information on these pages is for preliminary planning — not a guide for use. Be sure to follow the manufacturer's directions, notwithstanding information contained here.

For personal protective equipment and EPA registration numbers, please turn to page 39.

Remarks: Dry bulb onions. Do not fumigate when soil temperature is below 50°F. For best results, fumigate when temperature is 60°-80°F at depth of 6". Refer to label for use directions.

State Restrictions: Brom-O-Gas not for use in Arizona and not for use in California for oak root fungus (Armillaria root rot) control.

- - - - - - - - - - - - - - - -

Signal Word/Toxicity Class: Danger/I.

Micro Flo — see copper compounds

Microsperse COC — see copper compounds

Microthiol Special — see sulfur

Nu-Cop — see copper compounds

Nu-Gro Isotox — see captan premix

Penncozeb — see mancozeb

Redball — see sulfur

Ridomil Gold — see mefenoxam

Ridomil Gold/Bravo — see mefenoxam premix

Ridomil Gold/Copper — see mefenoxam premix

Ridomil Gold MZ — see mefenoxam premix

Ronilan — see vinclozolin

Rovral — see iprodione

Seed Mate — see captan premix

SULFUR

Drexel Suffa (6FL), Sulfa 80 (80DF) or Sulfur (90W) *(Drexel)*
Kumulus DF *(BASF)*
Thiolux (DF) *(Platte)*

Rate: 4.5-9 lb. ai/A
6-12 pt. 6FL/A
5-10 lb. 90W/A
2.4-8 lb. ai/A
3-10 lb. 80DF, DF/A

Disease:
powdery mildew

Time: Apply before disease appears; repeat as necessary.

Remarks: Do not use within 2 weeks of oil application (4 weeks 6FL). Some plants are susceptible to injury from sulfur under certain climatic conditions. Do not make sulfur applications at excessively high temperatures. Refer to label for use directions and precautions.

- - - - - - - - - - - - - - - -

Signal Word/Toxicity Class: Caution.
REI: 24 hr.

Micro Flo Sulfur (6L), Dual Spraying & Dusting Sulfur (DS) or Microsperse Wettable Sulfur (WS) *(Micro Flo)*
Terra Sulfur (6L) *(Terra)*

Rate: 3-8 lb. ai/A
4-10.6 pt. 6L/A
5.4 lb. ai/A
6 lb. DS, WS/A

Disease:
powdery mildew

Time: Apply when disease first appears; repeat as necessary.

Remarks: Onions (including dry), shallots. Do not apply DS and WS through any type of irrigation system. Refer to label for use directions.

- - - - - - - - - - - - - - - -

Signal Word/Toxicity Class: Caution.
REI: 24 hr.

Microthiol Special (80WP) *(Elf Atochem)*

Rate: 2.4-8 lb. ai/A
3-10 lb. 80WP/A
California:
4.8 lb. ai/A
6 lb. 80WP/A

Disease:
powdery mildew

Time: Apply at early leaf stage; repeat at 14-day intervals or as needed.

Remarks: Dry onions, dry shallots. Refer to label for use directions.

- - - - - - - - - - - - - - - -

Signal Word/Toxicity Class: Caution.
REI: 24 hr.

Redball EM-53 (6L), Superfine Wettable Sulphur (WS) or Commercial Dusting Sulphur (DS) *(International Sulphur)*

Rate: 2-12 pt. 6L/A
5-10 lb. WS/100 gal. water
10-30 lb. DS/A

Disease
Botrytis blight (WS, DS) — downy mildew (WS, DS) — powdery mildew

Time: Apply at first sign of disease; repeat as necessary.

Remarks: Apply in sufficient water for thorough coverage. Do not apply when temperatures exceed 90°F. Refer to label for use directions.

- - - - - - - - - - - - - - - -

Signal Word/Toxicity Class: Caution.
REI: 12 hr.

SULFUR (Premix)

Top Cop with Sulfur (6.25 lb./gal. sulfur & 1 lb./gal. tribasic copper sulfate) *(Stoller)*

Rate: 2-3 qt./A

Disease:
downy mildew — purple blotch

Time: Apply when plants are 4-6"; repeat at 7- to 10-day intervals.

Remarks: Refer to label for use directions.

- - - - - - - - - - - - - - - -

Signal Word/Toxicity Class: Caution.
REI: 24 hr.

TELONE (Premix)

Restricted Use Pesticide.

Telone C-17 (8.2 lb./gal. dichloropropene & 1.7 lb./gal. chloropicrin) *(Dow AgroSciences)*

Rate: 10.8-30 gal. C-17/A

Disease:
pink rot

Time: Preplant soil treatment. Apply whenever soil conditions permit.

Remarks: Rate varies by soil type. Do not use in extremely heavy clay or muck soils. Do not apply this product through any type of irrigation system. Refer to label for use directions, precautions, and restrictions.

- - - - - - - - - - - - - - - -

Signal Word/Toxicity Class: Danger/I.
REI: 72 hr.

Terranil — see chlorothalonil

◆ - **new product** • ai-active ingredient • DF-dry flowable • E/EC-emulsifiable concentrate • EW-emulsion in water
F/FL-flowable • G-granule • L/LC-liquid • SP/WSP-soluble packet • W/WP-wettable powder • WSB-water soluble bag

TERR-O-GAS (Premix)

Classified as an ozone depleter under the Clean Air Act. Use to be eliminated starting Jan. 1, 2001. Restricted Use Pesticide.

Terr-O-Gas (67% methyl bromide & 33% chloropicrin) *(Great Lakes Chemical)*

Rate: Nontarp:
270-355 lb./A

Disease:
Fusarium, Pythium, Sclerotinia, Phytophthora, Rhizoctonia, Sclerotium, Plasmodiophora

Time: Apply up to 2 weeks prior to planting when soil moisture is good and temperature is 60°-85°F.

Remarks: Dry bulb onions. Work upwind of product and never fumigate without 2 qualified applicators. Do not fumigate when there is little or no air movement. Tarping is not used after field application. Refer to label for use directions.

Signal Word/Toxicity Class: Danger/I.

Thiolux — see sulfur

THIOPHANATE-METHYL

Topsin M (WSB, 70W) *(Elf Atochem)*

Rate: 0.23-0.35 lb. ai/1000 ft. row
0.33-0.5 lb. WSB, 70W/1000 ft. row

Disease:
white rot

Time: Apply by spraying directly into open furrow at time of planting seed, sets, or bulbs.

Remarks: Not for this use through any type of irrigation system. Refer to label for use directions.

Signal Word/Toxicity Class: Caution.
REI: 12 hr.

THIRAM

42-S Thiram *(Gustafson)*
Thiram Granuflo (WDG) *(UCB Agrochemicals)*

Rate: 6 fl. oz. 42-S/100 lb. seed
4 oz. WDG/100 lb. seed

Disease:
damping-off, seed decay, seedling blight

Time: Seed treatment.

Remarks: Use with EPA-approved dye that imparts unnatural color to seed. Do not use treated seed for food, feed, or oil purposes. Do not apply Thiram Granuflo through any type of irrigation system. Refer to label for use directions.

State Restrictions: Thiram Granuflo not for use in California.

Signal Word/Toxicity Class: Caution.
REI: 24 hr. (Thiram Granuflo).

TOP COP — see copper compounds

Top Cop with Sulfur — see sulfur premix

Topsin M — see thiophanate-methyl

VINCLOZOLIN

Ronilan FL, DF, EG *(BASF)*

Rate: 0.75-1 lb. ai/A
1.5-2 pt. FL/A
1.5-2 lb. DF, EG/A

Disease:
leaf blight, purple blotch, white rot, neck rot, watery soft rot

Time: Apply at planting; repeat 4- to 6-weeks after crop emergence. For leaf blight and purple blotch, repeat 2 weeks after first application.

Remarks: Dry bulb onions. Use high rate for neck rot, soft rot, and white rot. Up to 3 additional sprays should be applied if conditions occur which favor disease. Refer to label for use directions. Do not apply more than 10 lb. ai/A in one crop season. Do not apply through any type of irrigation system. Do not apply to direct seeded onions for 4 weeks after emergence.

Days To Harvest: 18.

Signal Word/Toxicity Class: Caution.
REI: 12 hr.

Peas

Allegiance — see metalaxyl

Apron — see metalaxyl

Apron XL — see mefenoxam

CAPTAN

Gustafson Captan 30-DD or 400 *(Gustafson)*

Rate: 2 fl. oz. 30-DD/100 lb. seed
2.5 fl. oz. 400/100 lb. seed

Disease:
damping-off, seed decay, seedling blight, mold

Time: Seed treatment.

Remarks: Do not combine colloidal suspensions, wettable powders, or emulsifiable concentrates unless previous experience assures compatibility. Use with EPA-approved dye that imparts unnatural color to seed. Do not use treated seed for food, feed, or oil purposes. Refer to label for use directions.

Signal Word/Toxicity Class: Danger/I.

Drexel Captan 4FL *(Drexel)*

Rate: 2.5 fl. oz. 4FL/100 lb. seed

Disease:
seed decay, seedling blight, soilborne disease

Time: Liquid or slurry seed treatment.

Remarks: Do not use treated seed for food, feed, or oil purposes. Refer to label for use directions.

Signal Word/Toxicity Class: Danger/I.

CAPTAN (Premixes)

Captan-Diazinon contains graphite.

Captan-Diazinon (36.67% captan & 25% diazinon) *(Trace)*
Kernel Guard (14.67% captan & 15% diazinon & 25% lindane) *(Trace)*
Nu-Gro Isotox (12.2% captan & 25% lindane) *(Wilbur-Ellis)*
Seed Mate Isotox (12.2% captan & 25% lindane) *(Loveland)*

Rate: 2 oz./bu. seed (Trace)
4 oz./100 lb. seed (Isotox)

Disease:
damping-off, seed decay, seedling blight

Time: Planter box seed treatment.

Remarks: Do not use treated seed for food, feed, or oil purposes. Refer to label for use directions and precautions.

Signal Word/Toxicity Class: Danger/I.
REI: 4 days (Isotox); 12 hr. (Kernel Guard).

Champ — see copper compounds

7 Vegetables

NOTICE	
The information on these pages is for preliminary planning — not a guide for use. Be sure to follow the manufacturer's directions, notwithstanding information contained here.	For personal protective equipment and EPA registration numbers, please turn to page 39.

Champion — see copper compounds

Clean Crop C-O-C-S — see copper compounds

COPPER COMPOUNDS

Champ (FL or Formula 2) or Champion (WP) (copper hydroxide) *(Agtrol)*
Kocide (101, DF, 2000, 4.5LF, LF) (copper hydroxide) *(Griffin)*
KOP-Hydroxide (FL, 50WP) (copper hydroxide) *(Drexel)*

Rate: 2-4 pt. FL, LF/A
1-2 pt. Formula 2, 4.5LF/A
1.5-3 lb. WP, 101, DF, 50WP/A
1.25-2.25 lb. 2000/A

Disease:
powdery mildew

Time: Apply when disease first appears; repeat at weekly intervals.
Remarks: Do not apply in a spray solution having a pH less than 6.5 as phytotoxicity could occur. Refer to label for use directions.

Signal Word/Toxicity Class: Varies by formulation.
REI: Varies by formulation.

Clean Crop C-O-C-S (WDG) (74.8% copper oxychloride & 14.2% basic copper sulfate) *(Platte)*

Rate: 1-2 lb. ai/A
2-4 lb. WDG/A

Disease:
Alternaria, Anthracnose, downy mildew, Phytophthora blight

Time: Apply when disease first threatens; repeat at 7- to 10-day intervals to protect new growth throughout growing season.
Remarks: Do not apply this product through any type of irrigation system. Refer to label for use directions.

Signal Word/Toxicity Class: Caution.
REI: 24 hr.

Microsperse COC 53WP (copper oxychloride) or Nu-Cop (3L, 50DF, 50WP) (copper hydroxide) *(Micro Flo)*

Rate: 1-4 pt. 3L/A
1.5-3 lb. 50DF, 50WP, 53WP/A

Disease:
bacterial blight (53WP), powdery mildew

Time: Apply when disease first appears; repeat at weekly intervals.
Remarks: Refer to label for use directions.

Signal Word/Toxicity Class: Danger/I (50DF, 50WP, 53WP); Warning/II (3L).
REI: 48 hr. (50DF, 50WP); 24 hr. (53WP, 3L).

Kernel Guard — see captan premix

Kocide — see copper compounds

KOP-Hydroxide — see copper compounds

Kumulus — see sulfur

MEFENOXAM

Apron XL LS *(Novartis)*

Rate: 0.32-1.28 fl. oz. LS/100 lb. seed

Disease:
damping off, downy mildew, Phytophthora (early season)

Time: Seed treatment at planting.
Remarks: Field, garden peas. Use with EPA-approved dye that imparts unnatural color to seed. Do not use treated seed for food, feed, or oil purposes. Refer to label for use directions.

Signal Word/Toxicity Class: Warning/II.

Ridomil Gold EC or WSP *(Novartis)*

Rate: 0.25-0.5 lb. ai/A
0.5-1 pt. EC/A
0.225-0.45 lb. ai/A
0.5-1 lb. WSP/A

Disease:
damping-off, root and stem rot

Time: Preplant incorporated or surface application just after planting.
Remarks: Field, garden, sugar peas. For banded applications, 7" band recommended. Do not use in greenhouse or on field grown vegetable bedding plants. May be tank mixed. Do not plant any crop which is not registered for use with mefenoxam in treated soil for 12 months.
Days To Harvest: 1.

Signal Word/Toxicity Class: Warning/II (EC); Caution (WSP).
REI: 48 hr.

METALAXYL

Allegiance or Apron (FL) *(Gustafson)*

Rate: 0.75 fl. oz. Allegiance/100 lb. seed
0.75-1.5 fl. oz. Apron/100 lb. seed

Disease:
damping-off, Phytophthora (early season)

Rate: 3.4 fl. oz. Apron/100 lb. seed

Disease:
downy mildew

Time: Seed treatment.
Remarks: Field, garden peas. Use with EPA-approved dye that imparts unnatural color to seed. Do not use treated seed for food, feed, or oil purposes. Refer to label for use directions.

Signal Word/Toxicity Class: Warning/II.

Apron Dry *(Trace)*

Rate: 4 oz./100 lb. seed

Disease:
damping-off, Phytophthora

Time: Seed treatment.
Remarks: Treat only those seeds needed for immediate use. Do not use treated seed for food, feed, or oil purposes. Refer to label for use directions.

Signal Word/Toxicity Class: Caution.
REI: 12 hr.

Microsperse — see copper compounds

Nu-Gro Isotox — see captan premix

PCNB

PCNB 2-E or 2LF *(Amvac)*

Rate: 3.35-6.70 fl. oz. 2-E/100 lb. seed
4-8 fl. oz. 2LF/100 lb. seed

Disease:
covered smut, Fusarium, Rhizoctonia

Time: Seed treatment.
Remarks: Do not apply through any type of irrigation system. Do not plant root crops within 12 months of application unless PCNB is registered for use on those crops.

Signal Word/Toxicity Class: Danger/I (2-E); Caution (2LF).
REI: 12 hr.

RTU-PCNB *(Gustafson)*

Rate: 3-6 fl. oz. RTU/100 lb. seed

Disease:
Fusarium, Rhizoctonia

Time: Seed treatment.
Remarks: Do not use treated seed for food, feed, or oil purposes. Do not feed treated seed to animals. Refer to label for use directions.
State Restrictions: Not for use in California.

Signal Word/Toxicity Class: Caution.

◆ - **new product** • ai-active ingredient • DF-dry flowable • E/EC-emulsifiable concentrate • EW-emulsion in water
F/FL-flowable • G-granule • L/LC-liquid • SP/WSP-soluble packet • W/WP-wettable powder • WSB-water soluble bag

Redball — see sulfur

Ridomil Gold — see mefenoxam

Seed Mate Isotox — see captan premix

SULFUR

Drexel Suffa (6FL), Sulfa 80 (80DF) or Sulfur (90W) *(Drexel)*
Thiolux (DF) *(Platte)*

Rate: 2.6-5.3 lb. ai/A
3.5-7 pt. 6FL/A
2-8 lb. ai/A
3-10 lb. 80DF, DF/A
2.7-5.4 lb. ai/A
3-6 lb. 90W/A

Disease:
leaf spot (6FL, 90W) powdery mildew rust

Time: Apply at first signs of infection; repeat at 7- to 14-day intervals as needed. For 80DF, DF, apply early leaf stage; repeat at 10- to 14-day intervals with or without pesticides immediately after rain.

Remarks: Do not use within 2 weeks of oil application (4 weeks 6FL). Some plants are susceptible to injury from sulfur under certain climatic conditions. Do not make sulfur applications at excessively high temperatures. Do not apply this product through any type of irrigation system. Refer to label for use directions.

Signal Word/Toxicity Class: Caution.
REI: 24 hr.

Dusting Sulfur (DS) *(Platte)*

Rate: 15-20 lb. DS/A

Disease:
powdery mildew

Time: Apply at 7- to 14-day intervals.

Remarks: Do not make applications within 2 weeks of oil spray. Refer to label for use directions.

Signal Word/Toxicity Class: Caution.
REI: 24 hr.

Kumulus DF *(BASF)*
Micro Flo Sulfur (6L) or Microsperse Wettable Sulfur (WS) *(Micro Flo)*
Redball EM-63 (6L) or Superfine Wettable Sulphur (WS) *(International Sulphur)*
Terra Sulfur (6L) *(Terra)*

Rate: 2.7-4.5 lb. ai/A
3-5 lb. DF, WS/A
2.25-8 lb. ai/A
3-10.66 pt. 6L/A

Disease:
powdery mildew Septoria leaf spot rust (6L)

Time: Apply at early leaf stage; repeat at 10- to 14-day intervals with or without other pesticides immediately after a rain.

Remarks: Do not apply through any type of irrigation system (except Micro Flo 6L). Refer to label for use directions.

Signal Word/Toxicity Class: Caution.
REI: 24 hr.

Liquid Sulfur Six *(Helena)*

Rate: 3-4 pt./A

Disease:
powdery mildew

Time: Apply at first sign of infection; repeat at 3-week intervals.

Remarks: Apply by air in 8 gal. water. Do not apply this product through any type of irrigation system. Refer to label for use directions.

Signal Word/Toxicity Class: Caution.
REI: 24 hr.

Redball Commercial Dusting Sulphur (DS) *(International Sulphur)*

Rate: 2-7 pt. 6L/A
6-16 lb. WS/A
10-25 lb. DS/A

Disease:
powdery mildew

Time: 6L: Apply in early leaf stage; repeat as necessary. WS, DS: Apply when disease first appears; repeat at 7- to 14-day intervals (WS), 7- to 10-day intervals (DS).

Remarks: Apply in sufficient water for thorough coverage. Do not apply when temperatures exceed 90°F. Refer to label for use directions.

Signal Word/Toxicity Class: Caution.
REI: 12 hr.

SULFUR (Premix)

Top Cop with Sulfur (6.25 lb./gal. sulfur & 1 lb./gal. tribasic copper sulfate) *(Stoller)*

Rate: 2 qt./A

Disease:
powdey mildew

Time: Apply at first sign of disease; repeat at 7-day intervals.

Remarks: Refer to label for use directions.

Signal Word/Toxicity Class: Caution.
REI: 24 hr.

SYSTEM3

System3 (0.1% *Bacillus subtilis* & 14.25% metalaxyl & 16.67% PCNB) *(Helena)*

Rate: 4-6 oz./bu.

Disease:
Phytophthora Pythium *Rhizoctonia solani*

Time: Apply at planting.

Remarks: Non-oil lubricant has been added for seeds to flow more easily and evenly from planter boxes. Incorporation of dye is recommended to distinguish seed as treated. Do not use treated seed for food, feed or oil purposes. Refer to label for use directions.

Signal Word/Toxicity Class: Caution.
REI: 12 hr.

Thiolux — see sulfur

THIRAM

42-S Thiram or Thiram 50WP Dyed *(Gustafson)*
Thiram Granuflo (WDG) *(UCB Agrochemicals)*

Rate: 3 fl. oz. 42-S/100 lb. seed
3 oz. 50WP/100 lb. seed
2 oz. WDG/100 lb. seed

Disease:
damping-off seed decay seedling blight

Time: Seed treatment.

Remarks: Use with EPA-approved dye that imparts unnatural color to seed. Do not use treated seed for food, feed, or oil purposes. Do not apply Thiram Granuflo through any type of irrigation system. Refer to label for use directions.

State Restrictions: Thiram Granuflo not for use in California.

Signal Word/Toxicity Class: Caution.
REI: 24 hr. (Thiram Granuflo).

Top Cop with Sulfur — see sulfur premix

Peppers

Basicop — see copper compounds

Brom-O-Gas — see methyl bromide

7 Vegetables

NOTICE	The information on these pages is for preliminary planning — not a guide for use. Be sure to follow the manufacturer's directions, notwithstanding information contained here.	For personal protective equipment and EPA registration numbers, please turn to page 39.

CAPTAN

Gustafson Captan 30-DD or 400 *(Gustafson)*

Rate: 2.33 fl. oz. 30-DD/100 lb. seed
2-3 fl. oz. 400/100 lb. seed

Disease:
damping-off, seed decay, seedling blight, mold

Time: Seed treatment.

Remarks: Do not combine colloidal suspensions, wettable powders, or emulsifiable concentrates unless previous experience assures compatibility. Use with EPA-approved dye that imparts unnatural color to seed. Do not use treated seed for food, feed, or oil purposes. Refer to label for use directions.

- - - - - - - - - - - - - - - -

Signal Word/Toxicity Class: Danger/I.

Drexel Captan 4FL *(Drexel)*

Rate: 3 fl. oz. 4FL/100 lb. seed

Disease:
seed decay, seedling blight, soilborne disease

Time: Liquid or slurry seed treatment.

Remarks: California wonder peppers. Do not use treated seed for food, feed, or oil purposes. Refer to label for use directions.

- - - - - - - - - - - - - - - -

Signal Word/Toxicity Class: Danger/I.

Champ — see copper compounds

Champion — see copper compounds

Clean Crop C-O-C-S — see copper compounds

COPPER COMPOUNDS

Basicop (WP) (basic copper sulfate) *(Griffin)*

Rate: 1-2 lb. ai/A
2-4 lb. WP/A

Disease:
Anthracnose, Cercospora leaf spot, early blight, bacterial spot, downy mildew, late blight

Time: Apply when disease appears; repeat at 7- to 10-day intervals.

Remarks: Use higher rate when disease is heavy. Do not apply through any type of irrigation system. Refer to label for use directions.

- - - - - - - - - - - - - - - -

Signal Word/Toxicity Class: Warning/II.
REI: 24 hr.

Champ (FL or Formula 2) or Champion (WP) (copper hydroxide) *(Agtrol)*
Kocide (101, DF, 2000, 4.5LF, LF) (copper hydroxide) *(Griffin)*
Nu-Cop (3L, 50DF, 50WP) (copper hydroxide) *(Micro Flo)*

Rate: 2.66-4 pt. FL, LF/A
1.33-2 pt. Formula 2, 4.5LF/A
2-3 lb. WP, 101, DF, 50DF, 50WP/A
1.5-2.25 lb. 2000/A
1.33-4 pt. 3L/A

Disease:
bacterial spot

Time: Apply when disease threatens; repeat at 5- to 10-day intervals as needed.

Remarks: Do not apply in a spray solution having a pH less than 6.5 as phytotoxicity could occur. Refer to label for use directions.

- - - - - - - - - - - - - - - -

Signal Word/Toxicity Class: Varies by formulation.
REI: Varies by formulation.

Clean Crop C-O-C-S (WDG) (74.8% copper oxychloride & 14.2% copper sulfate, basic) *(Platte)*

Rate: 1-2 lb. ai/A
2-4 lb. WDG/A

Disease:
Alternaria, downy mildew, Phytophthora blight, Anthracnose

Time: Apply when disease first threatens; repeat at 7- to 10-day intervals to protect new growth throughout growing season.

Remarks: Do not apply this product through any type of irrigation system. Refer to label for use directions.

- - - - - - - - - - - - - - - -

Signal Word/Toxicity Class: Warning/II.
REI: 24 hr.

Copper-Count-N (L) (copper ammonium) *(Mineral Research)*

Rate: 3-6 pt. L/A

Disease:
bacterial spot, Cercospora leaf spot

Time: Apply at 7- to 10-day intervals. In plant beds, apply when plants emerge; in fields, immediately after transplanting or emergence. In areas or season when disease is severe, spray at 4- to 5-day intervals.

Remarks: Disease control is critical during fruiting. Apply by ground or air; refer to label for use directions. Do not apply through any type of irrigation system unless supplemental labeling on chemigation is followed.

- - - - - - - - - - - - - - - -

Signal Word/Toxicity Class: Caution.
REI: 12 hr.

Drexel Basic Copper Sulfate (WP) or KOP 300 (3FL) (basic copper sulfate) *(Drexel)*

Rate: 3-4 lb. WP/A
2.6-4 pt. 3FL/A

Disease:
Anthracnose, Cercospora leaf spot, early blight, bacterial spot, downy mildew, late blight

Time: Apply in seedbeds; continue in field at 7-day intervals.

Remarks: Do not apply through any type of irrigation system. Refer to label for use directions.

- - - - - - - - - - - - - - - -

Signal Word/Toxicity Class: Warning/II.
REI: 24 hr. (WP); 12 hr. (3FL).

KOP-Hydroxide (FL, 50WP) (copper hydroxide) *(Drexel)*

Rate: 2.66-4 pt. FL/A
2-3 lb. 50WP/A

Disease:
bacterial spot

Time: Apply when disease threatens; repeat at 7- to 14-day intervals (5- to 10-day intervals for WP).

Remarks: KOP-Hydroxide 50WP should not be applied in spray solution having pH less than 6.5 or greater than 9.0 as phytotoxicity may occur. Refer to label for use directions.

- - - - - - - - - - - - - - - -

Signal Word/Toxicity Class: Danger/I.
REI: 48 hr.

KOP Oxy-85 (WP) (copper oxychloride) *(Drexel)*

Rate: 4 lb. WP/100 gal. water

Disease:
damping-off

Time: Apply at 4- to 7-day intervals, beginning when plants break through the ground by spraying both plants and soil surrounding them.

Remarks: Cold frames, greenhouses. Refer to label for use directions.

- - - - - - - - - - - - - - - -

Signal Word/Toxicity Class: Danger/I.
REI: 24 hr.

Micro Flo Basic Copper "53" (basic copper sulfate) or Microsperse COC (53WP) (copper oxychloride) *(Micro Flo)*

Rate: 3-4 lb. 53, 53WP/A

Disease:
Anthracnose, Cercospora leaf spot, early blight, bacterial spot, downy mildew, late blight

Time: Apply in seedbed or field before disease first appears and right after transplanting; repeat at 7- to 10-day intervals (5- to 10-day for 53WP) as needed especially during fruiting stages.

Rate: Cold frames, greenhouses, etc.:
4-5 lb. 53WP/100 gal. water

Disease:
damping-off

Time: Apply before seedlings emerge, continue when plants emerge; repeat at 4- to 7-day intervals until transplanting time.

Remarks: Refer to label for use directions.

- - - - - - - - - - - - - - - -

Signal Word/Toxicity Class: Danger/I (53WP); Warning/II (53).
REI: 24 hr.

◆ - **new product** • ai-active ingredient • DF-dry flowable • E/EC-emulsifiable concentrate • EW-emulsion in water
F/FL-flowable • G-granule • L/LC-liquid • SP/WSP-soluble packet • W/WP-wettable powder • WSB-water soluble bag

TOP COP Tri-Basic (FL) (basic copper sulfate) *(Stoller)*
Rate: 3-4 pt. FL/A
Disease:
Anthracnose, Cercospora leaf spot, downy mildew, bacterial spot
Time: Start sprays in seed bed; repeat in field at 7-day intervals.
Remarks: During fruiting, use high rate. Refer to label for use directions.

Signal Word/Toxicity Class: Caution.
REI: 24 hr.

Copper-Count-N — see copper compounds

Kocide — see copper compounds

KOP-Hydroxide — see copper compounds

KOP Oxy — see copper compounds

Kumulus — see sulfur

MANEB

Maneb 75DF or 80WP *(Elf Atochem)*
Manex (FL) *(Griffin)*
Rate: East of the Mississippi:
1.12-2.25 lb. ai/A
1.5-3 lb. 75DF/A
1.2-2.4 lb. ai/A
1.5-3 lb. 80WP/A
1.2-2.4 qt. FL/A
West of the Mississippi:
1.12-1.5 lb. ai/A
1.5-2 lb. 75DF/A
1.2-1.6 lb. ai/A
1.5-2 lb. 80WP/A
1.2-1.6 qt. FL/A
Disease:
Anthracnose, Phytophthora blight, ripe rot, Cercospora leaf spot
Time: Begin when disease threatens; repeat at 7- to 10-day intervals (7-days east of the Mississippi for 75DF).
Remarks: Do not use any other product containing a different EBDC ai during the same growing season. Refer to label for use directions.
Days To Harvest: 7.

Signal Word/Toxicity Class: Caution.
REI: 24 hr.

Manex — see maneb

MEFENOXAM

Ridomil Gold EC or WSP *(Novartis)*
Rate: 0.5 lb. ai/A
1 pt. EC/A
0.45 lb. ai/A
1 lb. WSP/A
Disease:
damping-off, root and stem rot
Time: Preventative. Apply at planting; up to 2 supplemental applications may be made at 30-day intervals.
Remarks: May cause some yellowing of pepper leaves. Do not apply more than 3 pt. EC/A or 3 lb. WSP/A per season. Not for use in greenhouse or on field grown vegetable bedding plants. May be tank mixed. Do not plant any crop which is not registered for use with mefenoxam in treated soil for 12 months. Refer to label for use directions.
Days To Harvest: 7.

Signal Word/Toxicity Class: Warning/II (EC); Caution (WSP).
REI: 48 hr.

MEFENOXAM (Premix)

Ridomil Gold/Copper (WSB) (5% mefenoxam & 60% copper hydroxide) *(Novartis)*
Rate: 1.6 lb ai/A
2.5 lb. WSB/A
Disease:
Phytophthora, Pythium
Time: May be applied 30 days following Ridomil Gold EC at planting application and an EC supplemental application. Apply 3-4 times at 10- to 14-day intervals.
Remarks: Do not apply more than 1.5 lb. mefenoxam/A per crop. Do not plant any crop which is not registered for use with mefenoxam in treated soil for 12 months. Refer to label for use directions.
Days To Harvest: 7.

Signal Word/Toxicity Class: Danger/I.
REI: 48 hr.

METHYL BROMIDE

2% Chloropicrin added as warning agent. Restricted Use Pesticide. Classified as an ozone depleter under the Clean Air Act. Use is to be completely eliminated starting Jan. 1, 2001.

Brom-O-Gas *(Great Lakes Chemical)*
Rate: Injected with chisels + covered with tarp:
180-240 lb./A
Raised tarp with topical application:
0.4-0.55 lb./100 ft.
Disease:
Fusarium, Plasmodiophora, Sclerotinia, oak root fungus, Pythium, Sclerotium, Phytophthora, Rhizoctonia
Time: Preplant.
Remarks: Do not fumigate when soil temperature is below 50°F. For best results, fumigate when temperature is 60°-80°F at depth of 6". Refer to label for use directions.
State Restrictions: Brom-O-Gas not for use in Arizona and not for use in California for oak root fungus (Armillaria root rot) control.

Signal Word/Toxicity Class: Danger/I.

Micro Flo — see copper compounds

Microsperse COC — see copper compounds

Microthiol Special — see sulfur

Nu-Cop — see copper compounds

PCNB

PCNB 75DG or 75WP *(Amvac)*
Rate: Transplants:
2.25-3.75 lb. ai/100 gal. water
3-5 lb. 75DG, 75WP/100 gal. water
In-furrow:
7.5 lb. ai/100 gal. water
10 lb. 75DG, 75WP/100 gal. water
Disease:
southern blight
Time: For transplant, apply at time of transplanting; for in-furrow, apply just prior to planting.
Remarks: Set plows as flat as possible to avoid getting untreated soil against stems of plants. Do not apply through any type of irrigation system. Do not plant root crops within 12 months of application unless PCNB is registered for use on those crops.

Signal Word/Toxicity Class: Caution.
REI: 12 hr.

Redball — see sulfur

7 Vegetables

NOTICE The information on these pages is for preliminary planning — not a guide for use. Be sure to follow the manufacturer's directions, notwithstanding information contained here. For personal protective equipment and EPA registration numbers, please turn to page 39.

Ridomil Gold — see mefenoxam

Ridomil Gold/Copper — see mefenoxam premix

SULFUR

Drexel Suffa (6FL), Sulfa 80 (80DF) or Sulfur (90W) *(Drexel)*

Rate: 3-4.5 lb. ai/A
4-6 pt. 6FL/A
2-8 lb. ai/A
3-10 lb. 80DF/A
2.7-4.5 lb. ai/A
3-5 lb. 90W/A

Disease:
leaf spot (90W) powdery mildew rust (80DF)

Time: Apply early leaf stage; repeat at 10- to 14-day intervals as needed, or immediately after rain.

Remarks: Do not use within 2 weeks of oil application (4 weeks 6FL). Some plants are susceptible to injury from sulfur under certain climatic conditions. Do not make sulfur applications at excessively high temperatures. Do not apply this product through any type of irrigation system. Refer to label for use directions.

Signal Word/Toxicity Class: Caution.
REI: 24 hr.

Kumulus DF *(BASF)*

Rate: 2.4-4 lb. ai/A
3-5 lb. DF/A

Disease:
powdery mildew Septoria leaf spot

Time: Apply at early leaf stage; repeat at 10- to 14-day intervals with or without other pesticides immediately after a rain.

Remarks: Do not apply if temperatures during or within 3 days after application are expected to exceed 90°F. Do not apply through any type of irrigation system. Refer to label for use directions.

Signal Word/Toxicity Class: Caution.
REI: 24 hr.

Liquid Sulfur Six *(Helena)*

Rate: 3-4 pt./A

Disease:
powdery mildew

Time: Apply at first sign of infection; repeat at 3-week intervals.

Remarks: Apply by air in 8 gal. water. Do not apply this product through any type of irrigation system. Refer to label for use directions.

Signal Word/Toxicity Class: Caution.
REI: 24 hr.

Micro Flo Sulfur (6L), Dual Spraying & Dusting Sulfur (DS) or Microsperse Wettable Sulfur (WS) *(Micro Flo)*
Terra Sulfur (6L) *(Terra)*
Thiolux (DF) *(Platte)*

Rate: 2.25-8 lb. ai/A
3-10.66 pt. 6L/A
4.5 lb. ai/A
5 lb. DS/A
2.7-4.5 lb. ai/A
3-5 lb. WS, DF/A

Disease:
powdery mildew rust (6L)

Time: Apply at early leaf stage; repeat at 10- to 14-day intervals (14-day intervals-DS) with or without pesticides immediately after rain.

Remarks: Do not use within 2 weeks of oil application. Some plants are susceptible to injury from sulfur under certain climatic conditions. Do not make sulfur applications at excessively high temperatures. Do not apply DS and WS through any type of irrigation system. Refer to label for use directions.

Signal Word/Toxicity Class: Caution.
REI: 24 hr.

Microthiol Special (80WP) *(Elf Atochem)*

Rate: 2.4-8 lb. ai/A
3-10 lb. 80WP/A

Disease:
powdery mildew

Time: Apply at early leaf stage; repeat at 14-day intervals as needed.

Remarks: Refer to label for use directions.

Signal Word/Toxicity Class: Caution.
REI: 24 hr.

Redball Superfine Wettable Sulphur (WS) or Commercial Dusting Sulphur (DS) *(International Sulphur)*

Rate: 6-16 lb. WS/A
10-25 lb. DS/A

Disease:
powdery mildew

Time: Apply when disease first appears; repeat at 7- to 14-day intervals (WS), 7- to 10-day intervals (DS).

Remarks: Apply in sufficient water for thorough coverage. Do not apply when temperatures exceed 90°F. Refer to label for use directions.

Signal Word/Toxicity Class: Caution.
REI: 12 hr.

SULFUR (Premix)

Top Cop with Sulfur (6.25 lb./gal. sulfur & 1 lb./gal. tribasic copper sulfate) *(Stoller)*

Rate: 2 qt./A

Disease:
Anthracnose bacterial spot Cercospora leaf spot

Time: Apply at first sign of disease; repeat at 7- to 10-day intervals until harvest begins.

Remarks: Refer to label for use directions.

Signal Word/Toxicity Class: Caution.
REI: 24 hr.

TERR-O-GAS (Premix)

Classified as an ozone depleter under the Clean Air Act. Use to be eliminated starting Jan. 1, 2001. Restricted Use Pesticide.

Terr-O-Gas (67% methyl bromide & 33% chloropicrin) *(Great Lakes Chemical)*

Rate: Nontarp:
270-355 lb./A

Disease:
Fusarium Pythium Sclerotinia
Phytophthora Rhizoctonia Sclerotium
Plasmodiophora

Time: Apply up to 2 weeks prior to planting when soil moisture is good and temperature is 60°-85°F.

Remarks: Work upwind of product and never fumigate without 2 qualified applicators. Do not fumigate when there is little or no air movement. Tarping is not used after field application. Refer to label for use directions.

Signal Word/Toxicity Class: Danger/I.

Thiolux — see sulfur

THIRAM

42-S Thiram or Thiram 50WP Dyed *(Gustafson)*
Thiram Granuflo (WDG) *(UCB Agrochemicals)*

Rate: 8 fl. oz. 42-S/100 lb. seed
8 oz. 50WP/100 lb. seed
5.3 oz. WDG/100 lb. seed

Disease:
damping-off seed decay seedling blight

Time: Seed treatment.

Remarks: Use with EPA-approved dye that imparts unnatural color to seed. Do not use treated seed for food, feed, or oil purposes. Do not apply Thiram Granuflo through any type of irrigation system. Refer to label for use directions.

State Restrictions: Thiram Granuflo not for use in California.

Signal Word/Toxicity Class: Caution.
REI: 24 hr. (Thiram Granuflo).

◆ - **new product** • ai-active ingredient • DF-dry flowable • E/EC-emulsifiable concentrate • EW-emulsion in water
F/FL-flowable • G-granule • L/LC-liquid • SP/WSP-soluble packet • W/WP-wettable powder • WSB-water soluble bag

TOP COP — see copper compounds

Top Cop with Sulfur — see sulfur premix

Potatoes

Basicop — see copper compounds

Blue Viking — see copper compounds

Botran — see dicloran

Bravo — see chlorothalonil

Bravo S — see sulfur premix

CAPTAN

Clean Crop Captan 5% or 7.5 Dust *(Platte)*

Rate: 1-2 lb. 5%/bu. cut seed
10-16 oz. 7.5/100 lb. cut seed pieces

Disease:
damping-off seed decay

Time: Cut pieces should be treated within 6 hr. of cutting.

Remarks: Tuber damage and/or decay could occur on sandy soils under moist soil conditions and higher than normal soil temperatures. Do not use treated seed pieces for feed or food. Refer to label for use directions.

State Restrictions: Not for use in Jefferson County, Idaho.

Signal Word/Toxicity Class: Danger/I.
REI: Varies.

Champ — see copper compounds

Champion — see copper compounds

CHLOROTHALONIL

Bravo 500, 720/Weather Stik, Ultrex (WDG) or Zn *(ZENECA)*
Echo 500, 720 or 90DF *(Sostram)*

Rate: Early-season:
1 pt. 500, Zn/A
0.75 pt. 720/A
0.7 lb. WDG/A
0.8-1.2 lb. 90DF/A
Late-season:
1.5-2.125 pt. 500, Zn/A
1-1.5 pt. 720/A
0.9-1.4 lb. WDG/A
0.65-1.2 lb. 90DF/A

Disease:
Botrytis vine rot early blight late blight

Time: Early-season: Apply as banded treatment directed over foliage beginning when vines are first exposed and leaf wetness occurs; repeat at 7- to 10-day intervals until vines close. Late-season: As vines close between rows, apply broadcast to increase water carrier volume to cover denser canopy; repeat at 7- to 10-day intervals. When disease conditions are severe, use high rate at weekly intervals.

Remarks: May be applied through sprinkler irrigation equipment. Refer to label for use directions and restrictions.

State Restrictions: SLN-Maine (Bravo 500); SLN-Florida (Bravo 720, WDG); SLN-Minnesota, North Dakota, South Dakota, Washington, Wisconsin (Bravo 720, Weather Stik, Zn); SLN-New York, Oregon (Bravo 720); SLN-Michigan (Weather Stik); SLN-Michigan (Zn).

Signal Word/Toxicity Class: Danger/I (WDG, 90DF); Warning/II (500, 720, Zn).
REI: 48 hr.

Terranil (6L, 90DF or 90DF WSP) or Terranil Zn *(Terra)*

Rate: 0.78-1.12 lb. ai/A
1.5-2.125 pt. Zn/A
0.75-1.5 pt. 6L/A
0.875-1.25 lb. 90DF, 90DF WSP/A

Disease:
Botrytis vine rot early blight late blight

Time: Apply low rate when vines are first exposed and leaf wetness occurs; repeat at 7- to 10-day intervals. Apply high rate when vines close within row, late blight forecasting measures 18 disease severity, or crop reaches 300 P-days; repeat at 5- to 10-day intervals.

Remarks: Under severe conditions, use high rate and short interval. Refer to label for use directions.

Days To Harvest: 14.

Signal Word/Toxicity Class: Danger/I (90DF, 90DF WSP); Warning/II (6L, Zn).
REI: 48 hr.

Clean Crop C-O-C-S — see copper compounds

Clean Crop Nemasol — see metam-sodium

COPPER COMPOUNDS

Basicop (WP) (basic copper sulfate) *(Griffin)*
Drexel Basic Copper Sulfate (WP) or KOP 300 (3FL) (basic copper sulfate) *(Drexel)*

Rate: 1.5-3 lb. ai/A
3-6 lb. WP/A
4-5.3 pt. 3FL/A

Disease:
early blight late blight

Time: Apply when disease appears; repeat at 7- to 10-day intervals.

Remarks: Use higher rate when disease is heavy. Do not apply through any type of irrigation system. Refer to label for use directions.

Signal Word/Toxicity Class: Warning/II.
REI: 24 hr. (WP); 12 hr. (3FL).

Blue Viking or Tennessee Brand (copper sulfate pentahydrate) *(Griffin)*

Rate: 10 lb./10-100 gal. water (ground)

Disease:
late blight (suppression)

Time: Apply up to harvest.

Remarks: Do not apply through any type of irrigation system. Refer to label for use directions.

Days To Harvest: 7.

Signal Word/Toxicity Class: Danger/I.
REI: 24 hr.

Champ (FL or Formula 2) or Champion (WP) (copper hydroxide) *(Agtrol)*
Kocide (101, DF, 2000, 4.5LF, LF) (copper hydroxide) *(Griffin)*

Rate: 1.33-5.33 pt. FL, LF/A
0.66-2.66 pt. Formula 2, 4.5LF/A
1-4 lb. WP, 101, DF/A
0.75-3 lb. 2000/A

Disease:
early blight late blight

Time: Apply when plants are 6" high; repeat at 7- to 10-day intervals.

Remarks: Do not apply in a spray solution having a pH less than 6.5 as phytotoxicity could occur. Refer to label for use directions.

Days To Harvest: 2 weeks.

Signal Word/Toxicity Class: Varies by formulation.
REI: Varies by formulation.

7 Vegetables

NOTICE The information on these pages is for preliminary planning — not a guide for use. Be sure to follow the manufacturer's directions, notwithstanding information contained here.

For personal protective equipment and EPA registration numbers, please turn to page 39.

Clean Crop C-O-C-S (WDG) (74.8% copper oxychloride & 14.2% copper sulfate, basic) *(Platte)*
Rate: 0.75-2 lb. ai/A
1.5-4 lb. WDG/A
Disease:
early blight late blight
Time: Apply when disease first threatens; repeat at 7- to 10-day intervals to protect new growth throughout growing season.
Remarks: Do not apply this product through any type of irrigation system. Refer to label for use directions.

Signal Word/Toxicity Class: Warning/II.
REI: 24 hr.

Copper-Count-N (L) (copper ammonium) *(Mineral Research)*
Rate: 3-6 pt. L/A
Disease:
early blight late blight
Time: Apply on first appearance of disease at 7- to 10-day intervals.
Remarks: Applications may be made by ground or air; refer to label for use directions. Do not apply through any type of irrigation system unless supplemental labeling on chemigation is followed.

Signal Word/Toxicity Class: Caution.
REI: 12 hr.

KOP-Hydroxide (FL, 50WP) (copper hydroxide) *(Drexel)*
Rate: Light disease:
1.33-2 pt. FL/A
1-1.5 lb. 50WP/A
Severe disease:
4-5.33 pt. FL/A
3-4 lb. 50WP/A
Disease:
early blight late blight
Time: Apply at 7- to 10-day intervals when plants are 6" tall until 2 weeks before harvest.
Remarks: KOP-Hydroxide 50WP should not be applied in spray solution having pH less than 6.5 or greater than 9.0 as phytotoxicity may occur. Refer to label for use directions.

Signal Word/Toxicity Class: Danger/I.
REI: 48 hr.

KOP Oxy-85 (WP) (copper oxychloride) *(Drexel)*
Rate: 4 lb. WP/A
Disease:
early blight late blight
Time: Apply when disease first appears; repeat at 7- to 10-day intervals as needed.
Remarks: If late blight infection is severe, increase rate to 7 lb./A for large plants. Refer to label for use directions.

Signal Word/Toxicity Class: Danger/I.
REI: 24 hr.

Micro Flo Basic Copper "53" (basic copper sulfate) or Microsperse COC (53WP) (copper oxychloride) *(Micro Flo)*
Rate: 3-6 lb. 53/A
Disease: 53:
early blight late blight
Time: Apply before disease first appears; repeat at 7- to 10-day intervals as needed.

Rate: 3-4 lb. 53WP/A
Disease: 53WP:
early blight late blight
Time: Apply when plants are 4-6" high or when disease first appears; repeat at 4- to 10-day intervals to harvest or as needed.
Remarks: Refer to label for use directions.

Signal Word/Toxicity Class: Danger/I (53WP); Warning/II (53).
REI: 24 hr.

Nu-Cop (3L, 50DF, 50WP) (copper hydroxide) *(Micro Flo)*
Rate: 0.66-4 pt. 3L/A
1-4 lb. 50DF, 50WP/A
Disease:
early blight late blight
Time: Apply at 7- to 10-day intervals when plants are 4-6" high; continue until harvest (2 weeks before harvest for 3L).
Remarks: Use 1-1.5 lb./A for light disease; 3-4 lb./A for severe disease. Refer to label for use directions.

Signal Word/Toxicity Class: Danger/I (50DF, 50WP); Warning/II (3L).
REI: 48 hr. (50DF, 50WP); 24 hr. (3L).

TOP COP Tri-Basic (FL) (basic copper sulfate) *(Stoller)*
Rate: 4-6 pt. FL/A
Disease:
early blight late blight
Time: Apply when plants emerge; repeat at 7- to 10-day intervals.
Remarks: Refer to label for use directions.

Signal Word/Toxicity Class: Caution.
REI: 24 hr.

Copper-Count-N — see copper compounds

DICLORAN

Botran 75-W *(Gowan)*
Rate: 1.5 lb. ai/A
2 lb. 75-W/A
Disease:
Botrytis blight white mold
Time: Apply at layby; repeat at 10- to 14-day intervals.
Remarks: Do not feed treated potatoes to livestock. Refer to label for use directions.
Days To Harvest: 14.
State Restrictions: For use in North Central states.

Signal Word/Toxicity Class: Caution.
REI: 12 hr.

Dithane — see mancozeb

DUSTRET (Premix)

Dustret A is a dust based potato seed piece treatment containing maneb, streptomycin, and bark.

Dustret A or LD (8% maneb & 0.01% streptomycin sulfate) *(AGSCO)*
Rate: 1 lb. A, LD/100 lb. seed
Disease:
blackleg Fusarium Rhizoctonia
Time: Dry seed treatment.
Remarks: Thoroughly dust whole or cut potato seed pieces. If seed potatoes have not sprouted, subject them to temperatures of 50-60°F for 10- to 15-days before handling. Treat warmed seed pieces immediately after cut. Hold cut and treated seed pieces for 1 week or more at cool temperatures and high humidity; may be planted immediately. The holding period allows cut surfaces to heal. Do not use treated seed pieces for food, feed, or fodder.

Signal Word/Toxicity Class: Caution.

REI: 12 hr.

Dustret Maxim — see fludioxonil

Dustret T — see thiophanate-methyl

Echo — see chlorothalonil

FLUDIOXONIL

Dry fungicide seed protectant designed to suppress certain fungal diseases. Toxic to fish and aquatic invertebrates.

Dustret Maxim LD *(AGSCO)*
Rate: 0.5 lb. LD/100 lb. seed
Disease:
Fusarium dry rot *Helminthosporium solani* *Rhizoctonia solani* seed decay

◆ - **new product** • ai-active ingredient • DF-dry flowable • E/EC-emulsifiable concentrate • EW-emulsion in water
F/FL-flowable • G-granule • L/LC-liquid • SP/WSP-soluble packet • W/WP-wettable powder • WSB-water soluble bag

Time: Dry seed treatment.
Remarks: Apply so that cut seed pieces are thoroughly covered. After cutting, treated seed pieces should be planted in soil above 45°F with adequate soil moisture required for planting. If cut seed needs to be stored or held (7 days or less) make sure there is adequate cool air (60°F) movement through pile of cut seed potatoes at relative humidity of 85-90%. Cut and treated seed should not be piled above 6 ft. in height. Do not use on seed to be planted for production of seed. Do not use treated seed pieces for food, feed, or fodder.

Signal Word/Toxicity Class: Caution.
REI: 12 hr.

Maxim 4FS *(Novartis)*
Rate: 8 fl. oz. 4FS/100 lb. tubers
Disease:
Fusarium seed rot   scurf
Time: Seed treatment.
Remarks: May be tank mixed. Do not use treated seed for food, feed, or oil purposes. Do not graze on forage until 30 days after planting. Use EPA-approved dye or colorant that imparts unnatural color to seed.

Signal Word/Toxicity Class: Caution.

IPRODIONE

Rovral (4FL, WG, WP) *(Rhone-Poulenc)*
Rate: 0.5-1 lb. ai/A
1-2 pt. 4FL/A
1-2 lb. WG, WP/A
Disease:
early blight   white mold
Time: Early blight: Apply when conditions favor disease development. Up to 3 subsequent applications can be made at 10- to 14-day intervals. White mold: Apply just prior to row closing, or at early first sign of disease; repeat at 14- to 21-day intervals if favorable conditions for disease development continues.
Remarks: Under severe disease conditions, use high rate for early blight. Maximum of 4 applications per season. Refer to label for use directions.
Days To Harvest: 14.
Crop Rotation: Beans, broccoli, carrots, garlic, lettuce, onions (dry bulb), peanuts, potatoes, and rice can be rotated after harvest.
State Restrictions: Do not apply by air for white mold except in California.

Signal Word/Toxicity Class: Warning/II (WG); Caution (4FL, WP).
REI: 12 hr.

Kocide — see copper compounds

KOP-Hydroxide — see copper compounds

KOP Oxy — see copper compounds

Kumulus — see sulfur

MANCOZEB

Dithane (DF, F-45, M-45, ST, WSP) *(Rohm and Haas)*
Manzate 200DF *(Griffin)*
Penncozeb 75DF or 80WP *(Elf Atochem)*
Rate: 0.5-2 lb. DF/A
0.8-3.2 pt. F-45/A
0.5-2 lb. M-45, WSP/A
1-2 lb. 200DF, 75DF/A
1 2 lb. 80WP/A
Disease:
early blight   late blight
Time: Begin when plants are 4-6" high; repeat at 5- to 10-day intervals, (more frequently under severe disease conditions) throughout the season.
Remarks: Addition of surfactant will improve performance. It is recommended that this product be used with an Integrated Pest Management program. Vine kill should occur 14 days before harvest.
Days To Harvest: 14; Connecticut, Delaware, Florida, Maine, Massachusetts, Michigan, New Hampshire, New York, Ohio, Pennsylvania, Rhode Island, Vermont, Wisconsin-3.

State Restrictions: Refer to label for California precautions.

Rate: 1.25 lb. DF/50 gal. water
1.25 lb. M-45, WSP/50 gal. water
2 pt. F-45, ST/50 gal. water
1.25 lb. 200DF/50 gal. water
Disease:
Fusarium decay   seedborne common scab
Time: Seed piece treatment. Plant as soon as possible following treatment.
Remarks: Dip whole or cut potato tubers. Spread treated seed pieces in cool place if held before planting. Do not use treated seed for food, feed, or oil purposes. Refer to label for use directions.
State Restrictions: Refer to label for California precautions.

Signal Word/Toxicity Class: Caution.
REI: 24 hr.

MANEB

Maneb 75DF or 80WP *(Elf Atochem)*
Manex (FL) *(Griffin)*
Rate: 1.12-1.5 lb. ai/A
1.5-2 lb. 75DF/A
1.2-1.6 lb. ai/A
1.5-2 lb. 80WP/A
0.8-1.6 lb. ai/A
0.8-1.6 qt. FL/A
Disease:
early blight   late blight
Time: Begin when plants are 2-6" high; repeat at 5- to 10-day intervals.

Rate: Seed piece treatment:
0.75 lb. ai/10 gal.
1 lb. 75DF/10 gal.
0.8 lb. ai/10 gal.
1 lb. 80WP/10 gal.
0.8 qt. FL/10 gal.
Disease:
Fusarium seed piece decay
Time: Seed treatment.
Remarks: Dip whole or cut tubers. Spread in cool place if held before planting. Do not use treated seed potatoes for food, feed, or oil purposes. Refer to label for use directions.
Days To Harvest: 14; Connecticut, Delaware, Florida, Maine, Massachusetts, Michigan, New Hampshire, New York, Ohio, Pennsylvania, Rhode Island, Vermont, Wisconsin-3.

Signal Word/Toxicity Class: Caution.
REI: 24 hr.

MANEB (Premix)

ManKocide (DF) (15% maneb & 46.1% copper hydroxide) *(Griffin)*
Rate: 1.5-5 lb. DF/A
Disease:
early blight   late blight
Time: Apply when plants are 6" high at 7- to 10-day intervals.
Remarks: Use low rate where disease is light, high rate for severe disease. It is recommended that this product be used in an Integrated Pest Management program. Refer to label for use directions.
Days To Harvest: 14; Connecticut, Delaware, Florida, Maine, Massachusetts, Michigan, New Hampshire, New York, Ohio, Pennsylvania, Rhode Island, Vermont, Wisconsin-3.

Signal Word/Toxicity Class: Danger/I.
REI: 48 hr.

Manex — see maneb

ManKocide — see maneb premix

Manzate — see mancozeb

Maxim — see fludioxonil

NOTICE The information on these pages is for preliminary planning — not a guide for use. Be sure to follow the manufacturer's directions, notwithstanding information contained here. For personal protective equipment and EPA registration numbers, please turn to page 39.

MEFENOXAM (Co-Pack)

Twin-Pak containing Ridomil Gold and 6 lb. of chlorothalonil. Bravo 720 is toxic to fish, aquatic invertebrates, and marine/estuarine organisms.

Ridomil Gold/Bravo Liquid (L): Ridomil Gold EC (mefenoxam) *(Novartis)* **& Bravo 720 (chlorothalonil)** *(ZENECA)*

Rate: 1 container/5 A
Disease:
early blight
late blight
pink rot
Pythium leak

Time: For early and late blight, apply when conditions are favorable for disease development (before infection), but no later than when plant foliage meets within the row. Apply at 14-day intervals. For pink rot and Pythium leak, make first application at flowering and second 14 days later. If field has history of pink rot or Pythium leak problems, apply third application.

Remarks: For early and late blight, up to 3 applications per crop may be applied; (4 applications in Florida). Do not plant any crop which is not registered for use with mefenoxam in treated soil for 12 months.

Days To Harvest: 14.

- - - - - - - - - - - - - - - - -

Signal Word/Toxicity Class: Warning/II.
REI: 48 hr.

Mertect — see thiabendazole

METAM-SODIUM

Clean Crop Nemasol (42%) *(Platte)*

Rate: 30-75 gal. 42%/A
Disease:
early maturity disease

Time: Fall postharvest or spring preplant.

Remarks: Can be applied where crop stubble or vegetation exists without prior tillage, provided there is adequate soil penetration. Soil temperature should be in 40°-90°F range. Refer to label for use directions.

State Restrictions: In California, application must be in compliance with Technical Information Bulletin "Guidelines for All Application Methods for Metam-sodium in California." This bulletin may be obtained from your local pesticide dealer or a metam-sodium registrant.

- - - - - - - - - - - - - - - - -

Signal Word/Toxicity Class: Danger/I.
REI: 48 hr.

Metam CLR (42%) *(UCB Agrochemicals)*

Rate: 37.25-74.5 gal. 42%/A
Oregon:
29.75 gal. 42%/A
Disease:
early maturity disease

Time: Fall application.

Remarks: Inject into sprinkler system and apply in a minimum 1" water. Can be applied where crop stubble or vegetation exists without prior tillage, provided there is adequate soil penetration. In Oregon, apply using thin shank injector rig with shanks spaced 5" apart. Soil temperature should be in 40°-90°F range. Refer to label for use directions.

State Restrictions: In California, application must be in compliance with Technical Information Bulletin "Guidelines for All Application Methods for Metam-sodium in California." This bulletin may be obtained from your local pesticide dealer or a metam-sodium registrant.

- - - - - - - - - - - - - - - - -

Signal Word/Toxicity Class: Danger/I.
REI: 48 hr.

Micro Flo — see copper compounds

Microsperse COC — see copper compounds

Microthiol Special — see sulfur

Nu-Cop — see copper compounds

PCNB

PCNB 10% *(Amvac)*

Rate: 25 lb. 10%/A
Disease:
Rhizoctonia solani

Time: At planting.

Remarks: Broadcast treatment. Do not apply through any type of irrigation system. Do not plant root crops within 12 months of application unless PCNB is registered for use on those crops.

- - - - - - - - - - - - - - - - -

Signal Word/Toxicity Class: Caution.
REI: 12 hr.

Terraclor (75WP) or Terraclor 400 (FL) *(Uniroyal)*

Rate: 4-8 oz. 75WP/100 gal. water
6-12 fl. oz. FL/100 gal. water
Disease:
damping-off
root and stem rot

Time: Soil drench. One repeat application may be made 4-6 weeks later if necessary.

Remarks: Bedding plants in greenhouse or nursery. Refer to label for use directions.

Crop Rotation: Do not plant root crops in PCNB treated fields within 12 months of broadcast and banding applications unless PCNB registered for use on those crops.

- - - - - - - - - - - - - - - - -

Signal Word/Toxicity Class: Caution.

Turfcide (10G) *(Uniroyal)*

Rate: 1.5 oz. 10G/100 ft. row
Disease:
root and stem rot

Rate: 8 oz. 10G/100 ft. row
Disease:
southern blight

Time: Apply at planting time.

Remarks: Bedding plants container grown. Direct material into planting row and cover soil. Refer to label for use directions.

Crop Rotation: Do not plant root crops in PCNB-treated fields within 12 months of broadcast and banding applications unless PCNB registered for use on those crops.

- - - - - - - - - - - - - - - - -

Signal Word/Toxicity Class: Caution.

Penncozeb — see mancozeb

POLYRAM

Contact fungicide. Toxic to fish.

Polyram (80DF) (metiram) *(BASF, Platte)*

Rate: 1.2-1.6 lb. ai/A
1.5-2 lb. 80DF/A
Disease:
early blight
late blight

Time: Apply when plants are 4-6" high or when disease warrants; repeat at 5- to 10-day intervals.

Remarks: Addition of surfactant that will cause no phytotoxic effect will improve performance. This product should be used with an Integrated Pest Management program. Vine kill should occur 14 days before harvest. Do not graze livestock in treated areas. Do not apply this product through any type of irrigation system. Refer to label for use directions.

Days To Harvest: 14; Connecticut, Delaware, Florida, Maine, Massachusetts, Michigan, New Hampshire, New York, Ohio, Pennsylvania, Rhode Island, Vermont, Wisconsin-3.

- - - - - - - - - - - - - - - - -

Signal Word/Toxicity Class: Caution.
REI: 12 hr. (BASF); 24 hr. (Platte).

◆ - **new product** • ai-active ingredient • DF-dry flowable • E/EC-emulsifiable concentrate • EW-emulsion in water
F/FL-flowable • G-granule • L/LC-liquid • SP/WSP-soluble packet • W/WP-wettable powder • WSB-water soluble bag

PRO-TEX (Premix)

Toxic to fish and wildlife. Restricted Use Pesticide.

Pro-Tex (3.5 lb./gal. maneb & 0.5 lb./gal. triphenyltin hydroxide) *(Griffin)*
Rate: 0.75-1.5 lb. ai/A
0.75-1.5 qt./A
Disease:
early blight late blight
Time: Apply with appearance of blight weather conditions; repeat at 7-day intervals.
Remarks: For ground application (closed cabs only), apply in at least 15 gal. water; aerial application in 3-10 gal. water. Use low rate early in season and high rate mid- to late-season or when blight infection is in area. Do not exceed 6 qt./A per season. Refer to label for use directions.
Days To Harvest: 21.

Signal Word/Toxicity Class: Dange-Poison/I.
REI: 48 hr.

Redball — see sulfur

Rovral — see iprodione

SULFUR

Drexel Suffa (6FL, 8FL), Sulfa 80 (80DF) or Sulfur (90W) *(Drexel)*
Rate: 2.25-3 lb. ai/A
3-4 pt. 6FL/A
2.25-3 pt. 8FL/A (aerial)
2-8 lb. ai/A
3-10 lb. 80DF/A
2.7-4.5 lb. ai/A
3-5 lb. 90W/A
Disease:
leaf spot (90W) powdery mildew rust (80DF)
Time: Apply when disease first appears; repeat at 3- to 4-week intervals. For Sulfa 80, apply early leaf stage; repeat at 10- to 14-day intervals as needed, or immediately after rain.
Remarks: Do not use within 2 weeks of oil application (4 weeks FL). Some plants are susceptible to injury from sulfur under certain climatic conditions. Do not make sulfur applications at excessively high temperatures. Do not apply this product through any type of irrigation system. Refer to label for use directions.

Signal Word/Toxicity Class: Caution.
REI: 24 hr.

Kumulus DF *(BASF)*
Rate: 2.4-4 lb. ai/A
3-5 lb. DF/A
Disease:
powdery mildew Septoria leaf spot
Time: Apply at early leaf stage; repeat at 10- to 14-day intervals with or without other pesticides immediately after a rain.
Remarks: Do not apply if temperatures during or within 3 days after application are expected to exceed 90°F. Do not apply through any type of irrigation system. Refer to label for use directions.

Signal Word/Toxicity Class: Caution.
REI: 24 hr.

Liquid Sulfur Six *(Helena)*
Rate: 3-4 pt./A
Disease:
powdery mildew
Time: Apply at first sign of infection; repeat at 3-week intervals.
Remarks: Apply by air in 8 gal. water. Do not apply this product through any type of irrigation system. Refer to label for use directions.

Signal Word/Toxicity Class: Caution.
REI: 24 hr.

Micro Flo Sulfur (6L) or Microsperse Wettable Sulfur (WS) *(Miicro Flo)*
Terra Sulfur (6L) *(Terra)*
Rate: 2.25-8 lb. ai/A
3-10.66 pt. 6L/A
2.7-4.5 lb. ai/A
3-5 lb. WS/A
Disease:
powdery mildew rust (6L)
Time: Apply early leaf stage; repeat at 10- to 14-day intervals as needed.
Remarks: Do not use within 2 weeks of oil application. Some plants are susceptible to injury from sulfur under certain climatic conditions. Do not make sulfur applications at excessively high temperatures. Do not apply WS through any type of irrigation system. Refer to label for use directions.

Signal Word/Toxicity Class: Caution.
REI: 24 hr.

Microthiol Special (80WP) *(Elf Atochem)*
Rate: 4 lb. ai/A
5 lb. 80WP/A
Disease:
powdery mildew
Time: Apply at early leaf stage; repeat at 14-day intervals as needed.
Remarks: Refer to label for use directions.

Signal Word/Toxicity Class: Caution.
REI: 24 hr.

Redball EM-53 (6L), Superfine Wettable Sulphur (WS) or Commercial Dusting Sulphur (DS) *(International Sulphur)*
Rate: 2-7 pt. 6L/A
6-16 lb. WS/A
10-25 lb. DS/A
Disease:
powdery mildew
Time: 6L: Apply in early leaf stage; repeat as necessary. WS, DS: Apply when disease first appears; repeat at 7- to 14-day intervals (WS), 7- to 10-day intervals (DS).
Remarks: Apply in sufficient water for thorough coverage. Do not apply when temperatures exceed 90°F. Refer to label for use directions.

Signal Word/Toxicity Class: Caution.
REI: 12 hr.

THAT (6FL) or THAT Big 8 (8FL) *(Stoller)*
Rate: 2.25-3 lb. ai/A
3-4 pt. 6FL/A
2-3 lb. ai/A
2-3 pt. 8FL/A
Disease:
powdery mildew
Time: Apply at first sign of infection; repeat at 2 week intervals as needed.
Remarks: Apply by air in 8 gal. water. Refer to label for use directions.

Signal Word/Toxicity Class: Caution.
REI: 24 hr.

Thiolux (DF) *(Platte)*
Rate: 3-10 lb. DF/A
Disease:
powdery mildew rust
Time: Apply early leaf stage; repeat at 10- to 14-day intervals with or without pesticides immediately after rain.
Remarks: Do not make applications within 2 weeks of oil spray. Refer to label for use directions.

Signal Word/Toxicity Class: Caution.
REI: 24 hr.

SULFUR (Premixes)

Bravo S (27.25% sulfur & 19.15% chlorothalonil) *(Helena)*
Rate: 3-4.33 pt./A
Dry land culture:
2.5-4.25 pt./A
Disease:
Botrytis vine rot late blight powdery mildew
early blight

7 Vegetables

NOTICE The information on these pages is for preliminary planning — not a guide for use. Be sure to follow the manufacturer's directions, notwithstanding information contained here.

For personal protective equipment and EPA registration numbers, please turn to page 39.

Time: Apply when plants are 6-8" tall, or when disease threatens; repeat at 7- to 10-day intervals as needed. Powdery mildew: Apply at first sign of disease; repeat at 3-week intervals.
Remarks: Under severe conditions, use high rate and short interval. Do not allow livestock to graze treated areas, or feed hay or threshings from treated fields to livestock. Refer to label for use directions.

Signal Word/Toxicity Class: Caution.
REI: 48 hr.

Terranil S (27.25% sulfur & 19.15% chlorothalonil) *(Terra)*
Rate: 2.5-4.5 pt. S/A
Disease:
Botrytis vine rot, late blight, powdery mildew (S), early blight
Time: Apply when disease threatens; repeat at 7- to 10-day intervals as needed. For powdery mildew, apply at first sign of infection; repeat at 3-week intervals.
Remarks: Under severe disease infection, use high rate and shorter interval. Refer to label for use directions.

Signal Word/Toxicity Class: Warning/II.
REI: 48 hr.

Top Cop with Sulfur (6.25 lb./gal. sulfur & 1 lb./gal. tribasic copper sulfate) *(Stoller)*
Rate: 2-3 qt./A
Disease:
early blight, late blight
Time: Apply every 7-10 days throughout season.
Remarks: Use higher rate as vines increase in size. Refer to label for use directions.

Signal Word/Toxicity Class: Caution.
REI: 24 hr.

SUPER TIN

Packaged in AgPak, a water-soluble pack. Toxic to fish and wildlife. Restricted Use Pesticide.

Super Tin (80WP) (triphenyltin hydroxide) *(Griffin)*
Rate: 2.5-3.75 oz. 80WP/A
Disease:
early blight, late blight
Time: Apply with appearance of blight weather conditions; repeat at 7-day intervals.
Remarks: For ground application (closed cabs only), apply in at least 15 gal. water; 3-10 gal. water by air. Use low rate early in season; high rate mid- to late- season or when blight infection is in area. When tank mixed with another fungicide registered for potatoes, 1.87 oz./A may be employed. Do not exceed 15 oz. 80WP/A per season. Do not mix with pesticides, micronutrients, or adjuvants unless testing or prior experience has shown mixture to be nonphytotoxic to crop.
Days To Harvest: 21.
State Restrictions: Not for use in California.

Signal Word/Toxicity Class: Danger-Poison/I.
REI: 48 hr.

Tennessee Brand — see copper compounds

Terraclor — see PCNB

Terranil — see chlorothalonil

Terranil S — see sulfur premixes

TERRANIL Cu (Premix)

Terranil Cu (24% chlorothalonil & 24% copper oxychloride) *(Terra)*
Rate: 2.25-3.4 pt. Cu/A
Disease:
Botrytis vine rot, late blight, powdery mildew (S), early blight
Time: Apply when disease threatens; repeat at 7- to 10-day intervals as needed. For powdery mildew, apply at first sign of infection; repeat at 3-week intervals.
Remarks: Under severe disease infection, use high rate and shorter interval. Refer to label for use directions.
Days To Harvest: 7.

Signal Word/Toxicity Class: Danger/I.
REI: 48 hr.

THAT or THAT Big 8 — see sulfur

THIABENDAZOLE

Flowable water-dispersible suspension fungicide. Toxic to fish.

Mertect 340-F *(Novartis)*
Rate: 0.42 fl. oz. 340-F/2000 lb. tubers in complete water coverage
Disease:
tuber rot
Time: Mist before storage.
Remarks: Do not treat seed potatoes after cutting. Do not apply through any type of irrigation system.

Signal Word/Toxicity Class: Caution.

Thiolux — see sulfur

THIOPHANATE-METHYL

Potato seed piece treatment fungicide. Tops MZ is toxic to fish.

Dustret T *(AGSCO)*
Tops 2.5D, 5 or MZ *(Gustafson)*
Tops 2.5-D or 5 *(Platte)*
Rate: 0.5 lb. 5/100 lb. cut seed pieces
1 lb. T, 2.5D, 2.5-D, MZ/100 lb. cut seed pieces
Disease:
Fusarium, Rhizoctonia
Time: Dry seed treatment.
Remarks: Apply so that cut seed pieces are thoroughly covered with dust. Do not use treated seed pieces for food, feed, or fodder. Refer to label for use directions.

Signal Word/Toxicity Class: Caution.
REI: 12 hr. (Dustret).

TOP COP — see copper compounds

Top Cop with Sulfur — see sulfur premix

Tops — see thiophanate-methyl

Turfcide — see PCNB

Radishes

CAPTAN

Gustafson Captan 30-DD or 400 *(Gustafson)*
Rate: 1.25 fl. oz. 30-DD/100 lb. seed
1-2 fl. oz. 400/100 lb. seed
Disease:
damping-off, seed decay, seedling blight, mold
Time: Seed treatment.

◆ - **new product** • ai-active ingredient • DF-dry flowable • E/EC-emulsifiable concentrate • EW-emulsion in water
F/FL-flowable • G-granule • L/LC-liquid • SP/WSP-soluble packet • W/WP-wettable powder • WSB-water soluble bag

Remarks: Do not combine colloidal suspensions, wettable powders, or emulsifiable concentrates unless previous experience assures compatibility. Use with EPA-approved dye that imparts unnatural color to seed. Do not use treated seed for food, feed, or oil purposes. Refer to label for use directions.

Signal Word/Toxicity Class: Danger/I.

CAPTAN (Premix)

Nu-Gro Isotox (12.2% captan & 25% lindane) *(Wilbur-Ellis)*
Seed Mate Isotox (12.2% captan & 25% lindane) *(Loveland)*

Rate: 2 oz./100 lb. seed
Disease:
damping-off seed decay seedling blight
Time: Planter box treatment.
Remarks: Do not use treated seed for food, feed, or oil purposes. Refer to label for use directions.

Signal Word/Toxicity Class: Danger/I.
REI: 4 days.

COPPER COMPOUNDS

TOP COP Tri-Basic (FL) (basic copper sulfate) *(Stoller)*

Rate: 4 pt. FL/A
Disease:
Anthracnose downy mildew white rust
Cercospora leaf spot
Time: Apply when disease first appears; repeat at 7- to 10-day intervals in plant bed and fields.
Remarks: Refer to label for use directions.

Signal Word/Toxicity Class: Caution.
REI: 24 hr.

MEFENOXAM

Ridomil Gold EC or WSP *(Novartis)*

Rate: 0.5-1 lb. ai/A
1-2 pt. EC/A
0.45-0.9 lb. ai/A
1-2 lb. WSP/A
Disease:
Phytophthora Pythium
Time: Preplant incorporated or surface application just after planting.
Remarks: For banded application, 7" band recommended. Not for use in greenhouse or on field grown vegetable bedding plants. Do not plant any crop which is not registered for use with mefenoxam in treated soil for 12 months.
Days To Harvest: 1.

Signal Word/Toxicity Class: Warning/II (EC); Caution (WSP).
REI: 48 hr.

MEFENOXAM (Premix)

Ridomil Gold/Copper (WSB) (5% mefenoxam & 60% copper hydroxide) *(Novartis)*

Rate: 1.3 lb. ai/A
2 lb. WSB/A
Disease:
white rust
Time: May be applied 40-50 days after application or at planting. Apply 2-4 times at 14-day intervals.
Remarks: For season-long control, preplant or at planting application recommended. Do not plant any crop which is not registered for use with mefenoxam in treated soil for 12 months.
Days To Harvest: 7.

Signal Word/Toxicity Class: Danger/I.
REI: 48 hr.

Nu-Gro Isotox — see captan premix

Ridomil Gold — see mefenoxam

Ridomil Gold/Copper — see mefenoxam premix

Seed Mate — see captan premix

SULFUR (Premix)

Top Cop with Sulfur (6.25 lb./gal. sulfur & 1 lb./gal. tribasic copper sulfate) *(Stoller)*

Rate: 2 qt./A
Disease:
black rot downy mildew leaf spot
Time: Apply when disease threatens. Repeat at 7- to 10-day intervals; 3-day intervals in plant beds.
Remarks: Refer to label for use directions.

Signal Word/Toxicity Class: Caution.
REI: 24 hr.

THIRAM

42-S Thiram or Thiram 50WP Dyed *(Gustafson)*
Thiram Granuflo (WDG) *(UCB Agrochemicals)*

Rate: 8 fl. oz. 42-S/100 lb. seed
8 oz. 50WP/100 lb. seed
5.3 oz. WDG/100 lb. seed
Disease:
damping-off seed decay seedling blight
Time: Seed treatment.
Remarks: Use with EPA-approved dye that imparts unnatural color to seed. Do not use treated seed for food, feed, or oil purposes. Do not apply Thiram Granuflo through any type of irrigation system. Refer to label for use directions.
State Restrictions: Thiram Granuflo not for use in California.

Signal Word/Toxicity Class: Caution.
REI: 24 hr. (Thiram Granuflo).

TOP COP — see copper compounds

Top Cop with Sulfur — see sulfur premix

Rhubarb

Aliette — see fosetyl-aluminum

Botran — see dicloran

DICLORAN

Botran 75-W *(Gowan)*

Rate: 1 lb. ai/A
1.33 lb. 75-W/A
Disease:
Botrytis leaf rot
Time: Apply when buds emerge from crowns; repeat at weekly intervals.
Remarks: **Greenhouse/hothouse.** Refer to label for use directions.
Days To Harvest: 3.

Signal Word/Toxicity Class: Caution.
REI: 12 hr.

For personal protective equipment and EPA registration numbers, please turn to page 39.

FOSETYL-ALUMINUM

Aliette (WDG, WSP) *(Rhone-Poulenc)*
Rate: 1.6-4 lb. ai/A
2-5 lb. WDG, WSP/A
Disease:
downy mildew
Time: Apply when conditions favor disease development (high moisture and moderate temperatures); repeat at 7- to 21-day intervals.
Remarks: Do not apply less than 10 gal./A or make more than 7 applications per season. Do not tank mix with copper compounds. With moderate disease pressure, high rate and shorter interval should be used. Refer to label for use directions.
Days To Harvest: 3.

Signal Word/Toxicity Class: Caution.
REI: 12 hr.

MEFENOXAM

Ridomil Gold EC, GR, or WSP *(Novartis)*
Rate: 0.5-1 lb. ai/A
1-2 pt. EC/A
20-40 lb. GR/A
0.45-0.9 lb. ai/A
1-2 lb. WSP/A
Disease:
damping-off
Time: Apply at planting.
Remarks: For banded application, 7" band recommended. Not for use in greenhouse or on field grown vegetable bedding plants. Do not plant any crop which is not registered for use with mefenoxam in treated soil for 12 months.

Signal Word/Toxicity Class: Warning/II (EC); Caution (GR, WSP).
REI: 48 hr.

Ridomil Gold — see mefenoxam

Southern Peas

(Blackeyed, Crowder, Pinkeyed)

Allegiance — see metalaxyl

Apron — see metalaxyl

Apron XL — see mefenoxam

CAPTAN

Gustafson Captan 30-DD or 400 *(Gustafson)*
Rate: 2.25 fl. oz. 30-DD/100 lb. seed
2-3 fl. oz. 400/100 lb. seed
Disease:
damping-off, seed decay, seedling blight, mold
Time: Seed treatment.
Remarks: Cowpeas. Do not combine colloidal suspensions, wettable powders, or emulsifiable concentrates unless previous experience assures compatibility. Use with EPA-approved dye that imparts unnatural color to seed. Do not use treated seed for food, feed, or oil purposes. Refer to label for use directions.

Signal Word/Toxicity Class: Danger/I.

Drexel Captan 4FL *(Drexel)*
Rate: 2.5 fl. oz. 4FL/100 lb. seed
Disease:
seed decay, seedling blight, soilborne disease
Time: Liquid or slurry seed treatment.
Remarks: Cowpeas. Do not use treated seed for food, feed, or oil purposes. Refer to label for use directions.

Signal Word/Toxicity Class: Danger/I.

CHLORONEB

Supplemental seed treatment; use in conjunction with a standard fungicide seed treatment.

Demosan (65W or FL) *(Kincaid)*
Rate: 2.6 oz. ai/100 lb. seed
4 oz. 65W/100 lb. seed
7 oz. FL/100 lb. seed
Disease:
damping-off, Rhizoctonia, seedling diseases, Pythium blight, Sclerotium blight, sore shin
Time: Seed treatment or hopper box treatment.
Remarks: Blackeyed peas, cowpeas, Southern peas. Refer to label for use directions. Do not use treated seed for food, feed, or oil purposes. Do not graze within 45 days of planting.

Signal Word/Toxicity Class: Caution.
REI: 12 hr.

CHLOROTHALONIL

Terranil (6L, 90DF or 90DF WSP) or Terranil Zn *(Terra)*
Rate: 1-1.6 lb. ai/A
2-3 pt. Zn/A
1.375-2 pt. 6L/A
1.125-1.75 lb. 90DF, 90DF WSP/A
Disease:
Anthracnose, downy mildew, rust, Cercospora leaf spot
Time: Apply during early bloom stage; repeat at 7- to 10-day intervals. For use only on beans harvested dry with pods removed.
Remarks: Blackeyed peas. Use in sufficient water to obtain adequate coverage. Do not graze treated areas or feed treated plant parts to livestock. Refer to label for use directions.
Days To Harvest: 6 weeks.

Signal Word/Toxicity Class: Danger/I (90DF, 90DF WSP); Warning/II (6L, Zn).
REI: 48 hr.

Demosan — see chloroneb

MEFENOXAM

Apron XL LS *(Novartis)*
Rate: 0.32-0.64 fl. oz. LS/100 lb. seed
Disease:
damping-off, Phytophthora (early season)
Time: Seed treatment at planting.
Remarks: Blackeyed peas, cowpeas. Use with EPA-approved dye that imparts unnatural color to seed is recommended. Do not use treated seed for food, feed, or oil purposes.

Signal Word/Toxicity Class: Warning/II.

Ridomil Gold EC or WSP *(Novartis)*
Rate: 0.5-1 lb. ai/A
1-2 pt. EC/A
0.225-0.45 lb. ai/A
0.5-1 lb. WSP/A
Disease:
damping-off, root and stem rot
Time: Preplant incorporated or surface application just after planting.
Remarks: Blackeyed peas, cowpeas, crowder peas. For banded application, 7" band recommended. Not for use in greenhouse or on field grown vegetable bedding plants. May be tank mixed. Do not plant any crop which is not registered for use with mefenoxam in treated soil for 12 months.
Days To Harvest: 1.

Signal Word/Toxicity Class: Warning/II (EC); Caution (WSP).
REI: 48 hr.

◆ - **new product** • ai-active ingredient • DF-dry flowable • E/EC-emulsifiable concentrate • EW-emulsion in water
F/FL-flowable • G-granule • L/LC-liquid • SP/WSP-soluble packet • W/WP-wettable powder • WSB-water soluble bag

METALAXYL

Allegiance or Apron (FL) *(Gustafson)*
Rate: 0.75 fl. oz. Allegiance/100 lb. seed
0.75-1.5 fl. oz. Apron/100 lb. seed
Disease:
damping-off Phytophthora (early season)

Rate: 3.4 fl. oz. Apron/100 lb. seed
Disease:
downy mildew
Time: Seed treatment.
Remarks: Blackeyed peas, cowpeas. Use with EPA-approved dye that imparts unnatural color to seed. Do not use treated seed for food, feed, or oil purposes. Refer to label for use directions.

Signal Word/Toxicity Class: Warning/II.

Apron Dry *(Trace)*
Rate: 4 oz./100 lb. seed
Disease:
damping-off Phytophthora
Time: Seed treatment.
Remarks: Do not use treated seed for food, feed, or oil purposes. Refer to label for use directions.

Signal Word/Toxicity Class: Caution.
REI: 12 hr.

Ridomil Gold — see mefenoxam

SULFUR

Micro Flo Sulfur (6L) *(Micro Flo)*
Terra Sulfur (6L) *(Terra)*
Rate: 6-24 lb. ai/A
1-4 gal. 6L/A
Disease:
rust
Time: Apply soon after seedlings emerge; repeat at 7- to 10-day intervals throughout season.
Remarks: Cowpeas. Do not use within 2 weeks of oil application. Some plants are susceptible to injury from sulfur under certain climatic conditions. Do not make sulfur applications at excessively high temperatures. Refer to label for use directions. Refer to label for use directions.

Signal Word/Toxicity Class: Caution.
REI: 24 hr.

Terranil — see chlorothalonil

TERRANIL (Premix)

Terranil Cu (24% chlorothalonil & 24% copper oxychloride) *(Terra)*
Rate: 3-4.5 pt. Cu/A
Disease:
Anthracnose downy mildew rust
Cercospora leaf spot
Time: Apply during early bloom stage; repeat at 7- to 10-day intervals.
Remarks: Blackeyed peas. For use only on beans harvested dry with pods removed. Do not graze treated areas or feed treated plant parts to livesotck. Refer to label for use directions.
Days To Harvest: Dry-6 weeks; snap-7.

Signal Word/Toxicity Class: Danger/I.
REI: 48 hr.

THIRAM

42-S Thiram or Thiram 50WP Dyed *(Gustafson)*
Thiram Granuflo (WDG) *(UCB Agrochemicals)*
Rate: 2 fl. oz. 42-S/100 lb. seed
2 oz. 50WP/100 lb. seed
1.3 oz. WDG/100 lb. seed
Disease:
damping-off seed decay seedling blight
Time: Seed treatment.
Remarks: Cowpeas. Use with EPA-approved dye that imparts unnatural color to seed. Do not use treated seed for food, feed, or oil purposes. Do not apply Thiram Granuflo through any type of irrigation system. Refer to label for use directions.
State Restrictions: Thiram Granuflo not for use in California.

Signal Word/Toxicity Class: Caution.
REI: 24 hr. (Thiram Granuflo).

Spinach

Aliette — see fosetyl-aluminum

Allegiance — see metalaxyl

Apron XL — see mefenoxam

Basicop — see copper compounds

BENLATE

Benlate (SP, WP) (benomyl) *(DuPont)*
Rate: 1-2 lb. SP, WP/100 lb. seed in 0.5-2 pt. water
Disease:
Fusarium wilt
Time: Slurry treatment.
Remarks: Use with EPA-approved dye that imparts unnatural color to seed. Allow seeds to dry before planting; seeds may be planted any time after being well cured. Do not tank mix or alternate with benzimidazole or thiophanate products. Do not use treated seed for food, feed, or oil purposes. Refer to label for use directions.
State Restrictions: Benlate SP not registered for use in California.

Signal Word/Toxicity Class: Caution.
REI: 24 hr.

CAPTAN

Gustafson Captan 30-DD or 400 *(Gustafson)*
Rate: 4.75 fl. oz. 30-DD/100 lb. seed
4-6 fl. oz. 400/100 lb. seed
Disease:
damping-off seed decay seedling blight
mold
Time: Seed treatment.
Remarks: Do not combine colloidal suspensions, wettable powders, or emulsifiable concentrates unless previous experience assures compatibility. Use with EPA-approved dye that imparts unnatural color to seed. Do not use treated seed for food, feed, or oil purposes. Refer to label for use directions.

Signal Word/Toxicity Class: Danger/I.

Drexel Captan 4FL *(Drexel)*
Rate: 6 fl. oz. 4FL/100 lb. seed
Disease:
seed decay seedling blight soilborne disease
Time: Liquid or slurry seed treatment.
Remarks: Do not use treated seed for food, feed, or oil purposes. Refer to label for use directions.

Signal Word/Toxicity Class: Danger/I.

NOTICE The information on these pages is for preliminary planning — not a guide for use. Be sure to follow the manufacturer's directions, notwithstanding information contained here. For personal protective equipment and EPA registration numbers, please turn to page 39.

CAPTAN (Premix)

Nu-Gro Isotox (12.2% captan & 25% lindane) *(Wilbur-Ellis)*
Seed Mate Isotox (12.2% captan & 25% lindane) *(Loveland)*

Rate: 4 oz./100 lb. seed
Disease:
damping-off seed decay seedling blight
Time: Planter box treatment.
Remarks: Do not use treated seed for food, feed, or oil purposes. Refer to label for use directions.

Signal Word/Toxicity Class: Danger/I.
REI: 4 days.

Clean Crop C-O-C-S — see copper compounds

COPPER COMPOUNDS

Basicop (WP) (basic copper sulfate) *(Griffin)*
Drexel Basic Copper Sulfate (WP) or KOP 300 (3FL) (basic copper sulfate) *(Drexel)*
Micro Flo Basic Copper "53" (basic copper sulfate) or Nu-Cop (3L, 50DF, 50WP) (copper hydroxide) *(Micro Flo)*
TOP COP Tri-Basic (FL) (basic copper sulfate) *(Stoller)*

Rate: 1-2 lb. ai/A
2-4 lb. WP, 53, 50DF, 50WP/A
2.6 pt. 3FL/A
1.33-2.66 pt. 3L/A
4 pt. FL/A
Disease:
Anthracnose downy mildew white rust
Cercospora leaf spot
Time: Apply when disease appears; repeat at 7- to 10-day intervals.
Remarks: Refer to label for use directions.

Signal Word/Toxicity Class: Varies by formulation.
REI: Varies by formulation.

Clean Crop C-O-C-S (WDG) (74.8% copper oxychloride & 14.2% copper sulfate, basic) *(Platte)*

Rate: 0.5-1.5 lb. ai/A
1-3 lb. WDG/A
Disease:
downy mildew leaf spot
Time: Apply when disease first threatens; repeat at 7- to 10-day intervals to protect new growth throughout growing season.
Remarks: Leaf flecking may occur under adverse weather conditions. Do not apply this product through any type of irrigation system. Refer to label for use directions.

Signal Word/Toxicity Class: Warning/II.
REI: 24 hr.

Copper-Count-N (L) (copper ammonium) *(Mineral Research)*

Rate: 3 pt. L/A
Disease:
Anthracnose Cercospora leaf spot downy mildew
Time: Apply on first appearance of disease and continue on 7- to 10-day intervals.
Remarks: Apply by ground or air. Refer to label for use directions.
State Restrictions: Not for use in California.

Signal Word/Toxicity Class: Caution.
REI: 12 hr.

Kocide (101, DF, 2000, 4.5LF, LF) (copper hydroxide) *(Griffin)*

Rate: 1-2 lb. 101/A
2-3 lb. DF/A
1.5-2.25 lb. 2000/A
1.33-2 pt. 4.5LF/A
2-4 pt. LF/A
Disease
Anthracnose Cercospora leaf spot white rust
black leaf spot (101) downy mildew (101, LF)
Time: Apply when disease first appears; repeat every 7- to 10-days.
Remarks: Flecking may occur on spinach leaves. Do not apply in a spray solution having a pH less than 6.5 as phytotoxicity may occur. Refer to label for use directions.

Signal Word/Toxicity Class: Danger/I (101, DF); Warning/II (2000); Caution (4.5LF, LF).
REI: 24 hr.

Copper-Count-N — see copper compounds

FOSETYL-ALUMINUM

Aliette (WDG, WSP) *(Rhone-Poulenc)*

Rate: 1.6-4 lb. ai/A
2-5 lb. WDG, WSP/A
Disease:
downy mildew white rust
Time: Apply when conditions favor disease development (high moisture and moderate temperatures); repeat at 7- to 21-day intervals.
Remarks: Do not apply less than 10 gal./A or make more than 7 applications per season. Do not tank mix with copper compounds. With moderate disease pressure, high rate and shorter interval should be used. Refer to label for use directions.
Days To Harvest: 3.

Signal Word/Toxicity Class: Caution.
REI: 12 hr.

Kocide — see copper compounds

MEFENOXAM

Apron XL LS *(Novartis)*

Rate: 0.32-0.64 fl. oz. LS/100 lb. seed
Disease:
damping-off
Time: Seed treatment at planting.
Remarks: Do not apply to spinach grown in greenhouse. Do not use treated seed for food, feed, or oil purposes.

Signal Word/Toxicity Class: Warning/II.

Ridomil Gold EC or WSP *(Novartis)*

Rate: 0.5-1 lb. ai/A
1-2 pt. EC/A
0.45-0.9 lb. ai/A
1-2 lb. WSP/A
Disease:
damping-off downy mildew white rust
Time: Apply at planting only. For downy mildew and white rust, make up to 2 supplemental applications of 0.25 pt. EC or 0.25 lb. WSP/A shanked in 21 day intervals.
Remarks: For banded application, 7" band recommended. Not for use in greenhouse or on field grown vegetable bedding plants. Do not use more than 2.75 pt. EC /A or 2.75 lb. WSP/A per season. Do not plant any crop which is not registered for use with mefenoxam in treated soil for 12 months.
Days To Harvest: 21.

Signal Word/Toxicity Class: Warning/II (EC); Caution (WSP).
REI: 48 hr.

MEFENOXAM (Premix)

Ridomil Gold/Copper (WSB) (5% mefenoxam & 60% copper hydroxide) *(Novartis)*

Rate: 1.3 lb. ai/A
2 lb. WSB/A
Disease:
white rust
Time: May be applied 21 days after at planting application. Apply 1-2 times at 14-day intervals.
Remarks: Avoid late season applications when plants reach full maturity or begin senescence. Do not plant any crop which is not registered for use with mefenoxam in treated soil for 12 months.
Days To Harvest: 21.

Signal Word/Toxicity Class: Danger/I.
REI: 48 hr.

◆ - **new product** • ai-active ingredient • DF-dry flowable • E/EC-emulsifiable concentrate • EW-emulsion in water
F/FL-flowable • G-granule • L/LC-liquid • SP/WSP-soluble packet • W/WP-wettable powder • WSB-water soluble bag

METALAXYL

Allegiance (FL) *(Gustafson)*
Rate: 0.75 fl. oz. FL/100 lb. seed
Disease:
damping-off Phytophthora (early season)
Time: Seed treatment.
Remarks: Do not apply to spinach grown in greenhouse. Use with EPA-approved dye that imparts unnatural color to seed. Do not use treated seed for food, feed, or oil purposes. Refer to label for use directions.
State Restrictions: Not for use in California.

Signal Word/Toxicity Class: Warning/II.

Micro Flo — see copper compounds

Nu-Cop — see copper compounds

Nu-Gro Isotox — see captan premix

Redball — see sulfur

Ridomil Gold — see mefenoxam

Ridomil Gold/Copper — see mefenoxam premix

Seed Mate — see captan premix

SULFUR

Micro Flo Sulfur (6L) *(Micro Flo)*
Terra Sulfur (6L) *(Terra)*
Rate: 8-30 lb. ai/A
1.33-5 gal. 6L/A
Disease:
powdery mildew rust
Time: Apply when disease first appears; repeat at 10- to 14-day intervals.
Remarks: Do not use within 2 weeks of oil application. Some plants are susceptible to injury from sulfur under certain climatic conditions. Do not make sulfur applications at excessively high temperatures. Refer to label for use directions.

Signal Word/Toxicity Class: Caution.
REI: 24 hr.

Redball Superfine Wettable Sulphur (WS) or Commercial Dusting Sulphur (DS) *(International Sulphur)*
Rate: 6-16 lb. WS/A
10-25 lb. DS/A
Disease:
powdery mildew
Time: Apply when disease first appears; repeat at 7- to 14-day intervals (WS), 7- to 10-day intervals (DS).
Remarks: Apply in sufficient water for thorough coverage. Do not apply when temperatures exceed 90°F. Refer to label for use directions.

Signal Word/Toxicity Class: Caution.
REI: 12 hr.

SULFUR (Premix)

Top Cop with Sulfur (6.25 lb./gal. sulfur & 1 lb./gal. tribasic copper sulfate) *(Stoller)*
Rate: 4-8 pt./A
Disease:
Cercospora leaf spot downy mildew white rust
Time: Apply at first sign of disease; repeat at 7- to 10-day intervals.
Remarks: Refer to label for use directions.

Signal Word/Toxicity Class: Caution.
REI: 24 hr.

THIRAM

42-S Thiram or Thiram 50WP Dyed *(Gustafson)*
Thiram Granuflo (WDG) *(UCB Agrochemicals)*
Rate: 8 fl. oz. 42-S/100 lb. seed
8 oz. 50WP/100 lb. seed
5.3 oz. WDG/100 lb. seed
Disease:
damping-off seed decay seedling blight
Time: Seed treatment.
Remarks: Use with EPA-approved dye that imparts unnatural color to seed. Do not use treated seed for food, feed, or oil purposes. Do not apply Thiram Granuflo through any type of irrigation system. Refer to label for use directions.
State Restrictions: Thiram Granuflo not for use in California.

Signal Word/Toxicity Class: Caution.
REI: 24 hr. (Thiram Granuflo).

TOP COP — see copper compounds

Top Cop with Sulfur — see sulfur premix

Sweet Corn

Allegiance — see metalaxyl

Apron — see metalaxyl

Apron XL — see mefenoxam

Bravo — see chlorothalonil

CAPTAN

Captan 30-DD contains the toxic inert Rhodamine B.

Gustafson Captan 30-DD or 400 *(Gustafson)*
Rate: 3.25 fl. oz. 30-DD/100 lb. seed
2-4 fl. oz. 400/100 lb. seed
Disease:
damping-off seed decay seedling blight
mold
Time: Seed treatment.
Remarks: Do not combine colloidal suspensions, wettable powders, or emulsifiable concentrates unless previous experience assures compatibility. Use with EPA-approved dye that imparts unnatural color to seed. Do not use treated seed for food, feed, or oil purposes. Refer to label for use directions.

Signal Word/Toxicity Class: Danger/I.

Drexel Captan 4FL *(Drexel)*
Rate: 3.5 fl. oz. 4FL/100 lb. seed
Disease:
seed decay seedling blight soilborne disease
Time: Liquid or slurry seed treatment.
Remarks: Sweet corn. Do not use treated seed for food, feed, or oil purposes. Refer to label for use directions.

Signal Word/Toxicity Class: Danger/I.

7 Vegetables

NOTICE The information on these pages is for preliminary planning — not a guide for use. Be sure to follow the manufacturer's directions, notwithstanding information contained here.

For personal protective equipment and EPA registration numbers, please turn to page 39.

CAPTAN (Premixes)

Premix combining a fungicide with an insecticide. Captan-Diazinon contains graphite.

Captan-Diazinon (36.67% captan & 25% diazinon) *(Trace)*
Kernel Guard (14.67% captan & 15% diazinon & 25% lindane) *(Trace)*

Rate: 2 oz./bu. seed
Disease:
damping-off   seed decay   seedling blight
Time: Planter box seed treatment.
Remarks: Do not use treated seed for food, feed, or oil purposes. Refer to label for use directions.

Signal Word/Toxicity Class: Danger/I.
REI: 12 hr. (Kernel Guard).

CARBOXIN (Premixes)

Germate Plus contains graphite. Vitaflo-280 is a combination of a systemic fungicide (Vitavax) and a contact fungicide (thiram). Toxic to fish. Germate Plus is also toxic to birds and other wildlife.

Germate Plus (14% carboxin & 15% diazinon & 25% lindane) *(Trace)*

Rate: 2 oz./bu. seed
Disease:
seedling diseases
Time: Planter box seed treatment.
Remarks: Do not use treated seed for food, feed, or oil purposes. Refer to label for use directions.

Signal Word/Toxicity Class: Caution.
REI: 12 hr.

Vitaflo-280 (14.9% carboxin & 13.2% thiram) *(Gustafson)*

Rate: 4.5 fl. oz./100 lb. seed
Corn head smut:
8.5-11.5 fl. oz./100 lb. seed
Disease:
corn head smut   damping-off   seed decay
Time: Seed treatment.
Remarks: Sweet corn. Thorough seed coverage is required. Use high rate under expected conditions of heavy disease pressure. Avoid planting corn under cold, wet conditions. Stress conditions during first few weeks after planting may increase incidence of head smut. Will not control soilborne head smut. Do not use treated seed for food, feed, or oil purposes. Do not graze or feed livestock on treated areas for 6 weeks after planting. Refer to label for use directions.

Signal Word/Toxicity Class: Caution.

CHLOROTHALONIL

Broad-spectrum contact fungicide. Toxic to fish, aquatic invertebrates and marine/estuarine organisms.

Bravo 500, 720/Weather Stik, Ultrex (WDG) *(ZENECA)*
Echo 500, 720 or 90DF *(Sostram)*

Rate: 1.125-2.75 pt. 500/A
0.7-1.8 lb. WDG/A
0.75-2 pt. 720/A
0.65-1.6 lb. 90DF/A
Disease:
Helminthosporium leaf blight   rust
Time: Apply when conditions favor disease development; repeat at 4- to 7-day intervals as necessary.
Remarks: Sweet corn (grown for seed). Do not apply to sweet corn to be processed. Do not graze treated fields or feed treated parts to livestock. Do not ensile treated corn or use as livestock forage. Refer to label for use directions and restrictions.
Days To Harvest: 14.

Signal Word/Toxicity Class: Danger/I (WDG, 90DF); Warning/II (500, 720).
REI: 48 hr.

Terranil (6L, 90DF or 90DF WSP) *(Terra)*

Rate: 0.75-2 pt. 6L/A
0.625-1.5 lb. 90DF, 90DF WSP/A
Disease:
Helminthosporium leaf blight   rust
Time: Apply when disease threatens; repeat at 4- to 7-day intervals.
Remarks: Sweet corn, corn grown for seed. Do not allow livestock to graze in treated fields. Do not ensile treated corn or use as livestock forage. Refer to label for use directions.
Days To Harvest: 14.

Signal Word/Toxicity Class: Danger/I (90DF, 90DF WSP); Warning/II (6L).
REI: 48 hr.

Dithane — see mancozeb

Echo — see chlorothalonil

FLUDIOXONIL

Contact fungicide for control of seedborne and soilborne fungi. Toxic to fish and aquatic invertebrates.

Maxim 4FS *(Novartis)*

Rate: 0.08 fl. oz. 4FS/100 lb. seed
Disease:
damping-off   seedling blight   soilborne disease
seed decay
Time: Seed treatment
Remarks: May be tank mixed. Do not use treated seed for food, feed, or oil purposes. Do not graze on forage until 30 days after planting. Use EPA-approved dye that imparts unnatural color to seed.

Signal Word/Toxicity Class: Caution.

Germate Plus — see carboxin premixes

Kernel Guard — see captan premixes

MANCOZEB

Dithiocarbamate fungicide with broad-spectrum activity with no resistance problems. Coordination product of zinc ion and manganese ethylene bisdithiocarbamate. Toxic to fish.

Dithane (DF, F-45, M-45, WSP) *(Rohm and Haas)*
Manzate 200DF *(Griffin)*
Penncozeb 75DF or 80WP *(Elf Atochem)*

Rate: 1.2 lb. ai/A
1.5 lb. DF, M-45, WSP, 200DF/A
2.4 pt. F-45/A
0.75-1.12 lb. ai/A
1-1.5 lb. 75DF/A
0.8-1.2 lb. ai/A
1-1.5 lb. 80WP/A
Disease:
common rust   gray leaf spot   Helminthosporium leaf blight
Time: Apply when symptoms first appear; repeat at 4- to 7-day intervals.
Remarks: Sweet corn, sweet corn used for seed production including hybrid. Addition of surfactant will improve performance. Do not feed treated forage to livestock. Refer to label for use directions.
Days To Harvest: 7.
State Restrictions: Manzate and Penncozeb: For use east of the Mississippi (including Arkansas and Louisiana); west of the Mississippi (except Arkansas and Louisiana). Refer to label for California precautions.

Signal Word/Toxicity Class: Caution.
REI: 24 hr.

◆ - **new product** • ai-active ingredient • DF-dry flowable • E/EC-emulsifiable concentrate • EW-emulsion in water
F/FL-flowable • G-granule • L/LC-liquid • SP/WSP-soluble packet • W/WP-wettable powder • WSB-water soluble bag

MANEB

Dithiocarbamate contact fungicide. Toxic to fish.

Maneb 75DF or 80WP *(Elf Atochem)*
Manex (FL) *(Griffin)*

Rate: 1.12 lb. ai/A
1.5 lb. 75DF/A
1.2 lb. ai/A
1.5 lb. 80WP/A
1.2 qt. FL/A

Disease:
common rust; Helminthosporium blight

Time: Begin when disease threatens; repeat at 3- to 10-day intervals.
Remarks: Sweet corn, sweet corn for seed production. Add surfactant or spreader/sticker for better coverage. Do not feed treated forage to livestock. Refer to label for use directions.
Days To Harvest: 7.
State Restrictions: For use east of the Mississippi (including Arkansas, Louisiana), and west of the Mississippi (except Arkansas, Louisiana).

Signal Word/Toxicity Class: Caution.
REI: 24 hr.

Manex — see maneb

Manzate — see mancozeb

Maxim — see fludioxonil

MEFENOXAM

Systemic fungicide having a specific mode of action.

Apron XL LS *(Novartis)*

Rate: 0.32-0.64 fl. oz. LS/100 lb. seed
Disease:
damping-off

Rate: 1.28 fl. oz. LS/100 lb. seed
Disease:
downy mildew

Time: Seed treatment at planting.
Remarks: Sweet corn. Use of EPA-approved dye is recommended to impart unnatural color to seed. Do not use treated seed for food, feed, or oil purposes. Refer to label for use directions.

Signal Word/Toxicity Class: Warning/II.

METALAXYL

Systemic fungicide having a specific mode of action.

Allegiance or Apron (FL) *(Gustafson)*

Rate: 0.375-0.75 fl. oz. Allegiance/100 lb. seed
0.75-1.5 fl. oz. Apron/100 lb. seed
Disease:
damping-off; Phytophthora (early season)

Rate: 3 fl. oz. Apron/100 lb. seed
Disease:
downy mildew

Time: Seed treatment.
Remarks: Use with EPA-approved dye that imparts unnatural color to seed. Do not use for feed, food, or oil purposes. Refer to label for use directions.

Signal Word/Toxicity Class: Warning/II.

Apron Dry *(Trace)*

Rate: 3-4 oz./100 lb. seed
Disease:
damping-off

Time: Seed treatment.
Remarks: Do not use treated seed for food, feed, or oil purposes. Refer to label for use directions.

Signal Word/Toxicity Class: Caution.
REI: 12 hr.

Penncozeb — see mancozeb

PROPICONAZOLE

Triazole sterol inhibitor. Toxic to fish.

Tilt (EC) *(Novartis)*

Rate: 2-4 fl. oz. EC/A
Disease:
eyespot; gray leaf spot; Helminthosporium leaf blight; rust

Time: Apply when disease first appears. Repeat at 7- to 14-day intervals for leaf blight and rust, 14-day intervals for eyespot and gray leaf spot.
Remarks: Do not apply more than 16 fl. oz./crop per season. Do not graze or feed livestock treated forage or cut green crop for hay or silage. After harvest, hay can be used for bedding or feed. Refer to label for use directions.
Days To Harvest: 14.
Signal Word/Toxicity Class: Warning/II.
REI: 24 hr.

Sulfa — see sulfur

SULFUR

Nonsystemic fungicide with protective action. Microthiol Special, Sulfa 80, and Thiolux are micronized sulfurs.

Sulfa 80 (80DF) *(Drexel)*

Rate: 6-10 lb. ai/A
7.5-12.5 lb. 80DF/A
Disease:
powdery mildew

Time: Apply when disease appears; repeat as necessary.
Remarks: Do not use within 2 weeks of oil application. Some plants are susceptible to injury from sulfur under certain climatic conditions. Do not make sulfur applications at excessively high temperatures. Do not apply this product through any type of irrigation system. Refer to label for use directions.

Signal Word/Toxicity Class: Caution.
REI: 24 hr.

Terranil — see chlorothalonil

THIRAM

Broad-spectrum fungicide. Toxic to fish. Thiram 50WP contains the toxic inert Rhodamine B and is also toxic to birds and wildlife.

42-S Thiram or Thiram 50WP Dyed *(Gustafson)*
Thiram Granuflo (WDG) *(UCB Agrochemicals)*

Rate: 5 fl. oz. 42-S/100 lb. seed
3 oz. 50WP/100 lb. seed
2 oz. WDG/100 lb. seed
Disease:
damping-off; seed decay; seedling blight

Time: Seed treatment.
Remarks: Sweet corn. Use with EPA-approved dye that imparts unnatural color to seed. Do not use treated seed for food, feed, or oil purposes. Do not apply Thiram Granuflo through any type of irrigation system. Refer to label for use directions.
State Restrictions: Thiram Granuflo not for use in California.

Signal Word/Toxicity Class: Caution.
REI: 24 hr. (Thiram Granuflo).

Tilt — see propiconazole

Vitaflo — see carboxin premixes

NOTICE The information on these pages is for preliminary planning — not a guide for use. Be sure to follow the manufacturer's directions, notwithstanding information contained here.

For personal protective equipment and EPA registration numbers, please turn to page 39.

Sweet Potatoes, Yams

BENLATE

Benlate (SP, WP) (benomyl) *(DuPont)*
Rate: 0.5-1 lb. ai/A
1-2 lb. SP, WP/A
Disease:
Anthracnose
Time: Apply when disease appears; repeat at 3- to 4-week intervals.
Remarks: Yams. Do not tank mix or alternate with benzimidazole or thiophanate products. Refer to label for use directions.
Days To Harvest: 90.
State Restrictions: Benlate WP for use in Puerto Rico; Benlate SP for use in Florida and Puerto Rico.

Signal Word/Toxicity Class: Caution.
REI: 24 hr.

Botran — see dicloran

CHLOR-O-PIC

Restricted Use Pesticide.

Chlor-O-Pic (chloropicrin) *(Great Lakes Chemical)*
Rate: Nontarp:
150-400 lb./A
Disease:
Cylindrocladium, Fusarium, Phytophthora, Plasmodiophora, Pseudomonas, Pythium, Rhizoctonia, Sclerotinia, Sclerotium, Verticillium
Time: Apply up to 2 weeks prior to planting when soil moisture is good and temperature is 60°-85°F.
Remarks: Work upwind of product and never fumigate without 2 qualified applicators. Do not fumigate when there is no or little air movement. Refer to label for use directions.

Signal Word/Toxicity Class: Danger/I.

DICLORAN

Botran 75-W *(Gowan)*
Rate: Seed dip:
0.75 lb. ai/A
1 lb. 75-W/A
Plantbed spray:
2.25-2.8 lb. ai/A
3-3.75 lb. 75-W/A
Disease:
Rhizopus rot, Sclerotinia blight, scurf
Time: Dip 10-15 seconds for seeds, 5-10 seconds for roots; drain and bed promptly. For plantbeds, apply before covering with soil.
Remarks: Refer to label for use directions.
Crop Rotation: Do not plant tomatoes as a follow-up crop.

Signal Word/Toxicity Class: Caution.
REI: 12 hr.

MEFENOXAM

Ridomil Gold EC or WSP *(Novartis)*
Rate: 0.5-1 lb. ai/A
1-2 pt. EC/A
0.45-0.9 lb. ai/A
1-2 lb. WSP/A
Disease:
Phytophthora, Pythium
Time: Preplant incorporated or surface application just after planting.
Remarks: Sweet potatoes, yams. For banded application, 7" band recommended. Not for use in greenhouse or on field grown vegetable bedding plants. Do not plant any crop which is not registered for use with mefenoxam in treated soil for 12 months.
Days To Harvest: 1.

Signal Word/Toxicity Class: Warning/II (EC); Caution (WSP).
REI: 48 hr.

Mertect — see thiabendazole

Ridomil Gold — see mefenoxam

TELONE (Premix)

Restricted Use Pesticide.

Telone C-17 (8.2 lb./gal. dichloropropene & 1.7 lb./gal. chloropicrin) *(Dow AgroSciences)*
Rate: 10.8-30 gal. C-17/A
Disease:
soil rot
Time: Preplant soil treatment. Apply whenever soil conditions are suitable, but preferably in late summer or fall or in spring up to 2-3 weeks before planting.
Remarks: Do not use in extremely heavy clay or muck soils. Do not apply this product through any type of irrigation system. Refer to label for further use directions and precautions.

Signal Word/Toxicity Class: Danger/I.
REI: 72 hr.

THIABENDAZOLE

Mertect 340-F *(Novartis)*
Rate: 8 fl. oz. 340-F/7.5 gal. water
Disease:
black rot, foot rot, scurf
Time: Treat seed roots 1-2 minutes and plant immediately.
Remarks: Do not use treated roots for food or feed. Do not apply through any type of irrigation system.

Signal Word/Toxicity Class: Caution.

Table Beets

(Red Beets)

Apron — see metalaxyl

Apron XL — see mefenoxam

Basicop — see copper compounds

CAPTAN

Gustafson Captan 30-DD or 400 *(Gustafson)*
Rate: 9.5 fl. oz. 30-DD/100 lb. seed
2-4 fl. oz. 400/100 lb. seed
Disease:
damping-off, seed decay, seedling blight, mold
Time: Seed treatment.
Remarks: Do not combine colloidal suspensions, wettable powders, or emulsifiable concentrates unless previous experience assures compatibility. Use with EPA-approved dye that imparts unnatural color to seed. Do not use treated seed for food, feed, or oil purposes. Refer to label for use directions.

Signal Word/Toxicity Class: Danger/I.

◆ - **new product** • ai-active ingredient • DF-dry flowable • E/EC-emulsifiable concentrate • EW-emulsion in water
F/FL-flowable • G-granule • L/LC-liquid • SP/WSP-soluble packet • W/WP-wettable powder • WSB-water soluble bag

Drexel Captan 4FL *(Drexel)*
Rate: 12 fl. oz. 4FL/100 lb. seed
Disease:
seed decay seedling blight soilborne disease
Time: Liquid or slurry seed treatment.
Remarks: Do not use treated seed for food, feed, or oil purposes. Refer to label for use directions.

Signal Word/Toxicity Class: Danger/I.

Clean Crop C-O-C-S — see copper compounds

COPPER COMPOUNDS

Basicop (WP) (basic copper sulfate) *(Griffin)*
Rate: 1.5 lb. ai/A
3 lb. WP/A
Disease:
downy mildew
Time: Apply when disease appears; repeat 7- to 10-days after.
Remarks: Do not apply through any type of irrigation system. Refer to label for use directions.

Signal Word/Toxicity Class: Warning/II.
REI: 24 hr.

Clean Crop C-O-C-S (WDG) (74.8% copper oxychloride & 14.2% copper sulfate, basic) *(Platte)*
Rate: 1.5-2 lb. ai/A
3-4 lb. WDG/A
Disease:
downy mildew leaf spot
Time: Apply prior to disease infection; repeat at 7- to 10-day intervals to protect new growth throughout growing season.
Remarks: Do not apply this product through any type of irrigation system. Refer to label for use directions.

Signal Word/Toxicity Class: Warning/II.
REI: 24 hr.

Drexel Basic Copper Sulfate (WP) or KOP 300 (3FL) (basic copper sulfate) *(Drexel)*
Micro Flo Basic Copper "53" (basic copper sulfate) or Microsperse COC (53WP) (copper oxychloride) *(Micro Flo)*
Rate: 2-3 lb. WP/A
2.6 pt. 3FL/A
2-4 lb. 53, 53WP/A
Disease:
downy mildew leaf blight leaf spot
Time: Apply when disease first appears; repeat at 7- to 10-day intervals.
Remarks: Refer to label for use directions.

Signal Word/Toxicity Class: Danger/I (53WP); Warning/II (WP, 3FL, 53).
REI: 24 hr. (WP, 53WP, 53); 12 hr. (3FL).

Kocide (101, DF, 2000, 4.5LF) (copper hydroxide) *(Griffin)*
Rate: 2-5 lb. 101, DF/A
1.5-3.75 lb. 2000/A
1.33-3.33 pt. 4.5LF/A (including tops)
Disease:
Cercospora leaf spot
Time: Apply when disease threatens; repeat every 10- to 14-days.

Rate: Beet tops:
1-2 lb. DF/A
Disease:
black leaf spot black rot downy mildew
Time: Apply at 7- to 10-day intervals after transplants are set in the field, shortly after emergence of field-seeded crops, or when conditions favor disease development.
Remarks: Addition of 2 qt./A agricultural spray oil is recommended for table beets. Do not apply in a spray solution having a pH less than 6.5 as phytotoxicity may occur. Refer to label for use directions.
State Restrictions: Kocide 4.5LF is not for use on beet tops in California.

Signal Word/Toxicity Class: Danger/I (101, DF); Warning/II (2000); Caution (4.5LF).
REI: 24 hr.

Nu-Cop (3L) (copper hydroxide) *(Micro Flo)*
Rate: 1.33-6.66 pt. 3L/A
Disease:
Cercospora leaf spot
Time: Apply when disease threatens; repeat at 10- to 14-day intervals depending on weather conditions.
Remarks: Refer to label for use directions.

Signal Word/Toxicity Class: Warning/II.
REI: 24 hr.

TOP COP Tri-Basic (FL) (basic copper sulfate) *(Stoller)*
Rate: 2-4 pt. FL/A
Disease:
downy mildew leaf blight Septoria leaf spot
gray mold
Time: Apply when disease first threatens; repeat at 7- to 10-day intervals.
Remarks: Refer to label for use directions.

Signal Word/Toxicity Class: Caution.
REI: 24 hr.

Kocide — see copper compounds

Kumulus — see sulfur

MEFENOXAM

Apron XL LS *(Novartis)*
Rate: 0.32-0.64 fl. oz. LS/100 lb. seed
Disease:
damping-off
Time: Seed treatment at planting.
Remarks: Do not use treated seed for food, feed, or oil purposes.

Signal Word/Toxicity Class: Warning/II.

Ridomil Gold EC or WSP *(Novartis)*
Rate: 0.5-1 lb. ai/A
1-2 pt. EC/A
0.45-0.9 lb. ai/A
1-2 lb. WSP/A
Disease:
Phytophthora Pythium
Time: Preplant incorporated or surface application just after planting.
Remarks: For banded application, 7" band recommended. Not for use in greenhouse or on field grown vegetable bedding plants. Do not plant any crop which is not registered for use with mefenoxam in treated soil for 12 months.
Days To Harvest: 1.

Signal Word/Toxicity Class: Warning/II (EC); Caution (WSP).
REI: 48 hr.

METALAXYL

Apron Dry *(Trace)*
Rate: 4 oz./100 lb. seed
Disease:
damping-off
Time: Seed treatment.
Remarks: Do not use treated seed for food, feed, or oil purposes. Refer to label for use directions.

Signal Word/Toxicity Class: Caution.
REI: 12 hr.

Micro Flo — see copper compounds

Microsperse COC — see copper compounds

7 Vegetables

NOTICE	The information on these pages is for preliminary planning — not a guide for use. Be sure to follow the manufacturer's directions, notwithstanding information contained here.	For personal protective equipment and EPA registration numbers, please turn to page 39.

Nu-Cop — see copper compounds

Redball — see sulfur

Ridomil Gold — see mefenoxam

Sulfa — see sulfur

SULFUR

Kumulus DF *(BASF)*
Rate: 2.4-4 lb. ai/A
3-5 lb. DF/A
Disease:
powdery mildew
Time: Apply when disease first appears with or without other pesticides and repeat as necessary.
Remarks: Do not apply if temperatures during or within 3 days after application are expected to exceed 90°F. Do not apply through any type of irrigation system. Refer to label for use directions.

Signal Word/Toxicity Class: Caution.
REI: 24 hr.

Sulfa 80 (80DF) *(Drexel)*
Rate: 4-8 lb. ai/A
5-10 lb. 80DF/A
Disease:
powdery mildew
Time: Apply when disease first appears; repeat at 10- to 30-day intervals.
Remarks: Do not use within 2 weeks of oil application. Some plants are susceptible to injury from sulfur under certain climatic conditions. Do not make sulfur applications at excessively high temperatures. Do not apply this product through any type of irrigation system. Refer to label for use directions.

Signal Word/Toxicity Class: Caution.
REI: 24 hr.

Redball EM-53 (6L), Superfine Wettable Sulphur (WS) or Commercial Dusting Sulphur (DS) *(International Sulphur)*
Rate: 8-16 pt. 6L/A
5-10 lb. WS/100 gal. water
5-15 lb. DS/A
Disease:
powdery mildew
Time: Apply at first sign of disease; repeat at 10- to 30-day intervals as needed throughout season.
Remarks: Apply in sufficient water for thorough coverage. Do not apply when temperatures exceed 90°F. Refer to label for use directions.
Days To Harvest: 0.

Signal Word/Toxicity Class: Caution.
REI: 12 hr.

Micro Flo Sulfur (6L) or Microsperse Wettable Sulfur (WS) *(Micro Flo)*
Terra Sulfur (6L) *(Terra)*
Rate: 4 lb. ai/A
5.3 pt. 6L/A
3.6 lb. ai/A
4 lb. WS/A
Disease:
powdery mildew
Time: Apply at 18-day intervals.
Remarks: Do not use within 2 weeks of oil application. Some plants are susceptible to injury from sulfur under certain climatic conditions. Do not make sulfur applications at excessively high temperatures. Do not apply WS through any type of irrigation system. Refer to label for use directions.

Signal Word/Toxicity Class: Caution.
REI: 24 hr.

Thiolux (DF) *(Platte)*
Rate: 3-5 lb. DF/A
Disease:
powdery mildew
Time: Apply at first sign of symptoms; repeat at 10- to 30-day intervals with or without pesticides immediately after rain.
Remarks: Thorough coverage is required. Do not make applications within 2 weeks of oil spray. Refer to label for use directions.

Signal Word/Toxicity Class: Caution.
REI: 24 hr.

SULFUR (Premix)

Top Cop with Sulfur (6.25 lb./gal. sulfur & 1 lb./gal. tribasic copper sulfate) *(Stoller)*
Rate: 2 qt./A
Disease:
downy mildew powdery mildew
Time: Apply at first sign of disease; repeat at 7- to 14-day intervals.
Remarks: Refer to label for use directions.

Signal Word/Toxicity Class: Caution.
REI: 24 hr.

Thiolux — see sulfur

THIRAM

42-S Thiram or Thiram 50WP Dyed *(Gustafson)*
Thiram Granuflo (WDG) *(UCB Agrochemicals)*
Rate: 8 fl. oz. 42-S/100 lb. seed
8 oz. 50WP/100 lb. seed
5.3 oz. WDG/100 lb. seed
Disease:
damping-off seed decay seedling blight
Time: Seed treatment.
Remarks: Use with EPA-approved dye that imparts unnatural color to seed. Do not use treated seed for food, feed, or oil purposes. Do not apply Thiram Granuflo through any type of irrigation system. Refer to label for use directions.
State Restrictions: Thiram Granuflo not for use in California.

Signal Word/Toxicity Class: Caution.
REI: 24 hr. (Thiram Granuflo).

TOP COP — see copper compounds

Top Cop with Sulfur — see sulfur premix

Tomatoes

Aliette — see fosetyl-aluminum

AZOXYSTROBIN

Broad-spectrum fungicide for control of plant diseases. Toxic to freshwater and estuarine/marine fish and aquatic invertebrates.

Quadris (FL) *(ZENECA)*
Rate: 5-6.2 fl. oz. FL/A
Disease:
Anthracnose early blight Septoria leaf spot
black mold late blight target spot
buckeye rot powdery mildew
Time: Apply at 7- to 14-day intervals. Do not apply until 21 days after transplanting or 35 days after seeding.

◆ - **new product** • ai-active ingredient • DF-dry flowable • E/EC-emulsifiable concentrate • EW-emulsion in water
F/FL-flowable • G-granule • L/LC-liquid • SP/WSP-soluble packet • W/WP-wettable powder • WSB-water soluble bag

Remarks: Apply by ground only. For heavy disease, use high rate and short intervals. Do not make more than 6 applications/A per year. Do not apply through any type of irrigation system or through any ultra-low volume spray system. Do not plant other crops within 45 days after last application. Do not apply with adjuvant due to potential for phytotoxicity. Refer to label for use directions.
Days To Harvest: 7.

Signal Word/Toxicity Class: Caution.
REI: 12 hr.

Basicop — see copper compounds

BENLATE

Benlate (SP, WP) (benomyl) *(DuPont)*
Rate: 0.25-0.5 lb. ai/A
0.5-1 lb. SP, WP/A
Disease:

Cercospora leaf spot	leaf mold	target spot
gray mold	Phoma leaf spot	white mold

Time: Apply when disease first appears; repeat at 7- to 14-day intervals.
Remarks: Do not tank mix or alternate with benzimidazole or thiophanate products. Refer to label for use directions.
State Restrictions: Benlate SP not registered for use in California.

Signal Word/Toxicity Class: Caution.
REI: 24 hr.

Botran — see dicloran

Bravo — see chlorothalonil

Bravo S — see sulfur premix

Brom-O-Gas — see methyl bromide

CAPTAN

Gustafson Captan 30-DD *(Gustafson)*
Rate: 2.33 fl. oz. 30-DD, 300/100 lb. seed
Disease:

damping-off	seed decay	seedling blight
mold		

Time: Seed treatment.
Remarks: Do not combine colloidal suspensions, wettable powders, or emulsifiable concentrates unless previous experience assures compatibility. Use with EPA-approved dye that imparts unnatural color to seed. Do not use treated seed for food, feed, or oil purposes. Refer to label for use directions.

Signal Word/Toxicity Class: Danger/I.

Champ — see copper compounds

Champion — see copper compounds

CHLOROTHALONIL

Bravo 500, 720/Weather Stik, Ultrex (WDG) *(ZENECA)*
Echo 500, 720 or 90DF *(Sostram)*
Terranil (6L, 90DF or 90DF WSP) *(Terra)*
Rate: Foliage:
2-3 pt. 500/A
1.375-2 pt. 720, 6L/A
1.3-1.8 lb. WDG/A
1.2-1.7 lb. 90DF, 90DF WSP/A
Disease:

early blight	gray mold	Septoria leaf spot
gray leaf spot	late blight	target spot

Time: Apply every 7- to 10-days or when dew or rain occurs and disease threatens.

Rate: Fruit:
3-4 pt. 500/A
2-3 pt. 720, 6L/A
1.8-2.75 lb. WDG/A
1.7-2.3 lb. 90DF, 90DF WSP/A
Disease:

Alternaria fruit rot	Botrytis gray mold	late blight
Anthracnose	fruit rot	Rhizoctonia fruit rot

Time: Apply every 7- to 14-days beginning at fruit set or when dew or rain occurs and disease threatens.
Remarks: May be combined in spray tank with registered pesticide products that claim copper as the ai and are labeled for control of bacterial disease of tomatoes. Check copper manufacturer's label for specific instructions, precautions, and limitations. Do not use with Copper-Count-N in concentrated spray suspensions. Under severe conditions, use high rate and short interval. Use in sufficient water to obtain adequate coverage.

Signal Word/Toxicity Class: Danger/I (WDG, 90DF); Warning/II (6L, 500, 720).
REI: 48 hr.

Clean Crop C-O-C-S — see copper compounds

COPPER COMPOUNDS

Basicop (WP) (basic copper sulfate) *(Griffin)*
Champ (FL or Formula 2) or Champion (WP) (copper hydroxide) *(Agtrol)*
Rate: 1-2 lb. ai/A
2-4 lb. WP/A
2.66-5.33 pt. FL/A
1.33-2.66 pt. Formula 2/A
Disease:

Anthracnose	bacterial spot	late blight
bacterial canker	early blight	Septoria leaf spot

Time: Apply when disease appears; repeat at 7- to 10-day intervals.
Remarks: Use higher rate when disease is heavy. Do not apply through any type of irrigation system. Refer to label for use directions.

Signal Word/Toxicity Class: Danger/I (Champion); Warning/II (Basicop); Caution (Champ).
REI: 24 hr.

Clean Crop C-O-C-6 (WDG) (74.0% copper oxychloride & 14.2% copper sulfate, basic) *(Platte)*
Rate: 1-2 lb. ai/A
2-4 lb. WDG/A
Disease:

early blight	late blight	leaf spot

Time: Apply when disease first threatens; repeat at 7- to 10-day intervals to protect new growth throughout growing season.

Rate: 1-3 lb. ai/A
2-6 lb. WDG/A
Disease:
bacterial spot
Time: Apply every 5-7 days after true leaves appear.
Remarks: If late blight is severe, increase rate to 7 lb./A on large plants. Do not apply this product through any type of irrigation system. Refer to label for use directions.

Signal Word/Toxicity Class: Warning/II.
REI: 24 hr.

Copper-Count-N (L) (copper ammonium) *(Mineral Research)*
Rate: 3-6 pt. L/A
Disease:

bacterial speck	early blight	late blight
bacterial spot		

Time: Apply at 7- to 10-day intervals. In areas or seasons when disease is severe, spray at 4- to 5-day intervals.
Remarks: Applications may be made by ground or air; refer to label for use directions. Do not apply through any type of irrigation system unless supplemental labeling on chemigation is followed.

Signal Word/Toxicity Class: Caution.
REI: 12 hr.

7 Vegetables

NOTICE The information on these pages is for preliminary planning — not a guide for use. Be sure to follow the manufacturer's directions, notwithstanding information contained here. For personal protective equipment and EPA registration numbers, please turn to page 39.

Drexel Basic Copper Sulfate (WP) or KOP 300 (3FL) (basic copper sulfate) *(Drexel)*

Rate: 2-4 lb. WP/A
2.6-4 pt. 3FL/A

Disease:

Anthracnose	bacterial spot	nailhead rust
bacterial canker (Southeastern states)	early blight	Septoria leaf spot
	late blight	Stemphylium leaf spot

Time: Apply in seedbed or field before disease appears; repeat at 7- to 10-day intervals.

Remarks: Do not apply through any type of irrigation system. Refer to label for use directions.

Signal Word/Toxicity Class: Warning/II.
REI: 24 hr. (WP); 12 hr. (3FL).

Kocide (101, DF, 2000, 4.5LF, LF) (copper hydroxide) *(Griffin)*

Rate: 2-4 lb. 101, DF/A
1.5-3 lb. 2000/A
1.33-2.66 4.5LF/A
2.66-5.33 pt. LF/A

Disease:

Anthracnose	early blight	late blight
bacterial speck	gray mold	Septoria leaf spot
bacterial spot		

Time: Apply when disease first threatens; repeat every 7- to 10-days (10-30 days for bacterial speck).

Remarks: May be used in greenhouses and shadehouses; refer to label. Do not apply in spray solution having pH less than 6.5 as phytotoxicity may occur. Refer to label for use directions.

Days To Harvest: 5.

Signal Word/Toxicity Class: Danger/I (101, DF); Warning/II (2000); Caution (4.5LF, LF).
REI: 24 hr.

KOP-Hydroxide (FL, 50WP) (copper hydroxide) *(Drexel)*

Rate: 2.66-5.33 pt. FL/A

Disease:

bacterial speck	bacterial spot	early blight

Rate: 2-4 lb. 50WP/A

Disease:

Anthracnose	early blight	late blight
bacterial speck	gray mold	Septoria leaf spot
bacterial spot		

Time: Apply when disease threatens; repeat at 7- to 10-day intervals (10- to 30-day intervals for bacterial speck).

Remarks: Do not apply KOP-Hydroxide 50WP in spray solution having pH less than 6.5 or greater than 9.0 as phytotoxicity may occur. Refer to label for use directions.

Days To Harvest: 5.

Signal Word/Toxicity Class: Danger/I.
REI: 48 hr.

KOP Oxy-85 (WP) (copper oxychloride) *(Drexel)*

Rate: 4 lb. WP/A

Disease:

early blight	late blight	Septoria leaf spot

Time: Apply when disease is reported to be present in area; repeat at 7- to 10-day intervals as needed.

Rate: 5-6 lb. WP/A

Disease:
bacterial spot

Time: Seedbed: Apply every 5-6 days after first leaves appear. Field: Apply at 5- to 7-day intervals if disease is known to occur in area. During periods of rainy weather, use shorter intervals.

Remarks: If late blight infection is severe, increase rate to 7 lb./A for large plants. Use in sufficient water and thoroughly wet plants. Refer to label for use directions.

Rate: 4 lb. WP/100 gal. water

Disease:
damping-off

Time: Apply at 4- to 7-day intervals, beginning when plants break through the ground by spraying both plants and soil surrounding them.

Remarks: Cold frames, greenhouses. Use in sufficient water and thoroughly wet the plants. Refer to label for use directions.

Signal Word/Toxicity Class: Danger/I.
REI: 24 hr.

Micro Flo Basic Copper "53" (basic copper sulfate) or Microsperse COC (53WP) (copper oxychloride) *(Micro Flo)*

Rate: 2-4 lb. 53/A

Disease:

Anthracnose	early blight	Septoria leaf spot
bacterial canker	late blight	Stemphylium leaf spot
bacterial speck	leaf mold	
bacterial spot	nailhead rust	

Time: Apply in seedbed or field before disease appears; repeat at 7- to 10-day intervals as needed.

Rate: 3-4 lb. 53WP/A

Disease:

Anthracnose	early blight	Septoria leaf spot
bacterial speck	gray mold	
bacterial spot	late blight	

Time: Seedbeds: Apply to beds; repeat at 5- to 7-day intervals after first leaves appear. Field: Apply after transplanting or when disease is first expected; repeat at 4- to 7-day intervals.

Rate: Cold frames, greenhouses, etc.:
4-5 lb. 53WP/100 gal. water

Disease:
damping-off

Time: Begin when plants emerge; repeat at 4- to 7-day intervals until transplanting time.

Remarks: Refer to label for use directions.

Signal Word/Toxicity Class: Danger/I (53WP); Warning/II (53).
REI: 24 hr.

Nu-Cop (3L, 50DF, 50WP) (copper hydroxide) *(Micro Flo)*

Rate: 1.33-5.33 pt. 3L/A
2-3 lb. 50DF, 50WP/A

Disease:

Anthracnose	early blight	late blight (3L)
bacterial speck	gray leaf spot	Septoria leaf spot (3L)
bacterial spot	gray mold (3L)	

Time: Apply when disease threatens; repeat at 7- to 10-day intervals as needed.

Remarks: Refer to label for use directions.

Signal Word/Toxicity Class: Danger/I (50DF, 50WP); Warning/II (3L).
REI: 48 hr. (50DF, 50WP); 24 hr. (3L).

TOP COP Tri-Basic (FL) (basic copper sulfate) *(Stoller)*

Rate: 4 pt. FL/A

Disease:

Anthracnose	early blight	nailhead rust
bacterial canker	late blight	Septoria leaf spot
bacterial spot	leaf mold	Stemphylium leaf spot

Time: Begin in seedbed or field before disease appears; repeat at 7- to 10-day intervals.

Remarks: Refer to label for use directions.

Signal Word/Toxicity Class: Caution.
REI: 24 hr.

Copper-Count-N — see copper compounds

DICLORAN

Botran 6 Dust *(Wilbur-Ellis)*

Rate: 1.8 lb. ai/A
30 lb. Dust/A

Disease:
stem canker

Time: Spray stem from ground level up to height of 18-24"; repeat weekly as necessary.

Remarks: Greenhouse. Seedlings or newly set transplants may be injured by drenching. Refer to label for use directions.

Signal Word/Toxicity Class: Caution.
REI: 12 hr.

◆ - **new product** • ai-active ingredient • DF-dry flowable • E/EC-emulsifiable concentrate • EW-emulsion in water
F/FL-flowable • G-granule • L/LC-liquid • SP/WSP-soluble packet • W/WP-wettable powder • WSB-water soluble bag

Botran 75-W *(Gowan)*
Rate: 0.75 lb. ai/A
1 lb. 75-W/A
Disease:
stem canker
Time: Spray stem from ground level up to height of 18-24"; repeat weekly as necessary.
Remarks: Greenhouse/hothouse. Seedlings or newly set transplants may be injured by drenching. Refer to label for use directions.
Days To Harvest: 14.

Signal Word/Toxicity Class: Caution.
REI: 12 hr.

Dithane — see mancozeb

Echo — see chlorothalonil

FOSETYL-ALUMINUM

Aliette (WDG, WSP) *(Rhone-Poulenc)*
Rate: 2-4 lb. ai/A
2.5-5 lb. WDG, WSP/A
Disease:
damping-off   root and stem rot
Time: Apply at 2-4 leaf stage growth for direct seeded or immediately after transplanting to field; repeat at 7- to 14-day intervals.
Remarks: Apply in sufficient water for good coverage. Do not apply in less than 10 gal./A. Do not apply more than 20 lb./A per season. With moderate disease pressure, use high rate and shorter interval. Do not tank mix with copper compounds. Refer to label for use directions.
Days To Harvest: 14.
State Restrictions: Not for use in certain counties in Alabama, California, Illinois, Indiana, Iowa, Kentucky, Louisiana, Maryland, Missouri, North Carolina, Ohio, and Tennessee; refer to label.

Signal Word/Toxicity Class: Caution.
REI: 12 hr.

Kocide — see copper compounds

KOP-Hydroxide — see copper compounds

KOP Oxy — see copper compounds

MANCOZEB

Dithane (DF, F-45, M-45, ST, WSP) *(Rohm and Haas)*
Manzate 200DF *(Griffin)*
Penncozeb 75DF or 80WP *(Elf Atochem)*
Rate: East of the Mississippi:
1.5-3 lb. DF, M-45, WSP, 200DF/A
2.4-4.8 pt. F-45/A
0.75-3 lb. 75DF/A
0.75-3 lb. 80WP/A
West of the Mississippi:
1.5-2 lb. DF, M-45, WSP/A
2.4-3.2 pt. F-45/A
1-2 lb. 200DF/A
0.75-2 lb. 75DF/A
0.75-2 lb. 80WP/A
Disease:
Anthracnose   early blight   leaf mold
bacterial speck (Dithane)   gray leaf spot   Septoria leaf spot
bacterial spot (Dithane)   late blight
Time: Apply when seedlings emerge or transplants are set; repeat at 7- to 10-day intervals throughout season.
Remarks: For bacterial speck and spot, tank mix with fixed copper fungicide; follow application intervals recommended on label. Refer to label for use directions. Do not apply more than 16.8 lb./A per crop east of the Mississippi River; 6.4 lb./A per crop west of the Mississippi River.
Days To Harvest: 5.

Rate: 8.5 oz. DF/100 lb. seed
8 oz. M-45, WSP, 200DF/100 lb. seed
12.8 fl. oz. F-45, ST/100 lb. seed
Disease:
damping-off   seed rot   seedling blight
Time: Seed treatment.
Remarks: Use of EPA-approved dye is recommended to impart unnatural color to seed. Do not use treated seed for food, feed, or oil purposes. Refer to label for use directions.

Signal Word/Toxicity Class: Caution.
REI: 24 hr.

MANEB

Maneb 75DF or 80WP *(Elf Atochem)*
Manex (FL) *(Griffin)*
Rate: East of the Mississippi:
1.5-3 lb. 75DF, 80WP/A
2.4-4.8 pt. FL/A
West of the Mississippi:
1.5-2 lb. 75DF, 80WP/A
2.4-3.2 pt. FL/A
Disease:
Anthracnose   early blight   late blight
Cladosporium leaf mold   gray leaf spot   Septoria leaf spot
Time: East of Mississippi: Apply at first fruit cluster; West of Mississippi: Apply when first fruit forms. Repeat at 7- to 10-day interval.
Remarks: Field and greenhouse. Use high rate under severe pressure. Do not use any other product containing a different EBDC ai during the same growing season if using Manex. Refer to label for use directions.
Days To Harvest: 5.

Signal Word/Toxicity Class: Caution.
REI: 24 hr.

Manex — see maneb

Manzate — see mancozeb

MEFENOXAM

Ridomil Gold EC, GR, or WSP *(Novartis)*
Rate: 1-2 pt. EC/A
20 lb. GR/A
1-2 lb. WSP/A
Disease:
damping-off   fruit rot   root and stem rot
Time: Apply 4-6 weeks after planting; 1-2 applications up to 4 weeks before harvest.
Remarks: For banded application, 7" band recommended. Do not use more than 3 pt. EC/A, 40 lb. GR/A, or 3 lb. WSP/A per season. Not for use in greenhouse or on field grown vegetable bedding plants. Do not plant any crop which is not registered for use with mefenoxam in treated soil for 12 months.
Days To Harvest: 1.

Signal Word/Toxicity Class: Warning/II (EC); Caution (GR, WSP).
REI: 48 hr.

MEFENOXAM (Premixes)

Ridomil Gold/Bravo (WP) (4.5% mefenoxam & 72% chlorothalonil) *(Novartis)*
Rate: 1.5 lb. ai/A
2 lb. WP/A
Disease:
early blight   gray leaf spot   late blight
fruit rot   gray mold   Septoria leaf spot
Time: Apply when conditions are favorable for disease development, but before infection.

Rate: 1.5-2.3 lb. ai/A
2-3 lb. WP/A
Disease:
Alternaria fruit rot   Botrytis gray mold   Rhizoctonia fruit rot
Antracnose

NOTICE	The information on these pages is for preliminary planning — not a guide for use. Be sure to follow the manufacturer's directions, notwithstanding information contained here.	For personal protective equipment and EPA registration numbers, please turn to page 39.

7 Vegetables

Time: Apply when conditions are favorable for disease development, but before infection.
Remarks: Up to 3 applications at 14-day intervals may be applied; (4 applications in Florida). The labeled rate of protectant fungicide should be applied. Do not plant any crop which is not registered for use with mefenoxam in treated soil for 12 months.
Days To Harvest: 14.

Signal Word/Toxicity Class: Danger/I.
REI: 48 hr.

Ridomil Gold/Copper (WSB) (5% mefenoxam & 60% copper hydroxide) *(Novartis)*
Rate: 1.3 lb. ai/A
2 lb. WSB/A
Disease:
late blight
Phytophthora fruit rot
Time: Apply when conditions are favorable for disease development (before infection), but no later than when foliage meets within row. Make up to 3 applications at 14-day intervals.
Remarks: Tank mix with 0.8 lb. ai/A of maneb, mancozeb, metiram, or chlorothalonil. Apply up to 3 applications per crop. Labeled rate of protectant fungicide should be applied between applications. If conditions for late blight development are still favorable after 3 applications, use other fungicides. Do not exceed 1.5 lb. mefenoxam/A per crop. Do not plant any crop which is not registered for use with mefenoxam in treated soil for 12 months.
Days To Harvest: 14.

Signal Word/Toxicity Class: Danger/I.
REI: 48 hr.

Ridomil Gold MZ (WP) (4% mefenoxam & 64% mancozeb) *(Novartis)*
Rate: 1.4 lb. ai/A
2 lb. WP/A
Disease:
late blight
Time: Apply when conditions are favorable for disease development (before infection). Make up to 3 applications at 14-day intervals.
Remarks: If conditions for late blight development are still favorable after 3 applications, use other fungicides for late blight control. If other registered fungicides containing EBDC are used during growing season, do not apply more than 16.8 lb. EBDC/A. If tank mixed with other EBDC products, apply no more than 1.6 lb. EBDC west of the Mississippi; 2.4 lb. east of the Mississippi. Do not plant any crop which is not registered for use with mefenoxam in treated soil for 12 months.
Days To Harvest: 5.

Signal Word/Toxicity Class: Caution.
REI: 48 hr.

METHYL BROMIDE

2% Chloropicrin added as warning agent. Restricted Use Pesticide. Classified as an ozone depleter under the Clean Air Act. Use is to be completely eliminated starting Jan. 1, 2001.

Brom-O-Gas *(Great Lakes Chemical)*
Rate: Injected with chisels + covered with tarp:
180-240 lb./A
Raised tarp with topical application:
0.4-0.55 lb./100 ft.
Disease:
Fusarium
oak root fungus
Phytophthora
Plasmodiophora
Pythium
Rhizoctonia
Sclerotinia
Sclerotium
Time: Preplant.
Remarks: Do not fumigate when soil temperature is below 50°F. For best results, fumigate when temperature is 60°-80°F at depth of 6". Refer to label for use directions.
State Restrictions: Brom-O-Gas not for use in Arizona and not for use in California for oak root fungus (Armillaria root rot) control.

Signal Word/Toxicity Class: Danger/I.

Micro Flo — see copper compounds

Microsperse COC — see copper compounds

Microthiol Special — see sulfur

Nu-Cop — see copper compounds

PCNB

PCNB 75DG or 75WP *(Amvac)*
Rate: Transplants:
2.25-3.75 lb. ai/100 gal. water
3-5 lb. 75DG, 75WP/100 gal. water
In-furrow:
7.5 lb. ai/100 gal. water
10 lb. 75DG, 75WP/100 gal. water
Disease:
southern blight
Time: For transplant, apply at time of transplanting; in-furrow just prior to planting.
Remarks: Use 72" row spacings for flat tomatoes, 48" for staked tomatoes. Set plows as flat as possible to avoid getting untreated soil against stems of plants. Do not apply through any type of irrigation system. Do not plant root crops within 12 months of application unless PCNB is registered for use on those crops. Refer to label for use directions.
State Restrictions: Not for use in California.

Signal Word/Toxicity Class: Caution.
REI: 12 hr.

Terraclor (75WP) or Terraclor 400 (FL) *(Uniroyal)*
Rate: 4-8 oz. 75WP/100 gal. water
6-12 fl. oz. FL/100 gal. water
Disease:
damping off
root and stem rot
Time: Soil drench. One repeat application may be made 4-6 weeks later if necessary.
Remarks: Bedding plants in greenhouse or nursery. Refer to label for use directions.
Crop Rotation: Do not plant root crops in PCNB treated fields within 12 months of broadcast and banding applications unless PCNB registered for use on those crops.

Signal Word/Toxicity Class: Caution.

Turfcide (10G) *(Uniroyal)*
Rate: 1.5 oz. 10G/100 ft. row
Disease:
root and stem rot

Rate: 11 oz. 10G/100 sq. ft.
Disease:
club root
Time: Apply at planting time.
Remarks: Direct material into planting row and cover soil. Refer to label for use directions.
Crop Rotation: Do not plant root crops in PCNB treated fields within 12 months of broadcast and banding applications unless PCNB registered for use on those crops.

Signal Word/Toxicity Class: Caution.

Penncozeb — see mancozeb

Quadris — see azoxystrobin

Ridomil Gold — see mefenoxam

Ridomil Gold/Bravo — see mefenoxam premixes

Ridomil Gold/Copper — see mefenoxam premixes

◆ - **new product** • ai-active ingredient • DF-dry flowable • E/EC-emulsifiable concentrate • EW-emulsion in water
F/FL-flowable • G-granule • L/LC-liquid • SP/WSP-soluble packet • W/WP-wettable powder • WSB-water soluble bag

Ridomil Gold MZ — see mefenoxam premixes

Sulfa — see sulfur

SULFUR

Micro Flo Sulfur (6L) *(Micro Flo)*
Terra Sulfur (6L) *(Terra)*
Rate: 3-8 lb. ai/A
4-10.66 pt. 6L/A
Disease:
powdery mildew
Time: Apply as necessary.
Remarks: Do not use within 2 weeks of oil application. Refer to label for use directions.

Signal Word/Toxicity Class: Caution.
REI: 24 hr.

Microthiol Special (80WP) *(Elf Atochem)*
Thiolux (DF) *(Platte)*
Rate: 4-8 lb. ai/A
5-10 lb. 80WP/A
4.8-16 lb. ai/A
6-20 lb. DF/A
Disease:
powdery mildew
Time: Begin application when symptoms first appear.
Remarks: Thorough coverage is required. Do not make applications within 2 weeks of oil spray. Refer to label for use directions.

Signal Word/Toxicity Class: Caution.
REI: 24 hr.

Sulfa 80 (80DF) *(Drexel)*
Rate: 2-8 lb. ai/A
3-10 lb. 80DF/A
Disease:
powdery mildew
Time: Apply when disease first appears; repeat at 3- to 4-week intervals.
Remarks: Do not use within 2 weeks of oil application. Do not apply this product through any type of irrigation system. Refer to label for use directions.

Signal Word/Toxicity Class: Caution.
REI: 24 hr.

SULFUR (Premixes)

Bravo S (27.25% sulfur & 19.15% chlorothalonil) *(Helena)*
Rate: 4.5-8.5 pt./A
Disease:

Anthracnose	gray leaf spot	powdery mildew
Botrytis gray mold	gray mold	Rhizoctonia fruit rot
early blight	late blight	Septoria leaf spot

Time: Apply when disease threatens; repeat at 7- to 10-day intervals as needed.
Remarks: Under severe conditions, use high rate and short interval. Refer to label for use directions.

Signal Word/Toxicity Class: Caution.
REI: 48 hr.

Terranil S (27.25% sulfur & 19.15% chlorothalonil) *(Terra)*
Rate: Foliage:
4.5-5.5 pt. S/A
Disease:

early blight	gray mold	Septoria leaf spot
gray leaf spot	late blight	target spot

Time: Apply every 7- to 10-days when dew or rain occurs and disease threatens.

Rate: Fruit:
5-8.5 pt. S/A
Disease:

Alternaria fruit rot	Botrytis fruit rot	Botrytis late blight
Anthracnose	Botrytis gray mold	Rhizoctonia fruit rot

Time: Apply every 7- to 14-days beginning at fruit set or when dew or rain occurs and disease threatens.
Remarks: Use high rate and shortest interval when disease conditions are severe. Use in sufficient water to obtain adequate coverage. Do not apply more than 3.6 pt./A during growing season. Refer to label for use directions.
Days To Harvest: 7.

Signal Word/Toxicity Class: Warning/II.
REI: 48 hr.

Top Cop with Sulfur (6.25 lb./gal. sulfur & 1 lb./gal. tribasic copper sulfate) *(Stoller)*
Rate: 2-3 qt./A
Disease:

Anthracnose	downy mildew	powdery mildew
bacterial speck	early blight	Septoria blight
bacterial spot	late blight	

Time: Apply when disease threatens; repeat at 5- to 10-day intervals.
Remarks: Refer to label for use directions.

Signal Word/Toxicity Class: Caution.
REI: 24 hr.

Terraclor — see PCNB

Terranil — see chlorothalonil

Terranil S — see sulfur premixes

TERRANIL Cu (Premix)

Terranil Cu (24% chlorothalonil & 24% copper oxychloride) *(Terra)*
Rate: Foliage:
3-4.5 pt. Cu/A
Disease:

early blight	gray mold	Septoria leaf spot
gray leaf spot	late blight	target spot

Time: Apply every 7- to 10-days when dew or rain occurs and disease threatens.

Rate: Fruit:
4.5-6.75 pt. Cu/A
Disease:

Alternaria fruit rot	Botrytis fruit rot	Botrytis late blight
Anthracnose	Botrytis gray mold	Rhizoctonia fruit rot

Time: Apply every 7- to 14-days beginning at fruit set or when dew or rain occurs and disease threatens.
Remarks: Use high rate and shortest interval when disease conditions are severe. Use in sufficient water to obtain adequate coverage. Do not apply more than 3.6 pt./A during growing season. Refer to label for use directions.
Days To Harvest: 7.

Signal Word/Toxicity Class: Danger/I.
REI: 48 hr.

TERR-O-GAS (Premix)

Classified as an ozone depleter under the Clean Air Act. Use to be eliminated starting Jan. 1, 2001. Restricted Use Pesticide.

Terr-O-Gas (67% methyl bromide & 33% chloropicrin) *(Great Lakes Chemical)*
Rate: Nontarp:
270-355 lb./A
Disease:

Fusarium	Pythium	Sclerotinia
Phytophthora	Rhizoctonia	Sclerotium
Plasmodiophora		

Time: Apply up to 2 weeks prior to planting when soil moisture is good and temperature is 60°-85°F.
Remarks: Work upwind of product and never fumigate without 2 qualified applicators. Do not fumigate when there is no or little air movement. Tarping is not used after field application. Refer to label for use directions.

Signal Word/Toxicity Class: Danger/I.

7 Vegetables

NOTICE The information on these pages is for preliminary planning — not a guide for use. Be sure to follow the manufacturer's directions, notwithstanding information contained here. For personal protective equipment and EPA registration numbers, please turn to page 39.

Thiolux — see sulfur

THIRAM

42-S Thiram or Thiram 50WP Dyed *(Gustafson)*
Thiram Granuflo (WDG) *(UCB Agrochemicals)*

Rate: 6 fl. oz. 42-S/100 lb. seed
6 oz. 50WP/100 lb. seed
4 oz. WDG/100 lb. seed

Disease:
damping-off — seed decay — seedling blight

Time: Seed treatment.

Remarks: Use with EPA-approved dye that imparts unnatural color to seed. Do not use treated seed for food, feed, or oil purposes. Do not apply Thiram Granuflo through any type of irrigation system. Refer to label for use directions.

State Restrictions: Thiram Granuflo not for use in California.

- - - - - - - - - - - - - - - - -

Signal Word/Toxicity Class: Caution.
REI: 24 hr. (Thiram Granuflo).

TOP COP — see copper compounds

Top Cop with Sulfur — see sulfur premix

Turfcide — see PCNB

ZIRAM

Ziram 4L or 76 *(Drexel)*

Rate: 1 lb. ai/100 gal.
2 pt. 4L/100 gal.
1.1-1.5 lb. ai/100 gal.
1.5-2 lb. 76/100 gal.

Disease:
Anthracnose — Botrytis gray mold (greenhouse) — early blight — Septoria leaf spot

Time: Anthracnose, early blight, Septoria leaf spot: Apply when plants are 2-6" tall; repeat at 7- to 10-day intervals as needed. Botrytis gray mold: Apply before disease appears; repeat at 3- to 10-day intervals as needed.

Remarks: Do not apply this product through any type of irrigation system. Refer to label for use directions.

State Restrictions: Not for use in California.

- - - - - - - - - - - - - - - - -

Signal Word/Toxicity Class: Danger/I.
REI: 48 hr.

Ziram 76DF *(Elf Atochem)*
Ziram Granuflo (WDG) *(UCB Agrochemicals)*

Rate: 2.3-3 lb. ai/A
3-4 lb. 76DF, WDG/A

Disease:
Anthracnose — early blight — Septoria leaf spot

Time: Begin applications at first sign of infection; repeat at 7- to 14-day intervals. Use shorter spray interval under high disease pressure.

Remarks: Do not use on cherry tomatoes. Do not apply this product through any type of irrigation system. Do not graze or feed cover crops from treated orchards. Refer to label for use directions.

Days To Harvest: 7.

State Restrictions: Not for use in California.

- - - - - - - - - - - - - - - - -

Signal Word/Toxicity Class: Danger/I.
REI: 48 hr.

Watercress

COPPER COMPOUNDS

Kocide (101, DF, 2000, 4.5LF, LF) (copper hydroxide) *(Griffin)*
Nu-Cop (3L) (copper hydroxide) *(Micro Flo)*

Rate: 2 lb. 101, DF/A
1.5 lb. 2000/A
1.33 pt. 4.5LF/A
4 pt. LF/A
2.66 pt. 3L/A

Disease:
Cercospora leaf spot

Time: Apply when plants are first established in field; repeat at 7- to 14-day intervals.

Remarks: Do not apply Kocide in a spray solution having a pH less than 6.5 as phytotoxicity may occur. Refer to label for use directions.

State Restrictions: Kocide 101 is not for use in California.

- - - - - - - - - - - - - - - - -

Signal Word/Toxicity Class: Danger/I (101, DF); Warning/II (2000, 3L); Caution (4.5LF, LF).
REI: *24* hr.

Kocide — see copper compounds

Nu-Cop — see copper compounds

◆ - **new product** • ai-active ingredient • DF-dry flowable • E/EC-emulsifiable concentrate • EW-emulsion in water
F/FL-flowable • G-granule • L/LC-liquid • SP/WSP-soluble packet • W/WP-wettable powder • WSB-water soluble bag

Section 8

Fruits & Nuts

Bt Chart 380

Insecticide Listings

Small Fruits. 381

Tree Fruits, Citrus 392

Tree Fruits, Deciduous . . . 401

Tree Fruits, Subtropical . . 418

Nuts 423

Fungicide Listings

Small Fruits. 433

Tree Fruits, Citrus 446

Tree Fruits, Deciduous . . . 452

Tree Fruits, Subtropical . . 470

Nuts 477

The information in this 1999 Insect and Disease Control Guide is updated to August 15, 1998.
Because manufacturers' product recommendations are constantly being updated and revised as additional product registrations and label changes are approved by EPA, there may be changes after this date. However, every effort has been made to give manufacturers the opportunity to include in the Guide their latest information. This Guide is for preliminary planning – not a use guide. Be sure to follow label directions, notwithstanding information contained herein.

Bt Products Chart

Fruits & Nuts

Find Bt products quickly and easily with this Bt products chart. Simply find the insect you want to control on the top of the chart, and follow down to the marked spot to find the Bt product you can use to control that pest.

	Amorbia moth	Armyworms	Banana skipper	Cankerworms	Caterpillars	Codling moth	Cutworms	Diamondback moth	Fruitworms	Grape berry moth	Grape leaffolder	Grapeleaf skeletonizer	Gypsy moth	Hornworms	Leafrollers	Loopers	Navel orangeworm	Orange tortix	Orangedog	Oriental fruit moth	Peach twig borer	Spanworm	Tobacco budworm	Tufted apple bud moth	Walnut caterpiller	Webworms
Agree		■					■						■		■		■		■			■		■		
Biobit	■	■	■	■	■	■	■		■	■	■	■	■	■	■	■	■	■	■		■	■	■	■	■	■
BMP 123	■	■	■	■	■	■	■			■		■	■	■	■	■	■	■	■		■	■	■	■	■	■
Bt. ssp. *aizawai*	■	■	■	■	■	■	■			■		■	■	■	■	■	■	■	■	■		■	■	■	■	■
Bt. ssp. *aizawai* + *Bt.* ssp. *kurstaki*		■			■	■	■	■				■			■	■					■					
Bt. ssp. *kurstaki*	■	■	■	■	■	■	■	■	■	■	■	■	■	■	■	■	■	■	■	■	■	■	■	■	■	■
Condor	■	■	■	■	■		■		■	■	■	■	■	■	■	■	■	■	■	■	■	■	■	■	■	■
Crymax	■	■	■	■	■		■		■	■	■	■	■	■	■	■	■	■	■	■	■	■	■	■	■	■
Cutlass	■	■	■	■	■		■		■	■	■	■	■	■	■	■	■	■	■	■	■	■	■	■	■	■
DiPel	■	■	■	■	■	■	■		■	■			■	■	■	■	■	■	■	■	■	■	■	■	■	■
Javelin	■	■	■		■	■	■		■		■	■	■		■	■	■	■	■	■	■	■		■		■
Mattch		■			■	■	■	■				■			■	■					■					
MVP II		■			■	■	■	■				■			■	■					■					
Troy-Bt				■	■		■	■	■				■	■	■	■						■				
XenTari	■	■	■	■	■	■	■			■		■	■	■	■	■		■	■	■		■	■	■	■	■

Manufacturers List

These Bt products are manufactured by the following companies. Look for their addresses, phone numbers, and fax numbers in the Suppliers Directory beginning on page 599.

Agree, Javelin – Thermo Trilogy
Biobit, DiPel, XenTari
– Abbott Laboratories
BMP 123 – Becker Microbial
Condor, Crymax, Cutlass – Ecogen
Mattch, MVP II – Mycogen
Troy-Bt – Troy Biosciences

Fruits & Nuts

Insecticides

Small Fruits

(Blueberries, Caneberries, Cranberries, Currants, Gooseberries, Grapes, Strawberries)

455 Oil — see petroleum oil

ACEPHATE

Organophosphate systemic insecticide with contact and stomach poison activity. Effective against many aphids, thrips, caterpillars, and plant bugs. Has moderate residual effects and is compatible with most pesticides.

Orthene 75S *(Valent)*

Rate: 1 lb. ai/A
1.33 lb. 75S/A

Pest:

cranberry blossomworm	fireworms	spanworms
false armyworm	gypsy moth	Sparganothis

Remarks: Cranberries. Apply in water by air, ground, or with sprinklers. Do not apply more than 1 application per growing season. Do not apply from start of bloom until all berries set.

Days To Harvest: 90.

- - - - - - - - - - - - - - - - -

Signal Word/Toxicity Class: Caution.
REI: 24 hr.

Ambush — see permethrin

AZADIRACHTIN

Extracted from seed of neem tree. Insect growth regulator that controls targeted insect larvae when they ingest or come in contact with it, by interfering with the insect's ability to molt and shutting off feeding behavior. Affected larvae slowly starve and die at next molt.

◆ **Azatin XL Plus** *(Thermo Trilogy)*

Rate: 5-21 oz. XL Plus/A

Pest:

aphids	grapeleaf skeletonizer	psyllids
armyworms	green peach aphid	rose chafer
beet armyworm	gypsy moth	thrips
beetles	Japanese beetle	tussock moth
black vine weevil	June beetle	webworms
caterpillars	leafhoppers	weevils
cutworms	leafminers	whiteflies
flea beetles	leafrollers	yellowstriped armyworm
fruit flies	mealybugs	

Remarks: Blueberries, caneberries (blackberries, boysenberries, dewberries, elderberries, loganberries, raspberries, youngberries), currants, gooseberries, huckleberries, grapes, strawberries. Rate varies by species. Apply when pests first appear; repeat at 7-day intervals. When pest populations are high, use higher rates. Do not apply more that 21 fl. oz./A.

Days To Harvest: 0.

- - - - - - - - - - - - - - - - -

Signal Word/Toxicity Class: Caution.
REI: 4 hr.

◆ **NEEMAZAD 4.5EC or NEEMIX 4.5** *(Thermo Trilogy)*

Rate: 0.125-1 pt. 4.5/A
High Volume:
2.25-4.5 oz. 4.5EC/100 gal. water
Low Volume:
0.5 pt. 4.5EC/A

Pest:

armyworms	grape leaffolder	loopers
blueberry leafroller	grapeleaf skeletonizer	obliquebanded leafroller
caterpillars	green fruitworm	omnivorous leafroller
cherry fruitworm	gypsy moth	thrips
fruittree leafroller	hornworm	tobacco budworm
grape berry moth		

Remarks: Blueberries, caneberries (blackberries, boysenberries, dewberries, loganberries, raspberries, youngberries), cranberries, currants, gooseberries, huckleberries, grapes, strawberries. For low volume, apply in sufficient water for adequate coverage. For low insect pressure, apply preventive at 7-day intervals or at first sign of insects; high insect pressure, apply at 3- to 4-day intervals.

- - - - - - - - - - - - - - - - -

Signal Word/Toxicity Class: Warning/II.
REI: 12 hr.

Azatin — see azadirachtin

AZINPHOS-METHYL

Restricted Use Pesticide (except Guthion Solupak).

Gowan Azinphos-M 50-W *(Gowan)*
Guthion 2L, Solupak (50WSP) *(Bayer)*
◆ **Micro Flo Azinphosmethyl 2EC or 50W** *(Micro Flo)*
Sniper 2-E, 50 PVA, 50W *(Platte)*

Rate: 0.75-1 lb. ai/A
3-4 pt. 2EC, 2L/A
1.5-2 lb. 50-W, 50WSP/A
0.25-0.5 lb. ai/A
1-2 pt 2EC, 2-E/A
0.5-1 lb. 50PVA, 50W/A

Pest:

grape berry moth	grapeleaf skeletonizer	mites (except Gowan)
grape cane girdler	leafhoppers	redbanded leafroller
grape mealybug		thrips

Remarks: Grapes. Apply as full coverage spray. Do not apply more than 3 times per season. Allow at least 14 days between applications (except 2-E, 50W).

Days To Harvest: Minimum rate-0; higher rate-10.
State Restrictions: Sniper 50PVA not registered in California.

- - - - - - - - - - - - - - - - -

Rate: 0.25-2 lb. ai/A
1.25-8 pt. 2EC, 2-E/A
0.5-4 lb. 50-W, 50PVA, 50W/A

Pest:

aphids	leafminers	obscure root weevil
leafhoppers	leafroller	raspberry crown borer

Remarks: Caneberries (blackberries, boysenberries, loganberries, raspberries). Apply as full coverage spray; repeat as needed. See label for use directions in special cases.

Days To Harvest: 14; borer and root weevil control-3.
State Restrictions: Sniper 50PVA not registered in California.

- - - - - - - - - - - - - - - - -

NOTICE	The information on these pages is for preliminary planning — not a guide for use. Be sure to follow the manufacturer's directions, notwithstanding information contained here.	For personal protective equipment and EPA registration numbers, please turn to page 39.

Rate: 0.5-0.75 lb. ai/A
2-3 pt. 2EC, 2-E, 2L/A
1-1.5 lb. 50-W, 50PVA, 50W, 50WSP/A

Pest:
blueberry maggot, lecanium scale, plum curculio, fruitworms

Remarks: Blueberries. Apply by air or ground using sufficient water for good coverage. Do not apply more than 3 times per season.
Days to Harvest: 7 (2-E-3; 50W-14).
State Restrictions: For use in Eastern and North Central states only.

- - - - - - - - - - - - - - - - -

Rate: 0.5-1 lb. ai/A
2-4 pt. 2EC, 2-E/A
1-2 lb. 50-W, 50PVA, 50W, 50WSP/A

Pest:
cranberry fruitworm, *Sparganothis sulfureana*, tipworm, fireworms

Remarks: Cranberries. Apply high rate for fireworms. Do not apply more than 3 times per season. Allow at least 14 days between applications.
Days To Harvest: 21.
State Restrictions: Sniper 50PVA not registered in California.

- - - - - - - - - - - - - - - - -

Rate: 0.5 lb. ai/A
2 pt. 2EC, 2-E, 2L/A
1 lb. 50-W, 50PVA, 50W, 50WSP/A

Pest:
aphids, obscure root weevil, strawberry clipper, meadow spittlebug, omnivorous leaftier, strawberry leafroller, obliquebanded leafroller, pea leaf weevil, whiteflies, small black weevil

Remarks: Strawberries. Apply as full coverage spray. Do not apply more than 4 times per season. Allow at least 5 days between applications.
Days To Harvest: 5.
State Restrictions: Sniper 50PVA not registered in California.

- - - - - - - - - - - - - - - - -

Signal Word/Toxicity Class: Danger/I.
REI: 48 hr.; 72 hr. if average rainfall less than 25" per year.

BT 320 SULFUR 25 or 50 DUST (Premix)

Bt 320 Sulfur 25 Dust (0.064% Bt & 25% sulfur) or Bt 320 Sulfur 50 Dust (0.064% Bt & 50% sulfur) *(Wilbur-Ellis)*

Rate: 20-50 lb. Dust/A
Pest:
grape leaffolder, omnivorous leaf roller (CA)

Remarks: Grapes. Apply by ground equipment as soon as possible after hatching and before larvae are protected by leaves.

- - - - - - - - - - - - - - - - -

Signal Word/Toxicity Class: Caution.
REI: 24 hr.

CALCIUM POLYSULFIDE

Lime-Sulfur Solution *(Best Sulfur)*

Rate: 4-10 gal. L/A
Pest:
mealybugs

Remarks: Grapes. Apply during dormant period in sufficient water for coverage.

- - - - - - - - - - - - - - - - -

Rate: 8 gal. L/100 gal. water
Pest:
blackberry mite, rose scale, San Jose scale, oyster scale

Remarks: Caneberries. Fall application, apply after old canes have been removed from the field. Delayed dormant application, apply when leaf buds first start to open. Do not apply through any type of irrigation system.

- - - - - - - - - - - - - - - - -

Signal Word/Toxicity Class: Danger/I.
REI: 48 hr.

Sulforix *(Best Sulfur)*

Rate: 3 gal. L/100 gal. water
Pest:
red berry mite

Remarks: Caneberries. Apply as delayed dormant spray. Do not apply through any type of irrigation system.

- - - - - - - - - - - - - - - - -

Signal Word/Toxicity Class: Danger/I.
REI: 48 hr.

CARBARYL

Carbamate insecticide highly toxic to bees.

Clean Crop Sevin 5 Bait *(Platte)*

Rate: Grapes:
2-2.4 lb. ai/A
40-48 lb. Bait/A
Strawberries:
2 lb. ai/A
40 lb. Bait/A

Pest:
ants, darkling beetles, grasshoppers, armyworms, field crickets, sowbugs, cutworms

Remarks: Grapes, strawberries. Apply with ground or air equipment.
Days To Harvest: Grapes-0; strawberries-1.

- - - - - - - - - - - - - - - - -

Signal Word/Toxicity Class: Caution/III.
REI: 12 hr.

Drexel Carbaryl 4L, 50W, or 80S *(Drexel)*
Sevin 4F, XLR Plus 80S or 80WSP *(Rhone-Poulenc)*

Rate: 1-2 lb. ai/A
2-4 pt. 4L, 4F, XLR/A
2-4 lb. 50W/A
1.25-2.5 lb. 80S, 80WSP/A

Pest:
cutworms, grape leafhopper, rose chafer, eightspotted forester, Japanese beetle, saltmarsh caterpillar, European fruit lecanium, June beetle, snowy tree cricket, leafhoppers, western grapeleaf skeletonizer, European raspberry aphid, leafrollers, omnivorous leafroller, western yellowstriped armyworm, grape berry moth, orange tortrix, grape leaffolder, redbanded leafroller

Remarks. Caneberries (blackberries, boysenberries, dewberries, logan berries, raspberries), grapes. Apply before first brood leaffolder larvae emerge from rolls. Rate varies by pest species, refer to label.
Days To Harvest: 7.
State Restrictions: California only for European raspberry aphid and omnivorous leafroller (high rate) on berries.

- - - - - - - - - - - - - - - - -

Rate: 3 lb. ai/A
6 pt. 4L/A
1.5-2 lb. ai/A
3-4 pt. 4F, XLR/A
3-4 lb. 50W/A
1.5 lb. ai/A
1.875 lb. 80S, 80WSP/100 gal. water (100-300 gal. dilute spray/A)

Pest:
blueberry maggot, European fruit lecanium, Japanese beetle, cherry fruitworm, cranberry fruitworm

Remarks: Blueberries. Apply 3 weeks before harvest; repeat as necessary.
Days To Harvest: 0.

- - - - - - - - - - - - - - - - -

Rate: 1.5-3 lb. ai/A
3-6 pt. 4L, 4F, XLR/A
3-6 lb. 50W/A
1.875-3.75 lb. 80S, 80WSP/100 gal. water (100-300 gal. dilute spray/A)

Pest:
cranberry fireworm, cutworms, leafhoppers, cranberry fruitworm, elm spanworm, rose chafer, cranberry twig girdler, Japanese beetle

Remarks: Cranberries. Apply late bloom and at 7- to 10-day intervals as needed. May kill shrimp and crabs. Do not use in areas where these are important resources.
Days To Harvest: 1.

- - - - - - - - - - - - - - - - -

Rate: 1-2 lb. ai/A
2-4 pt. 4L, 4F, XLR/A
2-4 lb. 50W/A
1.25-2.5 lb. 80S, 80WSP/100 gal. water (100-300 gal. dilute spray/A)

Pest:
flea beetles, strawberry bud weevil, strawberry fruitworm, meadow spittlebug, strawberry leafroller, omnivorous leaftier, strawberry clipper, strawberry weevil

Remarks: Strawberries. May injure Early Dawn and Sunrise varieties.
Days To Harvest: 1.

- - - - - - - - - - - - - - - - -

Signal Word/Toxicity Class: Caution (4F, 4L, XLR); Warning/II (50W, 80S, 80WSP).
REI: 12 hr.

◆ - **new product** • ai-active ingredient • DF-dry flowable • E/EC-emulsifiable concentrate • EW-emulsion in water
F/FL-flowable • G-granule • L/LC-liquid • SP/WSP-soluble packet • W/WP-wettable powder • WSB-water soluble bag

Gowan Sevin Bait *(Gowan)*

Rate: 2 lb. ai/A
40 lb. Bait/A

Pest:

armyworms	cutworms	grasshoppers
crickets	darkling beetle	sowbugs

Remarks: Strawberries. Apply when pests appear in damaging numbers; repeat at 7- to 14-day intervals if necessary.

Days To Harvest: 1.

State Restrictions: For use in Arizona and California.

Signal Word/Toxicity Class: Caution/III.

REI: 12 hr.

◆ **Wilbur-Ellis Sevin 5 or 10 Dust** *(Wilbur-Ellis)*

Rate: Grapes:
2-2.5 lb. ai/A
40-50 lb. 5/A
20-25 lb. 10/A

Pest:

grape berry moth	grape leafhopper	Japanese beetle
grape leaffolder		

Rate: Strawberries:
2 lb. ai/A
40 lb. 5/A
20 lb. 10/A

Pest:

leafrollers	meadow spittlebug

Remarks: Do not feed treated vines to livestock.

Days To Harvest: Grapes-0; strawberries-1.

Signal Word/Toxicity Class: Caution.

REI: 12 hr.

CHLORPYRIFOS

Lorsban-4E *(Dow AgroSciences)*

◆ **Micro Flo Chlorpyrifos 4E** *(Micro Flo)*

Rate: 2.25 lb. ai/100 gal. water
4.5 pt. 4E/100 gal. water

Pest:

grape root borer

Remarks: Grapes. Apply 4 pt./15 sq. ft. diluted spray mixture to soil surface around base of each vine. Do not apply more than once per season or allow spray to contact fruit or foliage. Refer to label for directions.

Days To Harvest: 35.

State Restrictions: For use east of the Rockies.

Rate: 1.5 lb. ai/A
3 pt. 4E/A

Pest:

brown spanworm	cutworms	Sparganothis fruitworm
cranberry fruitworm	fireworms	
cranberry weevil		

Remarks: Cranberries. Apply when insects become a problem. For weevils, apply at flower bud development, repeat after 100% bloom. To avoid pesticide contamination of flood waters, do not treat when bogs are flooded. Do not apply more than twice a year. Refer to label for directions.

Days To Harvest: 60.

Rate: 1 lb. ai/A
2 pt. 4E/A in minimum 40 gal. spray

Pest:

strawberry bud weevil

Remarks: Strawberries. Prebloom use only. Apply when buds first appear; repeat 10-14 days later. Do not apply when berries start to form or are present. Phytotoxicity may occur if plants are experiencing high temperatures or drought. Do not apply more than twice per season. Do not mix with other materials unless prior use has shown combinations to be noninjurious.

Days To Harvest: 21.

Signal Word/Toxicity Class: Warning/II.

REI: 24 hr.

CRYOLITE

Kryocide (96%) *(Elf Atochem)*

Rate: 3.8-9.6 lb. ai/A
4-10 lb. 96%/A

Pest:

climbing cutworms	grape leaffolder	western grapeleaf skeletonizer
flea beetles	omnivorous leafroller	
grape berry moth (except CA)	orange tortrix	yellowstriped armyworm

Remarks: Grapes. Rate varies by pest species. Apply at 21-day intervals as needed. To control later broods, repeat after fruit is 1/4" or larger in diameter or as needed. For grapes grown for table use after fruit is 1/4" or larger in diameter, use of concentrate sprays will reduce visible residue spots from use of dilute applications at that stage. Do not exceed 20 lb./A per season. Do not use in combination with lime or compounds containing free lime. Do not chemigate in California.

Days To Harvest: 30.

Rate: Postharvest:
4.8-7.7 lb. ai/A
5-8 lb. 96%/A

Pest:

cutworms	grape leaffolder	western grapeleaf skeletonizer
grape berry moth (except CA)	omnivorous leafroller	
	orange tortrix	

Remarks: Grapevines. Controls leaf-feeding insects after all harvestable fruit or raisins have been removed from vineyards. Apply before leaf drop while insects are actively feeding. Do not exceed 8 lb./A per season. Do not use in combination with lime or compounds containing free lime.

Signal Word/Toxicity Class: Caution.

REI: 12 hr.

Prokil Cryolite 96 *(Gowan)*

Rate: 3.8-9.6 lb. ai/A
4-10 lb. 96/A

Pest:

climbing cutworm	grape leaffolder	grapeleaf skeletonizer
flea beetles	omnivorous leafroller	yellowstriped armyworm
grape berry moth	orange tortrix	

Rate: Multiple (2-3) low-rate applications:
2.9-4.8 lb. ai/A
3-5 lb. 96/A

Pest:

omnivorous leafroller

Remarks: Grapes grown for raisins, wine, and table-fresh market. Apply at 14-day intervals as needed. Ground application preferred. Do not exceed 20 lb./A per season. Do not use in combination with lime or compounds containing free lime. Do not apply through any type of irrigation system.

Days To Harvest: 30.

Rate: Postharvest:
4.8-7.7 lb. ai/A
5-8 lb. 96%/A

Pest:

cutworm	grape leaffolder	western grapeleaf skeletonizer
grape berry moth (except CA)	omnivorous leafroller	
	orange tortrix	

Remarks: Apply after fruit or raisins have been removed from vineyards. Apply before fall leaf drop while insects are actively feeding. Do not exceed 28 lb./A for combination of preharvest and postharvest per crop year. Do not use in combination with lime or compounds containing free lime.

Signal Word/Toxicity Class: Caution.

REI: 12 hr.

Damoil — see petroleum oil

DANITOL

Restricted Use Pesticide.

Danitol 2.4EC (fenpropathrin) *(Valent)*

Rate: 0.2 lb. ai/A
10.66 fl. oz. 2.4EC/A

Pest:

Lygus bugs	spittle bug	tarnished plant bug

NOTICE The information on these pages is for preliminary planning — not a guide for use. Be sure to follow the manufacturer's directions, notwithstanding information contained here.

For personal protective equipment and EPA registration numbers, please turn to page 39.

Rate: 0.3-0.4 lb. ai/A
16-21.33 fl. oz. 2.4EC/A

Pest:
twospotted spider mite

Remarks: Strawberries. Apply before mite count exceeds 20/leaflet; repeat in 30 days if needed. Do not make more than 2 applications exceeding 2.66 pt./A to same planting in 12 consecutive months.

Days To Harvest: 2.

- - - - - - - - - - - - - - - - -

Signal Word/Toxicity Class: Danger/I.
REI: 24 hr.

DEADLINE

Packaged as Bullets, M-Ps (mini-pellets), and granules. Resistant to rain and watering.

Deadline Bullets, M-Ps, Granules (4%) (metaldehyde) *(Pace)*

Rate: 0.4-1 lb. ai/A
10-40 lb. 4%/A

Pest:
slugs — snails

Remarks: Blueberries, caneberries (blackberries, boysenberries, loganberries, raspberries), grapes, strawberries. Apply every 3-4 weeks during growing season or as needed. Do not apply directly to edible portion of plant.

- - - - - - - - - - - - - - - - -

Signal Word/Toxicity Class: Warning/II (granule); Caution (bullets, M-P).
REI: 12 hr.

Diatect — see pyrethrins

DIAZINON

Restricted Use Pesticide due to avian and aquatic toxicity.

D•z•n diazinon AG600 WBC, 50W *(Novartis)*
Drexel Diazinon Insecticide (4EC) *(Drexel)*
Gowan Diazinon 4E or 50WSB *(Gowan)*
Helena Diazinon AG500 *(Helena)*
Micro Flo Diazinon AG500 or 50W *(Micro Flo)*
Prentox Diazinon AG500 or 50W *(Prentiss)*
◆ **Wilbur-Ellis Diazinon 4 Spray (4E)** *(Wilbur-Ellis)*

Rate: 0.19 lb. ai/A
6 fl. oz. AG500, 4E, 4EC/A in 100-300 gal. water
6 oz. 50W, 50WSB/A in 100-300 gal. water
4.75 fl. oz. AG600/A in 100-300 gal. water

Pest:
Drosophila

Rate: 0.5-1 lb. ai/A
1-2 pt. AG500, 4E, 4EC/A in min. 100 gal. water
1-2 lb. 50W, 50WSB/A in min. 100 gal. water
12.75-25.5 fl. oz. AG600/A in min. 100 gal. water

Pest:
aphids, grape berry moth, grape leaffolder, leafhoppers, omnivorous leafroller, Pacific spider mite, spider mites

Remarks: Grapes. Apply as thorough cover spray at 7-day intervals. Do not apply more than 1 lb. ai/A per application or more than 5 lb. ai/A per season.

Days To Harvest: 28.

- - - - - - - - - - - - - - - - -

Rate: 1-2 lb. ai/A
2-4 pt. AG500, 4E, 4EC/A in min. 100 gal. water
2-4 lb. 50W, 50WSB/A in min. 100 gal. water
25.5-51 fl. oz. AG600/A in min. 100 gal. water

Pest:
aphids, dryberry mite, leafhoppers, raspberry crown borer, raspberry fruitworm, raspberry sawflies, thrips, twospotted spider mite

Remarks: Caneberries (blackberries, boysenberries, dewberries, loganberries, raspberries). Apply as insects occur; repeat after 14 days if necessary. For raspberry fruitworm, apply when blossom buds separate; make second application immediately before blossoms open. For dryberry mite, apply as dormant spray in early spring when canes are still on the ground and buds are closed; make second application when buds are well developed but still closed. For raspberry crown borer, apply single application as a drench to crown and lower canes in the spring before buds break. Rate varies by pest species; refer to label. Do not apply more than 5 times per season.

Days To Harvest: 7.
State Restrictions: For use in California, Oregon, and Washington only.

- - - - - - - - - - - - - - - - -

Rate: 2-3 lb. ai/A
4-6 pt. AG500/A
4-6 lb. 50W; 50WSB/A
51-76.5 fl. oz. AG600/A

Pest:
blackheaded fireworm, cranberry fruit worm

Remarks: Cranberries. Apply when pests appear; repeat after 14 days if necessary. Up to 6 applications per season permitted for fireworm; 4 applications per season for fruitworm. May be applied through sprinkler irrigation. Refer to label for specific chemigation use directions.

Days To Harvest: 7.

- - - - - - - - - - - - - - - - -

Rate: 0.375-1 lb. ai/A
0.75-2 pt. AG500, 4E, 4EC/A in 100 gal. water
0.75-2 lb. 50W, 50WSB/A in 100 gal. water
9.5-25.5 fl. oz. AG600/A in 100 gal. water

Pest:
aphids, cyclamen mite, mole crickets, strawberry leafroller, twospotted spider mite

Remarks: Strawberries. For aphids and mites, apply as pests occur and repeat after 7 days if necessary. For cyclamen mites, direct spray to plant crowns (plants should be sufficiently agitated to assure thorough foliage coverage); application may be made to plants after harvest. For mole crickets, apply 1-2 days before transplanting and immediately incorporate into top 1-2" soil. For leafrollers, apply 200 gal. spray/A when blossoms show color. Do not apply more than 3 times per season.

Days To Harvest: 5.

- - - - - - - - - - - - - - - - -

Signal Word/Toxicity Class: Caution/III.
REI: 24 hr.

DIBROM

Dibrom 8 (8E) (naled) *(Amvac)*

Rate: 0.5-0.66 lb. ai/100 gal. water
0.5-0.66 pt. 8E/100 gal water

Pest:
Drosophila, leafhoppers

Remarks: Grapes. Apply by ground at 7-day intervals in 100-130 gal. diluted spray/A, depending on size of vines. Use low rate on Muscat varieties. Do not apply more than 6 pt./A per season or when temperatures exceed 90°F. Do not apply through any type of irrigation system.

Days To Harvest: 3.

- - - - - - - - - - - - - - - - -

Rate: Airblast application:
1 lb. ai/A
1 pt. 8E/A

Pest: Prebloom:
leafhoppers (overwintering), thrips
Postbloom:
Drosophila, grapeleaf skeletonizer, leafhoppers

Remarks: Grapes. Apply at 7-day intervals in minimum 20 gal. water/A. Do not use on Muscat varieties as fruit injury may occur. Postbloom applications can cause fruit russeting. Do not apply more than 6 pt./A per season or when temperatures exceed 90°F. Do not apply through any type of irrigation system.

Days To Harvest: 3.
State Restrictions: For use in California only.

- - - - - - - - - - - - - - - - -

Rate: 1 lb. ai/A
1 pt. 8E/A

Pest:
aphids, leafrollers, Lygus bugs, omnivorous leaftier, spider mites, spittlebugs, thrips

Remarks: Strawberries. Apply at 7-day intervals. Do not apply more than 5 pt./A per season or when temperatures exceed 90°F. Do not apply through any type of irrigation system.

Days To Harvest: 1.

- - - - - - - - - - - - - - - - -

Signal Word/Toxicity Class: Danger/I.
REI: 48 hr.

DICOFOL

◆ **Griffin Dicofol 4E** *(Griffin)*
Kelthane MF (4E), 35 or 50 *(Rohm and Haas)*

Rate: 0.67-1.17 lb. ai/A
1.33-2.33 pt. 4E/A
0.5-1.2 lb. ai/A
1.5-3.5 lb. 35/A
0.5 lb. ai/A
2.5 lb. 50/A

◆ - **new product** • ai-active ingredient • DF-dry flowable • E/EC-emulsifiable concentrate • EW-emulsion in water
F/FL-flowable • G-granule • L/LC-liquid • SP/WSP-soluble packet • W/WP-wettable powder • WSB-water soluble bag

Pest:
brown almond mite, McDaniel mite, Willamette mite, clover mite, Pacific mite, yellow mite, European red mite, twospotted spider mite

Remarks: Grapes. Apply just after bloom or whenever mites appear; repeat if necessary. Do not apply through any type of irrigation system.
Days To Harvest: 7.

- - - - - - - - - - - - - - - - -

Rate: 1 lb. ai/A
2 pt. 4E/A
0.35-1 lb. ai/A
1-3 lb. 35/A
0.38-1 lb. ai/A
0.75-2 lb. 50/A

Pest:
sixspotted spider mite, strawberry mite, twospotted spider mite

Rate: 2 lb. ai/A
4 pt. 4E/A
1.4-2.4 lb. ai/A
4-6.75 lb. 35/A
0.75-2.4 lb. ai/A
1.5-4.75 lb. 50/A

Pest:
cyclamen mite

Remarks: Strawberries. Apply when mites appear; repeat as necessary. For cyclamen mite, apply 2-3 sprays at 10- to 20-day intervals when new growth starts in spring, or use as spot treatment in infested areas during production season. Do not apply 4E if temperature exceeds 90°F.
Days To Harvest: 2.

- - - - - - - - - - - - - - - - -

Signal Word/Toxicity Class: Warning/II (35); Caution (4E, 50).
REI: 12 hr.

DI-SYSTON

Restricted Use Pesticide.

Di-Syston 8 or 15% (disulfoton) *(Bayer)*

Rate: 8 lb. ai/A
53.3 lb. 15%/A

Pest:
aphids

Remarks: Raspberries. Nonbearing nursery stock. Band application: Apply in 24" band over row to new or established planting. Work granules 2-3" into soil using cultivator or other suitable equipment. Second application may be made to soil later in season. Consult state agricultural experiment station or extension service for proper timing of applications.
State Restrictions: For use in Northeastern states.

- - - - - - - - - - - - - - - - -

Rate: 2-4 lb. ai/A
2-4 pt. 8/A (42" row spacing)
13.3-26.7 lb. 15%/A (42" row spacing)

Pest:
aphids, mites (except cyclamen)

Remarks: Strawberries. Propagating plants. Apply in water emulsion or with liquid fertilizer as soil injection on each side of row as needed. Do not use fruit from treated plants for food purposes.

- - - - - - - - - - - - - - - - -

Signal Word/Toxicity Class: Danger/I.
REI: 48 hr.

D•z•n — see diazinon

ENDOSULFAN

Drexel Endosulfan 3EC or 50WP *(Drexel)*
Gowan Endosulfan 3EC or 50WSB *(Gowan)*
◆ **Micro Flo Endosulfan 3EC or 50WP** *(Micro Flo)*
Phaser 3EC or 50WSB *(AgrEvo USA)*
Thiodan 3EC or 50WP *(FMC)*

Rate: 1-1.5 lb. ai/A
1.33-2 qt. 3EC/A
2-3 lb. 50WP, 50WSB/A

Pest:
grape leafhopper, grape phylloxera, rose chafer

Remarks: Grapes. Ground application preferred. Apply when insects appear or feeding is noticed. Do not apply to Concord variety as severe injury is likely to occur. Do not exceed 3 lb. ai/A per year or 3 applications per year.
Days To Harvest: 7.

- - - - - - - - - - - - - - - - -

Rate: 1.5 lb. ai/300 gal. water
2 qt. 3EC/300 gal. water
3 lb. 50WP, 50WSB/300 gal. water

Pest:
blueberry bud mite

Remarks: Blueberries. Apply immediately after harvest and repeat 6-8 weeks later. Do not apply after buds are well formed. Do not exceed 2 applications per year.

- - - - - - - - - - - - - - - - -

Rate: 1 lb. ai/A
1.33 qt. 3EC/A
2 lb. 50WP, 50WSB/A

Pest:
meadow spittlebug, strawberry aphid, tarnished plant bug

Rate: 2 lb. ai/A
2.66 qt. 3EC/A in 400 gal. water
4 lb. 50 50WP, 50WSB/A in 400 gal. water

Pest:
cyclamen mite

Remarks: Strawberries. Do not reapply within 15 days or more than twice within a 35-day period when fruit is present.
Days To Harvest: 4.

- - - - - - - - - - - - - - - - -

Rate: 1 lb. ai/100 gal. water
1.33 qt. 3EC/100 gal. water
2 lb. 50WP, 50WSB/100 gal. water

Pest:
garden symphylans

Remarks: Strawberries. Mix thoroughly; dip entire plant. Wear rubber gloves during dipping operation.
State Restrictions: For use in the Northwest.

- - - - - - - - - - - - - - - - -

Signal Word/Toxicity Class: Danger-Poison/I.
REI: 24 hr.

FENOXYCARB

Insect growth regulator.

Logic *(Novartis)*

Rate: Mound treatment:
1-3 Tbs./mound
Broadcast or aerial application:
1-1.5 lb./A

Pest:
fire ants

Remarks: Nonbearing blueberries. Mound treatment: Uniformly distribute 3-4 ft. around mound. Broadcast or aerial application: Apply uniformly with equipment calibrated to give correct dosage. Do not apply by air where weather conditions favor drift from treated areas. Retreat after 3-4 months, if needed. Apply when ants are actively foraging, usually when soil temperatures are above 60°F. Avoid application during excessively hot periods or when grass is wet. Do not apply to nonbearing crops more than twice per year or sooner than 1 year before first harvest. Do not graze treated areas.

- - - - - - - - - - - - - - - - -

Signal Word/Toxicity Class: Caution.
REI: 12 hr.

FURADAN

Carbamate insecticide/nematicide toxic to fish and other wildlife. Restricted Use Pesticide.

Furadan 4F (carbofuran) *(FMC)*

Rate: 10 lb. ai/A
2.5 gal. 4F/A

Pest:
dagger nematode, grape phylloxera, root-knot nematode

Remarks: Grapes. Broadcast treatment. Apply to soil surface between vine rows and immediately incorporate. Remove dense weed growth prior to treatment. Do not use on soils of pH 8.0 or greater. Do not make more than 1 application per crop year. Do not apply through any type of irrigation system.
Days To Harvest: 200.
State Restrictions: For use in California.

- - - - - - - - - - - - - - - - -

For personal protective equipment and EPA registration numbers, please turn to page 39.

Rate: 1-2 lb. ai/A
2-4 pt. 4F/A
Pest:
root weevil
Remarks: Strawberries. Apply as 10-12" band over row after last harvest but before Oct. 1. Do not apply if berries are present or make more than 1 application per season. Do not apply through any type of irrigation system.
State Restrictions: For use in Oregon, Washington.

Signal Word/Toxicity Class: Danger-Poison/I.
REI: 48 hr.

Furadan 15G (carbofuran) *(FMC)*
Rate: 2 lb. ai/A
13.3 lb. 15G/A
Pest:
black vine weevil
Remarks: Cranberries. For cranberry bogs that are sprinkler irrigated and are not flooded for purposes of irrigation or harvest. Spread granules uniformly over soil surface using rotary spreader. Make application about June 15-25 and repeat 1 month later. Incorporate granules by following each application with sprinkler irrigation.
Days To Harvest: 60.
State Restrictions: For use in Washington.

Signal Word/Toxicity Class: Warning/II.
REI: 48 hr.

Fyfanon — see malathion

Guthion — see azinphos-methyl

IMIDACLOPRID

Provado Solupak *(Bayer)*
Rate: 0.03-0.05 lb. ai/A
0.75-1 oz. Solupak/A
Pest:
leafhoppers mealybugs
Remarks: Grapes. Apply in sufficient water to provide thorough coverage. Use high rate for heavy infestations of mealybugs. Allow 14 days between applications. Do not apply more than 2 oz./yr.
Days To Harvest: 0.

Signal Word/Toxicity Class: Caution.
REI: 12 hr.

IMIDAN

Imidan 70-W (phosmet) *(Gowan)*
Rate: 1 lb. ai/A
1.33 lb. 70-W/A
Pest:
blueberry maggot, cherry fruitworm, cranberry fruitworm, flea beetles, grasshoppers, Japanese beetle, obliquebanded leafroller, plum curculio, redbanded leafroller, redstriped fireworm, rose chafer, sawfly spanworm, strawberry root weevil
Remarks: Blueberries. Apply as foliar spray. Second application may be made when indicated by insect infestations and local or state spray programs.
Days To Harvest: 3.
State Restrictions: For use in the Northeast only.

Rate: 1-1.5 lb. ai/A
1.33-2.13 lb. 70-W/A
Pest:
flea beetles, grape berry moth, grape cane borer, grape cane girdler, grape leafhopper, grape mealybug, Japanese beetle, redbanded leafroller, rose chafer
Remarks: Grapes. Apply as necessary to maintain control in accordance with insect infestations and local or state spray programs. Spray both sides of each row, also tops of vines to assure adequate coverage of fruit and foliage. For grape berry moth, apply prebloom, postbloom, first, and late cover sprays as needed. For grape leafhopper, apply when most nymphs hatch (generally coincides with berry moth). Use higher rate for Japanese beetle.
Days To Harvest: 14.
State Restrictions: For use east of the Rockies.

Rate: 1 lb. ai/A
1.33 lb. 70-W/A
Pest:
grape leaffolder, grape mealybug, omnivorous leafroller, vine mealybug, western grapeleaf skeletonizer

Rate: Delayed dormant:
1.5 lb. ai/A
2.13 lb. 70-W/A
Pest:
grape mealybug, vine mealybug
Remarks: Grapes. Adequately cover fruit and foliage when insects are present. Apply as early as first sizing; repeat at 10- to 14- day intervals as needed. For delayed dormant application, apply prior to bud break in combination with oil or spreader sticker; use adequate volume to ensure thorough coverage.
Days To Harvest: 7.
State Restrictions: For use west of the Rockies.

Signal Word/Toxicity Class: Warning/II.
REI: 24 hr.; California-5 days.

Kelthane — see dicofol

Kryocide — see cryolite

Kumulus — see sulfur

LANNATE

Toxic to fish and wildlife. Restricted Use Pesticides due to high acute toxicity to humans.

Lannate SP or LV (methomyl) *(DuPont)*
Rate: 0.45-0.9 lb. ai/A
0.5-1 lb. SP/A
1.5-3 pt. LV/A
Pest: All areas:
grape leaffolder, omnivorous leafroller, orange tortrix
East of the Rockies:
grape berry moth, leafhoppers
West of the Rockies:
climbing cutworm, leafhoppers, saltmarsh caterpillar
California:
grapeleaf skeletonizer, thrips
Remarks: Grapes. Apply in sufficient water to obtain thorough, uniform coverage when insects first appear at 5- to 7-day intervals or as needed. For climbing cutworm, repeat at 7- to 14-day intervals as needed at 0.9 lb. ai rate only. For grape berry moth, apply pre- and postbloom; repeat at 10- to 14-day intervals or as needed. Ground application for leafhoppers and skeletonizer. Do not apply more than 5 applications or 4.5 lb. ai/A per crop.
Days To Harvest: Fresh and raisin-1; wine grapes-14.

Rate: 0.225-0.9 lb. ai/A
0.25-1 lb. SP/A
0.75-3 pt. LV/A
Pest:
aphids, blueberry leafhopper, blueberry leafroller, blueberry maggot, cherry fruitworm, cranberry fruitworm, flea beetle (larvae), sawfly (larvae), sharpnosed leafhopper, tussock moth, weevils
Remarks: Blueberries. Ground application only for fruitworms. Do not apply during bloom. Do not apply more than 4 applications or 3.6 lb. ai/A per crop.

Rate: 0.45-0.9 lb. ai/A
0.5-1 lb. SP/A
1.5-3 pt. LV/A
Pest:
aphids, beet armyworm, Lygus bugs, omnivorous leaftier, thrips

Rate: 0.67-0.9 lb. ai/A
0.75-1 lb. SP/A
2.25-3 pt. LV/A
Pest:
armyworms

◆ - **new product** • ai-active ingredient • DF-dry flowable • E/EC-emulsifiable concentrate • EW-emulsion in water
F/FL-flowable • G-granule • L/LC-liquid • SP/WSP-soluble packet • W/WP-wettable powder • WSB-water soluble bag

Remarks: Strawberries. Apply in sufficient water to obtain thorough, uniform coverage at 5- to 7-day intervals or as needed. Do not apply more than 10 applications or 4.5 lb. ai/A per crop.

Days To Harvest: Blueberries-3; strawberries: fresh-3, processing fruit-10.

Signal Word/Toxicity Class: Danger-Poison/I.
REI: Blueberries, strawberries-48 hr.; grapes-7 days.

Lime-Sulfur Solution — see calcium polysulfide

Logic — see fenoxycarb

Lorsban — see chlorpyrifos

MALATHION

Clean Crop Malathion 57EC *(Platte)*

Rate: Blueberries:
0.94-2 lb. ai/A
1.5-3.2 pt. 57EC/A

Pest:
cherry fruitworm, cranberry fruitworm, Japanese beetle, plum curculio, sharpnosed leafhopper

Rate: Blackberries, boysenberries, dewberries, loganberries:
0.94 lb. ai/100 gal.
1.5 pt. 57EC/100 gal.

Pest:
aphids, Japanese beetle, leafhoppers, mites

Rate: Blackberries, boysenberries, dewberries, loganberries:
1.9 lb. ai/A
3 pt. 57EC/A

Pest:
rose scale, thrips

Rate: Currants, gooseberries:
1.1-2.2 lb. ai/100 gal.
1.6-3.2 pt. 57EC/100 gal.

Pest:
currant aphid, imported currantworm, Japanese beetle, mites

Rate: Grapes:
0.94 lb. ai/A
1.5 pt. 5EC/A

Pest:
mealybugs

Rate: Raspberries:
0.94-1.9 lb. ai/A
1.5-3 pt. 57EC/A

Pest:
aphids, Japanese beetle, leafhoppers, mites, rose scale, sap beetle, thrips

Rate: Strawberries:
0.94-1.9 lb. ai/A
1.5-3 pt. 57EC/A

Pest:
aphids, field crickets, Lygus bugs, potato leafhopper, spider mites, spittlebugs, strawberry leafroller, strawberry root weevil, thrips, whiteflies

Remarks: Blueberries, caneberries, (blackberries, boysenberries, dewberries, loganberries, raspberries), cranberries, currants, gooseberries, grapes, strawberries. May injure foliage on some grapes varieties. For strawberry root weevil, apply to soil surface before planting; work into top 6-8" soil.

Days To Harvest: Blueberries, caneberries-1; currants, cranberries, gooseberries, grapes, strawberries-3.

Signal Word/Toxicity Class: Caution/III.
REI: 12 hr.

◆ Clean Crop Malathion 8EC *(Platte)*

Rate: 1.5-2 lb. ai/A
1.5-2 pt. 8EC/A

Pest:
aphids, blackheaded fireworm, leafhoppers, leafrollers, Lygus bugs, mites, rose scale, spittlebugs, thrips, whiteflies

Remarks: Caneberries, (blackberries, boysenberries, dewberries, loganberries, raspberries), cranberries, strawberries.

Days To Harvest: Caneberries-1; cranberries, strawberries-3.

Signal Word/Toxicity Class: Caution/III.
REI: 12 hr.

Drexel Malathion 5EC *(Drexel)*
◆ Micro Flo Malathion 5EC *(Micro Flo)*
Prentox 5 lb. Malathion Spray (5EC) *(Prentiss)*

Rate: 0.94-1.25 lb. ai/A
1.5-2 pt. 5EC/A

Pest:
cherry fruitworm (high rate), cranberry fruitworm (high rate), Japanese beetle (low rate)

Rate: Northeast:
0.94 lb. ai/100 gal. water
1 pt. 5EC/100 gal. water + 1.5 qt. Staley's Sauce Base #7

Pest:
blueberry maggot

Remarks: Blueberries. For cranberry fruitworm, apply at egg hatch; repeat at 4- to 5-day intervals (total 4 applications). For Japanese beetles, apply when first berries turn blue; repeat at 10-day intervals as long as necessary.

Days To Harvest: 1; Northeast-8 hr.

Rate: Caneberries:
1.9 lb. ai/A
3 pt. 5EC/A

Pest:
aphids, Japanese beetle, leafhoppers, mites, rose scale, thrips

Rate: Currants, gooseberries:
1.25-1.9 lb. ai/A
2-3 pt. 5EC/A

Pest:
currant aphid, imported cabbageworm, Japanese beetle, mites

Rate: Grapes:
2-2.5 lb. ai/100 gal. water
1.5 pt. 5EC/100 gal. water

Pest:
European fruit lecanium

Rate: Grapes:
2-2.5 lb. ai/A
1.5-3 pt. 5EC/A

Pest:
leafhoppers, mealybugs, spider mites

Rate: Strawberries:
2-2.5 lb. ai/A
1.5-3 pt. 5EC/A

Pest:
aphids (high rate), field crickets, Lygus bugs, potato leafhopper, spider mites, spittlebugs, strawberry leafroller, strawberry root weevil, thrips, whiteflies

Remarks: Blueberries, caneberries, (blackberries, boysenberries, dewberries, loganberries, raspberries), cranberries, currants, gooseberries, grapes, strawberries.

Days To Harvest: Caneberries-1; currants, cranberries, gooseberries, grapes, strawberries-3.

Signal Word/Toxicity Class: Warning/II.
REI: 12 hr.

8 Fruits & Nuts

NOTICE	The information on these pages is for preliminary planning — not a guide for use. Be sure to follow the manufacturer's directions, notwithstanding information contained here.	For personal protective equipment and EPA registration numbers, please turn to page 39.

◆Drexel Malathion ULV *(Drexel)*
Fyfanon ULV *(Cheminova)*
◆ Micro Flo Malathion ULV *(Micro Flo)*

Rate: 10 fl. oz. ULV/A

Pest:
blueberry maggot

Remarks: Blueberries.
Days To Harvest: 0.

Signal Word/Toxicity Class: Caution.
REI: 12 hr.

◆ Fyfanon (8E) *(Helena)*

Rate: Blueberries, caneberries:
2 lb. ai/A
2 pt. 8E/A

Pest:

aphids	rose leafhopper	sap beetles
Japanese beetle	rose scale	twospotted spider mites

Rate: Cranberries:
1 lb. ai/A
1 pt. 8E/A

Pest:

blackheaded fireworm	leafhoppers	spittlebugs (nymph)
cranberry fruitworm		

Rate: Grapes:
1 lb. ai/A
1 pt. 8E/A

Pest:

European fruit lecanium	grape leafhopper	spider mites
	mealybugs	

Rate: Strawberries:
1.5-2 lb. ai/A
1.5-2 pt. 8E/A

Pest:

aphids (high rate)	spider mites	strawberry root weevil
field crickets	spittlebugs	thrips
Lygus bugs	strawberry leafroller	whiteflies
potato leafhopper		

Remarks: Blueberries, caneberries, (blackberries, boysenberries, dewberries, loganberries, raspberries), cranberries, strawberries. Injury may occur to Almeria, Cardinal, and Ribier grape varieties when applied after clusters appear.
Days To Harvest: Blueberries, caneberries-0; cranberries, grapes, strawberries-3.

Signal Word/Toxicity Class: Caution.
REI: 12 hr.

◆ Gowan Malathion 8 (8E) *(Gowan)*

Rate: Blueberries:
1.5-2.5 lb. ai/A
1.5-2.5 pt. 8E/A

Pest:

aphids	cherry fruitworm	plum curculio
blueberry maggot	cranberry fruitworm	sharpnosed leafhopper
blueberry tip borer	Japanese beetle	white tussock moth

Rate: Caneberries:
1-4 lb. ai/A
1-4 pt. 8E/A

Pest:

aphids	leafhoppers	rose scale
Japanese beetle	mites	thrips

Rate: Currants, gooseberries:
1-2 lb. ai/A
1-2 pt. 8E/A

Pest:

Japanese beetle	mites	rose chafer

Rate: Cranberries:
1-2.5 lb. ai/A
1-2.5 pt. 8E/A

Pest:

blackheaded fireworm	cranberry fruitworm	spittlebug (nymph)
	leafhoppers	

Rate: Grapes:
2-2.5 lb. ai/A
2-2.5 pt. 8E/A

Pest:

Drosophila	grape leafhopper	mealybugs
European fruit lecanium	Japanese beetle	spider mites
	leafhoppers	terrapin scale

Rate: Strawberries:
1.5-2 lb. ai/A
1.5-2 pt. 8E/A

Pest:

aphids	spider mites	strawberry root weevil
field crickets	spittlebugs	thrips
Lygus bugs	strawberry leafroller	whiteflies
potato leafhopper		

Remarks: Blueberries, caneberries, (blackberries, boysenberries, dewberries, loganberries, raspberries), cranberries, currants, gooseberries, grapes, strawberries. Injury may occur to grapes when applications are made after bloom. For strawberry root weevil, apply to soil surface before planting; work into top 6-8" soil.
Days To Harvest: Blueberries, caneberries, currants-1; cranberries, gooseberries, grapes, strawberries-3.
State Restrictions: Not for use on cranberries in California.

Signal Word/Toxicity Class: Caution/III.
REI: 12 hr.

◆ Wilbur-Ellis Malathion 8 *(Wilbur-Ellis)*

Rate: Caneberries:
1.5-2 lb. ai/A
1.5-2 pt. 8/A

Pest:

aphids	Lygus bugs	strawberry root weevil
blackheaded fireworm	rose scale	
leafhoppers	spider mites	thrips
leafrollers	spittlebugs	whiteflies

Rate: Grapes:
1.5-2.5 lb. ai/A
1.5-2.5 pt. 8/A

Pest:

grape leafhopper	mealybugs	spider mites

Remarks: Caneberries, grapes. Apply in minimum 5 gal. by air; 10 gal. by ground. For grapes, injury may occur to Ribier, Italia, Cardinal, and Almenia varieties when malathion is applied after clusters appear.
Days To Harvest: Caneberries-1; grapes-3.

Signal Word/Toxicity Class: Caution.
REI: 12.

Marlate — see methoxychlor

METASYSTOX-R

Toxic to fish and wildlife. Restricted Use Pesticide.

Metasystox-R (SC) (oxydemeton-methyl) *(Gowan)*

Rate: 0.25-0.375 lb. ai/100 gal. water
1-1.5 pt. SC/100 gal. water

Pest:

aphids	mites

Remarks: Grapes. Apply as full coverage spray. Do not apply more than 3 times per season. Do not apply to vines that will bear fruit within next 12 months.

Signal Word/Toxicity Class: Warning/II.
REI: 48 hr.; 72 hr. if average rainfall less than 25" per year.

METHOXYCHLOR

Marlate 50 (WP) *(Kincaid)*
Prentox Methoxychlor 2EC *(Prentiss)*

Rate: 1.3-1.7 lb. ai/100 gal. water
75-100 fl. oz. 2EC/100 gal. water
2 Tbs. 50WP/gal. water

Pest:

cranberry fruitworm	leafhoppers	San Jose scale (crawler)
flea beetles	omnivorous leaftier	
grape berry moth	rednecked borer	spittlebugs
grapeleaf skeletonizer	rose chafer	strawberry rootworm
Japanese beetle		strawberry weevil

◆ - **new product** • ai-active ingredient • DF-dry flowable • E/EC-emulsifiable concentrate • EW-emulsion in water
F/FL-flowable • G-granule • L/LC-liquid • SP/WSP-soluble packet • W/WP-wettable powder • WSB-water soluble bag

Remarks: **Blueberries, caneberries (blackberries, loganberries, raspberries, etc.), cranberries, currants, gooseberries, grapes, strawberries.** Apply at petal fall; repeat at 7- to 14-day intervals as needed.
Days To Harvest: 14.

- - - - - - - - - - - - - - - - - -

Signal Word/Toxicity Class: Caution.
REI: 12 hr.

METHYL PARATHION

Penncap-M is a microencapsulated formulation. Restricted Use Pesticide due to residual effects to avian species and hazard to bees.

Penncap-M (2FL) *(Elf Atochem)*

Rate: 1-3 lb. ai/A
4-12 pt. 2FL/A
0.5-1.5 lb. ai/100 gal. dilute spray
2-6 pt. 2FL/100 gal. dilute spray

Pest:
grape berry moth, grape blossom midge, grapevine aphid, Japanese beetle, mealybug, redbanded leafroller, rose chafer, western grape leafhopper

Rate: 0.5-1 lb. ai/A
2-4 pt. 2FL/A
0.25-0.5 lb. ai/100 gal. dilute spray
1-2 pt. 2FL/100 gal. dilute spray

Pest:
eastern grape leafhopper, potato leafhopper

Remarks: Rate varies by pest species; refer to label. Use only as postharvest dormant or delayed dormant prebloom applications in areas where average rainfall is less than 25" per year. Apply at 7-day intervals in areas where average rainfall exceeds 25". From bloom to harvest, do not apply more than 2 applications or more than 4 pt./A per application. Do not make more than 4 applications per year.
Days To Harvest: 40.
State Restrictions: Not for use in California.

- - - - - - - - - - - - - - - - - -

Rate: 1 lb. ai/A
4 pt. 2FLA
0.5 lb. ai/100 gal. dilute spray
2 pt. 2FL/100 gal. dilute spray

Pest:
eastern grape leafhopper, grape blossom midge, grapevine aphid, potato leafhopper

Rate: 1-2 lb. ai/A
4-8 pt.2FL/A
0.5-1 lb. ai/100 gal. dilute spray
2-4 pt. 2FL/100 gal. dilute spray

Pest:
grape berry moth

Rate: 1.5-2 lb. ai/A
6-8 pt.2FL/A
0.75-1 lb. ai/100 gal. dilute spray
3-4 pt. 2FL/100 gal. dilute spray

Pest:
rose chafer

Rate: 2-3 lb. ai/A
8-12 pt. 2FL/A
1-1.5 lb. ai/100 gal. dilute spray
4-6 pt. 2FL/100 gal. dilute spray

Pest:
western grape leafhopper

Remarks: Postharvest dormant or delayed dormant prebloom application only. Apply at 7- day intervals. Do not make more than 2 applications per year.
State Restrictions: For use in California only.

- - - - - - - - - - - - - - - - - -

Signal Word/Toxicity Class: Warning/II.
REI: 48 hr.

Microthiol Special — see sulfur

MORESTAN

Morestan Solupak 25 (oxythioquinox) *(Bayer)*

Rate: 0.125-0.25 lb. ai/100 gal. water
0.5-1 lb. 25/100 gal. water

Pest:
mites (except cyclamen)

Remarks: **Strawberries.** Apply as prebloom only. Spray all foliage surfaces including underside of leaves for complete coverage; repeat as necessary. Apply to strawberries in propagating beds and in the field during the year of setting new plants, but not after first bloom.

- - - - - - - - - - - - - - - - - -

Rate: 0.5 lb. ai/A
2 lb. 25/A

Pest:
mite eggs, spider mites

Remarks: **Red raspberries.** Refer to label for use directions and precautions.
State Restrictions: SLN-Oregon.

- - - - - - - - - - - - - - - - - -

Signal Word/Toxicity Class: Warning/II.
REI: 24 hr.

M-PEDE

M-Pede (potassium salts of fatty acids) *(Mycogen)*

Rate: 1-2% v/v solution

Pest:
aphids, leafhoppers, mealybugs, scale crawlers, spider mites, whiteflies

Remarks: **Blueberries, caneberries (blackberries, boysenberries, raspberries), cranberries, currants, grapes, strawberries.** Do not apply to Calmeria or Italia grape varieties. May alter waxy bloom and/or cause marking on grape surface which can affect quality rating of table grapes at harvest. Effective for late season and preharvest suppression of leafhoppers in wine and raisin grapes. May be tank mixed with other insecticides. Refer to both labels for use directions and precautions.
Days To Harvest: 0.

- - - - - - - - - - - - - - - - - -

Signal Word/Toxicity Class: Warning/II.
REI: 12 hr.

NEEMAZAD — see azadirachtin

NEEMIX — see azadirachtin

NEMACUR

Restricted Use Pesticide.

Nemacur 3 (fenamiphos) *(Bayer)*

Rate: Band:
9 lb. ai/A
3 gal. 3/A
Low pressure irrigation:
0.75-3 lb. ai/A
1 qt.-1 gal. 3/A

Pest:
nematodes

Remarks: **Grapes.** Apply as a band to soil and incorporate immediately or apply by low pressure irrigation. Refer to label for further use information.
Days To Harvest: 2.

- - - - - - - - - - - - - - - - - -

Rate: 3-6 lb. ai/A
1-2 gal. 3/A

Pest:
nematodes

Remarks: **Raspberries.** Apply as a band (covering feeder roots) and immediately incorporate. Do not apply more than once per year. Refer to label for further use information.
Days To Harvest: 6 months.
State Restrictions: Not for use in California.

- - - - - - - - - - - - - - - - - -

Signal Word/Toxicity Class: Danger/I.
REI: 48 hr.

8 Fruits & Nuts

NOTICE	The information on these pages is for preliminary planning — not a guide for use. Be sure to follow the manufacturer's directions, notwithstanding information contained here.	For personal protective equipment and EPA registration numbers, please turn to page 39.

Nemacur 3 or 15% (fenamiphos) *(Bayer)*
Rate: 1.8-2.7 lb. ai/A
2.4-3.6 qt. 3/A (40" rows)
5.9-8.8 fl. oz. 3/1000 ft. row (any row spacing)
12-18 lb. 15%/A (40" rows)
14.7-22 oz. 15%/1000 ft. row (any row spacing)
Pest:
nematodes
Remarks: Strawberries. Apply in 12-18" band over row and incorporate. Do not make more than 1 application per season. Refer to label for further use information.
Days To Harvest: 110.

Rate: 2 lb. ai/A
13.8 lb. 15%/A (40" rows)
17 oz. 15%/1000 ft. row (any row spacing)
Pest:
nematodes
Remarks: Nonbearing strawberries (nursery stock). Apply in 12" band over row immediately after transplanting and incorporate into soil. Make second application 8 weeks after first application. Refer to label for further use information.

Signal Word/Toxicity Class: Danger/I.
REI: 48 hr.

Omite — see propargite

Omni Supreme Spray — see petroleum oil

Orthene — see acephate

Penncap-M — see methyl parathion

PERMETHRIN

Ambush 2E or 25W *(ZENECA)*
Rate: 0.2 lb. ai/A
12.8 fl. oz. 2E/A
12.8 oz. 25W/A
Pest:
root weevil (adults)
Remarks: Raspberries. Apply with ground equipment in minimum 100 gal. water/A. Do not make more than 2 applications per season. Do not make second application within 11 days of the first.
Days To Harvest: 3.
State Restrictions: For use in Oregon and Washington.

Signal Word/Toxicity Class: Warning/II.

PETROLEUM OIL

◆ **455 Oil 98.8 or Damoil** *(Drexel)*
Omni Supreme Spray *(Helena)*
SunSpray 6E, 9E, or 11E *(Sun)*
Volck Supreme Spray *(Valent)*
Rate: Blueberries:
2 gal. 455/100 gal. water
2-3 gal. 9E, 11E/100 gal. water
Pest:
Putnam scale San Jose scale

Rate: Blueberries:
3 gal. 6E/100 gal. water
Pest:
mites scales

Rate: Grapes:
2 gal. 455/100 gal. water
4-6 gal./A (Damoil, Volck)
1-2 gal./A (Omni)
Pest:
mealybugs

Remarks: Blueberries, grapes. Apply dormant. Refer to label for directions, precautions, and restrictions.

Signal Word/Toxicity Class: Caution.
REI: Damoil, Volck-12 hr.; 455, Omni, SunSpray-4 hr.

SAF-T-SIDE *(Brandt Consolidated)*
◆ **Sumulsion** *(Drexel)*
Rate: 1-2 gal./100 gal. water
Pest: Grapes:
leafhoppers mealybugs scales
leafminers mites whiteflies
Blueberries, strawberries:
aphids sawflies scales
mites
Remarks: Blueberries, grapes, strawberries. Will kill other immature forms if spray covers insect. For grapes, apply dormant, summer, or postharvest for grapes; for blueberries and strawberries, apply as needed. Time application to fit local growth and application conditions. Compatible with most commonly used insecticides and fungicides. Do not use in combination with or immediately before or after spraying with fungicides such as captan, Dyrene, folpet, Morestan, or any sulfur products. Do not use with carbaryl or dimethoate.

Signal Word/Toxicity Class: Caution.
REI: 4 hr.

◆ **SunSpray Ultra-Fine** *(Sun)*
Rate: Blueberries:
2-3 gal./100 gal. water
Pest:
mites sawflies (eggs) scales

Rate: Strawberries:
1 gal./100 gal. water
Pest:
aphids mites
Remarks: Blueberries, strawberries. Apply dormant and foliage. Refer to label for directions, precautions, and restrictions.

Rate: Dilute spray:
2 gal./100 gal. water
Pest:
mites

Rate: Dilute spray:
4-6 gal./200-300 gal. water/A
Pest:
mealybugs

Rate: Concentrate spray:
4 gal./A (40-100 gal. spray/A)
Pest:
leafhoppers mealybugs
Remarks: Grapes. For mites, apply at bud swell or shoot growth of 5-7 cm in spring or postharvest. Refer to label for directions, precautions, and restrictions.

Signal Word/Toxicity Class: Caution.
REI: 4 hr.

Phaser — see endosulfan

PROPARGITE

Omite 30W or 30WS *(Uniroyal)*
Rate: 1.5-2.7 lb. ai/A
5-9 lb. 30W, 30WS/A
Pest:
Pacific spider mite, strawberry spider mite, twospotted spider mite, Willamette spider mite
Remarks: Grapes. For table grapes, do not apply dilute sprays after berries reach 1/4" in diameter. Do not apply more than twice per year. Do not apply through any type of irrigation system.
Days To Harvest: 21.
State Restrictions: For use west of the Rockies.

Signal Word/Toxicity Class: Danger/I.
REI: Table grapes-21 days. After 48 hr., workers can enter treated fields to perform hand labor tasks if they wear early-entry personal protective equipment.

◆ **- new product** • ai-active ingredient • DF-dry flowable • E/EC-emulsifiable concentrate • EW-emulsion in water
F/FL-flowable • G-granule • L/LC-liquid • SP/WSP-soluble packet • W/WP-wettable powder • WSB-water soluble bag

Provado — see imidacloprid

PYRELLIN (Premix)

Premix containing pyrethrins and rotenone.

Pyrellin E.C. (0.042 lb./gal. pyrethrins & 0.035 lb./gal. rotenone) *(Webb Wright)*

Rate: 1-2 pt. E.C./A

Pest:

aphids	Japanese beetle	rose chafer
blueberry maggot	leafhoppers	strawberry rootworm
gooseberry fruitworm	mites	strawberry leafroller
imported currantworm	raspberry fruitworm	thrips

Remarks: Blueberries, caneberries (blackberries, boysenberries, dewberries, elderberries, loganberries, raspberries), cranberries, currants, gooseberries, grapes, strawberries. Apply at 7-day intervals or less; repeat as needed. For boysenberries, dewberries, elderberries, loganberries, and strawberries, apply as needed.

Days To Harvest: 0.

- - - - - - - - - - - - - - - - -

Signal Word/Toxicity Class: Caution/III.

REI: 12 hr.

Pyrenone — see pyrethrins

PYRETHRINS

Contains piperonyl butoxide as a synergist. Diatect also contains silicon dioxide (diatomaceous earth). May be combined in tank mix with other insecticides for quicker and more complete kill where insect resistance may be a problem, and as an "exciter" to flush insects out of hiding and into contact with other spray residues.

Diatect Multipurpose Insecticide II *(Diatect)*
Pyrenone (0.5EC) *(AgrEvo Environmental Health)*
Py-Rin 60-6 EC *(Wilbur-Ellis)*

Rate: 1-6 lb. II/A
1-12 fl. oz. 0.5EC/A
2-16 oz. 6-60 EC/A

Pest: Including but not limited to:

aphids	gooseberry fruitworm	raspberry fruitworm
blueberry maggot	Japanese beetle	rednecked borer
caterpillars	imported currantworm	rose chafer
cutworms	leafhoppers	stink bugs
fireworms	leafrollers	strawberry leafroller
fruit flies	mites	thrips
fruitworms		weevils

Remarks: Blueberries, caneberries (blackberries, boysenberries, dewberries, elderberries, loganberries, raspberries, youngberries), cranberries, currants, grapes, huckleberries, strawberries, etc. Apply in sufficient water for thorough coverage; repeat as needed. Apply Diatect as dust or wettable powder.

Days To Harvest: 0.

- - - - - - - - - - - - - - - - -

Signal Word/Toxicity Class: Caution.

REI: 12 hr.

Py-Rin — see pyrethrins

SAF-T-SIDE — see petroleum oil

Sevin — see carbaryl

Sniper — see azinphos-methyl

Sulforix — see calcium polysulfide

SULFUR

Microthiol and Thiolux are micronized sulfurs.

◆ **Kumulus DF** *(BASF)*

Rate: 1.6-4 lb. ai/A
2-5 lb. DF/A

Pest:

blister mite	bud mite	red spider mite

Remarks: Grapes. Apply late dormancy, or with cutworm spray, repeat early shoot period and through season at 14-day intervals or with other pesticides. Do not apply if temperatures during or within 3 days after application are expected to exceed 90°F. Do not apply through any type of irrigation system. Refer to label for use directions.

- - - - - - - - - - - - - - - - -

Rate: 4-8 lb. ai/A
5-10 lb. DF/A

Pest:
red spider mite

Remarks: Strawberries. Apply early leaf stage, repeat at 10- to 14-day intervals with or without other pesticides. Do not apply if temperatures during or within 3 days after application are expected to exceed 90°F. Do not apply through any type of irrigation system. Refer to label for use directions.

- - - - - - - - - - - - - - - - -

Signal Word/Toxicity Class: Caution.

REI: 24 hr.

Microthiol Special (80W) *(Elf Atochem)*
Thiolux (DF) *(Platte)*

Rate: 2.4-8 lb. ai/A
3-10 lb. 80W/A
4.8 lb. ai/A
6 lb. DF/A

Pest:

blister mite	bud mite	red spider mite

Remarks: Grapes. Apply throughout season. Use with caution on Concord or Labrusca type grapes as injury can occur after bloom.

- - - - - - - - - - - - - - - - -

Rate: 4-8 lb. ai/A
5-10 lb. 80W, DF/A

Pest:

red spider mite	twospotted spider mite (80W) (except CA)

Remarks: Strawberries. Apply at early leaf stage; repeat as needed. Do not use on sulfur-sensitive varieties. Do not chemigate for Microthiol Special in California.

- - - - - - - - - - - - - - - - -

Signal Word/Toxicity Class: Caution.

REI: 24 hr.

Sumulsion — see petroleum oil

SunSpray — see petroleum oil

Thiodan — see endosulfan

Thiolux — see sulfur

VENDEX

Restricted Use Pesticide

Vendex 50WP (fenbutatin-oxide) *(Griffin)*

Rate: 0.5-1.25 lb. ai/A
1-2.5 lb. 50WP/A

Pest:

European red mite	Pacific mite	twospotted spider mite

Remarks: Grapes. Apply when mites first appear; repeat at 21 day intervals. Do not apply more than 2 lb. ai/A per year or more than 2 applications per season.

Days To Harvest: 28.

- - - - - - - - - - - - - - - - -

NOTICE The information on these pages is for preliminary planning — not a guide for use. Be sure to follow the manufacturer's directions, notwithstanding information contained here.

For personal protective equipment and EPA registration numbers, please turn to page 39.

Rate: 0.75-1 lb. ai/A
1.5-2 lb. 50WP/A
California:
0.75-1.5 lb. ai/A
1.5-3 lb. 50WP/A

Pest:
twospotted spider mite

Remarks: Strawberries. Apply in sufficient water to fully cover crop. Adjust spray volume and nozzle placement for good coverage of top and underside of leaves. Do not apply more than 2 lb. ai/A per season or more than 2 applications per season. In California, do not apply more than 4.5 lb. ai/A per season or more than 3 times per season.

Days To Harvest: 1.

- - - - - - - - - - - - - - - -

Signal Word/Toxicity Class: Danger/I.
REI: 48 hr.

Volck Supreme Spray — see petroleum oil

Tree Fruits, Citrus

435 or 455 Oil — see petroleum oil

ABAMECTIN

Natural product produced by soil microorganism, Streptomyces avermitilis. Unrelated to other miticides. Efficacy on citrus is extended significantly when combined with petroleum oil. Not recommended unless combined with petroleum oil. Toxic to fish, mammals, aquatic organisms, and highly toxic to honeybees. Restricted Use Pesticide.

Agri-Mek 0.15EC *(Novartis)*

Rate: Dilute sprays:
0.5-2 fl. oz. 0.15EC/100 gal.
Concentrate sprays:
5-20 fl. oz. 0.15EC/A

Pest:
citrus rust mite

Rate: Dilute sprays:
1-2 fl. oz. 0.15EC/100 gal.
Concentrate sprays:
10-20 fl. oz. 0.15EC/A

Pest:
broad mite, citrus bud mite, citrus leafminer, twospotted spider mite

Remarks: Not for use in citrus nurseries. Apply when mites appear during spring, summer, and/or fall in minimum volume 0.2% dilute spray or 1 gal./A concentrate paraffinic spray oil. Thorough coverage essential for good control. Do not apply more than 3 times, within 30 days of last treatment, or 40 fl. oz./A in any growing season. Do not apply by air or through any type of irrigation system. For Florida, refer to Citrus Pest Management Guide for instructions and precautions.

Days To Harvest: 7.

- - - - - - - - - - - - - - - -

Signal Word/Toxicity Class: Warning/II.
REI: 12 hr.

ACEPHATE

Organophosphate systemic insecticide with contact and stomach poison activity. May increase citrus red mite populations. Has moderate residual effects; compatible with most pesticides.

Orthene 75S *(Valent)*

Rate: 0.5 lb. ai/A
0.66 lb. 75S, 75WSP/A

Pest:
aphids, grasshoppers, katydid, mealybugs, orangedog, plant bugs, thrips, whiteflies

Rate: 0.5-0.75 lb. ai/A
0.66-1 lb. 75S/A

Pest:
citrus blackfly

Rate: Foliar:
0.75 lb. ai/A
1 lb. 75S/A
Mound treatment—drench method:
0.05 lb. ai/5 gal. water
1 oz. 75S/5 gal. water

Pest:
imported fire ants

Remarks: Nonbearing citrus (including citrus nursery seedlings). Length of residual activity depends on amount of moisture following application. Apply as needed for control of existing population. For citrus blackfly, apply when eggs or adults first appear at 7- to 10-day intervals or as needed to maintain control. For fire ants, apply 1 gal. mix/mound area by sprinkling mound and surrounding area until wet. Do not apply through any type of irrigation system. Do not graze treated areas.

Days To Harvest: 1 year.
State Restrictions: Except California for Orthene 75S.

- - - - - - - - - - - - - - - -

Rate: 2-4 lb. ai/100 gal. water
2.66-5.32 lb. 75S/100 gal. water

Pest:
apopka weevil, blue-green root weevil, Fuller rose weevil, little leaf notcher

Rate: Mound treatment—drench method:
0.05 lb. ai/5 gal. water
1 oz. 75S/5 gal. water

Pest:
imported fire ants

Remarks: Nonbearing. Apply 1 gal. mix/mound area by sprinkling mound and surrounding area until wet. For best results, apply mix in early morning or late afternoon when ants are most active. Applications made under prolonged hot and dry conditions may be ineffective due to location of ants deep within nest. To make ants more active, lightly disturb tops of mounds. To avoid mound relocation, do not severely disturb mounds. Do not apply through any type of irrigation system. Do not graze treated areas. Refer to Florida Citrus Pest Management Guide for instructions and precautions for the state.

Days To Harvest: 1 year.
State Restrictions: SLN-Florida.

- - - - - - - - - - - - - - - -

Rate: 0.5-1 lb. ai/A
0.66-1.33 lb. 75S/A

Pest:
citrus blackfly

Remarks: Nonbearing nursery stock. Do not apply to commercial groves. Apply by ground in sufficient water to assure thorough coverage. Use high rate when heavy infestations are present. Apply at 7- to 10-day intervals or as needed when eggs or adults first appear. Do not apply through any type of irrigation system. Do not graze treated areas. Consult extension service or Texas A & M entomologist before applying product when beneficial insects are present.

State Restrictions: SLN-Texas.

- - - - - - - - - - - - - - - -

Signal Word/Toxicity Class: Caution.
REI: 24 hr.

Adios — see carbaryl

Agri-Mek — see abamectin

AZADIRACHTIN

Extracted from seed of neem tree. Insect growth regulator that controls targeted insect larvae when they ingest or come in contact with it, by interfering with the insect's ability to molt and shutting off feeding behavior. Affected larvae slowly starve and die at next molt.

◆ **Azatin XL Plus** *(Thermo Trilogy)*

Rate: 5-21 oz. XL Plus/A

Pest:
aphids, beet armyworm, beetles, caterpillars, citrus cutworm, citrus leafminer, citrus thrips, cutworms, flea beetles, fruit flies, leafhoppers, leafminers, leafrollers, mealybugs, psyllids, thrips, webworms, whiteflies

◆ - **new product** • ai-active ingredient • DF-dry flowable • E/EC-emulsifiable concentrate • EW-emulsion in water
F/FL-flowable • G-granule • L/LC-liquid • SP/WSP-soluble packet • W/WP-wettable powder • WSB-water soluble bag

Remarks: Calamandin, citrus citron, grapefruit, kumquat, lemon, limes, mandarin, pummelo, satsuma mandarin, sour orange, sweet orange. Rate varies by species. Apply when pests first appear; repeat at 7-day intervals. When pest populations are high, use higher rates. Do not apply more that 21 fl. oz./A.
Days To Harvest: 0.

Signal Word/Toxicity Class: Caution.
REI: 4 hr.

◆ **NEEMAZAD 4.5EC or NEEMIX 4.5** *(Thermo Trilogy)*
Rate: 0.125-1 pt. 4.5/A
High Volume:
2.25-4.5 oz. 4.5EC/100 gal. water
Low Volume:
0.5 pt. 4.5EC/A
Pest:

aphids	greenhouse whitefly	psyllids
caterpillars	gypsy moth	sweetpotato whitefly
fall webworm	leafhoppers	thrips
fruit flies	leafminers	weevils
grasshoppers	mealybugs	

Remarks: Grapefruits, kumquats, lemons, limes, oranges. For low volume, apply in sufficient water for adequate coverage. For low insect pressure, apply preventive at 7-day intervals or at first sign of insects; high insect pressure, apply at 3- to 4-day intervals.

Signal Word/Toxicity Class: Warning/II.
REI: 12 hr.

Azatin — see azadirachtin

AZINPHOS-METHYL

Organophosphate insecticide. May increase spider mite, citrus red mite, and Texas citrus mite populations; highly toxic to honey bees. Restricted Use Pesticide (except Guthion Solupak).

Gowan Azinphos-M 50-W *(Gowan)*
Guthion 2L, Solupak (50WSP) *(Bayer)*
◆ **Micro Flo Azinphosmethyl 2EC or 50W** *(Micro Flo)*
Sniper 2-E, 50PVA, 50W *(Platte)*
Rate: 1.25-2 lb. ai/A
5-8 pt. 2EC, 2L/A
2.5-4 lb. 50WSP/A
0.25-0.38 lb. ai/100 gal. water
1-1.5 pt. 2-E/100 gal. water
0.5-0.75 lb. 50W/100 gal. water (Platte)
2-3 lb. ai/A
4-6 lb. 50-W, 50PVA/A
1.25-3 lb. ai/A
2.5-6 lb. 50W/A (Micro Flo)
Pest:

aphids	citrus root weevil	Glover scale
black scale	citrus thrips	orange tortrix
brown soft scale	cottonycushion scale	purple scale
California red scale	European brown snail	snow scale
chaff scale	Florida red scale	western tussock moth
citricola scale	fruittree leafroller	whiteflies
citrus mealybug	Fuller rose beetle	yellow scale

Remarks: Apply as full coverage spray. Do not apply more than twice per fruit year. Refer to California's UCIPM Pest Management Guidelines or Florida Citrus Pest Management Guide for instructions and precautions for these states.
Days To Harvest: 7; 2 applications/yr.-28.
State Restrictions: Sniper 50 PVA not registered in California.

Signal Word/Toxicity Class: Danger/I.
REI: Gowan-48 hr.; Guthion, Micro Flo, Sniper-7 days.

CALCIUM POLYSULFIDE

Lime-Sulfur Solution *(Best Sulfur)*
Rate: 2 gal. L/100 gal. water
Pest:

citricola scale	red spider mite	thrips

Rate: 7.5-12 pt. L/100 gal. water
Pest:
rust mite

Remarks: Oranges (navel, Valencia), tangerines, etc. Rate varies by time of application. For tangerines, apply late winter and early spring. Early and mid-season oranges may be injured by Lime-Sulfur sprays during summer and early fall months. For spiders, apply at first appearance. Do not apply through any type of irrigation system.

Signal Word/Toxicity Class: Danger/I.
REI: 48 hr.

CARBARYL

Carbamate insecticide for use on bearing and nonbearing citrus trees. May increase citrus red mite populations. Extremely toxic to aquatic and estuarine invertebrates; highly toxic to bees exposed to direct treatment or residues on blooming crops or weeds.

Adios (13%) *(Micro Flo)*
Rate: 0.065-0.1 lb. ai/A
0.5-0.75 lb. 13%/A
Pest:
cucumber beetle
Remarks: Apricots, nectarines, peaches. Do not apply through any type of irrigation system. Do not apply more than 3.75 lb./A per season.
Days To Harvest: Apricots, nectarines-3; peaches-1.
State Restrictions: Not for use in California.

Signal Word/Toxicity Class: Caution/III.
REI: 12 hr.

Chipco Sevin SL, 80WSP or Sevin 4F, 80S, 80WSP *(Rhone-Poulenc)*
Drexel Carbaryl 4L, 50W, 80S *(Drexel)*
Rate: 0.75-1 lb. ai/100 gal. water
1.5-2 pt. SL, 4L, 4F/100 gal. water
1.5-2 lb. 50W/A
1-1.25 lb. 80S, 80WSP/100 gal. water
Pest:

black scale	California red scale	citrus snow scale
brown soft scale	citricola scale	yellow scale

Rate: 1 lb. ai/100 gal. water
2 pt. SL, 4L, 4F/100 gal. water
2 lb. 50W/A
1.25 lb. 80S, 80WSP/100 gal. water
Pest:

avocado leafroller	citrus snow scale	West Indian sugarcane borer
California orangedog	fruittree leafroller	western tussock moth
citrus cutworm	orange tortrix	
citrus root weevil		

Remarks: Citrus citron, grapefruit, kumquats, lemons, limes, oranges, tangelos, tangerines. Rate varies by pest species; refer to label. Do not apply less than 10 gal. dilute spray mixture/mature tree. May be mixed with petroleum oils commonly used on citrus. Refer to California's UCIPM Pest Management Guidelines or Florida Citrus Pest Management Guide for instructions and precautions for these states.
Days To Harvest: 5.

Signal Word/Toxicity Class: Caution/III (4F, 4L); Warning/II (SL, 50W, 80S, 80WSP).
REI: 12 hr.

Clean Crop Sevin 5 Bait *(Platte)*
Rate: 2-2.4 lb. ai/A
40-48 lb. Bait/A
Pest:

ants	darkling beetles	grasshoppers
armyworms	field crickets	sowbugs
cutworms		

Remarks: Apply with ground or air equipment.
Days To Harvest: 5.

Signal Word/Toxicity Class: Caution/III.
REI: 12 hr.

CARZOL

Toxic to wildlife and toxic to bees exposed to direct treatment. May increase citrus red mite populations.

Carzol (SP) (formetanate hydrochloride) *(AgrEvo USA)*
Rate: 0.46-1.38 lb. ai/A
0.5-1.5 lb. SP/A
Pest:
thrips

8 Fruits & Nuts

NOTICE The information on these pages is for preliminary planning — not a guide for use. Be sure to follow the manufacturer's directions, notwithstanding information contained here.

For personal protective equipment and EPA registration numbers, please turn to page 39.

Remarks: Grapefruit, lemons, limes (CA only), oranges, tangerines, tangelos. Apply as outside coverage spray when thrips appear. For aerial application, use minimum 10 gal. water/A. Allow 14 days between applications. Do not apply more than 5 lb./A per 12 month period. Do not apply more than twice per season in California. Do not apply through any type of irrigation system.
Days To Harvest: 7.
State Restrictions: For use in Arizona, California (refer to UCIPM Pest Management Guidelines).

Rate: 0.92 lb. ai/A
1 lb. SP/A
Pest:
Apopka weevil, blue-green weevil, citrus root weevil, Fuller rose beetle, little leaf notcher

Rate: 0.46-0.92 lb. ai/A
0.5-1 lb. SP/A
Pest:
rust mite
Remarks: Grapefruit, lemons, limes, oranges, tangerines. Apply when mites appear; repeat as needed. Allow 30 days between applications. Apply up to 5 lb./A as needed per 12 month period. Do not apply through any type of irrigation system.
Days To Harvest: 7.
State Restrictions: For use in Florida (refer to Citrus Pest Management Guide).

Rate: 0.115-0.23 lb. ai/100 gal. water
0.125-0.25 lb. SP/100 gal. water
Pest:
rust mite
Remarks: Grapefruit, lemons, oranges, tangerines. Apply when mites appear; repeat as needed. For full cover dilute spray, apply 500 gal. water/A. Allow 30 days between applications. Apply up to 5 lb./A as needed per 12 month period. Do not apply through any type of irrigation system.
Days To Harvest: 7.
State Restrictions: For use in Texas.

Signal Word/Toxicity Class: Danger-Poison/I.
REI: 48 hr.

Chipco Sevin — see carbaryl

Chlorfos — see chlorpyrifos

CHLORPYRIFOS

Organophosphate insecticide registered on citrus. May increase spider mite populations. Highly toxic to honeybees.

Lorsban-4E *(Dow AgroSciences)*
Lorsban 50-W *(Griffin)*
◆ **Micro Flo Chlorpyrifos 4E** *(Micro Flo)*
Rate: Arizona, California:
1-3.5 lb. ai/A
2-7 pt. 4E/A
1-2 lb. ai/A
2-4 lb. 50-W
Pest:
aphids, avocado leafroller, cutworms, fruittree leafroller, katydids, orange tortrix, western tussock moth

Rate: Arizona, California:
3-6 lb. ai/A
6-12 pt. 4E/A
6-12 lb. 50-W
Pest:
mealybugs, thrips (suppression)

Rate: Arizona, California:
4-6 lb. ai/A
8-12 pt. 4E/A
8-12 lb. 50-W
Pest:
black scale, brown scale, California red scale

Rate: Florida:
1-3.5 lb. ai/A
2-7 pt. 4E/A
2-7 lb. 50-W
Pest:
aphids, brown citrus aphid, citrus rust mite, grasshoppers, mealybugs, orangedog, scales

Rate: Florida:
2-3.5 lb. ai/A
4-7 pt. 4E/A
4-7 lb. 50-W
Pest:
citrus rust mite

Rate: Texas:
2-3.5 lb. ai/A
4-7 pt. 4E/A
4-7 lb. 50-W/A
Pest:
aphids, brown citrus aphid, cutworms, katydids, mealybugs, scales
Remarks: Grapefruit, lemons, oranges, etc. For small transplanted citrus trees in Texas, apply 1 oz./1 gal. water (max. 7 pt./lb.) with backpack sprayer to runoff. In Arizona and California, do not apply during Dec., Jan., or Feb., or use in combination with spray oil when temperatures are expected to exceed 95°F for several consecutive days after application. Do not apply more than twice per year or more than 15 pt./lb. per year. Do not apply second foliar application within 30 days of first application. Do not allow livestock to graze treated areas. Refer to label and to California's UCIPM Pest Management Guidelines or Florida Citrus Pest Management Guide for instructions and precautions.
Days To Harvest: Up to 7 pt.-21; over 7 pt.-35.

Rate: 0.75-1 lb. ai/A
1.5-2 pt. 4E/A
Pest:
ants, red imported fire ant
Remarks: Orchard floor. Apply minimum 25 gal. water/A by ground uniformly to orchard floor. Foliar application may be made in addition to orchard floor. Do not tank mix with Evik herbicide. Do not apply more than 20 pt./A per season (6 pt./A in Florida). Do not allow livestock to graze treated areas. Refer to label and to California's UCIPM Pest Management Guidelines or Florida Citrus Pest Management Guide for instructions and precautions.
Days To Harvest: 28.

Signal Word/Toxicity Class: Warning/II.
REI: 24 hr.

◆ **Chlorfos 15G** *(Griffin)*
Lorsban 15G *(Dow AgroSciences)*
◆ **Micro Flo Chlorpyrifos 15G** *(Micro Flo)*
Rate: 1 lb. ai/A
6.7 lb. 15G/A
Pest:
ants, red imported fire ant
Remarks: Orchard floor. For foraging ants and to suppress mounds, apply by ground; repeat as needed. Do not use where weed growth or other obstructions would impede uniform coverage of orchard floor. Do not apply more than 67 lb./A per season (20.1 lb./A in Florida). Do not allow livestock to graze treated areas. Refer to label and to California's UCIPM Pest Management Guidelines or Florida Citrus Pest Management Guide for instructions and precautions.
Days To Harvest: 28.

Signal Word/Toxicity Class: Caution/III.
REI: 12 hr.

Citri-Oil — see petroleum oil

Clean Crop Sevin — see carbaryl

Comite — see propargite

◆ - **new product** • ai-active ingredient • DF-dry flowable • E/EC-emulsifiable concentrate • EW-emulsion in water
F/FL-flowable • G-granule • L/LC-liquid • SP/WSP-soluble packet • W/WP-wettable powder • WSB-water soluble bag

CRYOLITE

Kryocide (96%) *(Elf Atochem)*

Rate: 7.7-19.8 lb. ai/A
8-20 lb. 96%/A

Pest:

blue-green weevil (except CA)	Fuller rose beetle	orange tortrix
citrus cutworm	garden tortrix	orangedog
fruittree leafroller	katydids	variegated cutworm

Remarks: Apply at 30-day intervals as needed. Do not exceed 90 lb./A per season. Do not use chemigation in California.

Days To Harvest: 15.

Signal Word/Toxicity Class: Caution.
REI: 12 hr.

Prokil Cryolite 96 *(Gowan)*

Rate: 24-29 lb. ai/A
25-30 lb. 96%/A

Pest:

citrus cutworm	garden tortrix	orange tortrix
fruittree leafroller	Halcocera	orangedog
Fuller rose beetle	katydids	variegated cutworm

Remarks: Grapefruit, lemons, oranges. Apply at 30-day intervals as needed. Do not exceed 90 lb./A per season. Do apply through any type of irrigation system.

Days To Harvest: 15.

Signal Word/Toxicity Class: Caution.
REI: 12 hr.

Damoil — see petroleum oil

DEADLINE

Packaged as Bullets, M-Ps (mini-pellets), and granules. Resistant to rain and watering.

Deadline Bullets, M-Ps, Granules (4%) (metaldehyde) *(Pace)*

Rate: 0.8-1 lb. ai/A
20-40 lb. 4%/A

Pest:

slugs snails

Remarks: Apply every 3-4 weeks during growing season or as needed. Do not apply directly to edible portion of plant.

Signal Word/Toxicity Class: Warning/II (granule); Caution (bullets, M-P).
REI: 12 hr.

Diatect — see pyrethrins

DIBROM

Short residual organophosphate insecticide. Performs well in dormant spray in combination with supreme oils. Toxic to fish, aquatic invertebrates, and wildlife. Highly toxic to bees.

Dibrom 8 (8E) (naled) *(Amvac)*

Rate: 1-2 lb. ai/A
1-2 pt. 8E/A

Pest:

aphids	leafrollers	spider mites
brown scale (crawler)	orange tortrix	thrips
citrus cutworm		

Remarks: Grapefruits, lemons, oranges, tangerines. Apply at first sign of insects; repeat at 7-day intervals. For ground application, apply in 100-500 gal. water on mature trees, 15 gal water on young fruit plantings; aerial application, in min. 10 gal. water. Do not apply more than 6 pt./A per season or more than 1 pt./A in Florida.

Days To Harvest: 7.

Signal Word/Toxicity Class: Danger/I.
REI: 48 hr.-1 pt./A; 72 hr.-over 1 pt./A.

DICOFOL

Chlorinated hydrocarbon miticide. May increase citrus snow scale populations if applied in 500 or more gal. water per acre. Toxic to fish.

◆ **Griffin Dicofol 4E** *(Griffin)*
Kelthane MF (4E) *(Rohm and Haas)*

Rate: 0.5 lb. ai/100 gal. dilute spray
1 pt. 4E/100 gal. dilute spray

Pest:

citrus flat mite	citrus rust mite	Texas citrus mite
citrus red mite	sixspotted spider mite	Yuma mite

Remarks: Grapefruit, kumquats, lemons, limes, oranges, tangelos, tangerines. Apply when mites appear in threatening numbers; repeat as needed. Do not apply through any type of irrigation system. Refer to California's UCIPM Pest Management Guidelines or Florida Citrus Pest Management Guide for instructions and precautions.

Days To Harvest: 7.

Signal Word/Toxicity Class: Caution.
REI: 12 hr.

Digon — see dimethoate

Dimate — see dimethoate

DIMETHOATE

Organophosphate insecticide registered for oranges, grapefruit, lemons, and tangerines. May increase citrus red mite populations. Highly toxic to honeybees.

Cheminova Dimethoate 4E *(Cheminova)*
Dimate 4E *(Terra)*
Drexel Dimethoate 2.67 EC or 4EC *(Drexel)*
Gowan Dimethoate E267 *(Gowan)*
Helena Dimethoate 4E *(Helena)*
◆ **Micro Flo Dimethoate 267 or 4E** *(Micro Flo)*
Wilbur-Ellis Dimethoate 267 or Digon 400 *(Wilbur-Ellis)*

Rate: 0.25-0.5 lb. ai/100 gal. water
0.75-1.5 pt. 2.67, 267, E267/100 gal. water
0.5-1 pt. 4E, 4EC, 400/100 gal. water

Pest:

aphids	thrips	whiteflies (high rate)
mites (except rust)		

Rate: 0.5-0.75 lb. ai/100 gal. water
1.5-2.25 pt. 2.67, 267, E267/100 gal. water
1-1.5 pt. 4E, 4EC, 400/100 gal. water
Dilute ground/aerial application:

Pest:

scales (except black, snow)

Remarks: Grapefruit, lemons, oranges, tangerines. Do not use on citrus seedlings or apply during bloom period. Do not make more than 2 applications to mature fruit. Do not graze livestock in treated orchards. Refer to California's UCIPM Pest Management Guidelines or Florida Citrus Pest Management Guide for instructions and precautions.

Days To Harvest: 15; scales-45.

Rate: Foliar spray:
0.5 lb. ai/100 gal. water
1.5 pt. 2.67, 267, E267/100 gal. water
1 pt. 4E, 4EC, 400/100 gal. water
Soil drench:
2 lb. ai/A
6 pt. 2.67, 267, E267/A
4 pt. 4E, 4EC, 400/A

Pest:

aphids thrips

Remarks: Nonbearing. For foliar, repeat as necessary. For soil drench (trees 1-3 yr. old), apply in furrow or soil ring around base of tree. Apply as soon as insect injury to new growth begins. Nursery stock application should not be made in year trees begin to bear fruit. Do not graze livestock in treated orchards.

State Restrictions: For use in Arizona, California (refer to UCIPM Pest Management Guidelines).

Signal Word/Toxicity Class: Warning/II.
REI: 4 days.

NOTICE The information on these pages is for preliminary planning — not a guide for use. Be sure to follow the manufacturer's directions, notwithstanding information contained here.

For personal protective equipment and EPA registration numbers, please turn to page 39.

Drexel Suffa — see sulfur

ENDOSULFAN

Drexel Endosulfan 3EC or 50WP *(Drexel)*
Gowan Endosulfan 3EC or 50WSB *(Gowan)*
◆ Micro Flo Endosulfan 3EC or 50WP *(Micro Flo)*
Phaser 3EC or 50WSB *(AgrEvo USA)*
Thiodan 3EC or 50WP *(FMC)*

Rate: 2.5 lb. ai/A
3.33 qt. 3EC/A
5 lb. 50WP, 50WSB/A

Pest:
citrus aphid

Remarks: Nonbearing trees and nursery stock. Do not apply to bearing trees or trees that will bear fruit within 12 months. Do not apply through any type of irrigation system. Do not exceed 2 applications per year or 3 lb. ai/A per year.

Signal Word/Toxicity Class: Danger-Poison/I.
REI: 24 hr.

ETHION

Organophosphate miticide/insecticide recommended alone or in combination with petroleum oil, scale, whitefly, spider mites, and sooty mold. Ethion + oil combination recommended for use in summer spray only. As an insecticide, recommended for citrus snow scale control as dilute spray only.

Ethion 4 Miscible *(FMC)*
Ethion 4EC *(Micro Flo)*

Rate: 2.5-3 lb. ai/A
5-6 pt. 4, 4EC/A

Pest:

citrus blackfly	Florida red scale	snow scale
citrus red mite	glover scale	Texas citrus mite
citrus rust mite	purple scale	yellow scale
citrus whitefly	sixspotted spider mite	

Rate: 2.5-3 lb. ai/A
5-6 pt. 4, 4EC/A + 0.5%-0.7% summer oil

Pest:

black scale	citrus red mite	purple scale
brown soft scale	citrus rust mite	sixspotted spider mite
California red scale	citrus whitefly	snow scale
chaff scale	Florida red scale	Texas citrus mite
citrus blackfly	glover scale	yellow scale

Remarks: Grapefruit, lemons, limes, oranges, tangelos, tangerines. Apply in sufficient water for thorough coverage when insects first appear; repeat as necessary. Coverage of upper and lower leaf surfaces essential for good control. Where snow scale is a problem, spray trunk, limbs, and twigs thoroughly. Do not use oil on navel oranges. Do not make applications less than 90 days apart. Do not apply more than 9 lb. ai/A per year. Observe use limitations. Refer to oil label for use directions and precautions.
Days To Harvest: Grapefruit, oranges-0; lemons, limes, tangelos, tangerines-15.
State Restrictions: For use in Florida and Texas.

Signal Word/Toxicity Class: Warning/II.
REI: 48 hr.

FENOXYCARB

Insect growth regulator for control of fire ants on nonbearing citrus trees.

Logic *(Novartis)*

Rate: Mound treatment:
1-3 Tbs./mound
Broadcast or aerial application:
1-1.5 lb./A

Pest:
fire ants

Remarks: Nonbearing. Mound treatment: Uniformly distribute material 3-4 ft. around mound. Broadcast or aerial application: Apply with equipment calibrated to give correct dosage. Do not apply by air where weather conditions favor drift from treated areas. Retreat after 3-4 months, if needed. Apply when ants are actively foraging, usually when soil temperatures are above 60°F. Avoid application during excessively hot periods or when grass is wet. Do not apply to nonbearing crops more than twice per year or sooner than 1 year before first harvest. Do not graze treated areas. Refer to Florida Citrus Pest Management Guide for instructions and precautions.

Signal Word/Toxicity Class: Caution.
REI: 12 hr.

Fyfanon — see malathion

Guthion — see azinphos-methyl

Kelthane — see dicofol

Kryocide — see cryolite

Kumulus — see sulfur

LANNATE

Carbamate insecticide for thrips and orangeworm control. Restricted Use Pesticide.

Lannate SP or LV (methomyl) *(DuPont)*

Rate: 0.45-0.9 lb. ai/A
0.5-1 lb. SP/A
1.5-3 pt. LV/A

Pest:

beet armyworm	fruittree leafroller (grapefruit, oranges)	thrips
citrus cutworm (oranges)	orange tortrix	western tussock moth

Remarks: Do not apply more than 2.7 lb. ai/A per crop or make more than 4 applications per crop.
Days To Harvest: 1; livestock grazing/feeding-10.
State Restrictions: For use in Arizona, California.

Signal Word/Toxicity Class: Danger-Poison/I.
REI: 72 hr.

Lime-Sulfur Solution — see calcium polysulfide

Logic — see fenoxycarb

Lorsban — see chlorpyrifos

MALATHION

Organophosphate insecticide highly toxic to honeybees.

Clean Crop Malathion 57EC *(Platte)*

Rate: 0.9 lb. ai/A
1.5 pt. 57EC/A

Pest:

black scale	citricola scale	soft scale
California red scale	purple scale	yellow scale

Rate: 0.78 lb. ai/A in 100 gal. water
1.25 pt. 57EC/A in 100 gal. water

Pest:

California red scale (light infestations)	purple scale (light infestations)	thrips

Remarks: Grapefruit, lemons, limes, oranges, tangerines. For moderate to heavy infestations of red and purple scale, up to 2 pt./100 gal. may be applied. Do not apply during full bloom.
Days To Harvest: 7.

Signal Word/Toxicity Class: Caution/III.
REI: 12 hr.

◆ - **new product** • ai-active ingredient • DF-dry flowable • E/EC-emulsifiable concentrate • EW-emulsion in water
F/FL-flowable • G-granule • L/LC-liquid • SP/WSP-soluble packet • W/WP-wettable powder • WSB-water soluble bag

Clean Crop Malathion 8EC *(Platte)*
◆ **Wilbur-Ellis Malathion 8** *(Wilbur-Ellis)*

Rate: 1-2 lb. ai/100 gal. water
1-2 pt. 8EC, 8/100 gal. water

Pest:
aphids scale crawlers

Rate: 2.5 lb. ai/200 gal. water
2.5 pt. 8EC, 8/200 gal. water

Pest:
thrips

Remarks: Grapefruit, kumquats, lemons, limes, oranges, tangelos, tangerines. Do not apply when trees are in bloom.

Days To Harvest: 7.

Signal Word/Toxicity Class: Caution/III.
REI: 12 hr.

Drexel Malathion 5EC *(Drexel)*
◆ **Micro Flo Malathion 5EC** *(Micro Flo)*
Prentox 5 lb. Malathion Spray (5EC) *(Prentiss)*

Rate: 0.63-1 lb. ai/100 gal.
1-1.5 pt. 5EC/100 gal.

Pest:
black scale (single, off-brooded)
California red scale
citricola scale
purple scale
soft scale
yellow scale

Rate: 1.25 lb. ai/100 gal.
2 pt. 5EC/100 gal.

Pest:
Florida purple scale Florida red scale

Rate: 1.6 lb. ai/A
2.5 pt. 5EC/A in 200 gal. water

Pest:
citrus red mites thrips

Remarks: Grapefruit, kumquats, lemons, limes, oranges, tangelos, tangerines. Do not apply when trees are in bloom. For mature trees, apply 500 gal./A dilute spray.

Days To Harvest: 7.

Signal Word/Toxicity Class: Warning/II.
REI: 12 hr.

Fyfanon (8E) *(Helena)*

Rate: 1 lb. ai/A
1 pt. 8E/A

Pest:
aphids
black scale (single, off-brooded)
California red scale
citricola scale
purple scale
soft scale
thrips
yellow scale

Remarks: Do not apply when trees are in bloom. In California, do not apply unless parasite activity in scale is at low level or absent.

Days To Harvest: 7.

Signal Word/Toxicity Class: Caution/III.
REI: 12 hr.

◆ **Gowan Malathion 8 (8E)** *(Gowan)*

Rate: 1-8 lb. ai/A
1-8 pt. 8E/A

Pest:
Mediterranean fruit fly

Rate: 7-25 lb. ai/A
7-25 pt. 8E/A

Pest:
aphids
black scale (single, off-brooded)
California red scale
citricola scale
orangeworm
purple scale
soft scale
thrips
yellow scale

Remarks: Grapefruit, kumquats, lemons, limes, oranges, tangelos, tangerines. Do not apply when trees are in bloom.

Days To Harvest: 7.

Signal Word/Toxicity Class: Caution/III.
REI: 24 hr.

MICROMITE

Will not kill adult weevils or adult stages of leafminers. Restricted Use Pesticide due to toxicity to aquatic invertebrates.

Micromite (25WS) (diflubenzuron) *(Uniroyal)*

Rate: 5 oz. ai/A
20 oz. 25WS/A

Pest:
citrus leafminer citrus root weevil citrus rust mite

Remarks: Grapefruit, oranges, tangerines. Apply by ground. For leafminers, apply when oviposition begins. For weevils, when adults are present or to newly expanded flush. For mites, apply when populations are low. Do not apply more than 20 oz./A. Do not harvest cover crops for animal feed or graze livestock in treated groves. Do not apply when weather conditions favor drift to surface water, within 110 ft. upwind of surface water, when wind speed is above 8 mph, or during a temperature inversion. Do not apply this product through any type of irrigation system.

Days To Harvest: 21.

State Restrictions: For use in Florida only. Do not apply east of U.S. Highway 1, south and east of State Rd. 846/832, or south of West Palm Beach Canal.

Signal Word/Toxicity Class: Caution/III.
REI: 48 hr.

Microthiol Special — see sulfur

MORESTAN

Heterocyclic compound with miticidal/fungicidal activity. Used on citrus to control citrus rust and spider mites during postbloom period. Toxic to fish.

Morestan Solupak 25 (oxythioquinox) *(Bayer)*

Rate: 0.25-0.375 lb. ai/100 gal. water
1-1.5 lb. Solupak 25/100 gal. water

Pest:
citrus red mites
citrus rust mites
mites (eggs)
Texas citrus mites

Remarks: Nonbearing. Spray all surfaces, including underside of leaves for complete coverage; repeat as necessary. Do not apply to trees that will bear fruit within 1 year of application. Refer to California's UCIPM Pest Management Guidelines or Florida Citrus Pest Management Guide for instructions and precautions.

Rate: 0.625-1.25 lb. ai/A
2.5-5 lb. Solupak 25/A

Pest:
citrus red mites
citrus rust mites
mites (eggs)
Texas citrus mites

Remarks: Bearing trees. Apply by concentrate or dilute ground spray using 20-500 gal. water/A. Schedule applications no closer than 90 days apart in Florida, Texas or 180 days apart in Arizona, California. Refer to California's UCIPM Pest Management Guidelines or Florida Citrus Pest Management Guide for instructions and precautions.

Days To Harvest: 21.

State Restrictions: For use in Arizona, California, Florida, Texas.

Rate: 1.25 lb. ai/A
5 lb. Solupak 25/A

Pest:
mites (eggs) mites

Remarks: Refer to label for use directions and precautions.

State Restrictions: SLN-California.

Rate: 0.625-1.25 lb. ai/A
2.5-5 lb. Solupak 25/A

Pest:
red spider mites

Remarks: Refer to label for use directions and precautions.

State Restrictions: SLN-Louisiana.

Signal Word/Toxicity Class: Warning/II.
REI: 24 hr.

8 Fruits & Nuts

NOTICE	The information on these pages is for preliminary planning — not a guide for use. Be sure to follow the manufacturer's directions, notwithstanding information contained here.	For personal protective equipment and EPA registration numbers, please turn to page 39.

M-PEDE

M-Pede (potassium salts of fatty acids) *(Mycogen)*

Rate: 1-2% v/v solution

Pest:

aphids, mealybugs, spider mites, citrus rust mite, scale crawlers, whiteflies, leafhoppers

Remarks: Citron, grapefruit, kumquats, lemons, limes, tangelos, tangerines. Apply to wet with air blast or other appropriate sprayer. May be tank mixed with other insecticides; refer to both labels.

Days To Harvest: 0.

Signal Word/Toxicity Class: Warning/II.
REI: 12 hr.

NEEMAZAD — see azadirachtin

NEEMIX — see azadirachtin

NEMACUR

Restricted Use Pesticide.

Nemacur 3 (fenamiphos) *(Bayer)*

Rate: Band:
5-10 lb. ai/A
1.66-3.33 gal. 3/A
Low pressure irrigation:
1.5-4.5 lb. ai/A
2-6 qt. 3/A

Pest:

citrus root weevil (suppression), Fuller rose beetle (suppression), nematodes

Remarks: For band application, incorporate immediately. Refer to label for further use information. Refer to California's UCIPM Pest Management Guidelines or Florida Citrus Pest Management Guide for instructions and precautions.

Days To Harvest: 30.

State Restrictions: In California, do not apply to kumquats, tangelos, or citrus/hybrids.

Signal Word/Toxicity Class: Danger/I.
REI: 48 hr.

Nexter — see pyridaben

Omite — see propargite

Omni Supreme Spray — see petroleum oil

Orthene — see acephate

PETROLEUM OIL

Oils meeting FC435-66 standards are recommended for use on Florida citrus to control certain mites, scale insects, and whiteflies, and to loosen sooty mold. Oils smother arthropod pests so thorough coverage is required to achieve optimal control.

◆ **435 or 455 Oil 98.8** *(Drexel)*
◆ **SunSpray 6E, 7E, or 9E** *(Sun)*

Rate: 1.5 gal. 435, 6E/100 gal. water
1 gal. 455, 7E, 9E/100 gal. water

Pest:

citrus rust mite, scales

Rate: 0.75 gal. 435, 6E/100 gal. water
0.5 gal. 455, 7E, 9E/100 gal. water

Pest:

spider mites

Remarks: Refer to label for application timing. Apply in enough water for thorough coverage.

Signal Word/Toxicity Class: Caution.
REI: 4 hr.

Citri-Oil *(Wilbur-Ellis)*

Rate: 5-8 gal./A

Pest:

California red scale, Florida red scale, Texas citrus mite, chaff scale, purple scale, whiteflies, citrus red mite

Remarks: Spring or summer spray. Do not apply within 30 days of applying sulfur. Do not apply to trees that are under moisture stress. Do not apply in periods of excessive temperature (90°F) or low relative humidity (20%).

Signal Word/Toxicity Class: Caution.
REI: 12 hr.

Damoil *(Drexel)*
Volck Supreme Spray *(Valent)*

Rate: Grapefruit, lemons:
1-1.25 gal./100 gal. dilute spray
Valencia oranges:
1 gal./100 gal. dilute spray
Navel, Valencia oranges:
3 qt. product + 1 lb. Malathion 25W

Pest:

black scale, red scale, yellow scale, purple scale, red spider mite

Remarks: Grapefruit, lemons, oranges (navel, Valenica). For grapefruit and lemons, spray during winter and spring months. For navel oranges, apply no later than Sept. 1; no later than Sept. 15 for Valencia oranges. Refer to California's UCIPM Pest Management Guidelines or Florida Citrus Pest Management Guide for instructions and precautions.

Days To Harvest: Lemons-8 weeks.

Signal Word/Toxicity Class: Caution.
REI: 12 hr.

Omni Supreme Spray *(Helena)*

Rate: 1.2-1.4 gal./A

Pest:

black scale, flat mite (lemons, limes), red scale, bud mite (lemons, limes), red spider mite, purple scale, yellow scale, citricola scale

Remarks: Citron, grapefruit, lemons, limes, oranges (mandarin, navel, Valenica). Refer to label for application timing. Apply in enough water for thorough coverage.

Days To Harvest: Lemons-8 weeks.

Signal Word/Toxicity Class: Caution.
REI: 4 hr.

SAF-T-SIDE *(Brandt Consolidated)*
◆ **Sumulsion** *(Drexel)*

Rate: 1-2 gal./100 gal. water

Pest:

blackfly, Glover scale, snow scale, brown scale, mites, soft scale, California red scale, purple scale, whiteflies, chaff scale, red scale, yellow scale

Remarks: Will kill other immature forms if spray covers insect. Apply fall, early winter, postbloom, March-April. Time application to fit local growth and climate conditions. Compatible with most commonly used insecticides and fungicides. Do not use in combination with or immediately before or after spraying with fungicides such as captan, Dyrene, folpet, Morestan, or any sulfur products. Do not use with carbaryl or dimethoate.

Signal Word/Toxicity Class: Caution.
REI: 4 hr.

SunSpray Ultra-Fine *(Sun)*

Rate: 1-1.6 gal./100 gal. water

Pest: Grapefruit, lemons:

California red scale, mites, scales

Oranges and other citrus varieties including nonbearing:

black scale, mites, spider mites, blackfly, rust mite, whiteflies, California red scale, scales

◆ - **new product** • ai-active ingredient • DF-dry flowable • E/EC-emulsifiable concentrate • EW-emulsion in water
F/FL-flowable • G-granule • L/LC-liquid • SP/WSP-soluble packet • W/WP-wettable powder • WSB-water soluble bag

Remarks: Grapefruit, lemons, oranges (navel, Valenica), and other citrus varieties including nonbearing. Refer to label for application timing.

Signal Word/Toxicity Class: Caution.
REI: 4 hr.

Phaser — see endosulfan

PROPARGITE

Sulfite compound for control of various mite pests.

Comite *(Uniroyal)*
Rate: 1.64-2.46 lb. ai/A
2-3 pt./A
Pest:

broad mite	citrus rust mite	sixspotted mite
citrus red mite	false spider mite	Texas citrus mite

Remarks: Grapefruit, oranges. Do not apply early in the season during leaf expansion. Do not apply within 2 weeks prior to or following oil application. Do not apply more than twice per year. Do not apply through any type of irrigation system. Refer to Florida Citrus Pest Management Guide for instructions and precautions.
Days To Harvest: 21.
State Restrictions: For use in Florida, Texas.

Signal Word/Toxicity Class: Danger/I.
REI: 48 hr.

Omite 30W or 30WS *(Uniroyal)*
Rate: 0.6-6.6 lb. ai/A
2-22 lb. 30W, 30WS/A
Pest:
spider mites
Remarks: Lemons. Do not apply through any type of irrigation system.
Days To Harvest: 10.5 lb./A or less-21; more than 10.5 lb./A-28.
State Restrictions: For use in Arizona, California.

Signal Word/Toxicity Class: Danger/I.
REI: 48 hr.

Omite-CR *(Uniroyal)*
Rate: Lemons-Arizona:
1.5-4.5 lb. ai/A
5-15 lb./A in 600-1500 gal. spray (ground)
Pest:
citrus red mite

Rate: Grapefruit, lemons, oranges-California:
3.15 lb. ai/A
10.5 lb./A in 1000 gal. spray (ground)
Pest:
citrus red mite — twospotted spider mite
Remarks: Do not apply through any type of irrigation system.
Days To Harvest: Arizona-28; California-42.

Signal Word/Toxicity Class: Danger/I.
REI: 48 hr.

PYRELLIN (Premix)

Premix containing pyrethrins and rotenone. May be combined with other insecticides to enhance knockdown or as an insect "exciter" to help facilitate contact with other spray residues.

Pyrellin E.C. (0.042 lb./gal. pyrethrins & 0.035 lb./gal. rotenone) *(Webb Wright)*
Rate: 1-2 pt. E.C./A
Pest:

aphids	Fuller rose beetle	whiteflies
caterpillars	thrips	

Remarks: Apply as necessary to maintain control.
Days To Harvest: 0.

Signal Word/Toxicity Class: Caution/III.
REI: 12 hr.

PYRIDABEN

Pyridoxone miticide/insecticide, a wettable powder in water-soluble bags, that provides knockdown and residual control in citrus groves. Toxic to bees, fish, and aquatic invertebrates.

Nexter (WP) *(BASF)*
Rate: 0.26-0.5 lb. ai/A
5.2-10.67 oz. WP/A (0.8 WSB/A)
Pest: Mites:

broad mite	false spider mite	Texas citrus mite
citrus red mite	pink citrus rust mite	twospotted spider mite
citrus rust mite	sixspotted spider mite	

Remarks: Do not make more than 2 applications per year or apply more than 21.3 oz./A per season. Do not apply by air or through any type of irrigation system. Refer to label for spray drift precautions.
Days To Harvest: 7.

Signal Word/Toxicity Class: Danger/I.
REI: 12 hr.

Pyrenone — see pyrethrins

PYRETHRINS

Contains piperonyl butoxide as a synergist. Diatect also contains silicon dioxide (diatomaceous earth). May be combined in tank mix with other insecticides for quicker and more complete kill where insect resistance may be a problem, and as an "exciter" to flush insects out of hiding and into contact with other spray residues.

Diatect Multipurpose Insecticide II *(Diatect)*
Pyrenone (0.5EC) *(AgrEvo Environmental Health)*
Py-Rin 60-6 EC *(Wilbur-Ellis)*
Rate: 1-6 lb. II/A
1-12 fl. oz. 0.5EC/A
2-16 oz. 6-60 EC/A
Pest: Including but not limited to:

aphids	fruit flies	thrips
caterpillars	mites	whiteflies

Remarks: Grapefruit, kumquats, lemons, limes, oranges, tangelos, tangerines, etc. Apply in sufficient water for thorough coverage; repeat as needed. Apply Diatect as dust or wettable powder.
Days To Harvest: 0.

Signal Word/Toxicity Class: Caution.
REI: 12 hr.

Py-Rin — see pyrethrins

SAF-T-SIDE — see petroleum oil

Sevin — see carbaryl

Sniper — see azinphos-methyl

SPINOSAD

Fermentation-derived insect control agent. Does not significantly impact predaceous and parasitic insect populations. Highly toxic to bees and aquatic invertebrates.

◆ **Success Naturalyte (SC) (spinosyn A + spinosyn D)** *(DowAgroSciences)*
Rate: 0.062-0.156 lb. ai/A
4-10 oz. SC/A
Pest:

avocado leafroller	cutworm	orange tortrix
citrus peelminer	fruittree leafroller	western tussock moth
citrus thrips	katydids	

* *Suppression*
Remarks: Grapefruit, lemons, limes, oranges, tangerines, etc. Do not apply to citrus nurseries or citrus in greenhouses. Do not make more than 2 applications per year or apply more than 29 oz./A per crop. Do not apply through any type of irrigation equipment.
Days To Harvest: 1.

Signal Word/Toxicity Class: Caution.
REI: 4 hr.

8 Fruits & Nuts

NOTICE The information on these pages is for preliminary planning — not a guide for use. Be sure to follow the manufacturer's directions, notwithstanding information contained here. For personal protective equipment and EPA registration numbers, please turn to page 39.

Success Naturalyte — see spinosad

SULFUR

Microthiol and Thiolux are micronized sulfurs. Elemental sulfur compounds are used primarily to control citrus rust mite. One application per season recommended and only as a supplemental spray between main spray applications. Repeated use of sulfur increases citrus red mite and armored scale populations. Not compatible with oil and cannot be applied within 3 weeks of an oil spray as fruit burn may result. Eye irritation to applicators and fruit harvesters can occur.

Drexel Suffa 6 or 8 or Sulfur 90W *(Drexel)*

Rate: 0.75-3 lb. ai/A
1-4 pt. 6/A
0.75-3 pt. 8/A
0.8-3.2 lb. 90W/A

Pest:
rust mite

Remarks: Grapefruit, lemons, oranges, tangelos, tangerines. Apply when insects appear. Refer to California's UCIPM Pest Management Guidelines or Florida Citrus Pest Management Guide for instructions and precautions.

Signal Word/Toxicity Class: Caution.
REI: 12 hr. (Suffa 8); 24 hr. (Suffa 6, Sulfur 90W).

◆ Kumulus DF *(BASF)*

Rate: 6.4-16 lb. ai/A
8-20 lb. DF/A

Pest:
flat mite, rust mite, silver mite, red spider mite

Remarks: Apply in any spray not containing oil. Do not apply if temperatures during or within 3 days after application are expected to exceed 90°F. Do not apply through any type of irrigation system. Refer to label for use directions.

Signal Word/Toxicity Class: Caution.
REI: 24 hr.

Microthiol Special (80W) *(Elf Atochem)*

Rate: 8-12 lb. ai/A
10-15 lb.80W/A

Pest:
broad mite*, flat mite, sixspotted spider mite*, bud mite*, red spider mite, thrips*, citrus rust mite*, silver mite, Yuma spider mite*

** Except California*

Remarks: Citrus. Apply as needed.

Signal Word/Toxicity Class: Caution.
REI: 24 hr.

THAT (6FL) or THAT Big 8 (8FL) *(Stoller)*

Rate: 0.75-3 lb. ai/100 gal. water
1-4 pt. 6FL/100 gal. water
1-3 lb. ai/100 gal. water
1-3 pt. 8FL/100 gal. water

Pest:
clover mite, rust mite

Rate: 1.88-2.8 lb. ai/100 gal. water
2.5-3.75 pt. 6FL/100 gal. water
2-3 lb. ai/100 gal. water
2-3 pt. 8FL/100 gal. water

Pest:
thrips

Remarks: Grapefruit, lemons, limes, oranges, tangelos, tangerines. For thrips, apply at spring flush at 3-4" growth and repeat if needed.

Signal Word/Toxicity Class: Caution.
REI: 24 hr.

Thiolux DF *(Platte)*

Rate: 6.4-16 lb. ai/A
8-20 lb. DF/A

Pest:
bud mite, rust mite, thrips, flat mite, silver mite, Yuma spider mite, red spider mite, sixspotted spider mite

Remarks: Apply in any spray not containing oil.

Signal Word/Toxicity Class: Caution.
REI: 24 hr.

Sumulsion — see petroleum oil

SunSpray — see petroleum oil

SUPRACIDE

Organophosphate insecticide used on citrus to control scale insects and whiteflies. May increase citrus rust mite and spider mite populations. Highly toxic to honeybees. Restricted Use Pesticide.

Supracide 25WP (methidathion) *(Gowan)*

Rate: Arizona, California:
0.25-0.5 lb. ai/100 gal. water
1-2 lb. 25WP/100 gal. water

Pest:
black scale, California red scale, mealybugs, brown soft scale, cottonycushion scale, purple scale

Rate: Florida—postbloom:
0.625-5 lb. ai/A in 250 gal. water
2.5-20 lb. 25WP/A in 250 gal. water (20 lb.—1000 gal. water)

Pest:
chaff scale, Glovers scale, snow scale, citrus blackfly, mealybugs, whiteflies, citrus leafminer, purple scale, yellow scale

Rate: Texas—summer:
1.5-5 lb. ai/A in 300 gal. water
6-20 lb. 25WP/A in 300 gal. water (20 lb.—1000 gal. water)

Pest:
brown soft scale, chaff scale, mealybugs, California red scale

Remarks: Apply any time except bloom period, preferably during immediate postbloom period before fruit becomes infested. Properly calibrated application equipment essential for uniform coverage. Use low rate for light infestations; high rate for moderate to heavy infestations. For heavy scale infestations, 1000 gal./A spray volume suggested. Do not make more than 2 applications per growing season (1 application for lemons when tank mixed with oil); severe infestations may require 2 applications. Allow at least 45 days between applications. Do not apply more than 40 lb. 25WP/A (10 lb. for lemons) in any one season. Do not graze treated areas. Do not apply through any type of irrigation system. In coastal areas of California and other areas with no distinct bloom period, notify beekeepers prior to application. Refer to California citrus spray guide for additional information.

Days To Harvest: 14; with oil-60.

State Restrictions: For use in Arizona, California, Florida, Texas. In California, Florida, and Texas, may be tank mixed with EPA-registered spray oils (Florida and Texas, not more than 8 gal. oil/A; California, tank mix rate must be reduced to $^3/_4$ of alone rate).

Signal Word/Toxicity Class: Warning/II.
REI: 48 hr.; 14 days if greater than 2 lb. ai.

TEMIK

Restricted Use Pesticide.

Temik 15G (aldicarb) *(Rhone-Poulenc)*

Rate: 5 lb. ai/A
33 lb. 15G/A

Pest:
aphids, citrus red mite, Texas citrus mite, citrus nematode (except Florida), citrus rust mite, whiteflies

◆ - **new product** • ai-active ingredient • DF-dry flowable • E/EC-emulsifiable concentrate • EW-emulsion in water
F/FL-flowable • G-granule • L/LC-liquid • SP/WSP-soluble packet • W/WP-wettable powder • WSB-water soluble bag

Remarks: Grapefruit, lemons, limes, oranges. Apply just prior to or during spring flush of foliage growth in a band along the dripline on both sides of tree row. Spread granules uniformly and work into soil (preferred method) or shank granules 2-3" deep using 4-6 shanks on 12" center (width of each band should equal 1/4 tree row spacing), or for aphids, mites, and whiteflies, apply in irrigation furrows using 2 shanks per furrow. Do not allow livestock to graze treated areas. Do not apply within 50 ft. of any drinking water well except Florida where more restrictive setbacks apply.
Days To Harvest: Lemons-30.
State Restrictions: California: Apply between Mar. 1-Sept. 1. Florida: Apply between Jan. 1-Apr. 30. Do not apply within 300 ft. of any drinking water well or 1000 ft. of drinking water well when soils have permeability rate greater than 20"/hour with available water capacity less than 0.06 in all layers to a depth of 80". These restrictions apply unless wells are either cased to 100 ft. below ground level or at least 30 ft. below water table. Refer to Florida Citrus Pest Management Guide for instructions and precautions.

Signal Word/Toxicity Class: Danger-Poison/I.
REI: 48 hr.

THAT or THAT Big 8 — see sulfur

Thiodan — see endosulfan

Thiolux — see sulfur

VENDEX

Organotin miticide highly effective against citrus rust mites and spider mites on citrus. Tank mixing with oil results in reduced residual activity. Toxic to fish. Restricted Use Pesticide.

Vendex 50WP (fenbutatin-oxide) *(Griffin)*
Rate: 1-2 lb. ai/A
2-4 lb. 50WP/A (AZ, CA)
1-1.5 lb. ai/A
2-3 lb. 50WP/A (FL, TX)
Pest:
citrus red mite, citrus rust mite, Texas citrus mite, twospotted spider mite, Yuma mite
Remarks: Apply when mites first appear and repeat as necessary to maintain control. Do not apply more than 2 sprays in any 12-month period. Do not apply at less than 60-day intervals (30 days for Arizona, California). Do not apply more than 6 lb./A per year in Arizona, California, and Texas; 4 lb./A per year in Florida. Refer to California's UCIPM Pest Management Guidelines or Florida Citrus Pest Management Guide for instructions and precautions.
Days To Harvest: 7.

Signal Word/Toxicity Class: Danger/I.
REI: 48 hr.

Volck Supreme Spray — see petroleum oil

VYDATE

Insecticide/nematicide carbamate compound. Toxic to bees. Restricted Use Pesticide.

Vydate L (oxamyl) *(DuPont)*
Rate: 0.063-0.25 lb. ai/100 gal. water
0.25-1 pt. L/100 gal. water
Pest:
citrus rust mite

Rate: 0.5-1 lb. ai/A
2-4 pt. L/A
Pest:
citrus thrips
Remarks: For mites, apply when significant infestations are found; repeat as needed. For light to moderate pressure, apply every 4-6 weeks; moderate to heavy pressure, every 2-3 weeks. Spray to runoff. Do not apply more than 4 pt./A or more than 6 applications per year. For thrips, apply in early spring before bloom when new growth is 3-4" long; repeat as needed. Applications recommended at petal fall to prevent fruit scarring and mid-summer to protect new growth on young trees. Do not graze livestock in treated orchards. Do not apply more than 24 pt./A per year.
Days To Harvest: 7.

Rate: 0.5-2 lb. ai/A
2-8 pt. L/A (14-day intervals)
1-2 lb. ai/A
4-8 pt. L/A (30-day intervals)
Pest:
citrus nematode
Remarks: Initiate treatment in spring when soil temperatures at 12" depth have reached 50°F; repeat until soil temperature drops below 50°F. Treatments in Apr., May, and June and continued through Aug., Sept., and Oct. usually give good response. Best results with drip irrigation occur when Vydate is introduced into irrigation water during last 1/3 of irrigation cycle. Adjust flow from injection equipment to use contents over period of not less than 1 hr. Do not apply more than 8 pt./per 30-day period. Do not graze livestock in treated orchards.
Days To Harvest: 7.
State Restrictions: For use in California.

Rate: Foliar:
0.5-1 lb. ai/100 gal.
2-4 pt. L/100 gal. (up to 200 gal./A)
Pest:
aphids, leafhoppers, leafminers, mites, thrips

Rate: Foliar alone or as supplement:
0.5-1 lb. ai/100 gal. water
2-4 pt. L/100 gal. water
Preplant soil incorporated:
6-8 lb. ai/A
3-4 gal. L/A
Pest:
burrowing nematode, lesion nematode, root-knot nematode (except Javanese), sting nematode
Remarks: Nonbearing citrus (trees that will not bear fruit within 12 months after application). Commercial plantings only. Foliar treatment: Apply on mites and insects as needed to maintain control. Foliar treatment alone or as supplement to soil treatment on nematodes: Apply at first full leaf or when plant is in active growth phase 4 times on 2- to 3-week schedule. Do not apply foliar to plants under water stress or not actively growing; include spreader/sticker. Preplant soil incorporated on nematodes: Apply in at least 20 gal. water/A within 24 hr. before tranplanting and thoroughly incorporate 4-8" into soil immediately after application. If band treatment, use proportionately less material. Test on small scale before proceeding to large-scale application. Varietal response may differ if Vydate L is mixed with other products. Do not apply more than 4 gal./A per season.

Signal Word/Toxicity Class: Danger-Poison/I.
REI: 48 hr.

Tree Fruits, Deciduous

435 or 455 Oil — see petroleum oil

Ambush — see permethrin

AMITRAZ

Formamidine insecticide-acaricide labeled on pears. Excellent for controlling summer generations of psylla that are resistant to organophosphates and pyrethroids. Not effective in cold weather; should be applied when temperatures are 60°F or higher.

Mitac W *(AgrEvo USA)*
Rate: 0.75-1.5 lb. ai/A
1.5-3 lb. W/A
Pest:
pear psylla

Rate: 1.5 lb. ai/A
3 lb. W/A
Pest:
grape mealybug (except CA)

8 Fruits & Nuts

NOTICE The information on these pages is for preliminary planning — not a guide for use. Be sure to follow the manufacturer's directions, notwithstanding information contained here.

For personal protective equipment and EPA registration numbers, please turn to page 39.

Remarks: Pears. For postbloom control, apply as needed. Rate should not exceed 1.5 lb. ai/A per application or more than 3 lb. ai/A per season. Best control will be obtained when majority of insects are in young nymphal stage of development. Consult university extension service for information concerning best timing for control of overwintering adults.
Days To Harvest: 14.

- - - - - - - - - - - - - - - - - -

Signal Word/Toxicity Class: Danger/I.
REI: 24 hr.

APOLLO

Provides control primarily by preventing mite eggs from hatching. Not an adulticide, and should be tank mixed with an adulticide when mite numbers are high at application.

Apollo SC (clofentezine) *(AgrEvo USA)*

Rate: **Apples, pears:**
4-8 fl. oz. SC/A
Apricot, cherries, nectarines, peaches:
2-8 fl. oz. SC/A

Pest:

European red mite	Pacific spider mite	yellow spider mite
McDaniel mite	twospotted spider mite	

Remarks: Apples, apricots, cherries, nectarines, peaches, pears. Apply at first sign of mites. For apples, do not apply after tight cluster growth stage. Applications below 4 oz. SC/A are recommended only in association with established IPM programs and only when adequate beneficial insects/mites are prevalent. Do not apply by air. Do not apply more than once per season. If additional applications are needed, use miticide with different mode of action. Do not apply through any type of irrigation system. Do not graze treated areas or harvest cover crop for livestock feed.
Days To Harvest: Apricots, cherries, nectarines, peaches, pears-21.

- - - - - - - - - - - - - - - - - -

Signal Word/Toxicity Class: Caution.
REI: 12 hr.

ASANA

Restricted Use Pesticide.

Asana XL (esfenvalerate) *(DuPont)*

Rate: 0.025-0.075 lb. ai/A
4.8-14.5 fl. oz. XL/A

Pest:

apple aphid	Oriental fruit moth	tentiform leafminer
apple maggot	periodical cicada	tarnished plant bug
codling moth	plum curculio	tufted apple bud moth
green fruitworm	redbanded leafroller	variegated leafroller
obliquebanded leafroller	rosy apple aphid	white apple leafhopper
lesser appleworm	San Jose scale	
	stink bugs	

Remarks: Apples. Repeat as necessary to maintain control. Do not exceed 0.525 lb. ai/A per season. For dilute sprays, apply 200-600 gal./A, do not exceed 14.5 fl. oz./A per treatment. For plant bugs and rosy apple aphids, time of application is critical in achieving control. Use prebloom and postbloom spray timings recommended by state extension service. Do not feed or graze livestock on treated orchard floors.
Days To Harvest: 21.

- - - - - - - - - - - - - - - - - -

Rate: 0.025-0.075 lb. ai/A
4.8-14.5 fl. oz. XL/A

Pest:

codling moth	pear psylla	periodical cicada
green fruitworm	pear slug	plum curculio
leafrollers		

Remarks: Pears. Repeat as necessary to maintain control. Do not exceed 0.375 lb. ai/A per season, or 0.225 lb. ai/A between bloom and harvest. For dilute spray, apply 200-600 gal./A, but do not exceed 14.5 fl. oz. XL/A per treatment. Do not feed or graze livestock on treated orchard floors.
Days To Harvest: 28.

- - - - - - - - - - - - - - - - - -

Rate: 0.05-0.1 lb. ai/A
9.6-19.2 fl. oz. XL/A

Pest:

pear psylla

Remarks: Pears (dormant). Apply during dormant to prebloom (white bud) stage only. Do not apply more than 0.2 lb. ai/A per season. For dilute spray, apply 150-250 gal./A, but do not exceed 19.2 fl. oz. XL/A per treatment. Do not graze livestock on treated orchard floors.

- - - - - - - - - - - - - - - - - -

Rate: 0.025-0.075 lb. ai/A
4.8-14.5 fl. oz. XL/A

Pest:

American plum borer	leafrollers	periodical cicada
black cherry aphid	lesser peachtree borer	plum curculio
cherry fruit fly	Oriental fruit moth	stink bugs
green fruitworm	peach tree borer	tarnished plant bug
leafhoppers	peach twig borer	western cherry fruit fly

Remarks: Apricots, cherries, nectarines, peaches, plums, prune plums. Repeat as necessary to maintain control. Do not exceed 0.375 lb. ai/A per season with no more than 0.3 lb. ai/A per season between bloom and harvest. For dilute sprays, apply 200-400 gal./A, but do not exceed 14.5 fl. oz./A per treatment. Do not graze livestock on treated orchard floors. Some use instructions vary by pest species; refer to label.
Days To Harvest: 14.

- - - - - - - - - - - - - - - - - -

Signal Word/Toxicity Class: Warning/II.
REI: 12 hr.

AZADIRACHTIN

Extracted from seed of neem tree. Insect growth regulator that controls targeted insect larvae when they ingest or come in contact with it, by interfering with the insect's ability to molt and shutting off feeding behavior. Affected larvae slowly starve and die at next molt.

◆ **Azatin XL Plus** *(Thermo Trilogy)*

Rate: 5-21 oz. XL Plus/A

Pest:

aphids	fruit flies	mealybugs
apple aphid	green peach aphid	peach twig borer
armyworms	gypsy moth	peachtree borer
bagworms	Japanese beetle	psyllids
beetles	June beetle	tent caterpillar
caterpillars	leafhoppers	thrips
codling moth	leafminers	webworms
cutworms	leafrollers	whiteflies
flea beetles		

Remarks: Apples, apricots, cherries (sour, sweet), crabapples, nectarines, peaches, pears, plums, prunes, quinces. Apply when pests first appear; repeat at 7-day intervals. When pest populations are high, use higher rates. Do not apply more that 21 fl. oz./A.
Days To Harvest: 0.

- - - - - - - - - - - - - - - - - -

Signal Word/Toxicity Class: Caution.
REI: 4 hr.

◆ **NEEMAZAD 4.5EC or NEEMIX 4.5** *(Thermo Trilogy)*

Rate: 0.125-1 pt. 4.5/A
High Volume:
2.25-4.5 oz. 4.5EC/100 gal. water
Low Volume:
0.5 pt. 4.5EC/A

Pest:

aphids	greenhouse whitefly	psyllids
caterpillars	gypsy moth	sweetpotato whitefly
fall webworm	leafhoppers	thrips
fruit flies	leafminers	weevils
grasshoppers	mealybugs	

Remarks: Apples, apricots, cherries, crabapples, nectarines, peaches, pears, plums, prunes, quinces. For low volume, apply in sufficient water for adequate coverage. For low insect pressure, apply preventive at 7-day intervals or at first sign of insects; high insect pressure, apply at 3- to 4-day intervals.

- - - - - - - - - - - - - - - - - -

Signal Word/Toxicity Class: Warning/II.
REI: 12 hr.

Azatin — see azadirachtin

AZINPHOS-METHYL

Organophosphate insecticide registered for use in pome and stone fruits. One of the most widely used insecticides in tree fruits, effective against variety of insect pests. Similar to parathion in toxicity to humans, but not toxic to predaceous mites in orchards. Restricted Use Pesticide (except Solupak).

Gowan Azinphos-M 50-W *(Gowan)*
Guthion Solupak (50WSP) *(Bayer)*
◆ **Micro Flo Azinphosmethyl 2EC or 50W** *(Micro Flo)*
Sniper 50PVA *(Platte)*

◆ - **new product** • ai-active ingredient • DF-dry flowable • E/EC-emulsifiable concentrate • EW-emulsion in water
F/FL-flowable • G-granule • L/LC-liquid • SP/WSP-soluble packet • W/WP-wettable powder • WSB-water soluble bag

Rate: 0.75-1.5 lb. ai/A
1.5-3 lb. 50-W, 50PVA, 50W/A
1-1.5 lb. ai/A
2-3 lb. 50WSP/A
0.5-1 lb. ai/A
2-4 pt. 2EC/A

Pest:

aphids	fruittree leafroller	plum curculio
apple maggot	green fruitworm	Putnam scale
codling moth	leafhoppers	redbanded leafroller
European apple sawfly	mealybugs	San Jose scale
eyespotted bud moth	orange tortrix	stink bugs
Forbes scale	pear midge	tarnished plant bug

Remarks: Apples, crabapples, pears (except 2EC), quinces (except 50WSP). Allow 7 days between applications. In some areas, certain pests may have developed resistance to organophosphate insecticides. Consult advisor or cooperative extension service for recommendations. Refer to label for directions and precautions.

Days To Harvest: 7.

State Restrictions: Sniper 50PVA not registered in California.

- - - - - - - - - - - - - - - -

Rate: Apricots, nectarines, peaches:
East of the Rockies:
0.88-1.125 lb. ai/A
1.75-2.25 lb. 50-W, 50PVA, 50WSB/A
3.5-4.5 pt. 2EC/A
West of the Rockies:
1.5-2 lb. ai/A
3-4 lb. 50-W, 50PVA, 50WSP
6-8 pt. 2EC/A

Pest:

codling moth	*Platynota flavidana* leafroller	**East of the Rockies:**
eyespotted bud moth	plum curculio	aphids
Forbes scale	redbanded leafroller	cottony peach scale
fruittree leafroller	stink bugs	European fruit lecanium scale
lesser peachtree borer	tarnished plant bug	San Jose scale
orange tortrix	thrips	terrapin scale
Oriental fruit moth	tussock moth	walnut scale
peach twig borer		white peach scale
peachtree borer		

Rate: Plums, prunes:
East of the Rockies:
0.875-1.25 lb. ai/A
1.75-2.5 lb. 50-W, 50PVA/A
West of the Rockies:
1-2 lb. ai/A
2-4 lb. 50-W, 50PVA/A

Pest:

codling moth	peachtree borer	**East of the Rockies:**
eyespotted bud moth	plum curculio	aphids
fruittree leafroller	redbanded leafroller	Forbes scale
lesser peachtree borer	stink bugs	San Jose scale
orange tortrix	tarnished plant bug	
peach twig borer	tussock moth	

Rate: Plums, prunes:
East of the Rockies:
1.5 lb. ai/A
3 lb. 50-W, 50PVA, 50W/A
West of the Rockies:
1.5-2 lb. ai/A
3-4 lb. 50-W, 50PVA/A

Pest:
American plum borer

Remarks: Apricots (Gowan), nectarines, peaches, plums, prunes. Allow at least 14 days between applications for apricots, nectarines, peaches; 10 days for plums, prunes. For borers, apply 2-3 sprays to trunk from ground to scaffold limbs, timed with moth flight. For scales, apply when crawlers are present. Compatible with dormant and summer oils, which may be added to sprays in accordance with local recommendations.

Days To Harvest: Apricots, nectarines, peaches-21; plums, prunes-15.

State Restrictions: Sniper 50 PVA not registered in California.

- - - - - - - - - - - - - - - -

Rate: 0.75 lb. ai/A
1.5 lb. 50-W, 50PVA, 50W, 50WSP/A
3 pt. 2EC/A
Northeast, Northcentral:
0.25 lb. ai./100 gal.
0.5 lb. 50-W, 50PVA/100 gal. (300 gal./A)

Pest:

cherry leafminer	fruit flies	mineola moth
eyespotted bud moth	fruittree leafroller	plum curculio
Forbes scale	lesser peachtree borer	San Jose scale

Remarks: Cherries. Allow at least 14 days between applications. Up to 6 lb./A per crop season may be applied.

Days To Harvest: 15.

State Restrictions: Not for use in California for 50-W, 50PVA. In California for 50WSP, apply postharvest only.

- - - - - - - - - - - - - - - -

Signal Word/Toxicity Class: Danger-Poison/I.
REI: 48 hr.

Guthion 2L *(Bayer)*

Rate: Nectarines, peaches:
East of the Rockies:
0.88-1.13 lb. ai/A
3.5-4.5 pt. 2L/A
West of the Rockies:
1.5-2 lb. ai/A
6-8 pt. 2L/A

Pest:

codling moth	*Platynota flavidana* leafroller	**East of the Rockies:**
eyespotted bud moth	plum curculio	aphids
Forbes scale	redbanded leafroller	cottony peach scale
fruittree leafroller	stink bugs	European fruit lecanium scale
lesser peachtree borer	tarnished plant bug	San Jose scale
orange tortrix	thrips	terrapin scale
Oriental fruit moth	tussock moth	walnut scale
peach twig borer		white peach scale
peachtree borer		

Rate: Plums, prunes:
East of the Rockies:
0.88-1.13 lb. ai/A
3.5-4.5 pt. 2L/A
West of the Rockies:
1-2 lb. ai/A
4-8 pt. 2L/A

Pest:

codling moth	peachtree borer	**East of the Rockies:**
eyespotted bud moth	plum curculio	aphids
fruittree leafroller	redbanded leafroller	Forbes scale
lesser peachtree borer	stink bugs	San Jose scale
orange tortrix	tarnished plant bug	
peach twig borer	tussock moth	

Rate: Plums, prunes:
East of the Rockies:
1.5 lb. ai/A
6 pt. 2L/A
West of the Rockies:
1.5-2 lb. ai/A
6-8 pt. 2L/A

Pest:
American plum borer

Remarks: Nectarines, peaches, plums, prunes. Allow at least 14 days between applications for nectarines, and peaches; 10 days for plums, prunes. For borers, apply 2-3 sprays to trunk from ground to scaffold limbs, timed with moth flight. For scales, apply when crawlers are present. Compatible with dormant and summer oils, which may be added to sprays in accordance with local recommendations.

Days To Harvest: Nectarines, peaches-21; plums, prunes-15.

- - - - - - - - - - - - - - - -

Signal Word/Toxicity Class: Danger-Poison/I.
REI: 48 hr.; 72 hr. if average rainfall less than 25" per year.

Sniper 2-E or 50W *(Platte)*

Rate: 0.25-0.31 lb. ai/100 gal. water
0.5-0.625 lb. 50W/100 gal. water

Pest:

aphids	Forbes scale	plum curculio
apple maggot	fruittree leafroller	Putnam scale
codling moth	green fruitworm	redbanded leafroller
European apple sawfly	leafhopper	San Jose scale
	mealybug	stink bugs
eyespotted bud moth	orange tortrix	tarnished plant bug

Remarks: Apples, crabapples, pears, quinces. Do not apply more than 1000 gal. finished spray/A or more than 8 times per season. Compatible with dormant and summer oils, which may be added to sprays in accordance with local recommendations. Do not use oils within 30 days of sulfur or lime-sulfur treatments.

Days To Harvest: 15.

- - - - - - - - - - - - - - - -

Rate: 0.25-0.31 lb. ai/100 gal. water
1-1.25 pt. 2-E/100 gal. water
0.5-0.625 lb. 50W/100 gal. water

NOTICE The information on these pages is for preliminary planning — not a guide for use. Be sure to follow the manufacturer's directions, notwithstanding information contained here.	For personal protective equipment and EPA registration numbers, please turn to page 39.

Pest:

aphids	lesser peachtree borer	stink bugs
codling moth	orange tortrix	tarnished plant bug
cottony peach scale	peachtree borer	terrapin scale
European fruit lecanium scale	*Platynota flavidana* leafroller	thrips (nectarines)
eyespotted bud moth	plum curculio	tussock moth
Forbes scale	redbanded leafroller	walnut scale
fruittree leafroller	San Jose scale	white peach scale

Rate: 0.25-0.38 lb. ai/100 gal. water
1-1.5 pt. 2-E/100 gal. water
0.5-0.75 lb. 50W/100 gal. water

Pest:

Oriental fruit moth	peach twig borer

Remarks: Apricots, nectarines, peaches. Do not apply more than 800 gal. finished spray/A or more than 8 times per season. For borers, apply 2-3 sprays to trunk from ground to scaffold limbs, timed with moth flight. For scales, apply when crawlers are present. Compatible with dormant and summer oils, which may be added to sprays in accordance with local recommendations.

Days To Harvest: 21.

- - - - - - - - - - - - - - - - -

Rate: Plums, prunes:
0.25-0.5 lb. ai/100 gal. water
1-2 lb.2-E/100 gal. water
0.5-1 lb. 50W/100 gal. water

Pest:

American plum borer (high rate)	fruittree leafroller	redbanded leafroller
aphids	lesser peachtree borer	San Jose scale
codling moth	orange tortrix	stink bugs
eyespotted bud moth	peach twig borer	tarnished plant bug
Forbes scale	peachtree borer	tussock moth
	plum curculio	

Rate: Cherries:
0.25-0.5 lb. ai/100 gal. water
0.5-1 lb. 50W/100 gal. water

Pest:

cherry leafminer	fruit flies	mineola moth
eyespotted bud moth	fruittree leafroller	plum curculio
Forbes scale	lesser peachtree borer	San Jose scale

Remarks: Cherries, plums, prunes. Do not apply more than 800 gal. finished spray/A or more than 8 times per season.

Days To Harvest: 15.

State Restrictions: Not for use on cherries in California.

- - - - - - - - - - - - - - - - -

Signal Word/Toxicity Class: Danger-Poison/I.

REI: 48 hr.; 72 hr. if average rainfall less than 25" per year.

CALCIUM POLYSULFIDE

Lime-Sulfur Solution *(Best Sulfur)*

Rate: Dormant:
6-11 gal. L/100 gal. water
3 gal. L/100 gal. water + 1.5 gal. oil spray
Delayed dormant:
3 gal. L/100 gal. water

Pest:

blister mite	European red mite	San Jose scale
brown mite	rust mite	

Remarks: Apples. Use on Delicious apples may cause injury. Do not apply through any type of irrigation system.

- - - - - - - - - - - - - - - - -

Rate: Dormant:
5-7 gal. L/100 gal. water

Pest:

pearleaf blister mite

Rate: Delayed dormant:
3 gal. L/100 gal. water + 1.5 gal. oil

Pest:

European red mite	pearleaf blister mite	rust mite
green apple aphid	red mite	San Jose scale
pear psylla	rosy aphid	

Rate: Postharvest (early to mid-Sept.):
11 gal. L/100 gal. water
4 gal. L/100 gal. water + 0.75 gal. oil
2 gal. L/100 gal. water + 4 lb. wettable sulfur

Pest:

pearleaf blister mite	rust mite

Rate: Postharvest (late fall, mid Oct.-Nov.):
3-5 gal. L/100 gal. water + 1.5-2 gal. oil

Pest:

aphid eggs	mite eggs	rust mite
bud mite (Pacific Coast states)	pear psylla	scales
	pearleaf blister mite	

Remarks: Pears. Do not use on d'Anjou, Comice, or Seckle varieties after delayed dormant. For bud mite, apply when mites first penetrate under bud scales. Do not allow postharvest sprays to drift to adjacent apple orchards as defoliation might occur. Do not apply through any type of irrigation system.

- - - - - - - - - - - - - - - - -

Rate: Dormant (except cherries), delayed dormant:
6-11 gal. L/100 gal. spray
3 gal. L/100 gal. spray + 1.5 gal. oil

Pest:

aphids (except cherries)	peach twig borer	scales
brown mite	red mite	silver mite

Rate: Postharvest (cherries):
1 gal. L/100 gal. spray + 4 lb. wettable sulfur

Pest:

brown mite	twospotted spider mite

Remarks: Cherries (sweet, tart), peaches, plums, prunes. Do not apply through any type of irrigation system.

- - - - - - - - - - - - - - - - -

Signal Word/Toxicity Class: Danger/I.

REI: 48 hr.

Sulforix *(Best Sulfur)*

Rate: Postharvest:
2.5 gal. L/A

Pest:

pear bud mite	blister mite

Rate: Delayed dormant:
2.5 gal. L/A + 1.5-2 gal. oil or 3-5 gal. Dormant Emulsion

Pest:

aphid eggs	pear psylla	scales
mite eggs	pearleaf blister mite	

Remarks: Pears. Apply in sufficient water for thorough coverage. Do not apply through any type of irrigation system.

- - - - - - - - - - - - - - - - -

Signal Word/Toxicity Class: Danger/I.

REI: 48 hr.

CARBARYL

Wide usage in the fruit market, due to relatively high degree of safety and short preharvest interval. Effective broad spectrum insecticide and fruit thinning agent especially for hard-to-thin apple varieties. Not a miticide and may cause buildup of mites and aphids. Compatible with most other pesticides, and may provide control of insects resistant to organophosphate insecticides.

Chipco Sevin SL, 80WSP or Sevin 4F, 80S, 80WSP *(Rhone-Poulenc)*
Drexel Carbaryl 4, 50W, 80S *(Drexel)*

Rate: 0.5-1 lb. ai/100 gal. water
1-2 pt. SL, 4L, 4F/100 gal. water
1-2 lb. 50W/100 gal. water
West of the Rockies:
0.75-1 lb. ai/100 gal. water
1-1.25 lb. 80S, 80WSP/100 gal. water
East of the Rockies:
0.5-1 lb. ai/100 gal. water
0.66-1.25 lb. 80S 80WSP /100 gal. water

Pest:

apple aphid	Forbes scale	pearleaf blister mite
apple maggot	fruittree leafroller	periodical cicada
apple mealybug	green fruitworm	plum curculio
apple rust mite	gypsy moth	redbanded leafroller
apple sucker	Japanese beetle	rosy apple aphid
bagworm	lecanium scale	San Jose scale
California pearslug (pear sawfly larvae)	lesser appleworm	tarnished plant bug
	Lygus bugs	tentiform leafminer
codling moth	orange tortrix	white apple leafhopper
eastern tent caterpillar	oystershell scale	woolly apple aphid
European apple sawfly	pear psylla	yellowheaded fireworm
eyespotted bud moth	pear rust mite	

Remarks: Apples, pears. Rate varies by pest species; refer to label. To avoid undesired apple thinning, delay at least 30 days after full bloom. For psylla control, apply when eggs hatch or young nymphs are present.

Days To Harvest: 1.

- - - - - - - - - - - - - - - - -

◆ - **new product** • ai-active ingredient • DF-dry flowable • E/EC-emulsifiable concentrate • EW-emulsion in water
F/FL-flowable • G-granule • L/LC-liquid • SP/WSP-soluble packet • W/WP-wettable powder • WSB-water soluble bag

Rate: Apricots, nectarines, peaches:
1 lb. ai/100 gal. water
2 pt. SL, 4L, 4F/100 gal. water
2 lb. 50W/100 gal. water
1.25 lb. 80S, 80WSP/100 gal. water

Pest:

apple pandemis	June beetle	periodical cicada
codling moth	lecanium scale	plum curculio
cucumber beetle	lesser peachtree borer	redbanded leafroller
European earwig	olive scale	San Jose scale
fruittree leafroller	orange tortrix	tarnished plant bug
gypsy moth	Oriental fruit moth	tussock moth
Japanese beetle	peach twig borer	variegated leafroller

Rate: Cherries, plums, prunes:
0.75-1 lb. ai/100 gal. water
1.5-2 pt. SL, 4L, 4F/100 gal. water
1-1.25 lb. 80S, 80WSP/100 gal. water

Pest:

black cherry aphid	fruittree leafroller	peach twig borer
brown soft scale	green fruitworm	plum curculio
cherry fruitworm	gypsy moth	prune leafhopper
cherry maggot	Japanese beetle	redbanded leafroller
codling moth	lecanium scale	rose chafer
eastern tent caterpillar	lesser peachtree borer	San Jose scale
eyespotted bud moth	mealy plum aphid	tussock moth
Forbes scale	orange tortrix	variegated leafroller
	oystershell scale	

Remarks: Apply when insects or their damage appear. Rate varies by pest species; refer to label. For optimum scale control, apply when crawlers are present. Thoroughly spray limbs and tree trunks weekly during moth flight.

Days To Harvest: Peaches, cherries, plums, prunes-1; apricots, nectarines-3.

- - - - - - - - - - - - - - - - -

Signal Word/Toxicity Class: Caution (4F, 4L, XLR); Warning/II (SL, 50W, 80S, 80WSP).

REI: 12 hr.

Clean Crop Sevin 5 Bait *(Platte)*

Rate: 2-2.4 lb. ai/A
40-48 lb. Bait/A

Pest:

ants	cutworms	grasshoppers
armyworms	field crickets	sowbugs

Remarks: Apples, apricots, cherries, nectarines, peaches, pears, plums. Apply with ground or air equipment.

Days To Harvest: Apples, pears, cherries, peaches, plums-1; nectarines-3; apricots-28.

- - - - - - - - - - - - - - - - -

Signal Word/Toxicity Class: Caution/III.

REI: 12 hr.

CARZOL

Carbamate miticide/insecticide registered on apples and pears. Most effective against sucking insects and mites, especially those resistant to organophosphates. Effective against mobile stages of pest, and will not affect eggs of pests such as European red mites. Not stable in alkaline mix-water. Highly toxic to predatory mites, and should be used carefully in integrated mite control programs. Toxic to wildlife and to bees exposed to direct treatment.

Carzol (SP) (formetanate hydrochloride) *(AgrEvo USA)*

Rate: 0.125-0.25 lb. SP/100 gal. water
Concentrate spray:
0.5-1 lb. SP/A in at least 50 gal. water

Pest:
white apple leafhopper

Rate: 0.25-0.31 lb. SP/100 gal. water
Concentrate spray:
1-1.25 lb. SP/A in at least 50 gal. water

Pest:
thrips

Remarks: Apples. Apply when leafhoppers appear. For thrips, apply at bloom and repeat through petal fall as needed. For full cover dilute spray, apply 400 gal. water/A for leafhoppers; 400-800 gal. water/A (depending on size of trees) for thrips. Do not apply more than 4 lb./A per season. For thrips, apply at night or while bees are not working in bloom. Do not apply through any type of irrigation system.

- - - - - - - - - - - - - - - - -

Rate: 0.25 lb. SP/100 gal. water
Concentrate spray:
1 lb. SP/A in at least 50 gal. water

Pest:
Campylomma

Remarks: Apples. Apply prebloom/bloom; repeat as necessary through petal fall. For full cover dilute spray, apply 400 gal. water/A. Do not apply more than 4 lb./A per season. Apply at night or while bees are not working in bloom. Do not apply through any type of irrigation system.

Days To Harvest: 7.

State Restrictions: Not for use in California.

- - - - - - - - - - - - - - - - -

Rate: Apples, pears:
0.25-0.5 lb. SP/100 gal. water
Concentrate spray:
1-4 lb. SP/A in at least 50 gal. water

Pest: Apples:
tentiform leafminer — thrips

Apples, pears:
European red mite — twospotted spider mite

Rate: Pears:
0.25 lb. SP/100 gal. water
Concentrate spray:
1-2 lb. SP/A in at least 50 gal. water

Pest:
pear rust mite

Remarks: Apples, pears. Apply when mites appear; repeat as needed. For pear rust mite, apply postbloom to harvest at first sign of infestation. For leafminers, apply at pink and petal fall. For full cover dilute spray, apply 400-800 gal. water/A depending on size of trees. Do not apply more than 4 lb./A per season. Do not apply through any type of irrigation system.

Days To Harvest: 7.

- - - - - - - - - - - - - - - - -

Rate: 0.25-0.375 lb. SP/100 gal. water
Concentrate/aerial spray:
1-2 lb. SP/A

Pest:
thrips

Remarks: Nectarines. Apply when thrips infest blossoms through petal fall; repeat as needed. For full cover dilute spray, apply 400-500 gal. water/A depending on size of trees. Do not apply more than 4 lb./A per season. Apply at night or while bees are not working in blooms. Do not apply through any type of irrigation system.

Days To Harvest: 14.

- - - - - - - - - - - - - - - - -

Rate: 0.25 lb. SP/100 gal. water
Concentrate spray:
1-1.25 lb. SP/A

Pest:
Lygus bugs — stink bugs

Rate: Concentrate spray:
1 lb. SP/A in 100 gal. water or less

Pest:
plum curculio

Remarks: Peaches. Apply at first sign of infestation; repeat as needed. For full cover dilute spray, apply 400-500 gal. water/A. Do not apply more than 2.5 lb./A per season. Do not apply through any type of irrigation system.

Days To Harvest: 21.

State Restrictions: Not for use in California for plum curculio.

- - - - - - - - - - - - - - - - -

Rate: 0.25 lb. SP/100 gal. water
Concentrate spray:
1-1.5 lb. SP/A

Pest:
European red mite — McDaniel mite — twospotted spider mite

Remarks: Plum, prunes. Apply when mites appear; repeat as needed. For full cover dilute spray, apply 400-600 gal. water/A depending on size of trees. Do not apply more than 3 lb./A per season. Do not apply through any type of irrigation system.

Days To Harvest: 7.

- - - - - - - - - - - - - - - - -

Signal Word/Toxicity Class: Danger-Poison/I.

REI: 48 hr.

Chipco Sevin — see carbaryl

NOTICE The information on these pages is for preliminary planning — not a guide for use. Be sure to follow the manufacturer's directions, notwithstanding information contained here.	For personal protective equipment and EPA registration numbers, please turn to page 39.

CHLORPYRIFOS

Organophosphate insecticide registered on pome and stone fruit (except sweet cherries). There are 3 major types of Lorsban applications: trunk spray of 4E for borer control, dormant spray of 4E with oil for insect egg control, and foliar spray of 50W. Broad-spectrum insecticide that provides some suppression of European red mite populations when used season-long. Particularly effective against sucking insects such as aphids, scale crawlers, and leafhoppers, except where organophosphate resistance occurs.

Lorsban-4E *(Dow AgroSciences)*
◆ **Micro Flo Chlorpyrifos 4E** *(Micro Flo)*

Rate: 0.25-0.5 lb. ai/100 gal. water
0.5-1 pt. 4E/100 gal. water + 1-2 gal. dormant oil

Pest: Apples:

climbing cutworm	obliquebanded leafroller	rosy apple aphid
Lygus bugs	pandemis leafroller	San Jose scale

Pears:

climbing cutworm	pear psylla	San Jose scale

Nectarines, peaches, plums, prunes:

climbing cutworm	mealy plum aphid (plums, prunes)	peach twig borer San Jose scale

Remarks: Apples, nectarines, peaches, pears, plums, prunes. Dormant or delayed dormant. Make only 1 application during dormant season. Do not graze treated areas. Refer to label for use directions and California restrictions.

- - - - - - - - - - - - - - - - -

Rate: 3 lb. ai/100 gal. water
6 pt. 4E/100 gal. water

Pest:
peachtree borer

Remarks: Nectarines, peaches. Make 1 application per season. Trunk spray application: Spray trunks of trees before newly-hatched borers enter trees; do not spray fruit. Preplant dip application: Dip trees several inches above grafting bud scar and plant immediately or allow to dry before returning to storage. Do not allow trees to remain in contact with solution. Do not graze treated orchards. Refer to label for use directions.

Days To Harvest: 14.

- - - - - - - - - - - - - - - - -

Rate: 1.5-3 lb. ai/100 gal. water
3-6 pt. 4E/100 gal. water

Pest:

American plum borer	lesser peachtree borer	peachtree borer

Remarks: Cherries. Trunk spray application. Spray trunks of trees; repeat in 2 weeks, make third application postharvest. Do not make more than 3 applications per year. Do not allow spray to contact foliage on sweet cherries. Do not graze treated orchards. Refer to label for use directions.

Days To Harvest: 6.

- - - - - - - - - - - - - - - - -

Signal Word/Toxicity Class: Warning/II.
REI: 24 hr.

Lorsban 50-W *(Gowan)*

Rate: 1.5 lb. ai/A
3 lb./A

Pest:

apple aphid	fruittree leafroller	plum curculio
apple maggot	green fruitworm	redbanded leafroller
climbing cutworm	lesser appleworm	rosy apple aphid
codling moth	obliquebanded leafroller	San Jose scale
dogwood borer	orange tortrix	tarnished plant bug
European apple sawfly	Oriental fruit moth	tufted apple bud moth
European corn borer	pandemis leafroller	variegated leafroller
eyespotted bud moth	periodical cicada	woolly apple aphid

Remarks: Apples. For apple aphid, do not use alone as rescue treatment. For dogwood borer, apply as trunk spray. Do not apply more than 8 applications per season or apply last 2 treatments closer than 21 days apart. In California, do not do work involving contact with trees within 2 days after treatment. Do not graze treated areas. Refer to label for use directions.

Days To Harvest: 28.

- - - - - - - - - - - - - - - - -

Rate: 1-1.5 lb. ai/A
2-3 lb. 50-W/A

Pest:

apple aphid	green fruitworm	rosy apple aphid
apple maggot	lesser appleworm	San Jose scale
codling moth	obliquebanded leafroller	spotted tentiform leafminer
dogwood borer	orange tortrix	
European apple sawfly	Oriental fruit moth	tarnished plant bug
European corn borer	periodical cicada	tufted apple bud moth
European red mite	plum curculio	variegated leafroller
eyespotted bud moth	potato leafhopper	white apple leafhopper
fruittree leafroller	redbanded leafroller	woolly apple aphid

Remarks: Apples. For apple aphid, do not use alone as rescue treatment. For dogwood borer, apply as trunk spray. Do not apply more than 8 applications per season or apply last 2 treatments closer than 21 days apart. Do not graze treated areas. May be tank mixed (except dogwood borer); refer to label for rates, specific pests, and products. Follow label directions, restrictions, and precautions.

Days To Harvest: 28.

State Restrictions: For use in Connecticut, Delaware, Maine, Maryland, Massachusetts, Michigan, New Hampshire, New Jersey, New York, North Carolina, Ohio, Pennsylvania, Rhode Island, South Carolina, Vermont, Virginia, and West Virginia.

- - - - - - - - - - - - - - - - -

Rate: 1.5-3 lb. ai/A
2-3 lb. 50-W/A

Pest:

American plum borer	lesser appleworm	peach twig borer
climbing cutworm	lesser peachtree borer	peachtree borer
European fruit lecanium	mineola moth	San Jose scale
	obliquebanded leafroller	shothole borer
eyespotted bud moth	Pacific flatheaded borer	stink bugs
fruittree leafroller	pandemis leafroller	tarnished plant bug
green fruitworm	redbanded leafroller	

Rate: 3 lb. ai/A
3 lb. 50-W/A

Pest:

black cherry aphid	plum curculio	rose chafer
cherry fruit fly		

Remarks: Sour cherries. Not for use on sweet cherries. Do not apply more than 8 applications per season. Do not graze treated orchards. In California, do not do work involving contact with trees within 2 days after treatment. Refer to label for geographic restrictions.

Days To Harvest: 14.

- - - - - - - - - - - - - - - - -

Signal Word/Toxicity Class: Warning/II.
REI: 24 hr.

Citri-Oil — see petroleum oil

Damoil — see petroleum oil

DEADLINE

Packaged as Bullets, M-Ps (mini-pellets), and granules. Resistant to rain and watering.

Deadline Bullets, M-Ps, Granules (4%) (metaldehyde) *(Pace)*

Rate: 0.8-1 lb. ai/A
20-40 lb. 4%/A

Pest:

slugs	snails

Remarks: Apples, apricots, cherries, peaches, pears. Apply every 3-4 weeks during growing season, or as needed. Do not apply directly to edible portion of plant.

- - - - - - - - - - - - - - - - -

Signal Word/Toxicity Class: Warning/II (granule); Caution (bullets, M-P).
REI: 12 hr.

Diatect — see pyrethrins

DIAZINON

Organophosphate insecticide registered for use on a variety of tree fruits and some small fruits. Broad-spectrum insecticide that is less toxic to humans than parathion. Restricted Use Pesticide due to avian and aquatic toxicity.

D•z•n diazinon AG600 WBC, 50W *(Novartis)*
Drexel Diazinon Insecticide (4EC) *(Drexel)*
Gowan Diazinon 4E or WSB *(Gowan)*
Helena Diazinon AG500 *(Helena)*
Micro Flo Diazinon AG500 or 50W *(Micro Flo)*
Prentox Diazinon AG500 or 50W *(Prentiss)*
◆ **Wilbur-Ellis Diazinon 4 Spray (4E)** *(Wilbur-Ellis)*

Rate: 0.25-0.5 lb. ai/100 gal. water
0.5-1 pt. AG500, 4E, 4EC/100 gal. water
0.5-1 lb. 50W, 50WSB/100 gal. water
6.5-12.75 fl. oz. AG600/100 gal. water

◆ - **new product** • ai-active ingredient • DF-dry flowable • E/EC-emulsifiable concentrate • EW-emulsion in water
F/FL-flowable • G-granule • L/LC-liquid • SP/WSP-soluble packet • W/WP-wettable powder • WSB-water soluble bag

Pest:
cherry fruit fly cherry rust mite

Rate: 0.5 lb. ai/100 gal. water
1 pt. AG500, 4E, 4EC/100 gal. water
1 lb. 50W, 50WSB/100 gal. water
12.75 fl. oz. AG600/100 gal. water

Pest:
black cherry aphid, eyespotted bud moth, fruittree leafroller, leafhoppers, San Jose scale (crawler)

Remarks: Cherries. For aphids, apply as cover spray when aphids appear; repeat after 10 days if necessary. For fruit flies, make 3-4 applications when adult flies begin to emerge; repeat at 10-day intervals. For mites, apply as single supplemental spray after harvest. For bud moth and fruittree leafroller, apply cover sprays as infestations occur beginning at petal fall; repeat after 7 days if necessary. For leafhoppers, make 3-4 applications beginning at petal fall and repeat at 7-day intervals through growing season. For San Jose scale, apply during pre-pink stage or when crawlers first appear. Do not apply more than 2 lb. ai/A per application or more than 6 lb. ai/A per season. Do not make more than 3 applications per season.

Days To Harvest: 21.

- - - - - - - - - - - - - - - - -

Rate: 0.25 lb. ai/100 gal. water
0.5 pt. AG500, 4E, 4EC/100 gal. water + 1.5 gal. horticultural oil
0.5 lb. 50W, 50WSB/100 gal. water + 1.5 gal. horticultural oil
6.5 fl. oz. AG600/100 gal. water + 1.5 gal. horticultural oil

Pest:
olive scale (crawler)

Rate: 0.5 lb. ai/100 gal. water
1 pt. AG500, 4E, 4EC/100 gal. water
1 lb. 50W, 50WSB/100 gal. water
12.75 fl. oz. AG600/100 gal. water

Pest:
aphids, apricot mealybug (apricots), brown mite (clover mite), leafhoppers (peaches), olive scale (crawler), Oriental fruit moth (except apricots), peach twig borer (except apricots), San Jose scale (crawler), twospotted spider mite

Remarks: Apricots, nectarines, peaches. Apply as infestations occur; apricot mealybug, as cover spray from petal fall to June; repeat after 14 days if necessary. For leafhoppers, make 3-4 applications at petal fall and repeat at 7-day intervals through growing season. For Oriental fruit moth, apply at 100% petal fall; make 2 applications at 10- to 15-day intervals and time last application to peak adult flight. For peach twig borer on peaches, apply as insects occur in May and June at 7-day intervals. Do not apply more than 2 lb. ai/A per application or more than 6 lb. ai/A per season.

Days To Harvest: 21.

- - - - - - - - - - - - - - - - -

Rate: 0.75-1 lb. ai/100 gal. water
1.5-2 pt. AG500, 4E, 4EC/100 gal. water
1.5-2 lb. 50W, 50WSB/100 gal. water
19-25.5 fl. oz. AG600/100 gal. water

Pest:
white peach scale

Remarks: Peaches. Apply postharvest timed to coincide with peak crawler and immature crawler activity. For heavy infestations, 2 applications, 7 days apart, may be needed.

- - - - - - - - - - - - - - - - -

Rate: 0.25-0.5 lb. ai/100 gal. water
0.5-1 pt. AG500, 4E, 4EC/100 gal. water
0.5-1 lb. 50W, 50WSB/100 gal. water
6.5-12.75 fl. oz. AG600/100 gal. water

Pest:
brown mite (clover mite), European red mite, leafcurl plum aphid, mealy plum aphid, thistle aphid

Remarks: Plums, prunes. Apply at 7-day intervals as needed. Do not apply more than 2 lb. ai/A per application or more than 6 lb. ai/A per season.

Days To Harvest: 21.

- - - - - - - - - - - - - - - - -

Signal Word/Toxicity Class: Caution/III.
REI: 24.

D•z•n diazinon 50W *(Novartis)*
Gowan Diazinon 50WSB *(Gowan)*
◆ **Micro Flo Diazinon 50W** *(Micro Flo)*
Prentox Diazinon 50W *(Prentiss)*

Rate: Apples, pears:
0.5 lb. ai/100 gal. water
1 lb. 50W, 50WSB/100 gal. water

Pest:
apple maggot, codling moth, eyespotted bud moth, Forbes scale crawler, fruittree leafroller, green apple aphid, leafhoppers, mealybugs, rosy apple aphid, San Jose scale crawler, spider mites, tentiform leafminer, woolly apple aphid

Rate: Pears:
0.25-0.5 lb. ai/100 gal. water
0.5-1 lb. 50W, 50WSB/100 gal. water

Pest:
pear psylla

Remarks: Apples, pears. Apply as dormant or delayed dormant spray for apples; dormant spray for pears. For leafhoppers, apply as infestations develop, beginning at petal fall. For scale crawlers, apply early in the season when crawlers first appear. For pear psylla, apply as cover spray at pre-pink stage; make 1-2 additional applications during first and second cover sprays if necessary. For other pests, make 3 cover sprays. Repeat applications in 14 days if necessary. Diazinon may not control organophosphate-resistant apple maggots. Do not apply more than 4 lb./A per application or more than 12 lb./A per season.

Days To Harvest: 21.

- - - - - - - - - - - - - - - - -

Rate: Apples, apricots, nectarines, peaches, pears, plums, prunes:
0.5 lb. ai/100 gal. water
1 lb. 50W, 50WSB/100 gal. water + dormant/superior type oil
Cherries:
0.625 lb. ai/100 gal. water
1.25 lb. 50W, 50WSB/100 gal water + dormant/superior type oil

Pest:
apple aphid (eggs), apricot scale (except nectarines), black cherry aphid (eggs), black scale, brown mite (eggs), brown scale (except nectarines), European red mite (eggs), mealy plum aphid (eggs), mealybugs, parlatoria scale, San Jose scale, twig borer

Remarks: Apples, apricots, cherries, nectarines, peaches, pears, plums, prunes. Apply as spray (dormant/delayed dormant for apples). For plums or prunes, if applied late April or May, mix with light medium summer horticultural oil/100 gal. water. Do not apply more than 4 lb./A per application or more than 12 lb./A per season.

- - - - - - - - - - - - - - - - -

Signal Word/Toxicity Class: Caution.
REI: 24 hr. Refer to label for California reentry restrictions for nectarines and peaches.

DIBROM

Short residual organophosphate insecticide. Performs well in dormant spray in combination with supreme oils. Toxic to fish, aquatic invertebrates, and wildlife. Highly toxic to bees.

Dibrom 8 (8E) (naled) *(Amvac)*

Rate: 2-3 lb. ai/A
2-3 pt. 8E/A

Pest:
peach twig borer

Remarks: Peaches. Dormant or delayed dormant period. Apply single application by ground only in 40-300 gal. water/A. Combine with dormant oil spray. Follow all label directions and precautions.

Days To Harvest: 7.

- - - - - - - - - - - - - - - - -

Signal Word/Toxicity Class: Danger/I.
REI: 72 hr.

DICOFOL

Chlorinated hydrocarbon that is an effective miticide in pome fruits, with registration pending on stone fruits. Will provide good control of adult twospotted and European red mites, but only one application per season is recommended due to rapid buildup of resistance.

◆ **Griffin Dicofol 4E** *(Griffin)*

Rate: 0.38-0.5 lb. ai/100 gal. dilute spray
0.75-1 pt./100 gal. dilute spray

Pest:
apple rust mite, clover mite, European red mite, McDaniel mite, Pacific mite, Schoene mite, twospotted spider mite, Willamette mite, yellow (carpini) mite

8 Fruits & Nuts

NOTICE The information on these pages is for preliminary planning — not a guide for use. Be sure to follow the manufacturer's directions, notwithstanding information contained here.

For personal protective equipment and EPA registration numbers, please turn to page 39.

Remarks: Apples, pears. Apply when mite population appears; repeat at 10- to 14-day intervals. Do not apply more than 1 gal./A. To avoid injury, do not use glyodin fungicide and dicofol in late cover sprays. Do not apply through any type of irrigation system.
Days To Harvest: 7.

Signal Word/Toxicity Class: Caution.
REI: 12 hr.

Kelthane 35 or 50 *(Rohm and Haas)*
Rate: 1.4-2.8 lb. ai/A
4-8 lb. 35/A
1.5-3 lb. ai/A
3-6 lb. 50/A
Pest:
apple rust mite, McDaniel mite, twospotted spider mite, clover mite, Pacific mite, Willamette mite, European red mite, Schoene mite, yellow (carpini) mite
Remarks: Apples, crabapples, pears, quinces. Use higher rate for larger trees or dense foliage. Apply at petal fall to second cover or whenever mite populations reach threshold levels. Do not make more than 2 applications per season. Do not use glyodin fungicide with Kelthane in late cover sprays. Do not apply through any type of irrigation system.
Days To Harvest: 7.

Signal Word/Toxicity Class: Warning/II (35); Caution (50).
REI: 12 hr.

Digon — see dimethoate

Dimate — see dimethoate

DIMETHOATE

Systemic, broad-spectrum insecticide labeled for apples and pears. Toxic to wildlife and aquatic invertebrates.

Cheminova Dimethoate 4E *(Cheminova)*
Dimate 4E *(Terra)*
Drexel Dimethoate 2.67 EC or 4EC *(Drexel)*
Gowan Dimethoate E267 *(Gowan)*
Helena Dimethoate 4E *(Helena)*
◆ Micro Flo Dimethoate 267 or 4E *(Micro Flo)*
Wilbur-Ellis Dimethoate 267 or Digon 400 *(Wilbur-Ellis)*
Rate: 0.25-0.5 lb. ai/100 gal. water
0.75-1.5 pt. 2.67, 267, E267/100 gal. water
0.5-1 pt. 4E, 4EC, 400/100 gal. water
Pest:
aphids, leafrollers, pear psylla, leafhoppers, mites (except rust)
Remarks: Apples, pears. Do not apply during bloom period. Do not graze treated orchards.
Days To Harvest: 28.

Rate: 0.5 lb. ai/100 gal. water
1.5 pt. E267/100 gal. water
1 pt. 4E, 4EC, 400/100 gal. water
Pest:
apple maggot, codling moth (Midwest, Eastern states)
Remarks: Apples. Apply at petal fall; repeat at 10- to 14-days intervals as necessary. Under heavy infestations of apple maggots, some sting injury may occur. Do not apply during bloom period. Do not graze treated orchards.
Days To Harvest: 28.

Signal Word/Toxicity Class: Warning/II.
REI: 48 hr.

DI-SYSTON

For nonbearing fruit trees only. Systemic action makes it an excellent choice for preventing injury to nursery stock. Restricted Use Pesticide.

Di-Syston 15% (disulfoton) *(Bayer)*
Rate: 2.5-27 oz. 15G/tree
2.5 oz. 15%/" trunk diameter
Pest:
aphids, leafhoppers, thrips, lace bug, mites, whiteflies
Remarks: Nonbearing: apples, apricots, cherries, crabapples, peaches, pears, plums, prunes. Spread granules uniformly from trunk to drip line on all sides. Work into soil and water thoroughly. Higher rate recommended for heavy organic or muck soils. For transplanting, equivalent quantities should be applied into planting hole. Do not apply to trees that will bear fruit during current crop year.

Signal Word/Toxicity Class: Danger/I.
REI: 48 hr.

D•z•n — see diazinon

ENDOSULFAN

One of the last cyclodiene insecticides remaining in use. Effective as control material for sucking insects in apples, peaches and pears, and as borer control material in cherries, peaches, and plums. Highly toxic.

Drexel Endosulfan 3EC or 50WP *(Drexel)*
Gowan Endosulfan 3EC or 50WSB *(Gowan)*
◆ Micro Flo Endosulfan 3EC or 50WP *(Micro Flo)*
Phaser 3EC or 50WSB *(AgrEvo USA)*
Thiodan 3EC or 50WP *(FMC)*
Rate: 2.5 lb. ai/A
3 qt. 3EC/A
5 lb. 50WP, 50WSB/A
Pest:
apple aphid, rosy apple aphid, white apple leafhopper, apple rust mite, tarnished plant bug, woolly apple aphid, green fruitworm
Remarks: Apples. For green fruitworm and tarnished plant bug, apply at pink and/or petal fall. For white apple leafhopper (1st generation), apply when nymphs first appear on leaves. Do not exceed 3 applications per year or 3 lb. ai/A per year. Do not feed pomace from treated apples to livestock, feed cull fruits to animals, or graze treated orchards.
Days To Harvest: 21.

Rate: 2-2.5 lb. ai/A
2.66-3.33 qt. 3EC/A
4-5 lb. 50WP, 50WSB/A
Pest:
consperse stink bug (foliar), green fruitworm, pear rust mite, pear psylla (high rate), tarnished plant bug

Rate: 0.5 lb. ai/100 gal. water
0.66 qt. 3EC/100 gal. water (200-400 gal./A)
1 lb. 50WP, 50WSB/100 gal. water (200-400 gal./A)
Pest:
consperse stink bug (soil)

Rate: Postharvest or dormant:
0.25-0.5 lb. ai/100 gal. water
0.33-0.66 qt. 3EC/100 gal. water
0.5-1 lb. 50WP, 50WSB/100 gal. water
Pest:
pearleaf blister mite
Remarks Pears. For foliar treatment, stink bugs must be wet by spray to obtain control; soil treatment, apply to orchard floor around trees prior to bloom. For green fruitworm and tarnished plant bug, apply at white bud or petal fall. For pear psylla, apply when adults are first observed or nymphs are small and repeat to maintain control. Do not exceed 2 applications per year or 3 lb. ai/A per year. Aerial application does not result in satisfactory control and should only be employed if impossible to apply by ground. Do not feed cull fruits to animals or graze livestock in treated orchards.
Days To Harvest: 7.

Rate: 0.75 lb. ai/100 gal. water
1 qt. 3EC/100 gal. water
1.5 lb. 50WP, 50WSB/100 gal. water
West Coast:
0.5-0.75 lb. ai/100 gal. water
0.66-1 qt. 3EC/100 gal. water
1-1.5 lb. 50WP, 50WSB/100 gal. water
Southeastern states-postharvest:
1.5-2.5 lb. ai/100 gal. water
2-3.33 qt. 3EC/100 gal. water
3-5 lb. 50WP, 50WSB/100 gal. water
Pest:
lesser peachtree borer, peachtree borer

◆ - **new product** • ai-active ingredient • DF-dry flowable • E/EC-emulsifiable concentrate • EW-emulsion in water
F/FL-flowable • G-granule • L/LC-liquid • SP/WSP-soluble packet • W/WP-wettable powder • WSB-water soluble bag

Rate: 2-2.5 lb. ai/A
2.66-3.33 qt. 3EC/A
4-5 lb. 50WP, 50WSB/A

Pest:

black cherry aphid	green peach aphid	rusty plum aphid
black peach aphid	peach silver mite	stink bugs
green fruitworm	peach twig borer	

Remarks: Apricots, nectarines, peaches. For borers, best control is obtained with single application during first week of Sept.; spray all bark areas from ground level to scaffold limbs. Do not exceed 2 applications per year or 3 lb. ai/A per year. Do not feed cull fruits to animals or graze livestock in treated orchards.

Days To Harvest: 30; aphids, borers-21.

- - - - - - - - - - - - - - - - -

Rate: 0.5-0.75 lb. ai/100 gal. water
0.66-1 qt. 3EC/100 gal. water
1-1.5 lb. 50WP, 50WSB/100 gal. water

Pest:

black cherry aphid	fruittree leafroller (Pacific Northwest)	nursery mite
eyespotted bud moth (Pacific Northwest)		plum rust mite

Rate: Michigan:
1 lb. ai/100 gal. water
1.33 qt. 3EC/100 gal. water
2 lb. 50WP, 50WSB/100 gal. water

Pest:
mineola moth

Remarks: Cherries. For borers, thoroughly wet all bark from ground to scaffold limbs. For bud moth, apply at popcorn stage. For leafrollers, apply during pre-pink stage. For mineola moth, apply in delayed dormant period.

Days To Harvest: 21.

- - - - - - - - - - - - - - - - -

Rate: 2-2.5 lb. ai/A
2.66 3.33 qt. 3EC/A
4-5 lb. 50WP, WSB/A

Pest:

fruittree leafroller (Pacific Northwest)	leafcurl plum aphid	plum rust mite (nursery mite)
hop aphid	lesser peachtree borer	thistle aphid

Rate: 0.75 lb. ai/100 gal. water
1 qt. 3EC/100 gal. water
1.5 lb. 50WP, 50WSB/100 gal. water
West Coast:
0.5-0.75 lb. ai/100 gal. water
0.66-1 qt. 3EC/100 gal. water
1-1.5 lb. 50WP, 50WSB/100 gal. water

Pest:
peachtree borer

Remarks: Plums, prunes. For aphids, apply when eggs hatch during prebloom or petal fall; summer applications should be made before leaves curl. For fruittree leafroller, apply during pre-pink stage of growth. For lesser peachtree borer, apply spray to thoroughly wet trunk and main branches. For peachtree borer, spray thoroughly to cover bark areas from ground to scaffold limbs. Do not exceed 2 applications per year or 3 lb. ai/A per year. Do not graze livestock on orchard crops or grasses in treated areas.

Days To Harvest: 7.

- - - - - - - - - - - - - - - - -

Rate: 2 lb. ai/40 gal. water
2.66 qt. 3EC/40 gal. water
4 lb. 50WP, 50WSB/40 gal. water

Pest:
peachtree borer

Remarks: Cherries, peaches, plums-nursery stock dip. Mix thoroughly. Immerse trees so that roots and crowns are covered well above grafting bud scar. Wear rubber gloves during dipping operation.

Days To Harvest: 7.

- - - - - - - - - - - - - - - - -

Signal Word/Toxicity Class: Danger-Poison/I.
REI: 24 hr.

FENOXYCARB

Insect growth regulator.

Logic *(Novartis)*

Rate: Mound treatment:
1-3 Tbs./mound
Broadcast or aerial application:
1-1.5 lb./A

Pest:
fire ants

Remarks: Nonbearing nectarines, peaches, plums. Mound treatment: Uniformly distribute material 3-4 ft. around mound. Broadcast or aerial application: Apply uniformly with equipment calibrated to give correct dosage. Do not apply by air where weather conditions favor drift from treated areas. Retreat after 3-4 months, if needed. Apply when ants are actively foraging, usually when soil temperatures are above 60°F. Avoid application during excessively hot periods or when grass is wet. Do not apply to nonbearing crops more than twice per year or sooner than 1 year before first harvest. Do not graze treated areas.

- - - - - - - - - - - - - - - - -

Signal Word/Toxicity Class: Caution.
REI: 12 hr.

FENPROPATHRIN

Highly toxic to fish, moderately toxic to insect predators.

Tame 2.4 EC *(Valent)*

Rate: 0.1 lb. ai/100 gal. water
5.33 fl. oz. 2.4 EC/100 gal. water

Pest:

codling moth	McDaniel mite	pandemis moth
European red mite	orange tortrix	tarnished plant bug
leafrollers		

Rate: 0.15-0.3 lb. ai/100 gal. water
8-16 fl. oz. 2.4 EC/100 gal. water

Pest:

southern red mite	twospotted spider mite

Remarks: Nonbearing apples, pears (greenhouse, lathhouse, shadehouse only). Apply after petal fall or as needed. May be tank mixed with 0.66 lb./100 gal. Orthene Turf, Tree & Ornamental Spray. Do not apply through any type of irrigation system Refer to label for spray drift precautions.

- - - - - - - - - - - - - - - - -

Signal Word/Toxicity Class: Danger/I.
REI: 24 hr.

Fyfanon — see malathion

Guthion — see azinphos-methyl

HEXYTHIAZOX

Savey 50-WP *(Gowan)*

Rate: Apples:
3 oz. 50-WP/A

Pest:
European red mite

Rate: Pears:
3-6 oz. 50-WP/A

Pest:

European red mite	McDaniel spider mite

Remarks: Apples, pears. In apples, apply after dormancy is broken, but no later than pink stage of flower bud development. Do not apply more than 1 application per growing season. Do not apply more than 3 oz./A per growing season on apples; 6 oz./A per growing season on pears. Do not use water from treated orchards for irrigating other crops. May be incompatible with some EC formulations; addition of compatibility agent may improve compatibility with those products. Do not graze or feed livestock on cover crops growing in treated areas.

Days To Harvest: Pears-28.

- - - - - - - - - - - - - - - - -

Signal Word/Toxicity Class: Caution.
REI: 12 hr.

IMIDACLOPRID

Systemic insecticide for control of aphids. Toxic to bees and aquatic invertebrates.

Provado 1.6 *(Bayer)*

Rate: Apples, crabapples, Oriental pears, quince:
0.1 lb. ai/A
8 fl. oz. 1.6F/A
0.025 lb. ai/A
2 fl. oz. 1.6F/100 gal. water

NOTICE The information on these pages is for preliminary planning — not a guide for use. Be sure to follow the manufacturer's directions, notwithstanding information contained here.

For personal protective equipment and EPA registration numbers, please turn to page 39.

Pest:

aphids (except woolly apple)	leafminers	San Jose scale

Rate: Apples, crabapples, Oriental pears, quince:
0.05-0.1 lb. ai/A
4-8 fl. oz. 1.6/A
0.013-0.025 lb. ai/A
1-2 fl. oz. 1.6/100 gal. water

Pest:
leafhoppers

Rate: Pears:
0.25 lb. ai/A
20 fl. oz. 1.6/A
0.0625 lb. ai/A
5 fl. oz. 1.6/100 gal. water

Pest:

aphids	pear psylla	San Jose scale
mealybugs		

Remarks: Apples, crabapples, Oriental pears, pears, quince. Postbloom. Apply as dilute or concentrate foliar spray as needed after pollination is complete. Allow 10 days between applications. Do not apply during bloom or when bees are present in orchard. Do not apply more than 8 fl. oz./A in a single application (20 fl. oz./A for pears) or a total of 40 fl. oz./A per year.
Days To Harvest: 7.

Signal Word/Toxicity Class: Caution.
REI: 12 hr.

IMIDAN

Organophosphate, broad-spectrum insecticide on pome and stone fruits, but not on sweet cherries because of phytotoxicity. Low mammalian toxicity.

Imidan 70-W (phosmet) *(Gowan)*

Rate: Apples:
1.5-3.7 lb. ai/A
2.13-5.33 lb. 70-W/A

Pest:

apple maggot	green fruitworm	Oriental fruit moth
codling moth	gypsy moth	peach twig borer
dock sawfly	Japanese beetle	plum curculio
elm spanworm	mealybugs	redbanded leafroller
European corn borer	obliquebanded leafroller	redhumped caterpillar
European sawfly		rose chafer
fruittree leafroller	orange tortrix	San Jose scale

Rate: Pears:
1.5-5 lb. ai/A
2.13-7.13 lb. 70-W/A

Pest:

apple maggot	gypsy moth	plum curculio
codling moth	Japanese beetle	redbanded leafroller
elm spanworm	mealybugs	rose chafer
fruittree leafroller		

Rate: Apricots, nectarines, peaches, plums, prunes:
1.5-3 lb. ai/A
2.13-4.25 lb. 70-W/A

Pest: Apricots:

apple maggot	orange tortrix	plum curculio
fruittree leafroller	Oriental fruit moth	redbanded leafroller
Japanese beetle	peach twig borer	rose chafer

Nectarines:

apple maggot	Oriental fruit moth	redbanded leafroller
Japanese beetle	peach twig borer	rose chafer
omnivorous leafroller	plum curculio	San Jose scale

Peaches:

Japanese beetle	peach twig borer	rose chafer
Oriental fruit moth	plum curculio	San Jose scale

Plums, prunes:

apple maggot	peach twig borer	redhumped caterpillar
codling moth	plum curculio	rose chafer
omnivorous leafroller	redbanded leafroller	San Jose scale
Oriental fruit moth		

Rate: Cherries (sour):
1.5-1.7 lb. ai/A
2.13-2.5 lb. 70-WP/A

Pest:

cherry fruit fly	peach twig borer	rose chafer
fruittree leafroller	plum curculio	San Jose scale
Japanese beetle		

Remarks: Apples, apricots, cherries (sour), nectarines, peaches, pears, plums, prunes. For west of the Rockies and heavy insect infestations, use higher rate. Repeat as necessary in accordance with infestations and local or state spray programs. Do not use on Early McIntosh or Wealthy apple varieties. Do not apply more than 30 lb./A per season on apples. Do not use in combination with Omite on late maturing nectarine varieties as fruit injury may result.
Days To Harvest: Apples, cherries, pears, plums, prunes-7; apricots, nectarines, peaches-14.

Signal Word/Toxicity Class: Warning/II.
REI: 24 hr.

Kelthane — see dicofol

Kumulus — see sulfur

LANNATE

Toxic to fish and wildlife. Restricted Use Pesticide due to high acute toxicity to humans.

Lannate SP or LV (methomyl) *(DuPont)*

Rate: 0.45-0.9 lb. ai/A
0.5-1 lb. SP/A
1.5-3 pt. LV/A

Pest:

apple aphid	lesser appleworm	tarnished plant bug
codling moth	obliquebanded leafroller	tentiform leafminer
cutworms		tufted apple bud moth
fruittree leafroller	redbanded leafroller	variegated leafroller
green fruitworm	rosy apple aphid	white apple leafhopper

Remarks: Apples. Ground application. Apply when insects first appear at 5- to 7-day intervals or as needed; 10- to 12-day intervals for codling moth. Rate varies by pest species; refer to label. Do not use on Early McIntosh or Wealthy varieties. Do not apply more than 4.5 lb. ai/A per crop or make more than 5 applications per crop.
Days To Harvest: 14; livestock grazing/feeding-10.

Rate: 0.45-0.9 lb. ai/A
0.5-1 lb. SP/A
1.5-3 pt. LV/A

Pest:
thrips

Remarks: Nectarines. Apply as needed. Do not apply more than 2.7 lb. ai/A per crop or make more than 3 applications per crop.
Days To Harvest: 1; livestock grazing/feeding-10.
State Restrictions: For use in Arizona, California.

Rate: 0.67-1.8 lb. ai/A
0.75-2 lb. SP/A
2.25-6 pt. LV/A

Pest:

green peach aphid	plant bugs	stink bugs
Oriental fruit moth		

Remarks: Peaches. Apply when insects first appear at 5- to 7-day intervals or as needed. Rate varies by pest species; refer to label. For catfacing insects, apply at petal fall and continue in cover sprays at 7- to 10-day intervals. For Oriental fruit moths, apply ground application at petal fall using trapping devices and frequent field inspection to determine need for treatment. Continue in cover sprays and alternate with residual type insecticides registered for this use. Do not apply more than 5.4 lb. ai/A per crop or make more than 6 applications per crop.
Days To Harvest: 4; livestock grazing/feeding-10.

Rate: 0.45-0.9 lb. ai/A
0.5-1 lb. SP/A
1.5-3 pt. LV/A

Pest:

green fruitworm	obliquebanded leafroller

Remarks: Pears. Apply as needed. Do not apply more than 1.8 lb. ai/A per crop or make more than 2 applications per crop.
Days To Harvest: 7; livestock grazing/feeding-10.
State Restrictions: For use in the Northeast only.

Signal Word/Toxicity Class: Danger-Poison/I.
REI: Pears-48 hr.; apples, nectarines-3 days.; peaches-4 days.

◆ - **new product** • ai-active ingredient • DF-dry flowable • E/EC-emulsifiable concentrate • EW-emulsion in water
F/FL-flowable • G-granule • L/LC-liquid • SP/WSP-soluble packet • W/WP-wettable powder • WSB-water soluble bag

Lime-Sulfur Solution — see calcium polysulfide

Logic — see fenoxycarb

Lorsban — see chlorpyrifos

MALATHION

Mild organophosphate that is only occasionally used in commercial tree fruit production. Low in toxicity to applicator. Short harvest interval.

Clean Crop Malathion 57EC *(Platte)*

Rate: Apples, pears:
0.63-1.25 lb. ai/100 gal. water
1-2 pt. 57EC/100 gal. water

Pest: Apples:

codling moth	green apple aphid	tent caterpillar
eyespotted bud moth	mealybugs	woolly apple aphid
Forbes scale	rosy apple aphid	

Pears:

codling moth	mealybugs	pear psylla
leafrollers		

Rate: Cherries:
0.63-1 lb. ai/100 gal. water
1-1.5 pt. 57EC/100 gal. water

Pest:

black cherry aphid	cherry fruit fly	fruittree leafroller
eyespotted bud moth	fruittree leafroller	

Rate: Apricots:
1.25 lb. ai/100 gal. water
2 pt. 57EC/100 gal. water

Pest:

orange tortrix	soft scale

Rate: Nectarines, peaches, quince:
0.63-1.25 lb. ai/100 gal. water
1-2 pt. 57EC100 gal. water

Pest:

codling moth	mites	plum curculio
Forbes scale	Oriental fruit moth	

Rate: Plums, prunes:
0.63 lb. ai/100 gal. water
1 pt. 57EC/100 gal. water

Pest:

eyespotted bud moth	mealy plum aphid

Remarks: Injury may occur on McIntosh apples, and in the Northwest, on Bosc pears and certain varieties of sweet cherries.

Days To Harvest: Apples, cherries, plums, prunes, quinces-3; pears-1; apricots, nectarines, peaches-7.

Signal Word/Toxicity Class: Caution/III.
REI: 12 hr.

Clean Crop Malathion (8EC) *(Platte)*
◆ **Wilbur-Ellis Malathion 8** *(Wilbur-Ellis)*

Rate: 1-1.5 lb. ai/A
1-1.5 pt. 8EC. 8/A

Pest:

aphids	eyespotted bud moth	scales (peaches)
cherry fruit fly	mealybugs	spider mites
codling moth	orange tortrix	thrips

Remarks: Apricots, cherries, nectarines, peaches. May cause injury on certain varieties of sweet cherries (particularly in the Northwest) and fruit spotting on nectarines. For scale on peaches, apply to completion of hatching (about last of June).

Days To Harvest: Cherries-3; pears-1; apricots, nectarines, peaches-7.

Signal Word/Toxicity Class: Caution/III.
REI: 12 hr.

Drexel Malathion 5EC *(Drexel)*
◆ **Micro Flo Malathion 5EC** *(Micro Flo)*
Prentox 5 lb. Malathion Spray (5EC) *(Prentiss)*

Rate: Cherries:
0.63-1 lb. ai/100 gal. water
1-1.5 pt. 5EC/100 gal. water

Pest:

black cherry aphid	cherry fruit fly	Japanese beetle
bud moth	fruittree leafroller	

Rate: Apricots:
1-1.25 lb. ai/100 gal. water
1.5-2 pt. 5EC/100 gal. water

Pest:

aphids	European fruit lecanium	soft brown scale
codling moth	orange tortrix	terrapin scale

Rate: Nectarines, peaches:
0.63-1.25 lb. ai/100 gal. water
1-2 pt. 5EC/100 gal. water

Pest: Nectarines:

plum curculio	spider mites

Peaches:

black cherry aphid	European red mite	plum curculio
black peach aphid	green peach aphid	rusty plum aphid
cottony peach scale	Japanese beetle	spider mites
European fruit lecanium	Oriental fruit moth	terrapin scale

Remarks: Apricots, cherries (sweet, tart), nectarines, peaches. Injury may occur on certain varieties of sweet cherries (particularly in the Northwest).

Days To Harvest: Cherries-3; apricots, nectarines, peaches-7.

Signal Word/Toxicity Class: Warning/II.
REI: 12 hr.

◆ **Drexel Malathion ULV** *(Drexel)*
Fyfanon ULV *(Cheminova)*
◆ **Micro Flo Malathion ULV** *(Micro Flo)*

Rate: 12-16 fl. oz. ULV/A

Pest:
cherry fruit fly

Remarks: Cherries. Apply aerial application when flies first appear. Use higher rate when foliage is heavy or infestation is severe.

Days To Harvest: 1.

Signal Word/Toxicity Class: Caution.
REI. 12 hr.

Fyfanon (8E) *(Helena)*

Rate: Apricots:
1.25 lb. ai/A
1.25 pt. 8E/A

Pest:

aphids	European fruit lecanium	lesser peachtree borer
codling moth	Forbes scale	San Jose scale

Rate: Cherries:
1 lb. ai/A
1 pt. 8E/A

Pest:

black cherry aphid	cherry fruit fly	orange tortrix
bud moth	fruittree leafroller	soft brown scale

Rate: Peaches:
0.75-1.25 lb. ai/A
0.75-1.25 pt. 8E/A

Pest:

black cherry aphid	European fruit lecanium	Oriental fruit moth
black peach aphid	green peach aphid	rusty plum aphid
cottony peach scale	Japanese beetle	terrapin scale

Remarks: Apricots, cherries, peaches. For peachtree borers, apply at emergence to trunk and scaffold limbs at 21-day intervals. For Forbes and San Jose scale, apply 3 weeks before and immediately after harvest. May cause injury on certain varieties of sweet cherries in the Northeast.

Days To Harvest: Cherries-3; apricots, peaches-7.

Signal Word/Toxicity Class: Caution/III.
REI: 12 hr.

NOTICE The information on these pages is for preliminary planning — not a guide for use. Be sure to follow the manufacturer's directions, notwithstanding information contained here.	For personal protective equipment and EPA registration numbers, please turn to page 39.

Gowan Malathion 8 (8E) *(Gowan)*

Rate: Apples, pears:
2-15 lb. ai/A
2-15 pt. 8E/A

Pest: Apples:

aphids	mealybugs	pear psylla
eyespotted bud moth	mites	tent caterpillar
Forbes scale		

Pears:

mealybugs	spider mites	woolly aphid
pear psylla		

Rate: Apples, pears:
Dormant, delayed dormant:
4-15 lb. ai/A
4-15 pt. 8E/A

Pest:

codling moth	plum curculio	San Jose scale
fruittree leafroller	redbanded leafroller	(apples)

Rate: Cherries:
2.5-8 lb. ai/A
2.5-8 pt. 8E/A

Pest:

black cherry aphid	Forbes scale	lesser peach twig borer
bud moth	fruittree leafroller	San Jose scale
cherry fruit fly		

Rate: Apricots:
4-10 lb. ai/A
4-10 pt. 8E/A

Pest:

aphids	European lecanium	soft brown scale
codling moth	orange tortrix	terrapin scale

Rate: Nectarines, peaches:
2.5-9 lb. ai/A
2.5-9 pt. 8E/A

Pest:

black cherry aphid	Japanese beetle	rusty plum aphid
black peach aphid	lesser peachtree borer	San Jose scale
cottony peach scale	Oriental fruit moth	terrapin scale
green peach aphid	plum curculio	

Rate: Plums, prunes:
5-10 lb. ai/A
5-10 pt. 8E/A

Pest:

aphids*	leafcurl plum aphid	San Jose scale*
European red mite	lesser peachtree borer	twospotted spider mite
eyespotted bud moth	mealy plum aphid	

****Dormant, delayed dormant***

Remarks: Apples, apricots, cherries, nectarines, peaches, pears, plums, prunes. For apples, do not apply in bloom period. May cause injury on McIntosh apples and certain varieties of sweet cherries (particularly in the Northeast). Spray oil may be used in dormant/delayed dormant sprays.

Days To Harvest: Apples, cherries, plums, prunes-3; pears-1; apricots, nectarines, peaches-7.

Signal Word/Toxicity Class: Caution/III.
REI: 12 hr.; peaches-24 hr.

Marlate — see methoxychlor

METASYSTOX-R

Systemic insecticide labeled for aphid and mite control on pears, and for control of these pests in nonbearing fruit trees. Restricted Use Pesticide.

Metasystox-R (SC) (oxydemeton-methyl) *(Gowan)*

Rate: 0.25-0.375 lb. ai/100 gal. water
1-1.5 pt. SC/100 gal. water

Pest:
aphids mites

Remarks: Nonbearing: apples, apricots, cherries, crabapples, nectarines, peaches, pears, plums, prunes, quinces. Use low rate for pears. Apply as full coverage or concentrate spray at appropriate dilute rate; repeat as necessary. Do not apply more than 3 times per season. Do not apply to trees that will bear fruit within next 12 months.

Signal Word/Toxicity Class: Warning/II.
REI: 48 hr.; 72 hr. if average rainfall less than 25" per year.

METHOXYCHLOR

Methoxychlor is the last of the chlorinated hydrocarbons available for use in fruit production.

Marlate 50 (WP) *(Kincaid)*
Prentox Methoxychlor 2EC *(Prentiss)*

Rate: 1.3-1.7 lb. ai/100 gal. water
75-100 fl. oz. 2EC/100 gal. water
2 Tbs. 50WP/gal. water

Pest:

apple maggot	codling moth	plum curculio
cankerworm	Japanese beetle	rose chafer
cherry fruit fly	leafhoppers	tent caterpillar
cherry fruitworm	pearslug	

Remarks: Apples, apricots, cherries, nectarines, peaches, pears, plums, prunes, quinces. Apply at petal fall and repeat at 7- to 14-day intervals as needed.

Days To Harvest: Apples, pears, cherries, plums, prunes, quinces-7; apricots, nectarines, peaches-21.

Signal Word/Toxicity Class: Caution.
REI: 12 hr.

METHYL PARATHION

Penncap-M is microencapsulated form of methyl parathion, which provides an increased safety factor for applicators and other nontarget organisms by releasing active ingredient over a long period of time. Restricted Use Pesticide due to residual effects to avian species and hazard to bees.

Penncap-M (2FL) *(Elf Atochem)*

Rate: Except California:
0.5-1 lb. ai/A
2-4 pt. 2FL/A
0.125-0.5 lb. ai/100 gal. dilute spray
0.5-2 pt. 2FL/100 gal. dilute spray
California only:
1-2 lb. ai/A
4-8 pt. 2FL/A
0.5-1 lb. ai/100 gal. dilute spray
1-2 pt. 2FL/100 gal. dilute spray

Pest:

apple maggot	Oriental fruit moth	tarnished plant bug
apple-and-thorn skeletonizer	pear plant bug (pears only)	tufted apple bud moth
codling moth	plum curculio	whitemarked tussock moth
green fruitworm	San Jose scale	
lesser appleworm		

Except California:

fruittree leafroller	oystershell scale	variegated leafroller
obliquebanded leafroller	pandemis leafroller	woolly apple aphid
	redbanded leafroller	

Remarks: Apples, pears. Under heavy insect pressure, use higher rate. Apply at 7-day intervals. Consult local agricultural experiment station for proper spray timing. Do not make initial application before total petal fall or make more than 5 applications from total petal fall to harvest. Do not tank mix Penncap-M with streptomycin since injury to fruit and foliage may result. Do not apply to d'Anjou pears as a concentrate or semi-concentrate spray since fruit injury is likely to occur. Do not apply when weeds or cover crops in orchard are in bloom and bees are actively visiting area to be treated.

Days To Harvest: Annual rainfall exceeds 25": up to 4 pt./A-14, over 4 pt./A-21; annual rainfall less than 25"-30.

Rate: 0.75-1.5 lb. ai/A
3-6 pt. 2FL/A
0.25-0.5 lb. ai/100 gal. dilute spray
1-2 pt. 2FL/100 gal. dilute spray

Pest:

black cherry aphids	fruit flies	leafrollers
cherry fruit fly	green fruitworms	plum curculio

Remarks: Cherries. Apply at 7-day intervals. Do not make initial application before total petal fall or apply when substantial number of weeds or cover crops in orchard are in bloom and bees are actively visiting area to be treated. From petal fall to harvest, do not apply more than 6 applications. Do not use in areas where annual rainfall is less than 25" per year. Do not graze or feed cover crops from treated orchards.

Days To Harvest: 14.
State Restrictions: Not for use in California.

◆ - **new product** • ai-active ingredient • DF-dry flowable • E/EC-emulsifiable concentrate • EW-emulsion in water
F/FL-flowable • G-granule • L/LC-liquid • SP/WSP-soluble packet • W/WP-wettable powder • WSB-water soluble bag

Rate: 1.125-1.5 lb. ai/A
4.5-6 pt. 2FL/A
0.375-0.5 lb. ai/100 gal. dilute spray
1.5-2 pt. 2FL/100 gal. dilute spray

Pest:
lesser peachtree borer, peachtree borer

Rate: Except California:
0.5-1.5 lb. ai/A
2-6 pt. 2FL/A
0.125-0.5 lb. ai/100 gal. dilute spray
0.5-2 pt. 2FL/100 gal. dilute spray
California only:
0.75-1.5 lb. ai/A
3-6 pt. 2FL/A
0.25-0.5 lb. ai/100 gal. dilute spray
1-2 pt. 2FL/100 gal. dilute spray

Pest:
catfacing insects, Oriental fruit moth, plum curculio

Rate: 2 lb. ai/A
8 pt. 2FL/A

Pest:
peach twig borer

Remarks: Nectarines, peaches, plums, prunes. Apply at 7-day intervals. Do not make initial application before total petal fall, or make more than 4 applications from total petal fall to harvest. Do not make more than 6 applications per year. Do not apply when substantial number of weeds or cover crops in orchards are in bloom and bees are visiting area to be treated. For peach twig borer, make dormant or delayed dormant application only and do not apply in less than 200 gal./A. Do not graze or feed cover crops from treated orchards.

Days To Harvest: Annual rainfall exceeds 25": up to 3 pt./A-21, over 3 pt./A-28; annual rainfall less than 25"-14.

- - - - - - - - - - - - - - - -

Signal Word/Toxicity Class: Warning/II.
REI: 48 hr.

Microthiol Special — see sulfur

Mitac — see amitraz

MORESTAN

Organophosphate dual-purpose insecticide controls powdery mildew and adults and eggs of certain insects and mites. For ovicidal action, should be applied prebloom.

Morestan Solupak 25 (oxythioquinox) *(Bayer)*

Rate: **Bearing: apples, pears (except NY)**
Nonbearing: apples, apricots, cherries, nectarines, peaches, pears, plums, prunes:
0.125-0.25 lb. ai/100 gal. water
0.5-1 lb. Solupak 25/100 gal. water

Pest:
mite eggs, mites (except cyclamen)

Rate: **Bearing: apples, pears:**
0.375 lb. ai/100 gal. water
1.5 lb. Solupak 25/100 gal. water

Pest:
pear psylla

Remarks: Apples, apricots, cherries, nectarines, peaches, pears, plums, prunes. Bearing fruit: Apply as prebloom or postharvest sprays (delayed-dormant, pre-pink, or pink). Do not apply after first bloom. Additional applications may be made to tree foliage after harvest. Nonbearing fruit: Spray all foliage surfaces including underside of leaves for complete coverage. Repeat as necessary.

- - - - - - - - - - - - - - - -

Rate: 1 lb. ai/A
4 lb. Solupak 25/A

Pest:
mite eggs, mites (except cyclamen)

Remarks: Apricots. Postharvest. Apply in minimum 10 gal. water/A by air; 40 gal. water/A by ground. Thorough coverage necessary. Make only 1 application between harvest and up to the time trees break dormancy. Do not allow animals to graze cover crops.

- - - - - - - - - - - - - - - -

Rate: 0.125-0.375 lb. ai/A
0.5-1.5 lb. Solupak 25/A

Pest:
mite eggs, mites, pear psylla

Remarks: Apples, pears. Refer to label for use directions and precautions.
State Restrictions: SLN-Washington.

- - - - - - - - - - - - - - - -

Signal Word/Toxicity Class: Warning/II.
REI: 24 hr.

M-PEDE

M-Pede (potassium salts of fatty acids) *(Mycogen)*

Rate: 1-2% v/v solution

Pest:
aphids, green apple aphid, leafhoppers, mealybugs, pear psylla, rosy apple aphid, scale crawlers, spider mites, whiteflies

Remarks: Apples, apricots, cherries, crabapples, nectarines, peaches, pears, plums, prunes, quinces. Do not apply after delayed dormant stage on fresh markets including d'Anjou and Asian varieties. Not for use on yellow-skinned nectarine varieties. May be tank mixed with other insecticides. Refer to labels for use directions and precautions.

Days To Harvest: 0.

- - - - - - - - - - - - - - - -

Signal Word/Toxicity Class: Warning/II.
REI: 12 hr.

NEMACUR

Restricted Use Pesticide.

Nemacur 3 (fenamiphos) *(Bayer)*

Rate: Band:
5-10 lb. ai/A
1.66-3.33 gal. 3/A
Low pressure irrigation:
1.5-4.5 lb. ai/A
2-6 qt. 3/A

Pest:
nematodes

Remarks: Apples, cherries, nectarines, peaches. Apply band to soil and incorporate immediately or apply 1-6 applications by low pressure irrigation with at least 14 days between applications. Refer to label for further use information.

Days To Harvest: Apples-72; cherries, nectarines, peaches-45.

- - - - - - - - - - - - - - - -

Signal Word/Toxicity Class: Danger/I.
REI: 48 hr.

NEEMAZAD — see azadirachtin

NEEMIX — see azadirachtin

Omite — see propargite

Omni Supreme Spray — see petroleum oil

Penncap-M — see methyl parathion

PERMETHRIN

Synthetic pyrethroid. Exhibits low mammalian toxicity while providing excellent control of many fruit insects. Acts both as a stomach and contact poison, and provides control of many insects resistant to organophosphates. Toxic to beneficial mites in the orchard system. Restricted Use Pesticide.

Ambush 2E or 25W *(ZENECA)*

Rate: 0.1-0.4 lb. ai/A
6.4-25.6 fl. oz. 2E/A
6.4-25.6 oz. 25W/A

8 Fruits & Nuts

NOTICE The information on these pages is for preliminary planning — not a guide for use. Be sure to follow the manufacturer's directions, notwithstanding information contained here.

For personal protective equipment and EPA registration numbers, please turn to page 39.

Pest:

apple aphid, obliquebanded leafroller, plum curculio, redbanded leafroller, spotted tentiform leafminer, tarnished plant bug, white apple leafhopper

Remarks: Apples. Apply in 25-400 gal. spray with ground equipment only. Spray to wet all foliage. Do not apply after petal fall.

Rate: 0.2-0.4 lb. ai/A
12.8-25.6 fl. oz. 2E/A
12.8-25.6 oz. 25W/A

Pest:

codling moth, green fruitworm, pear psylla

Remarks: Pears. Prebloom/seasonal: apply 2 prebloom sprays followed by 3 seasonal sprays. Prebloom can be applied from dormant through budburst; seasonal sprays from petal fall up to 14 days before harvest. Do not apply more than 2 lb. ai/A per season. Dormant: For pear psylla, apply during dormant-delayed dormant growth periods in 25-400 gal. spray by ground; 3-20 gal. spray/A by air. May be combined with 2-8 gal. spray oil/A. Do not apply more than 0.8 lb. ai/A per season. Do not allow animals to graze orchard floor.

Days To Harvest: 14.

Rate: 0.025-0.05 lb. ai/A
1.6-3.2 fl. oz. 2E/A
1.6-3.2 oz. 25W/A

Pest:

green fruitworm, lesser peachtree borer, plum curculio, redbanded leafroller, rose chafer, tarnished plant bug

Remarks: Cherries. Apply in minimum 10 gal./A by air; 25-400 gal. water/A by ground when insects appear. Do not exceed 400 gal./A. In the East, do not exceed 6 applications per season or 4 applications after petal fall. In the West, do not exceed 4 applications per season or 3 applications after petal fall. Do not allow livestock to graze treated orchards. Do not feed forage to livestock.

Days To Harvest: 3.

Rate: 0.1-0.4 lb. ai/A
6.4-25.6 fl. oz. 2E/A
6.4-25.6 oz. 25W/A

Pest:

green fruitworm, lesser peachtree borer, Oriental fruit moth, peach twig borer, plum curculio, rose chafer, tarnished plant bug

Remarks: Peaches. Spray to wet all foliage. Apply in minimum 10 gal./A by air; 25-400 gal. water/A by ground. Do not apply more than 3 lb. ai/A per season. Do not allow livestock to graze treated orchards.

Days To Harvest: 7.

Signal Word/Toxicity Class: Warning/II.
REI: 24 hr.

Pounce 3.2 EC, 25WP or WSB *(FMC)*

Rate: 0.1-0.2 lb. ai/A
4-8 oz. 3.2 EC/A
6.4-12.8 oz. 25WP/A
1-2 bags WSB/A

Pest:

green fruitworm, obliquebanded leafroller, plum curculio, redbanded leafroller, rosy apple aphid, spotted tentiform leafminer, tarnished plant bug, white apple leafhopper

Remarks: Apples. Do not apply more than 0.6 lb. ai/A per season. Do not apply after petal fall.

Rate: Dormant through delayed dormant:
0.2-0.4 lb. ai/A
8-16 fl. oz. 3.2 EC/A
12.8-25.6 oz. 25WP/A
2-4 bags WSB/A

Pest:

pear psylla

Remarks: Pears. May be combined with 2-8 gal. spray oil/A. Do not apply more than 0.8 lb. ai/A per season. Do not graze orchard floor.

Rate: Summer:
0.2-0.4 lb. ai/A
8-16 fl. oz. 3.2 EC/A
12.8-25.6 oz. 25WP/A
2-4 bags WSB/A

Pest:

codling moth, green fruitworm, pear psylla

Remarks: Pears. Use in sufficient water to achieve runoff and apply by ground equipment. Do not apply more than 1.2 lb. ai/A per season.

Days To Harvest: 14.

Rate: 0.025-0.05 lb. ai/100 gal. water
1-2 oz. 3.2 EC/100 gal. water
1.6-3.2 oz. 25WP/100 gal. water
1 bag WSB/100 gal. water

Pest:

green fruitworm, lesser peachtree borer, plum curculio, redbanded leafroller, rose chafer, tarnished plant bug

Remarks: Cherries. Apply when insects appear. East of the Rockies: Do not make more than 6 applications per season or more than 4 applications after petal fall. West of the Rockies: Do not make more than 4 applications per season or more than 3 applications after petal fall. Do not allow livestock to graze or feed crop forage from treated orchards.

Days To Harvest: 3.

Rate: 0.05-0.1 lb. ai/100 gal. water
2-4 fl. oz. 3.2 EC/100 gal. water
3.2-6.4 oz. 25WP/100 gal. water
1 bag WSB/100 gal. water

Pest:

green fruitworm, lesser peachtree borer, Oriental fruit moth, peach twig borer, plum curculio, rose chafer, tarnished plant bug

Remarks: Peaches. Do not make more than 8 applications per season. Do not allow livestock to graze treated orchards.

Days To Harvest: 7.

Signal Word/Toxicity Class: Caution (3.2 EC); Warning/II (25WP, WSB).
REI: 24 hr.

PETROLEUM OIL

Petroleum based spray oil that may be used at delayed dormant or at reduced rates through summer. Application timing is critical for good control.

◆ **435 or 455 Oil 98.8** *(Drexel)*
◆ **SunSpray 6E, 7E, 9E, or 11E** *(Sun)*

Rate: 2-3 gal. 435,455, 6E, 7E/100 gal. water

Pest:

apple red bug, European fruit lecanium scale, fruittree leafroller, European red mite, San Jose scale, scurfy scale

Remarks: Apples. For apple red bug, lecanium scale, and fruittree leafroller, apply dormant. For San Jose and scurfy scale, apply delayed dormant. For European red mite, apply from green tip to delayed dormant. Do not apply in combination with or immediately preceding or following application of captan, carbaryl, Dyrene, folpet, Morestan, propargite, and sulfur compounds.

Rate: Cherries, nectarines:
2 gal. 455, 9E, 11E/100 gal. water

Pest:

San Jose scale

Rate: Peaches:
3 gal. 435, 455, 6E, 7E, 9E, 11E/100 gal. water

Pest:

cottony peach scale

Rate: Pears:
2 gal. 435, 455, 6E, 7E, 9E, 11E/100 gal. water

Pest:

fruittree leafroller, pearleaf blister mite, pear psylla

Rate: Plums, prunes:
2 gal. 455, 6E, 9E, 11E/100 gal. water

Pest:

European fruit lecanium scale, European red mite

Remarks: Cherries, nectarines, peaches, pears, plums, prunes. Apply during dormant stage. For European red mite on plums and prunes, apply from dormant to delayed dormant. For lecanium scale on plums and prunes, apply at 1/2" green. Do not apply in combination with or immediately preceding or following application of captan, carbaryl, Dyrene, folpet, Morestan, propargite, and sulfur compounds.

Signal Word/Toxicity Class: Caution.
REI: 4 hr.

◆ - **new product** • ai-active ingredient • DF-dry flowable • E/EC-emulsifiable concentrate • EW-emulsion in water
F/FL-flowable • G-granule • L/LC-liquid • SP/WSP-soluble packet • W/WP-wettable powder • WSB-water soluble bag

Citri-Oil *(Wilbur-Ellis)*
Rate: **Peaches, plums:**
Dormant spray
3-4 gal./100 gal. water
Pest:

San Jose scale	white peach scale

Remarks: Apply when temperature is between 50°-70°F. In case of very severe infestations, 2 applications may be needed.

Signal Word/Toxicity Class: Caution.
REI: 12 hr.

Damoil *(Drexel)*
Omni Supreme Spray *(Helena)*
Volck Supreme Spray *(Valent)*
Rate: **Apples, pears:**
Dormant/delayed dormant:
1.5-2 gal./100 gal. dilute spray
Delayed dormant/fall (blister mite):
2 gal./100 gal. dilute spray
Summer (pear psylla, rust mite):
1-1.5 gal./100 gal. dilute spray
Pest:

aphids (eggs)	pear psylla	twospotted spider mites
mites	pear scale	
Pacific spider mite		

Rate: **Apricots, cherries, nectarines, peaches, plums, prunes:**
Dormant or delayed dormant:
1.5-2 gal./100 gal. dilute spray
Summer stages (brown apricot scale, San Jose scale):
1 gal./100 gal. dilute spray
Pest:

aphids (eggs)	frosted scale	Putnam scale
black scale	Italian pear scale	red spider mite
brown apricot scale	oleander scale	San Jose Scale
European fruit lecanium	olive scale	soft brown scale
	parlatoria scale	

Rate: **Apricots, cherries, nectarines, peaches, pears (CA), plums, prunes:**
Summer:
3-4 qt./100 gal. dilute spray
Pest:
red spider mite
Remarks: Refer to label for directions, cautions, and limitations.

Signal Word/Toxicity Class: Caution.
REI: 12 hr.; Omni-4 hr.

SAF-T-SIDE *(Brandt Consolidated)*
◆ **Sumulsion** *(Drexel)*
Rate: 1-2 gal./100 gal. water
Pest: Apples:

aphids	fruittree leafroller	scurfy scale
apple red bug	mites	whiteflies
European red mite	scales	

Apricots, cherries, nectarines, peaches, pears:

aphids	pear psylla	scales
fruittree leafroller	pearleaf blister mite	whiteflies

Remarks: Apply dormant; late dormant for pear psylla. Time application to fit local growth and application conditions. Compatible with most commonly used insecticides and fungicides. Do not use in combination with or immediately before or after spraying with fungicides such as captan, Dyrene, folpet, Morestan, or any sulfur products. Do not use with carbaryl or dimethoate.

Signal Word/Toxicity Class: Caution.
REI: 4 hr.

SunSpray (6E) *(Sun)*
Rate: 2-3 gal. 6E/100 gal. water
Pest: Apples:

apple red bug	European red mite	San Jose scale
European fruit lecanium	fruittree leafroller	scurfy scale

Peaches:
cottony peach scale
Pears:

fruittree leafroller	pear psylla	pearleaf blister mite

Plums, prunes:

European fruit lecanium	European red mite

Remarks: Apples, peaches, pears, plums, prunes. Dormant/delayed dormant. Refer to label for use directions and precautions.

Signal Word/Toxicity Class: Caution.
REI: 4 hr.

SunSpray Ultra-Fine *(Sun)*
Rate: 1-3 gal./100 gal. water
Pest: Apples:

aphids	European red mite	San Jose scale
apple red bug	fruittree leafroller	scales
European fruit lecanium	mites	scurfy scale

Apricots, cherries:

aphids	mites	scales
fruittree leafroller	San Jose scale	tent caterpillars

Nectarines, peaches:

aphids	mites	San Jose scale
cottony peach scale	peach twig borer	scales
fruittree leafroller		

Pears:

European red mite	pear psylla	scales
fruittree leafroller	pearleaf blister mite	

Plums, prunes:

aphids	European red mite	peach twig borer
European fruit lecanium	mites	scales

Remarks: Apples, apricots, cherries, nectarines, peaches, pears, plums, prunes. Dormant/delayed dormant, summer, and postharvest. Refer to label for use directions and precautions.

Signal Word/Toxicity Class: Caution.
REI: 4 hr.

Phaser — see endosulfan

Pounce — see permethrin

PROPARGITE

Provides good control of European red mites and spider mites during the summer, as it does not volatilize readily. Sulfite ester compound; provides control of mites resistant to organophosphates and chlorinated hydrocarbons. Compound is an adulticide, and should be used when temperatures are above 70°F. Minimal effects on beneficial mites in orchard systems.

Omite 30W or 30WS *(Uniroyal)*
Rate: **Cherries (postharvest):**
1.8 lb. ai/A
6 lb. 30W, WS/A
Pest:

European red mite	Pacific spider mite	twospotted spider mite
McDaniel mite		

Remarks: Cherries. Do not make more than 3 applications per year for apples. Do not use on Asian pears as leaf injury may occur. Do not feed cover crops to livestock. Do not apply this product through any type of irrigation system.

Rate: **Nectarines:**
1.5-2.7 lb. ai/A
5-9 lb. 30W, 30WS/A
Pest:

brown almond mite	Pacific spider mite	twospotted spider mite
European red mite	peach silver mite	
McDaniel mite	strawberry spider mite	

Remarks: Nectarines. Do not make more than 2 applications per year. Do not apply this product through any type of irrigation system.
Days To Harvest: 14.
State Restrictions: For use west of the Rockies.

Signal Word/Toxicity Class: Danger/I.
REI: 48 hr.

Provado — see imidacloprid

Pyramite — see pyridaben

NOTICE The information on these pages is for preliminary planning — not a guide for use. Be sure to follow the manufacturer's directions, notwithstanding information contained here.	For personal protective equipment and EPA registration numbers, please turn to page 39.

PYRELLIN (Premix)

Premix containing pyrethrins and rotenone. May be combined with other insecticides to enhance knockdown or as an insect "exciter" to help facilitate contact with other spray residues.

Pyrellin E.C. (0.042 lb./gal. pyrethrins & 0.035 lb./gal. rotenone) *(Webb Wright)*

Rate: 1-2 pt. E.C./A

Pest: Including but not limited to:

aphids, leafhoppers, rose chafer, cherry fruit fly, mites, thrips, Japanese beetles

Remarks: Apples, cherries, peaches, pears, plums. Apply in intervals of 7 days or less; repeat as necessary. Refer to label for complete list of fruit crops covered.

Days To Harvest: 0.

Signal Word/Toxicity Class: Caution/III.
REI: 12 hr.

Pyrenone — see pyrethrins

PYRETHRINS

Contains piperonyl butoxide as a synergist. Diatect also contains silicon dioxide (diatomaceous earth). May be combined in tank mix with other insecticides for quicker and more complete kill where insect resistance may be a problem, and as an "exciter" to flush insects out of hiding and into contact with other spray residues.

Diatect Multipurpose Insecticide II *(Diatect)*
Pyrenone (0.5EC) *(AgrEvo Environmental Health)*
Py-Rin 60-6 EC *(Wilbur-Ellis)*

Rate: 1-6 lb. II/A
1-12 fl. oz. 0.5EC/A
2-16 oz. 6-60 EC/A

Pest: Including but not limited to:

aphids, leafhoppers, peach borer, cherry fruit fly, leafrollers, thrips, fruit flies, mites, whiteflies

Remarks: Apples, apricots, cherries, crabapples, nectarines, peaches, pears, plums, prunes, quince, etc. Apply in sufficient water for thorough coverage; repeat as needed. Apply Diatect as dust or wettable powder.

Days To Harvest: 0.

Signal Word/Toxicity Class: Caution.
REI: 12 hr.

PYRIDABEN

Pyridoxone miticide/insecticide, a wettable powder in water-soluble bags, that provides knockdown and residual control. Toxic to bees, fish, and aquatic invertebrates.

Pyramite (WP) *(BASF)*

Rate: 4.4-13.2 oz. WP/A (1.5-3 WSB)

Pest: Mites:

apple rust mite, Pacific spider mite, pear rust mite, European red mite, pear psylla, twospotted spider mite, McDaniel spider mite

Remarks: Apples, pears. Use high rate to ensure adequate concentration in full size trees with dense foliage. Do not make more than 2 applications per year or apply more than 26.4 oz./A per season. Do not apply by air or through any type of irrigation system. Refer to label for spray drift precautions.

Days To Harvest: Apples-25; pears-7.

Signal Word/Toxicity Class: Danger/I.
REI: 12 hr.

Py-Rin — see pyrethrins

SAF-T-SIDE — see petroleum oil

Savey — see hexythiazox

Sevin — see carbaryl

Sniper — see azinphos-methyl

SPINOSAD

Fermentation-derived insect control agent. Does not significantly impact predaceous and parasitic insect populations. Highly toxic to bees and aquatic invertebrates.

◆ **Success Naturalyte (SC) (spinosyn A + spinosyn D)** *(Dow AgroSciences)*

Rate: 0.062-0.156 lb. ai/A
4-10 oz. SC/A

Pest:

spotted tentiform leafminer, western tentiform leafminer

Rate: 0.094-0.156 lb. ai/A
6-10 oz. SC/A

Pest:

obliquebanded leafroller, pandemis leafroller

Remarks: Apples. Do not apply more than 3 times for leafrollers. Do not apply more than 29 oz. SC/A per crop. Do not apply through any type of irrigation equipment.

Days To Harvest: 7.

Signal Word/Toxicity Class: Caution.
REI: 4 hr.

Success Naturalyte — see spinosad

Sulforix — see calcium polysulfide

SULFUR

Microthiol and Thiolux are micronized sulfurs. Elemental sulfur compounds primarily provide disease control, but also will suppress mite populations when used during the season. Not nearly as effective as other miticidal compounds, and must be applied in frequent intervals.

◆ **Kumulus DF** *(BASF)*

Rate: 8-16 lb. ai/A
10-20 lb. DF/A

Pest:

blister mite (apples, pears), flat mite (cherries, plums, prunes), red spider mite, silver mite

Remarks: Apples, cherries, nectarines, peaches, pears, plums, prunes. Do not apply to d'Anjou pears or to sensitive apple varieties such as Ontario or Cox orange. For apples and pears, apply prebloom and petal fall with cover sprays throughout the season at 10- to 14-day intervals. For stone fruits, apply at bloom or early petal fall; repeat as necessary, usually 10- to 14-day intervals or after wet weather. Do not apply if temperatures during or within 3 days after application are expected to exceed 90°F. Do not apply through any type of irrigation system. Refer to label for use directions.

Signal Word/Toxicity Class: Caution.
REI: 24 hr.

Microthiol Special (80W) *(Elf Atochem)*
Thiolux (DF) *(Platte)*

Rate: 8-16 lb. ai/A
10-20 lb. 80W/A
8-24 lb. ai/A
10-30 lb. DF/A

Pest: Apples, pears:

blister mite, red spider mite, twospotted spider mite (80W) (except CA)

Cherries, nectarines, peaches, plums, prunes:

flat mite, red spider mite, silver mite

◆ - **new product** • ai-active ingredient • DF-dry flowable • E/EC-emulsifiable concentrate • EW-emulsion in water
F/FL-flowable • G-granule • L/LC-liquid • SP/WSP-soluble packet • W/WP-wettable powder • WSB-water soluble bag

Remarks: Apples, cherries, nectarines, peaches, pears, plums, prunes. For pome fruits, apply prebloom and petal fall. Dormant applications with oil may be made to apples and d'Anjou pears. For stone fruits, apply at bloom or early petal fall; repeat as needed at 10- to 14-day intervals. Refer to label for use directions and precautions.

Signal Word/Toxicity Class: Caution.
REI: 24 hr.

THAT (6FL) or THAT Big 8 (8FL) *(Stoller)*

Rate: 1.5-2.8 lb. ai/100 gal. water
2-3.75 pt. 6FL/100 gal. water
1.5-3 lb. ai/100 gal. water
1.5-3 pt. 8FL/100 gal. water

Pest:
rust mite

Remarks: Apples, pears. Dormant, delayed dormant, or postharvest. Apply with 1-1.5 gal. superior type dormant oil.
Days To Harvest: 0.

Signal Word/Toxicity Class: Caution.
REI: 24 hr.

Sumulsion — see petroleum oil

SunSpray — see petroleum oil

SUPRACIDE

Restricted Use Pesticide.

Supracide 25WP (methidathion) *(Gowan)*

Rate: Apples, apricots, cherries, nectarines, peaches, pears, plums, prunes:
1-3 lb. ai/A in 50 gal. water
4-12 lb. 25WP/A in 50 gal. water

Pest: Apples:
rosy apple aphid — San Jose scale

Pears:
San Jose scale

Apricots, cherries, nectarines, peaches, plums, prunes:
peach twig borer — San Jose scale (except apricots)

Rate: Pears:
2-3 lb. ai/A
8-12 lb. 25WP/A + 4-8 gal. oil (min. 100 gal. total spray/A)

Pest:
pear psylla

Rate: Nectarines, peaches, plums, prunes:
2 lb. ai/A
8 lb. 25WP/A + 4-8 gal. oil

Pest:
green peach aphid

Remarks: Apples, apricots, cherries, nectarines, peaches, pears, plums, prunes. Dormant/delayed dormant. Apply before blossoms open. Use low rate for light infestations; high rate for moderate to heavy infestations. Properly calibrated application equipment essential for uniform coverage. Use on Red Beaut plums only as dormant application; do not apply delayed dormant stage. EPA-registered spray oils may be added dormant/delayed dormant period only. For dormant/delayed dormant applications at rates less than 6 lb. 25WP/A, at least 2 gal. oil/A must be added to spray mixture. Do not apply more than 8 gal. oil/A by ground; 5 gal. oil/A by air. Do not graze treated areas or feed cover crops grown in treated areas to livestock. Do not apply through any type of irrigation system. Refer to label for further use directions.

Signal Word/Toxicity Class: Warning/II.
REI: 48 hr.; 14 days if greater than 2 lb. ai.

Tame — see fenpropathrin

THAT or THAT Big 8 — see sulfur

Thiodan — see endosulfan

Thiolux — see sulfur

THIRETHRIN (Premix)

Thirethrin (2.9L) (2 lb./gal. endosulfan & 0.45 lb./gal. piperonyl butoxide & 0.045 lb./gal. pyrethrins) *(Terra)*

Rate: 4 qt. 2.9L/A

Pest:

black cherry aphid	green peach aphid	rusty plum aphid
catfacing insects	peach twig borers	

Rate: 0.5-1.5 qt. 2.9L/100 gal water

Pest:
lesser peach tree borer — peach silver mites — peach tree borer

Remarks: Peaches. Do not make more than 2 applications during fruiting period. Thorough wetting of trunks and main branches is important.
Days To Harvest: 30.

Signal Word/Toxicity Class: Warning/II.
REI: 24 hr.

VENDEX

This miticide is an organotin compound, and has good control activity against both spider and rust mites. Low toxicity to predator mites and has not shown any phytotoxic effects to apples or pears. Restricted Use Pesticide.

Vendex 50WP (fenbutatin-oxide) *(Griffin)*

Rate: 0.5-1 lb. ai/A
1-2 lb. 50WP/A

Pest:

apple rust mite	European red mite	pear rust mite
Carpini spider mite (apples)	McDaniel spider mite	twospotted spider mite

Remarks: Apples, pears. Apply when mites first appear and repeat as necessary to maintain control. Do not apply more than 400 gal. dilute spray/A. Do not apply more than 2 times per season. Apply no more than 4 lb./A per season.
Days To Harvest: 14.

Rate: Cherries (sour, sweet):
Concentrate:
0.75-1.5 lb. ai/A
1.5-3 lb. 50WP/A

Pest:
cherry rust mite — European red mite — McDaniel spider mite

Rate: Nectarines, peaches, plums, prunes:
0.5-1 lb. ai/A
1-2 lb. 50WP/A

Pest:

bigbeaked plum mite	Pacific spider mite	plum nursery mite
European red mite	peach silver mite	twospotted spider mite
McDaniel spider		

Remarks: Cherries, nectarines, peaches, plums, prunes. Apply when mites first appear and repeat as necessary to maintain control. Do not apply more than 1.5 lb. ai/A per year for nectarines, peaches, plums, prunes; 2.25 lb. ai/A per year for cherries. Do not apply more than 2 times per season.
Days To Harvest: 14.

Signal Word/Toxicity Class: Danger/I.
REI: 48 hr.

Volck Supreme Spray — see petroleum oil

8 Fruits & Nuts

NOTICE The information on these pages is for preliminary planning — not a guide for use. Be sure to follow the manufacturer's directions, notwithstanding information contained here.

For personal protective equipment and EPA registration numbers, please turn to page 39.

VYDATE

Carbamate insecticide-acaricide-nematicide formulation that has both contact and systemic activity. Gives excellent control of aphids, leafhoppers, mites, and leafminers, but highly toxic to mite predators and pollinators. Restricted Use Pesticide.

Vydate L (oxamyl) *(DuPont)*

Rate: 0.5-1 lb. ai/A
2-4 pt. L/A

Pest:
European red mite, spotted tentiform leafminer, twospotted spider mite, white apple leafhopper

Rate: 1-2 lb. ai/A
4-8 pt. L/A

Pest:
apple aphid, rosy apple aphid

Remarks: Apples. For 1st brood leafminer, apply at 1/2" green stage to early pink stage of bud development. Do not apply after blossom clusters have separated. For 2nd brood leafminer, apply when an average of 2 or more larvae/leaf are present in sap-feeding stage. For best results, apply before larvae enter tissue-feeding stage. If needed, repeat 7-14 days after first application. For mites, apply when populations reach 2-4 mites/leaf and repeat at 7- to 14-day intervals as needed. For white apple leafhopper, apply as needed at 10- to 14-day intervals. For apple aphid, apply when 50% of terminals are infested. For rosy apple aphid, apply at pink (before bloom and no petals open) when aphids are present in significant numbers. Do not apply at bloom or within 30 days after bloom, or fruit thinning may occur. Do not apply more than 8 pt./A per growing season. Do not graze treated orchards.

Days To Harvest: 14.

State Restrictions: Do not use in Suffolk and Nassau counties, Long Island, New York.

- - - - - - - - - - - - - - - - - -

Rate: 1.5-2 lb. ai/A
6-8 pt. L/A in 100-600 water by ground

Pest:
European red mite, McDaniel mite, pear rust mite, twospotted spider mite

Remarks: Pears. Apply when mites first appear; repeat as needed to maintain control. Use low rate on light infestations; high rate on heavy infestations. Do not apply at bloom or within 30 days after bloom, or fruit thinning may occur. Do not graze treated orchards. Product has been tested on Bartlett and d'Anjou varieties without russeting; use on other varieties should be on small scale until possibility of russeting has been evaluated. Do not apply more than 8 pt./A per season.

Days To Harvest: 14.

State Restrictions: Do not use in California or in Suffolk and Nassau counties, Long Island, New York.

- - - - - - - - - - - - - - - - - -

Rate: Foliar:
0.5-1 lb. ai/100 gal.
2-4 pt. L/100 gal. (up to 200 gal./A)

Pest:
aphids, leafhoppers, leafminers, mites, thrips

Rate: Foliar alone or as supplement:
0.5-1 lb. ai/100 gal. water
2-4 pt. L/100 gal. water
Preplant soil incorporated:
6-8 lb. ai/A
3-4 gal. L/A

Pest:
burrowing nematode, lesion nematode, root-knot nematode (except Javanese), sting nematode

Remarks: Nonbearing apples, cherries, peaches, pears. Do not use on trees that will bear fruit within 12 months after application. Commercial plantings only. Foliar on mites, insects: Apply as needed to maintain control. Foliar alone or as supplement to soil treatment on nematodes: Apply at first full leaf or when plant is in active growth phase 4 times on 2- to 3-week schedule. Do not apply foliar to plants under water stress or not actively growing; include spreader/sticker. Preplant soil incorporated on nematodes: Apply in at least 20 gal. water/A within 24 hr. before tranplanting; thoroughly incorporate 4-8" into soil immediately after application. If band treatment, use proportionately less material. Test on small scale before proceeding to large-scale application. Varietal response may differ if Vydate L is mixed with other products. Do not apply more than 4 gal./A per season.

State Restrictions: Do not use in Suffolk and Nassau counties, Long Island, New York.

- - - - - - - - - - - - - - - - - -

Signal Word/Toxicity Class: Danger-Poison/I.
REI: 48 hr.

Tree Fruits, Subtropical

455 Oil — see petroleum oil

Ambush — see permethrin

AZADIRACHTIN

Extracted from seed of neem tree. Insect growth regulator that controls targeted insect larvae when they ingest or come in contact with it, by interfering with the insect's ability to molt and shutting off feeding behavior. Affected larvae slowly starve and die at next molt.

◆ **Azatin XL Plus** *(Thermo Trilogy)*

Rate: 5-21 oz. XL Plus/A

Pest:
aphids, beetles, caterpillars, cutworms, fruits flies, leafhoppers, leafminers, leafrollers, loopers, mealybugs, thrips, whiteflies

Remarks: Avocado, cacao, coffee, figs, kiwifruit, olives, papaya, persimmon, pineapple, tea. Rate varies by species. Apply when pests first appear; repeat at 7-day intervals. When pest populations are high, use higher rates. Do not apply more that 21 fl. oz./A.

Days To Harvest: 0.

- - - - - - - - - - - - - - - - - -

Signal Word/Toxicity Class: Caution.
REI: 4 hr.

Azatin — see azadirachtin

CARBARYL

Chipco Sevin SL or 80WSP *(Rhone-Poulenc)*
Drexel Carbaryl 4L or 50W *(Drexel)*
Sevin 4F or 80S *(Rhone-Poulenc)*

Rate: 0.75-1 lb. ai/100 gal. water
1.5-2 pt. SL, 4L, 4F/100 gal. water
1.5-2 lb. 50W/ 100 gal. water
3-8 lb. ai/100 gal. water
1-1.25 lb. 80S, 80WSP/100 gal. water

Pest:
olive scale

Remarks: Olives. For optimum scale control, add 1.5 gal. summer oil and apply when crawlers are present. Do not exceed 2 applications per year.

Days To Harvest: 0.

- - - - - - - - - - - - - - - - - -

Signal Word/Toxicity Class: Caution (4F, 4L); Warning/II (SL, 50W, 80S, 80WSP).
REI: 12 hr.

Chipco Sevin — see carbaryl

CHLORPYRIFOS

Lorsban-4E *(Dow AgroSciences)*
◆ **Micro Flo Chlorpyrifos 4E** *(Micro Flo)*

Rate: 2 lb. ai/A
4 pt. 4E/A

Pest:
dried fruit beetle

Remarks: Figs. Apply to soil and incorporate 3". Apply once per year as dormant application in late winter prior to beetle emergence and leaf formation. Refer to label for complete use directions.

Days To Harvest: 7 months.

State Restrictions: For use in California only.

- - - - - - - - - - - - - - - - - -

Signal Word/Toxicity Class: Warning/II.
REI: 24 hr.

◆ **- new product** • ai-active ingredient • DF-dry flowable • E/EC-emulsifiable concentrate • EW-emulsion in water
F/FL-flowable • G-granule • L/LC-liquid • SP/WSP-soluble packet • W/WP-wettable powder • WSB-water soluble bag

CRYOLITE

Kryocide (96%) *(Elf Atochem)*
Prokil Cryolite 96 *(Gowan)*
Rate: 9.6 lb. ai/A
10 lb. 96%, 96/A
Pest:
omnivorous leafroller
Remarks: Kiwifruit. Apply with ground equipment only, using sufficient water to obtain thorough coverage, up to 200 gal. spray/A. Apply at 15-day intervals as needed.
Days To Harvest: 30.
State Restrictions: For use in California only.

Signal Word/Toxicity Class: Caution.
REI: 12 hr.

Damoil — see petroleum oil

DEADLINE

Packaged as Bullets, M-Ps (mini-pellets), and granules. Resistant to rain and watering.

Deadline Bullets, M-Ps, Granules (4%) (metaldehyde) *(Pace)*
Rate: 0.8-1 lb. ai/A
20-40 lb. 4%/A
Pest:
slugs snails
Remarks: Avocados. Apply every 3-4 weeks during growing season, or as needed. Do not apply directly to edible portion of plant.

Signal Word/Toxicity Class: Warning/II (granule); Caution (bullets, M-P).
REI: 12 hr.

Diatect — see pyrethrins

DIAZINON

Restricted Use Pesticide.

D•z•n diazinon 50W *(Novartis)*
Gowan Diazinon 50WSB *(Gowan)*
◆ **Micro Flo Diazinon 50W** *(Micro Flo)*
Prontox Diazinon 50W *(Prentiss)*
Rate: 1 lb. ai/A
2 lb. 50W, 50WSB/A
Pest:
pineapple mealybug
Remarks: Pineapple. Apply in 200-500 gal. water; repeat at 28-day intervals as necessary (maximum 8 applications). Pineapple forage and refuse may not be fed to livestock. Refer to label for directions and precautions.
Days To Harvest: 7.

Signal Word/Toxicity Class: Caution.
REI: 24 hr.

DI-SYSTON

Restricted Use Pesticide.

Di-Syston 15% (disulfoton) *(Bayer)*
Rate: 2-4 g 15%/ft. tree height
Pest:
leafminers
Remarks: Coffee. Distribute granules uniformly under tree canopy. Do not apply more than twice per year. Make no more than 1 preharvest and 1 postharvest application.
Days To Harvest: 90.
State Restrictions: For use in Puerto Rico.

Signal Word/Toxicity Class: Danger/I.
REI: 48 hr.

D•z•n — see diazinon

ENDOSULFAN

Drexel Endosulfan 3EC or 50WP *(Drexel)*
Gowan Endosulfan 3EC or 50WSB *(Gowan)*
◆ **Micro Flo Endosulfan 3EC or 50WP** *(Micro Flo)*
Thiodan 3EC or 50WP *(FMC)*
Rate: 1.5-2 lb. ai/A
2-2.66 qt. 3EC/A
3-4 lb. 50WP, 50WSB/A
Pest:
pineapple fruit mite
Remarks: Pineapple. Fresh market. Apply at 7- to 10-day intervals if necessary, particularly during 40-day period of blooming. Do not exceed 3 lb. ai/A per year or 2 applications per season. Do not feed treated forage or pineapple by-products to livestock.
Days To Harvest: 7.

Signal Word/Toxicity Class: Danger-Poison/I.
REI: 24 hr.

IMIDACLOPRID

Provado 1.6 *(Bayer)*
Rate: 0.0825 lb. ai/A
6.6 fl. oz. 1.6/A
Pest:
thrips
Remarks: Mangoes. Foliar. Apply as dilute or concentrate spray as needed. Do not apply during bloom or when bees are present in orchard. Allow at least 7 days between applications. Do not apply more than 6.6 fl. oz./A in a single application or a total of 40 fl. oz./A per year.
Days To Harvest: 30.

Signal Word/Toxicity Class: Caution.
REI: 12 hr.

Kryocide — see cryolite

Kumulus — see sulfur

LANNATE

Toxic to fish and wildlife. Restricted Use Pesticides due to high acute toxicity to humans.

Lannate SP or LV (methomyl) *(DuPont)*
Rate: 0.45-0.9 lb. ai/A
0.5-1 lb. SP/A
1.5-3 pt. LV/A
Pest:
omnivorous looper western avocado leafroller
Remarks: Avocados. Apply in sufficient water to obtain thorough, uniform coverage; repeat at 5-7 day intervals or as needed. Do not apply more than 0.9 lb. ai/A per crop or make more than 2 applications per crop.
Days To Harvest: 1; livestock feeding/grazing-10.

Rate: 0.9 lb. ai/A
1 lb. SP/A
3 pt. LV/A
Pest:
omnivorous leafroller
Remarks: Pomegranates. Apply in sufficient water to obtain thorough, uniform coverage; repeat at 5-7 day intervals or as needed. Do not apply more than 1.8 lb. ai/A per crop or make more than 2 applications per crop.
Days To Harvest: 14.

Signal Word/Toxicity Class: Danger-Poison/I.
REI: 48 hr.

Lorsban — see chlorpyrifos

For personal protective equipment and EPA registration numbers, please turn to page 39.

MALATHION

Clean Crop Malathion 57EC *(Platte)*

Rate: 2.5 lb. ai/A
4 pt. 57EC/A in 300 gal. water

Pest:
vinegar fly

Remarks: Figs. Add 1-2 gal./A sulfured molasses.
Days To Harvest: 3.

Signal Word/Toxicity Class: Caution/III.
REI: 12 hr.

Clean Crop Malathion 8EC *(Platte)*
◆ Wilbur-Ellis Malathion 8 *(Wilbur-Ellis)*

Rate: **Avocados:**
1-1.5 lb. ai/A
1-1.5 pt. 8EC, 8/A

Pest:
aphids, mealybugs, spider mites, codling moth, orange tortrix, thrips

Rate: **Dates (except California):**
1 lb. ai/A
1 pt. 8EC/A

Pest:
nitidulid beetles

Rate: **Figs:**
2.5 lb. ai/A
2.5 pt. 8EC, 8/A

Pest:
dried fruit beetle, vinegar fly

Remarks: For figs, add 1-2 gal./A sulfured molasses.
Days To Harvest: Figs-3; avocados, dates-7.

Signal Word/Toxicity Class: Caution/III.
REI: 12 hr.

Drexel Malathion 5EC *(Drexel)*
◆ Micro Flo Malathion 5EC *(Micro Flo)*
Prentox 5 lb. Malathion Spray (5EC) *(Prentiss)*

Rate: **Avocados:**
1 lb. ai/100 gal. water
1.5 pt. 5EC/100 gal. water
4.7 lb. ai/A
7.5 pt. 5EC/A

Pest:
greenhouse thrips, omnivorous looper, soft brown scale, latania scale, orange tortrix

Rate: **Figs:**
2.5 lb. ai/1009 gal.water
4 pt. product/100 gal. water

Pest:
dried fruit beetle, vinegar fly

Rate: **Papaya:**
0.9-1.25 lb. ai/A
1.5-2 pt. 5EC/A

Pest:
aphids, mealybugs

Remarks: For figs, add 1-2 gal./A sulfured molasses.
Days To Harvest: Papayas-0; figs-3; avocados-7.

Signal Word/Toxicity Class: Warning/II.
REI: 12 hr.

Gowan Malathion 8 (8E) *(Gowan)*

Rate: **Avocados:**
4-9 lb. ai/A
4-9 pt. 8E/A

Pest:
greenhouse thrips, omnivorous looper, soft brown scale, latania scale, orange tortrix

Rate: **Figs:**
2.5 lb. ai/A
2.5 pt. 8E/A

Pest:
dried fruit beetle, vinegar fly

Rate: **Guava, mangoes, passion fruit:**
0.75 lb. ai/A
0.75 pt. 8E/A

Pest:
fruit flies

Rate: **Pineapple:**
5 lb. ai/A
5 pt. 8E/A

Pest:
mealybugs

Remarks: For figs, add 1-2 gal./A sulfured molasses. For mangoes, passion fruit, and guava, apply with 1 lb. partially hydrolyzed yeast protein or enzymatic yeast hydrolyzate.
Days To Harvest: Guava, mangoes, passion fruit-2; figs-3; avocados, pineapple-7.

Signal Word/Toxicity Class: Caution/III.
REI: 12 hr.

Microthiol Special — see sulfur

MOCAP

Toxic to aquatic organisms. Mocap EC is a Restricted Use Pesticide.

Mocap EC or 10%G (ethoprop) *(Rhone-Poulenc)*

Rate: 8 ml EC/stem
60 gm10%G/stem

Pest:
banana root borer, nematodes

Remarks: Bananas, plantains. Apply in radius of $^3/_4$ meter around each producing stem. Repeat in 6 months. Before application, remove ground litter from area to be treated. For best results, apply just before irrigation or rainy period.

Rate: 6-12 lb. ai/A
1-2 gal. EC/A

Pest:
nematodes

Remarks: Pineapple. Postplant drip irrigation systems. Apply in 50-100 gal. water before injection. Plant crop: Apply immediately after planting. Do not apply more than 16 gal./A per crop. Ratoon crop: Apply immediately after crop harvest. Do not apply more than 10 gal./A per crop. Apply additional treatments every 2 months.
Days To Harvest: 120.
State Restrictions: For use in Hawaii only.

Signal Word/Toxicity Class: Danger-Poison/I (EC); Warning/II (10%G).
REI: 48 hr.; 72 hr. if average rainfall less than 25" per year.

M-PEDE

M-Pede (potassium salts of fatty acids) *(Mycogen)*

Rate: 1-2% v/v solution

Pest:
aphids, plant bugs, thrips, leafhoppers, scales, whiteflies, mealybugs

Remarks: Bananas, coffee, pineapple. Apply at first sign of infestation; repeat at 7- to 14-day intervals. May be tank mixed with other insecticides. Refer to both labels for use directions and precautions.
Days To Harvest: 0.
State Restrictions: Not for use in California.

Signal Word/Toxicity Class: Warning/II.
REI: 12 hr.

NEMACUR

Restricted Use Pesticide.

Nemacur 3 or 15% (fenamiphos) *(Bayer)*

Rate: Hawaii:
Preplant:
20 lb. ai/A
6.66 gal. 3/A
Postplant:
0.5-3 lb. ai/A
1.33-8 pt. 3/A

◆ - **new product** • ai-active ingredient • DF-dry flowable • E/EC-emulsifiable concentrate • EW-emulsion in water
F/FL-flowable • G-granule • L/LC-liquid • SP/WSP-soluble packet • W/WP-wettable powder • WSB-water soluble bag

Puerto Rico:
Postplant:
5-10 lb. ai/A
1.66-3.33 gal. 3/A

Pest:
nematodes

Remarks: Pineapple. Hawaii: Apply preplant by broadcast and incorporation; postplant by broadcast or drip irrigation; additional applications may be made at 1-3 month intervals. Puerto Rico: Apply postplant by broadcast applications; additional applications may be made at 3-6 month intervals. Refer to label for further use information.

Days To Harvest: Hawaii-30 days; Puerto Rico-7 1/2 months.

State Restrictions: For use in Hawaii and Puerto Rico.

Rate: 3.33-6.66 gal. 3/A
66.66-133.33 lb. 15%/A

Pest:
nematodes

Remarks: Pineapple.

State Restrictions: SLN-Puerto Rico.

Rate: 1.5-3 lb. ai/A
2-4 qt. 3/A

Pest:
nematodes

Remarks: Kiwifruit. Up to 3 gal./A may be applied preharvest; maximum 1 gal./A postharvest. Do not apply to vineyards less than 1 year old. Refer to label for further use information

Days To Harvest: 31.

State Restrictions: For use in California.

Rate: 2.20-6.66 qt. 3/A
11.1-33.33 lb. 15%/A

Pest:
banana root borer, nematodes, root-knot nematode, burrowing nematode

Remarks: Bananas.

State Restrictions: SLN-Hawaii.

Rate: 20-30 g 15%/unit

Pest:
banana root borer, nematodes

Remarks: Bananas, plantains.

State Restrictions: SLN-Puerto Rico.

Signal Word/Toxicity Class: Danger/I.

REI: 48 hr.

Omni Supreme Spray — see petroleum oil

PERMETHRIN

Restricted Use Pesticide.

Ambush 2E or 25W *(ZENECA)*

Rate: Avocados:
0.2 lb. ai/100 gal. water
12.8 fl. oz. 2E/100 gal. water
12.8 oz. 25W/100 gal. water

Pest:
leafhoppers, mirid bugs, thrips, lepidopterous larvae

Rate: Papaya:
0.1 lb. ai/100 gal. water
6.4 fl. oz. 2E/100 gal. water
6.4 oz. 25W/100 gal. water

Pest:
fruit fly

Remarks: Avocados, papaya. Apply with ground equipment when insects appear and repeat at 7-10 day intervals as needed. Spray to wet all foliage. Do not make more than 6 applications per season. Do not graze livestock in treated areas or feed cover crops from treated areas.

Days To Harvest: 7.

State Restrictions: For use on papaya in Florida only.

Signal Word/Toxicity Class: Warning/II.

REI: 24 hr.

PETROLEUM OIL

Mode of action is through suffocation of eggs, larvae, and nymphs of insects and soft-bodied insects.

◆ **455 Oil 98.8 or Damoil** *(Drexel)*
Omni Supreme Spray *(Helena)*
Volck Supreme Spray *(Valent)*

Rate: Figs:
3 gal./100 gal. dilute spray

Pest:
fig scale

Rate: Olives:
1.5 gal./100 gal. water

Pest:
black scale, olive scale, spider mites, oleander scale, parlatoria scale

Remarks: Figs, olives. For figs, apply dormant/delayed dormant; olives, prebloom or postbloom up to buckshot. Refer to label for directions, precautions, and limitations.

Signal Word/Toxicity Class: Caution.

REI: 12 hr.; OMNI, 455-4 hr.

SAF-T-SIDE *(Brandt Consolidated)*

Rate: 1-2 gal./100 gal. water

Pest:
leafhoppers, scales, thrips, plant bugs, Sigatoka

Remarks: Avocados, bananas, figs, olives, mangoes, papaya. Apply as needed. Time application to fit local growth and application conditions. Compatible with most commonly used insecticides and fungicides. Do not use in combination with or immediately before or after spraying with fungicides such as captan, Dyrene, folpet, Morestan, or any sulfur products. Do not use with carbaryl or dimethoate.

Signal Word/Toxicity Class: Caution.

REI: 4 hr.

◆ **Sumulsion** *(Drexel)*

Rate: 1-2 gal./100 gal. water

Pest: Avocados, bananas:
leafhoppers, scales, thrips, leafrollers, Sigatoka

Figs:
aphids, mites, whiteflies, fruittree leafroller, scales

Olives:
black scale, parlatoria scale, spider mites, oleander scale

Remarks: Avocados, bananas, figs, olives. For avocados, bananas, apply at dormant stage. For figs, apply at dormant or delayed dormant stage. For olives, apply prebloom or postbloom to buckshot. Do not use in combination with or immediately before or after spraying with fungicides such as captan, Dyrene, folpet, Morestan, or any sulfur products. Do not use with carbaryl or dimethoate.

Signal Word/Toxicity Class: Caution.

REI: 4 hr.

SunSpray (6E) *(Sun)*

Rate: Figs:
3 gal. 6E/A
Olives:
1.5 gal. 6E/100 gal. water

Pest:
scales

Remarks: Figs, olives. For figs, apply dormant/delayed dormant; olives, prebloom or postbloom to buckshot. Refer to label for use directions and precautions.

Signal Word/Toxicity Class: Caution.

REI: 4 hr.

Provado — see imidacloprid

NOTICE The information on these pages is for preliminary planning — not a guide for use. Be sure to follow the manufacturer's directions, notwithstanding information contained here. | For personal protective equipment and EPA registration numbers, please turn to page 39.

PYRELLIN (Premix)

Premix containing pyrethrins and rotenone. May be combined with other insecticides to enhance knockdown or as an insect "exciter" to help facilitate contact with other spray residues.

Pyrellin E.C. (0.042 lb./gal. pyrethrins & 0.035 lb./gal. rotenone) *(Webb Wright)*

Rate: 1-2 pt. E.C./A

Pest:

aphids	fruit fly	thrips
caterpillars	leafhoppers	weevils
fireworms	mites	whiteflies

Remarks: Avocados, dates, guava, kiwifruit, mangoes, papaya, passion fruit, persimmon, pineapple. Apply as necessary to maintain control.

Days To Harvest: 0.

- - - - - - - - - - - - - - - - -

Signal Word/Toxicity Class: Caution/III.

REI: 12 hr.

Pyrenone — see pyrethrins

PYRETHRINS

Contains piperonyl butoxide as a synergist. Diatect also contains silicon dioxide (diatomaceous earth). May be combined in tank mix with other insecticides for quicker and more complete kill where insect resistance may be a problem, and as an "exciter" to flush insects out of hiding and into contact with other spray residues.

Diatect Multipurpose Insecticide II *(Diatect)*

Pyrenone (0.5EC) *(AgrEvo Environmental Health)*

Rate: 1-6 lb.II/A
1-12 fl. oz. 0.5EC/A

Pest: Including but not limited to:

aphids	leafhoppers	thrips
fruit flies	mites	whiteflies

Remarks: Avocados, bananas, dates, figs, guava, kiwifruit, mangoes, papaya, passion fruit, persimmon, pineapple, pomegranates. Apply in sufficient water for thorough coverage; repeat as needed. Apply Diatect as dust or wettable powder.

Days To Harvest: 0.

- - - - - - - - - - - - - - - - -

Signal Word/Toxicity Class: Caution.

REI: 12 hr.

SAF-T-SIDE — see petroleum oil

Sevin — see carbaryl

SULFUR

Microthiol and Thiolux are micronized sulfurs.

◆ **Kumulus DF** *(BASF)*

Rate: Avocados:
8-24 lb. ai/A
10-30 lb. DF/A

Pest:
brown mite

Rate: Figs:
Dilute spray:
3.2-4.8 lb. ai/water
4-6 lb. DF/100 gal. water
Air or concentrate:
12-24 lb. ai/A
15-30 lb. DF/A

Pest:
fig rust mite

Remarks: Avocados, figs. For avocados, apply when mites are first noticed. For figs, apply in spring and early summer when russeting appears. Do not apply if temperatures during or within 3 days after application are expected to exceed 90°F. Do not apply through any type of irrigation system. Refer to label for use directions.

- - - - - - - - - - - - - - - - -

Signal Word/Toxicity Class: Caution.

REI: 24 hr.

Microthiol Special (80W) *(Elf Atochem)*

Rate: Avocados:
8-16 lb. ai/A
10-20 lb. 80W/A

Pest:
brown mite

Rate: Figs:
4 lb. ai/A
5 lb. 80W/A

Pest:
eriophyrid mite   fig mite

Rate: Pomegranates:
2.4-8 lb. ai/A
3-10 lb. 80W/A

Pest:
mites

Remarks: Avocados, figs, pomegranates. For avocados, apply as needed. For figs, apply beginning in April; repeat at 3- to 4-week intervals through June. For pomegranates, apply in May or June; repeat at 3- to 4-week intervals or as needed. Use higher rates if past mite damage has been high. If temperature exceeds 95°F, lower rates and more frequent applications are advised.

- - - - - - - - - - - - - - - - -

Signal Word/Toxicity Class: Caution.

REI: 24 hr.

THAT Big 8 (8FL) *(Stoller)*

Rate: 16 lb. ai/A
2 gal. 8FL/A

Pest:
avocado brown mite   sixspotted spider mite

Remarks: Avocados. Do not apply within 4 weeks of oil application.

- - - - - - - - - - - - - - - - -

Signal Word/Toxicity Class: Caution.

REI: 24 hr.

Thiolux (DF) *(Novartis)*

Rate: 8-24 lb. ai/A
10-30 lb. DF/A

Pest:
brown mite

Remarks: Avocados. Apply as needed.

- - - - - - - - - - - - - - - - -

Signal Word/Toxicity Class: Caution.

REI: 24 hr.

Sumulsion — see petroleum oil

SunSpray — see petroleum oil

SUPRACIDE

Restricted Use Pesticide.

Supracide 25WP (methidathion) *(Gowan)*

Rate: 2-3 lb. ai/400 gal. water
8-12 lb. 25WP/400 gal. water

Pest:
black scale

Remarks: Olives. Apply once during postharvest or prebloom period. Do not apply more than 12 lb. 25WP/A per year. Do not apply after buds break. Do not graze treated areas or feed cover crops grown in treated areas to livestock. Do not apply through any type of irrigation system.

- - - - - - - - - - - - - - - - -

Signal Word/Toxicity Class: Warning/II.

REI: 48 hr.

THAT Big 8 — see sulfur

Thiodan — see endosulfan

◆ - **new product** • ai-active ingredient • DF-dry flowable • E/EC-emulsifiable concentrate • EW-emulsion in water
F/FL-flowable • G-granule • L/LC-liquid • SP/WSP-soluble packet • W/WP-wettable powder • WSB-water soluble bag

Thiolux — see sulfur

VENDEX

Restricted Use Pesticide.

Vendex 50WP (fenbutatin-oxide) *(Griffin)*

Rate: 1 lb. ai/ A
2 lb. 50WP/A

Pest:
broad mite, citrus red mite, Texas citrus mite, Carmine mite, red and black flat mite

Remarks: Papaya. Apply only as dilute spray when mites first appear. Repeat as necessary. Do not apply more than once per month to bearing trees, more than 4 times per season, or more than 4 lb. ai/A per year.
Days To Harvest: 7.
State Restrictions: Not for use in California.

Signal Word/Toxicity Class: Danger/I.
REI: 48 hr.

Volck Supreme Spray — see petroleum oil

VYDATE

Restricted Use Pesticide.

Vydate L (oxamyl) *(DuPont)*

Rate: Preplant:
4 lb. ai/A
2 gal. L/A
At planting—broadcast:
2-4 lb. ai/A
1-2 gal./A
At planting—drip irrigation:
1-4 lb. ai/A
0.5-2 gal./A

Pest:
reniform nematode, root-knot nematode

Remarks: Pineapple. Preplant: Apply within 1 week prior to planting and incorporate 4-6". At planting: Apply within 1 week after planting. Subsequent treatments: Apply 0.5-1 gal./A foliar (ground) at 2- to 4-week intervals; 0.25-1 gal./A by drip irrigation at 2-, 4- or 8-week intervals. Do not apply more than 6 gal./A per season.
Days To Harvest: Harvest or grazing-30.
State Restrictions: Not for use in California.

Rate: 5-10 ml undiluted/corm or seed

Pest:
banana corm borer, nematodes

Remarks: Bananas, plantains. At planting: Apply in planting hole; cover treated corm with soil; repeat 2-3 months after planting. If developing pseudostem is 1 ft. or shorter, apply directly over the top, wetting leaves and leaf axils; if higher, direct treatment to soil and apply as close as possible to developing pseudostem in semi-circular pattern. Use higher rate and shorter intervals when nematode or corm borer population levels are high. Do not apply more than 2 gal./A per year. Do not graze or forage treated fields.
Days To Harvest: 1.
State Restrictions: For use in Puerto Rico only.

Signal Word/Toxicity Class: Danger-Poison/I.
REI: 48 hr.

Nuts

435 or 455 Oil — see petroleum oil

Ambush — see permethrin

Ammo — see cypermethrin

APOLLO

Provides control primarily by causing female mites to lay sterile eggs. Not an adulticide, and should be tank mixed with an adulticide when mite numbers are high at application.

Apollo SC (clofentezine) *(AgrEvo USA)*

Rate: 2-8 fl. oz. SC/A

Pest:
European red mite, Pacific spider mite, yellow spider mite, McDaniel mite, twospotted spider mite

Remarks: Almonds, walnuts. Apply at first sign of mites. Applications below 4 oz. SC/A are recommended only in association with established IPM programs and only when adequate beneficial insects/mites are prevalent. Use at least 4 oz. SC/A south of CA Highway 152. Do not apply by air. Do not apply more than once per season. If additional applications are needed, use a miticide with a different mode of action. Do not apply this product through any type of irrigation system. Do not graze treated areas or harvest cover crop for livestock feed.
Days To Harvest: 30.

Signal Word/Toxicity Class: Caution.
REI: 12 hr.

ASANA

Fourth-generation synthetic pyrethroid active against a number of pecan pests. Long residual, but use on almonds and walnuts results in spider mite outbreaks. Toxic to wildlife, extremely toxic to fish, and highly toxic to bees. Low toxicity to humans and other mammals. Restricted Use Pesticide.

Asana XL (esfenvalerate) *(DuPont)*

Rate: 0.05-0.1 lb. ai/A
9.6-19.2 fl. oz. XL/A

Pest:
navel orangeworm, peach twig borer

Remarks: Almonds. Repeat as necessary to maintain control. Do not exceed 0.2 lb. ai/A per season. Do not feed or graze livestock on treated orchard floors.
Days To Harvest: 21.

Rate: 0.05-0.1 lb. ai/A
9.6-19.2 fl. oz. XL/A

Pest:
filbertworm, obliquebanded leafroller

Remarks: Filberts. Make first application after emergence of filbertworm moths in early summer. Do not apply a second treatment within 3 weeks after the first. Do not apply over 0.2 lb. ai/A per season. For dilute spray, apply 200-400 gal./A, but do not exceed 19.2 fl. oz./A per treatment. Do not feed or graze livestock on treated orchard floors.
Days To Harvest: 21.

Rate: 0.025-0.075 lb. ai/A
4.8-14.5 fl. oz. XL/A

Pest:
hickory shuckworm, pecan nut casebearer, pecan stem phylloxera, pecan aphid, pecan spittlebug, pecan leaf phylloxera, pecan weevil

Remarks: Pecans. Repeat as necessary to maintain control. Do not exceed 0.3 lb. ai/A per season. For dilute sprays, apply 200-600 gal./A, but do not exceed 14.5 fl. oz./A per treatment. Correct timing of spray applications is critical in achieving optimum control of leaf and stem phylloxera. Consult local spray recommendations for correct times of application. Do not feed or graze livestock on treated orchard floors.
Days To Harvest: 21.

Rate: 0.05-0.1 lb. ai/A
9.6-19.2 fl. oz. XL/A

Pest:
codling moth, walnut aphid, walnut huskfly, navel orangeworm

Remarks: Walnuts. Repeat as necessary to maintain control. Do not exceed 0.2 lb. ai/A per season. For dilute sprays, apply 200-400 gal./A. Use of baits in walnut husk fly sprays is recommended where endorsed by local agricultural extension service. For concentrate spray, apply in no less than 30 gal. water/A by ground or 5 gal. water/A by air. Do not feed or graze livestock on treated orchard floors.
Days To Harvest: 21.

Signal Word/Toxicity Class: Warning/II.
REI: 12 hr.

For personal protective equipment and EPA registration numbers, please turn to page 39.

AZADIRACHTIN

Extracted from seed of neem tree. Insect growth regulator that controls targeted insect larvae when they ingest or come in contact with it, by interfering with the insect's ability to molt and shutting off feeding behavior. Affected larvae slowly starve and die at next molt.

◆ **Azatin XL Plus** *(Thermo Trilogy)*

Rate: 5-21 oz. XL Plus/A

Pest:

aphids	fall webworms	mealybugs
beetles	leafhoppers	moths
caterpillar	leafminers	navel orangeworm
codling moth	leafrollers	peachtree borer
cutworms	loopers	whiteflies

Remarks: Almonds, beechnuts, Brazil, butternuts, cashews, chestnuts, chinquapins, filberts. Rate varies by species. Apply when pests first appear; repeat at 7-day intervals. When pest populations are high, use higher rates. Do not apply more that 21 fl. oz./A.

Days To Harvest: 0.

- - - - - - - - - - - - - - - -

Signal Word/Toxicity Class: Caution.

REI: 4 hr.

◆ **NEEMAZAD 4.5EC or NEEMIX 4.5** *(Thermo Trilogy)*

Rate: 0.125-1 pt. 4.5/A
High Volume:
2.25-4.5 oz. 4.5EC/100 gal. water
Low Volume:
0.5 pt. 4.5EC/A

Pest:

aphids	filbertworm	pecan nut casebearer
black margined aphid	frosted scale	pecan spittlebug
calico scale	fruittree leafroller	pecan stem phylloxera
chestnut weevil	hickory shuckworm	
codling moth	lesser webworm	pecan weevil
fall webworm	navel orangeworm	San Jose scale
filbert aphid	peach twig borer	twig girdler
filbert leafroller	pecan leaf phylloxera	walnut caterpillars

Remarks: Almonds, beechnuts, Brazil, butternuts, cashews, chestnuts, filberts, hickory, macadamias, pecans, pistachios, walnuts. For low volume, apply in sufficient water for adequate coverage. For low insect pressure, apply preventive at 7-day intervals or at first sign of insects; high insect pressure, apply at 3- to 4-day intervals.

- - - - - - - - - - - - - - - -

Signal Word/Toxicity Class: Warning/II.

REI: 12 hr.

Azatin — see azadirachtin

AZINPHOS-METHYL

Long residual, very broad spectrum organophosphate that is effective on lepidopterous pests. Will not control mites in the West. Can be disruptive to Trioxys parasite of walnut aphids. Extremely toxic to fish and wildlife; highly toxic to bees. Similar to parathion in toxicity to humans and other mammals. Guthion 2L and Gowan 50-W are Restricted Use Pesticides.

Gowan Azinphos-M 50-W *(Gowan)*
Guthion 2L, Solupak (50WSP) *(Bayer)*
◆ **Micro Flo Azinphosmethyl 2EC or 50W** *(Micro Flo)*

Rate: 1.5-2 lb. ai/A
6-8 pt. 2EC/A
3-4 lb. 50-W, 50 WSP, 50W/A

Pest:

navel orangeworm	peach twig borer

Remarks: Almonds. Apply as full coverage spray. Do not apply more than 1 dormant application or 2 postbloom applications. Allow 120 days between dormant and first postbloom application and 28 days for subsequent application. Do not apply after husk split. Do not graze livestock in treated groves for 21 days after treatment.

Days To Harvest: 28.

- - - - - - - - - - - - - - - -

Rate: 1.5-2 lb. ai/A
6-8 pt. 2EC, 2L/A
3-4 lb. 50-W, 50WSP, 50W/A

Pest:

apple mealybug	filbert leafroller	filbertworm
filbert aphid		

Remarks: Filberts. Apply as full coverage spray. Do not apply more than 3 times per season. Allow 14 days between applications.

Days To Harvest: Gowan, Micro Flo-30, Guthion-45.

State Restrictions: For use in the Pacific Northwest.

- - - - - - - - - - - - - - - -

Rate: 1.5-2 lb. ai/A
6-8 pt. 2EC, 2L/A
3-4 lb. 50-W, 50WSP, 50W/A

Pest:

aphids	May beetle	spittlebugs
fall webworm	pecan casebearer	twig girdler
hickory shuckworm	southern green stink bug	walnut caterpillar
leafminers		

Remarks: Pecans. Apply as full coverage spray. Do not apply more than 3 times per season. Do not apply on pecans after shuck split; walnuts after husk split.

Days To Harvest: Harvest-45 (Guthion); grazing-21.

- - - - - - - - - - - - - - - -

Rate: 1-2 lb. ai/A
6-8 pt. 2EC, 2L/A
2-4 lb. 50-W, 50W/A
1.5-2 lb. ai/A
3-4 lb. 50WSP/A

Pest:

codling moth	navel orangeworm	walnut huskfly*
filbertworm	redhumped caterpillar*	

**Guthion*

Remarks: Walnuts. Apply as full coverage spray. Do not apply more than 3 times per season. May by applied as split application; refer to label.

Days To Harvest: Harvest/grazing-21.

- - - - - - - - - - - - - - - -

Rate: 2.5 lb. ai/A
5 lb. 50-W, 50W/A

Pest:

navel orangeworm

Remarks: Pistachios. Apply as full coverage spray. Make 1 application from bud break to 10% hullsplit.

Days To Harvest: Harvest/grazing-21.

State Restrictions: For use in California only.

- - - - - - - - - - - - - - - -

Signal Word/Toxicity Class: Danger/I.

REI: 48 hr.

Sniper 50W *(Platte)*

Rate: 0.375-0.5 lb. ai/100 gal. water
0.75-1 lb. 50W/100 gal. water

Pest:

peachtwig borer

Remarks: Almonds. Apply as full coverage spray. Do not apply more than twice per season. Allow 30 days between applications.

Days To Harvest: 60.

- - - - - - - - - - - - - - - -

Rate: 0.25-0.75 lb. ai/100 gal. water
0.5-1.5 lb. 50W/100 gal. water

Pest:

apple mealybug	filbert leafroller	filbertworm
filbert aphid		

Remarks: Filberts. Apply as full coverage spray. Repeat as necessary.

Days To Harvest: Harvest-30; grazing 21.

State Restrictions: For use in the Pacific Northwest.

- - - - - - - - - - - - - - - -

Rate: Pecans:
0.38-0.56 lb. ai/100 gal. water
0.75-1.125 lb. 50W/100 gal. water

Pest:

aphids	May beetle	spittlebugs
fall webworm	pecan casebearer	twig girdler
hickory shuckworm	southern green stink bug	walnut caterpillar
leafminers		

Rate: Walnuts:
0.38-0.94 lb. ai/100 gal. water
0.75-1.88 lb. 50W/100 gal. water

Pest:

codling moth	filbertworm	navel orangeworm

Remarks: Pecans, walnuts. Apply as full coverage spray. Do not apply after husk split.

Days To Harvest: 21.

- - - - - - - - - - - - - - - -

Signal Word/Toxicity Class: Danger-Poison/I.

REI: 48 hr.; 72 hr. if average rainfall less than 25" per year.

◆ - **new product** • ai-active ingredient • DF-dry flowable • E/EC-emulsifiable concentrate • EW-emulsion in water
F/FL-flowable • G-granule • L/LC-liquid • SP/WSP-soluble packet • W/WP-wettable powder • WSB-water soluble bag

CALCIUM POLYSULFIDE

Lime-Sulfur Solution *(Best Sulfur)*

Rate: 1 gal. L/A
Concentrate, mature trees:
2-4 gal. L/A

Pest:
mites yellow pecan aphid

Remarks: Pecans. For yellow aphids, spray as needed to prevent excessive honeydew buildup. For mites, spray when infestation is first noticed. Do not apply through any type of irrigation system.

- - - - - - - - - - - - - - - - - -

Signal Word/Toxicity Class: Danger/I.
REI: 48 hr.

CARBARYL

Carbamate insecticide with activity against a wide variety of pests. Quite effective on lepidopterous pests. In almonds and walnuts, use will result in spider mite problems during year of application. Highly toxic to bees; extremely toxic to aquatic and estuarine invertebrates.

Chipco Sevin SL or 80WSP *(Rhone-Poulenc)*
Drexel Carbaryl 4L, 50W, 80S *(Drexel)*
Sevin 4F, 80S or 80WSP *(Rhone-Poulenc)*

Rate: 1 lb. ai/100 gal. water
2 pt. SL, 4L, 4F/100 gal. water
2 lb. 50W/100 gal. water
1.25 lb. 80S, 80WSP/100 gal. water

Pest:
fruittree leafroller, navel orangeworm, peach twig borer, San Jose scale

Remarks: Almonds. For navel orangeworm, time early- and mid-season applications to correspond with egg laying peaks. Make late-season application at initiation of hull split or up to 10% hull split. For other pests, apply in popcorn or petal fall stages, repeat when May brood of peach twig borer begins to hatch or as needed. For dormant or delayed dormant, apply in combination with a recommended dormant oil. Refer to oil label for additional use directions and restrictions.
Days To Harvest: 28.

- - - - - - - - - - - - - - - - - -

Rate: 2-3 lb. ai/100 gal. water
4-6 pt. SL, 4L, 4F/100 gal. water dilute spray
2.5-3.75 lb. 80S, 80WSP/100 gal. water

Pest:
chestnut weevil

Remarks: Chestnuts. Make 4 applications at weekly intervals beginning in late July. Last application should be made prior to shuck split.
Days To Harvest: 0.
State Restrictions: Not for use in California.

- - - - - - - - - - - - - - - - - -

Rate: 1 lb. ai/A
2 pt. SL, 4L, 4F/A
2 lb. 50W/100 gal. water
1.25 lb. 80S, 80WSP/100 gal. water

Pest:
filbert aphid, filbert leafroller, filbert moth, filbertworm

Remarks: Filberts. Apply when leafroller eggs are hatching. Repeat with first appearance of adult filbert moths and again 3-4 weeks later. Apply dilute sprays in 300-400 gal. water/A.
Days To Harvest: 0.

- - - - - - - - - - - - - - - - - -

Rate: 1-2.5 lb. ai/100 gal. water
2-5 pt. SL, 4L, 4F/100 gal. water
1.25-2.875 lb. 80S, 80WSP/100 gal. water

Pest:
black margined aphid, fall webworm, hickory shuckworm, lesser webworm, pecan leaf phylloxera, pecan nut casebearer, pecan spittlebug, pecan stem phylloxera, pecan weevil, twig girdler, walnut caterpillar

Remarks: Pecans. Do not apply more than 7.2 lb. ai/A per application. Apply dilute sprays in 200-400 gal. water/A.
Days To Harvest: 0.

- - - - - - - - - - - - - - - - - -

Rate: 0.5-2 lb. ai/A
1-4 pt. SL, 4L, 4F/A
0.66-2.5 lb. 80S, 80WSP/100 gal. water

Pest:
navel orangeworm

Remarks: Pistachios. Apply dilute volumes of 150-300 gal. mixed spray/A for full coverage.
Days To Harvest: 14.

- - - - - - - - - - - - - - - - - -

Rate: 0.5 lb. ai/A
1 pt. SL, 4L, 4F/A
1 lb. 50W/100 gal. water
0.66 lb. 80S, 80WSP/100 gal. water

Pest:
calico scale, codling moth, European fruit lecanium, filbertworm, frosted scale, fruittree leafroller

Rate: 2 lb. ai/A
4 pt. SL, 4L, 4F/A
4 lb. 520W/100 gal. water
2.5 lb. 80S, 80WSP/100 gal. water

Pest:
European earwig

Remarks: Walnuts. For codling moth, apply first spray when average cross-sectional diameters of developing nuts are $1/2$-$3/4$". Repeat during middle or late June as needed. Apply dilute sprays in 200-500 gal. water/A. For European earwig, spray tree trunks to point of runoff. For other pests, apply dilute sprays in 1000 gal. water/A for mature trees.
Days To Harvest: 0; except codling moth treatments-6.

- - - - - - - - - - - - - - - - - -

Signal Word/Toxicity Class: Caution (4F, 4L); Warning/II (SL, 50W, 80S, 80WSP).
REI: 12 hr.

Chipco Sevin — see carbaryl

CHLORPYRIFOS

Organophosphate insecticide/miticide with broad-spectrum activity on pecan and other nut pests. Quite effective in controlling ants on almond and walnut orchard floors. Toxic to birds and wildlife and extremely toxic to fish and aquatic organisms. Highly toxic to bees.

Lorsban-4E *(Dow AgroSciences)*
◆ **Micro Flo Chlorpyrifos 4E** *(Micro Flo)*

Rate: Almonds:
2 lb. ai/A
4 pt. 4E/A
4 lb. 50-W/A

Pest:
navel orangeworm, peach twig borer, San Jose scale

Rate: Filberts:
1.5-2 lb. ai/A
3-4 pt. 4E/A
3-4 50-W

Pest:
eyespotted bud moth, filbert aphid, filbert leafroller, filbertworm, obliquebanded leafroller, omnivorous leaftier, winter moth

Rate: Walnuts:
2 lb. ai/A
4 pt. 4E/A
4 lb. 50-W/A

Pest:
codling moth, walnut huskfly, walnut scale

Remarks: Almonds, filberts, walnuts. Make no more than 2 foliar applications per season for walnuts, 3 for almonds and filberts. For 50-W, may apply 1 dormant/delayed dormant application for almonds and walnuts. Do not allow livestock to graze treated orchards. Refer to label for use directions and precautions.
Days To Harvest: 14.
State Restrictions: Do not apply 50-W as dormant spray on almonds in the following California counties: Butte, Colusa, Glenn, Solano, Sutter, Tehema, Yolo, and Yuba.

- - - - - - - - - - - - - - - - - -

Rate: 0.5-2 lb. ai/A
1-4 pt. 4E/A
0.5-1 lb. ai/100 gal.
1-2 lb. 50-W/100 gal.

Pest:
black pecan aphid, fall webworm, hickory shuckworm, pecan leaf scorch mite (suppression), pecan nut casebearer, Phylloxera, spittlebugs

8

Fruits & Nuts

NOTICE	The information on these pages is for preliminary planning — not a guide for use. Be sure to follow the manufacturer's directions, notwithstanding information contained here.	For personal protective equipment and EPA registration numbers, please turn to page 39.

Remarks: Pecans. Make no more than 5 applications per season. Do not allow livestock to graze treated orchards. Refer to label for use directions and precautions.
Days To Harvest: 28.

- - - - - - - - - - - - - - - -

Rate: 2-4 lb. ai/A
4-8 pt. 4E/A
Pest:
pavement ants, southern fire ants
Remarks: Almonds, walnuts. Orchard floors. Apply with ground equipment that will uniformly apply spray when ant activity becomes evident and when temperature is below 90°F. Do not apply more than 16 pt./year to orchard floor. Do not graze treated orchards.
Days To Harvest: 14.

- - - - - - - - - - - - - - - -

Signal Word/Toxicity Class: Warning/II.
REI: 24 hr.

Citri-Oil — see petroleum oil

CONFIRM

Selective insecticide does not stimulate mite reproduction.

Confirm 2F (tebufenozide) *(Rohm and Haas)*
Rate: 0.25 lb. ai/A
16 oz. 2F/A
Pest:
codling moth, fall webworm, navel orangeworm, redhumped caterpillar
Remarks: Walnuts. For codling moth and navel orangeworm, apply at initiation of egg hatch. For fall webworm and redhumped caterpillar, apply when larvae first appear. Do not apply more than 16 oz./application or 64 oz./season. Do not graze livestock in treated areas or feed cover crops grown in treated areas to livestock.
Days To Harvest: 30.

- - - - - - - - - - - - - - - -

Signal Word/Toxicity Class: Caution.
REI: 12 hr.

CYPERMETHRIN

Broad spectrum synthetic pyrethroid insecticide with activity against several pecan pests. Long residual, but use on many crops will result in spider mite outbreaks, often the second year after use. Highly toxic to fish and bees and minor toxicity to humans. Restricted Use Pesticide.

Ammo 2.5 EC or WSB *(FMC)*
Rate: 0.06-0.1 lb. ai/A
3-5 fl. oz. 2.5 EC/A
0.05-0.1 lb. ai/A
1-2 bags WSB/A
Pest:
black pecan aphid, hickory shuckworm, pecan nut casebearer, pecan weevil, yellow pecan aphid
Remarks: Pecans. Apply lower rate when pest populations are low; increase rate as pest pressure increases. Apply by ground equipment to point of drip. Use 100 gal. dilute spray/A for smaller trees; 200-300 gal. water for larger trees. Up to 0.6 lb. ai/A per crop season may be applied. Do not graze treated orchards or cut treated cover crops for feed.
Days To Harvest: 21.
Crop Rotation: Not within 30 days after last application.

- - - - - - - - - - - - - - - -

Signal Word/Toxicity Class: Caution (2.5 EC); Warning/II (WSB).
REI: 12 hr.

zeta-CYPERMETHRIN

Synthetic pyrethroid with broad spectrum activity. Extremely toxic to fish and aquatic invertebrates; highly toxic to bees. Restricted Use Pesticide.

Mustang 1.5 EW *(FMC)*
Rate: 0.033-0.05 lb. ai/A
2.82-4.26 fl. oz. 1.5 EW/A
Pest:
black pecan aphid, hickory shuckworm, pecan nut casebearer, pecan weevil, yellow pecan aphid
Remarks: Pecans. Apply by ground equipment to point of drip. Use 100 gal. dilute spray/A for smaller trees. For larger trees, apply in 200-300 gal. water. Rate varies by insect pressure, timing of spray, and field scouting. Up to 0.3 lb. ai/A may be applied per season. Do not graze livestock in treated orchards or cut treated cover crops for feed.
Days To Harvest: 21.
Crop Rotation: Not within 30 days after last application.

- - - - - - - - - - - - - - - -

Signal Word/Toxicity Class: Warning/II.
REI: 12 hr.

Damoil — see petroleum oil

Diatect — see pyrethrins

DIAZINON

Organophosphate insecticide with broad spectrum activity on many nut pests. Excellent control of peach twig borer during growing season and as dormant. No longer effective on webspinning spider mite. Toxic to birds, fish, and wildlife including waterfowl. Restricted Use Pesticide due to avian and aquatic toxicity.

D•z•n diazinon AG600 WBC, 50W *(Novartis)*
Drexel Diazinon Insecticide (4EC) *(Drexel)*
Gowan Diazinon 4E or 50WSB *(Gowan)*
Helena Diazinon AG500 *(Helena)*
Micro Flo Diazinon AG500 *(Micro Flo)*
Prentox Diazinon AG500 or 50W *(Prentiss)*
◆ **Wilbur-Ellis Diazinon 4 Spray (4E)** *(Wilbur-Ellis)*
Rate: 0.5-0.75 lb. ai/100 gal. water
1-1.5 pt. AG500, 4E, 4EC/100 gal. water
1-1.5 lb. 50W, 50WSB/100 gal. water
12.75-19 fl. oz. AG600/100 gal. water
Pest:
parlatoria scale, San Jose scale
50W only:
aphid (eggs), apricot scale, black scale, brown scale, mealybugs, mite (eggs), twig borer
Remarks: Almonds. Apply as dormant spray. Add dormant or superior oil.
State Restrictions: For use in California only.

- - - - - - - - - - - - - - - -

Rate: 1 lb. ai/A
2 pt. 4E, 4EC/A in 100 gal. water
25.5 fl. oz. AG600/A in 100 gal. water
Pest:
aphids, spider mites

Rate: 3 lb. ai/A
6 pt. 4E, 4EC/A
76.5 fl. oz.AG600/A
Pest:
codling moth, scale crawler, walnut caterpillar
Remarks: Walnuts. Apply as foliar spray. Repeat after 14 days if necessary. Do not apply after husks open or illegal residues may result. Do not make more than 3 applications per season.
Days To Harvest: 45.
State Restrictions: For use in California only.

- - - - - - - - - - - - - - - -

Signal Word/Toxicity Class: Caution/III.
REI: 24 hr.

DIBROM

Short residual organophosphate insecticide. Performs well in dormant spray in combination with supreme oils. Toxic to fish, aquatic invertebrates, and wildlife. Highly toxic to bees.

Dibrom 8 (8E) (naled) *(Amvac)*
Rate: 2-5 lb. ai/A
2-5 pt. 8E/A
Pest:
peach twig borer
Remarks: Almonds. Apply single application during dormant or delayed dormant period. Apply in combination with dormant oil spray; follow instructions on spray oil label and observe all precautions. Do not apply through any type of irrigation system.

- - - - - - - - - - - - - - - -

◆ - **new product** • ai-active ingredient • DF-dry flowable • E/EC-emulsifiable concentrate • EW-emulsion in water
F/FL-flowable • G-granule • L/LC-liquid • SP/WSP-soluble packet • W/WP-wettable powder • WSB-water soluble bag

Rate: 0.5-1 lb. ai/A
0.5-1 pt. 8E/A in 100 gal. water

Pest:
aphids

Rate: 1-2 lb. ai/A
1-2 pt. 8E/A in 100-200 gal. water

Pest:
walnut husk fly

Rate: 2 lb. ai/A
2 pt. 8E/A in 100-200 gal. water

Pest:
walnut husk maggot

Rate: California-aerial:
1.5-2 lb. ai/A
1.5-2 pt. 8E/A in 10-20 gal. water

Pest:
aphids redhumped caterpillar

Remarks: Walnuts. Apply at first sign of insects; repeat at 7-day intervals as necessary. Do not use with highly alkaline materials such as lime or Bordeaux. Do not apply more than 4 pt./A per season. Do not graze treated orchards. Do not apply through any type of irrigation system.
Days To Harvest: 10.

Signal Word/Toxicity Class: Danger/I.
REI: 48 hr.-1 pt./A; 72 hr.-over 1 pt./A.

DICOFOL

Highly selective hydrocarbon miticide used to control pecan leaf scorch mite on pecans and several mites on other nut trees. Toxic to western predator mite.

◆ **Griffin Dicofol 4E** *(Griffin)*
Kelthane MF (4E) *(Rohm and Haas)*

Rate: 0.38-0.5 lb. ai/100 gal. dilute spray
0.75-1 pt. MF/100 gal. dilute spray

Pest:
brown almond mite, clover mite, European red mite, McDaniel mite, Pacific mite, twospotted spider mite, Willamette mite, yellow mite

Remarks: Pecans (Griffin), walnuts. Apply just after bloom or whenever mites appear and repeat if necessary. Do not apply through any type of irrigation system.
Days To Harvest: 14.

Signal Word/Toxicity Class: Caution/III.
REI: 48 hr.

Digon — see dimethoate

Dimate — see dimethoate

DIMETHOATE

Organophosphate insecticide/miticide with foliar systemic activity. Toxic to wildlife and aquatic invertebrates. Highly toxic to bees.

Cheminova Dimethoate 4E *(Cheminova)*
Dimate 4E *(Terra)*
Drexel Dimethoate 2.67 or 4EC *(Drexel)*
Gowan Dimethoate E267 *(Gowan)*
Helena Dimethoate 4E *(Helena)*
◆ **Micro Flo Dimethoate 267 or 4E** *(Micro Flo)*
Wilbur-Ellis Dimethoate 267 or Digon 400 *(Wilbur-Ellis)*

Rate: 0.33 lb. ai/A
1 pt. 2.67, 267, E267/A
0.66 pt. 4E, 4EC, 400/A

Pest:
aphids leafhoppers mites

Remarks: Pecans. Do not graze livestock in treated groves.
Days To Harvest: 21.

Signal Word/Toxicity Class: Warning/II.
REI: 48 hr.

DI-SYSTON

Organophosphate insecticide/miticide with foliar systemic and soil systemic activity. Toxic to fish, wildlife, and bees. Hazardous to humans and domestic animals. Restricted Use Pesticide.

Di-Syston 8 or 15% (disulfoton) *(Bayer)*

Rate: Foliar:
0.25-0.375 lb. ai/100 gal. water
0.25-0.375 pt. 8/100 gal. water
Aerial:
0.75-1 lb. ai/A
0.75-1 pt. 8/A

Pest:
aphids

Remarks: Pecans. For foliar application, apply as full coverage spray using not more than 400 gal. finished spray/A; for aerial application, apply in sufficient water to obtain thorough coverage but not less than 5 gal. total volume/A. Do not apply more than 3 times per season. Do not graze grass under treated trees.
Days To Harvest: 30.

Rate: 1.5-3 lb. ai/A
1.5-3 pt. 8/A
10-20 lb. 15%/A

Pest:
aphids

Remarks: Pecans. Apply by treating 6 ft. bands of soil on both sides of the trees. Bands should be located under main drip area of trees. Work into upper 2-3" of soil. On irrigated pecans, apply water as soon as possible after treatment. Make only 1 application per season. Early spring treatment recommended. Do not graze grass under treated trees.
Days To Harvest: 80.
State Restrictions: For use in South Central, Southwestern U.S.

Signal Word/Toxicity Class: Danger/I.
REI: 48 hr.

D•z•n — see diazinon

ENDOSULFAN

Chlorinated bicyclic sulfite insecticide. Toxic to fish, birds, bees, and wildlife. Hazardous to humans and domestic animals.

Drexel Endosulfan 3EC or 50WP *(Drexel)*
Gowan Endosulfan 3EC or 50WSB *(Gowan)*
◆ **Micro Flo Endosulfan 3EC or 50WP** *(Micro Flo)*
Phaser 3EC or 50WSB *(AgrEvo USA)*
Thiodan 3EC or 50WP *(FMC)*

Rate: 0.5-0.75 lb. ai/100 gal. water
0.66-1 qt. 3EC/100 gal. water
1-1.5 lb. 50WP, 50WSB/100 gal. water

Pest:
pecan leaf phylloxera

Rate: 0.75 lb. ai/100 gal. water
1 qt. 3EC/100 gal. water
1.5 lb. 50WP, 50WSB/100 gal. water

Pest:
black pecan aphid pecan nut casebearer spittlebug

Remarks: Pecans. For pecan leaf phylloxera, apply when nymphs appear and before they are enclosed in plant tissue; use higher rate for high populations. For casebearer, apply when eggs of first generation appear on tips of young nuts; another application may be required after second generation eggs are deposited. Do not apply after shuck split. Do not graze livestock on orchard crops or grasses in treated areas. Do not exceed 3 lb.ai/A per year or 2 applications per year.

Rate: 1.5-2.5 lb. ai/A
2-3.33 qt. 3EC/A
3-4 lb. 50WP, 50WSB/A

Pest:
walnut aphid

Remarks: Walnuts. Apply when insects appear or feeding is noticed. Do not apply after husk split. Do not graze livestock on orchard crops or grasses in treated areas. Do not exceed 3 lb.ai/A per year or 2 applications per year.

Signal Word/Toxicity Class: Danger-Poison/I.
REI: 24 hr.

NOTICE The information on these pages is for preliminary planning — not a guide for use. Be sure to follow the manufacturer's directions, notwithstanding information contained here. For personal protective equipment and EPA registration numbers, please turn to page 39.

Fyfanon— see malathion

Guthion — see azinphos-methyl

IMIDAN

Organophosphate insecticide that can be used near residential areas to control lepidopterous pests. Has low mammalian toxicity, but is extremely toxic to fish and highly toxic to bees.

Imidan 70-W (phosmet) *(Gowan)*

Rate: 3 lb. ai/A
4.33 lb. 70-W/A

Pest:
peach twig borer

Rate: California:
3-3.7 lb. ai/A
4.33-5.33 lb. 70-W/A

Pest:
navel orangeworm peach twig borer

Remarks: Almonds. Do not make more than 2 foliar applications per season.
Days To Harvest: 30.

- - - - - - - - - - - - - - - - -

Rate: 3-6 lb. ai/A
4.33-8.5 lb. 70-W/A

Pest:
codling moth navel orangeworm walnut huskfly

Remarks: Beechnuts, Brazil, cashews, chestnuts, filberts, hickory, macadamias, walnuts. Do not apply after hull split. For heavy infestations, apply high rate. Repeat as necessary in accordance with insect infestations and local or state programs. Do not apply more than 8.5 lb./A or more than 5 applications per season.
Days To Harvest: 14.

- - - - - - - - - - - - - - - - -

Rate: 1.4-2 lb. ai/A
2-3.13 lb. 70-W/A

Pest:
black pecan aphid, fall webworm, hickory shuckworm, pecan nut casebearer, pecan weevil, southern green stink bug, spittlebug

Remarks: Pecans. Apply as full coverage spray when infestation starts. For pecan weevil, repeat at 7-day intervals. Adequate control may not be achieved with heavy populations of fall webworm, pecan weevil, and stink bug. Do not graze or feed livestock on cover crops grown in treated groves.
Days To Harvest: 14.

- - - - - - - - - - - - - - - - -

Signal Word/Toxicity Class: Warning/II.
REI: 24 hr.

Kelthane — see dicofol

Kumulus — see sulfur

Lime-Sulfur Solution — see calcium polysulfide

Lorsban — see chlorpyrifos

MALATHION

Organophosphate insecticide that can be used on lepidopterous pests near residential areas. Toxic to fish and aquatic invertebrates; highly toxic to bees. Low toxicity to applicators.

Clean Crop Malathion 57EC *(Platte)*
Gowan Malathion 8 (8E) *(Gowan)*

Rate: Almonds:
0.63-1.25 lb. ai/100 gal. water
1-2 pt. 57EC/100 gal. water
2.5-8 lb. ai/A
2.5-8 pt. 8E/A

Pest:
aphids peach twig borer red spider mites

Rate: Filberts:
0.63 lb. ai/100 gal. water
1 pt. 57EC/100 gal. water
2.5-3 lb. ai/A
2.5-3 pt. 8E/A

Pest:
apple mealybug, eyespotted bud moth, filbert aphid, scale crawlers, tingids

Rate: Macadamias:
3-15 lb. ai/A
3-15 pt. 8E

Pest:
green stink bug

Rate: Pecans:
0.625-1.25 lb. ai/A
1-2 pt. 57EC/A
2.5-12.5 lb. ai/A
2.5-12.5 pt. 8E/A

Pest:
aphids, mites, pecan bud moth, pecan leaf casebearer, pecan nut casebearer, pecan phylloxera

Rate: Walnuts:
0.125-0.375 lb. ai/100 gal. water
0.2-0.6 pt. 57EC/100 gal. water
4-12 lb. ai/A
4-12.5 pt. 8E/A

Pest:
aphids mites walnut huskfly

Remarks: Rate varies by pest species.
Days To Harvest: 0.

- - - - - - - - - - - - - - - - -

Signal Word/Toxicity Class: Caution/III.
REI: 12 hr.

Clean Crop Malathion 8EC *(Platte)*

Rate: Filberts:
1 lb. ai/100 gal. water
1 pt. 8EC/100 gal. water

Pest:
apple mealybug, eyespotted bud moth, filbert aphid, scale crawlers, tingids

Rate: Pecans:
1-1.5 lb. ai/A
1-1.5 pt. 8EC/A

Pest:
aphids, codling moth, eyespotted bud moth, mealybugs, orange tortrix, spider mites, thrips

Remarks: Rate varies by pest species.
Days To Harvest: 0.

- - - - - - - - - - - - - - - - -

Signal Word/Toxicity Class: Caution/III.
REI: 12 hr.

Drexel Malathion 5EC *(Drexel)*
◆ Micro Flo Malathion 5EC *(Micro Flo)*
Prentox 5 lb. Malathion Spray (5EC) *(Prentiss)*

Rate: Macadamias:
1 lb. ai/100 gal. water
1.5 pt. 5EC/100 gal. water

Pest:
green stink bug

Rate: Pecans:
0.625-1.25 lb. ai/A
1-2 pt. 5EC/A

Pest:
aphids, mites, pecan bud moth, pecan leaf casebearer, pecan nut casebearer, pecan phylloxera

Rate: Walnuts:
0.125-0.375 lb. ai/100 gal. water
0.2-0.6 pt. 5EC/100 gal. water

Pest:
aphids mites walnut huskfly

Remarks: Rate varies by pest species.
Days To Harvest: 0.

- - - - - - - - - - - - - - - - -

Signal Word/Toxicity Class: Warning/II.
REI: 12 hr.

◆ - **new product** • ai-active ingredient • DF-dry flowable • E/EC-emulsifiable concentrate • EW-emulsion in water
F/FL-flowable • G-granule • L/LC-liquid • SP/WSP-soluble packet • W/WP-wettable powder • WSB-water soluble bag

Fyfanon (5EC) *(Helena)*
Rate: 0.63-1.25 lb. ai/100 gal. water
1-2 pt. 5EC/100 gal. water
Pest: Almonds:
aphids peach twig borer red spider mites
Pecans:
aphids pecan leaf casebearer pecan phylloxera
pecan bud moth pecan nut casebearer spider mites
Remarks: Almonds, pecans. Rate varies by pest species.
Days To Harvest: 0.

Signal Word/Toxicity Class: Caution/III.
REI: 12 hr.

Wilbur-Ellis Malathion 8 *(Wilbur-Ellis)*
Rate: Pecans:
1-1.5 lb. ai/100 gal. water
1-1.5 pt. 8/100 gal. water
Pest:
aphids leafrollers scale crawlers
codling moth mealybugs spider mites
eyespotted moth orange tortrix thrips

Rate: Walnuts:
4-12.5 lb. ai/100 gal. water
4-12.5 pt. 8/100 gal. water
Pest:
aphids mites walnut husk fly
Remarks: Pecans, walnuts. Apply in min. 5 gal. by air; 10 gal. by ground.
Days To Harvest: 0.

Signal Word/Toxicity Class: Caution/III.
REI: 12 hr.

METASYSTOX-R

Systemic organophosphate insecticide. Toxic to fish, wildlife, and bees. Restricted Use Pesticide.

Metasystox-R (SC) (oxydemeton-methyl) *(Gowan)*
Rate: 0.25 lb. ai/10 trees
1 pt. SC/10 trees (4 1/2" diameter)
Pest:
aphids
Remarks: Filberts. Apply in band around trunk with paint brush or low pressure sprayer, either undiluted or diluted with equal volume of water. For undiluted application, apply 0.33 fl. oz./" trunk diameter; diluted application, apply 0.66 fl. oz./" trunk diameter. On trees with extremely short trunks, it may be necessary to treat bases of scaffold limbs in order to apply correct rate; width of treated band will vary with trunk diameter. Do not apply more than once per season. Do not use on heavily stressed trees or young trees with trunk diameters less than 2".
Days To Harvest: 105.
State Restrictions: For use in Oregon, Washington.

Rate: 0.375 lb. ai/100 gal. water
1.5 pt. SC/100 gal. water
Pest:
aphid
Remarks: Walnuts. Apply as full coverage or concentrate spray at appropriate dilute rate. Do not apply more than once per season.
Days To Harvest: 30.

Signal Word/Toxicity Class: Warning/II.
REI: 48 hr.; 72 hr. if average rainfall less than 25" per year.

METHYL PARATHION

Penncap-M is microencapsulated form of methyl parathion, which provides an increased safety factor for applicators and other nontarget organisms by releasing active ingredient over a long period of time. Restricted Use Pesticide due to residual effects to avian species and hazard to bees.

Penncap-M (2FL) *(Elf Atochem)*
Rate: 0.5-2 lb. ai/A
2-8 pt. 2FL/A
Pest:
brown stink bug leaffooted bug southern green stink bug
dusky stink bug

Rate: 0.75-2 lb. ai/A
3-8 pt. 2FL/A
Pest:
hickory shuckworm pecan weevil
Remarks: Pecans. Do not apply after shuck split. Do not apply when weed or cover crops in orchard are in bloom and bees are foraging in areas to be treated. Consult local spray recommendations for correct timing of applications for optimal control. Do not apply more than 64 pt. per year. Do not graze or feed cover crops within 15 days after application.

Signal Word/Toxicity Class: Warning/II.
REI: 48 hr.

Microthiol Special — see sulfur

MORESTAN

Organophosphate insecticide/miticide/fungicide. Toxic to western predator mite and fish.

Morestan Solupak 25 (oxythioquinox) *(Bayer)*
Rate: 0.25-0.375 lb. ai/100 gal. water
1-1.5 lb. Solupak 25/100 gal. water
Pest:
aphids mite eggs mites
Remarks: Walnuts. Apply as full coverage spray using up to 250 gal. finished spray/A depending on number and size of trees. Spray all surfaces, including undersides of leaves, for complete coverage. Do not apply more than 2 times per season.
Days To Harvest: 30.

Rate: 0.25-0.5 lb. ai/100 gal. water
1-2 lb. Solupak 25/100 gal. water
Pest:
mites
Remarks: Macadamias. Apply as full coverage spray using up to 400 gal. finished spray/A depending on number and size of trees. Spray all surfaces, including undersides of leaves, for complete coverage. Three applications per season may be made.
Days To Harvest: 0.

Signal Word/Toxicity Class: Warning/II.
REI: 24 hr.

M-PEDE

M-Pede (potassium salts of fatty acids) *(Mycogen)*
Rate: 1-2% v/v solution
Pest:
aphids mealybugs spider mites
leafhoppers scale crawlers whiteflies
Remarks: Almonds, chestnuts, filberts, pecans, pistachios, walnuts (black, English). Apply to wet with an air blast or other appropriate sprayer. May be tank mixed with other insecticides. Refer to both labels for use directions and precautions.
Days To Harvest: 0.

Signal Word/Toxicity Class: Warning/II.
REI: 12 hr.

Mustang — see *zeta*-cypermethrin

NEEMAZAD — see azadirachtin

NEEMIX — see azadirachtin

Omite — see propargite

Omni Supreme Spray — see petroleum oil

Penncap-M — see methyl parathion

NOTICE	The information on these pages is for preliminary planning — not a guide for use. Be sure to follow the manufacturer's directions, notwithstanding information contained here.	For personal protective equipment and EPA registration numbers, please turn to page 39.

PERMETHRIN

Broad-spectrum synthetic pyrethroid insecticide that is readily broken down in most soils. Results in spider mite outbreaks second year after use. Highly toxic to fish and bees, but has low toxicity to mammals and birds. Restricted Use Pesticide.

Ambush 2E or 25W *(ZENECA)*
Rate: 0.2-0.4 lb. ai/A
12.8-25.6 fl. oz. 2E/A
12.8-25.6 oz. 25W/A
Pest:
navel orangeworm peach twig borer
Remarks: Almonds. Apply by ground equipment using sufficient water to obtain uniform coverage. Do not apply more than 2 lb. ai/A per season. Do not graze livestock in treated orchards.
Days To Harvest: 7.

- - - - - - - - - - - - - - - -

Rate: 0.2-0.4 lb. ai/A
12.8-25.6 fl. oz. 2E/A
12.8-25.6 oz. 25W/A
Pest:
filbertworm obliquebanded leafroller
Remarks: Filberts. Apply when insects appear. Apply 2E in minimum 15 gal. finished spray/A by air or 20 gal. finished spray/A by ground; 25W in a minimum 15 gal. finished spray/A by ground. Do not apply more than 1.6 lb. ai/A per season. Do not graze livestock in treated orchards.
Days To Harvest: 14.

- - - - - - - - - - - - - - - -

Rate: 0.2-0.4 lb. ai/A
12.8-25.6 fl. oz. 2E/A
12.8-25.6 oz. 25W/A
Pest:
chinch bug navel orangeworm peach twig borer
Remarks: Pistachios. Use sufficient water to obtain full coverage. Do not apply more than 0.8 lb. ai/A per season. Do not apply after 10% hull split. Do not graze livestock in treated orchards.
Days To Harvest: 0.

- - - - - - - - - - - - - - - -

Rate: 0.2-0.4 lb. ai/A
12.8-25.6 fl. oz. 2E/A
12.8-25.6 oz. 25W/A
Pest:
codling moth navel orangeworm walnut huskfly
Remarks: Walnuts. Apply when insects appear, in minimum 20 gal. finished spray/A by air or 100 gal. finished spray/A by ground. Do not apply more than 1.6 lb. ai/A per season. Do not graze livestock in treated orchards.
Days To Harvest: 1.

- - - - - - - - - - - - - - - -

Signal Word/Toxicity Class: Warning/II.
REI: 24 hr.

Pounce 1.5G, 3.2 EC or WSB *(FMC)*
Rate: 0.4 lb. ai/A
26 lb. 1.5G/A
16 oz. 3.2 EC/A
2-4 bags WSB/A
Pest:
ants
Remarks: Almonds, pistachios. Apply 1.5G as a broadcast treatment by ground or air; 3.2 EC by ground equipment in a minimum of 15 gal. finished spray/A. Do not apply more than 52 lb. 1.5G/A per crop season. Do not apply after 10% hull split. Overhead moisture following application will enhance activity. Do not graze livestock in treated orchards.
Days To Harvest: 0.

- - - - - - - - - - - - - - - -

Signal Word/Toxicity Class: Caution (1.5G, 3.2 EC); Warning/II (WSB).
REI: 24 hr.

Pounce 3.2 EC, 25WP or WSB *(FMC)*
Rate: 0.2-0.4 lb. ai/A
8-16 oz. 3.2 EC/A
12.8-25.6 oz. 25WP/A
2-4 bags WSB/A
Pest:
navel orangeworm peach twig borer
Remarks: Almonds, pistachios. Apply when insects appear. For almonds, do not make more than 2 applications during hull split. For pistachios, do not apply after 10% hull split. Do not graze livestock in treated orchards.
Days To Harvest: Almonds-7; pistachios-0.

- - - - - - - - - - - - - - - -

Rate: 0.2-0.4 lb. ai/A
8-16 oz. 3.2 EC/A
12.8-25.6 oz. 25WP/A
2-4 bags WSB/A
Full coverage application:
0.05-0.1 lb. ai/100 gal. (400 gal. finished spray/A)
Low volume:
0.2-0.4 lb. ai/A (50-200 gal. finished spray)/A
Pest: Filberts:
filbertworm obliquebanded leafroller
Walnuts:
codling moth navel orangeworm walnut huskfly
Remarks: Filberts, walnuts. Apply by air when insects appear. Do not apply more than 1.6 lb. ai/A per season. For full coverage application, spray to runoff. Do not graze livestock in treated orchards or feed cover crops from treated orchards to livestock.
Days To Harvest: Filberts-14; walnuts-1.

- - - - - - - - - - - - - - - -

Signal Word/Toxicity Class: Caution (3.2 EC); Warning/II (25WP, WSB).
REI: 24 hr.

PETROLEUM OIL

Short residual nondisruptive petroleum based spray oil commonly used for control of scale insects, certain mites, and eggs of certain insects. Time of application critical for good control and to prevent plant injury.

◆ **435 or 455 Oil 98.8** *(Drexel)*
◆ **SunSpray 6E or 7E** *(Sun)*
Rate: 1 gal. 455, 6E/100 gal. water
Pest:
aphids (eggs) scales

Rate: 1.5 gal. 455, 6E/100 gal. water
Pest:
spider mites
Remarks: Almonds. For aphid eggs and scales, apply in summer. For spider mites, apply during dormant stage. Do not apply in combination with or immediately preceding or following application of captan, carbaryl, Dyrene, folpet, Morestan, propargite, and sulfur compounds.

- - - - - - - - - - - - - - - -

Rate: 3 gal. 435, 455, 6E, 7E/100 gal. water
Pest:
obscure scale
Remarks: Pecans. Dormant. Do not apply in combination with or immediately preceding or following application of captan, carbaryl, Dyrene, folpet, Morestan, propargite, and sulfur compounds.

- - - - - - - - - - - - - - - -

Signal Word/Toxicity Class: Caution.
REI: 4 hr.

Citri-Oil *(Wilbur-Ellis)*
Rate: 3-3.5 gal./100 gal. water
Pest:
obscure scale
Remarks: Pecans. Apply during free dormancy, when temperature is between 50°-70°F.

- - - - - - - - - - - - - - - -

Signal Word/Toxicity Class: Caution.
REI: 12 hr.

Damoil *(Drexel)*
Omni Supreme Spray *(Helena)*
Volck Supreme Spray *(Valent)*
Rate: Summer:
3-4 qt./100 gal. dilute spray
Pest:
red spider mites

Rate: Summer:
1 gal./100 gal. dilute spray
Pest:
brown apricot scale San Jose scale

Rate: Dormant/delayed dormant:
1-1.5 gal./100 gal. dilute spray

◆ - **new product** • ai-active ingredient • DF-dry flowable • E/EC-emulsifiable concentrate • EW-emulsion in water
F/FL-flowable • G-granule • L/LC-liquid • SP/WSP-soluble packet • W/WP-wettable powder • WSB-water soluble bag

Pest:

aphid eggs	frosted scale	Putnam scale
black scale	Italian pear scale	red spider mites
brown apricot scale	oleander scale	San Jose scale
brown mites	olive scale	soft brown scale
European fruit lecanium	parlatoria scale	twospotted spider mites (summer)

Remarks: Almonds. Refer to label for use directions and precautions.

- - - - - - - - - - - - - - - - -

Rate: Late spring to midsummer:
2-4 qt./A

Pest:

calico scale	European fruit scale	San Jose scale
red spider mites (eggs)	Frosted scale	walnut aphid (eggs)

Remarks: Walnuts. Do not apply after husk split. Addition of Trithion or ethion at labeled rates aids in scale control.

- - - - - - - - - - - - - - - - -

Signal Word/Toxicity Class: Caution.
REI: 12 hr.; OMNI-4 hr.

SAF-T-SIDE *(Brandt Consolidated)*
Rate: 1-2 gal./100 gal. water
Pest:

aphids	mites	thrips
fruittree leafroller	plant bugs	whiteflies
leafhoppers	scales	

Remarks: Almonds, macadamias, pecans. For almonds and pecans, apply dormant, summer, or postharvest; for macadamias, apply as needed. Time application to fit local growth and application conditions. Compatible with most commonly used insecticides and fungicides. Do not use in combination with or immediately before or after spraying with fungicides such as captan, Dyrene, folpet, Morestan, or any sulfur products. Do not use with carbaryl or dimethoate.

- - - - - - - - - - - - - - - - -

Signal Word/Toxicity Class: Caution.
REI: 4 hr.

◆ **Sumulsion** *(Drexel)*
Rate: 1-2 gal./100 gal. water
Pest: Almonds:

aphids	mites	whiteflies
fruittree leafroller	scales	

Macadamias:

leafhoppers	scales	thrips
leafrollers	Sigatoka	

Walnuts:

aphids	mites	scales

Remarks: For almonds and macadamias, apply at dormant stage. For walnuts, apply in late spring to mid-summer, do not apply after husk splits. Do not use in combination with or immediately before or after spraying with fungicides such as captan, Dyrene, folpet, Morestan, or any sulfur products. Do not use with carbaryl or dimethoate.

- - - - - - - - - - - - - - - - -

Signal Word/Toxicity Class: Caution.
REI: 4 hr.

SunSpray Ultra-Fine (6E) *(Sun)*
Rate: Almonds:
1-3 gal. 6E/100 gal. water
Pest:

aphids eggs	San Jose scale	spider mites
fruittree leafroller	scales	tent caterpillar

Rate: Pecans:
3 gal. 6E/100 gal. water
Pest:

aphids	mites	scales

Remarks: Almonds, pecans. For use throughout growing season. Refer to label for use directions and precautions.

- - - - - - - - - - - - - - - - -

Signal Word/Toxicity Class: Caution.
REI: 4 hr.

Phaser — see endosulfan

Pounce — see permethrin

PROPARGITE

Organic cyclohexyl miticide with long residual effects. Controls mites through contact. Selective to western predator mite and insect predators. Requires minimal exposure training. Should not be combined with liquid formulation insecticide. Toxic to fish.

Omite 6E, 30W or 30WS *(Uniroyal)*
Rate: 2.25-3 lb. ai/A
3-4 pt. 6E/A
1.5-4.5 lb. ai/A
5-15 lb. 30W, 30WS/A
Pest: Mites

clover mite	Pacific spider mite	twospotted spider mite
European red mite	strawberry spider mite	

Remarks: Almonds. Omite 6E will cause phytotoxicity to almond leaves in the presence of summer spray oils. Apply no more than twice per year. Do not apply through any type of irrigation system.
Days To Harvest: 28.
State Restrictions: For use in Arizona, California.

- - - - - - - - - - - - - - - - -

Rate: 1.5-6.5 lb. ai/A
2-9 pt. 6E/A
2.1-6.5 lb. ai/A
7-22 lb. 30W, 30WS/A
Pest:

European red mite	twospotted spider mite

Remarks: Walnuts. Apply no more than twice per year. Do not apply this product through any type of irrigation system.
Days To Harvest: 21.
State Restrictions: For use in California only for 30W, 30WS.

- - - - - - - - - - - - - - - - -

Signal Word/Toxicity Class: Warning/II (6E); Danger/I (30W, 30WS).
REI: 48 hr.

Pyramite — see pyridaben

PYRELLIN (Premix)

Premix containing pyrethrins and rotenone. Short residual botanical.

Pyrellin E.C. (0.042 lb./gal. pyrethrins & 0.035 lb./gal. rotenone) *(Webb Wright)*
Rate: 1-2 pt. E.C./A
Pest:

aphids	thrips	whiteflies
mites		

Remarks: Almonds, cashews, chestnuts, filberts, macadamias, pecans, pistachios, walnuts. Apply as needed to maintain control.
Days To Harvest: 0.

- - - - - - - - - - - - - - - - -

Signal Word/Toxicity Class: Caution/III.
REI: 12 hr.

Pyrenone — see pyrethrins

PYRETHRINS

Contains piperonyl butoxide as a synergist. Diatect also contains silicon dioxide (diatomaceous earth). May be combined in tank mix with other insecticides for quicker and more complete kill where insect resistance may be a problem, and as an "exciter" to flush insects out of hiding and into contact with other spray residues.

Diatect Multipurpose Insecticide II *(Diatect)*
Pyrenone (0.5EC) *(AgrEvo Environmental Health)*
Rate: 1-6 lb. II/A
1-12 fl. oz. 0.5EC/A
Pest: Including but not limited to:

aphids	mites	thrips
caterpillars	pecan weevil	

Remarks: Almonds, beechnuts, Brazil, cashews, chestnuts, filberts, hickory, macadamias, pecans, pistachios, walnuts, etc. Apply in sufficient water for thorough coverage; repeat as needed. Apply Diatect as dust or wettable powder.
Days To Harvest: 0.

- - - - - - - - - - - - - - - - -

Signal Word/Toxicity Class: Caution.
REI: 12 hr.

8 Fruits & Nuts

NOTICE The information on these pages is for preliminary planning — not a guide for use. Be sure to follow the manufacturer's directions, notwithstanding information contained here. | For personal protective equipment and EPA registration numbers, please turn to page 39.

PYRIDABEN

Pyridoxone miticide/insecticide, a wettable powder in water-soluble bags, that provides knockdown and residual control. Toxic to bees, fish, and aquatic invertebrates.

Pyramite (WP) *(BASF)*
Rate: 4.4-13.2 oz. WP/A (1.5-3 WSB)
Pest: Mites:
apple rust mite, European red mite, McDaniel spider mite, Pacific spider mite, pear psylla, pear rust mite, twospotted spider mite
Remarks: Almonds. Use high rate to ensure adequate concentration in full size trees with dense foliage. Do not make more than 2 applications per year or apply more than 26.4 oz./A per season. Do not apply by air or through any type of irrigation system. Refer to label for spray drift precautions.
Days To Harvest: 7.

Signal Word/Toxicity Class: Danger/I.
REI: 12 hr.

SAF-T-SIDE — see petroleum oil

Sevin — see carbaryl

Sniper — see azinphos-methyl

SPINOSAD

Fermentation-derived insect control agent. Does not significantly impact predaceous and parasitic insect populations. Highly toxic to bees and aquatic invertebrates.

◆ **Success Naturalyte (SC) (spinosyn A + spinosyn D)** *(Dow AgroSciences)*
Rate: 0.062-0.156 lb. ai/A
4-10 oz. SC/A
Pest:
mites (overwintering)*, navel orangeworm, peach twig borer, redhumped caterpillar, scales*
***Suppression**
Remarks: Almonds. Apply at 14-day intervals. For mites and scales, add crop oils labeled for agricultural use added to dormant spray; consult specific oil labels and University of California recommendations for precautions and restrictions. Do not apply more than 29 oz./A per crop. Do not apply through any type of irrigation equipment.
Days To Harvest: 14.

Signal Word/Toxicity Class: Caution.
REI: 4 hr.

Success Naturalyte — see spinosad

SULFUR

Microthiol and Thiolux are micronized sulfurs.

◆ **Kumulus DF** *(BASF)*
Rate: 8-16 lb. ai/A
10-20 lb. DF/A
Pest: Almonds, pecans:
flat mite, red spider mite, silver mite
Walnuts:
almond mite, European red mite, Pacific mite, red spider mite, twospotted spider mite
Remarks: Almonds, pecans, walnuts. For almonds and pecans, apply at bloom or early petal fall; repeat as needed, usually at 10- to 14-day intervals or after wet weather. For walnuts, apply with mite sprays during season to increase effective of miticides. Certain walnut varieties are sensitive to sulfur spray under certain conditions; apply only to tolerant varieties. Do not apply if temperatures during or within 3 days after application are expected to exceed 90°F. Do not apply through any type of irrigation system. Refer to label for use directions.

Signal Word/Toxicity Class: Caution.
REI: 24 hr.

Microthiol Special (80W) *(Elf Atochem)*
Rate: 8-16 lb. ai/A
10-20 lb. 80W/A
Pest: Almonds:
flat mite, red spider mite, silver mite
Macadamias, walnuts:
almond mite, broad mite (except CA), Pacific mite, red spider mite, twospotted spider mite (except CA)
Pistachios:
mites

Rate: 4-8 lb. ai/A
5-10 lb. 80W/A
Pest: Pecans:
eriophyid mite, red spider mite, twospotted spider mite
Remarks: Almonds, macadamias, pecans, pistachios, walnuts. For almonds, apply at bloom or early petal fall; repeat at 10- to 14-day intervals. For macadamias, pecans, and walnuts, apply throughout season as needed. Do not apply unless varieties are known to be sulfur tolerant. For pistachios, apply in June-July; repeat as needed. When temperatures exceed 95°F, lower rates and more frequent applications advised.
State Restrictions: Not for use in California on macadamias, pecans.

Signal Word/Toxicity Class: Caution.
REI: 24 hr.

Thiolux (DF) *(Platte)*
Rate: Almonds:
8-24 lb. ai/A
10-30 lb. DF/A
Pest:
flat mite, red spider mite, silver mite

Rate: Walnuts:
8-16 lb. ai/A
10-20 lb. DF/A
Pest:
almond mite, Pacific mite, red spider mite, twospotted spider mite
Remarks: Almonds, walnuts. For walnuts, apply with miticide sprays to increase effectiveness. Certain walnut varieties are sensitive to sulfur; do not apply unless variety is known to be sulfur tolerant.

Signal Word/Toxicity Class: Caution.
REI: 24 hr.

Sumulsion — see petroleum oil

SunSpray — see petroleum oil

SUPRACIDE

Broad-spectrum organophosphate insecticide/miticide. Toxic to fish, wildlife, and mammals. Highly toxic to bees. Restricted Use Pesticide.

Supracide 25WP (methidathion) *(Gowan)*
Rate: 1-3 lb. ai/A in 50 gal. water
4-12 lb. 25WP/A in 50 gal. water
Aerial:
1.5-3 lb. ai/A in 20 gal. water
6-12 lb. 25WP/A in 20 gal. water
Pest:
brown almond mite, peach twig borer, San Jose scale
Remarks: Almonds. Apply during dormant/delayed dormant period or as cover spray. EPA-registered oils may be added to spray mixture in dormant/delayed dormant period. Do not make more than 1 application or more than 1 cover spray per season. Do not graze treated areas or feed cover crops grown in treated areas to livestock.
Days To Harvest: 80.

Rate: 2-4 lb. ai/A
8-16 lb. 25WP/A
Pest:
hickory shuckworm, serpentine leafminer, yellow pecan aphid

◆ - **new product** • ai-active ingredient • DF-dry flowable • E/EC-emulsifiable concentrate • EW-emulsion in water
F/FL-flowable • G-granule • L/LC-liquid • SP/WSP-soluble packet • W/WP-wettable powder • WSB-water soluble bag

Remarks: Pecans. Apply as full cover spray in sufficient water for thorough coverage using minimum spray volume of 200 gal./A (400 gal. for 16 lb. rate) when insects or their damage first appear. Repeat as necessary at 14- to 21-day intervals. Apply high rate for heavy infestations; low rate for yellow pecan aphid. Make no more than 2 applications per season. Do not tank mix with Cyprex. Do not graze treated areas.

Days To Harvest: 60.

State Restrictions: For use in southern U.S.

Rate: 1-3 lb. ai/A in 50 gal. water
4-12 lb. 25WP/A in 50 gal. water
Aerial:
1.5-3 lb. ai/A in 20 gal. water
6-12 lb. 25WP/A in 20 gal. water

Pest:
codling moth, frosted scale, Italian pear scale, walnut aphid, walnut scale

Remarks: Walnuts. Apply during dormant/delayed dormant period or as cover spray when insects appear. Use low rate for light infestations; high rate for moderate to heavy infestations. Properly calibrated application equipment essential for uniform coverage. EPA-registered oils may be added in dormant/delayed dormant period only. For dormant/delayed dormant applications at rates less than 6 lb. 25WP/A, at least 2 gal. oil/A must be added to spray mixture. Do not apply more than 8 gal. oil/A by ground; 5 gal. oil/A by air. Make no more than 1 application during dormant/delayed dormant period or more than 2 cover sprays per season. For walnuts, do not add oil to sprays in southern or central California, or to trees that may be subject to stress due to drought, drying winds, disease, or severe insect infestations. Do not graze treated areas or feed cover crops grown in treated areas to livestock. Do not apply through any type of irrigation system. Refer to label for further use directions.

Days To Harvest: 7.

Signal Word/Toxicity Class: Warning/II.
REI: 48 hr.

TEMIK

Systemic carbamate type insecticide/miticide. Applied to the soil to control aphids and mites. Extremely toxic to humans and other mammals, birds, fish, and wildlife. Restricted Use Pesticide.

Temik 15G (aldicarb) *(Rhone-Poulenc)*

Rate: Except Arizona, Florida:
5-10 lb. ai/A
33-67 lb. 15G/A

Pest:
aphids, mites

Rate: Arizona:
5-7 lb. ai/A
33-47 lb. 15G/A

Pest:
aphids

Rate: Florida:
5 lb. ai/A
33 lb. 15G/A

Pest:
pecan leaf phylloxera (suppression)

Remarks: Pecans. Producing trees. For aphids and mites, apply from bud break to nut set; for phylloxera, apply 1-2 weeks before bud break. Apply as 4-6 ft. band along dripline on both sides of tree row by spreading granules uniformly and working into soil or by shanking 2-3" into soil on 12" centers. Do not make more than 1 application per year. Do not exceed 47 lb./A in Arizona, 33 lb./A in Florida, 67 lb./A in other states. Do not harvest forage or hay, or graze in treated areas. Refer to label for environmental precautions and soil restrictions table.

State Restrictions: For use in the Southeast and Arizona. In Florida, apply between Jan.1-Apr. 30 only.

Rate: 21-84 grams ai/tree
5-20 oz. 15G/tree

Pest:
aphids, bud moth, mites, pecan leaf phylloxera (suppression)

Remarks: Pecans. Newly transplanted trees (1-5 yr. old). For aphids and mites, apply during bud break. For bud moth and phylloxera, apply 1-2 weeks before bud break. Apply as sidedress by spreading granules uniformly around tree and work into soil to depth of 2-3". Do not make more than 1 application per year. Do not harvest forage or hay, or graze in treated areas. Refer to label for environmental precautions and soil restrictions table.

State Restrictions: For use in the Southeast and Arizona. In Florida, apply between Jan.1-Apr. 30 only.

Signal Word/Toxicity Class: Danger-Poison/I.
REI: 48 hr.

Thiodan — see endosulfan

Thiolux — see sulfur

VENDEX

Selective fenbutatin oxide miticide which can be combined with oils on almonds. Not toxic to insect predators and moderately toxic to western predator mites. Toxic to fish.

Vendex 50WP (fenbutatin-oxide) *(Griffin)*

Rate: 0.5-1.25 lb. ai/A
1-2.5 lb. 50WP/A

Pest: Almonds, walnuts:
European red mite, hickory mite (almonds), Pacific spider mite, twospotted spider mite

Pecans:
hickory mite, pecan mite

Romarks: Almonds, pecans, walnuts. Apply when mites first appear. Do not apply more than 2 times per season or more than 2 lb. ai/A per year.

Days To Harvest: 14.

Signal Word/Toxicity Class: Danger/I.
REI: 48 hr.

Volck Supreme Spray — see petroleum oil

Fungicides

Small Fruits

(Blueberries, Caneberries, Cranberries, Currants, Gooseberries, Grapes, Strawberries)

Abound — see azoxystrobin

Aliette — see fosetyl-aluminum

AZOXYSTROBIN

Flowable broad-spectrum fungicide with preventive, systemic, and acurative properties. Toxic to freshwater and estuarine/marine fish and aquatic invertebrates.

Abound (FL) *(ZENECA)*

Rate: 11-15.4 fl. oz. FL/A

Disease:
black rot, downy mildew, Phomopsis cane spot, Phomopsis leaf spot, powdery mildew

Time: Apply prior to or in early stages of disease development.

NOTICE	The information on these pages is for preliminary planning — not a guide for use. Be sure to follow the manufacturer's directions, notwithstanding information contained here.	For personal protective equipment and EPA registration numbers, please turn to page 39.

Remarks: Grapes. Do not apply more than 2.88 qt./A per yr. or more than 6 applications per yr. Do not apply more than 2 sequential sprays before alternating with a fungicide that has different mode of action. Do not apply through any type of irrigation system or through any ultra-low volume spray system. Refer to label for use directions.
Days To Harvest: 14.

Signal Word/Toxicity Class: Caution.
REI: 12 hr.

BASAMID

Granular soil fumigant. Toxic to fish.

Basamid Granular (dazomet) *(BASF)*
Rate: 6-8 lb./1000 sq ft.
Time: Preplant.
Disease:
soilborne diseases
Remarks: Nonbearing: blackberries, blueberries, cranberries, elderberries, gooseberries, raspberries, currants, grapes, strawberries. Till to desired depth and follow suggested cultural practices. Do not use if soil temperature is below 43°F. Refer to label for use directions.
Days To Harvest: 1 year.

Signal Word/Toxicity Class: Warning/II.
REI: 24 hr.

Basicop — see copper compounds

Bayleton — see triadimefon

BENLATE

Benzimidazole systemic fungicide. Toxic to fish.

Benlate (SP, WP) (benomyl) *(DuPont)*
Rate: Grapes:
0.37-0.75 lb. ai/A
0.75-1.5 lb. SP, WP/A
Disease: East of the Rockies:
bitter rot | black rot | powdery mildew
Time: Apply at foliar emergence; repeat at 14- to 21-day intervals.

Rate: Grapes:
0.5-0.75 lb. ai/A
1-1.5 lb. SP, WP/A
Disease:
Anthracnose | Botrytis bunch rot | Isariopsis leaf spot
Time: Botrytis bunch rot: Apply at first bloom; repeat 14-days later if severe disease conditions persist. Anthracnose, Isariopsis leaf spot: Apply at 4-10" shoot growth; repeat at 10- to 14-day intervals. After harvest, apply to vines at 4-week intervals.

Rate: Grapes:
3.2 oz. SP, WP/gal.
Disease:
Eutypa dieback
Time: Paint or spray on immediately after pruning, before rain, dew, and spores come in contact with fresh wood.

Rate: Blueberries:
0.5 lb. ai/A
1 lb. SP, WP/A
Disease:
Anthracnose leaf spot | Botrytis blossom blight | mummy berry
Time: Anthracnose: Apply when disease first appears; repeat 14-days later. After harvest make 2 applications at 14-day intervals as needed. For Botrytis or mummy berry: Apply at green tip; repeat at 7- to 10-day intervals through petal fall.

Rate: Caneberries:
0.37 lb. ai/A
0.75 lb. SP, WP/A
Disease:
Botrytis blossom blight | powdery mildew
Time: Apply at 5-10% bloom and full bloom; repeat at 14-day intervals.

Rate: Currants:
0.37 lb. ai/A
0.75 lb. SP, WP/A
Disease:
powdery mildew
Time: Apply at early bloom and full bloom; repeat at 7- to 14-day intervals.

Rate: Strawberries:
0.25-0.5 lb. ai/A
0.5-1 lb. SP, WP/A
Disease:
Anthracnose | leaf scorch | powdery mildew
leaf blight | leaf spot
Time: Apply at 10% bloom and full bloom; repeat at 10- to 14-day intervals using 0.5 lb./A. Anthracnose: Apply when plants are established; repeat at 7-day intervals.
Remarks: Blueberries, caneberries (blackberries, boysenberries, dewberries, loganberries, raspberries), currants, grapes, strawberries. Do not tank mix or alternate with benzimidazole or thiophanate products. Refer to label for use directions.
Days To Harvest: Strawberries-1; caneberries-3; blueberries, currants-21; grapes-50 (except Eutypa dieback).
State Restrictions: Benlate SP not registered for use in California.

Signal Word/Toxicity Class: Caution.
REI: 24 hr.

Blue Viking — see copper compounds

Botran — see dicloran

Botran 6 Sulfur 25 — see dicloran premix

Bravo — see chlorothalonil

Brom-O-Gas — see methyl bromide

BT 320 SULFUR 25 OR 50 DUST (Premix)

Microbial insecticide/fungicide containing Bacillus thuringiensis and sulfur.

Bt 320 Sulfur 25 Dust (0.064% Bt & 25% sulfur) or Bt 320 Sulfur 50 Dust (0.064% Bt & 50% sulfur) *(Wilbur-Ellis)*
Rate: 20-50 lb. Dust/A
Disease:
powdery mildew
Time: Apply at 7- to 10-day intervals.
Remarks: Grapes. Refer to label for use directions.

Signal Word/Toxicity Class: Caution.
REI: 24 hr.

CALCIUM POLYSULFIDE

Balanced fungicide, insecticide, miticide composition containing special effective wetting agents.

Lime-Sulfur Solution *(Best Sulfur)*
Rate: Blueberries:
5-6 gal./100-150 gal. water
Disease:
Phomopsis cane | *Sclerotinia urnula* | twig blight
Time: Apply delayed dormant stage after leaf buds begin to break. For *Sclerotinia urnula*, apply postharvest or early dormant using 8 gal./100 gal. water.

Rate: Caneberries (blackberries, boysenberries, raspberries):
6-12 gal./100 gal. water
Disease:
Anthracnose | powdery mildew | yellow rust
cane blight | rust
Time: Dormant or delayed dormant. For spur blight, apply before buds break and show silver. For yellow rust, apply delayed dormant.

◆ - **new product** • ai-active ingredient • DF-dry flowable • E/EC-emulsifiable concentrate • EW-emulsion in water
F/FL-flowable • G-granule • L/LC-liquid • SP/WSP-soluble packet • W/WP-wettable powder • WSB-water soluble bag

Rate: Currants, gooseberries:
3-20 pt./100 gal. water
Grapes:
1 pt./in sufficient water (mature foliage)
4 pt./in sufficient water (new shoots)

Disease:
Anthracnose powdery mildew

Time: For currants and gooseberries, apply at bud break; repeat 10- to 15-days after initial application. Apply 3 pt. at 10-day intervals after second spray if needed. For grapes, apply during growing season when new shoots are 4-6" long.

Rate: Grapes:
4-10 gal./A

Disease:
powdery mildew

Time: Dormant.

Remarks: Blueberries, caneberries (blackberries, boysenberries, raspberries), currants, gooseberries, grapes. For grapes, do not apply after 5 a.m. or during the day when maximum temperatures are expected to exceed 85°F or sulfur burn may occur. For mature foliage, mix with 4 lb. wettable sulfur. During late growing season may be mixed with wettable sulfur and Sulforix Foliage spray. Do not apply this product through any type of irrigation system. Refer to label for use directions.

- - - - - - - - - - - - - - - - -

Signal Word/Toxicity Class: Danger/I.
REI: 48 hr.

Sulforix *(Best Sulfur)*

Rate: 3 gal./100 gal. water

Disease:
Anthracnose leaf spot yellow rust
cane blight

Time: Delayed dormant spray.

Remarks: Caneberries (blackberries, boysenberries, raspberries). Do not apply this product through any type of irrigation system. Refer to label for use directions.

- - - - - - - - - - - - - - - - -

Rate: 1 pt./100 gal. water

Disease:
powdery mildew

Time: Begin applications as early shoot growth occurs.

Remarks: Grapes. Mix with 4 lb. wettable sulfur. Do not apply this product through any type of irrigation system. Refer to label for use directions.

- - - - - - - - - - - - - - - - -

Signal Word/Toxicity Class: Danger/I.
REI: 48 hr.

CAPTAN

Toxic to fish.

Drexel Captan (4L, 50W, 80W) *(Drexel)*

Rate: Grapes-Eastern U.S.:
4 pt. 4L/A
2-4 lb. 50W/A
2.5 lb. 80W/A

Disease:
black rot downy mildew leaf spot (50W)
cane spot (50W)

Time: Apply before bloom, after bloom; repeat at 10- to 14-day intervals depending on disease conditions.

Rate: 2-4 pt. 4L/100 gal. water
1.25-2.5 lb. 80W/A

Disease:
dead-arm

Time: Apply when shoots are 1/2-1" long; repeat when shoots are 4-6" long.

Rate: Grapes-California:
2 pt. 4L/A
4 lb. 50W/A
2.5 lb. 80W/A

Disease:
bunch rot

Time: Make 2 applications before bloom and 1 just after bloom; repeat periodically making 3 cover applications before bunches close.

Rate: 4 pt. 4L/100 gal. water
3-4 lb. 50W/A
2-2.5 lb. 80W/A

Disease:
cane spot (50W) dead-arm (4L, 80W) leaf spot (50W)

Time: Apply first spray when green tissue begins to show but before shoots are 1" long; second spray when shoots are 6-8" long.

Rate: Blueberries:
Eastern:
5 lb. 50W/A
3.125 lb. 80W/A
Western:
2-5 lb. 50W/A
1.5 lb. 80W/A

Disease:
berry rot gray mold mummy berry

Time: Eastern: Apply when buds swell or when buds have loose scales; repeat at 7-day intervals through blossom period, continue at 7- to 10-day intervals (7- to 20-day intervals for 80W) from late bloom. Western: Apply midbloom; repeat at 7- to 10-day intervals until maturity.

Rate: Strawberries:
3 qt. 4L/A
3-6 lb. 50W/A
3.75 lb. 80W/A

Disease:
botrytis fruit rot (4L) gray mold (50W, 80W) leaf spot (50W, 80W)

Time: Apply when new growth starts in spring before fruit starts to form; repeat at weekly intervals up to harvest.

Remarks: Blueberries, grapes, strawberries. Do not apply through any type of irrigation system. Refer to label for use directions.

- - - - - - - - - - - - - - - - -

Signal Word/Toxicity Class: Danger/I.
REI: 4 days; 24 hr. (strawberries-50W).

Micro Flo Captan (50WP, 80WP) or Captec (4L) *(Micro Flo)*

Rate: Grapes:
2-4 lb. 50WP/A
1.25-2.5 lb. 80WP/A
0.75-1 qt. 4L/100 gal.

Disease:
black rot (suppression) cane spot leaf spot
bunch rot (CA) downy mildew

Time: U.S.: Apply when shoots are 1/2-1 1/2" long, 3-5" long, and 9-12" long; repeat just before bloom, immediately after bloom and continue at 10- to 14-day intervals as needed. California: Make 2 applications before bloom, 1 immediately after bloom, and 3 cover applications before bunches close. For cane and leaf spot: Apply when green tissue begins to show; repeat when shoots are 6-8" long.

Rate: Blueberries:
Eastern U.S.
5 lb. 50WP/A
3.125 lb. 80WP/A
0.75-1 qt. 4L/100 gal.
Western U.S.
2-5 lb. 50WP/A
1.25-3.125 lb. 80WP/A
0.75-1 qt. 4L/100 gal. (excluding CA)

Disease:
berry rot Botrytis gray mold mummy berry

Time: Eastern U.S.: Apply when buds swell or when buds have loose scales; repeat at 7-day intervals through blossom period; 7- to 10-day intervals from late bloom. Western U.S.: Apply at midbloom; repeat at 7- to 10-day intervals until maturity.

Rate: Strawberries:
3-6 lb. 50WP/A
1.88-3.75 lb. 80WP/A
1.5-3 qt. 4L/100 gal. (1.5-2.5 qt. for CA)

Disease:
Botrytis gray mold leaf spot

Time: Apply when new growth starts in spring and before fruit starts to form; repeat at 7- to 14-day intervals. Under conditions favorable for fruit rot, continue applications through harvest, treating immediately after each picking.

Remarks: Blueberries, grapes, strawberries. Do not apply this product through any type of irrigation system. Refer to label for use directions.

- - - - - - - - - - - - - - - - -

Signal Word/Toxicity Class: Danger/I.
REI: 96 hr.; strawberries: 24 hr.

Champ — see copper compounds

8 Fruits & Nuts

NOTICE	The information on these pages is for preliminary planning — not a guide for use. Be sure to follow the manufacturer's directions, notwithstanding information contained here.	For personal protective equipment and EPA registration numbers, please turn to page 39.

Champion — see copper compounds

CHLOROTHALONIL

Broad-spectrum contact fungicide. Toxic to fish, aquatic invertebrates, and marine/estuarine organisms.

Bravo 500, 720/Weather Stik, Ultrex (WDG) *(ZENECA)*
Echo 500, 720 or 90DF *(Sostram)*
Terranil (6L, 90DF or 90DF WSP) *(Terra)*

Rate: 2-3 lb. ai/A
3-4 pt. 720/A
2.7-3.6 lb. WDG/A

Disease:
Anthracnose mummy berry

Time: Apply at bud break; repeat through bloom at 7-day intervals using low rate, 10-day intervals using high rate. Do not apply after full bloom.
Remarks: Blueberries. Do not use more than 15 pt./A during growing season. Refer to label for use directions.
Days To Harvest: 42.
State Restrictions: SLN-Connecticut, Maine, New Jersey, Oregon, Washington (Bravo 720, WDG); SLN-New York, Vermont (Bravo 720); SLN-Michigan (Bravo 720, Weather Stik); SLN-Washington (Weather Stik).

- - - - - - - - - - - - - - - -

Rate: 3-5 lb. ai/A
6-10 pt. 500/A
4-7 pt. 720, 6L/A
3.8-6.3 lb. WDG/A
3.5-5.75 lb. 90DF, 90DF WSP/A

Disease:
fruit rot leaf blight twig blight

Time: Apply at early bloom; repeat at 10- to 14-day intervals. Under severe disease conditions, use high rate on 10-day schedule.
Remarks: Cranberries. Do not use more than 3 times per season. Do not apply to bogs when flooded or allow release of irrigation water from bogs for at least 3 days following application. Can be applied through sprinkler irrigation equipment (solid set systems only); refer to label for use directions.
Days To Harvest: 50.
State Restrictions: SLN-Massachusetts (Bravo 720, WDG), Wisconsin (Bravo 720).

- - - - - - - - - - - - - - - -

Rate: 1-1.4 lb. ai/A
2-2.75 lb. 500/A
1.375-3 pt. 720/A

Disease:
Ramularia leaf spot

Time: Apply when conditions favor leaf spot development, usually following rainy weather or sprinkler irrigation; repeat at 10- to 14-day intervals until runners are dug.
Remarks: Nonbearing strawberries. Can be applied through sprinkler irrigation equipment; refer to label for use directions.
State Restrictions: SLN-California (Bravo 500); SLN-Washington (Bravo 720).

- - - - - - - - - - - - - - - -

Signal Word/Toxicity Class: Danger/I (90DF, 90DF WSP); Warning/II (6L, 500, 720, WDG).
REI: 48 hr.

Clean Crop C-O-C-S — see copper compounds

COPPER COMPOUNDS

Fungicide/bactericide.

Basicop (WP) (basic copper sulfate) *(Griffin)*

Rate: Grapes:
1 lb. ai/100 gal. water
2 lb. WP/100 gal. water

Disease:
black rot downy mildew

Time: Apply when first leaves appear; repeat just before blossom opens, after petal fall, and every 2 weeks up to 2 weeks before harvest.

Rate: Caneberries:
1 lb. ai/100 gal. water
2 lb. WP/100 gal. water

Disease:
Anthracnose leaf spot yellow rust
cane spot

Time: Apply when leaf buds begin to open; repeat when flower buds show white at 10- to 14-day intervals.

Rate: Strawberries:
1-1.5 lb. ai/A
2-3 lb. WP/A

Disease:
leaf spot

Time: Apply at prebloom and postharvest.

Rate: Currants, gooseberries:
3 lb. ai/100 gal. water
6 lb. WP/100 gal. water

Disease:
Anthracnose leaf spot

Time: Make 3 applications; at full bloom, 2 weeks later, and after harvest.
Remarks: Caneberries (blackberries, boysenberries, dewberries, loganberries, raspberries) currants, gooseberries, grapes, strawberries. For strawberries, use high rate when disease is heavy. Do not apply through any type of irrigation system. Refer to label for use directions.

- - - - - - - - - - - - - - - -

Signal Word/Toxicity Class: Warning/II.
REI: 24 hr.

Blue Viking or Tennessee Brand (copper sulfate pentahydrate) *(Griffin)*

Rate: 4-8 lb./100 gal.

Disease:
powdery mildew

Time: Dormant: Apply in spring before bud swell, and before any green tissue is present.

Rate: 6 lb. + 2 lb. hydrated lime/100 gal.

Disease:
downy mildew

Time: Apply at first sign of disease; repeat as necessary.
Remarks: Grapes. Thoroughly wet vine, especially bark of trunk, head or cordons. Do not apply through any type of irrigation system. Refer to label for use directions.
State Restrictions: Not for use on downy mildew in California.

- - - - - - - - - - - - - - - -

Signal Word/Toxicity Class: Danger/I.
REI: 24 hr.

Champ (FL or Formula 2) or Champion (WP) (copper hydroxide) *(Agtrol)*

Rate: Grapes:
2.66 pt. FL/A
1.33-2.66 pt. Formula 2/A
2 lb. WP/A

Disease:
black rot downy mildew powdery mildew

Time: Apply at bud break; repeat through season depending on disease severity.

Rate: Blackberries, raspberries:
5.33 pt. FL/A
2.66 pt. Formula 2/A
4 lb. WP/A

Disease:
cane spot leaf spot

Time: Delayed dormant spray after training in spring, fall after harvest.

Rate: Blueberries:
0.66-1.33 gal. FL/A
0.33-0.66 pt. Formula 2/A
4-8 lb. WP/A

Disease:
bacterial canker

Time: Make 2 applications. First before fall rains; second 4 weeks later.

Rate: Cranberries:
1.33 gal. FL/A
0.66 gal. Formula 2/A
8 lb. WP/A

Disease:
dieback (Formula 2) fruit rot

Time: Dieback: Apply prebloom; second application 10- to 14-days later if necessary. Fruit rot: Make first application late bloom; 1-2 additional applications at 10- to 14-day intervals depending on disease severity.

Rate: Currants, gooseberries:
1.66 gal. FL/A
0.83 gal. Formula 2/A
10 lb. WP/A

◆ - **new product** • ai-active ingredient • DF-dry flowable • E/EC-emulsifiable concentrate • EW-emulsion in water
F/FL-flowable • G-granule • L/LC-liquid • SP/WSP-soluble packet • W/WP-wettable powder • WSB-water soluble bag

Disease:
leaf spot

Time: Make 3 applications. First after harvest, second before bloom, third after petal fall.

Rate: Strawberries:
2.66-4 pt. FL/A
1.33-2 pt. Formula 2/A
2-3 lb. WP/A

Disease:
leaf blight — leaf spot

Time: Apply when plants are established; repeat at weekly intervals throughout season. Discontinue if signs of crop injury appear.

Remarks: Blackberries (Auroras, Boysens, Cascades, Chehalems, Marions, Santiams, Thornless Evergreens), blueberries, cranberries, currants, gooseberries, grapes, raspberries, strawberries. On grapes, foliage injury may occur on copper-sensitive varieties such as Concord, Delaware, Niagara, and Rosette; either test for sensitivity or add hydrated lime. For blackberries and raspberries, add 1 qt. crop oil/A. Refer to label for use directions.

State Restrictions: Not for use on blueberries, raspberries in California.

- - - - - - - - - - - - - - - -

Signal Word/Toxicity Class: Danger/I (Champion); Caution (Champ).
REI: 24 hr.

Clean Crop C-O-C-S (WDG) (74.8% copper oxychloride & 14.2% copper sulfate, basic) *(Platte)*

Rate: Grapes:
1 lb. ai/A
2 lb. WDG/A

Disease:
black rot — downy mildew — powdery mildew

Time: For black rot, apply when buds break; repeat when shoots are 10" long. For mildew, apply at bud break; repeat as needed.

Rate: Caneberries:
2-3 lb. ai/A
4-6 lb. WDG/A

Disease:
cane spot — leaf spot — yellow rust

Time: For cane and leaf spot, apply when new laterals are leafing out; repeat when flowers begin to open. For yellow rust, apply when disease first threatens; repeat at 7- to 10-day intervals.

Remarks: Caneberries (blackberries, boysenberries, dewberries, loganberries, raspberries, youngberries), grapes. Postharvest spray after pruning is beneficial for caneberries. Where danger of copper injury is severe, add hydrated lime. Do not apply this product through any type of irrigation system. Refer to label for use directions.

- - - - - - - - - - - - - - - -

Signal Word/Toxicity Class: Warning/II.
REI: 24 hr.

Copper-Count-N (L) (copper ammonium) *(Mineral Research)*

Rate: Grapes:
2 qt. + 4 lb. hydrated lime L/A

Disease:
Anthracnose — downy mildew — powdery mildew
black rot

Time: Apply as last 1-2 late summer applications following early-season application of another fungicide.

Rate: Blackberries:
4 qt. L/A

Disease:
cane spot — leaf spot

Time: Apply delayed dormant spray after training in spring, again in late spring (2 qt./A), and make fall application after harvest (4 qt./A).

Rate: Cranberries:
8 qt. L/A

Disease:
fruit rot

Time: Apply late bloom; repeat 1-2 applications at 10- to 14-day intervals.

Rate: Strawberries:
1.5-2 qt. L/A

Disease:
leaf spot — scorch

Time: Apply at 7- to 10-day intervals from time new growth starts until harvest.

Remarks: Blackberries (Auroras, Boysens, Cascades, Chehalems, Logans, Marions, Santians, Thornless Evergreens), cranberries, grapes, strawberries. On grapes, foliage injury may occur on copper-sensitive varieties such as Concord, Delaware, Niagara, and Rossetti. Applications may be made by ground or air; refer to label for use directions. Do not apply through any type of irrigation system unless supplemental labeling on chemigation is followed.

State Restrictions: Not for use on grapes, blackberries in California.

- - - - - - - - - - - - - - - -

Signal Word/Toxicity Class: Caution.
REI: 12 hr.

Drexel Basic Copper Sulfate (WP) or KOP 300 (3FL) (basic copper sulfate) *(Drexel)*
TOP COP Tri-Basic (FL) (basic copper sulfate) *(Stoller)*

Rate: Grapes:
4-5 lb. WP/100 gal. water
2.66-4 pt. 3FL/A
2-3 pt. FL/A

Disease:
Anthracnose — black rot — downy mildew

Time: Apply when new growth is 1/2"; repeat at 10- to 14-day intervals.

Rate: Caneberries:
2 lb. WP/100 gal. water
2.66 pt. 3FL/100 gal. water
2-3 pt. FL/l00 gal. spray

Disease:
Anthracnose — leaf spot — yellow rust
cane spot

Time: Apply when leaf buds begin to open; repeat when flower buds show white at 10- to 14-day intervals. Postharvest spray after pruning, but before fall rains.

Rate: Strawberries:
2-3 lb. WP/100 gal. water
2.66-4 pt. 3FL/100 gal. water
4 pt. FL/100 gal. water

Disease:
leaf spot

Time: Apply prebloom and postbloom.

Remarks: Caneberries (blackberries, boysenberries, dewberries, loganberries, raspberries), grapes, strawberries. Do not apply WP or 3FL through any type of irrigation system. Refer to label for use directions.

- - - - - - - - - - - - - - - -

Signal Word/Toxicity Class: Warning/II (WP, 3FL); Caution (FL).
REI: 24 hr. (WP, FL); 12 hr. (3FL).

Kocide (101, DF, 2000, 4.5LF, LF) (copper hydroxide) *(Griffin)*

Rate: Grapes:
2 lb. 101, DF/A
1.5 lb. 2000/A
1.33 pt. 4.5LF/A
2.66 pt. LF/A

Disease:
black rot — downy mildew — powdery mildew

Time: Apply late dormant up to bud break with subsequent applications throughout season depending upon disease severity.

Rate: Blackberries, raspberries:
4 lb. 101, DF/A
5.33 pt. LF/A

Disease:
cane spot — leaf spot

Time: Apply delayed dormant spray after training in the spring; make fall application after harvest.

Rate: Blackberries, raspberries:
1.5-3 lb. 2000/A
1.33-2.66 pt. 4.5LF/A

Disease:
Anthracnose — leaf spot — yellow rust
cane spot — Pseudomonas blight (high rate)

Time: Low rate: Apply when leaf buds begin to open; repeat when flower buds show white. High rate: Apply delayed dormant spray after training in spring; make fall application after harvest.

Rate: Blueberries:
3.5 lb. 101/A
4.5 lb. 2000/A
4 pt. 4.5LF/A
6 lb. DF/A

NOTICE	The information on these pages is for preliminary planning — not a guide for use. Be sure to follow the manufacturer's directions, notwithstanding information contained here.	For personal protective equipment and EPA registration numbers, please turn to page 39.

Disease:
bacterial canker
Time: Make first application before fall rains; second application 4 weeks later.

Rate: Cranberries:
8 lb. 101, DF/A
10.66 pt. LF/A
Disease:
bacterial stem canker (DF) fruit rot
Time: Bacterial stem canker: Apply postharvest and again in spring before bud burst. Fruit rot: Make first application in late bloom. Make 1-2 additional applications at 10- to 14-day intervals depending on disease severity.

Rate: Cranberries:
6 lb. 2000/A
5.33 pt. 4.5LF/A
Disease:
bacterial stem canker, leaf blight, rose bloom, fruit rot, leaf spot
Time: Bacterial stem canker: Apply postharvest and again in spring at bud swell. Fruit rot: Make first application in late bloom. Rose bloom: Apply 3 sprays as soon as symptoms are observed. Make 1-2 additional applications at 10- to 14-day intervals. Leaf blight, red leaf spot, stem blight, tip blight: Apply delayed dormant in spring; repeat at 10- to 14-day intervals through prebloom.

Rate: Currants, gooseberries:
10 lb. 101, DF/A
7.5 lb. 2000/A
6.66 pt. 4.5LF/A
13.33 pt. LF/A
Disease:
Anthracnose (2000, 4.5LF) leaf spot
Time: Anthracnose: Apply after first leaves have expanded; continue on a 10- to 14-day schedule during wet conditions in spring. Make additional applications after harvest. Leaf spot: Apply after harvest, before bloom, after petal fall.

Rate: Strawberries:
2-3 lb. 101, DF/A
1.5-2.75 lb. 2000/A
1.33-2 pt. 4.5LF/A
2.66-4 pt. LF/A
Disease:
Angular leaf spot (4.5LF), leaf blight, leaf scorch (4.5LF), leaf spot
Time: Apply when plants are established and continue on weekly schedule throughout season. Discontinue if signs of crop injury appear.
Remarks: Blackberries (Auroras, Boysens, Cascades, Chehalems, Marions, Santiams, Thornless Evergreens), blueberries, cranberries, currants, gooseberries, grapes, raspberries, strawberries. On grapes, foliage injury may occur on copper-sensitive varieties such as Concord, Delaware, Niagara, and Rosette; either test for sensitivity or add hydrated lime. For blackberries and raspberries, add 1 qt. crop oil/A. Refer to label for use directions.
State Restrictions: Not for use on blueberries in California.

- - - - - - - - - - - - - - - -

Signal Word/Toxicity Class: Danger/I (101, DF,); Warning/II (2000); Caution (4.5LF, LF).
REI: 24 hr.

KOP-Hydroxide (FL or 50WP) (copper hydroxide) *(Drexel)*

Rate: Grapes:
2.66 pt. FL/A
2 lb. 50WP/A
Disease:
black rot, downy mildew, powdery mildew
Time: FL: Apply as late summer application. WP: Apply at bud break; repeat through season as needed.

Rate: Blackberries:
5.33 gal. FL/100 gal. water
Blackberries, raspberries:
4 lb. 50WP/A
Disease:
cane spot, leaf spot
Time: Apply delayed dormant after training in spring, and as full spray application after harvest.

Rate: Cranberries:
10.66 gal. FL/A
8 lb. 50WP/A
Disease:
fruit rot
Time: Apply in late bloom; 1-2 additional applications at 10- to 14-day intervals may be needed in severe cases.

Rate: Currants, gooseberries:
13.33 gal. FL/100 gal. water
10 lb. 50WP/A
Disease:
leaf spot
Time: Apply after harvest, second before bloom, third at petal fall.

Rate: Strawberries:
2.66-4 pt. FL/100 gal. water
2-3 lb. 50WP/A
Disease:
leaf blight, leaf spot
Time: Apply when plants are established; repeat on weekly intervals throughout season.
Remarks: Blackberries (Auroras, Boysens, Cascades, Cheha-lems, Logans, Marions, Santiams, Thornless Evergreen), cranberries, currants, gooseberries, grapes, strawberries. Slight to severe foliage injury may occur on copper sensitive grape varieties such as Concord, Delaware, Niagara, and Rosette. Do not apply through any type of irrigation system. KOP-Hydroxide 50 should not be applied in spray solution having pH less than 6.5 or greater than 9.0 as phytotoxicity may occur. Refer to label for use directions.
State Restrictions: KOP-Hydroxide 50WP not for use in California on raspberries.

- - - - - - - - - - - - - - - -

Signal Word/Toxicity Class: Danger/I.
REI: 48 hr.

Micro Flo Basic Copper "53" (basic copper sulfate) or Microsperse COC (53WP) (copper oxychloride) *(Micro Flo)*

Rate: Grapes:
4-5 lb. 53/A
Disease:
Anthracnose, black rot (suppression), downy mildew
Time: Apply when new growth is 1/2" long; repeat at 10- to 14-day intervals as needed.

Rate: Caneberries:
4-5 lb. 53, 53WP/A
Disease:
Anthracnose, leaf spot, yellow rust, cane spot
Time: Apply when leaf buds begin to open, repeat when flower buds show white. For Anthracnose, repeat at 10- to 14-day intervals. Postharvest spray after pruning, but before fall rains begin, use 12-15 lb./A combined with spreader/sticker.

Rate: Currants, gooseberries:
6 lb. 53/100 gal.
Disease:
Anthracnose, leaf spot
Time: Apply full bloom, 2 weeks later, after harvest.

Rate: Strawberries:
2-3 lb. 53, 53WP/100 gal.
Disease:
downy mildew, leaf spot
Time: Apply after leaves form; repeat at 10- to 14-day intervals.
Remarks: Caneberries (blackberries, boysenberries, dewberries, loganberries, raspberries), currants, gooseberries, grapes, strawberries. On grapes, foliage injury may occur on copper-sensitive varieties such as Concord, Delaware, Niagara, and Rosette. Refer to label for use directions.

- - - - - - - - - - - - - - - -

Signal Word/Toxicity Class: Danger/I (53WP); Warning/II (53).
REI: 24 hr.

Nu-Cop (3L, 50DF, 50WP) (copper hydroxide) *(Micro Flo)*

Rate: Grapes:
1.33-2.66 pt. 3L/A
2 lb. 50DF, 50WP/A + 1-3 lb. hydrated lime
Disease:
black rot, downy mildew, powdery mildew
Time: 50DF, 50WP: Apply as the last one or two late summer applications following early-season application of another fungicide. 3L: Apply at bud break with additional applications throughout rainy season.

Rate: Blackberries:
2.66-5.33 pt. 3L/100 gal. + 1 qt. superior-type oil
4 lb. 50DF, 50WP/100 gal. + 1 qt. superior-type oil
Raspberries:
2.66-5.33 pt. 3L/A + 1 qt. crop oil
4 lb. 50DF, 50WP/A + 1 qt. crop oil

◆ - **new product** • ai-active ingredient • DF-dry flowable • E/EC-emulsifiable concentrate • EW-emulsion in water
F/FL-flowable • G-granule • L/LC-liquid • SP/WSP-soluble packet • W/WP-wettable powder • WSB-water soluble bag

Disease:
cane spot leaf spot
Time: Blackberries: Apply delayed dormant, late spring, fall after harvest. Raspberries: Apply delayed dormant, fall after harvest.

Rate: Blueberries:
2.66-5.33 pt. 3L/A
Disease:
bacterial canker
Time: Apply before fall rains; second 4 weeks later.

Rate: Cranberries:
5.33-10.66 pt. 3L/A
8 lb. 50DF, 50WP/A
Disease:
fruit rot
Time: Apply late bloom (mid-July); 2 additional applications at 10- to 14-day intervals.

Rate: Currants, gooseberries:
6.66-10.66 pt. 3L/A
10 lb. 50DF, 50WP/A
Disease:
Anthracnose (3L) leaf spot
Time: Apply after harvest, before bloom, after petal fall.

Rate: Strawberries:
1.33-4 pt. 3L/100 gal.
2-3 lb. 50DF, 50WP/100 gal.
Disease:
downy mildew leaf blight leaf spot
Time: Apply when plants are established; continue weekly throughout season.
Remarks: Blackberries, blueberries, cranberries, currants, gooseberries, grapes, raspberries, strawberries. On grapes, slight to severe foliage injury may occur on copper sensitive varieties. Discontinue if signs of phytotoxicity appears on strawberries. Refer to label for use directions.
State Restrictions: Not for use on raspberries in California.

- - - - - - - - - - - - - - - - -

Signal Word/Toxicity Class: Danger/I (50DF, 50WP); Warning/II (3L).
REI: 48 hr. (50DF, 50WP); 24 hr. (3L).

Copper-Count-N — see copper compounds

DICLORAN
DICLORAN (Premix)

Protective fungicide. Toxic to fish.

Botran 6 Dust *(Wilbur-Ellis)*
Botran 6 Sulfur 25 Dust (dicloran & sulfur) *(Wilbur-Ellis)*
Rate: 30 lb. Dust/A
Disease:
Botrytis bunch rot powdery mildew storage rot
Time: Apply while clusters are still loose; repeat at 2-week intervals.
Remarks: Grapes. Refer to label for use directions.

- - - - - - - - - - - - - - - - -

Signal Word/Toxicity Class: Warning/II (6 Sulfur 25 Dust); Caution.
REI: 12 hr.

Botran 75-W *(Gowan)*
Rate: 1.5-3.5 lb. ai/A
2-4.66 lb. 75-W/A
Disease:
Botrytis rot bunch rot storage rot
Botrytis stem rot
Time: Apply when disease appears; repeat at 7-day intervals as needed.
Remarks: Grapes. Up to 5.33 lb./A may be applied per season. Refer to label for use directions.
State Restrictions: For use west of the Rockies only.

- - - - - - - - - - - - - - - - -

Signal Word/Toxicity Class: Caution.
REI: 12 hr.

Dithane — see mancozeb

DODINE

Protective fungicide in a wettable powder formulation recommended for the control of major fungal diseases in fruits and nuts. Not compatible with nonionic surfactants.

Syllit (65W) *(Platte)*
Rate: 1-1.3 lb. ai/A
1.5-2 lb. 65W/A
Disease:
leaf blight leaf scorch leaf spot
Time: Apply as soon as new growth starts; repeat at weekly intervals.
Remarks: Strawberries. Refer to label for use directions.
Days To Harvest: 14.

- - - - - - - - - - - - - - - - -

Signal Word/Toxicity Class: Danger/I.
REI: 48 hr.

Echo — see chlorothalonil

FENARIMOL

Triazole sterol inhibitor systemic fungicide. Toxic to fish.

Rubigan E.C. *(Dow AgroSciences)*
Rate: 2-6 fl. oz. E.C./A
Disease:
powdery mildew
Time: Apply 2-3 fl. oz./A 2 weeks prior to bloom, 4 fl. oz./A at full bloom, 6 fl. oz./A at first and second postbloom; repeat at 14- to 18-day intervals.
Remarks: Grapes. Apply with ground equipment only. Do not apply more than 6 fl. oz./A per application or more than 19 fl. oz. per season. Can be tank mixed with protectant fungicide; refer to label. Do not apply through any type of irrigation system. Do not graze cover crops or harvest forage for silage or hay.
Days To Harvest: 30.

- - - - - - - - - - - - - - - - -

Signal Word/Toxicity Class: Warning/II.
REI: 12 hr.

FERBAM

Broad-spectrum fungicide. Toxic to fish.

Ferbam Granuflo (WDG) *(UBC Agrochemicals)*
Rate: Grapes:
1.5 lb. ai/100 gal. water
2 lb. WDG/100 gal. water
Disease:
black rot
Time: Apply just before bloom, after bloom, and as disease conditions warrant.

Rate: Cranberries:
1.5 lb. ai/100 gal. water
2 lb. WDG/100 gal. water
Disease:
fruit rot Lophodermium twig blight
Time: Apply early blossoming period; repeat at 2-week intervals for total of 5 applications. Do not apply within 28 days after mid-bloom.

Rate: 6.8 lb. ai/100 gal. water
9 lb. WDG/100 gal. water (1 gal. mixture/1 ft. sq. area)
Disease:
fairy ring
Time: Apply in fall immediately after harvest.
Remarks: Cranberries, grapes. Do not make more than 3 applications per season on grapes. Do not use water from treated cranberry bogs for irrigating other crops. For fairy ring, treat area 3 ft. beyond advancing line of dead vines and 2 ft. within line. When used in combination with sulfur, do not apply within 10-14 days of oil spray or just prior to or during periods of excessively high temperatures. Refer to label for use directions.
Days To Harvest: Grapes-7.

- - - - - - - - - - - - - - - - -

Signal Word/Toxicity Class: Caution.
REI: 24 hr.

NOTICE The information on these pages is for preliminary planning — not a guide for use. Be sure to follow the manufacturer's directions, notwithstanding information contained here.	For personal protective equipment and EPA registration numbers, please turn to page 39.

FOSETYL-ALUMINUM

Systemic fungicide/bactericide. Toxic to estuarine invertebrates.

Aliette (WDG, WSP) *(Rhone-Poulenc)*

Rate: 4 lb. ai/A
5 lb. WDG, WSP/A

Disease:
Alternaria fruit rot (suppression) | Anthracnose fruit rot | Phomopsis canker (suppression) | Phytophthora root rot

Time: Apply in spring at pink bud stage; repeat at 14- to 21-day intervals.

Remarks: Blueberries. Apply in sufficient water to thoroughly wet foliage. Do not exceed 4 applications per year. Do not apply in less than 10 gal./A or exceed 20 lb./A per year. Refer to label for use directions.

Days To Harvest: 0.

State Restrictions: Not registered in California.

- - - - - - - - - - - - - - - - - -

Rate: 4 lb. ai/A
5 lb. WDG, WSP/A

Disease:
Phytophthora root rot

Time: West of the Rockies: For fall application, apply when conditions favor disease development; repeat in 3-4 weeks if necessary. For spring application, apply first spray after bud break (1-3" new growth); second spray 3-4 weeks later. East of the Rockies: Apply in spring after bud break (1-3" new growth); repeat on 45-60 day schedule.

Remarks: Caneberries (blackberries, loganberries, raspberries). Do not exceed 4 applications per year. The last fall application should be applied at least 30 days prior to leaf drop. Refer to label for use directions.

Days To Harvest: 60.

- - - - - - - - - - - - - - - - - -

Rate: Dip:
2 lb. ai/100 gal.
2.5 lb. WDG/100 gal.
0.8 lb. ai/40 gal.
1 lb. WSP/40 gal.
Foliar:
2-4 lb. ai/A
2.5-5 lb. WDG, WSP/A

Disease:
leather rot | red stele

Time: Leather rot: Apply between 10% bloom and early fruit set; repeat at 7- to 14-day intervals as long as conditions are favorable for disease development. Red stele: For dip, apply preplant to roots and crowns for 15-30 minutes; for foliar on annuals, apply 14-21 days after planting; for perennials, apply in spring when plants start active growth. Repeat at 30- to 60-day intervals if disease conditions continue.

Remarks: Strawberries. Do not apply by fixed wing or rotary aircraft. Do not exceed 30 lb./A per season. Refer to label for use directions.

Days To Harvest: 0.

State Restrictions: In certain counties of Arkansas, Illinois, Indiana, Iowa, Kentucky, Louisiana, Maryland, Minnesota, Mississippi, Missouri, North Carolina, New Hampshire, Ohio, Wisconsin, rate is limited to maximum 3.75 lb. product/A; refer to label for specific counties.

- - - - - - - - - - - - - - - - - -

Signal Word/Toxicity Class: Caution.
REI: 12 hr.

GALLEX (Premix)

Ready-to-use paint-on emulsion.

Gallex (2,4-xylenol & meta-cresol) *(AgBioChem)*

Rate: Apply undiluted.

Disease:
crown gall

Time: Apply first application before breaking of dormancy or during spring and early summer.

Remarks: Blueberries, caneberries (blackberries, boysenberries, raspberries, youngberries), grapes. Paint gall and cut surfaces + 1/2" surrounding healthy bark. Generally 1 application will give control; repeat if needed. Do not dilute, mix, or combine with other pesticides. Complete girdling of trunk by galls affects water-conducting capacity of tree. To prevent sudden loss of water movement, do not treat more than half of galled trunk at any one time. Treat remaining diseased tissue 4-6 months later when examination of initial treatment is made.

- - - - - - - - - - - - - - - - - -

Signal Word/Toxicity Class: Warning/II.

IPRODIONE

Dicarboximide fungicide. Toxic to aquatic invertebrates.

Rovral (4FL, WG, WP) *(Rhone-Poulenc)*

Rate: 0.5-1 lb. ai/A
1-2 pt. 4FL/A
1-2 lb. WG, WP/A

Disease: Grapes:
bunch rot

Time: Apply early to midbloom, prior to bunch closing, beginning of fruit ripening, and prior to harvest as needed.

Disease: Blueberries, caneberries, currants, gooseberries, huckleberries:
Botrytis fruit rot

Time: Apply at early bloom and again at full bloom; 2 additional applications can be made at 14-day intervals if needed.

Rate: Strawberries:
Foliar:
0.75-1 lb. ai/A
1.5-2 pt. 4FL/A
1.5-2 lb. WG, WP/A
Dip:
1 lb. ai/100 gal.
2 pt. 4FL/100 gal.
2 lb. WG, WP/100 gal.

Disease: Foliar:
Anthracnose | leaf spot | stem end rot
gray mold | Phomopsis soft rot

Dip:
Botrytis crown rot

Time: Foliar: Apply no later than 10% bloom; repeat at 10- to 14-day intervals. Dip: Apply in solution for 1-5 minutes and plant immediately.

Remarks: Blueberries, caneberries (blackberries, loganberries, raspberries), currants, gooseberries, grapes, huckleberries, strawberries. For grapes, do not graze treated orchards or feed cover crops grown in treated orchards to livestock. Under severe disease conditions, use high rate. Maximum of 4 applications per season (1 application for dip treatment on strawberries). Refer to label for use directions.

Days To Harvest: 0; grapes-7.

State Restrictions: Chemigation may be made except in New York.

- - - - - - - - - - - - - - - - - -

Signal Word/Toxicity Class: Warning/II (WG); Caution (4FL, WP).
REI: 12 hr.

Kocide — see copper compounds

KOP-Hydroxide — see copper compounds

Kumulus — see sulfur

MANCOZEB

Carbamate fungicide with broad-spectrum activity. Coordination product of zinc ion and manganese ethylene bisdithiocarbamate. Toxic to fish.

Dithane (DF, F-45, M-45, WSP) *(Rohm and Haas)*
Manzate 200DF *(Griffin)*
Penncozeb 75DF or 80WP *(Elf Atochem)*

Rate: East of the Rockies:
1.12-3 lb. ai/A
1.5-4 lb. DF, M-45, WSP, 200DF/A, 75DF
1.2-3.2 lb. ai/A
2.4-6.4 pt. F-45/A
1.5-4 lb. 80WP/A
West of the Rockies:
1.12-1.9 lb. ai/A
1.5-2.5 lb. DF, M-45, WSP, 200DF/A
1.2-2 lb. ai/A
2.4-4 pt. F-45/A
1.1-1.9 lb. ai/A
1.5-2.5 lb. 75DF/A
1.2-2 lb. ai/A
1.5-2.5 lb. 80WP/A

Disease:
black rot dead-arm downy mildew
bunch rot

Time: Apply when new shoots are 1/2-1 1/2" long, 3-5" long, 8-10" long; repeat at 7- to 10-day intervals until fruit is set.
Remarks: Grapes. In California, do not apply after bloom. Apply in sufficient water to provide thorough coverage. For late season control of black rot, dead-arm, and downy mildew, use of other approved and recommended fungicides is suggested. Dithane: Do not apply more than 19.2 lb. ai/A per season east of the Mississippi; 6 lb. ai/A west of the Mississippi. Refer to label for use directions.
Days To Harvest: 66.

Rate: 2.25-4.5 lb. ai/A
3-6 lb. DF, M-45, WSP, 200DF, 75DF/A
2.4-4.8 lb. ai/A
4.8-9.6 pt. F-45/A
3-6 lb. 80WP/A
Disease:
fruit rot
Time: Apply early bloom; repeat at 7- to 10-day intervals as required.
Remarks: Cranberries. Refer to label for use directions.
Days To Harvest: 30.
State Restrictions: Refer to label for California precautions.

Signal Word/Toxicity Class: Caution.
REI: 24 hr.

MANEB

Dithiocarbamate contact fungicide. Toxic to fish.

Maneb 75DF or 80WP *(Elf Atochem)*
Manex (FL) *(Griffin)*
Rate: East of the Rockies:
1.1-3 lb. ai/A
1.5-4 lb. 75DF/A
1.2-3.2 lb. ai/A
1.5-4 lb. 80WP/A
1.2-3.2 qt. FL/A
West of the Rockies:
1.1-1.9 lb. ai/A
1.5-2.5 lb. 75DF/A
1.2-2 lb. ai/A
1.5-2.5 lb. 80WP/A
1.2-2 qt. FL/A
Disease: Grapes:
black rot downy mildew Phomopsis
bunch rot
Time: Apply when shoots are 1/2 to 1 1/2"; repeat at 7- to 10-day intervals.

Rate: 3.6-4.5 lb. ai/A
4.8-6 lb. 75DF/A
3.8-4.8 lb. ai/A
4.8-6 lb. 80WP/A
2.8-4.8 lb. ai/A
2.8-4.8 qt. FL/A
Disease: Cranberries:
fruit rot
Time: Begin applications at mid-bloom; repeat at 14- to 21-day intervals (7- to 10-day for 80WP and FL).
Remarks: Cranberries, grapes. Refer to label for use directions.
Days To Harvest: Cranberries-30; grapes-66.
State Restrictions: Do not apply after bloom on grapes in California. For Manex, do not apply more than 19.2 lb. ai/A per season east of the Rockies; 6 lb. ai/A per season west of the Rockies.

Signal Word/Toxicity Class: Caution.
REI: 24 hr.

MANEB (Premix)

Fungicide/bactericide. Toxic to fish, birds, and other wildlife.

ManKocide (DF) (15% maneb & 46.1% copper hydroxide) *(Griffin)*
Rate: Grapes:
2.5 lb. DF/A
Disease:
black rot powdery mildew Phomopsis
downy mildew
Time: Apply late dormant or bud break; repeat when shoots are 1/2-1 1/2", 3-5", and 8-10" long, then at 7- to 10-day intervals until fruit is set.

Rate: Cranberries:
10.5 lb. DF/A
Disease:
fruit rot
Time: Apply mid-bloom; repeat at 7- to 10-day intervals.
Remarks: Cranberries, grapes. Foliage injury may occur on copper sensitive grape varieties such as Concord, Delaware, Niagara, and Rosette, either treat for sensitivity or add 1-3 lb. of hydrated lime per lb. of DF. On grapes, do not exceed 128 lb./A per season east of the Rockies, 40 lb./A per season west of the Rockies. On cranberries, do not exceed 96 lb./A per season. Refer to label for use directions.
Days To Harvest: Cranberries-30; grapes-66. Do not apply after bloom in California.

Signal Word/Toxicity Class: Danger/I.
REI: 48 hr.

Manex — see maneb

ManKocide — see maneb premix

Manzate — see mancozeb

MEFENOXAM

Systemic fungicide having a specific mode of action.

Ridomil Gold EC, GR, or WSP *(Novartis)*
Rate: Blueberries:
Established plantings:
0.125 lb. ai/1000 ft. lin. row
0.25 pt. EC/1000 lin. ft. row
0.11 lb. ai/1000 lin. ft. row
0.25 lb. WSP/1000 lin. ft. row
New plantings:
1.8 lb. ai/A
3.6 pt. EC/A
1.6 lb. ai/A
3.6 lb. WSP/A
Disease:
Phytophthora root rot
Time: Established plantings: Apply before plant starts to grow in spring. New plantings: Apply at planting. Additional application may be made to coincide with periods favorable for disease development.
Remarks: For banded application, 18" band recommended. On new plantings, do not apply more than 0.9 gal. EC/A broadcast during 12 months before bearing harvestable fruit. Do not plant any crop which is not registered for use with mefenoxam in treated soil for 12 months.

Rate: Cranberries:
0.5-0.875 ai/A
1-1.75 pt. EC/A
20-35 lb. GR/A
0.45-0.675 lb. ai/A
1-1.5 lb. WSP/A
Disease:
Phytophthora root rot
Time: Make 3 applications; in fall after harvest, in spring, and at least 45 days before harvest.
Remarks: Do not apply by air. Use in minimum 20 gal. water. Do not apply more than 2.6 lb. ai/A per growing season. Do not plant any crop which is not registered for use with mefenoxam in treated soil for 12 months.
Days To Harvest: 45.

Rate: Raspberries:
0.125 lb. ai/1000 ft. lin. row
0.25 pt. EC/1000 lin. ft. row
5 lb. GR/1000 lin. ft. row
0.11 lb. ai/1000 lin. ft. row
0.25 lb. WSP/1000 lin. ft. row
Disease:
Phytophthora root rot
Time: Apply in spring and fall.
Remarks: Apply as soil surface spray. Do not plant any crop which is not registered for use with mefenoxam in treated soil for 12 months.
Days To Harvest: 45.

For personal protective equipment and EPA registration numbers, please turn to page 39.

Rate: Strawberries:
Established and new plantings:
0.5 lb. ai/A
1 pt. EC/A
0.45 lb. ai/A
1 lb. WSP/A

Disease:
leather rot red stele

Time: Established plantings: Apply in spring after ground thaws and before first bloom; second application may be made after harvest in fall. New plantings: Apply at transplant; repeat 30 days before harvest.

Remarks: Do not apply more than 1.5 qt. EC/A or 3 lb. WSP/A per year. For low annual rainfall areas, surface applications need to be moved into root zone by rainfall, overhead irrigation, or mechanical incorporation. Do not plant any crop which is not registered for use with mefenoxam in treated soil for 12 months.

- - - - - - - - - - - - - - - - -

Signal Word/Toxicity Class: Warning/II (EC); Caution (GR, WSP).
REI: 48 hr.

MEFENOXAM (Premix)

Systemic fungicide having a specific mode of action. Packaged in Accu-Pak, a wettable powder in a water-soluble bag. Toxic to fish and aquatic organisms.

Ridomil Gold/Copper (WSB) (5% mefenoxam & 60% copper hydroxide) *(Novartis)*

Rate: 1.3 lb. ai/A
2 lb. WSB/A

Disease:
downy mildew

Time: Make up to 4 applications beginning before bloom.

Remarks: Grapes. Do not apply more than 0.4 lb. ai/A. For late season downy mildew, use other registered fungicides. Do not use on copper sensitive grape varieties. Do not plant any crop which is not registered for use with mefenoxam in treated soil for 12 months.

Days To Harvest: 66.

- - - - - - - - - - - - - - - - -

Signal Word/Toxicity Class: Danger/I.
REI: 48 hr.

METHYL BROMIDE

2% Chloropicrin added as warning agent. Restricted Use Pesticide. Classified as an ozone depleter under the Clean Air Act. Use is to be completely eliminated starting Jan. 1, 2001.

Brom-O-Gas *(Great Lakes Chemical)*

Rate: Injected with chisels + covered with tarp:
180-240 lb./A
Raised tarp:
0.4-0.55 1-2 lb./100 sq. ft.

Disease:
Fusarium
oak root fungus
Phytophthora
Plasmodiophora
Pythium
Rhizoctonia
Sclerotinia

Time: Preplant.

Remarks: Strawberries. Do not fumigate when soil temperature is below 50°F. For best results, fumigate when temperature is 60°-80°F at depth of 6". Refer to label for use directions.

State Restrictions: Brom-O-Gas not for use in Arizona and not for use in California for oak root fungus (Armillaria root rot) control.

- - - - - - - - - - - - - - - - -

Signal Word/Toxicity Class: Danger/I.

Micro Flo — see copper compounds

Microsperse COC — see copper compounds

Microthiol Special — see sulfur

M-PEDE

Effective contact insecticide, miticide, and fungicide. Toxic to aquatic invertebrates.

M-Pede (postassium salts of fatty acids) *(Mycogen)*

Rate: 1-2% v/v solution

Disease:
powdery mildew

Time: Apply at bud break; repeat at 7- to 10-day intervals.

Remarks: Grapes. May be tank mixed. Complete coverage of clusters as well as leaf surfaces essential for effective control. Do not apply through any type of irrigation system. Refer to label for use directions.

- - - - - - - - - - - - - - - - -

Signal Word/Toxicity Class: Warning/II.
REI: 12 hr.

MYCLOBUTANIL

Triazole sterol inhibitor. Systemic, protectant, and curative fungicide.

Nova or Rally (40W) *(Rohm and Haas)*

Rate: 1.2-2 oz. ai/A
3-5 oz. 40W/A

Disease:
Anthracnose black rot powdery mildew

Time: Apply when new shoots are 1-3"; repeat at 14-day intervals. For powdery mildew, apply prebloom (12-18" shoots); repeat at 21-day intervals.

Remarks: Grapes. Use high rate and shorter interval on susceptible varieties or under heavy disease pressure. Do not apply more than 1.5 lb./A per yr. Refer to label for use directions.

Days To Harvest: 14.

State Restrictions: Nova: Not for use in Alaska, Arizona, California, Hawaii, Idaho, Montana, Nevada, Oregon, Utah, Washington, Wyoming. Rally: For use only in Alaska, Arizona, California, Hawaii, Idaho, Montana, Nevada, Oregon, Utah, Washington, Wyoming; not for anthracnose in Arizona, California, Hawaii.

- - - - - - - - - - - - - - - - -

Signal Word/Toxicity Class: Nova: Caution; Rally: Warning/II or Caution (varies by region).
REI: 24 hr.

Nova — see myclobutanil

Nu-Cop — see copper compounds

Penncozeb — see mancozeb

Rally — see myclobutanil

Redball — see sulfur

Ridomil Gold — see mefenoxam

Ridomil Gold/Copper — see mefenoxam premix

RIDOMIL GOLD MZ (Premix)

Premix containing the active ingredients of Dithane and Manzate. Packaged in Accu-Pak, a wettable powder in a water-soluble bag. Toxic to fish.

Ridomil Gold MZ (WP) (4% mefenoxam & 64% mancozeb) *(Novartis)*

Rate: 1.7 lb. ai/A
2.5 lb. WP/A

Disease:
downy mildew

◆ - **new product** • ai-active ingredient • DF-dry flowable • E/EC-emulsifiable concentrate • EW-emulsion in water
F/FL-flowable • G-granule • L/LC-liquid • SP/WSP-soluble packet • W/WP-wettable powder • WSB-water soluble bag

Time: Make up to 4 applications beginning before bloom.
Remarks: Grapes. Do not apply more than 0.4 lb. ai/A. For late season downy mildew, use other registered fungicides. Do not use on copper sensitive grape varieties. Do not plant any crop which is not registered for use with mefenoxam in treated soil for 12 months.
Days To Harvest: 66.

- - - - - - - - - - - - - - - - -

Signal Word/Toxicity Class: Caution.

REI: 48 hr.

Ronilan — see vinclozolin

Rovral — see iprodione

Rubigan — see fenarimol

Sulfa 80 — see sulfur

Sulforix — see calcium polysulfide

SULFUR

Nonsystemic fungicide with protective action. Microthiol Special, Sulfa 80, and Thiolux are micronized sulfurs.

Drexel Suffa (6FL, 8FL), Sulfa 80 (80DF) or Sulfur (90W) *(Drexel)*

Rate: Grapes:
1.5-4.5 lb. ai/A
2-6 pt. 6FL/A (ground)
4.8 lb. ai/A
6 lb. 80DF/A
1.8-4.5 lb. ai/A
2-5 lb. 90W/A (ground)
California only:
0.75-1.5 lb. ai/100 gal. water
0.75-1.5 pt. 8FL/100 gal. water

Time: Apply when shoots are 6-8" long; repeat at 7- to 10-day intervals. For Sulfa 80, apply late dormant or early shoot period; repeat at 14-day intervals through season.

Rate: Caneberries:
6-13.5 lb. ai/A
8-18 pt. 6FL/A
5.4-13.5 lb. ai/A
6-15 lb. 90W/A
Blueberries, caneberries, currants, gooseberries, huckleberries:
4.8-12 lb. ai/A
6-15 lb. 80DF/A

Time: Apply when shoots are 6" long and before blossoms open; repeat at 7- to 10-day intervals as necessary. For raspberries-80DF: Spray each week starting at first bloom color and continue until all fruit is set; repeat as weather requires.

Rate: Strawberries:
4.5 lb. ai/A
6 pt. 6FL/A
4-8 lb. ai/A
5-10 lb. 80DF/A
1.8-4.5 lb. ai/A
5-10 lb. 90W/A
California only:
0.75 lb. ai/100 gal. water
0.75 pt. 8FL/100 gal. water

Disease:
powdery mildew

Time: Apply at first signs of disease; repeat at 3 week intervals.
Remarks: Blueberries, caneberries (blackberries, boysenberries, dewberries, loganberries, raspberries), currants, gooseberries, grapes, huckleberries, strawberries. For strawberries that will be canned, discontinue well before harvest in accordance with canners' recommendations. Concord and other Labrusca varieties of grapes and strawberries are susceptible to injury from sulfur under certain climatic conditions. Do not use within 2 weeks of oil application (4 weeks FL). Some plants are susceptible to injury from sulfur under certain climatic conditions. Do not make sulfur applications at excessively high temperatures. Do not apply this product through any type of irrigation system. Refer to label for use directions.

- - - - - - - - - - - - - - - - -

Signal Word/Toxicity Class: Caution.
REI: 24 hr.

Dusting Sulfur (DS) or Wettable Sulfur (WS) *(Platte)*

Rate: Grapes:
8-15 lb. DS/A
2-5 lb. WS/100 gal. water
Caneberries, gooseberries, strawberries:
10-20 lb. DS/A
2 lb. WS/100 gal. water

Disease:
powdery mildew

Time: For grapes, apply when shoots are 6-8" long; caneberries, strawberries, when disease first appears. Repeat at 7- to 10-day intervals.
Remarks: Caneberries (blackberries, boysenberries, loganberries, raspberries), gooseberries, grapes, strawberries. Concord and other Labrusca type grapes may be injured. Do not make applications within 2 weeks of oil spray. Refer to label for use directions.

- - - - - - - - - - - - - - - - -

Signal Word/Toxicity Class: Caution.
REI: 24 hr.

Kumulus DF (sulfur) *(BASF)*

Rate: Grapes:
1.6-4 lb. ai/A
2-5 lb. DF/A

Time: Apply in late dormancy.

Rate: Caneberries:
4.8-9.6 lb. ai/A
6-12 lb. DF/A

Time: Apply when new shoots are 6" long and before blossoms open; repeat at 10-day intervals or as necessary.

Rate: Blueberries, currants, gooseberries, huckleberries:
4.8-12 lb. ai/A
6-15 lb. DF/A

Time: Apply when disease first appears; repeat at 7- to 10-day intervals or as necessary.

Rate: Strawberries:
4-8 lb. ai/A
5-10 lb. DF/A

Time: Apply at first sign of disease and repeat as necessary.
Disease:
powdery mildew

Remarks: Blueberries, caneberries (blackberries, boysenberries, dewberris, loganberries, raspberries), currants, gooseberries, grapes, huckleberries, strawberries. Do not apply if temperatures during or within 3 days after application are expected to exceed 90°F. Do not apply through any type of irrigation system. Refer to label for use directions.

- - - - - - - - - - - - - - - - -

Signal Word/Toxicity Class: Caution.
REI: 24 hr.

Liquid Sulfur Six (6L) *(Helena)*
Redball EM-53 (6L) *(International Sulphur)*

Rate: Grapes:
1-2 pt. 6L/100 gal. water
California (Helena):
2-3 qt. 6L/100 gal. water

Disease:
powdery mildew

Time: Apply first application when shoots are 6-8" long; second when shoots are 12-15" long; third at bloom stage.

Rate: Strawberries:
2 pt. 6L/100 gal. water

Disease:
powdery mildew

Time: Apply at first sign of infections; repeat at 3-week intervals.
Remarks: Grapes, strawberries. Timing, rates, and number of applications vary with local conditions. Consult state Agricultural Extension Service for specific recommendations. For strawberries that will be canned, discontinue well before harvest in accordance with local canners' recommendations. Do not apply this product through any type of irrigation system. Refer to label for use directions.

- - - - - - - - - - - - - - - - -

Signal Word/Toxicity Class: Caution.
REI: 24 hr. (Liquid Sulfur Six); 12 hr. (Redball).

8 Fruits & Nuts

NOTICE	The information on these pages is for preliminary planning — not a guide for use. Be sure to follow the manufacturer's directions, notwithstanding information contained here.	For personal protective equipment and EPA registration numbers, please turn to page 39.

Micro Flo Sulfur (6L), Dual Spraying & Dusting Sulfur (DS) or Microsperse Wettable Sulfur (WS) *(Micro Flo)*

Rate: Grapes:
5.3-6 lb. ai/A
7-8 pt. 6L/A
4.5-5.4 lb. ai/A
5-6 lb. DS, WS/A

Disease:
powdery mildew

Time: 6L: Apply throughout season at 7- to 14-day intervals. DS, WS: Apply throughout season at 14-day intervals according to stage of development.

Rate: Caneberries, currants, gooseberries, huckleberries:
15 lb. ai/A
20 pt. 6L/A
13.5 lb. ai/A
15 lb. DS, WS/A

Disease:
powdery mildew

Time: Apply before blossom; repeat at 10-day intervals as necessary. Raspberries: Apply each week from first bloom to fruit set.

Rate: Strawberries:
4-9.75 lb. ai/A
5.33-13 pt. 6L/A
9 lb. ai/A
10 lb. DS, WS/A

Disease:
powdery mildew

Time: Apply at early leaf stage; repeat as necessary.

Remarks: Blueberries, caneberries (blackberries, boysenberries, dewberries, loganberries, raspberries), currants, gooseberries, grapes, huckleberries, strawberries. For grapes, use low rate before flowering; high rate after flowering. Do not apply DS and WS through any type of irrigation system. Refer to label for use directions.

Signal Word/Toxicity Class: Caution.
REI: 24 hr.

Microthiol Special (80WP) *(Elf Atochem)*

Rate: 2.4-8 lb. ai/A
3-10 lb. 80WP/A

Disease:
Phomopsis powdery mildew

Time: Apply throughout the season at 7- to 14-day intervals as needed.

Remarks: Grapes. Use caution on Concord or Labrusca grapes as injury can occur after bloom. Refer to label for use directions.

Rate: 4-8 lb. ai/A
5-10 lb. 80WP/A

Disease:
powdery mildew

Time: Begin application at early leaf stage; repeat as needed.

Remarks: Strawberries. Do not use on sulfur sensitive varieties. Refer to label for use directions.

Rate: 4.8-12 lb. ai/A
6-15 lb. 80WP/A

Disease:
powdery mildew

Time: Begin application before blossoms open; repeat at 10-day intervals as needed. Raspberries: Apply each week from first bloom to fruit set; repeat as weather requires.

Remarks: Blueberries, caneberries (blackberries, boysenberries, dewberries, loganberries, raspberries), currants, gooseberries, huckleberries. Refer to label for use directions.

Signal Word/Toxicity Class: Caution.
REI: 24 hr.

Redball Superfine Wettable Sulphur (WS) or Commercial Dusting Sulphur (DS) *(International Sulphur)*

Rate: 5-10 lb. WS/100 gal. water
5-15 lb. DS/A

Disease:
powdery mildew

Time: Apply when shoots are 6-8" long; repeat at 12-15" long, and bloom stage.

Remarks: Grapes. Sulfur may injure Concord and other New York varieties. Do not apply when temperatures exceed 90°F. Refer to label for use directions.

Rate: 5-10 lb. WS/100 gal. water
10-20 lb. DS/A

Disease:
powdery mildew

Time: Apply at first sign of infection; repeat at 3-week intervals (WS), as necessary (DS).

Remarks: Blackberries, blueberries, currants, raspberries, strawberries. Do not apply when temperatures exceed 90°F. Refer to label for use directions.

Signal Word/Toxicity Class: Caution.
REI: 12 hr.

Terra Sulfur (6L) *(Terra)*

Rate: Grapes:
Before flowering:
7 pt. 6L/A
After flowering:
8 pt. 6L/A

Disease:
Phomopsis powdery mildew

Time: Apply at 7- to 14-day intervals through season as needed.

Rate: Blueberries, caneberries, currants, gooseberries, huckleberries:
20 pt. 6L/A

Disease:
powdery mildew

Time: Apply before bloom; repeat at 10-day intervals as needed. For raspberries, apply each week from first bloom to fruit set.

Rate: Strawberries:
5.33-13 pt. 6L/A

Disease:
powdery mildew

Time: Apply at early leaf stage; repeat as necessary.

Remarks: Blueberries, caneberries (blackberries, boysenberries, dewberries, loganberries), currants, gooseberries, grapes, huckleberries, raspberries, strawberries. Do not use within 2 weeks of oil application. Some plants are susceptible to injury from sulfur under certain climatic conditions. Do not make sulfur applications at excessively high temperatures. Refer to label for use directions.

Signal Word/Toxicity Class: Caution.
REI: 24 hr.

THAT (6FL) or THAT Big 8 (8FL) *(Stoller)*

Rate: 0.75-1.5 lb. ai/100 gal. water
1-2 pt. 6FL/100 gal. water
0.66-1.33 lb. ai/100 gal. water
0.66-1.33 pt. 8FL/100 gal. water

Disease:
powdery mildew

Time: For first application, apply when shoots are 6-8" long; second application when shoots are 12-15" long; third application bloom stage.

Remarks: Grapes. Timing, rate, and number of applications will vary with local conditions; consult state agricultural experiment station. Refer to label for use directions.

Rate: 0.75 lb. ai/A
1 pt. 6FL/A
0.66 lb. ai/A
0.66 pt. 8FL/A

Disease:
powdery mildew

Time: Apply at first sign of infection; repeat at 2-week intervals.

Remarks: Strawberries. For strawberries that will be canned, discontinue sulfur application well before harvest, in accordance with local canners' recommendations. Refer to label for use directions.

Signal Word/Toxicity Class: Caution.
REI: 24 hr.

Thiolux (DF) *(Platte)*

Rate: Grapes:
6 lb. DF/A
Blueberries, currants, dewberries, huckleberries, strawberries:
5-10 lb. DF/A
Blackberries, boysenberries, gooseberries, loganberries, raspberries:
6-15 lb. DF/A

Disease:
powdery mildew

◆ - **new product** • ai-active ingredient • DF-dry flowable • E/EC-emulsifiable concentrate • EW-emulsion in water
F/FL-flowable • G-granule • L/LC-liquid • SP/WSP-soluble packet • W/WP-wettable powder • WSB-water soluble bag

Time: For grapes, apply late dormant, or early shoot period; repeat at 14-day intervals through season. For berries and currants, apply when new shoots are 6" long and before blossoms open or when disease first appears; repeat at 7- to 10-day intervals. For strawberries, apply at first sign of disease; repeat as necessary. For raspberries, apply weekly, starting at first bloom color and continue through fruit set; repeat as weather requires.
Remarks: Blueberries, caneberries (blackberries, boysenberries, dewberries, loganberries, raspberries), currants, gooseberries, grapes, huckleberries, strawberries. Do not use on sulfur sensitive varieties. Do not make applications within 2 weeks of oil spray. Refer to label for use directions.

Signal Word/Toxicity Class: Caution.
REI: 24 hr.

SULFUR (Premix)

Fungicide/bactericide in flowable formulation. Toxic to fish and aquatic organisms.

Top Cop with Sulfur (6.25 lb./gal. sulfur & 1 lb./gal. tribasic copper sulfate) *(Stoller)*
Rate: 2 qt./A
Disease: Grapes:
bunch rot, downy mildew, powdery mildew
Time: Apply as foliar spray when new growth is 1/2" long; repeat at 10- to 14-day intervals.

Disease: Caneberries:
Anthracnose
Time: Apply as foliar spray when leaf buds begin to open; repeat when flower buds show white at 10- to 14-day intervals.

Disease: Cranberries:
fruit rot
Time: Apply as foliar spray at midbloom; repeat at 7- to 10-day intervals.

Disease: Strawberries:
downy mildew
Time: Apply delayed dormant, after leaves form, at 10- to 14-day intervals.
Remarks: Caneberries (boysenberries, dewberries, loganberries, raspberries), cranberries, grapes, strawberries. Refer to label for use directions.

Signal Word/Toxicity Class: Caution.
REI: 24 hr.

Syllit — see dodine

Tennessee Brand — see copper compounds

Terranil — see chlorothalonil

TERR-O-GAS (Premix)

Restricted Use Pesticide. Classified as an ozone depleter under the Clean Air Act. Use is to be completely eliminated starting Jan. 1, 2001.

Terr-O-Gas (67% methyl bromide & 33% chloropicrin) *(Great Lakes Chemical)*
Rate: Nontarp:
270-355 lb./A
Disease:
Fusarium, Plasmodiophora, Rhizoctonia, Phytophthora, Pythium, Sclerotinia
Time: Apply up to 2 weeks prior to planting when soil moisture is good and temperature is 60°-85°F.
Remarks: Strawberries. Work upwind of product and never fumigate without 2 qualified applicators. Do not fumigate when there is little or no air movement. Tarping is not used after field application. Refer to label for use directions.

Signal Word/Toxicity Class: Danger/I.

THAT or THAT Big 8 — see sulfur

Thiolux — see sulfur

THIOPHANATE-METHYL

Benzimidazole systemic fungicide with broad-spectrum control. Toxic to fish.

Topsin M (WSB, 70W) *(Elf Atochem)*
Rate: 0.5-0.7 lb. ai/A
0.75-1 lb. WSB, 70W/A
Disease:
fruit rot, leaf blight, leaf scorch
Time: Begin applications at early bloom; repeat at 7- to 10-day intervals.
Remarks: Strawberries. Use high rate east of the Rockies and under conditions of severe disease pressure. Refer to label for use directions.
Days To Harvest: 1.

Signal Word/Toxicity Class: Caution.
REI: 12 hr.

THIRAM

Broad-spectrum fungicide. Toxic to fish.

Thiram 65 (WP) or Thiram Granuflo (WDG) *(UCB Agrochemicals)*
Rate: Strawberries:
2-2.5 lb. WP/100 gal. spray
1.7-2.2 lb. WDG/100 gal. spray
Disease:
gray mold
Time: Apply at early blossom stage; broadcast 3-5 applications at 10-day intervals per crop cycle.
Remarks: Strawberries. Good coverage of buds, blossoms, and developing fruit required. Do not apply within 3 days of picking. Do not apply through any type of irrigation system. Do not graze or feed cover crops from treated orchards. Refer to label for use directions.
Crop Rotation: Do not rotate treated crops with crops other than those registered for Thiram.

Signal Word/Toxicity Class: Caution.
REI: 24 hr.

TOP COP — see copper compounds

Top Cop with Sulfur — see sulfur premix

Topsin M — see thiophanate-methyl

TRIADIMEFON

Tirazole sterol inhibitor systemic fungicide compatible with many insecticides, miticides, and fungicides.

Bayleton 50% DF *(Bayer)*
Rate: Powdery mildew:
2-6 oz. 50% DF/A
Black rot:
2-3 oz. 50% DF/A (preventive)
3-4 oz. 50% DF/A (curative)
Disease:
black rot, powdery mildew
Time: For powdery mildew, apply prebloom; repeat at 14- to 21-day intervals as needed. For black rot preventive, apply at 10" green shoot; repeat at 7- to 14-day intervals until veraison is complete. Under severe disease conditions and/or on highly susceptible varieties, apply at 7-day intervals using high rate. For black rot curative, apply during bloom within 72 hr. after beginning of infection. Do not make spray applications closer than 7-day intervals; after bloom, continue using preventive rates.
Remarks: Grapes. Do not apply this product through any type of irrigation system. Refer to label for use directions.
Days To Harvest: 14.
Crop Rotation: Small grains, corn, sorghum, soybeans, beans, peas, and cucurbits may be planted 35 days after last application; forage or vines from these crops may not be used for food or feed. Root crops may be planted 120 days after last application but tops must not be used for food or feed. All crops may be planted 12 months or later after last application without restrictions.

For personal protective equipment and EPA registration numbers, please turn to page 39.

- - - - - - - - - - - - - - - - -

Rate: 4 oz./A in 20 gal. water (ground)

Disease:
powdery mildew

Time: Apply when disease symptoms first appear; repeat at 4-6 week intervals as needed.

Remarks: Raspberries. Do not make more than 7 applications per year or more than 2 applications within same 30-day period. Do not apply this product through any type of irrigation system.

Days To Harvest: 1.

Crop Rotation: Small grains, corn, sorghum, soybeans, beans, peas, and cucurbits may be planted 35 days after last application; forage or vines from these crops may not be used for food or feed. Root crops may be planted 120 days after last application but tops must not be used for food or feed. All crops may be planted 12 months or later after last application without restrictions.

State Restrictions: For use in California only.

- - - - - - - - - - - - - - - - -

Signal Word/Toxicity Class: Caution.
REI: 12 hr.

VANGARD

Broad-spectrum fungicide. Toxic to fish and aquatic invertebrates.

Vangard (WG) (cyprodinil) *(Novartis)*

Rate: 10 oz./A
Tank mix:
5-10 oz./A

Disease:
Botrytis bunch rot

Time: Apply early bloom; repeat at berry touch, verasion, or preharvest.

Remarks: Grapes. Apply by ground in sufficient water to provide thorough coverage. Under severe disease pressure, use shorter intervals. May be used alone or in tank mix. Use high rate for severe disease conditions. Do not apply more than 20 oz. WG/A per crop per season. Refer to label for use directions.

Days To Harvest: 7.

- - - - - - - - - - - - - - - - -

Signal Word/Toxicity Class: Caution.
REI: 12 hr.

VINCLOZOLIN

Dicarboximide fungicide. Toxic to fish.

Ronilan DF, EG *(BASF)*

Rate: 0.5-1 lb. ai/A
1-2 lb. DF, EG/A

Disease:
Botrytis fruit rot   gray mold

Time: First application should be made no later than 10% primary bloom; interval between subsequent applications will vary according to weather conditions and disease pressure.

Remarks: Raspberries. Do not apply more than 4 lb. ai/A. Do not use as a plant dip or injury may occur. Do not apply during rain; wait until spray will dry on plants. Do not apply this product through any type of irrigation system. Refer to label for use directions.

Days To Harvest: 9.

Crop Rotation: Lettuce and dry bulb onions may be planted after strawberries; refer to label for other restrictions.

- - - - - - - - - - - - - - - - -

Signal Word/Toxicity Class: Caution.
REI: 12 hr.

ZIRAM

Toxic to fish and wildlife.

Ziram 4L or 76 *(Drexel)*

Rate: Grapes:
0.16 lb. ai/12 gal.
0.33 pt. 4L/12 gal.
0.15 lb. ai/12 gal.
0.2 lb. 76/12 gal.

Disease:
black rot   powdery mildew

Time: Apply before buds swell; repeat after blossoming, but before fruit forms.

Rate: Blueberries:
2 lb. ai/100 gal.
4 pt. 4L/100 gal.
3 lb. 76/100 gal.

Disease:
Botrytis blossom blight   mummy berry

Time: Botrytis: Apply just before bloom, full bloom; repeat 2-3 weeks later. Mummy berry: Apply at loose bud scale stage; repeat 7-days later.

Rate: Cranberries:
2 lb. ai/100 gal.
4 pt. 4L/100 gal.

Disease:
fruit rot

Time: Apply 3 times at 2-week intervals beginning at mid bloom.

Rate: Strawberries:
1-1.5 lb. ai/100 gal.
2-3 pt. 4L/100 gal.
1.1-1.5 lb. ai/100 gal.
1.5-2 lb. 76/100 gal.

Disease:
Botrytis gray mold   leaf spot

Time: Botrytis: Apply early blossom stage; repeat 2-4 times at 5- to 10-day intervals. Leaf spot: Apply when blossom buds first appear; repeat when first blooms open.

Remarks: Blueberries, cranberries, grapes, strawberries. Do not apply this product through any type of irrigation system. Refer to label for use directions.

State Restrictions: Not for use in California.

- - - - - - - - - - - - - - - - -

Signal Word/Toxicity Class: Danger/I.
REI: 48 hr.

Ziram Granuflo (WDG) *(UCB Agrochemicals)*
Ziram 76DF *(Elf Atochem)*

Rate: Blueberries:
2.3 lb. ai/A
3 lb. 76DF/A

Disease:
Anthracnose fruit rot   mummy berry   Phomopsis

Time: Apply at loose bud scale stage; repeat 7-days later. Do not apply later than 3 weeks after full bloom.

Rate: Grapes:
2.3-3 lb. ai/A
3-4 lb. WDG, 76DF/A

Disease:
black rot   downy mildew   Phomopsis
Botrytis bunch rot

Time: Apply before buds swell; repeat after blossoming but before fruit forms.

Remarks: Blueberries, grapes. Do not apply this product through any type of irrigation system. Do not graze or feed cover crops from treated orchards. Refer to label for use directions.

Days To Harvest: 7 (except WDG).

State Restrictions: Not for use on blueberries in California.

- - - - - - - - - - - - - - - - -

Signal Word/Toxicity Class: Danger/I.
REI: 48 hr.

Tree Fruits, Citrus

435 or 455 Oil — see petroleum oil

Aliette — see fosetyl-aluminum

Banner — see propiconazole

Basicop — see copper compounds

◆ - **new product** • ai-active ingredient • DF-dry flowable • E/EC-emulsifiable concentrate • EW-emulsion in water
F/FL-flowable • G-granule • L/LC-liquid • SP/WSP-soluble packet • W/WP-wettable powder • WSB-water soluble bag

BENLATE

Benzimidazole systemic fungicide. Toxic to fish.

Benlate (SP, WP) (benomyl) *(DuPont)*

Rate: 0.5-1 lb. ai/A
1-2 lb. SP, WP/A

Disease:
fruit decay

Rate: 0.75-1.5 lb./A
1.5-3 lb. SP, WP/A

Disease:
greasy spot scab

Time: Scab: Apply once at 2/3 petal fall. Greasy spot: Apply once from mid-June to mid-July. Fruit decay: Make single application from 3 weeks to 2 days before harvest.

Remarks: Do not tank mix or alternate Benlate with benzimidazole or thiophanate products. Do not graze livestock in treated groves. Refer to label for use directions.

State Restrictions: Benlate SP not registered for use in California.

Signal Word/Toxicity Class: Caution.
REI: 24 hr.

Blue Viking — see copper compounds

Break — see propiconazole

Brom-O-Gas — see methyl bromide

CALCIUM POLYSULFIDE

Lime-Sulfur Solution *(Best Sulfur)*

Rate: 3.5 gal./100 gal. water

Disease:
overwintering spores

Time: Dormant.
Remarks: Bearing.

Rate: 2.5 qt./100 gal. water

Disease:
blotch powdery mildew scab

Time: Cluster bud, prebloom.
Remarks: Nonbearing. Refer to label for use directions.

Signal Word/Toxicity Class: Danger/I.
REI: 48 hr.

Champ — see copper compounds

Champion — see copper compounds

Clean Crop C-O-C-S — see copper compounds

COPPER COMPOUNDS

Fungicide/bactericide.

Basicop (WP) (basic copper sulfate) *(Griffin)*
Drexel Basic Copper Sulfate (WP) or KOP 300 (3FL) (basic copper sulfate) *(Drexel)*

Rate: 0.5 lb. ai/100 gal. water
1 lb. WP/100 gal. water
1.33 pt. 3FL/100 gal. water
Florida:
1 lb. ai/100 gal. water
2 lb. WP/100 gal. water

Disease:
brown rot

Time: Apply at first sign of rain or disease; repeat as necessary.

Rate: 0.75-1.5 lb. ai/100 gal. water
1.5-3 lb. WP/100 gal. water
1-2 pt. 3FL/100 gal. water

Disease:
greasy spot pink pitting scab
melanose

Time: For greasy spot and pink pitting, apply as summer spray (mid-July). For melanose, apply 1-3 weeks after petal fall; repeat 4 weeks later if necessary. For scab, apply immediately before flush; repeat at 2/3 petal fall.

Remarks: Do not apply through any type of irrigation system. Refer to label for use directions.

Signal Word/Toxicity Class: Warning/II.
REI: 24 hr. (WP); 12 hr. (3FL).

Blue Viking or Tennessee Brand (copper sulfate pentahydrate) *(Griffin)*

Rate: 10-10-100 Bordeaux spray

Disease:
bacterial blast

Time: Apply late Oct.-early Nov. or before fall rains.

Remarks: Apply 10-25 gal./mature tree. Do not apply through any type of irrigation system. Refer to label for use directions.

Rate: 3-2-6-100 Bordeaux spray

Disease:
brown rot Phytophthora Septoria fruit spot
leaf spot brown rot

Time: Apply Nov.-Dec. before or just after first rains. For severe brown rot, repeat in Jan.-Feb.

Remarks: Lemons, oranges, grapefruit. Apply 10-25 gal./tree to tree skirt, trunk, and ground under tree. Do not apply through any type of irrigation system. Refer to label for use directions.

Signal Word/Toxicity Class: Danger/I.
REI: 24 hr.

Champ (FL or Formula 2) or Champion (WP) (copper hydroxide) *(Agtrol)*

Rate: 0.66-1.33 gal. FL/A
0.33-0.66 gal. Formula 2/A
4-8 lb. WP/A

Disease:
brown rot

Time: Apply in fall to tree skirt to height of at least 4 ft.; also to bare ground 1 ft. beyond skirt; repeat as needed.

Rate: 0.66-2 gal. FL/A
0.33-1 gal. Formula 2/A
4-12 lb. WP/A

Disease:
melanose pink pitting scab

Rate: 0.33-1 gal. FL/A
0.16-0.5 gal. Formula 2/A
2-6 lb. WP/A

Disease:
greasy spot

Time: Apply prebloom and postbloom.

Rate: 2 gal. FL/A
1 gal. Formula 2/A
12 lb. WP/A

Disease:
citrus canker (suppression)

Time: Spray flushes 7-14 days after shoots begin to grow. Young fruit may require additional application. Number and timing of application will depend on disease pressure. Under heavy pressure, each flush of new growth should be sprayed.

Remarks: Grapefruit, lemons, limes, oranges, tangelos, tangerines. Refer to label for use directions.

State Restrictions: California for brown rot: In areas subject to copper injury, add 0.33-1 lb. high quality lime.

Rate: 1.33 pt. FL/1 gal. water
0.66 pt. Formula 2/1 gal. water
1 lb. WP/1 gal. water

Disease:
Phytophthora

Time: Apply in May prior to summer rains and/or in the fall prior to wrapping trees for freeze protection.

NOTICE The information on these pages is for preliminary planning — not a guide for use. Be sure to follow the manufacturer's directions, notwithstanding information contained here.

For personal protective equipment and EPA registration numbers, please turn to page 39.

Remarks: Grapefruit, lemons, limes, oranges, tangelos, tangerines. Paint trunks from soil surface to lowest scaffold limbs. Treatment serves for protection for up to 1 year, but does not cure existing infections. Refer to label for use directions.

- - - - - - - - - - - - - - - - -

Signal Word/Toxicity Class: Danger/I (Champion); Caution (Champ).
REI: 24 hr.

Clean Crop C-O-C-S (WDG) (74.8% copper oxychloride & 14.2% copper sulfate, basic) *(Platte)*

Rate: 2-5 lb. ai/A
4-10 lb. WDG/A
Disease:
brown rot
Time: Apply in fall before or just after first rains. Repeat if necessary.
Remarks: Add 0.33-1 lb. hydrated lime/1 lb. product. Spray ground and skirts of trees. Treat inside as well as outside, covering fruit, trunks, and skirts thoroughly. Do not disturb soil after treatment. Do not apply in areas where copper injury is known to occur. Do not apply this product through any type of irrigation system. Refer to label for use directions.
State Restrictions: For use in Arizona and California.

- - - - - - - - - - - - - - - - -

Rate: 2-6 lb. ai/A
4-12 lb. WDG/A
Disease:
greasy spot melanose scab
Time: Apply as prebloom and postbloom. For melanose and scab, repeat 4 weeks later. Do not apply after fruit is 3/4" in diameter.
Remarks: Citrus citron, grapefruit, kumquats, lemons, limes, oranges, tangelos, tangerines. Do not apply in areas where copper injury is known to occur. Do not apply this product through any type of irrigation system. Refer to label for use directions.
State Restrictions: Not for use in California on melanose, scab, and greasy spot.

- - - - - - - - - - - - - - - - -

Signal Word/Toxicity Class: Warning/II.
REI: 24 hr.

Copper-Count-N (L) (copper ammonium) *(Mineral Research)*

Rate: 2-6 qt. L/A
Disease:
brown rot
Time: Apply in fall before or just after heavy rain.
Remarks: In areas of skirt sprays, apply to a height of 4 ft. Applications may be made by ground or air; refer to label for use directions. Do not apply through any type of irrigation system unless supplemental labeling on chemigation is followed.

- - - - - - - - - - - - - - - - -

Rate: 0.75-2 qt. L/100 gal.
Disease:
greasy spot pink pitting scab
melanose
Time: Apply as prebloom and postbloom sprays.
Remarks: May be used in concentrate sprays at equivalent rates. For aerial application, use 6-8 qt. L/A in 10 gal. Do not apply through any type of irrigation system unless supplemental labeling on chemigation is followed.

- - - - - - - - - - - - - - - - -

Rate: Florida:
8 qt. L/A
Disease:
greasy spot
Time: Apply in June and continue at monthly intervals through Aug.
Remarks: Limes. Applications may be made by ground or air; refer to label for use directions. Do not apply through any type of irrigation system unless supplemental labeling on chemigation is followed.

- - - - - - - - - - - - - - - - -

Signal Word/Toxicity Class: Caution.
REI: 12 hr.

Kocide (101, DF, 20/20, 2000, 4.5LF, LF) (copper hydroxide) *(Griffin)*

Rate: 1 lb. 101, DF, 2000/A
0.66 pt. 4.5LF/A
Disease:
Phytophthora
Time: Apply in May prior to summer rains and/or in the fall prior to wrapping trees for freeze protection.
Remarks: Grapefruit, lemons, limes, oranges, tangelos, tangerines. Paint trunks from soil surface to lowest scaffold limbs. Treatment serves for protection for up to 1 year, but does not cure existing infections. Do not apply in a spray solution having a pH of less than 6.5 as phytotoxicity may occur. Refer to label for use directions.

- - - - - - - - - - - - - - - - -

Rate: 2-6 lb. 101, DF/A
1.5-4.5 lb. 2000/A
1.33-4 pt. 4.5LF/A
2.66-8 pt. LF/A
Disease:
greasy spot pink pitting (DF, 2000, 4.5LF)
Time: Apply in summer on expanded new flush; repeat on subsequent flushes where disease pressure is severe.

Rate: 4-8 lb. 101, DF/A
10-25 lb. 20/20/A
3-6 lb. 2000/A
2.66-5.33 pt. 4.5LF/A
5.33-10.66 pt. LF/A
Disease:
brown rot Septoria spot (20/20, 4.5LF)
Time: Brown rot: Apply in fall to tree skirt to height of at least 4 ft. or to bare ground 1 ft. beyond skirt; repeat as needed. Septoria spot: Apply in fall before or just after first rain; repeat as needed.

Rate: 4-12 lb. 101, DF/A
3-9 lb. 2000/A
2.66-8 pt. 4.5LF/A
6-16 pt. LF/A
Disease:
melanose pink pitting (101, LF) scab
Time: Apply as prebloom and postbloom sprays.

Rate: 8-10 lb. 101/A
6-7.5 lb. 2000/A
5.33-6.66 pt. 4.5LF/A
Disease:
Alternaria brown spot (suppression)
Time: Apply when first spring flush appears and each flush thereafter. After 2/3 petals have fallen, repeat on 21-day schedule.

Rate: 12 lb. 101, DF/A
9 lb. 2000/A
8 pt. 4.5LF/A
16 pt. LF/A
Disease:
citrus canker (suppression only)
Time: Spray flushes 7- to 14-days after shoots begin to grow. Young fruit may require additional application. Number and timing of application will depend on disease pressure. Under heavy pressure, each flush of new growth should be sprayed.

Rate: Field nursery grown:
2 lb. 101/A
3-6 lb. 2000/A
1.33 pt. 4.5LF/A
Disease:
brown rot greasy spot pink pitting
citrus canker (suppression) melanose scab
Time: Apply at 28-day intervals or as needed depending on disease severity.

Rate: Greenhouse and shadehouse:
6 Tbs. 101/1000 sq. ft.
Disease:
brown rot greasy spot pink pitting
citrus canker melanose scab
Time: Apply when disease threatens; repeat at 30-day intervals as needed.
Remarks: Nonbearing grapefruit, lemons, limes, oranges, tangelos, tangerines. Do not use Kocide DF or 4.5LF on citrus seedlings less than 2 yr. old grown in greenhouses or shadehouses. Refer to label for use directions.
State Restrictions: California for brown rot (DF, LF); greasy spot (LF); Alternaria brown spot (101): In areas subject to copper injury, add 0.33-1 lb. high quality lime. Kocide 101 not for use in California on greenhouse and shadehouse crops.

- - - - - - - - - - - - - - - - -

Signal Word/Toxicity Class: Danger/I (101, DF, 20/20); Warning/II (2000); Caution (4.5LF, LF).
REI: 24 hr.

KOP-Hydroxide (FL, 50WP) (copper hydroxide) *(Drexel)*

Rate: 1-2 pt. FL/A
2-6 lb. 50WP/A
Disease:
greasy spot pink pitting scab
melanose
Time: Apply as prebloom and postbloom.

◆ - **new product** • ai-active ingredient • DF-dry flowable • E/EC-emulsifiable concentrate • EW-emulsion in water
F/FL-flowable • G-granule • L/LC-liquid • SP/WSP-soluble packet • W/WP-wettable powder • WSB-water soluble bag

Rate: 0.66-1.33 pt. FL/100 gal. water
4-8 lb. 50WP/A
Disease:
brown rot
Time: Apply in fall before or just after first heavy rains. Apply to skirts of trees to height of at least 4 ft. and to bare ground 1 ft. beyond skirt.

Rate: 1 lb. 50WP/A
Disease:
Phytophthora
Time: Apply in May prior to summer rains and/or in fall prior to wrapping trees for freeze protection by painting tree trunks from soil surface to lowest scaffold limbs.
Remarks: In California, add 0.33-1 lb. lime per 1.33 pt./product in areas subject to copper injury. KOP-Hydroxide 50WP should not be applied in spray solution having pH less than 6.5 or greater than 9.0 as phytotoxicity may occur. Refer to label for use directions.
State Restrictions: KOP-Hydroxide 50WP for use in California only.

Signal Word/Toxicity Class: Danger/I.
REI: 48 hr.

KOP Oxy-85 (WP) (copper oxychloride) *(Drexel)*
Rate: 6 lb. WP/500 gal. water
Disease:
greasy spot melanose
Time: Apply dormant before spring growth appears; repeat 1- to 3-weeks after petal fall. For melanose scab, repeat 4-weeks later.
Remarks: Grapefruit, lemons, limes, oranges, tangelos, tangerines. Refer to label for use directions.

Signal Word/Toxicity Class: Danger/I.
REI: 24 hr.

Micro Flo Basic Copper "53" (basic copper sulfate) *(Micro Flo)*
Rate: 1-3 lb. 53/100 gal.
Florida:
10 lb. 53/500 gal.
Disease:
brown rot
Time: Apply at first appearance; repeat as needed to maintain protection during wet weather.

Rate: 7.5 lb. 53/500 gal.
Disease:
bacterial canker
Time: Spray to point of runoff monthly, or as needed.

Rate: 7-15 lb. 53/500 gal.
Florida:
2 lb. 53/100 gal.
Disease:
Alternaria melanose scab (suppression)
brown spot pink pitting
greasy spot
Time: Alternaria, brown spot: Apply in April; repeat 2-3 times at monthly intervals as needed. Greasy spot, pink pitting: Apply as summer spray (July 15). Melanose: Apply 1-3 weeks after petal fall; repeat 2-4 weeks later if necessary. Scab: Apply just before trees begin to flush; repeat at 2/3 petal fall.

Rate: Northern California:
10-20 lb. 53/500 gal.
Disease:
bacterial blast
Time: Spray entire tree Oct.-Nov. before first rains.
Remarks: Bearing, nonbearing trees. Do not apply to mandarins until after fruit has been picked. Refer to label for use directions.
State Restrictions: "Shot bag" mixes for Florida use only. Not for use in California on bacterial canker.

Signal Word/Toxicity Class: Warning/II.
REI: 24 hr.

Microsperse COC 53WP (copper oxychloride) *(Micro Flo)*
Rate: 7.5 lb. 53WP/A
Disease:
greasy spot
Time: Apply late June-early Aug. For control on summer flushes grown out after July; repeat in Aug. or Sept.

Rate: 15 lb. 53WP/A
Disease: Fresh market:
melanose scab
Time: Apply 2-3 weeks after petal fall; second spray 3-4 weeks later.
Remarks: Do not use 53WP in areas where copper injury is known to have occurred. Refer to label for use directions.

Signal Word/Toxicity Class: Danger/I.
REI: 24 hr.

Nu-Cop (3L, 50DF, 50WP) (copper hydroxide) *(Micro Flo)*
Rate: 0.75-4 pt. 3L/A
2-6 lb. 50DF, 50WP/A
Disease:
greasy spot

Rate: 2.66-5.33 pt. 3L/A
4-8 lb. 50DF, 50WP/A
Disease:
brown rot

Rate: 0.33-1 gal. 3L/A
4-12 lb. 50DF, 50WP/A
Disease:
citrus canker (suppression) melanose scab
pink pitting
Time: Apply prebloom, postbloom. For brown rot, apply in fall and continue as needed to tree height of at least 4 ft. and to bare ground 1 ft. beyond skirt. For citrus canker, apply 7- to 14-days after shoots begin to grow; additional applications may be necessary.
Remarks: Do not use 53WP in areas where copper injury is known to have occurred. Refer to label for use directions.
State Restrictions: California: For areas subject to copper injury, add 0.33-1 lb. high quality lime/1 lb. or gal. product.

Signal Word/Toxicity Class: Danger/I (50DF, 50WP, 53WP); Warning/II (3L).
REI: 48 hr. (50DF, 50WP); 24 hr. (53WP, 3L).

TOP COP Tri-Basic (FL) (basic copper sulfate) *(Stoller)*
Rate: 2-4 gal. FL/A
Disease:
melanose
Time: Apply 1-3 weeks after petal fall; repeat 2-4 weeks later. Where there is history of disease problems, wet spring, late or scattered bloom, apply second spray 4 weeks later.
Remarks: Grapefruit, lemons, limes. Compatible with oil sprays used on citrus. Copper and oil combinations should be avoided after fruit has attained a 3/4" diameter and during hot temperatures. Refer to label for use directions.

Rate: 1-4 gal. FL/A
Disease:
brown rot greasy spot scab
Time: For brown rot, apply when disease occurs or immediately on appearance of affected fruit. For greasy spot, apply during 2 months after expansion of early spring flush and within 1 month after expansion of late spring or other flushes. For greasy spot and pink pitting, make summer spray. For scab, apply just before flush and again at 2/3 petal fall. Under high moisture conditions, spray summer and fall flushes when new growth begins to appear.
Remarks: Oranges, tangelos, tangerines. Do not use on Valencias. If scab is likely to be severe, use double rate in first spay. Compatible with oil sprays used on citrus. Copper and oil combinations should be avoided after fruit has attained 3/4" in diameter and during hot temperatures. Refer to label for use directions.

Signal Word/Toxicity Class: Caution.
REI: 24 hr.

Copper-Count-N — see copper compounds

FERBAM

Broad-spectrum fungicide. Toxic to fish.

Ferbam Granuflo (WDG) *(UBC Agrochemicals)*
Rate: 1.1 lb. ai/100 gal. water
1.5 lb. WDG/100 gal. water
Disease:
Anthracnose scab

NOTICE The information on these pages is for preliminary planning — not a guide for use. Be sure to follow the manufacturer's directions, notwithstanding information contained here.	For personal protective equipment and EPA registration numbers, please turn to page 39.

Time: Apply during prebloom period and when 2/3 petals have fallen. May be applied in late summer or early fall if heavy flush of growth appears. For rough lemon nursery stock, up to 2.6 lb./100 gal. may be used at 7- to 10-day intervals and after heavy rains during growing period.

Remarks: Grapefruit, lemons, limes, oranges, tangerines, tangelos. Do not apply more than 1000 gal. spray/A. When used in combination with sulfur, do not apply within 10-14 days of oil spray or just prior to or during periods of excessively high temperatures. Refer to label for use directions.

Days To Harvest: 0.

- - - - - - - - - - - - - - - - -

Signal Word/Toxicity Class: Caution.
REI: 24 hr.

FOSETYL-ALUMINUM

Systemic fungicide/bactericide. Toxic to estuarine invertebrates.

Aliette (WDG, WSP) *(Rhone-Poulenc)*

Rate: Bearing, nonbearing foliar:
4 lb. ai/A
5 lb. WDG, WSP/A
Nonbearing dip:
2.4 lb. ai/100 gal.
3 lb. WDG, WSP/100 gal.

Disease:
brown rot (bearing) Phytophthora foot rot Phytophthora root rot

Time: Foliar: Apply when conditions favor disease development. Dip: Dip entire root system for 30-60 minutes prior to planting in field.

Remarks: Bearing, nonbearing citrus. Nonbearing trees will not produce marketable fruit for 12 months after last application. Do not exceed 4 applications or 20 lb. ai/A per year. Do not exceed 100 gal./A per application for nonbearing; 500 gal./A for bearing. Do not graze livestock in treated groves or feed forage from groves. Refer to label for use directions.

Days To Harvest: Bearing-30.

- - - - - - - - - - - - - - - - -

Rate: 2-4 lb. ai/5 gal.
2.5-5 lb. WDG, WSP/5 gal.

Disease:
Gummosis Phytophthora foot rot (bearing) Phytophthora root rot

Time: Nursery tree resets and new plantings at time of planting.

Remarks: Bearing, nonbearing citrus. Nonbearing trees will not produce marketable fruit for 12 months after last application. Apply as trunk spray or paint. Apply to trunk lesion in sufficient volume to thoroughly wet entire lesion. If no lesion is present, application should be made from soil line up the trunk approximately 2 ft. Under severe conditions (trunk lesions), higher rate should be used. Do not exceed a total of 4 applications or 20 lb. ai/A per year. Do not graze livestock in treated groves or feed forage from groves. Refer to label for use directions.

- - - - - - - - - - - - - - - - -

Rate: Foliar (California, Florida):
4 lb. ai/A
5 lb. WDG, WSP/A

Disease:
Alternaria brown spot (suppression) (FL) ice-nucleating bacteria (CA)

Time: Alternaria brown spot: Apply in spring at first sign of new growth; 2nd and 3rd applications should follow at 30- to 45-day intervals. Ice-nucleting bacteria: Apply as spray mid-Nov.

Remarks: Bearing citrus. Do not exceed 4 applications or 20 lb. ai/A per year. Do not graze livestock in treated groves or feed forage from groves. Refer to label for use directions.

- - - - - - - - - - - - - - - - -

Signal Word/Toxicity Class: Caution.
REI: 12 hr.

Kocide — see copper compounds

KOP-Hydroxide — see copper compounds

KOP Oxy — see copper compounds

Lime-Sulfur Solution — see calcium polysulfide

MEFENOXAM

Systemic fungicide having a specific mode of action.

Ridomil Gold EC, GR, or WSP *(Novartis)*

Rate: Nurseries (AZ, CA, FL, PR):
Soil Drench:
1-1.5 fl. oz. EC/100 gal. water
1-1.5 oz. WSP/100 gal. water
Soil Surface:
0.5-1 lb. ai/A
2-4 pt. EC/A
1-1.5 lb. ai/A
40-80 lb. GR/A
0.9-1.8 lb. ai/A
2-4 lb. WSP/A

Disease:
citrus foot rot root rot trunk canker

Time: Apply at planting; repeat at 3-month intervals during period when plants are actively growing.

Rate: Citrus Resets or New Plantings (AZ, CA, FL, PR):
Water Ring Drench:
1-1.5 fl. oz. EC/100 gal. water
1-1.5 oz. GR in 5 gal. water/tree
1-1.5 oz. WSP/100 gal. water
Soil Surface (AZ, CA):
0.5-1 lb. ai/A
2-4 pt. EC/A
1-1.5 lb. ai/A
40-80 lb. GR/A
0.9 lb. ai/A
2 lb. WSP/A
Soil Surface (FL, PR):
0.5 lb. ai/A
1 qt. EC/A
1-1.5 lb. ai/A
40-80 lb. GR/A
0.9 lb. ai/A
2 lb. WSP/A

Disease:
citrus foot rot root rot trunk canker

Time: Apply at transplanting; up to 3 additional applications may be made at 3-month intervals or when root flushes occur.

Rate: Established Plantings:
Soil Application (FL, PR):
0.5-1 lb. ai/A
2-4 pt. EC/A
1-1.5 lb. ai/A
40-80 lb. GR/A
0.45 lb. ai/A
1 lb. WSP/A
Soil Surface (AZ, CA):
0.5-1 lb. ai/A
2-4 pt. EC/A
1-3 lb. ai/A
80-120 lb. GR/A
0.9-1.8 lb. ai/A
2-4 lb. WSP/A

Disease:
citrus foot rot root rot trunk canker

Time: For soil applications, 2-3 applications may be made a year. For surface sprays, apply during spring root flush. Make 1 or 2 additional applications at 3-month intervals.

Rate: 0.5 lb. ai/3 gal. water
2 pt. EC/3 gal. water
0.9 lb. ai/3 gal. water
2 lb. WSP/3 gal. water

Disease:
gummosis

Time: Apply as trunk spray up to 3 times per year.

Remarks: Do not use in greenhouse nurseries or apply to bare roots. Do not make trunk and soil applications to same tree in same cropping season. Do not apply more than 1.5 gal. EC/A or 12 lb. WSP/A per year. For soil applications, treatment width should be wide enough to cover root systems of plants. Do not plant any crop which is not registered for use with mefenoxam in treated soil for 12 months.

Days To Harvest: 1.

- - - - - - - - - - - - - - - - -

Signal Word/Toxicity Class: Warning/II (EC); Caution (GR, WSP).
REI: 48 hr.

◆ - **new product** • ai-active ingredient • DF-dry flowable • E/EC-emulsifiable concentrate • EW-emulsion in water
F/FL-flowable • G-granule • L/LC-liquid • SP/WSP-soluble packet • W/WP-wettable powder • WSB-water soluble bag

METHYL BROMIDE

2% Chloropicrin added as warning agent. Restricted Use Pesticide. Classified as an ozone depleter under the Clean Air Act. Use is to be completely eliminated starting Jan. 1, 2001.

Brom-O-Gas *(Great Lakes Chemical)*

Rate: Injected with chisels + covered with tarp:
400-870 lb./A
Nontarp deep injections, auger-probe
1-2 lb./100 ft.
Florida-Injected with chisels + covered with tarp:
1 lb./100 ft.

Disease:

Fusarium	Plasmodiophora	Rhizoctonia
oak root fungus	Pythium	Sclerotinia
Phytophthora		

Time: Preplant.

Remarks: Do not fumigate when soil temperature is below 50°F. For best results, fumigate when temperature is 60°-80°F at depth of 6". Refer to label for use directions.

State Restrictions: Brom-O-Gas not for use in Arizona and not for use in California for oak root fungus (Armillaria root rot) control.

Signal Word/Toxicity Class: Danger/I.

Micro Flo — see copper compounds

Microsperse — see copper compounds

Nu-Cop — see copper compounds

Orbit — see propiconazole

PETROLEUM OIL

Oils recommended for use on Florida citrus. Toxic to fish.

435 or 455 Oil 98.8 *(Drexel)*
SunSpray 6E, 7E, or 9E *(Sun)*

Rate: 1.5 gal. 435, 6E/100 gal. water
1 gal. 455, 7E, 9E/100 gal. water

Disease:
greasy spot

Rate: 0.75 gal. 435, 6E/100 gal. water
0.5 gal. 455, 7E, 9E/100 gal. water

Disease:
sooty mold

Time: Timing of application varies by climatic conditions.

Remarks: Oils applied after Oct. may increase cold damage to trees; applied in Aug. and Sept. may affect solid contents and retard color. Refer to label for use directions.

Signal Word/Toxicity Class: Caution.
REI: 4 hr.

Sumulsion *(Drexel)*

Rate: 1-2 gal./A

Disease:
greasy spot   sooty mold

Time: Timing of application varies by climatic conditions.

Remarks: Oils applied after Oct. may increase cold damage to trees; applied in Aug. and Sept. may affect solid contents and retard color. Refer to label for use directions.

Signal Word/Toxicity Class: Caution.
REI: 4 hr.

SunSpray Ultra-Fine *(Sun)*

Rate: 1-1.3 gal./100 gal. water

Disease:
sooty mold

Remarks: Oranges (Navel, Valencia), other citrus varieties including nonbearing. For low volume treatments in spring, do not apply when temperature exceeds 90°F or when humidity goes down to 20% or below. Oils applied after Oct. may increase cold damage to trees; applied in Aug. and Sept. may affect solid contents and retard color. Refer to label for use directions.

Days To Harvest: 0.

Signal Word/Toxicity Class: Caution.
REI: 4 hr.

PROPICONAZOLE

Triazole sterol inhibitor. Toxic to fish.

Banner MAXX *(Novartis)*
Break or Orbit (EC) *(Novartis)*

Rate: 20-24 fl. oz. MAXX/100 gal. water
6-8 fl. oz. EC/A

Disease:
greasy spot

Time: Apply during June and August.

Remarks: Nonbearing. Apply in sufficient water for thorough coverage. Do not exceed 4 applications per season. Do not mix Break EC with Cyprex. Do not apply this product through any type of irrigation system. Do not graze livestock in treated areas. Break for use in California only. Refer to label for use directions.

Signal Word/Toxicity Class: Warning/II.
REI: 24 hr.

Ridomil Gold — see mefenoxam

SULFUR (Premix)

Fungicide/bactericide in flowable formulation. Toxic to fish and aquatic organisms.

Top Cop with Sulfur (6.25 lb./gal. sulfur & 1 lb./gal. tribasic copper sulfate) *(Stoller)*

Rate: 4 qt./100 gal. water

Disease:
brown rot

Time: Apply in fall, just before or just after first heavy rains.

Rate: 6 qt./100 gal. water

Disease:
melanose   scab

Time: For melanose, apply 1-3 weeks after petal fall; repeat 4 weeks later if necessary. For scab, apply before trees begin to flush and at 2/3 petal fall.

Remarks: Refer to label for use directions.

Signal Word/Toxicity Class: Caution.
REI: 24 hr.

Sumulsion — see petroleum oil

SunSpray — see petroleum oil

Tennessee Brand — see copper compounds

TERR-O-GAS (Premix)

Restricted Use Pesticide. Classified as an ozone depleter under the Clean Air Act. Use is to be completely eliminated starting Jan. 1, 2001.

Terr-O-Gas (67% methyl bromide & 33% chloropicrin) *(Great Lakes Chemical)*

Rate: 1.5-2 lb./100 ft.

Disease:

Fusarium	Plasmodiophora	Rhizoctonia
Phytophthora	Pythium	Sclerotinia

Time: Apply up to 2 weeks prior to planting when soil moisture is good and temperature is 60°-85°F.

NOTICE The information on these pages is for preliminary planning — not a guide for use. Be sure to follow the manufacturer's directions, notwithstanding information contained here.

For personal protective equipment and EPA registration numbers, please turn to page 39.

Remarks: Work upwind of product and never fumigate without 2 qualified applicators. Do not fumigate when there is no or little air movement. Tarping is not used after field application. Refer to label for use directions.
State Restrictions: For use in Florida.

Signal Word/Toxicity Class: Danger/I.
REI: 96 hr.

TOP COP — see copper compounds

Tree Fruits, Deciduous

Abound — see azoxystrobin

Aliette — see fosetyl-aluminum

AZOXYSTROBIN

Flowable broad-spectrum fungicide with preventive, systemic, and acurative properties. Toxic to freshwater and estuarine/marine fish and aquatic invertebrates.

Abound (FL) *(ZENECA)*
Rate: 6.2-9.2 fl. oz. FL/A
Disease:
scab
Time: Apply at petal fall; repeat at 12- to 14-day intervals.
Remarks: Peaches. Applications by ground or air. For heavy disease pressure, use high rate and short interval. Do not apply more than 1.73 qt./A per yr. or more than 6 applications per yr. Refer to label for use directions.
Days To Harvest: 14.

Signal Word/Toxicity Class: Caution.
REI: 12 hr.

Banner — see propiconazole

BASAMID

Granular soil fumigant. Toxic to fish.

Basamid Granular (dazomet) *(BASF)*
Rate: 6-8 lb./1000 sq ft.
Time: Preplant.
Disease:
soilborne diseases
Remarks: Nonbearing: apples, apricots, cherries, nectarines, peaches, pears, plums, prunes. Till to desired depth and follow suggested cultural practices. Do not use if soil temperature is below 43°F.
Days To Harvest: 1 year.

Signal Word/Toxicity Class: Warning/II.
REI: 24 hr.

Basicop — see copper compounds

Bayleton — see triadimefon

BENLATE

Benzimidazole systemic fungicide. Toxic to fish.

Benlate (SP, WP) (benomyl) *(DuPont)*
Rate: Apples:
0.19-0.375 lb. ai/A
0.375-0.75 lb. SP, WP/A
Disease:
black rot, fly speck, powdery mildew, scab, sooty blotch

Rate: Apples:
0.19 lb. ai/100 gal.
0.375 lb. SP, WP/100 gal.
Disease:
fruit rot

Rate: Pears:
0.375-0.75 ai/A
0.75-1.5 lb. SP, WP/A
Disease:
powdery mildew, scab

Rate: Pears:
0.125-0.19 lb. ai/100 gal. water
0.25-0.375 lb. SP, WP/100 gal. water
Disease:
fly speck, fruit rot, sooty blotch
Time: Scab, powdery mildew: Apply 1/2" green tip; repeat at 7- to 14-day intervals. Fly speck, sooty blotch: Apply at petal fall or when disease threatens; repeat at 14- to 21-day intervals. Fruit rot: Apply once 2-3 weeks before harvest.
Remarks: Apples, pears. Do not tank mix or alternate with benzimidazole or thiophanate products. Do not graze livestock in treated orchards. Refer to label for use directions.
Days To Harvest: 14.
State Restrictions: Benlate SP not registered for use in California.

Rate: Apricots, cherries, nectarines, peaches, plums, prunes:
East of Rockies:
0.5-1 lb. ai/A
1-2 lb. SP, WP/A
West of Rockies:
0.75-1 lb. ai/A
1.5-2 lb. SP, WP/A
Disease:
brown rot blossom blight, cherry leaf spot, fruit brown rot, peach scab, powdery mildew
Time: Brown rot blossom blight, peach scab, powdery mildew: Apply early bloom; repeat at full bloom or 10 days later. Fruit brown rot: Apply 3- to 21-days before harvest. Cherry leaf spot: Apply at 10- to 14-day intervals; make one application 2-3 weeks after harvest.

Rate: Plums, prunes:
0.375-0.75 lb. ai/A
0.75-1.5 lb. SP, WP/A
Disease:
black knot
Time: Apply at early bloom; repeat at 7- to 10-day intervals through mid-June.
Remarks: Apricots, cherries, nectarines, peaches, plums, prunes. Does not control peach leaf curl, shot hole, bacterial blast, or fruit rots caused by Rhizopus and Alternaria. Do not graze livestock in treated orchards. Refer to label for use directions.
Days To Harvest: 3 (except Cytospora canker).
State Restrictions: Benlate SP not registered for use in California.

Signal Word/Toxicity Class: Caution.
REI: 24 hr.

Blue Viking — see copper compounds

Botran — see dicloran

Botran 6 Sulfur 25 — see dicloran premix

Bravo — see chlorothalonil

Bravo S — see sulfur premixes

Break — see propiconazole

◆ - **new product** • ai-active ingredient • DF-dry flowable • E/EC-emulsifiable concentrate • EW-emulsion in water
F/FL-flowable • G-granule • L/LC-liquid • SP/WSP-soluble packet • W/WP-wettable powder • WSB-water soluble bag

CALCIUM POLYSULFIDE

Balanced fungicide, insecticide, miticide composition containing special effective wetting agents.

Lime-Sulfur Solution *(Best Sulfur)*

Rate: Apples:
Bearing:
3.5 gal./100 gal. water
Disease:
overwintering spores
Time: Dormant.

Rate: Apples:
Nonbearing:
2.5 qt./100 gal. water
Disease:
blotch powdery mildew scab
Time: Cluster bud, prebloom.

Rate: Apples:
1.5-10 gal./100 gal. water
Disease:
apple scab, powdery mildew, scab, blotch, leaf spot
Time: Apply delayed dormant; repeat in prepink and pink stages; blotch, apply prebloom, petal fall; repeat in 10-days.

Rate: Quince:
1 gal./100 gal. water
Disease:
scab
Time: Apply at 10-day intervals during primary infection period.

Rate: Pears:
2-11 gal./100 gal. water
Disease:
pear scab powdery mildew scab
Time: For pear scab, apply dormant, delayed dormant; powdery mildew and scab, apply at cracked bud, cluster bud, finger bud, green tip to finger bud, or petal fall.
Remarks: Apples, pears, quince. Use on Delicious apples may cause injury. Do not use on d'Anjou, Comice, or Seckle pear varieties after delayed dormant. For pears, do not apply oil after lime-sulfur sprays or if oil has been applied after delayed dormant. Allow minimum of 10 days between delayed dormant oil application and later lime-sulfur application; 30 days between oil and lime-sulfur sprays in the growing season. Do not apply lime-sulfur or oils to trees in weakened condition. Refer to label for use directions and precautions.

Rate: Sweet cherries:
7.5 gal./100 gal. water
Disease:
powdery mildew
Time: Delayed dormant.

Rate: Tart cherries:
1.5 gal./100 gal. water
Disease:
blossom blight, leaf spot, scab, brown rot
Time: Apply up to 3 sprays, prebloom, early bloom, full bloom.

Rate: Tart cherries:
0.5 gal./100 gal. water
Disease:
brown rot leaf spot powdery mildew
Time: Late season. Apply 3-5 times at weekly intervals until harvest.

Rate: Nectarines:
2-24 qt./100 gal. water
Disease:
brown rot, leaf spot, shot hole, leaf curl, powdery mildew
Time: For growing season, apply 3-5 times at weekly intervals until harvest. For leaf curl, apply early winter, late dormant.

Rate: Peaches:
0.5-12 gal./100 gal. water
Disease:
brown rot blossom blight, peach leaf curl, scab, powdery mildew, shot hole, Cornyneum blight
Time: For brown rot blossom blight, apply prebloom, early bloom, or full bloom. For brown rot and leaf spot, apply in summer 3-5 times at weekly intervals until harvest; all others, apply dormant (early and delayed).
Remarks: Cherries (sweet, tart), nectarines, peaches. Peaches: For severe leaf curl in dormant stage, 2 applications may be needed. Avoid spraying during hot periods of the day or hot humid conditions. Spray only to drip point, not to runoff. Do not apply dormant to peaches immediately after or during periods of 5 days or more of unseasonably high temperatures if a sufficient number of dormant cold hours have occurred. Refer to label for use directions.

Rate: Plums, prunes:
2-25 qt./100 gal. water
Disease:
black knot, brown rot, leaf spot, blotch, fruit rot, powdery mildew
Time: Apply late season. For black knot, apply at green tip stage.

Rate: Plums:
24-48 qt./100 gal. water
Disease:
Coryneum blight leaf curl
Time: Dormant, delayed dormant. For late season, apply weekly beginning 3-5 weeks before harvest.
Remarks: Plums, prunes. For brown rot and fruit rot, apply 3 qt./100 gal. water at 20, 10, and 2 days before harvest. Refer to label for use directions.

Signal Word/Toxicity Class: Danger/I.
REI: 48 hr.

Sulforix *(Best Sulfur)*

Rate: 2 qt./100 gal. water
Disease:
powdery mildew scab
Time: Apply in prepink, pink, and calyx periods.
Remarks: Apples, pears. Do not use on d'Anjou pears after calyx period. For powdery mildew, continued spraying is required during growing season. Do not apply this product through any type of irrigation system. Refer to label for use directions.

Rate: 2 qt./100 gal. water
Disease:
brown rot leaf spot powdery mildew
Time: For brown rot and leaf spot, apply 3-5 weeks before harvest; repeating at weekly intervals until just before picking. For powdery mildew, apply when disease appears.
Remarks: Cherries, nectarines, plums, prunes. For powdery mildew, use $1/2$ pt. + 4 lb. wettable sulfur. Do not apply this product through any type of irrigation system. Refer to label for use directions.

Rate: Dilute:
1 pt./100 gal. water
Concentrate:
2-4 pt./A
Disease:
powdery mildew
Time: Apply after all leaves have fallen and before buds begin to swell. If leaf curl has been severe, 2 applications may be needed, the first in early dormant and the second in delayed dormant.

Rate: 4 gal./100 gal. water
Disease:
leaf curl
Time: Apply after all leaves have fallen and before buds begin to swell. If leaf curl has been severe, 2 applications may be needed, the first in early dormant and the second in delayed dormant.
Remarks: Peaches. For powdery mildew, add 5-10 lb. wettable sulfur dilute; 15-32 lb. concentrate. Do not apply this product through any type of irrigation system. Refer to label for use directions.

Signal Word/Toxicity Class: Danger/I.
REI: 48 hr.

CAPTAN

Toxic to fish.

Drexel Captan (4L, 50W, 80W) *(Drexel)*

Rate: Apples:
Eastern U.S.:
8 pt. 4L/A
8 lb. 50W/A
5 lb. 80W/A

NOTICE	The information on these pages is for preliminary planning — not a guide for use. Be sure to follow the manufacturer's directions, notwithstanding information contained here.	For personal protective equipment and EPA registration numbers, please turn to page 39.

8 Fruits & Nuts

Disease:

black rot	Botrytis	scab
blossom end rot		

Time: Apply at 5- to 7-day intervals as needed in prebloom, bloom, petal fall, and first cover sprays.

Rate: 4-8 pt. 4L/A
4-8 lb. 50W/A
2.5-5 lb. 80W/A

Disease:

bitter rot	Brooks fruit spot	scab
black pox	fly speck	sooty blotch
black rot		

Time: Apply at 10- to 14-day intervals in second and later cover sprays up to start of harvest.

Rate: Western U.S.:
8-10 pt. 4L/A
4-8 lb. 50W/A
2.5-5 lb. 80W/A

Disease:

apple scab	Botrytis rot (Pacific NW)	bull's eye rot (Pacific NW)

Time: Apple scab: Apply as preblossom spray using high rate; mid-summer reduce to low rate. Botrytis rot, bull's eye rot: Make 1-2 applications with late cover sprays and 1-2 preharvest applications.

Rate: Apricots:
1 qt. 4L/A
3-5 lb. 50W/A
3.125 lb. 80W/A

Disease:

brown rot	jacket rot	twig blight

Time: Apply at red bud, bloom; repeat at 75% petal fall.

Rate: Cherries:
4 pt. 4L/A
2.5 lb. 80W/A
Eastern U.S.:
4 lb. 50W/A
Western U.S.:
3-4 lb. 50W/A

Disease: Eastern U.S.:

blossom blight	brown rot	leaf spot
Botrytis rot		

Western U.S.:

brown rot blossom blight	brown rot	leaf spot

Time: Apply prebloom, bloom, petal fall, shuck, cover and preharvest sprays. For leaf spot: Apply immediately after harvest; repeat in 10- to 14-days. Eastern U.S.: Applications at 3- to 4-day intervals may be necessary during bloom to control blossom blight; repeat at 7- to 10-day intervals as needed.

Rate: Nectarines, peaches:
2 pt. 4L/100 gal. water
4-8 lb. 50W/A
5 lb. 80W/A

Disease:

blossom blight	Coryneum blight	scab
brown rot		

Time: Brown rot, scab: Apply full pink bloom, petal fall, shuck, cover and preharvest sprays. Blossom blight: Applications at 3- to 4-day intervals may be necessary during bloom; repeat at 7- to 14-day intervals as needed. Coryneum blight: Apply pink bud, full bloom, petal fall and cover periods as necessary.

Rate: Plums, prunes:
2 pt. 4L/A
6 lb. 50W/A
3.75 lb. 80W/A
Western U.S.:
4-6 lb. 50W/A

Disease:
brown rot

Time: Eastern: Apply at full pink, bloom and petal fall sprays; repeat at 7- to 14-day intervals as needed. Western: Apply at green bud, popcorn, bloom and petal fall stages; repeat in cover sprays as conditions warrant. In California: Apply at full bloom for russet scab on prunes.

Remarks: Apples, apricots, cherries, nectarines, peaches, plums, prunes. For powdery mildew in apples, cherries, nectarines, peaches, sulfur may be added. Do not use in combination with or closely following or in alternation with wettable sulfur on sulfur sensitive apple varieties such as Red Delicious, Staymen, Baldwin, King, as severe injury and defoliation may occur. Do not apply through any type of irrigation system. Refer to label for use directions.

- - - - - - - - - - - - - - - -

Signal Word/Toxicity Class: Warning/II.
REI: 24 hr.

Micro Flo Captan (50WP, 80WP) or Captec (4L) *(Micro Flo)*

Rate: Apples:
4-8 lb. 50WP/A
2.5-5 lb. 80WP/A
0.5-1 qt. 4L/100 gal.

Disease: Eastern U.S.:

bitter rot	blossom end rot	fly speck
black pox	Brooks fruit spot	sooty blotch
black rot		

Time: Apply at 5- to 7-day intervals as needed in prebloom, bloom, petal fall and first cover sprays for blossom end rot, black rot, primary scab. For all other diseases, apply at 10- to 14-day intervals in second and later cover sprays.

Disease: Western U.S.:

Botrytis rot	bull's-eye rot	scab

Time: Bull's-eye rot, Botrytis rot (Pacific Northwest): Make 1-2 applications with late cover sprays and 1 final spray prior to harvest. Scab: Apply as midsummer cover spray.

Rate: Apricots:
3-5 lb. 50WP/A
1.88-3.125 lb. 80WP/A
0.75-1 qt. 4L/100 gal.

Disease:

brown rot	jacket rot	twig blight

Time: Apply at red bud, bloom, 75% petal fall, cover spray. California: Do not apply 50WP after 75% petal fall.

Rate: Cherries:
Eastern U.S.
4 lb. 50WP/A
2.5 lb. 80WP/A
0.75-1 qt. 4L/100 gal.
Western U.S.
3-4 lb. 50WP/A
1.88-2.5 lb. 80WP/A
0.75-1 qt. 4L/100 gal.

Disease:

Botrytis rot	brown rot blossom blight	leaf spot
brown rot		

Time: Apply prebloom, bloom, petal fall, shuck, cover, preharvest sprays; repeat at 7- to 20-day intervals as needed. For leaf spot postharvest, apply immediately after harvest; repeat in 10- to 14-days.

Rate: Nectarines, peaches (U.S.):
4-8 lb. 50WP/A
2.5-5 lb. 80WP/A
0.75-1 qt. 4L/100 gal.

Disease:

brown rot	Coryneum blight	scab

Time: Brown rot, scab: Apply in full pink, bloom, petal fall, shuck, cover, and preharvest sprays; repeat at 7- to 14-day intervals as needed. Coryneum blight: Apply pink bud, full bloom, petal fall, cover sprays as necessary and as postharvest spray (but before leaves drop).

Rate: Plums, prunes (fresh):
Eastern U.S.
6 lb. 50WP/A
3.75 lb. 80WP/A
1-3 qt. 4L/100 gal.
Western U.S.
4-6 lb. 50WP/A
2.5-3.75 lb. 80WP/A
1-3 qt. 4L/100 gal. (except CA)

Disease:

brown rot	lacy scab (Western U.S.)

Time: Eastern U.S.: Apply full pink, bloom, petal fall; repeat at 7- to 14-day intervals as needed. Western U.S.: Apply at green bud, popcorn, bloom, petal fall; repeat cover sprays as needed. For lacy scab, apply at full bloom.

Remarks: Apples, apricots, cherries, nectarines, peaches, plums, prunes. Refer to label for rates, directions, and precautions if sulfur is added to control powdery mildew on apples, cherries, nectarines, and peaches. Do not apply this product through any type of irrigation system.

- - - - - - - - - - - - - - - -

Signal Word/Toxicity Class: Danger/I.
REI: 96 hr.

◆ - **new product** • ai-active ingredient • DF-dry flowable • E/EC-emulsifiable concentrate • EW-emulsion in water
F/FL-flowable • G-granule • L/LC-liquid • SP/WSP-soluble packet • W/WP-wettable powder • WSB-water soluble bag

Champ — see copper compounds

Champion — see copper compounds

CHLOROTHALONIL

Broad-spectrum contact fungicide. Toxic to fish, aquatic invertebrates and marine/estuarine organisms.

Bravo 500, 720/Weather Stik, Ultrex (WDG) *(ZENECA)*
Echo 500, 720 or 90DF *(Sostram)*

Rate: 2.25-4 lb. ai/A
4.5-8 pt. 500/A
3.125-5.5 pt. 720/A
2.8-5 lb. WDG/A
2.6-4.6 lb. 90DF/A

Disease:
brown rot blossom blight; lacy scab (plums, prunes)

Time: Make 1 application at popcorn and another application at full bloom. If weather conditions favor disease development, make additional application at petal fall.

Remarks: Apricots, cherries, nectarines, peaches, plums, prunes. Use low rate on smaller trees and higher rate on trees taller than 20 ft. Refer to label for use directions and restrictions.

- - - - - - - - - - - - - - - - -

Rate: 2.25-3 lb. ai/A
4.5-6 pt. 500/A
3.125-4.125 pt. 720/A
2.8-3.8 lb. WDG/A
2.6-3.4 lb. 90DF/A

Disease:
cherry leaf spot; leaf curl; Coryneum blight; peach scab (apricots, nectarines)

Time: Scab and leaf spot: In addition to bloom applications, make 1 application at shuck split. Do not apply after shuck split and before harvest. If additional control is needed before harvest, use another registered fungicide. For leaf spot postharvest, make 1 application to foliage within 7 days after fruit is removed. In orchards with a history of high leaf spot incidence, repeat 10-14 days later. Coryneum blight and leaf curl: Apply at leaf fall in late autumn. When conditions favor high disease levels, use high rate and apply 1-2 applications in mid- to late-winter before bud swell. For leaf curl, if fall application is not practical, apply any time prior to budswell the following spring. Where Coryneum blight occurs, also apply at budbreak to protect newly emerging leaves and at shuck split to prevent fruit infections.

Remarks: Apricots, cherries, nectarines, peaches, plums, prunes. Refer to label for use directions and restrictions.

State Restrictions: Peaches: SLN-Alabama, Arkansas, Florida, Louisiana, Mississippi, South Carolina, Texas, Virginia (Bravo 720); SLN-New Jersey (Bravo WDG, Weather Stik).

- - - - - - - - - - - - - - - - -

Signal Word/Toxicity Class: Danger/I (WDG, 90DF); Warning/II (500, 720).
REI: 48 hr.

Terranil (6L, 90DF or 90DF WSP) *(Terra)*

Rate: 2.25-4 lb. ai/A
3.125-5.5 pt. 6L/A
2.5-4.5 lb. 90DF, 90DF WSP/A

Disease:
brown rot blossom blight; Coryneum blight; lacy scab (apricot, nectarines, peaches); cherry leaf spot; leaf curl

Time: For Coryneum blight and leaf curl, apply in late autumn. For brown rot blossom blight, and lacy scab, make 1 application at popcorn and second application at full bloom. If weather conditions favor disease development, make additional application at petal fall. In addition to bloom application, make 1 application at shuck split. Do not apply after shuck split and before harvest. For control of cherry leaf spot after harvest, make 1 application to foliage within 7 days after fruit is removed. In orchards with history of high leaf spot incidence, repeat 10-14 days later.

Remarks: Apricots, cherries, nectarines, peaches, plums, prunes. Use in sufficient water to obtain adequate coverage. Use 2.5-3.1 lb. ai on small trees, 3.1-4 lb. ai on trees taller than 20 ft. For Coryneum blight and leaf curl, when conditions favor high disease levels, use high rate and apply once or twice more in mid- to late-winter before buds swell. If fall application is not practical, apply anytime prior to bud swell the following spring. When Coryneum blight occurs, also apply at budbreak to protect newly emerging leaves and at shuck split to prevent fruit infections. Do not allow livestock to graze treated areas. Refer to label for use directions.

- - - - - - - - - - - - - - - - -

Signal Word/Toxicity Class: Danger/I (90DF, 90DF WSP); Warning/II (6L).
REI: 48 hr.

Clean Crop C-O-C-S — see copper compounds

COPPER COMPOUNDS

Fungicide/bactericide.

Basicop (WP) (basic copper sulfate) *(Griffin)*

Rate: Apples:
0.25-0.5 lb. ai/100 gal. water
0.5-1 lb. WP/100 gal. water

Disease:
fire blight; scab

Time: For fire blight, apply at 10% bloom; repeat at 5-day intervals until late bloom is over. For scab, apply delayed dormant. In California do not apply to apples during bloom.

Rate: Pears:
1-2.5 lb. ai/A
2-5 lb. WP/A

Disease:
blossom blast; blossom blight; fire blight

Time: For blossom blast and blight, apply before fall rains and again during dormancy before spring growth. For fire blight, apply at 10% bloom; repeat at 5- to 7-day intervals during bloom.

Rate: Cherries:
1.5-2.5 lb. ai/100 gal. water
3-5 lb. WP/100 gal. water

Disease:
Coryneum blight; dead bud; gummosis

Time: Apply in fall before heavy rains, repeat late dormant.

Rate: Sour cherries:
1-2.5 lb. ai/100 gal. water
2-5 lb. WP/100 gal. water

Disease:
blossom blight; brown rot; leaf spot

Time: For blossom blight and brown rot, apply during popcorn to full bloom and petal fall. For leaf spot, apply at petal fall, including 1-2 times after petal fall.

Rate: Apricots, plums, prunes:
2-2.5 lb. ai/100 gal. water
4-5 lb. WP/100 gal. water

Disease:
blossom blight; brown rot; shot hole

Time: Apply as dormant or bud spray (from bud swell to popcorn). Do not spray when trees are in leaf.

Rate: Nectarines, peaches:
2-2.5 lb. ai/100 gal. water
4-5 lb. WP/100 gal. water

Disease:
blossom blight; leaf curl; peach blight; brown rot

Time: For blossom blight and brown rot, apply in pink bud. For leaf curl, and peach blight, apply dormant. Do not use when trees are in leaf.

Remarks: Apples, apricots, cherries (sour, sweet), nectarines, peaches, pears, plums, prunes. In California, if dormant spray is made by air, apply 16-20 lb./A in 20 gal. water. Use on d'Anjou, Comice, or Seckel pear varieties may result in fruit russet. Do not apply to sweet cherries or English Morelo variety. For sour cherries with leaf spot, add 4 lb. hydrated lime; use low rate at leaf fall or when disease is light, high rate for dormant or when disease is moderate to heavy. Do not apply through any type of irrigation system. Refer to label for use directions.

- - - - - - - - - - - - - - - - -

Signal Word/Toxicity Class: Warning/II.
REI: 24 hr.

Blue Viking or Tennessee Brand (copper sulfate pentahydrate) *(Griffin)*

Rate: 5 lb./100 gal. water

Disease:
fire blight

Time: Apply dormant only at silver tip stage.

Remarks: Apples. After silver tip, severe burn will occur on any exposed green tissue. Do not mix lime to make Bordeaux spray for this treatment. Do not apply through any type of irrigation system. Refer to label for use directions.

- - - - - - - - - - - - - - - - -

NOTICE The information on these pages is for preliminary planning — not a guide for use. Be sure to follow the manufacturer's directions, notwithstanding information contained here.	For personal protective equipment and EPA registration numbers, please turn to page 39.

Rate: 10-10-100 Bordeaux spray
Disease:
Coryneum blight leaf curl leaf spot
Time: For Coryneum blight and leaf curl, apply as dormant spray in late fall and early spring before buds begin to swell. For leaf spot, apply as full cover spray after petal fall.
Remarks: Apricots, nectarines, peaches, plums, prunes, sour cherries. Do not apply through any type of irrigation system. Refer to label for use directions.

– – – – – – – – – – – – – – – – –

Rate: 12-12-100 Bordeaux spray
Disease:
bacterial canker dead bud
Time: Apply at leaf fall; repeat late winter before buds begin to swell. In wet, cool Northwest winters, third spray may be needed.
Remarks: Sweet cherries. Do not apply through any type of irrigation system. Refer to label for use directions.

– – – – – – – – – – – – – – – – –

Signal Word/Toxicity Class: Danger/I.
REI: 24 hr.

Champ (FL or Formula 2) or Champion (WP) (copper hydroxide) *(Agtrol)*
Kocide (101, DF, 20/20, 2000, 4.5LF, LF) (copper hydroxide) *(Griffin)*

Rate: Apples:
5.33 pt. FL, LF/100 gal. water
2.66 pt. Formula 2, 4.5LF/100 gal. water
3 lb. 2000/100 gal. water
4 lb. 101, DF, WP/100 gal. water
Disease:
collar rot crown rot
Time: Apply early spring or late fall after harvest. Apply 4 gal. suspension as a drench on lower trunk area of tree.

Rate: Apples:
10.66-21.33 pt. FL, LF/A
5.33-10.66 pt. Formula 2, 4.5LF/A
6-12 lb. 2000/A
8-16 lb. 101, DF, WP/A
Disease:
fire blight
Time: Apply as full cover spray between silver tip and green tip. Crop injury may occur from late application. Discontinue when green tip reaches 1/2".

Rate: Apples:
16-21.33 pt. FL, LF/A
8-10.66 pt. Formula 2, 4.5LF/A
9-12 lb. 2000/A
12-16 lb. 101, DF, WP/A
Disease:
Anthracnose European canker Pseudomonas
Time: Apply before fall rains.

Rate: Pears, quince:
1.33 pt. FL, LF/A
0.66 pt. Formula 2, 4.5LF/A
0.75 lb. 2000/A
2.5 lb. 20/20/A
Pears:
1 lb. 101, DF, WP/A
Quince:
8-12 lb. 101/A
Disease:
fire blight
Time: Apply at 5-day intervals throughout bloom period.

Rate: Pears:
16-21 pt. FL, LF/A
8-10.66 pt. Formula 2, 4.5LF/A
9-12 lb. 2000/A
12-16 lb. 101, DF, WP/A
22.5-40 lb. 20/20/A
Disease:
Pseudomonas blight
Time: Apply before fall rains and again during dormancy before spring growth starts.

Rate: Cherries:
10.66-16 pt. FL, LF/A
5.33-8 pt. Formula 2, 4.5LF/A
8-12 lb. 101, DF, WP/A
6-9 lb. 2000/A
Disease:
brown rot blossom blight Coryneum blight dead bud
Time: For Coryneum blight and dead bud, apply in fall before heavy rains and at late dormant. In orchards where disease is severe, apply shortly after harvest. For brown rot, apply full cover spray at popcorn stage and full bloom.

Rate: Apricots:
8-16 lb. 101/A
8-12 lb. DF, WP/A
15-30 lb. 20/20/A
6-9 lb. 2000/A
5.33-8 pt. Formula 2, 4.5LF/A
10.66-16 pt. FL, LF/A
Disease:
bacterial blast (101, 4.5LF) blossom brown rot Coryneum blight
Time: Apply at popcorn to full bloom. Application after bloom will result in crop injury. For bacterial blast, apply at dormant to early pink bud.

Rate: Nectarines, peaches, plums, prunes:
10.66-21.33 pt. FL, LF/A
5.33-10.66 pt. Formula 2, 4.5LF/A
8-16 lb. 101, DF, WP/A
6-12 lb. 2000/A
15-40 lb. 20/20/A
Disease:
bacterial blast (4.5LF) blossom blight Coryneum blight
bacterial canker (4.5LF) brown rot leaf curl (nectarines, peaches)
Time: Nectarines, peaches: For brown rot or blossom blight, apply at pink bud. For blight and leaf curl, apply at leaf fall. For 4.5LF, apply first application before fall rains, second at late dormant (late dormant must be made before leaf bud swell for leaf curl). Plums, prunes: For blight, apply as dormant spray. For brown rot or blossom blight, apply at pink, red, or early white bud stage.

Rate: Nectarines, peaches:
Postbloom:
1 lb. 101, DF, WP/A
0.75 lb. 2000/A
0.66 pt. Formula 2, 4.5LF/A
1.33 pt. FL, LF/A
Dormant:
8-16 lb. 101/A
8 lb. DF, WP/A
1.33 pt. Formula 2/A
10.66 pt. FL, LF/A
15 lb. 20/20/A
Disease:
bacterial spot
Time: Apply postbloom at first and second cover spray or dormant.
Remarks: Apples, apricots, cherries, nectarines, peaches, pears, plums, prunes, quince. Use on yellow apple varieties may cause discoloration. Excessive dosages may cause fruit russet on pears. For apricots, use high rates when conditions favor disease. May be used with agricultural spray oil for nectarines and peaches. Kocide 20/20 not for use on plums, prunes. Refer to label for use directions.
Days To Harvest: Nectarines, peaches-3 weeks.

– – – – – – – – – – – – – – – – –

Signal Word/Toxicity Class: Varies by formulation.
REI: Varies by formulation.

Clean Crop C-O-C-S (WDG) (74.8% copper oxychloride & 14.2% basic copper sulfate) *(Platte)*

Rate: Apples, pears, quince:
1-2 lb. ai/100 gal. water
2-4 lb. WDG/100 gal. water
Disease:
fire blight
Time: For apples, make 1 application between silver tip and green tip. Discontinue when green tip exceeds 1/2" as phytotoxicity may occur. Use 0.25 lb. when 20% of blossoms are open and repeat when 75% are open (in California, follow 0.25 lb. directions only). For pears and quince (including California), apply dormant to bud swell as full cover spray. Apply 0.25-0.5 lb. at 10% bloom; repeat at 3- to 5-day intervals until late bloom is over.

Rate: Apples, pears, quince:
2-2.5 lb. ai/100 gal. water
4-5 lb. WDG/100 gal. water
Disease:
Anthracnose (pears, quince) European canker (apples) pear blight (pears, quince) Pseudomonas (pears, quince)

◆ - **new product** • ai-active ingredient • DF-dry flowable • E/EC-emulsifiable concentrate • EW-emulsion in water
F/FL-flowable • G-granule • L/LC-liquid • SP/WSP-soluble packet • W/WP-wettable powder • WSB-water soluble bag

Time: Apply after harvest and before fall rains; repeat dormant to delayed dormant. For European canker, apply postharvest in fall. For pear blight, apply during bloom and other periods favorable to infection.

Rate: Apricots, nectarines, peaches, plums, prunes:
2 lb. ai/100 gal. water
4 lb. WDG/100 gal. water
Disease: Apricots, plums, prunes:
brown rot Coryneum blight
Nectarines, peaches:
bacterial spot peach blight peach leaf curl
brown rot
Time: For brown rot, apply preblossom, blossom, petal fall. For Coryneum blight, apply dormant up to red bud. For bacterial spot, leaf curl, and peach blight, apply as dormant spray up to bud swell.

Remarks: Apples, apricots, nectarines, peaches, pears, plums, prunes, quince. Apply low rate in California for apples, pears, quince. Dormant oil may be added. Do not apply this product through any type of irrigation system. Refer to label for use directions.

Rate: Cherries:
2-3 lb. ai/100 gal. water
4-6 lb. WDG/100 gal. water
Disease:
brown rot dead bud leaf spot

Rate: Tart cherries:
0.75 lb. ai/100 gal. water
1.5 lb. WDG/100 gal. water
Disease:
cherry leaf spot
Time: For brown rot and leaf spot, apply popcorn, bloom, and petal fall. For dead bud, apply in fall before heavy fall rains, and again before bud swell in spring.
Remarks: Cherries. For cherry leaf spot, add hydrated lime. For dead bud, use high rate + superior oil. Do not apply this product through any type of irrigation system. Refer to label for use directions.
State Restrictions: Not for use in California for cherry leaf spot or dead bud.

Signal Word/Toxicity Class: Warning/II.
REI: 24 hr.

Copper 3 Dust (basic copper sulfate) *(Wilbur-Ellis)*
Rate: 2-3 lb. ai/A
35-50 lb. Dust/A
Disease: Pears:
fire blight
Apricots, nectarines, peaches:
blossom blight brown rot Coryneum blight
Time: Apply at 7- to 10-day intervals. For fire blight, apply when trees are in full bloom or when cankers ooze; repeat at 2-week intervals, or sooner in case of heavy rain or hail. In rainy season, keep trees dusted as long as blooms continue to develop. For brown rot, apply early spring when trees are in red bud to popcorn. For Coryneum blight on nectarines and peaches, apply late Jan.
Remarks: Apricots, nectarines, peaches, pears. Copper can russet d'Anjou pears. Thorough coverage should be made on all parts of plant with particular attention to new growth. Refer to label for use directions.
Days To Harvest: 7.
State Restrictions: SLN-Washington.

Signal Word/Toxicity Class: Caution.
REI: 12 hr.

Copper-Count-N (L) (copper ammonium) *(Mineral Research)*
Rate: Apples, pears:
3-4 qt. L/100 gal.
Disease:
Anthracnose (apples) European canker (apples) Pseudomonas blight (pears)
Time: Apply before fall rain. For pears, repeat at dormant before spring growth starts.

Rate: Apples:
2-4 qt. L/A
Pears:
1 qt. L/A
Disease:
fire blight
Time: For apples, apply between silver tip and green tip. For pears, apply at 5-day intervals through bloom period.

Rate: Apricots:
8-12 qt. L/A
Disease:
brown rot blossom blight shot hole
Time: For brown rot blossom blight, apply at red bud to jacket fall stage. For shot hole, apply at leaf fall to protect buds and shoots from infection during rainy periods. Reapply up to late bud swell. Do not apply after full bloom.

Rate: Cherries:
2-3 qt. L/100 gal.
Disease:
brown rot blossom blight
Time: Apply as full cover spray at popcorn and full bloom.

Rate: Cherries:
6 qt. L/100 gal.
Disease:
Coryneum blight dead bud
Time: Apply before heavy fall rains (Oct.) and again in Jan. If disease pressure is severe, also apply in Aug.

Rate: Peaches, nectarines:
8-12 qt. L/A
Disease:
blossom brown rot leaf curl shot hole
Time: For blossom brown rot, apply as dormant, or delayed dormant spray. Can use with dormant spray oil. Do not apply at or after full bloom. For leaf curl and shot hole, apply at leaf fall to protect buds and shoots from infection during rainy periods. Reapply up until late bud swell. Do not apply after full bloom.

Rate: Peaches, nectarines:
0.5-4 pt. L/100 gal.
Disease:
bacterial spot
Time: For postbloom, apply low rate at first and second cover sprays. Do not spray later than 3 weeks prior to harvest. For dormant application, apply high rate.
Remarks: Apples, apricots, cherries, nectarines, peaches, pears. Can cause discoloration on yellow apple varieties; pick before spraying; discontinue when green tip is 1/2". Excessive dosages may cause fruit russet on pears. Slight defoliation and spotting of leaves may occur from use in cover sprays on peaches and nectarines. Applications may be made by ground or air; refer to label for use directions. Do not apply through any type of irrigation system unless supplemental labeling on chemigation is followed.
State Restrictions: Not for use on apples, pears, or for bacterial spot on peaches or nectarines in California.

Signal Word/Toxicity Class: Caution.
REI: 12 hr.

Drexel Basic Copper Sulfate (WP) or KOP 300 (3FL) (basic copper sulfate) *(Drexel)*
Rate: Pears:
Dilute:
0.5 lb. WP/100 gal. water
5.33 oz. 3FL/100 gal. water
Concentrate:
2 lb. WP/A
1.3 pt. 3FL/A
Disease:
fire blight
Time: Apply at 10% bloom; repeat at 5- to 7-day intervals during bloom.

Rate: Apricots:
Dilute:
4-5 lb. WP/100 gal. water
0.33-1.5 gal. 3FL/100 gal. water
Concentrate:
10-12 lb. WP/A
1-1.17 gal. 3FL/A
Disease:
brown rot bud blight shot hole
Time: Apply dormant or bud spray (from bud swell to popcorn). Do not spray when trees are in leaf.

Rate: Nectarines, peaches:
Dilute:
4-5 lb. WP/100 gal. water
0.33-0.5 gal. 3FL/100 gal. water
Concentrate:
12-16 lb. WP/A
1-1.33 gal. 3FL/A

NOTICE The information on these pages is for preliminary planning — not a guide for use. Be sure to follow the manufacturer's directions, notwithstanding information contained here.

For personal protective equipment and EPA registration numbers, please turn to page 39.

Disease:
brown rot leaf curl peach blight

Time: For brown rot, apply before bud swell and in full pink bud stage. Do not apply after pink bud or after trees are in leaf. For leaf curl and peach blight, apply as dormant spray before fall rains begin (Nov. 15-Dec. 15).

Rate: Nectarines, peaches-California:
16-20 lb. WP/A in 20 gal. water
1.33-1.66 gal. 3FL/A

Disease:
brown rot leaf curl peach blight

Rate: Nectarines, peaches-California:
3 lb. WP/100 gal. water
0.33 gal. 3FL/A

Disease:
bacterial disease

Time: Apply dormant application by air.

Rate: Plums, prunes:
Dilute:
4-5 lb. WP/100 gal. water
0.33-0.5 gal. 3FL/100 gal. water
Concentrate:
12 lb. WP/A
1 gal. 3FL/A
California:
16-20 lb. WP/A
1.33-1.66 gal. 3FL/A

Disease:
brown rot shot hole

Time: Brown rot: Apply at early green bud and full popcorn stages. Shot hole: Apply as dormant spray before heavy fall rains begin. For California, apply only dormant application by air.

Rate: Sour cherries:
5.3-6.6 pt. 3FL/100 gal. water
Dilute:
4-5 lb. WP/100 gal. water
Concentrate:
2 lb. WP/A

Disease:
blossom blight leaf spot shot hole
brown rot

Time: Blossom blight, brown rot: Apply at popcorn and late bloom stages. Leaf spot: Apply at petal fall. Shot hole: Apply as dormany spray.

Remarks: Apricots, cherries (sour), nectarines, peaches, pears, plums, prunes. Do not use on d'Anjou, Comice, or Seckel pear varieties. Do not apply through any type of irrigation system. Refer to label for use directions and precautions.

\- - - - - - - - - - - - - - - - - -

Signal Word/Toxicity Class: Warning/II.
REI: 24 hr. (WP); 12 hr. (3FL).

KOP-Hydroxide (FL, 50WP) (copper hydroxide) *(Drexel)*

Rate: Apples:
2.66-5.33 pt. FL/100 gal. water
4-16 lb. 50WP/A

Disease:
Anthracnose crown rot fire blight
collar rot European canker Pseudomonas

Time: Apply before fall rains. For fire blight, apply between silver tip and green tip; discontinue when green tip is 1/2". For collar and crown rot, apply as drench on lower trunk area of tree either early spring or fall after harvest.

Rate: Pears:
0.33 pt. FL/100 gal. water
1 lb. 50WP/A

Disease:
fire blight

Rate: 4-5.33 pt. FL/100 gal. water
12-16 lb. 50WP/A

Disease:
Pseudomonas blight

Time: Fire blight: Apply at 5-day intervals through bloom. Pseudomonas: Apply before fall rains, repeat at dormant before spring growth starts.

Rate: Apricots:
2.66-4 pt. FL/100 gal. water
8-12 lb. 50WP/A

Disease:
blossom brown rot Coryneum blight

Time: Apply from popcorn to full bloom. Do not apply after bloom.

Rate: Cherries:
8 pt. FL/100 gal. water
8-12 lb. 50WP/A

Disease:
Coryneum blight dead bud

Time: Apply before heavy fall rains (Oct.); repeat in Jan. For severe disease, apply in Aug. also.

Rate: 2.66-4 pt. FL/100 gal. water
8-12 lb. 50WP/A

Disease:
brown rot blossom blight

Time: Apply as full cover spray at popcorn and full bloom.

Rate: Nectarines, peaches:
2.66-5.33 pt. FL/100 gal. water
8-16 lb. 50WP/A

Disease:
brown rot blossom blight Coryneum blight leaf curl

Time: For brown rot, apply at pink bud. For Coryneum blight and leaf curl, apply dormant at leaf fall.

Rate: Nectarines, peaches:
Dormant:
2.66 pt. FL/100 gal. water
8 lb. 50WP/A
Postbloom:
0.33 pt. FL/100 gal. water
1 lb. 50WP/A

Disease:
bacterial spot

Time: Apply at first and second cover sprays. Do not spray later than 3 weeks prior to harvest.

Rate: Plums, prunes:
8-16 lb. 50WP/A

Disease:
blossom blight brown rot Coryneum blight

Time: For Coryneum blight, apply as dormant spray. For blossom blight and brown rot, apply at pink, red or early white stage.

Remarks: Apples, apricots, nectarines, peaches, pears, plums, prunes. To avoid discoloration on yellow apple varieties, pick before spraying. Excessive dosage may cause fruit russet on pears. For nectarines, peaches, plums, prunes, apply high rate when rainfall or disease is heavy. Slight defoliation and spotting of leaves may occur from use in cover sprays on nectarines and peaches. Refer to label for use directions and precautions.

Days To Harvest: Nectarines, peaches-3 weeks.
State Restrictions: Not for use in California on apples.

\- - - - - - - - - - - - - - - - - -

Signal Word/Toxicity Class: Danger/I.
REI: 48 hr.

KOP Oxy-85 (WP) (copper oxychloride) *(Drexel)*

Rate: 4 lb. WP/A

Disease:
leaf curl

Time: Apply prebloom and petal fall.

Remarks: Peaches. Use in sufficient water and thoroughly wet plants. Refer to label for use directions.

\- - - - - - - - - - - - - - - - - -

Signal Word/Toxicity Class: Danger/I.
REI: 24 hr.

Micro Flo Basic Copper "53" (basic copper sulfate) *(Micro Flo)*

Rate: 0.33-0.5 lb. 53/100 gal.

Disease:
fire blight

Time: Apply at 10% bloom; repeat at 5-day intervals through bloom.

Rate: 0.5-1.5 lb. 53/100 gal. + hydrated lime

Disease:
bitter rot blotch powdery mildew

Time: Apply at petal fall; repeat through fourth cover spray.

Rate: 1-1.25 lb. 53/100 gal. + hydrated lime

Disease:
Brooks fruit spot scab sooty blotch

◆ - **new product** • ai-active ingredient • DF-dry flowable • E/EC-emulsifiable concentrate • EW-emulsion in water
F/FL-flowable • G-granule • L/LC-liquid • SP/WSP-soluble packet • W/WP-wettable powder • WSB-water soluble bag

Time: Fruit spot, blotch: Apply late cover spray. Scab: Apply delayed dormant stage.

Rate: 3-4 lb. 53/100 gal.
Disease:
Anthracnose
Time: Red varieties: Apply to foliage before or after harvest. Yellow varieties: Apply to foliage after harvest once every 2-3 years as necessary.

Rate: 4 lb. 53/100 gal. + spray oil
Disease:
bull's-eye rot
Time: Apply after harvest.
Remarks: Apples. Where copper injury is likely to occur, add lime. Refer to label for use directions.

Rate: 0.5 lb. 53/100 gal.
Disease:
fire blight
Time: Apply at 10% bloom; repeat at 5- to 7-day intervals during bloom.
Remarks: Pears. Do not use on D'anjou, Comice, or Seckel pear varieties. Refer to label for use directions.

Rate: Cherries, sour cherries:
13-16 lb. 53/A
Disease:
Coryneum blight | gummosis | Pseudomonas canker
Time: Apply late leaf fall.

Rate: Cherries, sour cherries:
12-16 lb. 53/A
Northwestern states:
20-25 lb. 53/A
Disease:
Coryneum blight
Time: Dormant spray.

Rate: Sour cherries (except Great Lakes):
12-16 lb. 53/A
Disease:
brown rot | blossom blight
Time: Red bud, popcorn, late bloom.
Remarks: Cherries (sour, sweet). Do not apply nondormant sprays to sweet cherries or English Morelo varieties. Refer to label for use directions.

Rate: Apricots:
14-20 lb. 53/A
Disease:
brown rot (suppression) | shot hole
Time: Brown rot: Apply low rate through swelling buds and popcorn stage. Do not spray trees in leaf. For bud blight, apply high rate after most leaves have fallen, usually Nov.-Dec. but ahead of fall rains. Shot hole: Apply as dormant spray in California.

Rate: Nectarines, peaches:
15-25 lb. 53/A
Disease:
bacterial diseases (peaches) | brown rot | shot hole
blossom blight | leaf curl | twig blight
Time: Bacterial disease, leaf curl, shot hole, blossom blight, twig blight: Apply dormant. Brown rot, shot hole: Apply before bud swell, full pink bud to popcorn stage before leaves emerge.

Rate: Plums, prunes:
2-3 lb. 53/100 gal. + hydrated lime
Disease:
black knot
Time: Apply at green tip, shuck shed, and 2 cover sprays if needed.

Rate: 4-5 lb. 53/100 gal. + sticker
Disease:
black knot | brown rot (suppression) | shot hole
Time: Brown rot: Apply early green bud and full bloom stages. Shot hole: Apply dormant before heavy fall rains begin.
Remarks: Apricots, nectarines, peaches, plums, prunes. Spray may injure tender foliage of apricots. Refer to label for use directions.

Signal Word/Toxicity Class: Warning/II.
REI: 24 hr.

Microsperse COC (53WP) (copper oxychloride) *(Micro Flo)*

Rate: Apples:
8-16 lb. 53WP/A
Disease:
Anthracnose | European canker | Pseudomonas
Time: Apply at 10% and 80% leaf fall before fall rains; repeat before winter pruning. If fire blight spray is not made, repeat at silver and green tip but not after green tip exceed 1/2".

Rate: Apples:
8-16 lb. 53WP/A
Dilute cover spray:
1-1.5 lb. 53WP/A
Disease:
fire blight
Time: For fire blight, apply between silver tip and green tip, not to exceed 1/2". For dilute cover spray, apply at 10% bloom; repeat at 5-day intervals through late bloom.

Rate: Apricots, cherries:
8-12 lb. 53WP/A
Disease: Apricots:
brown rot blossom blight | Coryneum blight | twig blight
Cherries:
blight | leaf spot | twig blight
brown rot
Time: For apricots, apply fall dormant; repeat at popcorn to full bloom; cherries, apply popcorn, full bloom, petal fall.

Rate: Nectarines, peaches:
8-16 lb. 53WP/A
Disease:
bacterial spot | Coryneum blight | peach blight
brown rot blossom blight | peach leaf curl | twig blight
Time: Brown rot blossom blight, twig blight: Apply before buds swell, at pink bud before leaves emerge. Bacterial spot: Apply dormant at bud swell. Coryneum blight, leaf curl, and peach blight: Apply fall dormant before fall rains begin; repeat in spring before buds begin to swell.

Rate: Plums, prunes:
4-16 lb. 53WP/A
Disease:
brown rot blossom blight | Coryneum blight | twig blight
Time: Brown rot blossom blight, twig blight: Apply at green bud, early pink to white bud. Coryneum blight: Apply dormant before heavy fall rains.
Remarks: Apples, apricots, cherries, nectarines, peaches, plums, prunes. Use on yellow apple varieties may cause fruit russeting; non-yellow varieties may differ in susceptibility to copper resulting in russeting or injury. Refer to label for use directions.

Signal Word/Toxicity Class: Danger/I.
REI: 24 hr.

Nu-Cop (3L, 50DF, 50WP) (copper hydroxide) *(Micro Flo)*

Rate: Apples:
0.66-1.3 gal. 3L/A
2-4 lb. 50DF, 50WP/100 gal. water
Disease:
Anthracnose | crown rot (except 3L) | fire blight
collar rot (except 3L) | European canker | Pseudomonas
Time: Anthracnose, European canker, Pseudomonas: Apply before fall rains. Fire blight: Apply at silver and green tip. Do not apply after green tips reach 1/2" high. Collar and crown rot: Apply as drench on lower trunk areas in early spring or late fall after harvest.

Rate: Pears:
0.66-1.33 pt. 3L/A
Western U.S.:
1 lb. 50DF, 50WP/A
Quince:
1.33 pt. 3L/A
Disease:
fire blight
Time: Apply at 5-day intervals through bloom.

Rate: Pears:
1-1.33 gal. 3L/A
12-16 lb. 50DF, 50WP/A

NOTICE The information on these pages is for preliminary planning — not a guide for use. Be sure to follow the manufacturer's directions, notwithstanding information contained here.

For personal protective equipment and EPA registration numbers, please turn to page 39.

Disease:
Pseudomonas blight

Time: Apply before fall rains; repeat during dormancy before spring growth begins.

Rate: Apricots:
1.33-8 pt. 3L/A
8-12 lb. 50DF, 50WP/A
Disease:
blossom brown rot Coryneum blight
Time: Apply at popcorn to full bloom.

Rate: Cherries:
2.66-4 pt. 3L/100 gal. water
2-3 lb. 50DF, 50WP/A
Disease:
brown rot blossom blight

Rate: 0.66-1 gal. 3L/100 gal. water
8-12 lb. 50DF, 50WP/A
Disease:
Coryneum blight Pseudomonas
Time: Brown rot blossom blight: Apply popcorn and full bloom. Coryneum blight, Pseudomonas: Apply before heavy rains and again late dormant.

Rate: Nectarines, peaches:
0.66-2.66 gal. 3L/A
8-20 lb. 50DF, 50WP/A
Disease:
bacterial spot Coryneum blight leaf curl
brown rot blossom blight
Time: Brown rot blossom blight: Apply at pink bud. Bacterial spot: Apply low rate dormant. Coryneum blight, leaf curl: Apply after leaf fall.

Rate: Plums, prunes:
0.66-2.66 gal. 3L/A
8-16 lb. 50DF, 50WP/A
Disease:
brown rot blossom blight Coryneum blight
Time: Brown rot blossom blight: Apply at pink, red, or early white bud. Coryneum blight: Apply dormant.
Remarks: Apples, apricots, cherries, nectarines, peaches, pears, plums, prunes, quince. Use on yellow apple varieties may cause discoloration. Excessive dosages may cause fruit russet on pears. Refer to label for use directions.
State Restrictions: California: For areas subject to copper injury, add 0.33-1 lb. high quality lime/1 lb. product. Nu-Cop 3L is for use only on brown rot blossom blight.

Signal Word/Toxicity Class: Danger/I (50DF, 50WP); Warning/II (3L).
REI: 48 hr. (50DF, 50WP); 24 hr. (3L).

TOP COP Tri-Basic (FL) (basic copper sulfate) *(Stoller)*
Rate: Apples, pears:
1.25 pt. FL/A
Disease:
fire blight
Time: Apply at 10% bloom; repeat at 5- to 7-day intervals.

Rate: Apricots:
8-12 pt. FL/A
Plums, prunes:
8 pt. FL/A
Disease:
brown rot (8 pt. rate) shot hole (12 pt. rate)
Time: For brown rot, apply through bud up to popcorn. Do not spray when trees are in leaf. Apply dormant before heavy fall rains.

Rate: Nectarines, peaches:
8-12 pt. FL/A
Disease:
bacterial disease leaf curl shot hole
brown rot peach blight
Time: Apply before bud swell and pink bud stage. Do not apply after pink bud or after trees are in leaf as injury may occur. Apply in fall before rains begin (Nov. 15-Dec. 15).

Rate: Sour cherries:
2-6 pt. FL/100 gal. water
Disease:
brown rot blossom blight leaf spot shot hole
Time: For blossom blight, apply at popcorn and late bloom; leaf spot, at petal fall; shot hole, apply dormant.
Remarks: Apricots, cherries (sour), nectarines, peaches, plums, prunes. Do not use on d'Anjou, Comice, or Seckel pears varieties. Overspraying may cause injury to tender foliage of apricots, peaches, and some Japanese plums. Add hydrated lime for leaf spot and brown rot (late bloom stage) on cherries. Refer to label for use directions.

Signal Word/Toxicity Class: Caution.
REI: 24 hr.

Copper-Count-N — see copper compounds

DICLORAN
DICLORAN (Premix)

Protective fungicide. Toxic to fish.

Botran 6 Dust *(Wilbur-Ellis)*
Botran 6 Sulfur 25 Dust (dicloran & sulfur) *(Wilbur-Ellis)*
Rate: Nectarines, peaches:
60 lb. Dust/A
Disease:
brown rot brown rot blossom blight fruit decay
Time: For brown rot blossom blight, apply at pink bud and full bloom. For brown rot fruit decay, apply 18, 10 and 1 day before harvest. For brown rot, apply 1 day before harvest.

Rate: Sweet cherries:
50 lb. Dust/A
Disease:
brown rot brown rot blossom blight Rhizopus fruit rot
Time: For brown rot blossom blight, apply at popcorn, bloom, full bloom and petal fall. For fruit decay and Rhizopus fruit rot, apply 10 days and 1 day prior to harvest.
Remarks: Cherries (sweet), nectarines, peaches. Do not allow animals to graze treated areas. Refer to label for use directions.

Signal Word/Toxicity Class: Warning/II (6 Sulfur 25 Dust); Caution.
REI: 12 hr.

Botran 75-W *(Gowan)*
Rate: 1-4 lb. ai/A
1.33-5.33 lb. 75-W/A
Disease:
blossom blight fruit decay
Time: For fruit decay on apricots, nectarines, and peaches, apply 18 days and 10 days before harvest. For blossom blight on nectarines and peaches, apply at pink bud and full bloom; on plums and prunes, apply at popcorn and full bloom; on sweet cherries, apply at popcorn, bloom, full bloom and petal fall.
Remarks: Apricots, cherries (sweet), nectarines, peaches, plums, prunes. Refer to label for use directions.
Days To Harvest: Apricots, cherries, nectarines, peaches-10.

Signal Word/Toxicity Class: Caution.
REI: 12 hr.

Dithane — see mancozeb

DODINE

Protective fungicide in a wettable powder formulation recommended for the control of major fungal diseases in fruits and nuts. Not compatible with nonionic surfactants.

Syllit (65W) *(Platte)*
Rate: 0.65-1.3 lb. ai/A
1-2 lb. 65W/A (ground)
0.98 lb. ai/A
1.5 lb. 65W/A in 5-7 gal. (aerial)
2 lb. ai/A
3 lb. 65W/A (curative)
Disease:
scab

◆ - **new product** • ai-active ingredient • DF-dry flowable • E/EC-emulsifiable concentrate • EW-emulsion in water
F/FL-flowable • G-granule • L/LC-liquid • SP/WSP-soluble packet • W/WP-wettable powder • WSB-water soluble bag

Time: Prebloom through first cover applications by ground or air. Can apply delayed dormant with 8 gal. 60-75 viscosity oil/A. Repeat at 5- to 7-day intervals. Curative: Apply after infection has occurred during or following rain (within 36 hr.) from prebloom through first cover, and later in cover sprays as necessary.

Remarks: Apples. May be used with wettable sulfur for powdery mildew control. Consult state extension service for proper rates and timing of sprays. Do not use treated apples in the manufacture of apple pomace for use in livestock feed. Do not use on Golden Delicious and Grimes Golden varieties during cool, slow-drying weather conditions. Do not exceed 2 lb./A immediately before, during, or after occurrence of freezing or near freezing temperatures, particularly when accompanied by slow-drying conditions. Refer to label for use directions.

Days To Harvest: 7.

Rate: Western States:
2 lb. ai/A
2-3 lb. 65W/A
Oregon (aerial):
1.3-2 lb. ai/A
2-3 lb. 65W/A

Disease:
scab

Time: Apply from prebloom through first cover at 5- to 7-day intervals; repeat subsequent cover sprays as needed.

Remarks: Pears. Do not graze cover crops in treated orchards. Do not apply through any type of irrigation system. Refer to label for use directions.

Days To Harvest: 7.

Rate: 0.65-1.3 lb. ai/A
1-2 lb. 65W/A

Disease:
blossom brown rot (Western States) leaf spot

Time: For leaf spot, apply at petal fall or when first leaves unfold; repeat at 7- to 10-day intervals as needed (can be applied through harvest if conditions favor disease). For blossom brown rot, apply high rate when petals begin to open; repeat at full bloom and again at petal fall.

Remarks: Cherries (sour, sweet). May cause spotting to foliage or fruit when temperatures and humidity are excessively high. Can be combined with wettable sulfur for powdery mildew control. Consult state extension service for proper rates and timing. Do not graze cover crops in treated orchards. Do not apply through any type of irrigation system. Refer to label for use directions.

Rate: Western States:
1.3-2.6 lb. ai/A
2-4 lb. 65W/A

Disease:
blossom brown rot leaf curl

Time: For blossom brown rot, apply when petals begin to open; repeat at full bloom and petal fall. For leaf curl, apply before buds swell in spring. If disease was severe previous season, repeat just prior to bud swell.

Remarks: Peaches. Do not apply after petal fall. Refer to label for use directions.

Rate: 0.65-1.3 lb. ai/A
1-2 lb. 65W/A

Disease:
bacterial spot

Time: Spray at shuck split; repeat at 7- to 10-day intervals as needed through cover sprays. Shorten intervals during wet periods.

Remarks: Peaches. Tank mix with 4 lb. 50% captan. Under light to moderate disease conditions, use high rate. Do not use concentrate sprays higher than 3x. Do not combine with solvent formulations of insecticides or sulfur. Do not apply with, immediately before, or closely following oil or zinc-sulfate-lime spray. Do not graze cover crops in treated orchards. Do not apply through any type of irrigation system. Refer to label for use directions.

Days To Harvest: 15.

Signal Word/Toxicity Class: Danger/I.
REI: 48 hr.

Echo — see chlorothalonil

Elite — see tebuconazole

FENARIMOL

Triazole sterol inhibitor systemic fungicide. Toxic to fish.

Rubigan E.C. *(Dow AgroSciences)*

Rate: Apples, pears:
8-12 fl. oz. E.C./A

Disease:
powdery mildew; rust (east of the Rockies); scab

Time: Apply from green tip to tight cluster; repeat intervals vary, refer to label.

Rate: Cherries:
6-12 fl. oz. E.C./A

Disease:
leaf spot; powdery mildew

Time: East of the Rockies: Apply at petal fall; repeat at 7-10 day intervals until harvest. West of the Rockies: Apply at late bloom to petal fall; repeat at 14-21 day intervals until harvest.

Remarks: Apples, cherries, pears. Do not apply to pears until after petal fall. Do not apply more than 12 fl. oz./A per application and more than 84 fl. oz./A per season. Apply with ground equipment only. Rate varies according to climatic condition, varietal susceptibilty, disease history, and current disease status of the orchard. May be used alone or in tank mix combinations for concentrate sprays. Do not apply through any type of irrigation system. Refer to label for use directions.

Days To Harvest: Apples, pears-30; cherries-0.

Signal Word/Toxicity Class: Warning/II.
REI: 12 hr.

FENBUCONAZOLE

Triazole sterol inhibitor. Systemic fungicide with protectant, curative, and eradicant properties. Translocated upwards in the plant. Toxic to fish.

Indar (75WSP) *(Rohm and Haas)*

Rate: 2 oz. 75WSP/A in 50 gal. spray

Disease:
blossom blight; leaf spot; scab; fruit rot

Time: For blossom blight, apply at red bud stage. If conditions for disease develop, apply again at full bloom and petal fall. For fruit brown rot, apply 2-3 weeks before harvest at 7- to 10-day intervals. For leaf spot, use blossom blight schedule and continue at 10-14 day intervals. Additional applications may be made after harvest. For scab, apply at shuck split. Make 2-3 applications at 10- to 14-day intervals.

Remarks: Apricots, cherries, nectarines, peaches. Addition of surfactant will improve performance. Do not apply more than 1 lb./A per season. Do not graze livestock in treated areas or feed cover crops grown in treated areas to livestock. Refer to label for use directions.

Signal Word/Toxicity Class: Caution.
REI: 12 hr.

FERBAM

Broad-spectrum fungicide. Toxic to fish.

Ferbam Granuflo (WDG) *(UBC Agrochemicals)*

Rate: 0.76-1.1 lb. ai/100 gal. water
1-1.5 lb. WDG/100 gal. water
or 0.5 lb. WDG + 3 lb. wettable sulfur/100 gal. water

Disease:
blotch; cedar rust; scab

Time: For early-season control, apply prebloom, calyx, and early cover sprays. For rust, make additional bloom application using 0.5 lb./100 gal. water.

Rate: 0.57 lb. ai/100 gal. water
0.75 lb. WDG/100 gal. water

Disease:
bitter rot; Brooks fruit spot; scab; black rot; fly speck; sooty blotch

Time: Late-season control.

Remarks: Apples. Where bitter rot and Brooks spot are severe problems, increase rate to 1.5-2 lb./100 gal. water. On Golden Delicious, Jonathan, Baldwin, and other apple varieties which are especially apt to russet, some injury may result if used in early-season or bloom applications. The best time is cover sprays. When used in combination with sulfur, do not apply within 10-14 days of oil spray or just prior to or during periods of excessively high temperatures. Refer to label for use directions.

Days To Harvest: 7.

8 Fruits & Nuts

NOTICE The information on these pages is for preliminary planning — not a guide for use. Be sure to follow the manufacturer's directions, notwithstanding information contained here.

For personal protective equipment and EPA registration numbers, please turn to page 39.

Rate: 0.57 lb. ai/100 gal. water
1-1.5 lb./100 gal. water

Disease: Cherries:
brown rot leaf spot

Time: Apply petal fall and cover sprays. For leaf spot, apply immediately after harvest but prior to leaf drop.

Disease: Nectarines, peaches:
leaf curl

Time: Apply during dormant period in fall after leaves drop or in the spring before buds begin to swell. If leaf curl has been severe, make 2 applications in fall and spring during dormant period.

Disease: Pears:
scab sooty blotch

Time: Apply high rate in pink, calyx, first, and second cover spray; low rate in summer sprays.

Remarks: Cherries, nectarines, peaches, pears. When used in combination with sulfur, do not apply within 10-14 days of oil spray or just prior to or during periods of excessively high temperatures. Refer to label for use directions.

Days To Harvest: Cherries-0; pears-7; nectarines, peaches-21.

Signal Word/Toxicity Class: Caution.
REI: 24 hr.

FOSETYL-ALUMINUM

Systemic fungicide/bactericide. Toxic to estuarine invertebrates.

Aliette (WDG, WSP) *(Rhone-Poulenc)*

Rate: Bearing:
Foliar:
2-4 lb. ai/A
2.5-5 lb. WDG, WSP/A
Nonbearing:
Dip:
2.4 lb. ai/100 gal.
3 lb. WDG, WSP/100 gal.
Foliar:
2-4 lb. ai/100 gal.
2.5-5 lb. WDG, WSP/100 gal.

Disease:
apple blister spot (bearing); fire blight; Phytophthora collar rot; Phytophthora root rot

Time: For apple blister spot, apply early in season when blossoms are at pink stage; repeat at 7-day intervals only as long as conditions favor development. For fire blight, apply in spring when conditions are favorable for disease development; repeat at 4- to 7-day intervals only as long as conditions favor development. For Phytophthora collar and root rot, apply at 30- to 60-day intervals when conditions favor disease development. Do not apply for collar and root rot control until 30-days after last fire blight or blister spot application.

Remarks: Bearing, nonbearing: apples, crabapples, pears, quince. Nonbearing trees will not produce marketable fruit for 12 months after last application. Apply as foliar spray. Do not apply within 2-3 weeks of leaf senescence on bearing trees. Do not apply more than 5 lb./A per application, or 20 lb./A per season. Do not graze livestock on floor of treated orchards. Refer to label for use directions.

Days To Harvest: Bearing-14.

State Restrictions: In certain counties of Illinois, Minnesota, Ohio, Tennessee, Virginia, Vermont, Wisconsin, West Virginia, rate is limited to maximum of 3.75 lb. product/A; refer to label for specific counties.

Rate: 4 lb. ai
5 lb. WDG, WSP/100 gal. water

Disease:
Phytophthora collar rot Phytophthora root rot

Time: Apply at 60-day intervals when conditions favor disease development. Nursery tree resets and new plantings should be treated after leaf emergence.

Remarks: Nonbearing: apricots, cherries, nectarines, peaches, plums, prunes. Apply as foliar spray. Do not apply more than 100 gal./A or 5 lb./A per application or more than 20 lb./A per year. Do not exceed 4 applications per year. Refer to label for use directions.

Signal Word/Toxicity Class: Caution.
REI: 12 hr.

GALLEX (Premix)

Ready-to-use paint-on emulsion.

Gallex (2,4-xylenol & meta-cresol) *(AgBioChem)*

Rate: Apply undiluted.

Disease:
crown gall

Time: Apply first application before breaking of dormancy or during spring and early summer.

Remarks: Apples, apricots, cherries, nectarines, peaches, pears, plums, prunes. Paint gall and cut surfaces + 1/2" of surrounding healthy bark. Generally 1 application will give control; retreatment may be necessary. Do not mix or combine with other pesticides. Complete girdling of trunk by galls affects water-conducting capacity of tree. To prevent sudden loss of water movement, do not treat more than half of galled trunk at any one time. Treat remaining diseased tissue 4-6 months later when examination of initial treatment is made.

Signal Word/Toxicity Class: Warning/II.

Indar — see fenbuconazole

IPRODIONE

Dicarboximide fungicide. Toxic to aquatic invertebrates.

Rovral (4FL, WG, WP) *(Rhone-Poulenc)*

Rate: 0.5-1 lb. ai/A
1-2 pt. 4FL/A
1-2 lb. WG, WP/A

Disease: Peaches:
brown rot blossom blight scab shot hole

Time: Apply when bud tissue is susceptible to disease (pink bud); if conditions favorable for disease development persist or recur, apply at full bloom and petal fall.

Disease: Apricots, cherries, nectarines, plums, prunes:
brown rot blossom blight; cherry leaf spot; fruit brown rot; scab; shot hole

Time: Brown rot blossom blight: Apply when bud tissue is susceptible to disease development (pink, red, or white bud); if conditions favorable for disease development persist or recur, apply at full bloom and petal fall. Fruit brown rot: Apply when temperatures and moisture conditions favor disease infection in 5-week period prior to harvest; repeat at 7- to 14-day intervals following first preharvest application if needed. Scab, shot hole: Apply one blossom spray at shuck split; second foliar spray 14 days later, followed by 2-3 preharvest sprays starting 5-weeks before harvest at 7- to 14-day intervals. Cherry leaf spot: Follow brown rot blossom blight and fruit rot spray program. Make one foliar spray between petal fall and shuck split, with additional spray applied 14 days after harvest.

Remarks: Apricots, cherries, nectarines, peaches, plums, prunes. Under severe disease conditions, use high rate. Maximum of 4 applications per season (3 for peaches). Do not graze animals in treated orchards or feed cover crops grown in treated orchards to livestock. Refer to label for use directions.

Days To Harvest: 7 (except peaches).

Signal Word/Toxicity Class: Warning/II (WG); Caution (4FL, WP).
REI: 12 hr.

Kocide — see copper compounds

KOP 300 — see copper compounds

KOP-Hydroxide — see copper compounds

KOP Oxy — see copper compounds

Kumulus — see sulfur

Lime-Sulfur Solution — see calcium polysulfide

◆ - **new product** • ai-active ingredient • DF-dry flowable • E/EC-emulsifiable concentrate • EW-emulsion in water
F/FL-flowable • G-granule • L/LC-liquid • SP/WSP-soluble packet • W/WP-wettable powder • WSB-water soluble bag

MANCOZEB

Carbamate fungicide with broad-spectrum activity with no resistance problems. Coordination product of zinc ion and manganese ethylene bisdithiocarbamate. Toxic to fish.

Dithane (DF, F-45, M-45, WSP) *(Rohm and Haas)*
Manzate 200DF *(Griffin)*
Penncozeb 75DF or 80WP *(Elf Atochem)*

Rate: Prebloom/bloom:
4.8 lb. ai/A
6 lb. M-45, WSP, 80WP/A
6.4 lb. DF/A
9.6 pt. F-45/A
4.5 lb. ai/A
6 lb. 200DF, 75DF/A
Extended application/tank mix:
2.4 lb. ai/A
3 lb. M-45, WSP, 80WP/A
3.2 lb. DF/A
4.8 pt. F-45/A
2.25 lb. ai/A
3 lb. 200DF, 75DF/A

Disease:
cedar apple rust, fly speck, scab, Fabraea leaf spot, rust, sooty blotch

Time: Apply at $^1/_4$-$^1/_2$" green tip; repeat at 7- to 10-day intervals through bloom. For prebloom, apply through bloom; for extended application/tank mix, through second cover spray.

Remarks: Apples, crabapples, pears, quince. Do not combine or integrate the 2 treatment schedules. Extended applications based on tree-row volume or for use as a resistance management tool. Do not graze livestock in treated areas. Refer to label for use directions.

Days To Harvest: Extended schedule-77.

State Restrictions: Refer to label for California precautions.

Signal Word/Toxicity Class: Caution.
REI: 24 hr.

MANEB

Dithiocarbamate contact fungicide. Toxic to fish.

Maneb 75DF or 80WP *(Elf Atochem)*
Manex (FL) *(Griffin)*

Rate: Prebloom:
4.5 lb. ai/A
6 lb. 75DF/A
4.8 lb. ai/A
6 lb. 80WP/A
4.8 qt. FL/A
Extended application:
2.25 lb. ai/A
3 lb. 75DF/A
2.4 lb. ai/A
3 lb. 80WP/A
2.4 qt. FL/A

Disease:
cedar apple rust, fly speck, scab, Fabraea leaf spot, rust, sooty blotch

Time: Apply at $^1/_4$ to $^1/_2$" green tip. For prebloom, repeat at 7- to 10-day intervals through bloom; extended application, through second cover spray.

Remarks: Apples. Do not combine or integrate the 2 treatment schedules. Extended applications based on tree-row volume or for use as a resistance management tool. Do not graze livestock in treated areas. Refer to label for use directions.

Days To Harvest: Extended application-77.

Signal Word/Toxicity Class: Caution.
REI: 24 hr.

MANEB (Premix)

Fungicide/bactericide. Toxic to fish, birds, and other wildlife.

ManKocide (DF) (15% maneb & 46.1% copper hydroxide) *(Griffin)*

Rate: Apples:
10.66-21.33 lb. DF/A

Disease:
fire blight (suppression)

Time: Apply single application between silver tip and green tip as full cover spray. Crop injury may occur from late application; discontinue when green tip reaches $^1/_2$".

Rate: Pears:
1.5 lb. DF/A

Disease:
fire blight

Time: Apply at 5-day intervals throughout bloom period. Do not apply after bloom.

Remarks: Apples, pears. It is recommended that this product be used in an integrated pest management program. Russeting may occur in copper-sensitive pear varieties. Excessive dosages may cause fruit russet. Do not graze livestock in treated areas. Refer to label for use directions.

Days To Harvest: Apples-26.

Signal Word/Toxicity Class: Danger/I.
REI: 48 hr.

Manex — see maneb

ManKocide — see maneb premix

Manzate — see mancozeb

MEFENOXAM

Systemic fungicide having a specific mode of action.

Ridomil Gold EC, GR, or WSP *(Novartis)*

Rate: Broadcast spray or banded:
1 lb. ai/A
2 qt. EC/A
1.5 lb. ai/A
80 lb. GR/A
1.8 lb. ai/A
4 lb. WSP/A
Drench (apples):
0.25 lb. ai/100 gal water
0.5 pt. EC/100 gal. water
0.45 lb. ai/100 gal. water
1 lb. WSP/100 gal. water

Disease:
collar rot, crown rot, root rot

Time: Apples: Apply before disease development in early spring. Make additional application in fall after harvest, but before ground freezes. For new planting, delay first application until 2 weeks after planting. Stone fruit: For established plantings, apply in spring before plants begin growth. For new plantings, make first application 2 weeks after planting; repeat at 2- to 3-month intervals.

Remarks: Apples, apricots, cherries (sweet, tart), nectarines, peaches, plums (Chicksaw, Damson, Japanese), prunes. Apply only on nonbearing trees unless labeled for bearing trees. Do not dip roots in solution, spray roots, or concentrate it around tree trunk. Do not apply to trees under stress. Do not graze or feed cover crops in treated orchards. Do not apply more than 12 lb. WSP/A or 240 lb. GR/A per year. Do not plant any crop which is not registered for use with mefenoxam in treated soil for 12 months.

State Restrictions: In California, do not apply to newly planted trees within 90 days of planting.

Signal Word/Toxicity Class: Warning/II (EC); Caution (GR, WSP).
REI: 48 hr.

Mertect — see thiabendazole

METAM-SODIUM

Water-soluble liquid when applied to properly prepared soil is converted into a gaseous fumigant.

Vapam (WS) or HL *(Amvac)*

Rate: 1 qt. WS/100 sq. ft.
0.75 qt. HL/100 sq. ft.

Disease:
oak root fungus, soilborne diseases

Time: Tree replant sites in commercial orchards.

Remarks: Make shallow basin over planting site after removing dead or diseased trees and as much of root system as possible. Add Vapam to stream of water while filling basin. Apply in sufficient water to penetrate at least 6 ft.

NOTICE	The information on these pages is for preliminary planning — not a guide for use. Be sure to follow the manufacturer's directions, notwithstanding information contained here.	For personal protective equipment and EPA registration numbers, please turn to page 39.

Signal Word/Toxicity Class: Danger/I.
REI: 48 hr.

Micro Flo — see copper compounds

Microsperse — see copper compounds

Microthiol Special — see sulfur

MYCLOBUTANIL

Triazole sterol inhibitor. Systemic, protectant, and curative fungicide.

Nova or Rally (40W) *(Rohm and Haas)*
Rate: 1.25-2 oz. 40W/100 gal.
Disease:
rust   scab

Rate: 1.25-2.5 oz. 40W/100 gal.
Disease:
powdery mildew

Time: For best results, apply on 7- to 10-day schedule. Powdery mildew: Apply at tight cluster and continue through second cover spray. Additional sprays beyond second cover may be needed on susceptible varieties or under heavy disease pressure. Rusts: Apply at pink stage and continue through second cover spray. Scab: Prebloom, bloom, postbloom, postinfection.
Remarks: Apples. Prebloom, apply alone or tank mix with protectant fungicide. Bloom, postbloom, for summer disease control, tank mix with protectant fungicide registered for use on apples. Postinfection provides 96-hour control or curative activity, follow with a standard preventive spray schedule. Do not apply more than 5 lb./A per season. Do not apply this product through any type of irrigation system. Refer to label for use directions.
Days To Harvest: 14.
State Restrictions: Nova: Not for use in Alaska, Arizona, California, Hawaii, Idaho, Montana, Nevada, Oregon, Utah, Washington, Wyoming. Not for use on stone fruits in New York. Rally: For use only in Alaska, Arizona, California, Hawaii, Idaho, Montana, Nevada, Oregon, Utah, Washington, Wyoming.

- - - - - - - - - - - - - - - -

Rate: 1-2.4 oz. 40W/A
2.5-6 oz. 40W/A
Disease: Apricots, cherries, nectarines, peaches:
brown rot   leaf spot (cherries)   shot hole (except peaches)
brown rot blossom blight   powdery mildew
rust (peaches)
Plums, prunes:
brown rot blossom blight   rust

Time: For brown rot blossom blight, apply at red bud stage for apricots before infection occurs (popcorn stage for cherries; early pink bud stage for nectarines, peaches; at green tip for plums, prunes); if conditions are favorable for disease development, apply again at full bloom and petal fall. For brown rot, apply whenever conditions favor disease development during month prior to harvest. For powdery mildew and shot hole, follow brown rot blossom blight schedule and continue application at 10- to 14-day intervals until terminal growth ceases (powdery mildew) or 7- to 10-day intervals or as long as needed (shot hole). For leaf spot, follow brown rot schedule and repeat at 7- to 10-day intervals; make additional applications after harvest. For rust, apply 8 weeks after flowering if conditions are favorable for disease development.
Remarks: Apricots, cherries, nectarines, peaches, plums, prunes. Do not apply more than 2.75 lb./A per season for apricots, plums, prunes; 3.25 lb./A per season for cherries, nectarines, peaches. Do not apply this product through any type of irrigation system. Refer to label for use directions.
Days To Harvest: 0.
State Restrictions: Nova: Not for use in Alaska, Arizona, California, Hawaii, Idaho, Montana, Nevada, Oregon, Utah, Washington, Wyoming; not for sale or use in Nassau or Suffolk counties, New York. Do not apply by air in New York. Rally: For use only in Alaska, Idaho, Montana, Nevada, Oregon, Utah, Washington, Wyoming.

- - - - - - - - - - - - - - - -

Signal Word/Toxicity Class: Nova: Caution; Rally: Warning/II or Caution (varies by region).
REI: 24 hr.

Nova — see myclobutanil

Nu-Cop — see copper compounds

Orbit — see propiconazole

Penncozeb — see mancozeb

POLYRAM

Contact fungicide. Toxic to fish.

Polyram 80DF (metiram) *(BASF, Platte)*
Rate: Extended:
2.4 lb. ai/A
3 lb. 80DF/A
Prebloom:
4.8 lb. ai/A
6 lb. 80DF/A
Disease:
rust   scab

Time: Apply at 1/4-1/2" green tip; repeat at 7- to 10-day intervals through second cover spray; through bloom for prebloom. Do not apply after bloom.
Remarks: Apples. Do not combine or integrate the 2 treatment schedules. Do not graze livestock in treated areas. Do not apply this product through any type of irrigation system. Refer to label for use directions.
Days To Harvest: Extended application-77.

- - - - - - - - - - - - - - - -

Signal Word/Toxicity Class: Caution.
REI: 12 hr. (BASF); 24 hr. (Platte).

PROPICONAZOLE

Triazole sterol inhibitor. Toxic to fish.

Banner MAXX *(Novartis)*
Rate: Apples:
2-4 fl. oz. MAXX/100 gal. water
Disease:
powdery mildew   rust   scab

Rate: Cherries:
2-4 fl. oz. MAXX/100 gal. water
Disease:
blossom blight   leaf spot   powdery mildew
Time: Apply at 14- to 21-day intervals.

Rate: Nectarines, peaches, plums:
2-4 fl. oz. MAXX/100 gal. water
Disease:
blossom blight
Time: Apply at 5-10% bloom and 70-100% bloom.
Remarks: Nonbearing. Do not apply this product through any type of irrigation system. Refer to label for use directions.

- - - - - - - - - - - - - - - -

Signal Word/Toxicity Class: Warning/II.
REI: 24 hr.

Break (EC) *(Novartis)*
Rate: 4 fl. oz. EC/A
Disease: Nectarines, peaches, plums:
brown rot blossom blight   fruit brown rot
Time: For brown rot blossom blight, apply at 5-10% bloom and at 80-100% bloom. For fruit brown rot, apply maximum of 2 preharvest sprays 10-14 days before harvest up to day of harvest.

Disease: Prunes:
brown rot blossom blight
Time: Apply at 5-10% bloom and 50-75% bloom. If blossoming is prolonged, apply at petal fall.
Remarks: Apply minimum 20 gal./A spray volume by air; 50 gal./A spray volume by ground.
State Restrictions: For use west of the Rockies.

- - - - - - - - - - - - - - - -

Signal Word/Toxicity Class: Warning/II.
REI: 24 hr.

◆ - **new product** • ai-active ingredient • DF-dry flowable • E/EC-emulsifiable concentrate • EW-emulsion in water
F/FL-flowable • G-granule • L/LC-liquid • SP/WSP-soluble packet • W/WP-wettable powder • WSB-water soluble bag

Orbit (EC) *(Novartis)*
Rate: 4 fl. oz. EC/A4
Disease:
brown rot blossom blight fruit brown rot
Time: For brown rot blossom blight, apply early bloom stage; make second application at 50-75% bloom. If blossoming is prolonged or conditions persist, make third application at petal fall. For fruit brown rot, make maximum of 2 preharvest sprays from 3 weeks before harvest through day of harvest.
Remarks: Apricots, cherries, nectarines, peaches, plums. Apply in sufficient water for thorough coverage. Do not exceed 4 applications per season. Do not mix with Cyprex. Do not apply this product through any type of irrigation system. Do not graze livestock in treated areas.
State Restrictions: For use east of the Rockies.

Signal Word/Toxicity Class: Warning/II.
REI: 24 hr.

Rally — see myclobutanil

Redball — see sulfur

Ridomil Gold — see mefenoxam

Rovral — see iprodione

Rubigan — see fenarimol

Sulfa 80 — see sulfur

Sulforix — see calcium polysulfide

SULFUR

Nonsystemic fungicide with protective action. Microthiol Special, Sulfa 80, and Thiolux are micronized sulfurs.

Drexel Suffa (6FL, 8FL), Sulfa 80 (80DF) or Sulfur (90W) *(Drexel)*
Rate: 1.1-2.6 lb. ai/100 gal. water
1.5-3.5 pt. 6FL/100 gal. water
0.75-1.5 lb. ai/100 gal. water
0.75-1.5 pt. 8FL/100 gal. water
8-24 lb. ai/A
10-30 lb. 80DF/A
1.8-5.4 lb. ai/100 gal. water
2-6 lb. 90W/100 gal. water
Disease:
powdery mildew scab
Time: Apply prebloom, bloom, petal fall, shuck, and cover spray throughout season.
Remarks: Apples, pears. Sensitive pear varieties such as d'Anjou and Comice may be injured by sulfur under certain climatic conditions. Do not use within 2 weeks of oil application (4 weeks FL). Some plants are susceptible to injury from sulfur under certain climatic conditions. Do not make sulfur applications at excessively high temperatures. Do not apply this product through any type of irrigation system. Refer to label for use directions.

Rate: **Cherries, peaches, plums, prunes:**
1.1-2.6 lb. ai
1.5-3.5 pt. 6FL/100 gal. water
1.8-5.4 lb. ai
2-6 lb. 90W/100 gal. water
Peaches:
0.75-1.33 lb. ai
0.75-1.33 pt. 8FL/100 gal. water
Cherries, plums, prunes:
0.5-1.33 lb. ai
0.5-1.33 pt. 8FL/100 gal. water
Disease:
blossom blight leaf spot rust
brown rot powdery mildew scab
Coryneum blight
Time: Apply prebloom, bloom, petal fall, and as cover spray at 7- to 14-day intervals for cherries, nectarines, and peaches; 7- to 10-day intervals for plums and prunes. Also as postharvest foliar treatment for leaf spot on cherries.

Rate: **Cherries, nectarines, peaches, plums, prunes:**
8-24 lb. ai/A
10-30 lb. 80DF/A
Disease:
brown rot blossom blight leaf spot scab
cherry leaf spot powdery mildew
Time: Apply at bloom or early petal fall; repeat as necessary, usually 10- to 14-days after periods of wet weather. For brown rot of fruit, apply when fruit starts to ripen.
Remarks: Cherries, nectarines, peaches, plums, prunes. Do not use within 2 weeks of oil application (4 weeks FL). Some plants are susceptible to injury from sulfur under certain climatic conditions. Do not make sulfur applications at excessively high temperatures. Do not apply this product through any type of irrigation system. Refer to label for use directions.

Signal Word/Toxicity Class: Caution.
REI: 24 hr.

Kumulus DF (sulfur) *(BASF)*
Microthiol Special (80WP) *(Elf Atochem)*
Rate: 8-16 lb. ai/A
10-20 lb. DF, 80WP/A
Disease: Apples, pears:
powdery mildew scab
Time: Apply prebloom and petal fall with cover sprays throughout the season at 10- to 14-day intervals.

Rate: 8-16 lb. ai/A
10-20 lb. DF, 80WP/A
Disease: Cherries, nectarines, peaches, plums, prunes:
brown rot blossom blight cherry leaf spot powdery mildew
Time: Apply at bloom or early petal fall; repeat as necessary, usually 10- to 14-days or after wet weather. For brown rot, apply when fruit starts to ripen.
Remarks: Apples, cherries, nectarines, peaches, pears, plums, prunes. Do not apply to d'Anjou pears, apricots, or sensitive apple varieties such as Ontario or Cox orange. Sulfur may burn foliage when temperature is excessively high. Do not apply if temperatures during or within 3 days after application are expected to exceed 90°F. Do not apply Kumulus through any type of irrigation system. Refer to label for use directions.

Signal Word/Toxicity Class: Caution.
REI: 24 hr.

Liquid Sulfur Six *(Helena)*
Rate: **Apples, pears:**
0.75-3.5 pt./100 gal. water
Disease:
powdery mildew scab
Time: Apply prebloom through calyx, or as cover spray.

Rate: **Nectarines, peaches:**
1.5-3.5 pt./100 gal. water
Disease:
brown rot powdery mildew scab
Time: For brown rot, apply when fruit starts to ripen. For pink and bloom sprays, apply lower rate; for petal fall, shuck split, shuck fall, apply higher rate and repeat at 10- to 14-day intervals as needed.

Rate: **Cherries, plums, prunes:**
0.66-3.5 pt./100 gal. water
Disease:
brown rot leaf spot
Time: For petal fall, shuck, and cover spray, apply lower rate; pink and bloom sprays, apply higher rate.
Remarks: Apples, cherries, nectarines, peaches, pears, plums, prunes. Rates vary by time of application. Application to mature nectarines may cause discoloration. Do not apply this product through any type of irrigation system. Refer to label for use directions.

Signal Word/Toxicity Class: Caution.
REI: 24 hr.

NOTICE The information on these pages is for preliminary planning — not a guide for use. Be sure to follow the manufacturer's directions, notwithstanding information contained here.	For personal protective equipment and EPA registration numbers, please turn to page 39.

Micro Flo Sulfur (6L), Dual Spraying & Dusting Sulfur (DS) or Microsperse Wettable Sulfur (WS) *(Micro Flo)*
Terra Sulfur (6L) *(Terra)*

Rate: Apples, pears:
15-30 lb. ai/A
2.5-5 gal. 6L/A
13.5-27 lb. ai/A
15-30 lb. DS, WS/A

Disease:
powdery mildew scab

Time: Apply lower rate before flowering, higher rate after flowering.

Rate: Cherries, nectarines, peaches, plums, prunes:
30 lb. ai/A
5 gal. 6L/A
27 lb. ai/A
30 lb. DS, WS/A

Disease:
brown rot leaf spot rust
Coryneum blight powdery mildew scab

Time: Apply at bloom; repeat at 12-day intervals as necessary.

Rate: Quince:
34-58.5 lb. ai/A
5.66-9.75 gal. 6L/A

Disease:
brown rot powdery mildew scab

Time: Apply before disease appears; repeat at 7- to 10-day intervals as needed.
Remarks: Apples, cherries, nectarines, peaches, pears, plums, prunes, quince. Do not use within 2 weeks of oil application. Do not apply DS and WS through any type of irrigation system. Refer to label for use directions.

Signal Word/Toxicity Class: Caution.
REI: 24 hr.

Redball EM-53 (6L) *(International Sulphur)*

Rate: Apples, pears:
0.75-3.5 pt. 6L/100 gal. water

Disease:
powdery mildew scab

Rate: Cherries, plums, prunes:
0.75-3.5 pt. 6L/100 gal. water

Disease:
brown rot leaf spot

Rate: Nectarines, peaches:
0.66-3.5 pt. 6L/100 gal. water

Disease:
brown rot powdery mildew scab

Time: Apples, pears: prebloom through calyx, cover sprays. Cherries, nectarines, peaches, plums, prunes: pink, bloom, petal fall, shuck, and cover sprays.
Remarks: Apples, cherries, nectarines, peaches, pears, plums, prunes. Applications to mature nectarines may cause discoloration. Apply in sufficient water for thorough coverage. Do not apply when temperatures exceed 90°F. Refer to label for use directions.

Signal Word/Toxicity Class: Caution.
REI: 12 hr.

Redball Superfine Wettable Sulphur (WS) or Commercial Dusting Sulphur (DS) *(International Sulphur)*

Rate: 5-8 lb. WS/100 gal. water
30-50 lb. DS/A

Disease: Apples, pears:
powdery mildew rust scab

Apricots, cherries, nectarines, peaches, plums, prunes:
blossom blight fruit brown rot powdery mildew
brown rot leaf spot

Time: Apples, pears: pink, bloom, petal fall, and cover sprays. Apricots, cherries, nectarines, peaches, plums, prunes: pink, bloom, and petal fall (blossom blight, brown rot); pink, bloom, petal fall, shuck, and cover sprays (leaf spot, powdery mildew); apply when fruit starts to ripen, continue until harvest (fruit brown rot).
Remarks: Apples, apricots, cherries, nectarines, peaches, pears, plums, prunes. Sulfur may injure certain varieties of apples, check tolerance prior to application. Do not apply after pears have begun to turn down; do not apply to d'Anjou pears. Do not apply when temperatures exceed 90°F. Refer to label for use directions.

Signal Word/Toxicity Class: Caution.
REI: 12 hr.

THAT (6FL) or THAT Big 8 (8FL) *(Stoller)*

Rate: Apples, pears:
0.75-2.6 lb. ai/100 gal. water
1-3.5 pt. 6FL/100 gal. water
1-2.33 lb. ai/100 gal. water
1-2.33 pt. 8FL/100 gal. water

Disease:
powdery mildew scab

Time: Prebloom through calyx, cover sprays.

Rate: Nectarines, peaches:
1.1-2.8 lb. ai/100 gal. water
1.5-3.75 pt. 6FL/100 gal. water
1-3 lb. ai/100 gal. water
1-3 pt. 8FL/100 gal. water

Disease:
brown rot powdery mildew scab
leaf spot

Time: Pink, bloom sprays, petal fall, shuck, cover sprays, and for scab, shuck-split through preharvest. Do not apply during periods of excessively high temperature.

Rate: Cherries, plums, prunes:
0.5-1.3 lb. ai/100 gal. water
0.66-1.75 pt. 6FL/100 gal. water
0.5-1 lb. ai/100 gal. water
0.5-1 pt. 8FL/100 gal. water

Disease:
brown rot leaf spot

Time: Pink, bloom sprays, petal fall, shuck, cover sprays.
Remarks: Apples, cherries, nectarines, peaches, pears, plums, prunes. Refer to label for use directions.

Signal Word/Toxicity Class: Caution.
REI: 24 hr.

Thiolux (DF) *(Platte)*

Rate: 10-30 lb. DF/A

Disease: Apples, pears:
powdery mildew scab

Time: Apply prebloom, petal fall, and with cover sprays throughout season.

Disease: Cherries, nectarines, peaches, plums, prunes:
brown rot blossom blight leaf spot rust
powdery mildew scab
cherry leaf spot

Time: Apply bloom or early petal fall; repeat at 10- to 14-day intervals or as necessary after periods of wet weather. For brown rot, apply when fruit starts to ripen. For rust on peaches and prunes, apply any time from late spring to fall in preventative program.
Remarks: Apples, cherries, nectarines, peaches, pears, plums, prunes. Dormant application with oil may be made to apples and pears. Apply only to dormant d'Anjou pears. Do not make applications within 2 weeks of oil spray. Refer to label for use directions.

Signal Word/Toxicity Class: Caution.
REI: 24 hr.

Wettable Sulfur (WS) *(Platte)*

Rate: 6-8 lb. WS/100 gal. water

Disease:
powdery mildew scab

Time: Apply prebloom, calyx, and early cover sprays.
Remarks: Apples, pears. May also use 4-6 lb. product + 2-3 lb. polysulphide compound or 2 gal. liquid lime sulfur/100 gal. dilute spray. Sulfur may injure certain varieties of apples, check specific varietal tolerance to sulfur prior to using. Do not make applications within 2 weeks of oil spray. Refer to label for use directions.

Rate: 5-10 lb. WS/100 gal. water

Disease:
brown rot cherry leaf spot powdery mildew

Time: Apply prebloom, bloom, and postbloom, or when disease threatens during rainy periods; repeat at 7- to 10-day intervals.
Remarks: Cherries, nectarines, peaches, plums, prunes. Do not make applications within 2 weeks of oil spray. Refer to label for use directions.

Signal Word/Toxicity Class: Caution.
REI: 24 hr.

◆ - **new product** • ai-active ingredient • DF-dry flowable • E/EC-emulsifiable concentrate • EW-emulsion in water
F/FL-flowable • G-granule • L/LC-liquid • SP/WSP-soluble packet • W/WP-wettable powder • WSB-water soluble bag

SULFUR (Premixes)

Fungicide/bactericide in flowable formulation. Toxic to fish and aquatic organisms. Terranil S is also toxic to marine/estuarine organisms.

Bravo S (27.25% sulfur & 19.15% chlorothalonil) *(Helena)*

Rate: 3-4 pt./100 gal. water
Disease:
Coryneum blight (nectarines, peaches); leaf curl
Time: Make 1 application in late autumn to early winter before hard freezing occurs; make 1-2 additional applications in mid-late winter before buds begin to swell. Coryneum blight: Apply once at petal fall or at shuck split.

Disease:
brown rot blossom blight; cherry leaf spot; powdery mildew (nectarines, peaches); scab (nectarines, peaches)
Time: Make 1 application at popcorn; second at full bloom. If conditions favor disease development, make additional application at petal fall. In addition to bloom application, make 1 application at shuck split. Cherry leaf spot after harvest: Make 1 application to foliage within 7 days after fruit is removed. In orchards with history of high incidence, make second application 10- to 14-days later.
Remarks: Cherries (sweet, tart), nectarines, peaches, plums, prunes. Apply 12-16 pt./A on trees taller than 20 ft.; 9-12 pt./A on smaller trees. Do not apply after shuck split and before harvest. Refer to label for use directions.

- - - - - - - - - - - - - - - -

Signal Word/Toxicity Class: Caution.
REI: 48 hr.

Terranil S (19.15% chlorothalonil & 27.25% sulfur) *(Terra)*

Rate: 3-4 pt./100 gal.
Disease:
brown rot blossom blight; Coryneum blight; leaf curl; leaf spot (cherries); powdery mildew (peaches); scab (apricot, nectarines, peaches)
Time: For Coryneum blight and leaf curl, apply at leaf fall in late autumn. For brown rot blossom blight, powdery mildew, and scab, make 1 application at popcorn (pink, red, or early white bud) and second application at full bloom. If weather conditions favor disease development, make additional application at petal fall. In addition to bloom application, make 1 application at shuck split. Do not apply after shuck split and before harvest. For control of cherry leaf spot after harvest, make 1 application to foliage within 7 days after fruit is removed. In orchards with history of high leaf spot incidence, make second application 10-14 days later.
Remarks: Apricots, cherries (sweet, tart), nectarines, peaches, plums, prunes. Use in sufficient water to obtain adequate coverage. Use 9-12 pt./A on small trees, 12-16 pt./A on trees taller than 20 ft. For Coryneum blight and leaf curl, when conditions favor high disease levels, use high rate and apply once or twice more in mid- to late-winter before buds swell. If leaf fall application is not practical, apply anytime prior to bud swell the following spring. When Coryneum blight occurs, also apply at bud break to protect newly emerging leaves and at shuck split to prevent fruit infections. Do not allow livestock to graze in treated areas. Refer to label for use directions.

- - - - - - - - - - - - - - - -

Signal Word/Toxicity Class: Warning/II.
REI: 48 hr.

Top Cop with Sulfur (6.25 lb./gal. sulfur & 1 lb./gal. tribasic copper sulfate) *(Stoller)*

Rate: 4 pt./100 gal. water
Disease:
bull's-eye rot; fire blight
Time: For bull's-eye rot, apply foliar before harvest. For fire blight, apply at 10% bloom; repeat at 5- to 7-day intervals through bloom.
Remarks: Pears. Add spray adjuvant as spreader for bull's-eye rot. Refer to label for use directions.

- - - - - - - - - - - - - - - -

Rate: 8-12 pt./100 gal. water
Disease:
brown rot; shot hole
Time: Apply as delayed dormant or swelling bud to popcorn stage.
Remarks: Apricots. Avoid spraying when in leaf, as injury may occur. Refer to label for use directions.

- - - - - - - - - - - - - - - -

Rate: 1-3 pt./100 gal. water
Disease:
brown rot; leaf spot; powdery mildew; scab
Time: For brown rot and leaf spot, apply petal fall, shuck and cover sprays. For powdery mildew, apply pink and bloom spray. For scab, apply shuck split through preharvest.
Remarks: Nectarines, peaches. Refer to label for use directions.

- - - - - - - - - - - - - - - -

Rate: 1-1.8 pt./100 gal. water
Disease:
brown rot; leaf spot
Time: For brown rot, apply pink and bloom spray. For leaf spot, apply petal fall shuck and cover sprays.
Remarks: Cherries, plums, prunes. Refer to label for use directions.

- - - - - - - - - - - - - - - -

Signal Word/Toxicity Class: Caution.
REI: 24 hr.

Syllit — see dodine

TEBUCONAZOLE

Foliar fungicide for control of specified diseases. Toxic to estuarine and marine invertebrates.

Elite 45DF *(Bayer)*

Rate: 4-8 oz. 45DF/A
Disease:
blossom blight; brown rot; fruit rot; leaf spot (cherry); powdery mildew (cherry)
Time: For blossom blight, apply at white bud on cherry or pink bud on nectarine and peach; reapply at 50% bloom and at petal fall if necessary. For fruit rot, apply 21, 14, and 7 days before harvest; additional cover sprays during early postbloom. For leaf spot, apply at petal fall or when first leaves unfold; repeat at 10- to 14-day intervals. For powdery mildew, follow leaf spot schedule until terminal growth ceases.
Remarks: Cherries, nectarines, peaches. Do not apply through any type of irrigation system. Refer to label for use directions.
Days To Harvest: 0.
Crop Rotation: Any crop not specified on this label may be planted into treated areas 120 days after last application.

- - - - - - - - - - - - - - - -

Signal Word/Toxicity Class: Warning/II.
REI: 12 hr.

TELONE

Liquid fumigant for preplant soil treatment. Restricted Use Pesticide.

Telone II (dicloropropene) *(Dow AgroSciences)*

Rate: 35 gal. II/A
Disease:
bacterial canker
Time: Preplant broadcast. Apply preferably in the fall (light sandy soils) when soil is warm and moist (55°-80°F at injection depth).
Remarks: Peaches. Inject fumigant at a depth of 12-14" with chisels mounted on 12" centers. Do not apply through any type of irrigation system. Refer to label for use directions.

- - - - - - - - - - - - - - - -

Signal Word/Toxicity Class: Warning/II.
REI: 72 hr.

Tennessee Brand — see copper compounds

Terranil — see chlorothalonil

Terranil S — see sulfur premixes

THAT or THAT Big 8 — see sulfur

For personal protective equipment and EPA registration numbers, please turn to page 39.

THIABENDAZOLE

Toxic to fish.

Mertect 340-F *(Novartis)*
Rate: 16 fl. oz. 340-F/100 gal. water
Disease:
bull's-eye rot gray mold stem end rot
Time: Apply by dipping, flooding, or spraying.
Remarks: Apples, pears. Treat apples before and after storage, pears only once. Do not apply through any type of irrigation system.

Signal Word/Toxicity Class: Caution.

Thiolux — see sulfur

THIOPHANATE-METHYL

Benzimidazole systemic fungicide. Toxic to fish.

Topsin M (WSB, 70W) *(Elf Atochem)*
Rate: 0.7-1 lb. ai/A
1-1.5 lb. WSB, 70W/A
Disease:
apple scab, Brooks fruit rot, powdery mildew, black rot, fly speck, sooty blotch
Time: Preharvest. Apply at 5- to 10-day intervals from green tip through petal fall; continue at 7- to 14-day intervals in cover sprays.
Remarks: Apples. Refer to label for use directions.

Rate: 1 lb. ai/A
1.5 lb. WSB, 70W/A
Peaches:
1-1.6 lb. ai/A
1.5-2.25 lb. WSB, 70W/A
Disease:
black knot (plums, prunes), cherry leaf spot (cherries), peach scab (peaches), brown rot blossom blight, fruit brown rot, powder mildew (cherries)
Time: Preharvest: Apply at early bloom; make second application at full bloom. For fruit brown rot, apply 1-2 sprays starting 3 weeks before harvest (repeat at 10- to 14-day intervals between full bloom and final preharvest sprays if needed). For black knot, apply at prebloom, petal fall, and at first, second, and third cover sprays at 10- to 14-day intervals. For leaf spot, apply at petal fall, shuck, and at first, second, and third cover sprays at 10- to 14-day intervals and one spray 14-21 days after harvest. For cherry leaf spot, apply at petal fall or before, when leaves first unfold, and at first, second, and third cover at 10- to 14-day intervals and one spray 14-21 days after harvest. For peach scab and powdery mildew, apply at shuck split and at first cover sprays.
Remarks: Apricots, cherries, nectarines, peaches, plums, prunes. Refer to label for use directions.
Days To Harvest: 1.

Signal Word/Toxicity Class: Caution.
REI: 12 hr.

THIRAM

Broad-spectrum fungicide. Toxic to fish.

Thiram 65 (WP) or Thiram Granuflo (WDG) *(UCB Agrochemicals)*
Rate: Apples:
0.75-2 lb. WP/100 gal. spray
1.3-1.7 lb. WDG/100 gal. spray
Disease:
apple blotch, Brooks fruit spot, gray mold, bitter rot, bull's-eye rot, scab, black pox, cedar apple rust, sooty blotch, black rot, fly speck
Time: Apply late cover sprays. For cedar apple rust, apple blotch, bitter rot, black pox, and scab, apply at prebloom, calyx, and early cover sprays.

Rate: Peaches:
1.5-2 lb. WP/100 gal. spray
1.3-1.7 lb. WDG/100 gal. spray
Disease:
brown rot blossom blight, fruit rot, Rhizopus rot, leaf curl, scab

Time: For brown rot blossom blight, apply at 3- to 4-day intervals during bloom to keep newly opened blossoms covered. For fruit and Rhizopus rot, apply at petal fall, repeat 2 weeks later, then make 1-2 additional applications at 7-day intervals about 1 month before harvest or earlier if rot is apparent on green fruit. For scab, apply at petal fall; repeat 3-6 times at 10- to 14-day intervals. For leaf curl, apply postharvest in late fall and/or in late winter prior to bud swell.
Remarks: Apples, peaches. Do not apply through any type of irrigation system. Do not graze or feed cover crops from treated orchards. Refer to label for use directions.
Days To Harvest: Peaches-7.

Signal Word/Toxicity Class: Caution.
REI: 24 hr.

TOP COP — see copper compounds

Top Cop with Sulfur — see sulfur premixes

Topsin M — see thiophanate-methyl

TRIADIMEFON

Triazole sterol inhibitor systemic fungicide compatible with many insecticides, miticides, and fungicides.

Bayleton 50% DF *(Bayer)*
Rate: 2-8 oz. 50% DF/A
Disease:
cedar-apple rust, powdery mildew
Time: For apples, make first application at green tip; for pears, at bud burst. Repeat at 7- to 14-day intervals.
Remarks: Apples, pears. Do not apply more than 24 oz./A per season. Do not graze livestock in treated orchards. Do not apply through any type of irrigation system. Refer to label for use directions.
Days To Harvest: 45.

Signal Word/Toxicity Class: Caution.
REI: 12 hr.

VANGARD

Broad-spectrum fungicide. Toxic to fish and aquatic invertebrates.

Vangard (WG) (cyprodinil) *(Novartis)*
Rate: 5 oz./A
Tank mix:
3 oz./A
Disease:
scab
Time: Repeat at 7- to 10-day intervals. Apply at green tip if used alone; prebloom, pink, bloom, postbloom for tank mixes.
Remarks: Apples, crabapples, pears. Under severe disease pressure, use shorter intervals. For pears, must be tank mixed. Use high rate for severe disease conditions. Do not apply more than 22 oz. WG/A per crop per season. Refer to label for use directions.
Days To Harvest: 72.

Rate: 5 oz./A
Disease:
brown rot blossom blight
Time: Apply at bloom stages (apricots at red bud, cherries at popcorn, nectarines, peaches at pink bud, plums and prunes at green tip); repeat at full bloom.

Rate: California:
10 oz./A
Disease:
fruit rot
Time: Apply maximum of 2 applications as needed at 1/2-rate from preharvest to 2 days before harvest.
Remarks: Apricots, cherries, nectarines, peaches, plums, prunes. May be used alone or in tank mix. Use high rate for severe disease conditions. Do not apply more than 10 oz. WG/A per crop per season for brown rot blossom blight; 20 oz. WG/A per crop per season for fruit brown rot. Refer to label for use directions.
Days To Harvest: 2.

Signal Word/Toxicity Class: Caution.
REI: 12 hr.

◆ - **new product** • ai-active ingredient • DF-dry flowable • E/EC-emulsifiable concentrate • EW-emulsion in water
F/FL-flowable • G-granule • L/LC-liquid • SP/WSP-soluble packet • W/WP-wettable powder • WSB-water soluble bag

Vapam — see metam-sodium

ZIRAM

Toxic to fish and wildlife.

Ziram 4L or 76 *(Drexel)*

Rate: Apples, pears:
Pacific NW only:
1 lb. ai/100 gal.
2 pt. 4L/100 gal.
0.76-1.1 lb. ai/100 gal.
1-1.5 lb. 76/100 gal.

Disease:
bull's-eye rot; scab

Time: Bull's-eye rot: Apply first cover spray and preharvest spray before fall rains begin; repeat as necessary, especially if excessive rainfall occurs. Scab: Apply at prebloom, calyx, and cover sprays. In Washington: Apply first cover spray and in early Sept. on Delicious apples, late Sept. on Winesap apples as protectant spray only.

Rate: Cherries:
1 lb. ai/100 gal.
2 pt. 4L/100 gal.
1.1 lb. ai/100 gal.
1.5 lb. 76/100 gal.

Disease:
brown rot; leaf spot

Time: Brown rot: Apply popcorn, full bloom, petal fall, and as necessary in cover applications. Leaf spot: Apply petal fall, shuck fall, and 2 weeks after shuck fall.

Rate: Apricots:
1.5-2 lb. ai/100 gal.
3-4 pt. 4L/100 gal.
1.1-1.5 lb. ai/100 gal.
1.5-2 lb. 76/100 gal.

Disease:
brown rot; shot hole

Time: Brown rot: Apply prebloom, blossom, petal fall stages, and early cover stages. Shot hole: Apply at prebloom, early bloom, and full bloom (under severe disease conditions or when weather is extremely wet, repeat every 5 days until full bloom). For postharvest application, apply fall before heavy rains.

Rate: Nectarines, peaches:
1-2 lb. ai/100 gal.
2-4 pt. 4L/100 gal.
1.1-2 lb. ai/100 gal.
1.5-3 lb. 76/100 gal.

Disease:
brown rot; peach leaf curl; shot hole

Time: Brown rot: Apply at pink bud, 25%-75% of full bloom, petal fall, and cover stages. Leaf curl and shot hole: Apply at dormant and popcorn stages. Peaches-West Coast: Leaf curl: Apply just before fall rains or in November; repeat in 3 months while trees are still dormant and again in popcorn stage.

Remarks: Apples, apricots, cherries, nectarines, peaches, pears. Do not apply this product through any type of irrigation system. Refer to label for use directions.

State Restrictions: Not for use on apples, pears, cherries in California.

Signal Word/Toxicity Class: Danger/I.

REI: 48 hr.

Ziram 76DF *(Elf Atochem)*

Rate: Apples, pears:
4.6-6 lb. ai/A
6-8 lb. 76DF/A

Disease: Apples:
apple scab; bitter rot; bull's-eye rot (Pacific NW only); cedar apple rust; fly speck; leaf blotch; quince rust; sooty blotch

Pears:
bull's-eye rot (Pacific NW only); Fabraea leaf spot; pear scab

Time: Apply from prebloom through cover sprays as needed; under severe disease pressure use higher rate. For bull's-eye rot, apply in first cover spray and in preharvest spray before fall rains begin; repeat as necessary, especially if excessive rainfall occurs.

Rate: Apricots:
Western U.S.:
4.6 lb. ai/A
6 lb. 76DF/A

Disease:
brown rot; shot hole

Time: Apply at popcorn, full bloom, petal fall and/or 5 weeks after petal fall, and in cover sprays as needed.

Rate: Cherries:
Eastern U.S.:
3.8-6 lb. ai/A
5-8 lb. 76DF/A
Western U.S.:
4.6 lb. ai/A
6 lb. 76DF/A

Disease:
brown rot; leaf spot

Time: Apply at prebloom through cover sprays as needed.

Rate: Nectarines, peaches:
Eastern U.S.:
3.4-6 lb. ai/A
4.5-8 lb. 76DF/A
Western U.S.:
4.6-6 lb. ai/A
6-8 lb. 76DF/A

Disease:
brown rot; peach leaf curl; peach scab

Time: Apply at prebloom through cover sprays or as needed; under severe disease pressure use higher rate. For leaf curl, make dormant applications after leaf drop and/or prior to bud swell.

Remarks: Apples, apricots, cherries, nectarines, peaches, pears. For apples and pears do not apply more than 56 lb./A per crop cycle in Eastern U.S; 32 lb./A per crop cycle in Western U.S. Do not apply this product through any type of irrigation system. Do not graze or feed cover crops from treated orchards. Refer to label for use directions.

Days To Harvest: Pears (Western U.S.)-5. Apples, pears & cherries (Eastern U.S.)-14. Apricots, cherries (Western U.S.)-30.

State Restrictions: California labels may vary; refer to label.

Ziram Granuflo (WDG) *(UCB Agrochemicals)*

Rate: 4.6-6 lb. ai/A
6-8 lb. WDG/A

Disease: Apples:
apple scab; bitter rot; bull's-eye rot (Pacific NW); cedar apple rust; fly speck; quince rust; sooty blotch

Time: Apply from prebloom through early cover sprays as needed. For bull's eye rot, apply 12-14 days after calyx stage; repeat as necessary in summer sprays.

Disease: Pears:
bitter rot; bull's-eye rot (Pacific NW); Fabraea leaf spot; fly speck; pear scab; quince rust; sooty blotch

Time: Apply from prebloom through cover sprays as needed. For bull's-eye rot, apply first cover spray and preharvest spray before fall rains begin; repeat as necessary, especially if excessive rainfall occurs.

Rate: Cherries:
Eastern U.S.:
3.8-6 lb. ai/A
5-8 lb. WDG/A
Western U.S.:
3-4.6 lb. ai/A
4-6 lb. WDG/A
California only:
3-3.8 lb. ai/A
4-5 lb. WDG/A

Disease:
brown rot blossom blight; fruit rot; leaf spot; twig blight

Time: Apply prebloom, bloom, petal fall, shuck stages, and approximately 2 weeks after shuck fall.

Rate: Apricots:
4.6-6 lb. ai/A
6-8 lb. WDG/A
California only:
6 lb. ai/A
8 lb. WDG/A

NOTICE The information on these pages is for preliminary planning — not a guide for use. Be sure to follow the manufacturer's directions, notwithstanding information contained here. For personal protective equipment and EPA registration numbers, please turn to page 39.

Disease:
brown rot blossom blight, fruit rot, shot hole, twig blight

Time: Apply prebloom, bloom, and petal fall through early cover sprays.

Rate: Nectarines, peaches:
Eastern U.S.:
2.8-6 lb. ai/A
3.75-8 lb. WDG/A
Western U.S.:
4.6-7.6 lb. ai/A
6-10 lb. WDG/A
California only:
6 lb. ai/A
8 lb. WDG/A

Disease:
brown rot blossom blight, fruit rot, leaf curl, peach blight, twig blight

Time: For brown rot blossom blight, fruit rot, and twig blight, apply at prebloom through cover sprays as needed. For peach blight, follow schedule above. In addition to spray program, apply high rate in fall before heavy rains. For leaf curl, apply as dormant application after leaf drop and/or prior to bud swell.

Remarks: Apples, apricots, cherries, nectarines, peaches, pears. Do not apply this product through any type of irrigation system. Refer to label for use directions.

Days To Harvest: Apples-14. Pears: Eastern U.S.-14, Western U.S.-5. Cherries: California-7, Eastern U.S.-14, Western U.S.-30. Apricots-30; nectarines, peaches: Eastern U.S.-14, Western U.S.-30.

- - - - - - - - - - - - - - - - -

Signal Word/Toxicity Class: Danger/I.
REI: 48 hr.

Tree Fruits, Subtropical

Abound — see azoxystrobin

Aliette — see fosetyl-aluminum

AZOXYSTROBIN

Flowable broad-spectrum fungicide with preventive, systemic, and acurative properties. Toxic to freshwater and estuarine/marine fish and aquatic invertebrates.

Abound (FL) *(ZENECA)*

Rate: 5.5-8.3 fl. oz. FL/A

Disease:
black Sigatoka, yellow Sigatoka

Time: Apply prior to or in early stages of disease development.

Remarks: Bananas. Do not apply more than 2.08 qt./A per yr. Do not apply through any type of irrigation system or ultra-low volume spray system. Refer to label for use directions.

Days To Harvest: 0.

- - - - - - - - - - - - - - - - -

Signal Word/Toxicity Class: Caution.
REI: 12 hr.

Basicop — see copper compounds

Bayleton — see triadimefon

BENLATE

Benzimidazole systemic fungicide. Toxic to fish.

Benlate (SP, WP) (benomyl) *(DuPont)*

Rate: 0.5-1 lb. ai/A
1-2 lb. SP, WP/A

Disease: Avocados:
Anthracnose, Cercospora spot, scab

Mangos:
Anthracnose

Papaya:
Anthracnose, powdery mildew

Time: Avocados: Apply at bud swell; repeat at 3- to 4-week intervals. Mangos: Apply at first appearance of panicles (2" long); repeat at 7-day intervals through fruit set; continue at 3-4 week intervals. Papaya: Apply at bud expansion; repeat at 3- to 4-week intervals.

Remarks: Avocados, mangos, papaya. Do not tank mix or alternate with benzimidazole or thiophanate products. Refer to label for use directions.

Days To Harvest: Mangos, papaya-14; avocados-30.

State Restrictions: Benlate WP for use on avocados in Florida and Puerto Rico. Benlate SP not registered for use in California.

- - - - - - - - - - - - - - - - -

Rate: 0.6 lb. ai/100 gal.
1.25 lb. SP, WP/100 gal.

Disease:
pineapple butt rot

Time: Dip treatment.

Remarks: Pineapple. Immerse seedpieces wetting thoroughly; remove and allow to drain. Do not tank mix or alternate Benlate with benzimidazole or thiophanate products. Refer to label for use directions.

State Restrictions: Benlate SP not registered for use in California.

- - - - - - - - - - - - - - - - -

Signal Word/Toxicity Class: Caution.
REI: 24 hr.

Blue Viking — see copper compounds

Bravo — see chlorothalonil

Brom-O-Gas — see methyl bromide

Champ — see copper compounds

Champion — see copper compounds

CHLOROTHALONIL

Broad-spectrum contact fungicide. Toxic to fish, aquatic invertebrates, and marine/estuarine organisms.

Bravo 500, 720/Weather Stik, Ultrex (WDG) *(ZENECA)*
Echo 500, 720 or 90DF *(Sostram)*
Terranil (6L, 90DF or 90DF WSP) *(Terra)*

Rate: Papaya:
3-5.75 pt. 500/A
2-4 pt. 720, 6L/A
1.7-3.3 lb. 90DF/A
1.4-2.8 lb. WDG/A
1.75-2.25 lb. 90DF WSP/A

Disease:
Alternaria fruit spot, Anthracnose, stem end rot

Rate: Passion fruit:
2.75 pt. 500/A
2 pt. 720, 6L/A
1.8 lb. WDG/A
1.5 lb. 90DF, 90DF WSP/A (Terranil)
1.6 lb. 90DF/A (Echo)

Disease:
Alternaria fruit spot, brown spot, leaf spot

Time: Apply when conditions favor disease development. For passion fruit, apply when fruit spots appear (April-July). Repeat at 14-day intervals until weather condition no longer favor development.

Remarks: Papaya, passion fruit. Apply with ground equipment only. Use in sufficient water to obtain adequate coverage of fruit and leaves. Do not graze treated areas or feed vines or processing by-products to livestock. Refer to label for use directions and restrictions.

State Restrictions: For use in Hawaii on passion fruit. SLN-Florida (Bravo 720).

- - - - - - - - - - - - - - - - -

Signal Word/Toxicity Class: Danger/I (WDG, 90DF); Warning/II (6L, 500, 720).
REI: 48 hr.

Clean Crop C-O-C-S — see copper compounds

◆ - **new product** • ai-active ingredient • DF-dry flowable • E/EC-emulsifiable concentrate • EW-emulsion in water
F/FL-flowable • G-granule • L/LC-liquid • SP/WSP-soluble packet • W/WP-wettable powder • WSB-water soluble bag

COPPER COMPOUNDS

Fungicide/bactericide.

Basicop (WP) (basic copper sulfate) *(Griffin)*

Rate: Avocado:
1.5 lb. ai/100 gal. water
3 lb. WP/100 gal. water

Disease:
Anthracnose blotch scab

Time: Apply when blossom buds open; repeat at 4-week intervals for up to 5 applications.

Rate: Olives:
2-3 lb. ai/100 gal. water
4-6 lb. WP/100 gal. water

Disease:
olive knot peacock spot

Time: Make first application before winter rains fall, second, in early spring.

Rate: Papaya:
0.75 lb. ai/100 gal. water
1.5 lb. WP/100 gal. water

Disease:
Anthracnose

Time: Apply when disease is expected; repeat at 10- to 14-day intervals (5- to 7-day during heavy rainfall).

Remarks: Avocado, olives, papaya. Use higher rate when disease is heavy. Do not apply through any type of irrigation system. Refer to label for use directions.

Signal Word/Toxicity Class: Warning/II.
REI: 24 hr.

Blue Viking or Tennessee Brand (copper sulfate pentahydrate) *(Griffin)*

Rate: 10-10-100 Bordeaux spray
Less than 10" annual rainfall:
5-5-100 Bordeaux spray

Disease:
olive knot olive leaf spot

Time: Apply late Oct.-early Nov. or before fall rains. In wet winters, a repeat spray may be needed mid-winter.

Remarks: Olives. Do not apply through any type of irrigation system. Refer to label for use directions.

Signal Word/Toxicity Class: Danger/I.
REI: 24 hr.

Champ (FL or Formula 2) or Champion (WP) (copper hydroxide) *(Agtrol)*
Microsperse COC 53WP (copper oxychloride) or
Nu-Cop (50DF, 50WP) (copper hydroxide) *(Micro Flo)*

Rate: Avocados:
1.33-2 gal. FL/A
0.66-1 gal. Formula 2/A
8-12 lb. WP/A
1.33-8 pt. 3L/A
8-12 lb. 50DF, 50WP, 53WP/A

Disease:
scab

Time: Apply when bloom buds begin to swell; repeat at monthly intervals for 5-6 applications.

Rate: Bananas:
2.66 pt. FL/A in 3 gal. water
1.33 pt. Formula 2/A in 3 gal. water
2 lb. WP/A in 3 gal. water
1.33-2.66 pt. 3L/A in 3 gal. water
2 lb. 50DF, 50WP/A in 3 gal. water
3-4 lb. 53WP/A in 3 gal. water

Disease:
Sigatoka

Time: Apply on 14-day schedule throughout wet season; 21-day schedule during dry period.

Rate: 5.33 pt. FL/100 gal. water
2.66 pt. Formula 2/100 gal. water
4 lb. WP/100 gal. water
2.66-5.33 pt. 3L/A
4 lb. 50DF, 50WP/100 gal.
4-6 lb. 53WP/100 gal.

Disease:
black pitting

Time: Apply during first and second week of fruit emergence.

Rate: Cacao:
2.66-11.33 pt. FL/A
1.33-5.66 pt. Formula 2/A
2-4.5 lb. WP/A
1.33-11.33 pt. 3L/A
2-8.5 lb. 50DF, 50WP/A
3-10 lb. 53WP/A

Disease:
black pod

Time: Apply at start of rainy season; continue while infection conditions persist. Apply every 14-21 days in high rainfall areas, depending on disease severity. For drier areas, use 2-4 applications at longer intervals.

Rate: Guava:
4 pt. 3L/A

Disease:
Anthracnose red alga

Time: Apply just before flowering; repeat weekly until just before harvest.

Rate: Kiwifruit:
2.66-4 pt. FL/A
1.33-2 pt. Formula 2/A
2-3 lb. WP/A
1.33 gal. 3L/A
8 lb. 50DF, 50WP/A

Disease:
blossom blight leaf spot Pseudomonas
Erwinia herbicola

Time: Make 2-3 applications during dormant season.

Rate: Mangos:
1.33-1.66 gal. FL/A
0.66-0.83 gal. Formula 2/A
8-10 lb. WP, 50DF, 50WP/A
5.33-13.33 pt. 3L/A
12-15 lb. 53WP/A

Disease:
Anthracnose

Time: Apply monthly after fruit set until harvest.

Rate: Olives:
1.33-2 gal. FL/A
0.66-1 gal. Formula 2/A
8-12 lb. WP/A
0.66-2 gal. 3L/A
2 lb. 50DF, 50WP/100 gal. water
5-6 lb. 53WP/A

Disease:
Anthracnose olive knot peacock spot

Time: Anthracnose: Apply before rains begin; repeat at 10- to 14-day intervals during periods of heavy rainfall. Olive knot, peacock spot: Apply before winter rains fall; make second application in early spring for severe disease.

Rate: Passion fruit:
1 gal. 3L/A

Disease:
Anthracnose

Time: Apply just before flowering; repeat weekly to just before harvest.

Remarks: Avocados, bananas, cacao, guava, kiwifruit, mangos, olives, passion fruit. For Sigatoka, add 0.5 gal. agricultural oil and apply by air. For black pitting, apply directly to fruit stem and include basal portion of leaf crown. Refer to label for use directions.

State Restrictions: California: For use on olives, papaya; Nu-Cop only for use on mangos. For use in Florida on mangos.

Rate: Coffee:
2.66-10.6 pt. FL/A
1.33-5.3 pt. Formula 2/A
1.33-7.33 pt. 3L/A
2-8 lb. WP, 50DF, 50WP/A
3-12 lb. 53WP/A

Disease:
blight (except 53WP) coffee berry disease (except 53WP) leaf rust
iron spot pink disease

Time: Iron spot, pink disease: Apply at start of wet season; repeat at monthly intervals for 3 applications. Coffee berry disease: Apply after flowering and before onset of long rains; repeat at 21- to 28-day intervals until picking. Blight, leaf rust: Apply before onset of long rains; repeat at 14- to 21-day intervals through rainy season.

NOTICE The information on these pages is for preliminary planning — not a guide for use. Be sure to follow the manufacturer's directions, notwithstanding information contained here.

For personal protective equipment and EPA registration numbers, please turn to page 39.

Remarks: Coffee. Use higher rates when rainfall is heavy and disease pressure is high. Depending on equipment used, volume/A will differ. Refer to label for use directions.

Rate: Papaya:
0.25-0.83 gal. Formula 2/A
4-10 lb. WP/A
2-6.66 pt. 3L/A
2 lb. 50DF, 50WP/100 gal. water
2 lb. 53WP/100 gal. water

Disease:
Anthracnose

Time: Apply before disease appears, repeat at 10- to 14-day intervals or 5- to 7-day intervals during heavy rainfall.
Remarks: Papaya. Use higher rates when conditions favor disease. Spreader-sticker may be used during periods of heavy rains. Refer to label for use directions.

Signal Word/Toxicity Class: Varies by formulation.
REI: Varies by formulation.

Clean Crop C-O-C-S (WDG) (74.8% copper oxychloride & 14.2% basic copper sulfate) *(Platte)*
Rate: 2 lb. ai/100 gal. water
4 lb. WDG/100 gal. water

Disease:
olive leaf spot

Time: Apply after harvest and prior to rainy season. Under heavy disease pressure, second application may be necessary.
Remarks: Olives. Do not apply this product through any type of irrigation system. Refer to label for use directions.

Signal Word/Toxicity Class: Warning/II.
REI: 24 hr.

Copper-Count-N (L) (copper ammonium) *(Mineral Research)*
Rate: Avocados, mangos:
8 qt. L/A

Disease:
Anthracnose

Time: For avocados, apply when bloom bud begins to swell. For mangos, apply weekly when pinicles are 2" in length until all fruits are set. Repeat monthly until Aug.

Rate: Olives:
8-12 qt. L/A

Disease:
peacock spot

Time: Apply before winter rains fall. Second application should be made in early spring if disease is severe.
Remarks: Avocados, mangos, olives. Applications may be made by ground or air; refer to label for use directions. Do not apply through any type of irrigation system unless supplemental labeling on chemigation is followed.
State Restrictions: Not for use on avocados in California.

Signal Word/Toxicity Class: Caution.
REI: 12 hr.

Drexel Basic Copper Sulfate (WP) or KOP 300 (3FL) (basic copper sulfate) *(Drexel)*
Rate: Avocados:
3 lb. WP/100 gal. water
0.33 gal. 3FL/100 gal. water

Disease:
Anthracnose   blotch   scab

Time: Apply when blossom buds open; repeat at 4-week intervals for total of 5 applications. Do not apply later than 140 days after bloom.

Rate: Olives:
Dilute:
4 lb. WP/100 gal. water
0.33-0.5 gal. 3FL/100 gal. water
Concentrate:
12 lb. WP/A
1 gal. 3FL/A

Disease:
peacock spot

Time: Apply late Oct. In areas with 10" or less rainfall per year, use 2 lb. WP/100 gal. water.
Remarks: Avocados, olives. Do not apply through any type of irrigation system. Refer to label for use directions.

Signal Word/Toxicity Class: Warning/II.
REI: 24 hr. (WP); 12 hr. (3FL).

Kocide (101, DF, 2000, 4.5LF, LF) (copper hydroxide) *(Griffin)*
Rate: Avocados:
8-12 lb. 101, DF/A
6-9 lb. 2000/A
5.33-8 pt. 4.5LF/A
10.66-16 pt. LF/A

Disease:
Anthracnose (2000, 4.5LF)   blotch (2000, 4.5LF)   scab

Time: Apply when bloom buds begin to swell; repeat at monthly intervals for 5-6 applications.

Rate: Bananas:
2-4 lb. 101, DF
1.5-3 lb. 2000/A
2.66 pt. 4.5LF/A
2.66 pt. LF/A

Disease:
Sigatoka

Time: Apply on a 14-day schedule through wet season; 21-day schedule during dry period.

Rate: 2-4 lb. 101, DF
1.5-3 lb. 2000/A
5.33 pt. LF/100 gal.
1.33 pt. 4.5LF/A

Disease:
black pitting

Time: Apply during first and second week of fruit emergence.

Rate: Cacao:
2-8.5 lb. 101, DF/A
1.5-6.5 lb. 2000/A
1.33-5.66 pt. 4.5LF/A
10.33-24 pt. LF/A

Disease:
black pod

Time: Apply at start of rainy season; continue while infection conditions persist. Apply every 14-21 days in high rainfall areas, depending on disease severity. For drier areas, use 2-4 applications at longer intervals.

Rate: Guava:
3 lb. 101, DF/A
2.25 lb. 2000/A
2 pt. 4.5LF/A
6 pt. LF/A

Disease:
Anthracnose   red alga

Time: Apply before flowering; repeat weekly until just before harvest.

Rate: Kiwifruit:
8 lb. 101, DF/A
6 lb. 2000/A
5.33 pt. 4.5LF/A
10.66 pt. LF/A

Disease:
Erwinia herbicola   Pseudomonas

Time: Apply monthly.

Rate: Olives-California:
8-12 lb. 101, DF/A
6-9 lb. 2000/A
5.33-8 pt. 4.5LF/A

Disease:
olive knot   peacock spot

Time: Apply before winter rains fall; second application in early spring should be made if disease is severe.

Rate: Papaya:
4-10 lb. 101, DF/A
3-7.5 lb. 2000/A
2.66-6.66 pt. 4.5LF/A

Disease:
Anthracnose

Time: Apply at 10- to 14-day intervals under light disease pressure, 5- to 7-day intervals under heavy disease pressure.

Rate: Passion fruit:
6 lb. 101, DF/A
4.5 lb. 2000/A
4 pt. 4.5LF/A
12 pt. LF/A

◆ - **new product** • ai-active ingredient • DF-dry flowable • E/EC-emulsifiable concentrate • EW-emulsion in water
F/FL-flowable • G-granule • L/LC-liquid • SP/WSP-soluble packet • W/WP-wettable powder • WSB-water soluble bag

Disease:
Anthracnose
Time: Apply just before flowering; repeat weekly until just before harvest.
Remarks: Avocados, bananas, cacao, guava, kiwifruit, olives, papaya, passion fruit. For Sigatoka, add 0.5 gal. agricultural oil; apply by air. For black pitting, apply directly to fruit stem and include basal portion of leaf crown. Do not apply in a spray solution having a pH less than 6.5 as phytotoxicity may occur. Refer to label for use directions.
State Restrictions: Not for use in California for guava, papaya, and passion fruit. For use in California for olives.

- - - - - - - - - - - - - - - - -

Rate: Mangos:
8 lb. 101, DF/A
6 lb. 2000/A
5.33 pt. 4.5LF/A
10.66-13.33 pt. LF/A
Disease:
Anthracnose
Time: Apply monthly after fruit set until harvest.
Remarks: Mangos. Use higher rates when rainfall is heavy and disease pressure is high. Do not apply in a spray solution having a pH less than 6.5 as phytotoxicity may occur. Refer to label for use directions.
State Restrictions: Kocide DF not for use in California. Kocide LF and 101 for use in Florida only.

- - - - - - - - - - - - - - - - -

Rate: Coffee:
2-4 lb. 101, DF/A
1.5-3 lb. 2000/A
1.33-2.66 pt. 4.5LF/A
2.66-5.33 pt. LF/A
Disease:
iron spot leaf rust pink disease
Time: Iron spot, pink disease: Apply at start of wet season; make 3 applications at monthly intervals. Leaf rust: Apply before onset of rain; repeat every 21 days while rains continue.

Rate: 6-8 lb. 101, DF/A
4.5-6 lb. 2000/A
4-5.33 pt. 4.5LF/A
8-16.66 pt. LF/A
Disease:
blight coffee berry disease
Time: Blight: Apply before onset of long rains; repeat throughout rainy season at 14- to 21-day intervals. Coffee berry disease: Apply first spray after flowering and before onset of long rains; repeat at 21- to 28-day intervals until picking.

Rate: Low density plantings:
4.66-7.33 pt. LF/A
High density plantings:
9.33-10.66 pt. LF/A
Disease:
leaf rust - Brazil
Time: Apply before onset of rain; repeat at 3- to 4-week intervals depending on disease severity and rainfall conditions.
Remarks: Coffee. Use higher rates when rainfall is heavy and disease pressure is high. Depending on equipment used, volume/A will differ. Do not apply in a spray solution having a pH less than 6.5 as phytotoxicity may occur. Refer to label for use directions.

- - - - - - - - - - - - - - - - -

Signal Word/Toxicity Class: Danger/I (101, DF); Warning/II (2000); Caution (4.5LF, LF).
REI: 24 hr.

KOP-Hydroxide (FL, 50WP) (copper hydroxide) *(Drexel)*

Rate: Avocados:
10.66-13.33 pt. FL/A
8-12 lb. 50WP/A
Disease:
scab
Time: Apply when bloom buds begin to swell; repeat at monthly intervals for 5-6 applications.

Rate: Bananas:
2.66 pt. FL/A
2 lb. 50WP/A
Disease:
Sigatoka
Time: Apply by air at 14-day intervals throughout the wet season; 21-day intervals during dry periods.

Rate: 5.33 pt. FL/A
4 lb. 50WP/A
Disease:
black pitting
Time: Apply directly to fruit stem and basal portion of leaf crown 1-2 weeks after emergence.

Rate: Cacao:
High rainfall areas:
2.66-6 pt. FL/A
2-4.5 lb. 50WP/A
Low rainfall areas:
8.66-11.33 pt. FL/A
8.5 lb. 50WP/A
Disease:
black pod
Time: Wet areas: Apply at beginning of rainy season and continue while infection persists at 14- to 21-day intervals. Dry areas: Make 2-4 applications during critical infection periods and at long intervals.

Rate: Coffee:
2.66 pt. FL/A
2 lb. 50WP/A
Disease:
iron spot pink disease
Time: Apply at start of wet season; repeat at monthly intervals for 3 applications.

Rate: Average density plantations:
4.66-7.33 pt. FL/A
2 lb. 50WP/A
High density plantations:
9.33-10.66 pt. FL/A
4 lb. 50WP/A
Disease:
leaf rust
Time: Apply at 3- to 4-week intervals, depending on disease and rainfall from Sept.-Mar.

Rate: 6-8 lb. 50WP/A
Disease:
blight coffee berry disease
Time: Apply after flowering and before onset of long rains; repeat at 21- to 28-day intervals until picking.

Rate: Mangos:
2.66 pt. FL/100 gal.
8 lb. 50WP/A
Disease:
Anthracnose
Time: Apply monthly after fruit set until harvest.

Rate: Olives:
2.66-4 pt. FL/100 gal.
8-12 lb. 50WP/A
Disease:
peacock spot
Time: Apply before winter rains fall; second application early spring if disease if severe.
Remarks: KOP-Hydroxide 50WP should not be applied in spray solution having pH less than 6.5 or greater than 9.0 as phytotoxicity may occur. Refer to label for use directions.
State Restrictions: For use in California on olives. For use in Florida on mangos.

- - - - - - - - - - - - - - - - -

Signal Word/Toxicity Class: Danger/I.
REI: 48 hr.

Micro Flo Basic Copper "53" (basic copper sulfate) *(Micro Flo)*

Rate: 12-20 lb. 53/A
Disease:
Anthracnose blotch scab
Time: Apply when flower buds open; repeat at 4-week intervals. Do not apply later than 140 days after bloom.
Remarks: Avocados. Refer to label for use directions.
State Restrictions: In Florida, addition of spreader/sticker is recommended.

- - - - - - - - - - - - - - - - -

Rate: 12-15 lb. 53/A
Disease:
Anthracnose

8 Fruits & Nuts

NOTICE The information on these pages is for preliminary planning — not a guide for use. Be sure to follow the manufacturer's directions, notwithstanding information contained here.	For personal protective equipment and EPA registration numbers, please turn to page 39.

Time: Apply when panicles are 2" long; repeat weekly until fruit set, then spray monthly through Sept. for total of 5-12 applications depending upon area.
Remarks: Mangos. Refer to label for use directions.
State Restrictions: Not for use in California.

- - - - - - - - - - - - - - - - -

Rate: 5-6 lb. 53/100 gal.
Disease:
leaf spot
Time: Apply in fall, usually Oct.-Nov.
Remarks: Olives. Refer to label for use directions.

- - - - - - - - - - - - - - - - -

Rate: 2 lb. 53/100 gal.
Disease:
Anthracnose
Time: Apply before rains when disease is expected; repeat at 10- to 14-day intervals during periods of heavy rainfall.
Remarks: Papayas. Addition of spreader/sticker is recommended. Refer to label for use directions.
State Restrictions: Not for use in California.

- - - - - - - - - - - - - - - - -

Signal Word/Toxicity Class: Warning/II.
REI: 24 hr.

TOP COP Tri-Basic (FL) (basic copper sulfate) *(Stoller)*
Rate: 3 pt. FL/100 gal. water
Disease:
Anthracnose blotch scab
Time: Apply when bloom buds open; repeat at 4-week intervals. Apply only within 140 days of bloom.
Remarks: Avocados. Do not apply more than 5 applications. Refer to label for use directions.

- - - - - - - - - - - - - - - - -

Rate: 4-8 pt. FL/A
Disease:
leaf spot
Time: Apply in fall.
Remarks: Olives. Use low rate where less than 10" rainfall is expected. Refer to label for use directions.

- - - - - - - - - - - - - - - - -

Signal Word/Toxicity Class: Caution.
REI: 24 hr.

Copper-Count-N — see copper compounds

Dithane — see mancozeb

Echo — see chlorothalonil

Elite — see tebuconazole

Enable — see fenbuconazole

FENBUCONAZOLE

Triazole sterol inhibitor. Systemic fungicide with protectant, curative, and eradicant properties. Translocated upwards in the plant. Toxic to fish.

Enable (2F) *(Rohm and Haas)*
Rate: 6 fl. oz. 2F/A
Disease:
Sigatoka
Time: Apply when leaves first appear; repeat at 14- to 21-day intervals as needed.
Remarks: Bananas. Do not apply more than 8 times per year. Do not graze livestock in treated areas. Do not cut cover crops for livestock food. Refer to label for use directions.

- - - - - - - - - - - - - - - - -

Signal Word/Toxicity Class: Caution.
REI: 12 hr.

FOSETYL-ALUMINUM

Systemic fungicide/bactericide. Toxic to estuarine invertebrates.

Aliette (WDG, WSP) *(Rhone-Poulenc)*
Rate: Bearing, nonbearing:
Trunk spray or paint:
2-4 lb. ai/5 gal.
2.5-5 lb. WDG, WSP/5 gal.
Foliar:
4 lb. ai/A
5 lb. WDG, WSP/A
Nonbearing:
Drench:
5 oz. WDG/10 gal.
0.8 lb. ai/32 gal.
1 lb. WSP/32 gal.
Disease:
Phytophthora canker Phytophthora root rot
Time: Trunk spray or paint: Nursery tree resets and new plantings should be treated at time of planting. Drench application: Apply 2-3 days prior to transplanting. Foliar application: Apply at start of growing season; up to 4 foliar sprays/year may be applied at 60-day intervals.
Remarks: Bearing, nonbearing avocados. Nonbearing trees will not produce marketable fruit for 12 months after last application. For trunk spray or paint, if lesions are present, use high rate. For drench application, apply 1 qt. solution to pot or sleeve of tree. For foliar, apply to point of runoff. Do not exceed 100 gal./A for nonbearing, 500 gal./A bearing. Do not allow livestock to graze on floor of treated groves. Do not cut cover crops for feed. Refer to label for use directions.

- - - - - - - - - - - - - - - - -

Rate: Dip:
2 lb. ai/100 gal.
2.5 lb. WDG/100 gal.
0.8 lb. ai/40 gal.
1 lb. WSP/40 gal.
Foliar:
3 lb. ai/100 gal.
3.75 lb. WDG/100 gal.
12 lb. ai/400 gal.
15 lb. WSP/400 gal.
Disease:
heart rot
Time: Dip treatment: Apply as preplant dip immediately prior to planting. Foliar spray: Apply to established plants when aggravating conditions (excessive rainfall) occur or are anticipated. Up to 6 sprays can be applied at 3-month intervals.
Remarks: Pineapple. Apply in sufficient water for thorough coverage. Do not exceed 400 gal./A. Do not tank mix with copper compounds. Refer to label for use directions.
Days To Harvest: 3 months.

- - - - - - - - - - - - - - - - -

Signal Word/Toxicity Class: Caution.
REI: 12 hr.

GALLEX (Premix)

Ready-to-use paint-on emulsion.

Gallex (2,4-xylenol & meta-cresol) *(AgBioChem)*
Rate: Apply undiluted.
Disease:
olive knot
Time: Apply first application before breaking of dormancy or during spring and early summer.
Remarks: Olives. Paint gall and cut surfaces + 1/2" of surrounding healthy bark. Generally 1 application will give control; retreatment may be necessary. Do not mix or combine with other pesticides. Complete girdling of trunk by galls affects water-conducting capacity of tree. To prevent sudden loss of water movement, do not treat more than 1/2 galled trunk at any one time. Treat remaining diseased tissue 4-6 months later when examination of initial treatment is made.

- - - - - - - - - - - - - - - - -

Signal Word/Toxicity Class: Warning/II.

Kocide — see copper compounds

KOP-Hydroxide — see copper compounds

◆ - **new product** • ai-active ingredient • DF-dry flowable • E/EC-emulsifiable concentrate • EW-emulsion in water
F/FL-flowable • G-granule • L/LC-liquid • SP/WSP-soluble packet • W/WP-wettable powder • WSB-water soluble bag

Kumulus — see sulfur

MANCOZEB

Carbamate fungicide with broad-spectrum activity with no resistance problems. Coordination product of zinc ion and manganese ethylene bisdithiocarbamate. Toxic to fish.

Dithane (DF, F-45, M-45, ST, WSP) *(Rohm and Haas)*
Manzate 200DF *(Griffin)*
Penncozeb 75DF or 80WP *(Elf Atochem)*

Rate: 1.5-2.25 lb. ai/A
2-3 lb. DF, M-45, WSP, 200DF, 75DF/A
1.6-2.4 lb. ai/A
3.2-4.8 pt. F-45/A
2-3 lb. 80WP/A

Disease:
Sigatoka

Time: Apply when leaves first appear; repeat at 14- to 21-day intervals as needed.
Remarks: Bananas. Addition of surfactant will improve performance. Apply in sufficient water to provide adequate coverage. Refer to label for use directions.
Days To Harvest: 0.
State Restrictions: Refer to label for California precautions.

Rate: 0.75 lb. ai/25 gal.
1 lb. DF, WSP/25 gal.
0.8 lb. ai/25 gal.
1 lb. M-45/25 gal.
1.6 pt. F-45, ST/25 gal.
3 lb. ai/100 gal.
4 lb. 200DF/100 gal.

Disease:
Endosepsis Fusarium mold

Remarks: Caprifig. Prepare mamme figs by making shallow cut through the eye and then hand-dividing to avoid wasp injury. Submerge mamme figs in suspension for a minimum of 15 minutes; stir frequently. Fresh dipping solution should be used after treating 4-5 batches of figs. After treatment, figs should be drained prior to placement in trees. Do not use mamme figs for food, feed, or oil purposes. Refer to label for use directions.
State Restrictions: Refer to label for California precautions.

Rate: 1.5-1.9 lb. ai/A
2-2.5 lb. DF, M-45, WSP, 200DF/A
1.6-2 lb. ai/A
3.2-4 pt. F-45/A
1.12-1.9 lb. ai/A
1.5-2.5 lb. 75DF/A
1.2-2 lb. ai/A
1.5-2.5 lb. 80WP/A

Disease:
Anthracnose black spot Phytophthora fruit rot

Time: Begin at flowering; repeat at 14- to 21-day intervals. Treat central column crown, blossom areas, and developing fruit.
Remarks: Papayas. Addition of surfactant will improve performance. Refer to label for use directions.
Days To Harvest: 0.
State Restrictions: Refer to label for California precautions.

Rate: 2.4 lb. ai/10 gal.
3.2 lb. DF, M-45, WSP, 200DF/10 gal.
2.6 lb. ai/10 gal.
5.2 pt. F-45, ST/10 gal.

Disease:
Phytophthora heart rot

Time: Seed piece treatment.
Remarks: Pineapple. Dip planting material in solution prior to planting; stir frequently. New solution should be prepared when at least 2/3 volume has been used or sooner if soil from plant material has noticeably discolored solution. Depending on cleanliness and size of planting stock, up to 100 gal. solution/A should be used to treat propagative materials. Do not use treated seed for food, feed, or oil purposes. Refer to label for use directions.
State Restrictions: Refer to label for California precautions.

Signal Word/Toxicity Class: Caution.
REI: 24 hr.

MANEB

Dithiocarbamate contact fungicide. Toxic to fish.

Maneb 75DF or 80WP *(Elf Atochem)*
Manex (FL) *(Griffin)*

Rate: Bananas:
1.5-2.25 lb. ai/A
2-3 lb. 75DF/A
1.6-2.4 lb. ai/A
2-3 lb. 80WP/A
1.6-2.4 qt. FL/A

Disease:
Sigatoka

Rate: Figs:
0.5 lb. ai/100 gal. water
0.75 lb. 75DF/100 gal. water
0.6 lb. ai/100 gal. water
0.75 lb. 80WP/100 gal. water
0.6 qt. FL/100 gal. water

Disease:
surface mold surface rot

Rate: Papayas:
1.5-1.9 lb. ai/A
2-2.5 lb. 75DF/A
1.6-2 lb. ai/A
2-2.5 lb. 80WP/A
1.6-2 qt. FL/A

Disease:
Anthracnose Cercospora black spot Phytophthora fruit rot

Time: On bananas, apply when leaves first appear; on figs and papayas when disease first threatens; repeat at 14- to 21-day intervals.
Remarks: Bananas, figs, papayas. Refer to label for use directions.
Days To Harvest: Bananas, papayas-0; figs-10.
State Restrictions: Not for use on Kadota figs in California.

Signal Word/Toxicity Class: Caution.
REI: 24 hr.

MANEB (Premix)

Fungicide/bactericide. Toxic to fish, birds, and other wildlife.

ManKocide (DF) (15% maneb & 46.1% copper hydroxide) *(Griffin)*

Rate: Bananas:
2.5 lb. DF/A

Disease:
Sigatoka

Time: Apply when leaves first appear; repeat at 14- to 21-day intervals.

Rate: Papayas:
5.5-13 lb. DF/A

Disease:
Anthracnose

Time: Apply before disease appears at 14- to 21-day intervals under light disease pressure; 5- to 7-day intervals under heavy disease pressure.
Remarks: Bananas, papayas. Addition of approved surfactant/spreader is desirable. Refer to label for use directions.
Days To Harvest: 0.

Signal Word/Toxicity Class: Danger/I.
REI: 48 hr.

Manex — see maneb

ManKocide — see maneb premix

Manzate — see mancozeb

NOTICE	The information on these pages is for preliminary planning — not a guide for use. Be sure to follow the manufacturer's directions, notwithstanding information contained here.	For personal protective equipment and EPA registration numbers, please turn to page 39.

MEFENOXAM

Systemic fungicide having a specific mode of action.

Ridomil Gold EC, GR, or WSP *(Novartis)*

Rate: Sleeve Drench:
1 lb. ai/A
2 qt. EC/A
Injection into Irrigation Water:
0.5-1 fl. oz. EC/1000 gal. water
0.5-1 oz. WSP/1000 gal. water
Sprinkler irrigated orchards:
0.75-1.5oz. GR/sq. yd.
Drip irrigated orchards:
1.5-28 oz. GR/tree

Disease:
root rot

Time: Apply at start of growing season or at transplanting; 2 additional applications can be made at 3-month intervals.

Remarks: Avocado. Do not make more than 1.5 gal. EC/A, 12 lb. WSP/A, or 240 lb. GR/A per year. Do not plant any crop which is not registered for use with mefenoxam in treated soil for 12 months.

Days To Harvest: 28.

- - - - - - - - - - - - - - - - -

Rate: Nurseries:
0.5-1 fl. oz. EC in 40 gal. water/800 sq. ft.
Field:
0.44-0.875 lb. ai/A
0.88-1.75 qt. EC/A
0.8-1.6 lb. ai/A
1.75-3.5 lb. WSP/A

Disease:
root rot

Time: In nurseries, apply to seedling 3-4 weeks after planting seed; repeat 1 month later if needed. In field, make first application at time of transplanting, repeat in 2-4 weeks if needed.

Remarks: Papaya. Do not apply more than 2 applications per year. Do not plant any crop which is not registered for use with mefenoxam in treated soil for 12 months.

Days To Harvest: 90.

State Restrictions: Hawaii only.

- - - - - - - - - - - - - - - - -

Rate: 0.5-1 pt. EC/100 gal.water
0.5-1 lb. WSP/100 gal. water

Disease:
heart rot

Time: Apply as crown dip before planting.

Remarks: Pineapple. If there is crop failure within 1 year of planting treated crowns, do not harvest plant material for animal feed. Do not plant any crop which is not registered for use with mefenoxam in treated soil for 12 months.

- - - - - - - - - - - - - - - - -

Signal Word/Toxicity Class: Warning/II (EC); Caution (GR, WSP).
REI: 48 hr.

METHYL BROMIDE

2% Chloropicrin added as warning agent. Restricted Use Pesticide. Classified as an ozone depleter under the Clean Air Act. Use is to be completely eliminated starting Jan. 1, 2001.

Brom-O-Gas *(Great Lakes Chemical)*

Rate: Injected with chisels + covered with tarp:
180-240 lb./A
Raised tarp:
0.4-0.55 lb./100 ft.

Disease:

Fusarium	Plasmodiophora	Sclerotinia
oak root fungus	Pythium	Sclerotium
Phytophthora	Rhizoctonia	

Time: Preplant.

Remarks: Pineapples. Do not fumigate when soil temperature is below 50°F. For best results, fumigate when temperature is 60°-80°F at depth of 6". Refer to label for use directions.

State Restrictions: Brom-O-Gas not for use in Arizona and not for use in California for oak root fungus (Armillaria root rot) control.

- - - - - - - - - - - - - - - - -

Signal Word/Toxicity Class: Danger/I.

Micro Flo — see copper compounds

Microsperse — see copper compounds

Microthiol Special — see sulfur

Nu-Cop — see copper compounds

Penncozeb — see mancozeb

Ridomil Gold — see mefenoxam

Sulfa — see sulfur

SULFUR

Nonsystemic fungicide with protective action. Microthiol Special, Sulfa 80, and Thiolux are micronized sulfurs.

Kumulus DF (sulfur) *(BASF)*
Microthiol Special (80WP) *(Elf Atochem)*
Sulfa 80 (80DF) *(Drexel)*
Thiolux (DF) *(Platte)*

Rate: 8-24 lb. ai/A
10-30 lb. DF, 80WP, 80DF/A

Disease:
powdery mildew

Time: Apply before bloom; repeat in bloom, after fruit set, and 3 weeks later or as necessary.

Remarks: Mangos. Do not apply this product through any type of irrigation system. Refer to label for use directions and precautions.

- - - - - - - - - - - - - - - - -

Signal Word/Toxicity Class: Caution.
REI: 24 hr.

Micro Flo Sulfur (6L), Dual Spraying & Dusting Sulfur (DS) or Microsperse Wettable Sulfur (WS) *(Micro Flo)*
Terra Sulfur (6L) *(Terra)*

Rate: 30 lb. ai/A
5 gal. 6L/A
27 lb. ai/A
30 lb. DS, WS/A

Disease:
powdery mildew

Time: Apply before flowering; repeat at 20-day intervals.

Remarks: Mangos. Do not apply DS and WS through any type of irrigation system. Refer to label for use directions and precautions.

- - - - - - - - - - - - - - - - -

Signal Word/Toxicity Class: Caution.
REI: 24 hr.

SULFUR (Premix)

Fungicide/bactericide in flowable formulation. Toxic to fish and aquatic organisms.

Top Cop with Sulfur (6.25 lb./gal. sulfur & 1 lb./gal. tribasic copper sulfate) *(Stoller)*

Rate: Avocados:
4 pt./100 gal. water

Disease:

Anthracnose	Cercospora fruit spot	scab

Time: Apply when blossom buds open; repeat at 4 week intervals for 5 applications.

Rate: Bananas:
6 pt./100 gal. water

Disease:
Cercospora leaf spot

Time: Apply foliar application at 14-day intervals during wet season, 21-day intervals during dry season.

◆ - **new product** • ai-active ingredient • DF-dry flowable • E/EC-emulsifiable concentrate • EW-emulsion in water
F/FL-flowable • G-granule • L/LC-liquid • SP/WSP-soluble packet • W/WP-wettable powder • WSB-water soluble bag

Rate: Mangos, papaya:
4 pt./100 gal. water
Disease:
Anthracnose
Time: For mangos, apply when first bloom clusters appear; repeat weekly until fruit set, then monthly for 5-12 applications depending on area. For papaya, apply when disease is expected.

Rate: Olives:
4-6 pt./100 gal. water
Disease:
peacock spot
Time: Apply in fall.
Remarks: Avocados, bananas, mangos, olives, papaya. For bananas, add adjuvant as spreader-sticker. Refer to label for use directions and precautions.

- - - - - - - - - - - - - - - - -

Signal Word/Toxicity Class: Caution.
REI: 24 hr.

TEBUCONAZOLE

Foliar fungicide for control of specified diseases. Toxic to estuarine and marine invertebrates.

Elite 45DF *(Bayer)*
Rate: 3.2 oz. 45DF/A
Disease:
black leaf streak yellow sigatoka
Time: Apply before disease is established; repeat at 14-day intervals.
Remarks: Bananas, plantains. Do not apply through any type of irrigation system. Refer to label for use directions.
Days To Harvest: 0.
Crop Rotation: Any crop not specified on this label may be planted into treated areas 120 days after last application.

- - - - - - - - - - - - - - - - -

Signal Word/Toxicity Class: Warning/II.
REI: 12 hr.

Tennessee Brand — see copper compounds

Terranil — see chlorothalonil

TERR-O-GAS (Premix)

Restricted Use Pesticide. Classified as an ozone depleter under the Clean Air Act. Use is to be completely eliminated starting Jan. 1, 2001.

Terr-O-Gas (67% methyl bromide & 33% chloropicrin) *(Great Lakes Chemical)*
Rate: Nontarp:
270-355 lb./A
Disease:
Fusarium Pythium Sclerotinia
Phytophthora Rhizoctonia Sclerotium
Plasmodiophora
Time: Apply up to 2 weeks prior to planting when soil moisture is good and temperature is 60°-85°F.
Remarks: Pineapple. Work upwind of product and never fumigate without 2 qualified applicators. Do not fumigate when there is no or little air movement. Tarping is not used after field application. Refer to label for use directions.

- - - - - - - - - - - - - - - - -

Signal Word/Toxicity Class: Danger/I.

Thiolux — see sulfur

TOP COP — see copper compounds

Top Cop with Sulfur — see sulfur premix

TRIADIMEFON

Triazole sterol inhibitor systemic fungicide compatible with many insecticides, miticides, and fungicides.

Bayleton 50% DF *(Bayer)*
Rate: 2.66-9 oz. 50% DF/A
Disease:
Ceratocystis paradoxal fresh fruit rot pineapple butt rot
Time: Preplant and postharvest treatment.
Remarks: Pineapples. For preplant, apply as a planting material dip and allow for thorough wetting; for postharvest, dip or spray fruit for thorough coverage and allow to drain. Limit dipping time to not more than 3 minutes. Do not apply through any type of irrigation system. Do not use green forage or fodder for animal feed. Cannery waste may be fed. Refer to label for use directions.
Crop Rotation: Small grains, corn, sorghum, soybeans, beans, peas, and cucurbits may be planted 35 days after last application; forage or vines from these crops may not be used for food or feed. Root crops may be planted 120 days after last application but tops must not be used for food or feed. All crops may be planted 12 months or later after last application without restrictions.

- - - - - - - - - - - - - - - - -

Signal Word/Toxicity Class: Caution.
REI: 12 hr.

Nuts

Abound — see azoxystrobin

Aliette — see fosetyl-aluminum

AZOXYSTROBIN

Flowable broad-spectrum fungicide with preventive, systemic, and acurative properties. Toxic to freshwater and estuarine/marine fish and aquatic invertebrates.

Abound (FL) *(ZENECA)*
Rate: 9.2-12.3 fl. oa. FL/A
Disease:
Anthracnose scab
Time: Seasonal program: For early-season, apply at bud break; repeat at 2-week intervals through pollination. For cover spray, apply 3 weeks after final early-season spray; repeat at 3-week intervals for remainder of season. Block program: For prepollination applications, apply at bud break; repeat at 2-week intervals. For cover sprays, 3-week intervals through season.
Remarks: Pecans. Applications by ground or air. Do not apply more than 2.31 qt./A per yr. or more than 6 applications per yr. Do not apply through any type of irrigation system or through any ultra-low volume spray system. Refer to label for use directions.
Days To Harvest: 45.

- - - - - - - - - - - - - - - - -

Signal Word/Toxicity Class: Caution.
REI: 12 hr.

Banner — see propiconazole

BASAMID

Granular soil fumigant. Toxic to fish.

Basamid Granular (dazomet) *(BASF)*
Rate: 6-8 lb./1000 sq ft.
Time: Preplant.
Disease:
soilborne diseases
Remarks: Nonbearing: filberts, walnuts. Till to desired depth and follow suggested cultural practices. Refer to label for use directions. Do not use if soil temperature is below 43°F.

8 Fruits & Nuts

NOTICE The information on these pages is for preliminary planning — not a guide for use. Be sure to follow the manufacturer's directions, notwithstanding information contained here.	For personal protective equipment and EPA registration numbers, please turn to page 39.

Days To Harvest: 1 year.

- - - - - - - - - - - - - - - - -

Signal Word/Toxicity Class: Warning/II.
REI: 24 hr.

Basicop — see copper compounds

BENLATE

Benzimidazole systemic fungicide. Toxic to fish.

Benlate (SP, WP) (benomyl) *(DuPont)*

Rate: 0.5-0.75 lb. ai/A
1-1.5 lb. SP, WP/A
Disease:
brown rot blossom blight
Time: Apply at pink bud; second application during half-to-full bloom if necessary.
Remarks: Almonds. Do not tank mix or alternate with benzimidazole or thiophanate products. Refer to label for use directions.
Days To Harvest: 50.

- - - - - - - - - - - - - - - - -

Rate: 0.87 lb. ai/A
1.75 lb. SP, WP/A
Disease:
Botrytis blossom blight
Time: Apply 1-2 weeks before bloom; repeat at 7- to 14-day intervals through bloom.
Remarks: Macadamias. Do not tank mix or alternate with benzimidazole or thiophanate products. Refer to label for use directions.
State Restrictions: For use in Hawaii only.

- - - - - - - - - - - - - - - - -

Rate: 0.25-0.5 lb. ai/A
0.5-1 lb. SP, WP/A
Disease:
brown leaf spot, downy spot, leaf scorch, liver spot, powdery mildew, scab, Zonate leaf spot
Time: Apply at bud break or leaf unfolding, nut formation; repeat at 3- to 4-week intervals. Do not apply after shuck split.
Remarks: Pecans. Use higher rate on trees over 30 ft. tall. Do not tank mix or alternate with benzimidazole or thiophanate products. Refer to label for use directions.
Days To Harvest: 15.
State Restrictions: Benlate SP not registered for use in California.

- - - - - - - - - - - - - - - - -

Signal Word/Toxicity Class: Caution.
REI: 24 hr.

Blue Viking — see copper compounds

BROM-O-GAS

Restricted Use Pesticide. Classified as an ozone depleter under the Clean Air Act. Use is to be completely eliminated starting Jan. 1, 2001.

Brom-O-Gas (2.0%) (98% methyl bromide & 2% chloropicrin) *(Great Lakes Chemical)*

Rate: Injected with chisels + covered with tarp:
400-870 lb./A
Nontarp deep injections, auger-probe
1-2 lb./100 sq. ft.
Disease:
soilborne diseases
Time: Preplant.
Remarks: Do not fumigate when soil temperature is below 50°F. For best results, fumigate when temperature is 60°-80°F at depth of 6". Refer to label for use directions.
State Restrictions: Brom-O-Gas not for use in Arizona.

CALCIUM POLYSULFIDE

Balanced fungicide, insecticide, miticide composition containing special effective wetting agents.

Lime-Sulfur Solution *(Best Sulfur)*

Rate: 0.75-16 gal./100 gal. water
Disease:
brown rot blossom blight, powdery mildew, scab, shot hole
Time: For brown rot blossom blight, apply prebloom, early bloom, or full bloom; powdery mildew and shot hole, apply dormant; scab, apply delayed dormant.
Remarks: Almonds. For dormant stages, apply 1-3 applications at 2-week intervals between Nov. 15-Dec. 15. Avoid applying when temperatures are above 85°F as injury to foliage may occur. Do not apply this product through any type of irrigation system. Refer to label for use directions.

- - - - - - - - - - - - - - - - -

Signal Word/Toxicity Class: Danger/I.
REI: 48 hr.

CAPTAN

Toxic to fish.

Drexel Captan (4L, 50W, 80W) *(Drexel)*

Rate: 9 pt. 4L/A
4-10.4 lb. 50W/A
5.5 lb. 80W/A
Disease:
blossom blight, brown rot twig blight, leaf blight, scab, shot hole
Time: Apply at popcorn, bloom, petal fall.
Remarks: Almonds. Hulls may be fed to livestock. Do not apply through any type of irrigation system. Refer to label for use directions.
Days To Harvest: 30.

- - - - - - - - - - - - - - - - -

Signal Word/Toxicity Class: Danger/I.
REI: 4 days.

Micro Flo Captan (50WP, 80WP) or Captec (4L) *(Micro Flo)*

Rate: 4-9 lb. 50WP/A
2.5-6.6 lb. 80WP/A
1.5-2 pt. 4L/100 gal.
Disease:
brown rot blossom blight, brown rot twig blight, leaf blight, scab, shot hole
Time: Apply at popcorn, bloom, petal fall up to 5 weeks after petal fall.
Remarks: Almonds. Hulls may be fed to livestock. Do not apply this product through any type of irrigation system. Refer to label for use directions.
Days To Harvest: 30.

- - - - - - - - - - - - - - - - -

Signal Word/Toxicity Class: Danger/I.
REI: 96 hr.

Champ — see copper compounds

Champion — see copper compounds

CHLOROTHALONIL

Broad-spectrum contact fungicide. Toxic to fish, aquatic invertebrates and marine/estuarine organisms.

Terranil (6L) *(Terra)*

Rate: 3.2 lb. ai/A
4 pt. 6L/A
Disease:
eastern filbert blight
Time: Apply at leaf bud break; repeat at 2- to 4-week intervals.
Remarks: Filberts. Do not make more than 3 applications per season. Do not apply within 1 week of an oil-based pesticide application. Do not apply this product through any type of irrigation system. Do not feed treated plant parts to livestock. Refer to label for use directions.
Days To Harvest: 120.
State Restrictions: For use in Oregon.

- - - - - - - - - - - - - - - - -

Signal Word/Toxicity Class: Warning/II.
REI: 48 hr.

Clean Crop C-O-C-S — see copper compounds

◆ - **new product** • ai-active ingredient • DF-dry flowable • E/EC-emulsifiable concentrate • EW-emulsion in water
F/FL-flowable • G-granule • L/LC-liquid • SP/WSP-soluble packet • W/WP-wettable powder • WSB-water soluble bag

COPPER COMPOUNDS

Fungicide/bactericide.

Basicop (WP) (basic copper sulfate) *(Griffin)*

Rate: Almonds:
2-2.5 lb. ai/100 gal. water
4-5 lb. WP/100 gal. water

Disease:
blossom blight — brown rot

Time: Apply as dormant or bud spray. For bud sprays, apply at pink bud through popcorn stage. Do not apply after 50% bloom or when trees are in leaf.

Rate: Pecans:
1 lb. ai/100 gal. water
2 lb. WP/100 gal. water

Disease:
scab

Time: Apply according to State Extension Service recommended spray schedule.

Rate: Walnuts:
2-2.5 lb. ai/100 gal. water
4-5 lb. WP/100 gal. water

Disease:
blight

Time: Make first application late dormant prior to when catkins expand; repeat during bloom and early nutlet stage or as needed if frequent rainfall occurs.

Remarks: Almonds, pecans, walnuts. For almonds, use high rate when disease is heavy. Do not apply through any type of irrigation system. Refer to label for use directions.

Signal Word/Toxicity Class: Warning/II.
REI: 24 hr.

Blue Viking or Tennessee Brand (copper sulfate pentahydrate) *(Griffin)*

Rate: Almonds:
10-10-100 Bordeaux spray

Disease:
brown rot blossom blight — Coryneum blight

Time: For brown rot blossom blight, apply at pink bud. For Coryneum blight, apply dormant spray in late fall and early spring before buds begin to swell.

Rate: Walnuts:
15-10-100 Bordeaux spray

Disease:
downy mildew

Time: Apply early prebloom when 10%-20% pistillate are showing before or after rains; repeat after rains if necessary.

Remarks: Almonds, walnuts. Use only if Bordeaux mixture has been shown to be nonphytotoxic in area. Do not apply through any type of irrigation system. Refer to label for use directions.

Signal Word/Toxicity Class: Danger/I.
REI: 24 hr.

Champ (FL or Formula 2) or Champion (WP) (copper hydroxide) *(Agtrol)*

Rate: Almonds:
1.33-2 gal. FL/A
0.66-1 gal. Formula 2/A
8-12 lb. WP/A
Sprinkler irrigated orchards:
1.33-4 pt. FL/A
0.66-2 pt. Formula 2/A
1-3 lb. WP/A

Disease:
bacterial blast — blossom brown rot — Coryneum blight

Time: For dormant application, apply before foliage buds swell. If frequent rainfall occurs, second application should be made during early bloom (popcorn). For blast control in sprinkler irrigated orchards, apply at 2 week postbloom intervals or just before sprinkling. To avoid plant injury, do not apply after full bloom.

Rate: Filberts:
2.66-4 gal. FL/A
1.33-2 gal. Formula 2/A
16-24 lb. WP/A

Disease:
blight — eastern filbert blight

Time: Apply as postharvest spray. In seasons of heavy rainfall, apply second spray when 3/4 of leaves have dropped. For eastern filbert blight, make first application at budswell to budbreak; additional application at 10- to 14-day intervals depending on disease severity.

Rate: Macadamias (except California):
6-12 pt. FL/A
3-6 pt. Formula 2/A
3-6 lb. WP/A

Disease:
blossom blight — Raceme blight

Time: Apply during peak raceme development and bloom periods.

Rate: Pistachios:
0.66-1.33 gal. FL/A
0.33-0.66 gal. Formula 2/A
4-8 lb. WP/A

Disease:
Botrytis blight — Septoria leaf blight — shoot blight — late blight

Time: Apply beginning at bud swell. Repeat at 14- to 28-day intervals depending on disease conditions. Use high rate and shorter interval for severe disease.

Rate: Walnuts:
1.33-2.1 gal. FL/A
0.66-1.06 gal. Formula 2/A
8-12.5 lb. WP/A

Disease:
walnut blight

Time: Apply first application at early prebloom prior to or when catkins are partially expanded. Make additional applications during bloom and early nutlet stage or as needed if frequent rainfall occurs.

Remarks: Almonds, filberts, macadamias, pistachios, walnuts. For almonds, injury may occur from postbloom sprays on Mission and Peerless varieties. For filberts, add 1 pt. superior type oil/100 gal. water. Refer to label for use directions.

Signal Word/Toxicity Class: Danger/I (Champion); Caution (Champ).
REI: 24 hr.

Clean Crop C-O-C-S (WDG) (74.8% copper oxychloride & 14.2% copper sulfate, basic) *(Platte)*

Rate: 2 lb. ai/100 gal. water
4 lb. WDG/100 gal. water

Disease: Almonds:
brown rot — Coryneum blight

Time: Apply as dormant/delayed dormant spray to bud swell. For brown rot, apply prebloom and petal fall.

Disease: Filberts:
blight

Time: Apply as postharvest spray. In seasons of heavy rainfall, apply second spray when 75% leaves have dropped.

Rate: Walnuts:
1.5-2 lb. ai/100 gal. water
3-4 lb. WDG/100 gal. water

Disease:
walnut blight

Time: Apply prebloom, bloom (1% and 10%-20% bloom), and postbloom. Additional applications made immediately before or after rains important.

Remarks: Almonds, filberts, walnuts. May add dormant oil/100 gal water for almonds and filberts. Do not apply this product through any type of irrigation system. Refer to label for use directions.

State Restrictions: Not for use on filberts in California.

Signal Word/Toxicity Class: Warning/II.
REI: 24 hr.

Copper-Count-N (L) (copper ammonium) *(Mineral Research)*

Rate: Almonds:
8-12 qt. L/A

Disease:
brown rot — shot hole

Time: For brown rot, apply at delayed dormant swell stage. For shot hole, apply at leaf fall to protect buds and shoots from infection during rainy periods. Reapply up to late bud swell. Do not apply after full bloom.

NOTICE The information on these pages is for preliminary planning — not a guide for use. Be sure to follow the manufacturer's directions, notwithstanding information contained here.

For personal protective equipment and EPA registration numbers, please turn to page 39.

Rate: Filberts:
4-6 qt. L/100 gal.
Disease:
blight
Time: Apply late Aug.-early Sept. In heavy rains, apply another spray when 3/4 of leaves have dropped.

Rate: Walnuts:
8-12 qt. L/A
Disease:
walnut blight
Time: Make first application at early prebloom; second application at late prebloom, and additional applications if disease conditions persist.
Remarks: Almonds, filberts, walnuts. Applications may be made by ground or air; refer to label for use directions. Do not apply through any type of irrigation system unless supplemental labeling on chemigation is followed.
State Restrictions: Not for use on filberts in California.

Signal Word/Toxicity Class: Caution.
REI: 12 hr.

Drexel Basic Copper Sulfate (WP) or KOP 300 (3FL) (basic copper sulfate) *(Drexel)*
Rate: Almonds:
4-5 lb. WP/100 gal. water
2.6-4 pt. 3FL/100 gal. water
Dormant (aerial)
16-20 lb. WP/A
10.6-18 pt. 3FL/A
Disease:
brown rot shot hole
Time: Apply at pink bud through popcorn stage. Do not apply after trees are in leaf.

Rate: Walnuts:
4-5 lb. WP/100 gal. water
2.6-4 pt. 3FL/100 gal. water
Disease:
blight
Time: Make first applications in early prebloom (1% pistillate not catkin blooms showing); second when 10%-20% pistillate not catkin blooms are showing. Repeat 3-4 times per growing season if necessary.
Remarks: Almonds, walnuts. Do not apply through any type of irrigation system. Refer to label for use directions.

Signal Word/Toxicity Class: Warning/II.
REI: 24 hr. (WP); 12 hr. (3FL).

Kocide (101, DF, 20/20, 2000, 4.5LF, LF) (copper hydroxide) *(Griffin)*
Rate: Almonds:
6-12 lb. 101, DF/A
15-30 lb. 20/20/A
4.5-6 lb. 2000/A
4-5.33 pt. 4.5LF/A
8-16 pt. LF/A
Disease:
blossom brown rot (101, DF, 20/20, LF) Coryneum blight
Time: For dormant application, apply before foliage buds begin to swell. For early bloom (popcorn) application, apply before full bloom. To avoid plant injury, do not apply after full bloom.

Rate: Almonds:
12-16 lb. 101, DF/A
30-40 lb. 20/20/A
6-9 lb. 2000/A
5.33-8 pt. 4.5LF/A
12-21.33 pt. LF/A
Disease:
bacterial blast bacterial canker (2000, 4.5LF)
Time: Apply at dormant to early pink bud. Injury may occur from postbloom sprays, especially on Neplus varieties.

Rate: Filberts:
16-24 lb. 101, DF/A
12-18 lb. 2000/A
10.66-16 pt. 4.5LF/A
21.33-32 pt. LF/A
Disease:
blight eastern filbert blight
Time: Apply as postharvest spray. In seasons of heavy rainfall, apply second spray when 3/4 of leaves have dropped. For eastern filbert blight, make first application after harvest in Oct. before heavy winter rains begin; second application in late Feb.-early Mar.; third application 1 month later.

Rate: Macadamias:
4.5-6 lb. 101, DF/A
3.5-4.5 lb. 2000/A
3-4 pt. 4.5LF/A
9-12 pt. LF/A
Disease:
Anthracnose Phytophthora blight raceme blight
Time: Anthracnose: Apply at first sign of flowering; repeat weekly until just before harvest. Phytophthora blight, raceme blight: Apply during raceme development and bloom periods.

Rate: Pecans:
2-4 lb. 101, DF/A
1.5-3 lb. 2000/A
1.33-2.66 pt. 4.5LF/A
2-4 qt. LF/A
Disease:
kernel rot shuck rot Zonate leaf spot
Time: For suppression, apply at 2-4 week intervals starting at kernel growth and continuing until shucks open. Use higher rate and shorter interval if frequent rainfall occurs.

Rate: Pistachios:
4-8 lb. 101, DF/A
3-6 lb. 2000/A
2.66-5.33 pt. 4.5LF/A
4-8 qt. LF/A
Disease:
Botrytis blight Septoria leaf blight shoot blight late blight
Time: Apply at bud swell; repeat at 14- to 28-day intervals. Use higher rate and shorter interval for severe disease conditions.

Rate: Walnuts:
8-12.5 lb. 101, DF/A
20 lb. 20/20/A
6-9.5 lb. 2000/A
5.33-8.33 pt. 4.5LF/A
10.66-16 pt. LF/A
Disease:
walnut blight
Time: Apply first application spray at early prebloom prior to or when catkins are partially expanded. Make additional applications during bloom and early nutlet stage or as needed if frequent rainfall occurs.
Remarks: Almonds, filberts, macadamias, pecans, pistachios, walnuts. For filberts, add 1 pt. superior type oil/100 gal. water. Do not apply in a spray solution having a pH less than 6.5 as phytotoxicity may occur. Refer to label for use directions.
State Restrictions: Not for use in California on macadamia nuts.

Signal Word/Toxicity Class: Danger/I (101, DF, 20/20); Warning/II (2000); Caution (4.5LF, LF).
REI: 24 hr.

KOP-Hydroxide (FL, 50WP) (copper hydroxide) *(Drexel)*
Rate: Almonds:
2.66-4 pt. FL/100 gal. water
6-12 lb. 50WP/A
Disease:
blossom brown rot Coryneum blight
Time: Apply late dormant before foliage buds swell if frequent rainfall occurs; or during early bloom stage (popcorn). Do not use after full bloom.

Rate: Almonds:
4-5.33 pt. FL/100 gal. water
12-16 lb. 50WP/A
Disease:
bacterial blast
Time: Apply dormant to early pink bud.

Rate: Filberts:
5.33-8 pt. FL/100 gal. water
4-6 lb. 50WP/A
Disease:
blight
Time: Apply late Aug.-early Sept. In seasons with heavy rainfall, apply again when 3/4 leaves have dropped.

◆ - **new product** • ai-active ingredient • DF-dry flowable • E/EC-emulsifiable concentrate • EW-emulsion in water
F/FL-flowable • G-granule • L/LC-liquid • SP/WSP-soluble packet • W/WP-wettable powder • WSB-water soluble bag

Rate: Walnuts:
8-12.5 lb. 50WP/A
Dilute:
2.66 pt. FL/100 gal. water
Concentrate:
2.66 pt. FL/A (in 50-100 gal. water)
Disease:
walnut blight
Time: Apply at early prebloom when catkins are partially expanded; make 3 additional applications during bloom and early nutlet stages at 7- to 10-day intervals.
Remarks: Almonds, filberts, walnuts. KOP-Hydroxide 50WP should not be applied in spray solution having pH less than 6.5 or greater than 9.0 as phytotoxicity may occur. Refer to label for use directions.

Signal Word/Toxicity Class: Danger/I.
REI: 48 hr.

Micro Flo Basic Copper "53" (basic copper sulfate) *(Micro Flo)*
Rate: Almonds:
12-20 lb. 53/A
Disease:
brown rot (suppression) shot hole
Time: Apply at pink bud to popcorn stages, full bloom and petal fall if trees are not in leaf. Apply higher rate as dormant spray (California).

Rate: Filberts:
6 lb. 53/100 gal. water
Walnuts:
12-25 lb. 53/A
Disease:
blight
Time: Filberts: Apply postharvest before first heavy rain; if heavy fall rains occur, repeat after 3/4 of leaves have dropped. Walnuts: Apply early prebloom (1% pistillate), second when 10%-20% pistillate blooms are showing; repeat application 3-4 times as needed.
Remarks: Almonds, filberts, walnuts. Spraying tender foliage may cause leaf injury on almonds. Refer to label for use directions.

Signal Word/Toxicity Class: Warning/II.
REI: 24 hr.

Microsperse COC (53WP) (copper oxychloride) *(Micro Flo)*
Rate: Almonds:
6-12 lb. 53WP/100 gal. water
Disease:
brown rot blossom blight Coryneum blight twig blight
Time: Apply prebloom, bud swell, petal fall; 8-12 lb. in dormant before buds begin to swell.

Rate: Filberts:
16-24 lb. concentrate 53WP/A
Disease:
blight
Time: Apply postharvest in late Aug.-early Sept. before first heavy fall rain. If heavy rain occurs, repeat after 3/4 leaves have dropped.

Rate: Pecans:
2-4 lb. 53WP/A
Disease:
kernel rot shuck rot Zonate leaf spot
Time: Apply at 2- to 4-week intervals starting at kernel growth and continuing until shucks open.

Rate: Walnuts:
8-14 lb. 53WP/A
Disease:
blight
Time: Apply early prebloom (1% pistillate), second application when 10%-20% pistillate blooms are showing; repeat 3-4 times as needed during bloom and nutlet development.
Remarks: Almonds, filberts, pecans, walnuts. Refer to label for use directions.

Signal Word/Toxicity Class: Danger/I.
REI: 24 hr.

Nu-Cop (3L, 50DF, 50WP) (copper hydroxide) *(Micro Flo)*
Rate: Almonds:
1.33-8 pt. 3L/A
6-12 lb. 50DF, 50WP/A
Disease:
bacterial blast (3L) blossom brown rot Coryneum blight
Time: 50DF, 50WP: Apply dormant, before buds begin to swell, early bloom popcorn, full bloom. 3L: Apply during early bloom (popcorn); late dormant before foliage buds swell may be necessary when frequent rainfall occurs. For bacterial blast, apply early pink bud.

Rate: Filberts:
10.6-16 pt. 3L/A
16-24 lb. 50DF, 50WP/100 gal.
Disease:
blight eastern filbert blight
Time: Blight: Apply postharvest late Aug.-early Sept. In seasons of heavy rainfall, apply another when 3/4 leaves have dropped. Eastern filbert blight: Apply after harvest (Oct.) before heavy winter rains begin; repeat late Feb.-early Mar., and again 4 weeks later.

Rate: Macadamias:
4-8 pt. 3L/A
3-6 lb. 50DF, 50WP/A
Disease:
Anthracnose blossom blight raceme blight
Time: Anthracnose: Apply at first sign of flowering; repeat weekly until just before harvest. Blossom blight, raceme blight: Apply during peak raceme development and bloom period.

Rate: Pecans:
2.66-5.33 pt. 3L/A
2-4 lb. 50DF, 50WP/A
Disease:
kernel rot shuck rot Zonate leaf spot
Time: Apply at 2- to 4-week intervals starting at kernel growth and continuing until shucks open.

Rate: Pistachios:
5.3-10.6 pt. 3L/A
Disease:
Botrytis blight Septoria leaf blight shoot blight late blight
Time: Apply at bud swell; repeat at 14- to 28-day intervals.

Rate: Walnuts:
5.33-17 pt. 3L/A
8-12.5 lb. 50DF, 50WP/100 gal.
Disease:
walnut blight
Time: Apply early prebloom when catkins are partially expanded; during bloom and early nutlet stage or as needed if frequent rainfall occurs.
Remarks: Almonds, filberts, macadamias, pecans, pistachios, walnuts. Refer to label for use directions.

Signal Word/Toxicity Class: Danger/I (50DF, 50WP); Warning/II (3L).
REI: 48 hr. (50DF, 50WP); 24 hr. (3L).

TOP COP Tri-Basic (FL) (basic copper sulfate) *(Stoller)*
Rate: Almonds:
4-7 qt. FL/A
Disease:
brown rot shot hole
Time: Apply at pink bud through popcorn stage at low rates; high rates for dormant applications. Spray before 50% bloom; do not spray when trees are in leaf.

Rate: Filberts:
4 qt. FL/100 gal. water
Disease:
blight
Time: Apply in early fall. In season of heavy rainfall; repeat when 3/4 of leaves have dropped.

Rate: Walnuts:
2-2.5 qt. FL/100 gal. water
Disease:
blight
Time: Apply in early prebloom (1% pistillate and catkins bloom showing); apply second application during 10%-20% pistillate and catkins bloom; repeat at 3- to 4-day intervals.
Remarks: Almonds, filberts, walnuts. Overspraying of almonds may injure tender foliage. For filberts, add 1 pt. superior type oil; spray to wet. Refer to label for use directions.

Signal Word/Toxicity Class: Caution.
REI: 24 hr.

NOTICE The information on these pages is for preliminary planning — not a guide for use. Be sure to follow the manufacturer's directions, notwithstanding information contained here. For personal protective equipment and EPA registration numbers, please turn to page 39.

Copper-Count-N — see copper compounds

DODINE

Protective fungicide in a wettable powder formulation recommended for the control of major fungal diseases in fruits and nuts. Not compatible with nonionic surfactants.

Syllit (65W) *(Platte)*
Rate: Black walnuts:
2.6 lb. ai/A
4 lb. 65W/A
Disease:
Anthracnose

Rate: Pecans:
1.3-2.6 lb. ai/A
2-4 lb. 65W/A (ground)
Disease:
brown leaf spot, downy mildew, downy spot, leaf blotch, liver spot, scab
Time: For black walnuts, apply when first leaves unfold; repeat at 10- to 14-day intervals as needed. For pecans, apply when first leaves are showing at 10- to 14-day intervals through first cover; repeat subsequent applications at 2-3 week intervals as needed. Do not apply after shucks have started to open.
Remarks: Black walnuts, pecans. Severe pecan foliage injury has been experienced on Moore and Dan Deman varieties in the Southeast. Do not graze meat or dairy animals in treated groves. Do not apply through any type of irrigation system. Refer to label for use directions.
Days To Harvest: Black walnut-14.

Signal Word/Toxicity Class: Danger/I.
REI: 48 hr.

Enable — see fenbuconazole

FENBUCONAZOLE

Triazole sterol inhibitor. Systemic fungicide with protectant, curative, and eradicant properties. Translocated upwards in the plant. Toxic to fish.

Enable (2F) *(Rohm and Haas)*
Rate: 8 fl. oz. 2F/A
Disease:
downy spot, leaf blotch, leaf scorch, leaf spot, powdery mildew, scab, vein spot
Time: Apply at bud break; repeat at 10- to 14-day intervals through pollination.
Remarks: Pecans. Addition of surfactant will improve performance. Do not apply more than 1.5 qt./A per season. Do not graze livestock in treated areas or feed crops grown in treated areas to livestock. Refer to label for use directions.
Days To Harvest: 28.

Signal Word/Toxicity Class: Caution.
REI: 12 hr.

FOSETYL-ALUMINUM

Systemic fungicide/bactericide. Toxic to estuarine invertebrates.

Aliette (WDG, WSP) *(Rhone-Poulenc)*
Rate: 2.4-4 lb. ai/100 gal.
3-5 lb. WDG, WSP/100 gal.
Disease:
Phytophthora
Time: Nonbearing.
Remarks: Nonbearing almonds. Do not apply more than 100 gal./A, 5 lb./A per application, or more than 20 lb./A per year. Apply as paint or spray to pruning wound area in sufficient volume to thoroughly wet entire wound surface. Do not exceed 4 applications per year. Refer to label for use directions.

Signal Word/Toxicity Class: Caution.
REI: 12 hr.

GALLEX (Premix)

Ready-to-use paint-on emulsion.

Gallex (2,4-xylenol & meta-cresol) *(AgBioChem)*
Rate: Apply undiluted.
Disease:
crown gall
Time: Apply first application before breaking of dormancy or during spring and early summer.
Remarks: Almonds, filberts, pecans, walnuts. Paint gall and cut surfaces + 1/2" of surrounding healthy bark. Generally 1 application will give control; retreatment may be necessary. Do not mix or combine with other pesticides. Complete girdling of trunk by galls affects water-conducting capacity of tree. To prevent sudden loss of water movement, do not treat more than 1/2 galled trunk at any one time. Treat remaining diseased tissue 4-6 months later when examination of initial treatment is made.

Signal Word/Toxicity Class: Warning/II.

IPRODIONE

Dicarboximide fungicide. Toxic to aquatic invertebrates.

Rovral (4FL, WG, WP) *(Rhone-Poulenc)*
Rate: 0.5 lb. ai/A
1 pt. 4FL/A
1 lb. WG, WP/A
Disease:
brown rot blossom blight, shot hole
Time: Apply at pink bud; if conditions favorable for disease development persist or recur up to 3 subsequent applications can be made at full bloom, petal fall, up to 5 weeks after petal fall.
Remarks: Almonds. Maximum of 4 applications per season. Do not graze animals in treated orchards or feed cover crops grown in treated orchards to livestock. Refer to label for use directions.

Signal Word/Toxicity Class: Warning/II (WG); Caution (4FL, WP).
REI: 12 hr.

Kocide — see copper compounds

KOP-Hydroxide — see copper compounds

Kumulus — see sulfur

Lime-Sulfur Solution — see calcium polysulfide

MANEB

Dithiocarbamate contact fungicide. Toxic to fish.

Maneb 75DF or 80WP *(Elf Atochem)*
Manex (FL) *(Griffin)*
Rate: 4.5-6 lb. ai/A
6-8 lb. 75DF/A
4.8-6.4 lb. ai/A
6-8 lb. 80WP/A
4.8-6.4 qt. FL/A
Disease:
brown rot blossom blight, brown rot twig blight, rust, scab, shot hole
Time: Apply at popcorn, full bloom, petal fall, or every 7- to 10-days if bloom is staggered. Omit petal fall spray if only brown rot is present. Do not apply later than 5 weeks after petal fall.
Remarks: Almonds. Do not allow livestock to graze orchards. Refer to label for use directions.
Days To Harvest: 145.

Signal Word/Toxicity Class: Caution.
REI: 24 hr.

Manex — see maneb

◆ - **new product** • ai-active ingredient • DF-dry flowable • E/EC-emulsifiable concentrate • EW-emulsion in water
F/FL-flowable • G-granule • L/LC-liquid • SP/WSP-soluble packet • W/WP-wettable powder • WSB-water soluble bag

MEFENOXAM

Systemic fungicide having a specific mode of action.

Ridomil Gold EC, GR, or WSP *(Novartis)*

Rate: 1 lb. ai/A
2 qt. EC/A
1.5 lb. ai/A
80 lb. GR/A
1.8 lb. ai/A
4 lb. WSP/A

Disease:
collar rot · crown rot · root rot

Time: Established plantings: Apply in spring before plants begin their growth. New plantings: Make first application 2 weeks after planting. Additional applications can be made at 2- to 3-month intervals.

Remarks: Almonds, walnuts. Do not dip roots in Ridomil Gold EC solution, spray roots, or concentrate it around tree trunks, or injury may occur. Do not apply to trees under stress. Do not graze or feed cover crops in treated orchards. Do not plant any crop which is not registered for use with mefenoxam in treated soil for 12 months.

Days To Harvest: 1.

Signal Word/Toxicity Class: Warning/II (EC); Caution (GR, WSP).
REI: 48 hr.

METALAXYL

Systemic fungicide having a specific mode of action.

Subdue 2E *(Novartis)*

Rate: 6-12 fl. oz. 2E/1000 sq. ft.

Disease:
Phytophthora collar rot · Phytophthora root rot · Pythium root rot · Phytophthora crown rot

Time: Apply as necessary at 3-month intervals during growing season.

Remarks: Nonbearing in nurseries. Do not apply to trees that will bear harvestable nuts within 12 months of application. Apply in sufficient water for thorough coverage. Use high rates in areas known to be infested with Phytophthora.

Signal Word/Toxicity Class: Warning/II.
REI: 12 hr.

Micro Flo — see copper compounds

Microsperse — see copper compounds

Microthiol Special — see sulfur

MYCLOBUTANIL

Triazole sterol inhibitor. Systemic, protectant, and curative fungicide.

Rally (40W) *(Rohm and Haas)*

Rate: 5-8 oz. 40W/A

Disease:
blossom blight · rust · shot hole

Time: Apply at pink bud stage (5% bloom). If conditions are favorable for disease development, repeat at full bloom and petal fall; for rust, apply 8 weeks after flowering.

Remarks: Almonds. Do not make more than 3 applications or apply more than 1.5 lb./A per season. Refer to label for use directions.

Days To Harvest: 90.

State Restrictions: For use only in Arizona, California, Hawaii.

Signal Word/Toxicity Class: Warning/II.
REI: 24 hr.

Nu-Cop — see copper compounds

ORBIT/SUPER TIN (Co-Pack)

Two fungicides in water-soluble packages which must be used together providing a broad spectrum control of diseases.

Orbit/Super Tin: Orbit 45WP (45% propiconazole) *(Novartis)* **& Super Tin 80WP (80% triphenyltin hydroxide)** *(Griffin)*

Rate: 20 oz. 45WP + 25 oz. 80WP/5A

Disease:
brown leaf spot · liver spot · sooty mold
downy spot · powdery mildew · vein spot
leaf blotch · scab · Zonate leaf spot
leaf spot

Time: Apply at prepollination stages when young leaves are unfolding; second application when small nuts are forming.

Remarks: Pecans (bearing, nonbearing). For ground application (closed cabs only), apply in sufficient water to provide full coverage. For aerial application, apply in minimum of 20 gal. water. Apply maximum of 8 treatments during single growing season at 2- to 4-week intervals as needed. Do not apply after shucks have started to open. Refer to label for use directions.

State Restrictions: Not for use in certain counties in Alabama, Arkansas, Kentucky, Mississippi, Missouri, and Tennessee; refer to label for specific counties.

Signal Word/Toxicity Class: Danger/I (Super Tin); Caution (Orbit).
REI: 48 hr.

PROPICONAZOLE

Triazole sterol inhibitor. Toxic to fish.

Banner MAXX *(Novartis)*

Rate: Pecans:
8-12 fl. oz. MAXX/100 gal. water

Disease:
brown spot · leaf scorch · scab
downy spot · powdery mildew · Zonate leaf spot

Time: Apply at beginning of bud break; repeat at 14-day intervals for 3 applications.

Rate: Walnuts:
8.5 fl. oz. MAXX/100 gal. water

Disease:
Anthracnose

Time: Apply at 7- to 14-day intervals.

Remarks: Nonbearing pecans, walnuts. Do not apply this product through any type of irrigation system.

Signal Word/Toxicity Class: Warning/II.
REI: 24 hr.

Rally — see myclobutanil

Ridomil Gold — see mefenoxam

Rovral — see iprodione

Subdue — see metalaxyl

Sulfa 80 — see sulfur

SULFUR

Nonsystemic fungicide with protective action. Microthiol Special, Sulfa 80, and Thiolux are micronized sulfurs.

Drexel Suffa (6FL), Sulfa 80 (80DF) or Sulfur (90W) *(Drexel)*

Rate: 1.1-2.6 lb. ai/100 gal. water
1.5-3.5 pt. 6FL/100 gal. water
8-24 lb. ai/A
10-30 lb. 80DF/A
1.8-5.4 lb. ai/100 gal. water
2-6 lb. 90W/100 gal. water

NOTICE The information on these pages is for preliminary planning — not a guide for use. Be sure to follow the manufacturer's directions, notwithstanding information contained here.

For personal protective equipment and EPA registration numbers, please turn to page 39.

Disease:
brown rot powdery mildew scab
leaf spot

Time: Apply prebloom, petal fall, and as cover spray at 10- to 14-day intervals or after periods of wet weather.

Remarks: Almonds. Do not use within 2 weeks of oil application (4 weeks 6FL). Some nuts are susceptible to injury from sulfur under certain climatic conditions. Do not make sulfur applications at excessively high temperatures. Do not apply this product through any type of irrigation system. Refer to label for use directions.

Signal Word/Toxicity Class: Caution.
REI: 24 hr.

Kumulus DF (sulfur) *(BASF)*
Rate: 8-16 lb. ai/A
10-20 lb. DF/A
Disease:
brown rot blossom blight leaf spot powdery mildew

Time: Apply at bloom or early petal fall; repeat as necessary, usually 10- to 14-days or after wet weather. For brown rot, apply when fruit starts to ripen.

Remarks: Almonds, pecans. Do not apply if temperatures during or within 3 days after application are expected to exceed 90°F. Do not apply through any type of irrigation system. Refer to label for use directions.

Signal Word/Toxicity Class: Caution.
REI: 24 hr.

Liquid Sulfur Six *(Helena)*
Rate: 0.66-3.5 pt./100 gal. water
Disease:
brown rot leaf spot

Time: For petal fall, shuck, and cover spray, apply lower rate; pink and bloom sprays, apply higher rate.

Remarks. Almonds. Rates vary by time of application. Do not apply this product through any type of irrigation system. Refer to label for use directions.

Signal Word/Toxicity Class: Caution.
REI: 24 hr.

Micro Flo Sulfur (6L), Dual Spraying & Dusting Sulfur (DS) or Microsperse Wettable Sulfur (WS) *(Micro Flo)*
Rate: 8-24 lb. ai/A
10.6-32 pt. 6L/A
27 lb. ai/A
30 lb. DS/A
9-27 lb. ai/A
10-30 lb. WS/A
Disease:
blossom blight (WS) leaf spot twig blight (WS)
brown rot

Time: DS: Apply at bloom; repeat at 12-day intervals as needed. WS: Apply at bloom or early petal fall; repeat at 10- to 14-day intervals or after a period of wet weather.

Rate: 4-16 lb. ai/A
5.33-21 pt. 6L/A
Disease:
leaf spot powdery mildew sooty mold

Time: Apply throughout season as needed.

Remarks: Almonds, pecans. Some pecans varieties are sensitive to sulfur, do not apply unless varieties are known to be tolerant. Do not apply DS and WS through any type of irrigation system. Refer to label for use directions.

Signal Word/Toxicity Class: Caution.
REI: 24 hr.

Microthiol Special (80WP) *(Elf Atochem)*
Rate: Almonds:
8-16 lb. ai/A
10-20 lb. 80WP/A
Disease:
brown rot leaf spot

Time: Apply at bloom or early petal fall; repeat at 10- to 14-day intervals or after periods of wet weather.

Rate: Pecans:
4-8 lb. ai/A
5-10 lb. 80WP/A
Disease:
powdery mildew sooty mold

Time: Apply throughout the season as needed.

Remarks: Almonds, pecans. Some pecans varieties are sensitive to sulfur under certain conditions. Do not apply unless varieties are known to be tolerant of sulfur. Refer to label for use directions.

State Restrictions: Not for use on pecans in California.

Signal Word/Toxicity Class: Caution.
REI: 24 hr.

Terra Sulfur (6L) *(Terra)*
Rate: Almonds:
8-24 lb. ai/A
1.33-4 gal. 6L/A
Disease:
blossom blight powdery mildew scab
brown rot rust twig blight
leaf spot

Time: Apply at bloom or early petal fall; repeat at 10- to 14-day intervals or after period of wet weather.

Rate: Pecans:
5-16 lb. ai/A
0.66-2.66 gal. 6L/A
Disease:
leaf spot powdery mildew sooty mold

Time: Apply throughout season as needed.

Remarks: Almonds, pecans. Do not use within 2 weeks of oil application. Some plants are susceptible to injury from sulfur under certain climatic conditions. Do not make sulfur applications at excessively high temperatures. Refer to label for use directions.

Signal Word/Toxicity Class: Caution.
REI: 24 hr.

Thiolux (DF) *(Platte)*
Rate: 10-30 lb. DF/A
Disease:
brown rot blossom blight leaf spot rust
powdery mildew scab

Time: Apply at bloom or early petal fall; repeat at 10- to 14-day intervals as necessary after periods of wet weather. For brown rot, apply when fruit starts to ripen.

Remarks: Almonds. Do not make applications within 2 weeks of oil spray.

Days To Harvest: 30.

Signal Word/Toxicity Class: Caution.
REI: 24 hr.

Wettable Sulfur (WS) *(Platte)*
Rate: 6 lb. WS/100 gal. water
Disease:
brown rot scab

Time: Apply prebloom, bloom, postbloom, or when disease threatens during rainy periods; repeat at 7- to 10-day intervals.

Remarks: Do not make applications within 2 weeks of oil spray.

Signal Word/Toxicity Class: Caution.
REI: 24 hr.

SULFUR (Premix)

Fungicide/bactericide in flowable formulation. Toxic to fish and aquatic organisms.

Top Cop with Sulfur (6.25 lb./gal. sulfur & 1 lb./gal. tribasic copper sulfate) *(Stoller)*
Rate: Almonds:
2-4 qt./100 gal. water
Disease:
brown rot shot hole

Time: Apply as dormant spray or during swelling bud stage to early bloom stage (popcorn). Spray before 50% bloom or after petal fall.

Rate: Pecans:
2-3 qt./100 gal. water
Disease:
scab

Time: Apply as foliar spray when catkins show; repeat 3-4 times at 3 week intervals.

◆ - **new product** • ai-active ingredient • DF-dry flowable • E/EC-emulsifiable concentrate • EW-emulsion in water
F/FL-flowable • G-granule • L/LC-liquid • SP/WSP-soluble packet • W/WP-wettable powder • WSB-water soluble bag

Rate: **Walnuts:**
1-3 qt./100 gal. water
Disease:
blight
Time: Apply early prebloom; repeat at 7- to 10-day intervals until fruit set.
Remarks: Almonds, pecans, walnuts. Overspraying of almonds may injure tender foliage. Refer to label for use directions.

- - - - - - - - - - - - - - - -

Signal Word/Toxicity Class: Caution.
REI: 24 hr.

SUPER TIN

Nonsystemic fungicide with mainly protective action, but also some curative action. Restricted Use Pesticide.

Super Tin (80WP) (triphenyltin hydroxide) *(Griffin)*
Rate: 5-7.5 oz. 80WP/A
Disease:

downy spot	liver spot	scab
leaf blotch	powdery mildew	sooty mold
leaf spot		

Time: Apply at prepollination stages when young leaves are unfolding; second application when small nuts are forming; repeat at 2-4 week intervals as necessary. Do not apply after shucks have started to open.
Remarks: Pecans. For ground application (closed cabs only), use sufficient water for full coverage; aerial application apply in minimum of 20 gal. water. Use low rate for first 2 applications, until disease becomes severe, and during dry weather. Use high rate during wet weather or during severe scab, powdery mildew, or other disease infections. Do not mix with pesticides, micronutrients, or adjuvants unless testing or prior experience has shown mixture to be nonphytotoxic to crop.

- - - - - - - - - - - - - - - -

Signal Word/Toxicity Class: Danger-Poison/I.
REI: 48 hr.

Syllit — see dodine

Tennessee Brand — see copper compounds

Terranil — see chlorothalonil

Thiolux — see sulfur

THIOPHANATE-METHYL

Benzimidazole systemic fungicide. Toxic to fish.

Topsin M (WSB, 70W) *(Elf Atochem)*
Rate: Almonds:
0.7-1.4 lb. ai/A
1-2 lb. WSB, 70W/A
Disease:
brown rot blossom blight
Time: Apply at early pink bud and early petal fall.

Rate: Pecans:
0.35-0.7 lb. ai/A
0.5-1 lb. WSB, 70W/A
Disease:

brown spot	powdery mildew	stem end rot
downy spot	scab	Zonate leaf spot
liver spot		

Time: Apply when first leaves are showing; repeat at 3-4 week intervals until shuck split. Do not apply after shuck split.
Remarks: Almonds, pecans. For pecans, use higher rate for trees over 30 ft. tall and for aerial applications in Arkansas, Louisiana, Mississippi, Oklahoma, and Texas. Refer to label for use directions.

- - - - - - - - - - - - - - - -

Signal Word/Toxicity Class: Caution.
REI: 12 hr.

TOP COP — see copper compounds

Top Cop with Sulfur — see sulfur premix

Topsin M — see thiophanate-methyl

VANGARD

Broad-spectrum fungicide. Toxic to fish and aquatic invertebrates.

Vangard (WG) (cyprodinil) *(Novartis)*
Rate: 5-10 oz./A
Tank mix:
5 oz./A
Disease:
brown rot blossom blight
Time: Apply at 5%-10% bloom. If conditions persist, make additional application at 50%-100% bloom, and petal fall.
Remarks: Almonds. May be used alone or in tank mix. Use high rate for severe disease conditions. Do not apply more than 30 oz. WG/A per crop per season. Refer to label for use directions.
Days To Harvest: 50.

- - - - - - - - - - - - - - - -

Signal Word/Toxicity Class: Caution.
REI: 12 hr.

ZIRAM

Toxic to fish and wildlife.

Ziram 4L or 76 *(Drexel)*
Rate: Almonds:
1-2 lb. ai/100 gal.
2-4 pt. 4L/100 gal.
1.1-1.5 lb. ai/100 gal.
1.5-2 lb. 76/100 gal.
Disease:

blossom blight	leaf blight	shot hole
brown rot	scab	twig blight

Time: Blossom blight, brown rot, and twig blight: Apply at popcorn and full bloom. Leaf blight and scab: Apply at petal fall and 5 weeks later. For shot hole: Apply at popcorn, full bloom, petal fall, and 5 weeks later.

Rate: Pecans:
1.5 lb. ai/100 gal.
3 pt. 4L/100 gal.
1.5 lb. ai/100 gal.
2 lb. 76/100 gal.
Disease:
scab
Time: Apply when leaves are 1/4-1/2" long; repeat after pollination is complete and up to 5 cover sprays at 3-4 week intervals through Aug.
Remarks: Almonds, pecans. Do not apply this product through any type of irrigation system. Refer to label for use directions.
State Restrictions: Not for use on pecans in California.

- - - - - - - - - - - - - - - -

Signal Word/Toxicity Class: Danger/I.
REI: 48 hr.

Ziram 76DF *(Elf Atochem)*
Ziram Granuflo (WDG) *(UCB Agrochemicals)*
Rate: Almonds:
6 lb. ai/A
8 lb. 76DF, WDG/A
Disease:

brown rot	scab	shot hole
leaf blight		

Time: Apply at popcorn, full bloom, petal fall or as needed, but no later than 5 weeks after petal fall.

Rate: Pecans:
4.5-6 lb. ai/A
6-8 lb. 76DF, WDG/A
Disease:
Anthracnose scab
Time: Begin applications when leaves are 1/4-1/2" long; repeat after pollination for up to 5 cover sprays at 3-4 week intervals.
Remarks: Almonds, pecans. Do not apply this product through any type of irrigation system. Do not graze or feed cover crops from treated orchards.
Days To Harvest: Pecans-55.

- - - - - - - - - - - - - - - -

Signal Word/Toxicity Class: Danger/I.
REI: 48 hr.

For personal protective equipment and EPA registration numbers, please turn to page 39.

Notes • Notes • Notes

Section 9

Turf & Ornamentals

Bt Chart 488

Insecticide Listings

Turf 489

Ornamentals 494

Fungicide Listings

Turf 509

Ornamentals 519

The information in this 1999 Insect and Disease Control Guide is updated to August 15, 1998.

Because manufacturers' product recommendations are constantly being updated and revised as additional product registrations and label changes are approved by EPA, there may be changes after this date. However, every effort has been made to give manufacturers the opportunity to include in the Guide their latest information. This Guide is for preliminary planning – not a use guide. Be sure to follow label directions, notwithstanding information contained herein.

Bt Products Chart

Turf & Ornamentals

Find Bt products quickly and easily with this Bt products chart. Simply find the insect you want to control on the top of the chart, and follow down to the marked spot to find the Bt product you can use to control that pest.

	Armyworms	Azalea moth	Bagworms	Browntail moth	Buck moth	Budworms	Cankerworms	Caterpillars	Coneworm	Diamondback moth	Ello moth	Elm spanworm	Fungus gnats	Greenstriped mapleworm	Gypsy moth	Heliothis	Io moth	Leaf beetles	Leafrollers	Loopers	Oakworms	Oleander moth	Pine butterfly	Satin moth	Sod webworm	Tortrix	Tussock moths	Webworms
Biobit	■	■	■	■	■	■	■	■	■	■	■	■			■	■	■		■	■	■	■	■	■	■	■	■	
BMP 123	■	■	■			■	■	■		■	■	■			■	■	■		■	■		■	■		■		■	■
Bt. ssp. *aizawai*	■					■		■		■	■						■								■			
Bt. ssp. *aizawai* + *Bt.* ssp. *kurstaki*	■				■	■		■		■									■	■						■		
Bt. ssp. *israelensis*													■															
Bt. ssp. *kurstaki*	■	■	■	■	■	■	■	■	■	■	■	■		■	■	■	■		■	■	■	■	■	■	■	■	■	■
Bt. ssp. *tenebriosis*																		■										
Condor	■	■	■			■	■	■		■	■	■			■	■	■		■	■		■	■		■		■	■
Crymax	■	■	■			■	■	■		■	■	■			■		■		■	■		■	■		■		■	■
Cutlass																									■			
DiPel	■	■	■			■	■	■		■	■	■		■	■	■	■		■	■	■	■	■		■		■	■
Foray	■	■	■	■	■	■	■	■	■	■	■	■			■	■	■		■	■	■	■	■	■		■	■	■
Gnatrol													■															
Javelin	■		■			■	■	■				■			■				■		■				■		■	■
Lepinox	■																								■			
Mattch	■				■	■		■		■									■	■						■		
MVP II	■				■	■		■		■									■	■						■		
Novodor																		■										
Troy-Bt	■		■			■	■	■		■					■					■								
XenTari	■					■		■			■						■		■	■		■			■			

Manufacturers List

These Bt products are manufactured by the following companies. Look for their addresses, phone numbers, and fax numbers in the Suppliers Directory beginning on page 599.

Javelin – Thermo Trilogy
Biobit, DiPel, Foray, Gnatrol, Novodor, XenTari – Abbott Laboratories
BMP 123 – Becker Microbial
Condor, Crymax, Cutlass, Lepinox – Ecogen
Mattch, MVP II – Mycogen
Troy-Bt – Troy Biosciences

Turf & Ornamentals

Insecticides

Lawns & Turf

ACEPHATE

Organophosphate with contact and systemic activity. Active on most turf surface-feeding insects. Product of choice for resistant greenbugs and for rapid kill of mole crickets in southern states. Seems to have little effect on earthworms and other beneficials, but is toxic to birds and highly toxic to bees. The strong odor may be objectionable in high public use areas.

Orthene Turf, Tree & Ornamental Spray (75S, 75WSP) *(Valent)*

Rate: 1 lb. ai/A
1.33 lb. 75S, 75WSP/A (0.5 oz./1000 sq. ft.)

Pest:
grasshoppers greenbug

Rate: 1-2 lb. ai/A
1.33-2.75 lb. 75S, 75WSP/A (0.5-1 oz./1000 sq. ft.)

Pest:
sod webworm

Rate: 2.1 lb. ai/A
2.75 lb. 75S, 75WSP/A (1 oz./1000 sq. ft.)

Pest:
leafhoppers

Remarks: Turfgrass. Apply when insects appear; repeat at 1-week intervals as necessary. For grasshoppers and greenbug, do not mow for at least 24 hr. after application. For sod webworm, use higher rate when quick knockdown is needed or with heavy infestations.

- - - - - - - - - - - - - - - - -

Rate: 1-2.4 lb. ai/A
1.33-3.25 lb. 75S, 75WSP/A (0.5-1.2 oz./1000 sq. ft.)

Pest:
fall armyworm southern armyworm yellowstriped armyworm

Rate: 2-5 lb. ai/A
3.25-6.66 lb. 75S, 75WSP/A (1.2-2.4 oz./1000 sq. ft.)

Pest:
chinch bug cutworms fleas

Remarks: Turfgrass. Apply when insects appear; repeat at 2-week intervals as necessary. For chinch bug and fleas, apply as needed for adult population knockdown.

- - - - - - - - - - - - - - - - -

Rate: 2-3.9 lb. ai/A
2.66-5.2 lb. 75S, 75WSP/A (1-1.9 oz./1000 sq. ft.)

Pest:
mole cricket
(except CA)

Remarks: Turfgrass. Apply late afternoon, early evening, and after irrigation when insects appear. Do not irrigate after application. For knockdown of existing populations, more than 1 application may be required throughout growing season. For heavier infestations, use higher rate.

- - - - - - - - - - - - - - - - -

Rate: 2-3.9 lb. ai/A
2.7-5.2 lb. 75S, 75WSP/A (1-1.9 oz./1000 sq. ft.)

Pest:
cutworms southern armyworm yellowstriped armyworm
flea beetles

Remarks: Dichondra. Apply when insects appear; repeat at 2-week intervals if necessary.

- - - - - - - - - - - - - - - - -

Rate: Mound treatment:
Dry method: Evenly distribute 1-2 tsp. 75S
Drench method: Mix 1 oz. 75S/5 gal. water
Drench method: Mix 0.33 lb. 75WSP/26.6 gal. water

Pest:
imported fire ant

Rate: SLN-Oklahoma—Mound treatment:
Dry method: Evenly distribute 4 Tbs. 75S over mound
Drench method: Mix 6 oz. 75S/5 gal. water

Pest:
harvester ants

Remarks: Apply 1 gal. mix/mound area by sprinkling mound until wet; treat 4 ft. diameter circle around mound. For best results, apply in early morning or late afternoon when ants are most active. Applications made after prolonged hot and dry conditions may be ineffective due to location of ants deep in nest. Do not treat mound more than once per season.

- - - - - - - - - - - - - - - - -

Signal Word/Toxicity Class: Caution/III.
REI: 24 hr.

Pinpoint or Velocity (15G) *(Valent)*

Rate: Mound treatment:
Evenly distribute 2 Tbs. 15G over mound

Pest:
ants imported fire ant

Remarks: Apply early morning or late afternoon when ants are most active. Applications made under prolonged hot and dry conditions may be ineffective due to location of ants deep in nest. Grass in treated area may be injured.

- - - - - - - - - - - - - - - - -

Signal Word/Toxicity Class: Caution.
REI: 12 hr.

Amdro — see hydramethylnon

Award — see fenoxycarb

CARBARYL

Carbamate insecticide with wide spectrum of activity on insects. Continuous use may lower earthworm populations. Lower rates effective on most surface insects while higher rates are needed for soil-dwelling pests. Susceptible to alkaline hydrolysis in high pH tank mixes. Most turf applications have irrigation and entry restrictions. Toxic to aquatic and estuarine invertebrates, highly toxic to honeybees.

Chipco Sevin SL or 80WSP *(Rhone-Poulenc)*

Rate: 2-4 lb. ai/A
4-8 pt. SL/A
2.5-5 lb. 80WSP/A

Pest:

ants	fall armyworm	mosquitoes (adult)
armyworms	fiery skipper	sowbugs
centipedes	grasshoppers	spittlebugs
chiggers	green June beetle	springtails
cutworms	June beetle	ticks
earwigs	leafhoppers	yellowstriped armyworm
Essex skipper	lucerne moth	
European chafer	millipedes	

9 Turf & Ornamentals

NOTICE The information on these pages is for preliminary planning — not a guide for use. Be sure to follow the manufacturer's directions, notwithstanding information contained here.	For personal protective equipment and EPA registration numbers, please turn to page 39.

Rate: 6-8 lb. ai/A
12-16 pt. SL/A
7.5-10 lb. 80WSP/A

Pest:

chinch bug	mole cricket	sod webworm
lawn moth		

Rate: 8 lb. ai/A
16 pt. SL/A
10 lb. 80WSP/A

Pest:

bluegrass billbug	fleas	white grubs
European crane fly		

Remarks: Apply in sufficient spray volume for thorough coverage and thatch penetration. Use higher rate on large pest populations, advanced instars, and dense vegetation. For mole crickets, apply afternoon or early evening following irrigation. For green June beetle and white grubs, apply when grubs are feeding near soil surface, usually late Mar.-May, July-early Sept., or as recommended by extension service agents. For chinch bug, treat entire area rather than just damaged areas. For European crane fly, apply early spring (Apr. 1-Apr. 15) or as recommended by extension service. For armyworms, cutworms, fall armyworm, mole crickets, and sod webworm, do not irrigate treated areas following insecticide application. For chinch bug, European crane fly, green June beetle, and white grubs, irrigation of turfgrass soon after treatment will aid in penetration of insecticide into thatch.

Signal Word/Toxicity Class: Warning/II.
REI: 12 hr.

Clean Crop Sevin Bait *(Platte)*
Rate: 1.5 lb. ai/A
30 lb. Bait/A

Pest:

ants	darkling beetle	grasshoppers
armyworms	field cricket	sowbugs
cutworms		

Remarks: Ground application only. For best results, cut lawn, remove clippings, and water thoroughly before treatment.

Signal Word/Toxicity Class: Caution/III.
REI: 12 hr.

Drexel Carbaryl 4L, 50W, 80S *(Drexel)*
Rate: 1 lb. ai/A
2 pt. 4L/A
2 lb. 50W/A
1.25 lb. 80S/A

Pest:

ants	fall armyworm	millipedes
armyworms	fiery skipper	mosquitoes
bluegrass billbug	fleas	sod webworm
centipedes	grasshoppers	sowbugs
chiggers	June beetle	springtails
cutworms	lawn moths	ticks
earwigs	leafhoppers	yellowstriped armyworm
Essex skipper	lucerne moth	
European chafer		

Remarks: Use 150-200 gal. spray mixture/5000 sq. ft. established lawn, recreational area, roadway, or path border. For best results, mow lawn and apply after wetting or rain to ensure good penetration of turf. Garden hose or pressure applications may be used. Repeat 2-3 weeks later if necessary.

Rate: 1.5 lb. ai/A
3 pt. 4L/A

Pest:

chinch bug	white grubs

Remarks: Apply in 150-200 gal. spray/5000 sq. ft.

Signal Word/Toxicity Class: Caution.
REI: 12 hr.

CHIPCO CHOICE

Phenylpyrazole insecticide whose mode of action is as an antagonist of the GABA regulated chloride channels. Toxic to birds, fish, and aquatic invertebrates.

Chipco Choice 0.1%G (fipronil) *(Rhone-Poulenc)*
Rate: 0.0125-0.025 lb. ai/A
12.5-25 lb. 0.1%G/A

Pest:
mole cricket

Remarks: Commercial grounds, golf turf. For control of overwintering adult mole crickets or at peak egg hatch for young nymphs. Use higher rate against heavy infestations or older insect life stages. In case of heavy infestation, second application may be necessary. Do not make more than 2 applications/yr. Do not apply more than 25 lb./A per application.

Signal Word/Toxicity Class: Caution.

CHIPCO MOCAP

Unpleasant odor due to mercaptan, an impurity and breakdown product of ethoprop. Toxic to aquatic organisms and wildlife. Birds feeding in treated areas may be killed.

Chipco Mocap 10G (ethoprop) *(Rhone-Poulenc)*
Rate: 5 lb. ai/A
50 lb. 10G/A

Pest:

black turfgrass ataenius beetle	chafer beetle	Japanese beetle
bluegrass billbug	chinch bug	sod webworm
	cutworms	white grubs

Rate: 7.5-10 lb. ai/A
75-100 lb. 10G/A

Pest:
mole cricket

Rate: 20 lb. ai/A
200 lb. 10G/A

Pest:
nematodes

Remarks: Commercial turfgrass such as golf courses and sod farms. Not for use on domestic turfgrass. Bentgrass and ryegrass are more sensitive than other species. Do not exceed 400 lb./A per year or use more than 200 lb./A at any single application.
State Restrictions: Do not apply in Long Island, New York.

Signal Word/Toxicity Class: Warning/II.
REI: 24 hr.

Chipco Sevin — see carbaryl

CHLORPYRIFOS

Organophosphate insecticide/miticide with broad spectrum activity against most turf pests through contact and stomach action. Is effective against surface-feeding insects. Tends to attach to organic matter such as thatch and, therefore, is a better surface insect control product.

◆ **Cyren 2E or 4E** *(Cheminova)*
Dursban 50W *(Dow AgroSciences)*
Rate: 1-4 lb. ai/A
2-8 qt. 2E/A (1.5-6 fl. oz./1000 sq. ft.)
1-4 qt. 4E/A (0.75-3 fl. oz./1000 sq. ft.)
2-8 lb. 50W/A (8-32 packets)

Pest: Including:

ants	crickets	Hyperodes weevil
armyworms	cutworms	mosquitoes
billbugs	earwigs	sod webworm
chiggers	fire ants	ticks
chinch bug	fleas	white grubs
clover mite	grasshoppers	

Rate: Commercial sod:
8 lb. ai/A
16 lb. 50W/A (64 packets)

Pest:
imported fire ant

Remarks: Rate varies by pest species. For billbugs, spray early in season when adults first appear; repeat as needed. For fire ants, apply as drench. For Hyperodes weevil, spray suspected problem areas in mid-April and again in mid-May. For sod webworm, delay watering or mowing treated area for 12-24 hr. after treatment. For white grubs, spray when grubs are young and actively feeding near soil surface, usually during late Jul.-Aug. For best results, irrigate treated area with 1/2-1" water immediately after spraying to wash insecticide into soil. Consult extension service specialist for application timing.

Signal Word/Toxicity Class: Warning/II.
REI: 12 hr.

◆ - **new product** • ai-active ingredient • DF-dry flowable • E/EC-emulsifiable concentrate • EW-emulsion in water
F/FL-flowable • G-granule • L/LC-liquid • SP/WSP-soluble packet • W/WP-wettable powder • WSB-water soluble bag

ProTurf Insecticide III *(Scotts)*

Rate: 28.12 lb. III/22,000 sq. ft.

Pest:

cutworms	flea beetles	vegetable weevil
earwigs	sod webworm	

Rate: 28.12 lb. III/11,000 sq. ft.

Pest:

ants	chinch bug	grasshoppers
brown dog tick	crickets	

Rate: 28.12 lb. III/8250 sq. ft.

Pest:

Hyperodes weevil

Remarks: Not for use on turf being grown for sale, commercial seed production, or other commercial use as sod. Do not mow or collect clippings for 24 hr. after treatment. Water thoroughly after application. For chinch bug control, apply any time activity or damage is noticed; prevention, as soon as daily temperatures reach 80°F. For Hyperodes weevil, apply as soon as grass greens in the spring and repeat in 1 month.

Signal Word/Toxicity Class: Caution.
REI: 12 hr.

Clean Crop Sevin — see carbaryl

Conserve — see spinosad

CYFLUTHRIN

Pyrethroid with contact and stomach activity against a wide variety of turf surface pests. Highly toxic to fish but has relatively low mammalian toxicity.

Tempo 2, 20WP or WSP *(Bayer)*

Rate: 5.9-8.75 fl. oz. 2/A
7.7-10.75 oz. 20WP/A
50 gram WSP/7143-10,000 sq. ft.

Pest:

ants	cutworms	Hyperodes weevil (adult)
armyworms	earwigs	Japanese beetle (adult)
bluegrass billbug (adult)	fire ant (aids in control)	mole cricket
chiggers	fleas	scorpions
chinch bug	grasshoppers	sod webworm
crickets		ticks

Remarks: Landscape and recreational turfgrass. Apply when pests first appear. Repeat when necessary based on pest reinfestations, not exceeding 6 applications per year. Delay watering or mowing for 24 hr. after treatment. Do not apply through any type of irrigation.

Signal Word/Toxicity Class: Warning/II (2); Caution (20WP, WSP).
REI: 12 hr.

lambda-CYHALOTHRIN

Synthetic pyrethroid insecticide with broad-spectrum activity. Restricted Use Pesticide due to toxicity to fish and aquatic organisms.

Scimitar CS, GC, or WP *(ZENECA Professional)*

Rate: 5-10 fl. oz. CS, GC/A
3-6 gram WP/1000 sq. ft.

Pest:

ants	fleas (adult)	mites
armyworms	grasshoppers	pillbugs
centipede	imported fire ant	sod webworm
crickets	Japanese beetle (adult)	ticks (including spp. for Lyme disease)
cutworms	millipedes	
earwigs		

Rate: 10 fl. oz. CS, GC/A
6 gram WP/1000 sq. ft.

Pest:

black turfgrass ataenius (adult)	chinch bug	mole cricket (20 fl. oz. GC)
bluegrass billbug (adult)	fleas (adult)	sowbugs
chiggers	Hyperodes weevil (adult)	

Remarks: Apply at 7-day intervals as needed when pests first appear. Delay watering for at least 24 hr. for optimum control of surface feeding insects. Do not apply within 25 ft. of any kind of body of water or when wind speed is 15 mph or greater.

Signal Word/Toxicity Class: Warning/II.
REI: 24 hr.

Cyren — see chlorpyrifos

DEADLINE

Deadline (4G) (metaldehyde) *(Pace)*

Rate: 0.02-0.8 lb. ai/1000 sq. ft.
0.5-2 lb. 4G/1000 sq. ft.

Pest:

slugs	snails

Remarks: Grass grown for sod, turfgrass (including Dichondra). Apply every 3-4 weeks during growing season, or as needed. Do not apply directly to edible portion of plant.

Signal Word/Toxicity Class: Warning/II.
REI: 12 hr.

Diatect — see pyrethrins

DIAZINON

Organophosphate insecticide with wide spectrum of activity for both surface and soil dwelling pests. Not allowed on golf courses or sod farms due to bird toxicity. Also toxic to bees, earthworms, and wildlife. Do not apply where waterfowl graze. Restricted Use Pesticide due to avian and aquatic toxicity.

D•z•n 50W *(Novartis)*
Drexel Diazinon Insecticide (4EC) *(Drexel)*
Micro Flo Diazinon AG500 or 50W *(Micro Flo)*
Prentox Diazinon AG500 or 50W *(Prentiss)*
◆ **Wilbur-Ellis Diazinon 4 Spray (4E)** *(Wilbur-Ellis)*

Rate: 0.06-0.09 lb. ai/1000 sq. ft.
2-3 fl. oz. 4EC, AG500/1000 sq. ft.
1.5-2.5 fl. oz. AG600/1000 sq. ft.
2-3 oz. 50W/1000 sq. ft.

Pest:

ants	earwigs	leafhoppers
armyworms	European chafer	millipedes
bermudagrass mite	fleas	rhodesgrass scale
brown dog tick	frit fly	sod webworm
chiggers	Hyperodes weevil	southern chafer
clover mite	Japanese beetle (larvae)	sowbugs (pillbugs)
crickets	lawn billbug	springtails (Collembola)
cutworms	lawn chinch bug	
digger wasp	lawn moth	white grubs

Remarks: Spot-spray ant hills and wasp ground nest openings. For digger wasp, apply at dusk when wasps are active. For frit fly, mow grass and water well before treatment, but delay application until grass is dry. For brown dog tick, spray grass and under shrubbery. For Hyperodes weevil, treat problem areas in mid-April and in mid-May. For rhodesgrass scale, apply when crawlers first emerge. For white grubs, such as Japanese beetle, European chafer, or southern chafer, apply any time between late July-early Oct. Refer to label for directions and precautions.
State Restrictions: Do not apply in Nassau County, New York between Nov. 1-May 20 because of migratory habits of certain waterfowl.

Signal Word/Toxicity Class: Caution/III.
REI: 24 hr.

DICOFOL

Chlorinated hydrocarbon miticide which is useful against most of the mites which attack turfgrasses. Tends to have strong odor in hot conditions.

◆ **Dicofol 4E** *(Griffin)*
Kelthane 35 or T/O *(Rohm and Haas)*

Rate: 0.3-0.4 lb. ai/100 gal. water
0.6-0.8 pt. 4E/100 gal. water
1-1.33 lb. 35/100 gal. water
0.25-0.5 lb. ai/100 gal. water
0.5-1 lb. T/O/100 gal. water

9 Turf & Ornamentals

NOTICE The information on these pages is for preliminary planning — not a guide for use. Be sure to follow the manufacturer's directions, notwithstanding information contained here. For personal protective equipment and EPA registration numbers, please turn to page 39.

Pest:
bermudagrass mite, brown mite, clover mite

Remarks: Apply 2.5 gal./1000 sq. ft. when mites appear. Kelthane not for use on residential home lawns.

Signal Word/Toxicity Class: Warning/II (35); Caution (4E, T/O).
REI: 12 hr.

DI-SYSTON

Restricted Use Pesticide.

Di-Syston 8 (disulfoton) *(Bayer)*

Rate: 1 pt. 8/A

Pest:
mealybugs, thrips, whiteflies

Remarks: Bermudagrass (seed crop).
State Restrictions: SLN-California.

Signal Word/Toxicity Class: Danger/I.

Dursban — see chlorpyrifos

Dylox — see trichlorfon

D•z•n — see diazinon

FENOXYCARB

Insect growth regulator.

Award *(Novartis)*

Rate: Broadcast or aerial application:
1-1.5 lb./A
Mound treatment:
1-3 Tbs./mound

Pest:
bigheaded ants, fire ants

Remarks: Sod farms, turf. Apply uniformly with ground or aerial equipment calibrated to give correct dosage when ants are actively foraging, usually when soil temperatures are above 60°F. Do not apply during excessively hot periods or when grass is wet. Do not apply by air where weather conditions favor drift from treated areas. For mound treatment, uniformly distribute material 3-4 ft. around mound. For bigheaded ants, apply by ground only; allow 4-6 weeks for reduced activity of colonies. Do not graze treated areas.
State Restrictions: In California, for home lawn use only.

Signal Word/Toxicity Class: Caution.
REI: Do not reenter until application is complete.

HYDRAMETHYLNON

Attractant bait that worker ants find and carry back to the mound where it is passed along to other workers and the queen. Toxic to fish.

Amdro or Siege *(American Cyanamid)*

Rate: Broadcast:
1-2 lb./A
Mound treatment:
5 level Tbs./mound

Pest:
bigheaded ants, fire ants, harvester ants

Remarks: Lawns, turf, golf courses (Siege), sod (Amdro). Apply when ants are active and soil temperature is greater than 60°F. Broadcast bait uniformly with ground equipment (spinning disk types) or aerial equipment. When treating individual mounds, do not disturb mound or apply directly to top of mound. Do not apply more than 1.5 lb./A. Use higher rate on bigheaded ants.

Signal Word/Toxicity Class: Caution.
REI: 12 hr.

IMIDACLOPRID

Chloronicotinyl insecticide, a new class of compounds, with unique action on insect nervous systems. Effective at low rates. Broad spectrum activity with systemic action. Low toxicity to mammals, birds, and fish. Highly toxic to bees and aquatic invertebrates.

Merit 75WP, 75WSP or 0.5G *(Bayer)*

Rate: 6.4-8.6 oz. 75WP/A
6.4 oz. 75WSP/A
60-80 lb. 0.5G/A

Pest: Larvae:
annual bluegrass weevil, Asiatic garden beetle, billbugs, black turfgrass ataenius beetle, European chafer, Japanese beetle, northern masked chafer, Oriental beetle, Phyllophaga, southern masked chafer

Remarks: For optimum control, apply prior to egg hatch of target pest. Consult state agricultural experiment station, or state extension turf specialists for more specific information regarding application timing.

Rate: 8.6 oz. 75WP/A
6.4 oz. 75WSP/A
80 lb. 0.5G/A

Pest:
mole crickets

Remarks: Apply prior to or during peak egg hatch period, when adults or large nymphs are present and actively tunneling. Consult state agricultural experiment station or state extension turf specialists for more specific information regarding timing of application.

Signal Word/Toxicity Class: Caution.

Kelthane — see dicofol

LANNATE

Restricted Use Pesticide.

Lannate LV or SP (methomyl) *(DuPont)*

Rate: 0.9-1.8 lb. ai/A
3-6 pt. LV/A
1-2 lb. SP/A

Pest:
sod webworm

Remarks: Commercial sod farms. Ground application only. After application, sprinkle irrigate for 15 minutes. Do not apply more than 3.6 lb. ai/A per crop or make more than 4 applications per crop.

Signal Word/Toxicity Class: Danger-Poison/I.
REI: 48 hr.

MACH 2

Mimics actions of a natural insect hormone which induces molting and metamorphosis process in insects.

◆ **Mach 2 (2L or 1.5G) (halofenozide)** *(RohMid L.L.C.)*

Rate: 1 lb. ai/A
2 qt. 2L/A
67 lb. 1.5G/A

Pest: Larvae:
armyworms, cutworms, fall armyworm, sod webworms

Remarks: Apply at first sign of pest damage; repeat once if needed. Do not apply through any type of irrigation system.

Rate: 1.5 lb. ai/A
3 qt. 2L/A
100 lb. 1.5G/A

Pest:
white grubs (larvae)

Remarks: Preventive or curative treatment. Make only 1 application per year. Do not apply through any type of irrigation system.
Days To Harvest: Sod farms-7.
State Restrictions: Not for use in Nassau and Suffolk counties, New York.

Signal Word/Toxicity Class: Caution.
REI: 12 hr.

◆ - **new product** • ai-active ingredient • DF-dry flowable • E/EC-emulsifiable concentrate • EW-emulsion in water
F/FL-flowable • G-granule • L/LC-liquid • SP/WSP-soluble packet • W/WP-wettable powder • WSB-water soluble bag

MAVRIK AQUAFLOW

Synthetic pyrethroid with contact and stomach activity against a wide variety of turf surface pests. Extremely low rates used to achieve control. Highly toxic to fish, low mammalian toxicity.

Mavrik Aquaflow (*tau*-fluvalinate) *(Wellmark)*

Rate: 5-10 fl. oz./A

Pest:

ants*	clover mite	leafhoppers
armyworms	crickets*	millipedes*
Bermuda mite*	cutworms	sod webworm
brown dog tick	earwigs*	sowbugs*
chiggers*	fleas	springtails*
chinch bug	lawn moth	ticks

*** Except California***

Remarks: Apply in sufficient water to penetrate to pest when pests are immature or at susceptible stage and populations are manageable, before crop damage occurs. Where heavy populations are present, repeat at 5- to 7-day intervals if necessary. Use lower rate for thatch inhabiters, such as fleas, ticks, and chiggers; repeat in 10 days; medium rate for grassfeeders, such as armyworms, Bermuda mite, chinch bug, or sod webworm; maximum rate for cutworms. Do not apply through any type of irrigation system.

Signal Word/Toxicity Class: Caution.
REI: 12 hr.

Merit — see imidacloprid

M-PEDE

M-Pede (potassium salts of fatty acids) *(Mycogen)*

Rate: 1-2% v/v solution

Pest:

chinch bug	greenbug	mole cricket
cutworms	grubs	sod webworm
fleas	mites	

Remarks: For mole cricket, irrigate turf prior to application to bring crickets to surface. May be tank mixed with other insecticides. Refer to labels for use directions and precautions.

Signal Word/Toxicity Class: Warning/II.
REI: 12 hr.

NEMACUR

Organophosphate used for nematode control. Toxic to fish and wildlife. Restricted Use Pesticide.

Nemacur 3 Turf or 10% T&O (fenamiphos) *(Bayer)*

Rate: 10 lb. ai/A
3.3 gal. 3/A
100 lb. 10%/A

Pest:
nematodes

Remarks: 3 Turf: golf courses (except tees and greens); 10% T&O: cemeteries, golf courses, and industrial grounds. Refer to label for use directions, restrictions, and precautions.
State Restrictions: In California, Nemacur 10% T&O for golf courses only.

Signal Word/Toxicity Class: Danger/I.
REI: 48 hr.

OFTANOL

Broad-spectrum organophosphate. Highly susceptible to accelerated microbial degradation so rotate with other insecticides from season to season.

Oftanol 2 (FL) or 5% G (isofenphos) *(Bayer)*

Rate: 1-2 lb. ai/A
2-4 qt. FL/A
2 lb. ai/A
40 lb. 5% G/A

Pest:

armyworms	fall armyworm	mole crickets
billbugs (larvae)	fire ants	sod webworm (larvae)
chinch bug	flea beetles	white grubs (larvae)
crane fly (larvae)	Hyperodes weevil (larvae)	
cutworms		

Remarks: For use by commercial applicators only.

Signal Word/Toxicity Class: Warning/II (Oftanol 2); Caution (5% G).

Orthene — see acephate

Pinpoint — see acephate

ProTurf Insecticide III — see chlorpyrifos

Proxol — see trichlorfon

Pyrenone — see pyrethrins

PYRETHRINS

Contains piperonyl butoxide as a synergist and silicon dioxide (diatomaceous earth).

Diatect Multipurpose Insecticide II *(Diatect)*
Pyrenone (0.5EC) *(AgrEvo Environmental Health)*

Rate: 1 lb. II/1000 sq. ft.
0.25-0.5 fl. oz. 0.5EC/1000 sq. ft.

Pest:

ants	caterpillars	cutworms
armyworms	chinch bug	fire ants

Remarks: Apply as dust or wettable powder. Repeat as needed.

Signal Word/Toxicity Class: Caution.
REI: 12 hr.

Scimitar — see *lambda*-cyhalothrin

Siege — see hydramethylnon

SPINOSAD

Fermentation-derived insect control agent. Does not significantly impact predaceous and parasitic insect populations. Highly toxic to bees and aquatic invertebrates.

◆ **Conserve (SC) (spinosyn A + spinosyn D)** *(Dow AgroSciences)*

Rate: 10-52 fl. oz. SC/A

Pest:

armyworms	cutworms	sod webworm
black cutworm	fall armyworm	tropical sod webworm

Remarks: Turfgrass. Apply at 7-day intervals. For fall armyworm, apply early morning or late afternoon. For black cutworm, sod webworm, and tropical sod webworm, apply early afternoon or late evening. Delay watering or mowing for 12-24 hr. after treatment.

Signal Word/Toxicity Class: Caution.
REI: 4 hr.

TALSTAR

Broad-spectrum pyrethroid insecticide/miticide. Restricted Use Pesticide due to toxicity to fish and aquatic organisms.

Talstar GC Flowable (FL) (bifenthrin) *(FMC)*

Rate: 0.05-0.1 lb. ai/A
10-20 fl. oz. GC/A

Pest:

annual bluegrass weevil	cutworms	mealybugs
ants	earwigs	millipedes
armyworms	fall armyworm	mites
billbugs	fleas	mole cricket
black turfgrass ataenius	grasshoppers	pillbugs
centipedes	imported fire ant	sod webworm
chinch bug	Japanese beetle	sowbugs
crickets	leafhoppers	ticks

For personal protective equipment and EPA registration numbers, please turn to page 39.

Rate: Large infestations—single application:
0.2 lb. ai/A
40 fl. oz. GC/A

Pest:

ants — imported fire ant — mole cricket

Remarks: Golf courses, parks, recreational and athletic fields, etc. For use by commercial applicators only. Not for use on sod farms or grass grown for seed. Apply as broadcast treatment.

Signal Word/Toxicity Class: Caution.
REI: 12 hr.

Talstar GC Granular (bifenthrin) *(FMC)*

Rate: 0.1-0.4 lb. ai/A
50-200 lb. GC/A

Pest:

annual bluegrass weevil	chinch bug	millipedes
ants	cutworms	mole cricket
armyworms	fleas	pillbugs
billbugs	imported fire ant	sod webworm
black turfgrass ataenius	leafhoppers	sowbugs
centipedes	mealybugs	ticks

Remarks: Golf courses, landscaped areas, parks, recreational areas, athletic fields, etc. For use by commercial applicators only. Not for use on sod farms, grass grown for seed, or commercial seed production.

Signal Word/Toxicity Class: Caution.
REI: 12 hr.

Talstar Nursery Flowable (FL) (bifenthrin) *(FMC)*

Rate: Broadcast:
0.05-0.1 lb. ai/A
10-20 fl. oz. FL/A

Pest:

ants	earwigs	Japanese beetle
armyworms	fall webworm	mites
billbugs	fleas	mole cricket
chinch bug	grasshoppers	sod webworm
crickets	imported fire ant	ticks
cutworms		

Remarks: Sod farms. For use by commercial applicators only. For armyworm, cutworm, and sod webworm, delay watering or mowing for 24 hr. after application. For overwintered mole cricket, use lower rate early spring. For adult mole cricket in late summer or early fall, use higher rate.

Signal Word/Toxicity Class: Caution.
REI: 12 hr.

Tempo — see cyfluthrin

TRICHLORFON

Fast acting organophosphate with moderate mammalian toxicity. Spray mixtures are susceptible to alkaline hydrolysis in high pH situations.

Dylox 6.2 or 80 *(Bayer)*

Rate: 2-3 lb. 6.2/1000 sq. ft.
2.5-3.75 oz. 80/1000 sq. ft.

Pest:

cutworms	mole cricket (6.2)	white grubs
lawn moth	sod webworm	

Remarks: Mow lawn and rake dead grass from damaged spots. Water lawn and allow grass to dry before application. For sod webworm and cutworms, do not irrigate after application. For white grubs, apply high rate and thoroughly irrigate following application. Consult state agricultural experiment station or state extension service specialists for more specific information regarding timing of applications. Do not use treated areas or clippings from treated areas for feed or forage for livestock. Refer to label for use directions.

Signal Word/Toxicity Class: Warning/II (Dylox 80); Caution (Dylox 6.2).

Proxol 80SP *(AgrEvo USA)*

Rate: 1.5-3.75 oz. 80SP/1000 sq. ft.

Pest:

armyworms — cutworms — sod webworm

Remarks: Apply when insect populations reach levels sufficient to warrant control measures.

Rate: 3.75 oz. 80SP/1000 sq. ft.

Pest:

European crane fly (larvae) — white grubs

Remarks: For European crane fly, apply when damage is first noticed in lawn or turf area. If necessary, irrigate lawn or turf after application to move Proxol 80SP into zone of crane fly larvae infestation. For white grubs, apply when grub population levels are sufficient to warrant control measures. Thoroughly irrigate turf immediately after application.

Signal Word/Toxicity Class: Warning/II.
REI: 24 hr.

TURCAM

Carbamate insecticide with broad spectrum activity through contact and stomach action. Lower rates are useful against many surface insects but higher rates are needed for control of soil dwelling pests. Controls many nuisance pests which invade homes after living in turf. Highly toxic to bees. Restricted Use Pesticide due to aquatic and avian toxicity.

Turcam WP or 2½G (bendiocarb) *(AgrEvo USA)*

Rate: 0.5-1.5 oz. WP/1000 sq. ft.
40-120 lb. 2½G/A

Pest:

billbugs	Japanese beetle	sod webworms
chinch bug	leafhoppers	white grubs
European crane fly (WP)	mole crickets	

Remarks: Do not apply more than 6 lb. ai/A per year; 3 lb. ai/A for spring applications. Refer to label for restrictions and precautions.

State Restrictions: Not for use in California. Do not use on golf courses or sod farms on islands of Hawaii and Maui.

Signal Word/Toxicity Class: Warning/II (WP); Caution (2½G).
REI: 12 hr.

Velocity — see acephate

Ornamentals

(Christmas Trees, Flowers, Forest Plantings, Nurseries, Shelterbelts)

455 Oil — see petroleum oil

ABAMECTIN

Insecticide/nematicide derived from naturally occurring soil microorganisms, registered on ornamental plants. Abamectin has translaminar activity on some plants.

Avid 0.15EC *(Novartis)*

Rate: 4 fl. oz. 0.15EC/100 gal. water

Pest:

spider mites

Rate: 8 fl. oz. 0.15EC/100 gal. water

Pest:

leafminers

Remarks: Field grown ornamentals, foliage plants, greenhouse, shadehouse, woody ornamentals. Not recommended for use on ferns or conifers. Apply as foliar spray in sufficient water to obtain uniform plant coverage. Do not apply by air. Do not apply through any type of irrigation system.

Signal Word/Toxicity Class: Warning/II.
REI: 12 hr.

◆ - **new product** • ai-active ingredient • DF-dry flowable • E/EC-emulsifiable concentrate • EW-emulsion in water
F/FL-flowable • G-granule • L/LC-liquid • SP/WSP-soluble packet • W/WP-wettable powder • WSB-water soluble bag

ACEPHATE

Systemic organophosphate insecticide registered as an aerosol formulation for greenhouse use, and as a foliar spray for greenhouse and outdoor use. Effective against a wide range of chewing and sucking pests.

Orthene Turf, Tree & Ornamental Spray (75S) *(Valent)*

Rate: Flowering crabapples, shrubs, trees:
0.25 lb. ai/100 gal. water
0.33 lb. 75S/100 gal. water

Pest:

aphids	birch leafminer	leafrollers
bagworm	lace bug	tent caterpillar

Rate: Shrubs, trees:
0.25-0.5 lb. ai/100 gal. water
0.33-0.66 lb. 75S/100 gal. water

Pest:

California oakworm	fall cankerworm	spring cankerworm

Rate: Shrubs, trees:
0.5 lb. ai/100 gal. water
0.66 lb. 75S/100 gal. water

Pest:

Douglas-fir tussock moth (larvae)	gypsy moth (larvae)	scale crawlers
grasshoppers	Ponderosa pine needleminer	webworms

Remarks: Flowering crabapples, shrubs, trees. Apply when insects begin to appear; repeat at 2-week intervals for continuous crawler production. Use higher amount when larger larvae are present. Addition of sticker improves control of gypsy moth. For flowering crabapples, do not apply more than 3 times per growing season at 4-week intervals. Phytotoxicity has occurred on following crabapple varieties: Hopa, Ichonoski, Malus floribunda, Pink Perfection, Red Wine, and Snow Cloud. Before treating large plantings, spray only a few plants and observe 2 weeks for phytotoxicity. Do not apply to Huckleberry, Balm of Gilead, Cottonwood, Lombardy poplar, and Viburnum suspensum.

- - - - - - - - - - - - - - - - -

Rate: 0.75 lb. ai/100 gal. water
1 lb. 75S/100 gal. water

Pest:

boxelder bug	Nantucket pine tip moth (larvae)	root weevil (adult)
budworms		sawflies
leafhoppers		

Remarks: Shrubs, trees. Apply with hydraulic sprayer when insects begin to appear. For root weevil, apply when first feeding damage occurs. Repeat at 4-week intervals until first heavy frost occurs. For Nantucket pine tip moth, time of application is important. Repeat applications will be required for subsequent generations.

- - - - - - - - - - - - - - - - -

Rate: 1 lb. ai/100 gal. water
1.33 lb. 75S/100 gal. water

Pest:

elm leaf beetle (larvae)	Japanese beetle

Remarks: Boston ivy (outdoor), roses, shrubs, trees. Apply with hydraulic sprayer as full coverage spray when Japanese beetle or larvae begin to appear. For Japanese beetle, repeat at 2-week intervals if necessary. Multiple applications may cause slight tip burn or marginal leaf necrosis on some varieties. Test on a few plants to determine varietal phytotoxicity. Will not prevent elm leaf beetle eggs from hatching.

- - - - - - - - - - - - - - - - -

Rate: 0.75 lb. ai/100 gal. solution
1 lb. 75S/100 gal. solution

Pest:

black vine weevil	imported fire ant	strawberry root weevil

Remarks: Container-grown nursery stock: arborvitae, azalea, camellia, rhododendron, roses, viburnum, yew. Thoroughly drench root system. Apply by mid-Sept. for greenhouse stock; mid-Oct. for outdoor stock.

- - - - - - - - - - - - - - - - -

Rate: 0.25 lb. ai/100 gal. water
0.33 lb. 75S/100 gal. water

Pest:

aphids (foliage plants)	sweetpotato whitefly (orchids) (except CA)

Rate: 0.5 lb. ai/100 gal. water
0.66 lb. 75S/100 gal. water

Pest:

aphids	mealybugs	thrips
leafrollers (roses)	scale crawlers	whiteflies
Lygus bugs		

Remarks: Greenhouse: carnations, chrysanthemums, foliage plants (anthuriums, cacti, orchids, poinsettia), roses. Outdoor: daisies, gladiolus, gypsophila, roses, statice, strawflower, yarrow carnations, chrysanthemums. Apply with hydraulic sprayer as full coverage spray when insects begin to appear. For leafrollers, use higher rate when large larvae are present. For mealybugs, whiteflies, and continuous crawler production, 2-week intervals may be necessary. Multiple applications may cause slight tip burn or marginal leaf necrosis on some outdoor varieties. Do not apply more than once every 28 days to carnations and chrysanthemums. For greenhouse, addition of wetting agent may be required on difficult-to-wet foliage. Before treating large plantings, spray only a few plants and observe 2 weeks for varietal phytotoxicity.

- - - - - - - - - - - - - - - - -

Rate: 3-4 lb. ai/A
4-5.33 lb. 75S/A

Pest:

coneborers (except FL)	coneworms	slash pine flower thrips
	seedbugs	

Remarks: Slash pine seed orchards (FL only), southern pine seed orchards. Apply with high volume sprayers to tree crown until foliage is wet. With low volume sprayers, apply 100 gal./A. Apply first application 6 days after peak pollen flight and repeat at 4-week intervals. For slash pine flower thrips, apply first application about 2 weeks before maximum flower receptivity to pollen. Do not apply through any type of irrigation system. Do not graze or harvest cover crops. Avoid contact with open water.

State Restrictions: SLN-Florida, Georgia, South Carolina, Virginia.

- - - - - - - - - - - - - - - - -

Rate: Alstromeria, daffodils, irises, lilies:
0.375-0.5 lb. ai/A
0.5-0.66 lb. 75S/A

Pest:
aphids

Rate: Gerberas (greenhouse):
0.75 lb. ai/A
1 lb. 75S/A

Pest:

leafminers	thrips	whiteflies

Remarks: Apply 100 gal. spray mix/A when insects first appear; repeat as needed.

State Restrictions: SLN-California.

- - - - - - - - - - - - - - - - -

Signal Word/Toxicity Class: Caution/III.
REI: 24 hr.

Pinpoint (15G) *(Valent)*

Rate: Broadcast application:
0.4 lb. 15G/1000 sq. ft.
Manual application:
0.5-0.75 tsp./pot

Pest:

ants	azalea lace bug	mealybugs
aphids	imported fire ant	

Remarks: Azalea, crape myrtle, holly, photinia, poinsettia, pyracantha. Apply over soil surface, irrigating after application as needed. Do not apply to freshly rooted cuttings.

- - - - - - - - - - - - - - - - -

Signal Word/Toxicity Class: Caution.
REI: 12 hr.

Ambush — see permethrin

Amdro — see hydramethylnon

ASANA

Restricted Use Pesticide.

Asana XL (esfenvalerate) *(DuPont)*

Rate: 0.03-0.05 lb. ai/A
5.8-9.6 fl. oz. XL/A

Pest:

balsam twig aphid	Nantucket pine tip moth	pine needle midge
balsam woolly adelgid	northern pine weevil	red pine sawfly
		redheaded pine sawfly
coneworms	pales weevil	seed chalcid
cranberry girdler (adult)	pine chafer	seedbugs
	pine conelet bug	spittlebugs
European pine sawfly	pine leaf chermid	spruce budworm

NOTICE The information on these pages is for preliminary planning — not a guide for use. Be sure to follow the manufacturer's directions, notwithstanding information contained here.

For personal protective equipment and EPA registration numbers, please turn to page 39.

Remarks: Christmas tree plantings, conifer seed orchards, forest tree nurseries. Apply as needed to maintain control. Spray to obtain good coverage of entire tree. For pales weevil and northern pine weevil, dilute formulation in kerosene rather than water, and spray tree stumps to control developing larvae. For coneworms, seedbugs, and seed chalcid, rate varies according to sprayer volume. Refer to label for further use directions. Do not graze or harvest cover crop. Avoid contact with open water.

- - - - - - - - - - - - - - - - -

Signal Word/Toxicity Class: Warning/II.
REI: 12 hr.

Astro — see permethrin

Avid — see abamectin

Award — see fenoxycarb

AZADIRACHTIN

Extracted from seed of neem tree. Insect growth regulator that controls targeted insect larvae when they ingest or come in contact with it, by interfering with the insect's ability to molt and shutting off feeding behavior. Affected larvae slowly starve and die at next molt.

◆ **Azatin EC or XL** *(Olympic)*
◆ **Azatin XL Plus** *(Thermo Trilogy)*

Rate: 8-21 oz. EC/A
5-21 oz. XL, XL Plus/A

Pest:

aphids	fall armyworm	pine tip moth
apple aphid	fall cankerworm	psyllids
armyworms	fall webworms	rose aphid
azalea leafminer	flea beetles	rose chafer
bagworms	gladiolus thrips	sawflies
beet armyworm	greenhouse leaftier	silverleaf whitefly
beetles	greenhouse whitefly	spring cankerworm
black vine weevil	gypsy moth	spruce budworm
budworms	holly leafminer	sweetpotato whitefly
cabbage looper	Japanese beetle	tent caterpillar
caterpillars	June beetle	thrips
chafers	leafhoppers	tobacco budworm
codling moth	leafminers	tussock moth
corn earworm	leafrollers	webworms
elm bark beetle	loopers	whiteflies
European pine shoot moth	mealybugs	yellowstriped armyworm
	peachtree borer	

Remarks: Bedding plants, flowers, ornamentals, potted plants, shrubs, trees. Rate varies by species. Apply when pests first appear; repeat at 7-day intervals. When pest populations are high, use higher rates. Do not apply more that 21 fl. oz./A.
Days To Harvest: 0.

- - - - - - - - - - - - - - - - -

Signal Word/Toxicity Class: Caution (XL); Warning/II (EC).
REI: 4 hr. (XL); 12 hr. (EC).

◆ **NEEMAZAD 0.25EC or 4.5EC** *(Thermo Trilogy)*

Rate: 1 gal. 0.25EC/A
0.5-2 gal. 4.5EC/A
High Volume:
2.5-5 pt. 0.25EC/100 gal. water
2.25-4.5 oz. 4.5EC/100 gal. water
Low Volume:
1 gal. 0.25EC/A
0.5 pt. 4.5EC/A

Pest: Bedding plants, potted plants, foliage plants, ornamentals, shrubs, and trees in and around greenhouses and commercial nurseries:

aphids	leafminers	thrips
beet armyworm	loopers	whiteflies
caterpillars	mealybugs	

Trees and shrubs on commercial landscape:

gypsy moth	psyllids	tent caterpillar
hornworms	spruce budworm	weevils
pine sawfly		

Remarks: For low volume, apply in sufficient water for adequate coverage. For low insect pressure, apply at 7-day intervals; high insect pressure, at 3- to 4-day intervals.

- - - - - - - - - - - - - - - - -

Rate: 1 gal. 0.25EC/A
Pest: Roses:
Japanese beetle

Remarks: Apply at first sign of beetle emergence. Spray to runoff, making sure to cover all plants including buds and flowers. Repeat weekly or after rainfall as long as beetles are present.

- - - - - - - - - - - - - - - - -

Signal Word/Toxicity Class: Caution (0.25EC), Warning/II (4.5EC).
REI: 12 hr.

Azatin — see azadirachtin

AZINPHOS-METHYL

Organophosphate insecticide. Restricted Use Pesticide (except Guthion Solupak).

Gowan Azinphos-M 50-W *(Gowan)*
Guthion 2L or Solupak (50WSP) *(Bayer)*
◆ **Micro Flo Azinphosmethyl 2EC or 50W** *(Micro Flo)*
Sniper 50W *(Platte)*

Rate: 0.38-0.75 lb. ai/A
1.5-3 pt. 2EC, 2L/A
0.75-1.5 lb. 50-W, 50W, 50WSP/A

Pest:

eastern pine shoot borer	Nantucket pine tip moth	sawflies
European pine shoot moth		scales

Remarks: Christmas trees. Apply in sufficient water to give complete coverage.
State Restrictions: Azinphos-M 50-W not registered in California.

- - - - - - - - - - - - - - - - -

Rate: 0.375-1 lb. ai/100 gal. water
1.5-4 pt. 2EC, 2L/100 gal. water
0.75-2 lb. 50-W, 50W, 50WSP/100 gal. water

Pest:

aphids	euonymus scale	olive scale
black pineleaf scale	European elm scale	oystershell scale
black vine weevil (Guthion)	juniper scale	Pulvinaria scale
	lace bug	Putnam scale
brown soft scale	leafhoppers	thrips
Cerococcus scale	mites	

Rate: 2-4 lb. ai/100 gal. water
8-16 pt. 2EC, 2L/100 gal. water
4-8 lb. 50-W, 50W, 50WSP/100 gal. water

Pest:

cone midge	cone moth

Remarks: Ornamentals, nursery plants, shade trees. Spray all foliage surfaces, including underside of leaves; repeat as necessary. Thorough coverage of cones necessary. Rate varies by pest species; refer to label. Injury to hawthorn or American linden may occur under some conditions.

- - - - - - - - - - - - - - - - -

Rate: Conventional hydraulic-type sprayers:
High volume spray (0.2% dilution):
6 pt. 2EC, 2L/100 gal. water
4 lb. 50-W, 50W, 50WSP/100 gal. water
Low volume spray (0.1% dilution):
3 pt. 2EC, 2L/10 gal. water
2.25 lb. 50-W, 50W, 50WSP/10 gal. water

Pest:

coneworms	seedworms

Remarks: Southern pine seed orchards. Apply first application within 30 days following conelet closure, repeat 3-5 times at 30-day intervals. Apply approximately 5-10 gal. 0.2% dilution with high volume sprayers; 1-2 gal. 0.1% dilution with low volume sprayers/tree. Thorough coverage of cones necessary.
State Restrictions: Azinphos-M 50-W not registered in California.

- - - - - - - - - - - - - - - - -

Signal Word/Toxicity Class: Danger/I.
REI: 48 hr.

Baythroid — see cyfluthrin

◆ - **new product** • ai-active ingredient • DF-dry flowable • E/EC-emulsifiable concentrate • EW-emulsion in water
F/FL-flowable • G-granule • L/LC-liquid • SP/WSP-soluble packet • W/WP-wettable powder • WSB-water soluble bag

CALCIUM POLYSULFIDE

Sulforix *(Best Sulfur)*

Rate: 1.5 pt. L/100 gal. water

Pest:

red spider mite

Remarks: Roses (nursery stock and field grown). Pick open flowers before spraying to avoid discoloration.

Signal Word/Toxicity Class: Danger/I.
Restricted Entry Interval: 48 hr.

CARBARYL

Broad-spectrum insecticide used on many ornamental plants, mostly outdoor ornamentals. Extremely toxic to aquatic invertebrates, highly toxic to bees.

Chipco Sevin SL or 80WSP *(Rhone-Poulenc)*
Drexel Carbaryl 4L, 50W, 80S *(Drexel)*

Rate: 1 lb. ai/100 gal. water
2 pt. 4L, SL/100 gal. water
2 lb. 50W/100 gal. water
1.25 lb. 80WSP/100 gal. water

Pest:

ants	hackberry nipplegall maker	plant bugs
apple aphid	holly bud moth	poinsettia hornworm
armyworms	holly leafminer	psyllids
azalea leafminer	Jack pine budworm	puss caterpillar
bagworm	Japanese beetle	redhumped oakworm
birch leafminer	Jeffrey pine needleminer	rose aphid
blister beetle	June beetle	rose chafer
boxelder bug	lace bugs	roseslug
boxwood leafminer	leafhoppers	saddled prominent caterpillar
browntail moth	leafrollers	sawflies (exposed)
cankerworms	locust borer	scales
catalpa sphinx chigger	maple leafcutter	sowbugs
cutworms	mealybugs	spiny elm caterpillar
cypress tip moth	mimosa webworm	springtails
eastern spruce gall aphid	Nantucket pine tip moth	spruce budworm
elm leaf aphid	oak leaf skeletonizer	spruce needleminer
elm leaf beetle	oak leafminer	subtropical pine tip moth
elm spanworm	oak moth	tent caterpillar
eriophyid mite	oakworm	thornbug
European pine shoot moth	oleander caterpillar	thrips (exposed)
fall armyworm	olive ash borer	ticks
flea beetles	orange tortrix	walnut caterpillar
Fuller rose beetle	orangestriped oakworm	webworms
gall midge	periodical cicada	western hemlock looper
gall wasp	pine looper	western spruce budworm
grasshoppers	pine sawflies	willow leaf beetle
greenstriped mapleworm	pine spittlebug	yellow-poplar weevil
gypsy moth	pitch pine tip moth	

Remarks: Apply dilute sprays to obtain thorough coverage of upper and lower leaf surfaces to control scales, and treat trunks, stems, and twigs in addition to plant foliage. For optimum worm control, treat at early instars. Addition of sticker may improve residual control. Do not apply to wet foliage or when rain or high humidity is expected during next 2 days. Observe plant response precautions.

Rate: 1-2 lb. ai/100 gal. water
2-4 pt. 4L, SL/100 gal. water

Pest:

balsam twig aphid	Douglas-fir tussock moth	western spruce budworm
Cooley spruce gall adelgid	spruce budworm	

Remarks: Use high rate when large larvae or heavy populations are present.

Rate: 16 lb. ai/100 gal. water
32 pt. 4L, SL/100 gal. water
20 lb. 80WSP/100 gal. water

Pest:

elm bark beetle	mountain pine beetle	spruce beetle
Ips engraver beetle	roundheaded pine beetle	western pine beetle

Remarks: Preventive treatment only. Repeat annually as required to prevent beetle attacks. Apply 1 gal. spray/50 sq. ft. bark in late May-early July, or prior to beetle attack. Treat tree trunks from ground level up until trunk diameter is less than 5". For elm bark beetle, apply 20-30 gal. spray/50 ft. of elm tree for thorough coverage of all bark surfaces.

Signal Word/Toxicity Class: Warning/II.
REI: 12 hr.

Clean Crop Sevin Bait *(Platte)*

Rate: 2-2.4 lb. ai/A
40-48 lb. Bait/A

Pest:

ants	darkling beetle	grasshoppers
armyworms	field crickets	sowbugs
cutworms		

Remarks: Annual plants, herbaceous plants and shrubs, trees, woody plants (ornamentals).

Signal Word/Toxicity Class: Caution/III.
REI: 12 hr.

Chipco Sevin — see carbaryl

CHLORPYRIFOS

Broad-spectrum contact/residual organophosphate insecticide that also has some vapor activity.

◆ **Chlorpyrifos Pro 2 or Pro 4** *(Micro Flo)*

Rate: 0.25-3 lb. ai/100 gal. water
1-12 pt. 2/100 gal. water
0.5-6 pt. 4/100 gal. water

Pest: Including:

aphids	leafhoppers	redhumped caterpillar
bagworm	mealybugs	scale crawlers
borers	mimosa webworm	spittlebugs
cutworms	mites	thrips (exposed)
Eastern tent caterpillar	orangestriped oakworm	whiteflies
fall webworm	peachtree borer	yellownecked caterpillar
grasshoppers		

Remarks: Flowers, shrubs, evergreens, vines, shade and flowering trees. Do not use on azaleas, camellias, poinsettias, roses, or variegated ivy. Apply wetting spray to upper and lower leaf surfaces and infested limb and trunk areas when pests first appear; repeat at 7-10 day intervals, if needed. For borers, apply to trunks and lower limbs of trees and shrubs when adults begin to emerge. For peachtree borers, apply to flowering trees and shrubs of genus *Prunus* in water as trunk spray before newly-hatched larvae enter trees. To prevent native elm bark beetles from overwintering in uninfested trees, apply in water to bottom 9 ft. of trunk wetting thoroughly but not to runoff.

Signal Word/Toxicity Class: Warning/II.
REI: 12 hr.

◆ **Cyren 2E or 4E** *(Cheminova)*
Dursban 50W *(Dow AgroSciences)*

Rate: 0.25 lb. ai/100 gal. water
1 pt. 2E/100 gal. water
0.5 pt. 4E/100 gal. water
0.25-0.5 lb. ai/100 gal. water
0.5-1 lb. 50W/100 gal. water (2-4 packets)

Pest: Including:

adelgids	Jack pine budworm	puss caterpillar
aphids	juniper webworm	sawflies (exposed)
armyworms	katydid	spring elm caterpillar
bagworm	lace bugs	spruce budworm
boxelder bug	leafhoppers	tent caterpillars
cankerworms	leafrollers	walnut caterpillar
catalpa sphinx	maple leafcutter	western spruce budworm
citrus mealybug	mites	yellownecked caterpillar
elm spanworms	oak skeletonizer	
fall webworm	poplar tentmaker	
greenstriped mapleworm		

Rate: 0.5 lb. ai/100 gal. water
2 pt. 2E/100 gal. water
1 pt. 4E/100 gal. water
1 lb. 50W/100 gal. water (4 packets)

Pest: Including:

beet armyworm	mealybugs	oakworm
cutworms	mimosa webworm	redhumped caterpillar
leafhoppers	moths	thrips (exposed)
leafminers	native elm bark beetle	weevils
mahogany webworm	needleminers	

Rate: 0.5-1 lb. ai/100 gal. water
1-2 pt. 4E/100 gal. water
1-2 lb. 50W/100 gal. water (4-8 packets)

Pest: Including:

cottonwood leaf beetle	flea beetles	willow leaf beetle
elm leaf beetle		

NOTICE The information on these pages is for preliminary planning — not a guide for use. Be sure to follow the manufacturer's directions, notwithstanding information contained here.

For personal protective equipment and EPA registration numbers, please turn to page 39.

Rate: 1 lb. ai/100 gal. water
4 pt. 2E/100 gal. water
2 pt. 4E/100 gal. water
2 lb. 50W/100 gal. water (4-8 packets)

Pest: Including:

borers (including peachtree borer)	leafminers needleminers	scale crawlers

Rate: 3 lb. ai/100 gal. water
12 pt. 2E/100 gal. water
6 pt. 4E/100 gal. water
6 lb. 50W/100 gal. water (24 packets)

Pest:
cottonwood borer   peachtree borer

Rate: 8 lb. ai/100 gal. water
16 pt. 4E/100 gal. water
16 lb. 50W/100 gal. water (64 packets)

Pest:
beetles

Remarks: Flowers, evergreens, shrubs, vines, and flowering and shade trees. Refer to label for species of beetles, borers, moths, needleminers, scale crawlers, and weevils controlled. Apply coarse spray to upper and lower leaf surfaces and infested limb and trunk areas when pests first appear; repeat at 7-10 day intervals, if needed. For bagworm, apply when insects are in crawler stage. For leafrollers, spray before leaves are tightly rolled. For maple leafcutters, apply spray to larvae as cases are being formed. Do not treat sugar maple trees intended for maple syrup production. When large number of spider mite eggs are present, apply second spray after 3-5 days in the South or 7-10 days in the North after initial treatment to control newly-hatched nymphs. For fall webworm, direct spray into web and adjacent foliage. To reduce twig and branch feeding by bark beetles, apply in spring or early summer. To control migrating and invading gypsy moth larvae, treat trunks and foliage. For borers, apply uniformly as coarse low-pressure spray to trunks and lower limbs of trees and shrubs when adults begin to emerge. For peachtree borers, apply to flowering trees and shrubs of the genus *Prunus* in water as trunk spray before newly hatched larvae enter trees; thoroughly wet all bark areas from ground level to scaffold limbs. For preventive treatment of beetles, apply spray to main tree trunk in early spring or when threat of attack exists from nearby infested trees; for remedial treatment, apply spray to main trunk of infested trees when damage occurs but before adult beetles begin to emerge. To prevent native elm bark beetles from overwintering in uninfested trees, apply in water to bottom 9 ft. of trunk wetting thoroughly but not to runoff.

Signal Word/Toxicity Class: Warning/II.
REI: 12 hr.

CITATION

Citation (WP) (cyromazine) *(Novartis)*
Rate: 2.66 oz. WP/A in 100 gal. water
Pest:
leafminers

Remarks: Chrysanthemums (container grown). Greenhouse. Apply to point of runoff when stippling first appears. Repeat at 7-day intervals or as needed. Do not apply through any type of irrigation system.

Signal Word/Toxicity Class: Caution.
REI: 12 hr.

Clean Crop Sevin — see carbaryl

Conserve — see spinosad

CRYOLITE

Kryocide (96%) *(Elf Atochem)*
Rate: 7.7-23 lb. ai/A
8-24 lb. 96%/A

Pest:

codling moth	gypsy moth	leafrollers
flea beetles	katydids	plum curculio
Fuller rose beetle		

Remarks: Ornamentals, shade trees. Use higher rate for severe infestations. Do not use in combination with lime or compounds containing free lime. Do not apply chemigation application in California.

Signal Word/Toxicity Class: Caution.
REI: 12 hr.

Prokil Cryolite 96 *(Gowan)*
Rate: 24-29 lb. ai/A
25-30 lb. 96/A

Pest:

codling moth	Fuller rose beetle	katydids
flea weevil	gypsy moth	plum curculio

Remarks: Ornamentals, shade trees. Do not apply more than 96 lb./A per season. Do not apply through any type of irrigation system.

Signal Word/Toxicity Class: Caution.
REI: 12 hr.

CYFLUTHRIN

Broad-spectrum pyrethroid insecticide with activity against several pest groups affecting ornamental plants. Baythroid 2 is Restricted Use Pesticide due to toxicity to fish and aquatic organisms.

Baythroid 2 (cyfluthrin) *(Bayer)*
Rate: 3.2 fl. oz. 2/A
Pest: Various insects.
Remarks: Christmas trees. Refer to label for spray drift precautions.
State Restrictions: SLN-Michigan.

Signal Word/Toxicity Class: Danger/I.
REI: 12 hr.

Decathlon 20WP or WSP *(Olympic)*
Rate: 1.3 oz. 20WP/100 gal. water
1 packet WSP/136 gal.

Pest:

armyworms	elm spanworm	sawflies (larvae)
azalea caterpillar	fungus gnat	sowbugs
bagworm	gypsy moth (larvae)	striped oakworm
bristly roseslug	lace bugs	tent caterpillar
California oakworm	oleander moth (larvae)	tussock moth (larvae)
cankerworms	pillbugs	webworm
caterpillars (leaffeeding)	pine shoot moth	yellownecked caterpillar
cutworms	pine tip moth	
	redhumped caterpillar	

Rate: 1.9 oz. 20WP/100 gal. water
1 packet WSP/92 gal.

Pest:

aphids	June beetle (adult)	plant bugs
boxelder bug	leafhoppers	scale crawlers
budworms	leafrollers	spittlebugs
casebearers	leafskeletonizers	striped beetle
clover mite	mealybugs	thrips
elm leaf beetle	orchid weevil	ticks
flea beetles	pear psylla	tussock moth (larvae)
grasshoppers	peppertree psyllid	whiteflies
Japanese beetle (adult)		

Remarks: Greenhouse nurseries (evergreens, foliage plants, shrubs, trees). For use by commercial applicators only. Apply prior to establishment of high pest population; repeat as necessary. Apply as dilute spray to point of runoff; good spray coverage necessary. Apply to flowering plants during period when pollinating insects are not present, such as early morning or late evening. Addition of spreader/sticker may enhance insect control on ornamentals having hard-to-wet foliage. Do not apply through any type of irrigation system.

Signal Word/Toxicity Class: Caution/III.
REI: 12 hr.

Tempo 2, 20WP or WSP *(Bayer)*
Rate: 1-1.5 fl. oz. 2/100 gal.
1.3-1.9 oz. 20WP/100 gal.
50 gram WSP/92.6-138 gal.
Pest: Most ornamental pests except mites and scales.
Remarks: Landscape ornamentals (evergreens, flowers, foliage plants, shrubs, trees). Refer to label for use directions and precautions.

Signal Word/Toxicity Class: Warning/II (2); Caution (20WP, WSP).

◆ - **new product** • ai-active ingredient • DF-dry flowable • E/EC-emulsifiable concentrate • EW-emulsion in water
F/FL-flowable • G-granule • L/LC-liquid • SP/WSP-soluble packet • W/WP-wettable powder • WSB-water soluble bag

lambda-CYHALOTHRIN

Synthetic pyrethroid insecticide with broad-spectrum activity. Restricted Use Pesticide due to toxicity to fish and aquatic organisms.

Scimitar CS, GC, or WP *(ZENECA Professional)*
◆ **Topcide O/S** *(Uniroyal)*

Rate: 1.5-5 fl. oz. CS, GC/100 gal. water
1.2-4.8 oz. O/S, WP/100 gal. water (1-4 packets)

Pest:

ants	elm leaf beetle	oleander moth (larvae)
aphids	European sawfly	pillbugs
armyworms	fall webworm	pine sawfly
azalea caterpillar	flea beetles	pine tip moth
bagworm	forest tent caterpillar	plant bugs
black vine weevil (adult)	gypsy moth (larvae)	root weevil
boxelder bug	imported fire ant	sawfly
budworms	Japanese beetle (adult)	scale crawlers
California oakworm	June beetle (adult)	spiders
cankerworms	lace bugs	spittlebugs
caterpillars (leaffeeding)	leafhoppers	striped beetle
crickets	leafminers (adult)	striped oakworm
cutworms	leafrollers	thrips
eastern tent caterpillar	leafskeletonizers	tip moth
	midges	tussock moth (larvae)
	mosquitoes	wasps

Rate: 3-5 fl. oz. CS, GC/100 gal. water
2.4-4.8 oz. O/S, WP/100 gal. water (2-4 packets)

Pest:

broad mite	clover mite	spider mites
brown soft scale	mealybugs	whiteflies
California red scale	pine needle scale	

Remarks: Apply prior to build-up of large pest populations. Use higher rate as pest pressure increases. Apply at 7-day intervals if treatment is necessary. Do not apply more than 0.36 lb. ai/A. Do not apply to any plants grown for food or feed.

Signal Word/Toxicity Class: Warning/II.
REI: 24 hr.

Cyren — see chlorpyrifos

Damoil — see petroleum oil

DEADLINE

Deadline (4G) (metaldehyde) *(Pace)*

Rate: 0.02-0.8 lb. ai/1000 sq. ft.
0.5-2 lb. 4G/1000 sq. ft.

Pest:

slugs	snails

Remarks: Commercial nurseries, greenhouses, groundcover. Apply every 3-4 weeks during growing season or as needed. Do not apply directly to edible portion of plant.

Signal Word/Toxicity Class: Warning/II.
REI: 12 hr.

Decathlon — see cyfluthrin

Diatect — see pyrethrins

DIAZINON

Organophosphate insecticide. KnoxOut GH is a flowable microencapsulated formulation. Restricted Use Pesticide due to avian and aquatic toxicity.

D·z·n diazinon 50W *(Novartis)*
Drexel Diazinon (4EC) *(Drexel)*
Micro Flo Diazinon AG500 or 50W *(Micro Flo)*
Prentox Diazinon AG500 or 50W *(Prentiss)*
◆ **Wilbur-Ellis Diazinon 4 Spray (4E)** *(Wilbur-Ellis)*

Rate: 0.5 lb. ai/100 gal. water
1 pt. 4EC, AG500/100 gal. water
12.75 fl. oz. AG600/100 gal. water
1 lb. 50W/100 gal. water

Pest:

aphids	European pine shoot moth	obscure root weevil
bagworm	European red mite	omnivorous leaftier
carnation bud mite	flea beetles	privet mite
carnation shoot mite	holly bud moth	San Jose scale crawler
clover mite	leafhoppers	soft scale
cottonycushion scale crawler	lecanium pine needle scale crawler	thrips
cyclamen mite	mimosa webworm	twospotted mite
dipterous leafminer		webworms
		whiteflies

Rate: 1.5 lb. ai/100 gal. water
3 pt. 4EC, AG500/100 gal. water
38.25 fl. oz. AG600/100 gal. water
3 lb. 50W/100 gal. water

Pest:

apple-and-thorn skeletonizer	fall webworm	obliquebanded leafroller
cotoneaster webworm	hemlock chermes	pear slug
	oak looper	tent caterpillar

Remarks: Arborvitae, azaleas, birch, boxwood, camellias, carnations, chrysanthemums, Douglas fir, elm, gladioli, hawthorn, holly, juniper, lilac, locust, maple, oak, pine, plum, poplar, rhododendron, rose, spruce, willow. Do not use on ferns, poinsettia, hibiscus, papaya, pilea, or gardenia, since injury to plants may occur. For mimosa webworm, apply as foliar spray when adults first appear. Make second and third application at 4- to 5-week intervals.

Signal Word/Toxicity Class: Caution/III.
REI: 12 hr.

◆ **KnoxOut GH** *(Cleary)*

Rate: 0.75-1.5 lb. ai/100 gal. water
3-6 pt. GH/100 gal. water

Pest:

aphids	leafminers	scales
fungus gnat (larvae)	mealybugs	thrips
	mites	whiteflies

Remarks: Greenhouses, interiorscapes, nurseries, and ornamentals in commercial landscapes. Do not use on Bella Palm, dracaena, hoya, maidenhair fern, neath, and prayer plant.

Signal Word/Toxicity Class: Caution.
REI: 12 hr.

DIBROM

Vapor-active organophosphate insecticide, closely related to dichlorvos. Can be vaporized or applied as a fog.

Dibrom 8 (8E) (naled) *(Amvac)*

Rate: 1 lb. ai/A
1 pt. 8E/A in 100 gal. water

Pest: Conifers:

aphids	spittlebugs	western hemlock bark borer
juniper webworm	spruce budworm	Zimmerman moth
red spider mite*	tussock moth	
Sierra fir bark borer	twospotted spider mite*	

Broadleaf trees:

aphids	fall webworm	spring cankerworm
California oakworm (CA only)	leafminers	thrips
elm leaf beetle	oak webworm	tussock moth
fall cankerworm	red spider mite*	twospotted spider mite*

* *Suppression*

Remarks: Forest, shade trees. Apply at first sign of insects and repeat as necessary. Do not use with highly alkaline materials such as lime or Bordeaux. Do not spray on ornamental cherries or Liquidambar as injury may occur. Do not apply through any type of irrigation system.

Rate: 1 fl. oz. 8E/10,000 cu. ft.

Pest:

aphids	mealybugs	whiteflies (adults)
leafrollers	spider mites	

Remarks: Roses and other ornamental plants. Greenhouse vapor treatment. Apply when plants are dry. For spider mites, make 3-4 treatments at 3- to 4-day intervals. For whiteflies, make 2 applications at 7-day intervals. Avoid overtreatment and direct application to plants. Do not spray on chrysanthemums as injury may occur.

Signal Word/Toxicity Class: Danger/I.
REI: 48 hr.

9 Turf & Ornamentals

NOTICE The information on these pages is for preliminary planning — not a guide for use. Be sure to follow the manufacturer's directions, notwithstanding information contained here.

For personal protective equipment and EPA registration numbers, please turn to page 39.

DIMILIN

Chitin synthesis inhibitor that is registered on some ornamental plants. This material disrupts molting process, so that adults are not formed. Outdoor use only. Restricted Use Pesticide.

Dimilin 4L or 25W (diflubenzuron) *(Uniroyal)*

Rate: 0.016-0.063 lb. ai/A
0.5-2 fl. oz. 4L/A
1-4 oz. 25W/A

Pest:
gypsy moth

Rate: 0.03-0.125 lb. ai/A
1-4 fl. oz. 4L/A
2-8 oz. 25W/A

Pest:
forest tent caterpillar

Rate: 0.063-0.125 lb. ai/A
2-4 fl. oz. 4L/A
4-8 oz. 25W/A

Pest:
terminal weevil (pine, spruce)

Remarks: Shrubs, trees. For gypsy moth and tent caterpillar, apply early instar, but prior to full leaf expansion. If single spray of 0.5-1 fl. oz. 4L or 1-2 oz. 25W does not result in adequate control of gypsy moth larvae, second application may be made 7-14 days later. For terminal weevil, apply to adults in spring as snow leaves prior to egg deposition. Add 1-2 gal./A emulsifiable paraffinic oil. Do not exceed more than 4 fl. oz. 4L or 8 oz. 25W/A per year.

- - - - - - - - - - - - - - - - -

Rate: 0.063 lb. ai/A
2 fl. oz. 4L/A
4 oz. 25W/A

Pest:
Nantucket pine tip moth

Remarks: Christmas tree plantations, pine tree nurseries. Apply in early larval stages of development, preferably at beginning of egg hatch of 2nd tip moth generation. Do not exceed more than 2 fl. oz. 4L or 4 oz. 25W/A per year.

- - - - - - - - - - - - - - - - -

Rate: 0.125-0.25 lb. ai/A
0.5-1 lb. 25W/A

Pest:
beet armyworm

Remarks: Chrysanthemums (field, greenhouse grown). Apply when larvae appear and repeat at weekly intervals as required. Apply in dilute spray not to exceed 200 gal./A.

- - - - - - - - - - - - - - - - -

Signal Word/Toxicity Class: Caution/III.

DISTANCE

Insect growth regulator which suppresses embryogenesis within insect egg and inhibits metamorphosis and adult emergence. No activity on adult insects, but hatching of eggs laid by treated adults will be suppressed.

◆ **Distance (pyriproxyfen)** *(Valent)*

Rate: 6-12 fl. oz./100 gal. water

Pest:
aphids (suppression), fungus gnat, greenhouse whitefly, silverleaf whitefly, sweetpotato whitefly

Rate: 8-12 fl. oz./100 gal. water

Pest:
black scale, California red scale, Florida wax scale, San Jose scale, snow scale, spotted tentiform leafminer

Remarks: Flowering and foliage plants, groundcovers, nonbearing fruit and nuts trees, ornamental trees. Apply when insects begin to appear. Apply uniformly to plant surfaces to point of runoff. If necessary, make second application 10-21 days after the first. May be mixed and/or alternated with Orthene and Tame. Do not apply to coral bell, ghost plant, salvia, or Schefflera. Do not apply to poinsettia after bract formation.

- - - - - - - - - - - - - - - - -

Signal Word/Toxicity Class: Caution.
REI: 12 hr.

DI-SYSTON

Restricted Use Pesticide.

Di-Syston 15% (disulfoton) *(Bayer)*

Rate: Shrubs:
3.75-7.5 gram 15%/ft. ht.
Trees (broadcast, soil implant, transplant):
2.5 oz. 15%/" trunk diameter

Pest:
aphids, birch leafminer, camellia scale, elm leaf beetle, European elm scale, holly leafminer, lace bugs, leafhoppers, mimosa webworm, mites, pine tip moth, soft scale, tea scale, thrips, whiteflies

Remarks: Shrubs, trees (including Christmas trees). For shrubs, apply low rate for 1-3 ft. tall, high rate for 3-6 ft. tall. Distribute granules uniformly under shrub canopy. Work into soil and water thoroughly. Repeat at 4- to 6-week intervals if necessary. For broadcast, spread granules uniformly from trunk to drip line on all sides; work into soil and water thoroughly. For transplanting, equivalent quantities should be applied into planting hole. For soil implant, dig 2 equally spaced holes 9-12" deep under drip line. Fill holes to surface with soil, compress, add additional soil to level. Keep treated area moist for 7-10 days.

- - - - - - - - - - - - - - - - -

Rate: Row treatment:
3.5-7.25 oz. 15%/100 ft. row
Broadcast:
7 oz. 15%/100 sq. ft.

Pest:
aphids, lace bug, leafhoppers, mites, thrips, whiteflies

Remarks: Flowers and groundcover. For row treatment, apply in band on each side of seed furrow or as band over row. Work into soil and water thoroughly. Use higher rate for heavy organic or muck soils. For Easter lilies and gladiolus, apply in furrow over bulbs after they are set but before covering with soil. Do not make in-furrow application to garden lily bulbs. For broadcast, apply into soil before planting or after plants are established; repeat at 4- to 6-week intervals if necessary.

- - - - - - - - - - - - - - - - -

Rate: 1-1.5 oz. 15%/tree

Pest:
aphids, mites, Nantucket pine tip moth, scales

Remarks: None.
State Restrictions: SLN-North Carolina.

- - - - - - - - - - - - - - - - -

Signal Word/Toxicity Class: Danger/I.
REI: 48 hr.

Dursban — see chlorpyrifos

Dylox — see trichlorfon

D•z•n — see diazinon

ENDOSULFAN

Chlorinated hydrocarbon insecticide that has been used for many years to control a broad range of pests on ornamental plants.

Drexel Endosulfan 3EC or 50WP *(Drexel)*
Gowan Endosulfan 3EC or 50WSB *(Gowan)*
◆ **Micro Flo Endosulfan 3EC or 50WP** *(Micro Flo)*
Thiodan 3EC or 50WP *(FMC)*

Rate: 0.5 lb. ai/100 gal. water
0.66 qt. 3EC/100 gal. water
1 lb. 50WP, 50WSB/100 gal. water

Pest:
aphids, Cooley spruce gall adelgid, cyclamen mite, Douglas-fir needle midge, leatherleaf fern borer, rose chafer, spruce gall aphid, taxus bud mite, whiteflies

◆ - **new product** • ai-active ingredient • DF-dry flowable • E/EC-emulsifiable concentrate • EW-emulsion in water
F/FL-flowable • G-granule • L/LC-liquid • SP/WSP-soluble packet • W/WP-wettable powder • WSB-water soluble bag

Remarks: Douglas fir, leatherleaf fern, ornamentals, shade trees (except birch), spruce, Taxus. On chrysanthemums, best results will be obtained if applied before plants flower. Do not apply to Bonnafon Deluxe, Fred Shoesmith, and White Knight chrysanthemums as injury may result. For bud mite, make 3-5 applications beginning in mid-May; thorough coverage of foliage, twigs, and bark essential. For gall adelgid, apply in late April-early May when aphids are present but before galls are formed. For gall aphid, apply when white cottony tufts appear. For needle midge, apply late April-early May just before buds open. For fern borer, apply when first larval feeding is observed in mid-vein area at base of leaflets; repeat at 2- to 3-week intervals as necessary.

State Restrictions: For use in Pacific Northwest on Douglas fir; in New England on spruce.

- - - - - - - - - - - - - - - - -

Rate: 0.5-1 lb. ai/100 gal. water
0.66-1.33 qt. 3EC/100 gal. water
1-2 lb. 50WP, 50WSB/100 gal. water

Pest:
dogwood borer   lilac borer

Remarks: Dogwood, lilac. Apply early June; repeat in 10-14 days. Drench all bark areas down to ground level. Do not exceed 3 lb. ai/A per year.

- - - - - - - - - - - - - - - - -

Rate: 0.75 lb. ai/100 gal. water
1 qt. 3EC/100 gal. water
1.5 lb. 50WP, 50WSB/100 gal. water

Pest:
Zimmerman pine moth

Remarks: Pines (Austrian, Jack, red, Scotch, white). Apply in mid-April and again, if necessary, in late fall; thoroughly wet bark and main stem, especially where branches join main stem. Do not exceed 3 lb. ai/A per year.

- - - - - - - - - - - - - - - - -

Rate: 1 lb. ai/100 gal. water
1.33 qt. 3EC/100 gal. water
2 lb. 50WP, 50WSB/100 gal. water

Pest:
black pine weevil

Remarks: Taxus. Spray thoroughly and drench soil under plants. Apply when beetles first appear; repeat in 14 days. Do not exceed 3 lb. ai/A per year.

- - - - - - - - - - - - - - - - -

Signal Word/Toxicity Class: Danger-Poison/I.
REI: 24 hr.

ENSTAR II

Insect growth regulator that can be used to control several pests of ornamental plants.

Enstar II (S-kinoprene) *(Wellmark)*

Rate: 0.5 oz. II/5 gal. water
Pest:
mealybugs

Rate: 5-10 oz. II/5 gal. water
Pest:
fungus gnat

Rate: Preventive:
5 oz. II/100 gal. water
Curative:
10 oz. II/100 gal. water

Pest:
aphids   soft scale   whiteflies
armored scale

Remarks: Greenhouse plants (azalea, chrysanthemum, dianthus, fuchsia, geranium, hydrangea, marigold, pelargonium, petunia, snapdragon). Application to poinsettia should be made prior to bract formation. Foliar damage on some sensitive varieties can result; test on small area before large scale use. Some rose varieties such as yellow-blooded show delayed damage. Preventive: Before insect populations reach damaging levels, apply to runoff every 2 weeks. Curative: For existing populations, make 2 applications 7 days apart, apply to runoff; begin preventive program 2 weeks later. When infestations of root mealybug are observed, entire pot containing infested plant should be drenched by submersion. Continue until no more air bubbles are seen escaping from fluid surface; repeat as necessary. Do not apply through any type of irrigation system.

- - - - - - - - - - - - - - - - -

Signal Word/Toxicity Class: Warning/II.
REI: 4 hr.

FENOXYCARB

Insect growth regulator.

Award *(Novartis)*

Rate: Mound treatment:
1-3 Tbs./mound
Broadcast or aerial application:
1-1.5 lb./A

Pest:
bigheaded ants   fire ants

Remarks: Mound treatment: Uniformly distribute material 3-4 ft. around mound. Broadcast or aerial application: Apply uniformly with ground or aerial equipment calibrated to give correct dosage. Do not apply by air where weather conditions favor drift from treated areas. Apply when ants are actively foraging, usually when soil temperatures are above 60°F. Avoid application during excessively hot periods or when grass is wet. Do not graze treated areas. For bigheaded ant, make ground application only; allow 4-6 weeks for insect growth regulator effect to reduce activity of colonies.

State Restrictions: In California for home lawn use only.

- - - - - - - - - - - - - - - - -

Signal Word/Toxicity Class: Caution.
REI: Do not reenter until application is complete.

Precision *(Novartis)*

Rate: 0.125-0.25 lb. ai/100 gal. water
2-4 oz. WP/100 gal. water

Pest:
fungus gnat   thrips   whiteflies
soft scales

Remarks: Landscape ornamentals, container grown ornamentals, Christmas tree plantations, greenhouses, etc. For scales, thrips, and whiteflies, apply when target insects are detected. For gnats, apply as spray or drench to all surfaces where insects may breed. Repeat at 7- to 14-day intervals for scales and whiteflies; 10- to 14-day intervals for gnats. Refer to label for restrictions and precautions.

- - - - - - - - - - - - - - - - -

Signal Word/Toxicity Class: Caution.
REI: 12 hr.

FENPROPATHRIN

Broad-spectrum pyrethroid insecticide/miticide for indoor ornamental plantings in greenhouses, lathhouse, shadehouses, interiorscapes, and on outdoor ornamental and nursery plantings in commercial nurseries and other landscaped areas. In New York, registered for indoor use only. Toxic to wildlife and bees. Restricted Use Pesticide due to toxicity to fish and aquatic organisms.

Tame 2.4 EC *(Valent)*

Rate: 0.1-0.3 lb. ai/A
5.33-16 fl. oz. 2.4EC/A

Pest:

aphids	greenhouse whitefly	southern red mite (low-moderate pressure)
apple aphid	Japanese beetle	spotted tentiform leafminer
beet armyworm	lace bug	thrips
black vine weevil	leafhoppers	twospotted spider mite
citrus mealybug (including immature stage)	leafminers	
European red mite	mealybugs	
	San Jose scale (crawlers)	

Remarks: Plants, shrubs, trees (conifers, deciduous). Rate varies by pest species; refer to label. For aphids, mealybugs, and sweetpotato and silverleaf whiteflies, tank mix with Orthene Turf, Tree & Ornamental Spray; refer to labels. Since ornamental varieties may react differently to Tame, test on small scale before making large-scale applications. High gallonage applications to certain chrysanthemum varieties may result in crop injury. Do not apply to chrysanthemums and roses with open flowers. Tame & Orthene tank mix on poinsettias after bract formation may result in phytotoxicity on certain varieties. Do not apply by air or through any type of irrigation system.

- - - - - - - - - - - - - - - - -

Signal Word/Toxicity Class: Danger/I.
REI: 24 hr.

NOTICE The information on these pages is for preliminary planning — not a guide for use. Be sure to follow the manufacturer's directions, notwithstanding information contained here.

For personal protective equipment and EPA registration numbers, please turn to page 39.

FURADAN

Restricted Use Pesticide.

Furadan 4F (carbofuran) *(FMC)*

Rate: 1.6 oz. 4F/0.5 gal. water + 2 lb. pulverized kaolin clay

Pest:

pales weevil — pitch-eating weevil

Remarks: Pine seedlings. Mix thoroughly. Apply slurry to roots of pine seedlings prior to transplanting. Refer to label for further use instructions.

State Restrictions: Not for use in Long Island, New York.

- - - - - - - - - - - - - - - - -

Rate: 1-2 fl. oz. 4F/100 gal. water

Pest:

root weevil (larvae)

Remarks: Container grown. Mix thoroughly and apply as soil drench in sufficient volume to saturate entire soil profile within each container. Make single application when larvae are present (July to mid-Oct. in outdoor growing areas). Early marginal necrosis or leaf drop may occur on hydrangea or birch. Not all species or varieties of ornamentals have been tested. Before treating large numbers of plants of a particular variety, treat a few plants and observe prior to full-scale application. Do not handle treated plants, soil, or containers that are wet with spray. Application through overhead sprinkler equipment is prohibited.

State Restrictions: Not for use in California and Long Island, New York.

- - - - - - - - - - - - - - - - -

Signal Word/Toxicity Class: Danger-Poison/I.

REI: 48 hr.

Fyfanon — see malathion

Guthion — see azinphos-methyl

Hexygon — see hexythiazox

HEXYTHIAZOX

Highly active mite ovicide and larvicide. Recommended for use as a foliar application in ornamental plants and nonbearing trees and vines. Not intended for greenhouse use. Toxic to fish and aquatic invertebrates.

Hexygon 50-WP *(Gowan)*

Rate: 1-2 oz. 50-WP/100 gal. water
4-6 oz. 50-WP/A

Pest:

European red mite — spruce spider mite — twospotted spider mite
Pacific spider mite — strawberry spider mite — Willamette mite

Remarks: Outdoor ornamental plants such as flowers, bedding plants, foliage plants, and container-grown and bare root nursery stock, and to nonbearing, fruit, nut, and vine plants. Apply lower rates on low mite egg infestation and higher rates on moderate to high egg infestation levels. Apply only once per crop cycle. Do not graze or feed livestock on cover crops growing in treated areas or use water from treated areas for irrigating other crops.

- - - - - - - - - - - - - - - - -

Signal Word/Toxicity Class: Caution.

REI: 12 hr.

Savey 50-WP *(Gowan)*

Rate: 3-6 oz. 50-WP/A

Pest:

Pacific spider mite — spruce spider mite — twospotted spider mite

Remarks: Christmas trees, Christmas tree plantations. Do not apply more than 1 application per year. Savey may be incompatible with some EC formulations; addition of compatibility agent may improve compatibility with those products. Do not graze or feed livestock on cover crops growing in treated areas or use water from treated orchards for irrigating other crops.

State Restrictions: Not for use in California.

- - - - - - - - - - - - - - - - -

Signal Word/Toxicity Class: Caution.

REI: 12 hr.

HYDRAMETHYLNON

Attractant bait that worker ants find and carry back to the mound where it is passed along to other workers and the queen. Toxic to fish.

Amdro or Siege *(American Cyanamid)*

Rate: Broadcast:
1-1.5 lb./A
Mound treatment:
5 level Tbs./mound

Pest:

bigheaded ants — fire ants — harvester ants

Remarks: Apply when ants are active and soil temperature is greater than 60°F. For broadcast treatments, broadcast bait uniformly with ground equipment (spinning disk types) or aerial equipment. When treating individual mounds, do not disturb mound or apply directly to top of mound. Do not apply more than 1.5 lb./A. Do not use on vegetable or food crops.

- - - - - - - - - - - - - - - - -

Signal Word/Toxicity Class: Caution.

REI: 12 hr.

IMIDACLOPRID

Marathon 1% Granular or 60WP *(Olympic)*

Rate: Flats, benches, beds:
0.4 lb. ai/A
15 oz. 1%G/1000 sq. ft.
1 packet 60WP/3000 sq. ft.
Bulk rate:
0.5-9 lb. 1%G/cu. yd.
Topical Application:
0.5-4.1 g. 1%G/pot
Drench and Irrigation:
1 packet 60WP/pot (number proportionate; refer to label)

Pest:

adelgids — thrips — whiteflies
aphids

Rate: Bulk rate:
1-5 lb. 1%G/cu. yd.
Topical Application:
5-37 g. 1%G/pot
Drench and Irrigation:
1 packet 60WP/pot (number proportionate; refer to label)

Pest:

Japanese beetle — mealybugs — scales
lace bug — root weevil — white grubs
leafminers

Rate: Field Nurseries:
2-2.75 lb. 1%G/1000 ft. row
1 packet 60WP/1000 ft. row

Pest:

white grub

Remarks: Greenhouse and nurseries.

- - - - - - - - - - - - - - - - -

Signal Word/Toxicity Class: Caution.

REI: 12 hr.

Merit 0.5G *(Bayer)*

Rate: Trees, shrubs:
1.4-1.8 lb. 0.5G/1000 sq. ft.

Pest:

Asiatic garden beetle — Japanese beetle — Phyllophaga
chafers — Oriental beetle

Remarks: Flowers, groundcovers, shrubs, trees. Apply broadcast treatment. Refer to label for specific directions.

- - - - - - - - - - - - - - - - -

Signal Word/Toxicity Class: Caution.

Merit 75WP or 75WSP *(Bayer)*

Rate: Trees, shrubs:
1-2 oz. 75WP/30" trunk diameter or 30 ft. ht.
1.6 oz. 75WSP/24-48" trunk diameter or 24-48 ft. ht.
Flowers, groundcovers:
3-4 tsp. 75WP/1000 sq. ft.
1.6 oz. 75WSP/8250-11,000 sq. ft.

◆ - **new product** • ai-active ingredient • DF-dry flowable • E/EC-emulsifiable concentrate • EW-emulsion in water
F/FL-flowable • G-granule • L/LC-liquid • SP/WSP-soluble packet • W/WP-wettable powder • WSB-water soluble bag

Pest:

adelgids	leafhoppers	scales
aphids	leafminers	thrips
elm leaf beetle	mealybugs	white grubs
Japanese beetle	pine tip moth	whiteflies
lace bug	sawflies (larvae)	

Remarks: Apply by soil injection or soil drenching. Refer to label for specific directions.

Signal Word/Toxicity Class: Caution.

Provado 1.6 *(Bayer)*

Rate: 4-8 fl. oz. 1.6/A

Pest:

adelgids — aphids — sawflies

Remarks: Christmas trees. For aphids and sawflies, apply as pests begin to build; 2 applications at 7- to 10-day intervals may be required. For adelgids, time applications to coincide with full bud-swell or first budbreak of earliest bud-breaking trees. Once galls form, spraying will be ineffective. Allow at least 7 days between applications. Do not apply more than 40 fl. oz./A per year.

Days To Harvest: 14.

Signal Word/Toxicity Class: Caution.
REI: 12 hr.

IMIDAN

Imidan 70-W (phosmet) *(Gowan)*

Rate: 0.5-0.75 lb. ai/100 gal. water
0.75-1 lb. 70-W/100 gal. water

Pest:

birch leafminer	elm spanworm	magnolia leafminer
eastern tent caterpillar	gypsy moth	mealybugs
	Japanese beetle	redhumped caterpillar
elm leaf beetle (larvae)	leafhoppers	spring cankerworm

Remarks: Deciduous shade and ornamental trees, woody evergreens. Thoroughly wet all parts of affected plants to point of runoff. For heavy infestations, use higher rate. Repeat as necessary to maintain control. Do not apply if rain is expected or before leaf surfaces are dry.

Rate: 1-1.1 lb. ai/A
1.33-1.5 lb. 70-W/A

Pest:

Eastern pine shoot moth	gypsy moth	pales weevil
	Nantucket pine tip moth	root collar weevil (adult)
European pine shoot moth	pitch-eating weevil	sawfly

Remarks: Christmas tree plantations. Apply when pest populations reach economic threshold levels as determined by local extension service, Forest Service, or other monitoring system.

Rate: 1-1.1 lb. ai/A
1.33-1.5 lb. 70-W/A

Pest:

European pine shoot moth

Remarks: Pine trees. Spray trees for thorough coverage of foliage. Repeat at 2-week intervals throughout moth flight period.

State Restrictions: For use in the Pacific Northwest only.

Rate: 4% Top dip solution:
1.65 lb. ai/5 gal.
2.5 lb. 70-W/5 gal.

Pest:

pales weevil — pitch-eating weevil

Remarks: Pine seedlings (white, slash, loblolly). Dip down to and including root collar only. Dip in bundles loose enough to allow solution to penetrate bundles. Avoid coverage of roots. Swish tops in solution for 10-15 seconds to assure adequate coverage of all top growth. Drain and allow seedlings to dry before planting. Some slight needle burn and first year growth reduction may occur following treatment of loblolly pine. Do not keep solution overnight; make fresh batch each day seedlings will be dipped.

Signal Word/Toxicity Class: Warning/II.
REI: 24 hr.

JOUST

Joust (oxythioquinox) *(Olympic)*

Rate: 0.125-0.25 lb. ai/100 gal. water
4-8 fl. oz. 4L/100 gal. water

Pest:

mite eggs — mites — whiteflies

Remarks: Outdoor use only. For best results, apply during good drying conditions before mites, mite eggs, and whiteflies are present in large numbers and foliar symptoms begin to appear. Do not use spreader or sticker. Spray all foliage surfaces including underside of leaves for complete coverage. Repeat as necessary.

Signal Word/Toxicity Class: Caution.
REI: 24 hr.

KELTHANE

Chlorinated hydrocarbon miticide used on many ornamental plants.

Kelthane 35, 50 or T/O (dicofol) *(Rohm and Haas)*

Rate: 0.35-0.47 lb. ai/100 gal. water
1-1.33 lb. 35/100 gal. water
0.25-0.5 lb. ai/100 gal. water
0.5-1 lb. 50, T/O/100 gal. water

Pest:

mites

Remarks: Flowers, nursery stock, ornamentals, shade trees. Apply when mites are present and repeat as necessary. Do not apply through any type of irrigation system.

Signal Word/Toxicity Class: Warning/II (35); Caution (50, T/O).
REI: 12 hr.

KnoxOut — see diazinon

Kryocide — see cryolite

MALATHION

Contact/residual organophosphate insecticide with broad spectrum activity. Many of its uses on ornamental crops will likely be lost during EPA's reregistration process.

Drexel Malathion 5EC *(Drexel)*
Fyfanon 8E *(Helena)*
◆ **Gowan Malathion 8** *(Gowan)*
◆ **Micro Flo Malathion 5EC** *(Micro Flo)*
Prentox 5 lb. Malathion Spray (5EC) *(Prentiss)*

Rate: 0.63-2.5 lb. ai/100 gal.
1-4 pt. 5EC/100 gal.
1-2 lb. ai/A
1-2 pt. 8E/100 gal.
1-2.5 lb. ai/100 gal.
1-2.5 pt. 8/100 gal.

Pest:

aphids	fourlined leaf bug	pine needle scale
azalea scale	Japanese beetle (adult)	potato leafhopper
bagworm		rose leafhopper
birch leafminer	Juniper scale	soft scale
black scale crawler	lace bugs	spider mites
boxwood leafminer	magnolia scale	tarnished plant bug
euonymus scale	mealybugs	tent caterpillar
European pine shoot moth	Monterey pine scale	thrips
	oak kermes	wax scale
Fletcher scale	oystershell scale	whiteflies
Florida red scale	pine leaf scale	

Remarks: Christmas tree plantations, flowers, ornamental flowering plants, nursery stock, pine seed orchards, shade tree, shrubs, woody plants. Rate varies by pest species; refer to label. For scales, apply when crawlers have settled on foliage. Injury may occur on ferns, hickory, viburnum, lantana, crassula, and canaerti juniper. Slight injury has also been reported on ferns (Boston, maidenhair, Pteris), junipers (Canaertii, Crassula), petunias, small leaf spirea, white pine, and maples.

Signal Word/Toxicity Class: Warning/II (5EC); Caution/III (8E, 8).
REI: 12 hr.

9 Turf & Ornamentals

NOTICE The information on these pages is for preliminary planning — not a guide for use. Be sure to follow the manufacturer's directions, notwithstanding information contained here.

For personal protective equipment and EPA registration numbers, please turn to page 39.

Wilbur-Ellis Malathion 8 *(Wilbur-Ellis)*

Rate: 1-1.75 lb. ai/A
1-1.75 pt. 8/A

Pest:

aphids	leafhoppers	soft scale
armyworms	mealybugs	spider mites
Fletcher scale	Monterey pine scale	tent caterpillar
Florida red scale	oak kermes	thrips
Juniper scale	pine leaf scale	whiteflies

Remarks: Apply full coverage when scales have settled on foliage.
State Restrictions: Not for use in California.

Signal Word/Toxicity Class: Caution/III.
REI: 12 hr.

Marathon — see imidacloprid

Marlate — see methoxychlor

MAVRIK AQUAFLOW

Pyrethroid insecticide/miticide registered for controlling many of the common pests of ornamental plants.

Mavrik Aquaflow (*tau*-fluvalinate) *(Wellmark)*

Rate: 4-10 fl. oz./100 gal. water

Pest:

aphids	elm leaf beetle	mites
caterpillars (leaffeeding)	flea beetles	plant bugs
	leafhoppers	thrips
cucumber beetle	Lygus bugs	whiteflies

Remarks: Conifers (includes bedding plants, Christmas trees, shade trees), flowering plants, foliage plants, herbaceous and woody ornamentals. Use lower rate for light-to-moderate insect pressure; higher rates for green peach aphid, unexposed thrips and heavy insect pressure. Treat when pests are immature or at susceptible stage and populations are manageable, before crop damage occurs. Where heavy populations are present, repeat as necessary at 5- to 7-day intervals. Do not apply through any type of irrigation system.

Rate: 6.4-10 fl. oz./100 gal. water

Pest:

black vine root weevil, obscure root weevil, strawberry root weevil

Remarks: Herbaceous and woody ornamentals. Apply spray on ground around base of tree or shrub and up main trunk or stem.

Rate: 5 fl. oz./100 gal. water

Pest:

aphids, mealybugs, twospotted spider mite

Remarks: Flower and foliage cuttings for dipping. Immerse basket of cuttings into diluted solution until thoroughly wet.

Signal Word/Toxicity Class: Caution.
REI: 12 hr.

Merit — see imidacloprid

MESUROL

Restricted Use Pesticide.

Mesurol (75-W) (methiocarb) *(Gowan)*

Rate: 0.38-0.75 lb. ai/A
0.5-1 lb. 75-W/A

Pest:

aphids, mites

Rate: 1.5 lb. ai/100 gal. water
2 lb. 75-W/100 gal. water

Pest:

snails, slugs

Remarks: Flowers, shrubs, trees. Nursery and greenhouse stock. Spray plants and foliage until thoroughly wet; repeat as necessary. Do not apply more than 8 lb./A or make more than 4 applications per year. Do not apply in conjunction with foliar fertilizer applications.

Signal Word/Toxicity Class: Warning/II.
REI: 24 hr.

METASYSTOX-R

Restricted Use Pesticide.

Metasystox-R (SC) or 2 (oxydemeton-methyl) *(Gowan)*

Rate: 0.25-0.5 lb. ai/A
1-2 pt. SC/A
1-1.5 fl. oz. 2/" trunk diameter

Pest:

aphids	pine needle scale	white pine weevil
mites	(high rate)	

Remarks: Christmas trees, field grown nursery stock. For Metasystox-R SC, apply as full coverage spray. Do not apply more than twice per season. Do not handle treated plants for 48 hr. after application unless chemical-resistant gloves are worn. For Metasystox-R 2, mix in equal or greater volumes of water as needed for equipment used. Applications to noncontainer trees, shrubs, and other ornamentals should be made around general area of drip line.

Signal Word/Toxicity Class: Warning/II.
REI: 48 hr.; 72 hr. if average rainfall less than 25" per year.

METHOXYCHLOR

Restricted Use Pesticide.

Marlate 50 (WP) *(Kincaid)*

Rate: 2 Tbs. 50WP/gal. water

Pest: Flowers, ornamentals:

blister beetle	fleahoppers	rose chafer
cankerworms	flower thrips	rose slug
cucumber beetle	Japanese beetle	sawflies
flea beetles	leafhoppers	

Shade trees and shrubs:

cankerworms, Japanese beetle, tent caterpillar

Remarks: Flowers, ornamentals, shade trees, shrubs. Spray thoroughly to runoff; repeat at 7- to 14-day intervals as needed.

Signal Word/Toxicity Class: Caution.
REI: 12 hr.

Prentox Methoxychlor 2EC *(Prentiss)*

Rate: 0.6-1.3 lb. ai/100 gal. water
37.5-75 fl. oz. 2EC/100 gal. water

Pest:

blister beetle	flea beetles	leafrollers
cankerworms	fleahoppers	May beetle
cucumber beetle	forest tent caterpillar	rose chafer
eastern tent caterpillar	gypsy moth	roseslug
	inchworm	sawflies
elm leaf beetle	Japanese beetle	spittlebugs
fall armyworm	lace bugs	tent caterpillar
fall webworm	leafhoppers	tussock moth

Remarks: Shade trees and ornamentals (dogwood, elm, holly, magnolia, maple, oak, pine, sycamore), evergreens and shrubs (arborvitae, boxwood, dogwood, juniper, taxus), flowers (andromedas, asters, azaleas, begonias, camellias, carnations, chrysanthemums, coleus, dahlias, delphiniums, euonymus, geraniums, gladioli, marigolds, nasturtiums, rhododendron, roses, snapdragons, zinnias). Spray thoroughly to cover all plant surfaces; repeat at 7- to 14-day intervals as needed. Do not apply to birch, Chinese elms, Japanese maple, red maple, or redbud as foliage may be injured. Do not use on crassula, ferns (Boston, maidenhair, Pteris), or petunias.

Rate: 2-3 lb. ai/100 gal. water
1-1.5 gal. 2EC/100 gal. water

Pest:

scales

Remarks: Dormant application only. Do not use on birch, Chinese elm, Japanese maple, red maple, or redbud as foliage may be injured.

Signal Word/Toxicity Class: Caution.
REI: 12 hr.

Wilbur-Ellis Methoxychlor 2 *(Wilbur-Ellis)*

Rate: 8 lb. ai/100 gal. water
4 gal. 2/100 gal. water

Pest:

cankerworms	leafhoppers	tent caterpillars
Japanese beetle	rose chafer	

◆ - **new product** • ai-active ingredient • DF-dry flowable • E/EC-emulsifiable concentrate • EW-emulsion in water
F/FL-flowable • G-granule • L/LC-liquid • SP/WSP-soluble packet • W/WP-wettable powder • WSB-water soluble bag

Rate: 1-1.5 lb. ai/100 gal. water
2-3 qt. 2/100 gal. water
Pest:
elm bark beetle
Remarks: Ornamentals, shade trees. For elm bark beetle, repeat at 7- to 14-day intervals. Do not apply within 14 days of sulfur or sulfur containing sprays.

Signal Word/Toxicity Class: Danger-Poison/I.
REI: 48 hr.

Microthiol Special — see sulfur

MORESTAN

Miticide that has residual activity against the eggs of some species. Toxic to fish.

Morestan 4 (oxythioquinox) *(Bayer)*
Rate: 4-8 fl. oz. 4/100 gal. water
Pest:
mite eggs, red mite, rust mite, twospotted spider mite, whiteflies (eggs and immature stages)
Remarks: Flowers, shrubs, trees. Apply dilute or concentrate. Use in shade houses and landscaped areas where ornamentals are grown. Read label for comments concerning phytotoxicity.

Signal Word/Toxicity Class: Caution.
REI: 24 hr.

M-PEDE

M-Pede (potassium salts of fatty acids) *(Mycogen)*
Rate: 1-2% v/v solution
Pest:
adelgids, balsam woolly adelgid, conifer scale, earwigs, leafminers (tank mix), pine needle scale, psyllid, scales, spider mites, spruce spider mite (conifers), whiteflies

Rate: 3% v/v solution
Pest:
tent caterpillar

Rate: 4% v/v solution
Pest:
Cooley spruce gall adelgid
Remarks: Bedding plants, Christmas trees, flowers (including roses), and ornamental landscape shrubs and trees. Spray to wet infested plant surfaces or crawlers. For balsam woolly aphid, spray trees thoroughly when nymphs are abundant in early spring and/or fall, make second application 10 days later. For earwigs, spray directly on insects after disturbing their hiding place. For caterpillars, spray directly on young larvae when congregated on outside of tent. Test a few flowering plants prior to large scale use.

Signal Word/Toxicity Class: Warning/II.
REI: 12 hr.

NEEMAZAD — see azadirachtin

NEMACUR

Restricted Use Pesticide.

Nemacur 3 (fenamiphos) *(Bayer)*
Rate: 2-4 gal. 3/A
Pest:
nematodes
Remarks: Iris, lily bulbs, narcissus bulbs.
State Restrictions: SLN-Oregon, Washington.

Signal Word/Toxicity Class: Danger/I.
REI: 48 hr.

OFTANOL

Oftanol 2 or 5% (isofenphos) *(Bayer)*
Rate: 1-2 lb. ai/A
2-4 qt. 2/A
2 lb. ai/A
40 lb. 5%/A
Pest:
armyworms, billbugs (larvae), cutworms, chinch bug, fall armyworm, fire ants, flea beetles, Hyperodes weevil (larvae), mole cricket, sod webworm (larvae), white grubs
Remarks: Ornamental beds. For use by commercial applicators only.

Signal Word/Toxicity Class: Warning/II (Oftanol 2); Caution (5%).

Omite — see propargite

Omni Supreme Spray — see petroleum oil

Ornamite — see propargite

Orthene — see acephate

PENTAC

Miticide effective against most phytophagous species. Toxic to fish.

Pentac Aquaflow (dienochlor) *(Novartis)*
Rate: 8 fl. oz. /100 gal. water
Pest:
mites
Remarks: Greenhouse. For heavy populations, second application may be made in 7 days.

Signal Word/Toxicity Class: Warning/II.
REI: 12 hr.

PERMETHRIN

Pyrethroid insecticide. Highly toxic to bees exposed to direct treatment. Restricted Use Pesticide due to toxicity to fish and aquatic organisms.

Ambush 2EC or 25WP *(ZENECA)*
Astro 3.2EC or 25WP *(FMC)*
Rate: 0.1-0.2 lb. ai/A
6.4-12.8 fl. oz. 2EC/A
4-8 fl. oz. 3.2EC/A
6.4-12.8 oz. 25WP/A
Pest: Chrysanthemums (Florida):
vegetable leafminer
Roses (field grown):
Heliothis
Conifers (container, field grown):
Nantucket pine tip moth

Rate: **Roses (greenhouse):**
2 lb. ai/100 gal. water
12 fl. oz. 2EC/100 gal. water
12 oz. 25WP/100 gal. water
Pest:
beet armyworm, cabbage looper, omnivorous leafroller
Remarks: Small number of plants should be treated to determine plant safety prior to commercial use. For pine tip moth, apply when adults appear and repeat at 5- to 7-day intervals or as needed.

Rate: 0.1-0.2 lb. ai/100 gal. water
6.4-12.8 fl. oz. 2EC/100 gal. water
4-8 fl. oz. 3.2EC/100 gal. water
6.4-12.8 oz. 25WP/100 gal. water
Pest:
beet armyworm, cabbage looper, citrus thrips, cutworms, Heliothis, leafhoppers, leafminers, whiteflies

NOTICE The information on these pages is for preliminary planning — not a guide for use. Be sure to follow the manufacturer's directions, notwithstanding information contained here.

For personal protective equipment and EPA registration numbers, please turn to page 39.

Remarks: Field grown nursery stock (except Christmas trees). Margin leaf burn may occur on Dieffenbachia, Salvia, and Pteris fern. Application to blooming plants may cause browning of petals. Before treating large number of plants of a particular variety, treat a few plants and observe prior to full scale application. Apply to nonbearing plants of fruiting species.

State Restrictions: Ambush 2EC or 25WP not for use on Christmas trees or ornamental nursery stock in California. Ambush 25WP not for use on conifers in California. For use on chrysanthemums in Florida only. Not for use on roses in California.

Rate: High volume sprayers:
13 fl. oz. 2EC/100 gal. water
8 oz. 3.2EC/100 gal. water
Low volume sprayers:
68 fl. oz. 2EC/100 gal.
42 oz. 3.2EC/100 gal.
Aerial applications:
48 fl. oz. 2EC/A in not less than 10 gal. water
30 oz. 3.2EC/A in not less than 10 gal. water

Pest:
coneworms, seed bugs

Remarks: Pine seed orchards. Apply within 30 days after female flower closure, 1 week of female flower closure, or peak pollen flight for webbing coneworm. Up to 6 applications may be made at 4-week intervals. Apply approximately 5-10 gal. 0.025% dilution/tree with high volume sprayers; 100 gal. 0.125% dilution/A with low volume sprayers. Do not graze or harvest cover crop. Avoid contact with open water.

Signal Word/Toxicity Class: Warning/II; Caution (Astro 3.2EC).
REI: 24 hr.

Pounce 3.2EC, 25WP or WSB *(FMC)*

Rate: 0.1-0.2 lb. ai/100 gal. water
4-8 oz. 3.2EC/100 gal. water
6.4-12.8 oz. 25WP/100 gal. water
1-2 bags WSB/100 gal. water

Pest:

bagworm	citrus thrips	leafhoppers
beet armyworm	Heliothis (roses)	leafminers
cabbage looper	lace bug	whiteflies

Remarks: Nursery stock-field grown. Refer to label for uses and cautions.

Rate: High volume:
13 oz. 3.2EC/100 gal. water
0.21 oz. 25WP/100 gal. water
1-2 bags WSB/100 gal. water
5-10 gal. finished spray/tree
Low volume:
68 oz. 3.2 EC/100 gal. water
104 oz. 25WP/100 gal. water
1-2 bags WSB/100 gal. water
Aerial:
48 oz. 3.2EC/A
77 oz. 25WP/A
1-2 bags WSB/A

Pest:
coneworms, seed bugs

Remarks: Pine seed orchards. For aerial application, apply in at least 10 gal. water/A. Apply within 30 days after flower closure; 1 week of flower closure or peak pollen flight for webbing coneworm. Up to 6 applications may be made at 4-week intervals. Avoid contact with open water. Do not graze or harvest cover crops.

Rate: Roses:
0.2 lb. ai/100 gal. water
8 oz. 3.2EC/100 gal. water
12.8 oz. 25WP/100 gal. water
2 bags WSB/100 gal. water

Pest:
beet armyworm, cabbage looper, omnivorous leafroller

Rate: Chrysanthemums:
0.5 lb. ai/100 gal. water
20 oz. 3.2EC/100 gal. water
32 oz. 25WP/100 gal. water
5 bags WSB/100 gal. water

Pest:
Liriomyza leafminer fly

Remarks: Greenhouse. For chrysanthemums, avoid spraying blooms. May be applied on weekly schedule. Refer to label for cautions.

Signal Word/Toxicity Class: Caution (3.2EC); Warning/II (25WP, WSB).
REI: 24 hr.

PETROLEUM OIL

◆ **455 Oil 98.8** *(Drexel)*
◆ **SunSpray 9E, 11E** *(Sun)*

Rate: 2 gal. 98.8, 9E, 11E/100 gal. water

Pest:
European red mite, scales, white peach scale

Remarks: Azalea, dogwood, euonymus, flowering cherry, flowering crab, hawthorn, holly, juniper, lilac, pine, privet, rhododendron, rose, spirea. Dormant to bud break. For white peach scale, apply up to 3 gal. 455/100 gal. dilute spray.

Signal Word/Toxicity Class: Caution.
REI: 4 hr.

◆ **Damoil** *(Drexel)*

Rate: 1 gal./100 gal. water
Dormant spray:
2 gal./100 gal. water (deciduous)

Pest:

black scale	red spider mite (including eggs)	San Jose scale
brown apricot scale		soft brown scale
oystershell scale		

Remarks: Birch, conifers, dogwood, flowering cherry, flowering peach, hawthorne, holly, magnolia, maple, oak, privet, pyracantha. May cause discoloration on blue spruce. Do not use on conifers or ferns not tolerant to oil sprays.

Signal Word/Toxicity Class: Caution.
REI: 12 hr.

Omni Supreme Spray *(Helena)*

Rate: Winter dormant
2-4 gal./100 gal. water
Summer (foliar or cover):
1-3 gal./100 gal. water

Pest: Eggs, larvae, and immature stages of:

adelgids	leaf beetle (larvae)	psyllids
aphids	leafminers	sawflies
caterpillars	mealybugs	scales
Eriophyid mite	mites	whiteflies
lace bug		

Remarks: Christmas trees, conifers, evergreens, flowers, foliage and bedding plants, ornamentals, shade trees, shrubs. Do not make more than 4 application per growing season. Do not spray when there is moisture deficit in leaves or when plant is under obvious stress.

Signal Word/Toxicity Class: Caution.
REI: 4 hr.

SAF-T-SIDE *(Brandt Consolidated)*
◆ **Sumulsion** *(Drexel)*
◆ **SunSpray Ultra-Fine** *(Sun)*

Rate: 1-2 gal./100 gal. water

Pest: Mites and insects in egg stage such as:

aphids	mealybugs	sawflies
caterpillars	mites	scales
fungus gnat	plant bugs	thrips
leafminers	psyllids	whiteflies

Remarks: Bedding and foliage plants, conifers, flowers (azalea, camellia, carnation, chrysanthemum, fuschia, gladiola, hibiscus, iris, lilies, poinsettia, rhododendron, roses), ornamentals, shade trees, shrubs. Compatible with most commonly used insecticides and fungicides. Do not use on conifers that are not tolerant to oil sprays. Do not use in combination with or immediately before or after spraying fungicides such as captan, Dyrene, folpet, Morestan, or any sulfur products. Do not use with carbaryl or dimethoate.

Signal Word/Toxicity Class: Caution.
REI: 4 hr.

Pinpoint — see acephate

Pounce — see permethrin

Precision — see fenoxycarb

◆ - **new product** • ai-active ingredient • DF-dry flowable • E/EC-emulsifiable concentrate • EW-emulsion in water
F/FL-flowable • G-granule • L/LC-liquid • SP/WSP-soluble packet • W/WP-wettable powder • WSB-water soluble bag

PROPARGITE

Toxic to fish.

Omite 30W or 30WS *(Uniroyal)*
Rate: 0.6-1.5 lb. ai/A
2-5 lb. 30W, 30WS/A (in 50 gal. spray by ground)
Pest:
Pacific spider mite, southern red mite, twospotted spider mite
Remarks: Roses (field grown). Do not apply through any type of irrigation system.

Signal Word/Toxicity Class: Danger/I.
REI: 7 days. After 48 hr., workers can enter treated areas to perform hand labor tasks if they wear early-entry personal protective equipment.

Omite-CR *(Uniroyal)*
Rate: 1.5-4.5 lb. ai/A
5-15 lb./A
Pest:
spider mites
Remarks: Christmas trees, conifers. Nursery and plantation use. Complete and uniform coverage necessary for optimum results. Do not apply through any type of irrigation system.
State Restrictions: Not for use in California.

Signal Word/Toxicity Class: Danger/I.
REI: 7 days. After 48 hr., workers can enter treated areas to perform hand labor tasks if they wear early-entry personal protective equipment.

Ornamite *(Uniroyal)*
Rate: 0.3 lb. ai/100 gal. water
1 lb./100 gal. water
Pest:
southern red mite, twospotted spider mite
Remarks: Field and nursery ornamentals. Initially treat only small portion of crop to confirm plant safety.

Rate: East of the Mississippi:
0.9-2.25 lb. ai/A
3-7.5 lb./A
West of the Mississippi:
1.35-2.25 lb. ai/A
4.5-7.5 lb./A
Pest:
spider mites
Remarks: Christmas trees, conifers. Plantation and nursery use.

Rate: 0.6-1.5 lb. ai/A
2-5 lb./A (in 50-300 gal. spray by ground)
Pest:
Pacific spider mite, southern red mite, twospotted spider mite
Remarks: Roses (field grown).

Signal Word/Toxicity Class: Danger/ I.
REI: 7 days. After 48 hr., workers can enter treated areas to perform hand labor tasks if they wear early-entry personal protective equipment.

Provado — see imidacloprid

Proxol — see trichlorfon

PYRELLIN (Premix)

Premix containing pyrethrins and rotenone. May be combined with other insecticides to enhance knockdown or as an insect "exciter" to help facilitate contact with other spray residues.

Pyrellin E.C. (0.042 lb./gal. pyrethrins & 0.035 lb./gal. rotenone) *(Webb Wright)*
Rate: 1-2 pt. E.C./A
Pest:

aphids	Japanese beetle	mealybugs
armyworms	leafhoppers	mites
caterpillars	leafminers	plant bugs
chinch bug	leafrollers	thrips
fleahoppers	loopers	whiteflies
fruit flies	Lygus bugs	

Remarks: Apply at 7-day intervals or less; repeat as needed.
Days To Harvest: 0.

Signal Word/Toxicity Class: Caution/III.
REI: 12 hr.

Pyrenone — see pyrethrins

PYRETHRINS

Contains piperonyl butoxide as a synergist. Diatect also contains silicon dioxide (diatomaceous earth). May be combined in tank mix with other insecticides for quicker and more complete kill where insect resistance may be a problem, and as an "exciter" to flush insects out of hiding and into contact with other spray residues.

Diatect Multipurpose Insecticide II *(Diatect)*
Pyrenone (0.5EC) *(AgrEvo Environmental Health)*
◆ **Py-Rin 60-6 EC** *(Wilbur-Ellis)*
Rate: 1-6 lb. II/A
12- 24 fl. oz. 0.5EC/100 gal. water
2-16 oz. 6-60 EC/A
Pest: Including but not limited to:

aphids	fruit flies	Lygus bugs
armyworms	Fuller rose beetle	mealybugs
caterpillars	Japanese beetle	mites
chinch bug	leafhoppers	plant bugs
flea beetles	leafminers	thrips
fleahoppers	leafrollers	webworms
flies	loopers	whiteflies

Remarks: Apply in sufficient water for thorough coverage; repeat as needed. Apply Diatect as dust or wettable powder.
Days To Harvest: 0.

Signal Word/Toxicity Class: Caution.
REI: 12 hr.

PYRIDABEN

Pyridoxone miticide/insecticide, a wettable powder in water-soluble bags, that provides knockdown and residual control. Toxic to bees, fish, and aquatic invertebrates.

Sanmite WP *(BASF)*
Rate: 0.1-0.2 lb. ai/100 gal. water
2-4 oz. WP/100 gal. water (2-4 WSB)
Pest:
broad mite, Tumid mite, twospotted spider mite, European red mite

Rate: 0.2-0.3 lb. ai/100 gal. water
4-6 oz. WP/100 gal. water 4-6 WSB)
Pest:
whiteflies
Remarks: Greenhouse grown and outdoor ornamental trees, shrubs, plants, flowers. Do not use in lathhouses. Apply in enough water to thoroughly cover foliage. Do not exceed 10.67 oz. WP/A per application indoors; 21.3 oz. WP/A per year outdoors. Before using, test to ensure phytotoxic response will not occur. Do not apply as smoke, mist, fog, or aerosol.

Signal Word/Toxicity Class: Danger/I.
REI: 12 hr.

Py-Rin — see pyrethrins

SAF-T-SIDE — see petroleum oil

Sanmite — see pyridaben

Savey — see hexythiazox

Scimitar — see *lambda*-cyhalothrin

NOTICE The information on these pages is for preliminary planning — not a guide for use. Be sure to follow the manufacturer's directions, notwithstanding information contained here.

For personal protective equipment and EPA registration numbers, please turn to page 39.

Siege — see hydramethylnon

Sniper — see azinphos-methyl

SPINOSAD

Fermentation-derived insect control agent. Does not significantly impact predaceous and parasitic insect populations. Highly toxic to bees and aquatic invertebrates.

◆ **Conserve (SC) (spinosyn A + spinosyn D)** *(Dow AgroSciences)*

Rate: 6 fl. oz. SC/100 gal. water

Pest:

bagworm	European pine sawfly	thrips
beet armyworm	fall webworm	western flower thrips
Cuban laurel thrips	gypsy moth	willow leaf beetle
eastern tent caterpillar	hickory tussock moth	yellownecked caterpillar
elm leaf beetle	spruce budworm	

Rate: 11 fl. oz. SC/100 gal. water

Pest:

gall midge; honeylocust pod gall midge

Rate: 22 fl. oz. SC/100 gal. water

Pest:

leafminers	spider mites	twospotted spider mite
serpentine leafminer	spruce spider mite	

Remarks: Herbaceous, woody ornamentals. Not for use in greenhouses or enclosed structures. Use in lath and shadehouses permitted. Apply at 7-day intervals.

Signal Word/Toxicity Class: Caution.
REI: 4 hr.

Sulforix — see calcium polysulfide

SULFUR

Micronized sulfur.

Microthiol Special (80W) *(Elf Atochem)*

Rate: 2.4-8 lb. ai/A
3-10 lb. 80W/A (ground)
2.4-16 lb. ai/A
3-20 lb. 80W/A (air)

Pest:

twospotted spider mite

Remarks: Ornamentals, roses. Apply when mites first appear; repeat at 5- to 10-day intervals.
State Restrictions: Not for use in California.

Signal Word/Toxicity Class: Caution.
REI: 24 hr.

Sumulsion — see petroleum oil

SunSpray — see petroleum oil

SUPRACIDE

Restricted Use Pesticide.

Supracide 2E (methidathion) *(Gowan)*

Rate: 0.5 lb. ai/100 gal. water
2 pt. 2E/100 gal. water

Pest: Northeast U.S. including Ohio:

mugho pine needle scale; red pine needle scale; scotch pine scale

Remarks: Nursery stock-professional use only. Apply as thorough cover spray to point of runoff. Apply once per season after scale crawlers have hatched. Apply either early spring (April-May) for spring-generation crawlers; or in summer (June-July) for second generation summer crawlers. Summer spray will also control pine tortoise scale if present.

Signal Word/Toxicity Class: Danger-Poison/I.
REI: 48 hr.

TALSTAR

Pyrethroid insecticide/miticide with broad spectrum activity. Talstar Flowable is Restricted Use Pesticide.

Talstar Nursery Flowable (bifenthrin) *(FMC)*

Rate: 0.04-0.1 lb. ai/A
8-20 fl. oz. FL/A

Pest:

ants	earwigs	pecan leaf scorch mite
aphids	elm leaf beetle	pillbugs
bagworm	European red mite	pine needle scale (crawler)
beet armyworm	fall webworm	plant bugs
black vine weevil	flea beetles	San Jose scale (crawlers)
broad mite	fungus gnat	sowbugs
brown soft scale	imported fire ant	spider mites
California red scale (crawlers)	Japanese beetle	tent caterpillar
caterpillars (leaffeeding)	lace bug	thrips
centipedes	leafhoppers	tip moth
citrus thrips	leafminers	twig borer
clover mite	leafrollers	weevils
cutworms	Lygus bug	whiteflies
Diaprepes	mealybugs	
	millipedes	
	Orchid weevil	

Remarks: Ornamentals in greenhouses, lath houses, shade house, and outdoor nurseries including nonbearing fruit and nut trees. Rate varies by pest species; refer to label. Apply by ground only. For bagworm, apply when larvae begin to hatch and spray larvae directly. For scale crawlers and twig borer, treat trunks, stems, and twigs in addition to foliage. For black vine weevil larvae, use only as rescue treatment of infested containerized plants. Certain cultivars may be sensitive to final spray solution; small number of plants should be treated before application to entire planting.

Signal Word/Toxicity Class: Caution.
REI: 12 hr.

Talstar Nursery Granular (bifenthrin) *(FMC)*

Rate: 1-2.5 lb. G/200 cu. yd.

Pest:

imported fire ant

Remarks: Ornamental Christmas trees, conifers, flowers, plants, shrubs, trees, and nonbearing fruit and nut trees. Quarantine treatment for balled and containerized nursery stocks use in accordance with USDA guidelines. For soil incorporation into potting media used in containerized plantings.

Rate: 0.5-2.5 lb. 2G/200 cu. yd.

Pest:

black vine weevil	imported fire ant	root weevil
fungus gnat	mealybugs	white grubs

Remarks: Ornamental plants outside scope of USDA guidelines. Use higher rates for extended residual control.

Signal Word/Toxicity Class: Caution.
REI: 12 hr.

Tame — see fenpropathrin

Tempo — see cyfluthrin

Thiodan — see endosulfan

TRICHLORFON

Dylox 80 *(Bayer)*
Proxol 80SP *(AgrEvo)*

Rate: 20-30 oz. 80/100 gal. water
2-3 oz. 80SP/10 gal. water

Pest:

armyworms	Nantucket pine tip moth	tobacco budworm
bagworm	narcissus bulb fly	webworms
climbing cutworm	stink bugs	Zimmerman pine moth
dipterous leafminer	tarnished plant bug	
Lygus bugs		

◆ - **new product** • ai-active ingredient • DF-dry flowable • E/EC-emulsifiable concentrate • EW-emulsion in water
F/FL-flowable • G-granule • L/LC-liquid • SP/WSP-soluble packet • W/WP-wettable powder • WSB-water soluble bag

Remarks: Flowers, shrubs, trees. Rate varies by pest species; refer to label. Injury from spray droplets may occur on leaves of certain trees. Direct stream to base of plants at beginning of adult fly activity (early May-June). Repeat annually. Phytotoxicity has occurred on certain varieties of carnations and zinnias. When large planting of these flowers are to be sprayed for the first time, treat only a few plants and observe for 4-5 days before spraying entire planting.

Signal Word/Toxicity Class: Warning/II.
REI: 24 hr.

Topcide — see *lambda*-cyhalothrin

TURCAM

Carbamate insecticide that has some activity against a wide range of insect pests, through contact and stomach action. There may also be systemic activity on some plants. Toxic to fish and wildlife. Restricted Use Pesticide.

Turcam WP or 2$^{1}/_{2}$G (bendiocarb) *(AgrEvo)*

Rate: 2.28-32 oz. ai/100 gal. water
3-42 oz. WP/100 gal. water
2-3 lb. ai/100 gal. water
80-120 lb. 2$^{1}/_{2}$G/100 gal. water

Pest: Including:

azalea leafminer	elm leaf beetle	pine spittlebug
bagworm	fall webworm	poplar tentmaker
billbugs	green apple aphid	spring cankerworm
black vine weevil	gypsy moth	tea scale
bronze birch borer	Japanese beetle	thrips
chrysanthemum aphid	mimosa webworm	whiteflies
eastern tent caterpillar	Nantucket pine tip moth	willow leaf beetle
	pales weevil	yellownecked caterpillar

Remarks: Greenhouse ornamentals, landscape trees. For Turcam WP, spray trees to point of runoff. For gypsy moths, apply to egg masses up to 3 weeks before eggs hatch. For residual larval control, apply to whole tree. For black vine weevil larvae in container production, mix 2-4 oz. WP/100 gal. water; apply as soil drench. Do not apply more than 6 lb. ai/A per year; spring applications must not exceed 3 lb. ai/A.
State Restrictions: Not for use in California.

Signal Word/Toxicity Class: Warning/II (WP); Caution (2$^{1}/_{2}$G).

Fungicides

Lawns & Turf

3336 — see thiophanate-methyl

ALIETTE

Systemic fungicide. Toxic to aquatic and estuarine invertebrates.

Chipco Aliette or Aliette Signature (WDG) (fosetyl-aluminum) *(Rhone-Poulenc)*

Rate: 4-8 oz. WDG/1000 sq. ft.
Disease:

Pythium blight	Rhizoctonia	yellow tuft
Pythium root rot		

Time: Apply when conditions favor disease development; repeat at 14-day intervals for low rate, 21-day intervals for high rate.
Remarks: Cool- and warm-season turfgrasses. Do not water or mow treated turf until foliage is dry. Do not graze animals on treated turf, or feed clipping from treated turf to livestock and poultry. May be tank mixed; refer to label for use directions and precautions.

Signal Word/Toxicity Class: Caution.
REI: 12 hr.

Allegiance — see metalaxyl

Apron — see metalaxyl

Apron XL — see mefenoxam

AZOXYSTROBIN

Broad-spectrum, preventive fungicide with systemic and curative properties.

Heritage *(ZENECA)*
Rate: 0.2-0.4 oz./1000 sq. ft.
Disease:

Anthracnose	leaf spot	red thread
brown patch	melting-out	summer patch
Fusarium patch		

Time: Apply when conditions are favorable for disease development; repeat at 14- to 28-day intervals (14- to 21-day intervals for leaf spot and melting-out).

Rate: 0.4 oz./1000 sq. ft.
Disease:

brown patch (cool weather)	pink snow mold	spring dead spot
	Pythium blight	take-all patch
gray snow mold	Pythium root rot	yellow patch
necrotic ring spot	Rhizoctonia large patch	Zoysia patch

Time: For brown patch, Rhizoctonia, dead spot, and yellow and Zoysia patch, apply 1-2 applications late fall before snow cover, or when conditions are favorable for disease development; repeat at 28-day intervals. For snow mold, make 2 applications, 14 days apart, applying late fall before snow cover. For Pythium, apply before disease is present; repeat at 10- to 14-day intervals. For take-all patch, apply 2 applications, 28 days apart, in spring and 2 applications, 28 days apart, in fall. For necrotic ring spot, apply when conditions are favorable for disease development; repeat at 14- to 28-day intervals.
Remarks: Golf courses, commercial sod farms. Do not apply on top of snow for brown patch (cool weather), snow mold, Rhizoctonia, spring dead spot, yellow patch, or Zoysia patch. Does not control dollar spot, but is compatible to tank mix with other fungicides. Do not apply through any type of irrigation system. Do not graze or feed clippings from treated areas to animals. Refer to label for use directions.

Signal Word/Toxicity Class: Caution.
REI: 12 hr.

Banner — see propiconazole

BANOL

Water miscible fungicide concentrate having preventive and curative properties.

Banol (propamocarb hydrochloride) *(AgrEvo USA)*
Rate: Preventive:
1.33-2 fl. oz./in 2-5 gal. water
Curative:
3-4 fl. oz./in 2-5 gal. water
Disease:

damping-off	Pythium blight	root rot

9 Turf & Ornamentals

NOTICE The information on these pages is for preliminary planning — not a guide for use. Be sure to follow the manufacturer's directions, notwithstanding information contained here.

For personal protective equipment and EPA registration numbers, please turn to page 39.

Time: Apply preventive during periods of high temperature and humidity favorable to development of disease; repeat at 7- to 21-day intervals if needed.
Remarks: Do not apply this product through any type of irrigation system. Do not graze or feed clippings from treated areas to livestock.

- - - - - - - - - - - - - - - - - -

Signal Word/Toxicity Class: Caution.
REI: 12 hr.

BASAMID

Granular soil fumigant. Toxic to fish.

Basamid Granular (dazomet) *(BASF)*
Rate: 6-8 lb./1000 sq. ft.
Disease:
soilborne diseases
Time: Apply in fall for early spring planting. Apply 5 days before seeding.
Remarks: Lawn and turf seed beds and renovation, after incorporation, immediately drench soil with 15 gal. water/100 sq. ft. Apply water only as fast as it can be absorbed without runoff. After 5- to 7-days, rake soil lightly, not deeper than 2". Do not use if soil temperature is below 43°F.

- - - - - - - - - - - - - - - - - -

Signal Word/Toxicity Class: Warning/II.
REI: 24 hr.

Bayleton — see triadimefon

Bravo — see chlorothalonil

CAPTAN

Gustafson Captan 30-DD contains the toxic inert Rhodamine B.

Drexel Captan 4FL *(Drexel)*
Rate: 8.25 fl. oz. 4FL/100 lb. seed
Disease:
seed decay   seedling blight   soilborne disease
Time: Liquid or slurry seed treatment.
Remarks: **Grasses including bluegrass.** Do not use treated seed for food, feed, or oil purposes. Refer to label for use directions.

- - - - - - - - - - - - - - - - - -

Signal Word/Toxicity Class: Danger/I.

Drexel Captan (50W, 80W) *(Drexel)*
Rate: 2 lb. 50W/100 gal. water
Disease:
damping-off   root rot   soilborne diseases
Time: Cultivate into upper 3-4" of soil before planting.

Disease: Dichondra-California:
white mold
Time: Make 2-3 applications at 7-day intervals. Preplant treatment for seedlings or transplants of lawn seedbeds.
Remarks: Lawn seedbeds, soil. Do not apply through any type of irrigation system. Refer to label for use directions.

- - - - - - - - - - - - - - - - - -

Rate: 1.25 lb. 80W/100 gal. water
Disease:
brown patch   leaf spot   root rot
damping-off
Time: Apply prior to infection; repeat at 7- to 10-day intervals as needed.
Remarks: Do not graze treated areas or feed clippings to livestock. Do not apply through any type of irrigation system. Refer to label for use directions.

- - - - - - - - - - - - - - - - - -

Signal Word/Toxicity Class: Danger/I.
REI: 4 days.

Gustafson Captan 30-DD or 400 *(Gustafson)*
Rate: 6.5 fl. oz. 30-DD/100 lb. seed
5-8.25 fl. oz. 400/100 lb. seed
Disease:
damping-off   seed decay   seedling blight
mold
Time: Seed treatment.
Remarks: Do not combine colloidal suspensions, wettable powders, or emulsifiable concentrates unless previous experience assures compatibility. Use with EPA-approved dye that imparts unnatural color to seed. Do not use for feed, food, or oil purposes. Refer to label for use directions.

- - - - - - - - - - - - - - - - - -

Signal Word/Toxicity Class: Danger/I.

Micro Flo Captan (50WP, 80WP) or Captec (4L) *(Micro Flo)*
Rate: 2 lb. 50WP/100 gal. water
1.25 lb. 80WP/100 gal. water
2 pt. 4L/100 gal. water
Disease:
damping-off   soilborne disease   white mold (dichondra)
root rot
Time: Apply into upper 3-4" soil before planting. For dichondra, make 2-3 applications at 7-day intervals.
Remarks: **Lawn seedbeds.** Do not graze treated areas or feed clippings to livestock. Do not apply this product through any type of irrigation system.
Days To Harvest: 30.
State Restrictions: For use on dichondra in California only.

- - - - - - - - - - - - - - - - - -

Signal Word/Toxicity Class: Danger/I.
REI: 96 hr.

Chipco — see iprodione

CHLORONEB

Systemic, foliar applied fungicide for broad-spectrum disease control in turf.

ProTurf Fungicide V *(Scotts)*
Rate: 32.75 lb./11,000 sq. ft. (1/4 acre)
Disease:
gray snow mold   Pythium blight
Time: For blight, apply to moist foliage when high temperatures and humidity favor disease development; repeat at 5- to 7-day intervals. On new seedings, apply low rate after sowing seed; repeat at 7- to 10-day intervals. For snow mold, apply to dry or moist foliage before first heavy snow; repeat mid-winter if turf is free of snow, and late winter as snow melts.
Remarks: For severe disease, apply double rate. In areas where snow cover is present entire winter, apply double rate before first snow and recommended rate when snow melts in spring. Refer to label for use directions.

- - - - - - - - - - - - - - - - - -

Signal Word/Toxicity Class: Caution.

Teremec SP (WP) *(PBI/Gordon)*
Terraneb SP (FL or WP) *(Kincaid)*
Rate: Cool-season lawns:
3-4 oz. WP/1000 sq. ft.
5.3-7.1 oz. FL/1000 sq. ft.
Warm-season lawns:
5 oz. WP/1000 sq. ft.
8.8 oz. FL/1000 sq. ft.
Disease:
brown patch   Rhizoctonia blight
Time: Apply in fall or spring at first signs of infection. For cool-season, repeat at 7- to 10-day intervals during warm, damp weather; warm-season, make second application in 3-4 weeks if disease is severe or reappears.

Rate: 4 oz. WP/1000 sq. ft.
7.1 oz. FL/1000 sq. ft.
Disease:
Pythium blight   *Sclerotium rolfsii*
Time: Apply when conditions are favorable for development of disease; repeat in 5- to 7-days if conditions persist. On new seedlings, apply following emergence. If conditions remain favorable to disease development; repeat in 7- to 10-days.

Rate: 6-9 oz. WP/1000 sq. ft.
10.5-15.9 oz. FL/1000 sq. ft.
Disease:
snow mold   Typhula blight
Time: Apply late in fall, prior to snowfall.
Remarks: Do not graze or feed clippings from treated areas to livestock. Do not apply Teremec SP through any type of irrigation system.

- - - - - - - - - - - - - - - - - -

Signal Word/Toxicity Class: Caution.
REI: 12 hr. (Terraneb).

◆ - **new product** • ai-active ingredient • DF-dry flowable • E/EC-emulsifiable concentrate • EW-emulsion in water
F/FL-flowable • G-granule • L/LC-liquid • SP/WSP-soluble packet • W/WP-wettable powder • WSB-water soluble bag

CHLORONEB (Premix)

Fast-acting, sustained disease control through root and foliage uptake, resulting in less susceptibility of product to weather and maintenance conditions.

ProTurf Fungicide IX (3.26% chloroneb & 1.63% thiophanate-methyl) *(Scotts)*

Rate: 1.92-5.76 lb./1000 sq. ft.

Disease:

brown patch, copper spot, dollar spot, gray snow mold, pink snow mold, Pythium blight

Time: For copper spot, dollar spot, and brown patch, apply when disease is noted or suspected; repeat at 2- to 3-week intervals. For blight, apply to moist foliage when high temperatures and humidity favor disease development; repeat at 5- to 7-day intervals. For new seeding, apply just after sowing seed; repeat at 7- to 10-day intervals. For snow mold, apply to dry foliage before first snow; repeat mid-winter if turf is free of snow, and again immediately after final snow melts in spring.

Remarks: In areas where snow cover is present entire winter, apply high rate before first snow and normal rate when snow melts in spring. Allow 1 week between applications of this and other products.

- - - - - - - - - - - - - - - - -

Signal Word/Toxicity Class: Caution.

CHLOROTHALONIL

Broad-spectrum contact fungicide. Toxic to fish, aquatic invertebrates, and marine/estuarine organisms.

Bravo 500, 720/Weather Stik, Ultrex (WDG) *(ZENECA)*
Echo 500, 720 or 90DF *(Sostram)*

Rate: 0.75-1 lb. ai/A
1.5-2 pt. 500/A
1-1.5 pt. 720/A
0.9-1.4 lb. WDG/A
0.8-1.2 lb. 90DF/A

Disease:

Bipolaris leaf spot, Dreschslera leaf spot, glume blotch, leaf rust, Septoria leaf spot, stem rust, stripe rust

Rate: 0.75-1.5 lb. ai/A
1.5-3 pt. 500/A
1-2 pt. 720/A
0.9-1.8 lb. WDG/A

Disease:

eyespot

Time: Apply during stem elongation when conditions favor disease development. Reapply at flag (top) leaf emergence; repeat at 14-day intervals.

Remarks: Grasses grown for seed. Use in sufficient water to obtain adequate coverage, 20-150 gal. water for dilute spray; 5-10 gal./A for concentrate. Do not graze treated areas or feed treated plant parts to livestock. Refer to label for use directions and restrictions.

Days To Harvest: 14.

- - - - - - - - - - - - - - - - -

Rate: 4.5-9 oz. 90DF/1000 sq. ft.

Disease:

Fusarium patch, gray snow mold

Time: Fusarium patch: Apply late autumn; repeat at 21- to 28-day intervals as needed. Gray snow mold: Apply before snow cover in autumn. Use high rate if turf layer remains frozen prior to snow cover; repeat at monthly intervals as needed.

Rate: 2.25-14 lb. 90DF/A

Disease:

Anthracnose, dollar spot, Helminthosporium leaf spot, Rhizoctonia brown patch, Scleratinia

Rate: Preventive:
1.75-5 oz. 90DF/1000 sq. ft.
Curative:
3.5-6.5 oz. 90DF/1000 sq. ft.

Disease:

Alternaria leaf spot (California), Anthracnose, copper spot, Curvularia leaf spot, dollar spot, gray leaf spot, Helminthosporium leaf spot, large brown patch, melting-out, red thread, stem rust

Time: Apply in adequate amount of water to provide complete coverage. Repeat at various intervals, refer to label.

Remarks: For turf, do not mow or water after treatment until spray is thoroughly dry. In areas where pink snow mold is likely to occur, use 4.5 oz./1000 sq. ft. + 4 oz./1000 sq. ft. of another fungicide. Refer to label for use directions and restrictions.

- - - - - - - - - - - - - - - - -

Signal Word/Toxicity Class: Danger/I (WDG, 90DF); Warning/II (500, 720).
REI: 48 hr.

Daconil 2787 (FL), Weather Stik (6FL), Ultrex (WDG) or Zn *(ZENECA)*

Rate: 8-14 fl. oz. FL, Zn/1000 sq. ft.
5.5-9.75 fl. oz. 6FL/1000 sq. ft.
13.6-24 lb. WDG/A

Disease:

Fusarium patch

Time: Apply late autumn for areas with intermittent or lacking snow cover during winter. Repeat at 21- to 28-day intervals as needed.

Rate: Turfgrass:
8-16 fl. oz. FL, Zn/1000 sq. ft.
5.5-11.125 fl. oz. 6FL/1000 sq. ft.
13.6-27.2 lb. WDG/A

Disease:

gray snow mold

Time: Apply before snow cover in autumn.

Rate: Golf course fairways, lawns, turfgrass:
4-24 pt. FL, Zn/A
2.75-16.66 pt. 6FL/A
2.5-15 lb. WDG/A

Disease:

Anthracnose, brown patch, dollar spot, gray leaf spot, leaf spot, melting-out, red thread

Time: Apply when conditions favor disease development; repeat as needed.

Rate: Golf course tees, greens; ornamental turfgrass:
Preventive:
3-9 fl. oz. FL, Zn/1000 sq. ft.
2.125-6.25 fl. oz. 6FL/1000 sq. ft.
5-15 lb. WDG/A
Curative:
6-11 fl. oz. FL, Zn/1000 sq. ft.
4.125-7.66 fl. oz. 6FL/1000 sq. ft.
10-20 lb. WDG/A

Disease:

Anthracnose, brown patch, coppor spot, dollar spot, gray leaf spot, leaf spot, melting-out, red thread, stem rust

Time: Under severe conditions, use high rate at 7-day intervals.

Remarks: Do not mow or water after treatment until spray on turfgrass is thoroughly dry. Refer to label for use directions and restrictions.

- - - - - - - - - - - - - - - - -

Signal Word/Toxicity Class: Danger/I (WDG); Warning/II (2787, 6FL, Zn).
REI: 48 hr.

Thalonil (4L, 90DF or 90DF WSP) *(Terra)*

Rate: Preventive:
3-6 fl. oz. 4L/1000 sq. ft.
1.75-3.5 oz. 90DF/1000 sq. ft.
1 packet 90DF WSP/5700-11,400 sq. ft.
Curative:
6-11 fl. oz. 4L/1000 sq. ft.
3.5-6.5 oz. 90DF/1000 sq. ft.
1 packet 90DF WSP/3000-5700 sq. ft.

Disease:

Curvularia leaf spot, dollar spot, gray leaf spot, Helminthosporium leaf spot, large brown patch, melting-out

Rate: Preventive:
5-9 fl. oz. 4L/1000 sq. ft.
2.5-5 oz. 90DF/1000 sq. ft.
1 packet 90DF WSP/4000-5700 sq. ft.
Curative:
9-11 fl. oz. 4L/1000 sq. ft.
5-6.5 oz. 90DF/1000 sq. ft.
1 packet 90DF WSP/3000-4000 sq. ft.

Disease:

Alternaria leaf spot (CA only), Anthracnose, copper spot, stem rust

NOTICE The information on these pages is for preliminary planning — not a guide for use. Be sure to follow the manufacturer's directions, notwithstanding information contained here.

For personal protective equipment and EPA registration numbers, please turn to page 39.

Rate: Preventive:
3-9 fl. oz. 4L/1000 sq. ft.
1.75-5 oz. 90DF/1000 sq. ft.
1 packet 90DF WSP/4000-11,400 sq. ft.
Curative:
9-11 fl. oz. 4L/1000 sq. ft.
5-6.5 oz. 90DF/1000 sq. ft.
1 packet 90DF WSP/3000-4000 sq. ft.

Disease:
red thread

Time: For preventive, intervals vary; refer to label. For curative, use 7-day intervals.

Rate: 8-16 fl. oz. 4L/1000 sq. ft.
4.5-9 oz. 90DF/1000 sq. ft.
1 packet 90DF WSP/2500-4400 sq. ft.

Disease:
Fusarium patch, gray snow mold

Time: For gray snow mold, apply before snow cover in autumn. Use high rate if turf layer remains frozen prior to snow cover. If snow cover is intermittent or lacking during winter, repeat using low rate at monthly intervals until conditions no longer prevail. For Fusarium patch, apply only in areas where snow cover is intermittent or lacking during winter by applying in late autumn; repeat at 21- to 28-day intervals as needed.

Remarks: Golf course tees and greens, ornamental turfgrass. Apply in sufficient water for complete coverage. Use high rate under severe conditions. Do not apply through sprinkler irrigation system on golf courses. Refer to label for use directions, precautions and restrictions.

- - - - - - - - - - - - - - - - -

Signal Word/Toxicity Class: Danger/I (90DF, 90DF WSP); Warning/II (4L).
REI: 48 hr.

Curalan — see vinclozolin

Daconil — see chlorothalonil

Dithane — see mancozeb

Eagle — see myclobutanil

Echo — see chlorothalonil

ETRIDIAZOLE

Preventive and curative fungicide. Koban is toxic to fish.

Koban (30) *(Scotts)*

Rate: Established turf:
2-4.5 oz. 30/1000 sq. ft. (in 5 gal. water)

Disease:
damping-off, grease spot, Pythium root rot, crown rot, Pythium blight

Time: Apply when disease or conditions favorable for disease exists.

Rate: 5-6 oz. 30/1000 sq. ft. (in 5 gal. water)

Disease:
crown rot, Pythium root rot

Time: Irrigate immediately following application. Retreat in 5-10 days if disease or conditions favorable for disease persist.

Rate: Newly seeded:
7-9 oz. 30/1000 sq. ft. (in 5-10 gal. water)

Disease:
damping-off

Time: Retreat at 5- to 7-day intervals at 2-4.5 oz. to maintain preventive program after germination.

Remarks: Do not use for feed, food, or oil purposes. Refer to label for use directions.

- - - - - - - - - - - - - - - - -

Signal Word/Toxicity Class: Warning/II.
REI: 12 hr.

Terrazole (35WP) *(Uniroyal)*

Rate: Established:
4-5 oz. 35WP/1000 sq. ft. (in 2-5 gal. water)
Newly seeded:
6-8 oz. 35WP/1000 sq. ft. (in 5-10 gal. water)

Disease:
cottony blight, grease spot, root rot, crown rot, Pythium blight, spot blight, damping-off

Time: Established: Apply at first sign of disease; repeat at 5- to 10-day intervals depending upon climatic conditions and severity of disease. Newly seeded: Apply immediately after seeding; repeat at 5- to 7-day intervals.

Remarks: Established or newly seeded turf. Use high rates and short intervals under heavy disease pressure. Do not allow livestock to graze treated areas or feed clippings from treated areas to livestock.

- - - - - - - - - - - - - - - - -

Signal Word/Toxicity Class: Warning/II.

Defend — see PCNB

FENARIMOL

Triazole sterol inhibitor. Toxic to fish.

Rubigan A.S. *(Dow AgroSciences)*

Rate: 0.75-1.5 fl. oz. A.S./1000 sq. ft.

Disease:
copper spot, large brown patch, Rhizoctonia blight, dollar spot

Time: For copper and dollar spot, apply at 10- to 28-day intervals. For brown patch and Rhizoctonia, apply at 7- to 14-day intervals.

Rate: 1.75-4 fl. oz. A.S./1000 sq. ft.

Disease:
Anthracnose, necrotic ring spot, summer patch, Fusarium blight, powdery mildew, take-all patch

Time: Repeat every 30 days. For Fusarium blight and necrotic ring spot, apply in April and May for greens and tees. For powdery mildew, make 1 application when symptoms are present; repeat at high rate until disease is controlled.

Rate: 4-6 fl. oz. A.S./1000 sq. ft.

Disease:
spring dead spot

Time: Apply low rate in Sept.; high rate Oct.-Nov. Irrigate with 1/2-1" water.

Rate: 8 fl. oz. A.S./1000 sq. ft.

Disease:
Fusarium patch, pink snow mold, Typhula blight, gray snow mold, red thread

Time: Apply 1-2 times in fall as needed.

Rate: 15 fl. oz. A.S./1000 sq. ft.

Disease:
stripe smut

Time: Apply single application in Oct. or early spring prior to turfgrass greenup. Irrigate with 1/2-1" water.

Remarks: Rates vary by disease and time of application. For brown patch (severe pressure), add contact fungicide. For snow mold, 8 oz. rate may eliminate Poa annua; for prolonged cool, wet conditions (western U.S.), apply 2 oz. rate every 3-4 weeks. Do not apply this product through any type of irrigation system. Refer to label for use directions.

- - - - - - - - - - - - - - - - -

Signal Word/Toxicity Class: Caution.
REI: 12 hr.

FLUTOLANIL

Systemic fungicide. Toxic to aquatic invertebrates.

ProStar 50WP *(AgrEvo USA)*

Rate: 2 oz./1000 sq. ft.

Disease:
pink patch, southern blight, yellow patch, red thread

Time: Apply prior to or after initial signs of disease appears; repeat after 21- to 28-days if needed.

◆ - **new product** • ai-active ingredient • DF-dry flowable • E/EC-emulsifiable concentrate • EW-emulsion in water
F/FL-flowable • G-granule • L/LC-liquid • SP/WSP-soluble packet • W/WP-wettable powder • WSB-water soluble bag

Rate: 3 oz./1000 sq. ft.
Disease:
large brown patch
Time: Apply prior to or after initial disease appears; repeat 30 days later if symptoms persist. Fall applications may suppress disease development the following spring.

Rate: 4-6 oz./1000 sq. ft.
Disease:
gray snow mold
Time: Apply to green or tee area in the fall prior to permanent snow cover.

Rate: Preventive:
3 oz./1000 sq. ft.
Curative:
6 oz./1000 sq. ft.
Disease:
fairy ring
Time: Preventive: Apply and immediately follow by irrigation with 1/4-1/2" water; second application may be made at 21- to 28-day intervals using same rate. Curative: Apply as spot treatment to affected area at first sign of development. If symptoms reoccur, a second application not less than 30 days after the first is suggested.

Rate: Preventive:
2-3 oz./1000 sq. ft.
Curative:
3 oz./1000 sq. ft.
Disease:
brown patch
Time: Preventive: Apply prior to initial signs of disease; repeat at 14- to 21-day intervals if conditions favor disease development. Curative: Apply to active disease; make second application 14- to 21-days later if conditions favor disease development.
Remarks: For fairy ring, treated areas should be irrigated prior to and after applications with sufficient water to maintain growth of turf. If turf has been damaged extensively, it may have to be reseeded. Apply no more than a total of 6 oz./1000 sq. ft. as a broadcast application during a single growing season. Do not apply this product through any type of irrigation system. Refer to label for use directions.

Signal Word/Toxicity Class: Caution.
REI: 12 hr.

Fore — see mancozeb

Formec — see mancozeb

Fungo — see thiophanate-methyl

Golden Eagle — see myclobutanil

Heritage — see azoxystrobin

IPRODIONE

Contact fungicide. Chipco 26019 FLO is toxic to aquatic invertebrates.

Chipco 26019 (WP or FLO) *(Rhone-Poulenc)*
Rate: 1.5-4 oz. WP/1000 ft.
3-8 oz. FLO/1000 ft.
Disease:

brown patch	Fusarium patch	leaf spot
Corticium red thread	gray snow mold	necrotic ring spot
dollar spot	large patch	pink snow mold
Fusarium blight		

Time: Apply when conditions favor disease development or when diseases first appear. For greens, tees, and Fusarium patch, repeat at 14- to 21-day intervals as needed; fairways and other turf areas, repeat at 14- to 28-day intervals. For large patch, apply in fall when conditions are favorable for disease development but no symptoms are visible; repeat in spring as needed at 14- to 21-day intervals. For Fusarium blight and ring spot, use only preventive foliar applications at high rate; repeat at 28-day intervals. For snow molds, 1 application before first permanent snow cover (if possible, another during midwinter thaw). For red thread, use low rate as preventive every 14 days as needed.

Remarks: Cool- and warm-season turfgrass. On fairways for dollar spot, use 1-2 oz. WP or 2-4 oz. FLO/1000 ft. Do not mow or irrigate treated areas until foliage is completely dry (usually 24 hr.). Do not mix with any sticker, extender, or wetting agent. Do not graze treated turf or feed clippings from treated turf to livestock or poultry. Refer to label for use directions.
State Restrictions: Not registered for use in California for control of necrotic ring spot or large patch. Pacific Northwest (west of the Cascades) only for Fusarium patch.

Signal Word/Toxicity Class: Caution.
REI: 12 hr.

ProTurf Fungicide X *(Scotts)*
Rate: 52.86 lb./7300-22,000 sq. ft.
Disease:

brown patch	leaf spot	red leaf spot
gray snow mold	pink snow mold	

Time: Apply to moist or dry foliage when disease is expected. For brown patch and red leaf spot, repeat at 2-week intervals; 2- to 3-week intervals for snow mold; 3- to 4-week intervals for leaf spot.
Remarks: Do not water or mow treated areas within 24 hr. of application. Allow 1 week between applications of this and other products.

Signal Word/Toxicity Class: Caution.
REI: 12 hr.

IPRODIONE (Premix)

Combines two synergistic active ingredients which provide a contact and systemic action. Toxic to catfish.

ProTurf Fluid Fungicide (19.65% iprodione & 19.65% thiophanate-methyl) *(Scotts)*
Rate: 1.1-2.1 fl. oz./M
Disease:

Anthracnose	dollar spot	melting-out
brown patch	leaf spot	pink snow mold (WA, OR, CA only)

Time: Apply when weather conditions are favorable for disease development; repeat at 10- to 14-day intervals for normal to mild disease infestations; 7- to 10-day intervals for moderate to severe.
Remarks: Use with 4-7 gal. water/1000 ft. for tees and greens (2 gal. water/1000 ft. for larger areas). Do not water or mow treated areas within 24 hr. of application. Except in severe cases, allow 1 week between applications of this and other products. Do not apply this product through any type of irrigation system. Do not contaminate feed or food stuff. Do not graze treated areas, or feed clippings to livestock. Refer to label for use directions.

Signal Word/Toxicity Class: Caution.
REI: 12 hr.

Koban — see etridiazole

MANCOZEB

Fungicide with protective action. Toxic to fish.

Dithane (DF, F-45, M-45, T&O, WF, WSP) *(Rohm and Haas)*
Fore (FL, FloXL, T&O, WSP) *(Rohm and Haas)*
Formec 80 *(PBI/Gordon)*
Protect T/O (WP) *(Cleary Chemical)*
Rate: 4-8 oz. DF, M-45, T&O, WSP, 80, WP/1000 sq. ft.
6.4-12.8 fl. oz. F-45, FL, FloXL, WF/1000 sq. ft.
Disease:

copper spot	melting-out	Rhizoctonia brown patch
dollar spot	powdery mildew	slime mold
Fusarium blight	Pythium blight	stem rust
Fusarium snow mold	red thread	stripe rust
leaf rust		

Time: Apply when grass greens-up in spring or when disease or algae first appear; repeat at 7- to 14-day intervals (Pythium 5-day, Rhizoctonia 7-day). For snow mold, apply during winter at 2- to 6-week intervals.
Remarks: Not for use by homeowners. Use sufficient water to provide full coverage. Do not apply Fore T&O, Fore WSP, or Formec through any type of irrigation system. Do not use on grasses grown for seed. Do not graze treated areas, use on grasses intended for grazing, or feed clippings to livestock. Refer to label for use directions, especially for small areas.

Signal Word/Toxicity Class: Caution.
REI: 24 hr. (Dithane, Fore, Protect).

For personal protective equipment and EPA registration numbers, please turn to page 39.

Manzate 200DF *(Griffin)*
Rate: 3-4 oz. 200DF/1000 sq. ft.
Disease:
Helminthosporium leaf spot — Rhizoctonia brown patch — stripe rust
Time: Apply when disease threatens; repeat at 7- to 10-day intervals.
Remarks: Do not use on grasses grown for seed. Do not graze treated areas or feed clippings to livestock. Refer to label for use directions.
State Restrictions: Refer to label for California precautions.

Signal Word/Toxicity Class: Caution.
REI: 24 hr.

MANEB

Dithiocarbamate contact fungicide. Toxic to fish.

Maneb 75DF or 80WP *(Elf Atochem)*
Manex (FL) *(Griffin)*
Rate: 0.15-0.375 lb. ai/A
0.2-0.5 lb. 75DF/A
0.16-0.4 lb. ai/A
0.2-0.5 lb. 80WP/A
4.8-12.8 fl. oz. FL/1000 sq. ft.
Disease:
brown patch — dollar spot — melting-out

Rate: 2.25 lb. ai/A
3 lb. 75DF/A
2.4 lb. ai/A
3 lb. 80WP/A
2.4 qt. FL/100 gal. water
Disease:
leaf rust — stem rust — stripe rust
Time: Begin at first sign of disease; repeat at 7- to 14-day intervals.
Remarks: Do not graze treated areas, or feed clippings to livestock. Do not use if grass is grown for seed. Refer to label for use directions.

Signal Word/Toxicity Class: Caution.
REI: 24 hr.

Manex — see maneb

Manzate — see mancozeb

MEFENOXAM

Systemic fungicide having a specific mode of action.

Apron XL LS *(Novartis)*
Rate: 0.64-1.28 fl. oz. LS/100 lb. seed
Disease:
damping-off
Time: Seed treatment at planting.
Remarks: Golf courses, reclamation and erosion projects. Do not use treated seed for food, feed, or oil purposes. For long-term control of Pythium in areas when using Apron XL treated seed, apply Subdue MAXX 7-10 days after seeding.

Signal Word/Toxicity Class: Warning/II.

Subdue GR, MAXX, or 2X WSP *(Novartis)*
Rate: 5-25 oz. GR/1000 sq. ft.
0.5-1 fl. oz. MAXX/1000 sq. ft.
0.11-0.56 oz. WSP/1000 sq. ft.
Disease:
damping-off — Pythium blight — yellow tuft — downy mildew
Time: Established turf: Apply as preventive treatment; repeat at 10- to 21-day intervals. Newly seeded areas: Apply immediately after seeding; repeat at 7- to 14-day intervals if conditions remain favorable for disease.
Remarks: Irrigate immediately with 1/4-1/2" water. Do not make more than 3 applications per season.

Signal Word/Toxicity Class: Caution.

METALAXYL

Systemic fungicide having a specific mode of action.

Allegiance or Apron (FL) *(Gustafson)*
Rate: 0.75-1.5 fl. oz. Allegiance/100 lb. seed
1.5 fl. oz. Apron/100 lb. seed
Disease:
damping-off
Time: Seed treatment.
Remarks: Use with EPA-approved dye that imparts unnatural color to seed. Do not use treated seed for feed, food, or oil purposes.

Signal Word/Toxicity Class: Warning/II.

Apron Dry *(Trace)*
Rate: 4-8 oz./100 lb. seed
Disease:
damping-off
Time: Seed treatment.
Remarks: Treat only those seeds needed for immediate use. Do not use treated seed for food, feed, or oil purposes. Do not store excess treated seeds beyond planting time. Refer to label for use directions.

Signal Word/Toxicity Class: Caution.
REI: 12 hr.

Subdue 2E *(Novartis)*
Rate: 1-2 fl. oz. 2E/1000 sq. ft.
Disease:
damping-off — Pythium blight — yellow tuft — downy mildew
Time: Established turf: Apply as preventive treatment; repeat at 10- to 14-day intervals. Newly seeded areas: Apply immediately after seeding; repeat at 7- to 14-day intervals if conditions remain favorable for disease.
Remarks: Irrigate immediately with 1/4-1/2" water. Do not make more than 3 applications per season.

Signal Word/Toxicity Class: Warning/II.
REI: 12 hr.

ProTurf Pythium Control *(Scotts)*
Rate: 28.25 lb./22,000 sq. ft. (1/2 acre)
Disease:
damping-off — Pythium blight
Time: Apply when weather conditions are favorable for disease development; repeat at 10-day intervals. For severe disease, use double rate at 10-day intervals. For newly seeded, apply immediately after seeding; repeat at 7- to 14-day intervals. Under severe disease conditions, apply double rate; repeat at 7-day intervals.
Remarks: Allow 1 week between applications of this and other products.

Signal Word/Toxicity Class: Caution.
REI: 24 hr.

METALAXYL (Premix)

Systemic fungicide having a specific mode of action.

ProTurf Fluid Fungicide II (16% metalaxyl & 16% triadimefon) *(Scotts)*
Rate: 0.7 fl. oz./M
Disease:
brown patch — dollar spot — Pythium blight — damping-off
Time: Apply when weather conditions are favorable for disease development. Repeat at 7- to 10-day intervals; 10-day intervals for Pythium. For severe disease, use double rate at 7- to 14-day intervals. Newly seeded: Apply immediately after seeding; repeat at 5- to 7-day intervals. Under severe disease conditions, apply double rate; repeat at 10- to 14-day intervals.
Remarks: Use with 4-7 gal. water/1000 ft. Low rates of water may result in turf injury, especially under hot and humid conditions. For newly seeded, irrigate with 1/4" water. Do not water or mow treated areas within 24 hr. of application. Except in severe cases, allow 1 week between applications of this and other products. Do not apply this product through any type of irrigation system. Refer to label for use directions.

Signal Word/Toxicity Class: Danger/I.

◆ - **new product** • ai-active ingredient • DF-dry flowable • E/EC-emulsifiable concentrate • EW-emulsion in water
F/FL-flowable • G-granule • L/LC-liquid • SP/WSP-soluble packet • W/WP-wettable powder • WSB-water soluble bag

Micro Flo — see sulfur

MYCLOBUTANIL

Systemic, protectant, and curative fungicide recommended for control of specific diseases.

Eagle (WSP) *(Rohm and Haas)*
Rate: 0.6-1.2 oz. WSP/1000 sq. ft. (1-2 WSP/5000 sq. ft.)
Disease:

Anthracnose	leaf spot	rust
brown patch	melting-out	Septoria leaf spot
copper spot	necrotic ring spot	spring dead spot
crown rot	powdery mildew	summer patch
dollar spot	red thread	Zonate leaf spot
leaf smut		

Time: Apply when conditions are favorable for disease development. For leaf smut, apply in fall after grass enters dormancy and/or in spring prior to initiation of growth. For necrotic ring spot and spring dead spot, apply at monthly intervals beginning in Aug. Make 2-3 applications depending on disease pressure. For summer patch, apply in spring when conditions are favorable for disease development. Application intervals vary by disease; refer to label.
Remarks: Optimum control is achieved when applied on regular scheduled preventive spray program. Use sufficient water to provide thorough coverage. For dollar spot, make no more than 3 consecutive applications before rotating to registered fungicide with different mode of action.
State Restrictions: Use is limited to 9 oz./5000 sq. ft. per yr. for Nassau and Suffolk counties in New York.

Signal Word/Toxicity Class: Warning/II.
REI: 24 hr.

Golden Eagle *(Scotts)*
Rate: 1.97 lb./1000 sq. ft.
Disease:

Anthracnose	necrotic ring spot	rust
brown patch	powdery mildew	summer patch
dollar spot		

Time: Apply at 14- to 21-day intervals.
Remarks: Refer to label for use directions.

Par-Flo — see PCNB

PCNB

Seed and soil contact fungicide. Toxic to fish; some formulations are also toxic to aquatic organisms.

Defend 10G or 75WP *(Cleary Chemical)*
Penstar (Flo, 10G) *(Scotts)*
Rate: Warm-season grasses:
7.5 lb. 10G/1000 sq. ft.
1 lb. 75WP/1000 sq. ft.
24 fl. oz. Flo/1000 sq. ft.
Cool-season grasses:
2-2.5 lb. 10G/1000 sq. ft.
3-4 oz. 75WP/1000 sq. ft.
4.5-6 fl. oz. Flo/1000 sq. ft.
Disease:
brown patch
Time: Warm-season: Apply once in spring or fall or when disease appears; repeat 3-4 weeks later if disease is severe or reappears. Cool-season grasses: Apply at first sign of disease; repeat every 7- to 10-days during warm damp weather.

Rate: 5-7.5 lb. 10G/1000 sq. ft.
7-10 oz. 75WP/1000 sq. ft.
10.5-15 fl. oz. Flo/1000 sq. ft.
Disease:

dollar spot	leaf spot	pink snow mold
gray snow mold	melting-out	

Time: Apply as preventive application in spring or fall of each year; retreat 3-4 weeks later if disease is severe or reappears. For snow mold, apply single application prior to first snowfall.
Remarks: For dollar spot and leaf spot, use high rate as soon as possible if disease is present. Under certain growing conditions, temporary discoloration of grass occasionally may occur, causing no harm and will disappear in a short time. Do not graze treated areas or feed grass clippings to livestock. Refer to label for use directions.

Crop Rotation: Do not plant root crops in PCNB-treated fields within 12 months of broadcast and banding applications unless PCNB is registered for use on those crops.
State Restrictions: Not for use in California on leaf spot or dollar spot.

Signal Word/Toxicity Class: Caution.
REI: 12 hr.

Par-Flo 4F or 6F *(Amvac)*
Rate: Southern grasses: (St. Augustine, Bermuda, dichondra)
16-24 fl. oz. 4F/1000 sq. ft.
10-16 fl. oz. 6F/1000 sq. ft.
Northern grasses: (Bluegrass, bent, fescue, rye)
5-6 fl. oz. 4F/1000 sq. ft
3-4 fl. oz. 6F/1000 sq. ft.
Disease:
brown patch — *Rhizoctonia solani*
Time: For southern grasses, apply once in spring or fall or when disease appears. If disease reappears, retreat 3- to 4-weeks later. For northern grasses, apply at first sign of disease, repeat at 7- to 10-day intervals during warm damp weather.
Remarks: Do not apply through any type of irrigation system. Do not allow livestock to graze or feed on foliage in treated fields. Do not plant root crops within 12 months of application unless PCNB is registered for use on those crops.

Signal Word/Toxicity Class: Caution.

PCNB 2-E, 2FL, 10%, 10G T&O, TurfPro WSP *(Amvac)*
Rate: Southern grasses: (St. Augustine, Bermuda, dichondra)
1 qt. 2-E/1000 sq. ft.
32-48 fl. oz. 2FL/1000 sq. ft.
7.5 lb. 10%, 10G/1000 sq. ft.
1 lb. WSP/1000 sq. ft.
Dichondra:
1.5 qt. 2-E/1000 sq. ft.
Northern grasses: (Bluegrass, bent, fescue, rye)
10-12 fl. oz. 2FL/1000 sq. ft.
2-2.5 lb. 10%, 10G/1000 sq. ft.
3-4 oz. WSP/1000 sq. ft.
Disease:
brown patch
Time: Apply in fall or spring at first sign of infection. Repeat in 3- to 4-weeks if heavy rains or flooding occurs for southern grasses; 7- to 10-day intervals during warm, damp weather for northern grasses.

Rate: 22-32 fl. oz. 2FL/1000 sq. ft.
7-10 oz. WSP/1000 sq. ft.
5-7.5 lb. 10%, 10G/1000 sq. ft.
Disease:

Holminthosporium	dollar spot*	pink snow mold
leaf spot*	gray snow mold	snow mold

***Not for use in California**
Time: For leaf spot, apply in spring or fall of each year. To control existing infection, apply high rate; repeat in 3- to 4-weeks if heavy rains or flooding occurs or if disease is severe. For snow mold, apply before first snow fall.
Remarks: Refer to label for use directions. Do not apply through any type of irrigation system. Do not graze treated turf nor feed clippings to livestock. Do not plant root crops within 12 months of application unless PCNB is registered for use on those crops.

Signal Word/Toxicity Class: Danger/I (2-E); Caution (2FL, 10%, 10G, WSP).
REI: 12 hr.

Terraclor (75WP) *(Uniroyal)*
Turfcide (10G) or Turfcide 400 (FL) *(Uniroyal)*
Rate: 7-10 oz. 75WP/1000 sq. ft.
5-10 lb. 10G/1000 sq. ft.
10.5-15 fl. oz. FL/1000 sq. ft.
Disease:

dollar spot	leaf spot	pink snow mold
gray snow mold	melting-out	

Rate: 16 oz. 75WP/1000 sq. ft.
2-7.5 lb. 10G/1000 sq. ft.
4.5-24 fl. oz. FL/1000 sq. ft.
Disease:
brown patch
Time: Warm season grasses and dichondra: Apply single application in spring or fall or at first sign of disease; retreat 3-4 weeks later if necessary. Cool season grasses: Apply at first sign of disease; repeat at 7- to 10-day intervals during warm, damp weather. For snow mold, apply single application prior to first snowfall or when temperatures remain below 60°F and extended wet conditions are expected.

For personal protective equipment and EPA registration numbers, please turn to page 39.

Remarks: **Warm season grasses (Bermuda, St. Augustine), dichondra, cool season grasses (bent, bluegrass, fescue, ryegrass).** Areas treated should be watered following application. If treated area is subjected to flooding or unusually heavy rainfall, or if disease is severe or reappears, treat area again. Do not graze treated areas or feed clippings to livestock.

Crop Rotation: Do not plant root crops in PCNB treated fields within 12 months of broadcast and banding applications unless PCNB registered for use on those crops.

State Restrictions: Not for use in California on dollar spot or leaf spot.

- - - - - - - - - - - - - - - -

Signal Word/Toxicity Class: Caution.

Penstar — see PCNB

PROPICONAZOLE

Triazole sterol inhibitor. Toxic to fish.

Banner MAXX *(Novartis)*

Rate: 22-88 fl. oz. MAXX/A

Disease:

dollar spot

Time: Apply when conditions are favorable for disease development. Use low rate at 7- to 14-day intervals, high rate at 14- to 28-day intervals.

Rate: 44-88 fl. oz. MAXX/A

Disease:

Anthracnose	rust	stripe smut
powdery mildew		

Time: Apply when conditions are favorable for disease development; repeat at 14- to 28-day intervals. For smut, make 1 application late fall or early spring.

Rate: 88 fl. oz. MAXX/A

Disease:

pink patch	red thread	rust

Time: Apply when conditions are favorable for disease development; repeat at 14- to 21-day intervals.

Rate: 44-176 fl. oz. MAXX/A

Disease:

brown patch	leaf spot	melting-out

Time: For brown patch, apply May-June, before disease is present; repeat at 10- to 21-day intervals for high rate; 14- to 21-day intervals for low rate. For leaf spot and melting-out, apply at 14-day intervals.

Rate: 88-176 fl. oz. MAXX/A

Disease:

Fusarium patch	pink snow mold	summer patch
gray snow mold	Poa patch	take-all patch

Time: For Fusarium patch, apply when conditions are favorable for disease development, fall or early spring. For snow mold, apply late fall before snow cover. For Poa and summer patch, apply beginning in April at 14- to 28-day intervals. For take-all patch, make 1-2 fall and spring applications.

Rate: 130-176 fl. oz. MAXX/A

Disease:

yellow patch	Zoysia patch

Time: For yellow patch, apply once late fall before snow cover. For Zoysia patch, make 1 application early fall prior to disease development.

Rate: 176 fl. oz. MAXX/A

Disease:

necrotic ring spot	spring dead spot

Time: For necrotic ring spot, apply fall and/or early spring. For spring dead spot, make 1-3 applications.

Remarks: Do not apply this product through any type of irrigation system.

- - - - - - - - - - - - - - - -

Signal Word/Toxicity Class: Warning/II.
REI: 24 hr.

ProStar — see flutolanil

PROSTAR PLUS (Co-Pack)

Systemic fungicide in water-soluble packaging. ProStar is toxic to aquatic invertebrates.

ProStar Plus: ProStar 50WP (50% flutolanil) *(AgrEvo USA)*
& Bayleton 50WP (50% triadimefon) *(Bayer)*

Rate: Preventive:
1 WSP/12,000 sq. ft.
Curative:
1 WSP/8000 sq. ft.

Disease:

brown patch	pink patch	rust
copper spot	powdery mildew	southern blight
dollar spot	red thread	yellow patch

Time: Preventive: Apply prior to initial signs of disease; repeat if conditions favor disease development. Curative: Repeat after 30 days if conditions favor disease development.

Rate: Preventive:
1 WSP/4000 sq. ft.

Disease:

gray snow mold	pink snow mold

Time: Apply in the fall prior to permanent snow cover; second application may be made as needed after a 60 day interval if turf breaks dormancy during winter months.

Remarks: For snow mold, do not apply over snow cover, or when turf is dormant. Do not irrigate until spray has dried; avoid mowing for 24 hr.

- - - - - - - - - - - - - - - -

Signal Word/Toxicity Class: Caution.
REI: 12 hr.

Protect — see mancozeb

ProTurf — see iprodione premix

ProTurf II — see metalaxyl premix

ProTurf V — see chloroneb

ProTurf VII — see triadimefon

ProTurf IX — see chloroneb premix

ProTurf X — see iprodione

ProTurf Pythium — see metalaxyl

ProTurf Systemic — see thiophanate-methyl

Rubigan — see fenarimol

SENTINEL

Broad-spectrum systemic fungicide for prevention and control of specific diseases on turf.

Sentinel 40WG (cyproconazole) *(Novartis)*

Rate: **Fairways, roughs, tees, and sod farms:**
0.25-0.33 oz. 40WG/1000 sq. ft.
Green height turf and closely maintained trees:
0.16-0.25 oz. 40WG/1000 sq. ft.

Disease:

Anthracnose	dollar spot	rust
brown patch	powdery mildew	southern blight
copper spot	red thread	

Time: Repeat at 21- to 28-day intervals when conditions favor disease development.

◆ - **new product** • ai-active ingredient • DF-dry flowable • E/EC-emulsifiable concentrate • EW-emulsion in water
F/FL-flowable • G-granule • L/LC-liquid • SP/WSP-soluble packet • W/WP-wettable powder • WSB-water soluble bag

Rate: **Fairways, roughs, tees, and sod farms:**
0.25-0.33 oz. 40WG/1000 sq. ft.
Green height turf and closely maintained trees:
0.25 oz. 40WG/1000 sq. ft.

Disease:

necrotic ring spot (fairways, etc.)	summer patch

Time: Begin applications in early spring when soil temperatures reach 60-65°F. Apply at 28-day intervals. Do not apply 0.25 oz. rate when greens are stressed due to temperature or PGR effects.

Rate: **Fairways, roughs, tees, and sod farms, Green height turf and closely maintained trees:**
0.33 oz. 40WG/1000 sq. ft.

Disease:

gray leaf spot (suppression)	necrotic ring spot (green height turf)	pink snow mold
gray snow mold		stripe smut

Time: For gray leaf spot preventive application. Repeat at 21- to 28-day intervals when conditions favor disease development. For snow mold, apply in late fall before snow cover; repeat early spring. For necrotic ring spot, apply in early spring; repeat at 28-day intervals for 2-3 applications. For stripe smut, apply once late fall or early spring.

Remarks: Most effective with preventive application. Length of control depends on weather conditions, disease pressure, and turf management practices. Apply in sufficient water for thorough coverage.

- - - - - - - - - - - - - - - - -

Signal Word/Toxicity Class: Caution.
REI: 12 hr.

Spotrete — see thiram

Subdue — see mefenoxam; metalaxyl

SULFUR

Nonsystemic fungicide with protective action.

Micro Flo Sulfur (6L) *(Micro Flo)*
Terra Sulfur (6L) *(Terra)*

Rate: 8-44 lb. ai/A
1.33-7.33 gal. 6L/A

Disease:

Fusarium patch	take-all patch

Time: Apply monthly Sept.-May.

Remarks: Use low rate when temperatures are above 80°F; high rate when temperatures are below 80°F. Refer to label for use directions.

- - - - - - - - - - - - - - - - -

Signal Word/Toxicity Class: Caution.
REI: 24 hr.

Teremec — see chloroneb

Terraclor — see PCNB

Terraneb — see chloroneb

Terrazole — see etridiazole

Thalonil — see chlorothalonil

THIOPHANATE-METHYL

Systemic fungicide with protective and curative action.

3336 (F, WP, or G) *(Cleary Chemical)*

Rate: 1-2 fl. oz. F/1000 sq. ft.
2 oz. WP/1000 sq. ft.
1.5-6 lb. G/1000 sq. ft.

Disease:

copper spot	large brown patch	pink snow mold
dollar spot	leaf spot (F)	red thread
Fusarium patch		

Time: Apply when disease first appears; repeat at 7- to 14-day intervals as needed. For pink snow mold, apply middle-late Nov. or just before grass has gone dormant.

Rate: 4-8 fl. oz. F/1000 sq. ft.
2 oz. WP/1000 sq. ft.
3-9 lb. G/1000 sq. ft.

Disease:
Anthracnose

Time: Apply when disease first appears; repeat at 7- to 14-day intervals as needed. For preventive application for granular, apply lower rate when soil temperatures reach 60°F using 2 applications at 14-day intervals.

Rate: 5-10 fl. oz. F/1000 sq. ft.
4-8 oz. WP/1000 sq. ft.
6-12 lb. G/1000 sq. ft.

Disease:

Fusarium blight	necrotic ring spot	summer patch
leaf spot (F, WP)	stripe smut	

Time: For blight and ring spot, apply 2 applications F and WP at 10- to 14-day intervals beginning when disease first appears. For ring spot, apply G at 14-day intervals as needed beginning when soil temperatures reach 55-60°F. For stripe smut, apply at 7- to 14-day intervals (14- to 21-days for F) when disease first appears; for prevention, apply in spring and fall. For summer patch, apply 3 applications at 21-day intervals as needed; for suppression, apply using 10- to 14-day intervals beginning when disease first appears.

Remarks: Apply after mowing or avoid mowing 12 hr. after application. Apply in sufficient water to obtain thorough coverage. Use high rate under conditions of severe disease. Refer to label for use directions.

- - - - - - - - - - - - - - - - -

Signal Word/Toxicity Class: Caution.
REI: 12 hr.

Fungo (50WSB) *(Scotts)*

Rate: 1-2 oz. 50WSB/1000 sq. ft.

Disease:

Anthracnose	dollar spot

Time: Apply when disease first appears; repeat at 7- to 14-day intervals as needed.

Rate: 2 oz. 50WSB/1000 sq. ft.

Disease:

brown patch	Fusarium patch	red thread
copper spot		

Time: Apply when disease first appears; repeat at 10- to 14-day intervals (5- to 7-day intervals in severe disease conditions).

Rate: 4-8 oz. 50WSB/1000 sq. ft.

Disease:

Fusarium blight	stripe smut	summer patch
necrotic ring spot		

Time: Apply 3 applications beginning early May; repeat at 21-day intervals. For smut, make 2 applications at 7- to 14-day intervals in fall or early spring before grass growth begins.

Remarks: For use on both warm and cool season grasses. Under severe conditions, apply every 5-7 days. Apply after mowing or avoid mowing for 12 hr. after application. Do not graze or feed clippings to livestock or poultry.

- - - - - - - - - - - - - - - - -

Signal Word/Toxicity Class: Warning/II.
REI: 12 hr.

ProTurf Systemic Fungicide *(Scotts)*

Rate: 29.75 lb./22,000 sq. ft. (1/2 acre)

Disease:

brown patch	dollar spot	pink snow mold
copper spot		

Time: Apply to moist or dry foliage; repeat at 2- to 3-week intervals.

Remarks: For suppression of snow mold in areas where snow cover occurs, apply to dry foliage at double rate prior to first snow; repeat midwinter when turf is free of snow, and again immediately after final snow melts in spring. Refer to label for use directions.

State Restrictions: For use on pink snow mold only in California, Oregon, and Washington.

- - - - - - - - - - - - - - - - -

Signal Word/Toxicity Class: Caution.

NOTICE	
The information on these pages is for preliminary planning — not a guide for use. Be sure to follow the manufacturer's directions, notwithstanding information contained here.	For personal protective equipment and EPA registration numbers, please turn to page 39.

THIRAM

Broad-spectrum fungicide. Toxic to fish. Thiram 50WP contains the toxic inert Rhodamine B and is also toxic to birds and wildlife.

Spotrete F or 75WDG *(Cleary Chemical)*
Thiram Granuflo (WDG) *(UCB Agrochemicals)*

Rate: Preventive:
3.75 fl. oz. F/1000 sq. ft.
2.5 oz. 75WDG/1000 sq. ft.
2.5-3 oz. WDG/1000 sq. ft.
Curative:
7.5 fl. oz. F/1000 sq. ft.
5 oz. 75WDG/1000 sq. ft.
4.5-5 oz. WDG/1000 sq. ft.
Disease:
brown patch, dollar spot
Time: Preventive: Apply early spring before disease appears; repeat at 7- to 10-day intervals. Curative: Apply at first sign of disease, then follow preventive schedule.

Rate: 12 fl. oz. F/1000 sq. ft.
8 oz. 75WDG/1000 sq. ft.
Disease:
snow mold
Time: Apply in fall prior to first snow fall; repeat in early spring. For 75WDG, spring application of 3 oz./1000 sq. ft. during spring thaw may be necessary.

Rate: Preventive:
8 oz. WDG/1000 sq. ft.
Curative:
3 oz. WDG/1000 sq. ft.
Disease:
Fusarium patch
Time: Apply late fall before expected permanent snow cover; follow with spring application at same rate when snow melts and runoff has ended.
Remarks: Do not apply Thiram Granuflo through any type of irrigation system. Do not graze or feed treated turf. Refer to label for use directions.

- - - - - - - - - - - - - - - - -

Signal Word/Toxicity Class: Caution.
REI: 24 hr. (Thiram Granuflo).

42-S Thiram or Thiram 50WP Dyed *(Gustafson)*
Rate: 8 fl. oz. 42-S/100 lb. seed
8 oz. 50WP/100 lb. seed
Disease:
damping-off, seed decay, seedling blight
Time: Seed treatment.
Remarks: Grasses. Use with EPA-approved dye that imparts unnatural color to seed. Do not use treated seed for feed, food, or oil purposes.

- - - - - - - - - - - - - - - - -

Signal Word/Toxicity Class: Caution.

TRIADIMEFON

Triazole sterol inhibitor with protective and curative action.

Bayleton 25 or 50 T&O *(Bayer)*
Rate: Preventive:
0.5-1 oz. 25/1000 sq. ft.
0.25-0.5 oz. 50/1000 sq. ft.
Curative:
2 oz. 25/1000 sq. ft.
1 oz. 50/1000 sq. ft.
Disease:
brown patch*, copper spot, dollar spot, powdery mildew, red thread, Rhizoctonia blight*, rust
****Suppression***
Time: Preventive: Apply low rate at 15- to 30-day intervals (30- to 60-day for dollar spot) when conditions favor disease. Curative: Apply high rate for first application; repeat with preventive rate and schedule.

Rate: 1-4 oz. 25/1000 sq. ft.
0.5-2 oz. 50/1000 sq. ft.
Disease:
southern blight
Time: Preventive: Apply at 14-day intervals for initial 2-3 treatments; apply subsequent applications at 14- to 28-day intervals. Curative: Apply high rate for initial 2-3 treatments; repeat with preventive rate and schedule.

Rate: 2 oz. 25/1000 sq. ft.
1 oz. 50/1000 sq. ft.
Disease:
Anthracnose, stripe smut
Time: Anthracnose: Apply at 30- to 45-day intervals. Stripe smut-Preventive: Apply once in spring as grass growth begins, second application just prior to summer and third application late summer or early fall.

Rate: 2-4 oz. 25/1000 sq. ft.
1-2 oz. 50/1000 s. ft.
Disease:
Fusarium blight, summer patch, Zoysia patch
Time: For Fusarium blight and summer patch, apply in the spring 30- to 60-days before symptoms appear; repeat at 30-day intervals. For zoysia patch, apply mid-Sept. to mid-Oct. prior to disease development; second application in early spring if disease is heavy.

Rate: Preventive:
2 oz. 25/1000 sq. ft.
1-2 oz. 50/1000 sq. ft.
Curative:
4 oz. 25/1000 sq. ft.
2 oz. 50/1000 sq. ft.
Disease:
take-all patch
Time: Preventive: Apply 2-4 weeks prior to disease appearance; repeat at 21- to 28-day intervals. Curative: Apply at 14- to 21-day intervals for initial 1-2 treatments; repeat 2 oz. at 21- to 28-day intervals.

Rate: 5 oz. 25/1000 sq. ft.
Preventive-gray snow mold:
4 oz. 25/1000 sq. ft.
2 oz. 50/1000 sq. ft.
Preventive-pink snow mold:
2-4 oz. 25/1000 sq. ft.
1-2 oz. 50/1000 sq. ft.
Disease:
gray snow mold, pink snow mold
Time: Apply single application in late fall before snow cover. Do not apply over snow cover. Use high rate in areas with history of severe snow mold damage. For gray snow mold preventive, apply in fall, 30 days before turf dormancy. For pink snow mold preventive, apply before conditions favorable for infection occur; repeat at 60- to 90-day intervals.
Remarks: Make all applications after mowing and allow foliage to dry thoroughly before irrigation. Do not use clippings for animal feed.
State Restrictions: California labels may vary; refer to label.

- - - - - - - - - - - - - - - - -

Signal Word/Toxicity Class: Caution.

Bayleton 50% DF *(Bayer)*
Rate: 2-4 oz. 50% DF/A
Disease:
powdery mildew
Time: Apply 1-2 applications at first signs of disease and when new symptoms reappear.

Rate: 4-8 oz. 50% DF/A
Disease:
rust
Time: Apply as soon as weather conditions are favorable for development, before rust pustules are present; repeat at 14- to 21-day intervals.
Remarks: Bentgrass, bermudagrass, bluegrass, crested wheatgrass, fine fescue, perennial ryegrass, orchardgrass, tall fescue. On highly susceptible grass varieties and/or prolonged conditions favorable for disease, use shorter interval between applications. Control is especially important from head emergence through last flowering. Do not apply through any type of irrigation system. Do not graze livestock in treated orchards.
State Restrictions: For use in Idaho, Oregon, and Washington.

- - - - - - - - - - - - - - - - -

Signal Word/Toxicity Class: Caution.
REI: 12 hr.

ProTurf Fungicide VII *(Scotts)*
Rate: 29.18 lb./22,000 sq. ft. (1/2 acre)
Double rate:
29.18 lb./11,000 sq. ft. (1/4 acre)
Heavy rate:
29.18 lb./5500 sq. ft. (1/8 acre)
Disease:
Anthracnose, brown patch, dollar spot, rust
Time: Apply to moist or dry foliage. For brown patch, apply at heavy rate every 2 weeks; when disease is under control, apply double rate. Repeat at 2-week intervals.
Remarks: Refer to label for use directions.

- - - - - - - - - - - - - - - - -

Signal Word/Toxicity Class: Caution.

◆ - **new product** • ai-active ingredient • DF-dry flowable • E/EC-emulsifiable concentrate • EW-emulsion in water
F/FL-flowable • G-granule • L/LC-liquid • SP/WSP-soluble packet • W/WP-wettable powder • WSB-water soluble bag

Turfcide — see PCNB

VINCLOZOLIN

Dicarboximide fungicide. Toxic to fish.

Curalan (FL, DF, EG) *(BASF)*
Vorlan (DF) *(Scotts)*

Rate: Preventive:
1-2 oz. DF, EG, FL/1000 sq. ft.
Curative:
2-4 oz. DF, EG, FL/1000 sq. ft.

Disease:

brown patch	gray snow mold	pink patch
dollar spot	leaf spot	pink snow mold
Fusarium patch	melting-out	red thread

Time: Rate and intervals vary; refer to label.

Remarks: Apply curative rate to control existing infections, subsequent treatments at preventive rate once control has been achieved. Do not apply this product through any type of irrigation system. Do not apply to areas likely to be grazed by livestock and do not feed clippings to livestock or poultry.

- - - - - - - - - - - - - - - - - -

Signal Word/Toxicity Class: Caution.

REI: 12 hr.

Vorlan — see vinclozolin

Ornamentals

(Christmas Trees, Flowers, Forest Plantings, Nurseries, Shelterbelts)

3336 — see thiophanate-methyl

Alamo — see propiconazole

ALIETTE

Systemic fungicide. Toxic to aquatic and estuarine invertebrates.

Chipco Aliette (WDG) (fosetyl-aluminum) *(Rhone-Poulenc)*

Rate: **Bedding plants:**
Foliar spray:
1-3.2 lb. ai/100 gal.
1.25-4 lb. WDG/100 gal.
Conifers:
Dip:
2 lb. ai/100 gal.
2.5 lb. WDG/100 gal.
Foliar spray:
2-4 lb. ai/100 gal.
2.5-5 lb. WDG/100 gal.
Ornamentals:
Foliar spray:
2-4 lb. ai/100 gal.
2.5-5 lb. WDG/100 gal.
Drench:
0.32-0.64 lb. ai/100 gal.
0.4-0.8 lb. WDG/100 gal.
Soil incorporation:
0.4-0.64 lb. ai/cu. yd. of soil
0.5-0.8 lb. WDG/cu. yd. of soil

Disease:

downy mildew (roses)	fire blight (suppression)	Phytophthora
		Pythium

Time: For foliar and drench applications, apply prior to disease development as needed but not to exceed once every 30 days (every 7 days for roses). For soil incorporation, apply immediately before potting. For dip treatments, dip immediately before transplanting by thoroughly wetting plant and root mass. For fire blight, apply at prebloom stage; repeat at 7-day intervals until bloom period ends.

Remarks: Bedding plants, conifers, ornamentals. Refer to label for use directions.

State Restrictions: Not registered for use in California on Phytophthora root rot.

- - - - - - - - - - - - - - - - -

Rate: 1.6-3.2 lb. ai/100 gal.
2-4 lb. WDG/100 gal.

Disease:
bacterial blight (suppression)

Time: Apply prior to disease development as needed but not to exceed once every 7 days.

Remarks: Anthurium, Dieffenbachia, English ivy, Ficus, Schefflera, Spathaphyllum, Syngonium. Refer to label for use directions.

State Restrictions: Not registered for use in California on bacterial blight.

- - - - - - - - - - - - - - - - -

Signal Word/Toxicity Class: Caution.

REI: 12 hr.

Arbotect — see thiabendazole

Banner — see propiconazole

BANOL

Water miscible fungicide concentrate having preventive and curative properties.

Banol (propamocarb hydrochloride) *(AgrEvo USA)*

Rate: Seeding, transplanting (seedlings):
3 fl. oz./in 10 gal. water
Potting, transplanting (cuttings):
2 fl. oz./in 10 gal. water
Woody ornamentals:
2.5 fl. oz./in 10 gal. water

Disease:

Phytophthora	Pythium (woody ornamentals)

Time: Ornamentals: Seeding; transplanting (seedlings, cuttings); potting. Woody ornamentals: Drench treatment-repeat at 3-4 month intervals.

Remarks: Ornamentals, woody ornamentals. Dip bare roots in solution before potting. Do not apply this product through any type of irrigation system. Do not graze or feed clippings from treated areas to livestock.

- - - - - - - - - - - - - - - - -

Signal Word/Toxicity Class: Caution.

REI: 12 hr.

BASAMID

Granular soil fumigant. Toxic to fish.

Basamid Granular (dazomet) *(BASF)*

Rate: 6-8 lb./1000 sq. ft.

Disease:
soilborne diseases

Time: Apply in fall for early spring planting.

Remarks: Ornamental production fields, landscape beds, conifer seedbeds, flower bulbs. After incorporation, immediately drench with 15-20 gal. water/100 sq. ft.; plastic tarp recommended. For flower bulbs, till to desired depth and follow suggested cultural practices. Do not apply within 3-4 ft. of growing plants or closer than drip line of trees and large shrubs. Do not use if soil temperature is below 43°F.

- - - - - - - - - - - - - - - - -

Signal Word/Toxicity Class: Warning/II.

REI: 24 hr.

Bayleton — see triadimefon

NOTICE The information on these pages is for preliminary planning — not a guide for use. Be sure to follow the manufacturer's directions, notwithstanding information contained here.	For personal protective equipment and EPA registration numbers, please turn to page 39.

BENLATE

Benzimidazole systemic fungicide. Toxic to fish.

Benlate (SP, WP) (benomyl) *(DuPont)*

Rate: 0.5 lb. ai/A
1 lb. SP, WP/A

Disease:
Rhabdocline needle cast; Swiss needle cast; tip blight

Time: Rhabdocline, Swiss needle cast: Apply at bud break; repeat 10- to 14-days later just before needles emerge from sheath and 10- to 14-days after needle emergence. Tip blight: Apply in early May; repeat at 4-week intervals.

Remarks: Conifer pines (Austrian, Red, Scots), Douglas fir. Add spreader/sticker to improve coverage. Do not tank mix or alternate Benlate with benzimidazole or thiophanate products. Do not graze livestock in treated areas.

State Restrictions: Benlate SP is not registered for use in California.

- - - - - - - - - - - - - - - -

Rate: 1 oz. SP, WP/9.5 oz. dry Kaolinite clay for seedling roots

Disease:
brown needle blight

Rate: 2 oz. SP, WP/50 oz. Kaolinite clay + water to make a slurry

Disease:
Fusarium root rot; Rhizoctonia root rot

Time: Brown needle blight: Wet seedling roots in clean water, apply mixture to wet roots. Fusarium and Rhizoctonia root rot: Thoroughly cover seedling roots with slurry.

Remarks: Conifer seedlings (loblolly, longleaf, slash). Do not apply mixture to seedling foliage. Avoid excessive drying of roots or exposure to temperatures greater than 90°F or less than 32°F. Do not tank mix or alternate with benzimidazole or thiophanate products.

State Restrictions: Benlate SP not registered for use in California.

- - - - - - - - - - - - - - - -

Signal Word/Toxicity Class: Caution.
REI: 24 hr.

Blue Viking — see copper compounds

Botran — see dicloran

Bravo — see chlorothalonil

CALCIUM POLYSULFIDE

Balanced fungicide, insecticide, miticide composition containing special effective wetting agents.

Sulforix *(Best Sulfur)*

Rate: 1.5-3 pt./100 gal. water

Disease:
black leaf spot; powdery mildew; rust

Time: Apply lower rate on tender new growth; higher rate if mildew becomes severe.

Remarks: Roses (nursery stock and field grown). Thorough coverage is necessary. Pick open flowers before spraying to avoid discoloration. Do not apply this product through any type of irrigation system.

- - - - - - - - - - - - - - - -

Signal Word/Toxicity Class: Danger/I.
REI: 48 hr.

CAPTAN

Protective and curative fungicide controlling a wide range of fungal diseases.

Drexel Captan 4FL *(Drexel)*

Rate: 2.25 fl. oz. 4FL/1 gal. water

Disease:
bulb rot

Time: Liquid or slurry seed treatment.

Remarks: Gladiola bulbs. Soak bulbs for 20-30 minutes. Do not use treated seed for food, feed, or oil purposes. Refer to label for use directions.

- - - - - - - - - - - - - - - -

Signal Word/Toxicity Class: Danger/I.

Drexel Captan (50W) *(Drexel)*
Micro Flo Captan (50WP, 80WP) or Captec (4L) *(Micro Flo)*

Rate: Azaleas:
2-4 lb. 50W, 50WP/100 gal. water
2.5 lb. 80WP/100 gal. water
1-2 qt. 4L/100 gal. water

Disease:
damping-off; flower blight

Time: Damping-off: Dip cuttings before planting. Flower blight: Apply to soil area around plants, spray flowers just before bloom; repeat at 7- to 14-day intervals through bloom.

Rate: Begonias (tuberous):
4 lb. 50W, 50WP/100 gal. water
2.5 lb. 80WP/100 gal. water
2 qt. 4L/100 gal. water

Disease:
damping-off; tuber rot

Time: Dip tubers for 30 minutes, drain and plant.

Rate: Camellias:
1 lb. 50W, 50WP/100 gal. water
0.63 lb. 80WP/100 gal. water
0.5 qt. 4L/100 gal. water

Disease:
flower blight

Time: Drench soil around plants when flowers start to open; repeat at 7- to 10-day intervals through bloom.

Rate: Carnations:
2-4 lb. 50W, 50WP/100 gal. water
1.25-2.5 lb. 80WP/100 gal. water
1 qt. 4L/100 gal. water

Disease:
Alternaria leaf spot; damping-off; rust

Time: Apply at first sign of disease; repeat at 7- to 10-day intervals (shorter intervals during frequent rains and heavy dews). For damping-off, dip cuttings before bedding.

Rate: Chrysanthemum:
1-4 lb. 50W/100 gal. water
2-4 lb. 50WP/100 gal. water
1.5-2.5 lb. 80WP/100 gal. water
1-2 qt. 4L/100 gal. water

Disease:
Botrytis flower blight; damping-off; Septoria leaf spot

Time: For flower blight and leaf spot, apply at first sign of disease; repeat at 7- to 10-day intervals. For damping-off, dip cuttings before bedding.

Rate: Gladiolus (corms):
0.5-1.5 lb. 50W, 50WP/10 gal. water
0.33-1 lb. 80WP/10 gal. water
0.25-0.75 qt. 4L/100 gal. water

Disease:
corm rot; damping-off; decay

Time: Dip corms 20-30 minutes, drain and plant.

Rate: Roses:
2 lb. 50W, 50WP/100 gal. water
1.25 lb. 80WP/100 gal. water
1 qt. 4L/100 gal. water

Disease:
black leaf spot; Botrytis blossom blight

Time: Apply at first growth or first sign of disease; repeat at 7- to 14-day intervals, more frequently during frequent rains and heavy dews.

Remarks: Azaleas, begonia (tuberous), camellias, carnations, chrysanthemums, gladiolus (corms), roses. Do not apply through any type of irrigation system. Refer to label for use directions.

- - - - - - - - - - - - - - - -

Rate: Bench treatment:
2 lb. 50W/100 gal. water

Disease:
damping-off; root rot

Time: Preplant treatment for seedlings or transplants. Cultivate into upper 3-4" of soil before planting.

Remarks: Roses, flowers, shrubs, trees. Do not apply through any type of irrigation system. Refer to label for use directions.

- - - - - - - - - - - - - - - -

Signal Word/Toxicity Class: Danger/I.
REI: 96 hr.

◆ - **new product** • ai-active ingredient • DF-dry flowable • E/EC-emulsifiable concentrate • EW-emulsion in water
F/FL-flowable • G-granule • L/LC-liquid • SP/WSP-soluble packet • W/WP-wettable powder • WSB-water soluble bag

Champ — see copper compounds

Champion — see copper compounds

Chipco — see iprodione

CHLORONEB

Systemic, foliar applied fungicide for broad-spectrum disease control in turf.

Terraneb SP (FL or WP) *(Kincaid)*

Rate: 2-3.8 lb. ai/40-50 gal. water
84.5-169 fl. oz. FL/40-50 gal. water
3-6 lb. WP/40-50 gal. water

Disease: Azaleas, roses, etc.
Cylindrocladium cutting rot
Ferns:
Rhizoctonia blight

Time: Use lower rate for maintenance, repeating in 1 month. During periods favoring development of fungi, repeat at 2-week intervals. Use higher rate for areas where severe disease outbreaks occur.

Remarks: Azaleas, ferns (including leatherleaf), miniature roses, and woody ornamentals. Refer to label for use directions. Do not graze or feed clippings from treated areas to livestock.

- - - - - - - - - - - - - - - - - -

Rate: 1 lb. ai/200 gal. water
43 fl. oz. FL/200 gal. water
1.5 lb. WP/200 gal. water

Disease:
Rhizoctonia solani

Time: Apply as a soil drench and repeat every 10-20 days as necessary.

Remarks: Container-grown bedding plants such as aster, carnation, impatiens, and vinca. Refer to label for use directions. Do not graze or feed clippings from treated areas to livestock.

- - - - - - - - - - - - - - - - - -

Signal Word/Toxicity Class: Caution.
REI: 12 hr.

CHLOROTHALONIL

Broad-spectrum contact fungicide. Toxic to fish, aquatic invertebrates and marine/estuarine organisms.

Daconil 2787 (FL), Weather Stik (6FL), Ultrex (WDG) or Zn *(ZENECA)*
Echo 90DF *(Sostram)*
Thalonil (4L, 90DF or 90DF WSP) *(Terra)*

Rate: 2 pt. FL, 4L, Zn/100 gal. water
1.375 pt. 6FL/100 gal. water
1.4 lb. WDG/100 gal. water
1.25 lb. 90DF, 90DF WSP/100 gal. water

Disease:

Alternaria	Fabraea leaf spot	Phytophthora
Anthracnose	Fusarium	powdery mildew
Ascochyta blight	Helminthosporium	Rhizoctonia
Bipolaris leaf spot	ink spot	rust
black leaf spot	leaf blotch	scab
Botrytis	leaf spot	Septoria leaf spot
cedar-apple rust	Marssonina leaf spot	Sphaeropsis leaf spot
Cercosporidium	Monilinia	Stagonspora leaf scorch
Cylindrocladium	Mycosphaerella ray blight	tan leaf spot
Cylindrosporium	Ovulinia flower blight	Taphrina blister
Dactylaria leaf spot		Volutella leaf blight
Didymellina		

Time: Apply as directed on label; repeat at various intervals until conditions are no longer favorable for disease.

Remarks: Bulbs and flowering plants, foliage plants, shrubs and trees. Apply in sufficient water for complete coverage. During periods when conditions favor severe disease incidence, generally cloudy or wet weather, use high rate and shortest interval between applications. Discoloration of blooms has been noted on certain varieties when applications are made during flowering. Refer to label for use directions, precautions, and restrictions.

- - - - - - - - - - - - - - - - - -

Signal Word/Toxicity Class: Varies by formulation.
REI: 48 hr.

Bravo 500, 720/Weather Stik, Ultrex (WDG) *(ZENECA)*
Daconil 2787 (FL), Weather Stik (6FL), Ultrex (WDG) or Zn *(ZENECA)*
Echo 500, 720 or 90DF *(Sostram)*
Terranil (6L, 90DF or 90DF WSP) *(Terra)*
Thalonil (4L, 90DF or 90DF WSP) *(Terra)*

Rate: 2-8 pt. 500, FL, Zn, 4L/A
1.5-5.5 pt. 720, 6L/A
1.25-5 lb. WDG/A
1.375-5.5 pt. 6FL/A
1.125-4.5 lb. 90DF/A (Echo)
1.25 lb. 90DF/100 gal. water
1 packet 90DF WSP/100 gal. water

Disease:

Botrytis seedling blight	Phoma twig blight	Scirrhia brown spot
Cyclaneusma needle cast	Rhabdocline needle cast	Scleroderris canker
Lophodermium needle cast	Rhizosphaera needle cast	Sirococcus tip blight
		Swiss needle cast

Time: Apply as directed on label; repeat at various intervals until conditions are no longer favorable for disease.

Remarks: Conifers. During periods when conditions favor severe disease incidence, generally cloudy or wet weather, use high rate and shortest interval between applications. Aerial application to conifers is permitted although ground applications give better coverage. If application with ground equipment is not feasible, apply aerially to forest stands in 10-20 gal. water and to Christmas trees in 10-50 gal. water. Refer to label for use directions, precautions and restrictions.

- - - - - - - - - - - - - - - - - -

Signal Word/Toxicity Class: Varies by formulation.
REI: 48 hr.

Clean Crop C-O-C-S — see copper compounds

Clean Crop Nemasol — see metam-sodium

COPPER COMPOUNDS

Fungicide/bactericide.

Blue Viking or Tennessee Brand (copper sulfate pentahydrate) *(Griffin)*

Rate: 10-10-100 Bordeaux spray

Disease:
Botrytis blight

Time: Apply thorough coverage at first sign of disease; repeat at 3- to 10-day intervals as needed.

Remarks: Easter lily, gladiolus, tulip. Use shorter interval during periods of frequent rains or when severe disease conditions persist. Do not apply through any type of irrigation system. Refer to label for use directions.

- - - - - - - - - - - - - - - - - -

Signal Word/Toxicity Class: Danger/I.
REI: 24 hr.

Champ (FL or Formula 2) or Champion (WP) (copper hydroxide) *(Agtrol)*
KOP-Hydroxide (FL, 50WP) (copper hydroxide) *(Drexel)*

Rate: 1.33 pt. FL/100 gal. water
0.66 pt. Formula 2/100 gal. water
1 lb. WP, 50WP/100 gal. water

Disease:

Alternaria blight	fire blight	scab
Anthracnose	Phomopsis stem blight	Septoria leaf spot
black leaf spot	Phytophthora dieback	Volutella leaf blight
Botrytis blight	powdery mildew	Xanthomonas leaf spot
Cercospora leaf spot	Pseudomonas leaf spot	

Time: Apply as full cover spray at first sign of disease; repeat at 7- to 14-day intervals (or shorter) depending on rainfall and disease severity.

Rate: Philodendron:
2 pt. FL/100 gal. water
1 pt. Formula 2/100 gal. water
1.5 lb. WP/100 gal. water

Disease:
bacterial leaf spot

Time: Apply weekly before disease appears.

Rate: Sycamore:
2.66-4 pt. FL/100 gal. water
1.33-2 pt. Formula 2/100 gal. water
2-3 lb. WP, 50WP/100 gal. water

NOTICE	The information on these pages is for preliminary planning — not a guide for use. Be sure to follow the manufacturer's directions, notwithstanding information contained here.	For personal protective equipment and EPA registration numbers, please turn to page 39.

Disease:
Anthracnose

Time: Make 2 applications as full cover spray. First application at bud crack; second 7-14 days later at 10% leaf expansion.

Remarks: Aralia, azalea, begonia, bulbs (Easter lily, tulip, gladiolus), carnation, chrysanthemum, cotoneaster, euonymus, ivy, pachysandra, periwinkle, pyracantha, rose, yucca (Adam's needle), philodendron, sycamore. Discoloration of foliage and/or blooms have been noted on some varieties. To prevent residues, do not spray ornamentals just before selling season. KOP-Hydroxide 50WP should not be applied in spray solution having pH less than 6.5 or greater than 9.0 as phytotoxicity may occur. Refer to label for use directions.

- - - - - - - - - - - - - - - - - -

Signal Word/Toxicity Class: Danger/I (Champion, KOP); Caution (Champ).
REI: 48 hr. (KOP); 24 hr. (Champ, Champion).

Clean Crop C-O-C-S (WDG) (74.8% copper oxychloride & 14.2% copper sulfate, basic) *(Platte)*

Rate: Flowering plants:
1.25 lb. ai/100 gal. water
2.5 lb. WDG/100 gal. water

Disease:

Anthracnose (peonies)	Botrytis blight (dahlias, peonies)	rust (roses)
black leaf spot (roses)	leaf spot (dahlias, nasturtiums, roses)	

Time: For black leaf spot on roses, apply when leaf buds swell in early spring. Apply before disease first appears; repeat at 7-10 day intervals and after rains.

Rate: Bushes, trees:
1.5-2 lb. ai/100 gal. water
3-4 lb. WDG/100 gal. water

Disease:

Anthracnose (oak, sycamore)	dieback (vinca)	tar leaf spot (maple)
bud blight (azalea)	leaf spot (pachysandra)	twig blight (azalea, pachysandra)
canker (pachysandra, vinca)	sycamore blight (sycamore)	

Time: Apply when disease first reported in area; repeat at 7-10 day intervals. For sycamore blight, apply when buds begin to swell; repeat at 7-14 days later at early leaf stage.

Remarks: For maple, white oak, and pachysandra, hydrated lime may be added. Do not apply this product through any type of irrigation system. Refer to label for use directions.

State Restrictions: Not for use in California on dahilas, nasturtiums, pachysandra, peonies, roses, vinca, and white oak.

- - - - - - - - - - - - - - - - - -

Signal Word/Toxicity Class: Warning/II.
REI: 24 hr.

Kocide (101, DF, 2000, 4.5LF, LF) (copper hydroxide) *(Griffin)*

Rate: 1 lb. 101, DF, 2000/100 gal. water
0.66 pt. 4.5LF/100 gal. water
1.33 pt. LF/A

Disease:

Alternaria	Cercospora leaf spot	Phytophthora dieback
Anthracnose	downy mildew	powdery mildew
bacterial leaf spot	Entomosporium leaf spot	Pseudomonas leaf spot
bacterial soft rot	fire blight	scab
bacterial spot	gray mold	Septoria leaf spot
black leaf spot	Pestalotia leaf spot	Volutella leaf blight
Botrytis blight	Phomopsis stem blight	Xanthomonas leaf spot
Cercospora leaf blight		

Time: Apply at first sign of disease; repeat at 7- to 14-day intervals as needed. Use shorter intervals during periods of frequent rains or when severe disease conditions persist.

Remarks: Greenhouse, field, and nursery ornamentals. Can be used as maintenance spray, or alone, or in combination with other fungicides such as dithiocarbamates. Discoloration of foliage and/or blooms have been noted on some varieties. To prevent residues on commercial plants, do not spray just before selling season. Do not apply in a spray solution having a pH less than 6.5 as phytotoxicity may occur. Refer to label for use directions.

State Restrictions: California labels may vary; refer to label.

- - - - - - - - - - - - - - - - - -

Rate: Douglas fir:
2 lb. 101, DF/A
1.5 lb. 2000/A
1.33 pt. 4.5LF/A
4 pt. LF/A

Disease:
Rhabdocline needle cast

Time: Apply at bud break; repeat at 3-4 week intervals.

Rate: Sycamore:
2-3 lb. 101, DF/A
1.5-2.25 lb. 2000/A
1.33-2 pt. 4.5LF/A
4-6 pt. LF/A

Disease:
Anthracnose

Time: Apply as full cover spray, making first application at bud crack; second 7- to 10-days later at 10% leaf expansion.

Remarks: Douglas fir, sycamore. Refer to label for use directions.

State Restrictions: Not for use on Douglas fir in California (except 2000).

- - - - - - - - - - - - - - - - - -

Signal Word/Toxicity Class: Danger/I (101, DF); Warning/II (2000); Caution (4.5LF, LF).
REI: 24 hr.

KOP Oxy-85 (WP) (copper oxychloride) *(Drexel)*

Rate: 4 lb. WP/100 gal. water

Disease:

bud blight	twig blight

Time: Apply prebloom and petal fall.

Remarks: Azaleas. Use in sufficient water and thoroughly wet plants. Refer to label for use directions.

- - - - - - - - - - - - - - - - - -

Signal Word/Toxicity Class: Danger/I.
REI: 24 hr.

Microsperse COC (53WP) (copper oxychloride) *(Micro Flo)*

Rate: Ornamental plants, hedges:
4 lb. 53WP/100 gal. water

Disease:

Anthracnose	downy mildew	powdery mildew
Botrytis blight	leaf spots	smut (palms)

Time: Apply before disease appears; repeat at 7- to 10-day intervals as needed and after each rain.

Rate: Pine:
1.5-3 lb. 53WP/A

Disease:
needle blight

Time: Apply as needles are just emerging; repeat at yearly intervals as needed (3 weeks later for heavy disease).

Rate: Poplar:
1.66-2 lb. 53WP/100 gal. water

Disease:
leaf rust

Time: Apply at first sign of rust pustules; repeat every 4 weeks as needed.

Remarks: Ornamental plants, hedges, pines, poplars. Refer to label for use directions.

- - - - - - - - - - - - - - - - - -

Signal Word/Toxicity Class: Danger/I.
REI: 24 hr.

Nu-Cop (3L, 50DF, 50WP) (copper hydroxide) *(Micro Flo)*

Rate: 1.33 pt. 3L/100 gal. water
1 lb. 50DF/100 gal. water

Disease:

Alternaria	downy mildew	Pseudomonas leaf spot
Anthracnose	fire blight	scab
bacterial leaf spot	Phomopsis	Septoria leaf spot
black leaf spot	Phytophthora	Volutella leaf blight
Botrytis	powdery mildew	Xanthomonas leaf spot
Cercospora		

Time: Apply at first sign of disesae; repeat at 7- to 14-day intervals as needed. Use shorter interval during periods of frequent rains or when severe disease conditions persist.

Remarks: Ornamental plants. Discoloration of foliage and/or blooms have been noted on some varieties. To prevent residues on commercial plants, do not spray before selling season. Refer to label for use directions.

- - - - - - - - - - - - - - - - - -

Rate: Douglas fir:
2.66 pt. 3L/A

Disease:
Rhabdocline needle cast

Time: Apply at bud break; repeat at 3-4 week intervals.

Rate: Sycamore:
1.33-4 pt. 3L/100 gal. water
2-3 lb. 50DF, 50WP/100 gal. water

◆ - **new product** • ai-active ingredient • DF-dry flowable • E/EC-emulsifiable concentrate • EW-emulsion in water
F/FL-flowable • G-granule • L/LC-liquid • SP/WSP-soluble packet • W/WP-wettable powder • WSB-water soluble bag

Disease:
Anthracnose
Time: Apply at bud crack; second application 7- to 14-days later at 10% leaf expansion.
Remarks: Douglas fir, sycamore. Refer to label for use directions.

Signal Word/Toxicity Class: Danger/I (50DF, 50WP); Warning/II (3L).
REI: 48 hr. (50DF, 50WP); 24 hr. (3L).

Curalan — see vinclozolin

Daconil — see chlorothalonil

Defend — see PCNB

DICLORAN

Protective fungicide. Toxic to fish.

Botran 75-W *(Gowan)*
Rate: 0.75 lb. ai/150-200 gal.
1 lb. 75-W/150-200 gal.
Disease: Chrysanthemums, geraniums, roses:
Botrytis blight
Time: Stock cuttings or greenhouse: Apply when disease first appears; repeat at 7- to 14-day intervals. During prolonged damp, cool periods, apply every 5- to 7-days.

Rate: 1 lb. ai/100 gal.
1.33 lb. 75-W/100 gal.
Disease: Roses, hydrangeas:
Botrytis blight
Time: Field: Apply prior to shipment or storage. Stored: Spray area thoroughly or dip plants; repeat during long storage periods.

Rate: Field, foliar-blight:
1 lb. ai/100 gal.
1.33 lb. 75-W/100 gal.
Bulb dip-dry rot:
0.75 lb. ai/75 gal.
1 lb. 75-W/75 gal.
Furrow-dry rot:
4 lb. ai/A
5.33 lb. 75-W/A
Disease: Gladiolus:
Botrytis blight   Stromatinia dry rot
Time: For field/foliar: Apply as a full coverage spray; Bulbs: Soak for 20-30 seconds; Furrow: Spray at planting time, so treated soil surrounds corms.
Remarks: Do not apply through any type of irrigation system.

Rate: 1-2 lb. ai/A
1.33-2.66 lb. 75-W/A
Disease: Conifer seedlings:
Botrytis blight
Time: Field: Apply as foliar spray when disease first appears; repeat at 10-to 14-day intervals.
Remarks: Bare rootstock, container, greenhouse, or nursery.

Signal Word/Toxicity Class: Caution.
REI: 12 hr.

Dithane — see mancozeb

Drexel Suffa — see sulfur

Eagle — see myclobutanil

Echo — see chlorothalonil

ETRIDIAZOLE

Preventive and curative fungicide.

Terrazole (35WP) *(Uniroyal)*
Rate: Dry soil mix:
1.5-3 oz. 35WP/cu. yd. soil media
Soil drench:
3.5-10 oz. 35WP/100 gal.
Disease:
crown rot   root rot   stem rot
damping-off
Time: Dry soil mix: Premix with 5 times its volume of dry soil before adding; repeat if needed. Soil drench: Repeat at 4- to 12-week intervals as necessary.
Remarks: Bedding plants, flowering plants, foliage plants, woody ornamentals. May be applied as dry soil mix or soil drench. Use high rates for peat and other high organic potting media, and for heavy disease pressure. Refer to label for use directions.

Signal Word/Toxicity Class: Warning/II.

FENARIMOL

Triazole sterol inhibitor. Toxic to fish.

Rubigan A.S. *(Dow AgroSciences)*
Rate: Ajuga, begonia, calendula, chrysanthemum, dahlia, delphinium, hydrangea, phlox, sweet pea, verbena, zinnia:
3-5 fl. oz. A.S./100 gal.
Crape myrtle, euonymus, photinia:
6-10 fl. oz. A.S./100 gal.
Disease:
powdery mildew
Time: Apply at earliest appearance of disease by thoroughly wetting new growth and plant areas; repeat at 10- to 14-day intervals.

Rate: 4-12 fl. oz. A.S./100 gal.
Disease: Field or landscape grown roses:
black leaf spot
Flowering crabapple:
powdery mildew   rust   scab
Time: Apply when disease first appears; repeat at 7- to 14-day intervals throughout growing season. For roses, apply according to percentage of leaves on individual plants infected.
Remarks: Do not apply this product through any type of irrigation system. Refer to label for use directions.

Signal Word/Toxicity Class: Caution.
REI: 12 hr.

Rubigan T&O (EC) *(SePRO)*
Rate: 0.02-0.08 lb. ai/100 gal. water
3-10 oz. EC/100 gal. water
Greenouse roses:
0.03 lb. ai/100 gal. water
4 oz. EC/100 gal. water
Disease:
powdery mildew
Time: Apply at earliest appearance of symptoms or when conditions are conducive for development. For preventive treatment, apply at 10- to 14-day intervals.
Remarks: Field or container grown ornamentals including begonia, dahlia, delphinium, hydrangea, roses, verbena, zinnia, and greenhouse roses. Use high rate if disease is present. Do not apply through any type of irrigation system. Refer to label for use directions.

Signal Word/Toxicity Class: Warning/II.
REI: 12 hr.

FERBAM

Broad-spectrum fungicide. Toxic to fish.

Ferbam Granuflo (WDG) *(UBC Agrochemicals)*
Rate: Cacti and succulents:
1.5-2.3 lb. ai/100 gal. water
2-3 lb. WDG/100 gal. water
Conifers, ornamentals:
0.76-1.5 lb. ai/100 gal. water
1-2 lb. WDG/100 gal. water

NOTICE The information on these pages is for preliminary planning — not a guide for use. Be sure to follow the manufacturer's directions, notwithstanding information contained here.

For personal protective equipment and EPA registration numbers, please turn to page 39.

Disease: Cacti, ornamentals:

Alternaria blight	Botrytis	leaf spot
Anthracnose	cedar rust	rust
black leaf spot	Drechslera stem rot	Septoria leaf spot
blossom blight	flower blight	Stemphylium leaf spot

Conifers:

Botrytis	Melamspora needle rust	southern cone rust
Fusiform rust	Puccineastrum needle rust	spruce cone rust
Lophodermium needle cast		

Time: Rate and intervals vary; refer to label.

Remarks: Conifers, cacti, ornamentals. For southern cone rust and Botrytis, add spreader-sticker. Black residue from use of this product may be undesirable on some ornamentals. When used in combination with sulfur, do not apply within 10-14 days of oil spray or just prior to or during periods of excessively high temperatures. Refer to label for use directions.

Signal Word/Toxicity Class: Caution.
REI: 24 hr.

Fore — see mancozeb

Formec — see mancozeb

Fungo — see thiophanate-methyl

GALLEX (Premix)

Ready-to-use paint-on emulsion.

Gallex (2,4-xylenol & meta-cresol) *(AgBioChem)*

Rate: Apply undiluted.

Disease:

crown gall; olive knot

Time: Apply first application before breaking of dormancy or during spring and early summer.

Remarks: Ornamentals including euonymus, flowering cherry, peach, and plum, oleander, poplar, roses, willows. Paint gall and cut surfaces + $^1/_2$" of surrounding healthy bark. Generally 1 application will give control; retreatment may be necessary. Do not mix or combine with other pesticides. Complete girdling of trunk by galls affects water-conducting capacity of tree. To prevent sudden loss of water movement, do not treat more than half of galled trunk at any one time. Treat remaining diseased tissue 4-6 months later when examination of initial treatment is made.

Signal Word/Toxicity Class: Warning/II.

IPRODIONE

Contact fungicide. Toxic to aquatic invertebrates.

Chipco 26019 (WP) *(Rhone-Poulenc)*

Rate: Drench:
6.5 oz. WP/100 gal.
Foliar or Dip:
0.5-1 lb. ai/100 gal.
1-2 lb. WP/100 gal.

Disease:

aerial web blight	Cylindrocladium	ink spot
Alternaria	daffodil leaf scorch	ray blight
blossom blight	Fusarium	Rhizoctonia
Botrytis	Helminthosporium	tulip fire

Time: Drench: Apply at seeding and/or after transplanting at 14-day intervals. Foliar: Apply prior to disease development; repeat at 7- to 14-day intervals. Dip treatment: For roses, apply bare root prior to cold storage; azalea and rhodendron, dip cuttings prior to planting; gladiolus, dip corms prior to storage.

Remarks: For use on over 80 ornamental flowering and foliage plant species; refer to label. Do not use on Spathiphyllum or as soil drench on impatiens and pothos. Refer to label for use directions.

State Restrictions: Not registered for use in California for control of blossom blight, Botrytis storage rot, Cylindrocladium blight and wilt, or daffodil leaf scorch.

Signal Word/Toxicity Class: Caution.
REI: 12 hr.

Kocide — see copper compounds

KOP Oxy — see copper compounds

KOP-Hydroxide — see copper compounds

Kumulus — see sulfur

M-PEDE

Effective contact insecticide, miticide, and fungicide. Toxic to aquatic invertebrates.

M-Pede (postassium salts of fatty acids) *(Mycogen)*

Rate: 1-2% v/v solution

Disease:

powdery mildew

Time: Apply at bud break; repeat at 7- to 10-day intervals.

Remarks: Roses (greenhouse grown). Do not apply through any type of irrigation system. Refer to label for use directions.

Signal Word/Toxicity Class: Warning/II.
REI: 12 hr.

MANCOZEB

Fungicide with protective action. Toxic to fish.

Dithane (DF, F-45, M-45, WSP) *(Rohm and Haas)*
Penncozeb 75DF or 80WP *(Elf Atochem)*

Rate: 1.6-3.2 lb. ai/A
2-4 lb. M-45, WSP, 80WP/A
2.13-4.27 lb. DF/A
3.2-6.4 pt. F-45/A
1.5-3 lb. ai/A
2-4 lb. 75DF/A

Disease: Christmas trees:

Lophodermium needle cast; pine gall rust; Scirrhia brown spot

Douglas fir:

Swiss needle cast

Time: Apply in spring or early summer before infection occurs; repeat after heavy rains and at 2-week intervals as needed.

Remarks: Christmas trees, Douglas fir. Refer to label for use directions.

Signal Word/Toxicity Class: Caution.
REI: 24 hr.

Dithane (T&O, WF) *(Rohm and Haas)*
Fore (FL, FloXL, T&O, WSP) *(Rohm and Haas)*
Formec 80 *(PBI/Gordon)*

Rate: 1.2 lb. ai/100 gal. dilute spray
1.5 lb. T&O, WSP/100 gal. dilute spray
2.4 pt. WF, FloXL/100 gal. dilute spray
1.2-1.6 lb. ai/100 gal. dilute spray
1.5-2 lb. 80/100 gal. dilute spray

Disease:

Actinopelte leaf spot	Entomosporium leaf spot	Phomopsis
Alternaria	Fabraea leaf spot	Phyllosticta leaf spot
Anthracnose	flower blight	Phytophthora
Ascochyta blight	frogeye leaf spot	pine gall rust
black leaf spot	Fusarium leaf spot	Puccinia rust
Botrytis	Fusicladium scab	purple spot
cedar-apple rust	Gloeosporium leaf spot	Ramularia leaf spot
Cephalosporium leaf spot	Guignardia leaf blotch	Rhizoctonia
Cercospora	hawthorn rust	rust
Curvularia leaf spot	Herpobasidium blight	scab
Cylindrocladium rot	Leptosphaeria brown spot	Scirrhia brown spot
Cylindrosporium leaf spot	Lophodermium needle cast	Septoria leaf spot
Dactylaria leaf spot	Monochaetia canker	smoulder
Didymellina leaf spot	Mystrosporium ink spot	Sphaceloma scab
Discosia leaf spot		Sphaeropsis leaf spot
downy mildew		Swiss needle cast
Elsinoe leaf spot		Taphrina leaf blister
		Volutella blight

◆ - **new product** • ai-active ingredient • DF-dry flowable • E/EC-emulsifiable concentrate • EW-emulsion in water
F/FL-flowable • G-granule • L/LC-liquid • SP/WSP-soluble packet • W/WP-wettable powder • WSB-water soluble bag

Time: For preventive, apply lower rate as full cover spray at 7- to 10-day intervals throughout season. For control, apply higher rate for 3 applications at 3- to 5-day intervals.
Remarks: Flowers, foliage plants, and ornamentals. Do not apply this product through any type of irrigation system. Do not use for food or feed purposes. Refer to label for use directions, restrictions, and precautions.

- - - - - - - - - - - - - - - - -

Signal Word/Toxicity Class: Caution.
REI: 24 hr. (Dithane, Fore).

Protect T/O (WP) *(Cleary Chemical)*
Rate: 1-2 lb. WP/100 gal. water (1-2 WSB)
Disease: Ornamental plants:

Alternaria	downy mildew	Ramularia
Anthracnose blight	Exobasidium blight	Rhizoctonia aerial web blight
Aschochyta blight	flower blight	Rhizoctonia web blight
black leaf spot	Fusarium	rust
Botryosphaeria blight	gray mold	scab
Botrytis blight	ink spot	shot hole
brown rot blight	Kabatina dieback	Stemphyllium blight
Cephalosporium leaf blight	leaf spot	tip blight
Cylindrocladium leaf blight	needle cast	Volutella blight
	Phytophthora dieback	

Azaleas, camellias, rhododendron:
flower blight
Chrysanthemum:
Botrytis petal spot
Flowering dogwood:
Anthracnose
Peony:
Botrytis blight Phytophthora blight

Time: For ornamentals, spray when plants are well leafed-out or at first sign of disease; repeat at 7- to 10-day intervals throughout the season; flower blight, spray 2-3 times a week while flowers are opening by spraying into flowers and ground under plants thoroughly. For Anthracnose, apply when buds begin to open, when bracts have fallen; 4-weeks later, and again in late summer after flower buds for next season are formed. For Botrytis blight and Phytophthora blight, apply early spring and early fall, drenching soil around plants as well as foliage. For Botrytis petal spot, apply twice weekly during blooming season as full cover spray.

Rate: Pachysandra:
4 lb. WP/100 gal. water (4 WSB)
Disease:
Volutella blight
Time: Apply as a drenching spray at first sign of disease, making at least 5 applications at 10- to 14-day intervals.
Remarks: Azaleas, camellias, chrysanthemum, flowering dogwood, ornamental plants, pachysandra, peony, rhododendron. Do not use on marigolds. Do not use any portions of any listed plant for food or feed purposes. Refer to label for use directions.

- - - - - - - - - - - - - - - - -

Signal Word/Toxicity Class: Caution.
REI: 24 hr.

MANEB

Dithiocarbamate contact fungicide. Toxic to fish.

Maneb 75DF or 80WP *(Elf Atochem)*
Rate: 0.75-1.12 lb. ai/A
1-1.5 lb. 75DF/A
0.8-1.2 lb. ai
1-1.5 lb. 80WP/A
Disease:

Alternaria	Cercospora leaf spot	Phytophthora blight
Anthracnose	Curvularia leaf spot	rust
black leaf spot	leaf spot	Stremphylium leaf spot
Botrytis blight		

Time: For carnations, dahlias, lilies, and pansies, apply when new growth starts; repeat weekly. For dogwood, apply when buds open; repeat when bracts fall 4 weeks later and in late summer. For gladiolus, apply when flower spikes are developing; repeat 2-3 times at weekly intervals. For peonies, apply to foliage and soil in early spring and fall; repeat at 7- to 10-day intervals during growing season. For snapdragons and zinnias, begin with emergence; repeat weekly. For roses, begin when first leaves unfold; repeat at 7-10 day intervals.
Remarks: Carnations, dahlias, dogwood, gladiolus, lilies, pansies, peonies, roses, snapdragons, zinnias. Refer to label for use directions.

- - - - - - - - - - - - - - - - -

Rate: 0.375 lb. ai/100 gal. water
0.5 lb. 75DF/100 gal. water
0.4 lb. ai/100 gal. water
0.5 lb. 80WP/100 gal. water
Disease:
Ascochyta ray blight Botrytis petal spot Septoria leaf spot
Time: Apply twice weekly during blooming season for petal spot and ray blight. For Septoria leaf spot, apply weekly throughout the season.
Remarks: Chrysanthemums. Refer to label for use directions.

- - - - - - - - - - - - - - - - -

Signal Word/Toxicity Class: Caution.
REI: 24 hr.

MEDALLION

Controls certain foliar and soilborne diseases grown in containers, greenhouses, and other enclosed structures. Toxic to fish and aquatic invertebrates.

Medallion (WP) (fludioxonil) *(Novartis)*
Rate: Begonia, celosia, impatiens, marigold, petunia, salvia, snapdragon, zinnia-at seeding:
1 packet WP/100 gal. water
Begonia, carnation, celosia, impatiens, marigold, poinsettia, pothos, salvia, snapdragon, Spathiphyllum, vinca, zinnia-transplants and cuttings:
2-4 packets WP/100 gal. water
Disease:
Rhizoctonia
Time: For transplants and cuttings repeat 21-28 days later if needed.
Remarks: Soil Drench. Apply sufficient solution to thoroughly wet growing media. Make only 1 application to seedling crop prior to transplanting. Under severe conditions, use high rate.

- - - - - - - - - - - - - - - - -

Rate: Gomphrena, poinsettia, pothos:
4-8 packets WP/100 gal. water
Disease:
Alternaria Rhizoctonia

Rate: African violet, chrysanthemum, impatiens, marigold:
12-16 packets WP/100 gal. water
Disease:
Botrytis
Time: Apply when disease first appears; repeat at 7- to 14-day intervals.
Remarks: Apply as foliar spray in greenhouse or other enclosed structure. Spray to point of runoff. Under severe conditions, use high rate. Do not apply to geraniums or to control seedling diseases of pansies.

- - - - - - - - - - - - - - - - -

Signal Word/Toxicity Class: Caution.
REI: 12 hr.

MEFENOXAM

Systemic fungicide having a specific mode of action.

Subdue GR, MAXX, or 2X WSP *(Novartis)*
Rate: Bedding/foliage plants, flowers:
Soil Drench:
0.3-1 fl. oz. MAXX/100 gal. water
0.17-0.35 oz. WSP/100 gal. water
Soil Surface:
6-60 oz. GR/1000 sq. ft.
1 fl. oz. MAXX/1000 sq. ft.
Soil Mix:
1.6-8 oz. GR/cu. yd.
0.13 fl. oz. MAXX/ cu. yd.
Bedding plants:
Soil Drench (seeding):
0.13-0.25 fl. oz. MAXX/100 gal. water
Soil Drench (transplanting):
0.5-1 fl. oz. MAXX/100 gal. water
0.1-0.5 oz. WSP/100 gal. water
Woody ornamentals:
Soil Drench:
1-2 fl. oz. MAXX/100 gal. water
0.6-1.1 oz. WSP/100 gal. water
Soil Surface:
26-125 oz. GR/1000 sq. ft.
1.25-2.5 fl. oz. MAXX/1000 sq. ft.
Soil Mix:
1.6-12.5 oz. GR/cu. yd.

NOTICE The information on these pages is for preliminary planning — not a guide for use. Be sure to follow the manufacturer's directions, notwithstanding information contained here.

For personal protective equipment and EPA registration numbers, please turn to page 39.

Disease:
damping-off, Phytophthora stem rot, Pythium stem rot, Phytophthora root rot, Pythium root rot

Time: Rate varies by plant species; refer to label. Apply at seeding or transplanting; repeat at 1- to 2-month intervals for bedding plants and flowers; 2- to 3-month intervals for foliage plants and woody ornamentals. Refer to label for use direction, restrictions, and precautions.

Rate: Azaleas:
Soil Drench:
0.63-1.25 fl. oz. MAXX/100 gal. water
0.35-0.7 oz. WSP/100 gal. water
Soil Surface:
14.8-75 oz. GR/1000 sq. ft.
1 fl. oz. MAXX/1000 sq. ft.
Soil Mix:
1.6-10 oz. GR/cu. yd.

Disease:
Phytophthora crown rot, Phytophthora root rot

Time: Apply at transplanting. For soil surface, and soil mix, repeat at 2- to 4-month intervals if necessary.

Remarks: Bedding and foliage plants, flowers, woody ornamentals (including azaleas). Irrigate after application. For woody ornamentals, after transplanting, irrigate with sufficient water to thoroughly wet root zone. For soil mix applications, Subdue GR can be reapplied after 6-8 weeks. Refer to label for use direction, restrictions, and precautions.

- - - - - - - - - - - - - - - - -

Rate: Seedbeds and plug-plantings:
6-30 lb. GR/A
1.25 pt. MAXX/A
0.35-0.7 oz. WSP/1000 sq. ft.
Transplants:
12-60 lb. GR/A
2.5 pt. MAXX/A in 50 gal. water
0.14-0.8 oz. WSP/1000 sq. ft.
Plantations:
50-250 lb. GR/A
0.63-1.25 gal. MAXX/A in 50 gal. water
1.5 lb. WSP/A in 50 gal. water

Disease:
Phytophthora root rot

Time: Apply in spring and fall.

Remarks: Conifers in nurseries and plantations (including Christmas trees). Best to apply 1/2- to 1" water after application if rain is not expected within 3 days. Do not apply as a foliar spray. Refer to label for use direction, restrictions, and precautions.

- - - - - - - - - - - - - - - - -

Signal Word/Toxicity Class: Caution.

Mertect — see thiabendazole

METALAXYL

Systemic fungicide having a specific mode of action.

ProTurf Pythium Control *(Scotts)*

Rate: Soil surface broadcast:
2-2.5 lb./1000 ft. bed area
Soil mix at potting:
0.75-1.5 lb./cu. yd.

Disease:
damping-off, root rot, stem rot

Time: Apply broadcast or at potting; repeat at 2- to 3-month intervals as needed.

Remarks: Woody ornamentals except azaleas. Apply 1/4-1/2" water to activate and move to plant root zone. Do not apply more than 4-5 lb./1000 ft. more than once every 10 weeks. Refer to label for use directions.

- - - - - - - - - - - - - - - - -

Signal Word/Toxicity Class: Caution.
REI: 24 hr.

Subdue 2E *(Novartis)*

Rate: Foliage plants:
Soil Drench:
0.5-1.25 fl. oz. 2E/100 gal. water
Soil Mix:
0.25-0.5 fl. oz. 2E/cu. yd.
Soil Surface Spray:
2 fl. oz. 2E/1000 sq. ft.
Bedding plants:
Drench at seeding:
0.25-0.5 fl. oz. 2E/100 gal. water
Drench at transplanting:
0.5-2 fl. oz. 2E/100 gal. water
Soil Mix:
0.25 fl. oz. 2E/cu. yd.
Flowers:
Soil Drench:
0.5-2 fl. oz. 2E/100 gal. water
Woody Ornamentals:
Drench:
1-4 fl. oz. 2E/100 gal. water

Disease:
damping-off, Phytophthora stem rot, Pythium stem rot, Phytophthora root rot, Pythium root rot

Time: Apply at seeding or transplanting; repeat at 2- to 3-month intervals if necessary.

Rate: Azaleas, rhododendrons:
Drench:
1-2.5 fl. oz. 2E/100 gal. water
Foliar spray:
1.25-2.5 fl. oz. 2E/100 gal. wate

Disease:
Phytophthora crown rot, Phytophthora root rot, Phytophthora shoot blight

Time: Apply at seeding or transplanting. For soil drench, soil surface, and soil mix, repeat at 2- to 4-month intervals if necessary. For foliar spray, repeat at 2- to 3-month intervals if necessary.

Remarks: Bedding and foliage plants, flowers, woody ornamentals (including azaleas, rhododendrons). Do not apply to English Ivy more than once every 6 months. For Easter lilies, make single application at planting. Do not apply to Euonymus. Refer to label for restrictions and precautions.

- - - - - - - - - - - - - - - - -

Rate: Seedbeds and plug-plantings:
2.5 pt. 2E/A
Transplants:
5 pt. 2E/A
Plantations:
1.25-2.5 gal. 2E/A

Disease:
Phytophthora root rot

Time: Apply in spring and fall.

Remarks: Conifers in nurseries and plantations (including Christmas trees). Apply in at least 50 gal. water/A. If rain is not expected within 3 days, apply 1/2-1" water after application. Do not apply as a foliar spray.

- - - - - - - - - - - - - - - - -

Signal Word/Toxicity Class: Warning/II.
REI: 12 hr.

METAM-SODIUM

Water-soluble liquid when applied to properly prepared soil is converted into a gaseous fumigant.

Clean Crop Nemasol (42%) *(Platte)*

Rate: 3 pt. 42%/100 sq. ft.

Disease:
oak root fungus

Time: At planting.

Remarks: Tree replant sites in orchards. Remove dead or disease trees and as much of root system as possible; make shallow basin over planting site. Add to stream of water while filling basin. Tarping of replant sites required when near 1/2 mile to populated areas such as schools, hospitals, commercial or office buildings, factories, residential areas. Trees immediately adjacent to treatment site may be injured.

State Restrictions: In California, application must be in compliance with Technical Information Bulletin "Guidelines for All Application Methods for Metam-sodium in California." This bulletin may be obtained from your local pesticide dealer or a metam-sodium registrant.

- - - - - - - - - - - - - - - - -

Signal Word/Toxicity Class: Danger/I.
REI: 48 hr.

Microsperse — see copper compounds

Microthiol Special — see sulfur

MYCLOBUTANIL

Systemic protectant, and curative fungicide recommended for control of specific diseases.

Eagle (WSP) *(Rohm and Haas)*
Systhane (WSP) *(Rohm and Haas)*
Rate: 3 oz. WSP/50 gal. spray solution (Eagle)
2 oz. WSP/50 gal. spray solution (Systhane)
Disease:

Anthracnose	Entomosporium leaf spot	needle rust
black leaf spot	Fabraea leaf spot	Ovulinia blight
blossom blight	flower blight	powdery mildew
Cephalosporium leaf spot	fruit blight	rust
Cercospora leaf blight	Fusicladium scab	scab
Cercospora leaf spot	Fusiform rust	twig blight
Didymellina leaf spot	leaf spot	white rust

Time: Apply at 10- to 14-day intervals. For flowering almond, apply prebloom, 50% bloom, and petal fall. For azalea, mountain laurel, and rhododendron, apply when flowers start to exhibit color. For Douglas fir, loblolly pine, and slash pine, apply early spring; repeat at 2- to 3-week intervals until threat of infection has passed. For roses, apply at 7- to 10-day intervals.
Remarks: Flowers, foliage plants, and ornamentals. Plants under nutritional or water stress will not respond as well to treatment. Addition of surfactant will improve performance. For Douglas fir, loblolly pine, and slash pine, spray adjuvants must be added to spray solutions. Nuts from treated trees may not be used for food purposes. Treated maple trees may not be used for syrup production. For chrysanthemums, Systhane may be applied as foliar spray or dip treatment (fully submerge cuttings until wet throughout for 2 minutes and then plant). Do not apply more than 6 oz./A per application or more than 5 lb./A per year. Do not use treated plant materials for food or feed. Refer to label for use directions, restrictions, and precautions.
State Restrictions: Not for use in Nassau and Suffolk counties, New York.

- - - - - - - - - - - - - - - - -

Signal Word/Toxicity Class: Warning/II.
REI: 24 hr.

Nova or Rally (40W) *(Rohm and Haas)*
Rate: 1.6-4 oz. ai/A
4-10 oz. 40W/A
Disease:
Fusiform rust — needle rust — rust
Time: Apply in early spring; repeat at 2- to 3-week intervals until threat of infection is past. For rust, apply at first sign of disease; repeat at 10- to 14-day intervals.
Remarks: Douglas fir, loblolly pine (nursery only); hybrid poplar (nursery, forested areas used for wood pulp production). Addition of surfactant will improve performance. Do not apply more than 1.5 lb./A per yr. Refer to label for use directions.
Days To Harvest: 14.
State Restrictions: Nova: Not for use in Alaska, Arizona, California, Hawaii, Idaho, Montana, Nevada, Oregon, Utah, Washington, and Wyoming. Rally: For use only in Alaska, Arizona, California, Hawaii, Idaho, Montana, Nevada, Oregon, Utah, Washington, and Wyoming; not for anthracnose in Arizona, California, and Hawaii.

- - - - - - - - - - - - - - - - -

Signal Word/Toxicity Class: Nova: Caution; Rally: Warning/II or Caution (varies by region).
REI: 24 hr.

Nova — see myclobutanil

Nu-Cop — see copper compounds

PCNB

Seed and soil contact fungicide. Toxic to fish; some formulations are also toxic to aquatic organisms.

Defend 10G or 75WP *(Cleary Chemical)*
Rate: Ornamental, bedding, and flowering plants:
15 lb. 10G/1000 sq. ft.
4 oz. 75WP/800 sq. ft.
Herbaceous and woody plants:
15-20 lb. 10G/1000 sq. ft.
4-8 oz. 75WP/400-800 sq. ft.
Time: Apply to flat or bench area at seeding or transplanting; mix into top 2-3" of soil surface.

Rate: Tropical foliage plants:
15 lb. 10G/1000 sq. ft.
4-8 oz. 75WP/400-800 sq. ft.
Time: Apply to bed or bench area and thoroughly mix into top 2" of soil surface; repeat 4-6 weeks later if necessary.
Disease: 10G:

black rot	crown rot	root rot
Botrytis storage rot	flower blight	Rhizoctonia
bulb dry rot	neck dry rot	stem rot
bulb rot	needle blight	

75WP:
Rhizoctonia
Remarks: Nursery and greenhouse. Do not apply to philodendron or pilea since injury may occur. Refer to label for use directions.

- - - - - - - - - - - - - - - - -

Rate: Preplant broadcast:
2.25-3.5 lb. 10G/100 sq. ft.
3.5-6.5 lb. 75WP/1000 sq. ft. (3.25-4.5 lb. for Dutch iris)
In-furrow at planting:
60 lb. 10G/14,500 sq. ft.
4.5-6 lb. 75WP/1000 lin. ft.
Band-Dutch iris:
3.75 lb. 10G/1000 lin. ft.
0.5 lb. 75WP/1000 lin. ft.
Dip treatment:
2 lb. 75WP/3-3.5 gal. water (4-6 lb. for lilies)
Disease: Easter lilies:
Rhizoctonia root rot
Iris, hyacinth, narcissus, tulips:
black rot — crown rot
Dutch iris:
bulb rot — stem rot
Time: For preplant, apply prior to planting and thoroughly incorporate 6-7" (Dutch iris 2-3"). For in-furrow, apply in 12" band centered on row at planting; repeat 3 applications at 15-day intervals.

Rate: Gladiolus:
Preplant:
2.75 lb. 10G/100 sq. ft.
3.5 lb. 75WP/1000 sq. ft.
Band:
42 lb. 10G/1000 lin. ft.
5.5 lb. 75WP/1000 lin. ft.
Disease:
bulb dry rot — neck dry rot
Time: For preplant, apply prior to planting; thoroughly incorporate 4-6". For band, apply in 12" band centered on row at planting; incorporate 4".

Rate: Calendula, larkspur, snapdragon, sweet pea:
2.25 lb. 10G/100 sq. ft.
3.5 lb. 75WP/1000 sq. ft.
Disease:
Rhizoctonia — stem rot
Time: Apply prior to planting; thoroughly incorporate 4".

Rate: Southern pine seedlings:
38 lb. 10G/1000 sq. ft.
50 lb. 75WP/A
Disease:
Rhizoctonia needle blight
Time: Apply 1½" water/A by irrigation prior to planting. Seed may be planted immediately after irrigation or within 1 week of that date.

Rate: Established azaleas, camellias:
7.5 lb. 10G/150 sq. ft.
1 lb. 75WP/150 sq. ft.
Disease:
flower blight
Time: Apply over soil beneath bushes and in plant beds prior to opening of buds; repeat every 3-4 weeks during bloom.

Rate: Roses:
2 lb. 75WP/100 gal. water
Disease:
Botrytis storage rot
Time: Dip or spray dormant plants before packaging or storage.
Remarks: Field grown ornamentals. For dip treatment, dip bulbs for 5 minutes (Easter lilies 5-15 minutes). Use lower rate for lighter soils and higher rate for heavier soils and black rot control. Refer to label for use directions.

- - - - - - - - - - - - - - - - -

Signal Word/Toxicity Class: Caution.
REI: 12 hr.

For personal protective equipment and EPA registration numbers, please turn to page 39.

PCNB 2-E or TurfPro WSP *(Amvac)*
Rate: 0.75 lb. ai
1.5 qt. 2-E/150 sq. ft.
1 lb. WSP/150 sq. ft.
Disease:
flower blight
Time: Apply to soil beneath plants prior to opening of buds and repeat every 3- to 4-weeks through bloom period.
Remarks: Azaleas, camellias. Refer to label for use directions. Do not graze treated areas or feed grass clippings to livestock. Do not plant root crops within 12 months of application unless PCNB is registered for use on those crops.

- - - - - - - - - - - - - - - - -

Signal Word/Toxicity Class: Danger/I (2-E); Caution (WSP).
REI: 12 hr.

PCNB 2FL, 75W, 10%, 10G T&O or TurfPro WSP *(Amvac)*
Rate: Bedding, tropical foliage, herbaceous & woody plants:
0.75-1.5 pt. 2FL/100 gal. water
4-8 oz. 75W, WSP/100 gal. water
15 lb. 10%, 10G/1000 sq. ft.
Bulbs, Easter lilies:
1.5-2.33 gal. 2FL/1000 sq. ft.
3.25-6.5 lb. 75W, WSP/1000 sq. ft.
1000-2000 lb. 10%, 10G/A
Disease:
black rot (bulbs) crown rot (bulbs) Rhizoctonia root rot

Rate: Calendulas, larkspur, snapdragon:
1.33 gal. 2FL/1000 sq. ft.
3.5 lb. 75W, WSP/1000 sq. ft.
1000 lb. 10%, 10G/A
Disease:
Sclerotinia stem rot

Rate: Dutch iris:
1.2-1.66 gal. 2FL/1000 sq. ft.
3.25-4.5 lb. 75W, WSP/1000 sq. ft.
1000-1500 lb. 10%, 10G/A
Disease:
bulb rot stem rot

Rate: Gladiolus:
1.33 gal. 2FL/1000 sq. ft.
3.66 lb. 75W, WSP/1000 sq. ft.
1200 lb. 10%, 10G/A
Disease:
bulb dry rot neck dry rot
Time: Preplant.
Remarks: Bedding plants include African violets, begonias, carnations, chrysanthemums, poinsettias, snapdragon, etc. Bulbs include hyacinth, iris, narcissus, tulips. For bedding plants, use low rate for 2FL and 75W. Do not apply to Philodendron or Pilea since injury may occur. Do not graze treated areas or feed grass clippings to livestock. Do not plant root crops within 12 months of application unless PCNB is registered for use on those crops.

- - - - - - - - - - - - - - - - -

Rate: 18.75 gal. 2FL/A
50 lb. 75W, WSP/A
Disease:
Rhizoctonia needle blight
Time: Immediately after irrigation or within 1 week.
Remarks: Southern pine seedlings. Apply uniform ground coverage; 1-1/2" water/A should be applied by irrigation prior to planting.

- - - - - - - - - - - - - - - - -

Signal Word/Toxicity Class: Caution.
REI: 12 hr.

Terraclor (75WP) or Terraclor 400 (FL) *(Uniroyal)*
Rate: Band:
0.75-8.25 pt. FL/1000 lin. ft.
0.5-5.5 lb. 75WP/1000 lin. ft.
Broadcast:
5.25-9.75 pt. FL/1000 sq. ft.
3.25-6.5 lb. 75WP/1000 sq. ft.
Bulb soak:
Hyacinth, iris, narcissus, tulip:
3 pt. FL/3.2 gal. water
2 lb. 75WP/3.2 gal. water
Lily:
6-9 pt. FL/100 gal. water
4-6 lb. 75WP/100 gal. water
Cut flowers:
1-3 pt. FL/100 gal. water
0.75-2 lb. 75WP/100 gal. water
Ground spray:
1.5 pt. FL/150 sq. ft.
1 lb. 75WP/150 sq. ft.
In furrow:
6.75-9 pt. FL/1000 lin. ft.
4.5-6 lb. 75WP/1000 lin. ft.
Soil drench:
6-12 fl. oz. FL/100 gal. water
4-8 oz. 75WP/100 gal. water
Disease:
black rot, dry rot, root rot, bulb rot, flower blight, stem rot, crown rot, leaf spot, storage rot, damping-off
Time: Rates and intervals vary; refer to label.
Remarks: Bedding plants, flowering plants, foliage plants, woody ornamentals. Use low rate on lighter or coarse textured soil or when disease pressure is expected to be low; high rates on heavy or fine textured soils in fields with history of disease problems, or when weather conditions are expected to be unfavorable for rapid germination and growth of seedlings.
Crop Rotation: Do not plant root crops in PCNB-treated fields within 12 months of broadcast and banding applications unless PCNB is registered for use on those crops.

- - - - - - - - - - - - - - - - -

Rate: 1.75 pt. FL/1000 sq. ft.
1.125 lb. 75WP/1000 sq. ft.
Disease:
damping-off, root rot, stem rot, needle blight
Time: Broadcast prior to planting.
Remarks: Southern pine seedlings. Additional 1/2" of water should be applied by irrigation following treatment. Seed may be planted immediately following irrigation or within 1 week. Refer to label for use directions.
Crop Rotation: Do not plant root crops in PCNB treated fields within 12 months of broadcast and banding applications unless PCNB registered for use on those crops.

- - - - - - - - - - - - - - - - -

Signal Word/Toxicity Class: Caution.

Turfcide (10G) *(Uniroyal)*
Rate: Bedding, flowering, and foliage plants, woody ornamentals:
Broadcast:
15-20 lb. 10G/1000 sq. ft.
Growing media:
1-1.5 lb. 10G/cu. yd.
Bulb crops:
Band:
60 lb. 10G/14,500 lin. ft.
Broadcast:
15-40 lb. 10G/1000 sq. ft.
Disease:
black rot, damping-off, root rot, bulb rot, dry bulb rot, stem rot, crown rot, neck rot
Time: Preventive.
Remarks: Bedding plants, flowering plants, foliage plants, woody ornamentals. Use low rate on lighter or coarse textured soil or when disease pressure is expected to be low; high rates on heavy or fine textured soils in fields with history of disease problems, or when weather conditions are expected to be unfavorable for rapid germination and growth of seedlings.
Crop Rotation: Do not plant root crops in PCNB treated fields within 12 months of broadcast and banding applications unless PCNB registered for use on those crops.

- - - - - - - - - - - - - - - - -

Signal Word/Toxicity Class: Caution.

Penncozeb — see mancozeb

PIPRON

Toxic to fish.

Pipron (LC) (piperalin) *(SePRO)*
Rate: 0.25-0.5 lb. ai/100 gal. water
4-8 oz. LC/100 gal. water
Disease:
powdery mildew
Time: Apply at first sign of disease; repeat at 7- to 10-day intervals as needed.

◆ - **new product** • ai-active ingredient • DF-dry flowable • E/EC-emulsifiable concentrate • EW-emulsion in water
F/FL-flowable • G-granule • L/LC-liquid • SP/WSP-soluble packet • W/WP-wettable powder • WSB-water soluble bag

Remarks: Greenhouse ornamentals including African violet, begonia, dahlia, Gerbera daisy, Kalanchoe, poinsettia, rose, verbena, and zinnia. Add 4.8 fl. oz. surfactant. Use high rate if disease is present. Before using on plant types and cultivars not listed, first test on small scale to confirm efficacy and plant safety. Do not apply through any type of irrigation system. Refer to label for use directions.

Signal Word/Toxicity Class: Warning/II.
REI: 12 hr.

PLANTVAX

Systemic fungicide with curative action.

Plantvax (75W) (oxycarboxin) *(Uniroyal)*
Rate: 1.6-2.4 oz. 75W/10 gal. water
Disease:
rust
Time: Apply when first signs of disease appear; repeat at 2 week intervals, using 2-4 sprays per season.
Remarks: Ornamentals. Refer to label for use directions.

Signal Word/Toxicity Class: Caution.
REI: 12 hr.

POLYRAM

Contact fungicide. Toxic to fish.

Polyram (80DF) (metiram) *(BASF, Platte)*
Rate: 0.8-1.2 lb. ai/A
1-1.5 lb. 80DF/A
Disease:
black leaf spot leaf spot rust
Time: Apply when first leaves unfold; repeat weekly as needed. During periods of continuous rainy weather, apply twice a week.
Remarks: Roses. Apply in sufficient water for thorough coverage. This product should be used with an Integrated Pest Management program. Do not apply through any type of irrigation system.

Signal Word/Toxicity Class: Caution.
REI: 12 hr. (BASF); 24 hr. (Platte).

PROPICONAZOLE

Triazole sterol inhibitor. Alamo is a trunk-injected systemic fungicide for control of certain wilt diseases in oaks and elms. Toxic to fish.

Alamo *(Novartis)*
Rate: Preventive:
6 ml. product/" DBH
Therapeutic:
10 ml. product/" DBH
Disease:
Dutch elm disease oak wilt
Time: Upper Midwest: Treat oaks after June 15. Wounds in oaks between May 15-June 15 attract insects that transmit oak wilt pathogen. For Dutch elm disease, consult local extension agent.
Remarks: American elm, oak, northern pin oak, northern red oak. Inject into flare roots in high volume of water to ensure distribution throughout vascular system of tree.

Signal Word/Toxicity Class: Warning/II.

Banner MAXX *(Novartis)*
Rate: Crepe myrtle:
5-8 fl. oz. MAXX/100 gal. water
Disease:
powdery mildew
Time: Apply as full coverage spray to point of drip at 21-day intervals.

Rate: Douglas fir, Shasta fir:
16 fl. oz. MAXX/100 gal. water
Disease:
needle rust
Time: For Douglas fir, apply once in May. For Shasta fir, apply May-June, prior to anticipated infection period.

Rate: Oaks:
16 fl. oz. MAXX/100 gal. water
Disease:
leaf spot
Time: Apply at 14- to 28-day intervals when conditions are favorable for disease development.

Rate: Red tip photinia:
5-8 fl. oz. MAXX/100 gal. water
Disease:
leaf spot
Time: Apply at 7- to 14-day intervals.

Rate: 5-8 fl. oz. MAXX/100 gal. water
Disease: Amelanchier, ash, calendula, delphininum, lilac, monarda, phlox, zinnia:
powdery mildew
Carnation, iris, snapdragon:
rust
Gomphrena, marigold, pyracantha, Sweet William:
leaf spot
Roses:
black leaf spot powdery mildew rust
Time: Apply as full coverage spray to point of drip. For black leaf spot, apply with registered contact fungicide.

Rate: Chrysanthemum:
8-12 fl. oz. MAXX/100 gal. water
Disease:
ray blight
Time: Apply at 30-day intervals when conditions are favorable for disease development.

Rate: 5-12 fl. oz. MAXX/100 gal. water
Disease: Crabapples:
leaf spot rust scab
powdery mildew
Dogwood:
flower blight
Time: Apply as full coverage spray to point of drip when conditions are favorable for disease development.

Rate: Hawthorn:
2-4 fl. oz. MAXX/100 gal. water
Disease:
rust
Time: Apply as full coverage spray at 14- to 21-day intervals during primary infection.

Rate: Juniper:
5-8 fl. oz. MAXX/100 gal. water
Disease:
phomopsis blight
Time: Apply as full coverage spray to point of drip. Make first application as soon as junipers start to grow; repeat at 14- to 21-day intervals during active growth.

Rate: Pines:
5-16 fl. oz. MAXX/100 gal. water
Disease:
Diplodia tip blight tip blight
Time: Apply as full coverage spray to point of drip. Initiate applications in mid- to late-winter. Apply at 2-month intervals for 3 applications.

Rate: Poplars:
16 fl. oz. MAXX/100 gal. water
Disease:
leaf spot
Time: Apply at 14- to 28-day intervals when conditions are favorable for disease development.

Rate: Rhododendron:
8-12 fl. oz. MAXX/100 gal. water
Disease:
ray blight
Time: Apply at 30-day intervals when conditions are favorable for disease development.
Remarks: Do not apply this product through any type of irrigation system.

Signal Word/Toxicity Class: Warning/II.
REI: 24 hr.

For personal protective equipment and EPA registration numbers, please turn to page 39.

Protect — see mancozeb

ProTurf Pythium — see metalaxyl

Rally — see myclobutanil

Redball — see sulfur

Rubigan — see fenarimol

SPORAX

Borax fungicide for control of annosus root disease.

Sporax (borax) *(Wilbur-Ellis)*

Rate: 1 lb./50 sq. ft. stump surface

Disease:

annosus root disease

Time: Apply to stump surface of cut tree as soon after felling as practical. For best results, apply immediately after cutting, but avoid delaying treatment more than one day.

Remarks: Conifers. Sprinkle material on freshly cut stump surface. Complete coverage, including exposed side areas and any splinters, is necessary; level rather than sloping stump surfaces are preferred, especially in areas of high rainfall. Moisture in exposed wood from freshly cut stumps, will dissolve product and leach it into wood. Contact local, state or federal forestry or plant pathology specialists to verify most economical and efficient methods and materials for your situation before using product. Refer to label for use directions.

- - - - - - - - - - - - - - - -

Signal Word/Toxicity Class: Danger/I.

Strike — see triadimefon

Subdue — see mefenoxam; metalaxyl

Sulfa 80 — see sulfur

Sulforix — see calcium polysulfide

SULFUR

Nonsystemic fungicide with protective action.

Drexel Suffa (6FL), Sulfa 80 (80DF) or Sulfur (90W) *(Drexel)*

Rate: 5.6 lb. ai/100 gal. water
7.5 pt. 6FL/100 gal. water
5.4 lb. ai/100 gal. water
6 lb. 90W/100 gal. water

Disease:

black leaf spot (roses)
Botrytis blight (hydrangea)
leaf spot
powdery mildew

Time: Apply at 5- to 7-day intervals, when disease first appears and continuing until all danger from disease is past. During periods of humid or rainy weather it may be necessary to make applications as often as every 2- to 3-days.

Rate: Ground:
2-8 lb. ai/A
3-10 lb. 80DF/A
Aerial:
2-16 lb. ai/A
3-20 lb. 80DF/A

Disease:

black leaf spot (roses)
leaf spot
powdery mildew

Time: Apply when disease first appears; repeat at 5- to 10-day intervals.

Remarks: Aster, calendula, carnations, cherry laurel, chrysanthemum, clematis, columbine, cosmos, crape myrtle, dahlia, daisy, delphinium, dogwood, English ivy, euonymus, foxglove, gladiolus, golden fleece, hibiscus, hollyhock, honeysuckle, hydrangea, lady's mantle, laurel, ligustrum, lilac, petunia, phlox, poplar, rose, sage, silver vine, smilax, snapdragon, spiraea, sunflower, sweet peas, verbena, violets, willow, zinnia. Apply as dust or spray. Do not use within 2 weeks of oil application (4 weeks 6FL). Some plants are susceptible to injury from sulfur under certain climatic conditions. Do not make sulfur applications at excessively high temperatures. Do not apply this product through any type of irrigation system. Refer to label for use directions.

- - - - - - - - - - - - - - - -

Signal Word/Toxicity Class: Caution.
REI: 24 hr.

Dusting Sulfur (DS) *(Platte)*

Rate: Apply to cover foliage.

Disease:

powdery mildew
rust

Time: Repeat at 5- to 10-day intervals.

Remarks: Flowers, shrubs, trees. Do not make applications within 2 weeks of oil spray. Refer to label for use directions.

- - - - - - - - - - - - - - - -

Signal Word/Toxicity Class: Caution.
REI: 24 hr.

Kumulus DF (sulfur) *(BASF)*

Rate: 2.4-4 lb. ai/A
3-5 lb. DF/A

Disease:

black leaf spot
leaf spot (roses)
powdery mildew

Time: Apply when disease first appears; repeat at 5- to 10-day intervals.

Remarks: Asters, calendula, carnations, cherry laurels, clematis, columbine, cosmos, crape myrtle, dahlia, delphinium, gladiolus, hollyhock, ivy (English), laurel, ligustrum, petunia, sage, smilax, sunflowers, sweet peas, violets, zinnias. Do not apply to or allow spray drift on sulfur-sensitive plants such as chrysanthemums; sulfur may burn foliage when temperature is excessively high. Do not apply if temperatures during or within 3 days after application are expected to exceed 90°F. Do not apply through any type of irrigation system. Refer to label for use directions.

- - - - - - - - - - - - - - - -

Signal Word/Toxicity Class: Caution.
REI: 24 hr.

Micro Flo Sulfur (6L), Dual Spraying & Dusting Sulfur (DS) or Microsperse Wettable Sulfur (WS) *(Micro Flo)*

Rate: 2.25-6 lb. ai/100 gal. water
3-8 pt. 6L/100 gal. water
5 lb. ai/100 gal. water
5.6 lb. DS/100 gal. water
2.4-4 lb. ai/100 gal. water
2.7-4.5 lb. WS/100 gal. water

Disease:

black leaf spot (6L)
leaf spot
powdery mildew
rust (6L)

Time: Apply when disease first appears; repeat at 5- to 10-day intervals. During periods of humid or rainy weather it may be necessary to apply as often as every 2-3 days.

Remarks: Do not apply DS and WS through any type of irrigation system.

- - - - - - - - - - - - - - - -

Signal Word/Toxicity Class: Caution.
REI: 24 hr.

Microthiol Special (80WP) *(Elf Atochem)*

Rate: Ground:
2.4-8 lb. ai/A
3-10 lb. 80WP/A
Air:
2.4-16 lb. ai/A
3-20 lb. 80WP/A

Disease:

leaf spot
powdery mildew

Time: Apply to ensure complete plant coverage when disease first appears; repeat at 5- to 10-day intervals.

Remarks: Ornamentals, roses. Refer to label for use directions.

- - - - - - - - - - - - - - - -

Signal Word/Toxicity Class: Caution.
REI: 24 hr.

◆ - **new product** • ai-active ingredient • DF-dry flowable • E/EC-emulsifiable concentrate • EW-emulsion in water
F/FL-flowable • G-granule • L/LC-liquid • SP/WSP-soluble packet • W/WP-wettable powder • WSB-water soluble bag

Redball EM-53 (6L), Superfine Wettable Sulphur (WS) or Commercial Dusting Sulphur (DS) *(International Sulphur)*

Rate: 4-8 pt. 6L/100 gal. water
6-8 lb. WS/100 gal. water
6-8 lb. DS/A

Disease: Ornamentals:
leaf spot — powdery mildew

Roses:
black leaf spot — leaf spot — rust (6L)
brown canker — powdery mildew

Time: Apply when disease first appears; repeat at 5- to 10-day intervals for 6L and DS, 7- to 14-days for WS.

Remarks: Aster, carnations, cherry laurel, chrysanthemums, crape myrtle, English ivy, gladiolus, hollyhock, laurel, petunia, roses, sage, sunflowers, sweet peas, violets, zinnias. Keep new growth protected. Apply for thorough coverage. Reapply after rainfall. Do not apply when temperatures exceed 90°F. Refer to label for use directions.

Signal Word/Toxicity Class: Caution.
REI: 12 hr.

Terra Sulfur (6L) *(Terra)*

Rate: 3-8 pt. 6L/100 gal. water

Disease:
black leaf spot (roses) — leaf spot — rust

Time: Apply when disease first appears; repeat at 5- to 10-day intervals. During periods of humid or rainy weather it may be necessary to make applications as often as every 2-3 days.

Remarks: Ornamentals, roses. Do not use within 2 weeks of oil application. Some plants are susceptible to injury from sulfur under certain climatic conditions. Do not make sulfur applications at excessively high temperatures. Refer to label for use directions.

Signal Word/Toxicity Class: Caution.
REI: 24 hr.

Thiolux (DF) *(Platte)*

Rate: Ground:
2.4-8 lb. ai/A
3-10 lb. DF/A
Air:
2.4-16 lb. ai/A
3-20 lb. DF/A

Disease:
black leaf spot (roses) — leaf spot — rust — powdery mildew

Time: Apply when disease first appears; repeat at 5- to 10-day intervals.

Remarks: Ornamentals. Do not make applications within 2 weeks of oil spray. Refer to label for use directions.

Signal Word/Toxicity Class: Caution.
REI: 24 hr.

Systhane — see myclobutanil

Tennessee Brand — see copper compounds

Terraclor — see PCNB

TERRAGUARD

Triazole sterol inhibitor fungicide with protective and curative action. Toxic to fish.

Terraguard (50W) (triflumizole) *(Uniroyal)*

Rate: 4-8 fl. oz. 50W/100 gal. water

Disease:
aerial blight — leaf spot — powdery mildew
black root rot — petiole rot — root rot
Cylindrocladium root rot

Time: Foliar spray: Apply at weekly intervals as needed. For powdery mildew on roses and poinsettias, apply 16 fl. oz. 50W/100 gal. water for initial application, then follow schedule above. Soil drench: Apply at 3- to 4-week intervals as needed. For black root rot, apply 1-4 fl. oz. 50W/100 gal. water as soil drench.

Rate: 6-16 fl. oz. 50W/100 gal. water

Disease:
crown rot — Cylindrocladium root rot — wilt

Time: Prestick cutting soak: Soak cutting for 10 minutes. Soil drench: Apply at 3- to 4-week intervals as needed, usually 2-4 applications sufficient.

Remarks: Ornamentals. May be applied as foliar spray or soil drench. Use high rates under heavy disease pressure. Refer to label for use directions.

Signal Word/Toxicity Class: Caution.

Terraneb — see chloroneb

Terranil — see chlorothalonil

Terrazole — see etridiazole

Thalonil — see chlorothalonil

THIABENDAZOLE

Systemic fungicide with protective and curative action. Mertect 340-F is toxic to fish.

Arbotect 20-S *(Novartis)*

Rate: Preventive 1 yr. treatment:
1-2 fl. oz. 20-S/5" trunk diameter
Therapeutic 1 yr. treatment:
2-4 fl. oz. 20-S/5" trunk diameter
Preventive/therapeutic 3 yr. treatment:
2 fl. oz. 20-S/5" trunk diameter

Disease:
Dutch elm disease

Time: Preventive: Apply in 40-80 fl. oz. water when leaves approach full size (late May-June). Therapeutic: Apply in 80-160 fl. oz. water as current year infections are seen (late June-August). Use high rate under heavy disease pressure. Preventive/therapeutic: Inject into exposed root flares, below ground, once every 3 years.

Remarks: Elm trees. For optimum disease control, preventive treatment recommended. Trees 5" or less in diameter at chest height should not be treated. For preventive and therapeutic 3 yr. treatment, do not use if trees are less than 10" diameter, apply in 1 gal. water.

Rate: 4 fl. oz. 20-S/5" trunk diameter

Disease:
Anthracnose

Time: Apply injections late summer or early fall, in 2 consecutive years; repeat if necessary.

Remarks: Sycamore trees. Inject at 3-10" intervals around trunk, maximum hole of 1/2" diameter, using minimum 3-4 injection points/tree. Apply in 80-160 fl. oz. water. Trees that are 5" or less in diameter at chest height should not be treated.

Signal Word/Toxicity Class: Caution.

Mertect 340-F *(Novartis)*

Rate: 30 fl. oz. 340-F/100 gal. water

Disease:
basal rot

Time: Dip bulbs for 15-30 minutes, corms for 15 minutes.

Remarks: Ornamental bulbs and corms. Submerge completely in dipping suspension. Clean and treat within 24-48 hr. of digging. Do not apply through any type of irrigation system.

Signal Word/Toxicity Class: Caution.

Thiolux — see sulfur

THIOPHANATE-METHYL

Systemic fungicide with protective and curative action.

3336 (F or WP) *(Cleary Chemical)*

Rate: 10-33 fl. oz. F/100 gal. water
8-24 oz. WP/100 gal. water

NOTICE	The information on these pages is for preliminary planning — not a guide for use. Be sure to follow the manufacturer's directions, notwithstanding information contained here.	For personal protective equipment and EPA registration numbers, please turn to page 39.

Disease:

Anthracnose	Ascochyta leaf blight

Time: Repeat at 7- to 10-day intervals as needed. For Anthracnose, apply as buds break.

Disease:

black leaf spot	Entomosporium leaf spot	leaf scab
Botrytis flower blight	iris leaf spot (Didymellina)	powdery mildew
Cercospora leaf spot		Ramularia leaf spot
Curvularia leaf blight		rust (crabapple) (WP)

Time: Apply mid-summer or at first sign of disease. For rust, apply every 10- to 14-days as needed.

Disease:
Ovulinia blight

Time: Apply as flowers open; repeat at 7- to 10-day intervals as needed.

Disease:
tip blight (pine)

Time: Apply in spring when new growth starts, make second application just before needles emerge from sheath, third application 10 days later.

Disease:
twig blight (juniper)
(Kabatina, Phomopsis)

Time: Repeat at 7- to 10-day intervals as needed. For Phomopsis, apply in spring; Kabatina, apply late summer.

Remarks: Foliar. Ground application used for prevention and control of disease; aerial when disease first appears; repeat at 10- to 14-day intervals during growing season. Shorten interval when conditions are unusually favorable for development of disease. Do not apply this product through any type of irrigation system. Refer to label for use directions.

- - - - - - - - - - - - - - - - -

Rate: 16-20 fl. oz. F/100 gal. water
12-16 oz. WP/100 gal. water

Disease:

black root rot	Cylindrocladium rot	stem rot
crown rot	root rot	

Time: Apply after transplanting into propagation beds or containers. Repeat at 21- to 28-day intervals to provide adequate crop protection.

Remarks: Soil drench. Does not control Pythium or Phytophthora. Refer to label for use directions.

- - - - - - - - - - - - - - - - -

Rate: Plants or cuttings:
20 fl. oz. F/100 gal. water
Bulbs:
33 fl. oz. F/100 gal. water
12-16 oz. WP/100 gal. water

Disease:

bulb dry rot	corm rot	Rhizome rot

Time: Preplant dip.

Remarks: Bulb treatment. Soak cleaned bulbs for up to 30 minutes in warm dip (80°-85°F) preferably within 48 hr. of digging. Dry bulbs after treatment before storing. If bulbs are for forcing, treat bulbs that have been heat-cured. Immerse plants or cuttings for 10-15 minutes; remove and allow to drain. Refer to label for use directions.

- - - - - - - - - - - - - - - - -

Signal Word/Toxicity Class: Caution.
REI: 12 hr.

3336 G *(Cleary Chemical)*

Rate: 22-30 lb. G/1000 sq. ft.

Disease:

black root rot	Cylindrocladium rot	stem rot
crown rot	root rot	

Time: Apply after transplanting to propagation bed, containers or pots, nursery or landscape beds when disease first appears; repeat at 14- to 28-day intervals as needed.

Remarks: Use higher rate under conditions of severe disease. Refer to label for use directions.

- - - - - - - - - - - - - - - - -

Signal Word/Toxicity Class: Caution.
REI: 12 hr.

Fungo (50WSB) *(Scotts)*

Rate: Preplant dip:
12-16 oz. 50WSB/100 gal. water

Disease:

Botrytis	Penicillium rot	Sclerotinia
Fusarium rot	Rhizoctonia	Thielaviopsis

Time: For cuttings and cane sections, immerse 10-15 minutes; remove and allow to drain. For bulbs, corms, and tubers, soak 15-30 minutes in warm dip; dry after treatment.

Rate: 8-16 oz. 50WSB/100 gal. water

Disease: Foliar:

Anthracnose	Curvularia leaf blight	Ovulinia blight
Ascochyta leaf blight	Didymellina lefa spot	powdery mildew
black leaf spot	Entomosporium leaf spot	tip blight
Botrytis blight		
Cercospora leaf spot		

Drench:

Botrytis	Rhizoctonia	Thielaviopsis
Fusarium	Sclerotinia	

Rate: 24-32 oz. 50WSB/100 gal. water

Disease:
twig blight

Time: Foliar spray: Apply when disease first appears; repeat at 10- to 14-day intervals during growing season until leaf tissue is fully expanded. Drench treatment: Apply after transplanting; repeat at 2-4 week intervals during periods favorable for disease.

Remarks: Flowering, woody, and herbaceous ornamental plants, tropical foliage, trees. Shorten intervals during humid, rainy weather. For diseases that become more severe throughout summer, repeat at monthly intervals after spring. Refer to label for use directions.

- - - - - - - - - - - - - - - - -

Signal Word/Toxicity Class: Warning/II.
REI: 12 hr.

THIRAM

Broad-spectrum fungicide. Toxic to fish. Thiram 50WP contains the toxic inert Rhodamine B and is also toxic to birds and wildlife.

42-S Thiram or Thiram 50WP Dyed *(Gustafson)*

Rate: **Conifer seed:**
4 pt. 42-S/100 lb. seed
Bulbs and tubers:
1.5 pt. 42-S/8 gal. water
Flower seeds:
8 fl. oz. 42-S/100 lb. seed
Flower seeds, palms, philodendrons, Scheffleras:
8 oz. 50WP/100 lb. seed

Disease:

basal rot	damping-off	decay

Time: Seed treatment.

Remarks: Bulbs, conifers, flower seeds, palms, philodendrons, Scheffleras, tubers. Use with EPA-approved dye that imparts unnatural color to seed. Do not use for feed, food, or oil purposes.

State Restrictions: Not for use in California on flower seeds.

- - - - - - - - - - - - - - - - -

Signal Word/Toxicity Class: Caution.

Thiram Granuflo (WDG) *(UCB Agrochemicals)*

Rate: 21.5 oz. WDG/100 lb. seed

Disease:

damping-off	seed decay	seedling blight

Time: Seed treatment.

Remarks: Conifers. Do not use treated seed for food, feed, or oil purposes. Treat only seeds needed for immediate use. Do not store excess treated seeds beyond planting time. Do not apply this product through any type of irrigation system.

State Restrictions: Not for use in California.

- - - - - - - - - - - - - - - - -

Signal Word/Toxicity Class: Caution.
REI: 24 hr.

TRIADIMEFON

Triazole sterol inhibitor with protectrive and curative action.

Bayleton 25 or 50 T&O *(Bayer)*

Rate: 11 oz. 25/in water
5.5 oz. 50/in water

Disease:

flower blight	leaf spot	rust
leaf blight	powdery mildew	tip blight

Time: Apply as full cover foliage spray to point of drip as needed.

◆ - **new product** • ai-active ingredient • DF-dry flowable • E/EC-emulsifiable concentrate • EW-emulsion in water
F/FL-flowable • G-granule • L/LC-liquid • SP/WSP-soluble packet • W/WP-wettable powder • WSB-water soluble bag

Remarks: Flowering and foliage plants (outdoor), shade trees, ornamental shrubs and trees. Do not use on crops grown for food or forage. Do not apply this product through any type of irrigation system. Refer to label for use directions.
State Restrictions: California labels may vary; refer to label.

Signal Word/Toxicity Class: Caution.

Bayleton 50% DF *(Bayer)*
Rate: Christmas trees:
8 oz. 50% DF/A
Disease:
cone rust gall; Lophodermium needle cast; stem rust; tip blight
Time: For rust, apply when needles break through the fascicle sheath; repeat at 14- to 21-day intervals until galls become pale to white in color. For tip blight, apply at bud break; make 2 additional applications at 14-day intervals. For needle cast, apply at spore release, beginning in mid-July and ending mid-Oct.; make applications at 21-day intervals. Extend interval to 28-days if spore release is light or dry weather is expected.

Rate: Pine seedlings:
4-16 oz. 50% DF/A
Disease:
pine rust
Time: Apply prior to infection; repeat at 2-3 week intervals.

Rate: Pine seed nurseries:
2-20 oz. 50% DF/A
Disease:
Fusiform rust
Time: For lower rate, apply to wetted seeds and mix for 10 minutes before additional treatment; thoroughly air dry before sowing. For higher rate, cover and soak at room temperature (75°F) for 24 hr., stirring occasionally. Thoroughly air dry before applying additional treatment or before planting.
Remarks: Not for use on Concolor firs. Use high rate in areas of severe disease pressure. Do not use treated seed for food or feed purposes. Do not apply through any type of irrigation system. Refer to label for use directions.
State Restrictions: Not for use on pine seedlings, or pine seed nurseries in California.

Signal Word/Toxicity Class: Caution.
REI: 12 hr.

Strike (25 WDG) *(Olympic)*
Rate: Nursery:
4-8 oz. 25 WDG/50 gal. water
Greenhouse:
1 oz. 25 WDG/50 gal. water (winter)
2 oz. 25 WDG/50 gal. water (summer)
Disease:
flower blight; leaf blight; leaf spot; powdery mildew; rust; tip blight
Time: Apply full coverage spray at recommended intervals when plants are fully established and actively growing.
Remarks: Flowering and foliage plants (greenhouse and outdoor), shade trees, ornamental shrubs and trees. Do not use on crops grown for food or forage. Do not apply this product through any type of irrigation system. Refer to label for use directions.
State Restrictions: California labels may vary; refer to label.

Signal Word/Toxicity Class: Caution.
REI: 12 hr.

Turfcide — see PCNB

TurfPro — see PCNB

VINCLOZOLIN

Dicarboximide fungicide. Toxic to fish.

Curalan FL, DF, EG *(BASF)*
Rate: 8-48 fl. oz. DF, EG, FL/100 gal. water
Disease:
Anthracnose; brown patch; dollar spot; Fusarium patch; gray mold; gray snow mold; Helminthosporium; leaf spots; melting-out; pink patch; pink snow mold; red thread
Time: Foliar spray or dip, preventive or curative.
Remarks: Herbaceous and woody ornamentals, field grown flowers, bulb and corm plants. Refer to label for specific recommendations for sprays and dips of foliage or root stocks, bulbs, and corms. Do not apply this product through any type of irrigation system.

Signal Word/Toxicity Class: Caution.
REI: 12 hr.

ZIRAM

Contact foliar fungicide with protective action. Toxic to fish and wildlife.

Ziram 4L or 76 *(Drexel)*
Rate: Asters:
1 lb. ai/12 gal.
2 pt. 4L/12 gal.
0.76 lb. ai/12 gal.
1 lb. 76/12 gal.
Disease:
rust
Time: Apply early-season; repeat at 7-day intervals until blossoms form.

Rate: Carnations, chrysanthemums, dahlias, delphiniums, gladiolus, golden glow, hollyhocks, marigolds, morningglories, phlox, roses, snapdragons:
1.25 lb. ai/100 gal.
2.5 pt. 4L/100 gal.
1.3 lb. ai/100 gal.
1.66 lb. 76/100 gal.
Disease:
black leaf spot (roses); leaf spot; powdery mildew; rust
Time: Apply at 7- to 10-day intervals as needed throughout growing season. For roses, apply when leaves unfold.

Rate: Azalea, camellias:
1.5-2 lb. ai/100 gal.
3-4 pt. 4L/100 gal.
1.1-1.5 lb. ai/100 gal.
1.5-2 lb. 76/100 gal.
Disease:
flower blight
Time: Apply at first bloom; repeat at 7- to 10-day intervals through bloom.

Rate: Pine seedlings (nursery beds):
2 lb. ai/100 gal.
4 pt. 4L/100 gal.
1.5 lb. ai/100 gal.
2 lb. 76/100 gal.
Disease:
Fusiform rust
Time: Apply at time of emergence; repeat at 3- to 5-day intervals to mid-June.
Remarks: Asters, azaleas, camellias, carnations, chrysanthemums, dahlias, delphiniums, gladiolus, golden glow, hollyhocks, marigolds, morningglories, phlox, roses, snapdragons, pine seedlings. For azaleas and camellias, apply as drench to mulch and litter beneath and around bushes. Do not apply this product through any type of irrigation system.

Signal Word/Toxicity Class: Danger/I.
REI: 48 hr.

Ziram 76DF *(Elf Atochem)*
Rate: 1.1-1.5 lb. ai/100 gal.
1.5-2 lb. 76DF/100 gal.
Disease: Asters:
rust
Time: Apply early-season; repeat at 7-day intervals until blossoms form.

Disease: Azaleas, camellias, cyclamens, gardenias:
Botrytis; flower blight; leaf spot
Time: Begin applications with first bloom; repeat at 7- to 10-day intervals through bloom.

Disease: Carnations, chrysanthemums, dahlias, gladiolus, marigolds, morningglories, snapdragons:
leaf spot; rust
Time: Apply at 7- to 10-day intervals as needed throughout growing season.

Disease: Dahlias, delphiniums, golden glow, hollyhocks, phlox, roses:
black leaf spot; powdery mildew

9 Turf & Ornamentals

NOTICE The information on these pages is for preliminary planning — not a guide for use. Be sure to follow the manufacturer's directions, notwithstanding information contained here. | For personal protective equipment and EPA registration numbers, please turn to page 39.

Time: Apply at 7- day intervals as needed throughout growing season. For roses, apply when leaves unfold; repeat at 7- to 10-day intervals throughout season.

Remarks: Ornamentals. For azaleas and camellias, apply as a drench to mulch and litter beneath and around bushes. Do not apply this product through any type of irrigation system. Do not graze or feed cover crops from treated orchards. Refer to label for use directions.

State Restrictions: Not for use in California.

- - - - - - - - - - - - - - - - -

Rate: 1.5 lb. ai/100 gal.
2 lb. 76DF/100 gal.

Disease:

Fusiform rust (pine seedlings)	Melampsora needle rust (Douglas fir)	Puccineastrum needle rust (Shasta fir)

Time: For pine seedlings, apply at time of emergence; repeat at 3- to 5-day intervals to mid-June. For Christmas trees, apply once in spring when new shoot growth is 1/2-2" in length.

Remarks: Conifers. Do not apply this product through any type of irrigation system. Refer to label for use directions.

State Restrictions: Not for use in California.

- - - - - - - - - - - - - - - - -

Signal Word/Toxicity Class: Danger/I.
REI: 48 hr.

Ziram Granuflo (WDG) *(UCB Agrochemicals)*

Rate: 1.1-1.5 lb. ai/100 gal. water
1.5-2 lb. WDG/100 gal. water

Disease: Asters, azaleas, camellias:
flower blight (azaleas, camellias) rust (asters)

Time: For flower blight, begin applications with first bloom; repeat at 3- to 5-day intervals through bloom. For rust, apply early-season; repeat at 7-day intervals until blossoms form.

Disease: Cyclamens, gardenias, lilies, peonies, poinsettias, tulips:
Botrytis blight leaf spot (cyclamens, gardenias)

Time: Apply at 7- to 10-day intervals throughout growing season.

Rate: Carnations, chrysanthemums, dahlias, delphinium, gladiolus, golden glow, hollyhocks, marigolds, morningglories, phlox, roses, snapdragons:
1.3-1.5 lb. ai/100 gal. water
1.66-2 lb. WDG/100 gal. water

Disease:
black leaf spot (roses) powdery mildew rust
leaf spot

Time: Apply at 7- to 10-day intervals as needed throughout growing season. For roses, begin when leaves unfold and follow schedule listed above.

Remarks: Ornamentals. For azaleas and camellias, apply as a drench to mulch and litter beneath and around bushes. Do not apply this product through any type of irrigation system. Refer to label for use directions.

State Restrictions: Not for use in California.

- - - - - - - - - - - - - - - - -

Rate: 1.5 lb. ai/100 gal. water
2 lb. WDG/100 gal. water

Disease:
Fusiform rust Melampsora needle rust Puccineastrum needle rust

Time: For pine seedlings, apply at emergence; repeat at 3- to 5-day intervals to mid-June. For Christmas trees, apply once in spring when new growth is 1/2-2" in length.

Remarks: Conifers. Do not apply this product through any type of irrigation system. Refer to label for use directions.

State Restrictions: Not for use in California.

- - - - - - - - - - - - - - - - -

Rate: 1.5-2.3 lb. ai/100 gal. water
2-3 lb. WDG/100 gal. water

Disease:

Anthracnose	leaf blister	shot hole
Catkin Hypertrophy	leaf curl	twig blight
European canker	plum pockets	witches' broom
Fusiform rust	shoot blight	yellow leaf blister

Time: Apply as thorough cover spray in dormant period prior to bud swell.

Remarks: Trees, nonbearing fruit trees. Do not apply this product through any type of irrigation system. Refer to label for use directions.

State Restrictions: Not for use in California.

- - - - - - - - - - - - - - - - -

Signal Word/Toxicity Class: Danger/I.
REI: 48 hr.

◆ - **new product** • ai-active ingredient • DF-dry flowable • E/EC-emulsifiable concentrate • EW-emulsion in water
F/FL-flowable • G-granule • L/LC-liquid • SP/WSP-soluble packet • W/WP-wettable powder • WSB-water soluble bag

Section 10

Other Uses

Bt Chart 536

Insecticide Listings

Turf 489
Forage (Alfalfa, etc.) 537
Forage (Grasses, Hay) . . . 543
Pastures & Rangeland . . . 544
Noncropland 546

Fungicide Listings

Forage (Alfalfa, etc.) 547
Forage (Grasses, Hay) . . . 549

The information in this 1999 Insect and Disease Control Guide is updated to August 15, 1998. *Because manufacturers' product recommendations are constantly being updated and revised as additional product registrations and label changes are approved by EPA, there may be changes after this date. However, every effort has been made to give manufacturers the opportunity to include in the Guide their latest information. This Guide is for preliminary planning – not a use guide. Be sure to follow label directions, notwithstanding information contained herein.*

Bt Products Chart

Other Uses

Find Bt and enhanced seed products quickly and easily with these charts. Simply find the insect you want to control on the top of the chart, and follow down to the marked spot to find the Bt product you can use to control that pest. Company address information can be found on page 599.

Bt Products	Alfalfa caterpillar	Armyworms	Loopers	Mosquitoes	Range caterpillar	Skippers	Webworms
Agree	■	■	■				
Bactimos				■			
Biobit	■	■	■			■	
BMP 123	■	■	■		■	■	
Bt. ssp. *aizawai*	■	■	■			■	
Bt. ssp. *israelensis*				■			
Bt. ssp. *kurstaki*	■	■	■		■	■	■
Condor	■	■	■			■	■
Crymax	■	■	■			■	■
Cutlass	■	■	■			■	■
DiPel	■	■	■		■	■	■
Javelin	■	■	■				
Lepinox	■	■	■			■	■
Troy-Bt		■	■		■	■	
VectoBac				■			
XenTari	■	■	■			■	

Manufacturers List

These Bt products are manufactured by the following companies. Look for their addresses, phone numbers, and fax numbers in the Suppliers Directory beginning on page 599.

Agree, Javelin – Thermo Trilogy

Bactimos, Biobit, DiPel, VectoBac, XenTari – Abbott Laboratories

BMP 123 – Becker Microbial

Condor, Crymax, Cutlass, Lepinox – Ecogen

Troy-Bt – Troy Biosciences

Enhanced Seed Products	AgriPro Seeds Inc.	Croplan Genetics	DEKALB Genetics Corp.	Garst Seed Co.	Hoffman Seeds Inc.	Interstate Payco Seed	Merschman Seeds	Novartis Seeds	Pioneer Hi-Bred
Alfalfa + leafhopper resistance/tolerance	■	■	■	■	■	■	■	■	■
Alfalfa + leafhopper resistance/tolerance + Phytophthora root rot resistance/tolerance + bacterial wilt resistance/tolerance + Verticillium wilt resistance/tolerance + Anthracnose resistance/tolerance						■			
Alfalfa + leafhopper resistance/tolerance + Phytophthora root rot resistance/tolerance + bacterial wilt resistance/tolerance + Verticillium wilt resistance/tolerance + Aphanomyces resistance/tolerance			■			■		■	
Alfalfa + Phytophthora root rot resistance/tolerance + Aphanomyces resistance/tolerance + Sclerotinia crown/stem rot resistance/tolerance			■						

Other Uses

Insecticides

Forage

(Alfalfa, Birdsfoot Trefoil, Clover, Lespedeza, Vetch)

Adios — see carbaryl

Ambush — see permethrin

AZINPHOS-METHYL

Restricted Use Pesticide (except Guthion Solupak).

Gowan Azinphos-M 50-W *(Gowan)*
Guthion Solupak (50WSP) *(Bayer)*
◆ **Micro Flo Azinphosmethyl 50W** *(Micro Flo)*
Sniper 50PVA, 50W *(Platte)*

Rate: 0.25-0.5 lb. ai/A
0.5-1 lb. 50PVA, 50WSP/A
0.25-0.75 lb. ai/A
0.5-1.5 lb. 50-W, 50W/A

Pest:

alfalfa plant bug	fleahoppers	Lygus bugs
alfalfa weevil	grasshoppers	spittlebugs
Egyptian alfalfa weevil	leafhoppers	

Remarks: Alfalfa, birdsfoot trefoil, clover. Apply in minimum 10 gal. water/A by ground; 1-5 gal. water/A by air. Rates and volume vary by pest species; refer to label. Apply only once per cutting regardless of rate, formulation, or application method used.
Days To Harvest: 14-21 days depending on rate and formulation; refer to label.
State Restrictions: For use east of the Mississippi only.

Signal Word/Toxicity Class: Danger-Poison/I.
REI: 48 hr.

Baythroid — see cyfluthrin

CARBARYL

Carbamate insecticide. Highly toxic to bees exposed to direct treatment or residues on blooming crops or weeds.

Adios (13%) *(Micro Flo)*
Rate: 0.065-0.1 lb. ai/A
0.5-0.75 lb. 13%/A
Pest:
cucumber beetle
Remarks: Alfalfa, clover. Do not apply through any type of irrigation system. Do not apply more than 3.75 lb./A per season.
Days To Harvest: 0.
State Restrictions: Not for use in California.

Signal Word/Toxicity Class: Caution/III.
REI: 12 hr.

Clean Crop Sevin 5 Bait *(Platte)*
Rate: 1.5 lb. ai/A
30 lb. Bait/A
Pest:

ants	darkling beetle	grasshoppers
armyworms	field crickets	sowbugs
cutworms		

Remarks: Alfalfa, birdsfoot trefoil (forage, hay), clover. Apply by ground or air. Do not make more than 2 applications after grain heads emerge from boot.
Days To Harvest: Birdsfoot trefoil (forage, hay), clover-0; alfalfa-3.

Signal Word/Toxicity Class: Caution/III.
REI: 12 hr.

Drexel Carbaryl 4L, 50W, or 80S *(Drexel)*
Sevin XLR Plus, 4F, 80S or 80WSP *(Rhone-Poulenc)*
Rate: 0.5-1.5 lb. ai/A
1-3 pt. 4L, 4F, XLR/A
1-3 lb. 50W/A
0.66-1.875 lb. 80S, 80WSP/A
Pest:

alfalfa blotch leafminer	Egyptian alfalfa weevil (larvae)	Lygus bugs
alfalfa caterpillar	Essex skipper	Mexican bean beetle
alfalfa weevil (larvae)	European alfalfa beetle	stink bugs
armyworms	fall armyworm	threecornered alfalfa hopper
bean leaf beetle	grasshoppers	thrips
blister beetle	green cloverworm	velvetbean caterpillar
clover head weevil	Japanese beetle	webworms
corn earworm	leafhoppers	yellowstriped armyworm
cutworms		

Remarks: Alfalfa, clover. Apply when insects or their damage appear. Do not apply to wet foliage or when rain or high humidity is expected during next 2 days. Rate varies by pest species and geographic area; refer to label. On dense growth, use 25-40 gal. water/A (20-40 gal. for 80S, 80WSP) with ground equipment. Apply only once per cutting. Observe plant response precautions.
Days To Harvest: Alfalfa-3; clover-0.

Signal Word/Toxicity Class: Caution (4F, 4L, XLR); Warning/II (50W, 80S, 80WSP).
REI: 12 hr.

Gowan Sevin Bait *(Gowan)*
Rate: 1.5 lb. ai/A
30 lb. Bait/A
Pest:

armyworms	cutworms	grasshoppers
crickets	darkling beetle	sowbugs

Remarks: Alfalfa. Apply when pests appear in damaging numbers; repeat at 7- to 14-day intervals if necessary.
Days To Harvest: 3.
State Restrictions: For use in Arizona and California.

Signal Word/Toxicity Class: Caution/III.
REI: 12 hr.

◆ **Wilbur-Ellis Sevin 5 or 10 Dust** *(Wilbur-Ellis)*
Rate: 1.5 lb. ai/A
30 lb. 5/A
15 lb. 10/A
Pest:

alfalfa caterpillars	Japanese beetle	threecornered alfalfa hopper
armyworms	leafhoppers	thrips
corn earworms	stink bugs	

 For personal protective equipment and EPA registration numbers, please turn to page 39.

Remarks: Do not feed treated vines to livestock.
Days To Harvest: 3.
State Restrictions: Sevin 5 not registered in California.

Signal Word/Toxicity Class: Caution.
REI: 12 hr.

CARZOL

Miticide/insecticide in water-soluble packaging. Toxic to wildlife and toxic to bees exposed to direct treatment.

Carzol (SP) (formetanate hydrochloride) *(AgrEvo USA)*

Rate: 0.46-0.92 lb. ai/A
0.5-1 lb. SP/A

Pest:

consperse stink bug (CA only)	Lygus bugs	twospotted spider mite

Remarks: Alfalfa (seed crop only). Apply high rate for consperse stink bug. Use only prebloom in areas except California.
State Restrictions: For use in California, Idaho, Nevada, Oregon, and Washington.

Signal Word/Toxicity Class: Danger-Poison/I.
REI: 48 hr.

CHLORPYRIFOS

Nonrestricted, organophosphate insecticide used as a foliar spray for pests on alfalfa. Gives excellent control of numerous insect pests. When applied at early bud stage on alfalfa, affords good control of grasshoppers and lygus. Toxic to birds and wildlife, extremely toxic to fish and aquatic organisms, and highly toxic to bees exposed to direct treatment on alfalfa. Do not apply if bees are foraging or clustered outside of hives.

Lock-On *(Dow AgroSciences)*

Rate: 0.375-0.5 lb. ai/A
1.5-2 pt./A

Pest:
alfalfa caterpillar

Rate: 0.5 lb. ai/A
2 pt./A

Pest:

armyworms	Egyptian alfalfa weevil	pea aphid
blue alfalfa aphid		

Remarks: Alfalfa. Apply in 3-10 gal. water/A. Do not make more than 4 applications per year or apply more than once per crop cutting. Do not cut or graze within 4 days after application.
State Restrictions: For use in Arizona, California.

Signal Word/Toxicity Class: Warning/II.
REI: 12 hr.

Lorsban-4E *(Dow AgroSciences)*
◆ **Micro Flo Chlorpyrifos 4E** *(Micro Flo)*
Nufos 4E *(Cheminova)*

Rate: 0.25-1 lb. ai/A
0.5-2 pt. 4E/A

Pest:

alfalfa blotch leafminer	corn rootworm (adult)	plant bugs
alfalfa looper	cutworms	spittlebugs
alfalfa weevil	Egyptian alfalfa weevil	spotted alfalfa aphid
aphids (suppression)	grasshoppers	spotted cucumber beetle
armyworms	leafhoppers	

Remarks: Alfalfa. Apply by ground, air, or through center pivot irrigation system. Rate varies by pest species and geographic area; refer to label. Stubble spray may be applied to control leafhoppers in the Northeast. Do not make more than 4 applications per year or apply more than once per crop cutting.
Days To Harvest: Cutting/grazing: 0.5 pt.-7; 1 pt.-14; over 1 pt.-21.

Signal Word/Toxicity Class: Warning/II.
REI: 24 hr.

Comite — see propargite

CYFLUTHRIN

Synthetic pyrethroid formulated as an emulsifiable concentrate. Toxic to fish and aquatic organisms. Restricted Use Pesticide.

Baythroid 2 *(Bayer)*

Rate: 0.8-2.8 fl. oz. 2/A

Pest:

alfalfa caterpillar	blue aphid	meadow spittlebug
alfalfa looper	corn earworm	pea aphid
alfalfa plant bug	cutworms	potato leafhopper
alfalfa webworm	Egyptian alfalfa weevil	redlegged grasshopper
alfalfa weevil	green cloverworm	tarnished plant bug
aster leafhopper	Lygus bugs	
blotch leafminer		

Remarks: Alfalfa (except grown for seed). Apply in sufficient water for complete coverage of foliage. Applications should be based on local economic thresholds. Use lower rate for low to moderate insect pressure (requires careful scouting); higher rate for moderate to heavy insect pressure. Apply only once per cutting. Total of 11.2 fl. oz./A may be applied per crop season. Refer to label for spray drift precautions.

Signal Word/Toxicity Class: Danger/I.
REI: 12 hr.

DEADLINE

Packaged as Bullets, M-Ps (mini-pellets), and granules.

Deadline Bullets, M-Ps, Granules (4%) (metaldehyde) *(Pace)*

Rate: 0.4-1 lb. ai/A
10-40 lb. 4%/A

Pest:

slugs	snails

Remarks: Apply every 3-4 weeks during growing season or as needed. Do not apply directly to edible portion of plant.

Signal Word/Toxicity Class: Warning/II (granule); Caution (bullets, M-P).
REI: 12 hr.

Declare — see methyl parathion

Diatect — see pyrethrins

DIBROM

Organophosphate insecticide formulated as an emulsifiable concentrate. Toxic to fish, aquatic invertebrates, and wildlife. Highly toxic to bees exposed to direct treatment.

Dibrom 8 (8E) (naled) *(Amvac)*

Rate: SLN-Oregon:
1 lb. ai/A
1 pt. 8E/A
SLN-Idaho, Nevada, Washington:
1-1.5 lb. ai/A
1-1.5 pt. 8E/A

Pest:
Lygus bugs

Rate: SLN-Idaho, Nevada, Washington:
0.5 lb. ai/A
0.5 pt. 8E/A

Pest:
thrips

Remarks: Alfalfa (seed only). For thrips, apply in 10-20 gal. water by ground; 3-5 gal. water by air. For Lygus bugs, apply at 7-day intervals. Do not use treated seed or crop parts for food, feed, or oil purposes. Do not graze treated alfalfa. Do not apply through any type of irrigation system. Refer to label for directions, restrictions, and precautions.
State Restrictions: SLN-Idaho, Nevada, Oregon, Washington. In Oregon, for use only on fields operated by members of Oregon Alfalfa Seed Growers Association or Garden Alfalfa Seed Growers Association, and located in Malheur, Umatilla, Josephine, Jackson or Morrow counties, Oregon.

Signal Word/Toxicity Class: Danger/I.
REI: 48 hr.

◆ - **new product** • ai-active ingredient • DF-dry flowable • E/EC-emulsifiable concentrate • EW-emulsion in water
F/FL-flowable • G-granule • L/LC-liquid • SP/WSP-soluble packet • W/WP-wettable powder • WSB-water soluble bag

DICOFOL

Emulsifiable formulation that gives high initial kill against most mite species. Toxic to fish.

◆ **Dicofol 4E** *(Griffin)*

Rate: 0.3-0.4 lb. ai/A
0.6-0.8 pt. 4E/A

Pest:
Banks grass mite
bermudagrass mite
brown mite
clover mite

Remarks: Bermudagrass grown for seed. Do not feed clippings or allow livestock to graze treated areas.

Signal Word/Toxicity Class: Caution.
REI: 12 hr.

Digon — see dimethoate

Dimate — see dimethoate

DIMETHOATE

Cheminova Dimethoate 4E *(Cheminova)*
Dimate 4E *(Terra)*
Drexel Dimethoate 2.67 or 4EC *(Drexel)*
Gowan Dimethoate E267 *(Gowan)*
Helena Dimethoate 4E *(Helena)*
◆ **Micro Flo Dimethoate 267 or 4E** *(Micro Flo)*
Wilbur-Ellis Dimethoate 267 or Digon 400 *(Wilbur-Ellis)*

Rate: 0.25-0.5 lb. ai/A
0.75-1.5 pt. 2.67, 267, E267/A
0.5-1 pt. 4E, 4EC, 400/A

Pest:
alfalfa weevil (larvae) (reduction)
aphids
grasshoppers
leafhoppers
Lygus bugs
plant bugs

Remarks: Alfalfa. Apply as infestation warrants. Do not apply to alfalfa in bloom period. For alfalfa grown for hay, apply only once per season. Effective only on cutting to which applied. Do not apply through any type of irrigation system.
Days To Harvest: 10.

Signal Word/Toxicity Class: Warning/II.
REI: 48 hr.

DI-SYSTON

Restricted Use Pesticide.

Di-Syston 15% (disulfoton) *(Bayer)*

Rate: 1 lb. ai/A
6.7 lb. 15%/A

Pest:
aphids
mites

Remarks: Alfalfa, clover. Make only 1 application per crop season.
Days To Harvest: Grazing or cutting for forage-28.

Signal Word/Toxicity Class: Danger-Poison/I.
REI: 48 hr.

ENDOSULFAN

Drexel Endosulfan 3EC or 50WP *(Drexel)*
Gowan Endosulfan 3EC or 50WSB *(Gowan)*
◆ **Micro Flo Endosulfan 3EC or 50WP** *(Micro Flo)*
Phaser 3EC or 50WSB *(AgrEvo USA)*
Thiodan 3EC or 50WP *(FMC)*

Rate: 0.25 lb. ai/A
0.66 qt. 3EC/A
0.5 lb. 50WP, 50WSB/A

Pest:
meadow spittlebug

Remarks: Alfalfa forage. Apply 1 week after eggs begin to hatch, or when first small masses of froth are produced in crown of plant (usually when alfalfa is 4-6" tall). Do not apply more than once per cutting or exceed 3 applications per year.
Days To Harvest: Grazing/cutting for forage: 21.

Signal Word/Toxicity Class: Danger-Poison/I.
REI: 24 hr.

FURADAN

Restricted Use Pesticide.

Furadan 4F (carbofuran) *(FMC)*

Rate: 0.125-1 lb. ai/A
0.25-2 pt. 4F/A

Pest:
alfalfa blotch leafminer
alfalfa weevil
blue alfalfa aphid
Egyptian alfalfa weevil
grasshoppers
Lygus bugs
pea aphid
potato leafhopper
snout beetle (New York)

Remarks: Apply when insects appear or feeding is noticed. Rate varies by pest species. Do not move bees to alfalfa fields within 7 days of application. Do not apply more than twice per season or once per cutting. Do not use more than 1 pt./A in second application. Do not apply through any type of irrigation system.
Days To Harvest: Up to 0.5 pt.-7; 1 pt.-14; 2 pt.-28.
Crop Rotation: Do not rotate with any crop on soil treated at greater than 3 lb. ai/A for at least 10 months or plant with any crop for which carbofuran treatment is not registered for at least 10 months.
State Restrictions: Not for use in Long Island, New York.

Signal Word/Toxicity Class: Danger-Poison/I.
REI: 48 hr.

Fyfanon — see malathion

Guthion — see azinphos-methyl

IMIDAN

Imidan 70-W (phosmet) *(Gowan)*

Rate: 0.75-1 lb. ai/A
1-1.33 lb. 70-W/A
Arizona, California, Nevada:
0.75 lb. ai/A
1 lb. 70-W/A

Pest:
alfalfa blotch leafminer
alfalfa plant bug
common alfalfa weevil
Egyptian alfalfa weevil
fleahoppers
grasshoppers
Lygus bugs
meadow spittlebug
pea aphid
potato leafhopper (Northeast, North Central states only)
spittlebugs

Remarks: Alfalfa. Spray when larvae are actively feeding. Application for adult weevils should be timed to periods of visible activity. For alfalfa blotch leafminer and meadow spittlebug, apply when first signs of infestation are visible. Consult county agent or extension service representative regarding proper application timing. Do not apply more than once per cutting. Do not apply to alfalfa in bloom period. For control of pea aphids, may be tank mixed with other insecticides registered for use on alfalfa; refer to both product labels.
Days To Harvest: Grazing, cutting for hay: 7 except AZ, CA, NV-14.

Signal Word/Toxicity Class: Warning/II.
REI: 24 hr.

Kumulus — see sulfur

LANNATE

Restricted Use Pesticides.

Lannate (90) or LV (methomyl) *(DuPont)*

Rate: 0.225-0.9 lb. ai/A
0.25-1 lb. 90/A
0.75-3 pt. LV/A

10 Other Uses

NOTICE The information on these pages is for preliminary planning — not a guide for use. Be sure to follow the manufacturer's directions, notwithstanding information contained here.

For personal protective equipment and EPA registration numbers, please turn to page 39.

Pest:

alfalfa caterpillar, alfalfa weevil (larvae), aphids, armyworms (bermudagrass), beet armyworm, blotch leafminer, Egyptian alfalfa weevil (larvae), fall armyworm (bermudagrass), loopers, Lygus bugs, pea aphid, variegated cutworm, western yellowstriped armyworm, yellowstriped armyworm

Remarks: Alfalfa, bermudagrass. Apply in sufficient water at 5- to 7-day intervals or as needed to obtain thorough, uniform coverage when insects first appear. Rate varies by pest species; refer to label. Do not apply to dormant or semi-dormant alfalfa when minimum daily temperature is 50°F or lower. For alfalfa, do not apply more than 3.6 lb. ai/A per crop or make more than 10 applications per crop. For bermudagrass, do not apply more than 0.9 lb. ai/A per crop or make more than 4 applications per crop.

Days To Harvest: Alfalfa: feeding/grazing-7. Bermudagrass: feeding/grazing/forage-7; dehydrated hay-3.

Signal Word/Toxicity Class: Danger-Poison/I.
REI: 48 hr.

Lock-On — see chlorpyrifos

Lorsban — see chlorpyrifos

MALATHION

Organophosphate insecticide toxic to fish, aquatic invertebrates, aquatic life stages of amphibians, and bees exposed to direct treatment on blooming crops or weeds.

Clean Crop Malathion (57EC) *(Platte)*

Rate: 1-1.5 lb. ai/A
1.5-2.25 pt. 57EC/A

Pest: Alfalfa:

aphids, armyworms, clover leaf weevil, grasshoppers, leafhoppers, Lygus bugs, spider mites, spittlebugs, spotted alfalfa aphid, weevil larvae

Rate: 1-1.3 lb. ai/A
1.5-2 pt. 57EC/A

Pest: Clover:

aphids, clover leaf weevil, leafhoppers, Lygus bugs

Lespedeza, lupines:

grasshoppers

Vetch:

omnivorous leaftier, pea aphid, vetch bruchid

Remarks: Apply to alfalfa in bloom in the evening or early morning when bees are not working in fields or hanging on outside of hives.

Days To Harvest: 0.

Signal Word/Toxicity Class: Caution/III.
REI: 12 hr.

Clean Crop Malathion 8EC *(Platte)*

Rate: 1.5 lb. ai/A
1.5 pt. 8EC/A

Pest:

alfalfa weevil (larvae), aphids, armyworms, grasshoppers, leafhoppers, Lygus bugs, spider mites, spittlebugs

Rate: Ground:
0.8-1 lb. ai/A
0.8-1 pt. 8EC/A
Air:
1-1.5 lb. ai/A
1-1.5 pt. 8EC/A

Pest:

spotted alfalfa aphid

Remarks: Alfalfa, clover. Do not apply to alfalfa and clover in bloom.

Days To Harvest: 0.

Signal Word/Toxicity Class: Caution/III.
REI: 12 hr.

Drexel Malathion 5EC *(Drexel)*
◆ Micro Flo Malathion 5EC *(Micro Flo)*
Prentox 5 lb. Malathion Spray (5EC) *(Prentiss)*

Rate: Alfalfa, clover:
1-1.6 lb. ai/A
1.5-2.5 pt. 5EC/A

Pest:

alfalfa weevil (larvae), aphids, armyworms, clover leaf weevil, grasshoppers, Lygus bugs, potato leafhoppers, spider mites, spittlebugs, stink bugs (alfalfa), vetch bruchid (alfalfa)

Rate: Vetch:
1-1.25 lb. ai/A
1.5-2 pt. 5EC/A

Pest:

omnivorous leaftier, pea aphid, vetch bruchid

Remarks: Apply to alfalfa in bloom in the evening or early morning when bees are not working in field or hanging on outside of hives. Rate varies by pest species; refer to label. Do not apply to clover in bloom. Do not apply through any type of irrigation system.

Days To Harvest: 0.

Signal Word/Toxicity Class: Warning/II.
REI: 12 hr.

Fyfanon (8E) *(Helena)*
◆ Wilbur-Ellis Malathion 8 *(Wilbur-Ellis)*

Rate: 1-2 lb. ai/A
1-2 pt. 8E, 8/A

Pest:

alfalfa weevil (larvae), aphids, armyworms, clover leaf weevil, grasshoppers, Lygus bugs, pea aphid, potato leafhopper, spider mites, spittlebugs, vetch bruchid

Remarks: Alfalfa, clover. Use higher rate for armyworms. Do not apply during full bloom.

Days To Harvest: Helena: Up to 1.5 pt.-0; over 2 pt.-7. Wilbur-Ellis: 7.

Signal Word/Toxicity Class: Caution/III.
REI: 12 hr.

Fyfanon ULV *(Cheminova)*
◆ Micro Flo Malathion ULV *(Micro Flo)*

Rate: 0.5-1 pt. ULV/A

Pest:

alfalfa caterpillar, alfalfa weevil (larvae), beet armyworm, grasshoppers, western yellowstriped armyworm

Remarks: Alfalfa. Rate varies by pest species; refer to label. Do not apply to seed alfalfa or alfalfa in bloom. For alfalfa weevil, apply when day temperatures are expected to exceed 65°F and when 50-75% of leaves show feeding damage.

Days To Harvest: 0.5 pt.-0; 0.75-1 pt.-5.

Rate: 0.5-0.75 pt. ULVA

Pest:

black grass bug, grasshoppers

Remarks: Clover. Do not apply to clover in bloom.

Days To Harvest: 0.

Signal Word/Toxicity Class: Caution.
REI: 12 hr.

◆ Gowan Malathion 8 (8E) *(Gowan)*

Rate: 1.25-1.5 lb. ai/A
1.25-1.5 pt. 8E/A

Pest:

alfalfa weevil (larvae), aphids, armyworms, clover leaf weevil, grasshoppers, Lygus bugs, pea aphid, potato leafhopper, spider mites, spittlebugs, spotted alfalfa aphid, stink bugs, vetch bruchid

Remarks: Alfalfa, clover. Apply to alfalfa in bloom in the evening or early morning when bees are not working in fields or hanging on outside of hives.

Days To Harvest: Up to 1.5 pt.-0; 2 pt. -7.

Signal Word/Toxicity Class: Caution/III.
REI: 12 hr.

METASYSTOX-R

Restricted Use Pesticide.

Metasystox-R SC (oxydemeton-methyl) *(Gowan)*

Rate: 0.375-0.5 lb. ai/A
1.5-2 pt. SC/A

Pest:

aphids	Lygus bugs	thrips
leafhoppers	mites	

Remarks: Alfalfa, clover (seed crop). Apply in sufficient water for complete coverage but not less than 1 gal./A. Chaff from seed crop may be used for feed or forage, but do not cut green crop for these purposes.
Days To Harvest: 21.
State Restrictions: Not registered in California.

Signal Word/Toxicity Class: Warning/II.
REI: 48 hr.; 72 hr. if average rainfall less than 25 " per year.

METHOXYCHLOR

Chlorinated hydrocarbon insecticide with good residual activity in field applications. Toxic to fish and bees.

Prentox Methoxychlor 2EC *(Prentiss)*

Rate: 0.5-2 lb. ai/A
1-4 qt. 2EC/A

Pest:

alfalfa caterpillar	cucumber beetle	pea weevil
alfalfa weevil (larvae)	fall armyworm	soybean caterpillar
armyworms	flea beetles	spittlebugs
blister beetle	Japanese beetle	velvetbean caterpillar
clover leaf weevil	leafhoppers	webworms
cowpea curculio	Mexican bean beetle	

Remarks: Alfalfa, clover. Apply at first sign of infestation; repeat at 7- to 14-day intervals or as needed. Rate varies by pest species; refer to label.
Days To Harvest: Harvest or grazing-7.

Signal Word/Toxicity Class: Caution.
REI: 12 hr.

Wilbur-Ellis Methoxychlor 2 *(Wilbur-Ellis)*

Rate: 1-1.5 lb. ai/A
2-3 qt. 2/A

Pest:

alfalfa webworm	fall armyworm	soybean caterpillar
alfalfa weevil	pea weevil	velvetbean caterpillar

Remarks: Alfalfa. Apply at first sign of infestation, repeat at 7- to 10-day intervals. Do not apply within 14 days of sulfur or sulfur-containing sprays.
Days To Harvest: 7.

Signal Word/Toxicity Class: Danger-Poison/I.
REI: 48 hr.

METHYL PARATHION

Penncap-M is microencapsulated form of methyl parathion. Restricted Use Pesticide due to residual effects to avian species and hazard to bees.

Cheminova Methyl 4EC *(Cheminova)*

Rate: 0.25-1 lb. ai/A
0.5-2 pt. 4EC/A

Pest:

alfalfa caterpillar	clover leaf weevil	Lygus bugs
alfalfa weevil	Egyptian alfalfa weevil	spider mites
aphids		spotted alfalfa aphid
armyworms	grasshoppers	webworms
climbing cutworm	leafhoppers	

Remarks: Alfalfa, clover. Rate varies by pest species; refer to label. Do not apply if crop or weeds in treatment area are in bloom. In California and Nevada, do not apply more than 0.75 pt. 4EC/A.
Days To Harvest: 15.

Rate: 0.5-1 lb. ai/A
1-2 pt. 4E/A
0.25-1.25 lb. ai/A
0.5-2.5 pt. 4EC/A

Pest:

alfalfa weevil (larvae)	climbing cutworm	Lygus bugs
aphids	leafhoppers	vetch bruchid
armyworms		

Remarks: Vetch. For alfalfa weevil, apply when 75% of terminals show feeding, when bugs and shoots appear ragged, or when larvae counts reach 25 per sweep.
Days To Harvest: Up to 2 pt.-15; 2-2.5 pt.-20.

Signal Word/Toxicity Class: Danger-Poison/I.
REI: 48 hr.; 72 hr. if average rainfall less than 25 " per year.

◆ **Declare (4EC)** *(Griffin)*
Helena 4 lb. Methyl Parathion (4EC) *(Helena)*

Rate: Alfalfa, clover, vetch:
0.23-1.18 lb. ai/A
0.5-2.5 pt. 4EC/A

Pest: Alfalfa:

alfalfa caterpillar	climbing cutworm	stink bugs
alfalfa seed chalcid	flea beetles	

Clover:

climbing cutworm	clover seed chalcid (seed crops only)

Vetch:

aphids	leafhoppers	mites
armyworms	Lygus bugs	vetch bruchid
climbing cutworm		

Rate: Alfalfa, clover:
0.23-0.94 lb. ai/A
0.5-2 pt. 4EC/A

Pest:

alfalfa weevil (adult)	leafhoppers	plant bugs (seed crops only)
armyworms	mites	
Egyptian alfalfa weevil	pea aphid (alfalfa)	spotted alfalfa aphid

Rate: Alfalfa, clover, vetch:
0.23-0.47 lb. ai/A
0.5-1 pt. 4EC/A

Pest:
alfalfa weevil (larvae)

Remarks: Rate varies by pest species; refer to label. For alfalfa weevil (larvae), apply when 75% of terminals show feeding, when buds and shoots appear ragged, or when larvae count reaches 25 per sweep. For aphids, apply in early spring.
Days To Harvest: Up to 2 pt.-15; more than 2 pt.-20.

Signal Word/Toxicity Class: Danger-Poison/I.
REI: 48 hr.

Penncap-M (2FL) *(Elf Atochem)*

Rate: 0.375-0.5 lb. ai/A
1.5-2 pt. 2FL/A

Pest:
blue alfalfa aphid

Rate: 0.5-0.75 lb. ai/A
2-3 pt. 2FL/A

Pest:

alfalfa blotch leafminer	grasshoppers	potato leafhopper
alfalfa plant bug	green cloverworm	spittlebugs
alfalfa weevil	pea aphid	tarnished plant bug

Rate: 0.5-1 lb. ai/A
2-4 pt. 2FL/A

Pest:

cutworms	Egyptian alfalfa weevil

Remarks: Alfalfa. Do not apply if alfalfa or weeds in treatment area are in bloom. Do not apply to alfalfa grown for seed within 7 days prior to bloom or while bloom is present in treatment area.
Days To Harvest: Harvest or grazing-15.

Signal Word/Toxicity Class: Warning/II.
REI: 48 hr.

Wilbur-Ellis Methyl Parathion 4 Spray *(Wilbur-Ellis)*

Rate: 0.5 lb. ai/A
1 pt. 4/A

Pest:

aphids	leafhoppers	spotted alfalfa aphid
armyworms	spider mites	

Remarks: Do not apply if the crop or weed in the treatment area is in bloom. Do not apply through any type of irrigation system.
Days To Harvest: 15.
State Restrictions: California: Do not apply more than 0.75 pt. /A.

Signal Word/Toxicity Class: Danger-Poison/I.
REI: 24 hr.

NOTICE The information on these pages is for preliminary planning — not a guide for use. Be sure to follow the manufacturer's directions, notwithstanding information contained here.

For personal protective equipment and EPA registration numbers, please turn to page 39.

Wilbur-Ellis Methyl Parathion 5 Spray *(Wilbur-Ellis)*
Rate: 1.3 lb. ai/A
2 pt. 5/A
Pest:
Lygus bugs
stink bugs
Remarks: Do not use crop residues for grazing or feeding.

Signal Word/Toxicity Class: Danger-Poison/I.
REI: 48 hr.

Wilbur-Ellis Methyl Parathion 7.5 Spray *(Wilbur-Ellis)*
Rate: 0.235-0.94 lb. ai/A
0.25-1 pt. 7.5/A
Pest:
alfalfa caterpillar
alfalfa weevil
aphids
armyworms
climbing cutworm
clover leaf weevil
Egyptian weevil
flea beetles
grasshoppers
leafhoppers
Lygus bugs
mites
pea aphid
spotted alfalfa aphid
stink bugs
thrips
webworms
Remarks: Rates vary by pest species.
Days To Harvest: 15.

Signal Word/Toxicity Class: Danger-Poison/I.
REI: 48 hr.; 72 hr. if annual rainfall is less than 25" per year.

Microthiol Special — see sulfur

M-PEDE

M-Pede (potassium salts of fatty acids) *(Mycogen)*
Rate: 4 pt./A in 3-25 gal. water
2% v/v spray solution/A in more than 25 gal. water
Pest:
aphids
caterpillars (tank mix only)
fleahoppers
greenbug
Lygus bugs
plant bugs
spider mites
stink bugs
thrips
whiteflies
Remarks: Alfalfa. Apply at first sign of infestation; repeat at 7- to 14-day intervals. May be tank mixed with other insecticides; refer to both labels for use directions and precautions.
Days To Harvest: 0.
State Restrictions: Not for use in California.

Signal Word/Toxicity Class: Warning/II.
REI: 12 hr.

Nufos — see chlorpyrifos

Penncap-M — see methyl parathion

PERMETHRIN

Synthetic pyrethroid with low mammalian toxicity. Highly toxic to bees exposed to direct treatment. Restricted Use Pesticide due to toxicity to fish and aquatic organisms.

Ambush 2EC or 25WP *(ZENECA)*
Pounce 3.2EC or 25WP *(FMC)*
Rate: 0.05-0.2 lb. ai/A
3.2-12.8 fl. oz. 2EC/A
2-8 oz. 3.2EC/A
3.2-12.8 oz. 25WP/A
Pest:
alfalfa caterpillar
alfalfa weevil
armyworms
blue alfalfa aphid
cucumber beetle
cutworms
Egyptian alfalfa weevil
green cloverworm
green peach aphid
loopers
meadow spittlebug
pea aphid
plant bugs
potato leafhopper
spotted alfalfa aphid
stink bugs
velvetbean caterpillar
webworms
Remarks: Alfalfa. Apply in minimum 10 gal. finished spray/A by ground; or 1 gal. by air. Rate and volume varies by pest species, population level, and plant density. Do not apply more than 0.2 lb. ai/A per cutting. Do not use within 20 yd. aquatic habitats for ground application; 100 yd. aquatic habitats for air. Do not graze or feed soybean forage.
Days To Harvest: Up to 0.1 lb. ai-0; over 0.1 lb. ai-14.

Signal Word/Toxicity Class: Caution (3.2EC); Warning/II (2EC, 25WP).
REI: 24 hr.

Phaser — see endosulfan

Pounce — see permethrin

PROPARGITE

Sulfite ester miticide with residual killing action, available as a liquid emulsifiable concentrate. Should not be mixed with alkaline materials or with materials containing large amounts of petroleum solvents.

Comite (EC) *(Uniroyal)*
Rate: 1.2-2.46 lb. ai/A
1.5-3 pt. EC/A
Pest:
twospotted spider mite
Remarks: Alfalfa grown for seed. Not hazardous to bees when used alone. However, bee hazards may be encountered when combined with Systox or Toxaphene. Observe state department of agriculture regulations on use of insecticides for seed alfalfa. Compatible with most insecticides. Refer to label for further use instructions. Do not apply through any type of irrigation system. Do not feed treated foliage, alfalfa trash, or seed screenings to livestock or graze treated fields.
State Restrictions: SLN-California, Idaho, Montana, Nevada, Oregon, Utah, and Washington.

Signal Word/Toxicity Class: Danger/I.
REI: 7 days. After 48 hr., workers can enter treated areas to perform hand labor tasks if they wear early-entry personal protective equipment.

PYRELLIN (Premix)

Premix containing pyrethrins and rotenone. May be combined with other insecticides to enhance knockdown or as an insect "exciter" to help facilitate contact with other spray residues.

Pyrellin E.C. (0.042 lb./gal. pyrethrins & 0.035 lb./gal. rotenone) *(Webb Wright)*
Rate: 1-2 pt. E.C./A
Pest:
aphids
armyworms
chinch bug
flies
mites
thrips
Remarks: Alfalfa, clover. Apply at 7-day intervals or less; repeat as needed.
Days To Harvest: 0.

Signal Word/Toxicity Class: Caution/III.
REI: 12 hr.

PYRETHRINS

Contains piperonyl butoxide as a synergist. Diatect also contains silicon dioxide (diatomaceous earth).

Diatect Multipurpose Insecticide II *(Diatect)*
Pyrenone (0.5EC) *(AgrEvo Environmental Health)*
Rate: 1-6 lb. II/A
1-12 fl. oz. 0.5EC/A
Pest:
aphids
armyworms
caterpillars
chinch bug
flies
mites
thrips
Remarks: Alfalfa, clover, lespedeza, trefoil, vetch. Repeat as needed. Apply Diatect as dust or wettable powder.

Signal Word/Toxicity Class: Caution.
REI: 12 hr.

Pyrenone — see pyrethrins

Sevin — see carbaryl

Sniper — see azinphos-methyl

◆ - **new product** • ai-active ingredient • DF-dry flowable • E/EC-emulsifiable concentrate • EW-emulsion in water
F/FL-flowable • G-granule • L/LC-liquid • SP/WSP-soluble packet • W/WP-wettable powder • WSB-water soluble bag

SULFUR

Microthiol and Thiolux are micronized sulfur.

◆ **Kumulus DF** *(BASF)*
Rate: Alfalfa:
9.6-20 lb. ai/A
12-25 lb. DF/A
Seed alfalfa:
2.4-4 lb. ai/A
3-5 lb. DF/A
Pest:
red spider mite
Remarks: Repeat at 1- to 2-week intervals for alfalfa. Do not apply if temperatures during or within 3 days after application are expected to exceed 90°F. Do not apply through any type of irrigation system. Refer to label for use directions.

Signal Word/Toxicity Class: Caution.
REI: 24 hr.

Microthiol Special (80W) *(Elf Atochem)*
Thiolux (DF) *(Platte)*
Rate: 6.4-8 lb. ai/A
8-10 lb. 80W/A
2.4-4 lb. ai/A
3-5 lb. DF/A
Pest:
Atlantic mite, red spider mite
Remarks: Apply as needed.
State Restrictions: Microthiol Special not for use in California.

Signal Word/Toxicity Class: Caution.
REI: 24 hr.

Thiodan — see endosulfan

Thiolux — see sulfur

Forage
(Grasses, Hay, etc.)

CARBARYL

Carbamate insecticide. Highly toxic to bees exposed to direct treatment or residues on blooming crops or weeds.

Clean Crop Sevin 5 Bait *(Platte)*
Rate: 1-2 lb. ai/A
20-40 lb. Bait/A
Pest:
ants, armyworms, cutworms, darkling beetle, field crickets, grasshoppers, sowbugs
Remarks: Forage grasses. Apply with ground or air equipment.
Days To Harvest: 0.

Signal Word/Toxicity Class: Caution/III.
REI: 12 hr.

Clean Crop Sevin — see carbaryl

lambda-CYHALOTHRIN

Synthetic pyrethroid insecticide with broad-spectrum activity, formulated as an emulsifiable concentrate. Restricted Use Pesticide due to toxicity to fish and aquatic organisms.

Karate or Warrior (1E) *(ZENECA)*
Rate: 0.015-0.03 lb. ai/A
1.92-3.84 fl. oz. 1E/A
Pest:
army cutworm, armyworms, birdcherry oat aphid, cereal leaf beetle, chinch bug, cutworms, English grain aphid, fall armyworm, flea beetles, grasshoppers, greenbug, Russian wheat aphid, stink bugs, yellowstriped armyworm
Remarks: Wheat hay. Apply as required by scouting, usually at 5-day intervals or more. For chinch bug, repeat at 3- to 5-day intervals if needed. For aphids and greenbug, apply before insects begin to roll leaves; once wheat has started to boot, may provide suppression only. Do not apply more than 0.48 pt./A per season. Refer to label for spray drift precautions and state and county restrictions.
Days To Harvest: 30.

Signal Word/Toxicity Class: Danger/I.
REI: 24 hr.

DEADLINE

Packaged as Bullets, M-Ps (mini-pellets), and granules.

Deadline Bullets, M-Ps, Granules (4%) (metaldehyde) *(Pace)*
Rate: 0.4-1 lb. ai/A
10-40 lb. 4%/A
Pest:
slugs, snails
Remarks: Apply every 3-4 weeks during growing season or as needed. Do not apply directly to edible portion of plant.

Signal Word/Toxicity Class: Warning/II (granule); Caution (bullets, M-P).
REI: 12 hr.

Declare — see methyl parathion

Diatect — see pyrethrins

Fyfanon — see malathion

Karate — see *lambda*-cyhalothrin

LINDANE

Lindane 30C contains toxic inert Rhodamine B. Toxic to fish, birds, and wildlife.

Lindane 30C *(Gustafson)*
◆ **Lindane ST 40** *(AGSCO)*
Rate: 1.1 fl. oz. 30C/100 lb. seed
0.8 fl. oz. 40/100 lb. seed
Pest:
seedcorn maggots, wireworms
Remarks: Sudangrass. Do not use treated seed for food, feed, or oil purposes. Exposed treated seed may be hazardous to birds and other wildlife. Dispose of excess seed by burial away from bodies of water.

Signal Word/Toxicity Class: Warning/II.
REI: 12 hr.

MALATHION

Organophosphate insecticide toxic to fish, aquatic invertebrates, aquatic life stages of amphibians, and bees exposed to direct treatment on blooming crops or weeds.

Clean Crop Malathion 57EC *(Platte)*
Drexel Malathion 5EC *(Drexel)*
◆ **Micro Flo Malathion 5EC** *(Micro Flo)*
Prentox 5 lb. Malathion Spray (5EC) *(Prentiss)*
Rate: 1-1.3 lb. ai/A
1.5-2 pt. 57E, 5EC/A
Pest:
aphids, armyworms, grasshoppers, leafhoppers
Remarks: Grass hay. Use high rate for armyworm.
Days To Harvest: 0.

Signal Word/Toxicity Class: Caution/III (57EC); Warning/II (5EC).
REI: 12 hr.

NOTICE The information on these pages is for preliminary planning — not a guide for use. Be sure to follow the manufacturer's directions, notwithstanding information contained here.

For personal protective equipment and EPA registration numbers, please turn to page 39.

Fyfanon ULV *(Cheminova)*
◆ **Micro Flo Malathion ULV** *(Micro Flo)*
Rate: 0.5-0.75 pt. ULV/A
Pest:
black grass bug
grasshoppers
Remarks: Grass hay.
Days To Harvest: 0.

Signal Word/Toxicity Class: Caution.
REI: 12 hr.

METHOXYCHLOR

Chlorinated hydrocarbon insecticide with good residual activity in field applications. Toxic to fish and bees.

Prentox Methoxychlor 2EC *(Prentiss)*
Rate: 0.5-2 lb. ai/A
1-4 qt. 2EC/A
Pest:
alfalfa caterpillar
alfalfa weevil (larvae)
armyworms
blister beetle
clover leaf weevil
cowpea curculio
cucumber beetle
fall armyworm
flea beetles
Japanese beetle
leafhoppers
Mexican bean beetle
pea weevil
soybean caterpillar
spittlebugs
velvetbean caterpillar
webworms
Remarks: Peanut forage. Apply at first sign of infestation and repeat at 7- to 14-day intervals or as needed. Rate varies by pest species; refer to label.
Days To Harvest: Harvest or grazing-7.

Signal Word/Toxicity Class: Caution.
REI: 12 hr.

Wilbur-Ellis Methoxychlor 2 *(Wilbur-Ellis)*
Rate: 1-1.5 lb. ai/A
2-3 qt. 2/A
Pest:
alfalfa webworm
alfalfa weevil
fall armyworm
pea weevil
soybean caterpillar
velvetbean caterpillar
Remarks: Grass, soybean forage. Apply at first sign of infestation, repeat at 7- to 10-day intervals. Do not apply within 14 days of sulfur or sulfur containing sprays.
Days To Harvest: 7.

Signal Word/Toxicity Class: Danger-Poison/I.
REI: 48 hr.

METHYL PARATHION

Restricted Use Pesticide.

Cheminova Methyl 4EC *(Cheminova)*
◆ **Declare (4EC)** *(Griffin)*
Rate: 0.5 lb. ai/A
1 pt. 4EC/A
Pest:
range caterpillar

Rate: 0.75 lb. ai/A
1.5 pt. 4EC/A
Pest:
armyworms
crested wheat bug
false chinch bug
grasshoppers
leafhoppers
Remarks: Apply when insects appear and repeat as needed.
Days To Harvest: 15.

Signal Word/Toxicity Class: Danger-Poison/I.
REI: 48 hr.; 72 hr. if average rainfall less than 25" per year.

Pyrenone — see pyrethrins

PYRETHRINS

Contains piperonyl butoxide as a synergist. Diatect also contains silicon dioxide (diatomaceous earth).

◆ **Diatect Multipurpose Insecticide II** *(Diatect)*
Pyrenone (0.5EC) *(AgrEvo Environmental Health)*
Rate: 1-6 lb. II/A
1-12 fl. oz. 0.5EC/A
Pest: Including but not limited to:
aphids
armyworms
caterpillars
chinch bug
fleahoppers
flies
thrips
Remarks: Grass, fodder, hay forage. Apply in sufficient water for thorough coverage; repeat as needed. Apply Diatect as dust or wettable powder.
Days To Harvest: 0.

Signal Word/Toxicity Class: Caution.
REI: 12 hr.

Warrior — see *lambda*-cyhalothrin

Pastures & Rangeland

Ambush — see permethrin

Amdro — see hydramethylnon

CARBARYL

Carbamate insecticide. Highly toxic to bees exposed to direct treatment or residues on blooming crops or weeds.

Clean Crop Sevin 5 Bait *(Platte)*
Rate: 1-2 lb. ai/A
20-40 lb. Bait/A
Pest:
ants
armyworms
cutworms
darkling beetle
field crickets
grasshoppers
sowbugs
Remarks: Apply with ground or air equipment.
Days To Harvest: 0.

Signal Word/Toxicity Class: Caution/III.
REI: 12 hr.

Drexel Carbaryl 4L or 80S *(Drexel)*
Sevin 4F, XLR Plus, 80S or 80WSP *(Rhone-Poulenc)*
Rate: 0.5-1.5 lb. ai/A
1-3 pt. 4L, 4F, XLR/A
0.66-1.875 lb. 80S, 80WSP/A
Pest: Pastures:
armyworms
black grass bug
chinch bug
Essex skipper
fall armyworm
green June beetle
grasshoppers
range caterpillar
range crane fly
striped grass looper
thrips
white grubs
Rangeland:
black grass bug
grasshoppers
Mormon cricket
range caterpillar
range crane fly
Remarks: For thrips in grasses grown for seed, use high spray pressure to improve penetration into boot. Apply up to 2 applications per year; allow at least 14 days between applications. Carefully mark swaths to avoid over-application.
Days To Harvest: Harvest, grazing, cutting or forage: aerial application-0; ground application-14.

Rate: 1.5 lb. ai/100 gal. water
3 pt. 4F, XLR/100 gal. water
1.6 lb. ai/100 gal. water
2 lb. 80S, 80WSP/100 gal. water
Pest:
imported fire ant
Remarks: Apply total of 2 gal. diluted solution over surface of each mound or at least 1 qt./6" mound diameter using a bucket, can, or other appropriate equipment. Thoroughly wet mound and surrounding area to 4 ft. diameter (12 sq. ft.). Do not disturb mounds prior to treatment. Pour solution from a height of about 3 ft. to give sufficient force to break mound apex and flow into ant tunnels. For best results, apply in cool weather, 65-80°F, or in early morning or late evening hours. Repeat application if mound activity resumes after 10 days. Treat new mounds as they appear. Pressurized sprays may disturb ants and cause migration, reducing product effectiveness.

Signal Word/Toxicity Class: Caution (4L, 4F, XLR); Warning/II (80S, 80WSP).
REI: 12 hr.

◆ - **new product** • ai-active ingredient • DF-dry flowable • E/EC-emulsifiable concentrate • EW-emulsion in water
F/FL-flowable • G-granule • L/LC-liquid • SP/WSP-soluble packet • W/WP-wettable powder • WSB-water soluble bag

Declare — see methyl parathion

Diatect — see pyrethrins

DIBROM

Organophosphate insecticide formulated as an emulsifiable concentrate. Toxic to fish, aquatic invertebrates, and wildlife. Highly toxic to bees exposed to direct treatment.

Dibrom 8 (8E) (naled) *(Amvac)*

Rate: Aerial:
East: 0.8-1.6 fl. oz. 8E/A diluted with water
West: 1.6 fl. oz. 8EA diluted with water
Mist or cold fog:
3-5 qt. 8E/100 gal. water

Pest:
mosquitoes (adult)

Remarks: Pastures. Apply 2-8 qt. dilute spray/A. Calibrate equipment to apply 0.1 lb. technical naled/A. Do not apply through any type of irrigation system.

- - - - - - - - - - - - - - - - -

Rate: 0.5-1 lb. ai/A
0.5-1 pt. 8E/A

Pest:
armyworms grasshoppers range caterpillar

Remarks: Rangeland. Apply in 30-250 gal. water by ground; 1-5 gal. water by air. Rate varies by pest species; refer to label. For young grasshoppers, use 0.5 pt./A. Apply when hopper hatch is completed and repeat as necessary. Animals may be present during treatment. Do not graze lactating dairy animals on treated areas. Do not apply through any type of irrigation system.

- - - - - - - - - - - - - - - - -

Signal Word/Toxicity Class: Danger/I.

HYDRAMETHYLNON

Attractant bait that worker ants find and carry back to the mound where it is passed along to other workers and the queen. Toxic to fish.

Amdro or Siege *(American Cyanamid)*

Rate: Broadcast:
1-1.5 lb./A
Mound treatment:
5 level Tbs./mound

Pest:
bigheaded ants fire ants harvester ants

Remarks: Apply when ants are active and soil temperature is greater than 60°F. Broadcast bait uniformly with ground equipment (spinning disc types) or aerial equipment. When treating individual mounds, do not disturb mound or apply directly to top of mound. Do not apply more than 1.5 lb./A. Do not use on vegetable or food crops.

- - - - - - - - - - - - - - - - -

Signal Word/Toxicity Class: Caution.
REI: 12 hr.

MALATHION

Organophosphate insecticide toxic to fish, aquatic invertebrates, aquatic life stages of amphibians, and bees exposed to direct treatment on blooming crops or weeds.

Clean Crop Malathion 57EC *(Platte)*
Drexel Malathion 5EC *(Drexel)*
◆ **Micro Flo Malathion 5EC** *(Micro Flo)*
Prentox 5 lb. Malathion Spray (5EC) *(Prentiss)*

Rate: 1-1.3 lb. ai/A
1.5-2 pt. 57EC, 5EC/A

Pest:
aphids grasshoppers leafhoppers
armyworms

Remarks: Use high rate for armyworms.
Days To Harvest: Harvest or grazing-0.

- - - - - - - - - - - - - - - - -

Signal Word/Toxicity Class: Caution/III (57EC); Warning/II (5EC).
REI: 12 hr.

Clean Crop Malathion 8EC *(Platte)*
Gowan Malathion 8 (8E) *(Gowan)*

Rate: 1.25 lb. ai/A
1.25 lb. 8EC/A
1-1.25 lb. ai/A
1-1.25 pt. 8E/A

Pest:
aphids grasshoppers leafhoppers

Remarks: May also use 1.25 pt./A in 1 gal. diesel fuel oil by airplane or turbine-blower type spray.
Days To Harvest: 8EC-0; 8E-1.

- - - - - - - - - - - - - - - - -

Signal Word/Toxicity Class: Caution/III.
REI: 12 hr.

METHYL PARATHION

Penncap-M is microencapsulated form of methyl parathion. Restricted Use Pesticide due to residual effects to avian species and hazard to bees.

◆ **Declare (4EC)** *(Griffin)*

Rate: 0.25 lb. ai/A
0.5 pt. 2EC/A

Pest:
mosquitoes

Remarks: Broadcast application. Refer to label.
Days To Harvest: 7.

- - - - - - - - - - - - - - - - -

Signal Word/Toxicity Class: Danger-Poison/I.
REI: 48 hr.

Penncap-M (2FL) *(Elf Atochem)*

Rate: 0.5 lb. ai/A
2 pt. 2FL/A

Pest:
thrips*

Rate: 0.5-0.75 lb. ai/A
2-3 pt. 2FL/A

Pest:
grasshoppers true armyworm*
****Except California***

Remarks: For grasshoppers, use higher rate if majority of insects are large or weather is cool.
Days To Harvest: Harvest or grazing-15.

- - - - - - - - - - - - - - - - -

Signal Word/Toxicity Class: Warning/II.
REI: 48 hr.

Wilbur-Ellis Methyl Parathion 5 Spray *(Wilbur-Ellis)*

Rate: 0.5 lb. ai/A
0.75 pt. 5/A

Pest:
European crane fly

Remarks: Do not apply after grass growth is over 6" high.
Days To Harvest: 21.
State Restrictions: California.

- - - - - - - - - - - - - - - - -

Signal Word/Toxicity Class: Danger-Poison/I.
REI: 48 hr.

Penncap-M — see methyl parathion

PERMETHRIN

Restricted Use Pesticide.

Ambush 2EC *(ZENECA)*
Pounce 3.2EC or 25WP *(FMC)*

Rate: 0.01 lb. ai/A
0.64 fl. oz. 2EC/A
0.4 oz. 3.2EC/A
0.64 oz. 25WP/A

Pest:
range caterpillar

Remarks: Rangeland. Do not apply more than once per year. Cattle may be present during application. Do not harvest or feed hay to livestock.

- - - - - - - - - - - - - - - - -

Signal Word/Toxicity Class: Caution (3.2EC); Warning/II (2EC, 25WP).
REI: 24 hr.

NOTICE	The information on these pages is for preliminary planning — not a guide for use. Be sure to follow the manufacturer's directions, notwithstanding information contained here.	For personal protective equipment and EPA registration numbers, please turn to page 39.

Pounce — see permethrin

Pyrenone — see pyrethrins

PYRETHRINS

Contains piperonyl butoxide as a synergist. Diatect also contains silicon dioxide (diatomaceous earth).

Diatect Multipurpose Insecticide II *(Diatect)*
Pyrenone (0.5EC) *(AgrEvo Environmental Health)*
Rate: 1-6 lb. II/A
1-12 fl. oz. 0.5EC/A
Pest: Including but not limited to:
aphids, caterpillars, thrips, armyworms, flies
Remarks: Apply in sufficient water for thorough coverage; repeat as needed. Apply Diatect as dust or wettable powder.
Days To Harvest: 0.

Signal Word/Toxicity Class: Caution.
REI: 12 hr.

Sevin — see carbaryl

Siege — see hydramethylnon

Noncropland

ACEPHATE

Orthene 75 S *(Valent)*
Rate: 0.25 lb. ai/A
0.33 lb. 75 S/A
Pest:
grasshoppers
Remarks: Field borders, fence rows, roadsides, borrow pits. Early- to mid-season application. Do not graze or feed vegetation cut from treated areas.

Signal Word/Toxicity Class: Caution.
REI: 24 hr.

Amdro — see hydramethylnon

ASANA

Restricted Use Pesticide.

Asana XL (esfenvalerate) *(DuPont)*
Rate: 0.015-0.05 lb. ai/A
2.9-9.6 fl. oz. XL/A
Pest:
army cutworm, chinch bug, saltmarsh caterpillar, armyworms, grasshoppers
Remarks: Rate varies by pest species; refer to label. For armyworms and grasshoppers, spray noncropland adjacent to tilled areas to control migrating insects. Repeat as necessary to maintain control, but do not exceed 0.5 lb. ai/A per year. Do not feed treated crop to livestock. Do not spray ditchbanks or areas adjacent to water.

Signal Word/Toxicity Class: Warning/II.
REI: 12 hr.

CARBARYL

Carbamate insecticide. Highly toxic to bees exposed to direct treatment or residues on blooming crops or weeds.

Clean Crop Sevin 5 Bait *(Platte)*
Rate: 1-2 lb. ai/A
20-40 lb. Bait/A
Pest:
ants, darkling beetle, grasshoppers, armyworms, field crickets, sowbugs, cutworms
Remarks: Fence rows, wastelands. Apply by ground or air.

Signal Word/Toxicity Class: Caution/III.
REI: 12 hr.

Sevin 4F, XLR Plus, 80S or 80WSP *(Rhone-Poulenc)*
Rate: 0.5-1 lb. ai/A
1-2 pt. 4F, XLR/A
0.66-1.25 lb. 80S, 80WSP/A
Pest:
black grass bug, range caterpillar, range crane fly, Mormon cricket

Rate: 0.5-1.5 lb. ai/A
1-3 pt. 4F, XLR/A
0.66-1.875 lb. 80S, 80WSP/A
Pest:
grasshoppers
Remarks: Ditchbanks, hedgerows, rights-of-way, roadsides, wasteland. Apply up to 2 applications per year. Allow at least 14 days between applications. Carefully mark swaths to avoid over-application.
Days To Harvest: Harvest, grazing, forage or cutting: aerial application-0; ground application-14.

Signal Word/Toxicity Class: Caution (4F, XLR); Warning (80S, 80WSP).
REI: 12 hr.

Declare — see methyl parathion

DIBROM

Organophosphate insecticide formulated as an emulsifiable concentrate. Toxic to fish, aquatic invertebrates, and wildlife. Highly toxic to bees exposed to direct treatment.

Dibrom 8 (8E) (naled) *(Amvac)*
Rate: Aerial:
East: 0.8-1.6 fl. oz. 8E/A diluted with water
West: 1.6 fl. oz. 8E/A diluted with water
Mist or cold fog:
3-5 qt. 8E/100 gal. water
Pest:
mosquitoes (adult)
Remarks: Calibrate equipment to apply 0.1 lb. technical naled/A. Do not apply to tidal or marsh waters which are important to shrimp producing areas. Do not apply through any type of irrigation system. Refer to label for directions, restrictions, and precautions. Do not make more than 1 application every 10 days to pastures or other areas where lactating dairy animals are being held.

Signal Word/Toxicity Class: Danger/I.

FENOXYCARB

Insect Growth Regulator.

Logic *(Novartis)*
Rate: Mound treatment:
1-3 Tbs./mound
Broadcast or aerial application:
1-1.5 lb./A
Pest:
fire ants
Remarks: Mound treatment: Distribute 3-4 ft. around mound. Broadcast or aerial application: Apply with ground or aerial equipment calibrated to give correct dosage. Do not apply by air where weather conditions favor drift from treated areas. Repeat after 3-4 months, if needed. Apply when ants are actively foraging, usually when soil temperatures are above 60°F. Avoid application during excessively hot periods or when grass is wet. Do not apply more than twice per year. Do not graze treated areas.

Rate: 1-1.5 lb./A
Pest:
bigheaded ants
Remarks: Broadcast application. Apply uniformly with ground or hand-held equipment calibrated to deliver correct dosage. Treat surrounding areas, adjacent borders, and fence lines to help prevent migration into treated area. Repeat after 4-6 months if needed. Do not apply to areas which may be grazed by cattle, sheep, or other domestic animals.

Signal Word/Toxicity Class: Caution.
REI: 12 hr.

◆ - **new product** • ai-active ingredient • DF-dry flowable • E/EC-emulsifiable concentrate • EW-emulsion in water
F/FL-flowable • G-granule • L/LC-liquid • SP/WSP-soluble packet • W/WP-wettable powder • WSB-water soluble bag

Fyfanon — see malathion

HYDRAMETHYLNON

Attractant bait that worker ants find and carry back to the mound where it is passed along to other workers and the queen. Toxic to fish.

Amdro or Siege *(American Cyanamid)*
Rate: Broadcast:
1-1.5 lb./A
Mound treatment:
5 level Tbs./mound
Pest:
bigheaded ants fire ants harvester ants
Remarks: Apply when ants are active and soil temperature is greater than 60°F. Broadcast bait uniformly with ground equipment (spinning disk types) or aerial equipment. When treating individual mounds, do not disturb mound or apply directly to top of mound. Do not apply more than 1.5 lb./A. Do not use on vegetable or food crops.
- - - - - - - - - - - - - - - - - -
Signal Word/Toxicity Class: Caution.
REI: 12 hr.

Logic — see fenoxycarb

MALATHION

Organophosphate insecticide toxic to fish, aquatic invertebrates, aquatic life stages of amphibians, and bees exposed to direct treatment on blooming crops or weeds.

Fyfanon ULV *(Cheminova)*
◆ **Micro Flo Malathion ULV** *(Micro Flo)*
Rate: Wild host plants:
0.5 pt. ULV/A
Pest:
beet leafhopper

Rate: 0.5-0.75 pt. ULV/A
Pest:
grasshoppers
Remarks: Roadsides, wastelands.
Days To Harvest: 0.
- - - - - - - - - - - - - - - - - -
Signal Word/Toxicity Class: Caution.
REI: 12 hr.

METHYL PARATHION

Penncap-M is microencapsulated form of methyl parathion. Restricted Use Pesticide due to residual effects to avian species and hazard to bees.

◆ **Declare (4EC)** *(Griffin)*
Rate: 0.125 pt. 2EC/A
Pest:
mosquitoes
Remarks: Broadcast application. Refer to label for use directions.
Days To Harvest: 7.
- - - - - - - - - - - - - - - - - -
Signal Word/Toxicity Class: Danger-Poison/I.
REI: 48 hr.

Penncap-M (2FL) *(Elf Atochem)*
Rate: 0.5 lb. ai/A
2 pt. 2FL/A
Pest:
thrips*

Rate: 0.5-0.75 lb. ai/A
2-3 pt. 2FL/A
Pest:
grasshoppers true armyworm*
****Except California***
Remarks: Roadside areas. For grasshoppers, use higher rate if majority of insects are large or weather is cool.
Days To Harvest: Harvest or grazing-15.
- - - - - - - - - - - - - - - - - -
Signal Word/Toxicity Class: Warning/II.
REI: 48 hr.

Orthene — see acephate

Penncap-M — see methyl parathion

Sevin — see carbaryl

Siege — see hydramethylnon

Fungicides

Forage

(Alfalfa, Birdsfoot Trefoil, Clover, Lespedeza, Vetch)

Apron XL — see mefenoxam

CALCIUM POLYSULFIDE

Balanced fungicide, insecticide, miticide composition containing special effective wetting agents.

Lime-Sulfur Solution *(Best Sulfur)*
Sulforix *(Best Sulfur)*
Rate: 1 gal./A in 50-100 gal. water
Disease:
powdery mildew

Time: Apply at bud stage.
Remarks: Alfalfa, red clover. Thorough coverage necessary. Do not apply this product through any type of irrigation system. Refer to label for use directions.
- - - - - - - - - - - - - - - - - -
Signal Word/Toxicity Class: Danger/I.
REI: 48 hr.

CAPTAN

Gustafson Captan 30-DD contains the toxic inert Rhodamine B. Toxic to fish.

Drexel Captan 4FL *(Drexel)*
Rate: 8.25 fl. oz. 4FL/100 lb. seed
Disease:
seed decay seedling blight soilborne disease
Time: Liquid or slurry seed treatment.
Remarks: Alfalfa, clover. Do not use treated seed for food, feed, or oil purposes. Refer to label for use directions.
- - - - - - - - - - - - - - - - - -
Signal Word/Toxicity Class: Danger/I.

NOTICE The information on these pages is for preliminary planning — not a guide for use. Be sure to follow the manufacturer's directions, notwithstanding information contained here.	For personal protective equipment and EPA registration numbers, please turn to page 39.

Gustafson Captan 30-DD or 400 *(Gustafson)*

Rate: 6.5 fl. oz. 30-DD/100 lb. seed
5-8 fl. oz. 400/100 lb. seed

Disease:
damping-off, seed decay, seedling blight, mold

Time: Seed treatment.

Remarks: Alfalfa, clover, lespedeza, trefoil. Do not combine colloidal suspensions, wettable powders, or emulsifiable concentrates unless previous experience assures compatibility. Use with EPA-approved dye that imparts unnatural color to seed. Do not use treated seed for feed, food, or oil purposes. Refer to label for use directions.

- - - - - - - - - - - - - - - - -

Signal Word/Toxicity Class: Danger/I.

CAPTAN (Premixes)

Premix combining a fungicide with an insecticide. Toxic to fish, birds, and wildlife.

Nu-Gro Isotox (12.2% captan & 25% lindane) *(Wilbur-Ellis)*
Seed Mate Isotox (12.2% captan & 25% lindane) *(Loveland)*

Rate: 2 oz./100 lb. seed

Disease:
damping-off, seed decay, seedling blight

Time: Planter box treatment.

Remarks: Alfalfa, clover. Do not use treated seed for food, feed, or oil purposes. Treat only those seeds needed for immediate use, minimizing the interval between treatment and planting. Do not store excess treated seeds beyond planting time. Refer to label for use directions.

- - - - - - - - - - - - - - - - -

Signal Word/Toxicity Class: Danger/I.
REI: 4 days.

Champ — see copper compounds

Champion — see copper compounds

COPPER COMPOUNDS

Fungicide/bactericide.

Champ (FL or Formula 2) or Champion (WP) (copper hydroxide) *(Agtrol)*
Kocide (101, DF, 2000, 4.5LF, LF) (copper hydroxide) *(Griffin)*
KOP-Hydroxide (FL, 50WP) (copper hydroxide) *(Drexel)*
Nu-Cop (3L, 50DF, 50WP) (copper hydroxide) *(Micro Flo)*

Rate: 2.66 pt. FL, LF/A
1.33 pt. Formula 2, 4.5LF/A
2 lb. WP, 50WP, 50DF/A
2 lb. 101, DF/A
1.5 lb. 2000/A
1.33-2.66 pt. 3L/A

Disease:
Cercospora leaf spot, Leptosphaerulina leaf spot

Time: Apply 10-14 days before harvest or earlier if disease threatens.

Remarks: Alfalfa. Spray injury may occur with sensitive varieties such as Lahontan. KOP-Hydroxide 50 should not be applied in spray solution having pH less than 6.5 or greater than 9.0 as phytotoxicity may occur. Refer to label for use directions.

- - - - - - - - - - - - - - - - -

Signal Word/Toxicity Class: Varies by formulation.
REI: Varies by formulation.

Kocide — see copper compounds

KOP-Hydroxide — see copper compounds

Lime-Sulfur Solution — see calcium polysulfide

MEFENOXAM

Systemic fungicide having a specific mode of action.

Apron XL LS *(Novartis)*

Rate: 0.64 fl. oz. LS/100 lb. seed

Disease:
damping-off, downy mildew, Phytophthora (early-season)

Time: Seed treatment at planting.

Remarks: Alfalfa, clover, lespedeza, vetch, etc. Do not use treated seed for food, feed, or oil purposes. Refer to label for use directions.

- - - - - - - - - - - - - - - - -

Signal Word/Toxicity Class: Warning/II.

Ridomil Gold EC or WSP *(Novartis)*

Rate: 0.125-0.25 lb. ai/A
0.25-0.5 pt. EC/A
0.11-0.22 lb. ai/A
0.25-0.5 lb. WSP/A

Disease:
damping-off, root rot

Time: Apply as broadcast spray at planting.

Remarks: Alfalfa, birdsfoot trefoil, clover. Apply in minimum 20 gal. water. Use high rate in areas where disease pressure is expected to be heavy. Do not feed green forage or cut hay for 60 days following application. Do not plant any crop which is not registered for use with mefenoxam in treated soil for 12 months. Refer to label for use directions.

- - - - - - - - - - - - - - - - -

Signal Word/Toxicity Class: Warning/II (EC); Caution (WSP).
REI: 48 hr.

Micro Flo — see sulfur

Nu-Cop — see copper compounds

Nu-Gro Isotox — see captan premix

Ridomil Gold — see mefenoxam

SULFUR

Nonsystemic fungicide with protective action.

Micro Flo Sulfur (6L) *(Micro Flo)*
Terra Sulfur (6L) *(Terra)*

Rate: Clover:
4-24 lb. ai/A
2.5-4 gal. 6L/A

Disease:
powdery mildew

Rate: Vetch:
24-39 lb. ai/A
4-6.5 gal. 6L/A

Disease:
rust

Time: Apply when disease first appears; repeat at 7- to 10-day intervals.

Remarks: Clover, vetch. Do not use within 2 weeks of oil application. Some plants are susceptible to injury from sulfur under certain climatic conditions. Do not make sulfur applications at excessively high temperatures. Refer to label for use directions.

- - - - - - - - - - - - - - - - -

Signal Word/Toxicity Class: Caution.
REI: 24 hr.

Seed Mate Isotox — see captan premixes

Sulforix — see calcium polysulfide

◆ - **new product** • ai-active ingredient • DF-dry flowable • E/EC-emulsifiable concentrate • EW-emulsion in water
F/FL-flowable • G-granule • L/LC-liquid • SP/WSP-soluble packet • W/WP-wettable powder • WSB-water soluble bag

SULFUR (Premix)

Fungicide/bactericide in flowable formulation. Toxic to fish and aquatic organisms.

Top Cop with Sulfur (6.25 lb./gal. sulfur & 1 lb./gal. tribasic copper sulfate) *(Stoller)*

Rate: 2 qt./A
Disease:
leaf spot

Time: Apply as foliar spray 10-14 days before harvest or earlier if disease threatens.
Remarks: Alfalfa. Refer to label for use directions.

- - - - - - - - - - - - - - - - -

Signal Word/Toxicity Class: Caution.
REI: 24 hr.

THIRAM

Broad-spectrum fungicide. Toxic to fish. Thiram 50WP contains the toxic inert Rhodamine B and is also toxic to birds and wildlife.

42-S Thiram or Thiram 50WP Dyed *(Gustafson)*

Rate: 8 fl. oz. 42-S/100 lb. seed
8 oz. 50WP/100 lb. seed
Disease:
damping-off   seed decay   seedling blight

Time: Seed treatment.
Remarks: Small-seeded legumes. Use with EPA-approved dye that imparts unnatural color to seed. Do not use treated seed for feed, food, or oil purposes. Refer to label for use directions.

- - - - - - - - - - - - - - - - -

Signal Word/Toxicity Class: Caution.

Top Cop with Sulfur — see sulfur premix

Forage

(Grasses, Hay, etc.)

Allegiance — see metalaxyl

Apron — see metalaxyl

Apron XL — see mefenoxam

MEFENOXAM

Systemic fungicide having a specific mode of action.

Apron XL LS *(Novartis)*

Rate: 0.32-0.64 fl. oz. LS/100 lb. seed
Disease:
damping-off   seed rot

Time: Seed treatment at planting.
Remarks: Grasses, hay. Do not use treated seed for food, feed, or oil purposes.

- - - - - - - - - - - - - - - - -

Signal Word/Toxicity Class: Warning/II.

Ridomil Gold EC or WSP *(Novartis)*

Rate: 0.125-0.25 lb. ai/A
0.25-0.5 pt. EC/A
0.11-0.22 lb. ai/A
0.25-0.5 lb. WSP/A
Disease:
Pythium

Time: Apply as broadcast spray at planting.
Remarks: Grasses, hay. Apply in minimum 20 gal. water. Use higher rates in areas where disease pressure is expected to be heavy. Do not feed green forage or cut hay for 60 days following application. Refer to label for use directions.

- - - - - - - - - - - - - - - - -

Signal Word/Toxicity Class: Warning/II (EC); Caution (WSP).
REI: 48 hr.

METALAXYL

Systemic fungicide having a specific mode of action.

Allegiance or Apron (FL) *(Gustafson)*

Rate: 0.75 fl. oz. Allegiance/100 lb. seed
0.75-1.5 fl. oz. Apron/100 lb. seed
Disease:
damping-off   Phytophthora (early-season)

Time: Seed treatment.
Remarks: Forage grasses, forage legumes. Use with EPA-approved dye that imparts unnatural color to seed. Do not use treated seed for feed, food, or oil purposes. Refer to label for use directions.

- - - - - - - - - - - - - - - - -

Signal Word/Toxicity Class: Warning/II.

Ridomil Gold — see mefenoxam

For personal protective equipment and EPA registration numbers, please turn to page 39.

Notes • Notes • Notes

Indexes

Crop Index *597*

Disease Index *590*

Pest Index *571*

Product Index *552*

The information in this 1999 Insect and Disease Control Guide is updated to August 15, 1998.
Because manufacturers' product recommendations are constantly being updated and revised as additional product registrations and label changes are approved by EPA, there may be changes after this date. However, every effort has been made to give manufacturers the opportunity to include in the Guide their latest information. This Guide is for preliminary planning – not a use guide. Be sure to follow label directions, notwithstanding information contained herein.

Product Index

- **435 Oil (petroleum oil)** ... **Drexel**
  - WPS chart ... 42
  - field corn, popcorn ... 83
  - nuts ... 430
  - sugar beets ... 164
  - sweet corn ... 293
  - tree fruits, citrus ... 398,451
  - tree fruits, deciduous ... 414
- **455 Oil (petroleum oil)** ... **Drexel**
  - WPS chart ... 42
  - field corn, popcorn ... 83
  - nuts ... 430
  - ornamentals ... 506
  - small fruits ... 390
  - sugar beets ... 164
  - sweet corn ... 293
  - tree fruits, citrus ... 398,451
  - tree fruits, deciduous ... 414
  - tree fruits, subtropical ... 421
- **3336 (thiophanate-methyl)** ... **Cleary**
  - WPS chart ... 42
  - lawns/turf ... 517
  - ornamentals ... 531

A

- abamectin (see Agri-Mek, Avid, Zephyr)
- **Abound (azoxystrobin)** ... **ZENECA**
  - WPS chart ... 51
  - nuts ... 477
  - peanuts ... 176
  - small fruits ... 433
  - tree fruits, deciduous ... 452
  - tree fruits, subtropical ... 470
- acephate (see Micro Flo Acephate, Orthene, Payload, Velocity)
- **Adept (diflubenzuron)** ... **Uniroyal**
  - WPS chart ... 50
- **Adios (carbaryl)** ... **Micro Flo**
  - WPS chart ... 45
  - beans ... 203
  - cucurbits ... 226
  - eggplant ... 232
  - field corn, popcorn ... 77
  - forage (alfalfa, etc.) ... 537
  - peanuts ... 151
  - peas ... 260
  - peppers ... 266
  - potatoes ... 271
  - sorghum ... 157
  - southern peas ... 282
  - soybeans ... 96
  - sweet corn ... 288
  - sweet potatoes, yams ... 295
  - tomatoes ... 301
  - tree fruits, deciduous ... 393
- **Admire (imidacloprid)** ... **Bayer**
  - WPS chart ... 41
  - cole crops ... 221
  - cotton ... 115
  - eggplant ... 233
  - greens ... 240
  - lettuce ... 251
  - peppers ... 268
  - potatoes ... 274
  - tobacco ... 171
  - tomatoes ... 304
- **Agree (*Bt.* spp. *aizawai*)** ... **Thermo Trilogy**
  - Bt chart ... 76,94,144,196,380,536
  - WPS chart ... 49
- **Agricultural Streptomycin (streptomycin sulfate)** ... **Gustafson**
  - WPS chart ... 44
- **Agri-Mek (abamectin)** ... **Novartis**
  - WPS chart ... 46
  - celery ... 213
  - lettuce ... 248
  - peppers ... 264
  - tomatoes ... 300
  - tree fruits, citrus ... 392
- **Agri-mycin 17 (streptomycin sulfate)** ... **Novartis**
  - WPS chart ... 46
  - biocontrols ... 54
- *Agrobacterium radiobacter* (see Galltrol)
- **Alamo (propiconazole)** ... **Novartis**
  - ornamentals ... 529
- aldicarb (see Temik)
- **Aliette (fosetyl-aluminum)** ... **Rhone-Poulenc**
  - WPS chart ... 47
  - asparagus ... 308
  - carrots, etc. ... 318
  - celery ... 321
  - cole crops ... 324
  - cucurbits ... 330
  - greens ... 339
  - herbs/spices ... 341
  - hops ... 174
  - lawns/turf ... 509
  - lettuce ... 343
  - nuts ... 482
  - onions ... 348
  - rhubarb ... 364
  - small fruits ... 440
  - spinach ... 366
  - tomatoes ... 375
  - tree fruits, citrus ... 450
  - tree fruits, deciduous ... 462
  - tree fruits, subtropical ... 474
- **Aliette/Maneb 2+2 (co-pack)** ... **Rhone-Poulenc & Elf Atochem**
  - WPS chart ... 47
  - cole crops ... 322
  - cucurbits ... 328
  - greens ... 338
  - lettuce ... 343
  - onions ... 346
- **Allegiance (metalaxyl)** ... **Gustafson**
  - WPS chart ... 44
  - beans ... 313
  - carrots, etc. ... 318
  - cotton ... 124
  - cucurbits ... 332
  - field corn, popcorn ... 87
  - forage (grasses, hay) ... 549
  - garbanzos ... 336
  - herbs/spices ... 341
  - lawns/turf ... 541
  - lentils ... 342
  - okra ... 346
  - peanuts ... 178
  - peas ... 352
  - rice ... 183
  - small grains ... 138
  - sorghum ... 186
  - southern peas ... 365
  - soybeans ... 103
  - spinach ... 367
  - sugar beets ... 189
  - sunflower ... 192
  - sweet corn ... 369
- **Ambush (permethrin)** ... **ZENECA**
  - WPS chart ... 51
  - artichokes ... 199
  - asparagus ... 201
  - celery ... 215
  - cole crops ... 223
  - cucurbits ... 230
  - eggplant ... 234
  - field corn, popcorn ... 82
  - forage (alfalfa, etc.) ... 542
  - garlic ... 237
  - greens ... 242
  - herbs/spices ... 244
  - horseradish ... 246
  - lettuce ... 253
  - mushrooms ... 254
  - nuts ... 430
  - onions ... 259
  - ornamentals ... 505
  - pastures/rangeland ... 545
  - peppers ... 269
  - potatoes ... 276
  - small fruits ... 390
  - soybeans ... 99
  - spinach ... 287
  - sweet corn ... 293
  - tomatoes ... 305
  - tree fruits, deciduous ... 413
  - tree fruits, subtropical ... 421
- ***Amblyesius cucumeris*** ... **Florikan**
  - biocontrols ... 66
- **Amdro (hydramethylnon)** ... **American Cyanamid**
  - WPS chart ... 40
  - lawns/turf ... 492
  - noncropland ... 547
  - ornamentals ... 502
  - pastures/rangeland ... 545
- amitraz (see Mitac, Ovasyn)
- **Ammo (cypermethrin)** ... **FMC**
  - WPS chart ... 43
  - cole crops ... 219
  - cotton ... 112
  - garlic ... 236
  - lettuce ... 249
  - nuts ... 426
  - onions ... 257
- *Ampelomyces quisqualis* (see AQ10)
- aphid parasite (see *Diaretiella rapae*)
- **Aphid Predators (*Aphidoletes aphidimyza*)** ... **Nature's Control**
  - biocontrols ... 62
- ***Aphidius colmani* (parasitic wasp)** ... **Florikan**
  - biocontrols ... 62
- ***Aphidius matricariae* (parasitic wasp)** ... **Hydro-Gardens**
  - biocontrols ... 62
- *Aphidius* spp. (see *Aphidius colmani, Aphidius matricariae*)
- ***Aphidoletes aphidimyza* (predator midge)** ... **Arbico, Hydro-Gardens, International Technology, IPM Laboratories**
  - biocontrols ... 62
- *Aphidoletes aphidimyza* (see Aphid Predators)
- ***Aphytis melinus* (scale parasite)** ... **Arbico, Caltec, Sespe Creek Insectary**
  - biocontrols ... 62
- **Apollo (clofentizine)** ... **AgrEvo USA**
  - WPS chart ... 40
  - nuts ... 423
  - tree fruits, deciduous ... 402
- **Apron (metalaxyl)** ... **Gustafson**
  - WPS chart ... 44
  - beans ... 313
  - carrots, etc. ... 318
  - cotton ... 124
  - field corn, popcorn ... 87
  - forage (grasses, hay) ... 549
  - garbanzos ... 336
  - herbs/spices ... 341
  - lawns/turf ... 514
  - lentils ... 342
  - okra ... 346
  - peas ... 352
  - rice ... 183
  - small grains ... 138
  - sorghum ... 186
  - southern peas ... 365
  - soybeans ... 103
  - sugar beets ... 189
  - sunflower ... 192
  - sweet corn ... 369
- **Apron Dry (metalaxyl)** ... **Trace**
  - WPS chart ... 49
  - beans ... 313
  - cotton ... 124
  - field corn, popcorn ... 87
  - garbanzos ... 336
  - herbs/spices ... 342
  - lawns/turf ... 514
  - lentils ... 342
  - okra ... 346
  - peas ... 352
  - rice ... 183
  - small grains ... 138
  - sorghum ... 186
  - southern peas ... 365
  - soybeans ... 104
  - sugar beets ... 189
  - sunflowers ... 192
  - sweet corn ... 369
  - table beets ... 371
- **Apron XL (mefenoxam)** ... **Novartis**
  - WPS chart ... 46
  - beans ... 313
  - carrots, etc. ... 318
  - cotton ... 124
  - field corn, popcorn ... 87
  - forage (alfalfa, etc.) ... 548
  - forage (grasses, hay) ... 549
  - garbanzos ... 336
  - lawns/turf ... 514
  - lentils ... 342
  - okra ... 346
  - peanuts ... 178
  - peas ... 352
  - rice ... 183
  - small grains ... 138
  - sorghum ... 186
  - southern peas ... 364
  - soybeans ... 103

spinach 366
sugar beets 189
sunflowers 192
sweet corn 369
table beets 371

AQ10 (*Ampelomyces quisqualis*) Ecogen
WPS chart 43
biocontrols 54

Arbotect (thiabendazole) Novartis
ornamentals 531

Asana XL (esfenvalerate) DuPont
WPS chart 42
artichokes 198
beans 202
carrots, etc. 209
cole crops 217
cotton 110
cucurbits 225
eggplant 231
field corn, popcorn 77
greens 238
noncropland 546
nuts 423
ornamentals 495
peanuts 150
peas 260
peppers 265
potatoes 271
radishes 278
soybeans 95
sugarcane 166
sunflower 166
sweet corn 288
tomatoes 301
tree fruits, deciduous 402

Astro (permethrin) FMC
WPS chart 43
ornamentals 505

Avid (abamectin) Novartis
WPS chart 46
ornamentals 494

Award (fenoxycarb) Novartis
WPS chart 46
lawns/turf 492
ornamentals 501

azadirachtin (see Azatin, NEEMAZAD, NEEMIX)

Azatin (azadirachtin) Olympic, Thermo-Trilogy
WPS chart 46,49
artichokes 198
asparagus 199
carrots, etc 209
celery 213
cole crops 217
cucurbits 226
eggplant 231
garlic 236
greens 238
herbs/spices 243
hops 146
horseradish 245
lentils 246
lettuce 248
mint 148
nuts 424
okra 255
onions 256
ornamentals 496
peppers 265
potatoes 271
radishes 278
rhubarb 281
small fruits 381
spinach 284
sugar beets 161
sugarcane 166
sweet potatoes, yams 294
table beets 297
tobacco 169
tomatoes 301
tree fruits, citrus 392
tree fruits, deciduous 402
tree fruits, subtropical 418
watercress 307

azinphos-methyl (see Guthion, Micro Flo Azinphocmethyl, Sniper)

azoxystrobin (see Abound, Heritage, Quadris)

Aztec (tebuprimphos & cyfluthrin) Bayer
WPS chart 41
field corn, popcorn 77
sweet corn 288

B

Bacillus berliner & sulfur (see Bt 320 Sulfur Dust)

Bacillus subtilis (see Kodiak)

Bacillus subtilis & metalaxyl & PCNB (see System[3])

Bacillus thuringiensis (see Agree, Bactimos, Biobit, BMP 123, Condor, CRYMAX, Cutlass, Design, DiPel, Foray, Gnatrol, Javelin, Larvo-Bt, Lepinox, Mattch, M-Peril, MVP, Novodor, Raven, Troy-Bt, Vault, VectoBac, XenTari)

Bactimos (*Bt.* spp. *israelensis*) Abbott
Bt chart 144,536

Banner (propiconazole) Novartis
WPS chart 46
lawns/turf 516
nuts 483
ornamentals 529
tree fruits, citrus 451
tree fruits, deciduous 464

Banol (propamocarb hydrochloride) AgrEvo USA
WPS chart 40
lawns/turf 509
ornamentals 519

Basamid Granular (dazomet) BASF
WPS chart 41
hops 173
lawns/turf 510
nuts 477
ornamentals 519
small fruits 434
tree fruits, deciduous 452

basic copper sulfate (see Basicop, Copper 3 Dust, Drexel Basic Copper Sulfate, KOP 300, Micro Flo Basic Copper "53", TOP COP Tri-Basic)

Basicop (basic copper sulfate) Griffin
WPS chart 44
beans 311
carrots, etc. 317
celery 320
cole crops 323
cucurbits 329
eggplant 334
nuts 479
peanuts 177
peppers 354
potatoes 357
small fruits 436
spinach 366
sugar beets 187
table beets 371
tomatoes 373
tree fruits, citrus 447
tree fruits, deciduous 455
tree fruits, subtropical 471

Bayleton (triadimefon) Bayer
WPS chart 41
cucurbits 334
lawns/turf 518
ornamentals 532
small fruits 445
small grains 141
sugar beets 191
tree fruits, deciduous 468
tree fruits, subtropical 477

Baytan (triadimenol) Gustafson
WPS chart 44
cotton 122
field corn, popcorn 85
small grains 134

Baythroid (cyfluthrin) Bayer
WPS chart 41
carrots, etc. 209
cotton 110
forage (alfalfa, etc.) 538
ornamentals 498
peppers 266
radishes 279
sorghum 157
sugarcane 166
sunflower 167
sweet corn 290
tomatoes 302

Bean beetle parasite (see *Pediobius foveolatus*)

Bean Guard (captan & carboxin) Trace
WPS chart 49
soybeans 102

Bean Treater (thiram) Terra
WPS chart 48

Beauveria bassiana (see BotaniGard, Mycotrol, Naturalis-O, Naturalis-T)

bendiocarb (see Turcam)

Beneficial insect attractant (see Good Bug Power Meal)

Beneficial Nematodes (*Steinernema* spp.) Hydro-Gardens, Nature's Control
biocontrols 63

Beneficial nematodes (see also Guardian nematodes)

Bene-Nem (*Steinernema* spp. & *Heterohabditis* spp.) Florikan
biocontrols 68

Benlate (benomyl) DuPont
WPS chart 42
beans 309
carrots, etc. 316
celery 319
cole crops 322
cucurbits 328
garbanzos 336
greens 338
mushrooms 345
nuts 478
ornamentals 520
peanuts 176
rice 181
small fruits 434
small grains 134
soybeans 101
spinach 365
sweet potatoes, yams 370
tomatoes 373
tree fruits, citrus 447
tree fruits, deciduous 452
tree fruits, subtropical 470

benomyl (see Benlate)

Bidrin (dicrotophos) Amvac
WPS chart 40
cotton 110

bifenthrin (see Brigade, Capture, Talstar)

Biobit (*Bt.* spp. *kurstaki*) Abbott
Bt chart 76,94,128,144,196,380,488,536
WPS chart 40

Bio-Trek (*Trichoderma harzianum*) Wilbur-Ellis
WPS chart 50
biocontrols 55

BioVector (*S. carpocapsae*) Thermo Trilogy
biocontrols 68

black scale parasite (see *Metaphycus helvolus*)

Blue Viking (copper sulfate pentahydrate) Griffin
WPS chart 44
nuts 479
ornamentals 521
potatoes 357
small fruits 436
tree fruits, citrus 447
tree fruits, deciduous 455
tree fruits, subtropical 471

BMP 123 (*Bt.* spp. *kurstaki*) Becker Microbial
Bt chart 76,94,108,128,144,196,380,488,536
WPS chart 41

Bolstar (sulprofos) Bayer
WPS chart 41
cotton 111

borax (see Sporax)

BotaniGard (*Beauveria bassiana*) Mycotech
WPS chart 46
biocontrols 54

Botran (dicloran) Gowan
WPS chart 44
beans 312
celery 321
cucurbits 330
garlic 337
herbs/spices 341
lettuce 343
onions 348
ornamentals 523
potatoes 358
rhubarb 363
small fruits 439
sweet potatoes, yams 370
tomatoes 375
tree fruits, deciduous 460

Botran (dicloran) Wilbur-Ellis
WPS chart 50
garlic 337
lettuce 343
onions 348
small fruits 439
tomatoes 374
tree fruits, deciduous 460

Botran 6 Sulfur (dicloran & sulfur) Wilbur-Ellis
WPS chart 50
small fruits 439
tree fruits, deciduous 460

Bravo (chlorothalonil) ZENECA
WPS chart 51
beans 311
carrots, etc. 316
celery 320
cole crops 323
cucurbits 329
field corn, popcorn 86
garlic 336
lawns/turf 511
mint 174
mushrooms 345
onions 347
ornamentals 521
peanuts 176
potatoes 357
small fruits 436
soybeans 103
sugar beets 187
sweet corn 368
tomatoes 373
tree fruits, deciduous 455
tree fruits, subtropical 470

Bravo S (sulfur & chlorothalonil) Helena
WPS chart 45
beans 208,315
cole crops 225,327
peanuts 180
potatoes 361
tomatoes 307,377
tree fruits, deciduous 467

Bravo Zn (sulfur) .... **Zeneca**
- WPS chart 51
- beans 311
- potatoes 357

Break (propiconazole) .... **Novartis**
- WPS chart 46
- tree fruits, citrus 451
- tree fruits, deciduous 464

Brigade (bifenthrin) .... **FMC**
- WPS chart 43
- hops 146

Brom-O-Gas (methyl bromide) .... **Great Lakes**
- WPS chart 44
- asparagus 308
- cole crops 325
- cucurbits 332
- eggplant 335
- lettuce 344
- nuts 478
- onions 349
- peppers 355
- small fruits 442
- tobacco 193
- tomatoes 376
- tree fruits, citrus 451
- tree fruits, subtropical 476

Bt 320 Sulfur Dust (*Bt. berliner* & sulfur) ..... **Wilbur-Ellis**
- WPS chart 50
- celery 214
- cole crops 218
- cotton 111
- greens 238
- lettuce 249
- potatoes 271
- radishes 278
- small fruits 382,434
- spinach 285
- sugar beets 161,187
- tomatoes 301

C

calcium polysulfide (see Lime-Sulfur Solution, Sulforix)
capsaicin (see Hot Pepper Wax)
captan (see Captan Moly, Captec, Clean Crop Captan, Drexel Captan, Drexel Captan Plus Hi Moly/Captan, Gustafson Captan, Hi Moly/Captan-D, Molybdenum, Micro Flo Captan, Seed Mate Soybean Mo, Soy Gro)

Captan-Diazinon (captan & diazinon) .... **Trace**
- WPS chart 49
- beans 310
- field corn, popcorn 79,85
- peas 261,351
- sweet corn 290,368

captan & carboxin (see Bean Guard)
captan & diazinon (see Captan-Diazinon)
captan & diazinon & lindane (see Kernel Guard)
captan & lindane (see Nu-Gro Isotox, Seed Mate Isotox, Sorghum Guard)

Captan Moly (captan) .... **Terra**
- WPS chart 48

captan & PCNB & carboxin (see Vitavax PC)
captan & PCNB & thiabendazole (see Rival)

Captec (captan) .... **Micro Flo**
- WPS chart 45
- lawns/turf 510
- nuts 478
- ornamentals 520
- small fruits 435
- tree fruits, deciduous 454

Capture (bifenthrin) .... **FMC**
- WPS chart 410
- cotton 111
- field corn, popcorn 77

carbaryl (see Adios, Chipco Sevin, Clean Crop Sevin, Drexel Carbaryl, Gowan Sevin, Sevin, Slam)
carbofuran (see Furadan)
carboxin (see Vitavax-30C, Vitavax-34)
carboxin & captan (see Enhance Vitavax-Captan 20-20, Seed Mate Vitavax-Captan)
carboxin & diazinon & lindane (see Germate Plus)
carboxin & imazalil & thiabendazole (see RTU-Vitavax-Extra)
carboxin & maneb & lindane (see DB-Green + Vitavax)
carboxin & PCNB (see Vitavax-PCNB)
carboxin & PCNB & metalaxyl (see Prevail)
carboxin & permethrin (see Kernel Guard Supreme)
carboxin & thiram (see RTU-Vitavax-Thiram, Vitaflo-280, Vitavax T-L)
carboxin & thiram & lindane (see Vitavax-Thiram-Lindane)

Carzol (formetanate hydrochloride) .... **AgrEvo USA**
- WPS chart 40
- forage (alfalfa, etc.) 538
- tree fruits, citrus 393
- tree fruits, deciduous 405

Caterpillar Parasites (*Trichogramma* spp.) ... **Nature's Control**
- biocontrols 69

Champ (copper compounds) .... **Agtrol**
- WPS chart 40
- beans 311
- carrots, etc. 317
- celery 320
- cole crops 323
- cucurbits 329
- eggplant 334
- forage (alfalfa, etc.) 548
- hops 173
- lettuce 343
- nuts 479
- onions 347
- ornamentals 521
- peanuts 177
- peas 352
- peppers 354
- potatoes 272,357
- rice 182
- small fruits 436
- small grains 136
- sugar beets 187
- tomatoes 373
- tree fruits, citrus 447
- tree fruits, deciduous 456
- tree fruits, subtropical 471

Champion (copper compounds) .... **Agtrol**
- WPS chart 40
- beans 311
- carrots, etc. 317
- celery 320
- cole crops 323
- cucurbits 329
- eggplant 334
- forage (alfalfa, etc.) 548
- hops 173
- lettuce 343
- nuts 479
- onions 347
- ornamentals 521
- peanuts 177
- peas 352
- peppers 354
- potatoes 272,357
- small fruits 436
- small grains 136
- sugar beets 187
- tomatoes 373
- tree fruits, citrus 447
- tree fruits, deciduous 456
- tree fruits, subtropical 471

Checkmate (pheromone) .... **Consep**
- biocontrols 59

Cheminova Dimethoate (dimethoate) .... **Cheminova**
- WPS chart 41
- beans 204
- cole crops 220
- cotton 114
- cucurbits 228
- field corn, popcorn 79
- forage (alfalfa, etc.) 539
- greens 240
- lentils 247
- lettuce 250
- nuts 427
- peas 262
- peppers 267
- potatoes 273
- radishes 280
- safflower 156
- small grains 130
- sorghum 158
- soybeans 97
- spinach 285
- table beets 298
- tomatoes 303
- tree fruits, citrus 395
- tree fruits, deciduous 408

Cheminova Methyl (methyl parathion) .... **Cheminova**
- WPS chart 41
- beans 206
- carrots, etc. 211
- celery 215
- cole crops 222
- cotton 118
- field corn, popcorn 82
- forage (alfalfa, etc.) 541
- forage (grasses, hay) 544
- greens 241
- lettuce 252
- oil seed crops 150
- onions 259
- peas 263
- potatoes 275
- rice 155
- small grains 132
- soybeans 99
- spinach 286
- sugar beets 164
- sunflower 168

Chipco 26019 (iprodione) .... **Rhone-Poulenc**
- WPS chart 47
- lawns/turf 513
- ornamentals 524

Chipco Aliette (fosetyl-aluminum) .... **Rhone-Poulenc**
- WPS chart 47
- lawns/turf 509
- ornamentals 519

Chipco Choice (fipronil) .... **Rhone-Poulenc**
- lawns/turf 490

Chipco Mocap (ethoprop) .... **Rhone-Poulenc**
- WPS chart 47
- lawns/turf 490

Chipco Sevin (carbaryl) .... **Rhone-Poulenc**
- WPS chart 47
- lawns/turf 489
- nuts 425
- ornamentals 497
- tree fruits, citrus 393
- tree fruits, deciduous 404
- tree fruits, subtropical 418

chitin-protein (see ClandoSan)
chlorethoxyfos (see Fortress)

Chlorfos (chlorpyrifos) .... **Griffin**
- WPS chart 44
- cole crops 219
- field corn, popcorn 78
- greens 239
- onions 257
- peanuts 151
- radishes 279
- sorghum 158
- soybeans 96
- sugar beets 162
- sunflower 167
- sweet corn 289
- sweet potatoes, yams 295
- tobacco 170
- tree fruits, citrus 394

Chlor-O-Pic (chloropicrin) .... **Great Lakes**
- WPS chart 44
- cucurbits 329
- eggplant 334
- onions 347
- sweet potatoes, yams 370
- tobacco 192

chloroneb (see Demosan, ProTurf Fungicide V, Toromoc, Terraneb)
chloroneb & thiophanate-methyl (see ProTurf Fungicide IX)
chloropicrin (see Chlor-O-Pic)
chlorothalonil (Bravo, Daconil, Echo, Ridomil Gold/Bravo, Terranil, Thalonil)
chlorothalonil & copper oxychloride (see Terranil Cu)

Chlorpyrifos Pro (chlorpyrifos) .... **Micro Flo**
- WPS chart 45
- ornamentals 497

chlorpyrifos (see Chlorfos, Chlorpyrifos Pro, Cyren, Lock-On, Lorsban, Micro Flo Chlorpyrios, ProTurf Insecticide III)

***Chrysopa carnea* (green lacewing)** ..... **A-1 Unique Insect Control, Beneficial Insectary, Caltec, Hydro-Gardens, Natural Pest Controls**
- biocontrols 64

***Chrysoperla carnea* (green lacewing)** .... **Arbico**
- biocontrols 64

***Chrysoperla rufilabris* (green lacewing)** .... **Arbico, Beneficial Insectary, Caltec, IPM Laboratories**
- biocontrols 64

***Chrysoperla* spp. (green lacewing)** .. **Florikan, M&R Durango**
- biocontrols 64

Citation (cyromazine) .... **Novartis**
- WPS chart 46
- ornamentals 498

Citri-Oil (petroleum oil) .... **Wilbur-Ellis**
- WPS chart 50
- nuts 430
- tree fruits, citrus 398
- tree fruits, deciduous 415

ClandoSan (chitin-protein) .... **Igene Biotechnology**
- biocontrols 55

Clean Crop Captan (captan) .... **Platte**
- WPS chart 47
- potatoes 357

Clean Crop C-O-C-S (copper oxychloride & basic copper sulfate) .... **Platte**
- WPS chart 47
- beans 213
- carrots, etc. 317
- celery 320
- cole crops 324
- cucurbits 329
- eggplant 334
- hops 173
- lettuce 343
- nuts 479
- onions 347
- ornamentals 522
- peas 352
- peppers 354
- potatoes 358
- small fruits 437
- small grains 136
- spinach 366
- sugar beets 188
- table beets 371
- tomatoes 373
- tree fruits, citrus 448
- tree fruits, deciduous 456
- tree fruits, subtropical 472

Clean Crop Dimethoate (dimethoate) .... **Platte**
WPS chart 47
beans 204
celery 214
cole crops 220
cotton 114
cucurbits 228
field corn, popcorn 79
greens 240
lentils 247
lettuce 250
peas 262
peppers 267
potatoes 273
radishes 280
safflower 156
small grains 130
sorghum 158
soybeans 97
spinach 285
table beets 298
tomatoes 303
Clean Crop Malathion (malathion) .... **Platte**
WPS chart 57
asparagus 200
beans 205
carrots, etc. 211
celery 215
cole crops 222
cotton 116
eggplant 233
forage (alfalfa, etc.) 540
forage (grasses, hay) 543
garlic 237
greens 241
herbs/spices 244
hops 147
horseradish 245
lentils 248
lettuce 251
mint 149
mushrooms 254
nuts 428
onions 258
pastures/rangeland 545
peanuts 152
peas 263
peppers 258
potatoes 274
radishes 280
rice 155
safflower 157
small fruits 387
small grains 132
sorghum 159
soybeans 98
spinach 286
sugar beets 163
sweet potatoes, yams 296
table beets 299
tobacco 171
tomatoes 304
tree fruits, citrus 396
tree fruits, deciduous 411
tree fruits, subtropical 420
watercress 307
Clean Crop Nemasol (metam-sodium) .... **Platte**
WPS chart 47
mint 174
ornamentals 526
peanuts 178
potatoes 360
tobacco 193
Clean Crop Phorate (phorate) .... **Platte**
WPS chart 47
beans 207
cotton 119
field corn, popcorn 83
peanuts 153
potatoes 276
small grains 133
sorghum 160
soybeans 99
sugar beets 169
sugarcane 166
sweet corn 294
Clean Crop Sevin (carbaryl) .... **Platte**
WPS chart 47
asparagus 200
beans 203
carrots, etc. 209
celery 214
cole crops 218
cotton 111
cucurbits 226
eggplant 232
field corn, popcorn 78
forage (alfalfa, etc.) 537
forage (grasses, hay) 543
greens 238
horseradish 245
lawns/turf 490
lettuce 249
noncropland 546
okra 255
ornamentals 497
pastures/rangeland 544
peanuts 151
peas 260
peppers 266
potatoes 271
radishes 278
rice 154
small fruits 382
small grains 129
sorghum 157
soybeans 96
spinach 285
sugar beets 162
sunflower 167
sweet corn 288
sweet potatoes, yams 295
table beets 298
tobacco 170
tomatoes 302
tree fruits, citrus 393
tree fruits, deciduous 405
clofentezine (see Apollo)
***Coleomegilla imaculata* (ladybird beetle)** .... **Arbico**
biocontrols 63
Comite (propargite) .... **Uniroyal**
WPS chart 411
beans 207
cotton 119
field corn, popcorn 83
forage (alfalfa, etc.) 507
peanuts 153
potatoes 277
sorghum 160
sweet corn 294
tree fruits, citrus 399
Condor (*Bt.* spp. *kurstaki*) .... **Ecogen**
Bt chart 76,94,108,128,144,196,380,488,536
WPS chart 43
Confirm (tebufenozide) .... **Rohm and Haas**
WPS chart 48
nuts 426
Conserve (spinosad) .... **Dow AgroSciences**
WPS chart 42
lawns/turf 493
ornamentals 508
Contesia spp. (see *Contesia marginiventris, Contesia plutella*)
Copper 3 Dust (basic copper sulfate) .... **Wilbur-Ellis**
WPS chart 50
tree fruits, deciduous 457
copper ammonium (see Copper-Count-N)
Copper-Count-N (copper ammonium) .... **Mineral Research**
WPS chart 46
beans 312
carrots, etc. 317
celery 320
cole crops 324
cucurbits 329
eggplant 334
nuts 479
onions 347
peanuts 177
peppers 354
potatoes 358
small fruits 437
spinach 366
sugar beets 188
tomatoes 373
tree fruits, citrus 448
tree fruits, deciduous 457
tree fruits, subtropical 472
copper hydroxide (see Champ, Champion, Kocide, KOP-Hydroxide, Nu-Cop)
copper oxychloride (see KOP Oxy-85, Microsperse COC)
copper oxychloride & basic copper sulfate (see Clean Crop C-O-C-S)
copper sulfate pentahydrate (see Blue Viking, Tennessee Brand)
***Cotesia marginiventris (Cotesia* spp.)** ...... **Arbico, Caltec**
biocontrols 63
***Cotesia plutella (Cotesia* spp.)** .... **Arbico, Caltec**
biocontrols 63
Counter (terbufos) .... **American Cyanamid**
WPS chart 40
field corn, popcorn 79
sorghum 158
sugar beets 162
sweet corn 290
Cruiser (*Heterorhabditis bacteriophora*) .... **Ecogen**
biocontrols 65
CRYMAX WDG (*Bt.* spp. *kurstaki*) .... **Ecogen**
Bt chart 76,94,108,128,144,196,380,488,536
WPS chart 412
cryolite (see Kryocide)
***Cryptolaemus montrouzieri* (predator)** ...... **Arbico, Caltect, Florikan, Hydro-Gardens, International Technology, IPM Laboratories, Natural Pest Controls**
biocontrols 63
Cryptolaemus montrouzieri (see Mealybug Predators)
Curacron (profenofos) .... **Novartis**
WPS chart 46
cotton 112
Curalan (vinclozolin) .... **BASF**
WPS chart 41
lawns/turf 519
ornamentals 533
Cutlass (*Bt.* spp. *kurstaki*) .... **Ecogen**
Bt chart 76,94,108,128,144,196,380,488,536
WPS chart 43
cyfluthrin (see Baythroid, Decathlon, Tempo)
lambda-cyhalothrin (see Karate, Scimitar, Topcide, Warrior)
cypermethrin (see Ammo)
zeta-cypermethrin (see Mustang)
cyproconazole (see Sentinel)
cyprodinil (see Vangard)
Cyren (chlorpyrifos) .... **Cheminova**
WPS chart 41
lawns/turf 490
ornamentals 497
cyromazine (see Citation, Trigard)

D

***Dacnusa sibirica & Diglyphus isaea* (parasites)** .. **Arbico, International Tecnology, Natural Pest Controls**
biocontrols 63
Daconil (chlorothalonil) .... **ZENECA**
WPS chart 51
lawns/turf 511
ornamentals 521
Damoil (petroleum oil) .... **Drexel**
WPS chart 42
nuts 430
ornamentals 506
small fruits 390
tree fruits, citrus 398
tree fruits, deciduous 415
tree fruits, subtropical 421
Danitol (fenpropathrin) .... **Valent**
WPS chart 50
cotton 113
peanuts 152
small fruits 383
tomatoes 302
dazomet (see Basamid Granular)
DB-Green (maneb & lindane) .... **AGSCO**
WPS chart 40
field corn, popcorn 81,87
small grains 131,138
soybeans 98,103
sunflower 192
DB-Green L (maneb & lindane) .... **AGSCO**
WPS chart 40
field corn, popcorn 81,87
small grains 131,138
soybeans 98,103
sunflower 192
DB-Green + Vitavax (carboxin & maneb & lindane) .... **AGSCO**
WPS chart 40
field corn, popcorn 87
small grains 138
Deadline (metaldehyde) .... **Pace International**
WPS chart 47
artichokes 198
asparagus 200
beans 203
carrots, etc. 210
celery 214
cole crops 220
cucurbits 227
eggplant 232
field corn, popcorn 79
forage (alfalfa, etc.) 538
forage (grasses, hay) 543
garlic 236
greens 239
horseradish 245
lawns/turf 491
lettuce 250
mint 148
okra 255
onions 257
ornamentals 499
peas 261
peppers 266
potatoes 272
radishes 279
rhubarb 281
small fruits 384
southern peas 283
spinach 285
sugar beets 162
sweet corn 290
sweet potatoes, yams 295
table beets 298
tomatoes 302
tree fruits, citrus 395
tree fruits, deciduous 406
tree fruits, subtropical 419

Product Index

Decathlon (cyfluthrin) .... **Olympic**
WPS chart 46
ornamentals 498
Decis (deltamethrin) .... **AgrEvo USA**
WPS chart 40
cotton 113
Declare (methyl parathion) .... **Griffin**
WPS chart 44
beans 206
carrots, etc. 211
celery 215
cole crops 222
cotton 118
field corn, popcorn 82
forage (alfalfa, etc.) 541
forage (grasses, hay) 544
greens 241
lettuce 252
noncropland 547
oil seed crops 150
onions 259
pastures/rangeland 545
peas 263
potatoes 275
rice 155
small grains 132
soybeans 99
spinach 286
sugar beets 164
sunflower 168
table beets 299
Decoy (pheromone) .... **Thermo-Trilogy**
biocontrols 60
Defend (PCNB) .... **Cleary**
WPS chart 42
lawns/turf 515
ornamentals 527
***Delphastus pusillus* (predatory beetle)** .... **Arbico, IPM Laboratories, Nature's Control**
biocontrols 63
deltamethrin (see Decis)
Demosan (chloroneb) .... **Kincaid**
WPS chart 45
beans 311
cotton 123
southern peas 364
soybeans 102
sugar beets 187
Devour (*Steinermema riobravis*) .... **Thermo-Trilogy**
biocontrols 68
***Diadigma insulare* (diamondback moth parasite)** .... **Arbico, Caltec**
biocontrols 63
diamondback moth parasite (see *Diadigma insulare*)
***Diaretiella rapae* (aphid parasite)** .... **Arbico**
biocontrols 63
Diatect Multipurpose Insecticide II (pyrethrins) .... **Diatect**
WPS chart 42
artichokes 199
asparagus 202
beans 208
carrots, etc. 212
celery 216
cole crops 224
cotton 120
cucurbits 230
eggplant 235
field corn, popcorn 84
forage (alfalfa, etc.) 542
forage (grasses, hay) 544
garlic 237
greens 242
herbs/spices 244
hops 147
horseradish 246
lawns/turf 493
lentils 248
lettuce 253
mint 149
mushrooms 255
nuts 431
okra 256
onions 260
ornamentals 507
pastures/rangeland 546
peas 264
peppers 270
potatoes 277
radishes 281
rhubarb 282
rice 156
small fruits 391
small grains 133
sorghum 160
southern peas 284
soybeans 100
spinach 287
sugar beets 165
sunflower 169
sweet potatoes, yams 297
table beets 300
tomatoes 306
tree fruits, citrus 399
tree fruits, deciduous 416
tree fruits, subtropical 422
Diatomaceous Earth .... **Nature's Control**
biocontrols 55
diazinon (see Drexel Diazinon, D•z•n, Gowan Diazinon, Helena Diazinon, KnoxOut, Micro Flo Diazinon, Prentox Diazinon, Riverside Diazinon, Wilbur-Ellis Diazinon)
Dibrom (naled) .... **Amvac**
beans 204
cole crops 220
cotton 113
cucurbits 227
eggplant 233
forage (alfalfa, etc.) 538
greens 239
hops 146
noncropland 546
nuts 426
ornamentals 499
pastures/rangeland 545
peas 261
peppers 267
safflower 156
small fruits 384
sugar beets 163
tree fruits, citrus 395
tree fruits, deciduous 407
dicloran (see Botran)
dicloran & sulfur (see Botran 6 Sulfur 25 Dust)
dicloropropene (see Telone II)
dicloropropene & chloropicrin (see Telone C-17)
dicofol (see Griffin Dicofol, Kelthane)
dicrotophos (see Bidrin)
dienochlor (see Pentac)
difenoconazole & mefenoxam (see Dividend XL)
diflubenzuron (see Adept, Dimilin)
Digon 400 (dimethoate) .... **Wilbur-Ellis**
WPS chart 50
beans 204
celery 214
cole crops 220
cotton 114
cucurbits 228
field corn, popcorn 79
forage (alfalfa, etc.) 539
greens 240
lentils 247
lettuce 250
nuts 427
peas 262
peppers 267
potatoes 273
radishes 280
safflower 156
small grains 130
sorghum 158
soybeans 97
spinach 285
table beets 298
tomatoes 303
tree fruits, citrus 395
tree fruits, deciduous 408
Dimate (dimethoate) .... **Terra**
WPS chart 48
beans 204
celery 214
cole crops 220
cotton 114
cucurbits 228
field corn, popcorn 79
forage (alfalfa, etc.) 539
greens 240
lentils 247
lettuce 250
nuts 427
peas 262
peppers 267
potatoes 273
radishes 280
safflower 156
small grains 130
sorghum 158
soybeans 97
spinach 285
table beets 298
tomatoes 303
tree fruits, citrus 395
tree fruits, deciduous 408
dimethoate (see Clean Crop Dimethoate, Digon 400, Dimate, Drexel Dimethoate, Gowan Dimethoate, Helena Dimethoate, Micro Flo Dimethoate)
Dimilin (diflubenzuron) .... **Uniroyal**
WPS chart 50
cotton 114
mushrooms 254
ornamentals 500
soybeans 97
DiPel (*Bt.* spp. *kurstaki*) .... **Abbott**
Bt chart 76,94,108,128,144,196,380,488,536
WPS chart 40
Disrupt (pheromone) .... **Ecogen**
biocontrols 60
Distance (pyriproxyfen) .... **Valent**
WPS chart 50
ornamentals 500
disulfoton (see Di-Syston)
Di-Syston (disulfoton) .... **Bayer**
WPS chart 41
asparagus 200
beans 205
cole crops 221
cotton 114
field corn, popcorn 80
forage (alfalfa, etc.) 539
lawns/turf 492
lentils 247
lettuce 250
nuts 427
ornamentals 500
peanuts 152
peas 262
peppers 267
potatoes 273
small fruits 385
small grains 130
sorghum 159
soybeans 97
tobacco 170
tomatoes 303
tree fruits, deciduous 408
tree fruits, subtropical 419
DiTera (*Myrothecium verrucaria*) .... **Abbott**
WPS chart 40
biocontrols 56
Dithane (mancozeb) .... **Rohm and Haas**
WPS chart 48
asparagus 308
cotton 123
cucurbits 331
field corn, popcorn 86
flax 173
herbs/spices 341
lawns/turf 513
onions 348
ornamentals 524
peanuts 178
potatoes 359
rice 182
safflower 184
small fruits 440
small grains 137
sorghum 185
sugar beets 188
sweet corn 368
tomatoes 375
tree fruits, deciduous 463
tree fruits, subtropical 475
Dividend XL (difenoconazole & mefenoxam) .... **Novartis**
WPS chart 46
small grains 136
dodine (see Syllit)
Double R (imazalil) .... **AGSCO**
WPS chart 40
small grains 137
Drexel Basic Copper Sulfate (basic copper sulfate) .... **Drexel**
WPS chart 42
beans 312
carrots, etc. 317
celery 320
cole crops 324
cucurbits 329
eggplant 334
nuts 480
onions 347
peanuts 177
peppers 354
potatoes 357
small fruits 437
spinach 366
sugar beets 187
table beets 371
tomatoes 374
tree fruits, citrus 447
tree fruits, deciduous 457
tree fruits, subtropical 472
Drexel Captan (captan) .... **Drexel**
WPS chart 42
beans 310
cole crops 323
cotton 122
cucurbits 328
field corn, popcorn 85
forage (alfalfa, etc.) 547
greens 339
lawns/turf 510
nuts 478
onions 346
ornamentals 520
peanuts 176
peas 351
peppers 354
rice 182
small fruits 435

small grains ... 134
sorghum ... 185
southern peas ... 364
soybeans ... 101
spinach ... 365
sugar beets ... 187
sweet corn ... 367
table beets ... 371
tree fruits, deciduous ... 453
Drexel Captan Plus Molybdenum (captan) ... Drexel
WPS chart ... 42
nuts ... 101
Drexel Carbaryl (carbaryl) ... Drexel
WPS chart ... 42
asparagus ... 200
beans ... 203
carrots, etc ... 209
celery ... 214
cole crops ... 218
cotton ... 111
cucurbits ... 226
field corn, popcorn ... 78
forage (alfalfa, etc.) ... 537
greens ... 238
horseradish ... 245
lawns/turf ... 490
lettuce ... 249
nuts ... 425
okra ... 255
ornamentals ... 497
pastures/rangeland ... 544
peanuts ... 151
peas ... 261
peppers ... 266
potatoes ... 272
radishes ... 279
rice ... 154
small fruits ... 382
small grains ... 129
southern peas ... 282
soybeans ... 96
spinach ... 285
sugar beets ... 162
sunflower ... 167
sweet corn ... 288
sweet potatoes, yams ... 295
table beets ... 298
tobacco ... 170
tomatoes ... 302
tree fruits, citrus ... 393
tree fruits, deciduous ... 404
tree fruits, subtropical ... 418
Drexel Diazinon (diazinon) ... Drexel
WPS chart ... 42
beans ... 204
carrots, etc. ... 210
celery ... 214
cole crops ... 220
cucurbits ... 227
greens ... 239
hops ... 146
lawns/turf ... 491
lettuce ... 250
nuts ... 426
onions ... 258
ornamentals ... 499
peas ... 261
peppers ... 267
potatoes ... 273
radishes ... 279
small fruits ... 384
spinach ... 285
sugar beets ... 163
sweet corn ... 290
sweet potatoes, yams ... 295
table beets ... 298
tomatoes ... 303
tree fruits, deciduous ... 406
Drexel Dimethoate (dimethoate) ... Drexel
WPS chart ... 42
beans ... 204
celery ... 214
cole crops ... 220
cotton ... 114
cucurbits ... 228
field corn, popcorn ... 79
forage (alfalfa, etc.) ... 539
greens ... 240
lettuce ... 250
nuts ... 427
peas ... 262
peppers ... 267
potatoes ... 273
radishes ... 280
safflower ... 156
small grains ... 130
sorghum ... 158
soybeans ... 97
spinach ... 285
table beets ... 298
tomatoes ... 303
tree fruits, citrus ... 395
tree fruits, deciduous ... 408
Drexel Endosulfan (endosulfan) ... Drexel
WPS chart ... 42
artichokes ... 198
beans ... 205
carrots, etc. ... 210
celery ... 214
cole crops ... 221
cotton ... 114
cucurbits ... 228
eggplant ... 233
forage (alfalfa, etc.) ... 539
greens ... 240
lettuce ... 251
nuts ... 427
ornamentals ... 500
peas ... 262
peppers ... 267
potatoes ... 273
safflower ... 156
small fruits ... 385
small grains ... 130
southern peas ... 283
spinach ... 285
sugar beets ... 163
sunflower ... 168
sweet corn ... 291
sweet potatoes, yams ... 295
tobacco ... 170
tomatoes ... 303
tree fruits, citrus ... 396
tree fruits, deciduous ... 408
tree fruits, subtropical ... 419
Drexel Malathion (malathion) ... Drexel
WPS chart ... 42
beans ... 205
carrots, etc. ... 211
celery ... 215
cole crops ... 222
cotton ... 117
cucurbits ... 229
eggplant ... 233
field corn, popcorn ... 81
forage (alfalfa, etc.) ... 540
forage (grasses, hay) ... 543
garlic ... 237
greens ... 241
herbs/spices ... 244
hops ... 147
horseradish ... 246
lentils ... 248
lettuce ... 252
mint ... 149
mushrooms ... 254
nuts ... 428
okra ... 256
onions ... 258
ornamentals ... 503
pastures/rangeland ... 545
peanuts ... 152
peas ... 263
peppers ... 268
potatoes ... 274
rice ... 155
small fruits ... 387
small grains ... 132
southern peas ... 283
soybeans ... 98
spinach ... 286
sugar beets ... 163
sweet corn ... 292
sweet potatoes, yams ... 296
table beets ... 299
tree fruits, citrus ... 397
tree fruits, deciduous ... 411
tree fruits, subtropical ... 420
Drexel Suffa (sulfur) ... Drexel
WPS chart ... 42
asparagus ... 309
beans ... 314
carrots, etc. ... 319
cole crops ... 326
cotton ... 120
cucurbits ... 332
field corn, popcorn ... 88
garlic ... 338
greens ... 340
nuts ... 483
onions ... 350
ornamentals ... 530
peanuts ... 179
peas ... 353
peppers ... 356
potatoes ... 361
small fruits ... 443
small grains ... 140
sorghum ... 186
soybeans ... 104
sugar beets ... 189
tomatoes ... 306
tree fruits, citrus ... 400
tree fruits, deciduous ... 465
Drexel Sulfa (sulfur) ... Drexel
WPS chart ... 42
asparagus ... 309
beans ... 314
carrots, etc. ... 319
celery ... 322
cole crops ... 326
cucurbits ... 332
field corn, popcorn ... 88
garlic ... 338
greens ... 340
lettuce ... 344
mint ... 175
nuts ... 483
onions ... 350
ornamentals ... 530
peanuts ... 179
peas ... 353
peppers ... 356
potatoes ... 361
rice ... 183
small fruits ... 443
sorghum ... 186
soybeans ... 104
sugar beets ... 189
sweet corn ... 369
table beets ... 372
tomatoes ... 377
tree fruits, deciduous ... 465
tree fruits, subtropical ... 476
Drexel Sulfur (sulfur) ... Drexel
WPS chart ... 42
asparagus ... 309
beans ... 314
carrots, etc. ... 319
cole crops ... 326
cotton ... 120
cucurbits ... 332
field corn, popcorn ... 88
garlic ... 338
greens ... 340
nuts ... 483
onions ... 350
ornamentals ... 530
peanuts ... 179
peas ... 353
peppers ... 356
potatoes ... 361
small fruits ... 443
small grains ... 140
sorghum ... 186
soybeans ... 104
sugar beets ... 189
tomatoes ... 306
tree fruits, citrus ... 400
tree fruits, deciduous ... 465
Dursban (chlorpyrifos) ... Dow AgroSciences
WPS chart ... 42
lawns/turf ... 480
ornamentals ... 497
Dusting Sulfur (sulfur) ... Platte
WPS chart ... 47
beans ... 314
carrots, etc. ... 319
cole crops ... 326
ornamentals ... 530
peas ... 353
small fruits ... 443
Dustret (maneb & streptomycin sulfate) ... AGSCO
WPS chart ... 40
potatoes ... 358
Dustret Maxim (fludioxonil) ... AGSCO
WPS chart ... 40
potatoes ... 358
Dustret T (thiophanate-methyl) ... AGSCO
WPS chart ... 40
potatoes ... 362
Dylox (trichlorfon) ... Bayer
lawns/turf ... 494
ornamentals ... 508
D•z•n (diazinon) ... Novartis
WPS chart ... 46
beans ... 204
carrots, etc. ... 210
cole crops ... 220
cucurbits ... 227
greens ... 239
hops ... 146
lawns/turf ... 491
lettuce ... 250
nuts ... 426
onions ... 258
ornamentals ... 499
peas ... 261
peppers ... 267
potatoes ... 272
radishes ... 279
small fruits ... 384
spinach ... 285
sugar beets ... 163
sweet corn ... 290

Product Index

sweet potatoes, yams....295
table beets....298
tomatoes....303
tree fruits, deciduous....406
tree fruits, subtropical....419

E

Eagle (myclobutanil)....Rohm and Haas
WPS chart....48
lawns/turf....515
ornamentals....527
Echo (chlorothalonil)....Sostram
WPS chart....48
beans....311
carrots, etc....316
celery....320
cole crops....323
cucurbits....329
garlic....336
lawns/turf....511
mint....174
onions....347
ornamentals....521
peanuts....176
potatoes....357
small fruits....436
soybeans....103
sweet corn....368
tomatoes....373
tree fruits, deciduous....455
tree fruits, subtropical....470
Ecomask (*Steinernema carpocapsae*)....Biologic
biocontrols....68
Ecosafe (eugenol)....Arbico
biocontrols....56
Elite (tebuconazole)....Bayer
WPS chart....41
tree fruits, deciduous....467
tree fruits, subtropical....477
Enable (fenbuconazole)....Rohm and Haas
WPS chart....48
nuts....482
tree fruits, subtropical....474
***Encarsia formosa*....A-1 Unique Insect Control, Arbico, Beneficial Insectary, Florikan, Harmony Farm Supply, Hydro-Gardens, International Technology, IPM Laboratories, Nature's Control**
biocontrols....69
Encarsia formosa (see En-Strip, Whitefly Parasites)
endosulfan (see Drexel Endosulfan, Golden Leaf, Gowan Endosulfan, Micro Flo Endosulfan, Phaser, Thiodan)
endosulfan & pyrethrins (see Thirethrin)
Enhance Plus (maneb & carboxin & lindane)....Trace
WPS chart....49
field corn, popcorn....81,87
small grains....131,138
sorghum....159
Enhance Vitavax-Captan (carboxin & captan)....Trace
WPS chart....49
small grains....135
soybeans....102
Enhanced Seed
cotton....108
field corn, popcorn....76
forage (alfalfa, etc.)....536
potatoes....159
soybeans....95
Enstar (S-kinoprene)....Wellmark International
WPS chart....50
ornamentals....501
En-Strip (*Encarsia formosa*)....Natural Pest Controls
biocontrols....69
Entice (insect feeding stimulant)....Custom Chemicides
biocontrols....56
Enzone (sodium tetrathiocarbonate)....Entek
WPS chart....43
***Eretmocerus californicus* Beneficial Insectary, Florikan, Hydro-Gardens, M & R Durangos**
biocontrols....69
Eretmocerus californicus (see Whitefly Parasites)
esfenvalerate (see Asana)
ethion (see Ethion 4EC, Ethion 4 Miscible)
Ethion (ethion)....FMC, Micro Flo
WPS chart....43,45
tree fruits, citrus....396
ethoprop (see Chipco Mocap, Mocap)
etridiazole (see Koban, Terrazole)
eugenol (see Ecosafe)

F

fenamiphos (see Nemacur)
fenamiphos & fensulfothion (see Nemacur-Dasanit)
fenarimol (see Rubigan)
fenbuconazole (see Enable, Indar)
fenbutatin-oxide (see Vendex)
fenoxycarb (see Award, Logic, Precision)
fenpropathrin (see Danitol, Tame)
fensulfothion (see Dasanit)
ferbam (see Ferbam Granuflo)
Ferbam Granuflo (ferbam)....UCB Agrochemicals
WPS chart....49
ornamentals....523
small fruits....439
tobacco....192
tree fruits, citrus....449
tree fruits, deciduous....461
fipronil (see Regent)
Fire mite....Caltec
biocontrols....64
Fireban (tefluthrin)....Uniroyal
WPS chart....50
Flo Pro IMZ (imazalil)....Gustafson
WPS chart....44
small grains....137
fludioxonil (see Dustret Maxim, Maxim, Medallion)
flutolanil (see Moncut, ProStar, ProStar Plus)
tau-fluvalinate (see Mavrik)
Fly Parasites....Beneficial Insectary, Hydro-Gardens
biocontrols....64
Fly Predators (*Muscidifurax Raptorellus, Spanglia Endius, Spanglia Cameroni*)....Spalding Laboratories
biocontrols....64
Folicur (tebuconazole)....Bayer
WPS chart....41
peanuts....180
fonofos (see Crusade)
Foray (*Bt.* spp. *kurstaki*)....Abbott
Bt chart....488
WPS chart....40
Force (tefluthrin)....ZENECA
WPS chart....51
field corn, popcorn....80
Fore (mancozeb)....Rohm and Haas
WPS chart....48
lawns/turf....513
ornamentals....524
Formec (mancozeb)....PBI Gordon
lawns/turf....513
ornamentals....524
formetanate hydrochloride (see Carzol)
Fortress (chlorethoxyfos)....DuPont
WPS chart....42
field corn, popcorn....80
sweet corn....291
fosetyl-aluminum (see Aliette, Aliette/Maneb 2+2, Chipco Aliette)
Frustrate (pheromone)....Thermo Trilogy
biocontrols....60
Fungo (thiophanate-methyl)....Scotts
WPS chart....48
lawns/turf....517
ornamentals....532
Fungus Gnat Predators (*Hypoaspis miles*)....Nature's Control
biocontrols....65
Furadan (carbofuran)....FMC
WPS chart....43
cotton....115
field corn, popcorn....80
forage (alfalfa, etc.)....539
ornamentals....502
potatoes....274
small fruits....385
small grains....130
soybeans....97
sugarcane....166
sunflower....168
sweet corn....291
tobacco....171
Fury (*zeta*-cypermethrin)....FMC
WPS chart....43
cole crops....219
cotton....113
lettuce....250
onions....257
Fyfanon (malathion)....Helena
WPS chart....45
beans....206
carrots, etc....211
cotton....117
cucurbits....229
eggplant....234
field corn, popcorn....81
forage (alfalfa, etc.)....540
horseradish....246
lettuce....252
mint....149
nuts....429
onions....258
ornamentals....503
peppers....268
potatoes....274
radishes....280
rice....155
small fruits....388
small grains....132
sweet corn....292
sweet potatoes, yams....296
tomatoes....304
tree fruits, citrus....397
tree fruits, deciduous....411
Fyfanon ULV (malathion)....Cheminova
beans....206
cotton....117
field corn, popcorn....81
forage (alfalfa, etc.)....540
forage (grasses, hay)....544
noncropland....547
peas....263
rice....155
safflower....157
small fruits....388
small grains....132
sorghum....160
soybeans....98
sugar beets....163
tree fruits, deciduous....411

G

***Galendromus helvelous* (predator mite)....Arbico**
biocontrols....64
***Galendromus occidentalis* (predator mite)....Arbico, Harmony Farm Supply**
biocontrols....64
Gallex (2,4-xylenol & meta-cresol)....AgBioChem
nuts....482
ornamentals....524
small fruits....440
tree fruits, deciduous....462
tree fruits, subtropical....474
Galltrol (*Agrobacterium radiobacter*)....AgBioChem
biocontrols....56
Garlic Gard (garlic juice)....Soil Technologies
WPS chart....48
biocontrols....56
garlic juice (see Garlic Gard)
Gaucho (imidacloprid)....Gustafson
cotton....115
small grains....131
sorghum....159
Gambusis affinis (see mosquito fish)
Gemstar (nuclear polyhedrosis virus)....Thermo Trilogy
biocontrols....59
Germate Plus (carboxin & diazinon & lindane)....Trace
WPS chart....49
beans....310
field corn, popcorn....79,86
sweet corn....290,368
Gliocladium virens (see SoilGard)
Gnatrol (*Bt.* spp. *israelensis*)....Abbott
Bt chart....196,488
WPS chart....40
Golden Eagle (myclobutanil)....Scotts
lawns/turf....515
Golden Leaf (endosulfan)....FMC
WPS chart....43
tobacco....170
Golden Natur'l Spray Oil (soybean oil)....Stoller
biocontrols....57
***Goniozus lengneri* (navel orangeworm parasite)....Caltec**
biocontrols....64
Good Bug Power Meal (beneficial insect attractant)....Arbico
biocontrols....57
Gowan Azinphos-M (azinphos-methyl)....Gowan
WPS chart....44
beans....203
celery....213
cole crops....218
cotton....110
cucurbits....226
eggplant....232
forage (alfalfa, etc.)....537
nuts....424
onions....257
ornamentals....496
peppers....265
potatoes....271
small fruits....381
small grains....129
southern peas....282
soybeans....96
spinach....284
tobacco....169
tomatoes....301
tree fruits, citrus....393
tree fruits, deciduous....402
Gowan Diazinon (diazinon)....Gowan
WPS chart....44
beans....204
carrots, etc....210
celery....214
cole crops....220
cucurbits....227
greens....239
hops....146
lettuce....250
nuts....426
onions....258
peas....261
peppers....267
potatoes....272
radishes....279
small fruits....384
spinach....285

sugar beets ... 163
sweet corn ... 290
sweet potatoes, yams ... 295
table beets ... 298
tomatoes ... 303
tree fruits, deciduous ... 406
tree fruits, subtropical ... 419

Gowan Dimethoate (dimethoate) ... Gowan
WPS chart ... 44
beans ... 204
cole crops ... 220
cotton ... 114
cucurbits ... 228
field corn, popcorn ... 79
forage (alfalfa, etc.) ... 539
greens ... 240
lentils ... 247
lettuce ... 250
nuts ... 427
peas ... 262
peppers ... 267
potatoes ... 273
radishes ... 280
safflower ... 156
small grains ... 130
sorghum ... 158
soybeans ... 97
spinach ... 285
table beets ... 298
tomatoes ... 303
tree fruits, citrus ... 395
tree fruits, deciduous ... 408

Gowan Endosulfan (endosulfan) ... Gowan
WPS chart ... 44
artichokes ... 198
beans ... 205
carrots, etc ... 210
celery ... 214
cole crops ... 221
cotton ... 114
cucurbits ... 228
eggplant ... 233
forage (alfalfa, etc.) ... 539
greens ... 240
lettuce ... 251
nuts ... 427
ornamentals ... 500
peas ... 262
peppers ... 267
potatoes ... 273
safflower ... 156
small fruits ... 385
small grains ... 130
southern peas ... 283
spinach ... 285
sugar beets ... 163
sunflower ... 168
sweet corn ... 291
sweet potatoes, yams ... 295
tobacco ... 170
tomatoes ... 303
tree fruits, citrus ... 396
tree fruits, deciduous ... 408
tree fruits, subtropical ... 419

Gowan Malathion (malathion) ... Gowan
WPS chart ... 44
asparagus ... 201
beans ... 206
carrots, etc. ... 211
celery ... 215
cole crops ... 222
cotton ... 117
cucurbits ... 229
eggplant ... 233
field corn, popcorn ... 82
forage (alfalfa, etc.) ... 540
garlic ... 237
greens ... 241
hops ... 147
horseradish ... 246
lentils ... 248
lettuce ... 252
mint ... 149
mushrooms ... 254
nuts ... 428
okra ... 256
onions ... 258
ornamentals ... 503
pastures/rangeland ... 545
peanuts ... 152
peas ... 263
peppers ... 268
potatoes ... 274
radishes ... 280
rice ... 155
small fruits ... 388
small grains ... 132
soybeans ... 98
spinach ... 286
sugar beets ... 164
sweet corn ... 292
sweet potatoes, yams ... 296
table beets ... 299
tobacco ... 171
tomatoes ... 304
tree fruits, citrus ... 397
tree fruits, deciduous ... 412
tree fruits, subtropical ... 420

Gowan Sevin (carbaryl) ... Gowan
WPS chart ... 44
asparagus ... 200
beans ... 203
carrots, etc. ... 209
cole crops ... 218
cotton ... 111
cucurbits ... 226
eggplant ... 232
field corn, popcorn ... 78
forage (alfalfa, etc.) ... 537
greens ... 238
horseradish ... 245
lettuce ... 249
okra ... 255
peas ... 261
peppers ... 266
potatoes ... 272
radishes ... 279
small fruits ... 383
spinach ... 285
sugar beets ... 162
sweet corn ... 289
table beets ... 298
tomatoes ... 302

Grain Guard (mancozeb) ... Trace
WPS chart ... 49
rice ... 183
safflower ... 184
small grains ... 138
sorghum ... 185

Grain Guard Plus (mancozeb & lindane) ... Trace
WPS chart ... 49
field corn, popcorn ... 81,86
small grains ... 131,137
sorghum ... 159,185

green lacewing (see *Chrysopa carna, Chrysopa carnea, Chrysopa rufilabris, Chrysoperla carnea, Chrysoperla rufilabris, Chrysoperla* spp.)

Griffin Dicofol (dicofol) ... Griffin
WPS chart ... 44
beans ... 204
cotton ... 113
cucurbits ... 228
forage (alfalfa, etc.) ... 539
hops ... 146
lawns/turf ... 491
mint ... 148
nuts ... 427
peppers ... 267
small fruits ... 384
tomatoes ... 303
tree fruits, citrus ... 395
tree fruits, deciduous ... 407

Guardian (*Steinernema* spp.) ... Hydro-Gardens
biocontrols ... 68

Gustafson Captan (captan) ... Gustafson
beans ... 310
cole crops ... 323
cotton ... 122
cucurbits ... 328
field corn, popcorn ... 85
flax ... 172
forage (alfalfa, etc.) ... 548
greens ... 339
lawns/turf ... 510
lentils ... 342
peanuts ... 176
peas ... 351
peppers ... 354
radishes ... 362
small grains ... 134
sorghum ... 185
southern peas ... 364
soybeans ... 101
spinach ... 365
sugar beets ... 187
sunflower ... 191
sweet corn ... 367
table beets ... 370
tomatoes ... 373

Gustafson LSP (thiabendazole) ... Gustafson
small grains ... 137

Guthion (azinphos-methyl) ... Bayer
WPS chart ... 41
cotton ... 110
cucurbits ... 226
forage (alfalfa, etc.) ... 537
nuts ... 424
onions ... 257
ornamentals ... 496
potatoes ... 271
small fruits ... 381
small grains ... 129
tomatoes ... 301
tree fruits, citrus ... 393
tree fruits, deciduous ... 402

H

Halofenozide (see Mach 2)

Helena Diazinon (diazinon) ... Helena
WPS chart ... 45
beans ... 204
carrots, etc. ... 210
cole crops ... 220
cucurbits ... 227
hops ... 146
nuts ... 426
onions ... 258
peas ... 261
peppers ... 267
potatoes ... 272
radishes ... 279
small fruits ... 384
sugar beets ... 163
sweet corn ... 290
tomatoes ... 303
tree fruits, deciduous ... 406

Helena Dimethoate (dimethoate) ... Helena
WPS chart ... 45
beans ... 204
cole crops ... 220
cotton ... 114
cucurbits ... 228
field corn, popcorn ... 79
forage (alfalfa, etc.) ... 539
lettuce ... 250
nuts ... 427
peas ... 262
peppers ... 267
potatoes ... 273
radishes ... 280
safflower ... 156
sorghum ... 158
spinach ... 285
table beets ... 298
tomatoes ... 303
tree fruits, citrus ... 395
tree fruits, deciduous ... 408

Helena Methyl Parathion (methyl parathion) ... Helena
WPS chart ... 45
beans ... 206
carrots, etc. ... 211
celery ... 215
cole crops ... 223
cotton ... 118
field corn ... 82
forage (alfalfa, etc.) ... 541
greens ... 241
lettuce ... 252
peas ... 263
potatoes ... 275
rice ... 155
small grains ... 132
spinach ... 286
sugar beets ... 164
sunflower ... 168

Heritage (azoxystrobin) ... ZENECA
WPS chart ... 509
lawns/turf ... 51

***Heterorhabditis bacteriophora* (*Heterorhabditis* spp.) ... Arbico, Harmony Farm Supply, M&R Durango**
biocontrols ... 65

Heterorhabditis bacteriophora (see Cruiser, Heteromask, Lawn Patrol)

Heteromask (*Heterorhabditis bacteriophora*) ... BioLogic
biocontrols ... 65

hexadecenal (see Chokegard)

Hexygon (hexythiazox) ... Gowan
WPS chart ... 44
ornamentals ... 502

hexythiazox (see Hexygon, Savey)

Hi Moly/Captan (captan) ... Trace
WPS chart ... 49
soybeans ... 101

Hot Pepper Wax (capsaicin) ... Hot Pepper Wax
WPS chart ... 45
biocontrols ... 57

Hippodamia convergens (see Ladybugs)

hydramethylnon (see Amdro, Siege)

***Hypoaspis miles* ... Florikan, Harmony Farm Supply**
biocontrols ... 65

Hypoaspis miles (see Fungus Gnat Predators)

I

imazalil (see Double R, Flo Pro IMZ)

imidacloprid (see Admire, Gaucho, Marathon, Merit, Provado)

Imidan (phosmet) ... Gowan
WPS chart ... 44
cotton ... 115
forage (alfalfa, etc.) ... 539
nuts ... 428
ornamentals ... 503

peas ... 262
potatoes ... 274
small fruits ... 286
tree fruits, deciduous ... 410
Indar (fenbuconazole) ... Rohm and Haas
WPS chart ... 48
tree fruits, deciduous ... 461
insect feeding stimulant (see Entice, Konsume)
iprodione (see Chipco 26019, ProTurf Fungicide X, Rovral)
iprodione & thiophanate-methyl (see ProTurf Fluid Fungicide)
isazofos (see Triumph)
Isomate (pheromone) ... Pacific Biocontrol
biocontrols ... 60
isofenphos (see Oftanol)

J

Javelin (*Bt.* spp. *kurstaki*) ... Thermo Trilogy
Bt chart ... 76,94,108,128,144,196,380,488,536
WPS chart ... 49
Joust (oxythioquinox) ... Olympic
WPS chart ... 46
ornamentals (flowers) ... 503

K

Karate (*lambda*-cyhalothrin) ... ZENECA
WPS chart ... 51
cole crops ... 219
cotton ... 112
field corn, popcorn ... 79
forage ... 543
garlic ... 236
lettuce ... 249
onions ... 257
peanuts ... 151
small grains ... 129
sorghum ... 158
soybeans ... 96
sunflower ... 167
sweet corn ... 290
tomatoes ... 302
Kelthane (dicofol) ... Rohm and Haas
WPS chart ... 48
beans ... 204
cotton ... 113
cucurbits ... 228
hops ... 146
lawns/turf ... 491
mint ... 148
nuts ... 427
ornamentals ... 503
peppers ... 267
small fruits ... 384
tomatoes ... 303
tree fruits, citrus ... 395
tree fruits, deciduous ... 408
Kernel Guard (captan & diazinon & lindane) ... Trace
WPS chart ... 49
beans ... 310
field corn, popcorn ... 79
peas ... 261,351
sweet corn ... 290,368
Kernel Guard Supreme (carboxin & permethrin) ... Trace
WPS chart ... 49
field corn, popcorn ... 83,86
soybeans ... 99,102
sweet corn ... 293
S-kinoprene (see Enstar)
Knack (pyriproxyfen) ... Valent
WPS chart ... 50
cotton ... 115
KnoxOut (diazinon) ... Cleary
WPS chart ... 42
ornamentals ... 499
Koban (etridiazole) ... Scotts
WPS chart ... 48
lawns/turf ... 512
Kocide (copper hydroxide) ... Griffin
WPS chart ... 44
beans ... 312
carrots, etc. ... 317
celery ... 320
cole crops ... 324
cucurbits ... 329
eggplant ... 334
forage (alfalfa, etc.) ... 248
greens ... 339
herbs/spices ... 341
hops ... 173
nuts ... 480
onions ... 347
ornamentals ... 522
peanuts ... 177
peas ... 352
peppers ... 354
potatoes ... 357
rice ... 182
small fruits ... 437
small grains ... 136
spinach ... 366
sugar beets ... 188
table beets ... 371
tomatoes ... 374
tree fruits, citrus ... 448
tree fruits, deciduous ... 456
tree fruits, subtropical ... 472
watercress ... 378
Kodiak (*Bacillus subtillus*) ... Gustafson, Trace
Bt chart ... 76,94,108,128,144,196
WPS chart ... 44
Konsume (insect feeding stimulant) ... Troy Biosciences
biocontrols ... 57
KOP 300 (basic copper sulfate) ... Drexel
WPS chart ... 42
beans ... 312
carrots, etc. ... 317
celery ... 320
cole crops ... 324
cucurbits ... 329
eggplant ... 334
nuts ... 480
onions ... 347
peanuts ... 177
peppers ... 354
potatoes ... 357
small fruits ... 437
spinach ... 366
sugar beets ... 187
table beets ... 371
tomatoes ... 374
tree fruits, citrus ... 447
tree fruits, deciduous ... 457
tree fruits, subtropical ... 472
KOP Oxy-85 (copper oxychloride) ... Drexel
WPS chart ... 42
carrots, etc. ... 317
celery ... 321
cucurbits ... 330
ornamentals ... 522
peppers ... 354
potatoes ... 358
sugar beets ... 188
tomatoes ... 374
tree fruits, citrus ... 449
tree fruits, deciduous ... 458
KOP-Hydroxide (copper hydroxide) ... Drexel
WPS chart ... 42
beans ... 312
celery ... 320
cole crops ... 324
cucurbits ... 329
eggplant ... 334
forage (alfalfa, etc.) ... 548
hops ... 174
lettuce ... 343
nuts ... 480
onions ... 347
ornamentals ... 521
peanuts ... 177
peas ... 352
peppers ... 354
potatoes ... 358
small fruits ... 438
small grains ... 136
sugar beets ... 187
tomatoes ... 374
tree fruits, citrus ... 448
tree fruits, deciduous ... 458
tree fruits, subtropical ... 473
Kryocide (cryolite) ... ELF Atochem
WPS chart ... 43
cole crops ... 219
cucurbits ... 227
greens ... 239
lettuce ... 249
ornamentals ... 498
peppers ... 266
small fruits ... 383
tomatoes ... 302
tree fruits, citrus ... 395
tree fruits, subtropical ... 419
Kumulus (sulfur) ... BASF
WPS chart ... 41
asparagus ... 309
beans ... 208,314
carrots, etc. ... 212,319
cole crops ... 225,326
cotton ... 120
cucurbits ... 333
field corn, popcorn ... 84,88
forage (alfalfa, etc.) ... 543
garlic ... 237,338
greens ... 243,340
hops ... 147
mint ... 175
nuts ... 432,484
onions ... 260,350
ornamentals ... 530
peanuts ... 154,179
peas ... 264,353
peppers ... 270,356
potatoes ... 277,361
small fruits ... 391,443
small grains ... 133,140
sorghum ... 161,186
soybeans ... 100,104
sugar beets ... 165,190
sweet corn ... 294
table beets ... 300,372
tomatoes ... 306
tree fruits, citrus ... 400
tree fruits, deciduous ... 416,465
tree fruits, subtropical ... 422,476

L

Lady beetle (see *Lindorus lophanthae*)
Ladybird beetle (see *Coleomegilla imaculata*)
Ladybugs (*Hippodamia convergens*) ... A-1 Unique Insect Control, Arbico, Caltec, Harmony Farm Supply, Hydro-Gardens, IPM Laboratories, Natural Pest Controls, Nature's Control
biocontrols ... 65
Ladybugs (see *Rhizibius lophanthae*, Spider Mite Destroyer)
Lannate (methomyl) ... DuPont
WPS chart ... 43
asparagus ... 200
beans ... 205
carrots, etc. ... 210
celery ... 214
cole crops ... 221
cotton ... 115
cucurbits ... 228
eggplant ... 233
field corn, popcorn ... 81
forage (alfalfa, etc.) ... 539
garlic ... 236
greens ... 240
herbs/spices ... 243
horseradish ... 245
lawns/turf ... 492
lentils ... 247
lettuce ... 251
mint ... 148
onions ... 258
peanuts ... 152
peas ... 262
peppers ... 268
potatoes ... 274
small fruits ... 386
small grains ... 131
sorghum ... 159
soybeans ... 97
spinach ... 286
sugar beets ... 163
sweet corn ... 291
table beets ... 299
tobacco ... 171
tomatoes ... 304
tree fruits, citrus ... 396
tree fruits, deciduous ... 410
tree fruits, subtropical ... 419
Larvin (thiodicarb) ... Rhone-Poulenc
WPS chart ... 47
carrots, etc. ... 210
celery ... 215
cole crops ... 221
cotton ... 116
greens ... 240
herbs/spices ... 243
lettuce ... 251
rhubarb ... 281
soybeans ... 98
spinach ... 286
sweet corn ... 292
Larvo-Bt (*Bt.* spp. *kurstaki*) ... Troy Biosciences
WPS chart ... 43
Lawn Patrol (*Heterorhabditis bacteriophora*) ... Hydro-Gardens
biocontrols ... 65
Lepinox (*Bt.* spp. *kurstaki*) ... Ecogen
Bt chart ... 76,94,108,128,144,196,488,536
WPS chart ... 43
***Leptomastix dactylopii* (parasitic wasp) ... Arbico, IPM Laboratories**
biocontrols ... 65
Lime-Sulfur Solution (calcium polysulfide) ... Best Sulfur
WPS chart ... 41
forage (alfalfa, etc.) ... 547
nuts ... 425,478
small fruits ... 382,434
tree fruits, citrus ... 393,447
tree fruits, deciduous ... 404,453
lindane (see Lindane 30C, Lindane ST 40)
Lindane 30C (lindane) ... Gustafson
WPS chart ... 44
beans ... 205
carrots, etc. ... 211
cole crops ... 221
cotton ... 116
cucurbits ... 228
forage (grasses, hay) ... 543

lentils....247
onions....258
peas....262
radishes....280
safflower....157
small grains....131
sunflower....168

Lindane ST 40 (lindane)....AGSCO
WPS chart....40
beans....205
carrots, etc....211
cole crops....221
cotton....116
cucurbits....228
forage (grasses, hay)....543
lentils....247
onions....258
peas....262
radishes....280
safflower....157
small grains....131
sunflower....168

***Lindorus lophanthae* (lady beetle)....Arbico**
biocontrols....65

Liquid Sulfur Six (sulfur)....Helena
WPS chart....45
beans....208,314
cole crops....225,326
nuts....484
peanuts....180
peas....264,353
peppers....270,356
potatoes....277,361
small fruits....443
sugar beets....190
tree fruits, deciduous....465

Lock-On (chlorpyrifos)....Dow AgroSciences
WPS chart....42
cotton....112
forage (alfalfa, etc.)....538

Logic (fenoxycarb)....Novartis
WPS chart....46
noncropland....546
small fruits....385
tree fruits, citrus....396
tree fruits, deciduous....409

Lorsban (chlorpyrifos)....Dow AgroSciences
WPS chart....42
asparagus....200
cole crops....218
cotton....112
field corn, popcorn....78
forage (alfalfa, etc.)....538
greens....239
mint....148
nuts....425
onions....257
peanuts....151
radishes....279
small fruits....383
small grains....129
sorghum....158
soybeans....96
sugar beets....162
sunflower....167
sweet corn....289
sweet potatoes, yams....295
tobacco....170
tree fruits, citrus....394
tree fruits, deciduous....406
tree fruits, subtropical....418

Lorsban (chlorpyrifos)....Gowan
WPS chart....44
cole crops....219
greens....239
tree fruits, citrus....394
tree fruits, deciduous....406

Lorsban (chlorpyrifos)....Gustafson
beans....203
cotton....112
cucurbits....227
peas....261
southern peas....283
sweet corn....289

M

malathion (see Clean Crop Malathion, Drexel Malathion, Fyfanon, Fyfanon ULV, Gowan Malathion, Hopkins Malathion, Micro Flo Malathion, Prentox Malathion, Riverside Malathion)
mancozeb (see Dithane, Fore, Formec, Grain Guard, Manzate, Penncozeb, Protect)
mancozeb & lindane (see Grain Guard Plus)
maneb (see Aliette/Maneb 2+2, Maneb, Manex)

Mach 2 (halofenozide)....RohMid L.L.C.
WPS chart....48
lawns/turf....492

Maneb (maneb)....Elf Atochem
WPS chart....43
beans....313
cole crops....325
cucurbits....331
eggplant....335
field corn, popcorn....87
greens....339
lettuce....514
lawns/turf....334
nuts....482
onions....349
ornamentals....525
peppers....355
potatoes....359
small fruits....441
sugar beets....188
sweet corn....369
tomatoes....375
tree fruits, deciduous....463
tree fruits, subtropical....475

maneb & carboxin & lindane (see Enhance Plus)
maneb & copper hydroxide (see ManKocide)
maneb & lindane (see DB-Green, DB-Green L, Seed Mate Maneb-Lindane)
maneb & streptomycin sulfate (see Dustret)
maneb & triphenyltin hydroxide (see Pro-Tex)

Manex (maneb)....Griffin
WPS chart....44
beans....313
cole crops....325
cucurbits....331
eggplant....335
field corn, popcorn....87
greens....339
lawns/turf....514
lettuce....344
nuts....482
onions....349
peppers....355
potatoes....359
small fruits....441
sugar beets....188
sweet corn....369
tomatoes....375
tree fruits, deciduous....463
tree fruits, subtropical....475

ManKocide (maneb & copper hydroxide)....Griffin
WPS chart....44
cucurbits....331
onions....349
peanuts....178
potatoes....359
small fruits....441
small grains....138
tree fruits, deciduous....463
tree fruits, subtropical....475

Manzate (mancozeb)....Griffin
WPS chart....44
asparagus....308
cotton....123
cucurbits....331
field corn, popcorn....86
flax....173
herbs/spices....341
lawns/turf....514
onions....348
peanuts....178
potatoes....359
rice....182
safflower....184
small fruits....440
small grains....137
sorghum....185
sugar beets....188
sweet corn....368
tomatoes....375
tree fruits, deciduous....463
tree fruits, subtropical....475

Marathon (imidacloprid)....Olympic
WPS chart....46
ornamentals....502

Marlate (methoxychlor)....Kincaid
WPS chart....206
beans....211
carrots, etc....222
cucurbits....229
eggplant....234
greens....241
lettuce....252
ornamentals....504
peas....263
peppers....268
potatoes....275
radishes....280
small fruits....388
southern peas....283
spinach....286
sweet corn....292
sweet potatoes, yams....296
table beets....299
tomatoes....305
tree fruits, deciduous....412

Mattch (*Bt*. spp. *kurstaki* & *aizawai*)....Mycogen
Bt chart....94,108,144,196,380,488
WPS chart....46

Mavrik (*tau*-fluvalinate)....Wellmark International
WPS chart....50
lawns/turf....493
ornamentals....504

Maxim (fludioxonil)....Novartis
field corn, popcorn....86
potatoes....359
sorghum....185
sweet corn....368

Medallion (fludioxonil)....Novartis
WPS chart....46
ornamentals....525

mefenoxam (see Apron XL, Dividend Apron XL, Ridomil Gold, Ridomil Gold/Bravo, Subdue 2X, Subdue GR, Subdue MAXX)
mefenoxam & chlorothalonil (see Ridomil Gold/Bravo)
mefenoxam & copper hydroxide (see Ridomil Gold/Copper)
mefenoxam & mancozeb (see Ridomil Gold MZ)
mefenoxam & PCNB (see Ridomil Gold PC)

Merit (imidacloprid)....Bayer
lawns/turf....492
ornamentals....502

Mertect (thiabendazole)....Novartis
carrots, etc....319
mushrooms....345
ornamentals....531
potatoes....362
sweet potatoes, yams....370
tobacco....193
tree fruits, deciduous....468

***Mesoseiulus longipes* (predator)....Arbico, Harmony Farm Supply, Nature's Control**
biocontrols....66

Mesurol (methiocarb)....Gowan
WPS chart....44
ornamentals....504

metalaxyl (see Allegiance, Apron, Apron Dry, ProTurf Pythium Control, Subdue)
metalaxyl & triadimefon (see ProTurf Fluid Fungicide II)

Metam CLR (metam-sodium)....UCB Agrochemicals
WPS chart....49
mint....149,174
peanuts....178
potatoes....275,360
small grains....138
tobacco....193

metam-sodium (see Clean Crop Nemasol, Metam CLR, Vapam)

***Metaphycus helvolus* (black scale parasite)....Arbico, Harmony Farm Supply, Natural Pest Controls, Sespe Creek Insectary**
biocontrols....66

Metasystox-R (oxydemeton-methyl)....Gowan
WPS chart....44
beans....206
cole crops....222
cotton....117
cucurbits....229
eggplant....234
forage (alfalfa, etc.)....541
lettuce....252
mint....149
nuts....429
ornamentals....504
peppers....268
small fruits....388
sugar beets....164
sweet corn....292
tree fruits, deciduous....412

methamidophos (see Monitor 4, Monitor 4 Spray)
methidathion (see Supracide)
methiocarb (see Mesurol)
methomyl (see Lannate, Riverside Methomyl)
methoxychlor (see Marlate, Prentox Methoxychlor)
methyl bromide (see Brom-O-Gas)
methyl bromide & chloropicrin (see Terr-O-Gas)
methyl parathion (see Cheminova Methyl, Declare, Helena Methyl Parathion, Penncap-M, Wilbur-Ellis Methyl Parathion)
methyl parathion & endosulfan (see Methyl Parathion 2 Thiodan)
metiram (see Polyram)

Methyl Parathion 2 Thiodan (methyl parathion & endosulfan)....FMC
cotton....118

Micro Flo Acephate (acephate)....Micro Flo
WPS chart....45
cotton....109
peanuts....150

Micro Flo Azinphosmethyl (azinphos-methyl)....Micro Flo
WPS chart....45
artichokes....198
beans....203
celery....213
cole crops....218
cotton....110
cucurbits....226
eggplant....232
forage (alfalfa, etc.)....537
nuts....424

onions ... 257
ornamentals ... 496
peppers ... 265
potatoes ... 271
small fruits ... 381
small grains ... 129
southern peas ... 282
soybeans ... 96
spinach ... 284
tobacco ... 169
tomatoes ... 301
tree fruits, citrus ... 393
tree fruits, deciduous ... 402

Micro Flo Basic Copper "53" (basic copper sulfate) ... Micro Flo
WPS chart ... 45
beans ... 312
carrots, etc. ... 317
celery ... 321
cole crops ... 323
cucurbits ... 329
eggplant ... 334
lettuce ... 343
nuts ... 481
onions ... 347
peanuts ... 177
peppers ... 354
potatoes ... 358
small fruits ... 438
spinach ... 366
sugar beets ... 188
table beets ... 371
tomatoes ... 374
tree fruits, citrus ... 449
tree fruits, deciduous ... 458
tree fruits, subtropical ... 473

Micro Flo Captan (captan) ... Micro Flo
WPS chart ... 45
lawns/turf ... 510
ornamentals ... 478
nuts ... 520
small fruits ... 435
tree fruits, deciduous ... 454

Micro Flo Chlorpyrifos (chlorpyrifos) ... Micro Flo
WPS chart ... 45
asparagus ... 200
cole crops ... 218
cotton ... 112
field corn, popcorn ... 78
forage (alfalfa, etc.) ... 538
greens ... 239
mint ... 148
nuts ... 425
onions ... 257
peanuts ... 151
radishes ... 279
small fruits ... 383
small grains ... 129
sorghum ... 158
soybeans ... 96
sugar beets ... 162
sunflower ... 167
sweet corn ... 289
sweet potatoes, yams ... 295
tobacco ... 170
tree fruits, citrus ... 394
tree fruits, deciduous ... 406
tree fruits, subtropical ... 418

Micro Flo Diazinon (diazinon) ... Micro Flo
WPS chart ... 45
beans ... 204
carrots, etc. ... 210
celery ... 214
cole crops ... 220
cucurbits ... 227
greens ... 239
hops ... 146
lawns/turf ... 491
lettuce ... 250
mushrooms ... 254
nuts ... 426
onions ... 258
ornamentals ... 499
peas ... 261
peppers ... 267
potatoes ... 272
radishes ... 279
small fruits ... 384
spinach ... 285
sugar beets ... 163
sweet corn ... 290
sweet potatoes, yams ... 295
table beets ... 298
tomatoes ... 303
tree fruits, deciduous ... 406
tree fruits, subtropical ... 419

Micro Flo Dimethoate (dimethoate) ... Micro Flo
WPS chart ... 45
beans ... 204
celery ... 214
cole crops ... 220
cotton ... 114
cucurbits ... 228
field corn, popcorn ... 79
forage (alfalfa, etc.) ... 539
greens ... 240
lettuce ... 250
nuts ... 427
peas ... 262
peppers ... 267
potatoes ... 273
radishes ... 280
safflower ... 156
small grains ... 130
sorghum ... 158
soybeans ... 97
spinach ... 285
table beets ... 298
tomatoes ... 303
tree fruits, deciduous ... 395
tree fruits, subtropical ... 408

Micro Flo Endosulfan (endosulfan) ... Micro Flo
WPS chart ... 45
artichokes ... 198
beans ... 205
carrots, etc. ... 210
celery ... 215
cole crops ... 221
cotton ... 114
cucurbits ... 228
eggplant ... 233
forage (alfalfa, etc.) ... 539
greens ... 240
lettuce ... 251
nuts ... 427
ornamentals ... 500
peas ... 262
peppers ... 267
potatoes ... 273
safflower ... 156
small fruits ... 385
small grains ... 130
southern peas ... 283
spinach ... 285
sugar beets ... 163
sunflower ... 168
sweet corn ... 291
sweet potatoes, yams ... 295
tobacco ... 170
tomatoes ... 303
tree fruits, citrus ... 396
tree fruits, deciduous ... 408
tree fruits, subtropical ... 415

Micro Flo Malathion (malathion) ... Micro Flo
WPS chart ... 45
beans ... 205
carrots, etc. ... 211
celery ... 215
cole crops ... 222
cotton ... 117
cucurbits ... 229
eggplant ... 233
field corn, popcorn ... 81
forage (alfalfa, etc.) ... 540
forage (grasses, hay) ... 543
garlic ... 237
greens ... 241
herbs/spices ... 244
hops ... 147
horseradish ... 246
lentils ... 248
lettuce ... 252
mint ... 149
mushrooms ... 254
noncropland ... 547
nuts ... 428
okra ... 256
onions ... 258
ornamentals ... 503
pastures/rangeland ... 545
peanuts ... 152
peas ... 263
peppers ... 268
potatoes ... 274
rice ... 155
safflower ... 157
small fruits ... 387
small grains ... 132
sorghum ... 160
southern peas ... 283
soybeans ... 98
spinach ... 286
sugar beets ... 163
sweet corn ... 292
sweet potatoes, yams ... 296
table beets ... 299
tree fruits, citrus ... 397
tree fruits, deciduous ... 411
tree fruits, subtropical ... 420

Micro Flo Sulfur (sulfur) ... Micro Flo
WPS chart ... 45
artichokes ... 308
asparagus ... 309
beans ... 314
carrots, etc. ... 319
cole crops ... 326
cucurbits ... 333
eggplant ... 335
field corn, popcorn ... 88
flax ... 173
forage (alfalfa, etc.) ... 548
garlic ... 338
greens ... 340
lawns/turf ... 517
lettuce ... 344
mint ... 175
nuts ... 484
okra ... 346
onions ... 350
ornamentals ... 530
peanuts ... 180
peas ... 353
peppers ... 356
potatoes ... 361
small fruits ... 444
small grains ... 140
sorghum ... 186
southern peas ... 365
soybeans ... 104
spinach ... 367
sugar beets ... 190
sugarcane ... 191
table beets ... 372
tomatoes ... 377
tree fruits, deciduous ... 466
tree fruits, subtropical ... 476

Micromite (diflubenzuron) ... Uniroyal
WPS chart ... 50
tree fruits, citrus ... 397

Microsperse (sulfur) ... Micro Flo
WPS chart ... 45
peas ... 353
potatoes ... 358
sugarcane ... 191
table beets ... 371

Microsperse COC (copper oxychloride) ... Micro Flo
WPS chart ... 45
beans ... 312
carrots, etc. ... 317
celery ... 321
cucurbits ... 330
eggplant ... 334
garlic ... 336
nuts ... 481
onions ... 347
ornamentals ... 522
peanuts ... 177
peas ... 352
peppers ... 354
potatoes ... 358
small fruits ... 438
small grains ... 136
sugar beets ... 188
table beets ... 371
tobacco ... 192
tomatoes ... 374
tree fruits, citrus ... 449
tree fruits, deciduous ... 459
tree fruits, subtropical ... 471

Microthiol Special (sulfur) ... Elf Atochem
WPS chart ... 43
beans ... 315
carrots, etc. ... 212,319
cole crops ... 225,327
cotton ... 120
cucurbits ... 333
field corn, popcorn ... 84
forage (alfalfa, etc.) ... 543
garlic ... 237,338
greens ... 243,340
hops ... 147
lettuce ... 254,344
mint ... 175
nuts ... 432,484
onions ... 260,350
ornamentals ... 508,530
peanuts ... 154,180
peppers ... 270,356
potatoes ... 277,361
small fruits ... 391,444
small grains ... 133
sorghum ... 161
soybeans ... 104
sugar beets ... 165,190
tomatoes ... 306,377
tree fruits, citrus ... 400
tree fruits, deciduous ... 416,465
tree fruits, subtropical ... 422,476

Millenium (*Steinermema carpocapse*) ... Thermo Trilogy
biocontrols ... 68

minute pirate bug (see *Orius insidiosus*)

Mitac (amitraz) ... AgrEvo USA
WPS chart ... 40
tree fruits, deciduous ... 401

Mocap (ethoprop) ... Rhone-Poulenc
WPS chart ... 47
beans ... 207
cole crops ... 223
cucurbits ... 229

field corn, popcorn 82
peanuts 152
potatoes 276
sugarcane 166
sweet corn 292
sweet potatoes, yams 296
tobacco 171
tree fruits, subtropical 420
Moncut (flutolanil) ..... Gowan
WPS chart 40
peanuts 177
rice 182
Monitor (methamidophos) ..... Bayer, Valent
WPS chart 41,50
cotton 117
potatoes 275
tomatoes 304
Morestan (oxythioquinox) ..... Bayer
cucurbits 229
hops 147
nuts 429
ornamentals 505
small fruits 389
tree fruits, citrus 397
tree fruits, deciduous 413
Mosquito Fish (*Gambusis affinis*) ..... Natural Pest Controls
biocontrols 66
M-Pede (potassium salts of fatty acids) ..... Mycogen
WPS chart 46
artichokes 199
asparagus 201
beans 207
carrots, etc 212
celery 215
cole crops 223
cotton 118
cucurbits 230,332
eggplant 234
forage (alfalfa, etc.) 542
garlic 237
greens 242
herbs/spices 244
hops 147
horseradish 246
lawns & turf 493
lettuce 252
nuts 429
oil seed crops 150
okra 256
onions 259
ornamentals 505,524
peanuts 152
peas 264
peppers 269
potatoes 276
radishes 280
small fruits 389,442
small grains 132
soybeans 99
spinach 287
sugar beets 164
sweet potatoes, yams 296
table beets 299
tobacco 171
tomatoes 305
tree fruits, citrus 398
tree fruits, deciduous 413
tree fruits, subtropical 420
watercress 307
M-Peril (*Bt.* spp. *kurstaki*) ..... Mycogen
Bt chart 76,144,196
WPS chart 46
Mustang (*zeta*-cypermethrin) ..... FMC
WPS chart 43
cole crops 219
cotton 113
lettuce 250
onions 257
nuts 426
MVP II (*Bt.* spp. *kurstaki*) ..... Mycogen
Bt chart 94,108,144,196,380,488
WPS chart 46
myclobutanil (see Eagle, Golden Eagle, Nova, Rally, Systhane)
Mycoshield (oxytetracycline-agricultural terramycin) ..... Novartis
WPS chart 46
biocontrols 58
Mycotrol (*Beauveria bassiana*) ..... Mycotech
WPS chart 46
biocontrols 55
Myrothecium verucaria (see DiTera)

N

naled (see Dibrom)
Naturalis-O (*Beauveria bassiana*) ..... Troy Biosciences
WPS chart 49
biocontrols 55
Naturalis-T (*Beauveria bassiana*) ..... Troy Biosciences
WPS chart 49
biocontrols 55
navel orangeworm parasite (see *Goniozus lengneri*, *Pentilitomastiz* spp.)
Neem Oil (see Triact, Trilogy)
NEEMAZAD (azadirachtin) ..... Thermo Trilogy
WPS chart 49
artichokes 198
beans 202
carrots, etc. 209
celery 213
cole crops 218
cotton 110
cucurbits 226
eggplant 232
garlic 236
greens 238
herbs/spices 243
horseradish 245
lentils 247
lettuce 249
mint 148
nuts 424
onions 256
ornamentals 496
peanuts 151
peas 260
peppers 265
potatoes 271
radishes 278
rhubarb 281
small fruits 381
soybeans 95
spinach 284
sweet potatoes, yams 295
table beets 297
tomatoes 301
tree fruits, citrus 393
tree fruits, deciduous 402
NEEMIX (azadirachtin) ..... Thermo Trilogy
WPS chart 49
artichokes 198
beans 202
carrots, etc. 209
celery 213
cole crops 218
cotton 110
cucurbits 226
eggplant 232
garlic 236
greens 238
herbs/spices 243
horseradish 245
lentils 247
lettuce 249
mint 148
nuts 424
onions 256
peanuts 151
peas 260
peppers 265
potatoes 271
radishes 278
rhubarb 281
small fruits 381
soybeans 95
spinach 284
sweet potatoes, yams 295
table beets 297
tomatoes 301
tree fruits, citrus 393
tree fruits, deciduous 402
Nemacur (fenamiphos) ..... Bayer
WPS chart 41
asparagus 201
cole crops 223
cotton 119
eggplant 234
garlic 237
lawns/turf 493
okra 256
ornamentals 505
peanuts 153
peppers 269
small fruits 289
table beets 300
tobacco 172
tree fruits, citrus 398
tree fruits, deciduous 413
tree fruits, subtropical 420
Nemastop (plant extracts) ..... Soil Technologies
WPS chart 48
biocontrols 61
***Neoseiulus barkeri* (*Neoseiulus* spp.) ..... IPM Laboratories**
biocontrols 66
***Neoseiulus californicus* (*Neoseiulus* spp.) ..... Arbico, Beneficial Insectary, Caltec, IPM Laboratories, Nature's Control**
biocontrols 66
***Neoseiulus cucumeris* (*Neoseiulus* spp.) ..... Arbico, Natural Pest Controls**
biocontrols 66
***Neoseiulus cucumeris & N. barkeri* (*Neoseiulus* spp.) ..... Hydro-Gardens**
biocontrols 66
Neoseiulus spp. (see *Amblyesius cucumeris, Neoseiulus barkeri, Neoseiulus californicus, Neoseiulus cucumeris, N. cucumeris & N. barkeri, Thrips Predator Mites*)
Nexter (pyridaben) ..... BASF
WPS chart 41
tree fruits, citrus 399
Nolo Bait (*Nosema locustae*) ..... M & R Durango
biocontrols 59
NoMate ..... Ecogen
biocontrols 61
***Nosema locustae* ..... Arbico, Harmony Farm Supply**
biocontrols 59
Nosema locustae (see Nolo Bait)
Nova (myclobutanil) ..... Rohm and Haas
WPS chart 48
ornamentals 527
small fruits 442
tree fruits, deciduous 464
Novodor (*Bt.* spp. *tenebrionis*) ..... Abbott
Bt chart 196,488
WPS chart 40
Nuclear polyhedrosis virus (see Gemstar, SPOD-X)
Nu-Cop (copper hydroxide) ..... Micro Flo
WPS chart 45
beans 312
carrots, etc. 317
celery 321
cole crops 324
cucurbits 330
eggplant 334
forage (alfalfa, etc.) 548
greens 339
herbs/spices 341
hops 174
lettuce 343
nuts 481
onions 347
ornamentals 522
peanuts 177
peas 352
peppers 354
potatoes 358
rice 182
small fruits 438
small grains 136
spinach 366
sugar beets 187
table beets 371
tomatoes 374
tree fruits, citrus 449
tree fruits, deciduous 459
tree fruits, subtropical 471
watercress 378
Nu-Gro Isotox (captan & lindane) ..... Wilbur-Ellis
WPS chart 50
beans 310
carrots, etc. 316
cole crops 323
cotton 122
cucurbits 328
field corn, popcorn 85
flax 173
forage (alfalfa, etc.) 548
okra 346
onions 347
peas 351
radishes 363
safflower 184
small grains 134
soybeans 102
spinach 366
sugar beets 187
sunflower 191
Nufos (chlorpyrifos) ..... Cheminova
WPS chart 42
cotton 112
field corn, popcorn 78
forage (alfalfa, etc.) 538
small grains 129
sorghum 158
sweet corn 289

O

Oftanol (isofenphos) ..... Bayer
lawns/turf 493
ornamentals 505
Omite (propargite) ..... Uniroyal
WPS chart 50
field corn, popcorn 84
hops 147
mint 149
nuts 431
ornamentals 507
peanuts 153
potatoes 277
small fruits 390
tree fruits, citrus 399
tree fruits, deciduous 415

Product Index

Omni (petroleum oil) ... **Helena**
- WPS chart ... 45
- asparagus ... 201
- beans ... 207
- cucurbits ... 230
- nuts ... 430
- ornamentals ... 506
- peppers ... 269
- small fruits ... 390
- sweet corn ... 293
- table beets ... 300
- tomatoes ... 305
- tree fruits, citrus ... 398
- tree fruits, deciduous ... 415
- tree fruits, subtropical ... 421

Orbit (propiconazole) ... **Novartis**
- WPS chart ... 46
- tree fruits, citrus ... 451
- tree fruits, deciduous ... 465

Orbit/Super Tin (co-pack) ... **Novartis & Griffin**
- WPS chart ... 44
- nuts ... 483

Orius insidiosus **(minute pirate bug)** ... **Florikan, Harmony Farm Supply, IPM Laboratories**
- biocontrols ... 66

Ornamite (propargite) ... **Uniroyal**
- WPS chart ... 50
- ornamentals ... 507

Orthene (acephate) ... **Valent**
- WPS chart ... 50
- beans ... 202
- celery ... 213
- cole crops ... 217
- cotton ... 109
- lawns/turf ... 489
- mint ... 148
- noncropland ... 546
- ornamentals ... 495
- peanuts ... 150
- peppers ... 265
- radishes ... 278
- small fruits ... 381
- soybeans ... 95
- tobacco ... 169
- tree fruits, citrus ... 392

Ovasyn (amitraz) ... **AgrEvo USA**
- WPS chart ... 40
- cotton ... 109

oxamyl (see Vydate)

oxycarboxin (see Plantvax)

oxydemeton-methyl (see Metasystox-R)

oxytetracycline-agricultural terramycin (see Mycoshield)

oxythioquinox (see Joust, Morestan)

P

parasite (see *Dacnusa sibirica, Diglyphus isaea*)

parasitic wasp (see *Aphidus colmani, Encarsia formosa, Leptomastix dactylopii*)

Par-Flo (PCNB) ... **Amvac**
- lawns/turf ... 515

Payload (acephate) ... **Valent**
- WPS chart ... 50
- cotton ... 109
- peanuts ... 150
- peppers ... 265

PB-Rope (pheromone) ... **Pacific Biocontrol**
- biocontrols ... 59

PCNB (see Defend, Par-Flo, PCNB, Penstar, RTU-PCNB, Terraclor, Terra-Coat, Turfcide, TurfPro, Win-Flo)

PCNB (PCNB) ... **Amvac**
- WPS chart ... 40
- beans ... 313
- cole crops ... 325
- cotton ... 124
- field corn, popcorn ... 87
- garlic ... 337
- lawns/turf ... 515
- ornamentals ... 528
- peanuts ... 179
- peas ... 352
- peppers ... 355
- potatoes ... 360
- rice ... 183
- safflower ... 184
- small grains ... 139
- sorghum ... 186
- soybeans ... 104
- sugar beets ... 189
- tomatoes ... 376

PCNB (PCNB) ... **Wilbur-Ellis**
- WPS chart ... 50
- beans ... 313
- cotton ... 124
- garlic ... 337

Pediobius foveolatus **(bean beetle parasite)** ... **Arbico**
- biocontrols ... 66

Penncap-M (methyl parathion) ... **ELF Atochem**
- WPS chart ... 43
- beans ... 206
- cotton ... 118
- field corn, popcorn ... 82
- forage (alfalfa) ... 541
- nuts ... 547
- onions ... 429
- pastures/rangeland ... 545
- peas ... 263
- potatoes ... 275
- rice ... 156
- small fruits ... 389
- small grains ... 132
- southern peas ... 283
- soybeans ... 99
- tomatoes ... 305
- tree fruits, deciduous ... 412

Penncozeb (mancozeb) ... **ELF Atochem**
- WPS chart ... 43
- asparagus ... 308
- cotton ... 123
- cucurbits ... 331
- field corn, popcorn ... 86
- flax ... 173
- herbs/spices ... 341
- onions ... 348
- ornamentals ... 524
- peanuts ... 178
- potatoes ... 359
- rice ... 182
- safflower ... 184
- small fruits ... 440
- small grains ... 137
- sorghum ... 185
- sugar beets ... 188
- sweet corn ... 368
- tomatoes ... 375
- tree fruits, deciduous ... 463
- tree fruits, subtropical ... 475

Penstar (PCNB) ... **Scotts**
- WPS chart ... 48
- lawns/turf ... 515

Pentac (dienochlor) ... **Novartis**
- ornamentals ... 505

Pentilitomastiz **spp. (navel orangeworm)** ... **Arbico, Caltec**
- biocontrols ... 67

permethrin (see Ambush, Astro, Pounce)

petroleum oil (see 435 Oil, 455 Oil, Citri-Oil, Damoil, OMNI, SAF-T-SIDE, Sumulsion, SunSpray)

Phaser (endosulfan) ... **AgrEvo USA**
- WPS chart ... 40
- artichokes ... 198
- beans ... 205
- carrots, etc. ... 210
- celery ... 214
- cole crops ... 221
- cotton ... 114
- cucurbits ... 228
- eggplant ... 233
- forage (alfalfa, etc.) ... 539
- greens ... 240
- lettuce ... 251
- nuts ... 427
- peas ... 262
- peppers ... 267
- potatoes ... 273
- safflower ... 156
- small fruits ... 385
- small grains ... 130
- southern peas ... 283
- spinach ... 285
- sugar beets ... 163
- sunflower ... 168
- sweet corn ... 291
- sweet potatoes, yams ... 295
- tobacco ... 170
- tomatoes ... 303
- tree fruits, citrus ... 396
- tree fruits, deciduous ... 408

pheromone (see CheckMate, Decoy, Disrupt, Frustrate, Isomate, NoMate, PB-Rope, Stirrup)

phorate (see Clean Crop Phorate, Riverside Phorate, Thimet)

phosmet (see Imidan)

Phytoseiulus persimilis **(*Phytoseiulus* spp.)** ... **Arbico, Beneficial Insectary, Caltec, Florikan, Harmony Farm Supply, Hydro-Gardens, International Technology, IPM Laboratories, Natural Pest Controls, Nature's Control**
- biocontrols ... 67

Phytoseiulus spp. (see *Phytoseiulus persimilis*)

Pinpoint (acephate) ... **Valent**
- WPS chart ... 50
- lawns/turf ... 489
- ornamentals ... 495

piperalin (see Pipron)

Pipron (piperalin) ... **SePRO**
- WPS chart ... 48
- ornamentals ... 528

pirimicarb (see Pirimor)

Pirimor (pirimicarb) ... **ZENECA**
- WPS chart ... 51

plant extracts (see Nemastop, Sincocin)

Plantvax (oxycarboxin) ... **Uniroyal**
- WPS chart ... 50
- ornamentals ... 529

Podisus maculiventris **(predator)** ... **Arbico**
- biocontrols ... 67

Polyram (metiram) ... **BASF, Platte**
- WPS chart ... 41,47
- ornamentals ... 529
- potatoes ... 360
- tree fruits, deciduous ... 464

potassium salts of fatty acids (see M-Pede)

Pounce (permethrin) ... **FMC**
- WPS chart ... 43
- asparagus ... 201
- carrots, etc. ... 212
- celery ... 215
- cole crops ... 224
- cucurbits ... 230
- eggplant ... 234
- field corn, popcorn ... 83
- forage (alfalfa, etc.) ... 542
- greens ... 242
- herbs/spices ... 244
- horseradish ... 246
- nuts ... 430
- ornamentals ... 506
- pastures/rangeland ... 545
- peppers ... 269
- potatoes ... 276
- rhubarb ... 282
- soybeans ... 99
- spinach ... 287
- sweet corn ... 293
- tomatoes ... 305
- tree fruits, deciduous ... 414

Precision (fenoxycarb) ... **Novartis**
- WPS chart ... 46
- ornamentals ... 501

predator (see *Cryptolaemus montrouzieri, Mesoseiulus longipe, Podisus maculiventris*)

predator midge (see *Aphidoletes aphidimyza*)

predator mite (see *Galendromus occidentalis*)

predatory beetle (see *Delphastus pusillus*)

predatory snail (see *Rumina decollata*)

PRED-FEED I.P.M. ... **Custom Chemicides**
- biocontrols ... 61

Prentox Diazinon (diazinon) ... **Prentiss**
- WPS chart ... 47
- beans ... 204
- carrots, etc. ... 210
- cole crops ... 220
- cucurbits ... 227
- greens ... 239
- hops ... 146
- lawns/turf ... 491
- lettuce ... 250
- mushrooms ... 254
- nuts ... 426
- onions ... 258
- ornamentals ... 499
- peas ... 261
- peppers ... 267
- potatoes ... 272
- radishes ... 279
- small fruits ... 384
- spinach ... 285
- sugar beets ... 163
- sweet corn ... 290
- sweet potatoes, yams ... 295
- table beets ... 298
- tomatoes ... 303
- tree fruits, deciduous ... 406
- tree fruits, subtropical ... 419

Prentox Fogging Insecticide (pyrethirns) ... **Prentiss**
- WPS chart ... 47
- mushrooms ... 255

Prentox Malathion (malathion) ... **Prentiss**
- WPS chart ... 47
- beans ... 205
- celery ... 215
- cole crops ... 222
- cotton ... 117
- cucurbits ... 229
- eggplant ... 233
- field corn, popcorn ... 81
- forage (alfalfa, etc.) ... 540
- forage (grasses, hay) ... 543
- garlic ... 237
- greens ... 241
- herbs/spices ... 244
- hops ... 147
- horseradish ... 246
- lentils ... 248
- lettuce ... 252
- mint ... 149
- mushrooms ... 254
- nuts ... 428
- okra ... 256
- onions ... 258
- ornamentals ... 503
- pastures/rangeland ... 545
- peas ... 263
- peppers ... 268
- potatoes ... 274
- rice ... 155
- small fruits ... 387
- small grains ... 132
- southern peas ... 283

spinach ... 286
sweet corn ... 292
sweet potatoes, yams ... 296
table beets ... 299
tree fruits, citrus ... 397
tree fruits, deciduous ... 411
tree fruits, subtropical ... 420
Prentox Methoxychlor (methoxychlor) ... Prentiss
WPS chart ... 47
asparagus ... 201
beans ... 206
carrots, etc. ... 211
cole crops ... 222
cucurbits ... 229
eggplant ... 234
forage (alfalfa, etc.) ... 541
forage (grasses, hay) ... 544
greens ... 241
lettuce ... 252
ornamentals ... 504
peas ... 263
peppers ... 268
potatoes ... 275
radishes ... 280
small fruits ... 388
southern peas ... 283
spinach ... 286
sweet corn ... 292
table beets ... 299
tomatoes ... 305
tree fruits, deciduous ... 412
Prevail (carboxin & PCNB & metalaxyl) ... Trace
WPS chart ... 49
beans ... 310
cotton ... 123
field corn, popcorn ... 86
peanuts ... 176
small grains ... 135
soybeans ... 102
profenofos (see Curacron)
Prokil Cryolite (cryolite) ... Gowan
WPS chart ... 44
cole crops ... 219
cucurbits ... 227
ornamentals ... 498
peppers ... 266
potatoes ... 272
small fruits ... 383
tree fruits, citrus ... 395
tree fruits, subtropical ... 419
propamocarb hydrochloride (see Banol)
propargite (see Comite, Omite, Ornamite)
propiconazole (see Alamo, Banner, Break, Orbit, Orbit/Super Tin, Tilt)
Prosper Nema ... Arbico
biocontrols ... 62
ProStar (flutolanil) ... AgrEvo USA
WPS chart ... 40
lawns/turf ... 512
ProStar Plus (co-pack): Prostar & Bayleton ... AgrEvo USA & Bayer
lawns/turf ... 516
Protect (mancozeb) ... Cleary
WPS chart ... 42
lawns/turf ... 513
ornamentals ... 525
Protector-L (thiram) ... Trace
WPS chart ... 49
soybeans ... 105
Pro-Tex (maneb & triphenyltin hydroxide) ... Griffin
WPS chart ... 44
potatoes ... 361
sugar beets ... 188
ProTurf Fluid Fungicide (iprodione & thiophanate-methyl) ... Scotts
WPS chart ... 48
lawns/turf ... 513
ProTurf Fluid Fungicide II (metalaxyl & triadimefon) ... Scotts
lawns/turf ... 514
ProTurf Fungicide V (chloroneb) ... Scotts
lawns/turf ... 510
ProTurf Fungicide VII (triadimefon) ... Scotts
lawns/turf ... 518
ProTurf Fungicide IX (chloroneb & thiophanate-methyl) ... Scotts
lawns/turf ... 511
ProTurf Fungicide X (iprodione) ... Scotts
lawns/turf ... 513
ProTurf Insecticide III (chlorpyrifos) ... Scotts
WPS chart ... 48
lawns/turf ... 491
ProTurf Pythium Control (metalaxyl) ... Scotts
WPS chart ... 48
lawns/turf ... 514
ornamentals ... 526
ProTurf Systemic Fungicide (thiophanate-methyl) ... Scotts
lawns/turf ... 517
Provado (imidacloprid) ... Bayer
WPS chart ... 41
cole crops ... 221
cotton ... 115
eggplant ... 233
greens ... 240
hops ... 146
lettuce ... 251
ornamentals ... 503
peppers ... 268
potatoes ... 274
small fruits ... 286
tomatoes ... 304
tree fruits, deciduous ... 409
tree fruits, subtropical ... 419
Proxol (trichlorfon) ... AgrEvo USA
WPS chart ... 423
lawns/turf ... 494
ornamentals ... 508
Pyramite (pyridaben) ... BASF
WPS chart ... 41
nuts ... 432
tree fruits, deciduous ... 416
Pyrellin (pyrethrins & rotenone) ... Webb Wright
WPS chart ... 50
artichokes ... 199
asparagus ... 202
beans ... 207
carrots, etc. ... 212
celery ... 216
cole crops ... 224
cotton ... 120
cucurbits ... 230
eggplant ... 235
field corn, popcorn ... 84
forage (alfalfa, etc.) ... 542
garlic ... 237
greens ... 242
herbs/spices ... 244
hops ... 147
horseradish ... 246
lentils ... 248
lettuce ... 253
mint ... 149
nuts ... 431
oil seed crops ... 150
okra ... 256
onions ... 259
ornamentals ... 507
peanuts ... 153
peas ... 264
peppers ... 269
potatoes ... 277
radishes ... 281
rhubarb ... 282
rice ... 156
small fruits ... 391
small grains ... 133
sorghum ... 160
southern peas ... 284
soybeans ... 100
spinach ... 287
sugar beets ... 165
sunflower ... 168
sweet corn ... 294
sweet potatoes, yams ... 296
table beets ... 300
tomatoes ... 306
tree fruits, citrus ... 399
tree fruits, deciduous ... 416
tree fruits, subtropical ... 422
Pyrenone (pyrethrins) ... AgrEvo Environmental
WPS chart ... 40
asparagus ... 202
beans ... 208
carrots, etc. ... 212
celery ... 216
cole crops ... 224
cotton ... 120
cucurbits ... 230
eggplant ... 235
field corn, popcorn ... 84
forage (alfalfa, etc.) ... 542
forage (grasses, hay) ... 544
garlic ... 237
greens ... 242
herbs/spices ... 244
horseradish ... 246
lawns/turf ... 493
lentils ... 248
lettuce ... 253
mushrooms ... 255
nuts ... 431
onions ... 260
ornamentals ... 507
pastures/rangeland ... 546
peas ... 264
peppers ... 270
potatoes ... 277
radishes ... 281
rhubarb ... 282
rice ... 156
small fruits ... 391
small grains ... 133
southern peas ... 284
soybeans ... 100
spinach ... 287
sugar beets ... 165
sweet potatoes, yams ... 297
table beets ... 300
tomatoes ... 306
tree fruits, citrus ... 399
tree fruits, deciduous ... 416
tree fruits, subtropical ... 422
pyrethrins (see Diatect, Prentox Fogging Insecticide, Pyrenone, Py-Rin)
pyrethrins & rotenone (see Pyrellin)
pyridaben (see Nexter, Pyramite, Sanmite)
Py-Rin (pyrethrins) ... Wilbur-Ellis
WPS chart ... 51
asparagus ... 202
beans ... 208
carrots, etc. ... 212
celery ... 216
cole crops ... 224
cucurbits ... 230
eggplant ... 235
greens ... 242
herbs/spices ... 244
horseradish ... 246
lentils ... 248
lettuce ... 253
mushrooms ... 255
onions ... 260
ornamentals ... 507
peas ... 264
peppers ... 270
potatoes ... 277
radishes ... 281
rhubarb ... 282
rice ... 156
small fruits ... 391
southern peas ... 284
soybeans ... 100
spinach ... 287
sugar beets ... 165
sweet potatoes, yams ... 297
table beets ... 300
tomatoes ... 306
tree fruits, citrus ... 399
tree fruits, deciduous ... 416
pyriproxyfen (see Distance, Knack)

Q

Quadris (azoxystrobin) ... ZENECA
WPS chart ... 51
tomatoes ... 372

R

Rally (myclobutanil) ... Rohm and Haas
WPS chart ... 48
nuts ... 483
ornamentals ... 527
small fruits ... 442
tree fruits, deciduous ... 464
Raven (*Bt.* spp. *kurstaki*) ... Ecogen
Bt chart ... 196
WPS chart ... 43
Raxil (tebuconazole) ... Gustafson
small grains ... 139
Raxil-Thiram (tebuconazole & thiram) ... Gustafson
WPS chart ... 44
small grains ... 139
Redball EM-53 (sulfur) ... International Sulphur
WPS chart ... 45
beans ... 315
carrots, etc. ... 319
cole crops ... 327
field corn, popcorn ... 88
onions ... 350
ornamentals ... 531
peanuts ... 180
peas ... 353
potatoes ... 361
small fruits ... 443
small grains ... 140
sorghum ... 186
soybeans ... 105
sugar beets ... 190
table beets ... 372
tree fruits, deciduous ... 466
Redball Ligthning (sulfur) ... International Sulphur
WPS chart ... 45
cotton ... 121
field corn, popcorn ... 84
small grains ... 133
sorghum ... 161
soybeans ... 100
Redball Sulfur (sulfur) ... International Sulphur
WPS chart ... 45
asparagus ... 309
beans ... 315
carrots, etc. ... 319
cole crops ... 327
field corn, popcorn ... 88

garlic ... 338
onions ... 350
ornamentals ... 531
peanuts ... 180
peas ... 353
peppers ... 356
potatoes ... 361
small fruits ... 444
small grains ... 140
spinach ... 367
sorghum ... 186
soybeans ... 105
sugar beets ... 190
table beets ... 372
tree fruits, deciduous ... 466

Regent (fipronil) ... Rhone-Poulenc
field corn, popcorn ... 84

***Rhizobius lophanthae* (ladybug) ... Harmony Farm Supply**
biocontrols ... 67

Ridomil Gold (mefenoxam) ... Novartis
WPS chart ... 423
asparagus ... 308
beans ... 313
carrots, etc. ... 318
celery ... 321
cole crops ... 325
cotton ... 124
cucurbits ... 331
eggplant ... 335
forage (alfalfa, etc.) ... 548
forage (grasses, hay) ... 549
garbanzos ... 336
garlic ... 337
greens ... 340
herbs/spices ... 341
hops ... 174
horseradish ... 342
lentils ... 342
lettuce ... 344
mung beans ... 175
nuts ... 483
onions ... 349
peanuts ... 178
peas ... 352
peppers ... 355
radishes ... 363
rhubarb ... 364
small fruits ... 441
southern peas ... 364
soybeans ... 103
spinach ... 366
sugar beets ... 189
sweet potatoes, yams ... 370
table beets ... 371
tobacco ... 193
tomatoes ... 375
tree fruits, citrus ... 450
tree fruits, deciduous ... 463
tree fruits, subtropical ... 476

Ridomil Gold MZ (mefenoxam & mancozeb) ... Novartis
WPS chart ... 46
cucurbits ... 332
garlic ... 337
onions ... 349
small fruits ... 442
sugar beets ... 189
tomatoes ... 376

Ridomil Gold PC (mefenoxam & PCNB) ... Novartis
WPS chart ... 46
beans ... 314
cotton ... 124
peanuts ... 179

Ridomil Gold/Bravo (mefenoxam & chlorothalonil) ... Novartis
WPS chart ... 46
carrots, etc. ... 318
cole crops ... 325
cucurbits ... 331
garlic ... 337
onions ... 349
tomatoes ... 375

Ridomil Gold/Bravo (co-pack) ... Novartis & ZENECA
WPS chart ... 46
potatoes ... 360

Ridomil Gold/Copper (mefenoxam & copper hydroxide) ... Novartis
WPS chart ... 46
carrots, etc. ... 318
cucurbits ... 331
garlic ... 337
onions ... 349
peppers ... 355
radishes ... 363
small fruits ... 442
spinach ... 366
tomatoes ... 376

Rival (captan & PCNB & thiabendazole) ... Gustafson
beans ... 310
small grains ... 135
soybeans ... 102

Riverside Diazinon (diazinon) ... Terra
WPS chart ... 49

Riverside Malathion (malathion) ... Terra
WPS chart ... 49

Riverside Methomyl (methomyl) ... Terra
WPS chart ... 49

Riverside Phorate (phorate) ... Terra
WPS chart ... 49
cotton ... 119
field corn, popcorn ... 83
peanuts ... 153
potatoes ... 276
small grains ... 133
sorghum ... 160
soybeans ... 99
sugar beets ... 164
sugarcane ... 166
sweet corn ... 294

Ronilan (vinclozolin) ... BASF
WPS chart ... 41
lettuce ... 345
onions ... 351
small fruits ... 446

Rovral (iprodione) ... Rhone-Poulenc
WPS chart ... 47
beans ... 312
carrots, etc. ... 318
cole crops ... 324
garlic ... 337
lettuce ... 344
nuts ... 482
onions ... 348
peanuts ... 177
potatoes ... 359
rice ... 182
small fruits ... 440
tree fruits, deciduous ... 432

RTU (thiram & thiabendazole) ... Gustafson
WPS chart ... 44
small grains ... 139
soybeans ... 104

RTU-Baytan-Thiram (triadimenol & thiram) ... Gustafson
cotton ... 125
small grains ... 140

RTU-PCNB (PCNB) ... Gustafson
beans ... 314
cotton ... 124
field corn, popcorn ... 87
peanuts ... 179
peas ... 352
small grains ... 135
sorghum ... 186
soybeans ... 104
sugar beets ... 189

RTU-Vitavax-Extra (carboxin & imazalil & thiabendazole) ... Gustafson
WPS chart ... 44
small grains ... 135

RTU-Vitavax-Thiram (carboxin & thiram) ... Gustafson
WPS chart ... 44
beans ... 310
cotton ... 123
rice ... 182
small grains ... 135
soybeans ... 102

Rubigan (fenarimol) ... Dow AgroSciences
WPS chart ... 42
lawns/turf ... 512
ornamentals ... 523
small fruits ... 439
tree fruits, deciduous ... 461

Rubigan (fenarimol) ... SePRO
WPS chart ... 48
ornamentals ... 523

***Rumina decollata* (predatory snail) ... Sespe Creek Insectary**
biocontrols ... 67

S

SAF-T-SIDE (petroleum oil) ... Brandt Consolidated
WPS chart ... 41
asparagus ... 201
beans ... 207
celery ... 216
cole crops ... 224
cotton ... 119
cucurbits ... 230
eggplant ... 234
field corn, popcorn ... 83
lettuce ... 253
nuts ... 431
onions ... 259
ornamentals ... 506
peanuts ... 153
peppers ... 269
potatoes ... 276
radishes ... 280
small fruits ... 390
sugar beets ... 164
sweet corn ... 293
sweet potatoes, yams ... 296
tobacco ... 172
tomatoes ... 306
tree fruits, citrus ... 398
tree fruits, deciduous ... 415
tree fruits, subtropical ... 421

Sanmite (pyridaben) ... BASF
WPS chart ... 41
ornamentals ... 517

Savey (hexythiazox) ... Gowan
WPS chart ... 44
ornamentals ... 502
tree fruits, deciduous ... 409

scale parasite (see *Aphytis melinus*)

Scanmask (*Steinernema feltiae*) ... BioLogic
biocontrols ... 68

Scimitar (*lambda*-cyhalothrin) ... ZENECA
lawns/turf ... 491
ornamentals ... 499

Scout X-TRA (tralomethrin) ... AgrEvo USA
WPS chart ... 40
cole crops ... 224
cotton ... 120
lettuce ... 253
soybeans ... 100
sunflower ... 169

Seed Mate Isotox (captan & lindane) ... Loveland
beans ... 310
carrots, etc. ... 316
cole crops ... 323
cotton ... 122
cucurbits ... 328
field corn, popcorn ... 85
flax ... 173
forage (alfalfa, etc.) ... 548
okra ... 346
onions ... 347
peas ... 351
radishes ... 363
safflower ... 184
small grains ... 134
soybeans ... 102
spinach ... 366
sugar beets ... 187
sunflower ... 191

Seed Mate Maneb-Lindane (maneb & lindane) ... Loveland
field corn, popcorn ... 87
small grains ... 138
soybeans ... 103

Seed Mate Soybean Mo (captan) ... Loveland
soybeans ... 101

Seed Mate Vitavax-Captan (carboxin & captan) ... Loveland
small grains ... 135
soybeans ... 102

Sentinel (cyproconazole) ... Novartis
WPS chart ... 46
lawns/turf ... 516

Sevin (carbaryl) ... Rhone-Poulenc
WPS chart ... 47
asparagus ... 200
beans ... 203
carrots, etc. ... 209
celery ... 214
cole crops ... 218
cotton ... 111
cucurbits ... 226
eggplant ... 232
field corn, popcorn ... 78
forage (alfalfa, etc.) ... 537
greens ... 238
horseradish ... 245
lentils ... 247
lettuce ... 249
noncropland ... 546
nuts ... 425
okra ... 255
pastures/rangeland ... 544
peanuts ... 151
peas ... 261
peppers ... 266
potatoes ... 272
radishes ... 279
rice ... 154
small fruits ... 382
small grains ... 129
sorghum ... 157
southern peas ... 282
soybeans ... 96
spinach ... 285
sugar beets ... 162
sunflower ... 167
sweet corn ... 288
sweet potatoes, yams ... 295
table beets ... 298
tobacco ... 170
tomatoes ... 302
tree fruits, citrus ... 393
tree fruits, deciduous ... 404
tree fruits, subtropical ... 418

Siege (hydramethylnon) ... American Cyanamid
lawns/turf ... 492
noncropland ... 547
ornamentals ... 502
pastures/rangeland ... 545

Sincocin (plant extracts) ... Agricultural Sciences (AGSCI)
WPS chart ... 40
biocontrols ... 61

Slam (carbaryl) .... Micro Flo
- WPS chart 45
- field corn, popcorn 77
- soybeans 96
- sweet corn 288

Sniper (azinphos-methyl) .... Platte
- WPS chart 47
- artichokes 198
- beans 203
- celery 213
- cole crops 218
- cucurbits 226
- eggplant 232
- forage (alfalfa, etc.) 537
- nuts 424
- onions 257
- ornamentals 496
- peppers 265
- potatoes 271
- small fruits 381
- small grains 129
- southern peas 282
- soybeans 96
- spinach 284
- tobacco 169
- tomatoes 301
- tree fruits, citrus 393
- tree fruits, deciduous 402

sodium tetrathiocarbonate (see Enzone)

SoilGard (*Gliocladium virens*) .... Thermo Trilogy
- WPS chart 49
- biocontrols 62

Sorghum Guard (captan & lindane) .... Trace
- WPS chart 49
- field corn, popcorn 81,85
- sorghum 159,185
- soybeans 98,102
- sunflowers 168,191

Soy Gro (captan) .... Drexel
- WPS chart 42
- soybeans 101

soybean oil (see Golden Natur'l Spray Oil)

Spider Mite Destroyer (*Stethorus punctillum*) .... Nature's Control
- biocontrols 67

spinosad (see Conserve, Success Naturalyte, Tracer Naturalyte)

SPOD-X LC (nuclear polyhedrosis virus)... Thermo Trilogy
- biocontrols 59

Sporax (borax) .... Wilbur-Ellis
- ornamentals 530

Spotrete (thiram) .... Cleary
- lawns/turf 518

Steinernema carpocapsae .... Harmony Farm Supply, M&R Durango
- biocontrols 68

Steinernema feltiae .... M&R Durango
- biocontrols 68

Stoinernema carpocapsae (see also BioVector, Ecomask, Millenium)

Steinernema feltiae (see Scanmask)

Steinernema riobravis (see Devour)

Steinernema spp. (see Beneficial nematodes, Guardian, *Steinernema carpocapsae*)

Steinernema spp. & *Hetehabditis* spp. (see Bene-Nem)

Stethorus punctillium (see Spider Mite Destroyer)

Stirrup (pheromone) .... Troy Biosciences
- WPS chart 49
- biocontrols 61

streptomycin sulfate (see Agricultural Streptomycin, Agri-mycin)

Strike (triadimefon) .... Olympic
- WPS chart 46
- ornamentals 533

Subdue (metalaxyl) .... Novartis
- WPS chart 46
- lawns/turf 514
- nuts 483
- ornamentals 526

Subdue 2X (mefenoxam) .... Novartis
- WPS chart 46
- lawns/turf 514
- ornamentals 525

Subdue GR (mefenoxam) .... Novartis
- lawns/turf 514
- ornamentals 626

Subdue MAXX (mefenoxam) .... Novartis
- lawns/turf 514
- ornamentals 525

Success Naturalyte (spinosad) .... Dow AgroSciences
- WPS chart 42
- nuts 432
- tree fruits, citrus 399
- tree fruits, deciduous 416

Sulforix (calcium polysulfide) .... Best Sulfur
- WPS chart 41
- forage (alfalfa, etc.) 547
- ornamentals 497,520
- small fruits 382,435
- tree fruits, deciduous 404,453

sulfur (see Drexel Suffa, Drexel Sulfa, Drexel Sulfur, Dusting Sulfur, Kumulus, Liquid Sulfur Six, Micro Flo Sulfur, Microsperse, Microthiol Special, Redball Lightning, Terra Sulfur, THAT, THAT Big 8, Thiolux, Wettable Sulfur)

sulfur & chlorothalonil (see Bravo S, Terranil S)

sulfur & tribasic copper sulfate (see Top Cop with Sulfur)

sulprofos (see Bolstar)

Sumulsion (petroleum oil) .... Drexel
- WPS chart 42
- asparagus 201
- beans 207
- celery 216
- cole crops 224
- cotton 119
- cucurbits 230
- eggplant 234
- field corn, popcorn 83
- lettuce 253
- nuts 431
- onions 259
- ornamentals 506
- peanuts 153
- peppers 269
- potatoes 276
- radishes 280
- small fruits 390
- sugar beets 164
- sweet corn 293
- sweet potatoes, yams 296
- tobacco 172
- tomatoes 306
- tree fruits, citrus 398,451
- tree fruits, deciduous 415
- tree fruits, subtropical 421

SunSpray (petroleum oil) .... Sun
- WPS chart 48
- asparagus 201
- beans 207
- celery 216
- cole crops 224
- cucurbits 230
- eggplant 234
- field corn, popcorn 83
- herbs/spices 244
- lettuce 253
- nuts 430
- ornamentals 506
- peppers 269
- potatoes 276
- radishes 280
- small fruits 390
- sugar beets 164
- sweet corn 293
- tobacco 172
- tomatoes 306
- tree fruits, citrus 398,451
- tree fruits, deciduous 414
- tree fruits, subtropical 421

Super Tin (triphenyltin hydroxide) .... Griffin
- WPS chart 44
- nuts 485
- potatoes 362
- sugar beets 190

Supracide (methidathion) .... Gowan
- WPS chart 44
- artichokes 199
- cotton 121
- nuts 432
- ornamentals 508
- safflower 157
- tree fruits, citrus 400
- tree fruits, deciduous 417
- tree fruits, subtropical 422

Syllit (dodine) .... Platte
- WPS chart 47
- nuts 482
- small fruits 439
- tree fruits, deciduous 460

System3 (*Bacillus subtilis* & metalaxyl & PCNB)... Helena
- WPS chart 45
- beans 315
- cotton 125
- field corn, popcorn 88
- peanuts 180
- peas 353
- rice 183
- small grains 141
- soybeans 105

Systhane (myclobutanil) .... Rohm and Haas
- WPS chart 48
- ornamentals 527

T

Talstar (bifenthrin) .... FMC
- WPS chart 43
- lawns/turf 493
- ornamentals 508

Tame (fenpropathrin) .... Valent
- WPS chart 50
- ornamentals 501
- tree fruits, deciduous 409

tebuconazole (see Elite, Folicur, Raxil)

tebuconazole & thiram (see Raxil-Thiram)

tebufenozide (see Confirm)

tefluthrin (see Force)

Telone C-17 (dicloropropene & chloropicrin) .... Dow AgroSciences
- WPS chart 42
- mint 175
- onions 350
- peanuts 181
- sweet potatoes, yams 370
- tobacco 193

Telone II (dicloropropene) .... Dow AgroSciences
- WPS chart 42
- cotton 125
- mint 175
- sugar beets 190
- tree fruits, deciduous 467

Temik (aldicarb) .... Rhone-Poulenc
- WPS chart 47
- beans 208
- cotton 121
- nuts 433
- peanuts 154
- sorghum 161
- sugar beets 165
- sugarcane 166
- sweet potatoes, yams 297
- tree fruits, citrus 400

Tempo (cyfluthrin) .... Bayer
- lawns/turf 491
- ornamentals 498

***Tenodera aridifolia sinensis* (praying mantis)** .... A-1 Unique Insect Control, Arbico, Natural Pest Controls, Nature's Control
- biocontrols 68

Tennessee Brand (copper sulfate pentahydrate)... Griffin
- WPS chart 44
- nuts 479
- ornamentals 521
- potatoes 357
- small fruits 436
- tree fruits, citrus 447
- tree fruits, deciduous 455
- tree fruits, subtropical 471

terbufos (see Counter)

Teremec (chloroneb) .... PBI Gordon
- lawns/turf 510

Terra Sulfur (sulfur) .... Terra
- WPS chart 49
- artichokes 308
- asparagus 309
- beans 314
- carrots, etc. 319
- cole crops 326
- cucurbits 333
- eggplant 335
- field corn, popcorn 88
- flax 173
- forage (alfalfa, etc.) 548
- garlic 338
- greens 340
- lawns/turf 517
- lettuce 344
- mint 175
- nuts 484
- okra 346
- onions 350
- ornamentals 531
- peanuts 180
- peas 353
- peppers 356
- potatoes 361
- small fruits 444
- small grains 140
- sorghum 186
- southern peas 365
- soybeans 105
- spinach 367
- sugar beets 190
- sugarcane 191
- table beets 372
- tomatoes 377
- tree fruits, deciduous 466
- tree fruits, subtropical 476

Terraclor (PCNB) .... Uniroyal
- WPS chart 50
- beans 314
- cole crops 326
- lawns/turf 528
- ornamentals 515
- potatoes 360
- tomatoes 376

Terra-Coat (PCNB) .... Gustafson
- cotton 124
- small grains 139

Terraguard (triflumizole) .... Uniroyal
- WPS chart 50
- ornamentals 531

Terraneb (chloroneb) .... Kincaid
- WPS chart 45
- lawns/turf 510
- ornamentals 521

Product Index

Terranil (chlorothalonil)....Terra
WPS chart....49
beans....311
carrots, etc.....316
celery....320
cole crops....323
cucurbits....329
garlic....336
mint....174
nuts....478
onions....347
ornamentals....521
peanuts....176
potatoes....357
small fruits....436
southern peas....364
soybeans....103
sweet corn....368
tomatoes....373
tree fruits, deciduous....455
tree fruits, subtropical....470
Terranil Cu (chlorothalonil & copper oxychloride) ... Terra
WPS chart....49
beans....315
carrots, etc.....319
celery....322
cucurbits....333
peanuts....181
potatoes....362
southern peas....365
tomatoes....377
Terranil S (sulfur & chlorothalonil)....Terra
WPS chart....49
beans....315
cole crops....327
peanuts....180
potatoes....362
tomatoes....377
tree fruits, deciduous....467
Terrazole (etridiazole)....Uniroyal
WPS chart....50
lawns/turf....512
ornamentals....523
Terr-O-Gas (methyl bromide & chloropicrin).... Great Lakes
asparagus....309
cole crops....327
cucurbits....333
eggplant....335
lettuce....344
onions....350
peppers....356
small fruits....445
tobacco....193
tomatoes....377
tree fruits, citrus....451
tree fruits, subtropical....477
Thalonil (chlorothalonil)....Terra
WPS chart....49
lawns/turf....511
ornamentals....521
THAT (sulfur)....Stoller
WPS chart....48
beans....315
cotton....121
field corn, popcorn....84
mint....175
peanuts....180
potatoes....361
small fruits....444
small grains....140
sorghum....161
soybeans....100
sugar beets....190
tomatoes....306
tree fruits, citrus....400
tree fruits, deciduous....417,466
THAT Big 8 (sulfur)....Stoller
WPS chart....48
cotton....121
field corn, popcorn....84
mint....175
peanuts....180
potatoes....361
small fruits....444
small grains....140
sorghum....161
soybeans....100
sugar beets....190
tomatoes....306
tree fruits, citrus....400
tree fruits, deciduous....417,466
thiabendazole (see Arbotect, Gustafson LSP, Mertect)
Thimet (phorate)....American Cyanamid
WPS chart....40
beans....207
cotton....119
field corn, popcorn....83
peanuts....153
potatoes....276
small grains....133
sorghum....160
soybeans....99
sugar beets....164
sugarcane....166
sweet corn....294
Thiodan (endosulfan)....FMC
WPS chart....43
artichokes....198
beans....205
carrots, etc.....210
celery....214
cole crops....221
cotton....114
cucurbits....228
eggplant....233
forage (alfalfa, etc.)....539
greens....240
lettuce....251
nuts....427
ornamentals....500
peas....262
peppers....267
potatoes....273
safflower....156
small fruits....385
small grains....130
southern peas....283
spinach....285
sugar beets....163
sunflower....168
sweet corn....291
sweet potatoes, yams....295
tobacco....170
tomatoes....303
tree fruits, citrus....396
tree fruits, deciduous....408
tree fruits, subtropical....419
thiodicarb (see Larvin)
Thiolux (sulfur)....Platte
WPS chart....47
asparagus....309
beans....208,214
carrots, etc.....212,319
cole crops....225,327
cotton....121
cucurbits....333
field corn....85
forage (grasses, hay)....543
garlic....237,338
greens....243,340
mint....175
nuts....432,484
onions....260,350
ornamentals....531
peanuts....154,180
peas....264,353
peppers....270,356
potatoes....277,361
small fruits....391,444
small grains....133
sorghum....161
soybeans....100,105
sugar beets....190
table beets....372
tomatoes....306,377
tree fruits, deciduous....416,466
tree fruits, subtropical....422,476
thiophanate-methyl (see 3336, Dustret T, Fungo, ProTurf Systemic Fungicide, Tops, Topsin M)
42-S Thiram (thiram)....Gustafson
WPS chart....44
beans....316
carrots, etc.....319
castor beans....172
cole crops....327
cotton....125
cucurbits....333
eggplant....335
field corn, popcorn....88
flax....173
forage (alfalfa, etc.)....549
greens....340
lawns/turf....518
lettuce....345
okra....346
onions....351
ornamentals....532
peanuts....181
peas....353
peppers....356
radishes....363
rice....184
safflower....184
small grains....141
sorghum....186
southern peas....365
soybeans....105
spinach....367
sugar beets....191
sunflower....192
sweet corn....369
table beets....372
tomatoes....378
thiram (see Bean Treater, Protector-L, Spotrete, 42-S Thiram, Thiram 65, Thiram Granuflo)
Thiram (thiram)....Gustafson
WPS chart....44
beans....316
carrots, etc.....319
castor beans....172
cole crops....327
cotton....125
cucurbits....333
eggplant....335
field corn, popcorn....88
flax....173
forage (alfalfa, etc.)....549
greens....340
lawns/turf....518
lettuce....345
okra....346
ornamentals....532
peanuts....181
peas....353
peppers....356
radishes....363
rice....184
safflower....184
small grains....141
sorghum....186
southern peas....365
soybeans....105
spinach....367
sugar beets....191
sweet corn....369
table beets....372
tomatoes....378
Thiram 65 (thiram)....UCB Agrochemicals
WPS chart....50
small fruits....165
tree fruits, deciduous....400
thiram & thiabendazole (see RTU)
Thiram Granuflo (thiram)....UCB Agrochemicals
WPS chart....50
beans....316
castor beans....172
cole crops....327
cotton....125
cucurbits....333
eggplant....335
field corn, popcorn....88
flax....173
greens....34
lawns/turf....518
lentils....342
lettuce....345
okra....346
onions....351
ornamentals....532
peanuts....181
peas....353
peppers....356
radishes....363
rice....184
safflower....184
small fruits....141
small grains....445
sorghum....186
southern peas....365
soybeans....105
spinach....367
sugar beets....191
sunflower....192
sweet corn....369
table beets....372
tomatoes....378
tree fruits, deciduous....468
Thirethrin (endosulfan & pyrethrins)....Terra
WPS chart....49
beans....208
carrots, etc.....212
celery....216
cole crops....225
cucurbits....231
eggplant....235
greens....243
lettuce....254
peppers....270
potatoes....277
spinach....287
tree fruits, deciduous....417
***Thripobius semileuteus* (thrips parasite)**
....Arbico, Nature's Control
biocontrols....69
Thrips parasite (see *Thripobius semileuteus*)
Thrips predator mites (see *Neoseiulus cucumeris*)
..... Harmony Farm Supply, International Technology Services, Nature's Control
biocontrols....66
Thrips predator mites (see *Neoseiulus cucumeris*)
Tilt (propiconazole)....Novartis
WPS chart....46
celery....321
field corn, popcorn....88
peanuts....179
rice....183

small grains....139
sugarcane....191
sweet corn....369
TOP COP Tri-Basic (basic copper sulfate) .... Stoller
WPS chart....48
beans....312
carrots, etc....318
celery....321
cole crops....324
cucurbits....330
eggplant....334
greens....339
hops....174
lettuce....343
nuts....481
onions....347
peanuts....177
peppers....355
potatoes....358
radishes....363
small fruits....437
small grains....136
spinach....366
sugar beets....188
table beets....371
tomatoes....374
tree fruits, citrus....449
tree fruits, deciduous....460
tree fruits, subtropical....474
Top Cop with Sulfur (sulfur & tribasic copper sulfate).... Stoller
WPS chart....48
asparagus....309
beans....315
carrots, etc....319
celery....322
cole crops....327
cotton....125
cucurbits....333
eggplant....335
field corn, popcorn....88
forage (alfalfa, etc.)....549
greens....340
hops....174
lettuce....344
nuts....484
onions....350
peanuts....180
peas....353
peppers....356
potatoes....362
radishes....363
rice....183
small fruits....445
small grains....140
soybeans....105
spinach....367
sugar beets....190
table beets....372
tomatoes....377
tree fruits, citrus....451
tree fruits, deciduous....467
tree fruits, subtropical....476
Topcide (*lambda*-cyhalothrin) .... Uniroyal
WPS chart....50
ornamentals....499
Tops (thiophanate-methyl).... Gustafson, Platte
WPS chart....44,47
potatoes....362
Topsin M (thiophanate-methyl).... Elf Atochem
WPS chart....43
beans....315
cucurbits....333
nuts....485
onions....351
peanuts....181
small fruits....445
small grains....141
soybeans....105
sugar beets....190
tree fruits, deciduous....468
Tracer Naturalyte (spinosad) .... Dow AgroSciences
WPS chart....42
cotton....120
tralomethrin (see Scout X-TRA)
Triact (neem oil).... Thermo Trilogy
WPS chart....58
biocontrols....49
triadimefon (see Bayleton, ProStar Plus, ProTurf Fungicide VII, Strike)
triadimenol (see Baytan 30)
triadimenol & thiram (see RTU-Baytan-Thiram)
Trichoderma harzianum (see Bio-Trek)
***Trichogramma* spp. .... A-1 Unique Insect Control, Arbico, Beneficial Insectary, Caltec, Harmony Farm Supply, M & R Durango, Natural Pest Controls**
biocontrols....69
Trichogramma spp. (see Caterpillar Parasites)
Trigard (cyromazine).... Novartis
WPS chart....46
celery....216
cucurbits....231
lettuce....254
peppers....370
spinach....288
tomatoes....304
Trilogy (neem oil).... Thermo Trilogy
WPS chart....49
biocontrols....58
trimethyl-dodecatriene (see Stirrup M)
triphenyltin hydroxide (see Orbit/Super Tin, Super Tin)
Triumph (isazofos).... Novartis
WPS chart....46
Troy-Bt (*Bt.* spp. *kurstaki*) .... Troy Biosciences
Bt chart....94,108,128,144,196,380,488,536
WPS chart....49
Turcam (bendiocarb).... AgrEvo USA
WPS chart....40
lawns/turf....494
ornamentals....509
Turfcide (PCNB).... Uniroyal
WPS chart....50
cole crops....326
lawns/turf....515
ornamentals....528
peppers....360
tomatoes....376
TurfPro (PCNB).... Amvac
WPS chart....41
ornamentals....528

V

Vangard (cyprodinil).... Novartis
WPS chart....46
nuts....485
small fruits....446
tree fruits, deciduous....468
Vapam (metam-sodium).... Amvac
WPS chart....41
mint....174
tobacco....193
tree fruits, deciduous....463
VectoBac (*Bt.* spp. *israelensis*)).... Abbott
Bt chart....144,536
Velocity (acephate).... Valent
WPS chart....50
lawns/turf....489
Vendex (fenbutatin-oxide).... Griffin
WPS chart....44
eggplant....235
nuts....433
small fruits....391
tree fruits, citrus....401
tree fruits, deciduous....417
tree fruits, subtropical....423
vinclozolin (see Curalan, Ronilan, Vorlan)
Vitaflo-280 (carboxin & thiram) .... Gustafson
beans....311
field corn, popcorn....86
small grains....135
soybeans....102
sweet corn....368
Vitavax-30C (carboxin).... Gustafson
beans....310
cotton....123
peanuts....176
Vitavax-34 (carboxin) .... Gustafson
beans....310
cotton....123
field corn, popcorn....85
peanuts....176
safflower....184
small grains....135
soybeans....102
Vitavax PC (captan & PCNB & carboxin) .... Gustafson
peanuts....176
Vitavax-PCNB (carboxin & PCNB) .... Gustafson
cotton....123
small grains....135
soybeans....102
Vitavax-Thiram-Lindane (carboxin & thiram & lindane) .... Gustafson
WPS chart....44
small grains....135
Vitavax T-L (carboxin & thiram) .... Trace
small grains....135
soybeans....102
Volck (petroleum oil).... Valent
WPS chart....50
nuts....430
small fruits....390
tree fruits, citrus....398
tree fruits, deciduous....415
tree fruits, subtropical....421
Vorlan (vinclozolin).... Scotts
WPS chart....48
lawns/turf....519
Vydate (oxamyl).... DuPont
WPS chart....43
carrots, etc....213
celery....317
cotton....122
cucurbits....231
eggplant....235
peanuts....154
peppers....270
potatoes....278
soybeans....101
sweet potatoes, yams....297
tobacco....172
tomatoes....307
tree fruits, citrus....401
tree fruits, deciduous....418
tree fruits, subtropical....423

W

Warrior (*lambda*-cyhalothrin) .... ZENECA
WPS chart....51
cole crops....219
field corn, popcorn....79
forage (grasses, hay)....543
garlic....235
lettuce....249
onions....257
peanuts....151
small grains....129
sorghum....158
soybeans....96
sunflower....167
sweet corn....290
tomatoes....302
Wettable Sulfur (sulfur).... Platte
WPS chart....47
nuts....484
tree fruits, deciduous....466
Whitefly parasites (see *Encarsia formosa, Eretmocerus californicus*)
Wilbur-Ellis Diazinon (diazinon).... Wilbur-Ellis
WPS chart....51
beans....204
carrots, etc....210
celery....214
cole crops....220
cucurbits....227
greens....239
hops....146
lawns/turf....491
lettuce....250
nuts....426
onions....258
ornamentals....499
peas....261
peppers....267
potatoes....273
radishes....279
small fruits....384
sorghum....285
sugar beets....163
sweet corn....290
sweet potatoes, yams....295
table beets....298
tomatoes....303
tree fruits, deciduous....406
Wilbur-Ellis Dimethoate (dimethoate).... Wilbur-Ellis
WPS chart....51
beans....204
cole crops....220
cotton....114
cucurbits....228
field corn, popcorn....79
forage (alfalfa, etc.)....539
greens....240
lettuce....250
nuts....427
peas....262
peppers....267
potatoes....273
safflower....156
small grains....130
sorghum....158
tomatoes....298
tree fruits, citrus....395
tree fruits, deciduous....408
Wilbur-Ellils Malathion (malathion) .... Wilbur-Ellis
WPS chart....51
beans....206
celery....215
cole crops....222
cotton....117
cucurbits....229
eggplant....233
forage (alfalfa, etc.)....540
garlic....237
greens....241
horseradish....246
lentils....248
lettuce....251
mint....149
mushrooms....254
nuts....429
ornamentals....504
peas....263
peppers....268

potatoes 275
radishes 280
rice 155
small fruits 388
small grains 132
spinach 286
table beets 299
tomatoes 304
tree fruits, citrus 397
tree fruits, deciduous 411
tree fruits, subtropical 420
watercress 307

Wilbur-Ellis Methoxychlor (methoxychlor) ... Wilbur-Ellis
forage (alfalfa, etc.) 541
forage (grasses, hay) 544
ornamentals 504
peas 263

Wilbur-Ellils Methyl Parathion (methyl parathion) .... Wilbur-Ellis
WPS chart 51
artichokes 199
beans 206
carrot, etc. 211
celery 215
cole crops 223
cotton 118
forage (alfalfa, etc.) 541
lettuce 252
pastures/rangeland 545
peas 263
potatoes 275
rice 156
small grains 132
sorghum 160
spinach 286
sugar beets 164
sunflower 168
table beets 299

Wilbur-Ellils Sevin (carbaryl) .... Wilbur-Ellis
WPS chart 51
beans 203
cotton 111
cucurbits 227
eggplant 232
field corn 78
forage (alfalfa, etc.) 537
peppers 266
potatoes 272
small fruit 383
tomatoes 302

Win-Flo (PCNB) .... Amvac
WPS chart 41
cotton 124

X

XenTari (*Bt* . spp. *aizawai*) .... Abbott
Bt chart.......... 76,94,108,128,144,196,380,488,536
WPS chart 40

2,4-xylenol & meta-cresol (see Gallex)

Z

Zephyr (abamectin) .... Novartis
cotton 109

ziram (see Ziram, Ziram Granuflo)

Ziram (ziram) .... Drexel
WPS chart 42
beans 316
celery 322
cucurbits 334
nuts 485
ornamentals 533
small fruits 446
tomatoes 378
tree fruits, deciduous 469

Ziram (ziram) .... Elf Atochem
WPS chart 43
nuts 485
ornamentals 533
small fruits 446
tomatoes 378
tree fruits, deciduous 469

Ziram Granuflo (ziram) .... UCB Agrochemicals
WPS chart 50
nuts 485
ornamentals 534
small fruits 446
tomatoes 378
tree fruits, deciduous 469

Pest Index

A

Acalitus essigi — redberry mite (blackberry mite)
Acalitus vaccinii — blueberry bud mite
Acalymma trivittatum — western striped cucumber beetle
Acalymma vittatum — striped cucumber beetle
Acari — mites, ticks
Achyra rantalis — garden webworm
Acleris minuta — yellowheaded fireworm
Acontia dacia — brown cotton leafworm
Acrididae — grasshoppers, locusts
Acrobasis indigenella — leaf crumpler (mineola moth)
Acrobasis juglandis — pecan leaf casebearer
Acrobasis nuxvorella — pecan nut casebearer
Acrobasis vaccinii — cranberry fruitworm
Acrosternum hilare — green stink bug
Aculops lycopersici — tomato russet mite
Aculops pelekassi — pink citrus rust mite
Aculus cornutus — peach silver mite (peachtree silver mite)
Aculus fockeui — plum rust mite (plum nursery mite)
Aculus schlechtendali — apple rust mite
Acyrthosiphon kondoi — blue alfalfa aphid (blue aphid)
Acyrthosiphon pisum — pea aphid
Adelges abietis — eastern spruce gall adelgid (eastern spruce gall aphid)
Adelges cooley — cooley spruce gall aphid
Adelges piceae — balsam woolly adelgid
Adelges tsugae — hemlock chermes
Adelgidae — woolly aphid (adelgids)
Adelgids *adelgidae:*
- biocontrols 57
- ornamentals 497,502,503,505,506

Adelphocoris lineolatus — alfalfa plant bug
Adelphocoris rapidus — rapid plant bug
Agrilus anxius — bronze birch borer
Agrius cingulatus — sweetpotato hornworm
Agromyza frontella — alfalfa blotch leafminer
Agrotis ipsilon — black cutworm (greasy cutworm)
Agrotis orthogonia — pale western cutworm
Alabama argillacea — cotton leafworm
Aleurocanthus woglumi — citrus blackfly
Aleyrodidae — whiteflies
Alfalfa blotch leafminer *Agromyza frontella:*
- forage (alfalfa, etc.) 537-539,541

Alfalfa caterpillar *Colias eurytheme:*
- Bt chart 144,196,536
- beans 203
- forage (alfalfa, etc.) 537,538,540-542
- forage (grasses, hay) 544
- lentils 247
- peanuts 151
- peas 260-262,264
- southern peas 282,284
- soybeans 96

Alfalfa looper *Autographa californica:*
- beans 203,204
- carrots, etc. 210-212
- celery 215
- cole crops 217,219,221,222,225
- cucurbits 234
- forage (alfalfa, etc.) 538
- greens 238,240-242
- herbs/spices 243,244
- lentils 247
- lettuce 249,251-253
- mint 148
- peas 260,261,263
- peppers 268
- radishes 280
- rhubarb 282
- southern peas 282
- spinach 286,287

Alfalfa plant bug *Adelphocoris lineolatus:*
- forage (alfalfa, etc.) 537-539,541

Alfalfa seed chalcid *Bruchophagus roddi:*
- forage (alfalfa, etc.) 541

Alfalfa webworm *Loxostege commixtalis:*
- beans 207
- carrots, etc. 212
- forage (alfalfa, etc.) 538,541
- forage (grasses, hay) 544
- potatoes 277
- southern peas 283
- sweet corn 294
- table beets 299

Alfalfa weevil *Hypera postica:*
- photo 10
- biocontrols 55
- forage (alfalfa, etc.) 537-542
- forage (grasses, hay) 544

Allonemobius fasciatus — striped ground cricket
Almond mite *Bryobia rubrioculus* (brown mite):
- nuts 432

Alsophila pometaria — fall cankerworm
Alticinae — flea beetles
Alypia octomaculata — eightspotted forester
American plum borer *Euzophera semifuneralis:*
- tree fruits, deciduous 402-404,406

Ametastegia glabrata — dock sawfly
Amorbia essigana — avocado brown mite, western avocado leafroller
Amorbia moth:
- Bt chart 380

Ampeloglypter ater — grape cane girdler
Amphorophora rubi — European raspberry aphid (raspberry aphid)
Amyelois transitella — navel orangeworm
Anacamptodes fragilaria — citrus looper
Anarsia lineatella — peach twig borer
Anasa tristis — squash bug
Ancylis comptana — strawberry leafroller
Anisota — oakworm complex
Anisota senatoria — orangestriped oakworm
Annual bluegrass weevil:
- biocontrols 65,68
- lawns/turf 492-494

Anomala orientalis — Oriental beetle
Anthomyiidae — root maggot
Anthonomus eugenii — pepper weevil
Anthonomus grandis — boll weevil (cotton boll weevil)
Anthonomus musculus — cranberry weevil
Anthonomus signatus — strawberry bud weevil (strawberry clipper, strawberry weevil)
***Anthonomus* spp.** — blossom weevil
Anticarsia gemmatalis — velvetbean caterpillar
Antonina graminis — rhodesgrass scale (rhodesgrass mealybug)
Ants *Formicidae:*
- asparagus 200
- beans 203
- biocontrols 55,56,64
- carrots, etc. 209
- celery 214
- cole crops 218
- cucurbits 226
- eggplant 232
- field corn, popcorn 78
- forage (alfalfa, etc.) 537
- forage (grasses, hay) 543
- greens 238
- horseradish 245
- lawns/turf 489-491,493,494
- lettuce 249
- noncropland 546
- nuts 430
- okra 255
- ornamentals 495,497,499,508
- pastures/rangeland 544
- peanuts 151
- peas 260
- peppers 266
- potatoes 271
- radishes 278
- rice 154
- small fruits 382
- sorghum 157
- soybeans 96
- spinach 285
- sugar beets 162
- sunflower 167
- sweet corn 288
- sweet potatoes, yams 295
- table beets 298
- tobacco 169,170
- tomatoes 302
- tree fruits, citrus 393,394
- tree fruits, deciduous 405

Aonidiella aurantii — California red scale
Aonidiella citrina — yellow scale
Apache cicada *Diceroprocta apache:*
- asparagus 200

Aphididae — aphids, root aphid, woolly aphid
Aphids *Aphididae:*
- photo 12
- artichokes 198,199
- asparagus 199,201,202
- beans 202-209
- biocontrols 54-58,61-66
- carrots, etc. 209-212
- celery 213-216
- cole crops 217-225
- cotton 109-122
- cucurbits 226-231
- eggplant 232-235
- field corn, popcorn 79-84
- forage (alfalfa, etc.) 538-543
- forage (grasses, hay) 543,544
- garlic 236,237
- greens 238-243
- herbs/spices 243,244
- hops 146,147
- horseradish 245,246
- lentils 246-248
- lettuce 248-253
- mint 148,149
- mushrooms 254
- nuts 423-433
- oil seed crops 150
- okra 255,256
- onions 256-260
- ornamentals 495-509
- pastures/rangeland 545,546
- peanuts 151-153
- peas 260-264
- peppers 265-270
- potatoes 271-278
- radishes 278-281
- rhubarb 281,282
- safflower 156,157
- small fruits 381,382,384-391
- small grains 129-133
- sorghum 158-160
- southern peas 283,284
- soybeans 96,97,99,100
- spinach 284-287
- sugar beets 161-165
- sunflower 168,169
- sweet corn 290-294
- sweet potatoes, yams 294-297
- table beets 297-300
- tobacco 169-172
- tomatoes 301-307
- tree fruits, citrus 392-400
- tree fruits, deciduous 401-418
- tree fruits, subtropical 418,420-422
- watercress 307

Aphis citricola — spirea aphid
Aphis craccivora — cowpea aphid
Aphis fabae — bean aphid
Aphis gossypii — cotton aphid, melon aphid
Aphis illinoisensis — grapevine aphid
Aphis nasturtii — buckthorn aphid
Aphis pomi — apple aphid (green aphid, green apple aphid)
Aphrophora parallela — pine spittlebug
Aphrophora saratogensis — Saratoga spittlebug
Apopka weevil *Diaprepes abbreviatus* (sugarcane rootstalk borer weevil):
- biocontrols 68
- tree fruits, citrus 392,394

Apple aphid *Aphis pomi:*
- ornamentals 496,497,501
- tree fruits, deciduous 402,404,406-408,410,412

Apple curculio *Tachypterellus quadrigibbus:*
- biocontrols 55

Apple grain aphid *Rhopalosiphum fitchii:*
- biocontrols 58

Apple maggot *Rhagoletis pomonella:*
- photo 12
- tree fruits, deciduous 402-404,406-408,410,412

Apple mealybug *Phenacoccus aceris:*
- nuts 424,428
- tree fruits, deciduous 404

Apple pandemis *Pandemis pyrusana:*
- tree fruits, deciduous 405

Apple red bug *Lygidea mendax:*
- biocontrols 57
- tree fruits, deciduous 414,415

Apple rust mite *Aculus schlechtendali:*
- nuts 432
- tree fruits, deciduous 404,407,408,416,417

Apple sucker *Psylla mali:*
- tree fruits, deciduous 404

Apple-and-thorn skeletonizer *Choreutis pariana:*
- ornamentals 499
- tree fruits, deciduous 412

Appleworms:
- biocontrols 56

Apricot mealybug:
- tree fruits, deciduous 407

Apricot scale:
- nuts 426
- tree fruits, deciduous 407

Araneae — spiders
Archips argyrospila — fruittree leafroller

Archips fervidana — oak webworm
Archips rosana — filbert leafroller
Arctiidae — woollybear caterpillar
Argyresthia cupressella — cypress tip moth
Argyrotaenia citrana — orange tortrix (tortrix moth)
Argyrotaenia velutinana — redbanded leafroller
Armored scale *Diaspididae:*
biocontrols ... 57,67
ornamentals ... 501
Army cutworm *Euxoa auxiliaris:*
forage (grasses, hay) ... 543
noncropland ... 546
small grains ... 129,130
Armyworm *Pseudaletia unipuncta:*
Bt chart ... 76,94,108,128,144,196,380,488,536
artichokes ... 198,199
asparagus ... 199,200
beans ... 202,203,205,206,208
biocontrols ... 55-57,63,65,68,69
carrots, etc. ... 209-212
celery ... 213-215
cole crops ... 218,219,221-225
cotton ... 109-114,118,120
cucurbits ... 226-228
eggplant ... 232
field corn, popcorn ... 77-84
forage (alfalfa, etc.) ... 537,538,540-542
forage (grasses, hay) ... 543,544
garlic ... 236,237
greens ... 238-242
herbs/spices ... 243
hops ... 146
horseradish ... 245
lawns/turf ... 489-494
lentils ... 246,247
lettuce ... 248-252
mint ... 148
noncropland ... 546,547
oil seed crops ... 150
okra ... 255
onions ... 256,257,259
ornamentals ... 496-499,504,505,507,508
pastures/rangeland ... 544-546
peanuts ... 151
peas ... 260-264
peppers ... 265,266,268
potatoes ... 271-273,275,277
radishes ... 278,279
rhubarb ... 281,282
rice ... 154-156
small fruits ... 381-383,386
small grains ... 129,131-133
sorghum ... 157-160
southern peas ... 282,284
soybeans ... 95,96,98-100
spinach ... 284-286
sugar beets ... 161-163,167
sunflower ... 167
sweet corn ... 288-293
sweet potatoes, yams ... 294,295
table beets ... 297-300
tobacco ... 169,170
tomatoes ... 301-305
tree fruits, citrus ... 393
tree fruits, deciduous ... 402,405,410
Artoga trialbamaculella — redstriped fireworm
Artichoke aphid *Capitophorus elaeagni:*
artichokes ... 198
Artichoke (plume) moth *Platyptilia carduidactyla:*
Bt chart ... 144,196
artichokes ... 198,199
biocontrols ... 68
carrots, etc. ... 209
horseradish ... 245
lentils ... 247
potatoes ... 271
radishes ... 278
sweet potatoes, yams ... 295
table beets ... 297
Artipus floridanus — little leaf notcher
Artogeia rapae — imported cabbageworm
Asiatic garden beetle *Maladera castanea:*
lawns/turf ... 492
ornamentals ... 502
Asparagus aphid *Brachycolus asparagi:*
asparagus ... 200,201
Asparagus beetle *Crioceris asparagi:*
asparagus ... 200-202
Aspidiotus nerii — ivy scale, oleander scale
Aster leafhopper (beetle) *Macrosteles fascifrons:*
artichokes ... 198
carrots, etc. ... 209,210
celery ... 214
cole crops ... 218
forage (alfalfa, etc.) ... 538
greens ... 238
horseradish ... 245
lentils ... 247
lettuce ... 249,251
potatoes ... 271,276
radishes ... 278,279
spinach ... 285
sweet corn ... 290,293
sweet potatoes, yams ... 295
table beets ... 297,299
Ataenius spretulus — blackturfgrass ataenius:
Athrips rancidella — cotoneaster webworm
Atlantic mite *Tetranychus turkestani* (strawberry spider mite):
cotton ... 120,121
field corn, popcorn ... 84
forage (alfalfa, etc.) ... 543
small grains ... 133
sorghum ... 161
Aulacaspis rosae — rose scale
Autographa californica — alfalfa looper
Automeris io — Io moth
Autoplusia egena — bean leafskeletonizer
Avocado brown mite *Amorbia essigana* (western avocado leafroller):
biocontrols ... 64
tree fruits, subtropical ... 422
Avocado leafroller *Idona minuenda:*
tree fruits, citrus ... 393,397,399
Azalea caterpillar *Datana major:*
ornamentals ... 498,499
Azalea lacebug:
ornamentals ... 495
Azalea leafminer *Caloptilia azaleella:*
ornamentals ... 496,497,509
Azalea moth:
ornamentals ... 503
Azalea scale:
Bt chart ... 488

B

Bagworm *Thyridopteryx ephemeraeformis:*
Bt chart ... 488
biocontrols ... 56
ornamentals ... 495-499,503,506,508,509
tree fruits, deciduous ... 402,404
Balsam twig aphid *Mindarus abietinus:*
ornamentals ... 495,497
Balsam woolly adelgid *Adelges piceae:*
ornamentals ... 495,505
Banana corm borer *Cosmopolites sordidus:*
tree fruits, subtropical ... 423
Banana moth:
biocontrols ... 63,68
Banana root borer *Cosmopolites sordidus:*
tree fruits, subtropical ... 420,421
Banana skipper *Pelopidas thrax:*
Bt chart ... 380
Banded cucumber beetle *Diabrotica balteata:*
potatoes ... 271,272
sweet corn ... 288
sweet potatoes, yams ... 296
tomatoes ... 301,303
Banded soybean thrips:
cotton ... 111,113
Banded sunflower moth *Conchylis hospes:*
Bt chart ... 144
sunflower ... 167-169
Bandedwing whitefly *Trialeurodes abutilonea:*
biocontrols ... 112,121
cotton ... 54,55
Banks grass mite *Oligonychus pratensis:*
field corn, popcorn ... 77,80,83-85
forage (alfalfa, etc.) ... 539
small grains ... 133
sorghum ... 158-161
soybeans ... 100
sugar beets ... 165
sweet corn ... 294
Baris lepidii — imported crucifer weevil
Barley thrips *Limothrips denticornis:*
small grains ... 132
Bean aphid *Aphis fabae:*
biocontrols ... 54,55
Bean beetles:
beans ... 207,208
carrots, etc. ... 212
celery ... 216
cole crops ... 224
cucurbits ... 230
eggplant ... 235
field corn, popcorn ... 80
lettuce ... 253
onions ... 259
peanuts ... 153
peppers ... 269
potatoes ... 277
soybeans ... 100
spinach ... 287
sugar beets ... 165
sunflower ... 168
sweet corn ... 294
table beets ... 300
tomatoes ... 306
Bean leaf beetle *Cerotoma trifurcata:*
beans ... 202,203,205,206
biocontrols ... 55
cole crops ... 222
eggplant ... 232
field corn, popcorn ... 79
forage (alfalfa, etc.) ... 537
lentils ... 247
peanuts ... 151
peas ... 260,263,264
peppers ... 265
southern peas ... 282-284
soybeans ... 95-99
table beets ... 299
tomatoes ... 301
Bean leafroller *Urbanus proteus:*
beans ... 202
Bean leafskeletonizer *Autoplusia egena:*
beans ... 205,209
southern peas ... 283
Bedellia somnulentella — morningglory leafminer
Beet armyworm *Spodoptera exigua:*
photo ... 15
Bt chart ... 76
asparagus ... 199,200
beans ... 202,205
biocontrols ... 57,59
carrots, etc. ... 209,210,212
celery ... 213-215
cole crops ... 217,219,221,224,225
cotton ... 109-114,116-118,120
cucurbits ... 226,228
eggplant ... 232,233
field corn, popcorn ... 79,81
forage (alfalfa, etc.) ... 540
garlic ... 236
greens ... 238-240,242
herbs/spices ... 243,244
lettuce ... 248,249,251,253
onions ... 256,258
ornamentals ... 496,497,500,501,505,506,508
peanuts ... 150-152
peas ... 260,262
peppers ... 265,266,268
potatoes ... 271,274,276
rhubarb ... 281,282
small fruits ... 381,386
sorghum ... 158,159
soybeans ... 95-100
spinach ... 284,286,287
sugar beets ... 161-163
sunflower ... 167
sweet corn ... 288-292
table beets ... 297,298
tomatoes ... 301 305
tree fruits, citrus ... 392,396
Beet leaf beetle *Erynephala puncticollis:*
sugar beets ... 162
Beet leafhopper *Circulifer tenellus:*
noncropland ... 547
sugar beets ... 163
Beet root maggot *Tetanops myopaeformis* (sugarbeet root maggot):
sugar beets ... 164,165
Beet webworm *Loxostege sticticalis:*
cole crops ... 224
cucurbits ... 230
eggplant ... 235
lettuce ... 253
potatoes ... 277
soybeans ... 100
sugar beets ... 163,165
sweet corn ... 294
table beets ... 300
Beetles *Coleoptera:*
asparagus ... 199,201
beans ... 207
biocontrols ... 56,57,65
carrots, etc. ... 209
celery ... 213
cole crops ... 217
cucurbits ... 226,230
eggplant ... 232,234
garlic ... 236
greens ... 238
horseradish ... 245
lentils ... 246
lettuce ... 248
nuts ... 424
okra ... 255
onions ... 256
ornamentals ... 496,498
peanuts ... 153
peppers ... 265,269
potatoes ... 271
radishes ... 278,280
rhubarb ... 281
small fruits ... 381
spinach ... 284
sugar beets ... 161
sugarcane ... 166
sweet corn ... 293
sweet potatoes, yams ... 294
table beets ... 297,300
tomatoes ... 301,305,306
tree fruits, citrus ... 392
tree fruits, deciduous ... 402
tree fruits, subtropical ... 418
Belonolaimus **spp.** — sting nematode
Bermuda mite:
lawns/turf ... 493
Bermudagrass mite *Eriophyes cynodoniensis:*
biocontrols ... 55
forage (alfalfa, etc.) ... 539
lawns/turf ... 491,492
Bigbeaked plum mite *Diptacusgigantorhynchus:*
tree fruits, deciduous ... 417

Bigheaded ant *Pheidole megacephala:*
lawns/turf ... 492
noncropland ... 546,547
ornamentals ... 501,502
pastures/rangeland ... 545
Billbugs *Sphenophorus* spp.:
biocontrols ... 55,65,68
field corn, popcorn ... 78,79
lawns/turf ... 490,492-494
ornamentals ... 505,509
sweet corn ... 289
Birch leafminer *Fenusa pusilla:*
ornamentals ... 495,497,500,503
Birdcherry oat aphid *Rhopalosiphum padi* (oat birdcherry aphid):
forage (grasses, hay) ... 543
Black bean aphid *Aphis fabae* (bean aphid):
beans ... 204,205,209
southern peas ... 283
Black cherry aphid *Myzus cerasi:*
biocontrols ... 58
tree fruits, deciduous ... 402,405-407, 409,411,412,417
Black currant borer:
biocontrols ... 68
Black cutworm *Agrotis ipsilon:*
photo ... 15
carrots, etc. ... 209
cotton ... 109,119
field corn, popcorn ... 80,82
horseradish ... 245
lawns/turf ... 493
lentils ... 246
lettuce ... 248
rhubarb ... 281
soybeans ... 95
spinach ... 284
sweet corn ... 290
tobacco ... 169
tomatoes ... 304,305
Black fleahopper *Spanogonicus albofasciatus* (whitemarked fleahopper):
cotton ... 110,114
Black flies:
tree fruits, citrus ... 298
Black grass bug *Irbisia solani:*
forage (alfalfa, etc.) ... 540
forage (grasses, hay) ... 544
noncropland ... 546
pastures/rangeland ... 544
small grains ... 132
Black margined aphid:
nuts ... 424,425
Black mite:
small grains ... 132
Black peach aphid *Brachycaudus persicae:*
tree fruits, deciduous ... 409,411,412
Black pecan aphid *Melanocallis caryaefoliae:*
nuts ... 425-428
Black pine weevil:
ornamentals ... 501
Black pineleaf scale *Nuculaspis californica:*
ornamentals ... 496
Black scale (crawlers) *Saissetia oleae:*
biocontrols ... 65
nuts ... 426,431
ornamentals ... 500,503,506
tree fruits, citrus ... 393,394,396-398,400
tree fruits, deciduous ... 407,415
tree fruits, subtropical ... 421,422
Black turfgrass ataenius beetle *Ataenius spretulus:*
biocontrols ... 55
lawns/turf ... 490-494
Black vine weevil *Otiorhynchus sulcatus:*
biocontrols ... 54,55,65,68
ornamentals ... 495,496,499,501,508,509
small fruits ... 381,386
Blackberry mite *Acalitus essigi* (redberry mite):
small fruits ... 382
Blackheaded fireworm *Rhopobota naevana:*
small fruits ... 384,387,388
Blissus leucopterus — chinch bug
***Blissus* spp.** — lawn chinch bug (chinch bug complex)
Blister beetle *Meloidae:*
beans ... 203,205
carrots, etc. ... 211
cole crops ... 222
eggplant ... 233-235
forage (alfalfa, etc.) ... 537,541
forage (grasses, hay) ... 544
greens ... 241,242
lentils ... 247
lettuce ... 252
ornamentals ... 497,504
peanuts ... 151
peppers ... 266-268
potatoes ... 274,275
radishes ... 280
southern peas ... 282,283
soybeans ... 96,99
spinach ... 286
table beets ... 299
tomatoes ... 302,303,305
Blister mite *Eriophyidae:*
small fruits ... 391
tree fruits, deciduous ... 404,416
Blotch leafminer:
forage (alfalfa, etc.) ... 538,540
Blue alfalfa aphid *Acyrthosiphonkondoi* (blue aphid):
forage (alfalfa, etc.) ... 538,539,541,542
Blueberry bud mite *Acalitus vaccinii:*
small fruits ... 385
Blueberry leafhopper *Scaphytopius magdalensis:*
small fruits ... 386
Blueberry leafroller *Sparganothis sulfureana* (Sparganothis fruitworm):
small fruits ... 381,386
Blueberry maggot *Rhagoletis mendax* (blueberry fruit fly):
small fruits ... 382,386-388,391
Blueberry tip moth *Oberea mypos:*
small fruits ... 388
Blue-green weevil:
biocontrols ... 68
tree fruits, citrus ... 392,394,395
Bluegrass billbug *Sphenophorus parvulus:*
lawns/turf ... 490,491
Boll weevil *Anthonomus grandis:*
photo ... 10
biocontrols ... 55,56
cotton ... 110-118,120-122
Bollworm *Helicoverpa zea:*
Bt chart ... 94,128,144
biocontrols ... 56,61,69
carrots, etc. ... 210
celery ... 215
cotton ... 109,111-113,115,116,118-121
greens ... 240
herbs/spices ... 243
lettuce ... 251
rhubarb ... 282
soybeans ... 98,99
spinach ... 286
Borers:
Bt chart ... 144
biocontrols ... 56
tree fruits, deciduous ... 497,498
Boxelder bug *Leptocoris trivittatus:*
photo ... 12
ornamentals ... 495,497-499
Boxwood leafminer *Monarthropalpus buxi:*
ornamentals ... 497,503
Brachycaudus cardui — thistle aphid
Brachycaudus helichrysi — leafcurl plum aphid
Brachycaudus persicae — black peach aphid
Brachycolus asparagi — asparagus aphid
Brevicoryne brassicae — cabbage aphid
Brevipalpus lewisi — citrus flat mite (flat mite)
Brevipalpus obovatus — privet mite
Brevipalpus phoenicis — red and black flat mite
Bristly roseslug *Cladius difformis:*
ornamentals ... 498
Broad mite *Polyphagotarsonemus latus:*
artichokes ... 199
asparagus ... 201
beans ... 207
biocontrols ... 66
carrots, etc. ... 212
celery ... 215
cole crops ... 223
cucurbits ... 230
eggplant ... 234
garlic ... 237
greens ... 242
herbs/spices ... 244
horseradish ... 246
lettuce ... 252
nuts ... 423,432
okra ... 256
onions ... 259
ornamentals ... 499,507,508
peas ... 264
peppers ... 269,270
potatoes ... 276
radishes ... 280
spinach ... 287
sugar beets ... 164
sweet potatoes, yams ... 296
table beets ... 299
tomatoes ... 305
tree fruits, citrus ... 392,399,400
watercress ... 307
Bronze birch borer *Agrilus anxius:*
ornamentals ... 509
Brown (almond) mite *Bryobia rubrioculus:*
nuts ... 427,432
small fruits ... 382
tree fruits, deciduous ... 415
Brown apricot scale *Parthenolecanium corni* (European fruit lecanium):
nuts ... 430,431
ornamentals ... 506
tree fruits, deciduous ... 415
Brown citrus aphid *Toxoptera citricida:*
tree fruits, citrus ... 394
Brown cotton leafworm *Acontia dacia:*
cotton ... 110,116,117
Brown dog tick *Rhipicephalus sanguineus:*
lawns/turf ... 491,493
Brown garden snail *Helix aspersa:*
biocontrols ... 67
Brown mite *Bryobia rubrioculus:*
forage (alfalfa, etc.) ... 539
lawns/turf ... 492
nuts ... 431
tree fruits, deciduous ... 404,407
tree fruits, subtropical ... 422
Brown (soft) scale *Coccus hesperidum:*
biocontrols ... 65
nuts ... 426
ornamentals ... 496,499,508
tree fruits, citrus ... 393-396,398,400
tree fruits, deciduous ... 405,407
Brown spanworm *Ematurga amitaria* (brown cranberry spanworm):
small fruits ... 383
Brown stink bug *Euschistus servus:*
cotton ... 112
nuts ... 429
Brown wheat mite *Petrobia latens:*
small grains ... 129,130,132
Browntail moth *Euproctis chrysorrhoea:*
Bt chart ... 488
ornamentals ... 497
Bruchidae — seed weevil
Bruchophagus platypterus — clover seed chalcid
Bruchophagus roddi — alfalfa seed chalcid
Bruchus brachialis — vetch bruchid
Bruchus pisorum — pea weevil
Bryobia rubrioculus — brown mite (brown almond mite)
Bucculatrix thurberiella — cotton leafperforator
Buck moth *Hemileuca maia:*
Bt chart ... 488
Buckthorn aphid *Aphis nasturtii:*
potatoes ... 271
Bud mite *Phytoptus pyri* (pearleaf blister mite):
small fruits ... 391
tree fruits, citrus ... 398,400
tree fruits, deciduous ... 404
Bud moth *Spilonota ocellana* (eyespotted bud moth):
biocontrols ... 56
nuts ... 433
tree fruits, deciduous ... 411,412
Budworms:
Bt chart ... 94,128,144,488
cucurbits ... 226
ornamentals ... 495,496,498,499
tobacco ... 169-171
Buffalograss mealybug:
biocontrols ... 54
Bulb nematode *Ditylenchus* spp.:
garlic ... 237
Burrowing nematode:
biocontrols ... 56
tree fruits, deciduous ... 401,418
tree fruits, subtropical ... 421
***Byturus* spp.** — raspberry fruitworm

C

Cabbage aphid *Brevicoryne brassicae:*
biocontrols ... 54,55
cole crops ... 219,221,224,225
greens ... 239,242
oil seed crops ... 150
radishes ... 278
Cabbage budworm:
Bt chart ... 196
Cabbage flea beetle *Phyllotreta cruciferae:*
cole crops ... 221
greens ... 240
Cabbage looper *Trichoplusia ni:*
artichokes ... 198
beans ... 202,206,208
biocontrols ... 63,67,69
carrots, etc. ... 209-212
celery ... 213-216
cole crops ... 217-219,221-225
cotton ... 109-118,120
cucurbits ... 225-228,230,231
eggplant ... 232,234
garlic ... 236
greens ... 238-243
herbs/spices ... 243,244
horseradish ... 245
lentils ... 247,248
lettuce ... 248,249,251-254
mushrooms ... 254
onions ... 257
ornamentals ... 496,505,506
peanuts ... 151
peas ... 260,263,264
peppers ... 265,266,269
potatoes ... 271,275,276
radishes ... 278,280,281
rhubarb ... 281,282
soybeans ... 95,96,98-100
spinach ... 284-287
sunflower ... 169
sweet potatoes, yams ... 294,295
table beets ... 297,299
tobacco ... 169,171
tomatoes ... 301-305
Cabbage (root) maggot *Delia radicum:*
biocontrols ... 68
cole crops ... 218

Cabbage seedpod weevil:
- oil seed crops ... 150

Cabbage webworm *Hellula rogatalis:*
- cole crops ... 219

Cabbageworm *Artogeia rapae* (imported cabbageworm):
- beans ... 207,209
- biocontrols ... 69
- carrots, etc. ... 212
- celery ... 216
- cole crops ... 224
- cucurbits ... 230,231
- eggplant ... 235
- greens ... 242
- lentils ... 248
- lettuce ... 253
- okra ... 256
- peppers ... 269
- potatoes ... 277
- spinach ... 287
- sweet corn ... 294
- table beets ... 300
- tomatoes ... 306

Cadra figulilella — raisin moth

Calico scale *Eulecanium cerasorum:*
- biocontrols ... 58
- nuts ... 424,425,431

California oakworm *Phryganidia californica:*
- ornamentals ... 495,498,499

California orangedog *Papilio zelicaon:*
- tree fruits, citrus ... 393

California pearslug *Pristiphora abbreviata* (California pear sawfly):
- tree fruits, deciduous ... 404

California red scale *Aonidiella aurantii:*
- biocontrols ... 62
- ornamentals ... 499,500,508
- tree fruits, citrus ... 393,394,396-398,400

Caliothrips fasciatus — bean thrips

Caliroa cerasi — pear sawfly (pearslug)

Calligrapha multipunctata — willow leaf beetle

Caloptilia azaleella — azalea leafminer

Cambium curculio *Conotrachelus anaglypticus:*
- biocontrols ... 56

Camellia scale *Lepidosaphes camelliae:*
- ornamentals ... 500

Cameraria — oak leafminer

Camplyomma spp.:
- tree fruits, deciduous ... 405

Cane borers:
- biocontrols ... 69

Cankerworms *Geometridae:*
- Bt chart ... 380,488
- biocontrols ... 57
- ornamentals ... 497-499,504
- tree fruits, deciduous ... 412

Capitophorus elaeagni — artichoke aphid

Carabidae — ground beetles

Carmine spider mite *Tetranychus cinnabarinus:*
- celery ... 214
- cotton ... 109,111
- field corn, popcorn ... 77
- nuts ... 423

Carnation bud mite *Eriophyes paradianthi:*
- ornamentals ... 499

Carnation shoot mite *Eriophyes dianthi:*
- ornamentals ... 499

Carpenterworm *Prionoxystus robiniae:*
- biocontrols ... 56

Carpini spider mite *Eotetranychus carpiniborealis* (yellow spider mite):
- tree fruits, deciduous ... 417

Carpophilus hemipterus — driedfruit beetle

***Carpophilus* spp.** — picnic beetle

Carrion beetles *Silphidae:*
- sugar beets ... 163

Carrot rust fly *Psila rosae:*
- carrots, etc. ... 212

Carrot weevil *Listronotus oregonensis:*
- biocontrols ... 68
- carrots, etc. ... 209,212,213
- celery ... 216,217

Casebearers *Coleophoridae:*
- biocontrols ... 56
- ornamentals ... 498

Cassidinae — tortoise beetle

Catalpa sphinx *Ceratomia catalpae:*
- ornamentals ... 497

Caterpillars *Lepidoptera:*
- Bt chart ... 94,144,380,488
- artichokes ... 198
- asparagus ... 199,201
- beans ... 202,203,205,207,208
- biocontrols ... 55-57,63,64,66,68
- carrots, etc. ... 209
- celery ... 213,216
- cole crops ... 217-220,222,224,225
- cotton ... 110-114,118-120
- cucurbits ... 226,230,231
- eggplant ... 232,235
- field corn, popcorn ... 84
- forage (alfalfa, etc.) ... 537,541,542
- forage (grasses, hay) ... 544
- garlic ... 236
- greens ... 238,239,241,242
- herbs/spices ... 244
- horseradish ... 245
- lawns/turf ... 493
- lentils ... 246,247
- lettuce ... 248,249,250,253
- mint ... 149
- noncropland ... 546
- nuts ... 424-427,431,432
- oil seed crops ... 150
- okra ... 255,256
- onions ... 256,257,259
- ornamentals ... 495-500,503-509
- pastures/rangeland ... 544-546
- peanuts ... 150-153
- peas ... 260,262,264
- peppers ... 265,269,270
- potatoes ... 271
- radishes ... 278
- rhubarb ... 281
- small fruits ... 381,382,386,391
- small grains ... 132,133
- sorghum ... 160
- southern peas ... 282,284
- soybeans ... 95-100
- spinach ... 284,287
- sugar beets ... 161,164,165
- sunflower ... 167,168
- sweet corn ... 293
- sweet potatoes, yams ... 294,295
- table beets ... 297,300
- tobacco ... 169,171
- tomatoes ... 301,305,306
- tree fruits, citrus ... 392,393,399
- tree fruits, deciduous ... 402,404,405,410-412,415
- tree fruits, subtropical ... 418,422

Catfacing insect:
- tree fruits, deciduous ... 413,417

Catocales:
- biocontrols ... 56

Cecidomyiidae — cone midges

Cecidophyopsis psilaspis — taxus bud mite

Celama sorghiella — sorghum webworm

Celery leaftier *Udea rubigalis:*
- Bt chart ... 196
- carrots, etc. ... 212
- celery ... 216
- cole crops ... 224
- cucurbits ... 230
- eggplant ... 235
- lettuce ... 253
- peppers ... 266,269
- spinach ... 287
- sweet corn ... 294
- table beets ... 300

Celery looper *Syngrapha falcifera:*
- peas ... 260

Centipedes *Chilopoda:*
- lawns/turf ... 489-491,493,494
- ornamentals ... 508

Ceratitis capitata — medfly, Mediterranean fruit fly

Ceratomia catalpae — catalpa sphinx

Cercopidae — spittlebugs

Cereal leaf beetle *Oulema melanopus:*
- biocontrols ... 55
- field corn, popcorn ... 79
- forage (grasses, hay) ... 543
- small grains ... 129-132

Cerococcus scale *Cerococcus* spp.:
- ornamentals ... 496

***Cerococcus* spp.** — Cerococcus scale

Ceroplastes floridensis. — Florida wax scale

***Ceroplastes* spp.** — wax scale

Cerotoma trifurcata — bean leaf beetle

Ceutorhyncus assimilis — cabbage seedpod weevil

Chaetocnema confinis — sweetpotato flea beetle

Chaetocnema pulicaria — corn flea beetle

Chaetosiphon fragaefolii — strawberry aphid

Chafers:
- biocontrols ... 56
- lawns/turf ... 490
- ornamentals ... 496,502

Chaff scale *Parlatoria pergandii:*
- tree fruits, citrus ... 393,396,398,400

Chalcodermus aeneus — cowpea curculio (pea curculio)

Cherry fruit fly *Rhagoletis cingulata:*
- tree fruits, deciduous ... 402,406,407,410-412,416

Cherry fruitworm *Grapholita packardi:*
- small fruits ... 381,382,386-388
- tree fruits, deciduous ... 405,412

Cherry leafminer:
- tree fruits, deciduous ... 402,404

Cherry maggot *Rhagoletis cingulata:*
- tree fruits, deciduous ... 405

Cherry rust mite:
- tree fruits, deciduous ... 407,417

Chestnut weevil:
- biocontrols ... 58
- nuts ... 424,425

Chiggers *Trombiculidae* (chigger mites):
- biocontrols ... 55
- lawns/turf ... 489-491,493

Chilopoda — centipedes

Chinch bug *Blissus leucopterus:*
- biocontrols ... 55,56,65,68
- carrots, etc. ... 212
- field corn, popcorn ... 77-79,83,84
- forage (alfalfa, etc.) ... 542
- forage (grasses, hay) ... 543,544
- lawns/turf ... 489-491,493,494
- noncropland ... 546
- nuts ... 430
- oil seed crops ... 150
- ornamentals ... 505,507
- pastures/rangeland ... 544
- rice ... 154,156
- small grains ... 129,132,133
- sorghum ... 157-161
- sweet corn ... 288-290

Chionaspis furfura — scurfy scale

Chionaspis pinifoliae — pine needle scale (pine leaf scale)

Chironomidae — midges

Chlorochroa sayi — Say stink bug

Choreutis pariana — apple-and-thorn skeletonizer

Choristoneura fumiferana — spruce budworm

Choristoneura occidentalis — western spruce budworm

Choristoneura pinus — Jack pine budworm

Choristoneura rosaceana — obliquebanded leafroller

Chromaphis juglandicola — walnut aphid

Chrysanthemum aphid *Macrosiphoniella sanborni:*
- ornamentals ... 509

Chrysanthemum leafminer *Phytomyza syngenesiae:*
- artichokes ... 199

Chrysobothris mali — Pacific flatheaded borer

Chrysomela scripta — cottonwood leaf beetle

Chrysomelidae — leaf beetles

Chrysomphalus aonidum — Florida red scale

Chrysomphalus dictyospermi — Dictyospermum scale

Chrysoteuchia topiaria — cranberry girdler

Cicadellidae — leafhoppers

Circulifer tenellus — beet leafhopper, sugarbeet leafhopper

Citricola scale *Coccus pseudomagnoliarum:*
- tree fruits, citrus ... 393,396-398

Citrus aphid:
- tree fruits, citrus ... 396

Citrus black scale:
- biocontrols ... 66

Citrus blackfly *Aleurocanthus woglumi:*
- biocontrols ... 54,55
- tree fruits, citrus ... 392,396,400

Citrus bud mite *Eriophyes sheldoni:*
- tree fruits, citrus ... 392

Citrus cutworm *Egira curialis:*
- tree fruits, citrus ... 392,393,395,396

Citrus flat mite *Brevipalpus lewisi:*
- tree fruits, citrus ... 395

Citrus leafminer *Phyllocnistis citrella:*
- tree fruits, citrus ... 392,397,400

Citrus mealybug *Planococcus citri:*
- photo ... 13
- biocontrols ... 54,55,65
- ornamentals ... 497,501
- tree fruits, citrus ... 393

Citrus nematode *Tylenchulus semipenetrans:*
- biocontrols ... 56
- tree fruits, citrus ... 400
- tree fruits, deciduous ... 401

Citrus peelminer *Mamara salictella:*
- tree fruits, citrus ... 399

Citrus red mite *Panonychus citri:*
- photo ... 10
- biocontrols ... 57
- nuts ... 423
- tree fruits, citrus ... 395-401

Citrus red scale:
- biocontrols ... 62

Citrus root weevil *Pachnaeus litus:*
- biocontrols ... 54,55,68
- tree fruits, citrus ... 393,394,397,398

Citrus rust mite *Phyllocoptruta oleivora:*
- tree fruits, citrus ... 392,394-401

Citrus snow scale *Unaspis citri:*
- tree fruits, citrus ... 393

Citrus thrips *Scirtothrips citri:*
- ornamentals ... 505,506,508
- tree fruits, citrus ... 392,393,399,401

Citrus whitefly *Dialeurodes citri:*
- biocontrols ... 54,55,58
- tree fruits, citrus ... 396

Citrus yellow scale:
- biocontrols ... 62

Cladius difformis — bristly roseslug

Clastoptera achatina — pecan spittlebug

Clearwing borer *Sesiidae:*
- biocontrols ... 68

Climbing cutworm *Noctuidae:*
- beans ... 206
- forage (alfalfa, etc.) ... 541,542
- ornamentals ... 508
- peas ... 263
- small fruits ... 383,386
- small grains ... 132
- soybeans ... 99
- tree fruits, deciduous ... 406

Clover head weevil *Hypera meles:*
- forage (alfalfa, etc.) ... 537

Clover leaf weevil *Hypera punctata:*
- forage (alfalfa, etc.) ... 540-542
- forage (grasses, hay) ... 544

Clover mite *Bryobia praetiosa:*
- eggplant ... 235
- forage (alfalfa, etc.) ... 539
- lawns/turf ... 490-493
- nuts ... 427,431
- ornamentals ... 498,499,508

- small fruits 385
- tree fruits, citrus 400
- tree fruits, deciduous 407,408

Clover seed chalcid *Bruchophagus platypterus:*
- forage (alfalfa, etc.) 541

Clytoleptus albofasciatus — grape cane borer
Cnephasia longana — omnivorous leaftier (strawberry fruitworm)
Coccoidea — scales (scale crawlers)
Coccus hesperidum — brown soft scale, soft scale
Coccus pseudomagnoliarum — citricola scale
Coccus viridis — green scale

Codling moth *Cydia pomonella:*
- photo 16
- Bt chart 380
- biocontrols 56,59,60,68,69
- nuts 423-426,428-430
- ornamentals 496,498
- tree fruits, deciduous 402,412,414
- tree fruits, subtropical 420

Colaspis brunnea — grape colaspis
Coleophora laricella — larch casebearer
Coleophoridae — casebearers
Coleoptera — beetles
***Coleotechnites* spp.** — Jeffrey pine needleminer, Ponderosa pine needleminer
Colias eurytheme — alfalfa caterpillar
Collembola — collembola, springtails

Colorado potato beetle *Lepinotarsa decemlineata:*
- Bt chart 196
- artichokes 198
- beans 202
- biocontrols 55,57
- carrots, etc. 209,212
- celery 213,216
- cole crops 217,222,224
- cucurbits 226,227,230
- eggplant 231-235
- garlic 236
- greens 238,242
- horseradish 245
- lentils 247
- lettuce 248,253
- onions 256,259
- peas 260,261
- peppers 265,266,268,270
- potatoes 271-278
- radishes 278
- rhubarb 281
- soybeans 95
- spinach 287
- sweet corn 294
- sweet potatoes, yams 294,295
- table beets 297,300
- tomatoes 300-307

Common alfalfa weevil *Hypera postica* (alfalfa weevil):
- forage (alfalfa, etc.) 539

Common stalk borer *Papaipema nebris* (stalk borer):
- field corn, popcorn 79,84
- sweet corn 289,290

Conchylis hospes — banded sunflower moth

Cone midges *Cecidomyiidae:*
- ornamentals 496

Cone moths:
- ornamentals 496

Coneborers:
- ornamentals 495

Coneworms:
- Bt chart 488
- ornamentals 495,496,506

Conifer scale *Diaspidiotus coniferarum:*
- ornamentals 505

Conoderus falli — southern potato wireworm
Conotrachelus anaglypticus — cambium curculio
Conotrachelus aratus — pecan shootcurculio (hickory shoot curculio)
Conotrachelus crataegi — quince curculio
Conotrachelus nenuphar — plum curculio

Consperse stink bug *Euschistus conspersus:*
- forage (alfalfa, etc.) 538
- tree fruits, deciduous 408

Contarinia constricta — Douglas-fir needle midge
Contarinia johnsoni — grape blossom midge
Contarinia schulzi — sunflower midge
Contarinia sorghicola — sorghum midge

Cooley spruce gall adelgid *Adelges cooley*
- ornamentals 497,500,505

Corn borers:
- Bt chart 196
- biocontrols 69
- field corn, popcorn 77-80,82,83
- sorghum 157,158
- sweet corn 288-290,293

Corn earworm (headworm) *Helicoverpa zea:*
- photo 16
- Bt chart 76,128,196
- beans 202,203,205,206
- biocontrols 57,59,68,69
- carrots, etc. 209,210,212
- celery 214,215
- cole crops 218,219,222,225
- cucurbits 225
- eggplant 231
- field corn, popcorn 77-79,81-84
- forage (alfalfa, etc.) 537,538
- greens 238-240,242
- herbs/spices 243,244
- horseradish 245
- lentils 246,247
- lettuce 249-251,253
- okra 255
- onions 256
- ornamentals 496
- peanuts 150-152
- peas 260
- peppers 265,266,269
- radishes 278,279
- rhubarb 281,282
- sorghum 157-160
- southern peas 282,283
- soybeans 95,96,98-100
- spinach 284-287
- sugar beets 164
- sunflower 169
- sweet corn 288-293
- sweet potatoes, yams 294,295
- table beets 297-299
- tomatoes 301

Corn flea beetle *Chaetocnema pulicaria:*
- field corn, popcorn 79
- sweet corn 290

Corn leaf aphid *Rhopalosiphum maidis:*
- field corn, popcorn 77,79,82
- sorghum 158-160
- sweet corn 288,290,291

Corn rootworm (beetle) *Diabrotica* spp.:
- beans 202
- biocontrols 55,57,65,68
- field corn, popcorn 77-84
- forage (alfalfa, etc.) 538
- peanuts 152
- peas 260
- sorghum 158
- sugar beets 164
- sweet corn 288-294

Cosmopolites sordidus — banana corm borer, banana root borer
Cotinis nitida — green June beetle

Cotoneaster webworm *Athrips rancidella:*
- ornamentals 499

Cotton aphid *Aphis gossypii:*
- cotton 109-113,116-118,122

Cotton boll weevil *Anthonomus grandis:*
- biocontrols 55

Cotton bollworm *Helicoverpa zea* (bollworm):
- Bt chart 108
- biocontrols 59
- cotton 110-113,116,120

Cotton fleahopper *Psuedatomoscelis seriatus:*
- cotton 110-113,115,117,118,120,122

Cotton leafhopper:
- cotton 110

Cotton leafperforator *Bucculatrix thurberiella:*
- Bt chart 108
- biocontrols 56
- cotton 110-113,115-118,120,122

Cotton leafworm *Alabama argillacea:*
- Bt chart 108
- cotton 110,111,113,116-118,120

Cottonwood borer *Plectrodera scalator:*
- ornamentals 498

Cottonwood leaf beetle *Chrysomela scripta:*
- ornamentals 497

Cottony peach scale *Pulvinaria amygdali:*
- biocontrols 57
- tree fruits, deciduous 403,404,411,412,414,415

Cottonycushion scale *Icerya purchasi:*
- photo 13
- ornamentals 499
- tree fruits, citrus 393,400

Cowpea aphid *Aphis craccivora:*
- lentils 54,55

Cowpea curculio *Chalcodermus aeneus:*
- beans 202,203,205,206,209
- forage (alfalfa, etc.) 541
- forage (grasses, hay) 544
- lentils 247
- peas 260,263
- southern peas 282,283

Crambinae — sod webworm (lawn moth)

Cranberry blossomworm *Epiglaea apiata:*
- small fruits 381

Cranberry fireworm:
- small fruits 382

Cranberry fruitworm *Acrobasis vaccinii:*
- small fruits 382-384,386-388

Cranberry (twig) girdler *Chrysoteuchia topiaria:*
- biocontrols 65
- ornamentals 495
- small fruits 382

Cranberry root grubs *Lichnanthe vulpina:*
- biocontrols 65

Cranberry weevil *Anthonomus musculus:*
- small fruits 383

Crane flies *Tipulidae:*
- lawns/turf 493

Crested wheat bug *Irbisia* spp. (grass bugs):
- forage (grasses, hay) 544

Crickets *Gryllidae:*
- asparagus 200
- beans 203
- biocontrols 59
- carrots, etc. 209
- cole crops 218,219
- cotton 111
- cucurbits 226
- eggplant 232
- field corn, popcorn 78
- forage (alfalfa, etc.) 537
- greens 238
- horseradish 245
- lawns/turf 490,491,493,494
- lettuce 249,250
- okra 255
- ornamentals 499
- peas 261
- peppers 266
- potatoes 272
- radishes 279
- small fruits 383
- spinach 285
- sugar beets 162
- sweet corn 289
- table beets 298
- tomatoes 302

Crioceris asparagi — asparagus beetle
Crioceris duodecimpunctata — spotted asparagus beetle

Cross-striped cabbageworm *Evergestis rimosalis:*
- Bt chart 144,196
- cole crops 221,222,225
- radishes 281

Crown borers:
- biocontrols 65

Crown gall:
- biocontrols 56

Crown mite *Tyrophagus dimidiatus:*
- spinach 286,287

Cryptomyzus ribis — currant aphid
Cryptophlebia illepida — Koa seedworm

Cuban laurel thrips *Gynaikothrips ficorum:*
- biocontrols 55
- ornamentals 508

Cucumber beetles *Galerucinae:*
- beans 202-206,209
- biocontrols 55,68
- carrots, etc. 212
- celery 216
- cole crops 219,224
- cucurbits 225-231
- eggplant 232,234,235
- forage (alfalfa, etc.) 537,541,542
- forage (grasses, hay) 544
- greens 242
- horseradish 246
- lentils 247
- lettuce 249,250,253
- okra 256
- ornamentals 504
- peanuts 151
- peas 260,264
- peppers 265,266,269
- potatoes 271
- radishes 281
- sorghum 157
- southern peas 282,283
- soybeans 96
- spinach 287
- sweet corn 290,294
- sweet potatoes, yams 295,297
- table beets 299,300
- tomatoes 301,306
- tree fruits, citrus 393
- tree fruits, deciduous 405

Culicidae — mosquitoes

Curculio:
- biocontrols 56

Curculio caryae — pecan weevil
Curculionidae — pine weevil, root weevil, snout beetle, weevils

Currant aphid *Cryptomyzus ribis:*
- small fruits 387

Cutworms *Noctuidae:*
- Bt chart 94,144,196,380
- asparagus 199-201
- beans 202-206
- biocontrols 56,57,63,65,68,69
- carrots, etc. 209-212
- celery 213-216
- cole crops 217-221,225
- cotton 109-113,116,118-120
- cucurbits 225-230
- eggplant 232,234,235
- field corn, popcorn 77-83
- forage (alfalfa, etc.) 537,538,540-542
- forage (grasses, hay) 543
- garlic 236,237
- greens 238-240,242
- herbs/spices 243,244
- hops 146
- horseradish 245
- lawns/turf 489-494
- lentils 246,247
- lettuce 248-251,253
- mint 148
- noncropland 546
- nuts 424
- okra 255
- onions 256-259
- ornamentals 497-499,505,508
- pastures/rangeland 544

- peanuts ... 150-152
- peas ... 260-264
- peppers ... 265-268,270
- potatoes ... 271-277
- radishes ... 278,279,281
- rhubarb ... 281,282
- rice ... 154
- small fruits ... 381-383,386,391
- small grains ... 129,130,132
- sorghum ... 157,158
- southern peas ... 282
- soybeans ... 95,96,98-100
- spinach ... 284-287
- sugar beets ... 161-164
- sunflower ... 167,169
- sweet corn ... 288-293
- sweet potatoes, yams ... 294,295,297
- table beets ... 297-300
- tobacco ... 169,170
- tomatoes ... 301-306
- tree fruits, citrus ... 392-395,399
- tree fruits, deciduous ... 402,405,406,410
- tree fruits, subtropical ... 418

Cyclamen mite *Steneotarsonemus pallidus:*
- biocontrols ... 66
- ornamentals ... 499,500
- small fruits ... 384,385

Cyclocephala immaculata — southernmasked chafer (southern chafer)
***Cyclocephala* spp.** — chafer
Cydia caryana — hickory shuckworm
Cydia pomonella — codling moth
Cylas formicarius — sweetpotato weevil
Cynipidae — gall wasp
Cypress tip moth *Argyresthia cupressella:*
- ornamentals ... 497

Cyrtopeltis notatus — suckfly (tobacco suckfly)
Cyst nematode:
- biocontrols ... 56
- soybeans ... 97,101
- table beets ... 300

D

Dagger nematode *Xiphinema* spp.:
- biocontrols ... 56
- field corn, popcorn ... 79
- small fruits ... 385
- soybeans ... 97
- sweet corn ... 290

Daktulosphaira vitifoliae — grape phylloxera
Darkling (ground) beetle *Tenebrionidae:*
- asparagus ... 200
- beans ... 203
- carrots, etc. ... 209
- celery ... 214
- cole crops ... 218
- cotton ... 111
- cucurbits ... 226,229
- eggplant ... 232
- field corn, popcorn ... 78
- forage (alfalfa, etc.) ... 537
- forage (grasses, hay) ... 545
- greens ... 238
- horseradish ... 245
- lawns/turf ... 490
- lettuce ... 249
- noncropland ... 546
- okra ... 255
- ornamentals ... 497
- pastures/rangeland ... 544
- peanuts ... 151
- peas ... 260,261
- peppers ... 266
- potatoes ... 271,272
- radishes ... 278,279
- rice ... 154
- small fruits ... 382,383
- small grains ... 129
- sorghum ... 157
- soybeans ... 96
- spinach ... 285
- sugar beets ... 162
- sunflower ... 167
- sweet corn ... 288,289
- sweet potatoes, yams ... 295
- table beets ... 298
- tobacco ... 170
- tomatoes ... 302
- tree fruits, citrus ... 393

Dasineura gleditchae — honeylocust podgall midge
Dasineura oxycoccana — cranberry tipworm midge (tipworm)
Datana integerrima — walnut caterpillar
Datana major — azalea caterpillar
Datana ministra — yellownecked caterpillar
Decanthus fultoni — snowy tree cricket
Delia antiqua — onion maggot
Delia platura — seedcorn maggot
Delia radicum — cabbage maggot
Dendroctonus adjunctus — roundheaded pine beetle
Dendroctonus brevicomis — western pine beetle
Dendroctonus frontalis — southern pine beetle
Dendroctonus ponderosae — mountain pine beetle
Dendroctonus rufipennis — spruce pine beetle
Dermaptera — earwig
Desert spider mite *Tetranychus desertorum:*
- cotton ... 113,117,118

Desmia funeralis — grape leaffolder (grape leafroller)
Diabrotica *Diabrotica* spp.:
- beans ... 206
- cole crops ... 219
- cucurbits ... 227
- greens ... 239
- southern peas ... 283

Diabrotica balteata — banded cucumber beetle
Diabrotica barberi — northern corn rootworm
***Diabrotica* spp.** — corn rootworm (corn rootworm beetle)
Diabrotica undecimpunctata — spotted cucumber beetle, southern corn rootworm
Diabrotica virgifera — Mexican corn rootworm, western corn rootworm
Dialeurodes citri — citrus whitefly
Diamondback moth *Plutella xylostella:*
- Bt chart ... 94,128,144,196,380,488
- artichokes ... 198
- biocontrols ... 55,60,63
- carrots, etc. ... 209,211,212
- celery ... 213
- cole crops ... 217-222,224,225
- cucurbits ... 226,231
- eggplant ... 232
- garlic ... 236
- greens ... 238-242
- horseradish ... 245,246
- lentils ... 246-248
- lettuce ... 248-251,253
- mushrooms ... 245
- oil seed crops ... 150
- okra ... 255,256
- onions ... 256,257
- peas ... 264
- peppers ... 265
- potatoes ... 271
- radishes ... 278,280,281
- rhubarb ... 281
- spinach ... 284
- sweet potatoes, yams ... 294,295
- table beets ... 297,299
- tomatoes ... 301

Diaphania hyalinata — melonworm
Diaphania nitidalis — pickleworm
Diaprepes:
- ornamentals ... 508

Diaprepes abbreviatus — sugarcane rootstalk borer weevil (Apopka weevil, West Indian sugarcane root borer)
Diaspididae — armored scale
Diaspidiotus ancylus — Putnam scale
Diaspidiotus coniferarum — conifer scale
Diaspis bromeliae — pineapple scale
Diatraea grandiosella — southwestern corn borer
Diatraea saccharalis — sugarcane borer
Diceroprocta apache — Apache cicada
Dichomeris marginella — juniper webworm
Dictyospermum scale *Chrysomphalus dictyospermi:*
- biocontrols ... 62

Digger wasps *Sphecidae:*
- lawns/turf ... 491

Dioryctria zimmermani — Zimmerman pine moth
Diplopoda — millipedes
Diprionidae — pine sawflies
Diptacus gigantorhynchus — bigbeaked plum mite
Diptera — flies, maggots
Dipterous leafminer:
- beans ... 204
- carrots, etc. ... 210
- cole crops ... 220
- cucurbits ... 227
- lettuce ... 250
- ornamentals ... 499,508
- peas ... 261
- potatoes ... 272
- radishes ... 279
- spinach ... 285
- sugar beets ... 163
- table beets ... 298
- tomatoes ... 302,303

***Ditylenchus* spp.** — bulb, stem nematodes
Diuraphis noxia — Russian wheat aphid
Dock sawfly *Ametastegia glabrata:*
- tree fruits, deciduous ... 410

Dogwood borer *Synanthedon scitula:*
- biocontrols ... 68
- ornamentals ... 501
- tree fruits, deciduous ... 406

Dolichotetranychus floridanus — pineapple false spider mite (pineapple fruit mite)
Douglas-fir needle midge *Contarinia constricta:*
- ornamentals ... 500

Douglas-fir tussock moth *Orgyia pseudotsugata:*
- ornamentals ... 495,497

Driedfruit beetle *Carpophilus hemipterus:*
- tree fruits, subtropical ... 418,420

Drosophila melanogaster — vinegar fly
Drosophila spp. *Drosophila* spp.:
- small fruits ... 284,288
- tomatoes ... 301,304

Dryberry mite *Phyllocoptes gracilis:*
- small fruits ... 384

Dryocampa rubicunda — greenstriped mapleworm
Dusky stink bug *Euschistus tristigmus:*
- nuts ... 429

Dysaphis plantaginea — rosy apple aphid, rosy aphid
Dysmicoccus brevipes — pineapple mealybug

E

Earwig *Dermaptera:*
- biocontrols ... 56
- lawns/turf ... 489-491,493,494
- ornamentals ... 505,508

Earworm *Helicoverpa zea* (corn earworm):
- Bt chart ... 94,144
- biocontrols ... 56,57
- field corn, popcorn ... 81,83
- sugar beets ... 164
- sweet corn ... 291,293

Eastern grape leafhopper *Erythroneura comes:*
- small fruits ... 389

Eastern pine shoot borer *Eucosma gloriola:*
- ornamentals ... 496

Eastern pine shoot:
- ornamentals ... 503

Eastern spruce gall aphid *Adelges abietis* (Eastern spruce gall adelgid):
- ornamentals ... 497

Eastern tent caterpillar *Malacosoma americanum:*
- ornamentals ... 497,499,503,504,508,509
- tree fruits, deciduous ... 404,405

Edwardsiana prunicola — prune leafhopper
Edwardsiana rosae — rose leafhopper
Egira curialis — citrus cutworm
Egyptian alfalfa weevil *Hypera brunnipennis:*
- forage (alfalfa, etc.) ... 537-542

Eightspotted forester *Alypia octomaculata:*
- small fruits ... 382

Elasmopalpus lignosellus — lesser cornstalk borer
Elateridae — wireworms
Ello moth:
- Bt chart ... 488

Elm bark beetle:
- ornamentals ... 496,497,505

Elm leaf aphid *Tinocallis ulmifolii:*
- ornamentals ... 497

Elm leaf beetle *Pyrrhalta luteola:*
- biocontrols ... 55
- ornamentals ... 495,497-500,503,504,508,509

Elm spanworm *Ennomos subsignarius:*
- Bt chart ... 488
- ornamentals ... 497,498,503
- small fruits ... 382
- tree fruits, deciduous ... 410

Ematurga amitaria — brown cranberry spanworm (brown spanworm)
Empoasca abrupta — western potato leafhopper (melon leafhopper)
Empoasca fabae — potato leafhopper (bean leafhopper)
Empoasca solana — southern garden leafhopper
Endelomyia aethiops — roseslug
Endopiza viteana — grape berry moth
Endothenia albolineana — spruce needleminer
English grain aphid *Macrosiphum avenae:*
- field corn, popcorn ... 79
- forage (grasses, hay) ... 543
- small grains ... 129,131,132
- sorghum ... 159

Ennomos subsignarius — elm spanworm
Eotetranychus carpini — yellow spider mite (Carpini spider mite)
Eotetranychus clitus — azalea white mite
Eotetranychus hicoriae — pecan leaf scorch mite
Eotetranychus sexmaculatus — sixspotted spider mite
Eotetranychus willamettei — Willamette spider mite
Eotetranychus yumensis — Yuma spider mite
Epargyreus clarus — silverspotted skipper
Ephydridae — shore flies
Epicauta vittata — striped blister beetle
Epidiaspis leperii — Italian pear scale
Epiglaea apiata — cranberry blossomworm
Epilachna borealis — squash beetle
Epilachna varivestis — Mexican bean beetle
Epitrimerus pyri — pear rust mite
Epitrix cucumeris — potato flea beetle
Epitrix hirtipennis — tobacco flea beetle
Epitrix tuberis — tuber flea beetle
Epochra canadensis — gooseberry fruit fly
Erinnyis ello — poinsettia hornworm
Eriophyes cynodoniensis — bermudagrass mite
Eriophyes dianthi — carnation shoot mite
Eriophyes ficus — fig mite
Eriophyes neocynodonis — bermudagrass stunt mite
Eriophyes paradianthi — carnation bud mite
Eriophyes sheldoni — citrus bud mite
Eriophyid mite *Eriophyidae:*
- nuts ... 432
- ornamentals ... 497,506
- tree fruits, subtropical ... 422

Eriophyidae — blister mite, Eriophyidae mite
Eriosoma lanigerum — woolly apple aphid
Erynephala puncticollis — beet leaf beetle
Erythroneura comes — eastern grape leafhopper
Erythroneura elegantula — grape leafhopper, western grape leafhopper
Erythroneura variabilis — variegated leafhopper
Erythroneura ziczac — Virginiacreeper leafhopper

Essex skipper *Thymelicus lineola* (European skipper):
forage (alfalfa, etc.) 537
lawns/turf 489,490
pastures/rangeland 544
Estigmene acrea — saltmarsh caterpillar
Eucosma gloriola — eastern pine shoot borer
Eulecanium cerasorum — calico scale
Euonymus scale *Unaspis euonymi:*
photo 13
ornamentals 496,503
Euproctis chrysorrhoea — browntail moth
European alfalfa beetle *Subcoccinella vigintiquatuorpunctata:*
forage (alfalfa, etc.) 537
European apple sawfly *Hoplocampa testudinea:*
tree fruits, deciduous 403,404,406
European brown snail *Helix aspersa* (brown garden snail):
tree fruits, citrus 393
European chafer *Rhizotrogus majalis:*
biocontrols 55,68
lawns/turf 489-492
European corn borer *Ostrinia nubilalis:*
photo 16
Bt chart 76,94
beans 202,203,205,206
biocontrols 65,68
carrots, etc. 212
celery 215,216
cole crops 224,225
cotton 110-113,120
cucurbits 230
eggplant 231,232,235
field corn, popcorn 77-80,82-84
greens 242
herbs/spices 244
lentils 247
lettuce 249,253
onions 259
peppers 265,266,269,270
potatoes 271-277
rhubarb 282
sorghum 158
southern peas 282,283
soybeans 96
spinach 287
sweet corn 288-294
table beets 300
tomatoes 301,302,305,306
tree fruits, deciduous 406,410
European crane fly *Tipula paludosa:*
lawns/turf 490,494
pastures/rangeland 545
European earwig *Forficula auricularia:*
nuts 425
tree fruits, deciduous 405
European elm scale *Gossyparia spuria:*
ornamentals 496,500
European fruit lecanium *Parthenolecanium corni:*
biocontrols 57
nuts 425,431
small fruits 382,387,388
tree fruits, deciduous 403,404,406,
411,412,414,415
European fruit scale *Quadraspidiotus ostreaeformis:*
nuts 431
European pine sawfly *Neodiprion sertifer:*
ornamentals 494,508
European pine shoot moth *Rhyacionia buoliana:*
ornamentals 496,497,499,503
European raspberry aphid *Amphorophora rubi:*
small fruits 382
European red mite *Panonychus ulmi:*
photo 10
biocontrols 57
nuts 423,427,431-433
ornamentals 499,501,502,506-508
small fruits 385,391
tree fruits, deciduous 402,404-409,
411,412,414-418
European sawfly:
ornamentals 499
tree fruits, deciduous 410
Eurytomidae — seed chalcid
Euschistus conspersus — consperse stink bug
Euschistus tristigmus — dusky stink bug
Eutetranychus banksi — Texas citrus mite
Euxoa auxiliaris — army cutworm
Euxoa ochrogaster — redbacked cutworm
Euxoa scandens — white cutworm
Euzophera semifuneralis — American plum borer
Evergestis rimosalis — cross-striped cabbageworm
Eyespotted bud moth *Spilonota ocellana:*
nuts 425,428,429
tree fruits, deciduous 403-407,409,411,412

F

Fall armyworm *Spodoptera frugiperda:*
photo 16
Bt chart 76
beans 203,205,206
biocontrols 57,63
carrots, etc. 209-212
celery 213-215
cole crops 217-219,221,222,225
cotton 110-114,116-118,120
cucurbits 226,228,229
eggplant 232,234
field corn, popcorn 78,81-83
forage (alfalfa, etc.) 537,540,541
forage (grasses, hay) 543,544
greens 238,240-242
herbs/spices 243,244
horseradish 245
lawns/turf 489,490,492,493
lentils 246,247
lettuce 248,249,251-253
ornamentals 496,497,504,505
pastures/rangeland 544
peanuts 150-152
peppers 265,266,268
potatoes 271,272,274,275
radishes 278-280
rhubarb 281,282
small grains 129,132
sorghum 157-159
southern peas 282,283
soybeans 96-100
spinach 284-287
sugar beets 161,162,164
sunflower 167
sweet corn 288-293
sweet potatoes, yams 296
table beets 297-299
tobacco 169-171
tomatoes 301-305
Fall cankerworm *Alsophila pometaria:*
ornamentals 495,496,499
Fall webworm *Hyphantria cunea:*
lawns/turf 494
nuts 424-426,428
ornamentals 496,497,499,504,508,509
tree fruits, citrus 393
tree fruits, deciduous 402
False armyworm *Xylena nupera:*
small fruits 381
False chinch bug *Nysius raphanus:*
cotton 118
forage (grasses, hay) 544
potatoes 273-275,277
small grains 132
sorghum 157
tomatoes 304
False spider mites *Tenuipalpidae:*
tree fruits, citrus 399
Feltia subterranea — granulate cutworm
Fenusa pusilla — birch leafminer
Field crickets *Gryllus* spp.:
asparagus 200
beans 203
carrots, etc. 209
celery 214
cole crops 218
cucurbits 226
eggplant 232
field corn, popcorn 78
forage (alfalfa, etc.) 537
forage (grasses, hay) 543
greens 238
horseradish 245
lawns/turf 490
lettuce 249
noncropland 546
okra 255
ornamentals 497
pastures/rangeland 544
peanuts 151
peas 260
peppers 266
potatoes 271
radishes 278
rice 154
small fruits 382,387,388
small grains 129
sorghum 157
soybeans 96
spinach 285
sugar beets 162
sunflower 167
sweet corn 288
sweet potatoes, yams 295
table beets 298
tobacco 170
tomatoes 302
tree fruits, citrus 393
tree fruits, deciduous 405
Fiery skipper *Hylephila phyleus:*
lawns/turf 489,490
Fig mite *Eriophyes ficus:*
tree fruits, subtropical 422
Fig scale *Lepidosaphes conchiformis:*
tree fruits, subtropical 421
Filbert aphid *Myzocallis coryli:*
nuts 424,425,428
Filbert leafroller *Archips rosana:*
nuts 424,425
Filbert moth *Melissopus latiferreanus* (filbertworm):
nuts 425
Filbertworm *Melissopus latiferreanus:*
nuts 423-425,430
Fiorinia theae — tea scale
Fire ants *Solenopsis geminata:*
biocontrols 55,64
lawns/turf 489-494
noncropland 546,547
ornamentals 495,499,501,502,504,508
pastures/rangeland 544,545
small fruits 385
sorghum 158,159
tobacco 169
tree fruits, citrus 392,396
tree fruits, deciduous 409
Fireworm:
small fruits 381-383,391
tree fruits, subtropical 422
Flat (citrus flat) mite *Brevipalpus lewisi:*
nuts 432
tree fruits, citrus 398,400
tree fruits, deciduous 416
Flea beetles *Alticinae:*
artichokes 198,199
asparagus 199
beans 202,203,205,206,208
biocontrols 55,65,68
carrots, etc. 209-212
celery 213-216
cole crops 217-225
cotton 110,111
cucurbits 226-231
eggplant 231-235
field corn, popcorn 77-84
forage (alfalfa, etc.) 541,542
forage (grasses, hay) 543,544
garlic 236
greens 238-242
herbs/spices 243
horseradish 245,246
lawns/turf 489,491,493
lentils 246-248
lettuce 248-250,252,253
mint 148,149
oil seed crops 150
okra 255,256
onions 256,259
ornamentals 496-499,504,505,507,508
peas 263,264
peppers 265-270
potatoes 271-278
radishes 278-281
rhubarb 281,282
small fruits 381-383,386,388
small grains 129
sorghum 158,160
southern peas 282-284
soybeans 100
spinach 284-287
sugar beets 161-164
sunflower 169
sweet corn 288-294
sweet potatoes, yams 294-297
table beets 297-300
tobacco 169-172
tomatoes 301-306
tree fruits, citrus 392
tree fruits, deciduous 402
Flea weevil *Rhynchaeninae:*
ornamentals 498
Fleahoppers *Miridae:*
beans 202
biocontrols 55,56
carrots 212
celery 216
cole crops 224
cotton 109-118,120-122
cucurbits 230,231
eggplant 235
field corn, popcorn 84
forage (alfalfa, etc.) 537,539,542
forage (grasses, hay) 544
greens 242
oil seed crops 150
ornamentals 504,507
peanuts 152
peas 264
peppers 269
potatoes 277
small grains 132,133
sorghum 161
soybeans 99,100
spinach 287
sweet corn 294
table beets 300
tobacco 171
tomatoes 306
Fleas *Siphonaptera:*
biocontrols 65
lawns/turf 489-491,493,494
Fletcher scale *Parthenolecanium fletcheri:*
ornamentals 503,504
Flies *Diptera:*
biocontrols 64,68
forage (alfalfa, etc.) 542
forage (grasses, hay) 544
ornamentals 507
pastures/rangeland 546
Florida red scale *Chrysomphalus aonidum:*
ornamentals 503,504
tree fruits, citrus 393,396-398

Florida wax scale *Ceroplastes floridensis:*
- ornamentals ... 500

Flower thrips *Frankliniella tritici:*
- garlic ... 236
- onions ... 257
- ornamentals ... 504

Forbes scale (crawler) *Quadraspidiotus forbesi:*
- tree fruits, deciduous ... 403-405,407,411,412

Forest tent caterpillar *Malacosoma disstria:*
- ornamentals ... 499,500,504

Formicidae — ants

Fourlined leaf bug *Poecilocapsus lineatus* (fourlined plant bug):
- ornamentals ... 503

Frankliniella fusca — tobacco thrips

Frankliniella occidentalis — western fruit flies

Frankliniella tritici — flower thrips

Frit fly *Oscinella frit:*
- lawns/turf ... 491

Frosted scale *Parthenolecanium:*
- biocontrols ... 58
- nuts ... 424,425,431,433
- tree fruits, deciduous ... 415

Fruit flies *Tephritidae:*
- biocontrols ... 56,58
- ornamentals ... 507
- small fruits ... 381,391
- tree fruits, citrus ... 392,393,397,399
- tree fruits, deciduous ... 402-404,406,407, 410-412,416
- tree fruit, subtropical ... 418,420-422

Fruit hornworm:
- tomatoes ... 305

Fruittree leafroller *Archips argyrospila:*
- biocontrols ... 57,58
- nuts ... 424,425,431
- small fruits ... 381
- tree fruits, citrus ... 393-396,399
- tree fruits, deciduous ... 403-407,409-412,414,415
- tree fruits, subtropical ... 421

Fruitworms:
- Bt chart ... 380
- biocontrols ... 56,69
- small fruits ... 381-384,386-388,391
- tomatoes ... 301,302,304,305
- tree fruits, deciduous ... 402-406,408-410,412,414

Fuller rose beetle *Pantomorus cervinus:*
- ornamentals ... 497,498,507
- tree fruits, citrus ... 393-395,398,399

Fuller rose weevil *Pantomorus cervinus:*
- biocontrols ... 55
- tree fruits, citrus ... 392

Fumibotys fumalis — mint root borer

Fungus gnats (fungus flies) *Mycetophilidae:*
- Bt chart ... 196,488
- biocontrols ... 57,63,65,68
- mushrooms ... 255
- ornamentals ... 498-501,506,508

G

Galerucinae — cucumber beetle

Gall midge:
- ornamentals ... 497,508

Gall mite:
- biocontrols ... 57

Gall wasp *Cynipidae:*
- biocontrols ... 56
- ornamentals ... 497

Garden fleahopper *Halticus bractatus:*
- carrots, etc. ... 212
- celery ... 216
- cole crops ... 224
- cucurbits ... 230,231
- eggplant ... 235
- garlic ... 269
- greens ... 242
- peas ... 264
- potatoes ... 277
- spinach ... 287
- sweet corn ... 294
- table beets ... 300
- tomatoes ... 306

Garden symphylans *Scutigerella immaculata:*
- beans ... 207
- field corn, popcorn ... 82
- potatoes ... 276
- small fruits ... 385
- sweet corn ... 292

Garden tortrix *Ptycholoma peritana:*
- tree fruits, citrus ... 395

Garden webworm *Achyra rantalis:*
- beans ... 202,206
- cotton ... 110,117,118
- eggplant ... 232
- peanuts ... 151
- peas ... 260
- peppers ... 265,266
- southern peas ... 283
- soybeans ... 95,99
- table beets ... 299
- tomatoes ... 301

Gastropoda — slugs, snails

Geometridae — cankerworms, geometrid moth (inchworm, spanworms), loopers

Giant palmetto weevil:
- biocontrols ... 68

Giant whitefly:
- biocontrols ... 54,55

Girdlers:
- biocontrols ... 68

Gladiolus thrips *Thrips simplex:*
- ornamentals ... 496

***Glischrochilus* spp.** — picnic beetle

Globodera tabacum — tobacco cyst nematode

Glover scale *Lepidosaphes gloveri:*
- tree fruits, citrus ... 393,396,398,400

Glyptoscelis squamulata — grape bud beetle

Gnopothrips fuscus — slash pine flower thrips

Gooseberry fruitworm *Zophodia convolutella:*
- small fruits ... 391

Gossyparia spuria — European elm scale

Gracillariidae — tentiform leafminer

Granulate cutworm *Feltia subterranea:*
- cucurbits ... 228
- peanuts ... 150,152
- sweet corn ... 290
- tomatoes ... 305

Grape berry moth *Endopiza viteana:*
- Bt chart ... 380
- biocontrols ... 60
- small fruits ... 381-384,386,388,389

Grape blossom midge *Contarinia johnsoni:*
- small fruits ... 389

Grape cane borer *Clytoleptus albofasciatus:*
- small fruits ... 386

Grape cane girdler *Ampeloglypter ater:*
- small fruits ... 381,386

Grape colaspis *Colaspis brunnea:*
- soybeans ... 96

Grape leaffolder *Desmia funeralis:*
- Bt chart ... 380
- small fruits ... 381-384,386,388

Grape leafhopper *Erythroneura elegantula:*
- biocontrols ... 54,55
- small fruits ... 382,383,385,386,388

Grape mealybug *Pseudococcus maritimus:*
- biocontrols ... 54,55
- small fruits ... 301,388
- tree fruits, deciduous ... 401

Grape phylloxera *Daktulosphaira vitifoliae:*
- small fruits ... 385

Grape root borer *Vitacea polistiformis:*
- small fruits ... 383

Grapeleaf skeletonizer *Harrisina americana:*
- Bt chart ... 380
- small fruits ... 381,383,384,386,388

Grapevine aphid *Aphis illinoisensis:*
- small fruits ... 389

Grapholita molesta — Oriental fruit moth

Grapholita packardi — cherry fruitworm

Grapholita prunivora — lesser appleworm

***Graphognathus* spp.** — whitefringed beetle (adults)

Grasshoppers *Acrididae:*
- asparagus ... 200
- beans ... 202,203,205
- biocontrols ... 55-57,59,68
- carrots, etc. ... 209
- celery ... 214
- cole crops ... 217-219,225
- cotton ... 110-112,117,118
- cucurbits ... 225,226
- eggplant ... 232
- field corn, popcorn ... 77-82
- forage (alfalfa, etc.) ... 537-542
- forage (grasses, hay) ... 543,544
- greens ... 238
- horseradish ... 245
- lawns/turf ... 489-491,493,494
- lentils ... 247
- lettuce ... 249,253
- noncropland ... 546,547
- okra ... 255
- ornamentals ... 407,495,498
- pastures/rangeland ... 544,545
- peanuts ... 150,151
- peas ... 260,261,263,264
- peppers ... 265,266
- potatoes ... 271-275
- radishes ... 278,279
- rice ... 154-156
- safflower ... 157
- small fruits ... 382,383,386
- small grains ... 129,130,132,133
- sorghum ... 157,158,160
- southern peas ... 282,284
- soybeans ... 95-100
- spinach ... 285
- sugar beets ... 162-164
- sunflower ... 167-169
- sweet corn ... 288-290,292
- sweet potatoes, yams ... 295
- table beets ... 298
- tobacco ... 169,170
- tomatoes ... 301,302,305
- tree fruits, citrus ... 392-394
- tree fruits, deciduous ... 402,405

Green apple aphid *Aphis pomi* (apple aphid):
- biocontrols ... 58
- ornamentals ... 509
- tree fruits, deciduous ... 404,407,411,413

Green cloverworm *Plathypena scabra:*
- Bt chart ... 94,108,128,144,196,380
- beans ... 202,203,206
- carrots, etc. ... 212
- celery ... 215
- cole crops ... 225
- field corn, popcorn ... 79
- forage (alfalfa, etc.) ... 537,538,541,542
- greens ... 242
- herbs/spices ... 244
- lentils ... 247
- lettuce ... 249,253
- peanuts ... 151,152
- peas ... 260,262,264
- rhubarb ... 282
- southern peas ... 282-284
- soybeans ... 95-100
- spinach ... 287

Green fruitworm *Lithophane antennata:*
- small fruits ... 381
- tree fruits, deciduous ... 402-406,408-410,412,414

Green June beetle *Cotinis nitida:*
- biocontrols ... 54,55
- lawns/turf ... 489
- pastures/rangeland ... 544
- tobacco ... 170

Green June bug:
- tobacco ... 170

Green peach aphid *Myzus persicae:*
- photo ... 13
- artichokes ... 198,199
- asparagus ... 199,201
- beans ... 207
- biocontrols ... 54,55
- carrots, etc. ... 209,210,212
- celery ... 213-216
- cole crops ... 217,223
- cotton ... 117
- cucurbits ... 226,230
- eggplant ... 232-235
- forage (alfalfa, etc.) ... 542
- garlic ... 236,237
- greens ... 238,242
- herbs/spices ... 243,244
- horseradish ... 245,246
- lentils ... 246
- lettuce ... 248,251,252
- mushrooms ... 254
- okra ... 255,256
- onions ... 256,259
- peas ... 264
- peppers ... 265,268-270
- potatoes ... 271,273,276,277
- radishes ... 278,280
- safflower ... 156
- small fruits ... 381
- spinach ... 284,286,287
- sugar beets ... 161,163,164
- sweet potatoes, yams ... 294,296
- table beets ... 297,299
- tobacco ... 169,171
- tomatoes ... 301,303,305,306
- tree fruits, deciduous ... 402,409-412,417
- watercress ... 307

Green stink bug *Acrosternum hilare:*
- beans ... 205,209
- cotton ... 112
- eggplant ... 233,235
- nuts ... 428
- potatoes ... 273,277
- southern peas ... 283
- tomatoes ... 301

Greenbug *Schizaphis graminum:*
- biocontrols ... 54,55
- cotton ... 118
- field corn, popcorn ... 79
- forage (alfalfa etc.) ... 542
- forage (grasses, hay) ... 543
- lawns/turf ... 489,493
- oil seed crops ... 150
- peanuts ... 152
- small grains ... 129-132
- sorghum ... 158-161
- soybeans ... 99
- tobacco ... 171

Greenhouse leaftier *Udea rubigalis:*
- biocontrols ... 54,55
- tree fruits, subtropical ... 420

Greenhouse thrips *Heliothrips haemorrhoidalis:*
- biocontrols ... 54,55
- tree fruits, subtropical ... 420

Greenhouse whitefly *Trialeurodes vaporariorum:*
- photo ... 13
- artichokes ... 198
- beans ... 202
- biocontrols ... 54,55,58,63,69
- carrots, etc. ... 209
- celery ... 213
- cole crops ... 218
- cucurbits ... 226
- eggplant ... 232
- garlic ... 236
- greens ... 238
- horseradish ... 245
- lentils ... 247
- lettuce ... 249
- onions ... 257

ornamentals 496,500,501
peas 260
peppers 265
potatoes 271
radishes 278
rhubarb 281
soybeans 95
spinach 284
sweet potatoes, yams 295
table beets 297
tomatoes 301
tree fruits, citrus 393
tree fruits, deciduous 402
Greenstriped mapleworm *Dryocampa rubicunda:*
Bt chart 488
ornamentals 497
Gretchena bolliana — pecan bud moth
Ground beetles *Carabidae:*
cole crops 219
peas 264
Grubs:
biocontrols 56,65
field corn, popcorn 78,84
lawns/turf 493
sweet corn 289
Gryllidae — crickets
Gryllotalpidae — mole crickets
***Gryllus* spp.** — field crickets
Gynaikothrips ficorum — Cuban laurel thrips
Gypsy moth (caterpillar) *Lymantria dispar:*
Bt chart 380,488
biocontrols 60,68
ornamentals 495-500,503,504,508,509
small fruits 381
tree fruits, citrus 393
tree fruits, deciduous 402,404,405,410

H

Hackberry nipplegall maker *Pachypsylla celtidismamma:*
ornamentals 497
Halticus bractatus — garden fleahopper
Halcocera:
tree fruits, citrus 395
Haplorhynchites aeneus — sunflower headcipper weevil
Harlequin bug *Murgantia histrionica:*
carrots, etc. 209,212
celery 216
cole crops 218,221,224,225
cucurbits 230,231
eggplant 235
greens 238-243
horseradish 245,246
lentils 248
lettuce 249,253
peppers 269
radishes 279,281
spinach 285,287
sweet corn 294
table beets 298
tomatoes 306
Harrisina americana — grapeleaf skeletonizer
Harrisina brillians — western grapeleaf skeletonizer
Harvester ants *Pogonomyrmex* spp.:
lawns/turf 489,492
noncropland 547
ornamentals 502
pastures/rangeland 545
tobacco 169
Headclipper weevil *Haplorhynchites aeneus:*
sunflower 167
Headworm:
Bt chart 76,144,196
sorghum 157
sunflower 167
Helicoverpa:
Bt chart 108,144
Helicoverpa zea — bollworm (cotton bollworm), corn earworm (earworm, headworm, podworm, sorghum headworm), tomato fruitworm
Heliothis *Heliothis* spp.:
Bt chart 128,144,196,488
biocontrols 63
ornamentals 505,506
sunflower 167
Heliothis virescens — tobacco budworm
Heliothrips haemorrhoidalis — greenhouse thrips
Helix aspersa — brown garden snail (European brown snail)
Hellula rogatalis — cabbage webworm
Hemiberlesia lataniae — latania scale
Hemileuca maia — buck moth
Hemileuca oliviae — range caterpillar
Hemlock chermes *Adelges tsugae:*
ornamentals 499
Herpetogramma phaeopteralis — tropical sod webworm
Hesperiidae — skippers
Hessian fly *Mayetiola destructor:*
small grains 130,131,133
Heterocampa guttivitta — saddled prominent caterpillar
Heterodera schachtii — sugarbeet cyst nematode
Hickory mite:
nuts 433
Hickory shuckworm *Cydia caryana:*
biocontrols 60
nuts 423-426,428,429,432
Hickory tussock moth *Lophocampa caryae:*
ornamentals 508
Holly bud moth *Rhopobota naevana* (blackheaded fireworm):
ornamentals 497,499
Holly leafminer *Phytomyza ilicis:*
ornamentals 496,497,500
Homadaula anisocentra — mimosa webworm
Homoeosoma electellum — sunflower moth (sunflower head moth):
Honeylocust podgall midge *Dasineura gleditchiae:*
ornamentals 508
Hop aphid *Phorodon humuli:*
biocontrols 54,55
hops 146
tree fruits, deciduous 409
Hop vine borer *Hydroecia immanis:*
field corn, popcorn 83
Hoplia:
biocontrols 83
Hoplocampa testudinea — European apple sawfly
***Hoplolaimus* spp.** — lance nematode
Horistonotus uhlerii — sand wireworm
Hornworms *Sphingidae:*
Bt chart 94,108,128,145,197,380
artichokes 198
beans 202
biocontrols 56
carrots, etc. 209
celery 213
cole crops 218
eggplant 232
garlic 236
greens 238
horseradish 245
lentils 247
lettuce 249
onions 257
ornamentals 496,497
peanuts 151
peas 260
peppers 265,266,268,270
potatoes 271
radishes 278
rhubarb 281
small fruits 381
soybeans 95
spinach 284
sweet potatoes, yams 295
table beets 297
tobacco 169-171
tomatoes 301-305
Hulstia undulatella — sugarbeet crown borer (beet crown borer)
Hyalopterus pruni — mealy plum aphid
Hydrellia griseola — rice leafminer
Hydroecia immanis — hop vine borer
Hylephila phyleus — fiery skipper
Hylobius pales — pales weevil
Hypena humuli — hop looper
Hypera brunnipennis — Egyptian alfalfa weevil
Hypera meles — clover head weevil
Hypera postica — alfalfa weevil
Hypera punctata — clover leaf weevil
Hyperodes weevil *Listronotus* spp.:
lawns/turf 490,491,493
ornamentals 505
Hyphantria cunea — fall webworm
Hysteroneura setariae — rusty plum aphid

I

Icerya purchasi — cottonycushion scale
Ichthyura inclusa — poplar tentmaker
Imported cabbageworm *Artogeia rapae:*
Bt chart 94,145,197
biocontrols 67
carrots, etc. 209,211
celery 213
cole crops 217-225
cucurbits 226
eggplant 232,234
field corn, popcorn 82
garlic 236
greens 238-243
herbs/spices 243
horseradish 245
lentils 246,248
lettuce 248-253
mushrooms 254
okra 255,256
onions 257
peas 260,264
peppers 265,268
potatoes 271
radishes 278-281
rhubarb 281
small fruits 387
spinach 284,286,287
sweet potatoes, yams 294
table beets 297,299
tomatoes 301
Imported crucifer weevil *Baris lepidii:*
horseradish 246
Imported currantworm *Nematus ribesii:*
small fruits 387,391
Imported fire ants:
lawns/turf 489-491,493,494
ornamentals 495,499,508
pastures/rangeland 544
tobacco 169
tree fruits, citrus 392
Inchworms *Geometridae* (geometrid moths):
ornamentals 504
Io moth *Automeris io:*
Bt chart 488
Ips engraver *Ips* spp. (Ips beetles):
ornamentals 497
***Ips* spp.** — Ips beetles (Ips engraver)
Irbisia solani — black grass bug
***Irbisia* spp.** — grass bugs (crested wheat bug)
Iris borer *Macronoctua onusta:*
biocontrols 68
Italian pear scale *Epidiaspis leperii:*
nuts 431,433
tree fruits, deciduous 415
Ivy scale *Aspidiotus nerii:*
biocontrols 62

J

Jack pine budworm *Choristoneura pinus:*
ornamentals 497
Japanese beetle *Popillia japonica:*
photo 11
asparagus 199,201
beans 203,205,206
biocontrols 55,56,65,68
carrots, etc. 209,211,212
celery 213,216
cole crops 217,219,222,224
cucurbits 226,230,231
eggplant 232,234,235
field corn, popcorn 78,79,82
forage (alfalfa, etc.) 537,541
forage (grasses, hay) 544
garlic 236
greens 238,241,242
horseradish 245
lawns/turf 490-494
lentils 246,247
lettuce 248,249,252
okra 255,256
onions 256
ornamentals 495-499,501-504,507-509
peanuts 151
peas 264
peppers 265,268
potatoes 271
radishes 278,280
rhubarb 281
small fruits 381-383,386-389,391
southern peas 282,283
soybeans 95,96,98,99
spinach 284,286
sunflower 167
sweet corn 288,290,292,294
sweet potatoes, yams 294
table beets 297,299
tobacco 169,170
tomatoes 301,305,306
tree fruits, deciduous 402,404,405,410-412,416
Jeffrey pine needleminer *Coleotechnites* spp.:
ornamentals 497
June beetle *Scarabaeidae:*
biocontrols 68
lawns/turf 490,499
onions 256
ornamentals 496-499
small fruits 381,382
tobacco 169,170
tree fruits, deciduous 402,405
Juniper scale *Carulaspis juniperi:*
ornamentals 496,503,504
Juniper webworm *Dichomeris marginella:*
ornamentals 497,499

K

Katydid *Tettigoniidae:*
ornamentals 497,498
tree fruits, citrus 392,394,395,399
Kermes pubescens — oak leaf kermes (oak kermes)
Koa seedworm *Cryptophlebia illepida:*
biocontrols 61

L

Lace bugs *Tingidae:*
biocontrols 57
eggplant 232-234
ornamentals 497-504,506,508
peppers 266
potatoes 272
tomatoes 302
tree fruits, deciduous 408

Lambdina fiscellaria — western hemlock looper (hemlock looper), western oak looper (oak looper)
Lance nematode *Hoplolaimus* spp.:
- biocontrols 56
- field corn, popcorn 79
- soybeans 97,101
- sweet corn 290

Large looper:
- beans 204

Lasioderma serricorne — cigarette beetle (tobacco beetle)
Latania scale *Hemiberlesia lataniae:*
- tree fruits, subtropical 420

Lawn billbug *Sphenophorus* spp.:
- lawns/turf 491

Lawn chinch bug *Blissus* spp. (chinch bug complex):
- lawns/turf 491

Lawn moth *Crambinae* (sod webworm):
- lawns/turf 490,491,493,494

Leaf beetles:
- Bt chart 488
- biocontrols 57
- ornamentals 506

Leafcurl aphid:
- biocontrols 58

Leafcurl plum aphid *Brachycaudus helichrysi:*
- tree fruits, deciduous 407,409,412

Leaffooted plant bug *Leptoglossus phyllopus:*
- nuts 429
- potatoes 273,277

Leafhoppers *Cicadellidae:*
- artichokes 198,199
- asparagus 199,201,202
- beans 202-208
- biocontrols 54-58,64
- carrots, etc. 209-212
- celery 213-216
- cole crops 217-219,221-225
- cotton 110,111,113,114,116-121
- cucurbits 225-231
- eggplant 232-235
- field corn, popcorn 78,80
- forage (alfalfa, etc.) 537-542
- forage (grasses, hay) 543,544
- garlic 236,237
- greens 238-242
- herbs/spices 243,244
- hops 146,147
- horseradish 245,246
- lawns/turf 489-491,493,494
- lentils 246-248
- lettuce 249,251,253
- mint 148,149
- nuts 424,427,429,431
- okra 255,256
- onions 256,259,260
- ornamentals 495-501,503-508
- pastures/rangeland 545
- peanuts 150-153
- peas 260-264
- peppers 265,266,268,269
- potatoes 271-278
- radishes 278-281
- rhubarb 281,282
- rice 154
- safflower 156
- small fruits 381,382,384,386-391
- small grains 132
- southern peas 282-284
- soybeans 95-97,99,100
- spinach 284-287
- sugar beets 161-165
- sunflower 167,169
- sweet corn 288,290-294
- sweet potatoes, yams 294-297
- table beets 297-300
- tobacco 169
- tomatoes 301-306
- tree fruits, citrus 392,393,398
- tree fruits, deciduous 401-408,410,412-414,416,418
- tree fruits, subtropical 418,420-422
- watercress 307

Leafminers:
- artichokes 198,199
- asparagus 199,201,202
- beans 202-205,207
- biocontrols 56-58,63
- carrots, etc. 209,210,212
- celery 213-216
- cole crops 217,218,220,222-224
- cotton 110,119,121
- cucurbits 226-231
- eggplant 232-235
- field corn, popcorn 83
- garlic 236,237
- greens 238,240-242
- herbs/spices 243,244
- hops 146,147
- horseradish 245,246
- lentils 246,248
- lettuce 248-250,252,253
- mushrooms 254
- nuts 424
- okra 255,256
- onions 256,257,259,260
- ornamentals 494-499,500-503,505-508
- peanuts 151,153
- peas 260,261,264
- peppers 265-267,269,270
- potatoes 271-273,276,277
- radishes 278-281
- rhubarb 281,282
- rice 155
- small fruits 381,390
- southern peas 282,284
- soybeans 95,96
- spinach 284,285,287,288
- sugar beets 161,163-165
- sweet corn 293,294
- sweet potatoes, yams 294,296,297
- table beets 297-300
- tobacco 169,172
- tomatoes 300-307
- tree fruits, citrus 392,393
- tree fruits, deciduous 401-407,410,418
- tree fruits, subtropical 418,419
- watercress 307

Leafperforators:
- cotton 110,111,113,121

Leafrollers:
- Bt chart 145,380,488
- asparagus 199
- beans 203
- biocontrols 56-59,61
- carrots, etc. 209
- celery 213
- cole crops 217
- cotton 110,111,118
- cucurbits 226,228,231
- eggplant 232
- greens 238
- lettuce 248
- nuts 424,429,431
- onions 256
- ornamentals 495-499,504-508
- peppers 265,266
- potatoes 271
- radishes 278
- small fruits 381-384,386-388,391
- soybeans 96
- spinach 284
- sweet potatoes, yams 294
- table beets 297
- tomatoes 301
- tree fruits, citrus 392,395
- tree fruits, deciduous 402-406,408-412,416
- tree fruits, subtropical 418,419,421

Leafskeletonizer:
- ornamentals 498,499

Leaftiers:
- forage (alfalfa, etc.) 540
- greens 242
- nuts 425
- ornamentals 499
- small fruits 382,384,386,388

Leafworms:
- biocontrols 69
- cotton 110,111

Leatherleaf fern borer *Undulambia polystichalis:*
- ornamentals 500

Lecanium pine needle scale crawler *Lecanium* spp.:
- ornamentals 499

Lecanium scale (crawler) *Lecanium* spp.:
- biocontrols 58
- small fruits 382
- tree fruits, deciduous 404,405

***Lecanium* spp.** — lecanium scale
Lema trilineata — threelined potato beetle
Lepidoptera — caterpillars (lepidopterous larvae, leafeating caterpillars)
Lepidopterous larvae *Lepidoptera* (caterpillars):
- tree fruits, subtropical 421

Lepidosaphes beckii — purple scale
Lepidosaphes camelliae — camellia scale
Lepidosaphes conchiformis — fig scale
Lepidosaphes gloveri — glover scale
Lepidosaphes ulmi — oystershell scale
Lepinotarsa decemlineata — Colorado potato beetle
Leptocoris trivittatus — boxelder bug
Leptoglossus phyllopus — leaffooted bug
Lesion nematode *Pratylenchus* spp.:
- biocontrols 56
- carrots, etc. 213
- cucurbits 231
- field corn, popcorn 79
- peanuts 154
- potatoes 276,278
- soybeans 97,101
- sweet corn 290
- tobacco 172
- tree fruits, deciduous 401,418

Lesser appleworm *Grapholita prunivora:*
- tree fruits, deciduous 402,404,406,410,412

Lesser cornstalk borer *Elasmopalpus lignosellus:*
- beans 204
- field corn, popcorn 78-80
- peanuts 150-152
- sorghum 158
- soybeans 96
- sweet corn 289,290

Lesser peachtree borer *Synanthedon pictipes:*
- tree fruits, deciduous 402-406,408,409, 411-414,417

Lesser webworm *Tetralopha subcanalis:*
- nuts 424,425

Lettuce root aphid:
- lettuce 251

Lichnenthe vulpina — cranberry root grubs
Lilac borer *Podosesia syringae:*
- ornamentals 501

Limabean pod borer *Etiella zinckenella:*
- beans 203
- southern peas 283

Limothrips denticornis — barley thrips
Liriomyza brassicae — serpentine leafminer
Liriomyza leafminer fly:
- ornamentals 506

Liriomyza sativae — vegetable leafminer
***Liriomyza* spp.** — melon leafminer
Listroderes costirostris — vegetable weevil
Listronotus oregonensis — carrot weevil
Lithophane antennata — green fruitworm
Little leaf notcher *artipus floridanus:*
- tree fruits, citrus 392,394

Locust borer *Megacyllene robiniae:*
- ornamentals 497

Locusts *Acrididae:*
- biocontrols 55,59,68

Longitarsus waterhousei — mint flea beetle
Longtailed mealybug *Pseudococcus longispinus:*
- biocontrols 55

Loopers *Geometridae, Noctuidae:*
- Bt chart 94,108,128,145,197,380,488,536
- artichokes 198,199
- asparagus 199,200
- beans 202,204,205,207,208
- biocontrols 56,57
- carrots, etc. 209,212
- celery 213,214,216
- cole crops 217,219-221,224
- cotton 110,113,119,120
- cucurbits 226-228,230,231
- eggplant 231,232,235
- forage (alfalfa, etc.) 540,542
- garlic 236,237
- greens 238-240,242
- herbs/spices 243,244
- hops 146,147
- horseradish 245,246
- lentils 246,248
- lettuce 248-251,253
- mint 148,149
- nuts 424
- okra 255,256
- onions 256,259
- ornamentals 496,507
- peanuts 150-153
- peas 260-262,264
- peppers 265,268-270
- potatoes 274,277
- radishes 278,281
- rhubarb 281,282
- small fruits 381
- southern peas 284
- soybeans 95,100
- spinach 284,286,287
- sugar beets 161,165
- sunflower 168,169
- sweet corn 294
- sweet potatoes, yams 294,296,297
- table beets 297,300
- tobacco 169,171
- tomatoes 301,304,306
- tree fruits, subtropical 418-420

Lophocampa caryae — hickory tussock moth
Loxagrotis albicosta — western bean cutworm
Loxostege commixtalis — alfalfa webworm
Loxostege sticticalis — beet webworm (sugarbeet webworm)
Lucerne moth *Nomophila nearctia:*
- lawns/turf 489,490

Lygidea mendax — apple red bug (red bug)
Lygocoris communis — pear plant bug
Lygus bugs *Lygus* spp.:
- artichokes 199
- asparagus 201
- beans 202-208
- biocontrols 55,56
- carrots, etc. 209,212
- celery 214-216
- cole crops 219,224
- cotton 109-118,120-122
- cucurbits 230,231
- eggplant 232,234,235
- forage (alfalfa, etc.) 537-542
- garlic 237
- greens 238,242
- hops 147
- horseradish 246
- lentils 247,248
- lettuce 249,250,253
- oil seed crops 150
- okra 256
- onions 259
- ornamentals 495,504,507,508
- peanuts 152
- peas 261,263,264
- peppers 266,269,270
- potatoes 275,277
- radishes 281
- rhubarb 282
- safflower 156,157
- small fruits 383,384,386-388

small grains 132
southern peas 283,284
soybeans 99
spinach 287
sugar beets 164
sweet corn 294
sweet potatoes, yams 296
table beets 299,300
tobacco 171
tomatoes 302,305,306
tree fruits, deciduous 404-406
Lygus hesperus — western lygus bug
Lygus lineolaris — tarnished plant bug
***Lygus* spp.** — lygus bugs
Lymantria dispar — gypsy moth
Lymantriidae — tussock moth

M

Macadamia nut borer:
biocontrols 61
Macalla thrysisalis — mahogany webworm
Macrodactylus subspinosus — rose chafer
Macrosiphoniella sanborni — chrysanthemum aphid
Macrosiphum avenae — English grain aphid
Macrosiphum euphorbiae — potato aphid
Macrosiphum rosae — rose aphid
Macrosteles fascifrons — aster leafhopper
Magicicada septendecim — periodical cicada
Maggots *Diptera:*
biocontrols 56,68
peppers 267
Magnolia leafminer *Phyllocnistis magnoliella:*
ornamentals 503
Magnolia scale *Neolecanium cornuparvum:*
ornamentals 503
Mahogany webworm *Macalla thrysisalis:*
ornamentals 497
Maize billbug *Sphenophorus maidis:*
field corn, popcorn 79
sweet corn 290
Malacosoma americanum — eastern tent caterpillar
Malacosoma disstria — forest tent caterpillar
***Malacosoma* spp.** — tent caterpillar
Maladera castanea — Asiatic garden beetle
Manduca quinquemaculata — tomato hornworm
Manduca sexta — tobacco hornworm
Maple leafcutter *Paraclemensia acerifoliella:*
ornamentals 497
Marmara salictella — citrus peelminer
Masked chafers *Cyclocephala* spp.:
biocontrols 65
Matsucoccus resinosae — red pine scale
May beetle *Scarabaedae:*
nuts 424
ornamentals 504
Mayetiola destructor — Hessian fly
McDaniel spider mite *Tetranychus McDanieli:*
nuts 423,427,432
small fruits 385
tree fruits, deciduous 402,405,407-409,415-418
Meadow spittlebug *Philaenus spumarius:*
cole crops 219
field corn, popcorn 79
forage (alfalfa etc.) 538,539,542
lettuce 249
small fruits 382,383,385
sugar beets 163
sunflower 167
tomatoes 302
Mealy plum aphid *Hyalopterus pruni:*
biocontrols 58
tree fruits, deciduous 405-407,411,412
Mealybugs *Pseudococcidae:*
biocontrols 54-58,63-65
hops 146,147
lawns/turf 492-494
nuts 424,426,428,429
ornamentals 495-499,501-504,506-508
potatoes 274
small fruits 381,382,386-390
tree fruits, citrus 392-394,398,400
tree fruits, deciduous 401,404,407,410-413
tree fruits, subtropical 418-420
Medfly *Ceratitis capitata* (Mediterranean fruit fly):
photo 12
Mediterranean fruit fly *Ceratitis capitata:*
tree fruits, citrus 397
Megacyllene robiniae — locust borer
Megalopyge opercularis — puss caterpillar
Melanispis obscura — obscure scale
Melanocallis caryaefoliae — black pecan aphid
Melanophila fulvoguttata — western hemlock bark borer (hemlock borer)
Melanoplus femurrubrum — redlegged grasshopper
Melissopus latiferreanus — filbert moth, filbertworm
Melittia cucurbitae — squash vine borer
Meloidae — blister beetle
***Meloidogyne* spp.** — root-knot nematode
Melon aphid *Aphis gossypii:*
biocontrols 54,55
cucurbits 226,228
Melonworm *Diaphania hyalinata:*
Bt chart 145,197
cucurbits 226-228,230,231
eggplant 234
Membracidae — treehoppers
Merhynchites bicolore — rose curculio
Merodon equestris — narcissus bulb fly
Mesolecanium nigrofasciatum — terrapin scale
Mexican bean beetle *Epilachna varivestis:*
beans 202-209
biocontrols 66,67
carrots, etc. 209
celery 213
cole crops 217,224
cucurbits 226,231
eggplant 232,233
forage (alfalfa, etc.) 437,541
forage (grasses, hay) 544
greens 238,241,242
lentils 247
lettuce 248
peanuts 151
peas 260,263
peppers 265,268
southern peas 282,283
soybeans 95-100
spinach 284
sugar beets 165
table beets 297,299
tomatoes 301
Midges *Chironomidae:*
ornamentals 499
Millipedes *Diplopoda:*
biocontrols 55
lawns/turf 489-491,493,494
Mimosa webworm *Homadaula anisocentra:*
ornamentals 497,499,500,509
Mindarus abietinus — balsam twig aphid
Mineola moth *Acrobasis indigenella* (leaf crumpler):
tree fruits, deciduous 403,404,406,409
Mint flea beetle *Longitarsus waterhousei:*
biocontrols 65
Mint root borer *Fumibotys fumalis:*
mint 148
Miridae — fleahoppers, plant bugs (mirids)
Mirids *Miridae* (plant bugs):
cotton 112
tree fruits, subtropical 421
Mites (mite eggs) *Acari:*
artichokes 199
asparagus 201,202
beans 203,205-208
biocontrols 55-58,61,63,65,66
carrots, etc. 211,212
celery 213-216
cole crops 221-224
cotton 109-114,117-121
cucurbits 229-231
eggplant 234,235
field corn, popcorn 77,80,82,84
forage (alfalfa, etc.) 539-543
garlic 237
greens 241,242
herbs/spices 244
hops 146,147
horseradish 246
lawns/turf 491-494
lentils 248
lettuce 250,252,253
mint 149
mushrooms 254
nuts 423,425,427-429,431-433
oil seed crops 150
okra 256
onions 259,260
ornamentals 496,497,499,500-508
peanuts 152,153
peas 263,264
peppers 265,269,270
potatoes 275-277
radishes 280,281
rhubarb 282
small fruits 381,385,387-391
small grains 130,132,133
sorghum 160,161
southern peas 284
soybeans 97,99,100
spinach 284,286,287
sugar beets 164,165
sunflower 168,169
sweet corn 292-294
sweet potatoes, yams 296,297
table beets 299,300
tobacco 172
tomatoes 303,305,306
tree fruits, citrus 392-395,397-401
tree fruits, deciduous 401,402,404-409,411-418
tree fruits, subtropical 421,422
watercress 307
***Mocis* spp.** — striped grass looper
Mole crickets *Gryllotalpidae:*
biocontrols 55,57,58
carrots, etc. 210
cole crops 220
greens 239
lawns/turf 489-494
lettuce 250
ornamentals 505
peppers 267
radishes 279
small fruits 384
table beets 298
tobacco 170,171
tomatoes 303
Monarthropalpus buxi — boxwood leafminer
***Monellia* spp.** — yellow pecan aphid
Monophadnoides geniculatus — raspberry sawfly
Monterey pine scale *Physokermes insignicola:*
ornamentals 503,504
Mormon crickets *Anabrus simplex:*
biocontrols 55,59
noncropland 546
pastures/rangeland 544
Morningglory leafminer *Bedellia somnulentella:*
sweet potatoes, yams 296
Mosquitoes *Culicidae:*
Bt chart 145,536
biocontrols 66,68
lawns/turf 489,490
noncropland 546,547
ornamentals 499
pastures/rangeland 545
Moth eggs:
biocontrols 69
Moths *Psychodidae:*
biocontrols 64
greens 238
herbs/spices 243
nuts 424
ornamentals 499
radishes 278
Mountain pine beetle *Dendroctonus ponderosae:*
ornamentals 497
Mugho (pine needle) scales:
ornamentals 508
Murgantia histrionica — harlequin bug
Mushroom flies *Mycetophilidae:*
biocontrols 255
mushrooms 254,255
Mycetophilidae — fungus gnats, mushroom flies
Myzocallis coryli — filbert aphid
Myzus cerasi — black cherry aphid
Myzus persicae — green peach aphid

N

Nantucket pine tip moth *Rhyacionia frustrana:*
ornamentals 495-497,500,503,505,508,509
Narcissus bulb fly *Merodon equestris:*
ornamentals 508
Native elm bark beetle *Hylurgopinus rufipes:*
ornamentals 497
Navel orangeworm *Amyelois transitella:*
Bt chart 380
biocontrols 64,67
nuts 423-426,428,430,432
Needle nematode:
biocontrols 56
Needle rust:
biocontrols 58
Needleminers:
ornamentals 497,498
Nematis ribesii — imported currantworm
Nematoda — nematodes
Nematodes *Nematoda:*
asparagus 201
beans 207,208
biocontrols 55,56,61,62,68
carrots, etc. 213
celery 217
cole crops 223
cotton 119,121
cucurbits 229,231
eggplant 234,235
field corn, popcorn 79,82
garlic 237
lawns/turf 490,493
nuts 423
okra 256
ornamentals 505
peanuts 152-154
peppers 209,270
potatoes 275,276,278
small fruits 385,389,390
sorghum 158,161
soybeans 97,101
sugar beets 162,165
sugarcane 166
sweet corn 290,292
sweet potatoes, yams 296,297
table beets 300
tobacco 170-172
tomatoes 307
tree fruits, citrus 398,400
tree fruits, deciduous 401,413,418
tree fruits, subtropical 420,421
Neodiprion nanulus — red pine sawfly
Neodiprion rugifrons — redheaded pine sawfly
Neodiprion sertifer — European pine sawfly
Neolecanium cornuparvum — magnolia scale

Nephasia longana — strawberry rootworm
Nezara viridula — southern green stink bug
Nitidulid beetle *Nitidulitae* (sap beetles):
  beans ... 204,206
  peas ... 261
  tree fruits, subtropical ... 420
Nitidulitae — sap beetles (Nitidulid beetle)
Noctuidae — cutworms, loopers
Nomophila nearctia — lucerne moth
Northern corn rootworm *Diabrotica barberi:*
  photo ... 11
  field corn, popcorn ... 77,78,80
  sweet corn ... 288,289,291
Northern masked chafer:
  biocontrols ... 55,68
  lawns/turf ... 492
Northern pine weevil *Pissodes approximatus:*
  ornamentals ... 495
Notostraca — tadpole shrimp
Nuculaspis californica — black pineleaf scale
Nursery mite:
  tree fruits, deciduous ... 409
Nymphalis antiopa — mourningcloak butterfly (spiny elm caterpillar):
Nysius raphanus — false chinch bug
***Nysius* spp.** — seed bugs

O

Oak kermes *Kermes pubescens* (oak leaf kermes):
  ornamentals ... 503,504
Oak leaf skeletonizer:
  ornamentals ... 497
Oak leafminer *Cameraria:*
  ornamentals ... 497
Oak looper *Lambdina fiscellaria* (western oak looper):
  ornamentals ... 499
Oak moth *Phryganidia californica* (California oakworm):
  ornamentals ... 497
Oak skeletonizer *Bucculatrix ainsliella:*
  ornamentals ... 497
Oak webworm *Archips fervidana:*
  ornamentals ... 499
Oakworm *Anisota:*
  Bt chart ... 488
  ornamentals ... 499
Oat birdcherry aphid *Rhopalosiphum padi:*
  field corn, popcorn ... 77,79
  small grains ... 129-131
  sweet corn ... 288
Oberea myops — blueberry tip borer
Obliquebanded leafroller *Choristoneura rosaceana:*
  nuts ... 423,425,430
  ornamentals ... 499
  small fruits ... 381,382,386
  tree fruits, deciduous ... 402,406,410,412,414,416
Obscure root weevil *Sciopithes obscurus:*
  biocontrols ... 55
  ornamentals ... 499,504
  small fruits ... 381,382
Obscure scale *Melanaspis obscura:*
  biocontrols ... 57
  nuts ... 430
Odontopus calceatus — yellow-poplar weevil
Oebalus pugnax — rice stink bug
Oecanthinae — tree crickets
Oleander caterpillar *Syntomeida epilais:*
  ornamentals ... 497
Oleander moth:
  Bt chart ... 488
  ornamentals ... 498,499
Oleander scale *Aspidiotus nerii:*
  biocontrols ... 62
  nuts ... 431
  tree fruits, deciduous ... 415
  tree fruits, subtropical ... 421
Olethreutinae — pine shoot moth, pine tip moth
Oligonychus ilicis — Southern red mite
Oligonychus pratensis — Banks grass mite
Oligonychus punicae — avocado brown mite
Oligonychus ununguis — spruce spider mite
Olive ash borer:
  ornamentals ... 497
Olive scale (crawler) *Parlatoria oleae:*
  nuts ... 431
  ornamentals ... 496
  tree fruits, deciduous ... 405,407,415
  tree fruits, subtropical ... 418,421
Omnivorous leafroller *Platynota stultana:*
  Bt chart ... 197
  biocontrols ... 59,61
  cucurbits ... 228,231
  ornamentals ... 505,506
  peppers ... 266
  small fruits ... 381-384,386
  tree fruits, deciduous ... 410
  tree fruits, subtropical ... 419
Omnivorous leaftier *Cnephasia longana:*
  Bt chart ... 145
  forage (alfalfa, etc.) ... 540
  nuts ... 425
  ornamentals ... 499
  small fruits ... 382,384,386,388
Omnivorous looper *Sabulodes aegrotata:*
  tree fruits, subtropical ... 419,420
Oncideres cingulata — twig girdler
Onion maggot *Delia antiqua:*
  biocontrols ... 68
  garlic ... 236,237
  greens ... 238
  lettuce ... 249
  onions ... 257-260
  spinach ... 284
Onion thrips *Thrips tabaci:*
  biocontrols ... 54,55
  cole crops ... 219
  garlic ... 236,237
  lettuce ... 249,250
  onions ... 257-259
Oniscoidea — pillbugs, sowbugs
Operophtera brumata — winter moth
Orange tortrix *Argyrotaenia citrana:*
  Bt chart ... 380
  nuts ... 428,429
  ornamentals ... 497
  small fruits ... 382,383,386
  tree fruits, citrus ... 393-396,399
  tree fruits, deciduous ... 403-406,409-412
  tree fruits, subtropical ... 420
Orangedog *Papilio cresphontes:*
  Bt chart ... 380
  tree fruits, citrus ... 420
Orangestriped oakworm *Anisota senatoria:*
  ornamentals ... 497
Orangeworm:
  tree fruits, citrus ... 397
Orchid weevil *Diorymerellus laevimargo:*
  ornamentals ... 498,508
Orgyia leucostigma — whitemarked tussock moth
Orgyia pseudotsugata — Douglas-fir tussock moth
Orgyia vetusta — western tussock moth
Oriental beetle:
  biocontrols ... 65,68
  lawns/turf ... 492
  ornamentals ... 502
Oriental fruit moth *Grapholita molesta:*
  Bt chart ... 380
  biocontrols ... 59-61
  tree fruits, deciduous ... 402-407,410-414
Oscinella frit — frit fly
Ostrinia nubilalis — European corn borer
Otiorhynchus ovatus — strawberry root weevil
Otiorhynchus sulcatus — black vine weevil (taxus weevil)
Oulema melanopus — cereal leaf beetle
Overwintering boll weevil:
  cotton ... 115
Oystershell scale *Lepidosaphes ulmi:*
  ornamentals ... 496,503,506
  small fruits ... 382
  tree fruits, deciduous ... 404,405,412

P

Pachnaeus litus — citrus root weevil
Pachylobius picivorus — pitch-eating weevil
Pachypsylla celtidismamma — hackberry nipplegall maker
Pacific flatheaded borer *Chrysobothris mali:*
  tree fruits, deciduous ... 406
Pacific spide mite *Tetranychus pacificus:*
  beans ... 204
  cotton ... 109,113,119
  cucurbits ... 228
  field corn, popcorn ... 84,85
  hops ... 146
  nuts ... 423,427,431-433
  ornamentals ... 502,507
  peppers ... 267
  small fruits ... 384,385,390,391
  small grains ... 133
  sorghum ... 161
  soybeans ... 100
  sugar beets ... 161,165
  sweet corn ... 294
  tomatoes ... 303
  tree fruits, deciduous ... 402,407,408,415-417
Painted lady butterfly *Vanessa cardui:*
  beans ... 202
  peas ... 260
  soybeans ... 96
  sunflower ... 167
Paleacrita vernata — spring cankerworm
Pales weevil *Hylobius pales:*
  ornamentals ... 495,502,503,509
Pandemis leafroller:
  tree fruits, deciduous ... 406,412,416
Pandemis moth:
  tree fruits, deciduous ... 409
Pandemis pyrusana — apple pandemis
Panonychus citri — citrus red mite (purple mite)
Panonychus ulmi — European red mite
Pantomorus cervinus — Fuller rose beetle
Papaipema nebris — stalk borer (common stalk borer)
Papilio cresphontes — orangedog
Papilio polyxenes — black swallowtail, parsleyworm (celeryworm)
Papilio zelicaon — California orangedog
Paraclemensia acerifoliella — maple leafcutter
Paratrioza cockerelli — potato psyllid, tomato psyllid
Parlatoria oleae — olive scale (parlatoria scale)
Parlatoria pergandii — chaff scale
Parlatoria scale *Parlatoria olea* (olive scale):
  nuts ... 426,431
  tree fruits, deciduous ... 407,415
  tree fruits, subtropical ... 421
Parthenolecanium — frosted scale
Parthenolecanium corni — European fruit lecanium (brown apricot scale)
Parthenolecanium fletcheri — Fletcher scale
Pavement ants *Tetramorium caespitum:*
  nuts ... 426
Pea aphid *Acyrthosiphon pisum:*
  beans ... 202
  biocontrols ... 54,55
  forage (alfalfa, etc.) ... 538-542
  lettuce ... 242
  peas ... 260,262,263
Pea leaf weevil *Sitona lineatus:*
  beans ... 206
  peas ... 261-263
  small fruits ... 382
Pea weevil *Bruchus pisorum:*
  beans ... 206
  forage (alfalfa, etc.) ... 541
  forage (grasses, hay) ... 544
  peas ... 261-263
Peach silver mite *Aculus cornutus* (peachtree silver mite):
  tree fruits, deciduous ... 409,415,417
Peach twig borer *Anarsia lineatella:*
  Bt chart ... 380
  biocontrols ... 59,60
  nuts ... 423-430,432
  tree fruits, deciduous ... 402-410,413-417
Peachtree borer *Synanthedon exitiosa:*
  biocontrols ... 61,68
  ornamentals ... 496-498
  tree fruits, deciduous... 402-404,406-409,413,416,417
Pear bud mite *Phytoptus pyri* (pearleaf blister mite):
  tree fruits, deciduous ... 404
Pear midge *Contarina pyrivora:*
  tree fruits, deciduous ... 403
Pear plant bug *Lygocoris communis:*
  tree fruits, deciduous ... 412
Pear psylla *Psylla pyricola:*
  biocontrols ... 55,57
  nuts ... 432
  ornamentals ... 498
  tree fruits, deciduous ... 401,402,404,406-408, 410-417
Pear rust mite *Epitrimerus pyri:*
  nuts ... 432
  tree fruits, deciduous ... 404,405,408,416-418
Pear scale*:*
  tree fruits, deciduous ... 415
Pear thrips *Taeniothrips inconsequens:*
  biocontrols ... 54,55
Pearleaf blister mite *Phytoptus pyri:*
  biocontrols ... 57
  tree fruits, deciduous ... 404,408,414,415
Pearslug *Caliroa cerasi* (pear sawfly):
  ornamentals ... 499
  tree fruits, deciduous ... 402,412
Pecan aphid:
  nuts ... 423
Pecan bud moth *Gretchena bolliana:*
  nuts ... 428,429
Pecan casebearer:
  nuts ... 424
Pecan leaf casebearer *Acrobasis juglandis:*
  nuts ... 428,429
Pecan leaf phylloxera *Pylloxera notabilis:*
  nuts ... 423-425,427,433
Pecan leaf scorch mite *Eotetranychus Hicoriae:*
  nuts ... 424
  ornamentals ... 508
Pecan mite:
  nuts ... 433
Pecan nut casebearer *Acrobasis nuxvorella:*
  nuts ... 423-429
Pecan phylloxera *Phylloxer devastrix:*
  nuts ... 428,429
Pecan spittlebug *Clastoptera achatina:*
  nuts ... 423-425
Pecan stem phylloxera:
  nuts ... 423-425
Pecan weevil *Curculio caryae:*
  biocontrols ... 55
  nuts ... 423,426,428,429,431
Pectinophora gossypiella — pink bollworm
Pelopidas thrax — banana skipper
Pemphigus populivenae — sugarbeet root aphid
Pentatomidae — stink bugs
Penthaleus major — winter grain mite
Pepper maggot *Zonosemata electa:*
  peppers ... 268,270
Pepper weevil *Anthonomus eugenii:*
  beans ... 202
  biocontrols ... 54
  eggplant ... 232
  peas ... 260
  peppers ... 265,266,269,270
  tomatoes ... 301
Peppertree psyllid*:*
  ornamentals ... 498
Peridroma saucia — variegated cutworm

Periodical cicada *Magicicada septendecim:*
ornamentals....497
tree fruits, deciduous....402,404-406
Petrobia latens — brown wheat mite
Petrobia mite *Petrobia* spp.:
carrots, etc. 212
garlic....237
onions....260
Phenacoccus aceris — apple mealybug
Pheidole megacephalaf — bigheaded ant
Philaenus spumarius — meadow spittlebug
Phorid mushroom flies *Phoridae:*
mushrooms....254
Phoridae — phorid mushroom flies
Phorodon humuli — hop aphid
Phryganidia californica — California oak moth, California oakworm
Phthorimaea operculella — potato tuberworm
Phyllocnistis citrella — citrus leafminer
Phyllocoptes gracilis — dryberry mite
Phyllocoptruta oleivora — citrus rust mite
Phyllonorycter blancardella — spotted tentiform leafminer
Phyllonorycter felinella — western tentiform leafminer
Phyllophaga spp.:
lawns/turf....492
ornamentals....502
Phyllotreta cruciferae — cabbage flea beetle
Phyllotreta striolata — striped flea beetle
Phylloxera devastrix — pecan phylloxera
Phylloxera notabilis — pecan leaf phylloxera
Phylloxera spp.:
biocontrols....56
nuts....425
Physokermes insignicola — Monterey pine scale
Phytomyza ilicis — holly leafminer
Phytomyza syngenesiae — chrysanthemum leafminer
Phytoptus pyri — pear bud mite, pearleaf blister mite
Pickleworm *Diaphania nitidalis:*
Bt chart....145,197
cucurbits....225-231
eggplant....234
Picnic beetle *Carpophilus* spp., *Glischrochilus* spp.:
field corn, popcorn....81
sweet corn....291
Pillbugs *Oniscoidea:*
lawns/turf....491,493,494
ornamentals....498,499,508
Pin nematode:
biocontrols....56
celery....217
Pine butterfly:
Bt chart....488
Pine chafer *Anomala oblivia:*
ornamentals....495
Pine conelet bug *Platylygus luridus:*
ornamentals....495
Pine leaf chermid *Pineus pinifoliae:*
ornamentals....495
Pine leaf scale *Chionaspis pinifoliae* (pine needle scale):
ornamentals....503,504
Pine looper:
ornamentals....497
Pine needle midge:
ornamentals....495
Pine needle scale *Chionaspis pinifoliae:*
ornamentals....499,503-505,508
Pine sawflies *Diprionidae:*
ornamentals....496,497,499
Pine shoot moth *Olethreutinae:*
ornamentals....498
Pine spittlebug *Aphrophora parallela:*
ornamentals....497,509
Pine tip moth *Olethreutinae:*
ornamentals....496,498-500,503
Pine weevils *Curculionidae:*
biocontrols....68
Pineapple fruit mite *Dolichotetranychus floridanus* (pineapple false spider mite):
tree fruits, subtropical....419
Pineapple mealybug *Dysmicoccus brevipes:*
tree fruits, subtropical....419
Pineus pinifoliae — pine leaf chermid
Pink bollworm *Pectinophora gossypiella:*
photo....15
biocontrols....59-61
cotton....109-113,116,118,120-122
Pink citrus rust mite *Aculops pelekassi:*
tree fruits, citrus....399
Pinworms *Keiferia lycopersicella* (tomato pinworm):
Bt chart....197
tomatoes....301,304,305
Pissodes approximatus — northern pine weevil
Pissodes strobi — white pine weevil
Pitch pine tip moth *Rhyacionia rigidana:*
ornamentals....497
Pitch-eating weevil *Pachylobius picivorus:*
ornamentals....502,503
Planococcus citri — citrus mealybug
Plant bugs *Miridae:*
artichokes....199
asparagus....201,202
beans....202,203,206-208
biocontrols....55-57
carrots, etc. 209,211,212
celery....213-216
cole crops....218,219,223,224
cotton....109-113,115-118,120-122
cucurbits....225,230,231
eggplant....232,234,235
forage (alfalfa, etc.)....537-539,541,542
garlic....236,237
greens....238,241,242
herbs/spices....244
hops....147
horseradish....245,246
lentils....247,248
lettuce....249,252,253
mint....149
nuts....431
oil seed crops....150
okra....256
onions....257,259
ornamentals....497-499,503,504,506-508
peanuts....152
peas....263,264
peppers....266,269,270
potatoes....271-273,275-278
radishes....279-281
rhubarb....282
safflower....156
small fruits....383,385
small grains....132
southern peas....282,284
soybeans....96,99
spinach....285-287
sugar beets....164
sweet corn....288,290,294
sweet potatoes, yams....296,297
table beets....298-300
tobacco....171
tomatoes....302,305,306
tree fruits, citrus....392
tree fruits, deciduous....402-406,408-410,412,414
tree fruits, subtropical....420,421
watercress....307
Planthoppers:
biocontrols....54
Plant-pathogenic nematode:
biocontrols....55
Plathypena scabra — green cloverworm
Platylygus luridus — pine conelet bug
Platynota flavedana:
tree fruits, deciduous....403,404
Platynota idaeusalis — tufted apple bud moth
Platynota stultana — omnivorous leafroller
Platyptilia carduidactyla — artichoke plume moth
Plectrodera scalator — cottonwood borer
Plum curculio *Conotrachelus nenuphar:*
photo....11
biocontrols....55
ornamentals....498
small fruits....382,386-388
tree fruits, deciduous....402-406,410-414
Plum nursery mite *Aculus fockeui* (plum rust mite):
tree fruits, deciduous....417
Plum rust mite *Aculus fockeui:*
tree fruits, deciduous....409
Plume moth *Pterophoridae:*
artichokes....198
Plutella xylostella — diamondback moth
Podosesia syringae — lilac borer
Podworms *Helicoverpa zea* (corn earworm):
Bt chart....94,145,197
soybeans....98
Poecilocapsus lineatus — fourlined plant bug (fourlined leaf bug)
***Pogonomyrmex* spp.** — harvester ants
Poinsettia hornworm *Erinnyis ello:*
ornamentals....497
Polyphagotarsonemus latus — broad mite
Ponderosa pine needleminer *Coleotechnites* spp.:
ornamentals....495
Pontia prodice — southern cabbageworm
Popillia japonica — Japanese beetle
Poplar clearwing borer:
biocontrols....68
Poplar tentmaker *Ichthyura inclusa:*
ornamentals....497,509
Potato aphid *Macrosiphum euphorbiae:*
biocontrols....54,55
cucurbits....226
potatoes....271,276
tomatoes....301,302
Potato flea beetle *Epitrix cucumeris:*
artichokes....198
carrots, etc. 209
horseradish....245
lentils....247
potatoes....271,273,274,276,277
radishes....278
sunflower....169
sweet potatoes, yams....295
table beets....297
Potato leafhopper *Empoasca fabae:*
photo....13
beans....202,205,206
biocontrols....54,55
forage (alfalfa, etc.)....538-542
ornamentals....503
peanuts....150-152
peas....260
peppers....266
potatoes....271,273-278
small fruits....387-389
southern peas....283
soybeans....95,96,99
table beets....299
tree fruits, deciduous....406
Potato psyllid *Paratrioza cockerelli:*
biocontrols....54,55
potatoes....271,273-277
Potato tuberworm *Phthorimaea operculella:*
potatoes....271,273-277
***Pratylenchus* spp.** — lesion nematode (root-lesion nematode)
Prinoxystus robiniae — carpenterworm
Pristiphora abbreviata — California pear sawfly (California pearslug)
Privet mite *Brevipalpus obovatus:*
ornamentals....499
Prune leafhopper *Edwardsiana prunicola:*
tree fruits, deciduous....405
Pseudaletia unipuncta — armyworm (true armyworm)
Psuedatomoscelis seriatus — cotton fleahopper
Pseudaulacaspis pentagona — white peach scale
Pseudococcidae — mealybugs
Pseudococcus longispinus — longtailed mealybug
Pseudococcus maritimus — grape mealybug (apricot mealybug)
Pseudoplusia includens — soybean looper (soybean caterpillar)
Psila rosae — carrot rust fly
Psorophora columbiae — dark ricefield mosquito
Psychodidae — moths
Psylla mali — apple sucker
Psylla pyricola — pear psylla
Psyllid *Psyllidae:*
artichokes....198
beans....202
biocontrols....55-58
carrots, etc. 209
cucurbits....226
eggplant....232,235
greens....238
horseradish....245
lentils....247
lettuce....248
ornamentals....496,497,506
peppers....265,270
potatoes....271,274,276
radishes....278
small fruits....381
soybeans....95
spinach....284
sweet potatoes, yams....294,295
table beets....297
tomatoes....301,306
tree fruits, citrus....392,393
tree fruits, deciduous....402
Psyllidae — psyllid
Ptycholoma peritana — garden tortrix
Pulvinaria amygdali — cottony peach scale
Pulvinaria scale *Pulvinaria* spp.:
ornamentals....496
Purple scale *Lepidosaphes beckii:*
tree fruits, citrus....393,396-398,400
Puss caterpillar *Megalopyge opercularis:*
ornamentals....497
Putnam scale *Diaspidiotus ancylus:*
nuts....431
ornamentals....496
small fruits....390
tree fruits, deciduous....403,415
Pyroderces rileyi — pink scavenger caterpillar
Pyrrhalta luteola — elm leaf beetle

Q

Quadraspidiotus forbesi — Forbes scale
Quadraspidiotus juglansregiae — walnut scale
Quadraspidiotus ostreaeformis — European fruit scale
Quadraspidiotus perniciosus — San Jose scale

R

Range caterpillar *Hemileuca oliviae:*
Bt chart....536
forage (grasses, hay)....544
noncropland....546
pastures/rangeland....544,545
Range crane fly *Tipula simplex:*
noncropland....546
pastures/rangeland....544
Rangeland caterpillar:
Bt chart....128
Rapid plant bug *Adelphocoris rapidus:*
cotton....110
Raspberry crown borer *Pennisetia marginata:*
biocontrols....68
small fruits....381,384
Raspberry fruitworm *Byturus* spp.:
small fruits....384,391

Raspberry sawfly *Monophadnoides geniculatus:*
small fruits ... 384
Red and black flat mite *Brevipalpus phoenicis:*
nuts ... 423
Red imported fire ant *Solenopsis invicta:*
tree fruits, citrus ... 394
Red pine (needle) scale *Matsucoccus resinosae:*
ornamentals ... 508
Red pine sawfly *Neodiprion nanulus:*
ornamentals ... 495
Red scale:
biocontrols ... 65
tree fruits, citrus ... 398
Red spider mite *Tetranychidae* (spider mites):
beans ... 205,208
biocontrols ... 65
cole crops ... 225
cotton ... 120,121
field corn, popcorn ... 84,85
forage (alfalfa, etc.) ... 543
greens ... 243
hops ... 147
mushrooms ... 254
nuts ... 428-432
ornamentals ... 497,499,505,506
peanuts ... 154
peas ... 264
peppers ... 270
potatoes ... 277
small fruits ... 391
small grains ... 133
sorghum ... 161
southern peas ... 283
soybeans ... 100
sugar beets ... 165
sweet corn ... 294
table beets ... 300
tree fruits, citrus ... 393,397,398,400
tree fruits, deciduous ... 404,415,416
Redbacked cutworm *Euxoa ochrogaster:*
asparagus ... 200
Redbanded leafroller *Argyrotaenia velutinana:*
small fruits ... 381,382,386,389
tree fruits, deciduous ... 402-406,410,412,414
Redberry mite *Acalitus essigi:*
small fruits ... 382
Redheaded pine sawfly *Neodiprion rugifrons:*
photo ... 15
ornamentals ... 495
Redhumped caterpillar *Schizura concinna:*
nuts ... 424,426,427,432
ornamentals ... 497,498,503
tree fruits, deciduous ... 410
Redhumped oakworm *Symmerista canicosta:*
ornamentals ... 497
Redlegged grasshopper *Melanoplus femurrubrum:*
forage (alfalfa, etc.) ... 538
Rednecked borer:
small fruits ... 388,391
Rednecked peanutworm *Stegasta bosqueella:*
peanuts ... 150,151
Redstriped fireworm *Aroga trialbamaculella:*
small fruits ... 386
Reniform nematode *Rotylenchulus reniformis:*
biocontrols ... 56
nuts ... 423
tomatoes ... 307
Rhagoletis cingulata — cherry fruit fly, cherry maggot
Rhagoletis completa — walnut huskfly
Rhagoletis indifferens — western cherry fruit fly
Rhagoletis mendax — blueberry fruit fly, blueberry maggot
Rhagoletis pomonella — apple maggot
Rhagoletis suavis —walnut husk maggot
Rhipicephalus sanguineus — brown dog tick
Rhizotrogus majalis — European chafer
Rhodesgrass scale *Antonina graminis* (rhodesgrass mealybug):
lawns/turf ... 491
Rhopalosiphum fitchii — apple grain aphid
Rhopalosiphum maidis — corn leaf aphid
Rhopalosiphum padi — oat birdcherry aphid
Rhopobota naevana — blackheaded fireworm (holly bud moth)
Rhyacionia buoliana — European pine shoot moth
Rhyacionia frustrana — Nantucket pine tip moth
Rhyacionia rigidana — pitch pine tip moth
Rhyacionia subtropica — subtropical pine tip moth
Rhynchaeninae — flea weevil
Rice borer:
sugarcane ... 166
Rice leafminer *Hydrellia griseola:*
rice ... 155,156
Rice stink bug *Oebalus pugnax:*
rice ... 155
Rindworms:
Bt chart ... 145
cucurbits ... 225,226,228,230,231
eggplant ... 234
Ring nematode:
biocontrols ... 56,68
cucurbits ... 231
peanuts ... 154
peppers ... 270
potatoes ... 276,278
soybeans ... 97,101
tomatoes ... 307
Root aphid *Aphididae:*
cole crops ... 221
lettuce ... 250
Root collar weevil:
ornamentals ... 503
Root maggot *Anthomyiidae:*
biocontrols ... 68
cole crops ... 218-220
greens ... 239
radishes ... 279
tobacco ... 170
Root weevil *Curculionidae:*
biocontrols ... 55
hops ... 146
ornamentals ... 495,499,502,508
small fruits ... 386,390
Root-knot nematode *Meloidogyne* spp.:
biocontrols ... 56,68
carrots, etc. ... 213
celery ... 217
cucurbits ... 231
field corn, popcorn ... 79
nuts ... 423
peanuts ... 154
peppers ... 270
potatoes ... 275,276,278
small fruits ... 385
soybeans ... 97,101
sweet corn ... 290
sweet potatoes, yams ... 297
tobacco ... 170-172
tomatoes ... 307
tree fruits, deciduous ... 401,418
tree fruits, subtropical ... 421
Rootworms:
biocontrols ... 56,65
field corn, popcorn ... 77,78,80,83
sugar beets ... 164
sweet corn ... 288,289,291,293
Rose aphid *Macrosiphum rosae:*
biocontrols ... 54,55
ornamentals ... 496,497
Rose chafer *Macrodactylus subspinosus:*
biocontrols ... 65
ornamentals ... 496,497,500,504
small fruits ... 381,382,385,386,388,389,391
tree fruits, deciduous ... 405,406,410,412,414,416
Rose curculio *Merhynchites bicolor:*
biocontrols ... 55
Rose leafhopper *Edwardsiana rosae:*
ornamentals ... 503
small fruits ... 388
Rose scale *Aulacaspis rosae:*
small fruits ... 382,387,388
Rose slug *Endelomyia aethiops:*
ornamentals ... 497,504
Rosy aphid *Dysaphis plantaginea:*
tree fruits, deciduous ... 404
Rosy apple aphid *Dysaphis plantaginea:*
biocontrols ... 58
tree fruits, deciduous ... 402,404,406-408,410,411,
... 413,414,417,418
Rotylenchulus reniformis — reniform nematode
Rotten sugarcane borer:
biocontrols ... 68
Roundheaded pine beetle *Dendroctonus adjunctus:*
ornamentals ... 497
Russet mite:
artichokes ... 199
asparagus ... 201
beans ... 207
carrots, etc. ... 212
celery ... 215
cole crops ... 223
cucurbits ... 230
eggplant ... 234
garlic ... 237
greens ... 242
herbs/spices ... 244
horseradish ... 246
lettuce ... 252
okra ... 256
onions ... 259
peas ... 264
peppers ... 269
potatoes ... 276
radishes ... 280
spinach ... 287
sugar beets ... 164
sweet potatoes, yams ... 296
table beets ... 299
tomatoes ... 305,306
watercress ... 307
Russian wheat aphid *Diuraphis noxia:*
biocontrols ... 54,55,64
forage (grasses, hay) ... 543
small grains ... 129,131
Rust flies:
biocontrols ... 56
Rust mite:
biocontrols ... 58
ornamentals ... 505
tree fruits, citrus ... 393,394,398,400
tree fruits, deciduous ... 404,417
Rusty plum aphid *Hysteroneura setariae:*
tree fruits, deciduous ... 409,411,412,417

S

Sabulodes aegrotata — omnivorous looper
Saddled prominent caterpillar *Heterocampa guttivitta:*
ornamentals ... 497
Saissetia oleae — black scale
Saltmarsh caterpillar *Estigmene acrea:*
Bt chart ... 94,108,128,145,197
beans ... 202,205
biocontrols ... 56
cole crops ... 219,225
cotton ... 110-114,118,120
lettuce ... 249,250,253
noncropland ... 546
peas ... 260,262
small fruits ... 382,386
soybeans ... 95,96,98,99
San Jose scale *Quadraspidiotus perniciosus:*
biocontrols ... 57,58,62
nuts ... 424-426,430-432
ornamentals ... 499-501,506,508
small fruits ... 382,388,390
tree fruits, deciduous ... 402-407,410-412,
... 414,415,417
Sandhill cutworm:
sweet corn ... 290
Sap beetles *Nitidulidae:*
field corn, popcorn ... 78,81
small fruits ... 387,388
sweet corn ... 288,290,292
Satin moth *Leucoma salicis:*
Bt chart ... 488
Sawflies:
biocontrols ... 56,57
ornamentals ... 495-499,503,504,506
small fruits ... 386,390
Sawfly spanworm:
small fruits ... 386
Say stink bug *Chlorochroa sayi:*
small grains ... 132
Scale crawlers *Coccoidea* (scales):
nuts ... 426,428,429
ornamentals ... 495,497-499
small fruits ... 389
tree fruits, citrus ... 397,398
tree fruits, deciduous ... 413
Scales *Coccoidea:*
biocontrols ... 56-58,62-67
hops ... 147
nuts ... 424-426,430-433
ornamentals ... 496,497,499-506,508
small fruits ... 382,388,390
tree fruits, citrus ... 393-398,400
tree fruits, deciduous ... 402-407,410-412,
... 414,415,417
tree fruits, subtropical ... 418,420,421
Scaphoideus luteolus — whitebanded elm leafhopper (elm leafhopper)
Scaphytopius acutus — sharpnosed leafhopper
Scaphytopius magdalensis — blueberry leafhopper
Scarabaeidae — dung beetle, May beetle, June beetle, white grubs
Schizaphis graminum — greenbug
Schizura concinna — redhumped caterpillar
Schoene spider mite *Tetranychus schoenei:*
tree fruits, deciduous ... 407,408
Sciarid larvae:
biocontrols ... 68
Sciarid mushroom flies *Sciaridae:*
biocontrols ... 65,68
mushrooms ... 254
Sciaridae — sciarid mushroom flies
Sciopithes obscurus — obscure root weevil
Scirtothrips citri — citrus thrips
Scleroracus vaccinii — bluntnosed leafhopper
Scolytus rugulosus — shothole borer (fruittree bark beetle)
Scorpions *Scorpiones:*
lawns/turf ... 491
Scopiones — Scorpions
Scotch pine scale:
ornamentals ... 508
Scurfy scale *Chionaspis furfura:*
biocontrols ... 57
tree fruit (deciduous) ... 414,415
Scutigerella — garden symphylans (garden centipede)
Scutellerinae — shieldbacked bugs (shield bugs)
Seed bugs *Nysius* spp.:
biocontrols ... 55
ornamentals ... 495,506
Seed chalcid *Eurytomidae:*
ornamentals ... 495
Seed weevil *Bruchidae:*
sunflower ... 167,168
Seedcorn beetle *Stenolophus lecontei:*
beans ... 203
cucurbits ... 227
field corn, popcorn ... 77-80,83,84
peas ... 261
southern peas ... 283
soybeans ... 99
sweet corn ... 288-290,293,294
Seedcorn maggot *Delia platura:*
beans ... 203,205,207,208
carrots, etc. ... 211
cole crops ... 222

cotton ... 109,112,116
cucurbits ... 227,228
field corn, popcorn ... 77-81,83,84
forage (grasses, hay) ... 543
garlic ... 236
lentils ... 247
onions ... 257,258
peas ... 261,262
radishes ... 280
safflower ... 157
small grains ... 131
southern peas ... 283
soybeans ... 98,99
spinach ... 286
sunflower ... 168
sweet corn ... 288-291,293,294
Seedworms:
ornamentals ... 496
Sericothrips variabilis — soybean thrips
Serpentine leafminer *Liriomyza brassicae:*
celery ... 213,217
cucurbits ... 226,227
nuts ... 432
ornamentals ... 508
peppers ... 267
tomatoes ... 301,307
Sharpnosed leafhopper *Scaphytopius acutus:*
small fruits ... 386-388
Shield bugs *Scutellerinae* (shieldbacked bugs):
potatoes ... 275
Shore flies *Ephydridae:*
biocontrols ... 63,68
Shothole borer *Scolytus rugulosus:*
tree fruits, deciduous ... 406
Shuckworms:
biocontrols ... 56
Sierra fir bark borer *Tetropium abietis* (roundheaded fir borer):
ornamentals ... 499
Sigatoka leafroller:
nuts ... 431
tree fruits, subtropical ... 421
Silphidae — carrion beetles
Silk fly:
field corn, popcorn ... 82
Silver mite:
nuts ... 432
tree fruits, citrus ... 400
tree fruits, deciduous ... 404,416
Silverleaf whiteflies:
biocontrols ... 54,55,58,69
cole crops ... 217,221
cotton ... 115
eggplant ... 232,233
greens ... 238,240
lettuce ... 248,251
ornamentals ... 496,500
peanuts ... 152
peppers ... 265,268
spinach ... 284
tomatoes ... 301,302,304
Silverspotted skipper *Epargyreus clarus:*
soybeans ... 96,98,99
Simuliidae — black flies
Sipha flava — yellow sugarcane aphid
Siphonaptera — fleas
Sitodiplosis mosellana — wheat midge
Sitona lineatus — pea leaf weevil
Sixspotted spider mite *Eotetranychus sexmaculatus:*
biocontrols ... 64
small fruits ... 385
tree fruits, citrus ... 395,396,399,400
tree fruits, subtropical ... 422
Skippers *Hesperiidae:*
Bt chart ... 128,536
Slash pine flower thrips *Gnopothrips fuscus:*
ornamentals ... 495
Slugs *Gastropoda:*
artichokes ... 198
asparagus ... 200
beans ... 203
biocontrols ... 56
carrots, etc. ... 210
celery ... 214
cole crops ... 220
cucubits ... 227
eggplant ... 232
field corn, popcorn ... 79
forage (alfalfa, etc.) ... 538
forage (grasses, hay) ... 543
garlic ... 236
greens ... 239
horseradish ... 245
lawns/turf ... 491
lettuce ... 250
mint ... 148
okra ... 255
onions ... 257
ornamentals ... 499,504
peas ... 261
peppers ... 266
potatoes ... 272
radishes ... 279
rhubarb ... 281
small fruits ... 384
southern peas ... 283
spinach ... 285
sugar beets ... 162
sweet corn ... 290
sweet potatoes, yams ... 295
table beets ... 298
tomatoes ... 302
tree fruits, citrus ... 395
tree fruits, deciduous ... 406
tree fruits, subtropical ... 419
Small black grass weevil *Trachyphloeus bifoveolatus* (crusted root weevil):
small fruits ... 382
Smicronyx fulvus — sunflower seed weevil
Snails *Gastropoda:*
artichokes ... 198
asparagus ... 200
beans ... 203
carrots, etc. ... 210
celery ... 214
cole crops ... 220
cucurbits ... 227
eggplant ... 232
field corn, popcorn ... 79
forage (alfalfa, etc.) ... 538
forage (grasses, hay) ... 543
garlic ... 236
greens ... 239
horseradish ... 245
lawns/turf ... 491
lettuce ... 250
mint ... 148
okra ... 255
onions ... 257
ornamentals ... 499,504
peas ... 261
peppers ... 266
potatoes ... 272
radishes ... 279
rhubarb ... 281
small fruits ... 384
southern peas ... 283
spinach ... 285
sugar beets ... 162
sweet corn ... 290
sweet potatoes, yams ... 295
table beets ... 298
tomatoes ... 302
tree fruits, citrus ... 395
tree fruits, deciduous ... 406
tree fruits, subtropical ... 419
Snout beetle *Curculionidae:*
forage (alfalfa, etc.) ... 539
Snow scale:
ornamentals ... 500
tree fruits, citrus ... 393,396,398,400
Snowy tree cricket *Decanthus fultoni:*
small fruits ... 382
Sod webworm *Crambinae:*
Bt chart ... 488
biocontrols ... 55,57,65,68
lawns/turf ... 489-494
ornamentals ... 505
Soft (brown) scale *Coccus hesperidum* (brown soft scale):
nuts ... 431
ornamentals ... 506
tree fruits, deciduous ... 411,412,415
tree fruits, subtropical ... 420
Soft scale *Coccus hesperidum:*
biocontrols ... 57
ornamentals ... 499-501,503,504
tree fruits, citrus ... 396-398
tree fruits, deciduous ... 411
Soft-bodied insects:
biocontrols ... 64
Solenopsis geminata — fire ants
Solenopsis invicta — red imported fire ant
Solenopsis xyloni — southern fire ant
Sorghum midge *Contarinia sorghicola:*
sorghum ... 157-160
Sorghum webworm *Celama sorghiella:*
sorghum ... 157,159
Southern armyworm *Spodoptera eridania:*
carrots, etc. ... 209,210
celery ... 213,215
cotton ... 114
greens ... 238,240
herbs/spices ... 243
lawns/turf ... 489
lettuce ... 248,249,251
peppers ... 265
potatoes ... 271,272
rhubarb ... 281,282
soybeans ... 98
spinach ... 284,286
sweet corn ... 290,293
tomatoes ... 301-305
Southern cabbageworm *Pontia protodice:*
cole crops ... 219
Southern chafer *Cyclocephala immaculata* (southern masked chafer):
lawns/turf ... 491
Southern corn billbug *Sphenophorus callosus:*
field corn, popcorn ... 79
sweet corn ... 290
Southern corn rootworm *Diabrotica undecimpunctata:*
field corn, popcorn ... 77,78,80
peanuts ... 151,153
sweet corn ... 288,289,291
Southern fire ant *Solenopsis xyloni:*
nuts ... 426
Southern garden leafhopper *Empoasca solana:*
cotton ... 110,111
Southern green stink bug *Nezara viridula:*
beans ... 203
cotton ... 112,118
nuts ... 424,428,429
soybeans ... 95,96
Southern masked chafer *Cyclocephala immaculata* (southern chafer):
lawns/turf ... 492
Southern potato wireworm *Conoderus falli:*
potatoes ... 272,273
Southern red spider mite *Oligonychus ilicis:*
ornamentals ... 501,507
tree fruits, deciduous ... 409
Southwestern corn borer *Diatraea grandiosella:*
Bt chart ... 76
field corn, popcorn ... 77-80,82,83
sorghum ... 157,158
sweet corn ... 288-290,293
Sowbugs *Oniscoidea:*
asparagus ... 200
beans ... 203
biocontrols ... 55,56
carrots, etc. ... 209
celery ... 214
cole crops ... 218
cotton ... 111
cucurbits ... 226
eggplant ... 232
field corn, popcorn ... 78
forage (alfalfa, etc.) ... 537
forage (grasses, hay) ... 543
greens ... 238
horseradish ... 245
lawns/turf ... 489-491,493,494
lettuce ... 249
noncropland ... 546
okra ... 255
ornamentals ... 497,498,508
pastures/rangeland ... 544
peanuts ... 151
peas ... 260,261
peppers ... 266
potatoes ... 271,272
radishes ... 278,279
rice ... 154
small fruits ... 382,383
small grains ... 129
sorghum ... 157
soybeans ... 96
spinach ... 285
sugar beets ... 162
sunflower ... 167
sweet corn ... 288,289
sweet potatoes, yams ... 295
table beets ... 298
tobacco ... 170
tomatoes ... 302
tree fruits, citrus ... 393
tree fruits, deciduous ... 405
Soybean caterpillar *Pseudoplusia includens* (soybean looper):
forage (alfalfa, etc.) ... 541
forage (grasses, hay) ... 544
Soybean looper *Pseudoplusia includens* (spotted cabbage looper):
photo ... 16
beans ... 202
cole crops ... 217
cotton ... 110,113,116,118,120
eggplant ... 232
peanuts ... 151
peas ... 260
peppers ... 265
soybeans ... 95-100
tomatoes ... 301
Soybean thrips *Sericothrips variabilis:*
cotton ... 112
Spanogonicus albofasciatus — whitemarked fleahopper (black fleahopper)
Spanworms *Geometridae* (geometrid moths):
Bt chart ... 380
small fruits ... 381
Sparganothis fruitworm *Sparganothis sulfureana* (blueberry leafroller):
small fruits ... 381-383
Sparganothis sulfureana — Sparganothis fruitworm (blueberry leafroller)
Sphecidae — digger wasps
Sphenophorus callosus — southern corn billbug
Sphenophorus maidis — maize billbug (corn billbug)
Sphenophorus parvulus — bluegrass billbug
***Sphenophorus* spp.** — billbugs, lawn billbug
Sphingidae — hornworm (ello moth)
Spider mites *Tetranychidae:*
artichokes ... 199
asparagus ... 201
beans ... 204-208
biocontrols ... 57,58,64-67
carrots, etc. ... 211,212
celery ... 214,215

cole crops ... 219,223,225
cotton ... 109,111-113,117-121
cucurbits ... 227-230
eggplant ... 233-235
field corn, popcorn ... 77,80,83-85
forage (alfalfa, etc.) ... 538,540-543
garlic ... 237
greens ... 241-243
herbs/spices ... 244
hops ... 146,147
horseradish ... 246
lettuce ... 249,252
mint ... 149
mushrooms ... 254
nuts ... 423,426-433
oil seed crops ... 150
okra ... 256
onions ... 259
ornamentals ... 494,497,499,502-508
peanuts ... 151,152,154
peas ... 261,263,264
peppers ... 267-270
potatoes ... 275-277
radishes ... 280
small fruits ... 384,385,387-391
small grains ... 132,133
sorghum ... 158,160,161
southern peas ... 283
soybeans ... 96,97,99,100
spinach ... 286,287
sugar beets ... 161-165
sunflower ... 167
sweet corn ... 290,294
sweet potatoes, yams ... 296
table beets ... 299,300
tobacco ... 171
tomatoes ... 300,303-305
tree fruits, citrus ... 393,395-400
tree fruits, deciduous ... 402,405,407-409, ... 411-413,415-418
tree fruits, subtropical ... 420-422
watercress ... 307

Spiders *Araneae:*
asparagus ... 201
beans ... 207
cucurbits ... 230
eggplant ... 234
ornamentals ... 499
peanuts ... 153
peppers ... 269
radishes ... 280
table beets ... 300
tomatoes ... 306

Spilonota ocellana — eyespotted bud moth (bud moth)

Spiny elm caterpillar *Nymphalis antiopa* (mourningcloak butterfly):
ornamentals ... 497

Spiral nematode:
biocontrols ... 56
carrots, etc. ... 213
field corn, popcorn ... 79
potatoes ... 276
soybeans ... 97
sweet corn ... 290
sweet potatoes, yams ... 297

Spissistilus festinus — threecornered alfalfa hopper

Spittlebugs *Cercopidae:*
biocontrols ... 56
carrots, etc. ... 209
celery ... 213,214
forage (alfalfa, etc.) ... 537-541
forage (grasses, hay) ... 544
greens ... 238
lawns/turf ... 489
lettuce ... 249
nuts ... 424,425,427,428
ornamentals ... 495,497-499,504
potatoes ... 271
small fruits ... 383,384,387,388

Spodoptera eridania — southern armyworm
Spodoptera exigua — beet armyworm (sugarbeet armyworm)
Spodoptera frugiperda — fall armyworm
Spodoptera ornithogalli — yellowstriped armyworm (cotton cutworm)
Spodoptera praefica — western yellowstriped armyworm

Spotted alfalfa aphid *Therioaphis maculata:*
biocontrols ... 54,55
forage (alfalfa, etc.) ... 538,540-542

Spotted asparagus beetle *Crioceris duodecimpunctata:*
asparagus ... 200

Spotted cabbage looper *Pseudoplusia includens:*
sunflower ... 167

Spotted cucumber beetle *Diabrotica undecimpunctata:*
photo ... 11
beans ... 203,205
cucurbits ... 226,231
forage (alfalfa, etc.) ... 538

Spotted tentiform leafminer *Phyllonorycter blancardella:*
ornamentals ... 500,501
tree fruits, deciduous ... 406,414,416,418

Spring cankerworm *Paleacrita vernata:*
ornamentals ... 495,496,499,503,509

Spring elm caterpillar:
ornamentals ... 497

Springtails *Collembola:*
biocontrols ... 65
lawns/turf ... 489-491,493
ornamentals ... 497

Spruce beetle *Dendroctonus rufipennis:*
ornamentals ... 497

Spruce budworm *Choristoneura fumiferana:*
ornamentals ... 495-497,499,508

Spruce gall aphid:
ornamentals ... 500

Spruce needleminer *Endothenia albolineana:*
ornamentals ... 497

Spruce spider mite *Oligonychus ununguis:*
ornamentals ... 502,505,508

Squash beetle *Epilachna borealis:*
cucurbits ... 228,231

Squash bug *Anasa tristis:*
biocontrols ... 58
cucurbits ... 225-228,230,231
eggplant ... 234

Squash vine borer *Melittia cucurbitae:*
photo ... 17
cucurbits ... 225,226,228-231
eggplant ... 234

Stalk borer *Papaipema nebris:*
field corn, popcorn ... 77,79,82,83
sweet corn ... 288,293

Stegasta bosqueella — rednecked peanutworm

Stem nematode *Ditylenchus* spp.:
garlic ... 237

Stem weevil:
sunflower ... 167,168

Steneotarsonemus pallidus — cyclamen mite
Stenolophus lecontei — seedcorn beetle

Sting nematode *Belonolaimus* spp.:
biocontrols ... 56,68
carrots, etc. ... 213
cucurbits ... 231
field corn, popcorn ... 79
peanuts ... 154
peppers ... 270
potatoes ... 276,278
soybeans ... 97
sweet corn ... 290
tomatoes ... 307
tree fruits, deciduous ... 401,418

Stink bugs (grubs) *Pentatomidae:*
beans ... 203,204,206-209
carrots, etc. ... 209,212
celery ... 214-216
cole crops ... 218,219,221-225
cotton ... 109-113,115,117,118,120
cucurbits ... 230,231
eggplant ... 232,234,235
field corn, popcorn ... 82
forage (alfalfa, etc.) ... 537,538,540-542
forage (grasses, hay) ... 543
garlic ... 236,237
greens ... 238,241,242
horseradish ... 245
lentils ... 247
lettuce ... 249,250,253
nuts ... 429
oil seed crops ... 150
okra ... 255,256
onions ... 257,259
ornamentals ... 508
peanuts ... 151,152
peas ... 263,264
peppers ... 266,269,270
potatoes ... 272,275,277
radishes ... 279,281
rice ... 154-156
small fruits ... 391
small grains ... 129,132
sorghum ... 157,158
southern peas ... 282-284
soybeans ... 95,96,98-100
spinach ... 285,287
sugar beets ... 164
sunflower ... 167
sweet corn ... 294
sweet potatoes, yams ... 297
table beets ... 298-300
tobacco ... 169,171
tomatoes ... 302-306
tree fruits, deciduous ... 402-406,408-410

Strauzia longipennis — sunflower maggot

Strawberry aphid *Chaetosiphon fragaefolii:*
small fruits ... 385

Strawberry clipper *Anthonomus signatus* (strawberry bud weevil):
small fruits ... 382

Strawberry fruitworm *Cnephasia longana* (omnivorous leaftier):
small fruits ... 382

Strawberry leafroller *Ancylis comptana:*
small fruits ... 382,384,387,388,391

Strawberry spider mite *Tetranychus turkestani:*
cotton ... 109,119
nuts ... 431
ornamentals ... 502
small fruits ... 385,390
sugar beets ... 161
tree fruits, deciduous ... 415

Strawberry root weevil *Otiorhynchus ovatus:*
biocontrols ... 55,65,68
mint ... 148
ornamentals ... 495,504
small fruits ... 386-388

Strawberry rootworm *Nephasia longana:*
small fruits ... 388,391

Strawberry (bud) weevil *Anthonomus signatus:*
small fruits ... 382,383,388

Striped beetle:
ornamentals ... 498,499

Striped blister beetle *Epicauta vittata:*
cotton ... 111

Striped cucumber beetle *Acalymma vittatum:*
photo ... 11
beans ... 203
cucurbits ... 226,227

Striped flea beetle *Phyllotreta striolata:*
cole crops ... 218,219
cucurbits ... 228,231
greens ... 239

Striped grass looper *Mocis* spp.:
pastures/rangeland ... 544

Striped oakworm:
ornamentals ... 498,499

Stubby-root nematode:
celery ... 217
field corn, popcorn ... 79
peppers ... 270
potatoes ... 276
soybeans ... 97
sweet corn ... 290
tomatoes ... 307

Stunt nematode:
biocontrols ... 56
carrots, etc. ... 213
cucurbits ... 231
field corn, popcorn ... 79
peppers ... 270
potatoes ... 276
soybeans ... 97,101
sweet corn ... 290
tomatoes ... 307

Subcoccinella vigintiquatuorpunctata — European alfalfa beetle

Subtropical pine tip moth *Rhyacionia subtropica:*
ornamentals ... 497

Suckfly *Cyrtopeltis notatus:*
tobacco ... 170

Sugarbeet cyst nematode *Heterodera schachtii:*
sugar beets ... 162

Sugarbeet leafhopper *Circulifer tenellus:*
sugar beets ... 162

Sugarbeet root aphid *Pemphigus populivenae:*
sugar beets ... 162

Sugarbeet root maggot *Tetanops myopaeformis:*
sugar beets ... 162-165

Sugarbeet webworm *Loxostege sticticalis:*
sugar beets ... 163,164

Sugarcane borer *Diatraea saccharalis:*
sugarcane ... 166

Suleima helianthana — sunflower bud moth

Sunflower beetle *Zygogramma exclamationis:*
sunflower ... 167-169

Sunflower bud moth *Suleima helianthana:*
sunflower ... 167

Sunflower headclipper weevil *Haplorhynchites aeneus:*
sunflower ... 167

Sunflower maggot *Strauzia longipennis:*
sunflower ... 167-169

Sunflower midge *Contarinia schulzi:*
sunflower ... 167

Sunflower moth *Homoeosoma electellum:*
Bt chart ... 145
sunflower ... 167-169

Sunflower seed weevil *Smicronyx fulvus:*
sunflower ... 167-169

Sunflower stem weevil:
sunflower ... 167-169

Surface (feeding) cutworm *Noctuidae:*
beans ... 206
peas ... 263

Sweetpotato flea beetle *Chaetocnema confinis:*
sweet potatoes, yams ... 295

Sweetpotato hornworm *Agrius cingulatus:*
sweet potatoes, yams ... 295

Sweetpotato weevil *Cylas formicarius:*
biocontrols ... 55
sweet potatoes, yams ... 295

Sweetpotato whitefly *Bemisia tabaci:*
photo ... 14
artichokes ... 198
beans ... 202
biocontrols ... 54,55,58,63,69
carrots, etc. ... 209
celery ... 213
cole crops ... 217,218,221
cotton ... 112,113,115
cucurbits ... 226
eggplant ... 232,233
garlic ... 236
greens ... 238,240
horseradish ... 245
lentils ... 247
lettuce ... 248,249,251
onions ... 257

ornamentals ... 495,496,500
peanuts ... 152
peas ... 260
peppers ... 265,268
potatoes ... 271
radishes ... 278
spinach ... 284
sweet potatoes, yams ... 295
table beets ... 297
tomatoes ... 301,302,304,305
tree fruits, citrus ... 393
tree fruits, deciduous ... 402
Symmerista canicosta — redhumped oakworm
Symphyla — symphylans, symphylids
Symphylans *Symphyla:*
biocontrols ... 56
cole crops ... 223
field corn, popcorn ... 78-80
sweet corn ... 289-291
Symphylids *Symphyla:*
mint ... 149
Synanthedon exitiosa — peachtree borer
Synanthedon pictipes — lesser peachtree borer
Synanthedon scitula — dogwood borer
Syngrapha falcifera — celery looper
Syntomeida epilais — oleander caterpillar
Systena blanda — palestriped flea beetle

T

Tachypterellus quadrigibbus — apple curculio
Tadpole shrimp *Notostraca:*
rice ... 154-156
Taeniothrips inconsequens — pear thrips
Tarnished plant bug *Lygus lineolaris:*
photo ... 12
asparagus ... 201
beans ... 203
biocontrols ... 55
carrots, etc. ... 209
celery ... 213,214,216
cole crops ... 218
cotton ... 110-113,117,122
eggplant ... 232,235
forage (alfalfa, etc.) ... 538,541
greens ... 238
horseradish ... 245
lentils ... 247
lettuce ... 249
ornamentals ... 503,508
peppers ... 266,270
potatoes ... 271,272,275,276,278
radishes ... 279
small fruits ... 383,385
southern peas ... 282
spinach ... 285
sweet corn ... 288,290
table beets ... 298
tomatoes ... 302,306
tree fruits, deciduous ... 402-406,408-410,412,414
Taxus bud mite *Cecidophyopsis psilaspis:*
ornamentals ... 500
Tea scale *Fiorinia theae:*
photo ... 14
ornamentals ... 500,509
Tenebrionidae — darkling beetle
Tent caterpillar *Malacosoma* spp.:
lettuce ... 248
nuts ... 431
ornamentals ... 495-499,503-505,508
tree fruits, deciduous ... 402,411,412,415
Tentiform leafminer *Gracillariidae:*
tree fruits, deciduous ... 402,404,405,407,410
Tenuipalpidae — false spider mites
Tephritidae — fruit flies
Terminal weevil:
ornamentals ... 500
Terrapin scale *Mesolecanium nigrofasciatum:*
small fruits ... 388
tree fruits, deciduous ... 403,404,411,412
Tetanops myopaeformis — sugarbeet root maggot (beet root maggot)
Tetralopha subcanalis — lesser webworm
Tetramorium caespitum — pavement ants
Tetranychid mites:
biocontrols ... 61,66
Tetranychidae — spider mites (red spider mites)
Tetranychus cinnabarinus — Carmine spider mite
Tetranychus desertorum — desert spider mite
Tetranychus marianae — tropical spider mite
Tetranychus McDanieli — McDaniel spider mite
Tetranychus pacificus — Pacific spider mite
Tetranychus schoenei — Schoene spider mite
Tetranychus tumidus — tumid spider mite
Tetranychus turkestani — strawberry spider mite (Atlantic mite)
Tetranychus urticae — twospotted spider mite
Tetropium abietis — roundheaded fir borer (Sierra fir bark borer)
Tettigoniidae — katydid
Texas citrus mite *Eutetranychus banksi:*
biocontrols ... 57
nuts ... 423
tree fruits, citrus ... 395-401
Therioaphis maculata — spotted alfalfa aphid
Therioaphis trifolii — yellow clover aphid
Thistle aphid *Brachycaudus cardui:*
tree fruits, deciduous ... 407,409
Thistle caterpillar *Vanessa cardui* (painted lady):
soybeans ... 96,99
sunflower ... 167
Thornbug *Umbona crassicornis:*
ornamentals ... 497
Threecornered alfalfa hopper *Spissistilus festinus:*
beans ... 203
cotton ... 110
forage (alfalfa, etc.) ... 537
lentils ... 247
peanuts ... 151
southern peas ... 282
soybeans ... 95,96,98-100
Threelined potato beetle *Lema trilineata:*
potatoes ... 273,277
Thrips *Thysanoptera:*
photo ... 17
artichokes ... 198,199
asparagus ... 199-202
beans ... 202,203,205-208
biocontrols ... 54-58,63-66,68
carrots, etc. ... 209,212
celery ... 213,215,216
cole crops ... 217-219,221-224
cotton ... 109-121
cucurbits ... 226-231
eggplant ... 232-235
field corn, popcorn ... 79-81,84
forage (alfalfa, etc.) ... 537,538,541,542
forage (grasses, hay) ... 544
garlic ... 236,237
greens ... 238,241,242
herbs/spices ... 243,244
hops ... 146,147
horseradish ... 245,246
lawns/turf ... 492
lentils ... 246-248
lettuce ... 248-253
mint ... 148,149
noncropland ... 547
nuts ... 428,429,431
oil seed crops ... 150
okra ... 255,256
onions ... 256-260
ornamentals ... 495-504,506-509
pastures/rangeland ... 544-546
peanuts ... 150-154
peas ... 263,264
peppers ... 265,266,268-270
potatoes ... 271,275-277
radishes ... 278,280,281
rhubarb ... 281,282
rice ... 156
safflower ... 156,157
small fruits ... 381,384,386-388,391
small grains ... 130,132,133
sorghum ... 160,161
southern peas ... 282,284
soybeans ... 95-100
spinach ... 284,287
sugar beets ... 161,164,165
sunflower ... 168,169
sweet corn ... 290,292,294
sweet potatoes, yams ... 294,296
table beets ... 297,299,300
tobacco ... 169,171
tomatoes ... 301,304-306
tree fruits, citrus ... 392-397,399,400
tree fruits, deciduous ... 401-405,408,410, 411,416,418
tree fruits, subtropical ... 418-422
watercress ... 307
Thrips palmi:
biocontrols ... 54,55
Thrips simplex — gladiolus thrips
Thrips tabaci — onion thrips
Thymelicus lineola — European skipper (Essex skipper)
Thyridopteryx ephemeraeformis — bagworm
Thysanoptera — thrips
Ticks *Acari:*
lawns/turf ... 489-491,493,494
ornamentals ... 497,498
Tingidae — lace bugs
Tingids *Tingidae* (lace bugs):
nuts ... 428
Tinocallis ulmifolii — elm leaf aphid
Tipula paludosa — European crane fly
Tipula simplex — range crane fly
Tipulidae — crane fly
Tip moth *Dasineura oxycoccana* (cranberry tipworm midge):
ornamentals ... 499,508
Tipworm *Dasineura oxycoccana* (cranberry tipworm midge):
small fruits ... 382
Tobacco aphid:
tobacco ... 169,171
Tobacco budworm *Heliothis virescens:*
Bt chart ... 108,145,197,380
biocontrols ... 57,59,68
cole crops ... 217,219
cotton ... 109-113,115,116,118,120
cucurbits ... 226,228
field corn, popcorn ... 79,82
lettuce ... 248-250
ornamentals ... 496,508
small fruits ... 381
soybeans ... 96,98
spinach ... 284
tobacco ... 169,171
tomatoes ... 302
Tobacco flea beetle *Epitrix hirtipennis:*
tobacco ... 169,170
Tobacco hornworm *Manduca sexta:*
photo ... 17
Bt chart ... 145
peppers ... 265
tobacco ... 169
tomatoes ... 301
Tobacco thrips *Frankliniella fusca:*
cotton ... 111-113
garlic ... 236
onions ... 257
peanuts ... 151
tobacco ... 169
Tomato fruitworm *Helicoverpa zea:*
Bt chart ... 145,197
beans ... 202
biocontrols ... 57,59
carrots, etc. ... 209,210
celery ... 213,215
eggplant ... 232,235
greens ... 238,240
herbs/spices ... 243
lettuce ... 248,251
peas ... 260
peppers ... 265,266,270
potatoes ... 272
rhubarb ... 281,282
spinach ... 284,286
tomatoes ... 301-306
Tomato hornworm *Manduca quinquemaculata:*
photo ... 17
biocontrols ... 69
cole crops ... 222
eggplant ... 232
peppers ... 266
potatoes ... 272
tomatoes ... 301,302
Tomato pinworm *Keiferia lycopersicella:*
beans ... 202
biocontrols ... 60,61
eggplant ... 232,233
peas ... 260
peppers ... 265,266
potatoes ... 272
tomatoes ... 300-302,304,305
Tomato psylla *Paratrioza cockerelli:*
biocontrols ... 54,55
Tomato russet mite *Aculops lycopersici:*
beans ... 204
cucurbits ... 228
peppers ... 267
tomatoes ... 300,303,304,306,307
Tortoise beetle *Cassidinae:*
sweet potatoes, yams ... 295
Tortrix moth *Argyrotaenia citrana* (orange tortix moth):
Bt chart ... 488
biocontrols ... 56
Toxoptera citricida — brown citrus aphid
Trachyphloeus bifoveolatus — crusted root weevil (small black grass weevil)
Tree crickets *Oecanthinae:*
biocontrols ... 56
Trialeurodes abutilonea — bandedwing whitefly
Trialeurodes vaporariorum — greenhouse whitefly
Trichoplusia ni — cabbage looper
Trimacha intricata — strawberry leaf beetle
Trombiculidae — chigger mites (chiggers)
Tropical spider mite *Tetranychus marianae:*
beans ... 204
cucurbits ... 228
hops ... 146
peppers ... 267
tomatoes ... 303
Tropical sod webworm *Herpetogramma phaeopteralis:*
biocontrols ... 65,68
lawns/turf ... 493
True armyworm *Pseudaletia unipuncta* (armyworm):
field corn, popcorn ... 77,82
noncropland ... 547
pasture/rangeland ... 545
small grains ... 132
sweet corn ... 290
tomatoes ... 305
Tuberworm:
biocontrols ... 56
potatoes ... 271,274
tomatoes ... 301
Tufted apple bud moth *Platynota idaeusalis:*
Bt chart ... 380
tree fruits, deciduous ... 402,406,410,412
Tumid spider mite *Tetranychus tumidus:*
ornamentals ... 507
Tussock moth *Lymantriidae:*
Bt chart ... 488
ornamentals ... 496,498,499,504
small fruits ... 381,386
tree fruits, deciduous ... 403-405
Twig borer:
nuts ... 426
ornamentals ... 508
tree fruits, deciduous ... 407

Twig girdler *Oncideres cingulata:*
biocontrols ... 56
nuts ... 424,425
Twospotted spider mite *Tetranychus urticae:*
beans ... 204,207
biocontrols ... 67
celery ... 214
cotton ... 109,111-113,118-120
cucurbits ... 228
eggplant ... 235
field corn, popcorn ... 77,80,83-85
forage (alfalfa, etc.) ... 538,542
hops ... 146,147
mint ... 148,149
nuts ... 423,427,431-433
ornamentals ... 499,501,502,504,505,507,508
peanuts ... 152-154
peppers ... 267
potatoes ... 277
small fruits ... 384,385,388,390,391
small grains ... 133
sorghum ... 160,161
soybeans ... 99,100
sugar beets ... 161,165
sweet corn ... 294
tomatoes ... 303,306
tree fruits, citrus ... 392,399,401
tree fruits, deciduous ... 402,404,405,
... 407-409,412,415-418
Tylenchulus semipenetrans — citrus nematode
Typhlocyba pomaria — white apple leafhopper
Tyrophagus dimidiatus — crown mite

U

Udea profundalis — false celery leaftier
Udea rubigalis — celery leaftier, greenhouse leaftier
Umbona crassicornis — thornbug
Unaspis citri — citrus snow scale
Unaspis euonymi — euonymus scale
Undulambia polystichalis — leatherleaf fern borer
Urbanus proteus — bean leafroller

V

Vanessa cardui — painted lady (thistle caterpillar)
Variegated cutworm *Peridroma saucia:*
Bt chart ... 76,128
asparagus ... 200
beans ... 205
carrots, etc. ... 210
celery ... 214
cole crops ... 221
cucurbits ... 228
field corn, popcorn ... 81
forage (alfalfa, etc.) ... 540
greens ... 240
lettuce ... 251
mint ... 148
onions ... 258
peas ... 262
peppers ... 268
potatoes ... 274
spinach ... 286
sugar beets ... 163
sweet corn ... 291
table beets ... 299
tomatoes ... 302,304,305
tree fruits, citrus ... 395
Variegated leafhopper *Erythroneura variabilis:*
biocontrols ... 54,55
Variegated leafroller *Platynota flavedana:*
tree fruits, deciduous ... 402,405,406,410,412
Vegetable leafminer *Liriomyza sativae:*
carrots, etc. ... 212
celery ... 215
eggplant ... 234
greens ... 242
herbs/spices ... 244
lettuce ... 253
ornamentals ... 505
peppers ... 269,270
spinach ... 287
tomatoes ... 301,305
Vegetable weevil *Listroderes costirostris:*
biocontrols ... 55
carrots, etc. ... 212
celery ... 216
cole crops ... 219,224
cucurbits ... 230,231
eggplant ... 235
greens ... 242
horseradish ... 246
lawns/turf ... 491
lettuce ... 249,253
okra ... 256
peanuts ... 151
peas ... 264
peppers ... 269,270
spinach ... 287
sweet corn ... 294
table beets ... 300
tobacco ... 169
tomatoes ... 306
Velvetbean caterpillar *Anticarsia gemmatalis:*
Bt chart ... 94,108,128,145,197
beans ... 202,203
forage (alfalfa, etc.) ... 537,541,542
forage (grasses, hay) ... 544
lentils ... 247
peanuts ... 150-152
peas ... 260
southern peas ... 282
soybeans ... 95-100
Vetch bruchid *Bruchus brachialis:*
forage (alfalfa, etc.) ... 540,541
Vine mealybug:
small fruits ... 386
Vinegar fly *Drosophila melanogaster:*
tomatoes ... 303
tree fruits, subtropical ... 420
Virginiacreeper leafhopper *Erythroneura ziczac:*
biocontrols ... 54
Vitacea polistiformis — grape root borer

W

Walnut aphid *Chromaphis juglandicola:*
nuts ... 423,427,431,433
Walnut caterpillar *Datana integerrima:*
Bt chart ... 380
biocontrols ... 56
nuts ... 424-426
ornamentals ... 497
Walnut husk maggot *Rhagoletis suavis:*
nuts ... 427
Walnut huskfly *Rhagoletis completa:*
nuts ... 423-425,427-430
Walnut scale *Quadraspidiotus juglansregiae:*
nuts ... 425,433
tree fruits, deciduous ... 403,404
Wasps *Apocrita:*
ornamentals ... 499
Wax scale *Ceroplastes* spp.:
ornamentals ... 503
Webworms:
Bt chart ... 76,128,145,197,380,488,536
artichokes ... 198
asparagus ... 199
beans ... 202,203,206
biocontrols ... 55-57,65,68,69
carrots, etc. ... 209
celery ... 213
cole crops ... 217,218
cotton ... 110,117,118
cucurbits ... 226
eggplant ... 232
field corn, popcorn ... 79
forage (alfalfa, etc.) ... 537,541,542
forage (grasses, hay) ... 544
garlic ... 236
greens ... 238,241,242
horseradish ... 245
lawns/turf ... 489-494
lentils ... 246,247
lettuce ... 248,249,253
nuts ... 424,425
okra ... 255
onions ... 256,257
ornamentals ... 495-500,505,507-509
peanuts ... 151
peas ... 260
peppers ... 265,266
potatoes ... 271
radishes ... 278
rhubarb ... 281
small fruits ... 381
sorghum ... 158
southern peas ... 282,283
soybeans ... 95,96,99
spinach ... 284
sugar beets ... 161,162,164
sweet corn ... 289,290
sweet potatoes, yams ... 294,295
table beets ... 297,299
tomatoes ... 301
tree fruits, citrus ... 392
tree fruits, deciduous ... 402
Weevil larvae:
forage (alfalfa, etc.) ... 540
Weevils *Curculionidae:*
beans ... 202
biocontrols ... 56,58
celery ... 213
cole crops ... 218
eggplant ... 232
garlic ... 236
greens ... 238
lettuce ... 249
onions ... 257
ornamentals ... 496,497,508
peanuts ... 151
peas ... 260
peppers ... 265
rhubarb ... 281
small fruits ... 381,386,391
soybeans ... 95
spinach ... 284
sweet potatoes, yams ... 297
tomatoes ... 301
tree fruits, citrus ... 393
tree fruits, deciduous ... 402
tree fruits, subtropical ... 422
West Indian sugarcane borer *Diaprepes abbreviatus*
(sugarcane rootstalk borer weevil):
tree fruits, citrus ... 393
Western avocado leafroller *Amorbia essigana:*
tree fruits, subtropical ... 419
Western bean cutworm *Loxagrotis albicosta:*
Bt chart ... 76
beans ... 202,203,205
field corn, popcorn ... 77-79,82,83
lentils ... 247
peas ... 260
southern peas ... 282,283
sweet corn ... 288-290,292,293
Western cherry fruit fly *Rhagoletis indifferens:*
tree fruits, deciduous ... 402
Western corn rootworm *Diabrotica virgifera:*
photo ... 11
field corn, popcorn ... 77,78,80
sweet corn ... 288,289,291
Western flower thrips *Frankliniella occidentalis:*
biocontrols ... 54,55
cotton ... 117,118
garlic ... 236
onions ... 257
ornamentals ... 508
tomatoes ... 302
Western grape leafhopper *Erythroneura elegantula:*
small fruits ... 389
Western grapeleaf skeletonizer *Harrisina brillians:*
small fruits ... 382,383,386
Western hemlock bark borer *Melanophila fulvoguttata*
(hemlock borer):
ornamentals ... 499
Western hemlock looper *Lambdina fiscellaria:*
ornamentals ... 497
Western pine beetle *Dendroctonus brevicomis:*
ornamentals ... 497
Western spruce budworm *Choristoneura occidentalis:*
ornamentals ... 497
Western striped cucumber beetle *Acalymma trivittatum:*
beans ... 203
cucurbits ... 226
Western tentiform leafminer *Phyllonorycter felinella:*
tree fruits, deciduous ... 416
Western tussock moth *Orgyia vetusta:*
tree fruits, citrus ... 393,394,396,399
Western wheat mite *Petrobia latens* (brown wheat mite):
small grains ... 132
Western yellowstriped armyworm *Spodoptera praefica:*
asparagus ... 200
beans ... 205
forage (alfalfa, etc.) ... 540
lentils ... 247
peppers ... 266
potatoes ... 271
small fruits ... 382
sugar beets ... 163
tomatoes ... 301,302
Wheat midge *Sitodiplosis mosellana:*
small grains ... 129
White apple leafhopper *Typhlocyba pomaria:*
tree fruits, deciduous ... 402,404-406,408,410,
... 414,418
White cutworm *Euxoa scandens:*
asparagus ... 200
White grubs *Scarabaeidae:*
biocontrols ... 54-56,65,68
field corn, popcorn ... 77,79,80,82,83
lawns/turf ... 490-494
ornamentals ... 502,503,505,508
pastures/rangeland ... 544
sorghum ... 158
sugar beets ... 162
sweet corn ... 288,290-292,294
sweet potatoes, yams ... 296
White peach scale *Pseudaulacaspis pentagona:*
photo ... 14
ornamentals ... 506
tree fruits, deciduous ... 403,404,407,415
White pine weevil *Pissodes strobi:*
ornamentals ... 504
White tussock moth *Orgyia leucostigma:*
small fruits ... 388
Whiteflies *Aleyrodidae:*
photo ... 14
artichokes ... 198,199
asparagus ... 199,201,202
beans ... 202,205,207,208
biocontrols ... 54-58,63-66,69
carrots, etc. ... 209,210,212
celery ... 213-216
cole crops ... 217-221,223,224
cotton ... 109-113,115-121
cucurbits ... 226,228,230,231
eggplant ... 232-235
field corn, popcorn ... 83
forage (alfalfa, etc.) ... 542
garlic ... 236,237
greens ... 238,240,242
herbs/spices ... 243,244
hops ... 146,147
horseradish ... 245,246

lawns/turf ... 492
lentils ... 247,248
lettuce ... 248,249,251-253
mint ... 148,149
nuts ... 424,429,431
oil seed crops ... 150
okra ... 255,256
onions ... 256,257,259
ornamentals ... 495-509
peanuts ... 152,153
peas ... 260,264
peppers ... 265,268-270
potatoes ... 271,273,276,277
radishes ... 278,280,281
rhubarb ... 281,282
safflower ... 156
small fruits ... 381,382,387-390
small grains ... 132
southern peas ... 283,284
soybeans ... 95,99
spinach ... 284,286,287
sugar beets ... 161,164
sweet corn ... 293,294
sweet potatoes, yams ... 294-297
table beets ... 297,299,300
tobacco ... 171,172
tomatoes ... 301-306
tree fruits, citrus ... 392,393,395,396,398-400
tree fruits, deciduous ... 402,408,413,415,416
tree fruits, subtropical ... 418,420-422
watercress ... 307

Whitefringed beetle (adults) *Graphognathus* spp.:
peanuts ... 151

Whitemarked tussock moth *Orgyia leucostigma:*
tree fruits, deciduous ... 412

Willamette spider mite *Eotetranychus willamettei:*
nuts ... 427
ornamentals ... 502
small fruits ... 385,390
tree fruits, deciduous ... 407,408

Willow leaf beetle *Calligrapha multipunctata:*
ornamentals ... 497,508,509

Winter grain mite *Penthaleus major:*
small grains ... 132

Winter moth *Operophtera brumata:*
nuts ... 425

Wireworms *Elateridae:*
beans ... 204,205
biocontrols ... 56,65,68
carrots, etc. ... 210,211
celery ... 214
cole crops ... 219,220,222
cotton ... 116
cucurbits ... 227,228
field corn, popcorn ... 77-81,83,84
forage (grasses, hay) ... 543
greens ... 239
lentils ... 247
lettuce ... 250
onions ... 258
peanuts ... 151
peas ... 261,262
peppers ... 267
potatoes ... 272,273,276
radishes ... 279,280
safflower ... 157
small grains ... 131
sorghum ... 158,159
soybeans ... 98,99
spinach ... 285
sugar beets ... 162,163
sugarcane ... 166
sunflower ... 168
sweet corn ... 288-291,293,294
sweet potatoes, yams ... 295,296
table beets ... 298
tobacco ... 170,171
tomatoes ... 303

Woolly aphid *Aphididae, Adelgidae:*
biocontrols ... 58
tree fruits, deciduous ... 412

Woolly apple aphid *Eriosoma lanigerum:*
tree fruits, deciduous ... 404,406-408,411,412

Woollybear caterpillar *Arctiidae:*
soybeans ... 95,96,98
sunflower ... 167

X

***Xiphinema* spp.** — dagger nematode
Xylena nupera — false armyworm

Y

Yellow pecan aphid *Monellia* spp.:
nuts ... 425,426,432

Yellow spider mite *Eotetranychus carpini:*
nuts ... 423,427
small fruits ... 385
tree fruits, deciduous ... 402,407,408

Yellow scale *Aonidiella citrina:*
biocontrols ... 62
tree fruits, citrus ... 393,396-398,400

Yellow sugarcane aphid *Sipha flava:*
sorghum ... 158,159

Yellowheaded fireworm *Acleris minuta:*
tree fruits, deciduous ... 404

Yellownecked caterpillar *Datana ministra:*
ornamentals ... 497,498,508,509

Yellow-poplar weevil *Odontopus calceatus:*
ornamentals ... 497

Yellowstriped armyworm *Spodoptera ornithogalli:*
beans ... 205
cole crops ... 219
cotton ... 110-114
cucurbits ... 226,228
forage (alfalfa, etc.) ... 537,540
forage (grasses, hay) ... 543
greens ... 238,239
lawns/turf ... 489,490
ornamentals ... 496
peas ... 261
small fruits ... 381,383
small grains ... 129
soybeans ... 96,98
sugar beets ... 162
sweet corn ... 290
sweet potatoes, yams ... 295
tomatoes ... 301-303

Yellowstriped caterpillar *Spodoptera ornithogalli:*
Bt chart ... 108

Yuma spider mite *Eotetranychus yumensis:*
tree fruits, citrus ... 395,400,401

Z

Zimmerman (pine) moth *Dioryctria zimmermani:*
ornamentals ... 499,501,508
Zonosemata electa — pepper maggot
Zophodia convolutella — gooseberry fruitworm
Zygogramma exclamationis — sunflower beetle

Disease Index

A

Achyla:
- rice 182

Acremonium crotocinigenum — Cephalosporium leaf spot
Actinopelte dryina — Actinopelte leaf spot
Actinopelte leaf spot *Actinopelte dryina:*
- ornamentals 524

Aerial blight:
- ornamentals 531
- soybeans 101,105

Aerial web blight *Rhizoctonia solani:*
- ornamentals 524

Agrobacerum tumefaciens — crown gall
Altenius:
- biocontrols 54

Alternaria:
- photo 28
- biocontrols 58
- cole crops 324
- cucurbits 329
- eggplant 334
- ornamentals 521,522,524,525
- peas 352
- peppers 354
- tree fruits, citrus 449

Alternaria alternata — black mold
Alternaria blight:
- carrots, etc. 316,318,319
- eggplant 334,335
- ornamentals 521,524

Alternaria brown spot:
- tree fruits, citrus 448,450

Alternaria fruit rot:
- small fruits 440
- tomatoes 373,375,377

Alternaria fruit spot:
- tree fruits, subtropical 470

Alternaria leaf blight:
- carrots, etc. 317-319
- cucurbits 329,333,334
- eggplant 334

Alternaria leaf spot:
- carrots, etc. 316,317
- cole crops 322,323,325,327
- cucurbits 328-331
- greens 338,339
- lawns/turf 511
- ornamentals 520

Alternaria porri — purple blotch
Alternaria solani — early blight
Angular leaf spot:
- beans 312
- cucurbits 328-331,334
- small fruits 438
- tobacco 192

Anisogramma anomala — eastern filbert blight
Annosus root disease:
- ornamentals 530

Anthracnose:
- photo 20,27,28
- beans 309,311-313,315,316
- biocontrols 58
- carrots, etc. 316
- cole crops 322
- cucurbits 328-334
- eggplant 334,335
- greens 338
- lawns/turf 509,511-513,515-518
- lettuce 344
- nuts 477,480-483,485
- ornamentals 521-525,527,531-534
- peas 352
- peppers 354-356
- radishes 363
- small fruits 434-440,442,445
- southern peas 364,365
- soybeans 101,103,105
- spinach 366
- sweet potatoes, yams 370
- tomatoes 372-375,377,378
- tree fruits, citrus 449
- tree fruits, deciduous 456-459
- tree fruits, subtropical 470-477

Anthracnose blight:
- ornamentals 525

Anthracnose fruit rot:
- small fruits 440
- tree fruits, citrus 446

Anthracnose leaf spot *Sphaceloma* spp.:
- small fruits 434

Apple blister:
- tree fruits, deciduous 462

Apple blotch *Phyllosticta solitaria:*
- tree fruits, deciduous 468

Apple scab *Venturia inaequalis:*
- tree fruits, deciduous 453,454,468,469

Armillaria mellea — oak root fungus
Ascochyta blight:
- beans 311
- ornamentals 521,524

Ascochyta chrysanthemi — Ascochyta ray blight
Ascochyta leaf blight:
- ornamentals 532

Ascochyta rabiei:
- garbanzos 336

Ascochyta ray blight *Ascochyta chrysanthemi:*
- ornamentals 525

Ascochyta web blotch:
- peanuts 176,181

Ascospora ruborum — cane spot
Asparagus spear slime:
- asparagus 308

Aspergillus:
- Bt chart 76
- peanuts 176

B

Bacterial blast *Pseudomonas syringae:*
- nuts 479-481
- tree fruits, citrus 447,449
- tree fruits, deciduous 456

Bacterial blight:
- beans 311,312
- biocontrols 54
- carrots, etc. 317
- celery 320,322
- onions 347,349
- ornamentals 519
- peas 352
- rice 182

Bacterial canker:
- nuts 480
- small fruits 436,438,439
- tomatoes 373,374
- tree fruits, citrus 449
- tree fruits, deciduous 456,467

Bacterial disease:
- tree fruits, deciduous 458-460

Bacterial fruit blotch:
- cucurbits 329-331

Bacterial leaf blight:
- small grains 136

Bacterial leaf rot:
- biocontrols 54

Bacterial leaf spot:
- ornamentals 521,522

Bacterial leaf streak:
- small grains 136

Bacterial soft rot:
- ornamentals 522

Bacterial speck:
- tomatoes 373-375,377

Bacterial spot:
- photo 26
- biocontrols 54,58
- ornamentals 522
- peppers 354-356
- tomatoes 373-375,377
- tree fruits, deciduous 456-461

Bacterial stem canker:
- small fruits 438

Bacterial stem rot:
- biocontrols 54

Bacterial wilt:
- photo 23
- biocontrols 54
- cucurbits 329,330,334

Barley leaf stripe *Pyrenophora graminea:*
- small grains 137,139

Barley rust *Puccinia gramminis:*
- small grains 134,139,140

Barley scald *Rhynchosporium secalis:*
- small grains 139

Basal plate rot:
- photo 23

Basal rot:
- carrots, etc. 320
- celery 320,322
- cole crops 325
- ornamentals 531,532

Belly rot *Rhizoctonia* spp.:
- cucurbits 329,333

Berry rot *Botrytis* spp.:
- small fruits 435

Bipolaris leaf spot *Bipolaris setariae:*
- lawns/turf 511
- ornamentals 521

Bipolaris setariae — Bipolaris leaf spot
Bird's eye rot:
- photo 21

Bitter rot *Glomerella cingulata, Melanconium fuligineum:*
- small fruits 434
- tree fruits, deciduous 454,458,461,468,469

Black crown rot *Centrospora acerina:*
- carrots, etc. 318

Black knot:
- tree fruits, deciduous 452,453,459,468

Black leaf spot:
- cole crops 323,324
- greens 339
- ornamentals 520-525,527,529-534
- spinach 366
- table beets 371

Black leaf streak:
- tree fruits, subtropical 477

Black mold *Alternaria alternata:*
- tomatoes 372

Black pitting *Pseudomonas syringae:*
- tree fruits, subtropical 471-473

Black pod:
- tree fruits, subtropical 471-473

Black pox *Helminthosporium papulosum:*
- tree fruits, deciduous 454,468

Black root rot *Rhizoctonia solani:*
- cotton 122,125
- ornamentals 531,532
- tobacco 193

Black rot:
- cole crops 323,324,326,327
- greens 339,340
- ornamentals 527,528
- radishes 363
- small fruits 433-439,441,442,445,446
- sweet potatoes, yams 370
- table beets 371
- tree fruits, citrus 446
- tree fruits, deciduous 452,454,461,468

Black shank *Phytophthora parasitica:*
- tobacco 193

Black Sigatoka:
- tree fruits, subtropical 470

Black spot:
- photo 25
- biocontrols 58
- tree fruits, subtropical 475

Black spot canker:
- photo 18

Blackhull:
- peanuts 176

Blackleg *Phoma lingam:*
- biocontrols 54
- cole crops 322,324
- greens 338
- potatoes 358

Blast:
- rice 181

Blight:
- photo 18,28
- biocontrols 58
- celery 320
- nuts 479-481,485
- tree fruits, deciduous 459
- tree fruits, subtropical 471,473

Blossom blast:
- tree fruits, deciduous 455

Blossom blight:
- photo 19
- tree fruits, subtropical 471

Blossom blight *Botrytis* spp.:
- nuts 478,479,481,483-485
- ornamentals 524,527

Blossom blight *Monilinia* spp.:
- tree fruits, deciduous 453-461,464-467

Blossom brown rot:
- nuts 479-481
- tree fruits, deciduous 456-458,460,461

Blossom end rot *Botryosphaeria* spp.:
- photo 24
- tree fruits, deciduous 454

Blotch:
- tree fruits, citrus 447
- tree fruits, deciduous 453,458,461

Blotch *Cercospora* spp.:
- tree fruits, subtropical 471-474

Blue mold:
- biocontrols 54
- tobacco 192,193

Boll rot:
- cotton 125

Botryosphaeria blight:
- ornamentals 525

***Botryosphaeria* spp.**— leaf blight, blossom end rot
Botryotinia narcissicola — smoulder
Botrytis:
- photo 21
- biocontrols 58
- ornamentals 521,522,524,525,532,533
- tree fruits, deciduous 454

Botrytis blight:
- photo 20,24,25
- beans 311,315
- carrots, etc. 316
- nuts 479-481
- onions 350
- ornamentals 521-523,525,530,532,534
- potatoes 358

Botrytis blossom blight:
- nuts 478
- ornamentals 520
- small fruits 434
- tree fruits, citrus 446

Botrytis bunch rot:
- small fruits 434,439,446
- tree fruits, citrus 446

Botrytis cinerea — Botrytis fruit rot, gray mold, leaf blight, vine rot

Botrytis crown rot:
- small fruits 440

Botrytis flower/petal blight:
- ornamentals 520,532

Botrytis fruit rot *Botrytis cinerea:*
- small fruits 435,440,446
- tomatoes 377

Botrytis gray mold:
- beans 309
- cole crops 322
- greens 338
- small fruits 435
- tomatoes 373,375,377,378
- tree fruits, citrus 446

Botrytis late blight:
- tomatoes 377

Botrytis leaf blight/blast *Botrytis squamosa:*
- garlic 336,337
- onions 346-349

Botrytis leaf rot:
- rhubarb 363

Botrytis neck rot:
- onions 347,348

Botrytis petal spot:
- ornamentals 525

Botrytis rot:
- garlic 337
- lettuce 343
- onions 348
- small fruits 439
- tree fruits, deciduous 454

Botrytis seedling blight:
- ornamentals 521

Botrytis squamosa — leaf blight/blast

Botrytis stem rot:
- small fruits 439

Botrytis storage rot:
- ornamentals 527

Botrytis vine rot *Botrytis cinerea:*
- potatoes 357,361,362

Bottom rot *Rhizoctonia solani:*
- photo 22
- carrots, etc. 316
- lettuce 344

Brooks fruit rot:
- tree fruits, deciduous 468

Brooks fruit spot:
- tree fruits, deciduous 454,458,461,468

Brown blotch:
- rice 183

Brown canker *Cryptosporella umbrina:*
- ornamentals 531

Brown leaf spot:
- nuts 478,482,483
- rice 183

Brown needle blight:
- ornamentals 520

Brown patch *Rhizoctonia solani:*
- lawns/turf 509-511,513-519
- ornamentals 533

Brown rot:
- tree fruits, citrus 447-451

Brown rot *Monilinia laxa:*
- nuts 479-481,484,485
- tree fruits, deciduous 453-460,462,464-467,469

Brown rot blight:
- ornamentals 525

Brown rot blossom blight:
- nuts 478,479,481,482,484,485
- tree fruits, deciduous 452-460,462,464-469
- tree fruits, subtropical 470

Brown rot twig blight *Monilinia* spp.:
- nuts 478,482

Brown spot:
- beans 312
- nuts 483,485
- soybeans 105
- tree fruits, citrus 449
- tree fruits, subtropical 470

Brown spot *Helminthosporium oryzae:*
- rice 182

Buckeye rot:
- tomatoes 372

Bud blight *Botrytis* spp.:
- ornamentals 522
- tree fruits, deciduous 457

Bulb dry rot *Sclerotium gladioli:*
- ornamentals 527,528,532

Bulb rot:
- ornamentals 520,527,528

Bull's-eye rot *Neofabraea malicorticus:*
- tree fruits, deciduous 454,459,467-469

Bunch rot *Botrytis* spp.:
- small fruits 435,439-441,445

Bunt *Tilletia tritici:*
- small grains 134,138,139,141

C

Cane blight:
- small fruits 434,435

Cane spot *Ascospora ruborum:*
- photo 19
- small fruits 435-439

Canker:
- ornamentals 522

Carrot blight:
- carrots, etc. 317

Catkin hypertrophy:
- ornamentals 534

Cavity spot:
- carrots, etc. 318

Cedar rust *Gymnosporangium* spp.:
- ornamentals 524
- tree fruits, deciduous 461

Cedar-apple rust *Phragmidium juniperi-virginiae:*
- ornamentals 521,524
- tree fruits, deciduous 463,468,469

Centrospora acerina — black crown rot

Cephalosporium leaf blight:
- ornamentals 525

Cephalosporium leaf spot *Acremonium crotocinigenum:*
- ornamentals 524,527

***Cephalosporium* spp.** — fruit rot

Cephalosporium wilt:
- cucurbits 328

Ceratocystis paradoxa — pineapple disease

Ceratocystis paradoxal:
- nuts 477

Cercospora:
- ornamentals 522,524

Cercospora black spot:
- tree fruits, subtropical 475

Cercospora blight:
- carrots, etc. 316-319
- celery 320,322

Cercospora coffeicola — iron spot

Cercospora fruit spot:
- tree fruits, subtropical 476

Cercospora kikuchii — purple seed stain

Cercospora leaf blight:
- carrots, etc. 317
- ornamentals 522,527

Cercospora leaf spot:
- asparagus 308
- beans 311,312
- carrots, etc. 317,318
- cucurbits 328-333
- forage (alfalfa, etc.) 548
- greens 339
- ornamentals 521,522,525,527,532
- peanuts 177,178,180,181
- peppers 354-356
- radishes 363
- southern peas 364,365
- soybeans 105
- spinach 366,367
- sugar beets 187,188,190
- table beets 371
- tomatoes 373
- tree fruits, subtropical 476
- watercress 378

Cercospora oryzae — narrow brown leaf spot

Cercospora spot:
- tree fruits, subtropical 470

Cercosporidium:
- ornamentals 521

Cercosporidium leaf spot:
- peanuts 180,181

***Cercosporidium* spp.** — late leaf spot

Cherry leaf spot *Coccomyces hiemalis:*
- tree fruits, deciduous 452,455,457,462,465-468

Christulariella pyramidalis — zonate leaf spot

Citrus canker *Xanthomonas campestris* pv *citrii:*
- tree fruits, citrus 447-449

Citrus foot rot *Phytophthora* spp.:
- tree fruits, citrus 450

Cladosporium carpophylum — lacy scab

Cladosporium fulvum — leaf mold

Cladosporium leaf mold:
- eggplant 335
- tomatoes 375

Clubroot *Plasmodiophora brassicae:*
- cole crops 326
- tomatoes 376

Coccomyces hiemalis — cherry leaf spot

Coffee berry disease *Colletotrichum coffeanum:*
- tree fruits, subtropical 471,473

Collar rot *Phytophthora* spp.:
- nuts 483
- tree fruits, deciduous 456,458,459,463

Colletotrichum coffeanum — coffee berry disease

Colletotrichum leaf spot:
- photo 24

Common blight *Xanthomonas campestris* pv *phaseoli:*
- beans 311,312

Common bunt:
- rice 182
- small grains 134-137,141

Common root rot:
- small grains 136,137,139

Common rot:
- photo 27

Common rust:
- field corn, popcorn 86,87
- sweet corn 368,369

Common smut:
- small grains 135,139,141

Cone rust:
- ornamentals 533

Copper spot *Gloeacericospora sorghi:*
- lawns/turf 511-513,515-518

Corm rot *Fusarium, Sclerotium, Stromatinia* spp.:
- ornamentals 520,532

Corn head smut:
- field corn, popcorn 86
- sweet corn 368

Corticium fuciforme — red thread

Corticium red thread *Corticium* spp.:
- lawns/turf 513

Corticium salmonicolor — pink disease

Coryneum blight (shot hole) *Stigmina carpophila:*
- nuts 479-481
- tree fruits, deciduous 453-460,465-467

Cottony blight:
- lawns/turf 512

Cottony leak:
- photo 21
- cucurbits 331

Covered kernel smut:
- sorghum 185,186

Covered smut:
- field corn, popcorn 87
- peas 352
- rice 183
- small grains 134,135,137-141
- sorghum 186
- soybeans 104

Cronartium fusiforme — Fusiform rust

Cronartium quercuum — pine gall rust

Crown and root rot:
- photo 28

Crown gall *Agrobacerum tumefaciens:*
- biocontrols 54
- nuts 482
- ornamentals 524
- small fruits 440
- tree fruits, deciduous 462

Crown rot:
- asparagus 308
- lawns/turf 512,515
- nuts 483
- ornamentals 523,527,528,531,532
- tree fruits, deciduous 456,458,459,463

Crown rust:
- small grains 134,139,140

Cryptosporella umbrina — brown canker

Cryptosporella viticola — dead-arm

Curvularia flower/leaf spot *Curvularia trifoli:*
- lawns/turf 511
- ornamentals 524,525

Curvularia leaf blight:
- ornamentals 532

Curvularia trifoli — Curvularia flower/leaf spot

Cyclaneusma needle cast:
- ornamentals 521

Cycloconium oleaginum — peacock spot

Cylindrocladium:
- cucurbits 329
- eggplant 334
- onions 347
- ornamentals 521,524
- sweet potatoes, yams 370
- tobacco 192

Cylindrocladium black rot *Cylindrocladium crotalariae:*
- peanuts 178

Cylindrocladium crotalariae — Cylindrocladium black rot

Cylindrocladium cutting rot:
- ornamentals 521

Cylindrocladium leaf blight:
- ornamentals 525

Cylindrocladium root rot:
- ornamentals 531

Cylindrocladium rot:
- ornamentals 524,532

Cylindrosporium:
- ornamentals 521

Cylindrosporium leaf spot:
- ornamentals 524

D

Dactylaria leaf spot:
- ornamentals 521,524

Daffodil leaf scorch *Stagonospora curtisii:*
- ornamentals 524

Damping-off:
- beans 310,311,313,314,316
- carrots, etc. 316,318,319
- castor beans 172
- celery 321
- cole crops 323,325-327
- cotton 122-125
- cucurbits 328,331-333
- eggplant 335
- field corn, popcorn 85-88
- flax 172,173
- forage (alfalfa, etc.) 548,549
- garbanzos 336
- garlic 337
- greens 339,340

herbs/spices ... 341,342
lawns/turf ... 509,510,512,514,518
lentils ... 342
lettuce ... 344,345
mung beans ... 175
okra ... 346
onions ... 347-349,351
ornamentals ... 520,523,526,528,532
peanuts ... 176,178,181
peas ... 351-353
peppers ... 354-356
potatoes ... 357,360
radishes ... 362,363
rhubarb ... 364
rice ... 182-184
safflower ... 184
small grains ... 134-139,141
sorghum ... 185,186
southern peas ... 364,365
soybeans ... 101-105
spinach ... 365-367
sugar beets ... 187,189,191
sunflower ... 191,192
sweet corn ... 367-369
table beets ... 370-372
tobacco ... 192,193
tomatoes ... 373-376,378

Dead-arm *Cryptosporella viticola:*
small fruits ... 435,441

Dead bud *Pseudomonas syringae:*
small grains ... 136
tree fruits, deciduous ... 455-458

Decay *Fusarium* spp.:
ornamentals ... 520,532

Diaporthe medusaea — Phomopsis stem blight
Diaporthe phaseolorum — pod and stem blight
***Diaporthe* spp.**— melanose
Didymella bryoniae — gummy stem blight

Didymellina:
ornamentals ... 521

Didymellina leaf spot *Didymellina macrospora:*
ornamentals ... 524,527,532

Didymellina macrospora — leaf spot
***Didymellina* spp.**— leaf spot

Dieback *Phytophthora* spp.:
ornamentals ... 522
small fruits ... 436

Diplodia pinea — Diplodia tip blight

Diplodia tip blight *Diplodia pinea:*
ornamentals ... 529

Discosia leaf spot:
ornamentals ... 524

Dollar spot *Sclerotinia homoecarpa:*
biocontrols ... 55
lawns/turf ... 511-519
ornamentals ... 533

Downy mildew:
photo ... 22,23,25
beans ... 311-313,315
biocontrols ... 58
carrots, etc. ... 316-319
celery ... 320,321
cole crops ... 322-325,327
cucurbits ... 328-334
eggplant ... 334
forage (alfalfa, etc.) ... 548
garlic ... 336-338
greens ... 338-340
herbs/spices ... 341
hops ... 173,174
lawns/turf ... 514
lettuce ... 343,344
nuts ... 479,482
onions ... 346-350
ornamentals ... 519,522,524,525
peas ... 352
peppers ... 354,355
radishes ... 363
rhubarb ... 364
small fruits ... 433,435-439,441,442,445
sorghum ... 186
southern peas ... 364,365
spinach ... 366,367
sugar beets ... 187-190
sunflower ... 192
sweet corn ... 369
table beets ... 371,372
tomatoes ... 377
tree fruits, citrus ... 446

Downy spot:
nuts ... 478,482,483,485

Drechslera leaf spot:
lawns/turf ... 511

Drechslera stem rot:
ornamentals ... 524

Drop rot:
lettuce ... 343

Dry bulb rot:
ornamentals ... 528

Dry rot:
ornamentals ... 528

Dutch elm disease:
ornamentals ... 529,531

Dwarf bunt *Tilletia controversa:*
small grains ... 136,137

E

Early blight:
photo ... 27
eggplant ... 335
herbs/spices ... 341
peppers ... 354

Early blight *Alternaria solani:*
potatoes ... 357-362
tomatoes ... 372-375,377,378

Early blight *Cercospora* spp.:
carrots, etc. ... 317,319
celery ... 320-322

Early leaf spot:
peanuts ... 176,177,179,180

Early maturity disease *Verticillium dahliae:*
potatoes ... 360

Eastern filbert blight *Anisogramma anomala:*
nuts ... 478-481

Elsinoe leaf spot:
ornamentals ... 524

Endosepsis *Fusarium moniliforme:*
tree fruits, subtropical ... 475

Entomosporium leaf spot:
ornamentals ... 522,524,527,532

***Entyloma* spp.**— leaf smut

Erwinia herbicola:
tree fruits, subtropical ... 471,472

European canker:
ornamentals ... 534
tree fruits, deciduous ... 456-459

Eutypa dieback:
small fruits ... 434

Exobasidium blight:
ornamentals ... 525

Eyespot:
field corn, popcorn ... 88
lawns/turf ... 511
small grains ... 141
sweet corn ... 369

F

Fabraea leaf spot:
ornamentals ... 521,524,527
tree fruits, deciduous ... 463,469

Fairy ring:
lawns/turf ... 513
small fruits ... 439

False loose smut *Ustilago nigra:*
small grains ... 137,138

Fire blight:
photo ... 10
biocontrols ... 54,58
ornamentals ... 519,521,522
tree fruits, deciduous ... 455-460,462,463,467

Flag smut *Urocystis agropyri:*
small grains ... 134-136,139,140

Flower blight:
biocontrols ... 58
ornamentals ... 520,524,525,527-529,532-534

Fly speck:
tree fruits, deciduous ... 452,454,461,463,468,469

Foliage blight:
herbs/spices ... 341

Foot (root) rot:
small grains ... 134,139-141
sweet potatoes, yams ... 370

Fresh fruit rot:
nuts ... 477

Frogeye leaf spot:
ornamentals ... 524
soybeans ... 101,103,105

Fruit blight:
ornamentals ... 527

Fruit brown rot *Monilinia* spp.:
tree fruits, deciduous ... 452,462,464-466,468

Fruit decay:
tree fruits, citrus ... 447
tree fruits, deciduous ... 460

Fruit drop:
biocontrols ... 58

Fruit rot:
small fruits ... 436-439,441,445
tomatoes ... 373,375
tree fruits, citrus ... 446
tree fruits, deciduous ... 452,453,461,467-469
tree fruits, subtropical ... 470

Fusarium:
Bt chart ... 76,94,108,128,144,196
asparagus ... 308,309
beans ... 310,314
biocontrols ... 55
cole crops ... 325,327
cotton ... 124
cucurbits ... 329,332,333
eggplant ... 334,335
field corn, popcorn ... 87
lettuce ... 344,345
onions ... 347,349,351
ornamentals ... 521,524,525,532
peanuts ... 179
peas ... 352
peppers ... 355,356
potatoes ... 358,362
rice ... 182,183
safflower ... 184
small fruits ... 442,445
small grains ... 137
sorghum ... 186
soybeans ... 102,104
sugar beets ... 189
sweet potatoes, yams ... 370
tobacco ... 192,193
tomatoes ... 376,377
tree fruits, citrus ... 451
tree fruits, subtropical ... 476,477

Fusarium blight:
lawns/turf ... 512,513,517,518

Fusarium crown rot:
small grains ... 136

Fusarium decay:
potatoes ... 359

Fusarium dry rot:
potatoes ... 358

Fusarium leaf spot:
ornamentals ... 524

Fusarium mold:
tree fruits, subtropical ... 475

Fusarium moniliforme — Endosepsis (figs)
Fusarium nivale — Fusarium patch, pink snow mold

Fusarium patch:
ornamentals ... 533

Fusarium patch *Fusarium nivale:*
lawns/turf ... 509,511-513,516-519

Fusarium root rot:
beans ... 309
ornamentals ... 520
small grains ... 136

Fusarium rot:
photo ... 20,23
ornamentals ... 532

Fusarium scab:
small grains ... 139

Fusarium seed piece decay:
potatoes ... 359

Fusarium seed rot:
potatoes ... 359

Fusarium seed scab:
small grains ... 136

Fusarium snow mold:
lawns/turf ... 513

***Fusarium* spp.**— corm rot, decay

Fusarium wilt:
cotton ... 125
spinach ... 365

Fusicladium scab:
ornamentals ... 524,527

Fusiform rust *Cronartium fusiforme:*
ornamentals ... 524,527,533,534

G

Gaeumannomyces gramminis — take-all

Gall:
ornamentals ... 533

Gloeacericospora sorghi — copper spot
Gloeodes pomigena — sooty blotch

Gloeosporium leaf spot:
ornamentals ... 524

Glomerella cingulata — bitter rot

Glume blotch *Septoria nodorum:*
lawns/turf ... 511
small grains ... 134-141

Gnomonia caryae — liver spot
Gnomonia nerviseda — vein spot

Granville bacterial wilt *Pseudomonas solanacearum:*
tobacco ... 192,193

Grape powdery mildew:
photo ... 21

Gray leaf spot *Cercospora* spp.:
field corn, popcorn ... 86,88
sweet corn ... 368,369

Gray leaf spot *Stemphylium* spp.:
eggplant ... 335
lawns/turf ... 511,517
tomatoes ... 373-375,377

Gray mold *Botrytis* spp.:
photo ... 18,19,22
beans ... 312,315
biocontrols ... 58
carrots, etc. ... 319
lettuce ... 343
ornamentals ... 522,525,533
small fruits ... 435,440,445,446
table beets ... 371
tomatoes ... 373-375,377
tree fruits, deciduous ... 468

Gray snow mold:
lawns/turf ... 509-513,515-519
ornamentals ... 533

Grease spot:
lawns/turf ... 512

Greasy spot:
photo ... 20
biocontrols ... 58
tree fruits, citrus ... 447-449,451

Guignardia leaf blotch:
ornamentals ... 524

Gummosis:
tree fruits, citrus ... 450
tree fruits, deciduous ... 455,459

Gummy stem blight *Didymella bryoniae, Mycosphaerella melonis:*
cucurbits ... 328-333

Gymnosporangium clavipes — quince rust
***Gymnosporangium* spp.**— cedar rust

H

Halo blight *Psuedomonas syringae* pv *phaseolicola:*
beans ... 311,312,315

Hawthorn rust:
ornamentals ... 524

Head blight:
photo ... 26

Head smut:
field corn, popcorn 85
Heart rot *Phytophthora* spp.:
tree fruits, subtropical 474,476
Helminthosporium:
beans 310
lawns/turf 511
ornamentals 521,524,533
small grains 137
soybeans 102
Helminthosporium blight:
sweet corn 369
Helminthosporium leaf blight:
field corn, popcorn 86-88
rice 183
small grains 134
sweet corn 368,369
Helminthosporium leaf blotch:
small grains 136
Helminthosporium leaf spot:
lawns/turf 511,514,515
small grains 137,138
Helminthosporium oryzae — brown spot
Helminthosporium papulosum — black pox
Helminthosporium solani:
potatoes 358
Helminthosporium spot blotch:
small grains 136,140
***Helminthosporium* spp.**— leaf blight, leaf spot
Herpobasidium blight:
ornamentals 524

I

Ice-nucleating bacteria:
tree fruits, citrus 450
Ink spot *Mystrosporium adjustum:*
ornamentals 521,524,525
Iris leaf spot:
ornamentals 532
Iron spot (eyespot/brown spot) *Cercospora coffeicola:*
tree fruits, subtropical 471,473
Isariopsis leaf spot:
small fruits 434

J

Jacket rot (green fruit rot) *Sclerotinia sclerotiorum:*
tree fruits, deciduous 454

K

Kabatina dieback:
ornamentals 525
Kernel rot:
nuts 480,481
Kernel smut:
sorghum 185

L

Lacy (russet) scab *Cladosporium carpophylum:*
tree fruits, deciduous 454,455
Large brown patch *Rhizoctonia* spp.:
lawns/turf 511-513,517
Large patch:
lawns/turf 513
Late blight:
photo 25
carrots, etc. 317,319
celery 320-322
eggplant 335
herbs/spices 341
nuts 479-481
peppers 354
potatoes 357-362
tomatoes 372-377
Late leaf spot *Cercosporidium* spp.:
peanuts 176,177,179,180
Leaf blight:
biocontrols 58
carrots, etc. 317
cucurbits 332
garlic 337
nuts 478,485
onions 348,351
ornamentals 532,533,534
small fruits 434,436-439,445
small grains 130
soybeans 103
table beets 371
Leaf blotch:
photo 18
nuts 482,483,485
ornamentals 521
small grains 141
tree fruits, deciduous 469
Leaf curl:
ornamentals 534
tree fruits, deciduous 453,455-462,467,468
tree fruits, subtropical 470
Leaf mold *Cladosporium fulvum:*
photo 27
tomatoes 373-375
Leaf rust:
lawns/turf 511,513,514
ornamentals 522
small grains 134,136,137,139,140
tree fruits, subtropical 471,473
Leaf scab:
ornamentals 532
Leaf scorch:
nuts 478,482,483
small fruits 434,438,439,445
Leaf smut *Entyloma* spp.:
lawns/turf 515
rice 181,183
Leaf spot:
photo 20,24,26
artichokes 308
beans 311,314,315
biocontrols 58
carrots, etc. 317
celery 320,322
cole crops 323,324,327
cucurbits 329,330
forage (alfalfa, etc.) 549
greens 338-340
lawns/turf 509-511,513,515-517,519
lettuce 344
nuts 482-485
ornamentals 521,522,524,525,527-534
peanuts 179-181
peas 353
peppers 356
potatoes 361
radishes 363
rice 183
small fruits 434-440
soybeans 104,105
spinach 366
sugar beets 188,190
table beets 371
tomatoes 373
tree fruits, citrus 446,447
tree fruits, deciduous 453-462,464-467,469
tree fruits, subtropical 470,471,474
Leather rot *Phytophthora cactorum:*
small fruits 440,442
Leptosphaeria brown spot *Leptosphaeria* spp.:
ornamentals 524
Leptosphaeria korrae — necrotic ring spot
Leptosphaeria nodorum — Septoria glume blotch
Leptosphaerulina leaf spot:
forage (alfalfa, etc.) 548
Lettuce drop:
lettuce 344
Limb rot:
peanuts 181
Lime anthracnose:
photo 20
Limonomyces roseipellis — pink patch
Liver spot *Gnomonia caryae:*
nuts 478,482,483,485
Loose smut:
small grains 134-137,139,140
Lophodermium needle cast:
ornamentals 521,524,533
Lophodermium twig blight:
small fruits 439

M

Magnaporthe poae — summer patch
Marssonina leaf spot *Marssonina populi:*
ornamentals 521
Marssonina populi — leaf spot
Melampsora needle rust:
ornamentals 524,534
Melanconium fuligineum — bitter rot
Melanose *Diaporthe* spp.:
tree fruits, citrus 447-449,451
Melting-out *Helminthosporium* spp.:
lawns/turf 509,511,513-516,519
ornamentals 533
Mold:
beans 310
biocontrols 58
cole crops 323
cotton 122
cucurbits 328
field corn, popcorn 85
flax 172
forage (alfalfa, etc.) 548
greens 339
lawns/turf 510
lentils 342
peanuts 176
peas 351
peppers 354
radishes 362
small grains 134
sorghum 185
southern peas 364
soybeans 101
spinach 365
sugar beets 187
sunflower 191
sweet corn 367
table beets 370
tobacco 193
tomatoes 373
Monilinia:
ornamentals 521
Monilinia laxa — brown rot
***Monilinia* spp.**— blossom/twig blight, brown rot, fruit brown rot
Monilinia vaccini — mummy berry
Monilochaetes infuscans — scurf
Monochaetia canker:
ornamentals 524
Mummy berry *Monilinia vaccini:*
small fruits 434-436
tree fruits, citrus 446
Mycosphaerella citri — pink pitting
Mycosphaerella melonis — gummy stem blight
Mycosphaerella musicola — Sigatoka
Mycosphaerella ray blight:
ornamentals 521
Mystrosporium adjustum — ink spot
Mystrosporium ink spot:
ornamentals 524

N

Nailhead rust:
tomatoes 374
Narrow brown leaf spot *Cercospora oryzae:*
rice 181-183
Neck dry rot *Stomatinia gladioli:*
ornamentals 527,528
Neck rot *Botrytis* spp.:
onions 346,348,349,351
ornamentals 528
Necrotic ring spot *Leptosphaeria korrae:*
lawns/turf 509,512,513,515-517
Needle blight:
ornamentals 522,527,528
Needle cast:
ornamentals 525
Needle rust:
ornamentals 527,529
Neofabraea malicorticus — bull's-eye rot
Net blotch *Pyrenophora teres:*
small grains 134,135,137,139,140

O

Oak root fungus *Armillaria mellea:*
asparagus 308
cole crops 325
cucurbits 332
eggplant 335
lettuce 344
onions 349
ornamentals 526
peppers 355
small fruits 442
tobacco 193
tomatoes 376
tree fruits, citrus 451
tree fruits, deciduous 463
tree fruits, subtropical 476
Oak wilt:
ornamentals 529
Oat smut *Ustilago* spp.:
small grains 135,137-139
Olive knot:
ornamentals 524
tree fruits, subtropical 471,472,474
Olive leaf spot (peacock spot):
tree fruits, subtropical 471,472
Overwintering spores:
tree fruits, citrus 447
tree fruits, deciduous 453
Ovulinia azaleae — Ovulinia flower blight
Ovulinia blight:
ornamentals 527,532
Ovulinia flower blight *Ovulinia azaleae:*
ornamentals 521

P

Peach Antracnose:
photo 25
Peach blight:
tree fruits, deciduous 455,457-460
tree fruits, subtropical 470
Peach leaf curl *Taphrina deformans:*
tree fruits, deciduous 453,457,459,469
Peach scab *Cladosporium carpophilum:*
tree fruits, deciduous 452,455,468,469
Peacock spot *Cycloconium oleaginum:*
tree fruits, subtropical 471-473,477
Pear blight:
tree fruits, deciduous 456
Pear scab *Venturia nashicola:*
tree fruits, deciduous 453,469
Penicillium rot:
ornamentals 532
Pestalotia leaf spot *Pestalotia rhododendri:*
ornamentals 522
Pestalotia rhododendri — pestalotia leaf spot
Petiole rot:
ornamentals 531
Phoma betae — Phoma leaf spot
Phoma leaf spot *Phoma betae:*
herbs/spices 341
tomatoes 373
Phoma lingam — blackleg
Phoma twig blight:
ornamentals 521
Phomopsis:
eggplant 334,335
ornamentals 522,524
small fruits 441,444
tree fruits, citrus 446
Phomopsis blight:
ornamentals 529

Disease Index

Phomopsis cane:
small fruits ... 434
Phomopsis cane spot:
small fruits ... 433
Phomopsis canker:
small fruits ... 440
Phomopsis leaf spot:
small fruits ... 433
Phomopsis obscurans — Phomopsis soft rot
Phomopsis soft rot *Phomopsis obscurans:*
small fruits ... 440
Phomopsis stem blight *Diaporthe medusaea:*
ornamentals ... 521,522
Phragmidium juniperi-virginiae — cedar-apple rust
Phyllosticta leaf spot:
ornamentals ... 524
Phyllosticta solitaria — apple blotch
Phytophthora:
photo ... 26
asparagus ... 308,309
beans ... 313
carrots, etc. ... 318
cole crops ... 325,327
cucurbits ... 329,332,333
eggplant ... 334,335
field corn, popcorn ... 87
forage (alfalfa, etc.) ... 548
forage (grasses, hay) ... 549
garbanzos ... 336
herbs/spices ... 341,342
horseradish ... 342
lentils ... 342
lettuce ... 344,345
nuts ... 482
okra ... 346
onions ... 347,349,351
ornamentals ... 519,521,522,524
peas ... 352,353
peppers ... 355,356
radishes ... 363
rice ... 183
small fruits ... 442,445
small grains ... 138
southern peas ... 364,365
soybeans ... 103-105
spinach ... 367
sweet corn ... 369
sweet potatoes, yams ... 370
table beets ... 371
tobacco ... 192,193
tomatoes ... 376,377
tree fruits, citrus ... 447-449,451
tree fruits, subtropical ... 476,477
Phytophthora blight:
eggplant ... 334
nuts ... 480
ornamentals ... 525
peas ... 352
peppers ... 354,355
Phytophthora brown rot:
tree fruits, citrus ... 447
Phytophthora canker:
tree fruits, subtropical ... 474
Phytophthora cactorum — leather rot
Phytophthora collar rot:
nuts ... 483
tree fruits, deciduous ... 462
Phytophthora crown rot:
nuts ... 483
ornamentals ... 526
Phytophthora dieback *Phytophthora nicotianae:*
ornamentals ... 521,522,525
Phytophthora foot rot:
tree fruits, citrus ... 450
Phytophthora fragariae — red stele
Phytophthora fruit rot:
cucurbits ... 330
tomatoes ... 376
tree fruits, subtropical ... 475
Phytophthora heart rot:
tree fruits, subtropical ... 475
Phytophthora nicotianae — Phytophthora dieback
Phytophthora parasitica nicotianae — black shank
Phytophthora root rot:
cucurbits ... 330
nuts ... 483
ornamentals ... 526
small fruits ... 440,441
soybeans ... 103
tree fruits, citrus ... 450
tree fruits, deciduous ... 462
tree fruits, subtropical ... 474
Phytophthora shoot blight:
ornamentals ... 526
***Phytophthora* spp.** — collar rot, crown rot, dieback, foot rot, heart rot
Phytophthora stem rot:
ornamentals ... 526
soybeans ... 103
Pine gall rust *Cronartium quercuum:*
ornamentals ... 524
Pine rust:
ornamentals ... 533
Pineapple butt rot *Thielaviopsis paradoxa:*
tree fruits, subtropical ... 470
Pineapple disease *Ceratocystis paradoxa:*
sugarcane ... 191
Pink disease *Corticium salmonicolor:*
tree fruits, subtropical ... 471,473
Pink patch *Limonomyces roseipellis:*
lawns/turf ... 512,516,519
Pink pitting *Mycosphaerella citri:*
tree fruits, citrus ... 447-449
Pink rot *Sclerotinia:*
celery ... 320,321
herbs/spices ... 341
onions ... 350
potatoes ... 360
Pink snow mold (Fusarium patch) *Fusarium nivale:*
lawns/turf ... 509,511-513,515-519
ornamentals ... 533
Plasmodiophora:
asparagus ... 308,309
cole crops ... 325,327
cucurbits ... 329,332,333
eggplant ... 334,335
lettuce ... 344,345
onions ... 347,349,351
peppers ... 355,356
small fruits ... 442,445
sweet potatoes, yams ... 370
tomatoes ... 376,377
tree fruits, citrus ... 451
tree fruits, deciduous ... 476,477
Plasmodiophora brassicae — clubroot
Plum pockets:
ornamentals ... 534
Poa patch:
lawns/turf ... 516
Pod blight:
beans ... 310
Pod and stem blight *Diaporthe phaseolorum:*
soybeans ... 101-105
Pod rot:
photo ... 24
peanuts ... 176-179,181
Powdery mildew:
photo ... 18,19,21,22,24,25
beans ... 311,312,314,315
biocontrols ... 54,58
carrots, etc. ... 319
celery ... 322
cole crops ... 326,327
cucurbits ... 328-330,332-334
eggplant ... 335
field corn, popcorn ... 88
flax ... 173
forage (alfalfa, etc.) ... 547,548
garlic ... 338
greens ... 338,340
hops ... 174
lawns/turf ... 512,513,515,516,518
lettuce ... 344
mint ... 175
nuts ... 478,482-485
okra ... 346
onions ... 350
ornamentals ... 520-524,527-534
peanuts ... 179,180
peas ... 352,353
peppers ... 356
potatoes ... 361,362
rice ... 183
small fruits ... 433-439,441-446
small grains ... 134,136,139-141
sorghum ... 186
soybeans ... 104,105
spinach ... 367
sugar beets ... 187,190,191
sweet corn ... 369
table beets ... 372
tomatoes ... 372,377
tree fruits, citrus ... 447
tree fruits, deciduous ... 452,453,458,461,464-468
tree fruits, subtropical ... 470,476
Pseudomonas:
cucurbits ... 329
eggplant ... 334
onions ... 347
sweet potatoes, yams ... 370
tree fruits, deciduous ... 456,458-460
tree fruits, subtropical ... 471,472
Pseudomonas blight:
small fruits ... 437
tree fruits, deciduous ... 456-458,460
Pseudomonas canker:
tree fruits, deciduous ... 459
Pseudomonas leaf spot:
ornamentals ... 521,522
Pseudomonas solanacearum — Granville bacterial wilt
Pseudomonas syringae — bacterial blast, black pitting, dead bud
Psuedomonas syringae* pv *phaseolicola — halo blight
Pucciniastrum needle rust:
ornamentals ... 524,534
Puccinia gramminis — barley/rye rust
Puccinia rust:
ornamentals ... 524
Puccinia striiformis — stripe rust
Purple blotch *Alternaria porri:*
garlic ... 336,337
onions ... 346-351
Purple seed stain *Cercospora kikuchii:*
soybeans ... 101,103,105
Purple spot:
ornamentals ... 524
Pyrenophora graminea — barley leaf stripe
***Pyrenophora* spp.** — tan spot
Pyrenophora teres — net blotch
Pythium:
photo ... 27
Bt chart ... 76
asparagus ... 308,309
beans ... 310,315
biocontrols ... 55
carrots, etc. ... 318
cole crops ... 325,327
cotton ... 125
cucurbits ... 329,332,333
eggplant ... 334,335
field corn, popcorn ... 88
forage (grasses, hay) ... 549
horseradish ... 342
lettuce ... 344,345
onions ... 347,349,351
ornamentals ... 519
peanuts ... 176,177,180
peas ... 353
peppers ... 355,356
radishes ... 363
rice ... 182,183
small fruits ... 442,445
small grains ... 141
soybeans ... 102,105
sugar beets ... 187,189
sweet potatoes, yams ... 370
table beets ... 371
tobacco ... 192,193
tomatoes ... 376,377
tree fruits, citrus ... 451
tree fruits, subtropical ... 476,477
Pythium blight (cottony blight):
beans ... 311
cotton ... 123
lawns/turf ... 509-514
southern peas ... 364
soybeans ... 102
Pythium fruit rot:
cucurbits ... 328,331
Pythium leak:
potatoes ... 360
Pythium root rot:
lawns/turf ... 509,512
nuts ... 483
ornamentals ... 526
Pythium rot:
small grains ... 140
***Pythium* spp.** — collar rot, cottony blight, damping-off, foot rot (crown rot), fruit rot
Pythium stem rot:
ornamentals ... 526

Q

Quince rust *Gymnosporangium clavipes:*
tree fruits, deciduous ... 469

R

Raceme blight:
nuts ... 479-481
Ramularia:
ornamentals ... 525
Ramularia beticola — leaf spot
Ramularia leaf spot *Ramularia beticola:*
ornamentals ... 524,532
small fruits ... 436
Ray blight:
ornamentals ... 524,529
Red alga:
tree fruits, subtropical ... 471,472
Red leaf spot:
lawns/turf ... 513
Red stele *Phytophthora fragariae:*
small fruits ... 440,442
Red thread *Corticium fuciforme:*
lawns/turf ... 509,511-513,515-519
ornamentals ... 533
Rhabdocline needle cast *Rhabdocline pseudotsugae:*
ornamentals ... 520-522
Rhabdocline pseudotsugae — needle cast (Douglas-fir)
Rhizobium:
Bt chart ... 197
Rhizoctonia:
photo ... 22
Bt chart ... 94,108,128,145,197
asparagus ... 308,309
beans ... 311,314
biocontrols ... 55
cole crops ... 325,327
cotton ... 122-125
cucurbits ... 329,332,333
eggplant ... 334,335
field corn, popcorn ... 87
herbs/spices ... 341
lawns/turf ... 509
lettuce ... 344,345
onions ... 347,349,351
ornamentals ... 521,524,525,527,532
peanuts ... 179
peas ... 352
peppers ... 355,356
potatoes ... 358,362
rice ... 183
safflower ... 184
small fruits ... 442,445
small grains ... 136,139
sorghum ... 186
southern peas ... 364

- soybeans 102,104
- sugar beets 187,189
- sweet potatoes, yams 370
- tobacco 192,193
- tomatoes 376,377
- tree fruits, citrus 451
- tree fruits, subtropical 476,477

Rhizoctonia aerial web blight *Rhizoctonia solani:*
- ornamentals 525

Rhizoctonia blight *Rhizoctonia solani:*
- lawns/turf 510,512,518
- ornamentals 521

Rhizoctonia bottom rot:
- lettuce 345

Rhizoctonia brown patch *Rhizoctonia solani:*
- lawns/turf 511,513,514

Rhizoctonia foliage blight:
- herbs/spices 341

Rhizoctonia fruit rot *Rhizoctonia solani:*
- tomatoes 373,375,377

Rhizoctonia large patch:
- lawns/turf 509

Rhizoctonia limb rot:
- peanuts 180

Rhizoctonia needle blight:
- ornamentals 527,528

Rhizoctonia oryzae — sheath spot

Rhizoctonia root rot:
- ornamentals 520,527,528

Rhizoctonia solani:
- beans 310,315
- cotton 123-125
- field corn, popcorn 85,88
- lawns/turf 515
- ornamentals 521
- peanuts 176,177,180
- peas 353
- potatoes 358,360
- rice 182,183
- small grains 135,141
- soybeans 102,105

Rhizoctonia solani — aerial web blight, black root rot, bottom rot, brown patch, fruit rot, seed piece decay, seedling disease, sheath blight, sore shin

Rhizoctonia web blight:
- ornamentals 525

Rhizomania:
- sugar beets 190

Rhizome rot:
- ornamentals 532

Rhizopus:
- peanuts 176

Rhizopus fruit rot:
- tree fruits, deciduous 460

Rhizopus rot *Rhizopus* spp.:
- sweet potatoes, yams 370
- tree fruits, deciduous 468

Rhizosphaera needle cast:
- ornamentals 521

Rhynchosporium secalis — barley scald

Ring spot:
- cole crops 322,323,327
- greens 338

Ripe rot:
- peppers 355

Root and stem rot:
- beans 313,314
- cole crops 326
- eggplant 335
- garbanzos 336
- lentils 342
- peas 352
- peppers 355
- potatoes 360
- southern peas 364
- tomatoes 375,376

Root rot:
- biocontrols 62
- forage (alfalfa, etc.) 548
- lawns/turf 509,510,512
- mung beans 175
- nuts 483
- ornamentals 520,523,526-528,531,532
- peanuts 179
- tobacco 192,193
- tree fruits, citrus 450
- tree fruits, deciduous 463
- tree fruits, subtropical 476

Rose bloom:
- small fruits 438

Rust:
- asparagus 308,309
- beans 311,313-316
- biocontrols 58
- celery 322
- cole crops 326,327
- cotton 123,125
- field corn, popcorn 86,88
- forage (alfalfa, etc.) 548
- greens 340
- lawns/turf 515,516,518
- lettuce 344
- mint 174
- nuts 482-484
- ornamentals 520-525,527,529-534
- peanuts 176,178-181
- peas 353
- peppers 356
- potatoes 361
- small fruits 434
- small grains 139,141
- southern peas 364,365
- spinach 367
- sugar beets 188
- sugarcane 191
- sweet corn 368,369
- tree fruits, deciduous 461,463-466

Rye rust *Puccinia gramminis:*
- small grains 134,140

S

Scab:
- biocontrols 58
- cucurbits 329-334
- nuts 477-479,482-485
- ornamentals 521-525,527,529
- small grains 140
- tree fruits, citrus 447-449,451
- tree fruits, deciduous 452-455,458,460-469
- tree fruits, subtropical 470-474,476

Sclerotium rolfsii:
- peanuts 177

Scirrhia acicola — brown spot

Scirrhia brown spot *Scirrhia acicola:*
- ornamentals 521,524

Scleroderris canker *Scleroderris* spp.:
- ornamentals 521

Sclerophthora macrospora — yellow tuft

Sclerotinia:
- asparagus 308,309
- cole crops 325,327
- cucurbits 329,332,333
- eggplant 334,335
- lawns/turf 511
- lettuce 344,345
- onions 347,349,351
- ornamentals 532
- peppers 355,356
- small fruits 442,445
- soybeans 105
- sweet potatoes, yams 370
- tobacco 192,193
- tomatoes 376,377
- tree fruits, citrus 451
- tree fruits, subtropical 476,477

Sclerotinia blight:
- lettuce 343
- sweet potatoes, yams 370

Sclerotinia blight *Sclerotinia minor:*
- peanuts 177

Sclerotinia drop:
- lettuce 345

Sclerotinia homoecarpa — dollar spot

Sclerotinia minor — Sclerotinia blight

Sclerotinia sclerotiorum — green fruit rot, jacket rot, watery soft rot, white mold

***Sclerotinia* spp.** — pink rot

Sclerotinia stem rot:
- ornamentals 528

Sclerotinia urnula:
- small fruits 434

Sclerotinia white mold:
- beans 309
- carrots, etc. 316
- cole crops 322
- greens 338

Sclerotinia wilt:
- lettuce 343

Sclerotium:
- asparagus 308,309
- carrots, etc. 319
- cole crops 325,327
- cucurbits 329,332,333
- eggplant 334,335
- lettuce 344,345
- onions 347,349,351
- peppers 355,356
- sweet potatoes, yams 370
- tobacco 192,193
- tomatoes 376,377
- tree fruits, subtropical 476,477

Sclerotium blight:
- beans 311
- cotton 123
- southern peas 364
- soybeans 102
- sugar beets 187

Sclerotium gladioli — bulb dry rot

Sclerotium rolfsii — southern blight

Sclerotium rolfsii:
- lawns/turf 510
- peanuts 177

Sclerotium stem rot:
- peanuts 180

Scorch:
- small fruits 437

Scurf *Monilochaetes infuscans:*
- potatoes 359
- sweet potatoes, yams 370

Seed decay:
- beans 310,311,316
- carrots, etc. 316,319
- castor beans 172
- cole crops 323,327
- cotton 122,123,125
- cucurbits 328,333
- eggplant 335
- field corn, popcorn 85-88
- flax 172,173
- forage (alfalfa, etc.) 547-549
- greens 339,340
- lawns/turf 510,518
- lentils 342
- lettuce 345
- okra 346
- onions 346,347,351
- ornamentals 532
- peanuts 176,181
- peas 351,353
- peppers 354,356
- potatoes 357,358
- radishes 362,363
- rice 182,184
- safflower 184
- small grains 134,135,137-139,141
- sorghum 185,186
- southern peas 364,365
- soybeans 101-105
- spinach 365-367
- sugar beets 187,191
- sunflower 191,192
- sweet corn 367-369
- table beets 370-372
- tomatoes 373,378

Seed rot:
- beans 310,314
- cotton 123,124
- field corn, popcorn 87
- flax 173
- forage (grasses, hay) 549
- hops 174
- onions 348
- peanuts 176,178,179
- rice 182,183
- safflower 184
- small grains 134-139
- sorghum 185
- soybeans 101,102
- tomatoes 375

Seedborne common scab:
- potatoes 359

Seedborne disease:
- field corn, popcorn 85,86
- safflower 184
- sorghum 185

Seedling blight:
- beans 310,316
- carrots, etc. 316,319
- castor beans 172
- cole crops 323,327
- cotton 122,123,125
- cucurbits 328,333
- eggplant 335
- field corn, popcorn 85,86,88
- flax 172,173
- forage (alfalfa, etc.) 547-549
- greens 339,340
- lawns/turf 510,518
- lentils 342
- lettuce 345
- okra 346
- onions 346-348,351
- ornamentals 532
- peanuts 176,178,181
- peas 351,353
- peppers 354,356
- radishes 362,363
- rice 182-184
- safflower 184
- small grains 134,135,137-139,141
- sorghum 185,186
- southern peas 364,365
- soybeans 101,102,104,105
- spinach 365-367
- sugar beets 187,191
- sunflower 191,192
- sweet corn 367-369
- table beets 370-372
- tomatoes 373,375,378

Seedling disease:
- beans 310,311
- cotton 122-125
- field corn, popcorn 85-88
- hops 174
- peanuts 176,178
- rice 182
- safflower 184
- small grains 135
- southern peas 364
- soybeans 101-105
- sugar beets 187
- sunflower 192
- sweet corn 368

Septoria:
- small grains 136,139

Septoria blight:
- celery 320,322
- tomatoes 377

Septoria brown spot:
- soybeans 101,103

Septoria fruit spot:
- tree fruits, citrus 447

Septoria glume blotch *Septoria nodorum:*
- small grains 137

Septoria late blight:
- celery 320

Septoria leaf blight:
- nuts 479-481

Septoria leaf blotch:
- small grains 134,136,140

Septoria leaf spot:
beans 314
cole crops 326
eggplant 335
greens 340
lawns/turf 511,515
mint 174
ornamentals 520-522,524,525
peas 353
peppers 356
potatoes 361
small grains 137,138
table beets 371
tomatoes 372-375,377,378
Septoria nodorum — glume blotch
Septoria spot:
tree fruits, citrus 448
***Septoria* spp.**— brown spot, late blight, leaf blotch
Sheath blight *Rhizoctonia solani:*
rice 181-183
Sheath spot *Rhizoctonia oryzae:*
rice 183
Shoot blight:
nuts 479-481
ornamentals 534
Shot hole (Coryneum blight) *Stigmina carpophila:*
photo 23
nuts 478-485
ornamentals 525,534
tree fruits, deciduous 453,455,457-460,462, 464,467,469
tree fruits, subtropical 470
Shuck rot:
nuts 480,481
Sigatoka *Mycosphaerella musicola:*
tree fruits, subtropical 471-475
Sirococcus strobilinus — Sirococcus tip blight
Sirococcus tip blight *Sirococcus strobilinus:*
ornamentals 521
Slime mold:
lawns/turf 513
Smoulder *Botryotinia narcissicola:*
ornamentals 524
Smut:
onions 348,349
ornamentals 522
Snow mold:
photo 28
lawns/turf 510,515,518
Soft rot:
biocontrols 54
Soil fungi:
small grains 138
Soil-line rot:
photo 23
Soil rot:
sweet potatoes, yams 370
Soilborne disease:
beans 310
cole crops 323
cotton 122
cucurbits 328
field corn, popcorn 85-87
forage (alfalfa, etc.) 547
greens 339
hops 173
lawns/turf 510
nuts 477,478
onions 346
ornamentals 519
peanuts 176
peas 351
peppers 354
rice 182
small fruits 434
small grains 134
sorghum 185
southern peas 364
soybeans 101-103
spinach 365
sugar beets 187
sunflower 192
sweet corn 367,368
table beets 371
tobacco 193
tree fruits, deciduous 452,463
Sooty blotch *Gloeodes pomigena:*
tree fruits, deciduous 452,454,458,461-463, 468,469
Sooty mold:
nuts 483-485
tree fruits, citrus 451
Sore shin *Rhizoctonia solani:*
beans 311
cotton 123
southern peas 364
soybeans 102
sugar beets 187
Southern blight *Sclerotium rolfsii:*
lawns/turf 512,516,518
peppers 355
potatoes 360
tomatoes 376
Southern cone rust:
ornamentals 524
Spear rot:
asparagus 308
Sphaceloma scab:
ornamentals 524
***Sphaceloma* spp.**— Anthracnose leaf spot
Sphacelotheca reiliana:
field corn, popcorn 85
Sphaeropsis leaf spot:
ornamentals 521,524
***Sphaeropsis* spp.**— leaf spot
Spot blight:
lawns/turf 512
Spring dead spot:
lawns/turf 509,512,515,516
Spruce cone rust:
ornamentals 524
Stagonospora curtisii — daffodil leaf scorch
Stagonospora leaf scorch:
ornamentals 521
Stem blight:
beans 310
small grains 135
soybeans 102-105
Stem canker:
soybeans 103
tomatoes 374,375
Stem end rot *Rhizoctonia solani:*
nuts 485
small fruits 440
tree fruits, deciduous 468
tree fruits, subtropical 470
Stem rot *Rhizoctonia* spp.:
ornamentals 523,526-528,532
peanuts 176,177
rice 181,183
Stem rust:
lawns/turf 511,513,514
ornamentals 533
Stemphylium blight:
ornamentals 525
Stemphylium leaf spot:
ornamentals 524,525
tomatoes 374
Stigmina carpophila — Coryneum blight/shot hole
Stinking smut:
small grains 134,135,137-140
Storage rot *Botrytis* spp.:
ornamentals 528
small fruits 439
Strawbreaker:
small grains 134,141
Stripe rust *Puccinia striiformis:*
lawns/turf 511,513,514
small grains 134,140
Stripe smut *Ustilago striiformis:*
lawns/turf 512,516-518
small grains 134,140
Stromatinia dry rot:
ornamentals 523
Stomatinia gladioli — neck dry rot
Summer black stem:
photo 26
Summer patch *Magnaporthe poae:*
lawns/turf 509,512,515-518
Surface mold:
tree fruits, subtropical 475
Surface rot:
tree fruits, subtropical 475
Swiss needle cast:
ornamentals 520,521,524
Sycamore blight:
ornamentals 522

T

Take-all *Gaeumannomyces gramminis:*
small grains 134,136,140
Take-all patch *Gaeumannomyces gramminis:*
lawns/turf 509,512,516-518
Tan leaf spot:
ornamentals 521
Tan spot *Pyrenophora* spp.:
small grains 137,139,141
Taphrina blister:
ornamentals 521
Taphrina leaf blister:
ornamentals 524
***Taphrina* spp.**— leaf blister, peach leaf curl
Tar leaf spot:
ornamentals 522
Target spot:
cucurbits 328,329,333
tomatoes 372,373,377
Thielaviopsis:
ornamentals 532
Thielaviopsis paradoxa — pineapple butt rot
Tilletia caries:
small grains 136
Tilletia controversa — dwarf bunt
Tilletia tritici — bunt/smut
Tip blight *Sirococcus:*
biocontrols 58
ornamentals 520,525,529,532,533
Trunk canker:
tree fruits, citrus 450
Tuber rot:
ornamentals 520
potatoes 362
Tulip fire:
ornamentals 524
Twig blight:
biocontrols 58
nuts 481,484,485
ornamentals 522,527,532,534
small fruits 434,436
tree fruits, deciduous 454,459,469
tree fruits, subtropical 470
Typhula blight (snow mold) *Typhula* spp.:
lawns/turf 510,512
Typhula rot:
small grains 134,140

U

Urocystis agropyri — flag smut (wheat, grass)
Ustilago nigra — false loose smut
***Ustilago* spp.**— smut
Ustilago striiformis — stripe smut

V

Vein spot *Gnomonia nerviseda:*
nuts 482,483
Venturia inaequalis — apple scab
Venturia nashicola — pear scab
Verticillium:
cucurbits 329
eggplant 334
mint 174,175
onions 347
sweet potatoes, yams 370
tobacco 192
Verticillium dahliae — early maturity disease
Verticillium spot:
mushrooms 345
Verticillium wilt:
photo 27
Volutella leaf blight:
ornamentals 521,522,524,525

W

Walnut blight *Xanthomonas juglandis:*
nuts 479-481
Water mold:
rice 182
Watery soft rot *Sclerotinia sclerotiorum:*
lettuce 343
onions 351
Web blotch:
peanuts 176,180,181
White mold *Sclerotinia sclerotiorum:*
beans 312,313,315
cucurbits 330
lawns/turf 510
peanuts 176,177,180
potatoes 358,359
tomatoes 373
White rot *Sclerotium* spp.:
garlic 337
onions 348,351
White rust:
biocontrols 58
ornamentals 527
radishes 363
spinach 366,367
Wildfire:
biocontrols 54
Wilt *Verticillium* spp.:
ornamentals 531
tobacco 193
Wire stem (Corticium black root) *Rhizoctonia* spp.:
cole crops 326
Witches' broom:
ornamentals 534

X

Xanthomonas:
cole crops 324
Xanthomonas campestris* pv *citri — citrus canker
Xanthomonas campestris* pv *phaseoli — common blight
Xanthomonas juglandis — walnut blight
Xanthomonas leaf spot:
ornamentals 521,522
Xanthomonas translucens:
small grains 136

Y

Yellow leaf blister:
ornamentals 534
Yellow patch:
lawns/turf 509,512,516
Yellow rust:
small fruits 434-438
Yellow Sigatoka:
tree fruits, subtropical 470,477
Yellow tuft *Sclerophthora macrospora:*
lawns/turf 509,514

Z

Zonate leaf spot *Christulariella pyramidalis:*
lawns/turf 515
nuts 478,480,481,483,485
Zoysia patch:
lawns/turf 509,516,518

Crop Index

A

Acorn squash (Cucurbits) 230
Alfalfa (Forage) 537-543
Almonds (Nuts) 423-426,428-433,478-485
Anise (Herbs & Spices) 243,244
Apples (Tree Fruits, Deciduous) 402-417, 452-466,468-470
Apricots (Tree Fruits, Deciduous) 402-413,415-417, 452,454-470
Artichokes 198,199,308
Asparagus 199-202,308,309
Avocados (Tree Fruits, Subtropical) 418-422, 470-474,476,477

B

Bananas (Tree Fruits, Subtropical) 420-423,470-477
Barley (Small Grains) 129-141
Basil (Herbs & Spices) 243,244
Beans 202-209, 309-316
Beans, Mung 175
Beechnuts (Nuts) 424,428,431
Beets, Sugar 161-165,187-191
Beets, Table 297-300,370-372
Bell peppers (Peppers) 265,268-270
Bermudagrass for seed (Forage, Alfalfa) 539,540
Birdsfoot trefoil (Forage, Alfalfa) 537,548
Black walnuts (Nuts) 429,482
Blackberries (Small Fruits) 381,382,384,387-389, 391,434-440,443-445
Blackeyed peas (Southern Peas) 282-284,364,365
Blueberries (Small Fruits) 381,382,384-391, 433-441,443-446
Bok choy (Cole Crops) 217,218,223,322,324
Boysenberries (Small Fruits) 381,382,384,387-389, 391,434-438,440,443-445
Brazil nuts (Nuts) 424,428,431
Broccoli (Cole Crops) 217-225,322-328
Brussels sprouts (Cole Crops) 217-225,322-328
Burley tobacco (Tobacco) 171
Butternut squash (Cucurbits) 230
Butternuts (Nuts) 424

C

Cabbage (Cole Crops) 217-225,322-328
Cacao (Tree Fruits, Subtropical) 418,471-473
Calamondin (Tree Fruits, Citrus) 393
Caneberries (Small Fruits) 381,382,384,387-389, 391,433-441,443-446
Canola (Oil Seed Crops) 150
Cantaloupe (Cucurbits) 226-231,328-334
Caprifig 475
Carrots 209-213,316-319
Casaba melon (Cucurbits) 226,227,229-231, 328,331,332,334
Cashews (Nuts) 424,428,431
Castor Beans 172
Cauliflower (Cole Crops) 217-225,322-328
Celery 213-217,319-322
Cherries (Tree Fruits, Deciduous) 402-418,452-470
Chestnuts (Nuts) 424,428,429,431
Chickpeas (Garbanzos) 336
Chicory (Lettuce) 251,345
Chili peppers (Peppers) 268
Chinese cabbage (Cole Crops) 217-224,322, 323,325,327
Chinquapins (Nuts) 424
Chive (Herbs & Spices) 244,341
Citron (Tree Fruits, Citrus) 393,398,448
Citrus, Tree Fruits 392-401,446-452
Clover (Forage, Alfalfa) 537,539-542,547,548
Coffee (Tree Fruits, Subtropical) 418-420,473
Cole Crops 217-225,322-328
Collards (Greens) 238-243,338-340
Corn, Field 77-88
Corn, Sweet 288-294,368-369
Cotton 109-125
Cowpeas (Southern Peas) 282-284,364,365
Crabapples (Tree Fruits, Deciduous) 402,403,408, 410,412,413,416,462,463,468
Cranberries (Small Fruits) 381-384,386-389,391, 433,434,436-439,441,445,446
Crenshaw melon (Cucurbits) 226,227,229-231, 328,331
Crowder peas (Southern Peas) 202,283,364
Cucumbers (Cucurbits) 225-231,328-334
Cucurbits 225-231,328-334
Currants (Small Fruits) 381,387-389,391, 434-440,443-445

D

Dandelion (Greens) 238,240-242,
Dates (Tree Fruits, Subtropical) 420,422
Deciduous, Tree Fruits 401-418,452-470
Dewberries (Small Fruits) 381,382,384,387,388, 391,434,436-438,443-445
Dill (Herbs & Spices) 243,244,341,342
Dry beans (Beans) 202-209,309-313,315,316

E

Eggplant 231-235,334-336
Elderberries (Small Fruits) 381,391
Endive (Lettuce) 248-253,343-345
English walnuts (Nuts) 429
Escarole (Lettuce) 249,250,252,253

F

Fall wheat (Small Grains) 130,141
Fennel (Herbs & Spices) 243,244,341
Field Corn 77-88
Figs (Tree Fruits, Subtropical) 418,420-422,475
Filberts (Nuts) 423-425,428-431,477-482
Flax 172,173
Forage (Alfalfa, etc.) 537-543
Forage (Grasses, Hay) 543,544
Forage corn (Field Corn, Popcorn) 78,81
Forage sorghum (Sorghum) 157,185

G

Garbanzos 336
Garlic 236,237,336-338
Gherkin (Cucurbits) 226,331
Ginger (Herbs & Spices) 244
Gooseberries (Small Fruits) 381,387-389,391, 434-440,443-445
Gourds (Cucurbits) 226
Grain corn (Field Corn, Popcorn) 77,81
Grain sorghum (Sorghum) 157-161,185
Grapefruit (Tree Fruits, Citrus) 393-401,447-450
Grapes (Small Fruits) 381-391,433-446
Green beans (Beans) 202-207,311-313,315
Green onions (Onions) 257-259,347,349
Greens 238-243,338-340
Guava (Tree Fruits, Subtropical) 420,422,471-473

H

Hanover salad (Greens) 238
Head lettuce (Lettuce) 248-254,343,344
Herb & Spices 243,244,341,342
Hickory (Nuts) 424,428,431
Honeyball melon (Cucurbits) 226,229,331,332,334
Honeydew melon (Cucurbits) 226-231,328-334
Hops 146,147,173,174
Horseradish 245,246,342

I

Irish potatoes (Potatoes) 271

J

Jojoba (Oil Seed Crops) 150

K

Kale (Greens) 238-243,338-340
Kidney beans (Beans) 203,205,206,208,311,313,315
Kiwifruit (Tree Fruits, Subtropical) 418,419,421, 422,471-473
Kohlrabi (Cole Crops) 217-224,322,324,325,327,328
Kumquats (Tree Fruits, Citrus) 393,395,397-399,448

L

Lawns & Turf....489-494,509-519
Leaf lettuce (Lettuce)....249-254,343,344
Leeks (Onions)....256-259,347,349
Lemons (Tree Fruits, Citrus)....393-401,447-450
Lentils....246-248,342
Lespedeza (Forage, Alfalfa)....542,548
Lettuce....248-254,343-345
Lima beans (Beans)....202-208,309-313,315,316
Limes (Tree Fruits, Citrus)....393-401,447-450
Loganberries (Small Fruits)....381,382,384,387-389, 391,434,436-438,440,443-445

M

Macadamias (Nuts)....424,428,429,431,432,479-481
Mandarins (Tree Fruits, Citrus)....393,398
Mangoes (Tree Fruits, Subtropical)....419-422, 470-474,476,477
Marjoram (Herbs & Spices)....244
Melons (Cucurbits)....225-231,328-334
Milo (Sorghum)....157-161,185
Mint....149,150,174,175
Mung Beans....175
Mushrooms....254,255,345
Muskmelon (Cucurbits)....226-231,328-334
Mustard greens (Greens)....238-243,338-340

N

Navel oranges (Tree Fruits, Citrus)....393,398,399
Navy beans (Beans)....202,203,205,206,208
Nectarines (Tree Fruits, Deciduous)....402-417,452-470
Noncropland....546,547
Nuts....423-433,477-485

O

Oats (Small Grains)....129-141
Oil Seed Crops....150
Okra....255,256,345,346
Olives (Tree Fruits, Subtropical)....418,421,422, 471-474,477
Onions....256-260,346-351
Oranges (Tree Fruits, Citrus)....393-401,447-451
Oriental pears (Tree Fruits, Deciduous)....410
Ornamentals....494-508,519-534

P

Papayas (Tree Fruits, Subtropical)....418,420-423, 470-477
Parsley (Carrots)....209-212,316-318
Parsnips (Carrots)....209-212,316,318
Passion fruit (Tree Fruits, Subtropical)....422,470-473
Pastures & Rangeland....544-546
Peaches (Tree Fruits, Deciduous)....402-418,452-470
Peanuts....150-154,175-181
Pears (Tree Fruits, Deciduous)....401-418,452,453, 455-463,465-470
Peas....260-264,351-353
Peas, Southern....282-284,364,365
Pecans (Nuts)....423-433,477-485
Peppermint (Mint)....148,149,174,175
Peppers....264-270,353-357
Persian melons (Cucurbits)....227,229-231,331,332,334
Persimmons (Tree Fruits, Subtropical)....418,422
Pimento (Peppers)....268
Pineapple (Tree Fruits, Subtropical)....418-423, 470,474-477
Pinto beans (Beans)....208,311,313,315
Pistachio (Nuts)....424,429-432,479-481
Plantains (Tree Fruits, Subtropical)....420,421,423,477
Plums (Tree Fruits, Deciduous)....402-417, 452-460,462-468
Pole beans (Beans)....204,311,313,314
Pomegranates (Tree Fruits, Subtropical)....419,422
Popcorn....77-88
Potatoes....271-278,357-362
Potatoes, Sweet....294-297,370
Prunes (Tree Fruits, Deciduous)....402-417, 452-460,462-468
Pummelo (Tree Fruits, Citrus)....393
Pumpkins (Cucurbits)....225-231,328-334

Q

Quince (Tree Fruits, Deciduous)....402,403,408, 410-413,416,453,456,460,462,463

R

Radishes....278-281,362,363
Rangeland, Pastures &....544-546
Rape (Greens)....240,339
Rapeseed (Oil Seed Crops)....150
Raspberries (Small Fruits)....381,382,384,385, 387-391,434-446
Red Beets (Table Beets)....297-300,370-372
Red clover (Forage, Alfalfa)....547
Rhubarb....281,282,363,364
Rice....154-156,181-184
Romaine (Lettuce)....253
Rosemary (Herbs & Spices)....243
Rutabagas (Cole Crops)....217-220,222,322,325-327
Rye (Small Grains)....129-134,137-141

S

Safflower....156,157,184,185
Sage (Herbs & Spices)....243,244
Salsify (Carrots)....209-212
Seed corn (Field Corn, Popcorn)....77-80,82-84,86,88
Shallots (Onions)....256-260,347-349
Silage corn (Field Corn, Popcorn)....77
Silage sorghum (Sorghum)....160
Small Fruits....381-392,443-446
Small Grains....129-141
Snap beans (Beans)....202-208,309-313,315,316
Sorghum....157-161,185,186
Sour cherries (Tree Fruits, Deciduous)....402,404,410, 411,453,455,458-461,463,467
Sour orange (Tree Fruits, Citrus)....393
Southern Peas....282-284,364,365
Soybeans....95-105
Spearmint (Mint)....148,149,174,175
Spinach....284-288,365-367
Spring wheat (Small Grains)....136,137
Squash (Cucurbits)....225-231,328-334
Strawberries (Small Fruits)....381-392,434-446
String beans (Beans)....203,311,313
Subtropical, Tree Fruits....418-423,470-477
Succulent beans (Beans)....202,205,206,209
Sudangrass (Forage, Grasses, Hay)....543,544
Sugar Beets....161-165,187-191
Sugarcane....166,191
Summer squash (Cucurbits)....226-231,328-334
Sunflower....166-169,191,192
Sweet cherries (Tree Fruits, Deciduous)....402,404, 411,453,455,459-461,463,467
Sweet Corn....288-294,368-369
Sweet orange (Tree Fruits, Citrus)....393
Sweet Potatoes....294-297,370
Sweet sorghum (Sorghum)....157
Swiss chard (Greens)....238-242,339,340

T

Table Beets....297-300,370-372
Tangelos (Tree Fruits, Citrus)....393-400,447-450
Tangerines (Tree Fruits, Citrus)....393-400,447-450
Tea (Tree Fruits, Subtropical)....418
Thyme (Herbs & Spices)....243
Tobacco....169-172,192,193
Tomatillo (Tomatoes)....301,304
Tomatoes....300-307,372-378
Tree Fruits, Citrus....392-401,446-452
Tree Fruits, Deciduous....401-418,452-470
Tree Fruits, Subtropical....418-423,470-477
Trefoil (Forage, Alfalfa)....542,548
Triticale (Small Grains)....129,130
Turf, Lawns &....489-494,509-519
Turnip greens (Greens)....238-243,338,339
Turnips (Cole Crops)....217-220,222-225,322-328

V

Valencia oranges (Tree Fruits, Citrus)....393,398,399
Vegetables....198-378
Vetch (Forage, Alfalfa)....541,542,548

W

Walnuts (Nuts)....423-433,477-485
Watercress....307,378
Watermelon (Cucurbits)....226-231,328-334
Wax beans (Beans)....202,203,205,206,208,311,313
Wheat (Small Grains)....129-141
Winter squash (Cucurbits)....226-231,328-334
Winter wheat (Small Grains)....136,137
Wintergreen (Mint)....148

Y

Yams....294-297,370
Youngberries (Small Fruits)....381,391,440

Z

Zucchini (Cucurbits)....230

Suppliers Directory

A-1 Unique Insect Control
5504 Sperry Dr.
Citrus Heights CA 95621
Telephone: 916-961-7945
Fax: 916-967-7082
E-Mail: ladybugs@a-1unique.com
www.a-1unique.com

Abbott Laboratories
1401 Sheridan Rd.
Dept. 44C, Bldg. A1
North Chicago IL 60064
Telephone: 847-937-7909
Fax: 847-937-3729
Toll Free: 800-323-9597

AgBioChem
3 Fleetwood Court
Orinda CA 94563
Telephone: 510-254-0789
Fax: 530-527-6288

AgrEvo Environmental Health
95 Chestnut Ridge Rd.
Montvale NJ 07645
Telephone: 201-307-9700
Fax: 201-307-3281

AgrEvo USA Company
Little Falls Centre One
2711 Centerville Rd.
Wilmington DE 19808
Telephone: 302-892-3000
Fax: 302-892-3013
www.na.agrevo.com

Agricultural Sciences Inc. (AGSCI)
3601 Garden Brook
Dallas TX 75234
Telephone: 972-243-8930
Fax: 972-406-1125
Toll Free: 800-274-8930
E-Mail: atldal@aol.com

AgriPro Seeds, Inc.
P.O. Box 2962
6700 Antioch
Shawnee Mission KS 66201-1362
Telephone: 913-384-4940
Fax: 913-384-0208
E-Mail: markj@agripro.com
www.agripro.com

AGSCO Inc.
P.O. Box 13458
Grand Forks ND 58208-3458
Telephone: 701-775-5325
Fax: 701-775-9587
Toll Free: 800-859-3047
E-Mail: agsco@corpcomm.net
www.agsco-agdepot.com

Agtrol International
7322 Southwest Freeway, Suite 1400
Houston TX 77074
Telephone: 713-995-0111
Fax: 713-995-9505

American Cyanamid Co.
One Campus Dr.
Parsippany NJ 07054-4492
Telephone: 973-683-2000
Fax: 973-683-4001
Toll Free: 800-327-4645

Amvac Chemical Corp.
4100 E. Washington Blvd.
Los Angeles CA 90023
Telephone: 213-264-3910
Fax: 213-887-9221

ARBICO Inc.
P.O. Box 4247 CRB
Tucson AZ 85738
Telephone: 520-825-9785
Fax: 520-825-2038
Toll Free: 800-SOS-BUGS
E-Mail: arbico@aol.com
www.arbico.com

Asgrow Seed Co.
P.O. Box 7570
Des Moines IA 50322
Telephone: 515-331-7100
Fax: 515-331-7110

BASF Corp.
P.O. Box 13528
26 Davis Dr.
Research Triangle Park NC 27709-3528
Telephone: 919-547-2000
Fax: 919-547-2420
Toll Free: 800-962-7830/326-1810

Bayer Corp.
P.O. Box 4913
8400 Hawthorn Rd.
Kansas City MO 64210-0013
Telephone: 816-242-2000
Fax: 816-242-2738
Toll Free: 800-842-8020

Becker Microbial Products, Inc.
9464 NW 11th St.
Plantation FL 33322
Telephone: 954-474-7590
Fax: 954-474-2463
E-Mail: tcouch@icanect.net

Beck's Superior Hybrids, Inc.
6767 E. 276 St.
Atlanta IN 46031
Telephone: 317-984-3508
Fax: 317-984-3500
Toll Free: 800-937-2325
E-Mail: sales@beckshybrids.com
www.beckshybrids.com

Beneficial Insectary
14751 Oak Run Rd.
Oak Run CA 96069
Telephone: 530-472-3715
Fax: 530-472-3523
Toll Free: 800-477-3715
E-Mail: bi@insectary.com
www.insectary.com

Bergmann-Taylor, Inc.
P.O. Box 317
10073 Ellis Rd.
St. Jacob IL 62281
Telephone: 618-644-5522
Fax: 618-644-9452
E-Mail: btag@hometel.com

Best Sulfur Products, Inc.
5427 E. Central Ave.
Fresno CA 93725
Telephone: 209-485-0114
Fax: 209-264-1715

BioLogic Co.
P.O. Box 177
Willow Hill PA 17271
Telephone: 717-349-2789/2922
E-Mail: pyealber@epix.net

Brandt Consolidated
ClawEL Div.
Route 125, Box 277
Pleasant Plains IL 62677
Telephone: 217-626-1031
Fax: 217-626-1563
Toll Free: 800-300-6559

Caltec Agri-Marketing Services
P.O. Box 576155
Modesto CA 95357
Telephone: 209-575-1295
Fax: 209-575-0366
www.caltecag.com

Campbell Seed, Inc.
1375 N. 800W
Tipton IN 46072
Fax: 765-963-2047
Toll Free: 800-788-5950
E-Mail: campbellseed@i.quest.net
www.campbellseed.com

Cargill Hybrid Seeds
P.O. Box 5645
Minneapolis MN 55440
Fax: 612-742-7233
Toll Free: 800-4-YIELDS
E-Mail: mark_winkle@cargill.com
www.seed.cargill.com

Chemgro Co.
P.O. Box 218
E. Petersburg PA 17520
Telephone: 717-569-3296
Fax: 717-560-0117

Cheminova, Inc.
Oak Hill Park
1700 Rt. 23, Suite 210
Wayne NJ 07470
Telephone: 973-305-6600
Fax: 973-305-1382
Toll Free: 800-548-6113

Cleary Chemical Corp.
178 Ridge Rd., Suite A
Dayton NJ 08810-1501
Telephone: 732-329-8399
Fax: 732-274-0894
Toll Free: 800-524-1662

Consep, Inc.
213 S.W. Columbia St.
Bend OR 97702
Telephone: 541-388-3688
Fax: 541-388-3705
Toll Free: 800-367-8727
www.consep.com

Countrymark Cooperative, Inc.
Growmark, Inc.
1701 Towanda Ave.
Bloomington IL 61701
Telephone: 309-557-6398/6399
Fax: 309-557-6860
www.growmark.com or
www.countrymark.com

Custom Chemicides
P.O. Box 11216
Fresno CA 93722
Telephone: 209-264-0441
Fax: 209-264-0829

Dairyland Seed Co., Inc.
P.O. Box 958
West Bend WI 53095
Fax: 414-626-2281
Toll Free: 800-236-0163

DEKALB Genetics Corp.
3100 Sycamore Rd.
Dekalb IL 60115
Telephone: 815-758-9273
Fax: 815-756-2672

Deltapine Seed
One Cotton Row
P.O. Box 157
Scott MS 38772
Telephone: 601-742-4000
Fax: 601-742-3472
Toll Free: 800-511-SEED

Diatect International Inc.
P.O. Box 367
Smith Center KS 66967
Telephone: 208-343-3451
Fax: 913-282-6607
Toll Free: 800-227-6616

Dow AgroSciences
9330 Zionsville Rd.
Indianapolis IN 46268-1054
Toll Free: 800-258-3033
Fax: 800-905-7326
E-Mail: info@dowagro.com
www.dowagro.com

Drexel Chemical Co.
1700 Channel Ave.
Memphis TN 38106-1412
Telephone: 901-774-4370
Fax: 901-774-4666
E-Mail: drexel_chemical@juno.com

DuPont Agricultural Products
P.O. Box 80038
Walker's Mill, Barley Mill Plaza
Wilmington DE 19880-0038
Fax: 302-992-2276
Toll Free: 800-638-7668

Ecogen Inc.
2005 Cabot Blvd. W
P.O. Box 3023
Langhorne PA 17271-3023
Telephone: 215-757-1590
Fax: 215-757-2956
Toll Free: 800-220-3326

Elf Atochem North America, Inc.
Three Parkway
2000 Market St., 21st. Fl.
Philadelphia PA 19103-3222
Telephone: 215-587-7409
Fax: 215-419-5012

Farm Science Genetics
P.O. Box 322
Battle Ground IN 47920
E-Mail: thunsley@growmark.com
www.growmark.com or www.countrymark.com

Florikan E.S.A. Corp.
1523 Edger Place
Sarasota FL 34240
Telephone: 941-377-8666
Fax: 941-377-3633
Toll Free: 800-322-8666
E-Mail: buglady@aol.com

FMC Corp.
1735 Market St.
Philadelphia PA 19103
Telephone: 215-299-6661
Fax: 215-299-6256
Toll Free: 800-321-1362

Garst Seed Co.
2369 330th St.
P.O. Box 500
Slater IA 50244
Telephone: 800-831-6630
Fax: 515-685-5080
www.garstseed.com

Golden Harvest Seeds, Inc.
220 Eldorado Rd.
Bloomington IL 61704-3544
Telephone: 309-664-0558
Fax: 309-664-0984
E-Mail: jrsherl7@gte.net
www.goldenharvestseeds.com

Gowan Co.
P.O. Box 5569
Yuma AZ 85366-5569
Telephone: 520-783-8844
Fax: 520-343-9255
E-Mail: gowan@primenet.com

Great Lakes Chemical Corp.
P.O. Box 2200
One Great Lakes Blvd.
West Lafayette IN 47906-0200
Telephone: 765-497-6204
Fax: 765-497-6666

Griffin L.L.C.
P.O. Box 1847
Rocky Ford Rd.
Valdosta GA 31601-1847
Telephone: 912-242-8635
Fax: 912-244-5813

Gustafson, Inc.
1400 Preston Rd.
Plano TX 75293
Telephone: 972-985-8877
Fax: 972-867-0816

Harmony Farm Supply
3244 Hwy. 116 No. F
Sebastopol CA 95472
Telephone: 707-823-9125
Fax: 707-823-1734
E-Mail: kate@harmonyfarm.com
www.harmony farm.com

Helena Chemical Co.
6075 Poplar Ave., Suite 500
Memphis TN 38119
Telephone: 901-761-0050
Fax: 901-756-9947

Hercon Environmental Corp.
P.O. Box 467
Aberdeen Rd.
Emigsville PA 17318-0467
Telephone: 717-764-1191
Fax: 717-767-1016

Hoegemeyer Hybrids
1755 Hoegemeyer Rd.
Hooper NE 68031
Telephone: 402-654-3399
Fax: 402-654-3342
E-Mail: hoeghyb@htchet.com
www.corn1.com

Hoffman Seeds, Inc.
167 Greenfield Rd.
P.O. Box 12400
Lancaster PA 17605-2400
Fax: 717293-0283
Toll Free: 800-776 7929
E-Mail: mark.mattingly@seeds.novartis.com

Hot Pepper Wax, Inc.
305 Third St.
Greenville PA 16125
Telephone: 888-667-3785/724-646-2300
Fax: 724-646-2302
E-Mail: lindag@hotpepperwax.com
www.hotpepperwax.com

Hydro-Gardens, Inc.
P.O. Box 25845
Colorado Springs CO 80932
Telephone: 719-495-2266
Fax: 719-531-0506

IGENE Biotechnology, Inc.
9110 Red Branch Rd.
Columbia MD 21045
Telephone: 410-997-2599
Fax: 410-730-0540

International Sulphur, Inc.
P.O. Box 611
Mt. Pleasant TX 75456-0611
Telephone: 903-577-5500
Fax: 903-577-5540

International Technology Services Inc.
P.O. Box 19227
Boulder CO 80308-2227
Telephone: 303-473-9141
Fax: 303-473-9143
E-Mail: intertechserv@worldnet.att.net

Interstate Payco Seed Co.
P.O. Box 338
W. Fargo ND 58078
Telephone: 701-282-7338
Fax: 701-282-8218

IPM Laboratories
P.O. Box 300
Locke NY 13092-0099
Telephone: 315-497-2063
Fax: 315-497-3129

Kincaid Enterprises, Inc.
P.O. Box 549
Nitro WV 25143
Telephone: 304-755-3377
Fax: 304-755-4547

Latham Seeds
131 180th St.
Alexander IA 50420
Telephone: 515-692-3258
Fax: 515-692-3250
E-Mail: lathamseeds@fiai.net
www.lathamseeds.com

Loveland Industries
P.O. Box 1289
Greeley CO 80632
Telephone: 970-356-8920
Fax: 970-356-8926

M & R Durango, Inc.
P.O. Box 886
Bayfield CO 81122
Telephone: 970-259-3521
Fax: 970-259-3857

Merschman Seeds
103 Ave. D
West Point IA 52656
Telephone: 319-837-6111
Fax: 319-837-6104
Toll Free: 800-848-7333
E-Mail: merseed@interl.net
www.merschmanseeds.com

Micro Flo Co.
P.O. Box 5948
Lakeland FL 33807
Telephone: 941-647-3608
Fax: 941-647-3412
Toll Free: 800-451-8461

Midwest Seed Genetics
23751 Hwy. 30 E.
Carroll IA 51401
Fax: 712-792-6725
Toll Free: 800-369-8218
E-Mail: terry@midwestseed.com
www.midwestseed.com

Mineral Research & Development
One Woodlawn Green
200 E. Woodlawn Rd., Suite 250
Charlotte NC 28217
Telephone: 704-525-2771
Fax: 704-527-8232
Toll Free: 800-334-0417
E-Mail: mrdc@mrdc.com
www.mrdc.com

Mycogen Crop Protection
5501 Oberlin Dr.
San Diego CA 92121
Telephone: 619-453-8030
Fax: 619-453-9089
Toll Free: 800-745-7476
E-Mail: soares@mycogen.com

Mycogen Seeds
1340 Corporate Center Curve
Eagan MN 55121
Telephone: 612-405-5800
Fax: 612-405-5957
E-Mail: minnichsof@mycogen.com
www.mycogen.com

Mycotech Corp.
P.O. Box 4109
Butte MT 59702-4109
Telephone: 406-782-2386
Fax: 406-782-9912
Toll Free: 800-383-4310
E-Mail: mycotech@montana.com

Natural Pest Controls
8864-Little Creek Dr.
Orangeville CA 95662
Telephone: 916-726-0855
Fax: 916-726-0855
E-Mail: natpestc@cwnet.com
www.natural-pest-control.com

NatureMark Potatoes
250 Bobwhite Court 3rd Fl.
Boise ID 83706
Telephone: 208-382--2236
Fax: 208-389-2280
E-Mail: jennifer.feldman@monsanto.com

Nature's Control
P.O. Box 35
Medford OR 97501
Telephone: 541-899-8318
Fax: 541-899-9121
Toll Free: 800-698-6250
E-Mail: bugsnc@teleport.com

NC+ Hybrids
3820 N. 56th St.
Lincoln NE 68504
Telephone: 402-467-2517
Fax: 402-467-4217
E-Mail: kanderson@nc-plus.com
www.nc-plus.com

Novartis Crop Protection
P.O. Box 18300
Greensboro NC 27419-8300
Telephone: 336-632-6000
Fax: 336-632-2861
Toll Free: 800-334-9481
www.cp.us.novartis.com

Novartis Seeds, Inc.
7500 Olson Memorial Hwy.
Golden Valley MN 55427
Telephone: 612-593-7189
Fax: 612-593-7203

Olympic Horticultural Products
P.O. Box 1885
Bradenton FL 34206-1885
Fax: 888-647-4329
Toll Free: 800-659-6745
E-Mail: olympic@hortnet.com
www.hortnet.com/olympic

Pace International LP
500 7th Ave. S.
Kirkland WA 98033
Telephone: 425-827-8711
Fax: 425-822-8261
Toll Free: 800-247-8711

Pacific Biocontrols Corp.
400 E. Evergreen Blvd., # 205
Vancouver WA 98660
Telephone: 360-693-2866
Fax: 360-693-3088

Patriot Seed, Inc.
208 S. Worrell
P.O. Box 97
Bowen IL 62316-0097
Telephone: 217-842-5612
Fax: 217-842-5209
Toll Free: 800-643-1518
E-Mail: patriot@patriotseed.com
www.patriotseed.com

PBI/Gordon Corp.
1217 W. 12th St.
Kansas City MO 64101
Telephone: 816-421-4070
Fax: 816-474-0462
Toll Free: 800-821-7925

Pioneer Hi-Bred International, Inc.
7100 NW 62nd Ave.
P.O. Box 1150
Johnston IA 50131-1150
Telephone: 515-334-6908
Fax: 515-334-6886
E-Mail: harringtonjr@phibred.com
www.pioneer.com

Platte Chemical Co.
419 18th St.
Greeley CO 80632
Telephone: 970-356-4400
Fax: 970-356-4418

Prentiss Incorporated
CB 2000
Floral Park NY 11001
Telephone: 516-326-1919
Fax: 516-326-2312
E-Mail: ramiller@prentiss.com
www.prentiss.com

PSA Genetics
661 510th St.
Alta IA 51002
Fax: 717-296-3663
Toll Free: 800-441-1610

Rhone-Poulenc Ag Co.
P.O. Box 12014
2 T.W. Alexander Dr.
Research Triangle Park NC 27709
Telephone: 919-549-2000/2084
Fax: 919-549-3924
Toll Free: 800-334-9745

Rohm and Haas Co.
100 Independence Mall West
Philadelphia PA 19106
Telephone: 215-592-3221
Fax: 215-592-2797
Toll Free: 800-523-0762
www.rohmhaas.com

RohMid L.L.C. (joint venture of American Cyanamid & Rohm And Haas)

Rupp Seeds, Inc.
17919 County Rd. B
Wauseon OH 43567
Telephone: 419-337-1841
Fax: 419-337-5491

Sands of Iowa
P.O. Box 648
Marcus IA 51035-0648
Telephone: 712-376-4135
Fax: 712-376-4140
E-Mail: soi@nwidt.com

The Scotts Co.
Professional Business Group
14111 Scottslawn Rd.
Marysville OH 43041
Telephone: 937-644-0011
Fax: 937-644-7679
Toll Free: 800-543-0006
E-Mail: mike.bandy@scottsco.com
www.scottscompany.com

SePRO Corp.
11550 N. Meridian St., Suite 180
Carmel IN 46032
Telephone: 317-580-8282
Fax: 317-580-8290
Toll Free: 800-419-7779
E-Mail: rogers@sepro.com
www.sepro.com

Sespe Creek Insectary
P.O. Box 176
Lindsay CA 93247
Telephone: 209-562-6464

Sieben Hybrids, Inc.
633 N. College Ave.
Geneseo IL 61254
Telephone: 309-944-5131
Fax: 309-944-6090
E-Mail: sieben@geneseo.net
www.siebenhybrids.com

Soil Technologies Corp.
2103 185th St.
Fairfield IA 52556
Telephone: 515-472-3963
Fax: 515-472-6189
Toll Free: 800-221-7645
E-Mail: soiltech@lisco.com
www.lisco.com/soiltech

Sostram Corp.
70 Mansell Ct., Suite 230
Roswell GA 30076
Telephone: 770-587-1032
Fax: 770-587-1115

Spalding Laboratories
760 Printz Rd.
Arroyo Grande CA 93420
Telephone: 800-845-2847
Fax: 805-489-0336

Stine Seed Co.
2225 Laredo Trail
Adel IA 50003
Fax: 515-677-2716
Toll Free: 800-362-2510
www.stine.com

Stoller Enterprises, Inc.
8582 Katy Freeway, Suite 200
Houston TX 77024
Telephone: 716-461-1493
Fax: 713-461-4467
Toll Free: 800-539-5283

Stoneville Pedigreed Seed Co.
6625 Lenox Park Dr., Suite 117
Memphis TN 38115
Telephone: 901-375-5800
Fax: 901-375-5860
E-Mail: dthreet@stoneville.com
www.stoneville.com

Sun Company Inc.
P.O. Box 1135
Marcus Hook PA 19061
Fax: 215-246-8452
Toll Free: 800-385-2786

Terra Industries
Terra Centre
600 Fourth St.
P.O. Box 6000
Sioux City IA 51102-6000
Telephone: 712-277-1340
Fax: 712-233-3648
Toll Free: 800-831-1002
E-Mail: hdavis@terraindustries.com
www.terraindustries.com

Thermo Trilogy Corp.
9145 Guilford Rd., Ste. 175
Columbia MD 21046
Telephone: 301-604-7340
Fax: 301-604-7015
Toll Free: 800-847-5620
www.thermotrilogy.com

Trace Chemicals, Inc.
839 Brenkman Dr.
Pekin IL 61554
Telephone: 309-347-2184
Fax: 309-347-5929
Toll Free: 800-846-2980

Trisler Seed Farms, Inc.
3274 E. 800 North Rd.
Fairmont IL 61841-6139
Telephone: 217-288-9301
Fax: 217-288-9095
E-Mail: trisler@prairienet.com

Troy Biosciences
2620 N. 37th Dr.
Phoenix AZ 85009
Telephone: 602-233-9047
Fax: 602-254-7989

UCB Agrochemicals
2000 Lake Park Dr.
Smyrna GA 30080
Telephone: 770-801-3202
Fax: 770-801-3238
Toll Free: 800-426-3820
E-Mail: Agro.group@ucb-group.com
www.ucb-group.com

Uniroyal Chemical Co., Inc.
Benson Rd.
Middlebury CT 06749
Telephone: 203-573-2000
Fax: 203-573-3394
Toll Free: 800-243-2850

Valent USA Corp.
1333 N. California Blvd., # 600
P.O. Box 8025
Walnut Creek CA 94596
Telephone: 510-256-2700
Fax: 510-256-2776
Toll Free: 800-682-5368
E-Mail: Valent@valent.com

Webb Wright Corp.
P.O. Box 1572
Ft. Myers FL 33902-1572
Telephone: 941-283-4400
Fax: 941-283-4844

Wellmark International
1000 Tower Lane
Suite 245
Bensenville IL 60108
Telephone: 800-842-3135
Fax: 630-227-6065

Wilbur-Ellis Co.
191 W. Shaw Ave., Suite 107
Fresno CA 93704-2876
Telephone: 209-226-1934
Fax: 209-226-7630

ZENECA Ag Products
1800 Concord Pike
Wilmington DE 19850-5458
Fax: 302-886-1660
Toll Free: 800-759-4500

ZENECA Professional Products
1800 Concord Pike
Wilmington DE 19850-5458
Telephone: 302-886-1000
Fax: 302-886-1644
Toll Free: 800-759-4500
www.zenecaprofprod.com